C0-AKI-680

FOURIER TRANSFORM SPECTROSCOPY

FOURIER TRANSFORM SPECTROSCOPY

Eleventh International Conference

Athens, GA August 1997

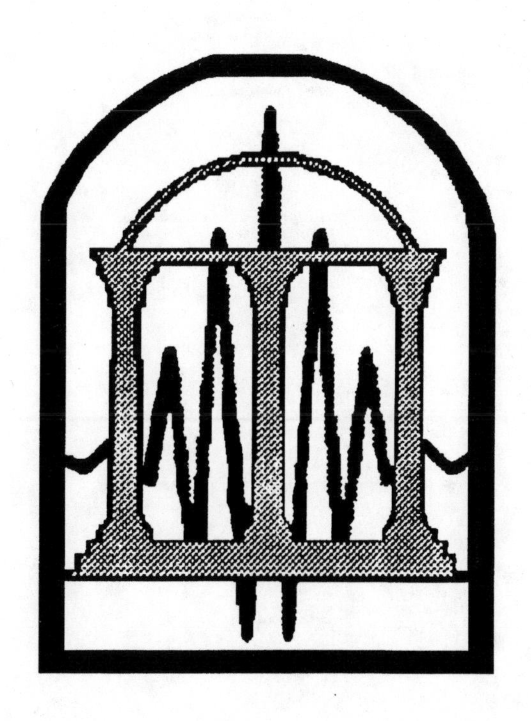

EDITOR
James A. de Haseth
University of Georgia, Athens

AIP CONFERENCE
PROCEEDINGS 430

American Institute of Physics

Woodbury, New York

Editor:

James A. de Haseth
Department of Chemistry
University of Georgia
Athens, GA 30602-2556

E-mail: dehaseth@dehsrv.chem.uga.edu

L.C. Catalog Card No. 98-70838
ISBN 1-56396-746-4
ISSN 0094-243X
DOE CONF- 970812

Printed in the United States of America

CONTENTS

1. KEYNOTE ADDRESS

2. PLENARY LECTURES

3. SUBMITTED PAPERS

3.1 ATMOSPHERIC STUDIES

3.2 CHEMOMETRICS

3.3 BIOCHEMICAL AND BIOMEDICAL STUDIES

3.4 INSTRUMENT DEVELOPMENT

3.5 FORENSICS

3.6 PROCESS CONTROL

3.7 METHODS DEVELOPMENT

3.8 POLYMERS

3.9 SURFACES

3.10 TEXTILES

3.11 MINERALS

3.12 STRUCTURAL ANALYSIS

3.13 THEORETICAL STUDIES

Preface

The Eleventh International Conference on Fourier Transform Spectroscopy (ICOFTS-11) took place August 10-15, 1997, at the Classic Center, Athens, Georgia, U.S.A. Athens is the northeast corner, or Piedmont area, of the State of Georgia. Athens is also home to the University of Georgia.

The Conference welcomed more than 300 attendees from five continents and 21 countries. The participants worked for industry, government agencies, and in academia. The Conference provided an atmosphere for people of diverse backgrounds to congregate and exchange information. The common thread for all participants was an interest in Fourier transform spectrometry, with an emphasis on optical applications. Most of the applications were in the mid and near infrared, and a few extended to the far infrared and visible regions of the electromagnetic spectrum.

As with other conferences in this series, the 11th ICOFTS was organized to encourage scientific exchange between the participants. The Conference began Sunday evening with a Keynote Address from Professor Alan G. Marshall, of the National High Magnetic Field Laboratory at Tallahassee, Florida. Professor Marshall introduced and discussed a technique related to the Conference focus, that is, Fourier Transform Ion Cyclotron Resonance. There was a total of sixteen Plenary Lectures. These lectures were scheduled in the mornings and late afternoons. This left the middle of the day open for poster presentations. More than 200 posters were presented, and each poster was available for viewing over a two-day period. Presenters were scheduled to be present at their posters for part of that time. This gave every participant, including the presenters, ample time to view and discuss all the presentations.

One morning was reserved for short presentations by young scientists. Several outstanding posters were chosen by the local program committee to be presented orally. This gave promising young scientists the opportunity to present their work to an international audience and highlight their contributions. This clearly increased the exposure of these young professionals, and provided many thought-provoking presentations.

The social program was also an important aspect of the Conference. The opening reception was sponsored by the Perkin-Elmer Corporation, and this took place immediately after the Keynote Address. Bruker Optics sponsored their traditional "German Evening" on Monday night. This reception was held at the State Botanical Gardens of Georgia, and was an excellent atmosphere to continue discussions. The Conference Banquet was held at Stone Mountain Park near Atlanta. After a Southern Barbeque Buffet a laser light show was presented. If nothing else, the laser technology was impressive! Both Nicolet Instrument Corporation and the Digilab Division of Bio-Rad Laboratories helped sponsor the Conference Banquet.

The instrument exhibit was an essential part of the Conference. More than twenty corporations presented their instrumentation or services at the Conference. The support of the instrument manufacturers is very important to the Fourier transform community as they build the instrumentation upon which much of the science is based. Their participation is an integral part of this series of conferences.

The primary financial sponsor of the Conference was the Coblentz Society. Without the support of the Coblentz Society this Conference would not have been possible. The Society provided the seed funds to begin the process of advertising and securing meeting facilities. The Coblentz Society has done much for the Fourier Transform Spectroscopic community, and their help is gratefully recognized. The Department of Chemistry at the University of Georgia assisted with the Conference both financially and with personnel. The financial support for social events from the instrument companies is gratefully acknowledged.

The Conference could not have taken place without the help of numerous people. First, Richard A. Dluhy must be recognized for his contributions as Program Chair. Rich organized the invited speaker program as well as the Young Investigator Symposium. He was a source of good ideas and many of his ideas were incorporated into the organization, scheduling, and program for the Conference. The help of the scientific reviewers whose careful comments led to the papers in this Proceedings is also gratefully acknowledged. More than twenty scientists from all over the world helped with the reviews. A special thanks must go to all the graduate and undergraduate students who helped with the Conference. Every student in Rich Dluhy's and my research groups spent many hours helping with registration, logistics, the manufacture of poster boards, etc. Students from Peter Griffiths' research group at the University of Idaho also assisted with some of the events. The Local Organizing Committee and the International Steering Committee members contributed greatly to the Conference.

Last, but not least, a special thanks goes to my wife, Leslie, who spent two years working for the Conference. Leslie arranged all the correspondence, registration, exhibits, abstracts, and social programs. Her organizational skills made the operation of the Conference much smoother.

The Twelfth International Conference on Fourier Transform Spectroscopy is to be held at Waseda University, Tokyo, Japan, August 23 through 27, 1999. The General Chairman of the 12th ICOFTS meeting is Koichi Itoh.

James A. de Haseth
General Chairman
University of Georgia

Dedication

This Proceedings of the
Eleventh International Conference on Fourier Transform Spectroscopy
is dedicated to the memory of:

Professor Dr. Robert Kellner
1945 - 1997

Robert was the Secretary General for ICOFTS-7
Vienna, Austria, 1987

and the

Program Chair for ICOFTS-10
Budapest, Hungary, 1995.

Robert's contributions to Fourier Transform Spectroscopy
were numerous and he will be missed by his many
friends and colleagues.

Scientific Program Committee

R. A. Dluhy, University of Georgia (USA) - *Chairman*
P. Bernath, University of Waterloo (Canada)
J. Bertie, University of Alberta (Canada)
B. Carli, CNR (Italy)
J. Chalmers, ICI (UK)
R. Corn, University of Wisconsin (USA)
L. Debouille, University of Liege (Belgium)
D. Haaland, Sandia National Laboratory (USA)
T. Keiderling, University of Illinois at Chicago (USA)
O. Kvalheim, University of Bergen (Norway)
C. Marcott, Procter & Gamble (USA)
J. Rabolt, University of Delaware (USA)
O. Schrems, Alfred Wegner Institute (Germany)
F. Siebert, University of Frieburg (Germany)
H. Strauss, University of California, Berleley (USA)
M. Tasumi, Saitama University (Japan)
J. van der Maas, University of Utrecht (Netherlands)
A. Vassallo, CSIRO (Australia)
J. Workman, Kimberly Clark (USA)

Local Organizing Committee

J. A. de Haseth, University of Georgia - *Chairman*
L. V. Azarraga, U.S. Environmental Protection Agency, Athens
R. T. Bishop, University of Georgia
T. W. Collette, U.S. Environmental Protection Agency, Athens
L. C. de Haseth - *Conference Administrator*
U. Happek, University of Georgia
T. A. Perenich, University of Georgia
R. A. Todebush, University of Georgia
J. Wilkin, University of Georgia

International Steering Committee

E. H. Korte, ISAS, Berlin (Germany) - *Chairman*	ICOFTS-8, L beck, 1991
H. Siesler, University of Essen (Germany)	
H. Wieser, University of Calgary (Canada)	ICOFTS-9, Calgary, 1993
J. E. Bertie, University of Edmonton (Canada)	
J. Mink, University of Veszprm (Hungary)	ICOFTS-10, Budapest, 1995

R. Kellner, Technical University of Vienna (Austria)
J. A. de Haseth, University of Georgia (USA) ICOFTS-11, Athens, 1997
R. A. Dluhy, University of Georgia (USA)
K. Itoh, Waseda University (Japan) ICOFTS-12, Tokyo, 1999
M. Tasumi, Saitama University (Japan)

1. KEYNOTE ADDRESS

Fourier Transform Ion Cyclotron Resonance Mass Spectrometry

Alan G. Marshall

Center for Interdisciplinary Magnetic Resonance, National High Magnetic Field Laboratory, Florida State University, 1800 East Paul Dirac Drive, Tallahassee, FL 32310

As for Fourier transform infrared (FT-IR) interferometry and nuclear magnetic resonance (NMR) spectroscopy, the introduction of pulsed Fourier transform techinques revolutionized ion cyclotron resonance mass spectrometry: increased speed (factor of 10,000), increased sensitivity (factor of 100), increased mass resolution (factor of 10,000—an improvement *not* shared by the introduction of FT techniques to IR or NMR spectroscopy), increased mass range (factor of 500), and automated operation. FT-ICR mass spectrometry is the most versatile technique for unscrambling and quantifying ion-molecule reaction kinetics and equilibria in the absence of solvent (i.e., the gas phase). In addition, FT-ICR MS has the following *analytically* important features: speed (~1 second per spectrum); ultrahigh mass resolution *and* ultrahigh mass accuracy for analysis of mixtures and polymers; attomole sensitivity; MS^n with one spectrometer, including two-dimensional FT/FT-ICR/MS; positive and/or negative ions; multiple ion sources (especially MALDI and electrospray); biomolecular molecular weight and sequencing; LC/MS; and single-molecule detection up to 10^8 Dalton. Here, some basic features and recent developments of FT-ICR mass spectrometry are reviewed, with applications ranging from crude oil to molecular biology.

INTRODUCTION

Various aspects of the Fourier transform relationship between an interferogram and a spectrum were known to Michelson and Rayleigh, among others. Modern FT optical interferometry may be said to have originated from P. Fellgett's 1951 thesis at Cambridge University (1), and practical implementation of FT-IR spectroscopy exploded almost immediately following the introduction of the Cooley-Tukey fast Fourier transform algorithm (2), which has become the most highly cited paper in all of mathematics. FT-NMR was introduced in 1966 by Ernst and Anderson (3). However, it was another nine years before Fourier transform methods could be implemented in ion cyclotron resonance (4, 5). FT-ICR MS evolution since then has paralleled that of FT-NMR (6), much as FT-NMR borrowed many concepts from FT-interferometry. The growth of FT-ICR MS is graphically evident from Figure 1.

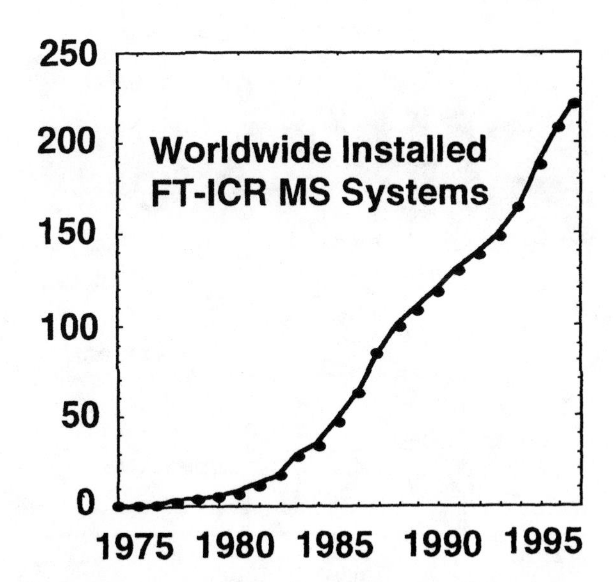

FIGURE 1. Following Comisarow/Marshall's first homebuilt instrument in December, 1973, the first commercial instrument (based on their 1976 patent) appeared in 1980.

CP430, *Fourier Transform Spectroscopy:* 11th International Conference
edited by J.A. de Haseth

ION EXCITATION AND DETECTION

An ion moving perpendicular to a spatially uniform magnetic field, B, executes "cyclotron" rotation at a frequency,

$$\omega_c = \frac{qB}{m} \qquad (1)$$

in which q and m are ion mass and charge. An ICR "interferogram" (like an NMR free induction decay) consists of the time-domain response to a pulsed excitation (Fig. 2, top). A broadband alternating voltage source excites the ion cyclotron cyclotron motion of ions. The excited ICR motion creates an alternating signal current, $I_S(t)$, and an alternating signal voltage, $V_S(t)$, in an external circuit (7). The excitation is applied perpendicular to the magnetic field $\boldsymbol{B}$.

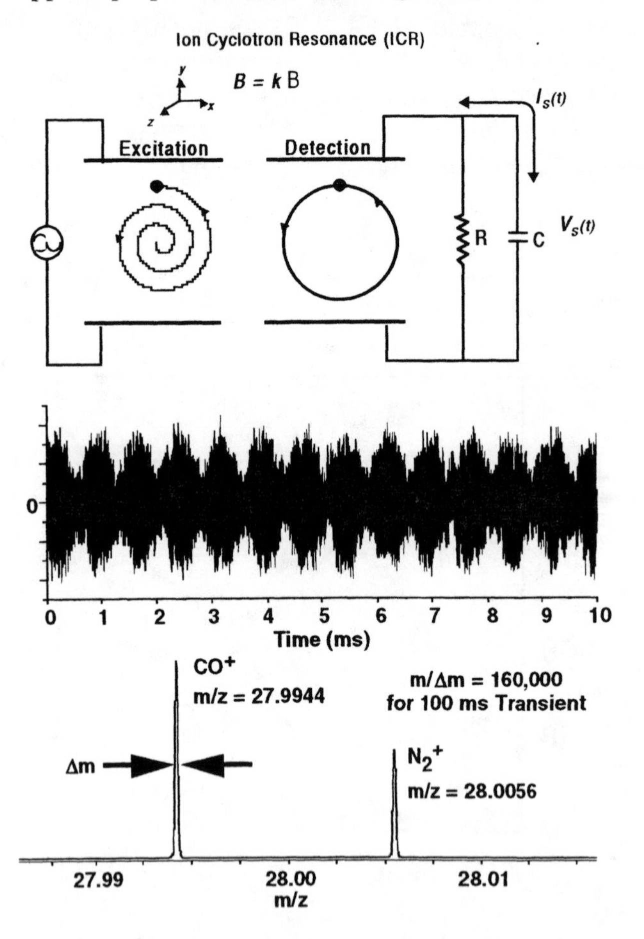

FIGURE 2. Top: FT-ICR excitation and detection (8). Bottom: Experimental time-domain signal and its Fourier-transformed and frequency-to-mass converted magnitude-mode mass spectrum.

Compared to FT-interferometry, FT-ICR (like FT-NMR) begins from a temporal (rather than spatial) one-sided (rather than two-sided) interferogram, as shown in Figure 2 (bottom).

Fortunately, ICR (unlike NMR) exhibits *linear* dipolar excitation: i.e., resonant excitation at twice the amplitude or for twice the duration doubles the detected ICR signal amplitude. Therefore, we do not need to compute the response to a given time-domain excitation waveform; the ICR response magnitude at a given frequency is simply proportional to me Fourier transform (at that frequency) of the excitation waveform (9-11). Figure 3 shows some representative excitation waveforms.

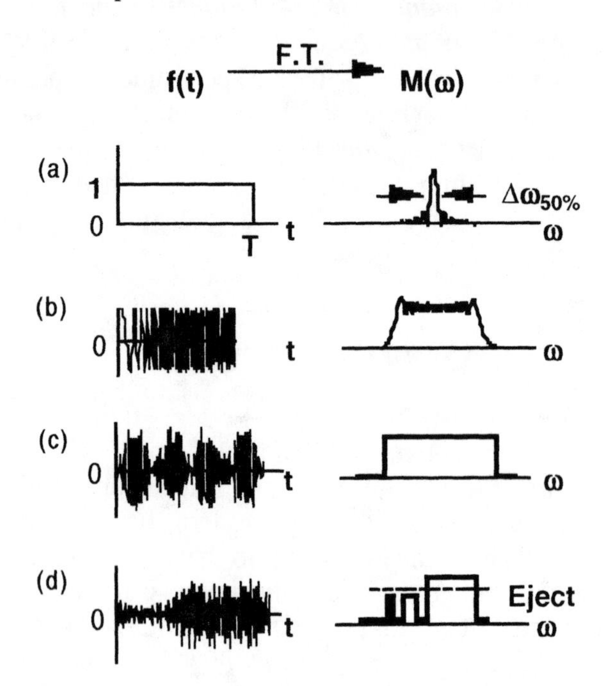

FIGURE 3. Time-domain (left) and frequency-domain (right) excitation waveforms for FT-ICR mass spectrometry (see text).

Fig. 3a gives the excitation spectrum for a simple rectangular pulse. Although suitable for FT-NMR, and in fact used for the very first (narrowband) FT-ICR experiment (4), such excitation is not appropriate for broadband FT-ICR, because the very wide spectral bandwidth (several MHz for the "chemical" mass range extending down to ~15 Da) would require a very short pulse (≤0.1 μs) with unreasonably high amplitude (>10 kV). Therefore, for the first ten years of FT-ICR MS, excitation was conducted with a frequency linear sweep (Fig. 3b) (5), followed by detection. Such "chirp" excitation made it possible to excite ions over a wide frequency range (and thus a wide mass range), but the excitation magnitude was uneven (leading to inaccurate ion relative abundances) and not particularly selective. In 1985, Marshall et al. introduced stored waveform inverse Fourier transform (SWIFT) excitation (Figs. 3c,d), which produces optimally flat and optimally mass-selective excitation or (at sufficiently high excitation amplitude and duration) radial ejection (when the ion cyclotron radius increases until ions strike the walls of the ion trap and are neutralized) (12). SWIFT excitation has recently been reviewed (13), and is now implemented on about 1/3 of all FT-ICR instruments, and provides maximum flexibility for MS/MS (see below).

FT MANIPULATIONS

FT-ICR shares with FT-IR and FT-NMR the usual features of discrete Fourier transform data handling (14): Nyquist sampling and foldover (15), fast Fourier transformation, zero-filling (16), apodization, deconvolution (9, 17, 18), oversampling (19), etc.

Moreover, various non-FT methods for obtaining a frequency-domain (and thus mass-domain) spectrum from a time-domain "interferogram" ICR signal include: the Hartley transform (a way of performing a Fourier transform on real-only data) (20, 21), Bayesian maximum entropy method (MEM) (22, 23), and linear prediction (24, 25).

However, several aspects of FT-ICR data reduction differ from those of FT-interferometry or FT-NMR spectroscopy. Notably, phasing of FT-ICR spectra over a broad frequency range is difficult (26); thus, spectra are usually reported in magnitude-mode rather than absorption-mode (27). As a result, resolving power is lower than for absorption-mode by a factor ranging from $\sim\sqrt{3}$ (Lorentzian peak shape) to 2 (sinc peak shape) (28), and new peak-fitting algorithms have been developed (29). Apodization of magnitude-mode spectra has been performed by means of magnitude-mode multiple-derivative techniques (30). Due to the need for massively large data sets (several megawords of time-domain data) to take advantage of potentially ultrahigh mass resolving power over a wide mass range, data clipping (to as low as 1 bit/word) has been used, with only modest distortion of the FT spectrum (31). Finally, the ICR time-domain signal damps exponentially only in the limit of very small ion cyclotron radius, corresponding to a Langevin ion-neutral collision model; under experimentally typical conditions (namely, hard-sphere ion-neutral collisions), the lineshape is distinctly non-Lorentzian (32); hence, it is common to apodize with a window function (e.g., Blackman-Harris) that damps the initial portion of the time-domain signal.

APPLICATIONS

Accurate Mass/Chemical Formula

The most unique advantage of FT-ICR over over other mass analysis methods is its ultrahigh resolving power, due to its inherently frequency-based measurement. What makes high resolution so useful in mass spectrometry is that mass measurement to ±0.001 Da or better often suffices to determine the chemical formula, $C_cH_hN_nO_o\cdot\cdot\cdot$, uniquely. Accurate mass measurement is particularly important for analysis of mixtures (see Figure 4), such as oil distillates (33), combinatorial drugs (34, 35), and environmental samples (36).

Ion Remeasurement

In most types of spectroscopy, we take it for granted that an experiment can be repeated or modified with a single sample of analyte(s). However, most mass analyses are destructive, so that ions are lost after one measurement. In FT-ICR, on the other hand, detection is based on the (non-destructive) induction of charge in detector electrodes. Thus, ions can in principle

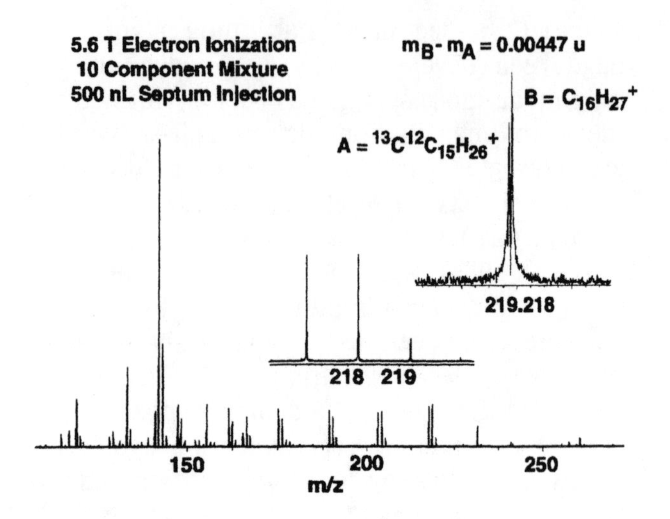

FIGURE 4. High-resolution FT-ICR mass spectrum of a 10-component hydrocarbon mixture (propyl–, pentyl-, hexyl-, heptyl-, octyl-, nonyl-, decyl-, and tridecylbenzene; fluorene, and 1-methylnaphthalene). Mass-scale expansion of the decylbenzene region shows resolution of $^{13}C_1$-decylbenzene and protonated decylbenzene.

be remeasured by FT-ICR MS (37). The problem is that ions tend to diffuse radially (i.e., in a plane perpendicular to the magnetic field direction), and are typically lost in a few seconds or less. Fortunately, by irradiation with an rf electric field of two-dimensional quadrupolar symmetry, it is possible to refocus ions back to the center of the ion trap (38-40), and then remeasure the same ions many times with virtually no loss of signal (41). Figure 5 shows an example of remeasurement. By use of remeasurement, as little as 8 attomole (total sample deposited on the probe tip) of the undecapeptide, substance P, has been detected by FT-ICR MS (42).

Tandem-in-Time Mass Spectrometry (MS^n)

As with FT-NMR, FT-ICR experiments are conducted by a series of time-separated events, during which ions are mass-selected, ejected, excited, detected, and/or allowed to collide and/or react with neutrals. This ability to control ions through a complex series of steps (see Figure 6) makes it possible to perform gas-phase ion-molecule chemistry in ways completely analogous to condensed-phase chemistry, with two big differences: (a) one typically proceeds from synthesis, purification, reaction, analysis, and cleanup in about 1 second in the gas-phase; and (b) reactions are conducted in the absence of solvent, making it possible to determine inherent reactivity of ions and molecules and to explore reactions that may be impossible in condensed phase.

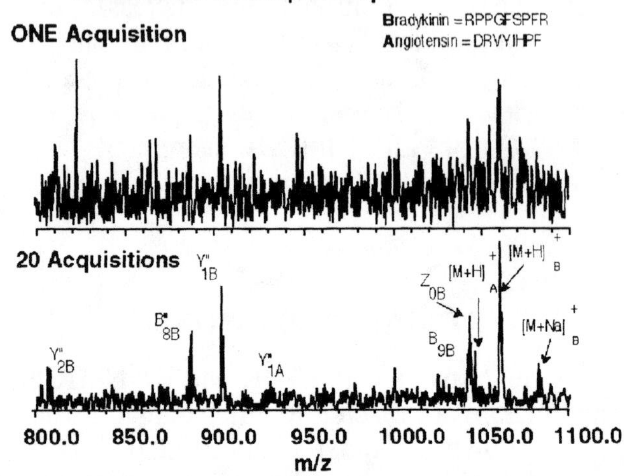

FIGURE 5. Enhancement in signal-to-noise ratio of a single batch of ions, formed by matrix-assisted laser desorption/ionization, by use of quadrupolar excitation and summation of remeasured signals. The labels denote the sites of peptide bond cleavage. (Data provided by T. Solouki.)

Experiments corresponding to "double-resonance" in NMR are known as MS/MS or "tandem" mass spectrometry, because they formerly required two mass spectrometers: one to isolate the "parent" ions of interest, then a reaction chamber, and then a second mass spectrometer to analyze the ion-molecule collision/reaction products. However, with FT-ICR, ions remain trapped throughout the experiment, and may be repeatedly remeasured, so that many stages of MS^n may be achieved from a single initial batch of ions.

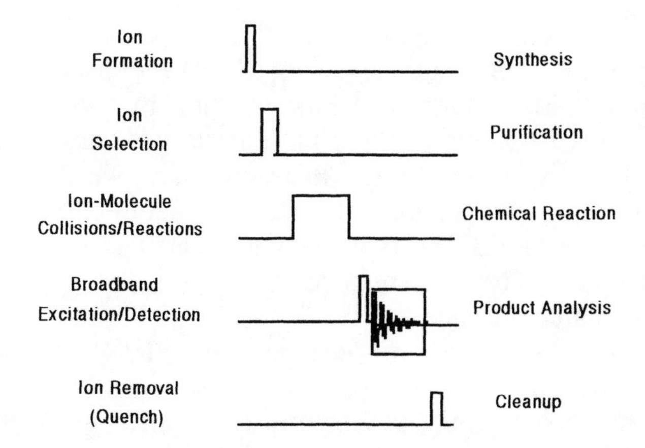

FIGURE 6. Stages of chemical analysis (right) and corresponding FT-ICR experimental event sequence (left) for an ion-molecule reaction.

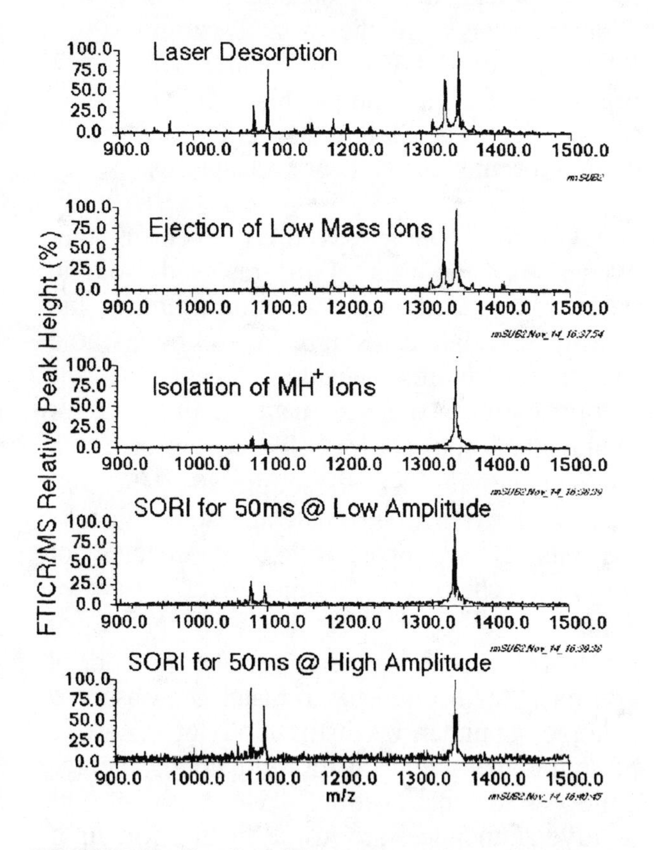

FIGURE 7. Peptide sequencing, based on multiple stage MS^n, conducted on a *single* batch of ions initially generated by matrix-assisted laser desorption/ionization.

Interactive Mass Spectrometry

The non-destructive nature of Fourier transform ion cyclotron resonance (FT-ICR) detection means that, following broadband detection, ions remain available for further manipulation and re-detection. Mass spectrometry may thus be performed interactively for the first time. Following each elementary experimental stage, such as ion generation, isolation, dissociation, or detection, the operator is free to choose and tailor the next stage without creating a fresh supply of ions. For example, one could test the effect of varying a single parameter over several values without having to repeat the entire experimental event sequence (and create a new batch of ions) each time (41), much like varying one letter or word in a sentence without having to rewrite the whole sentence. Such interactive control promises to speed development of complex experimental event sequences, as for optimizing the sequencing and structural analysis of tiny amounts (e.g., femtomoles or less) of biomacromolecules (peptides, nucleic acids, oligosaccharides) (43). Figure 7 shows an example of interactive MS^n. Proceeding from top to bottom in that Figure, ions are first formed by MALDI and observed (top spectrum); broadband mass-selective chirp ejection then removes low-mass ions, and broadband detection confirms the result (next spectrum). SWIFT ejection then removes ions whose masses are just above or below the species of interest, and the result is confirmed by broadband detection (next spectrum). Sustained off-resonance irradiation (SORI, which causes periodic ion heating) then accelerates ions for collisional dissociation, and the ions are detected again (next spectrum). Finally, because the previous spectrum revealed incomplete dissociation, SORI was repeated at higher amplitude, with the result shown in the lowermost spectrum. All of these manipulations were conducted on the *same* single batch of initially formed ions.

Biological Macromolecular Ions

In molecules with just a few atoms, it is usually not necessary to consider the various "rare" isotopes (^{13}C, ^{15}N, ^{18}O, ^{34}S, etc.). For example, carbon monoxide consists of about 99% $^{12}C^{16}O_2$ and about 1% $^{13}C^{16}O_2$ (Figure 8, top). However, the so-called "rare" isotopes become more prominent in larger molecules: e.g., in a fullerene with 70 carbons, the probability that one of the carbons is ^{13}C rather than ^{12}C is ~70 x 1%, or ~70% (Fig. 8, middle). For even a small protein, such as ubiquitin (8560 Da), species containing several ^{13}C are actually much more abundant than the all-^{12}C molecules (Fig. 8, bottom).

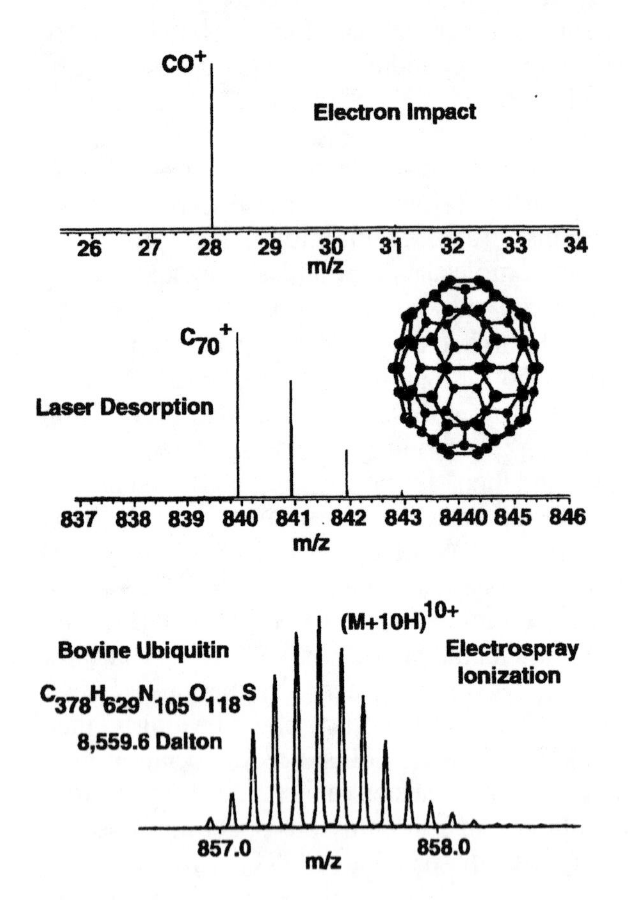

FIGURE 8. Isotopic distributions for small (top), medium (middle), and large (bottom) molecules. Note that the distribution widens and shifts to the right with increasing number of atoms per molecule, and that only the leftmost "monoisotopic" peak has a unique chemical formula (see text).

Several related problems arise from the presence of multiple isotopes in biological macromolecular ions. First, starting in about 1990, it became possible to generate gas-phase ions by electrospray ionization, in which multiply-charged ions are formed by addition of (e.g.) protons. A typical protein can take on about 1 proton per kDa, by protonation of available arginine and lysine residues. Because mass analyzers in general (and ICR in particular) separate ions based on mass-to-charge *ratio* (m/z, in which m is ion mass in u and z is charge in multiples of the elementary charge), the immediate problem then becomes how to determine charge independent from mass. Fortunately, if the isotopic envelope can be resolved to better than 1 u (as in Fig. 8, bottom), adjacent peaks differ by m/z = 1/z; ergo, the ion charge may be obtained simply as the reciprocal of the spacing between two adjacent peaks in the m/z spectrum (44). Because only FT-ICR MS can resolve such multiplets for macromolecular ions of more than a few kDa, FT-ICR MS is the method of choice for mass analysis of such species.

A second (more severe) problem is that even if m/z for each of the resolved isotopic peaks can be determined to ppm accuracy, the molecular weight can still be in error by a whole Dalton or more, because (except for the "monoisotopic" species—namely, the one for which all carbons are ^{12}C, all nitrogens are ^{14}N, all oxygens are ^{16}O, all sulfurs are ^{32}S, etc.) each of the remaining abundant isotopic peaks consists of a superposition of species of different chemical formula (e.g., species containing $^{13}C_2{}^{14}N^{32}S$ or $^{12}C_2{}^{14}N^{34}S$ or $^{13}C^{15}N^{32}S$, etc.). Therefore, the best prior approach has been to try to match the observed isotopic abundance distribution of (say) a protein to that predicted for a protein of average amino acid composition. However, if the relative abundances are just a bit in error, or if the unknown protein differs in composition from the average protein in the database, then the estimated molecular weight can be in error by 1 or more Da.

A simple solution to the above problem is to express a protein from a minimal medium containing ^{13}C-depleted glucose and ^{15}N-depleted ammonium sulfate. The result is shown in Figure 9. Both simulated and experimental electrospray FT-ICR mass spectra show that double-depletion of ^{13}C and ^{15}N effectively shifts the isotopic distribution to the left, so that the monoisotopic species is now prominent and easily identified (45). This single idea promises to extend the upper mass limit for protein mass spectrometry by about an order of magnitude.

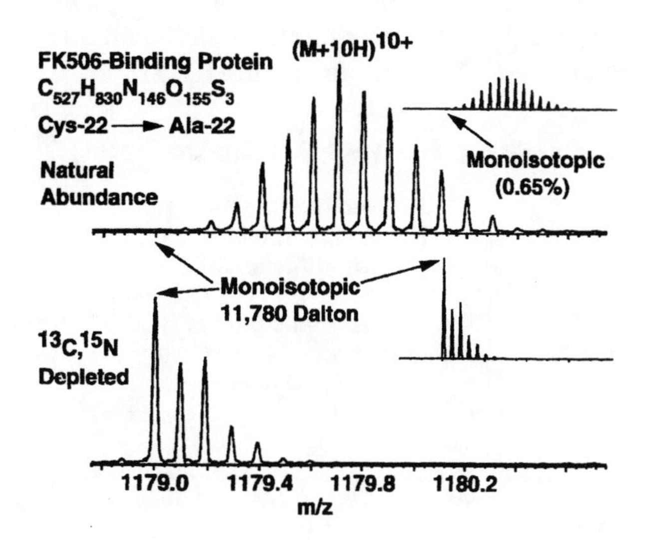

FIGURE 9. Electrospray ionization Fourier transform ion cyclotron resonance mass spectra (9.4 tesla) of a mutant (C22A) FK506-binding protein. Top: Natural-abundance isotopic distribution (~98.89% ^{12}C; ~99.63% ^{14}N. Bottom: Isotopic distribution for the same protein grown on a medium with 99.95% ^{12}C and 99.99% ^{14}N. Insets: Isotopic distributions calculated (same vertical scale) from the chemical formula for natural-abundance (top) and ^{13}C, ^{15}N doubly-depleted (bottom) FK506 binding protein. Reproduced, with permission, from (45). ©1997 American Chemical Society.

The advantages of being able to measure protein molecular weight to within 1 Da are manifold: (a) count the number of disulfide bridges (–S–S– is 2 Da lighter than 2 –SH); (b) identify deamidation (–NH_2 is 1 Da lighter than –OH); (c) identify such post-translational modifications as phosphorylation and glycosylation; (d) resolve and identify adducts; (e) identify variant amino acid sequences; etc. Figure 9 suggests several additional advantages (45). First, the monoisotopic species, present at only 0.65% at natural abundance, becomes the largest peak in the mass spectrum of the ^{13}C,^{15}N doubly-depleted protein! The molecular weight of the neutral protein is thus determined immediately and unambiguously. For a similarly doubly-depleted protein of 80 kDa molecular weight, the monoisotopic peak would still be 1% abundant, and thus easily identifiable. Second, depletion of rare isotopes increases mass spectral signal-to-noise ratio (because the same number of ions now exhibit fewer isotopic variants); mass spectral sensitivity and detection limit thus improve accordingly. Third, space charage distortions are reduced, because a mass spectrum of a given peak-height-to-noise ratio requires fewer total ions. Fourth, MS^n experiments are improved, because the narrow m/z distribution makes it easier to isolate the desired parent ions and facilitates identification of fragments (*e.g.*, one Da difference between loss of H_2O *vs.* NH_3 or glutamic Acid (129 Da) vs. glutamine (128 Da)). Fifth, depletion narrows all isotopic distributions, including any adducts (e.g., $(M+nH+mNa)^{(n+m)+}$), and thus dramatically increases the upper molecular weight limit before mass assignment is affected due to isotopic overlap of such impurities. Sixth, a narrower protein isotopic distribution makes it easier to observe and characterize non-covalent binding (protein:protein, protein:nucleic acid, enzyme:inhibitor, etc.). Seventh, identification of surface-accessible residues by H/D exchange is simpler, because of simpler deconvolution to yield the deuterium number distribution. Eighth, isotopically-depeleted proteins provide a good mass calibrant, whose own isotopic distribution is narrower than other natural-abundance proteins of similar molecular weight.

LC Electrospray FT-ICR MS

As for FT-IR, a rapidly growing application for FT-ICR mass spectrometry is its interface to a liquid chromatograph. The obvious problem is that FT-ICR is an inherently pulsed mass analyzer, whereas high-performance liquid chromatography (HPLC) is an inherently continuous sample source. A nice solution to this problem is to collect and accumulate ions in an octupole ion trap external to the magnet, and then eject the ions into the ICR ion trap for excitation and detection. Figure 10 shows such an arrangement, in which an rf-powered octupole ion guide consists of a first segment supplied with end caps to which d.c. voltage can be applied to trap and accumulate ions (46).

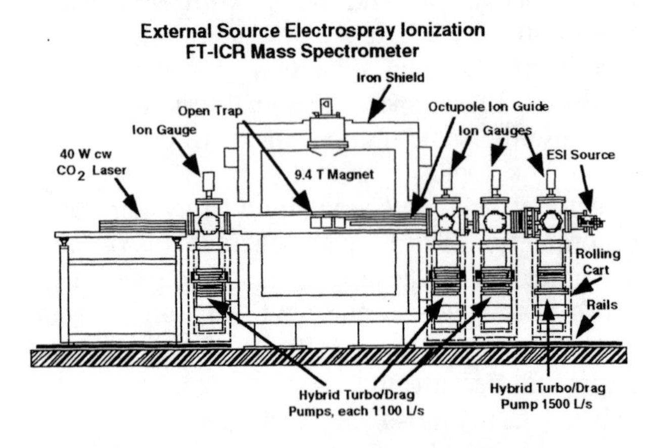

FIGURE 10. Schematic diagram of the electrospray 9.4 T FT-ICR mass spectrometer at the National High Magnetic Field Laboratory in Tallahassee, FL.

With this arrangement, micro-ESI (i.e., electrospray conducted with a 50 μ diameter capillary) has achieved remarkable sensitivity, as shown in Figure 11: 100 fmole/μL (i.e., 0.1 μM), from a total of 10 μL loaded (1 pmole total sample loaded) of a peptide hormone.

CONCLUSION

FT-ICR mass spectrometry shares many common conceptual and data reduction features with other types of FT spectroscopy, most notably FT-NMR (6). The most unique advantage of FT-ICR as a mass analyzer is that ion mass-to-charge ratio is experimentally manifested as a frequency. Because frequency can be measured more accurately than any other experimental parameter, ICR MS therefore offers inherently higher resolution (and thus mass accuracy) of any mass measurement. The introduction of FT techniques to ICR MS brought not only the Fellgett (multichannel, opening the exit slit) advantages of increased speed (factor of 10,000), or increased sensitivity (factor of 100), but also the advantages of fixed magnetic field rather than swept field, namely, increased mass resolution (factor of 10,000), and increased mass range (factor of 500). Applications deriving from these advantages include determination of chemical formulas, particularly in complex mixtures; detection limit in the attomole range; and multistage MS^n.

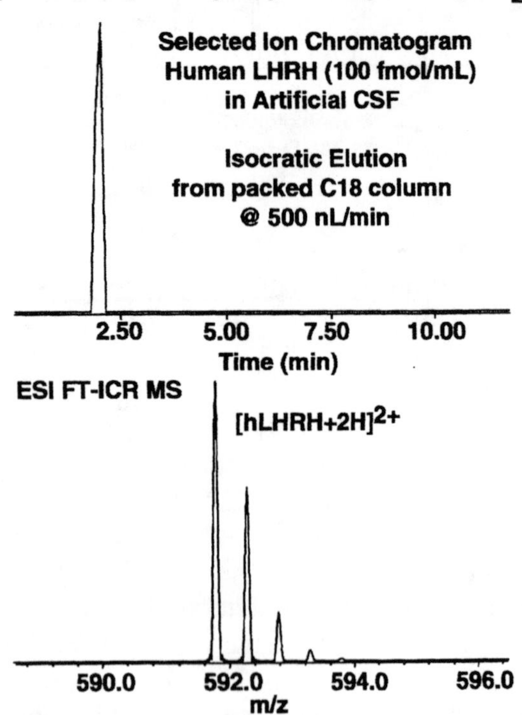

FIGURE 11. HPLC (isocratic elution with 70% MeOH, 0.25% HOAc at 500 nL/min) of human luteinizing hormone releasing hormone, LHRH, electrospray ionization FT-ICR mass spectrometry. Top: Mass chromatogram for molecular ion. Bottom: High-resolution mass spectrum. Figure kindly provided by F. M. White and M. R. Emmett, Florida State Univ.

More than 220 Fourier transform ion cyclotron resonance (FT-ICR) mass spectrometer systems have been installed worldwide, including national FT-ICR MS facilities in the U.S.A. (Tallahassee, Florida) and England (U. Warwick). Although FT-ICR instruments comprise less than one-half of one per cent of all mass spectrometers worldwide, ~10% of the (rapidly growing) total number of abstracts at the American Society for Mass Spectrometry Annual Conference on Mass Spectrometry & Allied Topics originate from FT-ICR. FT-ICR MS has been the sole or principal subject of three books (14, 47, 48), four refereed journal special issues (49-52), and more than 60 review articles. Listed here are a few reviews of early (8, 53, 54) and later developments (6, 40, 44, 55-61); other reviews are cited therein.

FT-ICR mass spectrometers are already displacing double-focusing magnetic sector instruments for high-performance mass analysis, because electrospray ionization (ESI) and matrix-assisted laser desorption ionization (MALDI) are much better suited to FT-ICR MS. For example, ESI FT-ICR sensitivity actually increases with increasing mass, because heavier ions tend to be more highly charged, and ICR signal strength is proportional to charge. Future directions for FT-ICR MS include: two-dimensional FT/FT-ICR MS (62) for automated analysis of complex mixtures; real-time interactive mass analysis and MS^n (43); synthetic and biological polymer sequencing, structure, and non-covalent adducts; LC ESI FT-ICR MS/MS, and optical spectroscopy of ions mass-selected and trapped by FT-ICR techniques (63). With such prospects, it seems safe to predict that the growth in number of FT-ICR MS systems will continue well into the next millenium.

ACKNOWLEDGMENTS

This work was supported by grants from NSF (CHE-93-22824), the NSF National High Field FT-ICR Mass Spectrometry Facility (CHE-94-13008), NIH (GM-31683), Florida State University, and the National High Magnetic Field Laboratory in Tallahassee, FL.

REFERENCES

1. Mertz, L. *Transformations in Optics* 1-116 (John Wiley & Sons, Inc., New York, 1965).
2. Cooley, J.W. & Tukey, J.W. *Math. Comput.* **19**, 297 (1965).
3. Ernst, R.R. & Anderson, W.A. *Rev. Sci. Instrum.* **37**, 93 (1966).
4. Comisarow, M.B. & Marshall, A.G. *Chem. Phys. Lett.* **25**, 282-283 (1974).
5. Comisarow, M.B. & Marshall, A.G. *Chem. Phys. Lett.* **26**, 489-490 (1974).
6. Marshall, A.G. *Acc. Chem. Res.* **9**, 307-316 (1996).
7. Comisarow, M.B. *J. Chem. Phys.* **69**, 4097-4104 (1978).
8. Comisarow, M.B. & Marshall, A.G. *J. Mass Spectrom.* **31**, 581-585 (1996).
9. Marshall, A.G. & Roe, D.C. *J. Chem. Phys.* **73**, 1581-1590 (1980).
10. Guan, S. *J. Am. Soc. Mass Spectrom.* **2**, 483-486 (1991).
11. Grosshans, P.B. & Marshall, A.G. *Anal. Chem.* **63**, 2057-2061 (1991).
12. Marshall, A.G., Wang, T.-C.L. & Ricca, T.L. *J. Amer. Chem. Soc.* **107**, 7893-7897 (1985).
13. Guan, S. & Marshall, A.G. *Int. J. Mass Spectrom. Ion Processes* **157/158**, 5-37 (1996).
14. Marshall, A.G. & Verdun, F.R. *Fourier Transforms in NMR, Optical, and Mass Spectrometry: A User's Handbook* 1-460 (Elsevier, Amsterdam, 1990).
15. Wang, M. & Marshall, A.G. *Anal. Chem.* **60**, 341-344 (1988).
16. Comisarow, M.B. & Melka, J. *Anal. Chem.* **51**, 2198-2203 (1979).
17. Marshall, A.G. *Chem. Phys. Lett.* **63**, 515-518 (1979).
18. Zhang, Z., Guan, S. & Marshall, A.G. *J. Am. Soc. Mass Spectrom.* **8**, 659-670 (1997).
19. Alber, G.M. & Marshall, A.G. *Appl. Spectrosc.* **44**, 1111-1116 (1990).
20. Williams, C.P. & Marshall, A.G. *Anal. Chem.* **61**, 428-431 (1989).
21. Williams, C.P. & Marshall, A.G. *Anal. Chem.* **64**, 916-923 (1992).

22. Meier, J.E. & Marshall, A.G. *Anal. Chem.* **62**, 201-208 (1990).
23. Meier, J.E. & Marshall, A.G. *Anal. Chem.* **63**, 551-560 (1991).
24. Loo, J.F., Krahling, M.D. & Farrar, T.C. *Rapid Commun. Mass Spectrom.* **4**, 297-299 (1990).
25. Guan, S. & Marshall, A.G. *Anal. Chem.* **69**, 1156-1162 (1997).
26. Craig, E.C., Santos, I. & Marshall, A.G. *Rapid Commun. Mass Spectrom.* **1**, 33-37 (1987).
27. Comisarow, M.B. & Marshall, A.G. *Can. J. Chem.* **52**, 1997-1999 (1974).
28. Marshall, A.G., Comisarow, M.B. & Parisod, G. *J. Chem. Phys.* **71**, 4434-4444 (1979).
29. Serreqi, A. & Comisarow, M.B. *Appl. Spectrosc.* **41**, 288-295 (1987).
30. Kim, H.S. & Marshall, A.G. *J. Mass Spectrom.* **30**, 1237-1244 (1995).
31. Hsu, A.T., Ricca, T.L. & Marshall, A.G. *Anal. Chim. Acta* **178**, 27-41 (1985).
32. Guan, S., Li, G.-Z. & Marshall, A.G. *Int. J. Mass Spectrom. Ion Processes* **000**, 0000-0000 (1997).
33. Guan, S., Marshall, A.G. & Scheppele, S.E. *Anal. Chem.* **68**, 46-71 (1996).
34. Cheng, X., *et al. J. Am. Chem. Soc.* **117**, 8859-8860 (1995).
35. Nawrocki, J.P., *et al. Rapid Commun. Mass Spectrom.* **10**, 1860-1864 (1996).
36. Fievre, A., Solouki, T., Marshall, A.G. & Cooper, W.T. *Energy & Fuels* **11**, 554-560 (1997).
37. Williams, E.R., Henry, K.D. & McLafferty, F.W. *J. Amer. Chem. Soc.* **112**, 6157-6162 (1990).
38. Savard, G., *et al. Phys. Lett. A* **158**, 247-252 (1991).
39. Schweikhard, L., Guan, S. & Marshall, A.G. *Int. J. Mass Spectrom. Ion Processes* **120**, 71-83 (1992).
40. Guan, S., *et al. Chem. Rev.* **8**, 2161-2182 (1994).
41. Speir, J.P., *et al. Anal. Chem.* **65**, 1746-1752 (1993).
42. Solouki, T., Marto, J.A., White, F.M., Guan, S. & Marshall, A.G. *Anal. Chem.* **67**, 4139-4144 (1995).
43. Guan, S. & Marshall, A.G. *Anal. Chem.* **69**, 1-4 (1997).
44. McLafferty, F.W. *Acc. Chem. Res.* **27**, 379-386 (1994).
45. Marshall, A.G., *et al. J. Am. Chem. Soc.* **119**, 433-434 (1997).
46. Senko, M.W., Hendrickson, C.L., Emmett, M.R., Shi, S.D.-H. & Marshall, A.G. *J. Am. Soc. Mass Spectrom.* **8**, 970-976 (1997).
47. Buchanan, M.V. in *ACS Symp. Series* 205 (American Chemical Society, Washington, D.C., 1987).
48. Asamoto, B. & Dunbar, R.C. *Analytical Applications of Fourier Transform Ion Cyclotron Resonance Mass Spectrometry* 1-306 (VCH, New York, 1991).
49. Pardue, H.L. *Analyt. Chim. Acta* **178**. 158 pp. (1985). (Special Issue: Fourier Transform Mass Spectrometry: Fundamental Aspects and Analytical Applications.)
50. Comisarow, M.B. & Nibbering, N.M.M. *Int. J. Mass Spectrom. Ion Proc.* **72**, 222 pp. (1986). (Special Issue: Fourier Transform Ion Cyclotron Resonance Mass Spectrometry)
51. Wilkins, C.L. *Trends in Analyt. Chem.* **13**, 223-251 (1994). (Special Issue: Fourier Transform Mass Spectrometry)
52. Marshall, A.G. *Int. J. Mass Spectrom. Ion Processes* **137/138**, 410 pp. (1996). (Special Issue: Fourier Transform Ion Cyclotron Resonance Mass Spectrometry)
53. Comisarow, M.B. in *Fourier, Hadamard, and Hilbert Transforms in Chemistry* (ed. Marshall, A.G.) 125-146 (Plenum, N.Y., 1982).
54. Marshall, A.G. *Acc. Chem. Res.* **18**, 316-322 (1985).
55. Freiser, B.S. in *Bonding Energetics in Organometallic Compounds* (ed. Marks, T.J.) 55-69 (Amer. Chem. Soc., Washington, D.C., 1990).
56. Marshall, A.G. & Schweikhard, L. *Int. J. Mass Spectrom. Ion Processes* **118/119**, 37-70 (1992).
57. Nibbering, N.M.M. *Analyst* **117**, 289-293 (1992).

58. Brenna, J.T., Creasy, W.R. & Zimmerman, J.A. *Amer. Chem. Soc. Symp. Ser.* **236**, 129-154 (1993).
59. Buchanan, M.V. & Hettich, R.L. *Anal. Chem.* **65**, 245A-259A (1993).
60. Guan, S. & Marshall, A.G. *Int. J. Mass Spectrom. Ion Processes* **146/147**, 261-296 (1995).
61. Marshall, A.G. in *Encyclopedia of Nuclear Magnetic Resonance* (eds. Grant, D.M. & Harris) 486-489 (Wiley, London, 1996).
62. Ross, C.W., III, Guan, S., Grosshans, P.B., Ricca, T.L. & Marshall, A.G. *J. Amer. Chem. Soc.* **115**, 7854-7861 (1993).
63. Li, G.-Z., Vining, B.A., Guan, S. & Marshall, A.G. *Rapid Commun. Mass Spectrom.* **10**, 1850-1854 (1996).

2. PLENARY LECTURES

Fourier Transform Infrared Spectroscopy as a Surface Science Technique

Hugo Celio and Michael Trenary

Department of Chemistry, University of Illinois at Chicago, 845 W Taylor St., Chicago, IL 60607-7061

A central goal of modern surface science is to obtain atomic and molecular level information about the structural and chemical properties of solid surfaces. For many, if not most, problems in surface science it is necessary to work under ultra high vacuum (UHV) conditions to obtain meaningful and reproducible results. A wide array of highly specialized and hence expensive UHV surface sensitive techniques have been developed to probe the gas-solid interface. Most of these techniques rely on the finite penetration depth of charge particles to achieve surface sensitivity. In contrast, surface sensitivity can also be achieved with reflection absorption infrared spectroscopy using unmodified low-cost commercial FTIR spectrometers. In this paper we show how a variety of problems in surface chemistry can be effectively addressed with FTIR spectroscopy.

INTRODUCTION

The chemical reactions that occur at the gas-solid interface are of central importance to a variety of technological areas. These areas include corrosion, lubrication, semiconductor processing, thin film growth, and heterogeneous catalysis. An atomic and molecular level understanding of the chemistry of gas-solid reactions has emerged in recent years through research employing the methods of modern surface science (1). Surface scientists seek a detailed understanding of the type of processes that can occur at the gas-solid interface by working under carefully controlled conditions using highly specialized techniques. Virtually all surface science techniques employ instrumentation that is developed and marketed solely for surface science studies. However, in recent years commercial FTIR spectrometers have been successfully incorporated into the arsenal of techniques used to study surface chemical reactions. In this paper, we consider the special requirements for surface science studies and how these requirements are being met with relatively low-cost general purpose FTIR spectrometers. We present several examples illustrating the capabilities and limitations of the technique. Finally, we discuss one method of potentially improving the sensitivity achievable with surface infrared spectroscopy.

Although much surface science research is motivated by important technological problems, the actual experiments are conducted under idealized conditions that are often quite remote from the underlying technological motivation. For example, the key chemical reactions associated with automobile catalytic converters occur under atmospheric pressures and at temperatures of a few hundred degrees on the surfaces of small platinum and rhodium particles embedded in an aluminum oxide matrix. Yet surface science studies of the chemical reactions related to catalytic converters are performed on large single crystals of platinum and rhodium under ultra high vacuum (pressures on the order of 1×10^{-10} torr). These conditions are needed in order to have control over the most important experimental variables, surface structure and composition. The use of single crystal samples oriented so as to expose particular crystallographic planes provides some control over the geometric arrangement of the surface atoms. Once a clean surface is prepared, UHV is needed in order to keep it free of contamination from the gas phase during the time period needed to study its properties. A useful rule of thumb is that at a gas pressure of 1×10^{-6} torr a surface will be covered with a monolayer of molecules in 1 second if the sticking coefficient is unity. Another source of surface impurities is segregation from the bulk, a process that can lead to a relatively high coverage of surface contaminants even when the bulk contaminants are present at undetectably low levels. The need to control surface structure and composition has led to the development of several specialized techniques. Low energy electron diffraction (LEED) is commonly used to verify that a single crystal surface is well-ordered with the expected periodicity. Auger electron spectroscopy (XPS), x-ray photoelectron spectroscopy (XPS), and ion scattering spectroscopy (ISS) are examples of techniques that can identify the presence of any of the elements on the surface, except hydrogen and helium. Techniques that provide information on surface structure and composition are necessary but not sufficient for investigating most chemical reactions, which usually involve a transformation from one molecular species to another with little change in the elemental composition. For example, the decomposition of ethylene on Pt(111) to surface carbon and hydrogen atoms occurs without any change in the coverage of carbon and hydrogen atoms but occurs through several molecular species containing both carbon and hydrogen. Vibrational spectroscopy offers one of the most effective ways of identifying molecular species on surfaces.

The two most common techniques used to obtain vibrational spectra of molecules absorbed on metal surfaces are high resolution electron energy loss spectroscopy (HREELS) (2) and reflection absorption infrared spectroscopy (RAIRS) (3). Although a variety of instrumental methods have been employed for RAIRS, currently most groups use commercial FTIR spectrometers. The strength and

CP430, *Fourier Transform Spectroscopy:* 11th International Conference
edited by J.A. de Haseth
 1-56396-746-4/98/$15.00

weaknesses of RAIRS are best illustrated by contrasting it with HREELS. In HREELS, a monoenergetic beam of electrons at a beam energy of a few eV is scattered from a sample and the scattered electrons are energy analyzed. Most electrons are scattered elastically but small peaks at lower energies are observed due to excitation of vibrations of adsorbed molecules or atoms. The resolution in HREELS is typically 40 cm^{-1} although technical refinements have led to the development of some instruments that are capable of resolutions of 10 cm^{-1}, or even slightly better. The spectral range with HREELS covers the region of 50-4000 cm^{-1} with a sensitivity that enables vibrations of most molecules to be measured at submonolayer coverages. Because the electrons exchange both energy and momentum with surface vibrations, HREELS can be used to measure the dispersion associated with surface phonons, a capability that RAIRS lacks. Like all techniques based on charged particles, HREELS can only operate under high vacuum conditions. And like most surface science instrumentation, HREELS spectrometers are built and marketed solely to the relatively small surface science community. This makes high performance HREELS spectrometers relatively expensive. In contrast, the use of FTIR spectrometers by surface scientists is a relatively small part of the FTIR market. Economies of scale and competition among the various manufacturers result in relatively low cost instruments as well as a much more broadly based development effort. The characteristics of surface vibrational spectroscopy with RAIRS are largely determined by the characteristics of FTIR spectroscopy in general. The instruments are capable of much higher resolution that is usually needed as most surface vibrational bands have intrinsic widths greater than 1 cm^{-1}. The spectral range is limited by the usual constraints of IR spectroscopy, i.e., by windows, IR optics, sources and detectors. The sensitivity is a crucial issue but most molecules have at least one vibrational band that is intense enough to detect at coverages between 0.1 and 1 monolayer. However, the sensitivity is still inadequate and even modest improvements in sensitivity would dramatically expand the type of surface problems that could be investigated with RAIRS. As an optical technique, RAIRS is not restricted to ultra high vacuum and many groups are attempting to investigate surface reactions under pressures approaching 1 atmosphere. For many applications, however, the most significant advantage of RAIRS over HREELS stems from the higher resolution, which is generally high enough that measured lineshapes are dominated by the intrinsic vibrational lineshapes rather than by the instrumental lineshape function. It is found that vibrational lineshapes vary dramatically with coverage and temperature and that the accurate measurement of such variations permits new surface phenomena to be studied that are inaccessible with other techniques. Another crucial advantage of high resolution is that it permits small shifts (less that 10 cm^{-1}) associated with isotopic substitution to be accurately measured. Such shifts can provide crucial information about the identity of surface intermediates formed in the course of a surface reaction.

The physical principles of the RAIRS technique were first established by Greenler (4). He showed that the change in reflectivity due to a thin layer of IR absorbing material on top of a metal substrate has a sharp maximum for grazing incidence, where the angle between the surface normal and the IR beam is about 85°. The magnitude of the expected reflectivity change as well as the optimum angle of incidence is dependent on which metal is used for the substrate. At grazing incidence the electric field of the IR beam that is parallel to the surface (perpendicular to the plane of incidence) is vanishingly small for metals and this leads to a highly restrictive surface IR selection rule. This rule states that only vibrational transitions that have a dynamic dipole moment with a component oriented along the surface normal will be IR active. This rule is best interpreted in terms of the symmetry properties of the molecule through the use of group theory (5). The rule is thus equivalent to stating that only vibrational modes that transform according to the molecules totally symmetric vibrations will have surface IR active fundamentals. For molecular adsorbates with a high degree of symmetry, this rule leads to considerably fewer IR active vibrations than associated with the corresponding gas phase molecule. A practical consequence of using grazing incidence with small single crystal samples is that the projection of the focused image of the IR source at the crystal is considerably larger than the sample itself. The sample thus acts like a throughput limiting aperture restricting the amount of infrared radiation that reaches the detector to only a few percent of the value it would otherwise have. This has several practical consequences. First, the expectation that globar sources operating near 1500 K will saturate sensitive IR detectors does not apply. Secondly, strategies for improving the sensitivity of FTIR that involve increasing the throughput of other components of the optical path will not be effective. Thirdly, although the absolute intensity of the infrared bands observed in RAIRS would ordinarily be high enough to achieve high signal-to-noise ratios, the much higher noise levels associated with severely restricted throughput seriously limits the sensitivity achievable. Finally, strategies for improving the signal-to-noise levels in RAIRS are somewhat different from standard FTIR usage. For example, we have found that a much higher temperature source leads to improved sensitivity in the RAIRS experiment although it would saturate the detector in most FTIR applications.

EXPERIMENTAL

Figure 1 shows a diagram of the apparatus used for the RAIRS experiments described in this paper. It consists of an ultrahigh vacuum chamber with a base pressure of less than 1×10^{-10} torr coupled to a commercial FTIR. The UHV chamber is equipped with instruments used with several surface science techniques. These include low energy electron diffraction, Auger electron spectroscopy, and temperature programmed desorption with a quadrupole mass spectrometer. The sample manipulator enables the single crystal samples to be heated to above 1100 K and cooled to 81 K. The sample temperature is measured with a type K thermocouple. The collimated IR beam from the FTIR is focused onto the crystal with an off-axis paraboloidal mirror with an effective focal length of 260 mm, recollimated with a second 260 mm focal length off-axis paraboloidal mirror and focused onto the detector with a third off-axis paraboloidal mirror. An IR polarizer is usually placed in front of the detector. The entire optical path is purge with dry air.

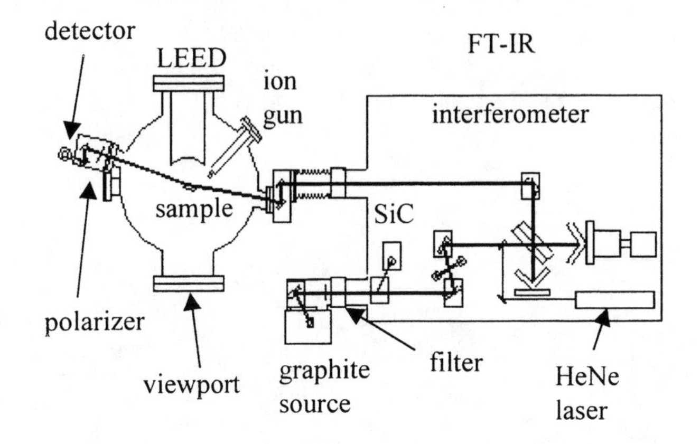

FIGURE 1. Apparatus for reflection absorption infrared spectroscopy consisting of a stainless steel ultra high vacuum chamber coupled to a commercial FTIR spectrometer.

The IR windows use salt crystals sealed with viton O-rings into special flanges that attach to the UHV chamber with copper gaskets. Although a variety of different salt crystals including KRS-5 (thalium bromoiodide), CsI and KBr have been used in the past, KBr is most commonly used. A variety of infrared detectors have been used including narrow and wide band mercury cadmium telluride detectors, an indium antimonide photovoltaic detector and a Si bolometer. Because the bolometer is a thermal detector it requires relatively slow mirror speeds whereas the photoconductive MCT and photovoltaic InSb detectors are best used with very fast mirror speeds. Typically a resolution of 2 cm^{-1} with triangle apodization was used and spectra were processed with zero-filling to yield eight data points per cm^{-1}. Gases were admitted to the UHV chamber through a leak valve and exposure amounts are given in units of Langmuir (L), with 1 L = $1x10^{-6}$ torr sec. In all cases the spectra are obtained by recording a background spectrum of the clean surface, exposing the surface to the gas of interest and immediately acquiring a sample spectrum using the same number of scans as for the background. The ratio of the sample and background spectra then yields the transmittance spectrum, which is equivalent to the reflectivity of the adsorbate/surface divided by the reflectivity of the clean surface. The ability to measure a spectrum of the clean and adsorbate covered surfaces under exactly the same conditions and as close in time as possible is essential for the successful measurement of the surface IR spectra.

RESULTS AND DISCUSSION

A central issue in RAIRS is the sensitivity achievable with the technique. However, this is highly dependent on the mode and molecule involved as well as on the type of information desired. The symmetric PF stretch of PF_3 chemisorbed on Pt(111) provides an example where extremely high signal-to-noise ratios can be achieved. The molecule also nicely illustrates the use of symmetry in interpreting the spectra in terms of the surface IR selection rule. We have explored the vibrational properties of PF_3 on Pt(111) in several previous studies (6-11). At a coverage of 0.33 monolayer and a temperature of 82 K we observe two bands at 951 cm^{-1} and 562 cm^{-1}. Both bands are quite sharp at 82 K and at a saturation coverage of 0.33 monolayer; the symmetric PF stretch at 951 cm^{-1} has a FWHM of 1.1 cm^{-1} and the symmetric PF_3 deformation at 562 cm^{-1} has a FWHM of 0.7 cm^{-1}. These FWHM were measured using an instrumental resolution of 0.25 cm^{-1}. Because the PF stretch band has such a high absolute IR intensity and because it is so sharp, when measured with p-polarized light, 0.25 cm^{-1} resolution, at a temperature of 81 K and at a coverage of 0.33 monolayer, it has a peak height equivalent to a transmittance of 40%. At low PF_3 coverages on Pt(111) the higher frequency band occurs at 899 cm^{-1} and the lower frequency band at 540 cm^{-1}. The frequencies of these fundamentals on Pt(111) can be compared with the fundamentals of gas phase PF_3 (12,13) and $Pt(PF_3)_4$ (13) molecules. Gas phase PF_3 is a pyramidal molecule with C_{3v} symmetry (12,13). It has a symmetric P-F stretch (892 cm^{-1}) and symmetric PF_3 bend (485 cm^{-1}) of A_1 symmetry and an asymmetric P-F stretch (860 cm^{-1}) and PF_3 bend (344 cm^{-1}) of E symmetry. The low coverage 899 cm^{-1} band on Pt(111) is much closer to the symmetric P-F stretch in $Pt(PF_3)_4$ at 903 cm^{-1} than to the asymmetric stretch at 864 cm^{-1}. Similarly, the 540 cm^{-1} band on Pt(111) is closer to the symmetric PF_3 bend at 513 cm^{-1} in $Pt(PF_3)_4$ than to the asymmetric bend at 384 cm^{-1}. Frequencies alone therefore support the conclusion that the two bands observed on Pt(111) are the symmetric stretch and bend of PF_3. The close correspondence in frequencies between PF_3/Pt(111) and $Pt(PF_3)_4$ indicates the same local bonding geometry, i.e., that PF_3 is bonded to a single platinum atom (as opposed to bridge bonded) through the phosphorus end of the molecule. The fact that only the symmetric vibrations are observed indicates that the molecule has local C_{3v} symmetry with the three-fold axis oriented along the surface normal.

A closer inspection of the P-F stretch region demonstrates just how completely the asymmetric vibrations are hidden from observation. Figure 2 shows the symmetric P-F stretch obtained with a signal-to-noise ratio of about 1000:1. The x10 expansion of the baseline reveals a peak which is at most a factor of 500 less than the symmetric stretch in intensity. This feature may or may not correspond to the asymmetric stretch. This 500 fold or more enhanced sensitivity for the symmetric stretch over the asymmetric stretch is in contrast to the gas phase infrared spectrum of $Pt(PF_3)_4$, where the asymmetric and symmetric P-F stretches are of equal intensity. This result demonstrates the very high discrimination against fundamentals which must, by symmetry, have a dynamic dipole moment oriented strictly parallel to the surface.

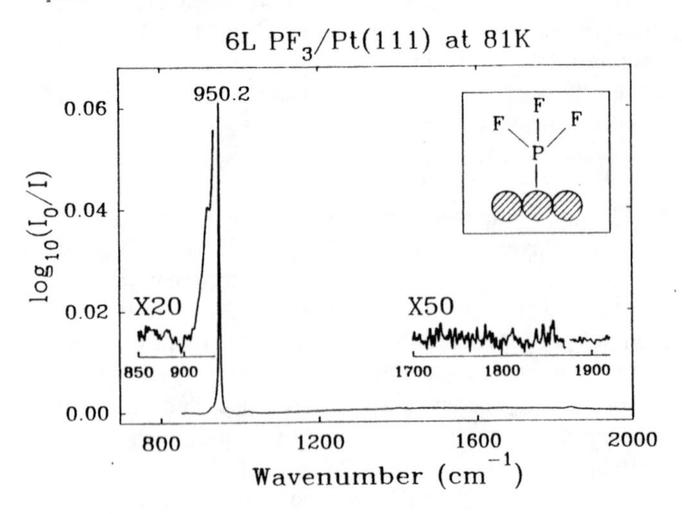

FIGURE 2. RAIRS spectrum of the PF symmetric stretch region obtained at 81 K and at a PF_3 coverage of 0.33 monolayer. The intensity scale is equivalent to absorbance.

The high SNR obtainable for PF_3/Pt(111) suggest that it may be possible to observe overtones or combination bands. The symmetric stretch overtone is of A_1 symmetry and the asymmetric stretch overtone has a component of A_1 and a component of E symmetry. Thus both overtones have the correct symmetry to be surface IR active. In gas phase PF_3, the symmetric stretch overtone is at 1781 cm^{-1}, 3 cm^{-1} less than twice the fundamental. The asymmetric stretch overtone is at 1713 cm^{-1}, 7 cm^{-1} less than twice the fundamental. For gas phase PF_3, the ratio of the intensities of these overtones relative to the intensity of the P-F symmetric stretch is 1:500 for the symmetric stretch overtone and 1:100 for the asymmetric overtone. Shown in Fig. 2 is an expanded scale in the region 1700-1920 cm^{-1} where we would expect to observe overtones of both the symmetric and asymmetric P-F stretch modes. The region below 1870 cm^{-1} was obtained with a narrow band mercury cadmium telluride detector while the region above 1870 was obtained with a InSb detector. The noise in this region is low enough that the overtones should be observable if they have the same intensity relative to the symmetric stretch fundamental as in gas phase PF_3. On Pt(111) the intensities of these overtones are reduced by at least factors of 5 and 2 relative to their gas phase intensities. Thus even for this unusually favorable case, only the first order transitions, the fundamentals of A_1 symmetry are observable with the current sensitivity limits of RAIRS.

The high intrinsic intensity of the symmetric PF stretch, the fact that its frequency matches the region of maximum sensitivity of an MCT detector and the fact that it forms an ordered overlayer that yields extremely sharp IR bands, are all factors that enable this band to be observed with very high signal-to-noise ratios. An example of a band that is extremely difficult to measure is provided by the Pt-C stretch of CO on Pt(111) shown in Fig. 3. This band occurs below the frequency range of MCT detectors and was therefore measured with a liquid helium cooled Si bolometer. The spectrum shown was obtained by signal averaging for over 61 minutes each for both the sample and the background. Not only is this vibration extremely weak, but is occurs in a difficult spectral region. Furthermore, the spectrum shown was obtained at a coverage and temperature where the band is narrowest and therefore easiest to measure. At both higher and lower coverages the band is considerably broader and just barely detectable above the noise level (14). Similarly, the band also broadens in response to increases in temperatures to the point where the SNR is only about 2:1 at 300 K.

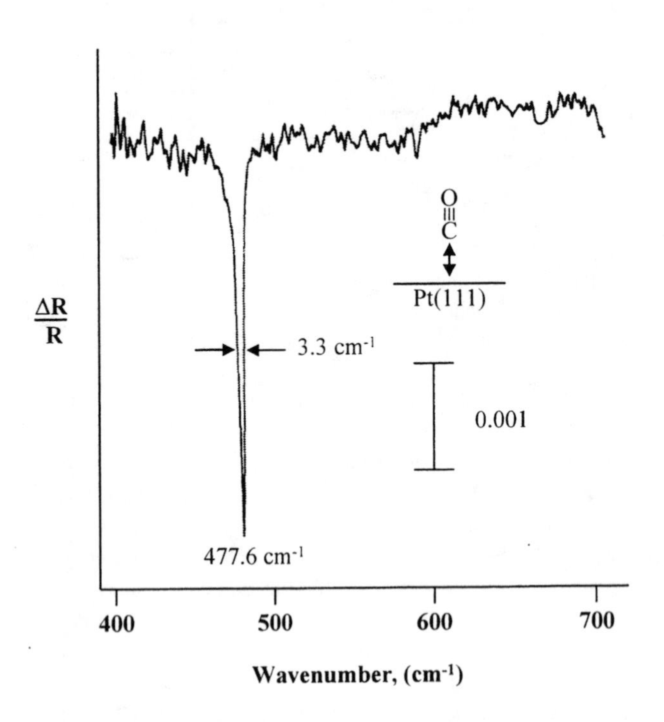

FIGURE 3. RAIRS spectrum of the C-Pt stretch for CO on Pt(111) at a temperature of 81 K and a CO coverage of 0.19 monolayer.

The examples of the symmetric PF stretch of PF_3 and the Pt-C stretch of CO on Pt(111) represent two extremes in terms of the sensitivity of the RAIRS technique. We have successfully studied a large number of systems where the sensitivity requirements fell between these two extremes. Yet every RAIRS study would benefit from higher sensitivity as it would permit observation of more allowed fundamentals as well as permit identification of surface species present at much lower coverages. In an effort to increase our sensitivity we have constructed a graphite source enclosed in a evacuable housing that can be operated at 2500-3000 K (15). This source is attached externally to the FTIR as shown in Fig. 1.

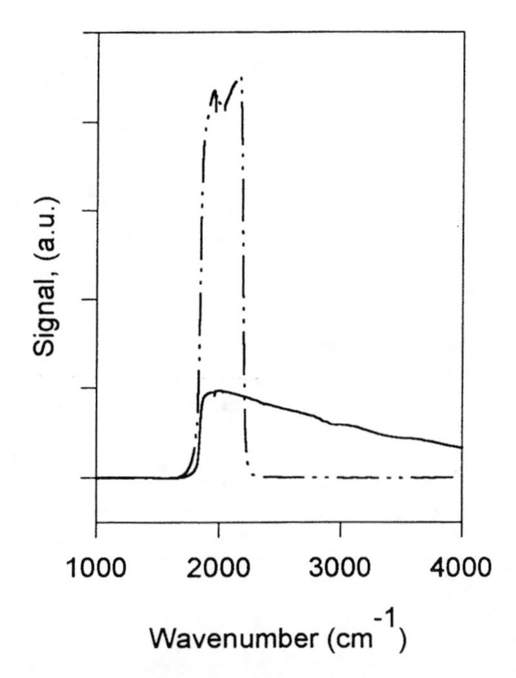

FIGURE 4. Single beam spectra comparing the IR intensity near 2000 cm^{-1} from a standard SiC source at 1500 K (solid line) and a graphite source (dashed line) at 2673 K. The low wavenumber cutoff at 1850 cm^{-1} is due to the InSb detector and a low pass filter was used with the graphite source.

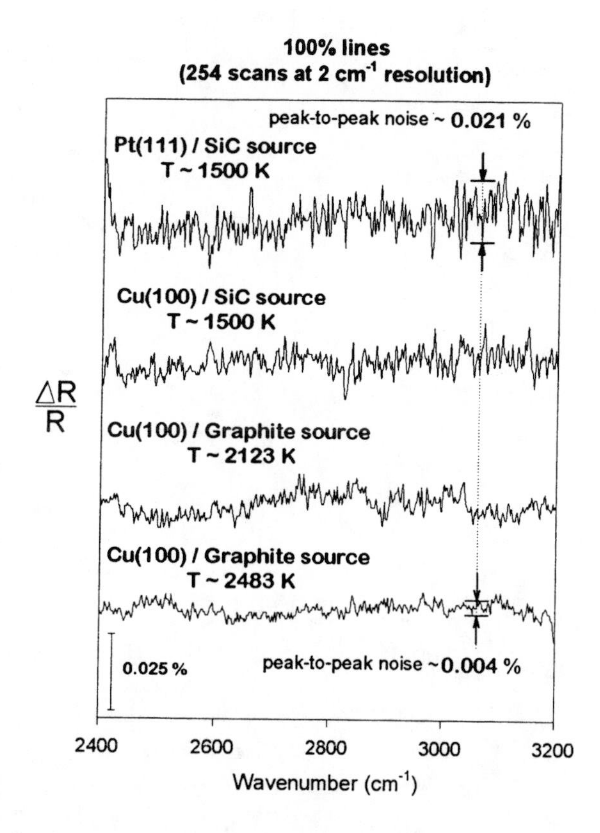

FIGURE 5. 100% lines comparing noise levels obtained with a Pt(111) sample and a larger Cu(100) sample using both a 1500 K SiC source and a graphite source at higher temperatures.

Figure 4 shows a comparison of single beam spectra obtained with both the graphite source and the standard SiC source operated at 1500 K. The sample is a Cu(100) crystal with a rectangular shape chosen to improve the throughput of the system. The spectra were obtained with an InSb detector with a low wavenumber cutoff of 1850 cm^{-1}. Because of detector saturation effects associated with the graphite source, a low

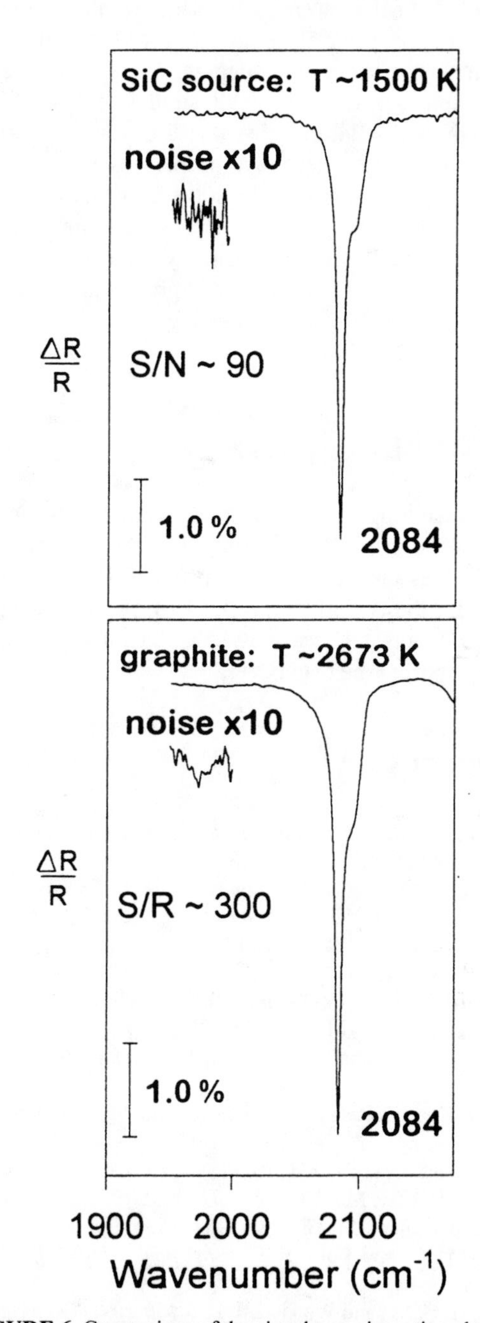

FIGURE 6. Comparison of the signal-to-noise ratios obtained with the graphite source and the SiC source for the C-O stretch of 0.5 monolayer of CO on Cu(100) obtained with 1 interferometer scan (0.25 sec) and 2 cm^{-1} resolution.

pass filter was used. Figure 4 demonstrates the considerably higher intensity of IR radiation available with the higher temperature source. Figure 5 shows a comparison of 100 %

lines showing how lower noise levels can be obtained by both using the higher temperature source as well as the Cu(100) sample which has a higher throughput than the Pt(111) crystal. The improved signal-to-noise level achievable with the higher temperature source is illustrated by the results shown in Fig. 6 for the C-O stretch of CO on the Cu(100) surface.These results were obtained with a single interferometer scan requiring about 0.25 sec. The high SNR achieved in such a short time suggests that this source is already suitable for time resolved studies. However, it currently shows less stability than the SiC source making signal averaging over extended time periods impractical. Once the stability problems can be solved, signal averaging for extended periods with the high temperature source should result in considerably higher sensitivity than is currently possible with RAIRS. This in turn will greatly improve the capabilities of RAIRS for investigating surface chemical phenomena.

ACKNOWLEDGEMENTS

This work was supported by a grant from the National Science Foundation (CHE 9616402). The PF_3 spectrum is from the PhD thesis work of Jingfu Fan and the Pt-CO stretch is from the PhD thesis work of Igor J. Malik performed at the University of Illinois at Chicago. We thank Craig L. Perkins for help in preparing the manuscript.

REFERENCES

1. Somorjai, G. A., *Introduction to Surface Science and Catalysis*, New York: Wiley, 1994.
2. Ibach, H. and Mills, D. L., *Electron Energy Loss Spectroscopy and Surface Vibrations*, New York: Academic Press, 1982.
3. Chabal, Y. J., *Surf. Sci. Rep.* 8, 214 (1988).
4. Greenler, R. G., *J. Chem. Phys.* 44, 310 (1966).
5. Fan, J., and Trenary, M., *Langmuir* 10, 649 (1994).
6. Liang, S. and Trenary, M., *J. Chem. Phys.* 89, 3323 (1988).
7. Agrawal, V. K. and Trenary, M., *J. Vac. Sci. Technol. A* **7**, 2235 (1989).
8. Agrawal, V. K. and Trenary, M., *J. Chem. Phys.* 95, 6962 (1991).
9. Zhou, Y., Jennings, G., Campuzano, J. C., Agrawal, V. K. and Trenary, M., *Surf. Sci.* 268, 378 (1992).
10. Fan, J., and Trenary, M., *J. Vac. Sci. Technol.* 10, 2576 (1992).
11. Fan, J., and Trenary, M., *Surf. Sci.* 282, 76 (1993).
12. Wilson, M. K., and Polo, S. R., *J. Chem. Phys.* 20, 1716 (1952).
13. Edwards, H. G. M., and Woodward, L. A., *Spectrochimca Acta*, 26A, 897, (1970).
14. Malik, I. J., and Trenary, M., *Surf. Sci.* 214, L237 (1989).
15. Celio, Hugo, PhD Thesis, University of Illinois at Chicago, 1997.

Emission Spectroscopy and Molecular Astronomy

P.F. Bernath

Department of Chemistry, University of Waterloo, Waterloo, Ontario, N2L 3G1, Canada
and
Department of Chemistry, University of Arizona, Tucson, Arizona, 85721, USA

Although emission spectroscopy is common in the visible and ultraviolet regions, the technique has been neglected in the infrared and far infrared. Fourier transform emission spectra of infrared electronic transitions, vibration-rotation bands and pure rotational transitions will be presented. The molecules of interest range from diatomics such as LiH and FeF to large molecules such as polycyclic aromatic hydrocarbons (PAHs), C_{60} and DNA bases. Even at long wavelengths in the far infrared region excellent spectra of hot molecules could be recorded. One of the primary applications of our laboratory emission spectra has been the assignment of astronomical spectra of objects such as the sun, sunspots, carbon stars and planetary nebulae. The discovery of hot water vapor in sunspots and the origin of the "unidentified infrared emission bands" will be discussed. Finally, some spectra obtained with a cryogenic infrared echelle spectrograph will be compared with spectra from a Fourier transform spectrometer.

INTRODUCTION

For the visible and ultraviolet regions, it is well-known that emission spectroscopy is more sensitive than absorption spectroscopy because emission spectra have "zero background." In an ideal absorption experiment, the noise arises mainly from shot-noise from the background continuum. By comparison, in an emission experiment, the noise is reduced because the continuum is absent. Of course the total signal level is also reduced in an emission experiment compared to an absorption experiment but the overall signal-to-noise ratio is increased because the noise has declined much more than the signal. For example, the violet color often seen in a flame is due to the easily detected $A^2\Delta \rightarrow X^2\Pi$ emission of the CH molecule. The detection of the corresponding absorption due to the CH molecule in a flame is possible but constitutes a much more difficult experiment (1).

What is surprising to many people is that the advantages of emission spectroscopy persist into the infrared. Infrared spectrometers are routinely operated in absorption, not emission. Infrared emission spectroscopy requires that the sample be at a different temperature (higher or lower) than the spectrometer. Higher temperatures are more favorable since the emitted power increases strongly with temperature. Infrared emission spectroscopy is plagued by thermal blackbody emission from the spectrometer itself as well as from the furnace or electrical discharge used to heat the sample. This blackbody emission is a continuum that provides nothing but noise. The Planck function (2) is a law of nature that can never be entirely avoided in the infrared. By limiting the field of view with cold apertures and by good optical design (e.g. by avoiding having the spectrometer look directly at a hot surface), as well as by limiting the spectral range of the detectors with cold filters, the effects of unwanted blackbody emission can be minimized.

We discovered the power of infrared emission spectroscopy by serendipity. At the request of the astronomer J. Keady we were trying to record an absorption spectrum of SiS(3) at 750 cm^{-1} to match molecular features in an infrared spectrum of a carbon star. The experimental plan was simple: heat Si powder and SiS_2 solid to 1000°C to make SiS by the reaction:

$$Si(s) + SiS_2(s) \rightarrow 2SiS(g)$$

The hot molecules would then be detected by absorption of infrared light from a glower. Much to our amazement a very high quality emission spectrum was recorded in only 14 minutes of data collection.

In 1989, when our SiS experiments were carried out with the McMath-Pierce Fourier transform spectrometer of the National Solar Observatory in Tucson, Arizona, there were scarcely any published examples of high resolution emission spectra at long wavelengths. Apart from a relatively low resolution emission spectrum of GeS(4), only the infrared emission spectrum of the FO molecule was available in the literature (5). The FO molecule was produced by the very exothermic reaction:

$$F + O_3 \rightarrow FO + O_2$$

and for some time we were convinced that an energetic reaction was necessary to see emission. Soon we recorded the thermal emission spectrum of BeF_2, proving that no chemical reactions are necessary to provide additional vibrational excitation (6). I have written a review article on the virtues of infrared emission spectroscopy (7).

Thermal emission spectra are, in fact, used to monitor atmospheric molecules such as O_3 in the stratosphere both from balloon platforms (8) and satellites (9). Thin films of solids and liquids also give excellent thermal emission spectra (10). Of course, spectra of materials in condensed phases are at low resolution.

CP430, *Fourier Transform Spectroscopy:* 11th International Conference
edited by J.A. de Haseth

The main application of our infrared emission spectroscopy is in molecular astronomy. In recent years astronomers have focussed their attention on "cool" sources (T=3-3000 K) such as dark molecular clouds, comets, planetary nebulae, brown dwarfs and stellar atmospheres. In contrast to "hot" stellar environments that are dominated by atoms, these cooler sources are abundantly endowed with molecules and solids (11). The temperatures achieved in our laboratory furnaces are appropriate for comparison with cool stellar atmospheres.

In astronomy, molecular observations provide a great deal of useful information about physical properties and chemical composition. Molecular spectra provide diagnostics that probe the temperature, pressure and composition.

Although the first detections of molecules in extraterrestrial environments were carried out in the visible region (12), the majority of recent discoveries have come in the millimeter wave region (13). The use of sensitive array detectors has, however, revitalized visible spectroscopy and infrared array detectors are starting to have an impact in infrared astronomy. Traditionally infrared astronomy has been carried out with Fourier transform spectrometers but these instruments are being replaced by cryogenic echelle spectrographs. As infrared and submillimeter wave astronomy expands, molecules will play an increasingly important role in astrophysics.

In this paper, recent work in our laboratory on the near infrared, infrared and far infrared spectroscopy of astrophysical molecules is discussed. The electronic spectra of CrH, TiO and FeF near 1 μm were analyzed. Vibration-rotation emission spectra of HF, SiS, SiO, H_2O and CH_4 were measured. For large molecules, the vibration-rotation bands of C_{60}, C_{70}, three DNA bases (uracil, thymine and adenine) and three PAHs (naphthalene, pyrene and chrysene) were recorded in the gas phase. The laboratory measurements for TiO and SiO were combined with data from sunspot spectra to provide improved spectroscopic constants. The hot H_2O emission spectra allowed the identification of H_2O absorption lines in the infrared spectra of sunspots. In the far infrared region pure rotation spectra of LiH and vibration-rotation spectra of PAHs and DNA bases were recorded.

Near Infrared Electronic Transitions

Most people find the idea of infrared electronic transition to be almost a contradiction, since most stable molecules typically have electronic transitions in the ultraviolet region. Many transient molecules such as ions or free radicals, however, have visible or infrared electronic transitions because of the presence of unpaired electrons in low-lying orbitals. While these reactive molecules cannot be purchased from a chemical supply company, they are, in fact, very common in energetic environments such as the atmospheres of stars (11).

The near infrared spectra of cool stars show strong absorption bands of metal hydrides and metal oxides. The CrH molecule has been identified in sunspots (14) in and S stars (15) through the $A^6\Sigma^+ - X^6\Sigma^+$ transition near 861 nm (16).

We have recorded an improved spectrum of the 0-0 band of the $A^6\Sigma^+ - X^6\Sigma^+$ transition of CrH (17) and CrD (18) using a hollow cathode lamp and a Fourier transform spectrometer. With these improved line positions, L. Wallace has looked in a new sunspot spectrum (19) and found no convincing evidence for the presence of CrH. In addition to this near infrared data, L. Wallace of Kitt Peak National Observatory has prepared new infrared sunspot and photospheric solar atlases (20).

In the course of hunting for CrH lines in the sunspot spectra, new lines of TiO were identified. The δ system of TiO ($b^1\Pi - a^1\Delta$) has a 0-0 band near 886 nm. This transition is well known in the spectra of M stars (21), but has not been previously noted in sunspots. In fact, the near infrared bands of TiO, VO and FeH are used in the detailed classification of M stars.

We measured new laboratory hollow cathode spectra of the 0-0 and 1-1 bands of the $b^1\Pi - a^1\Delta$ transition of TiO. The laboratory data and the sunspot lines were combined to provide improved molecular constants (19). In addition to the δ system of TiO, a few weak lines of the φ system ($b^1\Pi - d^1\Sigma^+$) were also identified in the sunspot spectra at 1.1 μm.

The surprising discovery of rotational transitions of AlCl, KCl, NaCl and AlF in the carbon-rich object IRC+10216 (22) suggests that metal halide molecules may be of astrophysical importance. Iron-containing compounds are particularly favorable for detection in stellar atmospheres because of the relatively high cosmic abundance of iron. The presence of the $F^4\Delta - X^4\Delta$ transition of FeH in sunspots and cool stellar spectra suggests that FeF might also be detectable in certain objects. Although ^{19}F is a fragile nucleus, it has an enhanced abundance in AGB stars (23).

There is a strong correspondence between the spectra of transition metal monofluorides and monohydrides so that a near infrared $^4\Delta - {}^4\Delta$ transition of FeF is expected, by analogy with FeH. We have recorded the 0-1, 0-0, 1-0 and 2-0 bands of the $g^4\Delta - a^4\Delta$ transition of FeF in the 9000 to 12000 cm^{-1} region (24). The FeF molecule was made in a carbon tube furnace by reaction of liquid Fe with CF_4 gas. Near infrared emission from the furnace was recorded with the Fourier transform spectrometer at the National Solar Observatory.

Vibration-Rotation Emission Spectroscopy

The element fluorine is difficult to detect in extraterrestrial sources. The ^{19}F nucleus is easily destroyed by nucleosynthetic reactions in stellar interiors so its abundance is low. The detection of fluorine is also difficult because most of the strong atomic lines, including the resonance lines, are found in the vacuum ultraviolet region.

In cooler sources the vibration-rotation transitions of HF near 2.5 μm are the most convenient and reliable lines for F abundance determinations. Recently the HF vibration-rotation lines were used by Jorissen *et al.* (23) to measure ^{19}F abundances in a large sample of cool stars. They found that AGB stars have up to 30 times the solar abundance of F, suggesting that ^{19}F is synthesized in He-burning shells.

Curiously there were no laboratory measurements for the highly-excited transitions of HF seen in stellar atmospheres. We measured the vibration-rotation as well as the pure rotational transitions of HF and DF at temperatures of up to 2600 K (25) in a furnace. These emission spectra were recorded by accident in the course of our work on the infrared spectra of metal fluorides.

Vibration-rotation emission spectroscopy above 1800 cm^{-1} is not particularly rare, even for transient molecules such as CH (26). This magic number of 1800 cm^{-1} (5.5 μm) is the band gap of the InSb detector. The InSb detector has nearly unit quantum efficiency and is the best infrared detector but it does not work for wavenumbers below 1800 cm^{-1}. The detection of emission at longer wavelengths is thus more difficult but excellent spectra can be recorded. For example, the emission due to AlH molecules were recorded at a resolution of 0.005 cm^{-1} with a signal-to-noise ratio of more than 150 for the strong lines (27). The measurement precision of 0.0002 cm^{-1} (6 MHz) is remarkable for a transient molecule.

Emission spectroscopy will work at even longer wavelengths as illustrated with the SiS spectrum (3). The 1-0 vibration-rotation lines of SiS have been detected in absorption in the source IRC+10216 (28). We recorded the laboratory vibration-rotation emission spectrum of SiS at 13 μm. The SiS molecules were made at 1300 K by the reaction of SiS_2 solid with Si powder in a tube furnace. Comparison of the laboratory and astronomical infrared spectra confirmed the SiS identification in IRC+10216.

The original discovery and preparation of the C_{60} molecule (29,30) was motivated by astronomical considerations. Now that laboratory spectra of C_{60} were not available, it is interesting to carry out an astronomical search for C_{60}. Although the infrared spectra of solid C_{60} are well known, the low temperature gas phase spectra are not available. We, therefore, recorded the infrared emission spectra of gaseous C_{60} and C_{70} as a function of temperature and extrapolated the band positions to low temperatures (31). For large molecules such as C_{60} excellent infrared emission spectra can be recorded, although the rotational structure is not resolved.

The strong C_{60} band at 1191 cm^{-1} is suitable for an astronomical search. We were unable to detect this band in IRC+10216 (32) at Kitt Peak. Other pure carbon moelcules such as C_3 (33) and C_5 (34) have been identified in IRC+10216 by infrared spectroscopy at 5 μm. However, the presence of large amounts of H_2 in this object may inhibit the chemical production of C_{60}. More favorable sources for C_{60} production may be R Coronae Borealis (R CrB) stars that are carbon rich but hydrogen deficient.. However, a search for C_{60} in three R CrB stars with a cryogenic echelle spectrograph also proved negative (35).

The sun and, particularly, sunspots are rich sources of astronomical spectra of molecules. For example in the 10 μm spectrum of a sunspot there is a dense forest of lines which can be assigned to SiO. In our laboratory a new SiO vibration-rotation emission spectrum was recorded by accident (36) during work on GaD. These laboratory measurements of SiO were combined with the sunspot data to provide a very extensive new set of line positions and molecular constants (37).

In the infrared sunspot atlas (20) numerous unidentified lines are present throughout the infrared region. We suspected that these lines were due to hot H_2O but the available laboratory data were inadequate to confirm this. We, therefore, recorded new laboratory emission spectra of hot water at 1800 K in the infrared region. Comparison of the laboratory emission spectra of H_2O and the sunspot absorption spectra identified most of the unassigned sunspot lines (38) as H_2O lines ("Water on the Sun").

Only a small fraction of the new water lines could be assigned quantum numbers. The water molecule is so light that the traditional power series expressions for the rotational energy levels are divergent. This massive breakdown of perturbation theory makes the assignment of quantum numbers to the lines very difficult. We have, therefore, begun a collaboration with the theoreticians O. Polyansky and J. Tennyson to apply more sophisticated approaches to the assignment of our spectra. Some preliminary assignments have already been made (39-41) and recently we have made a breakthrough in the prediction of the spectra of hot water.

The hot water lines in the laboratory and in the sunspot have been assigned by direct calculation of the infrared spectra starting with an *ab initio* potential energy surface (42). This surface was partially corrected for the breakdown of the Born-Oppenheimer approximation and used in a variational calculation of the energy levels. Although the predicted line positions were still in error, they were close enough to make the necessary assignments (43).

The spectrum of hot water is of enormous importance in molecular astronomy. Water is the main source of infrared opacity in all oxygen-rich cool objects. It is found in M stars and is particularly prominent in Mira variables (44). New results from the ISO satellite (45) are showing widespread water adsorption and emission in many objects. One result of our work will be improved infrared opacities (46) for modelling cool stellar atmospheres.

Hot water is also a primary product of the combustion of hydrocarbons. It can be detected remotely in the Fourier transform emission spectra of forest fires (47) and, for example, from an oxyacetylene torch (48). Other applications include the simulation of rocket plumes.

In the spectra of substellar objects such as brown dwarfs (e.g., Gl 229B) the infrared spectra of both H_2O and CH_4 are present (49). We have also recorded a new spectrum of hot CH_4 in the 3 μm region at the request of T. Oka in order to assign lines detected in the impact of comet Shoemaker-Levy with Jupiter.

The infrared spectra of planetary nebulae such as NGC 7027 and H II regions show unidentified infrared emission bands (UIRs) at 3.3 μm, 6.2 μm, 7.7 μm, 8.7 μm and 11.3 μm. The PAH hypothesis (50) attributes these bands to infrared emission from PAH molecules excited by strong UV radiation. These PAHs may be in the form of a gaseous, partly-ionized, partly-dehydrogenated mixture of species. The PAH hypothesis was based on the similarity between the characteristic infrared absorption bands of solid PAH molecules in KBr pellets and the UIRs. A more appropriate comparison would be between the emission spectra of gaseous PAH molecules and the UIRs.

Kurtz (51) and Joblin *et al.* (52) have recorded some PAH emission spectra and we have extended these measurments into the far infrared region (53). We find that even in the far infrared excellent emission spectra can be recorded for naphthalene, pyrene and chrysene. These experimental measurements are in excellent agreement with the *ab initio* predictions of Langhoff (54). In contrast to the mid-infrared spectra of PAH molecules, the far infrared spectra are unique for each PAH molecule. In fact the far infrared region provides a "spectroscopic fingerprint" for each of the different PAHs. Similar spectra were recorded for the DNA bases uracil, thymine and adenine (55).

Rotational Emission Spectra

Small gas phase molecules with large rotational constants have far infrared or even mid-infrared pure rotational spectra. For example, the hot water spectrum at 10 µm is essentially a pure rotational spectrum (38) with only a few lines from the ν_2 bending mode. Light transient molecules such as LiH and LiD have far infrared rotational spectra that can be measured in absorption (56) or emission (57).

The LiH and LiD molecules were made in a stainless steel tube furnace by the reaction of Li vapor with H_2 or D_2 at 1050°C. Excellent emission spectra were recorded (57) and the range of v and J was extended compared to the previous absorption measurements (56). Even the ^{6}Li-containing isotopomers could be detected in natural abundance. These spectra were recorded both to show feasability and to explore the effects of Born-Oppenheimer breakdown.

Future Prognosis

The improvement in sensitivity obtained by working in emission rather than in absorption with high-temperature molecules, ions and free radicals is typically a factor of 10 in the mid-infrared region. At the moment, the instrument of choice is the high-resolution Fourier transform spectrometer. For astronomical applications for which the light levels are very low, the Fourier transform spectrometer is inferior in sensitivity to the cryogenic echelle spectrograph.

A cryogenic echelle spectrograph is a compact spectrometer that uses an echelle grating in high orders to obtain high resolution spectra. The entire instrument is cooled to 50 K with a helium refrigerator to eliminate most of the background thermal emission from the spectrometer itself. The infrared radiation is measured with a large format (1024x1024) array of InSb detectors. Many of these spectrometers are under construction or have already been built at observatories around the world. The anticipated performance for an instrument called Phoenix at Kitt Peak National Observatory is spectacular. Phoenix has a theoretical resolving power in excess of 100,000 at 2000 cm^{-1} (a resolution of 0.02 cm^{-1}) with a sensitivity 100 times higher than that of a Fourier transform spectrometer. This quantum leap in performance promises to revolutionize both infrared astronomy and laboratory emission spectroscopy.

Phoenix has just become operational and we have used it, for example, to record spectra of the comet Hale-Bopp and the carbon star, IRC+10216. In Hale-Bopp the vibration-rotation emission lines of CH, CN, C_2H_6 and OCS could be identified.

The measured resolving power of Phoenix is about 70,000 and the sensitivity is very high compared to a Fourier transform spectrometer. For example the infrared emission spectrum of the free radical NH was recorded with both spectrometers at 3000 cm^{-1}. Phoenix achieved the same signal-to-noise ratio in one minute of integration as was obtained in 90 minutes with a Fourier transform spectrometer. Although this result is an unfair comparison for many reasons, it is clear that Phoenix is much more sensitive than an FTS.

ACKNOWLEDGEMENTS

The work described in this article was supported by grants from the Natural Sciences and Engineering Research Council of Canada, the Petroleum Research Fund of the American Chemical Society, and the NASA laboratory astrophysics program. The National Optical Astronomy Observatories are operated by the Association of Universities for Research in Astronomy under cooperative agreement with the National Science Foundation.

REFERENCES

1. Lynds L. and Woody, B.A., *Appl. Optics* **27**, 1225 (1988).
2. Bernath, P.F., *Spectra of Atoms and Molecules*, New York: Oxford University Press, 1995, ch. 1.
3. Frum, C.I., Engleman, R. and Bernath, P.F., *J. Chem. Phys.* **93**, 5457 (1990).
4. Uehara, H., Horiai, K., Sueoka, K. and Nakagawa, K., *Chem. Phys. Lett.* **161**, 149 (1989).
5. Hammer, P.D., Sinha, A., Burkholder, J.B. and Howard, C.J., *J. Mol. Spectrosc.* **129**, 99 (1988).
6. Frum, C.I., Engleman, R. and Bernath, P.F., *J. Chem. Phys.* **95**, 1435 (1991).
7. Bernath, P.F., *Chem. Soc. Rev.* **25**, 111 (1996).
8. Carli, B. and Carlotti, M., in *Spectroscopy of the Earth's Atmosphere and Interstellar Medium*, ed. K.N. Rao and A. Weber, San Diego: Academic Press, 1992, 1.
9. Roche, A.E., Kumer, J.B., Mergenthaler, J.L., Fly, G.A., Uplinger, W.G., Potter, J.F., James, T.C. and Sterritt, L.W., *J. Geophys. Res.* **98**, 10763 (1993).
10. Bernath, P.F., Sinquefield, S.A., Baxter, L.L., Sclippa, G., Rohlfing, C. and Barfield, M., *Vib. Spectrosc.* (submitted).
11. Jorgensen, U.G., in *Molecules in the Stellar Environment*, ed.

U.G. Jorgensen, Berlin: Springer-Verlag, 1994, pg. 29.
12. Black, J.H., in *Spectroscopy of Astrophysical Plasmas*, eds. A. Dalgarno and D. Layzer, Cambridge: Cambridge University Press, 1987, pg. 279.
13. Herbst, E., *Ann. Rev. Phys. Chem.* **46**, 27 (1995).
14. Engvold, O. Wöhl, H. and Brault, J.W., *Astron. Astrophys. Suppl.* **42**, 209 (1980).
15. Lindgren, B. and Olofsson, G.S., *Astron. Astrophys.* **84**, 300 (1980).
16. Kleman, B. and Uhler, U., *Can. J. Phys.* **37**, 537 (1959).
17. Ram, R.S., Jarman, C.N. and Bernath, P.F., *J. Mol. Spectrosc.* **161**, 445 (1993).
18. Ram, R.S. and Bernath, P.F., *J. Mol. Spectrosc.* **172**, 91 (1995).
19. Ram, R.S., Bernath, P.F. and Wallace, L., *Astrophys. J. Suppl.* **107**, 443 (1996).
20. Wallace, L., Livingston, W., Hinkle, K. and Bernath, P.F., *Astrophys. J. Suppl.* **106**, 165 (1996).
21. Kirkpatrick, J.D., Henry, T.J. and McCarthy, D.W., *Astrophys. J. Suppl.* **77**, 417 (1991).
22. Cernicharo, J. and Guélin, M., *Astron. Astrophys.* **183**, L10 (1987).
23. Jorissen, A., Smith, V.V. and Lambert, D.L., *Astron. Astrophys.* **261**, 164 (1992).
24. Ram, R.S., Bernath, P.F. and Davis, S.P., *J. Mol. Spectrosc.* **179**, 282 (1996).
25. Ram, R.S., Morbi, Z., Guo, B., Zhang, K.-Q., Bernath, P.F., Van der Auwera, J., Johns, J.W.C. and Davis, S.P., *Astrophys. J. Suppl.* **103**, 247 (1996).
26. Bernath, P.F., *J. Chem. Phys.* **86**, 4838 (1987).
27. White, J.B., Dulick, M. and Bernath, P.F., *J. Chem. Phys.* **99**, 8371 (1993).
28. Boyle, R.J., Keady, J.J., Jennings, D.E., Hirsch, K.L. and Wiedemann, G.R., *Astrophys. J.* **420**, 863 (1994).
29. Kroto, H.W., Heath, J.R., O'Brien, S.C., Curl, R.F. and Smalley, R.E., *Nature* **318**, 162 (1985).
30. Krätschmer, W., Lamb, L.D., Fostiropoulos, K. and Huffman, D.R., *Nature* **347**, 354 (1990).
31. Nemes, L., Ram, R.S., Bernath, P.F., Tinker, F.A., Zumwalt, M.C., Lamb, L.D. and Huffman, D.R., *Chem. Phys. Lett.* **218**, 295 (1994).
32. Hinkle, K. and Bernath, P.F. (unpublished).
33. Hinkle, K.H., Keady, J.J. and Bernath, P.F., *Science* **241**, 1319 (1988).
34. Bernath, P.F., Hinkle, K.H. and Keady, J.J., *Science* **244**, 562 (1989).
35. Clayton, G.C., Kelly, D.M., Lacy, J.H., Little-Marenin, I.R., Feldman, P.A. and Bernath, P.F., *Astron. J.* **109**, 2096 (1995).
36. Campbell, J.M., Dulick, M., Klapstein, D., White, J.B. and Bernath, P.F., *J. Chem. Phys.* **99**, 8379 (1993).
37. Campbell, J.M., Klapstein, D., Dulick, M., Bernath, P.F. and Wallace, L., *Astrophys. J. Suppl.* **101**, 237 (1995).
38. Wallace, L., Bernath, P.F., Livingston, W., Hinkle, K., Busler, J., Guo, B. and Zhang, K.-Q., *Science* **268**, 1155 (1995).
39. Polyansky, O.L., Busler, J.R., Guo, B., Zhang, K.-Q. and Bernath, P.F., *J. Mol. Spectrosc.* **176**, 305 (1996).
40. Polyansky, O.L., Zobov, N.F., Tennyson, J., Lotoski, J.A. and Bernath, P.F., *J. Mol. Spectrosc.* (in press).
41. Polyansky, O.L., Tennyson, J. and Bernath, P.F., *J. Mol. Spectrosc.* (submitted).
42. Partridge, H. and Schwenke, D., *J. Chem. Phys.* **106**, 4618 (1997).
43. Polyansky, O.L., Zobov, N.F., Viti, S., Tennyson, J., Bernath, P.F. and Wallace, L., *Science* **277**, 346 (1997).
44. Hinkle, K.H. and Barnes, T.G., *Astrophys. J.* **227**, 923 (1979).
45. Nov. issue of *Astron. Astrophys.* **315** (1996).
46. Allard, F., Hauschildt, P.H., Miller, S. and Tennyson, J., *Astrophys. J.* **426**, L39 (1994).
47. Worden, H., Beer, R. and Rinsland, C.P. *J. Geophys. Res.* **102**, 1287 (1997).
48. Fland, J.M., Camy-Peyret, C. and Maillard, J.P., *Mol. Phys.* **32**, 499 (1976).
49. Oppenheimer, B.R., Kulkarni, S.R., Mathews, K. and Nakajima, T., *Science* **270**, 1478 (1995).
50. Allamandola, L.J., Tielens, A.G.G.M. and Barker, J.R., *Astron. Astrophys.* **281** 923 (1995).
51. Kurtz, J., *Astron. Astrophys.* **255**, L1 (1992).
52. Joblin, C., D'Hendecourt, L., Léger, A. and Défourneau, D., *Astron. Astrophys.* **281**, 923 (1995).
53. Zhang, K.-Q., Guo, B., Colarusso, P. and Bernath, P.F., *Science* **274**, 582 (1996).
54. Langhoff, S., *J. Phys. Chem.* **100**, 2819 (1996).
55. Colarusso, P., Zhang, K.-Q., Guo, B. and Bernath, P.F., *Chem. Phys. Lett.* **269**, 39 (1997).
56. Maki, A.G., Olson, W.B. and Thompson, G., *J. Mol. Spectrosc.* **144**, 257 (1990).
57. Dulick, M., Zhang, K.-Q., Guo, B. and Bernath, P.F., *J. Mol. Spectrosc.* (in press).

Nanosecond Step-Scan FT-Infrared Absorption Spectroscopy in Photochemistry and Catalysis

H. Frei

Laboratory of Chemical Biodynamics, MS Calvin Laboratory, Lawrence Berkeley National Laboratory, University of California Berkeley, CA 94720

Time-resolved step-scan FT-IR absorption spectroscopy has been expanded to a resolution of 20 nanosecond. Following a description of the experimental set-up, applications in four research areas are presented. In the first project, we discuss a reversible isomerization, namely the bacteriorhodopsin photocycle. Main results are the discovery of 2 processes with distinct kinetics on the nanosecond time scale not detected by previous spectroscopic techniques, and observation of an instantaneous response of the protein environment to chromophore dynamics within the nanosecond laser pulse duration. In a second project, alkane C-H bond activation by a transition metal complex in room temperature solution is investigated and the first measurement of the formation of a C-H insertion product reported (alkyl hydride). Then, a nanosecond study of a pericyclic reaction, the ring-opening of cyclohexadiene, is discussed. The fourth example describes the first observation of a transient molecule in a zeolite matrix, a triplet excited quinone, by time-resolved infrared spectroscopy.

INTRODUCTION

Time-resolved vibrational spectroscopy is a powerful approach for elucidating reaction mechanisms because it furnishes information on the structure of chemical intermediates. This holds especially for methods that give access to the entire mid-infrared range (4000-400 cm^{-1}) and the time domain from nanoseconds to milliseconds. Of the two main techniques available, time-resolved infrared and resonance Raman spectroscopy, the latter is not suitable for chemistry of small molecules because reactants and transients typically lack the chromophores required for resonance enhancement. Our research interest is mainly in photochemistry and catalysis involving small polyatomics, hence infrared methods are preferred. Step-scan Fourier transform infrared absorption spectroscopy is an ideal method for broad-band infrared monitoring of condensed phase processes in the nano- and microsecond regime.

Until very recently, the time resolution of step-scan FT-IR absorption spectroscopy was limited to the microsec range (1-3), with 500 nanosec as the limit (4-6). Higher resolution (50 nanosec) was reported for step-scan emission measurements (7). However, the signal processing method used in step-scan emission spectroscopy is not suitable for transient absorption measurements in the nanosec range for reasons discussed in the next section. Alternative infrared methods for monitoring condensed phase reactions are based either on dispersive IR spectrometers or on cw laser sources. Dispersive IR spectrometers have been extensively used in transient spectroscopy of transition metal complexes for monitoring of CO and CN ligands (8-11) and for recording of nano- and microsecond processes of organic and biological molecules (12,13). The dispersive infrared method lacks the multiplex and throughput advantages of the FT-IR method and is therefore much less efficient. This becomes especially a problem for spectroscopy in the fingerprint region where bending and skeletal stretching modes often have small absorption cross sections. For laser spectroscopy, tunable infrared emitting semiconductor diodes and fixed frequency CO lasers have been used. The high intensity of these monochromatic sources allows recording even in the low nanosec regime (9-11, 14-16). Unfortunately, the laser methods do not allow spectroscopy over a range of more than a few hundred cm^{-1} typically. By contrast, the step-scan FT-IR method gives access to the entire infrared region.

Recording of spectra on the nanosec time scale and over the full infrared range is crucial for capturing chemical intermediates of interest in photochemistry and catalysis. For this reason, we have expanded the resolution of step-scan absorption spectroscopy to 20 nanosec. In this paper, we will first describe details of the experimental setup. Then, application of nanosec step-scan FT-IR spectroscopy for the study of four processes will be discussed; a photobiological process (bacteriorhodopsin photocycle), activation of alkane C-H bonds by transition metal complexes in solution, photochemical ring opening, and photochemistry in zeolites. Another example from our laboratory, not discussed in this paper, is nanosecond FT-IR spectroscopy of hemoglobin dynamics (17). Step-scan FT-IR experiments with nanosec resolution were also reported in the past months by several other research groups, namely bacteriorhodopsin studies by the groups of Braiman (18) and Gerwert (19), and excited state spectroscopy of transition metal complexes by the groups of Palmer (20) and Palmer and Meyer (21).

NANOSECOND STEP-SCAN FT-IR SYSTEM

In time-resolved step-scan FT-IR spectroscopy, the time course of the interferometer signal is recorded while holding the movable mirror fixed at each optical path difference (Fig. 1). At a given mirror position, a laser pulse excites the sample and the decay is measured and digitized. Typically tens of decays are recorded and averaged at each mirror position for signal improvement. After moving the mirror to the next position, decays are again measured and averaged. This process is repeated until the mirror has stepped through all positions required for the

CP430, *Fourier Transform Spectroscopy:* 11th International Conference
edited by J.A. de Haseth

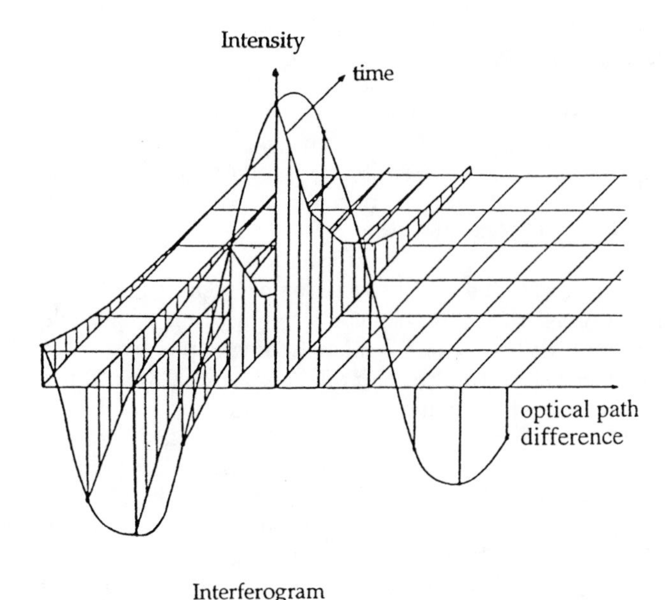

FIGURE 1. Principle of step-scan FT-infrared spectroscopy. The 3 axis represent optical path difference, interferometric signal, and time evolution, respectively. Adapted from Uhmann et al. (Ref. 1).

desired spectral resolution and undersampling. From this set of decay curves, an interferogram for each time delay with respect to the laser excitation pulse is computed. Fourier transformation furnishes a single beam infrared spectrum for each time point. Step-scan FT-IR spectroscopy on the time scale of nanoseconds has become feasible in recent years due to the availability of infrared photon detectors with nanosecond response time, of >100 MHz digitizers, and because of the development of methods that can hold the interferometer mirror steady within less than 1 nm.

The time-resolved step-scan system of our laboratory consists of a Bruker Model IFS88 spectrometer, a 486 PC with OPUS software, and auxiliary electronics for matching nanosecond infrared signals to a 200 MHz 8 bit digitizer (PAD82, sampling interval 10 nanosec) or a 40 MHz 12 bit digitizer (PAD1232, sampling interval 25 nanosec). For excitation of the sample at visible or UV wavelengths with pulses of 5-10 nanosec duration, a Quanta-Ray system consisting of a Nd:YAG laser Model DCR-2A (with GCR-3 optics, 10 Hz repetition rate), a dye laser Model PDL-1 and a WEX-1 wavelength extender was used.

A crucial requirement for the transient absorption measurements in the nanosec regime at high sensitivity is recording of the laser induced change of the interferometric signal while suppressing the orders of magnitude larger, static signal. Such recording and processing of AC-coupled infrared detector signals in step-scan FT-IR spectroscopy was first implemented by Siebert in a microsecond resolution instrument (1). Phase correction upon Fourier transformation of the AC-coupled interferogram requires a phase spectrum derived from a DC-coupled step-scan interferogram. Therefore, both AC and DC interferometric signals have to be recorded. Our spectrometer is equipped with photovoltaic infrared detectors featuring simultaneous AC and DC outputs (Kolmar Technologies HgCdTe detectors Model KMPV50-0.5-J2 and Model KMPV11-1-J2, and an InSb detector Model KISDP-1-J2). The FWHM of the response to a 10 nanosec 1.064 micron Nd:YAG excitation pulse is 20 and 50 nanosec for the HgCdTe detectors, and 42 nanosec for the InSb detector. The RC decay constant of the AC coupler is 3 millisec. The digitizer allowed simultaneous recording and processing of the AC and DC-coupled interferometric signal. To assure full use of the dynamic range of the 8 or 12 bit digitizer (± 1 V full range), the AC-coupled signal was amplified by a factor of 100 (CAL-AV Laboratories Model 7930 500 MHz amplifier) before transfer to the digitizer. The DC output of the detector was first adjusted to ground and then amplified (LeCroy Model 6103 programmable 150 MHz and CAL-AV amplifiers) to reach close to ± 1 V peak-to-peak interferogram amplitude before entering the digitizer.

Measurements in the microsec time range were done with a 200 kHz digitizer. A DC-coupled step-scan run of the sample was taken first in order to determine the phase spectrum since no simultaneous processing capability for AC and DC-coupled signals is available for this digitizer. The phase was stored for subsequent use in the Fourier transformation of the AC-coupled interferograms. The peak-to-peak noise of the HeNe reference laser with the movable mirror held in fixed position indicated a typical mirror stability of ±1.4 nm.

The visible or UV photolysis beam was aligned collinear, or at a small angle, with respect to the infrared probe beam. Excitation pulse energies were typically around 4 mJ cm^{-2} (80 μJ mm^{-2}). In order to prevent photolysis light from entering the detector or interferometer compartment, both ports were protected by a 2 inch germanium plate featuring a dielectric coating that maximized the infrared transmittance (95%, International Scientific). In addition, a dielectrically coated filter (Optical Coating Laboratory, Inc.) was mounted in front of the infrared detector in order to tailor the folding limits to the spectral region of interest for maximum sensitivity. Choice of the filter depended on the specific application. For triggering of the digitizer, a small fraction of the laser pulse was exposed to an EG&G silicon photodiode Model SGD-444.

APPLICATIONS

Bacteriorhodopsin Photocycle

Bacteriorhodopsin (BR) is a membrane protein with all-trans retinal as the light-sensitive chromophore (Fig. 2). The function of this protein of halophilic bacteria is to convert visible light energy into chemical energy by pumping protons across the cell membrane, thereby generating a proton gradient (22). The retinal is covalently bound at the Schiff base end to a lysine residue (Lys 216) of

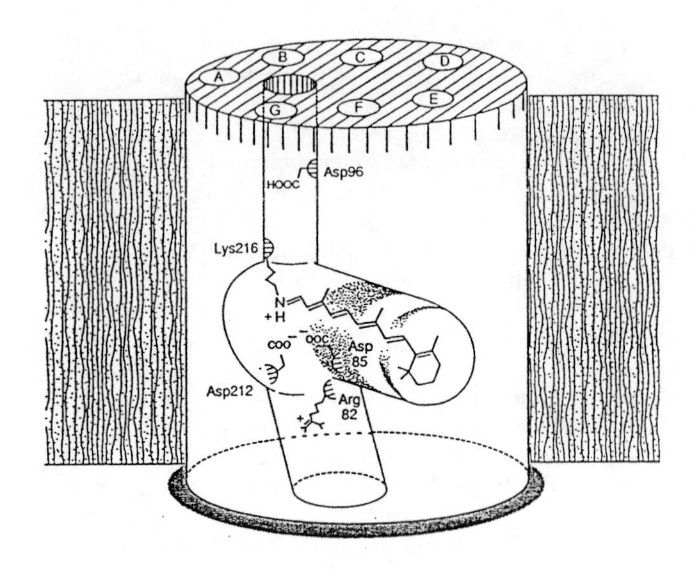

FIGURE 2. Cartoon of bacteriorhodopsin showing membrane, retinal chromophore, and key amino acid residues involved in the proton pump (vertical channel). Asp = aspartic acid, Arg = arginine, Lys = lysine. A-G designate the 7 α-helices of the protein. Intracellular space is at top. Adapted from Henderson et al. (Ref. 23).

one of the α-helices that constitute the protein (Fig. 2). The retinal-protein complex exhibits a broad visible absorption band with a maximum at 568 nm, which gives rise to its characteristic purple color. Upon absorption of a photon, the chromophore undergoes a rapid trans-to-cis isomerization about the C_{13}=C_{14} double bond (Fig. 3). BR then proceeds through a series of structurally distinct intermediates labeled J, K, L, M, N, and O before returning to its initial state after several milliseconds (22). The first step of the cycle, J formation, occurs in the subpicosec regime and is followed by conversion to K in 3 picosec. Upon further relaxation, the L intermediate is formed in about 2 microsec. The K to L transition is a relaxation

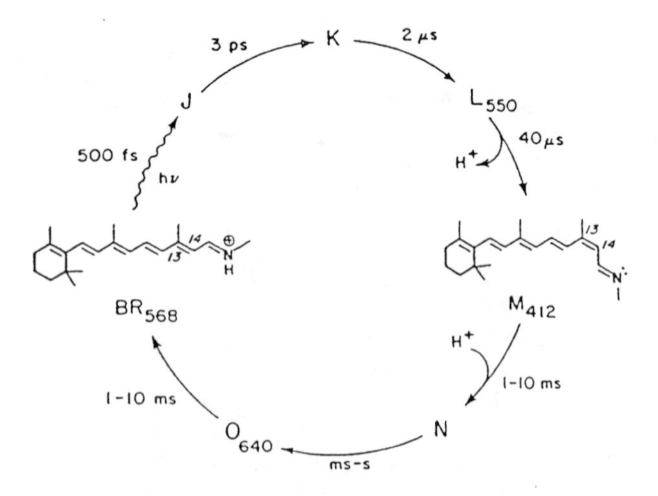

FIGURE 3. Photocycle of bacteriorhodopsin. Three-digit subscripts indicate visible absorption maxima (in nm) of the intermediates. Adapted from Fodor et al. (Ref. 24).

process of the C_{13}-cis retinal in the protein binding site in which energy stored in the distorted chromophore is transferred to the protein. It has been proposed by some groups to occur in 2 steps, namely relaxation to a species KL in 10 nanosec followed by conversion of KL to L within 2 microsec (25,26). Upon proton transfer to the protein in the L to M transition, reprotonation and reverse isomerization of the retinal in the M → N → O → BR transition completes the cycle.

Elucidation of the conformational changes of the chromophore and the protein environment during the K to L transition is crucial for the understanding of the photocycle because L is the key intermediate leading to proton transfer form the Schiff base to the protein. Our goal was to use the higher resolution and sensitivity of the step-scan FT-IR method compared to previous work with a dispersive infrared instrument to obtain improved spectral and kinetic information on the nanosec processes of the K to L conversion. Moreover, we prepared aqueous suspensions of BR in order to remove any ambiguity about the full state of hydration of the protein. Previous groups have used hydrated films throughout. Aqueous suspensions put additional demands on the sensitivity of the measurements.

Figure 4 shows transient absorption spectra of a suspension of BR in H_2O (22°C, pH = 7.4, 10 mM HEPES buffer) upon excitation at 532 nm with an 8 nanosec laser pulse (27). Data were taken with a sampling interval of 10 nanosec and a laser pulse repetition rate of 10 Hz. The time resolution is 20 nanosec (= FWHM of HgCdTe detector), the spectral resolution is 4 cm^{-1}. With an upper folding limit of 2250 cm^{-1}, 1141 mirror positions were required for each step-scan run. Thirty laser-induced decays were recorded at each mirror position. The spectra of Fig. 4 represent the average of the absorbance difference spectra of 17 such runs, each taken with a fresh sample spot. The absorbance difference spectra were computed as

$$\Delta A = -\log\left(\frac{S+\Delta S}{S}\right)$$

where S stands for static single beam spectrum (Fourier transform of the DC-coupled interferogram) and ΔS the laser-induced spectrum (Fourier transform of the AC-coupled interferogram normalized for amplification). The negative features of the spectra are due to ground state depletion of the complex, the positive bands originate from transient intermediates. The top trace (a) shows the average of the first 2 time slices taken with delays of 10 and 20 nanosec after the laser pulse. This is the first infrared spectrum of BR recorded with 20 nanosec resolution. Traces b-d are averages of the first 5, 10, and 20 ten nanosec slices, respectively, showing progressively higher signal-to-noise due to the more extensive averaging. The regions 1700-1540 cm^{-1} and <900 cm^{-1} cannot be used for analysis because of complete absorption of infrared light by liquid water.

Two new results emerge from this study. One is the observation of 2 distinct kinetics in the nanosecond regime, which can easily be discerned from the series of spectraltime slices shown in Fig. 5. These are centered at 100,

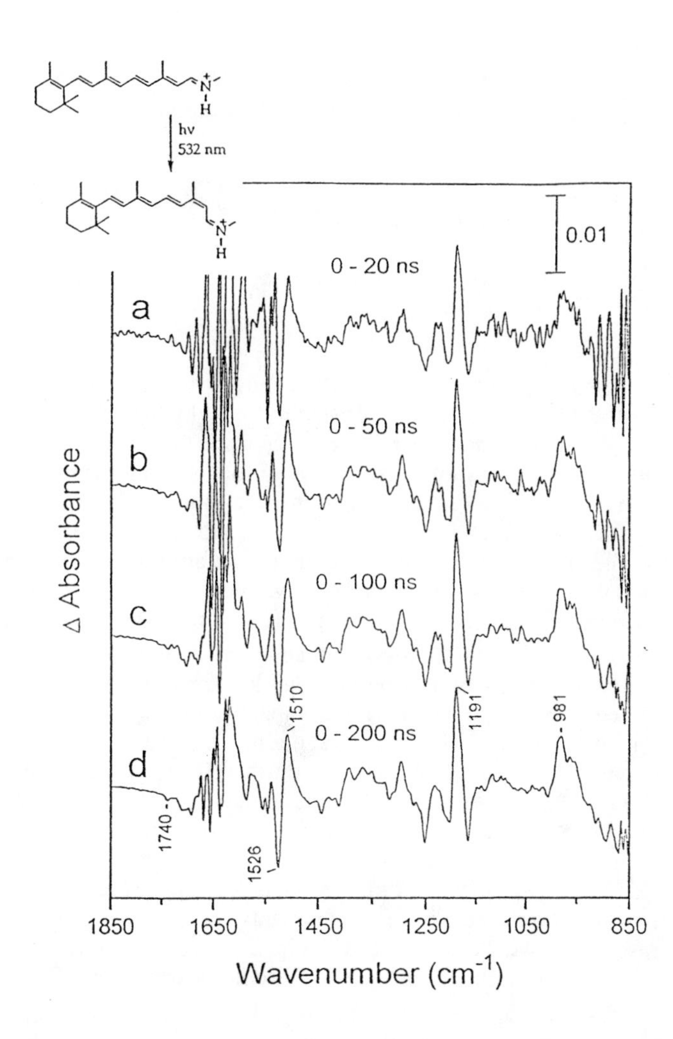

FIGURE 4. Nanosecond FT-IR spectra of aqueous suspension of bacteriorhodopsin. HgCdTe detector, FWHM = 20 nanosec, sampling interval 10 nanosec. BaF_2 sample cell with 10 μm path length. The spectra represent the average of the first 2 (trace a), 5 (trace b), 10 (trace c), and 20 (trace) ten nanosec time slices after the laser excitation pulse.

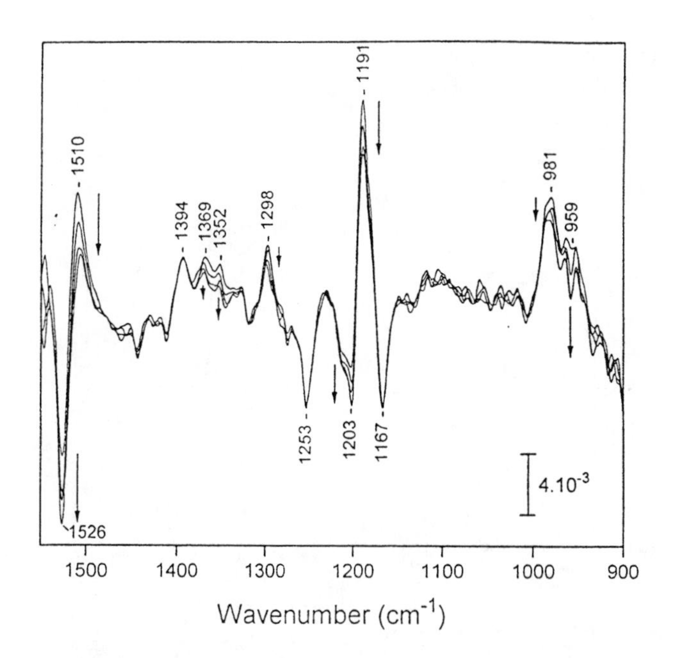

FIGURE 5. Time-resolved FT-IR spectra of BR taken from 0-200, 200-400, 400-600, and 600-800 nanosec. The arrows indicate the direction of spectral changes with time, with the length of the arrow reflecting the magnitude of the intensity change.

300, 500, and 700 nanosec after the laser excitation pulse. There are 2 sets of absorptions with distinct kinetic behavior. Bands at 1510, 1191, and 959 cm^{-1} decay with a constant of 400 nanosec (including the negative feature at 1526 cm^{-1}). By contrast, absorptions at 1369, 1352, 1298, and 981 cm^{-1} decay more slowly with a rate constant around 2 microsec. These different kinetics were not noted in previous work on BR by Sasaki et al. using a dispersive nanosec IR instrument (26), presumably because of a lack of sensitivity, but were confirmed in a recent step-scan FT-IR study by Braiman (18).

Spectral changes of Fig. 5 are due to structural relaxation of the C_{13}-cis retinal molecule in the protein binding pocket, which results from the fact that on the tens of nanosecond time scale the photoisomerized chromophore is still conformationally strained and theprotein not yet adjusted to its new structure. The nanosec FT-IR spectra reveal that the relaxation involves 2 distinct processes. The nature of the structural changes associated with each process can be determined by analyzing the vibrational modes involved in slow and fast kinetics. According to our interpretation, the 400 nanosecond process is mainly due to relaxation of the β-ionone end and the polyene chain because it mostly involves vibrations of these moieties. The 2 microsec process is mainly associated with modes of the Schiff base terminus and amino acids H-bonded to it. Braiman reached a somewhat different interpretation of the 2 processes, although spectral features and kinetics are in agreement with our measurements (18).

The second new result obtained by this study is the detection of a protein response to the chromophore photoisomerization in as little as 20 nanosec. Fig. 7 shows the evolution of spectral features around 1740 cm^{-1}, a region characteristic for the C=O stretch absorption of aspartic acid residues Asp 96 and 115 (28). At 20 nanosec after the excitation pulse, there is depletion around 1740 cm^{-1} and growth at 1735 cm^{-1} (dashed trace (a)). This means that one or both amino acids respond to the trans-cis isomerization of the chromophore within this short time interval. Kinetic analysis of the subsequent time slices at 50, 250, 450, and 900 nanosec after the pulse shows that the decay of the 1735 cm^{-1} peak and the bleach at 1740 cm^{-1} have both a rate constant of about 400 nanosec. Hence, the response of these amino acids is synchronous with the ast process of the chromophore. This reflects the proximity of the Asp 115 side chain and the β-ionone moiety (23), and may signal a direct coupling of the chromophore with this carboxylic acid residue. In a previous study on a hydrated film of BR using ultrafast lasers, Diller et al. did not detect any spectral changes

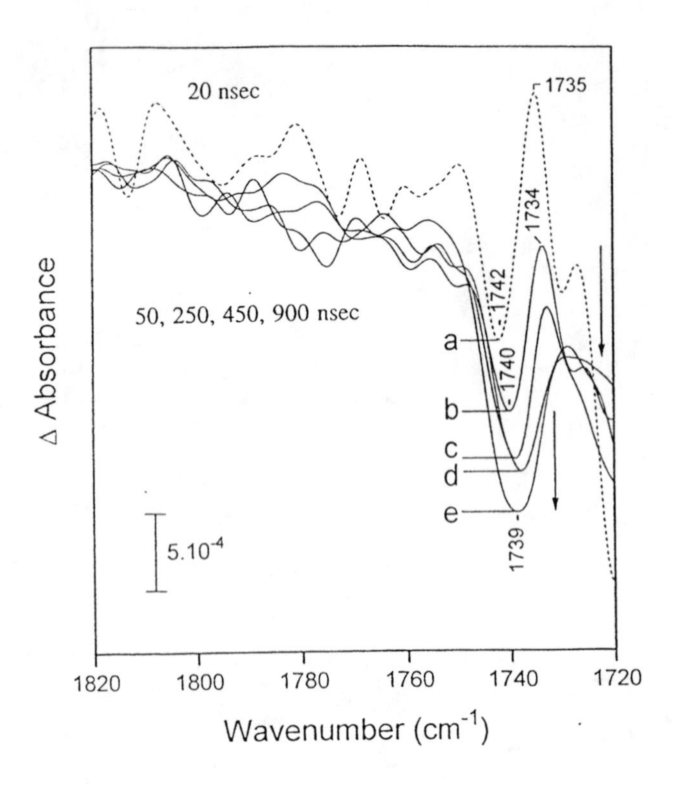

FIGURE 6. Expanded region 1720-1820 cm^{-1} showing spectral changes in the time intervals 0-20 nanosec (a), 0-100 nanosec (b), 200-300 nanosec (c), 400-500 nanosec (d), and 840-940 nanosec (e). The 20 nanosec trace (a) is shown with dashed line for clarity.

around 1740 cm^{-1} within 10 nanosec (29). We believe that this is due to insufficient hydration of the film, resulting in blocking or substantial slowdown of the protein response to the chromophore dynamics. This result points to the importance of conducting time-resolved studies under physiological sample conditions.

Alkane C-H Bond Activation

A key challenge in current catalysis research is to establish ways for the direct functionalization of small alkanes to useful products (30,31). The reason is that alkanes are inexpensive and highly abundant chemicals found in natural gas (methane, ethane, propane) and volatile fractions of petroleum. Yet, their use as feedstocks for industrial intermediates and organic building blocks is currently very limited. A breakthrough in fundamental studies of alkane activation occurred in the early 1980s when it was discovered that complexes of Ir or Rh react with C-H bonds to form an alkyl hydride insertion compound (32-34). A typical reaction is shown in the scheme below. Reaction is initiated by expulsion of a CO ligand under UV light, leading to a coordinatively unsaturated metal center that inserts into an alkane (RH) carbon-hydrogen bond in room temperature solution. The desired functionalized alkane is obtained by treatment of the alkyl hydride with an appropriate organic.

$$\mathrm{CpM(CO)_2} \xrightarrow[-\mathrm{CO}]{\mathrm{RH}} \mathrm{CpM(CO)(H)(R)} + \mathrm{CO}$$

(M = Ir, Rh)

alkyl hydride

A central research goal is to make metal-mediated alkane C-H bond activation catalytic, which requires a detailed understanding of the mechanism of the reaction. Infrared spectroscopy is an ideal tool for monitoring transition metal carbonyl chemistry because the CO frequency is very sensitive to the nature and stereochemistry of the ligands at the metal center. In addition, the IR absorption of CO ligands is very intense and lies in a spectral region free of most other functional group modes (1900-2100 cm^{-1}). Time-resolved infrared studies of C-H activation by transition metal complexes have focused in the past on gas phase and low temperature reactions in the microsec regime (35-38) and on the photodissociation dynamics of the metal complex on the picosec time scale (39). C-H bond insertion, the crucial step of alkane activation, has not been observed thus far in room temperature solution.

We have employed step-scan FT-IR spectroscopy to monitor the kinetics of alkane activation with a time resolution of 40 nanosec (40). Fig. 7 shows the course of the reaction upon excitation of the complex $Tp^*Rh(CO)_2$ (Tp^* = HB-(3,5-dimethyl pyrazolyl)$_3$) with a 5 nanosec pulse at 355 nm in liquid cyclohexane at room temperature. Photoelimination of a CO ligand, indicated by loss of the asymmetric and symmetric CO stretch absorption of the parent complex at 1981 and 2054 cm^{-1}, and growth of a monocarbonyl species at 1990 cm^{-1} is accomplished within the 40 nanosec response time of the InSb detector. The transient species with the CO ligand at 1990 cm^{-1} has been identified as a weak complex of the dissociated metal complex with the alkane (σ-complex). The frequency of the CO band indicates that one arm of the tri-dentate ligand is detached from the Rh center in this complex, rendering the metal more reactive towards the alkane (40). Conversion of the σ-complex to the alkyl hydride is signaled by the decay of the 1990 cm^{-1} band and concurrent growth of a CO stretch at 2032 cm^{-1} (Fig. 7). Single exponential fits give rise (decay) constant of around 230 nanosec (Fig. 8). The fact that the two rate constants are identical within uncertainties shows that the σ-complex is indeed the immediate precursor of the alkyl hydride. Using simple transition state theory, a free energy barrier of 8.3 kcal mol^{-1} for the C-H insertion step was derived from the rate constant. This constitutes the first measurement of the formation of a C-H insertion product, thereby establishing the time scale for C-H bond activation in room temperature alkane solution.

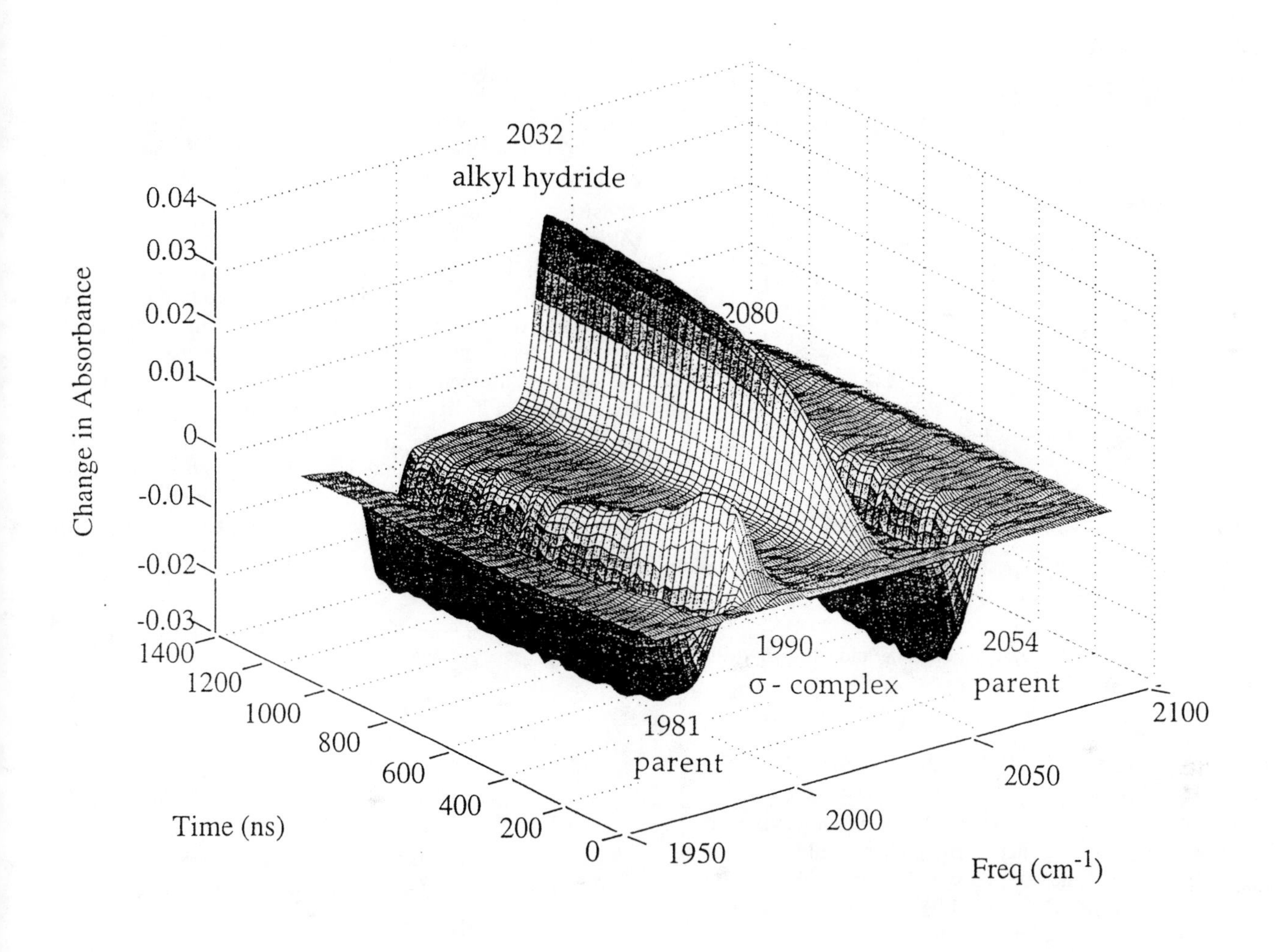

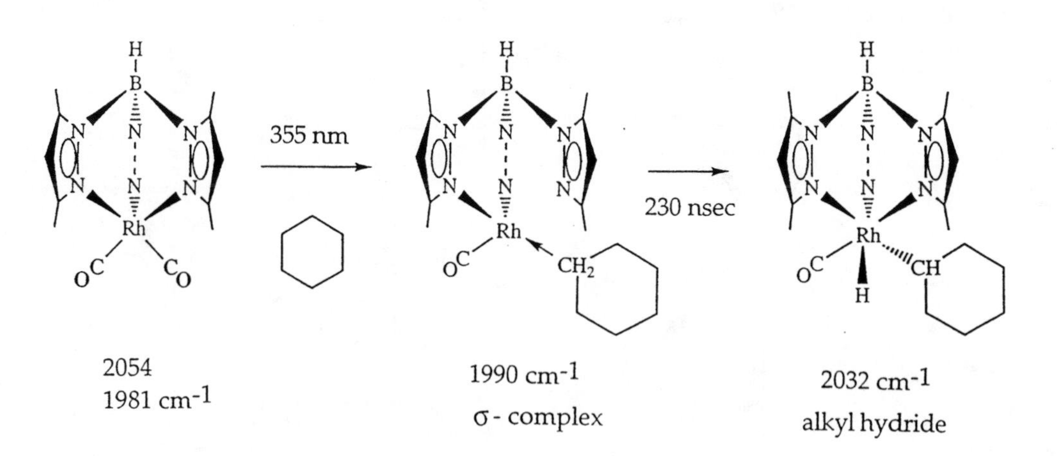

FIGURE 7. Nanosecond FT-IR spectrum of cyclohexane C-H bond activation by $Tp^{*}Rh(CO)_2$ following initiation by a UV laser pulse. InSb detector, FWHM = 40 nanosec, sampling interval 25 nanosec. Flow system, 1 mm path CaF_2 cell. The flow rate was adjusted such that each laser pulse encountered a fresh sample at a repetition rate of 10 Hz.

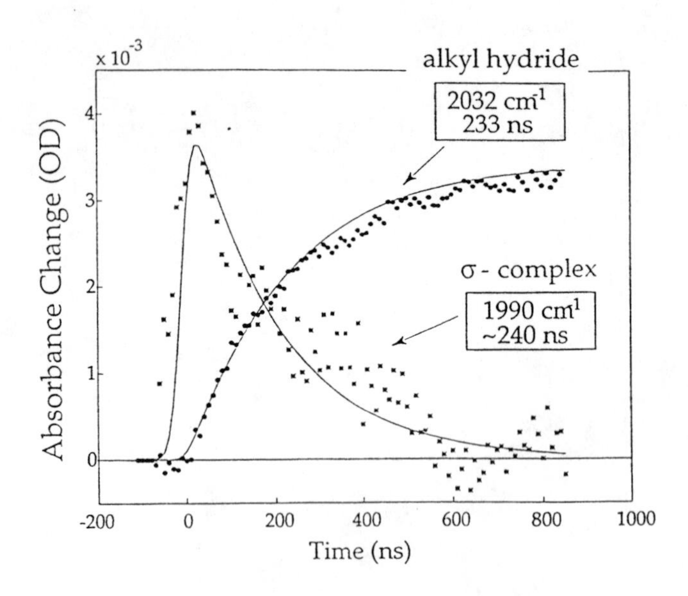

FIGURE 8. Kinetic curves for σ-complex and alkyl hydride product.

Photochemical Ring-Opening

UV light-induced ring-opening of 1,3-cyclohexadiene is a prototypical pericyclic reaction of fundamental interest in organic chemistry. It has been studied on the picosec time scale by resonance Raman (41-43) and by optical spectroscopy (44). These ultrafast techniques revealed that primary product of ring-opening, the cZc conformer of 1,3,5-hexatriene, is formed within less than a few picosec.

hν

cZc

dark

dark

tZt

cZt

The cZc conformer was observed to convert to a secondary product with a rise time of 7 picosec. The latter is assigned to the cZt conformer of hexatriene (42). According to the resonance Raman studies, there is no sign of the final product tZt-1,3,5-hexatriene at the longest delay time accessible by the technique, which is 3 nanosec. The stable tZt conformer is separated from the cZt form by a 4 kcal mol^{-1} internal rotation barrier (45-47). Simple transition state theory predicts for such a barrier that cZt → tZt interconversion should occur on the nanosec time scale.

In search of this interconversion, we have employed nanosec step-scan FT-IR spectroscopy to monitor ring-opening products (47). Cyclohexadiene in cyclohexane solution was excited at 292 nm with 8 mJ pulses. Flow of the solution allowed photolysis of a fresh sample with each laser pulse. FT-IR absorbance difference spectra in the region 1900-900 cm^{-1} are shown in Figs. 9 and 10. Data were recorded with a 10 nanosec sampling rate, and spectral averages of 5 consecutive time slices are shown. These are centered at 25 (a), 75 (b), and 125 nanosec (c) after the pulse. Trace (d) shows a 200 nanosec average, and

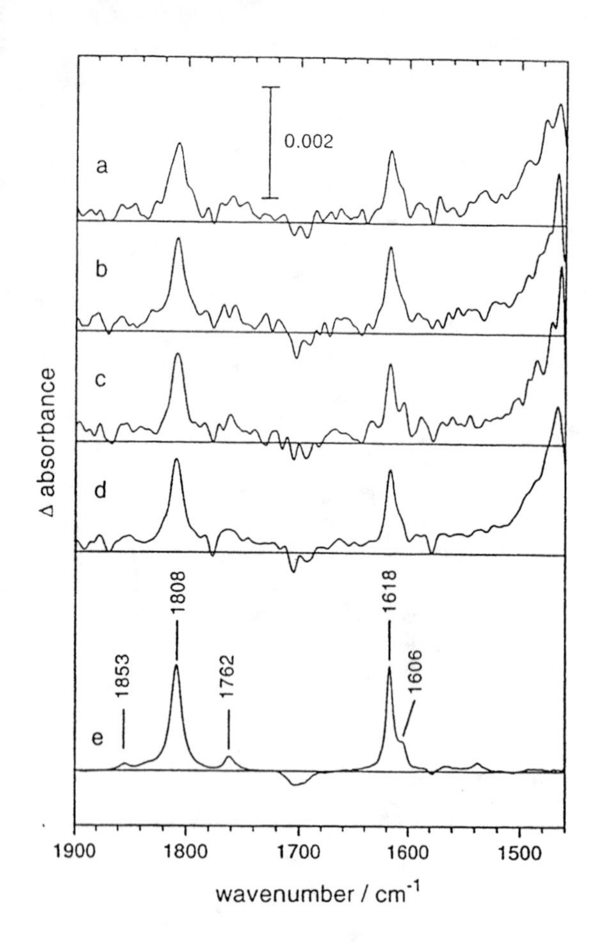

FIGURE 9. Nanosecond FT-IR absorbance difference spectra of cyclohexadiene photolysis. BaF_2 flow cell with 100 μm path length. The spectra were taken with a sampling internal of 10 nanosec and represent averages 0-50 nanosec (a), 50-100 nanosec (b), 100-150 nanosec (c), and 200-400 nanosec (d). Trace (e) is the static absorbance difference spectrum upon 4 min photolysis.

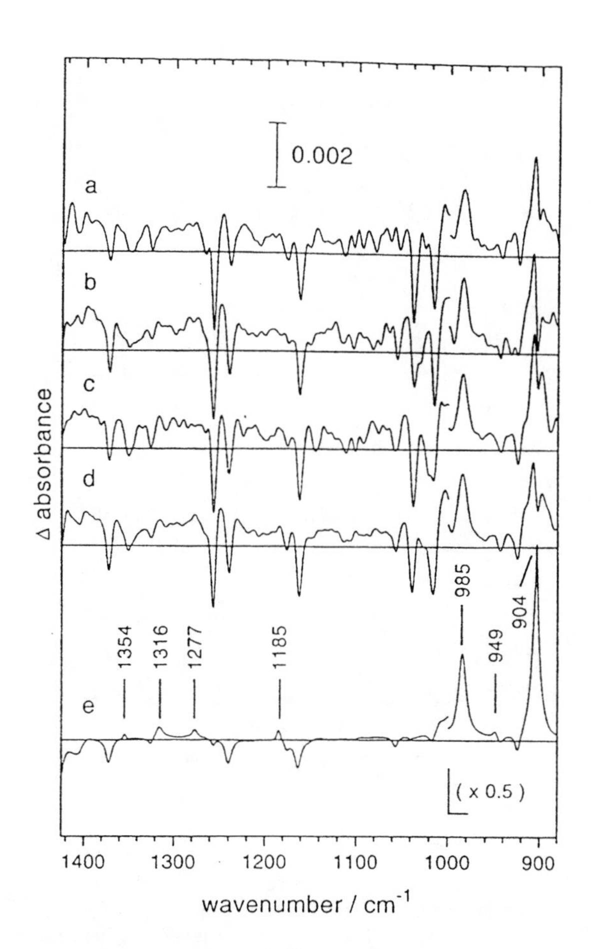

FIGURE 10. Nanosecond FT-IR absorbance difference spectra of cyclohexadiene photolysis. For explanation, see caption of Fig. 9.

spectrum (e) is a static absorbance difference spectrum upon 4 min photolysis. The product absorptions in the latter spectrum are those of the stable tZt-1,3,5-hexatriene product. Comparison of the static tZt spectrum with the nanosec spectra reveals that even at the shortest delay (spectrum (a)), tZt is the only conformer present (differences between nanosec and static spectra are due to thermal effects on the strong cyclohexane solvent absorptions caused by the laser-induced temperature rise of about 1 K (47)). Infrared absorptions of the unstable cZt conformer are expected at 1808, 1618, and 921 cm^{-1} (48,49), but no transient features are observed at these frequencies. If we take into account estimated intensities of cZt bands, the detection limit in spectral regions where these are expected to absorb, and consider the fact that the decay kinetics of cZt bands would be first order, we estimate that the cZt to tZt interconversion must occur in less than 20 nanosec (47). Since no tZt conformer was detected at a delay of 3 nanosec in the resonance Raman experiment, we conclude that the cZt to tZt interconversion takes place in the time interval between 10 and 20 nanosec.

Photochemistry in Zeolites

Nanopore matrices such as zeolites offer a unique environment for carrying out photoreactions due to their geometrical constraints, polarity, and, therefore, their ability to control reaction pathways (50,51). In order to explore the feasibility of time-resolved step-scan FT-infrared spectroscopy in zeolites as a mechanistic tool, we have attempted to record the infrared spectrum of an electronically excited organic, namely triplet duroquinone occluded in cation-exchanged zeolite Y. This zeolite consists of a three-dimensional network of spherical cages (13 Å diameter) interconnected by 7 Å windows, as shown in Fig. 11. The wall of each cage carries a negative charge of 7, which is counterbalanced by cations such as Na^+ (52). Aromatic carbonyl compounds in general, and quinones in particular, have two low-lying triplet states with $n\pi^*$ and $\pi\pi^*$ character, respectively (53). In the case of the $n\pi^*$ state, a non-bonding electron of the carbonyl oxygen is promoted to an antibonding π^* orbital, while a π electron is excited to the π^* orbital in the case of the $\pi\pi^*$ state. The energetics of these two states is influenced by the polarity of the solvent. This is due to the fact that the $\pi\pi^*$ state has a larger dipole moment than the $n\pi^*$ state (53). In fact, the relative ordering of the two states may depend on the solvent. The importance of this is that $n\pi^*$ states of carbonyls are more reactive than the $\pi\pi^*$ states. Hence, the chemical behavior of the lowest triplet state depends on whether it has $n\pi^*$ or $\pi\pi^*$ character. Vibrational spectroscopy is a reliable diagnostic tool for determining the nature of the triplet state; in the case of a $\pi\pi^*$ state, the CO group retains in essence its double bond, and the red shift of the mode relative to ground state is only about 100 cm^{-1}

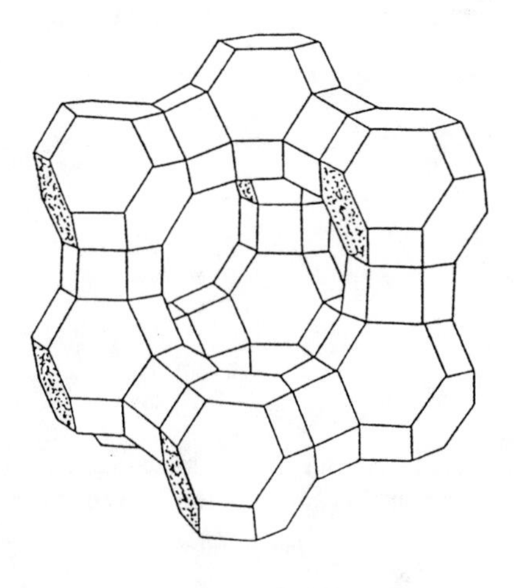

FIGURE 11. Large spherical cage (supercage) of zeolite Y. Each corner represents an Al or Si bridged by an O atom (not shown).

as has been shown for 4-phenyl benzophenone by dispersive infrared (54) and resonance Raman spectroscopy (55). By contrast, the CO group of $n\pi^*$ triplet states has single bond character and lies, therefore, at much lower frequencies (1200-1300 cm^{-1}). This has been observed in the case of the triplet $n\pi^*$ state of benzophenone by time-resolved resonance Raman (56) and phosphorescence excitation spectroscopy (57).

A static infrared spectrum of duroquinone occluded in zeolite NaY at -50°C is shown in Fig. 12. Step-scan experiments were conducted at a spectral resolution of 4 cm^{-1}. With the folding limits at 2250 and 1130 cm^{-1}, this resulted in 570 mirror positions per step-scan experiment. At each mirror position, 25 laser-induced decays were averaged, resulting in 30 min measurement time per experiment. Two spectral time slices are shown in Fig. 13a: One taken 5 µsec after the laser pulse, the other 55 µsec after the pulse. A glance at the two time slices reveals two distinct types of kinetic behavior: (i) positive bands such as those at 1542, 1354, and 1286 cm^{-1} that completely vanish within 50 µsec; (ii) negative features at 1642, 1442, 1380, 1313, and 1259 cm^{-1} which decrease only partially over the first 50 µsec. These latter bands overlap with those of ground state duroquinone, but the peaks are displaced by a few cm^{-1}. In addition, some of these features (1640, 1310, 1260 cm^{-1}) exhibit dispersive shapes. This phenomenon is characteristic for a small transient temperature change induced by the laser pulse and is well described in the literature (58).

When repeating the same step-scan experiment with a duroquinone-loaded zeolite, but this time by exposing the pellet to 1 atm of O_2 gas, only bands coinciding with type (ii) features of Fig. 13a were observed. No absorptions of set (i) appeared. This strongly supports our explanation of type (ii) features in terms of a

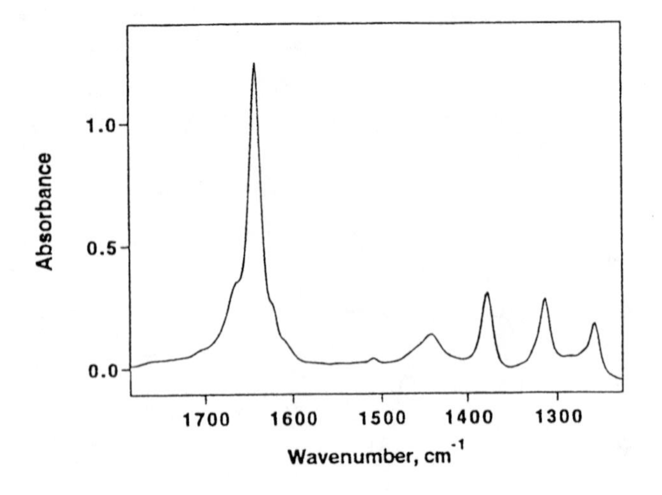

FIGURE 12. FT-infrared difference spectrum before and after vacuum loading of duroquinone into zeolite NaY.

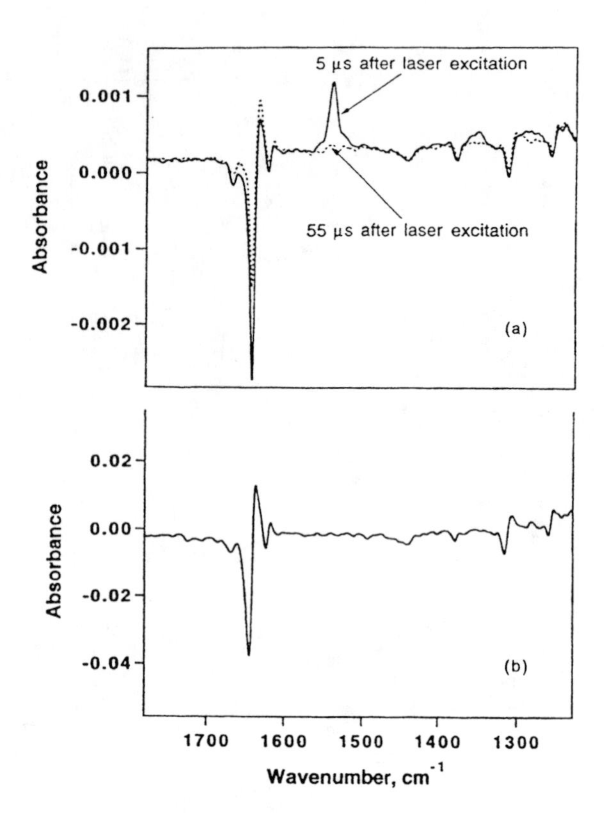

FIGURE 13. (a) Transient FT-infrared spectra recorded 5 and 55 ˉsec after laser excitation of duroquinone in NaY at -50°C. (b) FT-infrared difference spectrum of duroquinone in NaY before and after temperature increase from -50 to -30°C.

transient thermal effect induced by the UV laser pulse. In fact, a spectrum resembling the 55 µsec trace of Fig. 13a was obtained when recording the infrared difference spectrum before and after a temperature rise of the NaY matrix. This is shown in Fig. 13b. Note that the increase of the bulk zeolite temperature by 20 degrees gives an approximately 20 times larger signal than in Fig. 13a. This implies that the laser heating effect is a few degrees at the most.

More importantly, the reversible, quantitative quenching of type (i) bands upon exposure of the zeolite matrix to O_2 gas indicates that these originate from triplet duroquinone. At -50°C, 1 atm of O_2 gas results in about two O_2 molecule per supercage on average. Hopping of a small diatomic like O_2 between cages takes only nanoseconds even at this low temperature (59). Hence, deactivation of triplet excited duroquinone is certainly complete on the nanosecond time scale (60), consistent with what we observe. Since the decay of the heat-induced bands is at least several msec while the triplet absorptions decay within 50 µsec, we can obtain the transient infrared spectrum of the $T_1 \leftarrow S_0$ interconversion by subtracting the 55 µsec time slice from any preceding time slice. The result for the 5 µsec spectrum is shown in Fig. 14. An insert shows the

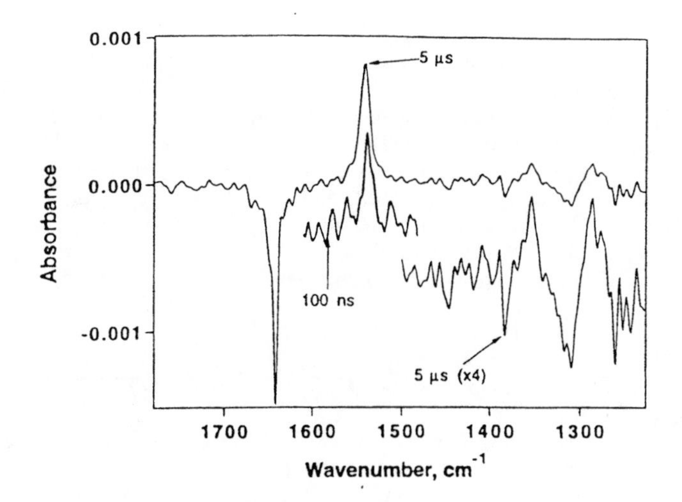

FIGURE 14. Transient FT-infrared spectrum of triplet duroquinone at 5 µsec after the laser pulse. It was obtained by subtracting the 55 µsec spectrum from the 5 µsec spectrum shown in Fig. 13a (see text). An insert shows the spectrum below 15 cm^{-1} expanded by a factor of 4. Another insert shows the ν(CO) band recorded during the first 100 nsec after the laser flash. All experiments were conducted with the FWHM=50 nsec resolution MCT detector.

spectrum in the region 1500-1200 cm^{-1} on an expanded absorbance scale. Transient product bands are readily observed at 1542, 1354, and 1286 cm^{-1}. Additionally, two bands were noted at 1408 and 1389 cm^{-1} with signal-to-noise barely above one, but strictly reproducible. Nanosecond spectra revealed a detector response limited rise of the transient spectrum. A 100 nsec time slice of the strongest absorption at 1542 cm^{-1} is shown in Fig. 14. The decay of this band is exponential, with a time constant of 19.6 ±1.0 µsec (Fig. 15). It is comparable with the lifetime of triplet duroquinone reported in various solvents (61,62).

^{18}O substitution of duroquinone results in a 25 cm^{-1} red shift of the 1542 cm^{-1} triplet state absorption. None of the other bands of the transient are affected significantly. This shift is close to the 29 cm^{-1} red shift of the ground state duroquinone C=O stretch at 1642 cm^{-1} (63) and is typical for a carbon-oxygen bond. The triplet state absorptions at 1354 and 1286 cm^{-1} are expected to be predominantly ring CC stretching modes by analogy with the infrared spectrum ground state duroquinone and of benzoquinone (64,65). The derivative feature around 1385 cm^{-1} signals a small (<10 cm^{-1}) blue shift of the triplet state mode relative to the ground state fundamental. It is assigned to the symmetric CH_3 bending mode which is indeed expected to be little affected by the electronic excitation. The small but reproducible triplet absorption at 1408 cm^{-1} is most likely a mixture of ν(CC) and asymmetric CH_3 bending modes (64,66).

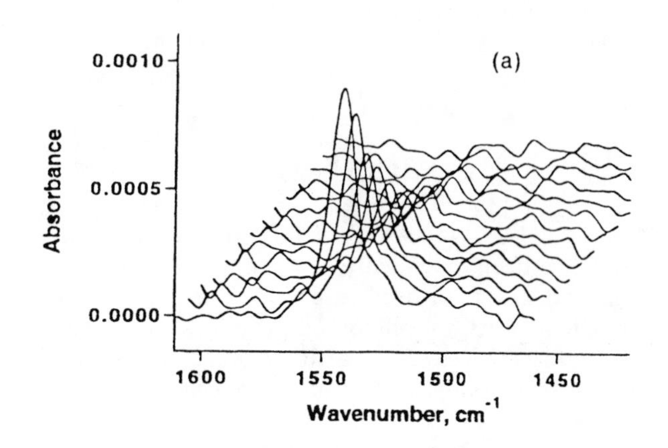

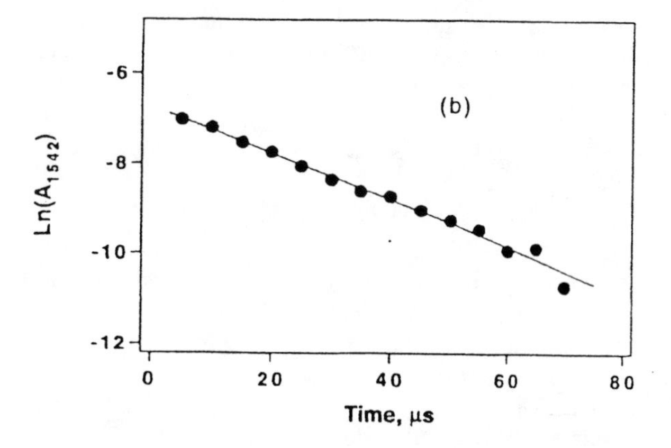

FIGURE 15. (a) Decay of the transient C=O stretch absorption of triplet duroquinone at 1542 cm^{-1}. (b) Fit of the absorbance decay yields a single exponential decay time of 19.6 µsec.

From the modest red shift of 100 cm^{-1} for the ν(CO) mode, we conclude that the CO group of triplet duroquinone has a double bond. Hence, the lowest triplet state of the molecule has predominantly $\pi\pi^*$ character. This is consistent with the proposal by Kemp and Porter that the T_1 state of duroquinone is $\pi\pi^*$ in polar solvents such as ethanol or water (62). It is to be expected that in the highly polar cages of zeolite NaY the duroquinone molecules will have the polar $\pi\pi^*$ state as the most stable triplet state, as observed. Strong stabilization of $\pi\pi^*$ triplet states in cation-exchanged zeolites has been proposed recently for a number of related aromatic ketones (67).

These measurements demonstrate the feasibility of broad band infrared absorption spectroscopy in zeolites on the nano-to-millisecond time scale in the 4000-1200 cm^{-1} region using the step-scan FT method (zeolites absorb strongly in most regions below 1200 cm^{-1}). It opens up detailed mechanistic studies of any chemical reaction or physical process that can be triggered with a nanosecond laser flash. For example, time-resolved studies of recently established reactions of hydrocarbons with O_2 under visible light in cation-exchanged zeolite Y can be studied by this method (68,69). Such work is in progress.

CONCLUSIONS

Time-resolved step-scan FT-IR absorption spectroscopy has been extended to 20 nanosecond resolution. The method has been used to elucidate mechanistic and dynamical aspects of photobiological systems, excited states of organics and transition metal complexes in solution and in solids, and of photo-induced organic and organometallic reactions in various condensed media. This work demonstrates the versatility of this new nanosecond spectroscopic tool for mechanistic studies in chemistry and biology. Particularly useful is the access to the entire fingerprint region of the infrared spectrum, which opens up structural information on nanosecond dynamics of chemical reactions previously beyond reach.

ACKNOWLEDGMENTS

The work in zeolites was conducted by postdoctoral fellows H. Sun and S. Vasenkov. The bacteriorhodopsin project was done in collaboration with Prof. R. A. Mathies, UC Berkeley Chemistry Department (Dr. W. Hage, Dr. M. Kim, E. Kauffmann). The alkane activation study was a collaboration with Profs. C. B. Harris and R. G. Bergman, UC Berkeley Chemistry Department (Dr. T. Lian, S. Bromberg, M. Asplund). This work was supported by the Director, Office of Energy Research, Office of Basic Energy Sciences, Chemical Sciences Division of the U. S. Department of Energy, under contract DE-AC03-76SF00098.

REFERENCES

1. Uhmann, W., Becker, A., Taran, C., and Siebert, F., *Appl. Spectrosc.* **45**, 390-397 (1991).
2. Palmer, R. A., Chao, J. L., Dittmar, R. M., Gregoriou, V. G., and Plunkett, S. E., *Appl. Spectrosc.* **47**, 1297-1310 (1993).
3. Plunkett, S. E., Chao, J. L., Tague, T. J., and Palmer, R. A., *Appl. Spectrosc.* **49**, 702-708 (1995).
4. Weidlich, O., and Siebert, F., *Appl. Spectrosc.* **47**, 1394-1400 (1993).
5. Nolker, K., and Siebert, F., In *Spectroscopy of Biological Molecules*, Hester, R. E., and Girling, R. B. (eds.), Cambridge: The Royal Society of Chemistry, 1991, pp. 205-206.
6. Schoonover, J. R., Strouse, G. F., Dyer, R. B., Bates, W. D., Chen, P., and Meyer, T. J., *Inorg. Chem.* **35**, 273-274 (1996).
7. Hartland, G. V., Xie, W., Dai, H. L., Simon, A., and Anderson, M. J., *Rev. Sci. Instrum.* **63**, 3261-3267 (1992).
8. Hermann, H., Grevels, F. W., Henne, A., and Schaffner, K., *J. Phys. Chem.* **86**, 5151-5154 (1982).
9. Poliakoff, M., and Weitz, E., *Adv. Organomet. Chem.* **25**, 277-316 (1986).
10. Turner, J. J., George, M. W., Johnson, F. P. A., and Westwell, J. R., *Coord. Chem. Rev.* **125**, 101-114 (1993).
11. George, M. W., Poliakoff, M., and Turner, J. J., *Analyst* **119**, 551-560 (1994).
12. Yuzawa, T., Kato, C., George, M. W., and Hamaguchi, H., *Appl. Spectrosc.* **48**, 684-690 (1994).
13. Iwata, K., and Hamaguchi, H., *Appl. Spectrosc.* **44**, 1431-1437 (1990).
14. Sluggett, G. W., Turro, C., George, M. W., Koptyug, I. V., and Turro, N. J., *J. Am. Chem. Soc.* **117**, 5148-5153 (1995).
15. Neville, A. G., Brown, C. E., Rayner, D. M., Lusztyk, J., and Ingold, K. U., *J. Am. Chem. Soc.* **113**, 1869-1870 (1991).
16. Ishikawa, Y., Hackett, P. A., and Rayner, D. M., *J. Am. Chem. Soc.* **109**, 6644-6650 (1987).
17. Hu, X., Frei, H., and Spiro, T. G., *Biochemistry* **35**, 13001-13005 (1996).
18. Dioumaev, A. K., and Braiman, M. S., *J. Phys. Chem. B* **101**, 1655-1662 (1997).
19. Rammelsberg, R., Hessling, B., Chorongiewski, H., and Gerwert, K., *Appl. Spectrosc.* **51**, 558-562 (1997).
20. Chen, P., and Palmer, R. A., *Appl. Spectrosc.* **51**, 580-583 (1997).
21. Chen, P., Omberg, K. M., Kavalinuas, D. A., Treadway, J. A., Palmer, R. A., and Meyer, T. J., *Inorg. Chem.* **36**, 954-955 (1997).
22. Mathies, R. A., Lin, S. W., Ames, J. B., Pollard, W. T., *Annu. Rev. Biophys. and Biophys. Chem.* **20**, 491-518 (1991).
23. Henderson, R., Baldwin, J. M., Ceska, T. A., Zemlin, F., Beckmann, E., and Downing, K. H., *J. Mol. Biol.* **213**, 899-929 (1990).
24. Fodor, S. P. A., Ames, J. B., Gebhard, R., Van den Berg, E. M. M., Stoeckenius, W., Lugtenburg, J., and Mathies, R. A., *Biochemistry* **27**, 7097-7101 (1988).
25. Shichida, Y., Matuoka, S., Hidaka, Y., Yoshizawa, T., *Biochim. Biophys. Acta* **723**, 240- (1983).
26. Sasaki, J., Yuzawa, T., Kandori, H., Maeda, A. Hamaguchi, H., *Biophys. J.* **68**, 2073-2080 (1995).
28. Hage, W., Kim, M., Frei, H., and Mathies, R. A., *J. Phys. Chem. B* **100**, 16026-16033 (1996).
28. Braiman, M. S., Mogi, T., Marti, T., Stern, L. J., Khorana, H. G., and Rothschild, K. J., *Biochemistry* **27**, 8516-8520 (1988).
29. Diller, R., Iannone, M., Cowen, B. R., Maiti, S., Bogomolni, R. A., and Hochstrasser, R. M., *Biochemistry* **31**, 5567-5572 (1992).
30. Hill, C. L. (ed.), *Activation and Functionalization of Alkanes,* New York: Wiley, 1989.
31. Lyons, J. E., and Parshall, G. W., *Catal. Today* **22**, 313-334 (1994).
32. Janowicz, A. H., Bergman, R. G., *J. Am. Chem. Soc.* **104**, 352- (1982).
33. Hoyano, W. A., Graham, G., *J. Am. Chem. Soc.* **104**, 3723- (1982).
34. Bergman, R. G. *Science* **223**, 902- (1984).
35. Perutz, R. N., *Pure Appl. Chem.* **62**, 1103-1106 (1990).
36. Wasserman, E. P., Moore, C. B., and Bergman, R. G., *Science* **255**, 315-318 (1992).
37. Bengali, A. A., Schultz, R. H., Moore, C. B., and Bergman, R. G., *J. Am. Chem. Soc.* **116**, 9585-9589 (1994).
38. Perutz, R. N., Belt, S. T., McCamley, A., and Whittlesey, M. K., *Pure Appl. Chem.* **62**, 1539-1545 (1990).

39. Lian, T., Bromberg, S. E., Yang, H., Bergman, R. G., and Harris, C. B. *J. Am. Chem. Soc.* **118**, 3769-3770 (1996).
40. Bromberg, S. E., Yang, H., Asplund, M. C., Lian, T., McNamara, B. K., Kotz, K. T., Yeston, J. S., Wilkens, M., Frei, H., Bergman, R. G., and Harris, C. B., *Science*, submitted.
41. Trulson, M. O., Dollinger, G. D., and Mathies, R. A., *J. Chem. Phys.* **90**, 4274-4281 (1989).
42. Reid, P. J., Doig, S. J., Wickam, S. D., and Mathies, R. A., *J. Am. Chem. Soc.* **115**, 4754-4763 (1993).
43. Lawless, M. K., Wickam, S. D., and Mathies, R. A., *Acc. Chem. Res.* **28**, 493-502 (1995).
44. Pullen, S., Walker, II, L. A., Donovan, B., and Sension, R. J ., *Chem. Phys. Lett.* **242**, 415-420 (1995).
45. Carreira, L. A., *J. Chem. Phys.* **62**, 3851-3854 (1975).
46. Ackerman, J. R., and Kohler, B. E., *J. Chem. Phys.* **80**, 45-50 (1984).
47. Kauffmann, E., Frei, H., and Mathies, R. A., *Chem. Phys. Lett.* **266**, 554-559 (1997).
48. Furukawa, Y., Takeuchi, H., Harada, I., and Tasumi, M., *J. Mol. Struct.* **100**, 341-350 (1983).
49. Yoshida, H., Furukawa, Y., and Tasumi, M., *J. Mol. Struct.* **194**, 279-299 (1989).
50. Ramamurthy, V., In *Photochemistry in Organized and Constrained Media*, Ramamurthy, V., (ed.), New York:VCH Publishers, 1991, 429-493.
51. Turro, N. J., *Pure Appl. Chem.* **58**, 1219-1229 (1986).
52. Breck, D. W., *Zeolite Molecular Sieves: Structure, Chemistry, and Use*, New York: Wiley, 1974.
53. Gilbert, A., and Baggott, J., *Essentials of Molecular Photochemistry*, Boca Raton: CRC Press, 1991, p. 288-294.
54. George, M. W., Kato, C., and Hamaguchi, H., *Chem. Lett.,* 873-876 (1993).
55. Tahara, T., Hamaguchi, H., and Tasumi, M., *J. Phys. Chem.* **94**, 170-178 (1990).
56. Tahara, T., Hamaguchi, H., and Tasumi, M., *J. Phys. Chem.* **91**, 5875-5880 (1987).
57. Ohmori, N., Susuki, T., and Ito, M., *J. Phys. Chem.* **92**, 1086-1093 (1988).
58. Yuzawa, T., Kato, C., George, M. W., and Hamaguchi, H., *Appl. Spectrosc.* **48**, 684-691 (1994).
59. Kärger, J., and Ruthven, D. M., *Diffusion in Zeolites*, New York: Wiley, 1992, Ch. 13.
60. Birks, J. B., *Photophysics of Aromatic Molecules*, London: Wiley, 1970, Ch. 10.
61. Amouyal, E., and Bensasson, R., *J. Chem. Soc., Farad. Trans. I* **72**, 1274-1287 (1976).
62. Kemp, D., and Porter, G., *Proc. Roy. Soc. London, Ser. A* **326**, 117-130 (1971).
63. Breton, J., Burie, J. R., Boullais, C., Berger, G., and Nabedryk, E., *Biochemistry* **33**, 12405-12415 (1994)
64. Baruah, G. D., Singh, S. N., and Jayaswal, M. G., *Indian J. Pure Appl. Phys.* **7**, 280-281 (1969).
65. Becker, E. D., Charney, E., and Anno, T., *J. Chem. Phys.* **42**, 942-949 (1965).
66. Flaig, W., and Salfeld, J. C., *Ann. Chem.* **626**, 215-224 (1959).
67. Ramamurthy, V., Corbin, D. R., and Johnston, L. J., *J. Am. Chem. Soc.* **114**, 3870-3877 (1992).
68. Sun, H., Blatter, F., and Frei, H., In *Heterogeneous Hydrocarbon Oxidation*, Oyama, S. T., and Warren, B. K., (eds.), Washington, D. C.: ACS Symposium Series No. 638, 1996, p. 409-427.
69. Frei, H., Blatter, F., and Sun, H., *CHEMTECH* **26**, 24-30 (1996).

New Quantitative Optical Depth Profiling Methods by Attenuated Total Reflectance Fourier Transform Infrared Spectroscopy: Multiple Angle / Single Frequency and Single Angle / Multiple Frequency Approaches

Hatsuo Ishida[1], Sanong Ekgasit[2], and Daniel P. Sanders[1]

[1]*Department of Macromolecular Science, Case Western Reserve University, Cleveland, Ohio 44106-7202, U.S.A.*
[2]*Thailand Institute of Scientific and Technological Research, Metal and Material Technology Department, Bangkok 10900, Thailand*

New analytical techniques for depth profiling using multiple-angle or single-angle attenuated total reflection Fourier transform infrared spectroscopy have been developed. Unlike the techniques traditionally used, these new approaches do not require prior knowledge of the complex refractive index profile for the depth profiling calculation, can be applied to strong absorption bands, and are able to utilize incident beams of arbitrary degree of polarization. The multiple-angle approach performs the depth profiling calculations on a set of angle-dependent absorptances at a single frequency, whereas the multiple-frequency approach uses a set of frequency-dependent absorptances from a single spectrum. Both techniques are based on a similar algorithm for the depth profiling calculations. The depth profiling analysis consists of three steps. First, the estimated complex refractive index profile is obtained by solving a set of linear equations of absorptance. This estimated profile is used as a trial profile in a linear fitting of absorptances calculated by exact optical theory to the experimental values. The reconstructed profile is obtained from the converged profile. Finally, the reconstructed profile is used in a non-linear fitting of reflectance. The converged profile from the non-linear fitting is the depth profile of the system. Many types of composition gradients can be successfully profiled without any *a priori* knowledge of the nature of the profile. The effect of noise on the quality of the profile is evaluated. The multiple-angle technique is appropriate for systems with slow changes in composition profile and those with significant intermolecular interactions, while the multiple-frequency approach is suitable for the study of systems with rapidly changing phenomenon where little intermolecular interactions or structural changes are present. The rapid spectral acquisition time of the multiple-frequency approach introduces the possibility of real-time depth profiling.

INTRODUCTION

Surface Characterization

The surface properties and composition of organic media have become the focus of intense interest in recent years. Many of the properties of organic systems arise from the specific chemical composition, structure, and orientation found at the surface or interphase between one medium and another. Properties such as catalysis, adhesion, diffusion, chemical reactivity, wettability, solubility, conductivity, adsorptivity, and degradative stability are directly influenced and, for the most part, determined by the chemical composition, molecular structure, and orientation near or at the surface. Often compositional, structural, and orientational gradients near the surface of organic systems are formed as a result of specific surface treatments and reactions, processing techniques, thermodynamic considerations, aging, and molecular design.

As the gap between macroscopic bulk processing methods and nanoscopic chemical syntheses becomes increasingly narrow, the surface and interphase in organic systems assume a more prominent role in determining vital material properties. Sensitive characterization techniques are required to carefully probe the composition and structure of molecularly-designed, macroscopic systems. Already, organic systems of surprisingly small scale such as polymeric waveguides, optical coatings, and selectively permeable membranes are contributing greatly not only in high-technology but also in day-to-day devices found in every household. If such miniaturization of existing technologies, manipulation of interfacial and surface phenomenon, and characterization of increasingly small systems is to continue, understanding of the specific chemical composition, structure, and orientation of the surface and interphase regions in organic systems must be gained in order to explain their observed phenomena, predict their behavior, and design their properties. Sensitive surface characterization techniques are required to probe near-surface and interphase regions in order to gain this understanding.

Recently, two new quantitative depth profiling techniques based on attenuated total reflectance (ATR) Fourier transform infrared (FTIR) spectroscopy have been developed in our laboratory (1-3). These two techniques overcome several major difficulties which have limited the applicability and widespread use of ATR as a quantitative depth profiling technique. Most importantly, they successfully eliminate the need for any *a priori* information

CP430, *Fourier Transform Spectroscopy:* 11th International Conference
edited by J.A. de Haseth

about the composition profile and remove operator bias inherent in such assumptions. The techniques have also been expanded to allow for the use of unpolarized incident radiation. In addition to reviewing the theoretical basis underlying the two new techniques, this paper will endeavor to compare and contrast these two techniques using spectrally simulated data in order to highlight the differences between these two techniques and their suitability for various systems.

BACKGROUND

Surface Characterization and Depth Profiling Techniques

The perfect surface characterization technique would be non-destructive, allow for quick, easy, in-situ sampling, and yield high accuracy, high resolution, quantitative information on a variety of scales (atomic, molecular, structural, and orientational) as a function of a deep sampling depth without requiring *a priori* knowledge of the distribution profile. At this point in time, no characterization technique can successfully claim to satisfy all of the above requirements. As a result, researchers must make sacrifices in sampling requirements, information scale, and depth of sampling when choosing a technique to investigate a particular system.

Many techniques have been used to study the depth dependent properties and phenomena in organic systems such as x-ray photoelectron spectroscopy (XPS), secondary ion mass spectrometry (SIMS), Rutherford backscattering, small-angle neutron scattering (SANS), forward recoil spectrometry (FRS), nuclear magnetic resonance spectroscopy (NMR), Raman spectroscopy, fluorescence spectroscopy, and a host of optical and Fourier transform infrared techniques.

Many of the aforementioned techniques, while offering exceptional sensitivity, are severely limited by their high cost/low availability, destructive nature, difficult sampling preparation, extreme testing conditions, and depth and scale of information obtained. An excellent review of depth profiling techniques used to characterize dielectric materials has been written by Bohn and Miller (4). A review of polymer-polymer interdiffusion with a portion on relevant characterization techniques was written by Jabbari and Peppas (5).

A host of FTIR techniques have been developed which fill the void left by the previously mentioned techniques. FTIR-based techniques are much less costly and therefore more available. Sampling requirements are generally less rigorous, the labeling of species of interest is not required, and the measurements are performed in a benign environment. Techniques such as diffuse reflectance, reflection-absorption, and external reflection have been used to gather chemical, structural, and orientational information from the surface of organic systems; however, they lack the ability to extract the depth dependence of the information obtained. The most common FTIR-based depth profiling techniques include ATR and photoacoustic spectroscopy (PAS). Several excellent reviews in the area of surface analysis and depth profiling via FTIR spectroscopic techniques have been written. Ishida (6) reviewed the quantitative surface analysis of polymers with FTIR. Urban and Koenig (7) and Fina (8) reviewed depth profiling via FTIR spectroscopy.

ATR offers a unique set of capabilities which overcome some of the disadvantages and restrictions inherent in the previously discussed techniques. ATR imposes very little sampling requirements other than good optical contact with the internal reflection element (IRE) and a wide variety of sampling arrangements are available. While ATR has a smaller depth of penetration than PAS (up to ten microns in the mid-IR), its probing range is still appropriate for a variety of applications. However, in order to perform depth profiling calculations with conventional ATR, prior knowledge of the composition profile must be known in order for the numerical fitting procedures to work. The two new quantitative ATR depth profiling techniques which have been developed in our laboratory eliminate the need for any *a priori* information about the composition profile. Before discussing the development of the two techniques, a brief review of pertinent ATR theory as well as a description of conventional ATR depth profiling procedures is necessary.

Attenuated Total Reflectance Fourier Transform Infrared Spectroscopy

ATR was independently developed by Harrick (9-10) and Fahrenfort (11) in the 1960's. ATR has since flourished as a spectroscopic technique, finding many applications in the areas of surface characterization (12-23), depth profiling (1, 2, 3, 24-30), and diffusion in polymer films (31-40). Readers interested in more complete descriptions of ATR theory and applications of ATR-FTIR spectroscopy should consult a number of excellent reviews found in the literature (6, 8, 41-44). In the following a brief overview of ATR theory, all quantities will be described in terms of incident angle and frequency to emphasize their dependence upon the choice of experimental conditions. This convention will help ensure that the subsequent discussion on the development of the new depth profiling approaches will be clear.

When an optically dense infrared (IR) transparent medium is placed in contact with an optically rare non-absorbing (i.e. the extinction coefficient, k, is zero) medium, total internal reflection will occur when an incident beam traveling through the optically dense medium strikes the optically rarer medium above the critical angle given by Snell's law (42)

$$\theta_c = \sin^{-1}\left(\frac{n_2(\upsilon)}{n_1(\upsilon)}\right) \tag{1}$$

where θ_c is the critical angle, $n_1(\upsilon)$ is the real portion of the complex refractive index of the optically dense

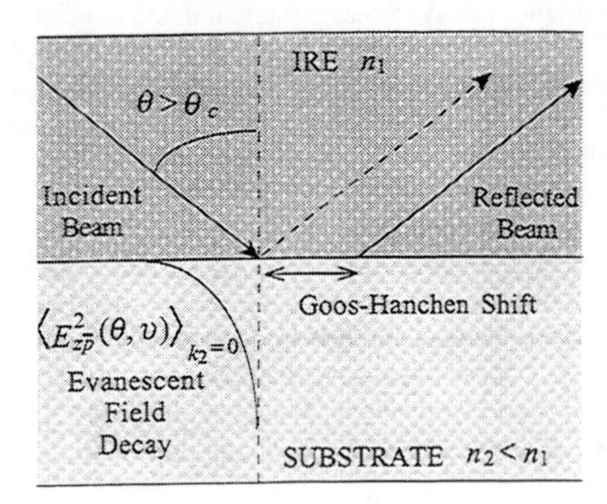

FIGURE 1. Fundamentals of total internal reflection. $\left\langle E^2_{z\bar{p}}(\theta, \upsilon)\right\rangle_{k_2=0}$ is the mean square evanescent field at depth z with an incident beam of degree of polarization $\bar{p}$ and frequency υ at an angle of incidence θ which is greater than the critical angle θ_c. n_1 and n_2 are the real portions of the complex refractive indices of the optically dense and rare mediums, respectively.

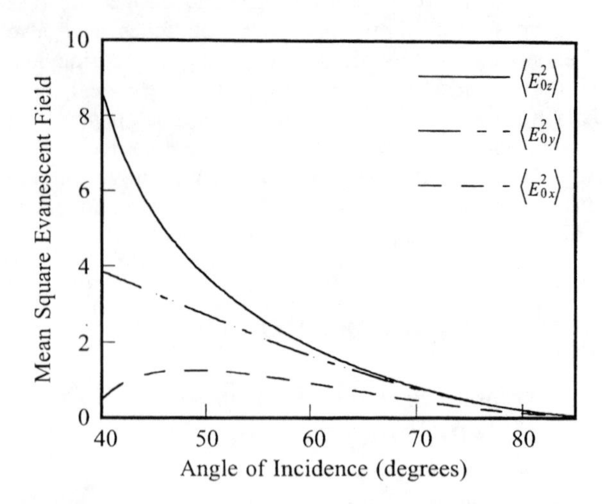

FIGURE 2. X-, y-, and z-components of the mean square evanescent field at the surface of the rarer medium in total internal reflection. Optical parameters: $n_1 = 2.40$, $n_2 = 1.50$.

medium, and $n_2(\upsilon)$ is the real portion of the complex refractive index of the optically rare medium (i.e. $n_2 < n_1$) at wavenumber υ. The complex refractive index of a medium is defined as:

$$\hat{n}(\upsilon) = n(\upsilon) + ik(\upsilon) \tag{2}$$

where i is equal to $\sqrt{-1}$, $\hat{n}(\upsilon)$ the complex refractive index, $n(\upsilon)$ the refractive index, and $k(\upsilon)$ the extinction coefficient at frequency υ. The optical conditions for total internal reflection are illustrated in Fig 1. Although no light actually penetrates into the rarer medium and no net energy flow crosses the boundary between the two media, an electric field is generated at the boundary and decays exponentially with depth into the rarer medium. This mean square electric field in a non-absorbing medium is termed the evanescent field. The decay function of the evanescent field is given by (1, 6, 8, 41-44):

$$\left\langle E^2_{zl}(\theta, \upsilon)\right\rangle_{k_2=0} = \left\langle E^2_{0l}(\theta, \upsilon)\right\rangle_{k_2=0} e^{-2z/d_p(\theta,\upsilon)} \tag{3}$$

where l indicates the polarization of the incident beam, θ the incident angle, υ the frequency of the incident beam, and $\left\langle E^2_{0l}(\theta, \upsilon)\right\rangle_{k_2=0}$ and $\left\langle E^2_{zl}(\theta, \upsilon)\right\rangle_{k_2=0}$ the mean square evanescent fields at the surface and depth z of the rarer medium respectively. The evanescent field at the surface of the rarer medium is determined by the angle of incidence, refractive indices of the two media, and the polarization of the incident radiation. The mean square evanescent wave at the surface for a variety of incident angles is shown in Fig 2. The decay constant, $d_p(\theta, \upsilon)$, is termed the penetration depth and is given by (1, 6, 8, 41-44):

$$d_p(\theta, \upsilon) = \frac{1}{2\pi\upsilon n_1(\upsilon)\sqrt{\sin^2\theta - \left(\dfrac{n_2(\upsilon)}{n_1(\upsilon)}\right)^2}}. \tag{4}$$

The penetration depth is the distance into the material where the mean square evanescent field decreases to $1/e^2$ (roughly 13%) of its value at the surface. One should note that $d_p(\theta, \upsilon)$ is independent of the polarization of the incident beam.

Goos and Hänchen discovered that the beam upon reflection is shifted by a distance that is dependent upon the angle of incidence, refractive indices of the system, and the frequency and polarization of the incident beam (45). While the amplitude of the evanescent field decays as it penetrates into the medium, the phase of the evanescent field carries the energy of the incident radiation parallel to the interface. After traveling the Goos-Hänchen shift, the radiation reflects back into the incident medium.

If the medium in contact with the incident medium is absorbing (i.e. $k_2 \neq 0$), total internal reflection no longer occurs and the infrared beam is attenuated as a result of absorption by the medium. The absorptance in the ATR configuration is related to reflectance by (1, 8, 29, 46-47):

$$A_l(\theta, \upsilon) = 1 - R_l(\theta, \upsilon) \tag{5}$$

where $A_l(\theta, \upsilon)$ and $R_l(\theta, \upsilon)$ are the absorptance and reflectance. The absorptance is given by (8, 41, 44, 48-51):

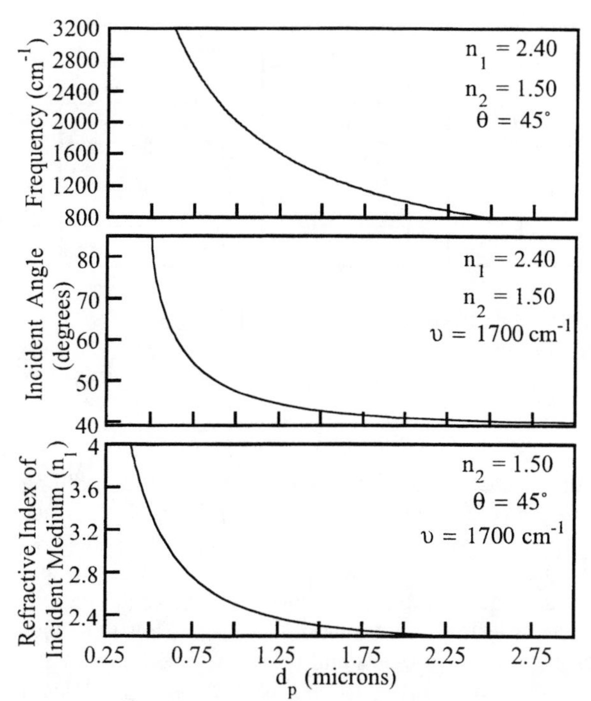

FIGURE 3. Effect of optical parameters on the penetration depth of the evanescent field. Default optical parameters: n_1 = 2.40, n_2 = 1.50, θ = 45°, and υ = 1700 cm^{-1}.

$$A_l(\theta,\upsilon) = \frac{1}{n_1(\upsilon)\cos\theta}\int_0^\infty n_{2z}(\upsilon)\alpha_{2z}(\upsilon)\left\langle E_{zl}^2(\theta,\upsilon)\right\rangle dz \quad (6)$$

where $n_{2z}(\upsilon)$ is the refractive index, $\alpha_{2z}(\upsilon)$ the absorption coefficient, and $\left\langle E_{zl}^2(\theta,\upsilon)\right\rangle$ the mean square electric field at depth z in the absorbing medium. Substitution of $4\pi k_{2z}(\upsilon)$ for $\alpha_{2z}(\upsilon)$ in equation 6, where $k_{2z}(\upsilon)$ is the extinction coefficient of the absorbing medium at depth z and frequency υ, yields (8):

$$A_l(\theta,\upsilon) = \frac{4\pi\upsilon}{n_1(\upsilon)\cos\theta}\int_0^\infty n_{2z}(\upsilon)k_{2z}(\upsilon)\left\langle E_{zl}^2(\theta,\upsilon)\right\rangle dz\,. \quad (7)$$

The absorptance can be seen as simply the product of the optical constants with the corresponding mean square electric field at each point with a cosine term in front of the integral to correct for the cross-sectional area difference between the beam in the incident medium and the area of the interface impinged upon.

Given the nature of the mean square electric field decay in the absorbing medium, the material closest to the interface has a much higher per thickness contribution to the total absorptance than material found further into the bulk. The material found within one-half of the penetration depth is responsible for 63% of an absorptance band in a homogeneous sample (52). ATR is in this manner surface selective, preferentially yielding information regarding the material found closest to the

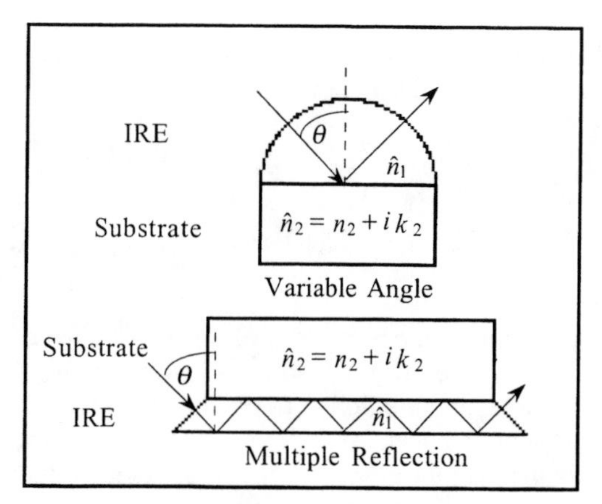

FIGURE 4. ATR configurations.

IRE. In fact, films as thin as 1.2 nm (53) will yield signals via ATR as absorption bands that can be detected.

Since there is appreciable mean square electric field beyond the penetration depth in the material, the material beyond this depth can still significantly affect the total absorptance of the system. Therefore, a working depth, $d_w(\theta,\upsilon)$, is defined as the point at which the mean square evanescent field decreases to 0.25% of its value at the surface. The working depth is approximately 3 times the penetration depth (43).

Depth Profiling Approaches with ATR FTIR

Through the examination of equation 4, one can see that the depth of penetration can be altered by changing the angle of incidence, incident medium, or frequency. The influence of changing the optical parameters on the penetration depth is shown in Fig. 3. Depth of penetration can be decreased by using an incident medium with a much greater refractive index than that of the rarer medium, a larger angle of incidence, or a higher frequency of incident radiation.

Of the many configurations available, two primary configurations are used in ATR FTIR depth profiling as shown in Fig. 4. The first uses hemicylindrical or hemispherical IREs in a variable angle configuration. The number of incident angles available is limited by the sophistication of the setup and the critical angle of the system. In the second configuration, a long planar IRE is used in which the incident beam can undergo multiple consecutive reflections off the sample at a single angle of incidence, thereby increasing the sensitivity of the technique.

Quantitative Depth Profiling Techniques

In the case of a weakly absorbing medium (k_2 << 0.1), the electric field is only slightly disrupted by the absorption. Therefore, the mean square electric field in the absorbing medium can be assumed to be identical to the mean square evanescent field in a non-absorbing medium of the same refractive index (1).

$$\left\langle E_{z\bar{p}}^{2}(\theta,\upsilon)\right\rangle_{k_2<<0.1} \cong \left\langle E_{z\bar{p}}^{2}(\theta,\upsilon)\right\rangle_{k_2=0} = \left\langle E_{0\bar{p}}^{2}(\theta,\upsilon)\right\rangle_{k_{2z}=0} e^{-2z/d_p(\theta,\upsilon)} \quad (8)$$

The equations in the remainder of this section have been modified from the original authors' expressions to be consistent with the notation convention established earlier in this paper. For a system in which the complex refractive index of the absorbing medium is not a function of depth with small extinction coefficients, Tompkins (54) originally derived equation 6 as:

$$A(\theta,\upsilon) = \frac{n_2(\upsilon)\left\langle E_{0\bar{p}}^{2}(\theta,\upsilon)\right\rangle_{k_2=0}}{n_1(\upsilon)\cos\theta} \int_0^\infty \alpha_{2z}(\upsilon)\, e^{-2z/d_p(\theta,\upsilon)}\, dz \quad (9)$$

Inverse Laplace Transformations

Hirschfeld (55) further generalized equation 9 for systems with an unknown profile in the extinction coefficient under the same limiting assumptions as Tompkins to yield:

$$A(\theta,\upsilon) \propto \int_0^\infty \alpha_{2z}(\upsilon)\, e^{-(\beta(\theta,\upsilon)z+\int_0^\infty \alpha_{2z}(\upsilon)dz)}\, dz \quad (10)$$

where

$$\beta(\theta,\upsilon) = \frac{4\pi\sqrt{n_1^2(\upsilon)\sin^2\theta - n_2^2(\theta,\upsilon)}}{\lambda} = \frac{2}{d_p(\theta,\upsilon)} \quad (11)$$

in which λ is the wavelength of the incident radiation. The additional integral in the exponential function in equation 11 takes the attenuation of the beam by the absorption process into account. When $\alpha_{2z}(\upsilon) << \beta(\theta,\upsilon)$, equation 10 can be simplified to:

$$A(\beta(\theta,\upsilon)) \propto \int_0^\infty \alpha_{2z}(\upsilon)\, e^{-\beta(\theta,\upsilon)z}\, dz \quad (12)$$

where the absorptance is described in terms of its dependence on the decay constant, $\beta(\theta,\upsilon)$. As a result of equation 12, a set of angle-dependent absorptances, $A(\beta(\theta,\upsilon))$, can be transformed into a set of depth-dependent absorption coefficients, $\alpha_{2z}(\upsilon)$, via an inverse Laplace transformation, thereby obtaining the depth profile. Other mathematical treatments of this problem have been offered by Rozanov and Zolotarev (24) and Stuchebryukov et al. (25). The technique was advanced by Fina and Chen (29) and, later, Shick et al. (47) who derived polarization dependent expressions for the Laplace transforms.

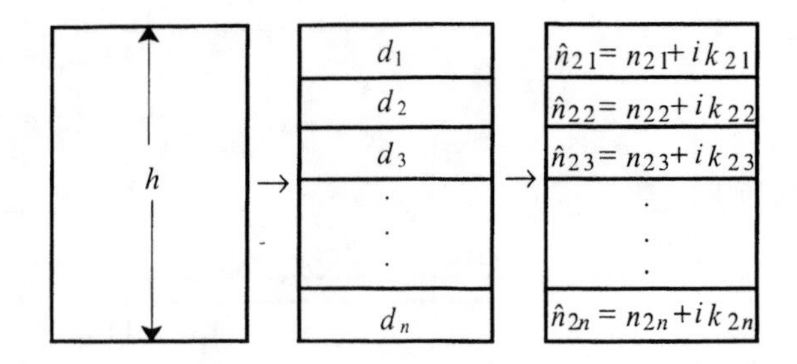

FIGURE 5. Definition of stratified medium. h is the thickness of the film. d_j is the thickness and $\hat{n}_j$ the complex refractive index of the of the j-th stratified layer.

Even when the extinction coefficients at the profiling frequency are restricted to very small values, the Laplace transformation is an "ill-conditioned" problem because there is more than one unique solution to the problem and not enough information exists in the experimental data to limit the solution to one possibility. As a result, small errors in experimental data and rounding deviations in the calculations lead to large errors in the calculated depth profile (50). In order to perform this transformation, a guessed profile of the absorption coefficient is required. This approach has been used by several groups (29, 47, 51, 56). The accuracy of profiles obtained by this approach is not good unless an extremely high signal-to-noise ratio is achieved and the guessed profile is fairly accurate (4). Although in some systems prior information about the profile is known, in many systems such information is not available. In addition, since the penetration depth range does not extend to infinity, a minimum distance exists below which the calculations cannot distinguish between any profile that conserves area (47).

Exact Optical Theory

The depth profiling methods based on exact optical theory discussed in this paper model a material of thickness, h, with depth-dependent properties as a stratified medium composed of a stack of n layers as illustrated in Fig. 5. Each layer is assumed to be homogeneous, isotropic, and parallel plane-bound with the adjacent layers. The complex refractive index of each layer, j, of thickness, d_j, is defined as (1-3):

$$\hat{n}_{2j}(\upsilon) = n_{2j}(\upsilon) + ik_{2j}(\upsilon). \quad (13)$$

The transmittance and reflectance of the stratified medium is identical to that of the film. It is important to remember that the assumption of isotropy in each layer requires that the complex refractive index is identical in the x, y, and z directions within each layer (i.e. $n_{2j_x}(\upsilon) = n_{2j_y}(\upsilon) = n_{2j_z}(\upsilon)$, $k_{2j_x}(\upsilon) = k_{2j_y}(\upsilon) = k_{2j_z}(\upsilon)$).
The optical configuration is shown in Fig 6. The plane of

incidence is defined by the plane perpendicular to the film that contains the incident and reflected beams. The incident medium is infrared transparent with a refractive index $n_1(\upsilon)$ while the infinitely thick substrate has a complex refractive index of $\hat{n}_3(\upsilon)$. The perpendicular and parallel polarized components of the incident radiation are given the notation s and p, respectively. For s polarization, the electric component of the electromagnetic wave is perpendicular to the plane of incidence and the magnetic component is parallel to the plane of incidence. For p polarization, the electric component of the electromagnetic wave is parallel to the plane of incidence and the magnetic component is perpendicular to the plane of incidence. Reflected and transmitted radiation are denoted by r and t, respectively.

Since the tangential components of both the electric and magnetic fields must be continuous across the interface between two layers, a characteristic 2x2 matrix M_j can be developed which relates the fields at the two boundaries of the j-th layer of the stratified medium. Matrix M_j is given by:

$$M_j = \begin{pmatrix} \cos\beta_j & \frac{-i}{p_j}\sin\beta_j \\ -i\,p_j\sin\beta_j & \cos\beta_j \end{pmatrix} \qquad (14)$$

where

$$\beta_j = 2\pi\upsilon\, d_j(\hat{n}^2_{2j}(\upsilon) - n_1^2(\upsilon)\sin^2\theta_1)^{1/2} \qquad (15)$$

and

$$p_j = (\hat{n}^2_{2j}(\upsilon) - n_1^2(\upsilon)\sin^2\theta_1)^{2} \qquad s\text{-polarization} \qquad (16)$$

$$p_j = \frac{(\hat{n}^2_{2j}(\upsilon) - n_1^2(\upsilon)\sin^2\theta_1)^{2}}{\hat{n}^2_{2j}(\upsilon)} \qquad p\text{-polarization} \qquad (17)$$

and θ_1 is the incident angle with respect to the surface, $\hat{n}_{2j}$ is the complex refractive index and d_j is the thickness of the j-th layer. Since the stratified layer consists of many adjacent layers, the total M matrix is the product of the consecutive component M_j matrices in the stratified medium.

$$M = \begin{bmatrix} m_{11} & m_{12} \\ m_{21} & m_{22} \end{bmatrix} = \prod_{j=1}^{N-1} M_j \qquad (18)$$

From this resulting matrix, the reflectivity at the surface of the stratified medium is determined via:

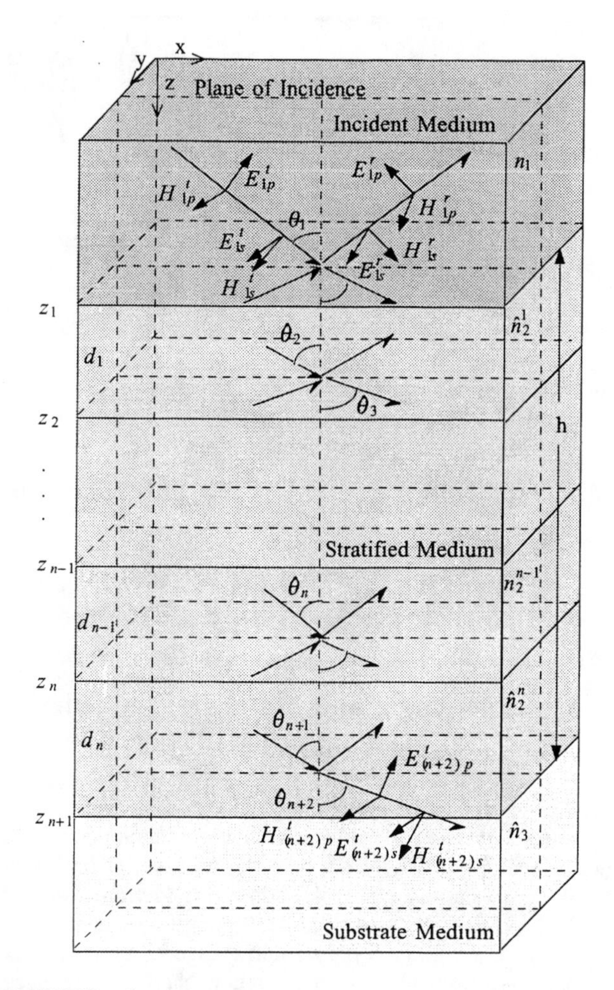

FIGURE 6. Interaction of electromagnetic radiation with a stratified medium.

$$\hat{r} = \frac{(m_{11} + m_{12}p_N)p_0 - (m_{21} + m_{22}p_N)}{(m_{11} + m_{12}p_N)p_0 + (m_{21} + m_{22}p_N)}. \qquad (19)$$

In the search for an improved depth profiling method, a technique based on exact optical theory was developed by Huang and Urban (50). When profiling frequencies were chosen where only one material absorbs, the volume fraction in the j-th layer could be determined and the refractive index calculated from the Kramers-Kronig relation. In this exact optical approach, a trial profile is used to fit calculated normalized reflectance intensities to experimental values. This approach overcomes the weak band limitation of the Laplace method and takes into account the effect of refractive index dispersion; however, it still requires a trial or guessed profile in order for the depth profiling calculations to work. Recently, two new depth profiling approaches (1-3) based on exact optical theory have been developed in our laboratory which, likewise, do not require small absorption coefficients, but also do not require a guessed profile.

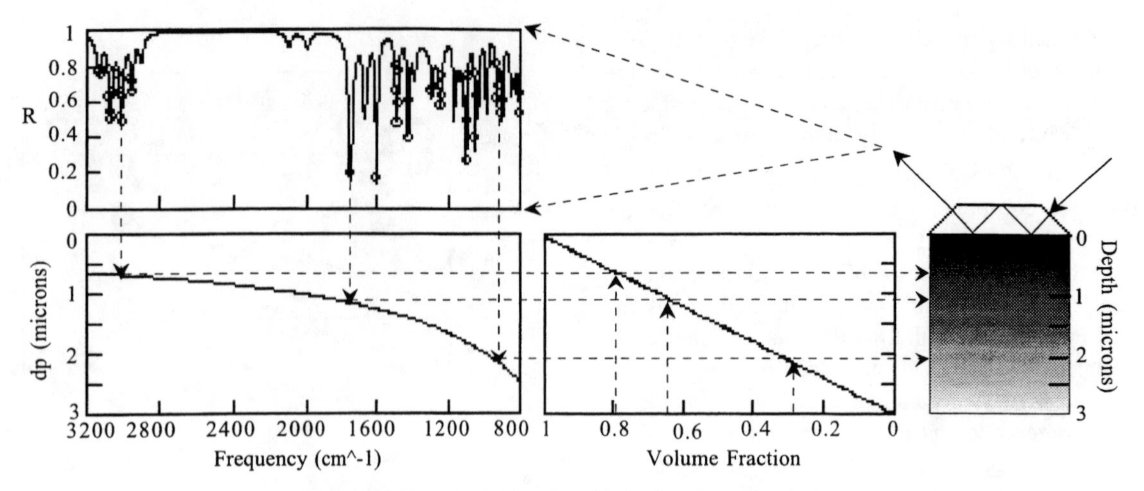

FIGURE 7. Principles of multiple-frequency analysis.

NEW ATR FTIR DEPTH PROFILING TECHNIQUES

Multiple-Angle and Multiple-Frequency Approaches

New depth profiling techniques taking advantage of both the variable-angle and multiple-reflection attachments have been developed. The first, hereafter termed the multiple-angle approach, calculates the depth profile from a set of angle-dependent absorptances at a single frequency. It does not require any *a priori* knowledge of the complex refractive index profile. In addition, it has been expanded to include the use of unpolarized or partially polarized incident beams.

The second approach, hereafter termed the multiple-frequency approach, calculates the depth profile from a set of frequency-dependent absorptances at a single angle of incidence. As discussed previously, the depth of penetration can be changed by using a different frequency as well as by changing the angle of incidence. Therefore, a single spectrum already contains information found at the entire range of depths throughout the medium up to the working depth of the lowest frequency used in the analysis. Only one composition profile will be able to generate the exact spectral intensity at each frequency chosen. An illustration of this concept is shown in Fig. 7. However, in order to accomplish this decoding of the composition profile from the reflectance spectrum, some assumptions need to be made in order to reduce the number of unknowns. It is assumed that one knows the complex refractive indices of the pure components in the system to be studied. The multiple-frequency approach does not require any prior knowledge of the composition profile and can be used with incident beams of arbitrary degree of polarization as well. Both techniques can be applied to strong absorption bands.

Theory

A short overview of the theoretical equations behind each of the techniques is needed before one can begin to understand their advantages, limitations, and appropriate uses. Those interested in the full derivation of all equations used in each technique should consult the original papers (1-3, 31).

By modeling a two component system (i.e. polymer/polymer, polymer/solvent, ...etc.) as a stratified medium, equation 7 can be rewritten as (1):

$$A_{\bar{p}}(\theta,\upsilon)=\frac{4\pi\upsilon}{n_1\cos\theta}\sum_{j=1}^{n}\left(n_{2j}(\upsilon)k_{2j}(\upsilon)\int_{z_j}^{z_{j+1}}\left\langle E_{\bar{zp}}^{2}(\theta,\upsilon)\right\rangle dz\right) \tag{20}$$

in which n is the number of stratified layers, z_j is the depth between the (j-1)-th and j-th layers, and z_{n+1} is the working depth. The absorptance of the stratified medium is simply the summation of the absorptance within each stratified layer.

One can either depth profile by measuring angle-dependent absorptance at a single frequency, thereby obtaining a set of linear equations in θ, or by measuring frequency-dependent absorptance at a single incident angle to obtain a set of linear equations in υ.

Multiple-Angle (1,2)

$$A_{\bar{p}}(\theta_1)=\frac{4\pi\upsilon}{n_1\cos\theta_1}\sum_{j=1}^{n}\left(n_{2j}k_{2j}\int_{z_j}^{z_{j+1}}\left\langle E_{\bar{zp}}^{2}(\theta_1)\right\rangle dz\right)$$

$$A_{\bar{p}}(\theta_2)=\frac{4\pi\upsilon}{n_1\cos\theta_2}\sum_{j=1}^{n}\left(n_{2j}k_{2j}\int_{z_j}^{z_{j+1}}\left\langle E_{\bar{zp}}^{2}(\theta_2)\right\rangle dz\right)$$

$$A_{\bar{p}}(\theta_3) = \frac{4\pi\upsilon}{n_1 \cos\theta_3} \sum_{j=1}^{n} \left(n_{2j} k_{2j} \int_{z_j}^{z_{j+1}} \left\langle E_{z\bar{p}}^2(\theta_3) \right\rangle dz \right) \quad (21)$$

$$\vdots$$

$$A_{\bar{p}}(\theta_m) = \frac{4\pi\upsilon}{n_1 \cos\theta_m} \sum_{j=1}^{n} \left(n_{2j} k_{2j} \int_{z_j}^{z_{j+1}} \left\langle E_{z\bar{p}}^2(\theta_m) \right\rangle dz \right)$$

where θ_i ($i = 1, 2, 3,..., m$) is the i-th incident angle.

Multiple-Frequency (3)

$$A_{\bar{p}}(\upsilon_1) = \frac{4\pi\upsilon_1}{n_1 \cos\theta} \sum_{j=1}^{n} \left(n_{2j}(\upsilon_1) k_{2j}(\upsilon_1) \int_{z_j}^{z_{j+1}} \left\langle E_{z\bar{p}}^2(\upsilon_1) \right\rangle dz \right)$$

$$A_{\bar{p}}(\upsilon_2) = \frac{4\pi\upsilon_2}{n_1 \cos\theta} \sum_{j=1}^{n} \left(n_{2j}(\upsilon_2) k_{2j}(\upsilon_2) \int_{z_j}^{z_{j+1}} \left\langle E_{z\bar{p}}^2(\upsilon_2) \right\rangle dz \right)$$

$$A_{\bar{p}}(\upsilon_3) = \frac{4\pi\upsilon_3}{n_1 \cos\theta} \sum_{j=1}^{n} \left(n_{2j}(\upsilon_3) k_{2j}(\upsilon_3) \int_{z_j}^{z_{j+1}} \left\langle E_{z\bar{p}}^2(\upsilon_3) \right\rangle dz \right)$$

$$\vdots \quad (22)$$

$$A_{\bar{p}}(\upsilon_m) = \frac{4\pi\upsilon_m}{n_1 \cos\theta} \sum_{j=1}^{n} \left(n_{2j}(\upsilon_m) k_{2j}(\upsilon_m) \int_{z_j}^{z_{j+1}} \left\langle E_{z\bar{p}}^2(\upsilon_m) \right\rangle dz \right)$$

where υ_i ($i = 1, 2, 3,..., m$) is the i-th frequency.

Both sets of linear equations can be written in a similar matrix form (1-2):

$$\mathbf{A} = \mathbf{E} \cdot \mathbf{X} \quad (23)$$

where

$$\mathbf{A} = \begin{bmatrix} A_{\bar{p}}(\theta_1 \text{ or } \upsilon_1) \\ A_{\bar{p}}(\theta_2 \text{ or } \upsilon_2) \\ A_{\bar{p}}(\theta_3 \text{ or } \upsilon_3) \\ \vdots \\ A_{\bar{p}}(\theta_m \text{ or } \upsilon_m) \end{bmatrix} \quad (24)$$

$$\mathbf{E} = \begin{bmatrix} e_{11} & e_{12} & e_{13} & \cdots & e_{1n} \\ e_{21} & e_{22} & e_{23} & \cdots & e_{2n} \\ e_{31} & e_{32} & e_{33} & \cdots & e_{3n} \\ \vdots & \vdots & \vdots & \cdots & \vdots \\ e_{m1} & e_{m2} & e_{m3} & \cdots & e_{mn} \end{bmatrix} \quad (25)$$

$$e_{ij} = C_i \int_{z_j}^{z_{j+1}} \left\langle E_{z\bar{p}}^2(\theta_i \text{ or } \upsilon_i) \right\rangle dz$$

$$(i = 1, 2, 3,..., m;\ j = 1, 2, 3,..., n) \quad (26)$$

$$\mathbf{X} = \begin{bmatrix} x_1 \\ x_2 \\ x_3 \\ \vdots \\ x_n \end{bmatrix}. \quad (27)$$

A is a column matrix of absorptances. **E** is a rectangular matrix of mean square electric fields integrated over the j-th layer. In this matrix, m is the number of profiling angles of incidence or frequencies, i represents the specific angle of incidence or frequency, n is the number of stratified layers, and j represents a specific stratified layer. **X** is a column matrix of optical constant dependent fitting parameters related to the depth profile of the material.

In order to solve these sets of linear equations, three problems must be solved. First, how can one calculate the mean square electric field without resorting to the weak absorption band limitation? Second, how can the polarization be taken into account so an incident beam of arbitrary degree of polarization can be used? Third, how can parameter **x** relate the optical constants of the system to those of the pure components and subsequently to the depth profile?

Estimation of the Mean Square Electric Field

While the aforementioned approximation of using the evanescent field ($k_2 = 0$) is acceptable for a weakly absorbing medium ($k_2 << 0.1$), in a moderately absorbing medium ($k_2 \geq 0.1$) the mean square electric field is too heavily attenuated to be considered identical to the evanescent field in a non-absorbing medium of the same refractive index. However, the mean square electric field can be estimated from the evanescent field along with the reflectance from the film (41).

$$\left\langle E_{z\bar{p}}^2(\theta, \upsilon) \right\rangle \cong \frac{1 + R_{\bar{p}}(\theta, \upsilon)}{2} \left\langle E_{z\bar{p}}^2(\theta, \upsilon) \right\rangle_{k_2 = 0} \quad (28)$$

This reflectance factor attempts to compensate for the absorption phenomenon and its effect on the electric field decay. At low extinction coefficients, the approximation yields excellent results. However, when higher extinction coefficients are used, the electric field decay is too strongly affected by the absorption process to be adequately modeled by the evanescent field even with the reflectance correction factor.

Determination of Polarized Mean Square Electric Field

Since the mean square electric field depends on the degree of polarization, $\bar{p}(\upsilon)$, the full expression for the substitution of the evanescent wave decay profile (equation 8) into equation 28 can be written as (2-3, 31):

$$\begin{aligned}\left\langle E_{z\bar{p}}^2(\theta,\upsilon)\right\rangle &= \frac{1}{4}\Big[\{1+\bar{p}(\upsilon)\}\{1+R_p(\theta,\upsilon)\}\left\langle E_{0p}^2(\theta,\upsilon)\right\rangle_{k_2=0} \\ &\quad + \{1-\bar{p}(\upsilon)\}\{1+R_s(\theta,\upsilon)\}\left\langle E_{0s}^2(\theta,\upsilon)\right\rangle_{k_2=0}\Big] \\ &\quad \cdot e^{-2z/d_p(\theta,\upsilon)} \\ &\equiv \beta(\theta,\upsilon)e^{-2z/d_p(\theta,\upsilon)}\end{aligned} \tag{29}$$

where $\left\langle E_{0p}^2(\upsilon)\right\rangle_{k_2=0}$ and $\left\langle E_{0s}^2(\upsilon)\right\rangle_{k_2=0}$ are p and s polarized mean square evanescent fields at the surface and $R_p(\theta,\upsilon)$ and $R_s(\theta,\upsilon)$, are the p and s polarized reflectances, respectively. The definition of $\bar{p}(\upsilon)$ is given by:

$$\bar{p}(\upsilon) = \frac{I_p(\upsilon) - I_s(\upsilon)}{I_p(\upsilon) + I_s(\upsilon)} \tag{30}$$

where $-1 \le \bar{p}(\upsilon) \le +1$ and $I_p(\upsilon)$ and $I_s(\upsilon)$ are the p and s polarized components of the incident beam at frequency υ, respectively. A value of -1 corresponds to an s polarized beam, +1 to a p polarized beam, and 0 to an unpolarized beam.

The relation given in equation 29 requires knowledge of the p and s polarization components of the reflectance. Ratioing the expressions for the p and s polarized absorptances ($A_p(\theta,\upsilon)$ and $A_s(\theta,\upsilon)$, respectively) and combining with the complete expression for $R_{\bar{p}}(\theta,\upsilon)$, one can obtain two quadratic equations which can be solved for the p and s polarization components of the reflectance ($R_p(\theta,\upsilon)$ and $R_s(\theta,\upsilon)$, respectively). The error in the s and p polarized reflectances determined by this technique compared to those calculated from exact optical theory as a function of angle and extinction coefficient is shown in Fig. 8. At small extinction coefficients, the error is negligible. Only at high extinction coefficients where significant modification of the electric field exists, do the reflectance values calculated by this approach differ substantially. As long as one knows the value of $\bar{p}(\upsilon)$, the mean square electric field can be estimated and may be used to calculate matrix **E** (2).

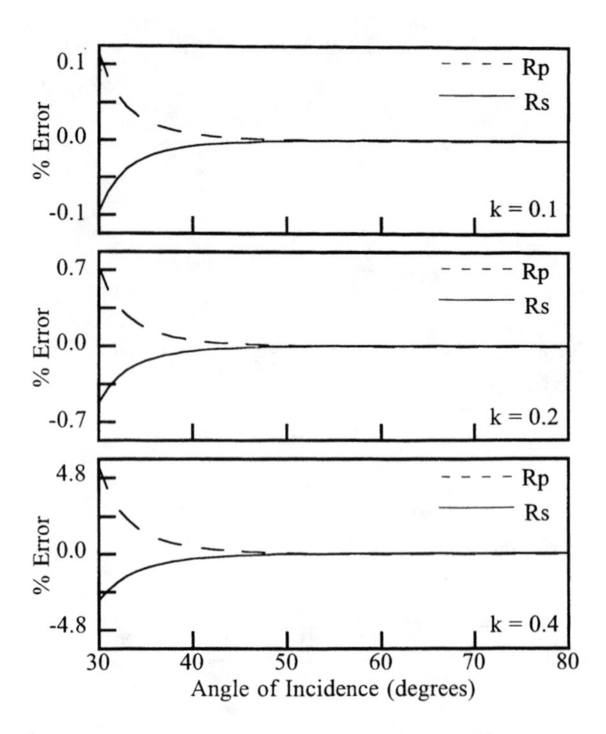

FIGURE 8. Accuracy of calculated p and s polarized reflectances as compared to reflectances from exact optical theory for an infinitely thick film. Optical parameters: $n_1 = 4.0$, $n_2 = 1.5$, $\upsilon = 1000\ \text{cm}^{-1}$, and $\bar{p} = 0$.

Relation of Optical Constants to Estimated Profile

If one assumes that no volume change occurs upon mixing, the following expression using the rule of mixtures for the optical constants of a binary system (i.e. film and solvent or polymer a and polymer b) can be used (1).

$$\begin{aligned} n_j^{\text{system}}(\upsilon)k_j^{\text{system}}(\upsilon) &= \\ \phi_j n_a(\upsilon)k_a(\upsilon) &+ (1-\varphi_j)n_b(\upsilon)k_b(\upsilon)\end{aligned} \tag{31}$$

where ϕ_j is the volume fraction in the j-th layer, n_j^{system} is the refractive index of the binary system, and k_j^{system} is the extinction coefficient of the binary system. In the case of the multiple-angle approach, the components of the **X** matrix are the product of the optical constants of each stratified layer (1).

$$\mathbf{X} = \begin{bmatrix} n_{21}k_{21} \\ n_{22}k_{22} \\ n_{23}k_{23} \\ \cdot \\ \cdot \\ \cdot \\ n_{2n}k_{2n} \end{bmatrix} \tag{32}$$

The j-th element of the matrix is the product of the refractive index and the extinction coefficient of the j-th layer, ($n_j^{system} k_j^{system}$), at the frequency of the depth profiling. This complex refractive index profile can be related to the physically meaningful composition depth profile in particular situations. At a frequency where one component absorbs weakly, the other does not absorb, and the refractive indices of the two components are similar, n_j^{system} for the binary mixture can be assumed to be constant and equal to n_a. The volume fraction in each layer j can be related to the extinction coefficient of the mixture by:

$$k_j^{system} = \phi_j k_a \tag{33}$$

where k_a is the extinction coefficient of the absorbing component. Since this method requires that one knows the refractive index of the film at each frequency of interest, the complex refractive index profile of the film yields the actual composition depth profile.

In the multiple-frequency approach, the complex refractive index spectra for both components of the system are required prior to the depth profiling calculation. Then, equation 31 can be rewritten as (3):

$$n_j^{system}(\upsilon) k_j^{system}(\upsilon) = n_a(\upsilon) k_a(\upsilon)\left[\{1-\gamma(\upsilon)\}\phi_j + \gamma(\upsilon)\right] \tag{34}$$

where

$$\gamma(\upsilon) = \frac{n_b(\upsilon) k_b(\upsilon)}{n_a(\upsilon) k_a(\upsilon)} \tag{35}$$

(i.e., $k_a(\upsilon) \neq 0$). When this relationship between the optical constants of the system and the volume fraction is substituted into the general absorptance equation for the stratified system (equation 20) along with the mean square electric field approximation, rearrangement yields (3):

$$\frac{A_{\bar{p}}(\upsilon) - A_{\bar{p}}^{\gamma}(\upsilon)}{\dfrac{4\pi\upsilon n_a(\upsilon) k_a(\upsilon)\{1-\gamma(\upsilon)\}}{n_1 \cos\theta}\beta(\upsilon)} = \sum_{j=1}^{n} \phi_j \int_{z_j}^{z_{j+1}} e^{-2z/d_p(\upsilon)} dz \tag{36}$$

where

$$A_{\bar{p}}^{\gamma}(\upsilon) = \frac{4\pi\upsilon n_a(\upsilon) k_a(\upsilon)\gamma(\upsilon)}{n_1 \cos\theta} \sum_{j=1}^{n} \int_{z_j}^{z_{j+1}} \left\langle E_{z\bar{p}}^2(\upsilon)\right\rangle dz \tag{37}$$

Equation 36 indicates that the components of the **X** matrix in the multiple-frequency approach are the volume fraction in each stratified layer. Thus, the **X** matrix directly yields the composition depth profile.

Therefore, the final form of the component matrices in equation 23 are as follows:

Multiple-Angle Approach (2)

$$\mathbf{A} = \begin{bmatrix} A_{\bar{p}}(\theta_1) \\ A_{\bar{p}}(\theta_2) \\ A_{\bar{p}}(\theta_3) \\ \cdot \\ \cdot \\ A_{\bar{p}}(\theta_m) \end{bmatrix}. \tag{38}$$

$$e_{ij} = \frac{2\pi\upsilon}{n_1 \cos\theta_i} \int_{z_j}^{z_{j+1}} \left\langle E_z^2(\theta_i)\right\rangle_{\bar{p}} dz$$

$$(i = 1, 2, 3, \ldots, m;\ j = 1, 2, 3, \ldots, n) \tag{39}$$

$$\mathbf{X} = \begin{bmatrix} n_{21}k_{21} \\ n_{22}k_{22} \\ n_{23}k_{23} \\ \cdot \\ \cdot \\ \cdot \\ n_{2n}k_{2n} \end{bmatrix} \tag{40}$$

Multiple-Frequency Approach (3)

$$\mathbf{A} = \frac{n_1\cos\theta}{4\pi} \begin{bmatrix} \dfrac{A_{\bar{p}}(\upsilon_1) - A_{\bar{p}}^{\gamma}(\upsilon_1)}{\upsilon_1 n_a(\upsilon_1) k_a(\upsilon_1)\{1-\gamma(\upsilon_1)\}\beta(\upsilon_1)} \\ \dfrac{A_{\bar{p}}(\upsilon_2) - A_{\bar{p}}^{\gamma}(\upsilon_2)}{\upsilon_2 n_a(\upsilon_2) k_a(\upsilon_2)\{1-\gamma(\upsilon_2)\}\beta(\upsilon_2)} \\ \dfrac{A_{\bar{p}}(\upsilon_3) - A_{\bar{p}}^{\gamma}(\upsilon_3)}{\upsilon_3 n_a(\upsilon_3) k_a(\upsilon_3)\{1-\gamma(\upsilon_3)\}\beta(\upsilon_3)} \\ \vdots \\ \dfrac{A_{\bar{p}}(\upsilon_m) - A_{\bar{p}}^{\gamma}(\upsilon_{m1})}{\upsilon_m n_a(\upsilon_m) k_a(\upsilon_m)\{1-\gamma(\upsilon_m)\}\beta(\upsilon_m)} \end{bmatrix} \tag{41}$$

$$e_{ij} = \frac{1}{2}\int_{z_j}^{z_{j+1}} e^{-2z/d_p(\upsilon_i)} dz$$

$$(i = 1, 2, 3, \ldots, m;\ j = 1, 2, 3, \ldots, n) \tag{42}$$

$$
\mathbf{X} = \begin{bmatrix} \phi_{21} \\ \phi_{22} \\ \phi_{23} \\ \cdot \\ \cdot \\ \cdot \\ \phi_{2n} \end{bmatrix} \tag{43}
$$

CALCULATIONS

Depth Profiling Procedure

The general form of the depth profiling procedure is nearly the same for the two approaches as illustrated in Figs. 9 and 10. First, the methods process the reflectance spectrum(a) and other data needed to set up the profiling calculations (i.e. profiling frequencies, optical constants of the system, incident angle(s), ...etc.). Next, the programs use this information to define the stratified medium (number of layers and thickness of each layer). The thickness of the stratified layers can be calculated in two manners. They can each be assigned an equal thickness, determined by the number of layers and the working depth, or they can each be assigned a thickness based on equal intervals of evanescent field decay at the lowest frequency and smallest incident angle (i.e. those with the largest d_w's) for the multiple-frequency and multiple-angle techniques, respectively. Since the evanescent field decays exponentially and the layers closer to the surface are sampled preferentially, it is advantageous to assign the stratified layer thickness via the exponential model. This method for defining the stratified medium is described elsewhere (1). However, this approach leads to difficulties in defining the spacial resolution. In the multiple angle method, the reflectances are interpolated via polynomial interpolation to equal the number of stratified layers and thereby convert **E** to a square matrix. In the multiple angle approach, frequencies with small extinction coefficients are selected. The *p* and *s* polarized components of the reflectances are then calculated. The programs are then able to generate matrix **E** from the estimated mean square electric fields. Once matrix **A** is generated from the absorptance spectra, the set of linear equations of absorptance can be solved by the singular value decomposition (SVD) method. Numerically, matrix **E** is very close to singularity. This would prevent usual numerical methods from solving the set of equations. Therefore, the SVD technique is used to overcome this problem. Details about this technique can be found elsewhere (1, 57). Solving the set of linear equations of absorptance yields the estimated profile. This estimated profile is subsequently used as a trial profile in a linear least square fitting of absorptance. Since the equations used in the linear fitting no longer require the evanescent wave assumption to calculate the mean square electric field, additional profiling frequencies with higher extinction coefficients are added to the calculations in the multiple frequency approach (3). The estimated profile is used to calculate the mean square electric fields. From these mean square electric fields, matrix **E** can be calculated. The new matrix **E** is used to calculate a new matrix **A**. Subsequently, solving the set of linear equations by the SVD method yields a new profile matrix **X**. The converged linear least square fitting yields a profile; however, since the actual relationship between absorptance and the optical constants of the stratified media are non-linear, this profile can only be regarded as an approximate profile. The profile from the linear least square fitting is used as a trial profile in a non-linear fitting using the Levenberg-Marquardt algorithm of reflectances calculated from exact optical theory to the experimental reflectances. The reflectances from exact optical theory are calculated using the matrix method, originally proposed by Abeles (58-59) and further developed by Hansen (46), which has been briefly mentioned earlier in this paper and is described in full detail elsewhere (1). The converged profile from the non-linear fitting yields the final profiles. More details on the depth profiling procedures can be found elsewhere (1-3).

Depth Profiling Computations

All programs discussed in this paper were originally written in FORTRAN 77. The depth profiling software programs used have been described previously (1-3). All of the calculations found in the previous papers and some of the calculations in this paper were performed on a Sun Sparc2 with Sun OS 4.2.1 running Open Windows v. 2. In order to take advantage of the rapidly increasing power of computer microprocessors, the code was updated to FORTRAN 90 to improve portability and recompiled using Microsoft FORTRAN Powerstation 4.0. Additional, depth profiling calculations were performed on a Digital Celebris XL 5120 with a 120 MHz Intel Pentium processor and 32 MB of RAM and a Gateway 2000 G6-200 with a 200 MHz Intel Pentium Pro processor and 128 MB of RAM. Both computers operate on a Microsoft Windows NT 4.0 platform and are typical of the computer capabilities available to regular research and development laboratories. Profiling results obtained on the newer PC-based systems were identical to those performed on the Sun Sparc2 but were performed in a fraction of the computation time.

Complex refractive indices of the film and substrate materials were determined via Kramers-Kronig analysis (3). Reflectances simulated using computer programs based on exact optical theory were used as inputs into the depth profiling programs. In this manner, reflectance from various experimental profiles can be easily obtained. This allows the quality of the depth profiling results to be determined. In testing the multiple-angle approach, a set of 15 profiling angles were used. In addition, a constant refractive index of 1.5 was assumed for the non-absorbing substrate. In the multiple-frequency approach, profiling

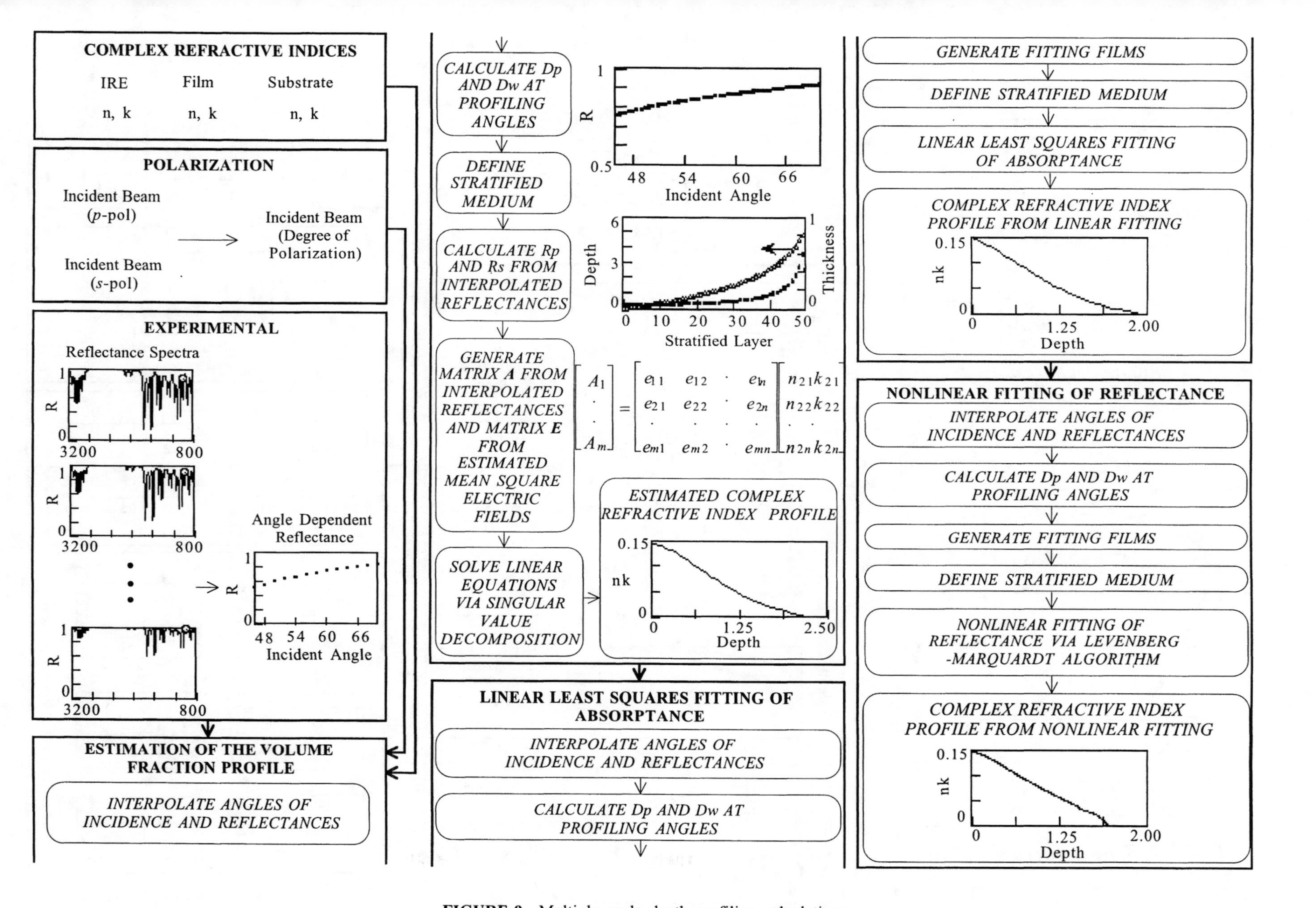

FIGURE 9. Multiple-angle depth profiling calculations.

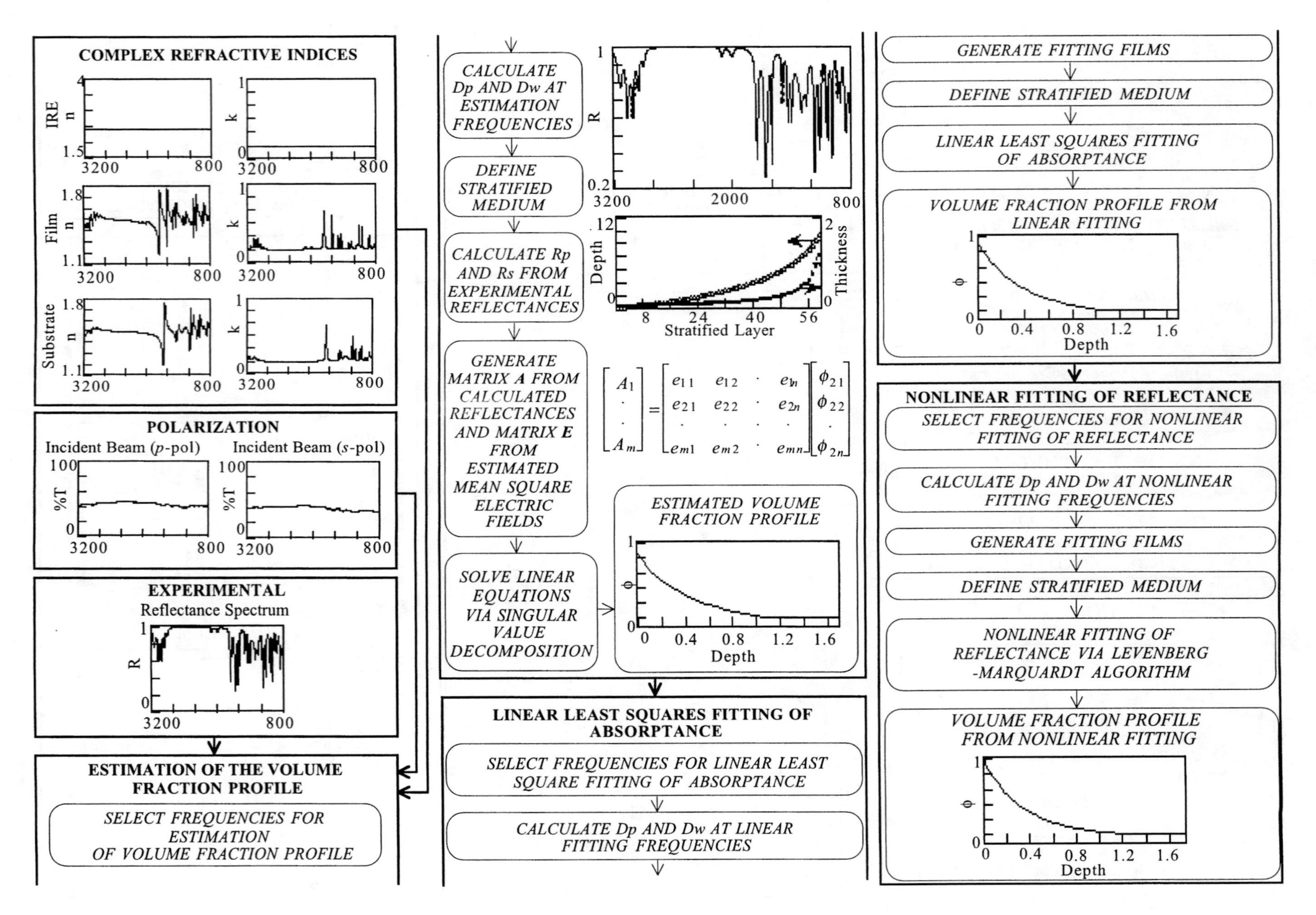

FIGURE 10. Multiple-frequency depth profiling calculations.

TABLE 1. Complex refractive index limits in the multiple-frequency approach

	Estimation		Linear Fitting		Non-Linear Fitting	
	minimum	*maximum*	*minimum*	*maximum*	*minimum*	*maximum*
n_{film}	1.4	1.7	1.4	1.7	1.4	1.7
k_{film}	0.05	0.2	0.05	0.6	0.05	0.6
$n_{substrate}$	1.4	1.7	1.4	1.7	1.4	1.7
$k_{substrate}$	0.0	0.03	0.0	0.1	0.0	0.1

frequencies were automatically selected by the computer according to preset complex refractive index limits at each stage of the calculation as shown in Table 1. In order to determine the effect of noise and more accurately reflect actual sampling conditions, white noise of varying magnitude from 10^{-5} to 10^{-3} absorptance units (a.u.) was added to the reflectance spectra prior to the depth profiling calculations.

DISCUSSION

Comparison with Previous Methods

When selecting an appropriate depth profiling technique, it is important to know several factors: What type of calculation is performed? What types of materials (i.e. bands) can the technique be used with? What are the experimental requirements to perform the profiling? A comparison between the multiple-angle and multiple-frequency approach, Laplace transform, and exact optical theory methods is shown in Table 2.

A traditional weakness of the Laplace transform method lies in the fact that it is only applicable to weakly absorbing bands. Both of the newly developed methods as well as those based on exact optical theory overcome this limitation. However, until now, exact optical theory approaches have not been able to overcome other weaknesses of the Laplace transform method. A *p* or *s* polarized incident beam is still required, necessitating the use of a polarizer each time data is collected. A guessed profile is still required in order to solve the system of equations. This introduces operator bias and uncertainty into the calculated profile. The converged profile is only as good as the guessed profile. Depending upon the nature of the fitting process, the techniques cannot distinguish between several calculated profiles if they have the same degree of "goodness" of fit to the test profile, which may be inaccurate. The assumption of a guessed profile fixes the possibilities of the possible calculated profiles to those similar to the guessed profile. It is very difficult to accurately model a profile unless very reliable *a priori* information about the system and profile is known. The numerical calculations may converge on a profile similar to the guessed profile when the actual profile is similar, yet significantly different enough to have a real impact on the system behavior.

Neither the multiple-angle nor the multiple-frequency technique require a guessed profile to perform the depth profiling. This reduces user bias and error. The multiple-frequency and multiple-angle approaches both estimate the profile and use this estimated profile as a trial profile in place of a guessed profile. This enables these approaches to distinguish very similar profiles accurately (1-3). These two new depth profiling approaches represent significant advances in the capabilities and ease of use of ATR depth profiling.

The *p* and *s* polarized components of the incident beam generate different surface fields at the interface and, therefore, interact differently with the absorbing media. In an FTIR spectrometer, polarization is a result of the light source, mirrors, beamsplitter materials, and reflection angles. Unless mirror or beamsplitter degradation is suspected, the degree of polarization can be measured once before a series of experiments, reducing the use of polarizers and the error due to misalignment or shifting of the polarizer throughout the course of the experiment. For FTIR spectrometers the ratio of polarizations is fairly close to one, and the effect of polarization is not as large as in grating spectrometers where a much larger range of polarization ratios is found as one traverses the mid-IR region (60). An incident beam of any polarization or an unpolarized incident beam may be used in the newly developed approaches without introducing significant error as long as one knows the degree of polarization at each frequency of interest. The ability to use beams with arbitrary degree of polarization offers a better signal-to-noise ratio than spectra taken with polarized incident beams and yield better depth profiling results (2-3). The effects of polarization and the necessary corrections have been discussed by Sperline (60).

In all of the quantitative depth profiling techniques, reproducibility of the spectra are essential. Excellent optical contact between the sample and the IRE is assumed. Air gaps will cause large deviations in the reflectance spectra which can only be accounted for if the gap thickness is known. Reproducibility of optical contact is essential, especially if the sample needs to occasionally be taken out of the fixture for heating or aging. Tests should be done to confirm that reproducibility of optical contact between samples to be studied exists.

Spectrometer stability and correct optical alignment are critical in all quantitative FTIR spectroscopic techniques. With the time consuming multiple-angle method, reference spectra must be taken at each angle of incidence. Care must be taken that purge times, cooling water flow rates, and other factors relating the spectrometer are identical between the reference and the sample spectra. In addition, the angular precision of the variable angle

TABLE 2. Comparison between ATR depth profiling techniques.

	Laplace Transform	Exact Optical Theory	Multiple-Angle	Multiple-Frequency
Calculation	Inverse Laplace transformation of absorptance	Kramers-Kronig analysis and linear fitting of reflectances	Linear fitting of absorptances and non-linear fitting of reflectances	Linear fitting of absorptances and non-linear fitting of reflectances
Fitting Parameters	Extinction coefficient	Complex refractive index	Complex refractive index or volume fraction	Volume fraction
Applicable Bands	Weakly absorbing bands	Any	Any	Any non-interacting bands
Required Guessed Profiles	Guessed extinction coefficient profile	Guessed complex refractive index profile	None	None
Polarization	*p* or *s* polarized incident beam	*p* or *s* polarized incident beam	Any degree of polarization or unpolarized incident beam	Any degree of polarization or unpolarized incident beam
ATR Fixture	Variable angle	Variable angle	Variable angle	Multiple reflection or variable angle
Experimental Requirements	Intensities at profiling frequencies at various incident angles	Intensities at profiling frequencies at various incident angles	Intensities at profiling frequencies at various incident angles	Intensity of entire spectrum at a single angle
	Optical contact between the IRE and film	Optical contact between the IRE and film	Optical contact between the IRE and film	Optical contact between the IRE and film
	Refractive index at the profiling frequency	Refractive index at infinite frequency	Refractive index at the profiling frequency	Complex refractive index spectra of the pure materials
	Polarizer	Polarizer	Degree of polarization at the profiling frequency	Degree of polarization at each frequency
Applicable systems	Systems with slow changes in composition profile and/or interacting bands	Systems with slow changes in composition profile and/or interacting bands	Systems with slow changes in composition profile and/or interacting bands	Systems with fast or slow changes in composition profile with few or no interacting bands
Additional Assumptions	Changes in complex refractive index of the material are due to composition changes and/or interactions	Changes in complex refractive index of the material are due to composition changes and/or interactions	Changes in complex refractive index of the material are due to composition changes and/or interactions	Changes in complex refractive index of the material are due to composition changes only

attachment is extremely important. Errors in the incident angle can have large effects on the resulting reflectance spectra, especially near the critical angle (48). The effect is more pronounced with lower refractive index materials like ZnSe ($n_1 = 2.4$) which are commonly used to probe more deeply into organic materials. A narrow aperture should be used to minimize beam divergence and its associated error. Kellner et al. (61) performed a set of experiments based on a chemometric experimental design to determine the effect of several experimental parameters on error levels in ATR-FTIR spectroscopy. Their results indicate that the relative importance of experimental conditions is as follows: refractive index > angle of incidence > aperture > time. The factors studied influence the penetration depth in the same manner. Factors which may cause error in the actual penetration depth compared to the nominal penetration depth can cause significant error in the reflectance spectra and, consequently, the quality of the depth profile. The angular imprecisions involved in the practical operation of the variable-angle ATR fixture introduce error which can significantly influence the quality of the resulting profile. The multiple-frequency approach can take advantage of

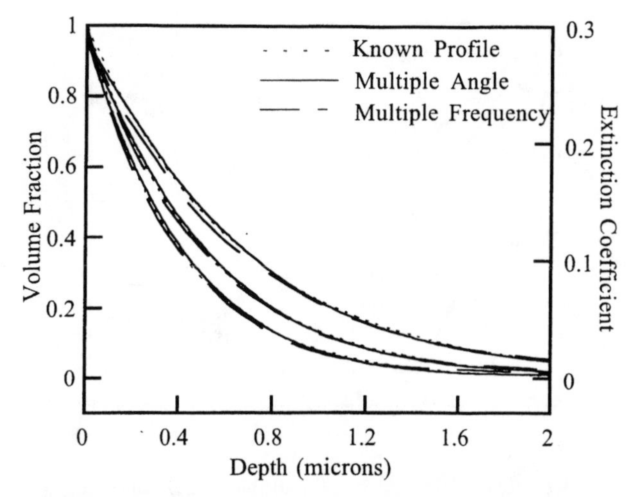

FIGURE 11. Comparison of multiple-angle and multiple-frequency fittings of exponentially decreasing profiles.

multiple-reflection attachments in which the angle of incidence is fixed, reducing a possible source of error. .

Removal of as much error as possible is required since unlike many other quantitative techniques, the multiple-angle and multiple-frequency techniques do not use normalized band intensities. It is often difficult to find a suitable reference component that can be assured of being uniformly distributed throughout the thickness of the material. It is also often difficult to find a reference band close to the band of interest. Even when one is found, corrections must be made for the difference in penetration depths between the two frequencies. In the multiple-frequency method, many absorption bands with different effective thicknesses are being used simultaneously. Since spectral acquisition time is very rapid, normalization should not introduce any significant improvement in short term tests in which the spectrometer conditions are not expected to change much. However, for longer term measurements with the multiple-frequency technique and for experiments using the multiple-angle technique, the quality of the depth profile will be affected to a degree dependent on the stability of the instrument and the ability of the user to correct for this.

Comparison between Multiple-Angle and Multiple-Frequency Methods

Simulated Depth Profiles

Since the electric field decays exponentially as it penetrates into the material, most of the information gained will be near the surface where the more intense electric field can interact with the material. Therefore, it is most desirable to monitor phenomena which happen near the surface of the IRE if possible. It has been shown that exponentially decreasing profiles are modeled best because they mimic the decay profile of the mean square electric field. Examples of fitting of exponentially decreasing profiles are shown in Fig 11. Many other profiles can be fitted as well. The depth profiling of step function profiles of various depths has been successfully performed using simulated data (3). They become increasingly difficult to profile accurately at the step front moves into the material because of the limited interaction between the material at the step interface with the mean square electric field which has decreased to a small value at that depth. These concerns arise for traditional ATR depth profiling techniques as well and are a result of the limitations imposed by the nature of the ATR phenomenon; however, they are important to take into consideration when using these or any other ATR depth profiling technique to study a system of interest.

As the fitting process progresses from the estimated profile stage through the final non-linear fitting procedure, the calculated volume fraction profile converges to the true depth profile. A comparison of the quality of fittings at each stage of the depth profiling procedure for the two techniques is shown in Fig. 12. Examination of the fittings shows the importance of the non-linear fitting process to the multiple-frequency approach. The multiple-angle method does not rely on the non-linear fitting as extensively as the multiple-frequency approach. The quality of the final profiles is, however, identical. To illustrate the effect of deeper profiles, a series of linear decreasing composition gradients have been profiled as well and are shown in Fig 13. Comparison between the results of the two methods shows that they are very similar in capabilities using near perfect data, although the multiple-frequency method seems to profile areas deeper in the material slightly more accurately with simulated data. This is likely due to the much larger number of independent data points available to the depth profiling calculations in the multiple-frequency approach. The multiple-frequency approach should be more accurate in practice as well due to a higher signal-to-noise ratio. This will only be the case provided the optical constants are accurate and no interaction peaks are selected for the profile. The effect of simulated noise on the accuracy of the depth profiling of similar profiles is shown in Fig. 14. Indeed, it is seen that the multiple-frequency approach is slightly more accurate than the multiple-angle approach. With a larger number of independent data points, the effect of random noise is more easily filtered out in the multiple-frequency approach as opposed to the multiple-angle approach where only a handful of independent data points are available. The effect of non-random noise has not been evaluated.

General Comparison

While the multiple-angle and multiple-frequency approaches both offer similar advantages over traditional ATR depth profiling approaches, each has a different set of capabilities and limitations that must be taken into consideration when choosing a depth profiling approach for a specific system. The multiple-frequency method operates best when there is a clear difference between the

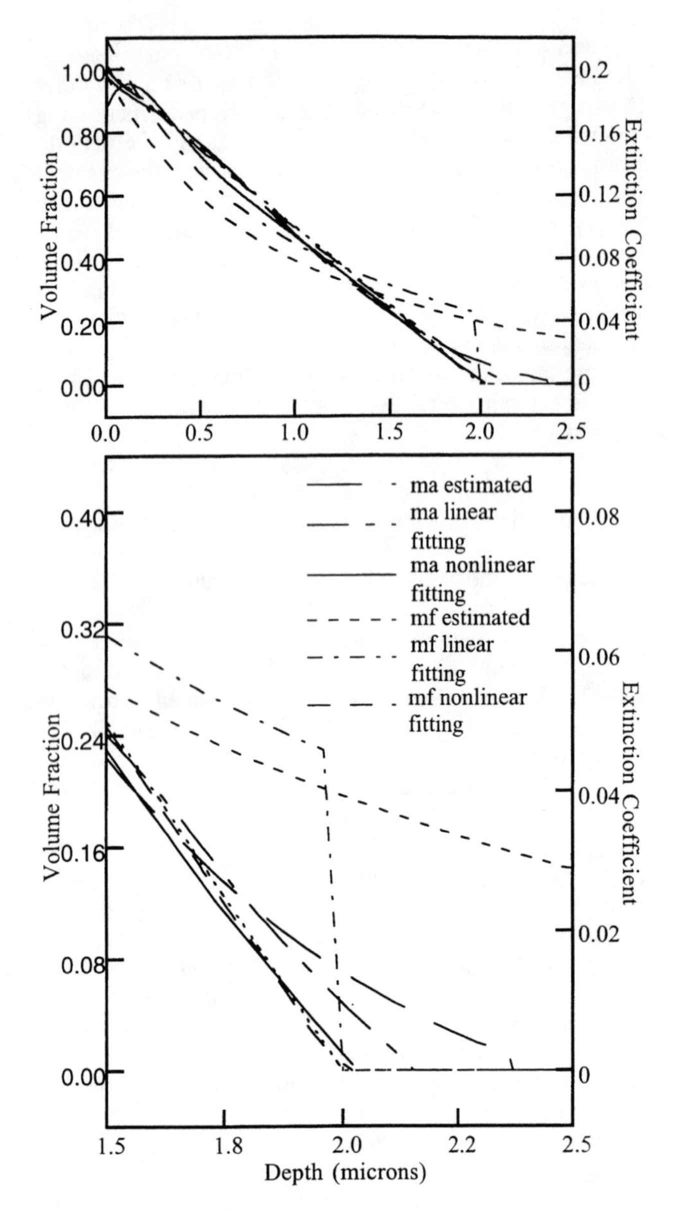

FIGURE 12. Comparison of the depth profiles after successive stages of the multiple-angle and multiple-frequency calculations.

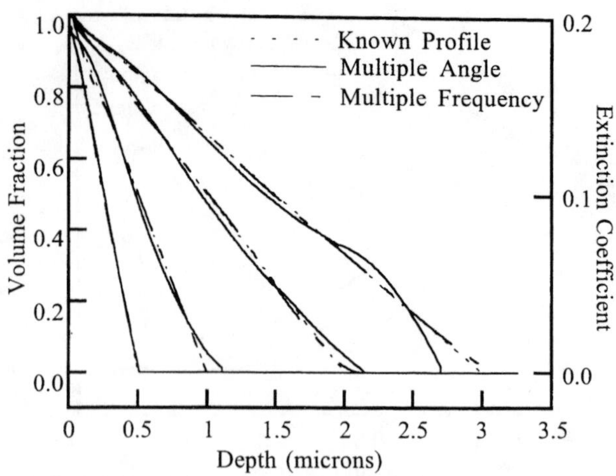

FIGURE 13. Comparison of multiple-angle and multiple-frequency fittings of linearly decreasing profiles.

extinction coefficients of the two materials at a particular frequency. Therefore, the accuracy of the calculated depth profile will be higher if the two materials have significantly different spectral features. The profiling of a chemically similar system (i.e. polystyrene and deuterated polystyrene) will suffer from a lack of accuracy due to the similarity of the extinction coefficients for nearly all the frequencies in the spectrum. The multiple-angle method can be used successfully here provided one utilizes a frequency where the two materials have significantly different extinction coefficients.

One of the assumptions inherent in the multiple-frequency method can limit its usefulness in some systems. In order to bring the complex refractive indices outside of the summation sign in equation 20, it is assumed that the complex refractive indices at all frequencies of the two materials are known and that the optical constants of the mixture of the two materials follows the additive property of the dielectric constant. Since there are many techniques for determining the optical constants of a material, the assumption of prior knowledge of the complex refractive indices of the two materials is not too burdensome. However, it is the second assumption which may cause some difficulties. All exact optical theory approaches discussed in this paper rely upon the stratified layer model to calculate the profile. The stratified layer model assumes that each layer is homogeneous with an isotropic complex refractive index. Most binary polymer systems tend to phase separate. Depending upon the size and nature of the phase separation, the stratified layer model breaks down. The material is not homogeneous in each stratified layer and the electric field is sufficiently perturbed that it does not decay in a same manner as the evanescent wave. Therefore, a miscible blend system is needed; however, in order for two chemically dissimilar polymers to be miscible on a small enough scale for the stratified layer theory to hold, specific molecular interactions are required. In high molecular weight polymeric systems, the combinatorial entropy of mixing, which is responsible for the stability of many small-molecule mixtures, becomes unimportant since its contribution to the free energy of mixing depends on the number of molecules. Therefore, the enthalpic contribution from the free volume and interaction terms are responsible for polymer miscibility. In order to achieve the negative Flory-Huggins interaction parameter, χ_{12}, leading to a negative enthalpy of mixing and consequently a negative free energy of mixing, specific interactions are necessary to counteract the positive contribution from the free volume term (62). Entropic considerations such as ordering and crystallization are not allowed either as these would tend to introduce anisotropy into the complex refractive indices of the layers and this breaks down the stratified layer model as well. Specific interactions such as hydrogen bonding, dipole-dipole

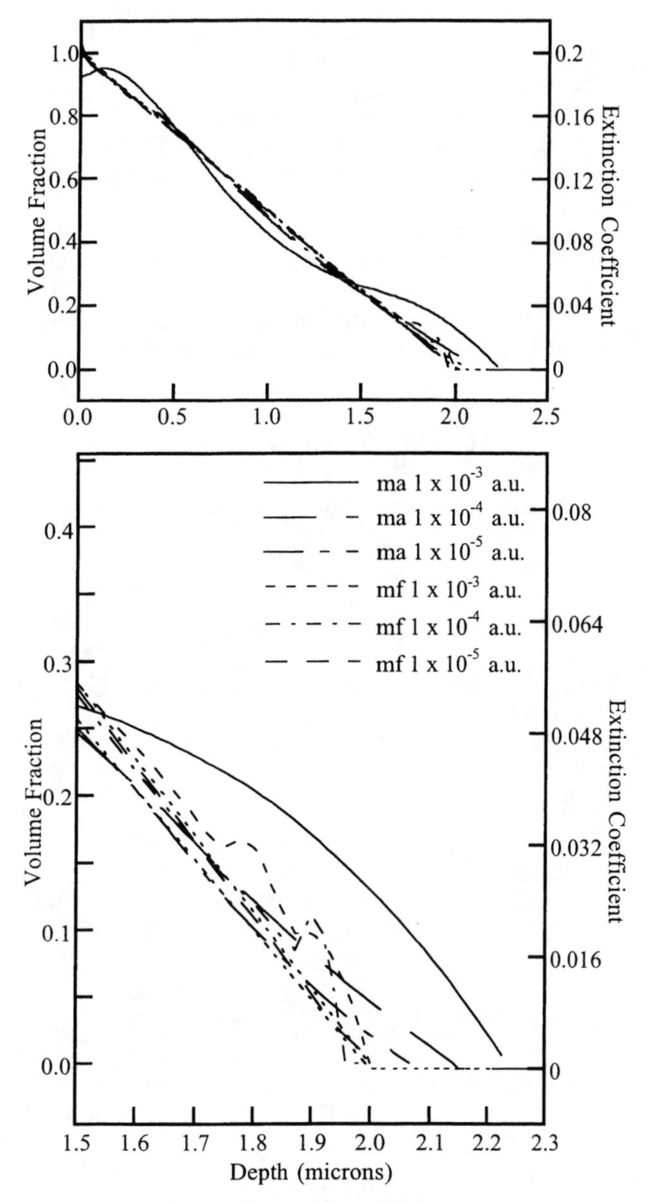

FIGURE 14. Evaluation of the effect of noise on the profiles calculated by the multiple-angle and multiple-frequency approaches.

interactions, steric factors, charge-transfer complex formation, and conformational effects all alter the complex refractive indices of the mixture and will be observed in the reflection spectrum though the appearance of interaction and hydrogen-bonding peaks, peak shifting, peak splitting, ...etc. The effects of specific interactions on the spectra of miscible polymer-polymer and polymer-solvent blends have been studied extensively (63-73).

The multiple-frequency approach cannot correctly deal with these effects. It will try to fit a simulated reflectance spectrum with the original optical constants of the material to the experimental data simple by altering the volume fraction. Since the multiple-frequency technique is based on exact optical theory and the optical constants of the pure materials are known, it can handle the effects of optical dispersion as described by Allara (74) and Huang and Urban (75). However, specific interactions change the complex refractive indices of the mixture such that it can no longer be successfully modeled by the normal additive property of the dielectric constant regardless of the volume fraction chosen. Unrealistic volume fractions (i.e. $\phi < 0$ or $\phi > 1$) will be selected by the program to model the absorptance in this region. In effect, the interaction effect will be treated as noise; however, since this "noise" is of the same magnitude as the reflectance data, the approach breaks down. The profile will never properly converge in this situation. One would reason that the multiple-frequency technique could handle interactions such as small peak shifts and peak magnitudes changing much better than the appearance of entirely new bands. Therefore, if one wishes to study a system in which these effects occur, it is important to know which spectral features are a result of interactions. It is recommended that comparison of experimental spectra and spectra from spectral simulation from the complex refractive indices of the two materials be used to identify interaction bands (74). These bands must be eliminated from consideration by the multiple-frequency program. This may, however, severely limit the number and desirability of the bands that remain eligible for the analysis. Distinct spectral features that are commonly used for material identification such as carboxyl, carbonyl, ether, ester, amide, and alcohol functional groups are commonly involved in intermolecular interactions. Therefore, some of the primary spectral features of the compounds which are the most useful for depth profiling cannot be used if large interactions are present. At this point in time, it is important to note that the effect of the magnitude of the interaction features on the accuracy of the resulting depth profile has not been determined, although work will continue in this direction in the future.

Since the multiple-angle method does not make this assumption, frequencies at which these interaction effects appear can still be used in the profiling. However, depending upon the magnitude of the interaction, physically meaningful interpretation of the resulting complex refractive index profile may prove to be difficult. Frequencies where large interaction effects occur should not be used for the depth profiling calculations if one desires to obtain a volume fraction profile since one does not know whether the magnitude of an interaction is linear with respect to composition of the system. If it is important to use frequencies with such interaction effects, the multiple-angle approach should be used. If the two compounds interact but have different enough spectral signatures that many other bands can be used, the multiple-frequency technique may be used provided the interaction regions are eliminated from consideration by the program.

Aside from the theoretical differences, one of the greatest practical differences between the two techniques is the experimental setup. The multiple-angle approach requires the traditional variable-angle ATR configuration with the collection of spectra at a series of incident angles. It takes a long time to take baselines at all the angles of interest, take the spectra at each angle, change the angle, wait for the instrument to purge if necessary,...etc. All approaches which use the variable-angle technique assume that the profile in the system is constant with respect to the duration of the experiment. Variable-angle techniques are applicable only to systems in which no diffusion, segregation, or reaction is taking place. If one would like to study such rapidly changing systems, an approach like the multiple-frequency approach, which requires only a single spectrum to perform the depth profiling calculations, can be used. This technique significantly shortens the constant profile assumption timescale. If the entire experiment is performed in the spectrometer (i.e. using a hot cell or a flow cell for example), the timescale is shortened even further. The multiple-frequency approach opens the door to real-time depth profiling studies with off-line data processing.

The multiple-frequency approach can use either the variable angle or multiple reflection ATR attachment. While variable-angle ATR accessories are less common and significantly more expensive, multiple-reflection accessories are found in many laboratories allowing the multiple-frequency approach to be used by a much wider audience. The technique may also be used with the variable-angle attachments although more scans will be necessary to achieve the same signal-to-noise ratio and the same accuracy in the converged depth profile. In addition, the angular resolution of the variable angle approach is limited to the precision of the attachment. In the multiple-frequency approach, a huge number of frequencies can be used, limited only by the resolution of the spectrum, the size of the spectral regions of interest, and the computing power of the microprocessors involved in the calculations. With the ever increasing power of computer microprocessors, on-line real-time depth profiling may be possible with a technique like the multiple-frequency approach.

CONCLUSIONS

Two new optical depth profiling techniques based on attenuated total reflection Fourier transform infrared spectroscopy have been developed. These techniques overcome several limitations which have hindered the widespread use of ATR depth profiling. No *a priori* information about the composition profile is required. Operator bias and influence associated with such guessed profiles is eliminated. In addition, the techniques can be used with incident radiation beams of arbitrary degree of polarization. A generalized form of the derivation of the two techniques has been presented. Depth profiling calculations on spectrally simulated reflectances from various composition profiles have been performed. The two techniques can successfully model a large range of profiles very accurately. The effect of noise has been evaluated by adding varying levels of white noise to the reflectance spectra prior to the depth profiling. The larger the signal-to-noise ratio, the more accurate the resulting profile. Finally, the capabilities and limitations of the two techniques have been outlined and compared with each other and various other methodologies.

The multiple-angle approach can be used to study systems with intermolecular interactions in which the profile changes only over a long timescale. The multiple-frequency approach, while unable to handle the shifts and peak distortions caused by interactions, eliminates the error associated with the angular settings of the variable angle fixture and offers the possibility of real-time depth profiling due to the short spectral acquisition time. The effect of interaction phenomena on the profiling capabilities of the multiple-frequency technique needs to be evaluated.

ACKNOWLEDGMENTS

This material is based upon work supported under a National Science Foundation Graduate Fellowship.

REFERENCES

1. Ekgasit, S., and Ishida, H., *Appl. Spectrosc.* **50**, 1187-1195 (1996).
2. Ekgasit, S., and Ishida, H., *Vib. Spectrosc.* **13**, 1-9 (1996).
3. Ekgasit, S., and Ishida, H., *Appl. Spectrosc.* (accepted).
4. Bohn, P.W., and Miller, D.R., *Crit. Rev. Anal. Chem.* **22**, 1-16 (1991).
5. Jabbari, E., and Peppas, N.A., *J.M.S-Rev. Macromol. Chem.Phys.*, **C34**, 205-241 (1994).
6. Ishida, H., *Rubber Chem. Technol.* **60**, 497-554 (1987).
7. Urban, M.W. and Koenig, J.L., *Vibrational Spectra and Structure 18: Applications of FT-IR Spectroscopy*, New York: Elsevier Science Publishing, 1990, pp. 127-181.
8. Fina, L.J., *Appl. Spectrosc. Rev.*, **29**, 309-365 (1994).
9. Harrick, N.J., *Phys. Rev. Lett.* **4**, 224-226 (1960).
10. Harrick, N.J., *J. Phys. Chem.* **64**, 1110-1114 (1960).
11. Fahrenfort, J., *Spectrochim. Acta* **17**, 698-709 (1961).
12. Carlsson, D.J., and Wiles, D.M., *Canad. Jrnl. Chem.* **48**, 2397-2406 (1970).
13. Carlsson, D.J., and Wiles, D.M., *Jrnl. Polym. Sci. Polym. Lett. Ed.* **8**, 419-424 (1970).
14. Carlsson, D.J., and Wiles, D.M., *Macromolecules* **4**, 174-179 (1971).
15. Webb, J.R., *Jrnl. Polym. Sci. A-1* **10**, 2335-2348 (1972).
16. Luoma, G.A., and Rowland, R.D., *Jrnl. Appl. Polym. Sci.* **32**, 5777-5790 (1986).
17. Kuhn, K.J., Hahn, B., Percec, V., and Urban, M.W., *Appl. Spectrosc.* **41**, 843-847 (1987).
18. Belali, R., Vigoreux, J.M., and Camelot, M., *Spectrochim. Acta* **10**, 1261-1267 (1987).
19. Cowie, J.M.G., Devlin, B.G., and McEwen, I.J., *Polymer* **34**, 501-504 (1993).
20. Cowie, J.M.G., Devlin, B.G., and McEwen, I.J., *Polymer* **34**, 4130-4134 (1993).
21. Cowie, J.M.G., Devlin, B.G., and McEwen, I.J., *Macromolecules* **26**, 5628-5632 (1993).
22. Niu, B.-J., Martin, L.R., Tebelius, L.K., and Urban, M.W., *Latex Film Formation at Surfaces and Interfaces* in *Film Formation in Waterborne Coatings*, Provder, T., Winnik, M.A., and Urban, M.W., Eds., Washington, D.C.: American Chemical Society, 1996, pp. 301-331.

23. Jang, W-H., and Miller, J.D., *J. Phys. Chem.* **99**, 10272-10279 (1995).
24. Rozanov, N.N., and V.M. Zolatarev, *Opt. Spectrosc.(USSR)* **49**, 506-510 (1980).
25. Stuchebryukov, S.D., Vavkushevskii, A.A., and Rudoi, V.M., *Sov. Phys. Dokl.* **27**, 931-932 (1982).
26. Hirayama, T., and Urban, M.W., *Prog. Org. Coat.* **20**, 81-96 (1992).
27. Zhao, C.L., Holl, Y., Pith, T., and Lambla, M., *Colloid & Polymer Sci.* **265**, 823-829 (1987).
28. Zerbi, G., Gallino, G., Del Fanti, N., and Baini, L., *Polymer* **30**, 2324-2327 (1989).
29. Fina, L.J., and Chen, G., *Vib. Spectrosc.* **1**, 353-361 (1991).
30. Pereira, M.R., and Yarwood, J., *Jrnl. Polym. Sci. Part B: Polym. Phys.*, **32**, 1881-1887 (1994).
31. Ekgasit, S., and Ishida, H., *Appl. Spectrosc.* **51**, 461-5 (1997).
32. Van Alsten, J.G., and Lustig, S.R., *Macromolecules* **25**, 5069-5073 (1992).
33. Fieldson, G.T., and Barbari, T.A., *Polymer* **34**, 1146-1153 (1993).
34. Jabbari, E. and Peppas, N.A., *Macromolecules* **26**, 2175-2186 (1993).
35. Jabbari, E. and Peppas, N.A., *Jrnl. Mat. Sci.* **29**, 3969-3978 (1994).
36. Jabbari, E. and Peppas, N.A., *Macromolecules* **28**, 6229-6237 (1995).
37. Farinas, K.C., Doh, L., Venkatraman, S., and Potts, R.O., *Macromolecules* **27**, 5220-5222 (1994).
38. Fieldson, G.T., and Barbari, T.A., *AIChE Jrnl.* **41**, 795-804 (1995).
39. Hellstern, U., and Hoffman, V., *Jrnl. Mol. Struct.* **349**, 329-332 (1995)
40. Nguyen, T., Bentz, D., and Byrd, E., *Jrnl. Coat. Tech.* **67**, 37-46 (1995).
41. Hansen, W.N., *Internal Reflection Spectroscopy in Electrochemistry* in *Advances in Electrochemistry and Electrochemical Engineering*, Muller, R.H., Ed., New York: Wiley, 1973, Vol. 9, pp. 1-60.
42. Harrick, N.J., *Internal Reflection Spectroscopy*, Ossining, NY: Harrick Scientific Corporation, 1978.
43. Mirabella Jr., F.M., *Appl. Spectrosc. Rev.* **21**, 45-178 (1985).
44. Urban, M.W., *Attenuated Total Reflectance Spectroscopy of Polymers*, Washington, D.C.: American Chemical Society, 1996.
45. Yamamoto, K., and Ishida, H., *Vib. Spectrosc.* **8**, 1-36 (1994).
46. Hansen, W.N., *J. Opt. Soc. Am.* **58**, 380-390 (1968).
47. Shick, R.A., Koenig, J.L., and Ishida, H., *Appl. Spectrosc.* **47**, 1237-1244 (1993).
48. Iwamoto, R. and Ohta, K., *Appl. Spectrosc.* **38**, 359-365 (1984).
49. Ishino, Y., and Ishida, H. *Appl. Spectrosc.* **42**, 1296-1302 (1988).
50. Huang, J., and Urban, M.W., *Appl. Spectrosc.* **47**, 973-981 (1993).
51. Ohta, K., and Ishida, H., *Appl. Opt.* **29**, 1952-1959 (1990).
52. Ohta, K., and Iwamoto, R., *Appl. Spectrosc.* **39**, 418-425 (1985).
53. Ohta, K., and Iwamoto, R., *Anal. Chem.* **57**, 2491-2499 (1985).
54. Tompkins, H.G., *Appl. Spectrosc.* **28**, 335-341 (1974).
55. Hirschfeld, T., *Appl. Spectrosc.* **31**, 289-292 (1977).
56. Yanagimachi, M., Toriumi, M., and Masuhara, H., *Appl. Spectrosc.* **46**, 832-840 (1992).
57. Press, W.H., Teukolsky, Vetterling, W.T., and Flannery, B.P., *Numerical Recipes in FORTRAN: The Art of Scientific Computing*, 2nd ed. New York: Cambridge University Press, 1992, pp. 51-63, 620-625.
58. Abeles, F. *J. Phys. (Paris)* **11**, 310 (1950).
59. Abeles, F., *Optics of Thin Films* in *Advanced Optical Techniques*, Van Heel, A.C.S., Ed., Amsterdam: North-Holland, 1967, pp. 144-188.
60. Sperline, R.P., *Appl. Spectrosc.* **45**, 677-681 (1991).
61. Kellner, R., Kashofer, F., and Simeonov, V., *Fresenius Z. Anal. Chem.* **325**, 258-262 (1986).
62. Patterson, D., and Robard, A., *Macromolecules* **11**, 690-695 (1978).
63. Wellinghoff, S.T., Koenig, J.L., and Baer, E., *Jrnl. Polym. Sci.: Polym. Phys. Ed.* **15**, 1913-1925 (1977).
64. Coleman, M.M., and Zarian, J., *Jrnl. Polym. Sci.: Polym. Phys. Ed.* **17**, 837-850 (1979).
65. Coleman, M.M., and Varnel, D.F., *Jrnl. Polym. Sci.: Polym. Phys. Ed.* **18**, 1403-1412 (1980).
66. Varnell, D.F., and Coleman, M.M., *Polymer* **22**, 1324-1328 (1981).
67. Varnell, D.F., Runt, J.P., and Coleman, M.M., *Macromolecules* **14**, 1350-1356 (1981).
68. Varnell, D.F., Runt, J.P., and Coleman, M.M., *Polymer* **24**, 37-42 (1983).
69. Coleman, M.M., and Moskala, E.J., *Polymer* **24**, 251-257 (1983).
70. Moskala, E.J., and Coleman, M.M. *Polym. Commun.* **24**, 206-208 (1983).
71. Varnell, D.F., Moskala, E.J., Painter, P.C., and Coleman, M.M., *Polym. Eng. Sci.* **23**, 658-662 (1983).
72. Coleman, M.M., Moskala, E.J., Painter, P.C, Walsh, D.J., and Rostami, S., *Polymer* **24**, 1410-1414 (1983).
73. Moskala, E.J., Howe, S.E., Painter, P.C., and Coleman, M.M., *Macromolecules* **17**, 1671-1678 (1984).
74. Allara, D.L., *Appl. Spectrosc.* **4**, 358-361 (1979).
75. Huang, J.B., and Urban, M.W., *Appl. Spectrosc.* **46**, 1014-1019 (1992).

Atmospheric FTS

Wesley A. Traub

Harvard-Smithsonian Center for Astrophysics
Cambridge, Massachusetts 02138

The chemical state of the upper atmosphere can be determined from FTS measurements of its infrared thermal emission spectrum. The atmosphere is self-luminous in this band, and can be measured in any direction, day or night. This paper focuses on results from the Smithsonian's far-infrared spectrometer FIRS-2. The techniques of emission and absorption spectroscopy are compared, and the unique features of FIRS-2 discussed. The driving force behind each FIRS-2 flight campaign has been the issue of ozone depletion in the stratosphere. In the Arctic winter vortex we measured large scale chemical change and vertical subsidence of the stratosphere. On a balloon platform we measured the abundances of radical and precursor species which control ozone loss in the middle stratosphere. We show evidence for two cases of "missing chemistry" in the stratosphere.

INTRODUCTION

Fourier transform methods are ideal for measuring properties of the Earth's stratosphere. Two major types of Fourier transform spectrometer (FTS) have been used to probe the stratosphere for nearly two decades now, emission spectroscopy and absorption spectroscopy. Emission FTS measurements use the intrinsic thermal emission spectrum of the stratosphere itself as the light source, and absorption FTS measurements use the sun as a light source with the atmosphere acting as an absorption cell. Both methods will be discussed here, with emphasis on emission FTS.

Both emission and absorption FTS measurements are forms of remote sensing, in which the spectrometer may be located on the ground, in an airplane, on a balloon platform, or on an orbiting satellite. It is the very remoteness of the stratosphere which makes these methods so useful.

Roughly speaking, the stratosphere constitutes about 10% of the mass of the whole atmosphere, ranging from an altitude of about 16 km (52,000 ft) to 50 km (164,000 ft). The boundaries are defined by temperature extremes, about 210 K at the bottom and 280 K at the top. The lower boundary varies with latitude and season from about 18 km (and 190 K) at the equator, to 8 km (and 220 K) at the poles. The increase of temperature with altitude is the opposite of what we experience in the lower atmosphere (troposphere), and is caused almost entirely by the absorption of solar ultra-violet (UV) photons by O_3 (ozone).

About 75% of the total mass of O_3 in the atmosphere is found in the stratosphere, and this is responsible for removing all but a few percent of the biologically harmful solar UV band known as UVB ($\lambda 280-320nm$). However the UVB flux which is transmitted depends critically on the instantaneous column abundance of O_3, so even small variations in O_3 can potentially have relatively large biological effects. The effect is non-linear: halving the ozone column results in an increase of 400–500% in the DNA-effective radiation reaching the surface.

Concern about O_3 loss is sufficient to drive a major worldwide scientific effort to understand the effects of chemistry and transport on the distribution of O_3 in the atmosphere[1]. Here we focus on the role of two particular FTS instruments which are contributing to this effort.

The major chemical groups with O_3-depleting power have been identified as HO_x (OH, HO_2), NO_x (NO, NO_2), Cl_x (Cl, ClO), and Br_x (Br, BrO). These gas-phase radical species destroy ozone catalytically by recycling themselves according to the general pattern:

$$X + O_3 \longrightarrow XO + O_2 \quad (1)$$

$$XO + O \longrightarrow X + O_2 \quad (2)$$

$$\text{net}: \quad O + O_3 \longrightarrow 2O_2 \quad (3)$$

where X = H, OH, NO, or Cl. In part because of the participation of the O atom, these four cycles are relatively most important in the upper stratosphere, where O is most abundant. In the lower stratosphere there are two additional cycles which contribute to O_3 loss:

$$OH + O_3 \longrightarrow HO_2 + O_2 \quad (4)$$

$$HO_2 + O_3 \longrightarrow OH + 2O_2 \quad (5)$$

$$\text{net}: \quad 2O_3 \longrightarrow 3O_2 \quad (6)$$

CP430, *Fourier Transform Spectroscopy:* 11th International Conference
edited by J.A. de Haseth

and

$$\begin{aligned} Br + O_3 &\longrightarrow BrO + O_2 && (7)\\ Cl + O_3 &\longrightarrow BrO + O_2 && (8)\\ BrO + ClO &\longrightarrow Br + Cl + 2O_2 && (9)\\ \text{net}: \quad 2O_3 &\longrightarrow 3O_2 && (10) \end{aligned}$$

In addition to the above gas-phase (homogeneous) reactions, there are particle-assisted (heterogeneous) reactions which are enabled by the presence of polar stratospheric clouds (PSCs) in the form of nitric acid trihydrate or water ice. The net result of these heterogeneous reactions is to convert stable species such as HCl and $ClONO_2$ into more reactive forms such as Cl_2 and HOCl. In the lower atmosphere, in the presence of a momentum conserving third molecule M, these feed another catalytic cycle which is responsible for about 75% of the loss[2] in the Antarctic ozone hole (and more moderate loss in the Arctic):

$$\begin{aligned} 2(Cl + O_3 &\longrightarrow ClO + O_2) && (11)\\ ClO + ClO + M &\longrightarrow Cl_2O_2 + M && (12)\\ Cl_2O_2 + h\nu &\longrightarrow Cl + ClOO && (13)\\ ClOO + M &\longrightarrow Cl + O_2 + M && (14)\\ \text{net}: \quad 2O_3 &\longrightarrow 3O_2 && (15) \end{aligned}$$

Outside the polar regions, sporadic volcanic eruptions produce SO_2 which then forms a ubiquitous and variable sulfuric acid aerosol layer in the lower stratosphere. This not only promotes the formation of Cl_2 and HOCl as noted above, but also tends to increase the relative amount of ClO and decrease the relative amount of NO_x, which in the lower stratosphere leads to more O_3 loss than expected from homogeneous chemistry alone.

These reactions and their results are the motivation behind regulatory actions on Cl_x, NO_x, and Br_x generating source species. The Antarctic ozone hole is a spring-time depletion of over 50% of column O_3 averaged over area throughout the month of October. Its first-time appearance in the early 1980s is now clearly linked to the presence of chlorofluorocarbons (CFCs), primarily freon-11 (CCl3F) and freon-12 (CCl_2F_2), both of which reached about one-half of their present concentrations around 1980, up from much smaller amounts in 1960.

The supersonic transport fleet which was proposed in the 1970s would have generated NO_x in the stratosphere which could have then depleted O_3, depending on the altitude; the current interest in high speed transport airplanes evokes similar issues. The widespread use of methyl bromide (CH_3Br) as a fungicide has prompted concerns that Br_x in the stratosphere will increase, thus depleting more O_3.

These are the issues which lie behind much of the research interest in stratospheric chemistry and transport. Many of the corresponding stratospheric measurements have been and will continue to be made by FTS techniques, to which we now turn our attention.

EMISSION OR ABSORPTION?

One of the most fundamental choices to be made in planning a spectroscopic experiment is to decide whether to consider the target as an emitting or an absorbing object. To be sure, in the majority of situations, there is no choice to be made, due to the equipment available or the nature of the source and available illumination. But in the case of the atmosphere, a choice is available, and each option brings with it certain advantages. In this section we present some basic considerations related to FTS observations of the atmosphere.

Our scientific priority is to understand how O_3 is produced and lost in the stratosphere; we accomplish this by measuring the abundances of those species which contribute to this process, and comparing these measurements with relevant theoretical models.

To measure these species spectroscopically, we need molecular transitions to observe. In the far-infrared and sub-millimeter regions there are numerous rotational transitions available, and in the mid-infrared there are ro-vibrational transitions as well. There are also some vibrational or electronic transitions available in the visible, but the selection is less rich than at longer wavelengths.

The reasons for choosing an FTS technique here are several: the infrared spectrum is rich with spectral lines from essentially all molecular species of interest, and all these lines can be measured at the same time and place; the FTS high spectral resolution capability is needed to separate the lines; the FTS multiplex advantage is tolerant of relatively noisy detectors; and the FTS large etendue accepts wide angle targets. To be sure, there are indeed other useful techniques (e.g., heterodyne spectroscopy, optical and radio wavelength lidars, filter photometry, in-situ spectroscopy, in-situ sampling, etc.) for investigating the stratosphere, but FTS has been and remains a very useful tool.

In absorption FTS the sun is the light source and the atmosphere acts as an absorbing path. For any given spectral transition, the absorption coefficient is determined by the population in the lower energy level and the intrinsic transition probability from that lower level to the corresponding upper level. To isolate the contributions from different altitudes, the sun is tracked as it rises or sets, and the rays sequentially pass through different tangent altitudes where the major contribution to the absorption is made. Subsequent mathematical modeling or inversion extracts the abundance variations as a function of altitude.

In emission FTS the atmosphere itself is the light

source, and the sun is in fact avoided. Most of the atmosphere is in local thermodynamic equilibrium (LTE), and the local kinetic temperature determines the population in each energy level of a given species, via collisional excitation with neighbors. Relaxation of the molecule from the collisionally excited upper level to a lower level produces a photon which can be detected. The intensity is determined by the population of the upper energy level and the transition probability to the lower level.

In both cases, absorption and emission are both occurring, because for example the emitted photon can of course be re-absorbed, and conversely the absorbed photon stream will never reach zero intensity because there will always be some emission contribution from molecules making the reverse transition.

For the moment we ignore these effects, and ask how much light is available in each technique. Obviously the sun is very bright, and we do not anticipate any lack of photons from it, but one may ask just how the flux from the atmosphere compares.

To estimate the solar absorption signal, we treat the sun as a black-body at temperature T_s, and note that the absorption spectrum will simply remove intensity at transition wavelengths. To estimate the emission spectrum, we treat the atmosphere as having potentially the intensity of a black-body at temperature T_a in the cores of strong spectral lines, and substantially less intensity (approaching zero) between spectral lines. So for both cases, we start with the relevant black-body flux, which is given by the Planck function B_σ for the specific intensity per unit wavenumber σ (cm^{-1}) and at temperature T (K).

$$B_\sigma = \frac{2hc^2\sigma^3}{e^{hc\sigma/kT}-1} \qquad \frac{erg}{cm^2\ s\ cm^{-1}\ ster} \tag{16}$$

An FTS instrument observing this source will have an internal collimated beam diameter d, so the area of collection will be taken to be that of the collimated beam inside the spectrometer, with area $\pi d^2/4$.

The collecting solid angle Ω is then determined by the resolution $R = \sigma/\Delta\sigma$ of the instrument, where $\Delta\sigma$ (cm^{-1}/resel) is the spectral width of a single resolution element (resel), independent of wavenumber. Technically Ω is always a fixed quantity, determined by the area of the entrance aperture divided by the square of the distance of that aperture from the collimating lens or mirror, and it should be determined by the largest wavenumber (smallest wavelength) in the spectrum. But for the purpose of this discussion let us allow Ω to vary with wavenumber, or alternatively think of it as representing a typical wavenumber in the band of interest. With this idealization, we have

$$\Omega = \frac{2\pi}{R} = 2\pi\frac{\Delta\sigma}{\sigma}. \tag{17}$$

We convert ergs to photons by multiplying by the unity expression $(1photon)/(hc\sigma\ erg)$. The quantity which is actually measured is not photons, but electrons, so for generality let us include an overall instrument efficiency factor η (elec/photon). This factor encapsulates all the transmission efficiencies of the optics, including beamsplitter, and it further includes the quantum efficiency of the detector. We end up with fundamental discrete quanta (electrons) of detected signal. For simplicity, let η be constant, independent of wavenumber.

Combining these quantities, we obtain the measured signal $M_{\Delta\sigma}$ as

$$M_{\Delta\sigma} = \frac{\eta c\sigma(\pi d\Delta\sigma)^2}{e^{hc\sigma/kT}-1} \qquad \frac{elec}{s\ resel}. \tag{18}$$

Now since $hc/k \simeq 1.44K/cm^{-1}$ is near unity, we can simplify the expression in the low-frequency (long-wave) limit to give

$$M_{\Delta\sigma} \simeq \frac{\eta k}{h}(\pi d\Delta\sigma)^2 T \qquad \frac{elec}{s\ resel} \tag{19}$$

in the regime

$$\sigma(cm^{-1}) < T(K). \tag{20}$$

Note that the detected signal is independent of wavenumber, and depends only on the collecting geometry, the efficiency factor, and the source temperature. This is certainly not a great surprise, particularly to radio astronomers, but it is nevertheless a remarkably simple result which helps us compare emission and absorption FTS instruments.

Let us choose some typical numerical values. Let the beam diameter in the FTS be set to $d = 10cm$. The efficiency includes a factor of 1/2 for the beamsplitter, roughly another factor of 1/2 for the optics and filters, and a quantum efficiency of a few percent, which is typical of some infrared detectors, for a nominal value of $\eta = 0.01$.

The resolution element is set by matching the instrumental width to that of a typical spectral line in the mid-stratosphere. At the long wavelengths of interest here, the line width is dominated by the pressure broadening width, not the doppler width. A nominal pressure broadening value is about 0.1 cm^{-1}/atmosphere, and a nominal pressure is 0.05 atmosphere, for a line width of 0.005 cm^{-1}.

Numerically, we then have the following count rate per resolution element in the spectrum, in the small wavenumber limit ($\sigma < T$):

$$M_{\Delta\sigma} \simeq 5\times 10^6 T \qquad \frac{elec}{s\ resel}. \tag{21}$$

For the sun ($T \simeq 5500K$) and the stratosphere ($T \simeq 250K$) the spectral distribution is shown in Fig. 1, illustrating: (a) on the red end of the spectrum, the flat

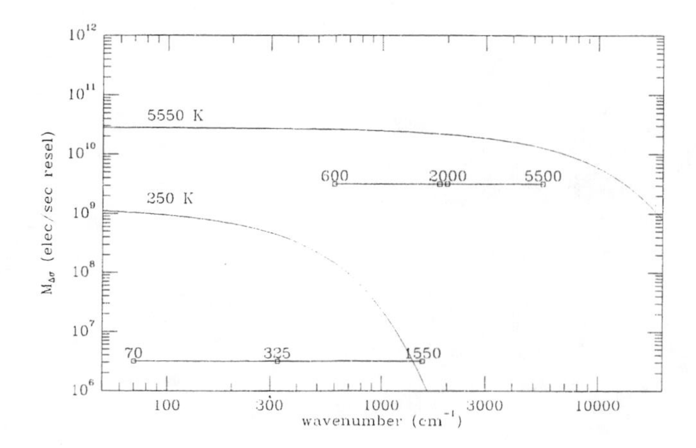

Figure 1: The expected detector signal is shown, in units of electrons per second per spectral resolution element, for idealized FTSs viewing black-body sources. The upper curve is for an idealized absorption FTS viewing the sun, and the lower curve is for an idealized emission FTS viewing the atmosphere. Experimentally realized bandpass ranges from filters and detectors are shown for the JPL Mk IV (upper insert) and the SAO FIRS-2 (lower insert).

count rate in the chosen FTS-appropriate units; (b) on the blue end of the spectrum, the exponential cutoff at $\sigma \sim (1-5)T$; and (c) the very high quantum count rates for either light source, and in particular the $T \sim 250K$ emission case.

We may now ask what signal to noise ratio we may expect in a spectrum for either of these sources, using a simplified experimental model. The number of spectral resolution elements and the number of interferogram samples are, by the Fourier transform connection, the same number n. If the maximum wavenumber is σ, the resolution element $\Delta\sigma$, the total interferogram time t, and the interferogram time per sample Δt, then

$$\begin{aligned} n &= \sigma/\Delta\sigma &(22)\\ &= t/\Delta t. &(23)\end{aligned}$$

Let us assume that the input spectrum is of the type shown in Fig. 1, i.e., an equal flux in essentially every spectral element. Then the interferogram signal level $S_{\Delta t}$ (before being ac-coupled to an amplifier) is the number of quanta per sample in the interferogram

$$\begin{aligned} S_{\Delta t} &= nM_{\Delta\sigma}\Delta t \quad elec &(24)\\ &= M_{\Delta\sigma}t \quad elec &(25)\end{aligned}$$

The quantum noise $N_{\Delta t}$ per interferogram sample is the square root of this number, assuming that purely Poisson statistics holds (i.e., that there are no other competing noise sources):

$$N_{\Delta t} = (M_{\Delta\sigma}t)^{1/2} \quad elec. \tag{26}$$

And the signal to noise ratio in the dc-coupled interferogram is the ratio of these quantities,

$$SNR_{\Delta t} = (M_{\Delta\sigma}t)^{1/2}. \tag{27}$$

It is this quantity which would be matched by the dynamic range of the analog to digital converter (ADC) which measured the interferogram.

In the FT process we project out of the interferogram the frequency component of interest, but here it is easier to envision the forward process in which the final spectral signal $S_{\Delta\sigma}$ is built up from the sum of n instantaneous spectral element contributions $M_{\Delta\sigma}$, so we can write

$$S_{\Delta\sigma} = M_{\Delta\sigma}t \quad \frac{elec}{resel}. \tag{28}$$

The noise per spectral resolution element $N_{\Delta\sigma}$ is the root sum square of the total measured noise each of n samples in the interferogram, giving

$$\begin{aligned} N_{\Delta\sigma} &= n^{1/2}N_{\Delta t} &(29)\\ &= (nM_{\Delta\sigma}t)^{1/2} \quad \frac{elec}{resel} &(30)\end{aligned}$$

The signal to noise ratio per spectral element $SNR_{\Delta\sigma}$ is then the ratio of these terms, which can be written in several interesting ways:

$$\begin{aligned} SNR_{\Delta\sigma} &\equiv \frac{S_{\Delta\sigma}}{N_{\Delta\sigma}} &(31)\\ &= (M_{\Delta\sigma}\Delta t)^{1/2} &(32)\\ &= \frac{SNR_{\Delta t}}{n^{1/2}} &(33)\end{aligned}$$

In particular, the SNR in the spectrum is reduced from the SNR in the interferogram by a factor $n^{1/2}$. Using this, we can now estimate the spectral quality in the emission and absorption cases discussed above.

For the emission case, we have $M_{\Delta\sigma} \sim 10^9$ elec/sec, and an interferogram time of $t \sim 10^2$ sec, so $SNR_{\Delta t} \sim 3 \times 10^5$ from quantum noise. As described below, the Smithsonian Astrophysical Observatory's (SAO) FIRS-2 FTS an effective dynamic range of 19 bits, or $2^{19} \sim 5 \times 10^5$. This is slightly better than needed for the example case at hand, so we know that the ADC noise will not add significantly to the quantum noise. There are on the order of $n \sim 10^5$ samples in the far-infrared spectrum, and about $n \sim 4 \times 10^5$ samples in the mid-infrared spectrum, giving an expected signal to noise ratio of about 1000 and 500 in the two bands, respectively. In practice we do indeed achieve SNR values within a factor of about 2 of these, validating the present crude calculation. A detailed comparison would involve substantial numerical simulation, but the basic dependencies would remain as outlined here.

For the absorption case, we have $M_{\Delta\sigma} \sim 2 \times 10^{10}$ elec/sec, and an interferogram time of $t \sim 10^2$ sec, so

$SNR_{\Delta t} \sim 1.4 \times 10^6$ from quantum noise. In this example, and assuming $n \sim 10^6$ samples, we would get $SNR_{\Delta\sigma} \sim 1500$. To compare with practice (see below) the Jet Propulsion Laboratory (JPL) Mk IV FTS has a beam diameter about 4 times smaller that our example 10 cm value, so the collected flux will be 16 times smaller, and the noise 4 times smaller. We thus expect $SNR_{\Delta t} \sim 4 \times 10^5$ from quantum noise. The JPL interferometer has an effective dynamic range of 19 bits, or 5×10^5, which is sufficient so as not to add appreciably to the quantum noise. There are on the order of $n \sim 10^6$ samples in the Mk IV spectrum, giving an expected signal to noise ratio of about 400 in the band. In practice the Mk IV does indeed achieve SNR values of about 400-500, depending upon the band chosen.

In summary, we see that although emission and absorption FTS instruments with similar collecting areas might be expected to give somewhat different signal to noise ratios, in practice two existing instruments are built with compensating collecting areas such that they both end up with roughly similar signal to noise ratios, both on the order of 500. With signal levels of this quality, it is possible to do significant and interesting science. We now turn to a quick overview of the emission spectra obtained by FIRS-2, followed by a brief description of the Mk IV and FIRS-2 instruments, and finally a sampling of scientific results.

EMISSION SPECTRA

To illustrate the quality of emission spectra which we have recently been able to achieve, we show in Fig. 2 the data from our recent (April 1997) FIRS-2 balloon flight from Fairbanks, Alaska.

The top panel shows a single scan (formed in our usual fashion from the sum of an "up" and a "down" scan), from a float altitude of 37 km looking down to a tangent altitude of 29 km, with the spectra binned for plotting (8 point binning from 0 to 320 cm^{-1} for the far-infrared detector, and 32 point binning above 320 cm^{-1} for the mid-infrared detector). Each detector segment has been normalized to unity for display; in practice each is normalized to a reference blackbody spectrum. The upper envelope to this data is formed by the product of the detector sensitivities, the beamsplitter efficiency (relatively flat, see next section), the modulation efficiency of the spectrometer, and the transmission of the optics.

The second panel shows an expanded version of the far-infrared spectrum, with the same 8 point binning, with an overall envelope shape which approaches zero at frequencies below 70 cm^{-1} mostly due to fall-off in the detector quantum efficiency, and is again nearly zero in the short region 260-275 cm^{-1} due to a quartz filter absorption.

The third panel shows an expanded portion of the sum of 5 apodized spectra, with no binning, and normalized

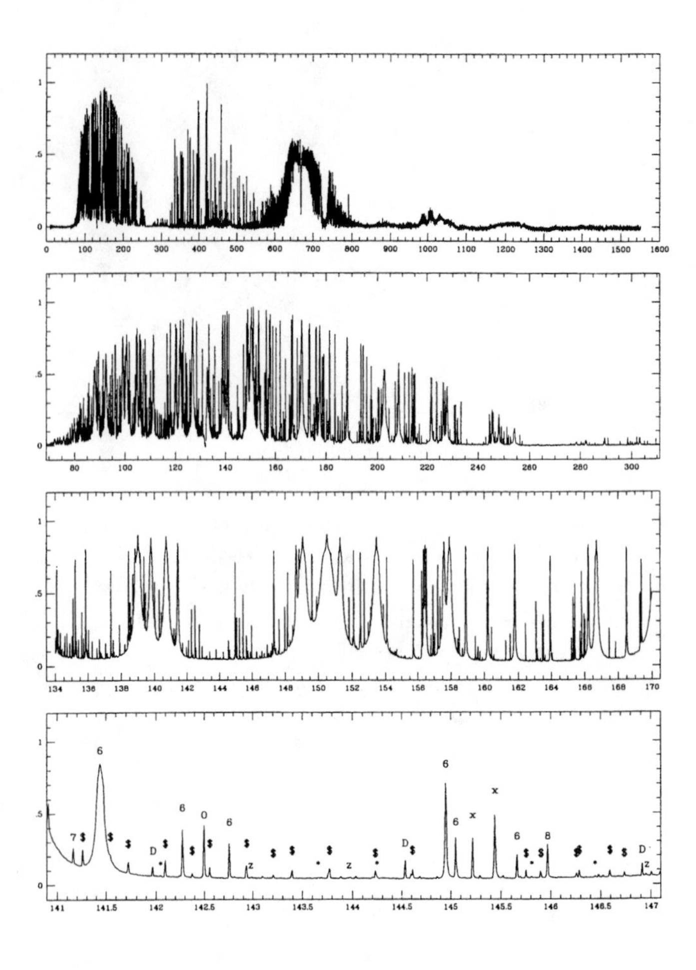

Figure 2: Atmospheric emission spectra, measured by FIRS-2 over Alaska in April 1997. Each lower panel is an expanded version of a section of the one above. The bottom panel key is O_3 ($), $H_2O^{16,17,18}$ (6, 7, 8), HDO (D), HOCl (Z), HO_2 (*), O_2 (O), HCl (x).

by the reference blackbody to a standard (277 K) blackbody spectrum. Broad features, typically the stronger H_2O lines, are seen, with much narrower features superposed, typically O_3 and other unsaturated lines. Some intermediate strength lines show broadening at their bases. One can also see the peculiar peaked tops of the strongest lines, in which the broad shoulder (at a relative intensity about 0.75) is formed in the lower, colder, higher pressure stratosphere, and the peaked center (at a relative intensity of about 0.85) is formed in the higher, warmer, lower pressure stratosphere.

The lower panel shows an expanded version of the one above, at a scale where the instrumental and/or intrinsic line widths begin to appear. The strength of each feature indicates its line of sight abundance. The altitude at which the refracted line of sight is closest to the earth is known as the tangent altitude; it is in this region where the maximum contribution to the spectrum usually occurs. The difference in such abundances between one tangent height and another indicates the abundance in the intervening layer. Analysis of these differences gives a vertical profile, abundance versus altitude, for each species.

THE Mk IV SPECTROMETER

The Mk IV spectrometer[3] is a double-passed Michelson interferometer which was designed and built at the Jet Propulsion Laboratory (JPL) for the purpose of measuring atmospheric absorption spectra in the 600 to 5500 cm^{-1} (1.8 to 16 μm) range, using the sun as a light source. It has operated on the ground at McMurdo Station in 1986, on the DC-8 in 1987, 1989, and 1992, and has been on many balloon flights from 1987 through 1997. The unapodized spectral resolution is 0.008 cm^{-1}, giving about 10^6 points per spectrum. The detectors are a HgCdTe for the 600-2000 cm^{-1} range, and an InSb photodiode for the 1850-5500 cm^{-1} range, operating simultaneously, and cooled with liquid nitrogen. Digitization is 19 bits, with a 12 bit ADC and 7 bits programmable. The beamsplitter is KBr, and the beam diameter is 2.5 cm. A quadrant detector controls the sun tracker to a position in the center of the solar disk. The data rate is 360 kbit/s, and both telemetry and onboard recording are used. Over 30 different gases can be measured, along with many isotopes, and pressure, temperature, and aerosol extinction can be measured. Data reduction and analysis are highly automated, and are similar in spirit but not in actual execution to those procedures followed for FIRS-2, which is described next in somewhat greater depth.

THE FIRS-2 SPECTROMETER

The FIRS-2 spectrometer[4,5,6] is a single-passed Michelson interferometer which was designed and built at the Smithsonian Astrophysical Observatory (SAO) for the purpose of measuring atmospheric emission spectra in the 70 to 1550 cm^{-1} (6.5 to 140 μm) range, using the atmosphere in thermal emission as a light source. It has operated on 10 balloon flights from 1987 to 1997, on the DC-8 in 1992, and recently on the ground in Fairbanks, Alaska in 1997. The unapodized spectral resolution is 0.004 cm^{-1}, giving up to about 0.4×10^6 points per spectrum. The detectors are a Ge:Ga photoconductor in the 70-350 cm^{-1} band, and a Ge:Cu photoconductor in the 350-1550 cm^{-1} range, operating in parallel, and both cooled to 4 K by liquid helium. Digitization is 19 bits, with a 14 bit ADC and an autorange single stage gain of $32 = 2^5$. The beamsplitter is Mylar with a Ge coating, discussed below. A reflecting telescope with commandable elevation control, and accurate to about 0.01 degree, independent of gondola tilt or motion, views the atmosphere in an elevation range from about $+30 to - 8$ degree. The data rate is 128 kbit/s, with telemetry to a ground station. Over 23 different gases can be measured, plus isotopes, and temperature, pressure, and aerosol extinction. Data reduction is described briefly below.

The FIRS-2 spectrometer as originally built operated in the 70-200 and 350-700 cm^{-1} bands, until the 1997 balloon flight from Alaska for which it was upgraded with a new beamsplitter[7], to allow operation in the full 70-1550 band. The previous beamsplitter was bare Mylar. The new beamsplitter was developed in cooperation with the National Research Council of Canada. The construction is a few-micron thick Mylar base with about one micron of germanium evaporated on one side. Early development issues were trying to get the Ge to adhere properly, keeping the Ge from crazing after evaporation, and preventing the substrate from melting. An even better beamsplitter could be made from polypropylene, but melting is still an unsolved problem. The advantage of polypropylene is that it has fewer absorption features, so the transmitted spectrum is cleaner. The efficiency of the beamsplitter, defined as usual for an FTS, is 4RT, where here R and T are the reflection and transmission of the beamsplitter. For the materials and thicknesses chosen, the theoretical efficiency is very high, greater than 90% over essentially all of the range that we use. This is to be compared with the former case of bare Mylar, where the peak efficiency was roughly 70% but varied sinusoidally with wavelength due to constructive and destructive interference between the two faces. The reason that a Ge coating works so well is that it breaks the symmetry, giving one face much higher reflectivity than the other, so interference between the faces is weak, and the sinusoidal pattern essentially disappears. The net result is that the detectors see much less unmodulated background light, which improves the signal to noise ratio, and all parts of the spectrum are roughly equally modulated, which improves the spectral coverage and gives more spectral lines and species for detection.

The telescope elevation control is based on a servo system which uses both a gyroscope for high frequency tilt information, and an inclinometer for low frequency tilt information. Since in thermal emission we do not have a target on which to track, and we must point near the horizon with an accuracy of a fraction of a kilometer at a distance of hundreds of kilometers, we need a pointing accuracy on the order of 0.01 degree, and this system delivers that. Azimuthal pointing is supplied by the gondola itself, with a nominal accuracy of about 1 degree, using a magnetometer measuring the earth's field direction as a reference. Both systems are commandable by telemetry.

The thermal emission spectra are calibrated[5] in intensity by occasionally viewing the cold sky at 30 degrees elevation, and a reference black body source which is passively allowed to equilibrate with the ambient temperature and is measured to an accuracy better than 1 K. The high sky spectra do contain some spectral lines, but these are quite narrow and mostly weak, so we remove them digitally, smooth the result to the intrinsic spectral content of the system, and use this for a zero-point calibration. Since the sky spectra are also used for data analysis of the overlying air mass, the calibration time is thus only that to view the blackbody and is only about 15% of the total, which is quite acceptable.

Individual spectra are obtained from mostly single-sided interferograms, and must therefore be phase-corrected[8] with care. Our method is basically to use the spectral lines themselves as test objects. We perform a standard complex Fourier transform on each suitably tapered interferogram, then search through the resulting complex spectrum in regions where we expect to find spectral lines, measure the local projection phase which best projects out a symmetric spectral line, pass a smooth phase function through all measured phases at all available wavenumbers, and use this smooth phase function to project out a real spectrum from the complex one over the full spectral range. This method is the only one we have found which works well with emission spectra in which the background can be weak, and sometimes changing sign throughout the spectrum, depending upon the nature of the two inputs to the spectrometer.

The data analysis uses a list of stratospheric transitions, both lines and continuous band structures, derived from the massive line compilation produced by the Air Force Phillips Lab and supplemented by our own ongoing updating work[9]. A recent addition to this is the capability to include the effects of line mixing, a subtle process which can significantly distort band-head structures (e.g., the sharp dip near 660 cm^{-1} in Fig. 2 below).

The radiative transfer code[5,6] works by splitting up the atmosphere along the line of sight into layers which are sufficiently homogeneous that they can be treated as each having a fixed temperature, pressure, and abundance in each species. The absorption in this layer is then calculated from the sum of the absorption coefficients of each of the constituents, including a semi-empirical continuum contribution from the water vapor (far wing) continuum. Doppler line broadening and pressure broadening are both included, the latter being temperature dependant. Where appropriate, line mixing is included as well. A stimulated emission factor is included, which is important at long wavelengths.

Starting with the most distant layer, the absorption is calculated, converted to an emissivity (1 - absorptivity), and multiplied by the local Planck function to generate an emitted flux. This flux is then transmitted through the next-nearest layer, where it is absorbed by that layer's absorption coefficient, and the local emission added. The process continues until the detector is reached.

The calculated spectrum is then compared to the measured spectrum, and the process iterated by varying the molecular abundances in the most sensitive layers until the process converges. By repeating this for each of a number of spectra taken at different elevation angles, chosen to give tangent heights separated by about 4 km ($\sim$ 1/2 scale height), we derive a series of mixing ratios for each species, as a function of altitude. A similar procedure is followed to determine temperature and pressure, using appropriate lines of CO_2 in the neighborhood of the 16 μm band. For each species, the results from different spectral lines are combined to produce a weighted mixing ratio profile, and appropriate error bars calculated based on the statistics of individual lines and their relative scatter. We carry out internal checks to ensure accuracy, including measuring the mixing ratio of O_2 and isotopes, and comparing our results with those of other instruments on those occasions when several instruments are observing the same air mass at the same time[10]. The resulting accuracies are quite high, with total errors in the range 2-10% for more than half of the species measured.

THE ARCTIC WINTER VORTEX

The polar vortex in the winter stratosphere is driven by reduced solar heat input to the absorbing ozone layer, allowing the stratosphere to cool and its pressure to drop. In this picture, a polar low-pressure system develops, outside air starts to move poleward, and the Coriolis force deflects the air to rotate in the same sense as the Earth, i.e., cyclonically. As the cooling air sinks and compresses, it is replaced by high-altitude air moving in from lower latitudes. This vertical compression of air, with replacement at the top, produces a downward transformation of mixing-ratio profiles.

During the Arctic Airborne Stratospheric Expedition (AASE II) in 1992, we used the FIRS-2 on board the NASA DC-8 aircraft to measure thermal emission from stratospheric hydrogen fluoride (HF), coupled with a subsidence model, to determine vertical displacement in the Arctic vortex during the northern hemisphere winter[11].

The FIRS-2 was adapted and used on the DC-8 as follows. The spectrometer views the stratosphere through a side port in the aircraft, sequentially recording thermal emission spectra at elevation angles of 32°, 16°, 8°, 4°, 2°, 1°, and 0°, followed by calibration scans of a hot and a cold blackbody source. This 700-s cycle is repeated while the aircraft is at or near cruise altitude, except for brief periods when the Sun is close to the line of sight. The viewing angles are selected by a tilting mirror, driven by a stepper motor, and controlled by a shaft encoder referenced to the aircraft roll angle as determined by the onboard inertial navigation system. The infrared radiation is focused by a fixed telescope mirror, then fed to the spectrometer. These two feed mirrors are rigidly mounted to the spectrometer, and the entire assembly is floated with respect to the aircraft on pneumatic isolators, to reduce vibration and potential interference fringe contrast losses. The aircraft optical window is removed; because of the difficulty of providing a suitable infrared-transmitting window, a pressure-tight enclosure is placed around the input optics, up to the spectrometer's vacuum tank, where a small-diameter polyethylene window admits light into the spectrometer.

Vertical subsidence

We use HF as a tracer of vertical motion in the stratosphere for 3 reasons. (1) HF is produced in the stratosphere, starting with photodissociation of chlorofluorocarbons and SF_6 and so has a positive vertical gradient in volume mixing ratio (VMR). If an air column is displaced downward, with replacement at the top by air from neighboring columns, then the total column abundance of HF will increase. If the nominal VMR profile is known, then the amount of vertical motion can be quantified from measurements of the change in column abundance. (2) HF is expected to be chemically unreactive, so abundance changes can be attributed to dynamical rather than chemical activity. (3) HF has an easily measured emission line in the FIRS-2 far-infrared spectrum.

For each spectrum obtained along the flight path, the overhead atmosphere is modeled with a nine-layer model; input parameters include atmospheric pressure at aircraft altitude and the atmospheric temperature profile.

The model VMR profile is taken to have a fixed amplitude but variable vertical scale. The atmosphere is assumed to contract by the factor $(1 + s)$, where s is independent of altitude but may vary with time; s is a dimensionless "subsidence" parameter. In other words, we assume that for a given parcel of air the quantity $(1+s)z$ is conserved with respect to changes in latitude, longitude, and time. With this scale transformation the VMR becomes

$$\mathrm{VMR}(z) = \mathrm{VMR}_0[(1 + s)z_0], \tag{34}$$

where $\mathrm{VMR}_0(z_0)$ is a midlatitude reference profile.

Data were obtained on 13 flights of the DC-8, each lasting about 10 hours, with a nominal cruise altitude of 11 km, which is above the tropopause at high northern latitudes. Spectra were recorded continuously. Each spectrum was analyzed in terms of a vertically scaled column profile of HF, in terms of the subsidence parameter s. The subsidence values were binned according to location inside the vortex, and month, and trends noted.

Using the subsidence trends, we calculated the corresponding vertical displacement by noting that for a parcel of air the value of $(1+s)z$ is conserved, so taking differentials, we obtain

$$\Delta z = -z\Delta s/(1 + s), \tag{35}$$

where Δz is a small displacement about a mean value z and Δs is the change in subsidence about a mean value s. The vertical velocity w is calculated by dividing the displacement by the elapsed time Δt, so

$$w = \Delta z/\Delta t. \tag{36}$$

We found that the derived vertical velocities were similar in both early and late winter, with average value

$$w = -0.052 \qquad cm\ s^{-1} \tag{37}$$

and an uncertainty of 25%. This suggests that the air in the descending regions is continuing to cool at about the same rate all winter.

From our measured winter-long average value of vertical velocity w and a nominal value of scale height H of 8 km, we can estimate the characteristic time to flush the stratosphere as

$$T = H/w \simeq 5.9 \qquad month, \tag{38}$$

at an altitude of about 18 km. This suggests that the air in the stratosphere is replaced relatively slowly, of the order of only once per winter season. The reinforces the concept that the vortex air is relatively isolated from lower-latitude air all winter, a condition which allows chlorine to accumulate and to be present in substantial quantities when photolysis begins again in spring, leading to O_3 loss.

Chemical change

We also measured column abundances[12,10] of HCl, O_3, HNO_3, and H_2O during the AASE II campaign. Using HF as a tracer, as discussed above, we removed the effects of subsidence from the measured column abundances; perturbations in the resulting column abundances are attributed to chemical processing.

The first stage of analysis is to determine the degree of stratospheric subsidence s. In the second stage of the analysis, we use this value to scale a standard mid-latitude mixing ratio profile for each of the other species, using the same type of relation as for HF, $VMR(z) = VMR_0[(1 + s)z]$, where $VMR(z)$ is the adjusted profile, and VMR_0 is the standard profile. The measured columns are divided by the corresponding theoretical columns for a subsided but chemically unchanged stratosphere, to produce dimensionless relative columns; the effect of this last operation is to remove residual effects due to variations of temperature or aircraft altitude.

The third stage is to take all measurements made within the span of a few days and sort them according to the value of potential vorticity (PV) at the point on the 440 K potential temperature (PT) surface through which the column density measurement passes. (The PV, PT coordinate system is frequently used to track air parcels because both are conserved along adiabatic frictionless trajectories.) High values of PV occur inside the rapidly rotating vortex.

The results are shown in Fig. 3 for HCl, O_3, HNO_3, and H_2O. Because HF is used to determine the subsidence, the corresponding HF points are all unity, by definition. The range of latitudes each month is nominally from 38°N (Moffett Field, California) to 90°N, except for one flight in February to 15°N (Puerto Rico). The January measurements at values of PV $> 3.5 \times 10^{-5}$ were made at latitudes having 24 hours of darkness in January.

In the HF subsidence study we found that the subsidence increases dramatically between the exterior and the interior of the vortex, and that the transition occurs in the region $(1.3 - 1.7) \times 10^{-5}$ $\mathrm{K\,m^2/kg\,s}$. We use this region as a gauge of the column abundances outside the vortex.

In panel (a) we show the normalized HCl column density as a function of PV. The column is depleted by about (55±10)% inside the vortex relative to the vortex exterior. If we assume that the loss of HCl occurs in the coldest region of the stratosphere (PT = (360 – 560) K, or altitudes of 14–22 km inside the vortex), then a nearly total loss of HCl in this altitude range is required to account for the observed loss in column density. In-situ measurements of HCl in the Arctic vortex near 20 km observed losses of > 95% in some air parcels in January, which were strongly correlated with increases in the ClO abundance. We observed some recovery of HCl inside the vortex in February, but overall the trends in HCl *vs.* PV changed little from January to March. The stability of the HCl gradient as a function of PV indicates that shuffling of vertical columns has little effect on our analysis.

The ozone columns in panel (b) show trends similar to HCl. Ozone inside the vortex is depleted by roughly (35±10)% relative to the exterior. The vortex interior remains relatively constant from January to March, but the depletion appears to spread to the vortex wall between January and February. The February and March measurements are quite similar. If, as for HCl, we assume that the losses are confined to the altitude range 14–22 km inside the vortex, then the ozone mixing ratio in this region must have decreased by about 70% relative to the subsided model profile. In-situ data showed losses relative to the vortex exterior in the 30–40% range in January around 17 km which is less than our estimate. However, since estimates of ozone loss depend on the choices of reference profile and altitude range, the difference between the two values is not surprising. The loss of ozone in the vortex wall in February may be due to dilution, by mixing with low altitude air from outside the vortex.

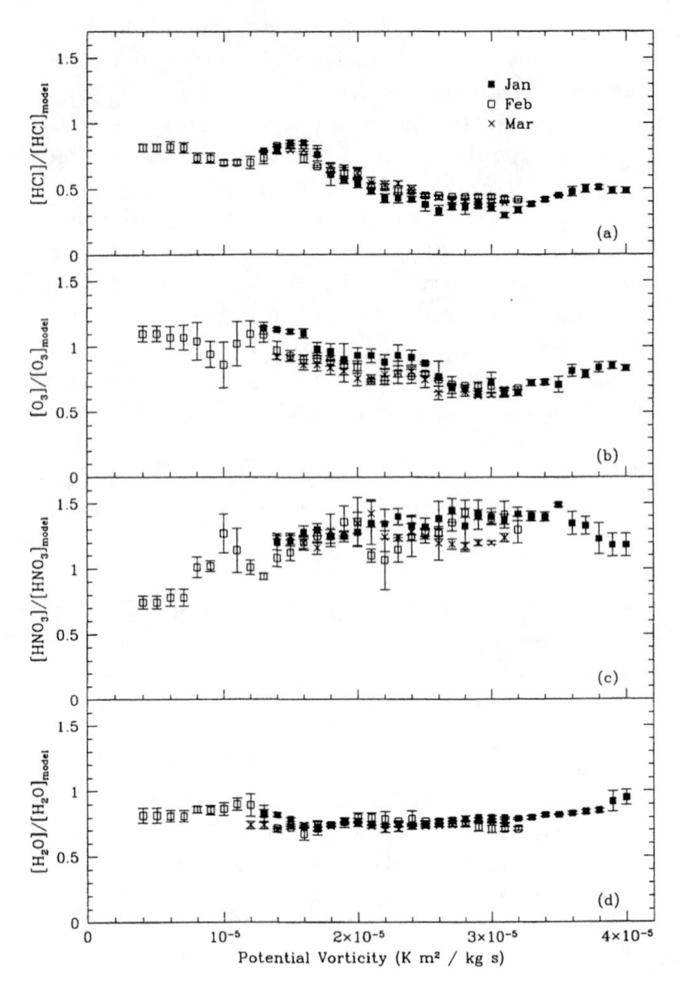

Figure 3: Monthly median values of column abundances of HCl, O_3, HNO_3, and H_2O, relative to a subsided model, showing chemical change as a function of time and distance into the Arctic vortex (increasing PV).

The nitric acid in panel (c) also show signs of perturbed chemistry in the vortex, with the column density elevated by (15±10)% in the vortex wall and interior in January. Measurements in February and March indicate a gradual return to values typical of those just outside the vortex.

The water in panel (d) shows little change, (0±10)%, across the vortex wall. This is consistent with the the absence of denitrification in panel (c) and also with the lack of extensive PSCs. Meteorological analysis of conditions in the Arctic vortex show that up to 30% of the 460 K surface inside the vortex was cold enough to form PSCs from late December to mid-January, but by the end of January temperatures were too warm for PSC formation.

Our data demonstrate that chemical change occurred inside the Arctic vortex. We find that in January, in the vortex, HCl is depleted by (55±10)%, O_3 is depleted by (35±10)%, HNO_3 is enhanced by (15±10)%, and H_2O is unchanged at (0±10)%. These results show that chlorine and nitrogen were repartitioned, O_3 was chemically destroyed, and no significant sedimentation of polar stratospheric clouds occurred.

OZONE PHOTOCHEMICAL BALANCE

Understanding the rate of removal of stratospheric ozone through catalytic cycles involving HO_x (HO_2 and OH), NO_x (NO_2 and NO), and halogen (ClO and BrO) radicals is essential for assessing the response of ozone to anthropogenic and natural perturbations such as industrial release of chlorofluorocarbons (CFCs) and halons, emission of nitrogen oxides from subsonic and supersonic aircraft, rising levels of N_2O and CH_4, and enhanced levels of sulfate aerosols following volcanic eruptions.

Recent calculations from models of the mid-latitude stratosphere give the following theoretical results: (a) catalytic cycles involving NO_x radicals are expected to dominate ozone destruction from about 22 to 45 km; (b) HO_x cycles dominate both above and below; and (c) modeled photochemical loss of ozone exceeds modeled production by 10 to 50% in the upper stratosphere. The latter has been coined the "ozone deficit" problem, because at these altitudes ozone is expected to be in photochemical equilibrium (production equalling loss), since the photochemical lifetime of ozone is much shorter than transport replacement times.

Using FIRS-2 on a balloon platform, we made simultaneous measurements of radicals and precursor species, to determine the ordering of the loss cycles of ozone in the middle stratosphere, and to study the ozone deficit problem[13,14,15,16].

We determined diurnally averaged photochemical removal rates of O_3 by each family of radicals are in two ways: from direct measurements of radical concentrations (radical method), and from radical concentrations calculated by a photochemical model constrained by measurements of radical precursors (precursor method). We used the radical method to derive, for the first time, empirical measures of the total loss rate of ozone as well as the relative removal rate by each family of radicals which regulate the abundance of ozone in the middle stratosphere (24-40 km). The precursor method provides a basis for comparing the empirical loss rates to theoretical rates constrained by appropriate environ-

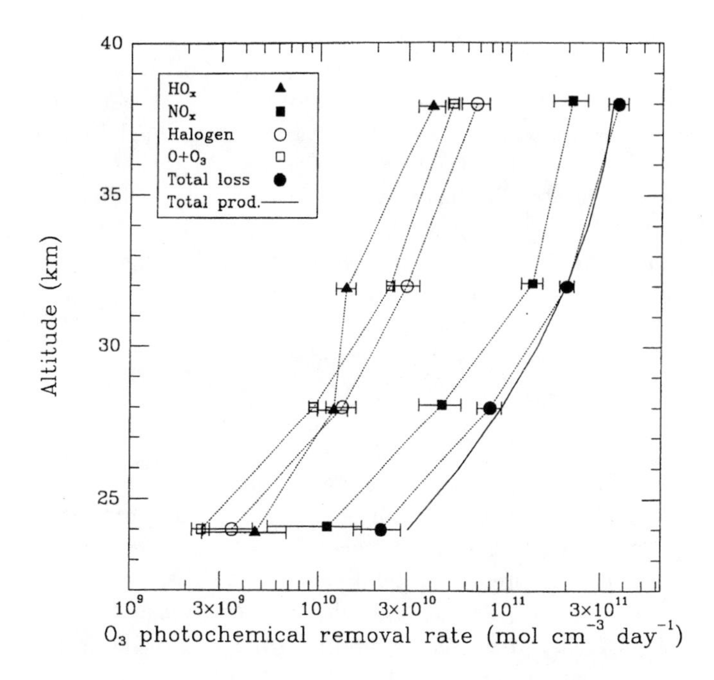

Figure 4: HO_x (solid triangles), NO_x (solid squares), and halogen (solid circles) catalyzed loss rates of O_3 inferred from radical method and calculated using the precursor method (curves).

mental conditions, represented by the observed concentrations of long lived radical precursors.

On an early FIRS-2 flight from Ft. Sumner, New Mexico we made the first simultaneous measurement of HO_x, Cl_x, and NO_x families throughout the middle atmosphere. The individual species measured were OH, HO_2, H_2O_2, H_2O, O_3, O_2, O^3P (in the mesosphere and thermosphere), HCl, HF, HOCl, HBr, CO_2, HNO_3, NO_2, N_2O, CO, HCN, and CO_2 (temperature and pressure profiles are derived from the 16 μm bands of CO_2). The gondola also carried the JPL BMLS microwave heterodyne spectrometer which measured ClO and O_3.

The strength of this set of measurements is the simultaneous observation of a large number of individual species, covering a large range of altitudes in the stratosphere and nearly one full diurnal cycle. The completeness of the data set provides a basis for examining our understanding of processes that regulate the partitioning of radicals using a photochemical model constrained by observations of the longer lived radical precursors.

Our constrained photochemical model balances the diurnally averaged production and loss of 35 reactive gases for the temperature, pressure, and latitude at which the observations were obtained. Standard reaction rates and absorption cross sections were used, except that photolysis cross sections for H_2O_2, HNO_2, and HNO_4 were extrapolated to longer wavelengths to accurately represent photolysis of these molecules at large solar zenith angles. A reaction probability of 0.1 was used for the heterogeneous hydrolysis of N_2O_5. The heterogeneous hydrolysis of $ClNO_3$ and $BrNO_3$ were included, but have a negligible effect on model results for the temperatures of these observations. The altitude profile for the surface area density of sulfate aerosols was adopted from SAGE II extinction measurements for this time period. Other inputs to the constrained photochemical model include the profiles of temperature, O_3, and H_2O measured by FIRS-2. Profiles of CH_4, odd nitrogen ($NO_y = NO + NO_2 + NO_3 + 2xN_2O_5 + HNO_2 + HNO_3 + HNO_4 + ClNO_3 + BrNO_3$), inorganic chlorine ($Cl_y = HCl + ClNO_3 + ClO + HOCl + Cl + OClO + 2xCl_2O_2 + ClOO$), and inorganic bromine ($Br_y = HBr + BrONO_2 + BrO + HOBr + BrCl$) have been inferred from the FIRS-2 measurement of N_2O using relations derived from previous satellite and in situ observations. The maximum Br_y at high altitude is assumed to be 21 ppt. The precursor method uses this photochemical model, constrained by our measurements of O_3, H_2O, N_2O, and temperature, to predict the abundances of O, HO_2, OH, NO_2, ClO, and BrO over a 24 hour period. These abundances are used to calculate the average rates O_3 loss.

The radical method uses direct measurements of HO_2, OH, NO_2, and ClO to calculate the rates of reactions of O_3 loss. The measurements are interpolated onto a uniform time grid by normalizing the abundances calculated by the constrained photochemical model to best fit the data.

Fig. 4 compares the removal rate of odd oxygen from the radical and precursor methods, partitioned into each radical family. An important result is that both methods agree well. Another result is that the general behavior agrees with theoretical expectations, in the sense that NO_x loss rates dominate between altitudes of about 22 and 38 km, and that HO_x loss rates dominate below about 22 km.

We have also calculated the production rate of O_3 from photolysis of O_2 (using a radiative transfer model that includes Rayleigh and aerosol scattering), and compared this with our measured rates[13]. We draw two conclusions from this comparison.

(a) Between 31 and 38 km, where the photochemical lifetime of ozone is short, we find that production and loss rates of odd oxygen calculated using both methods balance to within the uncertainty of the measurements of the radicals (approximately 10%), in contrast to previous studies (which lacked simultaneous measurements of HO_2, NO_2, and ClO) that reported loss rates up to 50% greater than production rates.

(b) Rates of production below 31 km exceed the loss rates calculated using both methods by 30 to 40%, indicating that this region is a net photochemical source region for O_3. This imbalance is consistent with results of 2-dimensional photochemical-transport models that show production and subsequent transport of O_3 to higher latitudes from this region.

Our finding that production and measured loss of of odd oxygen balance for altitudes between 31 and 38 km differs from the results of previous studies for two reasons: (i) we observe lower concentrations of ClO at these altitudes than are predicted by models which assume that production of HCl occurs only by reaction of Cl with hydrocarbons (see following section on "missing chemistry"); and (ii) we use a formulation for photolysis of O_2 that results in deeper penetration of ultraviolet radiation, and better agreement with measured transmittances, than is found using the standard formulation.

Our measurements demonstrate that production and loss rates of ozone (found using both methods) balance to within the uncertainty of measurement (10%) for altitudes between 31 and 38 km. This finding reconciles the long standing "ozone deficit" discrepancy in stratospheric chemistry, in this altitude range.

MISSING CHEMISTRY

ClO abundance

In the course of the photochemical balance work described above, we found that the standard chemical reaction rates were not successful in predicting the measured abundances of the Cl species HCl, HOCl, and ClO. Suspecting from this (and other independently reported inconsistencies) that the reactions $ClO + OH \rightarrow Cl + HO_2$ or $ClO + HO_2 \rightarrow HOCl + O_2$ might have another channel that produces HCl, we investigated[13] the predictions of various branching ratios giving small percentages of the total rate to $ClO + OH \rightarrow HCl + OH$ or $ClO + HO_2 \rightarrow HCl + O_3$.

Fig. 5 shows our measurements of three Cl_y species along with model calculations, corresponding to the average SZA of the measurements, assuming different branching ratios for the production of HCl from the reaction of ClO with both OH. The figure shows results from models incorporating a 0% branching ratio for the $ClO + HO_2$ reaction, and each panel shows three model calculations, assuming branching ratios for the ClO + OH reaction of 0%, 5%, and 10%. Assuming the total Cl_y used in the model is correct, these plots suggest that there must be some production of HCl at the expense of ClO and HOCl, possibly through one or both of these mechanisms. If the ClO + OH branch is responsible, then Fig. 5 suggests that the branching ratio is in the range 5-10%. We found that inclusion of these alternate branching ratios has negligible effect on HO_x and NO_x, as determined from direct comparison of model results.

A recent laboratory measurement[17] of the ClO +

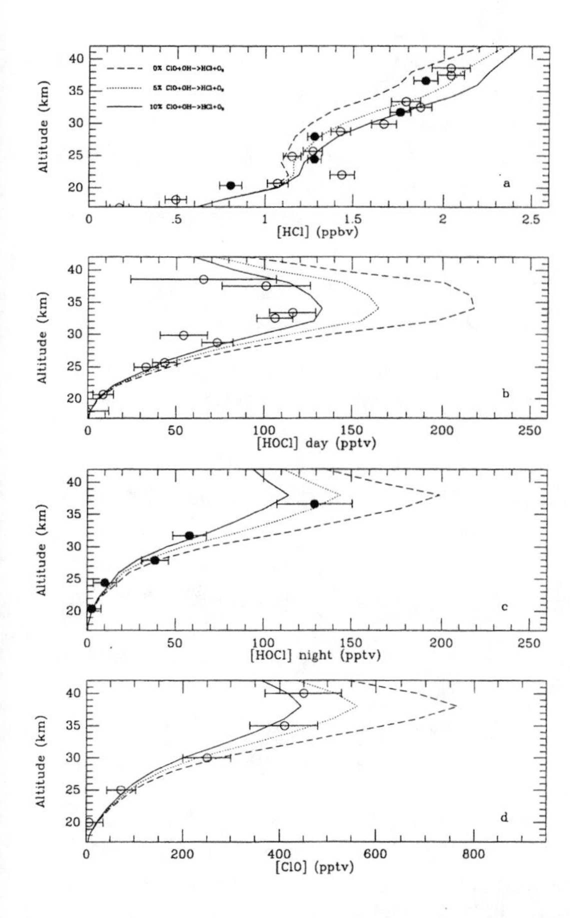

Figure 5: FIRS-2 and BMLS balloon measurements of Cl_y species and corresponding model calculations using varying branching ratios for the reactions of ClO with OH to produce HCl. A branching ratio in the range 5-10% produces a significantly better overall agreement than does 0%.

$OD \rightarrow DCl + O_2$ reaction, which should be a good proxy for the corresponding $ClO + OH \rightarrow HCl + O_2$ reaction, finds a branching ratio of $(6 \pm 2)\%$, which is in good agreement with the 5-10% range suggested by FIRS-2 and BMLS data.

The predicted decrease in ozone near 40 km resulting from the build up of halogens during the past several decades is mitigated by 50% if an additional pathway for production of HCl is included in models, resulting in closer agreement between observed and predicted trends in O_3 at these altitudes.

H_2O_2 abundance

Hydrogen peroxide (H_2O_2) is a significant reservoir for HO_x radicals (OH and HO_2) in both the stratosphere and troposphere, and is thought to undergo very simple photochemistry. Tropospheric measurements of H_2O_2 are used as a proxy for HO_2, which is thought to be directly involved in both the destruction and production of tropospheric ozone. The potential also exists for H_2O_2 to be mapped in the upper troposphere from space. As a result, the photochemistry of H_2O_2 ought to be well characterized for its measurements to be correctly applied to tropospheric ozone chemistry.

The primary production reaction of H_2O_2 is

$$HO_2 + HO_2 \rightarrow H_2O_2 + O_2. \qquad (39)$$

The significant chemical loss reactions of H_2O_2 are the reactions

$$H_2O_2 + OH \rightarrow H_2O + HO_2, \qquad (40)$$
$$H_2O_2 + h\nu \rightarrow OH + OH, \qquad (41)$$
$$H_2O_2 + O \rightarrow OH + HO_2. \qquad (42)$$

We ahve measured all of these reactants simultaneously in the stratosphere[18,19,20], and can therefore rigorously test the balance between production and loss of H_2O_2. Because the photochemistry of H_2O_2 is slow and occurs primarily during daylight, stratospheric concentrations of H_2O_2 are thought to vary only slightly over one diurnal cycle and can be determined from

$$[H_2O_2] = \frac{k_1[HO_2]^2}{j_3 + k_2[OH]} \qquad (43)$$

where [OH], $[HO_2]^2$, and j_3 are averaged over 24 hours.

In our study, we use measurements of profiles of [OH], $[HO_2]$, $[H_2O_2]$, $[O_3]$, and temperature between 20 and 38 km obtained during five balloon flights of the FIRS-2 spectrometer and profiles of $[H_2O_2]$, $[O_3]$, [CO], and temperature between 8 and 40 km from seven balloon flights of the Mk IV spectrometer during late September and late May/early June at 34°N.

The results are shown in Fig. 6, where the predicted H_2O_2 is compared to the measured H_2O_2 from both the FIRS-2 and Mk IV FTSs. The predicted abundances clearly exceed the measured ones by a significant margin. This overprediction suggests either the production in the model is too fast or that the loss terms are too slow, especially below 32 km where the model exceed the measurements by more than a factor of two.

If we assume that the loss rate is in error, then the missing loss term for H_2O_2 can be represented as

$$k_x[X] = \frac{k_1[HO_2]^2}{[H_2O_2]} - k_2[OH] - j_3 \qquad (44)$$

where the subscripts refer to the appropriate reactions above, k_x is the rate constant for the unknown reaction of H_2O_2 with molecule X, and the concentrations are averages over a 24 hour cycle. Starting with the list of known reactions, and eliminating those which were unlikely for one reason or another, we ultimately eliminated them all.

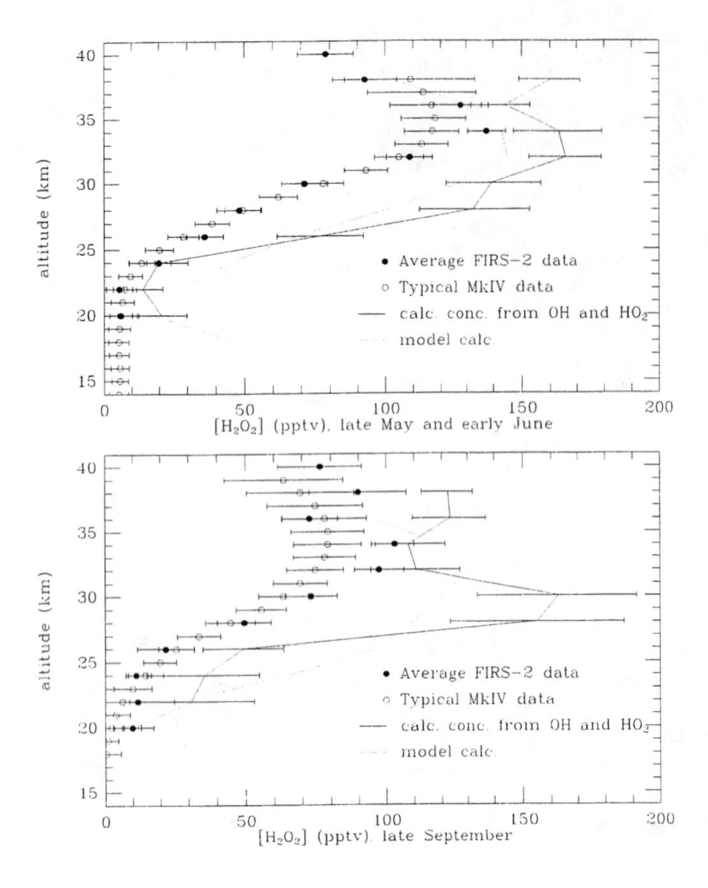

Figure 6: FIRS-2 (solid circles) and Mk IV (open circles) retrievals of stratospheric H_2O_2 concentration profiles during late May/early June and late September at 35°. Also plotted are calculated profiles of $[H_2O_2]$ from a photochemical model which is constrained by measured OH and HO_2 abundances from FIRS-2.

An alternate explanation is an uncatalogued loss reaction. As an illustration, one such potential reaction is

$$H_2O_2 + HO_2 \quad \rightarrow \quad H_2O + O_2 + OH, \qquad (45)$$

which is not included in the standard list of stratospheric reactions. This reaction requires multiple breaking and forming of bonds, which most likely would result in very slow rates with high activation energies and relatively large positive temperature dependences.

No laboratory studies have been performed for this reaction. It would be difficult to study because the self reaction of HO_2 creates H_2O_2, and if OH were produced, it could react with both HO_2 and H_2O_2. The concentration profile shape of $[HO_2]$ in the stratosphere is somewhat consistent with the profile shape of the H_2O_2 imbalance curve.

If the cause of the H_2O_2 discrepancy is indeed a missing reaction, and not an error in photolysis rates, then it most likely will be a significant sink of HO_x in the middle and lower troposphere. Thus it is important that the cause of this discrepancy be identified. As a possible useful application, we note that it is possible to monitor upper and middle tropospheric concentrations of H_2O_2 from space using mid-infrared ro-vibrational transitions. Assuming that the source of the stratospheric H_2O_2 discrepancy can be identified, such measurements could be used as a proxy to map tropospheric HO_x levels with space based measurements

HYDROGEN BROMIDE

Anthropogenic emission of methyl bromide and the halons has the potential to cause greater ozone losses than emissions of chlorofluorocarbons and hydrochlorofluorocarbons, in part because nearly 50% of the bromine released into the stratosphere is believed to exist in reactive forms, as compared to a few percent of the chlorine(1). Already, at a concentration of only 0.5% that of chlorine, catalytic cycles involving bromine account for part of the ozone loss in the Antarctic ozone hole and a few percent of the total ozone loss rate at mid-latitudes.

Accurate assessment of the ozone depletion potential (ODP) of brominated compounds requires an understanding of the partitioning of stratospheric bromine between reactive species (Br, BrO) and reservoir species (HBr, $BrNO_3$, and HOBr). Their low abundance has made measurement of bromine species difficult, which is unfortunate because there is still considerable uncertainty in the reaction rates used in photochemical models and comparison with observations would build confidence in model predictions.

In a recent FIRS-2 study(22) we expanded on earlier work(21) in which we combined measurements from 3 balloon flights to produce an upper limit for HBr of 4 parts per trillion (ppt) at an altitude of 32 km. The more recent work combines data from 7 balloon flights to produce mixing ratio profiles of HBr and HOBr from 22–38 km. After analyzing all these spectra for the very weak HBr features, we find an average mixing ratio of HBr in the range 22–34 km to be 2.0 ± 0.8 ppt, which is a a small fraction of the assumed total 19 ppt of total inorganic bromine. This is consistent with a model which includes 2% production of HBr through the reaction of BrO with HO_2, but is highly inconsistent with a 5 or 10% production of HBr.

ACKNOWLEDGEMENTS

This work was supported by NASA's Upper Atmosphere Research Program, Mission to Planet Earth, under grant NSG 5175 to the SAO. The 1997 Alaska balloon campaign, which included FIRS-2, Mk IV, and other instruments, was generously supported by the National Space Development Agency of Japan (NASDA), for the purpose of obtaining validation data for the Advanced Earth Observing Satellite (ADEOS). I am indebted to my colleagues for their ongoing collaboration, and for their contributions to this review: Dave Johnson, Ken Jucks, Kelly Chance, Geoff Toon, Bob Stachnik, Jim Margitan, Ross Salawitch, Hope Michelson, Paola Ciarpallini, and Kazuo Shibasaki. We together wish to thank the NASA JPL Balloon Group for providing gondolas and technical support for our balloon flights, the NASA National Scientific Balloon Facility for balloon launch and support service, and the NASA Ames ground and flight crews for assistance on all DC-8 flights.

REFERENCES

1. World Meteorological Organization, **Report No. 25**, 1991.
2. Anderson, J.G. in *Progress and Problems in Atmospheric Chemistry*, Ed. J.R. Barker, World Scientific, pp. 744-770, 1995.
3. Toon, G.C., *Optics and Photonics News*, **2**, pp. 298-307, 1991.
4. Traub, W.A., Chance, K.V., Johnson, D.G., and Jucks, K.W., *SPIE*, **1491**, pp.298-307, 1991.
5. Johnson, D.G., K.W. Jucks, W.A. Traub, and K.V. Chance, *JGR*, **100**, pp. 3091-3106, 1995.
6. Traub, W.A., and Stier, M.T., *Applied Optics*, **15**, pp. 364-377, 1976.
7. Dobrowolski, J.A., and W.A. Traub, *Applied Optics*, **35**, pp. 2934–2946, 1996.
8. Johnson, D.G., W.A. Traub, and K.W. Jucks, *Applied Optics*, **35**, pp. 2955–2959, 1996.
9. Chance, K., K.W. Jucks, D.G. Johnson, and W.A. Traub, *JQSRT* **52**, pp. 447-457, 1994.
10. Traub, W.A., K.W. Jucks, D.G.Johnson, M.T. Coffey, W.G. Mankin, and G.C. Toon, *GRL* **21**, pp. 2591-2594, 1994.
11. Traub, W.A., K.W. Jucks, D.G. Johnson, and K.V. Chance, *JGR*, **100**, pp. 11262-11267, 1995.
12. Traub, W.A., K.W. Jucks, D.G. Johnson, and K.V. Chance, *GRL*, **21**, pp. 2595-2598, 1994.
13. Jucks, K.W., D.G. Johnson, K.V. Chance, W.A. Traub, R.J. Salawitch, and R.A. Stachnik, *JGR*, **101**, pp. 28785-28792, 1996.
14. Chance, K., W.A. Traub, D.G. Johnson, K.W. Jucks, P. Ciarpallini, R.A. Stachnik, R.J. Salawitch, and H.A. Michelsen, *JGR*, **101**, pp. 9031–9043, 1996.
15. Johnson, D.G., W.A. Traub, K.V. Chance, K.W. Jucks, and R.A. Stachnik, *GRL*, **22**, pp. 1867–1871, 1995.
16. Johnson, D.G., J. Orphal, G.C. Toon, K.V. Chance, W.A. Traub, K.W. Jucks, G. Guelachvili, and M. Morillon-Chapey, *GRL*, **23**, pp. 1745–1748, 1996.
17. Lipson, J.B., Elrod, M.J., Beiderhase, T.W., Molina, L.T., and Molina, M.J., Submitted to *Faraday Transactions*, March 1997.
18. Jucks, K.W., D.G. Johnson, K.V. Chance, W.A. Traub, R.J. Salawitch, and C.G. Toon, to be submitted to *JGR*, 1997.
19. Chance, K.V., Johnson, D.G., Traub, W.A., and Jucks, K.W., *GRL* **18**, pp. 1003-1006, 1991.
20. Traub, W.A., D.G. Johnson, and K.V. Chance, *Science* **247**, pp. 446-449, 1990.
21. Traub, W.A., D.G. Johnson, K.W. Jucks, and K.V. Chance, *GRL* **19**, pp. 1651-1654, 1992.
22. Johnson, D.G., W.A. Traub, K.V. Chance, and K.W. Jucks, *GRL*, **22**, pp. 1373–1376, 1995.

Digital Signal Processing (DSP) Applications in FT-IR. Implementation Examples for Rapid and Step Scan Systems

Raúl Curbelo

Bio-Rad Digilab Division, 237 Putnam Av., Cambridge, MA 02139

Signal processing has always been a critical aspect of interferometric spectroscopy. The main value of Fourier Transform Spectroscopy (FTS) is to provide the maximum information per unit time. This means that careful signal processing is required to maximize the signal to noise ratio of the result. Digital signal processing (DSP) is not new in FTS. DSP for apodization, Fourier transform, phase correction, and some digital filtering have been used for many years. The first difficulty encountered in using DSP is the required transformation of the continuous time signal to a sequence of finite precision numbers, that is a discrete time, discrete magnitude representation of the continuous signal. Today, sampling and digitization with conventional techniques are reasonably straightforward, with digitization noise of the order of 2^{-16}. Advances in several aspects of FTS technology, including step scanning, have created the need to lower the digitization noise.

The processing of signals from multiple modulation FTS experiments, specially step-scan signals, has presented serious limitations until recently, due primarily to the performance restrictions of the available hardware, and the fundamental limitations of the analog systems used. The application of DSP techniques to reduce the sampling and digitization errors are reviewed, and practical solutions that can provide digitization noise lower than 2^{-24} with low cost systems are described. Some of the causes and effects of sampling error are reviewed, to clarify common misunderstandings. DSP applications in the detection of multiple modulation step-scan experiments and photoacoustic spectroscopy are described in detail. These applications allow the measurement of the signal during the entire step time, eliminate interference of secondary effects of phase modulation, and integrate the entire experiment in the data system, eliminating all the analog adjustments (knobs) needed when lock-in amplifiers are used.

INTRODUCTION

Discrete signal processing may be a more general definition for DSP. As such, it has roots in the 17th & 18th century. Discrete samples have been used in numerical methods to solve equations when analytical solutions are not possible. Newton used finite difference methods that are special cases of discrete time systems used today. Gauss discovered the fundamental principle of the fast Fourier transform around 1805.

Signal processing has always been a critical aspect of interferometric spectroscopy. The estimation of the desired spectrum presented enough difficulties that it eluded the workers in the field for many years: consider the time elapsed from Michelson to Fellgett (1891 to 1951). Even more important, (after Fellgett) the main value of Fourier Transform Spectroscopy (FTS) has been to provide the maximum information per unit time. This meant that careful signal processing from the detector to the end result was required to maximize the signal to noise ratio of the result.

Digital signal processing requires that the signals used be discrete amplitude, discrete time-sampled representations of continuous time, continuous amplitude (analog) signals

Digital signal processing is not new in FTS. When available, it was used by the early workers in the field (Fellgett, Gebbie and Strong) primarily for the Fourier transform, to circumvent the multiple limitations of analog spectrum analyzers. DSP for apodization, Fourier transform phase correction, and some digital filtering, have been used for many years.

SAMPLING AND DIGITIZATION

The first difficulty encountered in using DSP to process FT-IR signals is the required transformation of the continuous time interferometer signal to a sequence of finite precision numbers that is a representation of the continuous signal.

Today, sampling and digitization with conventional techniques are reasonably straightforward, with digitization noise of the order of 2^{-16}. Advances in several aspects of the technology, including step scanning, have created the need to lower the digitization noise. In some experiments, the spectrometer signal dynamic range can be larger than 2^{20}.

The processing of signals from multiple modulation FT-IR experiments, specially step-scan signals, has presented serious limitations until recently, due primarily to the performance restrictions of the available hardware, and the fundamental limitations of the analog systems used.

The ideal digitizer introduces a digitization error (noise) e_n that with reasonable assumptions [1], can be estimated as shown in equation 1.

CP430, *Fourier Transform Spectroscopy:* 11th International Conference
edited by J.A. de Haseth

$$e_n = LSB/\sqrt{12} \quad (1)$$

where LSB is the value of the least significant bit of the output. In any hardware implementation of an analog to digital converter (A/D), additional noise is introduced by the electronics. For a non-ideal digitizer, the word length N for the equivalent ideal digitizer (number of bits) is:

$$N = -\log_2 (e_n \cdot \sqrt{12}) \quad (2)$$

For a moderate cost 16 bit high end, 500 Ksamples/sec successive approximation A/D converter, (Analogic ADC4345) the RMS digitization noise is 5.2×10^{-6} of full scale (FS), which makes it a true 16 bit converter. This A/D was developed for medical imaging systems. Figure 1 shows the output of this A/D for a slow ramp input. The different steps correspond to successive output codes, and the effect of the small additional noise is seen in the transition between codes.

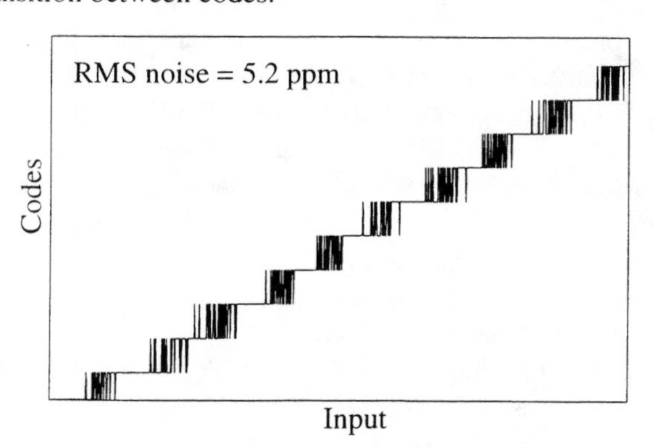

Figure 1. 16 bit A/D output for a ramp input

For a "20", bit 48 KSamples/sec Oversampling Sigma-Delta A/D (Crystal Semiconductor 5390) used in the audio industry, the RMS digitization noise is about 2×10^{-6} FS, which makes it slightly better than an ideal 17 bit converter. The output of this A/D for the same ramp input is shown in Fig. 2.

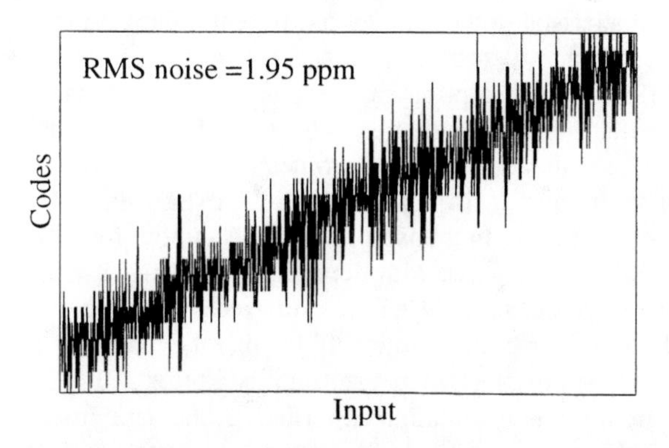

Figure 2. 20 bit Delta Sigma A/D output for a ramp input.

DSP techniques applied to FT-IR signals can improve the effective resolution of the A/D converter, and provide digitization noise smaller than 2^{-24} of full scale. A block diagram of an implementation of this technique is shown in Fig. 3.

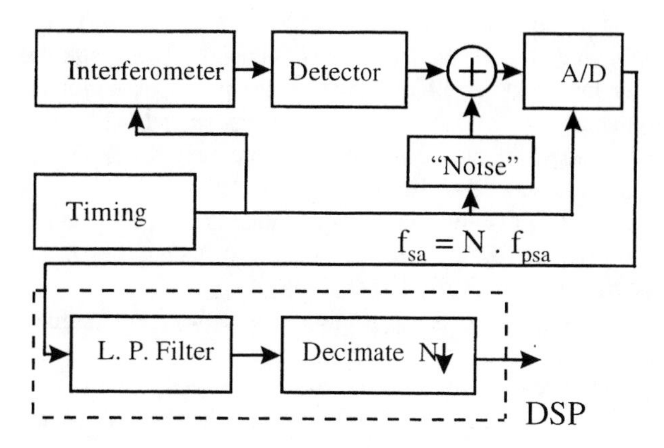

Figure 3. Improvement of the A/D resolution by adding noise, oversampling, digital filtering, and decimation.

This implementation is useful in both rapid scan and step-scan experiments. In step-scan experiments, it allows the measurement of the signal during the entire sample time, or step time; therefore, it uses the entire spectrometer signal channel capacity [2]. It also allows the reduction of the errors inherent in the step scanning process of multiple modulation experiments, and integrates the entire experiment in the data system, eliminating all the hardware instrumentation external to the spectrometer that is needed when lock-in amplifiers (LIA) are used.

In FT-IR, digitization noise limits the signal to noise ratio (S/N) for low spectral resolution features, which are generated by the interferogram centerburst. Coadding multiple scans does not improve the results when the detector noise is smaller than the digitization noise. Gain ranging is no help, as it only affects the smaller signal amplitudes of the interferogram.

We can solve the coadding problem by adding noise to the signal, larger than the digitization noise. If we require that the spectrum of the additive noise be outside the frequency range of the signal of interest, the additive noise can be filtered out in the digital domain. To reduce the effective digitization noise, we take N times more samples than the minimum required by the sampling theorem, therefore spreading the digitization noise power over a bandwidth (BW) N times larger. When the signal plus noise is filtered, the digitization noise in the BW of the signal is reduced by $\sqrt{N}$. After the filter, the BW of the signal and noise has been reduced by the same factor N, and only one of every N samples are needed. See Fig. 3.

A good choice for the "additive noise" is a triangular signal at the principal sampling frequency. This only has spectral components at multiples of the principal sampling frequency. For this case, the most efficient filter is very simple, and the filter and decimate process can be combined into one. The filter is a boxcar average, which gives the optimum S/N. The decimation is done by taking only one of every N samples for the output. Therefore, the digital filtering and decimation are performed by averaging successive sets of N samples. The resulting filter has zeros at all multiples of the principal sampling frequency f_{psa} . With an over sampling factor $N=2^{16}$, the A/D resolution can be improved theoretically by 2^8 at the same time that the digitization noise is reduced by the same factor, resulting in an equivalent 22 bit A/D, when a true 16 bit A/D is used. [3] Larger oversampling rates provide proportionally larger gains.

The sampling process itself is also a very demanding task when we try to minimize errors from second order effects. The sample position error, or the velocity error in rapid scan, needs to be minimized when we want to minimize second order errors [4].

The effects of sampling errors can in general be understood by using linear superposition and simple examples. The phase modulation (PM) from small position errors is approximately linear because the modulation index is much smaller than 1. The effect on broad band noise from a narrow band position error is to broaden the noise spectrum distribution, without changing the total noise power, because it is a PM. [5]. For a narrow band signal, the PM from the position error will create sidebands around the signal proportional to the amplitude of the signal, and with the spectrum of the position error. The PM of a broad band signal by a narrow band position error results in sidebands of the broad band signal separated by the frequency of the position error. The experimental result of an imposed sinusoidal position error of 10 nm at 0.2 Hz on the spectrum of polystyrene, collected at 10 Hz step rate and 400 Hz PM, is show in Fig. 4, where the spectrum from the disturbed system is superimposed with the spectrum from the undisturbed system.

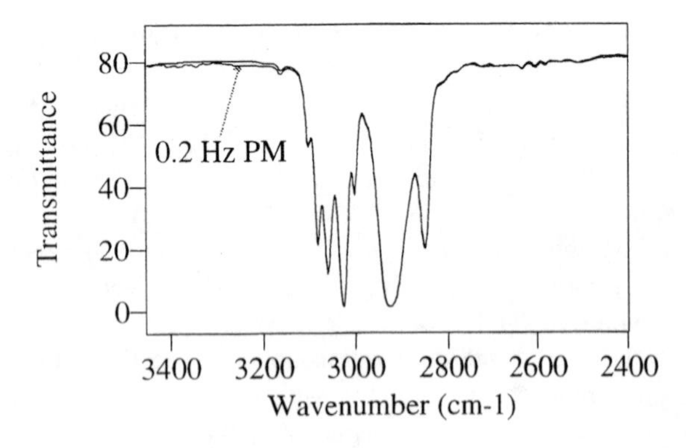

Figure 4. Polystyrene % T, with and without imposed position error.

Figure 5 shows the difference between the two spectra, illustrating the PM sidebands due to the induced position error.

In a modern FT-IR system, the main cause of position error is due to the effect of external vibrations on the moving mirror assembly. The sensitivity to these external excitations is determined by the control system driving the mirror, and in turn by the system dominant time constant.

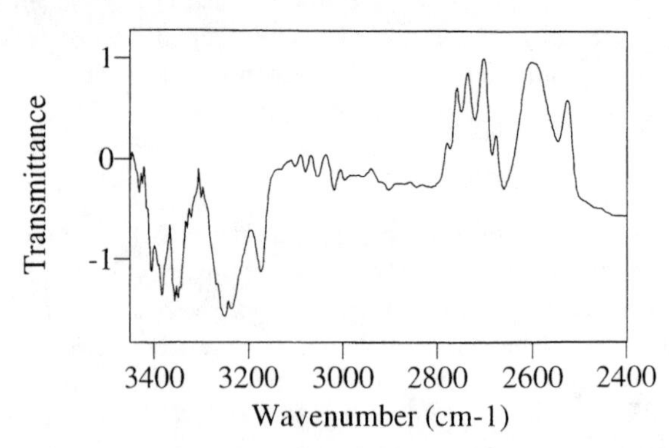

Figure 5. Difference spectrum showing the PM sidebands.

For a moving mirror and linear bearing system with a reasonable voice coil motor, the dominant time constant is of the order of 40 msec. In step-scan, this control system operating on a laboratory bench, will have a position error of about 2 nm RMS For a mirror driven by a piezo-electric transducer which is anchored rigidly to the interferometer frame, the time constant is about 40 µsec. With the combination of these two drives, the RMS position error under the same conditions is of the order of 0.2 nm., and the spectral content of this error below 25 Hz is about one tenth of that. Comparable values of velocity error are present in rapid scan experiments.

MULTIPLE MODULATION EXPERIMENTS USING A PHOTO-ELASTIC MODULATOR

The Photo-elastic Modulator (PEM) is used to modulate the polarization of the spectrometer beam, to allow the measurement of the difference in the spectral characteristic of the sample to different polarizations.

Circular dichroism (CD), measures the spectral differential absorption of the sample for left and right circular polarized radiation. The desired signal is obtained by demodulating the spectrometer signal at the PEM drive frequency. [6] Linear dichroism (LD) measures the spectral differential absorption of the sample for different linear polarizations, with demodulation at twice the PEM drive frequency. Dynamic infrared linear dichroism (DIRLD), measures the effect on the sample linear dichroism from a dynamic strain modulation. [7]

In multiple modulation experiments, the data processing system used until recently employed multiple Lock-in Amplifiers (LIA). The implementation of a DSP solution in this case is similar to the multiple modulation DSP previously described [8], with one more modulation as shown in Fig. 6. The interferometer step frequency and phase modulation frequencies are f_{St} and f_{PM} respectively.

The sample modulation frequency is f_S, and the sampling frequency is f_{sa}. The additional complexity in this case is that the PEM drive frequency f_M cannot be locked to the system clock. To demodulate the PEM carrier signal with a DSP process we need to know not only the exact frequency of the PEM drive at the time the data was collected, but also its phase relative to the sampling clock.

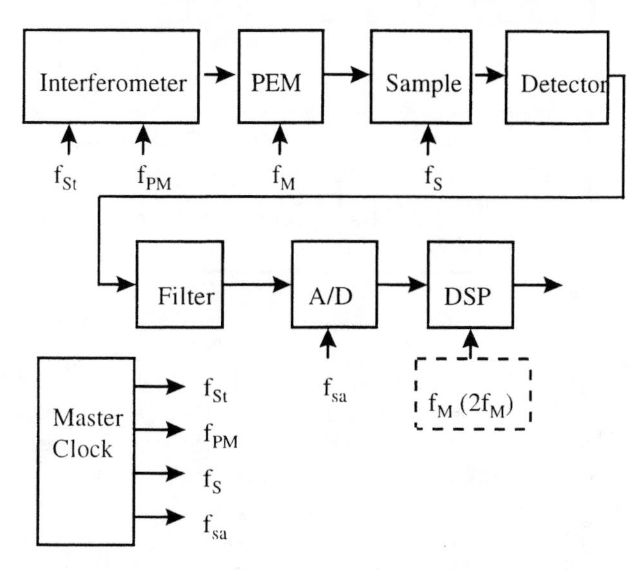

Figure 6. DSP demodulation frequencies. f_{St} is the interferometer stepping frequency.

Locking the PEM drive to the system master clock is not practical, as the bandwidth of the PEM is comparable to its resonant frequency drift, and would result in a changing PEM modulation index during an experiment.

One solution would be to collect the sinusoidal reference signal from the PEM drive with a second digitizer channel in the spectrometer, so that the frequency and phase information can be extracted in the DSP process using one of the methods developed for communication systems. [9, 10]. This choice involves more hardware capability than is required to collect only the signal of interest.

Another solution is to measure the frequency and phase of the PEM drive from the data collected for the actual experiment. In most cases, the PEM carrier is present in the signal due to residual polarization in the optical train. For the case when the residual polarization has been substantially eliminated, a small amount of the PEM drive can be added to the signal, as a pilot carrier, to be used to recover the frequency and phase of the PEM drive.

For circular dichroism, the polarization signal of interest is present in the side bands of the carrier at the frequency of the PEM drive f_M, and for linear dichroism, the signal is in the side bands of the carrier at the second harmonic of the PEM drive. In either case, the signals generated, although widely spread in the frequency domain, have quite small actual bandwidth as shown in Fig. 7 for a sample modulation f_S of 25 Hz, and a phase modulation f_{PM} of 400 Hz.

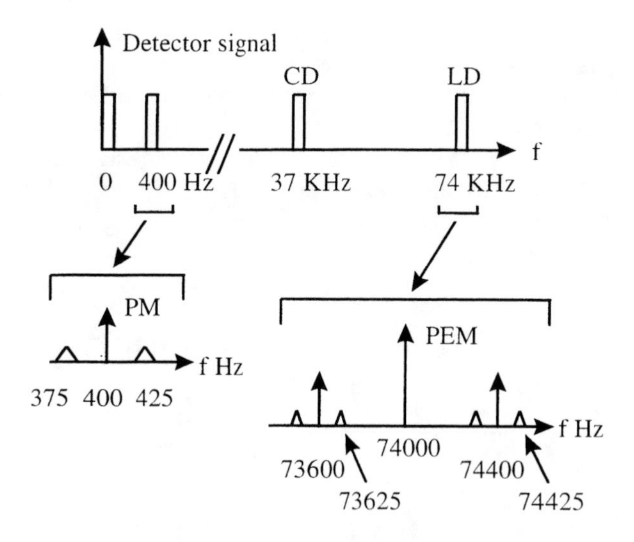

Figure 7. Spectra of DIRLD signals.

For the case of the DIRLD experiment the signals of interest due to f_S are near the PM frequency f_{PM}, and near $2f_M$. A nominal value for the PEM drive frequency f_M is 37 KHz with a tolerance of +/-20 Hz. To sample these signals applying the concept of low pass sampling would require a sampling frequency higher that 148 KHz. This rate would generate more than 300 Mb of data for a low resolution experiment (1000 steps of 1.3μ at 1 Hz for 8 cm^{-1} resolution). Considering that the data will require in the order of N^k (with k>1) processing operations, it is desirable to reduce the number N of samples to be processed for a given experiment.

Sampling can be viewed as a modulation process in which the spectrum of the sampled signal contains periodic repetitions of the base band spectrum of the input signal, with a period equal to the sampling frequency f_{sa}. See Fig. 8 .[1]

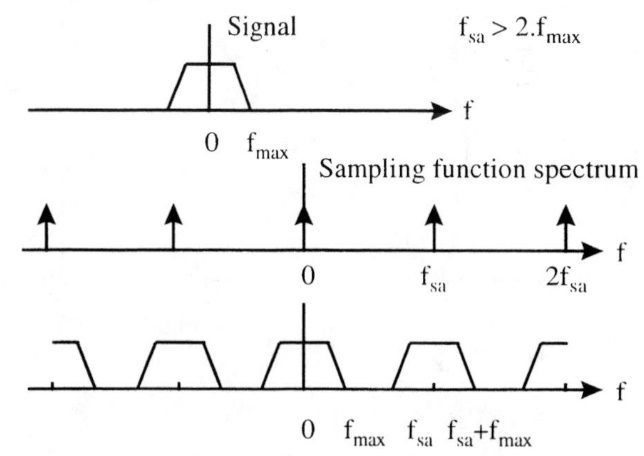

Figure 8. Spectra in low pass sampling implementation.

Because the actual signal bandwidth in this experiment is much smaller than the separation of the spectral regions of interest, a fairly simple anti-aliasing filter can be built that will allow sampling at a lower rate and prevents folding of noise over the signal bandwidth. For the DIRLD experiment, the filter pass bands are the trapezoids shown in Fig. 9. With this filter the signals can be sampled at 20 KHz, and signals near 74 KHz will be translated down to 6 KHz. Only the positive frequencies are shown in the spectrum of the sampled signal for simplicity. All sampling products at frequencies higher than 10 KHz will be filtered out in the DSP process at no cost.

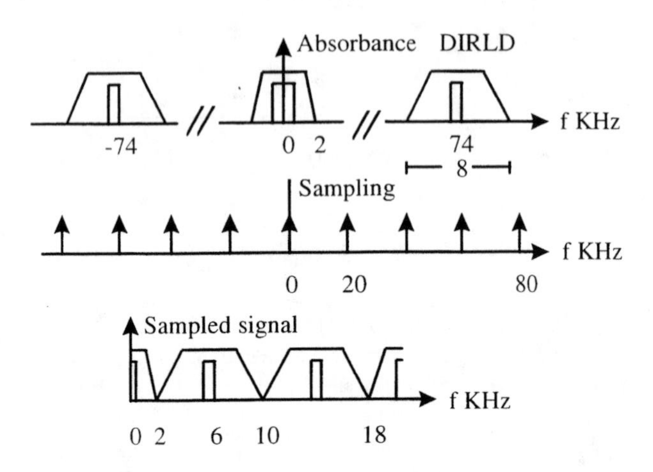

Figure 9. Filtering and undersampling.

For the case of circular dichroism, the modulated signals are near 37 KHz and can be sampled at 14 KHz using a similar anti-aliasing filter.

A portion of the sampled DIRLD signal from a step near the interferogram center burst is shown in Fig. 10 . The PEM carrier signal and the base band PM signal(distorted square wave) can be recognized.

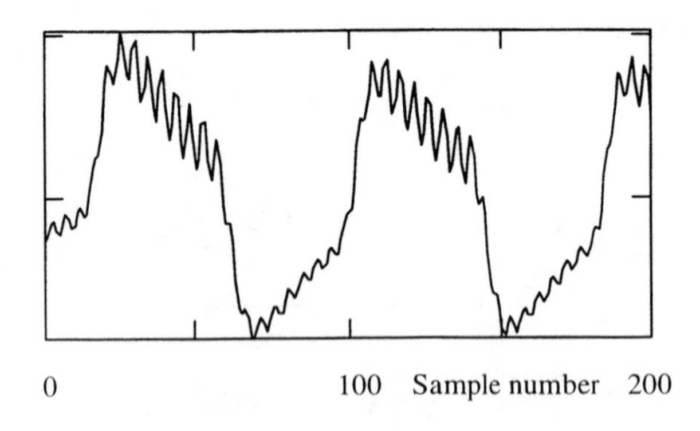

Figure 10. DIRLD digitized signal.

In general it is desirable to minimize the A/D digitization noise using its full channel capacity by over sampling the signal by a factor D. Using the same anti-aliasing filter as above, a digital low pass filter will reduce the digitization noise, and decimate the filtered signal by the factor D, as shown in Figure 11. In this case the gain will be limited to the reduction of the digitization noise, unless some of the signals provide the equivalent of the "noise " added in Fig 3. In that case we can also get an improvement in the A/D resolution.

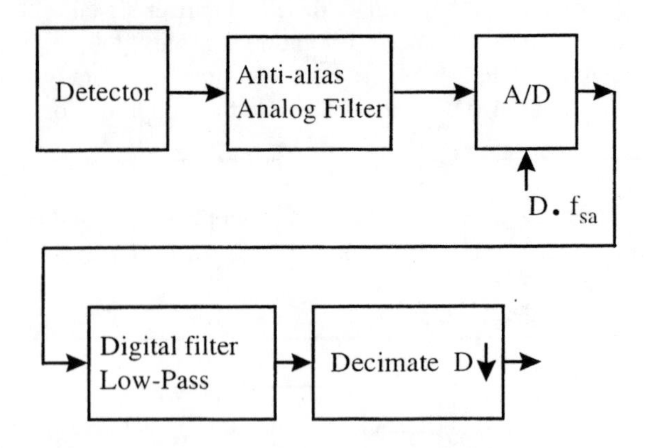

Figure 11. Oversampling, filtering and decimation.

The PEM carrier frequency could be measured at the maximum of the magnitude spectrum of the collected signal in a region centered on the nominal PEM frequency. This method is severely limited by the signal to noise ratio of the signal, and in general the measured frequency will have large errors.

A better method to determine the PEM carrier frequency is to measure the phase of the signal at different times in the signal from a step with respect to the phase of the nominal PEM drive frequency f_0 (ω_0 in radians/sec), and assuming constant frequency during the step, then calculate the average frequency.

We select three sets of M samples out of the N samples from the first step of the interferometer scan as shown in Fig. 12 and apodize each set with a triangular window function.

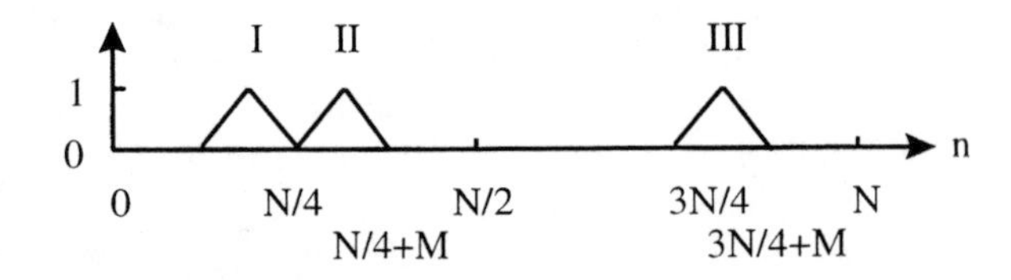

Figure 12. Apodization to compute the PEM frequency.

We call these samples x_n where "n" is the sample number in the step. We determine the phase error with respect to the nominal PEM frequency ω_0 for the set of x_n samples between N/4-M and N/4, (interval I) by computing the argument (phase angle) p1 of the discrete Fourier transform (DFT) of x_n at ω_0 .

$$p1=\arg\left[\sum_{N/4-M}^{N/4} x_n . e^{j\omega o \cdot n} \right] \quad (3)$$

Figure 13 shows the phase of the unknown PEM drive and the phase of the nominal PEM drive frequency as a function of sample number, for one step.

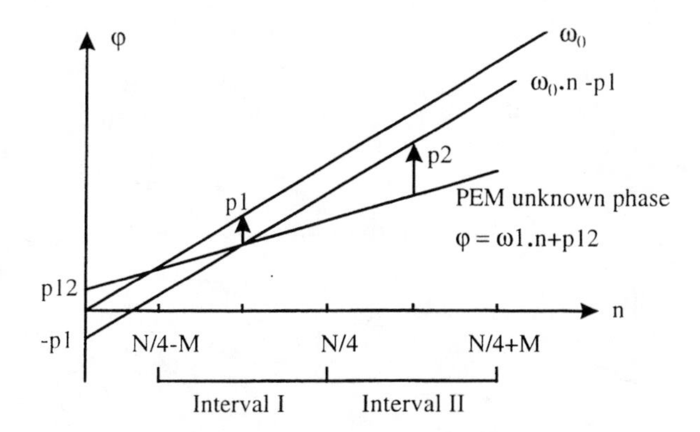

Figure 13. PEM drive phase vs. sample number.

If we recompute the DFT for the same samples in interval I with $e^{j(\omega o.n-p1)}$, the resulting argument would be zero, that is, ω_0 with phase -p1 is in phase with the PEM carrier at the center of the interval.

Now we compute the argument p2 of the DFT of x_n in interval II using $e^{j(\omega o.n-p1)}$.

$$p2=\arg\left[\sum_{N/4-M}^{N/4} x_n.e^{j\omega o \cdot n-p1}\right] \quad (4)$$

With the value of p2, we can compute an estimate ω_1 of the PEM frequency:

$$\omega_1 = \omega_0 - p2 / M \quad (5)$$

Using ω_1 we compute the phase p12 of ω_1 for zero error in interval I. With p12 we compute the phase error p3, in interval 3 using $e^{j(\omega 1.n-p12)}$. Equation 4 uses p3 to refine the estimate of the PEM frequency. (All equations unless indicated, use ω in radians, which is the abscissa of the DFT).

$$\omega_3 = \omega_1 - p3 / (N/2+M) \quad (6)$$

Another iteration may be required depending on the value of M used. The order of magnitude of the remnant error can be estimated evaluating $|\omega_3 - \omega_1|$.

M must be small enough so $p2< \pi$, and large enough so $p3< \pi$, so that equations 5 and 6 are valid. The first condition is met if:

$$|\omega_m - \omega_0| . M< \pi \text{ , then } M< f_{sa} / 2 . |f_m - f_0| \quad (7)$$

where f_{sa} is the sampling frequency. The second condition is met if:

$$|\omega_m - \omega_1| . N/2< \pi \quad (8)$$

If the second condition is not met, a second iteration in equation 5 is required with a larger interval length M', to obtain a better estimate of ω_1. At the same time, to optimize the convergence of the measurement of ω_m, M should be selected so the spectrum of the apodized signal has a minimal contribution at the PM side band frequency. The spectrum of a signal apodized with a triangular of a time interval M/ f_{sa} has zeros at $2.k_1.f_{sa}/ M$, where k_1 is an integer. For the PM frequency f_φ we can select M:

$$M = 2.k_1.f_{sa}/ f_\varphi \quad (9)$$

For example, for f_{sa} =20 KHz, f_m =6 KHz, $|f_m - f_0|$<20 HZ, and f_φ =400 Hz, M can be a multiple of 100, but no larger than 500.

Taking $\omega_m = \omega_3$, the phase error pm for the entire array of this step is computed as above using an apodization function A_n selected to provide filtering in the frequency domain, when demodulating the side band information from the PEM and PM modulations.

The apodization function A_n should provide substantial attenuation to the sidelobes of the different components of the signal, and in particular at ω_s and at or near ω_φ. A simple first choice is a triangular apodization. The Fourier transform of the apodization function is of the form:

$$A(\omega)|= C.\sin^2(k_{N'}. \omega)/(k_{N'}. \omega)^2 \quad (10)$$

For successive steps, the values of ω_m and pm must be measured using the process above with $\omega_1 = \omega_m$ from the previous step. Now only the last iteration will be required, as ω_m would have drifted only a small fraction of a Hertz in the time of one step.

Thereafter the DSP for the each step in the interferogram can proceed as shown in Fig. 14 where all the demodulating frequencies and the phase pm of ω_m are known.

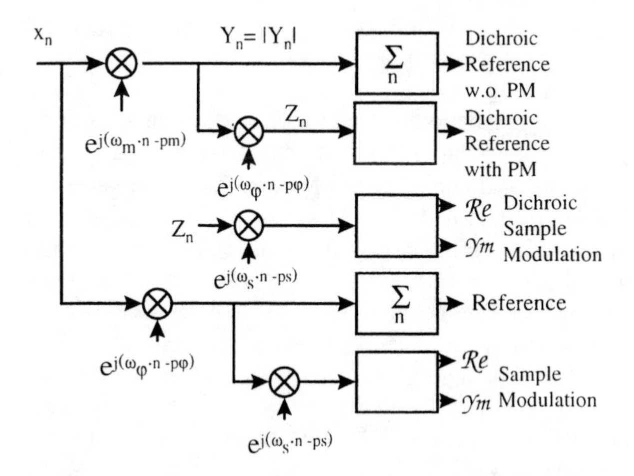

Figure 14. DSP for DIRLD signals

To determine the phase of ω_φ and ω_s to demodulate, we use a reference sample that introduces a zero or known phase shift to the modulation frequencies of interest. Assuming a zero phase reference for the phase modulation (open beam), we need to measure pφ , so the spectrum of the computed interferogram is positive real, which is the in-phase component, and has a zero quadrature component.

To be able to phase correct the resulting interferogram, we demodulate Y_n with ω_φ for each step to obtain the magnitude of the transform for that step, and the argument pφ1 for the step with the maximum magnitude in the scan. The rotation of the magnitude interferogram with pφ1 gives a pair of interferograms. The spectrum of the "real" interferogram is computed with Mertz phase correction and the phase array is stored to compute the "imaginary" interferogram using the same peak location. See Fig. 15.

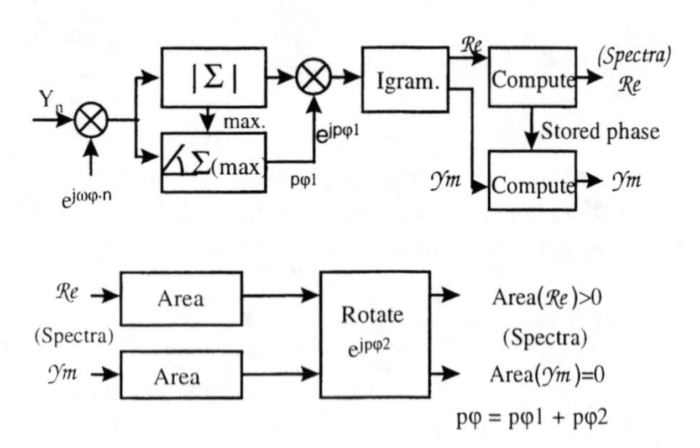

Figure 15. Rotation angle from reference sample.

Next, the area under the two spectra is computed over the spectral region of interest, and a second rotation pφ2 is applied so that the area under the quadrature component is zero. Therefore, the phase for the phase demodulation of any sample is:

$$p\varphi = p\varphi 1 + p\varphi 2\,. \qquad (11)$$

The demodulation of the sample modulation follows the same as above to determine the phase rotation ps. An additional difficulty appears because ω_s is in general a low frequency, and the PM sub-carrier is more than an order of magnitude larger amplitude, and thus it generates a large baseline shift at the sample modulation frequency. This effect can be minimized by a baseline correction method equivalent to one previously described. [11] In this case we can also assume that the baseline is a function that varies slowly with frequency as shown in Fig. 16. Here the transform for one step is shown already rotated for clarity.

The solution here is to fit a complex function to the baseline in regions at lower and higher frequencies than the sample modulation frequency, because the argument varies with frequency. This function can be a straight line or a higher order polynomial. At the sample modulation frequency, the value of the function obtained is subtracted from the demodulated value of the signal, and the argument of the function is subtracted from the demodulated value of the signal, as shown in Figure. 17.

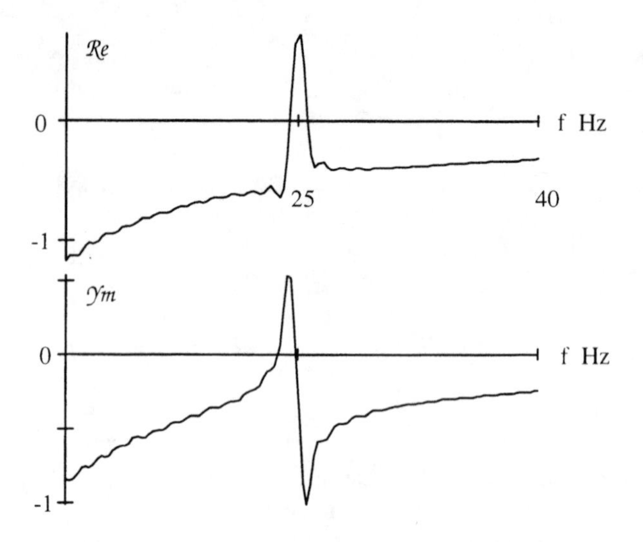

Figure 16. Signal at sample modulation, showing baseline offset.

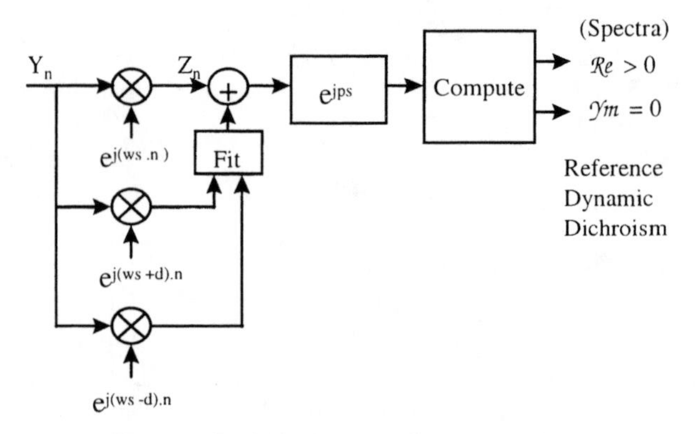

Figure 17. Complex baseline correction.

If a straight line is used for the baseline correction, it can be defined by two points in the spectrum at either side of the frequency of interest. To reduce the noise contribution from the baseline correction, these values could each be the average of several spectra values or the values of spectral elements obtained at a lower spectral resolution.

The above baseline correction is applied to the data at each step for the reference sample, and the phase ps for the sample frequency demodulation is determined. With all phases known, the demodulation of the data from a sample can be implemented as in Fig. 14. When demodulating the sample modulation, the baseline correction has to be applied to obtain the correct value of the interferogram at each step.

Figure 18 shows the transform of the data from one step around the frequency of the sample modulation after baseline correction and rotation. An LIA measuring the amplitude at the sample modulation frequency will report the total amplitude at that frequency, that includes the baseline offset, and the apparent frequency shift due to the different arguments (phase angles) between the baseline and the sample modulation signal at 25 Hz .

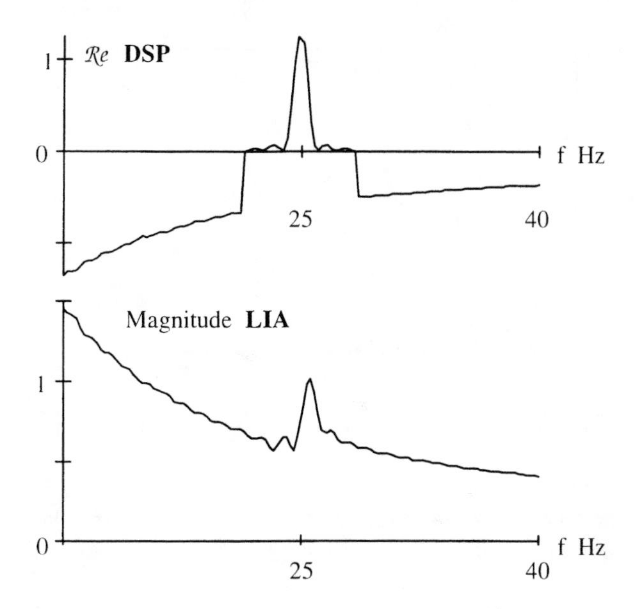

Figure 18. Baseline corrected demodulation from DSP, and magnitude spectrum as would be measured by a LIA.

PHOTOACOUSTIC SPECTROSCOPY IMPULSE RESPONSE

Step Scan Phase Modulated (PM) Photoacoustic Spectroscopy (PAS) experiments using DSP have been previously reported. [12], [13] These works used continuous PM with one or several discrete frequencies for sample excitation. From the resulting in-phase and quadrature data of the PM at each step, interferograms were created for different rotation angles between the two components to compute spectra at effectively different delays with respect to the excitation, giving the desired spectral depth profile information.

Alternative approaches were shown by Budevska. [14] The first uses amplitude modulation (AM) with a shutter to generate a pulse of infrared light from the interferometer at each step and collect the PAS time response. The second method uses a slow rise time (12 ms) PM step for the sample excitation. The AM method has the same limitations in pulse mode as in continuous modulation, in that it modulates the total of infrared light from the interferometer, where the average value is much larger than interferogram signal of interest, thereby reducing the resulting signal to noise ratio (S/N). The PM method proposed does not address the removal of the system function.

The AM method has the additional limitation of having a very low effective pulse power when the pulse length is set to achieve a useful resolution of a fraction of a millisecond, resulting in even lower S/N in the output result.

We have extended the method above, using pulse PM to obtain the PAS sample time response of a sample to a unit impulse. In practice, the excitation has to be a pulse of finite duration, and in general it can be any pulse p(t). In the general case, the system output at each step o(t) can be written:

$$o_S(t) = p(t)*s(t)*h(t) \quad (12)$$

The output signal $o_S(t)$ for each step for a linear time invariant system can be expressed as the convolution of the excitation pulse p(t), the sample impulse response s(t), and the detector system impulse response h(t), as shown in equation 12. The subscript S refers to the data from the sample of interest. In practice, the interferometer will be driven by a pulse p'(t), and p(t) is the PM pulse of infrared energy that excites the sample.

For the case of a PAS experiment, the detector system includes the PAS sample cell, its microphone and the signal channel electronics. Equation 12 can also be written as in Equation 13 by changing the order of the terms.

$$o_S(t) = p(t)*h(t)*s(t) \quad (13)$$

In general none of the three terms are known. Therefore we use for a reference a sample such that its impulse response s(t) is known, and better yet, it approximates an impulse.

It is common practice in PAS to use carbon black as a reference material, because it has very high absorptivity across the infrared spectrum and approximates total absorption at the surface and has negligible thermal mass [15]. Therefore its response approximates an impulse. Then the reference response can be equated to $u_0(t)$, the unit impulse, as shown in equation 14.

The subscript R refers to the data from the reference material. The term $u_0(t)$ drops out because the convolution with the unit impulse does not change the function.

$$o_R(t) = p(t)*h(t)*u_0(t) = p(t)*h(t) \quad (14)$$

Then from equation 13, we have:

$$o_S(t) = o_R(t)*s(t) \quad (15)$$

For equation 15 to be valid, h(t) cannot change from the reference to the sample measurement, which requires that the PAS cell must have the same volume to have the same response for both cases. This can be achieved with the carbon black backed by a substrate to match the volume of the sample.

Taking the Fourier transform of 15; in the frequency domain we have:

$$O_S(\omega)=[O_R(\omega)]\,[S(\omega)] \quad (16)$$

and solving for $S(\omega)$

$$S(\omega)=O_S(\omega)/O_R(\omega) \qquad (17)$$

Taking the inverse Fourier transform of the ratio of the complex spectrum $S(\omega)$ gives $s(t)$

$$s(t)=F^{-1}[S(\omega)]=F^{-1}[O_S(\omega)/O_R(\omega)] \qquad (18)$$

where $s(t)$ is the impulse response of the sample for this step of the interferogram.

Repeating this process for each step of retardation "d", we have an array $I(d,t)$ of data from which we can extract interferograms $i_t(d)$. From each of these interferograms we can compute (apodize, Fourier transform, phase correct) the corresponding spectra $S_t(\nu)$ for each desired value of "t", where t is the time delay from a theoretical impulse excitation to the computed spectrum.

$$S_t(\nu) = C\ [i_t(d)] \qquad (19)$$

The desired result is a set of infrared photoacoustic spectra of the sample that represent the spectral photoacoustic impulse response of the sample. That is, the spectra corresponding to different time delays after the excitation with an ideal impulse. For the case of a solid with uniform thermal properties and varying spectral characteristics as a function of depth, the time delays of the spectral response are related to the distance δ from the sample surface, and $S_t(\nu)$ is the photoacoustic spectrum of the sample at a depth δ from the surface. [16]

The above results hold for any pulse shape $p(t)$. The choice of excitation pulse shape can be used to optimize the S/N of the result by maximizing the pulse energy distribution in the time domain of interest.

An easy first choice for the pulse, can be the PM step generated in a step scanning spectrometer as suggested by Budevska, especially when the rise time of the spectrometer step is short enough to provide the time resolution required. Typical rise times for an interferometer with piezo-electric transducers driving a mirror are of the order of 100 μsec.

The following figures show the different functions for the data resulting from the PM step for a step near the centerburst of the interferogram. Figure 19 shows the discrete time transient response $o_R(t)$ generated by the reference sample, and its spectrum $O_R(f)$. Equivalent functions $o_S(t)$, and $O_S(f)$ are obtained for an unknown sample.

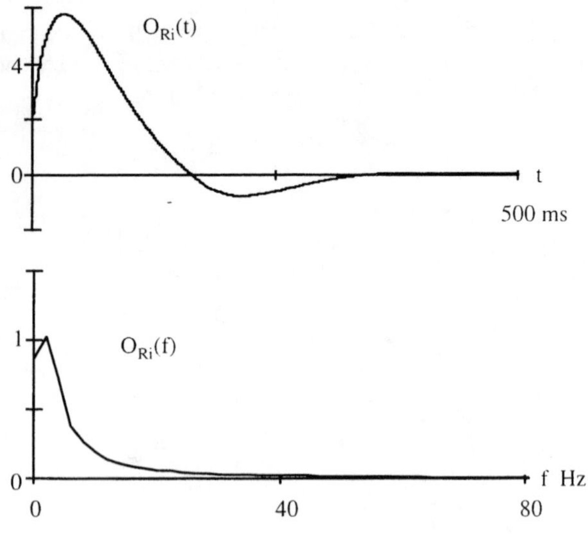

Figure 19. Reference signal and its spectrum for step i.

Figure 20 show the spectral response and the impulse response of the sample, corrected with the carbon black reference for the same step in the interferogram.

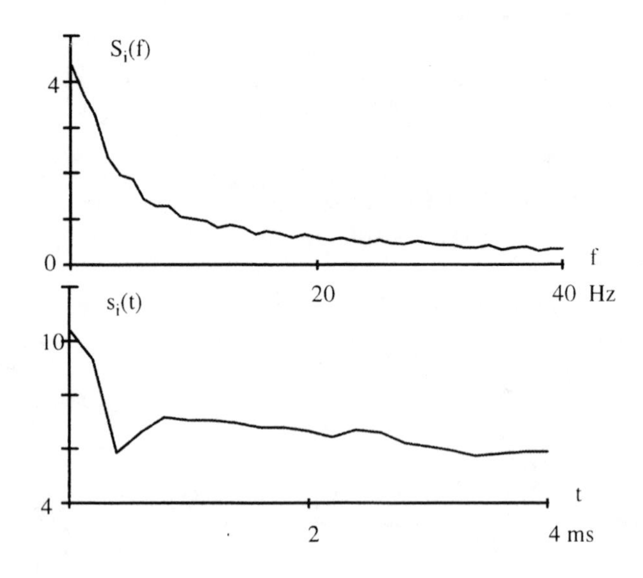

Figure 20. Ratio of sample to reference spectra for step i, and its inverse transform, the sample impulse response for that step.

Other possible generalized pulse signals that will have a better spectral power distribution are spread spectrum signals including chirps and pseudorandom unit value sequences. An example of a unit amplitude chirp and its spectrum are shown in Fig. 21.

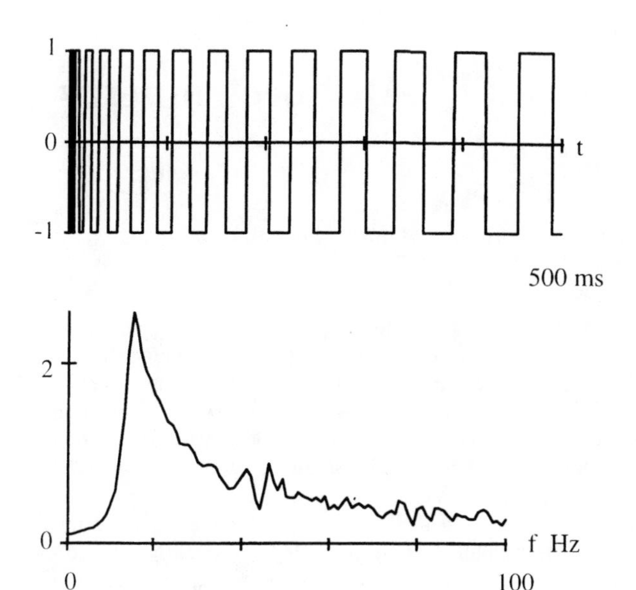

Figure 21. Unit value chirp and its spectrum.

This pulse provides much larger total energy, and a much higher spectral power density at high temporal frequencies, that will result in higher S/N for shallow depth PAS experiments.

REFERENCES

1. A. Oppenheim and R. Schafer, *Signal Processing,* Prentice Hall 1989.
2. C. Shannon, *The Mathematical Theory of Communications,* Bell System Technical Journal, Oct. 1948.
3. R. Curbelo, W. Howell, *Technique for Improving the Resolution of an A/D Converter,* U. S. Patent No. 5,265,039.
4. R. Curbelo, D. B. Johnson, R. A. Crocombe, *Optimizing Signal to Noise in Step-Scan Fourier Transform Spectrometers,* AIRS II Symposium Durham, NC, 1996.
5. Reference Data for Radio Engineers, page 21-8, H. P. Westman, *Editor,* Howard W. Sams & Co., Inc. 1970
6. P. Griffiths, J. de Haseth, *Fourier Transform Infrared Spectroscopy,* J. Wiley and Sons, 1986.
7. I. Noda, A. Dowrey, and C. Marcott, *A Spectrometer for Measuring Time-Resolved Linear Dichroism Induced by a Small-Amplitude Oscillatory Strain,* Appl. Spect. **42**, 2, 1988.
8. D. Drapcho, R. Curbelo, R. A. Crocombe, L. Zhang, D. Johnston, *Advances in Digital Signal Processing for Multiple Modulation FT-IR Measurements,* Pittsburgh Conference, March 1997.
9. R. Tervo and R. Enriquez, *Analysis of digital tanlock loop with adaptive filtering*, IEEE Pacific Rim Conference on Communications, Computers and Signal Processing, p. 5-8, vol. 1, 1993.
10. P. Y. Kam, *Tanlock carrier phase recovery without a divider and VCO*, Electronic Letters, vol. 30 No. 23 (Nov 1994).
11. R. Curbelo, *Digital Signal Processing Technique for a FT-IR Spectrometer Using Multiple Modulations,* U. S. Patent No. 5,612,784, 1997.
12. C. Manning, P. Griffiths, *Step-Scanning Interferometer with Digital Signal Processing,* Appl. Spect. **47**, 97, 1993.
13. D. Drapcho, R. Curbelo, E. Jianng, R. Crocombe, W McCarthy, *Digital Signal Processing for Step-Scan Fourier Transform Infrared Photoacoustic Spectroscopy,* Appl. Spect. **51**, 4, 1997.
14. O. Budevska and C. J. Manning, *Time-resolved Impulse Photoacoustic Measurement by Step-Scan FT-IR Spectroscopy*, Appl. Spect. **50**, 7, 1996.
15. J. McClelland, Private communication, August, 1997
16. E. Jiang, R. Palmer, J. Chao, *Development and applications of a photoacoustic phase theory for multilayer materials: The phase difference approach,* J. Appl. Phys. **78** (1), 1 July 1995.

Smoke And Mirrors: Ultra-Rapid-Scan FT-IR Spectrometry

C. J. Manning

Manning Applied Technology, 121 Sweet Avenue, Moscow, Idaho 83843

Fourier transform-infrared spectrometers have dominated the marketplace and the experimental literature of vibrational spectroscopy for almost three decades. These versatile instruments have been applied to a wide variety of measurements in both industrial and research settings. There has been, however, an ongoing need for enhanced time resolution. Limitations of time resolution in FT-IR measurements arise from the modulation frequencies intrinsic to the spectral multiplexing. Events which are slower than the minimum scan time, about 40 milliseconds at 4-cm^{-1} resolution, can be readily monitored with conventional instrumentation. For shorter transients, various step-scan, stroboscopic and asynchronous methods have been demonstrated to provide excellent time resolution, down to nanoseconds, but these approaches are limited to events which can be repeated many times with minimal variations. Some of these methods are also susceptible to low-frequency noise sources.

The instrinsic scan time of conventional FT-IR spectrometers is limited by the force that can be applied to the moving mirror. In commercial systems the moving mirror is invariably driven by a voice coil linear motor. The maximum force that can be exerted by the voice coil is sharply limited to a few Newtons. It is desirable to decrease the scan time by a large factor, but the required force scales as the square of the scan rate, while the voltage applied to the coil must scale as the cube of the rate.

A more suitable approach to very-rapid-scan FT-IR spectrometry may be the use of rotating optical components which do not have to turn around at the end of travel. There is, however, an apparent symmetry mismatch between rotating elements and the nominally planar wavefronts in a Michelson interferometer. In spite of the mismatch, numerous interferometer designs based on rotating elements have been proposed and demonstrated. Some of these designs are suitable for operation with scan times from tens of milliseconds to milliseconds, and perhaps faster, at 4-cm^{-1} resolution.

A novel interferometer design utilizing a single-sided precessing disk mirror allows a complete interferogram to be measured in 1 millisecond or less. A prototype instrument of this design has been constructed and tested. One application reported here is the measurement of a transient combustion event. While combustion reactions can be conveniently repeated under some circumstances, such as with gas-phase reactants, the shot-to-shot variation is unacceptably large for step-scan measurements. Preliminary data, illustrating operation and performance of the system, are presented. It is thought that the high modulation frequencies have resulted in superior rejection of multiplicative noise.

INTRODUCTION

Fourier transform infrared (FT-IR) spectrometers are versatile and powerful tools for measuring infrared spectra. Their success results largely from the well-known multiplex, throughput and registration advantages of interferometric spectrometry. In the usual rapid-scan mode of operation of FT-IR spectrometers, a collimated beam of radiation is intensity modulated by scanning one of the interferometer mirrors to produce a constant optical velocity in the range of 0.06 to 6 cm/s. The resulting modulation signal (an interferogram), as modified by interaction with a sample, is recorded from the output of an infrared detector. If the system under study varies with time, the time-scale of the spectral changes determines whether or not conventional rapid-scan FT-IR spectrometry may be used for the measurement. Typical FT-IR spectrometers require between 40 ms and 1 s to sweep an interferogram to 4-cm^{-1} resolution. If the spectral information varies with frequency components lower than 10 Hz, the time-scale of the spectral variations is longer than the time-scale of the spectral measurement (depending on the scan rate), *i.e.*, the system is varying slowly with time. Under these conditions, the conventional rapid-scan mode of operation of FT-IR spectrometers may be conveniently used to generate a time-resolved sequence of spectra (interferograms). By increasing the scan rate of the interferometer mirror, the time required for spectral measurement can be shortened to increase the range of application.

One of the fundamental limitations, however, of conventional FT-IR instruments is the rate at which a spectrum (*i.e.*, interferogram) can be scanned. For all commercial FT-IR instruments, the limitation arises principally from the voice coil linear motor used to drive interferometer scanning, together with the mass of the moving mirror assembly. In most systems, operating at 4-cm^{-1} resolution, the shortest possible scan time is approximately 40 ms. Under these conditions, the time required to turn around the mirror is comparable to the data acquisition time, *i.e.*, the duty cycle efficiency is low, and drops further if the scan rate is increased. A unique interferometer, based not on linear translation but on precession of a single plane mirror, is introduced. This design allows both high and constant duty cycle efficiency, so that scan times of 1 ms or less can be achieved.

CP430, *Fourier Transform Spectroscopy:* 11th International Conference
edited by J.A. de Haseth

$$x = A\cos(\omega t) \qquad A = 1.25 \text{ mm}$$

$$a = -A\omega^2\cos(\omega t) \quad a = \text{acceleration}$$

$$\omega = 2\pi f = 6.28 \times 500$$

$$\omega^2 = 9.86 \times 10^6$$

$$a_{peak} = 12{,}500 \text{ m/s}^2 \qquad F = ma,\ m = 0.03 \text{ kg}$$

$$F_{peak} = 375 \text{ N}$$

FIGURE 1. Derivation of force requirement.

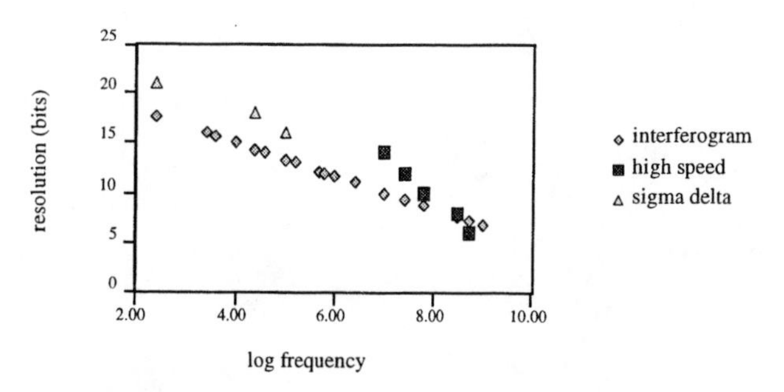

FIGURE 3. Plot of ADC performance *vs.* Frequency.

A brief review of conventional FT-IR spectrometry is necessary to understand why limited scan rates have been tolerated for so long. As noted above, in conventional instruments the interferometer mirror is translated by a voice coil drive. The mirror sweep encodes the radiation intensity so that all wavelengths can be measured simultaneously (the Fellgett advantage). A Fourier transform is then necessary to decode the interferogram and recover the spectrum. In step-scan operation the time-dependence of the encoding is essentially removed by stopping the mirror, thus simplifying a variety of time-resolved spectroscopic measurements. The step-scan and stroboscopic approaches are applicable only to processes that can be repeated many times and are not suitable for real-time measurements. A very useful alternative to these approaches is very-rapid-scan operation in which the mirror is moved very quickly, so that a system under study remains in a nearly constant state for the duration of one scan. This approach has the added benefits of producing data in real time and avoiding susceptibility to low-frequency noise sources that adversely affect step-scan and stroboscopic FT-IR measurements.

The force developed by a voice coil mirror drive is simply the product of the magnetic field strength, the length of the winding, and the current in the winding. For a conventional FT-IR system the maximum force that can be developed is on the order of 1 or 2 N. The mass of the moving mirror and the moving portion of the bearing can be taken to be 30 g, but in some systems these have a mass greater than 500 g. This is meant to be a best case view, so the smaller figure will be used. Figure 1 shows that from the equation for sinusoidal motion the acceleration required for any amplitude and frequency of motion can be readily derived. A typical FT-IR system is barely capable of driving the mirror through an amplitude suitable for phase modulation at 1 kHz. This is only a few fringes of the reference laser, much less than the 4000 fringe sweep required for 4-cm^{-1} resolution. The plot in Fig. 2 is typical of voice coil response. Individual systems vary in mass, voltage and other details, but not by large factors. On the left side of the plot the response drops as $1/f^2$ for purely mechanical reasons. At higher frequencies the coil inductance becomes significant and the response drops as $1/f^3$. The same equations can be worked backwards with the goal of measuring a 4-cm^{-1} spectrum in 1 ms. The required mirror acceleration is 12,500 m/s^2, and a force of about 375 N must be applied (to a 30 g mirror), with a resulting optical velocity of 250 cm/s. It might be possible to gain a factor of 100 in voltage (with a substantial high-voltage power supply and amplifier), and a factor of 10 in magnetic field strength, but not many mirror/bearing assemblies can withstand forces of this magnitude. Even if they could, the reaction force would be very likely to disturb other optical components.

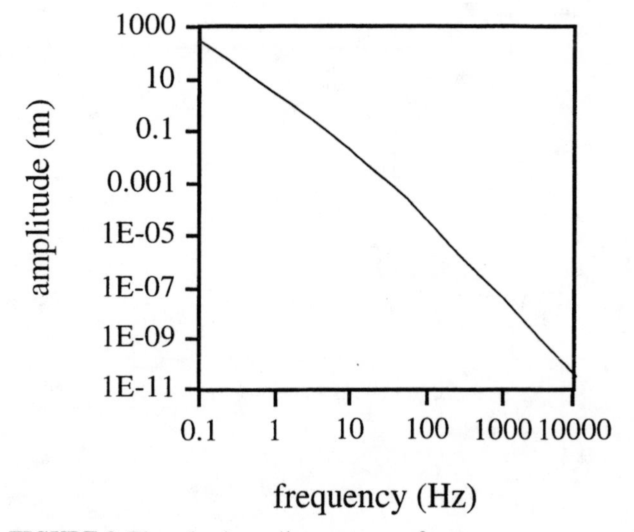

FIGURE 2. Plot of voice coil response *vs.* frequency.

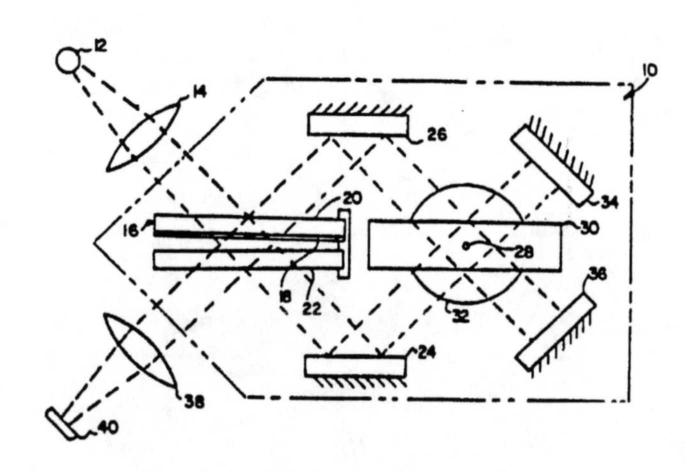

FIGURE 4. Dybwad - US Patent 4,654,530.

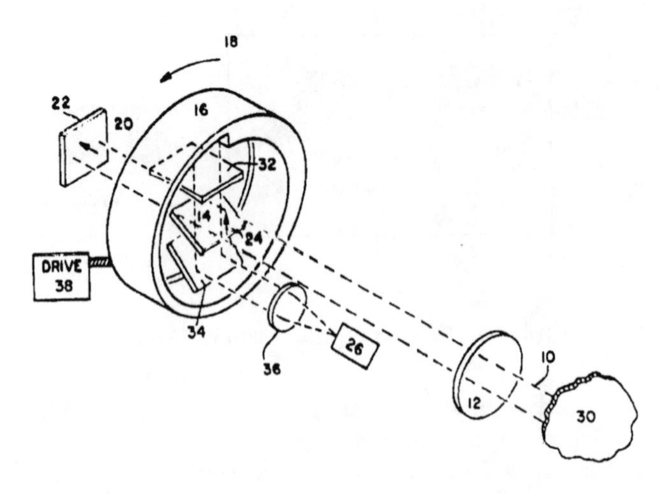

FIGURE 5. Smith - US Patent 4,179,219.

Were it possible to move a mirror with the velocities and accelerations suggested above, very-rapid-scan operation would combine many of the advantages of conventional FT-IR with the advantages of step-scan operation. It would be applicable to both repeatable and non-repeatable events on the millisecond time-scale. This approach would fail for transients which are too fast, and is generally not suitable for photoacoustic or photothermal measurements which are dependent on the modulation (Fourier) frequencies of the spectral multiplexing.

An optical retardation velocity appropriate for use with a pyroelectric (TGS) detector is about 0.3 cm/s, corresponding to a modulation frequency of 5 kHz for reference laser radiation at 15,804 cm^{-1}. At this speed a single 4-cm^{-1} scan requires 840 ms. For a mercury-cadmium-telluride (MCT) detector, a much higher optical velocity, typically 6 cm/s, produces better results. The reference laser is then modulated at 100 kHz, but the scan is still 42 times longer than the target 1 ms. If a 4-cm^{-1} interferogram is scanned at constant velocity in 1 ms, the optical velocity will be about 250 cm/s and the laser will be modulated at 4 MHz. This is 1000 times higher scan rate than is appropriate for a TGS detector, and a million times higher than is commonly used in step-scan measurements. The Fourier frequencies for the mid-infrared will be in the range from 100 kHz to 1 Mhz under these conditions.

As the mirror velocity changes, the dynamic range of the interferogram also changes. The amplitude of the optical

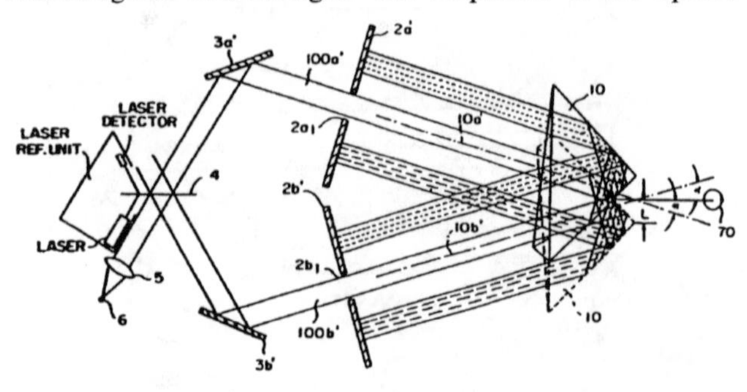

FIGURE 6. Tank - US Patent 5,341,207.

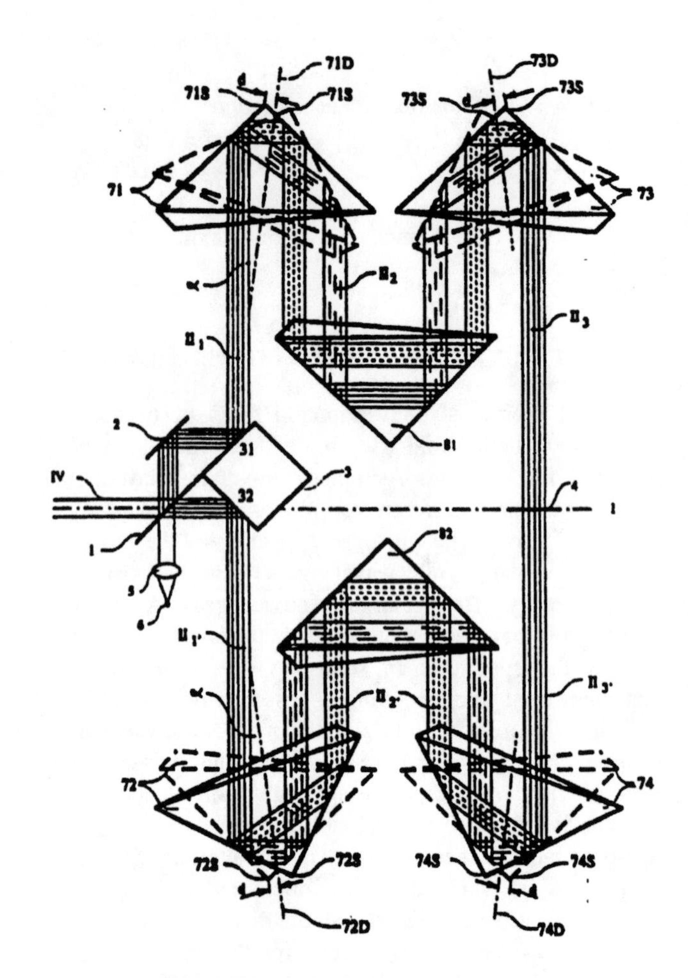

FIGURE 7. Tank - US Patent 5,148,235.

interference signal remains constant, independent of scan velocity. If the detector response is flat to the highest frequency present, then the electronic interferogram amplitude is also independent of scan speed. As the

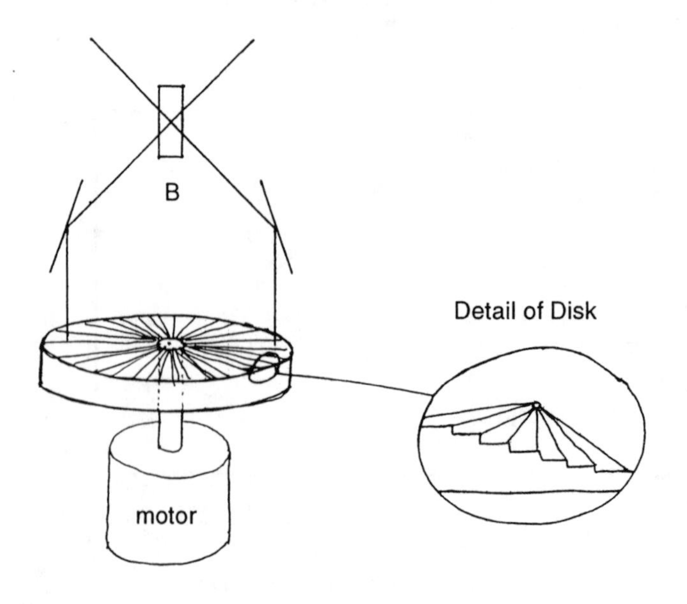

FIGURE 8. Manning.

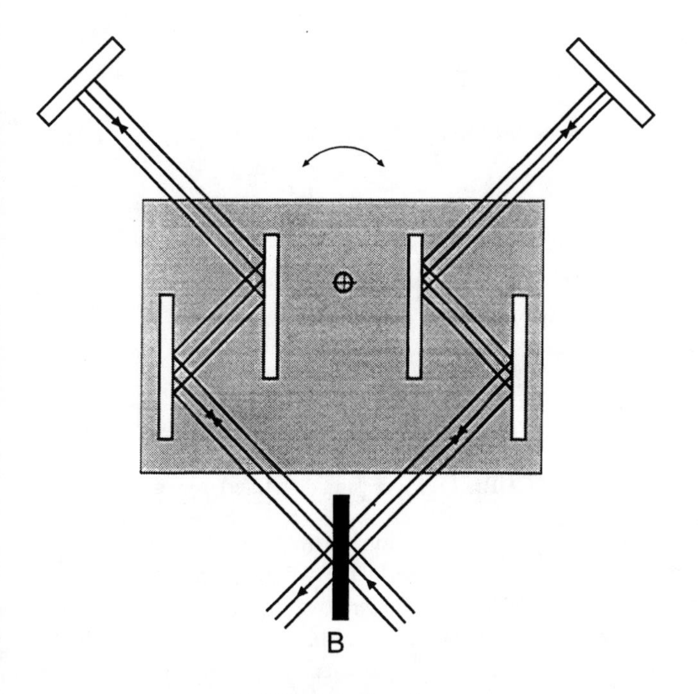

FIGURE 9. Perkin-Elmer.

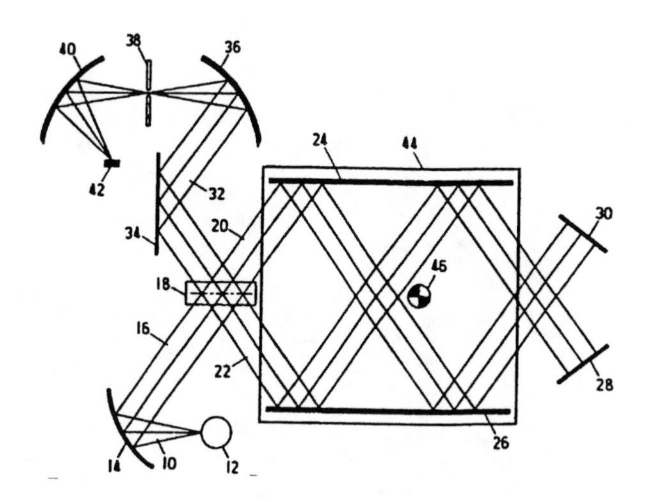

FIGURE 11. Brierley - US Patent 4,915,502.

detector bandwidth is opened to accommodate a broader range of modulation frequencies, more of the detector noise is passed. The net result is that the SNR per scan goes down as the square root of the scan time, while the SNR per measurement time remains constant. The performance of various commercial analog-to-digital converters (ADC) are presented in Fig. 3. The diamond points indicate approximately the number of bits of resolution required to accurately represent the interferogram for different mirror velocities. The exact vertical offset of the diamond points depends on source temperature, throughput and other experimental variables. The important point is the availability of high speed ADC's, indicated by the square points, which can digitize with sufficient resolution to allow interferogram recording out to 100 MHz, were it possible to scan a mirror that rapidly.

The use of voice coils to directly drive reciprocating motion for high scan rates has been ruled out because extreme measures would be required to make a workable system. A continuous rotary motion may be a better approach to generating rapid path difference variation. There is, however, an apparent symmetry mismatch between rotating optical elements and the nominally planar wavefronts in a Michelson interferometer. To meet a target of 1 ms per scan, the rotating elements will probably have to spin at 30,000 rpm. Under a stress of up to 50,000 g's they will have to distort less than approximately 250 nm, a tenth-wave of 2.5 μm radiation. At this rotation rate a 10-cm diameter aluminum disk of 1.25-cm thickness will store a respectable 1.8 kJ.

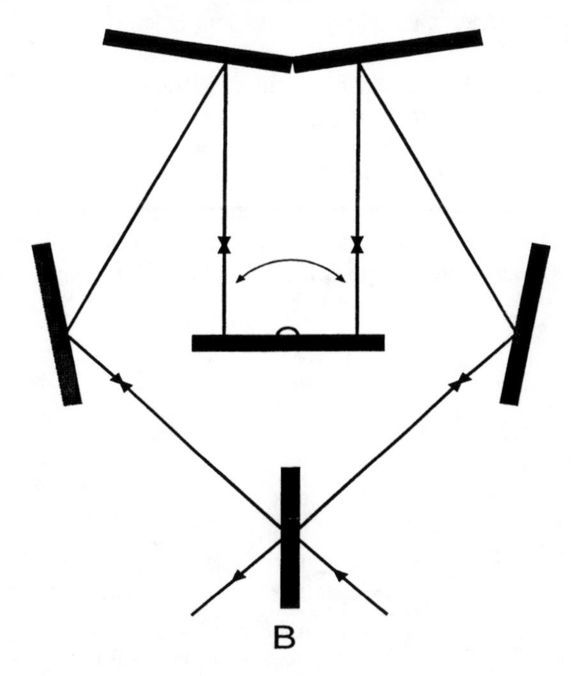

FIGURE 10. Kauppinen.

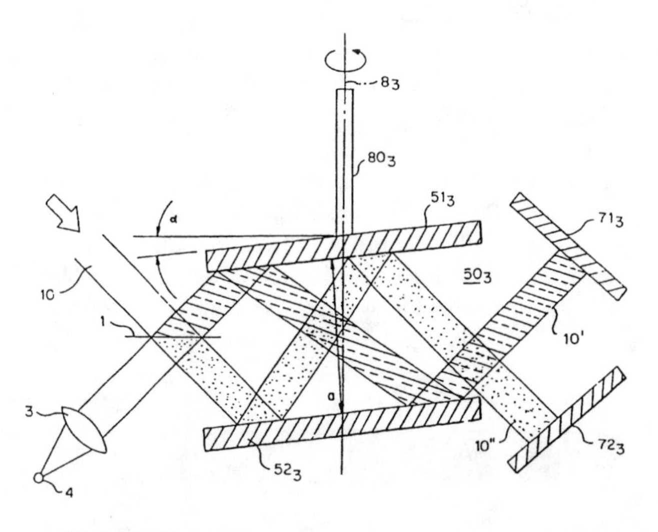

FIGURE 12. Tank - US Patent 5,457,529 - Illustration 6.

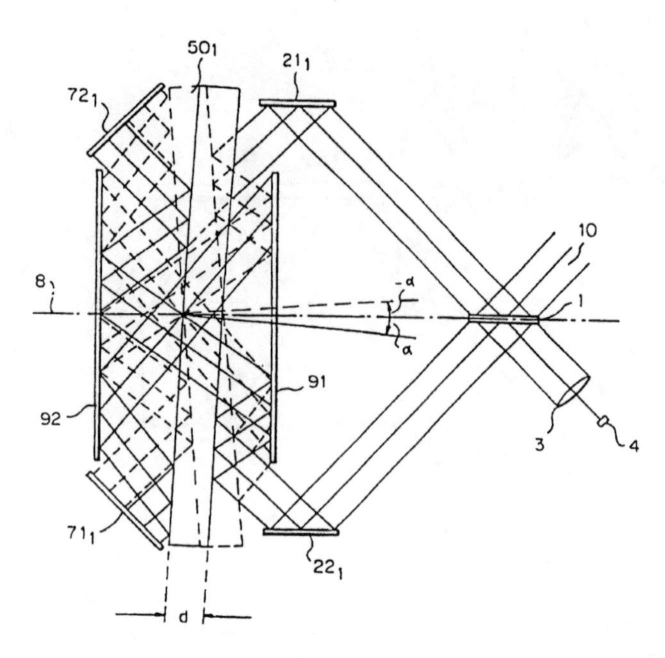

FIGURE 13. Tank - US Patent 5,457,529 - Illustration 3.

LITERATURE REVIEW

In spite of the apparent symmetry mismatch, numerous rotating interferometer designs have been proposed and demonstrated. Each of these designs exhibits some tradeoffs. Figure 4 shows Dybwad's design, disclosed in US Patent 4,654,530, which uses a rotating prism to vary optical path difference. It has the advantages of simplicity, compactness, and generation of 4 scans per revolution. Disadvantages include a duty cycle well below 100% because of the limited range of angle which produces useful modulation, and dispersion in the prism which distorts the wavenumber axis. There may be cause for concern about the mechanical strength of suitable IR transparent materials spinning at the 15,000 rpm required to achieve 1000 scans per second. The airflow around the flat surfaces may cause unacceptable drag and turbulence if the system is not evacuated.

Another approach to the symmetry problem is to use optical components of circular or spherical symmetry. A confocal interferometer design has been described (1) in which only a minute range of retardation can be used. Because the wavefronts approaching the detector are converging, the foci from the two arms of the interferometer will only match sufficiently well to produce useful interference for a very limited range of retardation. The key is the spherical symmetry of the reflectors which allows a large divergence angle in the input beam. The

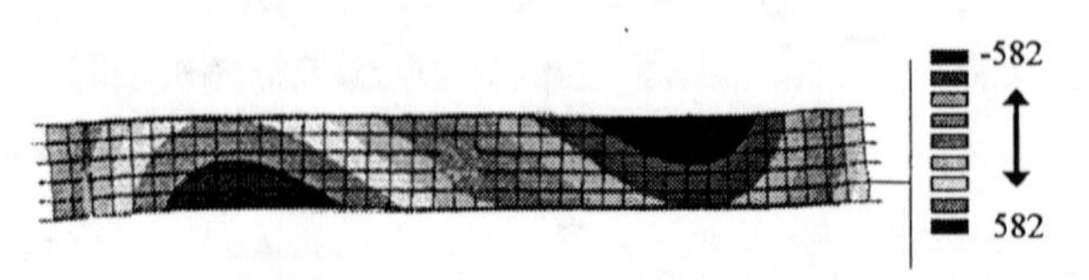

FIGURE 14. Deflection of tilted beryllium disk (nm).

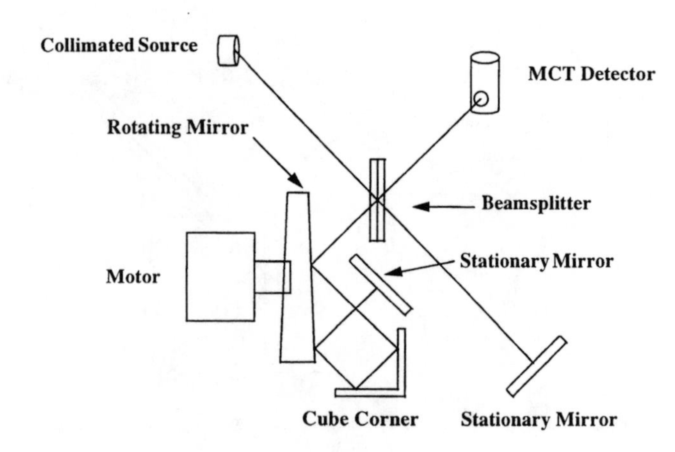

FIGURE 15. Precessing disk interferometer.

spherical symmetry also means that the reflector could be a portion of a toroidal reflector which is rotated about its center. Figure 5 shows an innovative design, described in US Patent 4,179,219 by Smith, in which a much larger range of retardation can be used. The convergence rate of the wavefronts is compensated by a varying focal length in the rotating reflector. The two key advantages of this design are compact size and large acceptance angle. It is, however, quite difficult to fabricate a mirror of the required figure. A second drawback of this design is that the mirror figure is a compromise between compensating the focal length for path variation and the slope of the surface required to vary the path difference.

Another approach to generating path difference by rotating optical elements is the use of spinning cube-corner retroreflectors mounted off axis such that the shear varies with rotation angle. Each shear distance produces a unique path difference. Any beam impinging on the reflector will be reflected antiparallel to the incident beam regardless of the rotation angle or shear offset. The design shown in Fig. 6 is disclosed by Tank, *et al.*, in US Patent 5,341,207. A variety of related configurations have been published, including the one illustrated in Fig. 7, also disclosed by Tank, *et al.* in US Patent 5,148,235. Some of these designs utilize multiple spinning retroreflectors to increase the rate of change and maximum excursion of path difference. The advantages of these designs include intrinsic tilt- and shear-

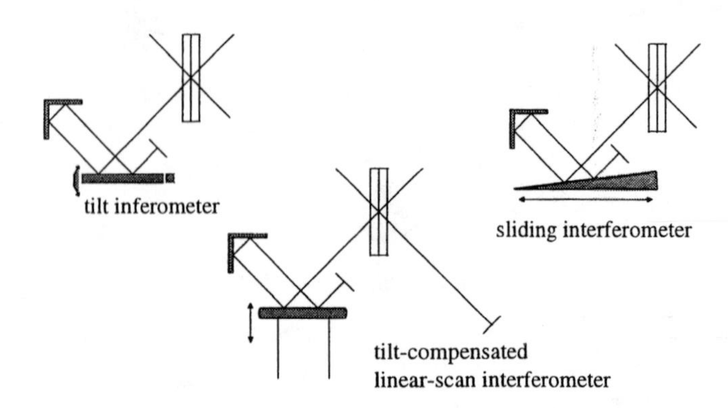

FIGURE 16. Variations of precessing disk interferometer.

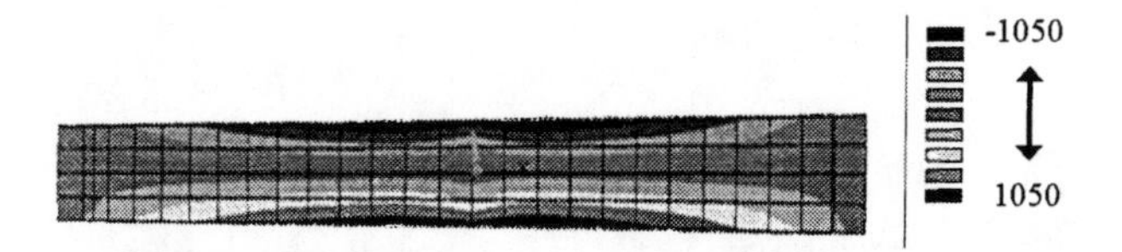

FIGURE 17. Deflection of double-wedged aluminum disk (nm).

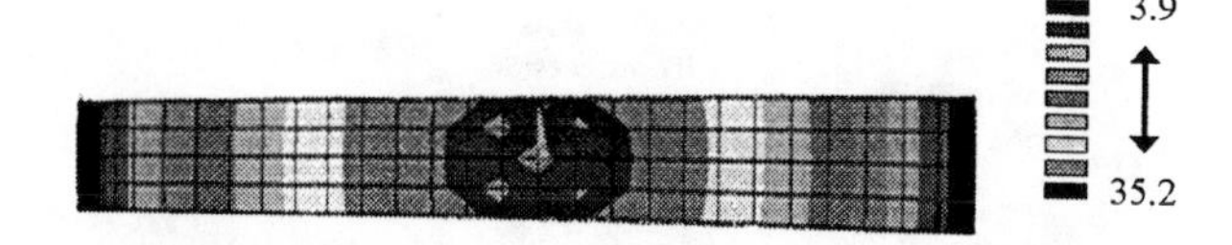

FIGURE 19. Stress in double-wedged aluminum disk (MPa).

compensation. Problems with this approach include the fact that off-axis spinning of retroreflectors is inherently unbalanced, and above about 6000 rpm distortion of the retroreflectors becomes significant. Large retroreflectors are fairly expensive.

The design diagrammed in Fig. 8 was an early attempt at solving the symmetry problem. It has a symmetrical rotor with the desired undersampling ratio cut into the surface as circular staircases. At any point across the interferometer aperture, the path difference between the two arms is the same because the images of the staircases match. It seemed a clever design until the cost of fabrication was considered; estimates ranged from US$50,000 to $100,000 with no guarantee of success. This design would be extremely susceptible to misalignment and there may be problems with dispersion because of the small facet size.

The literature held deeper insight, in Kauppinen's excellent paper (2), which discusses a series of related interferometers using the same approach to solve the symmetry problem: complementary reflections from two flat mirrors in each arm of the interferometer insure that a final reflection from a plane mirror is at normal incidence. This sends the beam exactly back to the beamsplitter *via* its original path. Fig. 9 shows one of Perkin-Elmer's designs, which is a variant of this general approach. Kauppinen's paper was particularly valuable because it showed enough variations on this theme to allow the common features to be extracted. It also clearly showed that exotic surfaces are not required to solve the symmetry problem, *i.e.*, convert rotation to planar retardation. Kauppinen has proposed and demonstrated a variant illustrated in Fig. 10 which also uses planar reflectors with rotational motion. It is optically equivalent to Fig. 9, but with the advantage that it is particularly compact and resistant to deformation by mechanical and thermal stresses. These two designs are not intended for complete rotation.

Another variation on this theme, Fig. 11, is taken from Brierley's US Patent 4,915,502 which was the point of departure for a series of novel interferometer designs. Brierley's design uses two parallel plane mirrors on a rotating platform. Although Brierley specifically excluded the case of complete rotation in his patent claims, the duty cycle for measuring 4-cm^{-1} interferograms using complete rotation of his apparatus is estimated to be 6%. Further, such a design would be difficult to balance for spinning at 500 revolutions per second, suffer serious distortion and have to be evacuated because the large flat surfaces would impede rotation in air. There is, however, a simple variation of this design that limits the sweep of retardation, while largely solving the duty cycle and air resistance problems. The idea is to rotate Brierley's parallel reflector assembly about a different axis. In Fig. 11, the axis of rotation is a line coming straight out of the paper. In Fig. 12, the rotation axis is a vertical line in the plane of the paper. As a consequence, the maximum retardation is set by the tilt of the reflector assembly when it is fixed to the rotating shaft.

This line of reasoning was followed by Tank, *et al.* in US Patent 5,457,529 which was the source of Fig. 12. The problem with this general design is that alignment between two fairly large mirrors spinning at 0.5 kHz must be maintained to 0.5 arcseconds. The retardation and optical velocity vary sinusoidally with rotation angle, but the resulting variations in the modulation frequencies are not a

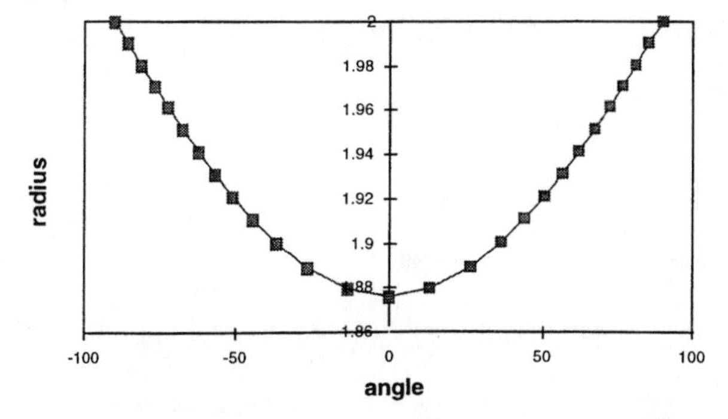

$$F(\theta) = 4\pi^2 \rho f^2 \int_0^r r^2 (t + 2r\cos\theta \sin\phi)\, dr\, d\theta$$

FIGURE 18. Plot of disk radius *vs.* angle.

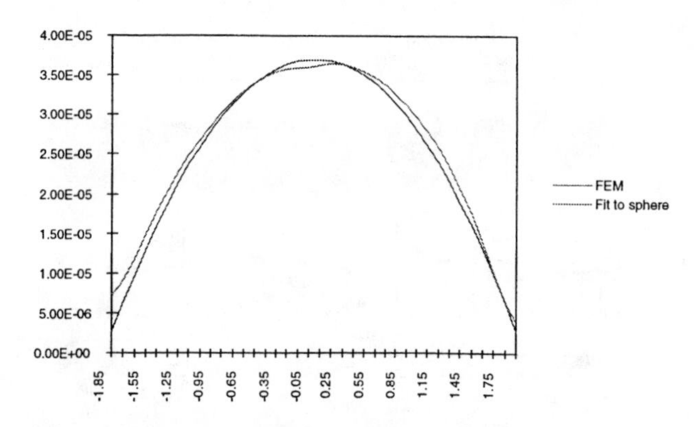

FIGURE 20. Plot of distortion of disk surface showing fit to spherical function.

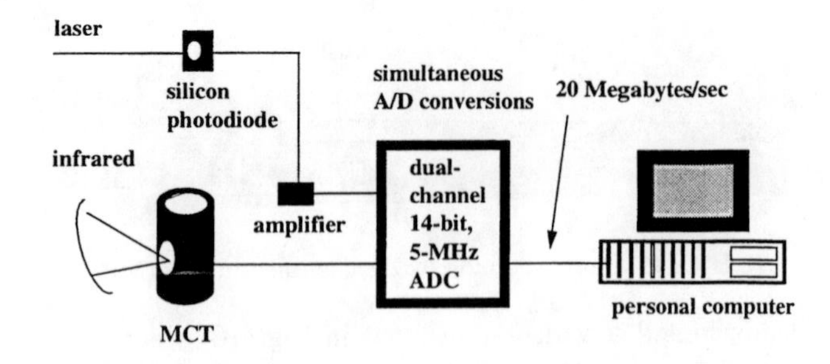

FIGURE 21. Schematic diagram of data acquisition system.

problem if the delays in the infrared and laser channels are matched (3,4). The design shown in Fig. 13, also by Tank, *et al.* represents a second innovation relative to Brierley. This approach turns the previous design inside-out, transforming the problem of aligning two separate rotating reflectors to the very much simpler problem of polishing two sides of a disk parallel. The penalty is the addition of folding mirrors, but it is much easier to adjust the alignment of several static mirrors than the alignment of even one spinning mirror. Tank places an aperture in the center of the disk which makes it more difficult to mount and spin the disk, but in compensation, has a considerable advantage over a solid disk design. Because the beam can pass through the center of the disk, the pathlength is greatly reduced; consequently a larger beam divergence can be tolerated.

One disadvantage of using a tilted disk to precess two parallel surfaces is that the tilt produces a strong bending moment in a centrifugal field. This is indicated in Fig. 14 which is the result of finite element modeling of the deformation of a beryllium disk mirror. The model assumed a 1.25-cm thickness and 10-cm diameter, tilted 1 degree relative to the shaft. The resulting deflection is approximately 600 nm. For an aluminum disk of equal dimension, the deflection would be about 6 times greater, or nearly 3600 nm. The criterion used in designing the prototype instrument was that deflection should be less than 250 nm.

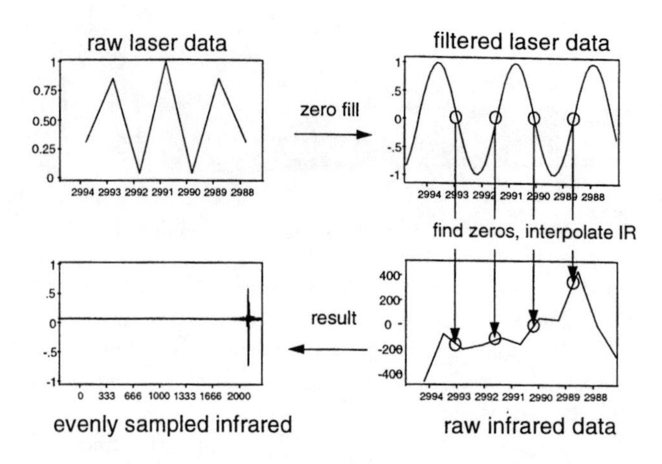

FIGURE 23. Schematic diagram of signal processing.

PROTOTYPE INSTRUMENT

Figure 15 is a diagram of a novel interferometer which represents a third innovation relative to Brierley. Instead of using two sides of the same disk to maintain alignment between complementary reflections, a cube-corner retroreflector is used to invert the image of a single side of the same disk. This greatly simplifies alignment; only two mirrors in each arm of the interferometer require angular adjustment. The design is intrinsically insensitive to alignment of the disk and retroreflector. This general approach is related to Steel's design (5) in which a moving cube-corner retroreflector is shear-compensated by reflection from a plane mirror. The advantage relative to Steel is that the moving element is not a cube-corner and hence can be made lighter and more resistant to deformation under the acceleration of motion. Further, fabrication and mounting of the disk mirror is also simplified because optical access to only one side of the disk is required.

The general disadvantages of this design are that the pathlength in each arm must be relatively large to accommodate the beam folding, the cube-corner reflector and disk mirror must have apertures at least twice as large as the interferometer aperture, and the resolution is limited by beam motion on the optical surfaces, as well as by the tilt angle. Throughput is limited by the beam divergence

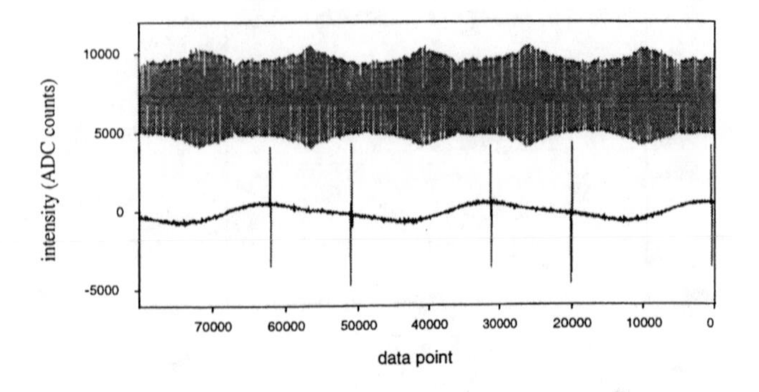

FIGURE 22. Plot of raw data from ADC; upper trace - laser, lower trace - infrared, 333 scans per second.

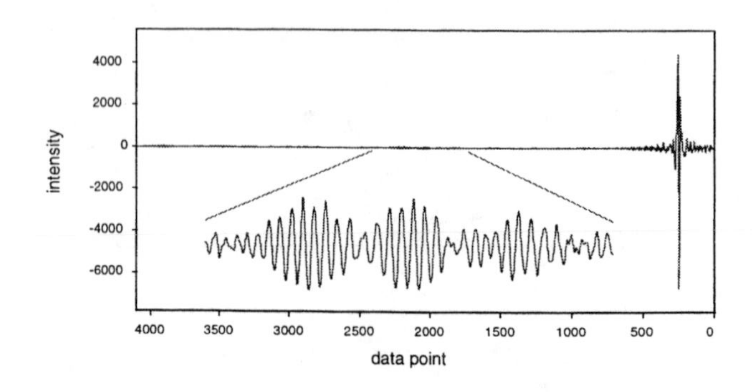

FIGURE 24. Interferogram acquired in 40 ms - after processing.

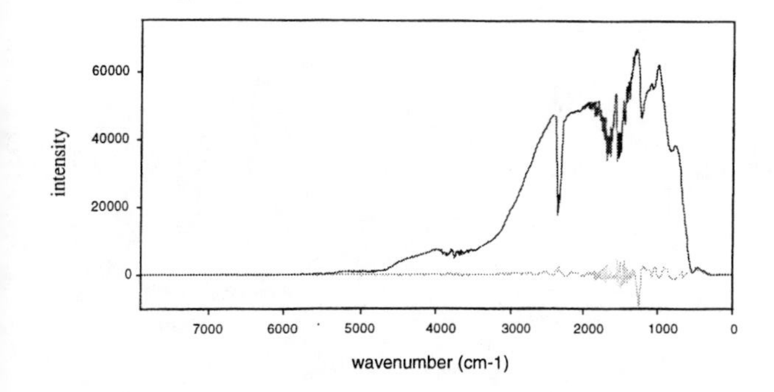

FIGURE 25. Spectrum computed from data acquired in 40 ms, resolution 8 cm^{-1}. Real (upper) and imaginary (lower) after Mertz phase correction.

which can be accommodated without the beams being clipped at the edges of the mirrors.

Referring to Fig. 15, the beam from a collimated source of radiation is split by the beamsplitter to produce a transmitted beam which enters the fixed arm to be returned by a flat mirror. The reflected beam from the beamsplitter enters the variable arm and impinges on the precessing disk. The reflected beam from the disk is returned by the cube-corner retroreflector to the disk. The beam returning from the cube-corner to the disk always makes a complementary reflection at the disk such that the beam arriving at the final plane mirror always has normal incidence. Thus the beam returns, exactly on its original path, to the beamsplitter and to the detector. A series of variations are shown in Fig. 16. These are various tilt-compensated interferometers which utilize more conventional mirror scanning means than the precessing disk, but share the tilt-compensation method of Fig. 15.

One of the major design concerns is distortion of the disk in the centrifugal field induced by rotation. Several different mathematical approaches were used to calculate distortion of a tilted flat disk and all indicate that, as demonstrated above, even a beryllium disk of these dimensions is marginally stiff. While the stiffness increases as the cube of the thickness, another important effect, radial stretching, is independent of thickness. Both beryllium and aluminum were considered for the disk material. Beryllium has approximately 2/3 the density of aluminum, and is

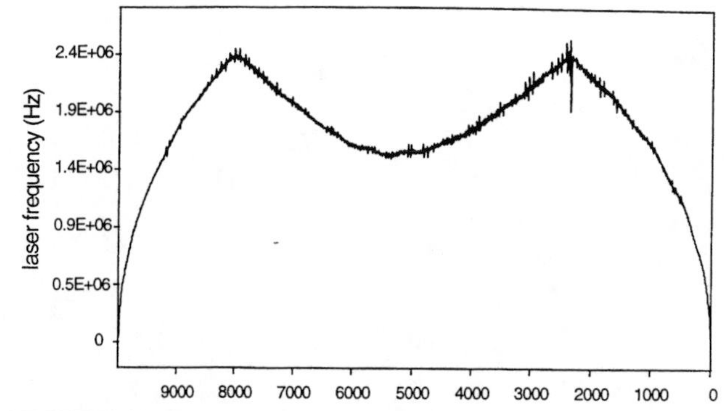

FIGURE 27. Laser frequency profile showing the modulation exceeding Nyquist limit.

about 4 times stiffer; together these attributes result in a 6x decrease of deflection for a given geometry. However, beryllium is rather expensive, both as a raw material and to machine. By using aluminum, these costs and the risks of exposure to beryllium have been avoided thus far.

Finding a suitable design for the disk mirror was quite difficult. At the design rotation speed of 30,000 rpm, even a flat untilted 10-cm disk of aluminum is radially stretched by some 2700 nm. For a disk of 1.25-cm thickness, the resulting spherical curvature of the surface is approximately 1800 nm from center to edge. If the disk is simply flat and tilted on the shaft, the distortion is far worse, as with the design shown in Fig. 12. By making the disk symmetric about the plane of rotation, a tilted surface can be precessed while completely avoiding bending distortion. This approach can be described as a double-wedge design as shown in cross-section in Fig. 17. It produces equal bending moments, from the two opposite faces of the disk mirror, which cancel. Because the tilt angle is small, the distortion of the surface is almost exactly spherical and can be precompensated with a spherical polish.

The balance equation for the radius of the disk as a function of angle was derived, by integrating the equation shown in Fig. 18, and fitted. Although the solution has cubic and quartic terms, it fits almost exactly to a circle minus a cosine term, as indicated by the plot in Fig. 18 (there are two superimposed sets of points and lines). The

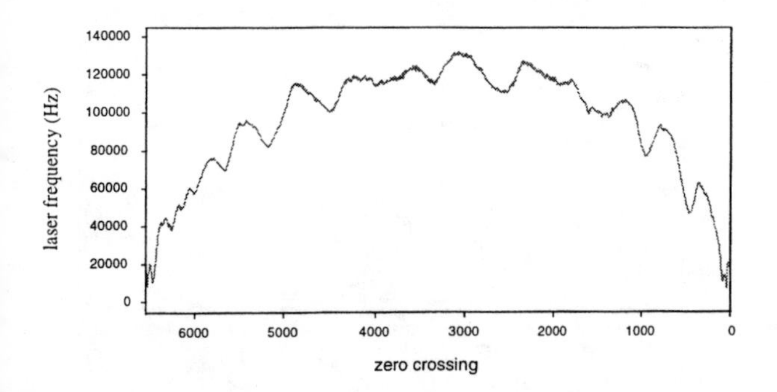

FIGURE 26. Laser frequency profile *vs.* Interferogram data point; note vibration.

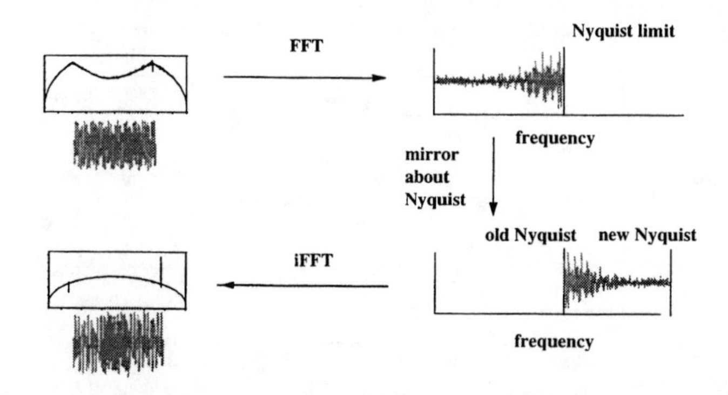

FIGURE 28. Method of reconstructing laser signal beyond Nyquist limit.

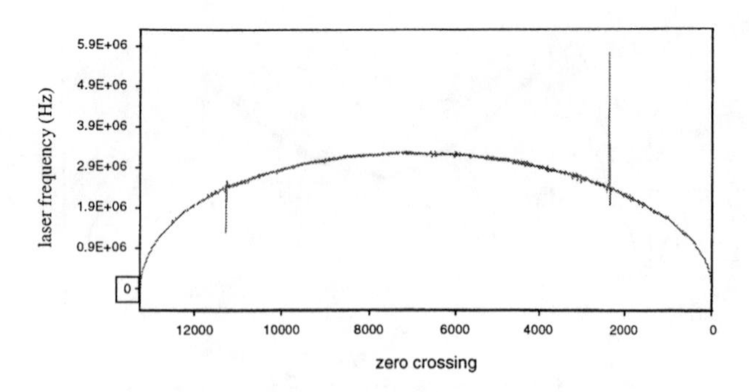

FIGURE 29. Laser frequency profile showing reconstruction of signal beyond Nyquist.

range of angles corresponds to the thick half of the disk, while the thin half was constrained to a constant radius of 5 cm. This result is intuitively reasonable, because the thickness of the disk around any constant radius varies sinusoidally. The force exerted by a differential of mass anywhere in the disk goes as its radius. The higher order terms in the balance equation arise because the mass in a particular differential of radius also varies radially. If the disk were of constant radius (as well as being doubly-wedged), it would be unbalanced by the extra material on the thick side. To balance the disk, the radius on the thick side has been made to vary as the function of angle plotted in Fig. 18. Alternatives include maintaining a constant radius, but drilling the edge of the rotor to remove mass from the thick side.

A series of disk designs were tested with the ANSYS 5 (Ansys, Inc., Cannonsburg, PA) finite element modeling program to estimate distortion. Figure 17 shows the results for the final design. The plot is vertical deflection as a function of position in the disk. These data were generated by a two-dimensional model of a slice through the center of the disk. Consequently, they indicate only half of the actual distortion which is produced by tensile forces acting in two axes. The slightly irregular trace is the result of modeling, which has distortions due to the constraints required by the program. The smooth trace is a spherical fit to the data, which is well within a 1/10 wave requirement at the shortest wavelength, 2.5 μm, intended for modulation in this interferometer. It would be quite economical to polish the disk to a spherical profile which precompensates the

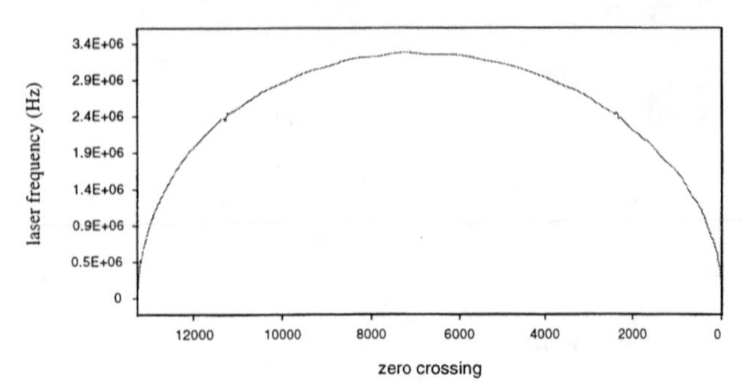

FIGURE 30. Laser frequency profile after filtering.

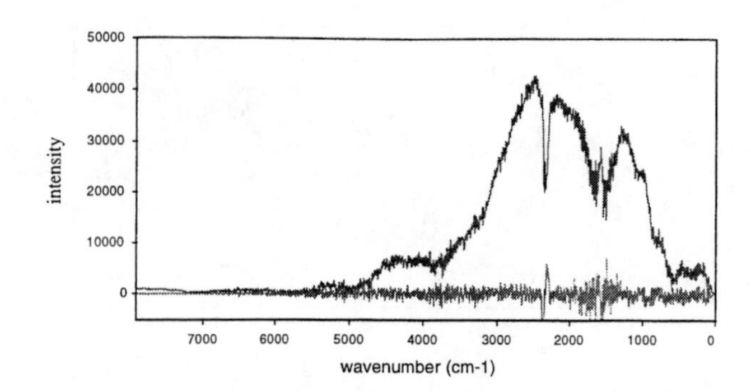

FIGURE 31. Spectrum computed from data acquired in 3 ms, resolution 4 cm^{-1}. Real (upper) and imaginary (lower) after Mertz phase correction.

stretching distortion, but this approach would result in a mirror which exhibits the desired surface flatness at only one rotational speed. The tensile stress in the disk, near the center, is approximately 35 MPa. The stress tapers off radially from the center as illustrated in Fig. 20. The data presented here were acquired with a disk mirror which is not compensated for the stretching induced by rotation. At the 667 scans per second rate of the fastest data presented here, the distortion is approximately one-half what it would be at 1000 scans per second.

One of the critical details of the high-speed rotating assembly is mounting of the disk to the motor shaft. Because the disk should remain undistorted to less than 250 nm, the mounting must be accomplished very carefully. Satisfactory results have been obtained in the prototype by fixing a steel hub into the aluminum disk with epoxy. The aluminum was lapped and coated with a bare gold mirror surface accurate to 1/6 wave at 632 nm.

The mirror drive motor is a standard brushless-DC design with good quality ball bearings. Runout and endplay in the bearings are quite small and would appear as scan-to-scan variations in the position of the centerburst. It should be possible to extract this data from the laser signal as a diagnostic of bearing condition. Better performance could be obtained with an airbearing motor but these are substantially more expensive.

Obvious concerns relating to the high-speed operation of the disk mirror arise. The first is the torque requirement

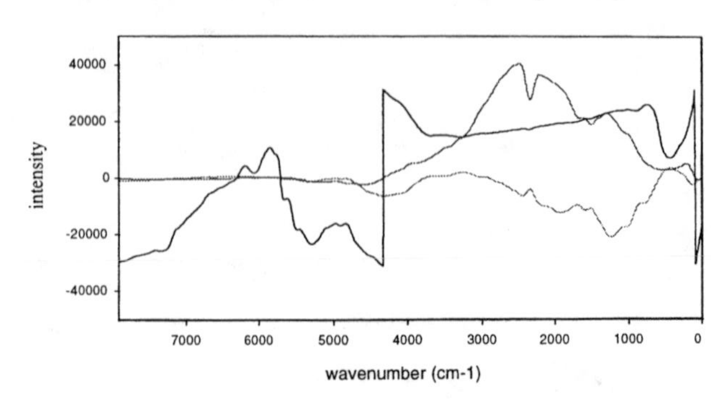

FIGURE 32. Phase of interferogram components (x10,000); real and imaginary low-resolution spectra.

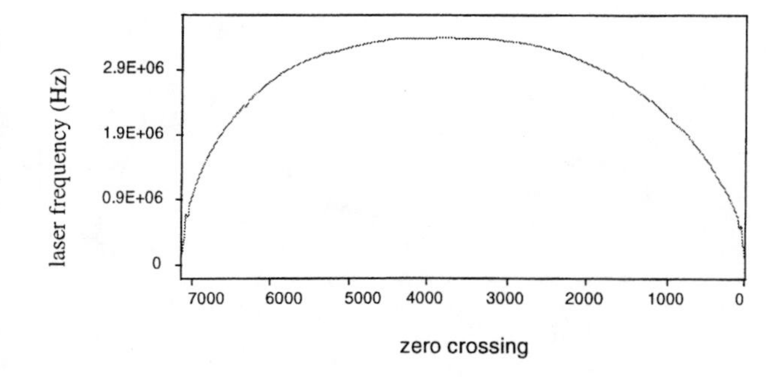

FIGURE 33. Laser frequency profile after filtering.

imposed by air drag. At 500 revolutions per second, even a flat disk generates considerable airflow. Measurements, however, indicate that the torque loading is well below the continuous rating of the motor. This is consistent with the fact that the motor temperature rise during extended operation is well below the maximum rating. The second concern is that turbulence generated by the drag will be a significant noise source, either because of vibrations that it induces or because of refractive index variations in the critical beam paths. The Reynold's number for the disk is quite large, on the order of 10^5 to 10^6, so the considerable turbulence observed during high-speed operation is in agreement with theoretical predictions. As long as any resulting signal fluctuations are small relative to the detector noise, they will disappear without effect. Because of the very large spread of Fourier frequencies, this system is much more resistant to multiplicative noise, including refractive index fluctuations caused by turbulence, than most FT-IR spectrometers. The frequency separation of adjacent spectral elements is approximately 1 to 2 kHz when operated at 4-cm^{-1} resolution and 30,000 rpm. This may be compared to the maximum separation of 26 Hz found in many commercial instruments. Only multiplicative variations with frequencies higher than 1 kHz will be a problem. It is likely, however, that the turbulence in front of the disk has frequency components well beyond 1 kHz. The edge velocity at speed is approximately 157 m/s. The third concern is that imbalance and acoustic noise from the disk/motor assembly will vibrate the system. Initial testing, however, shows that the disk balance is good even before fine balancing by drilling the periphery. The vibrations induced in the optical bench are fairly significant. Although they should not appear as noise, the resulting modulation of intensity can lead to photometric inaccuracy. For very sensitive measurements it may be appropriate to evacuate the system, in which case these problems will be obviated.

The other components of the breadboard system include a standard silicon carbide element (0206RC, The Carborundum Company, Niagara Falls, NY) mounted in a water-cooled jacket (removed from Polaris, ATI Instruments, Madison, WI), source collimating mirror (25.4-mm diameter, 50-mm FL, A8037-111, Janos Technology, Inc., Townshend, VT) as well as a photoconductive MCT detector (Graseby Infrared, Orlando, FL), cube-corner retroreflector (removed from Polaris) and a detector-focusing mirror (50-mm FL, removed from Polaris). These components are mounted on a 46-cm x 91-cm optical breadboard (HC-1836, Thorlabs, Inc., Newton, NJ) using standard hardware (Thorlabs, Inc.). The data acquisition subsystem is schematically diagrammed in Fig. 21. The laser interference signal is detected by a high-speed silicon photodiode and amplified. The resulting signal is passed to one channel of a dual-channel 14-bit 5-MHz ADC which simultaneously converts the infrared signal to produce time-matched pairs of samples. The infrared interference signal is detected by a standard photoconductive MCT element and the resulting electrical signal preamplified and digitized by the second channel of the ADC board. When the disk mirror is spinning at 166 revolutions per second, the stream of data generated by the acquisition system consists of a continuous series of scans as shown in Fig. 22. Each scan is 3 ms. The periodic decreases in the laser modulation amplitude are caused at least partly by capacitive loading of the preamplifier output by coaxial cable. The variation in the infrared level is caused by clipping of the beam at the edge of the mirror, which causes the energy reaching the detector to vary. The sampling times are completely asynchronous to the laser signal, so that the data must be post-processed to reconstruct an interference signal which is sampled at even intervals of retardation. The price of digital signal processors capable of 1 billion floating point operations per second has recently become quite reasonable. It is appropriate to consider processing the interferogram data in real time in the near future.

The method used to process the data reported here is similar to one described by Mertz (6), and is diagrammed in Fig. 23. The continuous stream of ADC samples are broken up into single-scan segments. That is, sets of matched infrared and laser data pairs are truncated at the turnaround points of retardation. The resulting single-scan segments are then digitally filtered to remove noise components outside the frequency range of the signals of interest. The laser data are zero-filled to produce at least 5 data points on each slope, and the resulting points are fitted to extract a best estimate of each zero-crossing position. The infrared signal is then interpolated to produce a set of data points matched to the zero-crossing positions. This is the reconstructed infrared interferogram consisting of samples at even intervals of retardation. This method relies on the delay in the infrared and laser channels being identical so that the effects of the sinusoidal velocity variation are intrinsically compensated (4). The reconstructed infrared interferogram may then be processed according to conventional practice.

EXPERIMENTAL RESULTS

Figure 24 shows an interferogram recorded in 40 ms while the interferometer mirror was operated at approximately 12 revolutions per second (24 scans per second). The corresponding spectrum appears in Fig. 25 and shows a signal-to-noise ratio (SNR) of more than 1000:1 (by the RMS method) at 8-cm^{-1} resolution. The laser frequency profile, indicating the optical velocity as a function of zero-crossing index, is shown in Fig. 26. The fluctuations of velocity indicate that the mirror was vibrating significantly, as well as rotating. At the time these data were recorded the high-speed disk mirror for the prototype instrument had not yet been constructed. The spinning mirror was a 10-cm Pyrex® flat (A2010-565, Janos Technology, Inc.) mounted on the spindle of an obsolete disk drive. Rotation was driven *via* a rubber band connected to a small induction motor. The imbalance of the Pyrex® flat, when tilted on the spindle for precession, caused the vibrations. The reasonable SNR of the resulting spectrum indicates that the velocity variations are effectively compensated, at least at these low velocities.

The laser frequency profile shown in Fig. 27 was obtained while the high-speed aluminum disk mirror was operated to 0.25-cm retardation at 166 revolutions per second, corresponding to 333 scans per second. The laser frequency exceeds the Nyquist sampling limit of the data acquisition subsystem for a portion of each scan. The frequency can be seen to increase smoothly up to 2.5 MHz and then it appears to decrease where the signal is aliased. This problem can be solved by the method shown in Fig. 28, in which the section of laser signal above the Nyquist sampling limit is clipped out of the array. Because this section of data contains no signal below the Nyquist sampling limit which could be confused with the aliased signal, it is possible to use a mirroring operation in the frequency domain to reconstruct the original laser signal. The reconstructed signal can then be spliced back into the original laser sample array to produce a data set like that shown in Fig. 29. The discontinuities at the splicing points, as well as some of the noise in the zero crossing positions, can be removed by filtering to produce a result like that shown in Fig. 30. It is important that this filtering not produce a relative phase shift between the laser and infrared interferograms. The infrared spectrum computed from these data is shown in Fig. 31 and has a SNR of approximately 15:1 computed by the peak-to-peak method. The phase array used in Mertz phase correction of this spectrum appears in Fig. 32, along with the real and imaginary parts of the low-resolution transform. The phase is quite flat throughout the spectral range, indicating that the delays in the infrared and laser signals are reasonably matched at this operating speed. The phase in radians was multiplied by 10,000 to bring it on scale with the other traces.

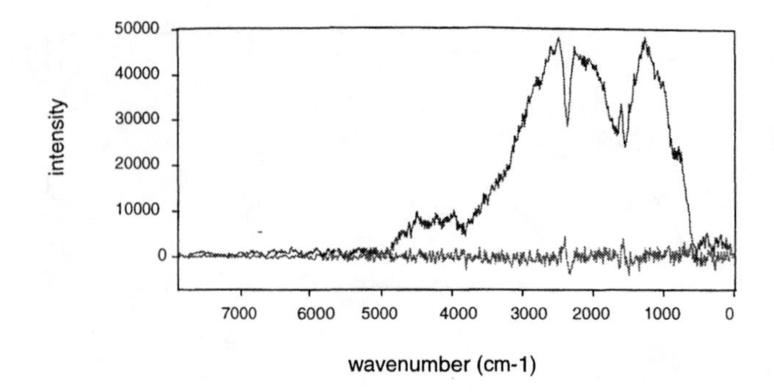

FIGURE 34. Spectrum computed from data acquired in 1.5 ms, resolution 8 cm^{-1}. Real (upper) and imaginary (lower) after Mertz phase correction.

The laser frequency profile (after Nyquist mirroring and filtering) obtained while the interferometer mirror was operated with 0.125-cm retardation at 333 revolutions per second, corresponding to 667 scans per second, is shown in Figure 33. An 8-cm^{-1} resolution spectrum computed from these data appears in Fig. 34 and exhibits a SNR of approximately 15:1 by the peak-to-peak method. This SNR is well below the expected performance based on the detector noise level. Some noise problems, such as electromagnetic interference from the mirror drive motor have been found. When operating at this speed, the turbulent airflow around the disk mirror is impressive. It is thought that the turbulence may induce density fluctuations, and consequently, refractive index fluctuations in the critical beam paths. Such fluctuations are inherently multiplicative and will degrade the SNR if they produce a higher noise level than the detector background. More study is needed to understand the nature and extent of the problem. It may prove necessary to evacuate either the entire instrument, or a compartment around the disk mirror.

The interferogram data shown in Fig. 35 were acquired with the silicon carbide source replaced by a model rocket igniter (Estes Industries, Penrose, CO), which consists of a heating wire with a pyrotechnic coating. The coating is thought to be black powder held with a binder. The igniter was fired at the beginning of data collection. The burn was somewhat erratic as indicated by the severe interferogram drift in Fig. 35. These noise components are multiplicative

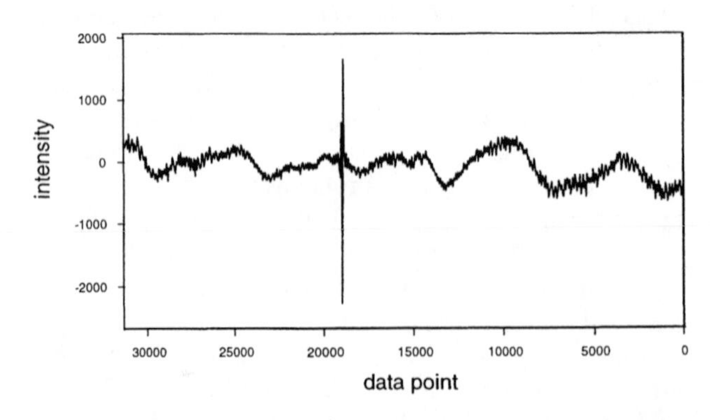

FIGURE 35. Interferogram of burning rocket igniter.

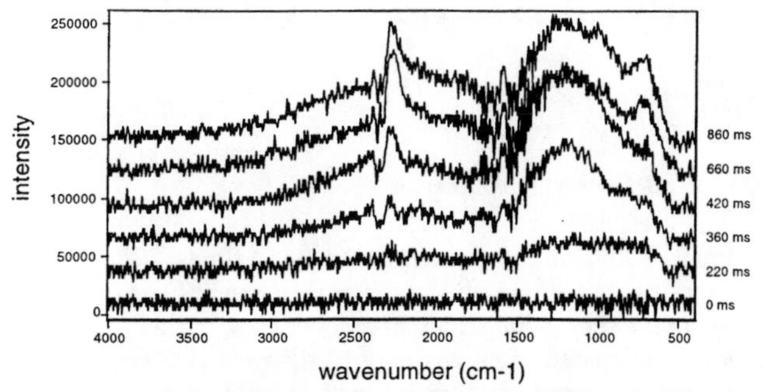

FIGURE 36. Time series of spectra of burning rocket igniter. Each spectrum computed from a single 6 ms scan.

in nature and have frequencies at least as high as 1 kHz; they would be a serious problem in a spectrometer operating with a small frequency spacing between adjacent spectral elements. Data were continuously acquired at a rate of 166 interferograms per second, or 6 ms per scan. A subset of the resulting time series is shown in Fig. 36. Each of the spectra is the result of a single scan. Emission bands, due to vibrationally excited carbon dioxide, are visible in the later spectra at 2300 and 700 cm^{-1}. The igniter is a weak and noisy emitter compared to the silicon carbide source. In spite of the low energy output, fluctuations and short scan time, the cold carbon dioxide absorption and even the rotational lines of water can be observed. The awater vapor may result from combustion of the binder. It appears that the noise level is independent of signal, at least at the low signal levels here. This indicates that multiplicative noise is not significant under these conditions.

CONCLUSIONS AND FUTURE DIRECTIONS

It is clear that the speed of spectral acquisition with current rapid-scan FT-IR technology is severely limited by voice coil performance. Work to date has only scratched the surface of interferometer design. A variety of interferometers that effect very rapid variation of path difference can be constructed. The advantages of very-rapid-scan operation include the ability to monitor transient events which are not repeatable, improved immunity to noise, particularly fluctuations in the events being monitored, and fast data acquisition. We anticipate that very-rapid-scan interferometers will find applications in spectral imaging, process monitoring, and time-resolved spectrometry.

ACKNOWLEDGMENTS

The people and organizations who made this work possible are gratefully acknowledged: DOE/INEL for funding the University of Idaho to purchase a prototype of the rotating interferometer described here. Roger Combs and Bob Kroutil of the US Army, Edgewood Arsenal for funding analysis of interferometer performance. Mike Itano for helping with the mechanical design of the disk mirror. Jeff Paul for helping with assembly of high speed op amp circuits. Blayne Hirsche for helping with instrument setup. Matt Williamson for help with figures. The advice and encouragement of Peter Griffiths, Richard Palmer and Bruce Chase are also greatly appreciated.

REFERENCES

1. M. Bottema and H. J. Bolle, Aspen Int. Conf. on Fourier Spectroscopy, 1970 (G. A. Vanasse, A. T. Stair, and D. J. Baker, eds.), AFCRL-71-0019, p. 293 (1971).

2. J. Kauppinen, I. K. Salomaa, and J.O Partanen, *Applied Optics* **34**, 6081 (1995).

3. A. S. Zachor, *Appl. Optics* **16**, 1412 (1977).

4. A. S. Zachor and S. M. Aaronson, *Appl. Optics* **18**, 1345 (1977).

5. W. H. Steel, Aspen Int. Conf. on Fourier Spectrosc., 1970 (G. A. Vanasse, A. T. Stair, and D. J. Baker, eds.), AFCRL-71-0019, p. 43 (1971).

6. L. Mertz, *Astron. J.* **70**, 548 (1965).

FT-IR and FT-NIR Raman Spectroscopy in Biomedical Research

D. Naumann

Robert Koch-Institute, Nordufer 20, 13353 Berlin, Germany

FT-IR and FT-NIR Raman spectra of intact microbial, plant animal or human cells, tissues, and body fluids are highly specific, fingerprint-like signatures which can be used to discriminate between diverse microbial species and strains, characterize growth-dependent phenomena and cell-drug interactions, and differentiate between various disease states. The spectral information potentially useful for biomedical characterizations may be distributed over the entire infrared region of the electromagnetic spectrum, i.e. over the near-, mid-, and far-infrared. It is therefore a key problem how the characteristic vibrational spectroscopic information can be systematically extracted from the infrared spectra of complex biological samples. In this report these questions are addressed by applying factor and cluster analysis treating the classification problem of microbial infrared spectra as a model task. Particularly interesting applications arise by means of a light microscope coupled to the FT-IR spectrometer. FT-IR spectra of single microcolonies of less than 40 µm in diameter can be obtained from colony replica applying a stamping technique that transfers the different, spatially separated microcolonies from the culture plate to a special IR-sample holder. Using a computer controlled x,y-stage together with mapping and video techniques, the fundamental tasks of microbiological analysis, namely detection, enumeration, and differentiation of micro-organisms can be integrated in one single apparatus. Since high quality, essentially fluorescence free Raman spectra may now be obtained in relatively short time intervals on previously intractable biological specimens, FT-IR and NIR-FT-Raman spectroscopy can be used in tandem to characterize biological samples. This approach seems to open up new horizons for biomedical characterizations of complex biological systems.

INTRODUCTION

It has frequently been shown in the literature that infrared and Raman spectra of complex biological matters such as intact microbial, plant or mamalian cells, tissues and body fluids are highly specific vibrational spectroscopic signatures that can be used to discriminate between e.g. diverse microbial species and strains, characterize particular cell structures in the cell cycle, describe cell-drug interactions, and identify various disease states in human [1-23]. While infrared spectroscopy as a tool for biodiagnostic purposes seems to be elaborated, and a first dedicated FT-IR instrumentation for microbial characterizations is already on the market, Raman spectroscopy, though potentially even more versatile, has not yet reached the stage of development at which routine biomedical applications are already possible. While most of the serious efforts made in the field of biomedical infrared spectroscopy of disease states in humans have been published by the spectroscopy group at the Institute for Biodiagnostics (NRC, Winnipeg, Canada) [8-15], a number of excellent papers have recently been reported on the application of biomedical microspectroscopy by the groups around Wetzel, LeVine, Lewis, and Levin [24-28].

This report will focus on the characterization of medically important microorganisms by FT-IR and FT-NIR Raman spectroscopy. Many of the experience collected in this context can be transferred to other areas of biomedical research. In this sense, intact microbial cells can be treated as model systems. An important point of this report will also be to clarify that the use of multivariate statistics is a main prerequisite to obtain meaningful results when FT-IR and FT-NIR Raman data from biological materials are collected and analysed.

EXPERIMENTAL

Strains and growth conditions

The strains used in this study were either from strain collections (ATCC = American Type Culture Collection, DSM = Deutsche Sammlung für Mikroorganismen, RKI = Robert Koch-Institute) or clinical isolates. Strains were streaked into agar plates using a four-quadrant streak pattern. *Enterobacteriaceae, Bacillus* and *Pseudomonas* strains were cultured on CASO- (casein peptone/soy meal peptone) agar plates purchased from Merck/Germany. *Streptococcus* strains were grown on Columbia blood agar plates from Becton Dickinson. Clostridia strains were also grown on Columbia agar in anaerobic jars. Yeast strains were cultivated on Sabouraud agar supplemented with 2% glucose (from Merck/Germany). All strains were grown for 24 ± 1 hrs. Temperature of growth was 37 ± 2° C except for the *Pseudomonas* strains (30 ± 2° C).

CP430, *Fourier Transform Spectroscopy:* 11th International Conference
edited by J.A. de Haseth
 1-56396-746-4/98/$15.00

Sample preparation and recording of spectra

Sample preparation

For IR measurements small amounts of the cells (≈ 10-60 µg dry wt) were carefully removed with a platinum loop from confluent colonies in the third quadrant of the agar surface and were suspended in 80 µl distilled water. An aliquot (30 µl) was transferred to a ZnSe (zinc selenide) optical plate and dried under moderate vacuum between 2.5 and 7.5 kPa to a transparent film suitable for absorbance/transmission FT-IR measurements. Details of sample preparation, cuvettes systems and data treatment techniques are described elsewhere [1, 3, 6].

For Raman measurements small amounts of the bacterial mass (≈ 10-60 µg dry wt) were harvested also with a platinum loop from regions of confluent colonies in the third quadrant of the agar surface and were transferred directly to steel cups with 2 mm half sphere holes suitable for Raman measurements. The biomass was dried under moderate vacuum similar as described for the FT-IR technique. In order to protect the surrounding atmosphere against contamination by microorganisms, the steel cups were closed by a CaF_2 window using small O-rings and special clamps as cuvette holders. The CaF_2 window could also be used as an internal standard by its Raman band at 322 cm^{-1}.

FT-IR measurements

FT-IR spectra were recorded on an IFS 28/B spectrometer (Bruker, Karlsruhe, Germany) especially designed for the measurement of microorganisms equipped with the FT-IR microscope model A590. For each FT-IR spectrum, 64 interferograms were co-added (time of measurement ≈ 70 s) and averaged, Fourier transformed using a Blackman-Harris 3 term apodization function, and a zerofilling factor of 4 to give a normal resolution of 6 cm^{-1} and an encoding interval of approximately 1 point per wavenumber.

FT-NIR spectroscopic measurements were performed on an IFS-66 spectrometer also purchased from Bruker Analytische Meßtechnik GmbH, Karlsruhe, Germany, using a CaF_2 beam splitter, a tungsten radiation source, and an InSb-detector. For each spectrum 512 interferograms were co-added, averaged, and Fourier transformed applying a Blackman-Harris 3 term apodization function. Nominal resolution used was 16 cm^{-1}.

NIR-FT-Raman measurements

The instrument used was a Bruker IFS66 spectrometer equipped with a FRA-106 Raman module and a liquid nitrogen cooled Ge-diode detector. The excitation source was an air-cooled, diode-pumped Nd:YAG laser (1064 nm) in the backscattering (180°) configuration. The diameter of the laser beam focus was approximately 100 µm, the spectral resolution 4 cm^{-1}, the zerofilling factor 4, the encoding interval approximately 1 point/cm^{-1}, and the apodization function Blackman Harris 4 term. At the sample, the laser power was 250-300 mW. The total number of scans co-added and averaged for each spectrum varied between 500 and 2000, the time of measurement needed was approximately 15 to 60 minutes.

Evaluation of spectral data

Evaluation of spectral data (calculation of derivatives, baseline correction, normalization etc.) was performed using the OPUS software version 3.0 (Bruker) running on a PC equipped with an Intel 486 processor and an operation system OS/2 Warp version 4.0 (IBM). To minimize problems from unavoidable baseline shifts and to enhance the resolution of superimposed bands, the first and second derivatives of the original IR spectra were calculated using a 9-point Savitzky-Golay filter. Calculation of the spectral distances and cluster analysis was performed using the OPUS option "Cluster Analysis". Part of the algorithms of this option are based on the ICARUS software package originally developed by Helm et al. [1]. Factorization of spectra was done using the OPUS option "Factorize". This factor analysis software is based on principal component analysis algorithms [29, 30].

DATA TREATMENT AND EVALUATION TECHNIQUES

When analysing and comparing hundreds, if not thousands of spectra collected from various different microorganisms, body fluids or tissues which all are encoded digitally by a defined number of data points (typically 2000 to 3000 data points in the mid-infrared region), the use of multivariate statistical techniques is a virtual necessity.

The arsenal of modern multivariate statistical techniques provides ample methodologies for the pretreatment, evaluation, and representation of huge and complex data structures. These techniques give at hand not only the possibility to get a survey over the data, but also allow the direct analysis and the interpretation of structures and inter-relationships within the "data cloud". While univariate statistical analysis considers only a single property of a given object, multivariate statistics evaluate several properties of the objects at the same time. In this way also the interrelationships between the properties can be taken into account.

Out of a large number of statistical techniques available for the analyst, four are of particular interest when considering infrared spectra of complex biological samples as the basis of biomedical diagnosis. These are factor analysis, hierarchical clustering, linear discriminant analysis and artifical neural networks [29-34].

The differences between these methods are that factor analysis is performed to extract the essential information from large and mixed data sets to achieve data reduction and to facilitate the recognition of patterns, while the hierarchical clustering technique, a so-called unsupervised classification method, attempts to find out intrinsic group structure within the data set without the *a priori* need of any class assignment or partitioning of the data into a training and test data set. The history of hierarchical cluster analysis process is generally represented by a minimal spanning tree, also called the "dendrogram" by which the merging process of classes can be visually followed. In contrast to hierarchical clustering, linear discriminant analysis and artifical neural networks are so-called supervised classifiers which need the class assignment of each individual (sample, spectra etc.) from the beginning. Partitioning of the whole data into a training and a test set is generally needed to ensure reliability of results.

In this report, factor analysis and hierarchical clustering is used to establish a classification scheme of microorganisms that is based on the infrared and Raman signals of microbial cell samples.

A particularly interesting application of pattern recognition techniques (factor analysis and artificial neuronal nets) to FT-IR microspectroscopic imaging of human melanoma thin sections has recently been described (see P. Lasch et al., this conference proceedings). This methodology may help to securely discriminate between healthy and malignant tissue regions and to enhance image contrast.

RESULTS AND DISCUSSION

FT-IR spectroscopy

The FT-IR technique establishes spectral fingerprints of the complex biological structures under investigation. These patterns comprise the vibrational spectroscopic characteristics of all constituents that are present in the cells, tissues or body fluids. From a simplified point of view these are primarily DNA/RNA, proteins, membranes, and cell-wall components. Consequently, FT-IR probes the total composition of a given biological sample in a single experiment. To understand FT-IR spectra of biological samples, some fundamental knowledge on cell composition and the particular structures of the building blocks in biological samples is essential.

Table 1 gives rough numbers of the gross composition of bacterial, yeast, and mammalian cells that may be helpful to better interprete FT-IR or Raman spectra of different biological samples. It is important to recognize that infrared or Raman spectra of complex biological materials do not only describe the total composition, but also give a number of specific bands that are sensitive to structural or

TABLE 1. Composition of Pro- and Eukaryotic Cells[1)]

	Bacteria	**Yeasts**	**Animal Cells**
Radius [µm]	1	10	100
Volume [µm^3]	1	1000	>10 000
Surface/volume [µm^{-1}]	1	0.1	tissue
Generation time [h]	0.2-10	1-10	20
DNA base-pairs	4 x 10^6	20 x 10^6	500-5000 x10^6
Number of "genes"	4000	20 000	>50 000
Size of ribosome	70 S	80 s	80 S
RNA % [w/w]	10-20	~ 3	3-4
DNA % [w/w]	3-4	~ 1	~ 1
Proteins % [w/w]	40-60	50-60	60
Lipids % [w/w]	10-15	~ 10	15-20
Polysaccharides % [w/w]	10-20	10-20	6-8

[1)] Numbers were taken from different sources

conformational changes. It is also a matter of fact that the physical state of the sample (hydration or aggregation state etc.) has a severe influence on FT-IR results and makes it necessary to standardize sampling, preparation, and data acquisation procedures rigorously.

Figure 1 A gives typical mid-infrared spectra of a microbial, body fluid, and tissue sample, each dehydrated to a thin film layer on ZnSe optical plates. Figure 1 B shows the infrared spectra of dehydrated film samples produced from cell suspensions of two different bacterial strains. While it is still impossible to obtain exact correlations between spectral features or bands and specific cellular structures, some preliminary assignments can be obtained when applying resolution enhancement techniques (see Fig. 1 C) in order to localize specific peaks under the typically broad band contours, and by referring to the numerous literature on infrared band assignments. Table 2 gives a number of tentative assignments that are based on a systematic, comparative analysis of infrared spectra of the main building blocks with representative numbers of different complex biological specimens.

Figure 2 shows FT-IR survey spectra of fully hydrated, and intact cells of *Escherichia coli* that span the entire mid- and near-infrared spectral range from 1000 to 10.000 wavenumbers. The spectrum of the left panel was obtained on a hydrated film sample with a thickness of about 8 µm, the right panel shows the spectrum recorded on a sample with a film thickness of about 250 µm. In both cases CaF_2 was used as the optical cuvette material. The insets of the two panels give the second derivatives as calculated from the original absorbance spectra. Obviously both, the mid-infrared and the near-infrared regions provide substantial amounts of spectral characteristics that may potentially be useful to analyse complex biological materials. Some rough band assignments are given at the top of the figure. It is especially noteworthy that even the spectral range between 1900 and 2800 cm^{-1}, a region that is generally thought to

TABLE 2. Tentative assignment of some bands frequently found in bacterial FT-IR spectra (peak frequencies have been deduced from the second derivatives and Fourier-deconvoluted spectra). Table 2 has been adapted from [3] and [6].

Band numbering (cf. Fig. 1C)	Frequency (cm^{-1})	Assignment[a)]
	~ 3500	O-H str of hydroxyl groups
	~ 3200	N-H str (amide A) of proteins
	2959	C-H str. (asym) of $-CH_3$ methyl
	2934	C-H str (asym) of $>CH_2$ methylene
	2921	C-H str. (asym) of $>CH_2$ methylene in fatty acids
	2898	C-H str of ≡C-H methine
	2872	C-H str (sym) of $-CH_3$ methyl
	2852	C-H str (sym) of $>CH_2$ methylene in fatty acids
1	1741	>C=O str of esters
	~ 1715	>C=O str of ester, protonated carboxyl and carbonyl groups
	1695 1685 1675	different amide I band components resulting from antiparallel pleated sheets and ß-turns of proteins
2	~ 1655	amide I of α-helical structures
3	~ 1637	amide I of ß-pleated sheet structures
4	1548	amide II band
5	1515	"tyrosine" band
6	1468	C-H sc of $>CH_2$ methylene
7	~ 1400	>C=O str (sym) of COO^-
	1310-1240	amide III band components of proteins
8	1250-1220	P=O str (asym) of $>PO_2^-$ phosphordiesters
	1088-1084	P=O str (sym) of $> PO_2^-$ phosphodiesters
9	1200 - 900	C-O-C, C-O dominated by the ring vibrations of carbohydrates, C-O-P, P-O-P
	720	C-H rocking of $>CH_2$ methylene
	900-600	"fingerprint region"

a) sc = scissoring; str = stretching; sym = symmetrical; asym = asymmetrical

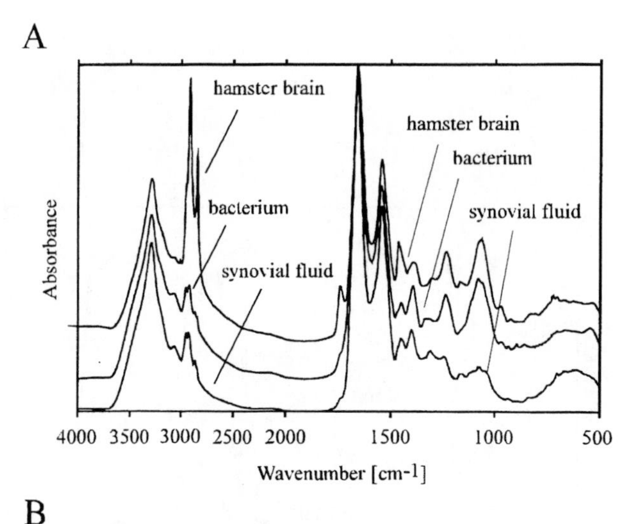

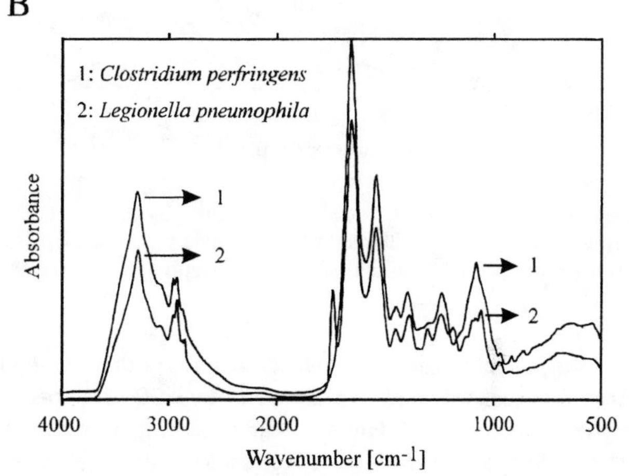

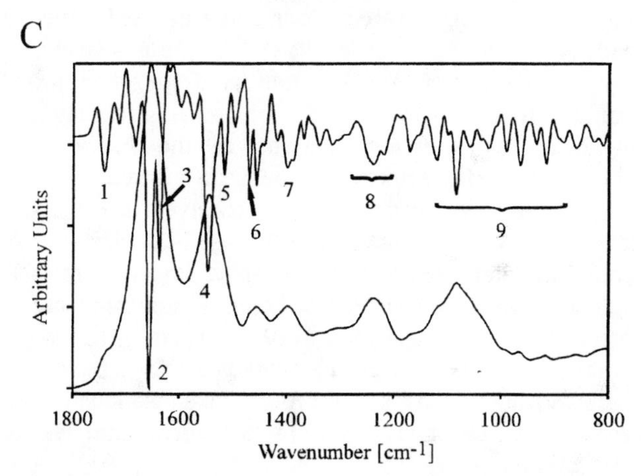

FIGURE 1. Representative mid-infrared spectra of microorganisms, tissues, and body fluids. (A) Infrared spectrum of a bacterial sample, small piece of hamster brain, and body fluid (synovial fluid from a patient suffering from rheumatoide arthritis). (B) Infrared spectra of different bacterial samples: *Legionella pneumophila* (1) and *Clostridium perfringens* (2). (C) Tentative band assignment of some bands observed in the 800 to 1000 wavenumber region. Bottom: Original absorbance spectrum. Top: Second derivative. For band assignment and band numbering cf. Table 1.

be free of any significant information, provides a number of spectral traits that, though minute in intensity, can be used to differentiate between biological substances. Up to now there is no relevant information available about which vibrational modes or chemical structures might stand behind these spectral features.

Figures 3 A and B show the results of classification trials on replicate measurements on a selection of

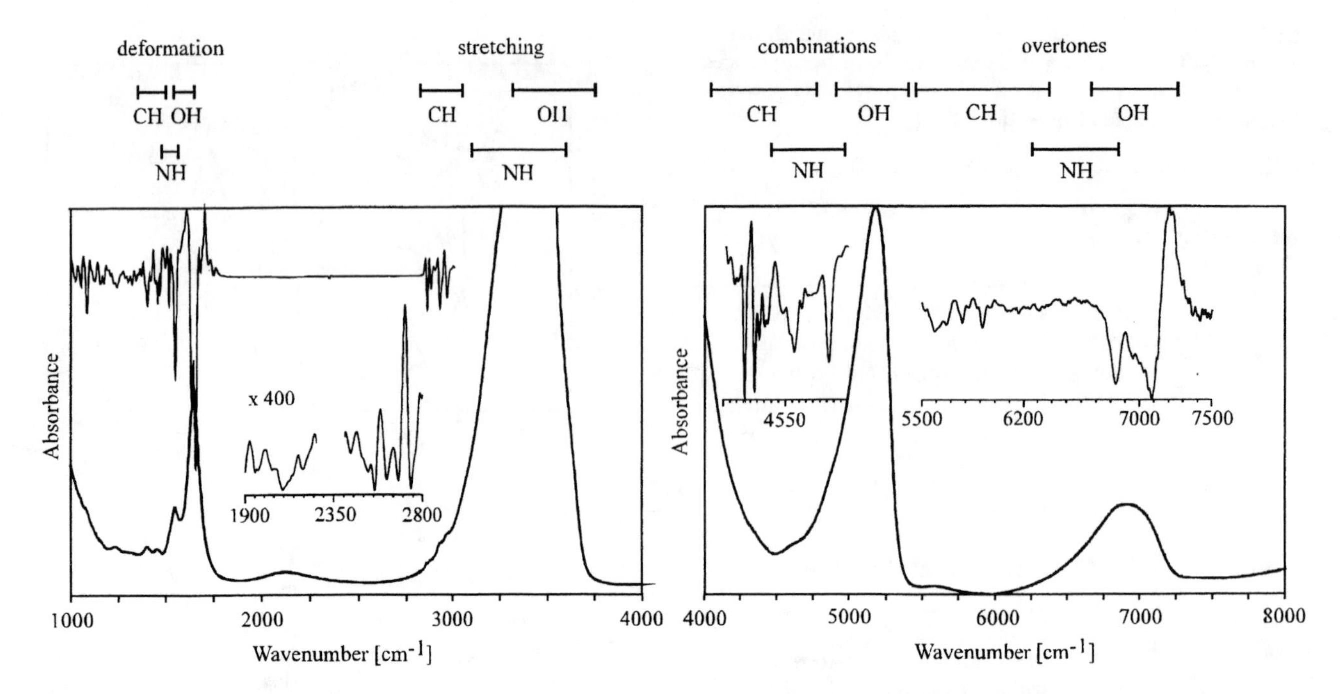

Figure 2. Infrared absorbance spectra of fully hydrated samples of *Escherichia coli* in the spectral range between 1000 and 10.000 cm^{-1} (mid-infrared, left panel and near-infrared, right panel). The insets show the second derivatives as calculated from the original absorbance spectra. Nominal physical resolution applied was 6 cm^{-1} for the mid-infrared and 16cm^{-1} for the near-infrared, respectively.

microorganisms using some spectral information available from the near-infrared and the 1900 to 2800 cm^{-1} spectral regions, respectively. Hierarchical clustering was applied in both cases to scrutinize spectroscopic similarities and spectral clustering. Ward's algorithm was used, since it tends to produce very dense clusters. The results of cluster analysis are shown in the familiar form of minimal spanning trees, the so-called dendrograms. Obviously also the overtones and combinations of the fundamental vibrational modes detected in the near-infrared, and the hitherto unknown spectral features observed in the 1900-2800 wavenumber regions can be used to characterize biological samples, provided the signal to noise ratio is high enough to make the inherently minute features detectable. The far-infrared (100-600 cm^{-1}), although principally also a good candidate for biological characterizations has not yet been taken seriously into account for investigations into biological samples of medical interest.

Differentiation and Classification of Medically Important Microorganisms

The FT-IR technique has a remarkably far-reaching differentiation capacity, it may provide results within a few minutes after obtaining single colonies of the pure culture and may be uniformly applied to all microorganisms which can be grown in culture. To demonstrate these potentials, Fig. 4 A shows the dendrograms of a hierarchical cluster analysis performed on 240 mid-infrared spectra of very diverse microorganisms, including different species and strains of Gram-positive and Gram-negative bacteria, and yeasts. It is particularly interesting that the technique is not restricted to the analysis of bacteria. Yeasts and fungi can also be analysed. This is in an exemplary manner demonstrated by Fig. 4 B which shows a successful, FT-IR based classification scheme of different species and strains of Candida.

As all cell components depend on the expression of smaller or larger parts of the genome, the FT-IR spectra of microorganisms display in a specific way a phenetic and a genetic fingerprint of the cells under study. This is why the specifity of the technique is rather high, allowing differentiations at very different taxonomic levels, even down to the subspecies, strain and/or serogroup/serotype level. The latter is demonstrated by the FT-IR based classification of different isolates of *Escherichia coli* (see Fig. 5). The purpose of this analysis was to group the 23 spectra according to the known *O*-antigenic properties of the E. coli isolates. *O*-antigen determination and serogroup analysis were performed by gel electrophoresis of the lipopolysaccharides (LPS) and by serological techniques. Since it was expected that spectral differences between the *O*-specific side chains of LPS (which are complex hetero-polysaccharides expressed at the cell surface) would preliminarly be detected in the spectral range where the polysaccharides dominate the observed spectral features, only the range between 700 and 1200 cm^{-1} was explored for cluster analysis. Using the Ward's algorithm, three distinct clusters are found containing the O-18, O-25, and O-114 strains, respectively (see Fig. 5 A). Complementary,

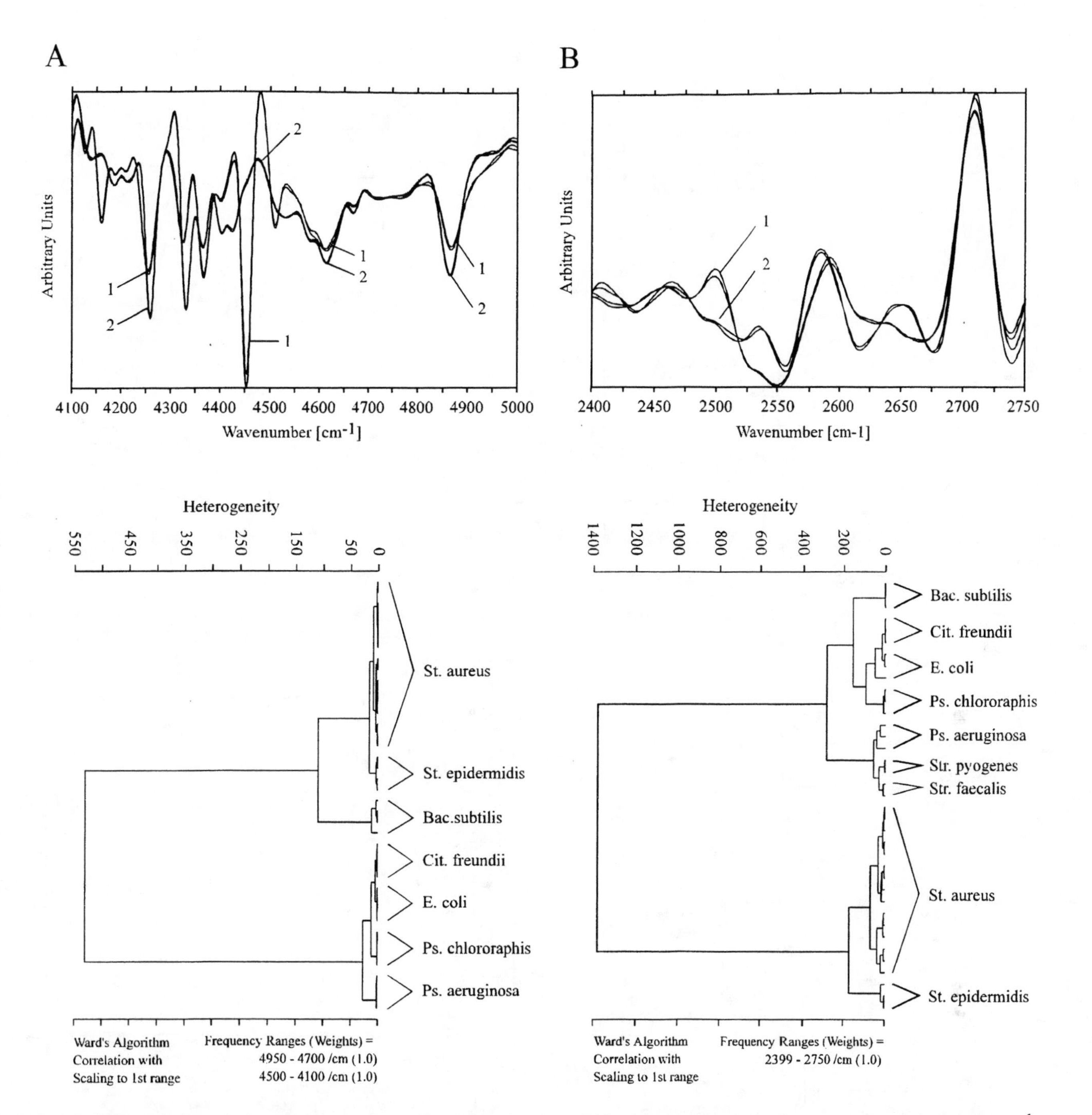

Figure 3. Differentiation of microorganims using spectral features observed in the near-infrared region between 4000 and 5000 cm^{-1} and in the mid-infrared region between 1900 and 2800 cm^{-1}. (A) Top: Second derivatives of two independent measurements on fully hydrated samples of *Staphylococcus aureus* (1) and *Escherichia coli* (2). Bottom: Dendrogram of the classification of near-infrared spectra obtained from a selection of Gram-positive and Gram-negative bacteria (strain labels have been neglected). Spectral ranges used for calculating the interspectral distance matrix were 4700-4950 and 4100-4500 cm^{-1}. (B) Top: Second derivatives of two independent measurements on samples of *Staphylococcus aureus* (1) and *Escherichia coli* (2). Bottom: Dendrogram of the classification of some Gram-positive and Gram-negative bacteria (strain labels have been neglected). Spectral range used for calculating the interspectral distance matrix was 2399-2750 cm^{-1}. Ward's algorithm was used in both cases. The distance measure used to calculate the distance matrix was the Pearson's momentum correlation coefficient as defined in [1].

factor analysis was performed on the same spectra. When using the factors 2 and 3 for two-dimensional projection of data, three different clusters can already be discriminated by eyes (see Fig. 5 B).

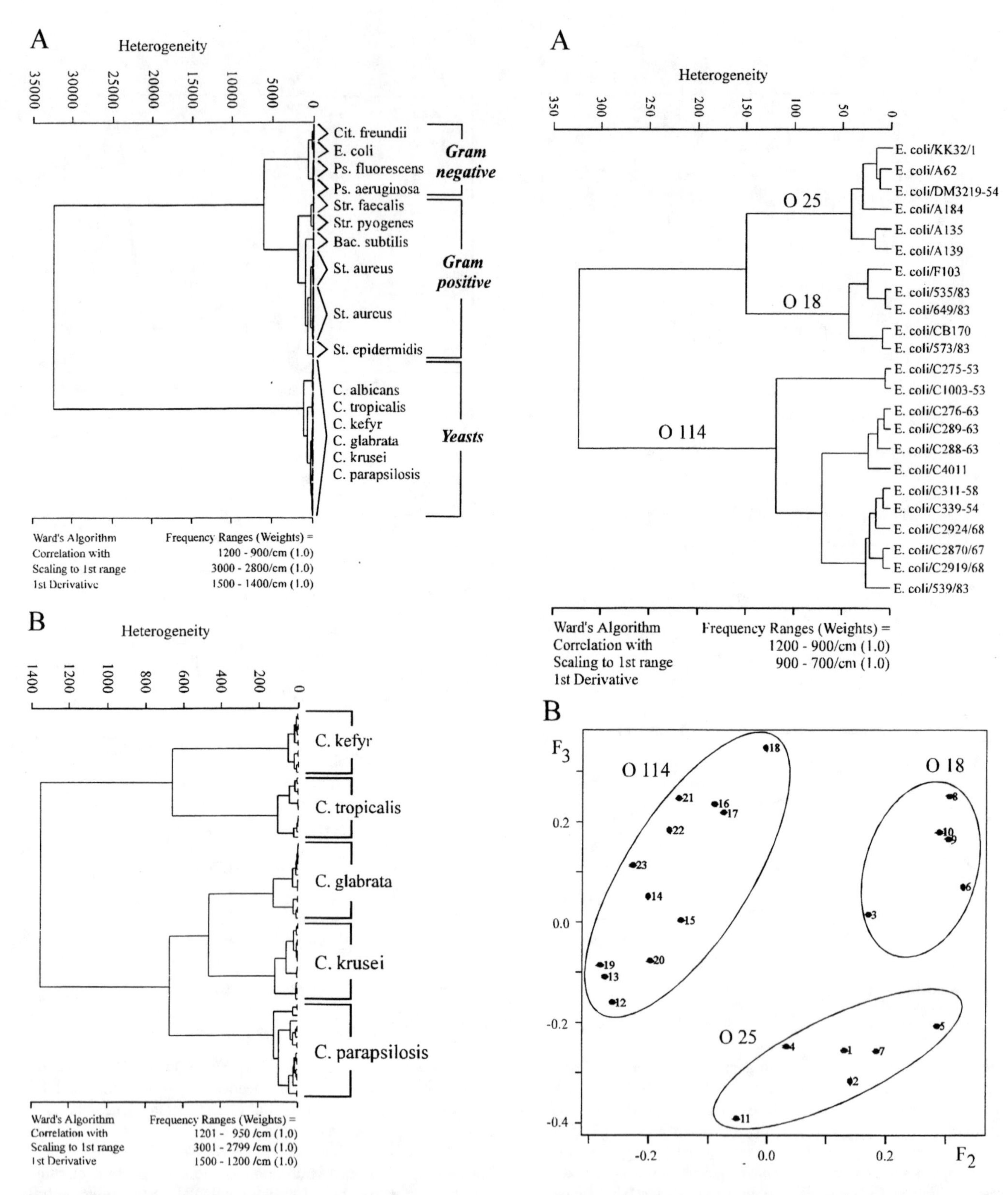

Figure 4. Dendrograms of classification trials based on the mid-infrared spectra of very diverse microorganisms. (A) Classification of a selection of different species and strains of Gram-negative, Gram-positive bacteria and of different species and strains belonging to the genus Candida. (B) Spectral classification scheme of different species and strains of Candida according to their species relationships. The spectral ranges used as input for both cluster analysis are given at the bottom of the dendrograms.

Figure 5. Spectral typing of closely related isolates of the species *Escherichia coli* according to their *O*-antigenic structure. (A) Spectral classification using the Ward's algorithm and the Pearson's correlation coefficient as the distance measure. (B) Factor analysis map using principal component analysis. Each point represents the spectrum of an isolate; for projection of data in factor space, the factorial coordinates 2 and 3 were used. In (A) and (B) the first derivatives of the spectra were used.

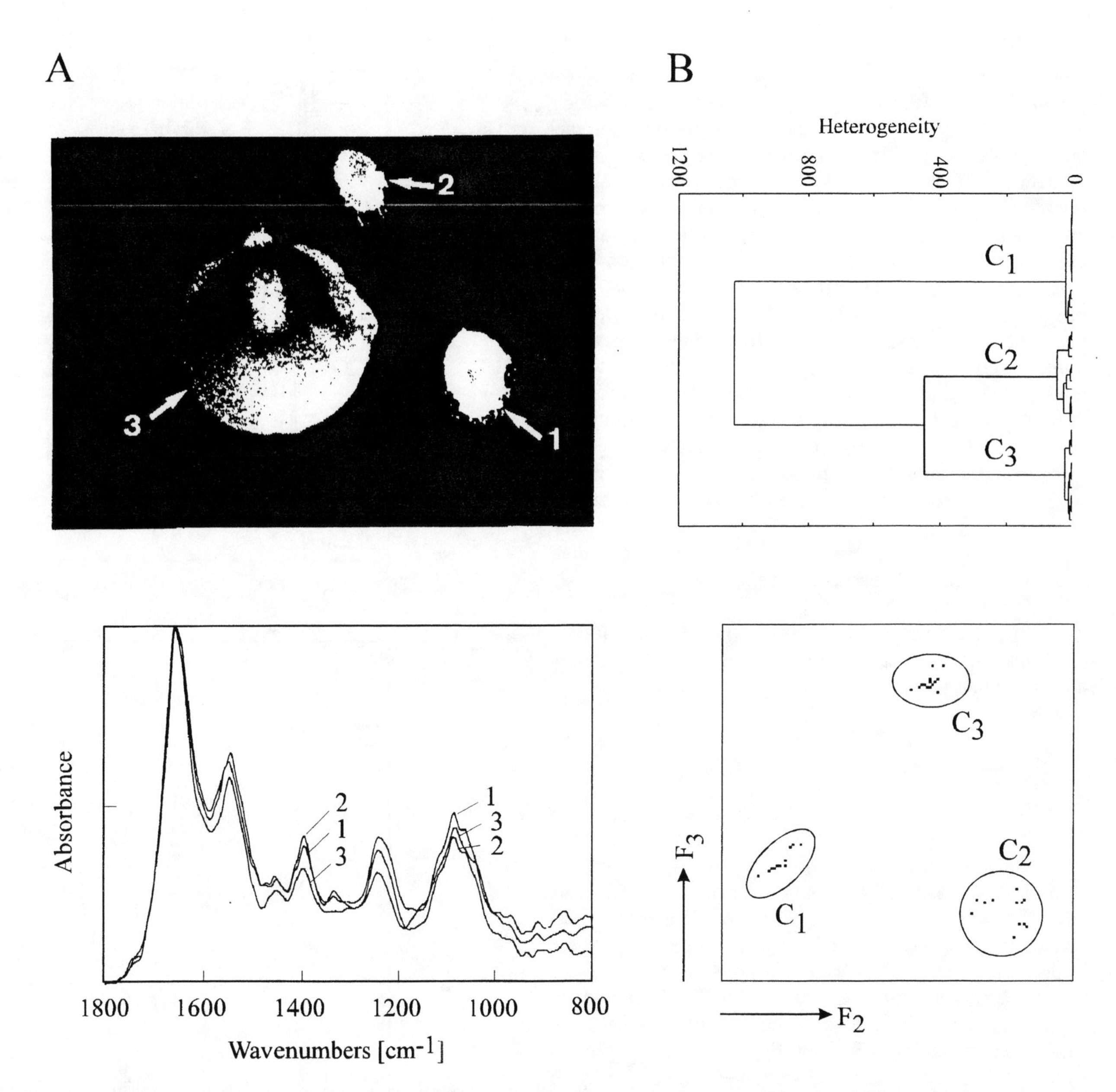

Figure 6. Detection, differentiation, and classification of different microbial microcolonies. (A) Upper panel: Micrograph (magnification approximately 150 x) of three different colony spots deposited on a BaF_2 window by a special stamping technique. Lower panel: Infrared spectra obtained from the colony spots shown above. (B) Left panel: Hierarchical cluster analysis performed on the measurement of approximately 40, randomly selected different colony spots. Right panel: Factor analysis performed on the same spectra. The clusters suggested by both classification techniques are: C_1: *Escherichia coli*; C_2 and C_3: two different strains of *Staphylococcus aureus*.

While it is a general prerequisite of the FT-IR technique to isolate pure cultures of the microorganisms in question, mixed cultures can be analysed by measuring microcolonies (40 to 100 µm in diameter) which grow separately on solid agar media by means of a light microscope coupled to the FT-IR spectrometer. Small amounts of the microcolony spots suitable for FT-IR microscopic measurements are obtained by applying a special replica technique that transfers spatially accurate the first two to three cell layers of the microcolonies from the solid agar plates to infrared transparent plates (e.g. made of BaF_2) [3, 6]. The air dried microbial spots deposited onto such IR-plates can then be measured microspectroscopically using a computer controlled x,y-stage. This may be done operator or computer controlled utilizing video and imaging techniques. The information accessible from the light-microscopic data (number of colonies, size, different shapes etc.) and from the FT-IR spectra of the microcolonies (cell composition, structural data, type-specific FT-IR fingerprints) can be used as input data to differentiate and classifiy very small amounts of microorganisms even from mixed cultures ($<10^3$ cells per colony spot), and to characterize colony growth. The combined use of both, the light-microscopic and the FT-IR spectroscopic information, may strongly enhance the potentials of FT-IR as applied to microorganisms which, however, will only be realized when imaging technologies and multivariate pattern recognition techniques are used in tandem. Only in this way, the fundamental tasks of microbiological analysis, namely detection, enumeration, and differentiation of microorganisms can be integrated in one single apparatus. Figure 6 gives examples of light-microscopic and microspectroscopic measurements on a mixed culture containing three different types of microorganisms. Figure 6 A, upper panel shows the micrographs from a selected area of a typical microcolony replica obtained by the imprinting technique. The presence of three different colonies can already be differentiated by eyes. The three FT-IR spectra obtained from the colony spots deposited onto the IR plate (see Fig. 6, lower panel) suggest that three different microorganisms are present in mixed culture. In a second step, after sampling representative numbers of FT-IR spectra of microcolonies, and after subjecting these spectra to multivariate statistical analysis including factor and hierarchical cluster analysis (see Fig. 6 B, upper and lower part), unequivocal classification and identification of the three different microorganisms can be achieved.

Detection and Characterization of Specific Cell Components and Cell Structures

Infrared spectroscopy can also be used to detect and identify *in situ* particular cell components present in intact microorganisms by some compound specific marker bands taking advantage of difference spectroscopic techniques. Endospore formation in bacteria, to give an example, can be detected by a number of marker bands diagnostic for dipicolinic acid which is a characteristic and essential compound in endospores (see Fig. 7 A), while poly-ß-hydroxybutyric acid, a typical storage material which is accumulated within the cytoplasm, may be identified using a number of infrared bands that are specified by Fig. 7 B. The difference spectrum shown in Fig. 7 A closely resembles spectra recorded from Ca^{2+} salts of dipicolinic acids (DPA). *In vivo* chelate binding of Ca^{2+} ions gives rise to symmetric and antisymmetric stretching vibration bands near 1378 and 1616 cm^{-1}, and the band near 1570 cm^{-1} may be due to C-N stretching vibration of the heterocyclic ring of DPA. Since the difference spectrum (c) of Fig. 7 A is clearly dominated by the spectral features of dipicolinate, monitoring of the multiphase process of spore formation *in situ* is possible. Similarly, the difference spectrum (c) shown in Fig. 7 B is nearly a perfect spectrum of PHB as could be verified by comparison with the FT-IR spectrum of the isolated pure compound. The most prominent difference band near 1740 cm^{-1} certainly represents the >C=O ester stretching band of the polyester compound. This particular intensive band can be used as a marker

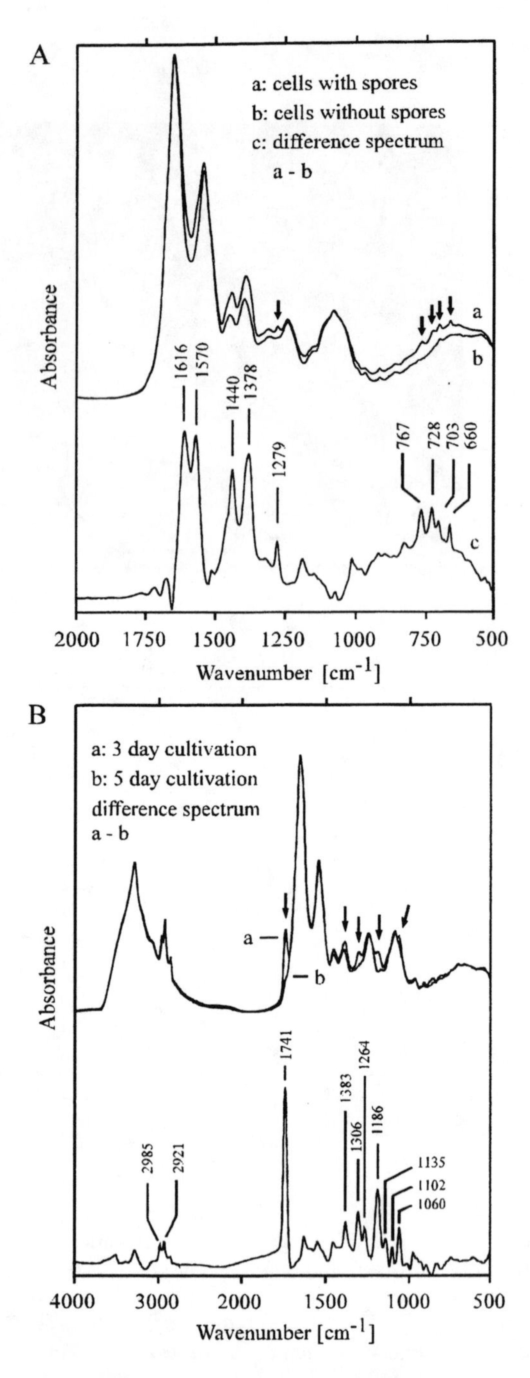

Figure 7. (A) FT-IR spectra of a strain of *Clostridium sordelli* which may produce endospores. Spectra were obtained from two populations with (a) and without (b) sporulating cells. The difference spectrum (a) minus (b) is shown by curve (c). (B) shows FT-IR spectra of a *Legionella pneumophila* strain that may produce the intracellular storage material poly-β-hydroxybutyric (PHB) acid as a function of time of growth. Spectra were obtained from cell populations with (a) and without (b) PHB. The difference spectrum (a) minus (b) is given by curve (c).
Some marker bands of the difference spectra in (A) and (B) are labeled. The difference spectra were obtained by one-to-one subtraction of vector normalized original absorbance spectra. Representations of the difference spectra are multiplied by a factor of 4.

band to monitor *in situ* intracellular production of PHB storage compounds.

The detection and quantitation of metabolically released CO_2 by bacteria and yeasts is an additionally interesting item of the FT-IR technique [15]. Intra-cellularly or elsewhere produced CO_2, which is usually detected as CO_2 hydrates in water, can be determined extremely sensitive, since the CO_2 band, located near 2342 cm^{-1}, is found in a spectral region where signal to noise ratio is optimal and which is usually devoid of overlapping spectral features.

FT-IR spectroscopy may also be a valuable tool for the *in situ* characterization of cell-drug interactions like e.g. the action of antibiotics. This very interesting possibility of the FT-IR technique has however still to be fully developed in the future [3].

NIR-FT Raman spectroscopy

Since high quality, essentially fluorescence free Raman spectra can now be obtained routinely on previously intractable biological samples using laser excitation radiation in the NIR range (e.g. with a Nd:YAG laser line at 1064 nm), it is possible to obtain the complementary vibrational spectroscopic information of a given, even highly coloured biological substance from its FT-IR and NIR-FT-Raman spectra [19, 35].

The nature of microbial FT-NIR Raman spectra, preliminary band assignments

Figure 8, lower panel shows typical FT-IR and NIR-FT Raman spectra obtained from a complex biological sample, in this case from intact cells of *Staphylococcus aureus* that have been dehydrated as described. The Raman spectra of microorganisms are characterized by the prominent C-H stretching bands in the range of 2700-3100 cm^{-1} (data not shown) and the C-H deformation band around 1450 cm^{-1} originating most likely from the abundantly present $-CH_3$, $>CH_2$ and C-H functional groups in lipids, amino acid side chains of the proteins and carbohydrates. In the 1200 to 1800 cm^{-1} region the amide I and amide III bands of the proteins also give rise to prominent bands around 1660 and 1250 cm^{-1}, respectively. Additionally, the Raman spectra show a series of well resolved and sharp bands of minor intensity, which are tentatively assigned to particular substructures like the RNA/DNA nucleotide base-ring vibrations of guanine (G), thymine (T), adenine (A), cytosine (C), uracil (U), and the amino acid side vibrations of tryptophane (Trp), tyrosine (Tyr), and phenylalanine (Phe) of the proteins. The FT-IR spectra of the same strain (Fig. 8, upper panel, curve 1), on the other side, reveals only broad superimposed spectral bands with the amide I and amide II bands being constantly the most prominent

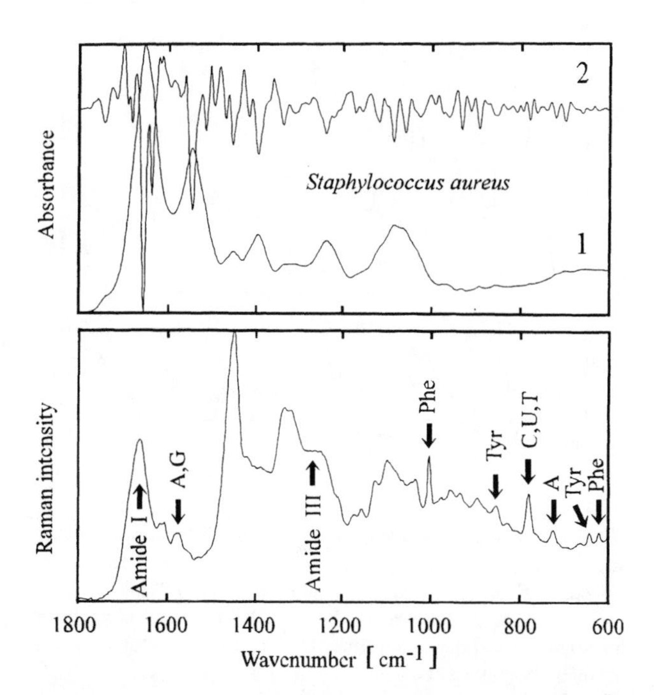

Figure 8. A typical Raman (lower panel) and infrared (upper panel) spectrum obtained from dried samples of Staphylococcus aureus PS 29 in the spectral region of 600 to 1800 cm^{-1}.
To enhance apparent resolution, the second derivative as calculated from the infrared absorbance spectrum (curve 1) is shown by curve 2 (see upper panel).
Abbreviations: C, U, T, A, G, Phe, and Tyr stand for cytosine, uracil, thymine, adenine, guanine, phenylalanine, and tyrosine, respectively.

features. Well resolved spectral bands can only be discriminated applying band narrowing techniques, e.g. by calculating the second derivatives as shown in Fig. 8, upper panel, curve 2. However, FT-IR spectra yield much better signal-to-noise ratios at comparable measurement times. While FT-IR gives typically S/N-ratios of better than 2000 at measuring times of less than a minute, about 45 minutes are needed for FT-NIR Raman spectroscopy to give a S/N ratio of only 400.

Figure 9 shows a typical FT-NIR Raman spectrum of *Escherichia coli* compared to Raman spectra of compounds which represent the main cellular components in bacteria, namely of a protein (RNase A), phospholipid (DMPC), and nucleotide (RNA). The striking similarity between the bacterial and the pure protein spectra merely reflects that the major portion of total cell mass is represented by the various cellular and membrane bound proteins, while the contribution of RNA/DNA-structures to bacterial Raman spectra is comparably small. The lipid compounds, although forming less then 10-15% of cell mass contribute markedly to the Raman spectra of bacteria due to their relatively high scattering coefficients.

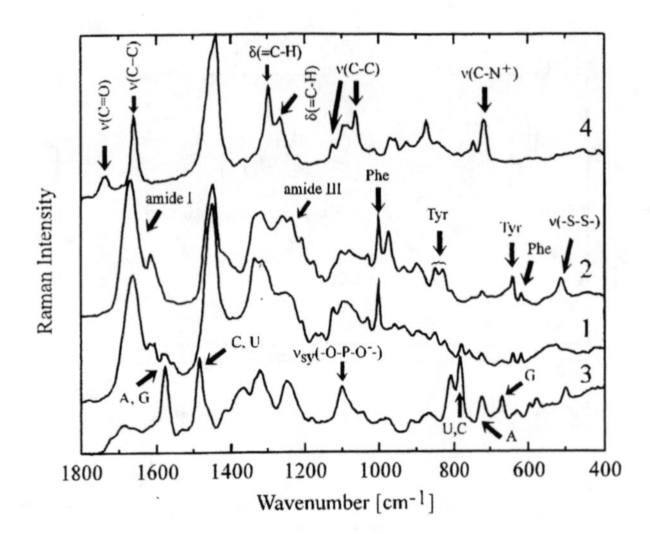

Figure 9. Raman spectrum of a representative sample of intact cells of *Escherichia coli* RKI/A139 (1). For spectral comparison, the spectra of the small, globular protein RNase A (2), a ribonucleic acid compound, RNA (3), and the phospholipid dimyristoylphosphatidylcholine, DMPC (4) are shown as well. For band assignments and abbreviations see legend to Fig. 8; ν = stretching vibration, δ = deformation, sy = symmetric.

Detection of particular cell structures

The FT-NIR Raman spectra of some microbial constituents like pigments of the carotenoid type may give rise to additonal peaks which, though being present in only small amounts, gain considerable in intensity due to a preresonance effect. Figure 10 A shows two spectra obtained from samples of a *Staphylococcus aureus* and a *Micrococcus roseus* strain that were golden (aureus) and deep rose (roseus) in colour, respectively. These spectra show two characteristic peaks near 1525 cm^{-1} and 1514 cm^{-1} (ν(C=C)) and 1159 cm^{-1} and 1155 cm^{-1} (νC-C)) for the *Staphylococcus aureus* and *Micrococcus roseus* samples, respectively, which were found to be markedly dependend on the time and temperature of growth, and which are quite diagnostic for the presence of carotenoid structures [36]. Fig. 10 B shows, for *Micrococcus roseus* as an example, that the amount of cellular pigment production strongly depends on the time of growth. The relative intensities of the diagnostic ν(C=C) and ν(C-C) bands, compared to the amide I band as an internal standard, mainly correlate with the time of growth, and thus demonstrate that the expression of this particular bacterial pigment is a growth dependent process.

Another example is the *in situ* Raman spectroscopic detection of relative amounts of protein and/or RNA as a function of the time of growth. Figure 11 gives FT-NIR Raman spectra of *Bacillus subtilis* cultivated for 48 (curve 1) and 120 hours (curve 2), respectively. The comparison of these two spectra strongly suggests a significant decrease in relative amount of proteins by the reduced relative peak intensity of the bands at 1679 cm^{-1} (amide I) and 1235 cm^{-1} (amide III), using the band intensity at

A

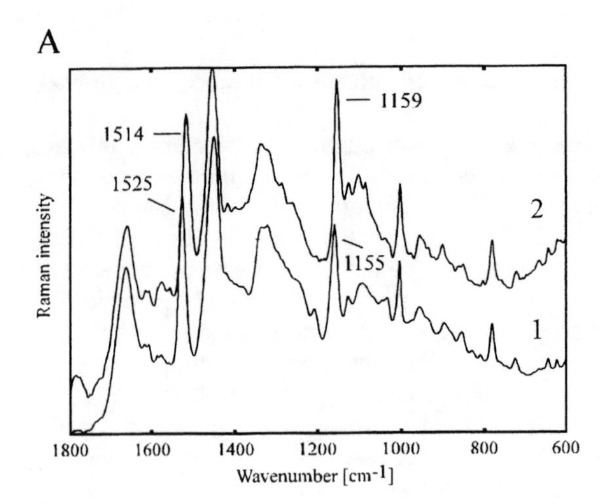

B

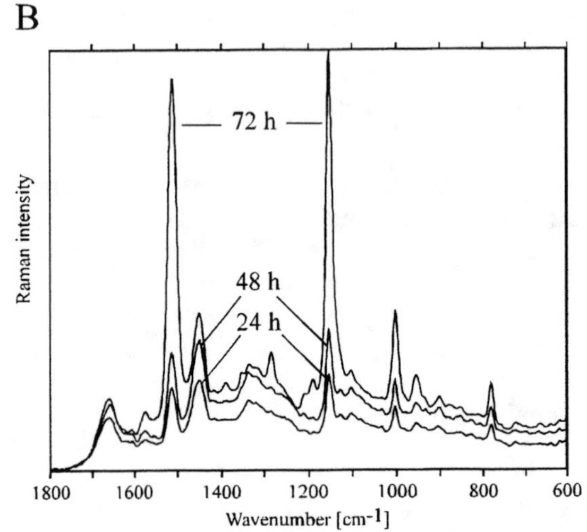

Figure 10. Raman spectra of bacteria exhibitting preresonance enhanced bands of carotenoids containing pigment compounds (A) *Staphylococcus aureus* DSM 20231 (1), *Micrococcus roseus* DSM 20447 (2); (B) Spectra obtained from *Micrococcus roseus* DSM 20447 after different times of growth.

1450 cm^{-1} (C-H deformation band) as an internal standard. The difference spectrum obtained by subtraction of spectrum 1 (48 hours time of growth) minus spectrum 2 (120 hours time of growth) (see Fig. 11, curve 3), suggests a strong decrease in the cellular amount of RNA as is indicated by the marker bands at 669 cm^{-1} (guanine), 726 cm^{-1} (adenine), 783 cm^{-1} (uracil and cytosine), 1095 cm^{-1} (symmetric stretching of $>PO_2^-$), 1481 cm^{-1} (cytosine and uracil), and 1576 cm^{-1} (adenine, guanine), respectively [37, 38]. For comparison, curve 4 gives the FT-NIR Raman spectrum of a RNA sample (see also Fig. 9). The observed band intensity changes may likely be assigned to the transition of the cells from the so-called logarithmic state of growth to the stationary state of growth. It is well known from the literature that the logarithmic state of growth is characterized by a much higher amount of RNA synthesis as compared to resting cells to guarantee high cell proliferation, i.e. high rates of protein biosynthesis.

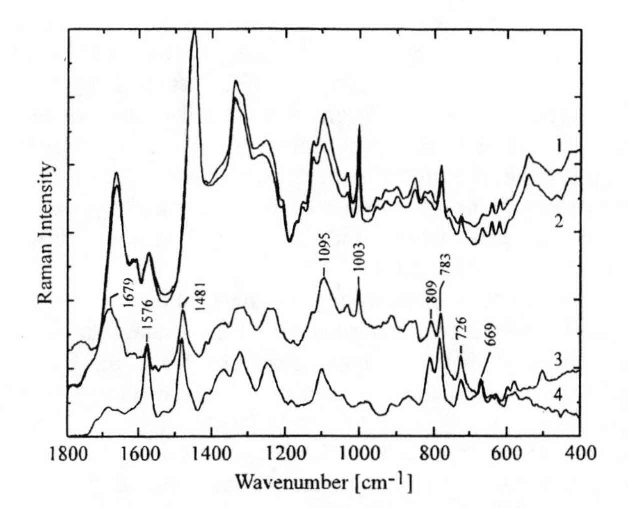

Figure 11. Spectral changes observed in samples of *Bacillus subtilis* RKI W2GR as a function of time of growth. 48 hours (1), 120 hours (2); difference spectrum (1) minus (2) (3) and - for comparison - the spectrum of RNA (4).

Differentiation of microorganisms

Just as FT-IR, microbial FT-NIR Raman spectra express the total composition and structure of all structures present, which in turn, may depend on various microbiological parameters such as time and temperature of growth, nutrient conditions etc.. Thus FT-NIR spectra of microorganisms can only be used for differentiation and classification purposes, if microbiological and sample preparation procedures are standardized rigorously. Figures 12 A and B show some Raman spectra of very diverse bacteria and yeasts. The intensity of the band at 787 cm^{-1} can be taken as a measure for the relative quantity of nucleic acids present, while useful information on the lipid/protein ratio can be calculated from the intensities of the bands at 1450 cm^{-1} (δ(C-H)) and 1660 cm^{-1} (amide I) or 1008 cm^{-1} (phenylalanine). These relationships may be helpful to quickly estimate gross compositions of microbial cell samples and allows the

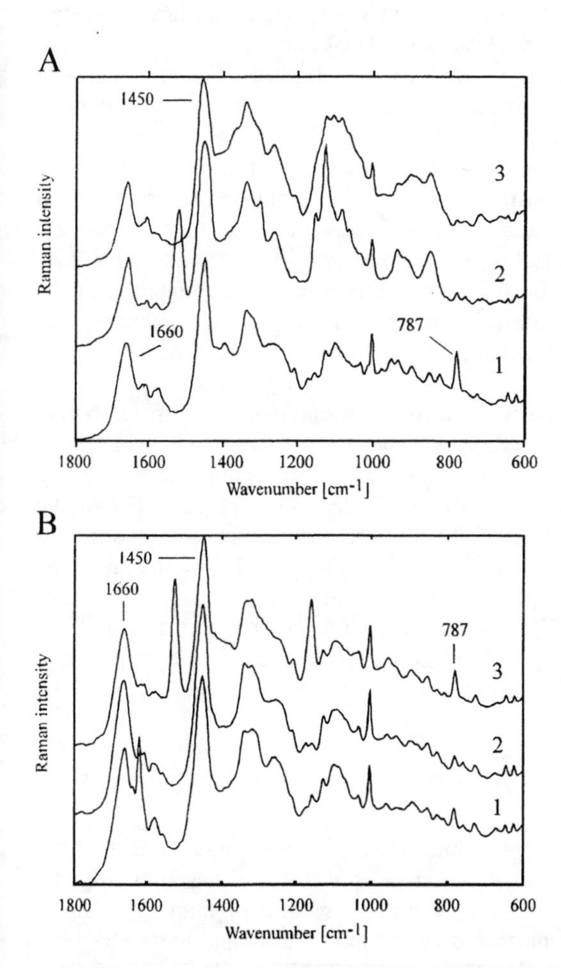

Figure 12. Survey FT-NIR Raman spectra of different microbial species and strains. (A) *Bacillus subtilis* RKI W2GR (1), *Mycobacterium smegmatis* HK 144 (2), *Candida albicans*, clinical isolate (3); (B) *Pseudomonas chlororaphis* ATCC 17809 (1), *Escherichia coli* RKI 139(2), *Staphylococcus aureus* DSM 20231 (3).

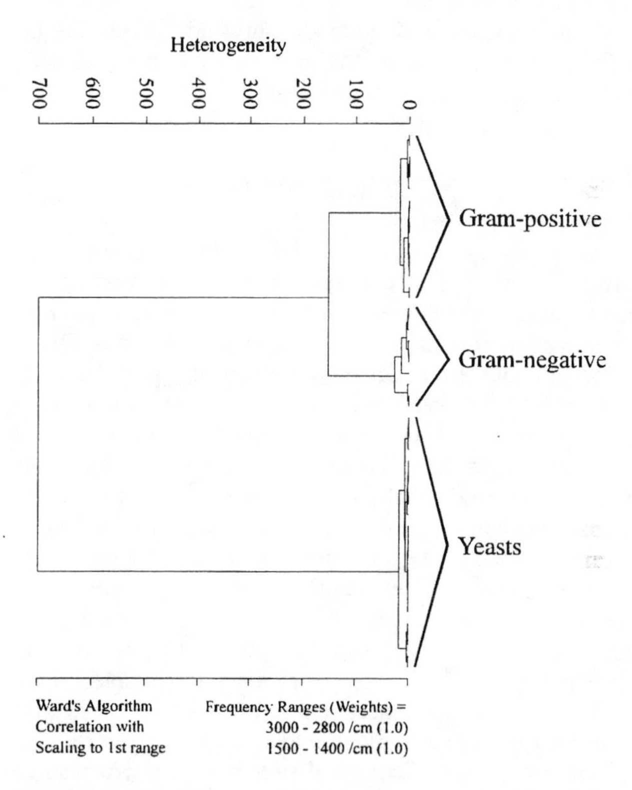

Figure 13. Classification trial on a selection of FT-NIR Raman spectra of Gram-positive and Gram-negative bacteria and yeasts. The Gram-postive, Gram-negative and yeast cluster comprise different strains of *Bacillus subtilis*, *Staphylococcus epidermidis*, *Staphylococcus aureus*, *Escherichia coli*, *Citrobacter freundii*, *Pseudomonas chlororaphis*, and *Pseudomonas aeruginosa* and of *Candida kefyr*, *krusei*, *tropicalis*, *parapsilosis*, and *glabrata*. Hierarchical clustering was applied using the Ward's algorithm. The original spectra in the spectral range of 2800-3000 and 1400-1500 cm^{-1} were used. The Pearson's momentum correlation coefficient as defined in [1] was used to calculate interspectral distances.

detection of changes of structural components as a function of e.g. time or temperature or growth. In principle the total amount of the spectral information provided by Raman spectra can be used to characterize the microbial species and strains under investigation.

For differentiation and identification purposes using large numbers of Raman spectra as a data basis, multivariate statistics have to be used as in the case of microbial FT-IR spectra. The combined use of factor and cluster analysis and the application of data pretreatment techniques, such as the calculation of derivatives and the use of a combination of smaller or larger regions of the spectrum that are most discriminative, may help to achieve optimal classifications. To give an example, Fig. 13 shows the dendrogram of a classification trial on a selection of Gram-positive and Gram-negative bacteria, and of different species and strains of Candida yeasts. As for FT-IR spectra on a similar collection of microorganisms (see Fig. 4 A), one obtains a distinct grouping just according to what a microbiologist would expect applying molecular genetic techniques, i.e. a grouping which nicely separates the Gram-positive bacteria away from the Gram-negative ones, and all bacteria quite distant from yeasts. At the same time all strains belonging to corresponding species and genus groups are grouped together.

CONCLUSIONS

The main advantages of FT-IR which constitute its attractiveness, are extreme rapidity compared to conventional techniques, uniform applicability to very diverse microorganisms, and a high specificity that allows differentiations even down to subspecies level. Last not least, the FT-IR technique requires only low amounts of consumables, it is computer compatible, and may thus promote that results and data bases are exchanged via data nets. The strength of the FT-IR technique is its ability to conduct epidemiological case studies and large screening experiments very quickly. Additional fields of applications are the detection of infection chains and the control of therapy, the maintenance of strain collections and the differentiation of microorganisms from the environment for which established systems are not yet available. In the food, water, and pharmaceutical industries FT-IR may also contribute to improve microbiological quality control. For the control of biotechnological processes it might also be an alternative or additional to already existing analytical tools.

The prospects of FT-IR microscopy for microbiological characterizations are very promising. This new technology may help to scale down the number of cells to less than 10^3, to analyse mixed cultures and to detect light-microscopic and spectroscopic features of microorganisms simultaneously. Perspectively, the development of a fully automated system of microbiological analysis that combines detection, enumeration, and identification of microorganisms can be addressed.

The sampling techniques, spectroscopic procedures and data evaluation strategies elaborated in the context of microbiological FT-IR can easily be carried over to characterize other microorganisms such as amoebae and viruses, and even plant or mammalian cells and tissues. Furthermore, the information density can be increased by the incorporation of additional "spectral traits" accessible from the near-infrared and far-infrared regions of the electromagnetic spectrum.

Drawing upon the knowledge obtained to date, the serial type of a dedicated instrument for FT-IR characterizations of microorganisms is now available from an FT-IR spectrometer producing company. A joint project was undertaken between 1986 and 1989 to develop the methodologies for the FT-IR analysis of intact microorganisms. This project, conducted under the auspices of the Ministry of Research and Technology and the Ministry of Health of Germany, was carried out by researchers at the Robert Koch-Institute in Berlin in collaboration with a manufacturer of FT-IR spectrometers. The prototype of a dedicated FT-IR system for microbial characterizations was then tested by several laboratories in a joint pilot study between 1992 and 1994.

FT-IR and FT-NIR Raman spectra of nearly identical samples of intact microorganisms can now be obtained with excellent reproducibilities. This makes it possible for the first time, to establish a combined use of infrared and Raman spectroscopy of complex biological samples, which constantly and notoriously give Raman spectra that are overwhelmed by the fluorescence background using laser excitation in the visible. Especially the exploitation of the complementarity of infrared and Raman spectroscopy may open new avenues for biomedical appliactions in the future.

A particularly intriguing challenge is to elaborate computer based pattern recognition techniques which give effective data reduction and optimal classification results from spectral data of biological samples collected in the near-, mid-, and far-infrared, or by combining different spectroscopic techniques such as FT-IR and FT-NIR spectroscopy. A neural network appraoch to this fundamental problem of information extraction from mixed spectral data bases is published by J. Schmitt in this conference proceedings.

ACKNOWLEDGMENTS

The excellent technical assistance of Angelika Brauer in preparing the manuscript is particularly acknowledged. Some of the FT-NIR Raman spectra of microorganisms have been measured by Thomas Löchte and Stefan Keller, Universität Essen, Germany. Their assistance in spectra interpretation and band assignment is gratefully acknowledged. The many persons who have contributed to this work throughout the years can be taken from the original literature and from the acknowledgments cited therein.

REFERENCES

1. Helm, D., Labischinski, H., Schallehn, G., Naumann, D., *J. Gen. Microbiol.* **137**, 69-79 (1991).
2. Naumann, D., Helm, D., Labischinski, H., *Nature* **351**, 81-82 (1991).
3. Naumann, D., Labischinski, H., Giesbrecht, P., In: *Instrumental Methods for Rapid Microbiological Analysis*, W. H. Nelson (ed.), New York, Deerfield Beach, Weinheim: VCH-Publisher, 1991, pp. 43-96.
4. Helm, D., Labischinski, H., Naumann, D., *J. Microbiol. Methods* **14**, 127-142 (1991).
5. Helm, D., and Naumann, D., *FEMS Microbiol. Lett.* **126**, 75-80 (1995).
6. Naumann, D., Schultz, C. P., Helm, D., In: *Infrared Spectroscopy of Biomolecules*, H. H. Mantsch and D. Chapman (eds.), New York: Wiley-Liss, Inc., 1996, pp. 279-310.
7. Malins, D. C., Polissar, N. L., Gunselman, S. J., *Proc. Natl. Acad. Sci USA* **94**, 259-264 (1997).
8. Eysel, H. H., Jackson, M., Nikulin, A., Somorjai, R. L., Thomson, G. T. D., Mantsch, H. H., *Biospectroscopy* **3**, 161-167 (1997).
9. Schultz, C. P., Liu, K.-Z., Johnston, J. B., Mantsch, H. H., *Leukemia Research* **20**, 649-655 (1996).
10. Choo, L.-P., Wetzel, D. L., Halliday, W. C., Jackson, M., LeVine, S. M., Mantsch, H. H., *Biophysical J.* **71**, 1672-1679 (1996).
11. Jackson, M., and Mantsch, H. H., In: *Infrared Spectroscopy of Biomolecules*, H. H. Mantsch and D. Chapman (eds.), New York: Wiley-Liss, Inc., 1996, pp. 311-340.
12. Jackson, M, and Mantsch, H. H., In: *Advances in Spectroscopy, Vol. 25: Biomedical Applications*, Clark, R. J. H., and Hester, R. E. (eds.), Chichester: John Wiley & Sons, Inc., 1996, pp. 185-215.
13. Choo, L.-P., Mansfield, J.R., Pizzi, N., Somorjai, R. L., Jackson, M., Halliday, W. C., Mantsch, H. H., *Biospectroscopy* **1**, 141-148 (1995).
14. Fabian, H., Jackson, M., Murphy, L., Watson, P. H., Fichtner, I., Mantsch, H. H., *Biospectroscopy* **1**, 37-46 (1995).
15. Schultz, C. P., Eysel, H. H., Mantsch, H. H., Jackson, M., *J. Phys. Chem.* **100**, 6845-6848 (1996).
16. Puppels, G. J., and Greve, J., In: *Advances in Spectroscopy, Vol. 25: Biomedical Applications*, Clark, R. J. H., and Hester, R. E. (eds.), Chichester: John Wiley & Sons, Inc., 1996, pp. 1-47.
17. Puppels, G. J., Garritsen, H. S. P., Segers-Nolten, G. M. J, de Mul, F. F. M., Greve, J., *Biophys. J.* **60**, 1046-1056 (1991).
18. Keller, S., Löchte, T., Dippel, B., Schrader, B., *Fresenius J. Anal. Chem.* **346**, 863 (1993).
19. Naumann, D., Keller, S., Helm, D., Schultz, C., Schrader, B., *J. Mol. Struct.* **347**, 399-406 (1995).
20. Manoharan, R., Ghiamati, E., Dalterio, R. A., Britton, K. A., Nelson, W. H., Sperry, J. F., *J. Microbiol. Meth.* **11**, 1-15 (1990).
21. Frank, C. J., McCreery, R. L., Redd, D. C. B., *Anal. Chem.* **67**, 777-783 (1995).
22. Redd, D. C. B., Feng, Z. C., Yue, K. T., Gansler, T. S., *Appl. Spectrosc.* **47**, 787 (1993).
23. Hawi, S. R., Campbell, W. B., Kajdacsy-Balla, A., Murphy, R., Adar, F., Nithipatikom, K., *Cancer Lett.* **110**, 35-40 (1996).
24. LeVine, S. M., and Wetzel, D. L. B., *Appl. Spectrosc. Rev.* **28**, 385-412 (1993).
25. LeVine, S. M., Wetzel, D. L., Eilert, A. J., *Int. J. Devl. Neuroscience* **12**, 275-288 (1994).
26. Estepa-Maurice, L., Hennequin, C., Marfisi, C., Bader, C., Lacour, B., Daudon, M., *Clin. Chem.* **105**, 576-582 (1996).
27. Lewis, E. N., Gorbach, A. M., Marcott, C., Levin, I. W., *Appl. Spectrosc.* **50**, 263-269 (1996).
28. Kidder, L. H., Kalasinsky, V. F., Luke, J. L., Levin, I. W., Lewis, E. N., *Nature Med.* **3**, 235-237 (1997).
29. Everitt, B. S., *Cluster Analysis*, Toronto: John Wiley, Inc., 1993.
30. Everitt, B. S., *Statistical Methods for Medical Investigations*, Toronto: John Wiley, Inc., 1994.
31. Manly, B. F. J., *Multivariate Statistical Methods. A Primer*, New York: Chapman & Hall, 1996.
32. McLachlan, G. J., *Discriminant Analysis and Statistical Pattern Recognition*, New York: Wiley & Sons, 1992.
33. Dayhoff, J. E., *Neuronal Network Architectures*, New York: Nostrand Reinhold, 1990.
34. Zupan, J., and Gasteiger, J., *Neuronal Networks for Chemists*, Weinheim: VCH, 1993.
35. Schrader, B., and Hoffmann, A., *Vibrational Spectroscopy* **1**, 239 (1991).
36. Inagaki, F., Tasumi, M., Miyazawa, T., *J. Raman Spectrosc.* **3**, 335 (1975).
37. Cao, A., Liquier, J., Taillandier, E., In: *Infrared and Raman Spectroscopy of Biomolecules*, Schrader, B. (ed.), Weinheim: Verlag Chemie, 1995, pp. 344.
38. Peticolas, W. L., Kubasek, W. L., Thomas, G. A., Tsuboi, M., In: *Biological Applications of Raman Spectroscopy*, Spiro, T. G. (ed.), New York, Chichester, Brisbane, Toronto, Singapure: John Wiley & Sons, 1987, pp. 81.

High Resolution Fourier Transform Infrared Spectroscopy of Short Lived Molecules.

Don McNaughton

Centre for High Resolution Spectroscopy and Optoelectronic Technology, Department of Chemistry Monash University, Clayton, Melbourne, Victoria 3168, Australia.

Fourier transform infrared spectroscopy at unapodized resolutions up to 0.0018 cm^{-1} has been used to probe the ro-vibrational energies, chemistry and structures of short lived molecular species. A number of methods for generation of transients, typically flow pyrolysis, photolysis and IR laser powered pyrolysis (IRLPP) have been coupled with a Bruker HR120 FTIR system and vibration-rotation spectra of molecules with lifetimes ranging from NCN ($t_{1/2} \approx 10^{-3}$ sec) through C_3O ($t_{1/2}$<1 sec), propadienone ($t_{1/2} \approx 10$ sec) and vinylamine($t_{1/2} \approx 10$ min) to difluoroacetylene ($t_{1/2} \approx 30$ min.) and chlorophosphaethyne ($t_{1/2} \approx 30$ min.) have been recorded. For asymmetric molecules in particular, assignment of a spectral band is often complicated because just a single band can contain thousands of resolved vibration-rotation lines. Computer assisted assignment of the spectra of species ranging from simple linear molecules, through near symmetric tops to highly asymmetric tops is discussed. In addition to using computer aided assignment to deal with the plethora of lines an experimental alternative has been used to reduce the number of lines. We have built a system based on a supersonic jet expansion coupled to our Bruker HR120 spectrometer and have used this to cool species down to ½ a rotational temperature of typically 20 - 60K. The prospects of using this system to study the spectra of transient molecules are discussed.

INTRODUCTION

High resolution gas phase spectroscopy provides rigorous experimental data on molecular structures, intramolecular force constants and molecular energy levels that together advance the understanding of the theoretical bases of molecular structure, chemical interactions and molecular dynamics. These experimental techniques are also of particular importance in the investigation of the highly reactive transient species that are mechanistically important reaction intermediates as well as possibly being molecules of astrophysical and/or atmospheric interest. Due to their inherent sensitivity, microwave spectroscopy and laser based spectroscopic techniques such as infrared diode laser systems have a long and highly successful association with the spectroscopic study and characterization of transient species. For the study of rotationally resolved vibrational spectra the laser based techniques suffer from very narrow band widths which make searching for new species a time consuming option. Since the introduction of commercial high resolution FTIR spectroscopy in the mid 80's it has provided an alternative technique capable of obtaining vibration-rotation spectra over a wide bandwidth but with a much reduced sensitivity when compared with microwave and laser techniques. For molecules with no dipole moment or a dipole moment close to zero, microwave spectroscopy is not generally successful and alternative techniques are required for structural determination and for exploration of the chemistry of such species. High resolution FTIR spectroscopy is thus capable of making a large contribution to the field of short lived molecules.

The work described in this paper has been carried out over the last 6 years in the department of chemistry at Monash university.

GENERAL EXPERIMENTAL CONSIDERATIONS

All spectra were recorded on a Bruker HR120 FTIR spectrometer with a maximum optical path difference of 5.4 m and a maximum unapodized resolution of *ca* 0.0018cm^{-1}. The instrument is equipped with a KBr beamsplitter and a MCT/InSb liquid nitrogen cooled sandwich detector for the mid IR region and 6

CP430, *Fourier Transform Spectroscopy:* 11th International Conference
edited by J.A. de Haseth

mylar beamsplitters and 2 helium cooled bolometers for the far-infrared region. The range of the instrument is thus 25 - 4000 cm^{-1}. Due to the large number of data points in a high resolution spectrum and the limitations resulting from the availability of only specific non aliased bandwidths, not all of this region can be covered at the highest resolution. In order to attain the highest resolution it is necessary to insert optical filters to remove the possibility of aliasing and also to reduce the interferogram to a workable size for Fourier transformation. A large number of expensive interference filters are required to cover the whole spectral region. Furthermore cold filters are required in the far IR to ensure that the sensitivity of the bolometer is not compromised by the heat produced from room temperature filters.

A further experimental factor that can affect resolution is the time required to obtain a spectrum. At high resolution a variable aperture must be closed down to ensure that the interference pattern focussed on the detector does not contain modulated radiation from spectral elements outside the desired resolution. This is wavelength dependent and to some extent negates the throughput advantage of FTIR. To obtain a reasonable S/N at the highest resolution usually requires anything from 20 to 100 coadded scans, depending on the type of detector and the spectral region. In addition, for the slow response of the Ge bolometer, below 350 cm^{-1} the scan speed must be greatly reduced. Typically collection times of 2 to 8 hours are required and for transient species where one is continually generating the species from a specifically designed precursor molecule, the time and cost of making precursors can be prohibitive. Most generation methods also produce large quantities of deposits that can quickly reduce the throughput of radiation, especially if the species are in contact with transfer mirrors as well as windows. At lower resolutions the scan times are much shorter and the S/N is improved and so spectra are often recorded at resolutions from 0.0019 cm^{-1} down to 0.01 cm^{-1}.

To retain the practical resolution at a value as low as possible, a "weak" apodization function (usually a 4 point function close to the boxcar) is used to minimize broadening, and sample pressures are kept well below one torr to ensure pressure broadening does not reduce the resolution. For the resultant narrow spectral lines the spectrum must also be zerofilled to eliminate inaccuracies in line centres emanating from the picket fence effect. Spectra are recorded with a zero fill factor of 2 (ZFF=2) and post zero filled to give a total ZFF=8. With a measurement accuracy of better than 0.0001 cm^{-1} it is also necessary to have a high quality temperature stabilized reference laser, to calibrate the instrument regularly and to preferentially calibrate each spectrum using well known lines of calibration species. For work on transients the latter is not usually a problem as a large number of additional simple gas species are often generated in the process of making the species of interest.

In order to increase the S/N by taking advantage of Beer's law, most experiments are carried out either using sample cells based on multiple passes of the probe infrared beam through the sample, or in cells up to 3 metres in length. The former solution generates problems with deposition on the mirrors of the White type cell with subsequent loss of signal throughput, whilst the latter requires a much larger throughput of sample gases. For a more expansive treatment of some of the above considerations see reference 1.

POST EXPERIMENTAL CONSIDERATIONS

Even for a simple linear triatomic molecule the spectral assignment can be long and tedious. This is mainly due to the many hundreds of lines that result from the hot bands arising from highly populated vibrational states and from bands due to the presence of isotopomers. For larger molecules, which have more complicated band systems and where there may be a greater number of populated vibrational states, there are often thousands of lines and the actual spectra show little in the way of patterns with which to begin an analysis. In our work, even for highly asymmetric molecules, we use a computer assisted assignment technique based on the Loomis-Wood (2) diagram for the initial assignment of all our spectra. Our program, based on an initial PC version obtained from the authors at Geissen University (3) uses the graphical power of a Macintosh system to aid in the spectral analysis.

Line centre wavenumber values and intensities are obtained by running the peakfind routine on the Bruker OPUS system and are then transferred to a Macintosh computer. The MacLoomis program is based on the Hamiltonian for a linear molecule from which is derived equation 1, an expression defining each individual vibration-rotation line in the P and R branches of a linear molecule.

$$\begin{aligned}\bar{\nu} = \bar{\nu}_0 &+ (B'' + B')m - (B'' - D'' - B' + D')m^2 \\ &- (2D'' - H'' + 2D' - H')m^3 \\ &+ (D'' - 3H'' - D' + 3H')m^4 \\ &+ 3(H'' + H')m^5 - (H'' - H')m^6 \qquad (1)\end{aligned}$$

For P branches $m = -J''$ and for R branches $m = J'' + 1$. The expression is thus a simple power series in m with the coefficients being combinations of the rotational and centrifugal distortion constants. To begin an analysis the operator first estimates the centre wavenumber of the band and an average B rotational constant (and thus the zero and first order coefficients of equation 1) by examination of the spectrum.

Initially the program replots the entire spectrum on the computer screen in horizontal segments $2B$ wide, vertically one under the other and centred on the chosen band centre with each peak represented as a triangle whose size depends on the peak intensity. For a linear or symmetric top molecule any spectral lines emanating from a molecule with a rotational spacing ($2B$) close to that chosen for the plot, will appear as a near vertical series of triangles or as a curved series of triangles, as in fig. 2. A series of lines is then chosen and assigned an m value by clicking with the mouse or dragging down the series. The chosen lines are then fitted to the polynomial expansion of equation 1 and the spectrum replotted using a chosen order of the polynomial. The centre and width of each horizontal segment for plotting is thus dictated by the coefficients of the polynomial. The program, which handles up to 20,000 lines and up to 100 chosen series, has many attributes eg. manipulating the screen, colour coding series, picking and unpicking lines, moving centre wavenumbers, weighting peaks, using spin statistics, handling Q branches, extrapolating series, hiding chosen lines, calculating combination differences, fitting combination differences and scaling of intensities.

The MacLoomis program has been used by us not only to assign spectra of linear and symmetric top species, where equation 1 is correct for each subband treated, but also to assign the spectra of species ranging from near prolate asymmetric tops to highly asymmetric tops eg. FNO (4), a number of HFC's (5,6), CFC's (7), HCFC's (8,9) and BCFC's (10), vinylamine (11) and thioketene (12). For asymmetric tops near the prolate limit, where the asymmetry parameter K_b is close to -1, each subband for a particular value of the K_a quantum number has lines spaced by *ca* $(B + C)$ so the subbands can be easily fitted to the polynomial expansion. As the asymmetry parameter decreases, the polynomial remains appropriate at high values of J and as one approaches the oblate limit the spacings reappear as *ca* $2C$. Consequently we have not yet found a molecule where an initial assignment of the spectrum cannot be made. For asymmetric species the chosen series are saved and manipulated into a file with the J, K_a, K_c assignments.

The data is then fitted using the appropriate Hamiltonian, normally Watson's S or A- reduced Hamiltonian (13), with the inclusion of interaction parameters to handle resonances. When there are lines that are not easily assigned using MacLoomis, line positions are predicted from the resultant constants and the assignment continued in the more traditional bootstrapping way. Where possible all assignments are checked using ground state combination differences.

The ν_3 band of N_2O in fig. 1, can be used to help

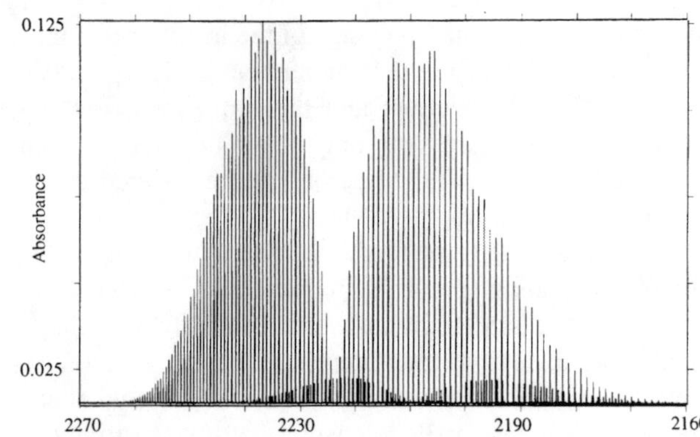

FIGURE 1. The high resolution spectrum (0.002cm^{-1}) of the ν_3 band of N_2O.

understand the way in which MacLoomis operates. This band consists of the cold band (Σ–Σ, a single intense P and R series), together with a number of weaker bands due to transitions from $\nu_2 = 1$ (Π–Σ, a doubled P and R series with a weak Q branch), $\nu_2 = 2$ (Σ–Σ, Δ–Δ, a tripled P and R series) and $\nu_3 = 1$ (Σ–Σ, a single P and R series). Hence there are seven easily discernible series.

If one begins an assignment by choosing the approximate centre wavenumber of the cold band as 2210 cm^{-1} and an approximate B value of 0.5 cm^{-1} (cf. B_0 = 0.419 cm^{-1}), the MacLoomis plot appears as in fig. 2. A number of curved series of lines are immediately apparent, even with the rough rotational. The related lines that appear as a single curved series in the lower left quadrant of fig. 2 is picked using the mouse. Upon a 4th order polynomial fit the lines of the series that have been chosen appear as a near vertical set of lines with the unpicked lines of the series curving off at higher and lower wavenumber. Upon picking all the lines in the series, moving the centre wavenumber to the central gap, and fitting to 4th order, the MacLoomis plot appears as in fig. 3.

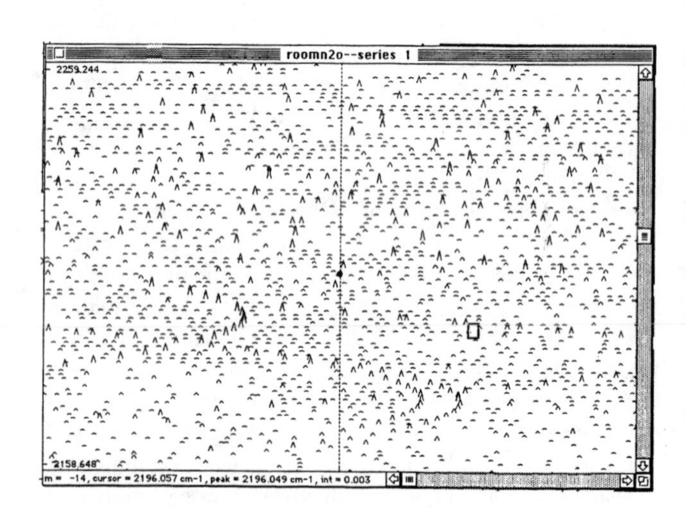

FIGURE 2. Initial MacLoomis plot of N_2O, ν_3.

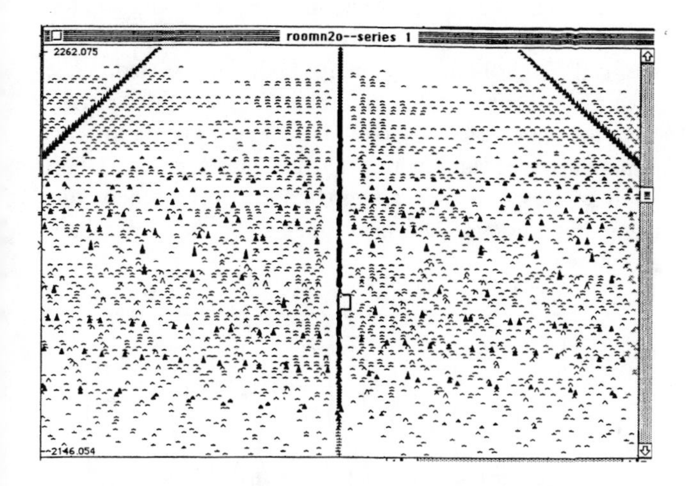

FIGURE 3. MacLoomis plot of N_2O, ν_3 after picking the main series and fitting to 4th order.

The series is now a vertical line and other series of lines can be seen diagonally across the screen. The series in the outer top quadrants are "ghosts" of the main series that appear due to wrap around effects because $B">B'$. In the spectrum this effect is observed as a closing up of the rotational spacing at high J values in the R branch.

After beginning a new series (taking the guesses for the coefficients from the cold band fit) and picking the lines of one of the doubled diagonal series with the mouse, the result of a 4th order polynomial fit is displayed in fig. 4. The vertical series is accompanied at an angle by the other part of the doubled series due to $\nu_2 = 1$ and the cold band now makes up a number of diagonal lines across the screen. The displays are normally colour coded and so previously fitted series are easily recognized on the screen. In fig. 4 the hot band, $\nu_3 = 1$, can now be seen as an additional curved set of low intensity peaks with its centre above that of the doubled $\nu_2 = 1$ series. For weak bands with no Q branch the assignment of a unique J numbering is difficult and assignments are usually made with the assistance of spin statistics and predictions using

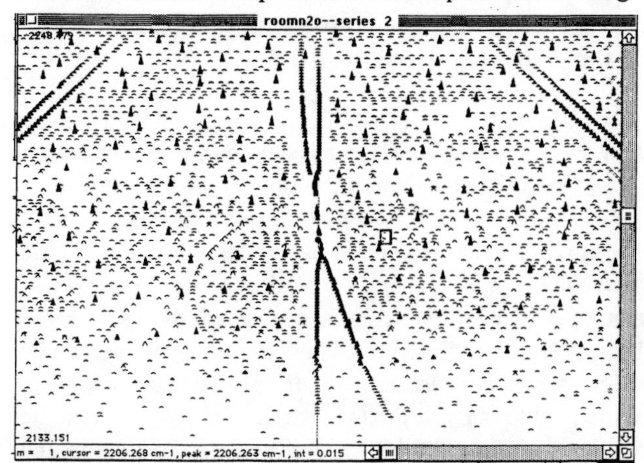

FIGURE 4. MacLoomis plot of N_2O, ν_3 with the doubled hot band series chosen and fitted to 4th order.

ab initio techniques or well known relationships between molecular parameter.

In this work all *ab initio* calculations were carried out using the Gaussian suite of programs (Gaussian 94 (14) or its predecessors) installed on VAX , IBM RISC6000 systems or DEC workstations.

GENERATION OF SPECIES USING FLOW PYROLYSIS AND PHOTOLYSIS

For species with half lives of the order of fractions of a second or more thermal or photolytic breakdown of a gas stream containing specifically designed precursor molecules has been successful in microwave and laser based experiments. In our experiments we have used this method to generate and observe high resolution spectra of the transient species: C_3O (15) , N_2S (16), ClC≡P (17), FC≡CF (18), $H_2C{=}S$ (19), $H_2C{=}C{=}S$ (12), $H_2C{=}CHNH_2$ (11), OCHCHS (19) and recently ClB=S and FB=S. It is also a useful method of generating small quantities of dangerous and difficult to synthesise species for spectral analysis and we have generated a number of explosive acetylenic species in this fashion eg. HC_6H (20), HC_4H(21), ClC≡CH (22), ClC≡CCl (23) and ClC_4H (15). For most pyrolysis experiments the vapour of a gaseous, liquid or solid precursor molecule is introduced by an appropriate method into a quartz tube that is externally heated by a furnace or a resistively heated wire wrapped around the tube. Figure 5 shows a typical flow pyrolysis setup for a volatile liquid sample. For solids a second furnace or heating element is included to vapourize the sample into the hot zone.

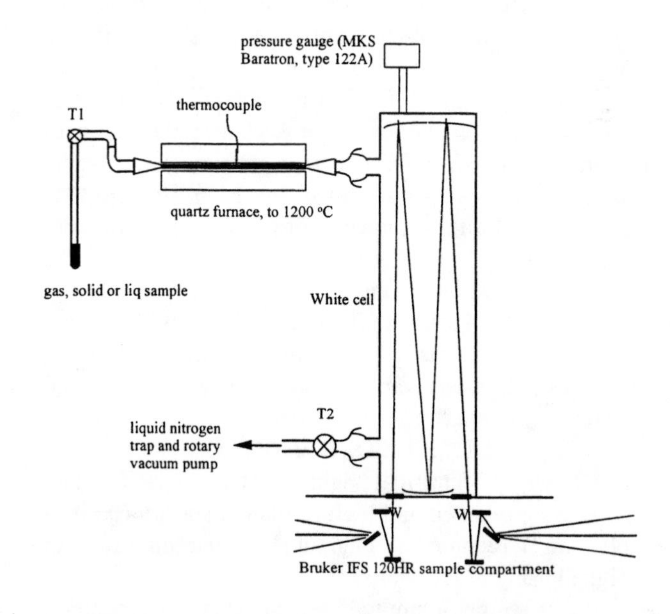

FIGURE 5. Pyrolyis flow system for a volatile liquid precursor.

The thermally degraded gas is continually pumped through a White cell and into a liquid nitrogen trap for later disposal. The production of the transient of interest and the quality of the final spectrum is affected by the following parameters, all of which must be systematically varied to maximize the yield and S/N.

a. The length and width of the quartz tube.
b. The temperature of the furnace.
c. The flow rate of the system.
d. The distance between the furnace and the White cell.
e. The surface of the absorption cell.

Tricarbon monoxide

Tricarbon monoxide is an interstellar species with a half life of <1 second that has been studied by microwave, millimetre-wave and matrix isolation IR spectroscopy using flow pyrolysis (24 - 29). After the initial preparation using specifically designed precursor molecules, fumaroyl dichloride was found to be the most efficient and effective starting material. In order to further characterize this interesting species and test the limits of flow pyrolysis in high resolution FTIR we have generated C_3O and recorded the band due to the cumulenone stretching mode.

Our initial experiments were carried out in a White cell set at a path length of 12 metres, at low resolution to conserve precursor, and at the pyrolysis temperature that achieved a maximum signal for C_3O in the microwave work ie. 1100°C . At this temperature a large number of bands from other species swamped the region of interest and no lines of C_3O were discernible. Even at slightly lower temperatures large numbers of broad features were observed. In order to ascertain if C_3O was being produced a number of spectra at a range of temperatures below 1100°C were recorded under fast flow pyrolysis conditions. After each run the cell was isolated and spectra of the species remaining after some 1,2 and 5 minutes were recorded. No features due to C_3O were observed after subtraction of the static spectra from the flow spectra. In order to determine the nature of the other species evolved in the experiment, two flow through traps were placed after the furnace and held at -196°C and -78°C. The contents of these traps were fractionally distilled by applying varying low temperature baths and low resolution spectra recorded. Spectra due to ClC≡CH, ClC≡CCl, ClC_4H, HCl, CO, CO_2 , HC≡CH and possibly ClC_4Cl were identified. Subsequently ClC≡CH, and ClC_4H were purified and high resolution spectra (0.002 - 0.005 cm^{-1}) recorded through the far and mid IR and assigned using MacLoomis.

With the failure to observe C_3O at low resolution the same procedure was carried out at a resolution of $0.01 cm^{-1}$. The upper trace in fig. 6 shows a spectrum of the flow pyrolysis at the optimum pyrolysis temperature, whilst the middle trace shows the spectrum after isolating the cell from both the pump and inlet system. The spectrum obtained after subtraction of the lower trace from the upper trace is shown at the bottom of fig. 6. This remaining spectrum, which must emanate from a short lived species, was then assignable to tricarbon monoxide using MacLoomis. The assignments for the major lines are shown.

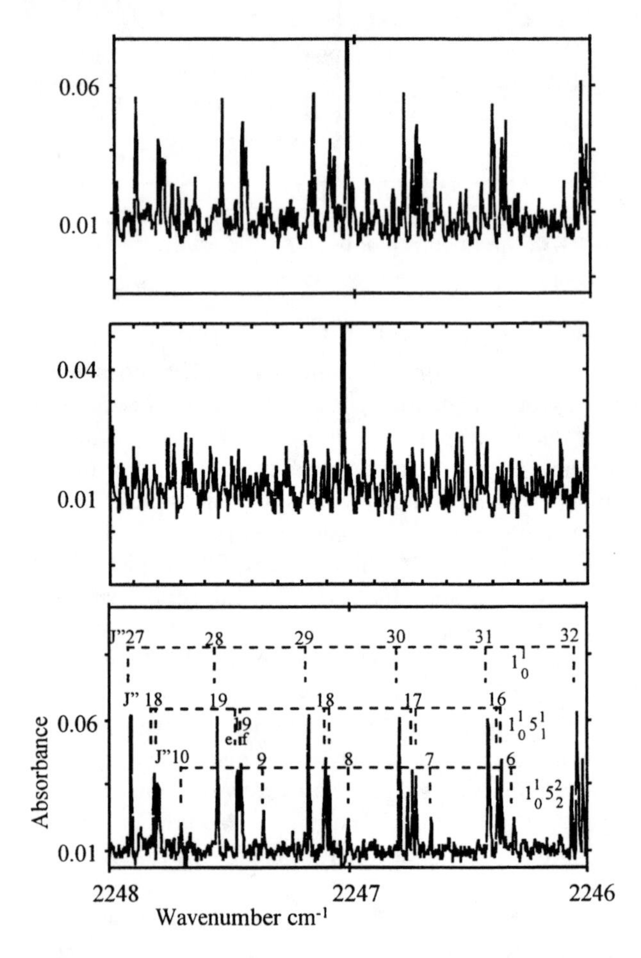

FIGURE 6. Spectra of pyrolysis products of fumaroyl choride (720°C) under flow conditions (top), static conditions (middle) and after subtraction (lower).

In order to determine the optimum temperature of production, pyrolysis runs at low and high resolution were carried out at a range of temperatures between 400 - 1100°C . During these runs a new spectrum, assignable to propenoyl chloride, appeared at low temperatures. Figure 7 contains a graphical representation of the evolution of each of the major species as measured by its infrared absorption intensity at increasing pyrolysis temperatures. The absorbance measurements of each species have been normalized to unity at the maximum of absorption and the individual points have been fitted to a polynomial expansion. Absorbance readings were kept below 0.1 in

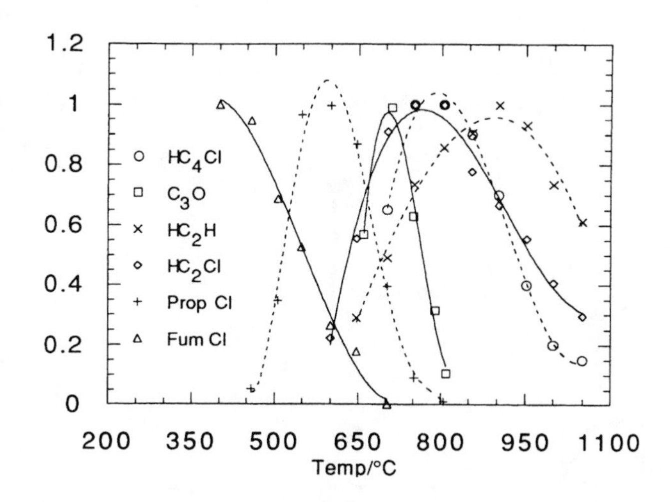

FIGURE 7. Production of pyrolysis products from fumaroyl chloride as a function of temperature. Absorbance maxima for all species are normalized to one.

order to ensure a linear relationship between absorbance and concentration.

The optimum temperature for production of C_3O is at 720°C, some 380°C below that of the microwave study, and it can be seen that over a very small range of temperature the spectrum is relatively free of most contaminants. At this temperature only small absorbances due to chloroacetylene obscure the spectrum as seen in fig. 6. Using spectra recorded under these conditions some 8 cold and hot subbands of C_3O were assigned, leading to molecular constants for a number of vibrational states in addition to the ground state. Table 1 below shows a comparison of the constants obtained for the ground states of the major species generated in the experiment, with those known from microwave studies. Although the microwave lines are some 4 orders of magnitude more accurately determined than the infrared lines, the large number of lines available results in rotational constants that approach those from the microwave study in accuracy. For the molecules where the microwave constants are derived from a very small number of lines the infrared derived constants are better determined.

From the range of products produced at a variety of temperatures it is possible to infer the process of the breakdown of the fumaroyl chloride. The two schemes described in scheme 1 were postulated by Brown *et al* (27) as possible pathways leading to tricarbon monoxide. From this work the lower of the two pathways is the correct one. The initial low temperature step was confirmed by trapping out essentially pure propenoyl chloride after pyrolysing fumaroyl chloride at 450°C. Pyrolysis of this propenoyl chloride leads to C_3O in a similar yield to that obtained from pyrolysis of fumaroyl chloride at similar temperatures.

Table 1 Comparison of Rotational constants obtained from infrared spectra with those obtained from microwave spectra. (MHz)

Species	B_{IR}	$B_{mw}{}^a$
C_3O	4810.913(20)	4810.889(2)
$HC{\equiv}C^{35}Cl$	5684.2065(30)	5684.216(5)
$HC{\equiv}C^{37}Cl$	5572.3600(36)	5572.355(5)
$HC_4{}^{35}Cl$	1373.112(7)	1373.10(1)
$HC_4{}^{37}Cl$	1341.973(19)	1342.04(4)

a. Taken from references 26, 30, 31. Numbers in parentheses are one standard deviation taken from the least squares fit

Attempts to generate C_3O in larger yield in the White cell and so record the weaker fundamentals were not successful. The ν_2 band of C_3O is predicted by *ab initio* calculations to be only some 10 times weaker than ν_1. Further experiments were carried out using a 3 metre gas cell coupled to the side port of the Bruker HR120 and evacuated by a $20M^3$ pump as described later in fig. 19. Despite the 25% reduction in path length and a reduction in the gas pressure to some 10% of the that of the White cell experiments, signal absorbances due to C_3O were essentially the same as those in the White cell experiments. Given this result, the large value of the dipole moment derivative for the ν_1 stretch predicted by the *ab initio* calculations, and the high sensitivity of the InSb detector, molecules with a lifetime shorter than that of C_3O are unlikely to be observed in this type of flow through experiment. For such short lived molecules attempts to work at low resolution are usually futile and even initial searching must be carried out at high resolution despite the long times and large amounts of precursor molecules required.

O=C=C=C=C=O
-HCl
-CO
ClOC(H)C=C(H)COCl
-HCl
O=C=C=C(H)COCl
lC=C=C=O
-CO
-HCl
HC≡C—COCl ← [lC=C(H)COCl]

Scheme 1. Proposed Pyrolytic breakdown of fumaroyl chloride via propenoyl chloride to give tricarbon monoxide (27). The lower of the two schemes is supported by our work.

Chlorophosphaethyne

In the late 1970's and early 1980's a number of short lived phosphaethyne, X-C≡P, species (X=CH_3, F, C≡N, C_3N, Ph) were generated and their microwave spectra recorded in the laboratory of H. W. Kroto (32) at the university of Sussex. This work opened up a new field of chemistry based on multiple bonded phosphorus species. Despite numerous attempts to observe the microwave spectrum of ClC≡P, its spectrum remained elusive, due to the very low dipole moment (estimated by *ab initio* calculations to be <0.1 D). Flow pyrolysis of trichloromethylphosphonous dichloride, Cl_3PCl_2, carried out at Sussex at 0.01cm^{-1} resolution using a Bomem DA-003 spectrometer, resulted in a weak spectrum that could not be analysed using traditional methods due to its weakness and the presence of perturbations. This spectrum was eventually understood using MacLoomis at Monash University to elucidate the strong perturbation that allowed a unique assignment (see below). The resultant ground state constants for both chlorine isotopomers were used in a successful search for the microwave spectrum. This microwave study subsequently determined the dipole moment to be 0.056(2)D (33) We have subsequently generated and studied ClC≡P in our laboratory.

Cl_3PCl_2, which can be made efficiently but tediously using very expensive starting materials, was made by us in small quantity by heating white phosphorus and CCl_4 in an autoclave. The Cl_3PCl_2 was obtained by vacuum distillation of the resultant mixture of PCl_3, red phosphorus and $POCl_3$. Gaseous Cl_3PCl_2 was flowed over granulated zinc, loosely packed in a heated quartz tube, and this resulted in spectra that could be assigned to ClC≡P, Cl_2CO, HC≡CCl, CCl_4, PCl_3 and HCl. It was found that the products could be cryogenically trapped at -196°C and separated by fractional distillation and that unlike many transient species ClC≡P could be successfully revapourized. This resulted in essentially contaminant free ClC≡P, which has a half life of *ca* 30min in our quartz White cell. High S/N spectra at resolutions between 0.003 and 0.005 cm^{-1} were obtained for all the fundamentals and a number of combination, overtone and difference bands. The spectrum of the ν_3 band is shown in fig. 8.

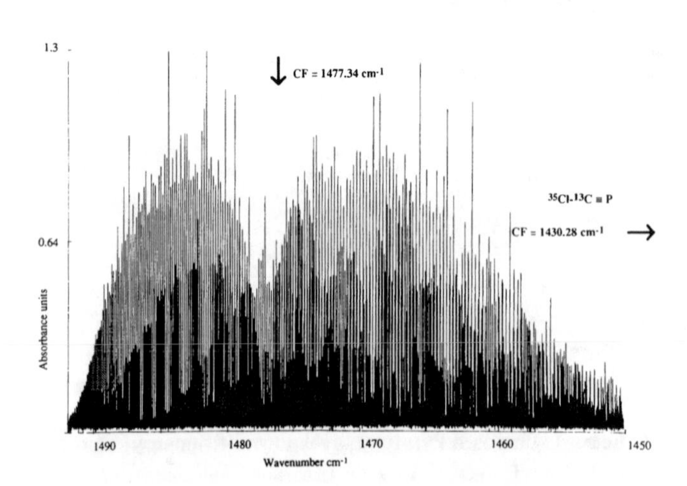

FIGURE 8. High resolution spectrum of the ν_3 band of ClC≡P at 0.0035cm^{-1}.

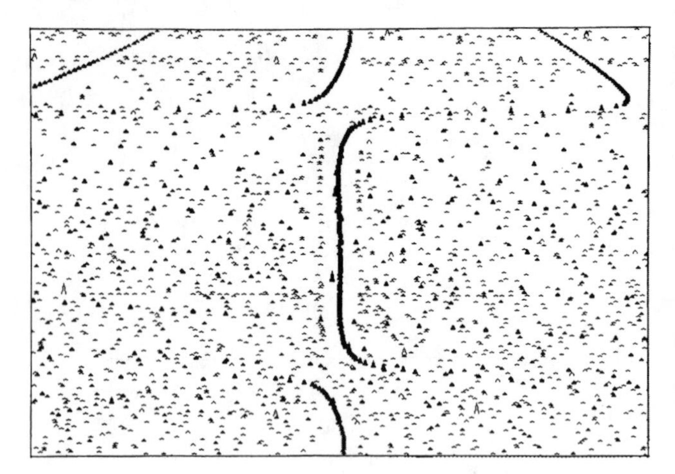

FIGURE 9. MacLoomis plot of the ν_3 band of ClC≡P with the cold band chosen and fitted to a 4th order polynomial.

The spectrum in fig. 8 is still difficult to assign due to the large number of hot bands, the presence of the two chlorine isotopes and a strong perturbation in the cold band. Some 13 subbands were uniquely assigned in this spectrum using MacLoomis. The MacLoomis plot is shown above in figure 9 with the cold band fitted to a 4th order polynomial.

The perturbation, which allowed us to derive our initial unique assignment is easily apparent in fig. 9. Such perturbations allow one to trial possible assignments in the MacLoomis program and fit ground state combination differences. Only when the assignment is correct will a good least squares fit to the combination differences be obtained. In the MacLoomis plot the effects of the strong apodization function used for high resolution spectra can be noticed. The weak subbands adjacent and parallel to the cold band are due to the residual feet in each strong line. For such semi-stable molecules that can be isolated and concentrated, the S/N of strong bands is high enough, even in the region where the less sensitive MCT detector is used, to observe and assign the spectra due to ^{13}C isotopomers, present in 1% natural abundance. In the case of ClC≡P the isotope shift for the ν_3, (C≡P str) mode shifts the centre of the $^{35}Cl^{13}C≡P$ band some 47 cm^{-1} lower and so removes the spectrum from the region cluttered with weak hot bands. Even in this region there are still a number of weak hot bands overlapping the weak lines due to the spectrum of the ^{13}C isotopomer.

In order to select the correct band and to achieve the correct assignment when a number of the low J components of the band are missing, it is necessary to obtain accurate estimates for the band centre and the expected B value. The band centre can be accurately predicted by *ab initio* techniques. Although the HF/6-31G* calculated wavenumber for ν_2 for the main isotopomer is predicted 200 cm^{-1} higher than that observed, the same calculation errors are included in the estimate of all isotopomers. The wavenumber shift upon isotopic substitution is thus much more accurately predicted than the fundamental wavenumber itself. The predicted shift is 47.4 cm^{-1} compared with the eventual experimental value of 47.06 cm^{-1}. An accurate estimate of the B_0 value can be obtained by calculating a B_0 value from those measured for the major isotopomers . In this case the predicted B_0 value is only 3×10^{-6} cm^{-1} higher than that obtained in the eventual fit of the assigned data. This technique has also been used successfully by us to obtain assignments for ^{13}C species in chloroacetylene (22) and dichloroacetylene (23).

A large number of constants can be calculated from the many thousands of assigned lines and in particular accurate B_V values can be obtained for all modes of both chlorine isotopomers after the effects of Fermi resonances have been taken into account. From these values accurate geometric parameters can be obtained (17). The equilibrium bond lengths are calculated to be:

$$r_{C\equiv P} = 155.27(29)\ \text{pm}$$
$$r_{C\text{-}Cl} = 163.41(27)\ \text{pm}$$

These values, with the number in parentheses being one standard deviation from the least squares fit, show the typical accuracy with which bond lengths can be calculated from high resolution infrared data. The standard deviations in the parameters are some 2 to 10 times larger than those typically obtained from microwave experiments on similar molecules even though the spectral resolution is some 3 or 4 orders of magnitude lower than that obtained in a microwave study. The large numbers of lines used to obtain the constants results in an accuracy for the rotational constants close to that obtained in a microwave study.

Cumulenones and Large Amplitude Motions

The generation and study by microwave spectroscopy of propadienone, H_2C_3O, a transient molecule with a half life of only a few seconds (34,35)

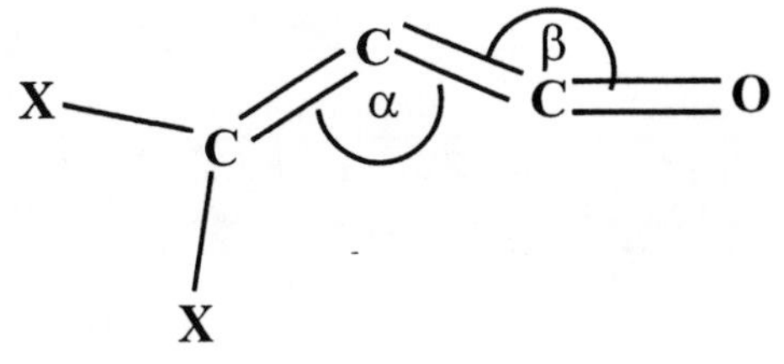

HF/6-31G** X = Cl; $\alpha = 151.1$ deg, $\beta = 173.4$ deg

HF/6-31G** X = F; $\alpha = 132.7$ deg, $\beta = 132.1$ deg

MP3/6-31G** X = H; $\alpha = 142.2$ deg, $\beta = 176.5$ deg

FIGURE 10. *Ab initio* predicted geometries of cumulenones.

proved to be a milestone in structural chemistry. In defiance of the predictions of what were considered to be high quality *ab initio* calculations using large basis sets (HF/6-31G**) (36), the molecule was proven to be kinked (37) as in the structure given in fig.10.

It was subsequently found that calculations incorporating electron correlation (MP3) and large basis sets could correctly predict the kinked structure of propadienone in fig. 10 (38). Subsequently *ab initio* calculations at the Hartree Fock level of theory predicted the kinked structures for difluoropropadienone and dichloropropadienone given in fig. 10, and we undertook a microwave study aimed at generating these species and determining their structural parameters.

Difluoromaleic anhydride and 3-bromo-3,3 - difluoropropanoic trifluoracetic acid anhydride, prepared as possible precursor molecules for the production of F_2C_3O, were pyrolysed at a range of temperatures but yielded no microwave lines assignable to F_2C_3O. In order to elucidate the chemistry and determine the optimum production conditions for difluoropropadienone the pyrolysis products were examined using FTIR spectroscopy. Pyrolysis of both species yielded low resolution spectra that were assignable to F_2C_3O by comparison with *ab initio* calculations and with the previous matrix isolation results (39,40). Although high resolution spectra of the breakdown products showed only a small amount of discernible fine structure, the band profiles could be simulated successfully by modelling with the rotational constants predicted from the *ab initio* calculations. The low resolution assignments of difluoropropadienone are shown in table 2. During the course of our work, which determined the presence of difluoropropadienone and the optimum conditions for its production, the group of Harmony, generated the difluoro species and confirmed its kinked nature (41) by microwave spectroscopy.

Table 2. Experimental and *ab initio* values for X_2C_3O species.

Difluoropropadienone		Difluorocyclopropenone		Dichloropropadienone		Dichlorocyclopropenone	
HF/6-31g* scaled cm^{-1}(KM mol^{-1})	Exp. cm^{-1}(rel.int)	HF/6-31g* scaled cm^{-1}(KM mol^{-1})	Exp. cm^{-1}(rel.int) cm^{-1}	HF/6-31g* scaled cm^{-1}(KM mol^{-1})	Exp. cm^{-1}	HF/6-31g* scaled cm^{-1}(KM mol^{-1})	Exp. cm^{-1}(rel.int)
94(2).		220(0.7)		40(2)		141(0.5)	
137(0.1)	258(0.6)		132(0.5)		164(4)		
479(6.4)	264(0.2)		277(0.7)		208(1)		
523(21)		624(0.0)		385(7)		480(10)	
614(41)		644(0.7)		487(40)		523(0)	
641(23)		709(0.7)		516(1)		640(20)	
693(85)	720(2.2)	695.0(5)		550(72)		654(11)	
811(98)	824(10)	876(3.7)		651(51)		689(35)	
1222(218)	1194(5)	1090(2.8)	1066.4(10)	869(330)		1010(0)	
1298(374)	1257(10)	1366(28)	1292(70)	1043(91)		1110(198)	
1770(989)	1764(55)	1803(49)	1788.9(100)	1773(275)		1681(332)	1631(50)
2149(1920)	2169(100)	1952(12)	1920.4(50)	2168(2817)	2148	1916(736)	1908(100)

In our work, where species with $\mu = 0$ are observable, we also obtained spectra, with intensities that diminished by 50% over 30 min., that could be assigned to difluoroethyne, FC≡CF. The bands were assigned to FC≡CF by observation of the alternating intensities of the J components (due to spin statistics arising from the equivalent fluorine atoms) and by comparison with the predicted *ab initio* wavenumber values and B value. The alternation in intensities can be seen in fig. 11. In an attempt to minimize the many by products and increase the yield of difluoroethyne the alternative breakdown method of flow through photolysis was investigated. This resulted in a cleaner decomposition and a good high resolution spectrum of FC≡CF (18) but yielded a different initial breakdown product in difluorocyclopropenone. The identity of this molecule was shown by comparison with *ab initio* predictions and with matrix isolation spectra (40) and by modelling the band profiles of the molecule using *ab initio* predicted rotational constants and a prediction/simulation program. Table 2 shows the results obtained and the reaction scheme in scheme 2 shows the two alternative breakdown routes from difluoromaleic anhydride.

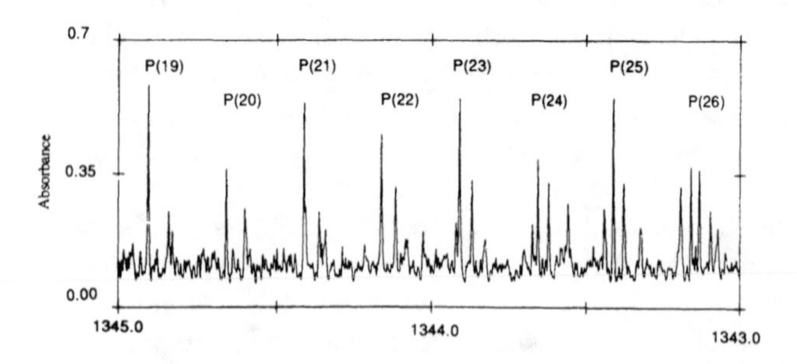

FIGURE 11. Spectrum of FC≡CF, ν_3 showing the alternation in intensities due to the effects of nuclear spin statistics.

Using dichloromaleic anhydride we have also generated by essentially the same pyrolysis and photolysis routes molecules whose spectra are consistent with dichloropropadienone, dichlorocylcopropenone and dichloroethyne. In this case no matrix IR results are available and apart from ClC≡CCl, where a full high resolution analysis was carried out (23), assignments were made by modelling band profiles and by comparison with *ab initio* predictions (see table 2). The lifetimes of the cumulenone (a few seconds) and the cyclopropenone (<1 hour) were also consistent with the expected lifetimes.

The spectrum of the $\nu_1 + \nu_3$ $(\Sigma\text{-}\Sigma)$ combination band of dichloroethyne is shown in fig. 12. The band at low resolution shows an unusual band profile with what appear to be prominent Q branch heads. At high resolution and with the use of MacLoomis, the band heads proved to be in the R branch of all the subbands.

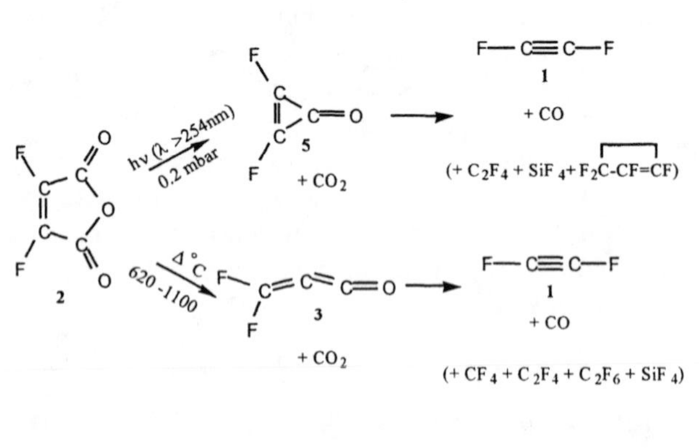

Scheme 2. Pyrolytic and photolytic breakdown of difluoromaleic anhydride.

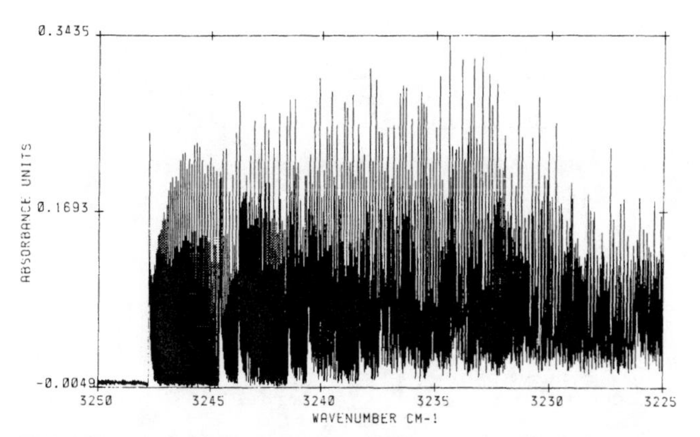

FIGURE 12. Infrared spectrum of the $\nu_1 + \nu_3$ *(Σ–Σ)* band of dichloroethyne at 0.0068 cm^{-1} resolution.

This type of band head, which is often observed in electronic spectra where the geometry change from one electronic state to another can be extremely large, is unusual in an infrared spectrum. The combination band consists of two stretching modes where the change in rotational constant between each state and the ground state is extremely large. The 2nd order coefficient in equation 1 thus is large. This band head is easily seen in the MacLoomis plot of the band in fig. 13 where the extremely large 2nd order coefficient leads to many "ghost subbands that converge in the MacLoomis plot. The large change in geometry for this combination mode also leads to an assignment for the ^{13}C isotopomer, which is located ~57 cm^{-1} lower and away from the congested spectra of the major isotopomers.

As part of our program aimed at investigating the kinked nature of the cumulenones and their potential functions we have also generated and recorded the far infrared spectrum of thioketene, a molecule with $t_{1/2} \approx 60$ mins in a quartz or glass cell. The mid infrared spectrum of this species has been studied previously (42,43) but our interest lies in understanding the low frequency modes.

In general, yields from pyrolysis are quite low due to the many reaction pathways that are available. Thioketene however was prepared in almost 90% yield by flow pyrolysis of 1,2,3 thiadiazole at 750°C. Spectra at 0.0023 and 0.0032 cm^{-1} resolution were recorded between 1200 - 300cm^{-1} . More than 20,000 individual vibration-rotation lines have now been assigned using MacLoomis and checked by lower state combination differences. A MacLoomis plot of part of the spectrum is shown in fig. 14. Many series can be seen and the selected series is essentially a vertical line. For thioketene, a near prolate rotor, the asymmetry parameter is very close to -1 and MacLoomis is an appropriate and simple way to assign bands in the congested spectrum.

Figure 15 shows some of the plethora of vibrational states in this spectral region that results in such spectral congestion and fig. 16 shows the spectrum in the region of the close lying ν_6 (B_1) and ν_9 (B_2) bands that are coupled by a strong *a*-axis Coriolis interaction. Without the effects of the perturbation the perpendicular type subbands are normally well separated with almost constant spacing. However in this case the wavefunctions are heavily mixed with the result that the ^rQ subband heads of ν_9 and the ^pQ subband heads of ν_6 tend to gain intensity and stack up, whilst the other Q subband heads lose intensity and spread out. The P and R structure of the subbands act in the same fashion to give overall a complex subband structure that is difficult to assign. The Q branch subband heads of the two modes are indicated in figure 16 to illustrate this effect. More than 3200 lines have been assigned in this region. Figure 17 shows a section of this spectrum in the region of the $^rQ_1$ subband head of ν_6 . This section of the spectrum is relatively straightforward and the major structure can be assigned by hand. Even so a number of hot band subbands are observable and assigned amongst the bands of the

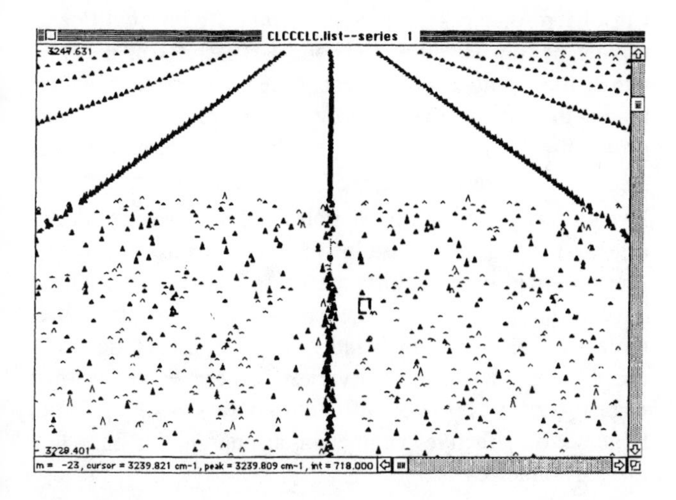

FIGURE 13. MacLoomis plot of the $\nu_1 + \nu_3$ *(Σ–Σ)* band of dichloroethyne.

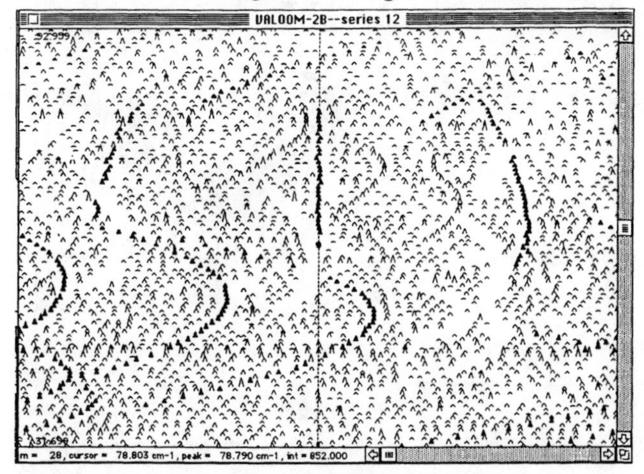

FIGURE 14. MacLoomis Plot of thioketene in the region 785.0 - 843.3 cm^{-1}. The $^qR_0$ $^qP_0$ series of ν_4 is displayed vertically.

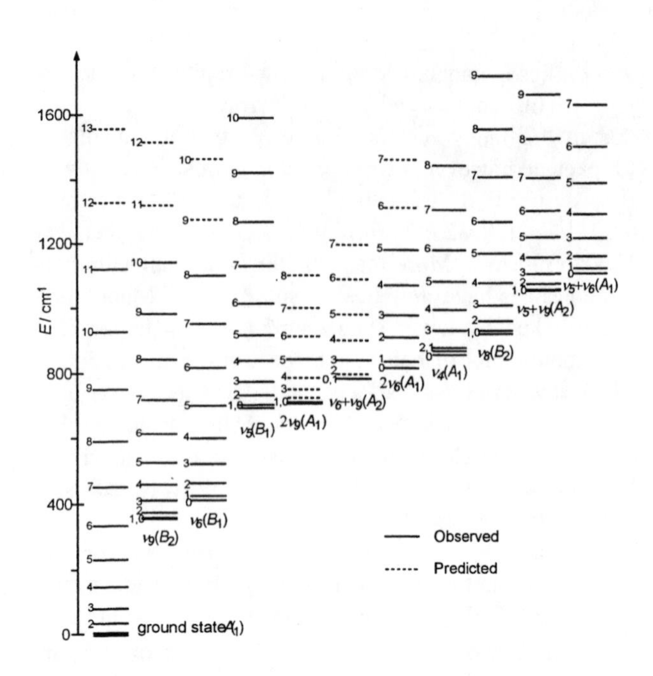

FIGURE 15. K_a sublevels (J=0 only). Of the low wavenumber modes of thioketene.

fundamental together with a subband of ν_5. Fitting this data to extract the desired information is an immense problem because the close lying ν_6 (B_1) and ν_9 (B_2) bands are also coupled to ν_5 (B_1) and ν_8 (B_2) by further a-axis resonances and and ν_6 (B_1) is coupled to $2\nu_9$ (A_1) by a b-axis resonance. To complicate matters further we also found that ν_9 is coupled to the ground state via a c-axis centrifugal distortion resonance of the type described by Urban and Yamada (44). These resonances are shown in figure 15.

A complete fit of all the interacting bands is out of the question due to the number of interacting energy levels, the high correlation of the parameters in any

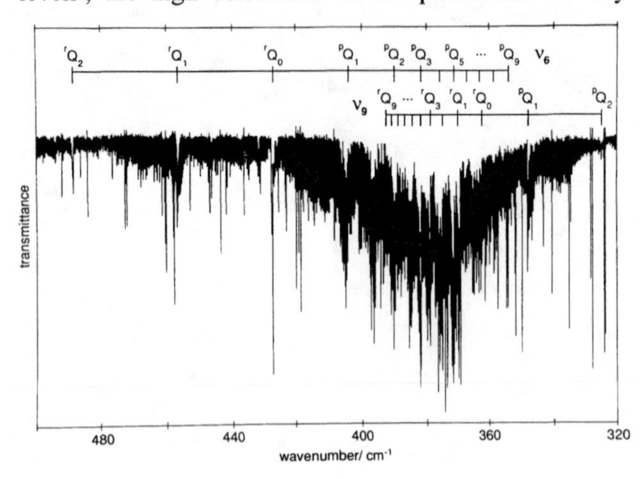

FIGURE 16. The ν_6/ν_9 region at reduced resolution (0.02cm^{-1}) with the Q branch subband origins marked.

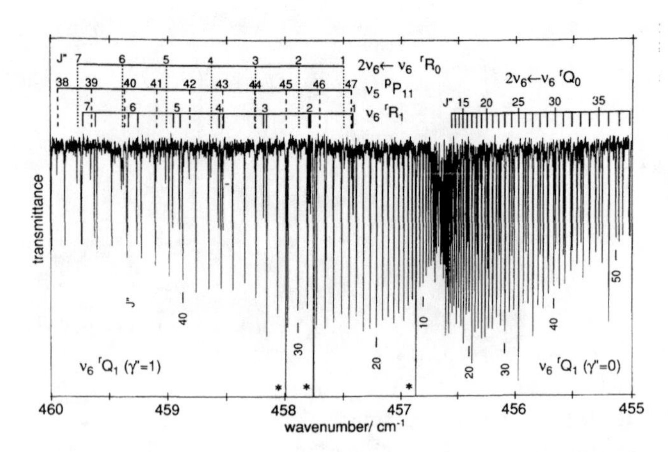

FIGURE 17. A section of the Far IR spectrum of thioketene at 0.0032cm^{-1}. .

effective Hamiltonian and the sheer scale of the problem. With the help of MP2/6-311G** *ab initio* predictions of the Coriolis coupling parameters we have successfully fitted the ground state with ν_6 and ν_9 to a Watson type Hamiltonian that includes the ground state interaction. This fit of 12165 transitions (12) yields a set of constants that describes the ground state well and a set of effective parameters for the two vibrational states. The ground state resonance shifts the ground state rotational levels by 3.4 cm^{-1} for J"=47, K_a"=11, the highest energy level included in an observed transition. It should be noted that all asymmetric molecules with light off a-axis atoms and low frequency modes are affected by such a centrifugal distortion resonance. Eg. For butatrienone , H_2C_4O, the effect is very large and the Watson Hamiltonian breaks down even for K_a" values ≥ 3.

A number of other interactions have been elucidated in the 10 modes of fig. 15. There are a number of Fermi type resonances and a,b and c - type Coriolis resonances effecting the spectrum. All the a - axis Coriolis resonances tend to be strong with significant mixing of the wavefunctions and large shifts in energy, whilst the b and c - axis resonances tend to be weaker unless the unperturbed energy levels are in very close proximity. Figure 18 shows the effect of a strong c-axis resonance between the K_a = 5 levels of ν_4 and the K_a = 4 levels of ν_8. The mixing of states "blurrs" the selection rules so that formally disallowed transitions are observed. The observed energy shifts are up to a wavenumber and have a large effect on the spectrum where lines are typically 0.003cm^{-1} in width. Because all modes are coupled together through a variety of Fermi and Coriolis resonances it has proved difficult to derive anything but a large set of effective parameters accounting for only the major resonances.

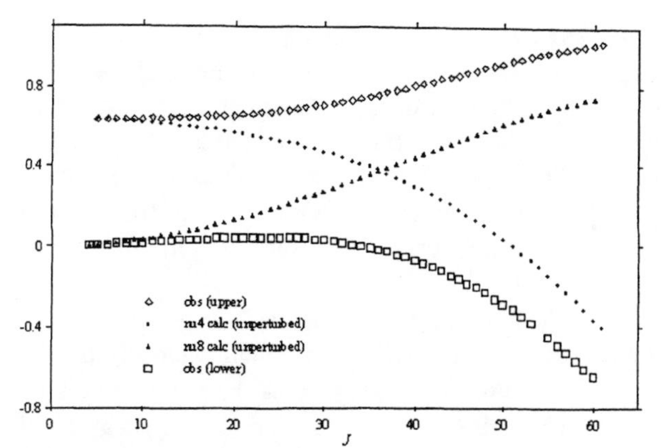

FIGURE 18. A strong *c*-type Coriolis resonance between the $K_a = 5$ levels of ν_4 and the $K_a = 4$ levels of ν_8.

Thioformaldehyde and Thioglyoxal

Both thioformaldehyde ($H_2C{=}S$) and thioglyoxal (O=CH-CH=S) are short lived species that polymerize easily. The high resolution infrared spectrum of thioformaldehyde has been studied previously by Turner *et al* (45) and Chedwell and Duxbury (46) but the ν_2 fundamental was obscured by lines from CS_2. Previous studies on glyoxal were restricted to matrix isolation studies of the photolysis products of 1,3-oxathiol-2-one by Torres *et al* (47). We have also reported the microwave spectrum of thioglyoxal (48)

In the matrix IR study of Torres *et al* (47) photolysis at 254nm produced a strong spectrum of thioglyoxal with very few by products. In order to produce thioglyoxal in high enough yield for a gas phase study we subjected the precurser molecule to flow pyrolysis. Figure 19 shows part of the low resolution spectrum that resulted from pyrolysis at the optimum temperature for production of thioglyoxal. A large number of by-products are apparent and are summarized in scheme 3.

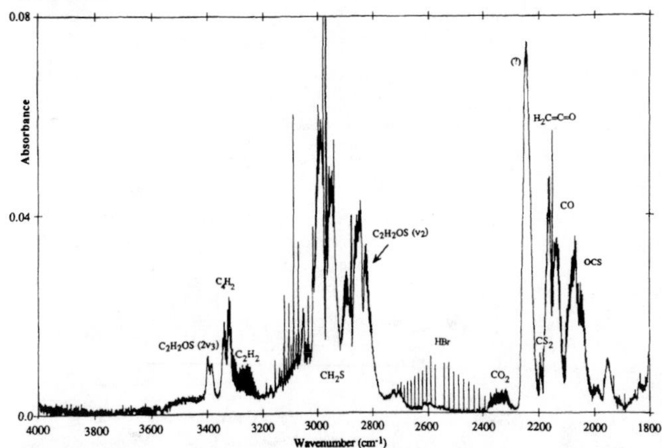

FIGURE 19. Spectrum obtained by flow pyrolysis of 1,3-oxathiol-2-one

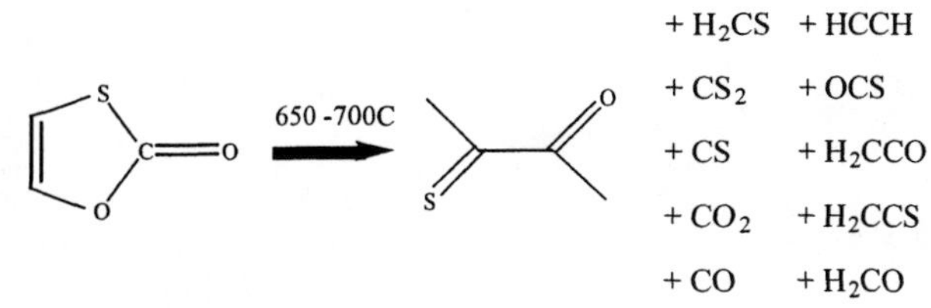

650 -700C → + H_2CS + HCCH + CS_2 + OCS + CS + H_2CCO + CO_2 + H_2CCS + CO + H_2CO

Scheme 3. Pyrolytic breakdown of 1,3-oxathiol-2-one

Bands due to thioglyoxal, which had a half life of *ca* 2 min in our apparatus, were identified by isolating the cell and monitoring the spectrum over time. Assignment of the low resolution spectrum was achieved by comparison with *ab initio* predictions and the matrix work (47). A high resolution spectrum of the ν_3 band was obtained at 0.005 cm^{-1} resolution. Although this band contained a lot of structure most of the lines overlapped and an assignment could not be obtained using MacLoomis. A set of rotational constants could be obtained only by holding the ground state constants to those of the ground state and simulating the band by varying the upper state constants. In this manner a good simulation was achieved with a set of sensible, but not necessarily unique constants.

Although the spectrum in fig 19 contains many by-products the amount of carbon disulphide is minimal and it was possible to obtain a fully resolved spectrum of the ν_2 band of thioformaldehyde at 0.005 cm^{-1} resolution. This has been analysed to give an accurate set of molecular constants for the ν_2 mode of thioformaldehyde.

In our high resolution spectrum lines thought to be due to the ^{34}S isotopomer were observed. In order to begin an assignment of these lines it was necessary to obtain a reliable set of molecular constants for this isotopomer. We thus recorded a pure rotational spectrum of thioformaldehyde in the far infrared region using our helium bolometer, a 3 metre gas cell and trimethylene sulphide as a precursor.

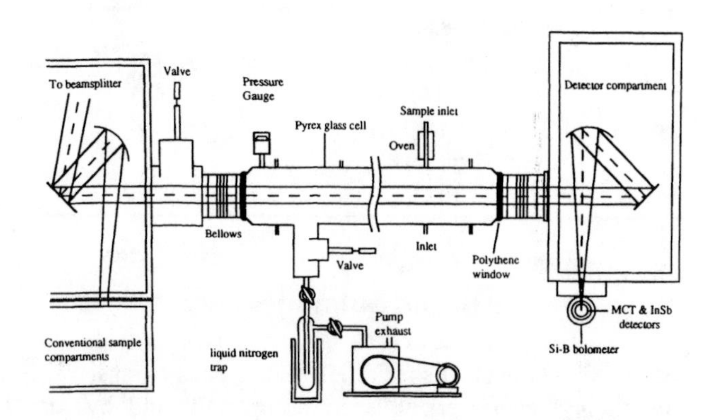

FIGURE 20. Experimental setup for far-IR spectrum of thioformaldehyde.

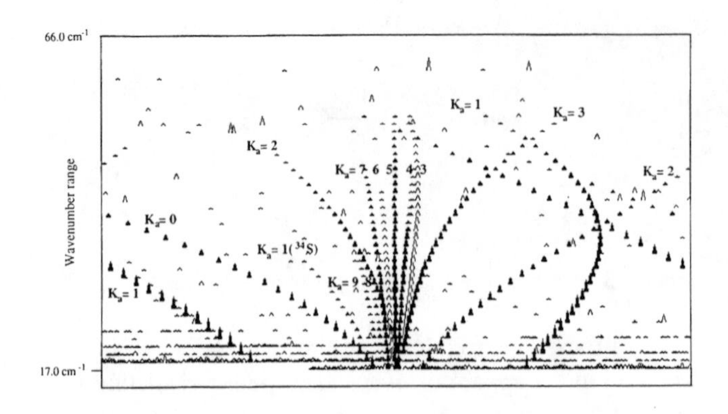

FIGURE 21. MacLoomis plot of the far-IR spectrum of thioformaldehyde.

The experimental setup is shown in fig. 20. Trimethylene sulphide was used to obtain a high yield of thioformaldehye. At low wavenumber the efficiency of the multipass cell is extremely low and in order to retain throughput of radiation and to eliminate the problem of polymer deposition on the White cell mirrors the long gas cell was used.

The pure rotational spectrum of thioformaldehyde was easily assigned using MacLoomis. The MacLoomis plot is shown in fig. 21 where all the lines other than those from the asymmetry split K_a=1 band essentially emanate from the same centre. One subband from $H_2C^{34}S$ is labelled in this plot. The spectrum of the $H_2C^{34}S$ presents the same pattern on a MacLoomis plot but with weaker lines. 126 transitions were assigned and this lead to a well determined set of constants. Table 3 compares our results with those previously determined from microwave measurements.

Table 3. Rotational and Centrifugal Distortion Constants of $H_2C^{34}S$ (MHz).

	Far IR	**Microwave**
A	291918.8(95)	
B	17388.426(16)	17388.31(44)
C	16377.344(13)	16377.80(48)
D_J	0.018387(13)	
D_{JK}	0.4985(17)	
d_I	-0.001072(4)	

Boron Sulphides

Although the transient species HB=S has been extensively studied and characterized spectroscopically and by *ab initio* molecular orbital calculations (49,50), much less is known about the halogenated thioxo-boranes XB=S (X=F,Cl,Br). Although their microwave spectra have been studied (51,52) and some photoelectron data is available, there have been no studies on their vibrational spectra at either low or high resolution. The microwave studies on these species have determined that XB=S molecules can be generated in good yield by passing X_2S_2 over Boron in a quartz furnace at temperatures ranging from 900 - 1250K. The XBS species then tend to form the cyclic trimers $(XBS)_3$.

We have recently generated ClB=S by passing S_2Cl_2 vapour over crystalline boron chips loosely packed along the length of a quartz tube heated to *ca.* 900-1000°C. The ClBS species was readily detected by FTIR spectroscopy by flowing the reaction products through a White cell set at 16 m pathlength at a sample pressure of approximately 0.35 Torr. At 1000°C all bands due to the precursor disappeared and bands with discernible rotational structure, evident in the mid infrared region, were attributed to ClB=S as well as to CO, CO_2, HCl, CS_2 and CS. In the far infrared region other bands that were identified, centred at 454.9 cm^{-1} and 474.3 cm^{-1} respectively, were assigned to $^{11}BCl_3$ and its isotopic partner, $^{10}BCl_3$. Following spectral assignment of the ro-vibrational structure of the four chlorine and boron isotopomers of ClB=S using MacLoomis, the data was fitted to the appropriate Hamiltonian for a linear molecule. Preliminary derived ground state constants agree well with the microwave results shown in table 4 and fitting of the data is ongoing.

Table 4. Preliminary Rotational Constants for ClBS /cm^{-1}

Species	B_{IR}	B_{MW}
35 11 32	0.09329055(31)	0.0932905 (2)
37 11 32	0.09082927(39)	0.0908294 (2)
35 10 32	0.09329301(35)	0.0932932 (5)
37 10 32	0.09083689(50)	0.0908365 (6)

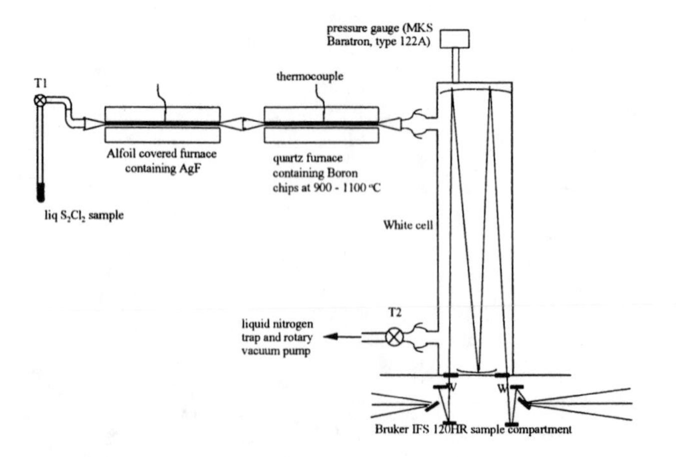

FIGURE 22. Experimental setup leading to the spectrum displayed in figure 23.

Recently we have also recorded some weak spectra of the ν_3 band of FBS using the experimental setup in fig. 22. S_2Cl_2 vapour was passed over AgF in a quartz tube heated to 100°C. The resultant F_2S_2 was then flowed directly over crystalline boron chips loosely packed along the length of a quartz tube and heated to *ca.* 900-1000°C. Under slow flow conditions, lines due to mainly BF_3 and a small amount of ClBS were observed. Under fast flow conditions we observe the spectrum shown in fig. 23 containing HBS, FBS and HBS. The bands labelled FBS have been confirmed as belonging to FBS by the *B* values derived via the MacLoomis assignment of the two bands. For $F^{11}BS$ the *B* value is 4953.837(13) MHz compared with 4953.853(3) Mhz from the microwave study(52). We are continuing this work in order to obtain better S/N spectra and thus fully characterize FBS. HBS which has been assigned only at a resolution of 0.1 cm^{-1} (50) will also be further studied.

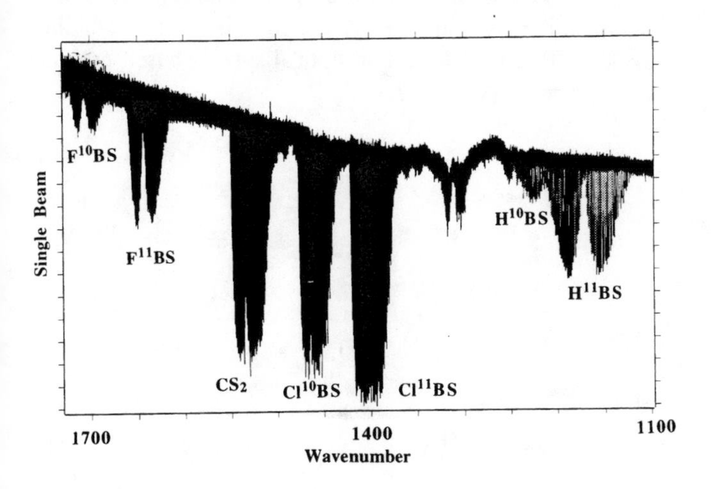

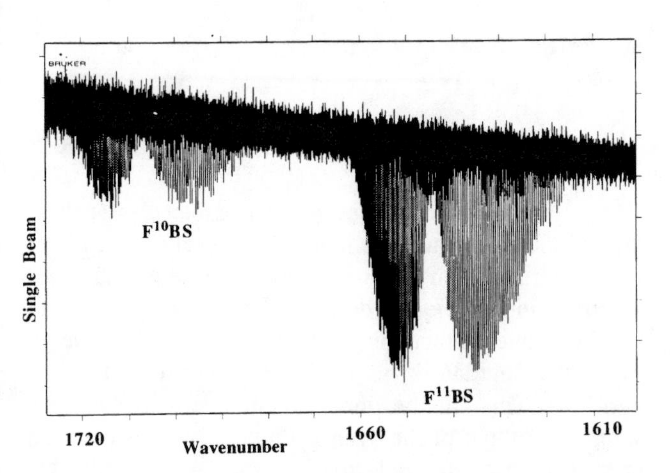

FIGURE 23. Spectrum under fast flow conditions using the setup in figure 18.

Infrared Laser Powered Pyrolysis (IRLPP)

Although flow through techniques using heated quartz tubes have proved successful in generating a number of short lived species, the method has a number of drawbacks. These are:

a. With the finite time required to introduce species into the absorption cell, molecules with $t_{1/2}$<1second cannot be observed.
b. A large number of secondary breakdown products are generated due to the large number of collisions that occur with the hot walls of the furnace tube.
c. Further chemical reactions in the hot zone lead to the formation of even more species.
d. The precursor may breakdown in a different fashion to that desired.

Photolysis often gives alternative products, as described above in the breakdown of the dihalomaleic anhydrides, but the yield is often very small. For very short lived species, where high pumping speeds are necessary, the retention time in the photolysis region is not long enough to build up a sufficient quantity for analysis.

In order to generate short lived species directly in the infrared probe beam we have used the method of infrared laser powered pyrolysis (IRLPP), used successfully by Russell (53) in a number of applications. In IRLPP the precursor molecule is introduced in to a gas cell along with a sensitizer molecule, typically SF_6, and irradiated with a high power CO_2 laser. The sensitizer molecule efficiently absorbs the laser radiation and the energy is collisionally transferred to the precursor molecule. Gas temperatures up to 1500°C have been reported in such systems and the species generated pyrolytically quickly migrate to the cold parts of the gas cell. In addition to the possibility of generating molecules directly in a beam of probe radiation, the hot wall collisions are eliminated in such a system.

After preliminary experiments where flowing mixtures of SF_6 and precursor species were irradiated by a SYNRAD 48-2 CO_2 laser in the centre of a 15 cm infrared absorption cell, we developed a system based around a 15cm long White cell. Figure 24 shows a typical setup for our experiments.

Initially we investigated a number of well known pyrolytic breakdowns and successfully generated a number of cumulenones. Propadienone spectra were obtained by bubbling SF_6 through acrylic anhydride and through the White cell. The optimum yield of propadienone was obtained with SF_6/anhydride ≈10:1, a laser power of 33 - 36 W and the pumping system restricted to slowly flow the gas stream through the cell.

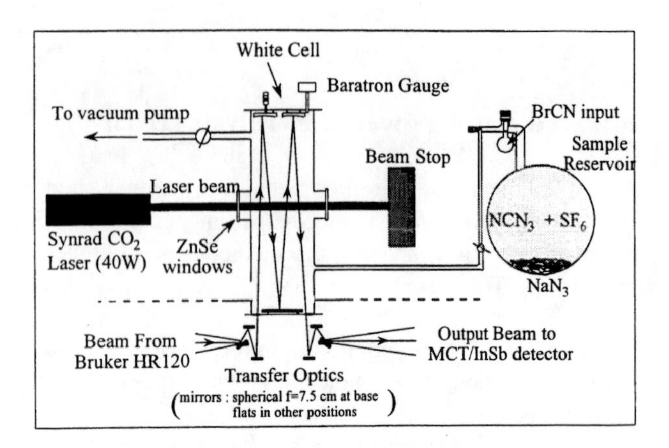

FIGURE 24. Schematic diagram of the experimental arrangement for the detection of NCN by IRLPP.

For propadienone, which is known to have a half-life of a few seconds we achieved spectra of almost the same S/N as spectra recorded by normal flow pyrolysis of pure acrylic anhydride. Given the different path lengths of the system, the yield of the short lived species in the absorption cell is some 10 times greater in the IRLPP system.

We have now generated and studied the free radical, NCN, in this system (54) and fig 24 shows the experimental system. Cyanogen azide (a dangerous compound that is highly explosive in the liquid or solid phase) was made in the gas phase (<11 torr) by the reaction of cyanogen bromide with sodium azide. SF_6 / NCN_3 mixtures were then flowed through the white cell and irradiated. The optimum conditions for observation of the spectrum of NCN were achieved with experimental conditions similar to that for propadienone.

Spectra at 0.01cm^{-1} resolution were recorded and fig. 25 shows a MacLoomis plot of the spectrum of the ν_3 band of NCN. NCN, which has a $^3\Sigma_g^-$ ground state shows a distinctive and expected spectrum with an intensity alternation N_{even}:N_{odd} = 2:1, due to the equivalent nitrogen nuclei and a doubling of each line due to the effects of spin rotation and spin-spin splitting. The MacLoomis plot shows these distinctive features. The components of the hot band appear centrally, with one component vertically displayed. A further doubling due to the coupling is apparent in the R branch. The cold band series appear diagonally across the screen and show a doubling and the expected intensity alternation. In addition to NCN, a spectrum that could be assigned to isocyanogen, N≡C-N≡C, recently observed by Stroth and Winnewisser (55), was observed.

During the IRLPP runs the region of the laser beam was accompanied by an intense red emission. We have investigated this emission by directing it through an optics system and monochromator on to a CCD array camera. The resultant restricted visible spectrum showed emission from the CN radical. The generation of NCNC is

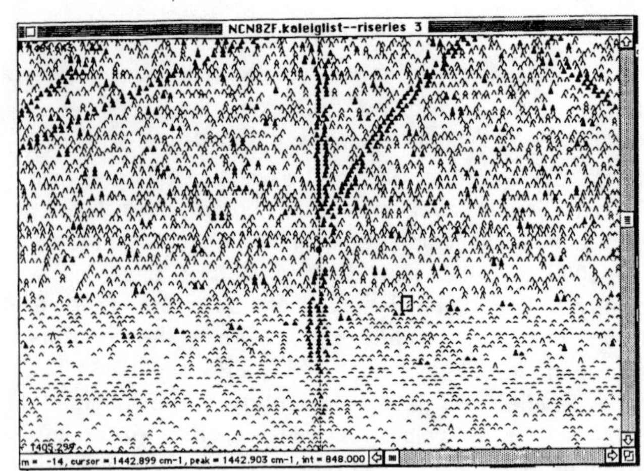

FIGURE 25. MacLoomis plot of the ν_3 band of NCN at 0.01cm^{-1} resolution. The selected series is the hot band emanating from the bending mode.

probably from a secondary process where the CN radicals attack further NCN_3 molecules and the pyrolysis breakdown is postulated in scheme 4.

We are currently carrying out experiments aimed at collection of both the IR and uv/vis using both the Bruker HR120 FTIR and an optical system based around a Peltier cooled CCD array.

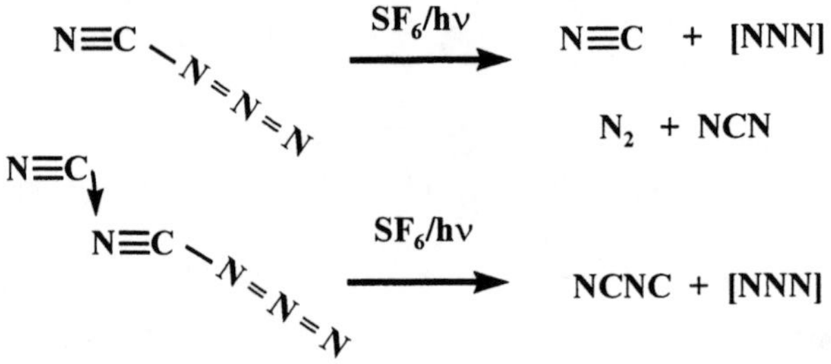

N≡C–N=N=N —(SF_6/hν)→ N≡C + [NNN]

N_2 + NCN

N≡C, N≡C–N=N=N —(SF_6/hν)→ NCNC + [NNN]

SCHEME 4. IRLPP of NCN_3/SF_6 with CO_2 laser.

Spectroscopy in a supersonic jet expansion.

For many of the asymmetric molecules such as the cumulenones, the large number of closely spaced lines in a spectrum taken at room temperature often make the spectrum intractable. In order to reduce the plethora of rotational lines emanating from heavily populated energy states the solution is to cool the molecules after the pyrolysis. This can be done by either effusive cooling where the output of the oven is "mixed" with a cold gas or by entraining the molecule in a supersonic expansion. The latter is commonly used in high resolution microwave and laser spectroscopy and has also been employed by Quack (56) in conjunction with FTIR spectroscopy. For transient species the method also may have the advantage of "trapping" the transient in the expansion before it breaks down further or reacts.

We have built a system (7) that couples a supersonic jet expansion apparatus to a Bruker HR120 FTIR. The apparatus, shown in fig. 26 consists of a vacuum chamber backed by a Varian HV12 cryopump that provides an inexpensive alternative to costly diffusion pumps with the restriction that helium cannot be used as a carrier gas. Our original system, coupled directly on to the HR120, allowed only a single pass of the focused infrared probe beam through the supersonic jet. expansion and was badly effected by vibrations from the cryopump compressor. Our current system is decoupled from the HR120 and allows 11 passes of a the beam waist (25mm. dia) of a nearly parallel beam through the supersonic expansion region and achieves *ca* 5 times the S/N of the original system.

Our expansion nozzle for observing the spectra of neat gases consists of a glass tube with a pinhole typically 100 - 300μm dia. with a heating jacket at the end capable of heating the nozzle to 200°C. With this system rotational temperatures of 6K for CO, 20K for N_2O and 40-60K for large asymmetric species have been achieved. This results in a considerable simplification of the spectrum as can be seen for N_2O in fig. 27. This spectrum was recorded in our single pass system and the top spectrum, which was achieved using neat N_2O should be compared with the room temperature spectrum of fig. 1.

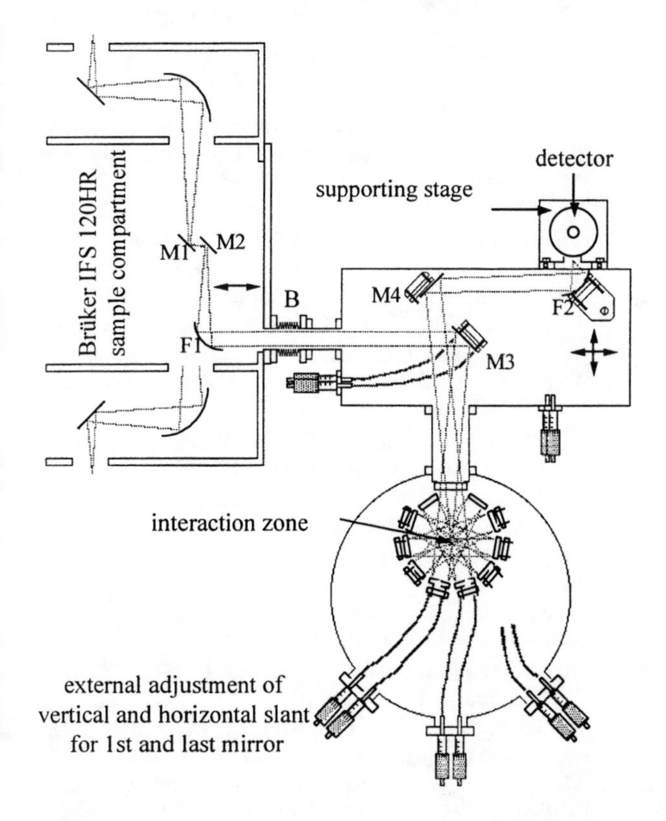

FIGURE 26a. Top view of jet expansion apparatus coupled to the Bruker HR120.

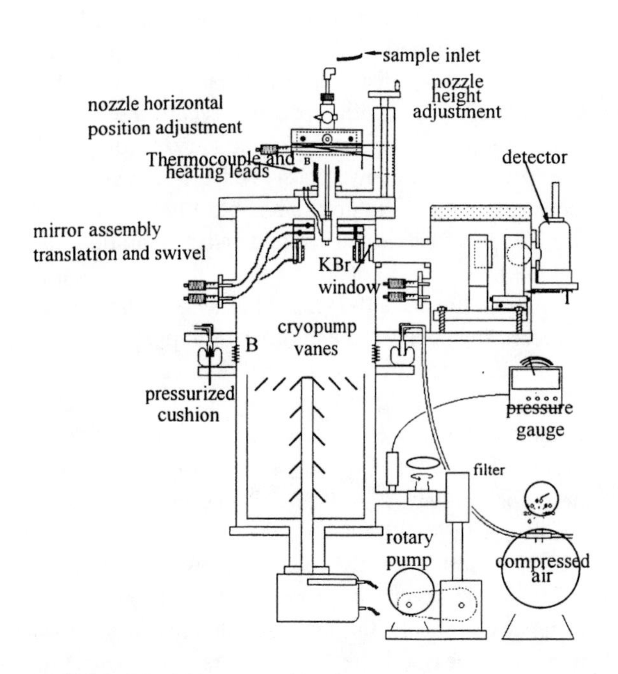

FIGURE 26b. Side view of jet expansion apparatus coupled to the Bruker HR120.

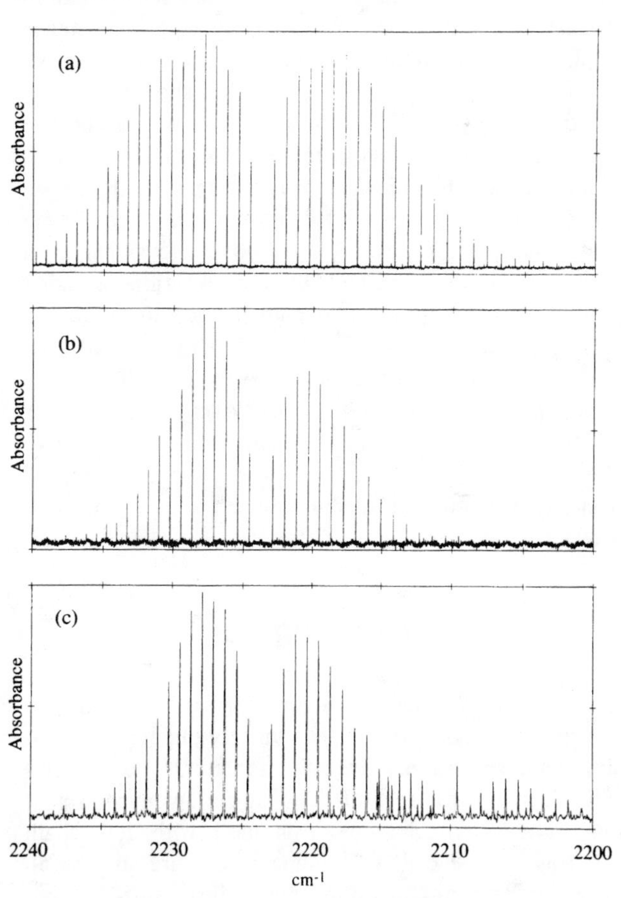

FIGURE 27. ν_3 band of N_2O. Top: neat N_2O at 300kPa, 180μm unheated nozzle, 5 scans at 0.01cm^{-1} resolution; center: as for top but diluted to 30% in Ar.; bottom: as for top with nozzle heated to 180°C, 5 scans at 0.01cm^{-1} resolution.

In addition to the rotational cooling, the hot band has almost disappeared and a vibrational temperature (assuming a Boltzman distribution of states) of *ca* 204K can be calculated. The middle spectrum where the N_2O is diluted in argon has a further reduction in rotational temperature and hence fewer observed lines, whilst the lower spectrum, where the nozzle has been heated, shows a similar degree of rotational cooling but is vibrationally hot (~490K). It is apparent that the hot band is also rotationally cold (~24K). This system has been used to record the spectra of a number of atmospherically significant species (5-10) at rotational temperatures of 40 - 60K. With the vastly simplified resultant spectra many of the bands have been successfully assigned using MacLoomis. Even for Freon 12, CCl_2F_2, where the asymmetry parameter K = -0.57, MacLoomis has been successfully used (7).

Using both flow pyrolysis and IRLPP we have recorded high S/N spectra of propadienone but due to the low barrier to inversion, the high populations in the low wavenumber modes and the large A rotational constant, the room temperature spectrum has proved intractable under analysis even with the assistance of MacLoomis. In an attempt to study a rotationally cooled and hence simplified, rovibrational spectrum of propadienone we have constructed a quartz nozzle with a short resistively heated furnace surrounding the end. Tests with an optical pyrometer showed temperatures of 700°C were achievable. Acrylic anhydride was entrained in dry argon at >100 kPa and expanded through the nozzle at temperatures ranging from 500 - 700°C. No signal from propadienone was observed. Similar experiments using acetic anhydride showed no evidence of lines due to ketene. Although these cumulenones would be significantly cooled in such a system the overall intensity of the band would still be distributed over a large number of states and it appears that a broad band technique like FTIR spectroscopy is not sensitive enough to record spectra of asymmetric molecules diluted to such an extent in a carrier gas.

CONCLUSION

High resolution Fourier transform infrared spectroscopy has been used to investigate the ro-vibrational spectra of transient molecules. For species with lifetimes of a second or greater flow pyrolysis or flow photolysis techniques combined with long path length absorption cells or multipass cells are successful. For shorter lifetimes, down to a few milliseconds, Infrared Laser Powered Pyrolysis (IRLPP) has been used to generate species directly in the infrared probe beam in a multipass cell. For all the species studied the spectra contain a plethora of ro-vibrational transitions and assignments have been made using a computer aided technique, MacLoomis. MacLoomis has proved useful in assigning the spectra of linear molecules, symmetric top molecules, near prolate asymmetric top molecules and totally asymmetric top molecules. In many cases the observed spectra are grossly effected by Fermi resonance, Coriolis resonances and in some cases by centrifugal distortion resonances. With the large number of transitions available from the infrared studies rotational constants, centrifugal distortion constants and structural parameters can be calculated with an accuracy approaching that obtained using microwave spectroscopy.

ACKNOWLEDGEMENTS

The author would like to thank the following people who contributed to the work described in this paper: Dr. P. S. Elmes, Dr. D. N. Bruget, Dr. G. Metha, Dr. E. G. Robertson, Dr. D. McGilvery, Ms. R. Tay-Lim and Mr. F. Shanks. I also acknowledge the Australian Research Council for funding part of this work. Scheme 1, scheme 2 and figures 7, 8, 9, 11, 12, 13, 16, 17, 21, 24, 25, 26 and 27 are reproduced from the references cited by kind permission of Elsevier Science BV , Academic Press Inc, The Royal Society of Chemistry and Plenum Publishing Corp.

REFERENCES

1. Herres, W., and Gronholz, J.,, "Understanding FTIR Data Processing", available from Bruker Analytische Messtechnik GmbH, Karlsruhe, Germany.
2 Loomis, F. W., and Wood, R. W., *Phys. Rev.* **32**, 223-236 (1928).
3. Winnewisser, B. P., Reinstadtler, J., Yamada, K.M.T., and Behrend, J., *J. Mol. Spec.* **136**, 12-16 (1989).
4. Evans, C., McNaughton, D., Dexter, P., and Lawrence, W., submitted to *J. Mol. Spec.*
5 McNaughton, D., and Evans, C., *J. Phys. Chem.* **100**, 8660-8664 (1996).
6. McNaughton, D., Evans , C., and Robertson, E G., *J. Chem. Soc. Faraday Trans* **91**, 1723-1728 (1995).
7. McNaughton, D., McGilvery , D., and Robertson, E. G., *J. Chem. Soc. Faraday Trans.* **90**, 1055-1071 (1994).
8. McNaughton, D., and Evans, C., *J. Mol. Spec.* **182**, 342-349 (1997).
9. McNaughton, D., Evans , C. and Robertson, E. G., *Mikrochim. Acta* **S14**, 543-546 (1997).
10. McNaughton, D., and Robertson, E. G., *Chemical Physics* **206**, 161-171 (1996).
11. McNaughton, D., and Robertson, E. R., *J. Mol. Spec.* **163**, 80-85 (1994).
12. McNaughton, D., Robertson, E.G., Hatherley, L.D., *J. Mol. Spec.* **175**, 377-385 (1996).
13. Watson, J. K. G., "Vibrational Spectra and Structure", Vol 6, 1-89, ed. J. R. Durig, Elsevier, 1977.

14. Gaussian 94, Revision B.1, Frisch, M. J., Trucks, G. W., Schlegel, H. B., Gill, P. M. W., Johnson, B. G., Robb, M. A., Cheeseman, J. R., Keith, T., Petersson, G. A., Montgomery, J. A., Raghavachari, K., Al-Laham, M. A., Zakrzewski, V. G., Ortiz, J. V., Foresman, J. B., Cioslowski, J., Stefanov, B. B., Nanayakkara, A., Challacombe, M., Peng, C. Y., Ayala, P. Y., Chen, W., Wong, M. W., Andres, J. L., Replogle, E. S., Gomperts, R., Martin, R. L., Fox, D. J., Binkley, J. S., Defrees, D. J., Baker, J., Stewart, J. P., Head-Gordon, M., Gonzalez, C., and Pople, J. A.,Gaussian, Inc., Pittsburgh PA, 1995.

15 McNaughton, D., McGilvery, D., and Shanks, F. S., *J. Mol. Spec.* **149**, 458–473 (1991).

16. Brown, R. D., Elmes , P.S., and McNaughton, D., *J. Mol. Spec.* **140**, 390 (1990).

17. McNaughton, D., and Bruget, D. N., *J. Mol. Spec.* **161**, 336–350 (1993).

18. McNaughton, D., and Elmes , P.S., *Spectrochim. Acta.* **48A**, 605–611 (1992).

19. McNaughton, D., and Bruget, D. N., *J. Mol. Spec.* **159**, 340–349 (1993) .

20. McNaughton, D., and Bruget, D. N., *J. Mol. Spec.* **150**, 620–634 (1991).

21. Tay, R., Metha, G., Shanks, F., and McNaughton, D., *Structural Chem.* **6(1)**, 47–55 (1995).

22. McNaughton, D., and Shallard, M., *J. Mol. Spec.* **165**, 185–194 (1994).

23. McNaughton, D., *Structural. Chem.* **3(4)**, 245–252 (1992).

24. Brown, R. D., Godfrey, P. D., Cragg, D. M., Rice, E. H. N., Irvine, W. M., Friberg, P., Suzuki, H., Ohishi, M., Kaifu, N., and Morimoto, M., *Astrophys. J.* **297**, 302–308 (1985).

25. Brown, R. D., Eastwood, F. W. , Elmes, P. S., and Godfrey, P. D., *J. Am.Chem.Soc.* **105**, 6496–6497 (1983).

26. Brown, R. D., Godfrey, P. D., Elmes, P. S., Rodler, M., and Tack, L. M., *J. Am.Chem.Soc.* **107**, 4112–4115 (1985).

27. Brown, R. D., Pullin, A. D. E., Rice, E. H. N., and Rodler, M., *J. Am.Chem.Soc.* **107**, 7877–7893 (1985).

28. Tang, T. B., Inokuchi, H., Saito, S., Yamada, C., and Hirota, E., *Chem. Phys. Lett.* **116**, 83–85 (1985).

29. Klebsch, W., Bester, M., Yamada, K. M. T., Winnewisser, G., Joentgen, W., Altenbach, H.J., and Vogel, E., *Astron.Astrophys.* **152**, L12–L13 (1985).

30. Jones, H., Takami, H., and Sheridan, J., *Z. Naturforsch.* **33A**, 156–163 (1978).

31. Bjorseth, A., Kloster-Jensen, E., Marstokk, K. M., and Mollendahl, H., *J. Mol. Struct.* **6**, 181–204 (1970).

32. Kroto, H. W., *Chem .Soc .Rev.* **11**, 435–491 (1982).

33. Firth, S., Khalaf, S., and Kroto, H. W., *J. Chem. Soc. Faraday Trans.* **88**, 3393–3395 (1992).

34. Blackman, G. L., Brown, R. D., Brown, R. F. C., Eastwood, F. W., and McMullen, *J. Mol. Spec.* **68**, 488–491 (1977).

35. Brown, R. D, Godfrey, P. D., Champion, R., and McNaughton, D., *J. Am. Chem. Soc.* **103**, 5711–5715 (1981).

36. Radom, L, *Aust . J. Chem.* **31**, 1–9 (1978).

37. Brown, R. D, Champion, R , Elmes, P. S., and Godfrey, P. D., *J.Am.Chem.Soc.* **107**, 4109–4112 (1985).

38 Farnell, L., and Radom, L., *Chem. Phys. Lett.* **91**, 373–377 (1982).

39. Brahms, J. C., and Dailey, W. P., *J. Am. Chem. Soc.* **111**, 8940 (1989).

40. Brahms, J. C., and Dailey, W. P., *J. Am. Chem. Soc.* **111**, 3071 (1989).

41 Tam, H. S., Harmony, M. D., Brahms, J. C., and Dailey, W.P., *J. Mol. Struct.* 223, 217–230 (1990).

42. Kroto, H. W., and McNaughton, D., *J, Mol. Spec.* **114**, 473–482 (1985).

43. Jarman, C.N., and Kroto, H. W., *J. Chem. Soc., Faraday Trans.*, **87**, 1815–1826 (1991).

44. Urban, S., and Yamada, K. M. T., *J. Mol. Spec.* **160**, 279–288 (1993).

45 Turner, H. , Halonen , L., and Mills, I. M. *J. Mol. Spec.,.,* **88**, 402-419 (1981).

46. Bedwell, D. J. , and G. Duxbury, *J. Mol. Spec.*, **84**, 531-558 (1980).

47. Torres, M. , Clement, A. , and O. P. Strausz, *Nouv. J. Chim.*, **7**, 269-270 (1983).

48. McNaughton, D., and Bruget, D. N. *J. Mol. Spec.*, **134**, 129-133 (1989).

49. Pearson, E. F., and McCormick, R.V., *J. Chem. Phys.* **58**,1619 (1973).

50. Turner, P. and Mills, I., *Mol. Phys.* **46**, 161 -170 (1981).

51. Kirby, C., and Kroto, H. W., *J. Mol. Spec.* 130–147 (1980)

52. Cooper, T. A., Firth, S., and Kroto, H. W., *J. Chem. Soc. Faraday Trans.* 87, 1499–1502 (1991).

53 Russell, D. K, *Chem. Soc. Rev.* **19**, 407–437 (1990).

54. McNaughton, D., Metha, G., and Tay, R., *Chemical Physics.* **198**, 107–117 (1995).

55. Stroth , F., and Winnewisser, M. W., *Chem. Phys. Lett.* **155**, 21–26 (1989).

56. Quack, M., *Annu. Rev. Phys. Chem.* **41**, 839 (1990).

Laboratory based FT Spectroscopy in support of Atmospheric Science

J. Ballard, D. A. Newnham and A. E. Heathfield

Space Science Department, Rutherford Appleton Laboratory, Chilton, Didcot, OX11 0QX, UK

Many aspects of atmospheric science rely on the availability of suitably accurate information which describes the interactions between electromagnetic radiation and atmospheric constituents. Examples include measurements of the atmosphere by remote sensing methods, calculations of the radiative forcing of the atmosphere by greenhouse gases, and studies of atmospheric processes such as ozone depletion. The number of gases whose spectral properties must be known is large and includes many unstable and reactive species. This, together with the requirement to understand the spectral properties of other atmospheric components such as aerosols, presents a difficult challenge to the spectroscopist.

The development of the high resolution Fourier transform spectrometer (FTS) has been central to meeting this challenge. FTS is widely used in the determination of many spectral parameters of atmospheric relevance, such as absolute absorption cross sections, line positions, line strengths and pressure broadening coefficients. Other developments in sample handling and sample containment have allowed realistic simulation in the laboratory of the physical conditions which pertain in terrestrial and planetary atmospheres.

In this paper some of the capabilities for and results from spectroscopic studies carried out at the Rutherford Appleton Laboratory (RAL), and elsewhere, in support of atmospheric science are described. Examples will include long optical path spectroscopy of weak absorbers such as molecular oxygen (relevant to satellite measurements of ozone), spectroscopy of greenhouse gases (relevant to radiative forcing and global warming potential calculations), and spectroscopy of gases and aerosol particles (relevant to ozone depletion). In addition to discussing the role that FTS plays in these investigations, experimental techniques which may be used to generate suitable sample conditions and a method for extracting accurate line parameters from measured spectra will also be described.

INTRODUCTION

Successful research in many branches of atmospheric science relies, *inter alia*, on good knowledge of interactions between electromagnetic radiation and the gases and particles contained in the atmosphere under study. A discussion of all aspects of spectroscopy which benefit atmospheric science is beyond the scope of this paper, which will concentrate on laboratory-based absorption spectroscopy using FTS. Such studies have application to remote sensing of atmospheric parameters such as temperature, pressure, and concentrations of gaseous and particulate species, and understanding of physical and chemical processes which occur in the atmosphere such as the greenhouse effect and ozone depletion.

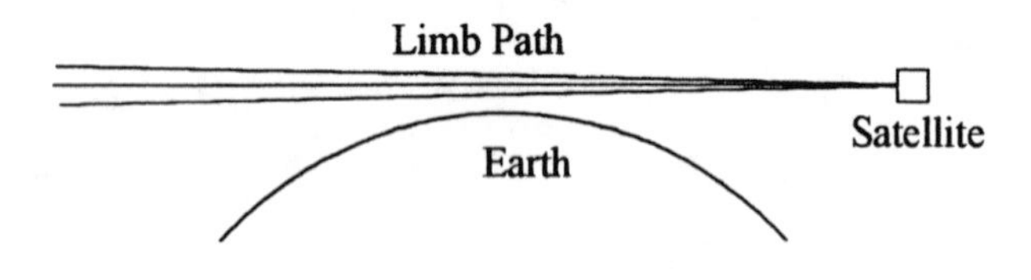

FIGURE 1: Illustrating a limb viewing geometry in a remote measurement of atmospheric parameters from space.

A typical example of a remote sensing measurement of atmospheric trace gases in the Earth's atmosphere is shown in Fig. 1. In this example an instrument on a satellite is designed so that its instantaneous field of view (FOV) is a relatively narrow pencil beam (typically < 0.02°) which passes through the limb of the Earth's atmosphere, without intersecting the Earth's surface. The instrument receives electromagnetic radiation from the FOV which is detected and is typically converted into an electrical signal for subsequent processing. The source of the radiation is typically either the sun or the atmosphere itself; the former case is often referred to as solar occultation. In both cases the radiation entering the instrument carries information pertaining to the composition and physical conditions in the atmosphere. Other experimental situations which have been commonly employed include instruments mounted at the Earth's surface or on balloon or aircraft platforms. Common to all these examples is that the basic quantity measured by the instrument is the intensity of the radiation emitted by the atmosphere, or transmitted through the atmosphere (with a varying degree of spectral resolution depending on the instrument), and from this basic measurement atmospheric parameters must be retrieved. In both the

CP430, *Fourier Transform Spectroscopy:* 11th International Conference
edited by J.A. de Haseth

solar occultation and thermal atmospheric emission cases the link between atmospheric emission (or transmission) and atmospheric parameters is the absorption cross section of atmospheric constituents as a function of optical wavenumber and physical conditions. The transmittance τ of the limb path shown in Fig. 1 at wavenumber $\tilde{\nu}$ can be written

$$\tau(\tilde{\nu}) = \exp - \left[\int_0^\infty k(\tilde{\nu}, x)\rho(x)dx \right] \quad (1)$$

where k is the molecular absorption cross section at the point x from the satellite and ρ is the density of absorbers at the same point. If k is known then ρ can in principle be deduced from the measurement of τ. For emission instruments the situation is slightly more complex in that both emission and transmission of the limb path has to be considered, and the radiance of the limb path can be written:

$$R \propto \int_0^\infty J(\tilde{\nu}, x)k(\tilde{\nu}, x)\rho(x)\tau(\tilde{\nu}, x)dx \quad (2)$$

where J is the source function, (the Planck or black body function for thermal emission), and τ is the transmittance as defined above. Again the density of absorbers (emitters) can in principle be deduced from knowledge of the intensity of emission provided that k is known. Since k can depend on temperature and pressure, a pre-requisite for composition measurements is that temperature and pressure along the limb path are known. Temperature is further needed in emission measurements since the Planck function is temperature dependent. Conversely, if ρ is known, atmospheric temperature can be measured, though again only if k is known. In some situations, particularly when measuring particles in atmospheres, the effects of scattering of radiation must be taken into consideration also.

Greenhouse gases are largely transparent to solar radiation in the visible spectral region, but absorb heavily the infra-red (IR) radiation emitted from the planetary surface. Consequently there is some trapping of IR radiation and a consequential warming of the surface compared to the temperature it would assume if the atmosphere did not contain gases with these properties. Again a key property of greenhouse gases is their absorption cross sections as a function of optical frequency and physical conditions. Clouds and aerosols also contribute to the greenhouse effect .

The range of optical frequencies at which absorption cross sections of atmospheric species must be known is large. Remote sensing instrumentation has been devised which operates through the entire electromagnetic spectrum from radio frequencies to the ultra-violet (UV). For example the Microwave Limb Sounder (1) on NASA's Upper Atmosphere Research Satellite is a very high resolution heterodyne receiver measuring emission from the Earth's atmosphere in the mm and sub-mm spectral region around 200GHz. The same platform carried the Improved Stratospheric and Mesospheric Sounder (2), a radiometer measuring atmospheric IR emission in 12 narrow intervals defined by filters in the wavelength range from 4.7 to 18 μm.

Fourier transform spectrometers (FTS) have also been used for atmospheric measurements from ground based, balloon, aircraft and space platforms. The Atmospheric Trace Molecule Spectroscopy programme (ATMOS) (3) involved deploying a relatively high resolution interferometer in solar occultation mode from the Space Shuttle, and an interferometer, MIPAS (4), observing Earth atmospheric emission will form part of the satellite instrument payload of the European Space Agency's (ESA) ENVISAT (5). Other space-borne FTS instruments are described in (6).

The UV spectral region is utilised by the Total Ozone Mapping Spectrometers (TOMS) (7) which have been flown for many years on several series of platforms, and also recently by the Global Ozone Monitoring Experiment (GOME) (8) which is currently in orbit on ESA's ERS-2 spacecraft. This spectral region is also employed by numerous ground-based instruments measuring trace gases in the stratosphere (9).

The range of physical conditions over which absorption cross sections must be known in connection with atmospheric science is also large. In the Earth's atmosphere pressures range from approximately 1000 hPa at the surface to near zero at high altitudes. Temperatures are around 300K at the surface and fall to below 200K in the lower stratosphere where processes such as ozone depletion take place. At higher altitudes, such as in the mesopause around 70km altitude temperatures can be as low as 160K. In some planetary atmospheres, such as Jupiter's, pressure and temperature ranges can be more extreme. Successful retrieval of atmospheric parameters from remotely sensed data, and studies involving radiative transfer calculations in atmospheres, therefore demand that absorption cross sections are known under these physical conditions.

There is also a broad range of atmospheric species whose absorption cross sections must be known. In radiative transfer studies connected with man-induced greenhouse warming of the Earth, not only do the spectral properties of naturally occuring gases such as carbon dioxide, water vapour and methane need to be known, but an increasing number of other species are important as well. These include the chlorofluoro carbons (CFCs) and their replacements such as HCFCs and HFCs. While the latter are less likely to destroy stratospheric ozone they are, molecule for molecule, very much more efficient greenhouse gases than carbon dioxide. Other species, such as perfluorinated hydrocarbons and ethers, have even larger global warming potentials, not only because of their spectral properties but also because of their extremely long lifetimes in the atmosphere.

Another very important process which is occuring in the Earth's atmosphere, ozone depletion, involves a large number of gaseous species. Full understanding of ozone depletion demands adequate knowledge of the distributions of many of these species, which in turn requires that absorption cross sections for many of them are known, at least for their measurement by remote sensing methods. Many of these species are highly reactive, which places additional demands on experimental determinations of absorption cross sections. Intensive studies of ozone depletion at high latitudes has highlighted the importance of polar stratospheric clouds (PSCs) (10), which form when temperatures fall below approximately 195K. The common occurence of sulphate aerosol in the stratosphere, has recently been shown to play an impotant role in determining trends and short-term fluctuations in stratospheric ozone at middle latitudes (11). Both PSC and sulphate aerosols can be detected by remote sensing methods (12, 13), but quantification again relies on availability of suitable spectral and other optical data.

SPECTRAL INFORMATION REQUIRED

The discussion above has indicated that a central spectroscopic property of atmospheric constituents required for many aspects of atmospheric science is the spectral absorption or scattering cross section over a range of physical conditions. It is not possible to include here a complete description of all the effects which determine a particular cross section, but in many cases gaseous absorption cross sections of gases can be calculated if certain basic line parameters are known.

Line Parameters

Gases are well known to absorb radiation in a number of reasonably well defined spectral regions which are specific to a given molecule. The absorptions correspond to the molecule going from one internal energy state to another and give rise to bands of spectral lines. Changes in rotational energy alone are primarily responsible for lines in the microwave and far-IR. Lines in the mid- and near-IR are caused primarily by changes in vibrational energy, and may show fine structure due to simultaneous changes in rotational energy. Visible and UV absorptions are caused mainly by changes in the electronic energy, with the possibility of fine structure due to simultaneous changes in vibrational and rotational energy.

Individual lines have a finite width, so can contribute to the absorption cross section at wavenumbers other than their line centre. The contribution from an individual line can be expressed

$$k(\tilde{\nu}-\tilde{\nu}_0) = SK(\tilde{\nu}-\tilde{\nu}_0) \tag{3}$$

where S is the line strength and K is a line shape factor, $\tilde{\nu}$ is the wavenumber and $\tilde{\nu}_0$ is the line centre wavenumber. Line strength is given by the expression

$$S \propto \frac{\nu_0}{Q(T)} |R|^2_{lu}\, g_l \exp[-E_l/kT] \times \left(1-\exp-[h\nu_0]/kT\right) \tag{4}$$

where R is the matrix element for transitions between the lower and upper states, T is the temperature, E_l is the energy of the lower state involved in the transition, g is a degeneracy factor and Q is the partition function. The line shape factor, K, is influenced by a number of physical effects, the most relevant to atmospheric science being the Doppler effect and intermolecular collisions.

Molecules in a gas at typical atmospheric temperatures move with a Maxwellian distribution of speeds characteristic of the temperature. Consequently there is a Doppler shift introduced by the motion into the frequency at which molecules absorb (or emit) radiation. This results in a line shape factor given by the expression

$$K(\tilde{\nu}-\tilde{\nu}_0) = \frac{1}{\sqrt{\pi}} \exp-\left[\frac{\alpha_D^2}{(\tilde{\nu}-\tilde{\nu}_0)^2}\right] \tag{5}$$

where α_D is the Doppler half-width, defined by the expression

$$\alpha_D = \frac{\tilde{\nu}_0}{c}\sqrt{\frac{2kT}{m}} \tag{6}$$

where m is the molecular mass, c the velocity of light, T the gas temperature and k the Boltzmann constant.

A further consequence of the molecular motion is that molecules collide with each other. The collisions perturb the ro-vibrational energy levels thereby contributing to the non-zero width of spectral lines. A simple form of the lineshape which is applicable to many atmospheric situations is the Lorentz lineshape, which is described by the expression

$$K(\tilde{\nu}-\tilde{\nu}_0) = \frac{1}{\pi}\frac{\alpha_L}{(\tilde{\nu}-\tilde{\nu}_0)^2-\alpha_L^2} \tag{7}$$

In this expression α_L is the Lorentz or collision broadened half-width. α_L depends on both pressure and temperature. Again much effort has been expended to establish these dependences for atmospheric applications. To a good approximation α_L varies with pressure and temperature according to the relation

$$\alpha_L(P,T) = \alpha_L(P_0,T_0) \times \frac{P}{P_0} \times \left[\frac{T}{T_0}\right]^{-n} \quad (8)$$

where n has a value between 0.5 and 1.0, and P_0 and T_0 are a standard pressure and temperature, usually 1 atm and 296K respectively. Both n and α_L can depend on the rotational states involved in the transition, but less so on the vibrational states.

In general, Doppler broadening is important near the line centre and collision broadening dominates in the line wings. α_L, typically has values between 0.05 and 0.1 cm^{-1} atm^{-1}, though collisions with some gases can result in significantly larger widths. A notable and important example in an atmospheric context is collisions between water molecules which can give line widths around 0.5 cm^{-1} atm^{-1}. Principally because of the pressure dependence, the Lorentz linewidth varies considerably at different locations in an atmosphere. For example, at the Earth's surface it is typically 0.1 cm^{-1} whereas at 20km altitude (lower stratosphere) it is typically 0.02 cm^{-1}, and becomes much smaller at higher altitudes still. For a given band in a particular molecule the Doppler width shows less variability in the Earth's atmosphere (since it is independent of pressure and depends only on the square root of temperature). However its dependence on optical frequency means that Doppler widths for spectral lines in the sub-mm and far IR spectral regions are typically $<10^{-5}$ cm^{-1} throughout most of the Earth's atmosphere so can be largely neglected in comparison to Lorentz widths. In the mid IR and visible spectral regions both Doppler broadening and pressure broadening have a similar magnitude through much of the atmosphere, so both effects must be taken into account. Several combined line shapes which do this have been published. The most widely used for studies of the Earth's atmosphere is the Voigt line shape, which is a simple convolution of the Doppler and Lorentz line shapes. It takes the form

$$\alpha_V = \frac{y}{\pi^{3/2}\alpha_D} \int_{-\infty}^{\infty} \frac{\exp-[t^2]}{(x-t)^2+y^2} dt \quad (9)$$

where $y = \alpha_L/\alpha_D$ and $x = (\tilde{\nu} - \tilde{\nu}_0)/\alpha_D$. Since there is no analytic function for the Voigt line shape it must be calculated numerically. Several schemes for doing so, with varying computational speed and accuracy, have been published (14, 15).

The absorption cross section at a given frequency in a ro-vibrational band is obtained in most cases with good accuracy by summing the contributions from all the lines in the band, treating each line as if it were isolated from the others. An important exception occurs when line mixing is present, and more sophisticated approaches must then be taken.

Broad - Band Effects

There are important atmospheric absorptions which cannot be described in terms of the summation of contributions from individual spectral lines. These include collision-induced absorptions and absorptions by molecules whose energy levels are so close together that the finite line width renders individual lines in their spectra unresolvable. Such absorptions can extend over several hundreds of cm^{-1} with few sharp features. In these cases direct measurements of absorption cross sections are required over the range of temperatures, pressures and spectral resolutions in which they will be applied.

Aerosols also give rise to broad-band effects. Among the spectral data required from laboratory studies are extinction measurements over a wide spectral range from the mid-IR to the UV, and the angular dependence of scattering. Measurements are needed on aerosols which mimic those found in the atmosphere, in terms of particle size distributions and chemical composition. Such laboratory data can then be applied to remote sensing measurements of aerosols, or to deduce the radiative properties of aerosols required for inclusion in radiative transfer models of the atmosphere or to understand their chemical properties.

APPLICABILITY OF FTS

The requirements for spectral data for atmospheric science support are very extensive. Many molecular absorptions spanning a large spectral range must be characterised at high spectral resolution and over a wide range of pressures and temperatures. This can only be achieved through a combination of measurements and theoretical modelling. The experimental aspects of this endeavour have become much more tractable in recent years owing to development of the FTS. The well-known attributes of the FTS (wide optical bandwidth, high spectral resolution, high energy throughput, efficiency of observations) makes it a very powerful tool for atmospheric spectroscopy when coupled to systems in which atmospheric conditions can be adequately simulated. For atmospheric studies the FTS is normally employed as an absorption spectrometer, though interesting atmospheric processes have been studied using an FTS in emission.

A typical experimental arrangement for measuring spectral line parameters and cross sections by absorption spectroscopy is shown schematically in Fig. 2, though other arrangements have the sample located between the source and the interferometer. The source of radiation is typically broad-band, for example a mercury lamp for the far-IR, a hot silicon carbide rod (eg globar) for the mid-IR, a quartz tungsten halogen lamp for the visible, and various discharge lamps (including xenon and deuterium

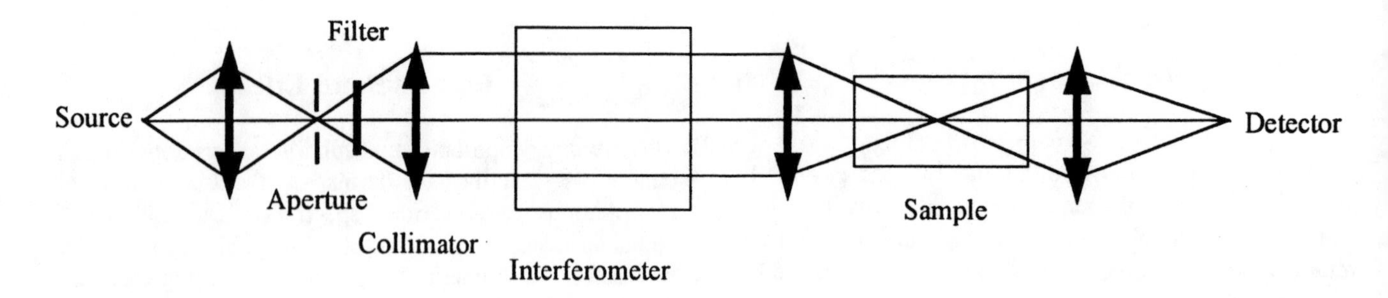

FIGURE 2: Schematic diagram of a typical experimental arrangement for FT absorption spectroscopy in support of atmospheric science.

arcs) for the UV. Radiation from the source is focussed onto an aperture which confines the angular spread of radiation passing through the interferometer to be no greater than that required to achieve the required spectral resolution. After being modulated by the interferometer, the radiation passes through the sample whose absorption spectrum is to be measured , and then is detected by a suitable detector. These include liquid helium cooled bolometers for the far-IR, liquid nitrogen cooled cadmium mercury telluride (CMT) and indium antimonide photoconductive and photovoltaic detectors for the mid- and near-IR, and silicon and gallium phosphide photodiodes and photomultipliers for the visible and UV. Transmittance spectra are obtained by ratioing measurements obtained with and without the sample present. Absorption cross sections can then be obtained by applying the Beer-Lambert law

$$k(\tilde{\nu}) = -\ln(\tau)/(absorber\ amount) \qquad (10)$$

or line parameters obtained by non-linear least-squares fitting, as will be discussed later. This straightforward-looking procedure hides a number of aspects which the experimentalist must be aware of if absorption cross sections and line parameters of sufficient accuracy for application to atmospheric problems are to be obtained. A catalogue of error sources when using an FTS for straightforward transmittance measurements of a plane parallel solid sample of a few mm thickness has recently been published (16). Some of the effects noted are beyond the control of the user and adequate performance relies on high quality support from the interferometer manufacturer. However there are a number of effects which are directly under the control of the user, three of which will be mentioned briefly here.

Sample Emission

This effect concerns radiation emitted by the sample itself and the gas cell windows. The following dicussion assumes a configuration shown in Fig. 2. The intensity and spectral distribution of the emitted radiation depends on the temperature of the sample and windows, and their emissivity. The radiation is emitted uniformly in all directions and therefore some enters the interferometer. Consequently the interferogram has contributions from both the source radiation (which generates an interferogram in transmission through the interferometer) and the sample radiation (which forms an interferogram in reflection from the interferometer). These contributions have an opposite phase with respect to optical path difference. This effect has been discussed in (17). Since the "sample emission" spectrum is not present when there is no sample present the sample transmittance is not simply given by the ratio of spectra recorded with and without the sample. The importance of sample emission in a given determination of absorption cross section depends on a number of factors and particularly needs to be considered when the radiation emitted by the sample, modulated by the interferometer and detected by the detector, is a significant fraction of that from the source. From considerations of the temperature and frequency dependence of the Planck function it is readily seen that particular care must be taken when working with relatively long wavelengths (far- and mid- IR), hot samples, or low brightness temperature sources. Furthermore the system throughput for radiation emitted by the sample may be significantly larger than that for source radiation if the throughput-defining elements for source radiation are on the source side of the interferometer and those for sample radiation are on the detector side. Sample emission can be corrected for by measuring spectra at two source brightnesses and subtracting them before ratioing against "no sample" spectra similarly recorded at the two source brightnesses and subtracted. If the radiation emitted by the sample does not change as a result of changing the source brightness it will be eliminated in the subtraction. However this does increase significantly the time required for the measurements to achieve a given signal to noise ratio in transmittance. An alternative approach is to attempt to model the sample emission and correct for it without further measurements (18). Not all designs of FTS used for quantitative measurements of absorption cross sections and spectral line parameters in the laboratory are similarly affected by sample emission. For example designs where the image of the detector produced by reflection in the interferometer is displaced laterally with respect to the detector are not, neither are "step and hold" inteferometers where radiation from the sample is not modulated. However, at least two commercially available high resolution interferometers widely used for spectroscopy in support of atmospheric science (the BOMEM DA8 and the Bruker IFS 120HR) have been shown to be potentially affected by this effect.

Signal Channel non-linearities

Another effect which can introduce errors into measured transmittances is non-linearity in the signal channel response. A common source of non-linearities arises from the photometric response of certain detectors which are commonly used in IR FTS, such as CMT photoconductive detectors. If excessive amounts of radiation is allowed to fall on these detectors, the detector response is not a linear function of the radiation intensity. This distorts the interferogram and consequently the spectrum. The effect can be minimized if the radiation is reduced to an acceptable level, preferably by optical filtering or by reducing the FOV rather than the source brightness for reasons discussed above. It may be necessary to work with an aperture which is significantly smaller than the maximum allowed from resolution considerations. Detector non-linearity can also be modelled in the data processing, as has been done (19) to derive a correction procedure for application to interferograms recorded by space-borne interferometers using CMT detectors in solar occultation mode.

Instrument Line Shape Distortion

Due account must be taken of the effect of the instrument line shape (ILS) in FTS measurements. As is well known the ILS for an FTS has two principal components due, respectively, to the finite optical path difference and the FOV. Accurate measurement of absorption cross sections and line parameters requires good knowledge of the ILS and effective treatment in the data processing. A related point concerns choice of measurement conditions to ensure appropriate depth of absorption in the measurements, particularly where the features being measured are not fully resolved.

SIMULATION OF ATMOSPHERIC CONDITIONS IN THE LABORATORY

The requirements for laboratory-based spectroscopic measurements in support of atmospheric science are such that absorption features covering a broad range of intensities must be measured over a wide range of temperatures, pressures, optical frequencies and spectral resolutions. In addition to the requirements placed on the spectrometer, which are met very effectively by the FTS, there are requirements placed on sample preparation and containment systems, and on systems for measuring the physical state of samples. Gas and vapour samples are usually contained in absorption cells through which source radiation passes before being focussed onto the detector. Cells are designed to enable samples to be held at appropriate temperatures and pressures. Many atmospheric applications involve measuring samples cooled to below room temperature, so designs involving a double-walled construction are often used. In these the inner volume contains the gas being measured while the outer volume contains a circulating fluid which can be cooled by an external heat exchanger. Condensation of water vapour in ambient air on the cooled surfaces can be prevented by enclosing the cell in an evacuated jacket, or by utilising the evacuated light path of the spectrometer. Alternatively the cell can be covered by a suitable insulating material (20). The pressure regime over which the cell will operate is determined by the details of its design and construction and systems must be provided to evacuate the cell to a low pressure and then fill it to the desired pressure of the gas under study. For high pressure operation precautions must be taken to reduce the risk of a catastrophic mechanical failure to an acceptable level. Radiation enters and leaves the cell through windows with suitable optical, thermal and mechanical properties. Suitable gas-tight seals around the windows have been devised using various materials. For operation down to approximately 200K, silicone rubber 'O' rings work well, but for lower temperatures other devices such as PTFE and indium wire seals are more suitable.

The optical pathlength through the absorption cell is an important parameter in setting up an experiment, and has a strong influence on the cell design. Cells through which the radiation makes a single pass have a pathlength which is typically similar to the physical length of the cell. It is often necessary, however, to use an optical pathlength which is very much greater than the physical dimensions of the cell. A number of optical arrangements have been devised to satisfy this requirement; perhaps the arrangement most often interfaced to FTS's is that due to White (21). This arrangement uses 3 spherical mirrors of equal radius of curvature. Two mirrors are placed adjacently at one end of the cell and the third is placed at the other end of the cell; the mirrors are separated by their radius of curvature. Optical paths which are many multiples of the mirror radius, and good matching to the f-number of the FTS are possible with this arrangement.

Non-equlibrium Cooling Techniques

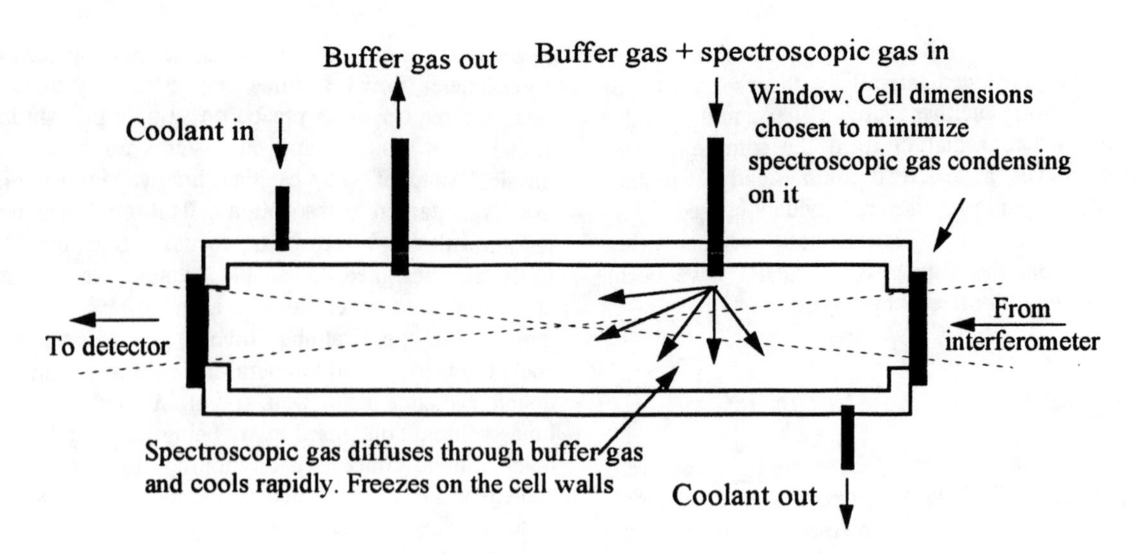

FIGURE 3: Schematic diagram of a collision cooling arrangement

The cooling arrangements described above utilize the conventional method for achieving a low temperature absorber sample. In this the gas or vapour reaches thermal equilibrium with the coolable cell via heat transfer resulting from collisions of the absorber gas molecules with the cold inner cell walls. However, for many atmospheric gases the lowest sample temperature is limited by the equilibrium vapour pressure of the gas, and continued cooling of the cell results in condensation of the vapour on the coldest parts of the cell. For readily condensible vapours such as H_2O and NO_2 the result is a rapid reduction in vapour pressure to negligible levels at temperatures significantly below their normal freezing points, making FTS measurements on vapour-phase samples in the laboratory at atmospheric conditions difficult or impossible. Measurements in the atmosphere, of course, are not so affected by SVP considerations because of the very long paths in the limb viewing geometry. To overcome SVP limitations in the laboratory, non-equilibrium cooling techniques have been developed, two of which are discussed briefly below.

Collision cooling

In the collision cooling technique the absorber gas is introduced into a cooled cell which contains a buffer gas, resulting in cooling of the absorber gas by collisions with the buffer gas molecules rather than by collisions with the cell wall. The buffer gas, typically helium or nitrogen, is chosen to have a high vapour pressure at the temperature of interest, and to have no absorption features which overlap those being studied in the absorber gas. The absorber gas molecules reach thermal equilibrium with the buffer gas after a few tens of collisions, but are diffusively trapped for tens of thousands of collisions, corresponding to tens of seconds residence time in the gas phase, as they diffuse to the cell walls where they freeze and are removed from the system. Continual replenishment of the absorber molecules allows a steady state concentration of low temperature sample to be achieved, as indicated in Fig. 3. The concentration of absorber gas and total pressure of absorber and buffer gas in the cell are important variables in the experiment, allowing generation of either supersaturated cold vapours or molecular clusters and aerosols, or mixtures of vapour, liquid, and solid phase components. For spectroscopic measurements, the gas cell is designed with suitable optical windows and mounts to allow radiation of the desired wavelength to be transmitted through the cooled gas sample, while maintaining a high vacuum seal at low temperatures.

The collision cooling technique was originally developed for application to microwave spectroscopy studies (22) of low temperature line broadening, for example providing information for water vapour broadened by nitrogen gas over the 80 to 600K temperature range (23). More recently, experiments have been described (24) in which cells coolable to temperatures as low as 5K have been used to obtain high resolution mid-IR spectra using tunable diode lasers. The method has also been used with FTS, but primarily for studies of molecular clusters and aerosol particles (25).

Supersonic jet cooling

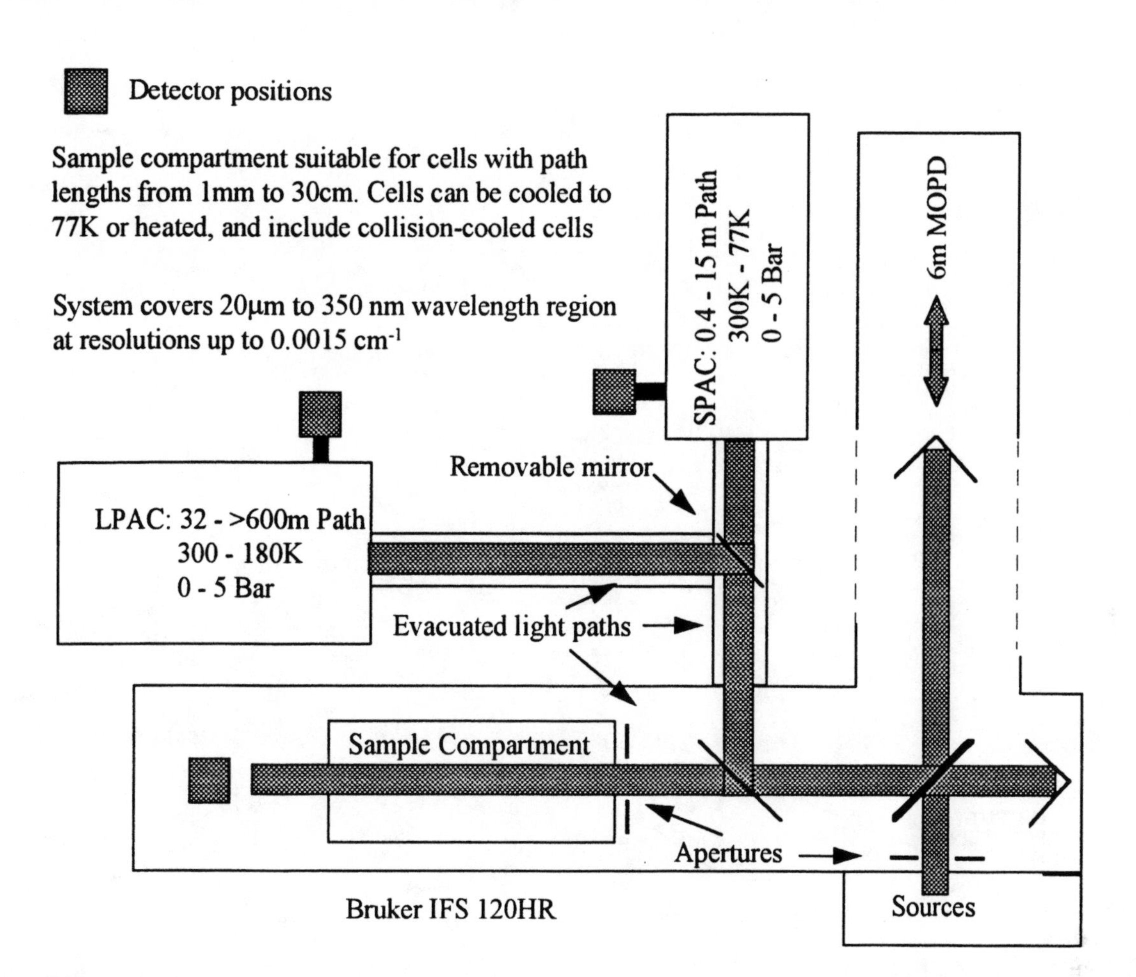

FIGURE 4: Schematic diagram of the optical arrangement used in the Molecular Spectroscopy Facility (MSF) at the Rutherford Appleton Laboratory (RAL)

Another non-equilibrium cooling technique involves forcing a gas through a small nozzle into a vacuum which cools, principally, the rotational degrees of freedom (26). Boltzmann population distributions in the rotational levels characteristic of a few K temperature are achievable under favourable conditions (seeded beams, high pumping speeds, relatively small molecules), though 30K is more common with unseeded beams of heavier molecules. Cooling of the vibrational degrees of freedom is however usually much less efficient. Absorber amounts in the light path tend to be relatively small, typically less than with collision cooling since the volume of cold gas in a typical molecular beam is only a few mm across, and absorber molecule densities in the seeded beam are very low. Consequently most molecular beam studies have been carried out with laser spectrometers where the sensitivity to absorptions can be significantly greater than with an FTS. However some work has been done on molecules of atmospheric interest using a molecular beam interfaced to a FTS (27, 28).

EXPERIMENTAL ARRANGEMEMENTS FOR FTS IN SUPPORT OF ATMOSPHERIC SCIENCE

Many laboratories possess some capabilities for conducting high quality research into the spectroscopy of atmospheric gases using FTS, and it is not possible to describe them all here. However the range of capabilities required is usefully discussed in the context of the Molecular Spectroscopy Facility (MSF) which has been established at the Rutherford Appleton Laboratory (RAL) specifically for laboratory-based atmospheric spectroscopy. This Facility comprises a high resolution FTS (Bruker model IFS 120HR) which is optically coupled to a number of absorption cells and other ancillary equipment.

Optical arrangement of the MSF

The FTS currently has a maximum optical path difference (MOPD) of 6m and continuous spectral coverage from around 500 cm^{-1} (20 μm) to 28,000 cm^{-1} (350 nm), through a number of beamsplitter, detector and source combinations. Extension of the spectral coverage into the far IR and UV (200 nm) are possible with

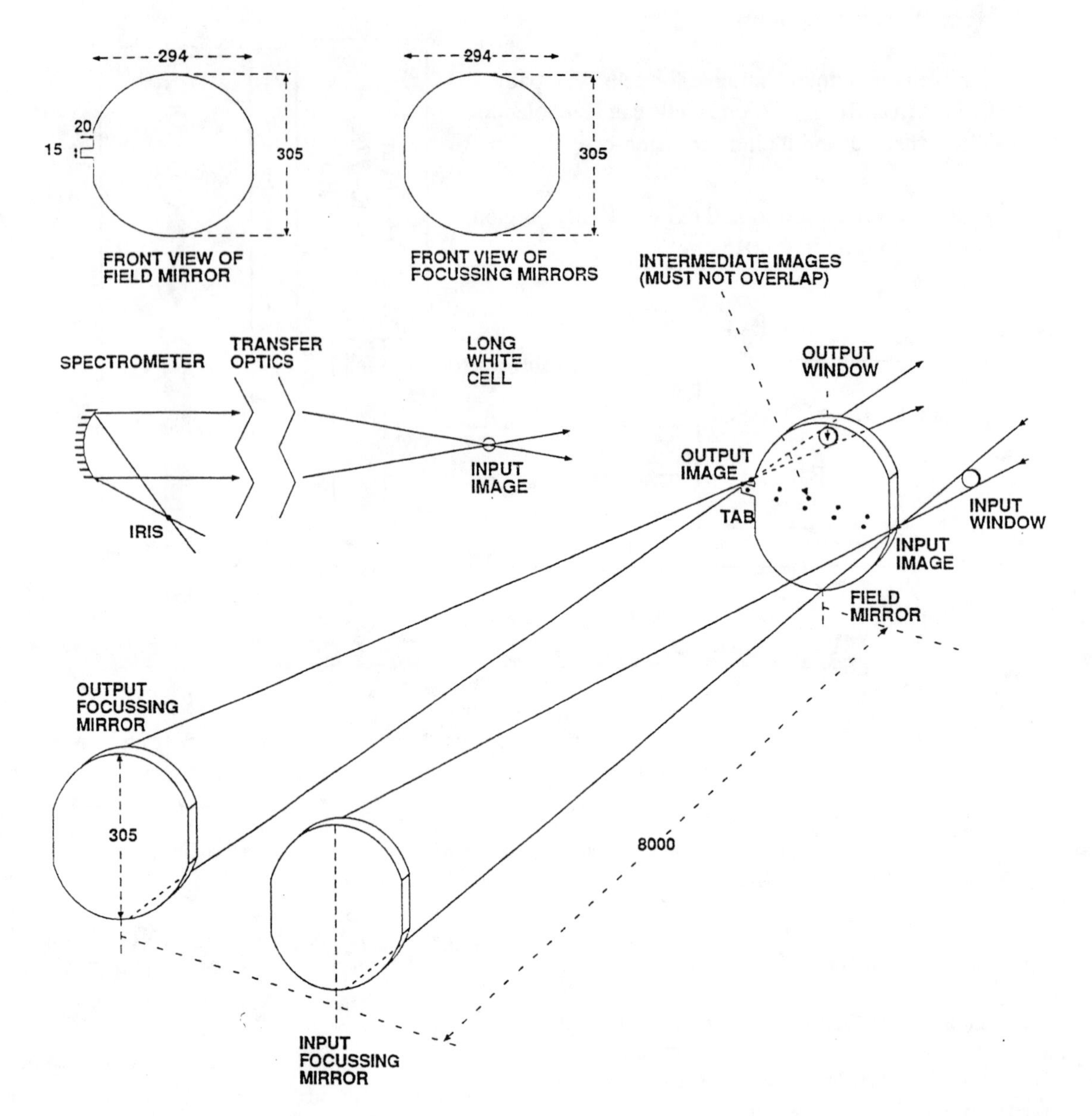

FIGURE 5: Schematic diagram of the optical arrangement used in the Long Path Absorption Cell at the MSF. Stated dimensions are mm. Reprinted from the Journal of Quantitative Spectroscopy and Radiative Transfer, Volume 52, J Ballard, K Strong, W B Johnston, M Page and J J Remedios, A coolable long path absorption cell for laboratory spectroscopy studies of gases, pages 677-691, Copyright (1994), with kind permission from Elsevier Science Ltd, The Boulevard, Langford Lane, Kidlington, OX5 1GB, UK.

commercially available add-ons. The optical arrangement is shown schematically in Fig. 4. Radiation from the source is focussed onto an variable aperture which defines the FOV. The modulated output beam from the interferometer is either allowed to pass through the spectrometer sample compartment, or is directed by a computer-controlled plane mirror through a port in the side of the spectrometer vacuum chamber. The former arrangement is used in experiments using single pass gas cells while the latter arrangement permits the source radiation to pass through longer pathlength cells via transfer optics which comprise reflective spherical and aspherical surfaces contained in evacuable tanks. Bandwidth-limiting optical filters and FOV-limiting apertures can be placed at appropriate positions in the light path.

Absorption cells

A number of absorption cells are provided which cover a wide range of pathlength, sample pressure, sample temperature and optical bandwidth, and ensure good sample stability. Optically the cells are either single or multi-pass. The former have pathlengths from a few mm to approximately 25 cm and are constructed from stainless steel or glass. They are typically coolable to temperatures below 200K with a cold fluid which is pumped through the jacket of the cell. Temperature control is achieved by regulating the flow of the coolant fluid. The details of one such cell are given in (29). The multipass cells are designed to give variable pathlengths between 1.6m and >600m. In total the range of pathlengths available spans over 5 orders of magnitude and permits a broad range of spectroscopic questions relevant to atmospheric science to be addressed. The long path absorption cell (LPAC) has been described in detail (30), but it is useful to recall some of the salient features here. Optically it is a White Cell which incorporates modifications described in (31). The spherical mirror optics are 300mm diameter and 8m radius of curvature (ie f/26.5) giving a minimum pathlength of 32m and a pathlength increment of 16m. Figure 5 shows the optical layout schematically. The cell is constructed from stainless steel and has a semi-mirrored finish on the inner surface which sample gases come into contact with. The absorption cell is wrapped in multi-layer super insulation and contained in a second stainless steel vessel which is essentially a vacuum jacket. This gives excellent thermal insulation when operating the absorption cell at temperatures below ambient.

Cooling of the cell is achieved by a heat exchanger which transfers heat from a cooling fluid to liquid nitrogen. The cooling fluid is pumped through channels on the outer surface of the absorption cell which are arranged in several circuits for optimizing uniformity of cooling. The temperature of the cell is controlled by varying the level of liquid nitrogen in the heat exchanger. Necessary signals for the temperature control are taken from a platinum resistance thermometer (PRT) which is also in contact with the cooling fluid. This system, shown schematically in Fig. 6, allows the cell to be cooled to any temperature between ambient and approximately 190K. Adjustment of the mirrors in the cell can be made when the cell is cold to compensate for small mechanical distortions which take place during cool-down.

The operating temperature of the cell and the sample gas contained therein is achieved using a number of PRTs distributed through the system. 16 PRTs protrude through the cell wall and are in good thermal contact with the gas itself and a further 16 are in thermal contact with the outer surface of the absorption cell.

The output from the interferometer can also be directed to a second multipass cell containing White-type optics of 40cm radius of curvature, thereby giving absorber pathlengths between 1.6m and approximately 20m. This system is also coolable and is well instrumented with PRTs and Baratron sensors for temperature and pressure measurements respectively. The current arrangement for

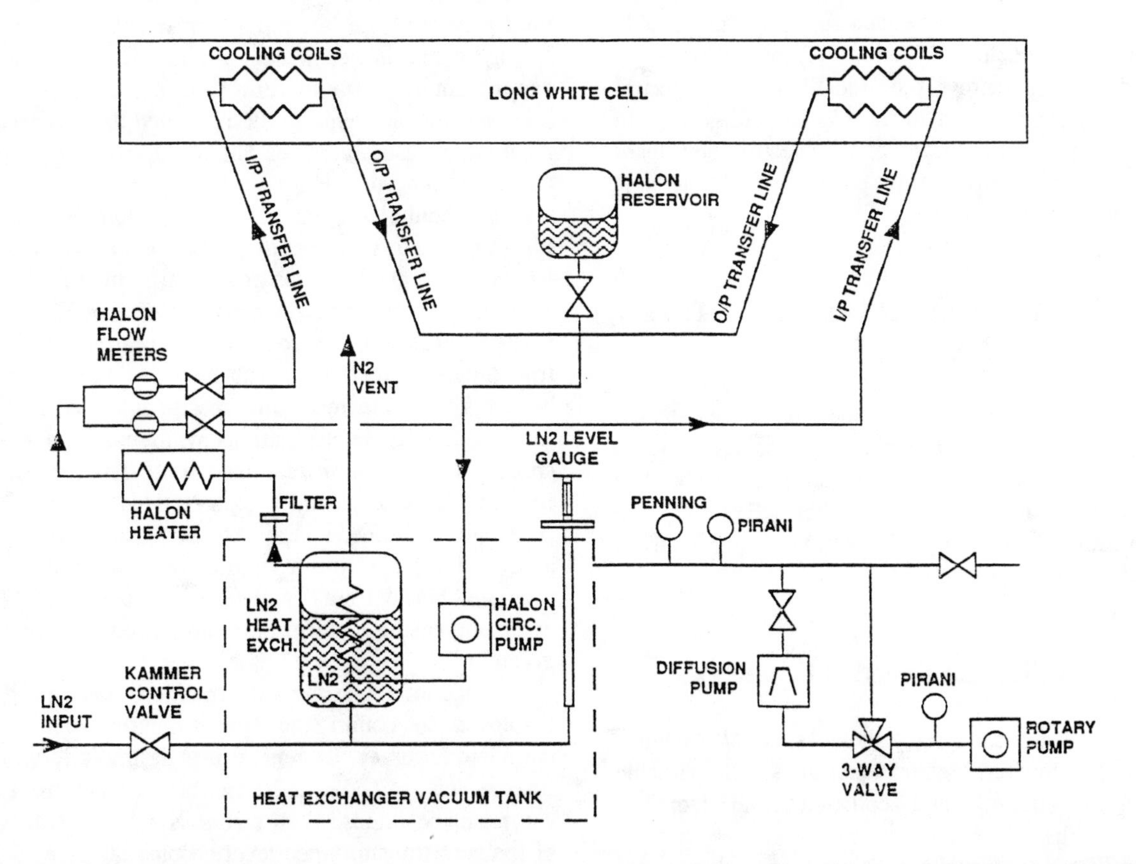

FIGURE 6: Schematic diagram of the cooling arrangement used in the Long Path Absorption Cell at the MSF. Reprinted from the Journal of Quantitative Spectroscopy and Radiative Transfer, Volume 52, J Ballard, K Strong, W B Johnston, M Page and J J Remedios, A coolable long path absorption cell for laboratory spectroscopy studies of gases, pages 677-691, Copyright (1994), with kind permission from Elsevier Science Ltd, The Boulevard, Langford Lane, Kidlington, OX5 1GB, UK.

cooling samples is somewhat different from that used in the LPAC, and has been developed by Oxford University (32). While the source of cooling is also liquid nitrogen, heat is extracted from the sample via a heat transfer gas, helium, which is contained in a vacuum tight volume between the sample volume and the liquid nitrogen-cooled surface. Temperatures between 77K and ambient are obtained by activating heaters distributed on the inner wall of the sample volume, and temperature uniformity is achieved through adjustment of the liquid nitrogen level and heater power. This system may also be operated at elevated pressures up to 5 atm absolute.

Comprehensive vacuum / gas handling systems are provided to allow filling of absorption cells to the required pressure of sample gas and its subsequent removal. Necessary precautions have been taken to permit the use of flammable gases in the LPAC, particularly so that large quantities of hydrogen and methane can be used in spectroscopic studies connected with science of Jupiter's atmosphere (33). Sample pressures are measured with calibrated Baratron capacitance manometers.

The MSF also incorporates a collision cooled cell which has been interfaced to the FTS, thereby offering extended FTS measurement capabilities for low temperature laboratory studies of vapours, with applications in atmospheric science, remote sensing, and fundamental molecular spectroscopy. Further developments of the gas handling techniques have extended the range of possible measurements to include FT spectroscopy of corrosive gases and aerosols across the mid-IR to UV spectral ranges, and at temperatures in the 180 to 300K range. In addition to applicability to gas phase spectroscopy, these techniques have application to studies of atmospheric aerosol mimics in the laboratory.

DATA ANALYSIS AND ILLUSTRATIVE RESULTS

A large number of molecular systems have been studied in the laboratory using FTS in an atmospheric science context, and it is only possible to refer to a few representative studies in any detail here. Examples are chosen to illustrate the range of experiments and analysis performed to obtain the required data.

Individual Spectral Lines

Many atmospheric gases show spectral features assignable to individual ro-vibrational lines with definable line position, strength and collision- and Doppler-broadened widths.

Line positions are usually measured at high spectral resolution and low pressure to minimise blending of adjacent spectral lines by the ILS and collision broadening. Absolute wavenumber calibration of the spectrum can be obtained, for example, by including a calibration gas in the optical light path. Due consideration must be taken of instrument effects which can introduce errors into measured line positions, such as asymmetries in the ILS. Measurement of line positions does not require accurate knowledge of the amount of sample in the light path. However the use of appropriate absorber amounts which avoid undesirable experimental conditions, such as lines which are excessively saturated, is necessary. Once empirical values have been obtained for the positions of a number of lines in a spectrum, an iterative process is typically employed in which theoretical line positions, calculated using a suitable model of the molecular energy levels, are fitted to the observed line positions. This process yields best estimates of the constants used in the model, which is then used to generate positions for the lines in the spectrum. The extent to which these calculated wavenumbers correspond to actual line positions depends on the sophistication of the model (ie how many and what type of interactions which can influence the energy levels are taken into account) and the extent and accuracy of the experimental data which is used to derive the constants in the model.

Unlike line positions, measurement of absolute line strengths requires knowledge of the absolute amount of absorber gas in the light path, as well as the parameters which determine the line shape, such as temperature and total pressure. The latter requirement arises because line strengths, and indeed widths, are mostly determined from experimental spectra by fitting a calculated spectrum with an assumed line shape to the measured data. As with line position measurements a high spectral resolution is usually employed.

Experimental determination of absolute line strengths places demands on the performance of an FTS in addition to those involved in position measurements alone. For example the photometric accuracy of the FTS becomes very important. Knowledge of the location of the "zero transmittance" level is critical, and this parameter should be checked before recording spectra intended to be used for line strength determinations by, for example, recording spectra which contain features having zero transmittance in the same spectral region. This can often be conveniently done simply by increasing the absorber amount until absorption features become so intense that they are "blacked out". A detailed discussion of potential error sources in determining line strengths by FTS is given in (34).

A range of experimental conditions may need to be employed to obtain the line strengths required with sufficient accuracy. In general it is desirable to work with lines which are between 20 and 80% transmitting at line centre; if too intense then errors associated with location of the zero transmittance level become large, while if too weak random noise may limit the measurement accuracy excessively. It may therefore be necessary to record a number of spectra with different absorber amounts (eg by varying the pathlength or pressure of absorber) so that variations of line strength across the spectral region of interest are compensated for.

Collision broadening can also be measured in the same way as intensities, and many of the same comments apply. Knowledge of the amount of the absorber gas in the light path is not required, nor is the partial pressure of the absorber except insofar as it influences the line shape through the difference between collisions between two absorber molecules and collisions between an absorber molecule and a broadening gas molecule. If "self broadening" is important relative to "foreign gas broadening" in determining the line shape then the partial pressure of the absorber gas must be known with appropriate accuracy. In any case the partial pressure of the broadening gas must be known so that the measured collision widths can be normalised to a standard pressure, usually 1 atmosphere. A common experimental practice is to measure collision line widths over a range of pressures, fitting a straight line to the resulting variation. The slope of the line gives an estimate of the broadening coefficient. Similarly the temperature dependence of collision broadening is often measured by fitting a straight line to the variation of $\ln(\alpha_L)$ with $\ln(T)$, which gives an estimate of the parameter n in equation (8). A particular challenge is measurement of line broadening in relatively crowded spectra, where the presence of line broadening significantly increases overlap of adjacent spectral lines.

Before describing some examples of line parameter determinations, the theoretical basis of the derivation of individual line parameters from measured spectra using non-linear least squares fitting will be described briefly. The observed single beam spectrum, $Y_{obs}(\nu)$, is a convolution of the true monochromatic single beam spectrum, $Y_{mono}(\tilde{\nu})$, and the ILS function, $f_{ILS}(\tilde{\nu})$:

$$Y_{obs}(\tilde{\nu}) = f_{ILS}(\tilde{\nu}) \otimes Y_{mono}(\tilde{\nu}) \tag{11}$$

The monochromatic single beam spectrum can be calculated from the line parameters of each line which produce absorption features in the wavenumber region of the fitted spectrum, using the expression:

$$Y_{mono}(\tilde{\nu}) = g(\tilde{\nu}) \prod_{i=1}^{n} \exp\left[-S_i N l K_{Voigt,i}(\tilde{\nu} - \tilde{\nu}_{0i})\right] \tag{12}$$

where $\tilde{\nu}$ is the actual wavenumber (in cm^{-1}), and $\tilde{\nu}_{0i}$ is the centre wavenumber of line i; S_i is the line strength (in units of $cm.molecule^{-1}$); N is the molecular concentration (in units of $molecule.cm^{-3}$); l is the optical pathlength through the absorber gas (in cm); $K_{Voigt,i}(\tilde{\nu} - \tilde{\nu}_{0,i})$ is the Voigt line shape, which, as shown above, has a Doppler and a collision broadening component, and $g(\tilde{\nu})$ is a function which describes the background (zero absorption) signal level. All lines with significant absorption at $\tilde{\nu}$ are included in the summation. This formalism assumes that effects such as line mixing are negligible. The line parameters (remembering that K_i depends on α_{Li}, the collision broadened width) and parameters describing $g(\tilde{\nu})$ (eg a quadratic in $\tilde{\nu}$) can be extracted from the measured spectra by performing a non-linear least squares fit of the calculated spectrum to the observed spectrum. The ILS is sometimes applied in Fourier space by multiplying the measured interferogram with the Fourier transform of the ILS.

For an ideal FTS with an infinitesimal field of view, the ILS, takes the form:

$$f(\tilde{\nu}) = \frac{\sin(2\pi\Delta\tilde{\nu})}{2\pi\Delta\tilde{\nu}} \tag{13}$$

where Δ is the maximum optical path difference (MOPD) of the interferometer. For interferograms recorded by such an instrument, the ILS corresponds to a boxcar function of length equal to the MOPD applied to an "infinite length" interferogram. Frequently an apodization function is applied to the interferogram prior to Fourier transformation to a spectrum. The finite aperture of the FTS also leads to additional 'self-apodization', $A_{\tilde{\nu}}^{\delta}$, of the recorded interferogram, which varies with the optical path difference δ, as:

$$A_{\tilde{\nu}}^{\delta} = \frac{\pi d^2}{4F^2} \frac{\sin\left(2\pi\tilde{\nu}\delta d^2/16F^2\right)}{2\pi\tilde{\nu}\delta d^2/16F^2} \tag{14}$$

where d is the diameter of the aperture, and F is the focal length of the collimating mirror. This contribution to the ILS should also be included in the calculation of the spectrum which is fitted to the observed spectrum. The sum of squared differences between calculated and observed spectra is computed, and calculation of the fitted parameters for the next iteration made using standard non-linear least-squares techniques based on, for example, the Levenberg-Marquardt algorithm (35). A judgement is made concerning the goodness of the fit of the calculated spectrum to the observed spectrum to decide whether further iteration of the parameters is required.

Two examples of line parameter determinations using FTS will be given, one concerning measurements of the red "atmospheric" bands of molecular oxygen, where individual lines are relatively well separated from each other, and the second concerning NO_2 vapour where a significant degree of line blending occurs even at low broadening pressures.

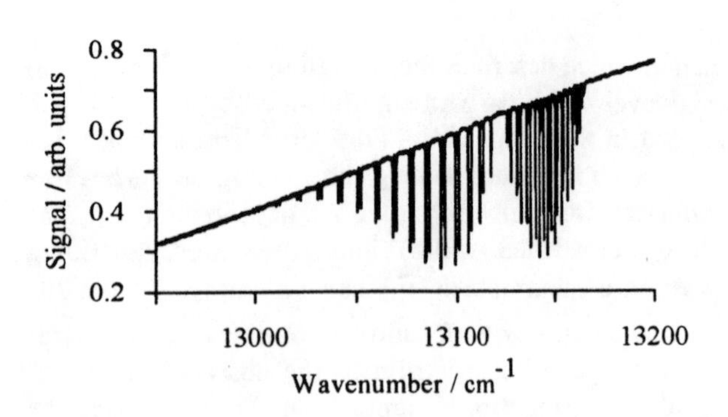

FIGURE 7: A spectrum of oxygen recorded at a resolution of 0.034 cm^{-1} through the LPAC at a 32.75m pathlength with 1000hPa of air in the cell at 283K

Oxygen Line Parameters

Oxygen (O_2) is a common atmospheric gas but has no permanent electric dipole, so does not exhibit a "normal" electric dipole spectrum. However its relatively weak spectral properties arising from, *inter alia*, magnetic dipole transitions are important in atmospheric science for a number of reasons. However the line data available in the current HITRAN database (36) is inadequate for some applications, particularly those requiring good knowledge of the collision line widths by air of the O_2 *A* and *B* bands, so new measurements have recently been made in the MSF at RAL.

Observation of these bands in the laboratory is difficult due to the weakness of the absorption at atmospheric conditions, so the LPAC was used at pathlengths up to 256m. Figure 7 shows a raw spectrum of oxygen in zero grade air (a mixture of oxygen and nitrogen of precisely known composition). This spectrum was recorded through the LPAC at a 32.75m pathlength with 1000hPa of air in the cell at 283K at a resolution of 0.034 cm^{-1}. The overall shape of the backround is determined by the optical components employed in the FTS (principally the source, beamsplitter, bandpass filter, cell windows and detector) and the sharp line features are due to absorption arising from O_2 $b^1\Sigma_g^+(v'=0)$ $\leftarrow$ $X^3\Sigma_g^-(v''=0)$ rovibrational transitions.

Spectral line parameters of O_2 in zero air were determined from measured spectra using a non-linear least squares fitting procedure similar to that described above. In this study, either single isolated lines were fitted using 6 parameters or up to 31 overlapping O_2 lines were fitted simultaneously to a total of up to 96 parameters. In all cases the fits included three spectroscopic parameters per line ($\tilde{\nu}_0$, S, α_L) and 3 parameters modelling an assumed quadratic baseline. The initial first guess positions and quantum number assignments of the lines identified in the spectra were taken from tables of literature values, and used to determine first guess line strength values. The line shape was assumed to be a Voigt profile, calculated using a computer code based on a standard algorithm (37). The Lorentzian component of the lineshape was calculated for each line from the air pressure broadening parameter (α_L) and the measured pressure.of zero air. The Doppler profile for each line was calculated from the temperature, molecular mass, and centre wavenumber of the line. The monochromatic spectrum $Y_{mono}(\tilde{\nu})$ was calculated from the line parameters, Fourier transformed and the ILS function was applied in Fourier space. An inverse Fourier transform was then applied. This spectrum was oversampled with respect to the experimental data to ensure correct application of the ILS. Tests showed that a sampling grid of at least 6 times finer spacing than the data-point spacing of the original experimental spectra was needed to minimize errors from this source. The actual number of data points in the calculated spectrum was 2^m, where m is an integer, allowing optimum

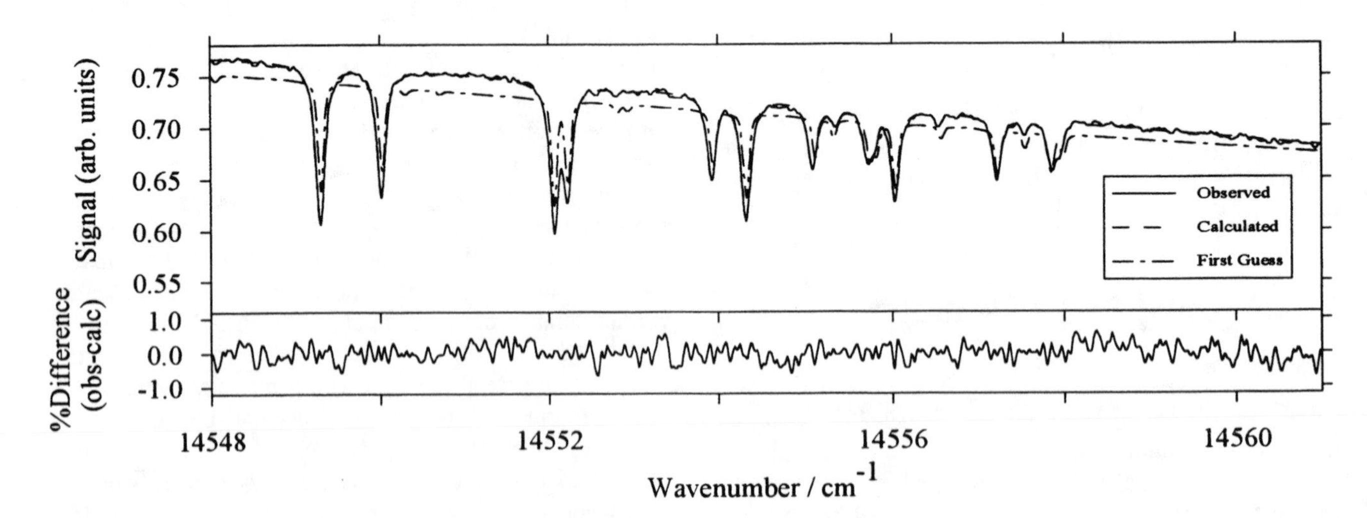

FIGURE 8: . The fitting procedure applied to a section of an O_2 B band spectrum. The spectrum was recorded at 0.04cm^{-1} resolution with 1000hPa of zero air in the LPAC at 128.75m path length and 283K The upper panel shows the measured spectrum as a solid line, the first guess calculated spectrum using initial values of the line and background parameters is shown as a dot-dash line, and the final (fitted) spectrum is shown as a dashed line. The difference between the experimental and fitted data is shown in the lower panel.

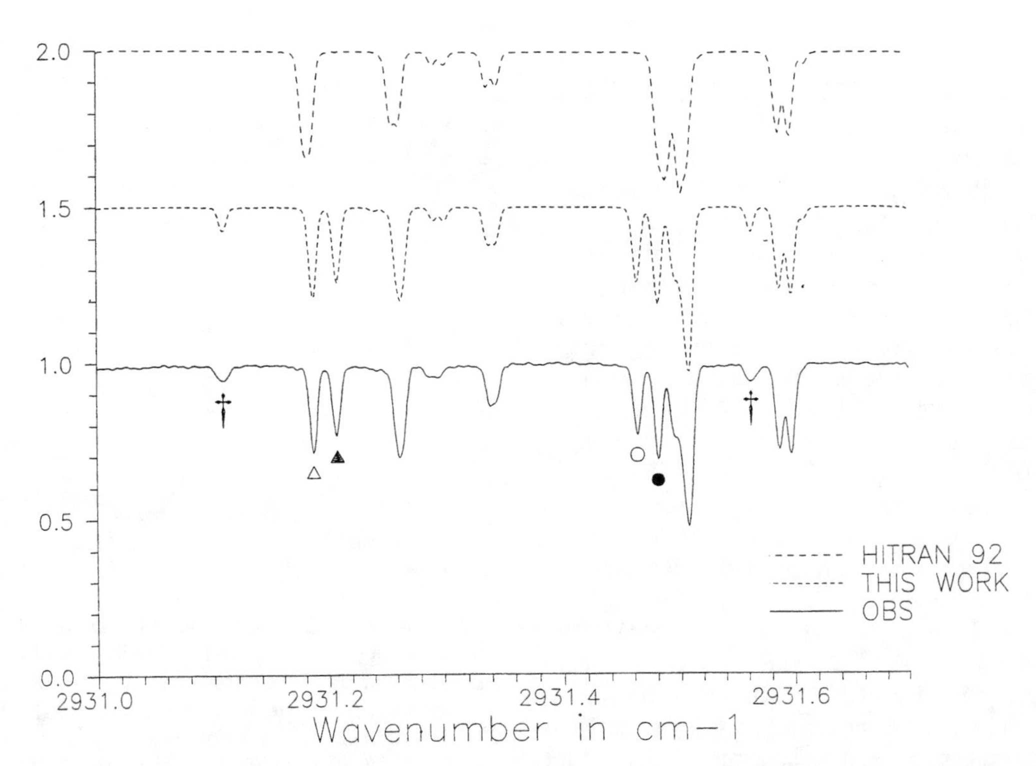

FIGURE 9: Improvement to NO_2 spectrum in the 3.4μm region. The lower trace is an experimental spectrum, the middle trace a spectrum calculated using line data available in the HITRAN-92 linelist (36) and the upper trace is a spectrum calculated using the results of (39). Certain transitions (indicated by daggers) were not represented at all on the HITRAN-92 linelist, and others (indicated by circles and triangles) are better represented due to an improved treatment of spin-rotation interactions. This figure is reproduced from (39) by kind permission from Academic Press

performance of the fast-Fourier transform (FFT) subroutine employed.

The general form of apodization functions employed in FTS is given by:

$$A(\delta) = \sum_{i=0}^{n} C_i \left[1 - \left(\frac{\delta}{\Delta}\right)^2\right]^i \tag{15}$$

where δ is the optical path difference and Δ is the MOPD. The coefficients C_i used to generate the apodization functions used in analysis of the O_2 spectra are given in Table 1.

The sum of squared differences, χ^2, between calculated and observed spectra was determined, and calculation of the fitted parameters for the next iteration made using standard non-linear least-squares techniques based on the Levenberg-Marquardt algorithm (35). This iterative procedure was continued automatically until the value of χ^2 was reduced to below a set level. Final output of the process was a listing of the fitted line parameters for each of the lines identified as contributing to the absorption features in the observed spectra. The fitting procedure applied to a section of an O_2 B band spectrum is illustrated in Fig. 8. The spectrum was recorded at 0.04cm^{-1} resolution with 1000hPa of zero air in the LPAC at 128.75m path length and 283K. In the upper panel of Fig. 8 the measured spectrum is shown as a solid line, the first guess calculated spectrum using initial values of the line and background parameters is shown as a dot-dash line, and the final (fitted) spectrum is shown as a dashed line. The difference between the experimental and fitted data is shown in the lower panel. A total of 96 parameters were fitted simultaneously; line position, strength and pressure-broadened width for each of 31 lines, and 3 parameters describing a quadratic background. The MOPD and self apodization contributions to the ILS were included.

TABLE 1. Apodization coefficients used in the O_2 analysis

Function	C_0	C_1	C_2	C_3
Boxcar	1.000000	0	0	0
Norton-Beer Strong	0.045335	0	0.554883	0.399782

Nitrogen Dioxide Line Parameters

NO_2 is also an important atmospheric gas despite its low concentration in the atmosphere. In the stratosphere it participates in one of the main catalytic cycles which destroy ozone, so much effort has been expended to develop techniques with which to measure NO_2 in the atmosphere. Among these are remote sensing techniques, for which spectral line parameters are required in the data reduction. For example NO_2 absorptions around 3.4μm wavelength have been used to measure atmospheric spectra from a ground-based FTS (38), and work has been carried out recently in the laboratory to determine appropriate line parameters.

Knowledge of these line parameters prior to this work was quite poor; for example the HITRAN data base contained the same value for the air broadened line width for all lines in all bands. The results have been reported in (39) and (40), which illustrate some of the difficulties which have to be overcome to make line parameter measurements in a molecule like NO_2 and the usefulness of the FTS in obtaining spectral data. The FTS enabled a high signal to noise (500) to be achieved at a high resolution (0.0033 cm^{-1}) over the spectral interval 2633 - 3511 cm^{-1} in a measurement time of 100 minutes. Broader band spectra (0 - 7900cm^{-1} , at 0.03 cm^{-1} resolution) could be recorded to check for impurities in the NO_2 sample, such as HNO_3, NO and CO_2. The pressure of the NO_2 vapour in the absorption cell was kept at a stable value during recording of spectra despite the relative high reactivity of NO_2. The actual NO_2 partial pressure had to be deduced from the total measured pressure since NO_2 exists in a pressure and temperature dependent equilibrium with its dimer, N_2O_4. The measured spectra were used to deduce band centres, and various constants in the energy level model employed (rotational, spin-rotation and coupling constants), as well as line strengths and line widths due to self, N_2 and O_2 broadening. Figure 9 illustrates the improvements in the representation of the NO_2 spectrum arising from this work.

Absorption Cross Sections

As has been observed already many atmospheric molecules do not lend themselves to a line-by-line analysis as described above. For these so-called "heavy molecules" it is necessary to adopt a purely empirical approach to obtain the absorption cross section data required. There have been many studies of absorption cross sections conducted by many groups; again it is only possible to give the basic principles and a couple of illustrative examples here.

The basic spectroscopic measurement is of transmittance of the sample at known physical conditions. Absorption cross sections are then derived using equation (10), often by combining a number of measurements at different absorber amounts. Deriving accurate cross sections therefore requires accurate transmittance measurements, and accurate knowledge of the absorber amount in the light path. The former is obtained by taking the ratio of spectra recorded with, and without, the sample in the cell, other factors being unchanged. This imposes further requirements on the photometric performance of

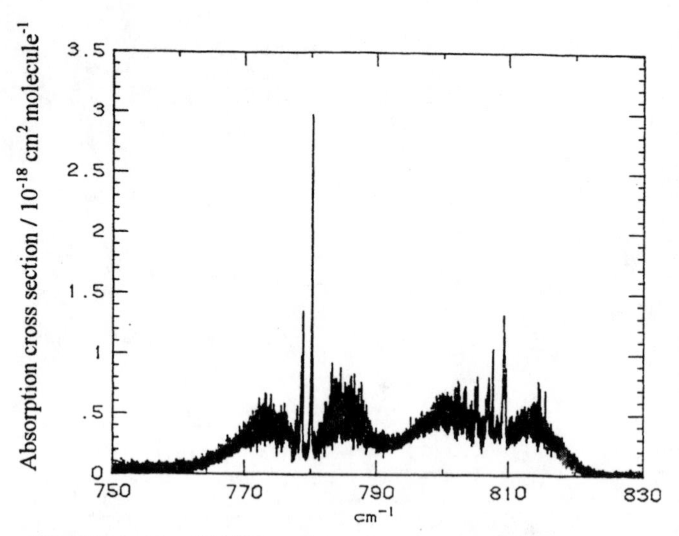

FIGURE 10: $ClONO_2$ absorption cross sections at 213K measured at 0.02 cm^{-1} resolution. Reproduced from J Ballard, W B Johnston, M R Gunson and P T Wassell, Journal of Geophysical Research, volume 93, page 1663, 1988, copyright by American Geophysical Union.

the system, such as reproducibility of the 100% transmittance signal level. It is particularly important that this ratio should be free of structure on the same wavelength scale as the absorption bands being measured. Cross section measurements are sometimes made at spectral resolutions which do not fully resolve spectral features, such as sharp Q branches. In this circumstance it is important that experimental conditions are used which ensure that the spectral features, were they fully resolved, do not approach saturation, if accurate cross sections at the measurement resolution are to be obtained. Stability of the sample in the absorption cell is very important too, particularly if there is a significant time delay between recording the background and the sample spectra. Effects which make samples unstable in a cell can be physical or chemical in origin. The latter can often be minimised by suitable choice of materials with which the sample comes into contact and by rendering such surfaces as dry as possible to prevent hydrolysis. Physical effects include thermal decomposition, photolysis and adsorption. Again choice of cell materials can be critical in minimising these.

A very important heavy molecule in atmospheric science is chlorine nitrate ($ClONO_2$), since it plays a significant role in ozone chemistry in the lower stratosphere. Its atmospheric concentration has only be measured spectroscopically, for example from space or ground-based FTS. $ClONO_2$ absorption cross sections were therefore measured at RAL using a 4.5cm pathlength coolable glass absorption cell and an FTS (41) and some results from this study are shown in Figs. 10 and 11.

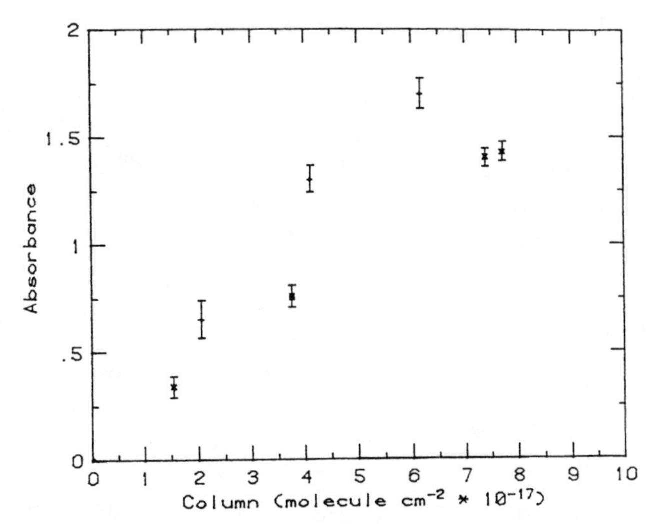

FIGURE 11: Absorbance as a function of absorber amount for the peak of the $ClONO_2$ ν_4 Q branch at 296K (crosses) and 213K (plus signs). Error bars are $\pm 1\sigma$ indicating random effects. Reproduced from J Ballard, W B Johnston, M R Gunson and P T Wassell, Journal of Geophysical Research, volume 93, page 1663, 1988, copyright by American Geophysical Union.

Another atmospherically important class of heavy molecules are HCFCs, HFCs and other fluorinated hydrocarbons and ethers which are potentially significant contributors to the enhanced greenhouse effect. Calculations of their global warming potential cannot proceed however unless adequate data on the spectral properties of these molecules is available. The FTS has become pre-eminent in obtaining the spectral data since these molecules often have several absorption bands in the wavelength region of interest to greenhouse warming (5μm to 20μm wavelength) and the broad coverage of the FTS is a distinct advantage.

A programme of measurements of IR absorption cross sections of selected greenhouse gases and assessment of their likely contribution to radiative forcing of the Earth's atmosphere is currently being carried out for the European Commission. This programme involves experimental spectroscopy groups in the UK (RAL and Strathclyde University), Italy (University of Bologna), Denmark (University of Copenhagen) and Belgium (Free University of Brussels) each of which is using an FTS to measure a number of CFC-substitutes or other fluorinated hydrocarbons. Associated radiative forcing calculations are being carried out at Reading University (UK). This project included an intercomparison of HCFC-22 cross sections measured by each group at the same experimental conditions. It was interesting to note that initial error estimates in most cases were unrealistically small and consequently measured values of the same parameter by different groups differed by more than the declared errors. Subsequent re-analysis taking account of all sources of error (including photometric, sample preparation and sample stability contributions) resulted in more realistic error estimates and a much greater degree of consistency in the measurements from the different groups. An example of the output coming from that project is given in Fig 12, which shows absorption cross sections for 2.1 hPa C_4F_8 vapour at 203K. The data were recorded at the MSF with a spectral resolution of $0.03 cm^{-1}$

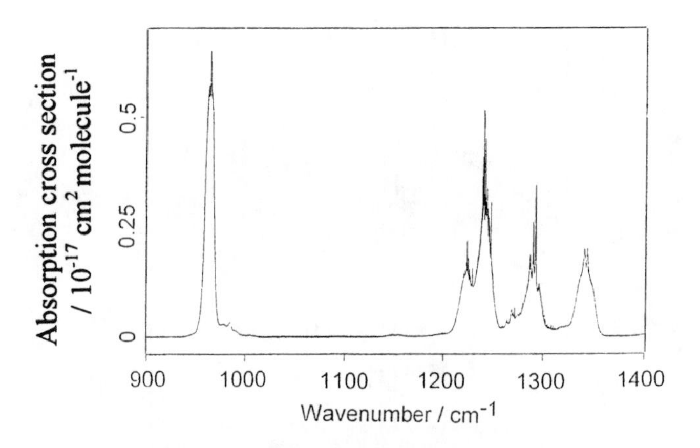

FIGURE 12: Absorption cross sections for 2.1 hPa C_4F_8 vapour at 203K. The data were recorded at the MSF with a spectral resolution of $0.03 cm^{-1}$

Collision cooling

As has been noted above, collision cooling is a technique for recording gas phase spectra at temperatures much colder than the sample's normal freezing point, and with sample pressures much larger than the normal SVP at the measurement temperature. It is therefore a very useful technique for simplifying spectra of heavy molecules, thereby aiding a line-by-line analysis of their absorption bands. Also this technique lends itself to measurements of line broadening by the buffer gas at cold temperatures, since for such measurements an estimate of the partial pressure of the absorber gas is not required. The RAL implementation of collision cooling has been applied to both these aspects of spectroscopy using an FTS.

Initial collision cooling experiments (42) utilized a gas handling system in which the cell was first filled with a fixed pressure of buffer gas, and pure absorber gas was continually flowed into the cell through an injector which was heated to prevent the sample freezing before reaching the buffer gas. At a total pressure of 10 hPa the absorber gas is rapidly cooled to the buffer gas temperature by collisions, and sufficient quantities of vapour could be generated at temperatures approaching 77K for high resolution FTS spectra to be recorded. Under these conditions, the spectra are simplified by efficient cooling of the rotational and vibrational degrees of freedom, which can be a useful aid in the assignment of congested

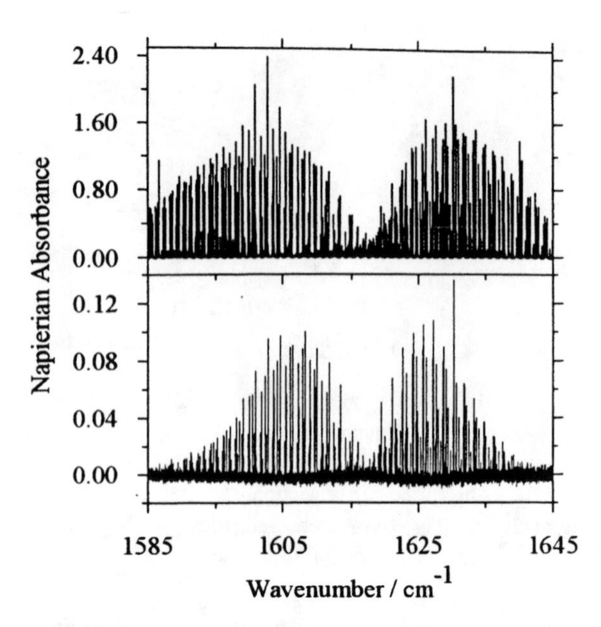

FIGURE 13: Illustrating spectral simplification by collision cooling of NO_2. Upper spectrum is of 0.1 hPa of vapour at 294K and 0.006 cm^{-1} resolution. Lower spectrum is collision cooled NO_2 vapour in 10 hPa helium to a temperature ≅ 110K. Reproduced from (42).

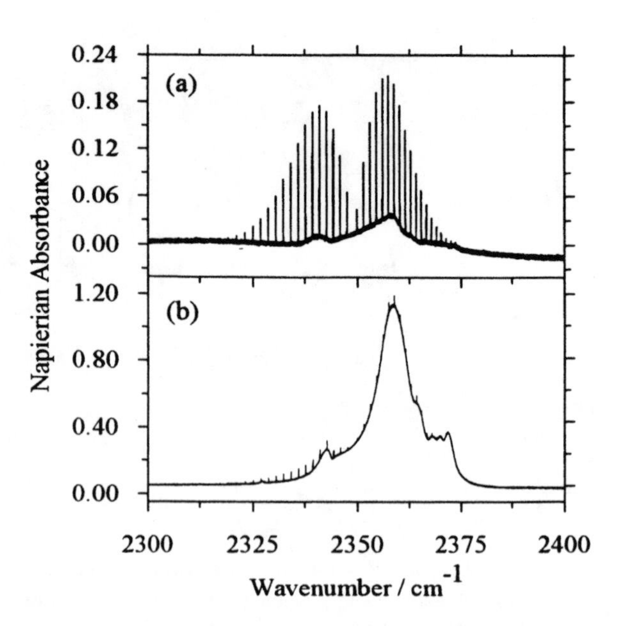

FIGURE 14: FTIR spectra recorded using a collision cooling cell with 1% CO_2 in 100 hPa of buffer gas. Argon was the buffer gas in spectrum (a), while the buffer gas was helium in spectrum (b). Data obtained at RAL MSF at 0.004 cm^{-1} resolution

and complex spectral features. Examples of molecules studied using this flow regime included CO_2, NO_2, HCFC-142b, and methyl silane (CH_3SiH_3). Figure 13(a) shows a spectrum of NO_2 at room temperature, and Fig. 13(b) shows an NO_2 spectrum recorded using the collision cooled cell cooled with liquid nitrogen and with 10 hPa He as the buffer gas. The temperature of the vapour in the latter is approximately 110K. Significant simplification of the spectrum occurs at the lower temperature, as expected. Figure 13 also illustrates the benefit of collision cooling in recording vapour phase spectra at temperatures significantly below the SVP of the sample at the measurement temperature. The SVP of NO_2 falls rapidly on cooling below room temperature, initially through conversion to N_2O_4 and then freezing at 263.8K. Conventional cooling to 110K would result in negligible vapour pressure of NO_2.

Measurements using the cell gas flow method described above with total gas pressures in excess of 100hPa resulted in clustering due to high localised concentration of low temperature absorber molecules. Although the molecular clusters generated are of considerable interest, for low temperature pressure broadening studies it is necessary to record FT spectra of the vapour in the buffer gas at total pressures of 100hPa and above. This has been achieved at RAL through the design and construction of a new collision cooling cell, in which a 1% mixture of the absorber gas in the desired broadening gas can be continually flowed through the cell. A system of gas handling has been developed which allows automatic control of the gas mixture flow and gas pressure to be achieved over extended measurement periods. Using this system, mid-IR spectra of CO_2 pressure broadened by 100 hPa of helium, argon, nitrogen,or air have been recorded at temperatures of 100K. Figure 14(a) shows a spectrum of CO_2, recorded with the cell cooled with liquid nitrogen and with 100 hPa of helium as the buffer gas. The broadening coefficient at the measurement temperature of 110K can be obtained by least squares fitting, as has been described previously. Figure 14(b) shows a spectrum of CO_2 recorded under the same conditions as (a) except that the buffer gas is argon. In this case very little of the CO_2 is in the gas phase. The cause of the increased clustering of CO_2 molecules when cooled with argon is unknown, but could be connected with the relative boiling points of helium and argon.

Aerosols

Research into spectral properties of aerosols is needed to elucidate their radiative properties in the atmosphere, and to assist in remote sensing of their composition and concentration in the atmosphere. Recent studies by a number of groups have concentrated on the mid-IR region for investigations into the freezing behaviour of sulphuric acid (43, 44, 45), nitic acid tri-hydrate (46, 47), and ternary solutions of nitric acid, sulphuric acid and water (48), as well as studies to retrieve complex refractive indices directly from extinction spectra of pure water (49), nitric acid tri-hydrate (50), and pure hydrazine (51) aerosols.

Experiments conducted at RAL have combined an FTS and a modification of the collision cooling experiment to generate aerosols. Nitrogen gas is bubbled through

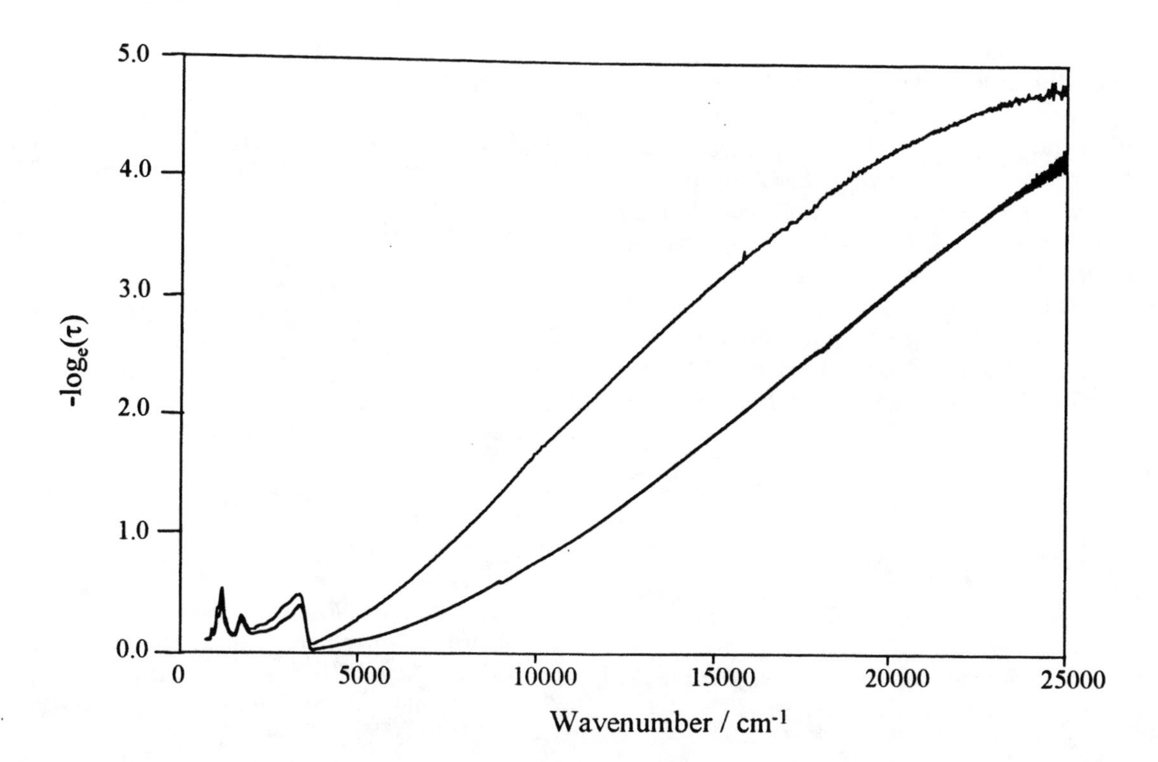

FIGURE 15: Preliminary spectra of sulphuric acid aerosol at 233K, recorded over a broad spectral interval at the RAL MSF. The absorption features in the mid-IR indicate that the aerosols in both spectra have the same composition (approx 65% weight of H_2SO_4), but the scattering extinction at short wavelengths indicates very different particle sizes.

separate liquid samples of oleum and water. The separate streams of H_2O and SO_3 vapour in N_2 are flowed into a cell where they react to form sulphuric acid aerosols. The temperature of the cell can be varied between 298 and 180K to represent temperatures found in the lower atmosphere, and the aerosols cool to the cell temperature by collision cooling. Corrosion of components from reactions with SO_3 and sulphuric acid is minimised by using glass, PFA and PTFE components in the gas transfer lines. The cell windows are made from acid-resistant BaF_2 which has a high transmittance from the mid-IR to the near UV, thereby permitting extinction spectra of the aerosols to be measured over a wide range of frequencies. To achieve complete spectral coverage between the infrared and the ultraviolet, measurements using four combinations of light source, beamsplitter and detector are required.

By carefully reproducing specific sets of flow conditions into and out of the cell, the same composition and size distribution of aerosols can be created with each of the different combinations of optical equipment in place, allowing the overlapping spectral intervals to be combined to give a single extinction ($-\log_e(\tau)$) spectrum for the entire wavenumber region of interest. Figure 15 is a preliminary example of two such extinction spectra of sulphuric acid aerosol at 233 K with the absorption being the primary cause of the bands present in the infrared region and the scattering causing the steadily increasing extinction towards higher wavenumbers.

The weight percent composition of the sulphuric acid in the aerosol has been estimated from the relative intensity of the group of sulphate absorption peaks (820-1470 cm^{-1}) and the absorption due to OH bond stretching (2409-3650 cm^{-1}) from a calibration curve reported in (43). Using this method the composition of the aerosol in the two spectra shown in figure n was deduced to be about 65 wt. % H_2SO_4 in both cases, and the differences in the extinction spectra can be attributed to the different size distributions of the droplets created by using different flow conditions in the two sets of experiments. The wide wavenumber range of the RAL data results in extinction spectra which contain information about the particle size distribution as well as the chemical composition, so are potentially useful in calculations of the short-wave radiative forcing due to sulphate aerosols. Furthermore, the spectra can be used in direct comparisons with remotely-sensed data at near-infrared and visible frequencies The main work on the sulphate aerosols at RAL has thus far concentrated on investigations into the composition and size of the droplets formed at different temperatures and development of techniques by which these two properties can be controlled and made as reproducible as possible. Calculating the size distribution of the particles has involved the use of Mie scattering and fitting procedures using reference data from Palmer and Williams (52).

ACKNOWLEDGEMENTS

The authors wish to thank the UK NERC for their support of the MSF at RAL, the European Space Agency for funding the oxygen spectroscopy carried out at the MSF under contract no. 11340/95/NL/CN; the European Commision for their funding of the Spectroscopy and Warming Potentials of Atmospheric Greenhouse Gases project under contract no. GT95-0069; Drs R Grainger and A Lambert (Oxford University) for use of their computer programs in analysing aerosol spectra; Mr R Knight (RAL) for measuring the C_4F_8 spectrum and supplying the associated figure; Dr W J Reburn (RAL) for developing the line fitting software used at the MSF

REFERENCES

1. Jarnot, R. F., Cofield, R. E., Waters, J. W., Flower, D. A., and Peckham, *J. Geophys. Res.* **101,** 9957-9982 (1996).
2. Taylor, F.W., Rodgers, C. D., Whitney, J. G., Werrett, S. T., Barnett, J. J., Peskett, G. D., Venters, P., Ballard, J., Palmer, C. W. P., Knight, R., Morris, P., Nightingale, T. J., and Dudhia, A., *J. Geophys. Res.* **98,** 10799-10814 (1993).
3. Gunson, M. R., Abbas, M. M., Abrams, M., C., Allen, M., Brown, L. R., Brown, T. L., Chang, A. Y., Goldman, A., Irion, F. W., Lowes, L. L., Mahieu, E., Manney, G. L., Michelson, H. A., Newchurch, M. J., Rinsland, C. P., Salawitch, R. J., Stiller, G. P., Toon, G. C., Yung, Y. L., and Zander, R, *Geophys. Res. Letts.* **23,** 2333-2336 (1996).
4. Endemann, M., and Fischer, H., *ESA Bulletin - European Space Agency* **76,** 47-52 (1993).
5. Readings, C. J., and Dubock, P. A., *ESA Bulletin - European Space Agency* **76,** 15-28 (1993).
6. Persky, M. J., *Rev. Sci. Instr.* **66,** 4763-4797 (1995).
7. Krueger, A., J., *Planet. Space Sci.* **37,** 1555-1565 (1989).
8. Hahne, A., Lefebvre, A., Callies, J., and Zobl, R., *ESA Bulletin - European Space Agency* **76,** 22-29 (1993).
9. Vaughan, G., Roscoe H. K., Bartlett L. M., O' Connor, F. M., Sarkissian, A., VanRoozendael, M., Lambert, J. C., Simon, P. C., Karlsen, K., Hoiskar, B. A. K., Fish, D. J., Jones, R. L., Freshwater, R. A., Pommereau, J. P., Goutail, F., Andersen, S. B., Drew, D. G., Hughes, P. A., Moore, D., Mellqvist, J., Hegels, E., Klupfel, T., Erle, F., Pfeilsticker, K., and Platt, U., *J. Geophys. Res.* **102,** 1411-1422 (1997).
10. Solomon, S., *Nature* **347,** 347-354 (1990).
11. Solomon, S., Portmann, R. W., Garcia, R. R., Thomason, L. W., Poole, L. R., and McCormick, M. P., *J. Geophys. Res.* **101,** 6713-6727 (1996).
12. Lambert, A., Grainger, R. G., Remedios, J. J., Rodgers, C. D., Corney, M., and Taylor, F. W., *Geophys. Res. Letts.* **20,** 1287-1290 (1993).
13. Wang, P. H., Kent, G. S., McCormick, M. P., Thomason, L. W., and Yue, G. K., *Appl. Opt.* **35,** 433-440 (1996).
14. Shippony, Z., and Read, W. G., *J Quant. Spectrosc. Radiat. Transfer* **50,** 635-646 (1993).
15. Schrier, F., *J Quant. Spectrosc. Radiat. Transfer* **48,** 743-762 (1993).
16. Birch, J. R., and Clarke, F. J. J., *Spectroscopy Europe* **7,** 16-22 (1995).
17. Ballard, J., Remedios, J. J., and Roscoe, H. K., *J Quant Spectrosc Radiat Transfer* **48,** 733 - 741 (1992).
18. Johns, J. W. C., Lu, Z., Weber, M., Sirota, J. M., and Reuter, D. C., *J. Molec. Spectrosc.* **177,** 203-210 (1996).
19. Abrams, M. C., Toon, G. C., and Schindler, R. A., *Appl. Opt.* **27,** 6307-6314 (1994).
20. Schermaul, R., Seibert, J. W. G., Mellau, J. C., and Winnewisser, M., Appl. Opt. 35, 2884-2890 (1996).
21. White, J. U., *J. Opt. Soc. Am.* **32,** 285 (1942).
22. Messer, J. K., and DeLucia, F. C., *Phys. Rev. Letts.* **53,** 2555-2558 (1984).
23. Goyette, G. M., and DeLucia, F. C., *J Molec. Spectrsoc.* **143,** 346-358 (1990).
24. Jin, P., Wang, H., Oatis, S., Hall, G. E., Sears, T.J., *J. Molec. Spectrosc.* **173,** 442-451 (1995).
25. Dunder, T., and Miller, R. E., *J. Chem. Phys.* **93,** 3693-3703 (1990).
26. Quack, M., *Ann. Rev. Phys. Chem.* **41,** 839 (1990).
27. Ballard, J., Newnham, D., and Page, M., ***Chem. Phys. Lett.*** **208,** 295-298 (1993).
28. McNaughton, D., and Evans, C., *J. Phys. Chem.* **100,** 8660-8664 (1996).
29. Ballard, J., Johnston, W. B., Moffat, P.H., and Llewellyn-Jones, D. T., *J. Quant Spectrosc Radiat Transfer,* **33,** 365-371 (1985).
30. Ballard, J., Johnston, W. B., Page, M., Strong, K., and Remedios, J. J., *J. Quant. Spectrosc. Radiat. Transfer* **52,** 677-691 (1994).
31. Bernstein, H. J., and Herzberg, G., *J. Chem. Phys.* **16,** 30-39 (1948).
32. Sihra, K. (Private Communication).
33. Strong, K., Taylor, F. W., Calcutt, S. B., Remedios, J. J., and Ballard, J., *J. Quant. Spectrosc. Radiat. Transfer.,* **50,** 363 - 429 (1993).
34. Birk, M., Hausamann, D, Wagner, G., and Johns, J. W., *Appl. Opt.* **35,** 2971-2985 (1996).
35. Press, W. H., Flannery, B. P., Teukolsky, S. P., and Vetterling, W. T., *Numerical Recipes - The Art of Scientific Computing,* Cambridge University Press, 1986
36. Rothman, L. S, Gamache, R. R., Tipping, R. H., Rinsland, C. P., Smith, M. A. H., Benner, D. C., Devi, V. M., Flaud, J. M., Camy-Peyret, C., Perrin, A., Goldman, A., Massie, S. T., Brown, L. R. and Toth, R. A., *J. Quant. Spectrosc. Radiat. Transfer* **48,** 645-651 (1992).
37. Armstrong, B. H., *J. Quant. Spectrosc. Radiat. Transfer* **7,** 61 (1967).
38. Nothold, J., and Schrems, O., *Geophys. Res. Letts.* **21,** 1355-1358 (1994).
39. Mandin, J.-Y., Dana, V., Perrin, A., Flaud, J.-M., Camy-Peyret, C., Regalia, L. and Barbe, A., *J. Molec. Spectrosc.* **181,** 379-388 (1997).
40. Dana, V., Mandin, J.-Y., Allout, M.-Y., Perrin, A., Regalia, L., Barbe, A., Plateaux, J.-J., and Thomas, X., *J. Quant. Spectrosc. Radiat. Transfer* **57,** 445-459 (1997).
41. Ballard, J., Johnston, W. B., Gunson, M. R., and Wassell, P. T., *J. Geophys. Res.* **93,** 1659-1665 (1988).
42. Newnham, D., Ballard, J. , and Page, M., *Rev. Sci. Instr.* **66,** 4475-4481 (1995).

43. Anthony, S. E., Tisdale, R. T., Disselkamp, R. S., Tolbert M. A., and Wilson J. C., *Geophys. Res.Lett.* **22,** 1105-1108 (1995).
44. Bertram, A. K., Patterson, D. D., and Sloan, J. J., *J. Phys. Chem.,* **100,** 2376-2383 (1996).
45. Clapp, M. L., Niedziela, R. F., Richwine, L. J., Dransfield, T., Miller, R. E., and Worsnop, D. R., *J. Geophys. Res.* **102,** 8899-8907 (1997).
46. Bertram, A. K., and Sloan, J. J., *J. Geophys. Res* In Press (1997)
47. Disselkamp, R. S., Anthony, S. E., Prenni, A. J., Onasch, T. B., and Tolbert, M. A., *J. Phys. Chem.* **100,** 9127-9137 (1996).
48. Anthony, S. E., Onasch, T. B., Tisdale, R. T., Disselkamp, R. S., Tolbert, M. A., and Wilson, J. C., *J. Geophys. Res.* **102,** 10777-10784 (1997).
49. Clapp, M. L., Miller, R. E., and Worsnop, D. R., *J. Phys. Chem.* **99,** 6317-6326 (1995).
50. Richwine, L. J., Clapp, M. L., Miller, R. E., and Worsnop D. R., *Geophys. Res. Lett.* **22,** 2625-2628 (1995).
51. Clapp M. L., and Miller, R. E., *Icarus* **123,** 396-403 (1996).
52. Palmer, K. F., and Williams, D., *Appl. Opt.* **1,** 208-219 (1975).

INFRARED SPECTROSCOPIC IMAGING MICROSCOPY: APPLICATIONS TO BIOLOGICAL SYSTEMS

Linda H. Kidder, Ira W. Levin, and E. Neil Lewis*

Laboratory of Chemical Physics, National Institutes of Diabetes and Digestive and Kidney Diseases, National Institutes of Health, Bethesda, MD 20892-0510

The coupling of imaging modalities with spectroscopic techniques adds additional dimensions to sample analysis in both the spectroscopic and spatial domains. The particular ability of infrared (IR) imaging to explore the spatial distribution of chemically distinct species on length scales ranging from microns to kilometers demonstrates the versatility and diversity of spectroscopic imaging. In this paper, we focus on the further development of our Fourier-transform (FT) based mid-IR spectroscopic imaging technique which combines the analytical capabilities of mid-IR spectroscopy with the morphological information obtained from optical imaging. The seamless combination of spectroscopy for molecular analysis with the power of visualization represents the future of infrared microscopy. Our spectroscopic imaging instrument integrates several infrared focal-plane arrays with a Michelson step-scan interferometer, generating high-fidelity and high spectral resolution mid-infrared spectroscopic images. The instrumentation produces multidimensional, chemically specific images, while simultaneously obtaining high resolution spectra for each detector pixel. The spatial resolution of the images approaches the diffraction limit for mid-infrared wavelengths, while the spectral resolution is determined by the interferometer, and can be 4 cm^{-1} or higher. Data derived from a variety of materials, particularly biological samples, illustrate the capabilities of the technique for readily visualizing chemical complexity and for providing statistical data on sample heterogeneity.

INTRODUCTION

Just as adding an imaging component to both fluorescence and NMR spectroscopies has broadened the utility of these methods (1-2), the coupling of infrared spectroscopy with sample visualization has analogously yielded a powerful new technique with implications in polymer research, materials science, biomedical diagnostics, and process control techniques. Infrared (IR) spectroscopy is particularly well suited for studying biomedical systems because of its ability to generate qualitative and quantitative information (chemical composition, structure, dynamics) for individual sample constituents. As such, it has been used as a probe to investigate a variety of biomedical problems, extending from the identification of foreign inclusions in biological material to the determination of diseased states. A far from exhaustive list of specific biomedical IR spectroscopic studies includes, for example, the analysis and classification of DNA from normal prostate tissue compared to DNA from adenocarcinoma (9), the investigation of stroke induced changes in protein and lipid composition of gerbil brains (10), the identification of foreign inclusions in human breast tissue (11), the determination of colorectal cancer (12), and the comparison of nucleic acid proteins in leukemic and normal lymphocytes (13).

The coupling of infrared spectroscopic techniques with imaging greatly extends the capabilities of conventional IR spectroscopy. From a biomedical perspective, existing pathological and histochemical protocols strongly depend on sample morphology and visualization. Therefore, the ability to maintain spatial integrity while accessing precise spectroscopic data intrinsic to a sample represents an ideal combination. We have developed an instrument which for the first time allows a practical means for simultaneously acquiring high-definition spectroscopic images and infrared spectra. Specifically, the imaging system incorporates a step-scan Fourier transform IR interferometer (FTIR), an infrared microscope (or appropriate image formation optics), and infrared sensitive focal plane arrays (FPAs) that encompass several different wavelength ranges. Using this technique we have undertaken a variety of studies (3-8).

Although infrared spectroscopy is widely used as an analytical tool in basic research environments, in this paper we will emphasize studies that reflect the application of this imaging technique to the study of biological and biomedical systems. The technique seamlessly combines IR spectroscopy and sample imaging by obtaining high

CP430, *Fourier Transform Spectroscopy:* 11th International Conference
edited by J.A. de Haseth

resolution IR spectra for each pixel on the array detector while generating chemically specific images. The spatial resolution of the images is limited by the optical resolution of the microscope and the wavelength of light used and, depending on the implementation, can vary from 3-11 microns. The spectral resolution, which is a function of interferometer settings, may be 4 cm^{-1} or higher. Image contrast for these IR spectroscopic images is derived from differences in the infrared absorption spectrum and is intrinsic to the sample. Therefore a researcher or clinician can use the technique to visualize a sample directly in terms of its chemical and structural heterogeneities without the addition of either stains or fluorescent tags. A further benefit of the technique is that it simultaneously records tens of thousands of independent spectra from different spatial locations within the sample. Thus, a variety of multivariate techniques can be exploited to characterize large sample areas and to provide robust statistics for the accurate classification of individual spectral signatures.

As an example of the use of an infrared imaging microscope in our laboratory, we have demonstrated the ability to rapidly screen and accurately determine the presence, size, and chemical composition of silicone gel inclusions in human breast tissue (6). Because relatively little is understood about the mechanism and extent of migration of implant material into either surrounding tissue or lymph nodes, one's ability to uniquely identify and determine the distribution of foreign material in human tissue has broad implications in the biomaterials/implant community (14). Since the identification of foreign materials is often performed by simply examining the morphology of the sample of interest (15, 16), this technique is complementary to existing methods of examination.

Additionally, because this imaging methodology depends on the intrinsic spectral properties of the samples being studied, it is remarkably versatile. We have exploited this versatility to distinguish heterogeneities in both the concentration and spatial distribution of integral constituents within tissue samples. For example, by investigating changes in the distribution of lipids and proteins in rat cerebellar tissue upon treatment with an antineoplastic drug, cytarabine, we have embarked on IR imaging studies of neurotoxicity. Similarly, we are also investigating the putative anomalous lipid distributions induced by the neurodegenerative storage disease Niemann Pick-C.

These successes in preliminary investigations of biological samples lead us to believe that infrared spectroscopic imaging will be broadly applied not only to compositional analysis in biological tissue, but also to the determination and analysis of diseased states. One of our goals is to utilize the changes in biochemical composition, distribution, and molecular organization caused by disease to rapidly and sensitively visualize these pathologies in tissue biopsies.

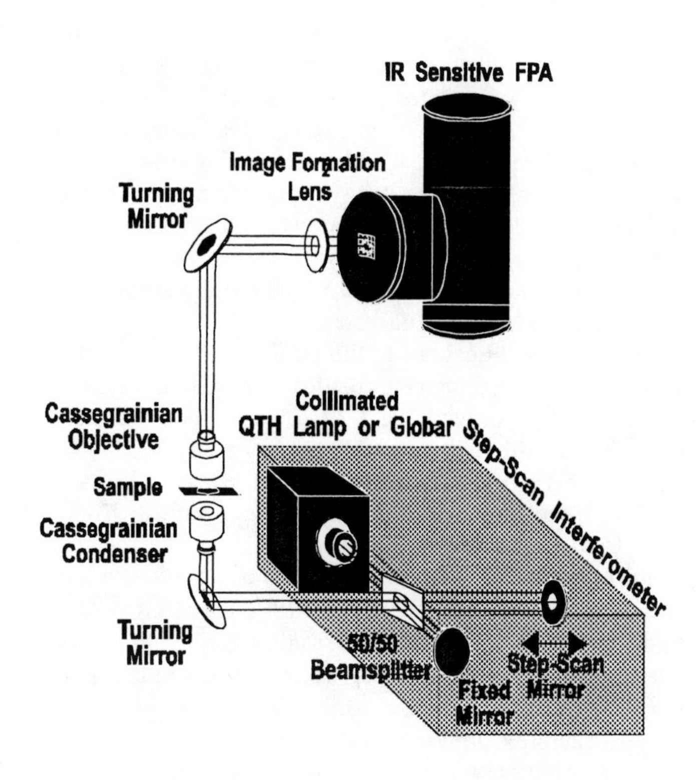

FIGURE 1. Step-scan infrared imaging system. An infrared microscope, coupled with a commercial FT-IR step-scan interferometer (outlined in gray), images samples onto an IR sensitive array. Samples are mounted on IR transparent slides., and analyzed in transmission mode. Depending on the implementation, the image formation lens is either BaF_2, ZnSe, or CaF_2. Typical focal lengths of the image formation lens also vary depending on the array, from 75-150 mm.

METHODS

Experimental

The Fourier transform imaging system developed in this laboratory (Fig. 1) comprises a commercially available step-scan interferometer (Bio-Rad FTS-6000) coupled to an infrared microscope (Bio-Rad UMA 500), or alternatively, appropriate image formation optics for imaging macroscopic samples. The sample image is focused onto one of several different types of infrared sensitive FPAs, depending on the wavelength region of interest. We have used four different IR sensitive arrays, that differ in array geometry as well as wavelength sensitivity. The original array implementation employed a 128x128 Indium Antimonide (InSb) detector, sensitive between 1-5.5 μm (4). The broadest wavelength detector used was an Arsenic doped Silicon (Si:As) detector with a spectral sensitivity between 1 and 25 μm (8). Recently, we have used two different Mercury Cadmium Telluride (MCT) detectors, both sensitive between 2 and 11 μm. We have previously presented results from the large format (256x256) MCT array (5), and will present imaging data sets obtained with the smaller format MCT (64x64) in this paper. These detectors, as will be discussed in the following section, were originally developed as components for defense related applications and have only recently been made commercially available and adapted for use in medical and scientific research.

Optical modifications to the commercial spectrometer system include an image-formation lens placed between the microscope objective and IR camera. For the most recent data collected with the small format MCT, the image formation lens is a 75 mm focal length CaF_2 lens. Optical filters within the dewar housing prevent visible light from impinging on the array. The source spectral bandwidth may also be reduced by blocking wavelengths shorter than 2.52 μm (3950 cm^{-1}) with a long-pass optical filter (OCLI). This spectral bandwidth filtering eliminates aliasing when employing a large interferometer mirror step size between image frames, while still conforming to the Nyquist sampling criterion for a given spectral resolution.

The primary, non-optical modification is the addition of a custom built circuit to synchronize the interferometer and collection of image frames by the FPA. As the interferometer mirror is stepped to a new position, the collection of image frames by the FPA is triggered. The data set is generated as a series of interferograms and image planes. Data collection is similar to that for conventional FTIR studies, except that entire image planes replace measurements from single detector elements. Each interferogram, corresponding to a pixel in the data set, is apodized, Fourier transformed, and then divided by an averaged air background spectrum to yield transmittance images and/or converted to absorbance (OD). Each complete data set consists of image planes that correspond to specific mid-IR frequencies and a separate IR spectrum for each of the pixels on the array. Figure 2 is a schematic representation of the data cube, showing how the spatial and spectral information are arranged, before and after the Fourier transform and conversion to absorbance.

Although each array implementation has variations in the experimental parameters, several broad concepts are presented below. Data sets are typically collected with spectral resolution between 16 and 32 cm^{-1}, with an interferometer step speed ranging from 1-25 sec. Typical averaged images consist of up to 128 co-added frames, with integration times ranging from 20-9000 μsec for each frame at every interferometer step. To obtain 16 cm^{-1} resolution over a 3975 cm^{-1} spectral range with an undersampling ratio (UDR) of 4, a data set will contain 512 image planes.

Biological sample preparation for the FTIR imaging system is essentially the same as for standard FTIR microspectroscopy. Samples are prepared by sectioning frozen tissue to 5-15 μm thickness. A section is immediately placed onto infrared transparent slides and then desiccated. Samples can also be prepared, as for standard histological examination, by embedding the tissue in paraffin, sectioning, rinsing in xylene to remove the paraffin, and then mounting on infrared transparent slides.

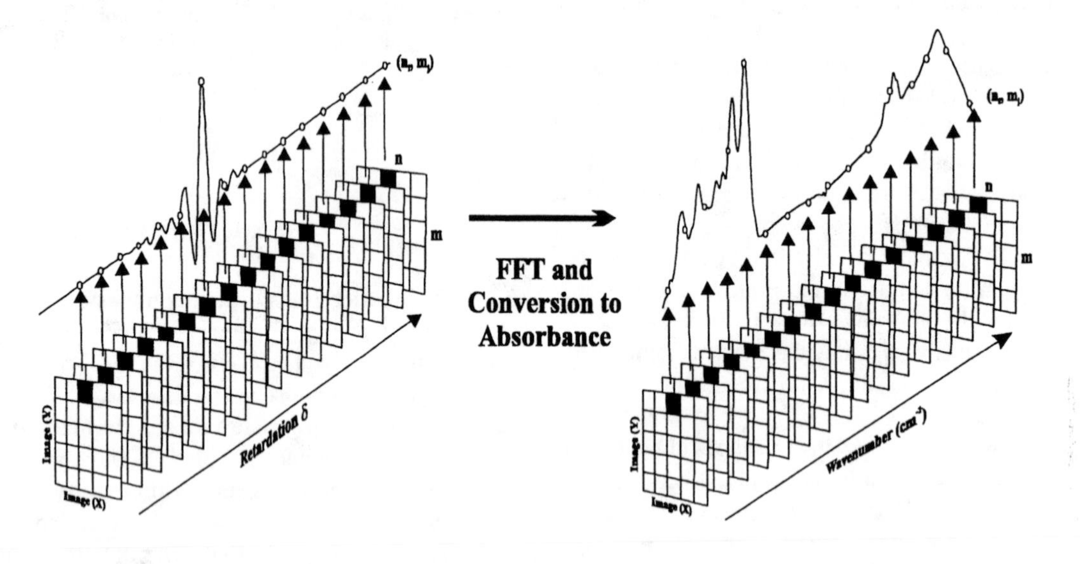

FIGURE 2. Schematic representation of a Fourier-transform infrared imaging data set. The data is acquired as an interferogram, with each pixel on the array having a response determined by its corresponding spatial location on the sample. Each delineated point on the interferogram represents a discrete interferometer mirror position. The resultant spectral data set is obtained by performing an FFT for each pixel on the array. After the FFT and conversion to absorbance, each pixel presents a unique infrared spectrum that gives chemical and spectral information pertaining to its spatial location on the sample.

TABLE 1. Properties of a Variety of Infrared Sensitive Focal Plane Arrays

	InSb	Small Format MCT	Large Format MCT	Small Format Si:As	Large Format Si:As
Pixel dimensions	50x50 Microns	61x61 microns	30x30 microns	100x100 microns	50x50 microns
Array geometry	128x128	64x64	256x256	20x64	240x320
A/D dynamic range	12 bits	14 bits	16 bits	15 bits	16 bits
Spectral Range	1-5.0 microns	2-10.5 microns	2.7-11 microns	1-25 microns	1-25 microns
Frame Rate	<240 Hz	<420 Hz	<120 Hz	800 frames/sec	50 frames/sec
Integration Time	50-400 μsec	78-1,500 μsec	15-9,000 μsec	1 μsec/pixel	1 μsec/pixel
Storage Capacity	$5x10^7$ electrons	$5x10^6$ electrons	$8x10^6$ electrons	$5x10^5$ electrons	$5x10^7$ electrons
Operating temperature	77 K	77 K	77 K	<10 K	<10 K

In terms of sample thickness, the technique has essentially the same constraints and limitations as standard optical microscopy based histopathological techniques; but because the technique is based on intrinsic sample chemical heterogeneities, there is no need to use dyes or stains to generate image contrast. In the future, we expect to be able to perform reflectance measurements, in which case the tissue samples need not be sectioned or mounted onto IR transparent slides. Finally, data files are visualized and processed using commercial software, (*ISIS*, Spectral Dimensions, and *Matlab 4.0,* The Mathworks).

Focal Plane Arrays

The impetus to develop infrared sensitive array detectors was fueled by the development of charge-coupled devices (CCDs). CCDs, although initially developed in 1969 as information storage devices, were found to be superb light sensitive detectors. Efforts soon turned toward making an equivalently useful infrared sensitive CCD for military uses such as missile guidance and general surveillance. Although direct charge injection input schemes were initially pursued, infrared sensitive materials are not suitable components from which to manufacture monolithic electronic circuits. Therefore, hybrid array types, in which an infrared sensitive material is bump-bonded to the silicon electronic readout circuitry using indium posts as connectors, became the most efficient array architecture.

After initial successes in building small linear arrays, expectations of developing large infrared sensitive focal plane arrays became high. Although a number of problems are associated with the physics and engineering of these detectors, several key factors have kept infrared focal plane array technology from advancing as rapidly as the earlier CCD technology. For example, because these infrared photon detectors sense room temperature as background noise, cryogenic techniques need to be incorporated into their design. Therefore, hybrid array structures must be able to withstand severe variations in temperature without the infrared sensitive material delaminating from the readout circuitry. Additionally, due to compromises related to charge depth limitations, the readout circuitry must be very fast, to prevent the array from saturating under conditions detecting terrestrial background radiation alone. These operating conditions necessitated the development of high speed readout electronics that not only worked well at cryogenic temperatures, but also maintained analog precision and linearity. Therefore, the development of large format IR sensitive arrays also depended heavily on simultaneous improvements in the supporting electronics systems used to read and process the data. Until now, the predominant use of these arrays has been in defense related applications, astronomical observations, and thermal and remote sensing. Although there are examples of thermal imaging in medicine and chemistry, the specific coupling of these arrays to laboratory based analytical instrumentation, such as tunable filters, spectrometers and microscopes, has opened new clinical and basic research applications.

Multivariate Techniques

Infrared spectroscopic imaging has demonstrated enormous potential for studying a variety of chemical and biological systems at both the microscopic and macroscopic levels. However, data sets acquired by these imaging systems are large and complicated, generally consisting of tens of thousands of pixels, each with an associated high-resolution IR spectrum. For a typical experiment, this leads to data sets ranging from 16-100 megabytes in size. Therefore, in order to realize the full potential of these spectral images, automated mechanisms are required to query the data in ways that reveal underlying structure and morphology.

Multivariate approaches to this challenge, in which each pixel is considered to be a single point in a multivariate (N-dimensional) space, can be employed as a way of highlighting important information contained within the data set. The variables (coordinates) of the point in N dimensions are reflected by the intensities of the N-point spectrum associated with the pixel. In this multivariate representation, pixels with similar spectra, i.e. similar coordinates, tend to cluster together. Although this representation appears abstract, it is utilized to definitively summarize large data sets in terms of small numbers of synthetic images, which, when combined with classical methods of spectral analysis and image segmentation, provides a powerful basis for data examination.

As an example, Principal Component Analysis (PCA) may be used to create spectral images and analyze data sets. PCA condenses information by projecting multivariate data to a smaller number of dimensions; the orientation of the projection is chosen to retain the maximum amount of information. The projections are used, in turn, to reconstruct synthetic images. For systems that consist of relatively small numbers of distinct components, minimal sets of principal component vectors can describe more than 98% of the data.

Other multivariate synthetic images can be created based on Euclidean distance calculations using either manually selected reference spectra from within the data set or externally derived single component spectra. In this method, each spectrum/pixel from the data set is assigned to a given representation by determining its closeness to one of the reference spectra. A synthetic image may then be constructed by digitally color coding the pixels according to their multispectral Euclidean distance classification. Because vibrational imaging systems typically acquire 16,000-64,000 spectra, it is a daunting, if not impossible task to analyze them individually. As a result, multivariate approaches are essential. The efficacy of multivariate classification techniques will be illustrated by analyzing vibrational spectroscopic imaging data from several systems.

RESULTS AND DISCUSSION

Spectral Characterization of Small Format MCT Array

Figure 3a shows a spectrum of polystyrene taken from a single pixel on the small format (64x64) MCT array; Fig. 3b displays the result of dividing the spectral response of two adjacent pixels. This imaging data set was acquired from a 5 μm thick polystyrene sheet. The data set was collected at 4 cm^{-1} spectral resolution, with an undersampling ratio of 4, corresponding to 2048 steps of the interferometer. The stepping rate of the interferometer was 1Hz, with the camera acquiring image frames at a rate of 210 Hz. With this array implementation and frame rate,

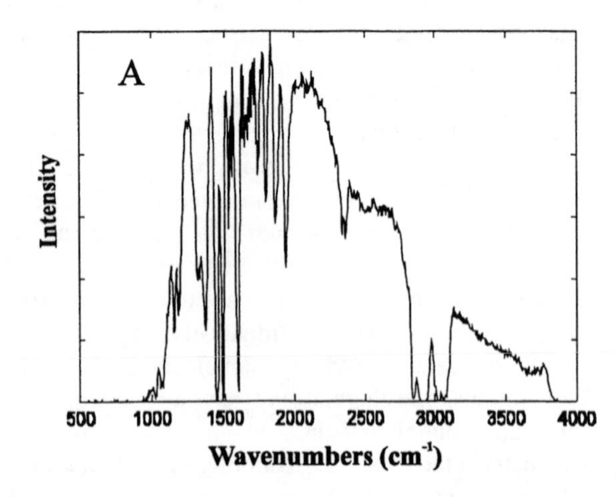

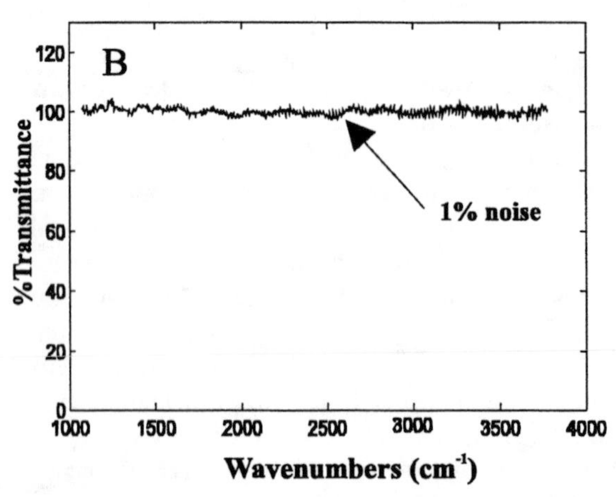

FIGURE 3. (A) Spectrum of polystyrene from a single pixel in the FTIR imaging data set taken with a 64 x 64 MCT array. (B) Plot showing the ratio of two spectra taken from adjacent pixels, giving an indication of the signal to noise characteristics of the array.

the integration time was fixed at 0.14 msec; 128 image frames were collected and averaged at each interferometer point. The total camera staring time to collect this data set containing 4096 infrared spectra was therefore 37 sec. This particular array, which digitizes the signal in the camera head, has extremely favorable noise characteristics. As can be seen from the figure, the resulting spectral noise over the range 2.5-10 μm is estimated to be ~1%. The broadband spectral response of the detector cannot be readily determined from this data set because both the high energy and low energy ends of the spectra are artificially truncated. On the low energy side, the CaF_2 image formation lens used does not transmit light below ~1000 cm^{-1}; whereas light above 4000 cm^{-1} is blocked by a bandpass filter in the interferometer housing. CO_2 absorption bands appearing at 2350 cm^{-1} also indicate that the instrument is not well purged.

Generating Image Contrast Based On Infrared Spectral Signatures

To demonstrate the ability of the infrared imaging system to distinguish between similar sample constituents on the basis of their infrared spectral signature, we prepared a sample comprised of two lipids whose mid-IR spectra differ only in the carbonyl stretching region. Specifically, we overlaid films of dipalmitoyl-phosphatidylcholine (DPPC) and dihexadecylphosphatidyl-choline (DHPC) lipids onto a BaF_2 window. The IR imaging data set was recorded on a 256x256 MCT array (in collaboration with E. J. Heilweil and V. D. Kleiman of the National Institute of Standards and Technology) with a spectral sensitivity from 900-3300 cm^{-1} (5). DHPC is the ether-linked analogue of DPPC, having acyl chains of the same length, but lacking a carbonyl moiety. The IR spectrum of DHPC, in contrast to that for DPPC, has no spectral intensity at 1737 cm^{-1} (the vibrational band assigned to a C=O stretching mode). The image data set, (16 cm^{-1} spectral resolution), consists of 512 image planes with 68 camera frames averaged at each interferometer step. As an example of using multivariate techniques to examine an imaging data set, Fig. 4 shows a composite "synthetic" image based on a Euclidean distance calculation (4a) and two spectra (4b) recorded from this sample. A Euclidean distance calculation is a "supervised" classification, where each pixel in the data set is compared to manually selected "pure component" spectra extracted from the data set. In this case, three pure component spectra were chosen, corresponding to DPPC, DHPC, or air. Using the following Euclidean distance metric, a comparison for each pixel was made. The image is a single binarized summary of this comparison, where each spectrum/pixel is determined to resemble one of the three

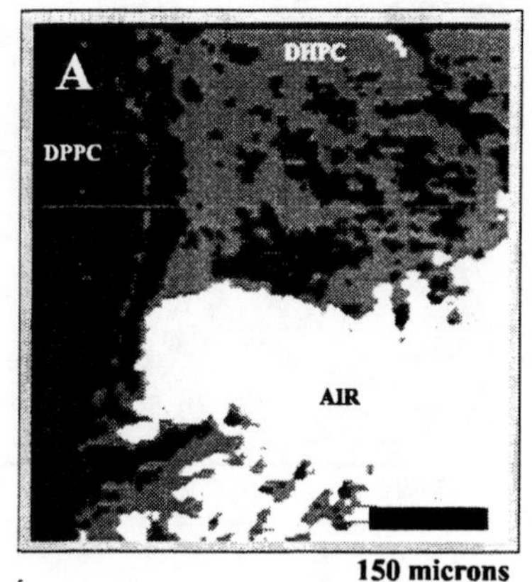

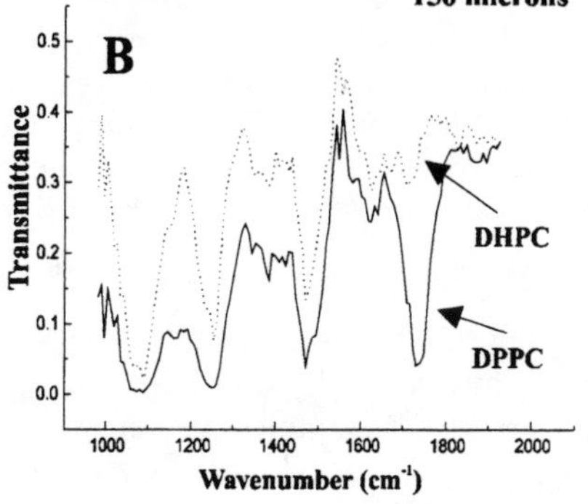

FIGURE 4. (A) Composite synthetic image and (B), two spectra from a sample that comprises a mixture of two lipids, DPPC and DHPC. The spectra are 40-pixel averages from the data set. The image is constructed by a multispectral Euclidean distance metric in which each spectrum is determined to resemble the spectrum of DPPC, of DHPC, or of air. This composite image is digitized such that the pixels that most closely resemble DPPC and DHPC are black and gray, respectively, and those for air or lack of sample are white.

pure component spectra, over the spectral range 1131-1847 cm^{-1}. This composite image is digitized such that the pixels that most closely resemble DPPC and DHPC are black and gray, respectively, and those for air or lack of sample are white. The images show clearly the ability of the technique to definitively distinguish between sample constituents on the basis of their mid-IR spectral differences.

Without using multivariate techniques, an image can be constructed to distinguish between the two lipid constituents. Because DHPC does not have a carbonyl absorption band at 1737 cm^{-1}, an extracted image that corresponds to this frequency will distinguish between the two lipids: DHPC will be dark, and DPPC will be bright. However, as the "air" component of the sample also does not have a carbonyl absorption band, it will be indistinguishable from the DHPC component. The Euclidean distance image in Fig. 4 clearly differentiates between the DHPC and air components. Even this simple example illustrates the importance of using all of the available spectral data and a multivariate analysis to

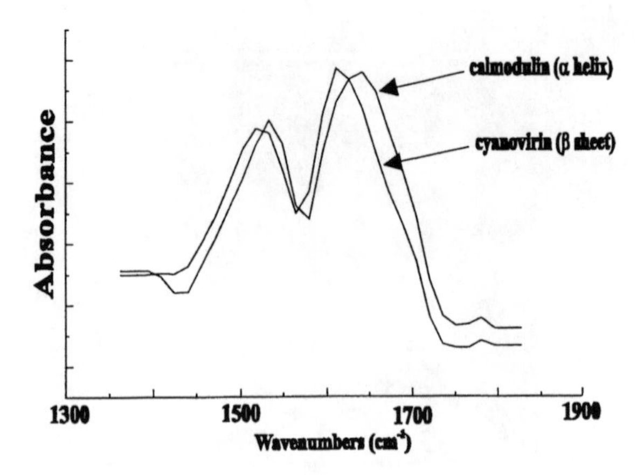

FIGURE 5. Spectra of calmodulin and cyanovirin extracted from single pixels in the imaging data set, comparing peak frequencies of the amide I band. This variation in the peak frequency is attributed to differences in protein secondary structure between the samples.

extract the most information from complex data sets.

Even more subtle spectral parameters can be used to distinguish between similar sample components. We have spectroscopically imaged a sample comprised of two proteins, calmodulin and cyanovirin. These two proteins have very different protein secondary structures: cyanovirin is in the β-sheet configuration, while calmodulin is α-helical. It is well known that these structural differences causes changes in the vibrational spectra of the two pure components; specifically in the frequency of the amide I band at 1650 cm^{-1}. Differences in spectra extracted from two separate spatial locations in the data, corresponding to the two protein types, are shown in Fig. 5.

The sample was prepared by dissolving the proteins separately in water. Then two films were cast, side by side, with a small region of overlap. To collect the spectroscopic imaging data set, the interferometer was run at a 1 Hz step scan rate, with 16 cm^{-1} spectral resolution and UDR 4. The small format MCT array was used to collect the data at a 210 Hz frame rate, collecting 128 image frames at each interferometer mirror position.

Figure 6a shows the image created by integrating the amide A absorption band, corresponding to an NH stretch centered around 3300 cm^{-1}. Because the amide A band does not distinguish between the two proteins, image contrast here is based only on sample thickness, not sample composition. However, Fig. 6b is derived from the center-of-gravity determination of the amide I band peak position: bright areas correspond to amide I peak frequencies centered around ~1650 cm^{-1} (calmodulin), dark areas correspond to peak frequencies centered at approximately 1625 cm^{-1} (cyanovirin). This image clearly delineates the distribution of the two proteins, and unlike the image in Fig. 6a, is relatively insensitive to the thickness of the material. Additionally, the distribution of the amide I peak frequencies into two distinct populations is further demonstrated by observing the peak position histogram in Fig. 6c. This figure shows that 97.8% of the data falls into two distinct chemical classes. The clarity with which infrared spectroscopic imaging can spatially resolve and identify sample components using both simple spectral analysis and multivariate techniques emphasizes its importance as an analytical and basic research tool.

Conclusive Identification of Foreign Inclusions

We have applied the infrared imaging technique (in collaboration with V F. Kalasinsky and J. L. Luke of the Armed Forces Institute of Pathology) to a common

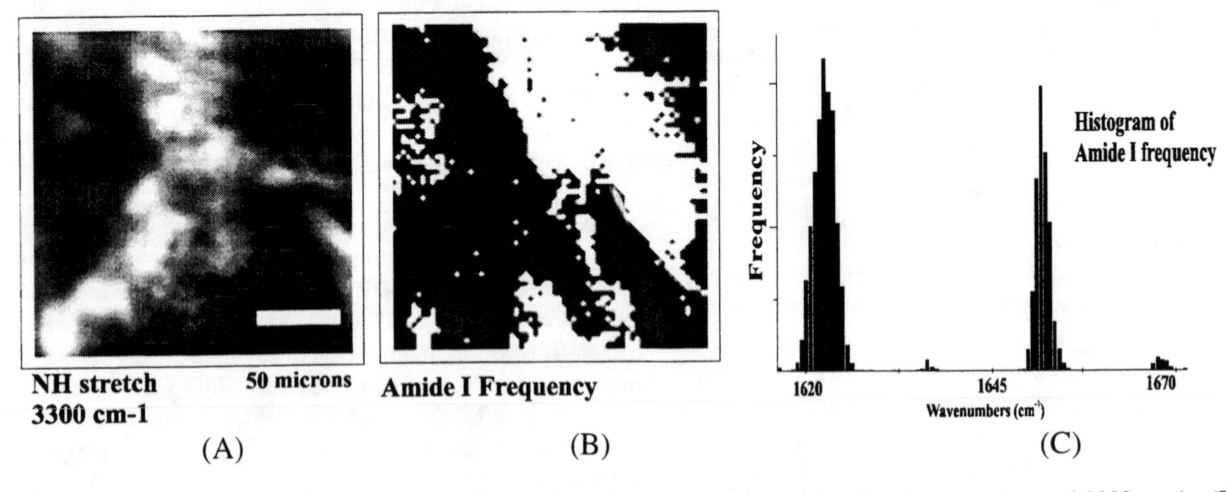

FIGURE 6. (A) Infrared spectroscopic image of the integrated intensity of the amide A band, centered around 3300 cm-1. (B) Infrared spectroscopic image of amide I peak frequencies (~1650 cm-1) that distinguishes between calmodulin and cyanovirin based on protein secondary structure. (C) Histogram of amide I peak frequencies contained within the imaging data set: 97.8% of the data are classified into two chemical classes.

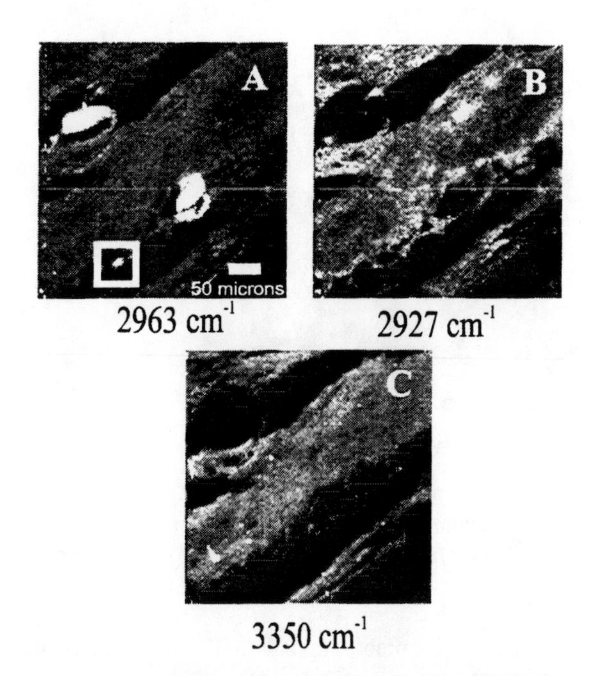

FIGURE 7. (A) Infrared image at 2963 cm^{-1} highlighting the presence of silicone gel inclusions in the human breast tissue. Image contrast is based on vibrational intensity. Species in the sample with a vibrational mode at this frequency, silicone gel for instance, will appear bright. (B) Infrared image at 2927 cm^{-1}, corresponding to the CH_2 and CH_3 stretching modes of lipid and paraffin. The inclusions are dark in this image frame, as silicone gel does not absorb at 2927 cm^{-1}. (C) Infrared image at 3350 cm^{-1} detailing the fairly homogeneous distribution of protein throughout the tissue sample. As in image (C) the inclusions are dark in comparison to the rest of the sample.

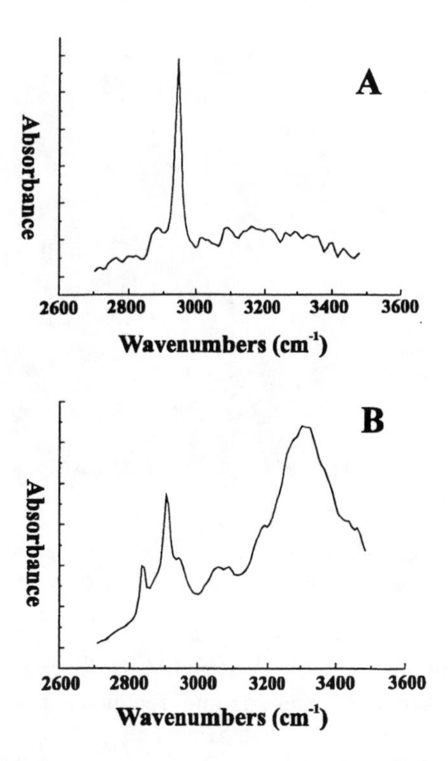

FIGURE 8. (A) Infrared absorption spectrum taken from a pixel within one of the inclusions. The spectrum corresponds to a 3x3 μm^2 portion of the sample, and clearly confirms the chemical identity of the inclusion as silicone gel. (B) Infrared absorption spectrum from a pixel in the human breast tissue. The lipid/paraffin CH_2 and CH_3 stretching modes give rise to peaks in the 2850-2950 cm^{-1} range; whereas the broad feature centered around 3350 cm^{-1} is characteristic of the NH stretching vibration for proteins.

pathological problem; namely that of locating and conclusively identifying foreign inclusions in human tissue (6). Although light microscopy is commonly used to visualize regions of tissue containing foreign substances, chemical identification is only possible using analytical techniques such as infrared or Raman microscopy. Accurate identification of foreign substances in human tissue has significant diagnostic and legal implications. For instance, the use of non-medical grade injection materials for tissue augmentation can confuse the diagnosis of silicone mastitis. Fourier transform infrared spectroscopy has been used to identify mineral oil and soy or olive oil as the principal organic oil component in breast biopsy specimens that had been clinically and histopathologically diagnosed as silicone mastitis (17).

Using infrared imaging spectroscopy, we have examined human breast tissue, containing silicone and polyester, from subjects with a history of silicone implant breast reconstructive surgery. This particular biopsy had been prepared for standard histological examination by embedding in paraffin, sectioning at 5 μm, and rinsing in xylene. Although the xylene rinse is designed to remove the embedding paraffin, it also changes the sample composition by removing lipids. However, this did not effect our ability to identify silicone inclusions in the biopsy. Figure 7 shows infrared spectroscopic images from this study. The images correspond to the silicone gel inclusion (A), the lipid component (B), and the protein component (C) and demonstrate the capability of the technique to determine the presence and distribution of the chemical constituents within the sample. Representative spectra derived from single pixels within regions of silicone gel and tissue respectively are shown in Fig. 8a and 8b. Image contrast (grayscale) based upon a particular chemical species is a powerful way to visualize a sample. Additionally, since the microscope system collects complete vibrational spectra at each pixel, the distribution of other components in the sample can be determined from vibrational signatures contained within the same data set.

Diseased Tissue Analysis

We have begun (in collaboration with D. S. Lester of the Food and Drug Administration, P. Pentchev of the National Institute of Child Health and Human

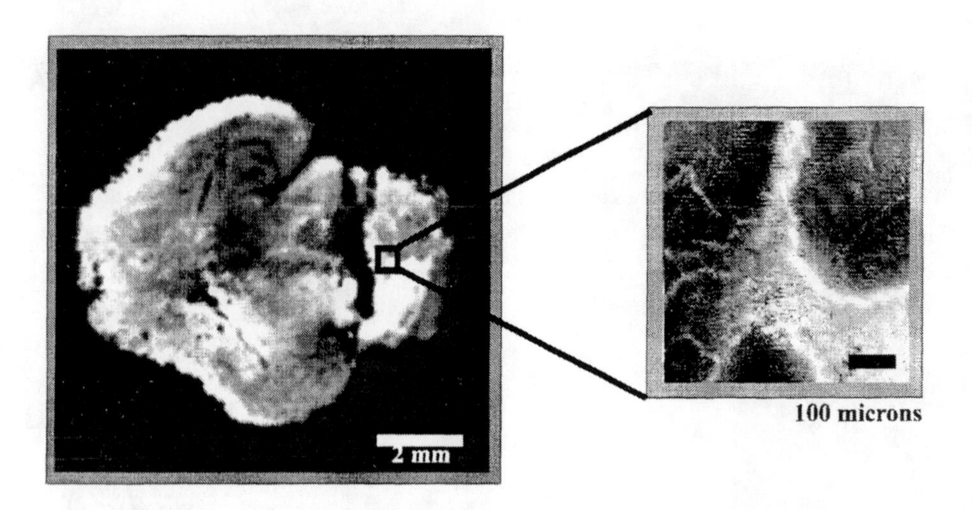

FIGURE 9. Macroscopic (A) and microscopic (B) images of a 10 µm mouse brain section showing spatial differences in the lipid and protein components, constructed from their respective infrared absorbance bands. The box in panel (A) shows the location of tissue that is imaged in panel (B).

Development, and E. J. Heilweil and V. D. Kleiman of the National Institute of Standards and Technology) an FTIR imaging study of the neurodegenerative genetic disease, Niemann-Pick C (NPC), by imaging intact brain sections from transgenic mice that express this human pathology. Using infrared imaging we hope to generate an understanding of the localization and changes in the distribution of cellular components as a result of the disease.

The samples were 10 µm thick horizontal sections of intact brains taken from both diseased and healthy mice. Figure 9a shows a macroscopic lipid to protein ratio image of an entire brain section (approximately 14 mm across) from a diseased mouse recorded with the 256x256 MCT array implementation (spectral sensitivity from 900-3300 cm^{-1}). Bright areas in the image are lipid rich, dark areas are richer in protein. Figure 9b shows a microscopic lipid to protein ratio image (600 µm field of view) from the same section, but focused on a small region located in the mouse cerebellum. The position of this area is indicated in Fig. 9a by a box enclosing the sample area. This data, in contrast to the MCT data, was collected on an InSb array, and contains spectral information in the 1975-3950 cm^{-1} region. These images indicate a possible anomalous lipid distribution in the diseased animal. Further studies are underway in our laboratory to asses this possibility, and a complete description of our results is forthcoming.

Examination of the Neurotoxic Effect of an Antineoplastic Drug

We have chemically imaged cerebellar tissue derived from Sprague Dawley rats that were treated with the drug cytarabine (in collaboration with D. S. Lester of the Food and Drug Administration) (18). This antineoplastic drug induces neurotoxic responses in the rat, which are morphologically expressed as significant Purkinje cell death. 10 µm thick cerebellar slices were prepared from rats exposed to either defined doses of cytarabine or control saline. Unstained slices were layered onto calcium fluoride windows for analysis by infrared imaging.

Figure 10 shows both an infrared brightfield image (A) and infrared spectroscopic image (B) of a brain slice extracted from a cytarabine treated rat cerebellum. Panel B has been processed by ratioing two images from the data set; one corresponding to a lipid absorption band at approximately 2927 cm^{-1} and the other corresponding to a protein absorption band at approximately 3350 cm^{-1}. The contrast in this resulting image is due entirely to

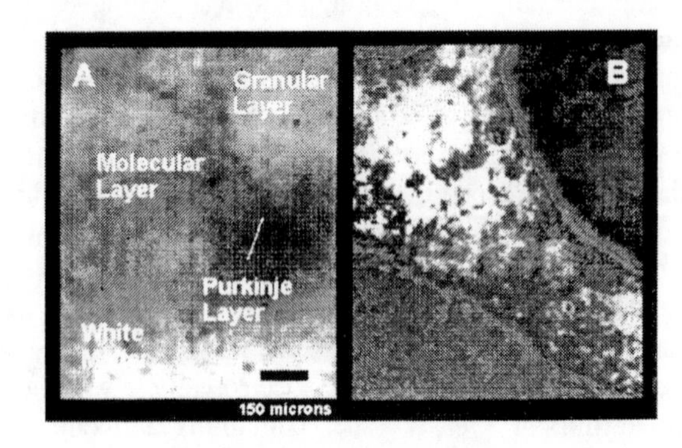

FIGURE 10. Infrared brightfield image (A) and spectroscopic image (B) of a cytarabine treated brain slice from a rat cerebellum. The infrared spectroscopic image was created from two images which highlighted both the lipid and protein fractions of the tissue. The resulting combination image shows the variation in the lipid/protein ratio in the various cell layers.

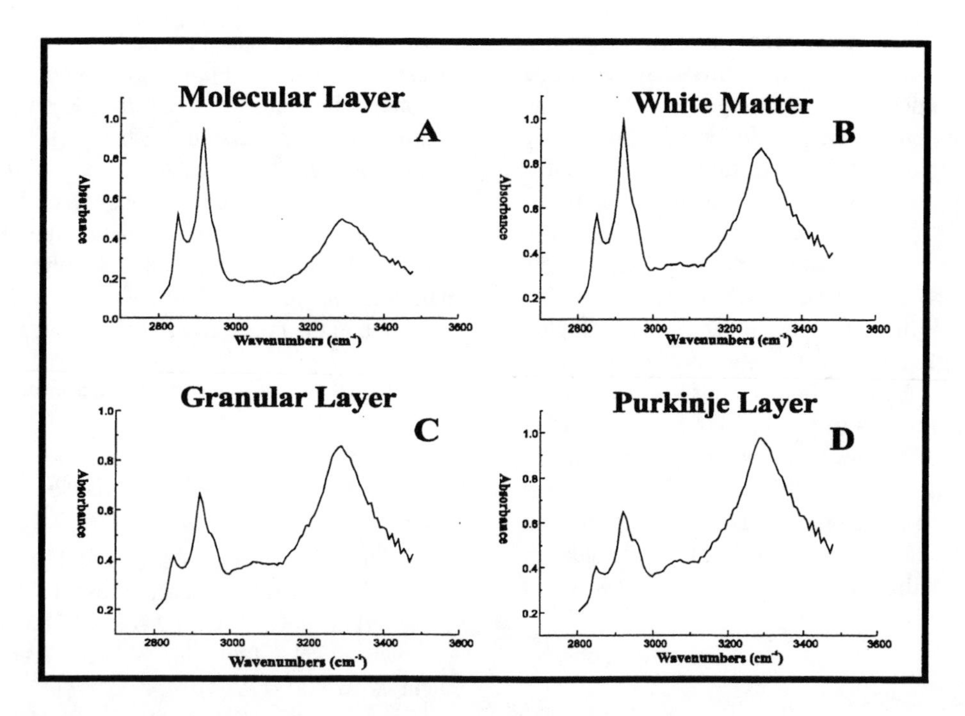

FIGURE 11. Infrared spectra extracted from the data set which provided the images for Figure 4. The panels show the variation of the spectral signature for each of the four distinct cellular regions in the cerebellum.

differences in the lipid/protein ratio in the different cell layers. The bright band running through the middle of the image in panel B corresponds to the molecular layer which contains a much higher percentage of lipid relative to the other layers. This chemically specific contrast is intrinsic to a specific sample and is obtained without the use of dyes or stains.

Figure 11 shows four infrared spectra extracted from the same data set corresponding to four distinct cellular regions within the section. From these spectra it can be clearly seen that the signatures are different and change dramatically from one cell layer to another. The cluster of peaks centered around 2920 cm^{-1} corresponds primarily to CH stretching modes of the hydrocarbon chains of the lipids in the sample, while the broad band centered around 3350 cm^{-1} arises from the protein species in the specimen. It is the differences in the relative intensities of these features that have been exploited to create the image in Fig. 11B. While the lipid/protein ratio varies dramatically from one cell layer to another, the highest relative amount of cellular protein occurs in the Purkinje cell layer (Fig. 11 panel D). This cell layer therefore appears the darkest in the spectroscopic image in Fig. 11B. These and other differences can be exploited to create different images of the same sample using the same data set to highlight a particular biochemical entity or phenomenon.

Very often, however, there are more subtle spectral changes that occur within the same chemical species or material as a result of changes in conformation or chemical bonding within the sample. These changes can result in, for example, a linewidth change or a frequency shift of a particular infrared absorption band. This more subtle effect can also be exploited in creating new images. For example, Fig. 12B shows an image created from the same sample and data set as shown in Fig. 10 and Fig. 11. This image, however, was created by measuring the frequency of the band centered at 2850 cm^{-1}, a feature which corresponds to the CH_2 symmetric stretching mode. Using an algorithm that determines the center of gravity of this peak, each of the 16,384 spectra in the data set were measured and a corresponding grayscale image was created which was either brighter or darker, in a continuous fashion, depending on whether the peak position for a

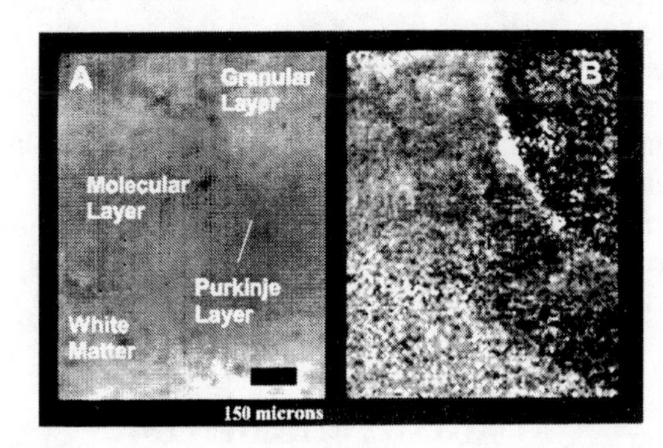

FIGURE 12. Infrared brightfield image (A) and spectroscopic image (B) of a cytarabine treated brain slice from a rat cerebellum. The infrared spectroscopic image was created by computing the frequency for the lipid chain CH_2 symmetric stretching mode for all 16,384 spectra and converting the results into grayscale values. The image provides information on the spatial variation of the lipid packing arrangements in the cell membrane for each of the layers in the sample.

particular pixel had shifted to higher or lower frequency. Although there is significant noise in this image resulting from the degree of uncertainty in fitting the peak position, several interesting features can be discerned. Firstly, the individual cell layers can still be identified, and, secondly, the Purkinje cell layer shows up as a brighter semicircular "band" that runs from the top of the image to the center of the right-hand-side. The brighter region suggests that the frequency of the CH_2 symmetric stretching mode is higher for this cell layer than for the others. This higher frequency is indicative of a greater ratio of *gauche* to *trans* isomers along the hydrocarbon chains of the lipids comprising the membranes in this cell layer. This finding is consistent with the images and spectra shown in Fig. 10 and Fig. 11 which indicated that the cell layer has a relatively higher protein content. This elevated protein contribution results in a disordered packing arrangement for the lipid acyl chains; that is, additional *gauche* chain isomers were induced.

CONCLUSIONS

Analogous to conventional FT-IR single point spectroscopy which has proven its ability to provide qualitative and quantitative information on biological materials, FT-IR spectroscopic imaging techniques also retain these same capabilities, but add a spatial dimensionality to the data. For example, in biological samples, it is the interaction of critical cellular components that determines biochemical function. Therefore, having the capability to resolve the spatial distribution of these components, while simultaneously probing subtle structural changes at the molecular level, goes beyond composition analyses and into the realm of determining biological function. As a result, we envisage FT-IR spectroscopic imaging as a powerful tool for probing the biochemistry of disease. As an example of this, we are currently undertaking infrared spectroscopic imaging studies on normal and malignant breast and prostate cancer cell lines to determine the efficacy of the technique to image and discriminate between individual cell types. Armed with this capability, we hope to generally image and distinguish diseased tissue *in vivo*, as well as *in vitro*.

ACKNOWLEDGMENTS

This work incorporates the results of collaborations with a variety of people at many different institutions. Although we have given due recognition throughout the paper, we would like to acknowledge our co-workers here according to their affiliations: Ted Heilweil and Valeria Kleiman at the National Institute of Standards and Technology; Vic Kalasinsky and James Luke at the Armed Forces Institute of Pathology; Peter Pentchev at the National Institute of Child Health and Human Development; and David Lester at the Food and Drug Administration. We would also like to thank Frank Delaglio at NIH for discussions of multivariate analysis techniques. Finally, we would like to express our appreciation to Abigail Haka for her contribution in characterizing the 64x64 MCT array, and for her diligent assistance in the preparation of this manuscript.

*Indicates author to whom correspondence should be addressed.

REFERENCES

1. Taylor, D. L. and Wang, Y. L., *Methods in Cell Biology*, New York: Academic Press, 1989, Vol. 30.
2. Lauterbur, P. C., *Nature* **242**, 191-192 (1972).
3. Lewis, E. N. , Levin. I. W., and Treado, P. J., USA Patent #5,377,003 (1994).
4. Lewis, E. N., Treado, P. J., Reeder, R. C., *et al.*, *Analytical Chemistry* **67**, 3377-3381 (1995).
5. Kidder, L. H., Levin, I. W., Lewis, E. N., *et al.*, *Optics Letters* **22**, 742-744 (1997).
6. Kidder, L. H., Kalasinsky, V. F., Luke, J. L., *et al.*, *Nature Medicine* **3**, 235-237 (1997).
7. Lewis, E. N., Gorbach, A. M., Marcott, C., *et al.*, *Applied Spectroscopy* **50**, 263-269 (1996).
8. Lewis, E. N., Kidder, L. H., Arens, J. F., *et al.*, *Applied Spectroscopy* **51**, 563-567 (1997).
9. Malins, D. C., Polissar, N. L., and Gunselman, S. J., *Proceedings of the National Academy of Science* **94**, 259-264 (1997).
10. LeVine, S. M. and Wetzel, D. L., *Neuroprotocols* **5**, 63-71 (1994).
11. Centano, J. A. and Johnson, F. B. *Applied Spectroscopy* **47**, 341-345 (1993).
12. Rigas, B., Morgello, S. N., Goldman, I. S., *et al.*, *Proceeding of the National Academy of Science* **87**, 8140-8144 (1990).
13. Benedetti, E., Bramanti, E., Papineschi, F., *et al.*, *Applied Spectroscopy* **51**, 792-797 (1997).
14. Hartman, L. C., Bessette, R. W., Baier, R. E., *et al.*, *Journal of Biomedical Materials Research* **22**, 475-484 (1988).
15. Dolwick, M. and Aufdemorte, T., *Oral Surgery, Oral Medicine, Oral Pathology* **59**, 449-552 (1985).
16. Eriksson, L. and Westesson, P.-L. *Oral Surgery, Oral Medicine, Oral Pathology* **62**, 2-6 (1986).
17. Kossovsky, N., Millet, D. E., and Wrobleski, D. A., *American Journal of Clinical Pathology* **97**, 34-39 (1992).
18. Lewis, E. N., Kidder, L. H., Levin, I. W., *et al.*, *Imaging Brain Structure and Function*, New York: Annals of the New York Academy of Sciences, 1997, pp. 234-247.

Continuous-Scan Time-Resolved FTIR Spectroscopy Measurements of High Energy Molecular Collisions

J.J. Sloan, and W.S. Neil

Department of Chemistry, University of Waterloo, Waterloo ON N2L 3G1 Canada

J. Roscoe

Department of Chemistry, Acadia University, Wolfeville NS B0P 1X0 Canada

F.-A. Kong

Institute of Chemistry, Academica Sinica, Beijing 100080 P.R.C.

We have studied the dynamics of gas phase collisions between high energy radicals and several small molecules under low pressure conditions, which enable us to resolve single collision events. The experiments are carried out in a fast-flow, low pressure chemiluminescence emission apparatus. Using nanosecond laser photofragmentation, we create pulses of high energy radicals in a flowing stream of a reagent molecule, then record a series of broad-band IR spectra of the emission from the products of the resulting reaction, using time-resolved FTIR spectroscopy (TRFTS). Our TRFTS instrument uses a continuous scan Michelson interferometer to which we have added a data acquisition system of our own design. The latter is based on VME architecture, and permits time resolutions between 1 μs and 3 ms at the full spectral bandwidth and maximum spectral resolution of the interferometer. Slower time resolutions, down to a few tens of milliseconds, can be obtained at reduced spectral bandwidths. The work to be reported uses a time resolution of a few tens of nanoseconds for the first spectrum, then from 1-3 μs for the second and subsequent spectra; typically, we record in excess of 20 spectra. The principal observable is the infrared emission from the vibrationally and rotationally excited products of the reaction. This is a very low-background experiment; the sensitivity is on the order of 10^{-15} moles/litre/quantum state. Where more than one reaction is possible, the relative intensities of the various spectra give the branching ratios for the different channels, while the time dependence of the emission gives the reaction kinetics

INTRODUCTION

Time-resolved Fourier transform spectroscopy (TRFTS) is conceptually simple and potentially very powerful, but its real value has not been fully exploited. The reason for this lies almost entirely in the technical problems involved in the instrumentation. As a result of these, there have been many attempts to implement it, but only a few have proven successful. This is especially true of early commercial efforts, which were often inflexible and difficult to support. Several more recent developments, however, have begun to change this situation, and commercial instrumentation is now quite practical for some applications.

In the following, I will introduce the subject without reference to specific commercial implementations, and compare the capabilities of the two major variants of the technique: step-scan and continuous-scan methods, with respect to important practical experimental considerations. After introducing the technique, I will discuss the current research in our laboratory, and indicate the present capability of our instrument. The development of this instrument has been evolutionary, building on the first successful system which was completed in 1982.

Techniques for Implementation of TRFTS

Presently, there are two distinct implementations of TRFTS, depending on whether a continuous-scan (CS) or step-scan (SS) Michelson interferometer is used. At the time of the original proposal for time-resolved FTS in the infrared (1) SS interferometers operating in this part of the spectrum were not available, so most of the early development was carried out with CS instruments. Recently, however, the capabilities of SS technology have become quite competitive, and workable instruments based on both methods are now available. We (2) and others (3) have discussed many of the general aspects of TRFTS previously, so I shall restrict the presentation here to a comparison of CS and SS implementations, and to some specific concepts which are necessary to understand the data to be presented later. The SS implementation is conceptually the simpler version, so I shall describe it first.

CP430, *Fourier Transform Spectroscopy:* 11th International Conference
edited by J.A. de Haseth
 1-56396-746-4/98/$15.00

For SS TRFTS, each measurement is carried out with the position of the moving mirror fixed at a specific location. This location is a known distance from the stationary phase point (the point of zero optical path difference, or ZPD) and is equal to an integer number (n) of whole or fractional reference laser fringes. With the mirror fixed at this location, a transient signal is initiated and recorded repeatedly, using standard transient recorder technology. The transient recorder measures M+1 points (0 to M) on the transient signal, digitizing these at regular intervals of the time delay (τ). Enough repetitions of the transient are averaged to achieve the desired signal to noise ratio, then the result is stored. The storage indices relating to each of these data points are the distance from ZPD, expressed as the number (n), and the delay time ($m\tau$) where $0 \leq m \leq M$. Following this, the mirror is stepped to the next reference laser fringe, locked there, and the signal averaging process is repeated. When the transient averaging and storage process has been carried out at each of a total of N mirror locations ($0 \leq n \leq N$; n=0 corresponds to the ZPD location) the resulting dataset consists of a two dimensional array, with N+1 columns and M+1 rows. Each column contains the transient recorded at a given location of the mirror, but different values of the time delay. Each row contains data recorded at the same time delay after the initiation of the transient, but different locations of the moving mirror. The spectrum of the transient at a time $m\tau$ after its initiation, is the Fourier transform of the mth row of this array.

CS TRFTS is conceptually identical to the SS version, but the hardware is quite different. For the case of a CS interferometer, the data must be recorded "on the fly" as the mirror reaches each required location, specified by a value of (n). On reaching such a location, the transient is initiated and a single set of ($m\tau$) data points are digitized, without stopping the mirror. These points are stored before the next - the (n+1)th - location of the moving mirror is reached. On completion of a single mirror scan, therefore, the entire data array of N+1 columns and M+1 rows has been recorded and stored, but only a single measurement of the transient is recorded at each value of (n). Signal averaging is achieved by scanning the mirror repeatedly, and co-adding the transients measured during each scan. With respect to signal averaging, therefore, continuous-scan TRFTS is identical to non-time-resolved FTS, in which a single data point is recorded at each mirror location (n).

Experimental Aspects of SS and CS TRFTS

Although the final data arrays produced by each implementation are identical, the differences in the methods by which they are obtained imply substantial differences in the experimental measurements to which the two techniques are suited, and also the potential noise sources to which they are subject. Broadly speaking, it is obvious that the major difficulties in SS technology revolve around the interferometer and the requirement that the mirror be held immobile at a known location for a relatively long period of time while the transient is signal averaged. The actual digitization of the transient is relatively less problematic because it involves the use of transient recorder technology, which is both a mature and widespread area of digital electronics. The difficulties are reversed for CS TRFTS, for which the interferometer is not the major problem; continuous-scan Michelson interferometers of a very high quality have existed for a long time. In the CS case, the challenge is in recording the transient, which requires digitizing the data points, co-adding them into the stored data array, and monitoring the operation of the interferometer, during the limited time between the occurrence of successive values of (n).

The nature of the hardware for the two implementations also dictates the kinds of measurements to which they are best suited. In the CS version, there is a maximum time between initiations of the transient, imposed by the fact that the mirror has a minimum velocity, below which its travel becomes unstable. Thus, the mirror arrives at successive values of (n) after a time interval which has a maximum value equal to $\delta x / v_{min}$ where δx is the distance between successive occurrences of (n) and v_{min} is the minimum velocity for stable operation of the moving mirror. For a typical system operating in the mid-IR and having a minimum mirror velocity of 10-4 m/s, to sample at each interference fringe of a He/Ne reference laser, the repetition rate of the transient must be greater than 315 Hz. If smaller spectral bandwidths are acceptable - for example 3950 cm^{-1} rather than 7900 cm^{-1} as in the above example - then the minimum repetition rate for the transient would be 158 Hz, and so on. For the SS version, there is no lower limit to the repetition rate of the transient, other than that imposed by the finite time to finish the experiment. This practical issue, however, still sets a genuine limit, on the experiment, due to the time required to move the mirror between successive measurement locations with SS interferometers. Movement requires the acceleration of the mirror from rest, then its deceleration and stopping at a precise location, where it must be held. The technology for this is evolving rapidly, and the times vary widely, depending on other characteristics of the instrument. With current technology, for example, 50-100 ms is about the shortest time for this cycle, and using the lower of these values, the time involved in accelerating, moving, and decelerating the mirror when recording a spectrum with an unapodized resolution of 0.12 cm^{-1} (65536 data points) is about one hour. The actual measurement time must be added to this.

From the preceding, it is clear that the main advantage of the SS technique is in measurements where the maximum repetition rate of the experiment is too low for observation by CS methods (e.g. less than 100 Hz). A practical caveat should be noted in this regard, however. To make a SS measurement of a transient with a lifetime of 0.01 s (or equivalently, a minimum repetition frequency of 100 Hz) 100 times at each mirror location would require 18 hours, and the total experiment time would be 19 hours (vide supra). If a CS apparatus could operate this slowly - something which is possible using a specialized mirror drive - the stop-start time, or about 1 hour, would be saved, but the measurement would still require 18 hours. Clearly, both the spectral resolution and the repetition frequency of the transient must be considered in the design of the experiment. It is generally true, however, that for short transients which can be repeated at high rates, CS methods are preferable, but this advantage is lost for transients which have lower repetition rates or slower time dependences.

An additional, but less obvious, consideration which must be weighed in such a comparison is the phase shift. In the SS method, there is no phase shift between successive time-delayed spectra, and the only phase shifts are those due to the optics and electronics. The CS method, however, suffers from the disadvantage that there is a phase shift between each of the successive data points, caused by the fact that the mirror moves a finite distance between the times at which the points are recorded. The consequences of this depend on the particular implementation of the CS technique. If all (m) time-delayed data points are recorded during the same reference laser fringe, as described above, the consequences are insignificant. In this case, there is a constant phase delay of $2\pi\tau v_{min}/\delta x$ between successive spectra, and each data point in a given interferogram is delayed by the same amount. Since the entire interferogram is shifted by the same phase angle with respect to the stationary phase point, the amplitudes of the Fourier components are unchanged; only their phases are altered (4). The normal correction for optical and electronic phase shifts is adequate to deal with this. Alternatively, if the phase of the signal is unimportant, as is often the case with pure emission spectroscopy, then the correct result is obtained by performing a power transform on the time-delayed interferograms. In principle, the same analysis can be used for the so-called "stroboscopic" or "interleaved" CS techniques(3,5), in which transients of longer duration than the time between reference laser fringes are recorded. It is beyond the scope of this article to discuss this method in detail, but it should be pointed out that due to the sorting necessary with this technique, the phase shifts are more complex and can lead to severe errors if not correctly taken into account.

Effects of Noise in SS and CS TRFTS

The final point which must be considered in comparing CS with SS methods for TRFTS, is the potential effect of noise and unwanted variations in the source intensity. This must be considered in light of the fact that time-resolved FTS requires a much wider bandwidth than the non-time-resolved procedure, because the transient to be measured has very high-frequency components. In principle, this eliminates one of the great advantages associated with FTS: the ability to convert a wide range of optical frequencies into a relatively narrow band of lower frequencies and achieve noise rejection by the application of narrow bandpass electronic filters. For TRFTS measurements having time resolutions in the megahertz range, the consequent loss of signal to noise ratio (SNR) is on the order of 100. Although this is the inescapable result of fast time resolution, it indicates the importance of extremely careful noise reduction. In some cases, post processing of the data with digital filters can ameliorate this loss somewhat.

Among the other noise reduction procedures, the most important is the elimination of variability in the excitation of the transient. We have pointed this out several times in the past (6,7) and it has been discussed more recently as well (8). In the gas phase chemical dynamics experiments which I shall describe later, the excitation source is a pulsed laser, which can have a pulse to pulse variability of 20%. To reduce the effects of this, which are very serious for the final SNR, the intensity of each laser pulse is separately recorded and the data points from the transient initiated by that pulse are divided by its intensity. It is essential that this procedure be carried out for each laser shot.

In addition to the noise from the wide electronic bandwidth and the variability of the excitation source, which are common to all time-resolved measurements, the CS and SS TRFTS measurements are each vulnerable to different kinds of experimental instabilities which can reduce their SNR. The most serious of these is long-term variability of the source, which causes a false variation in the intensity of the interferogram during the mirror scan. For successful operation, all FTS measurements require that the source parameters (for example, the density of reagents in the observation zone) remain stable on the timescale of one mirror scan. Variations which have a period longer than this are averaged out. The CS method collects one time-resolved dataset in each scan, and co-adds many scans for signal averaging. The SS technique collects the entire signal-averaged dataset in a single scan, hence the source intensity must remain stable for a much longer time in the SS experiment than in the CS case. In the example given previously, the CS instrument would require about 10 minutes for a single scan, whereas the SS instrument would require nearly 20 hours.

The next section will give relevant details of the experiments on gas phase, laser-ignited reactions which use our implementation of the CS technique. Following that, the results of several measurements using this method will be reported.

EXPERIMENT

CS TRFTS Data Acquisition System

The measurements are carried out using the CS TRFTS instrument which we have developed. It consists of a data acquisition system, with timing and control circuits designed specifically for CS TRFTS, but it will operate with either CS or SS interferometers. It was designed using CAD methods to maximize the efficiency, minimize the size, and ensure ease of fabrication.

The design includes many programmable logic devices (PLDs) which improve reliability and reduce propagation times. The main hardware is on one 6U height (9) ten-layer motherboard. This carries a single half-height mezzanine board which contains the analog circuits and analog to digital converter (A/D) chips. The architecture is based on the VME bus and the system is operated by a commercial single-board computer (SBC).

The software for the entire TRFTS experiment is resident in EEPROMS, addressed as user-programmable memory on the SBC. Parameters for the experiment (time-resolution, number of spectra to collect, etc. are downloaded into latches on the mother board by the SBC before the beginning of data collection, and thereafter the hardware controls the timing, digitizes the data points and stores them in fast buffers. The SBC, operating under interrupt control, downloads batches of data during gaps in the data acquisition process and performs basic arithmetic tasks such as normalizing the data for laser intensity variations and computing the storage addresses.

Experimental Apparatus

The basic configuration of the current experimental apparatus has been reported recently (10). The apparatus is designed to satisfy the criteria for CS TRFTS, as described in the previous sections. The goal of the research is the measurement of the dynamics and kinetics of high energy, laser-initiated chemical reactions and energy transfer processes occurring in the gas phase. The experiment is carried out in a tubular reactor having a window on one side to admit the beam of an excimer laser. The reactor contains a Welsh cell with its axis perpendicular to that of the excimer laser beam; the latter passes through the geometric centre of the Welsh cell. The observation zone defined by the Welsh cell mirrors has a 10 cm x 10 cm cross section; the excimer beam has a 1 cm x 2 cm cross section.

Premixed reagent gases flow through the reactor in the direction perpendicular to the plane containing the optical axes of the laser beam and the Welsh cell. The flow velocity of the gases is high enough to ensure the complete replacement of the reagents and removal of the products in the optical cell, between laser shots. This satisfies the basic requirement for TRFTS that each transient have the same baseline.

Laser photodissociation of one of the reagent molecules initiates the process by creating the reactive species (an atom or radical). Either the photofragments themselves, or their reactions with the other reagents are followed by observation of the IR chemiluminescence from the reaction zone. Usually, both the photofragment and the reaction products are highly excited, with energy distributions approximately corresponding to equivalent temperatures of many thousands of degrees. Under these circumstances, the chemiluminescence consists of many hundreds or thousands of transitions in the high temperature emission spectra of these excited species.

The simplest sequence of steps in a reactive study can be represented as follows:

Initiation:

$$AR \xrightarrow{h\nu_{las}} A + R \qquad (1)$$

Reaction:

$$A + MB \xrightarrow{k_{rxn}^{v,J}} AB(v,J) + M \qquad (2)$$

Observation:

$$AB(v,J) \xrightarrow{A_{v',J'}^{v,J}} AB(v',J') + h\nu_{IR} \qquad (3)$$

where $h\nu_{las}$ is the photofragmentation laser photon, $k_{rxn}^{v,J}$ is the detailed rate constant for reaction into the internal product state (v,J) and $A_{v',J'}^{v,J}$ is the Einstein transition probability for the emission of an IR photon.

Initiation by excimer laser photolysis (pulse width ~ 8 ns) is instantaneous on the 1 μs timescale of most of the measurements, so this establishes an accurate zero for the time axis. Simple kinetic information is obtained by measurement of the risetime of the total chemiluminescence from the excited state product - process (3) above. Because the chemiluminescent spectrum is frequency-resolved, dynamical information is also obtained. The populations of the product internal states created by the reaction are calculated by dividing the measured intensity of the (usually spectrally resolved) IR transitions by the appropriate Einstein transition probability. The dynamics of the reaction are inferred from the resulting populations of the products' internal states, which are created by the reaction.

The basic timescale of the process being examined is determined by the gas phase collision times of the reagents.

Within the limits of adequate SNR, the collisional timescale can be made long enough to permit the resolution of the important aspects of the process. In the typical experiments which we have carried out, some of which will be described below, the total reagent pressures are approximately 0.05 Torr. For 300 K thermal translational energies, this corresponds to average (gas kinetic) collision times of 1 - 3 μs. Thus, a 1 μs time resolution, which is achievable by most present TRFTS systems, is capable of resolving the processes on a sub-single collision basis. Since the rate constants for most interesting processes correspond to collision probabilities which are less than unity, the pressure and time resolution are customarily set such that a few collisions occur, on the average, between observations. This ensures that a measurable amount of the process occurs during the observation time.

The emission spectra recorded in these experiments have a very low background. Typically, the walls of the flow chamber are surrounded by cooled baffles, so the only background on the detector is the room temperature radiation emitted by the optics and their surroundings. As a consequence, the measurement has a very high sensitivity. Model kinetic calculations using known rate constants for reactions yielding spectrally-resolved products (e.g. hydrogen halides, CO, etc.) indicate that the minimum detectable time-resolved signal corresponds to about 10^6 - 10^7 molecules per cubic centimeter per quantum state in the observation volume (about 1000 cm^3). This corresponds to a detectivity of about 10^{-15} mole/liter. The following sections show examples of data which illustrate these points, and outline the dynamics of the processes which are presented

RESULTS

H Atom Reactions

In many cases, the laser initiation, process (1) yields fragments which have residual translational energy. This is the case for many of our recent studies involving the reactions of hydrogen atoms, in which the reagent in process (1) is usually H_2S, and the atomic hydrogen produced by photofragmentation at λ_{las}=193 nm has a translational energy of 2.3 eV. This is a convenient source for the study of the reactions of very high temperature H atoms. (A thermal kinetic energy distribution in which the average energy is 2.3 eV would have a temperature of about 40000 K.)

The high initial reagent energy, however, complicates the dynamics of the collisions. If the reaction does not occur at every collision, some H atoms recoil from the initial inelastic collisions with reduced kinetic energy, and have the possibility for a collision with a second reagent molecule. The second collision may be reactive or inelastic as well. In this scenario, the reagent relative kinetic energy in the first collision has a single value which is known very accurately, but thereafter, it has a distribution, which in general, is not known. For cases where we need to know the relative kinetic energy in order to interpret the experimental results, we add an excess of a "moderator" gas to thermalize the H atoms at a known rate. Depending on the identity of the moderator gas, and its pressure, this approach can thermalize the atoms either instantly, or at a rate which is under the control of the experimenter. For the hydrogen atom experiments, the moderator must be a rare gas, for which the energy transfer probabilities in collisions with high energy H atoms have been measured directly (11).

Using these measurements, the energy of the H atoms as a function of time can be modeled quite accurately. Some examples of the results obtainable from the application of CS TRFTS to the study of these energetic hydrogen atom reactions will be given in the remaining parts of this section. Then, some spectra obtained from reactions other radicals will be discussed.

The H + N_2O Reaction

We have recently completed a detailed analysis (12) of the dynamics and kinetics of the reaction:

$$H + N_2O \rightarrow OH + N_2 \qquad (4a)$$

in which H represents H atoms having high translational energies. In addition to (4a) the following processes can also occur in this experiment:

$$\boldsymbol{H} + N_2O \rightarrow OH(\ddagger) + N_2 \qquad (4b)$$

$$\boldsymbol{H} + N_2O \rightarrow NH + NO \qquad (4c)$$

$$\boldsymbol{H} + N_2O \rightarrow H + N_2O(\dagger) \qquad (5)$$

$$H + N_2O(\dagger) \rightarrow OH + N_2 \qquad (6)$$

$$H + N_2O(\dagger) \rightarrow N_2O \qquad (7)$$

in which (‡) represents electronic- and (†) represents vibrational- excitation. Although not indicated, the OH and N_2 products in the above equations may also be vibrationally excited. The H + N_2O system has been examined by several techniques over more than two decades, because it is important in fuel-lean combustion. A thorough review of past work is given in reference 12, including references to work in which the products of all of the reactive processes (4) were observed directly. Here we shall limit ourselves to reporting our time-resolved FTS measurements, and the information which can be derived from them.

The emission observed in our time-resolved FTS measurements of this reaction is shown in Fig. 1. The spectra are separated by equal time intervals of 6.2 μs, beginning 2.2 μs before the photolysis laser pulse. There are a total of 15 spectra, covering a total time interval of 86.8 μs. The first spectrum, recorded at -2.2 μs, is just visible along the front (frequency) axis. This shows the size of the background noise in the measurement. The

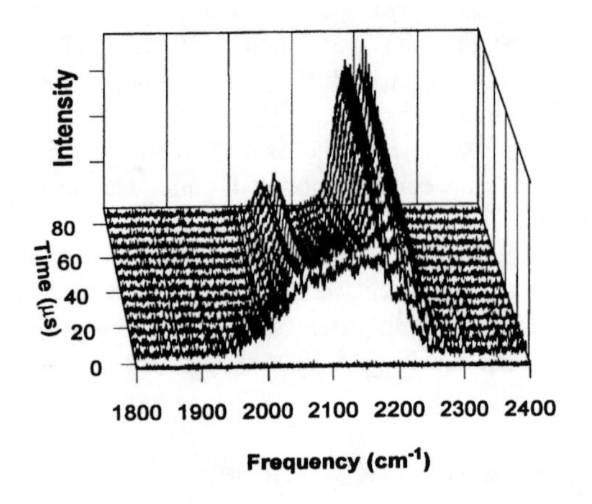

FIGURE 1. Time-resolved IR emission from vibrationally excited N_2O caused by collisions with high energy H atoms.

second spectrum, recorded at 4.0 μs after the laser pulse is a broad, nearly unstructured, band extending from about 1950 cm^{-1} to 2280 cm^{-1}.

The experiment used to record these bands was described in general terms in the Experimental section. The moderator gas used in this case was Argon. The pressures of the reagents, H_2S and N_2O, were adjusted such that the mean collision time of the average (300 K thermal) component with a reagent molecule was about 3 μs, and with Ar, about 300 nanoseconds. At this Ar collision frequency, the H atoms are thermalized in less than 10 μs, i.e. before the second N_2O spectrum is recorded. The experiment used a liquid nitrogen cooled InSb detector, having a maximum sensitivity between about 1800 cm^{-1} and 5000 cm^{-1}. In the frequency region to which the detector is sensitive, no other emission bands appeared at an intensity above 5% of the bands shown in Fig. 1.

The emission originates from the $N_2O(v_1,v_2,\Delta v_3=-1)$ transitions, where v_1, v_2 and v_3 are the symmetric stretch, bend, and asymmetric stretch modes, respectively. Under the emission envelope, there are a large number of bands, corresponding to different levels of excitation in the three modes.

The high wavenumber limit in the spectrum corresponds to the N_2O (0,0,1) → (0,0,0) transition. All of the transitions share the common characteristic that they are the fundamental bands in the asymmetric stretch quantum number ($\Delta v_3 = -1$). This assignment was confirmed by spectral simulation. The simulation for the $\tau = 35.0$ μs spectrum is shown in Fig. 2. This retrieves all of the significant features of the bands except a shoulder at about 2260 cm^{-1} and a small band centered at 2060 cm^{-1}. The former is due to absorption of the (0,0,1) → (0,0,0) emission by the background (300 K thermal) N_2O reagent; the latter is due to excited reagents and is not relevant to the analysis.

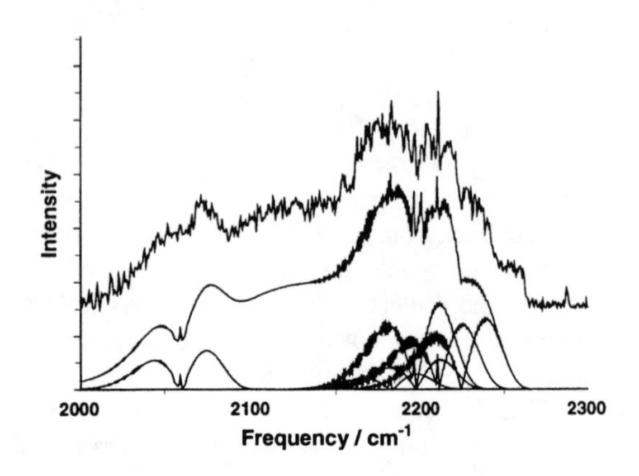

FIGURE 2. Simulation of the $N_2O(v_1,v_2,\Delta v_3=-1)$ bands. The upper trace is the spectrum measured at 35.0 μs, displaced for clarity. The middle curve is the simulation, which is the sum of the individual bands shown at the bottom.

These time-resolved spectra show that the most probable outcome of the $\boldsymbol{H}$ + N_2O encounter is process (5), the excitation of N_2O into a wide range of vibrational levels. Emission from OH, which might be expected from processes (4) and (6) was not observed under the conditions of the experiment shown in Fig. 1, although OH was observed in experiments using higher reagent pressures.

During the observation time, the emission in the central part of the excited band decreases, while that in the lower part is virtually unchanged (it increases slightly). The lifetimes of the excited states of N_2O are a few tens of milliseconds, so the disappearance of the emission is not due to infrared radiative relaxation. Furthermore, it is not due to collisional relaxation, since only a few tens of collisions occur during the observation time, and vibrational deactivation cannot occur at such a rapid rate.

To understand the time-dependence of the spectra in Fig. 1, it is necessary to consider the electronic surfaces controlling the interaction. These are sketched in Fig. 3. Only the zero point energies of the indicated species are shown in the figure; translational and vibrational energies must be added to these.

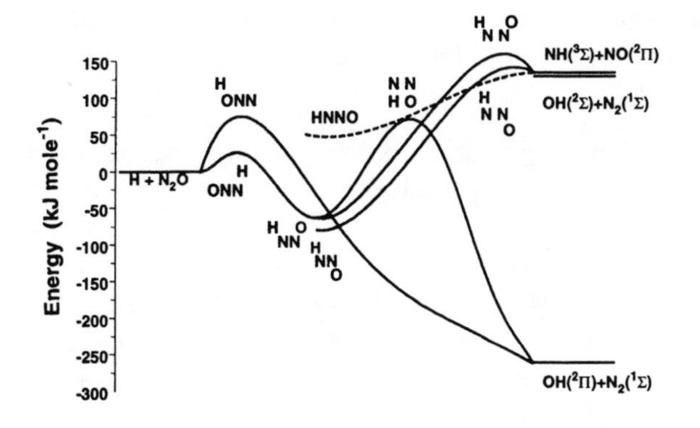

FIGURE 3. Potential energy surfaces for the HNNO system

When the relative translational energy in the ***H*** + N_2O collision, 221 kJ/mole, is added to the energy of the separated reagents (which is the zero of energy in Fig. 3) it raises the total energy of the colliding pair above all of the energy barriers for reaction on the ground states and the thresholds for formation of the excited state products, $OH(A^2\Sigma) + N_2$ and $NH(^3\Sigma) + NO(^2\Pi)$ which are nearly isoenergetic. If all of the N_2O internal states energetically accessible at this total energy were populated, the N_2O emission bands shown in Fig. 1, would extend much further to the left - well beyond the detector cutoff at about 1850 cm^{-1}. The observed low-frequency cutoff at about 1950 cm^{-1}, on the other hand, is the limit which would be observed if the collisions with ***H*** do not form any N_2O states with energies above the thresholds for the reactions yielding $OH(A^2\Sigma) + N_2$ and $NH(^3\Sigma) + NO(^2\Pi)$. We conclude from this that the latter reactions occur in the high energy collisions, in preference to N_2O vibrational excitation, process (5), and thus that the sum of the cross sections for processes (4b) and (4c) is larger than that of process (5).

Similar energetic considerations also suggest an explanation for the rapid disappearance of the emission below about 2150 cm^{-1}. The upper states of the transitions responsible for this emission have energies greater than 68 kJ/mole, which is the height of the barriers to reaction on the ground state surface, whereas the upper states of the transitions responsible for the emission below 2150 cm^{-1} are all below these energy barriers. This implies that the states above the barrier to reaction on the ground state surface disappear because they are removed by reaction (6). This must be predominantly a reaction between low-energy H atoms and N_2O in vibrational levels above the barrier because the translationally excited (***H***) atoms have been thermalized by the time the second spectrum is recorded, while the emission in this region continues to decrease during the entire 87 μs observation time.

We confirmed the preceding conclusion by writing a simple numerical model of the system, including all processes (4-7) and separately identifying the N_2O with vibrational excitation less than the ground state reaction barriers - denoted $N_2O(\dagger)$ - from that with vibrational excitation greater than these barriers - denoted $N_2O(\dagger\dagger)$. Thus, the energy transfer process (5) was separated into

$$\boldsymbol{H} + N_2O \rightarrow H + N_2O(\dagger) \tag{5a}$$

and

$$\boldsymbol{H} + N_2O \rightarrow H + N_2O(\dagger\dagger) \tag{5b}$$

The reaction with low-energy H atoms, reaction (6), was written as:

$$H + N_2O(\dagger\dagger) \rightarrow OH + N_2 \tag{6'}$$

while the reaction $H + N_2O(\dagger) \rightarrow OH + N_2$ was not considered because the reagents have insufficient energy to overcome the reaction barriers.

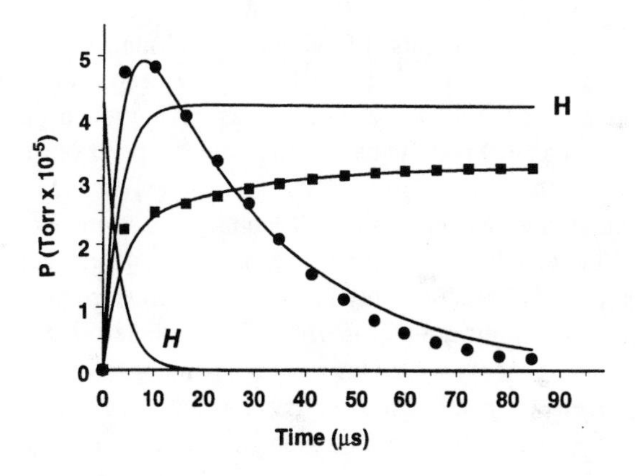

Figure 4 Kinetic model calculations of the concentrations of $N_2O(\dagger)$ - squares - and $N_2O(\dagger\dagger)$ - circles. The absolute concentrations of translationally excited and low-energy H atoms (***H*** and H respectively) have been reduced by a factor of 1000.

The predicted and measured time dependences of the emission from $N_2O(\dagger)$ and $N_2O(\dagger\dagger)$ are compared in. Fig. 4, along with the concentrations of ***H*** and H. The latter are calculated from the energy transfer probabilities determined in the direct measurements mentioned earlier (11). The excellent agreement of the model predictions with the observed time dependences of

$N_2O(\dagger)$and $N_2O(\dagger\dagger)$ emphasizes the validity of the model. The time dependence of $N_2O(\dagger\dagger)$ in the model allows the rate constant for reaction (6) to be determined. The value obtained is $1.7 \pm 0.8 \times 10^{-10}$ cm^3 molec^{-1} s^{-1}, which is about seven orders of magnitude larger than the room temperature rate constant. This value corresponds to a reaction probability of about 0.3 to 0.5 per gas kinetic collision, and shows that vibrational excitation of the N_2O to levels above the barrier to reaction on the ground state is a very effective way to increase the rate of this reaction.

The H + CO_2 Reaction

The reaction:

$$\boldsymbol{H} + CO_2 \rightarrow OH + CO \quad (7a)$$

was studied under very similar conditions to those in the ***H*** + N_2O study. Although the possible outcomes for the ***H*** + CO_2 reaction are very similar to those for the N_2O reaction, the details of the potential energy surfaces differ significantly. Reaction (7a) is endothermic (ΔH°_0=104 kJ/mole) and its activation energy is 110 kJ/mole (13), indicating that it has an intrinsic energy barrier of about 6 kJ/mole, in addition to the endothermicity. Between the reagents and products, the HCO_2 potential energy surface contains the intermediate HOCO, which is bound with respect to the reagents by about 60 kJ/mole. This intermediate, therefore, is located in a very deep potential minimum, between the barrier to reagent approach (the activation barrier) and the barrier to product separation (the endothermicity).

The initial kinetic energy of the ***H*** atoms in reaction (7a) is 221 kJ/mole, substantially in excess of the endothermicity, so reaction is possible only in the first collision or so, before the ***H*** atoms are thermalized. The endothermicity of the primary process precludes a reaction analogous to (4b) in which electronically excited OH is formed. The possible reactive processes (and the corresponding values of ΔH°_0 in units of kJ/mole) are:

$$\boldsymbol{H} + CO_2 \rightarrow CH + O_2 \quad +184 \quad (7b)$$

$$\boldsymbol{H} + CO_2 \rightarrow HCO + O \quad +112 \quad (7c)$$

The T-V (translational to vibrational) energy transfer is, of course, possible as well:

$$\boldsymbol{H} + CO_2 \rightarrow H + CO_2(\dagger\dagger) \quad (8)$$

but the subsequent reaction:

$$H + CO_2(\dagger\dagger) \rightarrow OH + CO \quad (9)$$

can only occur if the $CO_2(\dagger\dagger)$ retains at least 110 kJ/mole of the total possible 221 kJ/mole deposited in vibration during the initial collision.

This reaction has been studied extensively (14) due to its importance to combustion, and a review of its entire previous literature is beyond the scope of this article. It has been determined that the cross section for reaction (7a) increases rapidly as the translational energy is raised above the activation barrier, but in this experiment, as in the H + N_2O study, the ***H*** atoms are moderated by Ar, and their energies drop below the reaction barrier before the second collision with CO_2.

The most intense emission observed in our TRFTS measurement of this reaction is shown in Fig. 5. Here, the time=0 axis is at the rear of the Figure, and time advances toward the front of the plot. The band origins for the fundamental bands: CO(v=1→0) and CO_2 (0,0,1)→(0,0,0) are shown as dashed lines for reference. Vibrationally

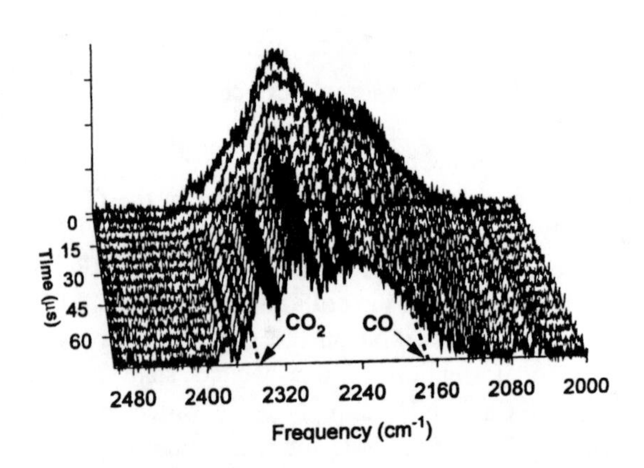

Figure 5. Time-resolved emission recorded from the ***H*** + CO_2 reaction.

excited ("hot") bands of each molecule are found to the lower frequency side of these lines.

The spectra show that the most intense emission is from highly vibrationally excited CO_2; it is possible that a small amount of CO emission is also present. Thus, like the ***H*** + N_2O reaction, the most probable event is T-V energy transfer, resulting in the vibrational excitation of the CO_2 molecule, process (8). The formation of CO and OH in reaction (7a) has been observed in other work on this reaction. We also see weak OH(v=1→0) emission with a SNR of about 2-3. The OH radical has very low Einstein transition probabilities in the infrared, while the Einstein A-factors of CO are about a factor of 2-3 larger than those of OH. Therefore, if a significant amount of vibrationally excited CO is formed in (7a) it would be detectable in our spectra. Since very little CO emission seems to be present, we conclude that this is created by the reaction mostly in the ground vibrational state.

Since the emission from the highly excited CO_2 formed

in the T-V process overlaps the region where the CO emission would be seen, it is very difficult to determine whether any vibrationally excited CO is produced under the conditions of our experiment. Detailed spectrum simulations, which we are undertaking at present, should permit the identification of the CO emission if it is present. There is no identifiable emission from CH or HCO from reactions (7b) and (7c) at the signal to noise of the present experiments.

The H + NO_2 Reaction

The reaction of 300 K thermal H atoms with NO_2 has been studied almost as extensively as the H + CO_2 reaction. Work on this reaction spans more than two decades, from the first IR chemiluminescence work (15) to the more recent and extensive laser induced fluorescence measurements (16). In this case as well, the complete literature of this reaction is beyond the scope of this article.

The major process which occurs is the reaction:

$$H + NO_2 \rightarrow OH + NO \qquad (10a)$$

At 300 K thermal reagent kinetic energies, reaction (10a) populates OH vibrational levels up to and including the energetic limit, v=3,. All energetically accessible rotational levels in each of the vibrational states are also populated.

In addition to (10a), a large number of other reactive and energy transfer processes are possible in this case:

$$H + NO_2 \rightarrow NH(\ddagger) + O_2 \qquad (10b)$$

$$\rightarrow NH + O_2(\ddagger) \qquad (10b)$$

$$\rightarrow HNO + O \qquad (10c)$$

$$\rightarrow HNO(\ddagger) + O \qquad (10d)$$

$$\rightarrow HO_2 + N \qquad (10e)$$

$$\rightarrow NO_2(\ddagger,\dagger) + H \qquad (11)$$

where the symbols (‡) and (†) indicate electronic and vibrational excitation as usual. Due to the close coupling between the vibrational levels of NO_2 and its low-lying electronic states, it is not possible to distinguish between these modes of excitation, and they are treated as a completely coupled set of modes in (11).

The IR emission bands from vibrationally excited OH, observed in our high energy H + NO_2 experiment, are shown in Fig. 6. The lowest panel shows the noise level recorded 1 μs before the photolysis laser is triggered. The times at which the successive spectra were recorded are indicated in each panel.

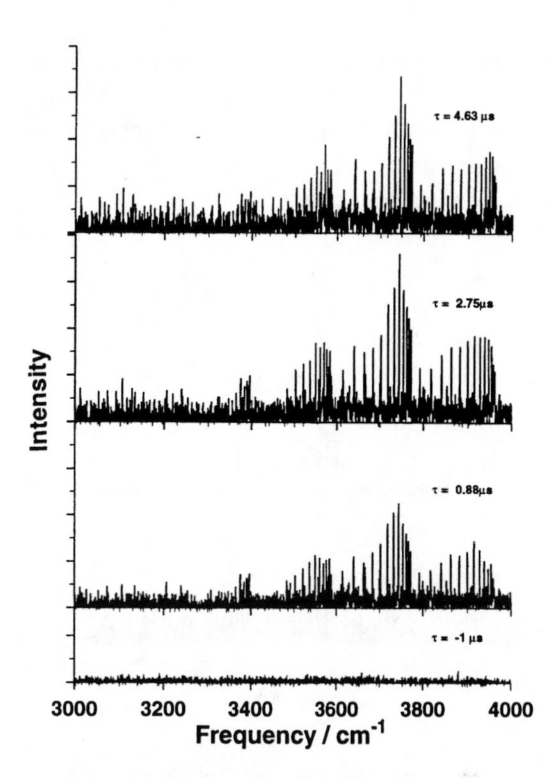

FIGURE 6 The earliest four time-resolved IR emission spectra from the reaction of ***H*** + NO_2.

The four groups of strong lines evident in each panel are the R-branch bandheads in the fundamental (Δv=-1) OH bands. The (1→0) bandhead is near 3960 cm^{-1}; the (2→1) bandhead is near 3770 cm^{-1}, the (3→2) near 3580 cm^{-1} and the (4→3) bandhead is near 3400 cm^{-1}. Although there is no emission from the P and Q branches evident in the spectra, these branches are indeed present, but at very low intensity.

The apparent absence of emission from the P and Q branches is due to the fact that their transition probabilities are strongly dependent on the OH rotational state. Specifically, the P and Q branch transition probabilities decrease very strongly with increasing rotational state, whereas those of the R-branch do not. The presence of strong R-branch emission in the early spectra therefore demonstrates directly that the major feature in the dynamics of the high-energy ***H*** + NO_2 reaction is the population of very high OH rotational states. The bandheads occur at rotational states N~19-21 but the R branches show that there is significant population out to values of N > 50.

The absence of emission from the P and Q

branches in the early spectra is evidence that the reaction not only produces very high rotational excitation, but that the distribution of rotational states is strongly inverted; there is virtually no initial population of rotational states below about N = 5. The consequence of reagent translational excitation in this case, therefore, is a dramatic increase in the observed rotational excitation of the OH product.

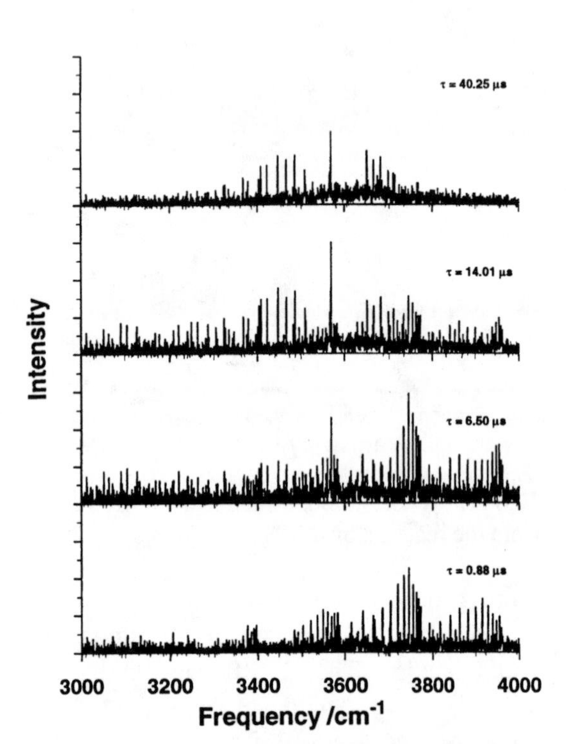

FIGURE 7. Time-resolved IR emission spectra from the reaction ***H*** + NO_2 showing the OH rotational relaxation.

The ***H*** + NO_2 experiment was carried out with no buffer gas, so there was very little relaxation of the OH rotation during the first few observations. At later times, however, relaxation becomes apparent. Fig. 7 shows the evolution of the OH spectrum over a longer time period, from the initial spectrum at 0.88 μs, to one at 40 μs. The loss of both the extremely high rotational excitation generated by the reaction, and most of the vibrational excitation as well, is indicated by the disappearance of the four prominent R-branch bandheads and the emergence of the P, Q and R-branches of the (v=1→0) transition. The prominent Q branch near 3560 cm^{-1} is the most characteristic feature of this transition.

In addition to the OH fundamental bands, these spectra also show emission from vibrationally excited NO and NO_2. These spectra are shown in Fig 8. The NO bands are below 2000 cm^{-1} and the NO_2 emission is centered near 2750 cm^{-1}. The NO is the expected second product of (10a), along with OH. The NO_2 is the result of T-V energy transfer. The ratio of the cross section for reaction (10a) to that for T-V energy transfer (11) is much larger for the ***H*** + NO_2 than for either the CO_2 or N_2O reactions. This is to be expected since the activation energy for the NO_2 reaction is less than 4 kJ/mole, while the others are much higher.

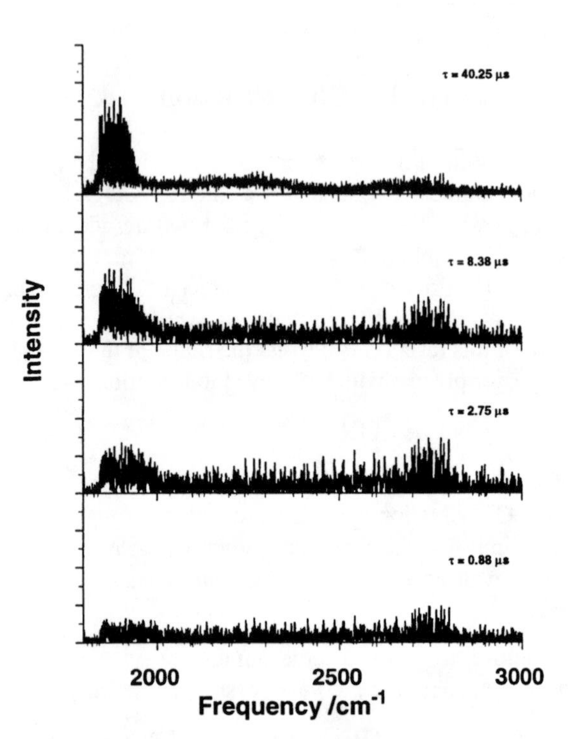

FIGURE 8 Emission from vibrationally excited NO (below 2000 cm^{-1}) and NO_2 (near 2750 cm^{-1}) from the reaction ***H*** + NO_2.

CONCLUSION

The data presented above show that TRFTS is capable of providing information on both the kinetics and dynamics of gas phase processes on a sub-single collision basis. It has adequate sensitivity to detect vibrationally excited molecules in the gas phase at concentrations of femtomoles per quantum state. Even at these concentrations, CS TRFTS can produce broad-band, high resolution infrared spectra in a reasonable time. The preceding spectra, for example, were recorded at spectral resolutions between 0.4 cm^{-1} and 0.12 cm^{-1}, and with time resolutions between a few μs and 1.3 μs. All cover the full bandwidth of the liquid nitrogen cooled InSb detector, and none required more than 3 hours of data collection time; most were recorded in 1 - 2 hours.

Since the technique has very high spectral resolution, it

can identify many different radicals and molecules and simultaneously measure their time-dependences. This information makes it possible to write very accurate kinetic models of the processes because the high sensitivity permits the concentrations of many of the components to be measured directly, while the high time resolution can be used to limit the complexity of the chemistry by eliminating contributions from secondary processes, if necessary.

ACKNOWLEDGMENTS

We thank Dustin Dickens for extensive assistance with data collection and processing, and J.-Y. Li and C. A. Carere for collection of the N_2O and NO_2 data. This work was supported by the Natural Sciences and Engineering Research Council of Canada.

REFERENCES

1 Murphy, R.E., and Sakai, H, in *Proceedings of Aspen International Conference on Fourier Spectroscopy*, G.A. Vanasse, A.T. Stair, and D. Baker, eds, 301 (1971).

2 Sloan, J.J. and Aker, P.M., "Time-resolved vibrational spectroscopy" *Springer Proceedings in Physics*, Vol 4, eds., A. Labereau, U. Stockburger, Springer-Verlag, Berlin (1985); Sloan, J.J., Aker, P.M., and Niefer, B.I. in *Laser Applications in Chemistry*, D.K. Evans, ed., Proc. SPIE, **669** 169 (1986); Sloan, J.J. and Kruus, E.J., *Advances in Spectroscopy* **18** (Time-resolved spectroscopy), R.J.H. Clark and R.E. Hester, eds, John Wiley & sons ltd, (1989) ch 5; Sloan, J.J. "Fourier Transform Methods in the Infrared", in *Atomic and Molecular Beam Methods*, Vol. 2, D.C. Laine and G. Scoles, eds.; Oxford University Press, (1992)

3 Leone, S.R., Acc. Chem Res, **22** 139 (1989); Heard, D.E., Brownsword, R.A., Weston, D.G. and Hancock, G., *Appl. Spectrosc.* **47** 1438 (1993)

4 Bracewell, R.N., *The Fourier transform and its applications*, 2nd ed. McGraw-Hill, New York, (1978) ch.6.

5 Sloan, J.J., *Proceedings of 7th International Conference on Fourier Spectroscopy*, Proc. SPIE, **1145**, 212 (1989)

6 Sloan, J.J., Kruus, E.J. and Neifer, B.I., *J. Chem. Phys.* **88** (1988) 985

7 P.A. Berg and J.J. Sloan, *Rev. Sci. Instrum.* **64** 2508 (1993)

8 Lindner, J., Lundberg, J.K., Williams, R.M. and Leone, S.R., *Rev. Sci. Instrum.* **66** 2812 (1995)

9 IEC subcommittee SC48D specification, compatible with DIN 41494

10 Sloan, J.J., Neil, W.S., and Carere, C.A., *Applied Optics* **35** 2857 (1996)

11 Park, J., Shafer, N., and Bersohn, R., *J. Chem. Phys.* **91**, 7861 (1989).

12 Sloan, J.J., Neil, W.S., Kong, F.-A., and Li, J.-Y., *J. Chem. Phys.* **22** Sept 1997, (in press)

13 Tsang, W. and Hampson, R.F., *J. Phys. Chem. Ref. Data*, **15** 1087 (1986)

14 See, for example, Rice, J.K. and Baronavski, A.P., *J. Chem. Phys.* **94** 1006 (1991), and references therein

15 Polanyi, J.C. and Sloan, J.J., *Int. J. Chem. Kinet. Symp. 1* **51** (1975).

16 Irvine, A.M., Smith, I.W.M., Tuckett, R.P. and Yang, X.F., *J. Chem. Phys.* **93** 3177 (1990); Irvine, A.M., Smith, I.W.M., and Tuckett, R.P., *ibid* pg. 3187.

LIFTIRS, The Livermore Imaging FTIR Spectrometer

Charles L. Bennett

Lawrence Livermore National Laboratory, L-43, Livermore, CA. 94550

The imaging FTIR spectrometer was invented 25 years ago. Only recently, however, with the development of infrared focal plane array technology and high speed microprocessors, has the imaging FTIR spectrometer become a practical instrument. Among the class of imaging spectrometer instruments, the imaging Fourier transform spectrometer enjoys a great advantage in terms of calibratibility, sensitivity, broad band coverage and resolution flexibility. Recent experience with the LIFTIRS instrument is summarized. As a concrete example of the acquisition, calibration, and comprehension of the data from an imaging Fourier transform spectrometer, the case history of a geological sample is discussed in great detail. In particular, the importance of principle component analysis to imaging spectroscopy is especially emphasized. It is shown how the various spatial/spectral constituents within a sample can be detected, located, identified and quantified.

INTRODUCTION

Imaging spectrometers produce a nearly overwhelming wealth of data. A relatively modest image format focal plane array (FPA), containing 128x128 pixels, for example, may produce 16,384 simultaneous spectra. It is beyond the capacity of the human sensory system to directly assimilate such data. Fortunately, there are sophisticated mathematical techniques which enable ready human appreciation of such "hyperspectral" data. The two most important of these techniques are principle component analysis (PCA), and matched filter analysis (MFA). Several of the intrinsic features of an Imaging Fourier Transform spectrometer (IFTS) are especially valuable in the application of these techniques. These features include the manifest spatial co-registration of all spectral images, and the strong preservation of channel to channel spectral correlations for a fixed pixel. Another more subtle, but quite valuable feature is the near perfect "whiteness" of the uncalibrated spectral data from an IFTS. In addition to these advantages for the utilization of hyperspectral data, an IFTS also offers a number of technical advantages, such as ease of calibration, high sensitivity, broad band coverage and resolution flexibility.

In this article, after a brief introduction to imaging spectrometers in general, and some of the terminology, a brief historical overview of the IFTS is presented. The remainder of the article is devoted to an illustration of the use of an IFTS with an archetypical imaging spectroscopy example.

IMAGING SPECTROMETERS

Imaging spectrometers acquire a "data cube" consisting of 2 spatial and 1 spectral dimension. There are a variety of means to this end, depending on whether spectra are dispersed in 0,1, or 2 dimensions, and whether spectra are temporally or interferometrically multiplexed. Any non-imaging spectrometer may have its field of view rastered across a scene to produce a data cube, but this is very inefficient. The availability of focal plane image detectors enables highly efficient acquisition of spectrally and spatially resolved data. A dispersive imaging spectrometer disperses light in 1 spatial dimension with either a grating or a prism, and provides imaging in an orthogonal direction. By concatenation of a succession of 1 dimensional slices of a scene, a 2 dimensional image may be reconstructed. A spatially multiplexed imaging interferometer produces a spatially extended interferogram in 1 dimension, and has imaging in an orthogonal direction. A variable filter spectrometer acquires a full 2-d image each frame, with successive frames viewed through successive filters. A temporally multiplexed imaging Fourier transform spectrometer (IFTS) acquires a full 2-d image per frame, with successive frames associated with different positions of a moving mirror in an interferometer. This is illustrated in figure 1, where a set of frames of a complete 2-d scene are shown in a stack.

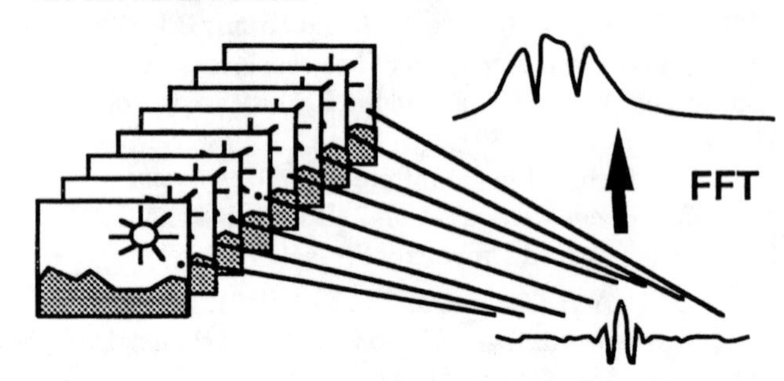

FIGURE 1. An overview of the data flow through an imaging FTIR spectrometer is sketched, illustrating the sequence of frames in the time domain at the left hand side, with a typical pixel's time dependent intensity in the lower right hand corner, and after fast Fourier transformation, the same pixel's spectrum is shown in the upper right hand corner.

The variation of light intensity for a single pixel in the scene is indicated by the curve to the lower right. The raw data cube for an IFTS consists of the interferogram in 1 dimension and the 2-d scene in the 2 orthogonal directions. Fourier transformation of the interferograms for each pixel in the field of view then produces the

CP430, *Fourier Transform Spectroscopy:* 11th International Conference
edited by J.A. de Haseth

spectral dimension of the data cube, as is illustrated by the curve in the upper right of Fig. 1.

There have been numerous comparisons of Fourier transform spectrometers with dispersive spectrometers over the years. One of the clearest of these comparisons is that of Schrader and Keller (1). The relative status of these two alternatives has depended on technology, and especially on the detector technology. For single element detectors in both a Fourier transform and a dispersive spectrometer, the need to scan over the spectrum with a dispersive spectrometer places it at a considerable disadvantage (2). However, with a linear array in the focal plane of a grating spectrometer, and a single element detector in the focal plane of a Fourier transform spectrometer, the classical Fellgett multiplex advantage of the Fourier transform spectrometer is offset by the linear array, for the purpose of measuring the spectrum of a single source. In the case that an extended subject area is being measured, the relative status of the Fourier transform and dispersive spectrometers is once again altered.

Two of the classical advantages of Fourier transform spectrometers are not relevant for imaging spectrometers. The Jacquinot throughput advantage (3) is made moot by the fact that the optical throughput per pixel for imagers which operate near the diffraction limit is determined solely by the mean wavelength of the observed light, and not by the spectral resolution, as is normally the case for a non-imaging spectrometer. Specifically, if the angular field of view of a pixel is given by the Rayleigh resolution criterion, $\theta=1.22\ \lambda/D$, then the throughput determining product $\theta \bullet D = 1.22\ \lambda$. The Fellgett multiplex advantage is also made moot by the fact that with an FPA in a dispersive spectrometer, all spectral channels are measured simultaneously. In effect, the temporal multiplexing characteristic of FT spectrometers is replaced by a spatial multiplexing in the dispersive spectrometers.

There is another aspect of Fourier transform spectrometer spectra, however, which comes into its own with an IFTS, and that is what may be called the "white noise" advantage. This aspect of an IFTS will be much discussed in the later sections, but in brief the recognition of distinctive spectral signatures, and the enumeration of the number of spectral constituents in a datacube are much facilitated by taking advantage of this spectral whiteness.

For **ideal** instruments making use of 2 dimensional detector FPA's, a remarkable parity holds for most imaging spectrometer designs, as was discussed in reference (4). In practice, however, the alternative designs are not equivalent. Currently available FPA's do not have sufficiently identical performance from pixel to pixel that any of the spatially multiplexed interferometric designs are practical for high performance imaging spectrometers. Residual, uncalibrated non-uniformities in real FPA's produce undesirable spectral aberrations and noise. These non-uniformities are found to be time dependent, and thus are practically impossible to calibrate away. Tunable filter imaging spectrometers and dispersive imaging spectrometers do both have their place in the broad field of imaging spectroscopy. The IFTS, however, has the distinction that it has the greatest flexibility in terms of spectral resolution, the least demanding requirements in terms of stray light (or in terms of cryogenic operation in the case of the thermal infrared spectral region), and the greatest fidelity in terms of spectral calibration of any of the imaging spectrometers, by virtue of the metrology used in determination of the optical path difference between the fixed and moving mirrors. (The classical Connes spectral advantage).

DEVELOPMENT HISTORY

An imaging Fourier transform spectrometer was first described in a 1972 patent by A.E. Potter, Jr (5). This concept was quite far ahead of its time then, however, and it was not until 1980 that results from a working instrument, with a rather modest 42 element array in the focal plane, were published (6). Truly "imaging" FTIR spectrometers required the development of relatively large format infrared FPA detectors. Also, in order to handle the computational demands in a reasonable time, relatively inexpensive, fast computers were needed.

In 1993, the first observations with the LIFTIRS instrument (7) were published. Since then, other researchers have begun to take advantage of imaging FTIR spectrometers as well (8), and at least one vendor is producing an imaging FTIR spectrometer commercially (9).

Over the past four years we have gained much experience with the LIFTIRS instrument (10), and have developed a number of techniques for the acquisition, reduction, calibration and exploitation of hyperspectral data. Some of these techniques are novel, while others are merely the application of classical mathematical analysis to the hyperspectral imaging problem. For the sake of being self contained, the complete train of steps for the analysis of a representative example are presented in the balance of this article.

REPRESENTATIVE EXAMPLE

A large number of fields of study stand to benefit by the application of imaging spectroscopy. Generally, any field for which spectroscopy is useful and for which there is a spatial variation to the spectra, can benefit. This covers a very large arena, and there isn't space here, nor is the author's imagination sufficient, to do justice to all of the possibilities. As an illustration of the type of data acquired with a hyperspectral instrument, on a subject of some familiarity, a small granite rock was used as the test subject for comprehensive analysis. The field of geochemistry (11) is concerned with the abundance and distribution of the elements in the earth. Hyperspectral data from granite, which is the most common rock on earth, is archetypal, not only of the application of hyperspectral imaging to geochemistry, but of the use of hyperspectral imaging in general.

Sample Case History

A typical data set taken with LIFTIRS is followed from the experimental setup to the raw data form, through the spectral computation, principle component analysis, and matched filter analysis. The steps discussed are representative of the nature of hyperspectral datacube analysis, and thus are generally useful, and independent of the subject matter.

Experimental Setup

In the example followed here, granite was the subject. Granite consists of a highly variable mixture of various feldspars, quartz, often mica and various other minor accessory minerals. Because of its great spatial variability and rich array of components, a granite sample provides an interesting subject for a hyperspectral imager.

A small sample of granite was heated with a hot air gun to approximately 60° C. It was then placed in the object plane of the instrument sketched in Fig. 2.

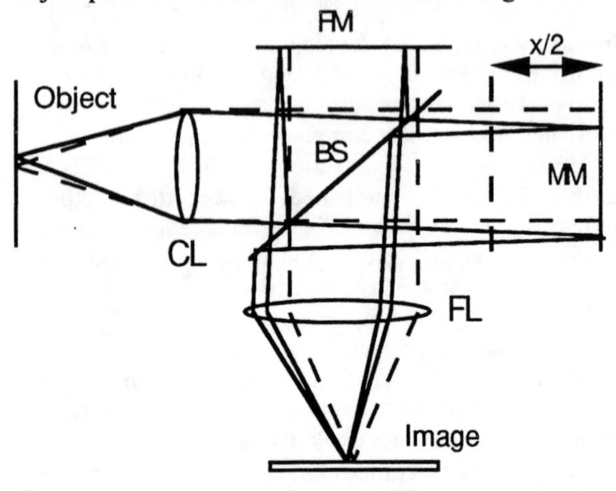

FIGURE 2. Optical layout of an imaging FT spectrometer.

Typical rays emerging from two representative points on the sample are drawn. One point is located on the optic axis, and it's rays are drawn dashed. The second point is displaced from the optic axis, and has solid emerging rays. In this setup the thermally emitted infrared light emerging from a point on the granite sample is collimated by the lens CL, passes through a Michelson interferometer, consisting of beam splitter, BS, fixed mirror, FM, and moving mirror, MM, and is then focused by the lens FL onto the image plane, where a thermal infrared sensitive FPA is located. In the data presented here, CL is an f/2 lens with a focal length of 100 mm, while FL is an f/6 lens with a focal length of 226 mm, thus producing a 2.26x magnification of the object. The cone of rays collected from points on the object has a solid angle of 0.1 steradian. The interferometer is the Bio-Rad model 896 interferometer, which has great flexibility in terms of scan rate, and can be easily interfaced to almost any framing FPA. The FPA used in the present measurements (manufactured by Rockwell), has a pixel pitch of 75 microns, and therefore the pixel spot size on the object was 33 microns. The FPA format is 128x128, which thus corresponds to a physical sample size of 4 x 4 mm.

The sample was held in place in front of the collimating lens with a 3 jaw chuck, and no attempt to carefully shield the sample from the laboratory thermal emission was made. Rather, the fact that the inside of an ordinary room has a radiation field which is a reasonable approximation to a black body cavity at room temperature was relied upon for the background "isothermal" bath. Because only 1.7% of the solid angle seen by the sample is filled by the effectively cold image of the FPA, almost all of the diffusely reflected light corresponds to the radiance expected for thermal emission at room temperature. As is apparent, the sample preparation for such measurements is trivial, and, provided only that the sample may be gently warmed, is non-destructive and non-intrusive.

The sample was positioned so as to bring the desired surface into focus, as judged by the sharpness of the fine features on the sample visible in a real time display of the FPA imagery. Since this same imagery is used to acquire the spectra, there is no question of a misalignment between the subject locating imagery and the spectral acquisition focus, as there is in a conventional FTIR microscope arrangement. Once properly positioned, a sequence of frames from the FPA are captured to volatile memory, and subsequently saved to magnetic disk storage. A series of such data cubes are acquired while the sample cools. Eventually, the temperature of the sample equilibrates with room temperature, and much less contrast is seen in the thermal imagery.

Raw Data Cubes

An example of the varied appearance of various frames seen through the interferometer is displayed in Fig. 3. The total number of frames acquired for this datacube was 2844, corresponding to an unapodized spectral resolution, that is the spectral channel spacing, of 2.2 cm^{-1}.

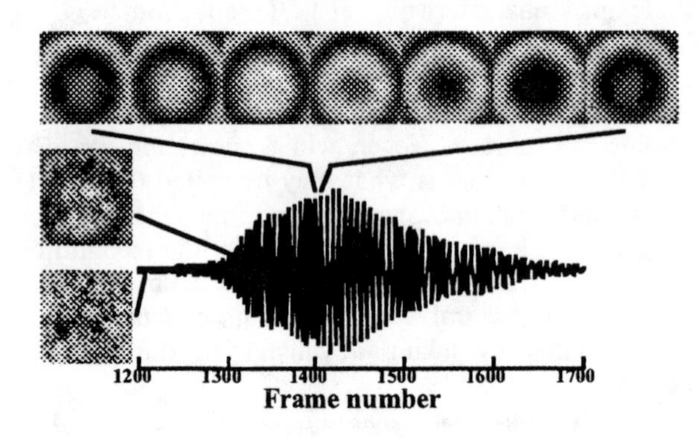

FIGURE 3. A composite display of a portion of a typical interferogram from the middle pixel in the array with images corresponding to the various indicated positions along the interferogram. The seven adjacent frames across the top correspond to the sequential frame numbers 1400 through 1406.

The interferogram shown in the figure associated with the central pixel in the FPA is quite representative, since the interferograms from pixels even at the corners of the image look very similar. For several positions along the interferogram, the images detected by the FPA are displayed, with the correspondence to the location in the interferogram indicated by the lines drawn in the figure. Note that for images taken from near the centerburst region, prominent interference rings are apparent. A sequence of seven successive frames are shown, corresponding approximately to a single cycle of the oscillating ring pattern. Also note that for images taken far from the centerburst, the interference rings are so subtle, that they are not apparent to visual inspection. When viewed in time sequence as a movie, the image detected in the FPA appears at first to have no interference fringes, followed by a period of increasingly prominent rings, which have the appearance of circular ripples in a pond which are continuously emerging from a center, and expanding out beyond the edges of the image. All through the centerburst region, the ripples appear to expand outwards from the center. As the centerburst intensity decreases on the right hand side of the plot, the rings become less and less pronounced, and soon have less contrast then the structures in the image.

Reduction to Spectra

For each pixel in the FPA, the time variation of the observed intensity comprises a standard interferogram. For each pixel's interferogram, the average level is subtracted, and the interferograms are optionally zero padded to make the total number of interferogram points equal a power of two. The substantial drawbacks of zero padding data for hyperspectral imaging purposes will be described later in the section ***the effect of zero padding***. The appearance of the uncalibrated hyperspectral datacube at this stage is illustrated in Fig. 4.

For simplicity of computation, the magnitude of the fast Fourier transform (FFT) of two sided interferograms is used to determine the uncalibrated spectrum. For high signal to noise ratio spectra, it is unnecessary to take the real part of the properly phase corrected complex spectrum, as is usually done, and for expediency in the processing of large quantities of data, omitting the phase correction stage saves significant time. The effect of simply taking the magnitude of the complex amplitude is that a small, positive definite noise generated baseline intensity is added to all spectra, but the rms random variations about the expected values are at the same level as would be the case with proper phase correction. Furthermore, the noise baseline is at a level, relative to the signal, of $1/SNR^2$, and is thus insignificant.

After this step, the datacube is reduced, but uncalibrated. Each pixels spectrum is the equivalent of what is commonly referred to as a "single beam" spectrum for non-imaging spectrometers. One particularly noteworthy aspect of these uncalibrated spectra is that the spectral noise is, to very good approximation, "white". This aspect of the uncalibrated spectra follows from the use of FPA's in imaging FTIR spectrometers, in that no "gain ranging" is employed away from the center burst, and the data is DC coupled, thus there is little variation in the expected rms noise level in the time domain, since all points in the interferogram have essentially the same gain and same signal level, and therefore approximately the same noise level. Since a constant rms noise level in the time domain transforms to a constant rms noise level in the frequency domain, the "whiteness" of the spectrum follows. It also follows that the relative spectral SNR is directly proportional to the uncalibrated spectral signal.

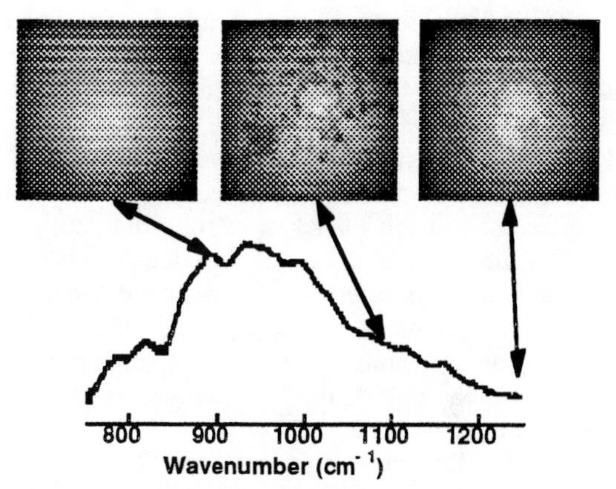

FIGURE 4. An example of a datacube transformed to the spectral domain, but uncalibrated. The sample spectrum from the center pixel is shown, and for a few representative spectral channels, the corresponding uncalibrated images are shown. Note that there are a number of pixel to pixel systematic non-uniformities apparent, including a horizontal striped pattern, and a circular vignetting pattern.

Calibration

Calibration of each pixels spectrum to spectral radiance units is made on the basis of linear interpolation between the radiance's observed from a pair of blackbodies at temperatures near, preferably just above and just below, that of the measured subject. Figure 5 displays the uncalibrated spectra from the central pixel of the data set for both a granite subject datacube and a pair of blackbody calibration datacubes.

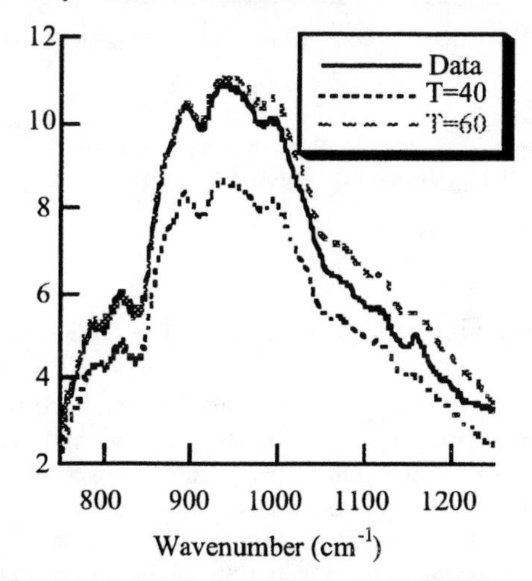

FIGURE 5. Representative uncalibrated spectra from the central pixel of the FPA are shown for three data sets, a blackbody at 40° C, (the lowest line) a blackbody at 60° C (the highest line) and the granite sample at approximately 60° C. The units on the ordinate for this data are arbitrary.

The explicit formula for calibration to spectral radiance units, for each pixel, is given by

$$S(k) = \frac{D(k) - C(k)}{H(k) - C(k)} * \Delta B(k) + B(k,T) \qquad (1)$$

where S(k) is the calibrated spectral radiance for the given pixel at wavenumber k. Similarly, D, C, H are the uncalibrated signals for the data, cold and hot blackbody references respectively. B(k,T) is the Planck function for blackbody spectral radiance at wavenumber k and the temperature T corresponding to the cold reference. $\Delta B(k)$ is the difference in the Planck spectral radiance function for blackbodies at the cold and hot temperatures. For convenience, the calibrated spectral radiance is expressed in terms of the brightness temperature, i.e. the brightness temperature at wavenumber k is the temperature of a blackbody which would produce the observed spectral radiance at wavenumber k.

Figure 6 illustrates this process for the same representative pixel treated in Fig. 5. After calibration to brightness temperature, the blackbody points are indicated by the horizontal lines, while the data for the granite sample are seen to have a wavenumber dependent brightness temperature.

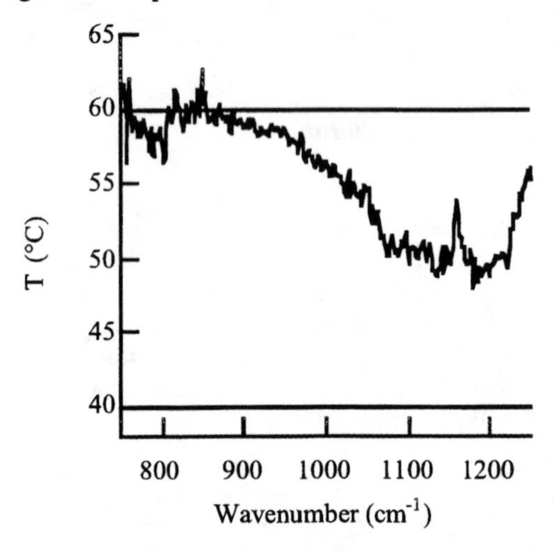

FIGURE 6. The data for the same pixel shown in Fig. 5 is displayed after calibration to brightness temperature units.

Spectral Data Cube

Once calibrated, the hyperspectral data cube contains a vast quantity of detailed information about the observed sample. Figure 7 displays a small set of spectra taken from various locations on the sample, and includes the mean spectrum computed from the average over all pixels. The locations were chosen on a "hunt and peck" basis, in order to provide a range of different spectra. On the basis of such casual inspection of the calibrated data cube, it is apparent that there is significant variation in the gradation of the spectra. That is, although adjacent pixels tend to have similar spectra, with a gradual and continuous change from one spectral shape to another; for other cases, an abrupt change between spectral shapes is found. This behavior is readily interpreted as being the result of a corresponding spatial dependence in the mineral composition within the sample. In some cases there is a mixture of two or more pure minerals within a single 33 micron pixel, and the relative admixtures of the pure components varies gradually from pixel to pixel. In other cases there is a much sharper boundary between the different mineral compositions. It is thus easy to understand that the mean spectrum for the entire 4 mm x 4 mm sample displayed in Fig. 7, which contains some stochastic mixture of the various pure mineral components, will not look like the spectrum for any one pure mineral.

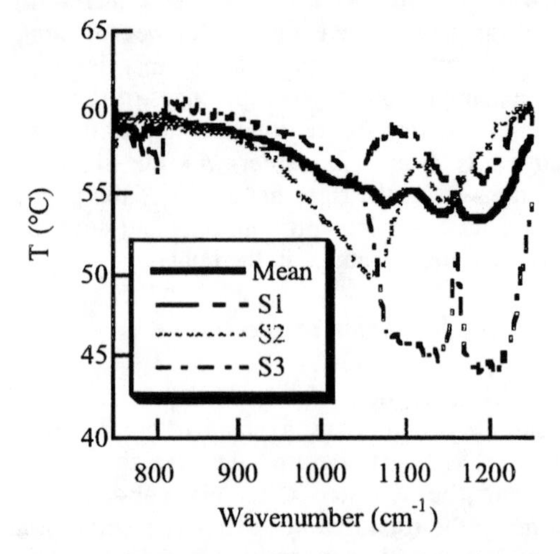

FIGURE 7. A selected set of brightness temperature calibrated spectra from various locations on the granite sample are shown. The solid curve corresponds to the mean over all pixels.

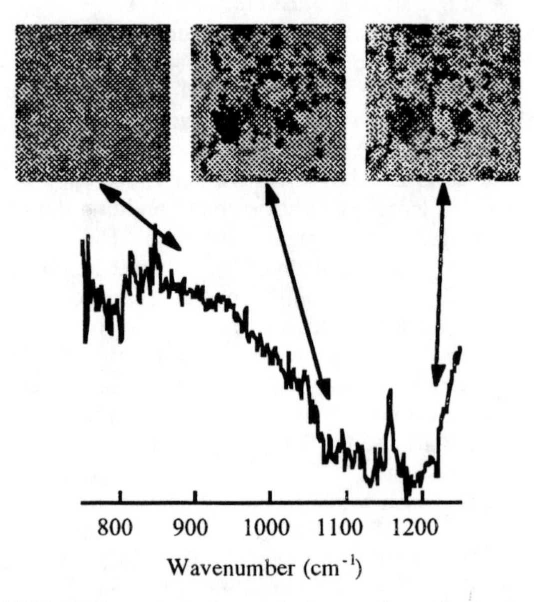

FIGURE 8. Three monochromatic images from the hypercube are displayed, corresponding to the same three frequencies as were shown in Fig. 4, viz. 900, 1100, and 1240 cm^{-1}. The curve shown is that of the brightness temperature calibrated spectrum for the central pixel in the image.

An indication of the spatial variability in composition and gradation can be gotten from the series of images corresponding to single spectral channels. Figure 8 displays three such monochromatic images. Such displays give a relatively crude indication of the spatial variability, since single spectral channels are often not very distinctive. For example, as can be seen for the three spectra shown in Fig. 7, at some frequencies, such as at about 900 cm^{-1}, there is virtually no difference between any of the three spectra. At other frequencies, such as at 1100 or 1240 cm^{-1}, there is a great deal of variability. The result of this difference can be seen in the images displayed in Fig. 8.

Spectral Interpretation

The brightness temperature spectrum, per se, is not an intrinsic property of the sample being observed. It is, however, simply related to the emissivity spectrum, and closely related to the spectral reflectance spectrum. As an example, for the most "colorful" of the spectra displayed in Fig. 7, labeled "S3", the conversion to both emissivity and reflectance is illustrated, and a comparison with a laboratory reference spectrum (12) for pure Quartz is shown in Fig. 9.

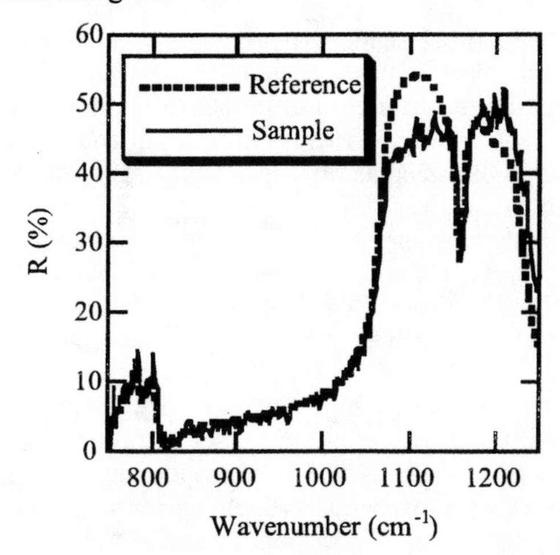

FIGURE 9. The reflectance spectrum computed for one of the sample location "S3" is compared with the library reference spectrum for the hemispherical reflectance spectrum of quartz, which is one of the major constituents of granite.

The relation between spectral radiance and emissivity, $\varepsilon(k)$, for any given pixel, is given to good approximation by

$$S(k) = \varepsilon(k) \cdot \Delta B(k) + B(k, T), \qquad (2)$$

where the notation is as in equation (1) with the exception that the cold temperature reference point is room temperature, and the hot reference temperature is that of the sample itself. This expression neglects the 1.7% of the solid angle seen by the sample which is directed back into the imaging spectrometer, and thus sees the cold FPA surface. For unit emissivity samples, this effect produces no error, while for highly reflective samples, this effect produces only a few percent error. This method is essentially that of Kember, et al. (13), and Chase (14).

The relation between emissivity and hemi-spherical reflectance is given by Kirchoff's law, $R(k)=1-\varepsilon(k)$. From this expression and equation (2), it follows that either emissivity or reflectivity reference spectra can be easily related to the brightness temperature calibrated data. For relatively modestly warmed samples, the difference between room temperature and sample temperature is small enough that the dependence of spectral radiance on temperature is linear, and all four quantities: spectral radiance, brightness temperature, emissivity and reflectance, are linearly related.

Hemispherical reflectance spectra are available for a large number of materials on the internet (15). Given any one pixel's spectrum, after conversion to reflectance, it is straightforward to compare against a library of possible spectra. It soon becomes apparent, however, that few of the pixel's spectra look exactly like any single library reference spectrum. Because the sample observed is obviously a mixture of several components, it is very plausible that even at the single pixel level, any one pixel has significant mixing of the individual spectra from the various pure components.

To disentangle the various pure components in the hyperspectral datacube, it is necessary to consider the variations in the spectra from pixel to pixel in order to pick out correlated spectral variations that may be attributed to the underlying basic spectral constituents.

Two Channel Spectral Correlations

As an introduction to the process of considering spectral correlations in the analysis of hyperspectral data, a two color case is treated. From the spectra displayed in Fig. 7, it can be seen that by comparison of spectral intensity near 900 cm^{-1} and 1100 cm^{-1} the "quartz like" pixels may be distinguished from "non-quartz like", for the reason that the quartz like spectrum has a substantial decrease in emissivity at 1100 cm^{-1} relative to the other spectra shown. A scatter plot of the brightness temperatures at these two frequencies for all 16,384 pixels in the sample is displayed in Fig. 10.

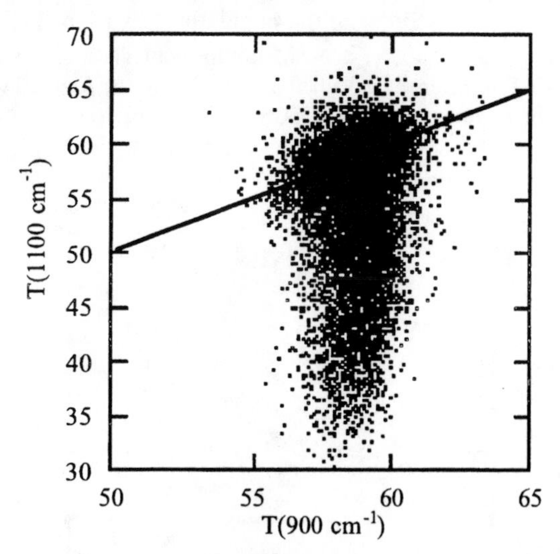

FIGURE 10. A scatter plot of the brightness temperature at 900 cm^{-1} vs. 1100 cm^{-1}. Greybody pixels, i.e. those whose brightness temperature is independent of wavelength, appear along the diagonal line shown, with position dependent on their brightness temperature. Pixels bearing an admixture of quartz like spectra are displaced downwards from this diagonal.

Note that the spread in brightness temperature for the 900 cm^{-1} channel is much less than the spread in brightness temperature for the 1100 cm^{-1} channel. Also note that there is not a clear separation between "quartz like" and "non-quartz like" pixels, since there is a nearly continuous distribution of points from the diagonal "greybody" line to the lowest points in the graph. Such a two channel scatter plot is clearly not very specific to the spectral nature of the constituents carrying the color difference. In order to have a greater degree of specificity, it is necessary to consider multi-channel correlations.

Multi-channel Spectral Correlations

It is of great use to generalize the two channel scatter plot analysis to M dimensions, where M is the number of spectral channels in the data. The basic quantities of interest are the mean spectrum, and the covariance matrix. The mean spectrum vector, $\overline{S}$, is defined by the expression for its components

$$\overline{S}_k = \frac{1}{N_{pixels}} \sum_{i,j}^{N_{pixels}} S_{ijk} \,, \tag{3}$$

where the summation is over all pixels within the datacube, and the indices specify the horizontal position (i), vertical position (j), and spectral channel (k) in the hyperspectral datacube. The covariance matrix, C, is defined by the expression for its matrix elements

$$C_{kk'} = \frac{1}{N_{pixels} - 1} \sum_{i,j}^{N_{pixels}} (S_{ijk} - \overline{S}_k)(S_{ijk'} - \overline{S}_{k'}), \tag{4}$$

where the normalizing factor involves $N_{pixels}-1$ rather than N_{pixels} in order that the sample estimate of the covariance be an unbiased estimate of the population covariance (16). The mean spectrum, $\overline{S}$, is an M component vector, where M is the number of spectral channels in the datacube, while the covariance matrix, C, is an MxM matrix.

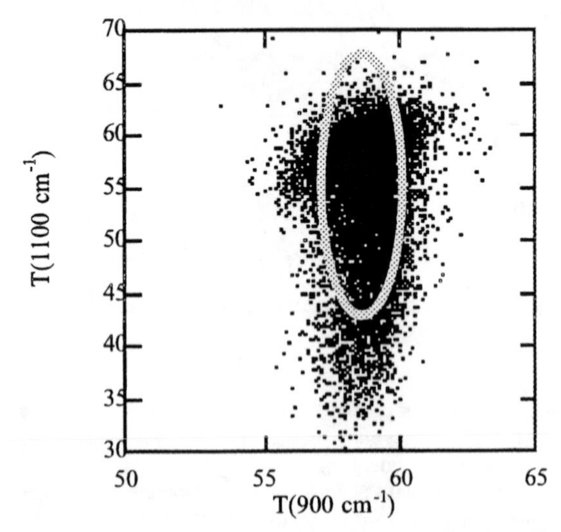

FIGURE 11. The same two channel scatter plot data displayed in figure 10 is replotted, with a superimposed principle component ellipse.

It is helpful to think of the covariance matrix in geometrical terms, and emphasize those aspects which are independent of the choice of axes for the coordinate system. Just as the length of a vector is doesn't change under arbitrary rotations, the moments about the principle axes of the covariance matrix also remain invariant under arbitrary unitary transformations. These moments are determined by the eigenvalues of the covariance matrix, while the directions of the principle axes are determined by the eigenvectors.

In the two dimensional scatter plot shown in Fig. 11, the principle axes of the covariance matrix for the data set shown are indicated by the sketched ellipse. The center of this ellipse is positioned to match the location defined by the mean spectrum values at 900 cm^{-1} and 1100 cm^{-1}. The axes of this ellipse coincide with the principle axes of the covariance matrix in the two dimensional subspace, while the lengths of the semi-major and semi-minor axes correspond to the 2 standard deviation spread along the two principle axes. In this particular case, because there is almost no "tilt" to the spread of points, the principle axes happen to be nearly perfectly horizontal and vertical respectively. In general, the principle axes may be inclined at any angle.

In three dimensions, it is almost as easy to imagine an ellipsoid, whose three principle axes would correspond to the direction of greatest, least, and intermediate spread of the points in a 3-d scatter plot. Beyond three dimensions, it becomes increasingly difficult to physically picture analogous hyper-ellipsoids, but in any number of dimensions, there will always be a set of principle axes, one for each dimension, with a corresponding spreading width for the set of points, which corresponds to the eigenvalue of the covariance matrix associated with the given principle axis.

Diagonalization of the covariance matrix (4) provides a set of M eigenvalues and eigenvectors. The eigenvectors form an orthonormal basis for the space of M spectral channels of the original datacube. By re-expressing the datacube in terms of the covariance matrix eigenvectors, a set of "eigenimages" corresponding to each of the eigenvalues of the covariance matrix may be formed. This principal component analysis (17) method, also known as factor analysis or the Karhunen-Loeve expansion (18), is quite sensitive to spectral variations correlated from pixel to pixel in the hyperspectral data.

Principle Component Analysis

The eigenvectors from the diagonalization of the covariance matrix are spectral vectors. The eigenspectra are given by the product of the eigenvectors, which have unit (dimensionless) normalization, with the square root of the corresponding eigenvalues. The eigenspectra thus have the same units as the calibrated datacube. Numbering the eigenspectra in order of decreasing magnitude of the eigenvalues, the first three eigenspectra are shown in Fig. 12. The dominant eigenspectrum, labeled PC1 in Fig. 12, can be seen to have a magnitude of nearly 6° C at 1100 cm^{-1}, and approximately 0° C at 900 cm^{-1}. Putting these magnitudes into perspective with respect to the spreading widths shown in Fig. 11, it may be noted that for the 1100 cm^{-1} channel, most of the spreading in the data may be attributed to contributions from the first principle component, which is primarily due to the presence of quartz.

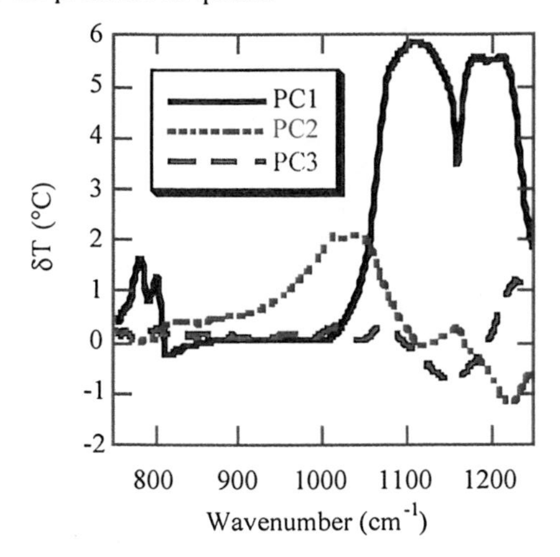

FIGURE 12. The first three most significant eigenspectra resulting from the principle component analysis are displayed.

The first nine eigenimages from the principle component analysis are displayed in Fig. 13. The first 7 of these eigenimages have correlated spatial structure which is attributable to true spectral/spatial variations in the datacube for this sample. From the 8th eigenimage on, the spatial structure is dominated by instrumental noise, and has very little significant pattern. Note that the sign of the contrast is not as significant as the magnitude. Although for some eigenvectors, the sign of the spectral features is unambiguous, as for example, with the first eigenvector, where it may be determined with confidence that the two bumps at 1100 cm^{-1} and 1200 cm^{-1} are associated with a greater degree of quartz reflectance, and correspondingly, a smaller emissivity, so that quartz features show up as relatively darker areas, for other eigenvectors the sign of the spectral features is ambiguous. The second eigenvector is a case in point, since it has both positive and negative spectral features. Furthermore, the same structure may show up with either positive or negative contrast, as is notably the case for the diamond structure in the lower right hand corner of the image, which shows up with positive contrast in the 4th and 5th principle component images, but shows up with negative contrast in the 6th principle component eigenimage.

By inter comparison of the various eigenimages displayed in Fig. 13, it can be seen that there are a great variety of spatial structures discernible in this particular granite sample. It is somewhat difficult, by eye, to make the proper associations between corresponding parts of the sample among the various images displayed in this figure. To facilitate such analysis, it is very helpful to construct false color images, which combine three of the individual eigenimages into a single color composite, and thus enable exploitation of the three color sensitivity of human vision. (Unfortunately, human vision is only sensitive to three linearly independent color combinations, and thus it is not humanly possible to see false color composites having more than three simultaneously contributing independent spectral dimensions). Figure 14 displays a false color image associated with the first three most significant principle components, whose eigenspectra are shown in Fig. 12. It can easily be seen that the "colorfullness" of the granite sample in the infrared is much more readily apparent on the basis of the false color image than on the basis of the separate gray scale images for each of the contributing components

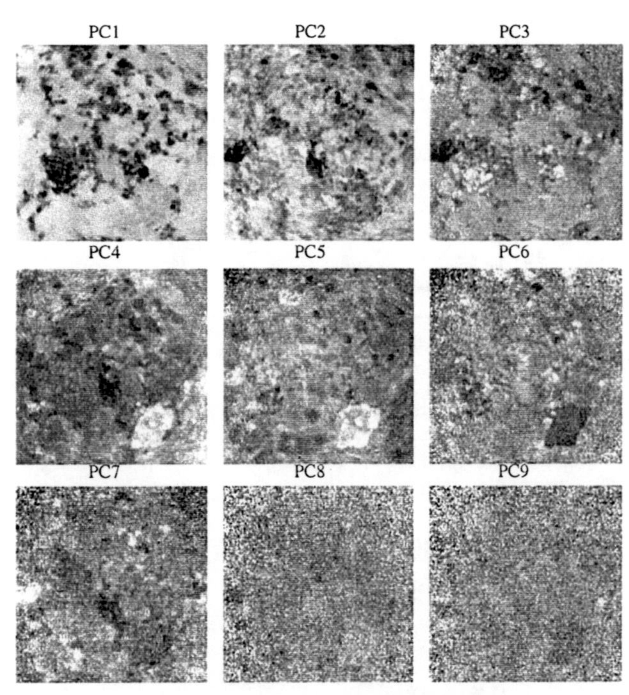

FIGURE 13. The first nine most significant eigenimages from the principle component analysis are displayed. The top three eigenimages correspond to the first three eigenspectra displayed in Fig. 12.

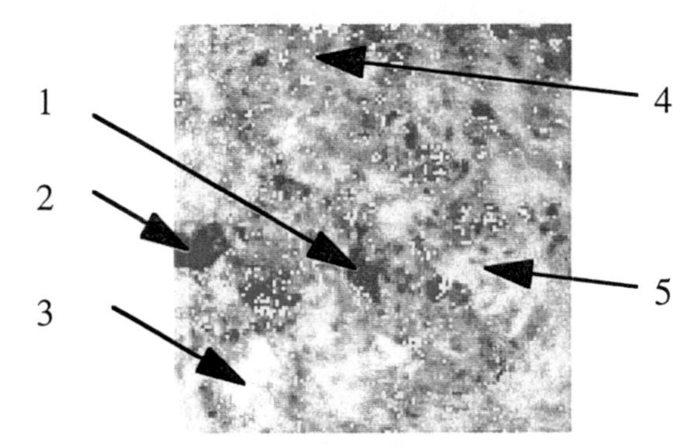

FIGURE 14. A false color image representing the contributions of the first three principle component eigenvectors to each pixel's spectrum. The intensity of the colors red, green, and blue correspond respectively to the contributions from the first, second and third principle components.

The spectra that correspond to each of the locations indicated in Fig. 14 are shown in Fig. 15. The fact that three colors are not sufficient to readily display the spectral diversity within this sample is indicated by the false color image shown in Fig. 16, based on mapping the

4th, 5th, and 6th principle components to the colors red, green and blue. The prominent yellow diamond, labeled with the number 6 in the figure, in the lower right hand corner, for example, is not apparent in the false color image of the first three principle components shown in Fig. 14.

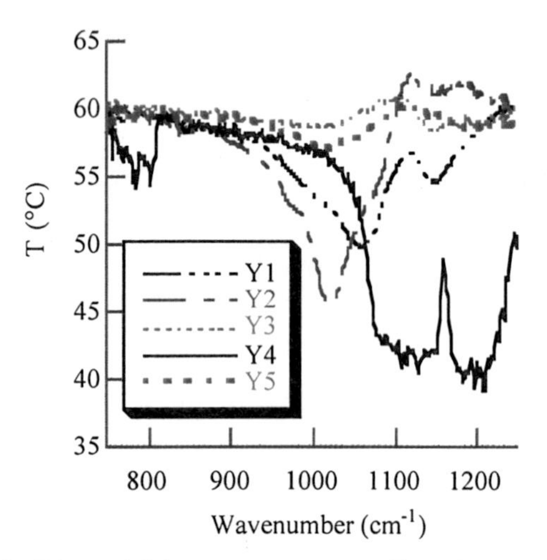

FIGURE 15. Brightness temperature calibrated spectra from the five locations shown in Fig. 14 are shown.

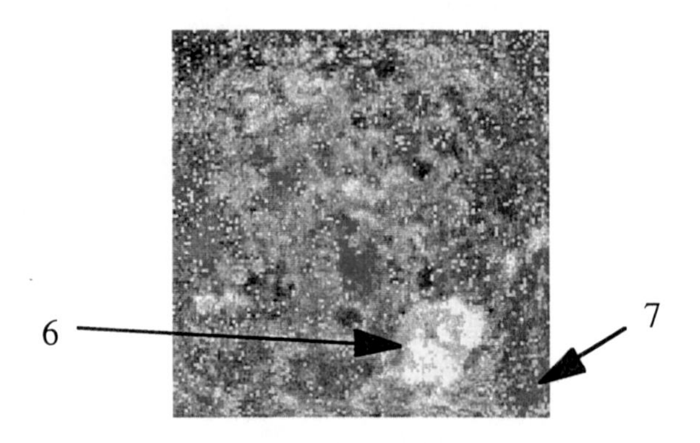

FIGURE 16. A false color image similar to that shown in Fig. 14, but representing instead the contributions of principle components 4,5, and 6. The locations of two more distinctive patches are indicated by arrows.

As can be seen in Fig. 17, the spectra for these two new locations are quite distinct from the spectra of the five locations displayed in Fig. 15. It is noteworthy that even though the spectral shapes corresponding to locations 6 and 7 are very similar, the main difference between them is the emissivity level. The yellow diamond area corresponds to a "shinier" patch on the sample than the purple corner, and in fact corresponds to a relatively pure crystal of albite.

The number of significant spectral constituents in a hyperspectral datacube is often of interest per se. In the case of the granite specimen, for example, the number of spectral constituents can be related to the number of different mineral species contained within the sample. Even without identification of the nature of the constituents, from the false color images, and the individual principle component images, it is clear that for the granite sample that there are at least 8 different spectral species.

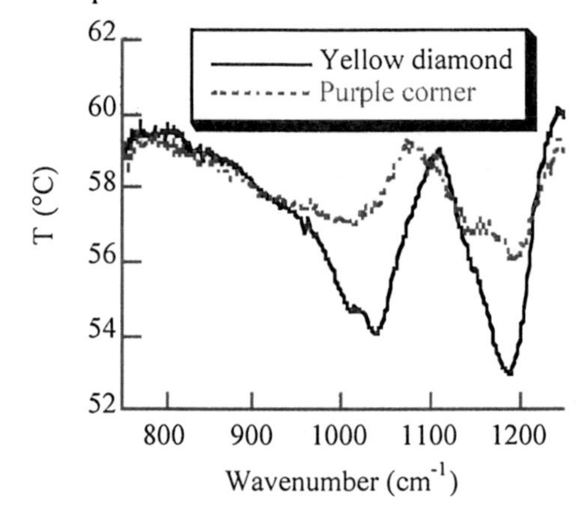

FIGURE 17. The spectra corresponding to the two locations indicated in Fig. 16 are displayed.

Eigenvalue Distributions

One of the most useful metrics for providing both a qualitative and a quantitative overview of the contents of a hyperspectral datacube is the distribution of the eigenvalues of the covariance matrix. An example is shown in Fig. 18 for the case of the granite sample.

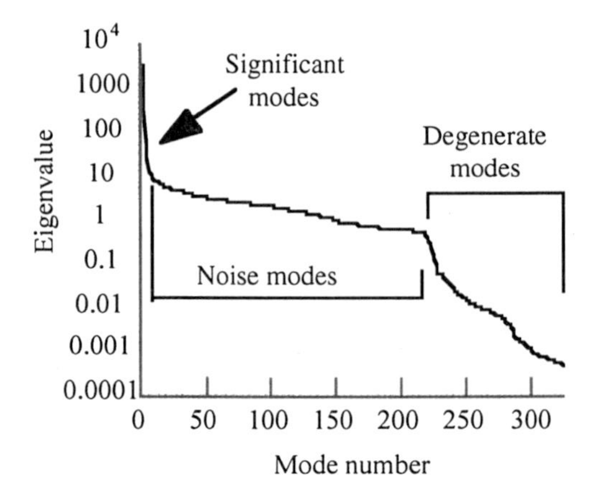

FIGURE 18. The eigenvalues of the covariance matrix for the granite datacube are plotted in decreasing numerical order. Three intervals are indicated in the plot, corresponding to ranges of eigenmodes having qualitatively different character.

The rapid drop off for the first few eigenvalues seen in this example frequently occurs in the principle component analysis of hyperspectral datacubes. In the general case this feature may be related to the relatively low entropy in typical spatial/spectral compositional mixtures. In the specific case of granite, which is produced by the gradual cooling of magma, within a small spatial region having nearly isothermal and isobaric conditions, local variations in the quantity of a particular mineral species formed will depend on the available concentrations of the various reacting compounds involved in the crystallization. As the freezing point for various species is reached, the available concentrations for reacting compounds will change in a spatially stochastic, time dependent manner. In a stochastic mixture of multiple components, it is highly unlikely that each component have the same abundance. Thus after eliminating the majority component, and considering a similar division of the remainder of the mixture, a similar argument can be carried out inductively. It is thus natural to find a succession of eigenmodes, with the eigenvalues dropping off exponentially with increasing mode number. This drop-off continues until the instrumental noise level is reached, at which point a Wigner distribution of modes is expected, as is discussed in the next section.

Finally, as can be seen in the figure, beyond the 217th mode (out of 325 total modes), there is another precipitous decrease in the eigenvalues. These modes, having very low variance, correspond to the degenerate modes that are introduced by zero-padding the interferograms from 2844 points to 4096 points before computation of spectra. Such zero-padding generates a higher density of spectral samples, but does not increase the spectral resolution. Zero-padding produces a particular type of spectral interpolation. The obvious fact that zero-padding does not increase the information content of an interferogram is reflected in the eigenvalue distribution by the steep drop off in eigenvalues, once the number of truly independent degrees of freedom has been exhausted. These eigenvalues do not becomes strictly zero, as they mathematically should, by virtue of the finite numerical precision involved in the computation. Indeed, the condition number for the covariance matrix for the granite sample datacube, given by the ratio of the smallest to the largest eigenvalue, is approximately 10^{-7}, which is just about the limit of the precision for the 23 significant bit floating point computations used in the data analysis.

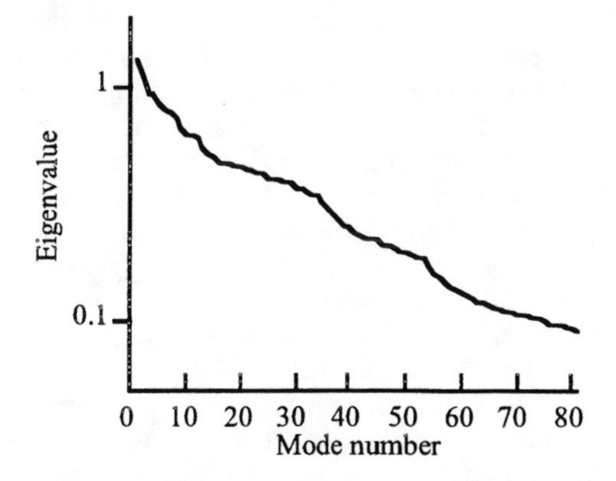

FIGURE 19. The eigenvalues of the covariance matrix for a calibrated blackbody reference datacube are plotted in order of decreasing magnitude.

On the basis of the above discussion, it is to be expected that for a completely spatially and spectrally unstructured datacube, with no zero padding prior to FFT computation, the distribution of eigenvalues should be determined by the instrumental noise alone. A good blackbody reference calibrated datacube should be both spatially and spectrally unstructured. Figure 19 displays the distribution of eigenvalues of the covariance matrix associated with a calibrated blackbody datacube, which was not zero padded prior to FFT computation of the spectra. This data was also acquired with a lower spectral resolution, thus fewer spectral channels and lower noise than that of Fig. 18. It is clear that the precipitous decrease in eigenvalues observed at either extreme of the curve shown in Fig. 18 is not present in Fig. 19.

The median eigenvalue, to good approximation, agrees with the spectral/spatial noise estimate based on a comparison of sequential measurements under nominally identical conditions. Further evidence in support of the identification of these modes with noise is apparent in the appearance of the eigenimages displayed in Fig. 20.

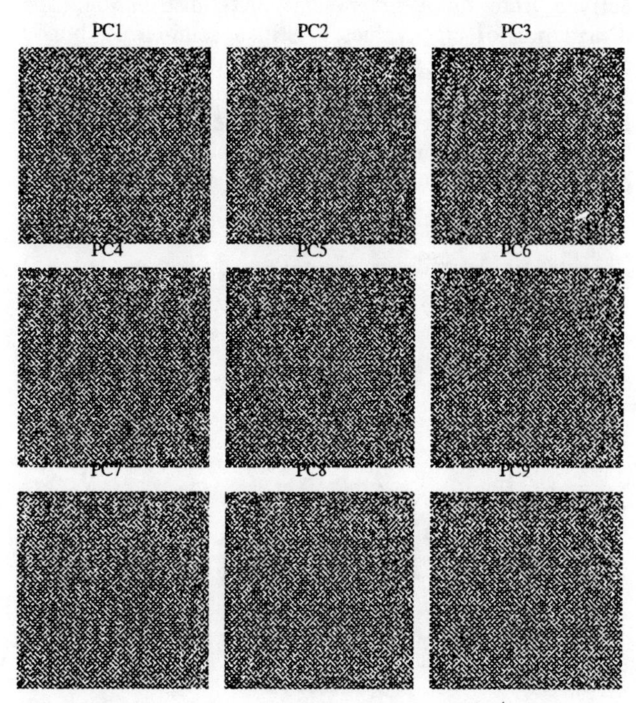

FIGURE 20. The first nine most significant eigenimages from the principle component analysis of a calibrated blackbody are displayed.

There is almost a complete lack of coherent spatial structure in these images. The very slight "rippling" or "striped" appearance in some cases, such as in PC7, is just at the rms spatial noise level, and is likely due to subtle instrumental artifacts which have not been completely calibrated out. The eigenimages for modes 10 and up are even less structured than the first 9, and most are consistent with random spatial noise.

The Wigner distribution

The eigenvalues of an NxN random matrix taken from a "Gaussian orthogonal ensemble", in the limit that N tends to infinity have a distribution known as the Wigner "semi-circle" distribution (19). Figure 21 displays a typical histogram of the eigenvalues of a 1000 x 1000 matrix of unit variance, zero mean random numbers. It can be seen that this histogram indeed has a semi-circular appearance, with the maximum eigenvalue magnitude given by $2\sqrt{N}$.

A hyperspectral datacube has purely white noise if every spatial/spectral cell has identical variance about the mean for that cell. The mean values are arbitrary. This is indeed a reasonable approximation to the raw, uncalibrated spectral datacubes from an imaging FTIR spectrometer, as was discussed in the previous section, ***reduction to spectra***. For such a white noise datacube, the expected value for the covariance matrix is proportional to the unit matrix, with coefficient equal to the variance of the noise for each cell of the datacube. Furthermore, the expected variances for each of the off diagonal elements of the covariance matrix are identical for a white noise datacube. The distribution of eigenvalues of such a random covariance matrix may therefore be simply derived from the Wigner semi-circle distribution. The distribution of eigenvalues is still a semi-circle, but is displaced along the abscissa. The radius of the displaced semi-circle, relative to the displacement of the center from the origin, is given in terms of the number of pixels, N_{pixels}, and the number of spectral channels, M, in the hyperspectral datacube as

$$\text{radius} = \frac{2\sqrt{M}}{\sqrt{N_{pixels}}} \quad . \tag{5}$$

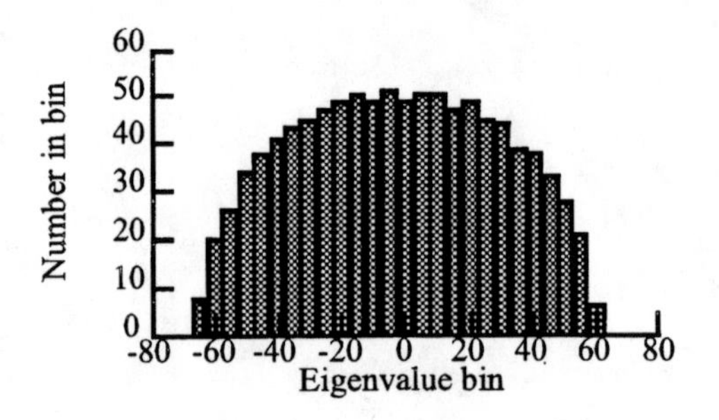

Figure 21. The histogram of the eigenvalues from a 1000x1000 random matrix has a Wigner semi-circle distribution.

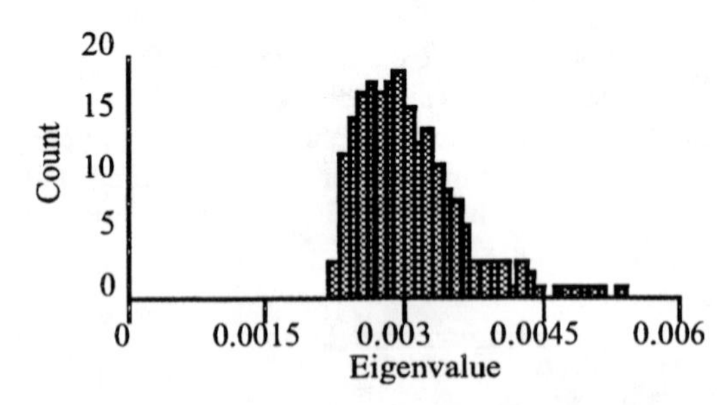

FIGURE 22. A histogram of the noise eigenvalues for an uncalibrated blackbody datacube is displayed.

This behavior is visible in the histogram plot, displayed in Fig. 22, for the eigenvalues of an uncalibrated blackbody reference datacube, which has been computed without zero padding of the interferograms. In this case the dimensionality of the data was 236, which is far from infinite, yet the shape of the histogram plot is already fairly close to semi-circular, albeit with an excess of larger eigenvalues. The first few large eigenmodes carry almost all of the non-noise structure present in the blackbody datacubes, such as the gradually varying efficiency attributable to vignetting, or the systematic variations in pixel to pixel response attributable to imperfect non-uniformity correction.

The Effect of Zero Padding

Using the same raw interferogram datacube as for the calculation shown in Fig. 22, but performing "zero padding" before computation of the spectra, leads to additional "degenerate" eigenmodes, which are apparent in Fig. 23. The number of modes after zero padding from 2844 points per interferogram to 4096 points increases from 236 to 338. All of the extra 102 modes introduced by zero padding show up in the two extra peaks in the histogram of eigenvalues shown in Fig. 23, one peak centered about the value 0, and one peak centered about the value 0.0015, which significantly is half the median eigenvalue.

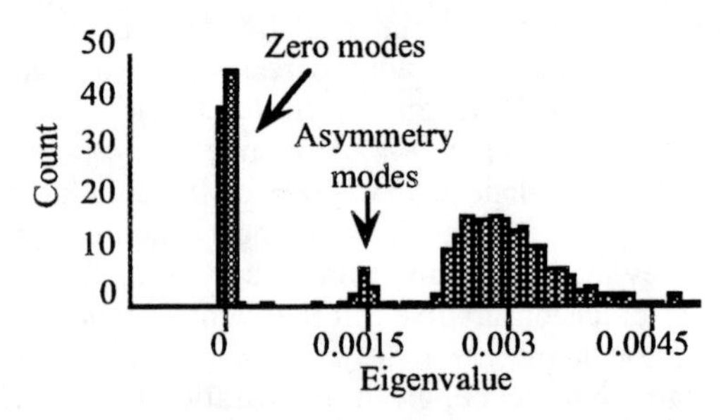

FIGURE 23. A histogram of the eigenvalues for an uncalibrated blackbody datacube calculated with zero padding of the interferogram is displayed.

Although the number of modes in each of the three groups is small, all three groups are plausibly consistent with Wigner semi-circle distributions. The semi-circular shape, which is dependent on the arbitrary scaling of the y axis, is only manifest in the plot for the main group centered about the value 0.003. The group centered about the value zero is readily accounted for as simply due to the additive roundoff error generated computational noise on the covariance matrix itself. In the absence of numerical roundoff error, and with perfectly symmetrical interferograms, the extra eigenvalues produced by zero padding would all be exactly zero. Within the null space corresponding to these zero modes, numerical roundoff error may plausibly be characterized as having zero mean and constant variance for each extra degree of freedom. This null subspace meets the criteria discussed in the section ***The Wigner Distribution*** to produce a semi-circle distribution of eigenvalues centered about the value zero. The group centered about the value 0.0015 has a more subtle explanation, related to mixing between the null space and its complement.

A two sided interferogram from a system with phase dispersion is not perfectly symmetric. In effect, each relatively narrow spectral band produces contributions to the interferogram with a different centerburst location. Thus the zero phase difference point is not perfectly centered for each frequency. For contributions to the interferogram which are not perfectly centered, a small number of large retardance points are under represented in the total interferogram by exactly a factor of two. By zero padding, this under representation is made manifest in the spectra, and a small number of the extra modes introduced end up with only half the proper strength. It is these modes which are apparent in Fig. 23 near the value 0.0015.

The zero modes indicated in Fig. 23, have non zero eigenvalues only by virtue of round off error in the computation. Their presence makes the inversion of the covariance matrix divergent, and they must be removed before a meaningful inversion of the covariance matrix is possible. Inverting the covariance matrix is an important ingredient for a number of signal processing steps, such as the generation of matched filters for the recognition of particular spectral fingerprints, as will be discussed in the section on ***Matched Filters***. Although it is possible to simply discard the portion of the hyperspectral space associated with small (or negative) eigenvalues, and proceed, it is very much better to avoid introducing degenerate modes in the first place, and thus to avoid zero padding. In practice, this means that it is highly advantageous to take two sided interferogram datacubes with the number of frames being an even power of two and to not perform zero padding.

The Effect of Apodization

Figure 24 displays the eigenvalue distribution for the same raw data as used in both Fig. 22 and 23, with the exception that the interferograms were apodized prior to Fourier transformation, using a cosine function which varied from 0 at the beginning, to unity in the middle, to zero at the end of each interferogram.

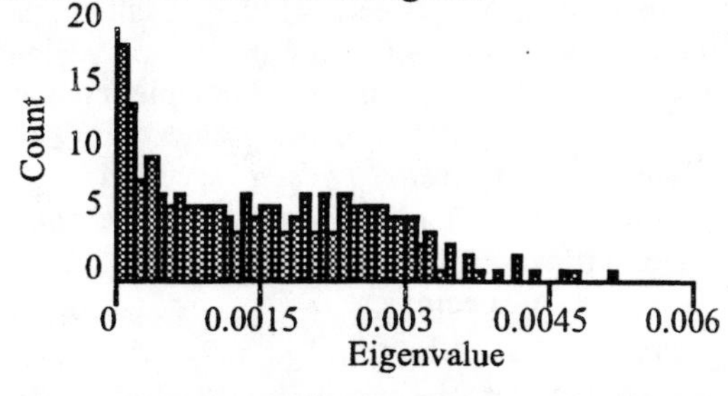

FIGURE 24. A histogram of the eigenvalues for an uncalibrated blackbody datacube is displayed. In this case, the spectra were calculated without zero padding of the interferogram, and were apodized using a cosine function.

Comparison of the distribution in Fig. 24 with that shown in Fig. 22, reveals that the eigenvalues for a large number of modes have been suppressed in magnitude, and a significant number have been approximately zeroed. Related to the discussion in the previous section, this suppression of the strength of certain modes introduces inaccuracies in signal processing algorithms which rely on the inversion of the covariance matrix, since near zero modes should be eliminated prior to the inversion. It is preferable to avoid this mode suppression at the outset, which implies that it is preferable to use unapodized interferograms to facilitate the signal processing of hyperspectral datacubes.

Calibration Cube Compression and Noise Reduction

The fact that most of the eigenmodes for an uncalibrated blackbody datacube are consistent with noise can be exploited both to reduce the noise in calibration data, and to compress the size of calibration datacubes without loss of fidelity of the data (20). The other key fact which enables this noise reduction and data compression is the a priori knowledge that a good blackbody reference source has very little spatial or spectral structure.

The datacubes taken with a blackbody subject are thus highly correlated. The spectra in each pixel are very similar, differing slightly only because there may be slight vignetting variations, angle dependent bandpass filter variations, possible field angle dependent responsivity variations in the FPA, or, more generally unknown sources of variation which are correlated from pixel to pixel. For this reason, in a principle component analysis, there are very few significant modes above the system noise level. Calibration datacubes can thus be very accurately represented by a small number of eigenmodes. The projection of datacubes onto the subspace corresponding to just those eigenmodes which are above the noise level, is denoted the Minimum Noise Representation (MNR). This representation differs from what is known, Boardman and Kruse (21), as the Minimum Noise Fraction (MNF). In the MNF, after diagonalization of the covariance matrix, an attempt is made to forcibly whiten the noise spectra by rescaling the principle component coordinate system to make all noise components have equal variance. In the case of data from an imaging FTIR spectrometer, in particular, it is not necessary to perform such forcible whitening, as discussed in the section on the ***Wigner Distribution***. Statistically, it is almost never found that all noise eigenvalues are equal, as is implicitly assumed in the use of rescaling in the MNF treatment. The MNR is thus both simpler, and of greater fidelity, than the MNF.

With a blackbody calibration source as a subject a great deal of the system noise may be eliminated from the calibration cubes by MNR. At the same time, the amount of data required to specify the calibration hypercube can be greatly compressed. This is accomplished by rotating the hypercube into the space of principle component eigenvectors, and keeping only the Minimum Noise Representation of the hyperspectral data as expressed in this basis. This is mathematically equivalent to performing a least squares fit of the calibration datacube to a linear combination of the mean spectrum and the non-noise eigenvectors, and keeping the coefficients involved in the linear combination as a representation of the calibration datacube.

As an example of the quality of this fit, the spectrum for a single pixel is compared with the corresponding least squares fit based on two components in figure 25. It can be seen that the least squares fit curve is much smoother, and indeed, the variations for the data points are due to the spectral noise in the blackbody datacube. By fitting to the smooth curve, a much less noisy curve is obtained for this particular pixel's spectrum (and all the others as well). Data taken with subjects that are not as spatially/spectrally simple as a blackbody calibration source may sometimes be represented in terms of the first few principle components alone, and for the dominant spatial/spectral features a similar level of data compression and noise reduction may be obtained as for blackbody calibration data. However, for spatially sparse and spectrally subtle features, it is necessary not only to retain more of the total hyperspace in the analysis, but to use a different analysis than PCA.

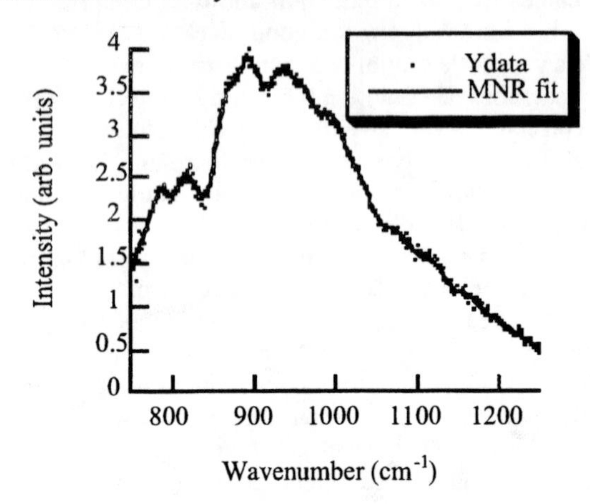

FIGURE 25. A Single pixel's spectrum from an uncalibrated blackbody reference datacube is compared with the Minimum Noise Representation (MNR).

Matched Filter Analysis

In order to image a particularly rare constituent within the complicated conglomeration of components, it is helpful to use the technique of matched filtering. A matched filter is defined as that for which the strength of a desired signal passing through the filter is maximized as much as possible while the strength of the underlying noise, or clutter variation, passing through the filter is minimized as much as possible. It is usually not possible to accomplish both at the same time, and so the figure of merit which is optimized by the matched filter is the signal to noise ratio. With a matched filter, it is possible to detect much weaker spectral signatures than are found with the PCA process described earlier. On the other hand, with a matched filter, it is necessary to specify the desired target spectrum before the filter may be constructed, in contrast to the case with PCA.

The effect of the matched filter on a hyperspectral datacube is to project each pixel's spectrum onto the direction of a matched filter vector, and thus to collapse the datacube into a single image. The expression for the matched filter vector, F, depends on the dot product of the desired target vector, T, with the covariance matrix for the hyperspectral datacube, C, in the following way:

$$F = T \cdot C^{-1}. \qquad (6)$$

The matched filter output image, I, is given by the dot product of the matched filter vector, F, with the hyperspectral datacube, S. In terms of the eigenvector basis derived from the diagonalization of the covariance matrix, C, the output image is written explicitly

$$I_{ij} = \sum_{k=1}^{M} \frac{T_k S_{ijk}}{\lambda_k} \quad , \qquad (7)$$

where the calibrated hyperspectral datacube elements are given by S_{ijk}, the image pixel value for the ith column and jth row is given by I_{ij}, the kth eigenvalue of the covariance matrix is λ_k, and the summation runs over the M spectral channels in the hyperspectral datacube.

It is the appearance of the inverse of the covariance matrix in expression (6), or equivalently the inverse eigenvalues in expression (7), which creates difficulties with zero modes. If the covariance matrix contains zero modes, then mathematically its inverse does not exist. With finite numerical precision, the probability that any eigenvalue be strictly zero is very small, and although the inverse of C may exist, it is very inaccurate unless the zero modes are removed prior to inversion. This may be accomplished by first recognizing the presence of anomalously small eigenvalues, aided by the discussion in the previous four sections, and second projection onto the subspace orthogonal to the approximately zero modes, which is equivalent to not summing over anomalously small eigenvalues in expression (7). Alternatively, it is preferable to avoid the zero modes in the first place, by not performing any zero padding of the interferograms, and by not performing any apodization either.

The output images corresponding to matched filters for three of the most common granite constituents, granite, biotite (22), and albite (23), are displayed in Fig. 26. By comparison of the quartz image in Fig. 26 with the image of the first principle component in Fig. 13, it is apparent that there is general similarity, although the quartz image tends to display only the central regions of the areas which show up in the principle component analysis. This is attributable to the greater purity of the central regions, and to the greater specificity of the matched filter image for pure quartz, since the reference target spectrum was taken from a reference library of highly pure mineral specimens.

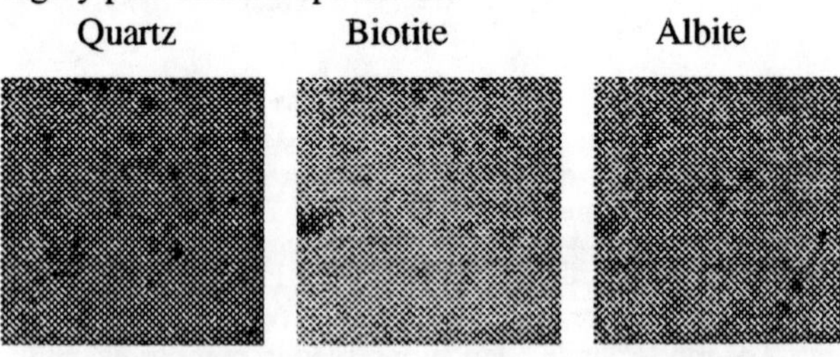

FIGURE 26. Representative matched filter output images for three filters constructed from reference spectra for three common granite constituents.

A similar comparison may be made between the biotite image and the second principle component image, and a similar explanation is plausible for the relatively smaller regions which show up in the matched filter image. On the other hand, the albite image shows a pattern of features which are not clearly seen in any of the first 9 principle component images, although there is a hint that relatively brighter components in the 5th principle component image do correspond to the darkest regions in the matched filter output image for albite.

It is interesting to compare the significance levels at which features become apparent in PCA with the significance levels at which features become apparent in MFA. To begin with, consider a feature confined to a single pixel. In order for the output of a matched filter to be significant for a single pixel out of 16,384 at the 65% confidence level, it is necessary that the single pixel signal differ by at least 4 standard deviations from the mean output of the filter. For a feature spread over n pixels in a cluster, the average difference of the cluster pixel's signal from the mean required to reach a given significance level varies approximately inversely with the square root of the number of pixels n. The detection of extended, weak clusters is, however, contingent on spatially filtering (or smoothing) the data on the size scale characteristic of the cluster. Such data manipulation is risky however, and once the mean significance of an individual member of an extended cluster drops below an SNR of approximately 1, such analysis should be viewed with caution.

On the other hand, in PCA, in order for a group of n pixels (clustered or not) to contribute strongly enough to the covariance matrix that they register in one of the principle components, that is, that they produce an eigenvalue outside the radius of the Wigner semi-circle in the white noise case, it is necessary that the rms average difference of the group's signal from the mean exceed the threshold given by expression (5). The rms signal to noise ratio required to detect a cluster of n pixels using PCA is compared with that required to detect a cluster of n pixels using MFA in Table 1, for the present case with 16,384 pixels and 236 channels.

TABLE 1. The dependence of the signal to noise ratio required to detect clusters of pixels on the number of pixels in the cluster is displayed for both principle component analysis (PCA) and matched filter analysis (MFA).

n_{pixels}	SNR_{PCA}	SNR_{MFA}
1	64	4
4	32	2
16	16	1
64	8	.5
256	4	.25
1024	2	.125
4096	1	.0625

The gain of the matched filter over straightforward principle component analysis for weak, spatially sparse features is thus approximately 16, as long as the correct filter vector is known. In practice, larger clusters of low significance pixels are more difficult to detect than indicated by the straightforward statistical analysis listed in the table, as discussed above, and the gain of MFA over PCA is less than 16 for the larger cluster numbers.

Endmember Detection

One of the drawbacks of matched filter analysis is the need to specify in advance, a particular target spectrum in order to generate the matched filter. A powerful technique for the determination of the pure constituent spectra whose mixtures provide the variations seen in a hyperspectral datacube exploits the geometric property of "convexity" (24). The set of points in a datacube are convex in hyperspectral space if the spectrum for every pixel is a linear combination of a given finite set of constituent spectra, with all positive fractions, and with the sum of the fractions of all constituents equal to unity. For the present case, and quite generally, this is a very good approximation. The points in a datacube thus lie within a simplex having P vertices. The number of vertices is determined by the number of endmembers. With P=3, for example, the spectra for all pixels may be written as linear sums of the spectra for 3 endmember spectra, the simplex is simply a triangle, and all of the points in the datacube should lie inside the boundary of the triangle. In the case of 4 endmembers, the simplex is a tetrahedron, and all points should plot within the boundary of the tetrahedron.

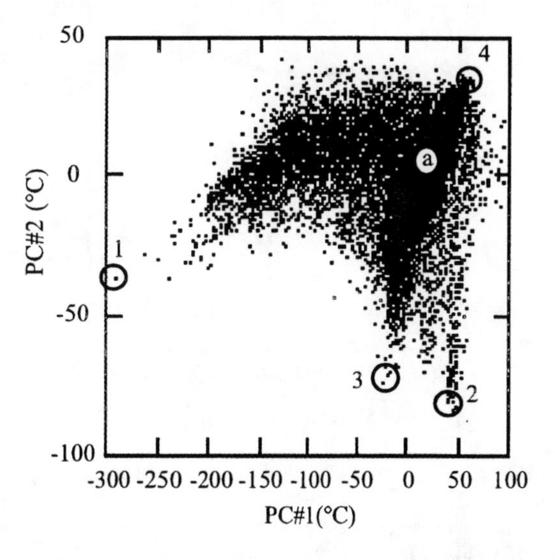

FIGURE 27. A scatter plot of the intensity of the first principle component vs. the intensity of the second principle component for every pixel in the hyperspectral datacube. Circles labeled 1,2,3, and 4 correspond to 4 endmembers of the data set. The circle labeled "a" corresponds to the location of the albite endmember, which is not distinguished in this particular two dimensional projection.

Such convex distributions, when plotted in a two dimensional projection, appear to have "spiky" fingers, which point towards the endmembers. An example from the granite data is shown in Fig. 27. In this figure, the intensity of the first principle component eigenimage (as shown in Fig. 13 as PC1) is plotted against the intensity of the second principle component eigenimage (as shown in Fig. 13 as PC2). The units shown (°C) reflect the apparent temperature deviation from the mean spectrum that would be seen in a particular direction in hyperspectral space (as specified by the eigenvector), and may have much larger excursions than the difference between the temperature of the sample and the temperature of the room since there are several spectral channels which contribute to the total length of the vector difference. As an example, for the "quartz" like eigenvector, there are approximately 100 spectral channels across the region of high quartz reflectance, as may be seen in Fig. 9.

A better intuitive understanding of the nature of the n-dimensional simplex structure of hyperspectral data is gotten by viewing the morphology of the 2-d scatter plot analogous to that shown in Fig. 27, as the directions chosen for the ordinate and abscissa vary continuously through the full hyperspace. Software which facilitates such visualization is provided in ENVI (25). ENVI also provides useful tools for the selection of particular pixel groups associated with endmembers.

Once candidate endmember pixels have been identified on the basis of their extremal behavior in hyperspace, their spectra may be compared with laboratory reference spectral libraries in order to identify the nature of the constituents within a sample. The spectra for the four endmembers identified in Fig. 27 are shown in Fig. 28.

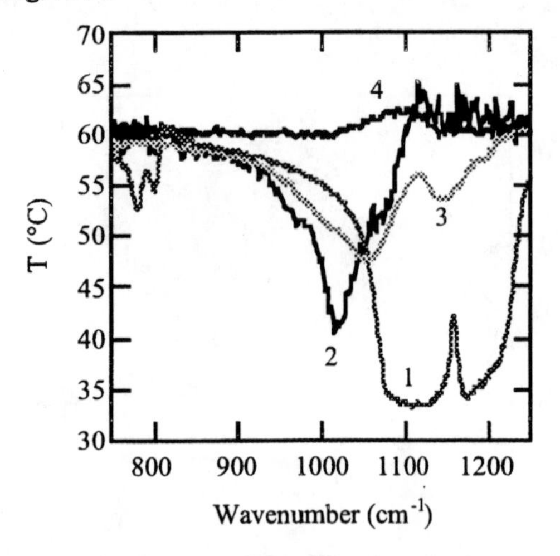

FIGURE 28. The spectra of the four endmembers identified in Fig. 27 by labeled circles are displayed, with the label number shown next to each curve corresponding to the position indicated in Fig. 27.

The identification of the endmember labeled 1 in both Fig. 27 and Fig. 28 is obviously quartz, as may be seen in Fig. 29, which compares the endmember reflectance spectrum, calculated from the brightness temperature spectrum according to the method described in the section ***Spectral Interpretation***, with that of a standard reference spectrum (26) for pure solid quartz. A similar reflectance spectrum for the endmember labeled 2 is shown in Fig. 30, and compared with a laboratory reference reflectance spectrum for pure, solid biotite. Note, however, the weakness of the biotite peak at 1100 cm^{-1}. This peak does occur in the spectrum labeled 3 in Fig. 28, and it is plausible that endmember 2 is not a pure endmember.

Conspicuously absent from the endmember spectra extracted from the scatter plot of the first principle component against the second principle component is anything like the albite spectrum. This is readily understandable in view of the previously noted point that the albite regions in the sample (the "yellow diamond" in Fig. 16) were only conspicuous in the 4th, 5th and 6th principle components.

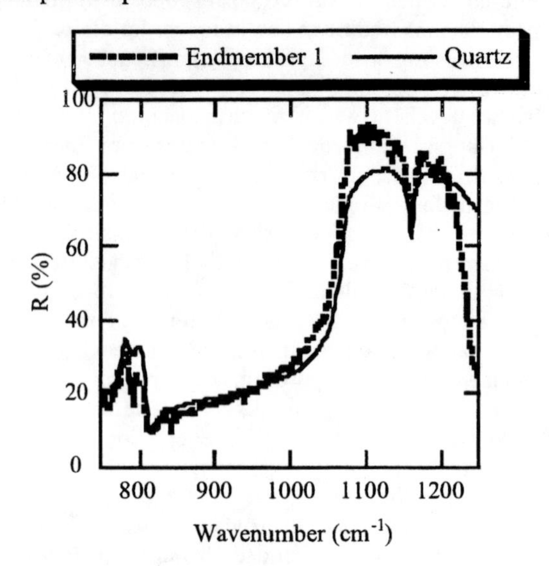

FIGURE 29. The reflectance spectrum corresponding to endmember 1 is compared with the laboratory reflectance spectrum for a solid sample of pure quartz.

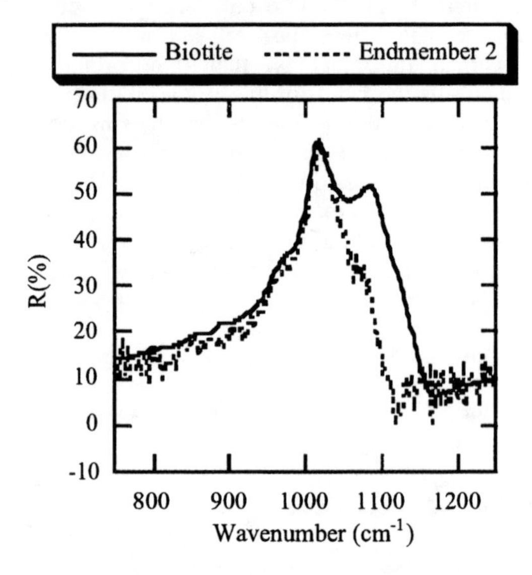

FIGURE 30. The reflectance spectrum corresponding to endmember 2 is compared with the laboratory reflectance spectrum for a solid sample of pure biotite.

The great degree of separation of the yellow diamond pixels which is so easily seen in the false color plot shown in Fig. 16 based on the 4th, 5th, and 6th principle components is manifest in the scatter plot shown in Fig. 31 by the great separation of the pixels which tend toward the circle labeled with the number 5. This is a good illustration of the great change in appearance of the hyperspectral data with change in projection direction.

The identification of the 5th endmember with albite is made on the basis of the comparison of its spectrum with the laboratory reference spectrum for a solid albite shown in Fig. 32. Although the magnitude of the reflectance of the observed endmember spectrum is not as great as for the laboratory reference spectrum, this can readily be attributed to a particle size effect. It is observed that the reflectance spectra of minerals has a strong dependence on the particle size (27). In general there is a decrease in the intensity of the reflectance peaks as the particle size decreases, and occasionally a change in the position and relative intensity of the peaks. The factor of 2 reduction in the observed reflectance for the endmember spectrum shown in Fig. 32 compared to the reference spectrum is not unusual.

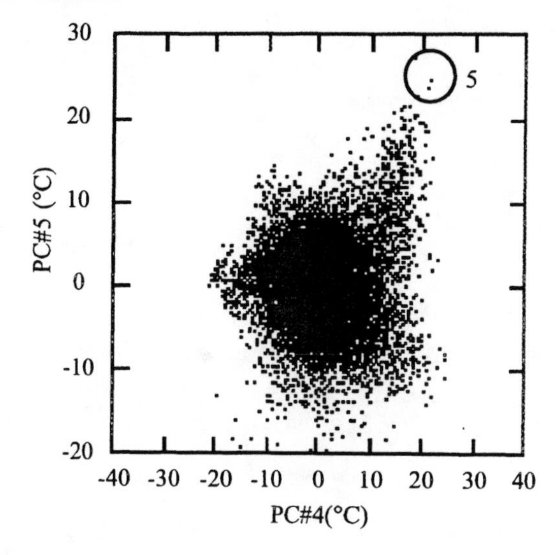

FIGURE 31. A scatter plot of the intensity of the fourth principle component vs. the intensity of the fifth principle component for every pixel in the hyperspectral datacube. The circle labeled 5 corresponds to the location of the albite endmember.

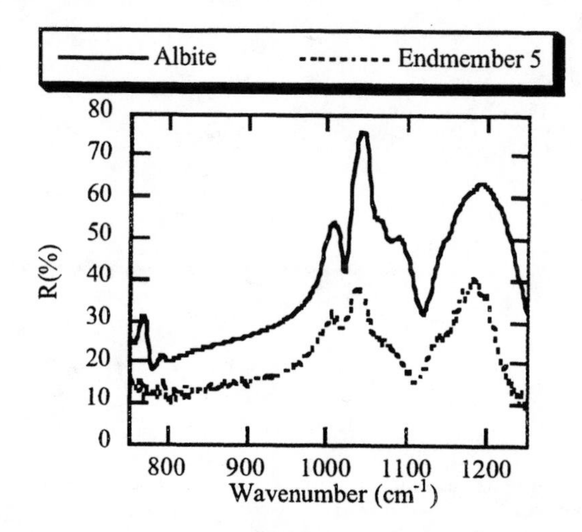

FIGURE 32. The reflectance spectrum corresponding to endmember 5 is compared with the laboratory reflectance spectrum for a solid sample of pure albite.

It is noteworthy that the total number of pixels in the present granite sample that may be considered pure is quite small, as can be seen especially clearly for the case of the quartz endmember in Fig. 27, and the albite endmember in Fig. 31. This may be explained by the typical crystal or grain size in the current sample relative to the area represented by a single pixel. If the grain size of the current sample were much smaller, then it is likely that all pixels would be mixed. Even in this case, it is still possible to identify the underlying endmembers by exploiting convexity. It is simply necessary to extrapolate in the direction that the "fingers of convexity" are pointing. In Fig. 31, for example, it would not have been necessary to have any pixels within the indicated circle in order to reconstruct the spectrum corresponding to the location of the circle. In this way, sub pixel mixtures can be disentangled on the basis of the geometry in hyperspace.

SUMMARY

A detailed chain of processing steps for the analysis of a typical hyperspectral datacube is carried out, with much reference to a specific example based on a sample of granite. The salient novel feature of this analysis is the use of the Wigner semi-circle distribution to determine the limits of the underlying noise in a hypercube. Other points of significance are the preference, for imaging Fourier transform spectroscopy purposes, to avoid zero padding, and to avoid apodization, if possible, since both of these computational processes have the effect of adding zero modes to the hyperspectral data. Data compression and noise reduction of hyperspectral datacubes is discussed, particularly with regard to calibration datacubes, for which there is a priori knowledge of the underlying spatial/spectral simplicity of the data. The comparison of principle component analysis and matched filter analysis for the revealing the constituents of hyperspectral datacubes is discussed. Finally, the location of endmembers based on extremal locations in hyperspace and the identification of the estimated endmember spectra by means of comparison with laboratory reference spectra is treated.

ACKNOWLEDGEMENTS

This work was performed under the auspices of the U.S. Department of Energy under Contract No W-7405-Eng-48. I thank W. Aimonetti, M. Carter, T. Hindley, R. Mitchell, P. Peaslee for their invaluable support in the development of the LIFTIRS instrument and program. I thank A. Stocker and W. Kendall for enlightening discussions concerning principal component analysis and matched filtering.

REFERENCES

1. Bernhard Schrader and Stefan Keller, "Raman: FT, or dispersive: is that the question?", p 30-39, in the proceedings of the *8th International Conference on Fourier Transform Spectroscopy*, Sept. 1991, Bubeck-Travemunde, ed. H.M. Heise, E.H. Korte, H.W. Siesler, SPIE vol 1575.

2. P.R. Griffiths and J.A. deHaseth, *Fourier Transform Infrared Spectrometry*, John Wiley & Sons, Inc., New York, 1986, pp. 274-283.

3. P. Jacquinot, "The Luminosity of Spectrometers with Prisms, Gratings, or Fabry-Perot Etalons", J. Opt. Soc. Am. 44, 761-765 (1954)
4. C.L. Bennett, M.R. Carter, D.J. Fields, "Hyperspectral Imaging in the Infrared Using LIFTIRS", in the proceedings of the conference on *Infrared Technology XXI*, San Diego, California, July 9-13, pp. 274-283, 1995, SPIE vol 1937
5. A.E. Potter, Jr., "Multispectral Imaging System", U.S. Patent # 3702735, Nov. 14, 1972.
6. C.W. Wells, A.E. Potter, T.H. Morgan, "Near-infrared spectral imaging Michelson interferometer for astronomical applications", in the proceedings of the conference: *Infrared Imaging Systems Technology*, April 10-11, 1980, Washington, D.C., ed. J. Zimmerman, W.L. Wolfe, SPIE vol 226.
7. C.L. Bennett, M. Carter, D. Fields, J. Hernandez, "Imaging Fourier Transform Spectrometer", p.191-200, in the proceedings of the *Imaging Spectrometry of the Terrestrial Environment*, April 14-15, 1993, Orlando, Fl., SPIE vol 1937.
8. E. Neil Lewis, "Fourier Transform Spectroscopic Imaging Using MCT Focal-Plane Arrays: A New Tool for Biological Imaging and Microscopy", in the proceedings of the *11th International Conference on Fourier Transform Spectroscopy*, August 10-15, 1997, Athens, Georgia, U.S.A.
9. The Bio-Rad FTS 6000 Stingray system, website address, http://www.biorad.com
10. M.R. Carter, C.L. Bennett, D.J. Fields, D. Lee, "Livermore Imaging Fourier Transform Infrared Spectrometer (LIFTIRS)", in the proceedings of *Imaging Spectrometry*, April 17-19, 1995, Orlando, Fl., SPIE vol 2480, pp. 380-386.
11. A.H. Brownlow, *Geochemistry*, Prentice-Hall, Inc., Englewood Cliffs, N.J., 1979.
12. Laboratory reference spectrum, quartz1c, obtained from the website address: http://asterweb.jpl.nasa.gov/speclib/
13. D. Kember, D.H. Chenery, N. Sheppard, and J. Fell, *Spectrochim. Acta*, vol **35A**, 455, 1979.
14. D.B. Chase, *Appl. Spectrosc.*, **35**, 77 (1981)
15. The website address http://asterweb.jpl.nasa.gov/speclib/ contains a convenient search and display page for a variety of rocks and minerals.
16. K. Fukunaga, *Introduction to Statistical Pattern Recognition*, p. 21, Academic Press, Inc. 1990.
17. E.R. Malinowski and D.G. Howery, *Factor Analysis in Chemistry*, John Wiley & Sons, Inc. 1980.
18. K. Fukunaga, op. cit., p. 403.
19. M.L. Mehta, *Random Matrices*, p. 75, Academic Press, Inc. 1991.
20. P.J. Ready and P.A. Wintz, "Information Extraction, SNR Improvement, and Data Compression in Multispectral Imagery", *IEEE Transactions on Communications, Vol. COM-21*, pp. 1123-1131, 1973.
21. J.W. Boardman, and F.A. Kruse, "Automated spectral analysis: a geological example using AVIRIS data, north Grapevine Mountains, Nevada", in the proceedings of the *ERIM Tenth Thematic Conference on Geologic Remote Sensing*, Environmental Research Institute of Michigan, Ann Arbor, MI, pp. I-407-I-418, 1994
22. Laboratory reference spectrum, biotite1s, obtained from the website address: http://asterweb.jpl.nasa.gov/speclib/
23. Laboratory reference spectrum, albite1s, obtained from the website address: http://asterweb.jpl.nasa.gov/speclib/
24. J.W. Boardman, "Automating Spectral Unmixing of AVIRIS Data Using Convex Geometry Concepts", in the *Summaries of the Fourth Annual JPL Airborne Geoscience Workshop*, **Vol. 1**, 1993, pp. 11-14.
25. ENVI, "The Environment for Visualizing Images", Research Systems, Inc., Boulder, CO, 1997.
26. Laboratory reference spectrum, quartz1s, obtained from the website address: http://asterweb.jpl.nasa.gov/speclib/
27. Internet address, http://asterweb.jpl.nasa.gov/archive/JHU/nicolet/minerals/minerals.txt

Structural Changes in the Ordering Processes of Macromolecular Compounds

M. Kobayashi and K. Tashiro*

Department of Macromolecular Science, Graduate School of Science, Osaka University, Toyonaka, Osaka 560, Japan

In order to clarify the microscopically-viewed relationship between the conformational ordering process and the aggregation process of the macromolecular chains in the phase transitions from melt to solid or from solution to gel, the time-resolved Fourier-transform infrared spectra and small-angle X-ray or neutron scattering data have been analyzed in an organized manner. Two concrete examples were presented. (1) In the gelation phenomenon of syndiotactic polystyrene-organic solvent system, the ordered TTGG conformation is formed and develops with time. This conformational ordering is accelerated by the aggregation of these chain segments, resulting in the formation of macroscopic gel network. (2) In the isothermal crystallization process from the melt of polyethylene, the following ordering mechanism was revealed. The conformationally-disordered short trans conformers appear at first in the random coils of the melt. These disordered trans sequences grow to longer and more regular trans sequences of the orthorhombic-type crystal and then the isolated lamellae are formed. Afterwards, the stacked lamellar structure is developed without change of lamellar thickness but with small decrease in the long period, indicating an insertion of new lamellae between the already produced lamellar layers.

INTRODUCTION

Macromolecular chains change their conformation drastically and sensitively between disordered and ordered forms by changing the external conditions such as temperature, stress, etc. The conformational change is accompanied always by the large change in the aggregation structure of chains. In order to clarify the detailed mechanism of disorder-to-order transitions of macromolecular compounds from the molecular level, therefore, we need to collect not only the information on conformational ordering process of each chain but also the information on the aggregation state of the chains and combine them in an organized manner.

Among the various types of experimental methods, a combination of vibrational spectroscopy with X-ray and neutron scattering technique is considered to be a powerful tool for revealing the essence of the transition mechanism. The vibrational spectroscopy may give us a detailed and concrete information on the conformational change of chains, and the X-ray and neutron data are useful for obtaining the information about the aggregation structure of chains. In particular, the recent development of the apparatus such as the rapid-scanning-type Fourier-transform spectrometer, the highly sensitive and quantitative X-ray (or neutron) detector etc. has allowed us to carry out the time-resolved measurements during the phase transitions.

In the present paper we will focus our attention to the two types of disorder-to-order phase transition. One is a gelation phenomenon. When a polymer solution is cooled from an elevated temperature, the fluid solution changes to nonfluid gel state due to the formation of cross-linkages between polymer chain segments (thermoreversible physical gel). In the solution, polymer chains take a form of so-called random coil but the unexpectedly regular chain conformation is attained in the cross-linking parts of the gel, although the details have not been clarified enough well. Another target is the crystallization phenomenon in the cooling process from the melt. As likely as in the solution, the polymer chains take random coil conformation in the melt. By cooling the sample, the crystalline regions of regular chain conformation are formed and coexist with the irregular amorphous phase. The molecular-level mechanism of transformation from random coil to regular crystalline structure has not yet been revealed clearly. These two phenomena, gelation and crystallization are apparently different from each other, but the essential features of the transition mechanisms may be quite similar, i.e., the regularization of the chain conformation and the associated ordering of the chain aggregation state, as will be discussed in the following sections.

CONFORMATIONAL ORDERING ON THE GELATION OF SYNDIOTACTIC POLYSTYRENE

Ordered Molecular Conformation and Critical Sequence Length

CP430, *Fourier Transform Spectroscopy:* 11th International Conference
edited by J.A. de Haseth
 1-56396-746-4/98/$15.00

In this paper the gelation mechanism of stereoregular syndiotactic polystyrene (SPS) will be discussed as a typical example (1). As shown in Fig. 1, SPS molecules take two kinds of conformation in the crystalline phase (2 - 9). One is the all-trans planar-zigzag type (TT) and the other is the helical form of TTGG regular sequence, where T and G denote, respectively, trans and gauche isomers. Another stereoregular polystyrene is isotactic species (IPS), which takes a (3/1) helical conformation consisted of a regular repetition of T and G isomers (10).

As shown in Fig. 2, the infrared spectra of SPS and IPS show many crystalline bands characteristic of these various types of conformation (11 - 13). In the molten or solution state as well as in the amorphous state, these bands disappear or become much broader. By heat treatment of the amorphous samples, these crystallization-sensitive bands begin to be observed. But the rate of increment of the infrared intensity differs from band to band. This is because of the difference in the sensitivity to the conformational order or the length of the regular helical segment. That is to say, the band is considered to have the characteristic limiting sequence length or the critical

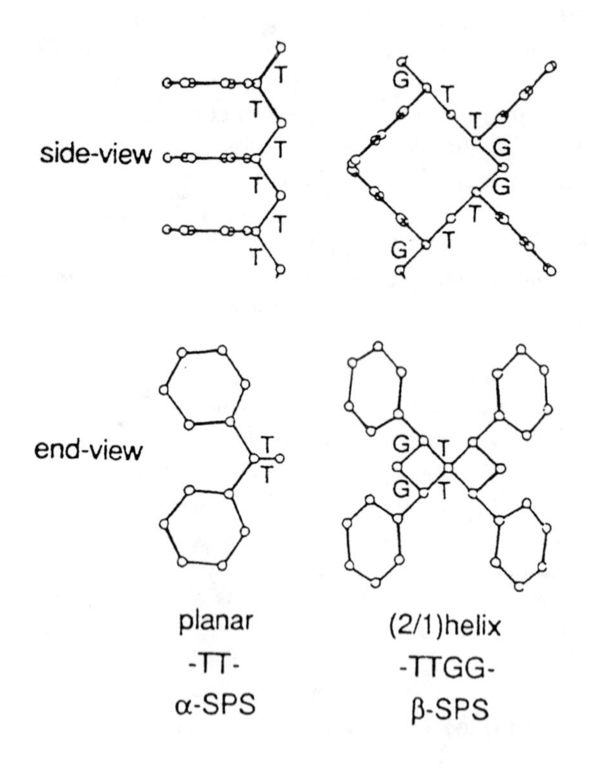

FIGURE 1. Molecular conformations of syndiotactic polystyrene taken in the crystalline region. (left) Planar-zigzag-type chain of crystal modification α. (right) (2/1) helical form of crystal modification β. T and G denote the trans and gauche isomers, respectively.

sequence length (CSL) necessary for the appearance of this band in the infrared (or Raman) spectrum (14, 15). This CSL is expressed by the number of monomeric residues m constructing the regular sequence and can be determined by the intramolecular isotope dilution technique. The deuterated and hydrogeneous monomers are copolymerized into one chain. The vibrational wave extending over the hydrogeneous (deuterated) monomeric sequence is cut by an invasion of deuterated (hydrogeneous) monomeric units. Let's consider the copolymer chain consisted of the D and H monomeric units with the molar fraction X of the H species. If the monomer arrangement in a chain obeys the Bernoulli statistics, the band intensity $I(X)$ for the finite chain segments longer than m is expressed by the equation

$$I(X)= X^m[m - (m -1)X]\ I(1.0) \qquad (1)$$

where $I(1.0)$ denotes the infrared intensity measured for the pure H polymer sample at highly regular conformational state. The band intensity is measured for a series of samples with different D/H content and plotted against the molar fraction X. By comparing the thus observed curve with that predicted by equation (1), we can determine the CSL m. One example is shown in Fig. 3, in which the 572 cm^{-1} band of the TT form (m = 13) and the 1124 cm^{-1} band of the TTGG form (m = 15) are illustrated (16).

Conformational Ordering Process on Gelation

SPS is soluble in various organic solvents at elevated temperature, and the hot solutions turn into transparent gels by allowing them to stand at room temperature (11 - 13). As seen in Fig. 4, where the case of SPS dissolved in carbon tetrachloride (CCl_4) is shown, the infrared spectra of the thus prepared gels exhibit many bands characteristic of the TTGG regular conformer, indicating that the highly ordered TTGG conformation is formed on gelation. Figure 5 (a) shows the time dependence of the infrared spectra taken at room temperature during the gelation of SPS/CCl_4 sample of the polymer concentration 1.97 wt %. With gelation time the TTGG bands at 572, 549, and 504 cm^{-1} increase in intensity, while the broad band centered at 500 cm^{-1} originated from the disordered conformation becomes weaker. Observation of the defined isobestic points shows this process of gelation is regarded as a two-component reaction from the disordered to the ordered state. The integrated intensities of a selected band pair of the ordered (I_{order}) and disordered ($I_{disorder}$) monomeric residues are expressed as

$$I_{order} = \varepsilon_{order}\ x_{order}\ c\ l \qquad (2)$$

$$I_{disorder} = \varepsilon_{disorder}\ x_{disorder}\ c\ l$$

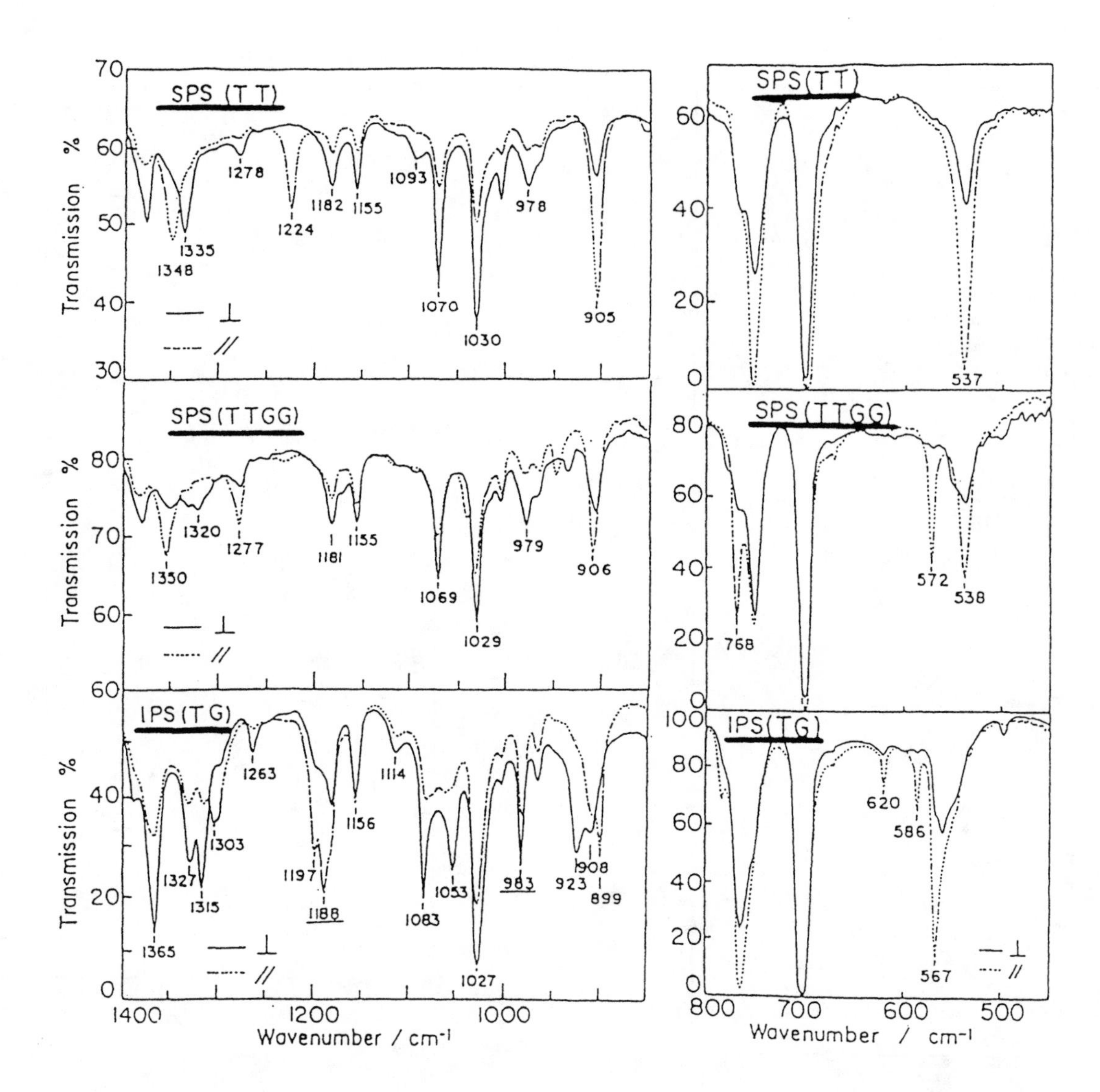

FIGURE 2. Polarized infrared spectra of oriented crystalline samples of syndiotactic polystyrene (SPS) and isotactic polystyrene (IPS) taken at room temperature. The solid and broken lines indicate the spectra taken with the electric vector of the incident infrared beam perpendicular and parallel to the orientation axis, respectively.

$$= \varepsilon_{disorder}\, c\, l - (\varepsilon_{disorder}/\varepsilon_{order})\, I_{order} \quad (3)$$

where ε_{order} and $\varepsilon_{disorder}$ denote the absorption coefficients, x_{order} and $x_{disorder}$ are the weight fractions of monomeric residues contained in the ordered and disordered conformations, respectively, c is the total concentration of the monomeric residues contained, and l is the optical path length. For the ordered and disordered states, respectively, the bands at 572 and 540 cm^{-1} were selected, as stated above. By plotting $I_{disorder}/cl$ against I_{order}/cl the quantities e_{order} and $e_{disorder}$ are known and so the fraction of the ordered state x_{order} can be evaluated in the absolute scale. Figure 5 (b) shows the linear relation between $I_{disorder}/cl$ and I_{order}/cl as predicted from equation (3). Figure 5 (c) shows the thus obtained time evolution of the orderliness parameter x_{order}. As mentioned above, if the Bernoulli statistics may be applied to the present system, the number average length L_n and the weight average length L_w of the ordered TTGG sequence are represented by the following equations (in a unit of number of monomeric residues), respectively.

$$L_n = 1/(1 - z) \quad (4)$$

$$L_w = (1 + z)/(1 - z) \quad (5)$$

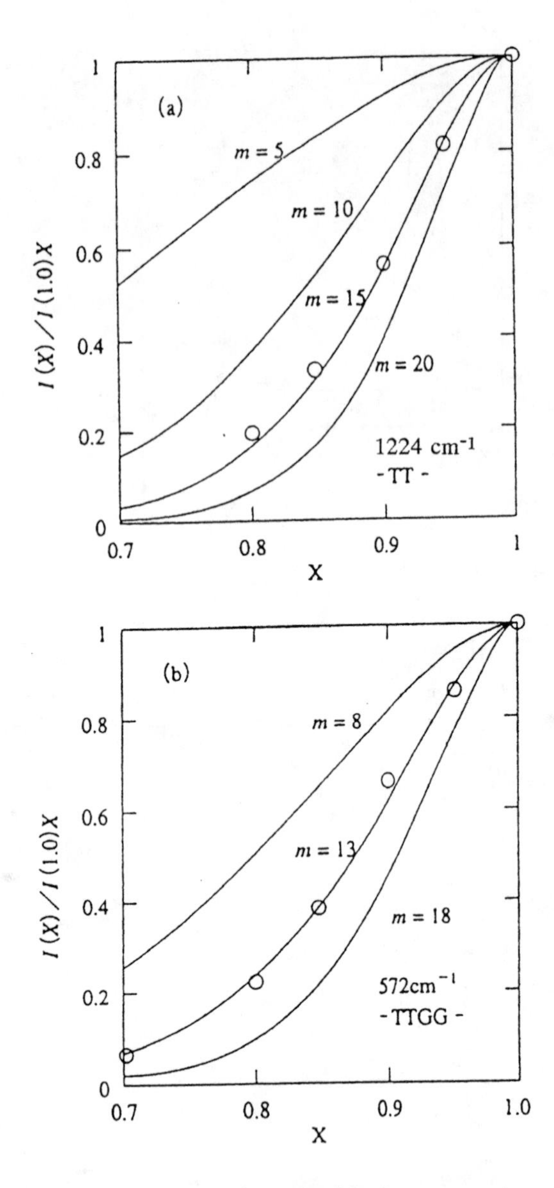

FIGURE 3. Determination of critical sequence lengths of infrared bands characteristic of the regular molecular conformations of syndiotactic polystyrene. (a) The band at 1224 cm^{-1} characteristic of all-trans-type conformation. The used samples were highly crystalline films. (b) The band at 572 cm^{-1} characteristic of TTGG-type conformation. The spectra were measured at -60°C for the gels prepared from the o-dichlorobenzene solution. The solid-line curves were calculated by the equation $I(X) = X^m[m - (m-1)X]\, I(1.0)$, where I(X) is the infrared band intensity expected for the sample with the molar fraction of the hydrogeneous monomeric residues X and m is the critical sequence length.

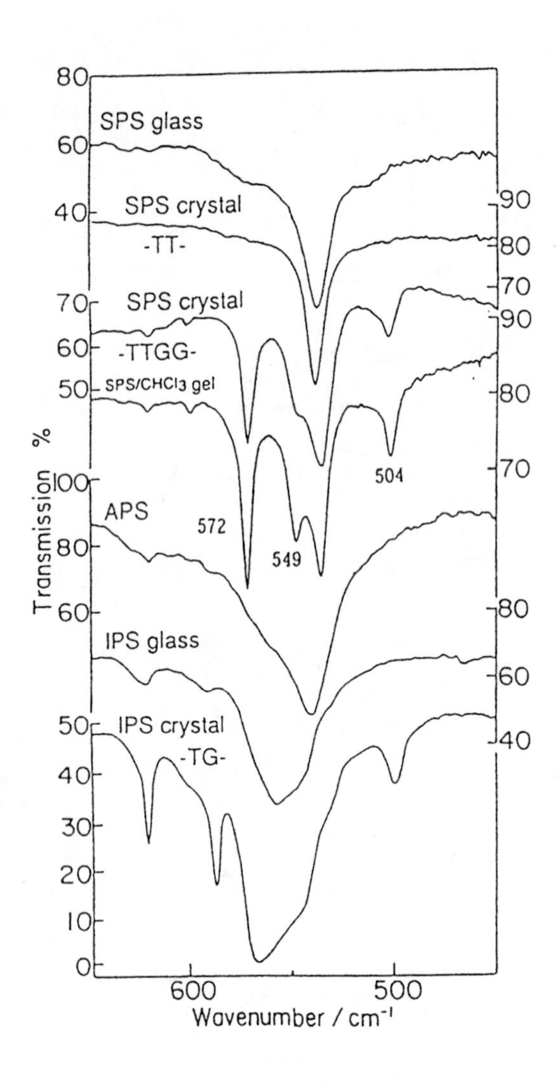

FIGURE 4. Infrared spectra of various polystyrene samples with different tacticity and aggregation state. SPS, APS, and IPS denote, respectively, syndiotactic, atactic, and isotactic polystyrenes. The glass samples were prepared by quenching the melt into ice-water bath. The crystal samples were obtained by annealing the glass samples.

where z is the probability of finding a monomeric residue in the ordered state. Since the z value is estimated from the relation $x_{order} = z^m[m - (m-1)z]$ if m is known, then L_n and L_w are calculated and plotted as a function of time in Fig. 5 (c).

Figure 6 shows the solvent effect on the conformational ordering process on gelation, where the solvents used were chloroform ($CHCl_3$) and carbon tetrachloride (CCl_4). The gels prepared from the different types of solvent exhibit essentially the same TTGG conformation as seen from the similar spectral pattern, but the formation rate is quite different. For example, in the case of SPS/CCl_4 (and SPS/benzene) system the gelation and the conformational ordering process accomplishes within several minutes at room temperature (at c = 1.97 wt %), while in the SPS/$CHCl_3$ system of the same concentration it proceeds very slowly with the time scale of several tens of hours or longer (1, 17).

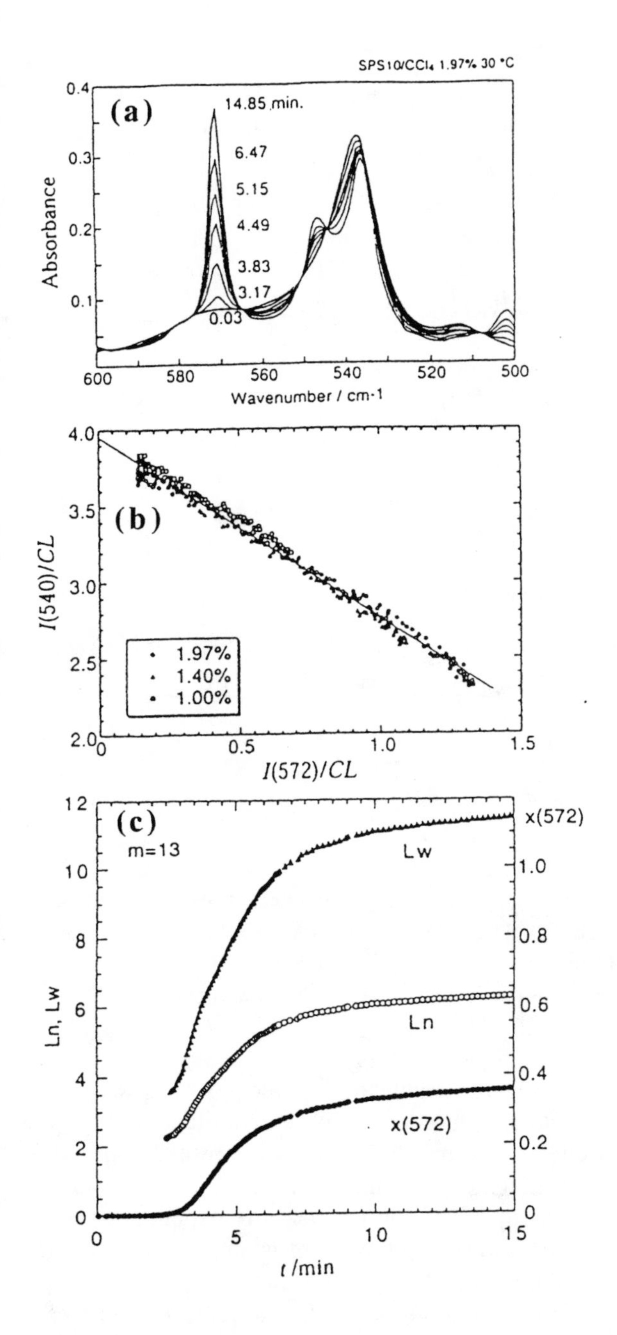

FIGURE 5. (a) Time dependence of infrared spectra taken at 30°C for the syndiotactic polystyrene dispersed in carbon tetrachloride (concentration 1.97 wt%). (b) Plot of the integrated intensity $I(540)$ of the infrared band at 540 cm^{-1} characteristic of the disordered conformation against that of the 572 cm^{-1} band characteristic of the ordered TTGG conformer. The intensities were reduced by the solution concentration c and the optical path length L. The data obtained for the samples at different concentrations were collected together. (c) Time evolution of the orderliness parameters determined by the method mentioned in the text. $x(572)$ is the weight fraction of monomeric residues contained in the ordered conformation. L_n and L_w are number-average and weight-average lengths of the conformationally-ordered segments, respectively.

Clustering of Conformationally-Ordered Segments on Gelation

In order to clarify the relationship between the conformational ordering process and the aggregation process of these ordered chain segments or the gel-network formation process, the time dependence of the small-angle neutron scattering (SANS) was measured in comparison with the results of infrared spectroscopy (18,19). For example, Fig. 7 shows the SANS data taken at 10 °C for the SPS/$CDCl_3$ system at 7.57 wt % concentration. The invariant Q defined as $Q = 4\pi \int I(q)\ q^2\ dq$ was evaluated and plotted against time in Fig. 8, where q is the momentum transfer and $I(q)$ is the corrected SANS intensity. The Q corresponds to the fraction of gel network structure. In this figure the above-mentioned conformational orderliness parameter x_{order} and the invariant Q are seen to change almost in parallel, indicating that the conformational ordering of the *molecular* chains and the clustering of the segments or the formation of the *macroscopic gel network* structure proceed simultaneously.

As seen in Fig. 8, the gelation process depends largely on the concentration of the system: the gelation rate increases remarkably with an increase of the concentration. This suggests that the conformational ordering process is accelerated by the aggregation of the polymer segments. We now assume that p of the conformationally-ordered segments (S_r), each consists of r monomeric residues, are aggregated together to form a cluster (Cr) as illustrated in Fig. 9. The rate equation corresponding to this aggregation process is expressed as follows (17).

$$p\ Sr \rightarrow Cr \qquad d[C_r]/dt = k\ [S_r]^p \qquad (6)$$

where [A] is the molar concentration of the state A, and k is the rate constant. Since the total number of monomeric residues included in the ordered conformational state longer than m is given by $p\ \Sigma_{r \geq m}\ r[C_r]$, where the summation is over the r larger than m, the infrared band intensity I_{order} of the ordered conformation (572 cm^{-1}) is expressed as $I_{order} = K\ p\ \Sigma r[C_r]$ (K is the proportionality constant). The initial slope of the curve I_{order} vs. time is given by

$$(dI_{order}/dt)_{t=0} = K p\ \ \Sigma\ r\ (d[C_r]/dt)_{t=0}$$

$$= K\ k\ p\ \Sigma\ r\ [Sr]^p{}_{t=0} \qquad (7)$$

At the initial stage of t = 0, the chain segment detected by the infrared measurement has the length r almost equal to m, and therefore equation (7) may be simplified in the following way ($\Sigma\ r\ [S_r] \sim m\ [S_m] \propto c$).

$$\ln\ (dI_{order}/dt)_{t=0} \propto p\ \ \ln\ c \qquad (8)$$

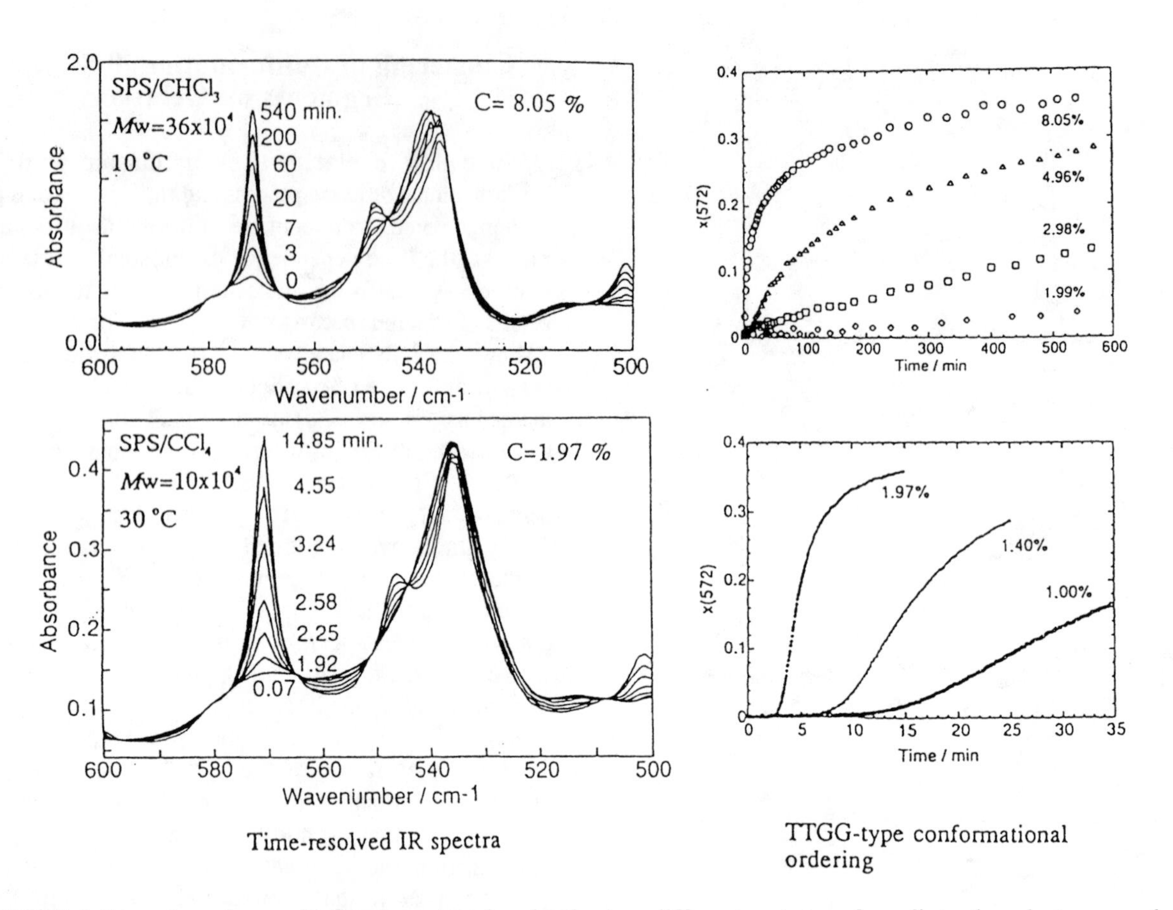

FIGURE 6. Time dependence of infrared spectra taken for the two different systems of syndiotactic polystyrene gels. (above) the chloroform system of concentration 8.05 wt % measured at 10°C. The orderliness parameter $x(572)$ obtained for the 572 cm^{-1} infrared band characteristic of the ordered TTGG conformation is plotted against time for the samples with the different concentrations. (below) the case of carbon tetrachloride system of concentration 1.97 wt% measured at 30°C.

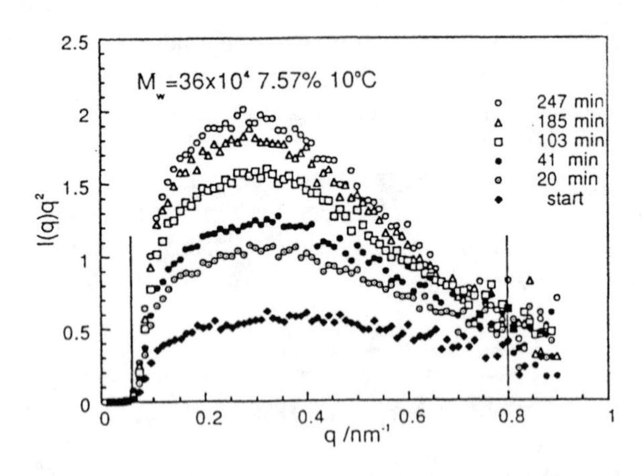

FIGURE 7. Time evolution of Lorentz-corrected small-angle neutron scattering intensity $I(q)$ measured during gelation of the syndiotactic polystyrene/chloroform system. The weight-averaged molecular weight M_w = 30 x 10^4, the concentration c = 7.57 wt%, and the measurement temperature = 10°C.

That is to say, the double logarithmic plot of $(dI_{order}/dt)_{t=0}$ against the concentration c should give the straight line of the slope p. Figure 10 (a) shows the results obtained for the SPS/$CHCl_3$ system at the various temperatures. Figure 10 (b) shows the temperature dependence of the number of chain segments p. For example, at 10°C the p is 5.2, indicating that about 5 chain segments, each of which is constructed by ca. 13 monomeric residues of TTGG regular conformation, are aggregated together at the initial stage of gelation. The value of p decreases with lowering temperature and is almost equal to unity below -15°C, indicating that at such a low temperature range the conformational ordering proceeds through a self-organization mechanism within one chain segment; i.e., the isolated chain can become ordered into the TTGG conformation without any supports by the surrounding chain segments. The crossover temperature, at which the transformation between the clustering and self-organization processes occurs, is dependent on the type of the used solvent. For example, for the SPS/o-dichlorobenzene system the p

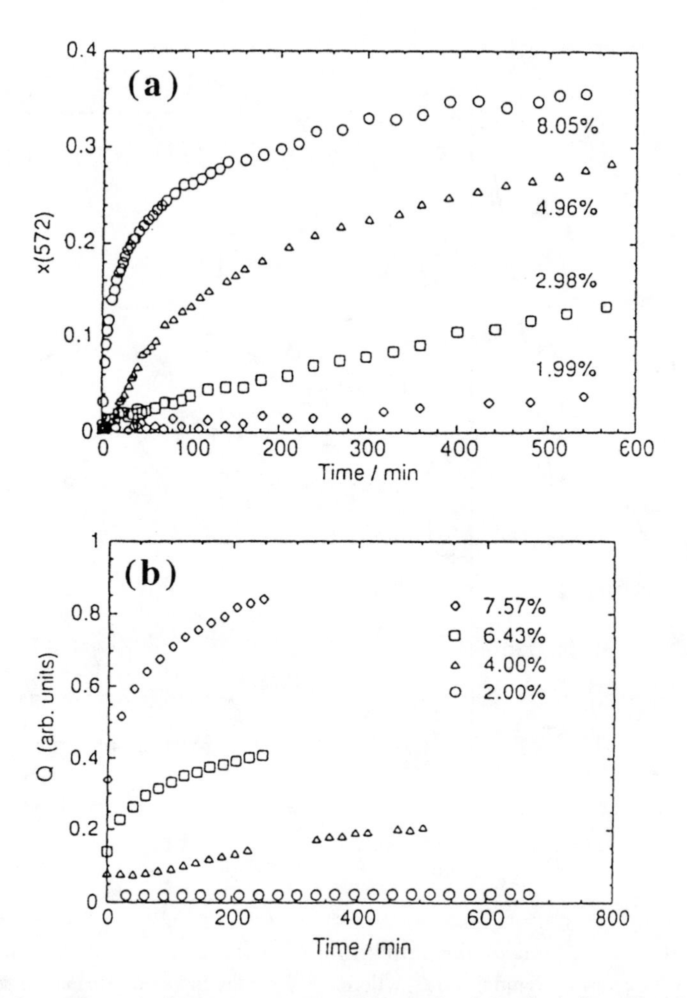

FIGURE 8. (a) Time evolution of the orderliness parameter $x(572)$ obtained for the infrared band at 572 cm^{-1} characteristic of the ordered TTGG conformer and (b) time evolution of the invariant Q evaluated from the SANS data shown in Figure 7. The data were collected at 10°C for the syndiotactic polystyrene/chloroform system with the various polymer concentrations.

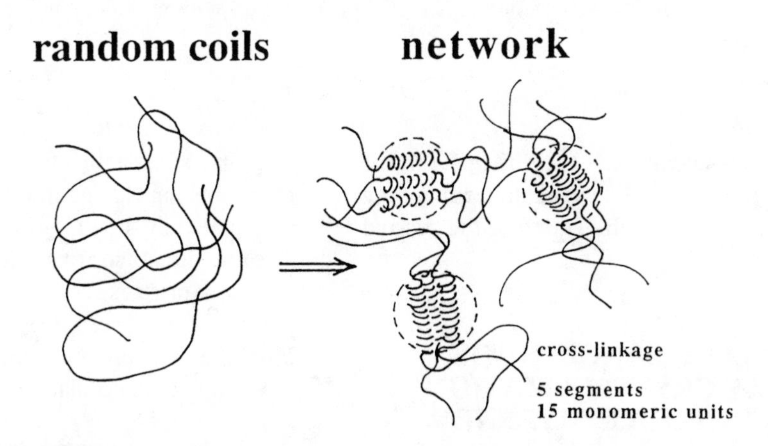

FIGURE 9. An illustration of the structural change from random coils to the aggregation state of the conformationally-ordered chain segments. The average number of aggregated chain segments is about 5, although it depends on temperature as well as on the type of solvent.

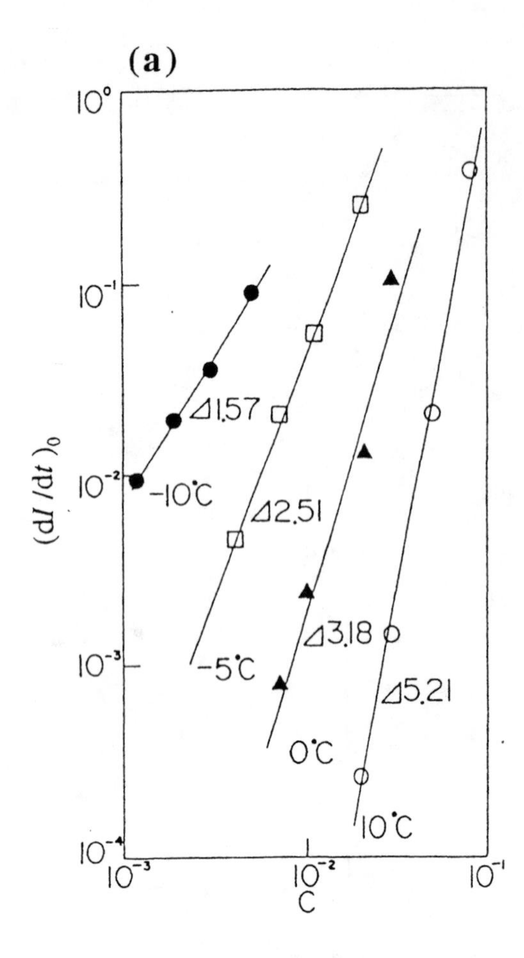

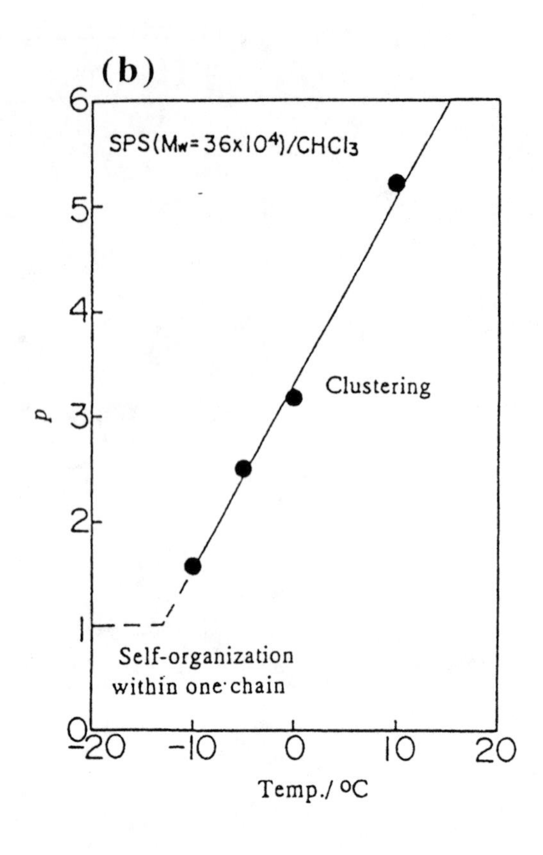

FIGURE 10. (a) Double logarithmic plots of the initial rate of increase in $I(572)$ vs c obtained for syndiotactic polystyrene/chloroform gels at the various temperatures, where $I(572)$ is the infrared intensity of the band at 572 cm^{-1} and c is the polymer concentration. The figure indicated for each linear line is the slope and corresponds to the value p. (b) Temperature dependence of the number of chain segments p. The p value larger than unity means the clustering of p chain segments coours in the gel. The value of p decreases with lowering temperature and is almost equal to unity below -15°C, indicating that the conformational ordering proceeds through a self-organization mechanism within one chain segment; i.e., the isolated chain can become ordered into the TTGG conformation without any supports by the surrounding chain segments.

is almost unity at room temperature. The difference in the interactions between polymer and solvent molecules may reflect on the difference in the stabilization of the ordered conformation in the gel.

STRUCTURAL CHANGE IN THE ISOTHERMAL CRYSTALLIZATION PROCESS OF POLYETHYLENE

When the molten sample of polyethylene (PE) is cooled below the crystallization temperature, the chains change their conformation from random coil to the regular planar-zigzag-type and form the crystalline lamella consisted of the parallel packing of these conformationally-ordered chain stems. How does such an ordering process occur during the course of crystallization ? In this section the molecular conformational change in the crystallization process will be investigated by the time-resolved FTIR measurement and at the same time the formation process of crystalline lamellae by the time-resolved small-angle X-ray scattering measurement. These two types of results are combined to clarify the crystallization mechanism of PE from both the view points of molecular level and crystallite level (20).

Temperature Jump

In general, PE crystallizes in a time scale of milliseconds to seconds depending on the degree of supercooling (ΔT), which is defined as $\Delta T = T^{\circ}m - Tc$, where the Tc is the predetermined crystallization temperature and the $T^{\circ}m$ is the equilibrium melting temperature. Therefore it is difficult

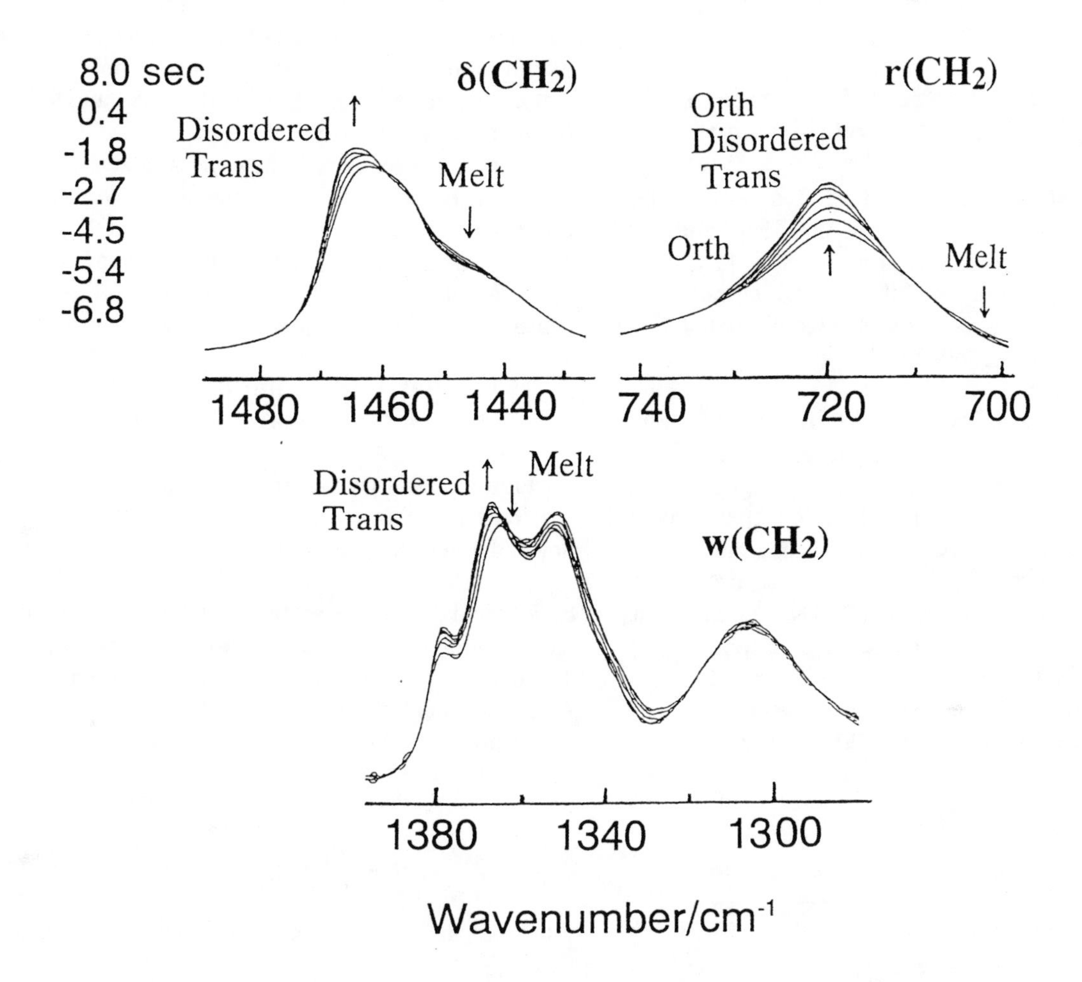

FIGURE 11. Time dependence of the infrared spectra measured during the isothermal crystallization process of linear low-density polyethylene sample (LLDPE(2)) at the supercooling $\Delta T = 4^{o}C$. The three sets of spectra shown in this figure correspond to the frequency regions of CH_2 bending mode (upper left), CH_2 rocking mode (upper right), and CH_2 wagging mode (lower), respectively. The times indicated in the figure were counted from the point where the temperature jump was finished.

to trace the structural change at a high resolution of time. In order to reduce the crystallization rate to the time scale treatable in the experiment without any loss of essential features of the structural change, a linear low-density PE sample with ethyl branching of ca. 17/1000 carbons [LLDPE(2)] was selected as a sample, which was supplied kindly from Exxon Chemicals Co., Ltd. The crystallization temperature of LLDPE(2) is 109 °C as estimated from the temperature-dependent measurement of the infrared spectra in the crystallization process from the melt (21).

In the isothermal crystallization experiments, it is important to cool the samples as fast as possible from the molten state to crystallization temperature Tc. Besides the temperature attained after this jump should be constant and stable. We constructed the temperature-jump apparatus specially designed for the time-resolved FTIR and SAXS measurements. The cooling rate was ca. 600 °C per minute for both the FTIR and SAXS experiments. After the jump was completed the temperature was stable with sufficiently small fluctuation (ca. ± 0.2 °C) (22, 23).

The FTIR spectra were measured by a Bio-Rad FTS-60A/896 Fourier-Transform infrared spectrometer equipped with an MCT detector in the rapid scan mode at a rate of 1 sec/spectrum with the spectral resolution of 2 cm^{-1}. The time-resolved SAXS measurements were performed by using a synchrotron-sourced X-ray beam on the line #10C of the Photon Factory, National Laboratory for High Energy Physics, Tsukuba, Ibaraki, Japan. The scattered X-ray signal was detected by using a one-dimensional position sensitive proportional counter (PSPC) at an interval of 5 - 6 sec.

Ordering Process of Chain Conformation During Isothermal Crystallization

In Fig. 11 is shown the time dependence of the IR spectra in the frequency regions of the CH_2 bending, wagging, and rocking modes measured at $\Delta T = 4$ °C. In the pervious study (24), we found the IR bands characteristic of the hexagonal phase or the CONDIS (conformationally-disordered) phase of PE as follows:
short and disordered trans segments = 1466, 719 cm^{-1}, kink (...TTGT$\bar{G}$TT...) contained in the disordered trans segments = 1368, 1306 cm^{-1}, local double-gauche defect (...TTGGTT...) 1352 cm^{-1}. As seen in Fig. 11, at the beginning of the isothermal crystallization, these bands obviously increase in intensity prior to the appearance of the bands intrinsic to the long and regular trans segments (the orthorhombic crystalline bands) at 1471 and 728 cm^{-1}. In order to make a quantitative analysis, the overlapped bands were separated into the components. The integrated intensities of the thus separated band components are plotted against time, as shown in Fig. 12. The starting point (t = 0) was assigned to the time when the temperature reached at the predetermined crystallization point (Tc). The intensity of the disordered trans band at 1368 cm^{-1}, for example, begins to increase at the initial stage of the temperature jump. After keeping the maximum for a while, this band intensity begins to decrease gradually. In this time region, the long and regular trans band at 728 cm^{-1} appears and increases the intensity. Such a tendency could be seen similarly for ΔT = 5, 6, 7 and 9 °C. Figure 12 indicates clearly the occurrence of the chain conformational ordering from the random coil to the regular planar-zigzag form *via the disordered trans form* .

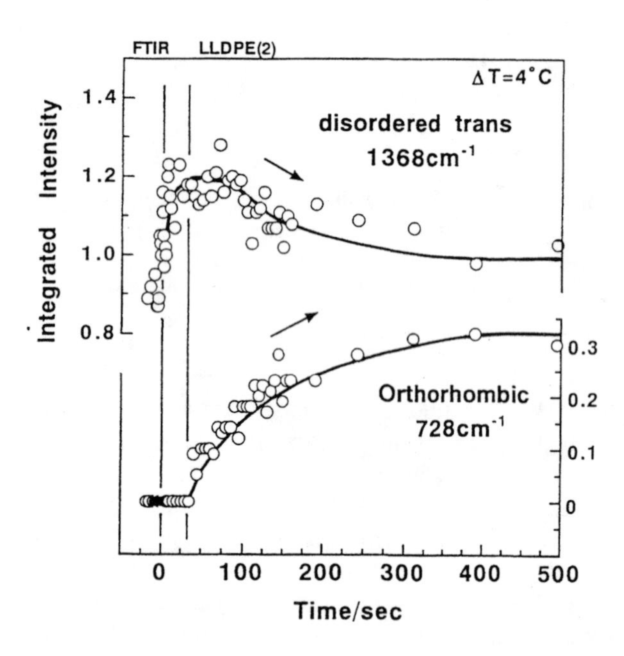

FIGURE 12. Time dependence of the integrated intensity of the infrared band at 1368 cm^{-1} characteristic of the conformationally-disordered trans sequence of polyethylene and that of the infrared band at 728 cm^{-1} characteristic of the regular trans sequence of the orthorhombic crystal. The sample used was linear low-density polyethylene (LLDPE(2)) and the supercooling $\Delta T = 4$°C.

Lamellar Formation in the Isothermal Crystallization Process

Figure 13 shows the time dependence of the Lorentz-corrected SAXS intensities $I(s)s^2$ [s: a scattering vector as defined later] measured for the LLDPE(2) during the isothermal crystallization at $\Delta T = 4$ °C, where the background due to the scattering from the window of the temperature jump cell was subtracted from the original data. The one-dimensional electron-density correlation function $K(z)$ was calculated from these data (25). Under the assumption of the two-phase model consisting of the alternately stacked structure of the crystalline and amorphous layers, the $K(z)$ is defined by

$$K(z)=\langle[\eta(z')-\langle\eta\rangle][\eta(z+z')-\langle\eta\rangle]\rangle \qquad (9)$$

where $\eta(z)$ and $\langle\eta\rangle$ are the electron density variation along

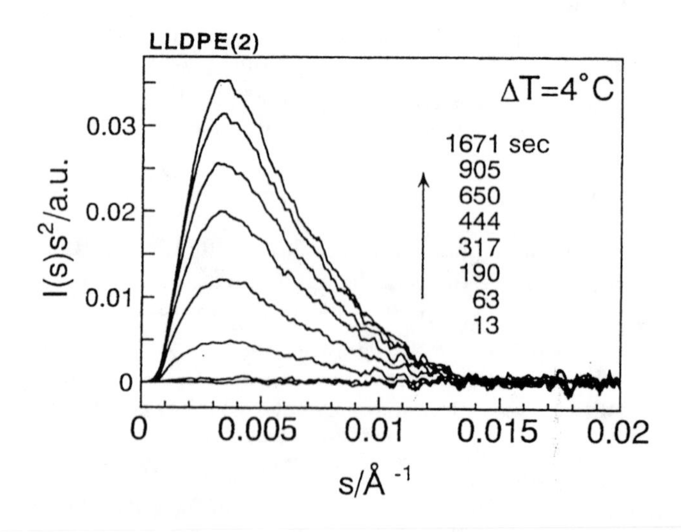

FIGURE 13. Time dependence of the Lorentz-corrected small-angle X-ray scattering profiles of the linear low-density polyethylene (LLDPE(2)) measured during the isothermal crystallization from the melt. The supercooling $\Delta T = 4$°C..

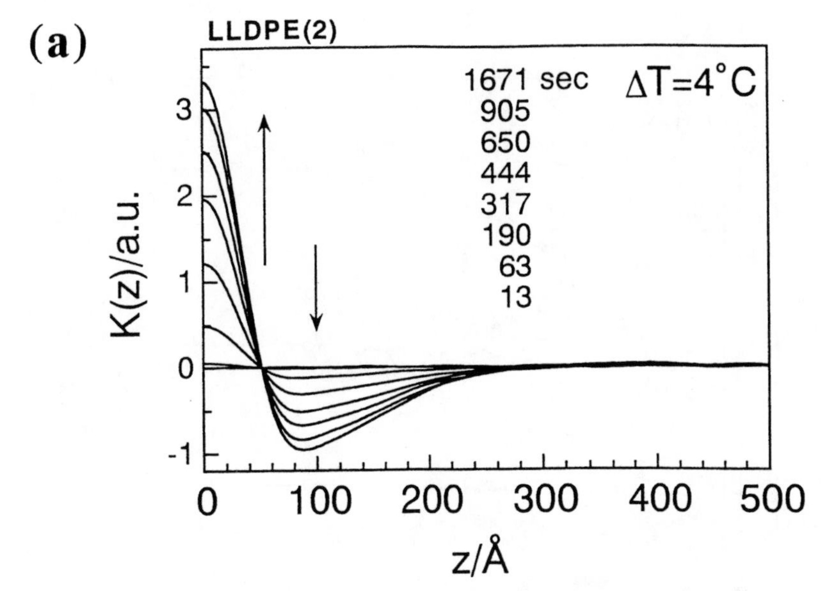

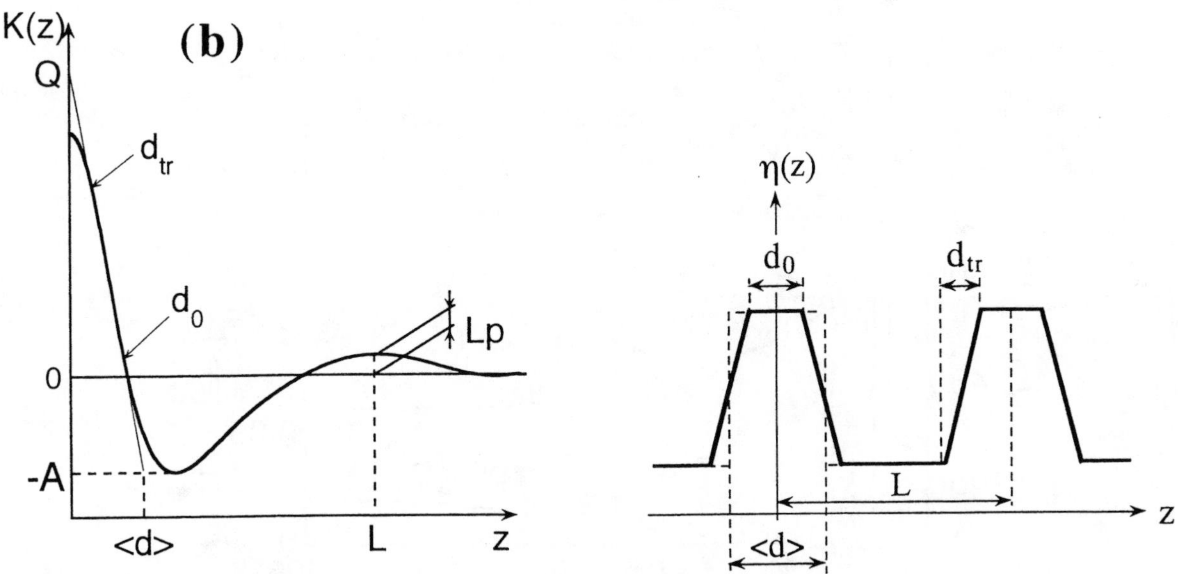

FIGURE 14. (a) Time evolution of one-dimensional correlation function $K(z)$ of the linear low-density polyethylene (LLDPE(2)) during the isothermal crystallization from the melt, which was calculated by using the data shown in Figure 13. (b) The indication of the physical meanings of the various points in the $K(z)$ function. z is the one-dimensional coordinate along the normal of the lamellae. Q is the invariant and A is the baseline at the bottom position of $K(z)$. The $\langle d \rangle$, d_{tr}, and d_0 are the averaged lamellar thickness, the thickness of the boundary, and the thickness of lamellar core, respectively. The L and Lp are the long spacing and the peak height of $K(z)$ at the point $z = L$, respectively.

the lamellar normal and the mean electron density, respectively. The $K(z)$ can be obtained by the Fourier transformation of the scattering curve.

$$K(z) = \int 4\pi\, s^2 I(s) \cos(2\pi s z)\, ds \qquad (10)$$

$$s = (2/\lambda) \sin\theta$$

where the wavelength of the incident X-ray beam is λ and the scattering angle is defined as 2θ.

The $K(z)$ thus obtained is shown in Fig. 14 (a), where the background coming from the molten state is corrected already at the point of Fourier transform calculation from the scattering intesity. At the early stage of the isothermal crystallization, the self-correlation triangle, as named by Strobl et al. (25), is observed in the $K(z)$ profile, implying

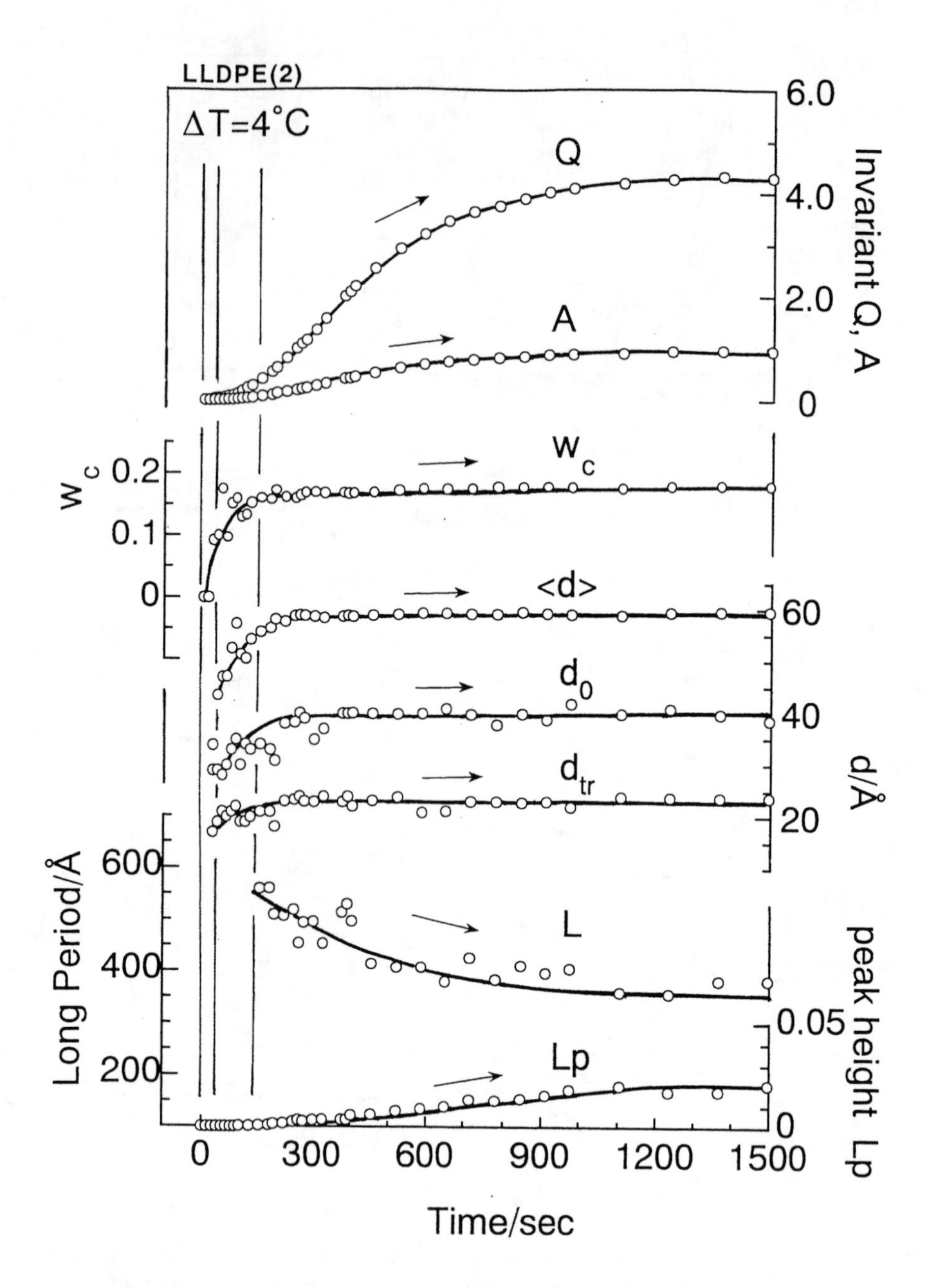

FIGURE 15. Time dependence of the structural parameters of the crystalline lamellae of the linear low-density polyethylene (LLDPE(2)) formed during the isothermal crystallization from the melt, which were evaluated from the one-dimensional correlation function $K(z)$ plotted in Figure 14.

the existence of single lamellae. As shown in Fig. 14 (b), the various points indicated in the $K(z)$ curve give the various physical parameters: the invariant $Q = \int I(s)\, s^2\, ds$, the degree of crystallinity $w_c = A/(Q + A)$, the mean lamellar thickness $\langle d \rangle$, the mean boundary thickness d_{tr}, the mean core thickness d_o, and the long spacing L. In Fig. 15 is shown the time dependence of the structural parameters of LLDPE(2) obtained from the $K(z)$ curves. In the early stage of the crystallization, the invariant Q and the baseline A increase gradually, reflecting on the increase of the crystallinity w_c. After ca. 150 sec, the growing rate of w_c decreases gradually. The d_{tr}, d_o and $\langle d \rangle$ begin to be observed around t = 40 sec and increase continuously to ca. 25, 40 and 60 Å, respectively, at t = 150 sec. At this

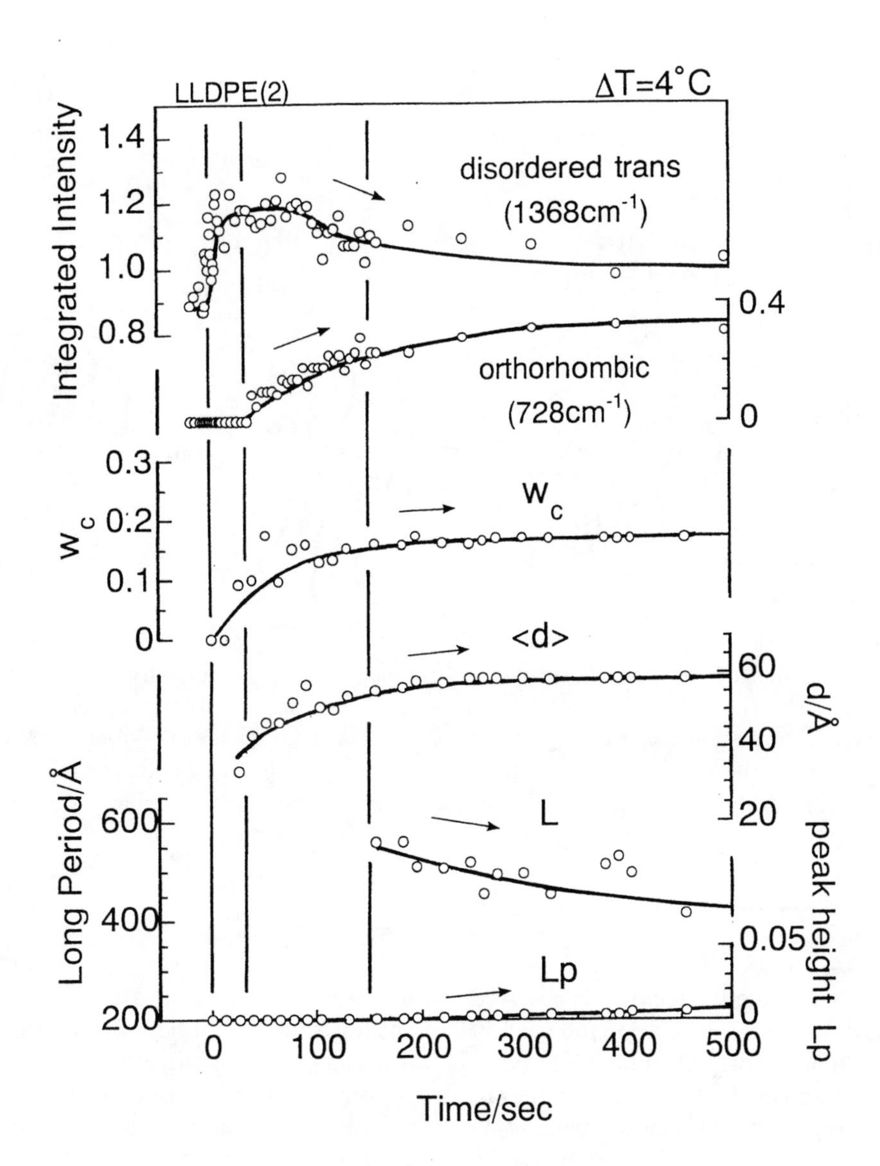

FIGURE 16. A comparison between the FTIR data and the SAXS data measured for the linear low-density polyethylene (LLDPE(2)) during the isothermal crystallization from the melt. Refer to the data shown in Figures 12 and 15.

stage, the peak of the long period (L) reflecting the lamellar stacking structure begins to appear and decreases the value from 500 Å to 350 Å. The peak height L_p evaluated from the $K(z)$ at $z = L$ position increases correspondingly, indicating the increase of the electron density difference between the crystalline lamella and amorphous phase.

Conformational Ordering and Aggregation Process

In Fig. 16 the time dependence of the structural parameters obtained from the SAXS data is compared with that of the IR band intensity, where the DT is 4 °C. This comparison can be made reasonably because the temperature jump in both the measurements was carried out at almost the same rate of ca. 600 °C/min. The structural model deduced from this comparison is illustrated in Fig. 17. In the early stage of crystallization the conformationally-disordered short trans sequences rapidly increase in population [Fig. 17 (b)]. In the time region of 40 - 150 sec, the disordered trans sequences decrease gradually and the regular trans sequences of orthorhombic type increase instead. The degree of crystallinity w_c also increases in parallel with this conformational transition. As seen in the

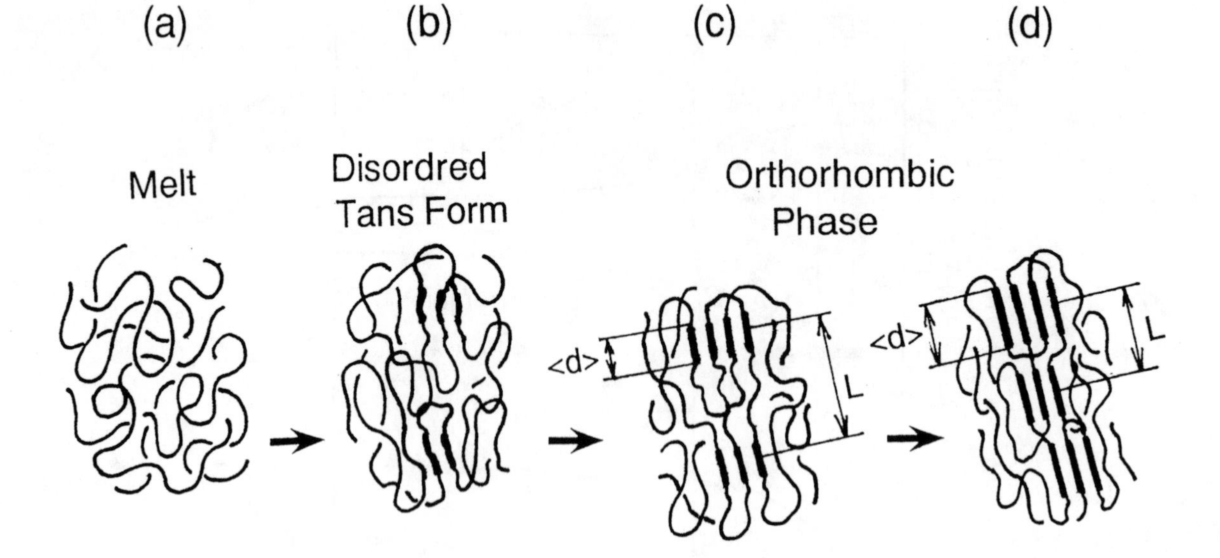

FIGURE 17. An illustration of the lamellar structural formation process in polyethylene during the isothermal crystallization from the melt. The $<d>$ and L denote the averaged thickness of a lamella and the long spacing between the stacked lamellae. (a) Starting from the random coil state in the melt, (b) the conformationally-disordered trans segments are formed as indicated by thick short lines. (c) These disordered trans segments change into the regular zigzag stems of orthorhombic crystal (straight thick lines) and form a sheet of lamella. The long spacing begins to be observed in the SAXS measurement. (d) With passage of time, new lamella is formed between the existing lamellae and the long spacing becomes shorter.

behavior of $<d>$, the isolated lamellae begin to be observed and increase the thickness up to ca. 60 Å. Around ca. 150 sec, when the formation of the orthorhombic-type crystal lamella is almost accomplished, the long spacing (L) reflecting the stacked lamellar structure begins to be detected [Fig. 17 (c)]. After that, the L becomes shorter but the lamellar thickness is almost constant. The regular trans sequences continue to increase in the population, though slowly, with the w_C. These changes suggest that new lamellae are formed between the existing isolated lamellae and at the same time the ordering occurs within the lamellae, as already pointed out by the several authors [Fig. 17 (d)] (26, 27).

CONCLUSIONS

In this paper we have investigated the ordering process of macromolecular compounds in the view-points of the chain conformation and the aggregation structure of chains by comparing the vibrational spectroscopic data with the X-ray or neutron scattering data. In both cases of the gelation of SPS and the isothermal crystallization of PE, the conformational ordering from random coil to regular form is observed at the transition point, although the systems are remarkably different from each other, one is the solution and another is the bulk solid. Such a conformational ordering is accompanied by the ordering in the chain aggregation structure, resulting in the formation of gel network or in the formation of crystalline lamellae. This type of structural change, that is, the process including both the intramolecular and intermolecular orderings is characteristic of macromolecular system and plays significantly important role in the structural development of gels and crystalline lamellae.

REFERENCES

1. M. Kobayashi, *Macromol. Symp.* **114**, 1 (1997).
2. A. Immirzi, F. de Candia, P. Ianneli, A. Zambelli, and V. Vittoria, *Makromol. Chem., Rapid Commun.* **9**, 761 (1988).
3. V. Vittoria, F. de Candia, P. Ianneli, and A. Immirzi, *Makromol. Chem., Rapid Commun.* **9**, 765 (1988).
4. M. Kobayashi, T. Nakaoki, and N. Ishihara, *Macromolecules* **22**, 4377 (1989).
5. Y. Chatani, Y. Fujii, Y. Shimane, and T. Ijitsu, *Polym. Prepr. Jpn. (Engl. Ed.)* **37**, E428 (1988).

6. G. Guerra, V. M. Vitagliano, C. De Rosa, V. Petraccone, and P. Corradini, *Macromolecules* **23**, 1539 (1990).
7. P. Corradini, R. Napolitano, and B. Pirozzi, *Eur. Polym. J.* **26**, 157 (1990).
8. O. Greis, Y. Xu, T. Asano, and J. Petermann, *Polymer* **30**, 590 (1989).
9. Y. Chatani, Y. Shimane, Y. Inoue, T. Inagaki, T. Ishioka, T. Ijitsu, and T. Yukinari, *Polymer* **33**, 488 (1992).
10. G. Natta, P. Pino, P. Corradini, F. Danusso, and E. Mantia, *J. Am. Chem. Soc.* **77**, 1700 (1955).
11. M. Kobayashi, T. Nakaoki, and N. Ishihara, *Macromolecules* **23**, 78 (1990).
12. T. Nakaoki and M. Kobayashi, *J. Mol. Struct.* **242**, 315 (1991).
13. T. Nakaoki and M. Kobayashi, *Rep. Progr. Polym. Phys. Jpn.* **33**, 91 (1990).
14. M. Kobayashi, K. Akita, and H. Tadokoro, *Makromol. Chem.* **118**, 324 (1968).
15. M. Kobayashi, K. Tsumura, and H. Tadokoro, *J. Polym. Sci., Polym. Phys. Ed.* **6**, 1493 (1968).
16. M. Kobayashi and Y. Ueno, *Macromolecules,* to be published.
17. M. Kobayashi and T. Kozasa, *Appl. Spectrosc.* **47**, 1417 (1993).
18. M. Kobayashi, T. Yoshioka, T. Kozasa, K. Tashiro, J. Suzuki, S. Funahashi, and Y. Izumi, *Macromolecules* **27**, 1349 (1994).
19. M. Kobayashi, T. Yoshioka, M. Imai, and Y. Itoh, *Macromolecules* **28**, 7376 (1995).
20. K. Tashiro, S. Sasaki, and M. Kobayashi, *Macromolecules,* to be published.
21. K. Tashiro, *Acta Polymer.* **46**, 100 (1995).
22. K. Tashiro, M. Izuchi, F. Kaneuchi, C. Jin, M. Kobayashi, and R. S. Stein, *Macromolecules* **27**, 1240 (1994).
23. K. Tashiro, K. Imanishi, Y. Izumi, M. Kobayashi, K. Kobayashi, M. Satoh, and R. S. Stein, *Macromolecules* **28**, 8477 (1995).
24. K. Tashiro, S. Sasaki, and M. Kobayashi, *Macromolecules* **29**, 7460 (1996).
25. G. Strobl and M. Schneider, *Macromolecules* **18**, 1343 (1980).
26. T. Albrecht and G. Strobl, *Macromolecules* **29**, 783 (1996).
27. R. S. Stein, J. Cronauer, and H. G. Zachmann, *J. Molecular Sci.* **383**, 19 (1996).

3. SUBMITTED PAPERS

3.1 ATMOSPHERIC STUDIES

Isotopic Analysis of Atmospheric Trace Gases by FTIR Spectroscopy

David W.T. Griffith, Michael B. Esler and Stephen R. Wilson

University of Wollongong, Department of Chemistry, Wollongong, NSW 2522, Australia

We have used FTIR absorption spectroscopy at both low ($1cm^{-1}$) and high ($0.004cm^{-1}$) resolution to measure small variations in the natural abundance of isotopes in CO_2. Quantitative analysis of the spectra was by Classic Least Squares using calculated calibration spectra. At low resolution, we achieve precision in the $^{13}C/^{12}C$ ratio of 0.15‰ for CO_2 in dried whole air samples, and demonstrate field measurements of the isotopic fractionation of carbon by plants in an actively growing pasture. Using high resolution, we discriminate the ^{13}C, ^{17}O and ^{18}O isotopomers of CO_2 and achieve a measurement precision for the oxygen isotope ratios of 2-4‰. A number of potential applications in atmospheric chemistry are described.

INTRODUCTION

The natural abundances of minor isotopes in atmospheric trace gases vary due to fractionation by physical, chemical and biological processes. Each process leaves its "signature" in the isotopic composition of the trace gases. The typical range of variations is normally only a few percent or less, but detailed study of these isotopic signatures provides a wealth of information on the sources and sinks of each gas and helps to constrain their global budgets. For examples, see Kaye [1], Ciais et al. [2], and Lowe et al. [3].

The technique most comonly used to determine isotopic ratios in nature is Isotope Ratio Mass Spectrometry (IRMS), which can with care achieve precision levels of better than 0.1‰ (ie 0.01%), and has been an extremely valuable technique in atmospheric trace gas chemistry and global budgeting. However, IRMS has some limitations because it discriminates between species based only on their mass, which leads to some ambiguities in certain applications. Species of the same nominal mass, such as $^{13}CO_2$ and $C^{17}OO$, $^{13}CH_4$ and CH_3D, or CO_2 and N_2O, are not distinguished and assumptions or corrections must be made when determining individual isotopic ratios. IRMS also cannot distinguish between symmetry isotopomers, such as $O^{18}OO$ and ^{18}OOO.

Infrared spectroscopy distinguishes species by both mass and structure and can potentially avoid these limitations. There are numerous atmospheric applications where these limitations are important. For example:

- Anomalously large and non-mass-dependent enrichment of heavy isotopes in stratospheric ozone have been observed since 1981 [4] but to date no satisfactory explanation has emerged. There is evidence from solar FTIR spectra that the enrichment is dependent on the symmetry of the isotopically substituted ozone. FTIR could be used to distinguish the position of isotopic substitution in ozone formation and shed light on the unexplained mechanism.
- $\delta^{13}C^*$ measurements of atmospheric CO_2 are a valuable tool in studies of the global carbon cycle [2]. IRMS analysis does not distinguish between ^{13}C and ^{17}O in CO_2 at mass 45, and a correction for ^{17}O to $\delta^{13}C$ measurements is usually made based on the determination of $\delta^{18}O$ at mass 46 and an assumed mass dependence of the oxygen isotope ratios. In stratospheric CO_2 this assumption may not hold because oxygen is believed to be exchanged via $O(^1D)$ atoms with isotopically anomalous ozone [5]. FTIR spectroscopy can be used to determine $\delta^{17}O$ and $\delta^{18}O$ in CO_2 directly.
- Hydrogen isotopes ^1H and ^2H (=D) are fractionated in nature by large amounts (up to 30%) because of the large relative mass difference between D and H. Thus analysis of atmospheric δD in CH_4 should provide valuable information on the atmospheric sources and sinks of methane. IRMS does not distinguish CH_3D from the more abundant $^{13}CH_4$ without laborious chemical separation, and FTIR could provide a valuable and direct measurement tool for D/H fractionation studies.

Infrared measurements of isotopic ratios by FTIR and tunable diode laser spectroscopy have recently been briefly reviewed [6]. In this paper we present initial results using FTIR spectroscopy at both low and high resolution on the

---

*Isotopic ratios are commonly quoted in the del notation, for example for ^{13}C:

$$\delta^{13}C = \frac{(^{13}C/^{12}C)_{sample} - (^{13}C/^{12}C)_{reference}}{(^{13}C/^{12}C)_{reference}} * 1000$$

CP430, *Fourier Transform Spectroscopy:* 11th International Conference
edited by J.A. de Haseth

determination of isotope ratios of ^{13}C, ^{17}O and ^{18}O in CO_2.

EXPERIMENTAL

Low resolution ($1cm^{-1}$) spectra of whole air samples dried to <10ppm H_2O were collected with a Bomem MB100 FTIR fitted with a 9.8m White cell and InSb detector. The spectrometer and cell were thermostatted to ±0.1°C to maintain stability, and cell temperature and pressure were measured and logged. Typically 128-256 spectra were coadded (8-16 minutes) giving noise levels in absorbance spectra of ca. $2x10^{-5}$ units. Air samples were either drawn directly from ambient air or from 2L flasks which had been filled at the sampling site. Regular analyses of an air standard traceable to NIST primary standards were interleaved with sample measurements. All gas handling, spectrometer control, temperature and pressure measurement and spectrum analysis were carried out automatically from Array Basic programs running under GRAMS-386 on the controlling computer. The spectrometer could either run samples of ambient air continuously or analyse batches of up to 8 flasks automatically without operator intervention.

High resolution spectra of pure CO_2 at 20 or 40mbar pressure in a 10cm cell were obtained with a Bomem DA8 high resolution FTIR spectrometer at 0.004 or $0.005cm^{-1}$ resolution using a globar source, KBr/Ge beamsplitter, narrow bandpass optical filter and InSb detector, typically taking 32 scans (20 minutes) per spectrum. CO_2 samples were introduced into the cell to the required pressure by liquid nitrogen trapping and expansion into fixed volumes to avoid possible isotopic fractionation when CO_2 is bled through small orifices.

All spectra were analysed by Classic Least Squares (CLS) using *calculated* calibration spectra in place of actual measured calibration spectra. The calibration spectra are calculated using the program MALT which is described in detail elsewhere [7]. This method provides convenient and fast calibration and quantitative analysis based on the HITRAN database of IR absorption line parameters [8] without the need for any calibration gases. In normal practice relative quantitative precision of <0.5% is obtainable, with absolute accuracy of <5% without the need for any calibration gases, and higher accuracy achieved by comparison with independent gas standards. Individual isotopomers may be analysed as independent species. As demonstrated in this work, relative precision of better than 0.2‰ (ie. 0.02%) can be achieved for isotopic ratios.

RESULTS

Figure 1 shows sections of the spectrum of the CO_2 ν_3 band at $1cm^{-1}$ and $0.004cm^{-1}$ resolution. At the lower resolution only the broader spectral features are resolved but the $^{13}CO_2$ P-branch is well separated from the $^{12}CO_2$ band, whilst at high resolution the individual rotational lines are well resolved.

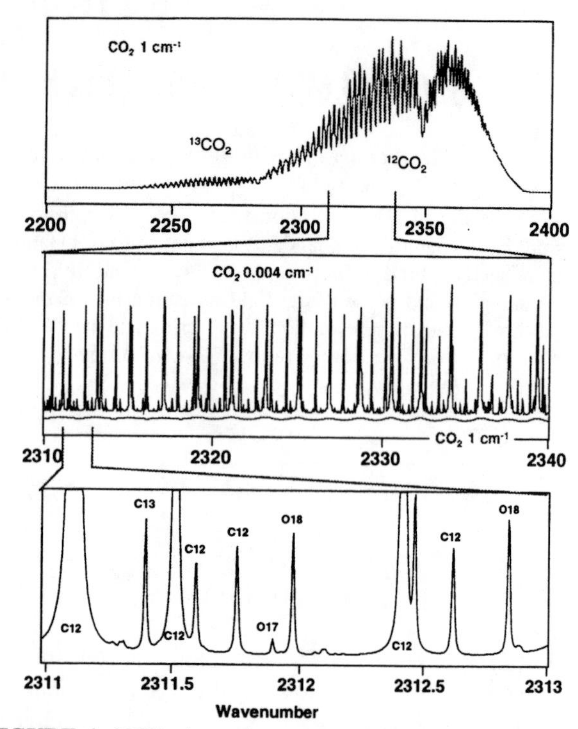

FIGURE 1. FTIR absorption spectra in the ν_3 band of CO_2 at $1cm^{-1}$ and $0.004cm^{-1}$ resolution.

Low resolution, $\delta^{13}CO_2$.

Figure 1 demonstrates that $1cm^{-1}$ resolution is sufficient to discriminate $^{13}CO_2$ from $^{12}CO_2$. MALT/CLS quantitative analysis of the CO_2 spectrum treating all isotopomers as independent species provides absolute concentrations of each isotopomer, from which isotopic ratios can be calculated. When ratioing the absolute concentrations, many potential sources of systematic error cancel and allow high precision for the ratios to be obtained. At $1cm^{-1}$ resolution, only the $^{12}CO_2$ and $^{13}CO_2$ isotopomers are resolved with any precision. Figure 2 shows a typical fit of a CO_2 spectrum used to determine isotopic ratios. The spectrum is fitted with MALT/CLS almost to within the spectrum noise level.

The precision of isotopic ratio retrieval may be assessed by repeated measurements of a single sample of air or of consecutive samples drawn from a single air cylinder. In a study of 32 consecutive measurements of a single sample the standard deviation of the measurements was ±0.15‰. The 95% confidence limit on the mean of the 32 measurements is less than 0.1‰, approaching the precision of IRMS. The ultimate precision may also be estimated by simulating the same experiment using an independent set of MALT spectra with realistic levels of random noise in place of the real measured spectra. Since the MALT spectra assume a spectrometer which is ideal and otherwise perfect measurement conditions, such a simulation provides an estimate of the best precision achievable for a given signal-to-noise ratio. The simulation indicates a precision of 0.1-0.2‰, comparable to that achieved in practice, and suggests that the experimental conditions and spectrometer performance are close to optimal.

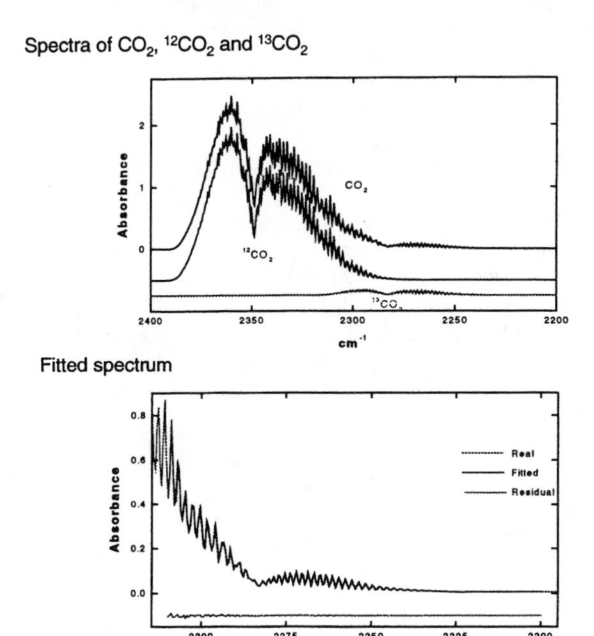

FIGURE 2. Low resolution spectra of $^{12}CO_2$, $^{13}CO_2$ and natural CO_2, (upper box), and a fitted spectrum of the CO_2 bands in an air sample (lower box). In the lower box, the actual and fitted spectrum are almost indistinguishable, and the lowest trace is the difference between the two spectra.

In October 1995 the spectrometer was used during OASIS, a coordinated campaign to measure fluxes of trace gases from an agricultural environment (see accompanying paper). In one experiment, vertical profiles of trace gases and $\delta^{13}CO_2$ were measured on a 22m tower located in a lucerne pasture. The profiles of CO_2 and $\delta^{13}CO_2$ obtained in the evening (22:30) when the plants and soil respire CO_2 are shown in Figure 3. The CO_2 mixing ratio increases strongly near the ground due to the strong ground-level source. The $\delta^{13}CO_2$ correspondingly decreases near the ground because the respired CO_2 is depleted in ^{13}C. The y-intercept of a plot of $\delta^{13}CO_2$ vs $1/CO_2$ gives the $\delta^{13}CO_2$ signature of the respired CO_2. The value obtained, -30‰, is in good agreement with the ^{13}C value expected for C3 plants such as lucerne, -28‰. This measurement suggests that FTIR spectroscopy could be a very useful tool for plant physiological studies since the measurements can be done rapidly in real time in the field with more than sufficient precision. The same instrument can also easily distinguish H_2O from HDO, allowing the possibility of monitoring both $^{13}C/^{12}C$ and D/H fractionation simultaneously.

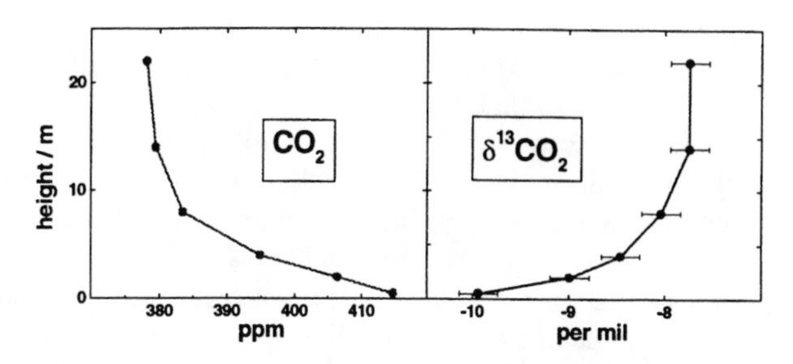

FIGURE 3. CO_2 and $\delta^{13}CO_2$ vertical profiles 0.5-22m above a lucerne pasture, Wagga Wagga, Oct 1995, 22:30 hrs.

High resolution, O isotopes in CO_2

Following the low resolution studies, we performed a pilot study to assess the feasibility of using high resolution FTIR to determine oxygen isotope ratios in CO_2 The $C^{17}OO$ and $C^{18}OO$ ν_3 bands fall beneath the parent CO_2 band and require high resolution spectra to be distinguished. Figure 1 suggests that there are many microwindows between the strong parent lines which may be used for minor isotope determinations. Figure 4 displays one such microwindow at 0.004cm^{-1} together with the MALT/CLS fit. The major contributions to the residual in this case are due to an imperfect instrument lineshape, and are strongly dependent on spectrometer alignment. To determine the oxygen isotopes in CO_2, 19 such microwindows were fitted between 2310 and 2340cm^{-1} and a weighted mean of the individual determinations was taken for each isotope

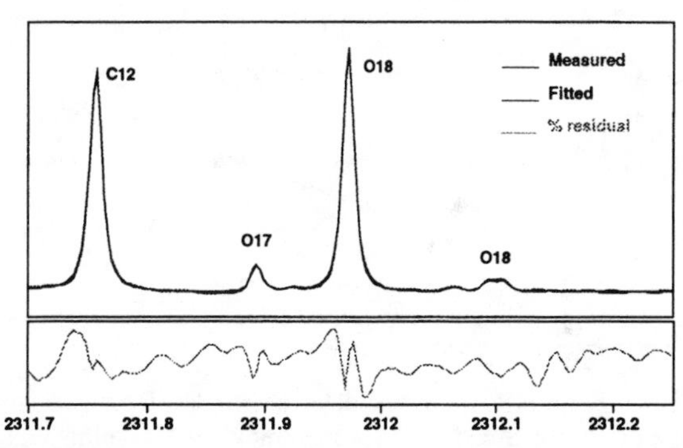

FIGURE 4. Example of a microwindow of high resolution CO_2 spectrum fitted by MALT/CLS. The upper frame shows the measured and fitted spectra, and the lower frame the residual (scale expanded).

Using this analysis method, 20 repeated measurements of a single CO_2 sample resulted in standard deviations of the retrieved values for $\delta^{17}O$ and $\delta^{18}O$ of 3.0 and 1.6‰ respectively. A simulation using MALT-calculated spectra as described above indicates that a theoretical precision of 0.2-0.4‰ should be achievable. We believe that the lower precision in the high resolution measurements is due partly to lower optical stability of the high resolution spectrometer, and partly to small changes in the environment during the measurements. By better control of the spectrometer environment, eg by maintaining constant temperature and sample pressure, we expect to realise the theoretical precision as we have done in the low resolution case.

A further study of O-isotope ratio determination was made in collaboration with Dr. S.K. Bhattacharrya (Physical Research Laboratory, India) and C.E. Allison (CSIRO Division of Atmospheric Research). A set of seven CO_2 samples were prepared in which the oxygen isotopic composition had been altered by exchange with H_2O enriched in ^{17}O or ^{18}O. The O-isotopic ratios were determined from the known water values and IRMS at CSIRO. The seven samples were then analysed by high resolution FTIR as described above. Figure 5 compares the $\delta^{17}O$ and $\delta^{18}O$ values determined by the two methods. After exclusion of the one outlier the standard errors of prediction for $\delta^{17}O$ and $\delta^{18}O$ were 2.6 and 4.0‰ respectively. The precision is marginally lower than that from the replicate measurement, but in this case the sample was changed between measurements so that any errors due to sample handling, pressure variations etc. will also be included in the observed scatter.

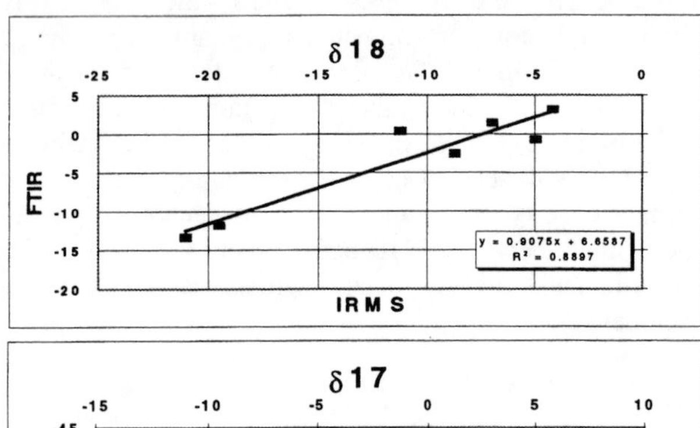

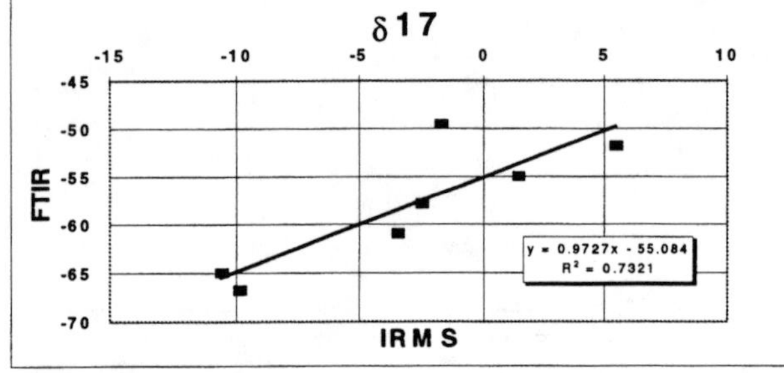

FIGURE 5. FTIR vs IRMS determinations of oxygen isotope ratios in CO_2 for a series of seven samples with synthetically altered isotopic compositions.

SUMMARY AND CONCLUSIONS

We have used FTIR absorption spectroscopy for isotopic analyses of CO_2 and demonstrated precisions for individual measurements of 0.15‰ for $\delta^{13}C$-CO_2 at low resolution and 2-4‰ for $\delta^{17}O$-CO_2 $\delta^{18}O$-CO_2 at high resolution. There are a number of atmospheric applications for which FTIR measurement of isotope ratios with this methodolgy and precision may prove useful, including field studies of C and H exchange in plants, stratospheric ozone formation, stratospheric CO_2 oxygen isotopic analysis, and deuterium in methane studies.

REFERENCES

(1) J.A. Kaye, *Rev. Geophys.* **28**, 1609-1658 (1987).

(2) P. Ciais, P.P. Tans, J.W.C. White, M. Trolier, R.J. Francey, J.A. Berry, D.R. Randall, P.J. Sellers, J.G. Collatz and D.S. Schimel, *J. Geophys. Res.* **D100**, 5051-5070 (1995).

(3) D.C. Lowe, C.A.M. Brenninkmeijer, G.W. Brailsford, K.R. Lassey, A.J. Gomez and E.G. Nesbit, *J. Geophys. Res.* **99**, 16,913-16,925 (1994).

(4) K. Mauersberger, *Geophys. Res. Lett.* **8**, 935-937 (1981).

(5) Y.L. Yung, W.B. DeMore and J.P. Pinto, *Geophys. Res. Lett.* **18**, 13-16 (1991).

(6) A. Kindness and I.L. Marr, *Appl. Spectrosc.* **51**, 17 (1997).

(7) D.W.T. Griffith, *Appl. Spectrosc.* **50**, 59-70 (1996).

(8) L.S. Rothman, R.R. Gamache, R.H. Tipping, C.P. Rinsland, M.A.H. Smith, D.C. Benner, V.M. Devi, J.-M. Flaud, C. Camy-Peyret, A. Perrin, A. Goldman, S.T. Massie, L.R. Brown and R.A. Toth, *J. Quant. Spectrosc. Radiat. Transfer* **48**, 469-507 (1992).

FTIR in the Paddock: Trace Gas Soil Flux Measurements using FTIR Spectroscopy

D.W.T. Griffith[1], I.M. Jamie[1], P.A. Beasley[1],
O.T. Denmead[2], R. Leuning[2], I.E. Galbally[3] and C.P. Meyer[3]

[1]University of Wollongong, Department of Chemistry, Wollongong, NSW 2522, Australia
[2]CSIRO Division of Land & Water, Environmental Mechanics Laboratory, Canberra, ACT 2600, Australia
[3]CSIRO Division of Atmospheric Research, Aspendale, VIC 3915, Australia

We have used FTIR spectroscopy for high precision trace gas analysis combined with micrometeorological flux-gradient and chamber methods to measure the fluxes of trace gases CO_2, CH_4, N_2O and CO between agricultural soils and the atmosphere. In flux gradient measurements, vertical profiles of the trace gases were measured every 30 minutes from the ground to 22 m. When combined with independent micrometeorological measurements of heat, water vapour and radiative fluxes, trace gas fluxes from the underlying surface could be determined. In chamber measurements, a closed chamber was placed over the soil surface and the air in the chamber monitored continuously by FTIR spectroscopy. Fluxes were calculated from the time rate of change of concentrations in the chamber after closure. The FTIR measurements were fully automated and ran reliably for several weeks, routinely obtaining precision of 0.1-0.5%.

INTRODUCTION

The earth's surface is a major source and sink of trace gases in the atmosphere. The measurement of surface-atmosphere fluxes of trace gases is a very active field of atmospheric research, and improved methods for trace gas flux measurement are an essential requirement to extend techniques. Chamber methods (1), which enclose a small surface area and measure the change in concentrations of trace gases in the enclosure, remain the most common because they are relatively simple to implement, but they sample only small areas and time spans, and may perturb the system being measured by changing the microclimate in the chambers. Micrometeorological methods (2) are to be preferred because they are non-disturbing, integrate over much larger areas and are more amenable to automation. However these techniques are significantly harder to implement and have stringent requirements on trace gas analysis; for eddy correlation measurements instruments must typically have 0.1s response times, and for gradient flux or eddy accumulation measurements relative precision of the order of 0.1% is required to resolve small but significant fluxes. Both chamber and micrometeorological methods would benefit greatly from improved precision, reliability and automation of trace gas concentration measurements.

The OASIS campaigns (Observations At Several Interacting Scales) (3) were carried out to measure and reconcile fluxes of energy, water vapour and trace gases in a heterogeneous agricultural environment near Wagga Wagga, SE Australia in October 1994 and 1995 over spatial scales from <$1m^2$ to over $100km^2$. During OASIS we used FTIR spectroscopy to make continuous, high precision, fully automated analyses of CO_2, CH_4, N_2O, CO and H_2O in air in both micrometeorological tower-based and chamber methods. The FTIR technique proved to be robust and reliable, providing simultaneous measurements of all five trace gases with 1-2 minute resolution and precision of the order of 0.1 - 0.5%. The trace gas measurements are being used to calculate surface fluxes of the trace gases. During the first OASIS campaign drought conditions prevailed and there was little growth or biological activity and consequently low trace gas fluxes. We report here techniques and preliminary results from the second OASIS campaign.

EXPERIMENTAL

FTIR spectroscopy

Both tower and chamber measurements were made with Bomem MB100-series spectrometers at $1cm^{-1}$ resolution. The tower instrument was a dual beam (MB104-2E) interferometer configured with a single globar source, two 57m White cells (one on each output beam) and a single InSb detector, housed in a thermostatted box in a caravan or demountable shed. In each 2-minute measurement period, one cell was evacuated and refilled with the next sample to

CP430, *Fourier Transform Spectroscopy:* 11th International Conference
edited by J.A. de Haseth

be analysed while the spectrum of the other cell was being collected; thus dead time was kept to a minimum. At the end of each 2 minute period, the FTIR beam was switched and roles of the 2 cells reversed. All sample handling, beam switching, spectrum collection, spectrum analysis and data logging and archiving were fully automated from a single Array Basic program running under GRAMS-386 on the spectrometer computer. Pressures and temperatures of the cells and enclosure were also logged by the program.

The spectrometer for chamber measurements was similarly fully automated and consisted of a standard MB104 model with a single 22m White cell and InSb detector.

Quantitative analysis - MALT/CLS

All spectra were analysed by Classic Least Squares (CLS) using *calculated* calibration spectra in place of actual measured calibration spectra. The calibration spectra are calculated using the program MALT which is described in detail elsewhere (4). This method provides convenient and fast calibration and quantitative analysis with relative precision of <0.5% and absolute accuracy <5% based on the HITRAN database of IR absorption line parameters (5) without the need for any calibration gases. To obtain absolute accuracy of <1%, a standard tank gas traceable to recognised standards was measured for comparison once per day. Spectra were fitted by CLS using pre-calculated calibrations and the analysed trace gas concentrations were displayed and archived in real time.

Figure 1 illustrates a fit of the 2160 - 2250 cm^{-1} spectral region of a typical clean air spectrum with MALT-calculated calibration spectra. The measured spectrum is fitted almost to within the noise level of the spectrum and the fit provides concentrations of CO_2, N_2O, CO and H_2O. CH_4 is determined by fitting the spectra near 2950-3000 cm^{-1}.

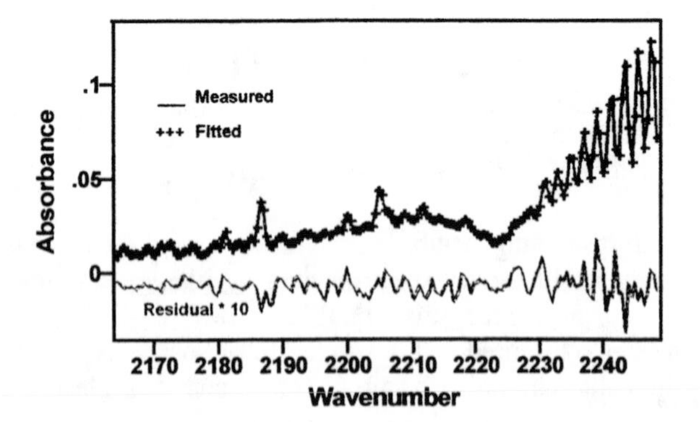

FIGURE 1. CLS fit of a typical clean air spectrum with MALT-calculated calibration spectra.

Tower Measurements

Vertical profiles of trace gas concentrations of CO_2, CH_4, N_2O, CO and H_2O were measured from a tower located in rapidly growing spring pasture (lucerne) about 100m from the FTIR caravan. Air was drawn through 100m of ½" polyethylene tubing from 7 inlets at heights of 0.5-22 m on the tower and the 7 inlets were sequentially analysed at 2 minutes per measurement. Two measurements from each height were averaged to provide a 7-point vertical profile of each trace gas every 30-minutes. The sampling protocol was controlled by solenoid valves switched by the Array Basic program. Measurements were made continuously for 19 days with a regular 1 hour stop near dusk each day for routine maintenance and standard gas calibration. Otherwise there were no periods of lost data (including during 2 days of heavy rain) and no apparent losses in data quality or precision.

Supporting micrometeorological measurements were made at the same 7 heights on the tower and included temperature, humidity, wind speed and direction, fluxes of heat, water vapour and CO_2 directly by eddy correlation, net radiation, and soil heat flux. These measurements are used *inter alia* to convert the measured trace gas vertical profiles to fluxes as described below.

Chamber measurements

Two automated chambers were located over the lucerne pasture and in an adjacent triticale wheat crop ca. 30m from the FTIR caravan. Air was circulated in a closed loop from one or the other chamber through a switching manifold and the FTIR absorption cell. Each chamber was connected in turn to the FTIR for 30 minutes, during which the chamber lid was closed after 5 minutes and reopened after 25 minutes to provide a 20-minute closure time during which the build-up or depletion of trace gases was measured. When closed, the chambers were dark so that photosynthesis was inhibited and only respiration fluxes were measured. Spectra were collected continuously and coadded in 1 minute (20 scans) groups. Sample line switching, chamber lid operation, spectrometer control, temperature and pressure logging and spectrum analysis were all controlled from a single Array Basic program, and each run was analysed in real time.

TOWER MEASUREMENTS

Vertical fluxes of heat, water vapour and trace gases are carried by turbulent diffusion and are proportional to the gradient of concentration with height:

$$F_x = -K\frac{dC_x}{dz} \qquad (1)$$

where F_x and C_x are respectively the flux and concentration of species x, K is the diffusion constant and z is height. To a reasonable approximation the diffusion constant K can be taken to be the same for all scalars (trace gases, heat, and water vapour). In the flux-gradient method, K is determined by independent direct measurements of the flux of some scalar and its vertical gradient. In the present work we have used water vapour gradients and the water vapour flux measured directly by eddy correlation to determine K for each 30 minute measurement period.

Figures 2 and 3 illustrate typical results. Figure 2 shows the average diurnal cycle in mixing ratio at all 7 levels for CO_2, N_2O and CH_4. CO measurements have not been interpreted and are not included here. Considering firstly CO_2, the gradients are negative (concentration decreasing upward) at night and positive during the day, reversing sign near dawn and dusk. Night gradients are often stronger due to low turbulent diffusion (small K in Eq. 1). There is an overall build-up of CO_2 at night above the local baseline level, and depletion below it during the day. This behaviour can also be seen in Figure 3, which shows vertical profiles of CO_2 from midnight to mid-morning. The reversal in sign of the vertical gradient after dawn can be clearly seen. The profiles may be interpreted in terms of net photosynthetic uptake of CO_2 during the day (surface sink, flux negative) and plant/soil respiration during the night (surface source).

Fluxes of CO_2 can be calculated from these gradients using Eq. 1 and water vapour to determine the diffusion coefficient. The peak daytime CO_2 flux due to photosynthetic uptake was typically around -1.0 mgCO_2 m^{-2} s^{-1}. Night values are more difficult to determine because (1) turbulence is often suppressed at night in the still, stable nocturnal boundary layer, and (2) supporting micrometeorological measurements were not usually made at night.

The average diurnal cycles of profiles of N_2O and CH_4 are less well resolved. The average daytime profile of N_2O varies by less than 0.5 ppbv from 0.5 to 22m, compared to measurement precision of about 0.2ppbv and a background level of 310ppbv. Despite the high measurement precision, flux calculations are subject to considerable error, and the mean N_2O flux is found to be 20±14 ngN m^{-2} s^{-1} over the measurement period. The soil is a N_2O source due to microbiological activity, but it is very weak in these ecosystems where soil nitrogen and moisture levels are low.

Methane similarly shows very weak gradients during the day, but a considerable buildup at night and overall levels significantly above baseline. In this case there is no significant source in the local (paddock) environment, but sheep and cattle in the general area provide a regional source. The flux gradient technique can barely resolve the surface flux in this case; we are pursuing nocturnal boundary layer budgeting methods to deduce the methane flux from the night-time buildup.

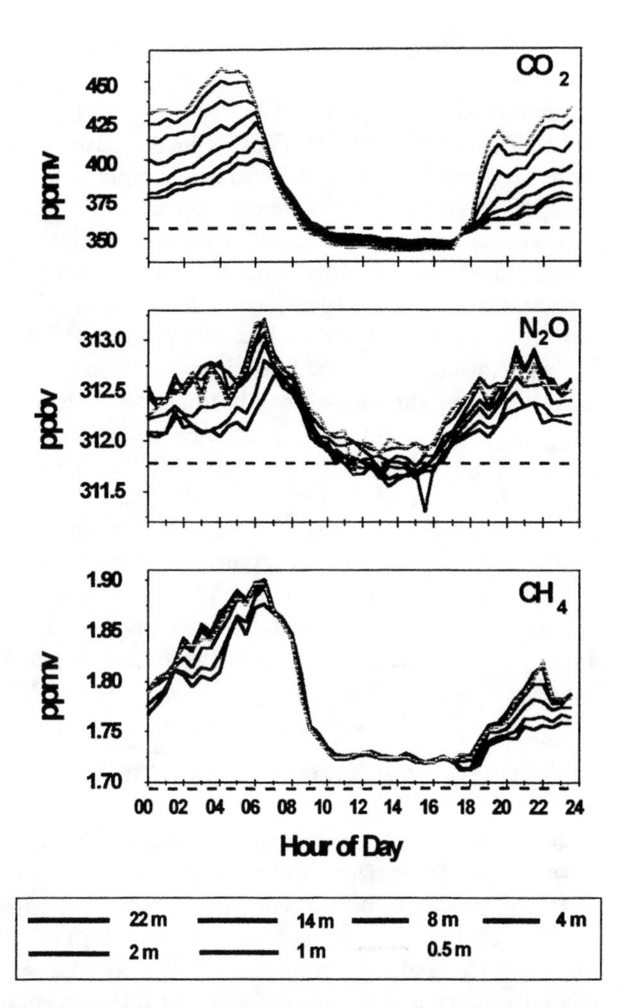

FIGURE 2. Diurnal mixing ratios of CO_2, N_2O and CH_4 at 7 tower levels 0.5 - 22 m averaged over 19 days. The dashed lines represent the clean air baseline level at Cape Grim, Tasmania at the time of the campaign.

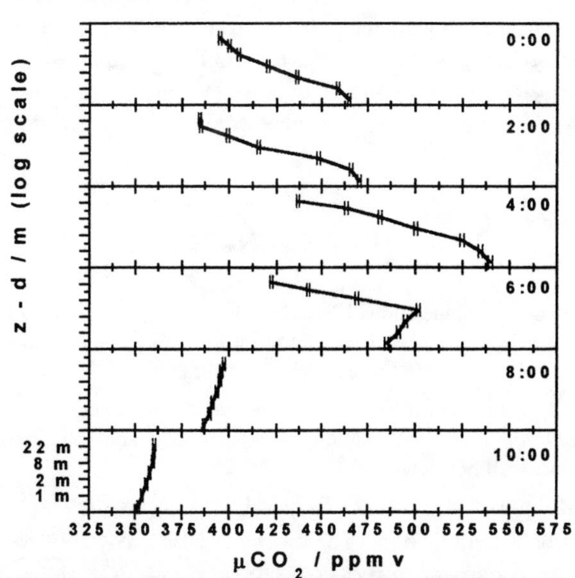

FIGURE 3. Typical vertical profiles of CO_2 from midnight until mid-morning, illustrating the reversal of gradient at dawn.

CHAMBER MEASUREMENTS

Chamber measurements of CO_2, N_2O and CH_4 fluxes were made continuously for 17 days in the pasture and crop environments. Figure 4 illustrates data for a typical 30 minute closure of the pasture chamber showing the linear increase in concentrations after chamber closure. The derived fluxes are calculated from the rate of change of concentration, the area covered and the volume of the chamber. Figure 5 collects all flux data from the 17 days of measurements. There is a clear diurnal cycle in the CO_2 respiration flux, attributable mainly to temperature. The diurnal variation in N_2O flux is less well pronounced, but there is a large increase in N_2O flux following ca. 30mm of rain which fell on 21-22 October. The "pulse" of N_2O following the rain can be seen to decrease to pre-rain values within 2-3 days of the rain event. A lesser amount of rain fell on 13 October, also leading to increased N_2O emissions. CH_4 fluxes were immeasurably small, with CH_4 mixing ratios changing by less than 10ppbv over a typical 20 minute closure.

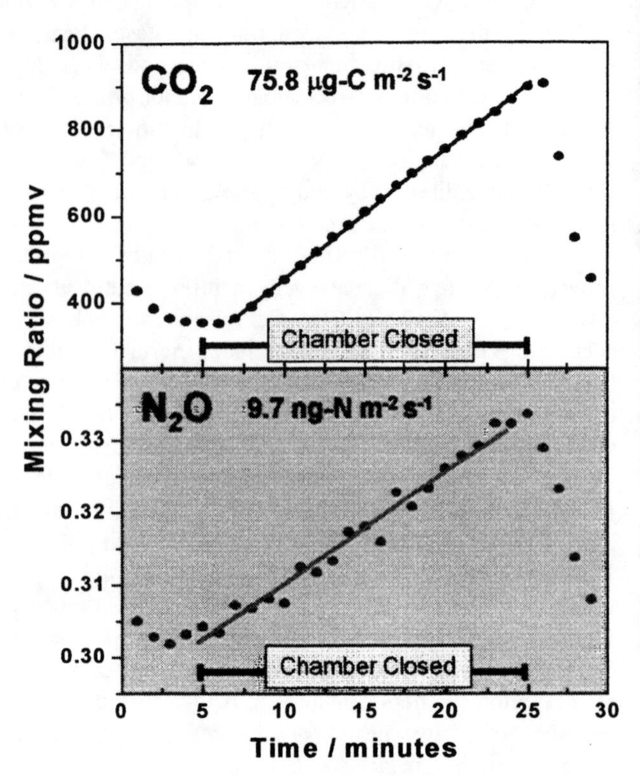

FIGURE 4. Changes in CO_2 and N_2O mixing ratios in the pasture chamber during a typical 20 minute closure

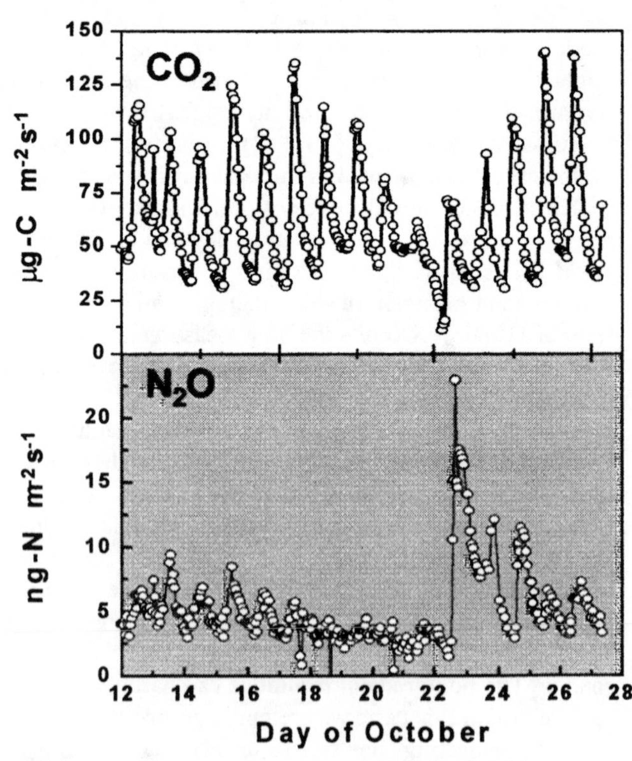

FIGURE 5. Chamber fluxes of CO_2 and N_2O, OASIS 95.

SUMMARY AND CONCLUSIONS

Automated FTIR determination of atmospheric trace gas concentrations for surface-atmosphere exchange measurements has proved to be a robust and reliable technique providing continuous simultaneous measurements of CO_2, CH_4, N_2O, CO and H_2O in air for 19 days without malfunction or data loss. High precision (0.1-0.5%) was obtained from 1 or 2 minute measurements for all species. This precision was sufficient to determine fluxes of CO_2 from both micrometeorological and chamber methods, and fluxes of N_2O from chambers. However vertical gradients were not sufficient for good micrometeorological determination of N_2O or CH_4 in the environment studied. Diurnal variations in N_2O and CH_4 however may be sufficient to estimate their fluxes from nocturnal boundary layer budgets.

REFERENCES

[1]G P Livingstone and G L Hutchinson, in *Biogenic Trace Gases: Measuring emissions from soil and water*, edited by P A Matson and R C Harriss (Blackwell Science, 1995).

[2]D H Lenschow, in *Biogenic Trace Gases: Measuring emissions from soil and water*, edited by P A Matson and R C Harriss (Blackwell Science, 1995).

[3]M.Raupach et al, CSIRO Technical Report no. 68, 1994.

[4]D.W.T. Griffith, *Applied Spectroscopy* **50** (1), 59-70 (1996).

[5]L.S. Rothman, R.R Gamache, R.H. Tipping *et al.*, *Journal of Quantitative Spectroscopy and Radiative Transfer* **48** (5/6), 469-507 (1992).

Measuring Air Pollutants in Presence of High Water Vapour Concentrations

Kai Wülbern

Technische Universität München, Lehrstuhl für Elektrische Meßtechnik, Arcisstraße 21, D-80290 München, Germany

In industrial emission monitoring applications sometimes very high water vapour concentrations can occur. In order to find out which accuracy a relatively simple FTS-based measuring system can achieve under such conditions, we performed NO measurements in presence of up to 60 vol.% water vapour. We used a Bruker IFS 66 with a spectral resolution of 1 cm^{-1} equipped with a pyroelectric DTGS-detector and a gas cell with 0.8 m path length. Concentrations were calculated from the measured spectra using the nonlinear NLS method. We found out that the loss of measuring effect caused by the reduction of path length is partially compensated by the absence of losses normally encountered with White cells. Futhermore, the capability of the NLS method to evaluate spectra with a low signal/noise ration made it possible to obtain sufficient accuracies for most industrial applications. The results make clear, that it is possible to build a relatively simple multicompound emission monitoring system based on an FTS.

INTRODUCTION

The utility of FTIR-spectrometry in extractive emission monitoring applications nowadays is undisputed. With an FTIR-spectrometer several air pollutants can be measured simultaneously. In addition, interfering compounds like water vapour (H_2O) and carbon dioxide (CO_2) can be included into analysis, which gives the opportunity to compensate their influence on the results of other compounds arithmetically. Therefore, no sample gas conditioning is required, i.e. the composition of the sample gas can be analysed unadulterated. Besides that, hazardous air pollutants (HAPs) such as HF and HCl or water soluble compounds such as NH_3 can be measured.

Recent measurements in a coal-fired power plant have demonstrated, that an emission monitoring system based on an FTIR-spectrometer can outperform today's standard emission analysers in terms of detection limit, linearity and cross-sensitivity (1). However, the maximum amount of water vapour encountered in the flue gas of this plant was not exceeding 15 vol.%. Depending on the kind of combustion respectively the flue gas cleaning equipment installed much higher concentrations can occur. If, for example, sewage sludge is burned, water vapour concentration can reach a level of 60 vol.%. The question discussed in this paper is, if in this case an FTIR-spectrometer is still capable to measure air pollutants without gas conditioning.

MEASURING SYSTEM

We first performed measurements with a heatable White cell. The absorption path length of this cell was adjustable from 0.8 to 8 m. Usually, a path length of several meters is used for gas analysis, but we found out, that it is disadvantageous at high water vapour concentrations. With a long path, water absorption bands become so intense, that many spectral regions of interest is opaque. Under those circumstances, analysis of air pollutants like NO, NO_2 and SO_2 turned out to be impossible. We concluded, that the path length must be reduced. As a consequence, a loss of measuring effect has to be accepted. On the other hand, a path length smaller than 1 m can be realised with a single pass cell, which is much easier to handle than the commonly used White cell. Especially in the field of emission monitoring, a White cell causes many problems because it has to be heated to 200 °C, which makes its optical adjustment quite difficult. Beside that, the mirrors inside the cells are in permanent contact with the sample gas, which may contain adsorptive or aggressive substances that pollute or damage the mirror surfaces. Since many reflections take place in a White cell, even a slight reduction of mirror reflectivity causes a considerable loss of signal.

The use of a single pass cell avoids these problems and faciliates the design of a FTIR-spectrometer-based emission monitoring system. But is it still possible to obtain sufficient accuracies for all legally mandated air pollutants with a path length below 1 m? In order to investigate this question we made laboratory measurements with a gas cell of 0.8 m path length. The cell was equipped with KBr-windows, had a volume of about 2 liters and was heated to 130 °C to prevent water vapour condensation. We used a Bruker IFS66 FTS with a Ge/KBr-beamsplitter and a Globar as infrared source. A relatively moderate spectral resolution of 1 cm^{-1} was chosen. This is a common value offered by commercially available FTIR-spectrometers designed for industrial applications. Since it is too costly to use a liquid nitrogen cooled MCT detector in industrial surroundings, we decided to apply a pyroelectric DTGS-

CP430, *Fourier Transform Spectroscopy:* 11th International Conference
edited by J.A. de Haseth

detector, which has only a low sensitivity but the valuable advantage that it can be operated at ambient temperature.

CALCULATION OF CONCENTRATIONS

Spectra were evaluated with the NLS method, a nonlinear extension of the widespread CLS-Method. CLS (Classical Least Squares) is a very efficient multivariate method for the evaluation of absorbance spectra (2). The calibration model represents Beer's law where the spectral absorbance *a* is described as a linear function of the component concentration *c*. Unfortunately, deviations from Beer's law occur if absorbance values are high or spectral resolution is bad. To better account for this effect, we extended the CLS calibration model by a second order term:

$$a = k_1 c + k_2 c^2$$

We called this method NLS (Nonlinear Least Squares). It has all advantages of CLS while yielding substantially better results in case of deviations from Beer's law. The algorithm is quite similar to CLS. In a calibration step the coefficients of the model function are least squares estimated from a set of calibration spectra of known composition. The calibration spectra are measured in the laboratory under similar conditions as the ones to be expected in field measurements. They must contain all components which may be present in the sample gas. In the prediction step of the analysis, the coefficients determined in the calibration are used to estimate component concentrations in unknown sample gas spectra.

From a practical point of view, the difference between CLS and NLS is, that in the prediction step a nonlinear least squares fit has to be performed. Therefore, NLS is significantly slower than CLS. However, experience shows, that calculation time is well below the time needed to measure a spectrum. So this disadvantage does not matter in this case. But there are some important advantages compared to other known methods. Due to the physical model, NLS offers the user a lot of qualitative information which helps him to solve even complex evaluation problems. Furthermore, a very efficient method for baseline correction can be integrated into the algorithm, a very important feature for online measurements. NLS has turned out to be a very stable, rugged and reliable evaluation method. We applied the method successfully for a couple of emission measurements in power plants, at car engines and helicopter turbines. A thorough description and investigation of the properties of NLS is given in (1).

EXPERIMENTAL

Figure 1 shows spectra of NO, NO_2, SO_2 and H_2O measured with a path length of 0.8 m. Even at a concentration

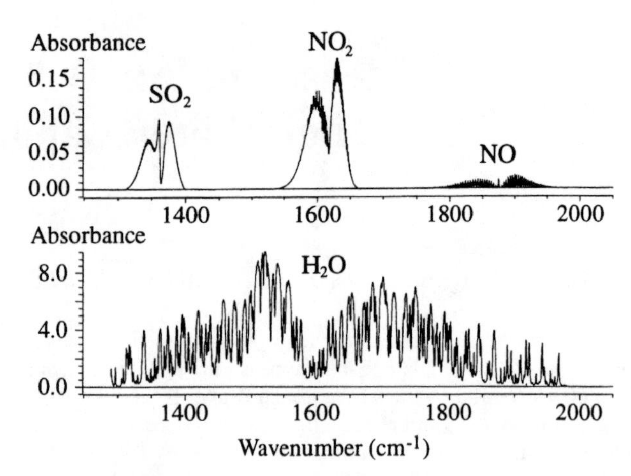

FIGURE 1. Spectra of air pollutants (100 ppm each) and water vapour (60 vol.%) at 0.8 m path length.

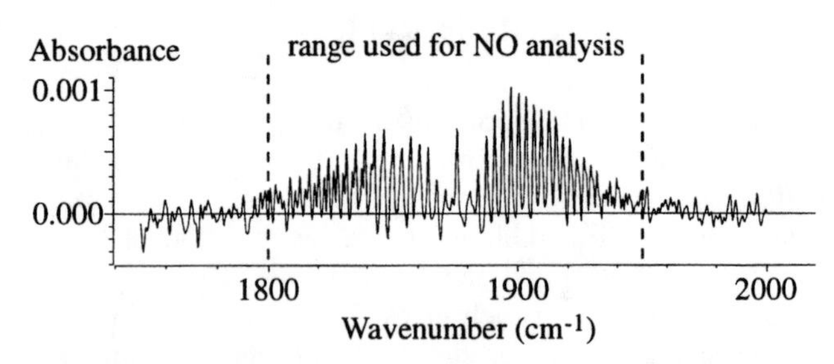

FIGURE 2. NO reference spectrum (5 ppm, measuring time 3 minutes).

of 60 vol.%, the maximum water vapour absorbance does not exceed 4.0 in the spectral regions of interest. This corresponds to a minimum transmission of about 2 %. For most data points, transmission is better than 15 %, which means that quantitative analysis should be possible. Of course, there is still a strong spectral interference between pollutants and water vapour absorption bands. A comparison of the spectra in Figure 1 shows, that the ratio of the intensities of water vapour and pollutant absorption lines is somewhere between 40:1 and 100:1. The situation for the analysis of NO is particulary difficult, because this gas has a very low specific absorbance. Therefore, it is recommendable, to investigate if NO can be analysed under the described conditions. If yes, other pollutants of interest can probably be measured, too.

For the investigation, an NO concentration range from 0 to 100 ppm was chosen. This is an appropriate value for emission monitoring systems installed in plants with modern flue gas cleaning equipment. The maximum H_2O concentration was 60 Vol.%, corresponding to a dew point temperature of about 85 °C. Several sets of test spectra were measured with the measuring system described above. Zerofilling was used for spectrum interpolation, a Happ-Genzel window for apodization.

For calibration, 23 pure-component spectra of were recorded, 11 NO spectra (5, 10, 20, ... 100 ppm) and 12 H_2O spectra (5, 10, ... 60 vol.%). We deliberately used pure component spectra, because pure components are much easier to handle than mixtures, especially if high water vapour concentrations are involved. To ensure a sufficient signal/noise-ratio, 60 scans were coadded resulting in a measuring time of about 3 minutes per spectrum. Figure 2 shows an NO reference spectrum at a concentration of 5 ppm. The noise amplitude under the conditions described above is distinctly smaller than most NO absorption lines.

For analysis, a spectral range from 1800 to 1950 cm^{-1} was chosen, containing 623 data points (see figure 2). Spectral values exceeding an absorbance value of 4.0 at maximum H_2O-concentration were excluded from analysis in advance. Then a iterative procedure was applied to determine the data points best suited for analysis. After having performed the calibration, a set of test spectra recorded at the same conditions as the reference spectra and covering the whole concentration range is analysed. Since the component concentration of these spectra is known, prediction errors can be calculated. To determine, which data points are to be included in the analysis, the spectral residuum calculated in the prediction step is analysed. The spectral residuum contains the deviations between model function and measurement for all data points and should ideally contain nothing but noise,. If for a particular datapoint, the relation between absorbance and concentration is not correctly described by the chosen model function, a high residuum value well beyond noise level will be observed. Such data points are excluded from analysis. Afterwards, a new calibration is performed with a reduced number of data points, prediction errors are calculated and the residuum for each spectrum is analysed again. This procedure is repeated until the prediction error reaches a minimum. For our test set, 177 from 623 data points remained for analysis.

RESULTS AND DISCUSSION

The first item investigated was the linearity of the measuring system. For this purpose, a set of 20 NO spectra (pure component with linearly increasing concentrations from 5 to 100 ppm) recorded under reference conditions (i.e. 3 minutes measuring time) was evaluated. Figure 3 shows the predicted versus the true concentration values. The linearity error is very small over the whole concentration range. This good result is clearly a consequence of the careful choice of data points.

In order to determine the detection limit, a measurement with a constant concentration of 10 ppm NO (without water vapour) was made. 10 scans were coadded per spectra, resulting in a measuring rate of 2 spectra per minute. This value is the minimum requirement demanded by German

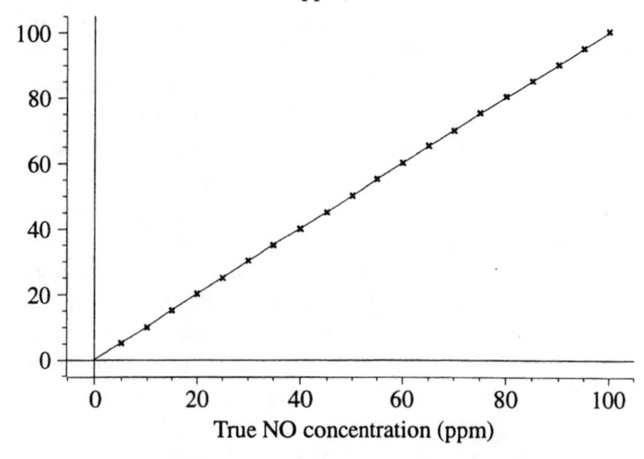

FIGURE 3. Linearity of the measuring system.

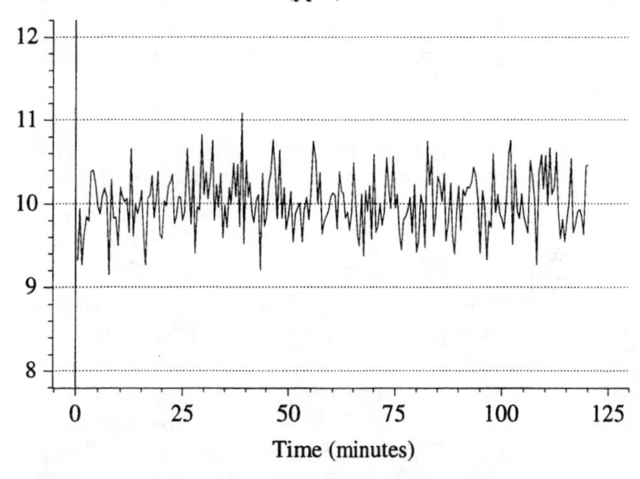

FIGURE 4. Determining the detection limit of the measuring system. The true NO concentration is 10 ppm. Measuring time is 30 seconds per spectrum (10 scans).

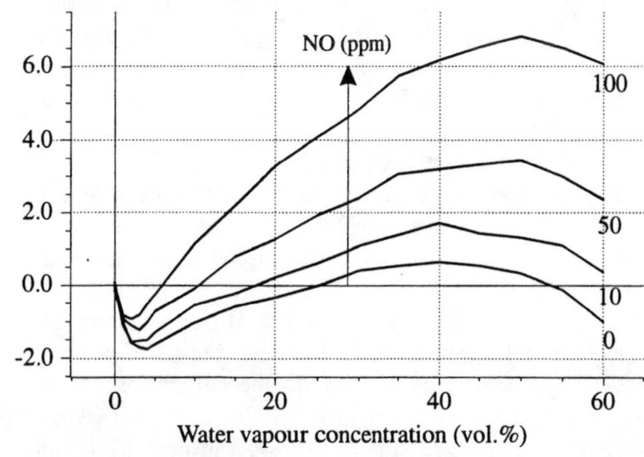

FIGURE 5. Cross sensitivity to water vapour for different NO concentrations. Since calibration was performed with pure component spectra, the prediction error is lowest, if pure H_2O-spectra are analysed (0 ppm NO).

regulations for on-line emission monitoring systems. Total measuring time was about 2 hours. The result can be seen in figure 4. A slight drift can be observed, but the dominant error is of statistical nature. Since only one background measurement was made to calculate the absorbance spectra, this is a clear evidence, that the integrated baseline correction works very well. The estimated concentration values fluctuate between 9.14 and 11.08 ppm. Standard deviation is 0.352 ppm. The 68 % confidence intervall calculated by the NLS-method is 0.361 ppm. Detection limit per definition is equivalent to the peak concentration noise amplitude, which is about 1 ppm in this case. This is a very good result, considering the small path length of only 0.8 m. It demonstrates, that multivariate evaluation methods permit quantitative analysis of spectra even if the signal/noise ratio is very low. This property has already been described in (1) and (3). It is a consequence of the fact, that many data points can simultaneously be included into the analysis, corresponding to averaging their results. Therefore, it is no longer necessary to use a path length of several meters to ensure that absorption bands of the compounds to be analysed are exceeding noise level.

The last but most important item to be investigated, is cross sensitivity to water vapour. If an analyser designed for the detection of a certain compound exhibits a response to other compounds which are present in the sample, this is called cross sensitivity. An FTIR-measuring system uses a complex evaluation procedure to calculate concentrations of all compounds present in the sample. Therefore, theoretically no cross sensitivity should occur. In practice however, no model can describe precisely the complicated relation between absorbance and component concentrations. So, even with an FTIR-spectrometer, a certain cross sensitivity can be observed. But if the reference spectra used for calibration are measured under similar conditions as expected in the field, cross sensitivity is small compared to standard analysers (1). For our investigation, we used pure components to faciliate the measurement of the reference spectra. Since we want to analyse mixture spectra, conditions for calibration and prediction are different and we have to expect an appreciable cross sensitivity of the NO prediction results to water vapour.

To determine this prediction error we analysed 4 sets of NO/H_2O-mixture spectra recorded in the laboratory. Every set contained 17 spectra, where NO concentration was constant (0, 10, 50 and 100 ppm) and H_2O-concentration increased from 0 to 60 vol.%. The results are presented in figure 5. Prediction errors are smallest, if no NO is present in the sample (0 ppm NO). In this case, pure H_2O-spectra are analysed, which corresponds to the calibration conditions. Nevertheless, prediction error fluctuates between -1.5 and +0.5 ppm, which is far from being negligible, but sufficient for most applications. If mixture spectra are evaluated, prediction error increases to a maximum value of +6.8.ppm (at 100 ppm NO, 50 Vol.% H_2O). This may be caused by molecular interactions taking place at higher water vapour concentrations. These interactions probably change the shape of the NO absorption band, which cannot be accounted for by the model used. There are two ways, in which prediction error due to cross sensitivity can be reduced to an acceptable level. Former investigations have shown, that the use of mixture spectra for calibration yields significantly better results (1). On the other hand, it is much more convenient to use pure component spectra, determine cross sensitivity for the range of interest and subsequently correct the predicted NO concentrations.

CONCLUSIONS

Our NO measurements at high water vapour concentration were carried out with a relatively simple measuring system, using an FTS with a moderate spectral resolution, a low sensitivity DTGS-detector and an absorption pathlength of 0.8 m. For calibration only pure component spectra were used. Due to the capability of multivariate evaluation methods to calculate component concentrations even from spectra with low signal/noise ratio, very good results were obtained. Linearity in the NO measuring range from 0 to 100 ppm was very good. For the detection limit a value of 1 ppm was determined, which is sufficient for most applications in industrial emission monitoring. NO prediction error due to water vapour cross sensitivity can reach values up to 7 ppm. Since H_2O concentration is calculated too, this error can subsequently be corrected for. The results show, that it is possible to build a simple FTS-based multicompound emission monitoring system using a single pass gas cell. However, further investigations have to be made to find out, if similar results can be obtained for NO_2 and SO_2, which are susceptible to water vapour cross sensitivity, too.

REFERENCES

1. K. Wülbern: *Prozeßgekoppelte Messung von Rauchgasen mit einem FTIR-Spektrometer*, Düsseldorf, VDI-Verlag, Fortschrittsbericht, Reihe 15, Nr. 154, 1996.
2. Haaland, D.M., Easterling, R.G., Vopicka, D.A., *Applied Spectroscopy* **39**, 73-83 (1985).
3. Haaland, D.M., Easterling R.G., *Applied Spectroscopy* **34**, 539 (1980).

Iterative Quantitative Evaluation of Atmospheric FTIR-Spectra Obtained in Open Path Monitoring

U. Müller and H.M. Heise

Institut für Spektrochemie und Angewandte Spektroskopie, Postfach 101352, D-44013 Dortmund, Germany

Open path monitoring using mobile FTIR-spectrometers is an efficient technique for air pollution analysis. Emphasis is laid on chemometric aspects for improved quantitative evaluation of mid-infrared atmospheric spectra. Multivariate techniques offer several advantages over univariate signal processing due to lower detection limits and less liability to interferences. Effects from atmospheric pressure and temperature variations, spectral wavenumber shifts and different resolutions are quantified using classical least squares regression. Iterative spectral exhaustion is considered relying on a set of reference spectra covering a range of concentrations to cope with non-linear absorbance dependencies. The spectral residuals due to model deficiencies are dependent on resolution, apodization and atmospheric measurement conditions and must be compared with possible residuals of unknown components not included in the modelling steps.

INTRODUCTION

The application of mobile spectrometers for in-situ determination of gaseous emissions has been the subject of many investigations in the past. While robust Fourier-transform infrared (FTIR) spectrometers for environmental monitoring are nowadays available (1), further development of the chemometric aspects leading to improved quantitative evaluation of atmospheric spectra is required. We have studied several aspects especially important in the multivariate analysis of long-path atmospheric spectra by using classical least-squares (CLS) methods. Multivariate techniques offer several advantages over univariate signal processing due to lower detection limits and less liability to interferences.

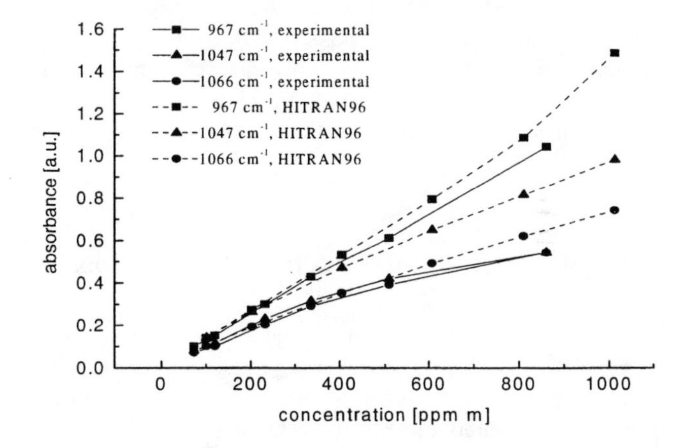

FIGURE 1. Univariate calibration using line maxima for NH_3 (boxcar apodization, 1 cm^{-1} nominal resolution).

NECESSITIES FOR AN EXPERT SYSTEM

Resolution and apodization

For pollutant measurements the mid-infrared spectral range is exploited. Here spectral features arising from water and carbon dioxide dominate which blank out wide spectral intervals due to strong absorption (2). To make use of spectral information under such conditions, a distinct advantage is offered by FTIR-systems capable of high spectral resolution (better than 0.2 cm^{-1}) for target compounds, showing a resolvable rotational fine structure.

Under such conditions deviations from linear calibration modelling are much smaller and the number of relevant spectral datapoints is increased compared to a resolution not adjusted to the true line halfwidth of atmospheric trace components (see also Figure 1 and 2).

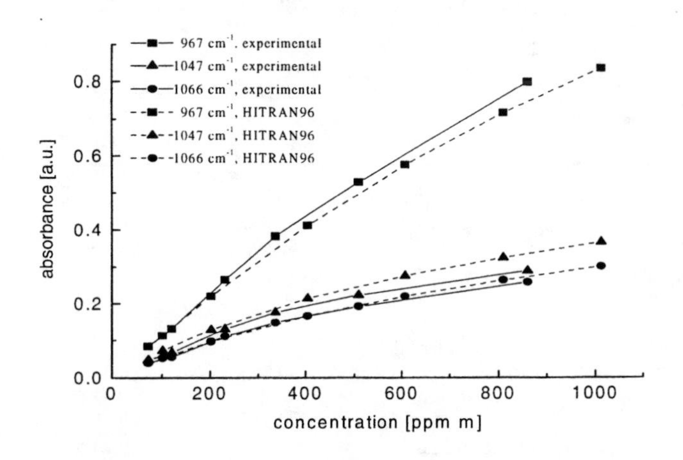

FIGURE 2. Univariate calibration using line maxima for NH_3 (triangular apodization, 1 cm^{-1} nominal resolution).

CP430, *Fourier Transform Spectroscopy:* 11th International Conference
edited by J.A. de Haseth

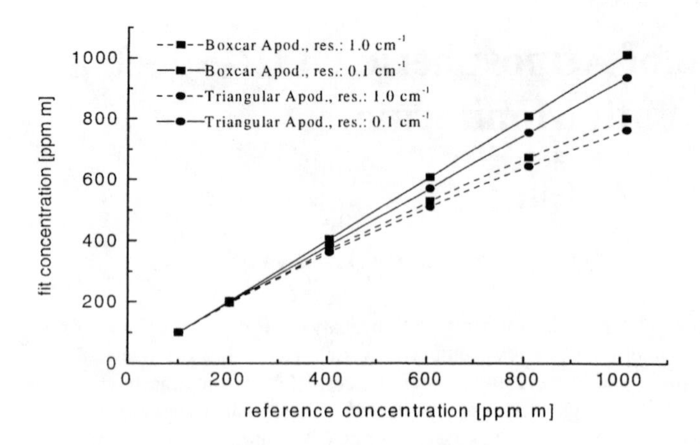

FIGURE 3. Multivariate calibration (500-2000 cm^{-1}) for NH_3 (absorbance limit: 1.0 a.u.). Results from HITRAN96 simulations.

Selection of optimum interferogram apodization can provide appropriate linear absorbance intervals dependent on concentration and pathlength (see, for example (3)). As a consequence, lowest spectral residuals will result when using multivariate calibration based on spectral fitting by CLS.

Nonlinearities in calibration curves can be significantly reduced by multivariate spectrum evaluation, especially when maximum line absorbance limits are considered (see Figure 3).

Deviations from standard reference conditions

The expert system has to account for changes in environmental variables such as temperature, pressure and humidity. Effects of atmospheric pressure and temperature differences compared to standard reference conditions are listed in Table I. Synthetic spectra were generated using the

TABLE I. Effect of differences in atmospheric pressure and temperature compared to standard reference conditions (1013 hPa, 20°C).

compound	resolution	$\Delta c_{rel}/\Delta P*10^3$ [hPa]	$\Delta c_{rel}/\Delta T*10^3$ [°C]
CH_3Cl	true linewidth	-0.77	-0.51
(652 ppm m)	0.2 cm^{-1}	-0.87	-0.40
	1.0 cm^{-1}	-0.99	-0.29
C_2H_6	true linewidth	-0.53	-1.13
(260 ppm m)	0.2 cm^{-1}	-0.70	-0.76
	1.0 cm^{-1}	-0.91	-0.29
CH_4	true linewidth	-0.75	-0.55
(128 ppm m)	0.2 cm^{-1}	-0.86	-0.38
	1.0 cm^{-1}	-0.95	-0.32
CO	true linewidth	-0.56	-0.41
(225 ppm m)	0.2 cm^{-1}	-0.78	0.40
	1.0 cm^{-1}	-0.96	0.92

Simulations were done with HITRAN96 database considering the pressure range of 960-1030 hPa and temperatures between 0-40 °C.

HITRAN96 spectral database. The successful implementation of such calibration spectra created from absorption line parameters was described recently (4,5).

Mismatch between open path and reference spectra due to wavenumber inaccuracy has to be corrected when broad spectral intervals are chosen for quantification.

TABLE II. Wavenumber accuracy for different Kayser-Threde K300- spectrometers of high resolution (0.2 cm^{-1}).

compound	ν[cm^{-1}] HITRAN96	Spectr. 1	$\Delta\nu/\nu \cdot 10^5$ Spectr. 2	Spectr. 3
H_2O	576.112	3.0	5.2	4.7
H_2O	1596.236	4.6	5.6	2.3
H_2O	1601.208	4.8	5.1	1.4
CO_2	2382.466	4.9	4.8	1.5
CO_2	2383.331	4.1	4.2	1.4
H_2O	3969.082	4.8	5.2	2.2

The line positions given in Table II were determined by using the centre of gravity algorithm applied to unperturbed symmetric lines (6). The spectral shifts considered in Table III were produced by interpolation using a cubic polynomial based on three datapoints each. Theoretically the shifts are proportional to wavenumber. By the developed CLS-software distinct limited spectral intervals are evaluated, so that a constant wavenumber shift correction can be applied (tolerances up to 0.003 cm^{-1}).

TABLE III. Relative concentration prediction error in % due to wavenumber shifts using CLS (spectral data simulated with nominal resolution of 0.2 cm^{-1}, triangular apodization).

compound	range [cm^{-1}]	conc. [ppm m]	0.003	0.01	$\Delta\nu$ [cm^{-1}] 0.05	0.09	0.18
Hexane	2900-2965	120	-0.02	-0.08	-0.35	0.60	-1.18
Toluene	960-1080	180	0.03	0.10	0.88	2.17	6.73
Ammonia	840-960	85	0.01	0.06	1.31	3.71	12.34
Methane	2900-2965	180	0.01	0.07	1.65	4.61	15.13

Calibration modelling

Instrumental effects from insufficient spectral resolution leading to significant deviations from Beer's law have to be appropriately modelled. The spectral residuals due to model deficiencies are dependent on resolution, apodization and atmospheric measurement conditions and must be rated against possible residuals of unknown components not included in the first modelling step. Different compensation methods are possible for removing main spectral features from water and carbon dioxide to enlarge the spectral range for monitoring, e.g., using orthogonalization against atmospheric factor spectra. In particular, this is necessary for our proposed generally applicable background procedure by recording a short-path single beam background spectrum. Iterative spectral exhaustion, including quadratic baseline components, is considered here relying on a set of reference spectra covering a wide range of concentrations. The closer the concentrations are chosen for the reference spectra compared to those of the atmospheric compound mixture, the smaller residues from linear model deficiencies can be achieved.

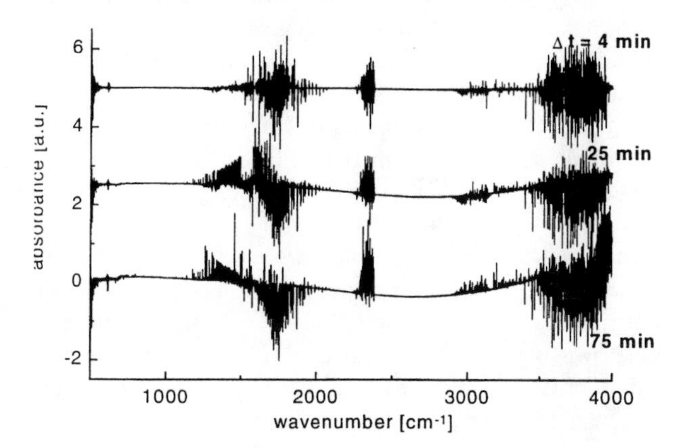

FIGURE 5. Baseline changes due to spectrometer instability. Spectra were measured with the Kayser-Threde K300-spectrometer.

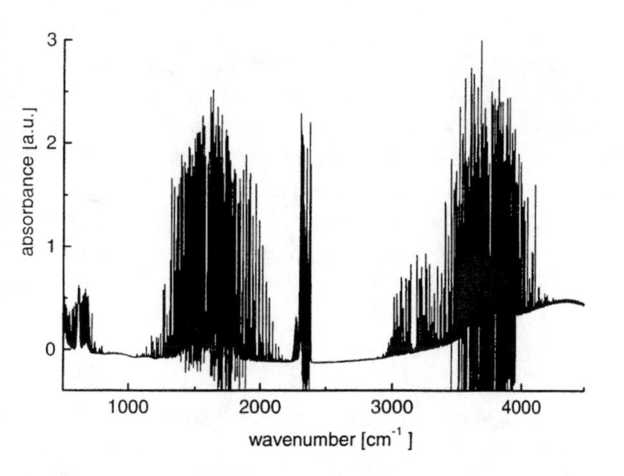

FIGURE 6. Atmospheric spectrum using short optical path background. Spectral resolution 0.2 cm^{-1}.

The generation of synthetic spectra, based on spectral parameter data, bears more flexibility, but is restricted to the availability of those data.

For reliable multivariate evaluation often a first qualitative analysis of atmospheric components from the measured spectrum is necessary for the selection of appropriate calibration reference spectra. Wavenumber selection is important in this context to optimize selectivity for special compounds and to eliminate systematic errors in concentration prediction due to modelling deficiencies arising from an insufficient calibration database.

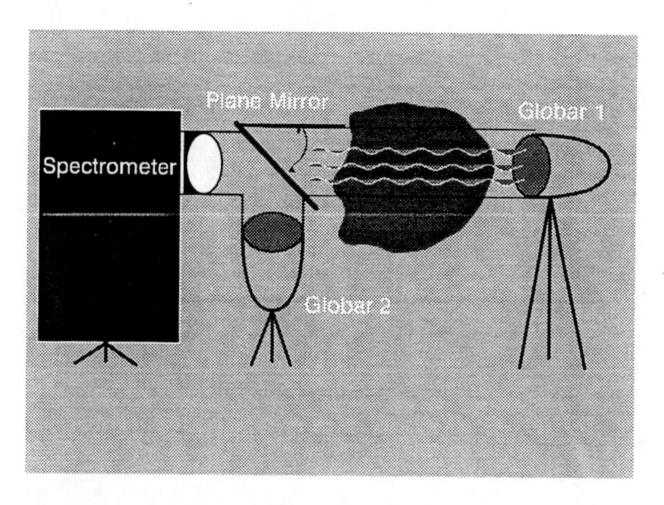

FIGURE 7. Experimental scheme for measuring atmospheric trace compounds from a diffuse emission source.

TABLE IV. Iterative concentration prediction using CLS with concentration adapted water reference spectra (spectral range 500-2000 cm^{-1}, absorbance limit 1.0, triangular apodization).

compound		concentration true [ppm m]	reference [ppm m]	nom. resol. 0.1 cm^{-1} Fit result * [%]	Iter. 1 [%]	nom. resol. 1.0 cm^{-1} Fit result * [%]	Iter. 2 [%]
Methane	Mix 1	600	101	0.5 (-1.4)	-1.2	11.7 (-6.0)	-3.4
	Mix 2	400		-0.7 (-0.8)	-0.7	11.5 (-3.5)	-0.9
Water	Mix 1	500 000	50 000	-27.5 (-28.4)	-3.4	-232 (-234)	-13.1
	Mix 2	300 000		-18.5 (-18.7)	-2.2	-146 (-145)	-8.1
Ammonia	Mix 1	800	101	-6.9 (-7.2)	-7.0	-22 (-25.7)	-3.9^{+}
	Mix 2	400		-4.5 (-4.9)	-4.8	-7.8 (-11.7)	-1.0^{+}
Sulf. diox.	Mix 1	800	101	0.5 (-0.5)	-0.2	5.4 (-2.7)	-0.6
	Mix 2	400		0.8 (-0.2)	0.1	7.1 (-1.1)	0.0
Spectral standard deviation	Mix 1			$3.7 \cdot 10^{-2}$	$6.8 \cdot 10^{-3}$	$6.9 \cdot 10^{-2}$	$9.1 \cdot 10^{-3}$
	Mix 2			$2.5 \cdot 10^{-2}$	$3.8 \cdot 10^{-3}$	$4.8 \cdot 10^{-2}$	$5.5 \cdot 10^{-3}$

* in brackets fitted results for regressing single component spectrum with given mixture concentration

$^{+}$ for these cases, also the NH_3 reference spectrum had been adapted

Recognition of unknown compounds by spectral cross correlation searching is presented (7). Compound specific intervals with unique spectral signatures with a minimum of possible interferences have been selected for cross correlation with atmospheric absorbance spectra. Principally, this procedure with normalized intensity spectra is insensitive to wavenumber shifts. In the cross correlation function the absolute maximum must occur around the zero position. The

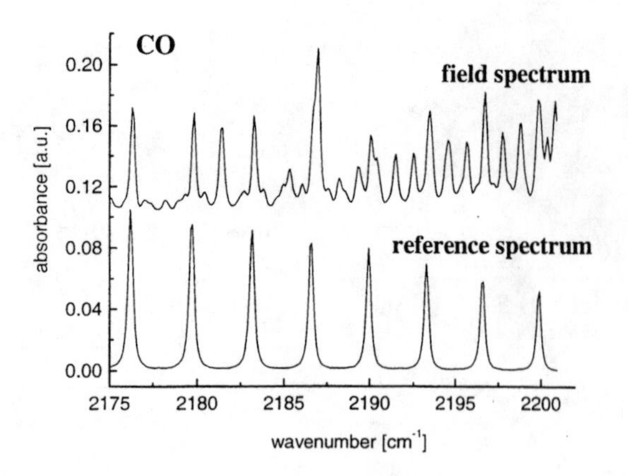

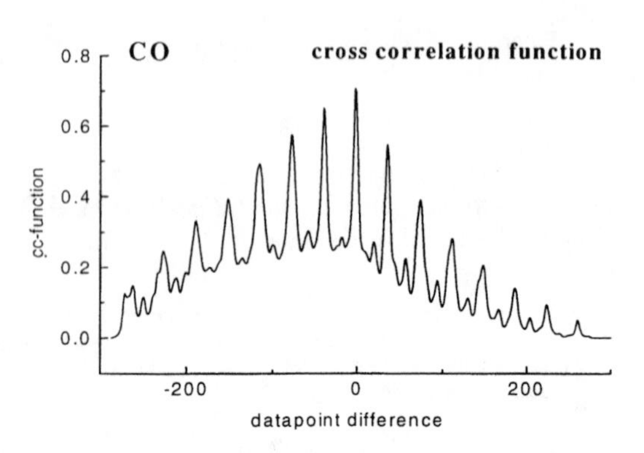

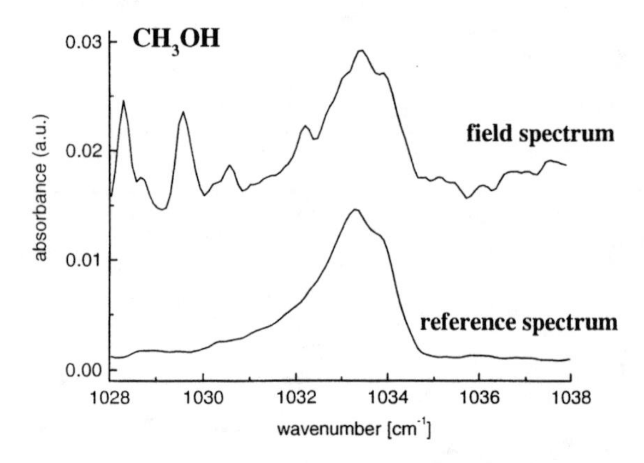

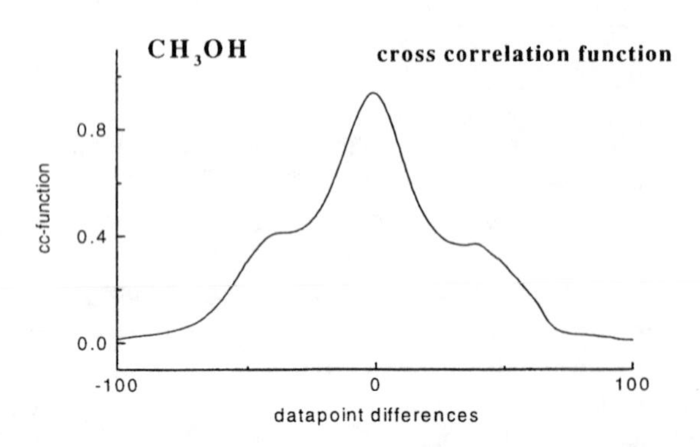

FIGURE 7. Normalized field- and reference spectra of CO_2 and CH_3OH and their cross correlation functions.

maximum value is tested for existence of the component in the mixture (for the ideal case it equals 1).

CONCLUSION

CLS has been successfully used in many applications, e.g. for chemical analysis of condensed phases. For evaluation of atmospheric IR-spectra a much greater complexity must be considered due to the rotational fine structure and strong absorptions occuring under such conditions. In addition, unknown and non-modelled components demand new strategies. In the absence of photometric errors and mismatch of wavenumber scales, iterative CLS-regression is straightforward and relies only on a set of reference spectra. This strategy copes with inevitable deviations from Beer's law which usually occur for spectral features with resolvable rotational fine structure under limited spectral resolution. Further significant spectral residuals are due to other model deficiencies, which can be caused by unknown components not included in the modelling steps. Identification of unknown compounds by spectral library searching (e.g., using cross correlation) can be based on the spectral residuals obtained from iterative least squares fitting.

ACKNOWLEDGEMENTS

We thank Kayser-Threde, München, the Landesumweltamt NRW, Essen and the Fachhochschule Düsseldorf, Labor für Umweltmeßtechnik, for fine cooperation. Financial support by the Bundesministerium für Bildung, Wissenschaft, Forschung und Technologie and the Ministerium für Wissenschaft und Forschung des Landes Nordrhein-Westfalen is gratefully acknowledged.

REFERENCES

1. Russwurm, G.M., Compendium of methods for the determination of toxic organic compounds in ambient air, "Compendium Method TO-16", U.S. Environmental Protection Agency, Cincinnati, OH 45268, 1997.
2. Hanst, P.L. , and Hanst, S.T., Gas measurement in the fundamental infrared region, in "Air monitoring by spectroscopic techniques", Sigrist M.W., ed.., New York: Wiley, 1994, pp. 335-470.
3. Zhu, C., Griffith, P.R., Proc. SPIE **2089**, 434-435 (1994).
4. Griffith, D.W.T., Appl. Spectrosc. **50**, 59-70 (1996).
5. Schäfer, K., Haus, R., Heland, J., Haak, A., Ber. Bunsenges. Phys. Chem. **99**, 405-411 (1995).
6. Cameron, D.G., Kauppinen, J.K., Appl. Spectrosc. **36**, 245-250 (1982).
7. Ehrentreich, F., Fresenius J. Anal. Chem. **359**, 56-60 (1997).

The $^{13}C/^{12}C$ Isotopic Ratio Determined by FTIR Spectroscopy

S. Söderholm[1,2], N. Meinander[3] and J. Kauppinen[1]

[1] Department of Applied Physics, Vesilinnantie 5, FIN-20014 University of Turku, Finland, email erik.soderholm@helsinki.fi
[2] Department of Food Technology, P.O.Box 27, FIN-00014 University of Helsinki, Finland, email erik.soderholm@helsinki.fi
[3] Department of Physics, P.O.Box 9, FIN-00014 University of Helsinki, Finland, email meinander@phcu.helsinki.fi

The isotopic ratio of carbon in atmospheric carbondioxide has been measured with the portable low-resolution GASMET™ FTIR spectrometer. The measured spectra have been analyzed with the Multicomponent analysis software provided with the instrument. The aim of the work has been to investigate how good a precision can be obtained. We have achieved a statistical uncertainty of 0.07% in the isotopic ratio. There is potential to further increase the precision of the method.

INTRODUCTION

When carbon moves between its large reservoirs (atmosphere, sea, biomass, fossil carbon) there is a small isotopic fractionation of the carbon. For example, the ^{13}C fraction in atmospheric CO_2 is around 0.8% lower than in CO_2 dissolved in the sea. The ^{13}C fraction in the biomass and the fossil deposits of C is around 2.5% lower. Thus long term changes in the isotopic composition of atmospheric CO_2 reflect the changes occurring in the distribution of carbon between its reservoirs. Short term fluctuations reveal the magnitude and time-evolution of local transport processes, such as the growth of the biomass.

The isotopic ratio is traditionally determined using mass-spectrometry, which achieves the precision of 2 parts in 100,000 necessary for transport studies. However, a lower precision of 1 part in 1,000 can be acceptable for many types of investigations. The isotopic ratio of a carbon sample is usually reported relative to a standard value, the isotopic ratio in the carbonate Pee Dee Belemnite and given in $\delta^{13}C$ units, where

$$\delta^{13}C = \left[\frac{\left([^{13}C] / [^{12}C] \right)}{\left([^{13}C] / [^{12}C] \right)_{PDB}} - 1 \right] \times 1000\,‰ \qquad (1)$$

All measurements of the isotopic ratio are done against a standard whose $\delta^{13}C$ value is known. Thus, if a set of standards covering the range of isotopic composition of interest is measured with the instrument, the precision obtainable with the instrument is determined by the reproducibility and statistical uncertainty of the measurements.

The instrument used in the present investigation is the GASMET FTIR (1) gas analyzer designed for high speed on site gas analyses in an industrial environment. The instrument is portable and has been designed for maximum ruggedness and insensitivity to changes in the ambient conditions. Low resolution and large apertures are used to maximize the signal to noise ratio (2) which determines the statistical uncertainty of the result.

EXPERIMENTAL

The complete gas analyzing system is shown in Fig. 1.

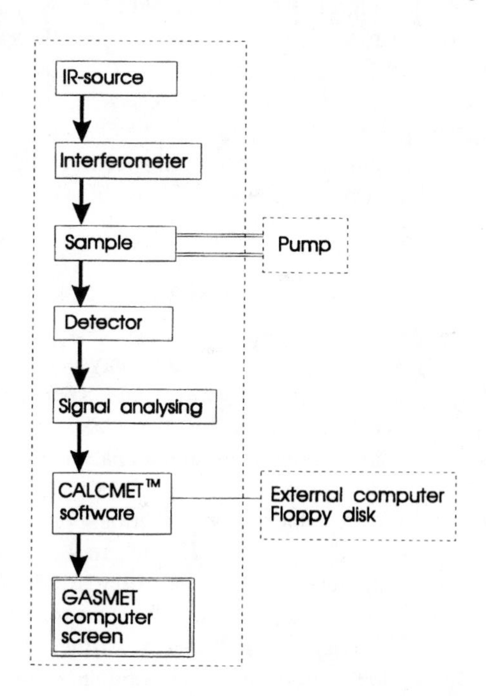

FIGURE 1. GASMET analyzing system.

CP430, *Fourier Transform Spectroscopy:* 11th International Conference
edited by J.A. de Haseth

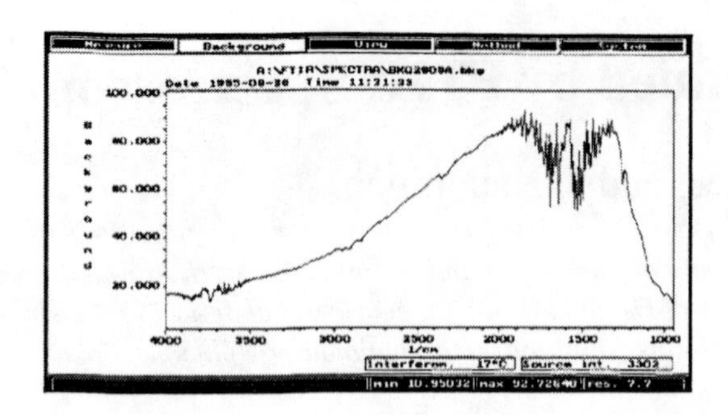

FIGURE 2. Background spectra measured with the GASMET analyzer.

The heat source is a cheramical SiC-crystal with a temperature of 1550 K. A typical background spectrum is shown in Fig. 2.

The interferometer is especially designed for the instrument and provides spectra recorded over the range of 850-4000 cm^{-1} with a resolution of 8 cm^{-1}. The aperture of the interferometer is approx. 2.5 cm and the maximum scanning rate is limited to 12 s^{-1}.

The sample compartment is a standard Vilk's cell with a volume of approx. 4400 l and an optical path length of 9 m.

For the purpose of handling the reference gases and controling the concentration in the sample compartment we constructed a gas cell with in-/outlets for sample gas and filling gas.

Using this cell we were able to calculate the partial pressure of the reference gas. We could also easily fill the sample compartment with a mixture of sample gas and neutral gas, usually N_2.

The reference gases, $^{12}CO_2$ and $^{13}CO_2$, were produced by oxidizing graphite with cupperoxide, CuO in sealed glass tube. Using our gas cell we measured spectra of $^{12}CO_2$ and $^{13}CO_2$ at different partial pressure.

THEORY

The theory for the multicomponent analysis (MCA) has been described in detail by Saarinen and Kauppinen (2). Their paper gives the basis for the choice of a low resolution approach to the quantitative analysis of gaseous samples.

MCA solves a linear optimization problem by minimizing the residual vector **s-Kx**, where **s** is a column vector containing the sample spectrum (dimension = number of data points), **K** is a rectangular matrix containing the library spectra, and **x** is the coefficient vector defining the composition of the sample.

The uncertainties in the coefficients obtained in the optimization depend on the S/N-ratio of the sample spectrum, the number of data points and the degree of overlapping of the library spectra. The noise in the library spectra is assumed to be small and not to affect the result of the analysis.

The effect on the uncertainties in the coefficient vector of decreasing in the signal to noise level by lowering the resolution is cancelled by the smaller number of data points in the spectra. The main advantage from lowering the resolution arises from the possibility to increase the aperture without further degrading the spectra, thus increasing the light throughput in the spectrometer, and from the shorter time it takes to make a single scan. These two advantages combined make the uncertainties in the coefficient vector approximately inversely proportional to the resolution.

The disadvantage of decreasing the resolution is that the library spectra become more linearly dependent, thus increasing the uncertainties in the coefficient vector. The effect of this is more than compensated for by the advantages, however. As a rule of thumb the number of datapoints should be around three times the number of library spectra needed in the analysis.

RESULTS AND DISCUSSION

The libray spectra were measured for 10-60 minutes and corrected for baseline errors. The final S/N was approx 8000. To further improve the analysis we have also calculated theoretical reference spectra using the HITRAN 1996 database (3). Measured $^{12}CO_2$ and $^{13}CO_2$ reference spectra are shown in Fig. 3a-b. Only the strong band in the 2300 cm^{-1} region is used in the analysis.

The airspectra we used for testing our methods were recorded for 15 minutes with a 25/75 air/N_2 mixture. This was necessary in order to obtain the same

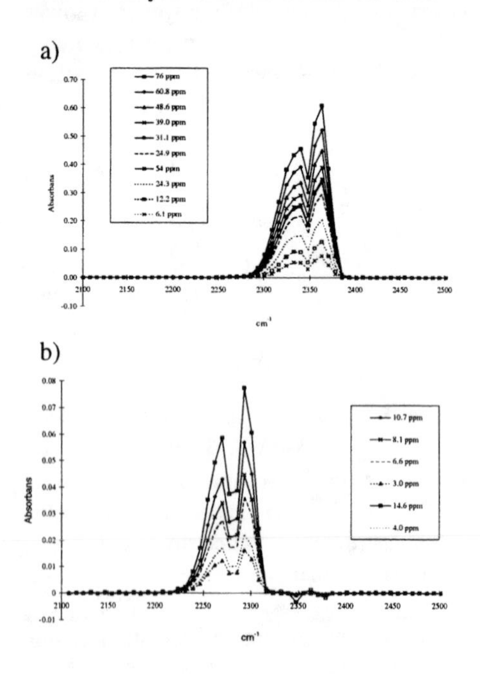

FIGURE 3a-b. Measured, $^{12}CO_2$ spectra (a) and $^{13}CO_2$ (b) spectra, at different partial pressures.

concentration of CO_2 as in the reference samples.

The factors affecting the analysis of the air spectra using the MCA software are the S/N-ratio, the shifting of the background level, the number of degrees of freedom in the analysis, the linear dependence of the $^{12}CO_2$ and $^{13}CO_2$ spectra, and non-linearity in the absorbance spectra.

The noise level in the measured spectra can be reduced by longer measurement times and hardware improvements.

We have investigated how the uncertainty in the coefficient vector are affected by noise in the reference spectra and in the air spectra. This was achieved by adding different amounts of white noise to theoretically calculated spectra. The S/N in the reference spectra will affect the absolute value of the isotopic ratio, but does not affect the reproducibility of the $\delta^{13}C$ ratio. The noise in the air spectra, however, is very critical for the final accuracy in $\delta^{13}C$ as is shown in Fig. 4.

One of the initial problems was a small but significant baseline drift (~ 10^{-3} absorption units). The origin of this drift has not yet been determined and different correction methods have been tested. The simplest correction is to use some set of correction functions (cosine, polynomial) as additional library spectra in the MCA. If the drift is linear this method should be successful. The baseline correction is done before the final MCA, but we still need to use some additional correction function in our final analysis.

We have calculated how much a baseline error affects the uncertainty in the analysis by adding different amount of some "error functions" . We have used 0-, 1-, and 2-order polynomials as error functions and the results are shown in Fig. 5.

The number of degrees of freedom depends on the number of library spectra used. Especially when using narrow bandwidths in the analysis the degrees of freedom is rather significant. Background correction should therefore also be done before the final MCA. As a simple test we have analyzed air spectra with different numbers of polynomial

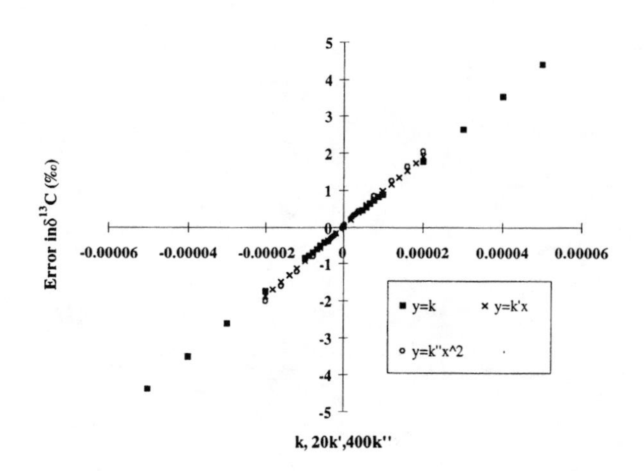

FIGURE 5. The shift in $\delta^{13}C$ as a function of k in y=k, 20k' in y=k'x, and k'' in $y=k''x^2$.

functions as library spectra.

The amount of information we used in our MCA is approx. 25 datapoints and therefore the number of library spectra is critical. This effect have been investigated by adding polynomial functions to the library set, see Fig. 6.

The $^{12}CO_2$ and $^{13}CO_2$ spectra are not completely independent due to band overlapping. This is not an ideal situation for the MCA and therefore we have investigated how the uncertainty in the coefficients are affected by the amount of band overlapping. The linear dependency have been varied by shifting the vibrational band of $^{13}CO_2$. The low resolution spectra from GASMET results in overlapping vibrational bands from $^{12}CO_2$ and $^{13}CO_2$ which becomes very significant when calculating the isotopic ratio. We have tested how much this effect affects the final uncertainty in $\delta^{13}C$ by shifting the vibrational band of $^{13}CO_2$ in relation to the band of $^{12}CO_2$. The result are shown in Fig. 7.

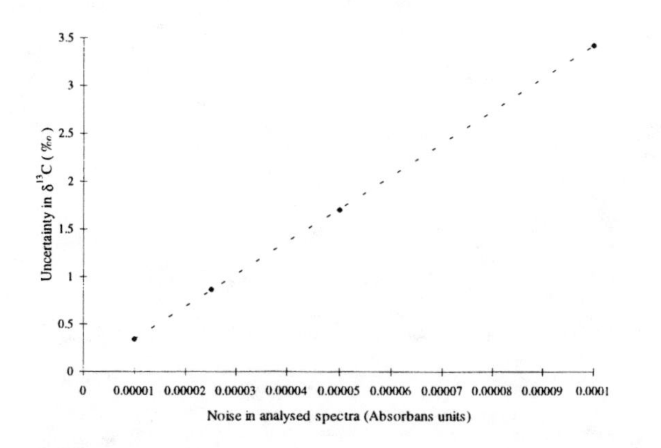

FIGURE 4. The uncertainty in $\delta^{13}C$ as a function of the noise level in the analyzed spectra (absorbance units).

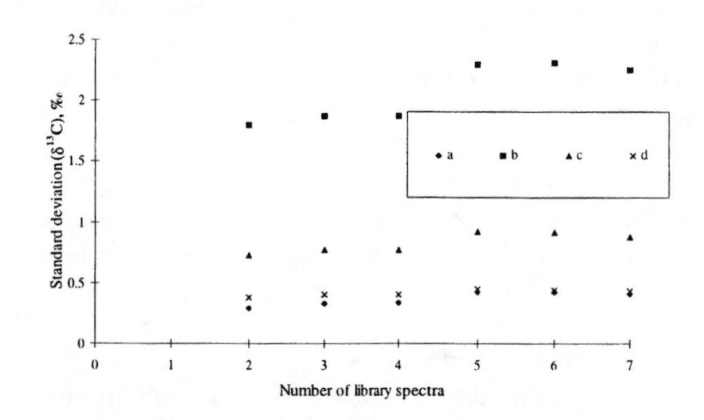

FIGURE 6. The uncertainty in $\delta^{13}C$ as a function of the number of library spectra. The noise in the used spectra was a) air spectrum: $1 \cdot 10^{-5}$, library spectra: $1 \cdot 10^{-4}$, b) air spectrum: $5 \cdot 10^{-5}$, library spectra: $1 \cdot 10^{-4}$, c) air spectrum: $2 \cdot 10^{-5}$, library spectra: $1 \cdot 10^{-4}$, d) air spectrum: $1 \cdot 10^{-5}$, library spectra: 0.

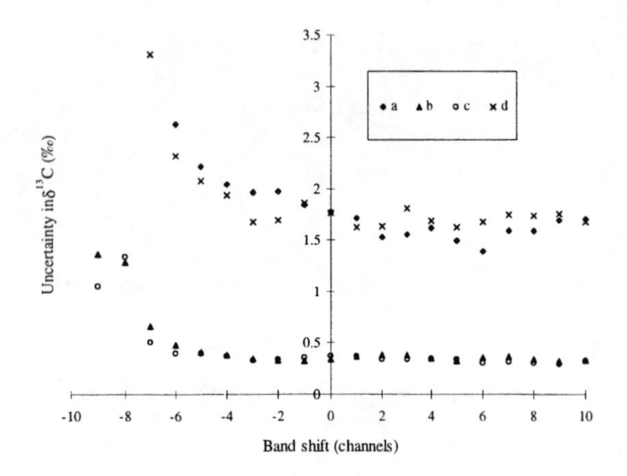

FIGURE 7. The uncertainty in $\delta^{13}C$ as a function of the $^{13}CO_2$ band shift for 2 different noise levels in the air spectra and in the library spectra. Negative x-values is defined as a shift towards more overlapping.

The nonlinearity problem is largely eliminated by using library spectra of reference gases with approximately the same CO_2 concentration as in the analyzed samples. However, the effect itself is of some interest and is investigated by analyzing a series of spectra measured from samples with different concentrations of CO_2. The nonlinearity in the absorption spectra is very critical to the final uncertainty of the analysis.

CONCLUSIONS

The analysis of the isotopic ratio of carbon in atmospheric CO_2 puts very high demands on the MCA. The abundance of the two isotopes differ by two orders of magnitude and the spectral bands of the two isotopomers partially overlap. On the one hand small errors in the computed baseline level will affect the computed intensity of the $^{13}CO_2$ band significantly. On the other hand the strong absorption bands of $^{12}CO_2$ are easily saturated causing changes in the band profile of the measured spectrum. This limits the CO_2 concentration that can be used in the sample.

We have investigated possible sources for increased uncertainty in the $\delta^{13}C$ coefficient and come to the conclusion that some important improvements have to be made in the MCA in order to achieve better accuracy in the analysis.

The reference spectra should be measured in the same concentrations as the air spectra to avoid non linearity effects.

The number of library spectra in the final MCA should not exceed 4, i.e. the 2 CO_2 spectra and a maximum of 2 additional baseline correction function. It is shown in figure 6 that when the number of library spectra increases further we loose accuracy.

A baseline shift is very critical for the analysis, but the form of the baseline correcting function is not so significant. This is shown in Fig. 7.

The overlapping vibrational bands are not crucial for the accuracy of the analysis as is shown in figure 7. A slight shift of the $^{13}CO_2$ band further towards the $^{12}CO_2$, however, would immediately result in poorer accuracy.

The overall limiting factor for the accuracy in determining the $\delta^{13}C$ value is the noise level in the measured air spectra. Figure 5 shows us that in order to achieve an uncertainty less than 0.5 ‰ in the $\delta^{13}C$ ratio we would have to measure the air spectra with a S/N ratio of approx. 100.000.

Using the above mentioned precautions when calculating the $\delta^{13}C$ for CO_2 in air we have improved the uncertainty in the analysis to approx. ± 0.7 ‰.

REFERENCES

1. Temet Instruments Oy, Asentajakatu 3, 00810 Helsinki, Finland.
2. Saarinen, P. and Kauppinen, J., *Appl. Spectr.* **45(6)** 953-963 (1991).
3. Ontar Corporation, HITRAN 1996.

Automated Detection of Ethanol Vapor by Passive Fourier Transform Infrared Remote Sensing Measurements

Patrick O. Idwasi and Gary W. Small

Center for Intelligent Chemical Instrumentation, Department of Chemistry, Clippinger Laboratories, Ohio University, Athens, OH 45701

Fourier transform infrared (FTIR) remote sensing measurements are used to implement an automated detection algorithm for ethanol vapor in the presence of ammonia, acetone, and isopropanol. This detection strategy is based on the direct analysis of interferogram data collected by the spectrometer. One or more bandpass digital filters are applied to the interferogram, and sections of the filtered data are combined to form a pattern for use with piecewise linear discriminant analysis (PLDA). The performance of PLDA with patterns constructed from the use of multiple filters is compared to the corresponding performance with patterns based on the use of a single filter. The use of multiple filters is observed to improve the ability of PLDA to detect small ethanol signatures in the presence of a large ammonia background.

INTRODUCTION

The detection of airborne volatile organic compounds (VOCs) is important in a variety of environmental monitoring applications. Passive Fourier transform infrared (FTIR) remote sensing measurements are useful in many of these monitoring scenarios due to the occurrence of characteristic infrared spectral absorption and emission features for most VOCs. The passive FTIR method utilizes an FTIR emission spectrometer and is configured to measure the absorption and emission features of VOCs against a naturally occurring infrared background emission. For example, the presence of VOCs in stack emissions can be detected by using the spectrometer to view the stack effluent against the infrared background of the sky.

A limitation of the passive remote sensing measurement is the use of an uncontrolled infrared background. This makes the collection of a representative background spectrum difficult, and complicates the subsequent analysis of the spectral data. Work in our laboratory is focusing on circumventing this problem through the development of methods to extract analyte information without the use of a separate measurement of the infrared background.

One application area of interest for FTIR remote sensing measurements is the implementation of an automated monitoring system for detecting the presence of one or more target VOCs. The work described here focuses on the development of such a system for the detection of ethanol vapor in the presence of acetone, ammonia, and isopropanol. Laboratory data designed to simulate remote sensing conditions is used in conjunction with actual field measurements of stack emissions to develop the detection methodology.

EXPERIMENTAL SECTION

The laboratory data used to simulate remote sensing conditions were collected with a Midac FTIR emission spectrometer (Midac Corp., Irvine, CA). The spectromeer used a liquid nitrogen-cooled Hg:Cd:Te detector which was designed to respond over the 800 to 1400 cm^{-1} range.

Data acquisition was performed with a Dell 486P/50 computer (Dell Computer, Austin, TX) and the MIDAS software package. The maximum spectral frequency collected was 1974.75 cm^{-1} with the interferogram points being sampled at every eighth zero crossing of the reference laser. The interferograms contained 1024 points, and the corresponding single-beam spectra had a nominal point spacing of 4 cm^{-1}.

The spectrometer was oriented to view the infrared energy from a 4×4 inch extended blackbody (Model SR-80, CI Systems, Agoura, CA). The source temperature ranged from 5 to 50 °C. A gas cell was placed between the source and spectrometer such that the cell filled the field-of-view. Interferograms were collected while the blackbody temperature was raised from 5 to 50 °C at intervals of approximately 5 °C. Various aqueous dilutions of ethanol, acetone, ammonium hydroxide, and mixtures of ethanol/acetone, ethanol/ammonium hydroxide, and acetone/ammonium hydroxide were placed in the gas cell to simulate different analyte and interference concentrations. The dilutions used for the one-component systems were 1 (pure analyte or interference) 1/2, 1/4, 1/8, 1/16, and 1/32. The two-component systems used dilution combinations of 1/8:1/8, 1/16:1/8, and 1/32:1/8.

The collection of the field data employed a similar Midac spectrometer to view the effluent from a plume generator operating at 200 °C. Pure ethanol, ammonia

CP430, *Fourier Transform Spectroscopy:* 11th International Conference
edited by J.A. de Haseth
 1-56396-746-4/98/$15.00

gas, pure isopropanaol, and mixtures of ethanol/ammonia were introduced into the plume generator and heated with hot air generated by a propane burner and attached blower. The spectrometer was oriented to view the stack effluent against a sky background. The data collection procedures and interferogram characteristics were identical to those described previously for the laboratory data.

The collected data were transferred to a cluster of four Silicon Graphics Indigo² R10000 workstations (Silicon Graphics, Mountain View, CA) operating under Irix (version 6.2). All computations were performed on these workstations with software written in FORTRAN 77.

RESULTS AND DISCUSSION

The data analysis strategy used in this work is based on the direct treatment of the interferogram data collected by the FTIR spectrometer. Sections of the interferogram are used as the input to piecewise linear discriminant analysis (PLDA), an automated pattern recognition procedure for numerical data. The PLDA algorithm produces a classification score which can be used to judge whether the analyte signature is present in the filtered interferogram.

Two strategies are used to enhance the selectivity of the method. First, bandpass digital filters are applied to the interferogram data to isolate the frequencies corresponding to a selected spectral band of the analyte. This adds spectral selectivity to the direct interferogram analysis. Second, a judicious choice of the interferogram segment used can add further selectivity by rejecting information on the basis of the widths of spectral features. For example, many of the components of the infrared background are either very broad or very narrow spectral features. The interferogram representation of a broad spectral feature is concentrated near the centerburst, while the representation of a very narrow feature is spread throughout the interferogram. Thus, by combining filtering with segment selection, analyte information can be discriminated from the infrared background on the basis of both frequency and width criteria.

This methodology has been used previously to identify the presence of VOCs in the atmosphere (1-4). In these previous studies, the filtering strategy has employed a single finite impulse response matrix (FIRM) filter (5) designed to isolate the frequencies associated with an analyte spectral band. The work presented here addresses the case in which the spectral information associated with an interfering species overlaps with that of the analyte. In this situation, information about both compounds passes through the filter, thus complicating the pattern recognition analysis. To address this problem, two filters are employed, one centered on the analyte band and the other centered on a band of the interference. Interferogram segments resulting from the application of both filters are combined and used to form a composite pattern.

In the present work, ethanol was the analyte, and ammonia served as the principal spectral interference. Figure 1 displays vapor-phase infrared spectra of ethanol, ammonia, and acetone. The ethanol filters used were centered in the region of the C-O stretch at 1066 cm^{-1}. The second filter was used to encode the ammonia signature. Ammonia filters were centered on the bands at 931 and 967 cm^{-1}. The filters were generated at varying positions and widths.

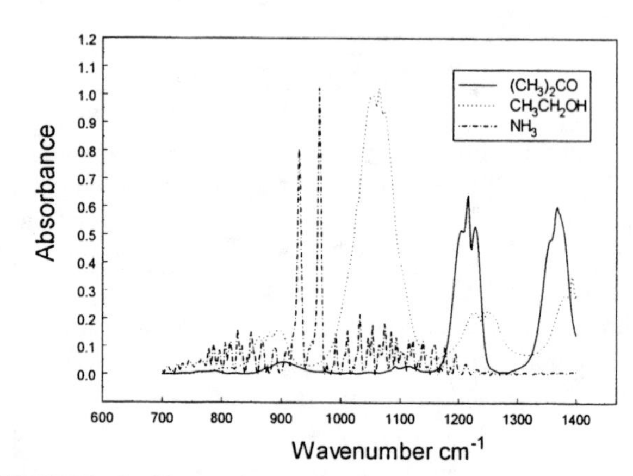

FIGURE 1. Vapor-phase absorbance spectra of acetone, ethanol, and ammonia.

Digital Filtering Methodology

The digital filters used in this work were modeled on Gaussian-shaped frequency response functions. The time domain Fourier pair of the frequency response was estimated by a time-varying finite impulse response method designed specifically for use with interferogram data (5). The resulting filters were applied directly to the interferogram data by computing the convolution sum of the interferogram and impulse response functions.

Pattern Recognition Methodology

The interferogram pattern used is represented as a point in m-dimensional space, where m is the number of points in the pattern. The use of the filter suppresses information that encodes the infrared background and spectral information from other compounds while it passes the analyte information. If the filter is effective in isolating the analyte information, the patterns will fall in groups corresponding to analyte-active and analyte-inactive clusters. The PLDA technique uses numerical optimization procedures to approximate a nonlinear discriminating boundary that separates the analyte-active and analyte-inactive clusters (1). The individual components of this boundary are termed discriminants.

Once the optimal position of the discrimants has been established, the classification of a pattern as analyte-active or analyte-inactive is based on the value of the discriminant score. The discriminant score encodes the side of the boundary on which the pattern lies and the distance to the boundary. By convention, the analyte-active patterns are those with positive discriminant scores. Patterns with small positive discriminant scores, i.e., those near the boundary, correspond to analyte signatures near the limit of detection.

Optimization of Detection Algorithm

The discriminants are placed as close to the analyte-inactive patterns as possible so that the limit of detection is kept as low as possible. From previous results we have observed that this works well in situations where the spectra are not complex (i.e., where the analyte spectral band of interest is not overlapped with bands due to other compounds). In situations where the spectrum contains overlapping features from more than one compound, increased difficulty is observed in achieving a low limit of detection. In this case, weak analyte signatures are obscured by the overlapping signatures of other compounds.

In the present work, the rotational bands of ammonia overlap the C-O stretch of ethanol. Our first investigation was to evaluate the effect of this band overlap by using a single filter centered in the region of the ethanol band. The laboratory data was used in this initial investigation due to the presence of a wide range of signal strengths. The data were divided into training and prediction data sets as shown in Table 1. The training data were used in optimizing the detection algorithm, while the prediction data were withheld from the optimization step and were used as an independent validation set for testing the performance of the developed algorithm.

From previous results, we used a 120-point interferogram segment as the input pattern for PLDA. Nine filters were used centered around the ethanol band at 1066 cm^{-1}. The starting point for the interferogram segment was varied from points 25 to 175, relative to the centerburst. The results of this study are shown in Table 2. Included in the table are the results produced by the

TABLE 1. Partitioning of the Data Sets

Data Category	Training set[a]	Prediction set[a]
Analyte-active	3494 (2900)	1501 (1100)
Analyte-inactive	6086 (5000)	24828 (14800)

[a]Number of interferograms (laboratory data in parentheses)

four best combinations of filter position, filter width, and interferogram segment location. Listed are classification percentages for the detection of ethanol in both the training and prediction data. Also listed are the false detection percentages for the prediction data. In the best case, the ethanol signature was recognized at a rate of approximately 90% in the prediction data. The false detection rate was less than one percent.

Investigation of the patterns that were misclassified revealed that the patterns affected most were those that were near the detection limit (i.e., those that had small temperature differences between the analyte and the blackbody and those that corresponded to small ethanol concentrations). The best results were exhibited by segments with starting positions about 50 to 75 points offset from the centerbust.

Segments further from the centerbust did not give very good results. This was attributed to the fact that since the ethanol band was wider than the narrow overlapping rotational bands of ammonia, most of the discriminating information regarding the band presence was located near the interferogram centerburst.

In a further study to observe if a shorter interferogram segment could give the same performance, 60-point segments were investigated with the same starting points as in the 120-point study. The results showed that the best 60-point segment produced classification rates that were not significantly different from those obtained with the 120-point segments. The four best 60-point segments were then selected for use in the second part of the study.

Next, the raw interferogram was filtered with two different digital filters, the first one tailored to pass ethanol information and the second one tailored to pass ammonia information. Segments from each of the filtered interferograms were concatenated, and the combined pattern was then used as the input in the PLDA analysis. The best 60-point segments from the interferograms filtered with the ethanol filters were augmented with a 60-point segment from the interferogram filtered using an ammonia filter. It was hoped that the introduction of the new information would add greater selectivity to the clustering of the patterns in the data space and thus lead to a better positioning of the discriminants.

Table 2 presents results corresponding to the four best combinations of ammonia and ethanol filters. The results for these filter combinations are indicated in parentheses.

A marked improvement in ethanol classification is observed along with a general reduction in the false alarm rate. In the best case, 97% of the ethanol signatures in the prediction set are correctly recognized, and the false alarm rate is less than 0.5%.

A similar study was performed with the field data. Pattern recognition analysis of field data is usually quite challenging due to the fact that the background is continuously changing. However, with the data used here, the ethanol signals were very strong due to a large ethanol flow rate and an extremely hot stack temperature. For this reason, training and prediction classification results exceeded 99% with a single ethanol filter. Results obtained with multiple filters were equally good, although no improvement was possible due to the essentially perfect classification rates. Future work will focus on the collection of stack emission data that is more challenging.

CONCLUSIONS

The multiple filtering protocol has been observed to improve the classification performance of PLDA by the introduction of new information. The choices of filter position, filter width, interferogram segment length, and segment starting position are important for optimal performance of the algorithm. The use of two filters increases the difficulty of this optimization by doubling the number of adjustable parameters. A more global method of optimization for these parameters is clearly needed, rather than limited scale optimization employed here. Current efforts in our laboratory are exploring the use of genetic algorithms for this purpose.

ACKNOWLEDGMENTS

Funding for this work was provided by the Department of the Army. Robert Kroutil, Roger Combs, and Robert Kroutil are acknowledged for providing the interferogram data. Charles Chaffin and Tim Marshall of AeroSurvey, Inc. are acknowledged for the construction and operation of the plume generator.

REFERENCES

1. Kaltenbach, T. F.; Small, G. W., *Anal. Chem.,* **63,** 936-944 (1991).
2. Shaffer, R. E.; Small, G. W., *Chemom. Intell. Lab. Syst.* , **32,** 95-109 (1996).
3. Shaffer, R. E.; Small, G. W.; Combs, R. J.; Knapp, R. B.; Kroutil, R. T., *Chemom. Intell. Lab. Syst.* , **29,** 89-108 (1995).
4. Bangalore, A. S.; Small, G. W.; Combs, R. J.; Knapp, R. B.; Kroutil, R. T.; Traynor, C. A.; Ko, J. D., *Anal. Chem.*, **69,** 118-129 (1997).
5. Small, G. W.; Harms, A. C.; Kroutil, R. T.; Dittilo, J. T.; Loerop, W. R. *Anal. Chem.*, **62,** 1768-1777 (1990).

TABLE 2. Pattern Recognition Results Obtained with Laboratory Interferogram Data

Filter position (cm^{-1})	Training (%) ethanol actives	Prediction (%) ethanol actives	Prediction (%) false detection	Prediction (%) total
1030 (105)[a]	92.14	90.45	0.41	98.96
1030 + 930[b]	(95.19)[c]	(94.08)	(0.45)	(99.14)
1030 (140)	90.62	90.91	0.58	98.83
1030 + 930	(90.16)	(94.42)	(0.32)	(99.29)
1090 (105)	90.52	87.55	0.36	98.81
1090 + 930	(96.04)	(97.08)	(0.45)	(99.37)
1090 (140)	90.48	88.64	0.61	98.64
1090 + 930	(92.59)	(95.00)	(0.22)	(99.42)

[a]fwhh in parentheses.
[b]Second entry indicates ammonia filter position.
[c]Multiple filtering results are indicated in parentheses.

Calibration Transfer Results for Automated Detection of Acetone and Sulfur Hexafluoride by FTIR Remote Sensing Measurements

Frederick W. Koehler IV and Gary W. Small

Center for Intelligent Chemical Instrumentation, Department of Chemistry, Clippinger Laboratories, Ohio University, Athens, OH 45701, U.S.A.

Fourier transform infrared (FTIR) remote sensing spectrometry is an environmental monitoring technique of increasing importance. A key impediment to applying FTIR remote sensing measurements on a large scale involves problems caused by the dynamic infrared background, with variations caused by fluctuations in the ambient environment as well as by instrument-specific signatures. Problems in data analysis and calibration transfer are explored in this work in the context of implementing an automated algorithm for detecting acetone or sulfur hexafluoride vapor.

INTRODUCTION

The use of FTIR remote sensing spectrometry for the detection of airborne pollutants is recently gaining in importance (1). This trend can be attributed to several advantages inherent to the technique. The sensitivity and specificity of FTIR spectrometry allows the determination of analytes in the presence of other, potentially interfering, species, which is critical for smokestack monitoring. In addition, the remote sensing aspect of the measurement provides a methodology which can monitor a wide area and a large volume of atmosphere without direct sample collection. This can be accomplished with a rugged portable instrument which can be deployed from ground-based as well as airborne data collection platforms.

However, two critical problems have been shown to hinder the widespread application of the technique. Traditionally in FTIR remote sensing, a reference background spectrum is collected and used to remove the background emission profile present in analyte spectra (1). Simple changes in the environment such as wind or temperature often prohibit stable, reproducible reference spectra from being measured. Analyte spectra obtained in this fashion contain widely varying baselines and can be difficult to analyze. In addition, a second important background problem is the large instrument-specific signatures which make the automated analysis of data from different instruments difficult.

The methodology developed in our laboratory seeks to overcome these challenges through signal processing and pattern recognition techniques applied directly to the raw interferogram data obtained from the passive remote sensing spectrometer, avoiding the need altogether of a separate background measurement. Digital filtering steps isolate the analyte signal from the background, and pattern recognition techniques are utilized to discriminate and characterize signals which contain analyte from those which do not in an automated fashion. In this paper, this methodology is extended with additional signal processing steps and with more strongly attenuating digital filtering techniques. It will be shown that instrument-specific background problems can be eliminated as well, allowing a successful transfer of qualitative calibration information between spectrometers.

EXPERIMENTAL SECTION

Calibration transfer issues in passive remote sensing were explored by collecting laboratory acetone and sulfur hexafluoride (SF_6) interferograms on a pair of similarly configured Midac Outfielder FTIR emission spectrometers, labeled units 120 and 145 (Midac Corp., Irvine, CA). These spectrometers employed liquid nitrogen-cooled Hg:Cd:Te detectors for use in the 800-1400 cm^{-1} spectral range.

These spectrometers were interfaced to a Dell system 486P/50 IBM PC compatible computer (Dell Computer, Austin, TX) operating under MSDOS (Microsoft, Redmond, WA). Data acquisition was performed with the MIDAS software package (2). A maximum spectral frequency of 1974.75 cm^{-1} was obtained with interferogram points being collected at every eighth zero crossing of the reference laser. Four cm^{-1} point spacing was obtained through the collection of 1024 interferogram points per scan.

A 4x4 inch extended blackbody (Model SR-80, CI Systems, Agoura, CA) provided a NIST traceable infrared source whose temperature was varied over 5 to 50 °C. The source temperature was accurate to 0.03 °C and precise to ± 0.01 °C. A sample gas cell with windows composed of low density polyethlyene (0.0005 in. thickness) was used. A thermocouple was utilized to monitor gas cell temperature.

CP430, *Fourier Transform Spectroscopy:* 11th International Conference
edited by J.A. de Haseth

TABLE 1. Partition of Acetone and SF_6 Data Sets

	Acetone		SF_6	
	Unit 120	Unit 145	Unit 120	Unit 145
Training	3170[a] (8782[b])	3239 (8284)	2640 (3940)	2041 (3942)
Prediction	2190 (8202)	2292 (7753)	2320 (3773)	2134 (3776)
Collected	6810 (16984)	6810 (16037)	8273 (7713)	8275 (7718)

[a] Analyte active
[b] Analyte inactive

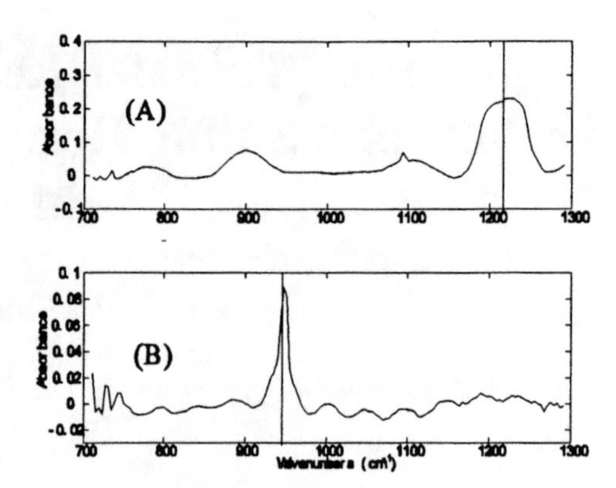

Figure 1. SF_6 and acetone absorbance spectra collected on Midac unit 120 under laboratory conditions. (A) Pure acetone spectrum at a blackbody temperature of 50 °C. The line at 1216 cm^{-1} highlights the acetone peak. (B) 1cc SF_6 at a blackbody temperature of 50 °C. The SF_6 peak at 945 cm^{-1} is highlighted.

Reagent grade acetone and sulfur hexafluoride were used as analytes.

For both acetone and SF_6 experiments, data collection for units 120 and 145 was performed alternately by moving the cell and blackbody in front of each instrument in turn. For the acetone data set, interferograms were collected with blackbody temperatures from 5 to 50 °C with steps at approximately 5 °C intervals for dilution factors with water of 1 (pure acetone), 1/2, 1/4, 1/8, 1/16, 1/32, and 1/64. Between 20 and 200 interferograms were collected at each level. For the SF_6 data set, interferograms were collected over the same temperature range with similar 5 °C steps with injected analyte volumes of 0.05, 0.02, 0.1, 0.2. 0.3, 0.5, and 1.0 cc. Between 20 and 150 interferograms were acquired at each level.

The collected interferograms were Fourier transformed and the resulting single-beam spectra were ratioed to corresponding background spectra collected when no analyte was present. After converting to absorbance units, the spectra were visually inspected to ensure that the analyte signal was clearly visible above the noise. Those which did not meet this criterion were removed from the data set. This led to the training and prediction sets for each analyte as listed in Table 1.

For data analysis, the collected data sets were transferred to a dual 180 MHZ Pentium Pro (Intel Corp., Santa Clara, CA) personal computer operating under the Linux operating system, version 2.0.14. The digital filtering and pattern recognition were performed on this system with original software written in FORTRAN 77 and C. Additional processing was performed with the aid of Matlab version 4.2c (The MathWorks, Natick, MA).

RESULTS AND DISCUSSION

Infrared signals measured through passive FTIR remote sensing experiments consist of analyte, background, and instrument-specific features superimposed (3). The lack of a stable background prevents the use of conventional data analysis methods such as the calculation of absorbance or difference spectra for an automated determination since they are unable to remove background and instrument features reliably through ratioing or subtraction. The purpose of the signal processing and pattern recognition steps outlined here is the extraction of analyte information and the suppression of interfering signals, thereby allowing an automated determination to be performed without the use of background measurements for ratioing or subtraction.

For analytes in this study, the features of interest are the 1216 cm^{-1} C-CO-C stretching band of acetone (49 cm^{-1} full width at half maximum (fwhm)) and the 945 cm^{-1} S-F stretching band of SF_6 (10 cm^{-1} fwhm). Figure 1 demonstrates the type of signal obtained through the calculation of absorbance spectra for interferograms collected from the laboratory acetone and SF_6 data sets. For the blackbody source temperature range covered in this study, both absorption and emission peaks were present in the data sets. Fine rotation features were absent in all spectra calculated from these data due to the 4 cm^{-1} spectral point spacing.

In order for our methodology to avoid the use of inactive backgrounds for ratioing or subtraction, signal processing and pattern recognition analysis are applied directly to the interferogram data. Direct interferogram analysis provides advantages by decomposing spectral features of different widths into different regions of the interferogram. This can be attributed to the fact that the interferogram representation of a narrow spectral feature dampens more slowly than the corresponding representation of a wide background feature. By optimal choice of the interferogram segment to use for analysis, a significant amount of background interference can be removed.

Once an optimal segment is isolated from the interferogram, digital filtering is used to enhance the analyte signal further. Time domain digital filtering involves the estimation of the convolution of the interferogram with the

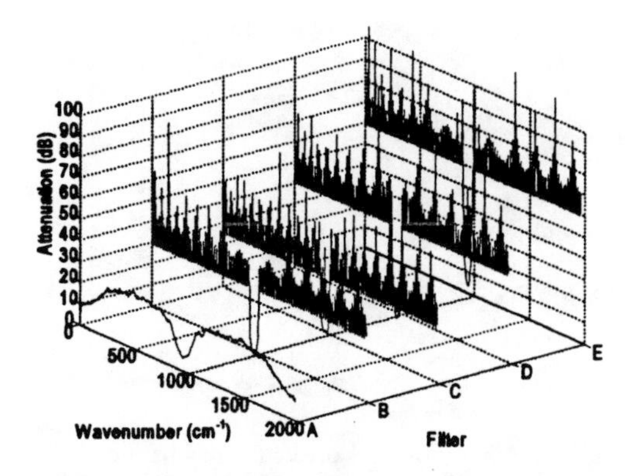

Figure 2. SF_6 FIRM and FIR filter frequency response plots demonstrating differences in attenuation and passband width. (A) FIRM filter with fwhm ~165 cm^{-1}. (B-E) FIR filters with fixed passband width of 72 cm^{-1}.

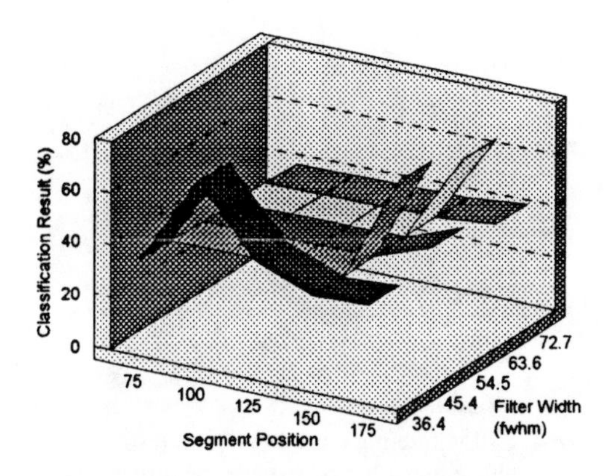

Figure 3. FIRM filtering cross prediction results for SF_6. Midac unit 120 was used as the primary instrument in predicting the unit 145 data set.

time domain representation of the filter frequency response function (4). Digital filtering provides a means of extracting frequency information due to the analyte from the problematic background frequencies while allowing the methodology to utilize key advantages found in signal processing data in the interferogram (time) domain.

Two types of digital filtering were used in this study, a time-varying finite impulse response matrix filter (FIRM) developed previously in our laboratory, and a standard FIR filter (4). FIRM filters sacrifice attenuation but offer high computational efficiency by having fewer coefficients. During filter generation, coefficients deemed statistically insignificant in the estimation of the convolution sum can be discarded. Standard FIR filters were calculated through the Remez exchange algorithm and provide exceptional out-of-band attenuation; however they contain nearly an order of magnitude more coefficients. Figure 2 shows frequency response plots for a representative SF_6 FIRM filter as well as several FIR filters utilized in this study. FIR filtering allows a closer approximation of the desired passband width to be attained. However, the FIRM filter attains approximately 25 decibels (dB) of attenuation with an average of 22 filter coefficients, whereas the FIR filters all contain 200 filter coefficients.

After filtering, a reliable pattern recognition step is required in the analysis to determine the presence or absence of analyte signal in the filtered data. Due to its high performance and simplicity in configuration, the nonlinear pattern recognition technique utilized for this methodology was piecewise linear discriminant analysis (PLDA). PLDA attempts to optimize the location of linear separating surfaces, termed discriminants, which divide the data space into analyte-active and inactive categories (5,6).

TABLE 2. FIRM Filter Parameters

Variable	SF_6	Acetone
Filter bandpass width (fwhm)	36.4[a] (81[b]), 45.4 (110), 54.5 (125), 63.6 (146), 72.7 (165) cm^{-1}	45.4 (85), 54.5 (103), 63.6 (150), 72.7 (167), 81.9 (201) cm^{-1}
Interferogram segment location[c]	75, 100, 125, 150, 175	50, 75, 100, 125, 150

[a] Specified fwhm during filter generation
[b] Measured fwhm
[c] Relative to centerburst

Previous work has demonstrated the most efficient means of optimizing the experimental parameters of FIRM filter passband center and width, interferogram segment starting position and length, as well as those of the PLDA pattern recognition algorithm (7). Using this protocol, and a subset of the overall experimental design used previously, FIRM filters were created with the same characteristics for SF_6. Acetone FIRM filters were also created, but with segment location and filter passband center optimized for its 1216 cm^{-1} peak. These filters were utilized to examine training and prediction as well as calibration transfer issues for acetone and SF_6. Table 2 summarizes the FIRM filter parameters used. Two values are indicated for FIRM filter width. The first is the width supplied to the filter generation algorithm, while the second width is the fwhm measured from the actual frequency response of the generated filter. Absolute values of the training and prediction interferograms were used in order to make the data space more robust for calibration transfer, and Forman phase correction was utilized. In all cases Midac unit 120 was used as a primary instrument, meaning that its interferograms were used during filter generation, as well as during pattern recognition training. Midac unit 145 was used as a secondary instrument to test calibration transfer. No unit 145 interferograms were included during training.

Results for FIRM filtering experiments from data collected on unit 120, and then utilized for both training and prediction were between 88.45 and 99.93% for both SF_6 and acetone. These results demonstrate that FIRM filtering performs well for same-instrument prediction for both analytes, as has been shown in the past. However, once these same discriminants were applied to data from a secondary instrument (unit 145), cross-prediction results decreased as seen in Figs. 3 and 4, particularly for SF_6. At ~40 cm^{-1}, the acetone spectral feature at 1216 cm^{-1}

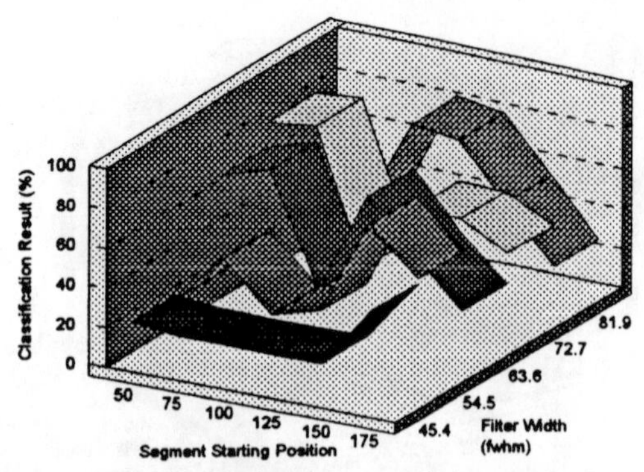

Figure 4. FIRM filtering cross-prediction results for acetone. Midac unit 120 was used as the primary instrument for predicting the unit 145 data set.

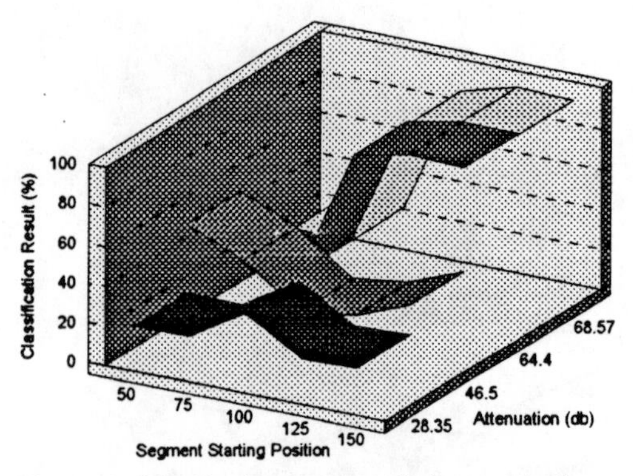

Figure 6. FIR filtering cross prediction results for acetone. Midac unit 120 was used as the primary instrument for predicting the unit 145 data set.

is approximately four times wider than SF_6. The typical FIRM passband more closely approximates the wider acetone peak, but lets a great deal of background information through for the narrow SF_6 peaks. Cross-prediction results appear to improve as acetone FIRM filter passband widths increase, however no clear trend is evident for the optimal segment location.

Although an extensive experimental study has yet to be performed with FIR filter parameters similar to that done with the FIRM study, four FIR filters were generated with constant passband width and varying attenuations for both acetone and SF_6. These filters were applied to the same interferogram segment positions used in the FIRM study. Frequency responses for these four filters for SF_6 can be seen in Fig. 2, with those of acetone being similar except for the passband center being located at 1216 cm^{-1}.

Results for same-instrument prediction for both analytes varied between 89.99 and 99.98%, and were similar to the results obtained with the FIRM filters. However, as seen in Figs. 5 and 6, cross-prediction scores for both compounds were markedly improved. The acetone and SF_6 predictions are observed to improve with increasing attenuation in the stopband, with the best results being observed for attenuations above 60 dB and segments located past point 125 (relative to the centerburst).

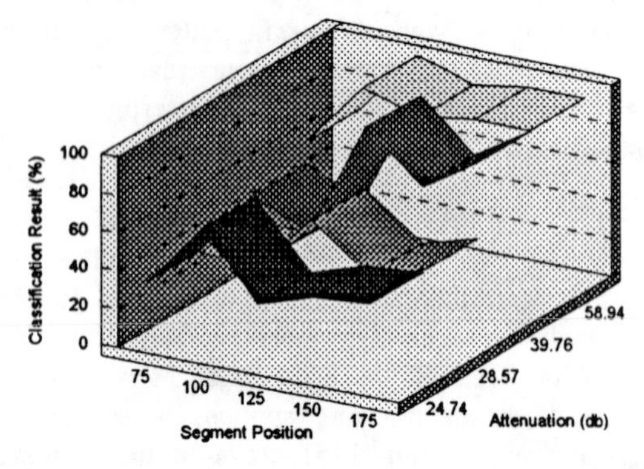

Figure 5. FIR filtering cross prediction results for SF6. Midac unit 120 was used as the primary instrument for predicting the unit 145 data set.

CONCLUSIONS

While FIRM filters provide sufficient performance for training and prediction on a single instrument, FIR filters with high degrees of stopband attenuation allow a successful transfer of qualitative calibration information across data spanning two spectrometers for both acetone and SF_6 analytes.

ACKNOWLEDGMENTS

Funding for this work was provided by the Department of the Army. Robert Kroutil, Roger Combs, and Robert Knapp are acknowledged for providing the interferogram data used in this work.

REFERENCES

1. Hammaker, R.M., Fateley, W.G., Chaffin, W.G., Marshall, T.C., Tucker, M.D., Makepeace, V.D., Poholarz, J.M., *Appl. Spectrosc.* **47**, 1471-1475 (1993).
2. Combs, R.J., Kroutil, R.T., Knapp, R.B., Unclassified Technical Report ERDEC-TR-298, Edgewood Research, Development, and Engineering Center: Aberdeen Proving Ground, MD, August 1995.
3. Kroutil, R.T., Ditillo, J.T., Small, G.W., Meuzelaar, H. (Ed.), *Computer-Enhanced Analytical Spectroscopy*, New York: Plenum, 1984, ch. 4.
4. Small, G.W., Harms, A.C., Kroutil, R.T, Ditillo, J.T., Loerop, W.R., *Anal. Chem.* **62**, 1768-1777 (1990).
5. Kaltenbach, T.F., Small, G.W., *Anal. Chem.* **63**, 936-944 (1991).
6. Shaffer, R.E., Small, G.W., *Chemom. Intell. Lab. Syst.* **32**, 95-109 (1996).
7. Shaffer, R.E., Small, G.W., Combs, R.J., Knapp, R.B., Kroutil, R.T., *Chemom. Intell. Lab. Syst.* **29**, 89-108 (1995).

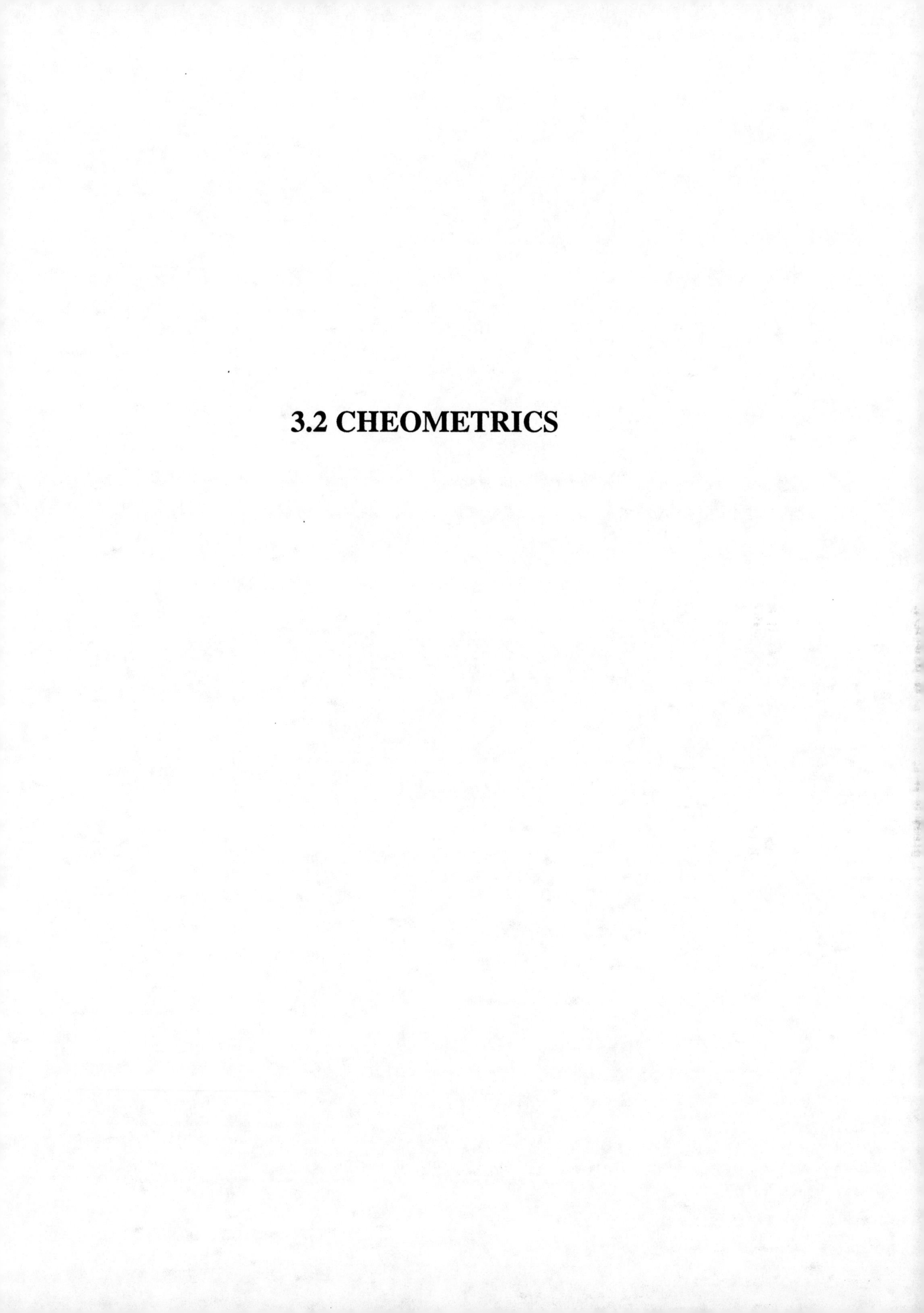

3.2 CHEOMETRICS

Genetic Algorithms Used for Spectral Window Selection in Open-Path Fourier Transform Infrared Spectrometry

R. James Berry and Peter R. Griffiths

Department of Chemistry, University of Idaho, Moscow, ID 83844-2343

Open-path Fourier transform infrared (OP/FT-IR) spectrometry is susceptible to interferences, especially those due to atmospheric water and carbon dioxide. To overcome these interferences, multivariate techniques such as classical least squares (CLS) and partial least squares (PLS) regression are used. The CLS method is very sensitive to variations in the background and the presence of lines due to water and carbon dioxide in the spectrum, and usually requires calculations to be made over small spectral windows to minimize these effects. The windows are usually chosen by the user and can reflect the user's bias. We have used a heuristic program called a genetic algorithm to select spectral windows that are independent of user bias and often yield results 2-4 times better than the generally accepted method of spectral window selection. We have also investigated the use of spectral window selection by a genetic algorithm for PLS using much larger spectral windows and found it to be advantageous.

INTRODUCTION

OP/FT-IR spectrometry is complicated by the absorption lines in the spectra of water and carbon dioxide (1). Fig. 1 shows the effect of these interferences for a 400-meter path spectrum, measured as 1 cm^{-1} resolution, ratioed against a short path length background. The features due to atmospheric species greatly complicate the calculation of concentration using a simple univariate application Beer's Law. Multivariate techniques are, therefore, used to model the OP/FT-IR spectra and to predict the concentrations of each analyte. The most rudimentary multivariate method is classical least squares (CLS) regression. In general, poor results are obtained via full-spectrum CLS. Instead, short spectral windows are used in the analysis.

The spectral windows that are chosen for CLS can be as short as 2 cm^{-1} to as wide as 50 cm^{-1}. The generally accepted method for selecting these windows is outlined in a draft document being prepared for the U.S. Environmental Protection Agency (EPA) that is known as the TO-16. This document outlines the procedure for selection of which peak(s) of the analyte to use as well as the width of the spectral window. However, it does not explicitly instruct on which peak(s) should be chosen for each individual compound, leaving the actual selection to the user.

An alternative method for selection of spectral windows is the half-height method. This is strictly an empirical technique where the user chooses a peak and uses only the portion of the peak at or above the half-height. Although

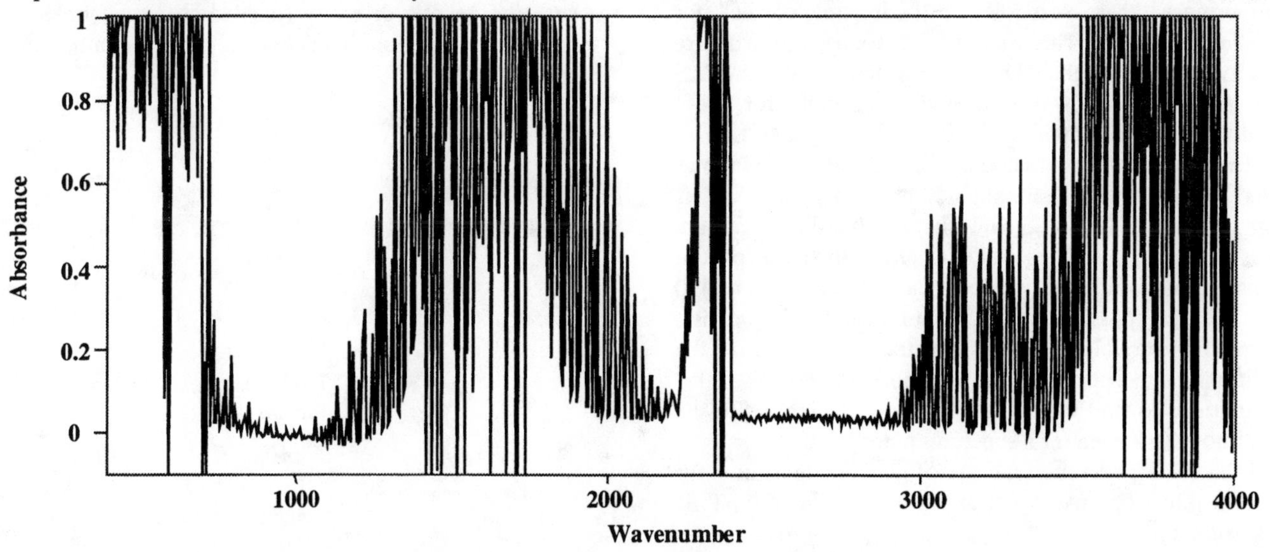

FIGURE 1. A typical open-path background

CP430, *Fourier Transform Spectroscopy:* 11th International Conference
edited by J.A. de Haseth

not a subjective as the previous method, this approach can give rise to problems of the various ways in which different users select the half-height points on the band. Furthermore, since the spectral windows are often much smaller than those in the TO-16 practice, the half-height method is much more susceptible to noise in the potion of the peak above the half-height points (but is immune to noise below the half-height points.

Partial least squares (PLS) regression is another multivariate technique that is gaining popularity in infrared analysis (2-4), although the OP/FT-IR community still strongly favors CLS. PLS is often regarded as a full spectrum method, *i.e.*, short spectral windows are rarely required. The more complicated the background, the more PLS factors will be required to model the background. While it is often believed that any background can be modeled with a sufficient number of factors, there is a stronger danger of overfitting spectra when too many factors are selected, *i.e.*, information on the analyte can be modeled into the background. In this case, the sensitivity and the prediction accuracy of PLS is reduced. It has been shown that by reducing the amount of extraneous data, the prediction accuracy can be improved (5). If this is indeed the case, the use of spectral windows for the PLS analysis of OP/FT-IR is called for.

The actual selection of spectral windows is not a trivial undertaking. The traditional windows that are used in OP/FT-IR are from 700-1300 cm^{-1}, 2000-2250 cm^{-1}, and 2400-3000 cm^{-1}. These windows eliminate many of the strong water and carbon dioxide interferences. Nonetheless, lines due to water and carbon dioxide are still present near the edges of the edges of the window, often with far greater peak absorbances than the bands due to trace analytes. Furthermore, these windows still do not correct for baseline variations. The windows that are used for PLS are usually much wider than the windows that are used in CLS and the selection of the wavenumber limits for these windows is just as important for PLS than is is for CLS.

However, assuming that only the traditional windows listed in the previous paragraph were chosen as the basis for a possible CLS or PLS analysis, and that the spectra are measured at quite low resolution (8 cm^{-1}), the total spectral domain from which the windows can be selected is 1450 cm^{-1}, with 363 possible wavenumbers from which to chose. From the possible 131,000 spectral windows in the traditional atmospheric windows, more than 1.889×10^{773} calculations to find the optimal window, which would be an arduous task even for a supercomputer.

In this paper, we report the use of genetic algorithms for selecting the optimal spectral windows for OP/FT-IR spectrometry.

THEORY

A genetic algorithm (GA) is a technique that uses Darwinian selection as a means for optimization. The genetic algorithm was first invented in 1975 and has enjoyed increasing popularity in many fields of endeavor, including analytical chemistry. (5-9) The GA uses 'genes' to search the information space. For our OP/FT-IR spectra, a gene would represent a spectral window and consists of two points (a first and last point of the spectral window).

A chromosome is a collection of genes. In its biological counterpart, a chromosome consists of many genes that express one thing, *i.e.* hair or eye color, length of fingers, etc.). A chromosome in the genetic algorithm can have one, but usually more than one gene. The chromosome is one possible answer to a problem. It is often called the 'parent' in the selection process.

The genetic algorithm improves its answer each successive iteration (or generation) by selecting the most fit parents (chromosomes). It takes the parents and has three choices, it can pass them directly into the next generation, mate it with another parent to produce offspring, or it can mutate the parent.

The mating of a parent is called crossover. In crossover, some of the genetic material (genes) from one parent is exchanged for some genetic material of the other parent. The hope is that one or both children made by the exchange will be better than their parents. Crossover is the primary method of search in a GA. Double crossover is an another type of crossover operator. Double crossover is the single crossover performed twice on the same pair of parents at different locations. Double (and other multiple) crossovers, tend to speed convergence.

Mutation also plays an important role in the GA. Mutation serves to keep the GA out of local minima. The mutation operator takes a parent and usually changes a single gene. Mutation helps to fine tune the GA and ensure the genetic diversity (the search space) broad and varied. The operation of a genetic algorithm can be seen in Figure 2.

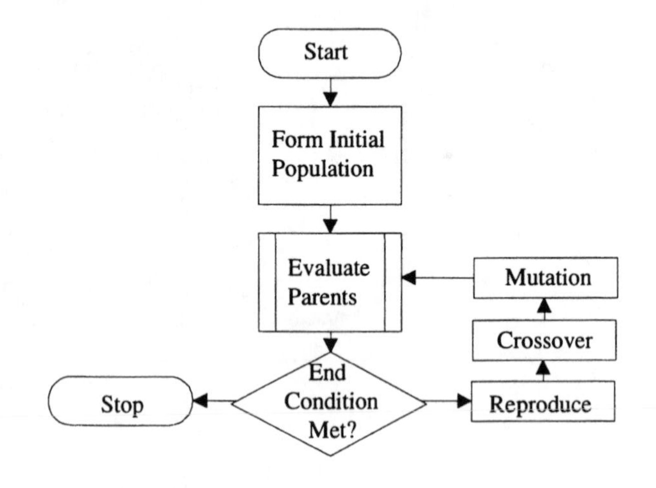

FIGURE 2. Flow chart of a Genetic Algorithm

A random initial population is chosen, the parents are evaluated and assigned a fitness by the fitness function. They are then sorted randomly based on their fitness, then passed directly into the next generation or operated on by either crossover or mutation. Then they are reevaluated, and the cycle repeated until the end condition (a certain level of error or preset number of generations has been achieved).

The fitness function is the metric by which the GA evaluates the goodness of each parent. The fitness function ensures that the GA gets progressively better each generation. The actual fitness function that is used varies widely with the problem that is to be solved. Any fitness function should satisfy the basic criteria of being a representative measure to achieve the solution.

EXPERIMETNAL

The backgrounds for the experiment were obtained at different pathlengths from 200 to 400 meters under a range of conditions of temperature, humidity, wind speeds, etc. on a Bomem BM-104 OP/FT-IR spectrometer. The spectra were ratioed against equidistant (used for the CLS regression) or short-path length (2-5 meter, used for the PLS regression) background spectrum and converted into absorbance. Several analyte spectra, stored linear on absorbance, were selected from the EPA vapor phase library (measured by Entropy Environmentalists, Inc., Research Triangle Park, NC, under EPA contract 68D90055), and were deresolved using the GRAMS (Galactic Software, Salem, NH) software package to have the same resolution as the background spectra. These spectra were base line corrected, scaled, and added to background spectra to yield synthetic spectra for analysis. They the resultant spectra were divided into a calibration set and a validation set.

All spectra treated by CLS regression were measured at a resolution of 1 cm^{-1}, which is the lowest resolution commonly used in OP/FT-IR spectrometry as currently practiced. Spectral windows were selected by the TO-16 and half-height methods and by using a genetic algorithm.

OP/FT-IR spectra treated by the PLS regression were measured at a lower resolution (8 cm^{-1}) than the previous set for several reasons. Firstly, the PLS algorithm is a computationally more intensive, and thus is much slower than the CLS regression. Secondly, PLS requires far less spectral information to achieve good results. Thirdly, we are developing a low resolution OP/FT-IR spectrometer in our laboratory and we are interested in a comparison of data obtained by PLS and CLS in order to decide which algorithm to use for the data acquired for this instrument.

The computations were carried out in the MATLAB 5.0 (MathWorks, Inc, South Natick, MA) using MS-Windows'95. The genetic algorithm was also written in MATLAB 5.0.

The fitness function of the GA depended on the root mean square error of prediction (RMSEP) of the path-integrated concentrations obtained using the validation set. The spectral windows of each chromosome were evaluated by the GA by using either PLS or CLS after which the RMSEP if the prediction set was found using the same spectral windows and the resulting matrix from the respective multivariate calibration. The actual fitness of each chromosome was the inverse square of the RMSEP.

The GA was limited to spectral windows of approximately 250 cm^{-1}; however, initially the GA was given enough spectral windows so that in theory it could encompass the entire spectrum (400-4000 cm^{-1}). This procedure allowed the GA to selectivity find the optimal spectral windows and the optimal number of spectral windows. The parameters for the GA were, mutation rate, 0.1; single crossover, 0.6; double crossover, 0.15; and default error, 0.005.

RESULTS

On hundred spectra were used as the data set for the CLS experiment. The training set for each analyte was composed of 75 spectra synthesized by adding the reference spectra of the analyte, each scaled by a different known coefficient, to 75 different backgrounds. The prediction set was composed of 25 corresponding spectra synthesized using different background spectra. In this way, we could know the exact concentration of the analyte for each run, which would be impossible for a 'real' field measurement. The results of this test are summarized in Table 1. The reported values for the TO-16 method are based on the best of all values attained. Similarly, the values for the half-height method were the best values that were obtained.

For each of the four analytes tested, the optimum results were always obtained using the genetic algorithm. For the CCl_2H_2, the differences in the results obtained by all three methods were statistically insignificant, while for the $CHCl_3$, the results obtained by the GA were twice as good as those obtained using the data between the half-height and more than three times better than the data obtained using the spectral windows obtained using the spectral windows recommended in the TO-16 practice.

For the PLS calibration, the spectra of 25 compounds were randomly chosen from the EPA vapor phase library

TABLE 1. RMSEPs of the CLS Results of the Various Methods of Spectral Window Selection

Analyte	GA	TO-16	Half-height
$CHCl_3$	0.0154	0.0324	0.0525
CCl_2H_2	0.0459	0.0490	0.0464
CH_3OH	0.0177	0.0355	0.0179
Hexane	0.0394	0.1085	0.0634

TABLE 2. Comparison of the average RMSEP for 25 analytes obtained by PLS using the traditional atmospheric windows with three (GA1) and two (GA2) spectral windows found by a GA

Method	RMSEP
Full-spectrum	0.0234
GA1	0.0018
GA2	0.0027

TABLE 3. Spectral Windows Chosen by the Genetic Algorithm for PLS (in cm^{-1})

	Window 1	Window 2	Window 3
GA1	712 - 1124	1136 - 1272	2756-2984
GA2	728 - 1256	2732 - 2980	------
Traditional	700 - 1300	2000 - 2250	2400 - 3000

of 100 different open-path background spectra. The prediction set of each compound was synthesized using 25 different open-path background spectra. The average RMSEP obtained with two and three windows selected by the GA are seen in Table 2, along with corresponding result obtained using the full spectrum. The window values selected by the GA are shown in Table 3. In general, the RMSEP obtained by the full spectrum PLS was comparable to the best RMSEP obtained by CLS. When the spectral windows for PLS were selected by a genetic algorithm, the RMSEP was reduced by a factor of ten, indicating the success of this approach.

We also experimented with allowing overlap of the spectral windows and forbidding overlap with respect to weighting of the PLS regression. In the training of the GA, we used a reduced number of factors during the training to increase the robustness of the PLS regression. This served to obtain the optimum window by severely stressing the ability of the PLS to accommodate the atmospheric water and carbon dioxide. In doing so, we ensured that the window the GA selected would contain the maximum amount of information.

DISCUSION AND CONCLUSIONS

The results of the CLS regression were positive. Since both the TO-16 and half-height methods use spectral windows that tend to be highly symmetric about the analyte peak, any noise that occurs on either side of the peak center change the prediction. Both methods are chosen using *a priori* knowledge, whereas the GA is a different type of method that makes no such assumptions. The results for the CLS indicate that the GA can select a spectral window that yields results that are slightly better than conventional methods. In some cases, there was an improvement by more than a factor of three, but in many cases the improvement was not as marked.

The results of the PLS experiment were more successful, as the selection of the optimum spectral windows allows for an improvement slightly greater than an order of magnitude. The windows selected by the GA roughly correspond to those of the 700-1300 cm^{-1} and the upper part of the 2250-3000 cm^{-1} window of the traditional usable spectral windows. It should also be noted that these results were obtained at a resolution of 8 cm^{-1}, which has been shown by a number of workers to give far poorer results using CLS than spectra measured as 8 cm^{-1} or better. Furthermore, this result indicates that there is far less need for obtaining spectra at high resolution than is currently recognized by the OP/FT-IR community.

ACKNOWLEDGEMENTS

The authors would like to that Dr. Erol Barbut in the Department of Mathematics for his assistance and the U.S. Department of Energy for their funding.

REFERENCES

1. Newman, A.R., *Anal. Chem.*, **69**, 43A-47A (1997).
2. Haarland, D. M., and Thomas, E. V., *Anal. Chem.*, **60**, 1193-1202 (1988).
3. Garriges, S., et al., *Ana. Chim. Acta*, **317**, 95-105 (1995).
4. Fredrickson, P. M., et al., *Appl. Spectrosc.*, **39**, 311-315 (1985).
5. Bangalore, A.S., et al., *Anal. Chem.*, **68**, 4200-4212 (1996).
6. Lucasius, C.B., and Kateman, G. *Trends in Anal. Chem.*, **10**, 254-261 (1991).
7. Paradkar, R. P., and Williams, R. R., *Appl. Spectrosc.*, **50**, 753-758 (1996).
8. Paradkar, R. P., and Williams, R. R., *Appl. Spectrosc.*, **51**, 92-100, (1997).
9. Massart, D., et al., *Anal. Chem.*, **67**,4295-4301 (1995).

Use of Partial Least Squares Regression for the Multivariate Calibration of Hazardous Air Pollutants in Open-Path FT-IR Spectrometry.

Brian K. Hart and Peter R. Griffiths

Department of Chemistry, University of Idaho, Moscow, ID, 83844-2343

Partial least squares (PLS) regression has been evaluated as a robust calibration technique for over 100 hazardous air pollutants (HAPs) measured by open path Fourier transform infrared (OP/FT-IR) spectrometry. PLS has the advantage over the current recommended calibration method of classical least squares (CLS), in that it can look at the whole useable spectrum (700-1300 cm^{-1}, 2000-2150 cm^{-1}, and 2400-3000 cm^{-1}), and detect several analytes simultaneously. Up to one hundred HAPs synthetically added to OP/FT-IR backgrounds have been simultaneously calibrated and detected using PLS. PLS also has the advantage in requiring less preprocessing of spectra than that which is required in CLS calibration schemes, allowing PLS to provide user independent real-time analysis of OP/FT-IR spectra.

INTRODUCTION

Open-path Fourier transform infrared (OP/FT-IR) spectrometry is an application of FT-IR spectrometry with many benefits to industry and regulatory agencies. OP/FT-IR allows the replacement of standard laboratory gas cells by a column of ambient air, typically 200-400 meters in length (although columns as long as 1 km or as short as 1 meter in length have been measured). This technique allows for the simultaneous detection and quantification over a wide spatial area of all compounds that are IR active and at sufficient concentration in the optical path (1). Since OP/FT-IR has the capability to perform real-time analysis of point and non-point source, without any sample collection and at a safe distance from the source, it has many advantages over competitive techniques. For example, OP/FT-IR has been used for the analysis of the atmosphere near Superfund sites and as a fence-line monitoring system at refineries and other industrial sites. It can provide early warning of any chemical release and serve to identify the type of gas being released to facilitate rapid and safe response. OP/FT-IR is not without its drawbacks. Weather conditions such as rain, snow, dust, or even wind can interfere with an analysis. The ubiquitous presence of water and CO_2 also limits the use of OP/FT-IR to the wavenumber ranges of 700-1300 cm^{-1}, 2000-2250 cm^{-1}, and 2400-3000 cm^{-1} (often referred to as the atmospheric windows).

Data reduction in OP/FT-IR spectrometry is usually done by a multivariate technique such as classical least squares (CLS) or partial least squares (PLS) regression. At present, CLS is the accepted calibration scheme for multivariate analysis by the OP/FT-IR community. In this approach, a suitable wavenumber range must be found that is free of any interference for each analyte of interest. Thus selection of the optimum spectral windows for CLS often requires an expert knowledge of spectroscopy on the part of the operator. As presently implemented for CLS, high-resolution spectra ($\Delta\nu \leq 1$ cm^{-1}) have been shown to yield the best predictions of concentration. Acquiring an appropriate background spectrum that is free of bands due to each analyte of interest is also important (and often time consuming) for CLS. An automated procedure capable of giving accurate results from low-resolution spectra would increase the acceptability of OP/FT-IR for environmental analysis.

In principle, PLS regression has few of the drawbacks listed above for CLS. The entire range of atmospheric windows may be used for the analysis of all compounds. A resolution of 8 cm^{-1} allows good results to be obtained by PLS, so that a relatively inexpensive, low-resolution instrument can be used for the analysis. Finally, as we will show in this paper, it is easier to select the appropriate background spectrum for PLS than for CLS. Ideally the background spectrum should be taken through a short (1-5 m) path, so that it can be acquired while the instrument is being set up, thereby minimizing both the installation time and logistical considerations. However, for CLS, it is necessary to compensate for the atmospheric absorption lines due to water and carbon dioxide prior to the data reduction step.

In this paper we will show that PLS can be easily automated to produce high quality results from low-resolution spectra even if the analyst knows little or nothing about spectroscopy. This system has been tested using a field test release of CH_2Cl_2 and $CHCl_3$ with excellent results.

THEORY

A basic assumption in any quantitative regression analysis is that the error primarily resides in the dependent

CP430, *Fourier Transform Spectroscopy:* 11th International Conference
edited by J.A. de Haseth

variable (2). CLS is based on the Lambert-Beer Law, which in its simplest terms is:

$$A = KC \quad (1)$$

where A is the optical signal (i.e., the absorbance at each wavelength of interest), K is a constant, and C is the concentration. In this notation A is the dependent variable implying that the main source of error is in the optical signal, which in FT-IR is rarely the case. PLS on the other hand is modeled on the inverse form of Beer's law:

$$C = PA \quad (2)$$

where P is a constant of proportionality. In this form of Beer's law, the primary source of error lies with the C term. Thus it can be easily argued that PLS is the more correct calibration scheme when working with spectroscopic data. A more thorough discussion of CLS and PLS is given in references (3)-(6).

The protocol for using OP/FT-IR in conjunction with CLS as the calibration routine is given in EPA Method TO-16 (7). In order to use CLS as the calibration routine, this protocol recommends that several steps be followed. These include selection of wavenumber regions for analysis in the presence of interfering species (TO-16, Section 8.3), which must be determined for each analyte combination separately. For the 100 analytes in the EPA vapor phase library, 100!, or 9.3×10^{157}, different possible sets of wavelengths can be selected. For PLS, on the other hand, we have used one set of wavelengths for every combination of analytes. Section 8.4 of TO-16 recommends that a background spectrum of the same optical path length used in the sampling procedure be used. This may require that the retroreflector, (or collection optics in a bistatic system) be moved upwind of the sample area and parallel to the wind direction, that the entire system be moved upwind of the sample area, that the operator is forced to wait until the analyte of interest has dissipated, or that a synthetic background be created from one of the sample spectra. For PLS, on the other hand, the reference spectrum can be a short path spectrum.

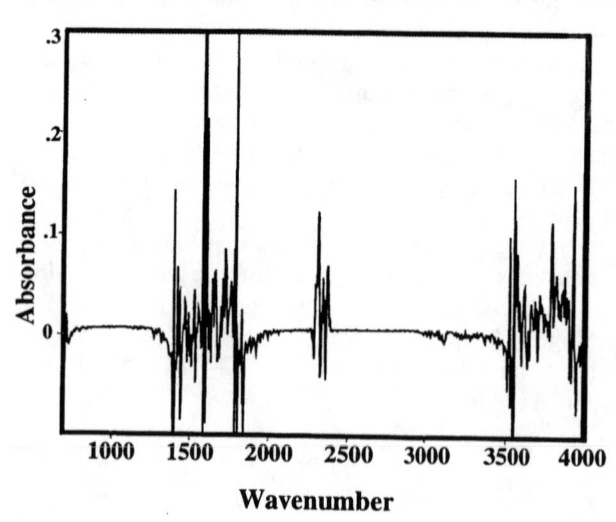

FIGURE 1. 600 meter OP/FT-IR spectrum with a 600 meter reference

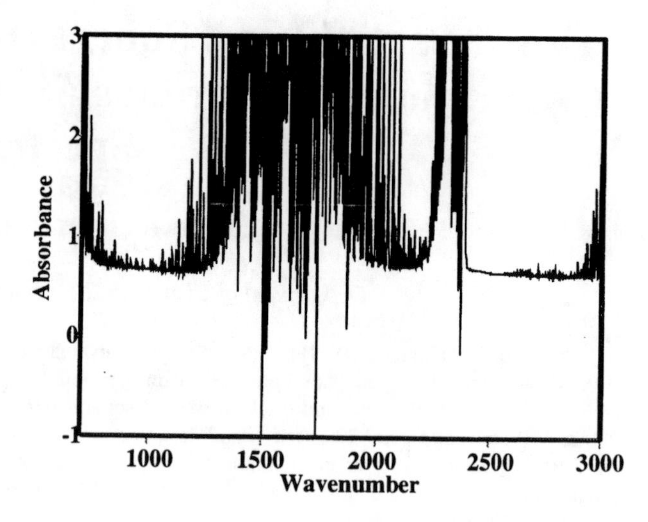

FIGURE 2. 600 meter OP/FT-IR spectrum with a short-path reference.

Figures 1 and 2 show the difference in the types of OP/FT-IR spectra used for the two types of calibration. Figure 1 is a typical background for an OP/FT-IR spectrum treated by CLS; note that the baseline is at zero and level, but that the amount of water vapor signal is greatly reduced from that found in Fig. 2. Figure 2 is an example of a background found in the OP/FT-IR spectra we have treated by PLS.

We have shown that PLS yields excellent analytical results, even when the spectra are measured at 8-cm^{-1}. Measuring the spectra at low resolution has the added benefit of greatly reducing the data acquisition for OP/FT-IR spectra.

EXPERIMENTAL

All calculations and regression analyses were performed with the use of MATLAB (The Mathworks, Inc., South Natick, MA, USA). Standards were all taken from the EPA Vapor Phase Library (prepared under EPA Contract 68D90055 by Entropy Environmentalists, Inc., Research Triangle Park, North Carolina, USA). Baseline correction and deresolution of the EPA vapor phase library were performed using GRAMS (Galactic Industries Corporation, Salem, NH, USA). The spectrometer used to acquire all OP/FT-IR spectra was a Bomem MB-104 capable of both monostatic and bistatic operation; for this study it was used exclusively in the monostatic configuration. The retro-reflector used for all measurements was an array of 44 individual 2-inch cube corners with an accuracy of 5 arc seconds (Opticon Inc., Billerica, MA, USA).

All spectra used for calibration and validation were synthetically created from over 500 OP/FT-IR background spectra and standard spectra from the EPA vapor-phase library. A typical calibration or training set consisted of

200 spectra containing between two and 100 analytes. To create the training set, M standard spectra were randomly selected from the list of 100 standards in a modified EPA vapor phase library, where the number of standards selected, M, is determined by the investigator. The modified EPA vapor phase library contains standards that have been background corrected, set to the appropriate resolution, and truncated to contain only the spectral region between 708-3000 cm^{-1}. The maximum absorbance due to the analyte was set to 0.3, with the intensities at all other wavelengths scaled accordingly. These standard spectra with maximum signal of 0.3 AU were then set to a concentration of 1.0 arbitrary concentration units (ACU).

A 1xN random number matrix was then generated for each of the M standards, where N is the number of spectra included in the training set. Random numbers were selected using MATLAB random number generator and bound between 1 and 0 inclusively. All the 1xN matrices were combined to create a MxN matrix, the **C** matrix. Each 1xN matrix was also multiplied by one of the standard spectra to create M LxN matrices, where L is the number of resolution elements in the standard spectra. A matrix of N OP/FT-IR spectra, truncated to the spectral region of 708-3000 cm^{-1} and at the proper resolution, would also be created to give a LxN matrix of backgrounds. All the LxN matrices would then be added together to create one LxN matrix, the **A** matrix. A validation set was created by the same method and used for testing of the calibration routine.

The PLS routine (PLS2) then took **A** and **C** from the training set and produced a set of matrices to be used in the PLS prediction routine. The **A** matrix from the validation set was then input into the prediction routine producing a matrix, **V**, the predicted concentrations of each of the analytes in each of the synthetic spectra.

The error in the calibration routine was then determined as the root mean squared error of prediction (RMSEP) according to:

$$RMSEP_M = \sqrt{\sum_{i=1}^{N} (V_{Mi} - C_{Mi})^2} \quad (3).$$

The error was also checked visually by plotting $\mathbf{V}_M$ vs. $\mathbf{C}_M$ for the validation set. This step was performed to ensure that the error was evenly distributed over the entire concentration range.

For CLS, the training and validation matrices were created in the same manner described above, with one important difference, in that all the OP/FT-IR spectra were baseline corrected before being used. CLS is very sensitive to changing baselines and if the baselines were left uncorrected, CLS gave meaningless results.

To test the success of each approach, a set of field spectra were also acquired. This set of OP/FT-IR spectra was acquired at 150 meters total pass, with a short-path background. Controlled releases of dichloromethane and trichloromethane were achieved by allowing these compounds to evaporate from four Petri dishes containing the pure analyte. Each Petri dish was placed adjacent to the retroreflector (75 meters from the spectrometer) and OP/FT-IR spectra were acquired at different resolutions. Calibration was performed using 200-400 meter OP/FT-IR with short-path backgrounds of the type shown in Fig. 2.

RESULTS

As a benchmark, the performance of CLS for various analytes was determined using EPA Method TO-16. For CLS, the RMSEP typically ranged between 0.05 and 0.02 ACU. The concentrations for all analytes varied from 0 to 1 in the test sets. For comparison, a RMSEP of 0.5 corresponds to an average error of 100%. Thus the average concentration error for CLS ranged from 10% to 4%. Our best result for CLS (a RMSEP of 0.017, corresponding to a concentration error of 3.4%), was acquired by using only the top 50% of the peak as compared to EPA Method TO-16 which uses the top 99% of the peak. This trend was not consistent and the actual result depended on the exact window used. Selection of windows became increasingly difficult as the number of analytes in the calibration and validation sets increased (with correspondingly higher RMSEPs).

The use of PLS was then tested under various conditions to investigate the effect of resolution and spectral preprocessing. The results of these tests showed that, unlike the case for CLS, the gain in accuracy resulting from any type of spectral preprocessing was minimal.

The selection of the correct number of factors for use in PLS is an area of constant debate. We decided to use the RMSEP as the criterion for selecting the number of factors with a RMSEP of 0.01 as our goal. Table 1 shows that this value was reached when using a X+8 formula for the number of factors, where X is the number of analytes in the calibration. In other words, the background in OP/FT-IR spectra measured over a wide range of pathlengths could be modeled with just 8 factors. This was further tested to include up to 100 analytes from the EPA Vapor Phase

TABLE 1. Number of PLS factors required to produce a RMSEP of less than 0.01.

# of Factors	Number of Analytes				
	1	**2**	**3**	**4**	**5**
7	0.0138	0.0141	0.0157	0.0282	0.1104
8	0.0105	0.0106	0.0137	0.0161	0.0276
9	***0.0097***	0.0105	0.0111	0.0163	0.0202
10	0.0088	***0.0096***	0.0109	0.0116	0.0173
11	0.0083	0.0091	***0.0097***	0.0112	0.0131
12	0.0079	0.0091	0.0091	***0.0096***	0.0126
13	0.0080	0.0090	0.0086	0.0087	***0.0098***
14	0.0088	0.0083	0.0088	0.0086	0.0093
15	0.0090	0.0082	0.0078	0.0076	0.0089
	RMSEP				

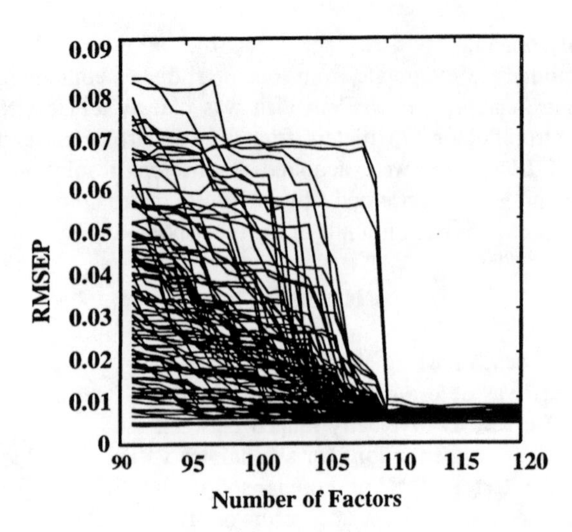

FIGURE 3. Simultaneous analysis of 100 analytes, showing that 110 factors are required to give a RMSEP of less than 0.01.

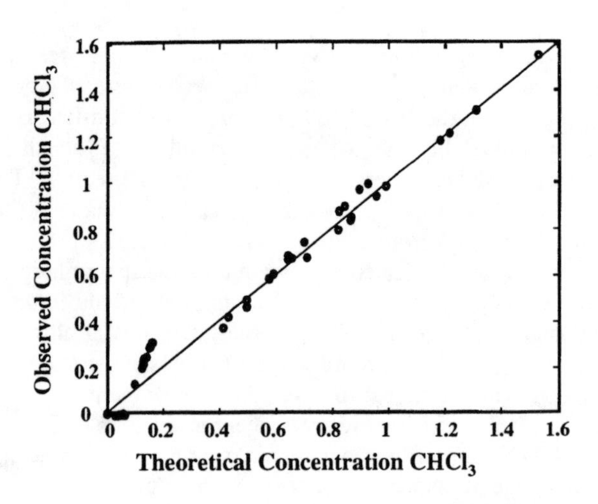

FIGURE 4. Results of the field test for $CHCl_3$

Library. As can be seen in Fig. 3, all 100 analytes could be determined to the 0.01 RMSEP level by using a X+10 factors formula. The need to add two more factors is attributed to the increased spectral overlap as more analytes are added to the test set. We have used the same factors for all synthetic and field data reported in this paper.

The results of the field test are shown in Fig. 4 for $CHCl_3$. The "theoretical" concentrations are the result of performing a univariate Beers-Law plot determination on each individual spectrum. The RMSEP for the PLS calibration using the atmospheric window was 0.0049 ACU.

DISCUSSION AND CONCLUSION

It has been shown that PLS calibration using low-resolution OP/FT-IR spectra offers many advantages over the conventional CLS calibration approach using high-resolution spectra. Using the full atmospheric windows, PLS was able to determine any set of analytes to an accuracy of greater than 0.01 RMSEP. This was accomplished without any spectral preprocessing, using short-path backgrounds (which are easier to obtain than any of the backgrounds currently recommended for CLS). Anywhere from 1 to 100 analytes could be determined simultaneously. The RMSEP was over an order of magnitude better than the best results we were able to attain with CLS under optimized conditions.

Measuring at lower resolution provides many advantages. The signal-to-noise ratio of the spectra is improved while the data acquisition time is reduced. Disk storage space and computational time are also greatly reduced as the number of resolution elements is reduced. Of even greater importance is the fact that the need for a trained spectroscopist is obviated by using an automated PLS calibration system without a loss in accuracy.

ACKNOWLEDGEMENTS

This work was in part funded by a grant from the University Research Consortium of the Idaho National Engineering and Environmental Laboratory.

We wish to thank Husheng Yang for his assistance in acquiring the OP/FT-IR spectra used in this study and James Berry for rewriting the PLS2 algorithm used for most of this study.

REFERENCES

1. Newman, A. R., *Analytical Chemistry*, **69,** 43A-47A, 1997
2. Mark, H, *Principles and Practice of Spectroscopic Calibration*, New York NY, USA, John Wiley & Sons Inc., 1991, Chapter 2.
3. Thomas, E. V., *Analytical Chemistry*, **66**, 795A-804A, 1994
4. Geladi, P., Kowalski, B. R., *Analytica Chimica Acta*, **185**, 1-17, 1986
5. Brown, P. J., *Measurement, Regression, and Calibration*, New York, Oxford University Press Inc., 1993
6. Lorber, A. and Kowalski, B. R., *Applied Spectroscopy*, **42**, 1572-1574, 1988
7. Draft EPA Method TO-16, Compendium of Methods for Organic Air Pollutants, U.S. Environmental Protection Agency, Washington, DC (1996).

Application of Neural Networks to Compound Identification in Open-Path FT-IR Spectrometry

Husheng Yang and Peter R. Griffiths

Department of Chemistry, University of Idaho, Moscow, ID 83844-2343

Neural networks have been applied in an attempt to determine the feasibility of recognizing whether or not a given analyte is present in an open-path Fourier transform infrared (OP/FT-IR) spectrum measured at low resolution. The neural network architecture used in this paper was a two layer feed-forward network trained by fast backpropagation. A hyperbolic tangent sigmoid transfer function was used in both layers. Each network has only one output and was trained to recognize only one compound. Synthesized open-path spectra, which were obtained by digitally adding randomly scaled reference spectra and open-path background spectra, were used to train the neural networks. Spectral windows containing only the absorption bands of the analyte were used as the neural network input. Trained neural networks were tested by experimentally measured OP/FT-IR spectra.

INTRODUCTION

Most gases and volatile organic compounds have characteristic absorption spectra in the mid-IR spectral region. OP/FT-IR spectrometry is considered a competitive method for outdoor environmental air monitoring [1-3]. One of the most important advantages of OP/FT-IR is that no sampling is required. But this advantage requires the capability of analyzing multiple components concurrently by the interpretation of the measured spectrum [3].

Classical least squares (CLS) and partial least squares (PLS) are the two most popularly used multivariate techniques in the analysis of OP/FT-IR data [1,4]. When using CLS and PLS, a prior knowledge of all the possible compounds whose absorption spectrum can interfere with the absorption feature being analyzed is required. If this process is done manually, it is usually time-consuming and requires the operator to be well trained in molecular spectroscopy [2]. Since OP/FT-IR spectra are measured in open air, many adverse conditions, such as a change in temperature and the strong absorption bands of water and carbon dioxide, can affect the quality of the spectrum. This also increases the difficulty of manual identification of compounds.

Artificial neural networks have been used extensively as a tool for pattern recognition and classification problems [5-7]. One advantage of neural networks is that once a network is well-trained, it can retain excellent performance even if degraded, noisy or missing data are applied [8]. In the area of infrared spectroscopy, neural networks have been used for spectral classification, structural feature recognition, and retrieval of compounds [6, 9-13].

We have tested the feasibility of using four types of networks including: a multilayer feed-forward network, a Kohonen self-organizing feature map, a bi-directional associative memory, and a Hopfield network for the data processing of OP/FT-IR spectrum. In this paper we report just the application of a two-layer feed-forward neural network for compound identification in an open-path spectrum. Training and testing methods are described. Trained networks were tested by field-measured OP/FT-IR spectra.

EXPERIMENTAL

Reference spectra used for the training of the neural networks came from the EPA reference library (prepared by Entropy Inc. under EPA contract 68D90055) and spectra measured in our laboratory. The EPA library is comprised of 385 spectra of 105 compounds which were measured at a nominal resolution of 0.25 cm^{-1}. 8-cm^{-1} resolution spectra were obtained by averaging the neighboring data points of high resolution spectra and were used in this research.

Reference spectra which were used in this paper but not included in the EPA library were measured in our laboratory. Vapor phase samples were generated using a static method by a procedure described previously [14]. FT-IR spectra were measured using a Perkin-Elmer Spectrum 2000 spectrometer (Beaconsfield, UK) equipped with a liquid nitrogen-cooled (non-factory installed) mid-band MCT detector (Graseby Infrared, Orlando, FL). The original data were recorded as unapodized interferograms. Data manipulation after collection of interferograms was performed using GRAMS/32 (Galactic Industries Corporation, Salem, NH).

Open-path background spectra were measured by a Bomem MB-104 OP/FT-IR spectrometer (Quebec, Canada) operating in the monostatic mode with a cube

CP430, *Fourier Transform Spectroscopy:* 11th International Conference
edited by J.A. de Haseth

corner array retroreflector. Interferograms were collected from a single-pass pathlength of 25 meters to 375 meters with a nominal resolution of 8 cm^{-1}. Single beam spectra were computed from the measured interferograms using medium Norton-Beer apodization. Absorbance spectra were obtained by ratioing a long-path single-beam spectrum with a short-path single-beam spectrum, with the resultant transmittance spectrum converted to absorbance. These absorbance spectra are referred to below as the open-path backgrounds.

Open path spectra of dichloromethane, chloroform, methanol, ethanol, 1-propanol, 2-propanol and 1-butanol were measured with the same spectrometer and retroreflector as the open-path background spectra. The pathlength was fixed at a single pathlength of 75 meters. All the original data were collected as interferograms at 8 cm^{-1} resolution. Absorbance spectra were obtained by using 2-meter pathlength single-beam spectra as the background.

All the neural network computations were conducted with MATLAB (The MathWorks, Inc., Natick, MA). Functions provided by the Neural Network Toolbox for Matlab (Howard Demuth and Mark Beale) were used if possible. The theory of neural networks is not covered in this paper. (For a description of neural network algorithms, the reader is referred to the reference [7] or [5]).

RESULTS AND DISCUSSION

If the infrared spectrum of a pure compound is measured in a laboratory, the compound can be easily identified by spectral searching. Unfortunately this is not the case in open-path FT-IR measurements. Currently there is not a simple method of resolving an OP/FT-IR mixture spectrum into pure compound spectra without prior knowledge of the compounds in question. In addition, the presence of high noise levels and poor baselines in an open-path spectrum cannot be corrected easily. As mentioned above, artificial neural networks have proven to have the ability of recognizing corrupted patterns. Thus, we believe it is possible to use neural networks to identify all the compounds in an OP/FT-IR spectrum.

Ideally we would like to have a neural network that can recognize all the compounds in a spectrum, for example, to train one neural network to recognize all the 105 compounds in the EPA library. Our preliminary studies have indicated that from a practical perspective it is impossible to train such a large-scale neural network, especially on a personal computer. Thus in this study, each neural network was trained to recognize only one compound; this has proved to be a more flexible and efficient method. In another attempt, we tried to train the neural network to give concentration information, but this has proven difficult to accomplish successfully with the network architecture used in this paper, and PLS regression is probably better for this application.

In a feed-forward multilayer neural network, several parameters, such as the number of hidden layers, the number of neurons and the transfer function used in each layer can be optimized for better performance. Currently there is not a simple method to determine these parameters. Most of the work is based on trial and error. Normally, a more complex neural network is a more powerful network; however, this may allow the neural network become easily overtrained. In practice, a rule of thumb is to use the simplest network that can adequately represent the training set.

Based on many tests by trial and error, we finally chose a two-layer neural network architecture which has ten neurons in the first (hidden) layer and one neuron in the second (output) layer. Both layers used a hyperbolic tangent sigmoid transfer function. The notation and representation used in this paper are the same as those in reference [5]. Fast backpropagation using momentum and an adaptive learning rate was used to train the network.

The FT-IR spectra we used as the training set and testing set have 856 data points in the frequency range of 4000-700 cm^{-1}. If the full spectrum is used to train the neural network, the network is too large and it takes a long time and a large amount of computer memory to train. For this reason a data reduction method is necessary. For this research, spectral windows containing only the absorption bands of the analyte were used as neural network inputs, instead of the full spectrum.

Backpropagation is a supervised training method which requires a target for each training spectrum. What we are interested in is a boundary to separate all spectra into two categories, those containing the compound that we want to identify and those not containing the compound. A natural choice of this boundary is zero. For example, we train all spectra containing the analyte being identified to give an output of greater than zero, and the spectra not containing the analyte of interest to give an output of less than zero. In practice, the exact target for each training spectrum is not known.

When the neural network was trained with a spectrum, the neural network output of the spectrum was calculated first. If the spectrum contained the analyte and gave an output greater than 0, the spectrum was not included in the training set; if a spectrum did not contain the analyte and gave an output less than 0, the spectrum was also skipped. Otherwise a spectrum is trained to a target of +0.1 or –0.1, which is a function of whether the spectrum contains the analyte or not.

Synthesized OP/FT-IR spectra were used to train and test the neural network. Compared to laboratory-measured spectra, open-path spectra have a significantly higher noise level and much poorer baseline. For example, the average RMS noise of 5 alcohol reference spectra is only 0.0007 absorbance unit, while the corresponding open-path spectra have an average RMS noise of 0.010, calculated from 2200 to 2100 cm^{-1}. The high noise level may interfere with the identification of a compound. Overlapped bands of other compounds may also interfere with the identification. The synthesized spectra should reflect both kinds of interference. To synthesize the OP/FT-IR spectra, reference spectra were scaled with randomly varying factors and added to open-path background spectra. The result was a series of simulated OP/FT-IR spectra, in which the analyte of interest was both present and absent. Normally half of the synthesized spectra were used as the training set and half as the testing set, with the training set and testing set being independent. In all spectra containing the analyte, the analyte absorbance was restricted to between 0.01 and 0.6. The lower absorbance limit was determined by the RMS noise of the open-path background spectra.

The open-path background has a significant influence on the training time of the neural network. When open-path spectra instead of pure reference spectra were used to train the neural network, the training time required usually increased significantly, indicating that the identification of a compound in an open-path spectrum is more difficult than in a laboratory-measured spectrum.

Two neural networks were trained to identify dichloromethane and chloroform in an open-path spectrum containing a total of ten components. Both neural networks were tested with synthesized open-path spectra, and more than 90% of the testing spectra were identified correctly. Field-measured open-path spectra of chloroform (12 spectra, peak absorbance 0.26-0.63), dichloromethane (12 spectra, peak absorbance 0.15-0.47), and mixtures of these two compounds (12 spectra, peak absorbance 0.20-0.51) were also used to test the two neural networks, and all spectra were identified correctly.

To further test this method, another five neural networks were trained to identify methanol, ethanol, 1-propanol, 2-propanol, and 1-butanol. Figure 1 shows the reference spectra of the five alcohols in the region of the C-O stretching bands. The structural and spectral similarity of these five compounds should be noted. Figure 2 shows the open-path spectra of these compounds. It can be seen that the peak shape and band position remains the same for the open-path and laboratory-measured spectra, but the noise level in the open-path spectra is much higher. These five neural networks were tested with field-measured spectra and the results are summarized in Table 1. It should be mentioned that these spectra were measured at relatively low concentrations; the highest peak absorbance was 0.115, the lowest peak absorbance was 0.005. All 92 spectra were identified correctly by the methanol, 1-propanol, and 2-propanol neural networks. There are four ethanol spectra (out of 24) not classified correctly by the ethanol neural network and eight 1-butanol spectra (out of 26) not classified correctly by the 1-butanol network. The high similarity of spectra of ethanol and 1-butanol can be seen from Figures 1 and 2. A careful inspection of the field-measured spectra showed that the incorrectly-identified spectra had the lowest absorbance values, with very low signal-to-noise ratios.

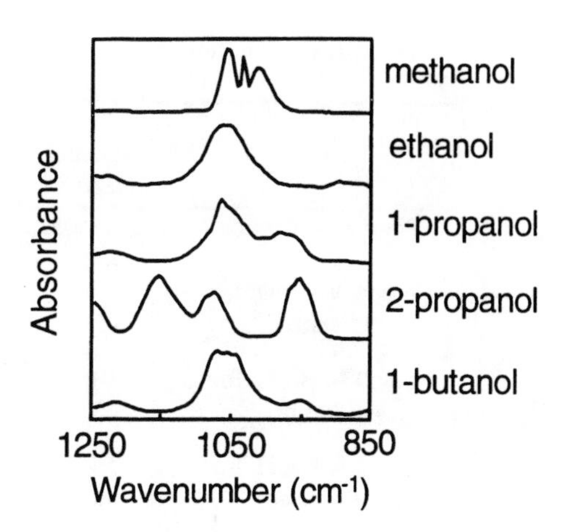

FIGURE 1. Reference spectra of five alcohol compounds in the wavenumber range of 1250-850 cm^{-1}.

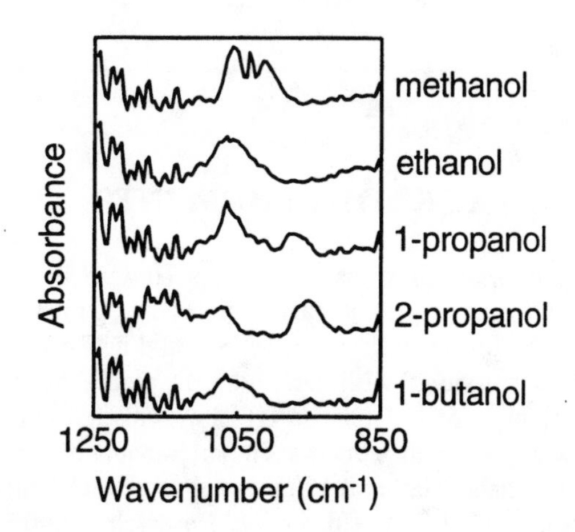

FIGURE 2. Field-measured open-path FT-IR spectra of the same compounds as in Fig. 1 in the wavenumber range of 1250-850 cm^{-1}. The noise level in the open-path spectra is much higher than in the reference spectra.

CONCLUSIONS

Two-layer feed-forward neural networks were successfully trained and tested for compound identification of open path FT-IR spectra. The neural networks can be trained by synthesized spectra which were obtained by digitally adding reference spectra and open-path background spectra. Using only those spectral windows containing absorption bands of the analyte as neural network input provided a convenient way for data reduction. The effectiveness of this method was proven by neural networks identifying five different alcohol compounds which have similar structure and reference spectra. A trained neural network can identify a compound in an open-path spectrum quickly and accurately. This provides a useful method for data processing and automation of OP/FT-IR spectrometry.

TABLE 1. Testing Results of the Five Alcohol Neural Networks by Field-Measured Open-Path Spectra

	22 Methanol Spectra	24 Ethanol Spectra	12 1-Propanol Spectra	8 2-Propanol Spectra	26 1-Butanol Spectra
Peak Absorbance	0.014-0.082	0.005-0.049	0.031-0.115	0.029-0.068	0.017-0.058
Neural Network for Methanol	22+ 0-	0+ 24-	0+ 12-	0+ 8-	0+ 26-
Neural Network for Ethanol	0+ 22-	20+ 4-	0+ 12-	0+ 8-	0+ 26-
Neural Network for 1-Propanol	0+ 22-	0+ 24-	12+ 0-	0+ 8-	0+ 26-
Neural Network for 2-Propanol	0+ 22-	0+ 24-	0+ 12-	8+ 0-	0+ 26-
Neural Network for 1-Butanol	0+ 22-	0+ 24-	0+ 12-	0+ 8-	18+ 8-

ACKNOWLEDGMENTS

The authors wish to thank Professor Howard B. Demuth of Department of Electrical Engineering, University of Idaho (UI), for his assistance in the neural network design. Useful discussions with Dr. John D. Jegla of the UI Department of Chemistry are also gratefully acknowledged. This work was in part supported by a grant from the Idaho National Engineering and Environmental Laboratory (INEEL) University Research Consortium. The INEEL is managed by Lockheed Martin Idaho Technologies Company for the U.S. Department of Energy, Idaho Operations Office under Contract No. DE-AC07-94ID13223.

REFERENCES

1. Russwurm, G. M. and Childers, J. W., *FT-IR Open-Path Monitoring Guidance Document,* Second Edition, Research Triangle Park, NC 27709: ManTech Environmental Technology, Inc., 1995.
2. Newman, A. R., *Analytical Chemistry* **69,** 43A-48A (1997).
3. McClenny, W. A., "Program Objectives and Status For the U.S.EPA Program on FTIR-Based Open-Path Monitoring", in Proceedings of *Optical Remote Sensing for Environmental and Process Monitoring*, Dallas, Texas, 1996.
4. Herget, W. F., "Effects of Spectral Resolution on Classical Least Squares Analysis of FTIR Spectra", in Proceedings of *Optical Remote Sensing for Environmental and Process Monitoring*, San Francisco, CA, 1995.
5. Hagan, M. T., Demuth, H. B., and Beale, M., *Neural Network Design*, Boston: PWS Publishing Co., 1996.
6. Weigel, U.-M. and Herges, R., *J. Chem. Inf. Comput. Sci.* **32,** 723-731 (1992).
7. Zupan, J. and Gasteiger, J., *Analytica Chimica Acta* **248,** 1-30 (1991).
8. Elling, J. W., Lahiri, S., Luck, J. P., Roberts, R. S., Hruska, S. I., Adair, K. L., Levis, A. P., Timpany, R. G., and Robinson, J. J., *Analytical Chemistry* **69,** 409A-415A (1997).
9. Meyer, M., Meyer, K., and Hobert, H., *Analytica Chimica Acta* **282,** 407-415 (1993).
10. Tanabe, K., Tamura, T., and Uesaka, H., *Applied Spectroscopy* **46,** 807-810 (1992).
11. Novic, M. and Zupan, J., *J. Chem. Inf. Comput. Sci.* **35,** 454-466 (1995).
12. Ricard, D., Cachet, C., and Cabrol-Bass, D., *J. Chem. Inf. Comput. Sci.* **33,** 202-210 (1993).
13. Klawun, C. and Wilkins, C. L., *J. Chem. Inf. Comput. Sci.* **36,** 69-81 (1996).
14. Richardson, R. L. and Griffiths, P. R., *Appl. Spectrosc.* (in press) (1997).

Investigation of the Effects of Resolution on Spectral and Interferogram-Based Analyses

Ndumiso A. Cingo and Gary W. Small

Center for Intelligent Chemical Instrumentation, Department of Chemistry, Clippinger Laboratories, Ohio University, Athens, OH 45701

The effects of resolution on spectral and interferogram-based analyses are investigated by comparing results obtained with multivariate calibration models built with spectral and interferogram data. The performance of the models in isolating information due to an analyte band that is overlapped to varying degrees with an interfering band is evaluated. The effects of changing spectral resolution on the spectral-based analyses is also studied, and the results obtained are compared to those from an analysis based on the direct use of interferogram data. A nonlinear multivariate calibration method is also applied to the interferogram-based analysis.

INTRODUCTION

Spectral resolution in infrared spectroscopy is an important issue when analyzing mixture samples characterized by closely spaced, overlapped spectral features. The successful resolution of these features may require a high resolution spectroscopic measurement. In Fourier transform infrared (FTIR) spectroscopy high resolution measurements increase the hardware requirements of the spectrometer, as the interferometer mirror drive must be able to maintain precise optical alignment over a longer distance (1).

Recent work performed in our laboratories has focused on developing data analysis methodology that would be compatible with a simplified spectrometer (2,3). The approach used implements an interferogram-based analysis by applying bandpass digital filters directly to interferogram data to isolate analyte features of interest, and performs qualitative and quantitative analyses on short segments of the resulting interferogram data

Although the feasibility of implementing this analysis has been established (3), an issue that needs to be addressed is how the concept of resolution in the spectral domain manifests itself in a quantitative analysis conducted in the time-domain with short interferogram segments displaced from the centerburst. In a conventional spectral analysis the last point collected in the interferogram will ultimately determine the resolution in the resulting spectra, whereas in an interferogram-based analysis additional factors such as the size and position of the segment may affect the effective spectral resolution. The work presented here explores the effect of spectral resolution when implementing a time-domain analysis, and compares the results obtained with those produced by a conventional analysis of spectral data.

EXPERIMENTAL SECTION

Simulated infrared spectra were used in this study. The spectra were constructed to simulate two components occurring at various concentration levels. The first component was the target analyte with a Gaussian peak centered at 950 cm^{-1}. The second component was an interference with a Gaussian peak centered at different distances from the analyte band to simulate varying degrees of overlap between the bands in different data sets. The mixture spectra were constructed by generating Gaussian transmittance spectra to simulate the analyte and interference peaks, and multiplying these by the spectral profile of a theoretical blackbody source operating at a temperature of 1500 ^{0}C. Gaussian-distributed random noise was then added to the resulting single-beam spectra at a level such that the absorbance signal-to-noise ratio at the highest analyte concentration was approximately 50.

The analyte and interference absorbances at the peak maxima were randomly set to 0.01-0.10 absorbance units (AU) at 100 different levels for each. This produced a total of 100 mixture samples. Assuming a path length and absorptivity product of unity, the concentrations were set at the same numerical values as the absorbances. The spectra corresponding to different samples were created in triplicate by using different random number seeds to generate the noise. Background spectra were created by adding noise to the blackbody spectral profile, and these were used to carry out an absorbance spectral analysis. The analyte and interference concentrations were randomly selected to minimize correlation between the two. The correlation coefficient between the two components was r = 0.0348. These spectra were inverse Fourier transformed into 4096 (4K) point interferograms with a sampling rate of 1 point collected at every eighth zero-crossing of the HeNe reference laser. This translated

CP430, *Fourier Transform Spectroscopy:* 11th International Conference
edited by J.A. de Haseth

TABLE 1. Partitioning of data.

Data Sets	No of Samples	No of Spectra
(1) **Calibration**:	**75**	**225**
Subsets		
Calibration:	*60*	*180*
Monitoring:	*15*	*45*
(2) **Prediction**:	**25**	**75**
Total = (1) + (2)	**100**	**300**

to a maximum digitized frequency of 1975.35 cm^{-1}, and a nominal spectral point spacing of 1 cm^{-1}.

Table 1 shows how the resulting data were partitioned. Partial least-squares (PLS) regression models were developed with the calibration subset, and these were used to explore different model sizes, spectral ranges, interferogram segment ranges and filter widths. These models were tested with the monitoring subset to determine the optimal parameters. The calibration and monitoring subsets were then recombined to form the main calibration set, and this was used to build models with the parameters determined to be optimal. These models were then used to predict the prediction data set, which had been withheld from the development process. Parameters explored were filter passband full-widths at half-height (fwhh) of 10 to 65 cm^{-1}, interferogram segments of lengths of 50 to 400 points, and segment starting points of 50 to 950, relative to the centerburst. Spectral ranges explored were 25 to 400 cm^{-1} wide, starting from 1200 and going to 800 cm^{-1}. Calibration model sizes investigated ranged from 1 to 5 PLS factors.

RESULTS AND DISCUSSION

Figure 1 shows different absorbance spectra with the analyte and interference peaks at different degrees of separation. The analyte peak is always centered at 950 cm^{-1}, and the interference peak is at (1) 952 cm^{-1}, (2) 959 cm^{-1} and (3) 970 cm^{-1} in different spectra. The fwhh of the analyte and interference bands are 15 and 50 cm^{-1},

TABLE 2. Effect of spectral overlap on (A) absorbance, (B) single-beam spectral analyses, and interferogram analyses using (C) linear and (D) nonlinear PLS.

Analysis type	Range	No. of PLS factors	SEC in arb. units $*10^{-4}$	SEP in arb. units $*10^{-4}$
A (cm^{-1})	1200-800	4	5.71	8.63
B (cm^{-1})	1200-825	3	13.7	16.7
C (points)	250-300	1	24.5	22.7
D (points)	150-250	5	7.12	7.94

respectively, while the concentrations are 0.035 and 0.094 arbitrary units, respectively, in each of the three plotted spectra. The appearance of the spectra clearly indicates the increasing overlap of the bands as the positions are brought closer together

Overview of Methodology

In the work presented here, the effect of spectral resolution on an analysis performed with interferogram data is addressed by comparing the results of quantitative analyses performed on data sets in which there were varying degrees of separation between the analyte and the interference bands. The results obtained were then compared to those produced by an analysis of both single-beam and absorbance spectral data.

Previous work in our laboratories has established the feasibility of performing a quantitative analysis on the basis of short segments of filtered interferograms (3). The analysis begins by applying bandpass digital filters directly to the analyte interferogram data. Calibration models are then developed from the resulting filtered interferogram data without the use of reference or background measurements. These models are used to extract quantitative information about the analyte from the interferogram.

Effect of Spectral Resolution on Calibration Model Performance

The issue of spectral resolution in the interferogram and spectral domains was studied by evaluating the performance of calibration models built using interferogram and spectral data sets in which the analyte and interference bands had varying degrees of overlap. The bands were separated by 2 cm^{-1}, 9 cm^{-1} and 20 cm^{-1}.

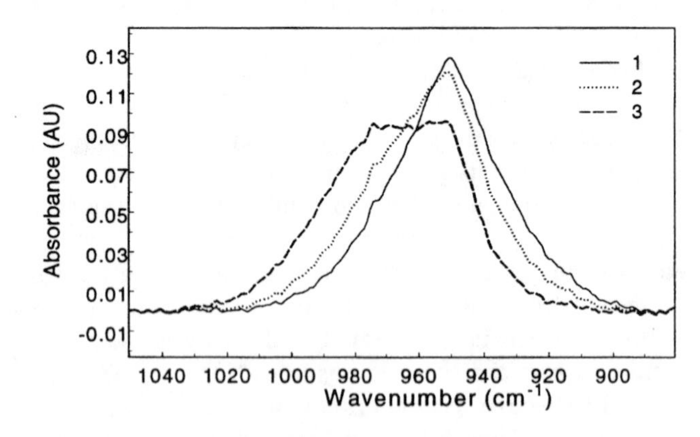

FIGURE 1. Spectra of target analyte band centered at 950 cm^{-1} and interference band centered at (1) 952 cm^{-1}, (2) 959 cm^{-1}, and (3) 970 cm^{-1}.

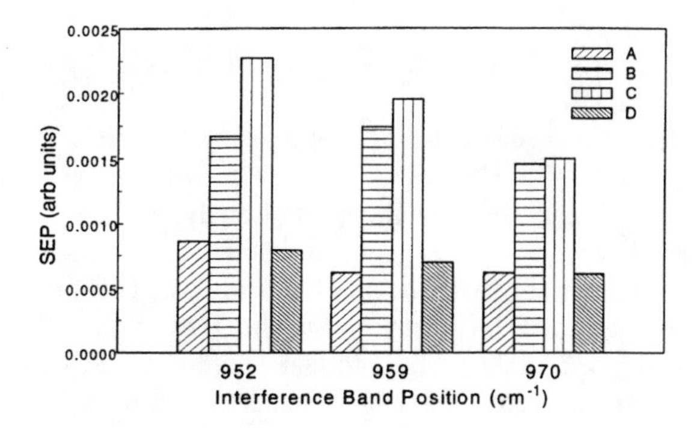

FIGURE 2. Effect of decreasing spectral overlap. The analyses were performed with (A) absorbance data, (B) single-beam data, (C) interferogram data and a linear PLS model and (D) interferogram data and a quadratic PLS model.

The fwhh values for the analyte and interference bands were 15 and 50 cm^{-1}, respectively.

Table 1 shows the results obtained when the analyses were performed with absorbance and single-beam spectral data, and interferogram data. The single-beam spectral analysis was implemented for comparison with the interferogram analysis since they both do not require the use of a background or reference measurement. In this case the interference band was centered at 952 cm^{-1}, only 2 cm^{-1} from the target analyte band. This data set presented the greatest challenge to the methods used in this study. For the absorbance and single-beam spectral data sets, the spectral resolution was 1 cm^{-1}, the highest used in this study. The absorbance analysis produced significantly better results than the single-beam analysis, most likely due to the linearity of absorbance with concentration described by the Beer-Lambert law. The contributions of the spectral background are also removed in the absorbance data. A comparison between the results obtained with an interferogram analysis implemented with a linear PLS algorithm, and those obtained with the spectral analyses reveals that the absorbance analysis gives the best results. While the single-beam analysis does not produce results that are as good as the absorbance case, it still gives significantly better results than the interferogram analysis. The interferogram analysis, however, is being performed with an interferogram segment of only 50 points, while the spectral analyses begin with interferograms of 4096 points each in each case.

A possible reason for this relatively poor performance of the interferogram analysis may lie in a key assumption that is made when building calibration models that relate analyte concentration to interferogram intensity data (3). This assumption is that the application of a bandpass digital filter to the raw interferogram will sufficiently suppress the frequencies outside of the analyte bandwidth and transmit those in which only the analyte is absorbing.

TABLE 3. Effect of spectral resolution on absorbance spectral analysis.

Resolution (cm^{-1})	Range (cm^{-1})	No. of PLS factors	SEC in arb units $*10^{-4}$	SEP in arb units $*10^{-4}$
1	1200-800	4	5.71	8.63
2	1200-800	4	6.11	8.76
4	1120-820	5	6.80	9.69
8	980-930	5	9.15	9.74
16	1120-920	2	13.5	14.1

If, however, there is more than one component contributing to variation in the spectral data, then the extent to which interference components are overlapped with the target analyte band will place more pressure on the filter to suppress frequencies other than those of the analyte. The closer the interference band is centered to the analyte band, the narrower the filter will need to be in order to transmit only analyte frequencies. A common characteristic of finite impulse response filters (FIR) is that attenuation characteristics outside the specified bandpass region become poorer as the bandpass width becomes narrower (4). In the current example, the interference band is centered only 2 cm^{-1} from the analyte, while the optimal FIR filter used in the analysis had a passband fwhh of 35 cm^{-1}. The assumption of only target analyte frequencies being transmitted is clearly being violated. This violation complicates the already nonlinear relationship between filtered interferogram intensity and concentration. The significantly degraded results from the interferogram analysis implemented with a linear PLS algorithm can thus be explained in this light.

To deal with the occurrence of this potential nonlinear effect, a quadratic PLS algorithm (5) was implemented with the interferogram analysis procedure. A comparison in Table 2 between the results obtained when this algorithm was implemented in the interferogram analysis and those obtained with other methods shows that the outcome of a nonlinear modeling of the data is a significant improvement over the linear approach, and indeed is better than the single-beam spectral analysis while being competitive with the absorbance approach. This result confirms the suspicion of the presence of nonlinearity in the filtered interferogram data.

Figure 2 shows the results obtained when the degree of overlap between the target analyte and interference band changes. The general trend observed is an improvement in the performance of all the methods employed as the degree of overlap decreases. This is expected since the ability to model the analyte increases with greater separation between the bands since there is less interference. Again, the results from the absorbance spectral analysis and nonlinear interferogram analysis are comparable, and are both significantly better than the

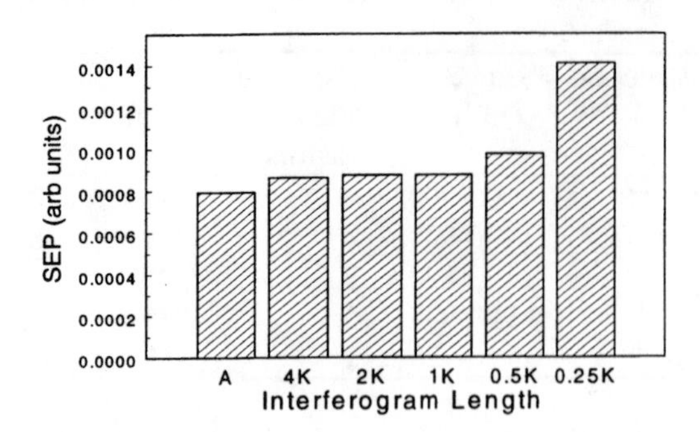

FIGURE 3. Effect of spectral resolution. Absorbance spectra of different resolutions were used for this analysis. The interference band was centered at 952 cm^{-1}. These results are compared with (A) nonlinear interferogram analysis results

single-beam spectral analysis and interferogram analysis implemented with a linear PLS algorithm. A closer look at the trend in the results of the interferogram analysis implemented with a linear PLS algorithm reveals that the results approach and nearly match those of the single-beam analysis as the separation increases. The filter used in this case had a passband fwhh of 25 cm^{-1}. This filter is moderately narrower than the previous one, and combined with decreasing contribution from the interference, reduces the extent of nonlinearity in the filtered interferogram data and helps bring the results to essentially the same level as those obtained with the single-beam spectral analysis.

Another approach taken in studying the effect of resolution on these analyses was to decrease the resolution of the spectra from 1 to 16 cm^{-1}. This was done by truncating the 4K interferograms before Fourier transforming them. This procedure yielded interferogram data sets with 2K, 1K, 0.5K and 0.25K interferograms, which, upon transformation yielded spectra with 2, 4, 8 and 16 cm^{-1} point spacing, respectively. The absorbance and single-beam spectral analyses were then performed with these lower resolution spectra. Table 3 shows the results obtained by implementing an absorbance spectral analysis with the interference peak centered at 952 cm^{-1}. Figure 3 compares these results to those obtained with the interferogram analysis implemented with the nonlinear PLS algorithm. These results show that the performance of the absorbance spectral analysis remains relatively constant as the resolution is changed from 1 to 4 cm^{-1}, and starts decreasing beyond that, being at its worst for the 16 cm^{-1} spectra. This is understandable since for the 8 and 16 cm^{-1} spectra there are fewer points available for modeling the relatively narrow analyte band. The results obtained with the interferogram analysis are somewhat better than even the best obtained with the highest and medium resolution spectra. This is extremely impressive given that the interferogram-based analysis uses a segment of only 101 points.

CONCLUSION

The interferogram-based analysis continues to produce good quantitative results, even when performed with interferogram segments as short as 50 points. A comparison of these results, however, with those from absorbance and single-beam spectral analyses shows that, while the results may be good on their own, they do not compare favorably with the spectral methods. This may be a consequence of the presence of nonlinearity in the filtered interferogram data. A formal nonlinear modeling procedure, implemented with a quadratic PLS algorithm was applied to the interferogram analysis, and produced results that were better than those of the single-beam spectral analysis, and comparable to those of the absorbance spectral analysis.

A comparison between results of the interferogram analysis implemented with a quadratic PLS model and those obtained with different resolution spectra shows that the interferogram analysis produces marginally better results than medium to high resolution spectral analyses. A combination of bandpass digital filtering, which ensures few sources of variation in the filtered data, judicious selection of the filtered interferogram segment for use in the analysis, and nonlinear multivariate calibration modeling combine to make the interferogram-based analysis competitive with spectral-based procedures that require the collection of significantly larger interferograms.

ACKNOWLEDGEMENTS

Funding for this work was provided by the Department of the Army.

REFERENCES

1. Griffiths, P. R.; de Haseth, J. A., *Fourier Transform Infrared Spectrometry*, New York: John Wiley & Sons, 1986, ch. 1.
2. Small, G. W.; Harms, A. C.; Kroutil, R. T.; Ditillo, J. T.; Loerop, W. R. *Anal. Chem.* **62**, 1768-1777 (1990).
3. Mattu, M. J.; Small, G. W. *Anal. Chem.* **67**, 2269-2278 (1995).
4. Wold, S, Kettaneh-Wold, N.; Skagerberg, B. *Chemom. Intell. Lab. Syst.* **7**, 53-65 (1990).
5. Oppenheim, A. V.; Schafer, R. W. *Discrete-Time Signal Processing*, Englewood Cliffs, NJ: Prentice Hall, 1989, ch. 7.

Weighted Partial Least Squares Method to Improve Calibration Precision for Spectroscopic Noise-Limited Data*

David M. Haaland and Howland D. T. Jones

Sandia National Laboratories, Albuquerque, NM 87185-0342

Multivariate calibration methods have been applied extensively to the quantitative analysis of Fourier transform infrared (FT-IR) spectral data. Partial least squares (PLS) methods have become the most widely used multivariate method for quantitative spectroscopic analyses. Most often these methods are limited by model error or the accuracy or precision of the reference methods. However, in some cases, the precision of the quantitative analysis is limited by the noise in the spectroscopic signal. In these situations, the precision of the PLS calibrations and predictions can be improved by the incorporation of weighting in the PLS algorithm. If the spectral noise of the system is known (e.g., in the case of detector-noise-limited cases), then appropriate weighting can be incorporated into the multivariate spectral calibrations and predictions. A weighted PLS (WPLS) algorithm was developed to improve the precision of the analyses in the case of spectral-noise-limited data. This new PLS algorithm was then tested with real and simulated data, and the results compared with the unweighted PLS algorithm. Using near-infrared (NIR) spectra obtained from a series of dilute aqueous solutions, the simulated data produced calibrations that demonstrate improved calibration precision when the WPLS algorithm was applied. The best WPLS method improved prediction precision for the analysis of one of the minor components by a factor of nearly 9 relative to the unweighted PLS algorithm.

INTRODUCTION

When multivariate spectral calibrations are limited by spectral noise and when the noise is nonuniform across the spectral region used in the analysis, the partial least squares (PLS) algorithm can be suboptimal. This suboptimal performance arises from the fact that PLS assumes that the spectral errors are normally distributed with zero mean and constant variance throughout the spectral region. For infrared spectra, the high-frequency spectral noise of transmittance spectra is often limited by detector noise that exhibits constant variance that is independent of the spectral intensity. However, once these transmittance spectra are converted to absorbance, the spectral noise variance is no longer constant. This nonuniform noise variance is accentuated in those cases where there are large variations in absorbance. In this case, the precision of the concentration prediction estimates should be improved using a WPLS algorithm relative to that of the traditional unweighted PLS algorithm. In this paper, we demonstrate one version of a weighted PLS (WPLS) algorithm and compare the prediction precisions for weighted and unweighted PLS analyses applied to both real and simulated data obtained from a set of dilute aqueous solutions.

*Sandia is a multiprogram laboratory operated by Sandia Corporation, a Lockheed Martin Company, for the United States Department of Energy under Contract DE-AC04-94AL85000.

EXPERIMENTAL PROCEDURES

The spectral data in this paper have been described previously (1). The data are from a set of spectra of urea, creatinine, and NaCl at 0 to 3000 mg/dL in water. 31 calibration samples were prepared in sealed cuvettes with each component varied at 13 concentration levels in a Latin hypercube experimental calibration design (1). The near-infrared spectra were obtained in 10-mm path-length cuvettes maintained at 23° C in a Nicolet 800 FT-IR spectrometer operated at 16 cm^{-1} resolution. The spectrometer was equipped with a liquid-nitrogen-cooled InSb detector, a quartz beam splitter, and a 75 W tungsten-halogen source. The run order of the samples was randomized. The spectra are exhibited in Fig. 1 for the analysis region (7500 to 11000 cm^{-1}). Figure 2 shows the same calibration spectra after mean centering.

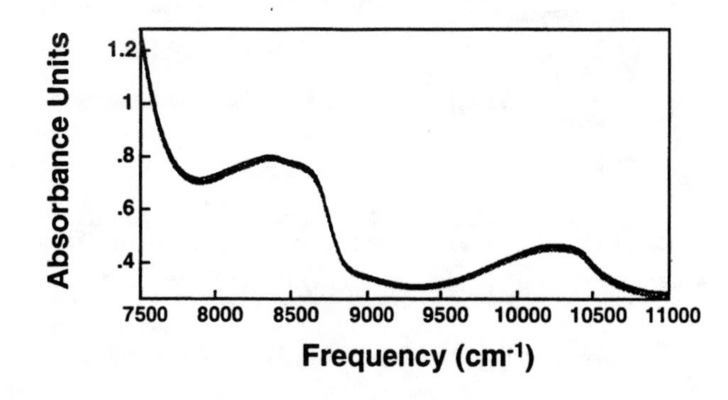

FIGURE 1. Spectra of 31 calibration samples.

CP430, *Fourier Transform Spectroscopy:* 11th International Conference
edited by J.A. de Haseth
1998 The American Institute of Physics 1-56396-746-4/98/$15.00

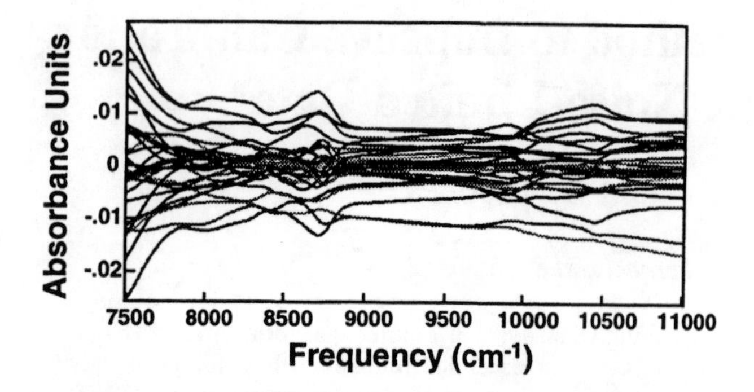

FIGURE 2. Mean-centered spectra of 31 calibration samples.

To further evaluate the advantages of the WPLS method, simulated spectra were generated to model the same system in a 25-mm path-length cell. These simulation spectra assured that the multivariate data analysis would be dominated by spectral noise. The simulation was accomplished by estimating the pure-component spectra from the 31 calibration spectra by CLS multivariate methods. Simulated absorbance spectra were generated by multiplying the pure-component spectra by 2.5 and then adding these path-length-corrected pure-component spectra according to their concentrations in each calibration sample. The baseline offset was also increased to further accentuate spectral noise differences across the spectrum. These simulated absorbance spectra were then converted to single-beam spectra and random noise (0 mean, normally distributed, and constant variance) was added. The resulting single-beam spectra were then converted back to absorbance. In this manner, the spectral noise mimics that expected for 25-mm path-length aqueous samples obtained on an FT-IR spectrometer with linear response. Figure 3 shows the mean-centered spectra of the 31 simulated samples. Note the greater noise in Fig. 3 relative to that in Fig. 2.

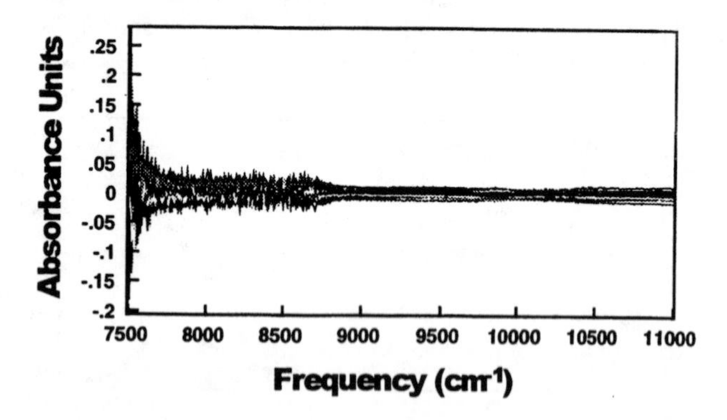

FIGURE 3. Mean-centered spectra of 31 simulated calibration samples.

All of the weighted and unweighted PLS analyses were performed using in-house software generated in Matlab 5.0. The cross-validated standard errors of prediction (CVSEP) for the two types of PLS calibrations were used for comparing the results.

THEORY

As reported by Haaland and Thomas (2), the PLS algorithm can be broken down into a series of CLS calibration and prediction steps and an inverse least squares (ILS) step. Haaland and coworkers (3,4) have presented the methods for performing CLS analyses in both the calibration and prediction phases. By analogy, the WPLS algorithm is readily constructed by adding weighting to the CLS portions of the PLS algorithm in the calculation of each PLS loading vector. Since the errors are presumed to be dominant in spectral rather in concentration space, weighting is not applied to the ILS steps of the PLS algorithm. Two basic unweighted PLS algorithms are currently available, one that generates orthogonal scores and one that generates orthogonal loading vectors (5). Without weighting, these two algorithms are equivalent. However, in the WPLS algorithms, it can be shown that the two PLS methods no longer yield identical results. We report in this paper results for the WPLS method that uses as its basis the unweighted PLS method with orthogonal loading vectors. The resulting WPLS algorithm has neither the scores nor the loading vectors orthogonal. However, orthogonal loading vectors can be generated during the WPLS algorithm simply by adding a Gram-Schmidt orthogonalization procedure after the second and subsequent loading vectors have been determined.

In Haaland and Easterling (3), the appropriate weighting function for detector-noise-limited FT-IR absorbance spectra is given. The proposed weighting factor was T_i^2 where T_i is the transmittance of the sample at frequency i. This weighting factor was determined by expanding the absorbance signal as a Taylor series about the transmittance and retaining only the first two terms. The variance of the transmittance noise was then found to be proportional to T_i^{-2}, and the appropriate weighting factor is inversely proportional to this variance. This weighting factor is then appropriate for the case presented in Ref. 3 where the analysis was performed over a small spectral window and the single-beam intensity of the spectrometer response was essentially constant with frequency.

A more general weighting factor can be determined for those cases where the spectrometer response is highly variable over the spectral region of interest. In this case, the Taylor series expansion of the absorbance is made about the single-beam intensity. Retaining the first two terms of this expansion shows that the appropriate weighting function is the square of the single-beam intensity spectrum (I_i^2). This weighting takes into account variations in spectral noise with sensitivity of the spectrometer as a function of frequency. In this paper, we used the square of the single-beam spectrum for weighting each step of the WPLS algorithm.

The WPLS algorithm is considerably slower than the standard PLS method. For the data presented in this paper, the WPLS method was a factor of seven slower than the traditional PLS method. Of the weighted steps in the PLS algorithm, by far the slowest is the weighted CLS calibration step of the WPLS method. However, in those cases where the spectral differences between samples are small relative to the range of absorbance variations within a sample, then this slow weighted CLS calibration step has relatively little influence on the predictions. Thus, in the case described above, the weighted portion of the CLS calibration step can be skipped. With this modification of the algorithm, the analyses time of the data presented in this paper differed by less than a factor of two between the WPLS and PLS algorithms.

RESULTS AND DISCUSSION

The cross-validated SEP values for the weighted and unweighted PLS analyses were not significantly different for the three components analyzed (CVSEP for the three components ranged from 13 to 24 mg/dL). The lack of significant difference between the two methods suggests that the spectral analysis of the original absorbance data is not dominated by spectral noise. Therefore, simulated spectra were generated as described above to assure that the analysis would be limited by spectral noise. The urea prediction results for the PLS and WPLS analyses are presented in Fig. 5 and Fig. 6, respectively. The results for the analyses of these simulated data are given in Table I for all three minor components in the calibration samples. The enormous benefit of WPLS over the normal PLS is observed in Figs. 4 and 5 and in Table I where the system is clearly limited by spectral noise and where the weighting function is highly variable over the spectral analysis region. Improvements in precision vary from factors of 3.8 to 8.9. In addition, the model sizes are approximately half the size for the WPLS models where the spectral noise is normalized. These results dramatically demonstrate the potential value of applying WPLS to systems where the errors are dominated by spectral noise.

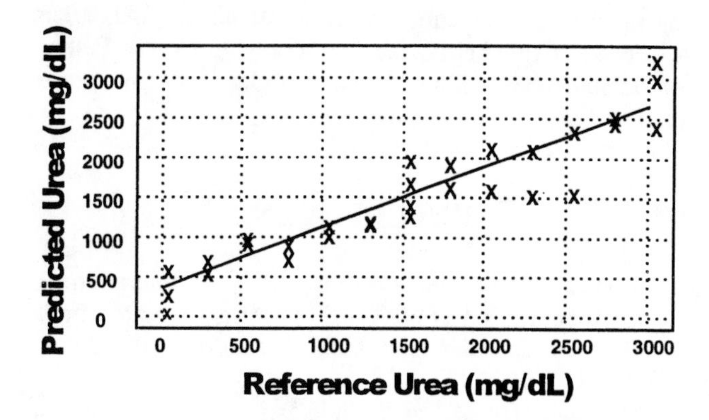

FIGURE 4. Prediction results for urea using the traditional PLS algorithm. The line is the linear least squares fit to the data.

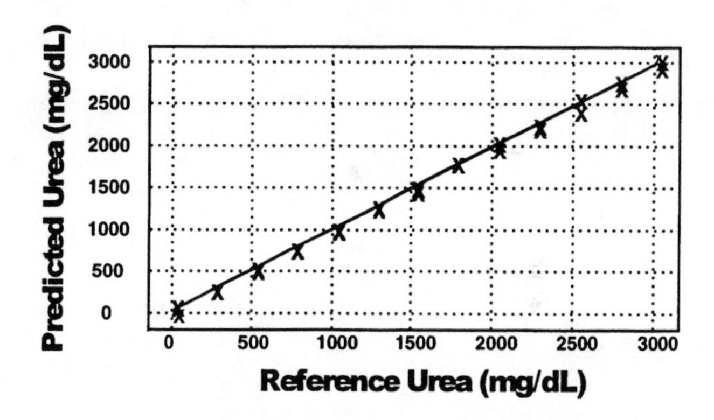

FIGURE 5. Prediction results for urea using the weighted PLS algorithm. The line is the linear least squares fit to the data.

Table I: Results of Applying PLS and WPLS to Simulated Data

Component	PLS			WPLS		
	CVSEP (mg/dL)	Factors	R^2	CVSEP (mg/dL)	Factors	R^2
Urea	350	7	0.8800	39.4	3	0.9986
Creatinine	216	8	0.9480	53.8	4	0.9967
NaCl	166	7	0.9689	43.5	4	0.9982

ACKNOWLEDGMENTS

The authors would like to acknowledge Edward Thomas for the Latin Hypercube experimental design and Brian Rohrback of Infometrix, Inc. for supplying the HP Peltier temperature controlling cuvette holder used in these studies.

REFERENCES

1. Haaland, D. M. and Jones, H. D. T., 9th International Conference on Fourier Transform Spectroscopy, Bertie, J. E. and Wisser, H., Editors, Proc. SPIE **2089**, 448-449 (1993).
2. Haaland, D. M. and Thomas, E. V., Analytical Chemistry **60**, 1193-1202 (1988).
3. Haaland, D. M. and Easterling, R. G., Applied Spectroscopy **36**, 665-673, (1982).
4. Haaland, D. M., Easterling, R. G., and Vopicka, D. A., Applied Spectroscopy **39**, 73-84, (1985).
5. H. Martens and T. Naes, Multivariate Calibration, New York: John Wiley & Sons, 1989, ch. 3, pp. 121-125.

Forensic Discrimination of Photocopy Toners by FT-Infrared Reflectance Spectroscopy

Edward G. Bartick, Rena A. Merrill
FBI Laboratory, Forensic Science Research and Training Center, FBI Academy, Quantico, VA 22135

William J. Egan, Brian K. Kochanowski, Stephen L. Morgan
Department of Chemistry & Biochemistry, The University of South Carolina, Columbia, SC 29208

Because of the speed, simplicity, and accessibility of photocopying, forensic examiners are encountering photocopies as often as original questioned documents. We investigated the ability of FT-infrared microscopy coupled with pattern recognition methods to discriminate among copy toner samples from a variety of manufacturers. Infrared microscopy is a preferred method due to its non-destructive nature, however, visual comparison is difficult because the observer may not be able to fully utilize the fine structure of the complex patterns. Principal component analysis and canonical variate analysis were used to visualize clustering of samples and to assess the statistical validity of the observed differences. The results illustrate the potential for computer-assisted data interpretation to provide decisive forensic identification of questioned samples.

INTRODUCTION

The use of office and personal photocopying machines has increased dramatically over the last 20 years. As a result, identifying the source of photocopied documents is not an easy task. Forensic scientists are often faced with having to identify photocopy machines involved in counterfeiting, fraud, false documents, anonymous letters, confidential materials, and acts of terrorism (1).

Techniques used for photocopy differentiation include optical microscopy, scanning electron microscopy (SEM), energy dispersive spectrometry (X-ray microprobe analysis), infrared (IR) spectroscopy, and pyrolysis gas chromatography/mass spectrometry (Py-GC/MS) (1-3). IR spectroscopy, in particular, provides information regarding the organic components of toners. Bartick and coworkers at the Forensic Science Research Institute (FBI Laboratory, FBI Academy, Quantico, VA) have performed systematic studies using microscopic infrared spectrometry (IR), diffuse reflectance, attenuated total internal reflectance, and reflection-absorption (4,5). Data from this work was summarized using a flow chart technique that enables discrimination of copy toners based on visual comparison of features in their IR spectra.

The present study is primarily concerned with a group of toners having a poly(styrene:acrylate) base component and which comprise a significant number of the commercial toners used in modern photocopy machines. Visual discrimination of the IR spectra among different samples within this group is difficult. These toners are often the hardest to differentiate by Py-GC/MS due to the complex, yet similar chromatographic patterns. FT-IR spectroscopy offers a powerful means for making comparative measurements of distinguishing characteristics for copy toners; however, an overwhelming amount of data is generated. An infrared spectrum may consist of measured light intensities at several thousand discrete wavelengths. Although an experienced analyst becomes expert at recognizing distinguishing features, the pattern recognition task is subjective and becomes quite difficult when numerous samples are compared. The objective of our preliminary work presented in this paper is the evaluation of multivariate statistical approaches for discriminating spectra of copy toners.

EXPERIMENTAL

Samples were prepared using a heat transfer process to move dry toner from documents to aluminum foil affixed to standard microscope slides as previously described (4,5). Although other materials provide a suitable reflective surface for the reflection-absorption technique, aluminum foil is readily available, inexpensive and permits the sample to be stored for further studies. The sample preparation is simple, fast and essentially non-destructive. The document is still legible after transferring the toner sample and only minimal destruction is visible microscopically. Care should be taken when sampling two-sided documents to avoid direct contact between the soldering iron and toner on the back of the document.

Spectra were collected using one of two systems: a Nicolet (Madison, WI) 20SXC infrared spectrometer with a Spectra Tech (Shelton, CT) IR-Plan microscope or a Nicolet 760 Magna infrared spectrometer with a Nicolet Nic-Plan microscope. Both microscopes utilize medium band Mercury Cadmium Telluride (MCT) detectors. Analyses were conducted at 4 cm^{-1} resolution. Between 128

CP430, *Fourier Transform Spectroscopy:* 11th International Conference
edited by J.A. de Haseth

and 256 scans were acquired for each spectrum over the range of 4000 to 650 cm^{-1}. Baseline adjustment was performed using the Nicolet OMNIC software as needed to flatten the baseline on each spectrum.

DATA ANALYSIS

Principal component analysis (PCA) is commonly used to reduce the dimensionality of data. The decomposition of the data matrix is efficiently accomplished by singular value decomposition (SVD) (6,7). SVD creates a linear combination of the correlated, original variables, forming a new set of variables which are orthogonal, *i.e.*, uncorrelated. This transformation also maximizes the variance of the original data explained by each of the new variables, which are called eigenvectors or PCs. Thus, as much information as possible is packed into the fewest number of unrelated variables. The decomposition of the data matrix allows the projection of points representing the samples into a two- or three-dimensional plot. The clustering of data points in this plot may then be used to base judgments concerning sample comparisons. Decisions about similarity of samples to one another reduce to comparison of distances between points on the plot. The clustering of similar samples can be assessed by comparison to the distances between samples judged different from one another. Whether this information is distributed among the PCs in a useful fashion, *e.g.*, permitting understandable display of the samples on the first 2-3 PCs when the samples had not been easily viewed on the many bi-plots of the original variables, is another question entirely.

Since the sample identities are known for the data presented here, the data was subjected to canonical variate analysis (CVA), also known as discriminant analysis (8). The implementation of CVA is quite similar to that of SVD and is based upon the eigen-decomposition of a matrix derived from the samples. CVA maximizes the ratio of between-group to within-group variance. Thus, CVA generates linear combinations of the original variables that best separate the groups specified by the analyst. The plot of the data points in the two- or three-dimensional space of the canonical variates permits the researcher to visualize clustering and similarity of the data. Recent work (8) has also shown the usefulness of calculating canonical variates that are orthogonal, a property not possessed by CVs in their usual formulation.

RESULTS AND DISCUSSION

FT-IR spectra were taken for each of the copy toners listed in Table 1. Representative spectra are shown in Figures 1 and 2. As can be seen, these spectra are particularly difficult to separate from one another by visual inspection of the spectra. The canonical variates analysis method requires that there be enough samples to make up a group; the 11 groups shown in Table 1 have at least 3 samples each.

TABLE 1. Copy toner samples.

Group	Number of samples
A (AB DICK)	5
B (Brother)	3
C (Citoh)	3
D (Canon)	18
E (Copystar)	5
F (HP)	6
G (Mannesmann)	3
H (Newgen)	5
I (Okidata)	5
J (Qume	4
K (TI)	3

Spectra were preprocessed using the Fast Fourier Transform (FFT). The magnitude of the modulus and the first half of the Fourier coefficients were used as the inputs to orthogonal canonical variate analysis (OCVA). OCVA attempts to find a representation of the data that minimizes the spread within a group and maximizes the spread between different groups. The resulting axes each explain different facets of the information contained in the data, *i.e.*, they are orthogonal. A projection of the samples into a three-dimensional space of the first three orthogonal CVs is shown in Figure 3.

The separation of groups by canonical variate axes are as follows:

Axis 1 & 2	A, B, E, I are well separated C, F, G, H, J are intermingled D and K are intermingled
Axis 1 & 3	E, H, K are well separated D and I are intermingled A, B, C, F, G, J are intermingled
Axis 2 & 3	A, B, E, I, K are well separated C, D, F, G, H, J are intermingled

It is apparent from these plots that samples in groups A, B, D, E, H, I, and K can be discriminated from one another. Replicate samples within these groups are closer together than the distances between groups. Samples in groups C, F, G, and J are not distinguishable by this method. In these cases, replicate samples from these groups overlap with data points representing samples from other groups. Complete differentiation of all the copy toners groups is not achieved, although several groups are clearly distinguishable. These groups include toners from AB Dick (A), Brother (B), Copystar (E), Okidata (I), Newgen (H), and Texas Instruments (K).

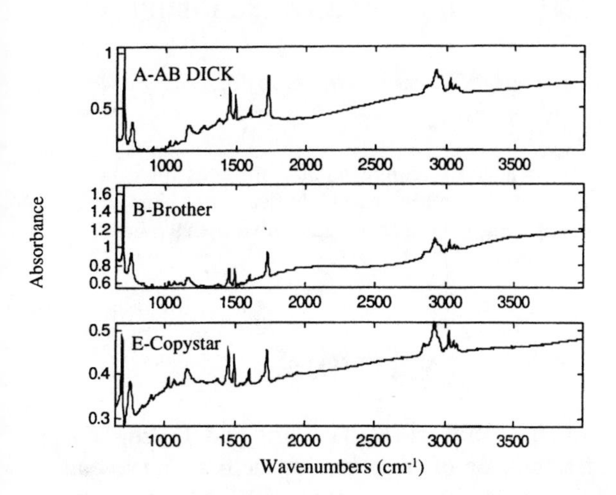

FIGURE 1. Representative FT-IR spectra from AB Dick, Brother, and Copystar toners.

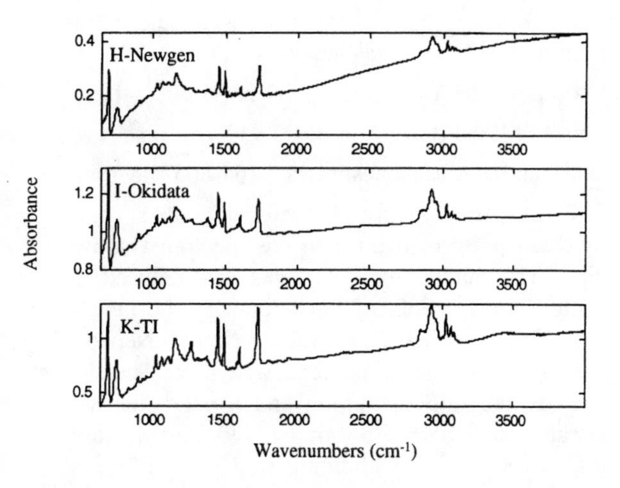

FIGURE 2. Representative FT-IR spectra from Newgen, Okidata, and Texas Instruments toners.

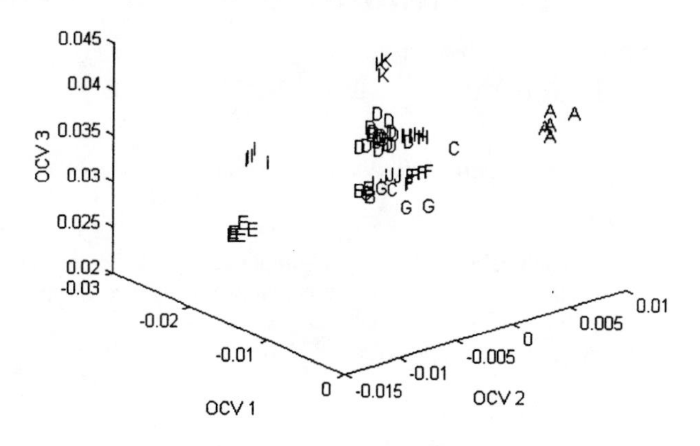

FIGURE 3. Projection of copy toner spectra on first three orthogonal canonical variate axes.

CONCLUSION

The focus of this work is development of automated, statistical-based, strategies for data handling that offer improvements in method validation and ease of interpretation. Although visual examination of these questioned materials by an expert using microscopy can be a powerful tool, such visual examination is subjective and time-consuming. The examples discussed illustrate the potential for computer-assisted data interpretation of forensic analytical data to provide decisive forensic identification of questioned samples. For each set of data, a visually interpretable map displaying the quantitative similarity of the IR spectra of forensic samples can be created. Statistical analysis of the spectra supported the discrimination ability of the IR method for several groups of poly(styrene:acrylate)-based toners.

REFERENCES

1. Totty, R. N., *Forensic Science Review*, **2**, 1-23 (1990).
2. Munson, T. O., *J. Forensic Sciences*, **34** (2), 352-365 (1989).
3. Chang, W.-T., Huang, C.-W., and Giang, Y.-S., *J. Forensic Sciences*, **38**, 843-863 (1993).
4. Bartick, E. G., Tungol, M .W., and Reffner, J. A., *Anal. Chim. Acta*, **288**, 35-42 (1994).
5. Merrill, R. A., Bartick, E. G., and Mazzella, W. D., *J. Forensic Sciences*, **41**, 264-271 (1996).
6. Press, W. H., Flannery, B. P., Teukolsky, S. A., and Vetterling, W. T. *Numerical Recipes: The Art of Scientific Computing*, New York: Cambridge University Press, 1986, pp. 52-64.
7. Jackson, J. E., *A User's Guide to Principal Components*, New York: John Wiley & Sons, Inc., 1991.
8. Krzanowski, W. J., *J. Chemom.*, **9**, 509-520 (1995).

Artificial Neural Networks Applied to FTIR and FT-Raman Spectra in Biomedical Applications

J. Schmitt*, T. Udelhoven+, T. Löchte**, H.-C. Flemming* and D. Naumann**

*Dept. Aquatic Microbiology, Spectroscopy Group, University of Duisburg, Germany, University of Trier FB VI
** Robert-Koch Institut, Berlin, Germany

Biomedical applications of vibrational spectroscopy developed for routine analysis reqiure reliable methods for data evaluation. Artificial neural networks open a new perspective for the spectra differentiation and identification of biologic samples with their small spectral variance. Spectral libraries based on different combined neural networks have been developed using FT-IR and FT-Raman spectra for bacteria and yeast identification.

INTRODUCTION

Biological and biomedical applications of vibrational spectroscopy have witnessed an enormous progress in the past. FTIR and Raman spectroscopy - to give some examples - are used to characterize and differentiate microorganisms, control bioreactors, test the susceptibilty of pharmaceuticals to bacteria and are used frequently to investigate mammalian cells, tissues and body fluids. The common spectral feature of all complex biological materials is their small spectral variance between the different objects. Especially because the transfer of elaborated spectroscopic techniques into routine analysis and other fields of application depends on its simplicity and reliability, new methods for data evaluation are required. Self-learning systems like artificial neural networks with appropriate techniques of spectral feature selection and data pretreatment are new and promising techniques. They have already been applied succesfully in several spectra interpretation applications for Raman and in FT-IR spectroscopy [1-3] Little has been published using neural networks in relatively large data sets with small spectral variances, which is typical in spectra of biological samples. In the present study the applicability of neural networks in spectra identification of bacteria and yeasts for practical purposes is illustrated. For the differentiation of 37 *Candida* strains both Raman and IR-spectra were used and the possibilities of a combination of both techniques are discussed. Also a neural library for 6 bacterial species with 355 strains based on FT-IR spectra is presented, which provides a data elucidation system for routine analysis.

METHODS

Network Architectures, Training and Testing. For the differentiation of single strains the first and second derivatives were selected instead of FT-IR absorbance spectra. This provides a higher differentiation depth of a given data set. No derivatives were used for Raman spectra. A smoothing function after Savitzky-Golay [4] with 9 smoothing points was applied prior to feature selection. Different spectral windows were selected as input units for the neural network. In general, the windows were the following: 3000-2800 cm^{-1}, 1750-1700 cm^{-1}, 1400-1205 cm^{-1} , 1200-950 cm^{-1} and 940-700 cm^{-1} and for Raman spectra 1164-805 cm^{-1} and 633-371 cm^{-1}, respectively. A vector normalization of the data was performed prior to the spectral windows selection. The number of datapoints was reduced by averaging 3 data points. For training, the neural network simulator SNNS (Stuttgart Neural Network Simulator) and own developements were used. The data set was divided into a training and a validation set. Each strain consisted of 3-5 training- and 2-3 validation spectra. In each case a validation step was performed after 20 learning cycles.

In the case of very similar vibrational spectra in relatively large datasets, as it occurs within the strain and substrain level of bacterial species, the seperation of individual classes requires long training phases. Secondly, if the number of single classes is too large no reliable seperation was possible. Therefore, we decided to establish a tree-like network with several linked subnets. The information flux in such a combined network is controlled by one heading network, whose single task is to seperate between different species. In the following, the further networks are optimized for the species and subspecies specific separation within a species.

CP430, *Fourier Transform Spectroscopy:* 11th International Conference
edited by J.A. de Haseth

At the beginning, training of the individual subnets was accomplished with different learning algorithms. The one with the best result was selected in each case. Besides standard backpropagation with flat spot elimination and weight decay the Quickprop algorithm [5,6] and Cascade correlation [7] were also utilised. Briefly, Quickprop assumes the error surface to be locally quadratic and attemps to jump in one step from the current position directly into the calculated minimum of a fitted parabola. After computing the first gradient during the learning step with regular backpropagation, a direct step to the error minimum is attempted by

$$\Delta(t+1)w_{ij} = \frac{S(t+1)}{S(t) - S(t+1)} \Delta(t)w_{ij}$$

where: w_{ij} weight between units i and j
$\Delta(t+1)$ actual weight change
$S(t+1)$ partial derivative of the error function by w_{ij}
$S(t)$ the last partial derivative

The Cascade-Correlation approach is not only a training algorithm but it also determines the topology of the network. The algorithm starts with a minimal network consisting only of an input and output layer. During the learning phase stepwise new appropiate hidden units from a set of candidate units are inserted to the hidden layer. The goal is to minimize the overall error of the net this way. By freezing the weights which is leading to the new hidden unit, a permanent feature detector for specific stuctures in the input layer is obtained. A new hidden unit is not only connected to the input and output layer also to each unit in the hidden layer. The addition of new hidden units is stopped if the overall error of the net falls below a given value. This leads to a compact but highly connected network topology [8].

RESULTS AND DISCUSSION

In order to establish an neural database for the identificaion of bacterial genera on the species/strain level, the following species were to be classified based on the FT-IR spectra.

TABLE 1:

Bacterial genera	Number of species
Pseudomonas	21
Bacillus	34
Aeromonas	10
Staphylococcus	10
Streptococcus	10
Nocardia	4

To solve the differentiation of these genera first, a feedforward network was trained. It constisted of 320 input neurons, a hidden layer with 20 neurons and an output layer with 6 neurons. Learning was performed with backpropagation. In dependence of the position of the highest activated output neuron, the subnet was activated (Fig. 1). This specialisation inreased the degree of the possible seperation of the individual species. The following table gives an overview over the topology and learning of the used sub-networks:

As it can be seen from table 2, considerable numbers of hidden units were required to separate the individual strains within one species. With exception of the Pseudomonads, only one hidden layer was used in the subnets . The numbers of learning cycles are quite high, especially for Pseudosomonas. Both properties are probably the result of the large similarities within the IR-spectra. The attempt to use Quickprop or Cascade-Correlation instead of Backpropagation in the case of the separation of Pseudosomonas strains failed because of chaotic fluctuations of the error curve during the learning phases. In the other cases learning with Quickprop or Cascade-Correlation was significantly

TABLE 2:

bacterial species	hidden units	learning function	No, of training cycles	% correct identified (in the validation phase)
Pseudosomonas	200*	Backprop	10000	92
Bacillus	100	Cascade-C.	1000	98
Aeromonas	120	Cascade-C.	3300	93
Staphylococcus	120	Cascade-C.	1600	95
Streptococcus	120	Cascade-C.	2000	94
Nocardia	50	Quickprop	500	100

* distributed within two hidden layers which consists of each 100 neurons.

TABLE 3:

37 yeast species	hidden units	learning function	No, of training cycles	% correct identified (in the validation phase)
Candida (FT-IR)	65	Quickprop	900	95
Candida (Raman)	65	Quickprop	1500	82
combined FT-IR /FT-Raman				98

faster and the generalisation properties of the networks better than with Backpropagation. It can also bee seen from table 2 that most of the spectra in the validation pattern set could be classified correctly. Therefore, IR-spectroscopy in combination with neural networks offers an alternative approach to traditional methods with large potential for a rapid and reliable identification of bacteria.

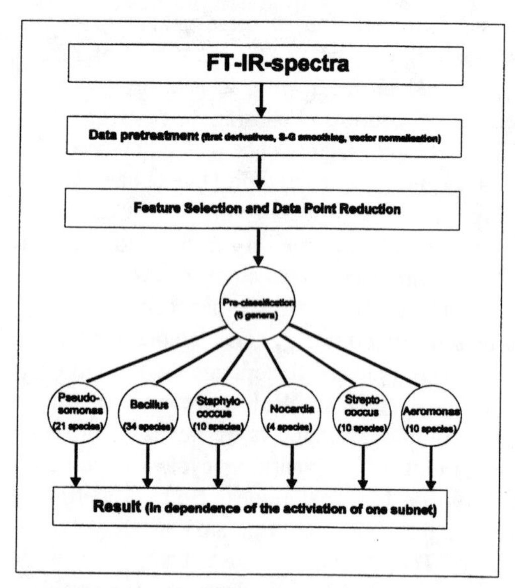

FIGURE 1. Combination of different neural networks after data pretreatment

In order to get additional information about the cells, Raman spectra can be integrated into the analysis. This is demonstrated in table 3, where 37 candida strains were seperated with neural networks.

It can be seen that an identification of the yeasts is possible on the basis of the IR spectra as well as with the Raman data. Despite a faster training and a better classification accuracy with FT-IR spectra, the Raman spectra can give valuable and complementary information with more distinct and well resolved bands. In contrast to the IR-spectra with dominantly amide I and II of the various proteins, the oligo/polysacchrides of the cell wall and fatty acids of the membrane [9,10] Raman spectra are more dominated by C-H stretching bands in the range of 2700-3100 cm^{-1} and the C-H deformation bands. They are originating from various amphiphilic compounds like phospholipids and amino side- chain groups (leucine, isoleucine and valine) of proteins. Also RNA/DNA oligo- and polynucleotides (base-ring vibrations of guanine, thymine, cytosine and uracils and the amino side-chain vibrations of tryptophane, tyrosine and phenylalanine of the proteins, as well as pigments and carotenoid structures can be assigned. Therefore, the classification results on the basis of both techniques are based on completely different properties of the cell surfaces of the yeast species.

CONCLUSION

IR-spectroscopy in combination with neural networks offers an alternative approach to traditional methods with large potentials for a rapid and reliable identifiation of bacteria. The range of the correct identification varied from 92-100% based on the validation data set and is therefore relatively high. On this background, it was possible to establish spectral libraries for the identification of bacterial species of Pseudomonas, Bacillus, Aeromonas, Staphylococcus, Streptococcus and Nocardia. Furthermore, the combination of FTIR and FT-Raman spectra of one and the same microbiologic sample provides the complete vibrational spectroscopic information available and enhances the specifity. These combined techniques seem to open the perspective of a methodology that can be used as an effective tool in diagnosis and therapy .

REFERENCES

1. Meyer M., Meyer K., Hobert H. *Analytica Chimica Acta*, 282, 407-415, 1993.
2. van Est Q.C., Schoenmakers P.J., Smits J.R., Nijssen W. *Vibrational Spectrsocopy*, Vol.4, 263-272, 1993.
3. Schulze G.H., Blades M., Bree A., Gorzalka B., Greek S., Turner F.B. *Appl. Spectrosc.*, Vol. 48, 50-57, 1994.
4. Savitzky A., Golay M.J. (1964): *Analytical Chemistry*, Vol.36,8, 1627-1639, 1964.
5. Fahlmann, S. E.: In: Touretzky, D., Hinton, G. and Sejnowsky, T. (Eds.): Proc. of the 1988 Connectionist Models Summer School, June 17-

26, 1988, Carnegie Mellon University, Morgan Kaufmann Publishers, Inc., 1989.
6. Fahlmann, S. E.: In: Touretzky, D. (Ed.): Advances in Neural Information Processing Systems 3, 190-198. Morgan Kaufmann Publishers, Inc., 1991.
7. Fahlmann, S. E. and Lebiere, C.: In: Touretzky, D. (Ed.): Advances in Neural Information Processing Systems 2, 524-532. Morgan Kaufmann Publishers, Inc., 1990.
8. Zell, A.: *Simulation neuronaler Netze.* 624 p., Stuttgart, Addison Wesley,1994.
9. Schmitt J., Nivens D.,White D.C., Flemming H.C: *Water Sci. & Technol.*, 32, 149-155, 1995.
10. Naumann D., Keller S., Helm D., Schultz Ch., Schrader B. *Journal Molecular Strucure*, 347, 399-406, 1995

Wavelength Selection for the Determination of Glucose in Human Serum by Near-Infrared Spectroscopy

Qing Ding and Gary W. Small

Center for Intelligent Chemical Instrumentation, Department of Chemistry, Clippinger Laboratories, Ohio University, Athens, OH 45701

A genetic algorithm (GA)-based wavelength selection procedure is developed to allow joint optimization of near-infrared wavelengths used and the number of latent variables employed in building partial least-squares calibration models for the determination of glucose in human serum samples. The random selection of a small number of initial wavelengths is found to decrease the number of final wavelengths selected dramatically. The effect of spectral resolution on the wavelength selection is also evaluated.

INTRODUCTION

Near-infrared (NIR) spectroscopy has been widely used in biological and clinical applications. The relatively low absorbances of analyte spectral bands and interference from overlapping bands in the NIR region cause the choice of data analysis strategies to be critical to the success of a NIR-based analysis. Although partial least-squares (PLS) regression is widely used to build calibration models from full spectra (1), recent research has demonstrated that wavelength selection can be used to enhance the performance of these PLS calibration models (2,3). An exhaustive search for the optimal set of wavelengths is time-prohibited, however, and the traditional steepest descent methods are easily trapped in local optima. To address these limitations, potentially more robust optimization techniques such as genetic algorithms (GAs) (4) have been used to search for the optimal set of wavelengths.

An area of research in our laboratory focuses on the development of techniques in NIR spectroscopy for the potentially noninvasive measurement of blood glucose (3,5,6). As part of this work, a GA method has been successfully implemented for jointly selecting the individual wavelengths used and the number of latent variables employed in developing a PLS calibration model (2). In this paper, this wavelength selection procedure is further investigated through the use of NIR spectra of human serum samples. The effect of spectral resolution on wavelength selection is also evaluated.

EXPERIMENTAL SECTION

The human serum data were collected at the University of Iowa with a Nicolet 740 Fourier transform spectrometer that employed a standard NIR configuration. A 2.5 mm Infrasil quartz transmission cell was employed, and the sample temperature was controlled to 37.0±0.2 °C by use of a water-jacketed cell holder. Interferograms of 16,384 (16K) points were collected, yielding spectra with a point spacing of 1.9 cm^{-1}.

Human serum samples were obtained from the Department of Pathology at the University of Iowa Hospitals and Clinics, and analyzed for glucose content by the hospital clinical chemistry laboratory. The serum samples were then frozen until just prior to collection of the spectral data. Two to four replicate spectra were collected for each sample. The samples spanned a range of 57-574 mg/dL (3.2-31.9 mM) in glucose concentration. The detailed spectral collection procedure was described in a previous paper (3).

The total of 235 serum samples (701 spectra) were used for this study. The partition of the data set is shown in Table 1. In this partitioning procedure, the PLS calibration models are built by use of the calibration subset, and then evaluated with the monitoring set. The prediction set is only used as an independent validation set to verify the final models.

RESULTS AND DISCUSSION

Figure 1A shows the NIR spectrum over the range of 4800-4200 cm^{-1} for glucose at a high concentration of 14,274 mg/dL in water. Three glucose absorption bands are clearly seen. In a previous study, the band centered at 4400 cm^{-1} was demonstrated to be most useful for extracting glucose information (4). Figure 1B shows the NIR spectra of serum samples from three different patients who happened to have the same blood glucose concentrations of 322 mg/dL. The characteristic glucose bands cannot be visually observed in the serum spectra. In addition, there are significant variations among the spectra even though they arise from the same glucose concentration. This is because

CP430, *Fourier Transform Spectroscopy:* 11th International Conference
edited by J.A. de Haseth

TABLE 1. Partition of Serum Data Set

Data Set	No. of Samples (Spectra)	
Calibration subset	151 (451)	
Monitoring set	37 (110)	
Subtotal (Calibration set)		188 (561)
Prediction set	47 (140)	
Total		235 (701)

the concentrations of other blood constituents such as proteins are different for the different patients and the features corresponding to the other constituents are overwhelming those of glucose. This suggests that the NIR-based determination of glucose in human serum is challenging and that the data analysis strategies employed are critical to the successful implementation of the method.

Overview of GA-Based Wavelength Selection

This study focuses on the development of a GA-based method to simultaneously select optimal sets of wavelengths in the NIR region and the number of latent variables used in building a PLS calibration model. An initial version of this algorithm has been reported previously (3).

GAs are numerical optimization methods derived from the concepts of genetics and natural selection. On the basis of their high efficency and ability to escape local optima, GA-based methods have become popular in selecting subsets of wavelengths for use in building multivariate calibration models (3,4).

The variables to be optimized in a GA are represented as genes on a chromosome. A group of chromosomes defines a population. After the initial population is generated, the fitness of each chromosome is evaluated. The fitness of the chromosome is judged through the use of an objective function (fitness function) that encodes the degree to which the variable settings are optimal. The optimization algorithm involves an iterative evolutionary process in which the chromosomes in the population recombine and mutate in search of a chromosome with the best fitness values. Each iteration of the algorithm is termed a generation, and the previous population is replaced by the current population with hopefully better fitness values. The algorithm terminates after a fixed number of generations or when a chromosome with a user-specified level of fitness is found.

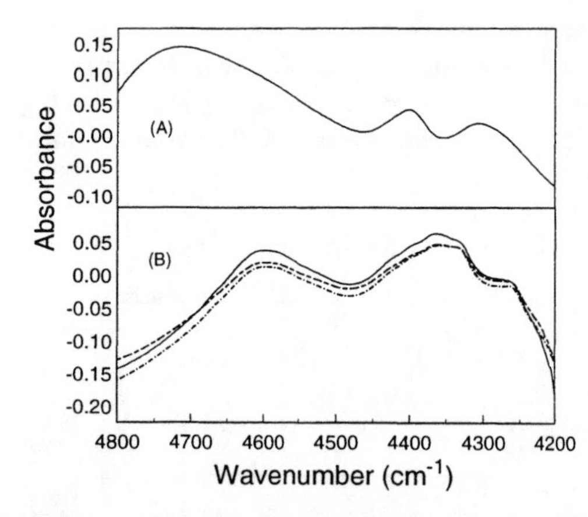

FIGURE 1. (A) NIR absorbance spectrum of glucose at a concentration of 14,274 mg/dL. (B) NIR absorbance spectra of three serum samples from three patients. The glucose concentration in each sample was 322 mg/dL.

1	2	...	101	102	...	399	400	...	...	519	520
0	0	1	0	1	1	0	0	1	0	0	15
5000 cm^{-1}	4998	...	4800	4798	...	4202	4200	...	...	4000 cm^{-1}	

FIGURE 2. Representation of a chromosome employed in optimizing the wavelengths and the number of PLS factors (sequence no. 520) used in building a calibration model.

As shown in Fig. 2, a chromosome in this work consisted of 520 genes, corresponding to 519 spectral points over the range of 5000-4000 cm^{-1}. The 520th gene encoded the number of PLS factors employed in the calibration model. The initial population was formed by randomly perturbing an initial starting chromosome.

The choice of an initial starting chromosome has a critical effect on the optimal set of wavelengths selected by a GA. In our original work with the serum data set (3), all wavelengths within a specified spectral window were initially selected. The optimal calibration models found through this procedure were based on approximately 230 wavelengths. In a subsequent study, we demonstrated that random selection of a small number of wavelengths within a limited range can dramatically decrease the number of wavelengths selected by the GA for building PLS calibration models. With this modified procedure, fewer than 80 wavelengths are selected by the GA with essentially the same performance of the calibration model (7). In the work reported here, the initial starting chromosome consisted of approximately 20 wavelengths randomly selected over the range of 4700-4300 cm^{-1} and 15 PLS factors.

The fitness function used in the GA optimization was

$$(MSE + MSME)^{-1} \qquad (1)$$

where MSE is the mean squared error in concentration of spectra in the calibration subset and MSME is the mean squared error in concentration of spectra in the monitoring set. The fitness function considered both calibration and monitoring errors to balance the performance of the calibration model. Also, to make the fitness function more robust, the fitness of each chromosome was taken as the

TABLE 2. Optimal Genetic Algorithm Configurations

Parameters	Setting
No. of generations	150
Population size	300
Initialization probability	0.1
Mutation probability	0.001
Recombination probability	0.9
Crossover scheme	single-point

mean value of eq. 1 computed across three different calibration subset and monitoring set combinations (i.e., three different rearrangements of the calibration set).

GA configuration parameters were optimized previously (3), and most of these parameters were used in the current work without further optimization. Table 2 lists the optimal GA configuration parameters used in this study.

Effect of Spectral Resolution and Interpolation by Zero-Filling on Wavelength Selection

Considering that the use of lower spectral resolution is somewhat analogous to wavelength selection, we sought to investigate the effect of resolution on the wavelength selection procedure. In addition, resolution is an important experimental parameter in NIR spectroscopy. Serum spectra were originally collected with a nominal 2 cm^{-1} spectral point spacing. The study of the effect of resolution on the glucose analysis is important in helping to define the characteristics of a dedicated spectrometer that might be used in the ultimate implementation of the method. The use of lower resolution could make the instrument more rugged and less costly. The effect of interpolation by zero-filling on wavelength selection was also studied.

The serum spectra with lower resolution were obtained by truncating the collected 16K interferograms to a length that corresponded to the desired lower resolution. For example, spectra with nominal point spacings of 4, 8, and 16 cm^{-1} were achieved by truncating the interferograms to 8K, 4K, and 2K points, respectively.

The spectra obtained for the different resolutions all contain one data point for each resolution element. By adding zeros to the end of the interferogram before computing the Fourier transform, the number of spectral points per resolution element can be increased. The points added are interpolations of the original linearly independent points. In the procedure, the applied apodization function determines the interpolation function. Triangular apodization was used in this study. The conventional purpose of this zero-filling procedure is to increase the photometric accuracy of the spectrum by smoothing the spectrum with the added interpolation points (8). The truncated 8K, 4K, and 2K interferograms were all zero-filled to 16K points. The spectra obtained from the zero-filled interferograms will be termed the 8Kzf, 4Kzf, and 2Kzf data sets, respectively.

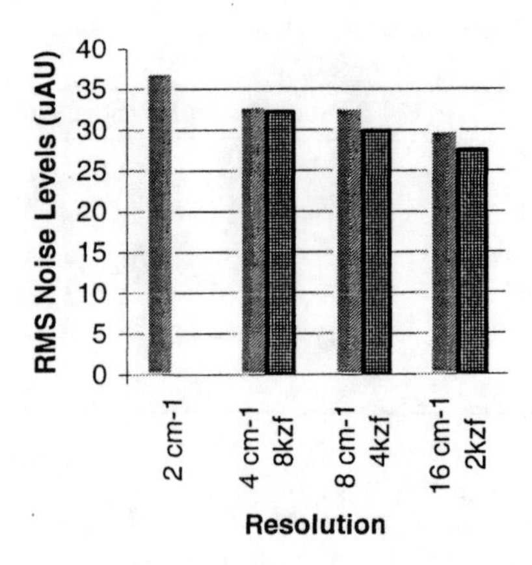

FIGURE 3. Pooled RMS noise levels of the data sets with different resolutions and levels of zero-filling.

The noise levels of the data sets constructed with different spectral resolutions and zero-fillings were estimated by employing the three replicate spectra for each sample to compute three noise absorbance spectra. All combinations of the replicate spectra were ratioed to each other and the resulting transmittance values were converted to absorbance. A total of 229 serum samples were used for computing the noise level. The noise level was computed as the simple root-mean-square (RMS) average of the noise values. Baseline artifacts such as drift and curvature were removed by computing the RMS noise about a second-order polynomial least-squares model fit to the noise spectrum. The RMS noise levels computed with the spectral range of 4700-4300 cm^{-1} for the data sets with different resolutions and levels of zero-filling are shown in Fig. 3. As expected, the noise levels decreased when the resolution decreased from 2 cm^{-1} to 16 cm^{-1}. However, the noise levels for 4 and 8 cm^{-1} were about the same. Interestingly, noise levels from spectra computed from zero-filled interferograms decreased compared to the corresponding spectra obtained without zero-filling.

The best results obtained from the optimal PLS calibration models for the data sets of different resolutions and levels of zero-filling are listed in Table 3. Also listed in the table are the number of wavelengths and the number of

TABLE 3. Effect of Resolution and Zero-Filling

Resolution (cm^{-1})	No. of wavelengths	No. of PLS factors	SEC (mg/dL)	SEP (mg/dL)
2	72	20	21.44	19.64
4	48	20	21.39	22.90
8	41	20	23.89	25.03
16	36	23	27.55	26.79
8kzf	70	21	21.47	21.66
4kzf	86	21	21.27	21.04
2kzf	86	22	21.72	22.25

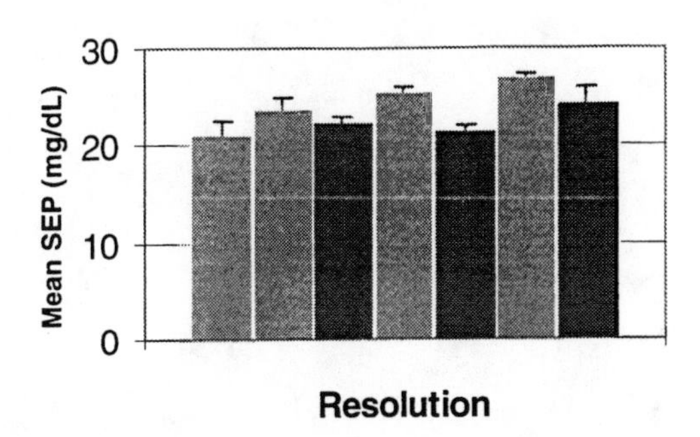

FIGURE 4. The means and standard deviations of the top five SEPs for (from left to right) 2 cm^{-1}, 4 cm^{-1}, 8Kzf, 8 cm^{-1}, 4Kzf, 16 cm^{-1}, and 2Kzf.

PLS factors present in the best chromosome selected by the GA along with the standard error of calibration (SEC) and standard error of prediction (SEP) of the model. For each data set, Fig. 4 shows the mean SEPs and standard deviations of the SEPs computed from the prediction results corresponding to the top five chromosomes selected by the GA. Figure 5 shows the optimal set of spectral points selected by the GA for the 2 cm^{-1}, 8 cm^{-1}, and 4Kzf data sets. Although the sets of wavelengths shown in Fig. 5 are not unique, they indicate that there are many spectral points relevant to glucose information (i.e., close to the 4700, 4400, and 4300 cm^{-1} glucose bands).

Without zero-filling, as the resolution decreased, the number of wavelengths selected decreased because the total number of wavelengths available was reduced. In terms of prediction results, there is a trend that SEP is increased when the resolution is decreased even though the noise levels at lower resolution were lower. That indicated that more spectral points relevant to glucose information may be needed to help explain the small variation of the glucose signal. With zero-filling (for 4, 8, and 16 cm^{-1}), the number of wavelengths selected was similar to that for 2 cm^{-1}

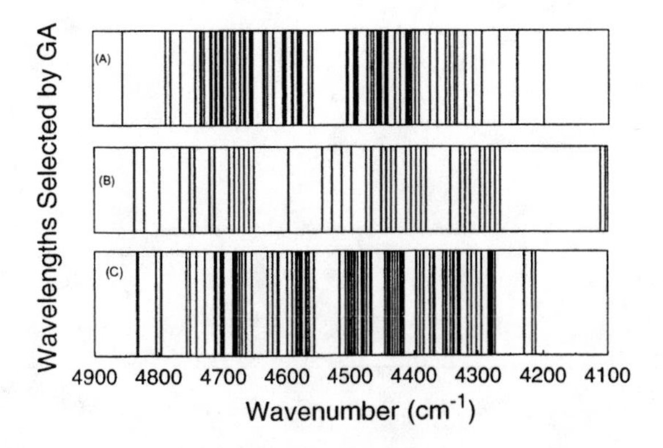

FIGURE 5. Wavelengths present in the best chromosome selected by GA for the data set: (A) 2 cm^{-1}, (B) 8 cm^{-1}, and (C) 4Kzf.

because the total number of wavelengths available was the same. The SEP was decreased with zero-filling compared to the corresponding results without zero-filling, and for 8Kzf and 4Kzf, they were about the same as that for 2 cm^{-1}. Comparing Fig. 4 to Fig. 3, the trend of prediction results with and without zero-filling is similar to that of the noise levels. The reason why zero-filling can reduce the noise level is not clear and remains under study.

CONCLUSIONS

The GA-based method is further demonstrated as an effective approach to wavelength selection. With random selection of wavelengths for the starting chromosome, the number of wavelengths in the optimal chromosomes has been significantly reduced. The use of lower resolution spectra can increase the GA optimization efficiency. However, the prediction performance of those calibration models may not be as good as that obtained with higher resolution spectra. Zero-filling is observed to improve the prediction results obtained with lower resolution spectra. Further work needs to be done to understand the reason for this observation.

ACKNOWLEDGMENTS

This work was supported by the National Institutes of Health under Grant DK 45126. Ron Shaffer and Arjun Bangalore are thanked for writing the original version of the GA software. Mark Arnold of the University of Iowa is acknowledged for providing the human serum data.

REFERENCES

1. Martens, H.; Næs, T., *Multivariate Calibration*, New York: Wiley, 1989, ch. 7, pp. 116-166.
2. Rimbaud, D. J.; Walczak, B.; Massart, D. L.; Last, I. R.; Prebble, K. A., *Anal. Chim. Acta* **304**, 285-295 (1995).
3. Bangalore, A. S.; Shaffer, R. E.; Small, G. W.; Arnold, M. A., *Anal. Chem.* **68**, 4200-4212 (1996).
4. Lucasius, C. B.; Beckers, M. L. M.; Kateman, G., *Anal. Chim. Acta* **286**, 135-153 (1994).
5. Arnold, M. A.; Small, G. W., *Anal. Chem.* **62**, 1457-1464 (1990).
6. Shaffer, R. E.; Small, G. W.; Arnold, M. A., *Anal. Chem.* **68,** 2663-2675 (1996).
7. Ding, Q.; Small, G. W., "Application of Genetic Algorithm-Based Wavelength Selection to Quantitative Near-Infrared Spectroscopy", presented at The Pittsburgh Conference on Analytical Chemistry and Applied Spectroscopy, Atlanta, GA, March 17, 1997.
8. Griffiths, P. R.; de Haseth, J. A., *Fourier Transform Infrared Spectrometry*, New York: John Willey & Sons, 1986, ch. 3, pp. 98-101.

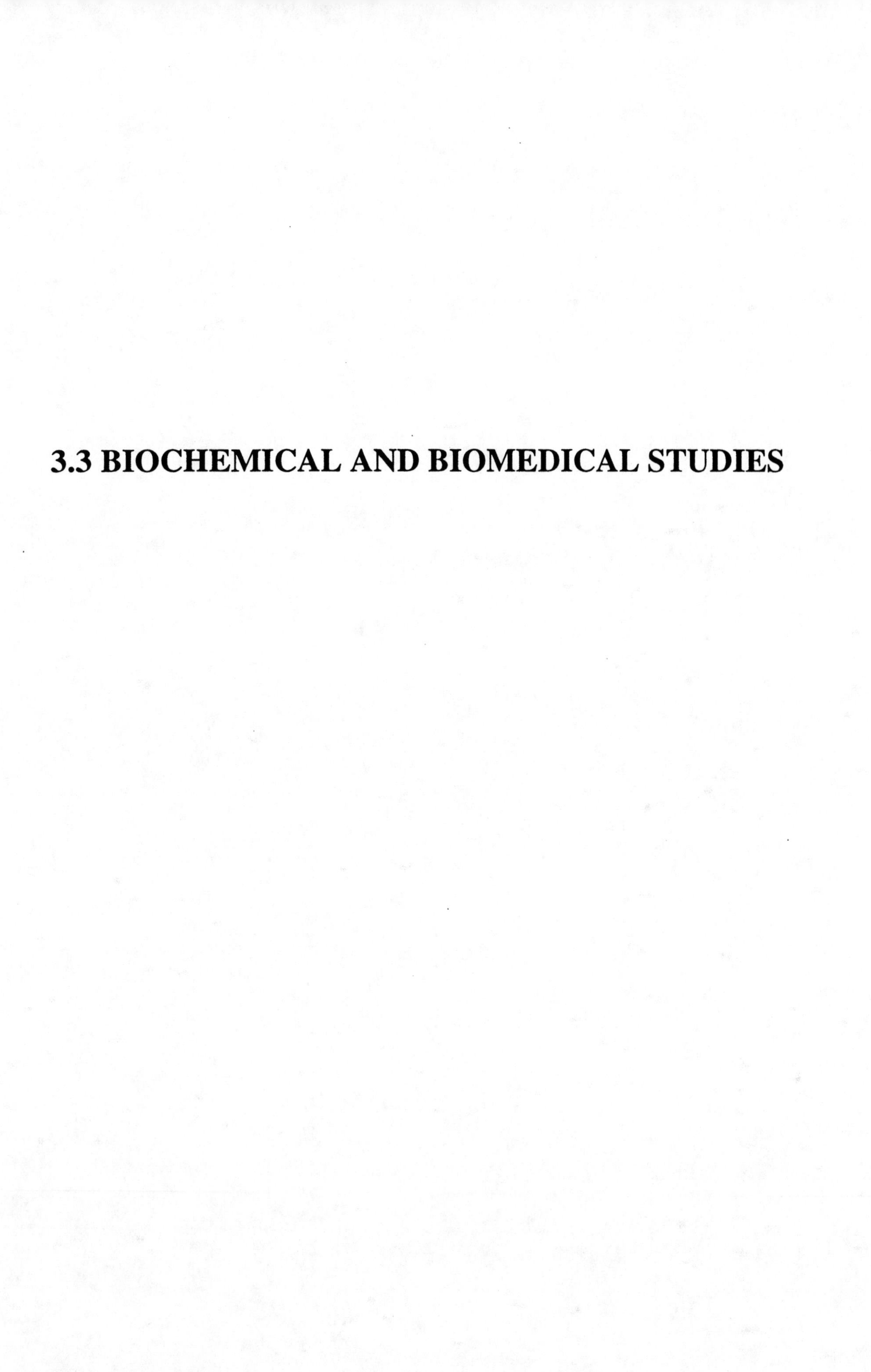

3.3 BIOCHEMICAL AND BIOMEDICAL STUDIES

Multivariate Determination of Hematocrit in Whole Blood by Attenuated Total Reflection Infrared Spectroscopy

S. Kostrewa[1], Ch. Paarmann[2], W. Goemann[2], and H.M. Heise[1]

[1] *Institut für Spektrochemie und Angewandte Spektroskopie, Bunsen-Kirchhoff-Str. 11, D-44139 Dortmund, Germany*

[2] *Eppendorf-Netheler-Hinz GmbH, D-22331 Hamburg, Germany*

A spectral analysis of whole blood was undertaken in the mid-infrared spectral range by using the attenuated total reflection technique. The reference hematocrit values of 109 blood samples were measured after cen-trifugation with a range between 30 % and 50 %. Multivariate calibration with the partial least-squares (PLS) algorithm was performed using baseline corrected absorbance spectra between 1600 and 1200 cm^{-1}. The relative prediction error achieved was 2.7 % based on average hematocrit values. The performance is comparable to that using centrifugation or conductivity measurements. The spectral effects from protein adsorption onto the ATR-crystal, as well as erythrocyte sedimentation have been investigated.

INTRODUCTION

Biofluid samples can be quantitatively characterized by infrared spectroscopy. In the last few years this method has been applied successfully to the determination of blood substrates, such as total protein or glucose in blood plasma (1, 2). Most clinical assays rely on blood plasma or serum, because the cellular components of whole blood such as the red blood cells hinder the photometric determination step routinely performed in the assays. An exception is the determination of glucose which can be carried out using hemolyzed blood. On the other hand, whole blood samples can be analyzed without any sample preparation by infrared spectroscopy (3).

Hematocrit is another important clinical parameter, which is determined as the relative volume ratio of the total erythrocytes as calculated versus a certain blood volume with its value usually given in percent. Deviations from normal ranges allow, for example, the recognition of fluid shifts between blood and the extravascular fluid space. For medical diagnostics, anemia is, in particular, a disease that is related to a loss or reduced production of red blood cells. Other pathological conditions affecting the hematocrit exist. Non-invasive spectroscopic methodology using near-infrared photoplethysmography has been proposed (4), similar to the techniques used in pulse oximetry, by which arterial hemoglobin concentration and oxygen saturation can also be measured, but this needs further clinical testing. In this investigation, we studied the mid-infrared spectra of whole blood recorded by using an horizontal attenuated total reflection (ATR) accessory and employed multivariate calibration modeling by partial least-squares (PLS) for the determination of hematocrit.

EXPERIMENTAL

Different blood samples were taken from 15 female and male volunteers by venous puncture using standard mono-vettes. Some specimens were randomly diluted with isotonic NaCl solutions containing hydroxyethylstarch (HES; Fresenius, Bad Homburg, Germany), which is regularly applied as plasma expander for the purpose of hemodilution. Thus a population of 109 samples was available spanning the hematocrit range between 30 and 50 %.

Several clinical methods exist for hematocrit measurement. The volume fraction of erythrocytes can be determined by automated cell counting or by the alternative spectrophotometric assay of total hemoglobin in lysed blood. Compact analysers suited to near-patient testing are available which estimate hematocrit by measuring the conductivity of undiluted blood (5). Furthermore, the determination of packed cell volume by centrifugation of blood filled capillary tubes is most frequently carried out, which was also used for this study to provide the reference values for the multivariate statistical calibration.

Mid infrared spectra were recorded by a Perkin Elmer FT-spectrometer Paragon 1000 PC equipped with globar, KBr-beamsplitter and a DTGS detector. A horizontal ATR accessory with a ZnSe-crystal and a trough for liquid sample containment was employed (internal reflection angle 45°, 13 reflections at sample interface). The measurements were carried out at room temperature. For the spectra 32 interferograms providing a resolution of 4 cm^{-1} were coadded. A strong Norton-Beer apodization function was applied before Fourier transformation. The spectrum measurement time including sample transfer was

CP430, *Fourier Transform Spectroscopy:* 11th International Conference
edited by J.A. de Haseth
 1-56396-746-4/98/$15.00

2 min. Due to cell sedimentation a strict time protocol was followed for spectrum recording. After discarding the sample, the ATR-crystal was cleaned by a solution of sodium-dodecylsulfate (Serva Heidelberg, Germany) and wiped with a soft tissue soaked with ethanol.

RESULTS AND DISCUSSION

Multivariate calibration using partial least-squares (PLS) has been very successful for the quantitative analysis of several parameters in many clinical blood assays (6). PLS-calibration was carried out using a software package written with MATLAB (The Mathworks, South Natick, U.S.A.); a detailed description has been published (7). Spectral data at 1800 and 960 cm^{-1} were used as anchor points for a linear baseline correction. The selection of a suitable spectral range is essential for achieving optimum results. Best predictions were obtained using spectral data in the interval between 1594 and 1200 cm^{-1}. Above the interval, larger spectral variances in the spectrum population are due to variations in water absorbances resulting from its intense absorption band at 1640 cm^{-1}. Below 1200 cm^{-1} we find significant absorptions from spiked HES or inherent blood glucose. Most prominent protein absorption bands can be found within the chosen interval, apart from the amide I band which is severely overlapped from the water band as discussed above. In addition, difference absorbance spectra against background spectra recorded with distilled water were used as input data for calibration. The population of sample spectra obtained by such preprocessing is shown in Fig. 1.

Crossvalidation of the different PLS-calibration models using 'leave-one-out' strategy yielded an optimum model based on 14 PLS-factors. The mean squared error for hematocrit prediction was 0.011, which gives a relative prediction error of 2.7 %, based on the hematocrit average value for the studied population. The prediction

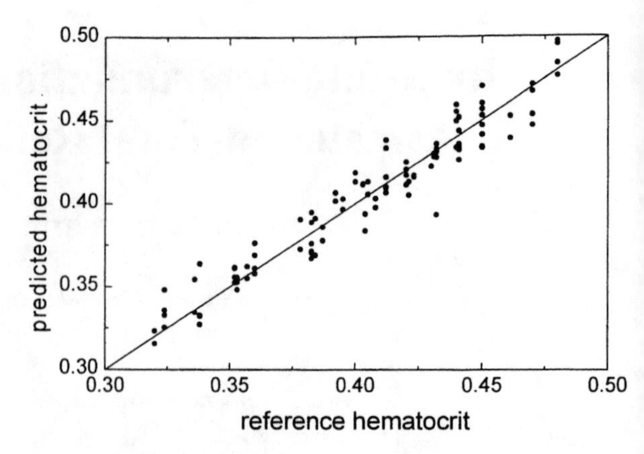

FIGURE 2. Scatter plot of hematocrit values calculated from IR-spectra versus reference values (predictions for whole blood samples originating from 15 patients with a calibration model based on 14 PLS-factors; for hemodilution, different volumes of isotonic HES-solutions were added).

performance is illustrated in Fig. 2, which gives a scatter plot of the predicted versus reference hematocrit values. In addition, the prediction quality was assessed by an *a posteriori* linear least-squares regression of the independent predictions versus the reference values ($c_{pred} = b_0 + b_1\ c_{ref}$), which yielded $b_0 = 0.024 \pm 0.010$ and $b_1 = 0.943 \pm 0.025$ with a standard error around the regression function of $s_{y/x} = 0.011$ and a correlation coefficient of $R = 0.965$.

As sedimentation of the cellular components can seriously influence the whole blood spectra, this critical assay parameter was further investigated. To separate from plasma protein adsorption effects which are well known, see for example Ref. (2), a dispersion of erythrocytes in isotonic NaCl solution was filled into the ATR-accessory trough, and the time dependent spectra were recorded within a period of 30 min. The series of spectra, obtained after scaled subtraction with a water absorbance spectrum, are displayed in Fig. 3 showing nicely the increase in

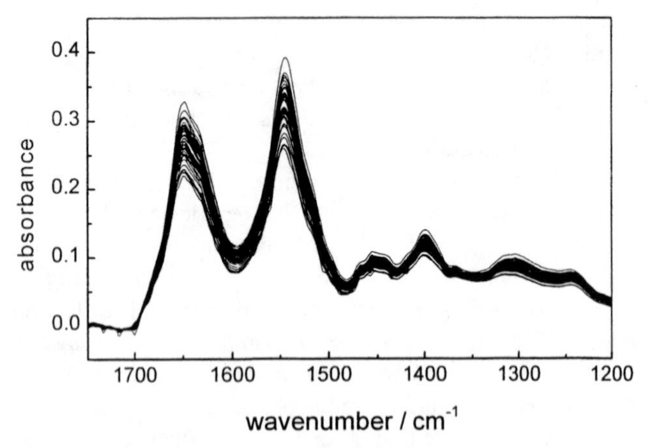

FIGURE 1. Baseline corrected absorbance ATR-spectra of whole blood samples calculated versus background spectra with a water filled cell, providing the spectral basis of the IR-spectrometric hematocrit assay.

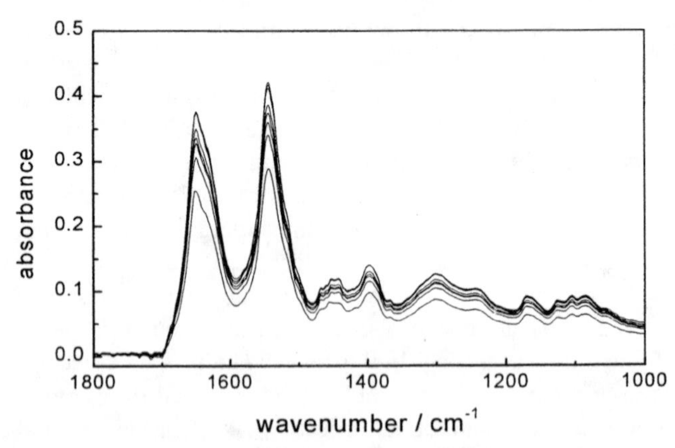

FIGURE 3. Sedimentation of washed erythrocytes in isotonic NaCl solution onto a horizontal ATR-crystal within a period of 30 min., manifested by their absorbance spectra obtained by scaled subtraction with a water spectrum (first spectrum with smallest absorbance values).

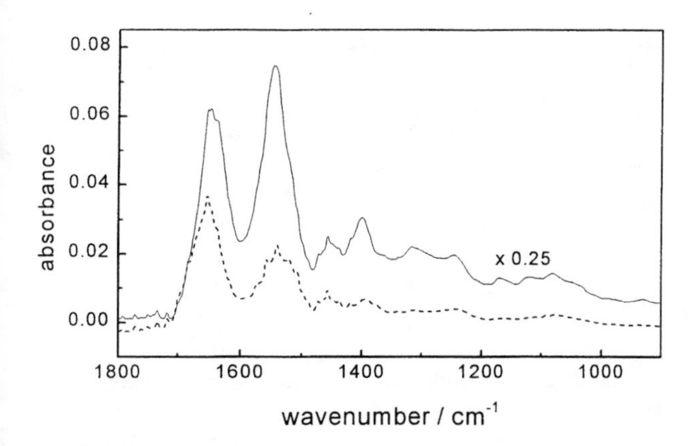

FIGURE 4. Absorbance ATR-spectra of blood plasma (solid curve, scaled by a factor of 0.25) and of the residual adsorbed and air dried protein film (dashed curve) after sample removal and a rinsing of the ATR-crystal by distilled water.

absorbances due to the time dependent growing ATR-crystal coverage.

The other effect which was studied in more detail was protein adsorption onto the crystal. After recording a blood plasma spectrum the sample was removed, and the crystal was rinsed with water only. An absorbance spectrum of the residual protein film, as calculated against the previously recorded background with the cleaned crystal, was obtained and compared with the spectrum of the blood plasma. These spectra are shown in Fig. 4. The amide II absorption band at 1540 cm^{-1} can be used for quantifying protein adsorption effects. After the usual spectrum recording time of 2 min., about 7.5 % of the protein signal can be related to the adsorbed fraction. This effect can certainly be calibrated when a tight time protocol is always taken into account.

A more serious influence comes from the sedimentation of the cellular components. A signal change of about 15 % is experienced after an additional time lapse of two further minutes after recording the first spectrum. Improvements due to independence of sedimentation can only be achieved with the application of a continuous homogenization by sample stirring or by rotating an ATR-cell type, which is principally possible, for example, with the popular Circle cell showing rotational symmetry. A shorter spectrum measurement time will also be useful in the context of reducing sedimentation effects. This can be obtained by using liquid nitrogen cooled semiconductive detectors from the mercury-cadmium-telluride type.

CONCLUSIONS

The overall performance of the IR-spectrometric hematocrit assay is comparable to that using centrifugation or conductivity measurements. This leaves opportunities for further improvements, if the reference values provided for calibration become more accurate. A strategy, as proposed recently by Faber and Kowalsky (8), can lead to improved and assay inherent prediction error estimates by correcting for the measurement error in the reference values.

A further advantage of the IR-spectroscopic assay presented here is that a variety of other blood substrates can be determined in addition to hematocrit. In this way, total protein, glucose, total cholesterol, triglycerides, urea and uric acid can also be estimated as previously shown (1, 2). For those investigations a micro-Circle cell with an internal volume of 50 μL was used, so that the necessary sample volume can be considerably reduced compared to the quantities of 0.5 mL being necessary with the horizontal ATR-accessory. Further assay improvements can be gained from simple ATR-measurements using flow-through microcells with IR-fibers based on silverhalides. The potential of fiber based accessories available for evanescent wave spectroscopy has recently been compared with conventional, but much more expensive ATR-accessories (9).

ACKNOWLEDGEMENTS

Financial support by the Ministerium für Wissenschaft und Forschung des Landes Nordrhein-Westfalen and the Bundesministerium für Bildung, Wissenschaft, Forschung und Technologie is gratefully acknowledged.

REFERENCES

1. Heise, H.M., and Bittner, A., J. Mol. Struct. **348**, 127-130 (1995)
2. Heise, H.M., Marbach, R., Koschinsky, Th., and Gries, F.A., Appl. Spectrosc. **48**, 85-95 (1994)
3. Heise, H.M., Marbach, R., Janatsch, G., and Kruse-Jarres, J.D., Anal. Chem. **61**, 2009-2015 (1989)
4. Schmitt, J.M., , Guan-Xiong, Z., and Miller, J., Proc. SPIE **1641**, 150-161 (1992)
5. Stott, R.A.W., Hortin, G.L., Wilhite, T.R., Miller, S.B., Smith, C.H., and Landt, M., Clin. Chem. **41**, 306-311 (1995)
6. Heise, H.M., Mikrochim. Acta [Suppl.] **14**, 67-77 (1997)
7. Marbach, R., and Heise, H.M., Chemom. Intell. Lab. Syst. **9**, 45-63 (1990)
8. Faber, K., and Kowalski, B.R., Appl. Spectrosc. **51**, 660-665 (1997)
9. Heise, H.M., Bittner, A., Küpper, L., and Butvina, L.N., J. Mol. Structure **410/411**, 521-525 (1997)

Essential Absorption Data for In-vitro and In-vivo Near Infrared Spectrometric Biotic Fluid Assays

H.M. Heise and A. Bittner

Institut für Spektrochemie und Angewandte Spektroskopie, Bunsen-Kirchhoff-Str. 11, D-44139 Dortmund, Germany

The performance of near infrared spectroscopic assays for the analysis of biotic samples within the frame of clinical chemistry has been the subject of several investigations. The determination of metabolites such as glucose can be carried out using multivariate calibration exploiting spectral data from different intervals. The reliability of such assays can be improved if special boundary conditions are fulfilled, as the existence of other components with similar spectra can lead to severe systematic errors for concentration prediction. Absorptivity data between 11000 and 3500 cm^{-1} is presented for the monosaccharides glucose, fructose and galactose, as well as for ethanol and hydroxyethylstarch used for hemodilution or as blood plasma expander. Pure compounds and aqueous solutions were studied. Since the hydrogen bonding network is considerably influenced by such solutes, also spectral information is provided on the variability of the spectra of aqueous samples due to temperature and electrolyte content of NaCl and KCl.

INTRODUCTION

Traditionally, infrared spectroscopy has been one of the most important physical methods in the chemical laboratory as it plays an important role in the elucidation of structures and the identification of organic compounds. Quantitative analysis has become another strength, in particular for process monitoring, which is dominated by near infrared spectrometry. Applications in clinical chemistry are growing because reagentless and fast spectrometric multicomponent assays are being developed. Promising results have been shown for near infrared spectrometry for several blood substrates (1-3). Furthermore, non-invasive diagnostic methods are desirable, as these would allow gentle and painless monitoring of important metabolites (4).

In this investigation important absorption data for the development of in-vitro and in-vivo near infrared spectrometric sensors are compiled for a broad spectral range from 11000 to 3500 cm^{-1}. Absorbance spectra are provided for different hexose sugars such as glucose, fructose and galactose, for ethanol and hydroxyethyl starch used for hemodilution or as blood plasma expander. For the sugars, spectra were recorded from crystalline powders, glassy samples prepared from sugar sirup and aqueous solutions. As the hydrogen bonding network of the water molecules within the aqueous solutions is affected by alcohol and carbohydrate solutes, other effects such as temperature variations and electrolyte concentrations (NaCl and KCl) on the spectrum of water are also investigated. A critical evaluation of the similarity between the spectra is given.

EXPERIMENTAL

Spectral measurements in the near infrared spectral range were carried out using a Bruker FT-spectrometer IFS 66 equipped with tungsten lamp, CaF_2 beamsplitter and InSb detector from Infrared Ass. (Suffolk, U.K.). For the aqueous solutions quartz transmission cells of 10 mm pathlength (short-wave near infrared) and of 0.5 mm (conventional near infrared) were used, which could be thermostatted to a temperature stability of ± 0.01 °C. Crystalline powders were measured using a diffuse reflectance accessory, which is attached via transfer optics to the spectrometer (5). The same InSb detector as for transmission measurements can be employed. Sample cups are adjustable in depth (5 mm was considered here). Glassy samples were prepared on diffusely reflecting gold-coated substrates of grade 1000 $inch^{-1}$ sandpaper from highly viscous sirup preparations and slow water evaporation by using a stream of warmed air. Such samples were also measured with the diffuse reflectance accessory. Usually, a spacer of 200 µm was used for pathlength setting, so that an average pathlength of 0.5 mm can be estimated for the transflection measurement, taking in account the solid angle seen by the detector and a Lambertian scattering characteristics for the substrates.

Glucose, fructose and galactose were for biochemical use and were supplied by Merck (Darmstadt, Germany). Ethanol, NaCl and KCl, each of grade *pro analysi*, were obtained from the same supplier. Hydroxyethyl starch (HES with average M_w 200000/ degree of substitution 0.5; 10 % in aqueous isotonic NaCl solution) was from Fresenius (Bad Homburg, Germany).

CP430, *Fourier Transform Spectroscopy:* 11th International Conference
edited by J.A. de Haseth

RESULTS AND DISCUSSION

The impetus for this investigation stems from our engagement in developing a non-invasive spectrometric blood glucose sensor for diabetic patients.This is desirable as monitoring is an important part of management of this disease. Frequent monitoring of blood glucose can help prevent complications from severe hypo- and hyper-glycemia. For this reason glucose is the key compound in this investigation. Fructose is a ketohexose used as caloric sweetener which is more than two times sweeter than glucose and can be fitted into the diet of diabetic patients, because it does not raise the blood sugar as fast as glucose (the carbohydrate glycemic index for fructose is 20 compared to 100 for glucose). The third monosaccharide studied is galactose, an aldohexose which shows, like glucose, α- and β-anomeric forms. Galactose is, besides glucose, part of the lactose disaccharide found in milk and is bound to most glycoproteins. Hydroxyethyl starch (HES) solutions have been reported to have good characteristics for hemodilution to improve micro-circulation of the blood or is used as blood substitute. Ethanol consumed as part of an alcoholic beverage can be found in similar concentrations in blood as glucose.

Quantitative absorption data on these compounds is still missing, although in some publications limited spectral

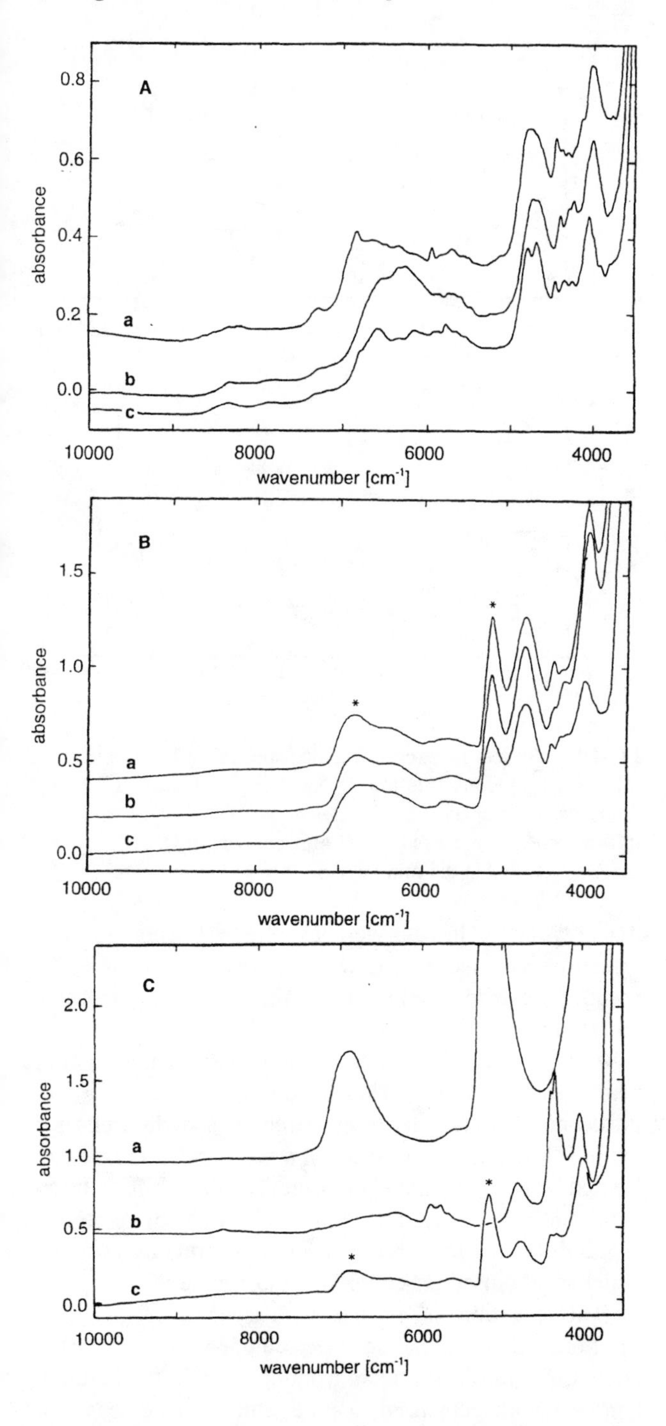

FIGURE 1. Comparison of pure component near infrared spectra: **A** Diffuse reflectance spectra of different crystalline monosaccharides: a) fructose, b) glucose and c) galactose; **B** spectra from glass-like sugar samples prepared from sirup preparations and measured in transflectance (same order as given in A); **C** absorbance spectrum of water (a) and ethanol (b), each recorded with 0.5 mm pathlength and at 25 °C, and transflectance spectrum of hydroxyethyl starch (c).

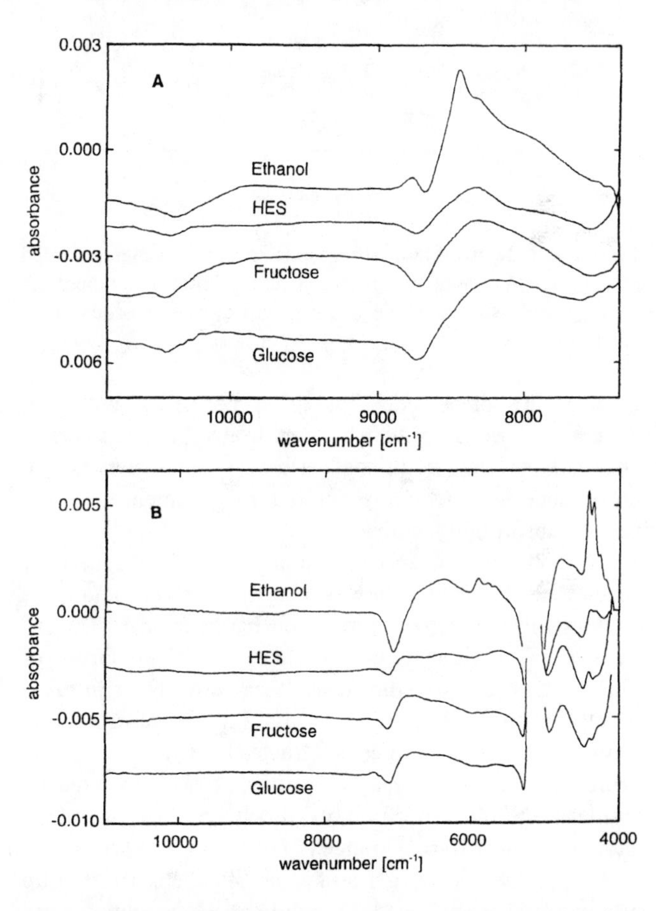

FIGURE 2. Difference spectra of some diluted aqueous solutions of compounds important for biotic assays with partial water absorbance compensation: **A** short-wave near infrared spectral range recorded from solutions of 0.5 % concentration each recorded at 30 °C with a 10 mm pathlength cell; **B** near infrared data from solutions of 1.5 % concentration recorded at 30 °C with a 0.5 mm pathlength cell.

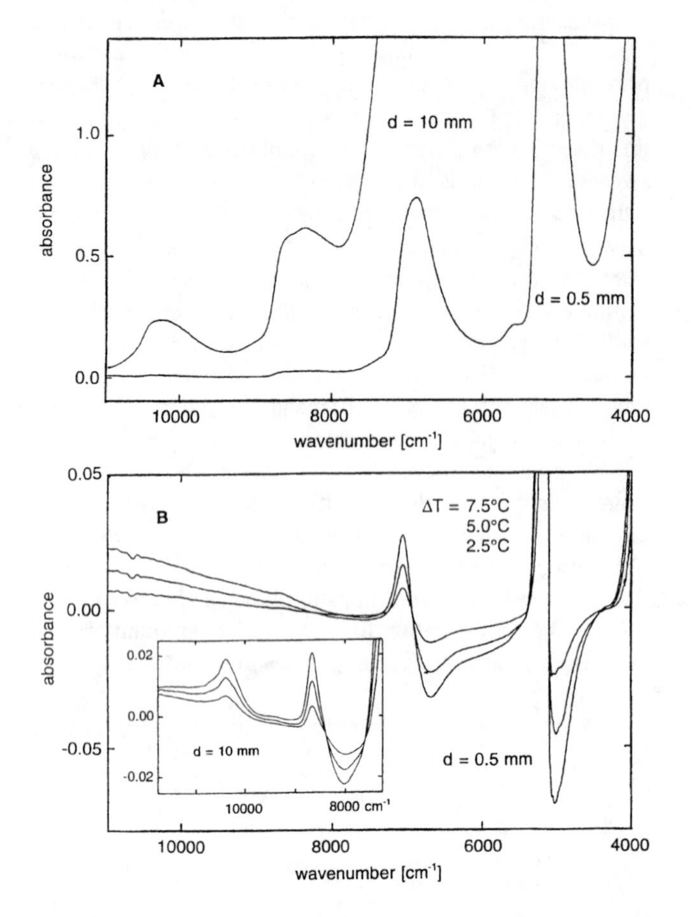

FIGURE 3. Near infrared absorbance spectra of water recorded at 25 °C (**A**), and difference spectra showing temperature dependencies (the reference water spectrum was recorded at 25 °C) (**B**).

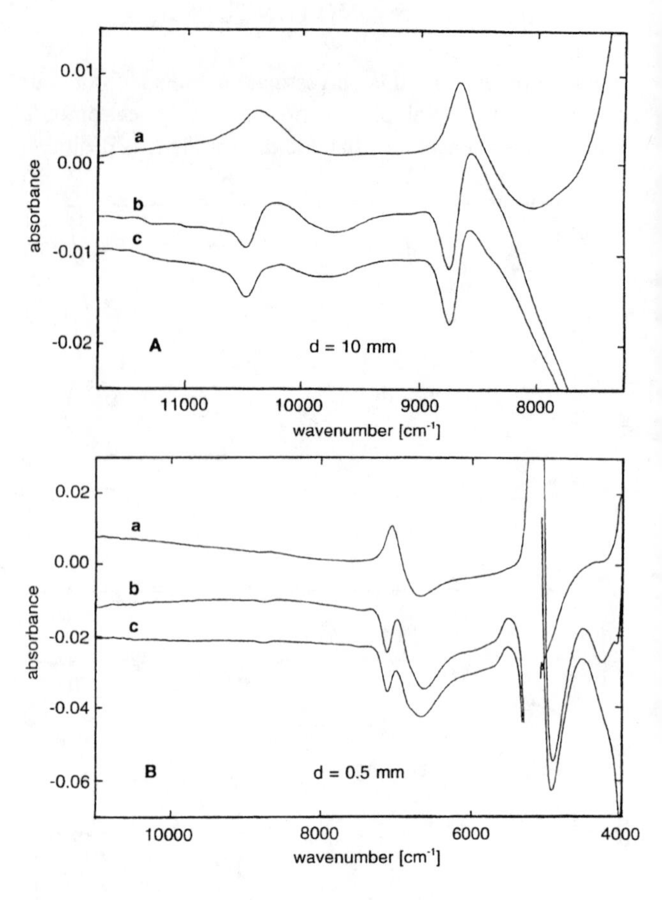

FIGURE 4. Near infrared difference spectra for water at 32.5 °C (a), for an aqueous solution of NaCl (b), and of KCl (c) (5 % each) versus a spectrum of water at 30 °C: **A** short-wave near infrared spectra recorded at 10 mm pathlength; **B** near infrared spectra recorded at 0.5 mm pathlength.

features are presented. This is in particular valid for aqueous solutions, for which the hydrogen bond network is changed, so that additional effects occur when scaled absorbance subtraction is applied for compensating the solvent absorption features.

As solids, the sugars show many sharp peaks in their spectra, see Figure 1A; however, with these compounds as solutions usually broad absorption bands appear. The influence of different aggregate states on the spectrum of glucose was recently studied by Reeves (6). Near infrared spectra of crystalline, molten, glass-like, freeze-dried, and saturated solution (water subtracted) of glucose were presented. The last three types of spectra show greatest similarity. Similar spectra have also been reported for the short-wave near infrared range (7).

We produced some glass-like modifications from sirup preparations which lead to solution equivalent spectra (compare Figures 1B, C and 2B). It is difficult to remove the remaining water without starting crystallization (residual water bands are marked by an asterisk). Water adsorbed to starch could, for example, only be removed after vacuum drying at 60 °C (8). For completeness Fig. 1C shows the spectra of water and ethanol recorded in a transmission cell, as well as the spectrum of HES measured as dried film. The comparison of the latter two spectra provide some clues for the assignment of vibrational bands.

There have been some efforts to quantify changes in the near infrared spectra of such compounds caused by the influence of water, pH, ionic strength and differences in the physical state using regression statistics, but these were limited to the long-wavelength interval between 4900 and 4100 cm^{-1}. Similarly, larger intervals could be tested, but especially in the latter region some fine structure is evident which can be used for compound discrimination.

The situation is similar with spectra from diluted solutions, for which some water compensation is applied. However, additional effects from changes in the hydrogen bond network compared to pure water can be seen from the first derivative features. Excluding the interval between 4500 and 4000 cm^{-1}, the spectral features for the monosaccharides and HES do not show significant differences. Using a multivariate approach, calibration models for ethanol will be less effected by spectral collinearity, although some interferences are still to be expected to the other compounds shown.

As we were interested in manifesting the O-H equilibrium structure perturbations, water spectra under different conditions were recorded. In Figure 3 the absorbance spectra of water at 25 °C for different pathlengths are shown. The effect from temperature changes is best illustrated by difference spectroscopy. One can see that the effects are linear with temperature change, so that temperature prediction can be based on such data; literature references on this subject can be found in (3).

The same can be stated for the effect that electrolytes have on the hydrogen bonding structure; for this reason only the spectrum of one NaCl and KCl solution is displayed in Figure 4. Spectral gaps above 5000 cm^{-1} are due to the high water absorbances seen even with a cell of 0.5 mm pathlength, so that the noise level becomes unacceptable. A much greater complexity of electrolyte samples has been recently studied by Brown and coworkers (9) who also give a good coverage of the literature available in this field. However, sodium and potassium ions are interesting from a physiological point of view because of some concentration variability.

Another parameter with influence on the spectrum is the pH value of water, which has recently been investigated by Molt and Cho (10). They followed a titration of an HCl-solution with NaOH. The resulting difference spectra show the same features as provided by Figure 4B (see trace b for features obtained for the NaCl-solution).

The expected large temperature influence on substrate concentration prediction has already been proved previously by us using partial least-squares calibration models from short-wave near infrared absorbance or logarithmized single beam spectra of blood plasma (11). When spectral differences against a water background spectrum recorded subsequently, the temperature impact on prediction could be eliminated. Improved difference spectra of biotic samples might be produced employing reference spectra of water with adjusted ionic strength.

CONCLUSIONS

To obtain reliable quantitative information on various analytes in tissue or blood, spectrometric methods are necessary which measure a broad spectral range allowing multivariate calibration modeling. Furthermore, certain behavior, such as avoiding fructose or alcohol, can assist in improving the reliability of glucose concentration prediction within diabetic patient self-monitoring. Restrictions for glucose have to be observed within clinical monitoring when hemodilution or the application of blood plasma expander with polysaccharide compounds has been considered for the patient. The same statements given for the in-vitro analysis of blood samples are valid for future non-invasive spectrometric in-vivo sensing systems.

ACKNOWLEDGEMENTS

Financial support by the Ministerium für Wissenschaft und Forschung des Landes Nordrhein-Westfalen and the Bundesministerium für Bildung, Wissenschaft, Forschung und Technologie is gratefully acknowledged. Further funds were granted by the Deutsche Forschungsgemeinschaft.

REFERENCES

1. Arnold, M.A., New developments and clinical impact of noninvasive monitoring' in: Handbook of Clinical Automation, Robotics, and Optimization, Kost, G.J., and Welsh, J., eds., New York: John Wiley & Sons, 1996, pp. 631-647
2. Heise, H.M., Mikrochim. Acta [Suppl.] **14**, 67-77 (1997)
3. Heise, H.M., Near-Infrared Spectrometry for in vivo Glucose Sensing, in: Biosensors in the Body: Continuous in vivo Monitoring, Fraser, D.M., ed., Chichester: John Wiley & Sons, 1997, ch. 3, pp. 79-116
4. Heise, H.M., Proc. IEEE Engineer. Med. & Biol., 18th Ann. Int. Conference, Amsterdam, Paper No. 1151, pp. 1-3 (1996).
5. Marbach, R., and Heise, H.M., Applied Optics **34**, 610-621 (1995)
6. Reeves, III, J.B., J. AOAC **77**, 814-820 (1994)
7. Reeves, III, J.B., J. Near Infrared Spectrosc. **2**, 199-212 (1994)
8. Delwiche, S.R., Norris, K.H., and Pitt, R.E., Appl. Spectrosc. **46**, 782-789 (1992)
9. Lin, J., Zhou, J., and Brown, C.W., Appl. Spectrosc. **50**, 444-448 (1996)
10. Molt, K., Cho, Y.J., J. Mol. Structure **349**, 345-348 (1995)
11. Bittner, A., Marbach, R., and Heise, H.M., J. Mol. Structure **349**, 341-344 (1995)

Dry Film Preparation from Whole Blood, Plasma and Serum for Quantitative Infrared Diffuse Reflectance Spectroscopy

A. Bittner and H.M. Heise

Institut für Spektrochemie und Angewandte Spektroskopie,
Bunsen-Kirchhoff-Str. 11, D-44139 Dortmund, Germany

The potential of infrared spectroscopy in the analysis of biotic fluids for the determination of important clinical parameters such as glucose and other blood substrates has been investigated. For this purpose dried films from whole blood, blood plasma and serum were prepared on diffusely reflecting gold-coated substrates from sandpaper of different grades. This enabled measurements in the mid and near infrared spectral ranges by using special diffuse reflectance accessories. The removal of water leads to a considerable enrichment of the fluid constituents. Due to the reduced sample complexity a considerable gain in spectral information is obtained. This is especially valid for measurements in the near infrared where the problems associated with variability in the spectra of aqueous samples due to several parameters, i.e. temperature, electrolyte content etc., are well known. Additionally, mid infrared studies were carried out into the stability of dried samples.

INTRODUCTION

An infrared spectroscopic multicomponent assay of biotic fluids such as whole blood, plasma or serum holds considerable promise for the clinical laboratory, being reagentless and fast. Most investigations were carried out with liquid samples using the attenuated total reflection technique (1), although other conventional measurement techniques have also been proposed (transmission, photoacoustic spectroscopy) (2). The removal of water leads to an enrichment of the blood constituents. As water dominates the absorption spectrum and is sensitive to temperature and the varying concentrations of electrolytes or proteins, removing it makes quantitative analysis much simpler. This is especially valid for the spectral regions covered by fundamental O-H stretching absorption and combination bands. Recently, samples of dried human serum were investigated for the near infrared spectrometric assay of urea (3). Spectra of dried samples of whole blood or serum on polyethylene foils were recorded in the mid-infrared spectral range by other researchers suggesting assays for glucose and cholesterol (4). Potassium thiocyanate served as internal spectral standard to improve the assay performance after spectrum normalization.

Spectra of high quality can be obtained from liquid films spread on a gold-coated, diffusely reflecting substrate, otherwise spectrum reproducibility suffers tremendously. We studied several effects during sample drying. The poor quality spectra of dried sample drops could be modelled by taking the inhomogeneous sample thicknesses into account. Improved spectra of thin films prepared from different biotic fluids, i.e. whole blood, plasma and serum, were recorded using diffuse reflectance accessories based on rotation-symmetric ellipsoidal mirrors and substrates of different roughness. Spectra in the near and mid-infrared are presented. Further studies were carried out into the stability of dried samples.

EXPERIMENTAL

Mid infrared spectra were recorded using a Perkin Elmer Modell 2000 FT-spectrometer with globar and KBr beamsplitter. The diffuse reflectance accessory in use here employs a rotational ellipsoidal mirror for collection of the diffusely reflected radiation, and for sample illumination a gold-coated light pipe (5). The accessory is completely purged with dry CO_2 free air. A dedicated liquid-nitrogen-cooled, narrow-range MCT detector from Santa Barbara Research Center (Goleta, CA, U.S.A.) is part of the accessory. Spectral measurements in the near infrared spectral range were carried out using a Bruker FT-spectrometer IFS 66 equipped with tungsten lamp and CaF_2 beamsplitter. The diffuse reflectance accessory housing with transfer optics (custom-made) and InSb detector with an element diameter size of 4 mm from Infrared Ass. (Suffolk, U.K.) (6) is attached to the spectrometer and can be purged by dried nitrogen (see also Figure 1). Instead of a light pipe, more efficient mirror optics for illumination are installed in the near infrared accessory. Spectral simulations on inhomogenous sample layer effects were carried out using the MATLAB software pakage (The Mathworks, South Natick, MA, U.S.A.).

CP430, *Fourier Transform Spectroscopy:* 11th International Conference
edited by J.A. de Haseth

RESULTS AND DISCUSSION

The diffuse reflectance (DR) technique is an appropriate technique for the measurement of scattering samples, although also high throughput optical accessories for transmission meassurements have been successfully applied (7). In this investigation a high throughput device for measuring DR-spectra of thin films from biotic fluids is used, by which also a local analysis on bulky samples can be performed. The near infrared accessory has been adapted to study also liquid samples, as well as powders or various skin tissue. The spot size studied can be varied by using different Jacquinot aperture stops, but the maximum outer diameter of the illuminated sample spot is 2 mm. The mid-infrared device has an irradiated area of about 3 mm diameter, but detector element size is 2 x 2 mm^2.

When starting this investigation we analyzed the reproducibility of measuring dried samples from blood plasma and serum, which had been filled into flat microcups of two mm inner diameter. Some spectra of dried specimens obtained from the same serum are displayed in Figure 2 illustrating the rather poor reproducibility for spectrum generation. This is the reason why an internal standard was considered by Budinova et al. for their transmission experiments with dried serum films.

Non-uniform samples have been characterized by FT-IR spectroscopy already in the past, see for example (8). We simulated the effects from inhomogeneous biotic films by considering simple layer models and integrating the sample transmittance over the cell lengths. The geometry for the individual cases is shown in Figure 3. Absorbances obtained by means of other cell types showing, e.g., wedge and lense shape have been analytically derived (9).

As basis for the absorption coefficients, scaled data from an absorbance spectrum of a dried plasma film with a minor water residue was used, which had been prepared between two plane disks of ZnSe pressed together. The

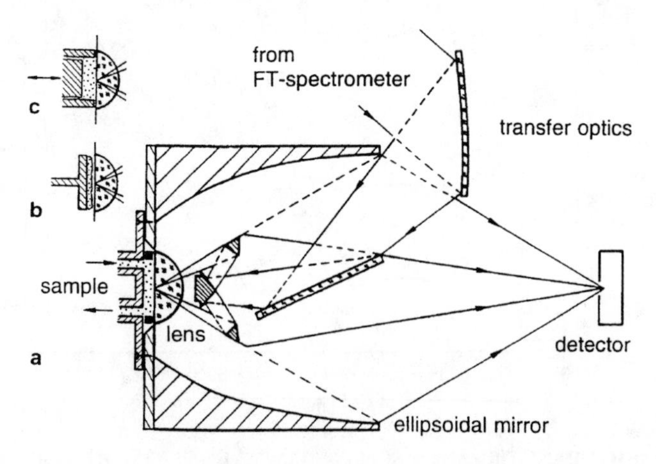

FIGURE 1. Diffuse reflectance accessory used for near infrared spectrometry of scattering samples: a) liquid samples, b) solid films and c) powders.

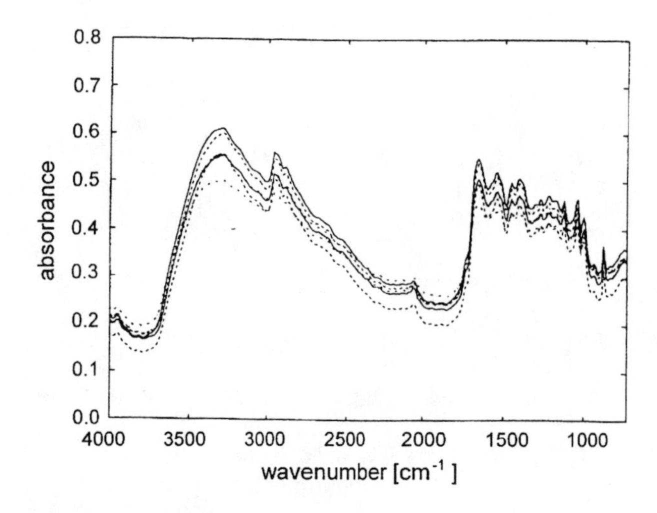

FIGURE 2. DR-spectra of dried serum samples of 1.8 μL volume prepared in gold-coated microcups.

different spectral traces are presented in Figure 4 which illustrate that the sample area should fill the illuminated and detected spot totally and the layer thicknesses have to be appropriately adjusted for optimum photometric performance.

For the following studies biotic fluid samples with a volume of about one μL were spread out onto a diffusely reflecting substrate by using a spatula allowing the reproducible production of fluid layers. These could be dried within short time by exposing the sample to atmospheric air. An example is provided in Figure 5 which shows the spectrum of a dried whole blood film prepared on a gold-coated rough substrate of 1000 $inch^{-1}$ grade sandpaper. Some features in the near infrared spectral range are also presented. However, such DR-spectra, but of better quality, could be recorded by using the near infrared accessory with InSb detector and the Bruker FT-spectrometer especially equipped for near infrared measurements.

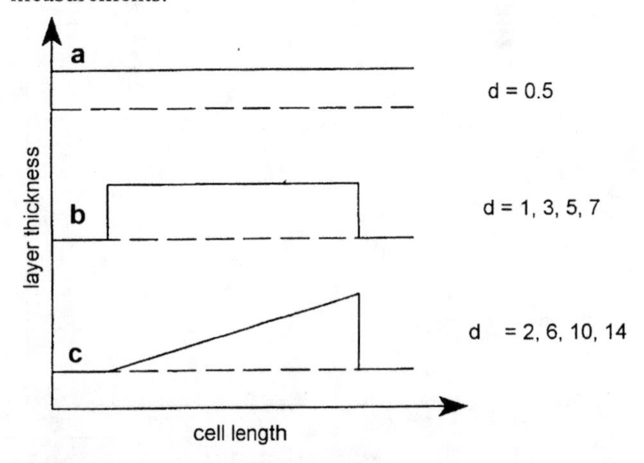

FIGURE 3. Different geometries of cell configurations used for the simulation of spectral effects from sample layer inhomogeneities: a) plane layer, b) plane layer sample not filling the probing aperture and c) wedged sample layer with incomplete coverage.

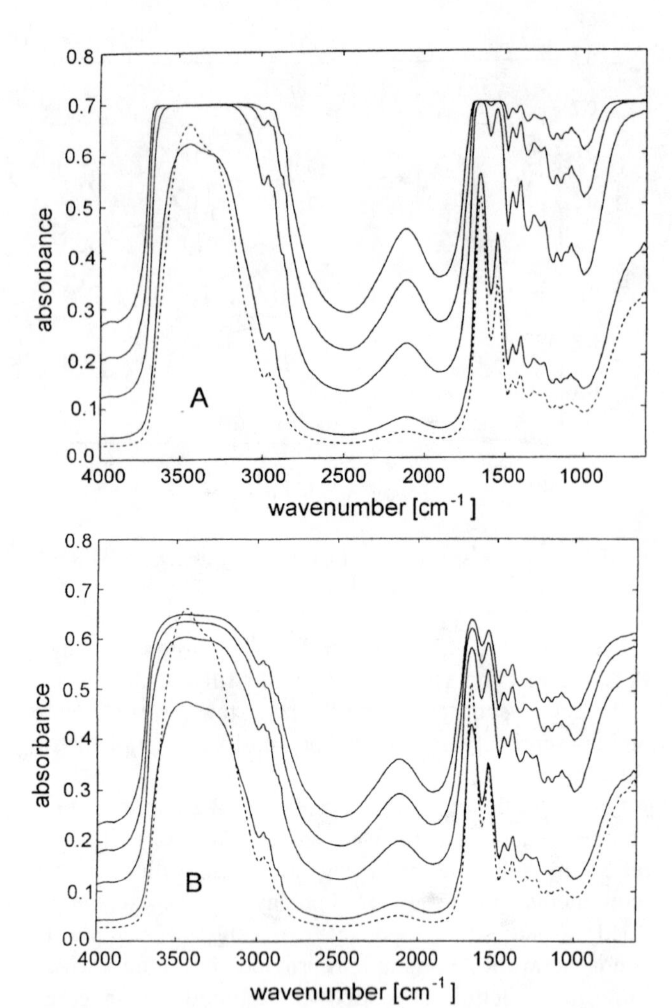

FIGURE 4. Simulated absorbance spectra of dried plasma films of different shapes and varying maximum thicknesses (dashed curves are for a homogeneous layer as shown in Fig. 3a): A stepped cell with the geometry and relative maximum layer thicknesses as shown in Fig. 3b with a sample coverage of 80 %; B wedged cell with same coverage as above, for the cell parameters, see Fig. 3c.

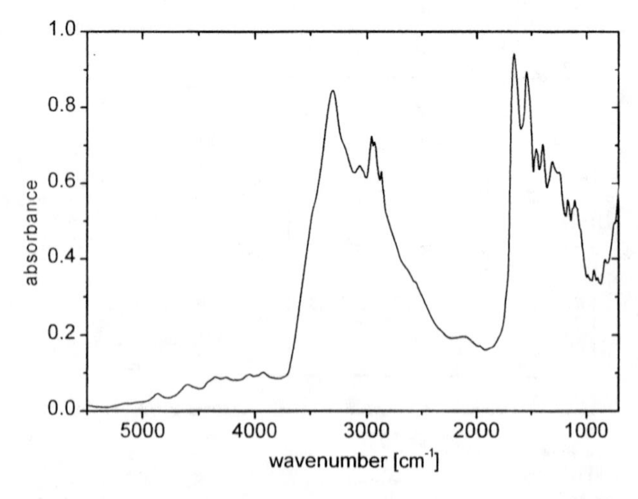

FIGURE 5. Mid infrared DR-spectrum of a dried whole blood film prepared on a gold-coated rough substrate of grade 1000 inch^{-1} sandpaper.

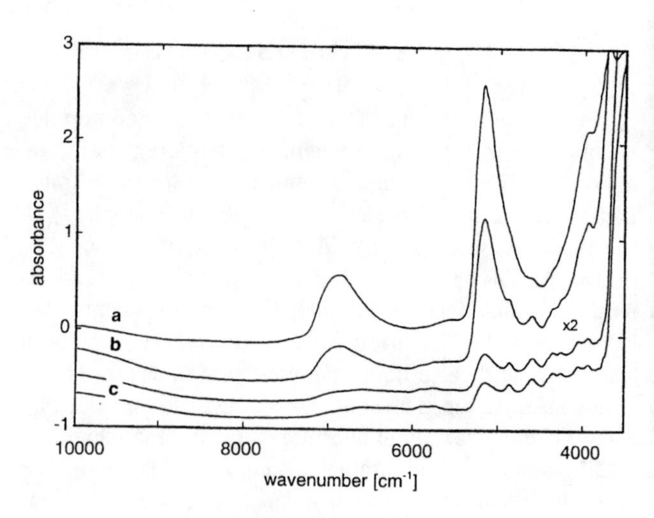

FIGURE 6. Absorbance spectra of whole capillary blood measured in a cell of 200 μm pathlength (the backing of the cell for the transflectance measurement is from a gold-coated substrate of 1000 grade sandpaper) (a); film from a 200 μm layer of whole blood after evaporation of about half of its water content (b); two spectra from dry blood films from the same sample as shown above, time differences between the recordings 2 minutes each (c). Traces b and c were enlarged by a factor of 2.

The advantages from the water evaporation are evident from Figure 6, where the DR-spectrum of a whole blood sample is shown along with the spectra taken during the water evaporation process. The spectral features of the protein compounds stand out prominently for the dry film preparation and are undisturbed by water, so that high quality spectra can be produced for quantitative analysis.

The preparation of films by means of rough substrates is easy, because the roughness determines the average sample pathlength of the attached film. The fluid has to brought to the substrate, immediately spread out and dried for quantitative spectrometry. In Figure 7 spectra of dried

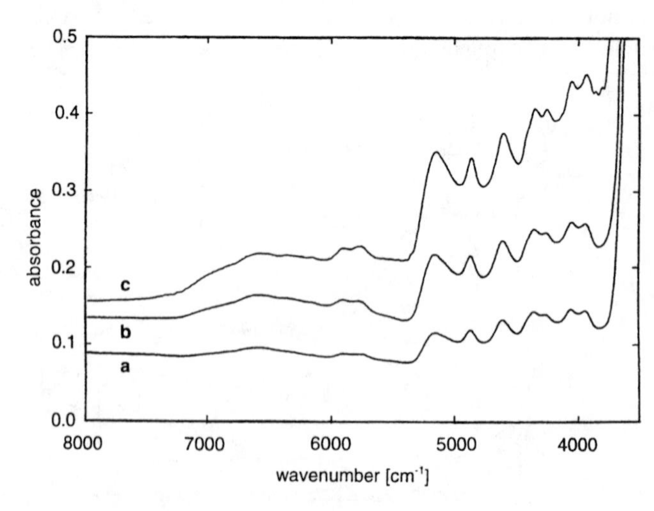

FIGURE 7. DR-spectra of blood films prepared on rough gold-coated substrates: grade 1000 inch^{-1} (a), grade 280 inch^{-1} (b), and grade 1000 inch^{-1} with an additional spacer on top of the substrate.

blood films produced by spreading whole blood specimens on substrates of different roughness and with the application of an additional cell spacer to increase the fluid film thickness are shown. If the original average sample thicknesses of the blood films are estimated from the dried protein band intensities and compared to those obtained from the 200 μm thick blood sample, values are about 115 μm for the 280 $inch^{-1}$ grade sand paper, which should theoretically have a particle size of about 90 μm. A scanning electron microscopic analysis of such a substrate shows particles actually about a factor of two smaller (10). On the other hand, the sample prepared on the 1000 $inch^{-1}$ grade paper stems from a liquid film of about 50 μm.

We also studied the influence from substrate inhomogenities which is noticeable for different smaller investigated areas and the coarser substrate. On the other hand, enlarging the film area to be studied for the less rough substrate by using a larger Jacquinot aperture stop does not influence the absorbance signals.

Stability of plasma and serum films was also studied. There are slight changes in intensity. These amount to about 3 % for dried plasma and to about 1 % for serum as noticed for the amide I and II bands in the mid infrared around 1600 cm^{-1}; the spectra were recorded after a period of one hour following water evaporation from the fluid film. Derivative features were noticed indicating a band shifting of about 30 cm^{-1}, see also (11). These facts have to be taken into consideration for repeat measurements. These can be considered, since dried biofilm preparations possess the capability of being stabilized for long-term archival. Further investigations have to be carried out covering also the near infrared spectra of dried biotic film specimens.

CONCLUSIONS

The preparation of dry films from aqueous biotic samples, such as whole blood or derived fluids, on inexpensive substrates is simple. The infrared spectroscopic measurement by using the diffuse reflectance technique is straightforward and reproducible. Much thicker film layers can be studied by this technique than by using the attenuated total reflection technique favoured so far for the analysis of biotic fluids (2, 12). Sedimentation, which exists for fluids with cellular components as for whole blood, leads to inhomogeneous samples, but does not affect the diffuse reflectance measurement integrating on the total sample layer thickness. This is not the case, for example, using the attenuated total reflection technique. Such a technique is more sensitive to sample gradients, since the probing depth is much shorter than experienced with the transflection measurements reported here. Sample volumes can be smaller than one μL, so that minimal invasive techniques can be applied to obtain body fluids for clinical testing. Infrared spectroscopic assays based on such a technique have to be tested on larger hospital sample populations such as previously described for liquid blood plasma (1). Since diffusely reflecting substrates can be easily produced and are inexpensive, disposable sampling strips can give the sample support for routine spectroscopic analysers.

ACKNOWLEDGEMENTS

Financial support by the Ministerium für Wissenschaft und Forschung des Landes Nordrhein-Westfalen and the Bundesministerium für Bildung, Wissenschaft, Forschung und Technologie is gratefully acknowledged. Further funds were granted by the Deutsche Forschungsgemeinschaft and Boehringer Mannheim GmbH (Mannheim, Germany).

REFERENCES

1. Heise, H.M., Marbach, R., Koschinsky, Th., and Gries, F.A., Appl. Spectrosc. **48**, 85-95 (1994)
2. Wang, J., Sowa, M., Mantsch, H.H., Bittner, A., and Heise, H.M., Trends Anal. Chem. **15**, 286-296 (1996)
3. Hall, J.W., and Pollard, A., J. Near Infrared Spectrosc. **1**, 127-132 (1993)
4. Budinova, G., Salva, J., and Volka, K., Appl. Spectrosc. **51**, 631-635 (1997)
5. Korte, E.H., and Otto, A., Appl. Spectrosc. **42**, 38-43 (1988)
6. Marbach, R., and Heise, H.M., Applied Optics **34**, 610-621 (1995)
7. Jakse, F.P., Friedman, R.M., and Freeman, J.J., Appl. Spectrosc. **38**, 700-705 (1984)
8. Hirschfeld, T., and Cody, C., Appl. Spectrosc. **31**, 551-552 (1977)
9. Hirschfeld, T., Appl. Spectrosc. **39**, 426-430 (1985)
10. Otto, A., Infrarotspektroskopie mit diffus reflektierter Strahlung: in-situ Messungen an schwach streuenden Proben, Fortschr. Ber. VDI Ser. 8, Vol. **146**, VDI Düsseldorf (1987)
11. Heise, H.M., Mikrochim. Acta [Suppl.] **14**, 67-77 (1997)
12. Heise, H.M., Bittner, A., Küpper, L., and Butvina, L.N., J. Mol. Structure, **410/411**, 521-525 (1997)

Near Infrared Spectrometric Investigation of Pulsatile Blood Flow for Non-Invasive Metabolite Monitoring

H.M. Heise and A. Bittner

Institut für Spektrochemie und Angewandte Spektroskopie,
Bunsen-Kirchhoff-Str. 11, D-44139 Dortmund, Germany

The non-invasive measurement of blood parameters is desirable for several reasons. A multi-wavelength near infrared approach based on the pulsatile blood flow - similar to pulse oximetry in the visible range - allows for a glucose assay with exclusive information probing from the vascular fluid space. Fast diffuse reflectance spectra were recorded from the human inner lip using a Fourier-Transform spectrometer equipped with an especially optimized custom-made accessory. The absorbance values of each spectral variable within the spectrum population obtained at least within a one-minute measurement period were Fourier transformed and the outstanding coefficient amplitudes were evaluated for the pulse frequency. These spectral Fourier coefficients can be composed to give the pulsatile near infrared blood spectrum (IR-plethysmography). An estimation of the experimental conditions necessary for pulse spectrometry with the goal of non-invasive metabolite monitoring is presented.

INTRODUCTION

The interest in the infrared spectral range for medical applications increases rapidly due to improvements in instrumentation and data processing. During the last few years near infrared spectrometry has been investigated as a non-invasive clinical tool for improved understanding of in-vivo processes by several researchers. The possibilities for the transcutaneous determination of important metabolites such as blood glucose were reviewed recently (1). Integral tissue probing suffers many limitations as described in detail (2). As most clinically relevant parameters are obtained by the analysis of whole blood or derived fluids, i.e. blood plasma or serum, it is highly desirable to have access to similar information by non-invasive spectrometric means. However, since blood represents only an unknown fraction of the total tissue, the corresponding signal changes due to the pulsatile blood flow are minimal when compared to the total tissue water signal, which varies according to the microvasculature and the tissue texture. Such a measurement principle has not yet been applied for practical metabolite measurements because of limitations in signal-to-noise ratio, although it is in use for pulse oximetry (3), the measurement of dyes injected as bolus to the vascular system (4) or as suggested for the determination of hemoglobin derivatives in whole blood (5).

Results of fast measurements on human oral mucosa using diffuse reflectance spectroscopy are reported. These allow the estimation of the water volume variations caused by changes in the arterial blood compartment associated with the cardiac cycle (near infrared plethysmography). Wavelength dependent penetration depth of the radiation provides information for different layer thicknesses of the skin tissue under investigation. An estimation of the optimum experimental conditions necessary for pulse spectrometry aiming at non-invasive metabolite monitoring is presented.

EXPERIMENTAL

Spectral measurements were carried out using a Bruker FT-spectrometer IFS 66 equipped with tungsten lamp and CaF_2 beamsplitter. The diffuse reflectance accessory housing with transfer optics (custom-made) and InSb detector from Infrared Ass. (Suffolk, U.K.) (6) was attached to the spectrometer for measuring skin tissue spectra. For fast measurements the spectrometer software for gas chromatography/FTIR coupling was employed. A total of nine interferograms with a spectral resolution of 32 cm^{-1} were averaged for each spectrum within 0.5 s. Diffuse reflectance spectra were recorded for a measurement time of at least two minutes. Subsequent background spectra could be recorded using Spectralon reflectance standard material from Labsphere (North Sutton, NH, U.S.A.). The spectra were transferred to a personal computer, and further data processing was carried out using the MATLAB software package (The Mathworks, South Natick, MA, U.S.A.). Savitzky-Golay smoothing was usually applied for noise reduction (quadratic polynomial with 25 data points) before Fourier analysis of the logarithmized single beam spectra.

CP430, *Fourier Transform Spectroscopy:* 11th International Conference
edited by J.A. de Haseth

RESULTS AND DISCUSSION

Several different reasons have shown the inner lip to be a good subject in several previous investigations. It is a rather homogenous tissue with the stratum corneum missing, so that good optical contact to the accessory immersion lens with refractive index matching can be guaranteed thus avoiding Fresnel reflection at the lip surface and large impact from scattering within the outmost skin layer. Furthermore, the lip is rich in capillary blood vessels and well thermostatted with the expectation of insignificant temperature gradients with increasing skin depth.

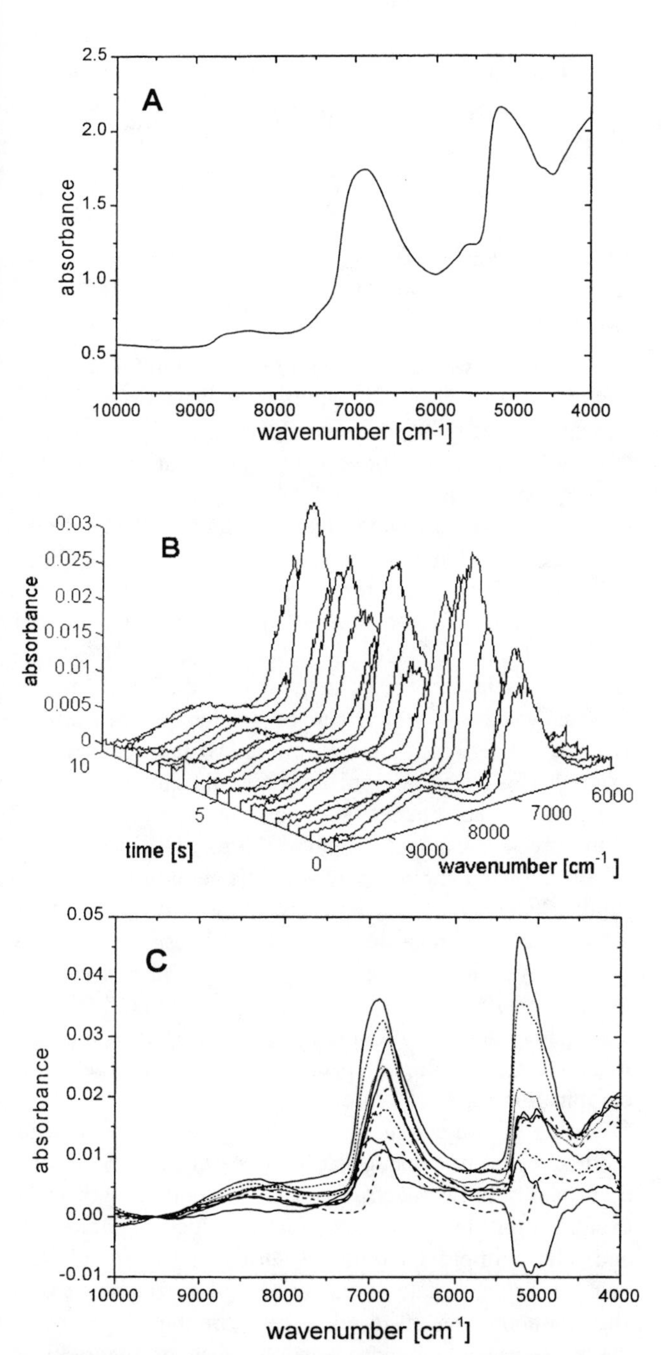

FIGURE 1. Mean of the diffuse lip spectra as calculated against a background spectrum using Spectralon (A) and some of the pulse resolved near infrared spectra shown as differences versus the first measured single beam lip spectrum (B); absorbance differences of the first 10 lip spectra as calculated versus the first recorded spectrum after baseline offset correction at 9500 cm^{-1} and Savitzky-Golay smoothing (C).

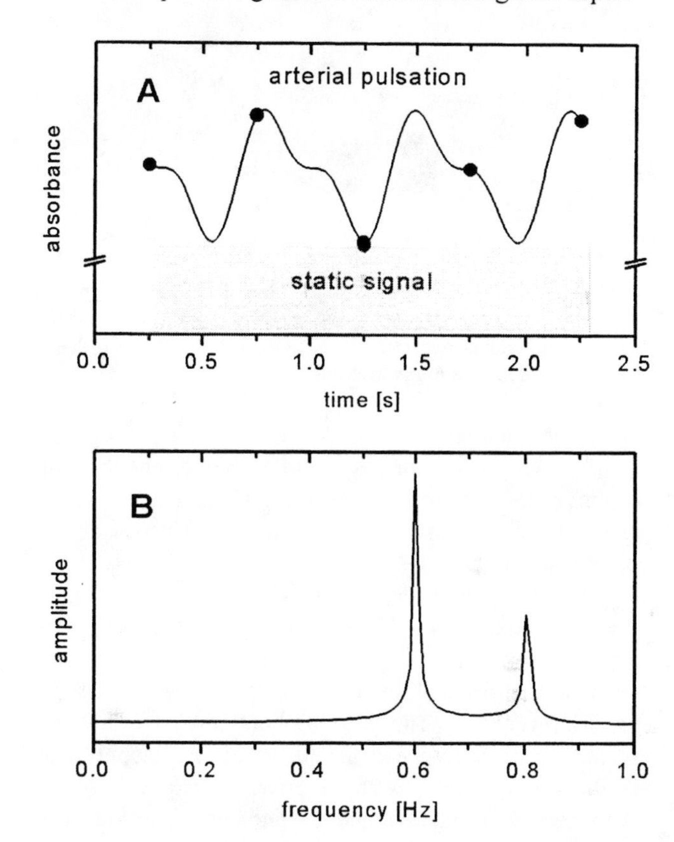

FIGURE 2. Pulsatile signal as known from pulse oximetry (pulse rate of 84 heart beats per minute) (A) and result from the Fourier analysis of the digitized trace during a time period of two minutes (two data points sampled per second; for aliasing see text) (B).

The near infrared average spectrum from a session of fast measurements and some time dependent difference spectra calculated versus the first spectrum recorded are shown in Fig. 1, which illustrates the rhythmic changes in the water content due to the pulsatile blood flow. The average spectrum from a two-minutes run is comparable to those from previous in-vivo tissue measurements when accumulating 1200 interferograms each necessary for signal-to-noise improvements (7). Positioning of the lip is rather critical for the time period of two minutes, because spectra are recorded from an area of about 1 mm^2. Heterogenity of the tissue due to position shifting can easily be documented by difference spectroscopy.

A closer look at the water band around 6800 cm^{-1}, after some slight baseline standardization, illustrates, as shown in Fig. 1C, a significant dynamic shifting of the absorption band maximum. This fact suggests a change in the water

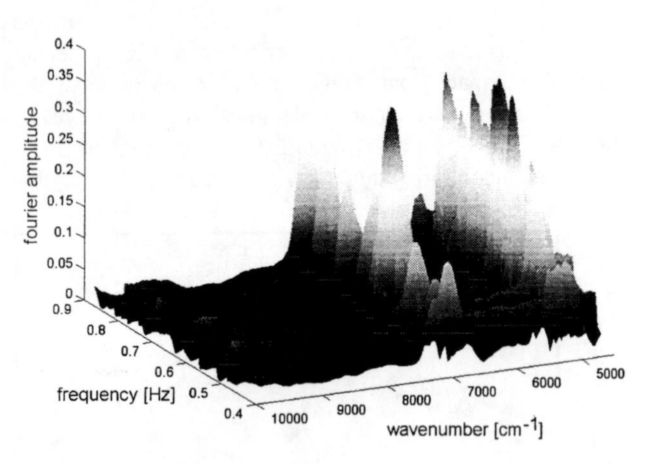

FIGURE 3. Three-dimensional plot of spectral Fourier amplitudes illustrating the pulsatile signals in the near infrared diffuse reflectance spectra of a human oral mucosa due to rhythmic blood pressure variations.

bonding structure which might be due to dynamic effects arising from the moving or standing blood, but this must be investigated further. Temperature fluctuations will produce different effects such as a first derivative absorption band feature.

The Fourier analysis of the pulsatile signals as known from pulse oximetry is shown in Fig. 2. Such a signal, with a cardiac cycle rate of 84 min^{-1}, was digitized in the same manner as provided by the repetition rate of the skin spectra recordings. The resulting Fourier transformation is presented in Figure 2B. It illustrates the inappropriate digitization frequency with the effect of frequency folding at the given Nyquist frequency limit which is known as aliasing. Also higher harmonics are folded into the frequency band resulting from the signal sampling. Nevertheless, a unique and corrected frequency assignment is still possible with the *a-priori* information about the existing pulse rates.

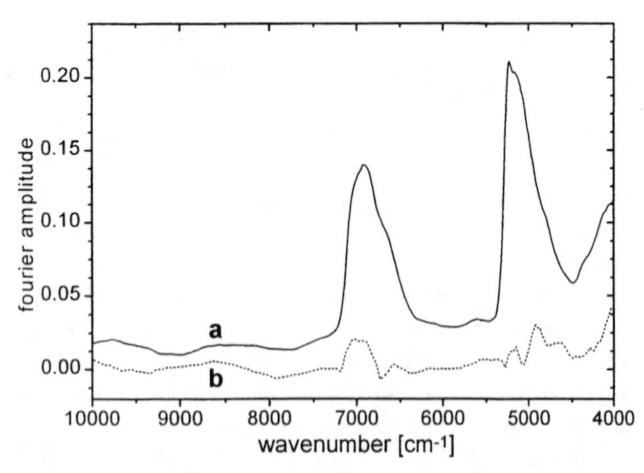

FIGURE 4. Averaged spectral Fourier amplitude coefficients: frequency interval between 0.59 and 0.65 Hz was considered (a), for the lower trace data were taken from the interval between 0.43 and 0.45 Hz (b).

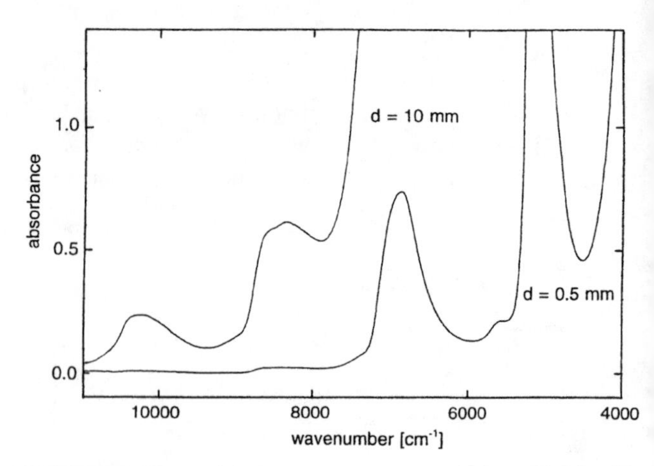

FIGURE 5. Water absorbance spectra measured at 25 °C with quartz transmission cells of 10 mm and 0.5 mm pathlength, respectively.

The three-dimensional diagram of the amplitude coefficients, obtained from the Fourier analysis of each time dependent spectral absorbance value traces, is provided in Figure 3 showing the prominent features around 0.6 Hz, which makes a pulse rate of 84 heart beats per minute plausible. The Savitzky-Golay smoothed data of the logarithmized single beam spectra were used for input; various procedures for baseline correction and standardization did not improve the results of the Fourier analysis. Coefficients for frequencies lower than 0.4 Hz are not displayed; these components arising from low frequency signals, which can be related to well known vasomotion effects of the microvasculature, see for example (8), can show significant contributions.

The water absorbance variations due to the rhythmic blood pressure changes are about 20 mA.U. for the overtone water band at 6900 cm^{-1} which is about a factor of 50 smaller compared to the signals we find for time integrative tissue measurements. The corresponding water sample layer thickness is about 15 μm which is about a factor of 70 smaller than usually applied for the quantitative analysis of blood plasma samples using data from the near infrared spectral range above the water combination band at 5175 cm^{-1}.

Further investigations have to be carried out into what kind of signal processing can be performed to produce a high quality pulse spectrum which is good enough to be evaluated for blood substrates such as glucose, cholesterol and others. Improved time resolution measurements can help to better isolate the cardiac pulse frequencies. A slightly poorer signal-to-noise ratio in the single beam tissue spectra will be produced because of the reduced number of interferograms accumulated which can certainly be tolerated after digital smoothing. In this context, also a degrading of the spectral resolution might be appropriate to increase the spectral signal-to-noise ratio. However, the available spectrometer software did not allow a further resolution degradation during the time of this investigation.

CONCLUSIONS

The overall performance of the near infrared spectrometric plethysmography of lip tissue, shown in this investigation, is promising for the development of future non-invasive multicomponent blood parameter assays. Since such a technique allows the probing of a part of the intravascular space, the perturbations from tissue can be excluded so that non-invasive blood analysis can become more reliable. This is especially valuable, because most diagnostic expert systems rely on the informations derived from blood specimens. For blood glucose self-monitoring for diabetic patients this might be the goal envisaged, because assaying the interstitial fluid as done by using microdialysis probes implanted in the subcutaneous fatty tissue or studying the aqueous tissue glucose by integrative skin tissue spectroscopy are monitoring means with many pitfalls included. However, further investigations with thorough calibration experiments have to be performed to prove patient independent calibration models such as have been shown to be possible for hospital blood plasma samples exploiting the full potential of clinical near infrared spectrometry (10).

ACKNOWLEDGEMENTS

Financial support by the Ministerium für Wissenschaft und Forschung des Landes Nordrhein-Westfalen and the Bundesministerium für Bildung, Wissenschaft, Forschung und Technologie is gratefully acknowledged. Further funds were granted by the Deutsche Forschungsgemeinschaft.

REFERENCES

1. Heise, H.M., Horm. Metab. Research **28**, 527-534 (1996)
2. Heise, H.M., Near-Infrared Spectrometry for in vivo Glucose Sensing, in "Biosensors in the Body: Continuous in vivo Monitoring", Fraser, D.M., ed., Chichester: John Wiley & Sons, 1997, ch. 3, pp. 79-116
3. de Kock, J.P., and Tarassenko, L., Med. Biol. Eng. & Comp. **31**, 291-300 (1993)
4. Nahm, W., and Gehring, H., Sensors and Actuators **B 29**, 174-179 (1995)
5. Mendelson, Y., Proc. IEEE Engineering in Medicine and Biology, 18th Ann. Int. Conference, Amsterdam, Paper No. 897, 1-3 (1996)
6. Marbach, R., and Heise, H.M., Applied Optics **34**, 610-621 (1995)
7. Bittner, A., Thomassen, S., and Heise, H.M., Mikrochim. Acta [Suppl.] **14**, 429-432 (1997)
8. Schechner, J.S., and Braverman, I.M., Microvascular Research **44**, 27-32 (1992)
9. Bertie, J.E., and Lan, Z., Appl. Spectrosc. **50**, 1047-1057 (1996)
10. Heise, H.M., Mikrochim. Acta [Suppl.] **14**, 67-77 (1997)

Chemical Analysis of Multiple Sclerosis Lesions by FT-IR Microspectroscopy

Steven M. LeVine[1] and David L. Wetzel[2]

[1]*Department of Molecular and Integrative Physiology and the Smith Mental Retardation and Human Development Center, University of Kansas Medical Center, Kansas City, Kansas and*
[2]*Microbeam Molecular Spectroscopy Laboratory, Kansas State University, Manhattan, Kansas*

Fourier transform infrared microspectroscopy can be used to collect infrared spectra from microscopic regions of tissue sections. If spectra are collected along a grid pattern, then maps of chemical functional groups can be produced and correlated to tissue histopathology. In the present study, white matter from multiple sclerosis and control brains were examined. Mapping experiments were designed such that 17 spectra were collected at 200 μm intervals along a line that was partially or wholly within a multiple sclerosis lesion site or within a representative white matter region of control tissue. Data analysis was based on earlier *in vitro* studies, which found that the carbonyl at 1740 cm^{-1} increases when lipids become oxidized, and the amide I peak at ~1660 cm^{-1} broadens when proteins become oxidized. The results indicated that the C=O to CH_2 ratio (1740 cm^{-1}:1468 cm^{-1}) was elevated at several collection points in lesion sites from each of five multiple sclerosis brains examined compared to values from white matter of four control brains. Inspection of the amide I peak at 1657 cm^{-1} revealed that it was broadened towards 1652 cm^{-1} in multiple sclerosis tissues but not control tissues. These results suggest that lipids and proteins are oxidized at active multiple sclerosis lesion sites. The localization of these products to lesion sites supports a role for free radicals in the pathogenesis of multiple sclerosis.

INTRODUCTION

Free radicals have been suggested to play a role in the pathogenesis of multiple sclerosis (MS), which is characterized pathologically by demyelination, reactive gliosis, and infiltration of inflammatory cells into the central nervous system. In the present study, we investigated whether products of free radical damage could be detected in active MS lesions. The localization of oxidation products at lesion sites would strengthen the hypothesis that free radicals have a pathogenic role in MS.

Most traditional biochemical techniques require tissue homogenization, which can dilute locally concentrated products beyond detection limits and disrupt their spatial distribution. Because oxidation products would be expected to be highly localized in active MS lesions, data from traditional techniques would not necessarily reflect the chemical changes that occur at sites of pathology. Unlike more traditional techniques, Fourier transform infrared (FT-IR) microspectroscopy can provide information about the *in situ* chemical features from microscopic regions of a tissue section (1,2). The advantages of this technique make it uniquely suited for the characterization of the chemical changes that are present in lesion sites of MS brain tissue. Earlier in vitro studies found that the carbonyl absorption at 1740 cm^{-1} increases when lipids become oxidized (3,4), and the amide I peak (CO stretch vibration of the amide group) at ~1660 cm^{-1} broadens when proteins become oxidized (5). We analyzed these peaks in spectra collected from lesion sites in MS brains and from white matter in control brains in order to determine if oxidation products could be identified in MS tissue.

MATERIALS AND METHODS

Tissue Preparation

Brain tissues from five MS and four control (one normal and three Alzheimer's disease) patients were fixed in buffered formalin. One of the Alzheimer's disease brains was from an individual that had severe disease, and the other two brains were from individuals that had mild to moderate disease. Areas of white matter that displayed features indicating potentially active lesion sites from MS brains and representative regions of white matter from control brains were dissected, frozen, and sectioned on a cryostat. Frozen sections, 10 μm thick, were thaw-mounted onto barium fluoride disks. Neighboring sections were mounted onto glass slides and stained with hematoxylin and eosin.

FT-IR Microspectroscopy

An IRμs™ microspectrometer (Spectra-Tech, Inc., Shelton, CT) was used to collect infrared spectra from microscopic

CP430, *Fourier Transform Spectroscopy:* 11th International Conference
edited by J.A. de Haseth
 1-56396-746-4/98/$15.00

regions. Spectra were collected using a 24 x 24 μm aperture, 256 coadded scans, and 4 cm^{-1} resolution. Experiments were designed such that 17 spectra were collected at 200 μm intervals along a line that was partially or wholly within an MS lesion site or within a representative white matter region of control tissue. Individual spectra were collected using an automated motorized stage that was programmed prior to each experiment (line map). A background spectrum was collected prior to each experiment using the conditions described above, except that 512 scans were coadded in place of 256 scans.

RESULTS

Infrared spectra from normal and MS tissue

Infrared spectra from white matter of control brains revealed absorption patterns similar to those described earlier (1,2). In particular, CH_2 absorptions at 2923 cm^{-1} and 1468 cm^{-1}, P=O at 1235 cm^{-1}, and HO-C-H at 1060 cm^{-1} were pronounced in these specimens (Fig. 1). These absorptions are found in lipids, phospholipids, and glycolipids, respectively, which are enriched as a percentage of tissue dry weight in white matter compared to gray matter. Spectra from normal areas of MS white matter had similar absorption profiles to spectra from control tissue, whereas spectra from MS lesion sites had absorptions that were significantly changed. For example, the CH_2 to NH (and OH) ratio (peak height of 2923 cm^{-1}:peak height of 3293 cm^{-1}) was substantially reduced in MS lesions compared to that in control tissue (Fig. 1A). Also, the other peaks characteristic of lipids were reduced in MS lesions compared to those in control tissue (Fig. 1B).

Additional spectroscopic changes were revealed when the data were analyzed for lipid and protein oxidation products. Because previous studies found that the carbonyl absorption at 1740 cm^{-1} increases when lipids become oxidized (3,4), and the amide I peak (CO stretch vibration of the amide group) at ~1660 cm^{-1} broadens when proteins become oxidized (5), these regions of the spectra were analyzed. The C=O to CH_2 ratio (peak area of 1740 cm^{-1}: peak height of 1468 cm^{-1}) was increased, and the peak at 1657 cm^{-1} was broadened in MS lesions compared to control tissue. In order to follow the spatial profile of these spectroscopic features, linear maps that were partially or wholly within a lesion site from MS brains or that were within representative areas of white matter from control tissues were prepared.

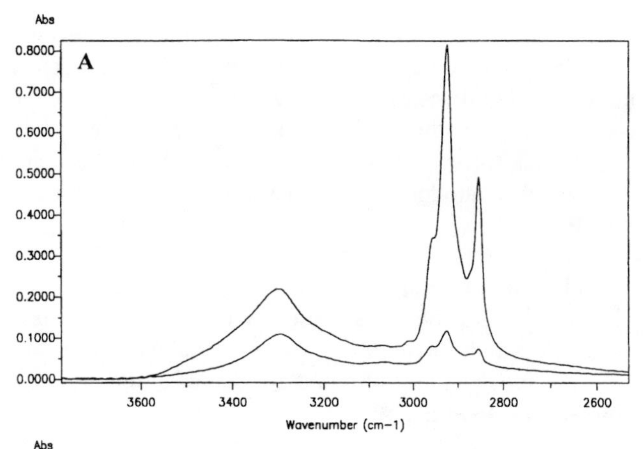

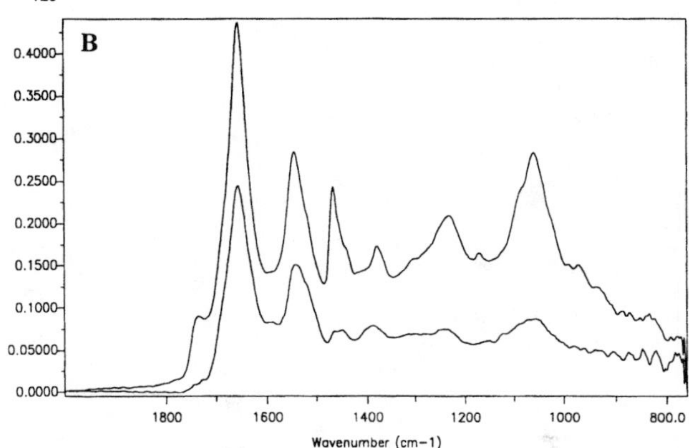

FIGURE 1 (*A*) The spectrum from the normal control (top) has a large ratio of 2923 cm^{-1} to 3293 cm^{-1} peak heights. This ratio is greatly reduced in the spectrum taken from an active lesion site of an MS brain (bottom). (*B*) The bands at 1468, 1395, 1235, and 1060 cm^{-1} are reduced in the spectrum from the MS brain (bottom) compared those for the normal brain (top). Spectra were smoothed (Savitsky-Golay 11 points) and baseline corrected (factor 1).

Maps

The data from line maps were analyzed using several absorption ratios, because ratios yield values that are independent from variations in section thickness. The CH_2 to amide II ratios (1468 cm^{-1}:1544 cm^{-1}) for the control tissues were tightly grouped with values generally ranging from 0.6 to 0.8. The MS tissues showed values ranging from less than 0.1 to 0.7. Examples of line maps depicting this ratio for normal and MS brains are presented in Fig. 2A.

In control patients, the C=O to amide II ratio (1740cm^{-1} :1544 cm^{-1}) generally ranged from 8.5 to 16. The ratios for several collection points from four of five MS patients were 8 or less. Examples of line maps depicting this ratio for normal and MS brains are presented in Fig. 2B. One MS patient had values in the control range. No MS patient had values exceeding the control range.

The C=O to CH_2 (1740 cm^{-1}:1468 cm^{-1}) values generally were grouped between 14 to 22 in the control tissues. Each of the five MS patients had several values that were above the control range, and three MS patients each had two collection points that were 32 or higher. Examples of line maps depicting this ratio for normal and MS brains are presented in Fig. 2C.

Plots of the first derivative at 1652 cm^{-1} revealed that the control values generally ranged from -0.025 to -0.07. Each of the five MS patients had values greater than -0.02 (less negative), with the majority of values falling in the -0.01 to -0.02 range. Examples of line maps depicting this ratio for normal and MS brains are presented in Fig. 2D.

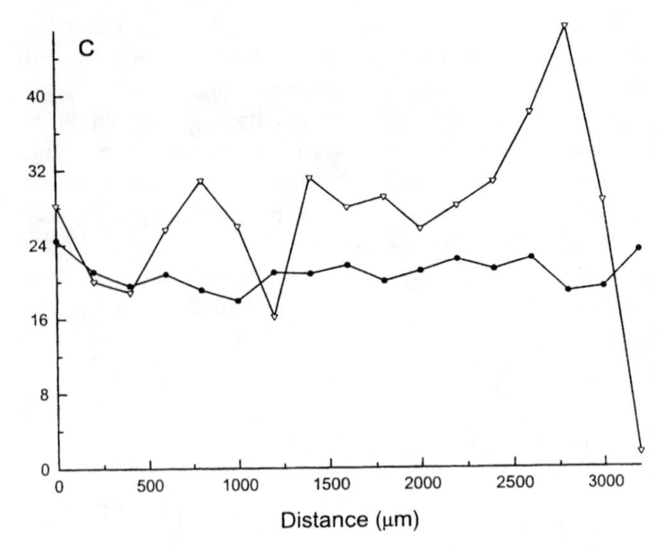

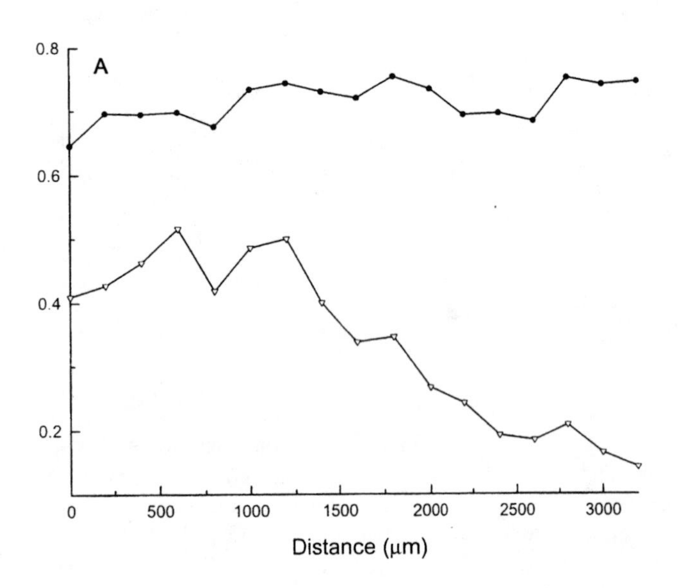

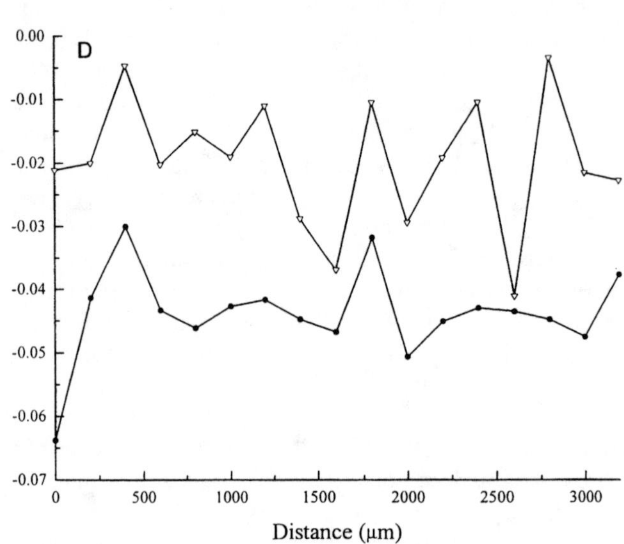

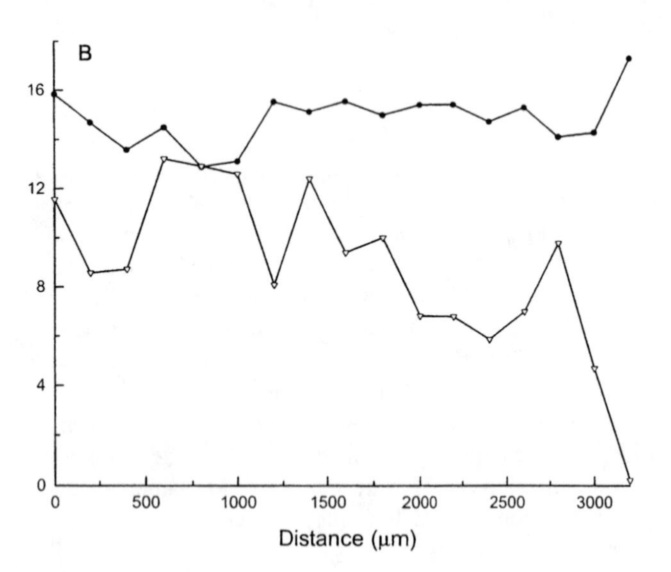

FIGURE 2: Linear maps for (*A*) 1468 cm^{-1} : 1544 cm^{-1}, (*B*) 1740 cm^{-1} :1544 cm^{-1}, (*C*) 1740 cm^{-1}:1468 cm^{-1}, and (*D*) the first derivative of 1652 cm^{-1}. The value for 1468 cm^{-1} was obtained from the height at 1468 cm^{-1} to a baseline at 1482 cm^{-1} to 1424 cm^{-1}. The value for 1544 cm^{-1} was obtained from the height at 1544 cm^{-1} to a baseline at 1578 cm^{-1} to 1484 cm^{-1}. The value for 1740 cm^{-1} was obtained from the peak area at 1716 cm^{-1} to 1765 cm^{-1} and a baseline at 1829 cm^{-1}. Solid circles=normal brain, open triangles=MS brain.

DISCUSSION

Oxidation to lipids causes the formation of carbonyl groups and a breakdown of long chains into smaller fragments, e.g., malondialdehyde. Earlier FT-IR studies have successfully detected oxidation products of lipids *in vitro* (3,4). In these studies, an increase in carbonyl groups centered around 1740 cm^{-1}, and a degradation of acyl chains were observed. In the present study, these findings have been extended to the identification of oxidation products *in situ* at lesion sites of tissue sections.

The values obtained from collection points within and between each map for the four control tissues were generally consistent. When there was variability in the controls, it was from an Alzheimer's tissue, which can show minor white matter changes. The higher carbonyl to CH_2 ratio that was detected in each of the MS patients suggests that lipids were being oxidized. Because oxidation can cause an increase in carbonyls and a degradation of lipids, both of these changes could have contributed to the elevated ratio.

Oxidation to proteins results in the production of additional carbonyls on some amino acid residues. Some of these carbonyls will reside adjacent to amines which, will give rise to a spectroscopic absorption for amide I. Protein oxidation was found to broaden the amide I peak at 1657 cm^{-1} (5). In the present study, several collection sites for each of the MS tissues had values for the first derivative of 1652 cm^{-1} that were greater than the values for 67 of the 68 collection sites from the four control tissues. This indicates that the peak at 1657 cm^{-1} had broadened towards 1652 cm^{-1} in MS tissue and that oxidation of proteins could be responsible for this shift.

In summary, the results presented in the present study are consistent with the presence of oxidation products at active MS lesion sites, which supports the hypothesis that free radicals are involved with the pathogenesis of MS.

ACKNOWLEDGMENTS

Tissue samples of four MS brains were obtained from the National Neurological Research Specimen Bank, VAMC, Los Angeles, CA 90073, which is sponsored by NINDS/NIMH, National Multiple Sclerosis Society, Hereditary Disease Foundation, and Veterans Health Services and Research Administration, Department of Veterans Affairs. The remaining tissue samples from MS, normal, and Alzheimer's diseased brains were supplied by the Department of Pathology at the University of Kansas Medical Center. This work was funded by NINDS (NS 33596) and the NICHD Mental Retardation Research Center Grant (HD 02528). We thank Joseph A. Sweat for his help with processing data.

REFERENCES

1. Wetzel, D. L. and LeVine, S. M., *Spectroscopy* **8**, 40-45, 1993.
2. LeVine, S. M. and Wetzel, D. L., *Appl. Spec. Rev.* **28**, 385-412, 1993.
3. Lamba, O. P., Borchman, D., and Garner, W. H., *Free Rad. Biol Med.* **16**, 591-601, 1994.
4. Lamba, O.P., Lal, S., Yappert, M. C., Lou, M. F., and Borchman, D. , *Biochim. Biophys. Acta.* **1081**, 181-187, 1991.
5. Signorini, C., Ferrali, M., Ciccoli, L., Sugherini, L.,Magnani, A., and Comporti, M., *FEBS Let.*, **362**, 165-170, 1995.

Diagnostics of Normal and Cancer Tissues by Fiberoptic Evanescent Wave Fourier Transform IR (FEW-FTIR) Spectroscopy

Natalia I. Afanasyeva

Institute of Spectroscopy, Russian Academy of Sciences, Troitsk, Moscow Region, 142092, Russia;
Current address: Department of Physics, University of Nevada-Reno, Reno NV 89557, USA

Fourier Transform Infrared (FTIR) Spectroscopy using optical fibers operated in the attenuated total reflection (ATR) regime in the mid-IR region in the range 850 to 4000 cm^{-1} has recently found an application in the noninvasive diagnostics of tissues *in vivo*. The method is suitable for nondestructive, nontoxic, fast (seconds), direct measurements of the spectra of normal and pathological tissues *in vitro, ex vivo, and in vivo* in real time. The aim of our studies is the express testing of various tumor tissues at the early stages of their development. The method is expected to be further developed for endoscopic and biopsy applications as well as for the research of different materials.

INTRODUCTION

Fourier Transform Infrared (FTIR) spectroscopy is traditionally used for analysis of the structure of biomolecules (1). In the past years this method has been widely applied for the diagnostics of tissues (2 - 6). It is a very important trend in diagnostics of cancer and precancer conditions. The normal, precancer, and cancer tissues of the different organs (2) have been investigated successfully by methods of fluorescent and autofluorescent, spectroscopy, elastic and Raman scattering (3). The nondestructive methods of vibrational spectroscopy (IR and Raman) enable us to diagnose tumors and cancer at an early stage on molecular level (2,4-6). Recently both methods of vibrational spectroscopy have been improved by using fiberoptic sensors (2,6). Fiberoptic technology is relatively inexpensive and can be adapted easily to any commercially available tabletop compact Fourier transform (FT) spectrometer (4-6).

The combination of fiber optical sensors and FT spectrometers can be applied to many fields: (i) noninvasive medical diagnostics of cancer and other different diseases *in vivo*, (ii) remote monitoring of biochemical processes and environment, (iii) surface diagnostics of numerous materials, (iv) minimally invasive bulk diagnostics of tissues and materials, (v) characterization of the quality of food, pharmacological products and cosmetics.

We suggested (5) that bare-core fibers should be used in the ATR regime with different configurations of the "probe" for fast, remote, noninvasive and nontoxic diagnostics of the cancer of kidney, stomach and lung *in vitro*, breast *ex vivo* during the operation and skin *in-* and *ex vivo* and in incisions. This technique could be used as minimally invasive for endoscopic and catheter applications.

In this paper we present IR spectra of normal and tumor tissues in the skin surface (*in vivo*) and breast precancer and cancer tissues (*ex vivo*) by using of FEW - FTIR method in the range of 1500 - 1800 cm^{-1}. The main changes in the spectra of tissues on molecular level are discussed here. The nondestructive, fast and noninvasive technique allows for the development of an appropriate method of cancer diagnostics.

MATERIALS AND METHOD

First we measured skin tissues *in vivo* (directly on patients) at the center of tumors and at 1 cm distance in the direction of the normal skin tissue noninvasively. After surgery the measurement procedure was repeated. Thereafter the samples were cut at the center of the tumors to measure the different layers of tumor and normal skin. The experiments have been performed in the operating rooms where fresh samples of tissue were provided from the organs of the patients and measured *ex vivo*. In this study mainly human breast tissue has been investigated. We have also studied lung, stomach and kidney tissues *ex vivo*. The results of these spectral measurements have been compared with available histological data.

We used polycrystalline $AgBr_xCl_{1-x}$ fibers (1 mm in diameter) in the spectral range 3-20μm with low optical losses (0.1 to 0.5 Db/m in the region of 10μm) and high flexibility ($R_{bending}$ > 10 to 100 fiber diameters). The self-made accessory had an optical system consisting of optical fibers to input and output the infrared radiation and a focusing spherical mirror to collect light onto a nitrogen-cooled MCT detector. The optical scheme of the accessory has been specifically designed for any commercial Fourier Transform (FT) spectrometer. The fiber has direct contact with the tissue similar to a prism in the attenuated total reflection method. The length of fiber interaction with the tissue varied from one millimeter to a few millimeters. Later we bent the fiber at a certain angle to obtain an optimal signal and enhanced

CP430, *Fourier Transform Spectroscopy:* 11th International Conference
edited by J.A. de Haseth

sensitivity for measurements at separate points (at about 1 mm) on tissue *in vivo*. We have used a tip probe for biopsy and endoscopic applications. The special tips allow us to collect or scatter the light for different tissue examinations . The contact between the fiber and the tissue has been optimized. All the spectra were measured with resolution of 4 cm^{-1} . The number of scans varied and the optimal recording time came to 40 s.

The spectra were treated with the help of a computer with a set of standard and special programs. All the absorption bands were interpreted and the regularities of conversion from normal to malignant tissues were defined.

RESULTS AND DISCUSSIONS

Our first fiberoptic evanescent wave Fourier transform infrared (FEW-FTIR) spectra measured *in vitro* (4-6) enabled us to select the main spectral ranges where the basic changes in: protein, lipid, and sugar systems, the hydrogen bond, and the process of disordering are most pronounced (4,5) In the papers mentioned above, the most attention was given to the characteristic sections of the IR spectrum in the range from 950 to 1500 cm^{-1} , associated with malignant tumors (4,5). After more detailed computer analysis we have further observed small but characteristic shifts of the most intense bands in the *mid*-IR spectrum associated with vibrations of amide groups (amide I and amide II) in the spectra of normal and tumor tissues of the kidney, stomach and lung. We indicate the largest intensity variations and shifts are observed for the amide II bands in the region of 1530-1560 cm^{-1}. In many cases both amide bands decrease in intensity and broadened. Furthermore the weak band in the region of around 1740 cm^{-1} was found to have a tendency to fully disappearance in the spectra of cancer tissue. These interesting facts have been confirmed in new studies of skin tissue *in vivo*, in particular, the spectra of breast tissue and small biopsy samples of stomach and lung tissues studied *ex vivo*. It became possible to test small samples of skin tissues *in vivo* using this new FEW-FTIR technique. These investigations have been performed in the region of the intense amide bands.

Figures 1-3 show characteristic spectra of tissue samples of tumors and skin of different pigmentation: nevus, melanoma, basaloma. Besides the main amide bands, we were interested in weaker absorption bands (1740-1745 cm^{-1} and ~1710 cm^{-1}) as shown, for example, in Figures 1-4. In particular we analyzed specific spectral regions where we could observe *in vivo* on the skin surface two bands simultaneously (Fig. 1, 3), one band (Fig. 2, 4) and no bands at all in the range 1700-1750 cm^{-1} (Fig. 2, 4). In addition we studied these bands and the changes of the main amide structures tissues after surgery. The tissue was cut in the center of the tumor or in the vicinity of a pigmentary spot. Using this method we were able to observe the changes in the spectra layer by layer and under the layer of the epidermis. In Fig. 1 and 2 we show spectra of nevus and melanoma. In Fig. 3 to 4 we have indicated spectra from basaloma and breast tissue at the points of normal and malignant tumors(cancer).

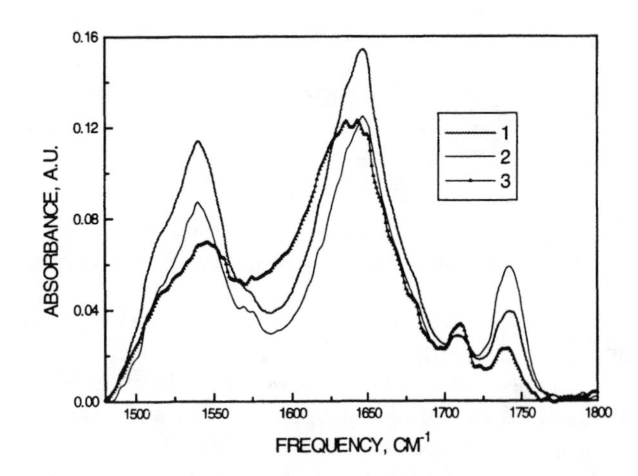

Figure 1. *In vivo* measurement of FEW-FTIR spectra of skin surface in central point of tumor in three cases of pigment nevus (noncancer) .

Fig. 1 show representative spectra of tumor tissue spots for skin on surface in case of nevus where the differences are most evident. In the spectra of real normal tissues around of pigment nevus we could observe four bands in the region of 1500-1800 cm^{-1} (Fig. 1). Two amide bands have the peak positions near 1648 cm^{-1} (amide I) and 1540 cm^{-1} (amide II). Two bands of hydrogen bonded ester carbonyl groups in lipids are located at 1710 and 1740 cm^{-1}. The gradual smoothing of the band 1710 cm^{-1} have been shown in the Fig. 1 as the beginning of disordering in the system of hydrogen bonds. In the Fig. 1 the typical spectra of the tumor (noncancer) tissues from three patients have been collected. The IR spectra in central points of these tumors shows good agreement of peak positions in two cases (curves 1, 2). Curve 3 in the Fig. 1 is different from the other curves in peak position of the main amide bands (1636 and 1545 cm^{-1}) and partial smoothing of the band 1740 cm^{-1}. The features appearing in these spectra were needed for adequate analysis. So in all our investigations we analyzed intensity ratios for three characteristic band parts: R_I (I_{1642}/I_{1545}), R_{II} (I_{1642}/I_{1742}) and R_{III} (I_{1742}/I_{1710}).

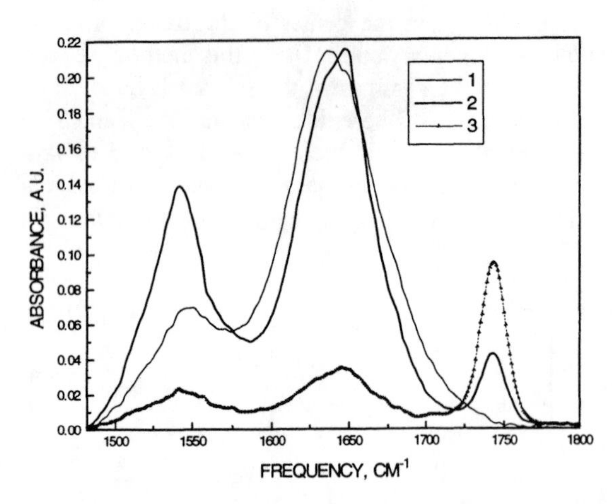

Figure 2. *In vivo* measurements of FEW-FTIR spectra of malignant skin surface tissues in three cases of melanoma.

We note that we have not only studied the changes in intensities and shifts of these bands ~ 1642 and ~ 1545 cm^{-1}, but also the distance between the peaks of the main amide bands (Δ) for almost all spectra. In particular, the parameter Δ was useful in analyzing the changes taking place in the spectra of malignant tissues. The distance Δ between the bands at 1642 and 1545 cm^{-1} mainly depends on the position of the 1545 cm^{-1} band structure. The average distance (Δ_{aver}) is 100 cm^{-1} for normal and noncancer tissues, although these differences can be much larger depending on the type of tumor or pigment. As for the spectra of malignant tissues this distance becomes less than 100 cm^{-1}.

The spectra of all normal tissues in cases of nevus, premelanoma and melanoma are similar. The intensity ratio of the chosen bands are the same. For example, R_I is 1.4 for all cases of normal tissue represented in Fig. 1-4. The intensity ratio R_{II} ranges from 2.0 to 2.5 for normal tissues and R_{III} is 1.0. But in the case of the tumor (Fig. 1, curves 1 and 3) R_{II} is 4.0 and 5.0. In the spectra of benign tumor (Fig. 1, curve 3) R_{II} changes immediately. This intensity ratio decreases in the case of premelanoma by 1.5 times. Spectrum 3 in Fig. 1 differs from other tumor spectra and is more similar to precancer. As can be seen the 1540 cm^{-1} band tends to shift up to 1553 cm^{-1}. In some cases it overlaps with a pronounced band feature at 1588 cm^{-1} (Fig. 1-4) producing the shoulder in the range of 1530 to 1540 cm^{-1}. Such displacements enable us to distinguish the α- and β-conformations of polypeptides. When passing from the α-helix to the β-sheet (7), the band of amide I shifts from about 1655 to 1630 cm^{-1} and for amide II from approximately 1550 to 1520 cm^{-1}. We can also differentiate parallel and antiparallel chain folding. For example, the weak band at 1685 cm^{-1} (Fig. 1) is only observed in a spectrum with β-conformation and antiparallel chain folding. On the other hand the 1655 cm^{-1} band characterizes parallel chain folding. However, the spectra structures in Fig. 1-4 are less suitable for the identification of clear α-helix fragments. The beginning of the structure rebuilding can be pointed out in almost all the cases presented above.

In certain cases of tumor and malignant tissues additional structures of amide bands can be observed. In the case of the additional component of the amide II band near 1515 cm^{-1}, "tyrosine" ring vibrational band (1) (or asymmetry of the band shape in that place) can be seen. Generally, tyrosine is found in the β-sheets of the chain. We can suppose very carefully that the base repairing can begin in many cases of precancerous and cancer phenomena. So FEW-FTIR spectra are sensitive to even small changes in the structure of tissues.

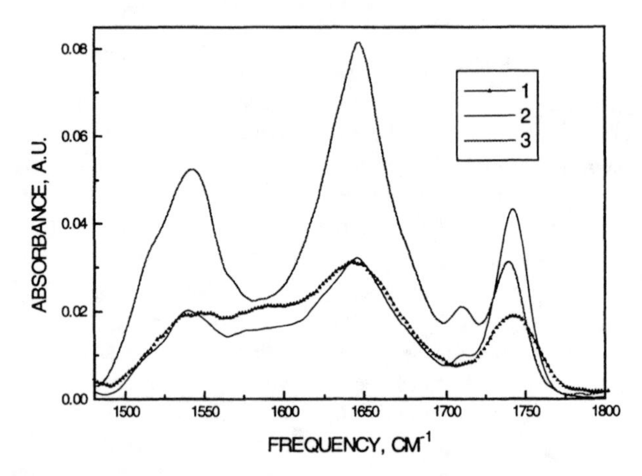

Figure 3. *In vivo* measurements of FEW-FTIR spectra of thre malignant skin surface tissues (basaloma)

In Fig. 2 the spectra of melanoma tissues (different types) on the surface are presented. In the first spectra (curve 1) the bands in the region 1710-1740 cm^{-1} are absent. Intensity ratio R_I (Fig. 2, curve 1) is 3. In the third spectra the intensity of amide bands is changed. The band near 1742 cm^{-1} belongs to the cyclic systems with hydrogen bonds (8). A second band structure around 1710 cm^{-1} is more likely characterized by hydrogen bonds in noncyclic system (8). The most interesting case we have observed shows a disappearance of the bands at 1744 and 1710 cm^{-1} in the spectra measured after operation. They are always absent in all cases of melanoma and nevus characterized by either one or two bands in the range 1700 - 1750 cm^{-1}. Thus, the hydrogen bond system is perturbed in both cyclic and noncyclic fragments of the main chain and the amino acid residuals (mainly proline or hydroxyproline) of rings (7,8).

Our spectral studies have also shown that basaloma greatly differs from nevus and melanoma in intensity and peak positions (Fig. 3). Our attention has been engaged by the emergence of a band at 1582 cm $^{-1}$ between two amide bands (Fig. 3 (1)). If we compare the spectra in Fig.

3 we can see that both amide and carbonyl bands decrease their intensities more drastically than in the spectrum 1. All four bands are conserved in spectra of basaloma Fig. 3 (2). For example, the spectra of basaloma on the surface show one 1740 cm^{-1} band (Fig. 3, curve 1), whereas this feature is not observed in the spectra in the cut. The distance between the peaks of the amide bands changed slightly. Basaloma are more variform than melanoma and nevus. Judging from their IR absorption spectra, one can say that several basaloma development stages (or species) exist.

The following figure 4 displays the spectra of breast noncancer, precancer and cancer measured *ex vivo*.

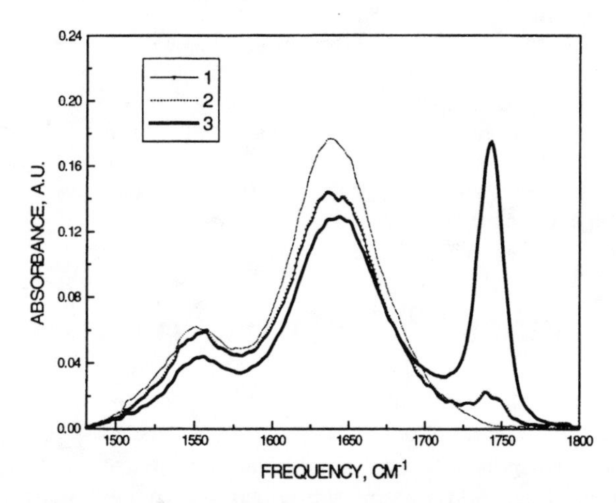

Figure 4. *Ex vivo* measurements of FEW-FTIR spectra of breast tissues: 1-precancer, 2-cancer, 3-tumor

In all cases of noncancer we can see the hydrogen bonded carbonyl band at 1740 cm^{-1} of the varied intensity. The spectra of precancer breast have a weak and as well as a nearly absent band in this range. In the case of real cancer the band is absent. It should be noted that this band is absent in spectra of breast tissues after radiation therapy and chemotherapy. And carbonyl band is also absent during the long period after these types of influences.

CONCLUSION

The main changes in the spectra have shown that in the spectra of normal tissue two carbonyl bands in the region of 1710 - 1740 cm^{-1} are present in more than 60% of cases. These bands are partially or completely absent in the spectra of precancer and cancer tissues. The changes in the peak position and the intensities of the main amide bands have occurred. So for tumor tissue the cooperative effect of cell reconstruction is typical. Basaloma can be detected from the skin surface directly. Melanoma can be registered as for surface as in the depth (bulk) of the skin. Nevus is detected in the deep layers of derma. But pigment nevus is well distinguished in the spectra of the skin surface. Such studies have enabled us to begin testing malignant tissues presently using the changes observed, in the basic and most intense absorption bands directly on the Fourier spectrometer monitor display. The methods of FTIR and Raman spectroscopy based on the usage of fiberoptic technique will be developed in a few years.

ACKNOWLEDGMENTS

We gratefully wish to thank Moscow department of Perkin-Elmer Inc. for giving us the spectrometer for our experiments and Institute of General Physics RAS- for optical fibers.

REFERENCES

1. Mantsch, H., Chapman, D., *Infrared Spectroscopy of Biomolecules*, , New York, Wiley-Liss, 1996, chapters 6, 7, 12.
2. B. Rigas, S. Morgello, I. S. Goldman, P. T. T. Wong, *Proc. Natl. Acad. Sci. USA*, **87**, 8140-8144, 1990.
3. Mahadevan-Jansen, A., Richards-Kortum, R. *J. Biomedical Optics*, **1** N1, 31-70, 1996.
4. Afanasyeva, N., Artjushenko, V., Lerman, A., Plotnichenko, V., Frank, G., Neuberger, W., *Macromol. Symp*., **94**, 269-272, 1995.
5. Afanasyeva, N., Artjushenko, V., Lerman, A., Plotnichenko, "Evanescent wave FTIR spectroscopy with MIR fibers", *Book of Abstracts, 9th Intern. Conf. on FTS*, Calgary, Canada, 223(PE48), 1993.
6. Artjushenko, V. G., Afanasyeva, N. I., Lerman, A. A., Ed. Wolfbeis, O. S., "Medical Applications of MIR-fiber spectroscopic probes", in: *Biomedical and Medical Sensors, Proc. of SPIE*, **2085**, 142-147, 1993.
7. Krimm, S., Ed. Durig, J.R., in: *Vibrational Spectra and Structure*, , Amsterdam: Elsevier, 1987, **16**, 1-2.
8. Maréchal, Y., Ed. Durig, J.R, in: *Vibrational Spectra and Structure*, , Amsterdam: Elsevier, 1987, **16**, 311-356.

Metabolically Deuterated Species Determined in Rat Cerebella by FT-IR Microspectroscopy as a Novel Probe of Brain Metabolism

David L. Wetzel[1], Steven M. LeVine[2], Daniel N. Slatkin[3], and Marta M. Nawrocky[3]

[1]*Microbeam Molecular Spectroscopy Laboratory, Kansas State University, Shellenberger Hall, Manhattan, KS 66506, USA*
[2]*Univeristy of Kansas Medical Center, Dept. of Mol. Int. Physiology, 3901 Rainbow Blvd., Kansas City, KS 66160, USA*
[3]*Brookhaven National Laboratory, Medical Dept., Bldg 490, Upton, NY 11973, USA*

Deuterated brain tissue in the cerebella of adult rats (from ingesting 40% D_2O prior to their being sacrificed) provides a means of studying brain metabolism without the use of radioisotopes. Microtomed frozen sections of the brains of adult rats were examined with FT-IR microspectroscopy. Corresponding brain sections of control animals (100% H_2O) were used for comparison. Multiple branches in the cerebella of several deuterated brains and control brains were mapped across the molecular layer, granular cell layer, and white matter to the opposite granular cell and molecular layers. Individual layers of the cerebella were compared for CD and OD or ND contents. Also, the absorbance of the CD band relative to that of other forms was compared among the molecular layer, the granular cell layer, and the white matter. Speculation concerning the relative metabolic uptake of deuterium was possible by comparing the absorbance of deuterated species with that of the ordinary hydrogen isotope form of the same functional group of the same tissue specimen.

INTRODUCTION

The spatial resolution of an optically efficient, integrated FT-IR microspectrometer allows the interrogation of selected portions of biological tissue with a sufficiently high signal-to-noise ratio so that it is possible to obtain the relative amount of materials present in small amounts in a localized area of the tissue. In this case, deuterated brain tissue in the cerebella of adult rats (from ingesting 40% D_2O prior to their being sacrificed) provides a means of studying brain metabolism without the use of radioisotopes. The isotopic effect on the fundamental vibrational frequency for the deuterated species provides absorption bands isolated from other absorption bands of the major components in the mammalian tissue. Therefore, spectroscopic interference is definitely not a problem. In previous work with heavy water used to study water migration in plant tissue it was necessary to use a 20 μm thick section of wheat in contrast to the usual 6-8 μm thick sections (1,2). Although scale expansion is necessary to see the stretching vibrations of OD, ND and CD, their measurement is still readily possible even from small portions of the microscopic field. In the present study, the relative concentrations of OD, ND, and CD were examined in the cerebella of adult rats that injested D_2O in their drinking water.

METABOLICALLY DEUTERATED TISSUE

Boron cancer therapy entails the introduction of boron usually in the form of porphorins. The boron nuclei within the cancerous tissue undergo neutron capture upon irradiation. The activated boron atoms subsequently provide localized radiation therapy to the cancerous tissue. Deuterated organic compounds are introduced within the tissue because, they provide a lower neutron capture cross section. Thus deuterated normal tissue is afforded some degree of protection. In support of these studies, adult rats had been treated by drinking water for 3½ weeks at select D_2O levels as part of ongoing work in the Medical Department at Brookhaven National Laboratory (3). We investigated the relative incorporation of deuterium in different regions of the cerebella in these rats.

EXPERIMENTATION AND RESULTS

Microtomed frozen sections of the cerebella of brains of adult rats were examined with FT-IR microspectroscopy. Sections taken near those used for the microspectroscopy were stained for examination by light microscopy. The experiments reported here were restricted to examination of individual branches of cerebella. Figure 1 shows the spectrum of major absorption bands from the three distinctly different layers in the cerebellum. These include the molecular layer, the

CP430, *Fourier Transform Spectroscopy:* 11th International Conference
edited by J.A. de Haseth

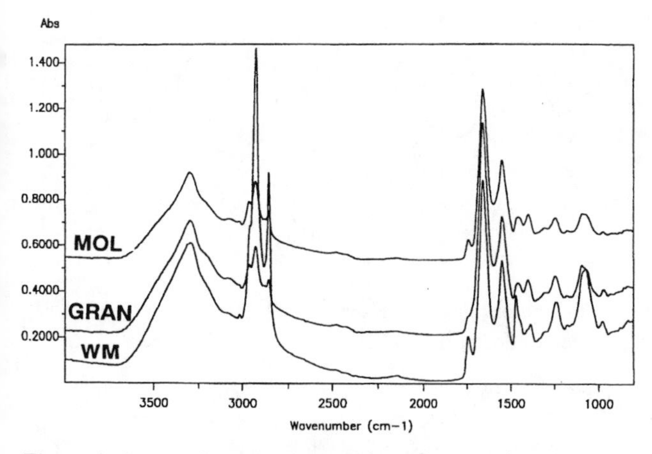

Figure 1. Spectra from the three regions (layers) of cerebellum

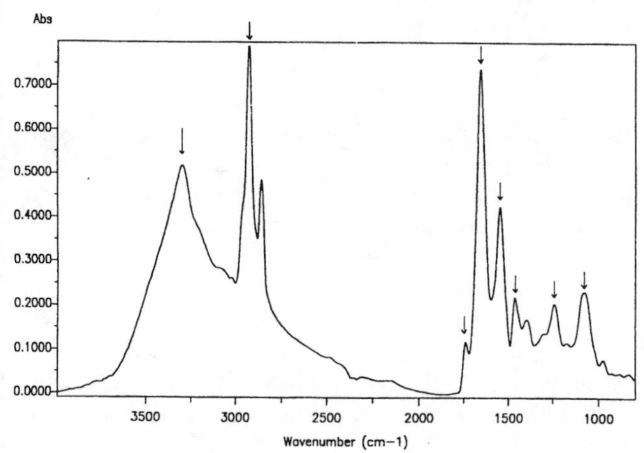

Figure 2. Major absorption bands of cerebellum white matter

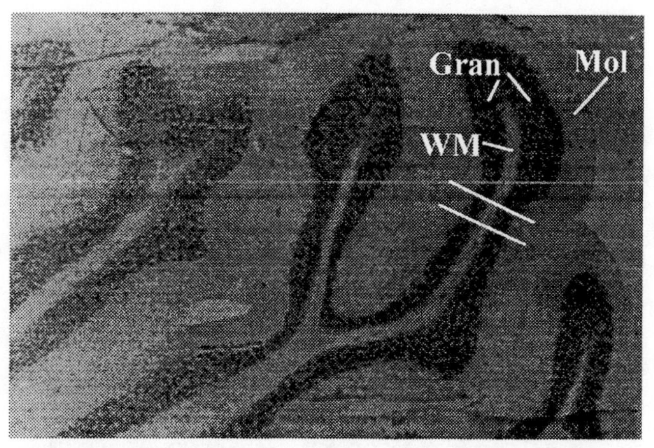

Figure 3. Photomicrograph of a stained cerebellum section indicating the three regions and the area mapped (between white lines)

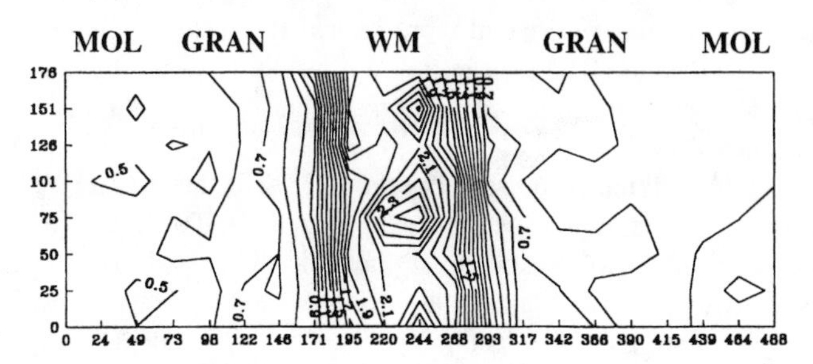

Figure 4. Contour map of peak areas at 1469 cm^{-1} (lipid)

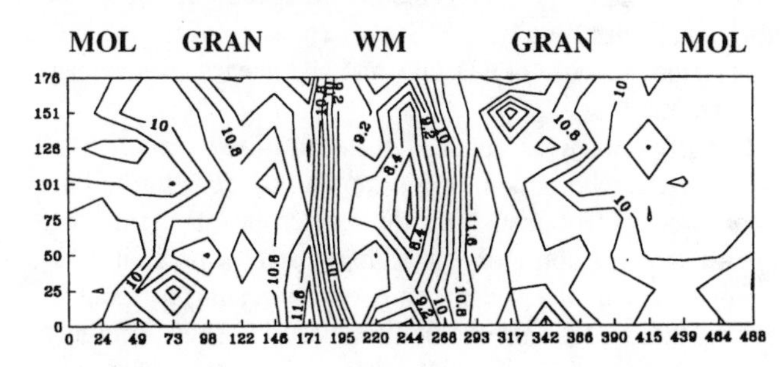

Figure 5. Contour map of peak area at 1550 cm^{-1} (protein)

granular cell layer, and the white matter. The white matter occurs at the center of each branch. It is bordered by the granular cell layer, and on the outside of that is the molecular layer. Differences in the composition of the major constituents are obvious from examination of these three spectra. Lipid is prevalent in the white matter. Absorption bands marked on Fig. 2, characteristic of white matter previously assigned from our earlier work involving mapping of the cerebrum (4,5), include 2927 cm^{-1} CH_2 stretch, 1740 cm^{-1} carbonyl, 1469 cm^{-1} CH_2 bend, 1235 cm^{-1} P=O, and 1085 cm^{-1} HCOH vibration indicative of carbohydrate. The lipids present in the white mater, in general, account for the carbonyl and the prominence of CH_2 bands. Long-chain compounds and cholesterol in brain tissue enhance the magnitude of this band and contribute to its prominence. If the lipids present are unsaturated, a very small band at 3015 cm^{-1} also may be exhibited. Phospholipids account for the prominence of the band at 1235 cm^{-1}, and galactocerebreside is a major contributor to the carbohydrate revealed at 1085 cm^{-1} in white matter tissue. The prominence of these same features in the white matter from the cerebellum shown in comparison to the granular and molecular layers is coincident with previous observations taken from white matter in the cerebrum (4).

A rectangular mapping experiment was done across the center of a branch of a cerebellum corresponding to the one marked (between white lines) on the photomicrograph (Fig. 3) of an adjacent stained section. Only the inner part of the molecular layer was included on either side of the granular cell layer. Contour maps were plotted from the peak areas of the distinguishing IR bands just cited. Two of these, 1469 cm^{-1}

indicative of white matter) and 1550 cm^{-1} (indicative of protein), are shown in Fig.4. Note that the white matter has the highest 1469 cm^{-1} peak area, and the molecular layer has the lowest. The distribution of other bands characteristic

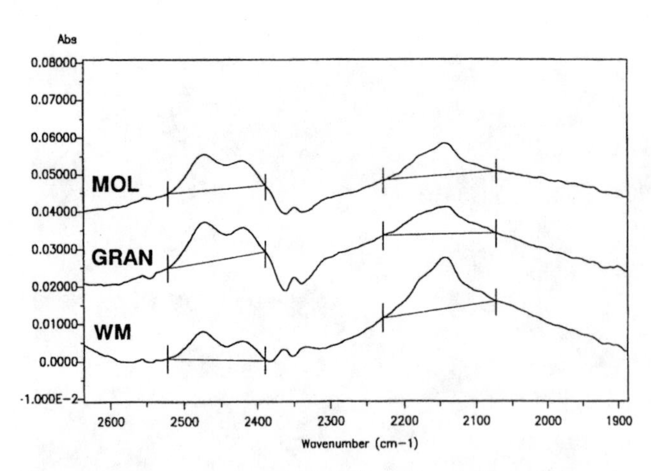

Figure 5. Spectral region of OD, ND (left) and CD (right) bands for three layers of cerebellum

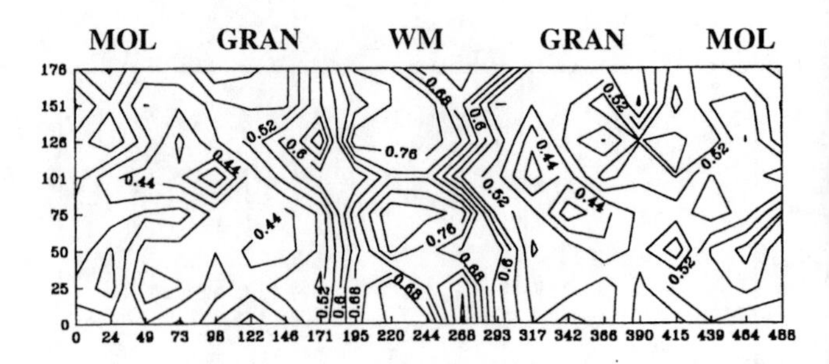

Figure 6. Contour map of peak areas at 2150 cm^{-1} showing localization of deuterium as CD

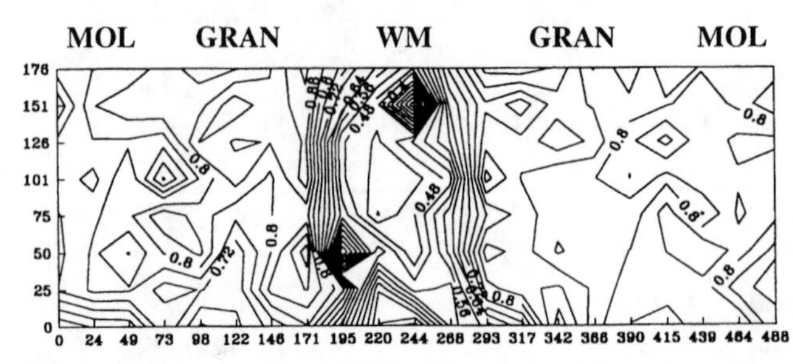

Figure 7. Contour map of peak areas at 2500-2400 cm^{-1} (combined ND and OD vibration)

of white matter responded more or less in concert with the 1469 cm^{-1} band. The concentration of protein defined by the amide II absorption at 1550 cm^{-1} is greatest for the granular cell layer.

High signal-to-noise operation of the IRμs™ microspectrometer (Spectra-Tech, Shelton, CT) using a conventional globar source at the Microbeam Molecular Spectroscopy Laboratory at Kansas State University (Manhattan, KS) and using a synchrotron source at the National Synchrotron Light Source of Brookhaven National Laboratory (Upton, NY) revealed good spectra of deuterated vibrational stretching bands. Scale expansion of these bands showed the presence of CD, OD, and ND in each targeted part of the tissue.

Initially data were obtained for each of the specimens at Brookhaven, where the tissue was obtained. The bulk of the data were produced at the Microbeam Molecular Spectroscopy Laboratory at Kansas State University. A cursory look at the spectra showed interesting differences among three types of cerebellum tissue. This difference persisted as other cerebella branches were examined. Multiple sections were examined, and corresponding tissues from the brains of other animals were subjected to the same analytical procedure. It is not within the scope of this brief preliminary report to present details or statistics for the entire study in which absorption band areas were calculated and averaged from a large number of spectra obtained from mapping experiments across individual branches.

The spectra shown in Figure 5 exemplify the relative population of CD vs OD, ND in each of the three types of tissues. As expected, the CD was most prominent in the fatty tissue (white matter). The OD, ND prominence relative to the CD was greater in the more proteinaceous granular cell layer and intermediate in the molecular layer.

Rectangular mapping across one branch of the cerebellum showed the difference in deuterated species among the different types of tissue. Only a representative part of the molecular layer was sampled on each side of the branch to allow less space between the samplings across the inner layers of the branch. In Fig. 6, the CD band at 2150 cm^{-1} is concentrated in the WM as would be anticipated, and the nominal 2500-2400 cm^{-1} term, embracing both OD and ND, is lowest in the WM.

DISCUSSION

The implications of these data and the ability to obtain quantitation even from relatively small amounts of deuterated species are profound. A ratio of the population of CD to the population of CH can give some indication of the metabolically incorporated deuterium in the hydrophobic part of the tissue. Similarly, a ratio of the OD, ND to OH, NH is indicative of the metabolic uptake for the hydrophilic type of organic compounds present. Because the deuterated species are present in small quantities compared to the same form of material with conventional hydrogen, scaling factors are necessary. However, it is readily possible to look at the CD to CH ratio in one tissue and compare it to the CD to CH ratio in the other two tissues. In such a case, whatever factor is used (to compare CD to CH) will cancel out. Similarly a comparison of the uptake of the hydrophilic OD, ND forms vs. their OH, NH counterparts can be made among the three tissues. In general, the occurrence of CD is greatest in the white matter, however, the ratio of CD to CH is lowest in the white matter. In the case of the granular cell layer, which is the most highly proteinaceous, the OD, ND occurrence is greater than that in the white matter. The OD, ND compared to OH, NH is reasonably high compared to the similar ratios in both the white matter and the molecular layer. The results of many measurements are summarized in Table 1.

Table 1. Significant Differences in Deuterated Ratios

CD/CH	OD,ND/OH,NH	CD/OD,ND
Mol>WM	Mol>WM	Mol<<WM
Mol>Gran	Mol>Gran	Mol>Gran
Gran>WM	Gran>WM	Gran<<WM

SUMMARY

We conclude from these observations, presented only in part in this brief preliminary report, that metabolically deuterated species can be determined readily *in situ* in mammaliam tissue. Furthermore, with the spatial resolution that is possible with an efficient FT-IR microspectrometer, data from these types of experiments can be used for a localized study of metabolism in the brain. This novel approach using deuterated tissue is accomplished without the use of radioisotopes.

ACKNOWLEDGMENT

This work was supported in part by the National Science Foundation EPSCoR Grant no. OSR 9255223 and the Kansas Agricultural Experiment Station. The National Synchrotron Light Source is supported by the United States Department of Energy contract no. DE-AC02-76CH00016.

REFERENCES

1. Wetzel, D.L. and Eilert, A.J., *Proc. SPIE - Int. Soc. Opt. Eng.*, **2089**, 464-465, 1994.
2. Sweat, J.A. Variations in tempering times for winter wheats by IRmicrospectroscopic tracking of D_2O [M.S. Thesis], Manhattan (KS): Kansas State University, 1995.
3. Coderre, J.A., Makar, M.S., Micca, P.L, Nawracky, M.M., Liu, H.B., Joel, D.D., Slatkin, D.N., and Amols, H.I., *J. Rad. Oncol. Biol. Phys.*, **27**(5), 1121-1129, 1993.
4. Wetzel, D.L. and LeVine, S.M., *Spectrosc.*, **8**, 40-45, 1993.
5. LeVine, S.M. and Wetzel, D.L., *Amer. J. Path.*, **145**(5), 1041-1047, 1994.

Contribution no. 98-55-B Kansas Agricultural Experiment Station, Manhattan.

Fourier Transform Infrared Spectroscopy Characterisation of Carotid Plaques

F. Alò[2], P. Bruni[1], C. Conti[1], E. Giorgini[1], C. Rubini[3], G. Tosi[1*]

[1]*Dipartimento di Scienze dei Materiali e della Terra, Università degli Studi di Ancona, via Brecce Bianche, 60131 ANCONA, Italy.* [2]*Clinica di Chirurgia Vascolare, Università degli Studi di Ancona, Ospedale Regionale Torrette, 60020 ANCONA, Italy.* [3]*Istituto di Anatomia Patologica, Università degli Studi di Ancona, Ospedale Regionale Torrette, 60020 ANCONA, Italy.*

Middle Fourier Transform Infrared determinations have been carried out in order to distinguish among the intima, homogeneous, heterogeneous and ulcerous carotid plaques through the analysis of morphology changes. In any specimen has been obtained a spectroscopic map that allowed to identify vibrational modes, that appear important to characterise the plaque.

INTRODUCTION

Fourier Transform Infrared Spectroscopy (FT-IR) can be considered an alternative tool to investigate biological materials because of its sensitivity and due to improvements in instrumentation and data handling. Any disease induces a clinical manifestation that is related with changes in the biochemistry of cells and tissues; consequently, morphological changes can be important to understand the nature of the disease. In particular, the study of plaque morphology is relevant to understand factors that can affect a cerebrovascular event. Duplex scanning is able to differentiate echogenic from heterogenic plaques (1). Recently, NIR and Middle Infrared Spectroscopy, associated with chemometric data handling, has shown that a non-invasive analysis of various part of human anatomy can be carried out (2a,b) in definite cases.

In particular, a topographic vibrational analysis on carotid artery and on brain or pulmonary tissues is able to detect changes in conformational disorder of membrane constituents and, among all, to evaluate the content of fatty esters and cholesterol (3,4).

In this work we report a FT-IR study that can distinguish among plaques from carotid endarterectomy, through the analysis of morphological changes.

EXPERIMENTAL

Samples of carotid arteries were collected from 30 patients (17 males aged 50-78 years and 13 females aged 60-75 years). Material consisted of endarterectomy specimens (n=30) of common carotid artery, frequently with bifurcation and an adjacent portion of the internal branch. One face of each plaque has been used for histopathological analysis and the other one for spectroscopic determinations.

Histological determinations. All selected cases had been routinely fixed in 10% neutral formalin (18-36 h), decalcified (10% formic acid; 12-72 h), processed and embedded in formalin. In each case 4 μm thick sections were stained with haematoxylin and eosin. In all cases additional stain were performed including Verhoeff for elastic tissue.

FT-IR determinations. The spectra have been obtained with a Nicolet 20-SX FT-IR spectrometer (DTGS detector) equipped with a Spectra Tech. Multiple Internal Reflectance (DRIFT) accessory for measurements in the solid state. Resolution 4 cm^{-1}, 250 scans. Treatment of the data has been achieved with a Galactic software package. Attribution of the bands has been done using literature data (5-7). On each sample, ten small parts out of each of thirty samples of carotid plaques, selected on the basis of surgical and histological characteristics, have been examined after grounding with KBr, that acts also as drying agent. With respect to KBr, lyophilization did not give better results.

----

* To whom correspondence should be addressed (E-mail: tosi@popcsi.unian.it; Fax: ++39-71-2204714)

RESULTS AND DISCUSSION

Results from histological determinations were interpreted on the basis of the following consideration: the most common disease of the vessels, atherosclerosis is recognisable in the gross examination. Atherosclerosis is a complex disease with a natural history and variation in appearance and constituents of the plaque during its evolution. In a few cases the plaques consist of florid

CP430, *Fourier Transform Spectroscopy:* 11th International Conference
edited by J.A. de Haseth

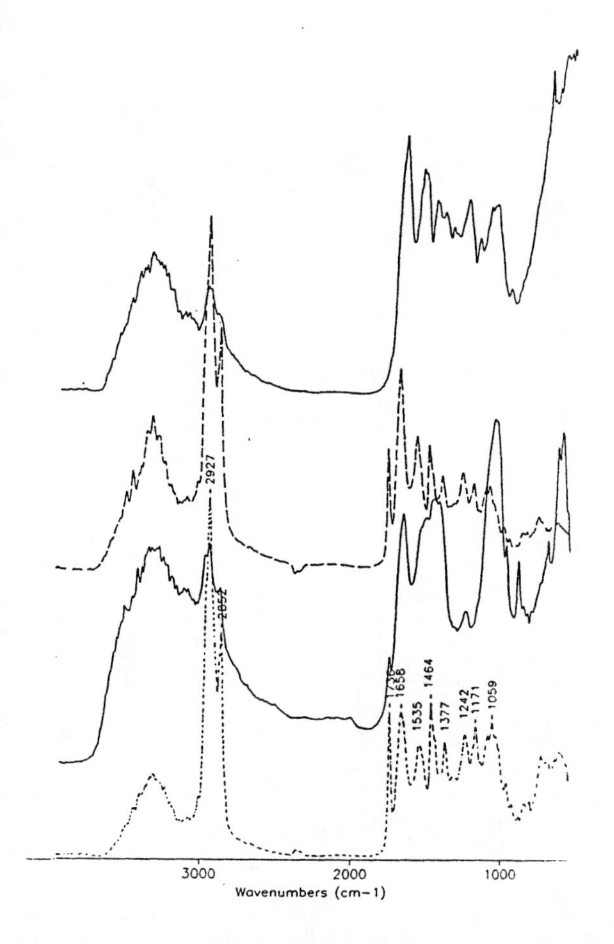

FIGURE 1. Infrared DRIFT spectra of specimens **A-D**: intima (**A**), homogeneous (**B**), dishomogeneous (**C**), ulcerated plaques (**D**).

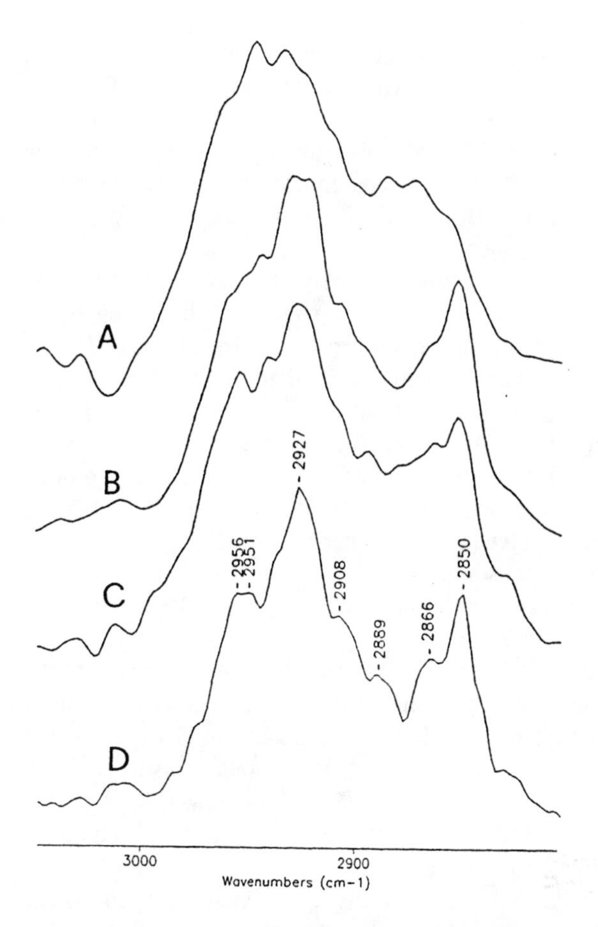

FIGURE 2. Deconvoluted spectra of CH stretching modes of specimens **A-D**: intima (**A**), homogeneous (**B**), dishomogeneous (**C**), ulcerated plaques (**D**).

granulation tissue. More commonly one sees dense fibrous tissue with varying amounts of glycosaminoglycans. These dense fibrous plaques break down with accumulation of various types and amounts of lipids. Moreover the fibrous cap overlying the plaque lipid core is actively weakened by inflammatory infiltrates. The inflammation may promote plaque rupture and thrombosis (8).

From the spectroscopic results, we can group the samples investigated so far into four categories: **A**, external 'common' part, the intima; **B**, homogeneous plaque; **C**, dishomogeneous plaque and **D**, ulcerated plaque. Fig. 1 shows the infrared spectra of representative specimens of each group.

In the 3000-2700 cm^{-1} region, CH_3 and CH_2 asymmetric and symmetric stretching modes are represented by a composite band with the most indicative maxima at *c.a.* 2955, 2924, 2866 and 2850 cm^{-1} (Fig. 2). In our samples, ν_{CH3} modes from proteins, are usually well distinguishable and absorb at 2873 and 2877 cm^{-1}. The overall band is much weaker in **A** than in the remaining samples. Band splitting on this region allowed to determine the intensity ratio CH_2/CH_3, related to lipid/protein content, that resulted 0.86 (**A**), 2.34 (**B**), 1.77 (**C**) and 1.72 (**D**) (9). In samples **A** (intima) and **B** (homogeneous plaque), we observe the highest content in proteins and the highest amount of lipids, respectively. Plaques **C** (dishomogeneous) and **D** (ulcerated) have almost the same CH_2/CH_3 ratio. These results are confirmed by the analysis of the intensity of the CH_3 stretching modes of proteins. With respect to **B-D**, a broadening of the whole profile is found in the spectrum of the intima (**A**). Broadening, related to a higher motional freedom and a higher interaction with the environment, can be induced by alkyl side chain groups of aminoacids in proteins. Inspection of the symmetric CH_2 stretching mode of lipids (*c.a.* 2850 cm^{-1}) evidenced a slight blue shift in sample **A** (2856 cm^{-1}) that can be related to lipid fluidity (10).

In the 1800-1500 cm^{-1} region the bands mainly arise from proteins, side-chains of aminoacids and interfacial zones of phospholipids. The vibration at 1735 cm^{-1}, due to the C=O stretching mode of phospholipids, appears as a single band even in second derivative. This means that the amount of water is not enough to give detectable hydrogen bonding with molecules of the sample and hence to give rise to an additional C=O band. In sample **D** the lipid content causes the intensity of this band to reach the maximum with respect with **A**, **B** and **C**.

The Amide I mode, centred in **A-D** at 1653, 1655, 1653 an 1658 cm^{-1}, respectively, mainly arises from proteins contributions and appears as a broad convoluted band (11). The overlapping of this band with OH bending modes of water, makes the assignment of secondary structures not easy. In the 1600-1400 cm^{-1} region, samples **A** and **B** show that band intensities decrease as one move to lower frequencies. In the dishomogeneous plaque (**C**), that can be considered a mixture containing mainly homogeneous plaque, intima and calcium phosphate, the bands have almost the same intensity and are convoluted with a characteristic profile. In the ulcerated plaque (**D**), the three main bands are again resolved with a proper pattern (Fig.1). The Amide II is present as a middle band at 1547 cm^{-1} (possibly coupled with contributes from tyrosine, 1517 cm^{-1}, and other components). The ratios Amide I/Amide II are 1.45, 3.49, 3.20, 3.37, respectively. Only the value in the intima **A** approaches the one in isolated proteins (roughly 1.4), while the higher ratio in **B-D** can be the consequence of a higher non-protein content or of a higher extinction coefficient of protein C=O an N-H moieties (12). The contribution of water to the Amide I, estimated from the ratio δ_{OH}/Amide, is 2.63, 3.40, 3.47, 3.97, in **A-D,** respectively. These values are consistent with the hypothesis that plaques with inflammation usually have a higher content of water (13). However, the slight higher content of water in **D** with respect to **C** does not allow to conclude regarding the inflammatory nature of plaque **D** and more data are needed.

Regarding the 1500-950 cm^{-1} region, we can make the following considerations: (i) the ratio CH_2/CH_3 scissoring vibrations at 1466 and 1456 cm^{-1}, is much lower in **A** than in other samples (where the CH_2 mode is predominant), thus confirming a higher lipid content; (ii) a similar trend is also observed between the convoluted band of **D** and **B** (centred at 1377 cm^{-1}) and the one of **A** (1398 cm^{-1}): the first band represents methyl symmetric bending while the second band, COO^- stretching and $(CH_3)_3N^+$ bending modes; (iii) in the dishomogeneous plaque **C**, the band at 1419 cm^{-1}, attributable to the trimethylammonium group of phosphatidylcholine, is particularly intense; (iv) the individuation of Amide III, that should be a weak band around 1310 cm^{-1}, is uncertain in all samples; (v) the blue shift of the maximum of $\delta_{PO_2^- asym}$ mode from 1234 (sample **A**) to 1243 cm^{-1} (sample **D**) appears interesting. Phosphate groups from lipids and DNA, mainly contribute to this absorption, with component bands at 1234 and 1241 cm^{-1}, respectively. Thus, the shift of the maximum to higher frequencies can be the consequence of an increased DNA contribution.

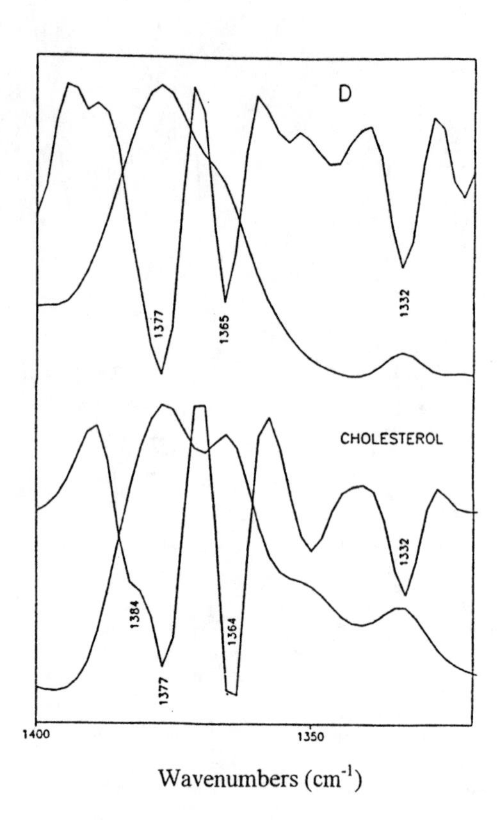

FIGURE 3. Second derivative DRIFT spectra of plaque **D** and cholesterol in the region 1400-1300 cm^{-1}.

The convoluted band around 1000 cm^{-1} is attributable to $\delta_{PO_2^- sym}$ modes (with contributions from C-O-C symmetric vibrations) and to the phosphate moiety of calcium phosphate (as a matter of fact, in plaque **C** this band is particularly intense).

Something can be said on cholesterol content. The second derivative DRIFT spectra, in the 1400-1300 cm^{-1} region, indicate that only the spectra of sample **C** and, mainly, **D**, have bands with profile and frequency similar to the ones of cholesterol at 1384, 1377, 1364 and 1332 cm^{-1} (Fig. 3) (4). Because also other modes contribute to these bands, as discussed above, the content of cholesterol (that increases from **A** to **D**) has been estimated with respect to the calibration curve derived for the band at 1364 cm^{-1}.

CONCLUSIONS

Our study has confirmed that reproducible differences can be observed among samples **A-D** (intima, homogeneous, dishomogeneous and ulcerated plaques). The lipid can be distinguished from protein as well as from other molecules (like DNA or water). The highest lipid content is found in heterogeneous plaques.

Further determinations are needed to affirm that ulcerous plaques have the highest cholesterol content. The analysis of relative intensities and profiles of bands in the 3100-2800 cm^{-1} region, of the intensity of the fatty ester band at 1735 cm^{-1} as well as of the water content, can represent a criterion to ascertain plaque dishomogeneity and inflammation.

From the comparison between histological and spectroscopic results, the plaques, studied so far, can be classified as homogeneous (n = 11), dishomogeneous (n = 19); six of the last plaques showed some ulceration. The agreeement between histolgical and FT-IR data, resulted higher than 90%.

REFERENCES

1. El-Barghouty, N., Nicolaides, A., Bacal, V., Geroulakos, G., Androulakis, A., *Eur. J. Vasc. Endovasc. Surg.,* **11**, 470-78 (1996).
2. a) Heise, H.M., *Mikrochim. Acta (Suppl.)*, **14**, 67-77 (1997). b) Applied Spectroscopy Research Group, University of Kentucky College of Pharmacy, Current Research (1994), private communication.
3. Williams, I.M., Mortimer, A.J., McCollum, C.N, *Eur. J. Vasc. Endovasc. Surg.*, **12**, 263-71 (1996).
4. Sakai, K., Yoshida, S., *Vibr. Spectroscopy*, **7**, 163-67 (1994).
5. Schrader, B., Ed., *Infrared and Raman Spectroscopy*, Weinheim, VCH Verlagsgesellschaft mbH, 1995.
6. Bruni, P., Cardellini, L., Giorgini, E., Iacussi, M., Maurelli, E., Tosi, G., *Mikrochim. Acta*, **14**, 169-72 (1997).
7. Bruni P., Iacussi, M.,Tosi, G., *J. Mol. Struct.*, 00, 000 (1997).
8. Hegyl, L., Skepper, J.N., Cary, N.R.B., Mitchinson, M.J., *J. Pathol*, **180**, 423-29 (1996); Jeziorska, M., McCollun, C., Woolley, D.E., *J. Pathol.*, **182**, 115-22 (1997); Boyle, J.J., *J. Pathol.*, **181**, 93-99 (1997).
9. Choo, L., Jackson, M., Halliday, W.C., Mantsh, H.H., *Biochim. Biophys. Acta*, **1182**, 333-37 (1993).
10. Cameron, D.G., Casal, H.L., Mantsch, H.H., *Biochemistry*, **19**, 3665-69 (1980).
11. Miyazawa, T., Fasman, G.G., Marcel Dekker, Ed., *Poly-α-Amino Acids*, New York, 1967.
12. Jackson, M., Mantsh, H.H., *Spectrochim. Acta Rev.*, **15**, 53-69 (1993).
13. Dubois-Dalcq, M., Armostrong, R., *Bioessay*, **12**, 569-74 (1990).

Localized (5 μm) Probing and Detailed Mapping of Hair with Synchrotron Powered FT-IR Microspectroscopy

David L. Wetzel[1] and Gwyn P. Williams[2]

[1]Microbeam Molecular Spectroscopy Laboratory, Kansas State University, Shellenberger Hall, Manhattan KS 66506, USA
[2]National Synchrotron Light Source, Bldg 725, Brookhaven National Laboratory, Upton, NY 11973, USA

The thickness and high absorptivity of single hairs typically result in the saturation of major infrared bands and their distortion. Single human hairs longitudinally microtomed and mounted on mirror slides were scanned routinely in the past with a 20 μm x 100 μm aperture that limited spatial resolution for localized probing and detailed mapping. Use of the nondivergent, bright, and low-noise synchrotron source for FT-IR microspectroscopy enables good S/N even at apertures as small as 5-6 μm. Functional group mapping as well as localized probing for extraneous materials illustrates the utility of this powerful probe.

INTRODUCTION

Human hair is a naturally occurring fiber that is of particular interest in pathology and in forensic science. FT-IR microspectroscopy of hair has provided information valuable in matching hair found at one location with the source of that hair. Material placed on the surface of the hair has been found by use of surface techniques such as attenuated total reflectance (ATR). In such cases, the hair composed largely of protein has provided a suitable substrate for absorbing or adsorbing other materials. Interesting presentations have been made of data produced from human hair subjected to a variety of cosmetics (1). Hair is also a very good adsorber for materials in the atmosphere such as solvents, smoke, and various odors. The probing discussed here is of the internal composition of the hair rather than materials that are found on the surface. When probing for the existence of foreign substances that have been placed in the human body, their uptake in hair is of concern. With this process, nature also provides a useful analytical concentration step.

PREVIOUS MICROSPECTROSCOPY

Traditionally, spectroscopists serving the area of pathology have separated the internal material from that found on the surface usually by a procedure of first washing the surface to remove contamination and their microtoming the sample in a block of paraffin to permit the usage of transmission reflection microspectroscopic techniques. Prior to using synchrotron powered FT-IR microspectroscopy, heroic efforts were made to obtain spectra of hair sliced in embedded paraffin that was placed on a reflecting microscope slide (2,3). Localization of chemicals making up the hair and potential localization of any foreign materials was limited by the signal-to-noise and thus a relatively large aperture was required. Typically, an aperture of 20 μm x 100 μm was used to obtain signal quantity and spectral quality that were dependable. Profiling of one particular hair was done by stepping the aperture in small increments across the width of the hair. Small (5 μm) steps of the 20 μm wide aperture did produce interesting information regarding the distribution of the materials that make up the hair. In particular, the protein bands were stronger in the center of the hair, near the medula, than they were out at the cortex (3). In cases where a foreign substance was present, it was necessary to use spectral subtraction of the spectra of control samples in order to locate or detect the presence of a foreign chemical substance.

SYNCHROTRON SOURCE

Localized probing and detailed mapping of hair was considered to be a good candidate for synchrotron powered FT-IR microspectroscopy. The critical requirement for these measurements is high signal to noise from a small illuminated spot on the sample, small being a few microns. The high signal to noise of the synchrotron source is apparent from Fig. 1. Infrared synchrotron radiation is a broadband source that is about 1000 times brighter than standard thermal sources (4). It is also in principle more stable, because no thermal fluctuations occur in this type of source. Light is generated by the electric field experienced as bunches of electrons circulate at relativistic energies in a storage ring. The intensity is proportional to the number of electrons, which is constant apart from a slowly varying exponential loss, so if the electron orbit is stable, the intensity is also constant. Brightness is crucial for microscopy where one is trying to illuminate as small an area as possible with as much light as possible. In the case of the synchrotron, the source size is roughly 1 mm by 1 mm, and the radiation is emitted into an angle of approximately 10 milliradians by 10 milliradians so

CP430, *Fourier Transform Spectroscopy:* 11th International Conference
edited by J.A. de Haseth
 1-56396-746-4/98/$15.00

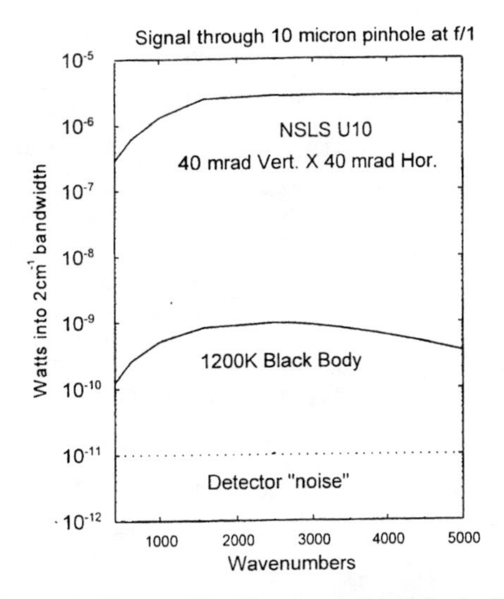

Figure 1. Signal of synchrotron vs. black body (5).

that the emittance is about 10^{-4} mm^2 steradians. In contrast, the emittance for a thermal source is more than 10^{-2} mm^2 steradians, so only 0.001% of the light from the thermal source would be available to illuminate a 10 μm sized spot on the sample. In practice with a thermal source, one requires illuminated areas of about 30 μm in order to obtain reasonable signal to noise ratio, and even then one has to average over many scans. A synchrotron illuminated infrared microscope facility was constructed at beamline U2B of the National Synchrotron Light Source at Bookhaven National Laboratory (6). Light from the synchrotron was collimated and introduced into a standard commercial FT-IR microspectrometer (IRμs, Spectra-Tech, Shelton, CT) by simply removing the collimating mirror for the thermal source. The overall optical system had some aberrations, so that the gain over the thermal source was a factor of about 50. This facility has served well both for material characterization and for use on biological substances or tissue. Localization is important not just to deal with small samples in the field of the microscope but to isolate subsamples within the microscopic field and to interrogate those without accidental sampling of neighboring tissue. The Spectra-Tech IRμs microspectrometer is doubly confocal, meaning that it has an aperture before the objective and another after the condenser, in the case of transmission, to limit the part of the field being interrogated and to reject the sampling of neighboring tissue brought about by diffraction. In practice, with the synchrotron source, we were able to obtain good signal to noise ratio when sampling areas whose size was of the order of the wavelength of the light being used, i.e. a few microns, in sampling times of less than a minute per sampled area.

EXPERIMENTATION AND RESULTS

With the synchrotron source and an IRμs microspectrometer (Spectra-Tech, Shelton, CT), probing of transmission reflection paraffin-microtomed samples of single human hairs on a mirror surface was attempted. Of the first six specimens with which this was attempted, only two were sufficiently thin to allow a double transmission. A rectangular mapping experiment with a 12 x 12 μm aperture (Fig. 2) showed maximum protein in the center of the hair as previously shown (2). Figure 3 shows a higher density of organic matter in the center and the effect of paraffin contamination on either side of the hair. Attempts to prepare thinner paraffin sections were wholly unsatisfactory. The localized probing reported here was done ultimately by removing thin sections of hair from the microtomed slab of paraffin and mounting them between barium fluoride disks 2 mm thick x 13 mm diameter in a microcompression cell. These specimens were probed using a transmission mode of operation with a 32X objective at a resolution of 4 or 8 cm^{-1} with 32-128 scans coadded. Apertures of 12 x 12 μm were used to obtain individual spectra. They also were used to step longitudinally along the fiber or across the fiber. The excellent spectrum in Fig. 4 was from a 12 x 12 μm aperture with 32 scans coadded. This represents an improvement of the previous requirement of a large 20 x 100 μm aperture. Figure 5 shows a series of spectra mapped across a hair also

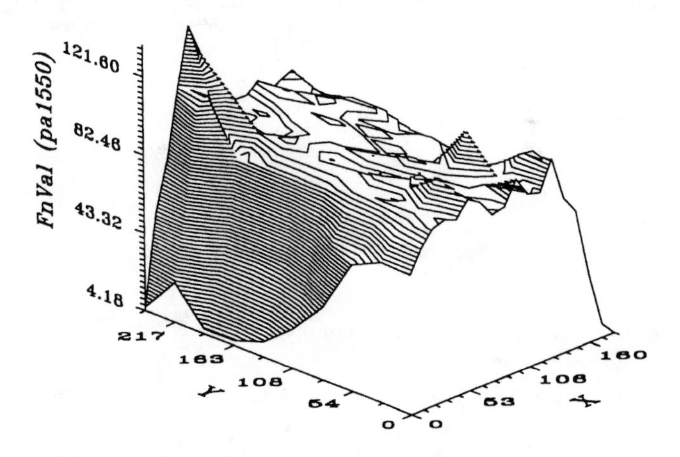

Figure 2. Peak area of 1550 cm^{-1} showing distribution of protein

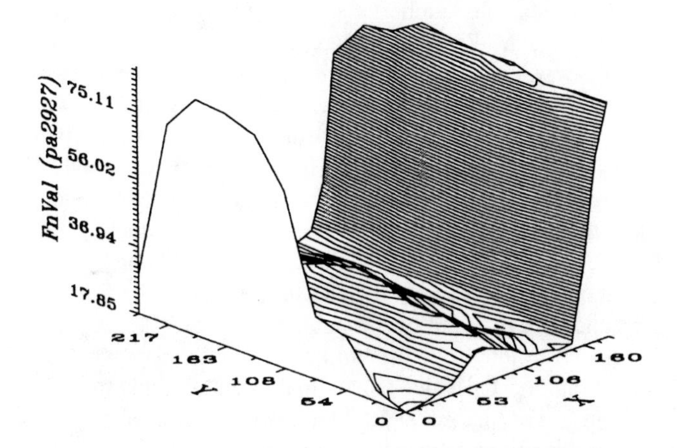

Figure 3. Peak area of 2927 cm-1 showing density of organic material

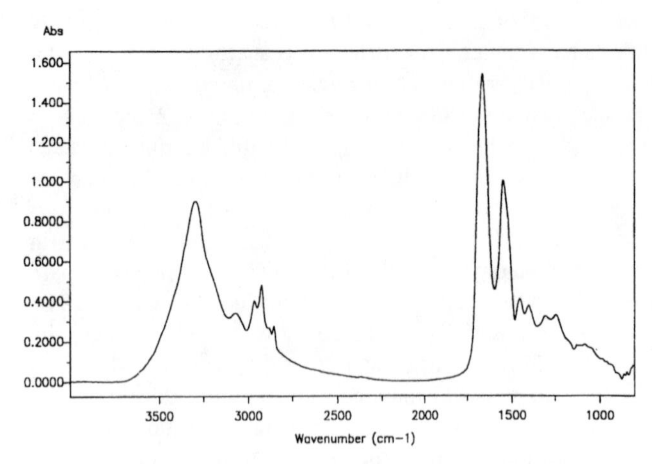

Figure 4. Spectrum obtained with 12 x 12 μm apertures

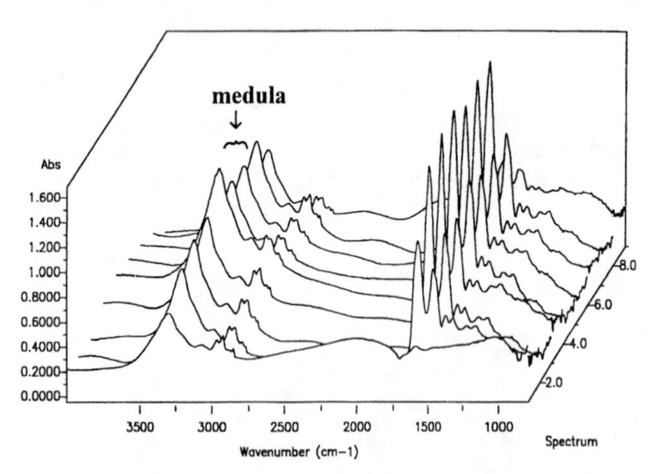

Figure 5. Spectra from a linear map across hair

with a 12 x 12 μm aperture. For specimens that were sufficiently thin, excellent spectra were produced in this manner. The most extreme case of localization that was done on hair specimens with the synchrotron powered system was that of an approximately 5 μm projected aperture that resulted from the use of a pin hole projected through a 32X objective. With a reasonable number of scans coadded, useful spectra shown resulted. With this aperture, different points along the medula were scanned, as shown in Fig. 6. At a point of interest, points across the hair were scanned. Figures 7 and 8 show typical spectra obtained in this manner from different places across the hair shown in the associated photos. Photographs at different parts of one particular hair specimen show the aperture projected at various sample points. Data from select points show the quality of spectra obtained. Note that at one particular point (Fig. 8), there is a hint of a foreign substance as evidenced by the carbonyl band at 1740 cm^{-1}. It should be pointed out that this particular specimen was a spiked sample. Once the longitudinal location of this foreign substance was located, several probes were taken across the hair in this region which showed a variance of the carbonyl (Fig. 9). The results of line mapping experiments from outside the medula across the medula to the opposite edge

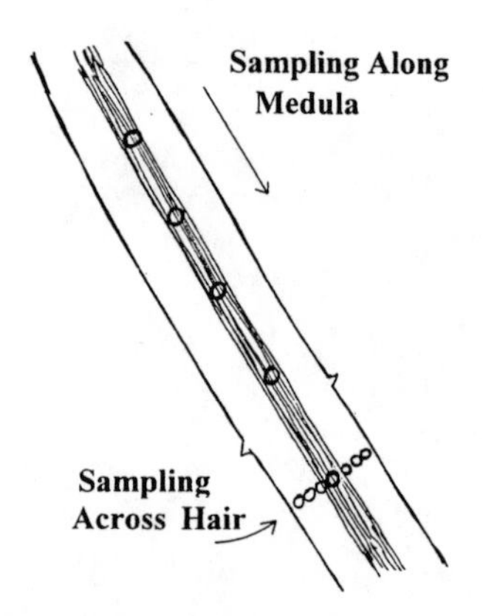

Figure 6. Sketch of individual points scanned

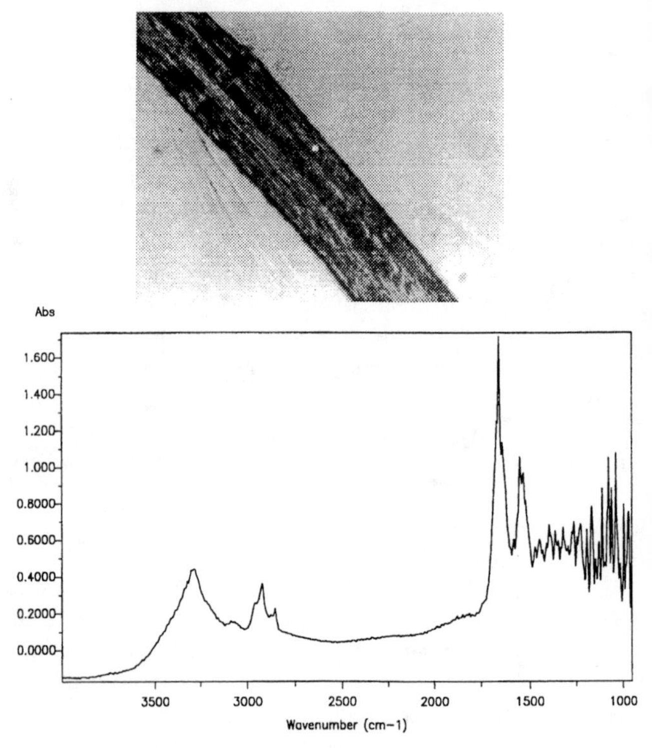

Figure 7. Photograph and corresponding spectrum at a point outside of the medula

with a 5 μm aperture showed maximum carbonyl peaks across at points 4 and 7 that straddled the two parts in the center of the medula (Fig. 10).

DISCUSSION AND CONCLUSIONS

The purpose of this work was to demonstrate the degree of localization achievable, the sampling techniques appropriate, and the improvement of localization in such a challenging task as mapping hair. Certainly the ability to change from a 20 μm x 100 μm aperture to a 5 μm circular aperture is a huge improvement in localization. It is also interesting to note that on the longitudinal axis at a typical growth rate for head hair, a 5 μm distance would represent approximately 25 minutes of

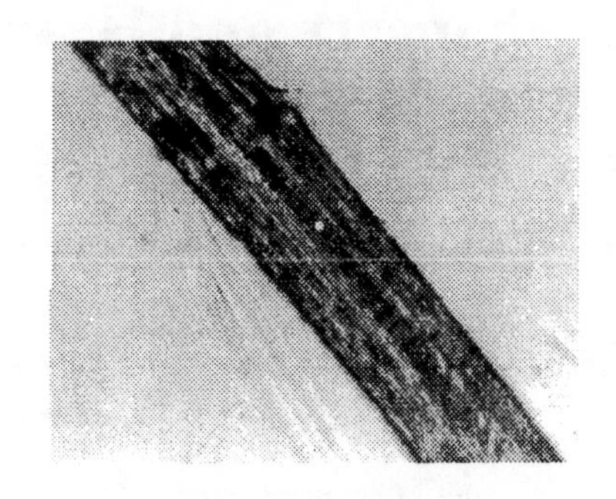

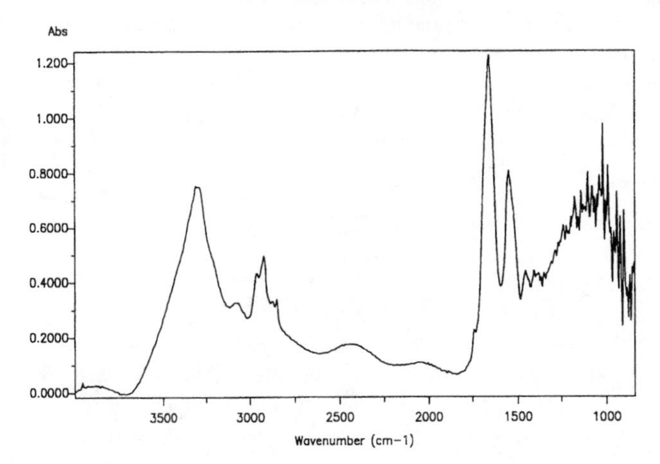

Figure 8. Photograph and corresponding spectrum of individual point at the edge of the medula

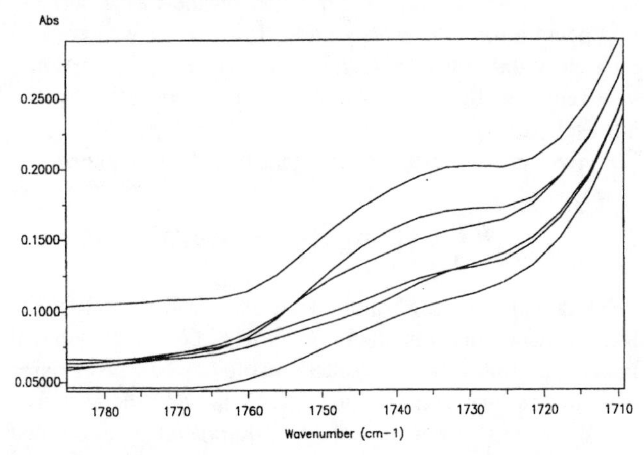

Figure 9. Spectral region at 1740 cm-1 showing variation across hair spiked with a drug metabolite

growth. Thus, where the actual hair serves as a recording medium, the microspectroscopy with this spatial resolution then can detect what is being laid down on that substrate in relative units of time. Localized probing of difficult to handle biological specimens is enhanced considerably with synchrotron powered FT-IR microspectroscopy. Excellent spectra resulted from spot sizes as small as 5 μm using a transmission mode for a 6 μm thick cross section between barium fluoride plates in a microcompression cell.

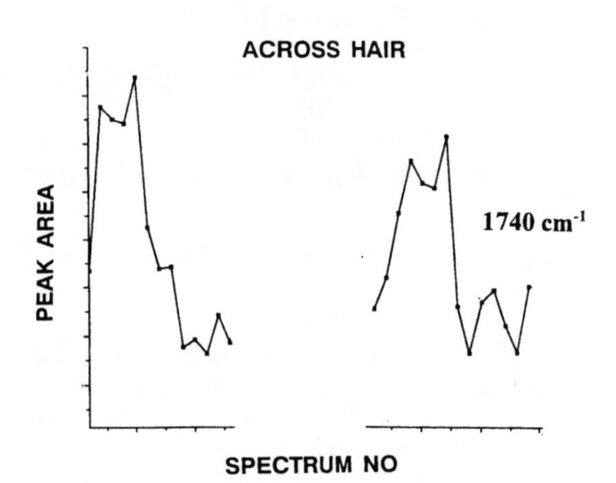

Figure 10. Distribution of carbonyl from two line maps across hair spiked with drug metabolite

ACKNOWLEDGMENTS

Paraffin-embedded microtomed control (drug free) hair specimens were provided by Dr. Kathryn S. Kalasinsky (Armed Forces Institute of Pathology, Washington, DC), whose long-term serious efforts to study the nature of the uptake and deposition of foreign substances in hair during growth inspired the analytical challenge that we accepted. A control hair specimen spiked with a carbonyl-containing major metabolyte of a drug was used in the small apertue search for foreign substance localization along the length and across the hair width. This work was supported in part by the National Science Foundation EPSCoR Grant no. OSR-9255223. The National Synchrotron Light Source is supported by the U.S. Dept. of Energy under contract no. DE-AC02-76CH00016.

REFERENCES

1. Martoglio, P.A., Reffner, J.A., and T., William, "Hair fiber differentiation by micro-ATR analysis", presented at the Pittsburgh Conference, Chicago, IL, Feb. 27-Mar. 4, 1994.
2. Kalasinsky, K.S., Magluilo, J., Jr., Schaefer, T., "Deposition and visulaization of drugs of abuse in hair by infrared microspectroscopy", in: Friel, J.J., (ed.), *Proceedings of the 28th Annual Microbeam Analysis Meeting*; 1994 Jul31-Aug5; New Orleans, LA. UCH Publishers, Inc., pp85-86.
3. Kalasinsky, K.S., Schaefer, T., and Smith, M.L., "Distribution and. visualization of drugs in hair by infrared microscopy", presented at the Pittsburgh Conference, New Orleans, LA, Mar. 5-10, 1995.
4. Williams, G.P., *Nuclear Instr. and Meth.*, **A291**,8, 1990.
5. Wetzel, D.L., Reffner, J.A., and Williams, G.P., *Mikrochim. Acta [Suppl.]*, **14**, 353-355, 1997.
6. Carr, G.L., Reffner, J.A., and Williams, G.P., *Rev. Sci. Instr.*, **66**, 1490, 1995.

Contribution no. 98-57 Kansas Agricultural Experiment Station, Manhattan.

The Interactions Between Penetration Enhancers and Human Skin Assessed by FT-Raman Spectroscopy

E. E. Lawson, A. C. Williams, H. G. M. Edwards and B. W. Barry

Drug Delivery Group, Postgraduate Studies in Pharmaceutical Technology, The School of Pharmacy, University of Bradford, Bradford BD7 1DP, U.K

FT-Raman spectra have been recorded from human stratum corneum *in vivo* and *in vitro* using a fibre optic coupling. Spectra from both tissues were near identical illustrating the validity of *in vitro* studies for examining the mechanisms of action of penetration enhancers *in vivo*. Treatment of excised human skin with the penetration enhancer 1,8-cineole showed that the permeation promoter induced increased order in the barrier bilayered lipids domains of the stratum corneum in most of the samples, but caused a decrease in order of these domains in some of the samples investigated. This may correlate with a mechanism of action where the enhancer exists as a discrete domain within the stratum corneum lipids which would provide defects in the barrier nature of the permeation limiting lipid domains

INTRODUCTION

Transdermal drug delivery offers potentially numerous advantages over conventional routes of drug administration. However, the skin is a very effective barrier and the range of drugs that can be administered to clinical levels via the transdermal route is limited. It is the skin's outermost layer, the stratum corneum or horny layer which is responsible for controlling the loss of water and other bodily constituents whilst preventing entry of noxious substances or drugs from the external environment. The stratum corneum comprises strata of protein (keratin) filled flattened cells embedded in a multiply bilayered lipid-rich intercellular matrix. The lipid domain is considered to be the major pathway by which drugs applied to the skin permeate through the tissue. Penetration enhancers are materials which temporarily and reversibly diminish the stratum corneum's barrier function thus encouraging percutaneous absorption of drugs through the skin. To date, the modes of action of penetration enhancers are unclear, although the most effective permeation promoters are able to interact with the highly organised lipid domains of the tissue. Here, the mechanisms by which a penetration enhancer, 1,8-cineole, interacts with stratum corneum lipids to disrupt their structure is investigated.

Previously, we have used FT-Raman spectroscopy to investigate the molecular interactions of penetration enhancers with isolated sheets of human stratum corneum; membranes prepared from post-mortem skin samples were soaked in an excess of terpene penetration enhancers. However, this methodology hindered spectroscopic investigation as the terpene bands masked the weaker skin bands (1). The current work addresses this problem and compares the utilization of *in vitro* experimental membranes with *in vivo* tissue.

METHODS

Membranes

Caucasian, full-thickness human abdominal skin was obtained post-mortem and was stored in evacuated polythene bags at -20°C until required (2). Prior to use, the membrane surface (stratum corneum) was swabbed briefly (10 sec) with acetone to remove any surface contamination such as fat which may be deposited on the tissue surface. For penetration enhancer studies, 1,8-cineole was applied by lightly dabbing it onto the skin; any excess remaining was removed by gentle swabbing and evaporation to the atmosphere.

In vivo studies utilized inner forearm Caucasian skin.

FT-Raman Spectroscopy

FT-Raman spectra were obtained using a Bruker 106 FT-Raman accessory mounted on an IFS 66 FT-IR optical bench. Samples were excited using a Nd:YAG laser providing a NIR exciting line at 1064 nm and the scattered radiation was collected using a liquid nitrogen cooled germanium diode detector with an extended spectral bandwidth. *In vivo* and *in vitro* Raman spectra were recorded from the skin by delivering the laser to the tissue via a 1 m long optical fibre and the scattered radiation was returned to the spectrometer via a fibre bundle. The probe head was mounted in contact with the skin with the focal point at the skin surface.

Skin spectra (both *in vivo* and *in vitro*) were recorded at a spectral resolution of 8 cm^{-1} for 1000 scans with an approximate laser power of 45 mW at the sample. FT-Raman spectra were recorded for normal, untreated skin (*in vivo* and *in vitro*) and for 1,8-cineole treated tissue.

CP430, *Fourier Transform Spectroscopy:* 11th International Conference
edited by J.A. de Haseth

RESULTS AND DISCUSSION

In vivo/In vitro Correlation

Spectra from human stratum corneum collected *in vivo* and *in vitro* were near identical with only minimal differences between these spectra - the spectrum recorded *in vivo* appeared to be slightly more noisy than that from the *in vitro* sample, possibly as a result of sweating or slight movement of the sample during scanning.

From elevated temperature studies of human stratum corneum (3) we have found that the relative position of the $\nu(CH_2)$ stretching bands at around 2930 and 2853 cm^{-1} are particularly sensitive to the degree of disorder within the stratum corneum lipid domain; band separation decreases with increasing disorder. This measure of disorder is especially valuable when probing penetration enhancer interactions with the tissue. From the spectra, the band separation of the $\nu(CH_2)$ stretching modes was similar for both *in vivo* and *in vitro* samples (76 - 77 cm^{-1}). The band separation for the *in vivo* sample was slightly, but not significantly, lower (i.e. indicating more disorder) possibly as the *in vivo* sample was measured at body temperature (32°C) whereas the *in vitro* sample was recorded at laboratory temperature (26°C).

The functionality regions of the spectra were also very similar with only minor alterations in the relative intensities of some bands. This is expected as human stratum corneum is known to exhibit considerable variability in diffusional properties between patient sites and between different patients, presumably arising from slight differences in tissue component contents.

Penetration Enhancer Studies

Spectra recorded from tissue (*in vitro*) treated with the enhancer showed no evidence for the terpene, although the enhancer decreased the intensity of the scattered Raman signal by approximately 25% compared to that of untreated tissue. This was a useful effect in that the treated areas could easily be located and the interference found with earlier studies from the terpene signal was avoided (1). The spectra showed that the $\nu(CH_2)$ stretching bands were markedly changed on application of the terpene. 35 post mortem samples were studied here before and after application of 1,8-cineole and the results were quite varied. The change in $\nu(CH_2)$ band separation upon terpene treatment was $\pm$ 5 cm^{-1} with the median value for this distribution being + 1.57 cm^{-1}. 86% of the samples showed an increase in band separation after treatment. This indicates an unexpected increase in lipid order on treatment with the enhancer. It would appear that the enhancer is able to partition into the tissue where it may exist as a separate domain (phase separated) which would cause a structural reorganization (ordering) of the endogenous lipids, possibly impelling the lipids to pack closer together as the enhancer pools are introduced into the lipid domains. This could account for the increase in order of the lipid derived modes whilst also explaining why drugs permeate across the tissue faster after enhancer treatment. Previous work has shown 1,8-cineole treatment of human epidermal membranes increases the transdermal flux of the model polar drug 5-fluorouracil nearly 100-fold (4,5). Thus, defects with a lower resistance to drug flux may have been introduced within the barrier lipid domains as separate "pools" of the penetration enhancer could exist within the lipid bilayers.

Other spectral features were altered in the terpene treated skin. Primarily, the keratotic modes around 1688 cm^{-1} altered with a suggestion that the α-helical nature of the keratin which predominates in untreated corneocytes alters somewhat to increase the proportion of β-sheets in terpene treated skin. Also, some *trans* to *cis* alterations of the lipid backbones were induced on terpene treatment.

CONCLUSIONS

It can be concluded that the use of human stratum corneum from cadavers is appropriate for investigating penetration enhancer effects on human stratum corneum *in vivo* as the spectra from *in vivo* and *in vitro* samples were essentially identical. Application of 1,8-cineole onto post-mortem skin altered the order/disorder equilibrium of the stratum corneum's lipid domains. Spectra showed some samples had a decrease in lipid order upon cineole treatment, but most samples showed an increase in order of the lipid domains. Unlike the elevated temperature studies of human stratum corneum, structural changes were also seen in the keratotic component of the tissue upon application of the cineole. It is suggested that 1,8-cineole performs its role as a penetration enhancer in part by ordering the lipid molecules thus creating space for solvent pathways in the barrier lipids, in addition to effects on the keratin component of the tissue.

REFERENCES

1. Anigbogu, A. N. C., Williams, A. C., Barry, B. W., and Edwards, H. G. M., *Int. J. Pharm.*, **125**, 265-282, 1995.
2. Harrison, S. M., Barry, B. W., and Dugard, P. H., *J. Pharm. Pharmacol.*, **36**, 261-262, 1984.
3. Lawson, E. E., Anigbogu, A. N. C., Williams, A. C., Barry, B. W., and Edwards, H.G.M., *Spectrochim. Acta*, submitted.
4. Williams, A. C. and Barry, B. W., *Int. J. Pharm.*, **74**, 157-168, 1991.
5. Yamane, M. A., Williams, A. C. and Barry, B. W., *J. Pharm. Pharmacol.*, **47**, 978-989, 1995.

FT-IR Microspectroscopic Imaging of Human Melanoma Thin Sections

P. Lasch*, W. Wäsche#, G. Müller# and D. Naumann*

*Robert Koch-Institut, F311, Nordufer 20, D-13353 Berlin, Germany
#Laser- und Medizin- Technologie GmbH, Krahmerstraße 6-10, D-1207 Berlin, Germany

FT-IR microscopic mapping techniques in combination with image construction methods have been used to characterize tissue thin sections from human melanoma. While IR imaging based on distinct spectral parameters (intensity, frequency, or half-width) often gives unsatisfactory results, pattern recognition analysis (e.g. by principal component analysis or Artificial Neural Networks) of the IR-data confirms standard histopathological techniques and turns out to be helpful to discriminate reliably between different tissues.

INTRODUCTION

Mid-infrared spectra of microorganisms, isolated cells, or tissues display the total chemical composition (proteins, membranes, nucleic acids, etc.) of the biological probes. Owing to the multitude of cellular compounds, broad and superimposed absorbance bands are observed throughout the entire spectral range. Some of the bands can be assigned to distinct functional groups of chemical substructures, but a comprehensive understanding of the spectral information is not available. Fortunately, some spectral ranges are dominated by particular chemical components and it is possible to subdivide the spectra into spectral 'windows' which are considered to contain more specific information (1). The IR-spectra recorded under standardized conditions of well-defined biological samples are known to be highly typical and reproducible. This property was utilized to develop a method for classifying and identifying microorganisms by FT-IR spectroscopy (2). For this purpose computer based multivariate statistical analysis (MSA) techniques (or other pattern recognition techniques) were found to be helpful to extract the relevant information from the mid-infrared spectra of the bacterial samples. Principally, these considerations should also be valid for data treatment of IR spectra obtained by FT-IR microspectroscopic mapping experiments on human tissue thin sections.

Here we present a study which demonstrates in an exemplary manner the potential of a new approach of FT-IR imaging: the combination of spatial resolved FT-IR microspectroscopy with pattern analysis techniques of the spectral data sets.

MATERIALS AND METHODS

Fresh tissue samples were cooled for cryo-sectioning rapidly in liquid nitrogen. The tissue specimens were then thaw mounted from the tissue sample as 8 µm thin slices for FT-IR microscopic measurements on CaF_2 windows. For comparative light microscopic investigations, slices of the same tissue sample were mounted in the same way on conventional microscope slides. These sections were analyzed after hemalum-eosin (HE) staining to confirm carcinomatosis and healthy regions. Samples for FT-IR microscopic measurements were stored at dry conditions. No fixation substances were added. In some cases for cryotoming it was necessary to embed the slices in a freezing medium prior to cryo-sectioning. It was confirmed that the IR properties of the tissue itself were not changed by the medium. Typically, the medium could be spectroscopically detected outside the tissue as an additional surrounding layer by specific IR bands. The IR samples could be stored for long periods of times under dry conditions at room temperatures.

Spectra were collected in transmission mode on a FT-IR microscope (Bruker Analytische Meßtechnik GmbH, Germany) which was equipped with a liquid nitrogen cooled MCT detector and connected to a IFS28B FT-IR spectrometer (Bruker). The microscope was linked to a computer controlled xy stage. Usually, rectangular regions of the sample were mapped with 80 µm steps in x and y direction using an aperture diameter of 100 µm. Background spectra were recorded at positions of the IR-window outside the tissue sample. Nominal spectral resolution used was 6 cm^{-1}. 22 scans were co-added per sample spectrum and apodized applying a Happ-Genzel apodization function for Fourier transformation. Interferograms were zero-filled using a zero-filling-factor of 6 to yield an encoding interval of approximately one data point per wavenumber.

The analysis of very large spectral data sets (up to 15.000 spectra per mapping experiment) was possible only with the help of a macro interpreter, integrated in the OPUS data collection software package (Bruker).

First steps of data evaluation included a 'quality test' of the raw spectral data (test for water vapor, test for tissue

CP430, *Fourier Transform Spectroscopy:* 11th International Conference
edited by J.A. de Haseth

freezing medium, and search for preparational artifacts like micro-rips or overlapping tissue thin section areas due to cryotoming, etc.). All spectra which have passed this test were subsequently offset corrected and normalized using the integral intensity between 1770 and 1100 cm^{-1} as an internal standard. From these spectra various spectral parameters (i.e. frequency of the $\nu_{sy}(PO_2)$ band, intensity at 1653 cm^{-1} amide I, or combinations like the 'lipid/protein ratio') were calculated and routinely plotted as a function of xy position. These kinds of IR images were denoted in the following as "conventional" IR-images.

MULTIVARIATE STATISTICAL ANALYSIS

Multivariate statistical analysis (MSA) are powerful tools for dealing with mixed populations of tissue spectra. In particular, principal component analysis (PCA) was used to extract relevant information from the spectral data sets. Each spectra can be thought as a point in a n-dimensional space, were n is the number of data points used. The entire set of spectra forms a cloud (or clouds) in this space. PCA forms a new, rotated coordinate system in which the first axis represents the direction of inter-spectra variance, the second axis the direction of remaining inter-spectra variance, and so on. The cloud of spectra can now be described with respect to this new coordinate system. More precisely, each of the aligned spectra can be expressed as a linear combination of the orthogonal and independent "eigenspectra". The first eigenspectrum points at the center of mass of the full data set. Consecutive eigenspectra describe decreasing amounts of the inter-spectral variance and, in our case, soon represent irrelevant information (noise). By disregarding these higher order components, and considering only the significant ones (typically in the order of 2 to 12), the spectra can be considered as points in a much smaller dimensional space.

Images were constructed by calculating the normalized distances $N_{(i,X)}$ of every point (spectra) to the centers of mass of the individual clouds and the assembly of these $N_{(i,X)}$ values with the original xy data to matrices (i is the number of classes or, in the present case, the number of spectral pattern, cf. Fig. 2). The matrix files which contain beside these distance values the original spatial information were further processed to produce i different color plots. The most intense color was assigned to points at the origins with $N_{(i,X)}= 0$, while white regions were characterized by high $N_{(I,X)}$ values ($N_{max} = 1000$, cf. Fig. 2). All i plots were superimposed to build up one, as demonstrated in Fig. 3. For principal component analysis and image construction, we have used first derivative spectra (Savitzky-Golay, 9 smoothing points) ranging from 900-1450 cm^{-1}. This spectral window was found to be particularly suitable for multivariate statistical analysis. It must be pointed out that the selection of certain spectral windows lead additionally to data reduction, which in turn significantly reduces calculation times.

Another aspect concerns the definition of the number of clouds (spectral classes) and the corresponding centers of mass. In our opinion, this step of data analysis is the most crucial one. To find an answer to this question, at present (without the help of elaborated tissue spectra libraries) it seems to be necessary to compare conventional HE stained specimens with the IR data and to define so-called "reference IR patterns" (e.g. by spectra averaging). It is clear, that the frequency of points at a given quadrant of principal component plots cannot be sufficient to define classes or even centers of mass. Therefore, the coordinates of the reference pattern in the PCA plots were used as mass centers which itself are essential to calculate the $N_{(i,X)}$ -values. For a better visualization of the results we have used only two principal components for PCA plot representation (vector 2 vs. vector 3- two-dimensional case). However, the approach presented here should be applicable for three or more dimensions, too.

It should be pointed out that the non supervised classification technique PCA alone is - in our opinion - not optimally suited to find out common properties of IR-spectra of defined tissues obtained from different patients. From our FT-IR mapping experiments we have learned that the biological variance of spectra from e.g. melanin containing parts of a malignant melanoma is relatively large. Since the mathematical algorithms of principal component analysis are mainly suited to find out the highest intra-spectra variance, large patient dependent biological variances will be included in principal component plots and can - in some cases - cover the relevant information. Therefore, to detect typical IR patterns of defined tissues and to include at the same time the biological variance, supervised classification techniques such as Artificial Neural Networks were applied to characterize tissue thin sections.

IR IMAGING USING ARTIFICIAL NEURAL NETWORKS (ANN)

Supervised pattern recognition analysis by Artificial Neural Networks (ANN) turned out to be helpful for tissue differentiation and identification and gives at hand an easy to use tool for rapid spectra analysis. We used a feed-forward backpropagation ANN for tissue identification and classification. In this ANN, the nodes are organized in an input layer, an output layer, and one hidden layer. The nodes are interconnected by weights and information propagates from one layer to the next through a sigmoidal activation function. The learning of the ANN was a supervised process in which known training cases are input to the ANN and the weights are adjusted with an iterative backpropagation procedure in order to achieve a desired input-output relationship. Detailed description of the backpropagation algorithm can be found in the literature

(3). In this study we used the Stuttgart Neuronale Netze Simulator (SNNS) for all calculations, running on a PC workstation (LINUX). The SNNS included a graphical user interface (gui) which allows visualization of e.g. the architecture of the network, the weights, or the errors as a function of increasing training epochs. Especially the graphical illustration of the error during the learning process was found to be useful to administrate the network training and to prevent "over-adaptation" of the net. Beside training data sets we have also defined validation data sets. During network training, the validation data were permanently tested and the error curves of both, the training data and validation data set was displayed on the screen. Spectra, suitable for network training and simultaneous internal validation were defined together on the basis of light microscopic investigations and randomly selected for the distinct data sets. Spectral data treatment was carried out as already described for PCA: Only spectra which had passed the quality test were offset corrected, normalized and derived for further processing.

The analysis of the FT-IR Microspectroscopic mapping data by ANN with supervised learning algorithms allows the easy combination of spectra recorded from different tissue thin sections (different patients). In the present study we have defined 5 distinct spectra pattern from 4 patients for network training and validation. Every pattern includes approximately 20 spectra per thin section for training and validation purposes, respectively. As already mentioned for PCA, the definition of spectral windows has significantly reduced calculation times and permitted network training with very large data sets.

Usually, after approximately 10.000 training cycles the error curves reached a constant level. If the error of the training and validation data set was comparable and constant, the learning process was stopped and the entire data set of a FT-IR microspectroscopic mapping experiment was tested. The output values of the ANN ranging from 0 (no activation) to 1 (full activation, pattern recognized) obtained from every spectra for every pre-defined pattern were utilized for image construction. As for principal component analysis, i matrices containing beside the output values the original spatial information were generated and converted to color images. Each image represented the spatial distribution of a distinct MIR pattern obtained by FT-IR microspectroscopic mapping measurements. These different color plots were again superimposed to one image (cf. Fig 4.). In contrast to Fig. 3, Fig 4. was generated using MIR data from 4 different melanoma thin sections. Obviously, typical spectral properties for all patterns were present in the IR data used.

DISCUSSION

Infrared microspectrometry can be used to map a particular functional group across a sample as a function of concentration using the absorbances or frequency of this group. This approach works at the best when e.g. „exotic“ substances in tissues are to be identified as shown by the published examples of silicone gel inclusions in human breast tissue (5), or the detection of Alzheimer plaques (β-amyloid peptide) in diseased human brain (6). In contrast to this „conventional“ functional group mapping, the spatial characterization of a tissues thin section requires techniques facilitating the analysis of spectral patterns which were distributed over the entire spectral range. In our opinion, the use of only one or two spectral parameters is not sufficient for differentiation and characterization of biological samples composed of hundreds of complex macromolecules. If one accepts the postulation that any diseased states in tissues involve complex (but reproducible) changes in intrinsic chemical composition and structural organization, a diseased state will produce unique infrared spectroscopic features as shown by the example of human colon carcinoma cell lines (7) or colorectal and cervical cancer (8, 9). In principal, one should be able to find out these signatures and to define spectral windows containing the relevant spectral information.

Comparing the two pattern recognition techniques used in this study, we have noted that the non-supervised classifier (PCA) is particularly meaningful if the input pattern is completely unknown or only partly defined. In contrast, ANN classifiers gave double or even multiple activation in such cases and required the knowledge of all IR pattern. It turned out that the combination of both techniques supply the best results. PCA was therefore mainly used to define the training and validation data sets for the ANN classifier. In the present study, we did not perform a systematic optimization of spectral windows, network architecture, and other important parameters. Many of the parameters were chosen based on our experience from other applications (identification of microorganisms). The goal of this study is to demonstrate the feasibility of pattern recognition techniques for FT-IR microspectroscopic mapping data and to present attempts for digital imaging of tissue thin sections by FT-IR spectroscopy. Our results indicate, that ample information is present in the mid-infrared region of tissue materials to define different spectral pattern suitable for tissue characterization. These patterns can be identified by computer based pattern recognition techniques and employed for further image construction. To include the significant biological variance and to facilitate the estimation of the discrimination power, the classifiers should be improved in future studies by optimization of the various parameters, mainly by including larger data sets and the elaboration of reference data bases. Additionally, integration of non-spectral data into network training such as age, sex, tumor grading, etc. are planned.

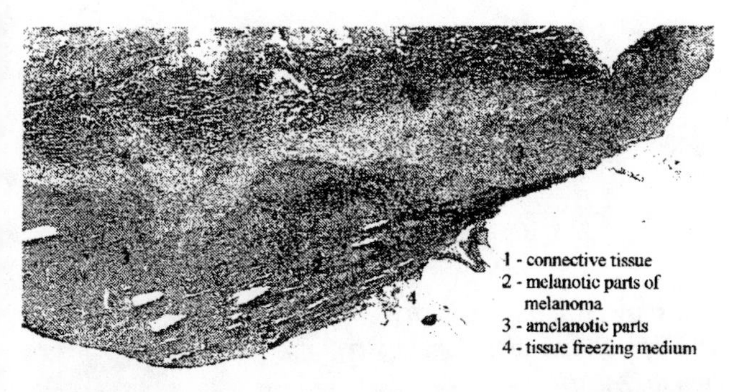

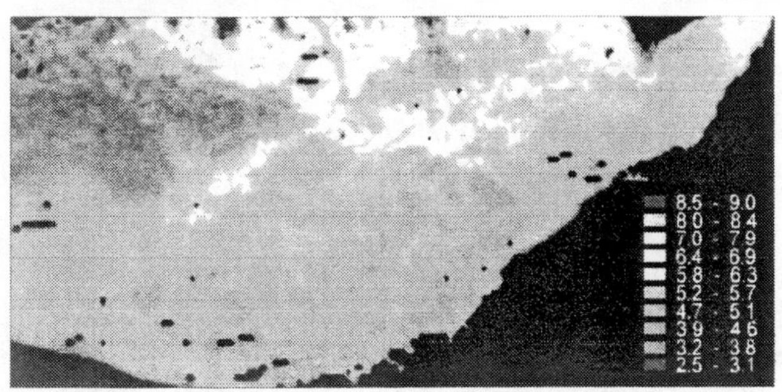

Figure 1. Comparison of images obtained on a human melanoma thin section in the visible (upper panel) and the IR range (lower panel, protein-lipid ratio). For this mapping experiment 176 (= 14 mm) spectra in x-direction, and 76 spectra (= 6 mm) in y-direction were recorded and analyzed. Areas of negative quality test results are indicated by black color.

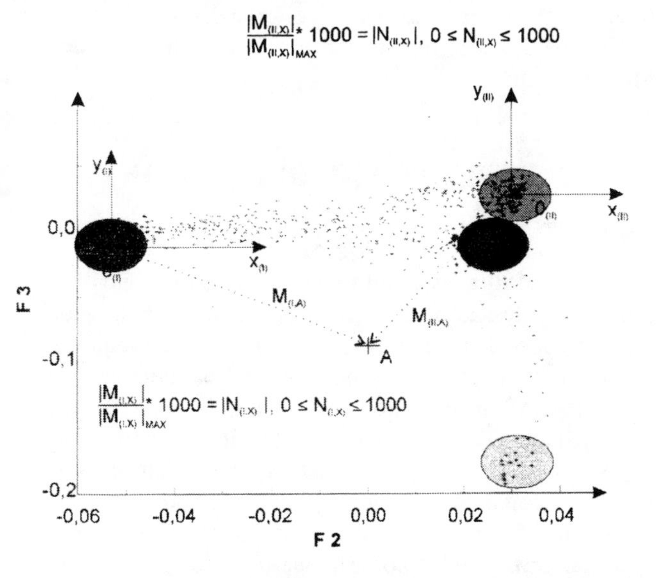

Figure 2. A two-dimensional PCA plot of human melanoma spectra. For projection of data, the second and the third principal component are used. Each of the spectra (normalized first derivatives in the spectral region from 1450-900 cm^{-1}) is depicted by a point. Four major populations of spectra are apparent. The normalized distances $N_{(i,X)}$ were utilized for image construction (see text).

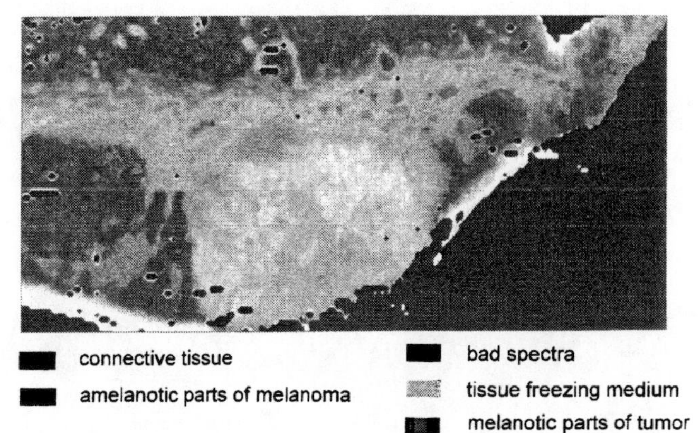

Figure 3. An example of enhanced image contrast using PCA for IR imaging. The tissue region corresponds exactly to that shown in Fig. 1. Patterns were defined as indicated.

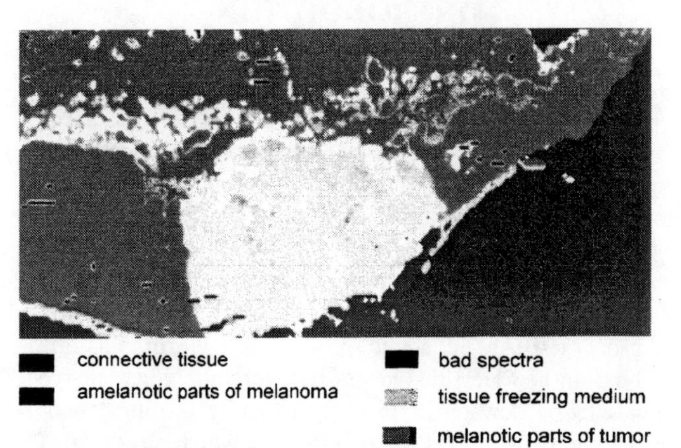

Figure 4. IR image constructed by ANN from FT-IR-mapping measurements data (again tissue thin section of a human melanoma, cf. Fig 1). Network training was performed using melanoma thin sections from 4 different patients. To reduce time needed for network training 4 spectral windows were defined (1392-1372, 1355-1240, 1233-1168, 1139-1084 cm^{-1}, respectively).

REFERENCES

1. Helm, D., Labischinski, H., and Naumann, D. *Journal Microbiological Methods* **14**. 127-141. (1991).
2. Naumann, D., Helm, D., and Labischinski, H., *Nature* **351**. 81. (1991)
3. Zell, A., Simulation Neuronaler Netze, Bonn: Addison-Wesley, 1994, ch. 8, pp. 105-106
4. Malins, D.C., Polissar, N.L., and Gunselman, S.J. *Proc. Natl. Acad. Sci. USA* **94**. 259-264. (1997)
5. Kidder, L.H., Kalasinsky, V.F., Luke, J.L., Levin, I.W., and Lewis, E.N. *Nature Medicine* **3**. 235-237. (1997)
6. Choo, L.P., Wetzel, D.L., Halliday, W.C., Jackson, M., LeVine, S.M., and Mantsch, H.H. *Biophysical J.* **71**. 1672-1679. (1996)
7. Rigas, B. and Wong, P.T.T., *Cancer Research* **52**. 84-88. (1992)
8. Rigas, B., Morgallo, I.S., Goldman, I.S., and Wong, P.T.T. *Proc. Natl. Acad. Sci. USA* **87,** 8140 (1990)
9. Wong, P.T.T. Lacelle, S., Yazdi, H.M., *Appl. Spectroscopy* **47**. 1830 (1993)

Structural and Temporal Behaviour of Biofilms Investigated by FTIR - ATR Spectroscopy

J. Schmitt*, U. P. Fringeli**, H.-C. Flemming*

** Dept. Aquatic Microbiology, Spectroscopy Group, University of Duisburg, IWW, Mülheim, Germany. ** Dept. Biophysical Chemistry, University of Vienna, Vienna, Austria*

The temporal and physiological behaviour of bacteria forming biofilms on a surface was investigated by FT-IR ATR spectroscopy. Time dependent spectra could be attributed to changes in biofilm properties. H-D exchange experiments offered insights in structural changes of biofilms after chemical treatment with chlorine, used as a desinfectant.

INTRODUCTION

Most bacteria on earth live in an immobilized form, embedded in extracellular polymeric substances (EPS) and adhere to surfaces in soils, sediments and many other habitats which provide interfaces, water and nutrients. This is referred to as biofilms. Their metabolic activity can change significantly with physico-chemical properties of the supporting material and the micro environment, e.g., influencing the kinetics of corrosion processes. An economic property of biofilms is the binding of water; this has to be overcome when sludge has to be dewatered. In order to study biofilms, non-destructive methods are required, in which structure of the matrix and concentration gradients are maintained. As a non-destructive, on-line and in situ method, FTIR-ATR is used for monitoring bacterial adherence, biofilm formation and development directly at the water/solid interface (1). In a specially designed ATR-optical device, biofilm studies can be performed over a period of days. Sample and reference spectra were obtained in quasi-real time in the single-beam, sample reference (SBSR) technique (2). Structural investigations of biofilms were made using H-D exchange.

METHODS

Biofilms were grown on the internal reflection element (IRE) of an ATR spectrometer cell after inoculation of the surface with bacteria in a continous flow through system. As test strains, a *Pseudomonas aeruginosa* SG22 and mixed cultures from a potable water system were used. Nutrients were supplied continously in the desired concentration (0.7, 1.5, 2 mg/l Caso bouillon). Based on the theoretical considerations about the electric field vector in the ATR-plate, sample and reference were measured on a single crystal without interference. This results in ATR-spectra, where long-term measurements can be performed without water/water vapour compensation and instrument instability. H-D exchange was performed in the fully hydrated biofilm through the gas phase, with an H_2O/air and D_2O/N_2 mixture at constant relative hymidity, temperature and the pH-value, the latter as far as possible. Measurements were performed on a Bruker IFS 66 FT-IR spectrometer (Bruker Karlsruhe) and an MCT-detector.

RESULTS AND DISCUSSION

Physiological studies

The IR spectrum of biofilms shows specific bands of proteins, polysaccharides, phosphoryl compounds and other groups of molecules (3,4). Biofilms are complex microbial systems and respond to nutritional conditions. Physiological changes are also possible in response to degraded and/or sorbed substances (1). It is of interest to know if and how much the activity may change and which concomitant physiological alterations occur. The detection of physiological and temporal changes of bacterial cells within biofilms due to different nutrient concentrations could be improved by the introduction of the SBSR-technique. No water and water vapour compensation is necessary in long term measurements,

CP430, *Fourier Transform Spectroscopy:* 11th International Conference
edited by J.A. de Haseth

which otherwise leads to unsatisfactory signal to noise ratios due to subtraction and further mathematical data treatment (e.g. derivatives) is limited.

Figure 1 gives a typical example of stacked spectra, representing biofilm growth. After only one hour the cells can be observed beginning to attach to the crystal surface. This is first indicated by the occurrence of the amide I and II bands. Within 24 hours a rapid increase in cell numbers on the surface takes place until a plateau phase is reached. The spectra demonstrate that the

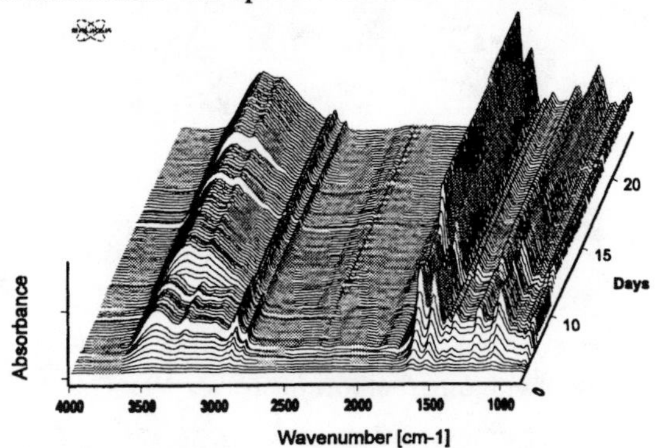

FIGURE 1. Stack plot of biofilm growth. ATR-FT-IR spectra aquired without any further data treatment (e.g.water subtraction, water vapour compensation)

intensity and composition of the bands between 1200 cm^{-1} and 900 cm^{-1} vary considerably during the measurement time of 120 hours according to the physiological response of the biofilm. These regions are mainly correlated with the formation of extracellular polymeric substances (EPS). This means that the physical and chemical properties of the EPS [5] may change more than expected during biofilm growth. Furthermore and with time, a decrease in EPS associated bands, a decrease in phosphate bands and differences in ratios and band shapes of amide I and II are observed. The decrease of the phospate bands is correlated by low cell activity. Changes in protein composition are manifested in morphological changes of the cells. They become smaller due to starvation.

H-D exchange in biofilms

It is well known that the spatial structure of biofilms change during growth. These changes were invesitgated by H-D exchange of the protons in the biofilm matrix, which consists significantly of bound water. Figure 2 shows the exchange kinetic of a mature biofilm.

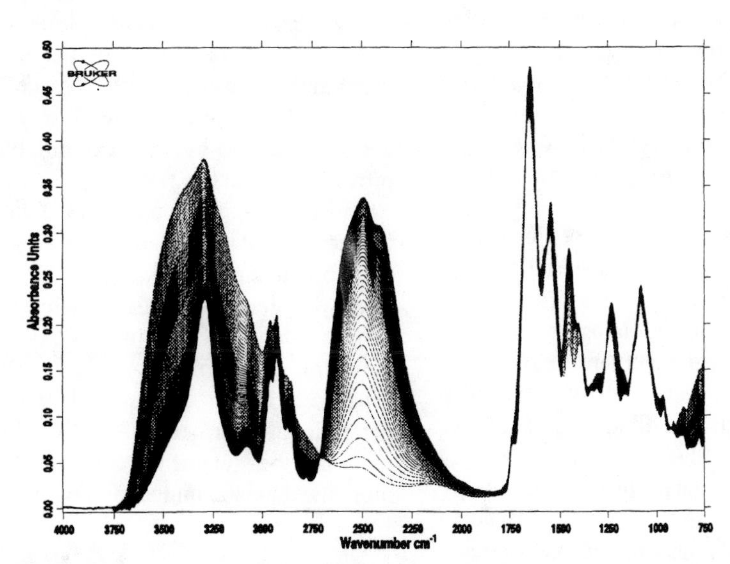

FIGURE 2. H-D exchange kinetic in a mature biofilm (5 days growth). Decrase in O-H (3400-3000 cm^{-1}) and increase in O-D bands (2700 -2000 cm^{-1})

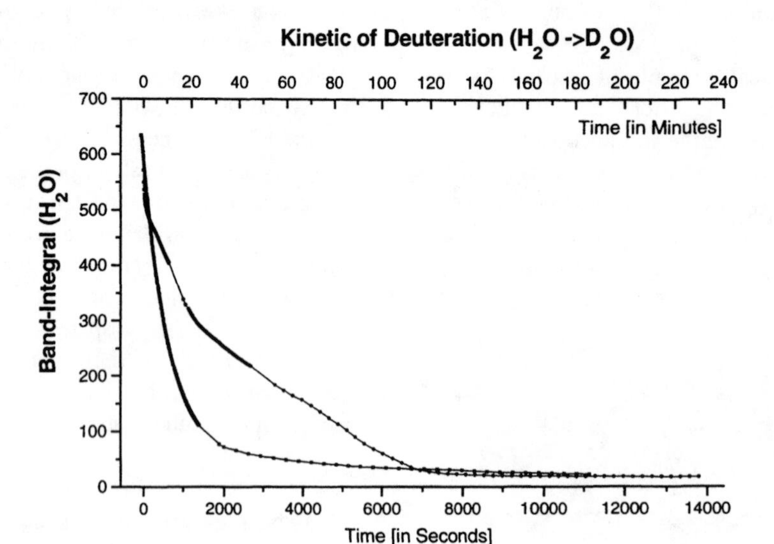

Figure.3. Exchange rates of biofilms treated with 6mg/l chlorine (below) and without (above).

The functional groups that are readily accessible are exchanged rapidly within the first 10 minutes of the deuteration process. The rate slows down between 15 and 30 minutes and finally falls off almost linearly. The exchange of the protons of interest is completed after 2 hours. Agents, which are suspected to alter the EPS matrix (6) show a significant change in the H-D exchange kinetics. Figure 3 shows the exchange rates of two biofilms, one treated with chlorine which is common in drinking water desinfection and without. The biofilms in drinking water, mixed culture biofilms, are grown under identical conditions for 2 days with 2 mg/l Caso bouillon.

It is obvious, that the exchange rate of the chlorinated biofilm is much faster, following an exponential decay. The untreated biofilm shows a multifunctional behaviour, with several underlying exchange processes. This suggests, that the chemical treatment of biofilms (e.g. chlorination) results in the destruction of coherence in the EPS matrix, as caused by the oxidative action of chlorine. The destruction of the EPS matrix again changes the physical and chemical properties within biofilms. This may result in different diffusion coefficients, in nutrient supply and in the mechanical resistance.

CONCLUSIONS

The complex matrix of biofilms was investigated by FT-IR / ATR spectroscopy biofilms in situ, on-line and non-destructively. The developement of a new ATR accessory enabled us to study the physiological behaviour of biofilms in long-term measurements, practically without limits of the spectrometer stability, water and water vapour interferences.

Spectral changes can be observed very sensitively and attributed to biochemical/mircobial reactions. This provides a tool for testing microorganisms attached to surfaces to chemicals used for cleaning, desinfection and in biomedical applications. Here, drug delivery and antibiotic testing is of interest.

The hydrogen/deuterium exchange in biofilms provides insight into the EPS matrix and hence in the spatial structure of biofilms. Different underlying exchange processes must be assumed which are still to be determined in further detail. Differences in the exchange rates can be quantified and attributed, resulting in a more mechanistic understanding of the biofilm matrix.

REFERENCES

1. Schmitt J., Nivens D.,White D.C., Flemming H.C: *Water Sci. & Technol.*, 32, 149-155, 1995.
2. Fringeli U.P, in Mirabella F.M. ed. *Internal Reflection Spectroscopy*, New York, Marcel Dekker, 1992, 255-323.
3. Schmitt J., Flemming H.C., in Heitz E., Flemming H.C., Sand W. (eds.*):Microbially Influenced Corrosion of Materials*, Berlin, Springer, 143-159.
4. Helm D., Naumann D.: *FEMS Microbiol. Letters*, 126,1995, 75-80
5. Christensen B.E., Charaklis W.G.,: in Characklis W.G., Marshall K.C. (eds.) Biofilms, NewYork, John Wiley, 1990,93-130.
6. Wingender J., Grobe S., Trüper H.G.: *J. Appl. Bacteriol.*, 79, 1995, 94-102.

Characterisation of Biomaterials using FT-Raman Spectroscopy

S.Söderholm[1], Y.H.Roos[1], N.Meinander[2] and M.Hotokka[3]

[1] *Department of Food Technology, P.O.Box 27, FIN-00014 University of Helsinki, Finland, email erik.soderholm@helsinki.fi*
[2] *Department of Physics, P.O.Box 9, FIN-00014 University of Helsinki, Finland, email meinander@phcu.helsinki.fi*
[3] *Department of Physical Chemistry, Åbo Akademi University, Porthansgatan 3, FIN-20500 Åbo, Finland, email matti.hotokka@abo.fi*

Carbohydrates play an important role in the quality and preservation of pharmaceutical and food materials. The storage temperature and water content is very critical in storage and, therefore, it is very important to understand how the physical state of carbohydrates is affected by water. Carbohydrates in foods and pharmaceuticals are usually present in the amorphous form even if other substances present affect the physical properties of carbohydrates it is mainly temperature and water content that determine the physical state. Amorphous carbohydrates show a second order phase transition, the glass transition, that is critical for stability. When carbohydrates are stored above their glass transition temperature they loose stability. Crystallization above the glass transition temperature may result in loss of quality. Raman spectroscopy offers a useful tool in the characterisation of phase transitions and effects of temperature and water content on material properties at a molecular level.

INTRODUCTION

Carbohydrates, such as sucrose, form an amorphous structure that has a second order phase transition where the molecular mobility is rapidly increased. This phase transition, i. e. the glass transition, is strongly sensitive to water content and temperature of the material. Traditionally, methods, such as TMA (thermal mechanical analysis), DEA (dielectric thermal analysis), DTA (differential thermal analysis), and DSC (differential scanning calorimetry), have been used by Roos (1) to study the physical state and the above mentioned phase transition of carbohydrates.

Raman spectroscopy is a well-known technique used as an analysis method of polymers (2). Raman spectra contain information of the overall molecular mobility and can therefore be used to detect the glass transition and the physical state of various materials. This is a rather new approach to understanding molecular changes in amorphous biomaterials and their phase transitions.

By analysing spectra of amorphous sugars with different water content and temperature we determined which spectral bands were affected by intermolecular interactions. These modes could function as indicators of the physical state of the material. They were also helpful in producing knowledge for understanding the "glass transition" - phenomenon in amorphous carbohydrates.

In the present study Raman spectra were determined for amorphous sucrose.

EXPERIMENTAL

Crystalline carbohydrates were obtained from Sigma. Amorphous sucrose was prepared by freeze-drying a 20% water solution in glass ampoules. The water content of the samples were then adjusted by storage over saturated salt solutions in vacuum desiccators. The salts used to obtain 11% and 23% relative storage humidity were LiCl and CH_3COOK. Dry samples were stored over P_2O_5. After storage the glass ampoules were sealed and transported to the Raman spectrometry laboratory for measurement.

For the purpose of temperature adjustment we constructed a sample holder/sample heater that consists of a heat sink, a Peltier heater/cooler, a glass ampoule holder and a controller unit. The heat sink is essentially a hollow compartment with inlet and outlet for water circulation. The purpose of this heat sink was to keep the backside of the Peltier element at a constant temperature at all times. The controller unit delivered the power needed for the Peltier device and kept the temperature at the desired level. Using this sample holder we measured Raman spectra directly through the walls of the sealed glass ampoules covering a temperature range of 5 to 95°C.

The spectrometer used was a Bruker IFS 66 with a FRA 106 FT-Raman Accessory unit that used the 1.06 μm laserline of a Nd:YAG laser working at a maximum output of 350 mW. The detector used was a Ge-diode. The spectra were all recorded using a resolution of 4 cm^{-1} over a spectral range of 50-4000 cm^{-1}.

The temperature gradient in the measurement was 0.25 °C / min and the spectra were recorded at intervals of 2°C. The laser power was kept at 90%, i. e. approx. 310 mW.

CP430, *Fourier Transform Spectroscopy:* 11th International Conference
edited by J.A. de Haseth

RESULTS

Raman spectra of amorphous, freeze-dried sucrose for three different temperatures are shown in Fig. 1a-c.

Due to the fact that the vibrational bands for a crystalline material experience much stricter selection rules the peaks became much narrower and showed large frequency shifts. In Fig. 1a-c the crystallization effect is clearly evident.

Analyzing the complete Raman spectrum of amorphous sucrose was a tedious task and we found only few earlier studies. Mathlouthi et al. (3) investigated sucrose-water interaction in a 60 % sucrose solution and identified the strongest vibrational bands. Due to the complex structure of the molecule the band assignments, however, were rather crude estimates. We should be able to locate and assign all vibrational modes more thoroughly using modern molecular modeling software. In this study, however, we used the vibrational assignments published by Mathlouthi et al.(3).

We made a fairly complete analysis of the Raman spectrum of sucrose and isolated over 50 vibrational bands. Some of these are shown in Fig. 2. The complete list of wavenumbers and relative intensities is shown in Table 1.

When analyzing the Raman spectra as a function of temperature and water content we have found some significant differences in the Raman intensities of well defined wavenumber intervals. The overall Raman intensity changes during the phase transition are shown in Fig. 3.

We located areas in the spectra which were especially sensitive to T_g and the glass transition. No large shifts in the peak wavenumbers were observed, but small changes in the relative intensities of various vibrational bands were observed. The relative intensities were calculated in relation to the overall Raman intensity. Changes in the relative intensities of CH_2 vibrations, i. e. wagging, bending, and rocking are shown in Fig. 4a-c.

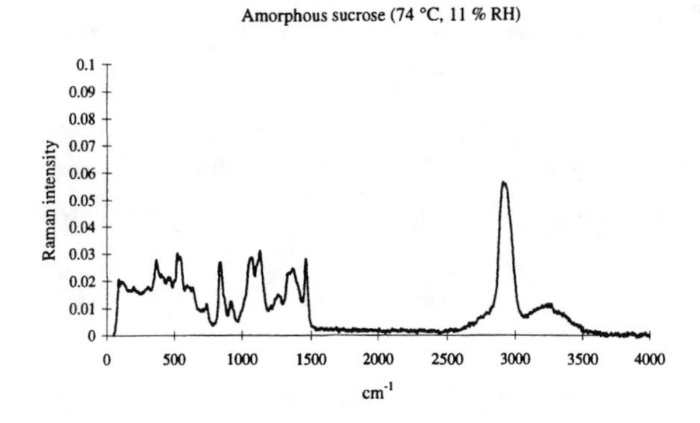

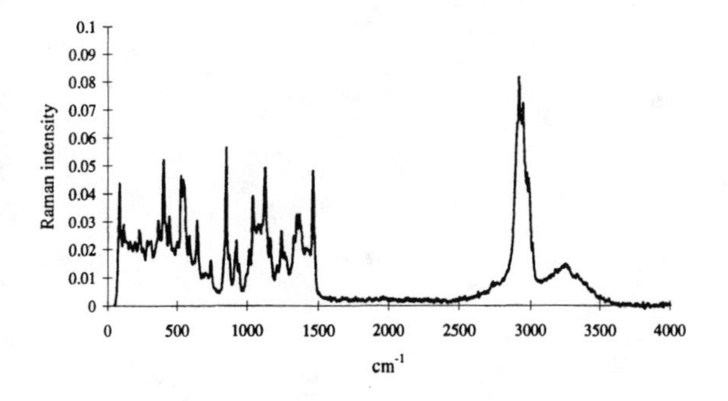

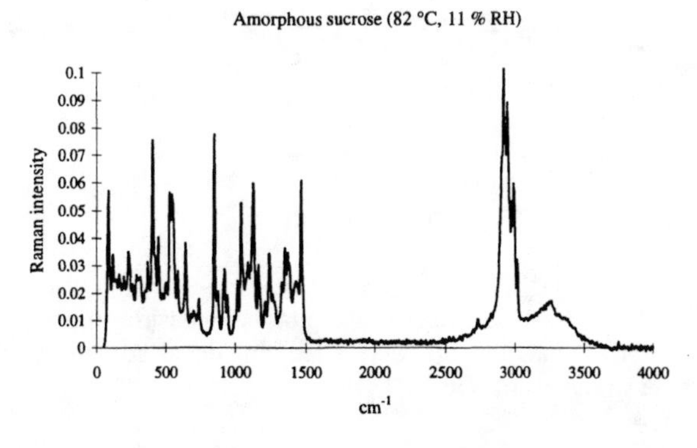

FIGURES 1a-c. Raman spectra of amorphous sucrose for the spectral range of 50-4000 cm^{-1}. The materials were stored over 11 % relative humidity at room temperature before sealing. The sample temperatures are 74, 78 and 82 °C.

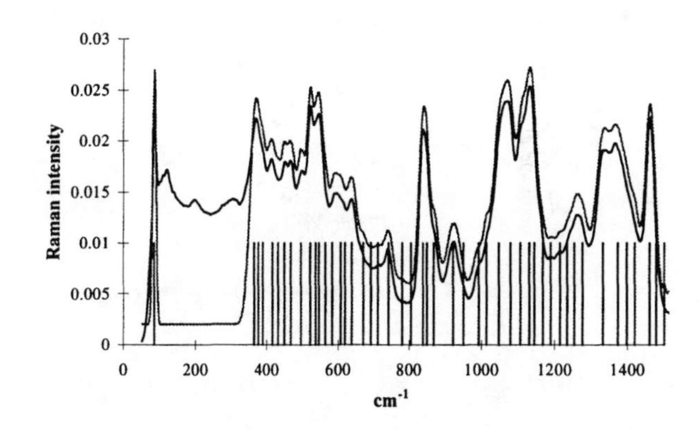

FIGURE 2. Calculated and measured Raman spectra with located vibrational modes. The calculated spectra was shifted slightly upwards for better visualization.

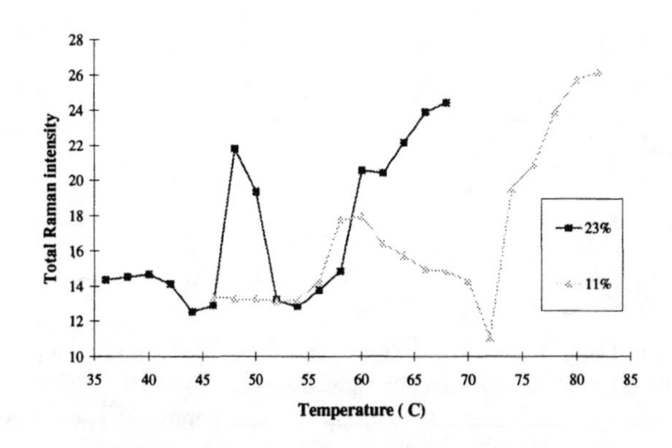

FIGURE 3. The glass transition could be observed from a strong increase in the overall Raman intensity at 46 °C (23%) and 56 °C (11%), respectively.

TABLE 1. Wavenumbers and relative intensities in the Raman spectra of freeze-dried amorphous sucrose.

cm^{-1}	Intensity	cm^{-1}	Intensity	cm^{-1}	Intensity
85.60	9.71	711.09	5.64	1214.92	7.49
363.48	14.26	742.33	11.84	1235.06	4.78
376.38	17.08	780.00	3.54	1254.00	8.40
388.74	1.43	805.00	3.31	1276.92	11.79
417.10	25.87	838.00	20.05	1332.48	34.58
433.22	0.55	850.00	1.17	1373.91	20.36
450.15	10.79	868.56	11.11	1398.00	7.88
469.23	9.88	921.68	17.45	1420.14	10.04
496.56	19.83	950.03	1.91	1459.79	24.42
523.18	13.81	992.18	10.25	1477.28	0.21
537.40	0.57	1013.09	2.35	1548.95	17.29
547.00	19.48	1047.71	30.38	1499.65	0.22
565.35	4.46	1079.14	21.31	2760.00	37.74
584.00	11.90	1106.04	4.57	2840.00	13.66
606.93	10.89	1131.12	40.44	2900.31	100.00
620.00	0.42	1145.02	0.62	2952.90	98.79
638.01	16.90	1168.00	4.13	2980.89	21.65
670.07	6.78	1190.00	6.94	3241.59	97.00
691.44	4.20				

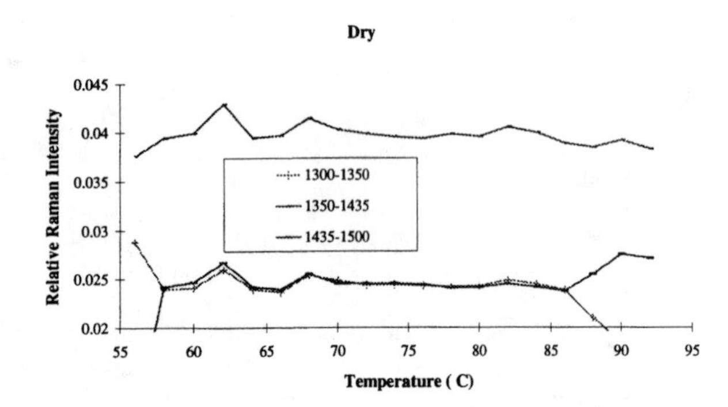

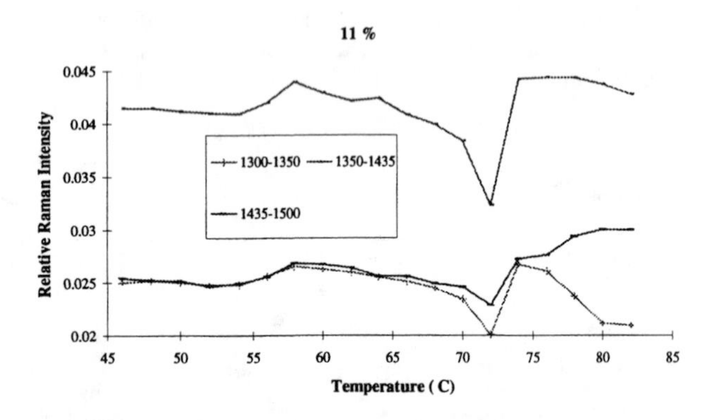

FIGURE 4 a-b. The relative intensities of the three areas in the spectra, which usually are assigned to CH_2 vibrations (wagging, rocking and stretching), for amorphous sucrose stored at 0% and 11% RH are shown to function as indicators of the glass transition at 54 and 66 °C.

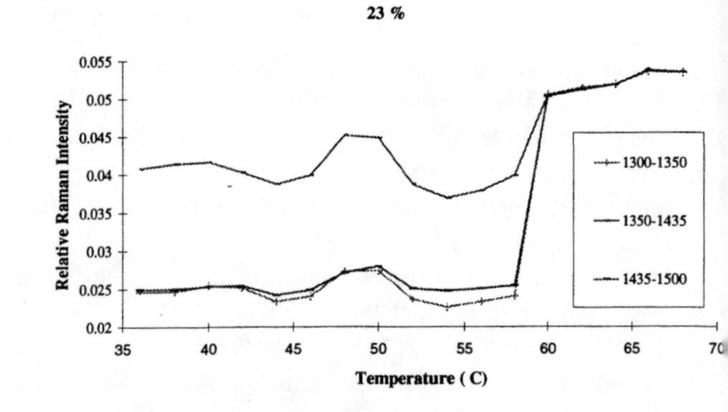

FIGURE 4 c. The relative intensities of the three areas in the spectra, which usually are assigned to CH_2 vibrations (wagging, rocking and stretching), for amorphous sucrose stored at 23% RH are shown to function as indicators of the glass transition at 46 °C.

The changes in the relative intensities of the vibrations which were more closely coupled to the -O-H groups in sucrose, such as δ(C-C-O), ν(C-O), are shown in Fig. 5a-c.

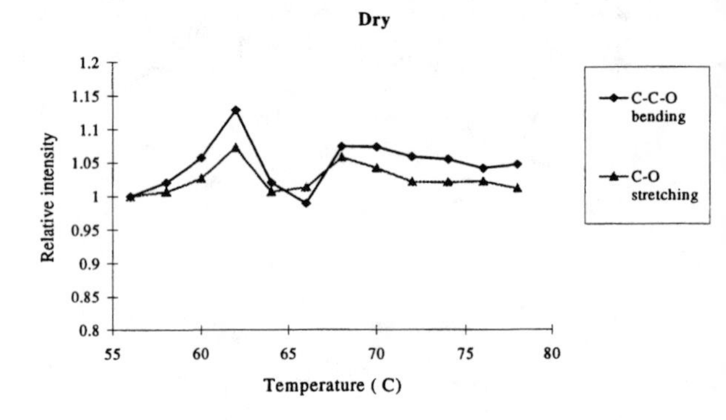

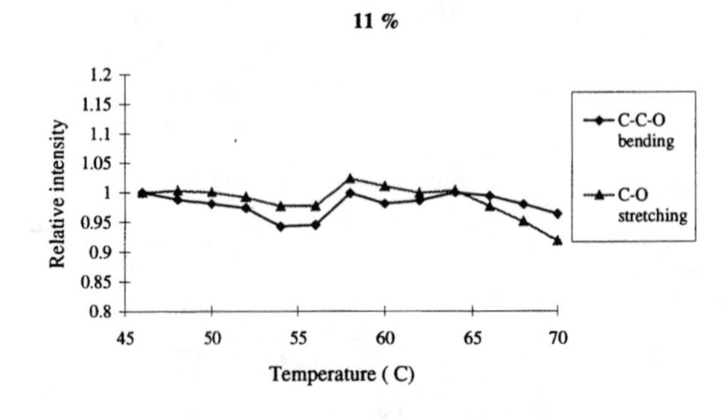

FIGURE 5 a-b. Changes in the relative intensities of the C-C-O bending band and the C-O stretching band for amorphous sucrose stored at 0% and 11% RH.

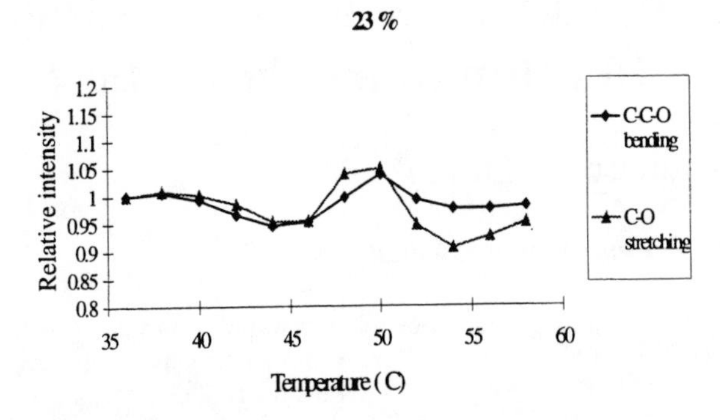

FIGURE 5 c. Changes in the relative intensities of the C-C-O bending band and the C-O stretching band for amorphous sucrose stored at 23% RH.

The Raman peak located at approximately 85 cm^{-1} also seemed to be a good indicator of the glass transition. When calculating the intensity of this peak relative to the total Raman intensity we observed a rapid change at the onset of the state transition and a remaining lower relative intensity of amorphous sucrose above T_g. This is shown in Fig. 6.

DISCUSSION

The second order state transition known as the glass transition in sucrose, occurs when the molecular mobility increases rapidly. This is due to changes in the inter-molecular bonding configurations. Below the glass transition temperature, T_g, the substance is a solid with the molecules "locked in place" and above T_g the substance becomes a syrup with rapidly decreasing viscosity. This occurs when the molecules can move more and more freely. One would expect that vibrational modes involving H atoms are those that will be most affected by changes in the molecular environment due to hydrogen bonding to neighboring molecules. Such modes are for example the C-C-O bending mode, C-O stretching mode.

These were the spectral regions that seemed to be most sensitive to the state transition. We have also noticed that the low wavenumber region is a good indicator of the state

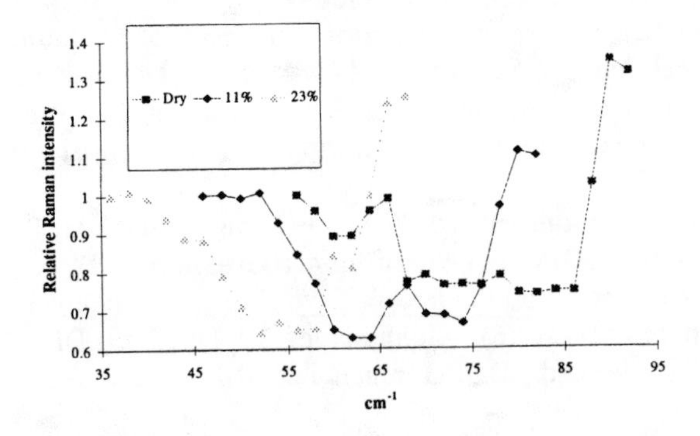

FIGURE 6. Intensity of Raman peak at 85 cm^{-1} relative to the total intensity.

transition. This was also expected because large cluster vibrations mainly contribute to this region.

The temperature referred to in all measurement was not necessarily the actual sample temperature, because of problems with heat conduction in the sample holder and the glass ampoule. The onset temperature of the glass transition of dry, 11% and 23% sucrose have been measured to be 56.6, 37.4 and 27.9 °C by Roos et al. (1,4). Depending on what we used as an indicator of the glass transition we can determine the corresponding temperatures to be 66, 54 and 46 °C.

We conclude that Raman spectroscopy is a promising method to determine the glass transition and investigating how it affects molecular vibrations and molecular mobility in carbohydrates. Much future work has still to be done in the field of molecular modeling to understand the actual processes of this state transition. Measuring this type of spectra on other types of carbohydrates will probably give more information. When a more complete understanding of the molecular changes are achieved we are planning to investigate more closely the glass separation that takes place in carbohydrate matrixes using the FT-Raman microscope.

REFERENCES

1. Roos, Y.H., *Carbohydr. Res.* **238**, 39-48 (1993).
2. Stuart, B.H., *Vibr. Spectr.* **10**, 79-87 (1996).
3. Mathlouthi, M., Luu, C., Meffroy-Biget, M.A. and Luu, D.V., *Carbohydr. Res.* **81**, 213-223 (1980).
4. Roos, Y.H. and Karel, M., *J. Food Sci.* **56(1)**, 38-43 (1991).

FTIR External Reflectance Studies of Lipid Monolayers at the Air-Water Interface: Applications to Pulmonary Surfactant

Jennifer M. Wilkin and Richard A. Dluhy

Department of Chemistry, University of Georgia, Athens, GA 30602

FTIR external reflectance spectra of monomolecular films of natural products and model mixtures relevant to pulmonary surfactant physiology were collected concurrently with surface measurements directly at the air-water interface. Films studied were calf lung surfactant extract (CLSE) and its phospholipid fraction (PPL) along with 2:1 DPPC-d_{62}:DPPG and 2:1 DPPC-d_{62}:DOPG containing 0, 1 or 2 wt% of the hydrophobic surfactant proteins SP-B and SP-C (SP-B+C). The CH_2 antisymmetric and symmetric stretching bands (~2920 and 2852 cm^{-1}) along with the analogous CD_2 stretching bands (~2194 and 2089 cm^{-1}) were analyzed, and band heights, integrated intensities and peak frequency positions were plotted as a function of measured surface pressure. Data suggest that 2:1 DPPC-d_{62}:DPPG + 2wt% SP-B+C is the most ordered and stable of the films and can be compressed to the highest sustainable surface pressure. Data from the model mixtures indicate that the surfactant protein interacts differently with each of the lipid components. Plots of the CH/CD intensity ratios versus surface pressure show an increase in this ratio upon the addition of SP-B+C as the protein apparently orders the CH component (DPPG or DOPG) and slightly disorders the CD component (DPPC-d_{62}).

INTRODUCTION

First proposed by von Neergaard in 1929 (1), pulmonary surfactant is a substance composed of lipids and proteins that lines the lungs preventing alveolar collapse during respiration and providing a first lung defense against air-borne infectious agents and irritants. Clements studied the surface properties of monolayer films of surfactant extracts on a Langmuir film balance, and attributed the stabilizing function of the surfactant to a lowering of the surface tension at the air-alveolar interface to values less than 10 dynes/cm (2). A lack of pulmonary surfactant is the chief cause of respiratory distress syndrome (RDS) found in preterm infants (3) and the related adult respiratory distress syndrome (ARDS). Surfactant replacement therapies are currently employed, and the development of more effective synthetic exogenous surfactants is dependent upon gaining a better understanding of how the individual surfactant components interact and function on a molecular level.

Surfactant is secreted from type II epithelial cells and is composed of 90 % lipid and 10% protein by weight. More than 80% of the lipid is phospholipid (4) with the most abundant single lipid being dipalmitoylphosphatidyl-choline (DPPC). Phosphoglycerols are the next largest phospholipid class comprising as much as 5 to 10% of the phospholipid population (5). Also found in mammalian pulmonary surfactant are four proteins specific to the lung: SP-A, SP-B, SP-C, and SP-D. SP-B and SP-C are low molecular weight (8.7 kDa and 4.2 kDa respectively) (6), hydrophobic proteins believed to help maintain low surface tensions in the film and to promote efficient spreading of the phospholipid components.

Monomolecular films of phospholipids at the air/water interface of a Langmuir film balance have long been used by researchers as model systems for studying pulmonary surfactant. To be effective, a model surfactant must be able to withstand high surface pressures (low surface tensions) during film compression and to respread quickly during film expansion. Films of DPPC alone will fulfill this first criterion, but do not respread quickly enough to function as effective exogenous surfactants in vivo. It is believed that unsaturated lipids, inorganic ions, and surfactant proteins SP-B and SP-C present in natural surfactant promote respreading of the DPPC component (7). Presented here are external reflectance FTIR spectroscopic data collected at the air/water interface which was used to investigate the effect these surfactant proteins have on binary phospholipid films of DPPC-d_{62} with DPPG and DOPG.

EXPERIMENTAL

DPPG, DOPG, and acyl chain perdeuterated DPPC-d_{62} were purchased from Avanti Polar Lipids (Alabaster, AL) and were used as received. Calf lung surfactant extract (CLSE), its total phospholipid fraction (PPL), and a mixture of the hydrophobic surfactant proteins B and C (SP-B+C) were generously provided by Dr. Robert H. Notter (Department of Pediatrics, University of Rochester Medical School). Subphase H_2O was obtained in-lab from a Barnstead (Dubuque, IA) ROpure/Nanopure reverse osmosis/deionization system having a nominal resistivity of 18 MΩ cm.

Stock solutions of DPPC-d_{62} (~1 mg/ml in $CHCl_3$), DPPG and DOPG (~1 mg/ml in 4:1 $CHCl_3$:MeOH) were prepared and concentrations verified by inorganic phosphorus assay (8). Solutions of 2:1 DPPC-d_{62}:DPPG and 2:1 DPPC-d_{62}:DOPG containing 0, 1, or 2 wt% SP-

CP430, *Fourier Transform Spectroscopy:* 11th International Conference
edited by J.A. de Haseth

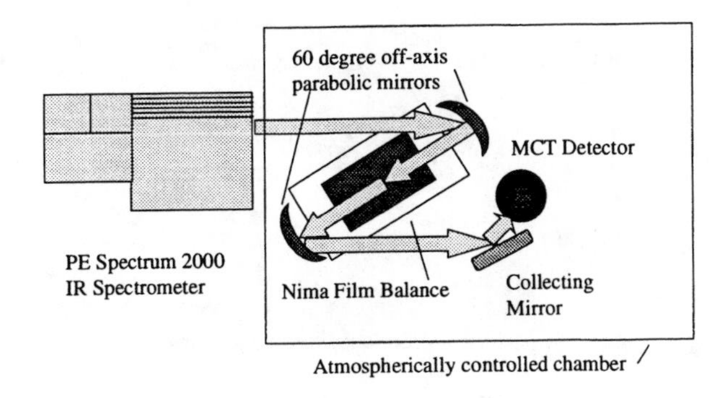

FIGURE 1. In-situ IR external reflection-absorption experimental design.

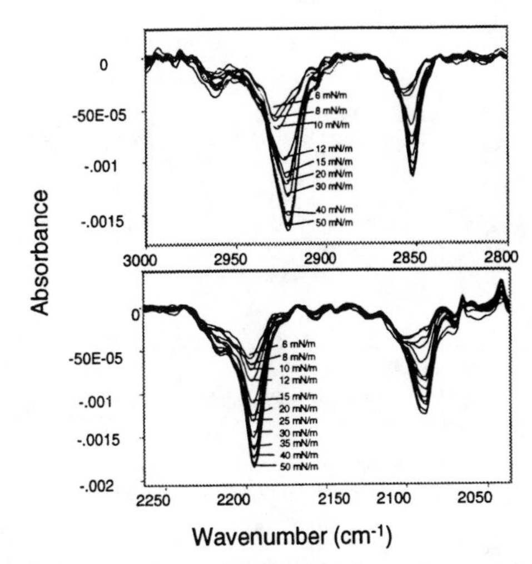

FIGURE 2. IR external reflection-absorption spectra of a 2:1 DPPC-d_{62}:DPPG binary monolayer film collected at various surface pressures. CH_2 (top) and CD_2 (bottom) stretching regions are illustrated.

B+C were prepared by mixing the appropriate amounts of the above lipid stock solutions and a 0.18 mg/ml stock solution of SP-B+C in 1:1 $CHCl_3$:MeOH. Aliquots of CLSE (2.5 mg/ml in $CHCl_3$) and PPL (11.5 mg/ml in $CHCl_3$) were diluted to ~1 mg/ml with $CHCl_3$ before use. The subphase used for all experiments was 150 mM NaCl in deionized H_2O (pH 5.6).

Figure 1 shows the details of the instrument design. Absorbance spectra are collected with a Perkin Elmer Spectrum 2000 FTIR spectrometer equipped with an external sample beam option. A sixty-degree, gold-coated, off-axis parabolic mirror (Janos Technology Inc., Townshend, VT) reflects the beam from the spectrometer onto the surface of a Nima 601M film balance (Coventry, England) at an incidence angle of 30°. The beam reflects off the subphase, sampling the film, and a second parabolic mirror collects the beam and directs it onto a collection mirror and then into a liquid N_2-cooled HgCdTe detector (EG&G Judson, Montgomeryville, PA). The film balance, optical components, and detector are housed in a sealed, Plexiglas chamber controlling the sensing environment and improving background subtraction of water vapor.

The subphase was first cleaned by aspiration and a single beam spectrum was collected for use as a background. The subphase temperature was held constant at 22 ± 1 °C by flowing thermostatted water through the hollow body of the trough. The temperature in the enclosed chamber was typically 24°C and the relative humidity remained fairly constant at 70%. Typically 5-10 µl of sample were spread via syringe onto the trough surface resulting in a surface pressure of 3-6 mN/m. The film was allowed to equilibrate for a period of 30 minutes and then was compressed intermittently and spectra collected over a range of surface pressures from ~5 mN/m to a maximum of 45-65 mN/m, depending on the nature of the film. Typically, 1024 scans were collected at 4 cm^{-1} resolution. Interferograms were apodized using a strong Norton-Beer function before being Fourier transformed. Spectra were not smoothed but were baseline corrected using GRAMS/32 (Galactic Industries Incorporated, Salem, NH) before peak positions and intensities were determined. Vibrational frequencies were calculated using a center of gravity algorithm (9).

RESULTS AND DISCUSSION

External reflectance spectra of the methylene stretching region (3000-2800 cm^{-1}) and the analogous CD_2 stretching region (2250-2050 cm^{-1}) for a monolayer of 2:1 DPPC-d_{62}:DPPG compressed over a range of surface pressures is shown in Fig. 2. As the film is compressed, the average area per molecule is reduced, resulting in an increase in the peak intensity. As the pressure of the film increases, an ordering of the acyl chains occurs and the frequencies of the peak maxima are observed to shift to lower wavenumbers. The values for both the CH_2 (2920 and 2852 cm^{-1}) and CD_2 (2194 and 2089 cm^{-1}) antisymmetric and symmetric stretching bands at a surface pressure of 50 mN/m indicate that both phospholipid species in the binary film exist in a relatively ordered state. The negative peaks seen in these absorbance spectra are theoretically predicted due to the low value for the attenuation constant of the complex refractive index for the water substrate (10). Similar spectra are seen for 2:1 DPPC-d_{62}:DOPG and for the natural products, CLSE and PPL.

Shown in Fig. 3 (top and bottom) are the antisymmetric CH_2 stretching vibration frequency and band height plotted against measured surface pressure for 2:1 DPPC-d_{62}:DOPG with and without surfactant protein added. Antisymmetric bands were chosen for analysis due to the higher signal to noise ratio than the corresponding symmetric bands. This was crucial for the low surface pressures where average molecular areas are high and peak intensities are inherently weak. Band height data was chosen over integrated intensity data due to the lesser influence of the partially overlapping CH_3/CD_3 stretching bands. Data were collected

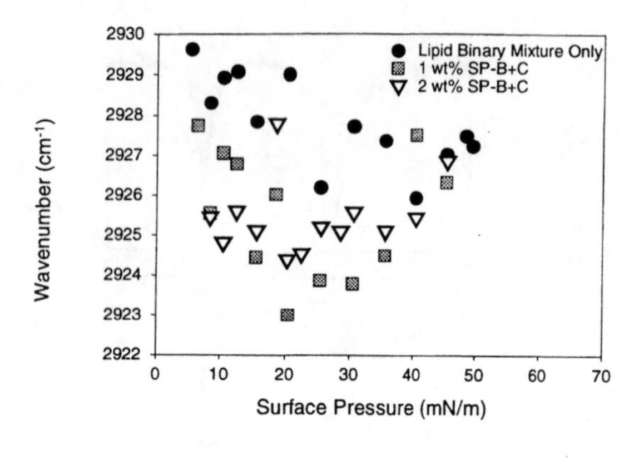

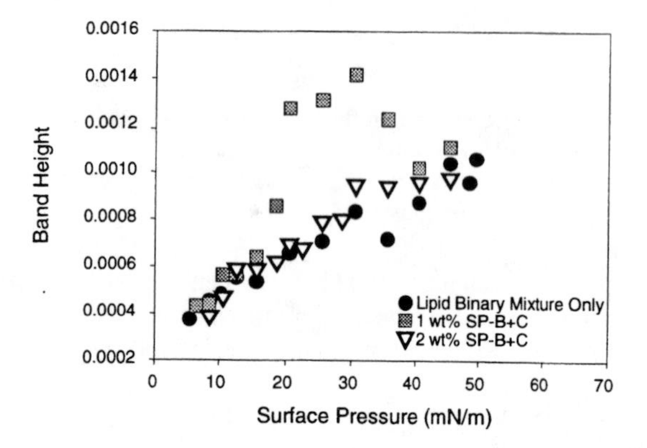

FIGURE 3. IR external reflection-absorption spectra band parameters for a 2:1 DPPC-d_{62}:DOPG binary monolayer film. Variation in frequency (top) and band height (bottom) for the CH_2 antisymmetric stretching band are plotted as a function of surface pressure.

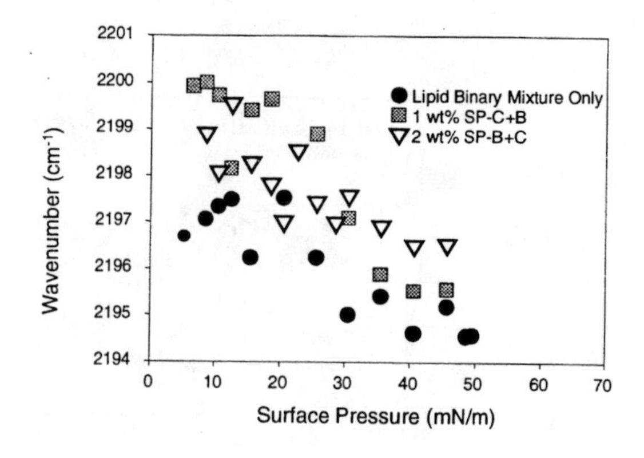

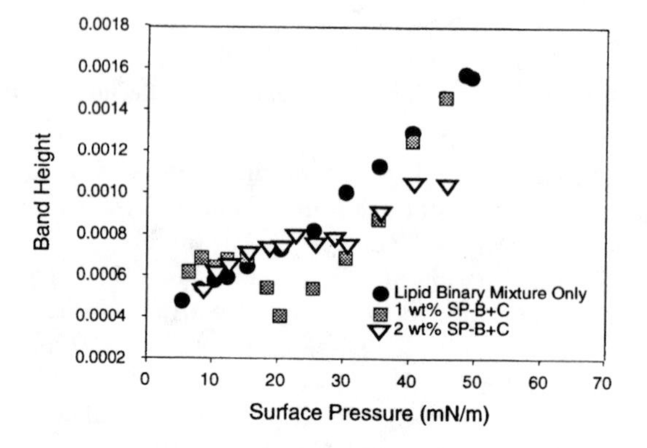

FIGURE 4. IR external reflection-absorption spectra band parameters for a 2:1 DPPC-d_{62}:DOPG binary monolayer film. Variation in frequency (top) and band height (bottom) for the CD_2 antisymmetric stretching band are plotted as a function of surface pressure.

for surface pressures from 5 to 50 mN/m (where the film begins to collapse due to the thermodynamic properties of the DOPG component). The lipid only film shows a steady lowering of the frequency value as the surface pressure increases and the DOPG acyl chains become more ordered (more all-trans configuration). This ordering, along with a decrease in average molecular area causes the band height to increase with pressure as is seen in Fig. 3 (bottom). Similarly, Fig. 4 (top and bottom) shows this same decrease in frequency and increase in band height with increased pressure for the antisymmetric CD_2 stretching band of the DPPC-d_{62} component of the binary film.

Adding surfactant protein appears to affect the CH and CD components of the film differently. The addition of 1 wt% protein shifts the antisymmetric CH_2 stretching peak to lower wavenumber values at all surface pressures, but with maximum displacements of 3 to 5 cm^{-1} occurring in the range from 15–35 mN/m. This effect is also clearly seen in Fig. 3 (bottom) as a marked increase in the band height for pressures in the 15–35 mN/m range. The addition of 2 wt% SP-(B+C) appears to also lower the frequency of the DOPG antisymmetric stretching band, but the magnitude of the decrease is constant over the entire range of pressures, as no maximum lowering is observed in either the wavenumber or band height plots. Overall, the addition of both 1 and 2 wt% protein acts to order the DOPG lipid acyl chains.

Addition of surfactant protein has the opposite effect on the deuterated lipid, DPPC-d_{62}, as is evidenced in Fig. 4. The addition of both 1 and 2 wt% SP-(B+C) causes approximately a 2 cm^{-1} increase in peak frequency over the entire range of surface pressures. The protein is also observed to shift the band heights to lower values than those seen for the lipid binary mixture only at pressures higher than 20 mN/m. A small dip in the band height data occurs at pressures from 15– 30 mN/m with the addition of 1 wt% protein, paralleling the trend seen in the DOPG data. The presence of surfactant protein appears to disorder the acyl chains of the DPPC-d_{62} component in the film, opposite of what was seen for the protiated component, DOPG.

Due to the presence of unsaturated sites in the acyl chains of DOPG, the thermodynamic properties of DPPC-d_{62}:DOPG monomolecular films are very different from those of films of DPPC-d_{62}:DPPG. Films of 2:1 DPPC-d_{62}:DPPG are more stable at high surface pressures, and were compressed to values near 70 mN/m before collapse. The exact nature of the plots of frequency and band height

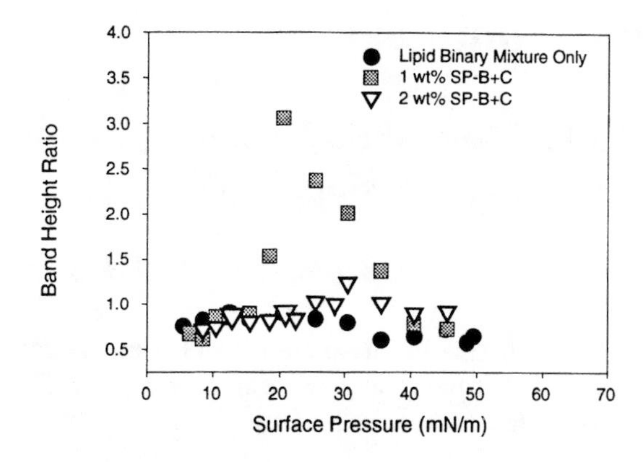

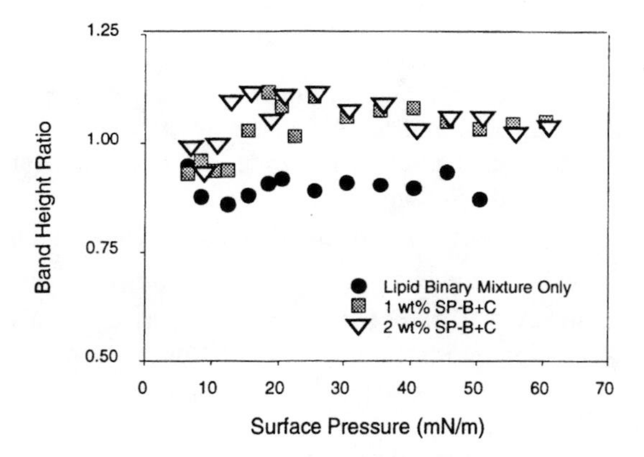

FIGURE 5. IR external reflection-absorption spectra band parameters for (top) a 2:1 DPPC-d_{62}:DOPG binary monolayer film, and (bottom) a 2:1 DPPC-d_{62}:DPPG binary monolayer film. Variation in band height ratio for the CH_2 / CD_2 antisymmetric stretching bands is plotted as a function of surface pressure.

versus surface pressure (data not shown) were not the same for the DPPG mixture as they were for the DOPG, but the overall trends upon addition of surfactant protein were alike, where a similar ordering of the CH component and slight disordering of the CD component occurred.

The ratio of the band heights for the CH_2 and CD_2 antisymmetric stretching vibration plotted as a function of applied surface pressure is shown in Fig. 5 for 2:1 DPPC-d_{62}:DOPG (top) and 2:1 DPPC-d_{62}:DPPG (bottom). For the lipid binary mixtures only, the CH/CD ratio remains constant at a value near one for all surface pressures. Because the extinction coefficient for protiated lipids is roughly twice that for deuterated ones, this value is as expected. When surfactant protein is added, however, the preferential ordering of the CH components and disordering of the CD components can be seen clearly as an increase in CH/CD ratio. This effect is large for the DOPG mixture upon addition of 1wt% protein and fades out when further protein is added, but both 1 and 2 wt% protein cause a marked increase for the DPPG mixture.

Figure 6 compares the behavior of natural surfactant products with the two model mixtures discussed above. From this plot of CH_2 antisymmetric stretching frequency

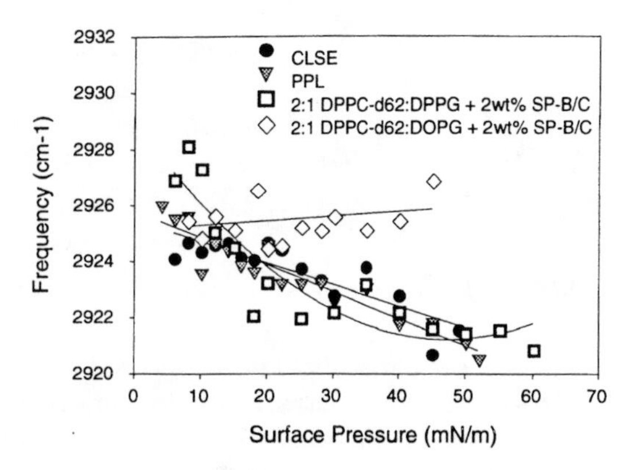

FIGURE 6. Frequency of antisymmetric CH_2 stretching vibration at 2920 cm^{-1} is plotted as a function of surface pressure for several different monolayer films of natural surfactant preparations and phospholipid model membranes.

versus surface pressure, it is noted that 2:1 DPPC-d_{62}:DPPG more resembles the natural products and causes the largest lowering of frequency. The DPPG model mixture was the most stable of the films and compressed to the highest surface pressure. It is noted that not much difference exists between the CLSE and PPL data. The absence of SP-B and SP-C in the PPL sample does not appear to cause any marked change in the appearance of the data and may possibly be due to incomplete extraction of the surfactant proteins.

ACKNOWLEDGEMENT

The work described here was supported by NIH grant GM40117 (R.A.D.)

REFERENCES

1. Von Neergaard, K. *Z. Gesamte. Exp. Med,.* **66**, 373-394, 1929.
2. Clements, J.A. *Proc. Soc. Exp. Biol. Med.,* **95**, 170-172, 1957.
3. Avery, M.E., Mead, J., *Am. J. Dis. Child,* **97**, 517-523, 1959.
4. King, R.J., Clements, J.A., In *Handbook of Physiology. Circulation and Nonrespiratory Functions,* Bethesda, MD, American Physiological Society, 1985, 309-336.
5. Notter, R.H., Finkelstein, *J. Appl. Physiol.* **57**, 1613-1624, 1984.
6. Curstedt, T., Johansson, J., Persson, P., Eklund, A., Robertson, B., Jöornvall, H. *Proc. Natl. Acad. Sci.USA.,* **87**, 2985-2989, 1990.
7. Wang, Z., Hall, S.B., Notter, R.H. *J. Lipid Res.,* **37**, 790-798, 1996.
8. Chen, P.S.; Toribara, T.Y.; Warner, H. *Anal. Chem.,* **28**, 1756-1758, 1956.
9. Cameron, D.G., Kauppinen, J.K.;,Moffat, D.J., Mantsch, H.H., *Appl. Spectrosc.,* **36**, 245-249, 1983.
10. Dluhy, R.A., *J. Phys. Chem.,* **90**, 1373-1379, 1986.

Synthesis and FT-IR Study of Ln-Glucose-Pyridine Complexes

Lianying Zheng, Ying Zhao, Ruifa Zong, Shifu Weng, Saijun Li, Yanmin Yang, Luqin Yang, Jinguang Wu

College of Chemistry and Molecular Engineering, Peking University, Beijing 100871, P.R. China

In this present work, $LnCl_3$-Glucose-Pyridine complexes (Ln=Sm, Eu, Er) have been synthesized and characterized by Fourier Transform Infrared Spectra. The results demonstrate that glucose and pyridine were coordinated to lanthanide ions successfully. In addition, the formula of the complex $[Sm(Glu)(Py)Cl_3]$ has been confirmed by elemental analysis.

INTRODUCTION

The chemistry and applications of rare earth have long been a keynote field in China, because China's rare earth resource accounts more than half of that of the world. In the last decades, great progress has been made in the application of rare earth, especially, as animal food additives and micro-fertilizer in agriculture. The results show that rare earth products can significantly increase the production of crops and meat as well as milk. However, these rare earth may enter human body through the food chain, and their effect on human health remains uncertain. On the other hand, saccharide is one of the most important substances in human body. Therefore, the study on the interaction between lanthanide ions and sugar will be great help to understand the bioactivity of rare earth.

Much work on the coordination of lanthanide-sugar complexes has been done in our previous work. Due to the complicated configurations of saccharide, the strong internal and external hydrogen bondings of the sugar, the syntheses of lanthanide-saccharide complexes in water or water-organic mixture solvents will be very difficult. Up to now, the preparation of single crystal of rare earth-sugar complexes has not been reported. FTIR spectra can provide much information on molecular structure. In this paper, the title complexes were synthesized and their FTIR spectra were studied to investigate the interaction between lanthanide ion and glucose.

EXPERIMENTS

Chemicals and Instruments

Sm_2O_3, Eu_2O_3, Er_2O_3 are 99.9% pure or better.
Hydrochloride acid: A.R; Glucose: C.P;
Pyridine: A.R; Alcohol: A.R
Nicolet Magna IR 750 Spectrometer.

Synthesis

First, lanthanide oxide was dissolved in 1:1 hydrochloric acid. Then the solution was concentrated and diluted with water respectively; and such operation had been repeated for several times until the pH of the solution was about 5-6. Hydrated lanthanide trichloride crystallized from the concentrated solution.

2g glucose in 20ml of pyridine was heated (less than 80 ℃) with continuous stirring for several hours to obtain solution A; 0.01mol of lanthanide trichloride was dissolved in 20ml of pyridine to obtain solution B. With vigorous stirring, solution A was added to solution B. The mixture was kept warm and stirred for two or three days until the precipitates were formed from it. The precipitates were washed with ethanol and desiccated in vacuum.

Physical Measurements

FT-IR spectra were performed on Nicolet Magna IR 750 Spectrometer. Mid-IR spectra were recorded in KBr pellets with a resolution of $4cm^{-1}$ and Far-IR spectra were recorded in cesium iodide pellet with a resolution of $4cm^{-1}$.

RESULTS AND DISCUSSIONS

General Properties

The complexes of Ln-Glucose-Pyridine are yellow-brown , sticky and more hygroscopic than glucose . Like saccharide, these complexes are very soluble in water, but hardly soluble in alcohol and ether. The composition of Sm-Glucose-Pyridine complex which is formulated as $[Sm(Glu)(Py)Cl_3]$ has been confirmed by elemental analysis.

Studies on FT-IR Spectra

The FT-IR spectra of the complexes resemble each other but differ from those of the ligands (glucose and pyridine).

In the spectra of complexes (Figure 1-3 and Table 1), the O-H vibration absorptions of glucose at 3448, 3390, 3322 and 3271 cm^{-1}, which were assigned to the O-H vibration of

CP430, *Fourier Transform Spectroscopy:* 11th International Conference
edited by J.A. de Haseth

and 3271 cm^{-1} , which were assigned to the O-H vibration of crystal water and the O-H absorption of glucose, can hardly be distinguished. This fact indicated that the H-bond network were weakened due to the lanthanide ions coordinated to glucose. The five C-H vibration absorption bands of glucose and pyridine in the range of 2800cm^{-1}-3000cm^{-1} were overlapped by the relatively strong O-H vibration absorptions in the complexes.

The three absorption bands of glucose in 1500-1300cm^{-1} were assigned to CH_2, COH, CCH, OCH bending vibrations and their coupling, while in the complexes, the shift and broadening of these bands showed the changes of the conformation of the sugar ring. The changes may be due to the coordination of glucose to lanthanide ions.

Several peaks in the spectrum of glucose in the range of 1300cm^{-1}-1100cm^{-1} can be observed. The two sharp absorption peaks at 1156cm^{-1} and 1111cm^{-1} were caused by two different C-O stretching vibrations. The former at 1156cm^{-1} may be assigned to the stretching vibration of C-O in the pyranose ring, and the latter (at 1111cm^{-1}) to the C-O stretching vibration of COH. The absorptions at 1210cm^{-1}, 1234cm^{-1} and 1250cm^{-1} were assigned to the deformation vibrations of OCH, while in the spectra of the title complexes, a broad band at about 1250cm^{-1} with a shoulder band at 1199cm^{-1} in this range can be seen. The shift to lower wavenumber and broadening of absorption bands revealed that the C-O stretching vibration in the complexes is not as free as that in the glucose.

It is very interesting to note that the interactions between lanthanide ions and glucose and pyridine can also be obviously shown in FIR (Figure 4 and Table 2). The far infrared spectra of the complexes only have a broad absorption band at 216cm^{-1}, which may be attributed to the Ln-O and Ln-N stretching vibrations. This means that the original crystal lattice vibration of glucose has vanished and the lanthanide ions are coordinated to the glucose.

The coordination of pyridine to lanthanide ions can also be confirmed by comparing the spectrum of pyridine (Figure 1) with those of the complexes (Figure 3) in the following two ranges. First, it is worth noticing that in the range of 1500-1700cm^{-1}, the former shows four peaks at 1685, 1633, 1597 and 1581m^{-1} of $\nu_{C\text{-}C}$,$\nu_{C\text{-}N}$, while the latter gives two or three peaks(see Table 1). secondly, in the range of 1400-1500cm^{-1} there are two absorption bands of $\nu_{C=C}$,$\nu_{C\text{-}N}$ at the 1481 and 1437cm^{-1} in the spectrum of pyridine ,while the band of 1437 cm^{-1} disappears in those of the complexes. On the basis of these changes between the spectra of pyridine and its complexes, it may be concluded that lanthanide ion is coordinated with pyridine

TABLE 1. IR Data of Glucose ,Pyridine and their Complexes(cm^{-1})

Glucose	ErGluPy	EuGluPy	SmGluPy	Pyridine
3448				3145
3388	3301	3354	3384	3078
3319	3222			3052
3265	3103		3106	3025
2972				3001
2962	2955	2956		2954
2936	2884	2885	2887	2929
2901	2853	2854		2910
2883				
				1685
	1633	1633	1634	1633
	1608	1610	1610	1597
		1572		1581
	1536	1534	1535	
	1485	1485	1485	1481
1432				1437
1375	1390	1397	1395	1375
1333	1334	1332	1333	-
1250	1250	1254	1253	1216
1234				
1210	1198		1199	
1156	1164			1146
1100	1101	1097		-
1094	1080	1081	1083	
1071				
1048	1056	1054	1051	1068
1031	1028	1029	1028	1030
1024				
1013				
915	908	900	900	
853				991
771	749	749	750	748
716				

TABLE 2. Far IR Data of Glucose and its Complexes(cm^{-1})

D-Glucose	556	519	460	421	402	321	273	251	170
Er-Glu-Py								216	

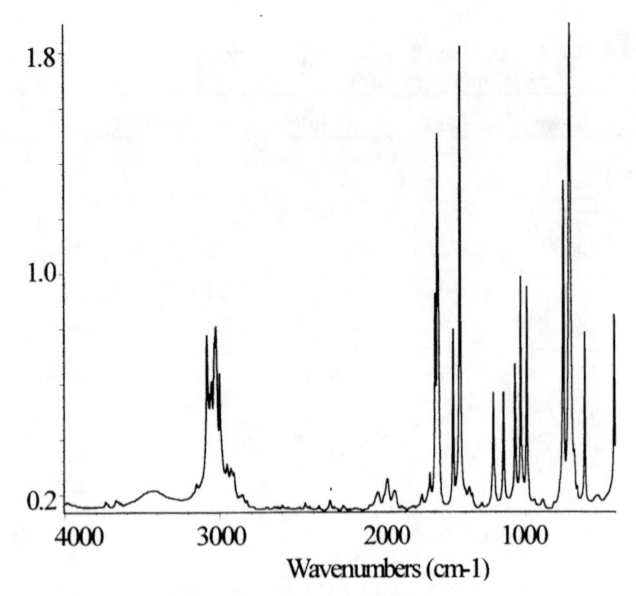

FIGURE. 1. Mid Infrared Spectra of Pyridine

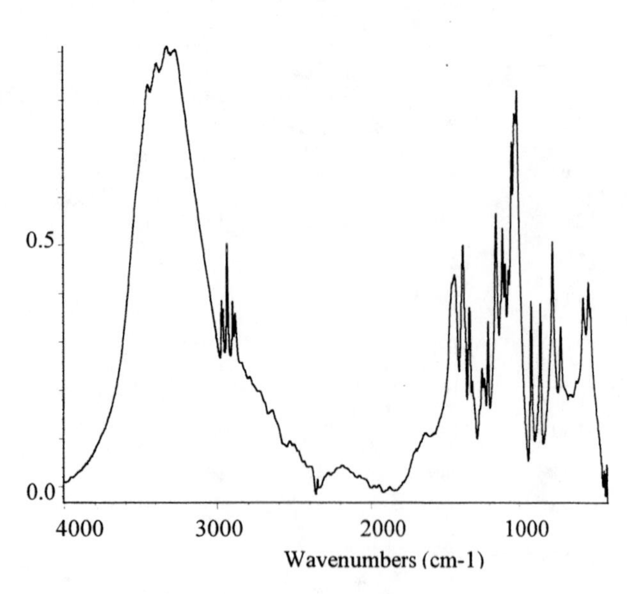

FIGURE 2. Mid Infrared Spectra of D-glucose

CONCLUSIONS

The title complexes have been successfully synthesized and characterized by infrared spectroscopic method. These studies show that strong interaction between lanthanide cation and glucose molecular exists in the complexes. The interaction between lanthanide ion and glucose have some effects on the conformation of the sugar, and different lanthanide ions have similar properties on the coordination with sugar.

ACKNOWLEDGMENTS

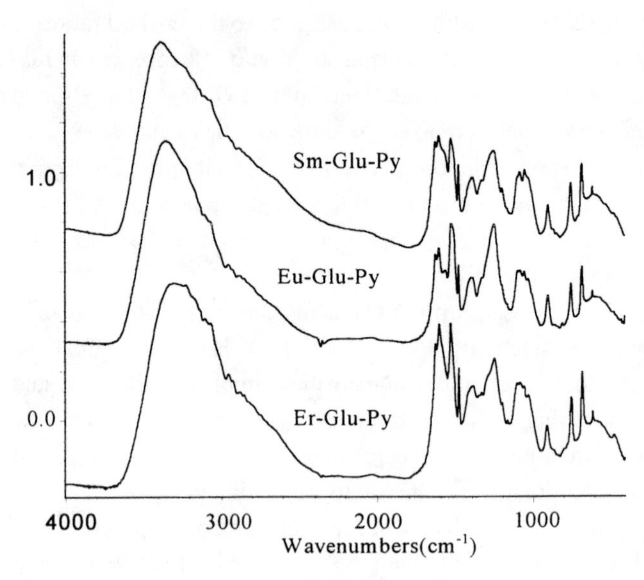

FIGURE 3. Mid Infrared Spectra of Title Complexes

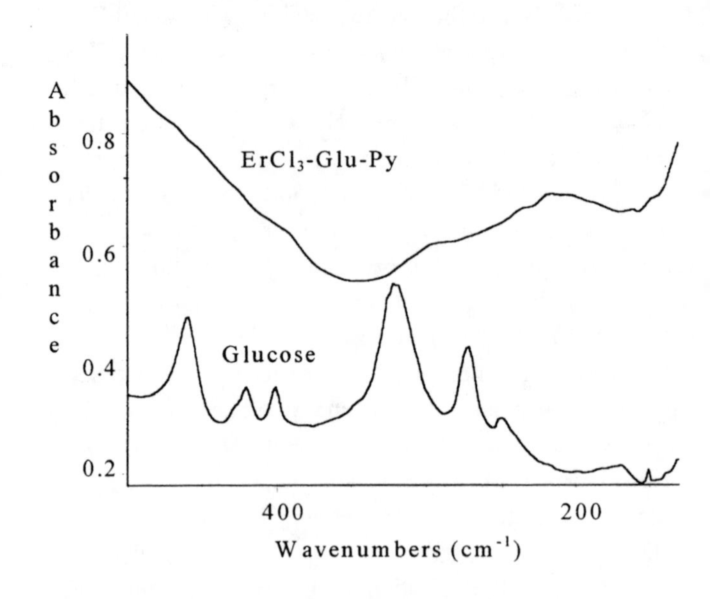

FIGURE 4. Far IR of D-glucose and $Er(Glu)(Py)Cl_3$

The support of the National Natural Sciences Foundation of China (NO. 29671002) is gratefully acknowledged.

REFERENCES

1. Masakazu Hineno .Carbohydr. Res., 1977, 56, 219-227
2. Zhou Qiao, Thesis, Peking University, 1994
3. Shigenobu Yano, Coordination Chemistry Reviews, 1988, 92, 113-156
4. H.A.Tajmir-Riahi, Carbohydr. Res., 1988, 172, 1-10

FTIR Spectroscopy and Sequence Prediction: Structure Of Human α_2-Macroglobulin

Rina K. Dukor[1], Michael N. Liebman[1], Anna I. Yuan[2] and Richard D. Feinman[2]

[1]*Vysis, Inc., 3100 Woodcreek Drive, Downers Grove, IL 60515;*
[2]*State University of New York Health Science Center at Brooklyn, Brooklyn, NY 11203*

The structure of a plasma proteinase inhibitor α_2-Macroglobulin (α_2m) is determined by FTIR spectroscopy and a number of sequence-structure prediction algorithms. In addition, α_2M dimers and complexes with methylamine and trypsin are examined. Our FTIR results estimate a helix content of 5-15% and a β-sheet content of 28-36%. None of the sequence prediction algorithms used in this study predicted values close to experimental data. Considerable differences in the FTIR spectra of α_2M dimer are observed and somewhat smaller changes are seen upon reaction of α_2M with methylamine and dithiodipyridine (DTP).

INTRODUCTION

α_2-macroglobulin (α_2m) is a major protease inhibitor in human blood plasma. It is a tetrameric protein of identical 185 kDa subunits where pairs of subunits are disulfide bonded (half-molecules) and half-molecules are non-covalently associated. When α_2m binds to proteinases, the resulting complex is rapidly cleared from the blood stream. Proteinases cleave one or more peptide bonds in the so-called 'bait' region of α_2m, resulting in tertiary structure changes and activation of an internal thioester that forms covalent bonds with proteinases (1). These thioesters are unique and have been found only in α_2m and complement proteins C3 and C4 (2). In addition to proteinases, α_2m has been shown to react with amines forming amides with the carboxyl of the thioesters and inducing a conformational change similar to that of proteinases. The structure of the α_2m and effects of binding of proteases and methylamine have been studied previously by circular dichroism (CD) (3,4), difference absorption spectroscopy (4), stopped-flow fluorescence spectroscopy (4) and small angle x-ray scattering (5). Percent secondary structure of α_2-macroglobulin and its complexes have been estimated based on the CD data, but reports are conflicting. All estimates agree on a high β-sheet content but not on the percent α-helix. In this study, Fourier Transform Infrared Spectroscopy (FTIR) is used to determine the structure of α_2m, its complex with methylamine, trypsin and α_2m dimers. Also, a number of sequence-to-structure prediction algorithms are used to estimate the secondary structure of all samples for comparison with experimental FTIR results.

MATERIALS AND METHODS

α_2-macroglobulin was purified from human blood as described previously (6). Samples were dissolved in sodium citrate buffer (pH=7) at a concentration in the range of 8-80 mg/ml. Solutions were placed in a cell with CaF_2 windows separated by a 6μm mylar spacer. FTIR data were recorded on a BOMEM MB-100 spectrophotometer, equipped with a DTGS detector at a resolution of 2 cm^{-1}. Data were analyzed with a software package called Prota™ (Vysis/Bomem). Prota™ includes a database of FTIR spectra of 37 proteins with known crystallographic structures and a factor analysis algorithm (7) that are used for estimation of secondary structure. Sequence predictions (Chou-Fasman and Garnier algorithms) were performed using the Cameleon program (Oxford Molecular) or by the methods of Solovyev and Strelets (available on the Internet).

RESULTS AND DISCUSSION

Figure 1 shows an FTIR absorption (bottom), Fourier-self deconvolution (middle) and second derivative (top) spectra of α_2-macroglobulin. The main features are bands at 1638 cm^{-1} and a sharp band at 1690 cm^{-1}, with shoulders at 1634, 1648, 1657, 1666 and 1680 cm^{-1}. If one were to use the frequency assignments alone, the peaks at 1634-1638 cm^{-1} would be assigned to a β-sheet structure and that would be consistent with the relatively high degree of β-structure determined by CD and earlier predictions from sequence (8). Table I summarizes factor analysis results based on FTIR spectroscopy, CD estimates and predictions from sequence. Our FTIR results estimate a helix content

CP430, *Fourier Transform Spectroscopy:* 11th International Conference
edited by J.A. de Haseth
 1-56396-746-4/98/$15.00

of 5-15% and a β-sheet of 28-36% and are consistent with some of the CD estimates (3) but not others (4). None of the sequence prediction algorithms predicted values close to the experimental data except for a composite (8). Upon reaction with methylamine, the spectrum changes little. This is consistent with other studies in the literature that the extreme changes in the tertiary structure observed upon methylamine binding are due to reorganization of intact secondary substructures rather than to a rearrangement of secondary structural motifs. Larger differences are observed in the spectrum of α_2m-methylamine treated with dithiodipyridine (DTP) indicating significant increase in the β-sheet structure and a decrease in the helix. Considerable differences are also observed in the spectrum of α_2m dimer. An unusual feature of the FTIR spectrum of α_2m is a very sharp peak at 1690 cm^{-1}. Although at the appropriate frequency for thioester, the peak persists in samples treated with methylamine and DTP, although both compounds cleave the thiol group. The 1690 cm^{-1} peak appears broader in the spectrum of dimers prepared by reduction and it is also apparent in the spectra of complexes with trypsin, suggesting that the structures that gives rise to the 1690 cm^{-1} absorption band are not part of the proteinase binding site.

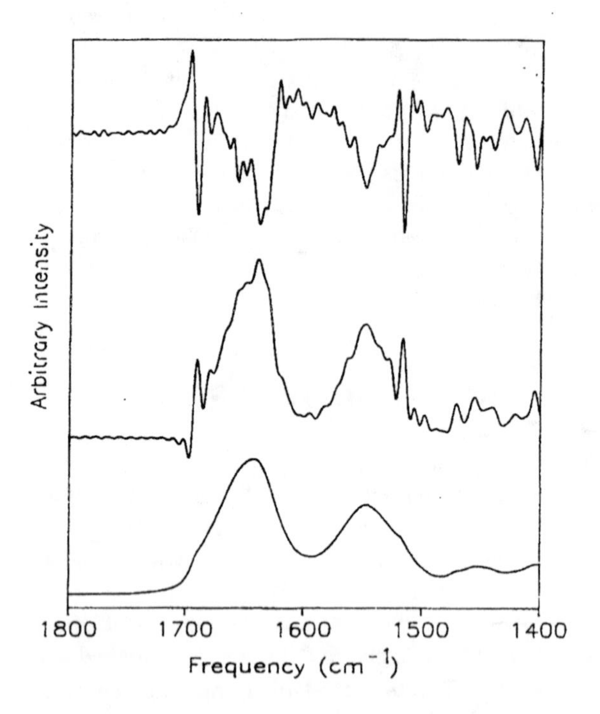

FIGURE 1. FTIR absorption (bottom), Fourier-self deconvolution (middle) and 2nd derivative (top) spectra of α_2-macroglobulin.

TABLE 1. Secondary Structure estimate for α_2-macroglobulin

	% helix	%sheet
Experimental Data		
FTIR	5-15	28-36
CD[3]	5	35-60
Sequence Prediction		
Chou & Fasman	32	46
Garnier et al	24	18
Solovyev	19	84
Strelets	31	11
Composite (Garnier & Lenstra)[8]	9	44

CONCLUSION

The importance of α_2-macroglobulin stems from its function in blood coagulation and other properties such as zinc binding capacity. Understanding its interaction with proteinases, and specifically the structural changes accompanying such an interaction, might shed some light on understanding the mechanism of fast clearance from the blood and α_2m's other functions. α_2m is a very large protein and its exact three dimensional structure has not been determined. Using FTIR spectroscopy combined with factor analysis, we have shown that the structure of α_2m consists of a high amount of β-sheet and a small amount of helix. This is in agreement with some previous studies and in contrast with others. The sequence prediction algorithms tested here do not agree with our FTIR results. However, these algorithms can be improved if the experimental data, such as FTIR or CD, are considered. These algorithms are presently being developed by several groups and hopefully soon will come to fruition, thus enabling structure determination that correlates with sequence.

REFERENCES

1. Feinman, R. D. *Annals N.Y. Acad. Sci 737*, 245 (1994)
2. Howard, J. B. *J. Biol. Chem. 255*, 7082 (1980)
3. Bjork, I. and Fish, W. W. *Biochem. J.* 207, 347 (1982)
4. Dangott, L. J., Puett, D., and Cunningham, L.W. *Biochemistry 22*, 3653 (1983)
5. Osterberg, R. and Malmensten, B. *Eur. J. Biochem. 143*, 541 (1984)
6. Chen, B., Wang, D., Yuan, A. I., and Feinman, R.D. *Biochemistry 31*, 8960 (1992)
7. Pancoska, P., Yasui, S.C., and Keiderling, T.A. *Biochemistry 30*, 5089 (1991)
8. Welinder, G., Mikkelsen, L. and Sottrup-Jensen, L. *J. Biol. Chem. 259*, 8328 (1984)

Electrochemically Induced FTIR Difference Spectra of the Cytochrome *c* Oxidase from *Paracoccus denitrificans*

Direct Evidence for the Protonation of Glu 278 upon Electron Transfer

P. Hellwig [a], C. Ostermeier [b], B. Ludwig [c], H. Michel [a], W. Mäntele [a]

[a] *Institut für Physikalische Chemie der Universität Erlangen, Egerlandstrasse 3, D-91058 Erlangen*
[b] *Max-Planck-Institut für Biophysik, Heinrich-Hoffmann-Strasse 7, D-60528 Frankfurt*
[c] *Institut für Biochemie, Marie-Curie-Strasse 9, D-60439 Frankfurt*

INTRODUCTION

The cytochrome *c* oxidase, the terminal enzyme of the respiratory chain, catalyses the stepwise reduction of oxygen to water. In this process electron and proton transfer is efficiently coupled to contribute to the formation of a proton gradient which drives ATP synthesis. There are four redox-active sites involved in the electron transfer: Cu_A accepts electrons from cytochrom *c*, heme *a* accepts the electrons from Cu_A and transfers them to the binuclear site, Cu_B and heme a_3, where O_2 is activated.

Two different types of protons are supposed to be involved in cytochrome *c* oxidase catalysis: "scalar" protons must be available to the binuclear center in order to form water; "vectorial" protons are pumped across the membrane to contribute to the generation of the proton gradient. For the translocation of the protons two separate pathways have been indicated from side-directed mutants in which proton translocation could apparently be prevented whilst still allowing oxygen reduction.

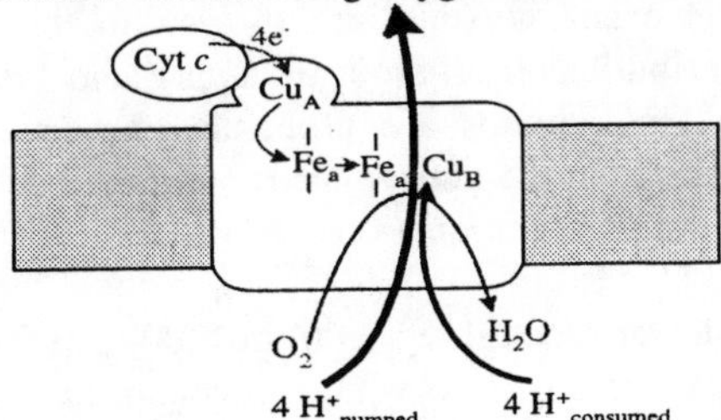

FIGURE 1: Schematic illustration of the cytochrome *c* oxidase

Recently the four-subunit cytochrome *c* oxidase from *Paracoccus denitrificans* has been crystallographically characterized (2) and the 13-subunit bovine heart cytochrome *c* oxidase has been crystallized (3) allowing a more detailed study of the mechanism of this redox-driven proton pump.

We have investigated the redox driven proton transfer of the cytochrome *c* oxidase from *Paracoccus denitrificans* by combination of FTIR difference spectroscopy and electrochemistry in a thin layer cell under anaerobic conditions. The highly sensitive redox induced FTIR difference spectra describe the reorganization of the cofactors and their protein site upon electron transfer, including conformational changes in the protein and protonation of amino acid side chain groups. Moreover, through combination of protein electrochemistry, UV/vis, and FTIR difference spectroscopy, individual cofactors can be adressed by selecting the appropriate electrode potential.

MATERIALS AND METHODS

Sample Preparation

The cytochrome *c* oxidase from *Paracoccus denitrificans* was prepared as reported previously.(1,2) For spectroelectrochemistry, a 0.5 mM solution of the sample in a 200 mM phosphate buffer (pH 6.9), 100 mM KCl as supporting electrolyte and n-decyl-β-D-maltopyranoside was used. For $^1H/^2H$-exchange the cytochrome *c* oxidase was repeatedly washed in ^2H_2O buffer. The exchange was found to be better than 80% as estimated from the shift of the amide II band in the FTIR absorbance spectrum.

Spectroelectrochemistry

The spectroelectrochemical cell for the UV/vis and IR range was used as described before.(4) With a pathlength of 6-9 μm, sufficient transmission in the range from 2000 - 1000 cm^{-1}, even for the strong water absorption at 1645 cm^{-1}, was achieved. The gold grid working electrode was chemically modified in a 2 mM cysteamine solution for one hour and then carefully washed with deionized water

CP430, *Fourier Transform Spectroscopy:* 11th International Conference
edited by J.A. de Haseth

(5). In order to accelerate the redox reactions, a mixture of several mediators (1) at a concentration of 40 μM for each mediator was added.

FTIR and UV/vis difference spectra as a function of the applied potential were obtained **simultaneously** combining IR and vis beams as previously described (1). For a difference spectrum, the sample was equilibrated at an initial potential, and a single beam spectrum of the protein was recorded. After equilibrating the sample at the final potential, another single beam spectrum was recorded and a difference spectrum was calculated from these two single beam spectra. Typically 128 interferogramms at 4 cm^{-1} resolution were coadded for each single beam spectrum. No smoothing or deconvolution procedures were applied. Potentials quoted with the data refer to the Ag/AgCl/3N KCl reference electrode; add 208 mV for SHE potentials.

RESULTS AND DISCUSSION

UV/Vis Difference Spectra

The two heme groups have several overlapping absorptions in the visible part of the spectrum. Whereas both hemes contribute nearly equally to the Soret band at 445 nm, heme *a* dominates the α-band at 605 mm. The titration of the soret band at 445 nm leads to a midpoint potential of - 0.035 V ± 10 mV for heme a_3 and 0.185 V ± 10 mV for heme *a*.

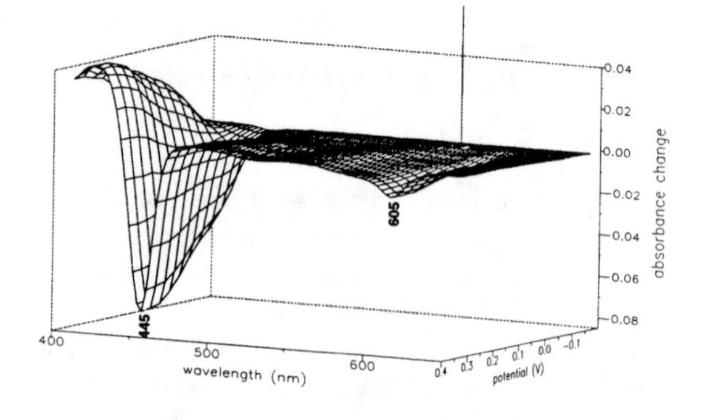

FIGURE 2: UV/vis difference spectra for a stepwise increase of the applied potential from -0.3 V to 0.4 V, taking the spectrum at -0.3 V as reference.

FTIR Difference Spectra

Reduced-minus-oxidized and oxidized-minus-reduced FTIR difference spectra of the cytochrome *c* oxidase in the mid-infrared (1800-1000 cm^{-1}) indicate **quantitative** and **fully reversible** electrochemistry. Figure 3 presents highly structured band patterns for the full potential step from + 0.5 V to - 0.5 V (thin line) and from - 0.5 V to + 0.5 V (bold line), where even minor smaller bands are perfectly reversible. An evaluation of the small amplitudes of the FTIR difference signals in the amide I region indicates that the redox process affects only a small number of peptide bonds.

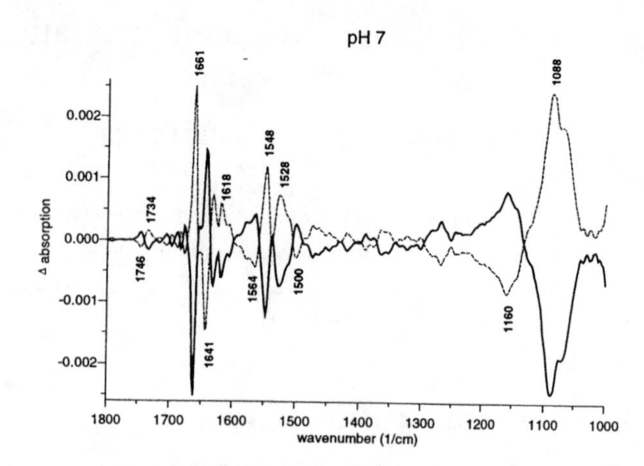

FIGURE 3: reduced minus oxidized (thin line) and oxidized minus reduced (bold line) FTIR difference spectra of the cytochrome *c* oxidase for a potential step from 0.5V to -0.5V (and *vice versa*) in phosphate buffer pH 6.9

In the amide I range (1680-1620 cm^{-1}) strong signals at 1661 cm^{-1}, 1641 cm^{-1}, 1632 cm^{-1} and 1618 cm^{-1} indicate absorbance changes of C=O groups from the polypeptide backbone. These difference signals can be interpreted in terms of small alterations in the structure upon the redox process. We cannot exclude contributions from the formyl group of the hemes and from amino acid side chains in this region.

In the spectral region from 1560 cm^{-1} to 1520 cm^{-1}, the amide II range, contributions from coupled CN stretching and NH bending modes are expected. An assignment for all these signals to the amide II mode is not very probable, since the bands do not show the expected strong shifts upon $^1H/^2H$ exchange (data not shown). In this spectral range, contributions from aromatic amino acid side chains and heme C=C groups are probable, too. In addition, contributions from the antisymmetric COO^- modes can be expected through protonation/deprotonation of COOH groups.

At the lower spectral region from 1200 to 1000 cm^{-1}, a broad difference structure with a minimum at 1160 cm^{-1} and a maximum at 1088 cm^{-1} upon reduction can be observed. As previously described [1], these bands can be assigned to PO modes caused by deprotonation of the phosphate buffer, correlating with the proton uptake of the protein and the mediators. If other buffers are used, these PO-signals are absent and characteristic modes for other buffers appear.

In the spectral region above 1680 cm^{-1}, contributions from the four heme propionates are expected, indicating proton uptake/release upon electron transfer. At higher frequencies a peak at 1736 cm^{-1} associated with the reduced form and a peak at 1746 cm^{-1} associated with the oxidized form was previously assigned to the C=O mode of Asp or Glu side chains, attributed to the proton transfer at the carboxylic group coupled with the redox reactions of the enzyme.(1)

Difference Spectra for a Stepwise Titration

A stepwise potential titration from the fully reduced to the fully oxidized state, shown in figure 4, resolved the signals at 1746/1734 cm^{-1} into several components. Performing a potential step from -0.5 V to 0 V does not result in a significant difference signal. At the potential range of 150 mV, a small positive component at 1738 cm^{-1} appears with a shoulder at 1744 cm^{-1} and a negative component at 1728 cm^{-1} develops. Increasing the potential to more positive values, the positive component at 1738 cm^{-1} disappears while the main component at 1746 cm^{-1} develops. The negative difference signal at 1728 cm^{-1} shifts to 1736 cm^{-1}.

This indicates that there is more than one ionizable residue involved in the observed redox-induced proton transfer. It cannot be excluded that part of these difference signals is caused by a completely protonated residue that reacts to the environment change upon the oxidation.

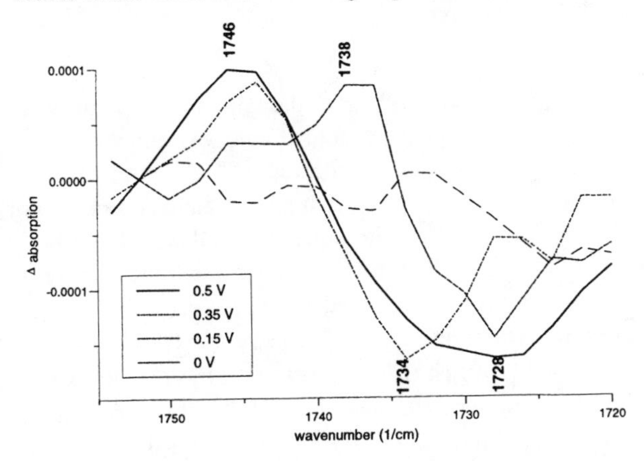

FIGURE 4: FTIR difference spectra of the cytochrome *c* oxidase for a stepwise increase of the applied potentials; the reference spectrum for each spectrum was recorded at a potential of -0.5 V

Difference Spectra of the Glu 278 Gln Mutant

Figure 5 shows an expanded view from 1760 to 1710 cm^{-1} of the FTIR difference spectra of the wild type cytochrome *c* oxidase and of the mutant enzyme where the ionizable residue glutamic acid 278 has been replaced by the non-ionizable residue glutamine. Protonation of E278 is proposed to be a step in the proton transfer through the enzyme coupled with electron transfer.(2) The relevance of E278 amino acid was already discussed on the basis of the activity and proton uptake of the E278Q mutant enzyme. The recently solved structure of the cytochrome *c* oxidase predicts its possible role in the proton pathway.(2)

In the FTIR difference spectra of the mutant enzyme and the wild type we could observe that the major component at 1746 cm^{-1} and at 1734 cm^{-1} are almost completely lost at the E278Q mutant enzyme **allowing the assignment of this signal to the protonation of the side chain of glutamic acid 278.**

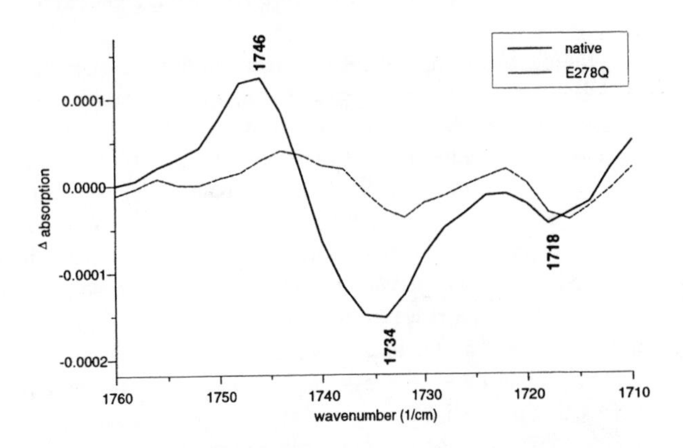

FIGURE 5: oxidized minus reduced FTIR difference spectra (-0.5V → 0.5V) of the native and E278Q mutant cytochrome *c* oxidase, the spectra were normalized at the α-band in the vis range

REFERENCES

1. Hellwig, P., Rost, B., Kaiser, U., Ostermeier, C., Michel, H. and Mäntele, W., *FEBS Lett.* **385**, 53-57 (1996).
2. Iwata, S., Ostermeier, C., Ludwig, B. and Michel, H., *Nature* **376**, 660-669 (1995).
3. Tsukihara, T., Aoyama, H., Yamashita, E., Tomizaki, T., Yamaguchi, H ., Shinzawa-Itoh, K., Nakashima, R., Yaono, R. and Yoshikawa, S., *Science* **269**, 1069-1074 (1995).
4. Baymann, F., Moss, D., A. and Mäntele, W. *Anal. Biochem.* **199**, 269-274 (1991).
5. Claudia Gries diploma Thesis Universität Erlangen, Germany (1997).
6. Sidhu, G., S., Hendler, R. W., *Biophysical Journal* **57**, 1125-1140 (1990).
7. Moody, J., A., Rich, P., R., *Biochimica et Biophysica Acta*, **1015** 205-215 (1990).

Effect of Sucrose on the Thermal Denaturation of a Protein: An FTIR Spectroscopic Study of a Monoclonal Antibody

D. M. Byler[1,2], D. L. Lee[2], T. D. Sokoloski[2], P. R. Dal Monte[2], & J. M. Baldoni[2]

[1]*Department of Chemistry, St. Joseph's University, Philadelphia, PA 19131-1395* and
[2]*SmithKline Beecham Pharmaceuticals, King of Prussia, PA 19406-0939*

The three-dimensional structure of many proteins and complex polypeptides is critically sensitive to the environment to which such a biopolymer is subjected. For example, in some cases lyophilization, which entails a complete cycle of freezing, drying *in vacuo*, and warming to ambient, brings about unpredictable changes to the secondary structure (conformation). Various research and development groups have suggested that the addition of sucrose or other oligosaccharides to protein solutions prior to lyophilization works to maintain the structural integrity of the native macromolecule. Less information is available, however, with regard to whether or not the presence of sugar inhibits thermal denaturation of proteins in solution. In the present study, FTIR spectroscopic examination of solutions of the monoclonal antibody in D_2O suggests that 6% (w/V) added sucrose produces little change in the observed mean temperature of thermal denaturation, $T_m \sim 68$ °C. Yet, when the protein solution is held at constant temperature in the range between about 60 and 75 °C, the presence of sucrose significantly alters the rate at which the protein unfolds. Based on the disappearance of the strong band near 1634 cm^{-1} associated with *intra*molecular β-structure of the native protein, the unfolding process was found to follow first-order kinetics. In the absence of sugar, the slope of a plot of ln (intensity) versus time yielded a first order rate constant of $k_1 = 2.65$ hr^{-1} at 69 °C. By contrast, when the protein solution contained 6% sucrose (w/V), $k_1 = 1.85$ hr^{-1} at the same temperature. Measuring the rate at different temperatures around T_m indicates that protein solutions containing sucrose unfold more slowly above ~ 68 °C than does the protein without the excipient. At lower temperatures, however, the opposite is true; the presence of the sugar enhances the rate of denaturation. Arrhenius plots of ln k_1 versus 1/T indicates that $E_a = \sim 600$ kJ/mol for the protein alone, but only ~330 kJ/mol with sucrose present.

INTRODUCTION

Proteins are under development as therapeutic agents for a wide variety of diseases. Many of the proteins under consideration are monoclonal antibodies [**Mab**] (immunoglobin G's) for specific antigens. The native state of each of these biopolymers has a complex three-dimensional conformational structure which is often stable over a relatively narrow range of critical factors, such as temperature, pH, ionic strength, etc. Sometimes the processing required ultimately to produce a suitable, effective protein formulation entails subjecting the protein to conditions that may lie outside of the envelope of stability. For example, lyophilization entails a complete cycle of freezing, drying *in vacuo*, and thawing, which in some cases induces unpredictable changes to the secondary structure (conformation) and hence the biological activity of the protein (1). In other instances, when the protein formulation is transported over long distances or stored in warehouses for extended times, situations may arise under which the sample may be subjected to above normal temperatures for varying times. Various research and development groups have found that addition of certain sugars, such as sucrose, frequently minimizes the loss of native structure when protein solutions are lyophilized. Less data exists, however, with regard to whether or not such excipients can enhance the structural stability of protein solutions exposed to temperature extremes.

The Fourier-transform infrared (FTIR) spectroscopic results reported here suggest that adding modest amounts of sucrose (~6% w/V) to an aqueous solution of antibody (~5% w/V) has little effect on the mean temperature at which the protein is observed to denature ($T_m \sim 68$ °C). Differential scanning microcalorimetry (DSC) discloses an increase in T_m of just 0.4 °C. Yet when the protein is held at constant temperature near T_m, FTIR shows that added sucrose significantly alters the rate at which the backbone of the protein unfolds and at which new *inter*molecular β-structures form. The latter lead to irreversible aggregation of the protein. The results reported here suggest that FTIR is an excellent method for probing the kinetics of such processes in proteins at temperatures close to T_m.

MATERIALS AND METHODS

Lyophilized samples of a purified **Mab** at pH 6.0 containing sucrose were obtained in house. Sucrose-free specimens were prepared by dialysis and then lyophilized. Each lyophile was dissolved in sufficient D_2O to yield clear, colorless solutions containing 51 mg/mL protein or 56 mg/mL protein plus 62 mg/mL sucrose (w/V). IR spectra were collected on a nitrogen-purged, Nicolet 510

CP430, *Fourier Transform Spectroscopy:* 11th International Conference
edited by J.A. de Haseth

FTIR spectrophotometer with a DTGS detector (2). The thermostatted, demountable IR cell had CaF_2 windows with a 50 μm spacer (2). Spectra typically consisted of 256 scans at 4 cm^{-1} resolution Fourier transformed with an extra level of zero filling. Second derivative data were calculated analytically (3). Residual water vapor lines were eliminated as described previously (2,4). Raw, unnormalized, unsmoothed peak intensities for selected second derivative bands were plotted versus temperature to determine the mean temperature of protein denaturation, T_m; for kinetic studies under isothermal conditions, graphs of peak intensity versus time and ln (intensity) vs. time were prepared.

Calorimetric were performed on a Microcal MC-2 differential scanning calorimeter (DSC) with deionized H_2O as a reference. Solutions were 18 mg/mL protein. The scan rate was ~1°C/min. DSC measures heat flow in J/min as a function of temperature. Heat flow divided by the scan rate in °C/min gives the heat capacity in J/°C. Normalization for protein concentration yields the molar heat capacity in kJ/mol·°C. T_m for protein denaturation corresponds to the peak maximum of the thermogram. The integrated area under the peak represents the total enthalpy (ΔH) for the denaturation process.

RESULTS

In the amide I' region (1700-1620 cm^{-1}), the second derivative IR spectrum of a D_2O solution of a **Mab** in its native state (Fig. 1), with or without sucrose, is dominated by two components: A strong band near 1638 cm^{-1} and a sharper, medium intensity peak at 1690 cm^{-1} flank two weaker features at 1662 and 1672 cm^{-1}. Subjecting a solution of the protein to small increments in temperature, these bands characteristically begin to weaken and then disappear over a relatively narrow temperature range centered around T_m ~ 68 °C (Fig. 2) as the backbone of the biopolymer unfolds. (A residual peak remains at 1641 cm^{-1} (w) even at the highest temperature.) Concomitantly, a new pair of bands wax near 1620 cm^{-1} (s) and 1685 (w) while the protein is observed to form an insoluble, irreversible aggregate. Whether solutions of this **Mab** contain added sucrose or not, the result is virtually identical. In particular, T_m remains almost unchanged.

DSC corroborates these FTIR results: thermal denaturation of the protein without sucrose produces a major endothermic event at 68.7 °C (T_m); a second minor endotherm, about 20% as large, occurs at 84.1 °C. The overall ΔH was 3480 kJ/mol. Corresponding values for protein plus sucrose were just slightly higher: 69.1 °C, 85.1 °C, and 3600 kJ/mol. FTIR discloses no obvious change near 84 °C.

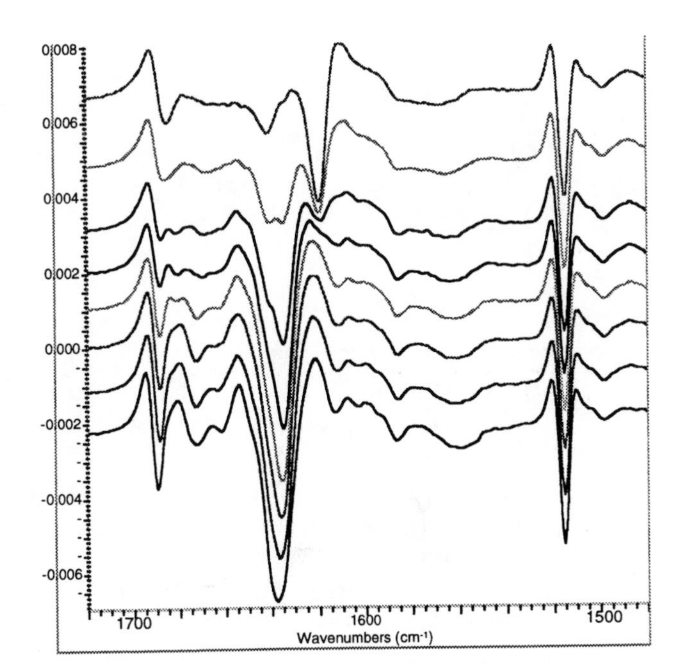

FIGURE 1. Second derivative IR spectra of sucrose-free **Mab** solutions at 29, 56, 61, 64, 66, 68, 69, and 83 °C (bottom to top).

By contrast, if the protein solution is held at fixed temperature a few degrees from T_m, the observed rate at which the amide I' IR bands change intensity is distinctly different when the carbohydrate is absent from solution than when it is present. For example, under isothermal conditions, the 1638 cm^{-1} band for solutions of the **Mab** without sucrose reveals an exponential loss in peak height with time. The rate of decrease becomes greater at higher temperatures (Fig. 3). As found above for the studies of temperature dependence, the appearance over time of a new absorption near 1620 cm^{-1} closely parallels the loss of intensity at 1638 cm^{-1}. The results for protein solutions with the sugar (not shown) are similar in appearance, but have distinctly different slopes.

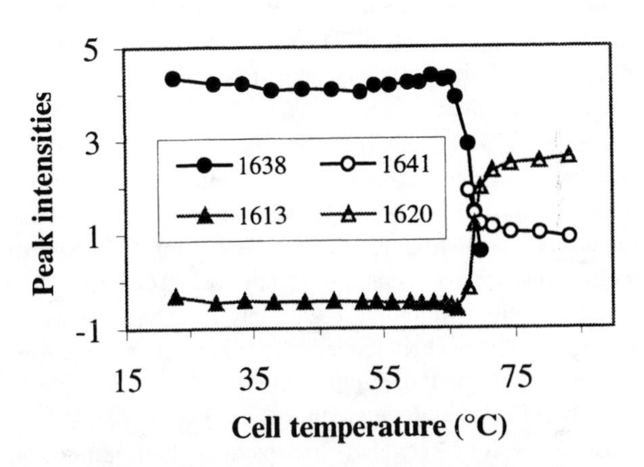

FIGURE 2. Change in intensity vs. cell temperature for selected second derivative IR bands of sucrose-free **Mab** in D_2O.

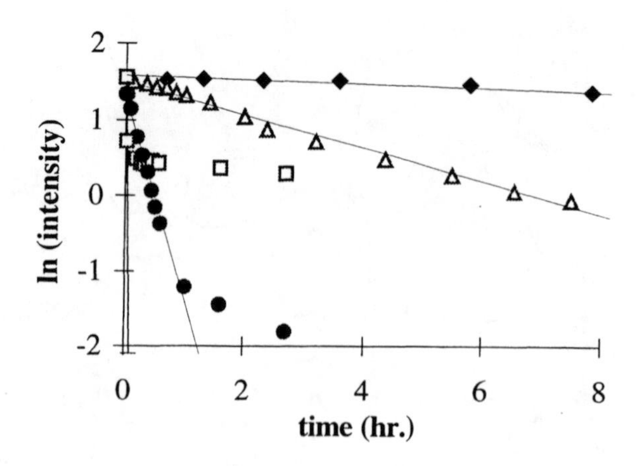

FIGURE 3. ln (1638 cm^{-1} intensity) vs. time at 62 (♦), 65 (Δ), 68 (•), and 74 °C (□) for D_2O solution of **Mab** without sucrose. At 62 °C, unfolding is not complete even after 50 hr. At 68 and 74 °C, the slope undergoes a marked change after the native protein conformation has totally disappeared.

DISCUSSION AND CONCLUSIONS

A typical **Mab** (immunoglobulin G) has two pairs of subunits. The two light and two heavy peptide chains are covalently linked by one or two disulfide bonds. As much as 75% of the secondary structure of the native protein takes the form of β-sheets (5). These consist of a series of anti-parallel β-strands anchored by turns. The former conformation gives distinct absorptions at 1638 (s) and 1690 cm^{-1} (m) in the amide I' region of resolution-enhanced IR spectra; the latter are probably responsible for the weaker bands near 1662 and 1672 cm^{-1} (3-6).

The relative intensities of these features in the second derivative IR spectra of the **Mab** in D_2O suffer little, if any, change as the protein is exposed to temperatures up to ~60 °C. This confirms that the conformation of this rather stable protein is largely unaffected by moderate thermal perturbation. (A small, but steady decrease in frequency of the strongest amide I' band from 1638 to 1634 cm^{-1} as the temperature approaches 60 °C probably results from the replacement of some residual hard-to-exchange hydrogens by deuterium atoms.)

Above 60 °C, however, further incremental increases in temperature induce gradual losses in intensity in all four amide I' bands (Figs. 1-2). These changes imply that the native secondary structure has begun to unfold. Almost simultaneously, new bands start to appear at 1685 and 1620 cm^{-1}. Such absorptions are commonly observed for many kinds of proteins when moderately concentrated aqueous solutions are heated to temperatures sufficiently high to cause unfolding and aggregation. These bands are attributed to long, regular, anti-parallel β-structures in which the unfolded peptide chain of one molecule interacts with that of another via strong *inter*molecular hydrogen bonds (6-8). This intermolecular coupling helps explain the observation for many proteins that thermal denaturation invariably leads to the irreversible formation of insoluble gels and aggregates. By about 69 °C, the amide I' spectrum is completely different from that of the native protein. Now the only bands are new ones at 1685, 1641, and 1620 cm^{-1}. The middle peak may represent residual structures resistant to unfolding.

From a plot of band intensity versus temperature (Fig. 2) one may estimate that the mean temperature of denaturation (T_m) for the **Mab** is ~68 °C. This value which corresponds to the midpoint of the narrow range of temperatures over which the protein completely unfolds and aggregates is consistent with T_m measured by DSC of 69 °C. Other workers have reported comparable values for other similar antibodies (8).

Now consider a solution of the **Mab** held for hours or even days at some fixed temperature between ambient and ~55 °C. The spectra show that all the protein in solution maintains its native conformation. Above T_m the opposite is true: The whole protein sample rapidly and irreversibly denatures and, at least for a concentration of 5% w/V, aggregates at a rate which increases with increasing temperature: In one hour or less, no IR bands due to native structures remain.

Unfortunately, one cannot so easily interpret the spectrum of the **Mab** at intermediate temperatures. Under such circumstances, bands characteristic of both the native and the denatured state appear together. On the basis of just the IR data, there is no simple means by which to determine if such a spectrum implies that every individual molecule is in some intermediate state, partly unfolded and partly denatured, or if the solution consists of a mixture (not necessarily at equilibrium) of completely folded and utterly denatured molecules with only a trace of other forms. Because the time scale for the interaction of an IR photon with the molecule is on the order of 10^{-13} s, intermediates with alternate conformations can be detected if they are present in sufficient concentration. Mathews and van Holde provide a clue: "...a protein tends to fold in a *cooperative* manner; in a partially folded mixture there are mainly wholly unfolded and wholly folded molecules, with few intermediate structures" (10). Apparently once a specific biomolecule has sufficient energy to begin to unfold, the complete process occurs quite rapidly.

Measurement over time of the the intensity of the 1638 cm^{-1} band for solutions of the **Mab** with and without sucrose held at four different constant temperatures close to T_m reveals an exponential decrease in the peak height. In fact, graphing the natural logarithm of the second derivative IR intensity against time gives a linear relationship implying that the unfolding protein follows first-order kinetics (Fig. 3). As expected, the higher the temperature, the greater the calculated rate constants, k_1,

Table 1.	First-Order Rate Constants and Half Lives for the Unfolding of the **Mab**			
	(without sucrose)		(with sucrose)	
T (°C)	k_1(hr^{-1})	$t_{1/2}$ (hr)	k_1(hr^{-1})	$t_{1/2}$ (hr)
57.1	-	-	**0.0316**	**21.9**
60.7	-	-	**0.112**	**6.22**
62.2	**0.0288**	**24.1**	*0.193*	*3.59*
64.3	*0.124*[a]	*5.59*	**0.418**	**1.66**
65.2	**0.220**	**3.16**	*0.545*	*1.27*
68.4	*1.61*	*0.431*	*1.61*	*0.431*
68.8	**2.65**	**0.261**	*1.85*	*0.375*
71.5	*10.9*	*0.064*	**4.46**	**0.155**
74.0	**42.**	**0.017**	-	-

[a]Values in italics are interpolated.

for the change in IR intensities with time (Table 1). Biphasic behavior is apparent for the data at higher temperatures. The individual spectra show that the change in slope occurs only once all the native secondary structure has been destroyed. These data presumably represent the rate of unfolding or denaturation of the **Mab** at the selected temperature. In contrast to the temperature profile of the IR intensities and the value of T_m reported above, the observed k_1's, take on markedly different values when 6% sucrose is added to the aqueous **Mab**.

Just as found with the studies of temperature dependence, a new band begins to appear in the isothermal spectra near 1620 cm^{-1} nearly concurrently with the onset of the loss of band intensity at 1638 cm^{-1}. Attempts to fit the logarithm of the increase of the former band with a linear trendline proved unsuccessful. Unlike the unfolding process, aggregate formation clearly does not follow a simple first-order rate law.

A plot of ln k_1 (for the intensity changes at 1638 cm^{-1} observed at different constant absolute temperatures) versus 1/T also gave straight lines (Fig. 4). Assuming one can represent the latter by the Arrhenius equation, $\ln k_1 = \ln A - (E_a/RT)$, one may estimate an activation energy, E_a, for the unfolding of the peptide chain. For solutions without added sugar, E_a is ~600 kJ/mol; when sugar is added, the value decreases to ~330 kJ/mol. The calculated intersection of lines which fit the two sets of values, solutions with and without sucrose, lies at 68.4 °C. It seems unclear whether this value is related to that for T_m or whether the similarity is just a coincidence. Nonetheless, the energy barrier to protein unfolding in the presence of 6% sucrose is much smaller than exists in the absence of this common excipient.

The preliminary FTIR results reported here provide glimpses of how **Mab**'s unfold in aqueous solution. Addition of 6 % sucrose has little effect on T_m, but does lower the barrier to unfolding. Further study is needed to verify these data and to uncover the details of the operative mechanism. Solutions of other proteins may display different behaviour. For example, Boye *et al.* (9) found changes of ~4 °C in T_m for bovine serum albumin when up to 50% sucrose was present. But they did not attempt any kinetic measurements. Now recent reports of time-resolved FTIR spectroscopy of the thermal and chemical denaturation of proteins (11) underscore the opportunities which this technology offers to researchers wishing to glean new insights into the nature protein of structure.

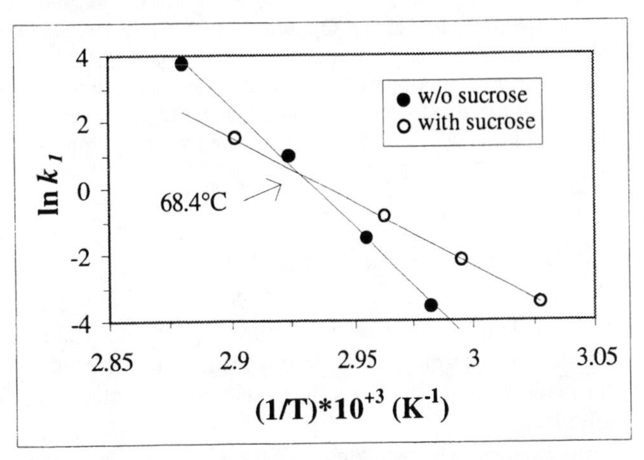

FIGURE 4. Arrhenius plots for rates of thermal denaturation of **Mab** in D_2O at various temperatures near T_m with sucrose [o], E_a = 326 kJ/mol and without sucrose [•], E_a = 600 kJ/mol.

REFERENCES

1. Prestrelski, S.J., Pikal, K.A., and Arakawa, T. *Pharm. Res.* **12**, 1250-1259 (1995).
2. Byler, D.M., Wilson, R.M., Randall, C.S., and Sokoloski, T.D., *Pharm. Res.*, **12**, 446-450 (1995).
3. Susi, H., and Byler, D.M., *Biochem. Biophys. Res. Commun.*, **115**, 391-397 (1983).
4. Prestrelski, S.J., Byler, D.M., and Liebman, M.N., *Proteins: Struct. Func. Gen.*, **14**, 440-450 (1992).
5. Byler, D.M., and Susi, H., *Biopolymers*, **25**, 469-487 (1986).
6. Surewicz, W.K., Mantsch, H.H. and Chapman, D. *Biochemistry*, **32**, 389-394 (1993).
7. Clark, A.H., Saunderson, D.H.P., and Suggett, A., *Int. J. Pept. Protein Res.*, **17**, 353-364 (1981).
8. van Stokkum, I.H.M., Linsdell, H., Hadden, J.M., Haris, P.I., Chapman, D., and Bloemendal, M., *Biochemistry*, **342**, 10508-10518 (1995).
9. Boye, J.I., Alli, I., and Ismail, A.A., *J. Food Agric. Chem.*, **44**, 996-1004 (1996).
10. Mathews, C.K., and van Holde, K.E., "Biochemistry," The Benjamin/Cummings Publishing Co., Inc., 1990, p. 193.
11. Reinstädler, D., Fabian, H., Backmann, J., and Naumann, D., *Biochemistry*, **35**, 15822-15830 (1996).

Interaction of Two Different Types of Membrane Proteins with Model Membranes Investigated by FTIR ATR Spectroscopy

M. Siam, G. Reiter, M. Schwarzott, D. Baurecht and U.P. Fringeli

Institute of Physical Chemistry, University of Vienna, Althanstrasse 14, A-1090 Vienna, Austria

Polarized FTIR ATR spectroscopy was used to investigate the interaction of mitochondrial creatine kinase (Mi-CK) and intestinal alkaline phosphatase (AP) with model membrane assemblies. Mi-CK was immobilized by adsorption to the negatively charged cardiolipin (CL) leaflet of a supported CL/DPPA bilayer. H-D-exchange of the enzyme and the stability under flowthrough conditions of the protein/membrane assembly were examined. AP, however, was bound to a DPPA Langmuir-Blodgett layer (LBL), followed by the completion of a bilayer-like structure by adsorption of POPC molecules from a vesicular solution. It turned out that the POPC adsorbate exhibited decreased molecular order compared to the POPC molecules on a supported POPC/DPPA bilayer. Enzymatic activity of immobilized AP was determined with p-nitrophenyl phosphate (p-NPP) as substrate and remained unchanged for at least 2 days.

INTRODUCTION

In vivo there are several ways how membrane proteins are attached to a lipid bilayer. We investigated two proteins with respect to their interactions with model membrane assemblies.

Mitochondrial creatine kinase (Mi-CK) is known to bind effectively to the surface of negatively charged bilayers like cardiolipin (CL) (1,2,3). Mi-CK catalyzes the phosphorylation of creatine by ATP and is therefore important for energy metabolism in cells of high and fluctuating energy requirements.

Alkaline phosphatase (AP) belongs to a wide group of enzymes that catalyze the non-specific hydrolysis of phosphate monoesters in an alkaline environment (4). AP is attached to the outer leaflet of the plasma membrane by the lipid moiety of a glycosyl-phosphatidylinositol (GPI) anchor (5). The hydrophobic hydrocarbon chains of this lipid moiety are responsible for the attachment of the enzyme to the lipid bilayer.

Our aim was to find experimental conditions that enable the formation of stable protein/membrane assemblies for both enzymes. The enzyme adsorption to the supported lipid matrices was monitored *in situ* by FTIR ATR spectroscopy, and the amount of bound protein was estimated.

MATERIAL AND METHODS

Chicken sarcomeric mitochondrial creatine kinase (Mi-CK, octamer with 340 kD) was obtained from T. Wallimann (Institute for Cell Biology, ETH-Hönggerberg, Zurich, Switzerland). Alkaline phosphatase (AP, dimer with 132kD) from bovine intestinal mucosa was provided by B. Roux (ICBMC, University Claude Bernard, Lyon, France). 1,2-Dipalmitoyl-*sn*-glycero-3-phosphoric acid (DPPA), Cardiolipin (CL), p-nitrophenylphosphate (p-NPP) and octyl-β-D-glucopyranoside (β-OG) were purchased from Fluka AG, and 1-palmitoyl-2-oleoyl-*sn*-glycero-3-phosphocholine (POPC) from Sigma.

When Mi-CK, when used as model membrane (Fig. 2, path 1: I - IV), a supported lipid bilayer was prepared using the Langmuir Blodgett (LB)/Vesicle-Method (6). With a film balance the first layer of DPPA was transferred to the surface of a clean germanium plate. The second layer of negatively charged CL was produced by spontaneous adsorption to the hydrophobic DPPA film from a vesicular solution. For the immobilization of Mi-CK a 0.55 mg/ml enzyme-solution in 10 mM phosphate buffer pH 7.0, 50 mM NaCl, was slowly pumped over the DPPA/CL bilayer at 25°C. The adsorption process could be monitored *in situ* because it occurred in an aligned cell in the sample compartment of the spectrometer by means of a peristaltic pump.

AP tends to precipitate in aqueous buffer solutions and needs some detergent for solubilization. Therefore, the transfer of AP from solution to the lipid bilayer is more critical than that of Mi-CK. At first, AP (50 μg/ml, 20 mM Tris buffer pH 7.4, 150 mM NaCl, 1 mM $MgCl_2$, 50 μM $ZnCl_2$) solubilized with β-OG was directly adsorbed to DPPA, then a POPC vesicle solution was circulated through the ATR cuvette to reconstitute the bilayer (Fig.2, path 2: I,V,VI). Afterwards, the activity of the immobilized AP was measured by pumping substrate solution (p-NPP) (7) through a flowthrough cuvette. Enzymatic activity was determined from the rate of p-nitrophenol production by ester hydrolysis as detected by VIS spectroscopy at 420 nm.

CP430, *Fourier Transform Spectroscopy:* 11th International Conference
edited by J.A. de Haseth

RESULTS AND DISCUSSION

Mi-CK rapidly binds to the CL/DPPA membrane: within 45 min 99% of the process took place (Fig.1). Data should be comparable to those of Stachowiak et al. (3), who found a fast process (k_1=0.11s^{-1}) and a slower one (k_2= 0.008s^{-1}) within the first 250s. In our experiment we can prove the slower process to have a rate constant of 0.01s^{-1}. Furthermore we detected a third process with a rate constant of 7.10^{-4} s^{-1} after observing the adsorption for about 50 min.

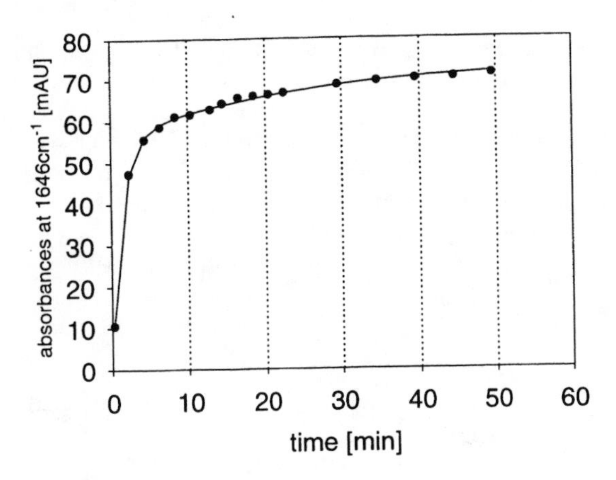

FIGURE 1. Absorbances at 1646 cm^{-1} during the adsorption of Mi-CK on a CL/DPPA bilayer: [•]... Data (A_{max} of amid I/I´-band) and [-]... kinetic fit of the adsorption of octameric Mi-CK on a DPPA/CL-bilayer.
Data were fitted with $f = A_0 - A_1 \cdot exp(-k_1 \cdot t) - A_2 \cdot exp(-k_2 \cdot t)$ using the Marquardt-Levenberg algorithm of Sigma-Plot. A_0 was estimated by extrapolation of a double reciprocal plot to be 74mAU. Results of the fit: A_1=55.4 ± 0.87 mAU, k_1=(1.23 ± 0.055)·10^{-2} s^{-1}, A_2=18.6 ± 0.62 mAU, k_2=(7.05 ± 0.34)·10^{-4} s^{-1} (15 iterations, Rsqr=0.999).

After 1 hour the medium was changed to D_2O (pH* 6.6) and the H-D-exchange was monitored. Analyzing the amide I/I´ and N-H stretching-bands indicates that Mi-CK is a rather stiff protein. About 20% of the H remain after 20 hours of D-exchange. The quantification of the amide I/I´-band (1650 cm^{-1}) (8) revealed a surface concentration of about $1 \cdot 10^{-12}$ mol·cm^{-2}, which corresponds to a density of coverage of 60%. Fig. 3 shows FTIR ATR spectra of the Mi-CK and the outer leaflet of the bilayer consisting of CL. A weak dichroism of the amide I/I´ band points to a slight distortion of the Mi-CK structure as determined by X-ray crystallography (point group 422) (9).

The protein/membrane assembly was checked for stability under flowthrough conditions and found to be stable for 3 days. Furthermore, Mi-CK seems to stabilize the CL/DPPA membrane.

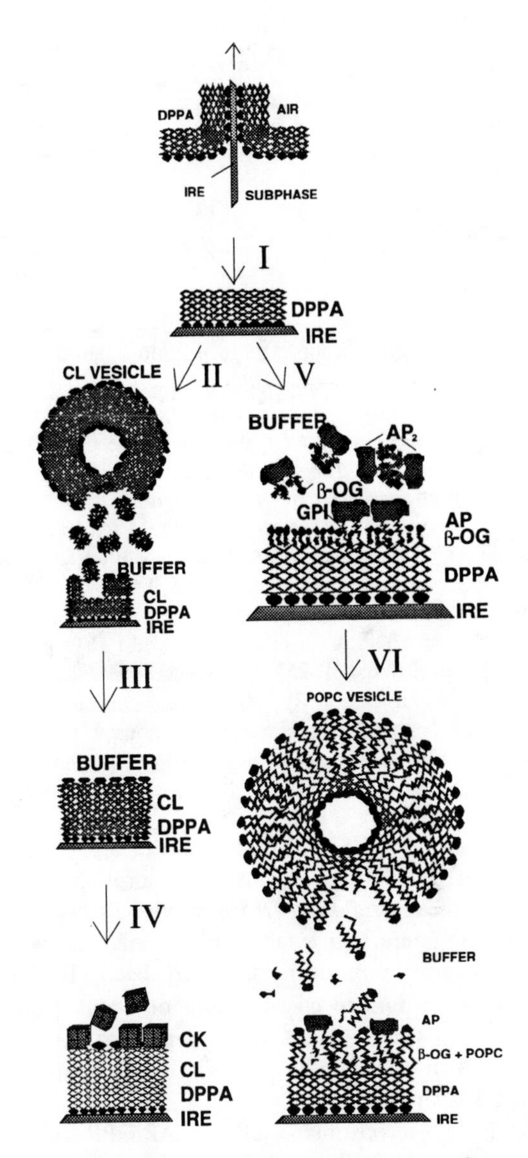

FIGURE 2. Schematic description of two pathways for immobilizing enzymes on lipid model membranes attached to an IRE-plate.
Path 1: Immobilization of Mi-CK. (I) Transfer of the inner IRE-attached DPPA-monolayer from the air/water interface of a film balance to an internal reflection element (IRE) by the Langmuir-Blodgett (LB) technique; (II) spontaneous adsorption of CL-lipids from vesicles energetically driven by the reduction of the unfavorable high energy of the hydrophobic surface of the DPPA monolayer in contact with the aqueous environment; (III) completed asymmetric CL/DPPA bilayer; (IV) adsorption of Mi-CK to the bilayer by electrostatic interactions.
Path 2: Immobilization of AP. (I) as described above; (V) spontaneous adsorption of AP (solubilized by b-OG) to the DPPA-Monolayer via its GPI anchor; (VI) reconstitution of a bilayer-like system by passing POPC vesicles over the AP-DPPA-assembly.

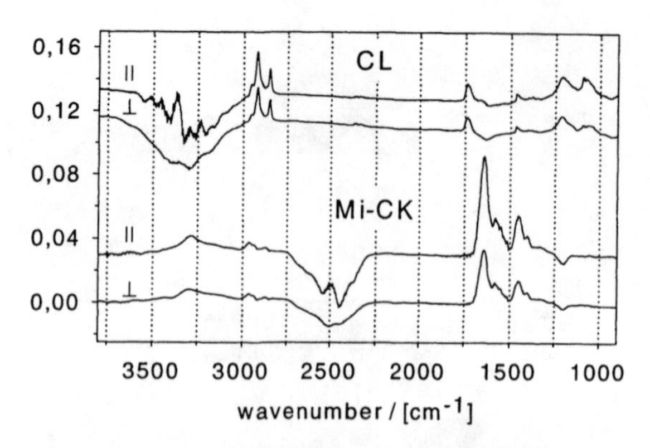

FIGURE 3. Polarized IR ATR absorbance spectra of CL and immobilized Mi-CK. *Top:* Cardiolipin from *E. Coli* (CL) assembled from a vesicle solution (0.67mg/ml CL) on a DPPA-layer. This bilayer was in contact with 20 mM phosphate buffer pH 7.0; T 18°C; reference, DPPA in phosphate buffer; dichroic ratio R, 1.16; surface concentration Γ = $1.68.10^{-10}$ mol cm^{-2}. *Bottom:* Polarized IR ATR absorbance spectra of Mi-CK immobilized on a DPPA/CL bilayer in 10 mM D_2O phosphate buffer solution pH* 6.6; T, 25°C; reference DPPA/CL bilayer in D_2O buffer; dichroic ratio R, 1.85; surface concentration Γ = $9.6.10^{-13}$ mol cm^{-2}; Both surface concentrations were estimated with the thin film approximation: angle of light incidence θ, 45°; number of active internal reflections N, 16.

In contrast to Mi-CK adsorbing to the negatively charged bilayer, AP exhibits a slower adsorption process. Saturation of adsorption to DPPA occurs after 7.5 h; the lipid-protein-interaction is taking place only via the GPI-anchor. Quantification of the amide I/I´-band (1650 cm^{-1}) (8) revealed a surface concentration of about $1.84 \cdot 10^{-12}$ $mol \cdot cm^{-2}$ (Fig.4) corresponding to a density of coverage of about 50%. It is supposed that AP is adsorbing to the DPPA LB layer as monomer (Fig. 2, V).

The POPC reconstitution of the AP-DPPA-assembly (Fig. 2, VI) yields a "bilayer"-like structure where the AP exhibits a constant catalytic specific activity of 30 U/mg for at least 2.5 days (compared to 60 U/mg in the AP solution before adsorption). Furthermore, the adsorbed POPC (Γ = $1.36 \cdot 10^{-10}$ mol cm^{-2}, density of coverage 50%) exhibits the same properties as a pure outer POPC layer adsorbed to DPPA, except for a diminished degree of order (Fig. 5). It seems that the presence of AP and/or of the tenside β-OG, which is also bound to DPPA, is disturbing the ordered adsorption of POPC molecules. Anyway, there must be care taken in the amount of tenside present in the protein sample: If it is too large, AP will remain solubilized and will not adsorb to the DPPA layer; if it is too small, AP will aggregate from the beginning.

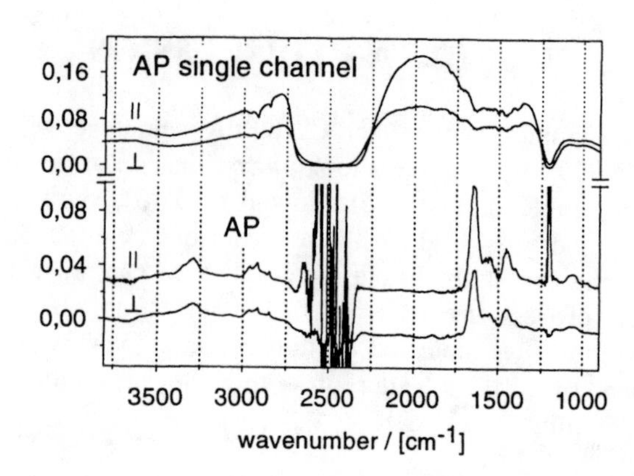

FIGURE 4. Polarized IR ATR spectra of AP. *Top*: Single channel spectra of AP immobilized on DPPA; 20 mM D_2O Tris buffer pH* 7.0; T 25°C. *Bottom:* Correspondent polarized IR ATR absorbance spectra; reference, DPPA in D_2O Tris buffer; dichroic ratio R, 1.62; surface concentration Γ = $1.84.10^{-12}$ mol cm^{-2};angle of light incidence θ, 45°; number of active internal reflections N, 35.

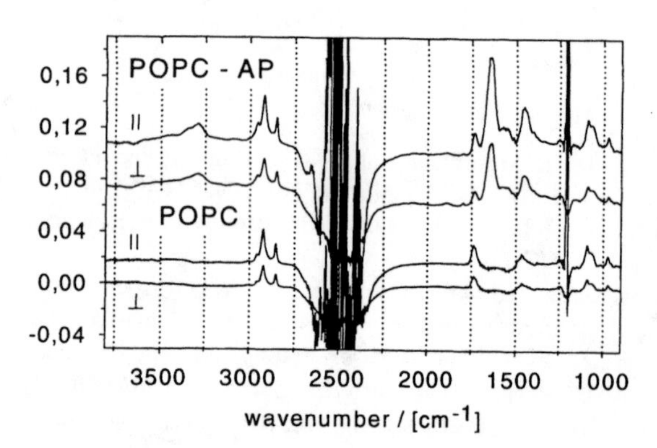

FIGURE 5. Polarized IR ATR absorbance spectra of POPC-AP and POPC. *Top:* POPC-AP assembled from a vesicular POPC-solution (0.67 mg/ml POPC) on a DPPA-layer with immobilized AP; 20 mM D_2O Tris buffer pH* 7.0; T 25°C; reference, DPPA in D_2O Tris buffer. *Bottom:* Polarized IR ATR absorbance spectra of POPC adsorbed on a AP-DPPA-assembly; T 25°C; reference AP-DPPA-assembly in D_2O Tris buffer; surface concentration $\Gamma = 1.36.10^{-10}$ mol cm^{-2}; dichroic ratio R, 1.59; angle of light incidence θ, 45°; number of active internal reflections N, 35.

Regardless, whether the interactions between enzymes and lipid membranes are of electrostatic or hydrophobic nature, our results show that it is possible to build up well-defined and stable protein/lipid-assemblies. The simultaneous application of FTIR ATR measurements with methods to determine the native enzymatic activity will give new insights into mechanisms of interaction between different types of membrane proteins and their lipid matrices.

ACKNOWLEDGEMENTS

We would like to thank Prof. T. Wallimann, O. Stachowiak and U. Schlattner (Institute for Cell Biology, ETH-Hönggerberg, Zurich, Switzerland) and Prof. B. Roux, M. Angrand and M. Bortolato (ICBMC, University Claude Bernard, Lyon, France) for kindly providing us with Mi-CK and AP, respectively.

REFERENCES

1. Wallimann T., Wyss M., Brdiczka D. and Nicolay K., *Biochem. J.* **281**, 21 (1992).
2. Rojo M., Hovius R., Demel R. and Wallimann T., *FEBS Letters*, **281**, 123 (1991).
3. Stachowiak O., Dolder M. and Wallimann T., *Biochemistry*, **35**, 15522 (1996).
4. MacComb R.B., Bowers G.N. and Posen S., *Alkaline Phosphatases*, New York: Plenum, 1979.
5. McConville M.J. and Ferguson M.A.J., *Biochem. J.* **294**, 305 (1993).
6. Wenzl P., Fringeli M., Goette J. and Fringeli U.P., *Langmuir*, **10**, 4253 (1994).
7. Garen A., and Levental C., *Biochim. Biophys. Acta* **38**, 470 (1960).
8. Fringeli U.P., Apell H.-J., Fringeli M. and Läuger P., *Biochim. Biophys. Acta* **984**, 301 (1989).
9. Fritz-Wolf K., Schnyder T., Wallimann T. and Kabsch W., *Nature*, **381**, 341 (1996).

2D-IR Investigation of the pH-Dependent Conformational Change in Cytochrome *c*

David A. Moss

Karlsruhe Research Center, Institute of Instrumental Analysis, P.O. Box 3640, D-76021 Karlsruhe, GERMANY

The reduced-minus-oxidized FTIR difference spectrum of cytochrome *c* was measured at a series of pH values between 6 and 11. The results obtained clearly document the increasing divergence in the conformations of reduced and oxidized cytochrome *c* with increasing pH. A parametric analysis of the data based on least squares fitting to a sum of Henderson-Hasselbalch functions showed that at least two pK values, 9.5 and 10.5, are required to describe the pH titration data. Simultaneously recorded spectral data in the visible region, by contrast, could be described with a single pK value of 9.5, in good agreement with the literature.The FTIR spectrum of the component titrating with a pK of 10.5 was strongly suggestive of a tyrosine deprotonation, while the spectrum of the pK 9.5 component was consistent with a carboxyl deprotonation. The pH dependency of the spectral data was also analyzed using the 2D correlation method. Two-dimensional correlation analysis of the FTIR data confirmed the presence of two components titrating at different pK values, and yielded spectra for these two components that were essential identical to those obtained by parametric least-squares fitting. Correlation analysis of the FTIR against the visible data was particularly useful in providing immediate access to the FTIR spectral component titrating at the same pK as the 695 nm band, without requiring any assumptions as to the total number of pH-dependent components present.

INTRODUCTION

It is well established that oxidized cytochrome *c* undergoes a reversible transition to a new, stable solution conformation when the pH is raised beyond ~9.5 (1). This conformational transition, which does not appear to occur in the reduced form of the protein, is characterized by the loss of an absorbance band at 695 nm. This is generally considered to indicate replacement of Met80-S as the axial haem ligand, presumably by a nitrogen atom. Despite extensive investigations with a variety of techniques, the identity of the axial haem ligand at alkaline pH has not yet been established. In addition, the deprotonation which triggers the conformational change has yet to be identified, although several possible assignments have been suggested.

In principle, a pH titration of the FTIR spectrum of oxidized cytochrome *c* would be expected to allow characterization of the deprotonation responsible for triggering the alkaline conformational transition. This approach is difficult to carry out in practice, since cytochrome *c* contains at least 20 amino acid residues that deprotonate in the pH range of interest and thus would severely complicate the interpretation of pH-dependent FTIR spectral features. Moreover, the spectral changes associated with the deprotonation and conformational change can be expected to be very small compared with the total IR absorbance of the protein, so that a sufficiently accurate background subtraction would be extremely difficult to achieve.

To circumvent these problems, a pH titration of the reduced-minus-oxidized FTIR difference spectrum, rather than of the absolute FTIR spectrum of cytochrome *c*, was performed. *In situ* reduction and oxidation of the protein by direct electrochemistry permitted a highly accurate elimination of all redox state-independent IR absorbance bands, as previously reported (2). This approach was thus expected to eliminate all pH-dependence common to both redox states, leaving only the redox-state dependent deprotonation involved in the alkaline conformational transition. The experimental data only partially fulfilled this expectation: the pH dependence of the FTIR difference spectrum was indeed greatly simplified, but not to the point where a single pK could adequately describe the data. Mathematical data analysis was thus required to extract the portion of FTIR spectral changes accompanying the loss of the 695 nm band.

The methods used to obtain the spectral data and for parametric non-linear fitting of the results have been presented in detail elsewhere (3) and will be summarized briefly here. The main subject of the present work is the use of 2D-IR correlation analysis to determine the number of titratable components present in the FTIR titration data, as well as the location of their pK values relative to the well-established single pK value obtained in titrations of the 695 nm band.

CP430, *Fourier Transform Spectroscopy:* 11th International Conference
edited by J.A. de Haseth
 1-56396-746-4/98/$15.00

MATERIALS AND METHODS

Horse heart cytochrome *c* was obtained from Sigma and rinsed repeatedly with buffer solution (50 mM PO_4, 50 mM glycine, 300 mM KCl, 2 mM dithiodipyridine) adjusted to the appropriate pH value before use. The final concentration of cytochrome *c* was between 3 and 11 mM. Eighteen such samples were prepared, covering the pH range 6 to 11. Reduced-minus-oxidized FTIR spectra were recorded with a Bruker IFS-25 spectrophotometer as the average of 32 scans at 4 cm^{-1} resolution, using the 10 μm pathlength spectroelectrochemical cell described previously (2). No data smoothing or resolution enhancement procedures were employed. A concentric visible measuring beam was used to obtain simultaneous difference spectra in the 500 - 750 nm region: these data were used to monitor the completeness of reduction/oxidation, and for subsequent scaling of the FTIR and visible data by normalization on the 550 nm band.

RESULTS

Figure 1 shows the reduced-minus-oxidized FTIR difference spectrum of cytochrome c at a selection of pH values. The data clearly show that the structural change associated with reduction and oxidation of the protein is small at neutral pH and becomes more pronounced at alkaline pH, i.e. the alkaline conformational transition involves a divergence of the reduced and oxidized structures.

While the titration of the 695 nm feature could be well

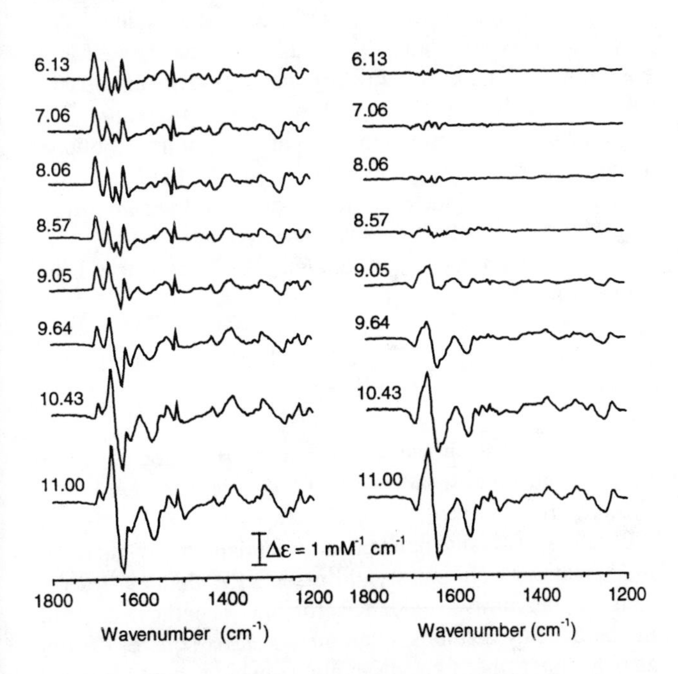

FIGURE 1. Reduced-minus-oxidized FTIR difference spectra of cytochrome c as a function of pH. Left panel, difference spectra at the pH indicated: right panel, change in the difference spectra with respect to the difference spectrum at neutral pH.

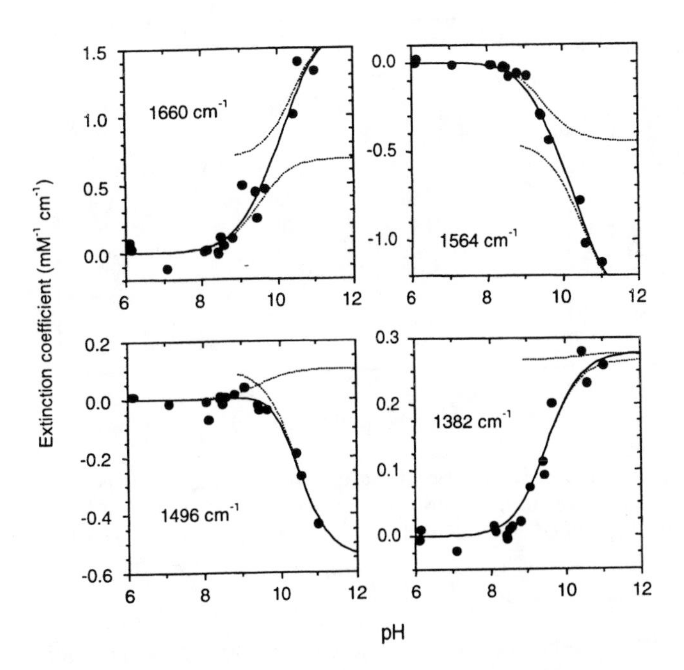

FIGURE 2. Parametric least squares fit of the data shown in Fig. 1 (right panel) at a selection of spectral frequencies. The solid lines show the results of non-linear fitting to the sum of two Henderson-Hasselbalch functions, with pK values of 9.48 and 10.46. The dotted lines show the individual components of the sum.

described with a single deprotonation at a pK value of 9.49 (data not shown), two deprotonations had to be assumed in order to model the FTIR data (Figure 2). The two pK values were determined by simultaneous non-linear least squares fitting of the data at 12 spectral frequencies, yielding values of 9.48 and 10.46. Subsequently, these pK values were treated as constants in the calculation of the two titration amplitudes for all spectral frequencies by singular value decomposition. The latter method has the advantage over other least-squares algorithms that it minimizes the values of any superfluous parameters in the model, and thus could be used to show that no more than two pK values are required to describe the data.

The results, shown in Figure 3, represent the FTIR spectral changes resulting from each deprotonation together with any associated changes in the structure of the protein. The two spectra are quite distinct, with the highest degree of similarity in the amide I region. The pK 10.46 deprotonation can be assigned with a high degree of certainty to tyrosine, while the data for pK 9.48 can be taken as supporting the assignment of this deprotonation to a heme propionic acid residue (6). The reasoning behind these assignments has been presented in detail elsewhere (3).

The generalized two-dimensional correlation method introduced by Noda (5) is an alternative method for the analysis of severely overlapping spectral data, in which phase differences in the dynamic spectral intensities resulting from an external perturbation of the sample are used as the basis of separation. The generalized form of the

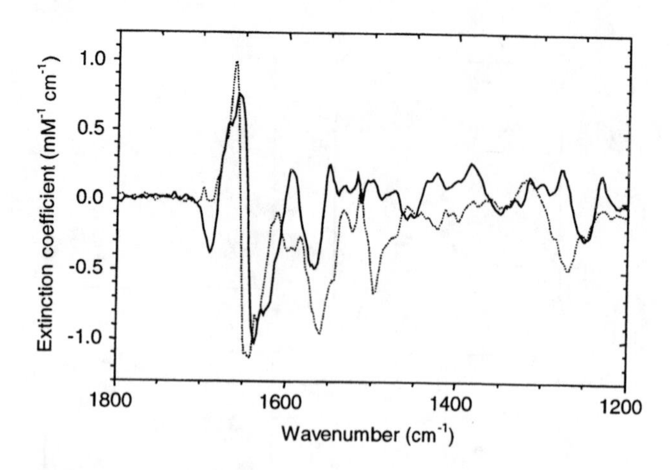

FIGURE 3. Resolution of the data of Fig. 1 into components with different pH dependencies by parametric least-squares fitting. Solid line, pK=9.48 component: dotted line, pK=10.46 component.

theory removes the restriction to sinusoidal perturbation functions. Although "dynamic" in this context usually refers to variations in the FTIR spectral intensity as a function of time, it has been pointed out that these variations may actually be presented as a function of any physical variable. The pH titration data of the present work should therefore be amenable to analysis with the 2D-IR correlation method, components with differing pK values being viewed as phase-shifted along the pH axis.

Synchronous and asynchronous correlation spectra were calculated with the Fourier transform method (5), which required the computation of the Fourier transform of the absorbance vs. pH data for each wavenumber. It should be noted that the sampling interval along the pH axis was by no means uniform, as is usually required for a discrete Fourier transform. This requirement is of no relevance here, because the generalized 2D-IR method makes no assumptions regarding the function describing the pH-dependency of the signal, i.e. an unevenly sampled

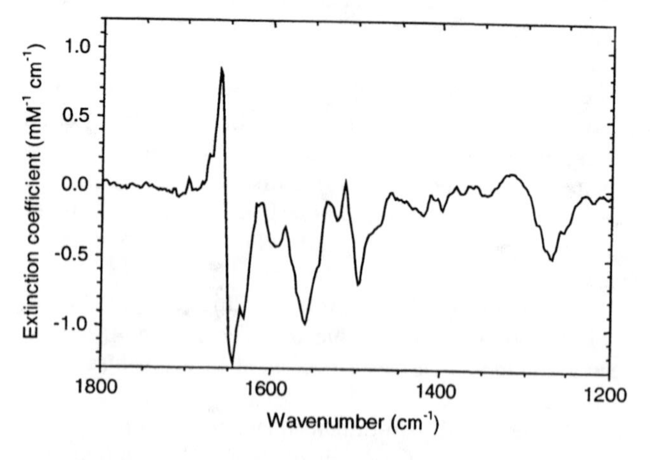

FIGURE 5. Asynchronous correlation spectrum calculated from the data of Fig. 1 with respect to the pH dependency at 695 nm. This spectrum represents that portion of the pH-dependent FTIR spectrum which titrates at a different pK value to that of the 695 nm band.

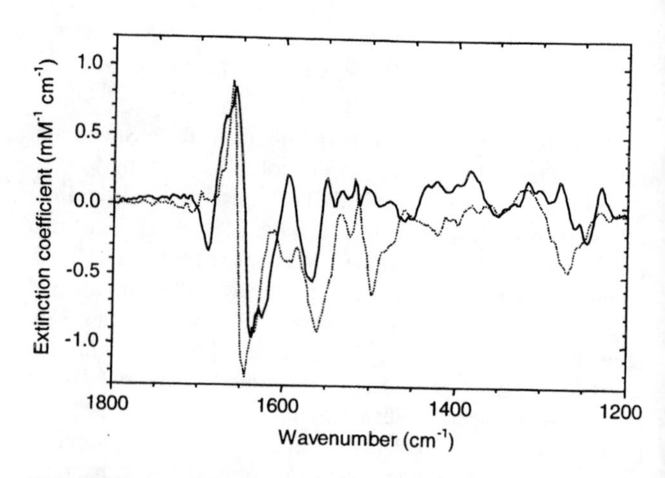

FIGURE 4. Resolution of the data of Fig. 1 into components with different pH dependencies by two-dimensional correlation analysis. Solid line, asynchronous correlation at 1492 cm^{-1}: dotted line, asynchronous correlation at 1384 cm^{-1} (the y axis has been scaled to correspond to that of Fig. 2).

Henderson-Hasselbalch function can be regarded as an evenly sampled, arbitrary function.

Although the synchronous correlation spectrum was found to be of limited value for the present analysis, the asynchronous correlation spectrum yielded an excellent separation of the FTIR titration data into its component spectra. The component spectra already determined by parametric least-squares fitting could be recovered nearly perfectly as the asynchronous correlation spectrum at a number of different spectral frequencies. The best matches to the parametric fitting results are shown in Figure 4.

As a further confirmation, the synchronous and asynchronous correlation intensities between the FTIR and visible spectral data were also calculated. The asynchronous correlation spectrum was the same for all visible wavelengths, confirming the homogeneous titration of the visible spectrum with a single pK value. The asynchronous correlation spectrum thus obtained (Figure 5) represents the sum of all FTIR bands that exhibit pH dependency but do <u>not</u> titrate at the same pK as the 695 nm band, and was once again in near-perfect agreement with the result obtained by parametric least-squares fitting.

DISCUSSION

This work has shown that when the pH dependency of the reduced-minus-oxidized FTIR difference spectrum of cytochrome *c* is resolved into components with different pK values, two independent and fundamentally different mathematical techniques yield essentially identical results. This is persuasive evidence for the accuracy of both mathematical methods. Parametric least-squares fitting also yielded the two pK values, which could not be obtained by 2D-IR correlation analysis. However, since the sign of the asynchronous correlation intensity indicates the direction of the phase shift, a comparison of the component spectra with the raw data can be used to determine the sequence of deprotonations. In addition, the 2D-IR analysis confirms that the lower of the two pK values for the FTIR data is identical with the single pK value found for the 695 nm band.

The analysis of the correlation intensities between the FTIR and visible range data was found to be particularly advantageous for the question originally addressed by this work, the determination of the change in the FTIR spectrum associated with the disappearance of the 695 nm band. Figure 5 shows that the asynchronous correlation spectrum of the FTIR data with the 695 nm band can be obtained in a single step: subtraction from the raw data would then yield the spectrum desired. This approach would work even if the asynchronous correlation spectrum of Figure 5 were the sum of many unresolved components.

In general, the 2D-IR correlation method was found to offer a number of significant advantages for the analysis of pH titration data from proteins:

1) In 2D-IR, the number of independent components can be determined by careful inspection of the asynchronous correlation map. Parametric fitting requires that the number of components be known in advance in order to set up the equation to be fitted, or - as in the present case - that the entire procedure is repeated with a variety of assumptions.

2) 2D-IR does not require a mathematical model. Parametric least-square fitting will produce meaningless data if the model used is incorrect: in the worst case, it may not be apparent that the results are meaningless.

3) 2D-IR automatically separates the dynamic and static components of the spectrum: i.e., the results obtained are identical regardless of whether the data from the left-hand or right-hand panel of Figure 1 are used. For least-squares fitting, it was necessary to extend the titration data down into the flat region below pH 8, in order to obtain and fit the data from the right-hand panel of Figure 1. Otherwise, an offset from zero would have had to be included as an additional parameter to be fitted, thus reducing the accuracy of fitting for the other parameters.

4) 2D-IR does not require any initial parameter estimates. The results obtained by parametric least-squares fitting can depend quite sensitively on the choice of initial estimates, which may allow some subjectivity to creep in.

Some degree of subjectivity may still remain when examining the asynchronous correlation map to determine the number of independent components, where genuinely unique spectra must be distinguished from linear combinations of two or more other spectra. A suitable mathematical algorithm would clearly be useful here.

In conclusion, this study has provided persuasive new evidence that the spectrum represented by the solid lines in Figures 3 and 4 is the FTIR difference spectrum corresponding to loss of the 695 nm band in oxidized cytochrome *c*. At the same time, the study has demonstrated that the 2D-IR correlation method is a useful tool for the analysis of pH titration data from proteins.

REFERENCES

1. Moore, G.R. and Pettigrew, G.W. (1990) *Cytochromes c: Evolutionary, Structural and Physicochemical Aspects*. Springer Verlag, Berlin
2. Moss, D.A., Nabedryk, E., Breton, J. and Mäntele, W. (1990) *Eur. J. Biochem.* **187**, 565-572
3. Moss, D.A., Baymann, F., Grzybek, S. and Mäntele, W. (1997) *Biochemistry* (submitted)
4. Noda, I. (1986) *Bull. Am. Phys. Soc.* **31**, 520
5. Noda, I. (1993) *Appl. Spectrosc.* **47**, 1329-1336
6. Tonge, P., Moore, G.R. & Wharton, C. (1989) *Biochem. J.* **258**, 599-605

Cylindrical Internal Reflectance FT-IR Interactions Between Components of Biological Molecules and Metal Ions - II

P. Bruni, C. Conti, E. Giorgini, M. Iacussi, E. Maurelli and G. Tosi*

Dipartimento di Scienze dei Materiali e della Terra, Università degli Studi di Ancona, via Brecce Bianche, 60131 ANCONA, ITALY

The interactions of metal ions with D-ribose and 2'-D-deoxyribose as well as 2'-D-deoxyadenosine and 2'-D-deoxyadenosine-5'-monophosphate have been investigated in solid, glassy state and in water solutions by comparing Diffuse Internal Reflectance, Cylindrical Internal Reflectance and Transmittance spectra. The metal ion mainly interacts with C-C, C-O bonds of sugars as well as with N(1)=C-NH_2 , N(7) and phosphate moieties of nucleotide. Cylindrical Reflectance, even in dilute solutions, constitutes an useful technique to study the behaviour of biological molecule fragments in water.

INTRODUCTION

An important contribute to the knowledge of the specific influence of each fragment on the biological activity of polynucleotides has been achieved through FT-IR spectroscopy: reports on interactions of D-glucose and D-galactose with some metal ions in the solid and glassy state have been recently published (1) and support X-ray determinations (2). For a better understanding on triple complexes between polynucleotides, vesicles and metal ions, the analysis of the role played by various fragments of the polynucleotide is fundamental (3). In this work, an infrared study of the interactions of biological molecules components with metal ions, is reported. The purpose of the present contribute is to outline the leading role of Cylindrical Diffuse Reflectance in the study of fragments in the puzzle of some complex biological molecules and to attempt a comparison with other spectroscopic techniques.

EXPERIMENTAL

Water solutions (neutral conditions) of D-ribose, 2'-D-deoxyribose with Mg, Ca, Mn, Cu, Zn chlorides as well as of 2'-D-deoxyadenosine and 2'-D-deoxyadenosine-5'-monophosphate with Ca chloride, have been used for liquid, glassy and solid state determinations (4). All compounds were Fluka reagents. The spectra have been obtained with a Nicolet 20-SX FT-IR spectrometer (DTGS detector) equipped with: (i) a Spectra Tech. Cylindrical Internal Reflectance (CIR) cell mounting a Zinc selenide rod for liquid phase determinations; (ii) a Spectra Tech. Multiple Internal Reflectance (DRIFT) accessory for measurements in the solid state; (iii) NaCl cells for determinations in glassy state (G).

Resolution 4 cm^{-1}, 250 or 64 scans for reflectance and transmittance spectra, respectively. Treatment of the data have been achieved with a Galactic software package.

RESULTS AND DISCUSSION

Vibrational modes due to alcohol, ether and ring moieties in sugars can be coupled each other and, in biological molecules, are often overlapped or obscured by strong absorptions from other groups (phosphate and nucleosides) (5). Spectroscopic data on sugars and on their interactions with metals have been reported on crystalline compounds, D_2O solutions and glassy state (1,6). Recently, CIR spectra of binary systems D-xylose, D-galactose and metal ions have been reported (7).

Important vibrational modes of the sugars can be found in the region 1500-700 cm^{-1}: the bands between 1500-1200 cm^{-1} are mainly due to C-C-H, C-O-H, O-C-H bending, those between 1200-900 cm^{-1} to C-O and C-C stretching and those between 900-700 cm^{-1} to skeletal vibrational modes (3, 8). The spectroscopic patterns of Ribose and 2-deoxyribose, that differ for a CH_2OH group, are rather dissimilar (Figs. 1a, 2a). DRIFT spectra, mainly in the region 1500-1200 cm^{-1}, evidence bands more resolved and intense than in glassy state or in solution (Figs. 1a, 1b).

----

* To whom correspondence should be addressed.
(E-mail: tosi@popcsi.unian.it; Fax: ++39-71-2204714)

CP430, *Fourier Transform Spectroscopy:* 11th International Conference
edited by J.A. de Haseth

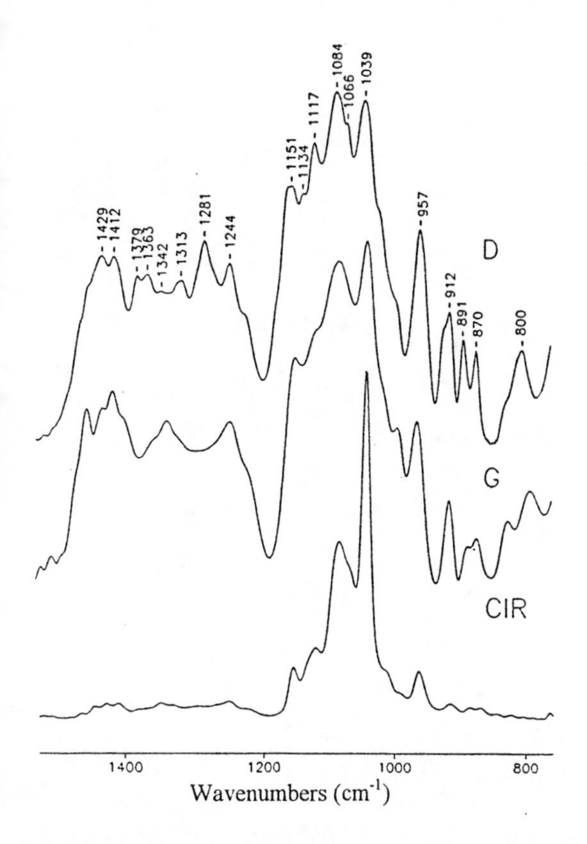

FIGURE 1a - From the top: DRIFT (D), Glassy State (G) and CIR spectra of D-Ribose in the 1500-800 cm^{-1} region.

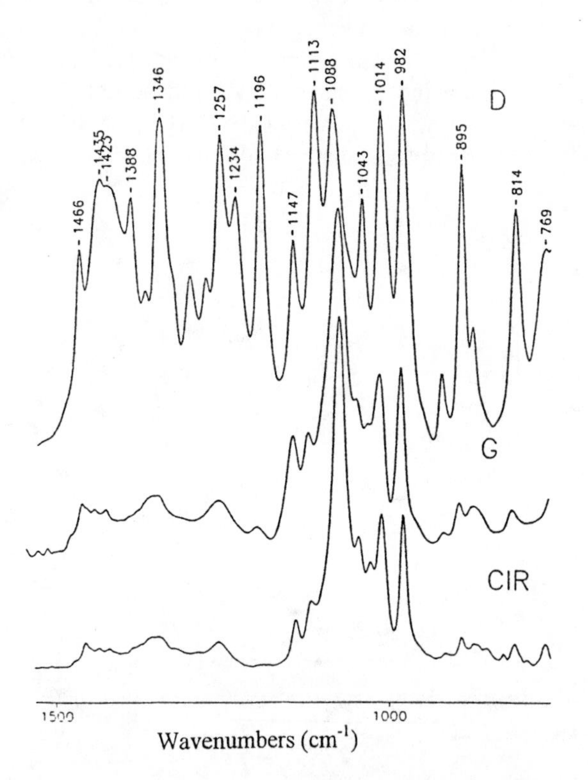

FIGURE 1b - From the top: DRIFT (D), Glassy State (G) and CIR spectra of 2'-D-deoxyribose in the 1500-800 cm^{-1} region.

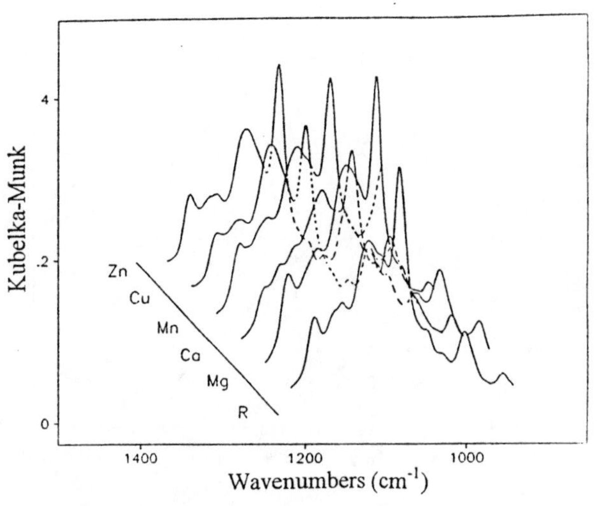

FIGURE 2 - 3D CIR spectra of D-Ribose (R) with Mg^{2+}, Ca^{2+}, Mn^{2+}, Cu^{2+}, Zn^{2+} between 1200-850 cm^{-1}.

The additional CH_2OH in ribose induces changes in $\nu_{C\text{-}O}$, ν_{CH2O} asym and ν_{CH2O} sym modes (1170-900 cm^{-1}) and in skeletal absorptions (850-700 cm^{-1}). In DRIFT spectra, additional, or stronger bands are found in 2'-D-deoxyeribose at 1196, 1088, 982 and 895 cm^{-1} while in D-ribose, the intensity of the absorptions (ν_{CH2O}, $\nu_{CH2Oring}$) at 1110-1040 cm^{-1} is more pronounced and convoluted. Differences between glassy and CIR spectra of the D-ribose can be noted around 1050 cm^{-1}.

Figure 2 shows the dependence of the vibrational modes of D-ribose from metal ions in the region 1200-900 cm^{-1}. The changes induced by the metal cation appear not relevant except for the modification of the intensity of some bands, first among all, the one at 1088 cm^{-1}. Calcium ion appears to induce more changes in the band profile. An analogous trend can be found in 2'-D-deoxyribose. At this regard, we attempted the evaluation of the degree of the interaction by examining the band at 1088 cm^{-1}.

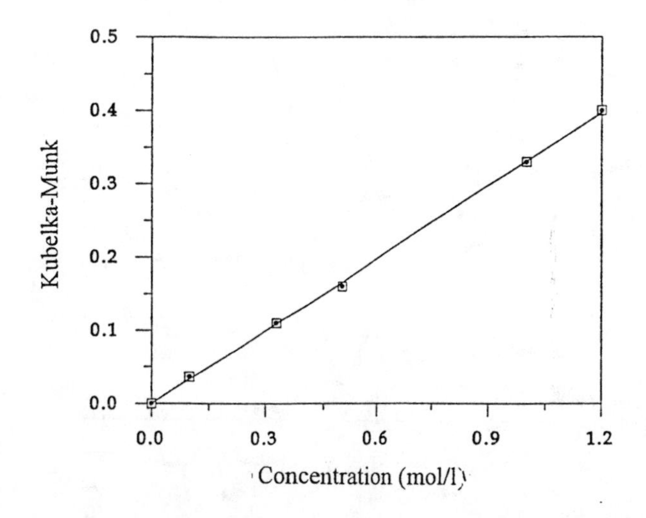

FIGURE 3. Calibration curve on the band at 1088 cm^{-1} of 2'-D-deoxyribose in water (CIR spectra).

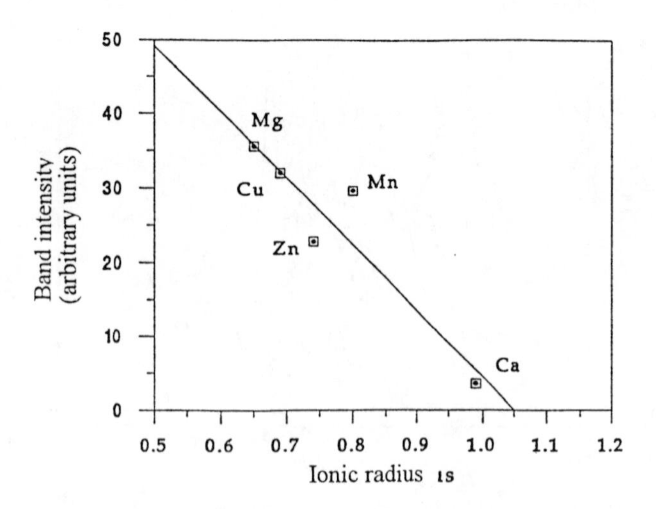

FIGURE 4. Dependence from the ionic radius of the subtraction band at 1088 cm^{-1} of 2'-D-deoxyribose

The calibration curve is reported in Figure 3. Figure 4 reports the dependence from the ionic radius of the metal of the subtraction band at 1088 cm^{-1}: on adding the metal ion, the intensity of the band of the sugar lowers and hence, subtracting the spectrum of 2'-D-deoyribose from the one of 2'-D-deoyribose/metal ion, results in a band (subtraction band) that is then arithmetically converted to obtain a positive value of the intensity. The lack of a satisfactory relation between the two parameters suggests that other facts, like hydration of the cation, must be operating. Either the intensity of some vibrational modes and, partially, the band profile of both sugars, depend on the concentration of the metal cation. Figure 5 shows the increase of the absorbance of subtraction bands of 2'-D-deoxyribose on increasing the amount of Zn^{+2}: an increase of the degree of the interaction lowers the intensity of the bands of the sugar in the complex and then, the intensity of the subtraction band increases.

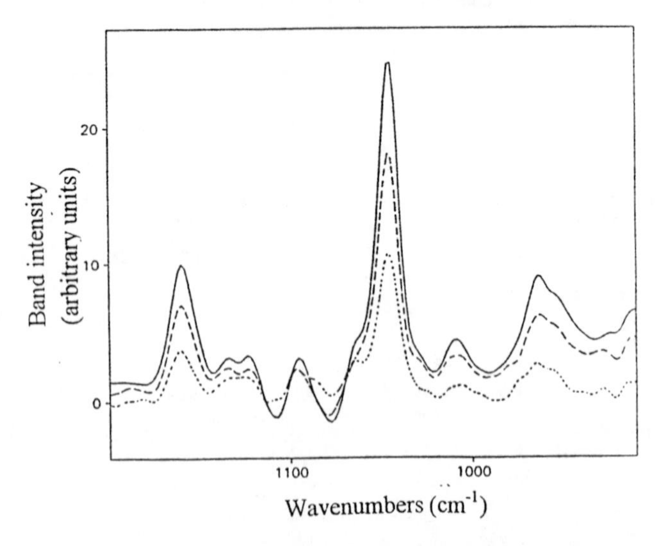

FIGURE 5. Increase of the intensity of the subtraction bands of 2'-D-deoxyribose between 1200-900 cm^{-1} from 2'-D-deoxyribose/ Zn^{+2} molar ratio: 1/1 (···), 1/2 (---), 1/3 (—).

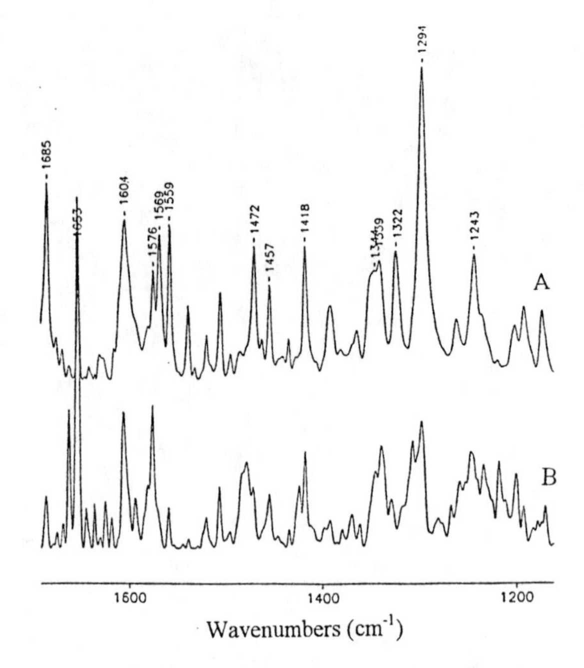

FIGURE 6. CIR spectra of 2'-D-deoxyadenosine (A) and 2'-D-deoxyadenosine/Ca^{+2} (B) in the 1700-1200 cm^{-1} region.

CIR spectra of 2-deoxyadenosine (A) and of equimolar 2-deoxyadenosine/Ca^{+2} (B) water solutions in the region 1700-980 cm^{-1}, are reported in Figure 6. Even if the solubility in water of the components is low, the spectra resulted satisfactorily resolved. On adding Ca^{+2}, the following considerations could be done: the intensity of the weak doublet of the NH_2-C(5)=N(6) at 1648, 1635 cm^{-1} is enhanced; also the band at 1486

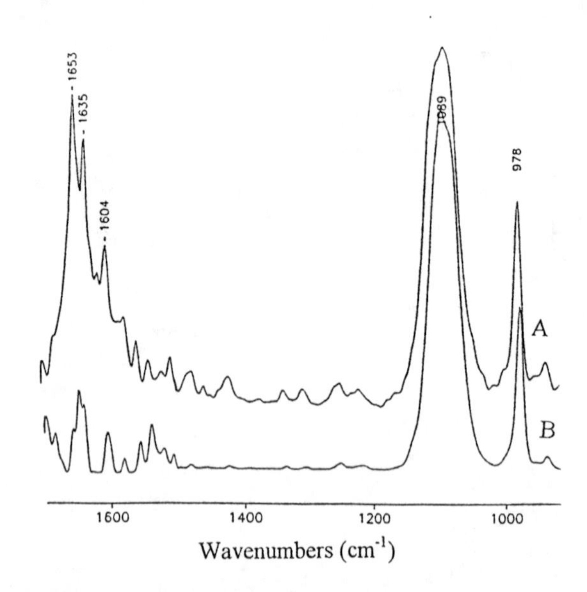

FIGURE 7. CIR spectra of 2'-D-deoxyadenosine-5'-phosphate (A) and 2'-D-deoxyadenosine-5'-phoshate/Ca^{+2} (B) in the 1700-1200 cm^{-1} region.

cm^{-1} (N(7)=C-) increases in intensity; the band at 1420 cm^{-1} is broadened; the strong and sharp band at 1294 cm^{-1} results in a doublet at 1304 and 1296 cm^{-1} in the complex; some changes in the 1240-1180 cm^{-1} evidence the contribute of the imidazole ring to the interaction; finally, as discussed above, modifications are induced in sugar vibration modes in the 1150-950 cm^{-1} region.

Figure 7 reports CIR spectra of solutions of 2'-D-deoxyadenosine-5'-phosphate (A) and equimolar 2'-D-deoxyadenosine-5'-phosphate/Ca^{+2} (B). Changes in intensity and band profile are found between 1700 and 1400 cm^{-1}. The interference of Ca^{+2} on the protonation of NH_2-C=N moiety, is responsible for changes in the band profile. On adding Ca^{+2}, the maximum of the convoluted band from phosphate and sugar contributions is 10 cm^{-1} blue shifted while the band at 978 cm^{-1} is red shifted.

The above results are in good agreement with determinations in DRIFT and glassy state (that have been cited only at the beginning for sake of shortness) and with the ones from Katsaros and Theophanides (9,5), concerning determinations in the crystalline state on the same or similar systems.

CONCLUSIONS

A comparative analysis with different techniques (DRIFT, Glassy state, X-ray, etc.), allows a recognition of spectral features of sugars and complex biological molecules and of their interactions with metals. From the results we can discriminate the influence of the metal ion on the molecule as a whole or on a specific group. If solubility problems are not crucial (enough revolved spectra are obtained by using 10^{-3} M solutions), Cylindrical Internal Reflectance can be considered a reliable technique to achieve information from water solutions of biological systems.

REFERENCES

1. Xi, N., Weng, S., Wu, J., Xu, G., *SPIE, Fourier Transform Spectroscopy*, 1145, 411-12 (1989).
2. Cook, J.W., Bugg, C.E., *J. Am. Chem. Soc.*, **95,** 6442 - 46 (1973).
3. Yu, B., Hansen, W.N., *SPIE, Fourier Transform Spectroscopy*, 1145. 409-10 (1989) ; Bruni, P., Giorgini, E., Maurelli, E., Iacussi, M., Tosi, G, *Mikrochimica Acta*, 00, 000 (1996).
4. Liang, H. Guo, H., Xu, D.F., Wu, J.G., *Mikrochim. Acta*, **I**, 215-17 (1988).
5. Tajmir-Riahi, H.A., Theophanides, T., *Inorg. Chim. Acta*, **80**, 183-190 (1983).
6. Yang, L., WU, J., Zhou, Q., Bian, J., Yang, Y., Xu, D., XU, G., *10th International Conference on Fourier Transform Spectroscopy (10th ICOFTS)*, Budapest (1995).
7. Bruni, P., Iacussi, M., Tosi, G., *J. Mol. Struct.*, 00,000 (1997).
8. Conley, R.T., *Infrared Spectroscopy*, Boston, Allyn-Bacon Inc., 1966; Schrader, B. Ed., VCH Verlagsgesellschaft mbH ,*Infrared and Raman Spectroscopy*, Weinheim, 1995.
9. Grigoratau, A., Katsaros, N., *J. Inorg. Biochem.*, **24**, 147-54 (1985).

An Investigation Of Insect Eggshell Using Fourier Transform Infrared Spectroscopy

J.M. Gentner[1], E. Wentrup-Byrne[1*] and B.W. Cribb[2]

[1]*Center for Instrument and Developmental Chemistry, Queensland University of Technology, P.O. Box 2434, Brisbane, Q4001, Australia.*
[2]*Department of Entomology and Center for Microscopy and Microanalysis, University of Queensland, St Lucia, Q4067, Australia.*

The moth *Helicoverpa amigera* is a major agricultural pest. The use of parasites such as wasps from the genus *Trichogramma* is a promising pest control strategy. *H. amigera* eggs exposed to this parasite undergo a colour change due to the deposition of a dark layer of material on the vitelline membrane. FT-IR analysis of spectra from this layer reveals that the deposit is proteinaceous. The location of the amide bands suggests that the secondary protein structure consists of β-turns, α-helix and irregular structures. There is evidence of oxidation changes to cystine residues due to an inflow of atmospheric oxygen during the development of the parasite. Absence of characteristic bands indicates the deposit is not melanotic.

INTRODUCTION

The biomaterials which make up the multilayer structure of the insect egg shell accommodate, protect and serve the physiological needs of the developing embryo. The architecture and chemistry of each layer influences their particular function. However, despite its importance, the chemistry of the insect eggshell is relatively poorly known.

Globally, the moth *Helicoverpa amigera* (Hubner) Lepidoptera: Noctuidae is considered one of the most significant agricultural pests. The caterpillar stage of this moth attacks more than 150 species of plants and causes substantial crop damage. Hence, an important aspect of research pertaining to this insect is the development of pest control strategies, particularly alternatives to chemical pesticides. One promising strategy is the use of predators and parasites, such as the members of the genus *Trichogramma*. These wasps kill their hosts in the egg stage before the caterpillar hatches and damages crops. Because these natural parasite populations occur sporadically, and only after crops have been damaged, entomologists have been searching for *in vitro* rearing and mass release techniques.

H. Amigera eggs exposed to *Trichogramma* undergo a colour change at prepupal stage of the parasite. This colour change has been attributed to the deposition of small dark granules on the host vitelline membrane. (1) The nature of this black layer may have consequences for successful *in vitro* parasitic culture.

EXPERIMENTAL

H. Amigera Eggs

Eggs of *H. amigera* were obtained from laboratory culture at the University of Queensland. They were laid onto paper towelling. Larvae were reared on bean-based artificial diet and maintained according to the protocol of Teakle and Jensen. (2) Adult *T. australicum* were obtained originally from Queensland Department of Primary Industries, Toowoomba, and then from laboratory culture at the University of Queensland. Wasps (*Trichogramma australicum* Girault (Hymenoptera: Trichogrammatidae) were reared in the eggs of *H. amigera*. Mated females were supplied with honey, exposed to 1-2 day old *H. amigera* eggs and allowed to oviposit in these eggs. Wasps were removed after 12 hours and retained. Parasitised eggs were sampled after emergence of the wasp. Unparasitised (control) eggs were not exposed to the wasps.

FT-IR microspectroscopy

Infrared transmission spectra were obtained from the egg shell (chorion plus vitelline membrane) of unparasitised eggs after hatching of the caterpillar and the chorion and vitelline membrane together with black deposit from parasitised eggs after the emergence of the wasp. The instrumentation included a Perkin Elmer IR

* To whom correspondence should be addressed.

CP430, *Fourier Transform Spectroscopy:* 11th International Conference
edited by J.A. de Haseth

microscope with a liquid nitrogen cooled MCT detector and PE series 2000 FT-IR spectrometer.

Spectra were recorded between 4000 - 700 cm^{-1} over 128 scans at 8 cm^{-1} resolution using 100 μm diameter aperture. Samples were placed on a 13mm x 2mm barium fluoride disc.

Transition Electron Microscopy (TEM)

Eggs were fixed in 3% glutaraldehyde in 0.1 M cacodylate buffer, pH 7.2, 4° C for at least 2 hours, washed in buffer and post fixed in 1% osmium tetroxide in cacodylate buffer. Tissue was then dehydrated in a graded series of acetone, infiltrated and embedded in Spurr resin. Semi-thin sections were cut with a glass knife and stained with toluidine blue before observation under an Olympus light microscope. Thin sections were obtained using a diamond knife and stained with uranyl acetate and lead citrate before viewing at 80kV with a JEOL 1010 transmission electron microscope.

RESULTS AND DISCUSSION

Previous studies by Orfanidou *et al* (3) have been carried out on the unparasitised eggshell of other Lepidoptera species. These studies have shown the insect chorion to be composed of proteins with a predominantly antiparallel β-sheet secondary structure.

A TEM micrograph of the cross-section of the parasitised egg shell is shown in Fig 1. This investigation revealed that the very thin vitelline membrane (40 nm) is still present in both the unparasitised and parasitised egg shells. Further, the black layer is actually continuous but changes markedly in thickness, ranging from 300 to 2500 nm, hence the granular appearance. Sometimes this layer appears to split or be two layers which then rejoin. The black deposit, together with the vitelline membrane, can be manually separated from the chorion. However, the vitelline membrane could not be separated from the chorion of the unparasitised egg shell.

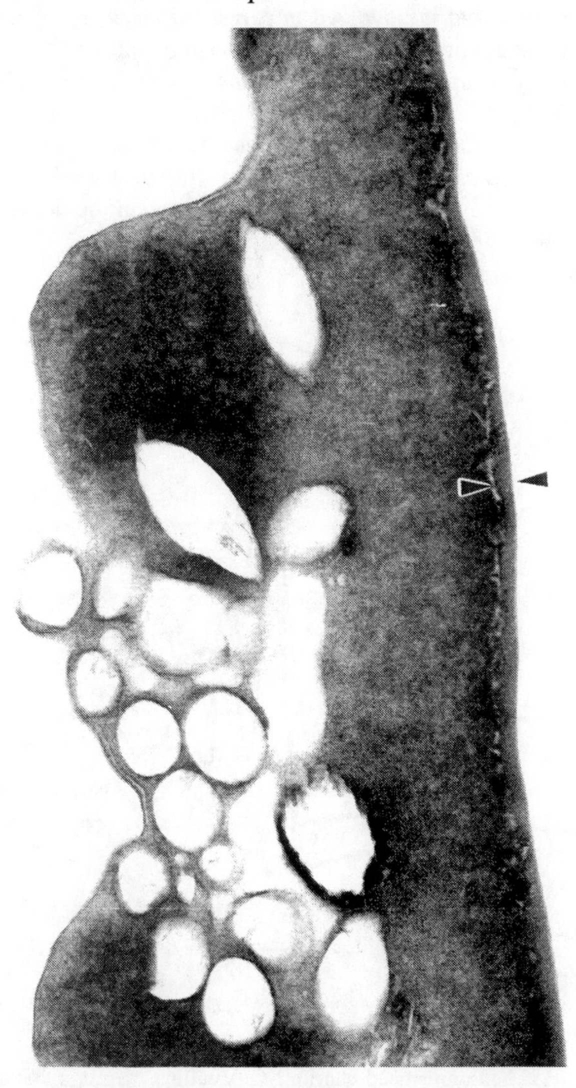

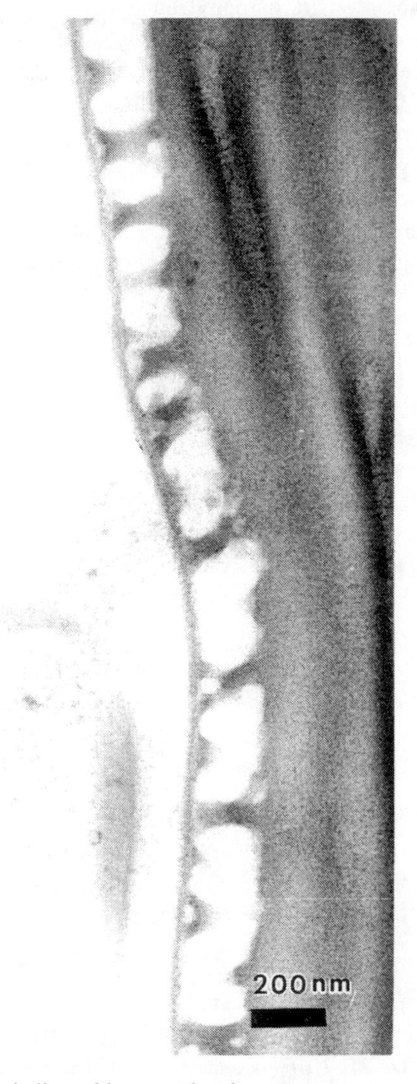

FIGURE 1. TEM micrograph of parasitised *H. amigera* egg shell. The vitelline membrane is indicated by arrowheads.

A representative infrared spectrum obtained from the hatched unparasitised egg shell is shown in Fig. 2a. The most intense feature in the amide I region, at 1634 cm^{-1}, indicates that the antiparallel β-pleated sheet structure is predominant.

In addition, shoulders at 1662 and 1692 cm^{-1} suggest the existence of β-turn structures in this proteinaceous material. This agrees with previous reports on the structure of other Lepidoptera egg shell structures. (3) Spectra from the chorion of the parasitised egg (Fig. 2b) closely resembled that of the unparasitised egg shell except for the intensities of the absorption bands between 2800 and 3000 cm^{-1}. These bands are assigned to the CH_3 (2959 and 2871 cm^{-1}) and CH_2 (2918 and 2850 cm^{-1}) asymmetric and symmetric stretching vibrations of the tissue components. In the absence of the vitelline membrane, the ratio of the CH_2: CH_3 bands are equal and typical of protein side-chain contributions. However, in the spectra of the unparasitised egg shell (chorion plus vitelline membrane) the CH_2 vibrations dominate this region of the spectrum. This is more typical of contributions from lipids, although a characteristic carbonyl band is absent, and would appear to be the only spectral contribution from the very thin vitelline membrane.

The spectra of the vitelline membrane and added black layer are shown in detail in Figs. 3 and 4. Given the negligible spectral contribution of the vitelline membrane and the comparative thickness of the these two layers, it can be assumed that these spectra are dominated by the composition of the added layer.

The first obvious property of the deposit is its proteinaceous nature. From Fig. 3 it can be seen that the amide I band is very broad. The dominant band at 1655 cm^{-1} reflects a significant proportion of α-helix and/or irregular secondary protein structure. From a further examination of the amide II (1544 cm^{-1}) and amide III regions (additional peak at 1284 cm^{-1}), it was concluded that both structures are present. The slightly elevated amide I/amide II intensity ratio, together with the broadening and shift to lower wavenumber of the N-H stretching vibration, suggests the black deposit may be hydrated. Further, the shoulder at 1624 cm^{-1} is characteristic of amide groups involved in stronger intermolecular hydrogen bonding. (4) It is unclear whether this intermolecular hydrogen bonding interaction is restricted to the black material or occurs between the black material and the vitelline membrane, or both.

A second clue was the absence of relatively intense bands at 1600, 1400 and 1045 cm^{-1}. These bands are characteristic of the natural pigment melanin. (5,6) In insects, melanin is found as a darkening agent in the cuticle and is involved in parasitoid encapsulation, which is a common defense mechanism employed by a host insect in response to invasion by foreign bodies. The host forms a capsule composed of host blood cells and melanin around the parasitoid egg, which may then die. (7) As these bands are not present in the spectra of the black layer, it is assumed that this is not a melanotic deposit.

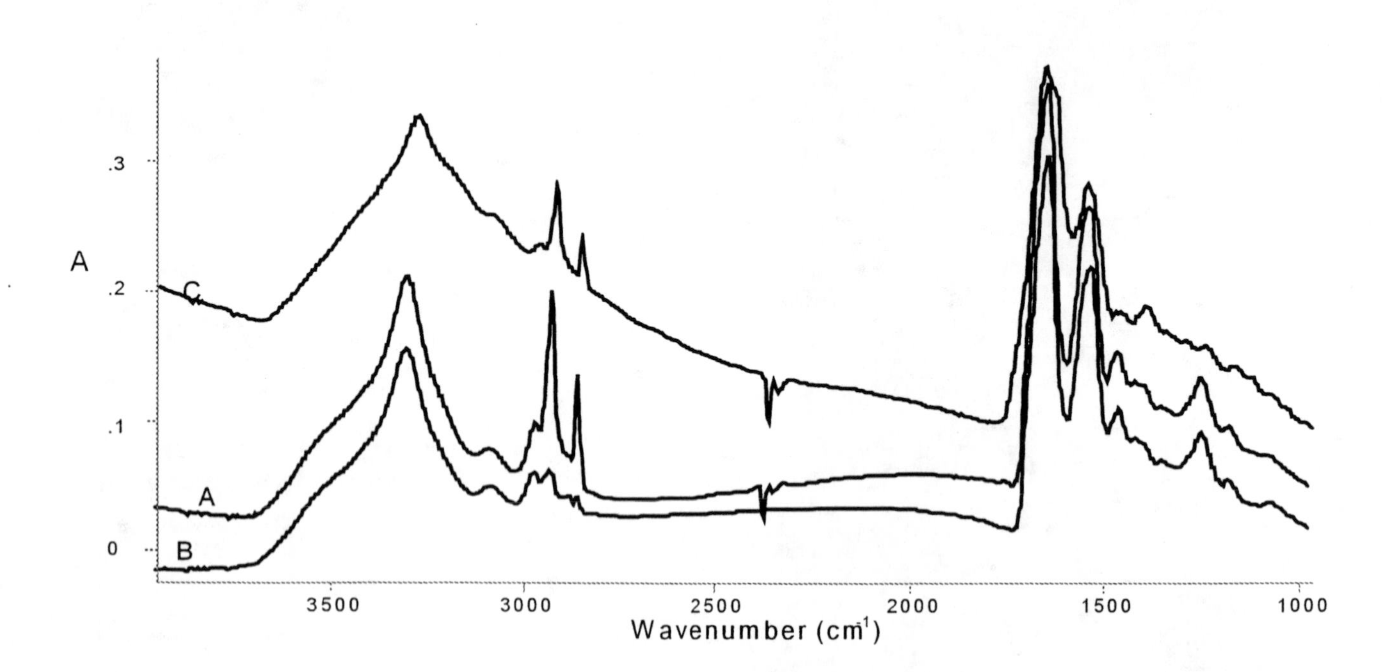

FIGURE 2. FT-IR spectra of *A*: Hatched unparasitised egg shell, *B*: Hatched parasitised chorion, *C*: Vitelline membrane and black deposit from parasitised hatched egg.

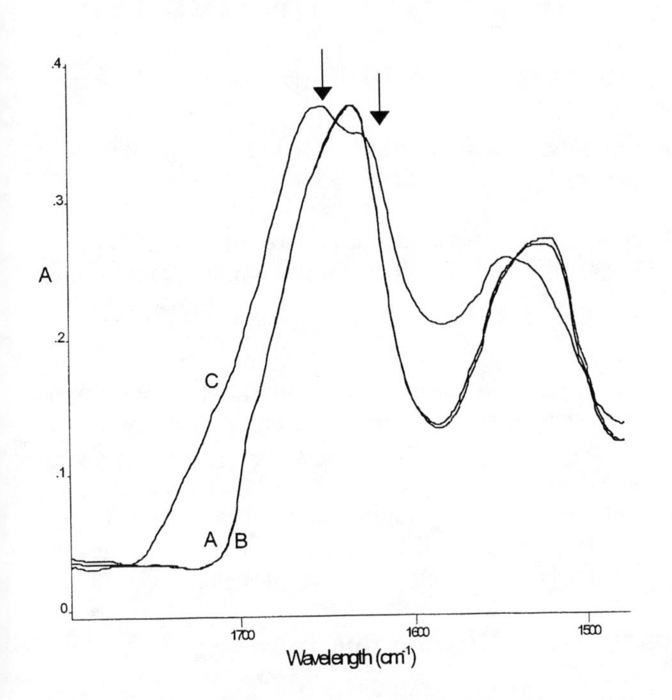

FIGURE 3. FT-IR spectra of *A*: Hatched unparasitised egg shell, *B*: Hatched parasitised chorion, *C*: Vitelline membrane and black deposit from parasitised hatched egg in the amide I and II region. Peaks at 1652 and 1624 cm^{-1} shown by arrows.

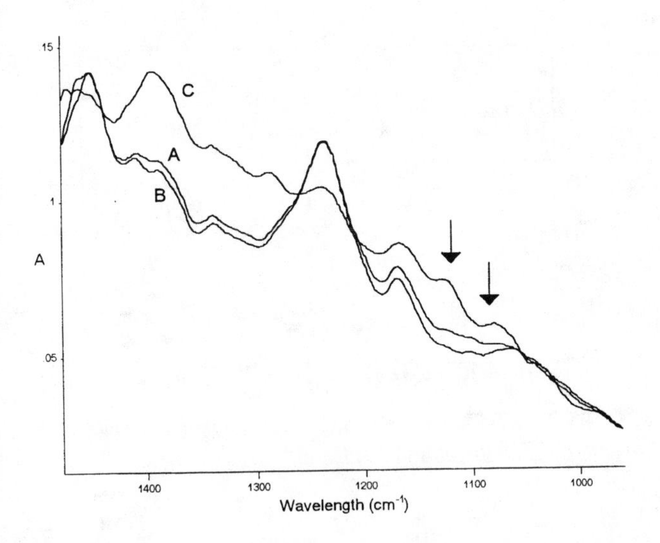

FIGURE 4. FT-IR spectra of *A*: Hatched unparasitised egg shell, *B*: Hatched parasitised chorion, *C*: Vitelline membrane and black deposit from parasitised hatched egg in the region 1500-1000 cm^{-1}. Peaks at 1121 and 1075 cm^{-1} shown by arrows.

Of further interest is the appearance of two bands at 1123 and 1075 cm^{-1} shown in detail in Fig. 4. These vibrations can be associated with the symmetric S-O stretching vibrations from the intermediate oxidation products of cystine residues, cystine monoxide and dioxide respectively. (8) Hutchinson *et al.* (1) observed that concurrent with the change in colour of the vitelline membrane, any unoccupied space in the parasitised egg fills with air. Natural oxidative degradation due in part to atmospheric oxygen has been observed in untreated hair. (9) A similar process is possibly the origin of these bands.

CONCLUSION

In this paper a multidisciplinary approach using FT-IR microspectroscopy and TEM has been employed as a first step to gaining a fuller understanding of the insect host-parasite interaction.

ACKNOWLEDGMENTS

The authors would like to thank Prof. Gordon Gordh for bringing together the proponents of the disparate sciences involved in this paper and Ms W. Amornsak for rearing and supply of *H. Amigera* and *T. australicum* adults. Both are from the Department of Entomology, The University of Queensland.

REFERENCES

1. Hutchinson, W.D., Moratorio, M., and Martin, J.M., *Annals Entomol. Soc. Am.* **83**, 46-54 (1990).
2. Teakle, R.E., and Jensen, J.M., in Singh, P. and Moore, R.F. (Eds), *Handbook of Insect Rearing Vol 2*, New York: Elsevier, 1985, pp. 313-322.
3. Orfanidou, C.C., Hamodrakas, S.J., Chryssikos, G.D., Kamitsos, E.I., Wellman, S.E., and Case, S.T., *Int. J. Bio. Macromol.*, **17**, 93-98 (1995).
4. Jackson, M., and Mantsch, H.H., *CRC Crit. Rev. Biochem. Mol. Biol.*, **30**, 95-120 (1995).
5. Zecca, L., Mecacci, C., Seraglia, R., and Parati, E., *Biochim. Biophys. Acta*, **1138**, 6-10 (1992).
6. Pierce, J.A., and Rast, D.M., *Phytochem.*, **39**, 49-55 (1995).
7. Blumberg, D., *Biological Control*, **8**, 225-236 (1997).
8. Joy, M., and Lewis, D.M. *Int. J. Cosmet. Sci.*, **13**, 249-261 (1991).
9. Strassburger, J., and Breuer, M.M., *J. Soc. Cosmet. Chem.*, **36**, 61-74 (1985).

FT-Raman Spectroscopy of Lichen Species from the Antarctic

J.M. Holder[1], H.G.M. Edwards[1], and D.D. Wynn-Williams[2].

1. *Chemistry and Chemical Technology, The University of Bradford, Bradford, West Yorkshire BD7 1DP. UK.*
2. *The British Antarctic Survey, High Cross, Madingley Road, Cambridge CB3 0ET. UK.*

In this investigation, FT-Raman spectroscopy has been utilised to characterize pigments and chemicals produced by Antarctic lichens, which have been exposed to increasing levels of UV-B radiation and other environmental conditions. The presence of calcium oxalate in some lichen species has also been confirmed spectroscopically.

INTRODUCTION

Many lichens are highly effective biodeteriorators of a range of rock substrata. The reaction of oxalic acid with calcium ions in rock substrata produces insoluble calcium oxalate. This plays a significant role in the weathering of calcareous rocks (1). If little or no oxalate is present, then other metabolic routes may be involved in the biodeterioration process. Lichens are known to produce secondary metabolic products which are usually specific to their symbiosis, and may prove to be important indicators of UV-radiation effects.

FT-Raman spectroscopy is currently being used to investigate the different strategies adopted by various lichen species to combat high levels of UV-radiation.

OBJECTIVES

- To investigate lichen pigment content in relation to exposure to ultraviolet radiation. The effects of UV are being studied on a latitudinal and altitudinal gradient, in the Antarctic, with the main species being *Xanthoria* and *Acarospora*.
- Other protective mechanisms developed by the lichens to minimise damage to their metabolic and reproductive systems by UV radiation eg calcium oxalate crystals.

EXPERIMENTAL PROCEDURE

Samples

Highly-pigmented lichen species from two different habitats have been compared. *Xanthoria parietina* (Figure 1a) from a maritime temperate location, has been compared with *Xanthoria mawsonii* (figure 1b) and *Xanthoria elegans* (Figure 1c) from climatically stressed Antarctic environments. Raman spectroscopy has also been used to investigate the pigments and chemicals produced during the biodeterioration process, for example, calcium oxalate crystals (Figure 2), in *X. elegans* and *Acarospora gwynii*. Extracts were obtained from these lichen species using recommended procedures from the literature (2). Identification of the lichen substances extracted has been confirmed with melting point tests and color reactions.

Raman Spectroscopy

Fourier-Transform Raman spectra of the lichen and extract samples were obtained using a Bruker IFS66 infra-red instrument with FRA 106 Raman module attachment. An Nd^{3+}/YAG laser operating at 1064 nm with a nominal source power of 100 mW was used to excite the spectra, which were accumulated from up to 4000 scans at 4 cm^{-1} spectral resolution for the lichen samples, and 1000 scans at 4 cm^{-1} spectral resolution for the extract samples, from a typical sample illumination area of 0.1 mm diameter.

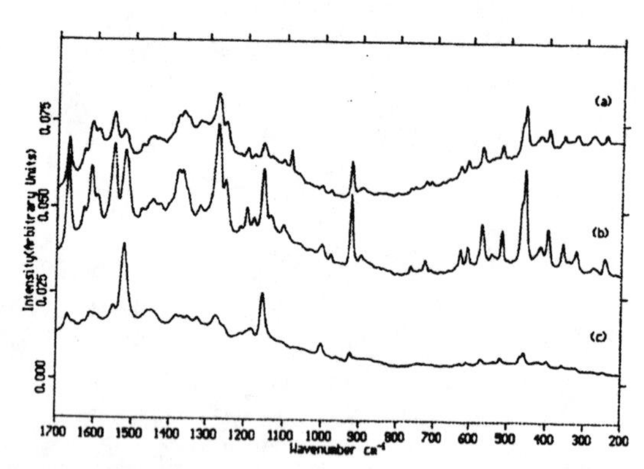

FIGURE 1. FT-Raman spectra of three *Xanthoria* species a) *X. parietina* b) *X. mawsonii* c) *X. elegans*.

CP430, *Fourier Transform Spectroscopy:* 11th International Conference
edited by J.A. de Haseth
 1-56396-746-4/98/$15.00

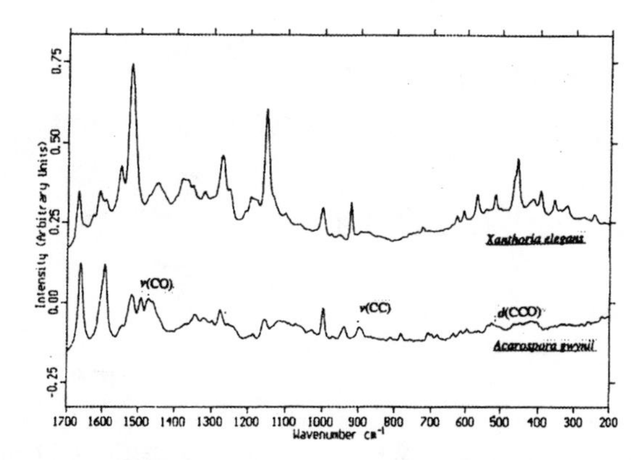

FIGURE 2. X. elegans and A. gwynii with the major oxalate bands indicated (1700 - 200 cm^{-1} region).

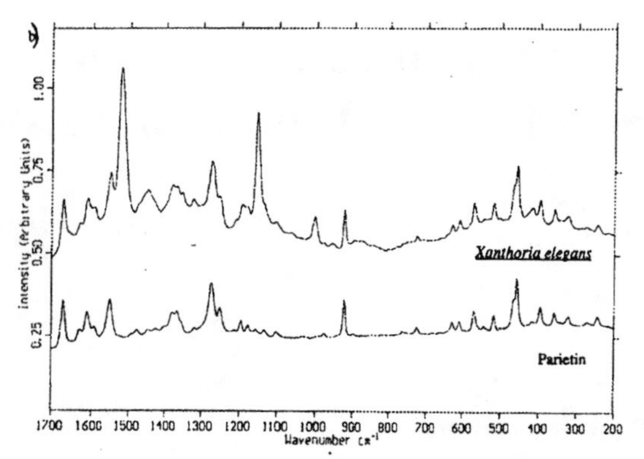

FIGURE 3. The stack plot of parietin and X. elegans (1700 - 200 cm^{-1} region).

RESULTS AND DISCUSSION

The environmental stresses which occur on the Xanthoria species from Victoria Land, Antarctica, are currently being correlated to pigment changes in the Raman spectra, particularly in the Raman bands at 1156 and 1525 cm^{-1} corresponding to carotenoids. Significant spectral differences occur at 1555, 1282 and 1004 cm^{-1}, between X. mawsonii and X. elegans from their differing Antarctic habitats, and the spectra indicate that there could be different survival strategies operating (Figures 1b and 1c).

Formation of calcium oxalate monohydrate occurs at a relatively high temperature (5 - 10^{0}C) in more acidic conditions than calcium oxalate dihydrate, which forms under colder, more humid conditions below 5^{0}C. A. gwynii has a much greater amount of oxalate present than X. elegans (Figure 2) which only has approximately 1 %.(3) Calcium oxalate monohydrate has Raman *V*(CO) bands at 1495 and 1465 cm^{-1}, whilst calcium oxalate dihydrate occurs at 1475 cm^{-1}.

The relationship between X. elegans and A. gwynii could possibly be a mutually beneficial one (symbiosis) in their respective oxalate and parietin contents. The oxalate could be a means of water storage in harsh climates, and so the A. gwynii may provide a water storage to the X. elegans, whilst the parietin may be a means of protection against high UV levels.

The extract believed to be parietin has been stack plotted with a number of lichen species, and is present in a few of these. Parietin was present in the largest quantity (Figure 3) in X. elegans. Vibrational assignments have been made for this extract using an anthraquinone model, which is thought to bestow the bright red-orange colour to any lichen in which it is present.

Model compounds based on lichen metabolic biproducts are now being tested as part of a comprehensive programme.

Wavenumber (1800 - 500 cm^{-1}) and vibrational assignments for the Raman spectra of X. elegans and parietin

Xanthoria elegans	Parietin	Proposed Assignment of vibrational mode
1672	1673	*v*(C=O) anthraquinone; parietin
1613	1614	*v*(CC) aromatic; parietin
1555	1555	*v*(CC) ring stretch, aromatic
1527		*v*(C=C) *B*-carotene
1452		*d*(CH_2)
1384	1381	*d*(CH_2)
1369	1370	*d*(CH_2)
1325	1322	*d*(CH_2); in-plane *d*(OH)
1279	1278	*d*(CH_2)
1257	1255	*d*(CH_2)
1155		*v*(C-C) *B*-carotene; anthraquinone
1003		aromatic; *v*(CC) ring breathing
925	924	*p*(CH_2)
631	629	*d*(CCO) ring
611	612	*d*(CCO) ring
572	571	*d*(CCO) ring
519	517	*d*(COC) glycosidic linkage

ACKNOWLEDGEMENTS

The authors would like to acknowledge the support of The British Antarctic Survey and NERC for aspects of the work presented here.

REFERENCES

1. C. Ascaso, J. Galvan and C. Rodriguez-Pascual, *Pedobiologia* **24**, 219 - 229, 1982.
2. Y. Asahina and S. Shibata, *Chemistry of lichen substances,* A .Asher and Co. Ltd Vaals-Amsterdam, 1971.
3. H.G.M. Edwards, N.C. Russell, M.R.D. Seaward and D. Slarke. *Spectrochimica Acta,* **51(A)**, 2091 - 2100, 1995.

Synchrotron Powered FT-IR Microspectroscopy Enhances Spatial Resolution for Probing and Mapping of Plant Materials

David L. Wetzel, Joseph A. Sweat, and Dia D. Panzer

Microbeam Molecular Spectroscopy Laboratory
Kansas State University, Shellenberger Hall, Manhattan, KS 66506, USA

Cross sections of grain kernels, leaves, other plant material, and their products have been examined routinely in our own laboratory with an integrated FT-IR microspectrometer equipped with a conventional (thermal) globar source. With plant material, scattering is often a problem. Representative (low density) mapping requires interpolation between spots on the tissue actually interrogated. High density (100%) mapping with a small pixel size is typically painstakingly done and requires coaddition of many scans. With the synchrotron source (National Synchrotron Light Source, Beamline U2B) of the U.S. Department of Energy's Brookhaven National Laboratory, Upton, New York, nearly all of these problems are solved. Low thermal noise and brightness of the beam provide high S/N. The non-divergence of the synchrotron microbeam allows the high S/N to be retained even with aperturing of 6 μm or 12 μm sizes. Diffraction influences the practical limit. Step sizes corresponding to the small aperture dimension reveal highly localized chemical differences between adjacent pixels of a tissue specimen.

INTRODUCTION

FT-IR microspectroscopy has been useful to us in the study of microchemical structure of plant materials for nearly a decade. The ability to analyze subsamples *in situ* in the microscope field of a 6 μm thick section of wheat and other biological tissues has been particularly exciting. Our earlier work was done using an infrared microscope accessory equipped with a front surface optics Cassegrainian objective and a similar condenser, dual remote project apertures and a sensitive detector miniaturized to coincide with the size of the beam (1). The microscope sampling accessory was mated to an interferometer bench with an optical interface containing a number of mirrors requiring multiple bounces. The ability to interrogate chemically different adjacent tissues was appealing. The infrared microspectrometer functions as a light microscope to permit viewing of the specimen and selection of the portion of the field to be interrogated by infrared. Not only were chemical differences observed between different botanical parts of the plant tissue, but different areas within the same botanical part such as wheat endosperm showed different chemical composition as did different parts of the wheat germ (2). Interest in interrogating small botanical features, that were surrounded by other types of plant tissue, led to the necessity for greater spatial resolution. This purpose was served with an integrated instrument (described later) in which the optics of the interferometer bench and the microscope are matched and signal is conserved. The object of this work was to achieve ultimate spatial resolution using the same optically efficient integrated instrument coupled to the "brightest light on earth" at the National Synchrotron Light Source. With the synchrotron brightness, radiant nondivergence, and absence of thermal noise, a high signal-to-noise ratio is achieved.

PREVIOUS MICROBEAM MOLECULAR SPECTROSCOPY OF PLANT TISSUES

Our need for improved spatial resolution was satisfied for use on a routine basis with an integrated FT-IR microspectrometer designed with the microscope and interferometer bench as a unit. It is optically efficient and as many as 8 mirror bounces are eliminated. This has been described as an infrared microscope with an embedded interferometer. With the IRμs™ (Spectra-Tech, Inc., Shelton, CT) instrument, 6 μm thick microtomed frozen sections of tissue are analyzed routinely. The tissue sections are mounted on 2 mm x 13 mm barium fluoride disks. A motorized stage with the IRμs instrument permits mapping of tissue particularly mapping across boundaries between different botanical parts such as kernel cross sections of cereal grains including wheat and corn (1-5). The transitions occurring from the pericarp to the aleurone cells and cell walls to subaleurone endosperm and finally to the central endosperm have shown interesting spectroscopic differences. The transitions from the germ to the endosperm have also been studied. The changes in chemical composition between subaleurone endosperm and central endosperm have been of particular interest in our work with wheat at Kansas State University. Careful sectioning of the germ has made it possible to examine the central root as well as the material around the root. Apart from the work with plant material, other biological materials such as mammalian tissue have been studies in cooperation with a neuroscientist at the University of Kansas Medical School.

CP430, *Fourier Transform Spectroscopy:* 11th International Conference
edited by J.A. de Haseth

In most of the mapping experiments, typically aperture sizes of 24 x 30 μm were used to selectively sample portions across a tissue. In general, 100% sampling was not attempted but maps were constructed by interpolation of peak areas from spectra obtained at regular intervals in an XY grid across the specimen being sampled. From spectra obtained in this manner, functional group maps were prepared from absorbance peak areas for bands of interest. The protein distribution was characterized from the amide II band at 1550 cm^{-1}. The 1025 cm^{-1} band was used for starch, the 1740 cm^{-1} small carbonyl peak was used for lipids and in cases where lipids were present in relatively large amounts CH_2 stretching vibrations at 2927 cm^{-1} and bending vibrations at 1469 cm^{-1} became of interest. The band at approximately 3300 cm^{-1} typically represented the OH in high carbohydrate containing specimens or the NH in primarily proteinaceous tissues. This broad band also served as a measure of the overall density of organic matter in the beam. In cereal processing, wheat milling for example is concerned with separating the starchy endosperm from the outer materials in the wheat and the ability to distinguish between structural carbohydrates such as cellulose and the edible starchy endosperm is of interest. Bands at 1420, 1370, and 1335 cm^{-1} were useful in this task. Furthermore, the carbohydrate band which for starch peaked at approximately 1025 cm^{-1} in a primarily cellulosic system would peak ordinarily at 1100 cm^{-1}. In the lipids from plant material, it was possible to detect unsaturation from a small peak at 3015 cm^{-1}.

The mapping of tissue sections is done routinely at the Microbeam Molecular Spectroscopy Laboratory at Kansas State University. It has also been possible using an aperture of 6 μm by 7 μm to perform a 100% map of a single aluerone cell of wheat (4,5). The cell was clearly defined from mapping the peak area of the amide II band and the cell walls were defined by the carbohydrate band area map. This achievement was accomplished painstakingly and required coaddition of many scans. Double aperturing was used before the objective and after the condenser to avoid accidental sampling of adjacent tissue by the process of diffraction. Without this precaution, spatial resolution is compromised. One particular challenge for which our system could not achieve sufficient spatial resolution was examination of the parenchyma bundle sheath in the vascular bundle of a cross section of grass. The chemical composition of this part of tissue is considered to influence the digestibility of forage materials by cattle. We wanted to use spectroscopic means to potentially predict this desirable feature of grasses being considered for future use in agriculture. All of these challenges led to the desire of seeking ultimate spatial resolution.

SYNCHROTRON SOURCE ADVANTAGES

High spatial resolution in microspectroscopy requires high signal-to-noise illumination of a spot of a few microns on the sample. Infrared synchrotron radiation is ideal to make this possible because this broad band source is very bright. It is not limited by thermal noise and the beam has a low divergence and is thus less subject to loss from small apertures used to restrict the target. Synchrotron radiation has been coupled to an integrated infrared microspectrometer (IRμS™) at Beamline U2B of the National Synchrotron Light Source at Brookhaven National Laboratory, Upton, New York. This was pioneered by Carr, Reffner, and Williams (6) in 1994.

In the synchrotron, electrons from an electron source are accelerated with a linear accelerator (LINAC) to an energy of *ca.* 75 million electron volts (MeV). Electrons from the LINAC enter a booster ring where they are further accelerated and from which they can enter either the X-ray storage ring or the vacuum ultraviolet (VUV) storage ring. Electrons entering the VUV storage ring are at an energy of approximately 750 meV. At the National Synchrotron Light Source, light is produced by accelerating bunches of electrons in either of the two closed orbit storage rings. It is well known that when electrons are accelerated they radiate energy in the form of electromagnetic waves. However, with the synchrotron, when the source of radiation (the bunches of electrons) is moving close to the speed of light a relativistic emission of radiation takes place where the radiation pattern becomes highly forward-directed. The intensity of light from the synchrotron beam is proportional to the number of electrons which is constant in the time required for a scan but varies slowly as the bunches of electrons in the storage ring declines. Intensity is constant when the electron orbit is stable. Unlike a conventional globar source that experiences thermal fluctuations, radiation from a synchrotron is characterized by the absence of thermal fluctuation and thus thermal noise. Spatial resolution in microspectroscopy requires a maximum of light illuminating the smallest possible area. Infrared

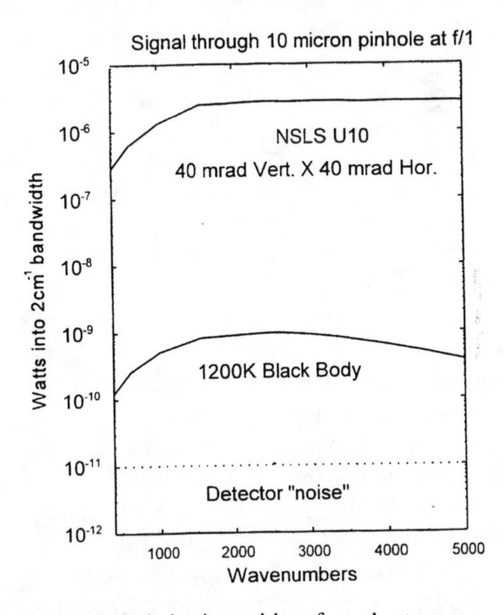

Figure 1. Relative intensities of synchrotron vs thermal source

synchrotron radiation is approximately 1000 times brighter than standard thermal sources (6). Figure 1 shows relative intensities of the synchrotron and globar sources in comparison to the noise that is typical for a miniature liquid nitrogen cooled mercury cadmium telluride detector. The high signal-to-noise of the synchrotron radiation has a great advantage. The synchrotron radiation is not only directional but concentrated within a narrow angle into a beam of high flux and losses of intensity are minimal for the radiation entering the interferometer and infrared microscope optics. Because a large percentage of the beam is concentrated into a small cross sectional area, aperturing of the field with a microscope does not discard a large percentage of the beam entering the microscope as it would for a more divergent conventional thermal source. As pointed out by Williams (7), in the case of the synchrotron, the radiation is emitted into an angle of approximately 10 milliradians by 10 milliradians. The emittance is approximately 10^{-4} mm^2 steradians. The emittance characteristic of the beam closely matches the throughput of the microscope for a properly illuminated 10 μm x 10 μm sample. This is a substantial improvement over a thermal source which has an emittance of more than 10^{-2} mm^2 steradians that provides only 0.001 % of light from the thermal source for illumination of a 10 μm sized spot on the sample. This is the reason that a conventional source equipped microspectrometer works well with 24-36 μm spots without coaddition of a large number of scans is acceptable. The commercial Spectra-Tech IRμs instrument identical to the one at the Microbeam Molecular Spectroscopy Laboratory was installed at a specially constructed beamline at the National Synchrotron Light Source at Brookhaven National Laboratory (5).

Radiation extracted from the VUV ring at the beamline contains soft X-rays and vacuum ultraviolet radiation in addition to that in the infrared region. The system used between the synchrotron beam port and the IR microspectrometer includes a first mirror that is a standard plain copper laser mirror. This mirror is water-cooled, absorbs X-rays and the VUV flux but reflects the infrared beam at right angles from the incident radiation to provide for Bremsstrahlung shielding. Other mirrors throughout the scheme direct radiation to the infrared microspectrometer. The storage ring operates at a high vacuum. The high vacuum is maintained through the optical arrangement by an intricate system of gate valves and window valves to a KBr window. The mirror chamber downbeam from the KBr windows is purged continually with nitrogen. An evacuated tube with KBr windows at each end is used to house the beam on its way to the microspectrometer. The columinated light was introduced into the standard IRμs microspectrometer by removing the columinating mirror normally used for the thermal source. At Brookhaven the facility has been used extensively for the characterization of materials. We have been privileged to use it on biological substances as well. Spatial resolution is important not just for microsampling, but for the isolation of subsamples within the microscopic field to avoid accidental sampling of neighboring tissue. The synchrotron powered Spectra-Tech IRμs microspectrometer operates near the diffraction limit and accidental sampling of neighboring tissue is minimized by the use of apertures after the condenser as well as before the objective to minimize the effect of diffraction on the sample area.

EXPERIMENTATION AND RESULTS

The facility described was used for obtaining single spectra from small spots selected by use of microscope viewing to find points of particular interest. A microscopist familiar with the morphology of the specimen may pick out individual areas of greater interest. By stepping aperture sized steps that are small in size, the result is 100% sampling. When this is done the pixel size that is used to produce an image from the mapping experiment provides greater detail. A section of wheat or corn endosperm may appear homogeneous to the eye looking through a microscope but not to the microspectrometer. Localized heterogeneities, heretofore unrecognized, may emerge. Figure 2 shows the localized heterogeneity of the lipids present in corn endosperm. This figure has the dimensions of 100 x 100 μm and heterogeneity is observed even within this small region. Previously without the use of the synchrotron at the Microbeam Molecular Spectroscopy Laboratory, we used 24 x 24 μm apertures and interpolation between separated data points to illustrate the heterogeneity in a map with the dimensions of 600 x 600 μm (3).

In addition to 100% mapping, the brightness of the synchrotron source and resulting high signal-to-noise enables large maps to be collected in a relatively short time due to less coaddition of scans. An example of a large map is shown in Fig. 3. A sorghum kernel was sectioned through the germ so that the section contained germ with endosperm on either side. By analyzing the peak area of the band at 1025 cm^{-1}, the distribution of carbohydrate in the form of starch reveals the location of the endosperm.

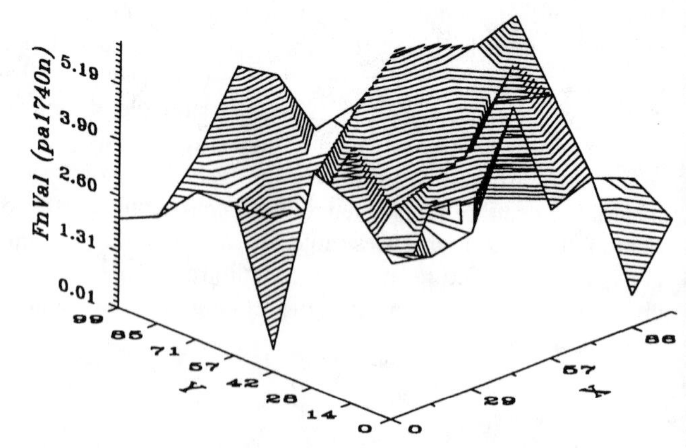

Figure 2. Peak area of 1740 cm-1 band for corn central endosperm 100% density map

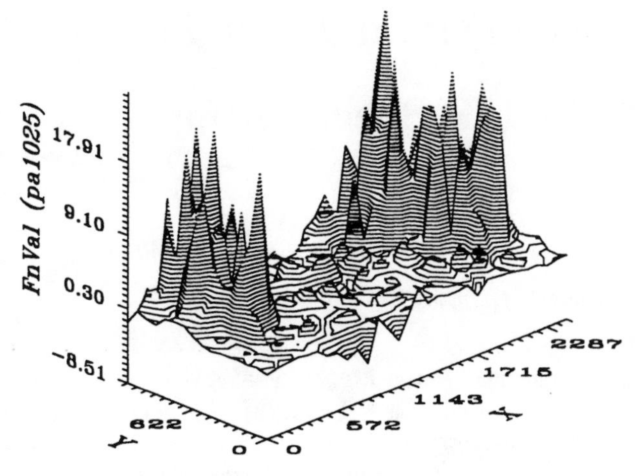

Figure 3. Peak area of band at 1025 cm-1 for a whole kernel section of sorghum

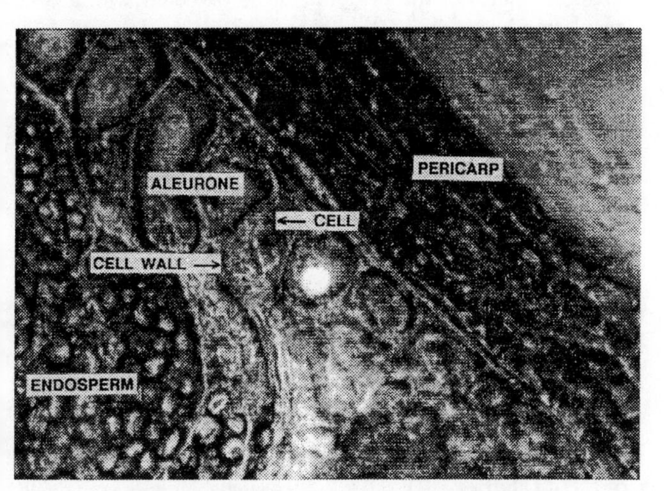

Figure 4. Photograph of a wheat section

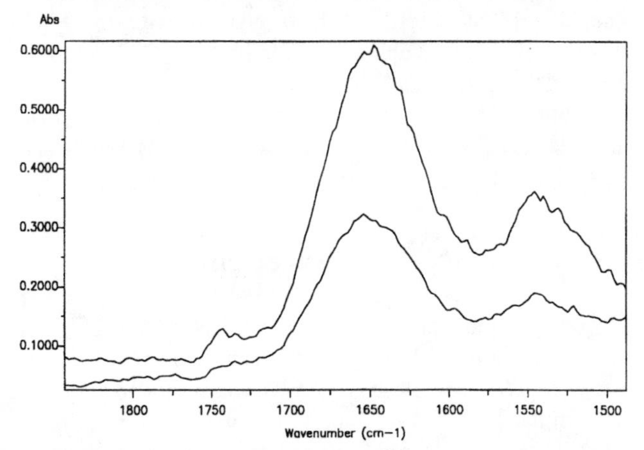

Figure 5. Spectra of aleurone cell (top) and an area outside of the cell (bottom)

Individual scans with an aperture of subcellular dimensions (5 μm) were also collected using the synchrotron source without the need for large numbers of coadditions. Figure 4 shows a photograph of the cross section of a wheat kernel with the various botanical part labeled. An illuminated image of the 5 μm aperture can be seen in the photograph positioned over the center of a single aleurone cell. The chemical contents of the single cell could be determined in this way. In Fig. 5, two spectra are shown for an aleurone cell and an area just outside of the cell. It can be observed that the cell contains larger relative amounts of protein (from the amide II band at 1550 cm^{-1}) and lipid (from the carbonyl band at 1740 cm^{-1}).

ACKNOWLEDGMENTS

This work was supported by the National Science Foundation, EPSCoR Grant No. OSR-9255223 and by the Microbeam Molecular Spectroscopy Laboratory at Kansas State University. The National Synchrotron Light Source is maintained by Brookhaven National Laboratory of the U.S. Department of Energy under project no. DE-AC02-76CH0016.

REFERENCES

1. Wetzel, D.L., Messerschmidt, R.G., and Fulcher, R.G., "Chemical Mapping of Wheat Kernels by FT-IR Microspectroscopy", presented at the Federation of Analytical Chemistry and Spectroscopy Societies 14th Annual Meeting, Detroit, MI, Oct., 1987.
2. Wetzel, D.L. and Fulcher, R.G. "Fourier transform infrared microspectroscopy of food ingredients", in: *Flavors and Off-Flavors*, Charalambous, G. (ed.), Amsterdam, Elsevier Science, p. 437, 1990.
3. Wetzel, D.L., "Molecular mapping of grain with a dedicated integrated Fourier transform infrared microspectrometer", in: *Food Flavors, Ingredients, and Composition*, Charalambous, G. (ed.), Amsterdam, Elsevier Science, pp. 679-728, 1993.
4. Wetzel, D.L. and Reffner, J.A., *Cereal Foods World*, **38**(1), 9-20, 1993
5. Wetzel, D.L., "Microbeam molecular spectroscopy of biological materials", in: *Generation, Analysis, Process Influences*, Charalambous, G. (ed.), Amsterdam, Elsevier Science, pp. 2039-2108, 1995.
6. Carr, G.L., Reffner, J.A., and Williams, G.P., *Rev. Sci. Instr.*, **66**, 1490, 1995.
7. Wetzel, D.L. and Williams, G.P., "Localized (5 μm) probing and detailed mapping of hair with synchrotron powered FT-IR microspectroscopy", in: *Proc. of 11th Int. Conf. of FT Spec.*, DeHaseth, J.A., American Inst. of Phyisics, 1998.

Contribution no. 98-56-B Kansas Agricultural Experiment Station, Manhattan, KS.

Studies on the mechanism of the *Sarcoplasmic Reticulum* Ca^{2+}-ATPase using Time-Resolved FTIR-Spectroscopy

F. v. Germar, A. Barth and W. Mäntele

Institut für Physikalische und Theoretische Chemie, Friedrich-Alexander-Universität Erlangen-Nürnberg
Egerlandstr. 3, D-91058 Erlangen, Germany

INTRODUCTION

We have studied changes in the vibrational spectrum of *sarcoplasmic reticulum* (SR) Ca^{2+}-ATPase upon nucleotide binding using ATP and the non-hydrolysable ATP-analogue, AMP-PNP (Adenylyl-imido-diphosphate), which does not allow progression of the reaction cycle beyond nucleotide binding.

Muscle contraction is triggered by an increase of the Ca^{2+}-concentration in skeletal muscle cells. For relaxation Ca^{2+} has to be transported back into the SR by the intrinsic membrane protein Ca^{2+}-ATPase, which couples active Ca^{2+}-transport to ATP hydrolysis. The reaction cycle is shown in a simplified form in Figure 1.

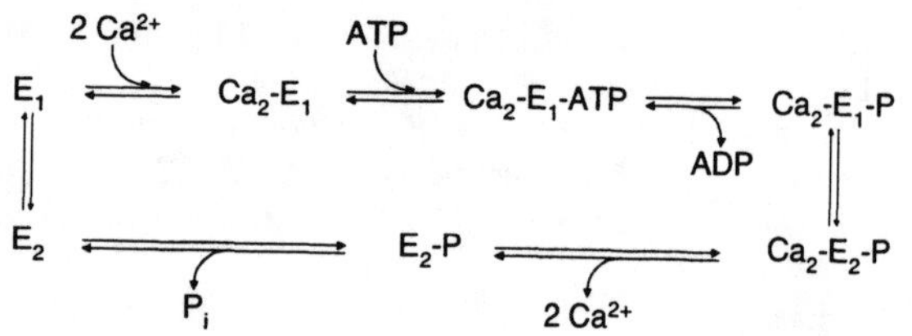

2 Ca^{2+} ; ATP
E_1 ⇌ Ca_2-E_1 ⇌ Ca_2-E_1-ATP ⇌ Ca_2-E_1-P (ADP)
E_2 ⇌ E_2-P ⇌ Ca_2-E_2-P
P_i ; 2 Ca^{2+}

FIGURE 1: Scheme of the reaction cycle of the *sarcoplasmic reticulum* Ca^{2+}-ATPase.

The model by deMeis and Vianna [1] is based on the assumption of two main functional conformation states, E_1 and E_2, of the protein. The transition from E_1 to E_2 is connected with a reorientation of the Ca^{2+}-binding sites from the cytoplasmic to the luminal side of the membrane. Two Ca^{2+} are bound from the cytoplasmic side to high affinity binding sites of the protein, which enables the Ca^{2+}-ATPase to hydrolyse ATP whereby the terminal phosphate-group is transferred to the amino acid side chain of Asp^{351}. After phosphorylation, the protein converts to the E_2-state where the Ca^{2+}-binding sites face the luminal side of the membrane. The decrease of the affinity for Ca^{2+} ions by a factor of 1000 leads to the release of both Ca^{2+} into the SR lumen. Next dephosphorylation of the enzyme completes the reaction cycle.

The reaction cycle can be triggered using "caged" substrates which are biologically inactive, UV-sensitive substrate-analogoues, i. e. "caged" ATP, "caged" AMP-PNP or "caged" Ca^{2+} (Figure 2). The FTIR technique detects conformational changes of the polypeptide backbone, of single amino acid side chains and of the functional groups of the substrate. Conversions between different states of the protein can be detected with a time resolution better than 50 milliseconds. Using "caged" ATP to initiate the reaction cycle, we can follow the reaction steps of nucleotide binding, Ca_2-E_1-P formation and the conversion to the E_2-P state. AMP-PNP is not hydrolysed by the Ca^{2+}-ATPase on the time scale of our experiments. Thus, only nucleotide binding is monitored with "caged" AMP-PNP.

MATERIALS AND METHODS

Sample preparation

Ca^{2+}-ATPase was prepared according to the method of Hasselbach and Makinose. After overnight dialysis in H_2O or ^2H_2O buffer, 10 µl of SR suspension were dried completely in a gentle stream of nitrogen and resuspended immediately in 0.6 µl of 12% glycerol/H_2O or ^2H_2O, respectively. The samples contained 1 mM Ca^{2+}-ATPase, 20 mM "caged" ATP or AMP-PNP, 0.5 mg/ml Ca^{2+} ionophor A23187, 2 mg/ml adenylate kinase, 100 mM imidazole HCl pH 7.0, 100 mM KCl, 10 mM $CaCl_2$ and 12% glycerol.

FTIR measurements

Measurements were carried out with a modified Bruker IFS 66 spectrometer equipped with a HgCdTe detector of selected sensitivity. The reaction was triggered by a Xenon flash tube. The flash energy was set to release approximately 2 mM nucleotide per flash. After the flash 70 spectra were recorded, 30 spectra with one scan each, 20 spectra with two scans and 20 spectra with fifty scans. Each scan consists of one complete forward and backward mirror movement, taking 55 ms. The cell had a pathlength of 5 µm and was thermostated at -7° C.

CP430, *Fourier Transform Spectroscopy:* 11th International Conference
edited by J.A. de Haseth

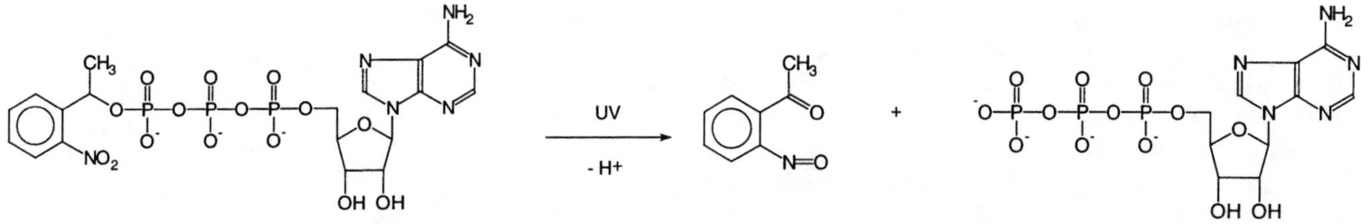

FIGURE 2: Release of ATP from "caged" ATP by an UV flash. In AMP-PNP, the oxygen between the β- and γ-phosphate is replaced by a NH-group.

Data processing

Difference spectra were obtained by calculating $-\log(I/I_0)$ from a spectrum after the flash (I) and a reference spectrum before the flash (I_0).
Positive bands indicate the state after the flash, while negative bands are characteristic of the state before the flash. For better results, each sample was probed 5 times and the spectra of 18 samples were averaged. For kinetic analysis, selected bands were integrated. Integration was performed by calculating the band area with a baseline that connects the two adjacent minima/maxima. If necessary, only half of the band area was integrated by using a horizontal baseline which starts from one of the adjacent minima/maxima.

RESULTS AND DISCUSSION

ATP and AMP-PNP release

The release of nucleotides in control samples without ATPase results in FTIR difference spectra (Figure 3) [(2)] which are discussed here because they superimpose on the protein absorbance changes in ATPase samples. The band pattern of the ATP- and AMP-PNP release spectra between 1800 and 1300 cm^{-1} are almost the same. These bands belong to changes that occur in the "caging" group. The negative bands at 1525 and 1345 cm^{-1} are assigned to the symmetric and antisymmetric stretching vibrations of the disappearing NO_2-group. The positive band at 1688 cm^{-1} is assigned to the C=O-group of the reaction products.

Bands in the region between 1300 and 900 cm^{-1} are caused by changes of the phosphate or the P-O-P-backbone vibrations. The transformation of the terminal PO_2^--group into a PO_3^{2-}-group can be seen at 1254 cm^{-1} and at 1120 cm^{-1}. In the AMP-PNP-spectrum, the influence of the NH-group shifts the 1254 cm^{-1} band to 1244 cm^{-1} At pH 7.0 the γ-phosphate is protonated in AMP-PNP, for that reason the PO_3^{2-}-band at 1120 cm^{-1} is missing.

The spectra of photolysis of "caged" ATP and AMP-PNP in ^2H_2O (not shown here) are very similar to the spectra in H_2O. The bands assigned to vibrations of the phosphate groups are shifting to smaller wavenumbers, but the band pattern is still the same.

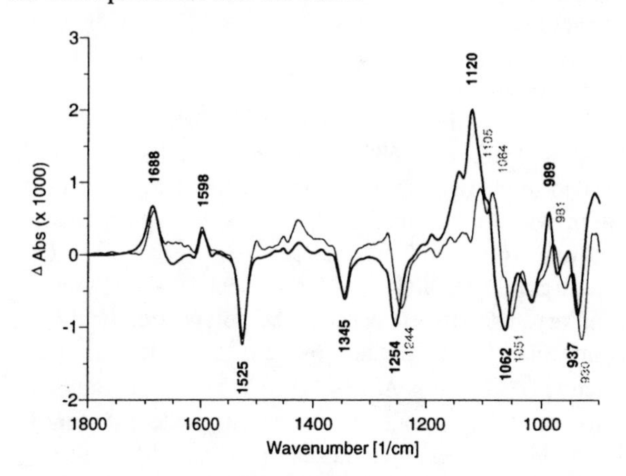

FIGURE 3: Infrared difference spectra of "caged" nucleotide photolysis in H_2O. ATP: bold line, AMP-PNP: thin line.

Nucleotide binding to the Ca^{2+}-ATPase in H_2O

We observe the "pure" nucleotide binding state of the protein between 1 and 2 seconds after the flash, because nucleotide release and binding are almost completely finished and the ATPase phosphorylation is still very low (Figure 4).

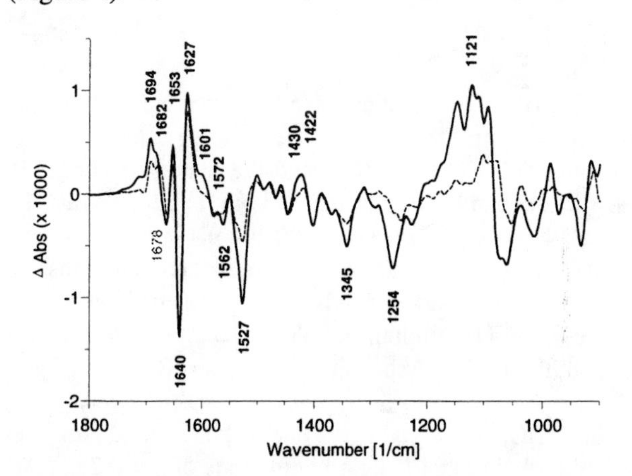

FIGURE 4: Difference spectra of ATP (bold line) and AMP-PNP binding (thin line)in H_2O.

The nucleotide binding spectra of ATP and AMP-PNP agree very well between 1800 and 1300 cm^{-1}. However, subtle differences between the ATP and the AMP-PNP binding spectra are observed caused by the imido group between the β- and γ-phosphate group of AMP-PNP. These can be seen most clearly at the positive shoulder at 1682 cm^{-1}, which is shifted to 1678 cm^{-1} in the AMP-PNP spectrum. Moreover, the positive bands at 1572 cm^{-1} and the negative band at 1562 cm^{-1} are missing in the AMP-PNP spectrum.

In the spectral region below 1300 cm^{-1}, the spectra are dominated by bands assigned to the photolytic release of the nucleotides, obscuring differences between the ATP- and AMP-PNP binding spectrum in this region, thus differences are not interpreted.[2]

Neither ATP nor AMP-PNP have vibrations that cause bands at 1682 or 1678 cm^{-1}, so these bands can be assigned to changes in the protein or to interactions of the protein with the nucleotide. Bands in the region between 1600 and 1690 cm^{-1} are characteristic for peptide C=O modes (amide I), and would represent small changes of secondary structure elements of the polypeptide backbone. In particular, bands around 1690 cm^{-1} indicate antiparallel β-sheets. ATP and AMP-PNP have almost the same configuration and should have the same nucleotide binding spectrum. However, there is one important difference that might account for the bandshift to 1678 cm^{-1}. The oxygen between the β- and γ-phosphate group of ATP is a hydrogen bonding acceptor, while the N-H group of AMP-PNP can serve as a hydrogen bonding donor. We thus take these spectra as strong evidence that the oxygen between the β- and γ-phosphate group in ATP binds to the polypeptide backbone of the SR Ca^{2+}-ATPase via a N-H•••O hydrogen bond.

Kinetic studies of the transition from the ATP binding state to the E_1-P state of the Ca^{2+}-ATPase

With a time resolution of 50 ms, we can follow the transition from the nucleotide binding state to the phosphorylated Ca^{2+}-ATPase and the hydrolysis of ATP. Most of the bands are affected by both. If we integrate the area of these bands, we obtain kinetic parameters that can be described as the sum of two first order reactions. In H_2O, only few marker bands show exclusively the time course of ATP binding (1466 cm^{-1}, $k = 3.2\ s^{-1}$) and phosphorylation of Asp^{351} (1719 cm^{-1}, $k = 0.2\ s^{-1}$) (Figure 5).

In ^2H_2O, there is only one marker band for the phosphorylation at 1711 cm^{-1} with a rate of $k = 0.2\ s^{-1}$. All other bands are affected by both of the two consecutive reactions. This can be seen most clearly by the 1663 cm^{-1} band, which increases intensity during the first two seconds (ATP binding) but decreases again while the Ca^{2+}-ATPase is phosphorylated (Figure 6). The rates are $k_1 = 9.4\ s^{-1}$ for the first and $k_2 = 0.25\ s^{-1}$ for the second reaction.

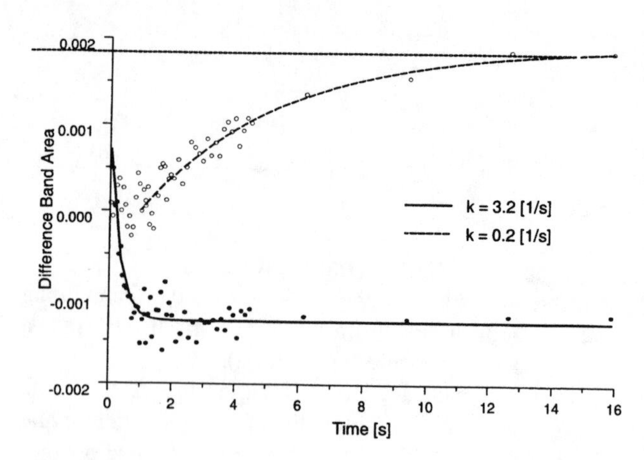

FIGURE 5: Kinetic traces for the 1466 cm^{-1} band (bold line, closed circles) monitoring ATP binding and of the 1719 cm^{-1} band (dashed line, open circles) monitoring phosphoenzyme formation in H_2O at -7° C.

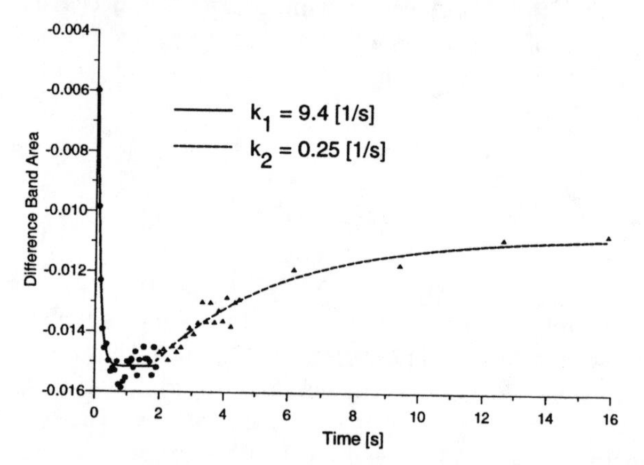

FIGURE 6: Kinetic trace for the 1663 cm^{-1} band, monitoring ATP binding (bold line, circles) and phosphoenzyme formation (dashed line, triangels) in ^2H_2O at -7° C

ACKNOWLEDGEMENTS

We are grateful to Prof. W. Hasselbach (Max-Planck-Institut, Heidelberg) for the gift of Ca^{2+}-ATPase and Dr. J. E. T. Corrie (National Institute for Medical Research, London) for the preparation of "caged" ATP and "caged" AMP-PNP.

REFERENCES

1. DeMeis, L. and Vianna, A., *Annu. Rev. Biochem.*, **48**, 275 - 292, 1979
2. Barth, A., Corrie, J. E. T., Gradwell, M. J., Maeda, Y., äntele, W., Meier, T. and Trnetham, D. R., *J. Am. Chem. Soc.*, **119 (18)**, 4149 - 4159, 1997
3. Barth, A., v. Germar, F. and Mäntele, W., *J. Biol. Chem.*, **271 (48)**, 30637 - 30646, 1996

Nanosecond Step-Scan FTIR Spectroscopy Applied to Photobiological Systems

C. Rödig, O. Weidlich, C. Hackmann, and F. Siebert

Institut für Biophysik und Strahlenbiologie, Albert-Ludwigs-Universität Freiburg, Albertstraße 23, 79104 Freiburg, Germany

Our improved step-scan FTIR instrument, capable of measuring spectra within 15 ns after the flash, is employed to measure flash-induced infrared difference spectra of bacteriorhodopsin, halorhodopsin and CO-myoglobin. For all three systems it is necessary to cover a large time range extending into several milliseconds. Therefore, the linear time base provided by the transient recorder board is converted to a quasi-logarithmic scale. Each of the three systems is characterized by several time constants extending over the large time range. For bacteriorhodopsin, it is shown that two spectral changes occur, one in the 20 and the other in the 100 ns time range. Furthermore, spectral differences between the two M states could be detected in the μs time range. For halorhodopsin, a clear batho intermediate with red-shifted ethylenic mode could be identified in the nanosecond time range. In addition, a transition corresponding to the N intermediate in bacteriorhodopsin was deduced. Further, it is shown that the millisecond time constant depends on Cl^- concentration, enabling the detection of the O intermediate. In the case of CO-myoglobin, spectral differences could be identified caused by mutations of the distal histidine of the heme binding pocket.

INTRODUCTION

Photobiological systems are characterized by photoreactions covering a large time range from femtoseconds to milliseconds or even seconds. It has been shown that time-resolved infrared difference spectroscopy is a powerful method for the study of the reaction mechanism of biological systems (e.g. (1)). Nevertheless, there will not be single infrared method available to investigate the complete temporal evolution. For the time range below 1 ns, the pump-probe or gated upconversion techniques have been developed (see articles in references (2,3)). In this contribution we demonstrate that time-resolved step-scan FTIR spectroscopy is well capable of covering the time-range from 15 ns onwards. We applied the technique to the investigation of the photoreactions of the light-driven proton pump bacteriorhodopsin (4,5), of the light-driven Cl^- pump halorhodopsin (6,7), and to CO rebinding to myoglobin after flash photolysis.

Bacteriorhodopsin contains as light-sensitive part the chromophore all-*trans* retinal which is bound to the protein via a protonated Schiff base. Absorption of light causes the all-*trans*→11-*cis* photoisomerization. In the M state the Schiff base deprotonates and all models of the pumping mechanism assume that it is this proton which is transported. The system has recently been studied by step-scan FTIR spectroscopy in the nanosecond time range (8,9). In this contribution, we also address this time range. In addition, taking advantage of the broad time range over that information is available, we searched for spectral differences between the two M states which have been postulated to explain the change in accessibility of the Schiff base from the extracellular to the cytosolic sides required for a vectorial proton transport (4,5). It is shown that for this task, although the transition between M_1 and M_2 is in the order of several μs, it is necessary to have information also on the 100 ns time range. The light-driven chloride pump is structurally very similar to bacteriorhodopsin. During the photoreaction the Schiff base does not deprotonate. Nevertheless it is thought that the Schiff base plays a crucial role in the pumping mechanism and that it has to change its accessibility. Therefore, a new intermediate has been postulated, which is, as in bacteriorhodopsin, called "N". This state has not yet been identified spectroscopically. A further aspect is the chromophore geometry in the next intermediate called "O". For the pumping mechanism it is important to know whether the chromophore has still the 13-*cis* geometry caused by the photoisomerization, or whether it has already thermally isomerized to the all-*trans* form. Both aspects are adressed by our step-scan FTIR investigations. Myoglobin is taken as a model system to investigate the binding of small ligands (CO) to proteins. This can be realized by photolyzing CO-myoglobin, by which the CO molecule is driven out of the heme pocket into the aqueous solvent. Subsequently, the recombination can be followed. Here, we address especially the conformational changes in the protein caused by the photolysis. It is shown that, by studying special mutants, they can be located to partly occur in the heme binding pocket.

CP430, *Fourier Transform Spectroscopy:* 11th International Conference
edited by J.A. de Haseth

EXPERIMENTAL

Bacteriorhodopsin from *Halobacterium salinarium* and halorhodopsin from *Natronobacterium pharaonis* were kindly provided by Dr. M. Engelhard (Dortmund). Recombinat human wild type myoglobin and myoglobin mutants were kindly provided by Dr. M. Ikeda-Saito (Cleveland). The instrument used for time-resolved step-scan measurements is described in the contribution by C. Rödig and F. Siebert in this volume, using our own software for data acquisition with quasilogarithmic time base (10) and for calculation of the spectra. The basic procedures are described in ref. (11). For photoexcitation, a frequency-doubled Nd:YAG laser was used, pulse energy was attenuated to approx. 1.5 mJ/cm^2.

RESULTS AND DISCUSSION

Bacteriorhodopsin

Fig. 1 shows the time time-resolved difference spectra in the ns-time range. They agree reasonably well with recently published data (8,9). The spectra are characterized mainly by chromophore bands. The band at 1640 cm^{-1} is partly caused by the C=N stretching vibration of the protonated Schiff base of the initial state. The band at 1511 cm^{-1}, the C=C stretching vibration of the retinal, shows that the photoproduct (KL) has a red-shifted absorption maximum. The corresponding band of the initial state is located at 1527 cm^{-1}. The band at 1190 cm^{-1} indicates the isomerization of the chromophore (C-C stretching mode coupled to CH bending). The range below 1000 cm^{-1} is characterized by hydrogen-out-of-plane bending vibrations. Protein bands can be seen in the amide-I (1690-1610 cm^{-1}) and amide-II (1570-1510 cm^{-1}) range. We especially agree with the analysis by Dioumaev and Braiman that two KL intermediates exist with a transition time in the order of 60 to 100 ns. The later intermediate has lower intensity of the C=C and C-C stretching modes at 1511 and 1190 cm^{-1}, respectively and of the HOOP mode at 960 cm^{-1}, indicating a relaxation process of the chromophore. However, we observe in addition very fast structural changes of the protein in the amide-I/II spectral range being completed after 30 ns. Such structural changes could explain the observed dependence of the photoproduct yield on the light pulse width (5 ns vs. 20 ns) (12). In addition, the negative band at 1558 has not been described in the recent nanosecond step-scan measurements, but it was found in our earlier step-scan investigations (13), and in time-resolved studies using a dispersive instrument (14). Thus, our data show important dynamic features of the chromophore and the protein which are not always directly correlated.

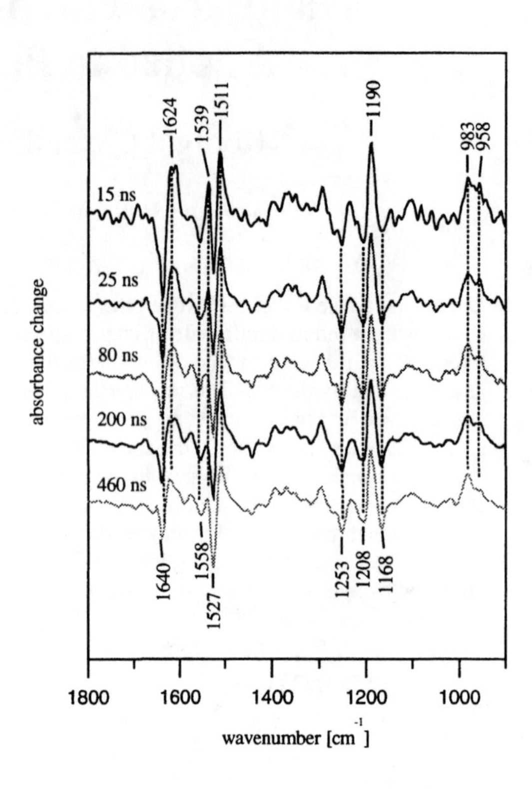

FIGURE 1. Nanosecond time-resolved step-scan FTIR difference spectra of bacteriorhodopsin

Fig. 2 shows corresponding spectra in the µs time range. Since the reaction between the intermediates also involves backreactions, the spectra of the pure states cannot be directly obtained. Thus, the spectrum of the L state (intermediate preceeding M) at 1.9 µs still contains contributions from the late KL state. By taking into account this contribution up to 75 µs, we were able to identify an early M state being in equlibrium with L, and the usual late M. The two states mainly differ in bands at 1650 (-), 1616(+) (amide-I), 1558(+) and 1535(-) (amide-II) cm^{-1} whose intensities increase with the formation of the late M state. These protein backbone changes could be responsible for the redirection of the accessibility of the Schiff base.

Halorhodopsin

Fig. 3 shows the time-resolved difference spectra of the chloride pump halorhodopsin. The spectra are lettered by the time-slice of the masurement and by the intermediate which dominates therein (pHR: *pharaonis* halorhodopsin, und the additional letters are adapted from the related photointermediates of bacteriorhodopsin). The first

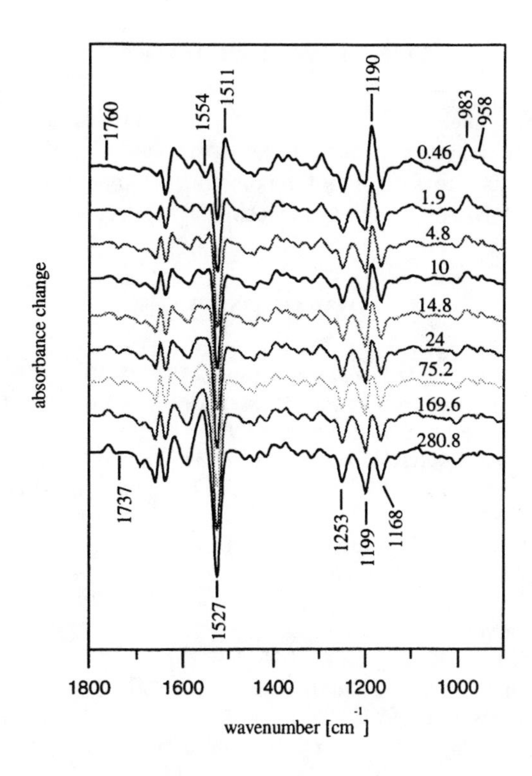

FIGURE 2. Microsecond time-resolved step-san FTIR difference spectra of bacteriorhodopsin

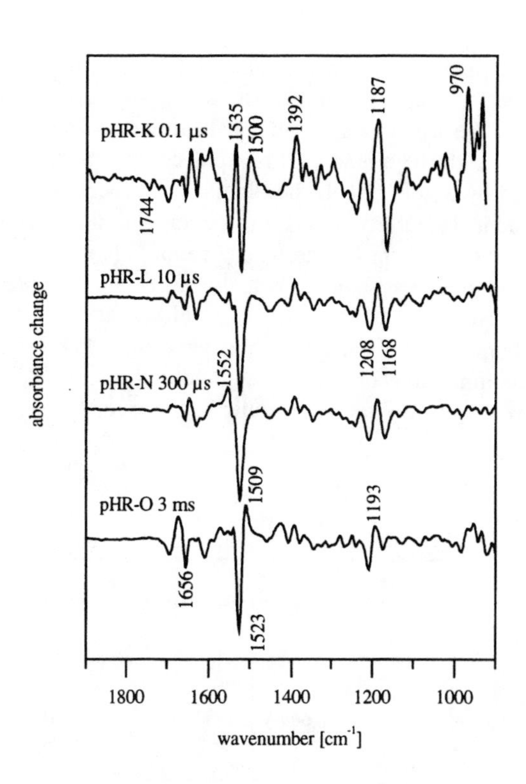

FIGURE 3. Time-resolved step-scan FTIR difference spectra of halorhodopsin

spectrum, pHR-K, is characterized by the ethylenic mode at 1500 cm^{-1}, indicating the red-shifted photoproduct. It is the first time that such an intermediate could be detected by infrared spectroscopy, since spectra of low-temperature measurements, stabilizing this intermediate, are lacking this mode (15). Again, the band at 1187 cm^{-1} indicates the isomerization of the chromophore and the strong HOOP mode its large distortion (16,17). Furthermore, rather strong bands are observed in the amide-I/II spectral range. The spectrum of the next intermediate (pHR-L) is very similar to the low-temperature spectrum (15). The large bands in the amide-II range and the large HOOP mode have disappeared, indicating a much more relaxed state. We provide first spectral evidence for the existence of an additional intermediate, labelled pHR-N. This species can also be deduced by a more careful kinetic analysis of the time-resolved spectra. With respect to the L state, differences can be seen in the amide-I/II spectral range and also small differeences in the chromophore bands between 1210 and 1165 cm^{-1}, indicating a very small change in the chromophore geometry. Interestingly, the amide-II intensity increase at 1552 cm^{-1} corresponds to the intensity increase observed at 1558 cm^{-1} during the M_1/M_2 transition in bactereiorhodopsin, which has been interpreted by the change in accessibility of the Schiff base. Finally, the last spectrum has been obtained in a seprate measurement under low Cl^- concentration. Again, the ethylenic mode at 1509 cm^{-1} indicates the red-shifted absorption maximum. The amide-II spectral changes have largely relaxed, and the disappearance of the negative band at 1168 cm^{-1} shows that the chromophore is thermally re-isomerized. A qualitatively similar spectrum, but with much more noise, has been obtained by static measurements (18). The HOOP mode, however, shows that the chromophore is still distorted. A kinetic analysis of the chloride dependence of the rate constants indicates that the rate of reformation of the intial state from the O species increases with increasing chloride concentration, in agreement with the idea that this step represents the Cl^- uptake (18). At high concentration the O species cannot be detected since it decays too fast.

CO-Myoglobin

Fig. 4 shows time-resolved difference spectra of the photolysis of CO-myoglobin for wild type and mutants H64I and H64G. The spectral changes have been taken at the maximum of their temporal evolution. The insert shows the spectral range of the CO stretching vibration of bound CO. In the mutations, the distal histidine, being in steric contact with the bound CO, is replaced by a bulkier group (I) and by a hole (G). If the binding influences the protein backbone of this region, corresponding amide-I changes should be different for the wild type and also different for the two mutants. This is actually seen. It confirms our hypothesis that binding of CO influences the protein backbone at the distal side of the heme ring. The influence of the mutation on the CO stretching band has been reported before based on static difference spectroscopy (19).

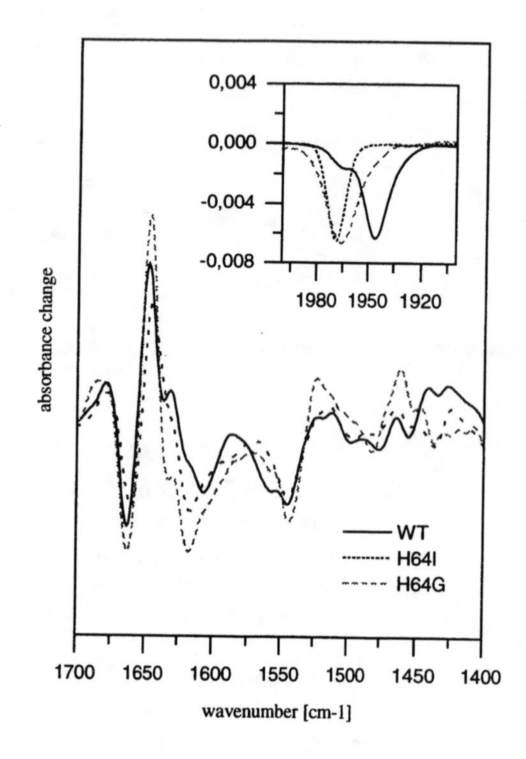

FIGURE 4. Time-resolved step-scan FTIR difference spectra of recombinant wild type and mutant CO-myoglobin.

Taken together, the examples presented demonstrate the strong potential of time-resolved step-scan FTIR difference spectroscopy for the study of biological reaction mechansims. The only, but essential, reqirement is that the reaction can be initiated several-thousand fold with high reproducibility.

ACKNOWLEDGEMENT

We thank Dr. M. Engelhard for providing us samples of halorhodopsin and bacteriorhodopsin, and Dr. M. Ikeda-Saito for the samples of recombinant human myoglobin, and the Bruker company for technical support.

REFERENCES

1. Siebert, F., *Mikrochim.Acta* **[Suppl.] 14**, 43-50 (1997).
2. *Time-resolved Vibrational Spectroscopy V*, Takahashi, H. ed., Berlin: Springer-Verlag, 1992.
3. *Time-Resolved Vibrational Spectroscopy VI*, Lau, A., Siebert, F., and Werncke, W. eds., Berlin: Springer-Verlag, 1994.
4. Oesterhelt, D., Tittor, J., and Bamberg, E., *J.Bioenerg.Biomembr.* **24**, 181-191 (1992).
5. Lanyi, J.K., *Biochim.Biophys.Acta* **1183**, 241-261 (1993).
6. Oesterhelt, D., *Israel J.Chem.* **35**, 475-494 (1995).
7. Lanyi, J.K., *Physiol.Rev.* **70**, 319-330 (1990).
8. Dioumaev, A.K. and Braiman, M.S., *J.Phys.Chem.B* **101**, 1655-1662 (1997).
9. Hage, W., Kim, M., Frei, H., and Mathies, R.A., *J.Phys.Chem.* **100**, 16026-16033 (1996).
10. Austin, R.H., Beeson, K.W., Chan, S.S., Debrunner, P.G., Downing, R., Eisenstein, L., Frauenfelder, H., and Nordlund, T.M., *Rev.Sci.Instrum.* **47**, 445-447 (1976).
11. Uhmann, W., Becker, A., Taran, Ch., and Siebert, F., *Appl.Spectrosc.* **45**, 390-397 (1991).
12. Weidlich, O., Friedman, N., Sheves, M., and Siebert, F., *Biochemistry* **34**, 13502-13510 (1995).
13. Weidlich, O. and Siebert, F., *Appl.Spectrosc.* **47**, 1394-1400 (1993).
14. Sasaki, J., Maeda, A., Kato, C., and Hamaguchi, H., *Biochemistry* **32**, 867-871 (1993).
15. Rothschild, K.J., Bousché, O., Braiman, M.S., Hasselbacher, C.A., and Spudich, J.L., *Biochemistry* **27**, 2420-2424 (1988).
16. Fahmy, K., Groβjean, M.F., Siebert, F., and Tavan, P., *J.Mol.Struct.* **214** , 257-288 (1989).
17. Fahmy, K., Siebert, F., and Tavan, P., *Biophys.J.* **60**, 989-1001 (1991).
18. Váró, G., Brown, L.S., Sasaki, J., Kandori, H., Maeda, A., Needleman, R., and Lanyi, J.K., *Biochemistry* **34**, 14490-14499 (1995).
19. Li, T., Quillin, M.L., Phillips, G.N., and Olson, J.S., *Biochemistry* **33**, 1433-1446 (1994).

3.4 INSTRUMENT DEVELOPMENT

Modification of Time-Resolved Step-Scan and Rapid-Scan FTIR Spectroscopy for Modulation Spectroscopy in the Frequency Range from Hz to kHz

D. Baurecht, W. Neuhäusser and U.P. Fringeli

Institute of Physical Chemistry, University of Vienna, Althanstrasse 14, A-1090 Vienna, Austria

Interleaved Rapid-Scan (continuous scan) and time-resolved Step-Scan techniques were modified in order to use these techniques for modulation experiments. The demodulation of the time-resolved spectra was done by software, replacing expensive Lock-In amplifiers. The new techniques lead to the same results compared to conventional Lock-In Step-Scan methods but show dramatically shortened measurement times with respect to the same signal-to-noise ratio.

INTRODUCTION

FTIR-Spectroscopy provides several techniques to study time-dependent phenomena (1, 2). Additionally, Modulation Spectroscopy can improve kinetic information and signal-to-noise ratio in experiments with reversible kinetics (3). Most of FTIR modulation experiments performed up to now were done with Step-Scan techniques in assistance of Hardware-Lock-In demodulation of the interferogram (LI-SS, Fig. 1).

On the other hand, all FTIR techniques capable of sampling time slices (sampling point spectra) synchronous with the modulation period can be used to analyze modulation experiments. In this case the demodulation of the time-resolved sampling point spectra is done by software (4) and allows to calculate both the fundamental tone and higher order harmonics after the data sampling. For the analysis of low frequency modulation (mHz), the sampling point spectra can be measured with the Fast-Scan technique (5).

Our aim was to modify time-resolved Interleaved Rapid-Scan technique IRS, which is a Continuous-Scan technique (1), and time-resolved Step-Scan techniques TR-SS (2) in order to use these techniques for Modulation Spectroscopy in the frequency range from Hz to kHz.

We compare the new methods with the conventional Lock-In Step-Scan technique regarding measurement time and signal-to-noise ratio.

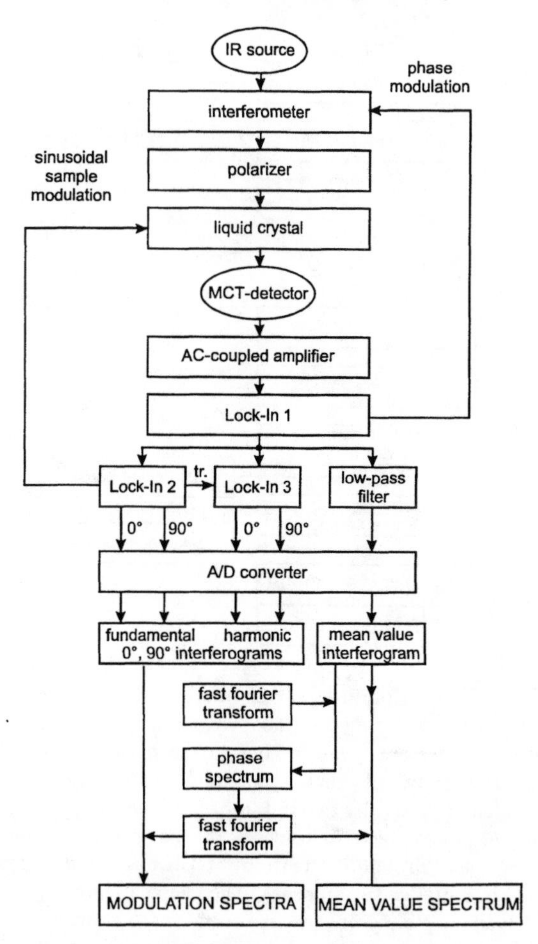

FIGURE 1. System setup for Step-Scan Lock-In modulation experiments. Phase modulation ("dithering" of the mirror) is used to modulate the IR beam (AC-coupled amplifier). Lock-In amplifier 1 demodulates the signal with respect to the phase modulation frequency. Lock-In amplifier 2 provides the sinusoidal sample modulation. Lock-In 2 and 3 demodulate the in-phase (0°) and quadrature (90°) elements of the fundamental tone and first harmonic, respectively. The output of the low-pass filter provides the interferogram of the mean value spectrum.

CP430, *Fourier Transform Spectroscopy:* 11th International Conference
edited by J.A. de Haseth

MATERIAL AND METHODS

Conventional Lock-In Step-Scan technique (LI-SS), modified Interleaved-Rapid-Scan technique (IRS, Fig. 2) and modified Time-Resolved Step-Scan technique (TR-SS, Fig. 3) on a Bruker IFS66 FTIR spectrometer are used to perform modulation experiments from 1 Hz to 6.25 kHz.

Nematic liquid crystals (LC) served as test samples. Polarized light and ATR-technique (Ge-plate, angle of incidence: 45°, number of active internal reflections: 12, n_{LC}=1.478, n_{Ge}=4.0) were used to study the dynamics of liquid crystals at the liquid crystal/Ge-interface (6).

Interleaved Rapid-Scan Technique (IRS)

In the IRS mode, the synchronization of sample modulation and data sampling is triggered by the zero crossing points of the laser interferogram (ZCP). The modulation frequency is determined by the time period between two rising ZCPs and therefore depends on the velocity of the movable mirror in the interferometer.

The sinusoidal modulation signal is generated by a Phase-Locked-Loop (PLL) circuit which is triggered by the ZCPs. The trigger impulses for data sampling are generated by another PLL loop that provides up to 16 sampling points per period which have to fit into one modulation period.

To produce different modulation frequencies the mirror velocity can be changed within a range depending on the control mechanism of the interferometer. To obtain very low modulation frequencies only the nth (n=1-32 on a Bruker IFS 66) ZCPs is used to trigger the PLL circuits. In this case n scans have to be performed to obtain a full interferogram, whereas the trigger impulse skips one rising ZCP each scan (interleaved mode).

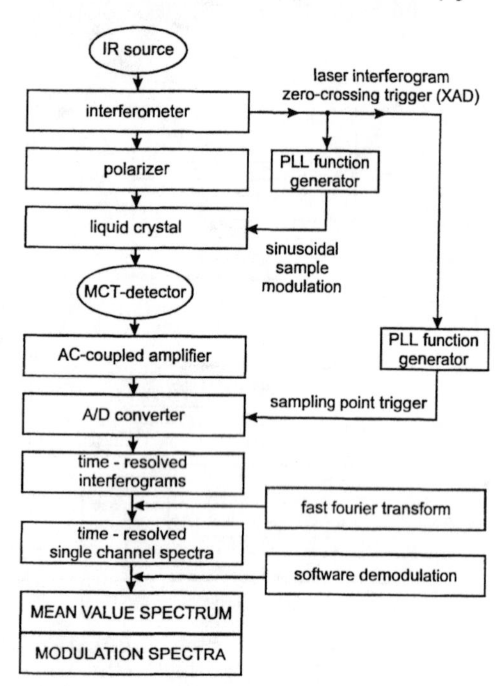

FIGURE 2. Modification of Interleaved Rapid-Scan (IRS) technique for modulation experiments. Time-resolved interferograms are measured during the mirror movement. The synchronization of data sampling and sinusoidal sample modulation is done by two PLL (phase-locked-loop) circuits. The sorting process and calculation of the corresponding single channel spectra (Fourier transform) takes place in the acquisition processor of the spectrometer. The demodulation of the time-resolved sampling point spectra is done by software and allows to calculate both the mean value spectrum and modulation spectra of the fundamental tone and harmonics after the data sampling.

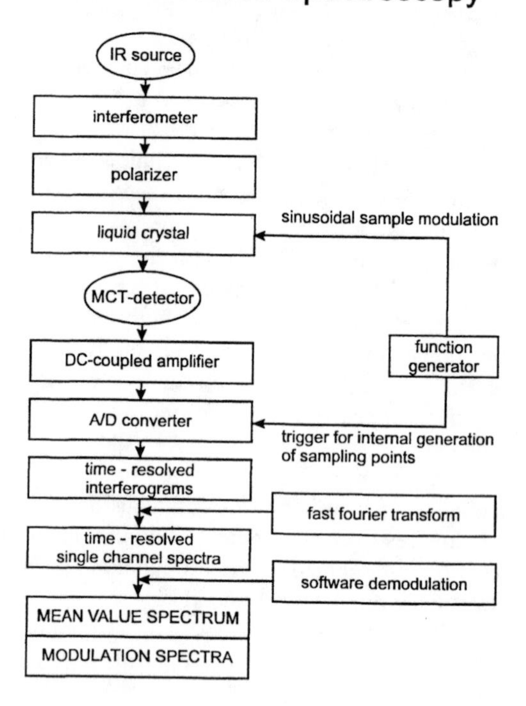

FIGURE 3. Modification of Step-Scan TRS technique for modulation experiments. A function generator provides the sinusoidal sample modulation and the trigger impulse for the internal generation of sampling points. Sampling points must fit to the period of the stimulus. Time-resolved interferograms are measured with the step scan technique (DC-coupled amplifier). The corresponding time-resolved single channel spectra are calculated by a FFT. The demodulation of the time-resolved sampling point spectra is done by software and allows to calculate both the mean value spectrum and modulation spectra of the fundamental tone and harmonics after the data sampling.

Time-Resolved Step-Scan Technique (TR-SS)

In the TR-SS mode, both the sinusoidal modulation signal and the trigger impulse for the beginning of data sampling are provided by an external function generator. The trigger impulses for data sampling are internally generated by the spectrometer electronics and must be tuned regarding the number and time gap to fit into one modulation period. The number of sampling points per period is only determined by the amount of memory of the data acquisition board of the spectrometer. The TR-SS technique provides therefore a better time-resolution of the sampling point spectra compared with the IRS technique.

Determination of modulation spectra

In the IRS and TR-SS mode, the demodulation of time-resolved spectra is done with the algorithm described by Fringeli (4). The software OPUS from Bruker was used to program the demodulation algorithm. In the case of IRS and TR-SS this leads directly to absolute phase and amplitude information of absorption bands.

The results of experiments performed with the LI-SS technique have to be corrected in amplitude and phase lag because of unavoidable gain and filter properties of the Lock-In amplifiers (used type: Stanford Research Systems SR 850). Due to the fact that the LI-SS technique demodulates interferograms or single channel spectra, a further correction has to be applied because of the logarithmic behavior of the absorption.

RESULTS AND DISCUSSION

Our results show that the measurement of time-resolved sampling point spectra with post-measurement demodulation is a more transparent method for determining absolute phase angles and modulation amplitudes of absorption bands in Modulation Spectroscopy. Both methods (TR-SS and IRS) provide full information in the frequency domain (i.e. for 2D-IR-correlation analysis) and lead to the same results compared with the LI-SS technique (Fig. 4). Although, the LI-SS method will be improved by digital demodulation of the interferograms (realized in the newest generation of FTIR-spectrometers), the techniques introduced in this paper are more transparent and better accessible for the user and provide additional time-dependent information (time domain, e.g. the time course of the absorbance during the stimulus period).

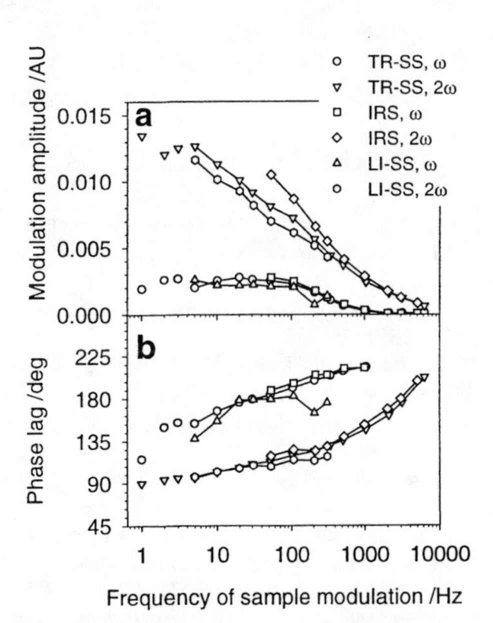

FIGURE 4a, b. Comparison of results obtained with Step-Scan Lock-In (LI-SS), Step-Scan Time-resolved (TR-SS) and Interleaved Rapid-Scan (IRS) spectroscopy used for ATR-FTIR modulation measurements. Frequency dependence of (*a*) absolute modulation amplitude and (*b*) phase lag for fundamental tone and first harmonic of a phenyl stretch vibration at 1611 cm^{-1} ν_{ar}(C=C). All three techniques lead to the same results.

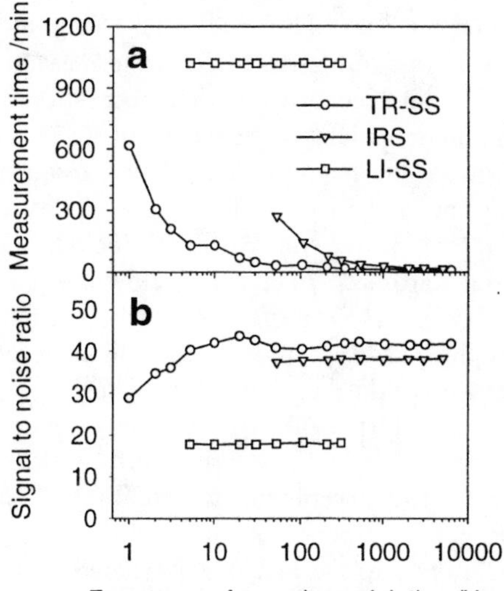

FIGURE 5a, b. Measurement time and signal-to-noise ratio of Step-Scan Lock-In (LI-SS), Step-Scan Time-resolved (TR-SS) and Interleaved Rapid-Scan (IRS) spectroscopy used for ATR-FTIR modulation measurements. *a* Measurement time of the used techniques. With the LI-SS technique no reduction of measurement time was possible (time constants of two Lock-In amplifiers, Tab. 1). *b* Comparison of the signal-to-noise ratio for mean value spectra of the modulated liquid crystal assembly (wavenumber range: 1900-2100 cm^{-1}). TR-SS and IRS technique show a significant higher S/N than LI-SS despite higher resolution and shorter measurement time. Resolution: 4 cm^{-1} (IRS, TR-SS), 8 cm^{-1} (LI-SS), high folding limit: 7899 cm^{-1}.

Table 1. Measurement parameters and results for modulation experiments performed with Interleaved Rapid-Scan Time-Resolved Spectroscopy (*IRS*), Step-Scan Time-Resolved Spectroscopy (*TR-SS*) and Lock-In Step-Scan Spectroscopy (*LI-SS*). *S/N* Signal-to-noise ratio measured in the mean value single channel spectra (wavenumber range 1900-2000 cm^{-1}), *ADC* analog to digital converter, *AQP* acquisition processor of the spectrometer. Resolution: 4 cm^{-1} (IRS, TR-SS), 8 cm^{-1}, high folding limit: 7899 cm^{-1}, interferogram size: 3665 (IRS, TR-SS), 1888 (LI-SS).

Method	LI-SS	IRS	TR-SS
frequency range	continuous up to ~ 320 Hz [a]	discrete 50 Hz - 12.5 kHz [b]	continuous 1 Hz - 12.5 kHz [b]
restriction for lower frequency limit	measurement time	slowest mirror velocity	spectral artifacts
restriction for upper frequency limit	phase modulation frequency	ADC sampling rate	ADC sampling rate
sampling point spectra / period	not necessary	up to 16 (Bruker IFS 66)	limited by memory of AQP
time-resolution	-----	5 µs [b]	5 µs [b]
No. of scans (coadditions)	1 (60000) [c]	100-3000 (1) [d]	1 (5-40)
modulation spectra	measured by Lock-In	calculated by software	calculated by software
mirror modulation	yes	no	no
DC-coupling	no	no	yes
measurement time	**17 h** [c]	**15 - 270 min**	**7 - 616 min**
S/N	**17.7 - 18.2**	**37.4 - 38.2**	**28.9 - 43.6**
drift of baseline	**no**	**yes**	**no**

[a] The mirror modulation frequency was set 10 times higher than the modulation frequency to allow analysis of higher order harmonics.
[b] ADC converter with a time-resolution of 5 µs
[c] The largest amount of measurement time is needed to get a stable value at the output of the LI amplifier 2 and 3 after each step of the mirror movement. The number of coadditions and the modulation frequency have therefore a negligible effect on the total measurement time.
[d] Depending on the mirror velocity of the interferometer and the modulation frequency, 1-30 scans are required to obtain a full interferogram.

The highest possible modulation frequency is only determined by the sampling rate of the ADC-converter of the spectrometer (IRS and TR-SS, Tab.1). Using an ADC converter with a time-resolution of 5 µs, the measurement of 16 sampling point spectra within one modulation period leads to a maximum modulation frequency of 12.5 kHz.

In the IRS-mode the minimum modulation frequency is determined by the slowest possible mirror velocity of the scanner (at present, 50 Hz with Bruker IFS66). In the SS-mode spectral artifacts increase below modulation frequencies of 3 Hz. These artifacts could be attributed to unstability of the mirror position and could therefore be improved using new digital control mechanisms.

The production of spectral artifacts due to long term drifts during the experiment is described for the IRS mode (7). Although our IRS experiments took up to 270 min, we did not find such a response in our modulation spectra. This is probably due to the fact that the demodulation of the time-resolved spectra will filter out all time-dependent signals which are not correlated with the modulation frequency.

Considering same signal-to-noise ratios, the Interleaved Rapid-Scan (IRS) and Time-Resolved Step-Scan (TR-SS) techniques lead to dramatically shortened measurement times compared to conventional Lock-In Step-Scan methods (Fig. 5). At a modulation frequency of 100 Hz we got measurement times of 17 h for LI-SS, 2.4 h for IRS and only 31 min for TR-SS technique, respectively.

ACKNOWLEDGEMENTS

We thank Dr. Martin Schadt from Rolic, Basel, Switzerland for kindly providing the liquid crystals. The DC-coupled amplifier was kindly provided by Bruker Analytische Messtechnik GmbH, Karlsruhe, BRD.

REFERENCES

1. Sloan, J. J. and Kruus, E. J., Time Resolved Spectroscopy, Chichester, England: John Wiley & Sons Ltd, 1989, ch. 5, pp. 219-253.
2. Johnson T. J., Simon A., Weil J. M. and Hassis G. W., Appl Spec **47**, 1376-1381 (1993).
3. Fringeli U. P., Internal Reflection Spectroscopy, New York: Marcel Dekker, 1992, ch. 10, pp. 255-324.
4. Fringeli U. P., Int. Patent Publication, PCT, WO 97/08598 (1997).
5. Müller M., Buchet R. and Fringeli U. P., J Phys Chem 100, pp. 10810-10825 (1996).
6. Neuhäusser W., Baurecht D., Schadt M. and Fringeli U. P., "Dynamics of a nematic liquid crystal in the Frequency range from 1 Hz to 6 kHz measured by ATR-FTIR Modulation Spectroscopy", presented at the 11th International Conference on Fourier Transform Spectroscopy, Athens, Georgia, USA, August 10-15, 1997.
7. Garrison A. A., Crocombe R. A., Mamantov G. and de Haseth J. A., Appl Spec **34**, 399-404 (1980).

The Design And Performance Of A Mid-Infrared FT-IR Spectroscopic Imaging System

N. A. Wright, R. A. Crocombe, D. L. Drapcho and W. J. McCarthy

Bio-Rad Digilab Division, 237 Putnam Avenue, Cambridge, MA 02139, U.S.A.

An integrated FT-IR spectroscopic imaging system is described. This consists of a step-scan FT-IR spectrometer, FT-IR microscope accessory and a 64 x 64 element MCT focal plane array, mounted on the microscope. Control of, and data acquisition from, both the spectrometer and the focal plane array is from a single PC. Typical data acquision times for an 8cm^{-1} resolution array are 1-4 minutes.

INTRODUCTION

This paper gives the first description of an integrated FT-IR spectroscopic imaging system, using an MCT array detector. FT-IR microscope imaging systems, employing InSb focal plane arrays (FPAs) have previously been described (1-3), as well as individual systems using Si:As (4) and MCT arrays (5). In addition, a macroscopic system, employed for gas plume imaging, has also been developed (6-7). Most of these employed separate data systems for interferometer control and FPA data acquisition, making them relatively unweidly.

SYSTEM DESCRIPTION

The integrated system consists of: a step-scan FT-IR spectrometer (Bio-Rad FTS 6000), operating in the mid-infrared with a ceramic source and KBr substrate beamsplitter; an FT-IR Microscope accessory (Bio-Rad UMA 500); and an MCT focal plane array (Santa Barbara Focal Plane) mounted on the microscope in the position usually occupied by the standard single element MCT. The final Cassegranian mirror assembly in the microscope was replaced by a ZnSe lens. The data system was a PC (200 MHz Pentium Pro) with 96 MB RAM, running Windows NT 4.0. Win-IR Pro software (Bio-Rad) controlled the FT-IR, and ImagIR software (Santa Barbara Focal Plane) controlled the FPA.

The MCT FPA had 64 x 64 (4096) total active detectors, and operated in photovoltaic mode. The pixels were on a 61 micron center-to-center spacing. The spectral range was 2.3 to 11 microns, and the detectivity is classified. A 3950 cm^{-1} long pass filter was placed in front of the FPA, so that and undersampling ratio of 4 could be used in FT-IR data collection.

In the FPA camera head was a 14-bit A/D converter, and it ran at up to 420 frames/second. Digital signals from the camera head were transferred to a digital frame grabber card in the PC (Imaging Technologies). This is in contrast to the arrangement in the InSb systems (1-3) which used two PCs, with the A/D in the PC image capture card set.

The optics were designed so that a 400 micron square sample was imaged onto the detector, giving a nominal spatial resolution of 6 microns. This resolution could be achieved at the short wavelength end of the optical frequency range used, but degraded at long infrared wavelengths.

The FT-IR spectrometer was operated in step-scan mode, with a typical dwell time of 0.2 to 1s per step. An undersampling ratio of 4 was used, resulting in a free-scanning spectral range of 3950 - 0 cm^{-1} . Spectroscopic resolutions of 4, 8 and 16 cm^{-1} were used. Under these conditions scan times ranged from 1 to 8 minutes.

The frame readout rate of the FPA can be set in software, and values from 200 - 316 Hz were used. Allowing for interferometer and electronic settling times, this allowed 20 frames to be co-added at 5 Hz step rate, 80 frames at 2.5 Hz step rate and 200 frames at 1Hz step rate.

METHOD

Spectra were acquired in both transmittance and reflectance. In both cases the arrays of sample spectra were referenced against background spectra. For transmittance the open beam is used for the reference and for reflectance a gold mirror.

Before a reference array is collected the FPA is 'calibrated'. This procedure corrects for non-uniform pixel response and non-uniform illumination. Ratioing the sample to the reference removes the optical function of the spectrometer.

The FPA data collection parameters are set up in the ImagIR software, which also provides the ability to monitor an interferogram-based infrared image. Data collection is initiated in Win-IR Pro, and the FTS 6000 provides are hardware-based signal to ImagIR indicating that the interferometer has moved to the next step. There is a settable delay time in ImagIR to allow the interferometer and electronics to settle at the new step.

CP430, *Fourier Transform Spectroscopy:* 11th International Conference
edited by J.A. de Haseth

Win-IR Pro computes the interferograms in the sample and reference arrays, and ratios them to produce an array of absorbance (or reflectance) spectra. This takes less than 30 seconds.

The data are initially presented as selectable images based on the absolute intensity at a particular infrared frequency. Clicking on a pixel displays the spectrum at that point in the image. Baseline corrected peaks can be defined on a spectrum and a new image created based on the height, area or position of the band maximum of that peak in the array of spectra. It is also possible to perform the usual FT-IR data manipulations on all the spectra in the array. Images can be viewed as flat, false color maps, or in 3D.

RESULTS

Some typical results are shown below. The sample was an epoxy/urethane sample. Different portions of it can be imaged at different infrared frequencies, and the images based on peaks at 1414 (Fig. 1) and 1467 cm^{-1} (Fig. 2) are shown. Where components have overlapped or blended spectral subtraction can be done to obtain spectra of the individual species (Fig. 3). These spectra can also be searched against standard databases for identification purposes.

ACKNOWLEDGEMENTS

The authors would like to thank Dr. Christoph Jansen (Bio-Rad Germany) for obtaining the data shown here, and Drs. E. N. Lewis and I. W. Levin (NIH) for helpful discussions.

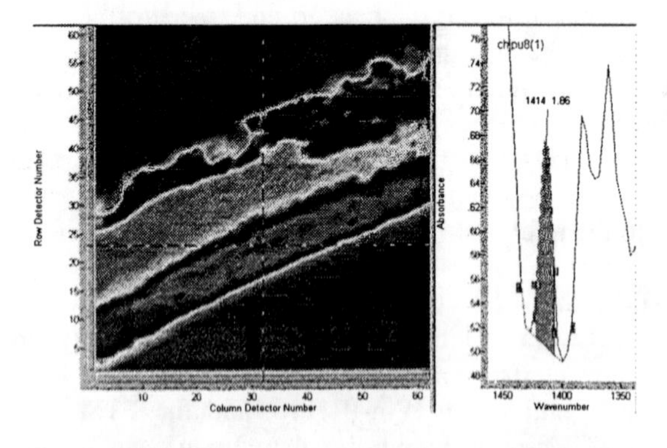

Figure 1. The infrared image of an epoxy/urethane sample (left), based on the area of a band centered at 1414 cm^{-1} (right).

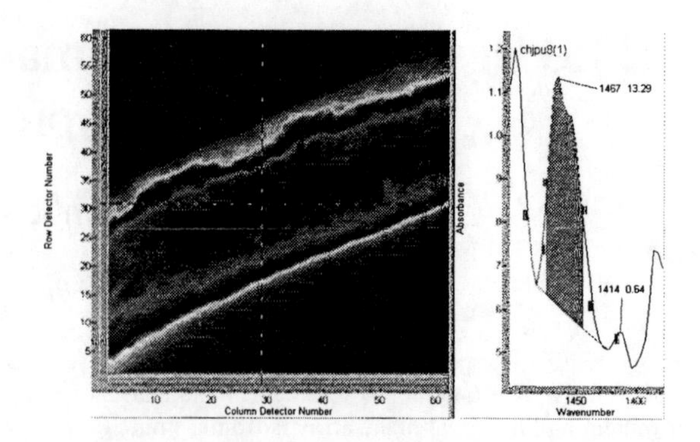

Figure 2. The infrared image of an epoxy/urethane sample (left) based on the integrated area of a band centered at 1467 cm^{-1} (right).

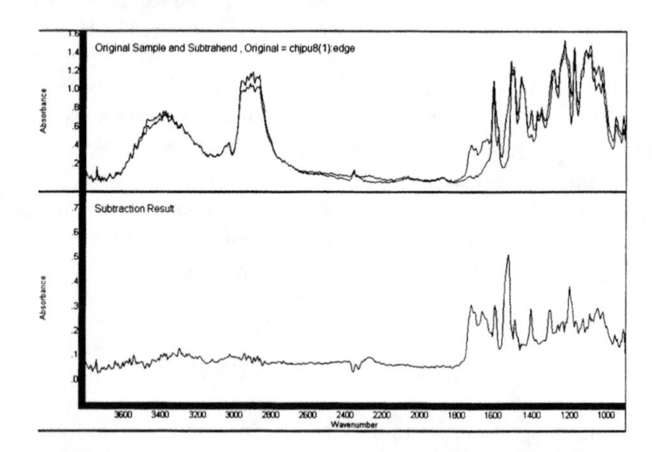

Figure 3. Pure component spectra (lower panel) are obtained by subtraction of spectra from individual pixels (upper panel).

REFERENCES

1. E. Neil Lewis, Patrick J. Treado, Robert C. Reeder, Gloria M. Story, Anthony E. Dowrey, Curtis Marcott and Ira W. Levin, *Anal. Chem.* **67**, 3377-3381 (1995)
2. E. Neil Lewis, Alexander M. Gorbach, Curtis Marcott and Ira W. Levin, *Appl. Spectrosc.* **50**, 263-269 (1996).
3. E. N. Lewis, I. W. Levin and R. A. Crocombe, *Mikrochim. Acta [Suppl.]* **14**, 589 (1997)
4. E. N. Lewis, L. H. Kidder, J. F. Arens, M. C. Peck and I. W. Levin, *Appl. Spectrosc.* **51**, 219 (1997)
5. MCT Focal-Plane Array Detection for Mid-Infrared Fourier Transform Spectroscopic Imaging, L. H. Kidder, I. W. Levin, E. N. Lewis, V. D. Kleiman and E. J. Heilweil, *Optics Letters* **22**, 742 (1997).
6. C. L. Bennett, M. Carter, D. Fields and J. Hernandez, *Proc. SPIE* **1937**, 191 (1993)
7. C. L. Bennett, *Proc. SPIE* **2266**, 25 (1994)

Visible Intracavity Laser Spectroscopy With A Step-Scan FTIR

T. J. Johnson+, K. Strong‡, and G.W. Harris

Centre for Atmospheric Chemistry/York University / North York, Ontario, M3J 1P3, Canada

+ Current address: Bruker Optics, 19 Fortune Drive, Billerica, MA 01821, USA
‡ Current address: Dept. of Physics, University of Toronto, 60 St. George Street, Toronto, Ontario M5S 1A7, Canada

INTRODUCTION

In order to determine the chemical composition of the atmosphere, techniques of high sensitivity are required since many gases are only present at trace levels. Intracavity Laser Spectroscopy (ILS) is such a technique and may be described as follows: A lasing medium with a broadband gain characteristic is discontinuously pumped by a chopped-CW source; the dye laser reaches threshold as several different cavity modes begin to oscillate, yielding a relatively broad emission. With time, mode competition sets in and certain modes within the cavity are enhanced, leading eventually to a narrow CW line. The total elapsed time between the start of broadband lasing and the stabilization of the spectral evolution of the developing CW emission is a function of many parameters, but typically there may be an initial pulse width ~100 cm^{-1} sharpening to ~5 cm^{-1} in the absence of a tuning element. For cavity lengths of ~25 cm, this takes place over one hundred to several hundred microseconds. ILS deals with the absorptions that occur when a species is present within the cavity whose absorption features lie in the emitted spectral region and are narrow enough to attenuate only a portion of the evolving broadband laser pulse. As the light resonates in the cavity, the path through the absorber increases, resulting in an absorption enhancement by factors as large as 10^7 versus a single pass (1). The cavity thus serves as a multipass cell where the total pathlength is proportional to the generation time t_g of the pulse, *i.e.* the time from when the dye laser reaches threshold to observation of the laser emission profile. Until recently, ILS has been performed using some variant of a trigger/scan scheme with a fixed time delay and a dispersive system (e.g. a monochromator with an OMA) recording the laser emission wavelength profile.

In contrast to dispersive systems, Fourier transform techniques are both broadband and exhibit a higher throughput, but FT spectrometers are limited in time resolution by the speed of the moving mirror. Step-scan FT spectrometers, however, obtain high temporal resolution of reproducible phenomena in a different manner (2). With these instruments, the mirror is translated (stepped) to a fixed position so that the optical path remains temporarily constant. An experiment is then initiated and the interferogram intensity at that mirror position is digitized at equi-spaced time intervals (*e.g.* every 5μs). The mirror is then stepped to the next position, the experiment triggered again, and a new set of fixed-retardation intensities is recorded. The process is repeated until the time-varying intensities are recorded for a complete set of mirror positions. At the conclusion, the two-dimensional matrix is rotated into a series of retardation-spaced intensities (*i.e.* interferograms) for specific times, each interferogram corresponding to a given time after trigger. The temporal resolution is thus limited only by the detector and ADC bandwidths. In step-scan TRS experiments, the full desired time resolution of the evolving pulse is obtained at essentially no additional cost in terms of the duration of the experiment. Moreover, one can use laser pulse averaging in-step to improve the signal-to-noise ratio (SNR). With such in-step averaging *N* Laser pulses are averaged while the mirror is fixed, thus increasing the SNR by $\sqrt{N}$, with only a small increase in the total experiment time. There is, of course, a limitation to this, namely when the number of in-step co-added laser pulses becomes comparable to the mirror step and stabilization time. In the present work we couple step-scan Fourier transform techniques to the ILS experiment, thus allowing, for example, the time evolution of the laser emission to be studied. This in turn enables the straightforward determination of the generation time t_g which optimizes intracavity absorption signal strength and linearity.

EXPERIMENTAL

Figure 1 depicts the experimental apparatus: A Coherent Innova 90 Ar^+ laser running 1.5 W all lines pumps the dye laser. The pump is intensity modulated by ~1 kHz acousto-optic modulator such that when switched the zero-order pump beam is transmitted at a slightly lower power to maintain the dye laser just below threshold. The AOM-modulated beam pumps a Spectra Physics 375B dye laser, yielding an emission maximum near ~615 nm. For the CH_4 experiments reported below, the 619 nm emitted beam was expanded and focused into the emission port of a Bruker IFS 66v FTIR equipped with a quartz beamsplitter and a Si-diode detector.

CP430, *Fourier Transform Spectroscopy:* 11th International Conference
edited by J.A. de Haseth

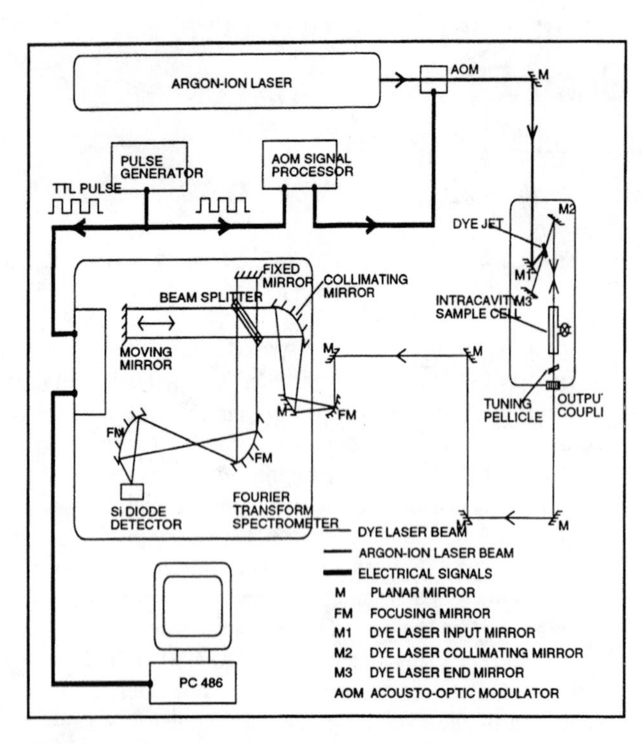

FIGURE 1. Schematic diagram of the step-scan FTIR experimental apparatus for intracavity laser spectroscopy.

In the triggering scheme, a square-wave produced by a pulse generator turns off the AOM such that the full Ar+ laser power appears in the zero order and simultaneously initiates the FTIR data acquisition at a given step. With the interferometer mirror held fixed, the interferogram intensity is recorded and digitized at 10 μs intervals. Typically 100 evolving laser pulses were digitized and averaged. After averaging, the mirror is moved to the next position and the cycle is repeated, until a complete two-dimensional matrix of interferogram intensities is collected. The data are then transposed and Fourier-transformed as described above.

THEORY AND RESULTS

Figure 2 illustrates the advantage of ILS via the step-scan technique, namely that extensive spectral and temporal information are obtained from a single experiment. This figure presents time-resolved emission spectra of the dye laser pumped by the Ar^+ laser at 0.90 W with the dye laser cavity purged by dry N_2. The spectral resolution is 4.0 cm^{-1} and the temporal resolution is 20 μs between the 50 slices displayed, for a total generation time of 1 ms; the pulse width narrows over the first few hundred microseconds as the pulse height grows. These spectra were obtained using 40 in-step averages because the dominant noise source originates from the dye laser (stochastic shot noise and pulse-to-pulse jitter).

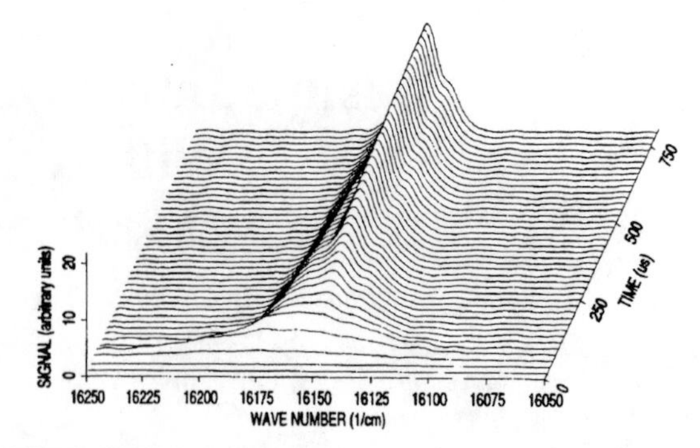

FIGURE 2. Time resolved spectra of the intracavity dye laser pulse for the N_2-purged cavity.

We can analyze the time evolution of both the peak height and spectral width. Stoeckel and Atkinson (3) have taken a model that assumes no wavelength-dependent absorber within the cavity and that the gain curve for the mode distribution is represented by an inverted parabola of finite width. Making these assumptions, they arrive at the following relation for the laser intensity as a function of wavenumber and generation time,

$$I\left(\tilde{v},t\right)=\frac{I'_o}{\Delta\tilde{v}_o}\left[\frac{\gamma t_g}{\pi}\right]^{\frac{1}{2}}\exp\left[-\left(\frac{\tilde{v}-\tilde{v}_o}{\Delta\tilde{v}_o}\right)^2\gamma t_g\right]$$

where I'_o = total laser intensity, t_g = generation time, γ = losses in the optical cavity, v = spectral frequency (cm^{-1}), v = frequency at line center of laser gain profile (cm^{-1}), and Δv_o = initial bandwidth (cm^{-1}). The second coefficient yields a peak height increasing in proportion to the square root of the generation time, $\sqrt{t_g}$, whilst the exponential term dictates a half-width that is decreasing as $1/\sqrt{t_g}$. Evaluating the above data we find that for the spectra between 80 and 500 μs the slope of the logarithm of the peak height versus time of 0.57 and a slope for the logarithm of the FWHM versus time of -0.76. For the peak height this is in reasonable agreement with the equation while the peak shape, and hence width, is less clearly defined. The large number of data points, which are conveniently obtained using the TRS-ILS technique, allows greater precision to the fit that has been obtained in previous studies. Further work has shown that linear relation between absorptance and generation time can be obtained. Work continues in this direction.

REFERENCES

1. Harris, S.J., *Applied Optics* **23**, 1311-1318, (1984).
2. Johnson, T.J., Simon, A., Weil, J.M., and Harris, G.W., *Applied Spectroscopy* **47**, 1376-1381, (1983).
3. Stoeckel, F., and Atkinson, G.H, *Applied Optics* **24**, 3591-3597, (1985).

Step-Scan Versus Rapid-Scan FTIR-VCD Spectroscopy Using HgCdTe and InSb Detectors

Fujin Long, Teresa B. Freedman and Laurence A. Nafie

Department of Chemistry, Syracuse University, Syracuse, New York 13244-4100 USA

A comparison of step-scan and rapid-scan Fourier transform vibrational circular dichroism (FT-VCD) spectroscopy in the hydrogen-stretching region has been carried out on Nicolet Magna 850 and Bruker IFS-55 FTIR spectrometers, each equipped with an accessory VCD bench. Single enantiomer VCD spectra of (+)- and (-)-camphor with flat baselines are obtained on each instrument, both in step-scan mode and rapid-scan mode. On both instruments, the RMS noise level is less than 1×10^{-6} (absorbance units) in less than two-hour collections, and the noise level obtained with the step-scan method is slightly lower than that for rapid-scan. The calibrated noise level of VCD spectra obtained on the Nicolet instrument is about twice that obtained on the Bruker instrument for a comparable collection time. Our preliminary results show that with an InSb detector on the Nicolet instrument, the noise level of VCD spectra is about a half of that obtained with an MCT detector.

INTRODUCTION

Since its introduction, FT-VCD has been mainly used in the mid-infrared region (1). For the hydrogen-stretching region, dispersive VCD instruments are mostly employed, since FT-VCD measurements are disadvantaged by the growing proximity between the VCD interferogram frequencies and those of the IR transmission interferogram. The step-scan technique provides a way to remove the major obstacle to extending FT-VCD measurements to the higher frequency region since there are no Fourier frequencies. Marcott et al. (2) and Keiderling et al. (3) have reported work in this area. In our studies of extending FT-VCD measurements to the higher spectral frequencies with the Bruker IFS 55, we find out from using both step-scan and rapid-scan techniques that we can obtain high quality VCD spectra, provided we use a small enough lock-in amplifier (LIA) time constant for the rapid-scan method (4). We also found about two times higher signal-to-noise ratios are achieved for the step-scan method on our Bruker-based instrument. In this work, we will present preliminary results of the comparison between Nicolet-based (5) and Bruker-based (4) VCD instruments and between using InSb and MCT detectors. The purpose is to further explore the limits of FT-VCD instrumentation.

EXPERIMENTAL

VCD measurements were undertaken on two VCD spectrometers, a Nicolet Magna 850 and a Bruker IFS-55. Both instruments use a tungsten lamp as an IR source for the hydrogen stretching region. The detailed instrument setup and measurement procedures were described previously (4, 5). Briefly, for the Nicolet instrument, a VCD bench is equipped with a ZnSe focusing lens, a KRS-5 grid polarizer, a ZnSe photoelastic modulator (PEM) oscillating at 37 kHz, a ZnSe f/1 lens and a 3×3 mm HgCdTe (MCT) detector ($D^* = 2.3 \times 10^{10}$ at 1 kHz), aligned in that order. In order to compare with the Bruker instrument, the InSb detector (and associated preamplifier) from the Bruker instrument was substituted for the MCT detector and carefully aligned. On the Bruker VCD bench, a BaF_2 grid polarizer, a CaF_2 PEM oscillating at 57 kHz, a ZnSe f/1 lens and a 2 mm diameter InSb detector ($D^* = 3.2 \times 10^{11}$ at 1 kHz) were placed in that order. On both instruments, the signal from the detectors is split into two pathways. One is for the normal IR single beam transmission, and the other is fed through an electronic filter, and an LIA tuned to the PEM frequency. The electronic filtering is somewhat different for the two instruments because of differences in instrument design. On the Nicolet, bandpath filters with variable bandwidth are used before and after the LIA, while on the Bruker, a high-pass filter is used before the LIA and a low-pass filter is used after. The filtering frequencies depend on the scanning mode and the moving mirror velocity.

Samples of (+)- and (-)-camphor, (+)- and (-)-α-pinene, albumin and CCl_4 were purchased from Aldrich and used without further purification. Samples of (+)- and (-)-propranolol were made from their hydrochloride salts purchased from Aldrich. Fixed pathlength cells with BaF_2 windows were used.

RESULTS AND DISCUSSION

As an example of a step-scan FT-VCD spectrum in the OH and NH regions, the VCD spectrum of (S)-propranolol is shown in Fig. 1. The spectrum was collected on the Bruker-based VCD instrument at 16 cm^{-1} resolution. The resolution is the same as in our earlier report of the FT-

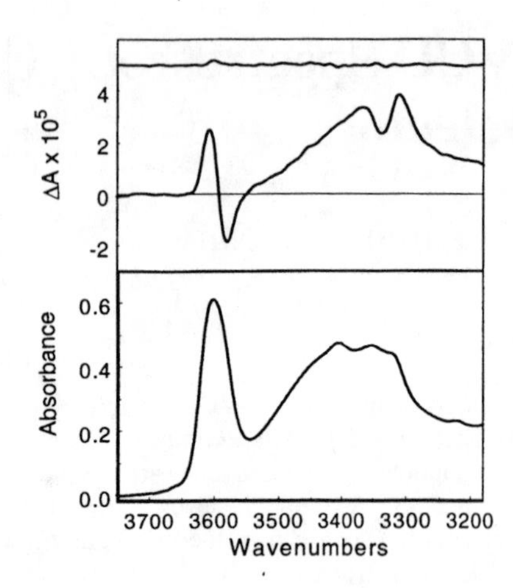

Figure 1. IR and step-scan FT-VCD spectra of (S)-propranolol, 0.1 M in $CDCl_3$, 16 cm^{-1} resolution, averaged for 6 hours.

VCD spectrum of camphor (4), which we labeled as 8 cm^{-1} resolution, because of a Bruker software error. For appropriate samples, all of the hydrogen stretching region can be measured at the same time. The noise level for a 16 cm^{-1} resolution spectrum is comparable to that achieved on earlier dispersive VCD instruments optimized in this region.

The comparison of step-scan and rapid-scan VCD on each instrument shows that step-scan gives better signal-to-noise ratio (S/N) than rapid-scan in the hydrogen stretching region. In the two experiments, the same samples are used and the collection time is the same. On the Nicolet instrument, we used the simultaneous rapid-scan mode, while on the Bruker instrument, sequential rapid-scan is used. Previously, we reported that in the mid-IR region, simultaneous rapid-scan mode gives the best S/N on the Nicolet-based instrument among step-scan, simultaneous rapid-scan and sequential rapid-scan modes (5). The reason for the difference is that with the increase of Fourier

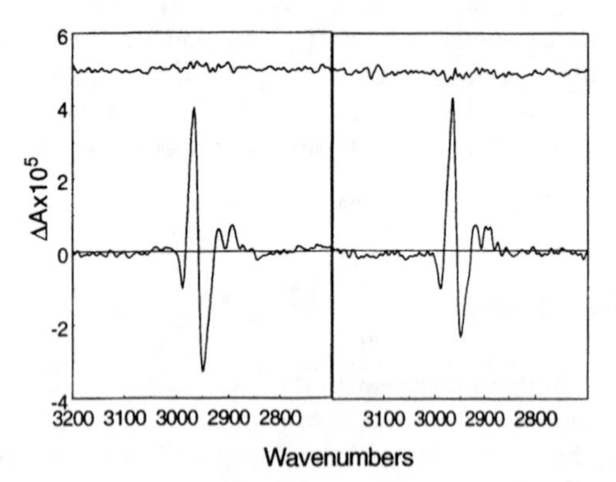

Figure 2. VCD and noise level of camphor collected on Nicolet-based VCD instrument under the same conditions, 8 cm^{-1} resolution, 2.4 hours, with InSb detector (left), and MCT detector (right).

frequency in the rapid-scan mode, the signal processing in the LIA is reduced, while in the step-scan mode, no such limitation is encountered.

It is expected that the S/N from the InSb detector would be much higher than from an MCT detector because of the significant difference in D* between the two detectors. However, the step-scan FT-VCD noise levels measured from the Bruker instrument and the Nicolet instrument are comparable. To eliminate the system difference, we moved the InSb detector and its associated preamplifier from the Bruker instrument to the Nicolet instrument. With all other conditions exactly the same, VCD spectra were collected. The noise level of the VCD spectra using the MCT detector is about twice as that using InSb detector (Fig. 2). Considering that in both cases the noise levels are below $3x10^{-6}$, we may have to consider this the noise limit of the LIA and FTIR system. Further investigation is needed. One way to reduce the noise level is to decrease the bandwidth of the bandpass filter before the LIA. The change from 20-80 kHz range to 31.5-40 kHz results in a noise level change from 2.7×10^{-6} to 2.0×10^{-6} for the same collection time. Further decrease in bandwidth requires a new electronic filter and such measurements will be undertaken in the near future.

CONCLUSION

FT-VCD spectra in the OH-, NH- and CH- stretching regions have been obtained. The noise level of the VCD spectra at 16 cm^{-1} resolution is comparable with dispersive VCD instruments. The difference between using MCT and InSb detectors is small, even though there is a large difference in D*. Narrowing the bandwidth of the electronic filter before the LIA reduces the VCD noise level.

ACKNOWLEDGMENTS

Support for this work from the National Institutes of Health (GM-23567) is gratefully acknowledged.

REFERENCES

1. Nafie, L .A. *Appl. Spectrosc.* **50(5)**, 14A-26A (1996).
2. Marcott C., Dowrey A. E. and Noda I., *Appl. Spectrosc.* **47**, 1324 (1993)
3. Wang, B. and Keiderling T. A. *Appl. Spectrosc.* **49**, 1347 (1995)
4. Long, F., Freedman T. B., Tague, T. J., and Nafie, L. A. *Appl. Spectrosc.* **51**, 508-511 (1997).
5. Long, F., Freedman T. B., Hapanowicz, R., and Nafie, L. A. *Appl. Spectrosc.* **51**, 504-507 (1997).

FT-IR Spectroscopic Imaging Microscopy Using an MCT Focal-Plane Array Detector

C. Marcott and R. C. Reeder

The Procter & Gamble Company, Miami Valley Laboratories, Cincinnati, OH 45253-8707

A 64 x 64 Mercury-Cadmium-Telluride (MCT) focal-plane array (FPA) detector coupled to a step-scanning Fourier transform infrared (FT-IR) microscope has been used to acquire spectroscopic images in the fingerprint region of the infrared spectrum (1800-900 cm^{-1}). FT-IR microspectroscopic imaging using this MCT FPA enabled us to chemically identify a 7.5-µm-thick adhesive tie layer in a polymer laminate packaging material. Infrared spectroscopic images of cross sections of human bone tissue obtained using the MCT FPA detector are also presented. The protein, carbonate, and phosphate concentrations change as a function of radial distance from an osteon.

INTRODUCTION

The combination of an infrared (IR) focal-plane array (FPA) detector and a Fourier transform infrared (FT-IR) microscope is a powerful one for obtaining spectroscopic images with unprecedented image fidelity.(1-5) Use of a step-scanning FT-IR spectrometer with an array detector placed at an image plane of the IR microscope enables full infrared spectra (or spectroscopic images) to be collected across the entire unapertured field of view of the sample in a single measurement. The first such measurements were demonstrated by Lewis, et al.,(1-2) using a 128 x 128 InSb focal-plane array detector. It has recently been demonstrated that a longer wavelength cut-off MCT FPA can also be used in conjunction with an FT-IR microscope to obtain spectroscopic images throughout the mid-infrared region of the spectrum (4000-900 cm^{-1}). These FPAs have recently become available at reasonable cost because they are a key component of the shoulder-fired Javelin anti-tank weapon system which is being mass produced for the U.S. Army.

EXPERIMENTAL

Spectroscopic images were recorded with a UMA 300A FT-IR microscope coupled to an FTS-60A step-scanning FT-IR spectrometer (both from Bio-Rad). A 64 x 64 MCT FPA detector was used in place of the usual single-element MCT detector. A 50-mm ZnSe image formation lens was substituted for the Cassegrainian mirror normally used for final focusing with a single-element MCT detector. A four-position cold filter wheel (Kadel) was added to the liquid-nitrogen dewar used to cool the FPA, and a 1800-cm^{-1} long-pass optical filter (Corion) was in place when the measurements were taken. The cold optical filter helps keep the individual detector array elements from saturating and removes thermal background noise originating outside the optical bandwidth of interest. Although the dewar, filter wheel, electronics, and FPA data acquisition software were all assembled and/or produced by Lockheed Martin Santa Barbara Focalplane, the Javelin 64 x 64 MCT FPA detector chip itself was manufactured by Hughes Santa Barbara Research Center.

RESULTS

Packaging Material Film

Spectroscopic images were collected on a commercial packaging material film (Paramount) consisting of a 12-µm-thick surface layer of poly(ethylene terephthalate) (PET), followed by a 7.5-µm-thick adhesive tie layer of unknown composition, a 15-µm-thick layer of ethylene vinyl alcohol copolymer (EVOH), another 7.5-µm-thick adhesive tie layer, and a 75-µm-thick layer of low-density polyethylene (LDPE). Figure 1 shows IR spectroscopic images representing the area under the peaks at 1740 cm^{-1} (ester vibration) and 1126 cm^{-1} (ethoxylate vibration). The 1740 cm^{-1} image clearly highlights the surface PET layer as well as the barrier EVOH layer. (EVOH usually contains a significant amount of ethylene vinyl acetate which absorbs strongly at 1740 cm^{-1}.) Each pixel represents a 5-µm x 5-µ m area of the sample. There is clearly a layer between the PET and EVOH layers that is slightly more than 5-µm wide. The 1126 cm^{-1} image clearly highlights this adhesive tie layer, since none of the other three components absorb strongly in this region. Figure 2 shows IR spectra of representative single pixels in the PET, EVOH, and adhesive tie layers of the image. The previously unidentified adhesive tie layer can now be identified as a polyethoxylate. These data demonstrate our spatial

CP430, *Fourier Transform Spectroscopy:* 11th International Conference
edited by J.A. de Haseth

resolution approaches 5 μm, as there is no evidence of any 1740 cm^{-1} absorbance from the adjacent PET and EVOH layers in the tie layer spectrum.

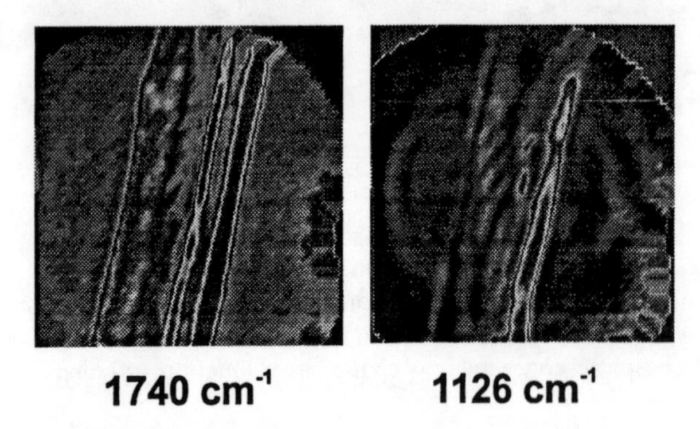

FIGURE 1. Spectroscopic image slices of the Paramount packaging material (5-μm cross section) representing the area under the IR band indicated.

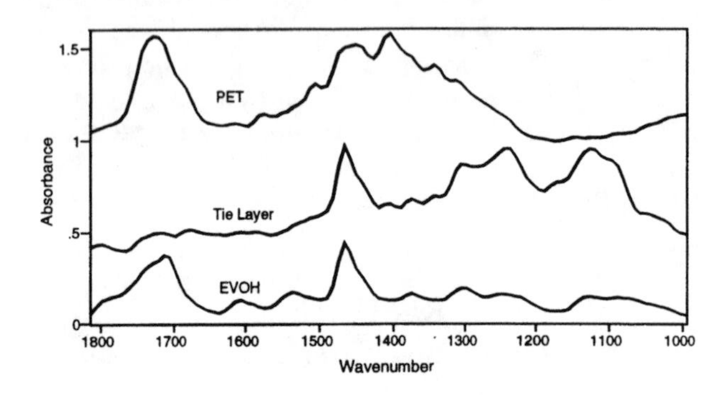

FIGURE 2. Single-pixel IR spectra (5 μm/pixel) taken from three different layers of the Paramount film (see Fig. 1).

Biomineralized Tissue

Infrared spectroscopy has been an important tool for studying bone mineralization.(6-7) FT-IR micro-spectroscopy, in particular, has been used to study site-to-site variations in mineral quality, providing information such as the relative amount of mineral present, the total amount of carbonate and phosphate, and the nature of the crystallinity in the apatite phase. In this study, the human bone tissue sections examined by FT-IR micro-spectroscopy were single osteons from normal human iliac crest biopsies obtained at necropsy. Figure 3 shows images representing the area under the IR peaks indicated of a 5-μ m-thick human bone section. Each pixel represents a 14-μ m x 14-μm area of the sample. The protein, carbonate, and phosphate concentrations change as a function of radial distance from an osteon.

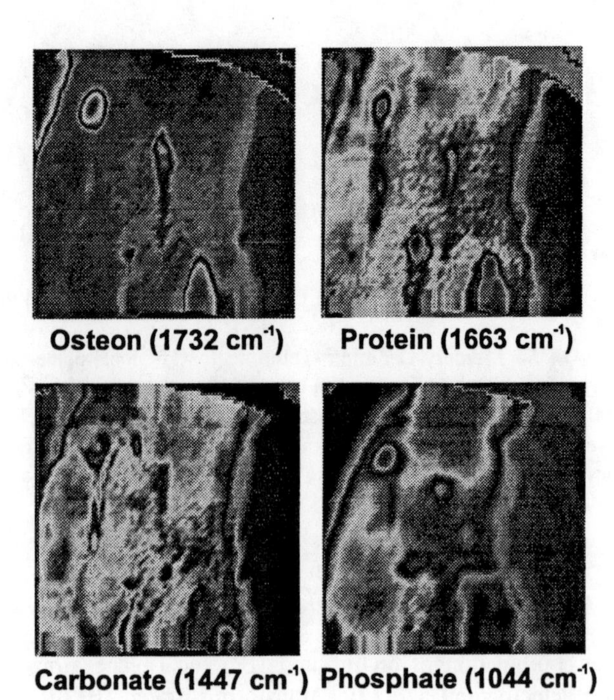

FIGURE 3. Spectroscopic image slices of a human bone tissue section representing the area under the IR band indicated.

ACKNOWLEDGMENTS

The authors wish to thank Dr. Douglas B. Zeik for providing the Paramount film and Dr. Matthew H. Chestnut for sectioning it. We also thank Dr. E. P. Pascalis and Dr. Adele L. Boskey of the Hospital for Special Surgery and Cornell Medical School, and Professor Richard Mendelsohn of Rutgers University, for providing all the sectioned bone tissue samples.

REFERENCES

1. Lewis, E. N., Treado, P. J., Reeder, R. C., Story, G. M., Dowrey, A. E., Marcott, C., and Levin, I. W., *Anal. Chem.* **67**, 3377-3381 (1995).
2. Lewis, E. N., Gorbach, A. M., Marcott, C., and Levin, I. W., *Appl. Spectrosc.* **50**, 263-269 (1996).
3. Marcott, C., and Reeder, R. C., *Microscopy and Microanalysis 1996*, Bailey, G. W., Corbett, J. M., Dimlich, R. V. W., Michael, J. R., and Zaluzec, N. J., Eds.: San Francisco Press, 1996, pp. 260-261.
4. Marcott, C., Story, G. M., Dowrey, A. E., Reeder, R. C., and Noda, I., *Mikrochim. Acta [Suppl.]* **14**, 157-163 (1996).
5. Lewis, E. N., Kidder, L. H., Levin, I. W., Kleiman, V. D., and Heilweil, E. J., *Optics Letters* **22**, 742-744 (1997).
6. Pleshko, N. L., Boskey, A. L., and Mendelsohn, R., *Biophys. J.* **60**, 786-793 (1991).
7. Paschalis, E. P., DiCarlo, E., Betts, F., Sherman, P., Mendelsohn, R., and Boskey, A. L., *Calcif. Tissue Int.* **59**, 480-487 (1996).

Pressure-Modulation Dynamic Attenuated-Total-Reflectance (ATR) FT-IR Spectroscopy

C. Marcott, G. M. Story, I. Noda, A. Bibby*, and C. J. Manning**

The Procter & Gamble Company, Miami Valley Laboratories, Cincinnati, OH 45253-8707
**Graseby Specac, 500 Technology Ct., Smyrna, GA 30082*
***Manning Applied Technology, 121 Sweet Ave., Moscow, ID 83843*

A single-reflectance attenuated-total-reflectance (ATR) accessory with a diamond internal-reflection element was modified by the addition of a piezoelectric transducer. Initial dynamic pressure-modulation experiments have been performed in the sample compartment of a step-scanning FT-IR spectrometer. A sinusoidal pressure modulation applied to samples of isotactic polypropylene and linear low density polyethylene resulted in dynamic responses which appear to be similar to those observed in previous dynamic 2D IR experiments. Preliminary pressure-modulation dynamic ATR results are also reported for a styrene-butadiene-styrene triblock copolymer. The new method has the advantages that a much wider variety of sample types and geometries can be studied and less sample preparation is required. Dynamic 2D IR experiments carried out by ATR no longer require thin films of large area and sufficient strength to withstand the dynamic strain applied by a rheometer. The ability to obtain dynamic IR spectroscopic information from a wider variety of sample types and thicknesses would greatly expand the amount of useful information that could be extracted from normally complicated, highly overlapped IR spectra.

INTRODUCTION

Many important properties of a material which differentiate it from similar materials may not show up in static ambient-condition measurements. Infrared (IR) peak frequencies and intensities, however, can be very sensitive to chemical or physical changes in a sample, such as phase transitions. Dynamic 2D IR spectroscopy of polymer films undergoing a small-amplitude oscillatory strain as a function of temperature is an example of a technique that can extract subtle spectroscopic differences which normally do not appear in static experiments at ambient conditions.(1-3) This technique, however, has so far been limited to thin polymer films that can be stretched in a rheometer. The ability to obtain dynamic IR spectroscopic information from a wider variety of sample types and thicknesses should greatly expand the amount of useful information that could be extracted from normally complicated, highly overlapped IR spectra.

In this paper we describe a new *pressure-modulation dynamic attenuated-total-reflectance* (ATR) accessory with a diamond internal-reflection element (IRE) capable of dynamic 2D IR measurements on a wide variety of sample types and geometries. Pressure-modulation results on isotactic polypropylene, linear low density polyethylene, and a styrene-butadiene-styrene triblock copolymer are presented and compared with previous dynamic 2D IR results on these same samples.(3)

EXPERIMENTAL

A commercial high-pressure diamond micro ATR accessory called the Golden Gate (Graseby Specac) has been modified with a piezoelectric pressure transducer attachment (Manning Applied Technology) for pressure modulation experiments. The modified ATR accessory was installed in the external sample compartment of a Bio-Rad FTS-6000 step-scanning FT-IR spectrometer equipped with a Mercury-Cadmium-Telluride (MCT) detector. A 400-Hz phase modulation was applied to the fixed mirror of the interferometer through the dynamic alignment piezoelectric devices. The phase modulation amplitude was set to twice the HeNe laser fringe wavelength (peak-to-peak). Full double-sided interferograms were collected at every eighth HeNe laser-fringe zero crossing, and a 1975-cm^{-1} long-pass optical filter was used to prevent aliasing. The interferometer was stepped at a rate of 2 s per step The piezoelectric pressure transducer was driven by a 10-Hz 80-volt sinusoidal signal synchronized with the clock in the spectrometer. This driving voltage leads to about a 15-μm mechanical displacement amplitude when there is no resistance on the piezoelectric transducer. All of the signal processing was done using Bio-Rad's Win-IR Pro DSP2 software with no lock-in amplifiers.(4)

Thin (ca. 25-μm-thick) sample films of isotactic polypropylene, linear low density polyethylene, and Kraton D1102 (Shell) were examined using the pressure-modulation dynamic ATR accessory. Kraton D1102 is a styrene-butadiene-styrene triblock copolymer containing 28% styrene. The necessary level of static pressure applied

CP430, *Fourier Transform Spectroscopy:* 11th International Conference
edited by J.A. de Haseth

to the sample (using a calibrated torque wrench) was determined by taking a series of rapid-scan measurements at increasing levels of static pressure until the normal IR absorbance signal no longer increased in size. All three samples were run at a spectral resolution of 8 cm^{-1}. A total of 16 scans were coadded for isotactic polypropylene, while 24 scans were used for linear low density polyethylene, and Kraton D1102.

RESULTS

The in-phase and quadrature pressure-modulation dynamic ATR and normal IR absorbance spectra of isotactic polypropylene are shown in Fig. 1. The dynamic raw data have been ratioed to the single-beam spectrum and plotted on the same scale. Due to the small penetration depths of ATR (ca. 1 μm), the absorbance peak heights we measure are about ten-times smaller than in our earlier dynamic infrared linear dichroism studies of isotactic polypropylene, which were performed by transmittance.(3) The y-axis scale shown is only correct for the absorbance spectrum, with the in-phase and quadrature spectra being substantially less intense. A large component of the negative in-phase signal in Fig. 1 looks essentially like the normal IR absorbance spectrum. This appears to be due to better optical contact of the sample with the diamond IRE, or because more sample is within the optical penetration depth when the sample is compressed. There are, however, some significant differences between the in-phase and absorbance spectra. The band at 1167 cm^{-1}, for example, is split into two bands in the in-phase spectrum, as it was in previous dynamic IR studies of isotactic polypropylene done in transmittance.(3) In addition, the band shapes of the peaks at 1458 and 1377 cm^{-1} in the in-phase, quadrature, and absorbance spectra are somewhat different. It was previously impossible to examine these bands by transmittance because they were too strongly absorbing.

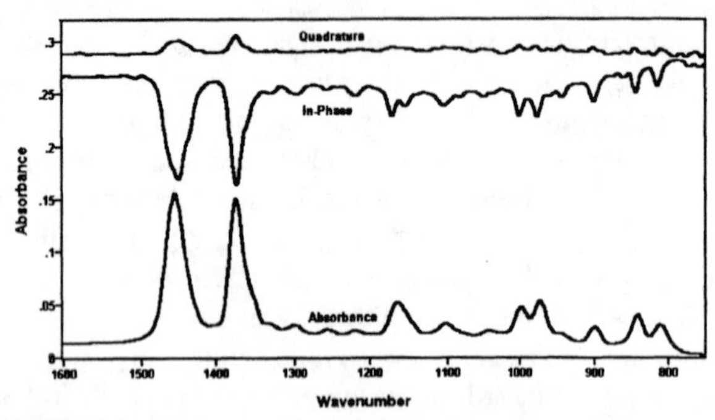

FIGURE 1. In-phase and quadrature dynamic pressure-modulation and normal IR absorbance spectra of isotactic polypropylene.

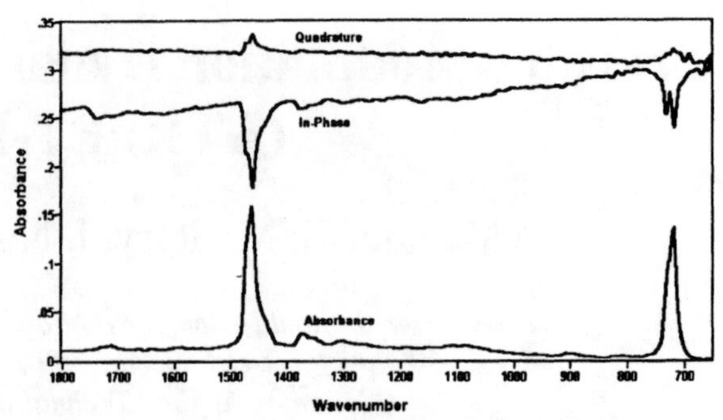

FIGURE 2. In-phase and quadrature dynamic pressure-modulation and normal IR absorbance spectra of linear low-density polyethylene.

The in-phase and quadrature pressure-modulation dynamic ATR and normal IR absorbance spectra of linear low density polyethylene is shown in Fig. 2. Notice that the dynamic signal appears to show better spectral resolution than the absorbance spectrum, even though the data were collected simultaneously.

Figure 3 shows the in-phase and quadrature pressure-modulation dynamic ATR and normal IR absorbance spectra of Kraton D1102. In this case there is not nearly as much of the absorbance spectrum superimposed negatively over the in-phase spectrum. This may be because the Kraton film was cast from toluene directly onto the diamond IRE and allowed to dry before the measurement, thereby providing more uniform contact less affected by the dynamic pressure perturbation.

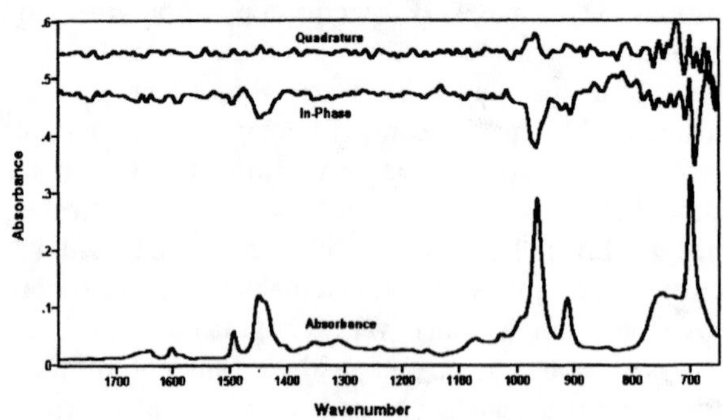

FIGURE 3. In-phase and quadrature dynamic pressure-modulation and normal IR absorbance spectra of Kraton D1102.

REFERENCES

1. Noda, I., *Appl. Spectrosc.* **44,** 550-561 (1990).
2. Marcott, C., Dowrey, A. E., and Noda, I., *Anal. Chem.* **66,** 1065A-1075A (1994).
3. Noda, I., Dowrey, A. E., and Marcott, C., *ACS Symposium Series on Characterization of Polymers Using Fourier Transform Infrared Spectroscopy Vol. 36*, Ishida, H. Ed., , New York: Plenum, 1987, pp. 33-59.
4. Drapcho, D. L., Curbelo, R., Jiang, E. J., Crocombe, R. A., and McCarthy, W. J., *Appl. Spectrosc.* **51,** 453-460 (1997).

FREQUENCY-RESOLVED, PHASE-RESOLVED AND TIME-RESOLVED STEP-SCAN FOURIER TRANSFORM INFRARED PHOTOACOUSTIC SPECTROSCOPY

E. Y. Jiang, D. L. Drapcho, W. J. McCarthy, and R. A. Crocombe

Bio-Rad Laboratories, Inc., Digilab Division, 237 Putnam Ave., Cambridge, MA, 02139

This paper demonstrates, compares and discusses frequency-resolved, phase-resolved and time-resolved step-scan Fourier transform infrared photoacoustic spectroscopic approaches used in depth profiling chemical analysis of layered polymeric samples.

INTRODUCTION

Step-scan Fourier transform infrared photoacoustic spectroscopy (S^2FTIR PAS) has been used in nondestructive depth profiling chemical analysis of thin layered / heterogeneous samples, due to the following major features of this technique: 1) uniform probing depth across the entire spectrum achieved by using phase modulation/demodulation, as compared to wavenumber dependent probing depth in rapid scan PAS; 2) easy extraction of photoacoustic phase from a multi-channel phase modulation experiment; 3) stepping modulation and easily accessible time-resolved spectroscopic capability offering an alternative depth profiling approach. In this paper, we will study three layered polymeric samples with phase modulation frequency-resolved, photoacoustic phase-resolved, and time-resolved stepping modulation approaches, respectively.

BACKGROUND

The inversely dependent relationship between photoacoustic probing depth (μ) and phase modulation frequency (f) offers an intuitive and straightforward modulation frequency-resolved depth profiling method in step-scan FTIR PAS. In this method the probing depths for a "homogeneous" sample at different modulation frequencies can be easily determined from the equation $\mu = (\alpha/\pi f)^{1/2}$, where α is the thermal diffusivity of the sample. This equation is reduced to $\mu = 180\ /f^{\ 1/2}$ µm for most organic polymers (with an average thermal diffusivity of 0.001 cm^2/sec). The effective probing depth for a layered sample, however, can be as deep as more than twice the thermal diffusion depth of the over layer, assuming each layer has a distinctive band.[1]

The use of photoacoustic signal phase, $\Phi = \tan^{-1}(Q/I)$, (I and Q are in-phase and quadrature photoacoustic responses, respectively), either alone or in combination with variation of the modulation frequency, provides the prospect of much greater details in depth profiling than use of the modulation frequency alone[2-3], because photoacoustic phase information offers a possibility to spatially resolve signals from within the probing depth defined by the modulation frequency. A phase lag of a photoacoustic signal with respect to the optical modulation is introduced mainly by the time scale difference between the slow because thermal diffusion (10^{-6} - 10^{-3} second) and the fast optical penetration (absorption) (10^{-15} - 10^{-13} second) within a sample. The phase of a photoacoustic signal characteristically represents the spatial origin of the signal. Photoacoustic signals from a deep layer of a sample are detected later than do those from a shallow layer. This relation has been quantitatively elucidated in a recently developed photoacoustic signal phase theory for multi-layered sample systems,[4] an extension to the classic Rosencwaig and Gersho theory for homogeneous solids.

Another approach to obtaining photoacoustic depth-resolved spectral information is to use time-resolved step-scan FTIR operation mode with either synchronized radiation pulse modulation (shuttle) or stepping modulation.[5] Time-resolved step-scan photoacoustic spectra are similar to phase-resolved ones in that a time delay of a photoacoustic signal from a deep layer can also be represented by a phase lag. In time-resolved "slow" (a few Hz or less) pulse or stepping modulation experiments, the quantitative relationship between time and probing depth may not be intuitively simple, because usually a stepping or pulse function is virtually a complicated

CP430, *Fourier Transform Spectroscopy:* 11th International Conference
edited by J.A. de Haseth

Fourier series. However, when the pulse or stepping modulation is substantially fast (tens or hundreds of Hz), the modulation frequency can be approximated to the Fourier frequency, $f = F = 2v\delta\sigma$, where v is the pulse or stepping frequency (Hz), δ the stepping size (cm) and σ the wavenumber of radiation. Therefore the wavenumber-dependent probing depth can be estimated using this Fourier frequency. However, use of high pulse or stepping frequencies is restricted in time-resolved photoacoustic measurements because of relatively slow photoacoustic signal response and long time constant of the photoacoustic cell preamplifier. A specific digital signal processing algorithm (DSP3) reported in this conference by our laboratory and a further modification of the photoacoustic cell electronics will improve the system performance on high stepping frequency time-resolved step-scan FTIR PAS experiments.

EXPERIMENTAL

All the data reported in this paper were collected in either step-scan phase modulation or step-scan time-resolved modes by using a Bio-Rad FTS 6000 spectrometer coupled with a helium purged MTEC 300 photoacoustic detector under room temperature. The recently developed software-based digital signal processing (DSP1) algorithm integrated in the Bio-Rad Win-IR Pro™ software package was used to simultaneously demodulate the phase modulation (fundamental) signal and its third, fifth, seventh and ninth harmonics. Phase modulation frequencies of 200, 100 Hz were used in the frequency-resolved and phase-resolved experiments, respectively. A stepping rate of 10 Hz and time resolution of 1 ms were used in the time-resolved experiment. An MTEC standard carbon black film was used to collect all reference spectra under the same condition for intensity normalization and a piece of carbon black-filled rubber was used as the surface reference for all phase modulation experiments.

RESULTS AND DISCUSSION

Frequency-resolved step-scan FTIR PAS. Fig. 1 overlays fundamental (200 Hz) and harmonic (600 Hz and 1000 Hz) photoacoustic magnitude [$M = (I^2 + Q^2)^{1/2}$] spectra calculated from the corresponding in-phase (I) and quadrature (Q) spectra collected from a single DSP1 scan for a two-layer sample, 10 μm polystyrene on 100 μm mylar. It can be seen from Fig. 1 that the intensity of the mylar band at 1725 cm^{-1} decreases much faster than that of the polystyrene band at 1680 cm^{-1} as modulation frequency increases from 200 Hz to 1000 Hz. This clearly indicates that the mylar layer is located deeper in the sample than the polystyrene layer. Note that at 1000 Hz, the theoretical probing depth for polystyrene is about 5 μm but the distinctive band of the substrate mylar is still detectable.

The frequency-resolved photoacoustic data can be further analyzed by using generalized two-dimensional (G2D) spectral correlation method.[7-8] Fig. 2 shows the asynchronous G2D correlation spectrum created by inputting the three magnitude spectra in Fig. 1. into the G2D algorithm available in the Bio-Rad Win-IR Pro™ software. The big negative contour at 1785/1725 cm^{-1} (X/Y) confirms the same results, i.e. the spatial origin of band at 1785 cm^{-1} is shallower than that of the band at 1725 cm^{-1}. More depth-related spectral details can be obtained by carefully analyzing the G2D spectra as suggested in literature.[8]

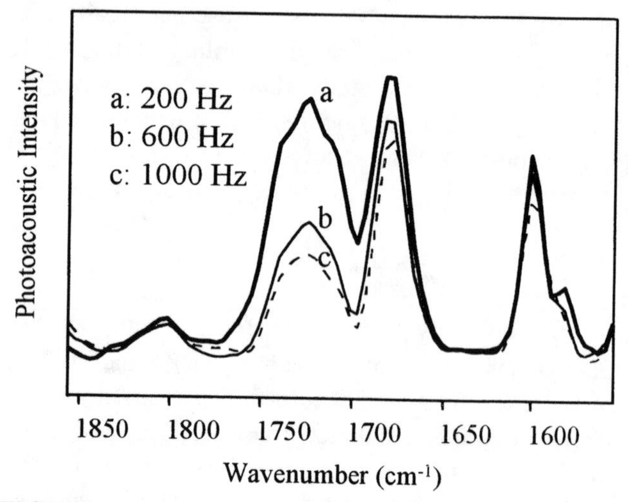

FIGURE 1. Step-scan DSP1 photoacoustic magnitude spectra of 1 μm polystyrene on 100 μm mylar collected at 200 Hz phase modulation frequency and 600 Hz and 1000 Hz harmonic frequencies.

Phase-resolved step-scan FTIR PAS. Fig. 3 illustrates step-scan DSP photoacoustic magnitude (M), and phase (Φ) spectra at 100 Hz phase modulation frequency for a two-layer sample, 2 μm silicone vacuum grease (VG) on 1 mm polyethylene (PE). It can be seen from the two spectra that the broad vacuum grease bands in the region of 1300-700 cm^{-1} correspond to very small phase lags while the PE band ~ 1460 cm^{-1} has a much greater phase lag. The phase lag

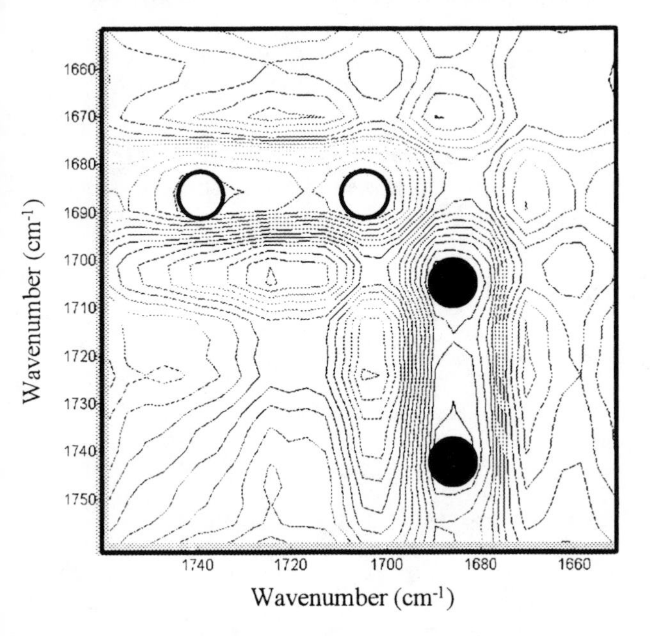

FIGURE 2. Generalized asynchronous 2D correlation spectra of 1 μm polystyrene on 100 μm mylar, generated from the spectra in Figure 1. O = A positive contour; ● = A negative contour.

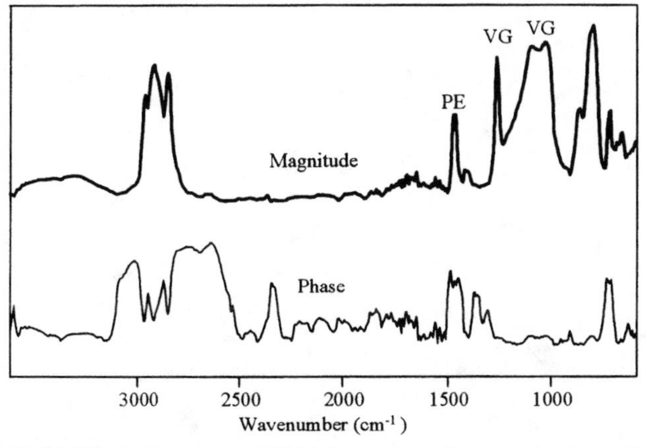

FIGURE 3. Step-scan DSP photoacoustic magnitude and phase spectra of 2 μm vacuum grease on 1 mm polyethylene (VG/PE) collected at 200 Hz phase modulation frequency.

difference between bands of different layers are obvious and can be used directly to differentiate signals spatially. There have been several other methods of using photoacoustic phase for depth profiling analysis, such as the continuos phase rotation plot (interpolation) method, in-phase and quadrature spectra method, and the conventional 2D spectral correlation method. The phase spectrum method illustrated here offers the most straightforward and yet very spatially informative analysis with the least spectral manipulations.

Time-resolved step-scan FTIR PAS. Fig. 4 (top) shows the three-dimensional view of step-scan time-resolved photoacoustic spectra of a four-layer sample, 10 μm polyethylene on 10 μm polypropylene on 6 μm polyethylene terephthalate on 6 mm polycarbonate (PE/PP/PET/PC) collected at stepping frequency of 10 Hz with time resolution of 1 ms. The probing depth range over the spectral region 1850-1200 cm^{-1} is approximately 165-205 μm, allowing

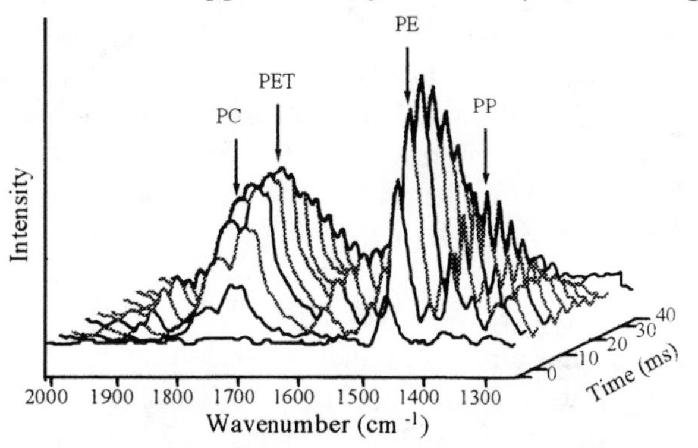

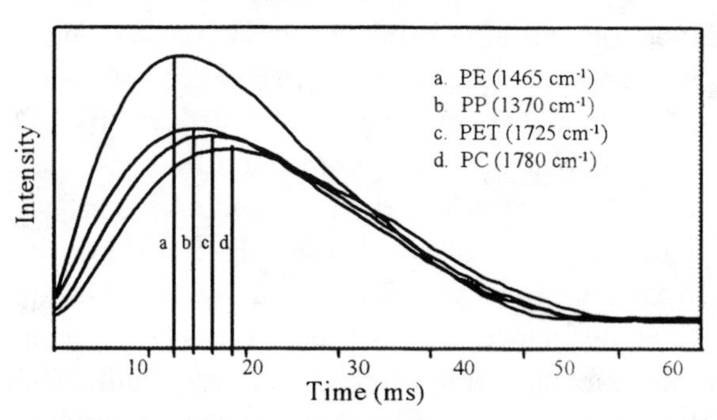

FIGURE 4. Three-dimensional step-scan time-resolved photoacoustic spectra (top) and the corresponding plot of characteristic band intensity vs. time (bottom) for 10 μm polyethylene on 10 μm polypropylene on 6 μm polyethylene terephthalate on 6 mm polycarbonate (PE/PP/PET/PC), collected at 10 Hz stepping rate and 1 ms time resolution.

signals from the bottom layer fully detectable. Fig. 4 (bottom) also shows photoacoustic intensity variation of the characteristic bands vs. time. It can be seen that the PE band (1465 cm^{-1}) reaches its maximum at an earlier time, and then sequentially, the PP (1370 cm^{-1}), PET (1725 cm^{-1}) and PC (1780 cm^{-1}) bands reach their maxima. The time difference between any two neighboring maxima is about 2 ms, indicating that the thicknesses of the top three layers are close to each other. The time difference (Δt) can be related to the phase difference ($\Delta\phi$) by $\Delta\phi = \omega\Delta t = 2\pi f\Delta t = 4\pi\nu\delta\sigma\ \Delta t$. Thus the calculated average phase difference corresponding to 2 ms time difference in this spectral region is about 0.7 degree (at 1500 cm^{-1}), giving a calculated layer thickness (d) of PE, PP, or PET of about 2.3 μm according to $d \approx \Delta\phi * \mu$ where the thermal diffusion length μ = 185 μm is used. The deviation factors of about 3 to 4 between the calculated thickness and the actual thickness of the over layers should be considered very good, given the approximations used in all the estimations and calculations. The detailed synchronous and asynchronous characteristics of the time-resolved photoacoustic spectra can also be further investigated by using generalized 2D spectral correlation analysis, which will be an interesting subject for future studies.

SUMMARY

In this paper we have demonstrated frequency-resolved, phase-resolved and time-resolved step-scan FTIR PAS depth profiling approaches. The frequency-resolved approach is the simplest, giving unambiguous probing depth information. The phase-resolved approach provides a detailed depth-resolved profile within the probing depth defined by the modulation frequency. The photoacoustic phase data can also be used for further quantitative analysis. Step-scan time-resolved FTIR PAS offers an alternative approach to depth profiling analysis and the time difference can be related easily to depth when the stepping frequency is relatively high. With much simpler analysis rules than the conventional 2D spectral correlation analysis, generalized 2D spectroscopy can be used to correlate a series of time-resolved photoacoustic spectra as well as a series of frequency-resolved photoacoustic spectra to unravel hidden spectral and spatial details of a heterogeneous sample.

REFERENCES

1. R. A. Palmer and R. M. Dittmar, *Thin Solid Films, Section A*, **223**, 31 (1993).

2. R. A. Palmer, E. Y. Jiang and J. L. Chao, *Proc. SPIE* 250 (1993).

3. R. A. Palmer and E. Y. Jiang, *J. Phys.* IV, 4 (July), C7-337 (1994).

4. E. Y. Jiang, R. A. Palmer and J. L. Chao, *J. Appl. Phys.*, 460 (1995).

5. O. Budevska and C. J. Manning, *Appl. Spectrosc.* **50**, 939 (1996).

6. D. L. Drapcho, R. Curbelo, E. Y. Jiang, R. A. Crocombe and W. J. McCarthy, *J. Appl. Spectrosc.* **51**, 453 (1997).

7. I. Noda, *Appl. Spectrosc.* **47**, 1329, (1993).

8. E. Y. Jiang, W. J. McCarthy, D. L. Drapcho, and R. A. Crocombe, *Appl. Spectrosc.* **51** (1997). (in press).

A Relaxation of Tilt Angle in a Ferroelectric Liquid Crystal Studied by Time-Resolved FT-IR

T. Matsumoto[1], K. Sakaguchi[2], A. Yasuda[3], and Y. Ozaki[1]

[1]*Department of Chemistry, School of Science, Kwansei-Gakuin University, Uegahara, Nishinomiya, 662 Japan.*
[2]*Department of Material Science, Faculty of Science, Osaka City University, Sugimoto-cho, Sumiyoshi-ku, Osaka 558 Japan.* [3]*SONY Corporation Research Center, 174 Fujitsuka-cho, Hodogaya-ku, Yokohama 240 Japan.*

Polarization angle dependences of infrared (IR) and time-resolved IR have been measured for a FLC mixture containing 20% cis-(2R,4R)-γ-butyrolactone 1 (YK230C) as a chiral dopant and 80% 5-octyl-2-(4-nonyloxyphenyl) pyrimidine (ONPP) as a nonchiral smectic base LC. These measurements and the measurements of the dichroic ratios of IR bands show that the apparent tilt angle and dichroic ratio in the dynamic state are larger than those in the static state. It seemed therefore that the order of the orientation is higher in the dynamic state than in the static state. In order to confirm the higher orientation order in the dynamic state, we performed time-resolved IR measurements of the FLC mixture for the delay time ranging from 0 to 500μs, which is much longer than the response time. The relaxation process was clearly observed after the response time.

INTRODUCTION

Liquid crystals (LC's) have been extensively employed for various kinds of devices, especially in display systems in which their ability to reorientate in the electric field is utilized (1). The mechanism of reorientation of LC's, however, is not well understood. During the recent part, time-resolved Fourier-transform infrared (FT-IR) spectroscopy has been developed and applied as a powerful tool for investigating the dynamics of LC's (2-7). We have been studying the mechanism of electric-field-induced switching of ferroelectric LC's (FLC's) (5-7) by use of an asynchronous time-resolved FT-IR technique (4). Recently, we have proposed a technique, the measurements of polarization angle dependence of time-resolved polarized IR spectra, to investigate the molecular orientation in the dynamic state of a FLC during the process of the electric-field induced reorientation.

In this communication, we report a time-resolved FT-IR study of a FLC mixture consisting of YK230C and ONNP (Figure 1). YK230C, which has a unique γ-lactone ring, shows large spontaneous polarization and fast response (8). For the FLC mixture, we have found a relaxation of tilt angle based upon the measurements of polarization angle dependence of IR and time-resolved IR spectra.

EXPERIMENTAL

YK230C was synthesized as previously reported (8). YK230C and the base LC were mixed at ratio of 1 to 4 or 1 to 49 by weight, and placed in a thin cell fabricated from two CaF_2 plates coated with layers of indium tin oxide and polyimide. The cell thickness was determined to be 1.6μm. The phase transition temperatures of the 1:4 and 1:49 mixtures are given in Fig.1.

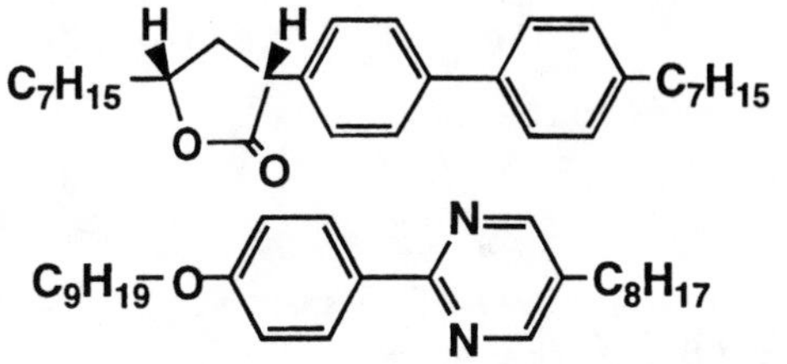

Mixing Ratio 1:4 (Chiral Dopant :Base LC)
Phase Transition/°C ;
I--(59.1~58.8)--N--(57.1~55.8)--Sa--(56.2~55.8)--Sc*--

Mixing Ratio 1:49 (Chiral Dopant :Base LC)
Phase Transition/°C ;
I--(66.8~66.6)--N--(63.6~63.4)--Sa--(59.1~58.3)--Sc*--

FIGURE 1. Structure of chiral dopant (YK230C) and nonchiral base LC (ONNP), and phase transition temperatures of the 1:4 and 1:49 mixtures.

The time-resolved FT-IR spectra were measured with a JEOL JIR-6500 FT-IR spectrophotometer working in the asynchronous mode equipped with a boxcar integrator consisting of a gate circuit and pulse delay circuit.

RESULTS AND DISCUSSION

Figure 2 shows time-resolved FT-IR spectra of the FLC mixture in the Sm-C* phase at 54°C under an electric field of ±2V and 5kHz repetition rate as a function of delay time from 0 to 100μs at intervals of 5μs. It is noted that bands at 2926 and 2854cm^{-1} due to the antisymmetric and symmetric CH_2 stretching modes of the alkyl groups, respectively, and a band at 1759cm^{-1} assigned to a C=O stretching mode of the γ-lactone ring decrease. Those at 1606, 1583, 1431, 1252, and 1165 cm^{-1} assigned to the

CP430, *Fourier Transform Spectroscopy:* 11th International Conference
edited by J.A. de Haseth

υ_{8a},υ_{8b},υ_{19} modes of the aromatic rings, Phe-O-C antisymmetric stretching mode, and ring stretching mode, respectively, increase with time. Analyzing the spectra in Fig. 2, we found that the molecule reorients nearly uniformly with the response time of 90μs.

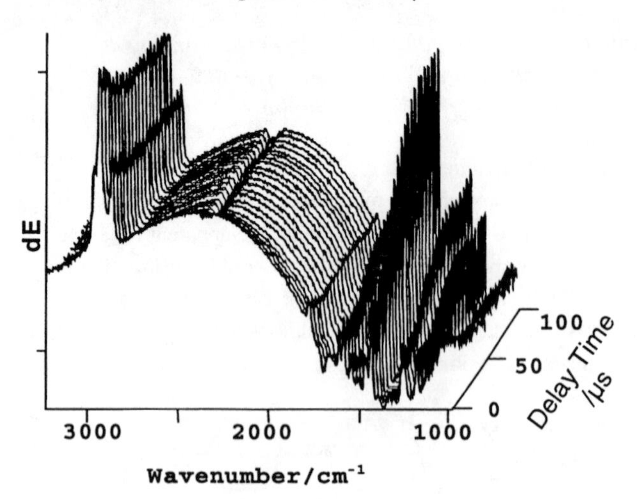

FIGURE 2. Time-resolved FT-IR spectra of the 1:4 FLC mixture in the Sm-C* phase at 54°C

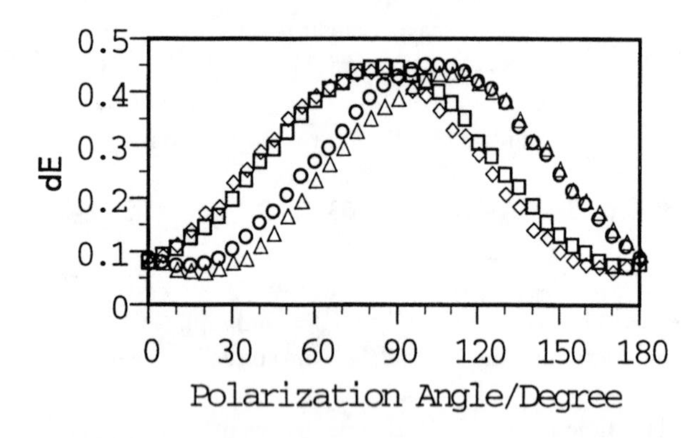

FIGURE 3. Polarized angle dependence of time-resolved FT-IR for the υ_{8a} band ; □ starting state (0μs), O final state (90μs). Polarized angle dependence of polarized FT-IR for the same band; ◊ DC +2V, Δ DC -2V

Figure 3 plots the absorption energy (dE) versus polarization angle at 0 (initial state) and 90 (final state) μs for the υ_{8a} band at 1606 cm^{-1} in the time-resolved polarized IR measurements of the 1:4 FLC mixture under an electric field of ±2V and 5kHz repetition rate. In the same figure is shown the absorption energy (dE) versus polarization angle for the same band in the polarized IR measurements of the same FLC mixture controlled under dc voltage of ±2V. In the polarized IR measurements the υ_{8a} band showed the maximum absorption energy at 85° and 105° when the positive and negative electric field was applied. Therefore, the apparent tilt angle was determined to be 10° in the static orientation state. The corresponding angle in the dynamic orientation state was estimated to be 15° because the υ_{8a} band showed the maximum absorption energy at 110° and 80° for the initial and final states, respectively. The time-resolved polarized and polarized IR measurements revealed that the apparent tilt angle is larger for the dynamic orientational state (15°) than the static orientational state (10°). In addition, the dichroic ratio (in the present study, the dichroic ratio was estimated by the ratio of absorption energy) for the υ_{8a} band is 7 and 6 for the dynamic and static states, respectively. It was thus concluded that the degree of the order in the orientation is higher for the dynamic state than for the static state. We inferred that the relaxation of the orientation occurs after the response time for the 1:4 FLC mixture.

In order to confirm the relaxation process, we carried out time-resolved IR measurements for the delay time ranging from 0 to 500μs at 54°C under an electric field of ±8V and 1kHz repetition rate. The existence of the relaxation in the orientation was clearly observed for all the IR bands investigated. Due to the limitation of the instrument used, we could not observe final points for the relaxation process. However, the final points may be estimated to be 800μs from the simulation based upon the changes in the tilt angle and absorption energy.

It seemed to us that the high mixing ratio of chiral dopant in the FLC mixture causes the relaxation in the orientation. Then, another FLC mixture with the ratio of 1:49 (chiral dopant : base LC) was prepared and the same series of IR measurements were carried out under the similar conditions. Time-resolved IR spectra of the 1:49 FLC mixture were measured at 55°C under an electric field of ±25V for the delay time from 0 to 100μs. From this experiment, the response time and the intermediate time were determined to be about 75 and 35μs, respectively.

We also measured polarized angle dependence of polarized IR spectra of the 1:49 FLC mixture at 55°C under ±25V and polarized angle dependence of time-resolved IR spectra for 0 (initial state) and 90μs (final state) to compare the apparent tilt angle between the static and dynamic states. In the polarized IR measurements the υ_{8a} band showed the maximum absorption energy at 80° and 105° when the positive and negative electric field was applied. Therefore, the apparent tilt angle was determined to be 12° in the static orientation state. The corresponding angle in the dynamic orientation state was estimated to be 12° because the υ_{8a} band showed the maximum absorption energy at 105° and 80° for the initial and final states, respectively. Of note is that the apparent tilt angle is the same for the static and dynamic states.

The time-resolved IR spectra of the 1:49 FLC mixture were measured also for the delay time of 0 to 500μs. However, the relaxation process was not observed. Moreover, the dichroic ratio for the υ_{8a} mode was found to be 8 and 7 for the static and dynamic orientation states, respectively. As a result, it turned out that the static orientation state has the higher order than the dynamic orientation state. This conclusion is sharply different from that for the FLC mixture with the higher mixing ratio of the chiral dopant.

CONCLUSION

Clear difference in the apparent tilt angle between the static and dynamic states observed for the FLC mixture with the high mixing ratio of the chiral dopant probably comes from the high content of the chiral dopant in the FLC mixture. More detailed discussion about this will be reported separately.

REFERENCES

1. J. W. Goodby, R. Blinc, N.A. Clark, S.T. Lagerwall, M. A. Osipov, S. A. Pikin, T. Sakurai, K. Yoshino, and B. Zeks, Ferroelectric Liquid Crystals: Principles, Properties and Applications (Gordon and Breach Science Publishers, Philadelphia).
2. H. Toriumi, H. Sugisawa, and H. Watanabe, Jpn. J. Appl. Phys. **27**. L935 (1988).
3. V. G. Gregoriou, J. L. Chao, H. Toriumi, and R. A. Palmer, Chem. Phys. Lett. **179**, 491, (1991).
4. K. Masutani, H. Sugisawa, A. Yokota, Y. Furukawa, and M. Tasumi, Appl. Spectrosc. **46**, 560 (1992).
5. M. Czarnecki, N. Katayama, M. Satoh, T. Watanabe, and Y. Ozaki, J. Phys. Chem. **99**, 14101 (1995).
6. A. L. Verma, B. Zhao, S. M. Jiang, J. C. Sheng, and Y. Ozaki, Phys. Rev. E, in press.
7. K. Taniike, Y. Ozaki, A. Yasuda, and T. Ikemoto, submitted for publication.
8. K. Sakaguchi, T. Kitamura, Y. Shiomi, M. Koden, and T. Kuratate, Chem. Lett. 1383, (1991).

Improvements in Signal Acquisition and Processing for Time-Resoved Step-scan FT-IR Spectroscopy

C. Rödig and F. Siebert

Institut für Biophysik und Strahlenbiologie, Albert-Ludwigs-Universität Freiburg, Albertstraße 23, 79104 Freiburg, Germany

A broadband amplifer with a short rise time was developed for the acquisition of the time dependent part of the interferogram in time-resolved step-scan measurements. Amplification of this portion of the interferogram by about 40 dB is necessary to digitize siganl and noise with adequate amplitude resolution. Therefore, the transient ac-signal from the detector has to be separated from the static dc-signal being larger by a factor of thousand. We describe here a way to avoid the currently used ac-coupling of the signal which is always related to the disadvantage of a low frequency limit for time-dependent signals. Since the recorded interferogram shows distorsions caused by nonlinear response of the MCT-detector, especially, when a photoconductive detector is used, we implemented a software noninearity correction method proposed by Keens and Simon (1). The effects of the nolinearity correction on the time-resolved step-scan difference spectra of biological samples are demonstrated.

INTRODUCTION

Time-resolved step-scan FT-IR spectroscopy is an excellent method for monitoring spectral changes in the time range from nanoseconds to seconds. The time resolution of this method is not intrinsically limited by the scanning velocity of the movable interferometer mirror. The interferogram I(x) is scanned by moving the mirror stepwise and additionally, at each mirror position x, the time course of changes of the interferogram intensity $\Delta I_x(t)$ after exciting the sample is recorded. Thus, time resolution of the system is limited only by the rise time of the detector, the bandwidth of the amplifiers and the sampling rate of the AD converter.

The step-scan technique is capable to detect very small absorption changes in the order of 10^{-4} OD against a large background absorption and has been successfully applied to different biological systems (2-6). It is necessary to achieve a high time resolution, because important steps in this kind of reactions occur in the early time domain. On the other hand, limited amount of sample material leads to the demand of recording the complete time course of a reaction, if needed, up to seconds, within the same measurement. Therefore, special attention has to be paid to the electronics amplifying the detector signal. The small transient intensity changes need a further amplification with a gain of 40 dB after the preamplifier to guarantee correct digitization. This is not the case for the large static interferogram values. Therefore dc- and ac- part of the detector signal have to be separated before entering the amplifier. The currently used ac-coupling of the signal for this purpose introduces a limitation of bandwidth towards low frequencies. Since for high time resolutions in the range of 20 ns usually low input impedance values are required (50 Ohms), capacitors used for the ac-coupling would amount to impractically high values. Thus, it would not be possible to process the slow time course of the interferogram signal with the fast electronics. Therefore, it is desireable to use an amplifier, which combines the advantages of ac- and dc-coupling, resulting in a large bandwidth and no low frequency limit.

Detectoion of small absorption changes with a high time resolution makes the use of photovoltaic or photoconductive MCT detectors indispensable. One problem in time-resolved step-scan FT-IR spetroscopy is related to the nonlinearity of the MCT detectors, particularly of photoconductive elements. The high throughput of the FT-IR spectrometer leads to very large interferogram values at the centerburst, resulting in a nonlinear detector response (7, 8). Several methods have been proposed for the elimination or reduction of detector nonlinearities (9, 10).

We implemented a software method for nonlinearity correction (NLC) of interferograms, based on the procedures patented by Bruker GmbH (1), to demonstrate the effects of detector nonlinearity on time-resolved step-scan difference spectra.

EXPERIMENTAL

The FT-IR instrument we used for all measurements is described in (2). For measurements with 20 ns time resolution, a photovoltaic MCT detector (Kolmar KV104) with a dc-coupled preamplifier (Kolmar KA020-A1) was employed, for measurements with 700 ns time resolution, we used a photoconductive MCT detector (Judson J15D16) with a dc-coupled preamplifier. Static interferogram values

CP430, *Fourier Transform Spectroscopy:* 11th International Conference
edited by J.A. de Haseth
 1-56396-746-4/98/$15.00

were digitized with a 13 bit AD-converter without further amplification, transient intensity changes were recorded wit a 8 bit/200 MHz transient recorder (Spectrum PAD82). The time dependent signals were amplified with the main aplifier described in the next section. For comparison, measurements were performed with ac-coupling the signal (1kHz cutoff) before amplification with two dc-coupled amplifiers (Comlinear).

The time dependent signals were transferred to a quasilogarithmic time-scale by coadding an increasing number of data points in order to reduce the amount of data and to improve the signal to noise ratio (11).

For contolling the measurement and data acquisition, including transformation of the data to a quasilogarithmic time scale, we developed a special computer software. Our implementation allows to communicate with the Bruker interferometer controller directly via serial interface RS232. A second serial interface is used to control the 13-bit AD converter, the amplifier settings (see next section) and the laser trigger. Optionally, this interface can be used to record other data (for example laser flash intensity) or to control other devices.

Time-resolved difference spectra were obtained as follows. In order to allow the application of the NLC, the procedure described in (2) was slightly altered. The change of the interferogram at a given time t after the laser flash $\Delta I_t(x)$, which is rearranged from the time-resolved signals $\Delta I_x(t)$, is not directly Fourier-transformed. Instead, the interferogram corresponding to a single beam spectrum at a given time after exciting the sample was reconstructed by coadding the static interferogram I(x) and the change of this interferogram. Both interferograms I(x) and $I(x)+\Delta I_t(x)$ were then Fourier-transformed and phase corrected as described in (2). The absorption change ΔA_t for the time t was calculated using the following formula:

$$\Delta A_t = -\lg[\, I(x)/(I(x)+\Delta I_t(x))\,].$$

NLC was then applied to both interferograms before phase correction and Fourier transformation. The details of the procedure of the NLC are given below. For further analysis of the difference spectra and extraction of the kinetics, the 3D-package of the OPUS software from Bruker was used.

RESULTS AND DISCUSSION

Step-scan Amplifier

As stated before, the ac part of the interferogram in step-scan measurements has to be further amplified after the preamplifier, whereas the dc part can be directly digitized. In order not to saturate the operational amplifiers of the main amplifier, the two signals have to be separated.

The principle of our amplifier is shown in Fig. 1. Two operational amplifiers mounted as inverting amplifiers generate a signal gain of 40 dB. An integrating feedback loop connects the output of the second stage with the inverting input of the first stage. The feedback loop can be opened and closed by the AC/DC switch, which is contolled by the software. When the feedback loop is closed, the amplifier is ac-coupled and the frequency limit is given by R and C.

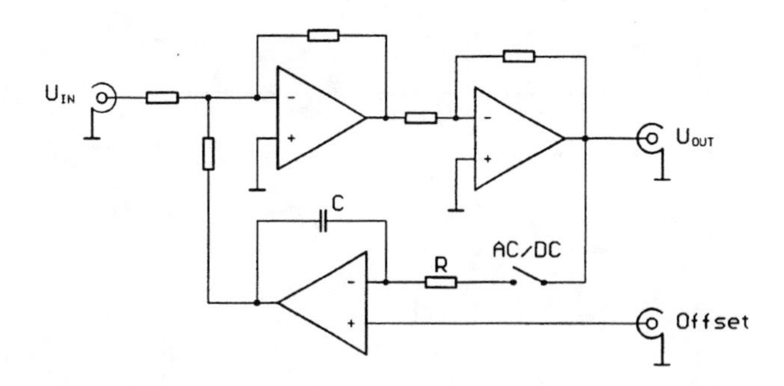

FIGURE 1. Schematics of the step-scan main ampifier.

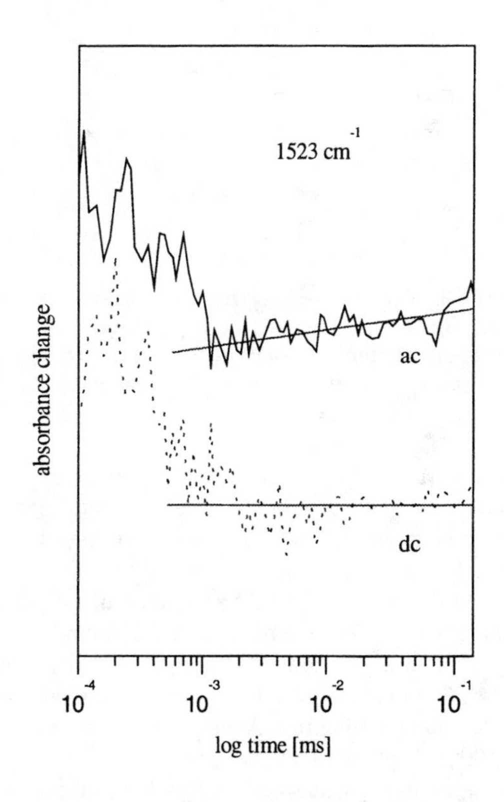

FIGURE 2. Time course of the halorhodopsin photocycle at 1523 cm^{-1} measured with an ac-coupled (top) and a dc-coupled main amplifier (bottom).

Upon opening the switch, the amplifier is quasi dc-coupled, while the input is compensated by the voltage over the capacitor C. This voltage corresponds to the interferogram level with the AC/DC switch closed. The only limit towards low frequencies is now the loss of capacitor voltage, which is very small (less than 1 mV/min). Aditionally an input is provided, which allows to set an external offset voltage into the feedback loop. This offset voltage, which is produced by a software-controlled DA-converter, ensures an optimal matching of the amplifier output to the input of the AD-converter. The bandwidth of the described amplifier is 50 MHz.

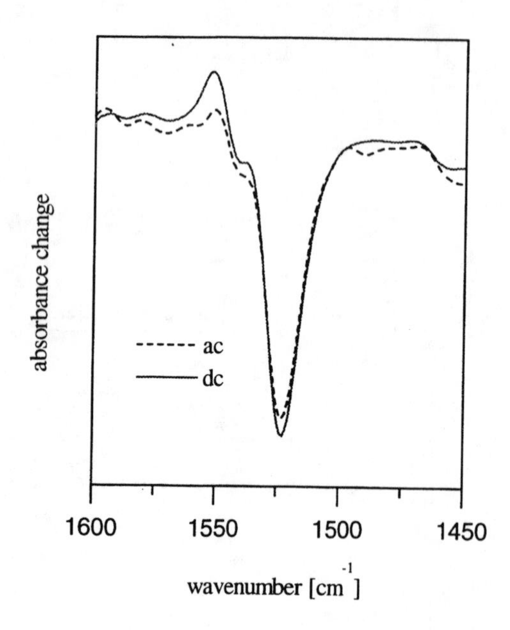

FIGURE 3. Part of the step-scan difference spectrum of halorhodopsin 100 μs after laser flash excitation. The solid line represents the spectrum measured with dc-coupling, the dashed line corresponds to the measurement with ac-coupling.

Figures 2 and 3 illustrate the differences between measurements with ac- and dc-coupled amplifiers. The measurements show infrared absorption changes during the photocycle of halorhodopsin from the nanosecond to microsecond time range. The time course of the ethylenic stretching band of the chromophore at 1523 cm^{-1} is plotted in Fig. 2. Top trace shows the measurement with the ac-coupled amplifier, the trace below was measured with the quasi dc-coupled amplifier. A linear least square fit from 10 to 100 μs is plotted for both curves. The damping of the signal becomes evident when the fits are compared. It is important to note that this effect is already observable after a few tens of microseconds. After 100 μs, the signal amplitude measured with ac coupling is reduced by about 10 %. This can be clearly seen in Fig. 3, where the difference spectra after 100 μs from ac- and dc-coupled measurements are compared in the spectral range around the ethylenic stretching band. In conclusion, it can be noted, that correct kinetical analysis of time resolved step-scan measurements over a large time range requires a careful choice of the signal amplifying electronics. The amplifier decribed above ensures correct representation of time-resolved signals without the limitations imposed by an ac-coupling.

Nonlinearity Correction

The nonlinearity correction method by Keens and Simon is based on a second order approximation of a corrected interferogram I_c of the following form:

$$I_c = \alpha\,(I + \beta\, I^2)\,,$$

where I is the measured interferogram. The correction coefficients α and β can be determined from the single beam spectrum after Fourier-transforming the measured interferogram. β is calculated by dividing the spectral intensity at zero frequency by the integral of the square of the spectrum above the detector cutoff frequency. The spectral intensity at zero frequency can be calculated from a linear fit of the spectrum below the detector cutoff. The second factor α is a function of β, the modulation efficiency of the interferometer and the integral of the spectrum above the cutoff frequency of the detector. A detailed derivation of the correction factors can be found in (1).

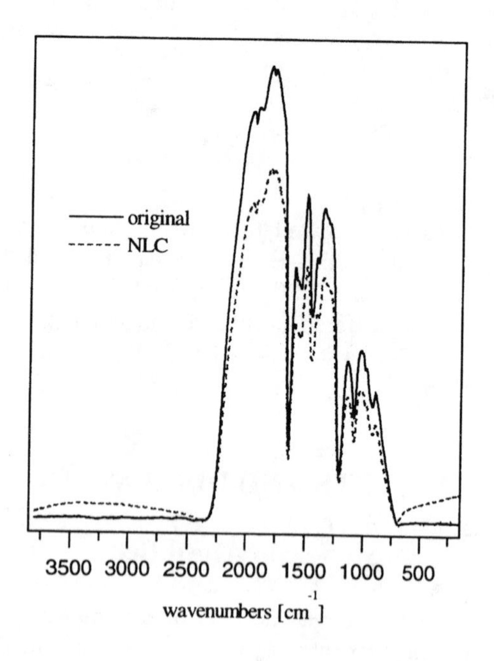

FIGURE 4. Single beam spectrum of carbonmonoxymyoglobin with nonlinearity correction (solid line) and without (dashed line).

Figure 4 shows the effect of the NLC on a single beam spectrum of carbonmonoxy myoglobin (MbCO) in D_2O. The spectral intensity below the detector cutoff and above the cutoff of the optical filter used in the experiment is completely removed. Additionally one observes large changes of the sigle beam intensity over the spectral range between 900 and 2300 wavenumbers.

The results of the nonlinearity correction of a step-scan difference spectrum are demonstrated in Fig. 5. The solid line represents the difference between unligated myoglobin and MbCO in D_2O 1.4 µs after laser flash photolysis, the dashed line shows the same spectrum with nonlinearity correction. The difference of corrected and non-corrected spectra is plotted below. For more clarity, only a small part of the spectral range is shown.

Two effects can be distinguished, when both spectra are compared. The NLC shifts the baseline of the difference spectrum slightly and produces claerly observable changes in the intensity of the difference bands. This variation of band intensities is particularly pronounced for the large difference bands (the CO-band at 1945 cm^{-1}, which is about five times larger than the amide bands, is not shown). These observations give rise to the conclusion, that this nonlinearity correction is necessary, when high accuracy of the intensities of the difference bands is required.

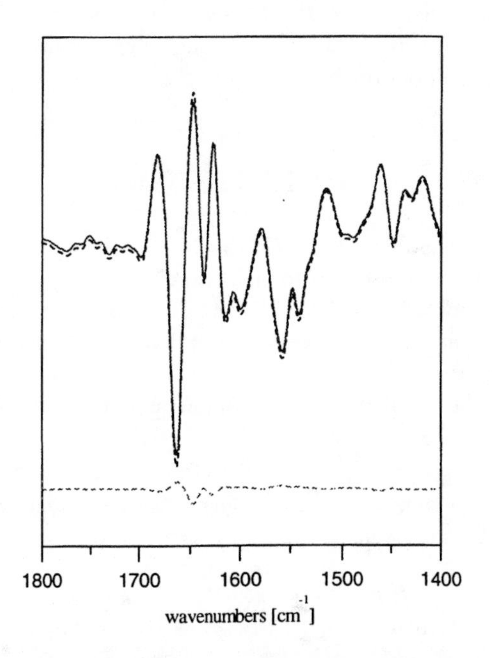

FIGURE 5. Difference spectra Mb/MbCO in D_2O 1.4 µs after laser flash photolysis (top). The spectrum drawn with a dashed line is corrected for detector nonlinearity, the solid line represents the spectrum without NLC. The dashed line on the bottom shows the differences between both spectra.

ACKNOWLEDGEMENT

We thank the Bruker Company for technical support.

REFERENCES

1. Keens, A., and Simon, A., *US Patent 4927269,* Bruker Analytische Meßtechnik GmbH, Rheinstetten, 1990
2. Uhmann, W., Becker, A., Taran, C., and Siebert, F., *Appl. Spectr.*, **45**, 390-397 (1991)
3. Rödig, C., Weidlich, O., and Siebert, F., *Time Resolved Vibrational Spectroscopy VI*, Lau, A., Siebert, F. and Werncke, W. eds., Berlin: Springer-Verlag, 1994, pp.227-230
4. Weidlich, O., and Siebert, F., *Appl. Spectr.*, **47**, 1394-1399 (1993)
5. Hu, X., Frei, H., and Spiro, T.G., *Biochemistry*, **35**, 13001-13005 (1996)
6. Dioumaev, A.K., and Braiman, M.S., *J. Phys. Chem B*, **101**, 1655-1662 (1997)
7. Chase, D.B., *Appl. Spectr.*, **38**, 491-494 (1984)
8. Zhang, Z.M., and Hanssen, L.M., *Mikrochim. Acta [Suppl.]*, **14**, 317-319 (1997)
9. Carter, R.O., Lindsay, N.E., and Bedhun, D., *Appl. Spectr.*, **44**, 1147-1151 (1990)
10. Bartoli, F., Allen, R., Esterowitz, L. and Kruer, M., *J. Appl. Phys.*, **45**, 2150-2154 (1974)
11. Austin, R.H., Beeson, K.W., Chan, S.S., Debrunner, P.G., Downing, R., Eisenstein, L., Frauenfelder, H., and Nordlund, T.M., *Rev. Sci. Instrum.*, **47**, 445-447 (1976)

Step-Scan FTIR Photoacoustic Spectroscopy: The Phase-Response of the Surface Reference Sample Carbon Black

M.W.C. Wahls*, J.L.Weisman#, W.J. Jesse*, and J.C. Leyte*

 # The College of William and Mary, C.S. Unit 1750,
200 Richmond Rd., Williamsburg, VA 23186-1750, USA

** Leiden Institute of Chemistry, Gorlaeus Laboratories*
Physical and Macromolecular Chemistry, University of Leiden
P.O. Box 9502, NL-2300 RA Leiden, The Netherlands

The Rosencwaig-Gersho theory of the photoacoustic effect is applied to simulate the PAS phase-shift of the surface reference sample carbon black. A layer thickness dependent PAS phase-response of carbon black for very thin layer thicknesses is calculated. First experimental results on various carbon black samples with different layer thicknesses confirm these simulations.

INTRODUCTION

In 1976, A.Rosencwaig and A.Gersho (1) introduced the thermal piston model of the photoacoustic experiment. The RG-theory relates the photoacoustic signal with magnitude M and phase Φ to the optical absorptivity β and thermal diffusity α of the investigated single layer with thickness l. In 1980, N.C.Fernelius (2,3) extended the Rosencwaig-Gersho theory for a double-layered polymer film.

Phase-resolved step-scan FTIR PAS is a convenient technique to obtain depth profiles from the simultaneously detected in-phase and quadrature interferograms. To compare phase spectra with theory, the phase shift due to the experimental set-up must be corrected for. To this end, the use of carbon black samples has been suggested as a reference. These samples are considered to represent the limiting case of optically opaque and thermally thin systems. They should produce the smallest phase lag expected theoretically, i.e. $+\pi/4$.

In 1979 Poulet, Chambron and Unterreiner (4) reported results of quantitative photoacoustic spectroscopy of thermally thick samples. They were the first to present a method to correct for the apparatus phase with a blackbody sample.

In 1996, we (5,6) presented a new method to correct the experimental signal phases Φ_{exp} of [PET]- single layers for the apparatus phase Φ_{app}.

As a phase reference, optically transparent and thermally thick polymer films were found to be more convenient and dependable than carbon black. The use of carbon black samples as phase reference had turned out to be ambiguous.

Here, simulated PAS-phase data of the surface reference sample carbon black are presented. They clearly indicate a dependency of $\Phi_{carbonblack}$ on its layer thickness, as predicted by the RG theory. First experimental PAS-phases of various carbon black samples with different film thickness confirm these simulations.

THE PAS PHASE Φ

Rosencwaig and Gersho (1,7) treated a one-dimensional model of the heat-flow in a sample caused by the absorbed modulated IR-light. They solved the thermal diffusion equation describing the heat transport in the sample.

The complex amplitude of the periodic temperature on the surface Q_0 is solved in terms of the optical (β), thermal (μ) and geometric (l) parameters of the sample.

The complex envelope Q of the sinusoidal pressure variation is given by:

$$Q = \frac{\beta I_0 \gamma P_0}{2\sqrt{2}\, T_0 \kappa l' a' (\beta^2 - \sigma^2)} \frac{(r-1)(b+1)e^{\sigma l} - (r+1)(b-1)e^{-\sigma l} + 2(b-r)e^{-\beta l}}{(g+1)(b+1)e^{\sigma l} - (g-1)(b-1)e^{-\sigma l}} \quad (1$$

β=optical absorptivity; I_0=incident monochromatic light flux; g=ratio of specific heats; P_0=ambient pressure; T_0=ambient temperature; κ=thermal conductivity; l=length;

CP430, *Fourier Transform Spectroscopy:* 11th International Conference
edited by J.A. de Haseth

a=thermal diffusion length; s=(1+i)a; b =(κ"a"/κa); g=(κ'a'/κa); r=(1-i)b/2a

Sample parameters are denoted by unprimed symbols, gas parameters by singly primed symbols and backing parameters by doubly primed symbols according to the notation used by Rosencwaig and Gersho.

Q_1 and Q_2 are respectively the real and imaginary parts of Q, as defined by the magnitude q and phase Φ of the acoustic signal:

$$Q = Q_1 + iQ_2 = qe^{-i\phi} \quad (2)$$

We used this solution for the evaluation of the signal-phase of the carbon black sample of the acoustic pressure variation in the cell caused by the photoacoustic effect .

The detectable acoustic pressure variation $\Delta p(t)$ at the microphone is:

$$\Delta p(t) = Q_1 \cos(\omega t - \frac{\pi}{4}) - Q_2 \sin(\omega t - \frac{\pi}{4}) \quad (3)$$

The calculated theoretical PAS phase Φ is obtained from the real [Q_1] and imaginary [Q_2] parts of [Q], as defined by Equations 1 and 2. The full equation 4 was used to evaluate the theoretical phase response without simplification or approximation.

$$\Phi = \tan^{-1}(Q_2 / Q_1) \quad (4)$$

The experimental signal phase is defined by ratioing the experimentally obtained in-**QU**adrature and **IN**-phase single-beam spectra:

$$\Phi = \tan^{-1}(QU / IN) \quad (5)$$

THE SIMULATED $\Phi_{carbonblack}$

The theoretical PAS phase $\Phi_{carbonblack}$ is simulated at 400 Hz phase modulation frequency as function of its layer-thickness with:

$\kappa = 6.85 * 10^{-4}$ (cal/(cm s °C)) [thermal conductivity]
$C_p = 0.365$ (cal/(g °C)) [heat capacity]
$\rho = 1.12$ (g / cm^3) [density] (Ref. no 8)

The optical absorption coefficient β is set to 1000000 cm^{-1}.

The image for the calculated PAS-phase of the carbon black sample $\Phi_{carbonblack}$ shows an approximately constant phase-shift to about π/4 for thicknesses $l_{carbonblack} > 40$ μm.

The samples having thicknesses lower than 40 μm show a clear dependency of the simulated PAS-phase $\Phi_{carbonblack}$ on the layer thickness. The phase-shift $\Phi_{carbonblack}$ increases towards the limit of π/2, while never reaching 90°, with decreasing $l_{carbonblack}$ and decreases to [again] π/4 for extremely thin carbon black samples with $l_{carbonblack}$ « 1 μm.

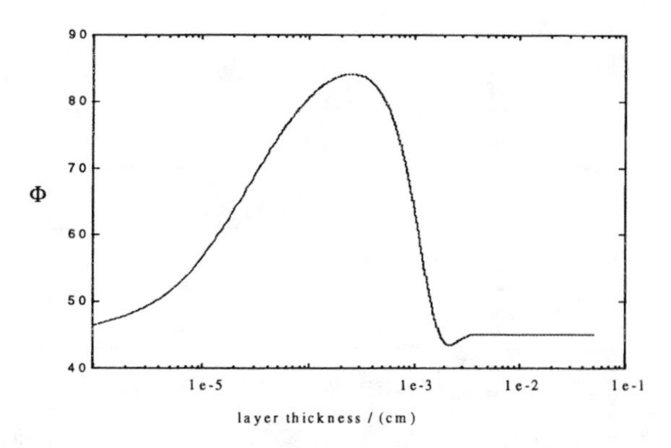

Figure 1.: Φ_{calc} of carbon black, at 400 Hz

THE EXPERIMENTAL $\Phi_{carbonblack}$

Carbon black films were made using different polymers with different percentages by weight of carbon black. The polymers, polyisoprene, polymethylmethacrylate and polystyrene, were dissolved in chloroform before carbon black was added. Films of 25% to 40% by weight carbon were pressed with the different polymers using a Graseby Specac film press. All films were heated and then pressed. Very thin films were made by spinning the carbon black dispersions on an aluminium disk.

A minimum concentration of 33% carbon was found to be sufficient to prevent IR absorbance of the polymeric material. These samples were compared to a 65% carbon filled rubber (kindly delivered by RA Palmer, thickness » 1.5mm) and the carbon black sample delivered with the MTEC PA cell, model 300, having unknown thickness.

The spectra were collected by a Bio-Rad FTS 60A step-scan interferometer coupled with an MTEC 300 PA cell. The orthogonal symmetric in-phase and quadrature interferograms with respect to the optical phase modulation were simultaneously detected and demodulated by a two-channel lock-in amplifier (Stanford SR 830 DSP).
[All data were obtained at a pmfreq. = 400Hz, 1Hz step-scan velocity, res.= 8cm^{-1}.]

Identical settings of the interferometer and the lock-in amplifier were maintained to prevent parameter dependent variations on $\Phi_{carbonblack}$. Φ_{app} is therefore assumed to be constant for all measurements presented here. No correction of $\Phi_{carbonblack}$ for Φ_{app} is applied due to the variance in $\Phi_{carbonblack}$.

Table 1. The experimental $\Phi_{carbonblack}$

The sample	l_{cb}	Φ_{cb}	Preparation method
carbon black, delivered with the MTEC PA cell, model 300	?	20°	---
65% carbon filled rubber (by RA Palmer)	>>1.5mm	58°	---
polyisoprene, 33% cb	100 μm	50°	Pressed
polyisoprene, 33% cb	70 μm	50°	Pressed
polyisoprene, 40% cb, 10% parafin	48 μm	50°	Pressed
polymethylmethacrylate, 25% cb	46 μm	51°	Pressed
polymethylmethacrylate, 25% cb	40 μm	49°	Pressed
polymethylmethacrylate, 35% cb	37 μm	32°	Spun
polymethylmethacrylate, 35% cb	26 μm	34°	Spun
polystyrene, 35% cb	30 μm	31°	Spun

CONCLUSIONS

To this end, all carbon black samples with thicknesses higher than 40 μm yield a constant experimental PAS phase shift [including Φ_{app}] of about 50°. Thus, the samples [still] made from different polymers with varying amounts of added carbon indicate an approximately constant PAS phase shift.

This may indicate those samples to be sufficiently thick carbon black samples, i.e. the PAS phase is independent on the sample thickness. Compared to the 65% carbon filled rubber, a small deviation is observed, maybe due to the differences in the amount of added carbon.

From the thinner carbon black samples with thicknesses of respectively 26, 30, and 37 μm, experimental PAS phase shifts in great disagreement to those already mentioned are observed. The obtained phase values may indicate an influence of the film thickness on $\Phi_{carbonblack}$, as indeed simulated from the RG theory (Figure I).

All experimental values of $\Phi_{carbonblack}$ include still Φ_{app}, so no comparison of absolute PAS phase shifts with simulated data (Figure I) can be made. Still, the relative variation in the experimental phase data may be related to the simulated changes in $\Phi_{carbonblack}$ at very low film thicknesses.

It might be concluded, that the photoacoustic surface reference sample carbon black may result in consistent experimental PAS phase shifts only if it is sufficiently thick.

REFERENCES

1. A.Rosencwaig, A.Gersho, J.Appl.Phys., 47, 64 (1976)
2.. N.C.Fernelius, J.Opt.Soc.Am., 70 (5), 480 (1980)
3. N.C.Fernelius, J.Appl.Phys., 51 (1), 650 (1980)
4. P.Poulet, J.Chambron, R.Unterreiner, J.Appl.Phys., 51, 1738 (1980)
5. M.W.C.Wahls, J.P.Toutenhoofd, L.H.Leyte-Zuiderweg, J.de Bleijser, J.C.Leyte, at the AIRS II conference, Durham, NC, USA, june 1996
6. M.W.C.Wahls, J.P.Toutenhoofd, L.H.Leyte-Zuiderweg, J. .de Bleijser, J.C.Leyte, J. Appl. Spectrosc. 51, 552 (1997)
7. A.Rosencwaig, in Photoacoustics and Photoacoustic Spectroscopy, R.E.Krieger Publishing Company, Inc. (1990)
8. Polymer Handbook, J.Brandrup and E.H.Immergut, Eds. (John Wiley and Sons, New York, 1975) 2nd ed.

Comparison of transform techniques for Event-Locked time-resolved Fourier transform spectroscopy

H. Weidner and R. E. Peale

Department of Physics, University of Central Florida, Orlando, FL 32816

A least-squares method of transforming interferograms into spectra is compared with the traditional fast Fourier transform (FFT) for varying degree of point-spacing unevenness. A Gaussian distribution is assumed for the deviations from even spacing. FFT gives inferior spectra quality above a critical unevenness, which decreases as more interferograms are averaged.

Introduction

Time-resolved Fourier spectroscopy is a powerful spectroscopic tool. Much of its success arises from the use of step-scan interferometers. Commercial implementations on continuously scanning interferometers are rare and have very limited capabilities. Several techniques have been developed to acquire time-resolved data under the conditions of continuously changing pathlength difference (1,2). The interleaved method accomodates the widest range of transient's time constants and repetition rates but suffers from noise and artifacts arising from scan speed variations (3,4). Event-Locked data acquisition (5) overcomes this problem by assuring accurate sampling times. We briefly describe the features of our fourth implementation of Event-Locked Fourier Spectroscopy (ELFS) and then compare two techniques for extracting spectral information from the unevenly spaced interferograms acquired by ELFS.

Event-Locked data acquisition is done externally to the unmodified interferometer. Our fourth improved implementation is based on home-made digital electronics and an ADC expansion card (1 MHz, ≥12 bits) both residing inside a personal computer (≥486DX2). The digital electronics follows the motion of the interferometer by passively observing the HeNe reference and a zero-pathlength signal. It accurately records the scan speed during data acquisition and creates trigger signals for the excitation source and the ADC. The speed recording is done with an accuracy of 25 ns assured by a quartz-stabilized time base. The counters, which trigger the transients and the ADC, use a subdivided frequency for a time resolution of 100 ns. ELFS can cover the full spectral range of the interferometer optics. Fringe skipping or subdivision by arbitrary integers minimizes the number of transients to the value set by the desired free spectral range.

Comparison of analysis techniques

The fast Fourier transform (FFT) is the main tool in analyzing common evenly spaced interferograms. Since FFT requires even sample spacing, a new method of transformation into spectra (5) was developed as part of ELFS. This new method requires measurement of the actual sampling positions and uses least-squares fitting of harmonic functions to the unevenly spaced interferograms. This technique will be referred to as "ELFS-style analysis" hereafter. ELFS-style analysis requires more computer memory and more computation time than a regular FFT procedure for spectra with the same resolution and frequency range. Hence, regular FFT analysis would be preferable in cases where ELFS style analysis does not result in better spectra quality. For almost equally spaced interferograms, it might be possible to ignore the unevenness and to use the more efficient FFT analysis without introducing significant noise or artifacts. This work quantitatively describes the effects of random sample-spacing unevenness on the quality of the spectra. Circumstances under which ELFS-style analysis is preferred to regular FFT are identified.

Since it is difficult to acquire interferograms with variable but known unevenness of the point spacing, numerical simulation is used. The experiment is replaced by a function $I(x)$, which gives an interferogram value for any pathlength difference x. The function used in our tests is the sum of three cosines with different amplitudes, frequencies, and phases apodized with a factor $\exp[(a\,x)^2]$. In the spectral domain, this function describes one isolated Gaussian peak at 11000 cm^{-1} and two partially overlapping Gaussian peaks near 12000 cm^{-1}. Discrete interferograms consisting of N points $\left(x_i', I\left(x_i'\right)\right)$ are computed with the index i assuming all integer values from 1 to N. Unevenly

CP430, *Fourier Transform Spectroscopy:* 11th International Conference
edited by J.A. de Haseth

spaced interferograms (indicated by primed variables) have points at

$$x'_i = i \cdot \Delta X + \Delta x_i - x_0 \, , \qquad (1)$$

where ΔX is the sampling interval, x_0 is some offset to give partially double-sided interferograms, and the Δx_i are slight individual offsets for the points. Evenly spaced points ($\Delta x_i = 0$), will be unprimed. For simplicity, a Gaussian distribution is assumed for the Δx_i with an rms value of $\overline{\Delta x}$. $\overline{\Delta x}$ is the unevenness of an interferogram.

An ELFS-style analysis routine, which includes apodization and phase correction procedures, is used to extract spectra from the interferograms. ELFS-style analysis reduces to FFT analysis in the limit of even sample spacing (5). This permits comparison of ELFS-style and regular FFT analysis using the same code on two slightly different sets of simulated input data. Using the same routine for both analysis styles has the advantage of using identical phase correction and apodization. For ELFS-style analysis, interferogram points $\left(x'_i, I\left(x'_i\right)\right)$ are provided as the input data. Using $\left(x_i, I\left(x'_i\right)\right)$ as the input simulates regular FFT analysis by assigning the interferogram values $I\left(x'_i\right)$, that are found at unevenly spaced points, to the evenly spaced x_i.

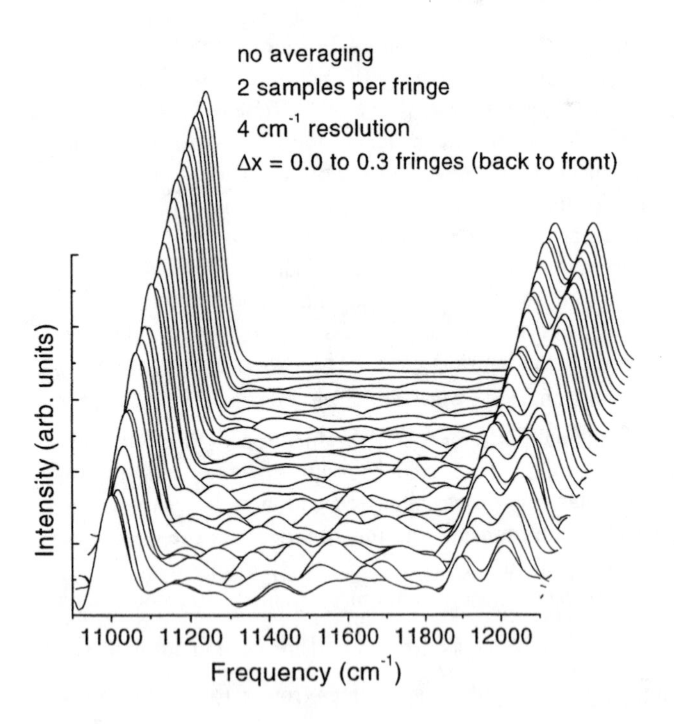

FIGURE 1. Simulated spectra obtained from regular FFT analysis with increasing (back to front in steps of 0.01 fringes) unevenness $\overline{\Delta x}$ of the data point spacing.

Interferograms with two samples per HeNe fringe ($\Delta X \approx 316\, nm$) up to a maximum pathlength difference of 2.5 mm were created. The results with regular FFT analysis for varying $\overline{\Delta x}$ are shown in Fig. 1. The ideal spectrum with $\overline{\Delta x} = 0$ is plotted in the background. As the value of $\overline{\Delta x}$ increases towards the front, noise with increasing amplitude appears. In addition, the heights of the three lines deviate more and more from their original values. Apparently, the lines lose strength to the predominantly positive noise.

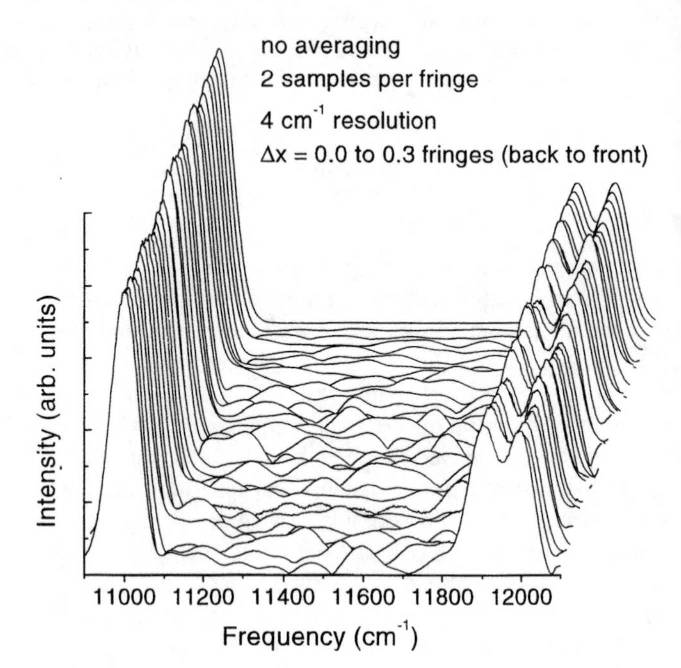

FIGURE 2. Simulated spectra obtained from ELFS-style analysis with increasing (back to front in steps of 0.01 fringes) unevenness of the data point spacing.

The results of ELFS-style analysis are shown in Fig. 2. Again, the only difference in the preparation of these spectra is the use of the primed pathlength coordinates ($\Delta x_i \neq 0$) in the interferograms. The noise appears to increase (at least initially) in a similar fashion as in Fig. 1. However, the line strengths vary only within the limits of the noise.

The differences between the spectral values $S'(\nu_i)$ and the ideal spectrum $S(\nu_i)$ may be used to quantify the spectra quality. A quality value Q is defined as the ratio of peak intensity $\hat{A}$ to the root-mean-squared (rms) value of such differences by

$$Q \equiv \hat{A}\left[\frac{1}{M}\sum_i^M \left(S'(\nu_i) - S(\nu_i)\right)^2\right]^{-\frac{1}{2}} \qquad (2)$$

where the summation is over the M spectral values in the range of interest. The peak intensity of the 11000 cm^{-1} line in the ideal spectrum is used for $\hat{A}$ in all further calculations. Larger Q values indicate better agreement with the ideal spectrum. Perfect agreement gives $Q = \infty$. In Fig. 3, Q is plotted against $\overline{\Delta x}$ for ELFS-style and regular FFT analyses for cases with and without averaging of interferograms. The spectra quality tends to decrease as the unevenness of the point spacing increases. Without averaging (one scan), the qualities for ELFS and FFT analyses are similar up to an unevenness of about 20% of the sampling interval ΔX. Above this limit, ELFS spectra have a nearly constant quality whereas FFT spectra continue to deteriorate until virtually no resemblance to the ideal spectrum remains for $\overline{\Delta x} > \Delta X$. The situation differs for averaged interferograms (64 scans). Only for very small $\overline{\Delta x}$ does FFT analysis perform better than ELFS. Above an unevenness of about 3% of the sampling interval, ELFS analysis performs better and reaches a constant level roughly three times higher than without averaging. In contrast, the quality value of FFT spectra decreases at an accelerated rate until it reaches the same low level as without averaging.

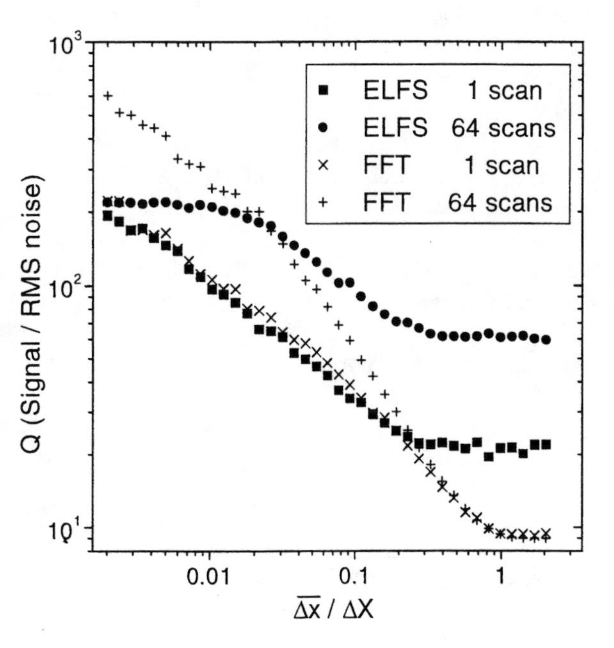

FIGURE 3. Quality value of spectra for ELFS-style (solid symbols) and usual FFT (crossed symbols) analyses as function of the sampling unevenness $\overline{\Delta x}$ (in units of the average sampling interval ΔX. A larger quality value means better agreement with the ideal spectrum.

The quality values for ELFS analysis of averaged interferograms is nearly constant for small $\overline{\Delta x}$. Apparently, there is at least one source of noise which is independent of $\overline{\Delta x}$. The approximate evaluation of the least squares formulae could be this source of noise as shown in Fig. 4. In ELFS-style analysis, the exact evaluation of harmonic functions at uneven intervals is replaced by an approximation using a certain number of neighboring points on an even grid and an interpolation technique (5). Six grid points/coefficients are used for the data in Fig. 3. Figure 4 shows the effect of decreasing the number of grid points on the quality of the spectra. Apparently, using fewer coefficients raises the noise floor (lower quality limit) as indicated by the numbered horizontal lines. For the values of $\overline{\Delta x}$ in Fig. 4, the spectra quality does not exceed these limits even with averaging. For medium and large values of $\overline{\Delta x}$, the noise contribution from the approximate evaluation becomes less important and the spectra quality for four and six coefficients are approximately equal.

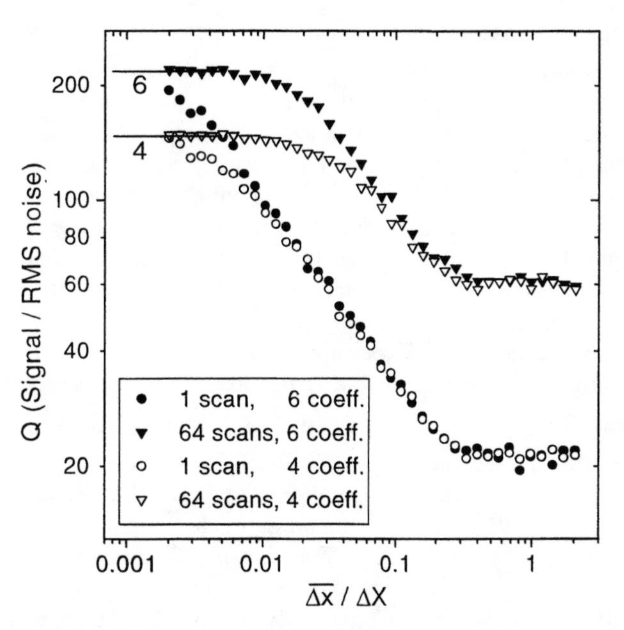

FIGURE 4. Quality of spectra obtained from ELFS-style analysis using 6 grid points (solid symbols) and 4 grid points (open symbols) for the approximate evaluation of the least-squares formulae. The horizontal lines indicate apparent quality limits for 4 and 6 grid points.

The quality dependence on $\overline{\Delta x}$ differs depending on the spectral content of the interferogram. High optical frequencies lead to steep slopes in the interferogram and to large changes of $I(x_i')$ with variations of Δx_i. This situation is shown in Fig. 5. The sampling positions at $\pm \Delta X$ lie on steep slopes, which leads to a large range of possible interferogram values and potentially to noise. By plotting the spectra quality versus $\overline{\Delta x}/\Delta X$ in Figs. 3 and 4, where ΔX is approximately half the inverse maximum optical frequency, the dependence of the curves on the spectral content of the spectra has been reduced.

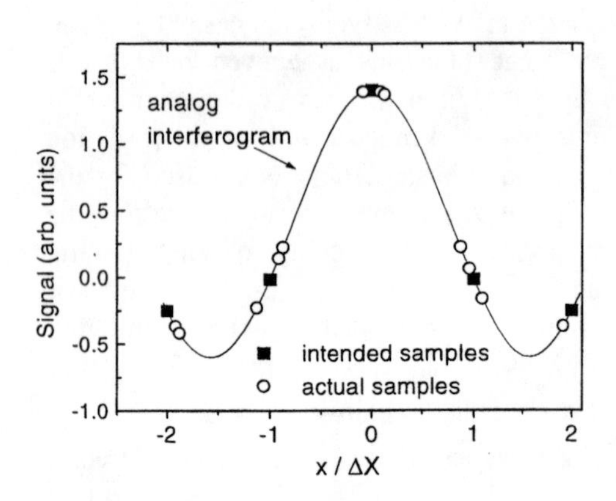

FIGURE 5. Analog interferogram with intended and actual samples.

Figure 5 also explains why line strengths decrease with $\overline{\Delta x}$ when using FFT analysis. The interferogram value assigned to zero pathlength difference determines the integrated strength of all lines. Sampling positions different from $x = 0$ give smaller interferogram values and hence lower line strengths. Averaging improves the situation only if actual samples are quite close to their intended position. ELFS analysis does not exhibit this problem since the actual sampling positions are used for calculations.

Practical conclusions may now be drawn. Depending on the number of averaged interferograms, one can extract from Fig. 3 a critical ratio $\overline{\Delta x}/\Delta X$, which defines a threshold above which one should use ELFS-style analysis. If the unevenness of the sampling positions is unknown, an approximate value may be found from the magnitude of mirror speed variations. For continuously scanning interferometers, an estimate may be obtained by observing the HeNe reference signal on an oscilloscope. Variations in the fringe period indicate speed changes. If the time delay between accurate knowledge of position and the actual sampling is t_d, and mirror speed variations are of size $\overline{\Delta v}$, the position error is $\overline{\Delta x} = \overline{\Delta v}\, t_d$. For a single interferogram (no averaging), the quality numbers for both analysis types are comparable for $\overline{\Delta x}/\Delta X \leq 0.2$. For 64 averaged interferograms, they are comparable for $\overline{\Delta x}/\Delta X \leq 0.03$. At a typical mirror speed of 0.5 cm/s with variations of 3%, it takes about 400 μs to reach the critical unevenness without averaging (60 μs for 64 averages). In other words, ELFS-style analysis will give better spectra for observation times (difference between the moment of accurate time and position knowledge and the sampling time) longer than 400 μs in TRFTS without coaddition.

The foregoing simulations assumed a random Δx_i distribution. Periodic changes in the mirror speed cause periodic changes in the sampled interferogram intensity. Artifacts are then expected when using regular FFT analysis (6). Applying both analysis techniques to a typical data set may help in deciding which technique to use for other spectra with similar spectral content.

The constant quality level for ELFS-style analysis at large values of $\overline{\Delta x}$ has an interesting implication. If $\overline{\Delta x}$ cannot be kept below 20% of the sampling interval, there is no longer an advantage in attempting to evenly distribute the interferogram samples. A random distribution of samples becomes acceptable (although accurate knowledge of the sampling positions x_i' is still required).

ACKNOWLEDGMENTS

This work was supported by AFOSR grant #F49620-961-0322 and by NSF grant #ECS-9531933.

References

1. J. J. Sloan and E. J. Kruus, "Time-Resolved Fourier Transform Spectroscopy", in *Time Resolved Spectroscopy*, R. J. H. Clark and R. E. Hester, (Wiley, New York, 1989), ch. 5, pp. 219-253.
2. Masutani, K., Sugisawa, H., Yokota, A., Furukawa, Y., and Tasumi, M., Appl. Spectrosc. **46**, 560-567 (1992).
3. Lindner, J., Lundberg, J. K., Williams, R. M., and Leone, S. R., Rev. Sci. Instrum **66**, 2812-2817 (1995).
4. Weidner, H. and Peale, R. E., Appl. Optics 16, 2849-2856 (1996).
5. Weidner, H. and Peale, R. E., "Event-Locked Time-Resolved Fourier Spectroscopy", Appl. Spectrosc., to appear 1997.
6. Learner, R. C. M., Thorne, A. P., and Brault, J. W., Appl. Optics **35**, 2947-2954 (1996).

Use of Generalized Two-Dimensional Correlation Spectra to Understand the Reaction Between Poly(Vinyl Alcohol) and Palladium (II) Chloride

Richard Wilson

Monsanto Company, 730 Worcester St., Springfield, MA 01151, USA

The use of two-dimensional correlation techniques in interpretation of time dependent IR spectra has proven to be a powerful tool. Extension from external sinusoidal perturbations of the sample to systems with a limited number of spectra along the time axis has opened the door to use of correlation spectra to understand chemical reaction pathways. Two-dimensional correlation spectra have been used to understand the reaction between poly(vinyl alcohol) and palladium (II) chloride. This chemical system is widely used in the field of electroless plating of metals. The palladium (II) chloride and poly(vinyl alcohol) are dissolved, spread on the area to be plated and dried. The poly(vinyl alcohol) is used as a binder. Heating results in reduction to palladium metal which is the catalyst for the electroless plating process. The heating step also results in chemical changes in the poly(vinyl alcohol) structure which are readily observed by Infrared spectroscopy. A series of spectra taken during the heating step were examined by using the generalized two-dimensional correlation technique and the results will be discussed.

INTRODUCTION

The electroless plating of non-conductive materials is almost as old an art as electroplating. The electroless plating process depends upon adhering catalytic amounts of a metal such as gold, platinum, or palladium to the substrate. The metal can then catalyze the oxidation of formaldehyde and the reduction of the metal to be plated, such as copper. The deposition process then proceeds autocatalytically.

Commercial systems usually rely on a palladium/tin colloid coating with subsequent activation by selectively dissolving the tin from the palladium particles. More recently, there have appeared systems utilizing a thin polymeric coating carrying palladium salts (1,2,3). These coatings usually require a heating step to reduce the Pd(II) to Pd(0) prior to the plating step. The interactions between the palladium and the polymeric carrier are the focus of this work.

EXPERIMENTAL

Solutions were prepared containing 2.5% by weight palladium (II) chloride (Aldrich Chemical Company) and 1.5% by weight poly(vinyl alcohol) dissolved in distilled water. The solution was cast as a thin film on a sodium chloride disk and allowed to dry at room temperature. The disk was then placed in a Spectra Tech variable temperature holder and heated to 130° C and held for thirty minutes. The IR spectra were obtained in transmission using a Bruker IFS 88 FT-IR spectrometer at a resolution of 4 cm^{-1}. A total of 30 interferograms were co-added for each spectrum obtained. Prior to computation, the spectra were deresolved to 8 cm^{-1}.

The correlation and disrelation spectra were calculated using an in-house developed Fortran program and were plotted using Psiplot version 5 plotting software.

THEORY

It has been shown (4,5) that when the time axis contains N data points, the 2-D synchronous correlation spectrum, $\underline{\Phi}(\gamma_1,\gamma_2)$, between any two frequencies, γ_1 and γ_2, is given by:

$$\underline{\Phi}(\gamma_1,\gamma_2) = \frac{1}{N-1}\sum_{i=1}^{N} y(\gamma_1,t_i)\cdot y(\gamma_2,t_i) \qquad (1)$$

and the 2-D disrelation spectrum, $\underline{\Lambda}(\gamma_1,\gamma_2)$, is given by the positive root of

$$\sqrt{\frac{1}{(N-1)^2}\sum_{i=1}^{N}\sum_{j>i}^{N}\left\{y(\gamma_1,t_i)\cdot y(\gamma_2,t_j) - y(\gamma_1,t_j)\cdot y(\gamma_2,t_i)\right\}^2} \quad ,(2)$$

where $y(\gamma,t)$ represents the intensity variation at frequency γ and time t. For this work, the deviation spectra, that is the absorbance at frequency γ and time t minus the time averaged absorbance at frequency γ, were used for $y(\gamma,t)$.

CP430, *Fourier Transform Spectroscopy:* 11th International Conference
edited by J.A. de Haseth

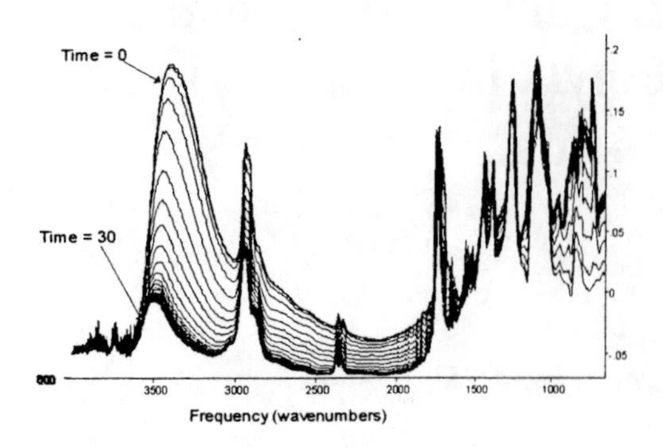

FIGURE 1. Stacked plot of absorbance spectra over time for poly(vinyl alcohol) / $PdCl_2$ heated to 130° C.

RESULTS AND DISCUSSIONS

From the spectra obtained, Fig. 1, there are several areas exhibiting marked changes during the thirty minutes of heating. Particularly prominent are the loss, as time passes, of the hydroxyl intensity at 3400 cm^{-1} and the growth in bands below 900 cm^{-1}. A sample of the same poly(vinyl alcohol), without palladium chloride, run under the same conditions, showed none of these spectral changes. These differences are even more obvious in the deviation spectra, Fig. 2. The deviation spectrum, at time t, is calculated as the absorbance spectrum at time t minus the time average spectrum. The deviation spectra eliminate bands unchanging with time while emphasizing bands which change with time. Those bands, due to reactants, which decrease with time appear as positive peaks in the initial deviation spectrum, while, conversely, bands due to products appear as negative peaks in the initial deviation spectrum. Thus, the hydroxyl bands are being consumed and the bands below 900 cm^{-1} are being generated as time progresses. The deviation spectra appears to show that a carbonyl band at 1710 cm^{-1} is converted to a carbonyl band at 1738 cm^{-1} during the reaction.

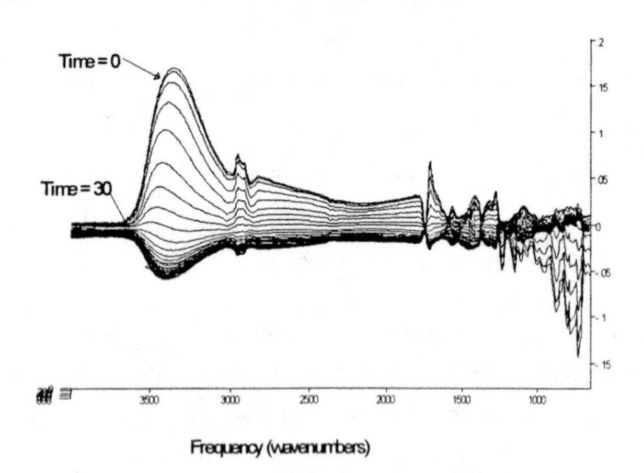

FIGURE 2. Stacked plot of deviation spectra over time for poly(vinyl alcohol) / PdC12 heated to 130° C.

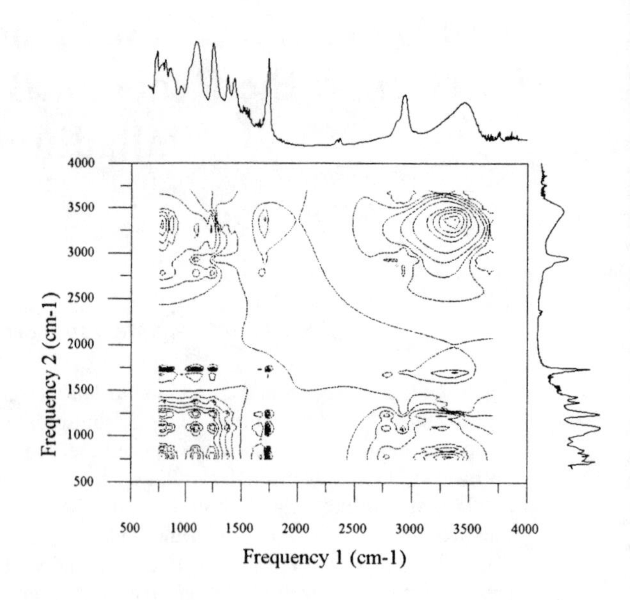

FIGURE 3. Contour plot of the correlation spectrum for the reaction of poly(vinyl alcohol) / $PdCl_2$ heated to 130° C. The time averaged spectrum is added along the x and y axis for comparison.

A contour plot of the 2-D correlation spectrum is shown in Fig. 3. The 2-D correlation spectrum is dominated, as expected by the large auto-correlation peak due to the hydroxyl groups centered at around 3400 cm^{-1}. The 2-D correlation spectrum also shows that there are correlation cross peaks between the hydroxyl bands and bands in the frequency region below 1800 cm^{-1}. Cross peaks in the 2-D correlation spectrum correspond to pairs of frequencies whose intensities change at the same rate

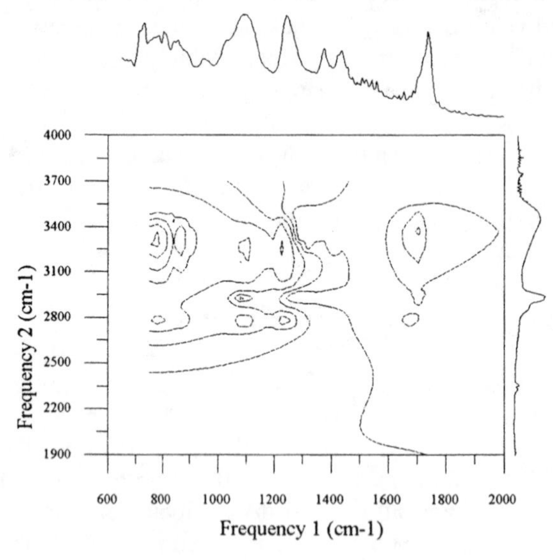

FIGURE 4. Expanded view of the contour plot of the correlation spectrum for the reaction of poly(vinyl alcohol) / $PdCl_2$ heated to 130° C. The time averaged spectrum is added along the x and y axis for comparison.

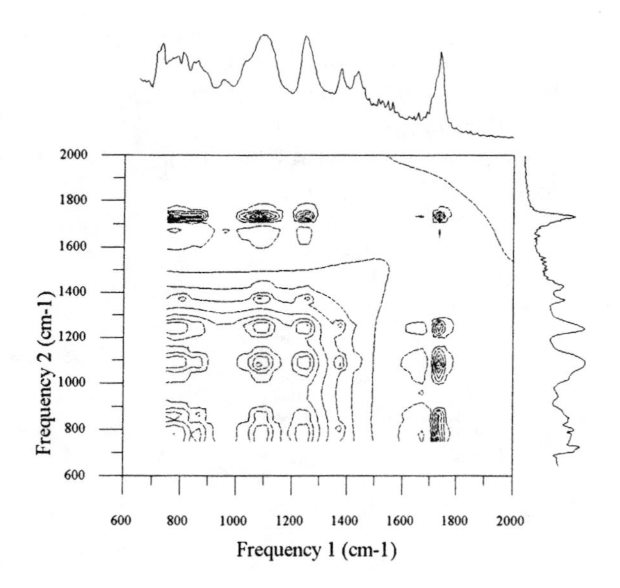

FIGURE 5. Expanded view of the contour plot of the correlation spectrum for the reaction of poly(vinyl alcohol) / $PdCl_2$ heated to 130° C. The time averaged spectrum is added along the x and y axis for comparison.

with time. This is shown more clearly in Fig. 4 which is an expanded view of the cross peaks between the decreasing hydroxyl band and the fingerprint region.

Closer examination of Fig. 4 shows that there are three sets of cross peaks due to the hydroxyl groups: one cross peak between 3380 and 1710 cm^{-1}; a pair of cross peaks between 3300 and the peaks 780 and 870 cm^{-1}; and a pair of cross peaks between 3260 and the peaks at 1100 and 1220 cm^{-1}.

Figure 5 shows an expanded view of the fingerprint region of the 2-D correlation spectrum contour plot.

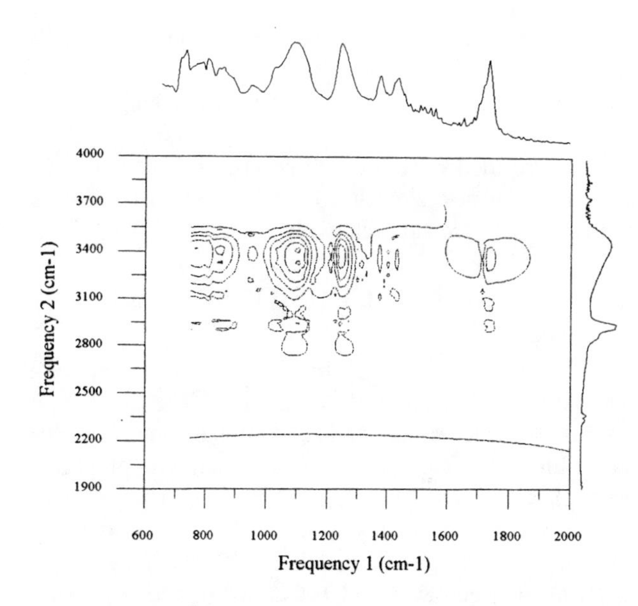

FIGURE 7. Expanded view of the contour plot of the disrelation spectrum for the reaction of poly(vinyl alcohol) / $PdCl_2$ heated to 130.° C. The time averaged spectrum is added along the x and y axis for comparison.

This region is dominated by the cross peaks due to the bands at 1262, 1090, 857 and 807 cm^{-1} and the residual acetate groups on the polymer at 1736 cm^{-1}. Each of these bands also has a weaker cross peak with a band at 1660 cm^{-1}.

Figure 6 is the contour plot of the 2-D disrelation spectrum. The 2-D disrelation spectrum contains cross peaks between frequencies changing at different rates with time. Not surprisingly, the largest cross peaks in the 2-D disrelation spectrum are between the hydroxyl band and the fingerprint region, Fig. 7. Here it is plain to see

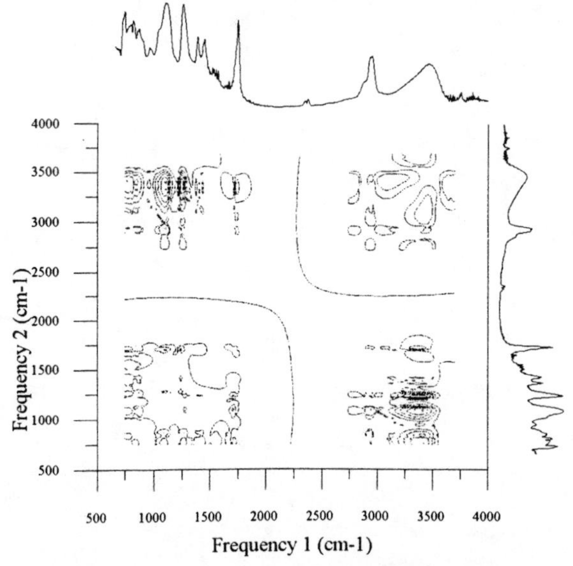

FIGURE 6. Contour plot of the disrelation spectrum for the reaction of poly(vinyl alcohol) / $PdCl_2$ heated to 130° C. The time averaged spectrum is added along the x and y axis for comparison.

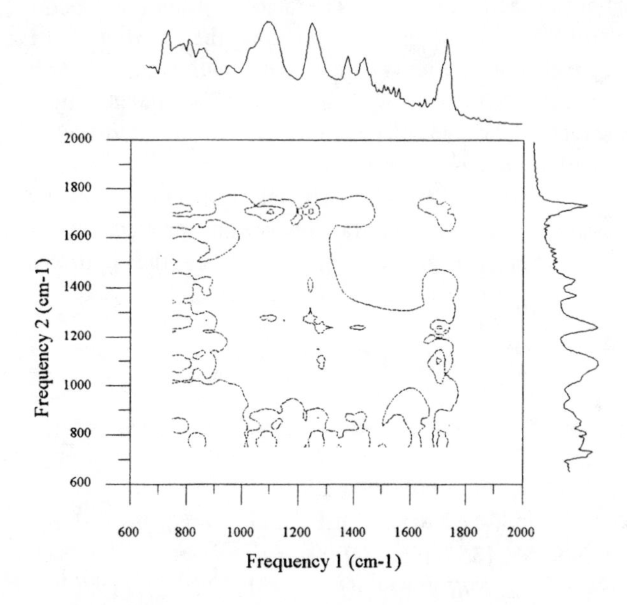

FIGURE 8. Expanded view of the contour plot of the disrelation spectrum for the reaction of poly(vinyl alcohol) / $PdCl_2$ heated to 130 C. The time averaged spectrum is added along the x and y axis for comparison.

two distinct cross peaks due to hydroxyl groups at 3390 cm^{-1} and 3280 cm^{-1}. There also are a series of cross peaks between bands at 1262 and 1090 cm^{-1} with a band at 3040 cm^{-1}.

An expanded view of the fingerprint region of the 2-D disrelation spectrum is shown in Fig. 8. All the cross peaks in this region are relatively small.

CONCLUSIONS

The band assignments for the infrared spectrum of poly(vinyl alcohol) are well known (6). The presence of residual acetate groups add additional bands to the spectrum. The carbonyl stretching band in poly(vinyl acetate) is a strong band at 1740 cm^{-1} (7). Hydrolysis of the acetate to alcohol results in a shift of the carbonyl stretching vibration to 1710 cm^{-1}, particularly evident in materials between 80% and 100% hydrolyzed (8). From the 2-D correlation spectrum, it is seen that the loss of this band at 1710 cm^{-1} correlates with the loss of hydroxyl intensity at 3380 cm^{-1}. The lack of a cross peak in the 2-D correlation spectrum between the 1710 cm^{-1} band and the 1740 cm^{-1} indicates that the two bands are not interconverting as might be assumed from inspection of the deviation spectra, Fig. 2.

Poly(vinyl alcohol) is known to dehydrate on heating to form a conjugate polyene structure (9). The presence of cross peaks in the 2-D correlation and disrelation spectra at frequencies of 1660 and 3040 cm^{-1} show dehydration is occurring under these experimental conditions.

From these experiments, it is clear that poly(vinyl alcohol) is far from acting as an inert binder during the palladium activation step. The alcohol groups are being hydrolyzed to form unsaturation along the polymer chain. Although not explicit in these data, palladium is likely functioning as a catalyst for many of the reactions observed. The lack of reactivity without the addition of the palladium clearly points in that direction.

The use of two-dimensional correlation spectra provides enhanced spectral resolution, particularly in broad, hydrogen bonded bands such as the hydroxyl groups in this experiment.

REFERENCES

1. Jackson, R.L., *J. Electrochem. Soc.* **135**, 95-101 (1990).
2. U. S. Patent 3,900,320.
3. Hirsch, T.J., Mirachy, R.F., and Lin, C., *Appl. Phys. Lett.* **57**, 1357-1359, (1990).
4. Noda, I., *Bull. Am. Phys. Soc.* **31**, 520-524, (1986).
5. Burie, J-R., *Applied Spectroscopy* **50**, 861-865, (1996).
6. Krimm, S., Liang, C.Y., and Sutherland, G.B.B.M, *J. Polymer Sci.* **22**, 227-247, (1956).
7. Thompson, H.W., and Torkington, P., *Trans. Faraday Soc.*, **41**, 246-260, (1945).
8. Blout, E.R. and Karplus, R., *J. Am. Chem. Soc.*, **70**, 862-864, (1948).
9. Finch, C.A., Polyvinyl Alcohol, London: John Wiley & Sons, 1973, ch. 8, pp. 172-174.

HYPHENATION OF SEQUENTIAL- AND FLOW INJECTION ANALYSIS WITH FTIR-SPECTROSCOPY FOR CHEMICAL ANALYSIS IN AQUEOUS SOLUTIONS

B. Lendl, R. Schindler and R. Kellner*

Institute for Analytical Chemistry, Vienna University of Technology Getreidemarkt 9/151, A-1060 Vienna, Austria (Europe)

A survey of the principles of sequential (SIA)-and flow injection analysis (FIA) systems with FTIR spectroscopic detection is presented to introduce these hyphenations as powerful techniques for performing chemical analysis in aqueous solution. The strength of FIA/SIA-FTIR systems lies in the possibility to perform highly reproducible and automated sample manipulations such as sample clean–up and/or chemical reactions prior to spectrum acquisition. It is shown that the hyphenation of FIA/SIA systems with an FTIR spectrometer enhances the problem solving capabilities of the FTIR spectrometer as also parameters which can not be measured directly (e.g. enzyme activities) can be determined. On the other hand application of FTIR spectroscopic detection in FIA or SIA is also of advantage as it allows to shorten conventional analysis procedures (e.g. sucrose or phosphate analysis) or to establish and apply a multivariate calibration model for simultaneous determinations (e.g. glucose, fructose and sucrose analysis). In addition to these examples two recent instrumental developments in miniaturized FIA/SIA-FTIR systems, a μ-Flow through cell based on IR fiber optics and a micromachined SI-enzyme reactor are presented in this paper.

INTRODUCTION

FTIR spectroscopy is widely accepted as technique for chemical analysis as nearly all molecules exhibit characteristic IR absorptions. However, the analysis of aqueous samples is made difficult by the strong water absorption in the mid-IR range and the need of water-resistant IR-transparent materials. Considering the fact that nearly all liquids originating from biological or environmental samples are water-based, still only limited use is made of the high information content of the mid-IR range. The development of routine high performance FTIR spectrometers and new auxiliary equipment causes this situation to change and to establish FTIR spectroscopy as a powerful tool for measurements also in the aqueous phase.

This work focuses on the hyphenation of automated flow manipulation systems like Flow Injection Analysis (FIA) and Sequential Injection Analysis (SIA) with FTIR spectrometers. Flow Injection Analysis (1975)[1] and Sequential Injection Analysis (1990)[2] were introduced by Ruzicka et al as means for automated analysis and have evolved during the last 20 years to important techniques in chemical analysis. The main idea of SIA and FIA is to abandon the principle of measurements in chemical and physical equilibria. It is no longer necessary to have equilibrated conditions, the only necessity is that the measurement conditions are highly reproducible. The sample is injected into a flowing stream, totally or partially mixed with reagents and transported to the detector. The analysis procedure can be repeated with a high stability of timing and flow rates, so that all samples are treated in exactly the same manner. If a sample is measured in an unequilibrated state, all successive samples will be measured in exactly the same state. Compared to classical chemical analysis it can be stated that by abolishing the principle of measurements in the equilibrated conditions, a great variety of fundamentally new techniques has been developed in FIA and SIA[1,2], including sample clean-up, separation, enrichment and reaction steps.

The hyphenation of FIA and SIA with FTIR spectrometers has evolved from simple sample introduction system to a very versatile analytical tool since the first applications in 1985[3,4]. Even if flow systems are solely used for sample introduction, significant improvements are gained in comparison with manual operation. As all samples are measured under the same conditions, problems due to temperature effects or pathlength difference are normally avoided. The relative standard deviation of subsequent absorbance measurements is typically below 2 %, even if highly absorbing water samples are measured. Furthermore, the whole system can be automated, speeding up the analysis procedure considerably. However, the full capabilities of the FIA/SIA-FTIR systems are only exploited if the advantages of a molecular specific, multidimensional detector are combined with automated chemical and physical sample treatment before measurement. The following paragraphs are aimed to illustrate firstly some instrumental developments and

CP430, *Fourier Transform Spectroscopy:* 11th International Conference
edited by J.A. de Haseth

secondly the possibilities to address complex analytical problems with FIA/SIA-FTIR.

INSTRUMENTAL DEVELOPMENTS

μ-Flow through transmission cell[5]:

As the typical sample volumes of FIA/SIA are in the low μl-range the dead volume of conventional IR liquid flow through cells (transmission as well as ATR cells) reduces the performance of the flow systems due to averaging and dilution effects. In our group a new μ-flow-through transmission cell was developed, having a detection volume of 30 nl and minimized dead volumina (Fig. 1).

Two fiber pieces ($AgCl_xBr_{1-x}$, diameter: 750μm) with plane-parallel faces were assembled coaxially to each other at an adjustable distance (23 μm for measurements in aqueous solutions) between the two fiber end faces. The cell body (PTFE) was inserted in a aluminum housing, mounted on a positioning device and connected with conventional FIA fittings to the flow system. The assembly was placed in the sample compartment of the FTIR spectrometer (Bruker IFS 88) and the IR radiation was focused onto the fiber tip and to the MCT detector by the use of KBr lenses. The optical performance of the cell (SNR, light throughput) is nearly the same as for standard transmission cells equipped with CaF_2 or ZnSe windows and considerably better than with standard ATR cells.

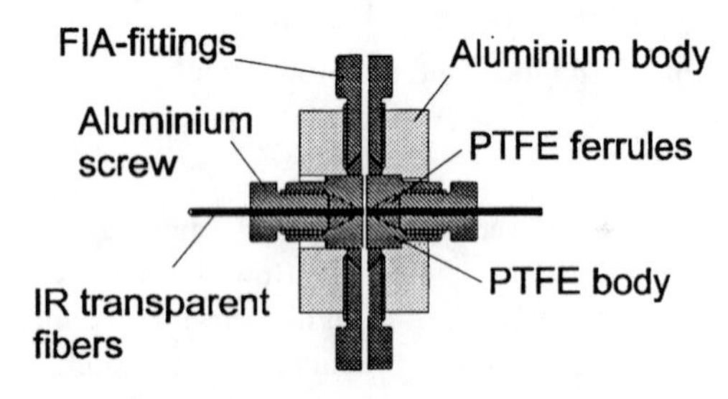

Fig. 1: Fiber optic flow through cell

Automated flow systems[6]:

FIA and SIA are well suited for automated operation, thus reducing possible human errors and increasing the long-term stability of the resulting analysis systems. Especially SIA is totally software controlled and operating procedures and conditions can be adjusted by simple changes of the controlling software rather than by manual reconfiguration of the manifold.

Representative FIA and SIA systems are depicted in Fig. 2. The individual parts are connected by PTFE tubings with inner diameters from 0.2 to 1 mm. Piston pumps and selection valves are normally used in SIA whereas peristaltic pumps and injection valves are preferred for FIA.

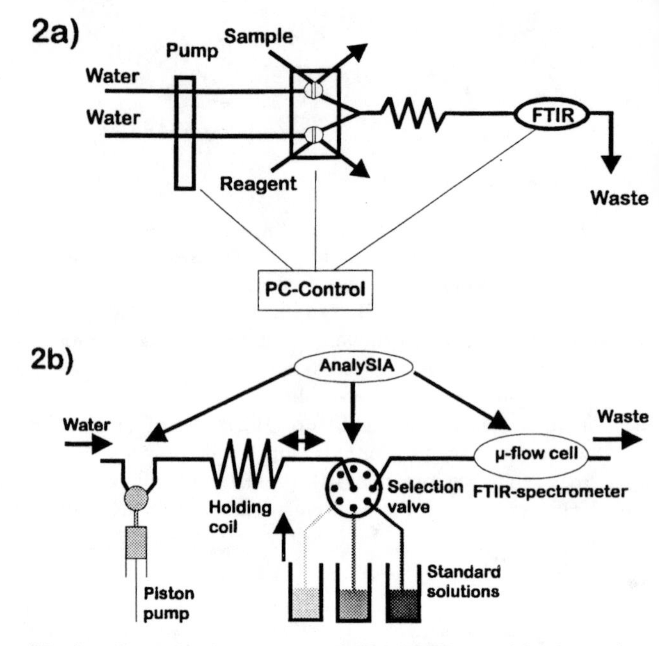

Fig.2a: Completely automated SIA/FTIR manifold (used for automated multi-standard preparation and miniaturized sucrose analysis).

Fig.2b: FIA/FTIR system (used for determination of phosphate and amyloglucosidase activities).

Miniaturized components[5]:

The achievements in the field of microengineering have triggered the development of miniaturized Total Analysis Systems (μ-TAS)7, which meant the incorporation of all analysis functions onto a small device, ideally a single chip. This miniaturization is not only a simple downscaling of the standard laboratory equipment but requires also new concepts and measurement principles. FTIR spectroscopy is of special interest here as concerning this technique no sensitive loss has to be faced upon miniaturization which is in contrast to UV/VIS absorption or fluorescence detection where the optical pathlength has to be reduced significantly upon miniaturization. All components of a μ-TAS system have to be selected to fit to the reduced dimensions regarding dead volume and flow conditions. In fig.3 a SEM picture of a miniaturized enzyme reactor made by silicon micromachining is shown, which was used in an FIA/FTIR system for sucrose analysis.

Fig. 3: SEM photograph of a micromachined enzyme reactor.

APPLICATION TO ANALYTICAL PROBLEMS

Miniaturized Total Analysis System for Sucrose analysis[5]:

A flow system containing a micromachined lamellae-type porous silicon reactor (Fig. 3) and the fiber optic flow through cell (Fig. 1) was applied to the enzymatic determination of sucrose in aqueous solutions such as soft drinks. By the application of difference spectra unspecific background absorption was eliminated successfully and the action of the • -fructosidase immobilized on the microreactor was monitored selectively. In the concentration range 10 and 100 mmol/l sucrose was measured using sample volumes less than 10 µl. The method was also successfully applied to the analysis of real samples.

Automated multicomponent standard preparation by SIA/FTIR[6]:

A new approach for automated preparation of multicomponent calibration standards with Sequential Injection Analysis (SIA) using one single stock solution for each component is proposed. The key idea is the generation of highly reproducible injection sequences in the flow system. Single standards of three different components are stacked in the holding coil in well defined sequences. Upon flow reversal and transport to the detector the stacked zones disperse into each other producing exactly defined mixtures of the three standards which can be recorded as a function of time. The exact dilutions of one stock solution due to dispersion can be found by the measurement of only one IR absorbing substance, which is injected in the same way as afterwards in the mixtures. Instead of the other stock solution water is used and so for each injection volume the dilution - time profile is recorded. If these profiles are established, they can be regarded as system properties afterwards and combined to give the desired component mixtures.

The method was successfully applied to the preparation of mixtures of glucose, fructose and sucrose, which were measured in the mid-IR range (900 - 1500 cm^{-1}) and used for the establishment of a Partial Least Squares (PLS) calibration model. The results for the analysis of soft drinks by measuring the FTIR spectrum and subsequent chemometric data evaluation were found to be in good agreement with the results of a reference method (HPLC with refractive index detection). Typical deviation from the results obtained by the reference method were in the order of 2.5, 4.4, 2.8 g/L (2.5, 4.1, 2.1 % of total sugar) for glucose, fructose and sucrose respectively.

Determination of phosphate in soft drinks[8]:

The different protonated phosphate ions (PO_4^{3-}, HPO_4^{2-}, $H_2PO_4^-$, H_3PO_4) present in aqueous solutions exhibit different IR absorption bands. A change in pH causing protonation or deprotonation of the phosphate ion is therefore accompanied by a change in the IR-spectrum. Influences from the strongly absorbing matrix of soft drinks can be easily removed by the evaluation of the difference spectrum obtained from the spectra before and after pH change because the matrix molecules (sugars, organic acids) do not show spectral changes in the spectral region of interest (950 - 1100 cm^{-1}) upon pH modulation. By implementation of this analysis principle in an automated FIA system it was shown that determination of phosphate in the technically relevant range from 100 to 1000 mg/l is possible with a precision better than 1 %. The introduction of a chemical reaction step has enhanced the selectivity and sensitivity of the FTIR measurement considerably and improved speed (sample frequency: 60 h^{-1}) and precision of the analysis.

Determination of enzyme activities[9]:

Amyloglucosidase (glucan 1,4-• -glucosidase) activity can be measured in aqueous solution by direct FTIR spectroscopic monitoring of the enzymatic starch hydrolysis. By the application of difference spectra unspecific background absorption is eliminated successfully and the method can be applied to determine enzyme activities in real samples, e.g. from fermentation processes. The changes in the starch spectrum caused by the enzyme action are directly related to the enzyme activity, so in this case no need exists for additional consecutive reaction steps or specialized substrates as it is common with UV/VIS or electrochemical detection.

The use of a FIA manifold guarantees optimum handling of the viscous substrate solution and a high reproducibility of the IR-measurements, which is absolutely necessary to get reproducible results for spectral changes of less than 1 mAU (absolute peak intensity: 350 mAU). Linear calibration curves from 50 to 2000 U/l amyloglucosidase activity were obtained and the method was successfully applied to the analysis of fermentation broth samples.

CONCLUSION

The above examples illustrate the possibilities of FTIR spectroscopy coupled with automated flow systems for solving of analytical problems. FTIR measurements in aqueous solutions can be performed successfully even in the transmission mode and due to the automation of the liquid management a high degree of reproducibility can be achieved. It is accepted that the proposed methods will find increasingly use, especially in the field of process analysis and process control.

1 Ruzicka, J.; Hansen, E.H.; Flow Injection Analysis, 2nd edn., Wiley, New York, 1988.

2 Ruzicka, J.; Marshall, G. D.; Christian, G.D. *Anal. Chem.* **1990,** *62,* 1861.

3 Morgan, D.K.; Danielson, N.D.; Katon, J.E. *Anal. Letters* **1985**, *18(A16)*, 1979.

4 Curran, D.J.; Collier, W.G. *Anal. Chim. Acta* **1985**, *177*, 259.

5 Lendl, B.; Schindler, R.; Frank, J.; Kellner, R.; Drott, J., Laurell, T. *Anal. Chem.* **1997,** *69,* 2877.

6 Schindler, R.; Watkins, M.; Vonach, R.; Lendl, B., Frank, J., Kellner, R.; Sara, R. submitted to *Anal. Chem.*

7 Manz, A.; Grabner, N.; Widmer, H.M. *Sens. Actuators B* **1990**, *1*, 244.

8 R. Vonach, B. Lendl, R. Kellner, *Analyst* **1997**, *122*, 525.

9 Schindler, R.; Lendl, B.; Kellner, R. *Analyst* **1997**, *122*, 531.

ACKNOWLEDGEMENT

Acknowledgment is given to the Austrian Science Foundation for the financial support provided within the project P11338 ÖCH.

A CE/FT-IR Spectrometric Interface: Preliminary Studies

Lin-Tao He and James A. de Haseth*

Department of Chemistry, University of Georgia, Athens, GA 30602-2556, USA

In this work, a CE/FT-IR spectrometric interface was investigated and has been proven feasible based on a solvent elimination approach. We have designed a metal nebulizer CE/FT-IR interface and have successfully collected spectra at the nanogram level. Standard buffers may pose a significant problem in CE/FT-IR spectrometry in that the buffer absorbances may mask the analyte spectrum. Alternate buffers have been investigated and have been found to be suitable.

INTRODUCTION

Capillary Electrophoresis (CE) is an electrically driven chromatography. It is a relatively new member in the chromatographic family. Compared to other pressure-driven chromatographic technologies, such as GC, HPLC, and SFC, CE can be broadly described as a high-efficiency separation technique with high analysis speed, low sample volumes, and wide applicability to polar and non-polar substances.

An interface between chromatography and FT-IR spectrometry is an important analytical technique. GC/FT-IR spectrometry has been commercially available for two decades. HPLC/FT-IR spectrometry and SFC/FT-IR spectrometry have been presented in a large number of publications, and interfaces are presently being commercialized. Two contrasting approaches, involving the use of flow cell (1, 2) and solvent elimination (3-17), have been used in the design of HPLC/FT-IR interfaces. Flow cell HPLC/FT-IR measurements have severe sensitivity limitations caused by the infrared absorption of the mobile phase. The solvent elimination approach has been proven to have great potential and many methods for solvent elimination, such as thermospray (7), electrospray (17), concentric flow nebulization (5, 8, 9, 17-20), ultrasonic nebulization (16, 21), and particle beam desolvation (13-15), have been developed for use in the HPLC/FT-IR interface.

CE/FT-IR spectrometry, however, has not been reported because CE/FT-IR is more difficult than other chromatography/FT-IR interfaces and other CE detection techniques such as CE/UV, CE/LIF (Laser Induced Fluorescence) (22), CE/Raman (23), CE/MS (24, 25), and CE/ICP (26). Basically, CE buffers have much less interference in the detection techniques other than FT-IR spectrometry. Furthermore, CE/UV, CE/LIF and CE/Raman can be built for on-column detection. CE/FT-IR spectrometry based on a solvent elimination approach has three major obstacles: 1) electrical contact for normal CE operation, 2) buffer interference, and 3) small sample deposition area for increased IR sensitivity to meet the low sample volumes in CE.

We have designed a metal nebulizer interface with the above considerations in mind and have successfully collected CE/FT-IR spectra in the presence of a volatile buffer system. The metal nebulizer CE/FT-IR spectrometric interface is applicable to qualitative and quantitative analysis, CE fraction preparation and, investigation of CE mechanisms and theory.

EXPERIMENTAL

The CE experiments were performed on a Beckman (Fullerton, California, USA) P/ACE 5000 Capillary Electrophoresis system with a UV detector. CE/FT-IR spectra were collected on a Perkin-Elmer (Norwalk, Connecticut, USA) Spectrum 2000 FT-IR spectrometer combined with an *i*-series FT-IR microscope. The uncoated silica capillary (75 μm i.d. and 360 μm o.d.) CE column was purchased from Polymicro Technologies Inc. (Phoenix, Arizona, USA).

RESULTS AND DISCUSSION

a. Design of CE/UV/FT-IR column and modification of a commercial CE instrument

For normal CE operation, two buffer vials and a high voltage supply are needed. The capillary column is filled with the required buffer solution and the capillary ends are dipped into reservoirs with high-voltage electrodes and the required buffer solutions.

CP430, *Fourier Transform Spectroscopy:* 11th International Conference
edited by J.A. de Haseth

In order to perform CE/FT-IR sample deposition, the outlet end buffer vial has to be removed and the CE column should be extended out of the commercial instrument. Second, the high voltage supply must be modified. We extended one high-voltage connector from the negative supply terminal. Lastly, electrical contact should be considered when the outlet buffer is removed. We used a commercial capillary cartridge and connected a 4 cm long stainless-steel capillary to the end of an extended CE column. The negative high-voltage terminal was connected to the stainless-steel capillary. Figure 1 shows the modified high-voltage system and CE/UV/FT-IR column that is 110 cm in length, and 60 cm extends beyond the UV detector.

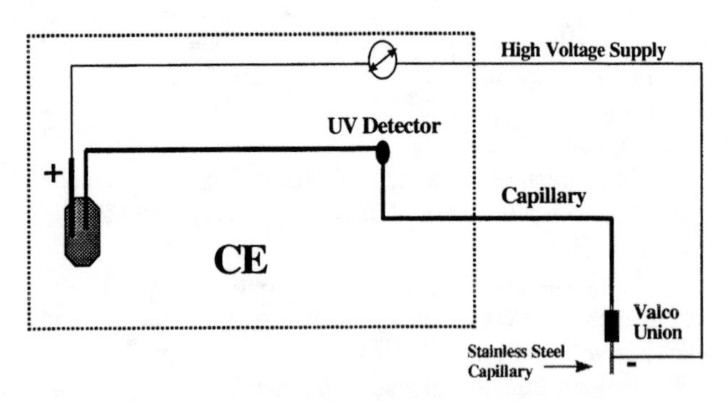

FIGURE 1. CE/UV/FT-IR column and the modified high voltage supply. Capillary: 75μm i.d., total column length: 110 cm, 50 cm between injection reservoir and UV detector.

In some commercial CE/MS interfaces (27), a sheath liquid is used for electrical contact. If the same technique is applied for CE/FT-IR spectrometry, the sheath liquid will complicate solvent elimination. The stainless-steel capillary is more practical than the sheath liquid for electrical contact in CE/FT-IR spectrometry.

b. Design of a metal nebulizer interface

CE has a flow rate of nanoliters per minute, which is much slower than HPLC. Consequently, solvent elimination in CE/FT-IR spectrometry is much easier than in HPLC/FT-IR. We have used a concentric flow nebulizer to eliminate the CE solvent. The metal nebulizer CE/FT-IR interface uses room-temperature helium sheath gas to nebulize the CE effluent. Figure 2 shows the metal nebulizer CE/FT-IR interface. Figure 3 shows a UV electropherogram of a four-compound mixture and a 30 kV CE current plot with the metal nebulizer CE/FT-IR interface. The CE current plot demonstrates that the electrical contact is constant, even though the helium sheath gas is applied to the interface.

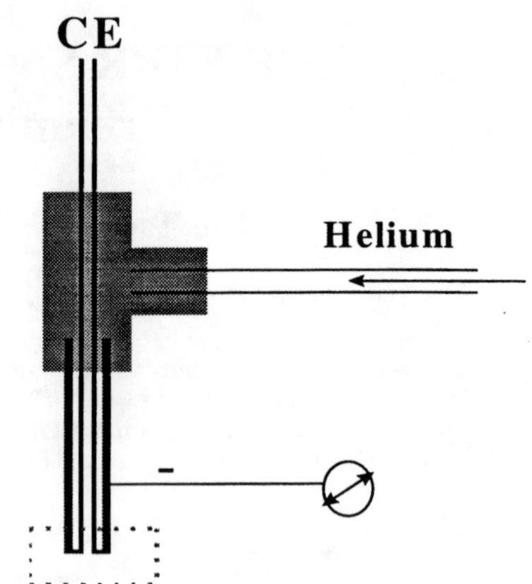

FIGURE 2. Metal nebulizer CE/FT-IR interface. The negative polarity of the high voltage supply is connected to the metal nebulizer.

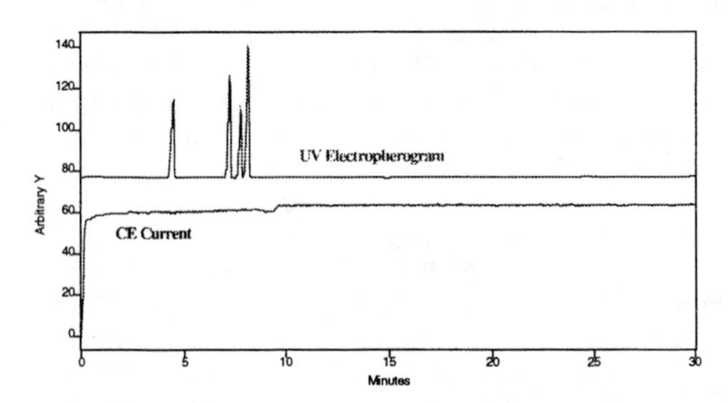

FIGURE 3. UV Electropherogram and CE current using the metal nebulizer CE/FT-IR interface. Sample: caffeine, salicylic acid, p-aminobenzoic acid, sodium benzoate, concentration: 1×10^{-3} M, voltage: 25kV, buffer: 0.05 M ammonium acetate, detection: 214 nm. Column: 75 μm i.d., 110 cm long, CE/UV/FT-IR column. Nebulization: 12 psi sheath helium.

c. Buffer considerations

Borate and phosphate are common CE buffers. They have strong IR absorbance, however, so it is necessary to choose other buffers or electrolytes with low interference for IR detection of analytes. Some IR transparent electrolytes, such as potassium chloride (KCl), were tested with the metal nebulizer interface. Unfortunately, a strong IR spectrum was obtained from potassium chloride solution after deposition on a ZnSe window. The spectrum was of potassium hydroxide (KOH). As more K^+ ions than Cl^- ions migrate from the inlet buffer (anode) to the metal nebulizer (cathode), they become KOH with OH^- from water. More successful buffers are volatile buffers, such as ammonium acetate.

d. CE/FT-IR sample deposition and IR measurement

As stated above, the metal nebulizer is used for CE electrical contact and solvent elimination with helium sheath gas, as well as for sample deposition onto a ZnSe crystal. About 12 to 20 psi of helium sheath gas is needed to deposit CE eluite onto the ZnSe crystal, which is located as closely as possible below the metal nebulizer. The deposition sizes are about 100 to 200 μm in diameter. The sample deposits were measured off-line with an infrared microscope. All CE/FT-IR spectra were obtained from 100 co-added scans at a resolution of 4 cm^{-1}. Figures 4 and 5 show CE/FT-IR spectra of caffeine and sodium benzoate, as well as their reference (normal) FT-IR spectra. In most cases, volatile buffers, such as ammonium acetate and ammonium hydroxide, have little effect on CE/FT-IR spectra, and spectral subtraction of the buffer is not required. In these preliminary results, high quality CE/FT-IR spectra can be obtained from about 50 ng of sample injected into a CE.

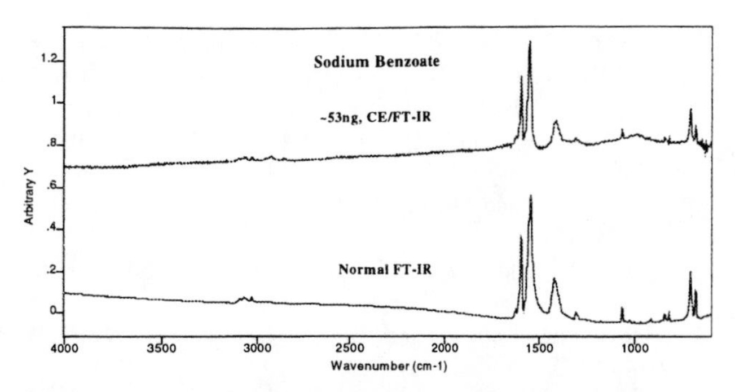

FIGURE 5. Sodium benzoate CE/FT-IR and reference (normal) FT-IR spectra. CE/FT-IR: 0.005M sodium benzoate. Pressure injection: 10 seconds, voltage: 15kV, buffer: 0.8M NH_4OH, column: 75 μm i.d., 78 cm long CE/UV/FT-IR column. Normal FT-IR: sodium benzoate solution was dropped on a ZnSe crystal and dried in air.

CONCLUSIONS

Though CE/FT-IR is a difficult technique, our preliminary research has shown that it is feasible. The metal CE/FT-IR interface has great potential for qualitative and quantitative analysis, CE fraction preparation, and investigation of CE mechanisms and theory.

Further studies on CE/FT-IR will focus the collection of more CE/FT-IR spectra, increased sensitivity, and investigation of other CE/FT-IR buffers.

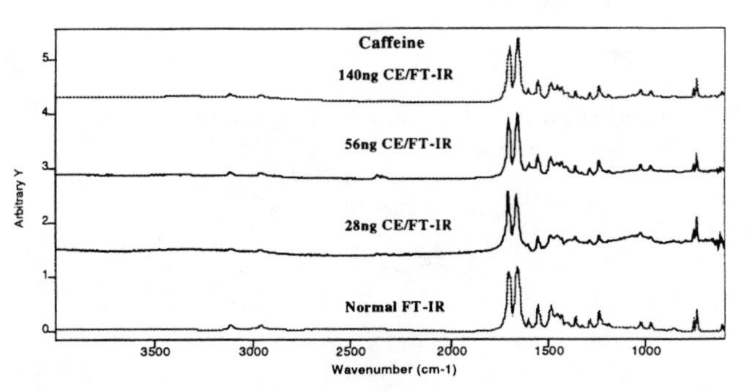

FIGURE 4. Caffeine CE/FT-IR and reference (normal) FT-IR spectra. Buffer: 0.05M NH_4Ac.

REFERENCES

1. Vidrine, D.W., and Mattson, D.R., *Appl. Spectrosc.* **32,** 502-506 (1978).
2. Johnson, C.C., and Taylor L.T., *Anal. Chem.* **56**, 2642-2647 (1984).
3. Kuehl, D., and Griffiths, P.R., *J. Chromatogr. Sci.* **17,** 471-476 (1979)
4. Conroy, C.M., Griffiths, P.R., Duff, P.J.,and Azarraga, L.V., *Anal. Chem.* **56,** 2636-2642 (1984).
5. Lange, A.J., Griffiths, P.R., and Fraser, D.J.J., *Anal. Chem.* **63,** 782-787 (1991).
6. Kalasinsky, V.F., Whitehead, K.G., Kenton, R.C., Smith, J.A.S., and Kalasinsky, K.S., *J. Chromatogr. Sci.* **25,** 273-280 (1987).
7. Robertson, A.M., Littlejohn, D., Brown, M., and Dowle, C.J., *J. Chromatogr.* **588,** 15-24 (1991).
8. Gagel, J.J., and Biemann, K., *Anal. Chem.* **58,** 2184-2189 (1986).
9. Gagel, J.J., and Biemann, K., *Anal. Chem.* **59,** 1266-1272 (1987).
10. Fujimoto, C., Jinno, K., and Hirata, Y., *J. Chromatogr.* **258,** 81-92 (1983).
11. Somsen, G.W., van Stee, L.P.P., Gooijer, C., Brinkman, U.A.Th., Velthorst, N.H., and Visser, T., *Anal. Chim. Acta* **290,** 269-276 (1994).
12. Somsen, G.W., Hooijschuur, E.W.J., Gooijer, C., Brinkman, U.A.Th., Velthorst, N.H., and Visser, T., *Anal. Chem.* **68,** 746-752 (1996).
13. Robertson, R.M., de Haseth, J.A., and Browner, R.F., *Appl. Spectrosc.* **44,** 8-13 (1990).
14. Turula, V.E., and de Haseth, J.A., *Appl. Spectrosc.* **48,** 1255- 1264 (1994).
15. Turula, V.E., and de Haseth, J.A., *Anal. Chem.* **68,** 629-638 (1996).
16. Dwyer, J.L., Champman, A.E., and Liu, X., *LC-GC* **13,** 240-250 (1995).
17. Raynor, M.W., Bartle, K.D., and Cook, B.W., *J. High Resol. Chromatogr.* **15,** 361-366 (1992).

18. Somsen, G.W., van de Nesse, R.J., Gooijer, C., Brinkman, U.A.Th., Velthorst, N.H., Visser, T., Kootstra, P.R., and de Jong, A.P.J.M., *J. Chromatogr.* **552,** 635-647 (1991).
19. Griffiths, P.R., and Lange, A.J., *J. Chromatogr. Sci.* **30,** 93-97 (1992).
20. Lange, A.J., and Griffiths, P.R., *Appl. Spectrosc.* **47,** 403-410 (1993).
21. Liu, M.X., and Dwyer, J.L., *Appl. Spectrosc.* **50,** 349-356 (1996).
22. Garner, T.W., and Yeung, E.S., *J. Chromatogr.* **515,** 639-644 (1990)
23. Li, H., and Morris M.D., Pittsburgh Conference on Analytical Chemistry and Applied Spectroscopy, Atlanta, Georgia, March 1997, Abstract No. 221.
24. Smith, R.D., Barinaga, C.J., and Udseth, H.R., *Anal. Chem.* **60,** 1948-1952 (1988).
25. Niessen, W.M.A., Tjaden, U.R., and van der Greef, J., *J. Chromatogr.* **636,** 3-19 (1993).
26. Olesik, J.W., Kinzer, J.A., and Olesik, S.V., *Anal. Chem.* **67,** 1-12 (1995).
27. Fisons Instruments/VG Bio Tech, Altrincham, Cheshire, UK, Application Note No. 209.

Universal On-Line HPLC/FT-IR and GC/FT-IR Direct Deposition Interface

Lin-Tao He and Peter R. Griffiths*

Department of Chemistry, University of Idaho, Moscow, ID 83844-2343

On-line micro- and semimicro-HPLC/FT-IR interfaces have been developed in our laboratory. These interfaces involve thermospray nebulization of the HPLC effluent and a modification of the Bio-Rad *Tracer* direct deposition GC/FT-IR interface. The detection limits for on-line narrow-bore HPLC/FT-IR measurements are at the low nanogram level. The performance of reversed-phase HPLC/FT-IR was tested at flow rates from 8 to 50 μl/min. The modified fused-silica and stainless-steel transfer lines developed for HPLC/FT-IR can also be used for direct deposition GC/FT-IR measurements. The feasibility of using the same instrument for on-line HPLC/FT-IR and GC/FT-IR direct deposition measurements has been demonstrated.

INTRODUCTION

The interfacing of high-performance liquid chromatography (HPLC) and Fourier transform infrared (FT-IR) spectrometry has been of interest since 1975[1] and many scientists have addressed the HPLC/FT-IR interface in the ensuing 22 years [1-22]. Two approaches have been used in the design of HPLC/FT-IR interface: the use of flow cells [1-3] and the solvent elimination approach [4-22].

The simple flow-cell HPLC/FT-IR interface is analogous to light-pipe gas chromatography (GC)/FT-IR interfaces. However, most solvents commonly used as the mobile phases in HPLC possess intense IR absorption bands across much of the mid-infrared spectrum, thus hindering the use of flow cells for the unequivocal identification of solutes. Solvent-elimination approaches have been proven to have great potential for combining HPLC with FT-IR because the absorption of radiation by the solvent is no longer a problem.

In typical solvent-elimination HPLC/FT-IR interfaces, the analyte is deposited on a suitable substrate while the solvent is eliminated. Many IR sampling techniques (*e.g.*, transmission, diffuse reflection (DR), reflection-absorption) have been used to obtain the spectrum of each deposited component. In first solvent-elimination HPLC/FT-IR interfaces [4,5,7,8], the analytes were condensed on a powered alkali halide substrate held in discrete cups, after which the cups were moved into the IR beam and DR spectra were measured off-line. Novel reflection and reflection-absorption measurement methods have also been applied in some off-line solvent-elimination HPLC/FT-IR interfaces [9,15-17]. In more recent developments, the IR analysis is accomplished by IR microscopy [6,12,14]. In this report, we demonstrate how FT-IR microscope optics can be used for on-line HPLC/FT
-IR measurements.

No solvent-elimination technique that has been described to date can be used for all samples and all mobile phases. For example, volatile analytes can evaporate along with the mobile phase. Ideally, all analytes should have higher melting points than the boiling point of the mobile phase at the time of elution. Another disadvantage of the solvent elimination approach is that buffered mobile phases are not easily used in HPLC/FT-IR [13,19]. Previous reports by Robertson et al. [13] and from our laboratory [19] demonstrated that it is possible to use a mobile phase incorporating a volatile buffer (ammonium acetate) with solvent-elimination HPLC/FT-IR interfaces. When non-volatile buffers such as phosphate are used, however, it becomes difficult to observe the absorption spectra of trace analytes over the very strong absorption bands of the buffer salts. In 1984, Conroy et al. showed that it was possible to measure DR HPLC/FT-IR spectra of polar molecules eluted using an aqueous mobile phase by continuously extracting the analytes into dichloromethane prior to eliminating the solvent and depositing the eluites on powdered KCl [4]. Recently, Somen et al. [12] applied an analogous on-line post-column liquid-liquid extraction (LLE) module in conjunction with a solvent elimination reversed-phase (RP) HPLC/FT-IR interface, where buffer salts are not extracted by post-column extractant (dichloromethane). The post-column LLE process extends the applicability of the use of buffered mobile phases in the solvent elimination HPLC/FT-IR interface.

In most solvent-elimination HPLC/FT-IR interfaces, the mobile phase is converted to an aerosol after emerging from the column. The two most common approaches are based on the thermospray [8] and concentric flow nebulization [6,18,19]. The fundamental difference between these two techniques is whether or not the column effluent is sheathed in a flowing gas. In the concentric flow nebulizer developed in our laboratory, warm helium or nitrogen gas is passed down the outer of two concentric

CP430, *Fourier Transform Spectroscopy:* 11th International Conference
edited by J.A. de Haseth

fused-silica tubes both to assist the conversion of the liquid solvent which is passing through the inner tube to an aerosol and to constrain this aerosol into a very narrow stream. The similar approach has been successfully adopted by several other groups [9,12,20,21]. Several other nebulization techniques have been applied for HPLC/FT-IR measurements. Raynor et al. [21] studied the feasibility of electrospray nebulization for RP-LC/FT-IR. The particle beam interface, originally developed for the HPLC-mass spectrometry interface, has been modified for HPLC/FT-IR by de Haseth and co-workers [13,14,22]. Finally, an ultrasonic nebulizer has been incorporated in the commercial off-line HPLC/FT-IR interface by Lab Connections, Inc. (Marlborough, MA) [16,17].

Even though at least 10 groups have worked on the development of a variety of HPLC/FT-IR interfaces for over 20 years, most devices based on the principle of solvent elimination have been used in an off-line mode and the development of on-line HPLC /FT-IR interfaces still presents a considerable challenge. *With the exception of the thermospray,* most of the interfaces summarized above have one inherent limitation for on-line measurements in that, to obtain the maximum sensitivity, the solutes should be deposited in as small an area as possible, necessitating the use of microscope-style optics to focus the beam onto the sample and to collect the transmitted or reflected radiation. Microscopes have intrinsically small working distances, however, and there is often simply not enough space in which to mount the nebulizer. The goal of this project was to develop an *on-line* microbore HPLC/FT-IR interface based on the commercial direct deposition (DD) GC/FT-IR interface known as the *Tracer* (Bio-Rad, Digilab Division, Cambridge, MA) [23] by modifying its transfer line and using thermospray nebulization. Since the modified transfer line can also be used for DD GC/FT-IR, we have also investigated the feasibility of fabricating a universal on-line HPLC/FT-IR and GC/FT-IR interface.

EXPERIMENTAL

Chromatography. A Carlo Erba (Milan, Italy) Phoenix 20 Micro-HPLC capable of flow rates from 1 to 4000 µl/min was used for these experiments. The system is equipped with a Valco (Houston, TX) 4-port internal sample injector, controlled by a Valco electronic valve actuator, and a 60-nl or 200-nl sample loop. HPLC separations were performed on a MetaChem (Torrance, CA) Intersil ODS-2 320-µm internal diameter (i.d.), 250-mm long micro-HPLC column and a Supelco (Bellefonte, PA) Supelcosil LC-ABZ 1.0-mm i.d., 300-mm long semimicro-HPLC column. In both cases, the diameter of the stationary phase was 5 µm. Mixtures of 50:50 acetonitrile (ACN) and H_2O (Milli-Q) or 60:40 methanol (MeOH) and H_2O were used as mobile phases. The flow rate was 10 µl/min for micro-HPLC and 30 or 50 µl/min for semimicro-HPLC. The actual conditions are given in the figure captions.

GC separations were performed on a Hewlett-Packard (Palo Alto, CA) Model 5890 gas chromatograph with an on-column injector and a Supelco SPB-608 30-m long, 0.25-mm i.d., fused silica capillary column with a 0.25-µm thick film of the stationary phase.

Spectroscopy. All data shown in this paper were obtained on-line with a Bio-Rad/Digilab (Cambridge, MA) FTS-40 FT-IR spectrometer with the *Tracer* interface [23] by averaging 4 scans at a resolution of 8 cm^{-1}. Functional group chromatograms were constructed by integrating the absorbance between 1670 and 1800 cm^{-1}.

On-Line DD HPLC/FT-IR Interface. The transfer line of the *Tracer* GC/FT-IR interface consists of four elements: a 200-µm i.d., 290-µm o.d., 1.5-m long fused silica capillary transfer line, a 50-µm i.d., ~8-cm long restrictor or deposition tip, a resistively-heated 0.030"-i.d., 1/16"-o.d., ~70-cm long stainless-steel guard tube, and a 25-W, 120-V heater (Watlow C1E14) for the deposition tip. We designed several new transfer lines for the DD HPLC/FT-IR interface based on the configuration of the *Tracer* GC/FT-IR interface. The following two were the most effective.

Transfer Line 1: a 100-µm i.d., 360-µm o.d., 1.2-m long fused-silica capillary transfer line was inserted into the same stainless-steel guard tubing used in the *Tracer* interface. This transfer line could also be heated by a 5-V power supply and the Watlow 25-W, 120-V heater.

Transfer Line 2: a 0.010" (~250-µm) i.d., 0.030" o.d., 70-cm long stainless-steel tube was used as the capillary transfer line, and a 0.004" (~100-µm) i.d., 0.008" o.d., 4 - 7 cm long stainless-steel capillary was used as restrictor or deposit tip. The heat elements are unchanged, see Figure 1.

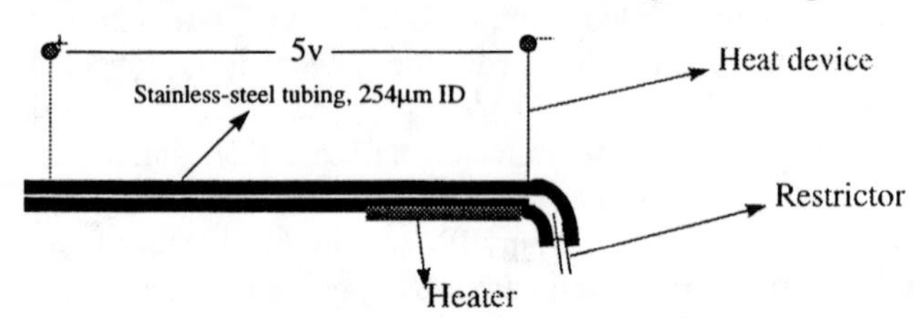

Figure 1. Diagram of transfer line 2 for on-line DD HPLC/FT-IR measurements.

RESULTS AND DISCUSSION

1. Solvent elimination and on-line FT-IR detection

As discussed above, the practical feasibility of on-line HPLC/FT-IR is most simple using a thermospray interface. In this paper, we report ways of modifying the transfer line of the *Tracer* DD GC/FT-IR interface and of using thermospray nebulization to eliminate the mobile phases under vacuum to convert this device to an on-line HPLC/FT-IR interface. As the solvent is evaporated, the analytes are deposited on a moving ZnSe slide operated by XY step-motor stage. The optimum thermospray temperature varies according to the composition and flow rate of the mobile phase. The speed at which the ZnSe slide moves is computer-controlled and also varies according to the chromatographic parameters. The deposition tip is located only 800 μm ahead of IR beam, so that the IR measurements can be made very shortly after deposition. For on-line detection the chamber must be evacuated or else the solvent vapor seriously interferes with the spectrum of each analyte.

2. On-line micro-HPLC/FT-IR by transfer line 1

The IR chromatogram for sequential injections of 60 ng caffeine onto the 320-μm i.d. ODS-2 column is shown in Figure 2A and a representative on-line micro-HPLC/FT-IR spectrum of 60 ng caffeine at the maximum of one of these peaks is shown in Figure 2B.

Caffeine is a fairly volatile compound and the intensity of its spectrum decreases fairly rapidly when the ZnSe slide is at ambient temperature under vacuum. When 6 ng aliquots of caffeine were injected, the peaks could not be detected in the IR chromatogram and no bands assignable to caffeine could be observed in the spectrum. When dry ice was added to the dewar used to cool the ZnSe slide, however, the efficiency at which the analyte is condensed was improved significantly. The detection of 6 ng and 15 ng caffeine by on-line micro-HPLC/FT-IR is demonstrated in Figure 3. Provided that the slide is not too cold to allow the mobile-phase to evaporate, decreasing the temperature of ZnSe slide always increased the sensitivity of IR detection.

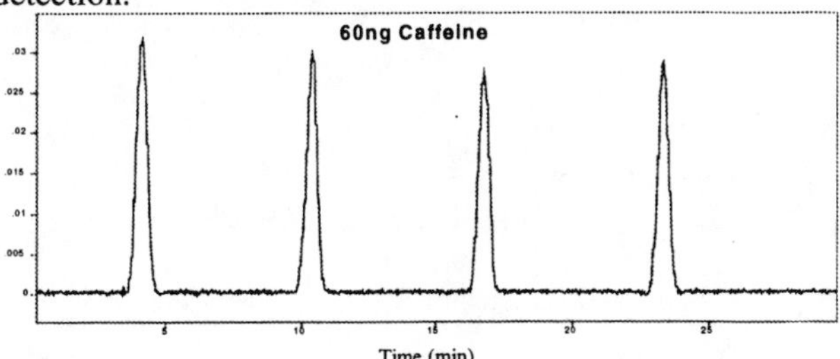

Figure 2A. Functional group chromatogram (1670-1800cm^{-1}) of sequential injections of 60 ng of caffeine. Mobile phase: ACN:H_2O=50:50; flow rate: 10μl/min; vacuum: 1.6x10^{-3} torr; thermospray: 180°C; ZnSe slide: 30°C.

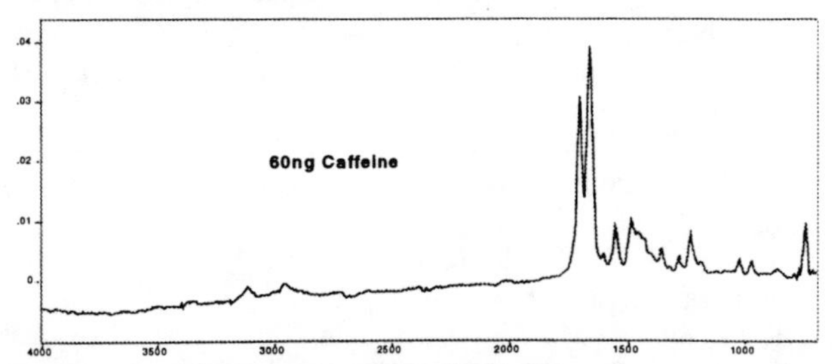

Figure 2B. On-line micro-HPLC/FT-IR spectrum of 60 ng caffeine at peak maximum of one peak in Figure 2A.

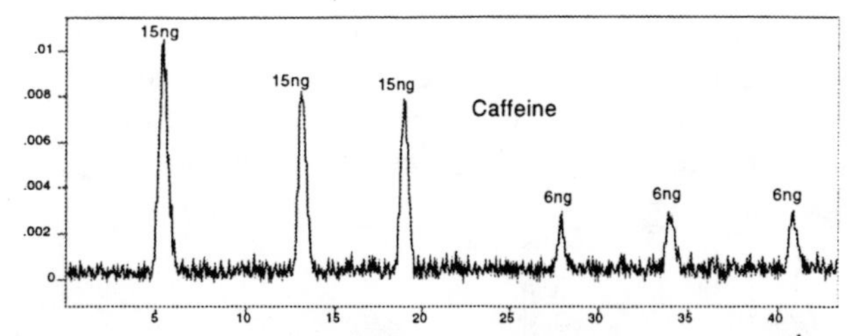

Figure 3A. Functional group chromatogram (1670-1800cm^{-1}) of sequential injections of 6 and 15 ng caffeine. MeOH:H_2O=60:40; flow rate: 8μl/min; vacuum: 3x10^{-1} torr; thermospray: 130°C; dry ice in *Tracer* dewar. X axis: Time (min).

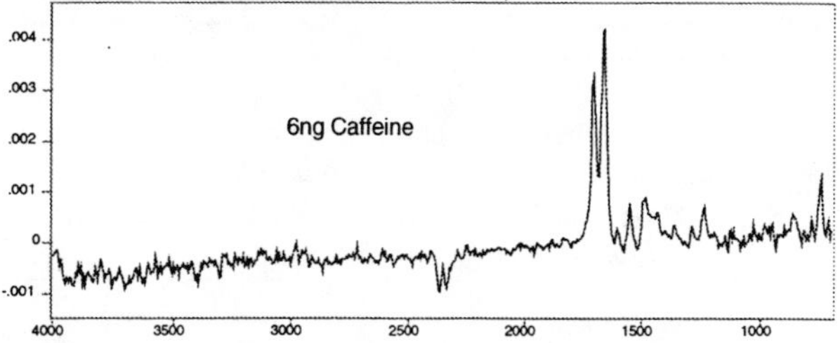

Figure 3B. On-line micro-HPLC/FT-IR detection of 6 ng caffeine from the maximum of one peak in Figure 3A.

3. On-line micro- and semimicro-HPLC/FT-IR by transfer line 2

Although transfer line 1 permitted micro-HPLC/FT-IR spectra to be measured, it was found to be quite fragile and frequently broke when the temperature of the thermospray was high. We therefore designed a new transfer line made of stainless-steel (Fig. 1), which can be used for both micro-HPLC (320-μm i.d. column) and semimicro-HPLC (1-mm i.d. column). Although detection limits were not as low as they were when transfer line 1 was used because of the difference in the diameter of the tubing, acceptable results were usually found. The IR chromatogram and spectrum of 50-ng injections of phenanthrenequinone on the 320-μm column obtained using the stainless-steel transfer line is shown in Figure 4.

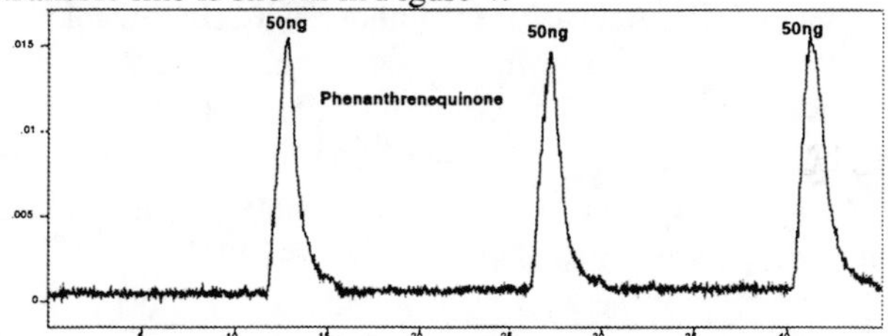

Figure 4A. Functional group chromatogram (1670-1800cm^{-1}) of 50 ng phenanthrenequinone by micro-HPLC via transfer line 2. Mobile phase: ACN:H_2O=50:50; flow rate: 10μl/min; vacuum: 2x10^{-3}torr; thermospray: 145°C, slide: 20°C. X axis: Time(min).

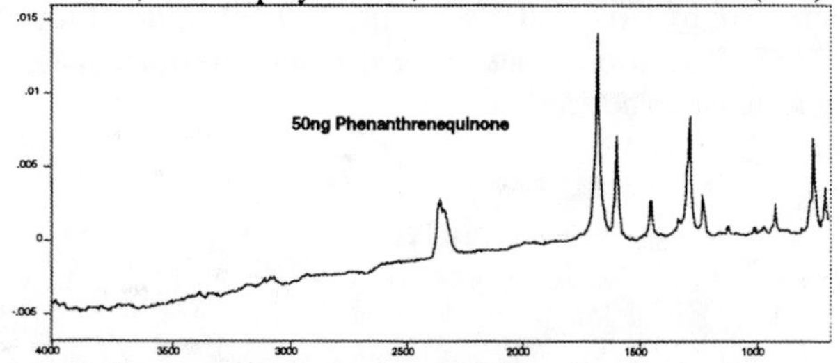

Figure 4B. On-line micro-HPLC/FT-IR spectrum of 50 ng phenanthrenequinone from the maximum of one peak in Fig. 4A.

Figure 5 demonstrates the separation of three barbiturates on a semimicro (1-mm i.d.) column. Despite the fact that barbiturates often decompose on stainless-steel GC columns, there is no evidence of decomposition in the spectra in Fig. 5.

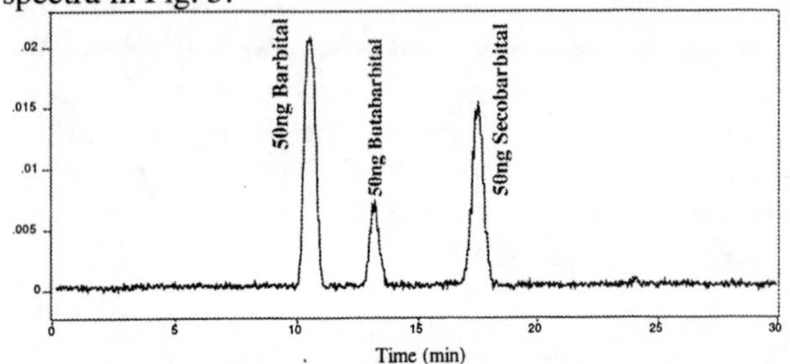

Figure 5A. Functional group chromatogram (1670-1800cm^{-1}) of three barbiturates (50ng) by semi-micro-HPLC. Column: 1.0mm LC-ABZ; mobile phase: ACN:H_2O=50:50; flow rate: 30μl/min; vacuum: 5x10^{-1}torr; thermospray: 250°C; slide: 28°C.

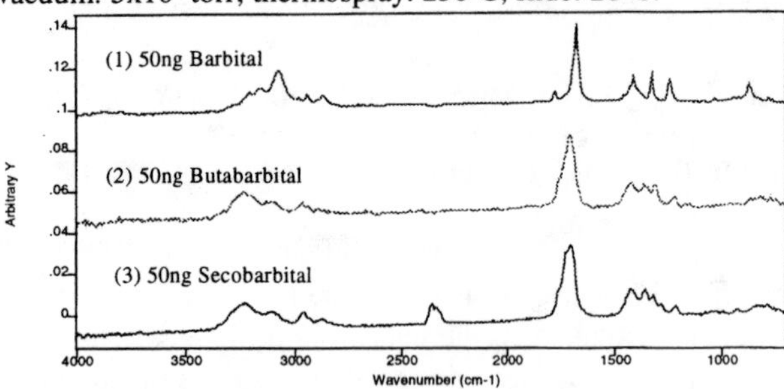

Figure 5B. On-line semi-micro-HPLC/FT-IR spectra of barbiturates in Figure 5A.

4. On-line DD GC/FT-IR by transfer line 2

The main differences between transfer line 2 and the transfer line on the Bio-Rad *Tracer* are the materials (stainless steel *vs.* fused silica) and the i.d. of deposition tips (100 μm *vs.* 50 μm). To test whether our stainless-steel transfer line is still adequate for GC/FT-IR, we first measured the deposition size by using transfer line 2 for DD GC/FT-IR. Acenaphthenequinone was chosen for this purpose as it was known to be non-volatile [24]. With transfer line 2, the full diameter of the deposit was found to be about 300 μm with the full-width at half height less than 150 μm. These numbers are considerably less than twice the corresponding diameters on the *Tracer* [25].

Assuming that the analytes can withstand a short time in contact with stainless steel, all of the three transfer lines could be used for both DD GC/FT-IR HPLC/FT-IR measurements. When transfer line 1 was used for DD GC/FT-IR, the column head pressure of GC had to be increased by 30%. Thus on-line DD HPLC/FT-IR and GC/FT-IR measurements using the same instrument have been shown to be feasible.

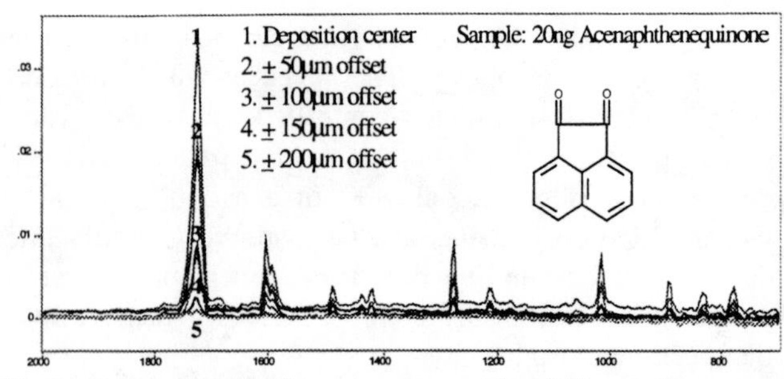

Figure 5. DD GC/FT-IR spectra measured at different distances from peak center obtained using transfer line 2.

References

1. Kizer, K. L., Mantz, A. W., Bonar, L. C., Am. Lab. **7**, 85 (1975)
2. Vidrine, D. W., and Mattson, D. R., Appl. Spectrosc. **32**, 502-506 (1978).
3. Johnson ,C. C., and Taylor, L T., Anal. Chem. **56**, 2642-2647 (1984).
4. Kuehl, D., and Griffiths, P. R., J. Chromatogr. Sci. **17**, 471-476 (1979).
5. Conroy, C. M., Griffiths, P. R., Duff, P. J., Azarraga, L. V., Anal. Chem. **56**, 2636-2642 (1984).
6. Lange, A. J., Griffiths, P. R., Fraser, D. J. J., Anal. Chem. **63**, 782-787 (1991).
7. Kalasinsky, V. F., Whitehead, K. G., Kenton, R. C., Smith, J. A. S., Kalasinsky, K. S., J. Chromatogr. Sci. **25**, 273-280 (1987).
8. Robertson, A. M., Littlejohn, D., Brown, M., Dowle, C. J., J. Chromatogr. **588**, 15-24 (1991).
9. Gagel, J. J., and Biemann, K., Anal. Chem. **59**, 1266-1272 (1987).
10. Fujimoto, C., Jinno, K., Hirata, Y., J. Chromatogr. **258**, 81-92 (1983).
11. Somsen, G. W., van Stee, L. P. P., Gooijer, C., Brinkman, U. A. Th., Velthorst, N. H., Anal. Chim. Acta **290**, 269-276 (1994).
12. Somsen, G. W., Hooijschuur, E. W. J., Gooijer, C., Brinkman, U. A. Th., Velthorst, N. H., Anal. Chem. **68**, 746- 752 (1996).
13. Robertson, R. M., de Haseth, J. A., Browner R. F., Appl. Spectrosc. **44**, 8-13 (1990).
14. Turula, V. E., and de Haseth, J. A., Anal. Chem. **68**, 629-638 (1996).
15. Gagel, J. J., and Biemann, K., Anal. Chem. **58**, 2184-2189 (1986).
16. Dwyer, J. L., Chapman, A. E., Liu, X., LC-GC **13**, 240-250 (1995).
17. Liu, M. X., and Dwyer, J. L., Appl. Spectrosc. **50**, 349-356 (1996).
18. Griffiths, P. R., and Lange, A. J., J. Chromatogr. Sci. **30**, 93-97 (1992).
19. Lange, A. J., and Griffiths, P. R., Appl. Spectrosc. **47**, 403-410 (1993).
20. Somen, G. W., van de Nesse, R. J., Gooijer, C., Brinkman, U. A. Th., Velthorst, N. H., J. Chromatogr. **552**, 635-647 (1991).
21. Raynor, M. W., Bartle, K. D., Cook, B. W., J. High Resol. Chromatogr. **15**, 361-366 (1992).
22. Robertson, R. M., de Haseth, J. A., Kirk, J. D., Browner, R. F., Appl. Spectrosc. **42**, 1365-1368 (1988).
23. Bourne, S., Haefner, A. M., Norton, K. L., Griffiths, P. R., Anal. Chem. **62**, 2448-2452 (1990).
24. Fuoco, R., Pentoney, S. L., Griffiths, P. R., Anal. Chem. **61**, 2212-2218 (1989).
25. *Tracer* Operator's Manual, p.284 and 311.

Miniature Fourier Transform Instrument for Radiation Thermometry

Franklin J. Dunmore and Leonard M. Hanssen

Optical Technology Division, National Institute of Standards and Technology (NIST), Gaithersburg MD 20899

A miniature Fourier transform (FT) spectrometer has been tested as a device for remotely measuring the temperature of a high stability/emissivity blackbody. The commercially manufactured device is based on the novel design of a polarizing Wollaston prism spatial domain interferometer, with a Si diode array detector, and without any moving parts. The measurement of temperature using Planck's law showed a consistent nonlinear effect. This results in an error of the order of 1% for measurement of temperatures 500 K and above. Planned calibration measurements should reduce the nonlinearity related error and improve the FT temperature measurement.

INTRODUCTION

Fourier transform spectrometry over a broad spectral range (using Planck's Radiation Law) shows promise for allowing accurate, contact-free temperature measurement and a Planck's Law based Kelvin Temperature scale(1). Advances in fast temporal scan FT spectroscopy promise much shorter measurement time compared to the standard dispersive methods of thermal radiometry. To further these possibilities a commercially manufactured (by Photonex Ltd. (2,3)) Wollaston prism polarizing type miniature Fourier transform spectrometer (MFTS) has been tested to measure the temperature of high stability/emissivity blackbodies (4). The MFTS in this report uses a Si diode array detector sensitive in the 200 nm to 1100 nm (9000 cm^{-1} to 50000 cm^{-1}) wavelength region. For blackbody sources at 2000 K, only millisecond data acquisition time is necessary. This makes the MFTS a potentially useful device where fast measurements are required.

THE INSTRUMENT

Shown in Fig. 1 is a schematic of the MFTS which is based on the design of Padgett, et al. (5). From left to right, the first element is a linear polarizer oriented at 45° with respect to the plane of the paper. In the center is a Wollaston prism which is an optical combination of two wedged birefringent calcite slabs with their optical axes oriented perpendicular to each other. This acts as a beam splitter for the light coming from the polarizer. For the polarization component parallel to the optical axis, the refractive index at 589 nm is n_o=1.658; for the perpendicular component it is n_e=1.486. For a ray crossing the prism at the center d=0, the induced optical path length difference between the two beams in the first wedge is compensated in the second, resulting in a net zero difference. For $d \neq 0$, the two beams will traverse differing optical path lengths. The focusing optics will recombine the rays at the 1024 element Si diode array.

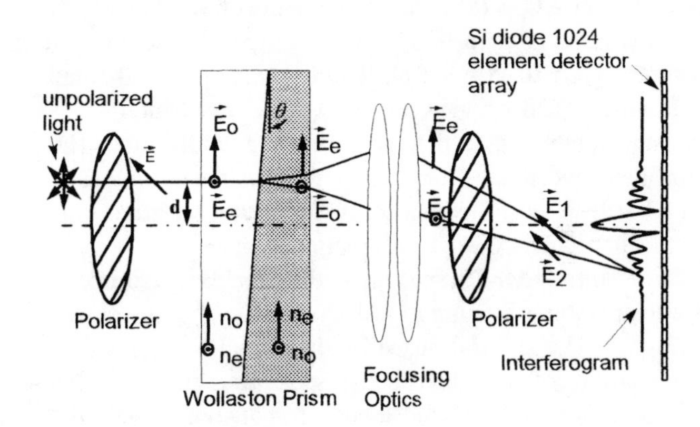

Figure 1. The optical layout of the MFTS.

All wavelengths interfere constructively for d=0 at the array leading to the central maximum of an interferogram in the spatial domain. Where $d \neq 0$ there will be modulation of the intensity according to

$$I(\nu) \propto |A(\nu)|^2 (1 + \cos(2\pi \Delta \nu)) \quad (1)$$

where A is the Electric field amplitude, ν is the wavenumber and

$$\Delta = 2d(n_e - n_o)\tan\theta \quad (2)$$

is the path length difference, with θ the prism wedge angle (5). The accompanying instrument software takes the Fourier transform of the spatial fringe pattern (interferogram) to give the instrument spectral signal $S(\nu)$ as a function of wavenumber. In addition, the software applies

CP430, *Fourier Transform Spectroscopy:* 11th International Conference
edited by J.A. de Haseth

corrections for the wavenumber variation of the refractive indices and polarizer efficiencies. The wavelength range of the polarizers limit the wavelengths that are modulated by the prism. For this instrument they cut off at 600 nm. The long wavelength cutoff of the Si diode array detector is at a wavelength less than 1100 nm. This results in an operational range of 9000 cm^{-1} to 16700 cm^{-1} (1100 nm to 600 nm) with a full width at half maximum resolution of 27 cm^{-1} for the instrument, determined primarily by the number of pixels of the Si diode array (5).

All expanded uncertainties in this report have a coverage factor k=2 (95% confidence) (6). The expanded uncertainty in the wavenumber, 10 cm^{-1}, is estimated to be less than half of resolution limit for this instrument. This uncertainty can be reduced by calibration with a stable laser source (5).

RESULTS AND ANALYSIS

To test its capability for radiometric temperature measurement the MFTS was set up to view a pyrolitic graphite cavity blackbody with an emissivity of 99.9% (7) made by VIINOFI (8). The temperature of the blackbody was measured using a NIST standard pyrometer, calibrated against a gold fixed point blackbody source. The 0.4 mrad (full angle) field of view of the MFTS was limited by a 0.2 mm aperture placed at the end of a baffle tube. The expanded uncertainty of the pyrometer measurement is 3.14 K (9). The following resullts are representative only for this particular device at this particular time.

The resulting MFTS spectra of the blackbody source at five temperatures ranging from 1500 K to 2500 K are shown in Fig. 2. The spectral shape of the signal S is due to a combination of the blackbody spectrum, the polarizer efficiencies and the Si diode responsivity. As the temperature increases, the signal increases rapidly,

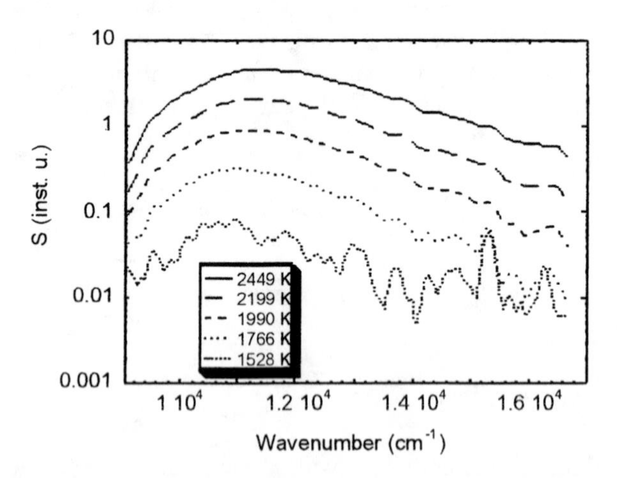

Figure 2. Spectral signal S in instrumental units, versus wavenumber for several blackbody source temperatures.

consistent with Planck's Law and the peak signal shifts to higher frequencies consistent with Wien's displacement Law. At the lower temperatures, a reproducible fine structure of small amplitude, common to all of the curves becomes apparent. This is due to a combination of the variability and nonlinearity of each pixel's responsivity.

Initially we assume linear response. Then a simple calibration procedure can be performed in order to measure temperature. First the instrument spectral signal $S(\nu,T)$ is measured for a blackbody source at several temperatures measured by the NIST standard pyrometer. Then the theoretical spectral flux $\Phi_\nu(\nu,T)=A\Omega\, L_\nu(\nu,T)$ is calculated at one calibration temperature T_0. Here A is the effective area of the blackbody source (determined by aperture size), Ω is the solid angle subtended by the detector array, and the Planckian spectral radiance L_ν is given by:

$$L_\nu(\nu,T)=\frac{c_1 n^2 \nu^3}{\exp(c_2\nu/T)-1} \qquad (\mathrm{W\ sr^{-1}\ m^{-1}})^{\dagger} \qquad (3)$$

where T is the true temperature measured by the NIST pyrometer. The constants are $c_1=1.191\times10^{-16}$ W sr^{-1} m^2, $c_2=1.439\times10^{-2}$ K m (10) and n=1.00028 is the refractive index of air at 1 μm (11). Then the instrument spectral responsivity function R is calculated at the calibration temperature T_0:

$$R(\nu)=\frac{S(\nu,T_0)}{\Phi_\nu(\nu,T_0)} \qquad (4)$$

Then the spectral flux $\Phi_{c\nu}$ calibrated to T_0 by Equation 4, at a different temperature T' is

$$\Phi_{c\nu}(\nu,T')=\frac{S(\nu,T')}{R(\nu)} \qquad (5)$$

where T' can be any other of the several temperatures of Fig. 2. Therefore the spectral flux can be deduced at any other temperature by using Eq. (4). Finally $\Phi_{c\nu}(\nu,T')$ is fitted to the Planckian theoretical spectral flux $\Phi_\nu(\nu,T')$ for the single parameter T'. It is expected that the deduced temperature T' should equal the true temperature T to within the expanded uncertainty of the pyrometer measurement, 3.14 K. However, it has not been the case and the reasons are discussed below.

In Fig. 3 the dashed lines are the Planckian spectral fluxes at the true temperatures, which are shown to the right, the dots are the calibrated spectral fluxes $\Phi_{c\nu}$, and the solid lines are the fits of Planck's Law to $\Phi_{c\nu}$ with the temperature values shown to the left. The differences between the fitted temperatures and the true temperatures are shown in the middle. In Fig. 3, 2449 K is the calibration temperature and

† As an aid to the reader the appropriate SI units in which a quantity should be expressed is indicated in parentheses when the quantity is first introduced.

the expanded uncertainty is less than 2 K for the temperature fits to Φ_{cv}. There is a systematic error in the measured temperature, on the order of 1% of the temperature, that decreases with decreasing temperature differences from the calibration temperature.

In order to ascertain the discrepancy between measured and deduced temperatures, we have measured the spectral responsivity function R at various temperatures of the blackbody. Shown in Fig. 4 is R versus wavenumber, derived from the data of Fig. 2, for several temperatures ranging from 2000 K to 2500 K. There is a significant deviation from linear spectral responsivity, which is apparent from the variation of R with temperature and therefore incident spectral flux. Interestingly, this nonlinear

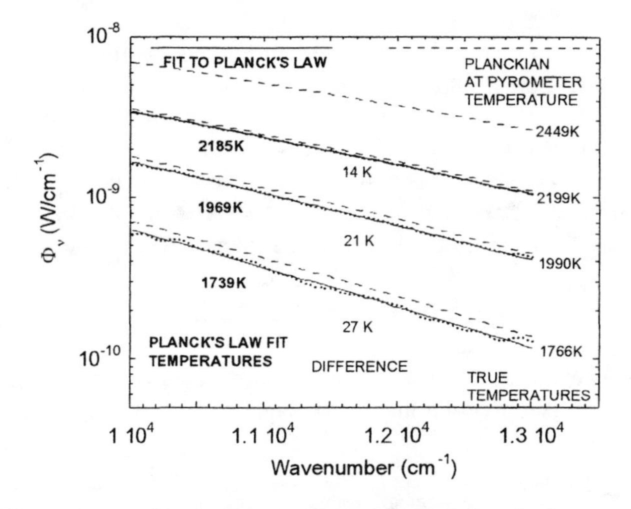

Figure 3. Logarithmic plot of calibrated spectral flux Φ_{cv} from data (dots), calculated spectral flux Φ_ν (thin line), and the Planck's Law fits to Φ_{cv} (solid line).

behavior is opposite to the more common decrease in responsivity as the incident flux reaches a saturation value. Instead, R increases as the flux increases. This is readily apparent in Fig. 5, where R versus Φ_ν is plotted for several frequencies. For linear responsivity, R should be a constant. For each curve the first data point to the left (at the lowest spectral flux) is T =1766 K, the second to the right is for 1990 K, the third is for 2199 K and the last is for 2449 K. The data, replotted in Fig. 6 between S and Φ_ν, is best represented as a cubic polynomial. The slope of the S versus Φ_ν curve is approximately the responsivity and is strongly wavenumber dependent.

The expanded uncertainty for S is less than 0.3%. For the Φ_ν and Φ_{cv} scales in Fig. 3, the Φ_ν scale in Figs. 5.and 6, and the R scale in Figs. 4 and 5, it is 5%. The uncertainty for the Φ_ν and Φ_{cv} scales is dominated by the uncertainty of the area of the aperture and does not contribute to the uncertainty of the measured temperatures derived below because it has no effect in the process of fitting $\Phi_{cv}(\nu, T')$ for T'. This is because the Φ_ν and Φ_{cv} scale uncertainties are independent of temperature and it is the shape of the curve, not the absolute spectral flux level that determines the fit for T'.

The cause of the instrumental nonlinear responsivity is most likely nonlinearity of the Si diode detector array. This nonlinear responsivity is due to stray capacitance in the sample and hold circuit of the array connecting each pixel to

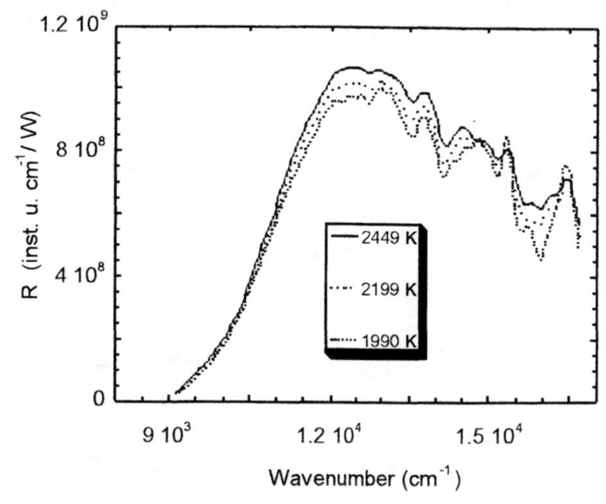

Figure 4. MFTS spectral responsivity R versus wavenumber for three blackbody temperatures.

the A/D converter; it is more prevalent in diode array detectors than in single diode detectors (12).

The MFTS temperature measurement is extremely sensitive to nonlinearity because, for the wavenumber and temperature ranges described in this report, the incident spectral flux has a very strong dependence on temperature. This can be seen in the following relation between the fractional changes of Φ_ν and T:

$$\frac{\Delta\Phi_\nu}{\Phi_\nu} = \frac{c_2\nu}{T}\left(1 - e^{-\frac{c_2\nu}{T}}\right)^{-1} \cong 8.6\frac{\Delta T}{T} \qquad (6)$$

This corresponds approximately to a ninth-power temperature dependence of the spectral flux in this wavenumber-temperature regime. The nonlinearity can be described by relating S to Φ_ν by

$$S = a + b\Phi_\nu + c\Phi_\nu^{\,2} + d\Phi_\nu^{\,3} \qquad (7)$$

The S offset, due to dark current in the Si diode pixels is $a \cong 0.01$. In the linear case $c=d=0$ and $R=(S-a)/\Phi$.

The data of Fig. 6 is fit to Eq. (7), and the result is shown in an inset to Fig. 6. The 12000 cm^{-1} to 13000 cm^{-1} wavenumber range is the region of maximum combined responsivity of the MFTS. Here the values of b, c, and d are relatively wavenumber independent. Eq. (7) can be inverted to give Φ_ν as a function of S for a single wavenumber. When data is calibrated this way, better agreement with the NIST pyrometer is found, where the differences between the true temperatures and the temperatures of the fits to Planck's

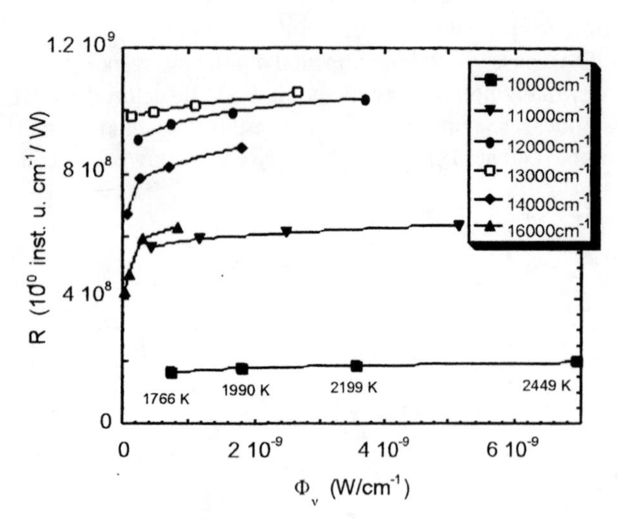

Figure 5. Spectral responsivity R versus incident spectral flux Φ_v at several wavenumbers spanning the spectral range of the MFTS.

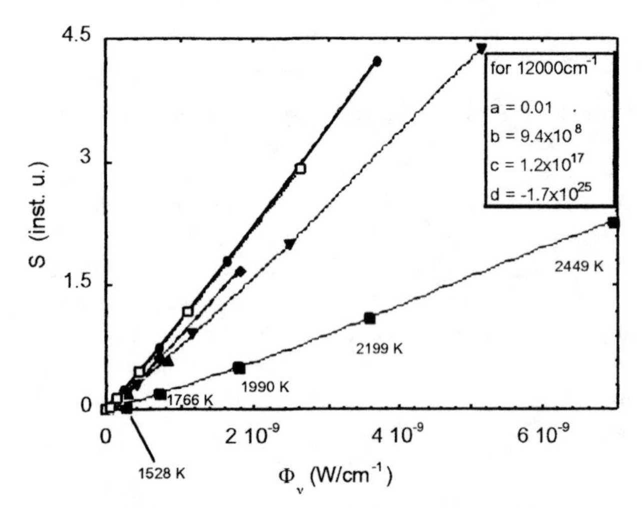

Figure 6. MFTS spectral signal S versus incident spectral flux Φ_v. The legend is the same as in Figure 5.

Law, as in Fig. 3, are 4 K for T=1766 K, 1 K for T=1990 K and 3 K for T=2199 K. The last two temperatures came within the expanded uncertainty of the NIST pyrometer, 3.14 K.

Attempts to recalibrate the spectrometer across its entire spectral range proved unsuccessful because the variation of the spectral responsivity has a strong wavenumber dependence as is seen for the sample of frequencies across the MFTS spectral range in Figs. 5 and 6. S at any one wavenumber is determined by the spectral responsivity at all wavenumbers of every pixel of the Si diode detector array. This makes detector linearization by recalibrating S over a broad spectral range difficult. It will be necessary to correct the interferogram directly prior to Fourier transform processing, in order to increase the radiometric accuracy of the MFTS temperature measurement.

CONCLUSIONS AND FUTURE WORK

The MFTS is a very useful device for spectral radiometry because of its speed and compact portable nature. However, its accuracy for temperature measurements is limited by the nonlinear behavior of its detector array. This may be remediable. An effective method for detector linearization in FT spectrometry has been developed for standard FT instruments used with nonlinear detectors, such as HgCdTe. The measured interferogram, is corrected via a calibrated responsivity curve obtained by comparison to a NIST transfer-standard linear Ge detector (13). For the case of the MFTS, each pixel of the bare Si diode array will have to be characterized with respect to a standard Si detector. When the interferogram is calibrated for individual pixel non-linearity, the resulting spectrum after Fourier transformation should have significantly reduced detector nonlinearity effects and the temperature measurement accuracy will be limited by other sources of error such as wavenumber scale uncertainty and deviations from ideal behavior of the polarizer and prism, which can be further investigated.

ACKNOWLEDGEMENTS

This work has been supported by the Air Force Calibration Coordination Group under the Rapid Blackbody Calibrator Project.

REFERENCES

1. H .A. Gebbie, R. A. Bohlander, and R.P. Futrelle, Nature **240**, 391 (1972): H .A. Gebbie, Infrared Phys. **34** , 575 (1993).
2. The mention of manufacturers and model names is intended solely for the purpose of providing technical information useful to the reader and in no way should be construed as an endorsement of the named manufacturer or product.
3. Photonex Ltd., 35 Cable Depot Road, Riverside Industrial Estate, Clydebank, Glasgow G81 1UY United Kingdom
4. F.J. Dunmore *et al.*, 25th Anniversary Conference of the Council for Optical and Radiation Measurements (1997).
5. M.J. Padgett et al, Applied Optics **33**, 6035 (1994).
6. B.N. Taylor and C.E. Kuyatt, *Guidelines for Evaluating and Expressing the Uncertainty of NIST Measurement Results,* NIST Technical Note 1297, 1994 Edition, pp 2-6.
7. V.I. Sapritsky and A.V. Prokhorov, Appl. Opt., **34**, 5645 (1995).
8. All Russian Institute for Optical and Physical Measurements, Moscow, Russia.
9. NIST Special Publication, 1997, (in press).
10. P.R. Griffiths and J. A. de Haseth, *Fourier Transform Infrared Spectroscopy*, New York: John Wiley & Sons, 1986, ch. 7, p. 248.
11. B. Edlen, Metrologia, **2**, 71 (1966).
12. Conversations with Bill Beckman, Mark Hollbrook and Andy Smith (Photonex LTD).
13. Z.M. Zhang, C.J. Zhu, and L.M. Hanssen, Applied Spectroscopy **51**, 576 (1997).

Making The Most Of The Throughput Advantage In Imaging Fourier Transform Spectrometers.

Jérôme Genest†, André Villemaire‡, Pierre Tremblay†

†- Centre d'optique photonique et laser,
Département de génie électrique et de génie informatique,
Université Laval, Québec, Canada G1K 7P4

‡-Bomem inc.
Special Projects in Radiometry division
450 Saint-Jean-Baptiste, Québec, Canada G2E 5S5

Some aspects of imaging Fourier transform spectrometers are still not generally understood. One particular point leading to confusion concerns the throughput of time-scanning imaging instruments and how it can be compared to the étendue of conventional Fourier transform infrared radiometers (FTIRs). This paper shows that a scanning imaging Fourier transform spectrometer can, in general, use more throughput than a similar non imaging instrument. It is important to understand that the overall instrument's throughputs must be compared. The pixel thoughput of the imaging instrument is not greater than the throughput of the non-imaging device. But many pixels side by side lead to a greater étendue. Examples of calculated throughput for various realistic configurations are shown to illustrate the throughput gain of imaging FTIRs. Another confusing aspect is the étendue of imaging spatially modulated interferometers (SMIs). The étendue of such devices is highly dependent on their particular design. It is shown that SMIs tilting collimated beams suffer from a throughput reduction with tilt angle. Experimental results confirm the theoretical expression derived for fringe visibility with respect to tilt angle. Configurations of imaging FTIRs with no moving parts that achieve higher étendue are presented and discussed.

INTRODUCTION

This paper deals with two features of imaging spectrometers not generally understood and accepted. The aim is to provide a good perspective to help the reach of an agreement of the community on these topics.

The first concerns the superior throughput advantage of time-scanning imaging interferometers. It is shown that by using many detectors the field of view of an FTIR can be widened. The throughput advantage of time-scanning imaging instrument is therefore greater than standard time-scanning interferometers. This is against various claims in the litterature[1] and we shall see why their basic assumption of using only the central lobe of the Haidinger fringes is incorrect.

The second aspect is also related to throughput. It is an attempt to clarify the question for spatially modulated interferometers. Many authors often assume that SMIs always benefit from the same -or larger- throughput than conventionnal time-scanning instruments[2,3,4]. This is almost never the case. Beam-tilting SMIs are much more resticted than standard FTIRs while source doubling and image tilting SMIs are not fundamentally limited in étendue[5]. These facts are not clear and confusion remains about the throughput of SMIs. The throughput limitation of beam tilting SMIs is calculated in the paper and is confirmed by experimental results. The absence of limitation of some other designs is explained.

ÉTENDUE OF TIME SCANNING FTIRs

The throughput of a time scanning interferometer is fixed by the self-apodization caused by the off-axis effect. Rays travelling at an angle θ experience indeed a path difference $\Delta X_o \cos\theta$ where ΔX_o is the path difference of the on-axis ray. Accepting rays from various angles reduces the visibility of the interferogram because of the ensemble of many effective path differences seen.

In imaging interferometers, the field of view is generally imaged on the detector so the radiance of the desired scene is obtained for many spectral bands. To use the precious pixels of the detector array and to collect all the energy entering in the instrument, it is clever to have a design with the detector matched to the field stop size. In that sense, the detector defines the field of view.

With all these considerations, a monochromatic interferogram seen by a detector anywhere in the focal plane of the imaging optics can, at first order, be expressed as equation 1[6,7] with the wavenumber υ and the solid angle element sustended by the detector $d\Omega$.

$$I = \int I_o [1 + \cos(2\pi\upsilon\Delta X_o \cos\theta)] d\Omega . \qquad \textbf{Eq. 1}$$

For circular, annular or pie-shaped detectors, equation 1 is easily integrated using $d\Omega = \sin\theta\, d\theta\, d\phi$. The result is equation 2. The interferogram is multiplied by a cardinal sine (*sinc*) window hence the name self-apodization. The parameter β is shown at equation 3.

CP430, *Fourier Transform Spectroscopy:* 11th International Conference
edited by J.A. de Haseth

$$I = I_o(\phi_{\max} - \phi_{\min})(\cos\theta_{\min} - \cos\theta_{\max}) \times \frac{\sin(\beta)}{\beta}\cos\left(2\pi\upsilon\Delta X_0 \frac{\cos\theta_{\min} + \cos\theta_{\max}}{2}\right)$$

Eq. 2

$$\beta = 2\pi\upsilon\Delta X_0 \frac{\cos\theta_{\min} - \cos\theta_{\max}}{2} \qquad \textbf{Eq. 3}$$

This *sinc* reduces the interferogram's modulation efficiency . The effect is obviously worst when β is higher, i.e. at maximal wavenumber $\upsilon_{\max}$, at maximal path difference $\Delta X_{\max}$ and for an important difference between the cosines of the extremal radial angles. If one wishes to keep a 80% modulation efficiency, one obtains the condition stated at equation 4. We shall use this 80% criterion to compare all the detectors and arrays presented.

$$\cos\theta_{\min} - \cos\theta_{\max} < \frac{1.131}{\pi\upsilon_{\max}\Delta X_{\max}} \qquad \textbf{Eq. 4}$$

For a circular centered detector $\theta_{min} = 0$ and equation 4 is then the 80% version of the well-known expression for the throughput limitation[8]. Under this condition, the detector size is matched to the central lobe of the Haidinger fringes. It is important to understand that this match applies only to this case and is by no way a general expression. As we shall see in the examples section, a smaller detector can be placed way outside this main lobe. It is therefore incorrect to assume the N pixels of a matrix must fit within the main lobe of the bull's-eye[1] .

If square or rectangular detectors are to be used, it is more convenient to integrate equation 1 in cartesian coordinates. One has therefore equation 5 to work with. The focal length f of the output optics is related to the radial angle of the field of view according to figure 1.

$$I = \int_{x_{\min}}^{x_{\max}}\int_{y_{\min}}^{y_{\max}} I_o[1+\cos\left(\frac{2\pi\upsilon\Delta X_o f}{\sqrt{f^2+x^2+y^2}}\right)]\frac{f\,dxdy}{\left(f^2+x^2+y^2\right)^{\frac{3}{2}}}$$

Eq. 5

√f2+x2+y2
√x2+y2
θ
f
Detector plane
Imaging lens

Figure 1: Output optics geometry.

Numerical integration of equation 5 will allow the calculation of fringe visibility for various cases of interest.

EXAMPLES

Let us first consider the common case of an on-axis circular detector. This will provide a comparison basis for further examples. For all cases, the parameters of table 1 will be used. In order to satisfy equation 4 and to keep a 80% fringe visibility, the detector radius is found to be 0.759 mm. The detector is shown at figure 2. This detector can obviously be cut in pie slices. Each slice is keeping the same visibility since it depends only on the radial angle. On can see the detector matches closely the bull's-eye main lobe (figure 3).

f	output optics focal length	200	mm
$\upsilon_{\max}$	maximal wavenumber	10000	cm^{-1}
$\Delta X_{\max}$	maximal path difference	5	cm

Table 1: Parameters for detector size calculations.

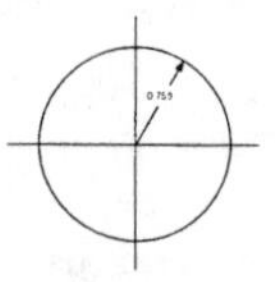

Figure 2: Circular on-axis detector.

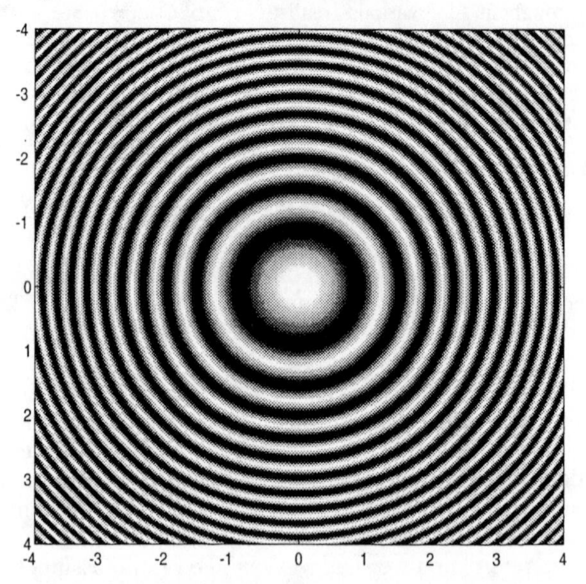

Figure 3: Haidinger fringes in the detector plane.

Let see now if it is possible to add annular layers to this on-axis circular detector. According to equation 4, it is. To preserve a 0.8 modulation efficiency, each layer will be thinner than the preceding one. The first few layers are shown at figure 4. The annuluses thicknesses are respectively: 0.759, 0.314, 0.241, 0.203, 0.179, 0.162 and 0.149 mm. These annuluses all have a 80% visibility. Layers can be added in this manner without limit. The example is not realistic but it makes clear that using multiple detectors can allow field widening.

Another illustrative example is multiple annular layers of constant size. This allows one to see the gain of reducing the size of detectors. Three cases are shown at figure 5. The thickness (r_{max} -r_{min}) of each layers is respectively the half, the quarter and the eighth the size of the circular, on-axis matched detector. For each case, layers are added until the last has a visibility just above 0.8. Annuluses toward the center have an higher visibility.

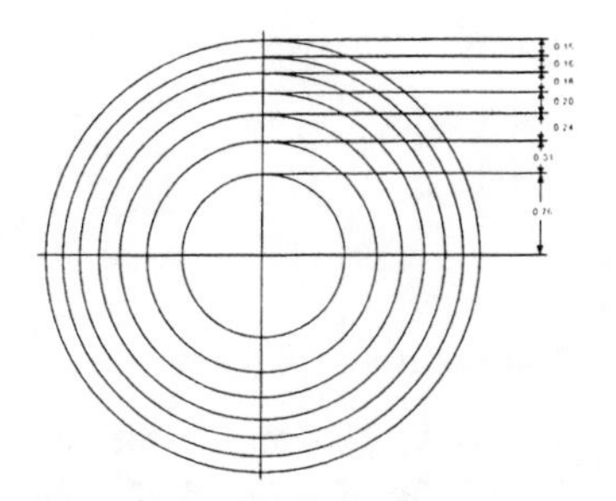

Figure 4: Matched annular detectors

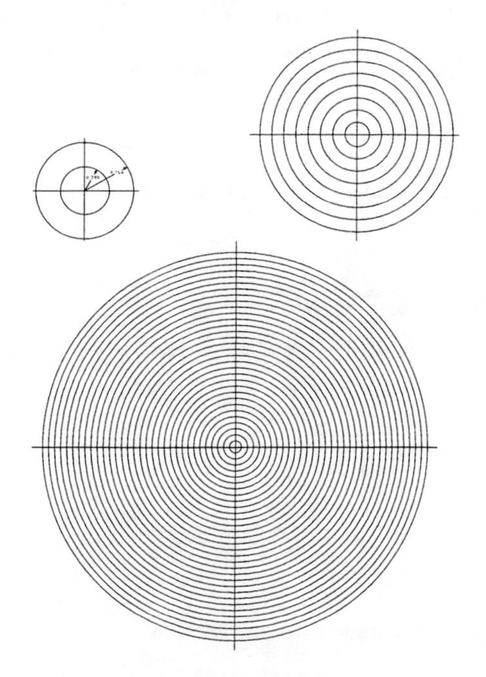

Figure 5: Annuluses of constant thickness t= 0.380, t= 0.190 and t=0.0949 mm.

These calculations with circular detectors are rather unrealistic but provide a formal treament that gives an idea of the superior throughput advantage of imaging time-scanning interferometers.

Using numerical integration, it is possible to repeat the same procedure for square or rectangular detectors. The single square detector having a 0.8 fringe visibility is shown first at figure 6. It has a 1.266 mm side. When comparing the detector with the circular one (figure 2), one sees that none of both is inside the other. It is therefore incorrect to assume the square detector must be bounded by the 0.8 modulation index circle[9]. It has an area 11% smaller and is thus suboptimal As the circular detector can be divided in pie slices, this single square detector can be divided in four 0.633 mm sided pixels, one in each quadrant and all having a 80% modulation efficiency. Shown at second, third and fourth in figure 6 are detector arrays with pixels respectively the half, the quarter and the eighth of 0.633 mm. Arrays are designed so the extrema corner pixels produce fringes with 0.8 visibility, all inner detectors having higher visibility. Again, the throughput gain of using detector arrays with small pixels is obvious.

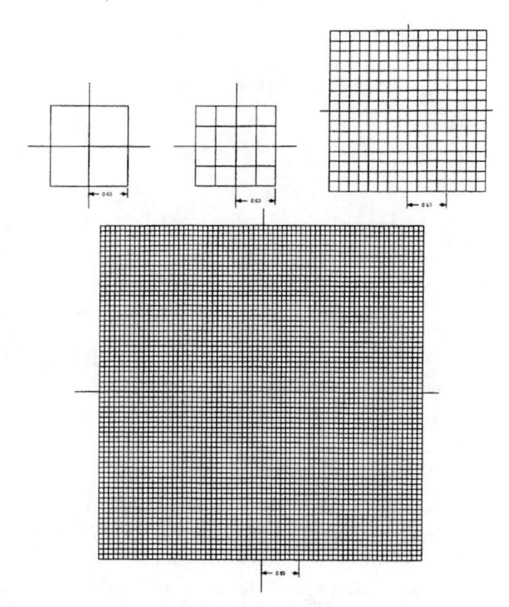

Figure 6: Square detectors of constant size a=0.633, a=0.317, a=0.158 and a=0.0791.

THROUGHPUT OF BEAM-TILTING SMIS

Beam-tilting SMIs are based on standard interferometers designs. They collimate the light source, pass it through the interferometer with one mirror tilted and detect the spatial interferogram. There are many configurations of such SMIs. Tilted Michelson, Sagnac and Mach-Zehnder interferometers may be used as well as wavefront division methods such as tilted mirror pairs, interlaced mirrors or mirror stairs. Figure 7 shows a tilted Michelson interferometer. The cylindrical lens may be used to image a slit in the direction perpendicular to the beam tilting. It is the preferable beam-tilting design because it allows the largest interference region and the smallest distance between beam tilting and detection[10].

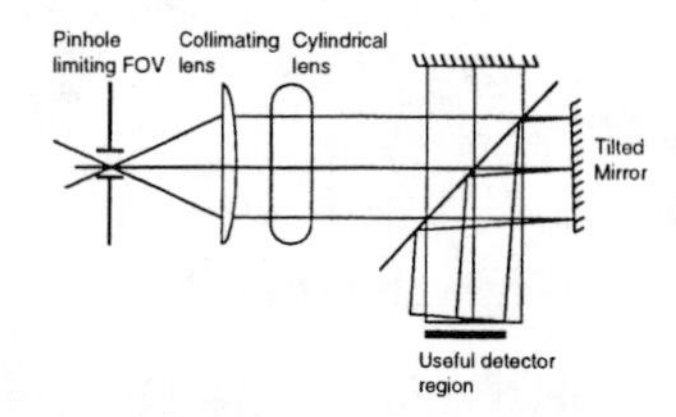

Figure 7: Tilted Michelson interferometer.

The throughput limitation of these SMIs comes from the detection of rays at several angles in the interferometer (see figure 8). The monochromatic interferogram can be expressed as equation 6.

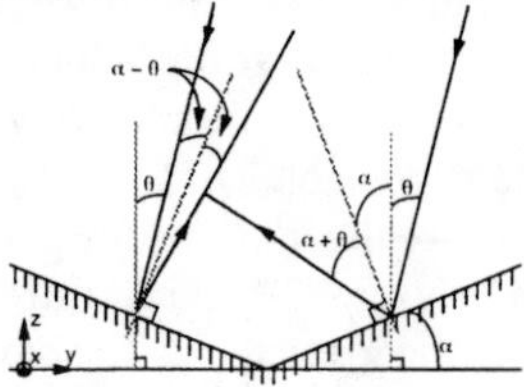

Figure 8: Collimated rays in tilted interferometer.

$$I = \int_{\theta_{min}}^{\theta_{max}} I_o \cos^2(\pi \upsilon[(\cos(2\alpha - \theta) - \cos(2\alpha + \theta))z + (\sin(2\alpha - \theta) - \cos(2\alpha + \theta))y])d\theta$$

Eq. 6

If both α and θ are small so $\sin(A) \sim A$ and $\cos(A) \sim 1$ and the field of view is symetric so that $\theta_{min} = -\theta_{max}$, integration of the first order approximation of equation 6 gives:

$$I = I_o \theta_{max} \left[1 + \frac{\sin(4\alpha 2\pi \upsilon \theta_{max} z)}{(4\alpha 2\pi \upsilon \theta_{max} z)} \cos(4\alpha 2\pi \upsilon y) \right]$$

Eq. 7

The spatial interferogram is killed by a cardinal sine term that depends on the tilt angle α, the distance z between the tilt application and detection, the field of view and the wavenumber υ. It is therefore obvious that z must be as small as possible; thus the advantage of Michelson interferometers to realize beam tilting SMIs.

Generally, the throughput limitation of beam-tilting SMIs is much more restrictive than the one for standard time scanning FTIRs. Beam tilting SMIs do not therefore benefit from the full throughput advantage of FTIRs.

EXPERIMENTAL RESULTS

An experimental setup similar to figure 7 was built to check the expression for the throughput of beam-tilting SMIs (from equation 7). Figure 9 shows the measured visibility with respect to tilt angle compared with the theoretical *sinc*. Only the amplitude of the theoretical curve is adjusted to take into account the poor quality of the beam splitter used. Results show good agreement.

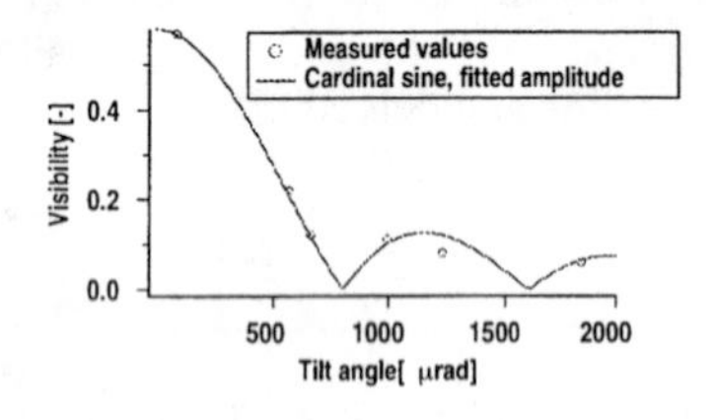

Figure 9: Visibility with tilt angle

ALTERNATE SMIS DESIGNS

Perhaps the most elegant SMI is the source doubling type (figure 10). An asymmetric Sagnac interferometer is used to produce to images of the source. These images are then placed in the focal plane of the output lens. The interferogram is detected at the other focal plane. As demonstrated by Okamoto[5], source-doubling SMIs have no interferometric étendue limitation. They benefit from the largest field of view possible. This is so because there is no collimation before the interferometer. The lens "sees" two images and the optical path difference for a given spatial position on the detector is the same for any pair of complementary points in the two images no matter how extended the sources are. The étendue of source-doubling SMIS is ultimately limited by the quality of the optics.

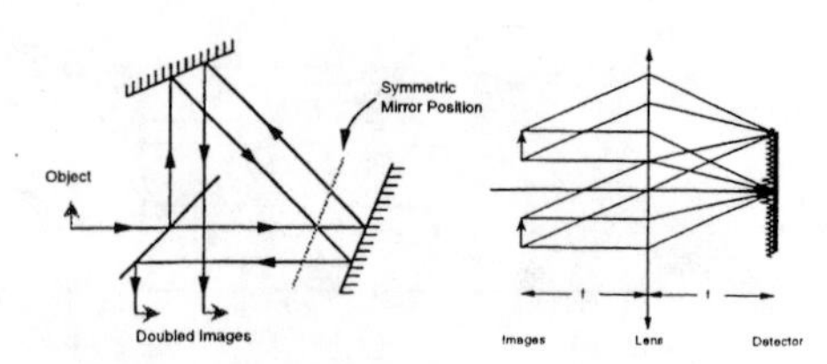

Figure 10: Source-doubling principle.

Recently proposed by Horton[11,12], the image-tilting SMI makes good use of the absence of collimation before the interferometer. Two images are formed as in figure 11. One is tilted in one interferometer arm. A modified Mach-Zehnder interferometer is necessary to have enough independance between both arms to tilt images and make sure they superpose correctly on the detector. The main difference between beam-tilting and image-tilting SMIs is the fact that beam tilting ones misalign images at infinity (collimated beams) while image tilting devices image converging beams forming an image at a finite distance. Image-tilting SMIs are a more complicated that source doubling ones and suffer from a major disadvantage: the data they measure need reordering to obtain the data cube usual in imaging spectrometry.

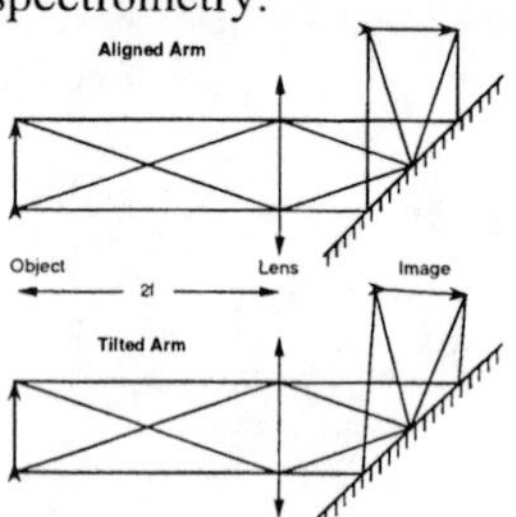

Figure 11: Image tilting principle.

CONCUSIONS

The superior throughput advantage of time-scanning imaging interferometers has been demonstrated and illustrated. It has also been shown that beam-tilting spatially modulated interferometers are severely limited in étendue and that alternate SMI designs are preferable, especially source-doubling SMIs.

REFERENCES

[1] Descour, M. R., 1996, proceedings of SPIE **2819**, 285.
[2] Aryamanya-Mugisha, H. *et al.*, 1985, Appl. Spect., **39**, 693.
[3] Barnes, T. H., 1985 Appl. Opt., **24**, 3702.
[4] Dierking, M. P. *et al.*, 1996, Appl. Opt., **35**, 84.
[5] Okamoto, T. *et al.*, 1984, Appl. Opt., **23**, 269.
[6] Connes, J., J., 1958, Phys. Radium, **19**, 197.
[7] Genest, J. *et al.*, for submission to Appl. Opt., 1997.
[8] Brault, J. W. 15th Advanced Course of the Swiss Society of Astronomy and Astronomy p. 1, 1985.
[9] Beer, R., Remote Sensing by Fourier Transform Spectrometry. Wiley, New-York, 1992.
[10] Genest J. *et al.*, Internal Bomem Report, 1997.
[11] Horton, R. F., 1996, proceedings of SPIE **2819**, 300.
[12] Horton, R. F. *et al*, 1997, proceedings of SPIE **3118**.

Design and Applications of a High-Throughput, Multi-Range FT-IR Microscope

[1]N. T. Kawai, [1]R. S. Jackson, [2]M. Jackson, and [2]H. Mantsch

[1]Bruker Optics, 19 Fortune Drive, Billerica, Massachusetts 01821
[2]Institute for Biodiagnostics, National Research Council of Canada, Winnipeg, Manitoba, R3B 1Y6

A new microscope design is discussed which incorporates viewing and measurement flexibility. Designed to accommodate dual detectors, the microscope can be optimized for measurements in multiple spectral ranges. When equipped with MCT and InSb detectors and an automated stage, chemical images across the near-IR and mid-IR ranges can be obtained. Measurement areas can be defined on the video image, and the resulting chemical images can be directly overlaid to provide an exact correspondence between areas of visual and chemical interest. The sensitivity and applicability of this technique is demonstrated by the measurement of biological tissue samples.

INTRODUCTION

The analysis of small samples by FT-IR microscopy has been exploited in the mid-infrared region (4,000 - 400 cm^{-1}) over the last ten years (1). Equipped with sensitive MCT detectors and reflective optics, commercial FT-IR microscopes combine visual and infrared optical paths to enable the location and measurement of samples down to the diffraction limits. Transmission FT-IR microscopy has high sensitivity provided that pathlengths are small; samples are ideally between 10 and 20 µm in thickness. The resulting spectra are severely degraded by diffraction when the sample or defining aperture is below 10µm.

NEAR-INFRARED MICROSCOPY

Spectra measured in the near-infrared region (10,000 - 4,000 cm^{-1}) are more difficult to interpret since they exhibit broad overtone and combination bands whose assignment to specific functional groups is not as straightforward as those of mid-infrared fundamentals. Absorptivities of near-infrared bands are also much weaker, typically >10 times less than mid-infrared peaks. However, the weaker absorptions in the near-infrared region provide an advantage to the microspectroscopist since transmission spectra can be obtained of species 100 µm or even 1 mm in thickness, minimizing sample preparation. Another benefit of near-infrared is that smaller samples can be measured before reaching the limits of diffraction. In practical terms, 5 µm sample sizes may be achievable with near-infrared microscopy.

OPTIMIZED MICROSCOPE DESIGN

The addition of visualization, spectral range, and measurement capabilities to any optical system is only justified if the overall throughput of the basic optical path is not compromised. A new commercial microscope combines flexibility of viewing and measurement while providing excellent signal-to-noise ratios. Fully equipped, the microscope can incorporate: binocular eyepieces and two camera ports for visual imaging; transmission and reflection optics for both viewing and measurement; simultaneous visible and infrared capabilities; dual detector channels for multi-range measurements; and a motorized stage for automated mapping measurements (*i.e.* chemical imaging).

EXPERIMENTAL

Mid-/Near-Infrared Microscope Configuration

Infrared spectra were measured with an IR ScopeII attached to an EQUINOX 55 FT-IR spectrometer (all from Bruker). The main optics bench was equipped with a broadband KBr beamsplitter and mid- and near-infrared sources to effectively cover the range from 10,000 - 370 cm^{-1}. Both a mid-band MCT detector (cutoff *ca.* 600 cm^{-1}) and an InSb detector (10,000 - 1,200 cm^{-1}) were mounted in the microscope, with detector switching controlled through software. The microscope also utilized a motorized stage and video camera for systematic mapping measurements. In this configuration, the IR ScopeII was able to provide chemical images in both mid- and near-infrared regions without any manual changes.

CP430, *Fourier Transform Spectroscopy:* 11th International Conference
edited by J.A. de Haseth

Biological Tissue Samples

The samples under investigation were 10μm slices obtained from a rabbit liver. The subject was given a diet high in cholesterol which resulted in high bilirubin levels (*i.e.* jaundice). The tissue samples were supported on 2mm BaF_2 windows and measured in transmission mode.

RESULTS AND DISCUSSION

Correspondence Between Chemical and Visual Images

Figure 1 shows the visual image of a liver tissue sample overlaid with the chemical image of the amide I band (1650 cm^{-1}). The large, darker area corresponds to higher concentrations of protein spectral contribution. The image of the C=O band at 1734 cm^{-1} also correlates to the darker area, but shows higher intensity in the lighter-coloured region (further to the left) as well.

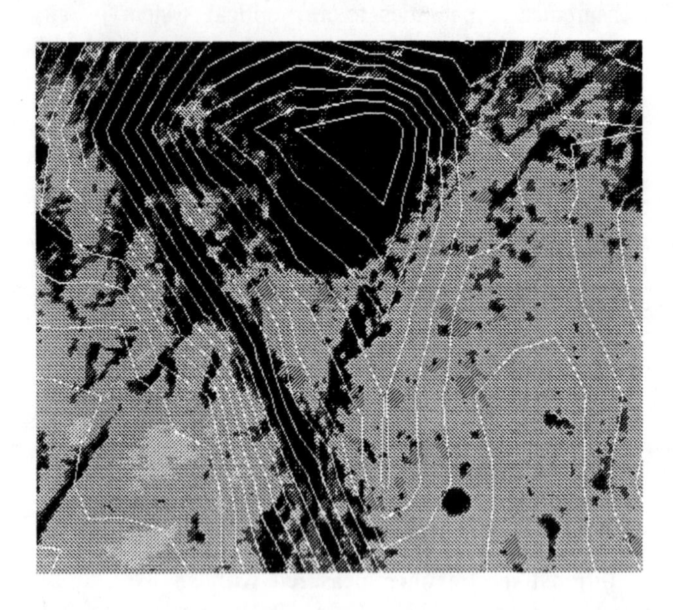

Figure 1. Visual and chemical image at 1650 cm^{-1} (amide I) of a rabbit liver tissue sample (280x205 μm). The infrared image was obtained with a 36 x 36 μm aperture.

Near-Infrared Chemical Imaging

A visual image and corresponding near-infrared chemical image of another tissue sample are shown in Figures 2 and 3. In this case, the chemical image at 4632 cm^{-1} (N-H stretch bend combination) corresponds to the yellow streak. For this sample, although the tissue slice is only 10 μm thick, the mid-infrared image of the amide I band shows peak intensities near 2.8 A.U., too strong for semi-quantitative analysis across the sample. However, many other features in the mid-infrared image are within 1.0 A.U. This last example effectively demonstrates the utility of dual-range chemical imaging on a single sample.

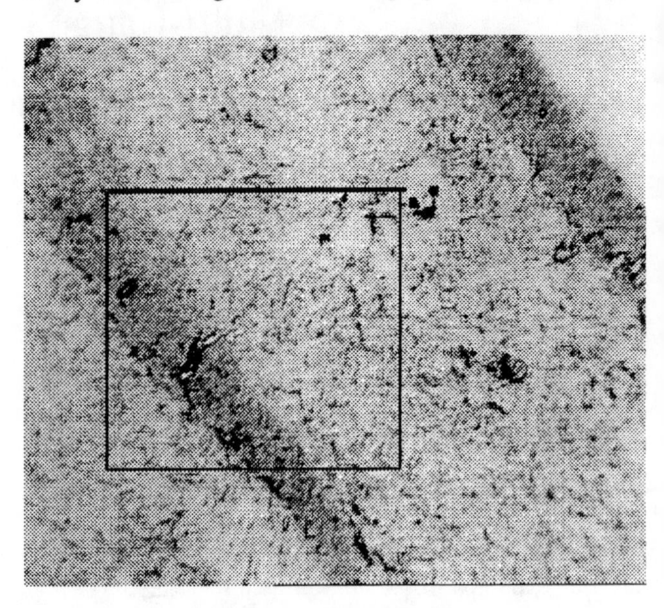

Figure 2. Visual image of a rabbit liver tissue sample showing measurement area (970 x 910 μm)..

Figure 3. Near-infrared chemical image at 4632 cm^{-1}, corresponding to the area indicated in Figure 2. The infrared image was obtained with a 47 x 42 μm aperture.

References

1. Katon, J. E., and Sommer, A. J., *Anal. Chem.*, **64**, 931A – 940A (1992).

Experimental Confirmation of Sample Area Definition in Infrared Microspectrometry

John A. Reffner, and Robert W. Hornlein

Spectra-Tech Inc. Shelton, CT. 06484

Using photolithography, thin polymer films with a controlled geometric pattern were produced as a test pattern for infrared microspectroscopy(IMS). Test samples were prepared on barium fluoride for transmission and gold mirrors for reflection measurements. The test pattern and methods for evaluating an IMS system are described. The purpose of this study is to provide a quantitative method for determining the ability of a microscope to define the infrared beam to a specific area and to determine limits of detection.

INTRODUCTION

In today's laboratory environment, it is important that the performance of analytical instruments be tested and validated. In the past, the testing of infrared microspectrometers was based solely upon measuring a signal-to-noise ratio (SNR) of the 100% line. This test is insufficient and misleading. Radiant energy from outside the sample area may contribute signal, but this spurious energy does not represent the sample. It is important to test how well the sample area is defined. In all current commercial infrared spectrometer systems, remote image plane masks are used to define specific areas for infrared spectral analysis. A special test specimen has been developed to evaluate the accuracy of the mask's ability to define the sample area.

CONSTRUCTION OF IMS RESOLUTION TEST SPECIMEN

IMS resolution test plates were constructed by photolithographic reproduction of a series geometrically shaped thin polymer films on both infrared transparent windows and gold mirrors. First the surface of the plate is coated with a thin, uniform layer of a photoresist polymer. The photoresist used is AZ 1518 Photoresist (Hoechst Celanese, Sommerville, New Jersey). This photoresist contains a dye (Basic Blue 81) to enhance the visibility of the photoresist film on the substrates. Then this surface is printed with geometrical patterns such as bands, discs and squares. In all cases the geometrical patterns are in several sizes and are of either the polymer or the background material (gold or barium fluoride). The width of these features varies from 10 μm to 100 μm. The various sized areas of polymer are used to test detection limits while the open areas surrounded by polymer are used to determine the sample accuracy definition.

EXPERIMENTAL

All spectra reported in this work were recorded using an IRμs infrared spectrometer (Nicolet Instrument Company). This system was used because it has remote image plane masks in positions before and after the sample. A variety of mask configurations were used to demonstrate variations in sample definition and detection limit. Unless specifically indicated, all spectra were collected at 4 wavenumber spectral resolution with the co-addition of 256 scans (2 min.).

EXPERIMENTS AND RESULTS

Measuring the accuracy of sample definition is best performed by recording the spectrum of a hole. A hole is defined as a clear area surrounded by the photoresist polymer film. The test plate provides rectangular, square and circular holes. A recommended test is to measure the spectrum of a 15 x 100 μm hole in the polymer film, masked to the same dimension. If the masking were perfectly accurate, then a featureless spectrum would be recorded. Because of diffraction effects and scatter (Schwarzschild-Villiger) (1) a remote image plane mask is not perfectly accurate and a residual spectrum will be recorded. For a quantitative evaluation of mask performance, the spectrum of a large area of the polymer film is recorded using the same remote image mask size and the residual spectrum is ratioed to this infinite area spectrum. The smaller the value of this ratio, the more accurate the definition of the specimen area.

CP430, *Fourier Transform Spectroscopy:* 11th International Conference
edited by J.A. de Haseth

This method is similar to hole counts used in AEM to evaluate energy dispersive x-ray systems.

An example illustrates the use of this method. Figure 1 contains data for the photoresist open slit test. Three spectra are reported; 1. A 15μm wide, open area, bound by photoresist recorded with dual masks. 2. The same hole recorded with a single mask (lower only). 3. A large area of photoresist. The mask size was 15 x 100 μm for all three spectral measurements. The spectrum of the large area of photoresist is used as a reference. The spectra of the holes show spectral features corresponding to the photoresist reference. There is a significant difference between the residual photoresist spectrum for the single and dual mask data. With a single mask the intensities are approximates 3 or more times as great as the corresponding features in the dual mask data. The ratios of residual to the reference spectra are reported in Table 1.

Figure 1 Photoresist Open Slit Test - 15x 100um

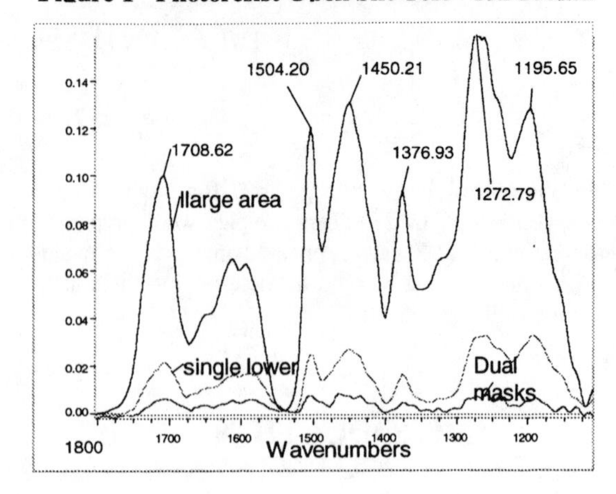

TABLE 1. Sample Definition Test Data for 15 X 100 μm Hole

Band (peak)		Photoresist	Lower Mask (single)		Dual Masks	
cm^{-1}	μm	Abs	Abs	%	Abs	%
1708	5.85	0.085	0.021	24.7	0.0069	8.1
1504	6.65	0.103	0.024	23.3	0.0074	7.2
1450	6.90	0.113	0.026	23.0	0.0079	7.0
1268	7.89	0.133	0.028	21.1	0.0065	4.9
1199	8.34	0.118	0.029	24.6	0.0074	6.3
972	10.29	0.030	0.024	80.0	0.0045	15.0

The evaluation of the detection limit is determined by recording spectra of small squares. The smallest square area that produces an analytical useful spectrum determines the detection limit. An analytically useful spectrum contains several peaks with a peak height to baseline noise ratio of 3:1 or greater. The squares of varying size are most useful for testing detection limit since rectangular adjustable masks are used.

An example of the evaluation of the detection limit for a 32x objective is illustrated by the spectra in Figure 2 and the data compiled in Table 2. These data show that for this objective on this spectrometer the detection limit in the reflection mode is slightly greater than 10 x 10 μm for the 1674 cm^{-1} absorption band. Examination of the data in Figure 1 clearly shows the spectral features of the photoresist in the 10 x 10 μm mask area. However, this data does not meet the 3:1 absorbance:noise requirement.

Figure 2. Detection limit test for 32x objective - Reflection mode

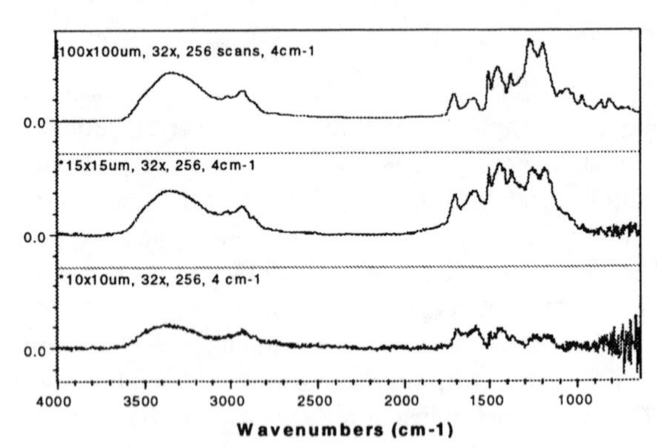

Table 2. Detection Limit Data

Area Size (μm)	Abs.	P-P Noise	Abs./P-P	Detection limit
10x10	0.11	0.038	2.89	> 10x10
15x15	0.23	0.011	20.91	< 15x15
100x100	0.134	0.00043	311.63	<< 100x100

CONCLUSION

The photoresist test pattern provides a convenient and direct means for evaluation and validation of infrared microspectrometer systems. Methods are proposed for testing both the ability of the system to define the sample area and to quantitatively assess a limit of detection.

REFERENCES

Piller, H., *Microscope Photometry,* Berlin, Springer-Verlag, 1977, Chapter 12

Factors Inducing and Correction of Photometric Error Introduced to FT-IR Spectrometers by a Nonlinear Detector Response

Robert L. Richardson, Husheng Yang, and Peter R. Griffiths

Department of Chemistry, University of Idaho, Moscow, ID 83844-2343

For strongly absorbing bands measured with a Fourier transform infrared (FT-IR) spectrometer, the effects of a nonlinear detector response must be eliminated before Beer's law linearity can be achieved. The effect of the non-linear response of mercury cadmium telluride (MCT) detectors has been evaluated on three commercial FT-IR spectrometers. The greater the photon flux, photon flux density, and the smaller the area of the detector on which the source image is focused, the greater are the effects of the nonlinearity. Detector nonlinearity is manifested by the generation of nonlinear Beer's law plots. A simple correction algorithm has been applied to Beer's law data acquired using both photoconductive MCT and pyroelectric detectors and found to work well.

INTRODUCTION

A nonlinear photometric response can be caused by the detector element and/or its associated hardware. Both software and hardware methods have been used to correct MCT detector nonlinearity. For software correction, a mathematical model is usually set up to simulate the nonlinear interferograms. Appropriate correction factors are calculated and applied to the measured interferogram before initiation of phase correction. Hardware correction can be achieved by using a linearizing circuit or using voltage-biased MCT detectors.

In the present work, the dependence of band intensities in spectra measured using MCT detector with a nonlinear response on photon flux, photon flux density and the area of the detector on which the source image is focused were studied. A simple method of correcting for the effect of a nonlinear detector response on both vapor- and liquid-phase FT-IR spectra is described. The validity of this method was verified on three spectrometers that utilize MCT detectors.

EXPERIMENTAL

FT-IR spectra of vapor phase mixtures of chloroform and/or dichloromethane in air free of CO_2 and H_2O (Balston, Lexington, MA) at ambient temperature and pressure were measured using a Perkin-Elmer (PE) Spectrum 2000 spectrometer (Beaconsfield, UK) equipped with a liquid nitrogen-cooled (non-factory installed) mid-band MCT (MB-MCT) detector (Graseby Infrared, Orlando, FL). Mixtures of dichloromethane and/or chloroform were generated using a static method by a procedure described previously[1]. The original data were recorded as unapodized interferograms. $CHCl_3$ and CH_2Cl_2 were chosen as analytes because the C-Cl stretching band of each molecule is broad, has a high absorptivity, and overlaps significantly with the absorption band of the other analyte between 780 and 740 cm^{-1}.

Liquid-phase samples of *t*-butyl ether in CCl_4 solutions (Aldrich, Milwaukee, WI) were measured on three FT-IR spectrometers: a PE Spectrum 2000 (Beaconsfield, England), a Bio-Rad FTS-175C (Cambridge, MA), and a Bomem MB-100 (Quebec, Canada). *t*-Butyl ether was selected because its spectrum is dominated by two strong, broad absorption bands (at ~1120 and 2960 cm^{-1}). Ten solutions, ranging in concentration from 0.0072 to 0.14 (v/v), were prepared volumetrically and loaded into a CaF_2 liquid cell (International Crystal Laboratories, Garfield, NJ) with a pathlength of 107.2 μm. All spectra were computed from double-sided interferograms collected at a nominal resolution of 2.0 cm^{-1} (~90 seconds acquisition time) and computed using medium Norton-Beer apodization. All absorbance spectra of *t*-butyl ether were calculated using the single-beam spectrum of the empty CaF_2 cell as the reference, and uncompensated absorption features due to CCl_4 were eliminated by spectral subtraction. Manipulation of data for both vapor and liquid phase samples after collection of interferograms was performed using GRAMS/32 (Galactic Industries Corporation, Salem, NH) and MATLAB (The MathWorks, Inc., Natick, MA) software.

In addition, the transmittance of a sheet of 50-μm thick polystyrene was used to further quantitatively evaluate the photometric performance of the Perkin-Elmer Spectrum 2000 spectrometer used in this study.

CP430, *Fourier Transform Spectroscopy:* 11th International Conference
edited by J.A. de Haseth

RESULTS AND DISCUSSION

On the PE Spectrum 2000 spectrometer, the size of the image of the source at the detector can be reduced by reducing the diameter of the Jacquinot stop (J-stop), d_J, while maintaining an approximately constant photon flux density in the region of the detector element illuminated. On the other hand, the total photon flux density can be decreased uniformly while the diameter of the source image on the detector remains the same by reducing the diameter of the beamsplitter stop (B-stop), d_B.

A series of ratios of the centerburst intensity measured using the NB-MCT detector, I_{MCT}, to that measured under the same conditions with a DLATGS detector, I_{TGS}, were measured on the PE Spectrum 2000 spectrometer. In these measurements, d_B was increased as d_J was decreased so that the total photon flux, as estimated by centerburst signal measured with the DLATGS detector, was kept constant to within ±5%. The ratio I_{MCT}/I_{TGS} is plotted against d_J in Figure 1. The fact that I_{MCT}/I_{TGS} is not constant verifies the well known fact that the response of the MCT detector varies nonlinearly with photon flux density. In each curve, the photon flux density decreases as d_J is increased (and d_B is increased).

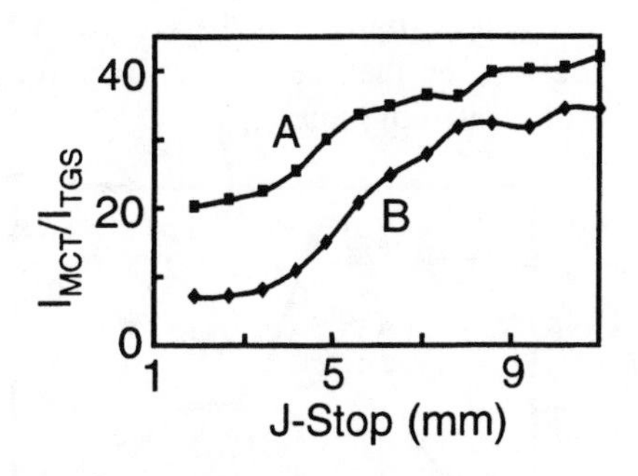

FIGURE 1. Variation of centerburst intensity ratio as a function of J-stop aperture measured on the PE Spectrum 2000 spectrometer. Curve A was measured with the beam only attenuated by the J and B stops. For curve B, the beam was further attenuated by a 12% screen.

One easily-measured indication of photometric accuracy is the transmittance measured at the peak of the 2919 cm^{-1} band in the spectrum of a 50-μm thick sheet of polystyrene. The transmittance is approximately 0.09% when measured with a pyroelectric detector. Extensive measurements in our laboratory have demonstrated this to be very close to the correct value. The values of the minimum transmittance of this band measured on the PE Spectrum 2000 equipped with a narrow-band (NB) MCT detector at three J-stop values are shown as a function of beam attenuation in Fig. 2. The highest transmittance (poorest photometric accuracy) was obtained with the smallest J-stop. This result was not unexpected because the PE linearizing circuit was calibrated when the detector was uniformly illuminated at d_J = 11.0 mm, corresponding to a nominal resolution of 4.0 cm^{-1}. In practice it appears to be a general finding that linearizing circuits should only be used when the J-stop and B-stop diameters are the same as they were when the circuit was calibrated.

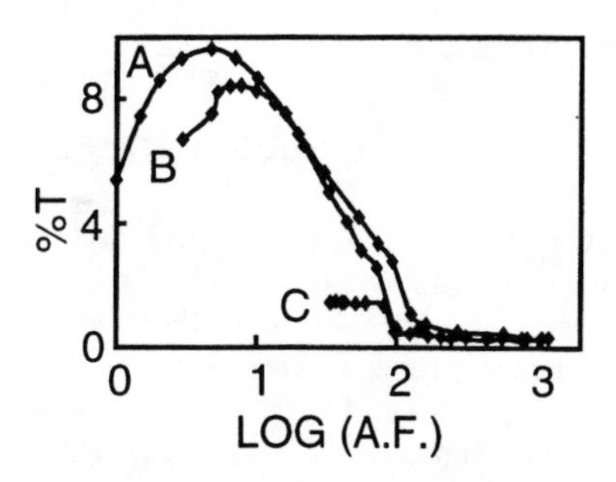

FIGURE 2. The minimum transmittance of the 2919-cm^{-1} band of a 50-μm thick polystyrene film as a function of incident signal intensity measured by the PE Spectrum 2000 spectrometer with a NBMCT detector. Beam attenuation was accomplished by reducing the B-stop aperture and/or inserting a 12%T screen. A.F. represents attenuation factor. Curves A, B, and C were measured at the J-stop value of 1.9, 3.4, and 7.8 mm, respectively.

Digital addition or subtraction of single-component spectra is frequently used for the quantitation and interpretation of multicomponent spectra[2]. Beer's law linearity is generally assumed for such manipulations, so that the digital addition of two or more overlapping spectra should accurately represent the band features and intensities of the physical mixture measured. However, we have found that detector nonlinearity may introduce significant photometric error in the measurement of reference spectra, particularly in the region of strongly absorbing bands.

An example is given in Figure 3, in which a measured absorbance spectrum of physically-mixed $CHCl_3$ and CH_2Cl_2, each at a concentration of 400 ppm, is plotted on the same scale as the spectrum generated by digitally adding the single-component spectra of each molecule at a concentration of 200 ppm and multiplying the resultant absorbance spectrum by a factor of two. It can be seen that the absorption features in the spectra of CH_2Cl_2 (where the maximum absorbance is less than 0.6 AU) correspond closely while the spectra in the region where $CHCl_3$ has its greatest absorption are significantly different. This result is caused by the effect of detector nonlinearity.

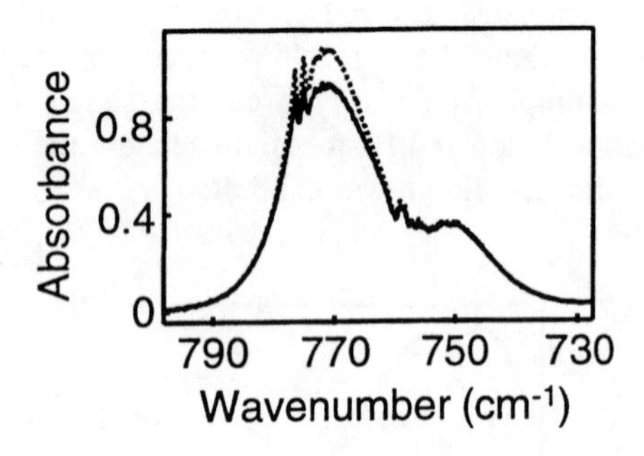

FIGURE 3. Comparison of the uncorrected spectra of physically mixed samples of 400 ppm chloroform and 400 ppm dichloromethane (solid line) with digitally added spectra of the individual compounds (broken line) in the region from 790 to 730 cm^{-1}. The digitally added spectrum was obtained by adding each analyte at a concentration of 200 ppm, then multiplying each data point by two.

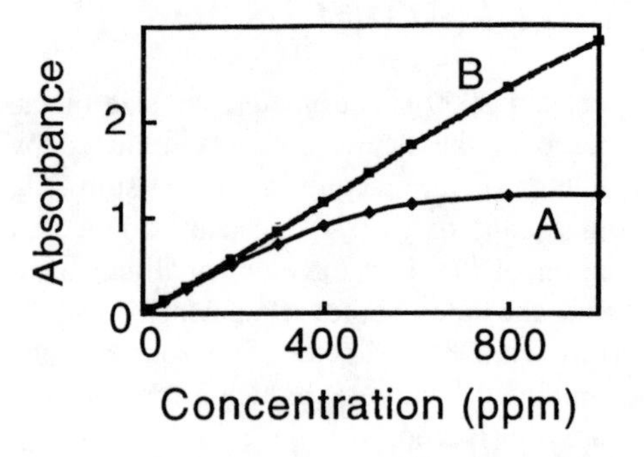

FIGURE 4. Beer's law plot for the strongly absorbing band of chloroform at 775.1 cm^{-1}. The correlation coefficient was 0.8699 before zero-offset correction (curve A) and 0.9998 after correction (curve B) with τ_{opt}=0.0563.

When the response of a detector of an FT-IR spectrometer, or its associated amplifier, is nonlinear, the single-beam spectrum usually shows a non-zero value at wavenumbers below the detector cut-off[3]. The non-zero baseline apparently extends into the spectrum, causing a photometric error that degrades the linearity of Beer's law plots. As we proposed recently[1], if this non-zero baseline continued across the entire spectrum, it could be corrected in a manner analogous to the typical stray light correction for dispersive spectrometers. If the true zero transmittance line is shifted to a transmittance τ by the effect of detector nonlinearity, a corrected transmittance, T_c, can be calculated using the following "zero-offset" equation:

$$T_c = (T_m - \tau)/(1 - \tau) \quad (1)$$

where T_m is the measured transmittance. The corrected absorbance, A_c, is then calculated as $-\log_{10}T_c$. The optimum value of τ, τ_{opt} is selected to give the highest value of R^2 in a Beer's law plot. This is usually done at the wavenumber of maximum absorption, where the effect of the non-zero baseline in the single-beam spectrum is greatest.

Beer's law plots for the wavelength of maximum absorptivity of the strongest band in the spectrum of chloroform are shown in Figure 4 both as measured (solid line), and after correction using τ_{opt} (broken line). This situation is analogous to a stray light of 0.056 transmittance (5.6%T) which, not surprisingly, is the value of τ_{opt} found to yield optimal Beer's law linearity for this band.

From a practical perspective, an estimate of τ_{opt} may be readily obtained by measuring the transmittance spectrum of any material giving rise to a totally absorbing band in the precise region of interest, such as the 700 cm^{-1} band in the spectrum of a 50-µm film of polystyrene, provided that the average transmittance over the rest of the spectrum is high enough that the photon flux, and hence the response non-linearity of the detector, is approximately the same. The improved similarity between the absorbance spectra of physically mixed and digitally added $CHCl_3$ and CH_2Cl_2 shown in Figure 3 after the "zero-offset" correction has been applied is illustrated in Figure 5.

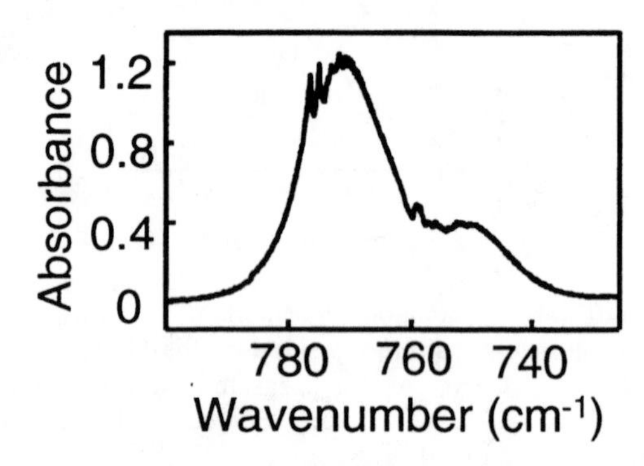

FIGURE 5. Corresponding data to Figure 3, after corrected for detector linearity using zero-offset method.

Beer's law plots for the 2960 cm^{-1} band of *t*-butyl ether measured on the Perkin-Elmer Spectrum 2000 with both DLATGS and NB-MCT detectors are shown in Figure 6. Reasonably good linearity was obtained for spectra measured using the DLATGS detector. Not unexpectedly, however, the absorbance values measured by the NB-MCT detector varied considerably with the diameter of the J-stop aperture used and substantial negative deviation from Beer's law behavior was observed. When the data shown in Fig. 6 are corrected using the zero-offset method, a substantial improvement of Beer's law linearity is obtained, as given by Figure 7.

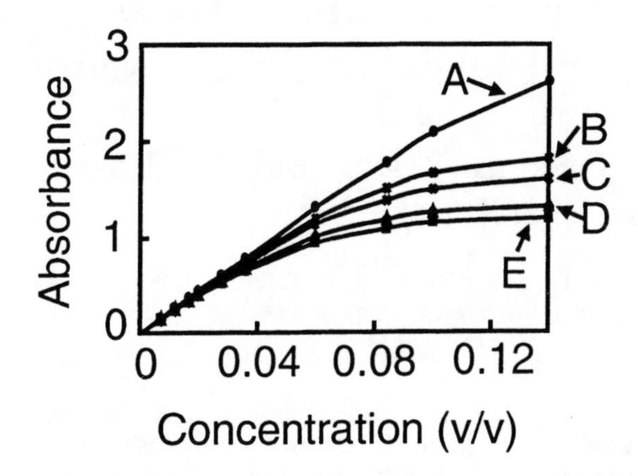

FIGURE 6. Beer's law plots for the 2960 cm^{-1} band of *t*-butyl ether in CCl_4 measured on the PE Spectrum 2000 spectrometer. Curve A was measured using the DLATGS detector with d_J=7.8 mm. Curve B (d_J=7.8 mm), C (d_J=6.3 mm), D (d_J=4.85 and 1.9 mm), and E (d_J=3.4 mm) was measured using the NBMCT detector.

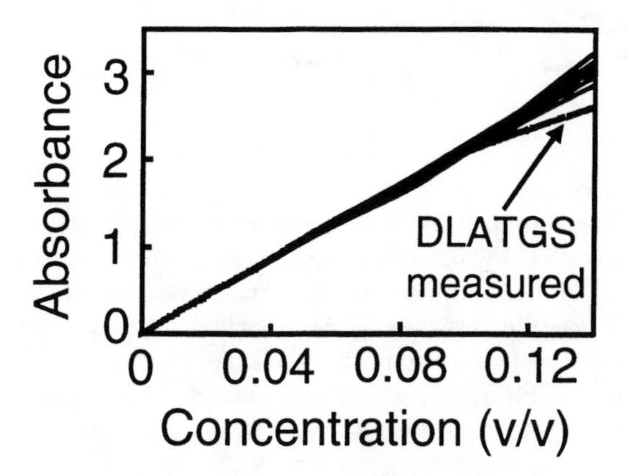

FIGURE 7. Corresponding data to Figure 6, after corrected for detector nonlinearity using zero-offset method. Uncorrected data measured with the DLATGS detector (broken line) is also shown.

The Beer's law plots for the 2960 cm^{-1} band of *t*-butyl ether were also measured by the Bio-Rad FTS-175C spectrometer with a DTGS and a WBMCT detector and a Bomem MB-100 spectrometer with a DTGS and a NBMCT detector. Beer's law plots measured by both DTGS detectors had fairly good linearity, but with the MCT detectors, significant deviations from Beer's law were observed. For all Beer's law plots measured by MCT detectors, a large improvement in photometric linearity is obtained using the zero-offset correction, with corrected absorbance values now very similar to those measured by the pyroelectric detectors. Pyroelectric detector linearity is also improved slightly, and in no instance was Beer's law linearity degraded.

The behavior of the C-O stretching band in the spectrum of *t*-butyl ether at 1120 cm^{-1} band was also evaluated. As with the stronger 2960-cm^{-1} band, correlation coefficients after the zero-offset correction was applied are typically between 0.999 to 0.9999 for all spectrometers and detectors under all conditions investigated.

CONCLUSIONS

The photometric accuracy of the MCT detectors has been determined to be strongly dependent upon both the intensity of incident signal as well as the diameter of the image for spectrometers equipped with linearizing circuits, whereas instruments equipped with pyroelectric detectors were relatively insensitive to these effects. The greater the photon flux, photon flux density, and the smaller the area of the detector on which the source image is focused, the greater are the effects of the MCT detector nonlinearity. A nonlinear response leads to nonlinear Beer's law plots on spectra measured on FT-IR spectrometers. The greatest effect is an offset of the zero line of the single-beam spectrum that can be corrected in a manner that is equivalent to correcting for stray light in a dispersive spectrometer. Using data acquired from three FT-IR spectrometers, we validated the zero-offset method as an accurate method of correcting for the effects of detector nonlinearity using Beer's law data.

ACKNOWLEDGMENTS

This work was in part supported by a grant from the Idaho National Engineering and Environmental Laboratory (INEEL) University Research Consortium. The INEEL is managed by Lockheed Martin Idaho Technologies Company for the U.S. Department of Energy, Idaho Operations Office under Contract No. DE-AC07-94ID13223. It was also partially supported by Cooperative Agreement 70NANB5H0046 from the National Institute for Standards and Technology.

REFERENCES

1. R. L. Richardson and P. R. Griffiths, "Evaluation of a System for Generating Quantitatively Accurate Vapor Phase Infrared Reference Spectra", *Appl. Spectrosc.*, in press (1997).
2. P. R. Griffiths and J. A. de Haseth, *Fourier Transform Infrared Spectroscopy*, Chemical Analysis Vol. 83, New York: John Wiley and Sons, Inc., 1986.
3. D. B. Chase, *Appl. Spectrosc.* **38**, 491 (1984).

The Determination of Enantiomeric Purity and Absolute Configuration by Vibrational Circular Dichroism Spectroscopy

Laurence A. Nafie[1], Fujin Long[1], Teresa B. Freedman[1], Henry Buijs[2], Allan Rilling[2], Jean-Rene Roy[2], and Rina K. Dukor[3]

[1]*Department of Chemistry, Syracuse University, Syracuse, NY, USA, 13244-4100*
[2]*Bomem Inc., 450 ave. St-Jean-Baptiste, Quebec, Quebec, CANADA, G2E 5S5*
[3]*Vysis Inc., Bioinformatics, Downers Grove, IL, USA, 60515*

There is an increasing need for new methods to determine percent enantiomeric excess (%ee) in chiral molecules. Four sets of determinations of %ee using Fourier transform infrared vibrational circular dichroism (FTIR-VCD) have been performed using three different instruments and several kinds of samples. These include measurements for neat α-pinene with two different FTIR spectrometers equipped with a mercury cadmium telluride (MCT) detector, measurements for lysine in H_2O using one of the MCT instruments, and measurements for 3-methylcyclohexanone in CCl_4 solution. We find that FT-VCD spectroscopy is capable of measuring %ee in the range of 1% or better for these samples using one to several hours of spectral collection time.

INTRODUCTION

The production of chiral pharmaceutical molecules that are optically pure is becoming increasingly important. Recent regulations from the Food and Drug Administration stress that chiral pharmaceuticals must be tested in optically pure form, or if the racemate is to be marketed, both enantiomers and the racemate must be tested separately. In order to insure that small amounts of the undesired enantiomer are not present in the final product, or anywhere along the process stream, new methods for the routine measurement of enantiomeric purity (enantiopurity) and absolute configuration are needed. Typically, enantiopurity is determined by separation of enantiomers using column chromatography, and absolute configuration is determined from x-ray crystallography. In practice, the development of a chiral column for a new compound can be costly in time and money, if it is successful at all. For absolute configuration, an impasse is reached if appropriate single crystals cannot be obtained.

Vibrational circular dichroism (VCD) provides alternatives to both of these problems (1). Percent enantiomeric excess (%ee) can be determined without separation or further derivatization if a sample of known optical purity is available without enantiomeric separation. Further, relative %ee can be determined for a set of samples even in the absence of a standard of known %ee. For absolute configuration, studies have recently been completed that demonstrate, through comparison of experiment to *ab initio* quantum mechanical calculations, that absolute configuration can be unequivocally established for molecules in the liquid or solution phase.

In recent years several studies have been published in which vibrational optical activity, either as VCD (2) or Raman optical activity (ROA) (1,3,4) have been used to determine enantiomeric purity in chiral samples. In this paper, we present new results demonstrating the capability of VCD to determine %ee.

EXPERIMENTAL

Measurements of %ee have been carried out with three different FT-VCD instruments. Two of these instruments are at Syracuse University and are adapted in-house with VCD sample benches (5,6) attached to commercially available FTIR spectrometers, a Nicolet Magna 850 (5) and a Bruker IFS-55 (6). The third instrument is located at Bomem, Inc. in Quebec, Canada and is a complete stand-alone commercially available VCD spectrometer, the Chiralir, available from Bomem and BioTools, Inc. The Nicolet and Chiralir instruments were operated in rapid-scan mode and were equipped with optics appropriate for mid-infrared measurements (800 to 2000 cm^{-1}) featuring ZnSe and BaF_2 optics and an MCT detector. The Bruker instrument was optimized for near infrared measurements (above 2000 cm^{-1}) and was operated in the step-scan mode with a tungsten source, CaF_2 optics and an InSb detector. Chiral samples and solvents were obtained from Aldrich and used without further purification.

RESULTS AND DISCUSSION

In this section, we present the results of several sets of measurements of %ee. The aim of these experiments is to determine an initial level of sensitivity of VCD to varying %ee for samples prepared with known enantiomeric dilution factors. In all cases, the commercially available

CP430, *Fourier Transform Spectroscopy:* 11th International Conference
edited by J.A. de Haseth

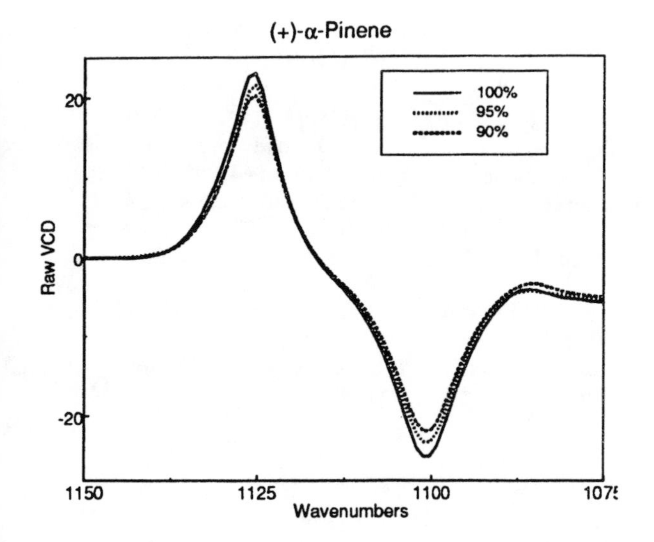

FIGURE 1. Comparison of raw VCD spectra for (+)-α-pinene, at 100% ee, 95% ee and 90% ee, 8 cm^{-1} resolution, 70 μm pathlength, 2 h collection.

single-enantiomer species was taken to be the 100% ee standard, even if the quoted commercial purity was less than that value by a percent or two. Samples were weighed on an electronic balance and dissolved in a known volume of solvent. Values of %ee were determined by mixing measured amounts of opposite pure enantiomers in the desired proportions. At this stage, no attempt to quantify the errors in the weighed %ee values was made, under the assumption that uncertainties in the VCD measurements in the range of 1%ee were greater.

In Fig. 1 we present VCD spectra using the Nicolet MCT instrument for neat (*R*)-(+)-α-pinene in the mid-infrared between 1150 and 1075 cm^{-1}. Samples of 100.0%, 94.4% and 90.3%ee were prepared, where the measurement time for each VCD spectrum was 2 h at 8 cm^{-1} resolution and a pathlength of 70 microns. The spectral region in Fig. 1 is expanded for clarity of presentation. The full region used for our VCD %ee measurements was 1350 to 950 cm^{-1}. Within the full region, we selected six major VCD peaks, some positive and some negative in intensity, and used approximately 7 data points near the vicinity of each peak maximum to generate a data set of 41 points. The statistical analysis of these points is listed in Table 1. The mean value of the VCD points was 88.7% for the 90.3% sample and 95.1% for the 94.4% sample. In these cases the departure of the VCD %ee from the weighed value is approximately 1%ee. The other statistical measures listed in the table carry their usual meanings.

In an alternate approach to the determination of %ee in neat α-pinene, we used the Chiralir MCT instrument and prepared samples at a wide range of weighed %ee, including a cluster of five above 90%ee. A partial least squares (PLS) analysis was used to determine the best value of %ee of the sample relative to the 100%ee reference spectrum over the range 1350 to 850 cm^{-1}. The measurement time was 50 min for each sample at 4 cm^{-1}

TABLE 1. Optical Purity Analysis by VCD for (R)-α-pinene

	Sample 1	Sample 2
Purity by weight (ee%)	90.3	94.4
Purity by VCD (ee%)	88.7	95.1
Variance	0.00032	0.00019
Standard Deviation	0.0179	0.0139
Standard Error	0.00280	0.0022
Total Observed. Data Points	41	41

with a 50 μm pathlength. The values of weighed and VCD pairs of %ee, in that order, are: (5.0, 4.8), (10.0, 8.2), (40.0, 41.6), (60.0, 60.2), (92.0, 92.5), (94.0, 93.3), (96.0, 95.8), (98.0, 98.7) and (99.0, 99.3). From these values it is clear that the VCD measurements are within 1%ee for all but two of the samples. This is consistent with the first study described above with the appearance of somewhat higher accuracy using roughly half the measurement time, twice the resolution and somewhat shorter pathlength. The application of PLS analysis to VCD %ee determinations looks promising and merits further study and comparison to the methodology of employing points in the vicinity of VCD peaks for %ee determinations.

As an example of %ee applied to a biological sample in aqueous solution, we report the results of VCD measurements of 1.7 M solutions of L-lysine in H_2O using 2 h of collection at 8 cm^{-1} resolution on the Nicolet MCT instrument with a 25 μm cell. The spectral region analyzed was from 1500 to 1200 cm^{-1}. Samples were prepared at 90.0 and 95.0 %ee and the VCD measured values, respectively, were 89.7 and 94.6 %ee. These results were closer in value to those reported above for this instrument for neat α-pinene samples.

Finally, we report the results of %ee VCD measurements in the CH stretching region for 3R-(+)-methylcyclohexanone using the Bruker-InSb instrument. The measurement conditions were 0.9M solutions at a pathlength of 50 μm using 16 cm^{-1} resolution and 3 h of collection time per sample. The values of %ee from the VCD measurements employed the principal peaks in the CH stretching region between 3000 and 2800 cm^{-1} using the methodology described above. The pairs of prepared and VCD-measured %ee values of each of five samples was (25.0, 23.6), (50.0, 50.1), (75.0, 74.5), (80.0, 81.3) and (90.0, 89.5). These variations are similar to those encountered in the mid-infrared although somewhat longer collection times were employed to compensate for somewhat lower signal to noise values in the spectra.

CONCLUSIONS

We have demonstrated in a variety of preliminary studies that VCD can measure %ee in the range of 1% for a variety of samples in different spectral regions without enantiomeric separation or chemical modification of the sample. The use of PLS as a straightforward method of statistical determination of %ee from a set of VCD spectra shows promise for future application and may give results that are more accurate than selective use of data points at principal VCD peaks. Future studies will entail more detailed considerations of sources of error in both the preparation of samples and VCD determinations, as well as improvements made possible by longer investments in VCD collection times per sample.

ACKNOWLEDGMENTS

Support for this work from the National Institutes of Health (GM-23567) is gratefully acknowledged.

REFERENCES

1. Nafie, L. A. *Appl. Spectrosc.* **50(5)**, 14A-26A (1996).
2. Spencer, K. M., Cianciosi, S. J., Baldwin, J. E., Freedman, T. B., and Nafie, L. A. *Appl. Spectrosc.* **44**, 235-238 (1990).
3. Spencer, K. M., Edwards, R. B., and Rauh, R. D. *Appl. Spectrosc.* **50**, 681-685 (1996).
4. Hecht, L., Phillips, A. L., and Barron, L. D. *J. Raman Spectrosc.* **26**, 727 (1995).
5. Long, F., Freedman, T. B., Hapanowicz, R. and Nafie, L. A. *Appl. Spectrosc.* **51**, 504-507 (1997).
6. Long, F., Freedman, T. B., Tague, T. J., and Nafie, L. A. *Appl. Spectrosc.* **51**, 508-511 (1997).

Achieving Accurate FTIR Measurements on High Performance Bandpass Filters

R Hunneman[1], R Sherwood[1], C Deeley[2] and R Spragg[2]

1. University of Reading, Infrared Multilayer Laboratory, Department of Cybernetics, Whiteknights, Reading, Berks RG6 2AY England, 2.Perkin-Elmer Ltd. Post Office Lane, Beaconsfield, Bucks HP9 1QA England

The sources of ordinate error in FTIR spectrometers are reviewed with reference to measuring small out-of-band features in the spectra of bandpass filters. Procedures for identifying instrumental artefacts are described. It is shown that features well below 0.01%T can be measured reliably.

INTRODUCTION

The optical industry has been slow to adopt Fourier Transform instrumentation for critical transmission measurements because of concerns about their ordinate accuracy. In this paper we look at the problems in characterizing a set of bandpass filters required for a satellite-borne instrument. Particular emphasis is placed on the measurement of weak features in the blocking regions of bandpass filters. The procedures we have adopted to identify any residual effects due to double modulation or non-linearities are reported. Out-of-band features can be reliably measured at levels well below 0.01%T.

PROBLEMS IN FT INSTRUMENTS

There are two main types of potential problem affecting these measurements. One is that of interreflections between surfaces in the optical train. Such problems are common to all optical spectrometers but their consequences in interferometers are different from those in dispersive instruments. The second is that of non-linearities in the detector and subsequent electronics. These problems are accentuated in FT instruments because of the high dynamic range resulting from the multiplex advantage. The consequences of the problems are seen both in inaccurate transmittance values and in the presence of spectral artefacts. The artefacts typically appear at multiples of the true wavenumber and so represent a significant problem in measuring the blocking region of bandpass filters.

In the example of Fig. 1 the filter passband starts at about 1500cm-1. There is apparent structure in the blocking region below 3000cm-1, Fig. 2, but most of this is due to the spectrometer rather than the sample. This can be demonstrated by measuring the spectrum in the presence of a sapphire window that is opaque below 1600cm-1.

Interreflections

Surfaces in the optical train are angled where practicable to deflect unwanted reflections. When interreflections involve the interferometer double modulation can occur, leading to artefacts outside the passband of the filter. The main problem is the reflection between the sample and interferometer. Masking can be used to intercept the reflections when tilting the sample is not acceptable. Doing this reduces the artefacts by about 80%, to below 0.2%T, Fig. 3. There are also reflections on the source side that can be removed by further masking. The resultant overall reduction in throughput of the spectrometer is little more than 50%. The residual features seen in Fig. 3 around 2400 to 2300cm-1 are genuine, also being observed when the passband is blocked using sapphire.

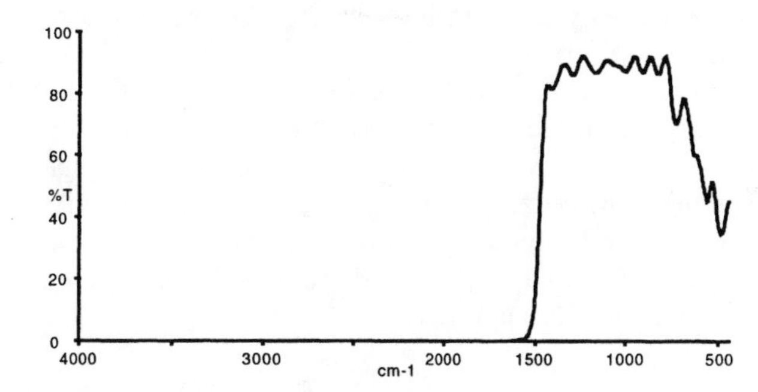

FIGURE 1. Spectrum of typical longpass filter.

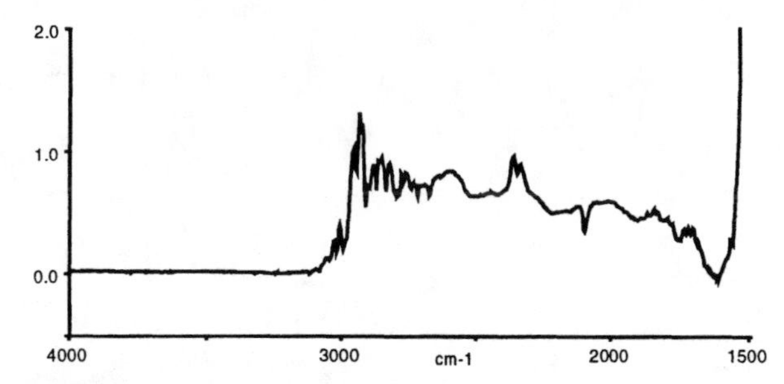

FIGURE 2. Spurious features observed in the blocking region.

CP430, *Fourier Transform Spectroscopy:* 11th International Conference
edited by J.A. de Haseth

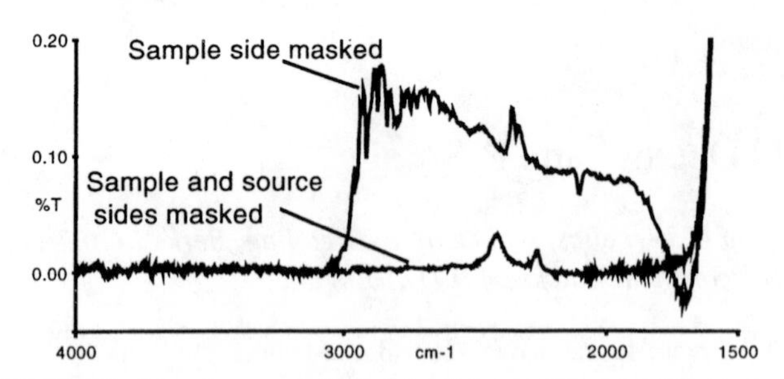

FIGURE 3. Spectra obtained with masking to block any interreflections involving the interferometer.

Residual Effects

With reflections eliminated lower level artefacts become apparent. The first of these is ripple caused by apodization. The commonly used Norton-Beer functions were originally identified as providing the best compromise between the amplitude of the first sidelobe and band broadening. However as the functions do not go to zero at the maximum OPD the sidelobe intensities decay rather slowly. In Fig. 3 the sidelobes resulting from stong Norton-Beer apodization are clearly visible around 2000cm-1. For these measurments a function providing more rapid decay of sidelobe intensity, such as Filler, may be more appropriate. Errors in the digitization of the interferogram can also introduce artefacts. If the number of bits used for digitization is inadequate the spectrum shows what looks like high frequency noise that is not reduced by signal averaging. The artefects caused by inaccuracies in the A/D conversion typically have broader structure. Figure 4. shows the effect of a single bit A/D conversion error on the spectrum of the filter in Fig. 1.

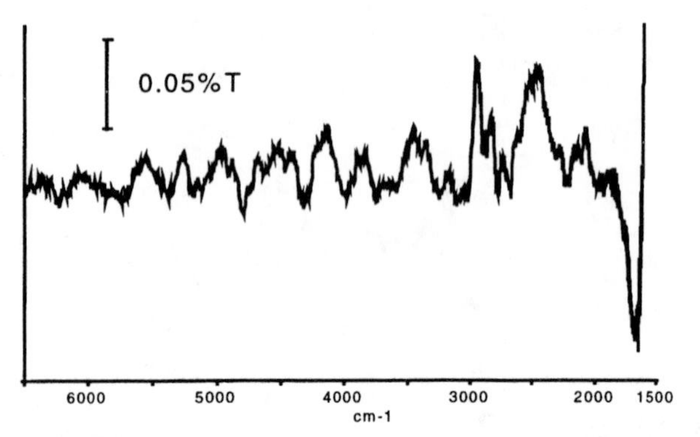

FIGURE 4. Out-of-band region measured with a single bit A/D conversion error.

TESTING PERFORMANCE

All the measurements are from Perkin-Elmer S2000 Optica instruments incorporating the features described earlier. Although very low noise levels are needed, DTGS detectors are used because of the extreme demands placed on linearity by these high dynamic range measurements. The overall ordinate accuracy is checked with samples of known refractive index, germanium and zinc selenide.

The spectrum from a high performance band pass filter is shown in Fig. 5. This was taken with 256 scans at 8cm-1 resolution in approximately 8 minutes. Transmission at 1596cm-1 is around 40%. The small features around 3200 and 4800cm-1 vanish when the region below 2000cm-1 is blocked with a glass slide, showing that these features are not genuine. The origin of the feature at 3200cm-1 could be optical or electronic. However tilting the sample has little effect, showing that interreflections with the sample are not significant.

The amplitude of these artefacts varies with the level of analog gain used in processing the detector signal. It is minimized by using the maximum gain, suggesting that they result from limitations in the digital signal processing.

This example demonstrates that any residual artefacts appear at or below the level of 0.01%T. These artefacts are found at multiples of the true wavenumbers. If it is possible to find a suitable material to block the passband measurements can be made at levels well below 0.01%T.

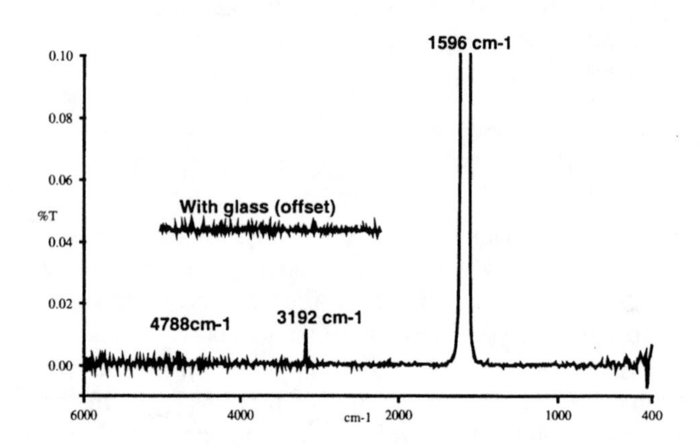

FIGURE 5. Spectrum of a high performance band pass filter showing the level of measurement artefacts.

THE HIRDLS INSTRUMENT

The High Resolution Dynamics Limb Sounder (HIRDLS) is a filter-based limb-viewing radiometer instrument scheduled to fly on the EOS CHEM satellite to be launched in 2002. The performance of the instrument is critically dependant on that of the filters used and the knowledge of their spectral characteristics. HIRDLS has 21 channels covering the spectral range from 6 to 17.5μm, with multilayer interference filters using Lead Telluride, Germanium, Zinc Selenide and Zinc Sulphide as layer materials. The channels are defined by a set of filters operating at 300K with bandwidths of between 1 and 8% (FWHM) with an additional set of bandpass filters operating at 65K at the cold focal plane which provide further blocking and to reduce cross-channel / height spectral contamination. Thus there are 42 sets of filter measurements needed, encompassing a transmission range of 0.01%T to nearly 100%.T, reflecting the required knowledge level of the filter blocking and that of the antireflection coating performance. Additionally in the case of the cold filters these measurements have to be made at 65K.

Measuring out-of-band Features

The filter used to illustrate low level measurements has a passband centered at about 580cm-1, Fig. 6. The out-of-band region of this filter has features with transmittance of up to 0.5%T. It should be pointed out that this leakage is not critical in the HIRDLS design because other elements in the system will remove it. It is clear from measurements incorporating a 2.5mm zinc sulfide window, that all the features seen at the 0.1% level in Fig. 7 are genuine. Closer examination in Fig. 8 shows artefacts around 1750 - 1600 and 1200 - 1100 cm-1 at the 0.01%T level. The artefact between 1200 and 1100 cm-1 overlaps with a genuine feature but its presence is evident when the spectra are overlaid. Another way of identifying this and other small artefacts is that their appearance is very sensitive to the phase calculation for the spectrum.

The observed spectrum is also compared with a calculated spectrum obtained using the FilmStar program[1] in Fig. 9. The agreement is extremely good at longer wavelengths. At shorter wavelengths the calculation is less satisfactory, with the calculated features displaced to shorter wavelengths than those observed. This is because the properties of lead telluride are less well modelled and the orders of interference are higher.

SUMMARY

These measurements demonstrate that spectral artefacts are typically smaller than 0.01%T even when the passband is not blocked.

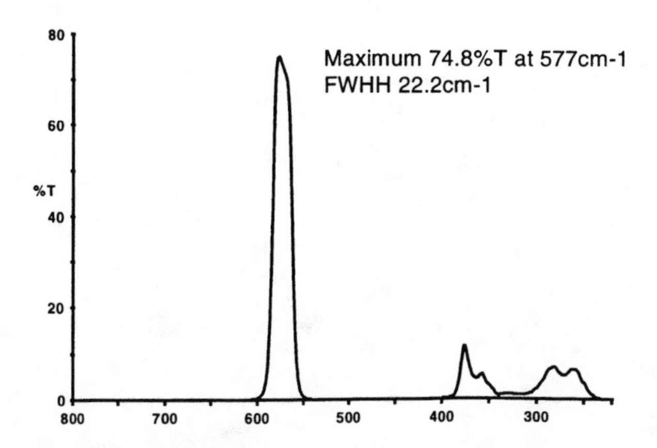

FIGURE 6. Passband of HIRDLS filter.

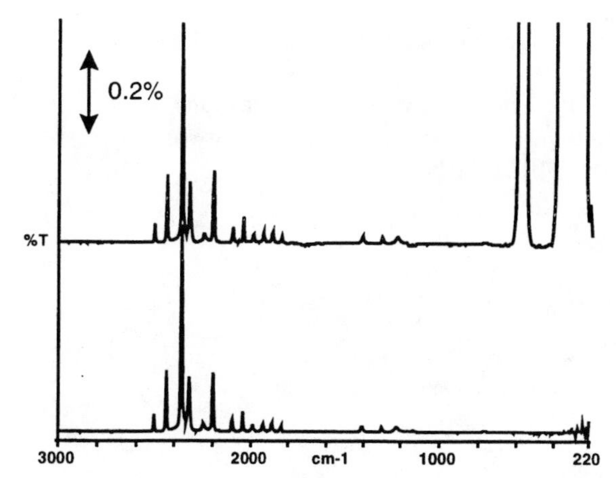

FIGURE 7. Out of band features of HIRDLS filter. Upper: full spectrum, lower: with ZnS blocking the passband.

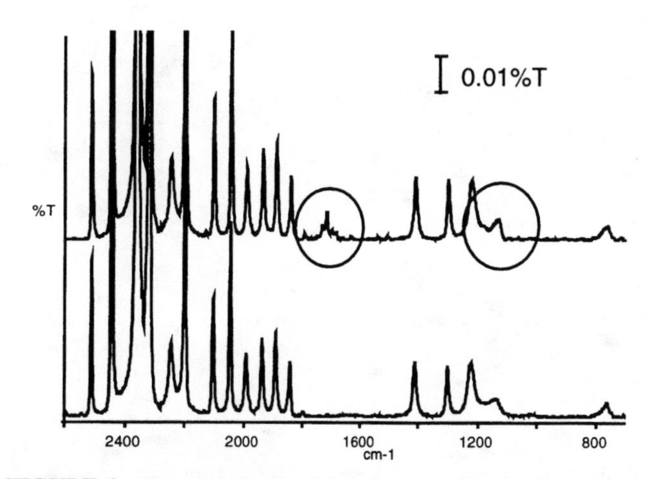

FIGURE 8. Exanded detail of the spectra of Fig.6. The upper spectrum shows artefacts near 1150 and 1750cm-1 that are absent in the lower spectrum with the passband blocked.

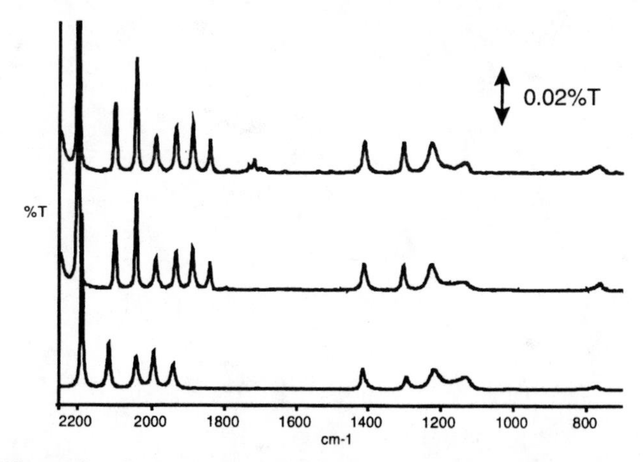

Figure 9. Comparison of observed and calculated out-of-band features. Top: full spectrum, center: observed with passband blocked, bottom: calculated.

ACKNOWLEDGEMENT

We are grateful to Dr Gary Hawkins of the Reading Infrared Multilayer Laboratory for the calculated spectrum of the HIRDLS filter.

REFERENCE

1. FTG Software Associates, Princeton NJ 08642

Observation of Picosecond Transient Raman Spectra by Asynchronous Fourier Transform Raman Spectroscopy

Akira Sakamoto[1], Hiromi Okamoto[2], and Mitsuo Tasumi[1]

[1] *Department of Chemistry, Faculty of Science, Saitama University, Urawa, Saitama 338, Japan*
[2] *Research Centre for Spectrochemistry, School of Science, The University of Tokyo, Bunkyo-ku, Tokyo 113, Japan*

Asynchronous Fourier transform (FT) Raman spectroscopy with 100 picosecond time resolution has been developed. An additional signal-processing assembly required for time-resolved and transient Raman measurements consists of a picosecond Nd:YLF laser system, a gate circuit, and a low-pass filter. This additional assembly can be attached to any conventional continuous-scan FT-Raman spectrophotometer. The principle of signal-processing employed in this method is almost the same as that of the asynchronous pulsed-laser-excited FT-Raman spectroscopy. This method does not require the synchronization between Raman excitation by probe laser pulses and the sampling by the A/D converter. Transient Raman spectra have been obtained from the first excited singlet state of anthracene in cyclohexane solution and photoexcited poly(*p*-phenylenevinylene) [$-(C_6H_4CH=CH)-_n$] by using 351 nm light (pulse width ≈ 70 ps) for photoexcitation and 1053 nm light (pulse width ≈ 100 ps) for Raman excitation.

INTRODUCTION

Since the development of Fourier transform (FT) Raman spectroscopy [1], near-infrared FT-Raman spectroscopy has become a powerful tool in a wide range of areas in science and technology in recent ten years [2]. In near-infrared FT-Raman spectroscopy, continuous-wave (CW) lasers have been mostly used for Raman excitation. Cutler *et al.* have used pulsed lasers in near-infrared FT-Raman spectroscopy by synchronizing the Raman excitation to the A/D converter sampling point [3]. In contrast with this synchronous method, we have developed an asynchronous pulsed-laser-excited FT-Raman spectrophotometer based on a conventional continuous-scan interferometer [4]. Our method does not require the synchronization between the Raman excitation and the sampling of the A/D converter. Therefore, the additional signal-processing assembly can be attached to any conventional FT-Raman spectrophotometer, without any further modifications in either hardware or software of the existing instrument. Asselin and Chase have recently carried out a thorough study on the effect of operating parameters (the repetition rate of the pulsed laser, the gate width of the gate circuit, the laser power, etc.) on the signal-to-noise (S/N) ratio in asynchronous pulsed-laser-excited FT-Raman spectroscopy [5]. In these studies [3-5], some advantages obtained by using pulsed Raman excitation and gate detection have been demonstrated.

Time-resolved and transient Raman spectroscopy has been successfully used to investigate the structure and dynamics of short-lived molecular species in condensed phase [6]. However, time-resolved and transient Raman measurements have been sometimes hampered by fluorescent background induced by the pump laser beam. Therefore, such measurements have been made either on samples with low fluorescence quantum yields or on samples whose excited-state Raman scattering generated by the probe laser beam is spectrally apart from the fluorescence induced by the pump laser beam. Since conventional near-infrared FT-Raman spectroscopy with CW excitation is now the most widely used method for overcoming the fluorescence problem in Raman spectroscopy, time-resolved FT-Raman spectroscopy with pulsed near-infrared excitation may be useful for obtaining Raman spectra from excited states of highly fluorescent samples. Recently, Jas *et al.* have reported picosecond time-resolved FT-Raman spectra of highly fluorescent 9,10-diphenylanthracene [7] and anthracene [8] in the first excited singlet state using a step-scan interferometer and picosecond laser pulses.

In the present study, we have developed asynchronous transient FT-Raman spectroscopy with picosecond time resolution. As an application of this method, we observe picosecond transient FT-Raman spectra of the first excited singlet state of anthracene and a transient FT-Raman spectrum from the photoexcited state of a photoconductive polymer, poly(*p*-phenylenevinylene) [$-(C_6H_4CH=CH)-_n$, PPV].

EXPERIMENTAL

1. Apparatus. A typical setup for the asynchronous transient FT-Raman measurement is schematically shown in Fig. 1. The FT-Raman spectrophotometer used in the present study was based on our original FT-Raman spectrophotometer (JEOL JIR-5500) reported previously [9]. Raman-scattered radiation collected with a 90° off-axis parabolic mirror was passed through three holographic notch filters (Kaiser) to reject Rayleigh scattering. A Ge

CP430, *Fourier Transform Spectroscopy:* 11th International Conference
edited by J.A. de Haseth

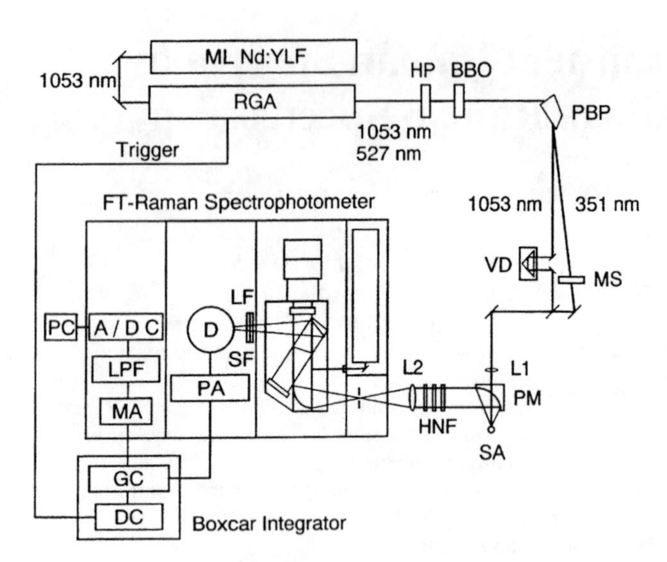

FIGURE 1. Schematic diagram of a typical setup for the asynchronous picosecond-transient FT-Raman measurement. **ML Nd:YLF**, CW mode-locked Nd:YLF laser ; **RGA**, CW Nd:YLF regenerative amplifier ; **HP**, half wavelength plate for 1053 nm ; **BBO**, β-barium borate crystal ; **PBP**, Pellin-Broca prism ; **VD**, variable optical delay line ; **MS**, mechanical shutter ; **L1** and **L2**, lens ; **PM**, parabolic mirror ; **SA**, sample ; **HNF**, holographic notch filter ; **LF**, long-wavelength-pass filter ; **SF**, short-wavelength-pass filter ; **D**, Ge detector ; **PA**, preamplifier ; **GC**, gate circuit ; **DC**, delay circuit ; **MA**, main amplifier ; **LPF**, low-pass filter ; **A/D C**, A/D converter ; **PC**, personal computer.

detector (North Coast EO-817P) was operated at liquid nitrogen temperature. A long-wavelength-pass filter and a short-wavelength-pass filter (Spectrogon) were placed before the detector to limit the observable spectral region. The third harmonic of a CW pumped Nd:YLF regenerative amplifier (Quantronix 4417RG, wavelength 351 nm, pulse width ≈ 70 ps, repetition rate 1 or 1.1 kHz) seeded by a CW mode-locked Nd:YLF laser (Quantronix 4216D) was used to excite the sample by the pump-probe technique. The 1053-nm (fundamental) line from the same regenerative amplifier (pulse width ≈ 100 ps, repetition rate 1 or 1.1 kHz) was used to probe the Raman scattering from the excited sample. After the probe beam was passed through a variable optical delay line, the pump and probe beams were overlapped collinearly and loosely focused on the sample. Additional signal-processing units required for transient Raman measurements were the gate circuit contained in a Boxcar integrator (SRS Model SR 250 & 280) and a low-pass filter. The gate width of the gate circuit was set at 200 ns. Interferograms were collected for 20 scans alternately with and without pump laser pulses, and each series of interferograms was accumulated for 1000 scans and averaged to obtain a satisfactory S/N ratio. The wavenumber resolution was 4 cm^{-1} for all the measurements performed in the present study.

2. Materials. Anthracene was purchased from the Tokyo Chemical Industry Co., Ltd. and used without further purification. HPLC-grade cyclohexane was used as the solvent. The sample solution (3×10^{-3} mol dm^{-3}) was circulated continuously through a quartz cell (pathlength 2 mm).

A poly(*p*-phenylenevinylene) (PPV) film deposited on a glass plate was prepared according to the method reported by Murase *et al* [10]. The PPV film on the glass plate was placed at the head of a cryostat (Oxford Instruments DN1754), and transient Raman measurements were made for such a film at 78 K.

SIGNAL PROCESSING

The principle of signal processing in asynchronous pulsed-laser-excited FT-Raman spectroscopy was described in our previous paper [4]. The concept of this method is based on asynchronous time-resolved FT-IR spectroscopy developed by Masutani *et al* [11].

In asynchronous FT-Raman spectroscopy, the smallest repetition frequency of the Raman excitation should be set by the requirement which is expressed by the Nyquist theorem as $\tau \leq 1/(2f_M) = 1/(4\,\upsilon\,\nu_M)$ (τ is the interval of the pump and probe laser pulses, f_M is the maximum modulation frequency, υ is the velocity of the moving mirror, and ν_M is the maximum wavenumber of the spectral region to be measured). Our present system does not satisfy this requirement strictly because of the higher repetition limit of the Pockels cell in a CW pumped Nd:YLF regenerative amplifier. However, we can practically handle this requirement by limiting the spectral region to be measured and choosing an appropriate combination of the moving-mirror velocity (υ) and the repetition frequency ($1/\tau$), in such a way that the original spectral component is not overlapped by any other higher-order components. This method for overcoming difficulties imposed by the Nyquist theorem is described in more detail in our paper [12].

RESULTS AND DISCUSSION

1. Anthracene. Transient FT-Raman spectra obtained from a 3×10^{-3} mol dm^{-3} solution of anthracene are shown in Fig. 2. The pumping and probing wavelengths were 351 and 1053 nm, respectively. The Raman bands due to anthracene in the ground-state (S_0) and the solvent (cyclohexane) were observed when the solution was irradiated by the probe laser only (not shown). With both the pump and probe lasers incident on the sample at a delay of 0 ps, several bands of the transient appeared and the intensities of the bands due to S_0 anthracene and the solvent decreased (Fig. 2a, asterisks in Fig. 2a indicate the bands due to S_0 anthracene and the solvent) compared with those obtained with the probe laser only. The reason for the decrease in the Raman intensities due to S_0 anthracene and the solvent will be discussed later. The weak

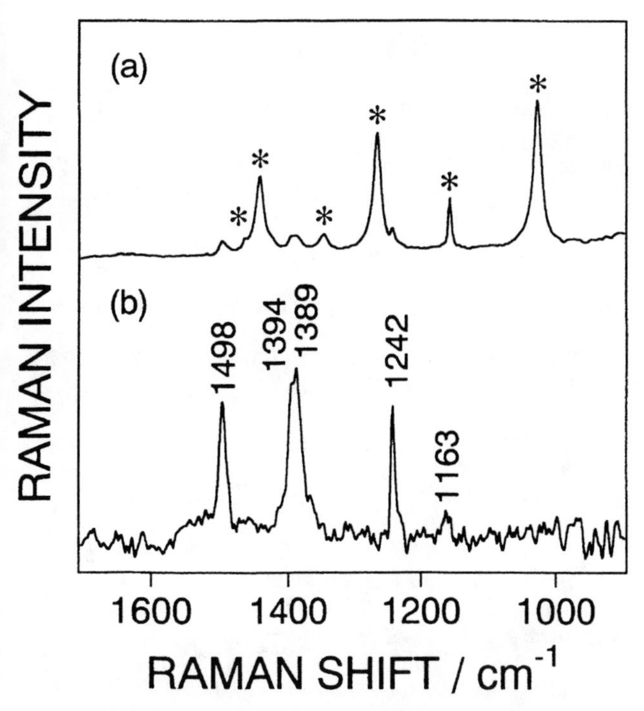

FIGURE 2. Transient FT-Raman spectra of anthracene in cyclohexane (3×10^{-3} mol dm^{-3}; pump laser 351 nm; probe laser 1053 nm). (**a**) Spectrum obtained with both the pump and probe lasers incident on the sample at a delay of 0 ps ; (**b**) difference spectrum.

fluorescence background of anthracene excited by the pump laser was also observed in Fig. 2a. Figure 2b shows the difference spectrum which was obtained by subtracting the two spectra observed with either one of the pump or probe lasers by appropriate factors from the spectrum observed with both the pump and probe lasers incident on the sample simultaneously (Fig. 2a). The factor for the pump only spectrum (fluorescence) was 1.0 and that for the probe only spectrum was 0.87. Five Raman bands of the transient were observed at 1498, 1394, 1389, 1242, and 1163 cm^{-1} in Fig. 2b.

We can assign these new bands to anthracene in the first excited singlet state (S_1) produced by the pump laser at 351 nm for the following reasons. (1) The lifetime of the S_1 anthracene is known to be about 5 ns [8]. Therefore, a sufficient number of anthracene molecules should be populated in the S_1 state by the picosecond ultraviolet excitation at a delay of 0 ps. (2) The frequencies of the observed Raman bands are in good agreement with the vibrational frequencies of anthracene in the S_1 state obtained from the absorption spectrum in a *n*-heptane matrix at 4 K (1503, 1389, 1247, and 1166 cm^{-1}) [13] and the fluorescence excitation spectrum of jet-cooled molecules (1501, 1389, 1380, 1247, and 1168 cm^{-1}) [14]. (3) In anthracene, the S_3 ($^1B_{3g}$) state has been reported to exist at about 8200 – 9100 cm^{-1} above the S_1 ($^1B_{1u}$) state [15]. Therefore, the probe laser at 1053 nm (9497 cm^{-1}) must be in resonance with the $S_3 \leftarrow S_1$ transition. The transient FT-Raman spectrum (Fig. 2b) is considered to be obtained by the resonance enhancement. This consideration rationalizes the observation that the intensities of the Raman bands due to S_0 anthracence and the solvent decreased in Fig. 2a (pump laser on) compared with those obtained with the probe laser only (pump laser off).

Recently, Jas *et al.* have reported the time-resolved FT-Raman spectra of S_1 anthracene using a step-scan interferometer with picosecond near-infrared laser pulses [8]. The band positions observed in Fig. 2b are in good agreement with the observation by Jas *et al.* Since Jas *et al.* have used a lock-in amplifier in order to selectively amplify the changes induced in the signal by pump pulses, they have simultaneously observed both positive Raman bands due to the transient and negative bands due to S_0 anthracene and the solvent. In contrast with their method, we obtained the Raman spectra only due to the transient (Fig. 2b) by adjusting the factor for subtraction.

2. Poly(*p*-phenylenevinylene). The transient FT-Raman spectrum of a poly(*p*-phenylenevinylene) (PPV) film on a glass plate at 78K is shown in Fig. 3a. This spectrum is obtained by subtracting two spectra observed with either one of the pump (351 nm) or probe (1053 nm) lasers from the spectrum observed with both the pump and probe lasers incident on the sample at a delay of 0 ps. Figure 3b shows FT-Raman spectrum of a neutral PPV film at 78K obtained with the probe laser (1053 nm) only.

PPV is a π-electron conjugated polymer with a nondegenerate ground state, which shows photoconductivities [16], nonlinear optical properties [17], electroluminescence [18], and high electrical conductivities upon doping [10]. Self-localized excitations (singlet excitons, triplet excitons, polarons, and bipolarons in the case of PPV) generated by photoexcitation or chemical doping are considered to play major roles in such interesting physical and chemical properties of π-electron conjugated polymers [19]. Therefore, it is important to characterize the self-localized excitations and to elucidate their dynamics for understanding such properties. Singlet and triplet excitons are the first excited singlet and triplet states, respectively, which are created in conjugated polymer chain upon photoexcitation. Polarons and bipolarons correspond, respectively, to radical ions and divalent ions, which are formed upon charge separation after photoexcitation or chemical doping. Such self-localized excitations extend over a certain number of repeating units and have structures different from the regular polymer chain.

In our previous studies, we demonstrated the usefulness of resonance Raman spectroscopy in the characterization of self-localized excitations (polarons and bipolarons) existing in the chemically doped PPV's [20-24]. In the present study, we extend such a study to photoexcited PPV. Since the picosecond photoinduced absorptions of PPV are observed in the region from near-infrared to infrared [25-28], picosecond transient Raman spectroscopy with near-infrared excitation is expected to give useful information

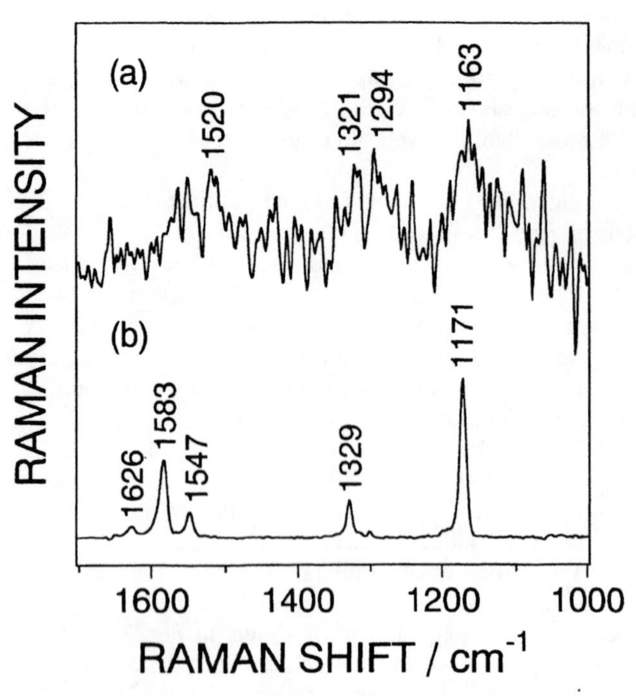

FIGURE 3. (**a**) Transient FT-Raman spectrum of a PPV film at 78 K (pump laser 351 nm; probe laser 1053 nm; delay time 0 ps) ; (**b**) FT-Raman spectrum of a neutral PPV film at 78 K.

on the self-localized excitations created in PPV upon photoexcitation.

The transient FT-Raman spectrum of the photoexcited PPV at 78K (Fig. 3a) is quite different from the Raman spectrum of neutral PPV (Fig. 3b). The Raman bands observed for photoexcited PPV (Fig. 3a) are lower in wavenumber than those of the corresponding bands of neutral PPV (Fig. 3b). In addition, the observed band widths of the photoexcited PPV (Fig. 3a) is broader than those of neutral PPV (Fig. 3b). These spectral features observed for photoexcited PPV are similar to those for chemically doped PPV [20-24], in which self-localized excitations (polarons and bipolarons) are clearly detected. Therefore, the observed transient FT-Raman spectrum (Fig. 3a) is attributable to self-localized excitations created in PPV upon photoexcitation. However, we cannot clarify at present the origin of the observed spectrum; in other words, which one of the singlet exciton, triplet exciton, polaron, and bipolaron is responsible for the observed spectrum is not clear. In the case of chemically doped PPV's, we utilized the resonance Raman spectra of the radical ions and divalent ions of the appropriate PPV oligomers for determining the types of the self-localized excitations in doped PPV's [20-24]. Therefore, observation of transient Raman spectra from the electronically excited singlet and triplet states of the oligomers must be useful for characterizing the observed transient Raman spectrum of photoexcited PPV.

REFERENCES

1. Hirschfeld, T. and Chase, B., *Appl. Spectrosc.* **40**, 133 (1986).
2. Hendra, P. J., Jones, C., and Warnes, G., *Fourier Transform Raman Spectroscopy: Instrumentation and Chemical Applications*, New York: Ellis Harwood, 1991.
3. Cutler, D. J. and Petty, C. J., *Spectrochim. Acta* **47A**, 1159 (1991).
4. Sakamoto, A., Furukawa, Y., Tasumi, M., and Masutani, K., *Appl. Spectrosc.* **47**, 1457 (1993).
5. Asselin, K. and Chase, B., *J. Mol. Struct.* **347**, 207 (1995).
6. Hamaguchi, H. and Gustafson, T. L., *Annu. Rev. Phys. Chem.* **45**, 593 (1994).
7. Jas, G. S., Wan, C., and Johnson, C. K., *Appl. Spectrosc.* **49**, 645 (1995).
8. Jas, G. S., Wan, C., Kuczera, K., and Johnson, C. K., *J. Phys. Chem.* **100**, 11857 (1996).
9. Furukawa, Y., Ohta, H., Sakamoto, A., and Tasumi, M., *Spectrochim. Acta* **47A**, 1367 (1991).
10. Murase, I., Ohnishi, T., Noguchi, T., and Hirooka, M., *Polym. Commun.* **25**, 327 (1984).
11. Masutani, K., Sugisawa, H., Yokota, A., Furukawa, Y., and Tasumi, M., *Appl. Spectrosc.* **46**, 560 (1992).
12. Sakamoto, A., Okamoto, H., and Tasumi, M., *Appl. Spectrosc.* in press.
13. Bree, A. V. and Katagiri, S., *J. Mol. Spectrosc.* **17**, 24 (1965).
14. Lambert, W. R., Felker, P. M., Syage, J. A., and Zewail, A. H., *J. Chem. Phys.* **81**, 2195 (1984).
15. Dick, B. and Hohlneicher, G., *Chem. Phys. Lett.* **83**, 615 (1981).
16. Tokito, S., Tsutsui, T., Tanaka, R., and Saito, S., *Jpn. J. Appl. Phys.* **25**, L680 (1986).
17. McBranch, D., Sinclair, M., Heeger, A. J., Patil, A. O., Shi, S., Askari, S., and Wudl, F., *Synth. Met.* **29**, E85 (1989).
18. Burroughes, J. H., Bradley, D. D. C., Brown, A. R., Marks, R. N., Mackay, K., Friend, R. H., Burns, P. L., and Holmes, A. B., *Nature* **347**, 539 (1990).
19. Heeger, A. J., Kivelson, S., Schrieffer, J. R., and Su, W. -P., *Rev. Mod. Phys.* **60**, 781 (1988).
20. Sakamoto, A., Furukawa, Y., and Tasumi, M., *J. Phys. Chem.* **96**, 3870 (1992).
21. Sakamoto, A., Furukawa, Y., and Tasumi, M., *Synth. Met.* **55**, 593 (1993).
22. Sakamoto, A., Furukawa, Y., and Tasumi, M., *J. Phys. Chem.* **98**, 4635 (1994).
23. Sakamoto, A., Furukawa, Y., and Tasumi, M., *J. Phys. Chem.* **101**, 1726 (1997).
24. Sakamoto, A., Furukawa, Y., Tasumi, M., Noguchi, T., and Ohnishi, T., *Synth. Met.* **69**, 439 (1995).
25. Sinclair, M. B., McBranch, D., Hagler, T. W., and Heeger, A. J., *Synth. Met.* **50**, 593 (1992).
26. Samuel, I. D. W., Raksi, F., Bradley, D. D. C., Friend, R. H., Burn, P. L., Holmes, A. B., Murata, H., Tsutsui, T., and Saito, S., *Synth. Met.* **55**, 15 (1993).
27. Hsu, J. W. P., Yan, M., Jedju, T. M., Rothberg, L. J., and Hsieh, B. R., *Phys. Rev. B* **49**, 712 (1994).
28. Leng, J. M., Jeglinski, S., Wei, X., Benner, R. E., Vardeny, Z. V., Guo, F., and Mazumdar, S., *Phys. Rev. Lett.* **72**, 156 (1994).

Chemical Mapping Using Two-Dimensional Hadamard Transform Raman Spectrometry

R. A. DeVerse, T. A. Mangold, R. M. Hammaker, and W. G. Fateley

Department of Chemistry, Kansas State University, Manhattan KS 66506-3701

A Raman chemical mapping investigation was conducted in the visible spectral region using an automated Spex 14018 double monochromator in concert with a moving two dimensional (2D) Hadamard encoding mask. The 2D Hadamard encoding mask combined with conventional Raman spectrometry was used to create chemical maps of heterogeneous liquid and solid samples. The 2D Hadamard encoding mask is based on the cyclic S_{255} matrix as a 15X17 array. It is mounted vertically and is moved from one encodement pattern to the next by an automated translation stage in concert with scans by the Spex monochromator. The moving 2D mechanical mask was used to spatially encode incident laser radiation or the Raman scattered radiation collected from a sample area illuminated by an argon ion laser operated at 514.5nm. The encoded radiation was decoded by a fast Hadamard transform that resulted in the assignment of a conventional Raman spectrum for each of the 255 "pixels" comprising the mask area. The Raman scattering intensity, plotted with respect to pixel coordinates, resulted in a chemical map of the illuminated sample area. The data demonstrates the utility of the moving 2D Hadamard encoding mask for Raman chemical mapping.

INTRODUCTION

Hadamard transform spectrometry (HTS) is a unique combination of dispersive and multiplexing spectrometries. As in a dispersive spectrometer, the radiation from a source is collected and separated into it's individual spectral resolution elements by a spectral separator and then is collected and focused for spatial presentation on a focal plane. Unlike a dispersive spectrometer, which possesses a single exit slit, the Hadamard transform (HT) spectrometer employs a multi-slit array (i.e. a mask) located on a focal plane. This arrangement allows for the simultaneous measurement of a multitude of spectral resolution elements at a single detector and produces a multiplexing spectrometer. To recover the spectrum of N resolution elements requires measurement of the detector response for N different encodements of (N+1)/2 open mask elements and (N-1)/2 closed mask elements. The primary data, recorded as a plot of detector response versus encodement number, is called an encodegram. Hadamard transformation of the encodegram yields the spectrum. Two choices for the Hadamard encoding mask are the moving(mechanical) mask and the stationary(electro-optic) mask. The moving mask for N resolution elements requires 2N-1 mask elements that are either open (transmittance of 1) or closed (transmittance of 0). The stationary mask for N resolution elements requires only N mask elements that are selected to be either transparent(transmittance of $T_t \leq 1$) or opaque (transmittance of $T_0 \geq 0$). The advantage and disadvantage of the moving mask are no optical transmission problems and potential moving parts problems, respectively. If the conventional one-dimensional(1D) Hadamard encoding mask were to be folded in some manner, the result would be the generation of a 2D Hadamard encoding mask. Similar to the 1D masks used for spectral multiplexing, 2D masks can be used for spatial multiplexing, For example the 1D mask represented by S_{15}: 00010 01101 01111 for one encodement converts to a 2D mask (3X5 array) represented for the corresponding encodement by:

0	0	0	1	0
0	1	1	0	1
0	1	1	1	1

BACKGROUND

For more than a decade we have utilized stationary Hadamard encoding masks based on liquid crystal technology in our application of Hadamard transform techniques in spectrometry, depth profiling and imaging, conveniently classified as one, two, three, or four dimensional spectrometry by the total number of spatial and spectral dimensions involved[1-3]. As a consequence of the nature of liquid crystal devices, investigations were restricted to the visible and near-infrared spectral regions. Upon a review of advances in translation device technology it was determined that the mask positioning reproducibility problems that impeded the initial

CP430, *Fourier Transform Spectroscopy:* 11th International Conference
edited by J.A. de Haseth

development of Hadamard transform techniques with moving (mechanical) Hadamard encoding masks could be solved. The use of a metal-etched mechanical Hadamard encoding mask with each mask element either completely open or completely closed makes imaging or mapping and spectrometry possible in any spectral region for which an appropriate spectral separator and detector are available.

EXPERIMENTAL

Three different experimental setups were investigated. The first two setups placed the sample in front of the mask and were only different in the lens system between the mask and the monochromator. It is important to note that in the first two setups, where the sample is in front of the encodement mask, the sample's Raman scattered radiation is encoded. A schematic representing the first two setups is shown in figure 1.

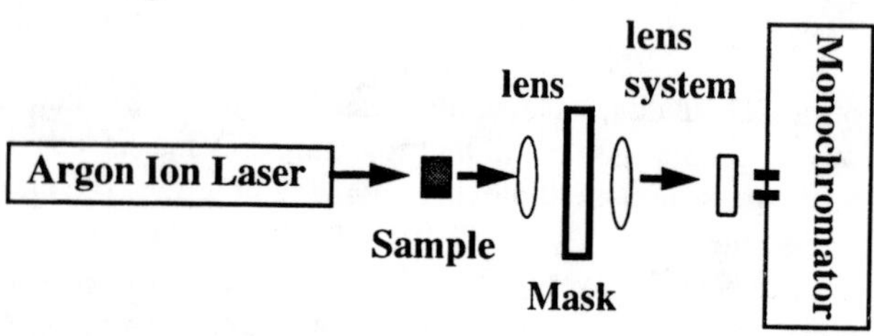

FIGURE 1. Representation of the first two setups investigated using pre-mask sample positioning where the Raman scattered radiation is encoded. The first two setups differed only in the post-mask lens system.

The setup used to collect the data presented here employed post-mask positioning of the sample. In this third configuration the incident radiation from the excitation laser is encoded and the resultant Raman scattered radiation is simply collected and measured as in a conventional Raman experiment. Figure 2 shows a schematic representing the third setup which was used to collect the data presented here. The benefits of the setup shown in figure 2 over the setup shown in figure 1 are simplified optics, simplified sample positioning and the potential employment of back scatter collection optics. The end result of encoding the incident radiation is better imaging or mapping capabilities and greater sensitivity.

The moving, etched-metal, 2D Hadamard encoding mask, which is based on the cyclic simplex S_{255} matrix as a 15X17 array, was mounted vertically on a precision Unidex 100 translation stage manufactured by Aerotech®. The precision translation stage was used to move the metal mask from one encodement to the next. Parts for mounting the translation device and mask were machined and assembled in-house.

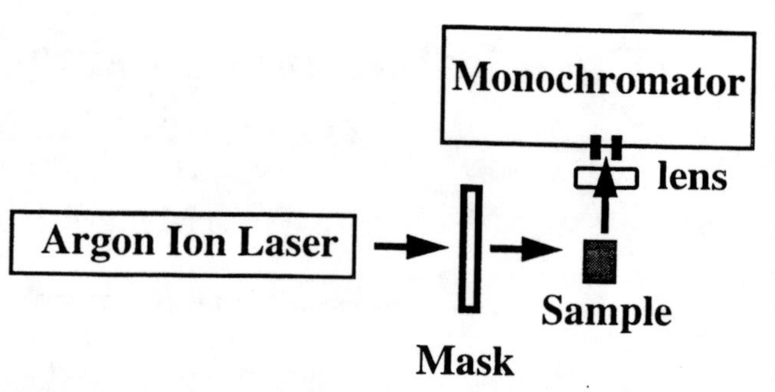

FIGURE 2. Third and final setup using post-mask sample positioning. The incident laser radiation is encoded by the mask.

RESULTS

In the course of investigating 2D Hadamard transform Raman chemical mapping, various liquid and solid samples were studied. Immiscible liquids nitrobenzene, water, and benzene in a cuvette were successfully mapped using this technique as well as heterogeneous solid samples of naphthalene with benzoic acid. As an example we present the mapping process of a capillary tube containing, in the illuminated volume, 81 femptograms of pararosaniline chloride in a silver colloid suspension. The success of this experiment with a moving mask demonstrates the sensitivity and chemical mapping potential of the Hadamard transform technique, as applied to Raman spectrometry. We also present an example of how a heterogeneous sample of immiscible liquids can be mapped using this technique. Figure 3 shows the enhanced Raman scattering spectrum of pararosaniline chloride illuminated by an argon ion laser operated at 514.5nm with no mask in place.

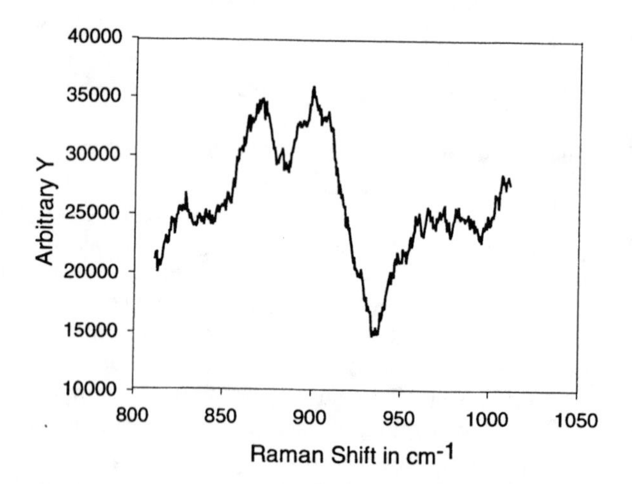

FIGURE 3. Enhanced Raman spectrum of 81 femptograms of pararosaniline chloride in a capillary tube. Raman shift of interest is 911cm^{-1} for mapping purposes.

A PC controlled Spex 14018 double monochromator was used to scan over the known Raman active pararosaniline chloride peak at 911cm^{-1}. The intensity at each data point of the scan was recorded by converting the output voltage of a hydroelectrically cooled photomultiplier tube connected to the exit slit of the monochromator to a digital value which was subsequently saved to a PC data file along with the corresponding wavenumber.

Figure 4 shows the sample position relative to mask position for mapping purposes. Figure 5 shows a full scale image or chemical map of the capillary tube containing, in the illuminated volume, 81 femptograms of pararosaniline chloride in a silver colloid solution.

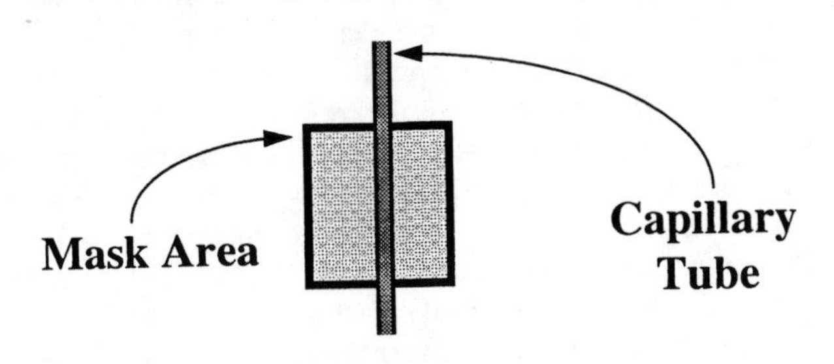

FIGURE 4. Schematic showing the relative position of the capillary tube containing pararosaniline chloride in a silver colloid solution and the encodement mask.

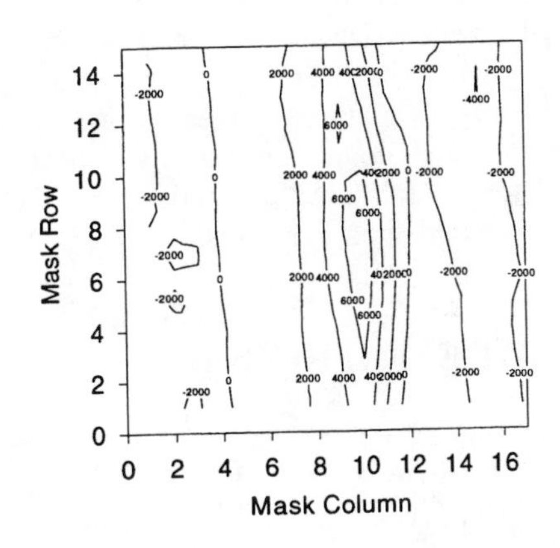

FIGURE 5. Full scale chemical map of the capillary tube containing, in the illuminated volume, 81 femptograms of pararosaniline chloride in a silver colloid solution. Raman scatter intensity at 911cm^{-1} v.s. pixel location. Numbers represent arbitrary intensity values.

The image data was obtained with the same monochromator and detector setup as the enhanced Raman spectrum and by employing the Hadamard encodement mask constructed out of a cyclic simplex S_{255} matrix in a 15X17 array. A single restricted scan of the Raman scattering was conducted for each of the 255 encodement patterns.

The intensity of each encodement pattern was collected and recorded in a data file by the PC. The mask was moved from one encodement to the next by a precision translation device. Once all of the encodement patterns had been scanned, the data was then transformed by a fast Hadamard transform into a file for each wavenumber scanned. For every wavenumber in which a data point was collected the Hadamard transform routine creates a file containing the intensity value for each "pixel" of the mask. This type of data file was used to create the chemical map shown in figure 5 by plotting the intensity value at the 911cm^{-1} data point for each "pixel" of the mask. The image is shown full scale with no noise floor removal. Figure 6 shows the same sample with the lower 40% of the full scale subtracted for clarity.

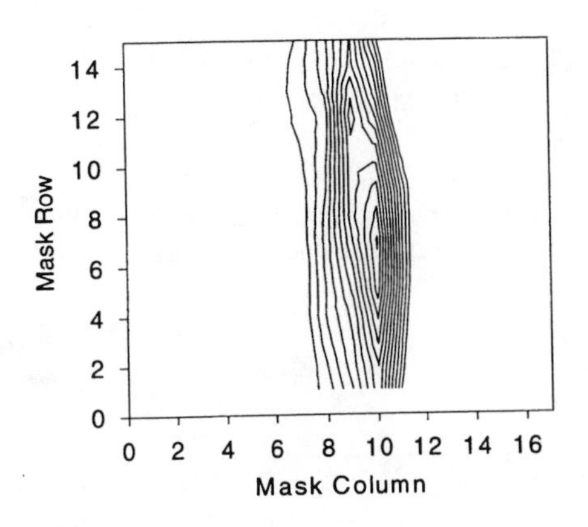

FIGURE 6. Scaled chemical map of the capillary tube containing, in the illuminated volume, 81 femptograms of pararosaniline chloride in a silver colloid solution. Image scaled for clarity by dropping the lower 40% of full scale.

Immiscable liquids were also mapped successfully using this technique and the scaled images are shown below. Only scaled images are presented here for reasons of space available and clarity.

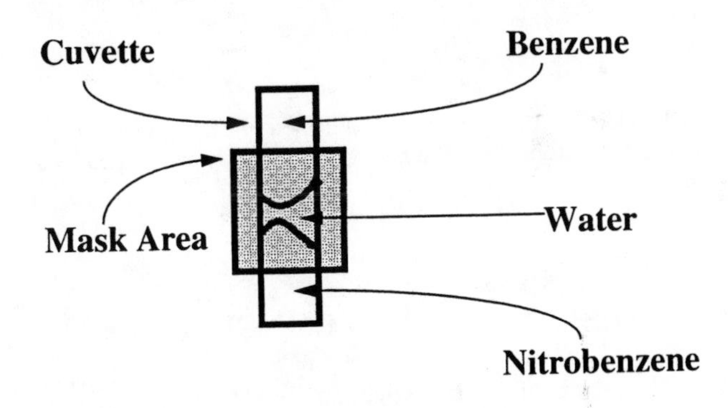

FIGURE 7. Immiscable liquids benzene, water, nitrobenzene in a cuvette. Schematic of sample position relative to encodement mask area.

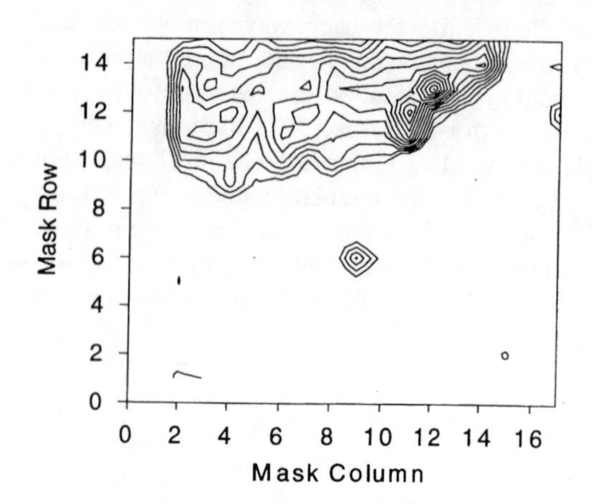

FIGURE 8. Chemical map of benzene in a cuvette with water and nitrobenzene. Raman scatter intensity at 996cm^{-1} v.s. pixel location.

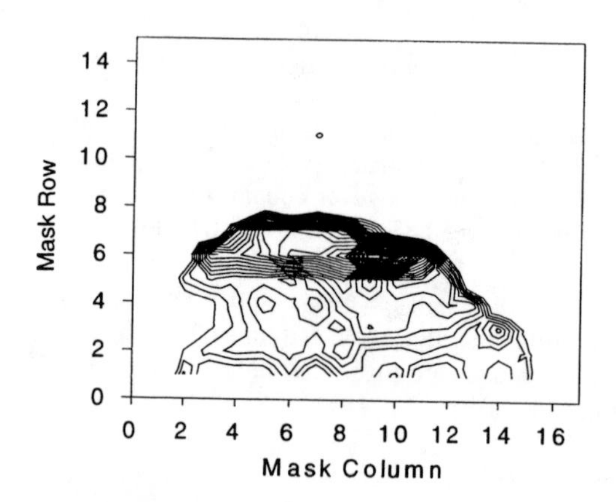

FIGURE 9. Chemical map of nitrobenzene in a cuvette with water and nitrobenzene. Raman scatter intensity at 1350cm^{-1} v.s. pixel location.

DISCUSSION AND CONCLUSIONS

The results of our investigation show that 2D Hadamard transform chemical mapping can be successful when used with conventional and enhanced Raman spectrometric techniques. The employment of a moving Hadamard encodement mask to accomplish chemical mapping over broad spectral regions is supported by this study in the visible portion of the spectrum. Chemical mapping in the mid- and near-IR spectral regions have also been successful using the etched-metal mechanical mask[4]. Technological advances in translation devices have made employing the moving mask in Hadamard transform chemical mapping possible by avoiding the reproducibility in positioning problems encountered in the past.

The conventional Raman spectrometric system can be used with little modification to perform mapping studies. The addition of an appropriate radiation encodement device (mask), data processing algorithms for the fast Hadamard transform, and laser beam expander to illuminate the mask area is all one needs to begin mapping studies of heterogeneous sample areas. As demonstrated by the data presented above, the location of the analyte and its distribution over the sampled area can be mapped using this technique. Provided that an analytes active Raman scattering peak is known and does not coincide precisely with other Raman active peaks from other constituents in the area of the sample being mapped, a chemical map can be successfully generated. Depending on the resolution of the spectral separator (i.e. monochromator) being used the Raman active peaks of the samples constituents can be very close in proximity without analyte spatial discrimination difficulties over the area being investigated.

ACKNOWLEDGMENTS

Some of the equipment used in this investigation was purchased by funds donated by the U.S. Department of Energy, Chemical Sciences Division, Grant Number DE-FG02-85ER13347. However, any opinions. findings, conclusions, or recommendations expressed herein are those of the authors and do not necessarily reflect the views of the DOE. Additional support from DOM Associates, Inc., Manhattan, Kansas, is gratefully acknowledged.

REFERENCES

1. Fateley, W. G., Hammaker, R. M., Paukstelis, J. V., Wright, S. L., Orr, E. A., Mortensen, A. N., and Latas, K. J., *Appl. Spectrosc.*,**47**, 1464-1470 (1993).

2. Fateley, W. G., Sobczynski, R., Paukstelis, J. V., Mortensen, A. N., Orr, E. A., and Hammaker, R. M., in *Spectrophotometry, Luminescence and Coulour; Science and Compliance, Volume 6 in Analytical Spectroscopy Library*, Burgess, C. and Jones, D. G., (eds.), Amsterdam, Elsevier, 1995, pp. 315-335.

3. Hammaker, R. M., Mortensen, A. N., Orr, E. A., Bellamy, M. K., Paukstelis, J. V., and Fateley, W. G., *J. Molec. Struct.*, **348**, 135-138 (1995).

4. Bellamy, M. K., Mortensen, A. N., Hammaker, R. M., and Fateley, W. G., *Appl. Spectrosc.*, **51**, 477-486 (1997).

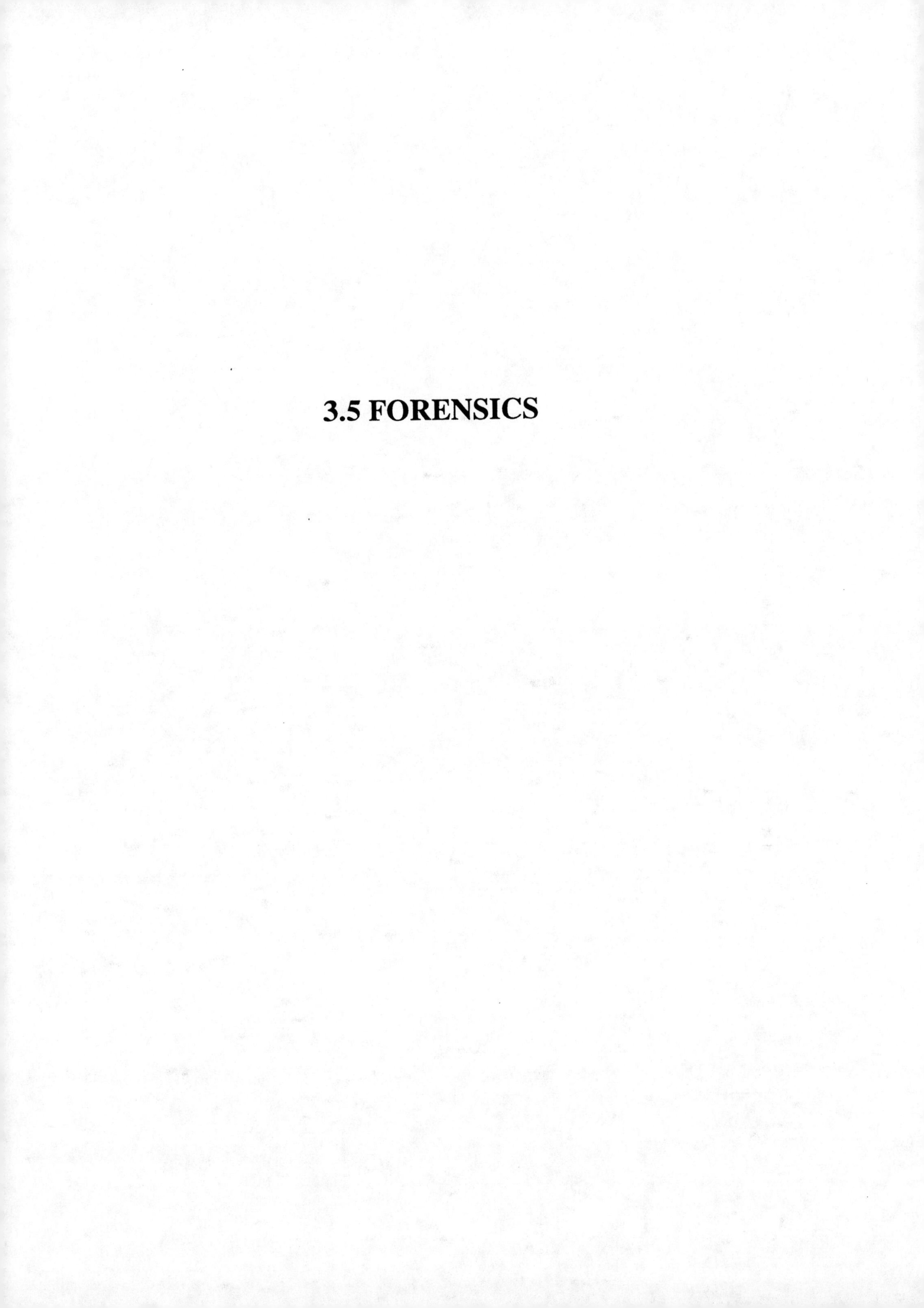

3.5 FORENSICS

GHB: Forensic Examination of a Dangerous Recreational Drug by FTIR Spectroscopy

J. P. Kindig, L. E. Ellis, T. W. Brueggemeyer, and R. D. Satzger

U.S. Food and Drug Administration, Forensic Chemistry Center, 1141 Central Parkway, Cincinnati, OH 45202, U.S.A.

Gamma-hydroxybutyric acid (GHB) is an illegal drug that has been abused for its intoxicating effects. However, GHB can also produce harmful physiological effects ranging from mild (nausea, drowsiness) to severe (coma, death). Because GHB is often produced by clandestine manufacture, its concentration, purity, and final form can be variable. Therefore, the analysis of suspected GHB samples using FTIR spectroscopy requires a variety of sample preparations and accessories, based on the sample matrix.

INTRODUCTION

Gamma-hydroxybutyric acid (GHB) occurs naturally in the human body as a metabolite of gamma-aminobutyric acid (GABA), a neurotransmitter (1). GHB acts as a depressant on the central nervous system (CNS) and was therefore investigated for use as an anesthetic. This usage was primarily abandoned because insufficient analgesia was produced and some patients experienced seizures (2). GHB has also been used medicinally to treat narcolepsy and in the treatment of alcohol and opiate withdrawal symptoms (3,4,5). In 1977, it was reported that GHB stimulated the release of human growth hormone, which led to the promotion of GHB as a steroid alternative for body-builders (6). The use of GHB as a hypnotic increased after the 1989 ban of L-tryptophan, a sleep-inducing nutritional supplement. GHB became popular for recreational use due to its intoxicating effects, which are similar to those produced by alcohol. However, the physiological effects of GHB are highly dose-sensitive, and overdose can result in symptoms as severe as coma, seizures, bradycardia, and respiratory depression. Additionally, the effects of GHB can be exacerbated by the concomitant use of other CNS depressants such as alcohol (7). In 1990, following an outbreak of 57 poisonings in nine states over a five-month period, the U. S. Food and Drug Administration (FDA) banned the production and sale of GHB (8). Currently, GHB is legally available in the U. S. only as an investigational new drug (IND) for sleep disorders.

Despite FDA's ban of GHB, its illicit manufacture and distribution have continued, and GHB has been promoted in the "night club" environment as a "legal" alternative to the federally scheduled hallucinogen "ecstasy" (MDMA: 3,4-methylenedioxymethamphetamine). There have also been reports of GHB being used as a "date rape" drug, in which GHB is added to a female's drink without her knowledge, rendering her semiconscious or unconscious (9). Meanwhile, reports have persisted of GHB-related illnesses, hospitalizations, and even deaths (10,11,12).

Procedures for the synthesis of GHB salts are widely available, particularly on the Internet, with the usual starting materials being gamma-butyrolactone (GBL) and sodium hydroxide (NaOH) (see Fig. 1). Because GHB is primarily synthesized in clandestine laboratories, the concentration, purity, and final form of the product can vary substantially. This intensifies public health concerns regarding GHB because those who procure the drug for recreational use may ingest product of unknown strength and composition, with potentially deadly consequences.

Most suspected GHB samples come into this laboratory as evidence seized by FDA's Office of Criminal Investigations (OCI). Samples have been encountered as solids and as solutions in various solvents, and they often contain mixtures of GHB and/or GBL with other components. Fourier transform infrared spectroscopy (FTIR) has been useful in the analysis of these samples, with a variety of sample preparations and accessories being required. FTIR results are presented from five cases in which GHB and GBL have been found in several different sample matrices.

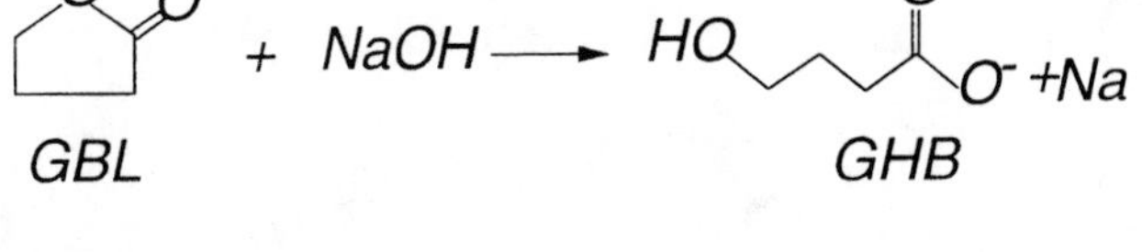

GBL + NaOH → GHB

FIGURE 1. GHB synthesis.

CP430, *Fourier Transform Spectroscopy:* 11th International Conference
edited by J.A. de Haseth

1-56396-746-4/98/$15.00

TABLE 1. Spectrometer Settings

	Main bench	Microscope
Resolution	4 cm^{-1}	4 cm^{-1}
# of scans	64	128
Apodization	Happ-Genzel	Happ-Genzel
Mirror velocity	0.6329 cm/s	1.8988 cm/s
Aperture	100	100
Gain	2	2

EXPERIMENTAL

FTIR spectra were measured using a Nicolet Magna-IR 550 spectrometer equipped with a Nic-Plan infrared microscope. Settings of the spectrometer are shown in Table 1. The software used for data collection and analysis was OMNIC, which is capable of performing a computerized spectral library search to aid in characterizing or identifying sample spectra. Solid samples were measured on a single face of a miniature diamond anvil cell (High Pressure Diamond Optics, Inc.) using the infrared microscope. Liquid samples were prepared in two ways. First, each was measured in the main bench of the spectrometer as a liquid film between two 25mm x 4mm barium fluoride (BaF_2) discs (Spectra-Tech). Second, liquid samples were deposited with a syringe on a BaF_2 disc and dried, and the solid residue was measured using the infrared microscope. Additional sample preparation was performed as necessary for certain samples, and is described where applicable.

A self-modeling curve resolution program was written to extract pure component spectra from a set of mixture spectra obtained through the microscope. The software was generated in this laboratory using MATLAB (version 4.2) and implemented on a 200 MHz Pentium-based microcomputer. The algorithm first uses principal components analysis to establish a spectral search space of suitable dimensionality. It then uses a fitness function to evaluate a grid of potential solutions found within this space.

GHB and GBL standards were purchased from Sigma Chemical Company.

RESULTS AND DISCUSSION

Case 1

The sample consisted of two items: one colorless liquid with a white solid precipitate, and a second colorless liquid. The white solid was identified as GHB. Both colorless liquids, when dried, left a residue that was identified as GHB. When analyzed by FTIR as liquid samples, one of the items appeared to be a solution of GHB in water, while the other was consistent with a solution of GHB in ethanol. This was

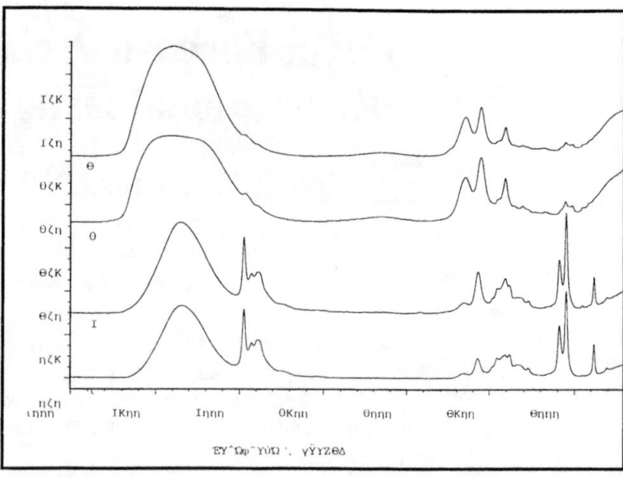

FIGURE 2. Spectra from Case 1. 1=Sample #1; 2=Control, GHB in water; 3=Sample #2; 4=Control, GHB in ethanol.

established by comparison with control solutions (see Fig. 2). This case demonstrates the value of FTIR analysis of suspected GHB solutions as liquids because information can be obtained regarding the solvent. A solution of GHB in ethanol is an unusual occurrence but, from a public health perspective, is of great concern due to the synergistic effects of GHB and alcohol on the body.

Case 2

The sample consisted of a colorless liquid. The FTIR spectrum of the liquid sample appeared to have spectral features of water, GBL, and GHB. For comparison, an "addition" spectrum was generated by combining the spectrum of a control solution of GHB in water with that of a control solution of GBL in water. The comparison with the addition spectrum (see Fig. 3) gave more certainty to the conclusion that both GHB and GBL were present in the aqueous solution. The fact that a GHB solution also had GBL present seems to indicate that, in the "manufacturing" process, insufficient base was used to complete the conversion of GBL to GHB.

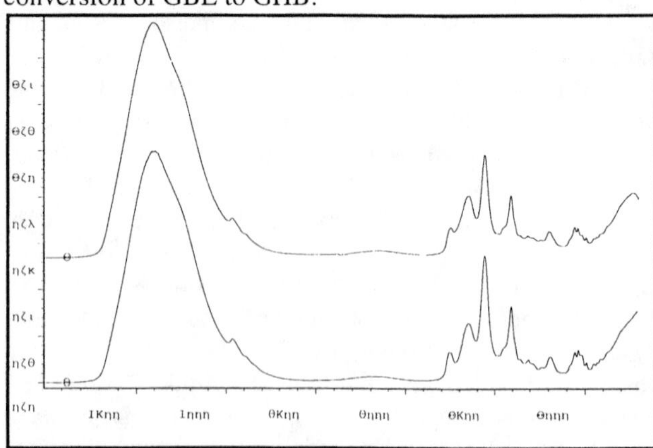

FIGURE 3. Spectra from Case 2. 1=Sample; 2=Control, addition spectrum of GHB in water and GBL in water.

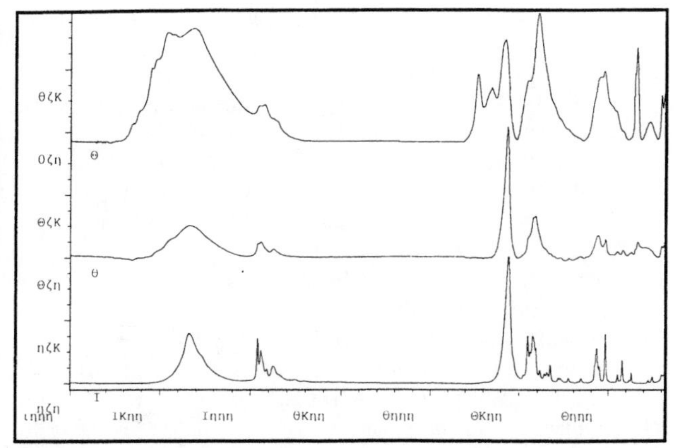

FIGURE 5. Spectra from Case 5. 1=Mixture spectrum; 2=Extracted component spectrum; 3=GHB standard.

Case 3

The first items of interest in the sample were two white solids which formed strongly basic solutions when dissolved in water. The FTIR spectra of the solids did not correspond well with the spectrum of GHB, but they did show some similarities including a strong band at the same frequency (~1560 cm^{-1}) as the major band in the spectrum of GHB. Furthermore, the sample spectra showed very little to no -OH character (lack of a broad band from ~3500-3000 cm^{-1}). Based primarily on these spectral characteristics, it was postulated that a GHB-like substance could be present in the sample which was deprotonated at both the acid *and* alcohol functionalities in the molecule (refer to Fig. 1). A compound such as this might be produced by treating GBL or GHB with an alkali metal or with an alkoxide salt, which acts as a stronger base than NaOH. This hypothesis was tested by dissolving each of the white solids in water, neutralizing the solutions with hydrochloric acid (HCl), and drying them on a BaF_2 disc. The resulting solid residues each had spectra that were consistent with that of GHB (see Fig. 4), supporting the idea that a GHB-like substance was present in the sample.

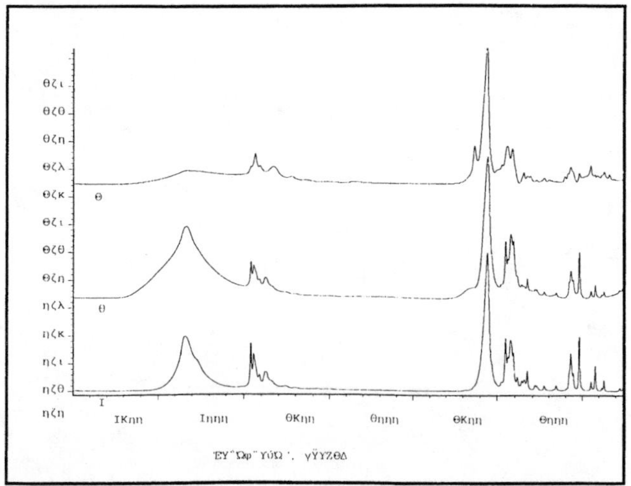

FIGURE 4. Spectra from Case 3. 1=Sample, solid; 2=Sample, residue after neutralization; 3=GHB standard.

Another item of interest in the sample contained paper towels that had been found in the suspect's garbage. To extract the paper towels for FTIR analysis, they were soaked in methanol, and the extract was removed and deposited on a BaF_2 disc. When dried, a residue was left that was identified as GHB. This is the only encounter with paper towels as a sample matrix.

Case 4

This sample was a GHB "kit" that was ordered by OCI from an illicit manufacturer. It consisted of a colorless liquid labeled to be GBL and a white solid labeled to be NaOH. The kit also included instructions for the preparation of GHB, which were followed in this laboratory to produce a colorless liquid product. The initial colorless liquid was identified as GBL, and the liquid product was consistent with a solution of GHB in water. When dried, the product left a residue that was identified as GHB.

A kit such as this is difficult to legally control because its two components, GBL and NaOH, can be lawfully possessed and distributed. Furthermore, the instructions included with the kit tout it as a chemistry experiment to demonstrate an exothermic reaction and state that the product is not intended for human consumption. However, the GHB produced from this kit is as dangerous as from any other source.

Case 5

The sample consisted of various off-white solids that had been scraped from surfaces in the kitchen of a suspected GHB manufacturer. Several FTIR spectra were obtained of the solids, but they all appeared to be of mixtures as none of them matched well with anything in the spectral libraries. However, the spectrum of one item in particular ("Item A") had a strong band at the same frequency (~1560 cm^{-1}) as the major band in the spectrum of GHB. Using the infrared microscope, seven spectra were measured from different regions of one preparation of Item A. These seven spectra were subjected to self-modeling curve resolution.

Self-modeling curve resolution is a multivariate technique which can be used to extract the pure component spectra from a set of mixture spectra (13). Success requires, first, that the concentrations of the chemical components vary independently of each other and, second, that the mixture spectra be linear combinations of the component spectra. It was shown that a set of FTIR spectra--each acquired with a microscope from a different region of a small sample--could be decomposed to the component spectra via a curve resolution technique (14). The approach capitalized on compositional heterogeneity found at the microscopic scale, with each mixture spectrum representing a slightly different blend of components.

Figure 5 shows one of the set of seven spectra of Item A.

A type of self-modeling curve resolution was used to mathematically extract a component spectrum from this mixture set. This extracted spectrum was observed to be similar in terms of band positions to the spectrum of a GHB standard (see Fig. 5).

CONCLUSION

GHB is a dangerous recreational drug, and its abuse does not appear to be subsiding, despite reports of its adverse health effects. The laboratory analysis of suspected GHB samples is a necessity in the prosecution of illegal manufacturers, and it is therefore essential that analytical methods be adapted to address complex sample matrices. The five cases outlined exemplify the variety of samples in which GHB has been found, and, more importantly, the approaches that have been taken using FTIR to obtain results. By implementing these techniques, FTIR spectroscopy can be very useful in the analysis of suspected GHB samples in many different matrices. The procedures used in these five cases will be applicable to future GHB cases, and may suggest means of treating other unusual samples.

ACKNOWLEDGMENTS

The authors wish to thank Mantai Z. Mesmer and Lora A. Lin, both of the FCC, who analyzed these samples using LC and/or LC/MS and provided preliminary information about their contents; Diane B. Fraser of the FCC, who provided invaluable reference materials on GHB; and Kirsten R. Stewart, graduate student at Michigan State University, who assisted with the FTIR analysis of some of the samples while interning at the FCC.

REFERENCES

1. Bessman, S. P., and Fishbein, W. N., *Nature* **200**, 1207-1208 (1963).
2. Dyer, J. E., *American Journal of Emergency Medicine* **9**, 321-324 (1991).
3. Mamelak, M., Scharf, M., and Woods, M., *Sleep* **9**, 285-289 (1986).
4. Gallimberti, L., Ferri, M., Ferrara, S. D., Falda, F., and Gessa, G. L., *Alcoholism: Clinical and Experimental Research* **16**, 673-676 (1992).
5. Gallimberti, L., Cibin, M., Pagnin, P., Sabbion, R., Pani, P. P., Piratsu, R., Ferrara, S. D., and Gessa, G. L., *Neuropsychopharmacology* **9**, 77-81 (1993).
6. Takahara, J., Yunoki, S., Yakushiji, W., Yamauchi, J., Yunane, Y., and Ofuji, T., *Journal of Clinical Endocrinology and Metabolism* **44**, 1014-1017 (1977).
7. Centers for Disease Control, *Morbidity and Mortality Weekly Report* **39**, 861-863 (1990).
8. Food and Drug Administration, "Gamma Hydroxybutyric Acid" [press release], Rockville, MD: Food and Drug Administration, November 8, 1990.
9. Andrews, K. M., "Historical Overview of GHB," presented in the GHB Workshop at the 49th Annual Meeting of the American Academy of Forensic Sciences, New York, NY, February 17-22, 1997.
10. Food and Drug Administration, "FDA Re-Issues Warning on GHB" [press release], Rockville, MD: Food and Drug Administration, February 18, 1997.
11. Ferrara, S.D., Tedeschi, L., Frison, G., and Rossi, A., *Journal of Forensic Sciences* **40**, 501-504 (1995).
12. Gorman, C., *Time*, September 30, 1996, p. 64.
13. Malinowski, E. R., *Factor Analysis in Chemistry*, 2nd Ed., New York: John Wiley and Sons, 1991.
14. Guilment, J., Markel, S., and Windig, W., *Applied Spectroscopy* **48**, 320-326 (1994).

RIBAVIRIN: The Analysis of a Polymorphic Substance by LC-MS and FTIR Spectroscopy

A. C. Machal, R. A. Flurer, T. W. Brueggemeyer, L. E. Ellis, R. D. Satzger, and K. R. Stewart

U.S. Food and Drug Administration, Forensic Chemistry Center (FCC), 1141 Central Parkway, Cincinnati, OH 45202, U.S.A.

The FTIR laboratory often has the task of identifying unknown pharmaceuticals. This case involves unknown capsules received at the Forensic Chemistry Center. Through extensive searching of pharmaceutical data bases, it was concluded that the capsules might contain ribavirin, which is classified as an anti-viral agent. Mass spectral analysis (LC-MS) concluded that the capsules contained ribavirin; however, the FTIR results did not agree with the mass spectral results. Additional experiments were performed and the results demonstrate the capabilities of FTIR to discern differences between polymorphic forms of a substance, such as ribavirin, when other techniques are unable to provide this information.

INTRODUCTION

A substance which may exist in two or more crystalline forms is referred to as polymorphic. Investigation of polymorphs has shown that approximately one-third of all organic compounds are polymorphic under normal pressure (1). Drug substances are frequently polymorphs with 67% of all steroids exhibiting multiple crystallinity (2).

The pharmaceutical industry is well aware of polymorphic substances, as different crystalline forms of a substance may possess different properties such as solubility, dissolution rate, melting point, and chemical reactivity. Furthermore, solubility of a polymorph may determine its bioavailability (1). In the case of a solid dosage form, tablet problems or even tablet failure may occur if the undesirable polymorph is used in the processing.

In the production of a specific pharmaceutical substance, the manufacturer is aware of the properties of that substance such as whether it is polymorphic. However, in a forensic setting, the pharmaceutical in question is frequently unknown, without markings to indicate the identities or properties of the substances it contains. If a pharmaceutical substance being examined is polymorphic, the analytical technique chosen is important for complete identification.

The methods routinely used in the examination of polymorphs are optical microscopy (hot-stage), infrared spectroscopy, thermal techniques such as DSC, and X-ray powder diffraction. Within FTIR spectroscopy, the methods of choice for examining polymorphs are diffuse reflectance (DRIFTS) and infrared microspectroscopy. Each of these techniques reduces the potential for interconversion between the polymorphs by reducing sample handling. Grinding is one sample preparation method which may cause polymorphic interconversion (1). With FTIR spectroscopy, the spectral differences observed are greater for some polymorphic substances than for others. In the case of ribavirin, the differences between the two polymorphs are evident.

Ribavirin, which is a synthetic nucleoside, is an anti-viral agent found to inhibit replication of a broad spectrum of viruses (3,4). The two crystalline forms of ribavirin have the same biological activity (3). However, whether these polymorphs have the same bioavailability remains to be determined.

EXPERIMENTAL

Case Study

The Forensic Chemistry Center (FCC) was asked to identify the contents of unmarked capsules. After extensive searching of pharmaceutical data bases, it was suspected these capsules contained ribavirin.

Melting Point Determination

Using a Thomas Melting Rate Programmer coupled with a Thomas Hoover Capillary Melting Point apparatus, the melting points of the suspect capsule contents and the

CP430, *Fourier Transform Spectroscopy:* 11th International Conference
edited by J.A. de Haseth

house reference standard were determined twice. The suspect sample melted at 170°C and 164°C, respectively. The standard melted at 177°C for both measurements.

LC-MS Analysis

Instrumentation:

A sample extract and the ribavirin house standard were analyzed by electrospray LC/MS/MS on a Finnigan MAT TSQ 700 mass spectrometer equipped with a Finnigan MAT API source and a Hewlett-Packard 1090L liquid chromatograph.

Suspect Sample and Standard Preparation:

The contents of one capsule were vortex mixed with 5 ml of DI water for 1 min. An aliquot was filtered through a 0.2 μm PTFE filter and diluted 1000 fold with DI water. The ribavirin standard was also prepared in DI water at a concentration of 10 μg/ml.

TABLE I. LC-MS Settings

LC Column	Hamilton PRP-X200,SCX, 2 mm x 150 mm, 10 μm particle size
Mobile Phase	0.1% (v/v) formic acid in water
Mobile Phase Flow Rate	200 μL/min
Column Oven Temperature	40° C
Injection Volume	10 μL for all
Heated Capillary Temp.	250° C
Sheath Gas Pressure	80 psi
AUX Gas Flow Rate	20 units
Spray Voltage	5 kV
Electron Multiplier Voltage	1200 V
Dynode Voltage	-15 kV
Precursor Ion m/z	245.0 (ribavirin MH^+)
Product Ion Scan Range	m/z 50 - m/z 250 in 1s
Collision Energy	20 eV
Target Gas Pressure	1.6 mTorr Ar

FTIR Analysis

Analyses using KBr Pellets and Diamond Cell

Instrumentation:

The FTIR spectra were measured using a Bio-Rad FTS-60A spectrometer interfaced with a Bio-Rad UMA 300A microscope. The software used for data collection and analysis was Bio-Rad 1.20.

TABLE II. Bio-Rad Spectrometer Settings

	Main Bench	Microscope
Resolution	4 cm^{-1}	4 cm^{-1}
# of Scans	64	256
Apodization	Triangular	Triangular
Scan Speed	5 KHz	20 KHz
Detector	DTGS	MCT
Aperture	max. res. 2 cm^{-1}	open
Gain	1	1

Suspect Sample and Standard Preparation:

A small amount of powder from one suspect capsule was prepared as a KBr pellet and the spectrum measured in the main sample compartment of the spectrometer.

Using a Nikon SMZ-U stereoscopic light microscope, the powder from the capsule was observed to consist of white clumps and small, white, needle-like crystals. Using a probe, a few of the needle-like crystals were removed from the matrix, pressed into a thin film using a diamond-anvil cell and the spectrum measured using a UMA 300A microscope. Similarly, a small amount of ribavirin house standard was prepared and analyzed as described above using both the KBr pellet and the diamond-anvil cell.

Analysis using a Melt Procedure

Instrumentation:

The FTIR spectra were measured using a Nicolet Magna-IR 550 spectrometer interfaced with a NicPlan microscope. The software used for data collection and analysis was Omnic 3.0. See TABLE III for instrument settings.

Suspect Sample and Standard Preparation:

Using a Nikon SMZ-U stereoscopic light microscope, a few of the needle-like crystals were removed from the powder matrix of the capsule using a probe, placed on a barium fluoride window, and the crystals melted using a hot plate. The melt was then analyzed using a NicPlan microscope. Similarly, a few crystals from the ribavirin house standard were prepared and analyzed as described above.

Additional Experiments

(The same instrumentation settings were used for the additional experiments as for the case study.)

A second ribavirin standard was received from a second source. Additional LC-MS and FTIR analyses were performed. The second ribavirin standard was prepared as

TABLE III. Nicolet Spectrometer Settings

	Microscope
Resolution	4 cm^{-1}
# of Scans	128
Apodization	Happ-Genzel
Scan Speed	1.8988 cm/sec
Detector	MCT/A
Gain	2

before for the LC-MS and FTIR analyses for the case study.

Due to possible polymorphic interconversion by using grinding in preparing pellets or pressure in using the diamond-anvil cell, both standards were measured using DRIFTS, and an unpressed crystal from each standard was measured using microspectroscopy. The diffuse reflectance data were obtained using the Bio-Rad spectrometer and the individual crystals measured using the NicPlan microscope.

In addition to the above examinations, the two ribavirin standards were subjected to various treatments to observe their behavior. Each standard was melted within a KBr pellet by increasing the temperature slowly using a GC oven; dissolved in various solvents, vortex mixed, spotted on a barium fluoride window and allowed to air dry; and dissolved in various solvents, vortex mixed, filtered using nylon 0.45 μm filters, spotted on a barium fluoride window and allowed to recrystallize at room temperature. Each melt, film, and crystal was measured using the NicPlan microscope.

RESULTS

Case Study

In the product ion spectrum of the standard, the ion at m/z 133 is produced by simple cleavage of the N-C bond indicated (Fig.1). The ion at m/z 113, which is the base peak, corresponds to cleavage of the N-C bond with simultaneous transfer of the proton from the carbon atom α to the N-C bond on the sugar subunit to the nitrogen atom. The ion at m/z 96 represents the loss of NH_3 from the amide to form the corresponding acylium ion. The product ion spectrum of the unknown capsule extract matches that of the ribavirin standard. Therefore, the mass spectrometric data are consistent with the presence of ribavirin in the unknown capsule. The data also demonstrate that mass spectrometric identification is not dependent on the crystalline form of the analyte. One would expect this since all memory of crystal structure is lost during sample dissolution.

Due to mass spectral results, the spectrum of the KBr pellet of the sample powder was compared to a spectrum of a KBr pellet of the ribavirin house standard. The spectra were not consistent with each other (Fig.2). The comparison of the melting points of the powder from the capsule to the melting point of the ribavirin house standard suggested that the powder from the capsule may be a mixture. Thus, a small number of the needle-like crystals were removed from the matrix and analyzed. The infrared absorbance spectrum of the needle-like crystals pressed into a film using a diamond-anvil cell was compared to a spectrum of the ribavirin house standard analyzed using the same method. The spectra were not consistent with each other (Fig.3). Since ribavirin exists in two polymorphic forms, further analysis was performed using a melt procedure in order to convert both the sample and standard to the same state for comparison.

The infrared absorbance spectrum of the melt of the needle-like crystals removed from the suspect sample matched the spectrum of the melt of the crystals from the ribavirin house standard (Fig.3). These spectra did not match with the original spectra of the sample or standard prepared as KBr pellets or by using the diamond-anvil cell. Because ribavirin can exist in two polymorphic forms, it was necessary to convert both the suspect sample and standard to the same state (non-crystalline) for comparison. According to the results of the melt analysis, the suspect sample is consistent with ribavirin.

Additional Experiments

After receiving a second ribavirin standard, additional LC-MS and FTIR analyses were performed. In the LC-MS experiment, the product ion spectrum of the second standard was very similar to the first two product ion spectra obtained in the case study (Fig.1).

The FTIR analyses confirmed the suspect capsule to contain ribavirin consistent with the second ribavirin standard, the other polymorph of ribavirin. The infrared absorbance spectra of the two reference standards were compared and found to be different from each other (Fig.4). In addition, the spectrum of each standard was added to a spectral library and then the spectra of the standards searched. Neither polymorph generated the other as a top fifteen match indicating that the spectral differences between the two are not minor.

Due to possible grinding or pressure interconversions, the two standards were compared by DRIFTS and by examining an unpressed crystal from each standard by microspectroscopy. Neither the grinding in preparing a KBr pellet nor the pressing using the diamond-anvil cell interconverts the ribavirin polymorphs (Fig.5).

In addition to the above experiments, each of the ribavirin polymorph standards were melted, dissolved, and recrystallized under various conditions. The spectra of the films of the melts and dissolutions for both standards were consistent with each other. This would be expected since crystalline form is lost when a polymorph is melted or dissolved.

The first polymorph standard (P1) was recrystallized to the second polymorph (P2) using distilled water, methanol, anhydrous ethanol, or 95% ethanol. The second polymorph standard (P2) did not recrystallize to the first polymorph (P1) using any of these solvents. Instead it converted back to its original crystalline form using either distilled water or methanol. However, during the recrystallization process using distilled water, P2 was observed to first recrystallize to what appeared to be diamond-shaped plates, with needle-like crystals forming later. Using a polarizing light microscope, both types of crystals were observed to be birefringent, as were the irregular-shaped crystals of the first polymorph (P1) standard and the needle-like crystals of the second polymorph (P2) standard, respectively.

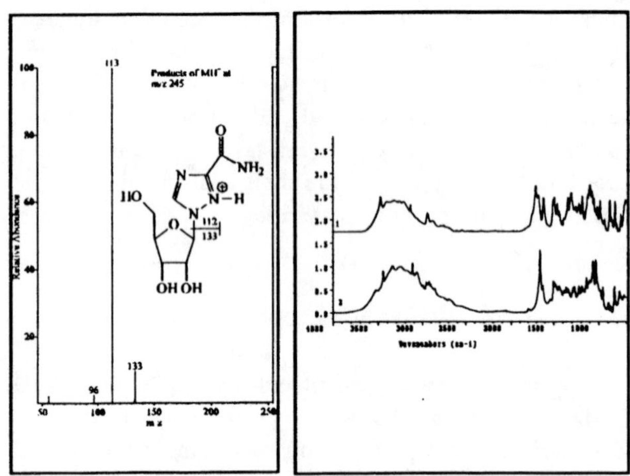

FIGURE 1. The product ion spectrum of ribavirin.

FIGURE 2. KBr: 1) ribavirin standard 1; 2) suspect sample

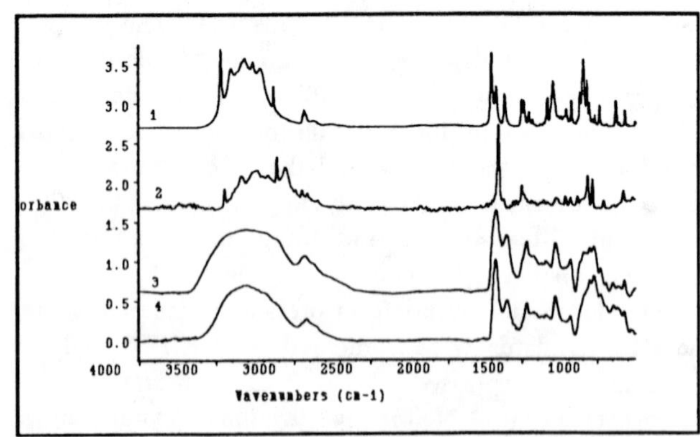

FIGURE 3. Diamond (upper two spectra); Melt (lower two) spectra 1,3: ribavirin std; spectra 2,4; spl crystals

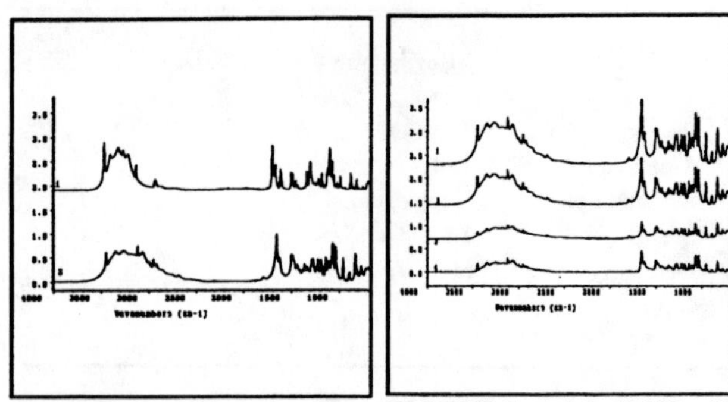

FIGURE 4. Diamond cell: 1) ribavirin std 1; 2) ribavirin std 2

FIGURE 5. std 2: 1) KBr; 2) diamond; 3) DRIFT; 4) crystal

CONCLUSION

Polymorphic substances react unpredictably under various conditions. Within FTIR spectroscopy, the sample preparation and technique chosen are important due to possible interconversion of polymorphs. If the technique chosen to examine a polymorphic substance dissolves the crystalline structure, as in the case of LC-MS, vital information may be lost. In a forensic setting, such information could aid in the identification of a substance made by an unapproved manufacturer.

The results of the ribavirin case study and the additional experiments performed illustrate the capability of FTIR to discern differences between polymorphic forms of a substance.

ACKNOWLEDGMENTS

The authors wish to thank the members of the Microscopy Laboratory at the Forensic Chemistry Center as well as Dr. Chris Brot from Sigma Chemical Company for the technical discussions.

REFERENCES

1. Threlfall, T. L., *Analyst* **120**, 2435-2460 (1995).
2. Aldridge, P. K., Evans, C. L., Ward, H. W. II, Colgan, S. T., Boyer, N., Gemperline, P. J., *Analytical Chemistry* **68,** 997-1002 (1996).
3. Sidwell, R. W., Robins, R. K., Hillyard, I. W., *Pharmacology and Therapeutics* **6**, 123-146 (1979).
4. *Chemical and Engineering News* **50**, 26-27 (1972).

Identification of VX Type Nerve Agents Using Cryodeposition GC–FTIR

Martin T. Söderström

Finnish Institute for Verification of the Chemical Weapons Convention,
P. O. Box 55, FIN-00014 University of Helsinki, Finland.

Analysis of VX type nerve agents with a gas chromatography–Fourier transform infrared spectroscopy (GC–FTIR) system using cryodeposition of the eluents is described. An interpretation system based on comparison of characteristic spectral features to the features in reference spectra, is used to characterize side chains in VX type nerve agents.

INTRODUCTION

VX type nerve agents are very toxic. They affect the transmission of nerve impulses by strongly inhibiting acetylcholine esterase. They are approximately ten times more toxic than e.g. nerve agent sarin.

The production, storage, and use of these O-alkyl S-[2-(dialkylamino)ethyl] alkylphosphonothiolates is forbidden by the Chemical Weapons Convention (CWC) (1), which entered into force on April 29, 1997. They are included in the CWC as a generic structure shown in Fig. 1, where $\mathbf{R^1}$ and $\mathbf{R^3}$ can be either methyl, ethyl, propyl, or isopropyl, and $\mathbf{R^2}$ can be either hydrogen or any aliphatic substituent containing from one to ten carbon atoms.

Mass spectrometry (MS), nuclear magnetic resonance spectroscopy (NMR), and infrared spectroscopy (IR) are used routinely for the analysis of this type of chemicals. Recommended operating procedures are based on the use of reference spectra. As it is virtually impossible to synthesize all possible reference chemicals in advance, spectral interpretation plays an important role in the analysis of this type of chemicals. This study concentrates on the interpretation of condensed phase IR spectra of VX type nerve agents for which direct library search cannot be applied due to the lack of reference spectra.

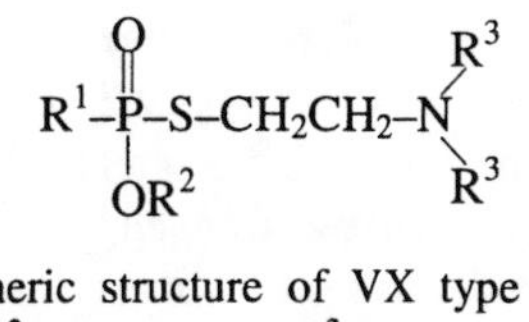

$$R^1-P(=O)(OR^2)-S-CH_2CH_2-N(R^3)_2$$

FIGURE 1. Generic structure of VX type nerve agents. Side chains: $\mathbf{R^1} \leq C_3$, $\mathbf{R^2}$ = H or $\leq C_{10}$, $\mathbf{R^3} \leq C_3$.

RESULTS AND DISCUSSION

Unambiguous Identification

The unambiguous identification of a chemical for the purposes of the CWC requires confirmation of the structure by two different spectroscopic techniques. For a reliable identification of the chemical from an environmental sample of unknown composition, reference spectra or reference chemicals are needed. If no reference material for the chemical exists, accurate interpretation of all available analytical data is required, so that the necessary chemicals can be synthesized quickly for reference purposes. The CWC allows only two weeks for analysing the samples including the final reporting of the results.

The most widely used analytical technique in the CWC related analysis is gas chromatography–electron ionization mass spectrometry (GC–EI/MS). For the VX type chemicals the spectral information given by GC–EI/MS is, however, quite limited: the base peak is normally produced by ion $CH_2N(\mathbf{R^3})_2^+$. Other fragment ions rarely give useful information. The information about $\mathbf{R^1}$ and $\mathbf{R^2}$ groups cannot always be deduced even after chemical ionization tandem mass spectrometric (GC–CI/MS/MS) analysis.

NMR is a very efficient method for identification of the alkyl groups unless the background of the sample overlaps with the resonances of interest.

The cryodeposition gas chromatography–Fourier transform infrared spectroscopy (GC–FTIR) (2) offers a complementary method for analysis. Infrared spectra of VX type chemicals provide information about alkyl groups $\mathbf{R^1}$, $\mathbf{R^2}$, and $\mathbf{R^3}$. The sensitivity is comparable to the GC–MS method because chemicals containing phosphorus-oxygen bonds are highly absorbing.

Cryodeposition GC–FTIR

Experimental

The system used was a Bio-Rad FTS-45 FTIR spectrometer equipped with a Tracer accessory. A Hewlett-Packard 5890 Series II gas chromatograph (GC) with a

CP430, *Fourier Transform Spectroscopy:* 11th International Conference
edited by J.A. de Haseth

split/splitless injector was used for sample introduction. The GC column used was a 25 m long Supelco PTE™-5 with 0.32 mm inner diameter and 0.25 μm film thickness.

The GC program was started from 40 °C (1 min) followed by heating 10 °C/min up to 250 °C for 10 minutes. The splitless time was 0.80 min. The transfer line between the GC and the Tracer unit was kept at 250 °C.

The "on-the-fly" data collection was performed with 4 scans at a resolution of 8 cm^{-1} giving a time resolution of 1 spectrum in 1.1 seconds. The post-run data collection was performed with 64 scans at a resolution of 4 cm^{-1}.

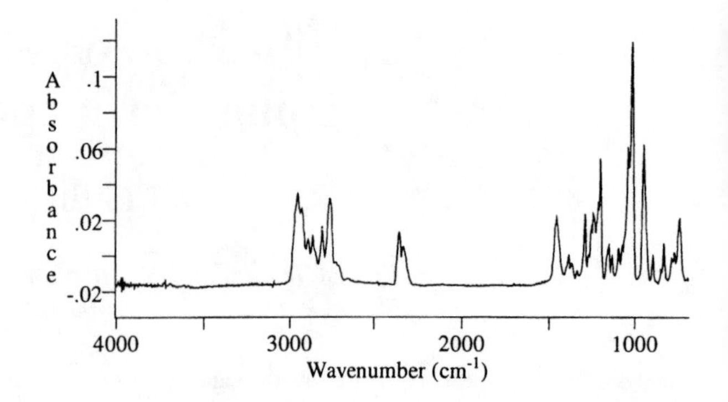

FIGURE 2. Spectrum of O-ethyl S-[2-(dimethylamino)ethyl] *n*-propylphosphonothiolate (chemical B in the chromatogram shown in Fig. 3). Signal-to-noise ratio of the spectrum is about one hundred.

GC–FTIR measurements

During a cryodeposition experiment the eluting chemicals are trapped on a moving ZnSe window, which is cooled by liquid nitrogen. The chemicals are deposited on the surface through a narrow transfer line with a 50 μm restrictor. After the deposition a focused infrared beam is passed through the ZnSe window to the detector.

During the GC run, the spectra are measured at 8 cm^{-1} and Gram-Schmidt and functional group chromatograms are calculated. After the run, the essential chromatographic peaks can be remeasured from the window using selected resolution and large number of scans.

Sensitivity

The sensitivity of the cryodeposition system, as of all infrared systems, depends on the absorptivity of the analytes. All nerve agents are good absorbers, because phosphorus-oxygen bonds have strong dipole moments. For VX type chemicals, the detection level in a chromatogram is between 500 pg and 5 ng depending on the sample background. A reasonable spectrum can be obtained at even lower quantities, because the system can accumulate hundreds of scans for a spectrum, thus increasing the signal-to-noise ratio. The signal-to-noise ratio in the spectrum of chemical B (Fig. 2), was slightly above one hundred using 64 scans for the measurement, but less than ten in the functional group chromatogram (Fig. 3). Sometimes it is even possible to measure an adequate spectrum from the correct retention time when the chemical is not visible at all in the chromatogram.

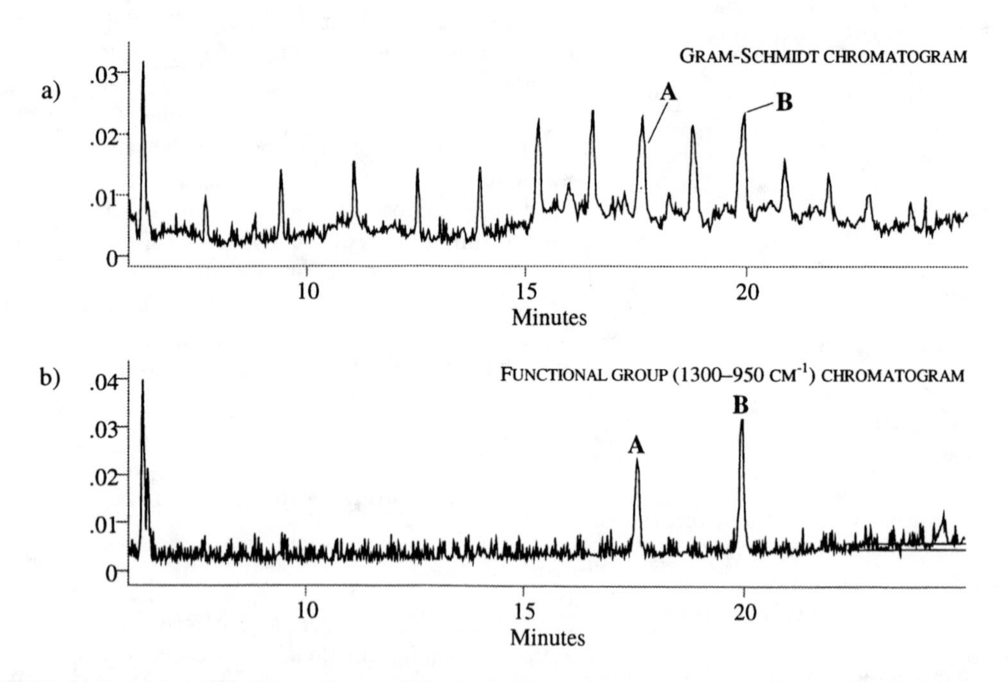

FIGURE 3. Two chromatograms of an organic liquid sample containing two VX type chemicals (A and B): a) Gram-Schmidt chromatogram and b) functional group chromatogram of region 1300–950 cm^{-1}. The spiking level of chemicals A and B was 10 ppm in a 500 ppm diesel fuel background added to the sample.

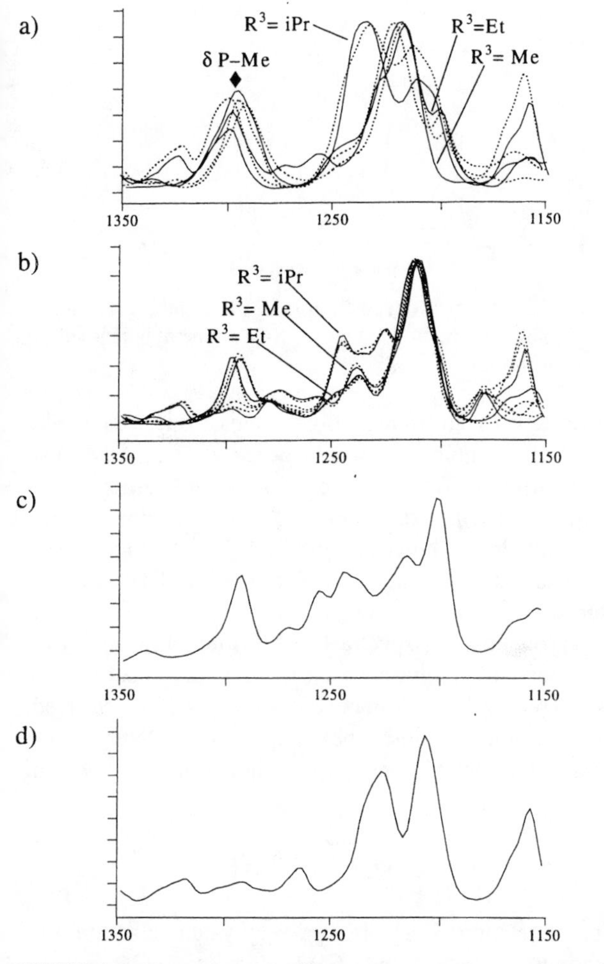

FIGURE 4. Examples of different pattern characteristics for different P-alkyl groups: a) P-methyl, b) P-ethyl, c) P-propyl, and d) P-isopropyl. Several VX type chemicals are shown overlaid in a) and b).

Identification

Type of Chemical

A characteristic group in the organophosphorus chemicals is the phosphorus-oxygen double bond. The frequency of the stretching of this group depends on the substituents on the phosphorus. Thomas and Chittenden (3) defined an equation for the calculation of the position of the P=O stretching, $\nu_{P=O}$:

$$\nu_{P=O} = 930 + 40\, \Sigma\, \pi. \qquad (1)$$

Empirical π constants (3, 4) relate to the electronegativity of the attached group. The calculated range, 1226–1208 cm^{-1}, fits well into the experimental range, 1220–1210 cm^{-1}, (5) and is well separated from the corresponding range in the other nerve agent types (sarin type 1278–1270 cm^{-1}; tabun type 1272–1265 cm^{-1}).

VX type chemicals, like all nerve agents, can be detected in the GC–FTIR from the functional group chromatogram for P=O and P–O bond vibrations. As can be seen in Fig. 3 the region 1300–950 cm^{-1} is useful.

Alkyl Group R^1

Group $\mathbf{R^1}$ can be characterized from the phosphorus-carbon deformation. The P-methyl can be identified from a sharp symmetric deformation near 1300 cm^{-1}. This is the only general P–C vibration found in the spectra (6, 7). P-ethyl produces two weak peaks at a lower wavenumber than P-methyl. These peaks are sometimes very difficult to identify. The $\mathbf{R^3}$ group has some influence on the pattern as seen in Fig. 4. Near 1300 cm^{-1} there are overlapping peaks, which are probably due to C–N stretchings.

The other alkyl groups have to be identified by comparing the peak pattern around the P=O stretching to those in known VX type chemicals. The overlap of P-alkyl (other than P-methyl) peaks with the P=O peak may split the P=O peak so that the interpretation is impossible leaving only pattern comparison. In Fig. 4 is shown examples of each P-alkyl pattern. More data is required for both P-propyl and P-isopropyl groups before final conclusions for these groups can be drawn.

Alkyl Groups R^2 and R^3

The identity of the $\mathbf{R^2}$ group can be determined from the relatively weak bands (region 1200–1050 cm^{-1}) due to the (P–)O–C–X group. Uncertainties arise from the N–C stretchings appearing in the same region. The pattern is always a combination of peaks produced by O–$\mathbf{R^2}$ and

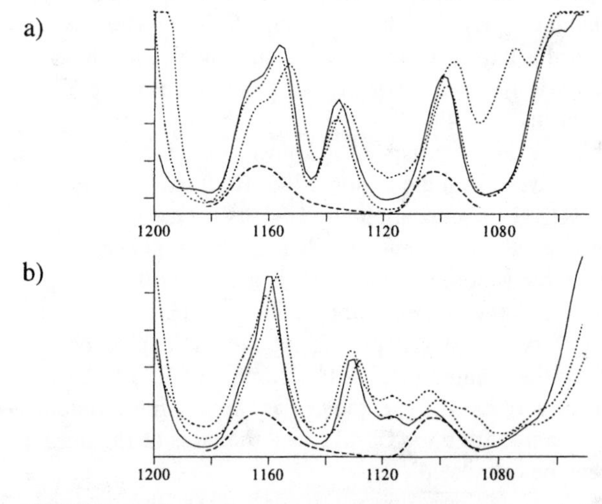

FIGURE 5. Examples of peaks characteristic to each $\mathbf{R^2}$–$\mathbf{R^3}$ combination: a) O-**ethyl** S-[2-(di**methyl**amino)ethyl] and b) O-**ethyl** S-[2-(di**isopropyl**amino)ethyl]. Several VX type chemicals are shown overlaid. The dashed line shows peaks produced by the O-ethyl group (from bis(O-ethyl) methylphosphonate).

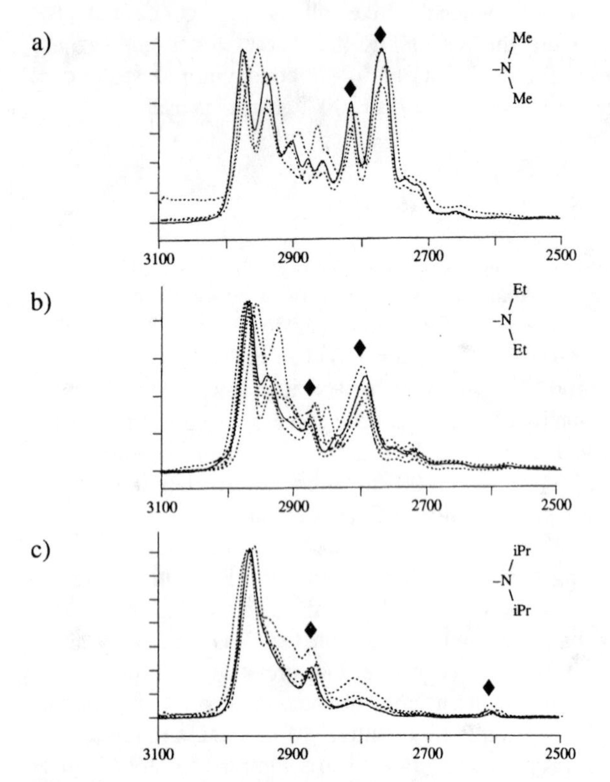

FIGURE 6. Carbon-hydrogen stretching vibrations characteristic for different N–$\mathbf{R^3}$ groups (marked with ♦): a) S-[2-(di**methyl**-amino)ethyl], b) S-[2-(di**ethyl**amino)ethyl], and c) S-[2-(di**iso-propyl**amino)ethyl]. Several VX type chemicals are shown overlaid.

N–$\mathbf{R^3}$ groups. In this context the combination is characteristic for each specific $\mathbf{R^2}$–$\mathbf{R^3}$ as shown in Fig. 5.

Carbon-hydrogen stretchings of carbons next to nitrogen are seen at lower wavenumbers than those of the other carbons (8). Alkyl group $\mathbf{R^3}$ can be identified from these vibrations, which seem to be characteristic for different dialkylamine groups as shown in Fig. 6. The assignment based on the IR spectrum for $\mathbf{R^3}$ group should match the ion $CH_2N(R^3)_2^+$ seen normally as the base peak in the EI/MS spectrum.

If the identity of group $\mathbf{R^3}$ is known, but group $\mathbf{R^2}$ can not be resolved from the region 1200–1050 cm^{-1}, it means that the $\mathbf{R^2}$–$\mathbf{R^3}$ combination is previously unseen. If the ester group $\mathbf{R^2}$ is present in other types of phosphorus chemicals for which reference data is available, this group can probably be identified (see Fig. 5 for comparison).

Even if only two groups, e.g. $\mathbf{R^1}$ and $\mathbf{R^3}$, have been identified, the number of possible isomers of $\mathbf{R^2}$ has been significantly reduced. The molecular weight information can be obtained from the CI–MS, and the mass of the third group can be calculated.

Need for Reference Data

This approach to identify VX type nerve agents requires information of some of the homologues. For the complete

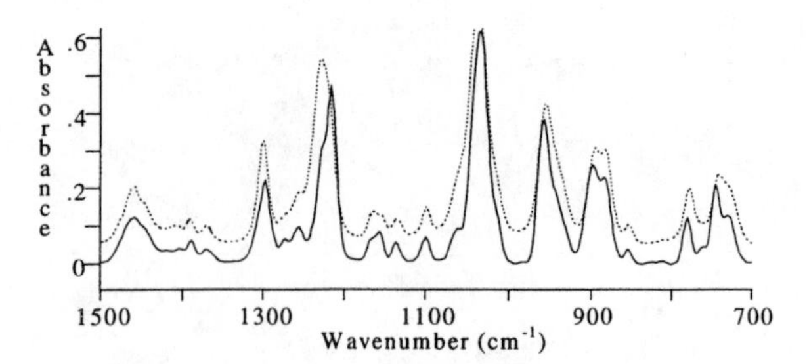

FIGURE 7. Part of cryodeposition (solid line) and liquid phase (dotted line) spectra of O-ethyl S-[2-(dimethylamino)ethyl] methylphosphonate

identification of all combinations at least one example of each $\mathbf{R^2}$–$\mathbf{R^3}$ combination would be necessary. Additional data for different $\mathbf{R^2}$ groups can, however, be applied to VX type chemicals from other phosphorus chemicals (e.g. dialkyl alkylphosphonates and alkyl alkylphosphonofluoridates), so that at least better candidates can be produced for synthetic work.

As cryodeposition spectra can be often compared with normal condensed phase spectra, reference spectra from other types of instruments can also be utilized. Cryodeposition and liquid phase spectra of O-ethyl S-[2-(dimethylamino)ethyl] methylphosphonate are shown in Fig. 7.

CONCLUSIONS

Cryodeposition GC–FTIR is a very valuable addition to a laboratory performing CWC analysis. Phosphorus chemicals are good analytes for this instrument: they are good absorbers and their spectral interpretation gives information on the substituents.

Interpretation of the IR spectra of VX type nerve agents found from unknown samples has in some cases made the structure elucidation more reliable, thus speeding up the synthesis of correct reference chemicals.

REFERENCES

1. *Convention on the Prohibition of the Development, Production, Stockpiling and Use of Chemical Weapons and on their Destruction*, 8 August 1994.
2. Bourne, S., Haefner, A. M., Norton, K. L., and Griffiths, P. R., *Anal. Chem.* **62**, 2448–2452 (1990).
3. Thomas, L. C., and Chittenden, R. A., *Spectrochim. Acta* **20**, 467–487 (1964).
4. Thomas, L. C., *Interpretation of the Infrared Spectra of Organophoshorus Compounds*, London: Heyden & Son Ltd., 1974, ch. 2, pp. 31–37.
5. Söderström, M. T., and Ketola, R. A., *Fresenius J. Chem.* **350**, 162–167 (1994).
6. Thomas, L. C., and Chittenden, R. A., *Spectrochim. Acta* **21**, 1905–1914 (1964).
7. Ref. 4, ch. 9, pp. 93–97.
8. Hill, R. D., and Meakins, G. D., *J. Chem. Soc.* 760 (1958).

3.6 PROCESS CONTROL

Wait and Watch – Monitoring Photoresist Thin Films During Heat Treatment

S.Hilbrich [a)], W.Theiß [a)], R.A.Carpio [b)]

a) I. Phys. Inst., Aachen University of Technology, D-52056 Aachen, Germany
b) SEMATECH, 2706 Montopolis Dr., Austin, TX 78741, USA

The changes of photoresist layers during heat treatment are studied by infrared reflection spectroscopy. A simulation approach is used to analyze the time-dependence of the reflectivity. Information on the development of the layer thickness and the chemical composition is obtained.

INTRODUCTION

Optical spectroscopy is widely used as a non-destructive analytical tool in process and product control. Typical applications in the field of semiconductor production are the determination of carrier concentrations or film thicknesses in multilayer films involving doped semiconductors and insulating oxides [1].

Here we present a method to investigate the changes of photoresist films during thermal treatment. These layers play the central role in photolithography which is used to produce the lateral structurization of semiconductor layer systems required for chip production. Photoresist baking steps are essential ingredients in the photolithography process.

In the first step of the investigation time series of reflectance spectra are recorded during the heat treatment of the photoresist layers deposited on silicon wafers. Then the recorded spectra are analyzed by an automated numerical simulation which fits the relevant model parameters such as the photoresist layer thickness or the strengths of absorption bands. Typical results will be shown.

EXPERIMENTAL

The experimental technique appropriate for photoresist inspection is reflection spectroscopy. During the heat treatment spectra were recorded with a repetition rate of about 0.8 s (Nicolet 560, 700 ... 6000 cm^{-1} , resolution 4 cm^{-1}, 15° angle of incidence, unpolarized radiation) and the time development was followed for about 100 s. This is a typical duration of photolithography postbake steps. The spectra were normalized to the reflectivity of a TiN coated mirror. All results shown in this article have been obtained on Shipley 510L photoresist.

Figure 1 shows one of the series of experimental spectra. The reflection spectra carry information on the layer thickness and the optical constants of the photoresist layers, as well.

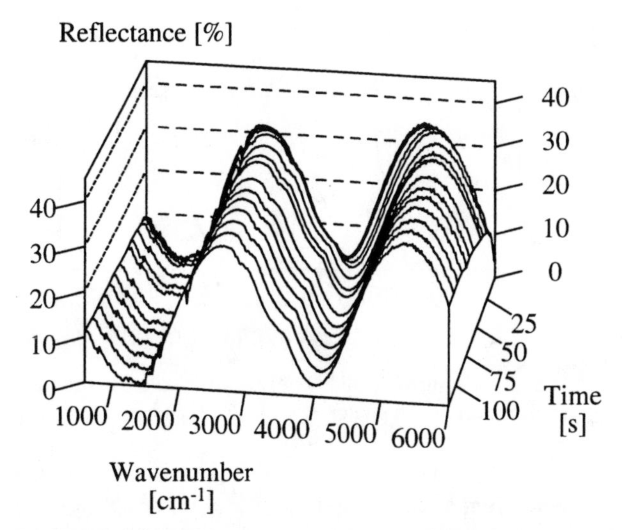

FIGURE 1. Time dependence of the reflectivity of a photoresist on a silicon wafer during an 80°C postbake treatment. The spectra are characterized by dominant thickness interference fringes and weak absorption bands.

DATA ANALYSIS

Optical spectra of thin films can be analyzed best by a comparison of the measured with simulated data [2].

Spectrum modelling is based on two steps. First, for each layer in the stack the frequency dependent optical constants have to be defined. One can use fixed imported literature data or flexible models. The latter must be used, of course, if parameters like oscillator frequencies or strengths are to be determined from the optical experiment. Besides the simple Drude dielectric function model for doped semiconductors and vibrational modes discussed in [2] we also employ the extension of the harmonic oscillator susceptibility suggested by Brendel and Bormann [3]. This

CP430, *Fourier Transform Spectroscopy:* 11th International Conference
edited by J.A. de Haseth

accounts for frequency variations due to local disorder in a material.

After defining the optical constants of each layer the propagation of light waves through the layer stack has been considered. For this work we have used the algorithm described in [4]. After computing the model spectrum for the experimental configuration the deviation of simulation and experiment is computed. In a least-square-fit the model parameters are adjusted to reproduce the measured spectra. A very good agreement can be obtained as demonstrated by the example shown in Figure 2.

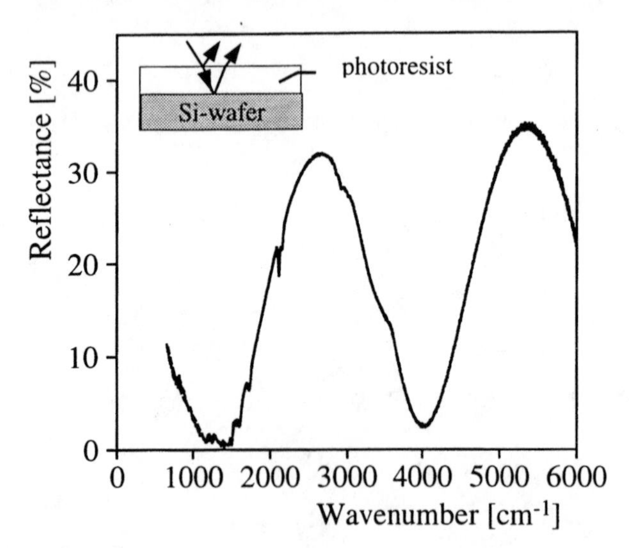

FIGURE 2. Measured (solid line) and calculated (dashed) reflectance spectrum of a photoresist film on a silicon wafer.

The parameters that are varied in the fit are the photoresist film thickness and the model parameters of the photoresist dielectric function (which is the square of the complex index of refraction). The optical constants determined from the fit example given in Figure 2 are displayed in Figure 3.

The dielectric function is composed of several contributions. First, there is a real and frequency independent constant (the so-called dielectric background) which is the high wavenumber limit of the real part (see figure 3). This constant is the sum of the low frequency polarizabilities of all the electronic excitations in the visible and ultraviolet spectral range.

In addition to the dielectric background there are many oscillator contributions representing local vibrational modes. The squared oscillator strength of a harmonic oscillator is proportional to the concentration of the corresponding molecule. Hence changes of the chemical composition of the photoresist during heat treatment can be followed by investigating the time dependence of the individual oscillator contributions.

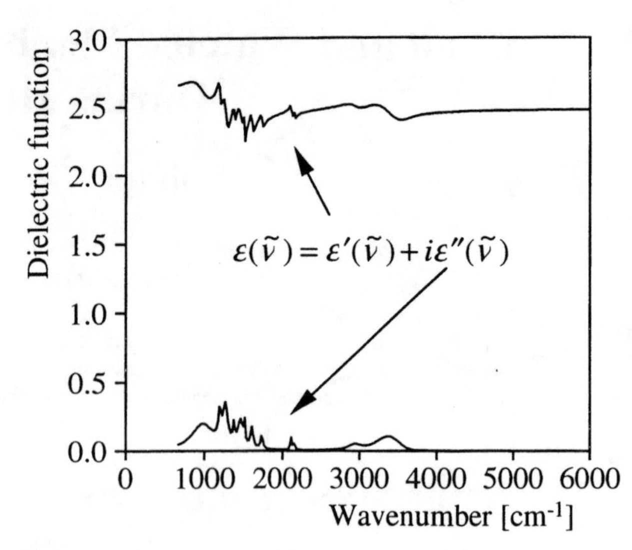

FIGURE 3. Dielectric function model for the photoresist layer obtained from the fit shown in Fig.2.

The simulation method described above has been realized in a software tool [5] which allows the batch processing of series of input spectra like the one shown in Figure 1. Each spectrum is fitted until a minimum of the deviation between model spectrum and measurement is found. If user-defined time or deviation thresholds are reached the fit is also stopped and the next experimental spectrum is loaded. The progress of the simulation analysis is monitored in report files which are generated and updated during the batch run. Tables of the obtained parameters and the achieved quality of the simulations (i.e. the mean squared deviation) are created. These can be used to easily inspect the time development of all parameters by automatically created graphs. The following section on results gives some examples.

RESULTS

After the deposition of a photoresist film on a wafer the material is densified by a post-apply bake step (PAB). This process takes about a minute and reduces significantly the solvent content of the photoresist. Heat treatments of this kind have been investigated with the method described in this work.

Figure 4 shows results obtained on several samples with photoresist layers deposited under the same initial conditions but baked at different temperatures.

First of all the expected behaviour concerning the 'shrinkage' speed is observed: The higher the temperature the faster the layer thickness decreases initially. After about 20 s the rapid changes merge into a slow decrease which ends after the 100 s observation time in different final thickness values. These depend significantly on the baking temperature.

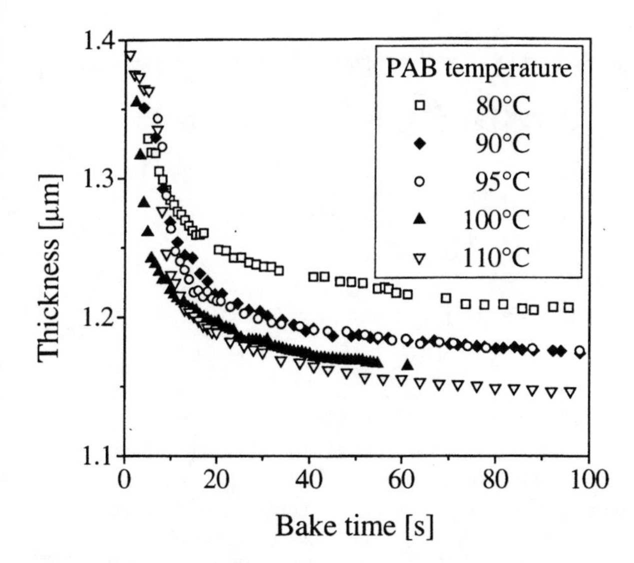

FIGURE 4. Time dependence of the geometrical thickness of a series of photoresist films. The post-apply bake temperature has been varied from sample to sample as indicated.

The optical constants that are obtained from the simulation analysis do also show systematic changes with bake time and temperature. The dielectric background increases during heat treatment (see Figure 5). After about 30 s the final value is reached which is about 4% above the initial level.

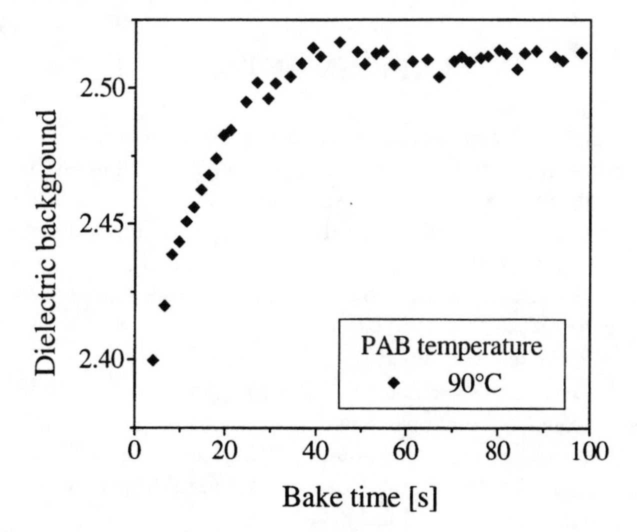

FIGURE 5. Time dependence of the dielectric background

This indicates a densification of the material (the corresponding refractive index increases from 1.55 initially to 1.58). Note that the dielectric background change is less pronounced than that of the geometrical thickness. This observation is consistent with the assumption that there is not only a photoresist densification but also a loss of material due to solvent evaporation during heating treatment.

The solvent loss can directly be detected looking at the squared oscillator strength of the carbonyl band at 1720 cm^{-1}. Results for various bake temperatures are given in Figure 6. The initial concentration is reduced by about 50% at 80° and by about 75% at the highest investigated temperature of 110°.

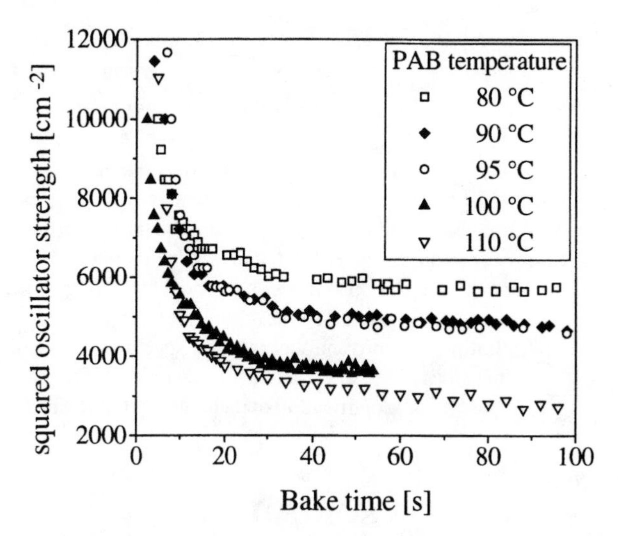

FIGURE 6. The squared oscillator strength of the carbonyl band (1720 cm^{-1}) decreases monotonically with bake time. After a rapid 'carbonyl loss' for about 20 s an almost time-independent concentration is reached which strongly depends on bake temperature.

CONCLUSION

The combination of time-resolved infrared reflection spectroscopy and automated data analysis by simulation has been shown to be a very useful tool to study thin film systems. In the example of photoresist bake steps information on the time dependence of the film thickness and the chemical composition could be achieved.

REFERENCES

[1] R.A.Carpio, B.W.Fowler, W.Theiß, Proc. SPIE Vol. 2638 (1995), 72

[2] P.Grosse, Proc. SPIE Vol. 1575 (1991), 169

[3] R.Brendel, D.Bormann, J. Appl. Phys. **71** (1992), 1

[4] B.Harbecke, Appl. Phys. B **39** (1986), 165

[5] SCOUT – SCientific Objects and UTilities, a Windows application for doing optical spectroscopy by computer, developed by one of the authors (W.Theiß)

Rapid Analysis of Wood Using Transient Infrared Spectroscopy and Photoacoustic Spectroscopy with PLS Regression

Stanley J. Bajic[1], Roger W. Jones[1], John F. McClelland[1], Bonnie R. Hames[2], and Robert R. Meglen[2]

[1]*Ames Laboratory, Iowa State University, Ames, IA 50011 U. S. A. and*
[2]*National Renewable Energy Laboratory, Golden, CO 80401 U. S. A.*

In the forest products industry, improved methods are needed for rapid analysis of wood and paper products. Currently, the best methods for determining chemical and physical properties of wood-based materials require considerable sample preparation and analysis time. Consequently, quantitative information is often not obtained on a time scale suitable for process monitoring, control, and quality assurance. The primary barriers to practical utilization of conventional infrared methods are the opaqueness and poor reflection properties of the wood-based materials. This paper demonstrates how photoacoustic and transient infrared spectroscopies have been combined with chemometric techniques to overcome the limitations of conventional infrared spectroscopies and to permit rapid chemical and physical characterization of wood chips. Both photoacoustic and transient infrared spectroscopic methods are examined as rapid at- and on-line techniques for feedstock identification and chemical composition analysis prior to processing.

INTRODUCTION

Methods for obtaining rapid, high quality and reproducible qualitative and quantitative data are needed in the paper and pulp industry to maintain process control. Currently, the best analytical methods for the determination of the chemical composition of wood require considerable sample size, elaborate sample preparation and several days of analysis time. Various chemical properties of wood can readily be determined from FTIR spectra of ground wood.(1-3) This paper demonstrates two methods, FTIR photoacoustic spectroscopy (FTIR-PAS) and transient infrared spectroscopy (TIRS), each combined with partial least squares (PLS) regression as faster and simpler alternatives for the characterization of feedstocks at and on line, respectively. Results demonstrating species identification (qualitative analysis) and chemical composition (quantitative analysis) for both analytical methods are presented.

FTIR-PAS can acquire both near infrared and mid-infrared spectra from small (submilliliter) samples of milled wood with sufficient precision for quantitative results when used with PLS regression. Transient infrared spectroscopy brings the capabilities of PAS to the process line. In TIRS, the surface of the moving material is heated with a jet of hot air. The warmed surface radiates in the mid-infrared region, and its emission spectrum is captured with an FTIR spectrometer. The heated surface is sufficiently thin that its emission is not overly affected by self-absorption, and it has the characteristic structure found in conventional spectra. Combining TIRS with PLS regression can provide real-time, on-line quantitative analysis of wood as well as other solid materials.(4)

EXPERIMENTAL

Eight different species of wood chips were acquired from several industrial collaborators and analyzed. Douglas fir, Southern yellow pines, acacia, eucalyptus, loblolly pine, ponderosa pine, white fir and a variety of mixed Southern hardwoods (yellow oak, hickory, blackgum and sweet gum) were harvested, chipped and milled for chemical analysis and spectroscopic characterization.

Photoacoustic wood data were measured from milled samples and collected on a Bio-Rad Digilab FTS-60A FT-IR spectrometer equipped with an MTEC model 200 photoacoustic cell. Two-hundred fifty six scans were co-added at 2.5 kHz (<5 mins) for each sample. All spectra were normalized to carbon black.

TIRS data were collected on a Bomem MB 100 FT-IR spectrometer equipped with an external focussing mirror and a wide-band MCT detector cooled with liquid nitrogen. The normal infrared source was replaced by a cold surface. A platter was positioned so that wood chips could be rotated within the spectrometer's field of view simulating a feedstock process line. A heated gas jet was also mounted to the

CP430, *Fourier Transform Spectroscopy:* 11th International Conference
edited by J.A. de Haseth

spectrometer so that it could be aimed onto the moving wood chips. Spectra were acquired while the chips were moving at 1.8 m/s (350 ft/min).

Calibration modeling was performed on all data sets using PLS. ASTM Standard Methods (5) for the chemical determination of ash, extractives, lignin, and carbohydrates were performed on all species to provide the "true" chemical composition or calibration standards.

DISCUSSION

Chemical variations between hard- and softwoods can easily be determined by their respective infrared spectra. For example, Figure 1 illustrates some differences between the FTIR-PAS spectra of a hardwood (acacia) and a softwood (white fir). Hardwoods typically have a higher content of xylan as evidenced by the more intense absorption band at 1730 cm^{-1} (band 1) in Figure 1. Hardwoods also show an absorption at 830 cm^{-1} (band 5) due to syringyl structures characteristic of hardwood lignin. In contrast, softwoods show enhanced absorptions at 870 cm^{-1} and 810 cm^{-1} (bands 3 and 4, respectively) characteristic of the benzene ring substitution of the guaiacyl moieties, the main structural unit of softwood lignins.

The PAS spectra from the different wood species were

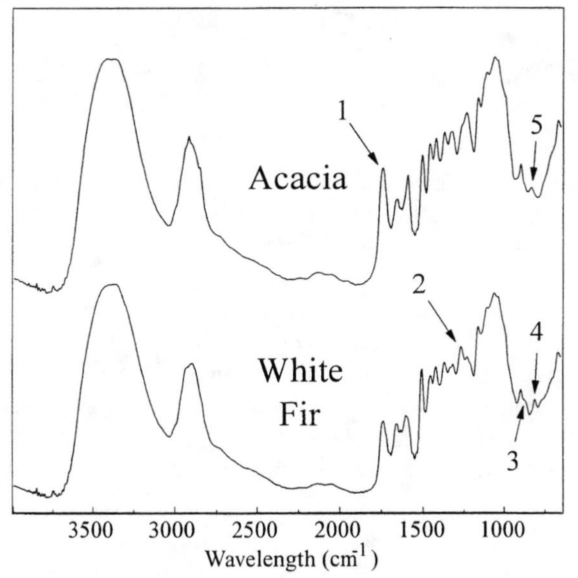

FIGURE 1. FTIR-PAS spectra of hard (acacia) and softwood (white fir). The labeled absorption bands correspond to: 1) C=O stretch in xylan; 2) guaiacyl nuclei in lignin; 3) 1,3,4-substituted benzene ring in softwood lignin; 4) 1,3,4-substituted benzene ring in softwood lignin; and 5) 1,3,4,5-substituted benzene ring in hardwood lignin.

subjected to principal component analysis to determine whether the wood species could be distinguished. The spectral range used for analysis in this study was 1800 to 800 cm^{-1}.

Figure 2 shows the two-vector principal component 1 vs principal component 2 score plot from the principal component analysis of the PAS data. This plot (or species map) shows the overall variation for the PAS spectra. The

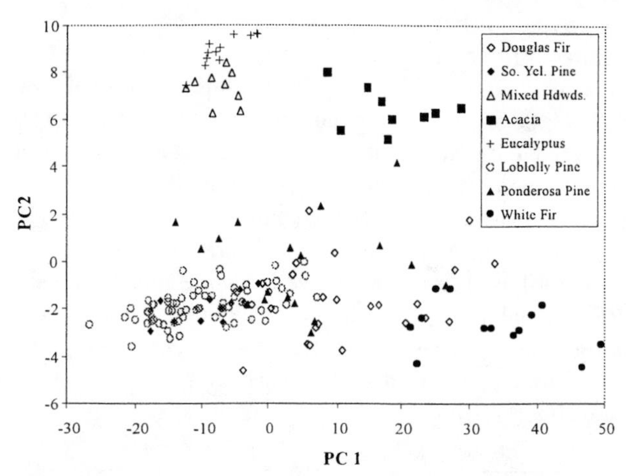

FIGURE 2. Species identification by principle component analysis applied to photoacoustic measurements of milled wood.

different wood species clearly cluster in different spectral space regions, reflecting their chemical similarities. This is demonstrated by noting that all of the hardwoods are located in the upper half of the species map and all softwoods in the lower portion. (While some of the species appear to be overlapped, additional orthogonal projections not shown are required to demonstrate the uniqueness of each species.) The species map also shows that the softwood and hardwood samples further cluster by species type. Note that the loblolly and Southern yellow pines samples occupy the same spectral space as expected, since loblolly pine is a major component of the Southern yellow pines mixture. The first principal factor of the score plot, which accounts for approximately **92%** of the variance in the PAS measurements, contains information about the chemical makeup of the wood samples. The ordinate shows that the second factor, which explains approximately **6%** of the remaining variance, quite clearly separates hard and soft woods.

The range of calibration was augmented by preparing "special" calibration samples. These samples were prepared by manually sorting wood chips into groups which contained primarily heartwood or sapwood, and samples with increased bark and no bark at all. These samples represented the extremes for each species and helped to delineate the species boundaries in qualitative analysis. The species maps confirm that the "extreme" samples span a broader range in spectral space than the unsorted, natural abundance species, as expected and actually formed the boundaries for the respective species.

An important aspect of this work was to demonstrate that quantitative chemical information can be extracted from the PAS infrared measurements. Chemical determination of extractives, lignin, and carbohydrates were performed on all samples using ASTM methods.(5) A full cross-validation PLS calibration model for all three chemical constituents was then developed using three softwood species (douglas fir, white fir and ponderosa pine).

Correlations between the true chemical compositions of the three softwoods and those predicted from the cross-validated PLS regression on PAS measurements for lignin, carbohydrate and extractives content were then made. In all cases, predictions were within the precisions (1%) of the wet chemical methods used to develop the model. Correlation coefficients ranged between 0.85 and 0.94.

While PAS has been demonstrated to provide a laboratory-based rapid alternative to wet chemical analysis, TIRS is capable of providing the same information in near real-time and *on line*. TIRS captures the "fingerprint region" mid-infrared spectra of solid materials moving along a process line. We have applied TIRS to the analysis of wood

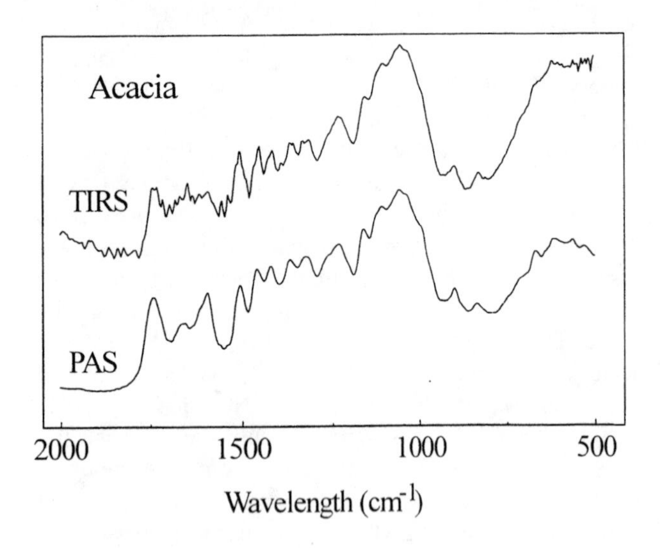

FIGURE 3. Comparison of normalized TIRS and PAS spectra.

chips in laboratory simulations of an on-line environment. As mentioned earlier, TIRS measurements of the moving chips produce similar infrared spectra to that of PAS. Figure 3 shows a comparison of normalized spectra obtained from TIRS and PAS measurements. The spectra are visually similar and PLS calibration modeling confirms that the chemical information content of PAS and TIRS spectra are equivalent. Water vapor bands observed in the TIRS spectrum are present because of the water vapor present in the ambient air between the spectrometer and wood chips. These bands, however, do not interfere with calibration modeling because they are uncorrelated with the chemical compositions of the samples.

TIRS data sets of all eight wood species were collected. A principal component analysis of the TIRS data yielded statistically similar species maps to those produced from PAS spectra. The TIRS principal component species maps also showed clear distinctions of hardwoods and softwoods, and that each wood species occupied a unique region of spectral space.

In order to determine if quantitative chemical information could be obtained from TIRS measurements on moving

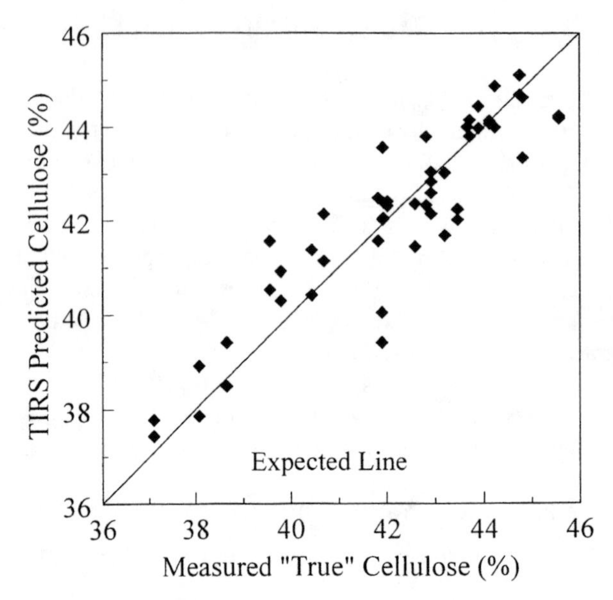

FIGURE 4. Correlation plot for the determination of cellulose based on a simulated on-line TIRS measurement of Douglas fir, ponderosa pine and white fir. RMS error = 0.90.

chips, PLS regression with full cross-validation was performed on the same three softwood species used in the PAS analysis.

As with PAS measurements, good correlations for predictions of lignin, carbohydrate and extractives content were also obtained from TIRS measurements. Figure 4 shows a typical analysis result from a simulated on-line TIRS measurement of wood chips moving at 350 ft/min. Each point in the plot represents a prediction from a model in which it was not a participant, namely full cross-validation modeling was used.

CONCLUSION

These data demonstrate that good quality mid- infrared spectra can rapidly be obtained by both PAS in the laboratory on milled wood and by TIRS on moving wood chips in a simulated on-line system. Chemometric methods such as principal component analysis and PLS modeling have demonstrated qualitative analysis of wood species. These infrared techniques can rapidly distinguish wood species at- and on-line. In addition, these rapid techniques are capable of quantitative analysis of extractives, lignin, and carbohydrates with accuracies comparable to wet chemical methods. The rapidity and low cost of these methods makes them suitable for use on-line in the paper and pulp industry.

ACKNOWLEDGMENTS

Ames Laboratory is operated for the U. S. Department of Energy by Iowa State University under contract No. W-7405-ENG-82. This work is supported by the U. S. Department of Energy, Office of Industrial Technology and is part of the Forest Products Agenda 2020 Initiative .

REFERENCES

1. Meder, R., Gallagher, S., Mackie, K., Bohler, H., and Meglen, R., *Holzforschung*, 1997 (submitted and accepted).
2. Meglen, Robert R., *J. of Chemometrics,* **5**, 163-179, (1991).
3. Kuo M.-L., McClelland, J. F., Chien, P.-L., Walker, R. D., and Hse, C.-Y., *Wood and Fiber Science*, **20**(1), 132-145 (1988).
4. Jones, R. W, and McClelland, J. F., *Process Control and Quality*, **4**, 253-260 (1993).
5. ASTM Standard Methods: E1757-95 wood preparation, E1755-95 ash content, E1756-95 total solids, E1690-95 ethanol extraction, E1721-95 acid insoluble residue, E1758-95 carbohydrates, and TAPPI UM250 acid soluble lignin.

FTIR Monitoring of Industrial Scale CVD Processes

V. Hopfe [1], H. Mosebach [2], M. Meyer [3], D. Sheel [4], W. Grählert [1], O. Throl [1] and B. Dresler [1]

1 Fraunhofer Institute-IWS, Winterbergstraße 28, D-01277 Dresden, Germany
2 Kayser-Threde GmbH, Wolfratshauser Str. 48, D-81379 Munich, Germany
3 Daimler - Benz AG, P.O. Box 80 04 65, D-81663 Munich, Germany
4 Pilkington plc, Hall Lane Lathom Ormskirk Lancashire L40 5UF, Great Britain

The goal is to improve chemical vapour deposition (CVD) and infiltration (CVI) process control by a multipurpose, knowledge based feedback system. For monitoring the CVD/CVI process in-situ FTIR spectroscopic data has been identified as input information. In the presentation, three commonly used, and distinctly different, types of industrial CVD/CVI processes are taken as test cases: (i) a thermal high capacity CVI batch process for manufacturing carbon fibre reinforced SiC composites for high temperature applications, (ii) a continuously driven CVD thermal process for coating float glass for energy protection, and (iii) a laser stimulated CVD process for continuously coating bundles of thin ceramic fibres. The feasibility of the concept with FTIR in-situ monitoring as a core technology has been demonstrated. FTIR monitoring sensibly reflects process conditions.

INTRODUCTION

Background

Chemical vapour deposition (CVD) as well as infiltration (CVI) processes are key technologies in many industrial sectors, including extensive use in micro electronics, surface protection of components, for energy efficient optical coatings on glass for buildings, and for manufacturing fibre reinforced ceramic composite materials. Gaseous precursors containing the elements to be deposited are pyrolysed at the heated substrate or, alternatively, by exciting the molecules with laser photons or by a plasma.

Although the techniques are used on a technological scale, the underlying chemistry is not completely understood. A typical reaction scheme covers pre-reactions in the gas phase which are linked with surface reactions and different mass transport processes (Fig. 1).

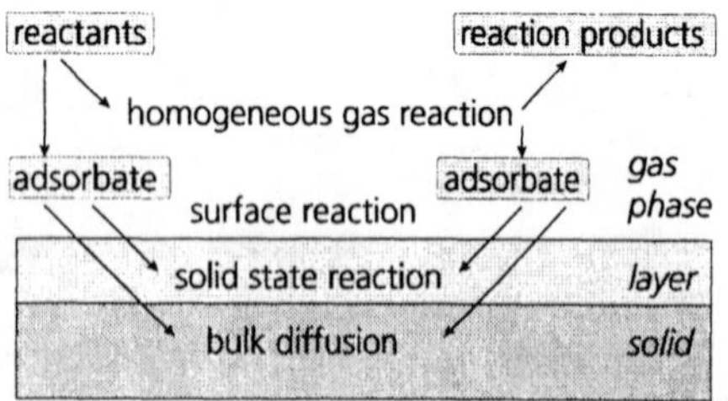

Fig. 1: Reaction scheme of CVD processes - reactions can be investigated by optical spectroscopy

Properties of the deposits depend on process chemistry; reaction intermediates are key components. Thus, deposition process and, as a result, the quality of the deposits can be controlled by monitoring the composition of the gas phase and/or the surface layer continuously.

Objectives

In order to maximize the potential of CVD technology, flexible, high yield processes are required. Critical in achieving this potential is an effective degree of process control. Reactor operation in an open loop with optimisation of process parameters based on characterisation of post processed materials is the conventional way of operation. The technical approach is characterised as

- CVD process control by a multipurpose, knowledge based feedback system
- FTIR spectroscopic data as input information
- In-situ spectroscopy of the gas phase near the surface of the growing layer (Fig. 2a); a direct monitoring of the surface (Fig. 2b) is not feasible in many industrial reactors
- extractive gas analysis is applicable in selected cases (quenching / diluting of the reactands)

FTIR based process sensors combine unique properties: multidetection capability, non invasive operation, robustness, short response time, which makes them core technology for industrial purposes

CP430, *Fourier Transform Spectroscopy:* 11th International Conference
edited by J.A. de Haseth

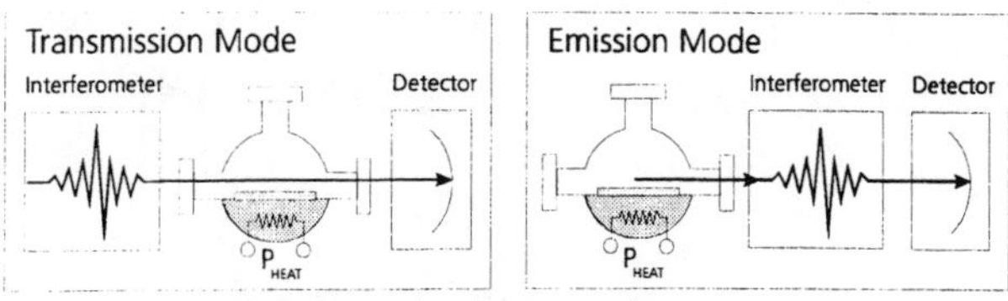

Fig. 2a: Set-up for monitoring gas phase:

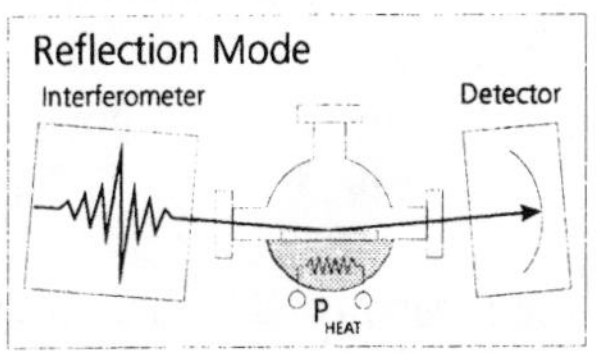

Fig. 2b: Set-up for monitoring surface and gas phase

Applications

Several applications are taken as test cases to check the feasibility of the process control approach. Three commonly used, and distinctly different, types of industrial CVD/CVI processes are taken as test cases: (i) a thermal high capacity CVI batch process for manufacturing carbon fibre reinforced SiC composites for high temperature applications, (ii) a continuous thermal CVD process for coating float glass for energy protection, and (iii) a laser stimulated CVD process for continuously coating bundles of thin ceramic fibres which currently runs in a R&D prototype coater. The presentation is focussed on first results concerning application (i) and (iii).

(i) FTIR MONITORING OF A CVI REACTOR FOR MANUFACTURING CERAMIC COMPOSITES

Industrial Target

Fibre reinforcement of ceramics result in damage tolerant materials with superior mechanical properties, e.g. high strength, stiffness and toughness, which make them candidate materials for high temperature applications: gas turbines, heat exchangers, nozzles or emergency brakes.

The fibre reinforced ceramic composite material is processed in the following way:

- Rovings of ceramic fibres are shaped into a free-standing structure (preform).
- The preform is inserted into the chemical vapour infiltration (CVI) reactor.
- The preform is densified by filling the pores between the fibres with ceramic particles; the particles are precipitated within the pores by pyrolysing a gaseous precursor. Caused by diffusion through long pores the overall rate is low.

Challenge of the Technical Process

The CVI process is characterised by the following properties:

- long infiltration time of typically some days to weeks
- low run-to-run reproducibility
- deviations from specified material properties are detected *post mortem*

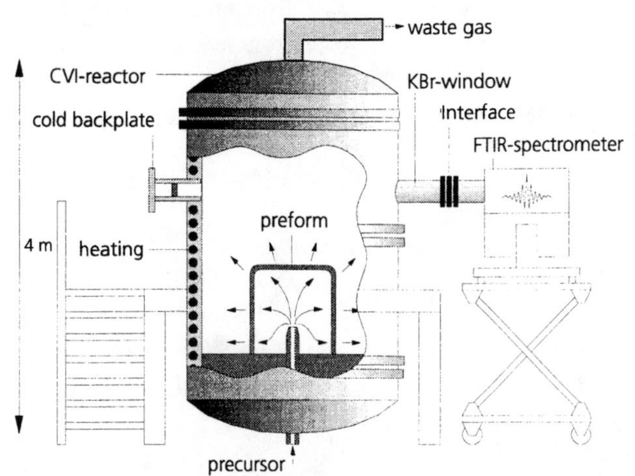

Fig. 3: Adaption of the FTIR spectrometer to the CVI reactor; Spectrometer: K300 Kayser-Threde GmbH; double-pendulum interferometer; resolution 0.01 cm^{-1}; KBr beamsplitter; MCT detector; Optical adaption: low f-number focussing mirror; two aligned gas purged windows; Reference measurements: by turning off precursor flow; Radiation background: cold backplate

Technical Approach: *in-situ* FTIR Emission

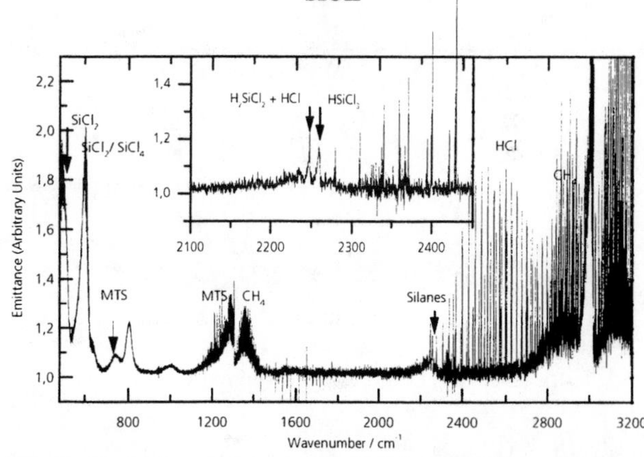

Fig. 4: Emission spectrum of a typical CVI run; Infiltration parameters: temperature 1220-1470 K; total pressure 10-50 mbar; precursor CH_3SiCl_3/H_2, CH_3SiCl_3/Ar

Process Chemistry: Species Identification

CH_3SiCl_3	←	CVI precursor
$SiCl_2$, $(SiCl_3)_{n=1,2}$, $SiCl_4$	←	silicon chlorides
$HSiCl_3$, H_2SiCl_2	←	chlorosilanes
CH_4, CH_3Cl, CH_3	←	hydrocarbons
SiC (s)	←	homogeneous nucleates
HCl	←	reaction product
CO, chlorosiloxanes	←	from precursor handling
NH_3	←	from reactor purge system

Mechanism of CH_3SiCl_3 Decomposition

after: M.D. Allendorf and C.F. Melius J.Phys. Chem. 97 (1993), 720-728

Primary reactions

$CH_3SiCl_3 \rightarrow CH_3 + SiCl_3$

$CH_3SiCl_3 \rightarrow CH_2SiCl_3 + H$

$CH_3SiCl_3 \rightarrow H_2C{=}SiCl_2 + HCl$

$CH_3SiCl_3 \rightarrow SiCl_2 + CH_3Cl$

Secondary reactions

with hydrogen

$CH_3 + H_2 \rightarrow CH_4 + H$

$SiCl_3 + H_2 \rightarrow HSiCl_3 + H$

without hydrogen

$SiCl_3 + CH_3 \rightarrow SiCl_2 + CH_3Cl$

$SiCl_3 \rightarrow SiCl_2 + Cl$

CONCLUSION: most species predicted by theory have been identified by FTIR.

In-situ Process Monitoring

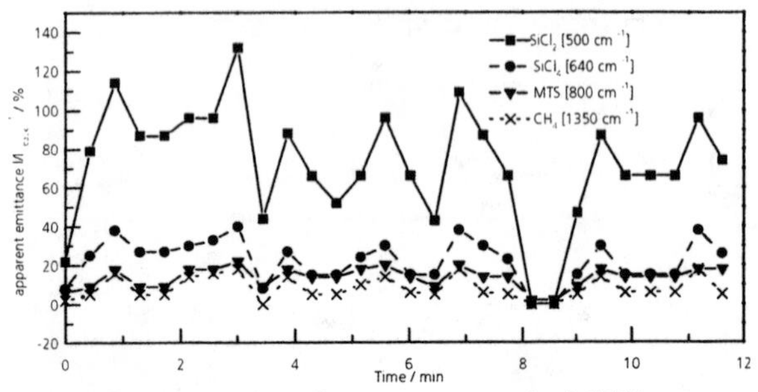

Fig. 5: Concentration changes on a typical CVI run

*Observations (*Fig. 5*)*

- Unexpected strong fluctuations of concentration of different species in the gas phase near the preform have been detected – the reason is not clear yet.
- The method is highly sensitive: common sensors (temperature, pressure, flow rate) could not detect these fluctuations.

(iii) DIAGNOSTICS OF A LASER CVD REACTOR FOR FIBRE COATING

Industrial Target

- Toughness and chemical stability of fibre reinforced ceramics at high temperatures are strongly influenced by the interface between fibre and matrix.
- The interface has to be engineered by coating the fibres with a thin ceramic layer (of BN, C, SiC, TiB_2, etc.).

A laser driven CVD process has been established for continuous high rate coating of fibre bundles at atmospheric pressure. The process runs in a prototype coater wich is based on an industrial cw-CO_2 laser with 6 kW photon power. The endless fibre bundles are coiled at ambient atmosphere and are continuously moved at atmospheric pressure through the open reactor via purged fibre gates (Fig. 6).

Challenge of the Technical Process

- stabilisation of the high rate deposition process; suppression of homogenous nucleation
- automatic control of (open) reactor operation; safety aspects
- quality assurance

Technical Approach: FTIR Monitoring

A multi-port extractive FTIR system has been established (Fig. 6) which realises a continuous measuring of (alternatively)

- purity of the precursor (port 4)
- decay rate of the precursor (ports 2, 4)
- composition and quantity of the gaseous reaction products (port 2)
- controlling the correct operation of the fibre gates; a lack of their inertness may damage reactor and fibres (ports 1, 3)

Different extraction lines are selected by computer control. The lines are heated to avoid condensation of products.

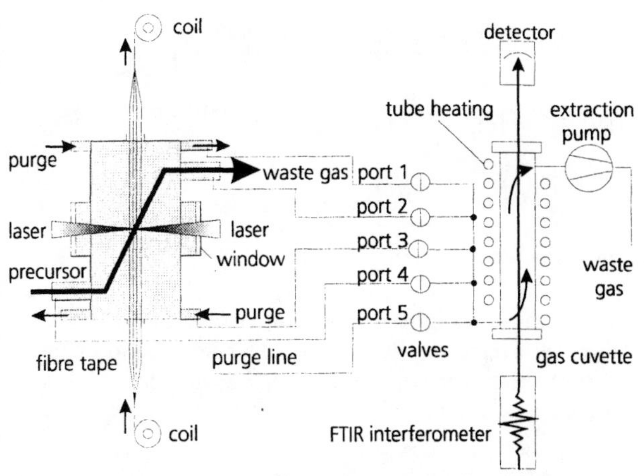

Fig. 6: Scheme of the multi-port extractive FTIR system Spectrometer: Bruker IFS 66, Michelson interferometer, air bearing, resolution 0.2 cm^{-1}, KBr beamsplitter, DTGS detector; Reference measurements: purged gas cuvette

Result: Process Chemistry

Fibre	Coating	Precursor	Products
C	py-C	CH_4	CH_4, C_2H_2, C_2H_4
C	py-C	C_2H_4	C_2H_4, C_2H_2, CH_4, C_6H_6
C	py-C	C_6H_6	C_6H_6, C_2H_2, C_2H_4, CH_4
C	py-C	C_2H_2	C_2H_2, C_2H_4, CH_4
SiC	py-C	C_2H_4	C_2H_4, C_2H_2, CH_4
C	SiC	CH_3SiCl_3	HCl, $HsiCl_3$, H_2SiCl_2, $SiCl_2$, CH_4, C_2H_2

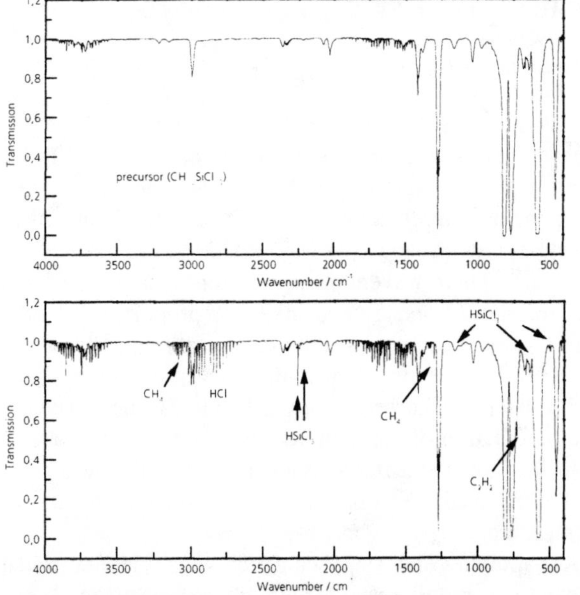

Fig. 7: Absorption spectra of a typical laser CVD run: deposition of SiC layer on carbon fibres

Process Monitoring

Observations(Fig. 7, 8):

- strong fluctuations of concentration of different species in the gas phase due to fluctuations of fibre quality, fluid dynamics and laser power
- product concentration is not correlated with fluctuations of precursor concentration

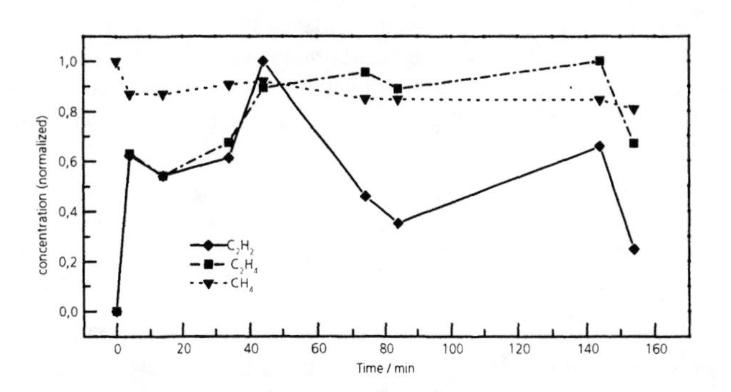

Fig. 8: Concentration changes on a typical laser CVD run: py-C deposition on carbon fibres, precursor CH_4, laser power 3800 W, laser switched on at time zero

Conclusion

- FTIR spectroscopy is feasible for monitoring and finally for controlling industrial scale CVD reactors.
- FTIR monitoring more sensibly reflects process conditions and thus is of considerably higher value for process control than common sensor systems.
- FTIR monitoring is "chemical sensitive", key species can be detected.

R&D Targets*

- development of novel, miniaturised, non-contact, multi-detection sensors for on-line monitoring in hostile environments
- advanced procedures for signal processing
- process control concepts which can be used in a wide range of industrial CVD processes

*) Objectives of an Consortium approach as part of an ongoing project supported by the Commission of the European Communities under contract BRPR-CT96-0322

FT Raman Investigations of Fast Moving Samples

R. Salzer, U. Roland, and R. Born*

Institut fuer Analytische Chemie, TU Dresden, D-01062 Dresden, Germany
** Institut fuer Werkstoffwissenschaft, TU Dresden, D-01062 Dresden, Germany*
Email: reiner.salzer@chemie.tu-dresden.de

Fast moving samples were investigated by conventional and FT Raman spectroscopy. Using near infrared excitation of a Nd-YAG laser instead of the visible excitation by gas-lasers the problem of fluorescence was overcome. The double modulation due to sample movement occurring in FT Raman spectroscopy was avoided by application of the step scan technique. Sample speeds up to 20 m/s were successfully applied. The fast movement of the sample inhibits the thermal degradation of the sample and allows quality control under draw.

INTRODUCTION

Many solids and liquids contain impurity chromophores or unsaturated groups. These moieties absorb visible photons, giving rise to fluorescence, but they do not absorb in the NIR region. In these cases Raman spectra are not available by visible excitation, because the fluorescence intensity is four to eight times higher than the Raman intensity. In most cases these spectra are observable using FT Raman spectroscopy and NIR excitation.

In conventional dispersive Raman spectroscopy sample movement (sample rotation) is a useful method to reduce the overheating and the thermal degradation of the sample, whereas in FT Raman spectroscopy any sample movement causes double modulation (1). This problem can be overcome by the step scan technique (2,3).

EXPERIMENTAL

The Raman experiments were carried out using a dispersive spectrometer DILOR XY 800 and a Fourier Transform spectrometer BRUKER RFS 100. In the DILOR system a CCD (charge coupled device) camera was employed for the detection of scattered light. The BRUKER RFS 100 spectrometer was equipped with a germanium detector. Excitation in the visible region was performed by an argon ion laser (514 nm) and a krypton ion laser (647 nm), in NIR by a Nd-YAG-laser with an excitation wavelength of 1064 nm. The BRUKER RFS 100 was equipped with the step scan option TRS I, which allows time resolutions between 5 µs and 65 ms. Sample movement was achieved by different spinning cells allowing sample speeds of up to 20 m/s.

RESULTS

Investigations of moving samples by FT Raman spectroscopy results in spectra severely distorted by additional bands due to double modulation (Fig. 1).

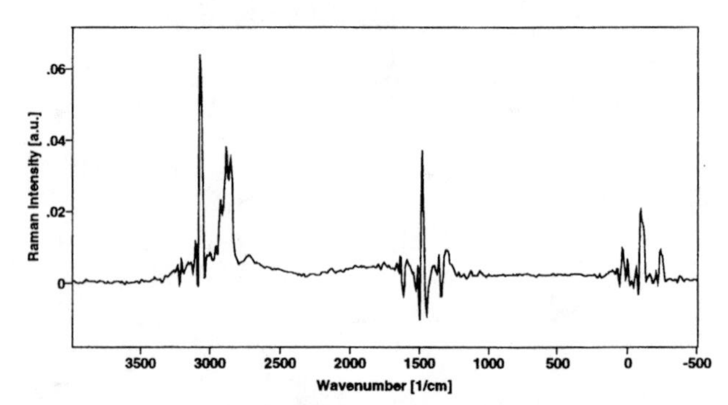

FIGURE 1. Distorted FT Raman spectrum of a fast moving (10 m/s) polyethylene sample recorded with a laser power of 350 mW and a resolution of 4 cm^{-1}

Mechanical reasons, scattered external light and the AC frequency can be excluded as the origin of the bands. In FT Raman spectroscopy additional bands due to sample movement or sample rotation can be separated into main and side spinning bands (4). The main spinning bands are equidistant. Their wavenumber difference is proportional to the rotation frequency and inversely proportional to the scanner velocity. The positions of the spinning bands do not depend on the position of the measuring point at the rotating sample. The origin of the bands can be attributed to the modulation of the scattered light due to roughness of the surface of the sample. Nevertheless, the additional bands prevent the evaluation of the Raman spectra of moving samples.

After applying the step scan technique the additional bands were never observed for moving samples (Fig. 2).

CP430, *Fourier Transform Spectroscopy:* 11th International Conference
edited by J.A. de Haseth

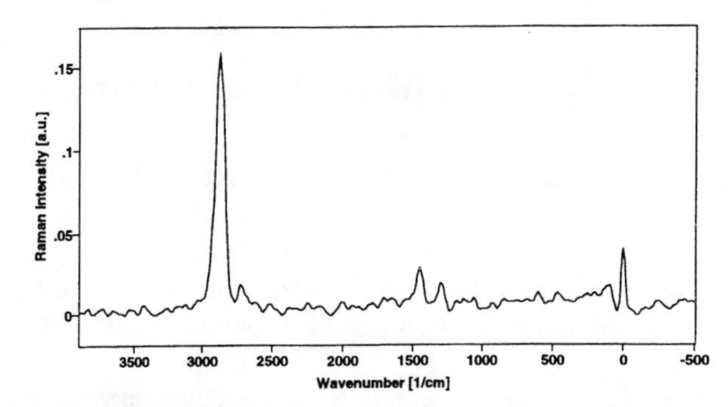

FIGURE 2. Undistorted step scan FT Raman spectrum of the polyethylene sample from fig. 1, again moving at 10 m/s

Sample movement can now be used in FT Raman spectroscopy. A large number of thermally sensitive samples become accessible to FT Raman overcoming the problem of sample burning and structural modification. This is especially important for biological specimens. The measurement of polymer samples under draw becomes possible, too. The thermal (black body) emission can be markedly reduced.

Apart from the fact that the technique allows the FT NIR Raman spectroscopic investigation under conditions where sample movement is essential, it also provides the possibility to obtain a surface image (Fig. 3 and 4) of rotating samples without readjustment during the measuring procedure (2,3).

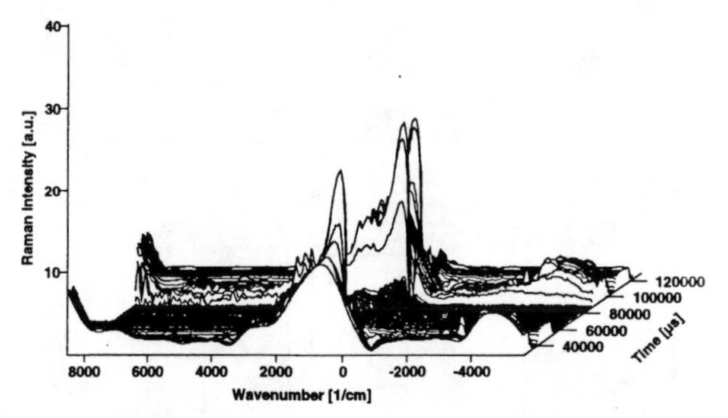

FIGURE 3. Step scan FT Raman spectrum of a rotating two component polymer sample

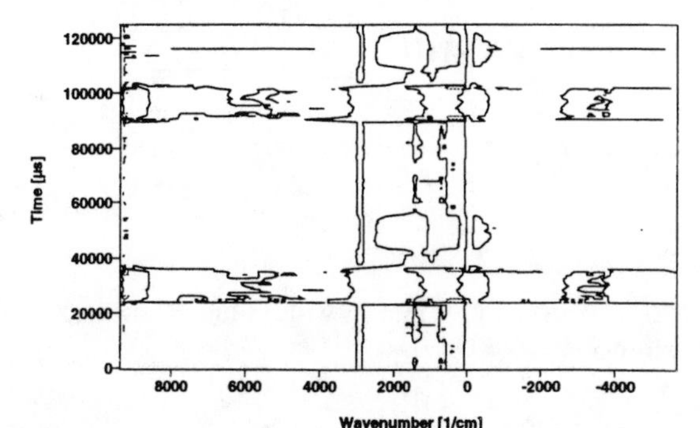

FIGURE 4. Contour plot according to Fig. 3

In order to apply the step scan technique to the surface imaging of inhomogeneous samples, the rotation frequency has to remain constant throughout the whole scan time and the measurement has to be restarted in a reversible manner (3).To fulfill the second condition an external trigger was used to start the measurement every time at the same position of the surface of the sample. As an example is shown in Fig. 3 a step scan FT Raman spectrum of a rotating two component polymer (frequency of rotation 30 Hz, resolution 64 cm^{-1}, time resolution 500 µs) and the according contour plot in Fig. 4.

REFERENCES

1 Bennett, R., *Spectrochim. Acta* 50A, 11, pp. 1813-1823 (1994)

2 Salzer, R., Roland, U., Born, R, in: *Proceedings of the XVth International Conference on Raman Spectroscopy*, (S. A. Asher, P. B. Stein, eds.), New York: John Wiley and sons, 1996, pp. 1200-1201

3 Salzer, R., Roland, U., Born, R, *Appl. spec.*, submitted

4 Salzer, R., Roland, U., *Microchim. Acta*, supplement 14, pp. 757-758 (1997)

Solution Properties and Spectroscopic Characterization of Polymeric Precursors to SiNCB and BN Ceramic Materials

E. Cortez[1], E. Remsen[1], V. Chlanda[1], T. Wideman[2], G. Zank[3], P. Carrol[2], and L. Sneddon[3]

Analytical Science Center, Monsanto Corporate Research, Monsanto Company, 800 North Lindbergh, St. Louis, MO 63167; Department of Chemistry and Laboratory for the Research on the Structure of Matter, University of Pennsylvania, Pennsylvania 19107-6323; The Advanced Ceramics Program, Dow Corning Corporation, Midland, Michigan, 48686-0995

Boron Nitride, BN, and composite SiNCB ceramic fibers are important structural materials because of their excellent thermal and oxidative stabilities. Consequently, polymeric materials as precursors to ceramic composites are receiving increasing attention. Characterization of these materials requires the ability to evaluate simultaneous molecular weight and compositional heterogeneity within the polymer. Size exclusion chromatography equipped with viscometric and refractive index detection as well as coupled to a LC-transform device for infrared absorption analysis has been employed to examine these heterogeneities. Using these combined approaches, the solution properties and the relative amounts of individual functional groups distributed through the molecular weight distribution of SiNCB and BN polymeric precursors were characterized.

INTRODUCTION

Boron nitride (BN) and composite SiNC ceramics are important structural material because of their wide range of attractive properties. Boron nitride exhibits high temperature stability and strength, a low dielectric constant, large thermal conductivity, hardness, and corrosion and oxidation resistance.[4] Composite SiNC materials are light weight and exhibit excellent thermal and oxidative stabilities.[5,6] Moreover, the addition of boron to this silicon-based material can result in greatly enhanced ceramic properties including reduced crystallinity and improved thermal and oxidative stabilities.[7] These attractive properties have led to increased efforts to develop polymeric precursors to BN and SiNCB ceramic materials.

The synthetic reactions which yield polymeric precursors to BN and SiNCB have been described previously.[7,8] In both cases, the resulting polymeric materials exhibit significant chemical complexities which present a challenge to conventional characterization tools because of the likelihood of simultaneous molecular weight and compositional heterogeneities within the polymer. This paper summarizes the use of multiple-detector size-exclusion chromatography employing viscometric, refractive index, and infrared absorption detection to evaluate the heterogeneities in hydridopolysilazane functionalized with pinacolborane (PIN-H) and polyborazylene (PBZ) modified with dipentylamine (DPA).

EXPERIMENTAL

Molecular weight distribution averages and weighted average intrinsic viscosities were determined by size-exculsion chromatograph employing in-line viscometric detection (SEC/VISC). Chromatograms were obtained with a Waters 150-CV SEC/VISC system operated at 35 °C. Figure 1 below depicts the experimental setup used in these analysis.

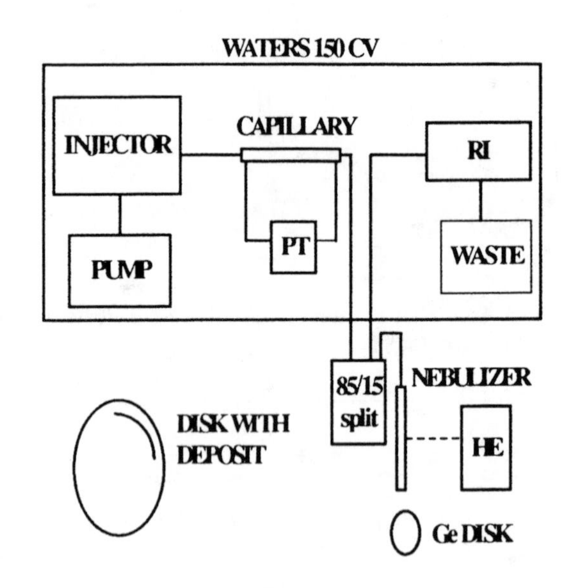

FIGURE 1. Waters 150 C system with in-line RI detector and LC-transform collection Ge disks.

CP430, *Fourier Transform Spectroscopy:* 11th International Conference
edited by J.A. de Haseth

The resulting Mark-Howink relationship for polystyrene in THF at 35 °C was:

$$[\eta]=1.2175 \times 10^{-4} M^{0.712} \quad (1)$$

where, M is the peak molecular weight of the calibrant.

Data acquisition and reduction were provided either by a micro pdp 11/23+ computer or a 486 desktop computer. Data acquisition performed with the 11/23+ computer employed a modified version of MOLWT3. Data acquisition perform by the desktop system used the TRISEC program. Universal calibration and molecular weight calculations made with the MOLTWT3 acquired data employed customized software. The same calculations performed with TRISEC acquired data employed calculation modules in the TRISEC software package. The data processing details have been described elsewhere.[7,8]

The variation of composition across a polymer molecular weight was determined by infrared spectroscopic analysis of collected SEC fractions. The approach employed generally followed methods previously described for the characterization of copolymer compositional heterogeneity. Specific conditions used in this study were as follows: a model LC-Transform was placed in-line with the SEC/VISC system and used to deposit eluting polymeric fractions onto a Ge disk rotating at a constant angular speed of 10 °/min. The deposition was performed by splitting of 15% of the eluting solution stream and devolatizing the stream with a 20 psi flow of He heated to 51 °C. The fractions were deposited around the circumference of the Ge disk as a continuous polymer film. The chromatographic conditions used with the LC-Transform were identical to those used for the SEC/VISC molecular weight analysis. The offset between the VISC, IR, and DRI chromatograms was corrected by overlaying the Gramm-Schmidt reconstructed IR chromatogram and the corresponding VISC and DRI chromatograms for a mondisperse polystyrene standard (M = 207,000 g/mol).

IR spectroscopy was conducted on either a Nicolet 800 or Magna 550 spectrometer. The Nicolet 800 system is equipped with a Ge coated KBr beamsplitter and a MCTB detector. Each spectra consist of 16 co-added sample scans recorded at 4 cm^{-1} resolution. The Nicolet Magna 550 system is equipped with a KBr beamsplitter and a DTGS detector. Each spectra consists of 5 co-added sample scans collected at 4 cm^{-1} resolution. The optics for the Model 100 LC-transform system were used in each system.

RESULTS AND DISCUSSIONS

Hydridopolysilazane (HPZ) is a well established precursor to SiNCB ceramic materials.[9] This polymer can be readily modified with borazine to form hybrid polymers (HPZ-B) in which pendent borazine rings are bonded to the polysilazane backbone via B-N linkages.[10] The resulting polymers are excellent precursors to SiNCB materials. However, the latent reactivity of the B-H groups on the borazine ring cause cross-linking of HPZ-B polymers which limits its applications to ceramic materials and/or coatings. The use of monofunctional boranes such as pinacolborane (PINH) eliminates the BH groups on the borane units and thus hinders cross-linking and allows for more advanced applications such as fibers.

The combined SEC/VISC/IR technique provides a means of characterizing the compositional heterogeneities in these polymeric systems. The example SEC/VISC chromatogram for HPZ and HPZ-PIN polymers depicted in Fig. 2 below are multi-modal which indicates heterogeneous chains and/or compositional heterogeneity.

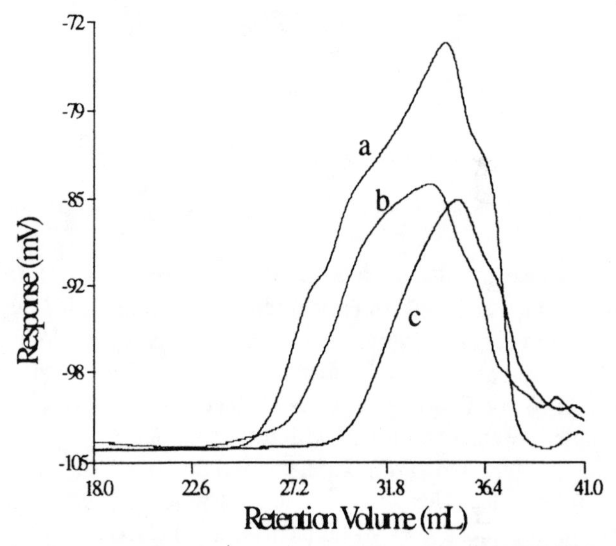

FIGURE 2. DRI chromatograms for (a) HPZ, (b) HPZ-PIN-1, (c) HPZ-PIN-3.

The molecular weight of the modified HPZ decreased with increased functionalization which was controlled by the duration of the reaction. Morevoer, the molecular weight for the different HPZ-PIN polymers are lower than the unmodified HPZ suggesting chain degradation. The infrared spectra for HPZ corresponding to molecular weights ranging from 680,800 to 4,900 showed little qualitative variation in the relative intensities of the characteristic vibrational bands such as υ_{N-H}, υ_{C-H}, and υ_{Si-H}, which suggests compositional homogeneity over this molecular range. The fingerprint region below 1000 cm^{-1} of the infrared spectra did show some intensity changes which suggest branching is possible. Functionalization of

HPZ with PIN, Fig. 3, resulted in an increase of the υ_{Si-H} to υ_{N-H} from 2.28 ± .04 for the unmodified HPZ to 3.24 ± .08 for HPZ-PIN-3. The increased standard deviation for the HPZ-PIN-3 polymer suggest more molecular weight variation than the parent compound.

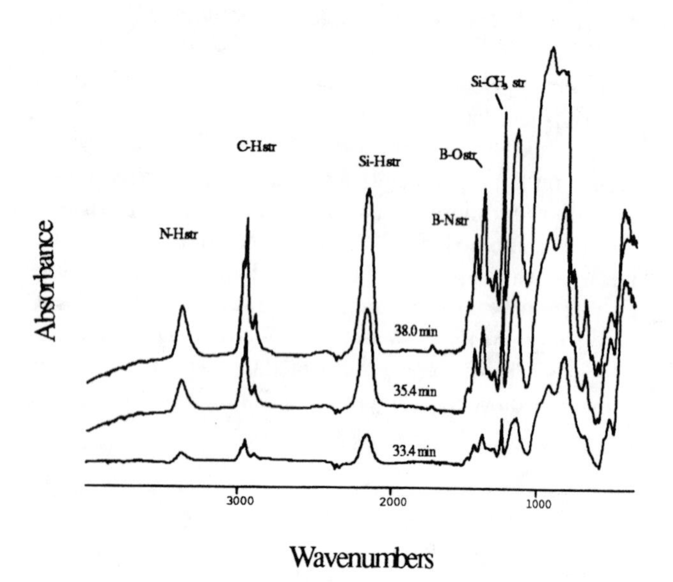

FIGURE 3. Stacked IR spectra for HPZ-PIN-3 at 33.4, 35.4, and 38.0 minutes.

The IR spectra also show a qualitative variation in υ_{B-O} to υ_{Si-H} which suggests an enrichment of PIN content in the lower molecular weight region of the SEC chromatogram. This observation is consistent with polymer modification via Si-N bond formation with chain cleavage producing smaller polymer fragments with terminal borane.

The combined molecular weight and infrared techniques used for SiNCB materials were also useful in examining precursors to boron nitride. Polyborazylene[11] is an excellent precursor to boron nitride; however, the reactive B-H and N-H sites present in the molecule cause polymeric cross-linking to occur. One way to inhibit the cross-linking reactions is to functionalize the polymer with secondary amines such as dipentlyamine. As with the HPZ-PIN materials described above, the degree of functionalization is controlled by varying the reaction time.

The molecular weight distribution for PBZ-DPA showed a shift to lower molecular weight compared to the unmodified PBZ, and produced an extra low molecular weight mode. These observations suggest polymer degradation, but, the mechanism for this degradation was not further elucidated. The infrared spectra at representative retention volumes provided evidence for compositional differences across the molecular weight distribution. The IR spectra for the higher molecular weight regions were dominated by the υ_{B-H} (2600-2400 cm^{-1}) and the characteristically sharp υ_{N-H} (3450-3300 cm^{-1}) of the borazine ring. The intensity of the υ_{C-H} modes (2700-2900 cm^{-1}) were comparable to those of υ_{N-H} in the lower molecular weight region of PBZ-DPA.

CONCLUSIONS

Combined molecular weight and infrared spectroscopic characterization studies of polymeric precursors to SiNCB and BN ceramic materials indicated functionalization of hydridosilizane and polyborazylene produces compositional heterogeneity. The PBZ-DPA polymer showed lower molecular weights than PBZ, with the highest amine concentrations in the lower molecular weight fractions, suggesting some backbone scission occurs during polymer modification. The molecular weight of the HPZ-PIN polymer also decreased compared to the parent polymer (HPZ), with the highest borane concentrations in the lower molecular weight fractions.

ACKNOWLEDGEMENTS

We thank the U.S. Department of Energy, Division of Chemical Sciences, Office of Basic Energy Sciences, Office of Energy Research, and the National Science Foundation

REFERENCES

1. Monsanto Company
2. University of Pennsylvania
3. Dow Corning Company
4. (a) Meller, A. *Gmelin Handbuch der Anorganische Chemie; Boron Compounds*; Springer Verlag: Berlin, 1983, 2nd Supplement, Vol. 1.; (b) Meller, A. *Gmelin Handbuch der Anorganische Chemie, Boron Compounds*; Springer-Verlag: Berlin, 1988, 3rd Supplement, Vol.3, and references therein.
5. (a) Mausikant, S. In *What Evey Engineer Should Know About Ceramics*; Marcel Dekker, Inc.; New York, 1991; (b) Messier, D.R.; Croft, W. J. In *Preparation and Properties of Solid State Materials*; Wilcox, W. R.; Ed.; Marcel Dekker, Inc.; New York, 1982, Vol.7, Chapter 2; (c) *Gmelin Handbook of Inorganic Chemistry*; Springer Verlag: Berlin, Silicon Supplement B2, 1984; B3, 1986 and references therein.

6. Laine, R.M.; Babonneau, F. *Chem. Mater.*, **5**, 260-279, (1993); (b) Birolt, .; Pillot, J. P.; Dunogues, J. *Chem. Rev.*, **95** , 1443-1477, (1995) and references therein.
7. Wideman, T.; Cortez, E.; Remsen, E.E.; Zank, G.A.; Carroll, P.J.; Sneddon, L.G. *Reactions of Monofuntional Borans with Hydridopolysilizane: Synthesis, Characterization and Ceramic Conversion Reactions of New Processible Precursors to SiNCB Ceramic Materials;* to be published and references therein.
8. Wideman, T.; Remsen, E. E.; Cortez, E.; Chlanda, V.L.; Sneddon, L.G. *Amine-Modified Polyborazylens: Second-Generation Precursors to Boron Nitride*; submitted March 1997 and references there in.
9. Legrow, G.E.; Lim, T.F.; Lipowitz, J.; Reaoch, R.S. *Am. Ceram. Soc. Bull.*, **66,** 363-367, (1987).
10. (a) Su, K.; Remsen, E.E.; Zank, G.A.; Sneddon, L.G. *Chem. Mater.*, **5**, 547-556, (1993); (b) Su, K.; Remsen, E.E.; Zank, G.A.; Sneddon, L.G. *Poly. Preprints*, **34**, 334-335, (1993); (c) Zank, G.A.; Sneddon, L.G.; Su, K. *U. S. Patent No. 5,252,684*, Ocober 12, 1993. ; (d) Zank, G.A.; Sneddon, L.G.; Su, K. *U. S. Patent No. 5,256,753*, Ocober 26, 1993. (e) Wideman, T.; Su, K; Remsen, E. E.; Zank, G. A.; Sneddon, L.G. *Chem. Mater.*, **7**, 2203-2212, (1995); (f) Wideman, T.; Su, K; Remsen, E. E.; Zank, G. A.; Sneddon, L.G. *Mat. Res. Soc. Symp. Proc.*, **410**, 417-423, (1995).
11. (a) Fazen, P..J.; Remsen, E.E.; Carroll, P.J.; Beck, J.S.; Sneddon, L.G. *Chem. Of Mater.*, **7**, 1942-1956, (1995); (b) Fazen, P.J.; Beck, J.S.; Lynch, A.T.; Remsen, E.E.; Sneddon, L.G. *Chem. Of Mater.*, **2**, 96-97, (1990).

IR Spectroscopic Studies in Microchannel Structures

A. E. Guber, W. Bier

Forschungszentrum Karlsruhe GmbH, Institut für Mikrostrukturtechnik
Postfach 3640, 76021 Karlsruhe, Germany

By means of the various microengineering methods available, microreaction systems can be produced among others. These microreactors consist of microchannels, where chemical reactions take place under defined conditions. For optimum process control, continuous online analytics is envisaged in the microchannels. For this purpose, a special analytical module has been developed. It may be applied for IR spectroscopic studies at any point of the microchannel.

INTRODUCTION

The various microengineering methods developed in the past years allow to produce smallest microchannel systems, micro heat exchangers, fluid systems and sensors. Increasing miniaturization of the above components meanwhile also allows the fabrication of the so-called microreactors, by means of which safety and efficiency will be improved in numerous chemical fields[1]. The currently existing microreactors contain one or several arrays of microchannels, where chemical substances react under defined conditions[2,3]. Precise spectroscopic control of the chemical reaction in a single microchannel can be achieved at a selected point by means of two transversely fixed fibers. Depending on the microreactor material, however, sophisticated sealing technology between the fiber and the channel wall is required. The Y mixer-based reaction system presented here allows to perform IR spectroscopic studies at nearly any point of the microreaction channel.

METAL MICROREACTORS

With the LIGA technique[4] and mechanical microengineering[5], two major microengineering methods have been developed at the Karlsruhe Research Center. The latter was employed by the Institute for Microstructure Technology and the Central Experimental Engineering Department to fabricate smallest metal micro heat exchangers[6]. Principle setup of the micro heat exchanger is represented schematically in FIGURE 1. At first, smallest microchannels are cut into thin metal foils. These are then cut into minute foil platelets, piled up in a crosswise manner and subjected to helium- and pressure-tight diffusion welding. The standard micro heat exchanger consists of 100 square foil platelets. 80 microchannels are available per foil layer. In general, channel dimensions can be varied within certain limits. Usually, a micro heat exchanger has about 4000 microchannels with a transmission area of 150 cm^2 and a transmission volume of 1 cm^3. Therefore, these micro heat exchangers are suited as microreactors for e.g. heterogeneously catalyzed reactions and rapid, highly exothermal gas-phase and liquid reactions.

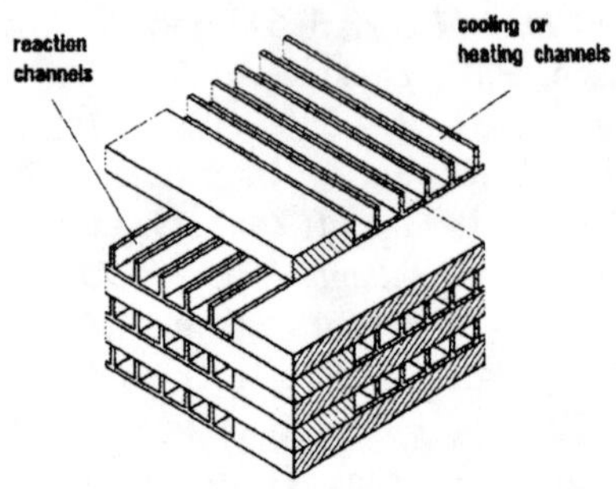

FIGURE 1. Schematic representation of crosswise piling of the foil platelets.

FIGURE 2 shows another schematic representation of a microreactor produced at the Karlsruhe Research Center. Via two separate medium inlets, the educts first enter a static micromixer. They are then passed on to the reactor of the cross-flow type. If necessary, transverse heating or cooling may be carried out.

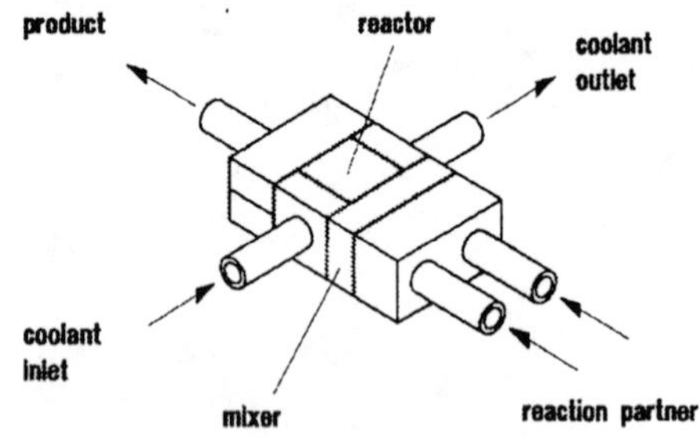

FIGURE 2. Schematic representation of a microreactor consisting of a mixer and a cross-flow type reactor.

These chemically rather interesting components allow an exact and rapid temperature cïntrol of chemical processes. At the same time, high product yields may be

CP430, *Fourier Transform Spectroscopy:* 11th International Conference
edited by J.A. de Haseth

expected, as the formation of undesired by-products can be reduced considerably.

For optimum operation of such microreactors, it is recommendable to have an integrated online analytics in at least one of the microchannels. As insertion and vacuum-tight fixation of very thin optical fibers in metal microreactors are impossible without constructional changes, the setup was optimized by producing a simple Y mixer consisting of two supply lines and a downstream reaction channel. In principle, such a component may be operated in parallel to the microreactor shown in FIGURE 2.

FABRICATION OF Y MIXERS BY THE μEDM TECHNIQUE

By means of the micro electrical discharge machining (μEDM) technique, a simple Y mixer is cut into a circular stainless steel platelet of about 1 mm thickness. Both supply channels are 200 μm wide and 7.5 mm long. The downstream reaction channel is 400 μm wide and 18 mm long. FIGURE 3 shows the design of the Y mixer.

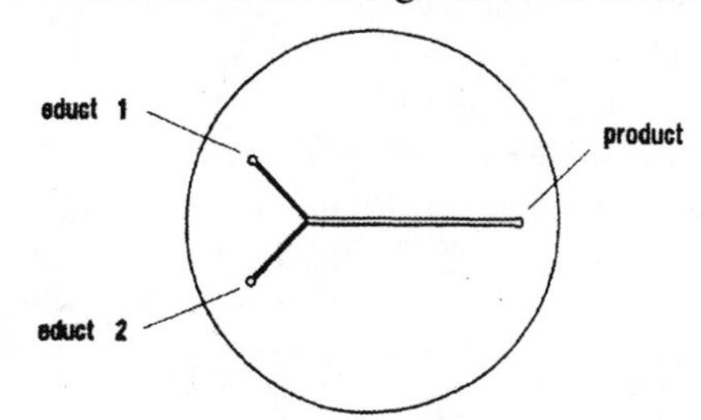

FIGURE 3. View of a straight Y mixer.

As known, μEDM technique allows high-precision processing of metal materials. Due to its flexibility, this method is often used in prototype and model construction[7]. The material is removed by a combination of electric, thermal and mechanical processes. FIGURE 4 illustrates the principle of wire erosion by way of the Y mixer example. Due to the very precise relative movement between the tool (cutting wire) and the work piece (Y mixer), the channel geometry desired may be generated by the CNC-EDM machine.

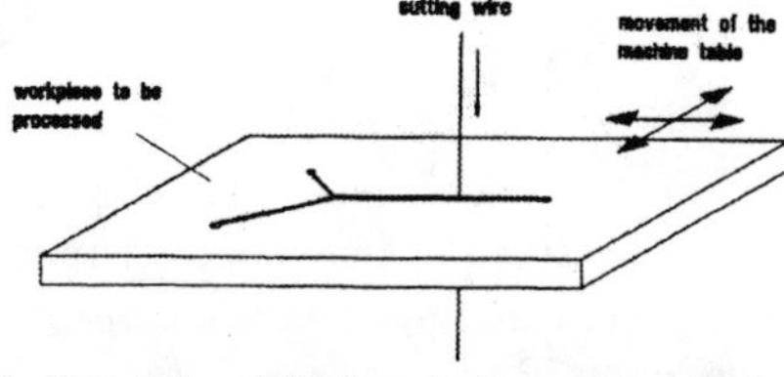

FIGURE 4. Principle μEDM technique. The work piece to be processed is moved relative to the cutting wire.

A thin tungsten cutting wire of about 100 μm thickness is used. Thus, gaps of about 200 μm can be cut. At present, minimum cutting wires of 30 μm in outer diameter only are available such that no difficulties arise when fabricating microchannels of 60 μm width[8]. As any shape is possible during cutting erosion, Y mixers with meandering reaction channels were generated in addition to the Y mixers with a straight reaction channel (cf. FIGURE 5). The supply lines are also 200 μm wide and 7.5 mm long. Both educts first pass a baffle before entering the 400 μm wide and 40 mm long meandering reaction channel.

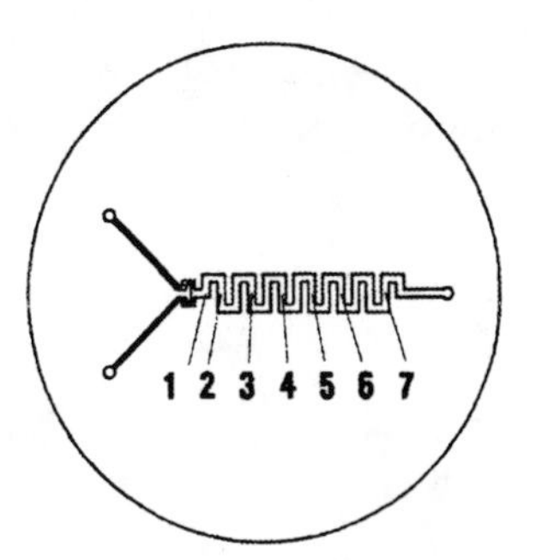

FIGURE 5. View of a meandering Y mixer

EXPERIMENTAL

The stainless steel platelets processed by means of the μEDM technique are inserted into a separately produced holding system of stainless steel and covered by IR transparent AgCl disks on both sides. Possible dead volumes directly below the Ag Cl disks can be neglected. The thus obtained microcuvette is helium-vacuum-tight and can be operated both in the low-pressure range and at high pressures (up to a maximum of 2 bar). The lines leaving the microcuvette are integrated into the test rig by means of a flange system. Via both supply lines, various substances can be introduced into the microcuvette for subsequent reaction. The reaction products generated leave the microcuvette via the product line. IR light is passed through the Y mixer under an IR microscope that is connected to a FTIR spectrometer of the BRUKER company (type IFS 88). Due to the fabrication technique applied, transmission may take place at any point of the Y mixer. By PC-supported control of the object carrier of the IR microscope, a number of measuring points can be reached along the reaction channels.

Up to now only in a first series of experiments the efficiency of both mixer types was tested. A gas reaction of halogen chemistry was selected, namely, the halogen exchange reaction between the gas $SiCl_4$ and pure fluorine gas:

$$SiCl_4 + 2F_2 \dashrightarrow SiF_4 + 2Cl_2$$

Starting at room temperature, SiF_4 and chlorine gas are formed[9,10]. From two storage tanks, various stoichiometric amounts of $SiCl_4$ gas and F_2 gas are metered into the Y mixer via both educt channels using two flow meters. The microcuvette is fixed under the IR microscope. IR transmission measurements are carried out at several measuring points along the reaction channel (cf. FIGURE 6).

The educt $SiCl_4$ and the reaction product SiF_4 are IR active. The main absorption bands are found at 619 cm^{-1} and 1029 cm^{-1} and, hence, can be identified in the IR spectrum. Possible intermediate products of the type $SiCl_xF_y$ are also IR active and located in the range of about 1020 and 920 cm^{-1}. Elementary Cl_2 gas and F_2 gas are not IR active and can only be identified by measurements in the UV range.

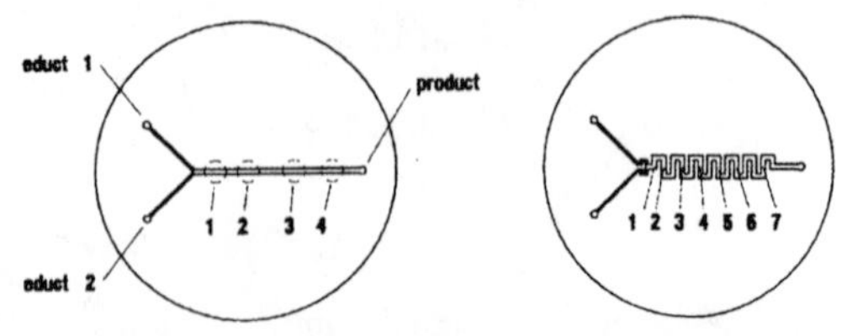

FIGURE 6. Measuring points along the Y mixer reached by the IR microscope.

RESULTS AND DISCUSSION

The result of a first experiment in the "straight Y mixer" is shown in FIGURE 7. The above chemical reaction was studied IR spectroscopically over a mixer length of 18 mm.

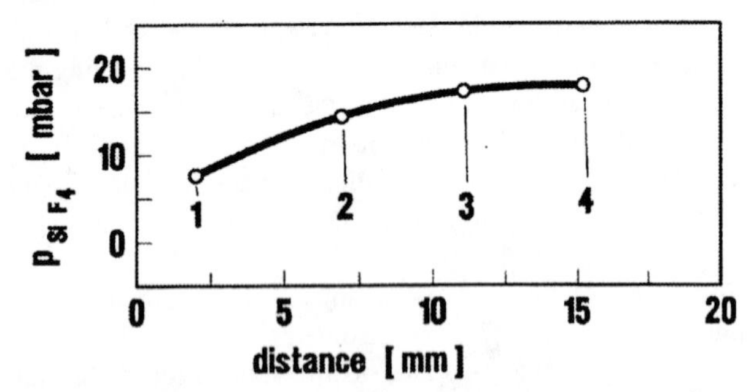

FIGURE 7. Development of the SiF_4 partial pressure along the reaction channel of the straight Y mixer.

Formation of SiF_4 could be detected after a short time already. It increases with increasing length of the reaction channel. The result obtained for the meandering Y mixer is represented in FIGURE 8. IR transmission measurements at a total of 7 measuring points were started at a distance of 2 mm downstream of the mixing point.

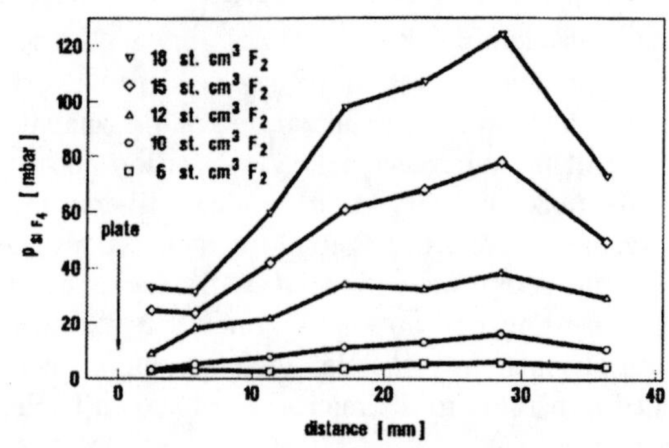

FIGURE 8. Development of the SiF_4 partial pressure along the meandering Y mixer at various stoichiometric $SiCl_4/F_2$ ratios.

In this case the $SiCl_4$ gas amount was 6 st. cm^3 / 60 s and the F_2 gas amount was varied over a large range (6 - 18 st. cm^3 / 60 s). The pressure in the end of the reaction channel was about 500 to 600 mbar. In the substoichiometric case, SiF_4 formation over the entire flow channel is relatively small. In case of stoichiometric and superstoichiometric supply, however, SiF_4 formation is increased significantly. Reduction of the SiF_4 signal at the last measuring point is probably caused by a pressure drop in the product channel and in the subsequent pipes. The IR spectrum recorded at a distance of about 36 mm from the mixing point is shown in FIGURE 9. SiF_4 formation during stoichiometric and superstoichiometric supply of F_2 gas can be noticed. Furthermore, the intermediate reaction products SiF_3Cl (1010 cm^{-1}), SiF_2Cl_2 (995 cm^{-1}) and $SiFCl_3$ (947 cm^{-1}) can be identified.

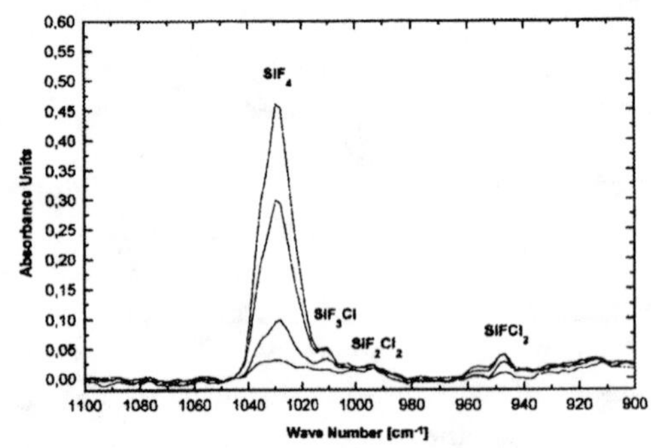

FIGURE 9. Part of the IR spectra recorded at a distance of 36 mm from the beginning of the mixing zone.

CONCLUSIONS

By means of the microcuvette developed, chemical reactions can be measured at any point of micro channel systems. For this purpose, the latter are subjected to IR transmission under an IR microscope. By microengineering methods based on the μEDM technique, any microchannel geometries can be cut into thin stainless steel platelets. In a first test phase, simple mixer types on the basis of a "straight" Y mixer and a meandering mixer were produced and tested successfully for the halogen exchange reaction selected. In the future these first results are to be completed by further physical and chemical boundary values.

ACKNOWLEDGMENTS

We would like to thank Prof. Dr. O.-F. Hagena and Dr. K. Schubert for the useful discussions and consistent support of this work. Our special thanks go to Messrs. P. Abaffy, A. Mayer and M. Schübel for the production of the various Y mixers, the assembly of the test rig, the performance of the measurements and the evaluation of a number of IR spectra.

REFERENCES

1. Ehrfeld,W., Hesse, V., Möbius, H., Richter, Th., Russow,K., "Potentials and realization of microreactors; in Micosystems Technology for Chemical and Biological Microreactors", DECHEMA Monographs, 1996, **132**, pp. 1-28.
2. Möbius, H., Ehrfeld, W., Hessel, V., Richter, Th., "Sensor controlled processes in chemical microreators", presented at TRANSDUCERS'95 and EUROSENSORS IX, Stockholm, Sweden, June 25-29, 1995.
3. Bier, W., Linder, G., Seidel, D., Schubert, K., *KfK-Nachr.* **23 (2-3)**, 165-173 (1991).
4. Bley, P., *Interdisciplinary Science Reviews* **18 (3),** 267-272 (1993).
5. Bier, W., Guber, A., Linder, G., Schaller, T., Schubert, K., *KfK-Bericht* **5238**, 132-137 (1993).
6. Bier, W., Keller, W., Linder, G., Seidel, D., Schubert, K., *Sensors and Actuators* **19**, 189-197 (1990).
7. König,W., *Fertigungsverfahren; Band 3: Abtragen*, VDI Verlag (1989).
8. Guber, A. E., Giordano, N., Loser, M., Wieneke, P., *F&M Feinwerktechnik, Mikrotechnik, Mikroelektronik* **104 (4)**, 247-251 (1997).
9. Gmelins Handbuch der anorganischen Chemie, *Silicium; Teil B*, 686, (1959).
10. Guber, A., Köhler, U., *Journal of Fluorine Chemistry* **58**, 320 (1992).

3.7 METHODS DEVELOPMENT

Polymer Standards for Testing Fourier Transform Infrared Spectrometers

Bryan T. Bowie and Peter R. Griffiths

Department of Chemistry, University of Idaho, Moscow, ID 83844-2343

An ideal sample for monitoring the photometric accuracy of Fourier transform infrared (FT-IR) spectrometers should be a self-supporting solid that is stable over a period of several years. Ideally, the spectrum of this sample should also have no interference fringes and the bands of interest should be stable over a wide temperature range. Two differently processed isotactic polypropylene samples, one stress-relieved and the other extruded, mounted between two BaF_2 windows were investigated for this purpose. The 841-cm^{-1} band of polypropylene has been investigated as a possible wavenumber and photometric standard, because it is sharp, is not overlapped by neighboring bands and absorbs in an atmospheric window. When measured at 2 cm^{-1} resolution, the band centers (as calculated by their center of mass) are 841.68 cm^{-1} and 840.08 cm^{-1} for the stress-relieved and extruded polymer respectively. When the sample temperature changes from 10 °C to 50 °C, the band shifts by 0.03 cm^{-1} and 0.014 cm^{-1} for the stress-relieved and extruded polymer respectively. The full width at half height (FWHH) of the band increases by 0.33 cm^{-1} for both films when the temperature is increased from 10°C to 50°C. Conversely, the absorbance decreases by 0.085 and 0.058 for the stress-relieved and extruded film respectively when the temperature is increased from 10°C to 50°C. The percent deviation of the band area is 4.69 and 2.96 for the stress-relieved and extruded polymer respectively. Another possible issue to be concerned with when using a polymer as a standard is the effect of dichroism. The intensity of the 841-cm^{-1} band of the stress relieved polymer varies by 4.5 % when the orientation of incident polarized radiation is varied by 90° about an axis normal to the plane of the polymer sample. When two layers of the polymer are overlaid such that the backbone chains of each layer are oriented orthogonally, the intensity of the analogous band for the extruded polymer changes by only 1.2 %. The results of studies on the effect of calculating the band center using a center of mass or a curve fit algorithm and the effect of band center shift due to small changes of sample position are also reported.

INTRODUCTION

An ideal sample for monitoring the photometric performance of FT-IR spectrometers should be a self-supporting solid that is stable over a period of several years. An isotropic polymer film of the appropriate thickness mounted between two windows to minimize the effect of interference fringing and to prevent the sample from being oxidized by the atmosphere would appear to have properties approaching the optimum. The polymer spectrum should contain at least one isolated (ideally Lorentzian) band with an absorbance of between 0.2 and 2 absorbance units (AU). The band should be broad enough that it can be readily measured with a resolution parameter, ρ[1,2] of 0.2 or less, but narrow enough that it is separated from other bands in the spectrum by at least 4 times its full width at half height (to allow it to be measured at a variety of different spectrometer resolutions).

Preliminary studies in our laboratory indicate that isotactic polypropylene films may fulfill these criteria. One particular band, at 841 cm^{-1}, in the spectrum of isotactic polypropylene (iPP) seems exceptionally promising as an FT-IR standard. We have, therefore, studied the temperature, polarization and resolution dependence of the center wavenumber and peak absorbance of this band.

EXPERIMENTAL

Two types of iPP were used in this investigation. The first type was a 50-μm thick stress relieved material acquired from Transilwrap (Chicago IL). The polymer chains in this material are fairly (but not totally) unoriented so the spectrum exhibited only a small amount of dichroism. The second type was a 6-μm thick extruded material acquired from Chemplex (Tuckahoe, NY). This material is more ordered and the spectrum exhibited greater dichroism.

It was determined that the window of choice would be a 2-mm thick BaF_2 plate from International Crystal Laboratories because of the index matching between the window and the

CP430, *Fourier Transform Spectroscopy:* 11th International Conference
edited by J.A. de Haseth
 1-56396-746-4/98/$15.00

polypropylene sample.

Spectra were obtained using a Bomem MB-100 FT-IR spectrometer with a DTGS detector that was known to produce a linear response. The sample was placed in a variable-temperature sample cell (Harrick Scientific, Ossining, NY) with a Fisher Isotemp refrigerated circulator to maintain and control the temperature.

RESULTS AND DISCUSSION

a. Sample Dichroism

Because of the properties of the materials used in the fabrication of beamsplitters, the beam in every FT-IR spectrometer is polarized to some extent. Thus any dichroism exhibited by a polymer film will reduce its desirability for use as a photometric standard for FT-IR spectrometry. In practice, both of the iPP samples that were tested exhibited some level of dichroism.

As expected, the stress-relieved isotactic polypropylene sample studied first had the lowest dichroism. A series of spectra was measured as the polarizer was rotated through 90°. The absorbance at 841 cm^{-1} varied by about ±4.5% for this sample. When the same experiment was performed using the extruded iPP, the absorbance varied by more than a factor of 2 as the polarizer was rotated through 90°. To minimize the effect of dichroism, 2 layers of the sample were mounted so that the polymer chains in each layer were mutually perpendicular. In this case, the absorbance at 841 cm^{-1} varied by about ±1.2%. Orientating the sample in this manner simulates the effect of orientation of the polymer chains that would be obtained if the sample was spin-cast on the window material, and appears to be a most promising means of eliminating the effect of chain orientation

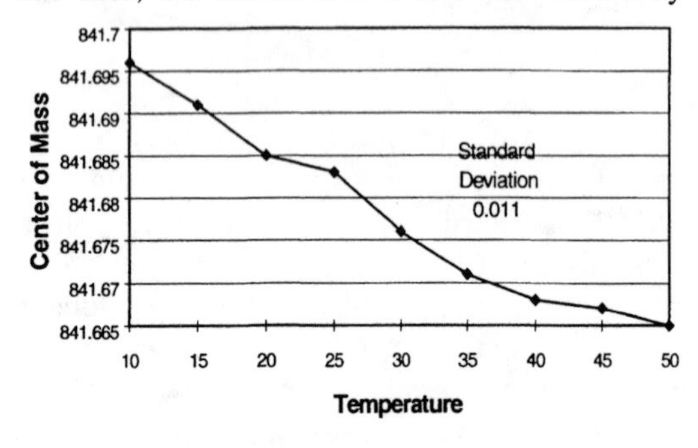

FIGURE 1. Center of Mass Calculated from 2 cm-1 Resolution Spectra of Stress-Relieved iPP

b. Reproducibility of Band Center Estimation

The position, FWHH and peak absorbance of a band can vary significantly as a function of temperature. Obviously, the bands in an ideal sample will exhibit no significant change in the center wavenumber, FWHH and peak absorbance of analytically useful bands over a wide temperature range. Three methods are commonly used for the determination of the center of a band. In the first, the center of mass (gravity) is calculated[3]. The second involves fitting an entire band or band multiplet to one or more synthetic bands with well defined shapes; this method is usually known as curve fitting. In the third method, known as either the cubic spline or peak-fitting methods, a cubic equation is fitted to the points around the top of the peak and the second derivative is calculated; the wavenumber at which the second derivative is zero is the maximum value of the peak. Different algorithms are typically recommended by different spectrometer manufacturers. The center-of-mass and curve-fitting methods, as incorporated in GRAMS/32 supplied by Galactic Industries Corporation, were tested in this study, with the center-of-mass algorithm typically having twice as good reproducibility as the curve-fitting algorithm.

To test the temperature dependence of the 841-cm^{-1} band of iPP, spectra were measured every 5°C from 10°C to 50°C. Spectra of the stress-relieved iPP sample were measured at 2-cm^{-1} resolution (unapodized); the standard deviation of the peak center determined by the center-of-mass algorithm was 0.011 cm^{-1}. Typical results are shown in Fig. 1. The same procedure used with the two-layer extruded iPP sample gave a standard deviation of 0.005 cm^{-1}, see Fig. 2.

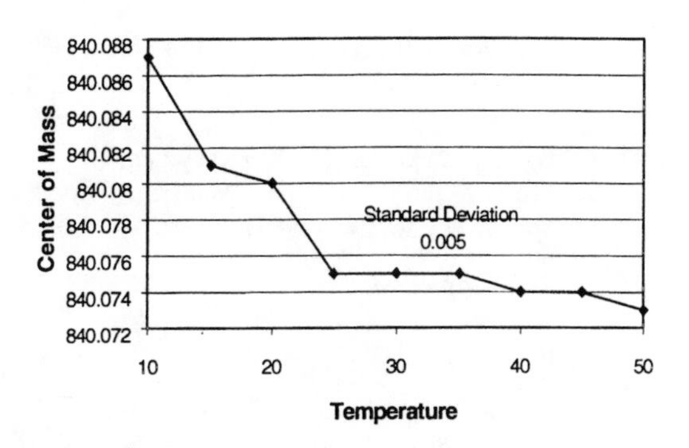

FIGURE 2. Center of Mass Calculated from 2 cm-1 Resolution Spectra of Extruded iPP

The standard deviation of the band center for the series of spectra of the stress-relieved iPP samples measured between 10 and 50°C obtained by curve fitting using a mixed Lorentzian/Gaussian band shape was 0.022 cm^{-1}, which was twice that of the center-of-mass operation. The corresponding standard deviation for the extruded iPP sample was 0.017 cm^{-1}. Thus it was concluded that the center-of-mass method gave the more reliable results. Perhaps of more importance was the fact that the actual value of the peak center reported using the two algorithms were different by about 0.2 cm^{-1}, possibly because of a small peak asymmetry. Thus to obtain consistent results, the same algorithm should always be used.

Two factors can affect the value of the peak position reported by the software, the reproducibility of the placement of the sample in the beam, discussed in *d* below, and the number of interpolated data points between each independent datum; in FT-IR spectrometers, the latter is usually determined by the extent of zero filling applied to the interferogram.

c. Variation of Band Width, Height and Area with Temperature

The fact that the center wavenumber changes by considerably less than 0.1 cm^{-1} as the temperature is varied by 40°C suggests that the 841-cm^{-1} band is a good one to use as a standard for testing the performance of FT-IR spectrometers. We were also interested in whether iPP could be used as a photometric standard for commercial FT-IR spectrometers. To be a good photometric standard, the height and width of the band should also remain constant as the temperature of the sample is changed. In practice, this is much more difficult to achieve as the FWHH of most samples increases with temperature. The FWHH of the 841 cm^{-1} band of extruded iPP, as estimated by curve-fitting a spectrum measured at 2 cm^{-1} resolution, is approximately 3.3 cm^{-1} and increases from 3.1 to 3.5 cm^{-1} as the temperature is increased from 10 to 50°C. When the iPP spectrum is measured at 8 and 16 cm^{-1}, the shape of the 841-cm^{-1} band changes from largely Lorentzian to that of the sinc function corresponding to the instrument line shape (ILS) function of the spectrometer[4]. (None of these spectra were apodized). For low resolution spectra, the measured FWHH is approximately equal to 0.61 divided by the maximum optical path difference, i.e., the width is about the same as the FWHH of the sinc ILS function of an unapodized spectrum. For such spectra, the 841-cm^{-1} band shows very strong side-lobes, as expected. The band in the spectrum of the (less oriented, and hence probably less crystalline) stress-relieved iPP is about 4.3 cm^{-1}, with slightly less temperature dependence than the extruded polymer. As would be expected, the peak absorbance of the band decreases with temperature for the extruded iPP and for stress-relieved iPP. Not surprisingly, the band area varies quite significantly with temperature for both samples. If one assumes that the temperature of most labs is held between 68 and 72°F (20-22°C), however, it is probable that the greatest variance in FWHH, peak height and band area will come from the amount by which the sample is heated by the infrared beam in the sample compartment.

d. Placement of Sample and Reproducibility

Small changes in the position at which the sample cell is mounted in the sample chamber can have large effects on the calculated band center if the diameter of the sample is smaller than the diameter of the beam in the sample chamber, *i.e.*, if *vignetting* occurs[5]. This effect can be seen in Figure 3, where the sample cell was bumped between the measurements of

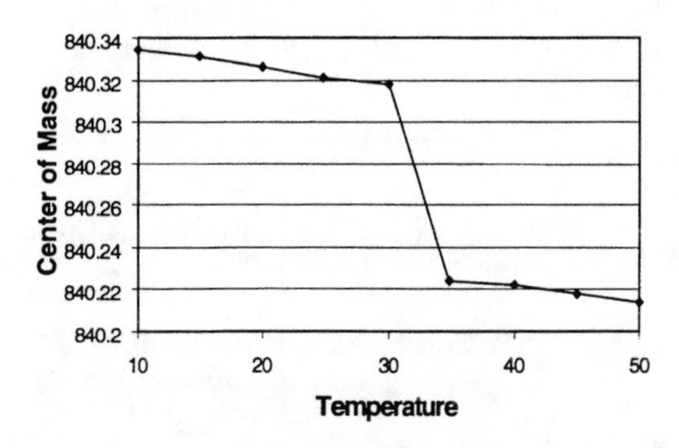

FIGURE 3. Center of Mass Calculated from 8 cm-1 Resolution Spectra of Extruded iPP

30 and 35°C by an amount that was enough to cause the discontinuity in the series. Ideally, the diameter of the sample should be large enough to avoid vignetting. The cell used for our measurements did not allow us to avoid vignetting, however. This leads to the question of how reproducible the series is from run to run instead of within the series itself. There is difference from run to run using the same sample, and sample position reproducibility is an important factor.

CONCLUSIONS

Extruded isotactic polypropylene is a reasonable choice for an a performance standard when two films are mounted at 90° to each other to avoid the effects of dichroism.

For the calculation of band centers, there appears to be a statistical advantage in using the center-of-mass algorithm over peak-fitting methods. This gives the additional advantage that the center of mass algorithm is more easily automated.

At least a 4x zero fill is required to determine the band center with any degree of statistical accuracy at resolutions of 2 cm^{-1} through 8 cm^{-1}. Because of the sharpness of the 841-cm^{-1} band (~3 cm^{-1} FWHH), the spectra measured at 8 cm^{-1} resolution using boxcar truncation do show the effects of side lobes and an appropriate apodization function should be applied.

The reproducibility of sample position appears to be more important than the effect of band shift due to temperature. Thus, either the sample must have a significantly larger diameter than the beam or great care must be taken during the positioning to guarantee reproducibility from run to run.

Both the width and height of the 841-cm^{-1} band of isotactic polypropylene are somewhat temperature dependent. Nonetheless, this band has many of the properties that are desirable of a wavenumber and photometric standard.

REFERENCES

1. R. J. Anderson and P. R. Griffiths, *Anal. Chem.*, **47**, 2339 (1975).

2. P. R. Griffiths and J. A. de Haseth, *Fourier Transform Infrared Spectrometry*, pp. 23-25, Wiley Interscience, New York, NY (1986).

3. D. G. Cameron, J. K. Kaupinen, D. J. Moffatt and H. H. Mantsch, *Appl. Spectrosc.*, **36**, 245 (1982).

4. Ref. 2, pp. 9-23.

5. Ref. 2, pp. 33-39.

Studies of a Polystyrene Wavenumber Standard for Infrared Spectrophotometry

Changjiang Zhu and Leonard M. Hanssen

Optical Technology Division, National Institute of Standards and Technology (NIST), Gaithersburg, MD 20899

Standard Reference Material (SRM) 1921 is a matte finish polystyrene film available from NIST for use in calibrating the wavenumber scale of spectrometers in the infrared spectral region from 545 cm^{-1} to 3082 cm^{-1}. New results of the dependence of the calibrated peak values on measurement resolution and peak determination method are presented. Appropriate zero filling of the interferogram was found to significantly decrease peak value dependence on resolution. For any resolution ≤ 4 cm^{-1}, as determined by the center of gravity method, the peak values differ by amounts less than 0.01 cm^{-1}, the least significant digit of the calibration values specified in Ref. [2]. A method of extrapolating the peak minimum for an absorption band in the transmission spectrum was carried out through a chi square fit on a series of peak values obtained at different peak fractions of the same band. The extrapolated minimum peak values are presented for the first three bands of polystyrene.

INTRODUCTION

The center of gravity (CG) method is a peak value determination algorithm described by D. G. Cameron, J. K. Kauppinen et al in reference [1]. For an absorption band in a transmission spectrum, its center of gravity, ν_{CG}, is given by

$$\nu_{CG} = \frac{\int_{\nu_i}^{\nu_k} \nu[T_f - T(\nu)]d\nu}{\int_{\nu_i}^{\nu_k} [T_f - T(\nu)]d\nu} \tag{1}$$

where f is a fraction of the peak height as determined by the smaller peak side, T(ν) is the transmittance, and ν_i and ν_k are the wavenumbers at which $T(\nu) = T_f$. The uncertainty associated with the use of the CG method to determine a peak value is at its minimum near f = 0.5 [1]. Since NIST first made available polystyrene as SRM 1921 in 1994 [2], others [3] have studied the dependence of the calibrated wavenumber values on various parameters such as resolution and temperature. We have expanded on those efforts with a view to make this standard more useful to the community. Here we present some of the results.

RESULTS AND ANALYSIS

First an extensive investigation was conducted on the influence of resolution (r) on peak position value for the CG algorithm. Twelve polystyrene samples were measured sequentially six times (72 measurements in all) under purge using a Bomem DA3 FT-IR spectrometer [4]. The instrumental conditions were as follows: r = 0.5 cm^{-1}, aperture = 0.1 mm, scan speed = 0.5 cm/s, gain = 4, scan number = 100, Hamming apodization, KBr beamsplitter, globar source, and MCT detector.

In order to eliminate the effects caused by measurement-to-measurement variation, the lower resolution (i.e. 1.0, 2.0 and 4.0 cm^{-1}) spectra were obtained by appropriate truncation of the interferograms acquired at a resolution of 0.5 cm^{-1}. The influence of resolution change on a representative polystyrene absorption band at 1028 cm^{-1} (FWHH = 11.1 cm^{-1}) can be seen in Fig. 1. Significant peak position shifts are observed. For example, the

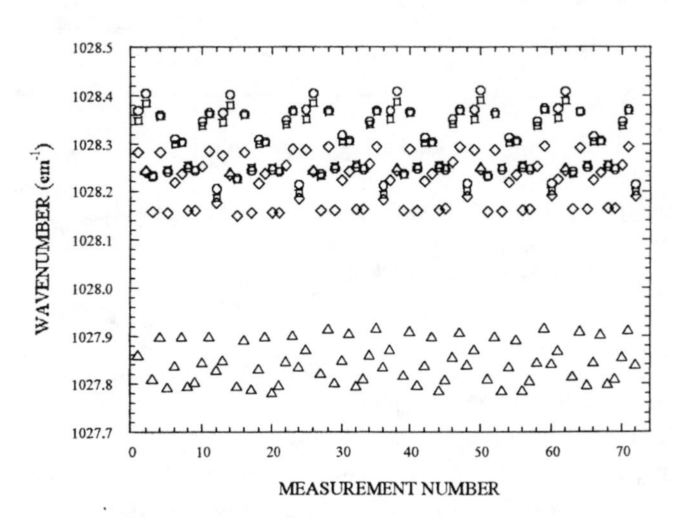

FIGURE 1. 1028 cm^{-1} peak values of polystyrene, measured at: ○, r = 0.5 cm^{-1}; □, r = 1.0 cm^{-1}; ◇, r = 2.0 cm^{-1} and △, r = 4.0 cm^{-1} without zero filling.

average peak value difference between that of r = 0.5 cm^{-1} and r = 4.0 cm^{-1} is about 0.42 cm^{-1} which surpasses the

CP430, *Fourier Transform Spectroscopy:* 11th International Conference
edited by J.A. de Haseth

maximum of the measurement-to-measurement variation (which is nearly all due to sample-to-sample variation) of ~ 0.2 cm^{-1} for this band.

This peak value shift of 0.42 cm^{-1} with resolution degradation is caused by the data point interval increase. Having fewer data points along the band contour results in a weight distribution shift when the CG method is applied. To examine this effect more carefully, the original interferograms were processed in such a way as to maintain the data point interval while degrading the spectral resolution. This was done by substituting zeros for the truncated portions of the interferograms. This corresponds to zero filling the lower resolution interferograms so that all the resulting spectra have the same data point interval. For the 1028 cm^{-1} band of polystyrene, the resulting peak values obtained from the different resolution spectra for any individual measurement are almost identical as shown in Fig. 2.

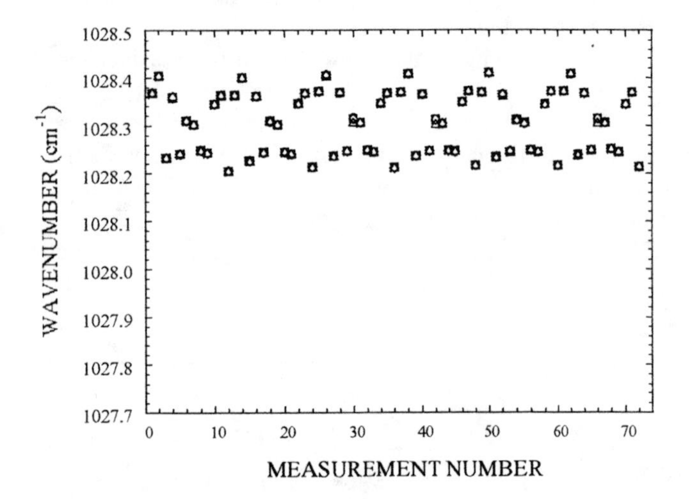

FIGURE 2. 1028 cm^{-1} peak values of polystyrene, measured at: ○, r = 0.5 cm^{-1}; □, r = 1.0 cm^{-1}; ◇, r = 2.0 cm^{-1} and △, r = 4.0 cm^{-1} with zero filling.

This significant result of the independence of peak value from resolution, when using the CG method, was observed for all 13 calibration bands of the polystyrene. The 13 peak mean ($\bar{\nu}$) differences between those measured for r = 0.5 cm^{-1} and r = 1.0, 2.0 or 4.0 cm^{-1} are summarized in Table 1. Each mean represents the average of all 72 measurements (12 samples measured 6 times).

To accommodate users who have access only to software which locates the peak minima, we have employed an extrapolated center of gravity method to produce peak minima values for SRM 1921. For each band, a series of peak values were obtained using the CG method with different peak fraction, f. These data were used to extrapolate to the band minimum at f = 0. Since the portions of the peaks used in the CG calculation for the smaller fraction contains fewer data points (than f = 0.5), a higher level of zero filling is required to produce good results. This can be seen in Fig. 3. Thus, all the spectra were zero filled to a level equivalent to that for spectrum of resolution 0.06 cm^{-1}.

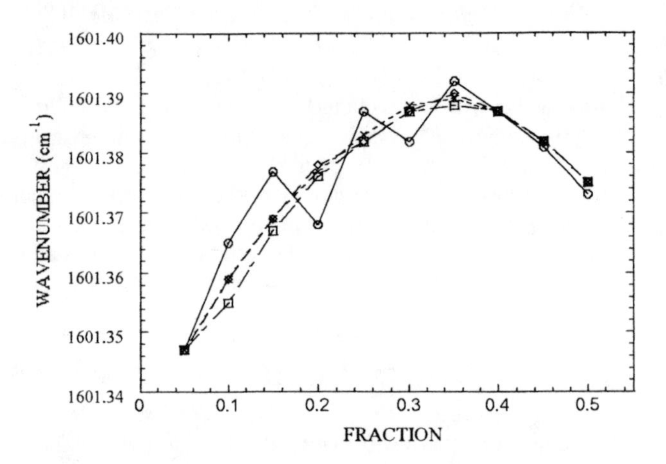

FIGURE 3. Effect of different levels of zero filling on the 1601 cm^{-1} band when r = 0.5 cm^{-1} and the peak fraction is changed from 0.05 to 0.5. The zero filling level is changed as: ○, 1 times; □, 2 times; ◇, 4 times, and ×, 8 times.

Peak value vs. Fraction plots for 4 bands of polystyrene from a single spectrum are shown in Fig. 4. One can observe the following:

(1) Each band has its own wavenumber vs. fraction curve pattern which is determined by the peak contour.

(2) The peak values associated with the different resolutions, tend to vary more as the CG fraction is reduced. The peak value at f = 0.5 is one of the stable points which are the least affected by resolution change.

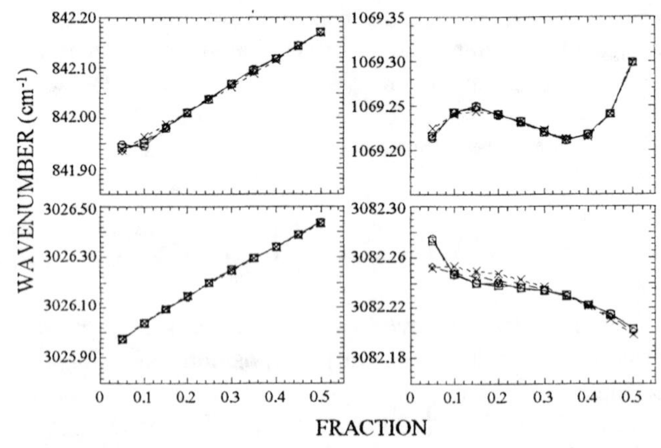

FIGURE 4. Dependence of peak values of 4 polystyrene bands on CG fraction and resolution. The different resolution results are shown as: ○, r = 0.5 cm^{-1}; □, r = 1.0 cm^{-1}; ◇, r = 2.0 cm^{-1}; and , ×, r = 4.0 cm^{-1}.

TABLE 1. 13 Peak Mean Differences Between That Measured at Resolution = 0.5 cm^{-1} and at Lower Resolutions

peak (cm^{-1})	without zero filling			with zero filling		
	$\bar{v}_{r=0.5}-\bar{v}_{r=1.0}$	$\bar{v}_{r=0.5}-\bar{v}_{r=2.0}$	$\bar{v}_{r=0.5}-\bar{v}_{r=4.0}$	$\bar{v}_{r=0.5}-\bar{v}_{r=1.0}$	$\bar{v}_{r=0.5}-\bar{v}_{r=2.0}$	$\bar{v}_{r=0.5}-\bar{v}_{r=4.0}$
545	0.0300	-0.3403	-1.3639	-0.0018	-0.0049	-0.0059
842	0.0403	0.0081	-0.1388	0.0005	0.0005	0.0006
906	0.0317	0.0159	-0.0222	0.0001	0.0002	0.0001
1028	0.0317	0.0865	0.4244	-0.0002	-0.0002	-0.0015
1069	0.0415	0.0199	0.0980	0.0002	0.0002	0.0000
1154	0.0195	-0.0215	-0.4097	0.0000	-0.0007	-0.0016
1583	0.0836	0.0593	-0.3846	-0.0001	-0.0009	-0.0035
1601	0.0067	0.0150	0.1545	-0.0002	-0.0001	0.0002
2850	-0.0547	0.0317	-0.0715	-0.0002	0.0002	-0.0002
3001	0.0081	0.0322	-0.3992	0.0002	-0.0002	-0.0002
3026	-0.0500	0.0276	0.1096	0.0000	0.0000	0.0005
3060	-0.0171	0.0005	0.0549	0.0002	-0.0002	0.0002
3028	-0.0146	0.1155	1:2263	0.0005	0.0002	0.0005

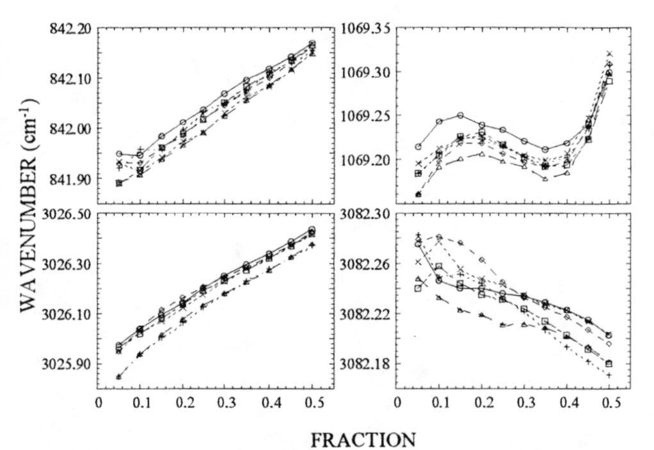

FIGURE 5. Six measurements of the same sample at r = 0.5 cm^{-1}. In the plot, the measurement number are as follows: ○, 1; □, 2; ◇, 3; ×, 4; +, 5; and △, 6.

The peak value vs. CG fraction for 6 measurements of the same sample with resolution of 0.5 cm^{-1} is shown in Fig. 5. In each case, the measurement-to-measurement variation of peak value is larger than that due to resolution change (Fig. 4).

Now we proceed to extrapolate the peak fraction results to find the peak minimum values. Since it is difficult to fit irregular curves such as some of these shown in Fig. 4, the extrapolated minimum operation was performed on a reduced range of CG fraction ($0.01 \leq f \leq 0.1$). The peak values obtained at this range of CG fraction for the 906 cm^{-1} of polystyrene are shown in Fig. 6. The dashed line in this plot represent the mean of 6 measurements of the same sample and error bars stand for the standard deviation of the mean. To make the extrapolation more accurate at lower fraction, chi square fit, a linear curve fit with weighting by the inverse of the standard deviation, was applied to the data. This fit is shown as a solid line in Fig. 6. Table 2 summarizes the chi square fitting results of the first three bands of polystyrene for r = 0.5, 1.0, 2.0 and 4.0 cm^{-1}.

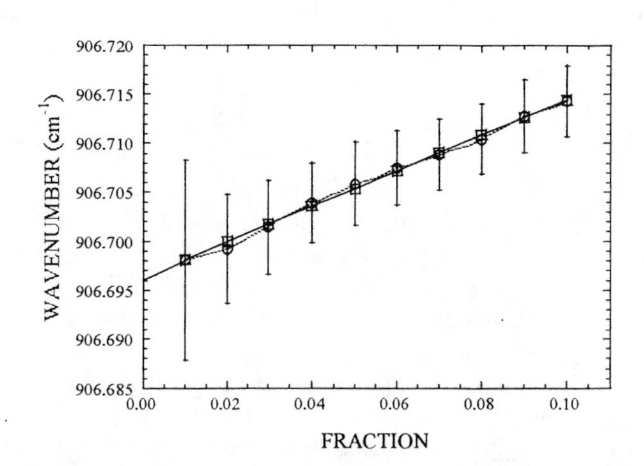

FIGURE 6. Chi square fitting on mean of 6 measurements with standard error as error bars at r = 0.5 cm^{-1}. In the plot, the dashed line with ○ represents the mean of 6 measurements and the solid line with □ stands for chi square fitting result.

CONCLUSIONS AND FUTURE WORKS

Appropriate zero filling is a very effective way to improve peak value determination for spectra measured at low resolution, especially for the CG method and specifically with f = 0.5.

The approach of the extrapolated CG method to obtain the peak minimum will be used to provide another set of reference data to SRM 1921. We will extend the

TABLE 2. Chi Square Fit on Mean Peak Value vs. Fraction Plots of Three Bands of Polystyrene, Comparison of Extrapolated Peak Minimum with Peak Value at f = 0.5

Peak (cm^{-1})	r (cm^{-1})	Slope (cm^{-1}/f)	Extrapolated Peak Minimum (cm^{-1})	Chi ^{2}	Peak Values (cm^{-1}) at f = 0.5
545	0.5	8.4487	539.855	0.17782	547.071
545	1	8.3836	539.872	0.19366	547.052
545	2	8.4154	539.875	0.09036	547.055
545	4	8.3842	539.879	0.05103	547.048
842	0.5	0.09729	841.916	0.17188	842.161
842	1	0.09714	841.917	0.32188	842.158
842	2	0.29563	841.901	0.09881	842.158
842	4	0.57009	841.875	0.00152	842.158
907	0.5	0.18028	906.696	0.01081	906.841
907	1	0.18208	906.699	0.01808	906.839
907	2	0.17629	906.696	0.01101	906.839
907	4	0.17671	906.697	0.02009	906.839

extrapolated CG method to all of our polystyrene data and generate calibrated minimum values for the SRM 1921.

Comparison of this method with other commercially available minimum find algorithms will be carried out. We will also extend the ongoing investigation of the effects of temperature change on peak value.

REFERENCES

1. Cameron D. G., Kauppinen J. K., Moffat D. J. and Mantsch H. H., *Appl. Spectrosc.*, **36**, 245-250, 1982
2. Gupta D., Wang L., Hanssen M. L., Hsia J. J. and Dalta R. U., *NIST Special Publication 260-122*, 1995
3. Spragg R. A. and Billingham M., *Spectrosc.* **10(1)**, 41-43, 1995
4. The use of a trade name or manufacturer in this publication is for identification purposes only and does not imply endorsement of the product by the National Institute of Standards and Technology.

Diffuse Reflectance Measurements at Nanogram Levels by Infrared Fourier Transform Spectroscopy

Guilan Tao, Qijia Fan, Tao Chen

Beijing Institute of Microchemistry, P. O. Box 1932, Beijing 100091, P. R. China

INTRODUCTION

Diffuse reflectance accessories have been widely used with infrared Fourier transform spectroscopy for almost 20 years. These accessories can in large part be attributed to their versatility for solids, liquids, powder, black body and some samples that are handled with difficulty in routine work. Also the measurement sensitivity is increased greatly. Especially in micro- and ultramicro analysis, the desired results can be obtained for diffuse reflectance spectrum measurements at microgram level by improving sample preparation methods.

In this report we describe how to measure the samples at nanogram levels and to obtain spectra with the desired signal-to-noise ratio in normal scanning times.

EXPERIMENTAL

Instrumentation A Digilab FTS-60 Fourier transform infrared spectrometer equipped with Globar source and TGS detector was used for the diffuse reflectance measurements. A Harrick Diffuse Reflectance Accessory was used. A microscope for sample preparation as well as a fine pointed probe (dissecting needle) and micro syringe and any other apparatus suitable for micro manipulation are also necessary. The solvents purified were provided. Powdered KBr (about 10-15μm in particle size) is prevented from contamination and moisture. Cleanliness throughout the preparation is imperative because of the small amount of sample involved in the analysis.

Sample preparation Generally powdered KBr was moved to sample cup without compression and was leveled using a spatula. Powdered samples were placed on a KBr surface of the cup by the aid of a microscope or samples dissolved in solvents were moved on the cup (solvents must evaporate easily). Crystallites or some difficult samples were shifted to the KBr surface by a pointed probe under the microscope. Liquid compounds were directly dropped to the surface of the cup and which kept the surface smooth and flat.

In this paper all diffuse reflectance spectra were recorded at 8 cm^{-1} resolution and 40 scans. The type of the spectra was absorbance spectrum of diffuse reflectance.

RESULTS AND DISCUSSION

The conventional method of sample preparation for diffuse reflection measurements at the nanogram level can not be used. Because of contamination and loss of samples, samples could not be prepared by the method of grinding the mixture of KBr and compounds. Impurities in solvents and contamination in KBr must be eliminated. Otherwise, the deviation of the determination will occur. A dust particle in the atmosphere is frequently 10-100 times the weight of the samples being considered. For the mentioned reasons, it is beneficial to use the solvent transform method for diffuse reflectance work at the ultramicro level.

For diffuse reflectance spectra the incident infrared beam can penetrate as much as 3-5 mm. So the sample cup depth is generally less than 5 mm. According to K-M equation, the greatest contribution to $f(R_\infty)$ originates in the first few layers of samples (1-2). For the microsample analysis of diffuse reflectance in our laboratory, we put samples on the KBr surface, so that the weakly incident infrared signal was enhanced and the sensitivity was raised greatly.

Figure 1 is a diffuse reflectance spectrum of 50ng of carbazole. From that we can see some bands belong to alcohol, as the sample was dissolved in it. For example, 2971, 2921, 2850, 1050, 888 cm^{-1} etc. belong to the alcohol. Although the sample amount is very low, the main peak of the compound can be distinguished.

CP430, *Fourier Transform Spectroscopy:* 11th International Conference
edited by J.A. de Haseth

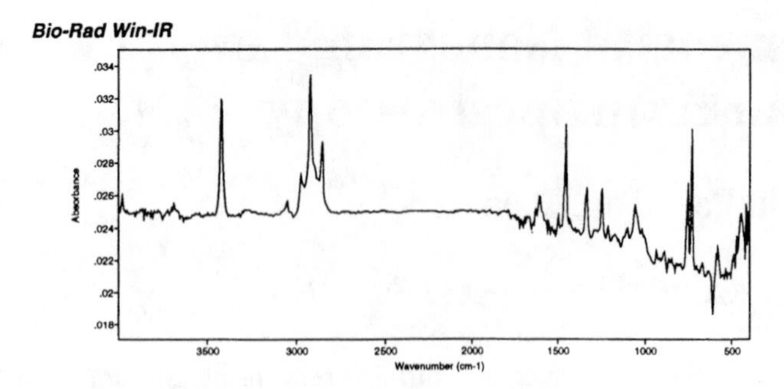

Figure 1. 50ng carbazole diffuse reflectance spectrum

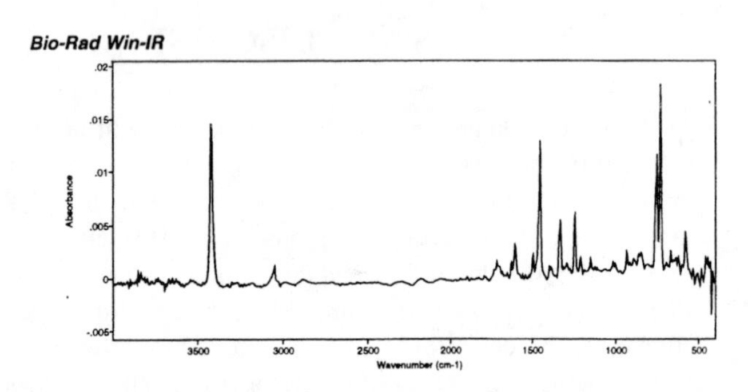

Figure 2. 100ng carbazole diffuse reflectance spectrum

Figures 2 and 3 are 100 ng and 500 ng carbazole diffuse reflectance spectra, respectively. It is very obvious that the spectra of the 500ng sample are sufficiently high quality that spectral subtractions will be feasible.

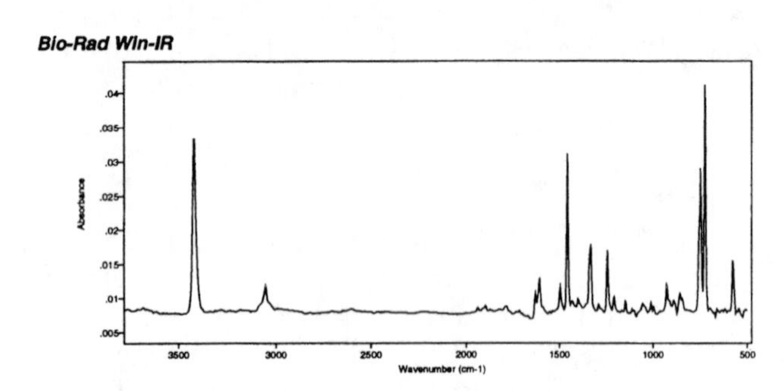

Figure 3. 500 n g carbazole diffuse reflectance spectrum

Figure 4 shows the diffuse reflectance spectrum of one tiny crystallite of Cu / C_{60} sample that is too small to be visible to naked eye. Under the microscope the particle was placed on the surface of the cup and measured. Results were comparable with IR microscopy. Whereas the wavenumber regions for the diffuse reflectance measurements is 4000 - 400 cm^{-1} and for the microscopy 4000-700 cm^{-1} (narrow range detector).

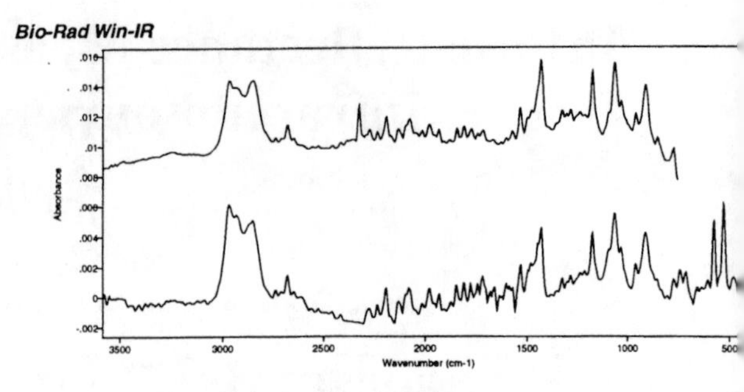

Figure 4. (below) Cu / C_{60} diffuse reflectance spectrum (above) Cu/C_{60} absorbance spectrum by microscopy

Figure 5 shows 500 ng caffeine diffuse reflectance spectrum. Figure 6 shows 50 ng and 80 ng PBBO sample diffuse reflectance spectrum.

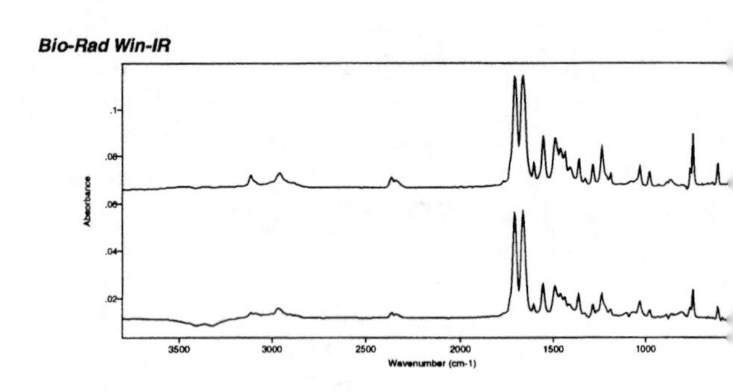

Figure 5. (below) 500ng caffeine diffuse reflectance spectrum (above) caffeine diffuse reflectance spectrum

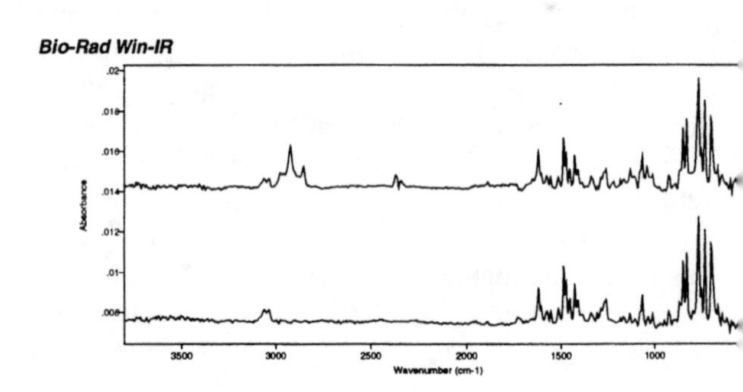

Figure 6. (below)80 ng PBBO diffuse reflectance spectrum (above) 50 ng PBBO diffuse reflectance spectrum. PBBO: 2-(4'-Biphenylyl)-6-phenylbenzoxazole.

The above data indicate that the sensitivity of the diffuse reflectance technique has reached the nanogram level. The microanalysis technique depends on improved microsample manipulation methods. These methods are so important that we can identify some micro amounts of material that could not be identified before.

REFERENCES

1. Fuller, M. P. and Griffiths, P. R., Anal. Chem. 50, 1906 (1978)
2. Kuehl, D. and Griffiths, P.R., J. Chromatogr. Sci. 17, 471 (1979)

Diffuse Reflectance mid-Infrared Fourier Transform Spectroscopy of Cellulose-Based Materials

J. E. Parks II[a], D. S. Blair[b], and G. L. Powell[a]

[a]*Oak Ridge Centers for Manufacturing Technologies*, P. O. Box 2009, Oak Ridge, TN 37831-8084*
[b]*Sandia National Laboratory, P. O. Box 5800, Albuquerque, NM 87185-0755*

Diffuse Reflectance mid-Infrared Fourier Transform (DRIFT) spectroscopy has been applied to the analysis of cellulose-based materials including wood, paper, and paper products. High-quality DRIFT spectra representing 1 mm spatial resolution on the substrate were obtained from different locations on the sample using a mechanical positioning technique. The analysis of the spectra results in spectral images showing the spatial variation in the chemistry of the sample. Chemical variations in cellulose-based samples have been observed due to variations in ink content, coating thickness, and concentration of phenyl and carbonyl groups associated with lignin.

INTRODUCTION

The production of paper and other cellulose-based products is an energy and material consuming process. Recycling is one method used by the forest products industry to conserve energy and materials. However, recycled stock can contain materials with a wide variety of characteristics that may or may not be suitable for a particular industrial process. In order to maximize the value of recycled materials, characterization of the recycled stock is desired. By characterizing the recycled stock according to qualities desired by the forest products industry, recycled stock can be selected for input into the manufacturing process so that process/machinery breakdown is minimized, energy use is minimized, pollution is minimized, and product quality is maximized.

Diffuse Reflectance mid-Infrared Fourier Transform (DRIFT) spectroscopy has been applied to the characterization of cellulose-based materials that are commonly found in recycled stock. DRIFT spectroscopy is sensitive to the chemical nature of the sample surface (with skin depths of ~1 μm), and cellulose materials are excellent samples for investigation in the diffuse reflectance geometry due to the surface structure of paper, cardboard, and wood. The spectral range of the work presented here is 4500 cm^{-1} to 500 cm^{-1}; thus, fundamental vibration modes are detected and lead to characterization of the chemical nature of the sample.

The presence of ink, coatings, and lignin on or in cellulose-based materials has been detected with DRIFT spectroscopy. Spectral images showing chemical variations in the sample have been generated by obtaining DRIFT spectra from multiple positions on the sample. The map data show a quantitative measurement of sample chemistry that can be used to characterize stock in terms useful for input into industrial processes.

*Managed for the U. S. Department of Energy by Lockheed Martin Energy Systems, Inc. under Contract No. DE-AC05-84OR21400.

EXPERIMENT DESIGN

Mid-infrared absorbance spectra were obtained with a Fourier transform interferometer-based device (Model SOC-400, Surface Optics Corporation, San Diego, CA 92127). The SOC-400 incorporates an ellipsoidal mirror for collection of the diffusely scattered light. The mirror extends a full 360° angle about the central axis of the ellipse; thus, a large amount of the diffusely scattered light is collected. The sample is positioned at one focal point of the ellipsoid, and the detector is positioned at the other focal point. Diffusely reflected light makes one reflection off of the mirror in route to the detector. The large collection solid angle and the single reflection surface combine to make an extremely efficient system for collecting diffusely scattered light.

The sample was scanned in two dimensions beneath the SOC-400 with translation stages driven by stepper motors. During spectrum acquisition the sample remained still. Array basic programs written in conjunction with the GRAMS software package (GRAMS/32, Galactic Industries Corporation, Salem, NH 03079) controlled positioning of the motors and spectra collection.

No sample preparation was required. However, care was taken to avoid contamination of samples since the technique is sensitive to surface oils and other contaminants. Purge gas (argon) was emitted from the sample aperture of the instrument onto the sample during spectral acquisition.

Spectra in the range of 4500 cm^{-1} to 500 cm^{-1} were collected with 8 cm^{-1} resolution. Typically, 16 interferometer scans were averaged to produce one spectrum. Absorbance spectra were computed with reference to a rough Au sample. Upon completion, an experiment yields a two dimensional array of spectra that correspond to positions in the two-dimensional spatial scan of the sample. The spectra are analyzed to obtain an image displaying the spatial distribution of a particular chemical

CP430, *Fourier Transform Spectroscopy:* 11th International Conference
edited by J.A. de Haseth
1998 The American Institute of Physics 1-56396-746-4/98/$15.00

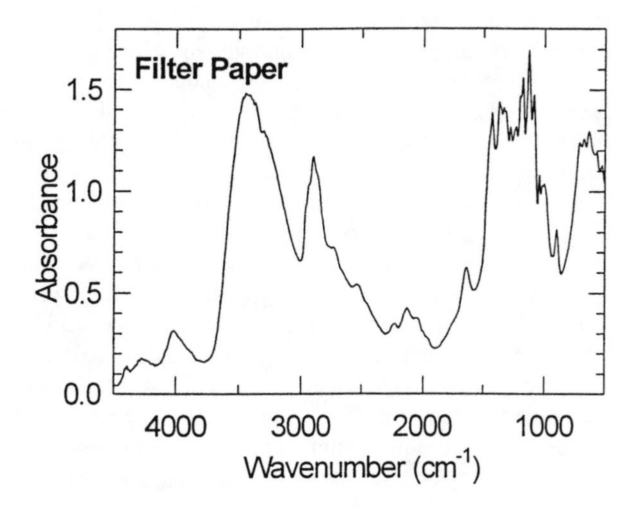

FIGURE 1. DRIFT spectrum of filter paper. The cellulose chemistry is evident in the O-H and C-H stretch vibrational mode bands.

moiety on the sample. The maximum spatial resolution obtained by the technique is limited by the 1 mm spot size of illumination on the sample.

SPECTRA AND SPECTRAL IMAGES

Figure 1 shows a DRIFT spectrum of filter paper in the mid-infrared region. Filter paper, which by necessity must be free of extractants and impurities, was chosen as a sample to illustrate the chemical nature of cellulose. Strong absorbance peaks occur at 3450 cm^{-1} and 2900 cm^{-1} due to absorption of light by stretching mode vibrations of O-H and C-H bonds in cellulose, respectively. Other strong absorbance features occur in the fingerprint region (1500-500 cm^{-1}). Since paper is an excellent diffuse scattering medium, no features from spectral reflecting light occur in the spectrum.

Ink Content in Paper

The presence of ink on paper is indicated in DRIFT spectra of paper by a baseline shift and the presence of absorbance peaks due to ink chemistry and/or ink interactions with paper. The baseline shift is due to absorption of the infrared light by elements in the ink; carbon is the primary absorbing element in black inks. Ink chemistry will vary greatly for the wide variety of available inks, and absorbance bands due to bonds between ink and paper will vary with the ink chemistry.

Figure 2 shows DRIFT data taken from the logo of the ICOFTS-11 Symposium on the call for papers announcement (Second Circular). The sample contained two inks (black and red) which affect the paper spectra uniquely. The black ink caused a large baseline shift in the spectrum of paper due to the strong absorbing properties of the carbon in the ink. Absorbance peaks characteristic of the inks occur at 1729 cm^{-1} for both inks and 1498 cm^{-1} for the red ink.

(a)

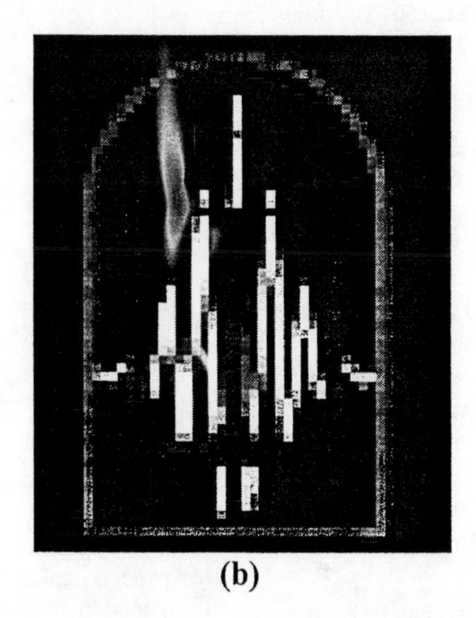

(b)

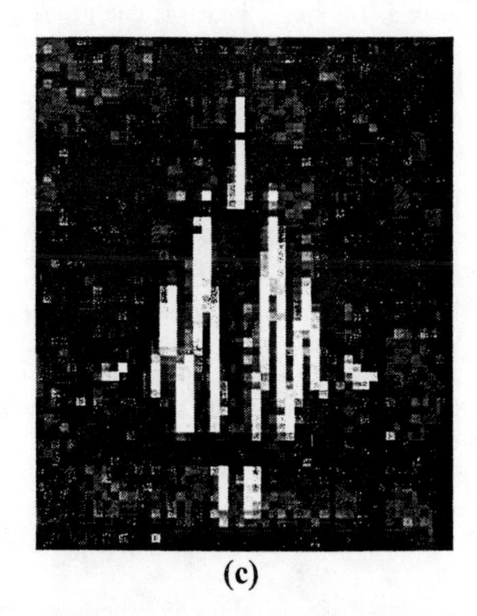

(c)

FIGURE 2. Spectral images of the ICOFTS-11 logo showing **(a)** the baseline shift at 4450 cm^{-1}, **(b)** the peak height of the carbonyl band at 1723 cm^{-1}, and **(c)** the height of the absorbance peak at 1498 cm^{-1}. The magnitudes of the grayscale images are 0.74, 0.25, and 0.15 absorbance units, respectively. The spectral images in **(a)**, **(b)**, and **(c)** indicate the presence of black ink, ink in general, and red ink in the paper sample, respectively. The pixel width in the images is 1 mm.

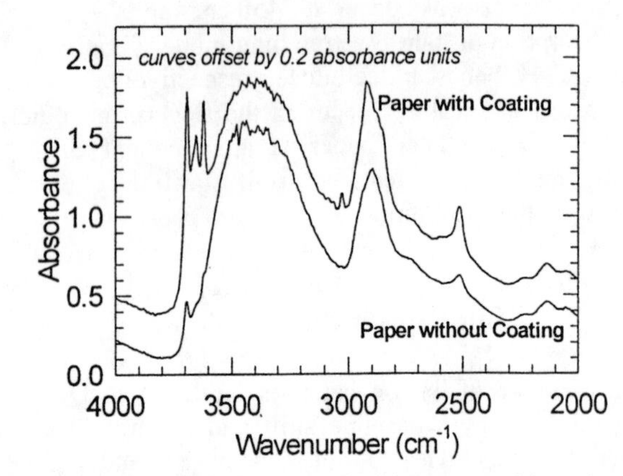

FIGURE 3. DRIFT spectra of white paper with a clay-based coating and white paper without a coating. The clay in the coating is evident by the Group II hydroxyl bands from 3600 to 3700 cm^{-1}.

Spectral images showing the distribution of inks on the sample have been generated by analyzing spectra obtained in a two-dimensional scan of the sample. 3000 spectra were obtained at 1 mm intervals. The maps were formed by plotting analysis results as grayscale colors in an image with the same spatial coordinates where the data was taken on the sample. Figure 2.a. is a spectral image showing the baseline shift at 4450 cm^{-1} which occurs due to absorption of the infrared light by the black ink. The Roman numeral three is a gray color and, therefore, has a weaker baseline shift due to the lower concentration of black ink. Figure 2.b. shows the peak height of the carbonyl band at 1729 cm^{-1} that corresponds to the presence of both black and red ink. Figure 2.c. shows the peak height of the absorbance peak at 1498 cm^{-1} which is due to the presence of red ink on the sample. For both Fig. 2.b. and Fig. 2.c., the spectra in the data set were normalized with multiplicative scattering correction; the correction helped to measure the relative concentration more accurately and account for variations in the data from differences in the overall signal level detected by the spectrometer.

The spectral images show the chemical variation in the sample due to the presence of different types and levels of ink in the paper. Each map was generated from the same set of data, and more information on the sample chemistry can be obtained from further analysis of the spectra. The maps shown represent basic analysis techniques of data obtained with a sensitive technique.

Clay-Based Coatings on Paper

Clay-based coatings on paper are detected easily by the DRIFT technique due to the strong absorbance features associated with the clay in the coatings. Figure 3 shows a spectrum of paper with a clay-based coating and a spectrum of uncoated paper for comparison. The absorbance peaks from 3600 to 3700 cm^{-1} are due to the Group II hydroxyl bonds associated with the clay in the coating material.

Figure 4 is a spectral image of a paper sample with a clay-based coating. The coating thickness varies across the sample. In the lower left corner of the sample, the coating has been removed entirely from the paper. Additionally, a circular region in the center of the sample has had the coating partially removed. The spectral region from 3580 to 3750 cm^{-1} was integrated to generate the spectral image. The spectra were normalized with multiplicative scattering correction prior to integration. The variation in the coating thickness is reflected by the grayscale magnitude of those regions in the image.

Lignin Chemistry in Woods

The primary constituents of wood are cellulose, hemicellulose, and lignin. The relatively complex nature of lignin with respect to cellulose allows detection of lignin with DRIFT spectrometry. The actual chemistry of the lignin varies with wood type; in addition, the lignin chemistry will vary from tree to tree and within the same tree as environmental factors affect the growth of a tree.

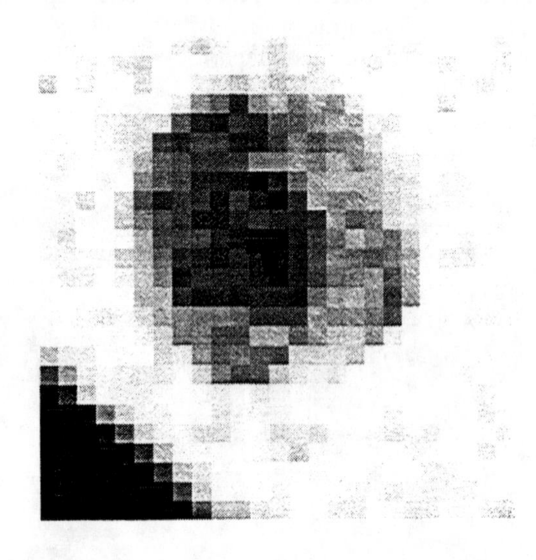

FIGURE 4. Spectral image of white paper with a clay-based coating. The grayscale corresponds to the integral of the 3580 to 3750 cm^{-1} spectral range and indicates the thickness of coating on the paper sample. The coating has been removed entirely from the paper in the lower left corner of the image. The shaded circular area in the center of the image indicates a region where the coating has been partially removed. The image is 32 mm by 32 mm.

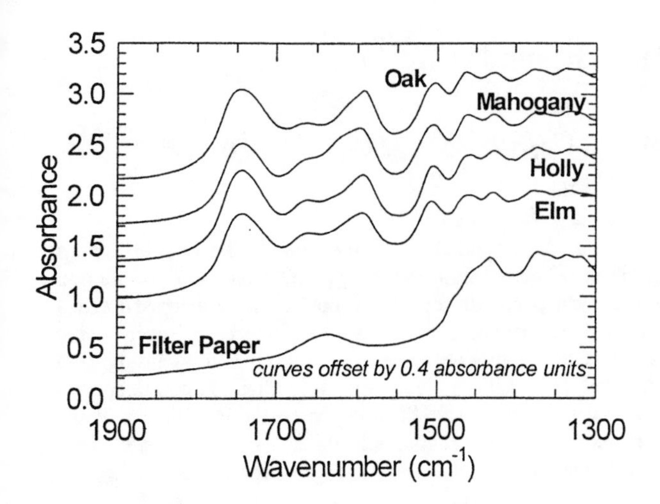

FIGURE 5. DRIFT spectra of elm, holly, mahogany, and oak wood samples, and a DRIFT spectrum of filter paper. Lignin in the wood samples is indicated by the absorbance bands from 1900 to 1500 cm^{-1}.

Figure 5 shows DRIFT spectra of four different wood types (elm, holly, mahogany, and oak); a spectrum of filter paper is shown as a reference of cellulose with no lignin. Lignin chemistries present in the spectra include an aliphatic ester carbonyl band at 1743 cm^{-1}, an aromatic ester band at 1663 cm^{-1}, and aromatic bands corresponding to phenyl vibration modes at 1594 cm^{-1} and 1505 cm^{-1}. The relative strengths of the lignin bands vary with wood type.

Figures 6.a., 6.b., and 6.c. show spectral images of a sample made up of the four wood types shown in Fig. 5. The sample is made up of 25 mm strips of wood; the types of woods from top to bottom in the images are elm, holly, mahogany, and oak, respectively. The grain pattern in each wood runs from top to bottom in the image. Figures 6.a., 6.b., and 6.c. were generated by measuring the peak heights of the 1743 cm^{-1}, 1663 cm^{-1}, and 1595 cm^{-1} bands, respectively. The lignin chemistry variation among the different types of wood is evident in the relative differences in the strengths of the absorbance bands analyzed. Variations in chemistry corresponding to grain pattern are also seen in the images.

CONCLUSION

Diffuse Reflectance mid-Infrared Fourier Transform (DRIFT) spectroscopy is a sensitive technique for analyzing the chemical nature of cellulose-based materials. The diffuse reflectance geometry is particularly useful for paper and wood due to the scattering properties of those materials. Macroscopic spectral images obtained by scanning the sample beneath the spectrometer allow visual identification of the spatial distribution of chemistries on the sample.

Ink content and chemistry, clay-based coating thickness, and lignin chemistry have been measured in cellulose-base materials with the DRIFT technique. Detection of the presence and chemistry of inks and lignin in feedstock for the forest products industry is useful for optimal control of processes to remove ink and lignin. In addition, the detection of lignin with the DRIFT technique indicates a potential for identification of types of wood by infrared spectroscopy. Such an identification capability would be useful for sorting wood stock and, thereby, selecting desired fiber properties. Clay-based coating uniformity is critical to the obtaining high quality printing; thus, the DRIFT technique could be used for quality control of coating applications.

ACKNOWLEDGEMENTS

The authors gratefully acknowledge the financial support of the Office of Industrial Technology, U. S. Department of Energy, and Surface Optics Corporation, San Diego, CA for providing the SOC-400 Surface Inspection Machine, Infrared used to obtain the spectra.

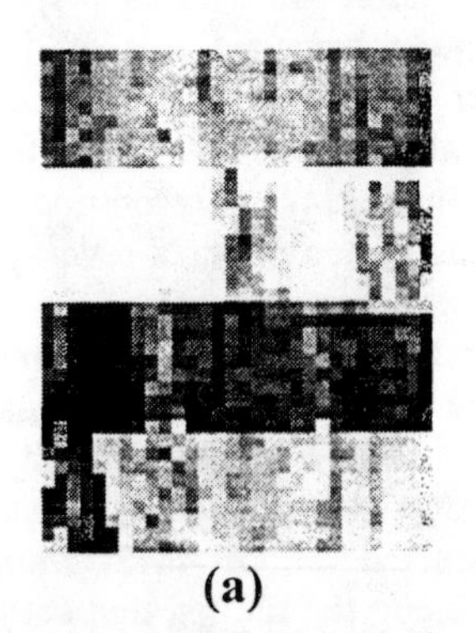

(a)

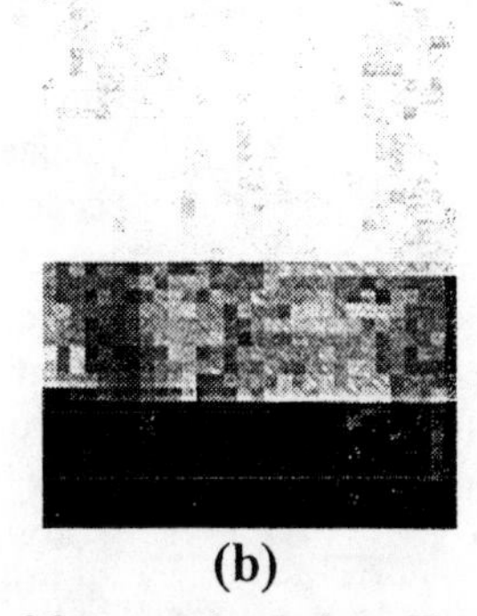

(b)

(c)

FIGURE 6. Spectral images of a sample composed of four wood samples. The woods were cut in 25 mm strips. The wood types from top to bottom are elm, holly, mahogany, and oak, respectively. The images are 76.2 mm by 101.6 mm. The grayscale images represent the magnitude of the peak heights of the **(a)** 1743 cm^{-1}, **(b)** 1663 cm^{-1}, and **(c)** 1595 cm^{-1} bands, respectively. The magnitudes of the grayscale images are 0.28, 0.06, and 0.21 absorbance units, respectively.

Diffuse Reflectance FTIR of Stains on Grit Blasted Metals

G. L. Powell, R. L. Hallman, Jr., and R. L. Cox

Oak Ridge Centers for Manufacturing Technologies, P. O. Box 2009, Oak Ridge, TN 37831-8096*

Diffuse reflectance mid-infrared Fourier transform (DRIFT) spectroscopy has been applied to the detection of oil contamination on grit-blasted metals. The object of this application is to detect and discriminate between silicone and hydrocarbon oil contamination at levels approaching 10 mg m^{-2}. A portable FTIR spectrometer with dedicated diffuse reflectance optics was developed for this purpose. Using translation devices positioned by instructions from the spectrometer operating system, images of macroscopic substrates were produced with millimeter spatial resolution. The pixels that comprise an image are each a full mid-infrared spectrum with excellent signal-to-noise, each determined as individual files and uniquely saved to disc. Reduced spectra amplitudes, based on peak height, area, or other chemometric techniques, mapped as a function of the spatial coordinates of the pixel are used to display the image. This paper demonstrates the application of the technique to the analysis of stains on grit-blasted metals, including the calibration of the method, the inspection of substrates, and the migration of oil contamination.

INTRODUCTION

The detection of oil contamination on grit-blasted metals, the arch-typical surfaces for bonding where there exists bond strength requirements, requires that silicone and hydrocarbon oil contamination be detected and discriminated at levels approaching 10 mg m^{-2}. For this purpose, a portable FTIR spectrometer with dedicated diffuse reflectance optics was developed through a collaboration between Lockheed Martin Energy Systems, Inc., National Aeronautics and Space Administration Marshal Space Flight Center, and Surface Optics Corporation. The instrument is used to inspect the interior walls of solid rocket motor cases after grit-blasting and cleaning, and prior to the application of a primer. This instrument has been coupled with translation devices positioned by instructions from the spectrometer operating system to produce images of macroscopic substrates with millimeter spatial resolution and unlimited substrate size. Preliminary work in this area has been reported.[1,2] This paper describes the application of general purpose spectral acquisition and data analysis techniques for producing images to the analysis of stains on grit-blasted metals including the calibration of the method, the inspection of substrates, and the migration of oil contamination on surfaces.

EXPERIMENTAL

The diffuse reflectance spectra were obtained using an SOC-400 Surface Inspection Machine/Infrared[2], a portable FTIR spectrometer with dedicated diffuse reflectance optics and sensitivities to efficient light scattering media approaching 1×10^{-4} absorbance units in one minute using a deuterated triglyceral sulfate detector and KBr windows and beam splitter. This instrument was coupled with linear translation devices (3 orthogonal 0.3 m travel VELMEX Unislides using an NF90 serial controller) positioned by instructions through serial communications from the MIDAC/GRAMS/32 operating system. The spectrometer was mounted such that the diffuse reflectance optics looked down on the specimen that was moved in the horizontal plane during mapping. The vertical Unislide manually located the plane of the substrate at the focal point of the diffuse reflectance optics.

The software for spectral collection and mapping was an Array Basic program that initially loaded a complete set of operational parameters and mapping grid, constructed a file list for building a GRAMS multifile sorted for optimum image conversion, and produced a text file containing the mapping parameters. The program then collected an individual spectrum with calibration and typically coadded 48 scans. The specimen was advanced the predetermined distance in the +X direction and another spectrum taken with the file name numerically incremented. Thus individual spectra were taken with appropriate calibration and with each spectrum saved as a closed file to protect against loss of data during power outages or other system crashes. This is important since the system typically collects 500 to 10,000 spectra per day, depending on wavenumber resolution and number of coadded scans, and often requires several days to collect a spectral map. The mapping motion is to move along the X direction the predetermined distance at predetermined step lengths, take one step in the Y direction, and then move in the opposite X direction resulting inthe same amount of time for each step. At the end of data collection, the software constructed a GRAMS Multifile from the individual files using the file list that sorted the spectra as though all spectra had been collected in the +X direction. A second Array Basic program operated on this GRAMS

*Managed for the U. S. Department of Energy by Lockheed Martin Energy Systems, Inc. under Contract No. DE-AC05-84OR21400.

CP430, *Fourier Transform Spectroscopy:* 11th International Conference
edited by J.A. de Haseth
1998 The American Institute of Physics 1-56396-746-4/98/$15.00

Multifile to read baseline corrected peak heights and areas as well as to pick values from the spectra and constructed other GRAMS Multifiles representing these reduced spectra as a course in the X direction.

Figure 1 demonstrates this mapping process and also demonstrates the spatial resolution of the SOC 400. A total of 2601 spectra were obtained from a 5-mm by 5-mm area on a grit-blasted gold substrate at 0.1-mm steps. In the center of this area two 0.1-mm diameter mono-filament fishing (Stren) lines formed a cross with the Y direction being parallel to the light entering the SOC 400's barrel ellipse. The amplitude in Fig. 1 is the peak height of the mono-filament line at 1675 cm^{-1} after linear baseline correction at 3800 cm^{-1} and 2000 cm^{-1}, each averaged over 50 cm^{-1}. The zigzag texture in the Y direction is the reversible whiplash (~0.1-mm) of the positioning device.

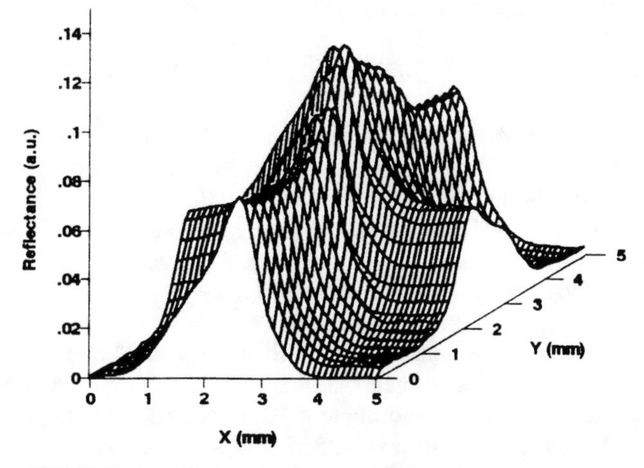

FIGURE 1. Reduced spectral image of 0.1-mm crossed filaments showing the spatial resolution of the SOC 400 as a GRAMS/32 Multifile.

The white region on each side of Fig. 1 marks the spatial resolution functions for the SOC 400 when crossing a linear boundary. This data, along with the spatial response function previously reported for the Z direction, demonstrates that the SOC 400 has a 2-mm diameter sphere in front of the focal point that is very robust with respect to sensitivity tolerance with respect to the sample position.

The stained grit-blasted steel substrates evaluated in this work consisted of two specimens of D6AC steel 112-mm by 112-mm by 12-mm step-plate contamination specimens prepared by the method of Booth[3] and provided by NASA-MSFC. The clean metal substrates were repeatedly masked with a witness foil and sprayed with non-volatile solutes in a volatile solvent to produce contamination steps on these plates that were 19-mm wide with each step more contaminated than the last by 108 mg m^{-2}. The two contaminants provided were a paraffin wax, chosen because it deposited as a solid and did not migrate, as well as being a very simple hydrocarbon, and (2) CRC silicone, because it was a simple silicone and deposited as a liquid that wet the substrate surface. A grit-blasted gold specimen 75-mm by 75-mm by 0.5-mm was also prepared using ta technique previously described[1]. This gold specimen was contaminated by placing a drop of Dow Corning DC-200 Silicone Oil (5cs.) at the center of this specimen using a microliter syringe while the specimen was being weighed with 0.01 mg resolution. The 0.70 mg of silicone oil spread rapidly and spectral mapping was begun 4 hrs. later.

The spectra were obtained at 16 cm^{-1} resolution using 48 coadded scans determined in absorbance units (a. u.). The steel specimens were mapped at increments of 2.5-mm resulting in a 44 X 44 pixel image. The gold specimens were mapped at 2.0-mm increments resulting in a 48 X 48 pixel image. The spectral reduction routine executed a linear baseline correction based on the average values over the range of 2700 cm^{-1} to 2725 cm^{-1} and 3150 cm^{-1} to 3175 cm^{-1}, constructed reduced spectral maps based on the peak height for the aliphatic C-H bands near 2920 cm^{-1} and 2960 cm^{-1} (aliphatic hydrocarbons and silicones), and aromatic C-H bands near 3020 cm^{-1} (polystyrene and other plastics). Reduced spectral maps were also produced using a linear baseline correction of the average values over the range of 2025 cm^{-1} to 2050 cm^{-1} and 3550 cm^{-1} to 3775 cm^{-1}, using the average value over 3375 cm^{-1} to 3400 cm^{-1} to detect O-H bands (fingerprints), and the peak height near 1260 cm^{-1} relative to the adjacent minima to detect the Si-C band (silicones). These peak height values were also stored as text in a table with the mapping coordinates and time.

RESULTS AND DISCUSSION

Figure 2 compares the first three contamination steps on a paraffin contaminated D6AC steel step-plate in terms of the main aliphatic and aromatic C-H bands. That the two bands represented either the paraffin or polystyrene was confirmed by inspection of the full individual infrared spectra. The "clean" step in the absence of polystyrene residue or particles, had an average value of 0.0002 a. u. ± 0.0004 a. u. This positive (greater than zero) bias results from comparing a maximum value for a peak height with an average value for the baseline and effectively measures the half width of the baseline noise. The first contamination step at 108 mg m^{-2} was at the 0.017 a. u. ± 0.005 a. u. level, where the variability was probably due to real contamination variations due to the contamination deposited as "splats" of millimeter or less size visible to the unaided eye. Analysis of a line map across the steps indicated a linear calibration factor of 3.6 g m^{-2} $a.u.^{-1}$ or detection limits below 2 mg m^{-2} (~ 2 nm film thickness). Previously reported work using safflower oil indicated a calibration factor twice that large (i.e., half the sensitivity), probably because the calibration is peak height determined and the safflower oil C-H bands are much less sharp. A calibration based on band area would probably bring the

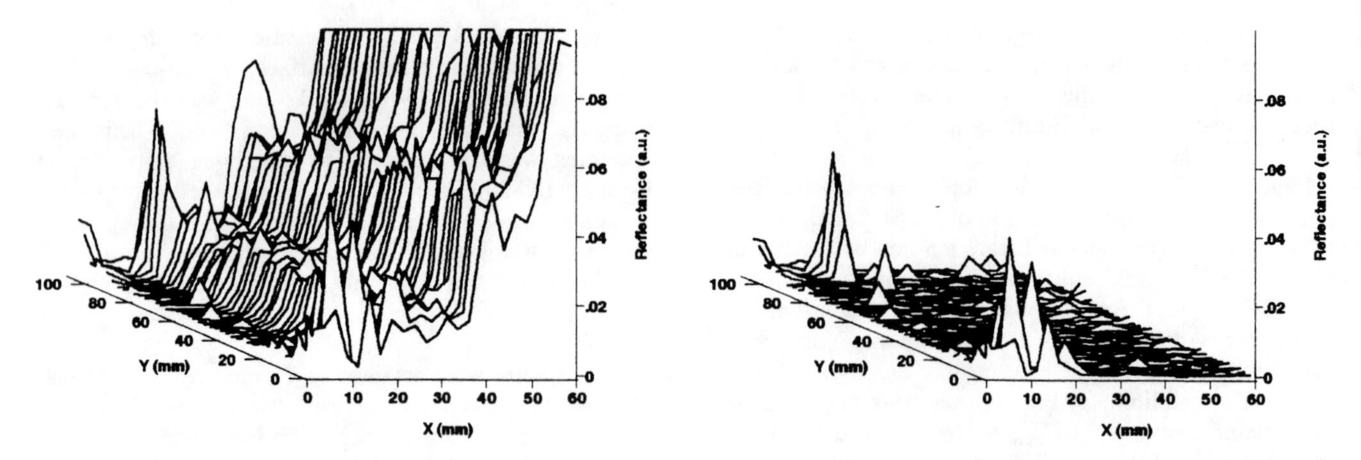

FIGURE 2. Reduced spectral maps (as GRAMS/32 hidden line 3-D displays) of a D6AC step-plate based on the main aliphatic (left) and aromatic (right) C-H bands. The step-plate was clamped at the corners with polystyrene contacts and some polystyrene dust migrated onto the rest of the step-plate since there is a one-to-one correspondence between the spikes in the aliphatic and aromatic C-H spectral maps. The high peak at the far end of the 0.06 a. u. step was due to a fingerprint.

two calibration factors into better agreement. The safflower oil spread as a liquid, but oxidatively cured to a solid film at ~ 1.7 g m^{-2}.

Figure 3 shows a step-plate map of the Si-C band on CRC silicone step-plate contamination standard. Note that, like the safflower oil, the silicone oil migrated, but unlike the safflower oil, it was not chemically fixed. The silicone oil not only migrated to erase the steps, but migrated off the front surface of the plate and also concentrated in the polystyrene at the edges. Thus, the silicone contaminated

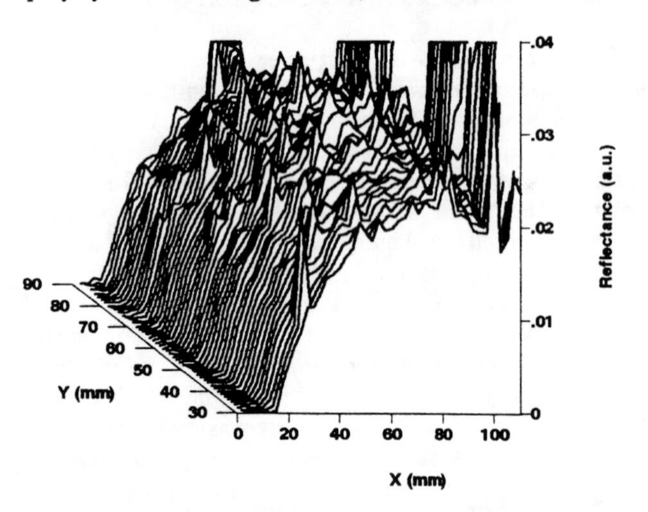

FIGURE 3. Reduced spectral may of a CRC silicone step-plate based on the peak height Si-C band on the silicone. The Y abscissa has been trimmed on each end so that the polystyrene effects at the corners does not distract from the observation of the silicone behavior. The spikes along the Y direction at the extreme X values represent silicone absorbed on the polystyrene.

step-plate was of little use in establishing quantitative limits and calibration factors. On the other hand the silicone did not migrate onto the first (clean) step, but rather migrated only where the spray process wet the substrate. The clean step region had an average peak height value of 0.0001 a. u. ± 0.0002 a. u.

Finding a contaminant that will wet the substrate, be applicable at levels near 100 mg m^{-2}, and be fixed sufficiently well to yield quantitative information led back to the oil drop spreading experiment[1]. Gold was used because it was capable of achieving constant weight at the 0.01 mg level sufficiently long to weigh a microliter of oil. The object was to allow the oil to migrate, but as it migrated, to map it over a region that included all the oil stain so that integration over the spectral map of the spot would yield a calibration factor. Figure 4 shows spectral maps taken of a DC 200 Silicone (5 cs.) stain taken over several days. Integration of these maps for 0.70 mg of silicone yielded calibration factors of 6.4 g m^{-2} for the highest concentration (first day after application) and 10.4 g m^{-2} for the map taken on the third day, and 12.2 g m^{-2} for the map taken on the sixth day. On the sixth day the stain was no longer visible to the unaided eye.

CONCLUSIONS

Diffuse reflectance mid-infrared Fourier transform (DRIFT) spectroscopy is capable of qualitatively and quantitatively detecting oil contamination on grit-blasted metals at levels below 10 mg m^{-2}. DRIFT measurements also give considerable insight into the processes associated with oils on metals, many of which are very complex. Portable DRIFT spectrometers with dedicated diffuse reflectance optics are capable of this task and of extending the task to manufacturing applications. Spatial mapping enhances this capability by using visual contrast to optimize reference spectra and decisions related to chemometric analysis.

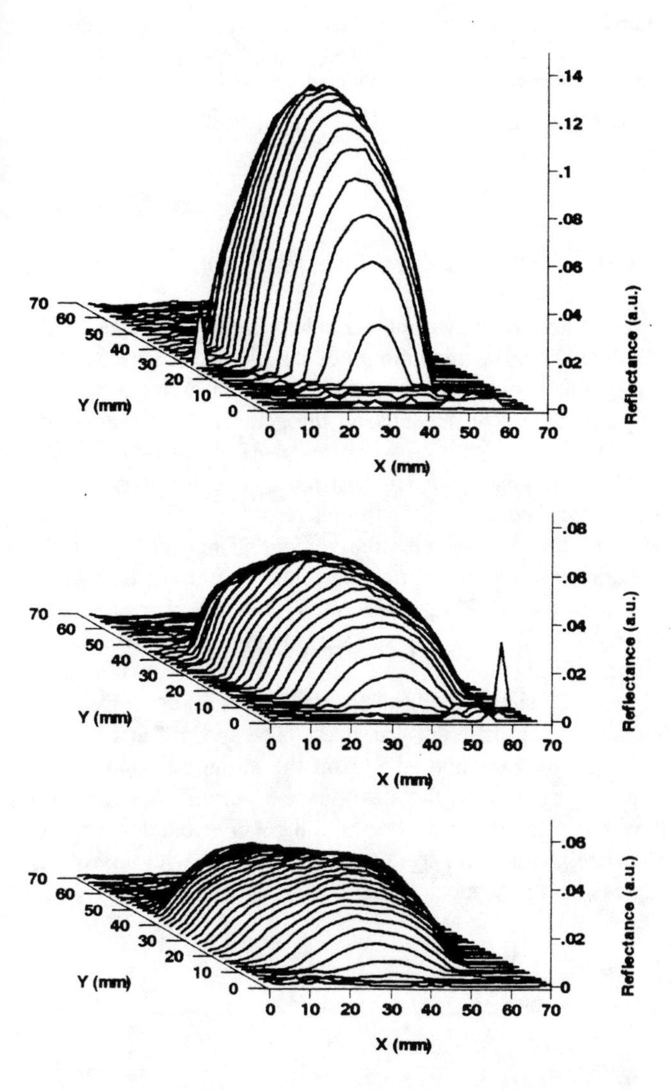

FIGURE 4. Images of the spread of a 0.7 mg silicone oil drop applied to a clean sandblasted gold surface based on the 1265 cm^{-1} silicone (Si-C) absorption band. Elapsed time: Upper map-first day, Middle map-third day, Lower map-sixth day.

ACKNOWLEDGEMENTS

The authors gratefully acknowledge the staff of Surface Optics Corporation, San Diego, CA for their diligence in developing this DRIFT spectrometer and to B. H. Nerren of NASA, MSFC, AL for his efforts to develop this technology and for providing the D6AC step-plate calibration specimens.

REFERENCES

1. 1. G. L. Powell, T. E. Barber, M. Marrero-Rivera, D. M. Williams, N. R. Smyrl, and J. T. Neu, **Mikrochim. Acta [Suppl.] 14**, 655 (1997).

2. G. L. Powell, T. E. Barber, and J. T. Neu, "Diffuse Reflectance Mid-Infrared Spectroscopy as a Tool for the Identification of Surface Contamination on Sandblasted Metals.," Aerospace Environmental Technology Conference, **NASA Conference Publication**, A. F. Whitaker, ed., MSFC, Alabama, (1996) in press.

3. R. E. Boothe, "Standardization of Surface Contamination Analysis Systems," Aerospace Environmental Technology Conference, **NASA Conference Publication 3298**, A. F. Whitaker, ed., MFSC, Alabama, 517-524 (1995).

Abrasive Sampling for Diffuse Reflectance Using Reflective Substrates

R A Hoult and R A Spragg

Perkin-Elmer Ltd, Post Office Lane, Beaconsfield HP9 1QA England

The measurement of diffuse reflection spectra from samples collected on an abrasive substrate is reviewed. Abrasives of 400 grit or finer are recommended for mid-IR spectra. Using a reflective metal-coated substrate is shown to improve the signal-to-noise ratio by up to a factor of five. An extremely robust diamond-in-metal sampling device can be used to obtain excellent spectra from very hard materials such as quartz.

INTRODUCTION

Abrasive sampling with silicon-carbide-coated paper was originally introduced by Sharp[1] as a means of taking a small sample of a large object for subsequent measurement in a KBr pellet. It has since been adopted in the simplified form of measuring diffuse reflection directly from the thin layer of material collected on the abrasive surface[2]. Here we review the technique and describe an enhancement involving reflective substrates. The abrasive materials commonly used consist of silicon carbide or diamond powder bonded to a flexible substrate. We illustrate the use of a diamond-in-metal abrasive device that can be used to obtain spectra from very hard objects.

REFLECTION MEASUREMENTS ON AN ABRASIVE SUBSTRATE.

The diffuse reflection mid-IR spectra of powders usually do not resemble transmission spectra of the same materials unless they are diluted in a non-absorbing matrix. However the spectra of neat samples prepared by abrasion have relative band intensities very similar to those obtained with dilution (Fig. 1). The explanation is that rays encounter few absorbing particles in a thin layer before emerging, just as happens with diluted powders.

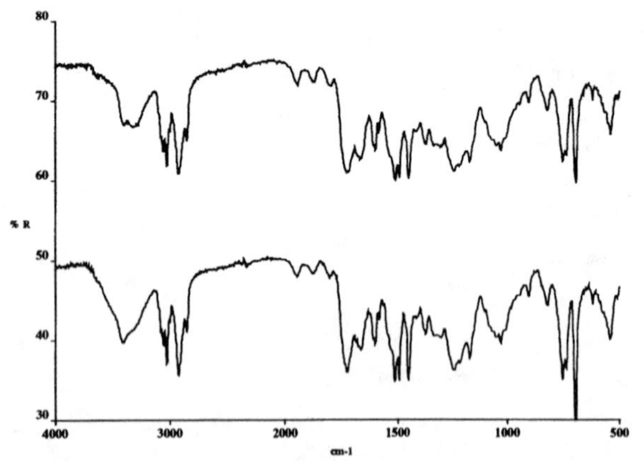

FIGURE 1. Diffuse reflection spectra of the same material, recorded as a thin film (upper) and diluted in KBr (lower).

Practical Considerations

There are two important variables, the particle size of the abraded material and amount of sample on the surface. The aim is to avoid strong absorptions within a single particle. The particle size depends on the grit size of the abrasive. With

surfaces coarser than 100 grit the band intensities become distorted because the particles are becoming too large. Even with fine particles the amount of material has to be restricted to avoid strong absorptions. 400 grit abrasive is suitable for most samples. With this grit size any loose powder visible on the surface suggests that there is too much sample.

Figure 2. shows spectra from a polystyrene coffee cup abraded with different grit sizes. The spectra, in Kubelka-Munk units, are normalized on the strongest band at 700 cm^{-1}. At the top is the absorbance spectrum of a thin film. As the grit becomes coarser the relative band intensities deviate increasingly from those of the film, with the weaker bands being enhanced.

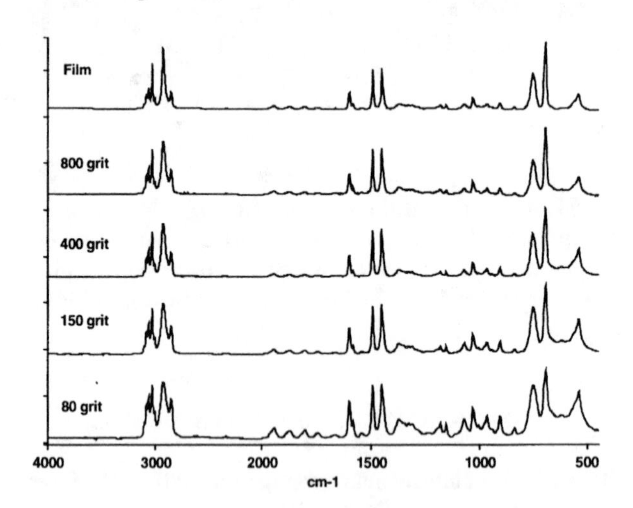

FIGURE 2. Spectra of polystyrene abraded with different grit sizes, compared with a transmission spectrum at the top.

CP430, *Fourier Transform Spectroscopy:* 11th International Conference
edited by J.A. de Haseth

COMPARISON OF REFLECTIVE AND CONVENTIONAL SUBSTRATES

We have recently investigated the use of abrasive surfaces with a reflective metal coating. These have two potential advantages; higher reflectivity, and freedom from spectral artefacts. As illustrated in Fig.3 the reflection from silicon-carbide-coated paper is enhanced by about a factor of twenty by coating with aluminium. Spectral features associated with the silicon carbide around 800cm^{-1} and with organic binders are eliminated.

The spectra in Fig. 4 from an acetaminophen tablet were obtained on uncoated and reflective substrates with the same grit size. The reflective substrate was used as the background. The signal to noise ratio for the sample on the reflective substrate is enhanced by about a factor of five relative to that on the uncoated surface. The distortion caused by the spectral contribution of the uncoated silicon carbide is evident below 1000cm^{-1}.

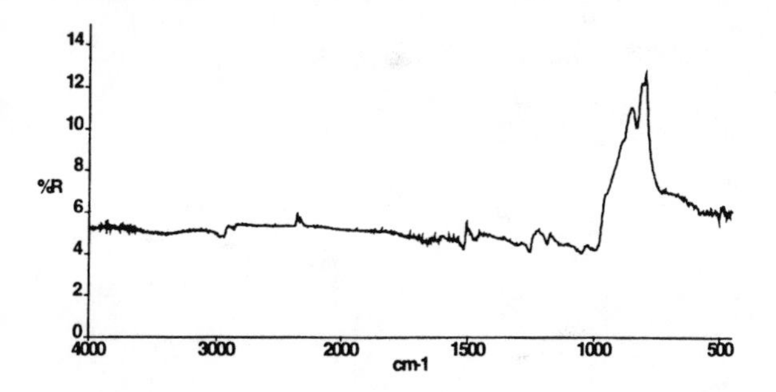

FIGURE 3. Spectrum of 800 grit silicon-carbide-coated paper with the aluminized material as reference.

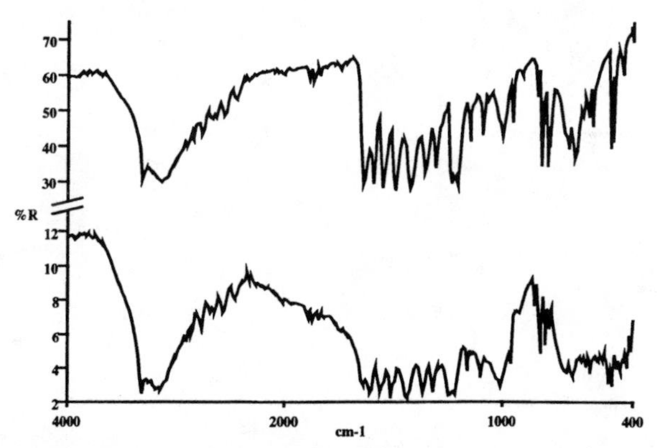

FIGURE 4. Spectra of acetaminophen on silicon-carbide-coated paper. Upper: aluminized, lower: conventional.

TYPICAL APPLICATIONS

This approach is perhaps most useful for samples that are too large or irregular in shape to be accommodated in conventional reflection or ATR accessories. These are typified by the soft drink bottles whose spectra appear in Fig. 5.

The abrasive-tipped metal rods can be used when only small surfaces are accessible. The abrasive is diamond powder embedded in the metal, providing an extremely robust device that can be used with very hard samples such as quartz as shown in Fig. 6.

LIMITATIONS

The sample has to be abraded to a fine powder so that soft, fibrous and rubbery materials are unlikely to give good spectra. If very hard materials such as some minerals are examined with unsuitable abrasive materials there is the possibility that the substrate will be damaged, leading to spurious features in the spectra.

The technique is not generally useful for quantitative analysis. One reason is that it is difficult to control the amount of sample, so that band intensities vary. Mechanical systems to control the abrasion have been devised, but these remove the basic simplicity of the approach. A further difficulty is that the spectra inevitably have a contribution from the abrasive surface itself. This takes the form of stray light from reflective substrates or spectral features from uncoated materials. Despite the stray light library searching works well although match qualities are poorer than with conventional spectra.

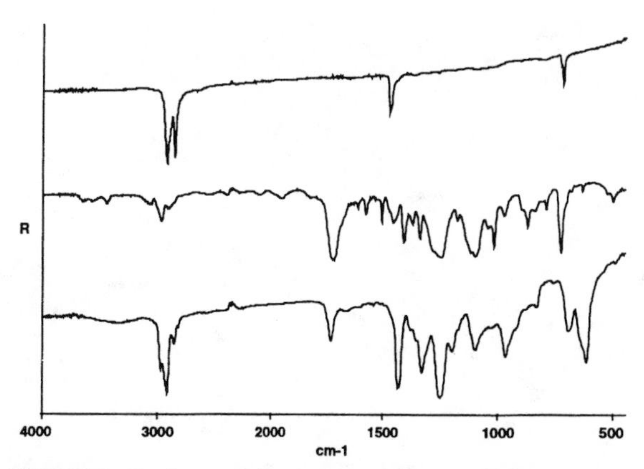

FIGURE 5. Spectra from soft drink containers abraded on aluminized silicon-carbide-coated paper. Top: polyethylene, center: PET, bottom: PVC.

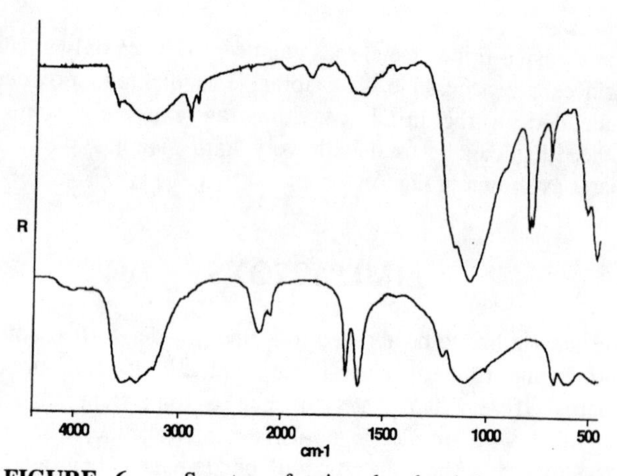

FIGURE 6. Spectra of minerals abraded on aluminized diamond-in-metal tipped rods. Upper: quartz, lower: gypsum.

REFERENCES

1. Sharp, J., *European Spectroscopy News* **43**, 4 (1982)
2. Spragg, R.A., *Applied Spectroscopy*, **38**, 604 (1984)

FTIR Spectroscopy as an Analytical Tool for Solid Phase Synthesis

D Clark and R A Spragg

Perkin-Elmer Ltd., Post Office Lane, Beaconsfield, Bucks, HP9 1QA, England

Methods for obtaining spectra of organic compounds bound to polymer supports for solid-phase synthesis are compared. ATR is the most convenient method for compounds on beads while diffuse reflection with abrasive sampling is preferred for crown supports. Microscopy is used for single beads or to investigate the distribution on crowns. FT Raman spectra can be obtained very easily but the data are perhaps less useful than IR for monitoring reactions.

INTRODUCTION

Rapidly growing interest in solid phase synthesis and combinatorial chemistry is posing new analytical challenges. The requirement is to follow the progress of multi-step reactions of organic molecules bound to the surface of porous solid polymers. Ideally this has to be done without cleaving the molecules from the surface. Vibrational spectroscopy is in general an excellent tool for following chemical reactions because of its ability to recognise specific functional groups. This poster reviews alternative ways of measuring the spectra of compounds on the polymer substrates used for this type of chemistry. In general the aim of the measurement is to look at a single reaction step through the appearance or disappearance of features associated with particular functional groups. At present vibrational spectroscopy cannot provide the rapid automated screening needed to handle the large numbers of compounds generated by parallel syntheses.

Two classes of substrate are in common use, both aimed at providing a large surface area. One is a powder with individual particles, called 'beads', ranging in size from about 50 to 200μm. The other substrates, 'crowns' or 'gears', are plastic mouldings with dimensions of several millimeters. The crowns are moulded from polyethylene or polypropylene with materials to provide reactive groups grafted on to them. Simultaneous reactions can be carried out on an array of these with a different compound bound to each crown.

The commonly used resin beads fall into two classes. One is based on polystyrene with suitable reactive groups on the surface. The others are 'gel' resins based on polyethylene-glycol-polystyrene graft copolymers. In these the reactive groups are linked to polyoxyethylene chains in order to facilitate measurements by NMR.

TYPES OF RESIN

In the normal resins the spectral features of polystyrene are prominent while the polyoxymethylene chains provide the strongest features in the spectra of the gel resins. The spectra below were obtained by ATR and corrected for the wavelength dependence of the penetration depth. Very similar spectra can be obtained by preparing KBr pellets but the need for rapid measurement makes reflection techniques more attractive.

DIFFUSE REFLECTION MEASUREMENTS

Aggregates of beads can be measured by diffuse reflection. Because of the relatively small particle size the required information can often be obtained directly from diffuse reflection spectra of the beads, especially in the important regions 1500-1800cm^{-1} and above 3200cm^{-1}. However as increasingly large beads are being used it may be necessary to grind them. Good quality spectra are obtained either by diluting the ground beads in KBr or by presenting them as a thin layer on a reflective substrate such as aluminized-silicon-carbide-coated paper.

Figure 1 shows spectra of a Wang resin with a compound believed to have three carbonyl groups, ester, carbamate and secondary amide. The spectrum of the unground beads is unsatisfactory because the stronger bands such as those of the carbonyl region are severely distorted. In the example

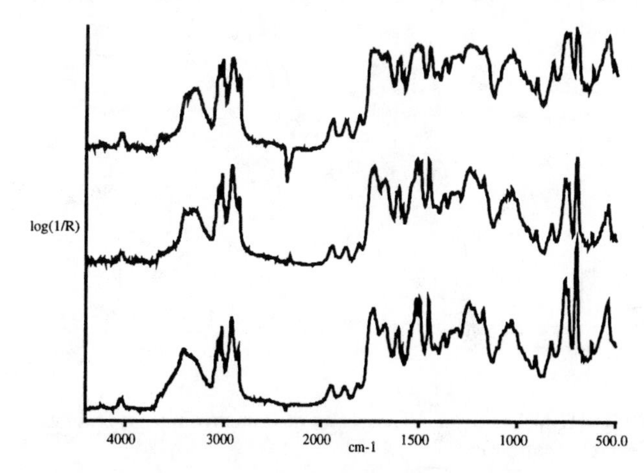

FIGURE 1. Diffuse reflection spectra of Wang resin beads. Top: unground, Center: thin layer of ground beads, Bottom: ground beads diluted in KBr.

CP430, *Fourier Transform Spectroscopy:* 11th International Conference
edited by J.A. de Haseth

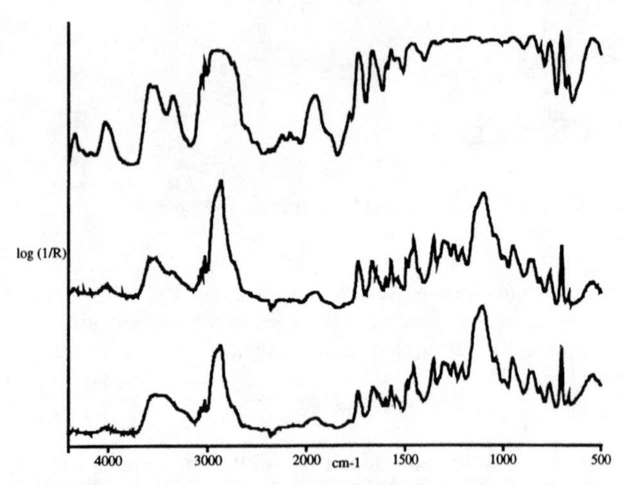

FIGURE 2. Diffuse reflection spectra of Tentagel beads. Top: unground, Center: thin layer of ground beads, Bottom: ground beads in KBr.

of Fig. 2 the presence of an ester group in a compound on a Tentagel resin can be confirmed from the spectrum of the unground beads.

ATR MEASUREMENTS

ATR spectra of the resins are obtained rather easily using an accessory with a small crystal and a clamping device under which the powder is trapped to ensure optical contact. In practice this is probably the simplest method of measuring these spectra. The spectra of the two resins previously measured by diffuse reflection are shown in Fig. 3. The intensities have been corrected for the wavelength dependence of the penetration depth.

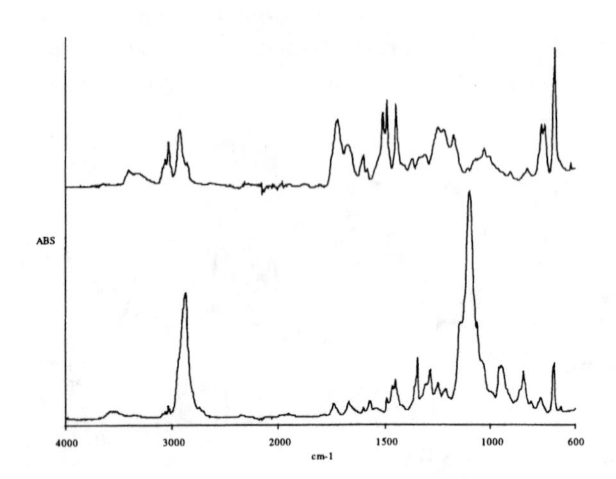

FIGURE 3. Diamond ATR spectra of Wang resin (top) and Tentagel (bottom)

The relative intensities of bands from the bound molecules and the polymer substrate are very similar in diffuse reflection and ATR spectra, as can be seen in Figs. 4 and 5. In general the greater convenience of the ATR method therefore makes this the preferred method. However diffuse reflection can be better for identifying weak OH bands because of the smaller ATR penetration depth at shorter wavelengths.

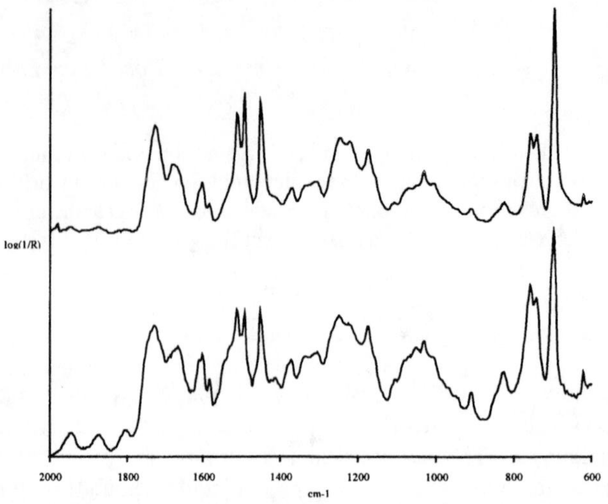

FIGURE 4. Spectra of Wang resin. Top: diffuse reflection, Bottom: diamond ATR.

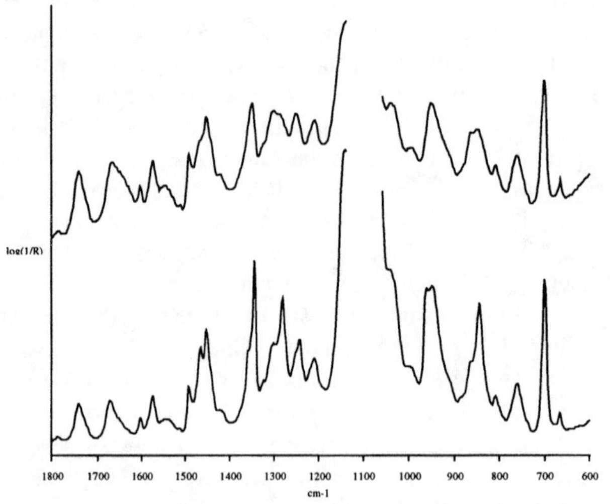

FIGURE 5. Spectra of Tentagel resin. Top: diffuse reflection, Bottom: diamond ATR.

MEASUREMENTS ON CROWNS

Measuring reflection spectra of crowns is inconvenient because of their irregular shapes. A simple alternative is to remove a small amount of material from the crown using an abrasive-tipped rod, and measure this in a diffuse reflection accessory. Such spectra, from a thin layer of powder, are generally much more similar to transmission spectra than are reflection spectra from bulk material. Typical spectra of material removed in this way from crowns are illustrated in Fig. 6.

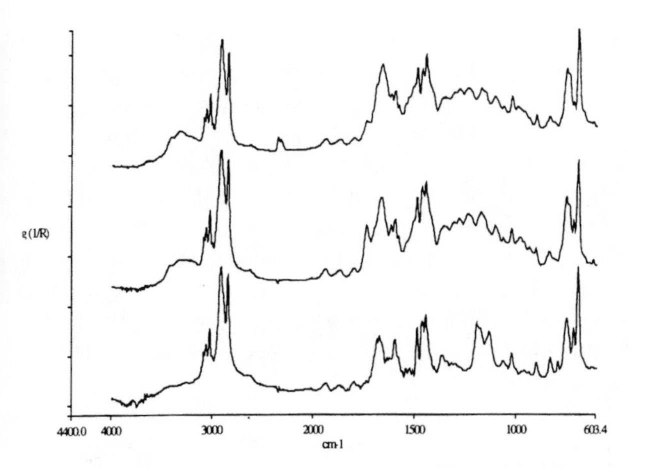

FIGURE 6. Diffuse reflection spectra of material abraded from the surface of crowns. The bottom spectrum is from an unreacted crown and the upper two are from crowns bearing different products.

APPLICATIONS OF MICROSPECTROSCOPY

Transmission Measurements on Single Beads

The spherical shape of beads distorts the relative band intensities in transmission spectra. However beads that have been swollen with solvent can be compressed between windows to provide a region with a uniform pathlength[1]. For the examples of Figs. 7 and 8 Tentagel and Wang resin beads, nominally 70µm in diameter, were trapped between the windows of a demountable liquid cell. Chloroform was added to swell the beads before compressing to about 30µm between calcium fluoride windows, giving regions more than 100µm across in contact with the windows. The spectra are of the beads in chloroform, with a chloroform spectrum as the background. The microscope aperture was 50 x 50µm. It has been shown[1] that this approach can be used to monitor reactions in single beads.

ATR Mapping

The manufacture of crowns involves grafting a polymer with reactive groups on to the initial moulding, and then linking molecules of interest to this. There is obvious interest in examining the uniformity of such surfaces. A microscope with an ATR objective can be used to map the surface. Figure 9 shows two spectra from a series recorded automatically at 100µm intervals along a section of a 'macro crown'. The surface is somewhat irregular so that differences could arise from variations in the contact of the ATR crystal with the sample. However it is clear from the spectra that there are changes in relative band intensities that result from differences in composition, not simply from changing contact. The spectra have almost identical intensities in the C-H stretching region but the relative intensities from ester and amide groups at 1730 and 1640cm^{-1} respectively are very different.

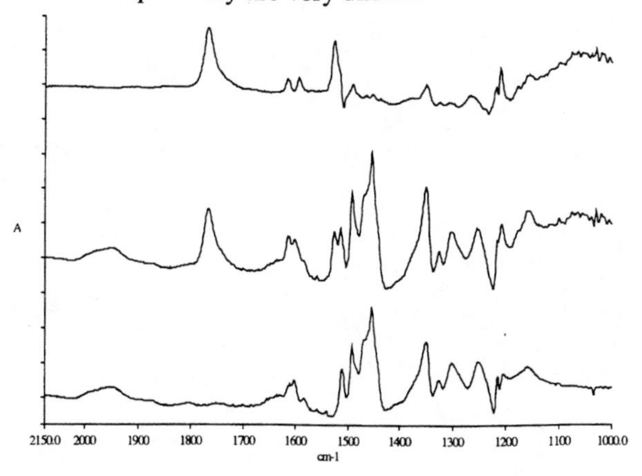

FIGURE 7. Transmission spectra of single Tentagel beads in chloroform. Bottom: Tentagel-PHB, Center: Tentagel-PHB bead with product, Top: difference.

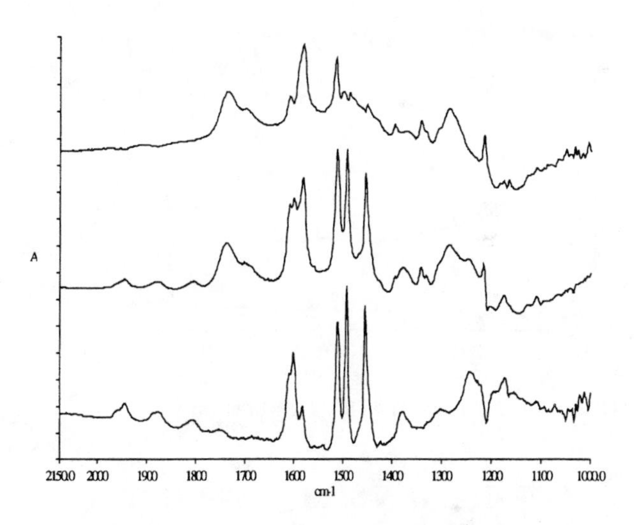

FIGURE 8. Transmission spectra of single Wang resin beads in chloroform. Bottom: unreacted bead, Center: bead with product, Top: difference.

Automated Measurements

Crowns are typically mounted in an array similar to a 96 well microplate. We have demonstrated the possibility of using an X-Y-Z stage to measure spectra sequentially from an array of crowns automatically. The 96 well plate was mounted on the sample stage of a Perkin-Elmer AutoIMAGE™ system. The ATR crystal is visible above the samples. The spectra of these crowns obtained from an automated measurement sequence are seen in Fig. 10. The bottom spectrum is of an ungrafted polypropylene crown.

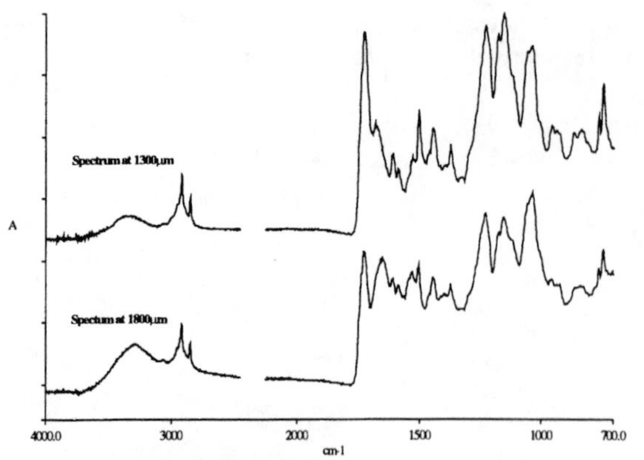

FIGURE 9. Microscope ATR spectra from different points on a line scan along a macro crown.

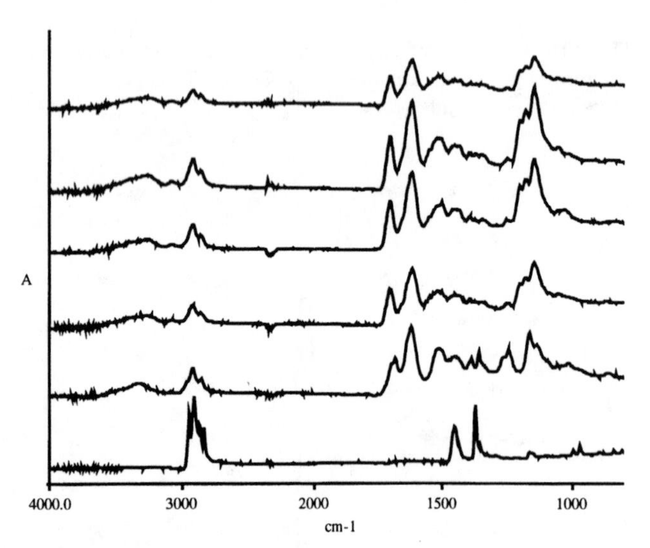

FIGURE 10. Microscope ATR spectra of crowns.

FT-RAMAN SPECTROSCOPY

Possibly the simplest method of obtaining spectra from single beads is by Raman spectroscopy. The example in Fig. 11 illustrates both the strengths and limitations of this method. The samples are the same Tentagel resin and product as for the IR spectra of Fig. 7.. The spectra are of excellent quality and allow a very clean subtraction. However the difference spectra give a quite different picture of the reacting groups because the carbonyl groups are not seen.

Comparison of IR and Raman Data

In Fig. 12 we compare the IR and Raman data from Tentagel beads differing by reaction with p-nitrophenyl chloroformate to protect a hydroxyl group. The IR data are from transmission of single beads swollen in chloroform. The Raman measurements were obtained directly. The data illustrate very clearly the complementary nature of IR and Raman data. They further illustrate how IR spectra tend to show more features associated with functional groups.

The most prominent band in the Raman spectrum is at 1336cm^{-1}, associated with a nitro group on an aromatic ring, is not seen in the IR spectrum. The shoulder at 1348 cm^{-1} corresponds to the nitro group band at 1350cm^{-1} in the IR spectrum. The other IR band from the nitro group at 1527cm^{-1} is not visible in the Raman spectrum. Both Raman and IR bands show characteristic aromatic ring bands close to 1600cm^{-1}. However the very strong carbonyl absorption at 1765cm^{-1} in the IR spectrum is not evident in the Raman spectrum.

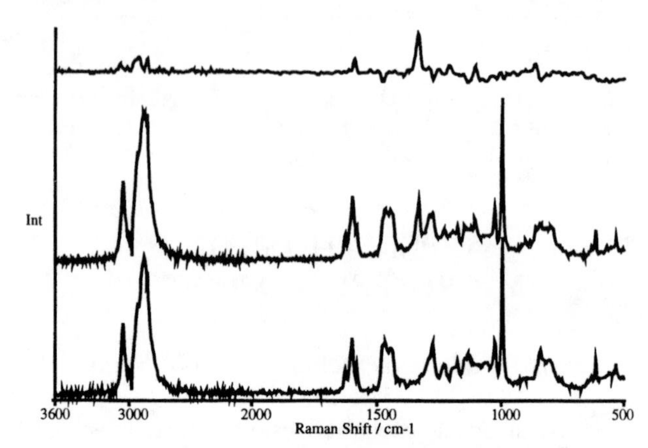

FIGURE 11. Raman spectra of single Tentagel beads. Bottom: Tentagel-PHB, Center: Tentagel-PHB bead with product, Top: difference.

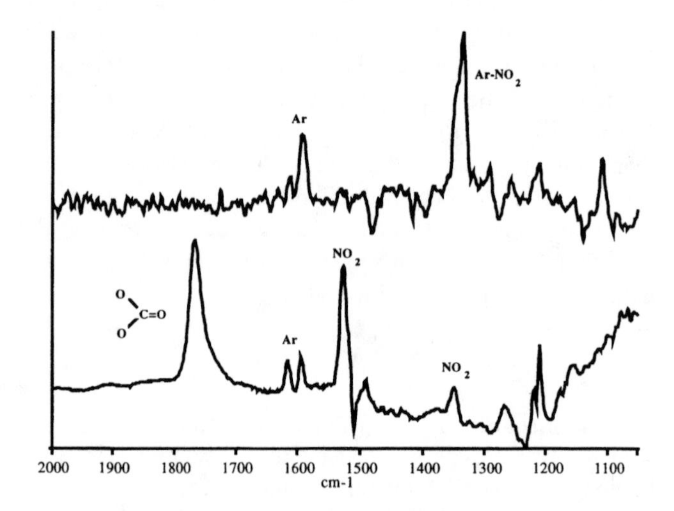

FIGURE 12. Difference spectra between Tentagel beads after reaction with p-nitrophenyl chloroformate and the original beads. Top: Raman, Bottom: IR.

ACKNOWLEDGEMENTS

We are extremely grateful to Dr Andrew Bray of Chiron Mimotopes Pty for providing samples and to Dr Robert Alexander of Perkin-Elmer Ltd for some of the data.

Uniform Depth Profiling of Multiple Sample Depth Ranges in a Single Step-Scanning FT-IR Photoacoustic Experiment

G. M. Story and C. Marcott

The Procter & Gamble Company, Miami Valley Laboratories, Cincinnati, OH 45253-8707

Photoacoustic spectroscopy (PAS) depth profiling results are consistent with the known layer structure of a Paramount laminate film and an adhesive label. There is a clear relationship between the measured phase angle differences and individual layer thicknesses. At low modulation frequencies, deeper sample depths are probed and the phase-angle differences are smaller than at higher frequencies. Digital signal processing enables multiple depth ranges to be probed in a single experiment. Comparing different modulation-frequency results helps to verify the layer thickness estimates. PAS depth profiling is a viable approach when the top layer is either: 1) thin; or 2) not strongly absorbing in key regions of the IR spectrum where deeper layers absorb.

INTRODUCTION

Over the past decade, step-scanning Fourier transform infrared (FT-IR) spectrometry has reemerged commercially and become particularly popular for uniform depth-profiling studies using photoacoustic spectroscopy (PAS) (1). The depth in the sample from which the photoacoustic signal originates depends on the modulation frequency of the incident light. A conventional rapid-scanning FT-IR spectrometer modulates each wavelength of IR radiation at a different frequency, leading to PAS spectra with convoluted depth information. In a step-scanning FT-IR experiment, on the other hand, the incident light is modulated the same way for each IR wavelength, making it straightforward to obtain PAS spectra from uniform depths in the sample across all wavelengths. Until recently, a cumbersome lock-in amplifier was required to extract the PAS spectrum, and only a single modulation frequency could be used. The use of powerful new PCs with digital-signal-processing (DSP) capability has greatly simplified the step-scanning FT-IR PAS experiment and reduced the cost of the instrumentation (2-3). In addition, the DSP approach permits the use of multiple modulation frequencies, thereby increasing the amount of depth-profiling information obtained from a single step-scanning PAS experiment. Examples will be presented, illustrating how the DSP approach can be used to probe multiple depth ranges uniformly in a single PAS experiment.

EXPERIMENTAL

All spectra were measured on a Bio-Rad FTS-6000 spectrometer operating in step-scan mode. A helium-purged MTEC Model 200 PAS cell was used. Phase modulation was applied to the fixed mirror through the dynamic alignment piezoelectric devices. The phase modulation amplitude was one HeNe laser fringe (peak-to-peak) for the near-infrared (NIR) experiments and two fringes for the mid-infrared work. Full double-sided interferograms were collected at every other HeNe laser-fringe zero crossing. The spectral resolution was 8 cm^{-1} and a scan rate of 1 Hz was used for the 25-Hz phase modulation frequency experiments and 2.5 Hz for all other experiments. An OCLI #L01278 long-pass optical filter was used for the NIR experiments to remove aliasing errors. A carbon black standard was used to assign zero phase using the DSP algorithm described in ref. 3. All sample interferograms were computed using the carbon black in-phase stored-phase array, which contains only instrument response function contributions (4-5). Finally, the computed PAS spectra were ratioed to the carbon black in-phase spectrum. The computed in-phase and quadrature sample spectra [S(I) and S(Q)] can be rotated in Cartesian space to calculate spectra at any desired phase angle $S(\theta)$ using the equation: $S(\theta) = S(I)\sin(\theta) + S(Q)\cos(\theta)$.

RESULTS

Packaging Material Film

A commercial packaging material film (Paramount) was examined in the near infrared region by PAS using multiple modulation frequencies. The sample consisted of a 12-µm-thick surface layer of poly(ethylene terephthalate) (PET), followed by a 7.5-µm-thick adhesive tie layer of unknown composition, a 15-mm-thick layer of ethylene vinyl alcohol copolymer (EVOH), another 7.5-µm-thick

CP430, *Fourier Transform Spectroscopy:* 11th International Conference
edited by J.A. de Haseth

Table 1. Phase angle where maximum signal is observed for 3 layers of a commercial polymer laminate at modulation frequencies of 25, 100, and 400 Hz. Phase angle differences between layers are given to the right.

	Odd Harmonics of PM Fundamental				
	25 Hz	75 Hz	125 Hz	175 Hz	225 Hz
PET	14	2	0	0	4
	22	50	68	82	74
EVOH	36	52	68	82	78
	58	58	52	38	22
PE	94	110	120	120	100

	Odd Harmonics of PM Fundamental				
	100 Hz	300 Hz	500 Hz	700 Hz	900 Hz
PET	2	0	2	0	0
	56	96	110	132	140
EVOH	58	96	112	132	140
	56	36	34	4	-
PE	114	132	146	136	-

	Odd Harmonics of PM Fundamental				
	400 Hz	1200 Hz	2000 Hz	2800 Hz	3600 Hz
PET	4	10	0	2	-
	112	146	158	158	-
EVOH	116	156	158	160	-
	38	10	-	-	-
PE	154	166	-	-	-

adhesive tie layer, and a 75-μm-thick layer of low-density polyethylene (LDPE). Separate experiments were performed at modulation frequencies of 25, 100, and 400 Hz. The PAS depth profiling results shown in Table 1 are consistent with the known layer structure of the Paramount laminate film. There is a clear relationship between the measured phase angle differences and individual layer thicknesses. At low modulation frequencies, deeper sample depths are probed and the phase-angle differences are smaller than at higher frequencies.

White Adhesive Label

A small, white adhesive label (Dennison) was examined on both sides by PAS using a fundamental phase-modulation frequency of 200 Hz. Figure 1, an overlay of 22° phase spectra at the fundamental, third, and fifth harmonic phase modulation frequencies, indicates the presence of a surface aliphatic adhesive layer. This layer is thin because the OH-stretching bands from the label are detectable as negative (deep) features until the most surface sensitive fifth-harmonic modulation frequency (1000 Hz). Figure 2, an overlay of -16° phase spectra at the fundamental, third, and fifth harmonic phase modulation frequencies, is dominated by the OH-stretching absorbances from the label. The aliphatic adhesive is not detectable even at the fundamental frequency, indicating the label is thick. However, a deep ester component, shown in the rectangular box in the fundamental PAS phase spectra of Figs. 1 and 2, is detected when sampling either side of the label. This feature was not detectable by more surface-sensitive techniques, such as attenuated total reflectance (ATR) or PAS at higher modulation frequencies.

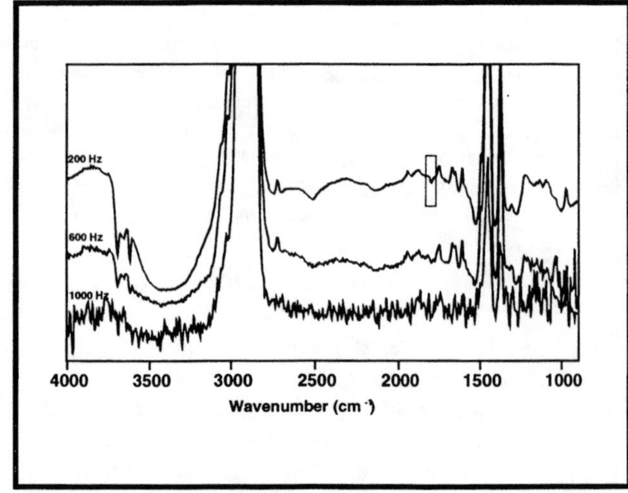

FIGURE 1. Fundamental, third, and fifth harmonic PAS spectra at a phase angle of 22° of a white adhesive label, adhesive side up.

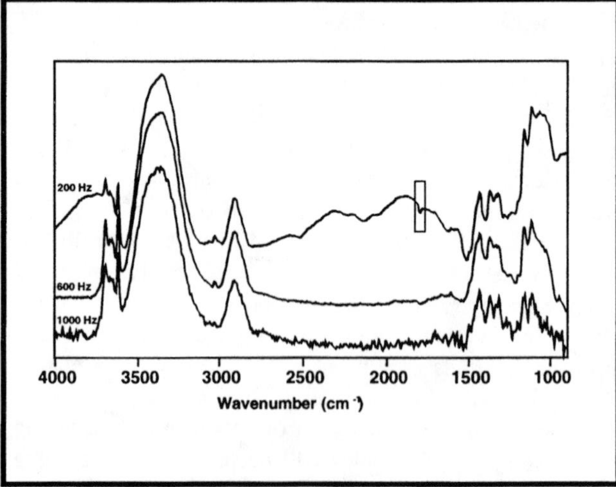

FIGURE 2. Fundamental, third, and fifth harmonic PAS spectra at a phase angle of -16° of a white adhesive label, label side up.

Drug Tablet in Blister Packaging

A generic decongestant tablet in a blister package was examined using a fundamental phase-modulation frequency of 200 Hz and compared to 200-Hz PAS spectra of an empty poly(vinylchloride) (PVC) blister package and a free tablet. The top spectrum in Fig. 3 shows evidence of the tablet in the sealed blister pack (OH stretch). There also appear to be negative features at both the aromatic CH-stretching region and the ester region of this spectrum, indicating a deeper layer of aromatic ester. ATR experiments indicated the presence of a phthalate plasticizer on both the inner and outer surfaces of the blister package as well as on the surface of the tablet. The PAS phase-angle data for this sample did not make sense, falsely suggesting that the surface layer was of the tablet, followed by the PVC of the blister pack. This observation appears to be an artifact due to the vibrating pocket of air trapped inside the blister pack (a drum effect). Although the tablet coating can be detected without destroying the blister package using PAS, the thermal phase information cannot be trusted due to this vibrating trapped air pocket, which shortcuts the propagation of the thermal wave through the PVC. Raman spectroscopy using a fiber-optic probe to focus through the blister pack and onto the tablet is perhaps a better way to study this system nondestructively.

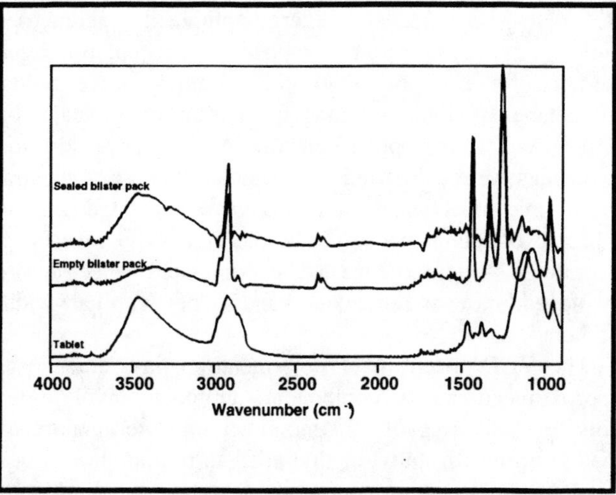

FIGURE 3. 200 Hz PAS spectra of a decongestant tablet (24°), blister package (24°), and the tablet inside the blister package (14°).

REFERENCES

1. Gregoriou, V. G., and Hapanowicz, R., *Spectroscopy* **12**, 37-41 (1997).
2. Manning, C. J., and Griffiths, P. R., *Appl. Spectrosc.* **47**, 1345-1349 (1993).
3. Drapcho, D. L., Curbelo, R., Jiang, E. J., Crocombe, R. A., and McCarthy, W. J., *Appl. Spectrosc.* **51**, 453-460 (1997).
4. Story, G. M., Marcott, C., and Noda, I., *Proc. of the 9th Int. Conf. on Fourier Transform Spectrosc., SPIE* **2089**, 242-243 (1993).
5. Marcott, C., Story, G. M., Dowrey, A. E., Reeder, R. C., and Noda, I., *Mikrochim. Acta [Suppl.]* **14**, 157-163 (1997).

Vibrational and VCD spectra of poly(menthyl vinyl ether)

J.L. McCann,[1] P. Bour,[2] and H. Wieser[1]

[1] *Department of Chemistry, University of Calgary, Calgary, AB, Canada, T2N 1N4*
[2] *Institute of Organic Chemistry and Biochemistry, Academy of Sciences of the Czech Republic, Flemingovo nam 2, 16610 Praha, CR*

The detailed assignments are reported for the vibrational and VCD spectra of (1*S*, 2*R*, 5*S*)-(+)-menthol. Energy minimized geometries, harmonic force fields, and atomic polar tensors were calculated at the Becke3LYP/6-31G** level, and atomic axial tensors with the vibronic coupling theory at the HF/6-31G level. The spectra consist of contributions mainly from two isomers (70%) distinguished only by conformation of the OH group. An attempt was made to simulate the absorption and VCD spectra of poly(methyl vinyl ether) using a component approach and invoking the excitation scheme with promising though not conclusive results at this stage.

INTRODUCTION

Vibrational circular dichroism (VCD) spectroscopy is now at the stage where the spectra of small and medium sized molecules can be measured readily and simulated theoretically with reasonable ease and confidence. Fruitful extensions of the technique in experimental as well as theoretical directions to determine structures and structural changes of polymeric materials are therefore desirable and should be feasible. While the VCD spectra of various biopolymers have been reported (1), few articles have been published dealing with VCD of synthetic chiral polymers (2). During the past four years we have explored correlations between VCD features and chiral polymers (3, 4, 5) targeting poly(menthyl vinyl ether) in particular. In the first instance we needed to understand the VCD spectrum of menthol, isomenthol and neomenthol. In this paper we summarize the assignment of the absorption and VCD spectra of menthol (6). In addition, we highlight progress with our ability to simulate VCD spectra of chiral polymers based on high level *ab initio* computations using poly(menthyl vinyl ether) as example.

EXPERIMENTAL AND COMPUTATIONAL DETAILS

(1*S*,2*R*,5*S*)-(+)-Menthol was obtained from Aldrich and was used without further purification. Menthyl vinyl ether was prepared from ethyl vinyl ether and menthol (7), and polymerized with $EtAlCl_2$ in toluene at -78°C to yield poly(menthyl vinyl ether). The spectra of the samples (in CCl_4 solution) were recorded at a resolution of 4 cm^{-1} with a Bomem MB100 interferometer (8). Typically 5000 ac and 500 dc scans were collected. The resulting VCD spectra were corrected for artifacts by subtracting the spectra of the opposite enantiomer.

The absorption and VCD spectra of menthol were computed as follows. With density functional theory at the B3LYP/6-31G** level, energy-optimized geometries yielded the equilibrium composition of ten rotational isomers, for each of which the harmonic force fields, frequencies of vibrational fundamentals uniformly scaled by 0.985, and atomic polar tensors (APTs) for absorption intensities were calculated. To simulate the VCD spectra, the atomic axial tensors (AATs) were generated via the vibronic coupling theory (9) at the HF/6-31G level with distributed origin gauge. The simulated spectra are displayed below as Lorentzian band shapes with half-width of 5 cm^{-1}.

The VCD spectrum of poly(menthyl vinyl ether) was approximated by a component approach involving a polymer consisting of 20 monomer units terminated by methyl groups; menthyl methyl ether to mimic the menthyl pendants; a 3-unit section of the polymer terminated by methyl groups; and an 8-carbon section of the polymer backbone, again terminated in all positions by methyl groups (3,5,7-trimethoxy-octane). Geometries, force fields, APTs and AATs were then calculated with a theory and at a level appropriate for the size of the component. Details of these calculations will be published elsewhere in due course.

RESULTS AND DISCUSSION

The observed VCD spectrum of (+)-menthol can be simulated successfully with only the two lowest energy conformers which have the same conformation of the ring and the same orientation of the isopropyl group, viz. chair and gauche conformations, respectively. They do differ, however, in the orientation of the OH group, being anti for the lowest (38% population) and gauche for the second

CP430, *Fourier Transform Spectroscopy:* 11th International Conference
edited by J.A. de Haseth

lowest (25% population) conformer. The calculated and observed VCD spectra are compared in Fig. 1.

Region I of the observed spectrum consists mostly of CH_3, CH_2, and "out-of-plane" CH rocking and ring stretching vibrations and is identically reproduced. Region II contains largely CO stretching modes and while it, too, is well reproduced, it shows some apparent discrepancies. Modes in Region III are extensively mixed and include CC stretching of the isopropyl group, "in-plane" CH rocking, CH_2 twisting, and CH_3 symmetric deformation modes. Bending of OH begins to contribute above 1200 cm^{-1} and gives rise to bands broadened by intermolecular hydrogen bonding. The asymmetric CH_3 and CH_2 bending modes in Region IV are generally well reproduced but are predicted too high in wavenumbers. The discrepancy in Region II, noted above, is believed to arise from slight differences between observed and calculated peak positions of close lying vibrations. The appearance of this group of bands is especially distinct and characteristic of menthol. Among vibrational modes that are characteristically different for the orientation of the OH group, the OH bend is most clearly distinguishable in the spectra. The VCD bands labeled **a** and **b** in Figure 1 characterize unambiguously the anti and gauche orientations of the OH group giving rise respectively to negative/positive and positive/negative features.

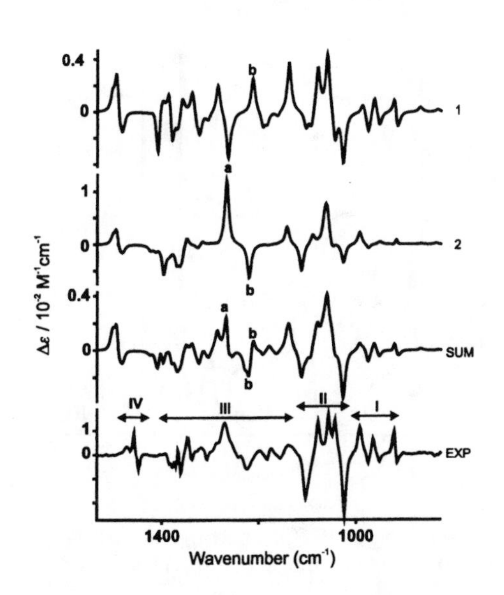

Figure 1. Calculated VCD spectra of Conformers 1 and 2 of menthol, the sum of all 10 conformers; and the observed spectrum of menthol (0.39 M).

Although the VCD spectra of the menthyl vinyl ether monomer and polymer have not yet been calculated and assigned, some qualitative comparisons can be made (Figure 2). Many bands of the monomer in Regions I and II are very similar to those of the parent compound, with the exception of the group of bands containing CO stretching vibrations. Noteworthy in particular is that the intensities of these bands, when normalized for concentration, are essentially equal. Bands in Region III of the monomer have sharpened considerably, likely due to the absence of hydrogen bonding, and are generally different from those of menthol. The VCD bands of the polymer in Regions III and IV appear to be quite different from those of either the monomer or menthol itself, while those in Region I are essentially identical. The most distinctive difference is the intense band in Region II in the polymer, which we attribute to conformational changes and complexities, and perhaps also intensity enhancements due to dipolar coupling in helical or near helical segments of the polymer.

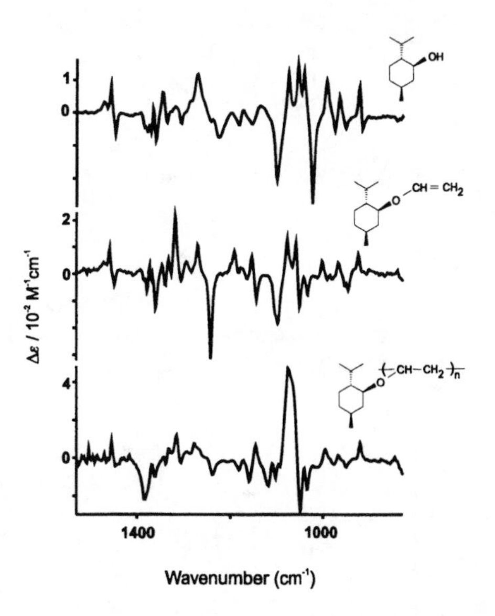

Figure 2. VCD spectra of menthol (0.39 M, top trace), menthyl vinyl ether (0.17 M, middle trace), and poly(menthyl vinyl ether) (0.031 g in 1 ml CCl_4, or 0.15 M monomer, bottom traces).

A segment of the computed helical structure of the polymer is displayed in Figure 3. It confirms the stacking of CO bonds in helical portions as one might have assumed intuitively. Distances between these bonds are of a magnitude that could lead to dipolar coupling, even though the characteristic conservative couplets are not evident in the VCD spectrum. The absence of these distinctive features may be due to conformational inhomogeneity which would further suggest considerable flexibility of the polymer.

The calculated absorption and VCD spectra of poly(menthyl vinyl ether) are compared in Figure 4. The calculated absorption bands resemble those observed even if they do not correspond in detail, indicating a degree of validity of the adopted procedure. Moreover, the calculated absorptions in Region II also resemble those of their counterparts in menthol (5). Comparison of the VCD features reveals some similar bands in Region III as indicated by arrows in Figure 4. The calculated bands in Region I are too weak for making any comparisons. While Region II is the most intense also in the calculated spectrum, there is little visual similarity. There may be several reasons for the discrepancies. One could be that the predicted wavenumber positions of the bands differ from those observed, which in VCD can alter the band contours substantially. Another very likely possibility could be that the calculations simulate only a helical conformation of the polymer, whereas in reality the polymer, likely being quite flexible, would adopt a number of conformations that can have quite have different VCD signatures.

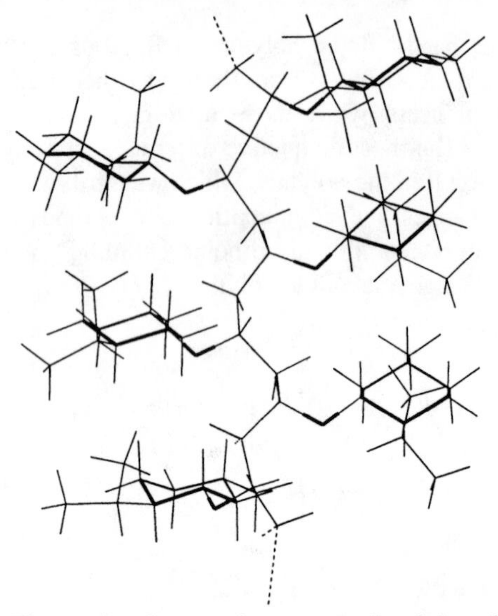

Figure 3. Computed structure of poly(menthyl vinyl ether) as determined by optimization and step-wise annealing of 20 repeat units using molecular mechanics.

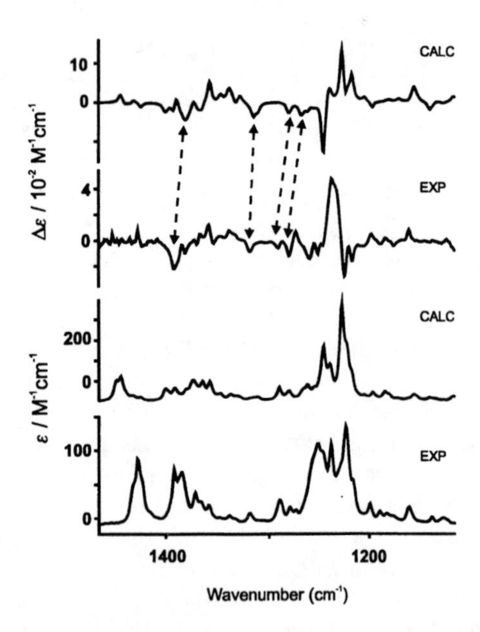

Figure 4. Observed and calculated absorption (bottom pair of traces) and VCD (top pair) spectra of poly(menthyl vinyl ether).

The argument suggesting that the VCD spectrum of poly(menthyl vinyl ether) differs from that calculated for a helical segment as a result of backbone flexibility is supported by the VCD spectrum of poly(triphenylmethyl methacrylate) (Figure 5). While this polymer has no chirality arising from its pendants, it is chiral by virtue of a stable helical conformation locked in place by the phenyl groups (10). The VCD spectrum displays sharp positive/negative peaks characteristic of dipolar coupling of oscillators in a regular helical environment Analysis of the VCD spectrum of this polymer is presently in progress.

CONCLUSION

Analysis of the absorption but especially of the VCD spectra of (+)-menthol based on high level *ab initio* simulations has led to a definitive vibrational assignment and identification of rotational isomers. The VCD features of the menthyl vinyl ether monomer that arise from the chiral centers of menthol itself remain essentially the same with nearly identical intensities, aside from a few bands the origin of which still needs to be identified. The VCD spectrum of the polymer is considerably different than that of either the pendant or the monomer, particularly in the CO stretching region. A preliminary attempt to simulate the VCD spectrum of the polymer in which a helical structure was assumed led to the tentative conclusion that the backbone of poly(menthyl vinyl ether) is reasonably flexible.

The results presented in this paper clearly indicate that VCD spectroscopy of polymeric materials, chiral either by virtue of asymmetric pendants or as a result of a regular helical structure, is not only feasible but also yields distinctive and unique structural information.

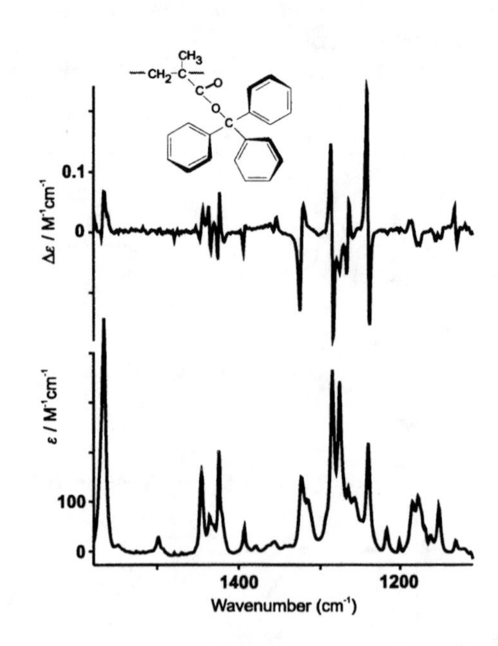

Figure 5. Absorption and VCD spectra of poly(triphenylmethyl methacrylate) (41.1 mg in 1 ml CCl_4, or 0.125 M monomer).

REFERENCES

1. Keiderling, T.A., and Pancoska, P, in *Advances in Spectroscopy Part B, Biomolecular Spectroscopy*, Vol. 21 (R.J.H. Clark and R. Hester, eds.), Chichester: Wiley, 1993, ch. X, pp. 267-315.
2. Nafie, L.A., Keiderling, T.A., Stephens, P.J., *J. Am. Chem. Soc.* **98**, 2715-2723 (1976).
3. Tsankov, D., Eggimann, E., Liu, G., and Wieser, H., *SPIE* **2089**, 178-179 (1993).
4. McCann, J., Tsankov, D., Hu, N., Liu, G., and Wieser, H., *J. Mol. Struct.* **349**, 309-312 (1994).
5. McCann, J.L., Schulte, B., Tsankov, D., and Wieser, H., *Mikrochim. Acta [Suppl.]* **14**, 809-810 (1997).

6. McCann, J.L., Rauk, A., and Wieser, H., *Can. J. Chem.*, submitted.
7. Watanabe, W.H., and Conlon, L.E., *J. Am. Chem. Soc.* **79**, 2828-2833 (1944).
8. Tsankov, D., Eggimann, E., and Wieser, H., *Appl. Spectrosc.* **49**, 132-138 (1995).
9. Nafie, L.A., and Friedman, T.B., *J. Chem. Phys.* **78**, 7108 (1983); Yang, D., and Rauk, A., *J. Chem. Phys.* **97**, 6517 (1992).
10. Cavallo, L., Corradini, P., and Vacatello, M., *Polymer Comm.* **30**, 236-238.

FTIR Spectroscopic Studies of Magnesium Impurities in Germanium

L. T. Ho

Insitute of Physics, Academia Sinica
Nankang, Taipei, Taiwan, Republic of China

From the FTIR measurement of magnesium-doped germanium, several absorption lines have been observed for the first time. The relative positions of these spectral lines are in good agreement with excitation spectra observed for other well-established acceptor impurities in germanium, which suggests that they are due to a magnesium-related acceptor impurity. Its ionization energy, estimated to be 118.91 meV, indicates that it is a deep acceptor impurity in germanium.

INTRODUCTION

It is well known that the group III acceptors in germanium can be considered as the solid-state analogues of the hydrogen atom(1). Due to their small binding energies, they are all shallow acceptor impurities in germanium(2,3).

Several group II acceptors in germanium have been studied as well(4~8). They are all deeper acceptor impurities. Due to a double acceptor, they can be considered as the solid-state analogues of the helium atom.Both the neutral and the singly ionized acceptor states for each of the group II impurities can be studied. For example,the excitation spectra for both neutral and singly ionized beryllium(4,7), both neutral and singly ionized zinc(5,8) in germanium have all been reported. For magnesium in germanium, however, so far only the excitation spectrum for neutral magnesium(7) has been studied and corresponding spectrum for singly ionized magnesium has not been observed yet.

The purpose of the present paper is to report the result of a spectroscopic study of magnesium impurities in germanium. In the course of our study, we have observed for the first time an unexpected series of spectral lines, which clearly indicates the existence of another magnesium-related acceptor impurity center.

EXPERIMENTAL PROCEDURE

Magnesium was thermally diffused into high-purity germanium by using a sandwich technique described elsewhere(9). During diffusion, the sample was heated at 900℃in vacuum, typically for 100 hours. After the heat treatment, the sample was cooled down by quenching into liquid nitrogen.

A Bomem DA3 Fourier-transform infrared spectrometer, equipped with appropriate beamsplitter and detector, was used for the infrared absorption measurement. Instrumental spectral resolution was typically set at 0.5 cm^{-1}. An Oxford continuous flow optical cryostat, equipped with CsI optical windows, was used for the low temperature measurements. The sample was mounted on the copper tailpiece of the cryostat using a strain-free mounting technique(10).

EXPERIMENTAL RESULT

From the excitation spectrum of neutral magnesium acceptor impurities in germanium previously reported(7), the excitation lines have the following positions in meV: 31.21(G line), 32.95(D line), 33.71(C line), and 34.34(B line). These spectral lines are due to transitions from the ground state to various excited states of the neutral magnesium acceptor. The ionization energy for this acceptor center has been calculated to be 35.83 meV.

Since spectrum of wide spectral range can be obtained in one FTIR measurement, it is easy to have the chance to search for other interesting features presented in the spectrum. This experimental technique is particularly useful for looking for the excitation spectrum of impurity center with unknown ionization energy. When measuring the magnesium-doped germanium we prepared , we have indeed observed for the first time a new series of spectral lines around 116 meV, as is shown in Fig.1. The spectral lines observed have the following positions in meV:114.19(G line), 115.93(D line), 116.63(C line), and 117.30(B line). Since this energy region is much higher than the ionization energy of neutral magnesium acceptor, it seems, therefore, that the most reasonable explanation for these extra spectral lines

CP430, *Fourier Transform Spectroscopy:* 11th International Conference
edited by J.A. de Haseth

would be the excitation lines for singly ionized magnesium acceptor.

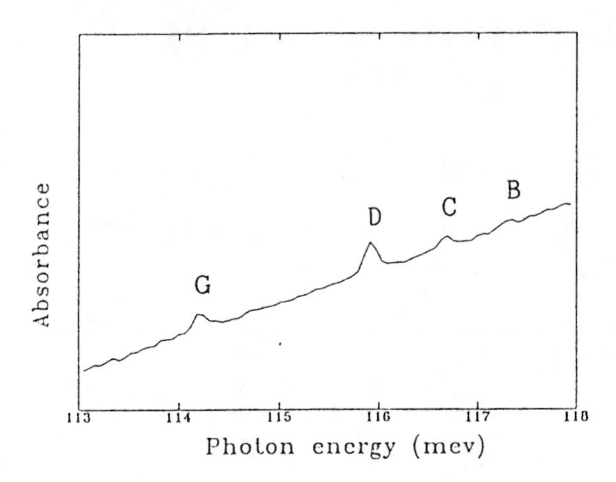

FIGURE 1. Excitation spectrum of magnesium-doped germanium. Liquid helium used as the coolant.

For a helium-like group II acceptor impurity in germanium, the binding energy of each energy state for the singly ionized acceptor should be 4 times that of the corresponding state for the neutral acceptor. This has been verified, e.g., in the case of zinc acceptor in germanium[8]. Shown in Table 1 is the comparison of the energy spacings between the corresponding lines of the extra lines in Fig.1 and those of the neutral magnesium acceptor. It is evident that the relative positions of the spectral lines shown in Fig.1 are strikingly similar to those obtained for neutral magnesium acceptor. The labeling of the spectral lines observed in Fig.1 is based on this similarity. Since the line spacings are quite close rather than 4 times, apparently the spectral lines observed in Fig.1 are not due to singly ionized magnesium acceptor. Most likely it is due to another magnesium-related acceptor impurity in germanium. It is not unusual for impurity atoms of a certain foreign element to produce more than one impurity centers in a semiconductor crystal. For example, beryllium has been found to form at least five different kinds of acceptor impurities in silicon[11]. In the present case, the ionization energy for the new acceptor center is estimated to be 118.91 meV clearly indicating that it is a much deeper acceptor impurity in germanium.

TABLE 1. Energy Spacings between Excitation Lines (in meV)

Label	Neutral Mg	Fig.1
G-D	1.74	1.74
G-C	2.50	2.44
G-B	3.13	3.11
D-C	0.76	0.70
D-B	1.39	1.37

In conclusion, from the FTIR measurement of magnesium-doped germanium, a new series of absorption lines have been observed for the first time. Evidence indicates that they are due to a much deeper magnesium-related acceptor impurity center in germanium.

Since spectrum and information of wide spectral range can be obtained in one measurement, the discovery of this new series of lines also demonstrates the advantage of FTIR spectroscopy in searching for new impurity centers in semiconductors.

ACKNOWLEDGMENT

This work was partially supported by the National Science Council of the Republic of China under contract number NSC86-2112-M-001-013.

REFERENCES

1. A. K. Ramdas and S. Rodriguez, Rep. Prog. Phys. 44, 1297(1981).
2. H. P. Soepangkat and P. Fisher, Phy. Rev. B 8, 870(1973).
3. R. L. Jones and P. Fisher, J. Phys. Chem. Solids 26, 1125(1965).
4. H. Shenker, E. M. Swiggard, and W. J. Moore, Trans. Metall Soc. AIME 239, 347(1967).
5. R. L. Jones, PhD Thesis, Purdue Univ.(1968).
6. R. A. Chapman and W. G. Hutchinson, Phys. Rev. 157, 615(1967).
7. J. W. Cross, L. T. Ho, A. K. Ramdas, R. Sauer, and E. E. Haller, Phys. Rev. B28, 6953(1983).
8. N. R. Butler and P. Fisher, Phys. Rev. B 13, 5465(1976).
9. L. T. Ho, Appl. Phys. Lett. 35, 409(1979).
10. C. Jagannath, Z. W. Grabowski, and A. K. Ramdas, Phys. Rev. B 23, 2080(1981).
11. L. T. Ho, F. Y. Lin, and W. J. Lin, Int. J. of IR & MM Waves 16, 339(1995).

A Comparative Study of the Dipole-Polarizability of the Metallocenes $Fe(C_5H_5)_2$, $Ru(C_5H_5)_2$ and $Os(C_5H_5)_2$ by Means of Dispersive Fourier Transform Spectroscopy in the Visible (DFTS-VIS)

U.Hohm and D.Goebel

Institut für Physikalische und Theoretische Chemie der TU Braunschweig, Hans-Sommer-Str.10, D-38106 Braunschweig, FRG

The gas-phase refractivity (n-1) of the metallocenes ferrocene, $Fe(C_5H_5)_2$, ruthenocene, $Ru(C_5H_5)_2$, and osmocene, $Os(C_5H_5)_2$, has been determined by dispersive Fourier transform spectroscopy and other traditional interferometric techniques in the wavenumber-range between $11500 cm^{-1}$ and $20000 cm^{-1}$. The results are used to calculate the frequency-dependence of the mean dipole-polarizability $\alpha(\omega)$ according to the Lorenz-Lorentz-relation. We observe an increase in the mean dipole-polarizability with increasing molecular weight. Extrapolation to zero frequency yields the electronic part of the static mean dipole-polarizability, $\alpha(0)$, which also increases in the sequence from ferrocene to osmocene as 126.13(66)au, 133.1(1.2)au, and 138.5(4.8)au.

INTRODUCTION

Since their discovery the d^6-metallocenes $FeCp_2$, $RuCp_2$, and $OsCp_2$ ($Cp=\eta^5-C_5H_5$) have served as a prototype of metal-organic compounds. However, very little is known from experiment and theory about multipole-moments, which reflect the static charge distribution, and electro-optical properties of the free metallocene molecules. So far, the available electro-optical and magnetical properties of $FeCp_2$ and $RuCp_2$ have been determined mostly in condensed phases (1,2), where intermolecular interactions may disturb significantly the properties of the free molecule. To the best of our knowledge, there are no experimental investigations of the electro-optical properties of osmocene. Only recently, we have reported high-level time-dependent DFT/B3-LYP calculations of the dipole-polarizability tensor of ferrocene which are supported by gas-phase measurements of the mean dipole-polarizability $\alpha(\omega)$ (3).

The main objective of this work is to make a comparative study of the frequency-dpendence of the mean dipole-polarizability $\alpha(\omega)$ of the d^6-metallocenes ferrocene, ruthenocene and osmocene by means of precise gas-phase measurements of the density- and frequency-dependence of the refractive index of these compounds. Additionally, combination with available polarizability-anisotropies will yield reasonable estimates of the dispersion of their dipole-polarizability anisotropy $\Delta\alpha(\omega)$. These data provide also valuable information on the isotropic and anisotropic dispersion-interaction energy and the isotropic diamagnetic susceptibility χ.

Throughout the whole paper atomic units (au) are used. For conversion into SI-units see Appendix.

THEORETICAL

If a molecule is placed into a uniform and weak electric field $\mathbf{E}$ it acquires a dipole moment $\mathbf{p}$ according to $\mathbf{p}=\alpha\mathbf{E}$, where α is the dipole-polarizability tensor. Now let z be the principal axis then a molecule with a five-fold rotation axis possesses only two different non-zero components of the dipole-polarizability tensor α, namely $\alpha_\perp(\omega) \equiv \alpha_{xx}(\omega) = \alpha_{yy}(\omega)$ and $\alpha_{||}(\omega) \equiv \alpha_{zz}(\omega)$. In what follows we will refer to these symmetry requirements. The two invariants of the dipole-polarizability tensor α are the mean dipole-polarizability $\alpha(\omega)$ and the dipole-polarizability anisotropy $\Delta\alpha(\omega)$, here given as:

$$\alpha(\omega) = [2\alpha_\perp(\omega)+\alpha_{||}(\omega)]/3 \quad (1)$$

$$\Delta\alpha(\omega) = \alpha_{||}(\omega)-\alpha_\perp(\omega) \quad (2)$$

For non-absorbing substances and for frequencies ω lower than the first electronic transition frequency ω_{01} the electronic part of $\alpha(\omega)$ and $\Delta\alpha(\omega)$ can be approximated with high accuracy by a two term Kramers-Heisenberg dispersion relation:

$$\alpha(\omega) = \frac{1}{3}\left(2\frac{f_\perp}{\omega_\perp^2-\omega^2}+\frac{f_\parallel}{\omega_\parallel^2-\omega^2}\right) \quad (3)$$

$$\Delta\alpha(\omega) = \frac{f_\parallel}{\omega_\parallel^2-\omega^2}-\frac{f_\perp}{\omega_\perp^2-\omega^2} \quad (4)$$

CP430, *Fourier Transform Spectroscopy:* 11th International Conference
edited by J.A. de Haseth

Here, $\omega_\perp$ and $\omega_{||}$ are effective transition frequencies with corresponding oscillator strengths $f_\perp$ and $f_{||}$. We have determined the mean dipole-polarizability by precise measurements of the gas-phase refractive-index $n(\omega,\rho,T)$, where ρ is the amount-of-substance density and T is the temperature. In the low density region n and α are related by:

$$n(\omega,\rho,T) - 1 = 2\,\pi N_A\, a_0^3 \rho\, \alpha(\omega,T) \quad (5)$$

where N_A is Avogadro's constant and a_0 is the atomic unit of length.

EXPERIMENTAL

The interferometric technique as well as the apparatus used are described elsewhere (5). To be brief MCp_2 of mass m (supplied by Strem Chemicals, purity better than 99%) is fed into an evacuated sample cell of length $l/2$ and volume V which is fused off thereafter. This sample cell is placed into the active arm of a high-temperature Michelson interferometer and is heated by means of a pipe-furnace until the sample is fully evaporated. Depending on the metallocene a maximum temperature of approximately 490K was used. The resulting amount-of-substance density of the vapor is given by $\rho = m/(MV)$, M being the molar mass. In all cases the maximum ρ is less than 2.62 mol/m^3. The interference fringe shift N, which is recorded during the evaporation of the sample, yields directly the refractivity $(n-1)=N\lambda/l$, where λ is the wavelength of the measuring light (λ_1=632.99nm (ω_1=0.07198116au), λ_2=543.516nm (ω_2=0.08383075au), λ_3=325.13nm (ω_3=0.1401388au), λ_3 only used in the case of $FeCp_2$). These absolute measurements have been performed at three different densities ρ and are repeated several times in order to reduce the uncertainty in the measured refractivity at a given density. The uncertainty in (n-1) is $\Delta(n-1) < 1\times10^{-6}$.

These non-isothermal measurements of the refractivity at discrete wavelengths have been augmented by dispersive Fourier transform spectroscopy (DFTS), in order to record an isothermal quasi-continuous refractive index spectrum between 0.05au and 0.091au, with a resolution of about 3.5×10^{-5}au. In the case of our quasi-continuous measurements in the visible wavelength range the recorded Fourier spectra are relative spectra and must be adapted to the results obtained at ω_1 and ω_2.

RESULTS AND DISCUSSION

The mean dipole-polarizabilities obtained in our absolute measurements are given in Table 1. From these results one immediately recognizes a decrease in the dispersion on going from ferrocene to ruthenocene and osmocene. This behaviour is also pointed out in Fig.1, where the quasi-continuously determined mean polarizability curves $\alpha(\omega)$ recorded in our DFTS measurements are presented.

In order to determine the frequency-dependence of the individual tensor components $\alpha_{||}(\omega)$ and $\alpha_\perp(\omega)$ we rely on the polarizability anisotropy $\Delta\alpha(\omega)$ of ferrocene and ruthenocene measured at 632.8nm (ω=0.07198au) by Ritchie *et al.* (1). These values are incorporated into a non-linear fit in order to determine the parameters in Eqs.(3,4). The details of this procedure are given in reference (4). Comparison of the mean dipole-polarizabilities and molar-volumes of all three metallocenes allows also for an estimate of the frequency-dependence of $\alpha_{||}(\omega)$ and $\alpha_\perp(\omega)$ of osmocene. In Table 2 the fit-parameters of Eqs.(3) and (4) are summarized.

Figure 2 presents the obtained frequency dependence of the individual components $\alpha_{xx}=\alpha_\perp$ and $\alpha_{zz}=\alpha_{||}$ of the polarizability tensor α of the metallocenes. For all three metallocenes, the parallel component $\alpha_{||}$ is larger in magnitude and shows the more pronounced frequency dependence.

The effective transition frequencies $\omega_\perp$ and $\omega_{||}$ are known to correlate with the energies of the ionization bands of the molecule. In the case of the metallocenes we observe that $\omega_{||}$ lies in the A region of the photoelectron spectrum, which is mostly due to ionization from the metal d orbitals. $\omega_\perp$ lies in the vicinity of the B-band region, which is dominated by ionization of essentially ligand electrons.

TABLE 1. Mean dipole-polarizability $\alpha(\omega)$ of the metallocenes obtained from monochromatic measurements of the refractivity (n-1). Rms-errors in parentheses

	$\alpha(\omega_1)$	$\alpha(\omega_2)$	$\alpha(\omega_3)$
$FeCp_2$	131.51(32)	133.74(34)	150.7(1.1)
$RuCp_2$	137.79(59)	139.70(25)	-
$OsCp_2$	142.46(25)	144.10(29)	-

TABLE 2. Effective oscillator strengths $f_\perp$ and $f_{||}$ and effective eigenfrequencies $\omega_\perp$ and $\omega_{||}$ of the metallocenes according to Eqs.(3,4). $\alpha(0)$ and $\Delta\alpha(0)$ are the static mean dipole-polarizability and dipole-polarizability anisotropy, $\Delta_{RMS}\alpha$ is the rms-error of the fit; all quantities in atomic units (au)

	$f_\perp$	$\omega_\perp$	$f_{||}$	$\omega_{||}$	$\alpha(0)$	$\Delta\alpha(0)$	$\Delta_{RMS}\alpha$
$FeCp_2$	20.43(12)	0.4218(12)	12.685(40)	0.29205(38)	126.13(66)	33.9(1.1)	0.072
$RuCp_2$	30.10(30)	0.4980(25)	14.870(73)	0.30830(74)	133.1(1.2)	35.1(2.0)	0.027
$OsCp_2$	47.9(1.9)	0.615(12)	17.15(24)	0.3251(22)	138.5(4.8)	35.6(7.7)	0.14

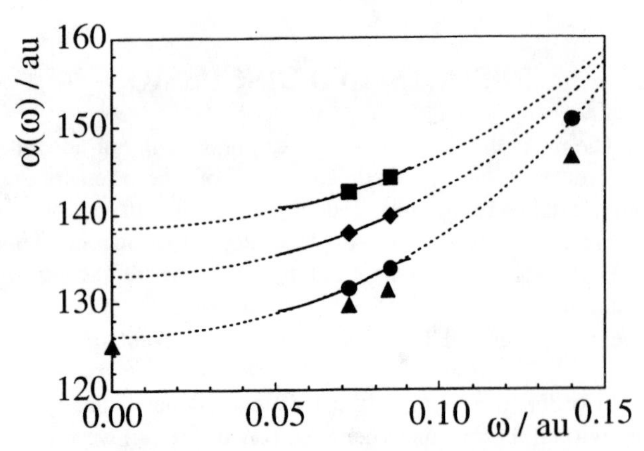

FIGURE 1. Frequency dependence of the mean dipole-polarizability $\alpha(\omega)$ of the metallocenes (lower curve ferrocene, middle curve ruthenocene, upper curve osmocene). The full lines were obtained by dispersive Fourier transform spectroscopy, the dashed lines are the fitted curves according to Eqs.(3,4). The full symbols were obtained from absolute measurements of the refractive index, except for (▲), which were obtained from time-dependent DFT/B3-LYP calculations (3).

For molecules in the ground state the oscillator strengths f_k should obey the Thomas-Reiche-Kuhn (TRK) sum-rule $\Sigma f_k = N_e$, where N_e is the total number of electrons in the molecule. The experimentally determined oscillator strengths $f_\perp$ and $f_{||}$ are obtained in a limited frequency range. Naturally, they do not obey the TRK sum-rule and we observe $F = 2f_\perp + f_{||} < N_e$. However, the increase of 53.5, 75.1, and 113.0 in F in the sequence $FeCp_2$, $RuCp_2$, and $OsCp_2$ nearly parallels the increase in N_e, which amounts to 96, 114, and 146.

Having established the frequency dependence of $\alpha(\omega)$ and $\Delta\alpha(\omega)$ of the metallocenes we can estimate further quantities of physico-chemical interest.

The Kramers-Heisenberg-type representation of $\alpha(\omega)$ and $\Delta\alpha(\omega)$ in Eqs.(3,4) is suitable for obtaining reliable estimates of the isotropic, C_6, and anisotropic, C_6' and C_6'', dispersion interaction-energy constants, which are defined in terms of integrals over the polarizability at imaginary frequencies: $C_6 = 3\int\alpha(i\omega)^2 d\omega/\pi$, $C_6' = \int\alpha(i\omega)\Delta\alpha(i\omega)d\omega/\pi$, and $C_6'' = \int\Delta\alpha(i\omega)^2 d\omega/(3\pi)$. The relation of these constants to the dispersion interaction-energy can be found e.g. in (6). The resulting coefficients are given in Table 3. In particular, the isotropic component C_6, which increases from ferrocene to osmocene, is very large. The magnitude of this coefficient reflects the strong intermolecular interactions between the metallocenes. It is in line with the observed concentration (of the order of 2%) of metallocene dimers in the vapor phase at elevated temperatures and with the experimentally obtained enthalpy of dimerization which is approximately 80 kJ mol^{-1}.

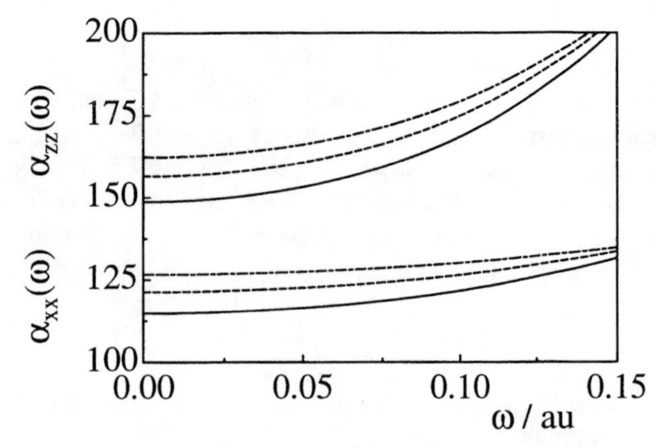

FIGURE 2. Frequency dependence of the individual components $\alpha_{xx} = \alpha_\perp$ (lower triple) and $\alpha_{zz} = \alpha_{||}$ (upper triple) of the polarizability tensor of the metallocenes (in au). (——) ferrocene, (- - -) ruthenocene, and (-···) osmocene.

The simple Drude model of electric and magnetic properties of matter allows for estimating further quantities of interest (7). According to Eq.(6) we can also get an estimate of the mean diamagnetic susceptibilty χ of these compounds.

$$-\chi = \frac{1}{4}\left(\frac{f_\perp}{\omega_\perp} + \frac{f_{||}}{\omega_{||}}\right) \tag{6}$$

In the sequence ferrocene, ruthenocene and osmocene we obtain -22.97au, -27.19au, and -32.66au. We compare these estimates with experimental results. Applying the 'spin-flip technique' on single crystals, Mulay and Mulay (2) measured the solid-state values of -26.0, -31.2, and -40.2au. The deviations between our estimates and the experimental values are between 11% and 19%. Of course, these differences may be due to the inadequacy of the Drude model. However, the measurements of the tensor components of χ do not obey the symmetry requirement that $\chi_{xx} = \chi_{yy} \neq \chi_{zz}$, so these experimental results may be in error.

TABLE 3. Isotropic and anisotropic dispersion interaction-energy constants C_6, C_6', and C_6'' of the metallocenes

	$10^{-3}xC_6$	$10^{-2}xC_6'$	C_6''
$FeCp_2$	4.356(43)	1.02(11)	19(1)
$RuCp_2$	5.484(95)	0.09(25)	35(1)
$OsCp_2$	6.92(48)	-2.0(1.3)	79(8)

CONCLUSIONS

Our gas-phase measurements of the mean dipole-polarizability $\alpha(\omega)$ of the homologous row $FeCp_2$, $RuCp_2$, and $OsCp_2$ yield a monotonous increase in $\alpha(0)$ and $\Delta\alpha(0)$, which is due to an increasing number of electrons in the metal atom. Relativistic effects, which are assumed to liminish α of osmium compared to ruthenium (which night be comparable to the dramatic decrease in α on going from cadmium to mercury (8,9)), are obviously nasked by bonding to the organic rings.

ACKNOWLEDGMENTS

Financial support by the Deutsche Forschungsgemeinschaft and Fonds der Chemischen Industrie are gratefully cknowledged.

APPENDIX

Conversion factors from atomic (au) to SI-units:
Length, $1a_0 = 5.291772x10^{-11}m$,
Energy, $1E_h = 4.369748x10^{-18}J$,
Frequency, ω, $1E_h/(h/2\pi) = 4.134137x10^{16}\ s^{-1}$,
Dipole-polarizability, α, $\Delta\alpha$: $1e^2a_0^2E_h^{-1} = 1.648778$ x $10^{-41}\ C^2m^2J^{-1}$,
Diamagnetic susceptibility, χ, $1e^2a_0^2m^{-1} = 7.98104$ x $10^{-29}\ JT^{-2}$,
Dispersion interaction-energy constant, C_6, $1a_0^6E_h = 9.573448x10^{-80}Jm^6$.

REFERENCES

1. Ritchie, G.L.D., Cooper, M.K., Calvert, R.L., Dennis, G.R., Phillips, L., and Vrbancich, J., *J.Am.Chem.Soc.* **105**, 5215-5219 (1983).
2. Mulay, L.N., and Mulay, I.L., *Anal.Chem.* **38**, 501R-512R (1966).
3. Hohm, U., Goebel, D., and Grimme, S., *Chem.Phys.Lett.* **272**, 328-334 (1997).
4. Goebel, D., and Hohm, U., *J.Chem.Soc. Faraday Trans.* (1997), in press.
5. Goebel, D., and Hohm, U., *J.Appl.Phys.* **29**, 3132-3136 (1996).
6. Hohm, U., *Chem.Phys.* **179**, 533-541 (1994).
7. Amos, R.T., *Int.J.Quant.Chem.* **60**, 67-74 (1996).
8. Goebel, D., and Hohm, U., *Phys.Rev. A* **52**, 3691-3694 (1995).
9. Goebel, D., and Hohm, U., *J.Phys.Chem.* **100**, 7710-7712 (1996).

Measurement of the Mid-Infrared Emission Spectrum of Diamond

G. L. Powell, R. L. Cox, and S. W. Allison

Oak Ridge Centers for Manufacturing Technologies, P. O. Box 2009, Oak Ridge, TN 37831-8096*

A method for measuring the emission spectra of diamond wafers, typically 0.1 to 1 mm thick, is described. The method uses a 40 W continuous CO laser (1800 cm^{-1} to 1900 cm^{-1}) to heat the diamond to ~500°C suspended in an evacuable cell in an emission collection accessory to a Fourier transform infrared spectrometer (FTIRS). A shutter protects the FTIRS from the intense laser light during the heating process and opens immediately after the laser is turned off to measure the emission as the specimen cools. The spectra are referenced to a "black body" source that uses the same evacuable cell window. Data from synthetic and natural diamonds are reported.

INTRODUCTION

Diamond has many applications as a mid-infrared transmitting window that has exceptional durability with respect to temperature, abrasion, and chemical inertness and synthetic diamonds hold the possibility of making windows large enough for many practical applications. The possibility that process related defects in synthetic diamonds may contribute to the emission of light in the mid-infrared region at temperatures above ambient could limit their application. Measuring the emission spectrum of diamond in the ambient-to-500°C temperature range is particularly challenging since diamond, with the exception of a weak, broad band near 2100 cm^{-1}, is transparent over most of the mid-infrared spectral range, where black body radiation from the surroundings obscures the emission from the diamond. To meet this challenge, the diamond specimen was suspended in vacuum at the focal point of an emission collection accessory of an FTIRS with a thermocouple affixed to the diamond as far as possible from this focal point. A continuous wave laser was then used to heat the diamond. The following is a description of the experimental system and results that define the performance that allowed for the coexistance of a powerful laser, an FTIRS emission collection system, and a thermally isolated diamond specimen sufficient to collect meaningful mid-infrared emissivities.

*Managed for the U. S. Department of Energy by Lockheed Martin Energy Systems, Inc. under Contract No. DE-AC05-84OR21400.

EXPERIMENTAL

Diamond specimens of approximately 10-mm by 10-mm by 0.5-mm, both natural and synthetic, were mounted on needles in an evacuable cell having a 32-mm OD/25-mm ID ZnSe dome that adapted to an enclosed custom emission collection accessory[1] (Harrick Scientific) attached to a BIO-RAD FTS-60 FTIRS. The emission accessory was designed so that a 40 W CW SYNRAD 57-6 CO Laser beam (1800 cm^{-1} to 1900 cm^{-1}) could heat the diamond in vacuum in such a manner that neither the reflected nor the refracted laser beam struck the emission collection optics. Scattered light from the dome entering the spectrometer during heating was blocked by a mechanical shutter (10 ms) inside the spectrometer near the emission focal point. Reflected light passed under the emission collection mirror to a black region behind that mirror. Refracted laser light entered a long steel cylinder underneath the specimen. The diamond specimen was heated to ~500°C, the WinIR spectral collection began collecting single-scan single-beam 8 cm^{-1} spectra at 0.86 s intervals, the laser was switched off, the shutter was opened during the return of the interferometer mirror, and emission spectra were collected using the spectrometer's DTGS detector. A fine-wire type K thermocouple was attached to the corner or edge of the diamond as far as possible from the focal point (>5 mm) using a small drop of silver paint, to monitor the temperature that dropped at a near exponential rate starting at 5°C s^{-1} immediately after the laser was turned off. Figure 1 shows a single-beam response of the spectrometer to a 0.647-mm thick synthetic diamond specimen 12.7 mm dia. at 500°C compared to the CO laser light scattered from the ZnSe dome while the laser was

CP430, *Fourier Transform Spectroscopy:* 11th International Conference
edited by J.A. de Haseth
1998 The American Institute of Physics 1-56396-746-4/98/$15.00

operated at low power. In all, eleven specimens were analyzed ranging in thickness from 0.13 mm to 0.65 mm. The thinner specimens were not as effectively heated by the laser and the 0.13 mm thick specimen achieved only 350°C because of low absorbance of light.

The reference spectra was from a "black body" consisting of a 3 mm diameter by 12 mm deep opening in a graphite cylinder (20-mm dia.) mounted on the reentrant heating element of an evacuable cell[2] having the ZnSe dome described above. Reference spectra were obtained in a cooling (kinetic) mode similar to the diamond emission spectra after power to the heater element was turned off. The graphite cooled at a much slower rate yielding more spectra over a given temperature range. The data, as multifiles, were parameterized with the observed cooling curves, co-averaged over a 1°C range and ratioed to similar spectra for diamond to yield diamond emissivity as a function of temperature as shown in Figure 2.

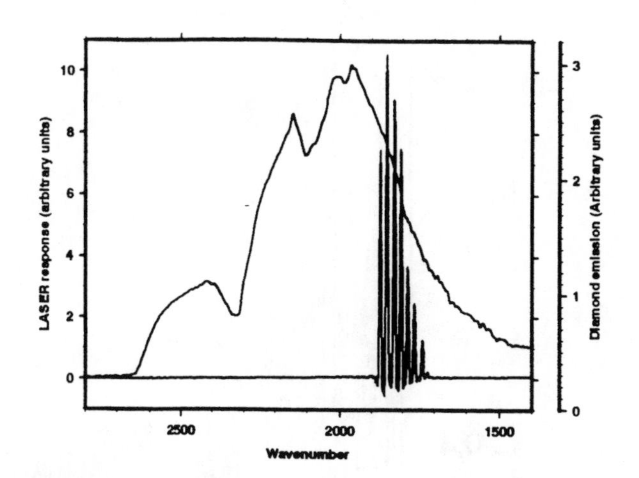

FIGURE 1. Single-beam emission spectra from a diamond 0.647 mm thick by 12.7 mm dia. (right ordinate) compared to that for scattered light from the CO laser (left ordinate).

RESULTS AND DISCUSSION

Figure 2 demonstrates that this experimental method not only yields good emissivity spectra, but

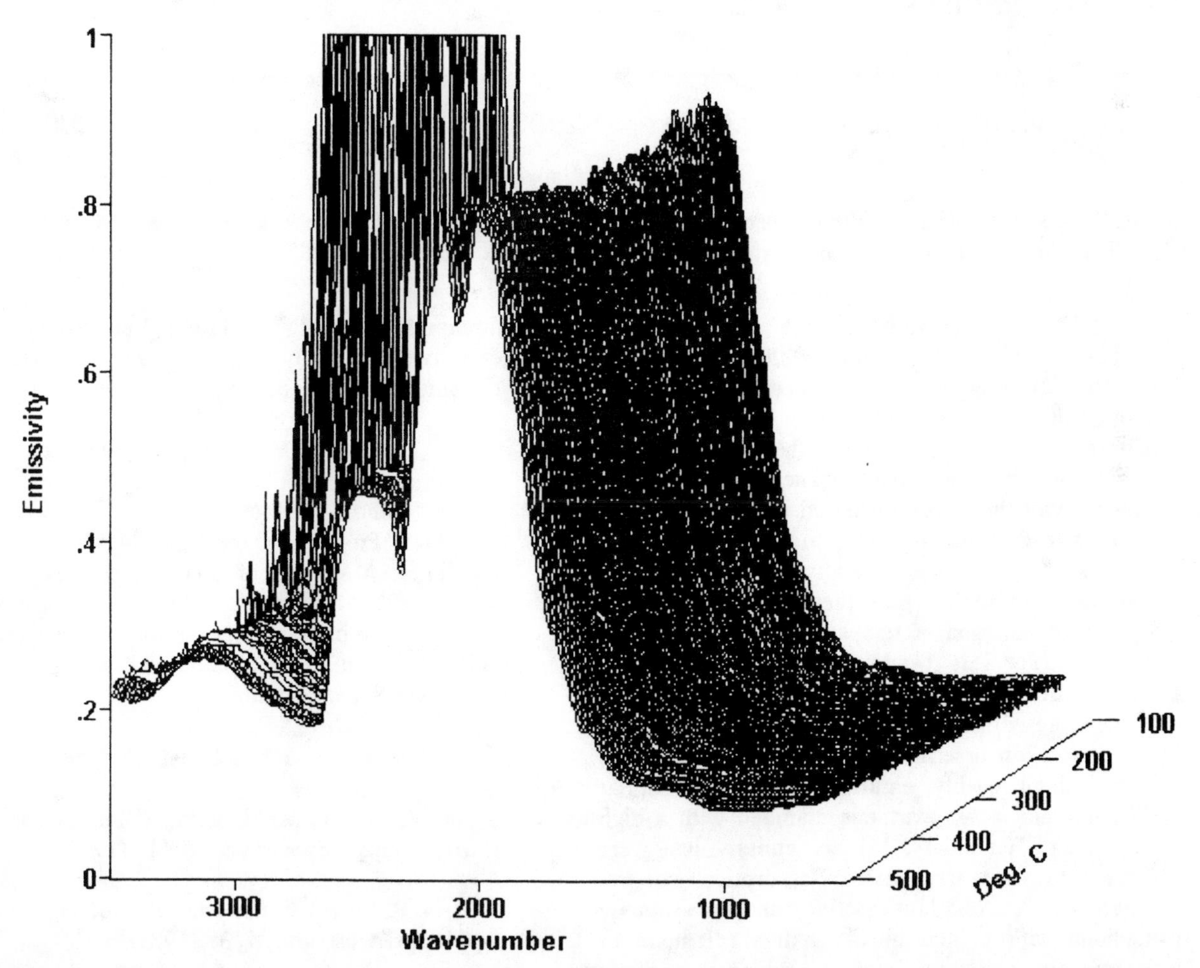

FIGURE 2. The temperature dependence of the emissivity spectra of a diamond 0.647 mm thick by 12.7 mm dia.

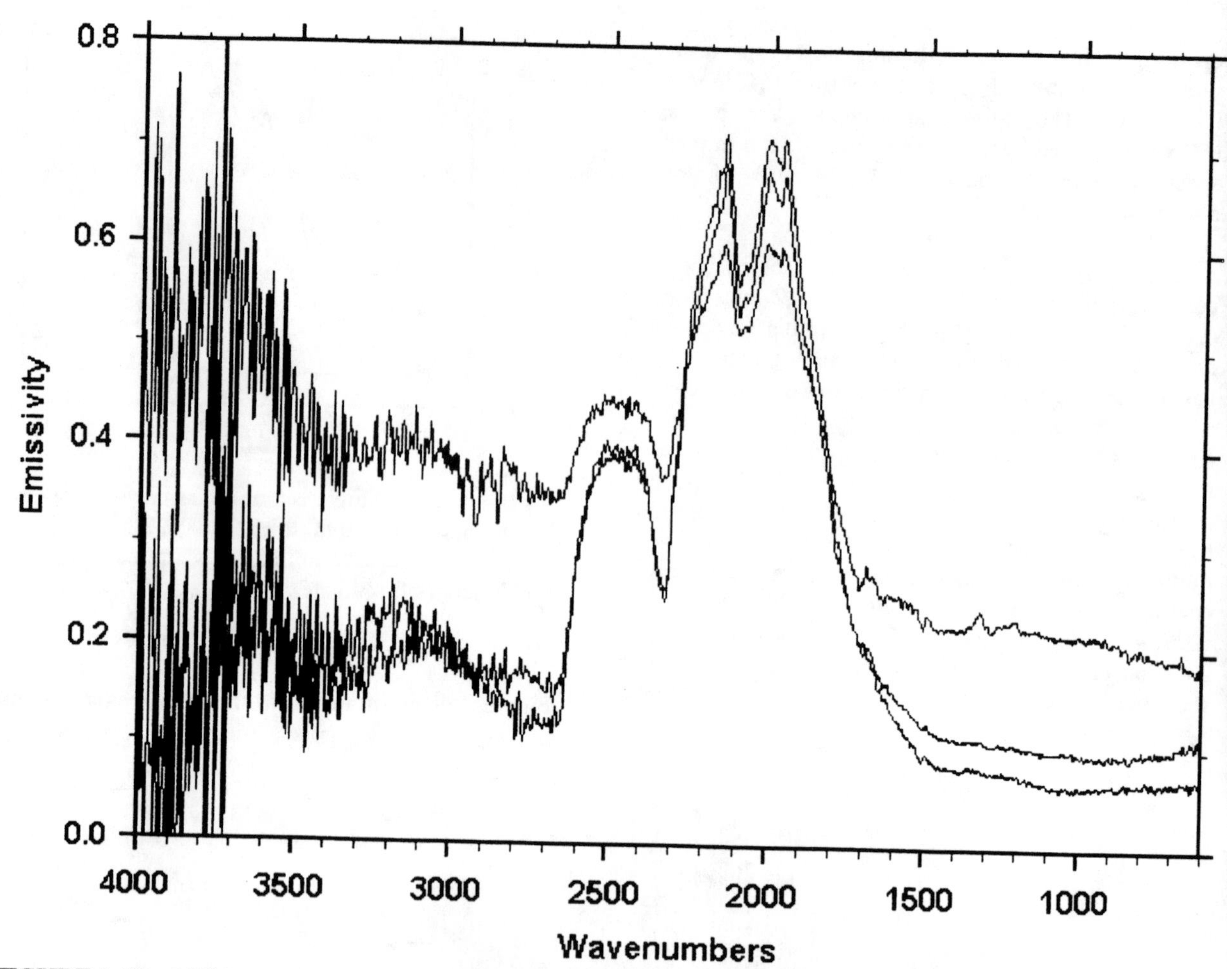

FIGURE 3. The 350°C emissivity spectra of 3 diamond specimens. At 1000 cm^{-1}: lower curve 0.647 mm synthetic, middle curve, 0.497 mm natural, 0.128 mm synthetic. At the 2020 cm^{-1} maximum, this order is inverted.

emissivity as a function of both frequency and temperature within the assumption that the reference is a true "black body". Of the specimens analyzed, ranging in thickness from 0.13 mm to 0.65 mm, the baseline corrected peak height of the 2020 cm^{-1} band was linear with specimen thickness. The baseline emissivity of these specimens did not correlate with thickness and ranged from <0.1 to >0.3. The baseline for these specimens was generally not very frequency dependent, indicating that there was not present sources of emission at temperatures different from the diamond. This would be expected for light emitted from objects other than the specimen at temperatures different from the specimen. Figure 3 compares a low baseline emission synthetic diamond with a high quality natural diamond of similar thickness and a thin synthetic diamond with a high background emissivity. These emissivities were similar to their transmission absorption spectra. Figure 3 compares a low baseline emission synthetic diamond with a high quality natural diamond of comparable thickness and a thinner synthetic diamond with a high background emissivity. These emissivities were qualitatively very similar to their transmission absorption spectra.

ACKNOWLEDGEMENTS

The authors acknowledge the Materials and Structures Program of the Ballistic Missile Defense for financial support of this project. The authors wish to acknowledge M. Milosivic, J. Lucania, and H. Tillenberg of Harrick Scientific for designing and building the infrared emission accessories that made these experiments possible.

REFERENCES

1. G. L. Powell, M. Milosevic, J. Lucania, and N. J. Harrick, **Appl. Spectrosc. 46**, 111 (1992).

2. N. R. Smyrl, E. L. Fuller, Jr., and G. L. Powell, **Appl. Spectroscopy, 37**, 38 (1983).

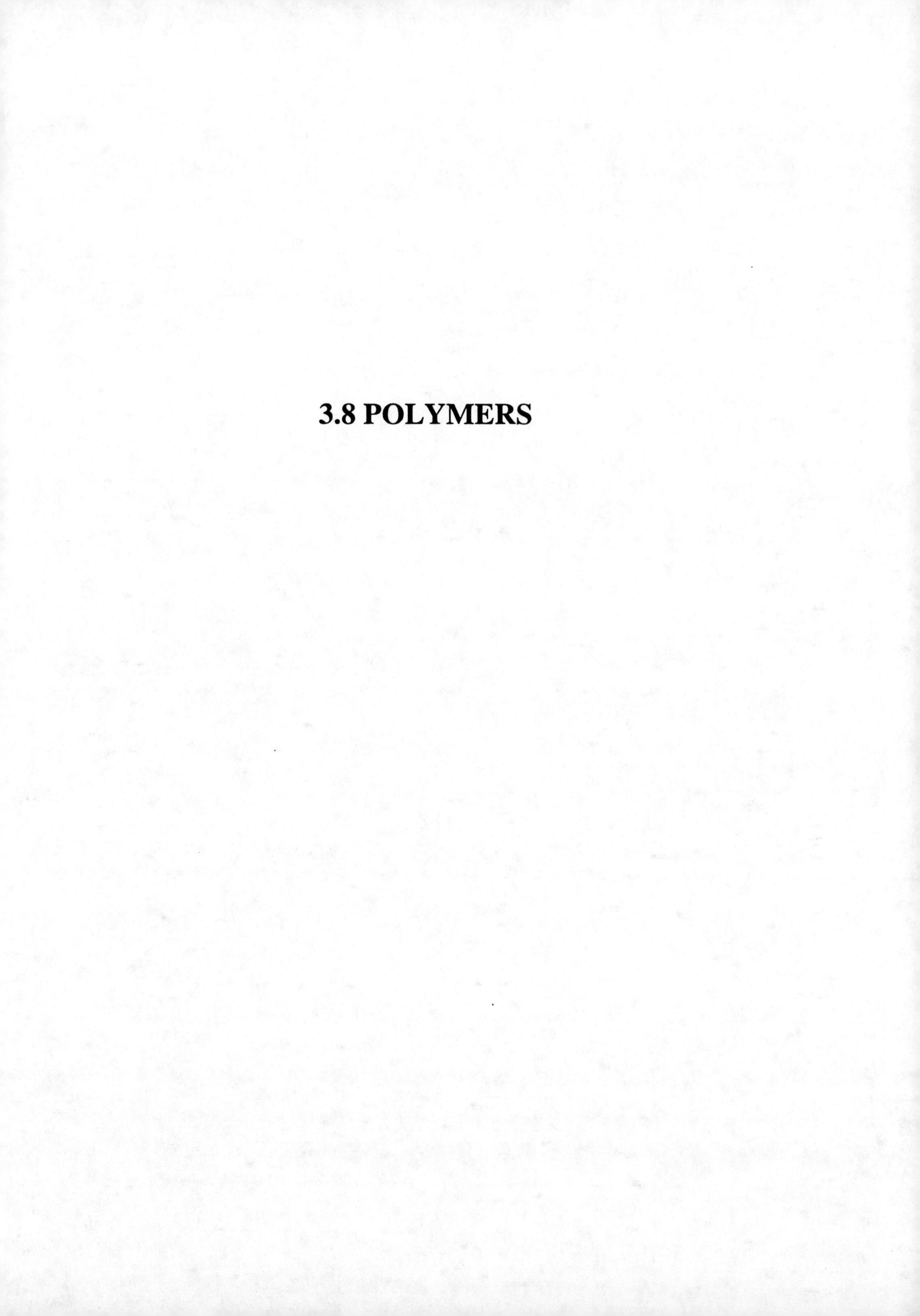

3.8 POLYMERS

Fourier-Transform Infrared Dichroism Investigation Of Molecular Orientation In Elastomeric Networks

L.Bokobza[1], T. Buffeteau[2] and B. Desbat[2]

[1]*Laboratoire de Physico-Chimie Structurale et Macromoléculaire, ESPCI, 10 rue Vauquelin, 75231 Paris, Cedex 05, France*

[2]*Laboratoire de Spectroscopie Moléculaire et Cristalline, Université de Bordeaux 1, 33405 Talence Cedex France*

Segmental orientation in uniaxially stretched poly(dimethylsiloxane) (PDMS) elastomeric networks is investigated by Fourier-transform infrared dichroism. Precise determination of the dichroic effects, in the mid- as well as in the near-infrared range, are obtained by polarization modulation of the incident electromagnetic field. Introduction of reinforcing filler such as silica in PDMS leads to an increase in orientation, attributed to the additional cross-links produced by the filler.

INTRODUCTION

Various spectroscopic techniques such as fluorescence polarization, nuclear magnetic resonance and infrared dichroism are increasingly used to study segmental orientation in deformed polymeric networks. These techniques directly probe the orientational behavior of network chains at a molecular level and are therefore superior to indirect estimates of segmental orientation based on macroscopic measurements of stress and strain. The careful analysis of segmental orientation in elastomeric networks leads to indispensable information in the physics of dense polymeric media. This information can be used to gain an understanding of the mechanisms of deformation and of the physical properties, especially mechanical properties.

Infrared linear dichroism, which measures directly the orientation of electric dipole-transition moments associated with particular molecular vibrations, is especially attractive for a detailed study of submolecular level orientations of materials such as polymers. On the other hand, further progress in FTIR spectroscopy by the introduction of modulation techniques renders this experimental field more challenging. By leading to a precise quantitative determination of the orientational behavior of chain segments, the polarization modulation FTIR spectroscopy is expected to allow extensive comparisons between experiment and theory.

In this work, the polarization modulation technique coupled to FTIR spectroscopy, is used in the mid- as well as in the near-infrared range, to investigate molecular orientation in poly(dimethylsiloxane) (PDMS) networks.

EXPERIMENTAL PART

Samples

The samples employed, kindly supplied by the Rhône-Poulenc Industries (Silicon Division), were cross-linked statistically by using peroxide cures. The filled network was blended, before cross-linking the polymers, with 40 parts by weight of pyrogenic silicas, previously treated for surface modification in order to reduce interactions between particles and the matrix. These samples have a number-average molecular weight between cross-links, M_C, of 17300.

Infrared dichroism measurements

Segmental orientation in a network submitted to uniaxial elongation may be conveniently described by the second Legendre polynomial (1) :

$$\langle P_2 (\cos \theta) \rangle = (3 \langle \cos^2\theta \rangle -1)/2$$

where θ is the angle between the direction of extension and the local chain axis of the polymer.

Infrared dichroism is based on the determination of the dichroic behavior of a selected absorption band of the infrared spectrum of a uniaxially stretched sample.

Parameters commonly used to characterize the degree of optical anisotropy in stretched polymers are the dichroic difference $\Delta A = A_{//} - A_{\perp}$ or the dichroic ratio $R = A_{//} / A_{\perp}$ ($A_{//}$ and $A_{\perp}$ being the absorbances of the investigated band, measured with radiation polarized parallel and perpendicular to the stretching direction, respectively).

The orientation function $\langle P_2(\cos\theta) \rangle$ is related to the dichroic ratio R by this expression :

$$\langle P_2 (\cos \theta) \rangle = \frac{2}{(3\cos^2 \beta -1)} \cdot \frac{(R-1)}{(R+2)} = F \frac{2}{(3\cos^2 \beta -1)}$$

where β is the angle between the transition moment vector of the vibrational mode considered and the local chain axis

CP430, *Fourier Transform Spectroscopy:* 11th International Conference
edited by J.A. de Haseth

of the polymer or any directional vector characteristic of a given chain segment and $F = (R-1)/(R+2)$ is the dichroic function.

$<P_2(\cos\theta)>$ is related to the structural absorbance A :

$$[A = (A_{//} + 2A_{\perp})/3]$$

by :

$$<P_2(\cos\theta)> = \frac{2}{(3\cos^2\beta - 1)} \frac{A_{//} - A_{\perp}}{A_{//} + 2A_{\perp}} = \frac{2}{(3\cos^2\beta - 1)} \frac{\Delta A}{3A}$$

It should be noted that orientational measurements by infrared technique require a knowledge of the respective transition moment direction relative to the chain axis. That means that the investigated absorption bands must be well assigned to normal vibrations of specified atomic groups. Such an assignment can be achieved by making a normal-coordinate analysis and experimentally, by looking at deuteration effects or dichroic behavior.

Standard dichroism measurements require two different spectra of the sample to get the polarized absorptions parallel and perpendicular to the stretching direction and form either the dichroic ratio R or the dichroic difference ΔA. $A//$ and $A\perp$ are subject to instrument and sample fluctuations occurring between the measurements. This static way of measuring infrared linear dichroism lacks sensitivity in the determination of small dichroic effects. Usually, it leads to significant measurements only for draw ratio $\alpha > 1.5$ ($\alpha = l / l_0$, l_0 and l being the undeformed and deformed lengths, respectively).

The polarization modulation approach, which consists basically of a fast modulation of the polarization state of the incident field between directions parallel and perpendicular to the stretching direction of the investigated polymer, can then be used to measure small dichroic effects with high sensitivity (2,3). By improving considerably the signal-to-noise ratio, this method is particularly recommended for characterizing molecular orientation in films under low deformation or in polymers exhibiting anisotropies nearly undetectable by the standard method.

RESULTS AND DISCUSSION

In previous studies (4,5), we have investigated the dichroic behaviors of the band located at 2500 cm^{-1}, and at 4164 cm^{-1} respectively ascribed to the overtone of the CH_3 symmetrical bending vibration located at 1260 cm^{-1} and to a combination of the symmetrical stretching mode of the methyl group at about 2905 cm^{-1} with the symmetrical bending mode. The transition moment associated with both vibrational modes lies along the CH_3-Si bond, which is a symmetry axis of the methyl group. As already mentioned, determination of the orientation function requires the definition of the angle β between the transition moment of the investigated band and a directional vector characteristic of a given chain segment. For the particular case of PDMS chains, the chosen directional vector is that joining two successive oxygen atoms. The angle β was thus expected to equal 90° (Fig.1).

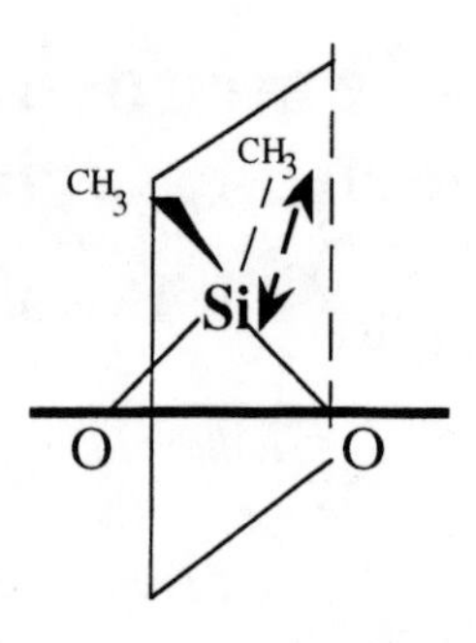

FIGURE 1. Definition of the transition moment vector.

In this work, in addition to the analysis of the band located at 2500 cm^{-1}, we have also characterized bands between 4000 and 6000 cm^{-1} located in the near infrared region. Let us point out that the near-infrared spectral region, which covers the interval between approximately 14000 and 4000 cm^{-1}, is dominated by absorption bands corresponding to overtones and combinations of fundamental C-H, O-H and N-H vibrations, because of the large anharmonicity of those vibrations involving the light hydrogen atoms. As the overtones and combinations are much weaker than the fundamental absorption bands (usually by a factor of 10 to 100), NIR spectroscopy allows the analysis of samples up to several millimeters thick._

The near-infrared spectrum of a PDMS film about 2 mm thick is shown in Fig.2. While the mid-infrared spectrum does not allow any determination of the dichroic effects on account of the very strong intensity of the absorption bands except for the band at 2500 cm^{-1}, the reduced intensity of the near-infrared absorptions makes a wide range of the spectrum available for the evaluation of anisotropy in polymeric materials.

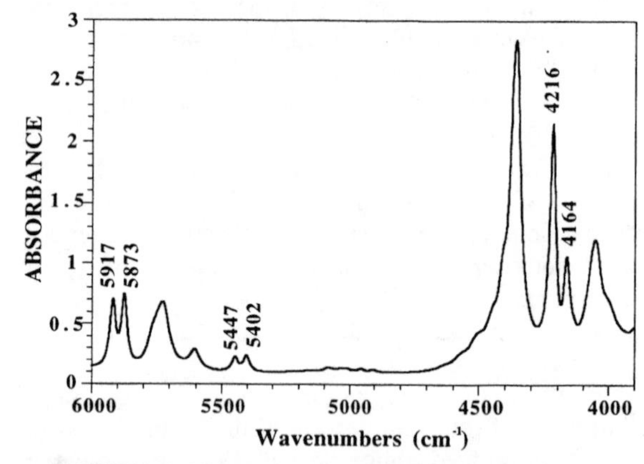

FIGURE 2. Near-infrared spectrum of a PDMS film.

Figure 3 represents the strain dependence of the dichroic difference, ΔA, for the spectral pattern between 4000 and 6000 cm^{-1}. The absolute value of ΔA for each band, and thus the anisotropy of the sample, increases with the draw ratio α.

On the other hand, as shown in Fig.3, one advanrtage of the polarization modulation technique is the possibility of separating the bands on account of their different dichroic behavior.

In Fig.4 , the dichroic functions F = (R-1)/(R+2) are plotted against what we call the strain function $(\alpha^2 - \alpha^{-1})$ for different near-infrared absorption bands. This way of plotting the data arises from the proportionality between the orientation function and the strain function, which generally appears in the theoretical models of rubber elasticity. On the other hand, plotting the dichroic function F instead of $\langle P_2(\cos\theta)\rangle$ means that we are only looking at the orientation of the transition moment vectors without any assumption concerning the local chain axis and thus the angles β.

The results presented in Fig. 4 show that the higher dichroic effect is obtained for the band at 5873 cm^{-1} which means that the angle between the transition moment associated with this vibrational mode and the local chain axis is more certainly close to 90°. This makes questionable our first evaluation of the second moment of the orientation function on the basis of an angle β of 90° for the band located at 2500 cm^{-1}. More probably, this mode is coupled with other vibrations leading to a lower value of the angle β. This results points out the main problem arising in the determination of molecular orientation by infrared dichroism and shows that near-infrared spectroscopy could be helpful for that purpose, since it allows the analysis of vibrational modes, probably less coupled that in the mid-infrared range.

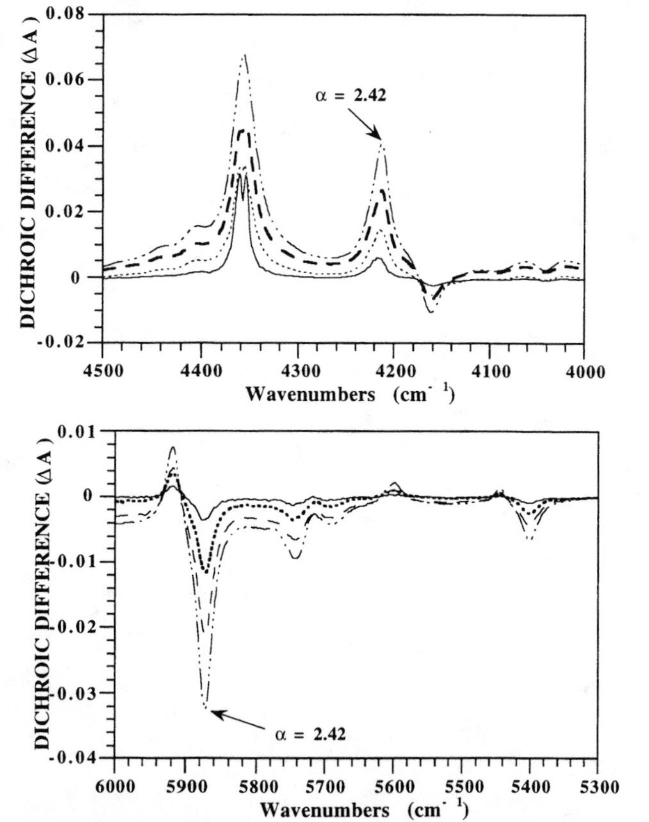

FIGURE 3. Strain dependence of the dichroic difference for the bands located between 4000 and 6000 cm^{-1}.

On the other hand, from the dichroic behaviors of the various bands, it could be possible to give a precise assignment of the observed absorptions and to correlate the two spectral regions, NIR and mid-IR.

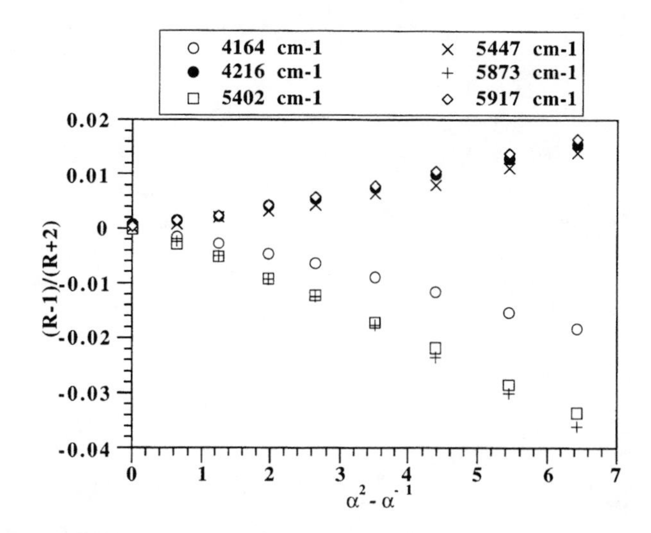

FIGURE 4. Dichroic functions versus $(\alpha^2-\alpha^{-1})$.

Incorporation of fillers into elastomers is of significant commercial importance due to the enhancement of the physical properties of the final materials (6,7). Analysis of the properties of filled elastomers under strain is an area of particular interest for understanding the molecular origin of the reinforcement effect.

The most important aspect of elastomer reinforcement is the nature of the bonding between the filler particles and the polymer chains. Typically, the network chains adsorb strongly onto the particle surfaces, and this increases the effective degree of cross-linking. This type of adsorption with permanent chemical bonding between filler particles and polymer chains will be especially strong if the particles contain some reactive surface groups.

In the silica-silicone rubber, the silanols present at the suface of the particle, interact with the oxygen atoms of the PDMS chains via hydrogen bonds.

Stress-strain curves for the unfilled and the silica-filled PDMS networks are represented in Fig.5 as plots of the true stress (force divided by the deformed area) against $(\alpha^2-\alpha^{-1})$. The vertical broken lines locate the rupture points. The data reveal clearly that the incorporation of the reinforcing particles significantly increases the modulus and the ultimate properties like the stress at rupture and the maximum extensibility.

Orientational analysis also shows that, at a given elongation, the orientation increases by a factor of about 1.8 by addition of the filler particles (Fig.6).

Such a behavior can be explained by the fact that the reinforcing particles can act as additional physical cross-links. The orientational order generated in uniaxially strained rubbers is determined by the average number of segments between chemical and physical cross-links.

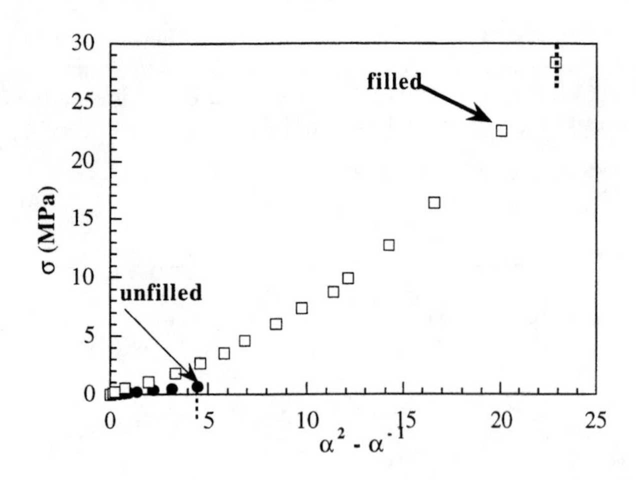

FIGURE 5. Stress-strain curves at room temperature.

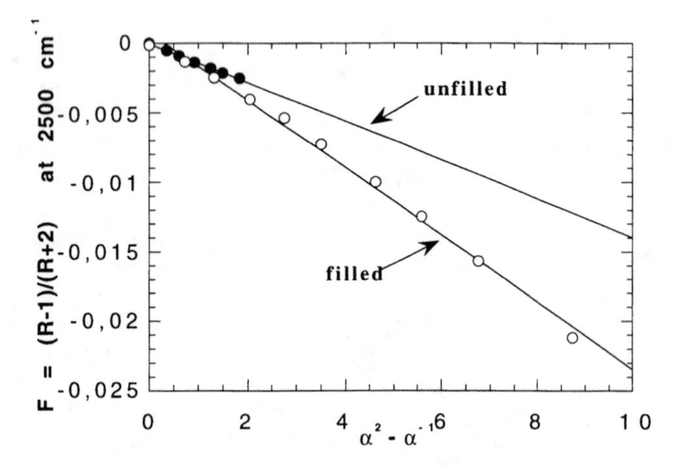

FIGURE 6. Dichroic functions for the unfilled and silica-filled networks versus (α^2-α^{-1}).

CONCLUSIONS

Infrared dichroism is a powerful tool for probing the chain segment orientation.

A better characterization of segmental orientation can be obtained by combining polarization-modulation with FT-IR spectroscopy. This technique should be exploited more in order to obtain reliable experimental results thus allowing more extensive comparisons between theory and experiment.

In the filled systems, infrared spectroscopy is able to identify the filler parameters responsible of the reinforcement thus leading to a better understanding of the molecular origin of the reinforcement effect.

REFERENCES

1. B. Jasse, J.L. Koenig, *J. Macromol. Sci., Rev. Macromol. Chem.,* **C17**, 61 (1979).
2. Buffeteau, T., Desbat, B. and Turlet, J.M. *Appl. Spectrosc.* **45**, 380 (1991).
3. Buffeteau, T., Desbat, B., Pezolet, M. and Turlet, J.M. *J. Chim. Phys.* **90**, 1467 (1993).
4. Besbes, S., Cermelli, I.,Bokobza, L., Monnerie, L., Bahar, I., Erman, B. and Herz, J. *Macromolecules* **25**, 1949 (1992).
5. Bokobza, L., Desbat, B. and Buffeteau, T. *Mikrochim. Acta* [Suppl.] **14**, 407 (1997).
6. Kraus, G. *Adv. Polym. Sci.* **8**, 155 (1971).
7. Wagner, M.P. *Rub. Chem. Tech.* **49**, 703 (1976).

Polarized Light-Induced Anisotropy of Azo Dyes Studied by Polarized FTIR Spectroscopy

Keiko Tawa[1], Kenji Kamada[2], Toru Sakaguchi[2], and Koji Ohta[2]

1. Fields and Reactions, PRESTO, JST and Osaka National Research Institute, 1-8-31Midorigaoka, Ikeda, Osaka 563, JAPAN, 2. Osaka National Research Institute, 1-8-31Midorigaoka, Ikeda, Osaka 563, JAPAN

The anisotropy induced in the poly(methyl methacrylate) (PMMA) films doped with Disperse Orange 3 (DO3, NO_2-phenyl-N=N-phenyl-NH_2) was investigated by polarized FTIR spectroscopy. Observed infrared absorption bands of DO3 in the polymers were assigned to symmetric (NO_2s, 1341cm^{-1}) and antisymmetric (NO_2as, 1523cm^{-1}) stretching modes of NO_2, and the C-N stretching mode of C-NH_2 (C-N, 1303cm^{-1}). By measuring the polarized IR spectra of DO3/PMMA, the infrared dichroism was observed in NO_2s and C-N bands. From these results, the orientation factors, K_{Zfn} (f=x, y, z), for each isomer were determined. The factors indicate that it is difficult for NO_2 group to move in the PMMA during trans-cis-trans isomerization and that phenyl group with NH_2 mainly moves in PMMA on isomerization. This study clarified that the anisotropy of DO3 in PMMA is induced by photoselection with the irradiation of linearly polarized light and is not induced by reorientation of DO3 molecules, since the entire DO3 molecule cannot rotate in PMMA during a series of isomerization process.

INTRODUCTION

An optical anisotropy[1, 2] has been previously observed in polymer films doped with photoresponsive molecules under the irradiation of the polarized light. Recently, new techniques of optical switching[3] and holography[4] by utilizing the polarized light-induced anisotropy of polymer films have been studied. Todorov et al.[5, 6] studied photoinduced anisotropy in rigid dye solutions for transient polarization holography which can be utilized as a real-time recording. However, the physical mechanisms of the induced anisotropy and its relaxation have not yet been clarified.

In this study, the photoinduced anisotropy of azo dyes in the polymer films was investigated by using polarized FTIR spectroscopy. Azo dyes have two spatial forms, trans and cis isomers. The trans form is more stable in the absence of light. Azo dyes are able to isomerize from the trans to cis form under irradiation. However, due to the instability of the cis form, it can isomerize back to the trans form thermally. The isomerization rates and the photoequilibrium ratio depend on the kind of azo dyes and polymer matrices. In observed polarized spectra, the absorbance of each functional group changed due to the photoisomerization and the photoinduced anisotropy of the azo dyes under the irradiation. The close relationship between the isomerization and the anisotropy will be discussed in terms of the orientation factors[7] obtained from analysis of several vibrational bands in FTIR spectra. This study shows that polarized FTIR spectroscopy is one of the most powerful methods available to analyze the physical mechanisms of photoinduced anisotropy.

EXPERIMENTS

We used Disperse Orange 3 (DO3, NO_2-phenyl-N=N-phenyl-NH_2), an azo dye, as the photoresponsive molecule and poly(methyl methacrylate) (PMMA) as the polymer matrix. The sample films were prepared by casting chloroform solution containing PMMA and DO3. The thickness of the films were less than 20 μm. The dye concentration was about 1-2 wt % in the PMMA films. A Nicolet 60SXR FTIR spectrometer was used for measuring infrared spectra. The FTIR spectra of the DO3/PMMA films were measured before irradiation and under irradiation of linearly polarized light from an Ar ion laser with a wavelength of 488 nm and an optical power of 20 mW/cm^2. The IR spectrum of PMMA was estimated as a standard. Polarized FTIR spectra were measured using the polarizer (KRS-5) in the parallel (Z) and perpendicular (Y) directions to the direction (Z) of polarized light of Ar ion laser. (Figure 1)

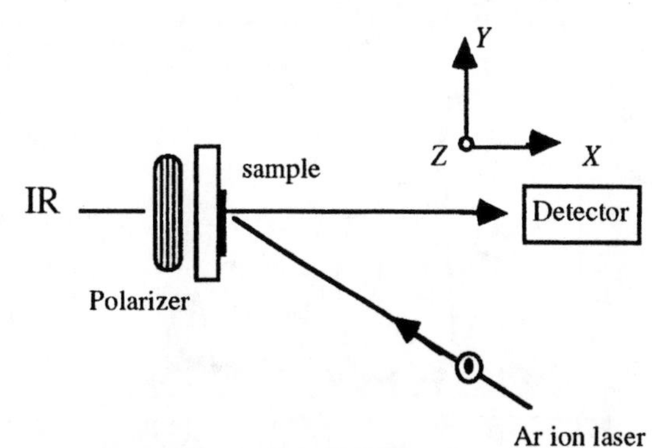

FIGURE 1. Optical alignment of experimental instruments.

CP430, *Fourier Transform Spectroscopy:* 11th International Conference
edited by J.A. de Haseth

RESULTS

Figure 2 shows the FTIR spectra of DO3/PMMA and PMMA. Subtraction of PMMA spectrum from DO3/PMMA spectrum clarifies the infrared absorption bands of DO3 in the polymer films as shown in Figure 3, where the FTIR spectra in the initial state (before irradiation) and in the photostationary state (under irradiation) are shown. The vibrational bands in the spectra were assigned[8,9] to symmetric (NO_2s, 1341cm^{-1}) and antisymmetric (NO_2as, 1523cm^{-1}) stretching modes of NO_2, and the C-N stretching mode of C-NH_2 (C-N, 1303cm^{-1}). By measuring the polarized IR spectrum of DO3/PMMA in the photostationary state, the infrared dichroism was observed in the NO_{2s} and C-N bands, where the absorbances in the *Z* direction decreased more than those in the *Y* direction. The absorbance of NO_{2as} band did not change in either the *Z* or the *Y* direction. Table 1 shows the absorbances of NO_2s, NO_2as, and C-N bands in the photostationary state. They are normalized to their initial absorbances before irradiation of polarized light.

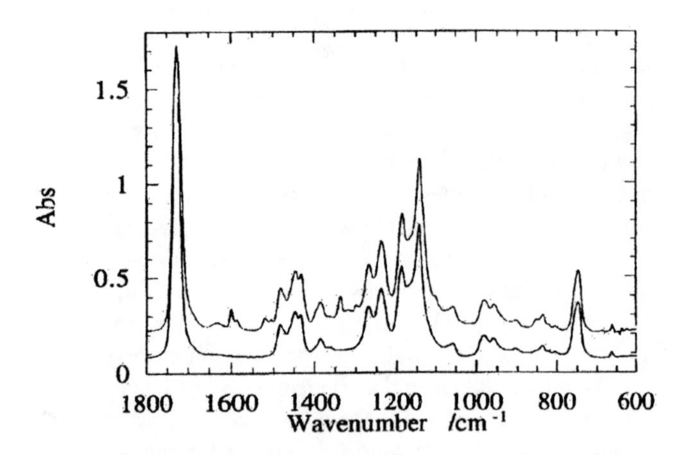

FIGURE 2. FTIR spectra of DO3/PMMA (upper) and PMMA (lower). (The spectrum of DO3/PMMA is indicated with shift.)

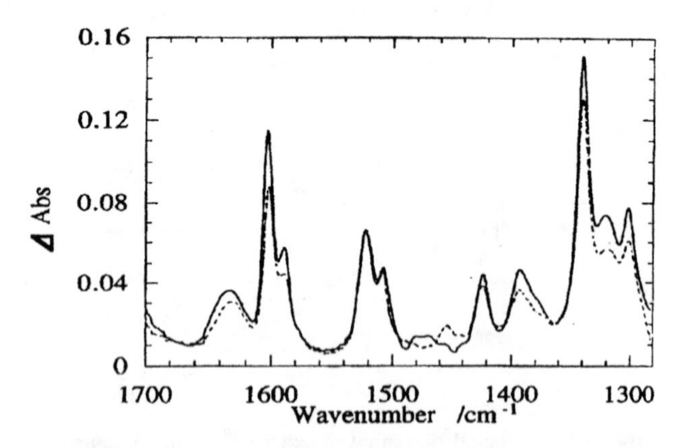

FIGURE 3. FTIR subtraction spectra between DO3/PMMA and PMMA in the initial state before irradiation (solid curve) and in the photostationary state under irradiation (dashed curve).

TABLE 1. The absorbances of NO_2s, NO_2as, and C-N bands of DO3 in PMMA in the photostationary state. They are normalized to the initial absorbances before irradiation.

a	$E_Z(a)$	$E_Y(a)$
NO_2s	0.910	0.985
NO_2as	1.00	0.990
C-N	0.700	0.865

DISCUSSION

The orientation factors, K_{Zf_n} (f=x,y,z, n=t (trans), c(cis)), for each isomer are introduced[7] in order to determine the degree of orientation of the molecules irradiated by *Z*-polarized light. In the bands of NO_2s, NO_2as, and C-N, each vibrational band for the cis isomer appears at the same wavenumber as that for the trans isomer. The absorbance of the *a*-vibrational band is the sum of the absorbance of the trans and cis isomers.

$$E(a) = \alpha E(a_t) + (1-\alpha)E(a_c). \quad (1)$$

$E(a)$ is the absorbance vector in the laboratory-fixed axes, *X,Y,Z*, and it is composed of $E_X(a)$, $E_Y(a)$, and $E_Z(a)$. α is the molar fraction of the trans isomer to the total number of molecules. $E(a_n)$ can be represented by the multiplication of the matrix of orientation factors, $K_Z^{(n)}$, and the absorbance vector, $A(a_n)$, in an arbitrary molecular framework, *x,y,z*.

$$E(a_n) = K_Z^{(n)} A(a_n). \quad (2)$$

The absorbance of the *a*-band is described by the combination of equations (1) and (2) as

$$E(a) = \alpha K_Z^{(t)} A(a_t) + (1-\alpha) K_Z^{(c)} A(a_c). \quad (3)$$

In this study, since the sample was irradiated by linearly polarized light along the Z axis, the sample can be treated as an uniaxial orientation by photoselection, $E_X(a)$=$E_Y(a)$. Therefore, the matrix of the orientation factor, $K_Z^{(n)}$, is described as

$$K_Z^{(n)} = \begin{pmatrix} \frac{1-K_{Zx_n}}{2} & \frac{1-K_{Zy_n}}{2} & \frac{1-K_{Zz_n}}{2} \\ \frac{1-K_{Zx_n}}{2} & \frac{1-K_{Zy_n}}{2} & \frac{1-K_{Zz_n}}{2} \\ K_{Zx_n} & K_{Zy_n} & K_{Zz_n} \end{pmatrix}. \quad (4)$$

In order to define K_{Zf_n}, we shall introduce the dot

product of the absorption probability of the linearly Z-polarized light, e_Z (unit vector), and an electric dipole transition moment, $M(a_n)$ as equation (5). The transition moment, $M(a_n)$, and the absorbance, $A(a_n)$, in equation (3) have the relationship $|M(a_n)|^2 \propto |A(a_n)|$. The absorption probability can be described in the simple form:

$$<(e_Z \cdot M(a_n))^2>= M_x^2(a_n)<\cos^2\theta_{x_n}>+ M_y^2(a_n)<\cos^2\theta_{y_n}>+M_z^2(a_n)<\cos^2\theta_{z_n}>, \quad (5)$$

where the other terms, such as $M_xM_y<\cos\theta_{x_n}\cos\theta_{y_n}>$, vanish by selecting the arbitrary system of molecular axes which coincide with the molecular orientation axes. The terms of $M_f^2(a_n)<\cos^2\theta_{f_n}>$ corresponds to the square of the projection of the f-component of the transition moment, $M(a_n)$, onto the Z axis as shown in Figure 4. The square of the cosines are defined as the orientation factors, K_{Zf_n}.

$$K_{Zf_n}=<\cos^2\theta_{f_n}>, \qquad f: x, y, z, \quad (6)$$

$$<(e_Z \cdot M(a_n))^2>= M_x^2(a_n)K_{Zx_n}+M_y^2(a_n)K_{Zy_n}+M_z^2(a_n)K_{Zz_n}. \quad (7)$$

Orientation factors K_{Zf_n} provide only a very incomplete specification of the orientation distribution function but still describe it sufficiently for many experiments. In the case of random distribution, such as the sample in its initial state before irradiation, the values of K_{Zx_t}, K_{Zy_t}, and K_{Zz_t} are equivalent to each other, and equal to 1/3.

We defined the molecular axes for the trans and cis isomers as shown in Figures 5 (a) and (b), respectively. The molecular axes for the trans isomer were chosen as the molecule takes a planar structure in y_t-z_t plane, and the $\pi\pi^*$ transition moment is assumed to be parallel to z_t axis.[10,11] In the trans isomer, the vibrational transition moments of NO_2s and C-N bands are almost parallel to the $\pi\pi^*$ transition moment and that of NO_2as is perpendicular to $\pi\pi^*$. The relationship $K_{Zx_t}=K_{Zy_t}$ can be assumed in this case since the $\pi\pi^*$ transition moment is assumed to be parallel to the z_t axis. In the molecular axes for the cis isomer, the C-NO_2 bond is oriented in the direction of z_c axis and the C-NH_2 bond forms an angle of 30° with the y_c axis. Both phenyl rings are tilted by 30° to the y_c-z_c plane in which the N=N bond lies.[10,12]

The K_{Zf_t} values for the two isomers were determined by solving the three equations which were represented for the three bands, NO_2s, NO_2as, and C-N by using the results in the photostationary state shown in Table 1, where the α value was determined by UV-VIS spectra.

The determined values of K_{Zf_t} are shown in Table 2. In the trans isomer, the K_{Zz_t} value is smaller than the K_{Zx_t} and K_{Zy_t} values. This means that the trans isomers were excited by the photoselection of the linearly Z-polarized light. In the cis isomer, the K_{Zz_c} value is larger than the K_{Zx_c} and K_{Zy_c} values. The larger K_{Zz_c} value indicates that there are a lot of cis molecules with a small angle between the z_c axis, in Figure 5(b), and the Z axis, or that more C-NO_2 bonds align to the Z axis. In other words, more trans molecules, where the C-NO_2 bonds align to the Z axis, are excited by photoselection and isomerize to cis molecules, where the C-NO_2 bonds tend to align to the Z axis. This indicates

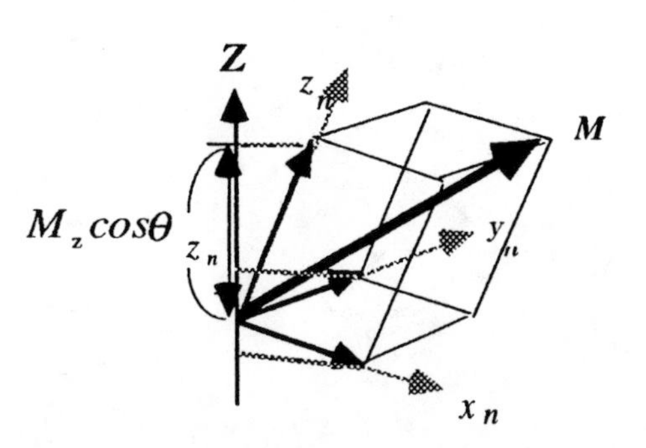

FIGURE 4. The projection of the z-component of an electric dipole transition moment, M, onto the Z axis.

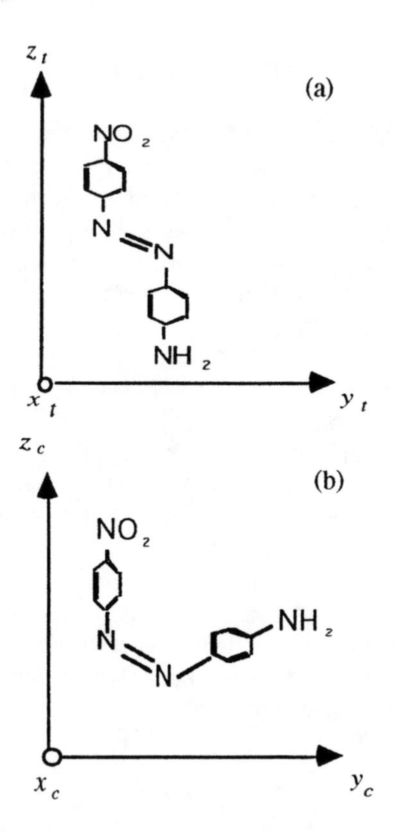

FIGURE 5. Molecular axes for trans (a) and cis (b) isomers.

TABLE 2. The values of K_{Zf_n} in the photostationary state.

	trans	cis
K_{Zx_n}	0.356	0.32
K_{Zy_n}	0.356	0.23
K_{Zz_n}	0.288	0.45

that it is difficult for the NO_2 group to move in the PMMA films during trans-cis-trans isomerization.

The physical mechanism of photoinduced anisotropy has been suggested to be induced by two processes, the photoselection and the reorientation. The latter means that the configuration of trans molecules rotates by 90° in the back isomerization process from cis to trans. However, this study indicated that the NO_2 group is difficult to move in the PMMA matrix or the entire DO3 molecule cannot rotate in PMMA during a series of isomerization process. Therefore, the physical mechanism of polarized light-induced anisotropy in PMMA doped with DO3 is mainly due to photoselection by the irradiation with linearly polarized light, and it is almost never induced by the reorientation of DO3 molecules. However, it is not clear why NO_2 group is difficult to move in the PMMA matrix. The NO_2 group as an electron acceptor may interact specifically with PMMA.

CONCLUSION

We investigated the anisotropy induced in the PMMA films doped with DO3 using polarized FTIR spectroscopy. In order to analyze the data from the dichroism, the orientation factors for each isomer were determined. The factors indicate that it is difficult for the NO_2 group to move in the PMMA matrix during trans-cis-trans isomerization. Polarized light-induced anisotropy in PMMA doped with DO3 depends strongly on photoselection by the irradiation with linearly polarized light, and is not induced by the reorientation of DO3 molecules. Since the NO_2 group is an electron acceptor, it probably interacts specifically with PMMA. In this study, it was indicated for the first time that the polarized FTIR spectroscopy is one of the most powerful methods available to analyze the physical mechanism of photoinduced anisotropy. We are studying the behavior of azo dyes in various polymers under the irradiation of polarized light in order to clarify on which interactions between dyes and polymer matrices the photoinduced anisotropy depends.

REFERENCES

1. Xie, S., Natansohn, A., and Rochon, P. *Chem. Mater.* **5**, 403-411 (1993).
2. Kumar, G. S. and Neckers, D. C. *Chem. Rev.* **89**, 1915-1925 (1989).
3. Ikeda, T., Sasaki, T., and Ichimura, K. *Nature* **361**, 428-430 (1993); Ikeda, T. and Tsutsumi, O. *Science* **268**, 1873-1875 (1995).
4. Todorov, T., Nikolova, L., Tomova, N., and Dragostinova, V. *IEEE J. Quantum Electron.* QE-**22**, 1262-1267 (1986).
5. Todorov, T., Nikolova, L., Tomova, N., and Dragostinova, V. *Opt. Quantum Electronics* **13**, 209-215 (1981).
6. Mateev, V., Markovsky, P., Nikolova, L., and Todorov, T. *J. Phys. Chem.* **96**, 3055-3058 (1992).
7. Michel, J. and Thulstrup, E. W. *Spectroscopy with Polarized Light*, New York: VCH Publishers, 1986.
8. Natansohn, A., Rochon, P., Pezolet, M., Audet, P., Brown, D., and To, S. *Macromolecules* **27**, 2580-2585 (1994).
9. Zemskov, A. V., Rodionova, G. N., Tuchin, Yu. G., and Karpov, V.V. *Zh. Prikl. Spektrosk.* **49**, 581-586 (1988).
10. Beveridge, D. L. and Jaffé, H. H. *J. Am. Chem. Soc.* **88**, 1948-1953 (1966).
11. Robin, M. B. and Simpson, W. T. *J. Chem. Phys.* **36**, 580-589 (1962).
12. Forber, C. L., Kelusky, E. C., Bunce, N. J., and Zerner, M. C. *J. Am. Chem. Soc.* **107**, 5584-5890 (1985).

Polyester (PET) Single Fiber FT-IR Dichroism: Potential Individualization

Liling Cho[1], David L. Wetzel[1], and John A. Reffner[2]

[1]*Microbeam Molecular Spectroscopy Laboratory, Kansas State University, Shellenberger Hall, Manhattan, KS, 66506, USA*
[2]*Spectra-Tech Inc., 2 Research Drive, Shelton, CT, 06484, USA*

Individualization of undyed fibers can be a problem in analyzing fiber evidence in forensic cases. In addition to the physical and optical microscopic features, the chemical composition information from single fiber FT-IR microspectroscopy may be useful. In the case of polyester, the most commonly used fiber, only a single generic class usually is recognized. Single fiber polarized FT-IR microspectroscopy provides a means of using the molecular orientation of the macromolecules in the fiber resulting from their manufacturing history to observe spectroscopic differences. Dichroic ratios for eight usable infrared bands for PET single fibers permit multidimensional discriminant analysis. The procedure described sorts PET fibers into 10 working subclasses and demonstrates the potential of this approach for single fiber individualization. This new dimension can be added to the traditional size, shape, and other distinguishing features.

INTRODUCTION

The individualization of single undyed fibers in fiber evidence for forensic cases presents a challenge. Physical and optical microscopic features commonly are used as the first line of attack. Single fibers that differ in chemical composition may benefit from analysis by FT-IR microspectroscopy (1,2). Unfortunately, the most commonly used fiber in the modern world is polyester. At present, only a single generic class is recognized for polyethylene terephthalate (PET). The intent of this work was to demonstrate the potential of single fiber individualization by using a pattern of dichroism at multiple wavelengths to possibly expand the PET generic class to subclasses.

PRELIMINARY WORK

Single fiber analysis by FT-IR microspectroscopy is in itself somewhat of a challenge. When one is not concerned with dichroism measurements but only with chemical composition, flattening of the fiber is permitted in order to make the specimen thinner and at the same time remove any lensing effect that may occur from the optical geometry of the specimen. Flattening is very definitely the single most advantageous sample preparation step for single fiber microspectroscopy. Unfortunately, when using polarized infrared radiation and studying dichroism, it is important not to disturb the fiber. In textile manufacturing, the fibers coming from the spinneret typically are heated and drawn over a series of rollers (see Fig. 1). The extent of a drawing of a fiber is determined by the draw ratio or the take-up speed, and the product is tailored for end use by maintaining good control of this step of the manufacturing process. The crystallinity of the fiber can be adjusted as well as the molecular orientation.

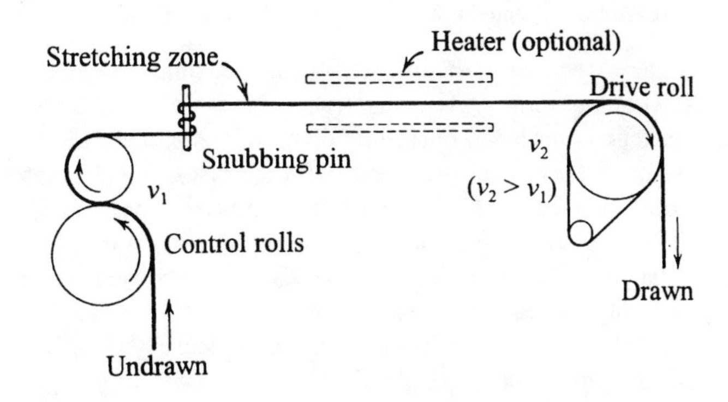

Figure 1. Diagram of stretching operation (from Ref. 3 with permission)

Dichroism particularly in the infrared region is a valuable technique to establish the orientation of the macromolecules that make up the fiber. In general, polarized FT-IR can be used to establish the manufacturing history of a polymer. In the case of single fibers, the experimentation is somewhat more of a challenge. Prior to this work, single fiber techniques were developed and a stretching device for use on the microscope stage of the microspectrometer was fabricated (4). Results from single fiber stretching experiments for PET and other materials were described previously (4,5,6). For PET single fibers, initially thin fibers (typically 3 denier) were used to establish which infrared bands were subject to changes in dichroic ratio upon stretching (7). Figure 2 shows the typical change in dichroic ratio with elongation of a single fiber. Subsequent stretching experiments with 6 denier fibers of the same composition (commonly used for apparel) expanded on the original findings and allowed stretching to a

CP430, *Fourier Transform Spectroscopy:* 11th International Conference
edited by J.A. de Haseth

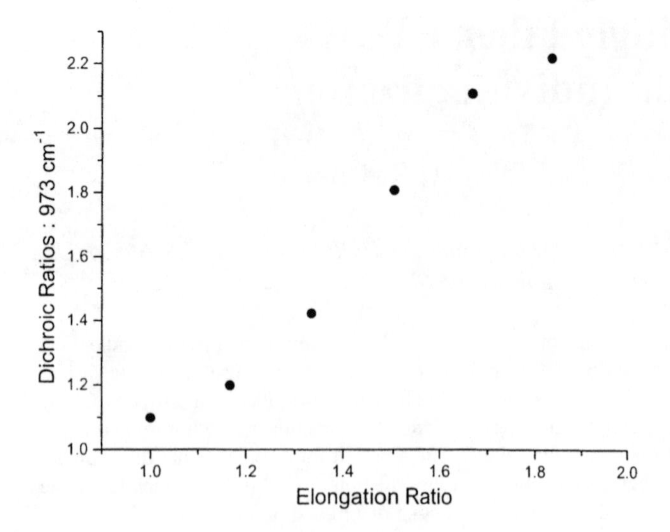

Figure 2. Change in dichroic ratio vs. elongation ratio (5).

greater extent. With the thicker fibers, stretching beyond the Newtonian limits of elongation was possible without breakage. This led to interesting observations (not shown) of abrupt breaking in the trend of dichroism change was observed once the Newtonian limit had been exceeded (8). The earlier stretching experiments showed that, for a typical unflattened fiber, some of the infrared absorption bands were off-scale or at least nonlinear due to high absorptivity (Fig. 3). Figure 4 shows a pair of parallel and perpedicularly polarized spectra (top) for a single PET fiber drawn to a high orientation. The other pair of spectra (bottom) are from a single fiber that was less oriented. However, in the infrared range from 4000 to 600 cm^{-1}, eight usable infrared bands were identified. They are listed in Table 1, which also indicates the direction of change and differences in dichroic ratios before and after stretching.

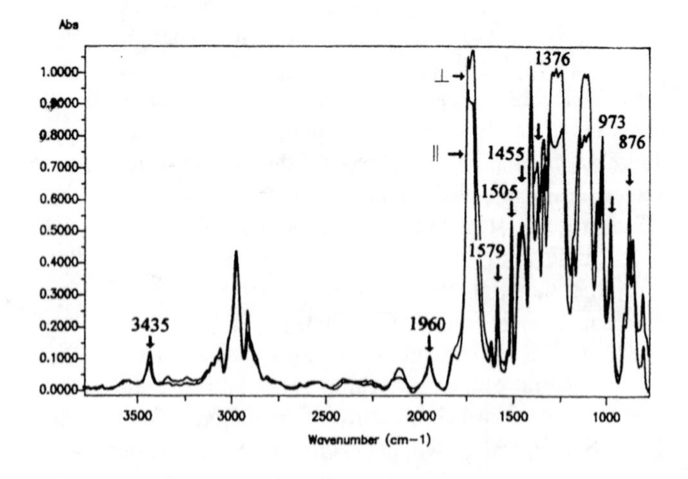

Figure 3. Parallel and perpendicular spectra of polyester fiber (5,6,8).

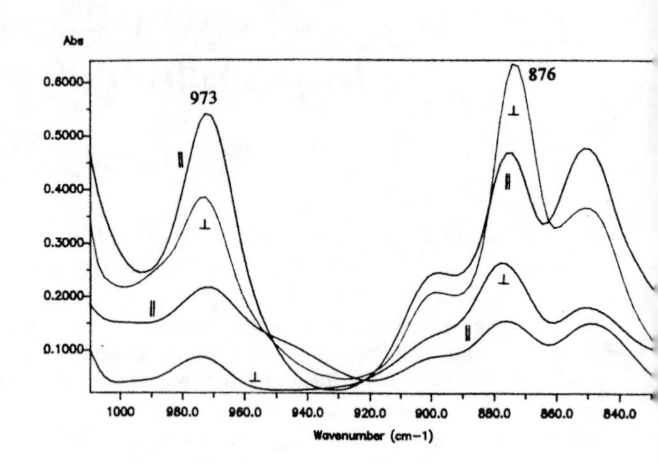

Figure 4. Parallel and perpendicular polarized spectra of single fibers at two different molecular orientation conditions (5).

TABLE 1. Polyester Absorption Bands (5).

Frequency (cm^{-1})	Dichroic Ratios Before/After Stretching	Assignment
3435	0.98/0.88	C=O stretch overtone
1960	1.06/1.04	1,4-substituted phenyl overtone
1579	1.04/1.41	phenyl C-C stretch
1505	1.04/1.33	phenyl C-C stretch (in-plane)
1455	1.02/1.16	CH_2 bending
1376	1.09/0.80	CH_2 wagging
973	1.09/2.11	C-O stretch of *trans* ethylene glycol unit
876	0.97/0.53	phenyl C-H band out-of-plane

EXPERIMENTATION AND RESULTS

The experimentation reported here involved assembling a collection of 33 PET fibers from five different manufacturers in North America. Single fibers were removed carefully from yarns obtained from each of these sources, and FT-IR microspectroscopy was performed for each of these fibers in five replications at both orientations of the polarizer. A grid-type polarizer suitable for mid-infrared use was placed in the microspectrometer between the beam splitter and the microscope optics. The polarizer was on a rack and pinion mount and could be adjusted between the two extreme positions using a circular vernier scale dial. In the one case, it was perpendicular to the axis of the fiber on the stage of the microscope, and in the other case, it was parallel to the axis of the fiber. Data were obtained for 10 replicates of each of the 33 fiber specimens at the eight infrared bands of interest. The 5280 peak heights were measured from spectra obtained with both parallel and perpendicular polarized radiation. From these data 2640, dichroic ratios were calculated in order to produce a multi-wavelength dichroic ratio pattern. Among the eight infrared bands previously identified, four of these emerged as being the best indicators for orientation difference through discriminant analysis procedures. Figure 5 shows the distribution of 165 dichroic ratios; five reps for 33 commercial fiber specimens. A Mahalanobis distance statistical routine was applied beginning with all eight wavelengths, and from this procedure, three were chosen for an initial separation.

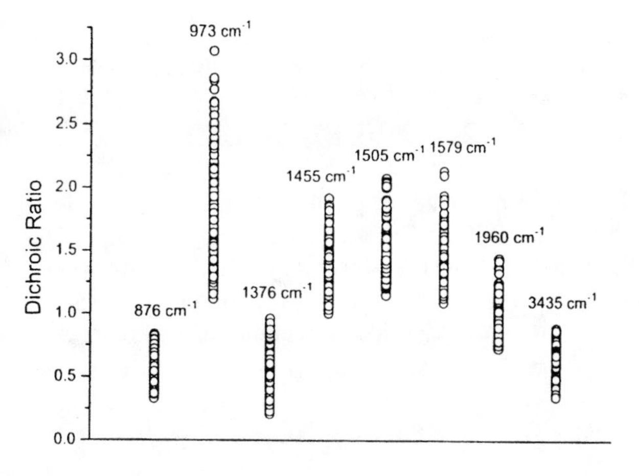

Figure 5. Dichroic ratio distribution for 5 replications of 33 fibers (4,7).

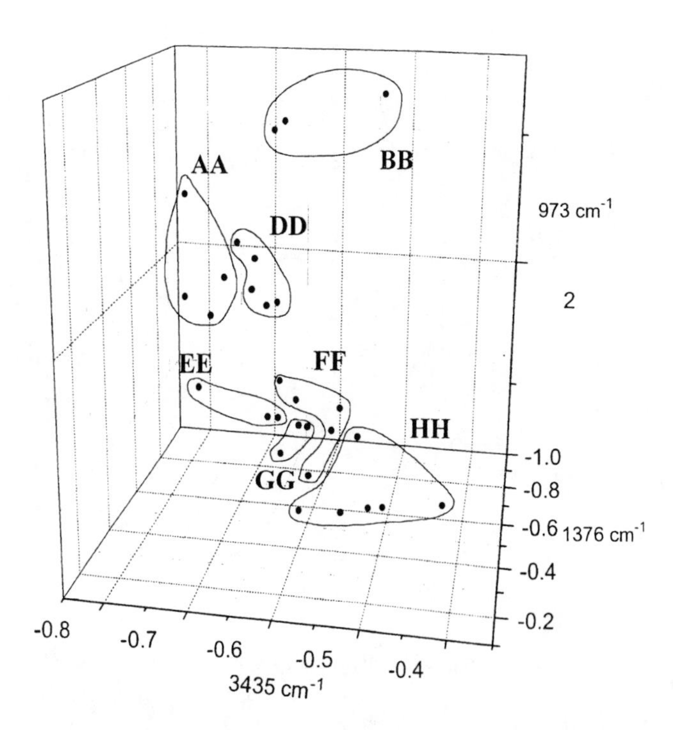

Figure 6. Seven groups classified from 3 band dichroic ratio Mahalanobis distance discriminant analysis routine (each dot is the average of 5 reps).

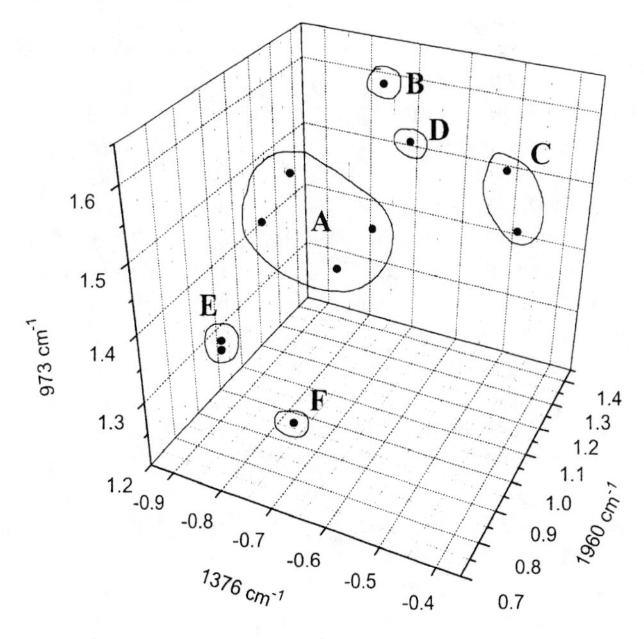

Figure 7. Six subgroups classified for 3 band dichroic ratio discriminant analysis routine applied to the fiber data pooled from groups EE, FF, and GG shown in Fig. 6.

DISCUSSION AND CONCLUSIONS

Unlike typical discriminate analysis procedures usually applied in the near-infrared or mid-infrared field, the lack of data describing the history of the fibers from 33 different sources made it necessary to use a bootstrap procedure to establish groupings of similar manufacturing conditions. Once three or four of the most useful wavelengths were established, two-dimensional plots were made where any two of these were put together to look for coincident similarity or differences. From these dichroic ratio empirical plots, preliminary grouping was assigned. Once the preliminary groups were assigned, a Mahalanobis distance discriminant analysis statistical routine was run in order to establish an expression to classify samples into their respective groups. The program also tested the expression to see that the proper classification was maintained for the learning set. A complete discussion of the stepwise procedure is beyond the scope of this brief report. Results of several iterations of discriminant analysis revealed a clean discrimination among four of seven groups (Fig. 6). The potential of overlap among the other three groups existed but was not actually experienced using averages of 5 reps. Dichroic ratio data

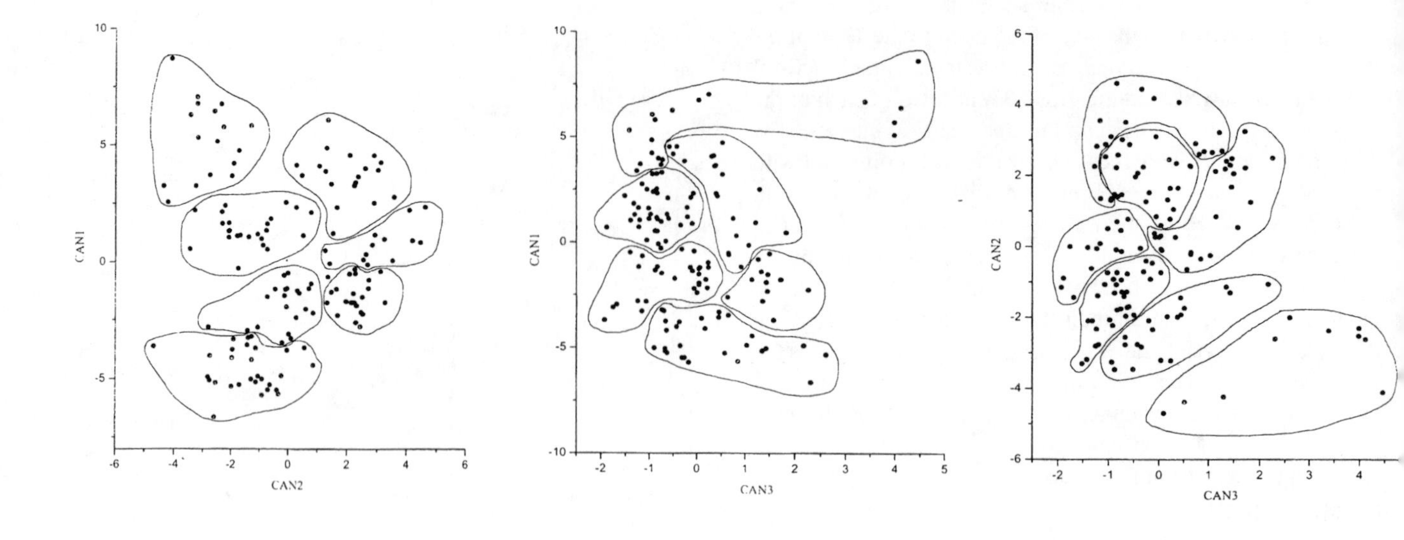

Figure 8. Two dimensional plots of 3 canonical variables derived from dichroic ratios at all 8 IR absorption bandslisted in Table 1.

from all reps of the three groups (EE, FF, and GG) were pooled to run a second-stage discriminant analysis . Figure 7 shows the second stage separation that results in six groups. Note that the absorption bands for the second stage were not all the same as those used for the first stage.

100% hits resulted when the two stage discriminant analysis protocol was tested. This was done with dichroic ratios calculated at each of the designated IR bands from single fiber polarization spectroscopic data not used in the method development. Twenty-nine PET fibers were sorted into 10 subclasses. The dichroic ratios used were calculated from five replicates of each specimen and averaged. This apparent success using dichroic data from four IR bands is supported by the same four being selected independently by running the SAS stepwise discriminant analysis routine. The subclass grouping selected for use in the 1st stage of the Mahalanobis distance portion of our protocol was tested with the SAS canonical variable discriminant analysis routine which incorporated dichroic ratios of all eight infrared bands for each of the specimens. A three dimensional plot (not shown) of the first three canonical variables calculated at all eight IR bands exhibited excellent separation with the subclass groupings. It illustrates the discrimination that is possible. Two dimensional plots of these data presented in Fig. 8 illustrate consistency of the subclass grouping selected. This work demonstrates that ten working subclasses can be assigned from infrared single fiber dichroic data instead of a single generic class for PET fibers. This new multiple IR band dichroic ratio approach added to the traditional size, shape, and other distinguishing features may be of assistance with forensic fiber evidence in the future.

ACKNOWLEDGMENTS

This work was supported primarily by the National Science Foundation, EPSCoR Grant No. OSR-9255223 and by the Microbeam Molecular Spectroscopy Laboratory at Kansas State University. The authors thank Joseph A. Sweat for computer routines for peak quantitation and for graphics.

REFERENCES

1. Grieve M.C., *Forensic Sci. Rev.,* **6**:59-80, 1994.
2. Tungol, M.W., Bartick E.G., and Montaser, A., *J. Forensic Sci.*, **36**(4):1027-43, 1991.
3. Riley, J.L., "Spinning and drawing fibers", in: *Polymer Processes*, Schildknecht, C.E., ed., New York, Interscience, 1956.
4. Cho L. and Wetzel, D.L., "Polarized microbeam FT-IR analysis of single fibers", In: Bailey G.W., Corbett, J.M., Dimlich, R.V.W., Michael, J.R., and Zaluzec, N.J., eds., *Proceedings of Microscopy and Microanalysis*; 1996 August 11-15; Minneapolis. San Fransico: San Fransico Press, Inc., pp. 206-207, 1996.
5. Wetzel D.L. and Cho, L., *Mikrochim. Acta [Suppl.]*, **14**:349-351, 1997
6. Cho L. Single fiber analysis by FT-IR microspectroscopy [PhD.dissertation]. Manhattan (KS): Kansas State University, 1997.
7. Cho, L., Reffner, J.A., and Wetzel, D.L., "Forensic clsssification of polyester fibers by infrared dichroic ratio pattern recognition", Submitted to *J. Forensic Sci.*
8. Cho, L., Gatewood, B., Reffner, J.A., and Wetzel, DL., "Polarization differentiation of PET single fibers with infrared microspectroscopy", Submitted to *J. Forensic Sci.*

Contribution no. 98-53, Kansas Agricultural Experiment Station, Manhattan.

Study of Ester Crosslinking Reactions on Aluminum Surfaces by Infrared Attenuated Total Reflectance Spectrometry

Sanmitra A. Bhat[1], Charles Q. Yang* and James A. de Haseth

Department of Chemistry, University of Georgia, Athens, Georgia 30602

Polycarboxylic acids are used as an alternative nonformaldehyde durable press finishing agents for cotton fabrics. Previous studies have shown that polycarboxylic acids esterify with cotton cellulose through intermediate formation of a cyclic anhydride. Cotton cellulose, due to the presence of hydroxyl groups, is a very active substrate. To understand the mechanism of ester formation, esterification reactions were studied on aluminum surfaces by infrared attenuated total reflectance (ATR) spectrometry. The infrared data showed that a five-membered cyclic anhydride is formed as an intermediate, that esterifies with the crosslinking agents. The data also demonstrated that formation of anhydride increases with temperature and also in the presence of a catalyst.

INTRODUCTION

In the textile industry, dimethyl dihydroxy ethylene urea (DMDHEU), an N-methylol reagent, has been used as a crosslinking agent for cotton cellulose to produce wrinkle resistant cotton fabrics [1,2]. Excellent laundering durability and a durable-press rating is obtained with DMDHEU. Formaldehyde is released during the production, storage and consumer use of DMDHEU treated fabrics. In 1987, formaldehyde was identified as a probable human carcinogen [3]. So in recent years, research has been geared towards an alternative non-formaldehyde crosslinking finishing agents. Polycarboxylic acids, such as butane tetracarboxylic acid, effectively crosslink to cellulose and was first reported by Welch in 1988 [4]. This crosslinking imparts a wrinkle resistant property to cotton fabrics. Welch also indicated that during the ester crosslinking, an anhydride is possibly formed as an intermediate. Fourier transform infrared spectrometry was used by Yang [5,6] to study the possible reaction mechanisms of polycarboxylic acids with cotton cellulose. The reaction takes place through the intermediate formation of a five membered cyclic anhydride. The cyclic anhydride reacts with the hydroxyl groups of cellulose. Cotton cellulose due to presence of hydroxyl groups, is a reactive substrate, so crosslinking reactions were studied on a non-interactive substrate.

In the present study, ester crosslinking reactions were studied on aluminum foil by Infrared Attenuated Total Reflectance (ATR) Spectrometry. Aluminum does not interfere with the reactions and does not absorb in the infrared region, thus making it an appropriate substrate. Poly(maleic acid) was used for the study. Polyvinyl alcohol and triethanol amine were used as crosslinking agents. Reaction of poly(maleic acid) and polyvinyl alcohol was carried out with and without a catalyst. Sodium hypophosphite (NaH_2PO_2) was used as a catalyst. The polymer film on aluminum foil was heated to different temperatures, and the ATR spectra were recorded at the corresponding temperatures to study the progress of ester crosslinking.

EXPERIMENTAL

Poly(maleic acid) (PMA) (33% solid in water) was supplied by the FMC Corporation (Princeton, NJ). Polyvinyl alcohol (PVA) was supplied by Scientific Polymer Product, Inc. (Ontario, NY). Heavy duty aluminum foil was used for the study. The aluminum foil was cut into 4 x 3 cm pieces. Two pieces were used for each experiment. Before each experiment, the two aluminum pieces were treated with 0.1M NaOH for 5 minutes to remove any dirt and aluminum oxide. The treated foil pieces were rinsed with water and later dried at 110 °C for 5 minutes. Each piece was dipped into the polymer solution (with/without crosslinking agents) for 5 minutes. The coated aluminum foil pieces were dried at 110 °C for 20 minutes. Subsequent heatings were carried out at 125, 140, 150, 170, 190 °C for 5 minutes each. The attenuated total reflectance spectrum (ATR) was recorded at each temperature. At the end of curing cycle, the cured film on each aluminum foil piece was treated with a mixture of dibasic and tribasic sodium phosphate solution of pH 11, and dried. The ATR spectrum was recorded again.

The ATR spectra were collected on a Perkin-Elmer System 2000 equipped with TGS detector. Resolution for all infrared spectra was 4 cm^{-1}, with 100 scans co-added for each spectrum. With the increase in the heating temperatures the coated film became hard, so nuprene rubber was used as a pressure device behind the aluminum foil to bring the film in intimate contact with the zinc selenide crystal. All spectra were normalized with respect

1. Current Address: STR Inc., Enfield, CT 06082

* Author to whom correspondence should be addressed

CP430, *Fourier Transform Spectroscopy:* 11th International Conference
edited by J.A. de Haseth

to the CH_3 and CH_2 bending mode region: 1480-1360 cm^{-1}, as the stretching region could not be used as it was overlapped by the broad OH peak of the carboxylic group.

RESULTS AND DISCUSSION

Initial studies were performed with the use of PMA (21%) solutions only. No crosslinking agents or catalysts were used. The ATR spectra of foils treated with PMA (21%) and heated to different temperatures are shown in Figure 1.

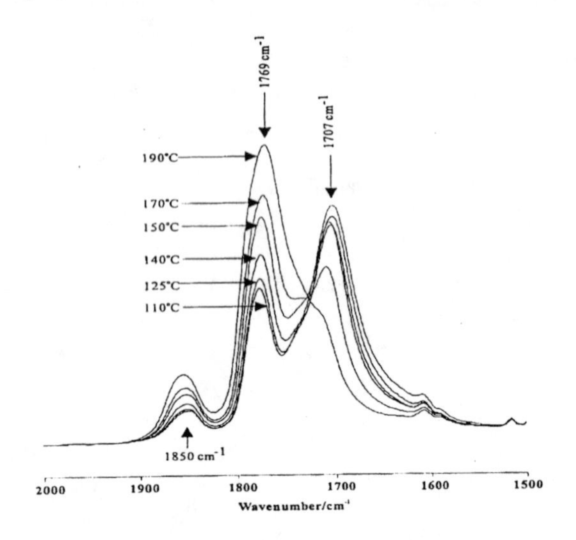

FIGURE 1: ATR Spectra of PMA at different temperatures

The intense band at 1707 cm^{-1} is due to the carbonyl stretching of carboxyl groups in PMA. With increasing temperature, the anhydride bands at 1769 and 1850 cm^{-1} increase, indicating that more anhydride is formed at higher temperatures. The anhydride is a five membered cyclic anhydride formed by dehydration of two carboxylic groups in PMA as shown below.

HC—COOH
HC—COOH

$\xrightarrow{-H_2O}$

HC—CO
  O
HC—CO

Poly(maleic acid)   Five membered cyclic anhydride

Due to the ring strain, the carbonyl band shifts from 1705 cm^{-1} to a higher wavenumber. The intensity ratios of anhydrides to free carboxyl is given in Table 1 for PMA. Nearly complete conversion of free carboxyl groups to five membered cyclic anhydrides takes place at 190 °C.

Earlier studies [5,6] have shown that the five membered cyclic anhydride formed is an effective intermediate that can undergo crosslinking reactions. To look at this aspect, studies were conducted with two different crosslinking agents: Triethanol Amine (TEA) and Polyvinyl Alcohol

TABLE 1: Intensity ratios for PMA

Temperature °C	Ratio of 1850/1707 cm^{-1}	Ratio of 1769/1707 cm^{-1}
110	0.138	0.652
125	0.148	0.725
140	0.178	0.867
150	0.221	1.026
170	0.291	1.436
190	0.535	2.360

(PVA). For TEA, the aluminum foils first were treated with 0.1M NaOH solution for five minutes. It was rinsed with distilled water and dried at 125 °C for five minutes. The treated foils were dipped in PMA/TEA solution (1:1 ratio) for five minutes and dried at 125 °C for 20 minutes. The ATR spectrum of the dried film on aluminum surfaces was recorded. Subsequent heatings were performed at higher temperatures for five minutes and the corresponding ATR spectra were recorded. The ATR spectra are shown in Figure 2.

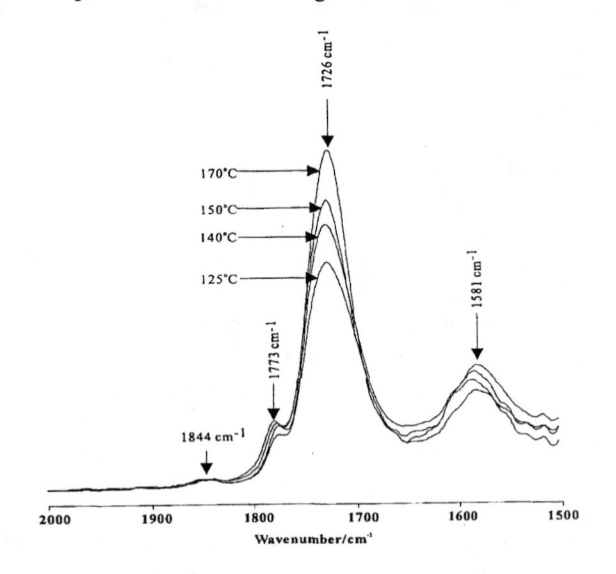

FIGURE 2: ATR Spectra of PMA/TEA system

The broad peak at 1725-1705 cm^{-1} is due to the overlapping of the ester carbonyl and the carboxyl carbonyl peaks. With an increase in temperature, more esters are formed as shown by a decrease in the intensity ratio of the anhydride to the broad peak at 1725-1705 cm^{-1}, as shown in Table 2. This decrease in the intensity ratio indicates that the esterification reaction between the cyclic anhydride and carboxyl groups of PMA takes place. During this esterification reaction between PMA and TEA, part of carboxyl groups of PMA exist as carboxylate ions as shown by the peak at 1581 cm^{-1}.

TABLE 2: Intensity ratios for PMA/TEA

Temperature °C	Ratio of 1773/1725-1705 cm^{-1}
125	0.236
140	0.234
150	0.222
170	0.202
Treated with soln. of pH 11	0.144

The dissociation constant (K_a) of carboxylic acid is given by

$$K_a = [COO^-]\,[H^+]\,/\,[COOH]$$

As more COOH groups form the ester linkages, the value of K_a increases due to the increase in $[COO^-]/[COOH]$. Hence a peak is observed at 1581 cm^{-1}. To separate the ester peak from the carboxyl peak, the cured films were treated with sodium phosphate solution at pH 11 to convert all the free carboxyl groups to carboxylate ions. The ATR spectrum of cured films treated with the pH 11 sodium phosphate solution is shown in Figure 3.

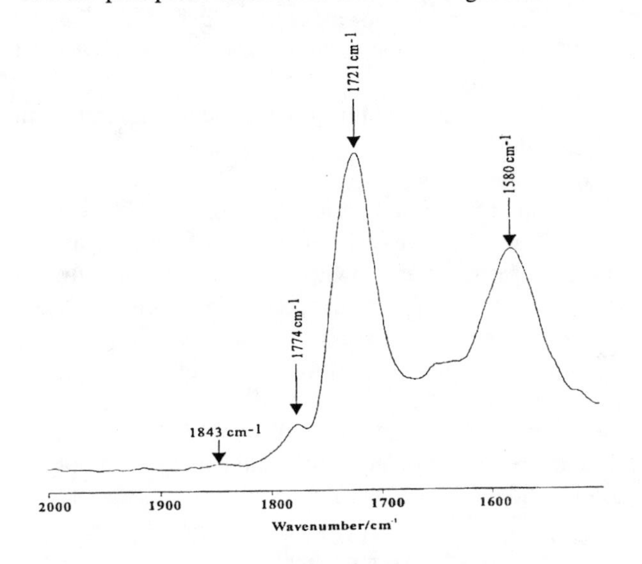

FIGURE 3: PMA/TEA cured film treated with pH 11 solution.

The peak at 1721 cm^{-1} is due only to the ester linkage between PMA and TEA. The intensity of this peak is 0.167 and that of 1580 cm^{-1} peak is 0.115. The ratio of these intensity values is 1.45, indicating a substantial amount of ester linkages in the cured films. A decrease in the intensities of the two anhydride bands indicated that part of the anhydride reacted with sodium phosphate to form sodium carboxylate.

Experiments were conducted with PVA at two different concentrations. The concentrations of PMA was kept constant, and only the concentration of PVA was varied. The film forming procedure on the aluminum surface was similar to one with TEA, except that the initial drying of aluminum foil after rinsing with the distilled water and the polymer film (PMA/PVA) was conducted at 110 °C instead of 125 °C. The ATR spectra of PMA-PVA (1:0.5) are shown in Figure 4.

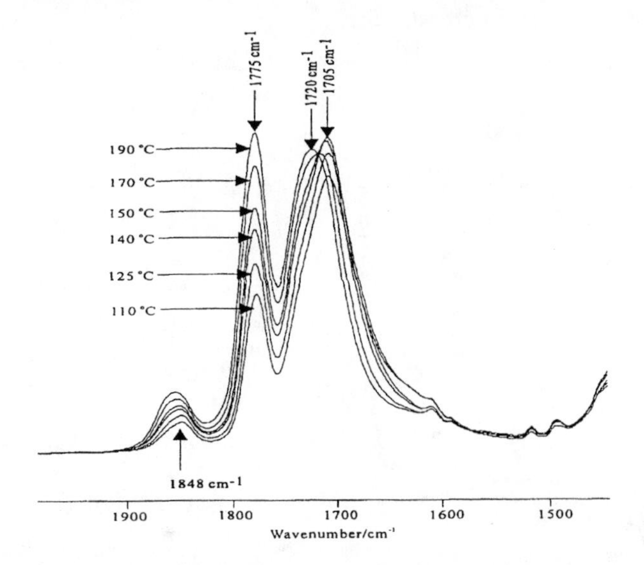

FIGURE 4: PMA/PVA (1:0.5) cured to different temperatures

The wavenumber of the 1705 cm^{-1} peak shifts to 1720 cm^{-1}, indicating esterification between PMA and PVA. Spectra also show intense anhydride peaks, suggesting fewer crosslinking sites available due to the low concentration of PVA. The cured/heated film was treated with sodium phosphate solution at pH 11 to convert all the free carboxyl group to carboxylate ion. The ratio of intensity of the ester peak at 1720 cm^{-1} to the carboxylate carbonyl peak was 0.54. The ratio suggests that a small amount of ester crosslinkages is formed, and was mainly due to the smaller number of hydroxyl groups present because of the low concentration of PVA. The reaction of the anhydride, formed by dehydration of two carboxyl groups on PMA, with PVA, proceeds as follows:

HC–CO / O / HC–CO (Five membered cyclic anhydride) + CH_2 / HO–CH / CH_2 / HO–CH (Polyvinyl Alcohol) → HO–CH / CH_2 / HC–CO–O–CH / CH_2 / HC–COOH

Five membered cyclic anhydride   Polyvinyl Alcohol

In the second part of the experiment, the concentration of PVA was increased two fold. The spectra of the cured PMA-PVA (1:1) film is shown in Figure 5. The anhydride bands in the spectra are not as intense as seen with the low concentration PVA (Figure 4). Anhydrides formed reacted with the hydroxyl groups to give esters. With high cm^{-1} to the carboxylate peak at 1581 cm^{-1} (after treating with pH

concentrations of PVA, the ratio of the ester peak at 1720 11 solution) was 0.88. Due to the increase in the PVA concentration, more crosslinking sites are available. Thus the ester linkages increases as more hydroxyl groups are available for crosslinking.

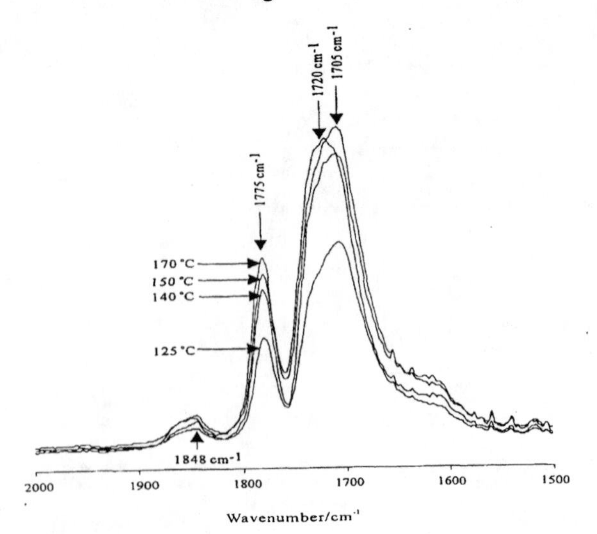

FIGURE 5: PMA/PVA (1:1) cured to different temperatures

The above study with PMA-PVA (1:1) also was conducted in the presence of a catalyst, sodium hydrophosphite. The cured film was treated with pH 11 solution to convert the free carboxyl group to carboxylate ion. The spectrum is shown in Figures 6. The intensity ratio of the ester to the carboxylate carbonyl peak was 1.14. This increase in the ratio from 0.88 to 1.14 shows that the catalyst accelerates the ester formation reaction through the intermediate formation of an anhydride.

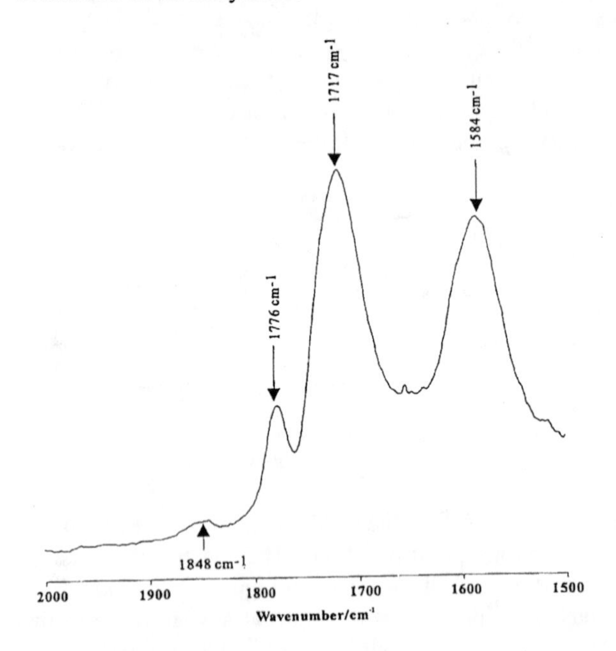

FIGURE 6: PMA/PVA (1:1) cured in presence of catalyst and treated with pH 11 solution

Infrared spectrometry data showed that, with PMA a five membered cyclic anhydride intermediate is formed which readily esterifies when crosslinking sites are available. The carboxylic acid, PMA, effectively crosslink with TEA and PVA to form a polymer network. A reactive cyclic anhydride is formed which then readily esterifies with the hydroxyl groups of TEA and PVA. The amount of ester formed increases with temperature, as more reactive anhydrides are formed at higher temperatures. The PMA-PVA system can be used as a finishing polymer to the hydrophobic substrates, particularly in the textile industry for synthetic fibers such as polyester, polypropylene and nylon. It also can be used as a binder during the chemical bonding of the substrate fibers in the nonwoven industry. The system introduces both the carboxyl and hydroxyl groups on the substrate. These functional groups can undergo chemical reactions with any other finishing agents applied to the substrate making it more durable.

CONCLUSION

The esterification reaction with both crosslinking agents goes in two steps:

1. The formation of the reactive cyclic anhydride by losing a water molecule from the two carboxyl groups.
2. The reaction of this reactive cyclic anhydride with the hydroxyl groups of crosslinking agents.

Formation of the anhydride by polycarboxylic acids increases with an increase in temperature. Thus, the amount of ester formed increases with temperature. The infrared spectra recorded support this mechanism. Ester formation increases with an increase in the concentration of crosslinking agents, as more crosslinking sites are available to form ester. Ester formation also increases in the presence of a catalyst, indicating that the reaction between the reactive anhydrides and hydroxyl groups are accelareted by the catalyst.

REFERENCES

1. Arceneanx, R. L., Frick, J. G., Jr., Reid, J. P., Gantreanx, G. A., Am. Dyest. Rep., **50**, 849, 1961.
2. Goldstein, H. B., Textile Chem. Color., **5**, 209, 1973.
3. Mckerron, C. D., Rich, L. A., Chem. Week, **140**, 15, 1987.
4. Welch, C. N., Textile Res. J., **58**, 480, 1988.
5. Yang, C. Q., J. Polym. Sci. Chem. Ed., **31**, 1187, 1993.
6. Yang, C. Q., J. Appl. Polym. Sci., **50**, 2047, 1993.

A Method for Checking Homogeneity of Subsurface Regions by Variable Angle ATR: Experiments on Polymers *vs.* Optical Modelling

I. Hopfe*, K.-J. Eichhorn*, V. Hopfe**, W. Grählert**

**Institute of Polymer research, PO Box 120 411, 01005 Dresden, Germany*
***Fraunhofer Institute Material and Beam Technology, Winterbergstraße 28, 01277 Dresden, Germany*

The subsurface structure of many technically applied materials is characterised by alterations of the composition (swelling or surface reactions on polymers, leaching of glasses etc.). Attenuated total reflection (ATR) spectroscopy is a powerful tool for analysing such surface regions. Because of problems with optical contact the ATR method is difficult to handle in practise. Based on variable angle ATR - FTIR spectroscopy a straightforward method has been established which can detect compositional inhomogeneities in subsurface regions or across layers. The method works as follows: (i) making ATR measurements at two different angles of incidence without changing the sample, (ii) normalising the ATR absorbance spectra by using an appropriate band of the substrate as an internal standard, (iii) calculating difference spectra of the normalised spectra. In the case of a homogeneous material the difference spectrum virtually vanishes whereas in the case of an inhomogeneous surface region the difference spectrum reflects the compositional gradient. The method has been tested at polyamide -12 foils: (a) chemically cleaned, (b) treated with initiator, and (c) treated with pluronic. The experimental findings have been supported by spectra modelling. Based on an optical multilayer model the ATR spectra of a homogeneous polymer and of the same material with a slightly altered surface layer have been calculated for different angles of incidence. By applying steps (ii) to (iii) the experimental results are confirmed.

INTRODUCTION

The subsurface structure of many widely used materials is characterised by alterations of composition (swelling or surface reactions on polymers, leaching of glasses, graded layers in coated materials etc.). As a consequence, in many cases we are faced with heterogeneities in composition and morphology on and near the surface. Attenuated total reflection (ATR) - or internal reflection spectroscopy (IRS) - is a powerful tool for analysing such surface regions. Because of the problems with optical contact the ATR method is difficult to handle in practice and the interpretation of the spectra have to be carried out with great care.

The traditional way of subsurface depth profiling is based on two or more ATR crystals in order to vary penetration depth by changing crystal refractive index and/or internal angle of incidence. The drawback of this method is that the sample contact and thus the sampling area changes after re-adjustment. Additionally, for making difference spectroscopy, the pure bulk material has to be measured as a reference. Thus, because of the different sampling steps, the reproducibility is low.

In this paper a straightforward method to overcome these problems based on variable angle ATR-FTIR spectroscopy is presented. The proposed method is evaluated by means of the detection of compositional inhomogeneities in subsurface regions of different modified polyamide-12 foils.

EXPERIMENTAL

Materials

- Polyamide-12 (PA12) , Hüls AG, foils d=500μm
- Initiator-Copolymer: (Poly-(propen-alt-(maleic anhydride/ maleic acid monotertiarybutyl-perester))
- Pluronics: Polyethyleneoxide-block-polypropyleneoxide-block-polyethyleneoxide-block

ATR

- FTIR spectrometer IFS66 (Bruker), purged with dry air
- Variable angle ATR (HARRICK): twin parallel mirror reflection attachment, range of optics angle 30° - 60°
- KRS-5 crystal (n=2.35), single-pass parallel plate, crystal end-face angle 45°
- reproducible contact is made between the sample and the KRS-5 crystal by specially designed pressure plates and torque screwdriver (range: 8 - 20 OZ/IN)

CP430, *Fourier Transform Spectroscopy:* 11th International Conference
edited by J.A. de Haseth

- Background and sample measurements at 45° and 60° optics angles (dial settings), resulting in internal angles of incidence of 45° and 51.3°
- 100 scans were co-added at a resolution of 4 cm^{-1}
- corrected absorbance spectra (ATR units) were baseline-corrected in the range 1850 - 850 cm^{-1}

THEORY AND METHOD

As Mirabella and Harrick pointed out, that there is "no accepted and tested theoretical treatment for experimental IRS data that can yield an absolute quantitative concentration depth profile of an absorbing species in a surface" (1). The problem is to describe the unknown concentration depth profile with an exponentially decaying probing field (of the evanescent wave). As many others in the following study the depth profile is based on the "penetration depth" d_p and "effective layer thickness" d_e. However it is known that the actual sample depth is larger than d_p, and d_e contains the total interaction between the radiation and sample (based on five factors!) and is also not a simple penetration distance (1). A new more realistic approach for determining the concentration profile of a surface by variable angle ATR has been introduced by Ishida (2). He used the inverse Laplace transform function of the absorbance A which depends on the decay constant $\gamma = 1/d_p$.

The following treatment remains on the d_e concept to provide the link between ATR and transmission or absorbance spectra. At first the internal reflection spectra (eq. 1) are transformed into absorbance spectra (eq. 2):

$$R = I/I_0 = \exp(-\alpha\, d_e) \tag{1}$$

$$A = \lg I_0/I = -\lg R = (1/2.303)\, \alpha\, d_e = \varepsilon\, d_e \tag{2}$$

α and ε = absorption coefficients

To make these spectra more "transmission-like" the common "ATR correction" is applied:
ATR unit = A ν/1000 (at the wavenumber ν).

For bulk materials, d_e is proportional to d_p (3):

$$d_e = (n_{21} E_0^2 / 2 \cos\theta)\, d_p \tag{3}$$

$$\text{with } d_p = \lambda / n_{12}(\sin^2\theta - n_{21}^2)^{1/2} \tag{4}$$

E_0^2 electric field amplitude of the incoming wave;
$n_{21} = n_2/n_1$ (refractive index ratio of the ATR crystal n_1 and polymer n_2 ~1.5);
λ wavelength, θ internal angle of incidence

The advantage of variable angle experiments is that only one parameter (θ) is changed, which results in the variation of d_e and A (see eq. 2). Note that the number of reflections N in the ATR plate is also changed with θ: N = (l/t) cos θ (l length, t thickness of the plate).
Multiple reflections have to be considered in eq. 2 for the absorbance spectra:.

$$A = N \varepsilon d_e \tag{5}$$

The proposed method for depth profiling works as follows:

(i) recording the spectra 1 and 2 at the two angles of incidence without changing the sample

(ii) normalising the ATR absorbance spectra by using an appropriate band of the substrate (bulk material) as an internal standard, this band must be within the linear absorbance range and should exhibit a low structural sensitivity (the PA12 band at 1121 cm^{-1} has been used)

(iii) making difference spectra of the normalised spectra

I. homogenous materials

spectrum1: $A_1 = N_1\, d_{e1}$
spectrum2: $A_2 = N_2\, d_{e2}$

After normalisation (n) the difference spectrum for two spectra at (slightly) different angles of incidence virtually vanishes because $A_{1n} \cong A_{2n}$.

II. heterogenous surface region (s) on homogenous bulk substrate (b) ($n_s \cong n_b$)

spectrum1: $A_{1b} + A_{1s} = N_1\, d_{e1}\, (\varepsilon_b + \varepsilon_s)$
spectrum2: $A_{2b} + A_{2s} = N_2\, d_{e2}\, (\varepsilon_b + \varepsilon_s)$

For both spectra the same surface region is penetrated ($A_{1s} = A_{2s}$), but different depths in the bulk are scanned ($A_{1b} \neq A_{2b}$). After normalising $A_{1bn} = A_{2bn}$ and $A_{1sn} \neq A_{2sn}$ the difference spectrum ΔA_{sn} nearly reflects the compositional homogeneity of the surface layer.

Clearly, the applicability of the method is restricted to samples and ATR configurations where the penetration depth for both angles of incidence is higher than layer thickness.

APPLICATION OF THE METHOD

From the experiments (Fig. 1-3) the following conclusions can be drawn:

(i) The difference spectrum (Fig. 1) of a chemically cleaned polymer sample vanishes. Hence the subsurface layer is identical with the bulk material; no swelling etc. can be detected.

(ii) The difference spectrum of chemically modified polymer samples clearly exhibits the spectrum of the heterogeneity within the subsurface layer (Fig. 2, 3)

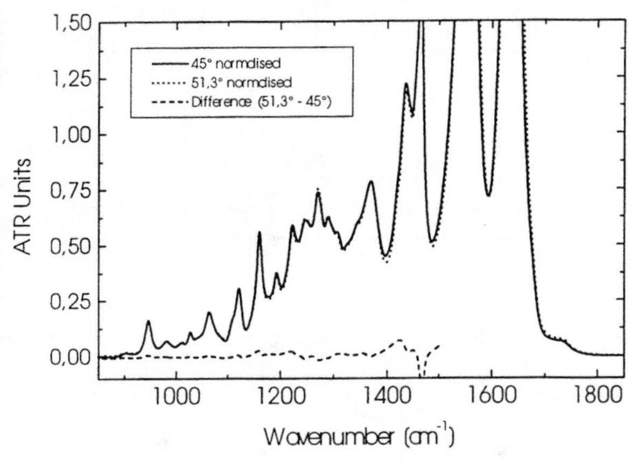

FIGURE 1. Chemically cleaned Polyamide 12

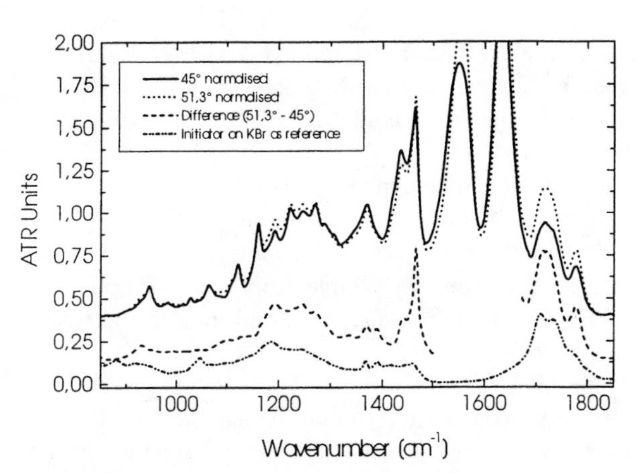

FIGURE 2. Initiator on polyamide 12

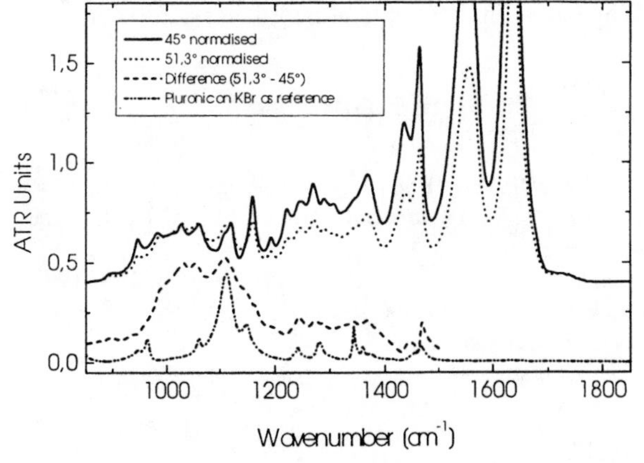

FIGURE 3. Pluronic on Polyamide 12

OPTICAL MODELLING

The experimental findings are supported by spectra modelling. Based on a optical multilayer model the ATR spectra of the homogenous polymer and of the material with a slightly altered surface layer have been calculated.

A set of damped Lorentzian oscillators is used for modelling the dielectric function of the polymer where oscillator parameters are extracted from experiments (Fig. 4). The heterogeneous subsurface region of the material is simulated by inserting two additional oscillators into the set of the "pure" polyamide. Within a monolayer optical model (Fig. 5) the thickness of the heterogeneous layer is varied (50 nm ... 2000 nm) and the ATR spectra are calculated for the two selected angles of incidence (Fig. 6, eq. 6). The further data treatment of the calculated spectra follows the above proposed procedure. The difference spectra clearly reflect the structure of the subsurface region.

Calculations

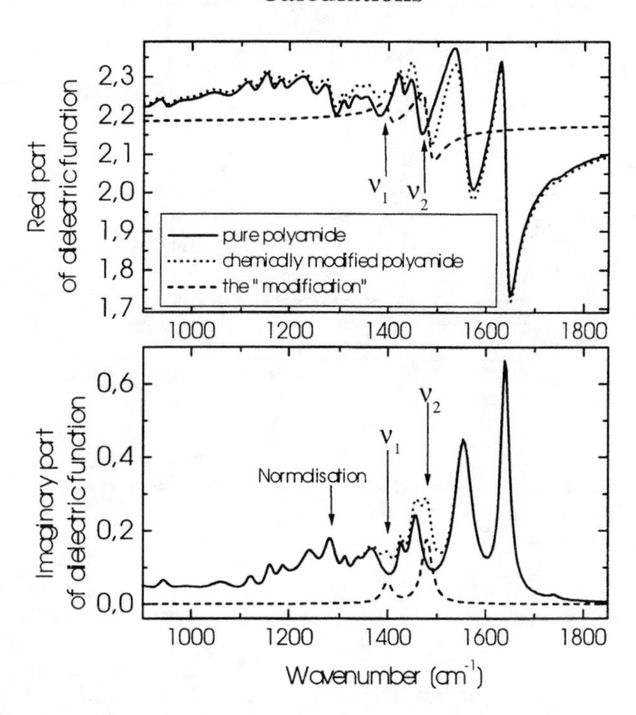

FIGURE 4. Comparison of dielectric function of polyamide and chemically modified polyamide. The modified material is characterized by additional oscillators at 1400 cm^{-1} and 1480 cm^{-1}.

For extracting signatures of the heterogeneity the absorbance spectra of the pure polyamide are subtracted at the two angles of incidence (Fig. 7). As expected from basic considerations the absorbance of the heterogeneity at first increases with increasing layer thickness and beyond a maximum around 500 nm the intensity decreases. Beyond this maximum additional spectral features appear at the high frequency side of the heterogeneity resonance which mainly coincide with the main refraction bands of the polyamide. This effect is caused either by a refraction induced influence of the heterogeneity resonances onto basic polymer or by restrictions of the penetration depth.

Optical model

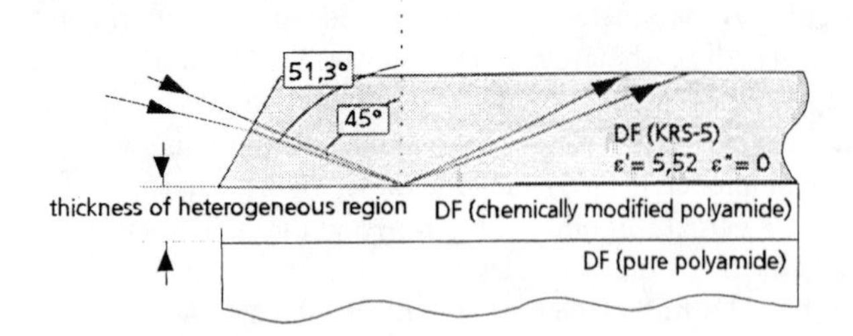

FIGURE 5. Optical model for calculation of variable angle ATR-spectra

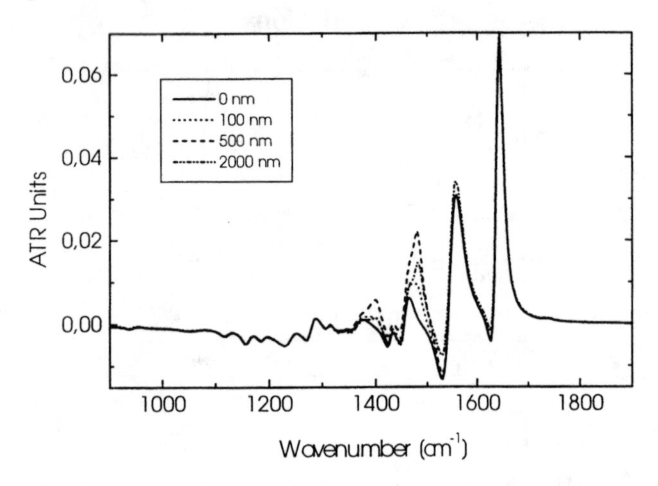

FIGURE 6. Influence of thickness of the heterogeneous region on normalised difference spectra between 51,3° and 45° ATR Spectra.

$$R = f(\hat{\varepsilon}_{KRS-5}, \hat{\varepsilon}_{PA}, \hat{\varepsilon}_{HPA}, d_{HPA}, \theta) \qquad (6)$$

$\varepsilon_{KRS-5,PA,HPA}$ dielectric function of KRS-5, polyamide and chemically modified polyamide respectively
d_{HPA} : thickness of heterogeneous subsurface region
θ : angle of incidence

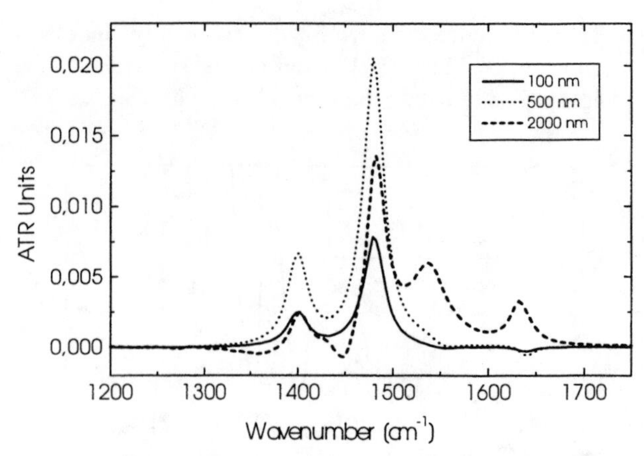

FIGURE 7. Resulting absorbance spectra of heterogeneous region after elimination of the influence of pure polyamide.

CONCLUSIONS

The advantages of the described variable angle ATR method in comparison with common ATR handling techniques for depth profiling to check surface homogeneity are:

- The method can be applied on one test specimen, there is no need for any reference material
- The method is carried out by changing only the angle of incidence, without any sample preparation or optical re-adjustment (probed surface area and sample contact are unchanged)

The proposed method should be characterised as an intuitive but very practicable attempt with proven applicability in the field of polymer spectroscopy. The experimental findings have been confirmed by spectra modelling based on exact optical multilayer approach.

REFERENCES

1. Mirabella Jr., F.M., Harrick, N.J., *Internal Reflection Spectroscopy: Review and Supplement*: Marcel Dekker, Inc., 1985.
2. Ishida, H., *Bull. Inst. Chem. Res. Kyoto Univ.* **71**, 198 (1993).
3. Harrick, N.J., *Internal Reflection Spectroscopy*, Harrick Scientific Corporation, 1987.

The effect of ionic strength variation in the orientation characteristics of ionic polymer multilayers on patterned self-assembled monolayers using infrared reflection absorption spectroscopy

Vasilis G. Gregoriou[1], Rick Hapanowicz[2], Sarah L. Clark[3] and Paula T. Hammond[3]

[1]*Polaroid Corporation, Analytical Research & Development. 1265 Main Street, W4-1D, Waltham, MA 0225;* [2]*Nicolet Instrument Corporation 5225-5 Verona Road, Madison, WI 53744 USA;* [3]*Department of Chemical Engineering, Massachusetts Institute of Technology, Cambridge MA, 02139.*

Infrared reflection absorption spectroscopy was used in the elucidation of the effect of NaCl concentration in the morphology and orientation behavior of polymeric thin films of alternating bilayers of sulfonatedpolystyrene /poly(diallyldimethylammonium chloride) (SPS/PDAC) fabricated by ionic multilayer assembly. It was found that the integrity of the underlined self-assembled monolayer depends on the ionic content of the initial solutions. Specifically, at higher salt concentrations (>1M) the loss of the lateral hydrogen bonding is observed, indicative of the disrupted nature of the alkanethiol packing. In addition, the presence of water was detected in the electrolyte samples, an indication of a high level of hydration in these films. Finally, an increase in the intensity of bands proportional to the salt content can be attributed to a corresponding increase in the film thickness due to charge-shielding along the polymer backbone.

INTRODUCTION

Reflectance infrared spectroscopy has become a mainstream analytical spectroscopic technique for studying the chemical and molecular orientation of thin films on metallic surfaces.[1-3] The sensitivity of the technique comes from the fact that the intensity of the p-polarized component of the reflected infrared radiation (electric field parallel to the plane of the metal surface) is enhanced at a metal surface at a high angle of incidence, the "so-called" grazing angle.

Recently, the technique of phase modulated infrared reflection absorption spectroscopy (PM-IRRAS) has appeared in the literature.[4-6] This new development offer several advantages when compared to the static measurements. Two of the most important of these advantages is the sensitivity enhancement attainable in a dynamic measurement along with the elimination of atmospheric interference.

A new technique for the assembly of ultrathin films involves the electrostatic adsorption of alternating layers of polyanions and polycations.[7] By using self-assembled monolayers (SAMs) patterned onto a gold surface, it has been demonstrated that layer-by-layer polyion adsorption can be selective.[8] The resulting thin films are patterned microstructures which may be of interest in optical device or sensor applications. The basis of this selective deposition is the difference in chemical functionality of the patterned surface. A carboxylic acid (COOH) functionalized SAM is used as an ionizable surface onto which polycation/polyanion bilayers are adsorbed. The COOH surface is patterned onto the surface, and the remaining gold regions are covered with a second SAM functionalized with oligoethyleneglycol units (EG) which prevent the adsorption of polymer. It has been found that the nature of deposition on these two different surfaces varies with ionic content in a manner which affects selectivity. Some of the reasons for these changes include specific interactions between the surfaces and the added electrolyte ions. PM-IRRAS has been used to investigate some of these changes on patterned samples adsorbed from various ionic content, and to confirm the proposed changes in adsorption behavior.[9] This spectroscopic tool allows the resolution of the underlying SAMs groups of 10 bilayer samples, and is sensitive to any orientation effects in or out of the plane of the surface.

CP430, *Fourier Transform Spectroscopy:* 11th International Conference
edited by J.A. de Haseth

EXPERIMENTAL

Sample preparation

Details of the sample preparation can be found in previous publications. [7,10] Gold substrates were prepared by thermally evaporating gold shot onto n-type test grade silicon wafers. Alkane thiols (R = COOH, $(OCH_2CH_2)_3OH$) were used to functionalize the gold surface using the microcontact printing process.[11] The stamps were inked using the supernatant of a saturated solution of $HS(CH_2)_{15}COOH$ (COOH in this paper) in hexadecane. The second alkane thiol $HS(CH_2)_{11}(OCH_2CH_2)_3OH$ (EG) was deposited on the surface from dilute solution in absolute ethanol. The patterned surface contained straight lines of 3.5 μm (COOH) with spaces of 2.5 μm between the lines (EG).

Patterned substrates were immersed in a dilute solution of polycation, in this case, polydiallyldimethylammonium chloride (PDAC) for 20 minutes. The substrate was then thoroughly rinsed with neutral Milli-Q water, dried with a nitrogen stream, and immersed in a second solution of polyanion, sulfonated polystyrene (SPS) for 20 minutes. Alternating layers of PDAC/SPS were created, with all samples consisting of 10 patterned bilayers.

Polydiallyldimethyl ammonium chloride (PDAC)	Sodium polystyrene sulfonate (SPS)

PM/IRRAS FT-IR spectroscopy

A Nicolet Magna-IR 850 FT-IR spectrometer, capable of simultaneously digitizing two channels of spectroscopic information was used in this study. A nitrogen purge box was used to house all of the additional hardware. A collimated infrared beam from the spectrometer was focused onto the sample using a parabolic mirror at an incident angle of 80°. A wire grid polarizer (Perkin Elmer Co.) and a Hinds Instruments 37KHz ZnSe photoelastic modulator (PEM) were inserted into the beam path between the parabolic mirror and the sample. The polarizer defines a p-polarized infrared beam (electric vector perpendicular to the plane of the metal surface) that was modulated between two linearly polarized states by the PEM at 74kHz. The peak modulation wavelength of the PEM was set to 7.5 microns, which maximized the modulation efficiency for the fingerprint region of the sample. After the sample, the reflected beam was refocused onto the narrow band mercury cadmium telluride (MCT) detector using a BaF_2 lens. Simultaneous spectral acquisition at 8 cm^{-1} spectral resolution of the differential and sum signals was performed using a 0.316 cm/sec optical mirror velocity. A lock-in amplifier performed the demodulation of the polarization modulation information. The sum spectral information is the result of the low-pass filtered (40kHz) intensity modulation information.

RESULTS & DISCUSSION

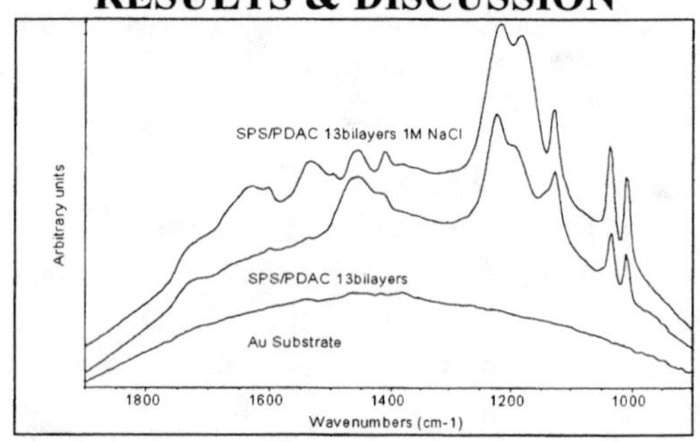

Figure 1

The PM-IRRAS spectra for the SPS/PDAC polyion system with and without the presence of NaCl can be found in Figure 1. The reflection absorption spectrum of the bare Au substrate is also plotted for comparison reasons. The latter spectrum is also used to correct the PM-IRRAS spectra from the Bessel function dependency introduced by the modulator. The data depicted in Figure 1 are the result of the application of the equation

$$\Delta R/R = (R_p - R_s)/(R_p + R_s)$$

where ΔR is the differential reflected R_s and R_p linearly polarized signals. The denominator is the sum of these two signals and is the low pass filtered intensity modulated signal from the infrared detector. The insensitivity of the measurement to the presence of water vapor is evident upon inspection of the 1800 - 1400 cm^{-1} region. This region is not easily accessible with a conventional IRRAS experimental set-up, due to the low intensity of the surface bands.

In a PM-IRRAS experiment, bands that show significant intensities have large components perpendicular to the surface plane. In contrast, if a transition moment lies along the surface plane, the corresponding infrared band should not be present. The differences in the appearance of the PM-IRRAS spectra for the electrolyte and non-electrolyte samples were explored in a different study.[9] A more random environment was found in the presence of the added electrolyte. The loss of planarity in the arrangement of the aromatic rings (the phenyl stretching bands at 1600 and 1500 cm^{-1} were used as markers) suggested that SPS adsorbs in a more planar arrangement along the surface in the absence of salt. This is consistent with models of highly extended, unshielded polyelectrolyte in solution, and the adsorption of polyions as two dimensional monolayers of polyion from solution to form thin monolayers (3-10 Å in thickness). In addition, the reduction of the PDAC peak at 1450 cm^{-1} on the addition of salt indicated that this polyion also absorbed from dilute solutions with the C-N stretching dipole moment pointing out of the plane of the film surface. The presence of the water band at 1640 cm^{-1} in all electrolyte samples is indicative of hydration found in these samples that is not apparent in the spectrum of the nonelectrolyte in Figure 1.

The variation of the ionic content from 0.001 M to 1 M NaCl produced the PM-IRRAS spectra depicted in Figure 2. All the spectra are normalized to the most intense band at 1221 cm^{-1} for a qualitative comparison.

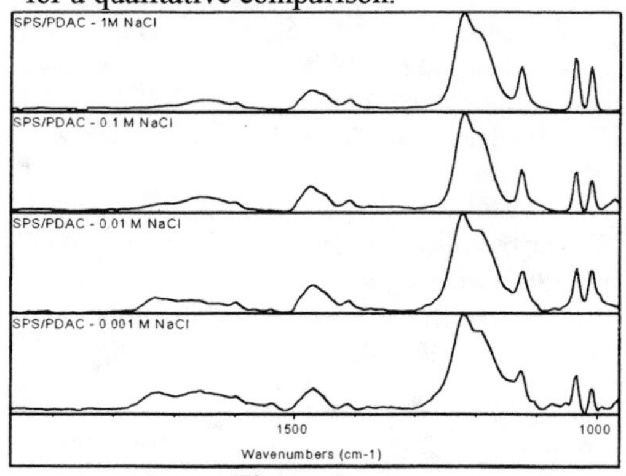

Figure 2

These spectra appear to be similar to each other and they are all distinctly different from the spectrum of the non-electrolyte sample shown in Figure 1. However, a closer inspection reveals subtle but real differences that are related to the effect of ion concentration on the multilayers. Figure 3 shows an expansion of the carbonyl spectral region of the PM-IRRAS spectra.

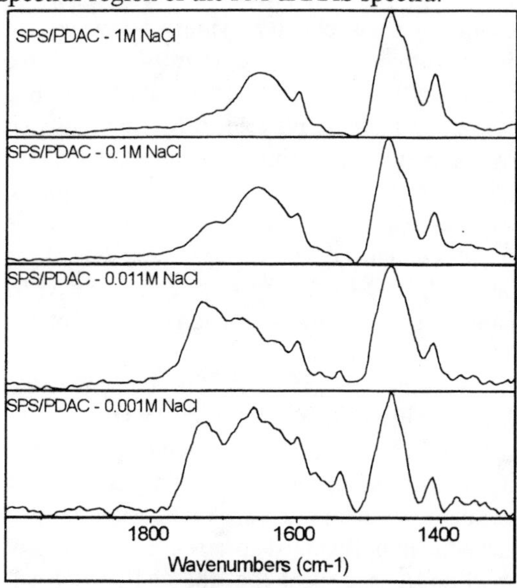

Figure 3

The band at 1732 cm^{-1} belongs to the C=O stretching mode of the COOH self-assembled monolayer on which the bilayers have adsorbed. The band is present in the spectra of the 0.001 M and 0.01 M NaCl solutions, but is diminished in intensity in the 0.1 M NaCl sample and it is almost absent in the 1 M NaCl sample. The presence of this band indicates the presence of hydrogen bonded COOH groups with a component of the bond stretch perpendicular to the substrate.

The frequency of this carbonyl stretch is higher than the frequency of the alkanethiol in the bulk by approximately 20 cm^{-1}. This is the result of the formation of linear hydrogen bonds along the surface plane, a well documented observation in the literature. Specifically, they have been reported as doublets in -COOH terminated SAMs at 1718 cm^{-1} and 1741 cm^{-1}.[13] All of the spectra in Figure 3 appear to have a shoulder band at 1711 cm^{-1}, an indication that there is also some free carbonyl present in the monolayer. However, as the salt content is increased, the presence of the hydrogen bonded peak at 1730 cm^{-1} is significantly diminished. We attribute the lessening of the intensity of the 1730 cm^{-1} band to the gradual weakening of intermolecular hydrogen bonding within the SAM with increased ionic strength due to competitive interactions between the added NaCl ions in solution and the COOH groups on the surface. Deposition of layer-by-layer films on COOH surfaces actually increases up to approximately 0.1 to 0.4 M salt concentration, due to the adsorption of shielded polyions. However, at higher salt concentrations, from 0.4 to 1 M, the amount of film deposited drops drastically due to screening of the polyelectrolyte charges and the competitive adsorption of small NaCl ions onto the COOH surface at high concentrations. The adsorption of the NaCl electrolyte is probably enhanced by specific interactions of the NaCl with the COOH groups, as evidenced by changes in the lateral hydrogen bonding of the COOH monolayer.

Since the thickness of the particular system is less than the wavelength of the incident radiation, a linear dependence between the intensity of the PM IRRAS signal and the thickness of the layer is expected. Figure 4 shows the unnormalized spectra for this series of samples. It is evident that the addition of salt results in a corresponding increase in the polyion film thickness. These findings are expected, and are in agreement with previous work;[10,12] the addition of electrolyte shields the polyions, allowing a more random conformation of the chains, and facilitating a loopier adsorption. The resulting films are thicker, with greater amounts of adsorbed polymer per layer. The spectra in Figure 4 illustrate this increased thickness, which has been confirmed using ellipsometry. At 1 M NaCl concentrations, the polyion that has adsorbed has actually adsorbed onto the EG surface, with the COOH surface remaining relatively bare, due to the competitive adsorption of salt ions as described above. The EG resist region is also affected by the presence of salt ions, and becomes a site for deposition at 1 M NaCl content. This phenomenon of reversed deposition has been described in a separate paper.[10]

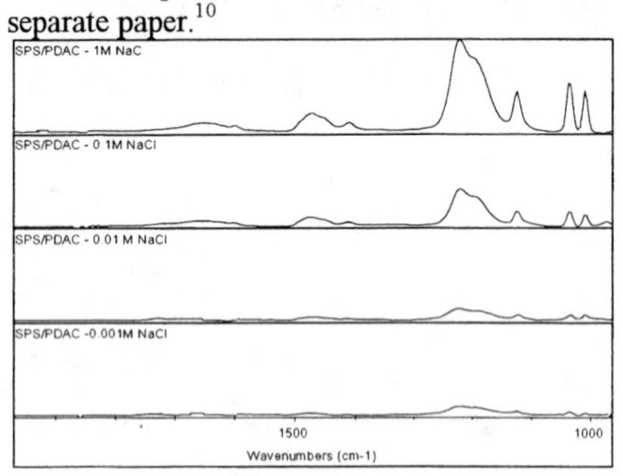

Figure 4

CONCLUSIONS

PM-IRRAS has been used to examine changes in polyion layer-by-layer films formed on SAMs surfaces. The addition of added electrolyte results in an increase in bilayer thickness, as observed by comparing relative absorption intensities. We have also confirmed that the COOH SAMs surface changes on exposure to polyion solutions of varying ionic content. At high NaCl content, specific interactions with salt ions interfere with hydrogen bonding within the SAM. These interactions result in the displacement of polyions during the adsorption process, and, at 1 M NaCl, the greatly reduced adsorption of polyion onto the COOH surface. This phenomenon is actually important to understanding the reversed deposition behavior observed in these systems at high salt content.

REFERENCES

1. R.G. Greenler, *J. Phys. Chem.* **44**, 310 (1966).
2. C.L. Hoffmann, and J.F. Rabolt, *Macromol.* **29**, 2543 (1996).
3. J.T. Young, F.J. Boerio, Z. Zhang, and T.L. Beck, *Langmuir* **12**, 1219 (1996).
4. T. Buffeteau, B. Desbat, and J.M. Turlet, *Appl. Spectr.* **45**, 380 (1991).
5. M.R. Anderson, M.N. Evaniak, and M. Zhang, *Langmuir* **12**, 2327 (1996).
6. H. Huehnerfuss, V. Neumann, and K.J. Stine, *Langmuir* **12**, 2561 (1996).
7. G. Decher, and J-D Hong *Mackromol. Chem., Macromol. Symp.*, **46**, 321 (1991).
8. S.L. Clark, M. Montague, and P.T. Hammond, *Supramolecular Science* **4**, 141 (1997).
9. V.G. Gregoriou, R. Hapanowicz, S. Clark, and P.T. Hammond, *Appl. Spectr.* **51**, 470 (1997).
10. S.L. Clark, M.F. Montague, P.T. Hammond, Macromolecules, 1997, accepted.
11. P.T. Hammond, and G.M. Whitesides *Macromol.* **28**, 7569 (1995).
12. S.L. Clark, M. F. Montague, P.T. Hammond, ACS Symposium Series, 1997, Organic Thin Films, Spring 1997, accepted.
13. R.G. Nuzzo, L.H. Dubois, and D. L. Allara, *J. Am. Chem. Soc.* **112**, 558, (1990).

Time-Domain Dynamic Opto-Rheology Study of Polymer Films Using Step-Scan FTIR Time-Resolved Spectroscopy (S^2FTIR TRS)

Haochuan Wang[1], Richard A. Palmer*[1], Christopher J. Manning[2], and Jon R. Schoonover[3]

[1]*Department of Chemistry, Duke University, Durham, NC 27708,* [2]*Manning Applied Technology, 121 Sweet Ave., Moscow, Idaho 83843,* [3]*Los Alamos National Lab, CST-4, Mail Stop J586, Los Alamos, NM 87545*

Step-scan Fourier transform infrared spectroscopy in conjunction with impulse stress on polymer films has been used to monitor dynamic rheological responses in "real time". A novel piezo-electrically-driven polymer microrheometer was employed to apply repetitive impulses to the polymer sample while time-domain spectra were recorded. Recent results include the study of both semi-crystalline polymers such as isotactic polypropylene (iPP) and elastomers such as Estane polyester/polyurethane copolymer and Kraton tri-block copolymer. The spectral changes of iPP are consistent with frequency-domain results. For iPP at room temperature, large differences in the response times of different absorption bands are not seen. However, the orientation response of the CH_3 rocking mode is slightly slower than the responses of the backbone modes. To our knowledge, this is the first reported successful step-scan FTIR time-domain dynamic polymer opto-rheology experiment. The advantages of the time-domain experiment over the frequency-domain experiment are also discussed briefly. This technique appears to be applicable to a variety of polymer samples, and examples from additional results are illustrated.

INTRODUCTION

In recent years, there has been a steadily increasing interest in the effect of mechanical stress on the IR spectra of polymers (1). This is the so-called IR opto-rheology or IR rheo-optics.

Since IR spectroscopy is sensitive to the orientation of transition dipoles, and since each functional group within a molecule generally has a different IR absorption frequency, the use of infrared spectra (particularly the dynamic infrared spectra) allows the responses to mechanical strain, of specific functional groups and of specific components of polymer blends, copolymers and composites to be determined and to be distinguished. This provides information on the molecular and submolecular origins of macroscopic rheological properties (2, 3).

The well-known advantages of interferometric (Fourier transform, or FT) methods for infrared spectroscopy make FTIR the choice for virtually all broad band infrared spectral measurements, compared with dispersive spectroscopic method (4). Application of time-resolved FTIR on dynamic rheo-optical study started with the rapid-scan FTIR methods more than a decade ago (5, 6). However, this approach was complicated by the difficulty of separating the time dependence of the sample response from that of the spectral multiplexing. Later, Noda *et al.* demonstrated mid-infrared dynamic opto-rheology measurements of isotactic polypropylene (iPP) using a monochromator (7). Although this work was very successful, the dispersive IR method required long measurement times, particularly for the study of broad spectral ranges.

In the last ten years, step-scan FTIR spectroscopy (S^2FTIR) has developed considerably. In a step-scan FTIR measurement, interferogram data are time-averaged retardation-by-retardation. The essential characteristic is that data are collected while the retardation is constant (or, in some cases, while the retardation is modulated about a fixed value). This greatly simplifies many types of dynamic spectroscopy by avoiding convolution of the time-dependence of the sample response with the time-dependence of the data collection process itself, as well as simplifying synchronization of the transient under study with the data collection process (2, 8).

Operationally, the application of S^2FT-IR to the study of time-dependent phenomena can be divided into time-domain and frequency-domain data collection modes (9). To date all the reported step-scan FT-IR/polymer rheo-optical measurements on the millisecond time-scale have been carried out in the frequency-domain, obtaining in-phase and quadrature components of the spectral response to sinusoidal perturbation (2, 7-12). Such results are often used to generate 2-D correlation spectra, which originated in

CP430, *Fourier Transform Spectroscopy:* 11th International Conference
edited by J.A. de Haseth

the work of Noda and Marcott (13-15). The stretchers used in much of the previous work utilize a cam drive mechanism, which can only generate smooth profiles of strain *vs.* time. As a result, use of time domain (time-resolved, or impulse/response) measurements in FTIR dynamic opto-rheology, in which the dynamic profile of the spectra change is recorded at a function of time, has not yet been reported in the open literature, although the first such report by our group is now in press (16). In that work, the rheo-optical properties of iPP was studied in time-domain, and a novel piezo-driven polymer stretcher (Polymer Modulator™, Manning Applied Technology, Moscow, ID) was employed to generate impulse stresses. The new device allows a variety of strain perturbation waveforms to be applied to samples, including repetitive impulse stretching. Hence, both frequency-domain and time-domain experimental approaches can be performed.

In this experiments reported here, the piezo-driven microrheometer, together with a step-scan FTIR spectrometer, is used again to measure dynamic impulse/response spectra of iPP film with much improved signal-to-noise ratio (SNR). Time-resolved data of Estane polyester/polyurethane copolymer are also measured. The dynamic IR spectra for both polymers are recorded with good SNR, and the spectral responses of iPP at selected absorption bands are studied in details. The time-domain data are also compared with the previous frequency-domain results.

EXPERIMENTAL

A Bruker IFS 66/DSP spectrometer (Bruker Optics, Billerica, MA) with a DC-coupled MCT detector were used for data collection. A personal computer was used to control the IFS 66/DSP, which is set to work in the step-scan time-resolved data collection mode.

The Polymer Modulator™ by Manning Applied Technology is use to generate impulse stresses on the polymer films. A square wave is used to drive the microrheometer to generate impulse stretching and relaxing on the polymer films. The stretcher responds fully to the frequency components of the applied waveform up to 70 Hz. The amplitude of stretching is set as the maximum, which is close to 75 μm.

A pulse generator controls the measurement, generating both the square waveforms for the stretcher and the signals necessary to trigger the spectrometer computer to sample the detector signals, as well as the pulse signaling the spectrometer when to step the interferometer mirror. During the experiments, when the scanning mirror has been stepped to a particular retardation, impulse signals (0~4V) are sent to the stretcher controller from the pulse generator. The controller amplifies the signals to 0~100V and applies them to the PZT to drive the stretching motion. Simultaneously, the detector signal is sampled by the internal spectrometer ADC. The stretching transient is repeated 500 times and the data points corresponding to the same time during each cycle are averaged to improve the SNR. When the data collection at a given retardation is completed, the interferometer mirror is stepped to the next retardation, where, after a short mirror stabilization delay time, the stretching sequence is repeated.

The sample of iPP film, approximately 30-μm thick, previously uniaxially stretched 4x1, was obtained from Procter & Gamble Co. The Estane sample is obtained from Los Alamo National Lab and is also prestretched to approximately 5X1. In both experiments, A 2000 cm^{-1} low-pass optical filter and a wire grid polarizer (PN 186-0243, Perkin-Elmer Co., Norwalk, CT) were mounted in front of the polymer sample. The low-pass filter allows decreased data collection time by preventing aliasing. An increased retardation between steps can then be used to decrease the total number of steps required for a particular resolution. The polymer spectral data reported here were collected to 8 cm^{-1} resolution with the polarizer oriented parallel to the stretching direction.

The square wave is set, so that the voltage jump from 100 V to 0 V on the PZT at 2nd ms, which increases the stress on the polymer film; the voltage jump from 0 V to 100 V at 11 ms, which decreases the stress back to the original state. At each step 1000 samples with a time resolution of 20 μs are recorded over the 20 ms stretching period. After data collection is complete, the interferograms are extracted from the 3-D data package and used to generate a stack plot of 1000 single beam spectra, or time "slices".

The first 75 time slices, corresponding to the 1.5 ms period before pulse-on, while the polymer remains in the more stretched state, were averaged, and the result was used as the reference spectrum ($I_{reference}$). Every 10 slices (corresponding to 0.2 ms time periods) of the iPP data and every 25 slices (corresponding to 0.5 ms time periods) of the Estane data were then averaged and used as the sample spectra at time *t* ($I_{sampled\ at\ time\ t}$). Equation 1 was then used to calculate the absorbance difference spectra at time *t* ($\Delta A_{at\ time\ t} = A_{sampled\ at\ time\ t} - A_{reference}$) during the pulse-on and pulse-off phases. Finally, the response curves of the selected absorption bands in the absorbance difference spectra were plotted.

$$\Delta A_{at\ time\ t} = \log (I_{reference} / I_{sampled\ at\ time\ t}) \quad (1)$$

RESULTS AND DISCUSSION

iPP:

The static absorbance spectrum of the iPP sample is shown in Figure 1. The spectral range 1250 cm^{-1} ~ 800 cm^{-1} was analyzed. This is the region where most of the backbone stretching and deformation vibrations absorb, which are known to be most affected by external deformation (17).

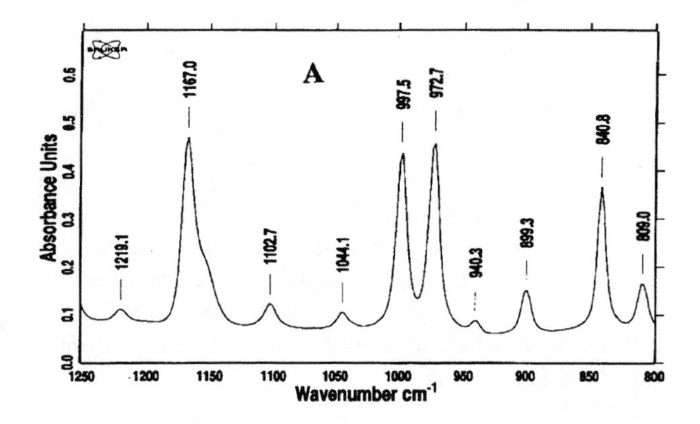

FIGURE 1. Absorption spectrum (between 1250 cm^{-1} and 800 cm^{-1}) of static iPP film in stretched state, obtained in the continuous-scan mode.

Figure 2 is the 3-D plot of the time-resolved absorbance difference spectrum. In this dynamic ΔA spectrum, between the 2 ms and 11 ms, only the 1167 cm^{-1}, 998 cm^{-1}, 973 cm`1, and 841 cm^{-1} bands appear, indicating that in the 800 cm^{-1} ~ 1250 cm^{-1} range only these four vibrational modes have measurable spectral changes in response to the change in stress. This is consistent with previous results. The 1167 cm^{-1}, 973 cm^{-1}, and 841 cm^{-1} bands are bisignate (derivative-like), with their negative components to higher energy. This indicates that these three absorbance bands shift to lower wavenumber when the polymer is more stretched. However, the 998 cm^{-1} absorbance band shows only an increase of intensity; it does not exhibit a frequency shift.

Theoretical calculation of the effect of strain on iPP has already been carried out and elaborated in detail by Tashiro *et al* (18). The absorption bands at 1167 cm^{-1}, 973 cm^{-1}, and 841 cm^{-1} are mostly associated with the skeletal backbone modes. When a tensile strain is applied along the polymer chain, this tensile strain will cause force field changes in the backbone modes, resulting in frequency shifts in these bands. Simultaneously, reorientation of the backbone chain occurs. Complete interpretation of the dynamic spectral change is not possible without a satisfactory separation of these two effects. On the other hand, the absorption band at 998 cm^{-1} consisting mostly of the rocking of the methyl side-group, exhibits almost no frequency shift, which can reasonably be attributed only to the orientation effect of the side chain.

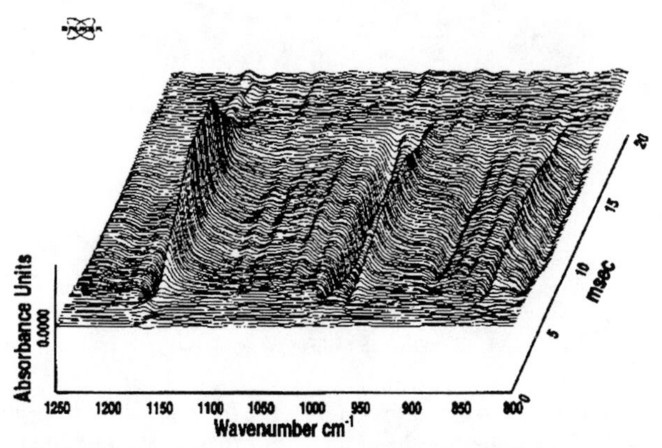

FIGURE 2. Time resolved absorption difference spectrum of iPP, obtained in the step-scan mode.

Considering the fact that the piezo element also has a response time on the millisecond time scale, prior to the dynamic measurements of polymers, a partial beam block is run to determine the instrument response curve, in effect, zero time. The partial beam block is clamped in the moving jaw of the stretcher, so as to chop the IR intensity with exactly the same speed as the jaw motion. To guard against the possibility that the additional load imposed by the polymer film might cause different delays, the polymer film is clamped in the stretcher jaws when the beam block measurement is made. The dynamic absorption difference spectra of the partial beam block experiment is analyzed in the same way as the polymer data.

The peak values of all the negative bands in the iPP stretching experiment (at 1170.3 cm^{-1}, 999.4 cm^{-1}, 975.3 cm^{-1}, and 843.1 cm^{-1}) are plotted as a function of time in Figure 3. The beam block data were treated identically, and these results are also plotted in Figure 3. From this comparison, it can be seen that the polypropylene sample responds to the change in stress so fast that it is difficult to separate the response curve of the sample from that of the stretcher. Additionally, as expected, significant differences in the response times for different absorbance bands are not seen. However, detailed study of the data does indicate that the 998 cm^{-1} band intensity change appears to lag behind the spectral change at 1170.3 cm^{-1} during the pulse-off (strain decrease) process.

Estane:

For Estane, the spectral range 1950 cm^{-1} ~ 650 cm^{-1} was analyzed. The experiment of Estane is performed in the same way as the iPP experiment and the final data of Estane is also analyzed is the same way as iPP data. The 3-D plot of the time-resolved absorbance difference spectrum of Estane is plotted in Figure 4.

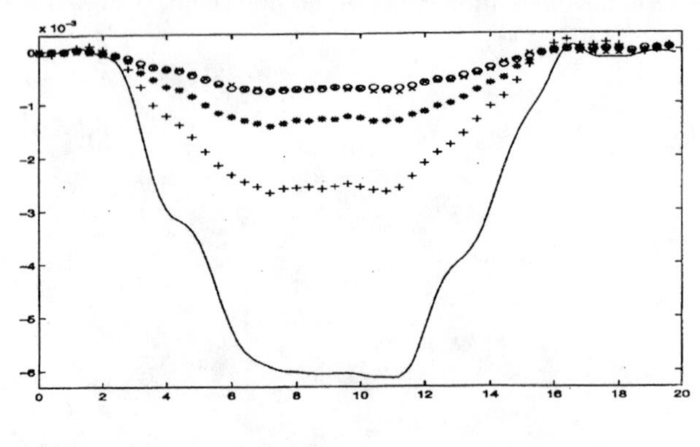

Time in Milliseconds

FIGURE 3. Time-resolved spectral response as absorption difference values for the beam block experiment (solid line) and for particular wavenumbers for the iPP stretching experiment: at 1170.3 cm^{-1} (+), 999.4 cm^{-1} (*), 975.3 cm^{-1} (o), 843.1 cm^{-1} (x) as a function of time.

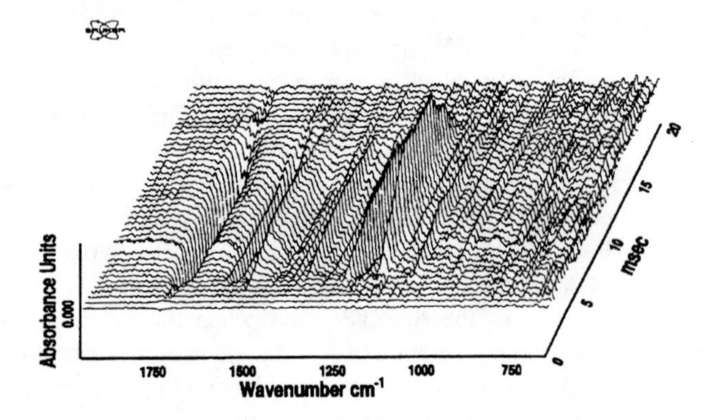

FIGURE 4. Time resolved absorption difference spectrum of Estane, obtained in the step-scan mode.

The SNR of the Estane data is not comparable with the iPP data and detailed study of the time-domain responses of different absorption band has not been completed. Experiments to improve the SNR are currently underway, so that both quantitative and qualitative analysis of the different responses can be carried out.

The objective of this investigation is to study the feasibility of the time-domain dynamic polymer opto-rheology using step-scan FTIR. The time domain data here, particularly the iPP time-domain data has been compared with the previous frequency-domain data. It is clear that the time-resolved data are consistent in this respect with the phase-resolved results. The shift of most backbone bands to lower wavenumber as the stress increases is also consistent with the earlier results of Siesler and with some simple analogies with strained aliphatic rings (1,19). The SNR of our time-resolved spectra is 21 peak-to-peak, comparable with the frequency-domain result for the same data collection time. Considering the fact that the time-resolved signal is spread over many spectra, rather than just in-phase and quadrature spectra, this SNR is quite good relative to the previous frequency-domain experiments.

The comparison of the time-domain experiment with previous frequency-domain method provides the conclusion that both methods have their advantages and disadvantages. Time domain measurement is conceptually simpler and is able to record both the exication and decay as a response to the impulse stretching and relaxing separately; whereas the use of lock-in amplifiers or digital signal processing (DSP) in frequency-domain measurement provides a better degree of noise rejection. Although upon inspection it appears that the frequency-domain method provides a better time resolution than the time-domain method. However, because of the underlying equivalence of the two methods, particularly in the limiting case of response to a sinusoidal perturbation recorded as a time-resolved sequence, there should be no difference between the two methods. In addition, the time-domain data obtain on Estane copolymer indicate that the time-domain technique is applicable to a variety of polymer samples.

ACKNOWLEDGEMENTS

Acknowledgements are made to the Lord Foundation of North Carolina, to the Bruker Instruments, Inc., to Manning Applied Technology, to Los Alamos National Laboratory, and to Duke University for the support of this work.

REFERENCES

1. H. W. Siesler and K. Holland-Moritz, *Infrared and Raman Spectroscopy of Polymers*, (Marcel Dekker, New York, 1980), p.266.
2. R. A. Palmer, James L. Chao, Rebecca M. Dittmar, Vasilis G. Gregoriou, and Susan E. Plunkett, Appl. Spec. **47**, 9 (1993).
3. C. J. Manning, Ph. D. Dissertation, Duke University, Durham, NC, (1991).
4. P. R. Griffiths and J. A. de Haseth, *Fourier Transform Infrared Spectrometry*, (New York: John Wiley and Sons, 1986).
5. W. G. Fateley and J. L. Koenig, J. Polym. Sci., Polym. Lett. Ed. **20**, 445 (1982).
6. J. A. Graham, W. M. Grim III, and W. G. Fateley, J. Mol Struct. **113**, 311 (1984).
7. I. Noda, A. E. Dowrey, and C. Marcott, Appl. Spectrosc. **42**, 203 (1988).
8. M. J. Smith, C. J. Manning, R. A. Palmer, and J. L. Chao. Appl. Spectrosc. **42**, 546 (1988).
9. R. A. Palmer, Spectroscopy, **8**, 2 (1993).
10. R. A. Palmer, C. J. Manning, J. L. Chao, I. Noda, A. E. Dowrey, and C. Marcott, Appl. Spectrosc. **45**, 12 (1991).

11. B. O. Budevska, C. J. Manning, P. R. Griffiths, and Robert T. Roginski, Appl. Spectrosc. **47**, 11 (1993).
12. V. G. Gregoriou, I. Noda, A. E. Dowrey, C. Marcott, J. L. Chao, and R. A. Palmer, J. Polym. Sci. B. Polym. Phys. **31**, 1769 (1993).
13. I. Noda, J. Am. Chem. Soc. **111**, 8116 (1989).
14. I. Noda, Appl. Spectrosc. **44**, 550 (1990
15. C. Marcott, paper presented at 1992 FACSS Symposium, Philadelphia (1992).
16. H. Wang, R. A. Palmer, C. J. Manning, Appl. Spectrosc. **51**, (1997), paper in press.
17. R. G. Snyder, and J. H. Schachtschneider, Spectrochim. Acta 20, 853 (1964).
18. K. Tashiro, S. Minami, G. Wu, and M. Kobayashi, Journal of Polymer Science: Part B: Polymer Physics, **30**, 1143 (1992).
19. D. Lin-Vein, N. Colthup, W. G. Fataley, and J. G. Grasselli, *Infrared and Raman Characteristic Frequencies of Organic Molecules* (Academic Press, New York, 1991), p. 19.

Infrared Spectroscopy of a Polyurethane Elastomer Under Thermal Stress

A. L. Marsh[1] and J. R. Schoonover[2]

[1]*Department of Chemistry, Hampden-Sydney College, Hampden-Sydney, Virginia 23943*
[2]*Chemical Science and Technology Division, Los Alamos National Laboratory, Los Alamos, New Mexico 87545*

FTIR spectroscopy was used to study changes in hydrogen bonding in estane, a polyurethane elastomer used as a binding agent in high explosive systems, as a function of temperature. Hydrogen bonding in estane has been observed to decrease with an increase in temperature.

INTRODUCTION

Infrared spectroscopy has been used to study morphology, particularly changes in hydrogen bonding, in polyurethanes and polyurethane elastomers. This technique is useful because certain absorption bands characteristic of polyurethanes are influenced by the environment in which the polymer is located. Specifically, changes in hydrogen bonding can be observed through the N-H and C=O stretching regions of the spectrum. Hydrogen bonding is known to occur between the N-H group of the urethane segment and the C=O group of the urethane or ester segment, and is also known to decrease with an increase in temperature. Several different types of polyurethanes have been studied by infrared spectroscopy. These types include poly(ether-urethanes) (1-15), poly(ester-urethanes) (3, 16), poly(urethane-urea)s (17, 18), and a simple polyurethane(19).

Srichatrapimuk and Cooper studied temperature dependent behavior of poly(ether-urethane) and poly(ester-urethane) elastomers as a function of hard and soft segment lengths (3). The IR absorptions in the N-H and C=O regions were monitored with temperature change in order to provide a quantitative measurement of phase separation. The enthalpy of hydrogen-bond dissociation was determined from the fraction of bonded groups at different temperatures. The H-bonded N-H stretch was located at 3320 cm^{-1}, and the non-bonded N-H stretch was a shoulder at 3420 cm^{-1}. The intensity of the bonded N-H stretch was observed to decrease, or shift to higher frequency, with an increase in temperature, indicating a decrease in hydrogen bonding. No splitting in the C=O region was observed in the poly(ester-urethanes). About 80% of N-H groups were calculated to be hydrogen-bonded at room temperature. More interurethane bonding occurred in polymers with longer hard and soft segments, and the hard-soft bonding was determined to dissociate first.

Siesler observed similar results as Srichatrapimuk and Cooper observed(16). He conducted rheo-optical FTIR experiments of poly(ester-urethanes) at different temperatures by stretching and relaxation along a single axis. Temperature dependence of hydrogen bonding and structural organization of the hard segments were also studied. The change in the N-H region was attributed to a dissociation of hydrogen bonds and the small change in the C=O region was attributed to the functional group being less displaced. A shift of the N-H bending and C-N stretching band to lower frequency was observed, and was attributed to the inverse effect of hydrogen bonding on deformation vibrations.

The present work is a study of estane, a poly(ester-urethane) used as a structural support in industrial applications and as a binding agent in various high-explosive systems. Over time, possibly due to the presence of certain environmental agents, changes develop in the bulk properties of estane and in the mechanical and sensitivity properties of the explosive composites. The relationship between the bulk properties and the molecular structure of estane are not well understood, and questions regarding the effects of aging of estane-based, high-explosive binding matrices have been difficult to answer. FTIR studies have been used to fundamentally relate molecular vibrations of estane, in its pure form at various temperatures, to its macroscopic structural and binding properties.

EXPERIMENTAL

Estane 5703, a polyurethane elastomer consisting of urethane hard segments and ester soft segments, was obtained from B.F. Goodrich. The repeating unit of this polymer is

CP430, *Fourier Transform Spectroscopy:* 11th International Conference
edited by J.A. de Haseth

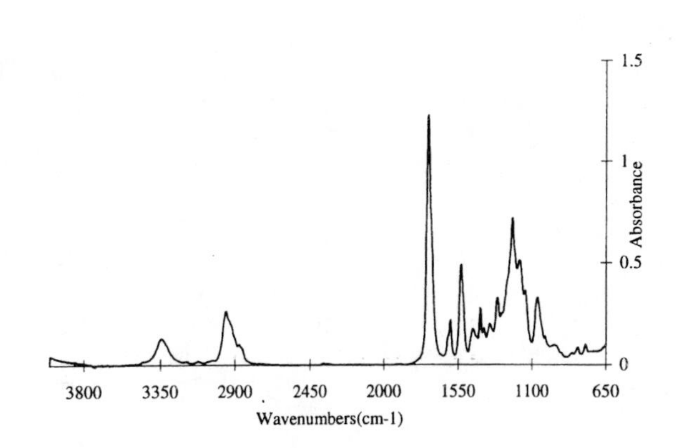

FIGURE 1. Infrared absorption spectrum of estane at 32.0 °C.

$$\{[O(CH_2)_4OOCNH(C_6H_4)CH_2(C_6H_4)NHCOO]_m\text{-}[(CH_2)_4OOC(CH_2)_4COO]_n\}$$

where m = 4-6 and n = 1-3. Samples were prepared by casting a dilute solution of the polymer onto 13mm x 2mm NaCl windows. Sample thickness was sufficient to yield an absorbance of under 1.5 absorbance units.

Samples were mounted between two NaCl windows in an aluminum block cell which contained a cartridge heater and was connected to a temperature controller. The temperature was monitored through a Type K thermocouple placed next to the sample in the temperature cell. Temperature readouts were to an accuracy of 0.1 °C. Spectra were taken with a Nicolet 20SXB FTIR spectrometer. 200 scans were measured with resolution of 8cm^{-1}. Difference spectra were calculated by subtracting the spectrum at 100.0 °C from the spectrum at 32.0 °C.

RESULTS AND DISCUSSION

Figure 1 is a spectrum of estane taken at 32.0 °C.

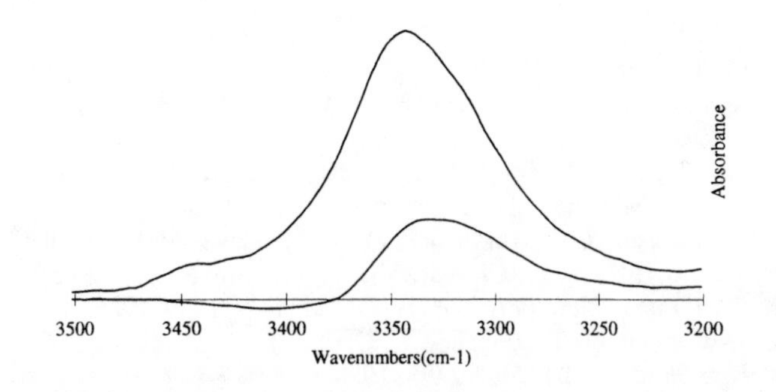

FIGURE 2. Relative change in N-H stretching region at 100.0 °C.

TABLE 1. Absorption Band Assignments

Frequency (cm-1)	Relative Intensity	Assignment
3440	weak, shoulder	free N-H stretch
3340	strong	bonded N-H stretch
3190	weak	CIS-TRANS bonded N-H stretch
3120	weak	overtone of 1531 cm^{-1}
2951	strong	CH_2 asymmetric stretch
1732	very strong	free and bonded C=O stretch in urethane and ester
1597	strong	C=C(benzene) stretch
1531	very strong	N-H bend/C-N stretch
1415	strong	C-C(benzene) stretch
1315	strong	N-H bend, C-N stretch, C-H bend
1223	strong	N-H bend/C-N stretch
1180	strong	C-O-C(ester) stretch
1068	strong	C-O-C(hard) stretch

Table 1 summarizes the band assignments for this spectrum as made by Srichatrapimuk and Cooper (3). The N-H region consists of two bands, one at 3340 cm^{-1} representing hydrogen-bonded N-H stretching and the other at 3440 cm^{-1} representing non-hydrogen-bonded, or "free," N-H stretching. The "free" N-H band appears as a shoulder on the higher frequency side of the H-bonded N-H band. A strong band at 1732 cm^{-1} represents "free" and H-bonded C=O stretch in both the urethane and ester segments. Two bands at 1531cm^{-1} and 1223 cm^{-1} represent both N-H bending and C-N stretching.

The relative change in the N-H stretching region at 100.0 °C is shown in Fig. 2. The difference spectrum shows a decrease in intensity of the bonded N-H stretch and an increase in intensity of the "free" N-H stretch. The N-H band is shifting to a higher intensity, indicating a decrease in hydrogen bonding.

The relative change in the C=O stretching region at 100.0 °C is shown in Fig. 3. The difference spectrum shows three bands are changing in intensity. Hydrogen bonding between N-H of the urethane segment and C=O of both the urethane and ester segments must be decreasing. The overall band is shifting to a higher frequency with the

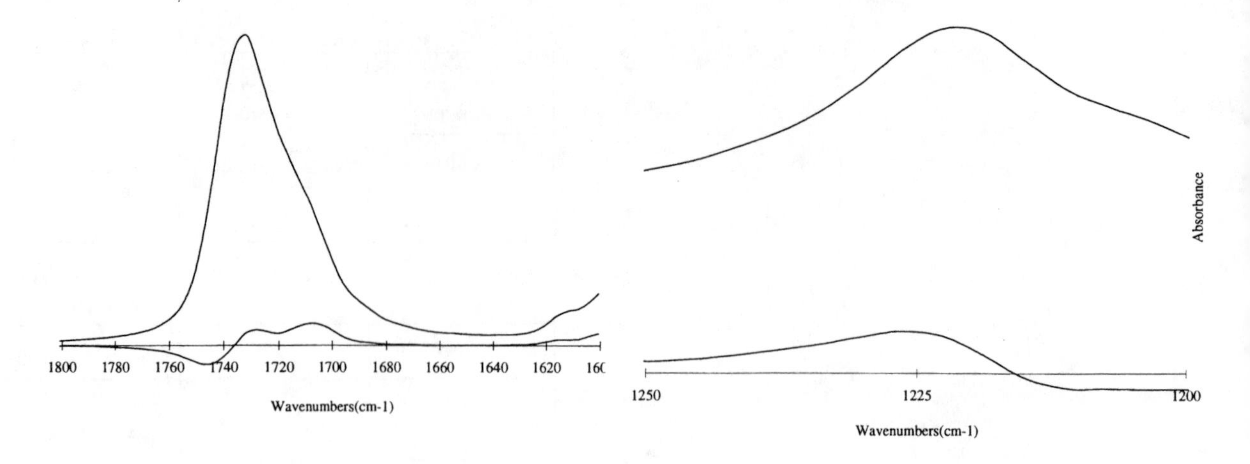

FIGURE 3. Relative change in C=O stretching region at 100.0 °C.

FIGURE 5. Relative band changes in the lower-frequency N-H bending and C-N stretching region at 100.0 °C.

measured temperature. The relative change in intensity, however, is small compared to the change for the N-H stretching region.

The relative changes in the N-H bending and C-N stretching regions are shown in Fig. 4 and Fig. 5. These bands are shifting to a lower frequency, due to the inverse effect as noted by Siesler (16). These relative changes are not small compared to the changes in the C=O stretching region.

CONCLUSION

Morphological changes in a poly(ester-urethane) were observed by infrared spectroscopy as changes in absorbance intensity of N-H, C=O, and C-N bands as a function of temperature. The N-H stretching and C=O stretching bands shifted towards higher frequency with an increase in temperature, while the two N-H bending and C-N stretching bands shifted towards lower frequency with an increase in temperature. Hydrogen bonding, both interurethane and hard-soft segment, decreases with temperature. The structural properties of estane, therefore, change with temperature. Two-dimensional cross-correlatin analysis will next be applied to the spectra to resolve overlapped bands and to view related changes with temperature.

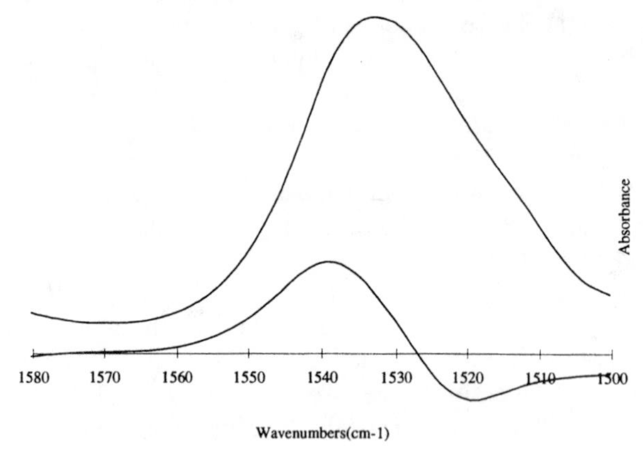

FIGURE 4. Relative band change in higher-frequency N-H bending and C-N stretching region at 100.0 °C.

ACKNOWLEDGMENTS

Funding for this experiment was provided through a Laboratory Directed Research and Development project at Los Alamos National Laboratory on polymer aging.

REFERENCES

1. Paik Sung, C. S.; Schneider, N. S., *Macromolecules*, **8**, 68-73, 1975.
2. Paik Sung, C. S.; Schneider, N. S., *Macromolecules*, **10**, 452-458, 1977.
3. Srichatrapimuk, V. W.; Cooper, S. L., *J. Macromol. Sci., Phys.*, **B15**, 267-311, 1978.
4. Senich, G. A.; MacKnight, W. J., *Macromolecules*, **13**, 106-110, 1980.
5. Brunett, C. M.; Hsu, S. L.; MacKnight, W. J., *Macromolecules*, **15**, 71-77, 1982.
6. Christenson, C. P.; Harthcock, M. A.; Meadows, M. D.; Spell, H. L.; Howard, W. L.; Creswick, M. W.; Guerra, R. E.; Turner, R. B., *J. Polym. Sci., Polym. Phys. Ed.*, **24**, 1401-1439, 1986.
7. Koberstein, J. T.; Gancurz, I.; Clarke, T. C., *J. Polym. Sci., Polym. Phys. Ed.*, **24**, 2487-2498, 1986.
8. Lee, H. S.; Wang, Y. K.; Hsu, S. L., *Macromolecules*, **20**, 2089-2095, 1987.

9. Coleman, M. M.; Skrovanek, D. J.; Hu, J.; Painter, P. C., *Macromolecules*, **21**, 59-65, 1988.
10. Meuse, C. W.; Yang, X.; Yang, D.; Hsu, S. L., *Macromolecules*, **25**, 925-932, 1992.
11. Hong, J. L.; Lillya, C. P.; Chien, J. C. W., *Polymer*, **33**, 4347-4351, 1992.
12. Wang, F. C.; Feve, M.; Lam, T. M.; Pascault, J. P., *J. Polym. Sci., Polym. Phys. Ed.*, **32**, 1305-1313, 1994.
13. Lee, H. S.; Hsu, S. L., *J. Polym. Sci., Polym. Phys. Ed.*, **32**, 2085-2098, 1994.
14. Reynolds, N.; Spiess, H. W.; Hayen, H.; Nefzger, H.; Eisenbach, C. D., *Macromol. Chem. Phys.*, **195**, 2855-2873, 1994.
15. Goddard, R. J.; Cooper, S. L., *Macromolecules*, **28**, 1390-1400, 1995.
16. Siesler, H. W., *Polym. Bull.*, **9**, 471-478, 1983.
17. Bummer, P. M.; Knutson, K., *Macromolecules*, **23**, 4357-4362, 1990.
18. Teo, L. S.; Chen, C. Y.; Kuo, J. F., *Macromolecules*, **30**, 1793- 1799, 1997.
19. Coleman, M. M.; Lee, K. H.; Skrovanek, D. J.; Painter, P. C., *Macromolecules*, **19**, 2149-2157, 1986.

Polymer Chain Disorder
Effect on Chemical and Spectroscopic Properties

P. Bruni[1], C. Conti[1], E. Giorgini[1], A. Mar'in[2], G. Tosi[1*]

[1] *Dipartimento di Scienze dei Materiali e della Terra, Università degli Studi di Ancona, Via Brecce Bianche, I-60131, ANCONA, Italy.*
[2] *United Institute of Chemical Physics Russian Academy of Sciences, 117334, MOSCOW, Russia*

The oxidation of high density polyethylene precipitated from chlorobenzene or decane, in presence of tetrakis[methylene-3-(3′,5′-di-*tert*-butyl-4′-hydroxy-phenyl)-propionate]methane, has been studied. Inhibition by the antioxidant is more efficient when the polymer has been obtained from dilute chlorobenzene solutions. FT-IR spectra indicate that a less ordered structure is obtained when decane is the solvent. The change of disorder in the amorphous state on drawing is also discussed.

INTRODUCTION

Reactant molecules, when dissolved in a polymer, show a low degree of mobility and of homogeneity in their spatial distribution (1). Most of a dissolved additive A is sorbed by the centres Z around chain entanglements, forming immobile complexes AZ, while only a negligible part is present as really dissolved (and hence it is mobile). A chemical reaction taking place on a polymer is then influenced by units of chain disorder. For example, retardation of polymer oxidation is based on the reaction of macroradicals and/or hydroperoxide groups, attached to the polymer chains with antioxidant molecules. Because the macroradicals are virtually immobile, only mobile antioxidants can participate in retardation of the process. The precipitation of the polymer from different solvents can be also used to change the polymer structure and the efficiency of antioxidants. Polymer packing in the solid and in melt should also depend on solvent and on concentration of the polymer in solution at precipitation: lower concentration of macromolecules means lower intermolecular interactions (2-4). In this work, we report a FT-IR analysis on high density Polyethylene (PE), precipitated from chlorobenzene (CBZ) and decane (DEC), submitted to oxidation and tensile stress. The experiments were performed on the polymer melt at 180°C in the presence of antioxidant and in solid state without antioxidant.

EXPERIMENTAL

Materials. High density powder PE (MW ≈ $3x10^5$, Russian production) was dissolved in chlorobenzene or decane by stirring under nitrogen at 130°C (2h) and then poured in cold water and filtered. PE films (140-50 μm) were prepared by pressing the powder at 140°C for two minutes. **Infrared determinations**. The spectra have been obtained with a Nicolet 20-SX FT-IR spectrometer (DTGS detector) equipped with a Spectra Tech. Multiple Internal Reflectance (DRIFT) accessory for measurements in the solid state. Resolution 4 cm^{-1}, 250 scans. Data handling has been achieved with a Galactic software package. The area of the peaks in the amorphous region (1400-1300 cm^{-1}), was calculated from the convoluted spectra and referred to the same sample thickness; in this way, samples obtained by different procedures can be compared. For **oxidation experiments**, PE powder was mixed with different amounts of purified tetrakis[methylene-3-(3′,5′-di-*tert*-butyl-4′-hydroxyphenyl) -propionate]methane, TMM (Irganox 1010), dissolved in benzene and dried under vacuum. The polymer oxidation was monitored by measuring oxygen consumption. For **sorption studies**, polymer powder was mixed with different amounts of 2,6-di-*tert*-butyl-4-methyl-phenol (BMP) dissolved in benzene and dried in air; the samples, sealed in glass tubes, were heated at 180°C for 2 hours and then rapidly immersed in liquid nitrogen: BMP condensed on the walls was extracted by cold heptane and its amount determined by UV spectrophotometry.

RESULTS AND DISCUSSION

We have determined oxygen consumption during oxidation of PE samples containing 0.15% of Irganox 1010 (TMM). and precipitated from chlorobenzene and decane with different concentrations of PE. The results indicate that the induction interval (time corresponding to the slow oxidation) is higher in samples precipitated from chlorobenzene and from dilute solutions: it means that samples precipitated from chlorobenzene are more stable to oxidation than the ones from decane (Fig. 1). The lower

----

* To whom correspondence should be addressed (E-mail: tosi@popcsi.unian.it; Fax: ++39-71-2204714)

CP430, *Fourier Transform Spectroscopy:* 11th International Conference
edited by J.A. de Haseth

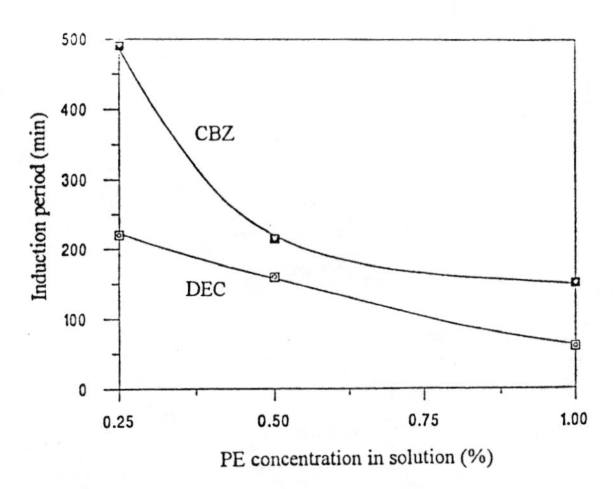

FIGURE 1. Induction period of the oxidation of PE at 180°C in the presence of TMM as a function of the concentration used for precipitation of PE from chlorobenzene (CBZ) and decane (DEC).

concentration of antioxidant needed to retard the oxidation, is higher in samples precipitated from decane and decreases by lowering PE concentration at precipitation (1). To verify these changes in the melt, the sorption of vapours of 2,6-di-tert-butyl-4-methylphenol (BMP) on PE was studied. Fig. 2 shows the sorption isotherms of BMP on PE samples in coordinates 1/[BMP]p - 1/[BMP]m. The linear correlation means that the sorption of the antioxidant follows the Langmuir law if one assumes the existence of centres responsible for the sorption of the additive. A higher concentration of sorption centres is obtained in samples prepared from decane solution, in more concentrated solutions and by increasing the rate of precipitation. Table 1 shows the peaks intensity in the region 1400-1300 cm^{-1} of the spectra of PE films prepared by different methods. The absorbances at 1303 and 1367 cm^{-1}, corresponding to *-GTG-*, *-GTG'* conformational defects (*G* and *G'* represent the two *gauche* and *T* the *trans*

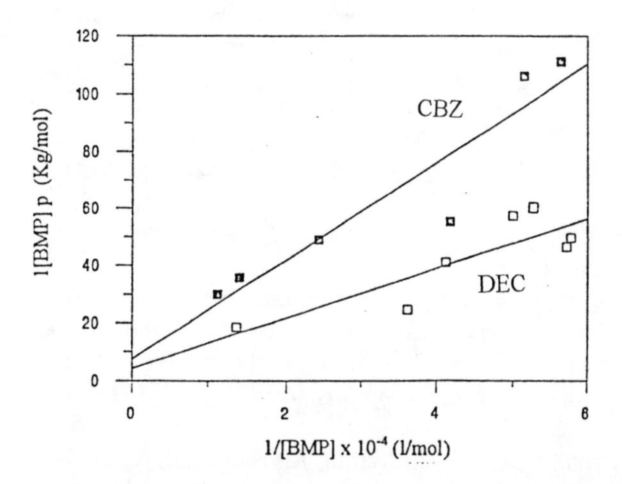

FIGURE 2. Sorption of 2,6-di-tert-butyl-4-methylphenol (BMP) on PE samples precipitated from chlorobenzene (CBZ) and decane (DEC).

TABLE 1. Relative peak intensities in the 1400-1300 cm^{-1} region and BMP solubility (%) in samples precipitated from chlorobenzene and decane.

Solvent	1367	1351	1301	Solubility
Chlorobenzene	3.01	2.34	4.94	0.628
Decane	4.18	3.14	6.17	0.702

conformations), and at 1352 cm^{-1} (*GG* conformation) are higher in samples precipitated from decane solution, that correspond to less ordered polymer structure. Orientation of PE precipitated from the two solvents at small drawing (20%) is accompanied by some increase in absorption at 1367,1351 and 1301 cm^{-1} (that represent the disorder in the amorphous state) with a similar behaviour (5); at higher drawing, these intensities continue to increase for samples prepared from decane and decrease for samples precipitated from chlorobenzene. Fig. 3 shows the correlation between the ratio band area/film thickness versus drawing percentage for the band at 1367 cm^{-1}.

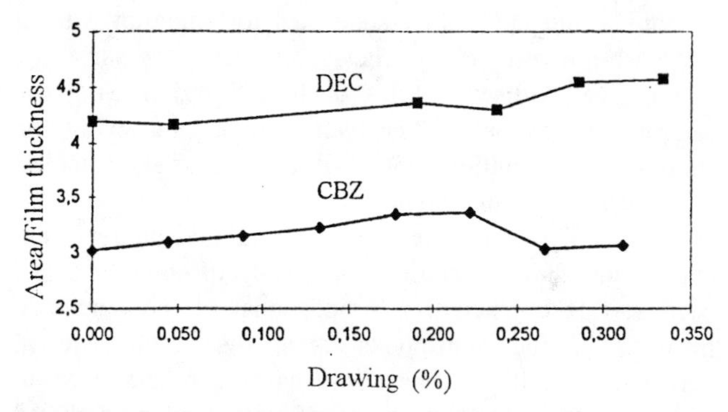

FIGURE 3. The dependence of the ratio band area/film thickness from drawing (%) in the band at 1367 cm^{-1}.

Figure 4 shows the change of the ratio (band area at 720 - band area at 730)/film thickness *vs* elongation in PE samples precipitated from chlorobenzene and decane, at a qualitative level. In samples of PE precipitated from chlorobenzene the area of the band at 720 changes more than the one at 730 cm^{-1} (6).

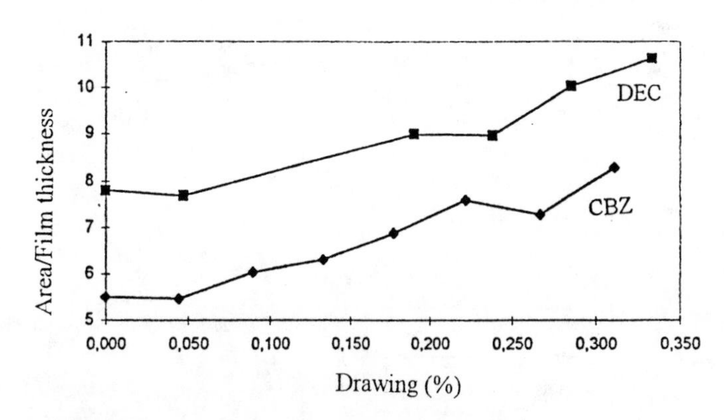

FIGURE 4. The dependence of the ratio (band area at 730 - band area at 720)/film thickness from drawing (%), in samples precipitated from chlorobenzene (CBZ) and decane (DEC).

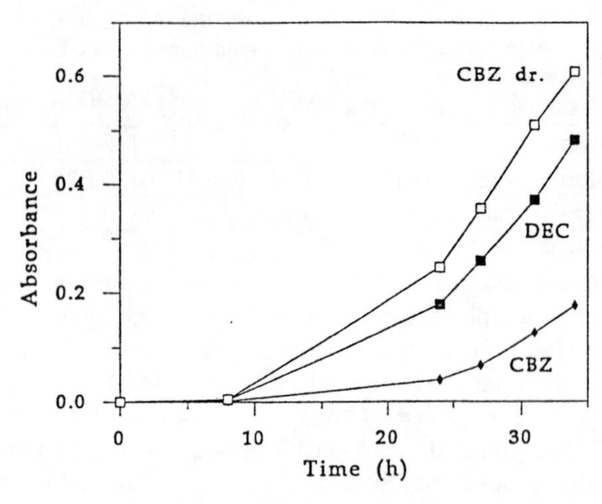

FIGURE 5. Time dependence of absorbance of the carbonyl band at 1720 cm^{-1} in PE samples precipitated from chlorobenzene (CBZ), decane (DEC) and chlorobenzene after drawing (CBZ dr.).

In Fig. 5, is shown the increase of the intensity of the carbonyl band in samples submitted to oxidation without and with drawing (15%). The solvent being the same, the amount of oxidised species is higher on drawing the sample. In addition, it is noteworthy that, on drawing, the solubility of the antioxidant, BMP in both type of samples, passes through a maximum (Fig.6).

The elasticity coefficient in samples precipitated from decane, is 15% higher than the one from chlorobenzene; because this coefficient is proportional to the degree of disorder in the amorphous state, the results are in agreement FT-IR findings. We have also attempted to determine the tensile strength of specimens precipitated from chlorobenzene and decane (Fig. 7). The elasticity of the films precipitated from decane, results 15% higher than the one of films precipitated from chlorobenzene. The difference can be explained assuming that the elasticity is proportional to the degree of disorder in the amorphous

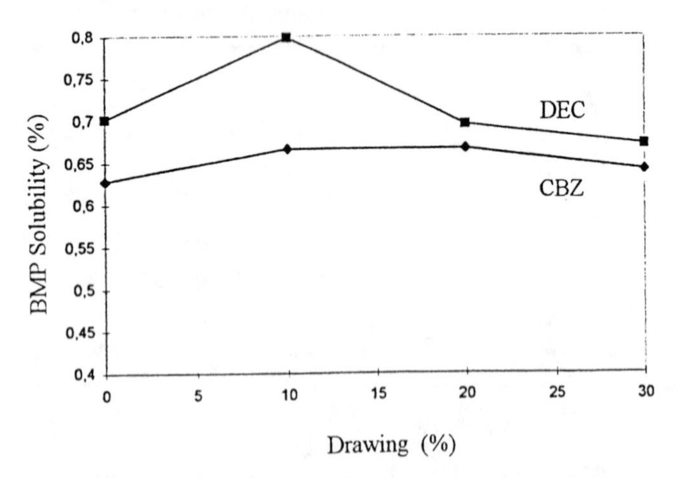

FIGURE 6. BMP solubility (%) *vs* drawing (%) in samples precipitated from chlorobenzene (CBZ) and decane (DEC).

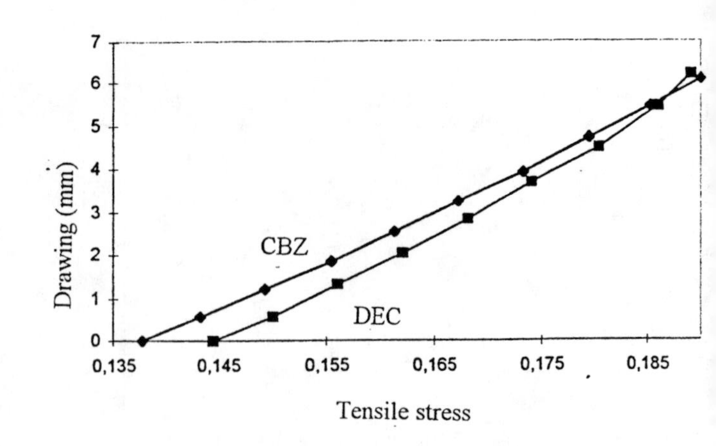

FIGURE 7. Drawing (mm) *vs* tensile stress (Newton) in samples precipitated from chlorobenzene (CBZ) and decane (DEC).

state. To a higher degree of disorder in PE precipitated from decane corresponds a higher elasticity.

CONCLUSIONS

It has been shown that changing the condition of polymer precipitation, i.e. using different solvents, it is possible to obtain polymer samples with different density of macromolecular packing and different concentration of chain entanglements. This difference in the physical structure affects the polymer stability to oxidation. FT-IR analysis allows to understand not only morphological changes in the structure but also the mechanical behaviour of polymers precipitated from various solvents.

ACKNOWLEDGMENTS

The Authors thank Prof. Francesco Branca and his group (University of Rome) for mechanical determinations.

REFERENCES

1. Mar'in, A., Shlyapnikov, Yu. A., *European Polymer J.*, **23**, 633-38 (1987); Bruni, P., Conti, C., Mar'in, A., Shlyapnikov, Yu. A., Tosi, G., *European Polymer J.* , in press.
2. Bruni, P., Mar'in, A., Maurelli, E., Tosi, G., *Polymer Degradation and Stability*, **46**, 151-57 (1994).
3. Bruni, P., Conti, C., Mar'in, A., Shlyapnikov, Yu. A., Tosi, G., *Polymer Preprints (J.A.C.S.)*, **37**, 101- 102 (1996).
4. Bruni, P., Conti, C., Mar'in, A., Shlyapnikov, Yu. A., Tosi, G., *European Polymer J.* , in press.
5. Zoppo, M., Zerbi, G., *Polymer*, **31**, 658-62 (1990).
6. Agosti, E., Zerbi, G., Ward, I. M., *Polymer*, **33**, 4219-29 (1992).

Synchrotron Powered FT-IR Microspectroscopic Incremental Probing of Photochemically Degraded Polymer Films

David L. Wetzel[1] and Roscoe O. Carter III[2]

[1]*Microbeam Molecular Spectroscopy Laboratory, Kansas State University, Shellenberger Hall, Manhattan, KS, 66506, USA*
National Synchrotron Light Source, Beamline U2B, Bldg. 725, Brookhaven National Laboratory, Upton, NY, 11973, USA
[2]*Ford Motor Co., Scientific Research Laboratory, PO Box 2053, MD 3061, Dearborn, MI 48121, USA*

An acrylic polymer automotive coating that had been subjected to Florida sun for 3 years was subsequently exposed to accelerated photochemical attack with Xenon lamps. Microtomed 6 µm-thick sections of the photochemically degraded polymer films were mounted between two 13 mm diameter x 2 mm thick barium fluoride disks in a compression cell. With dual apertures 6 µm wide and 36 µm long, a line mapping procedure was performed by stepping the motorized stage in 5 µm increments. The chemical composition was mapped from the outermost edge through the degraded and washed out area into the pristine part of the clear coat, the base coat, and finally the primer. The results of incremental probing of the exposed acrylic polymer coating was compared to a retained specimen of the same material that had been protected from exposure to ultraviolet radiation. Previous attempts with photoacoustic infrared spectroscopy had established the destruction of some absorption bands and the appearance of new broad bands of oxidation products. The depth of the photochemical action was revealed by transmission probing as described here including 1 µm increment line mapping across the clear coat. Interdiffusion of adjacent clear and base coats was also evident. Other polymers subjected to impingement of O^{+4} at different levels of flux showed oxidation by ATR microspectroscopy of the exposed surface in comparison to spectra obtained by the same means from the unexposed back side of the 0.25 in-thick specimen of polypropylene.

INTRODUCTION

FT-IR microspectroscopy has long been known for its ability to deal with polymer laminates. In such a case, each layer in the laminated material will have a finite thickness, and with appropriate aperturing, a good spectrum is obtained from each layer. From this, a qualitative analysis can be run usually with the assistance of a spectral library. Furthermore, various spectroscopic surface techniques such as ATR or photoacoustic spectroscopy have been useful in detecting and measuring differences in the chemical composition of the surface versus the body of the polymer or other material. In this work, polymer degradation by oxidation or photochemical reaction has been shown to exist on the surface of polymer specimens by these two techniques, respectively. However, in order to measure the depth of penetration of the incident UV photons and the extent of their destructive force at each depth, it is necessary to use highly spatially resolved FT-IR microspectroscopic mapping of cross sections in a transmission mode. Previous infrared spectroscopic probing surface preference sampling has been utilized previously to concentrate the measurement of chemical change in the region where it has taken place, namely, the surface. Whereas a transmission spectrum will show primarily the spectral features of the body of material it is difficult to notice relatively small changes superimposed upon the spectrum that is on the transmission spectrum of the parent material. When the spectrum of a retained sample is subtracted from the surface preferenced sampling of the same material that has been subjected to an oxidative or photochemical degradation process, the spectrum of the reaction products is then enhanced so that it is possible to obtain a frequency response and perhaps determine the class of compounds being produced. Surface preferenced sampling may be accomplished with attenuated total reflectance where the depth of penetration can be controlled somewhat by the choice of the ATR crystal in terms of its index refraction. Photoacoustic detection of FT-IR spectroscopy is also used to obtain surface preferenced sampling. This procedure also has some degree of selectivity in the depth of penetration. Examining a microtomed cross section by a transmission FT-IR microspectroscopy is also a useful procedure, however, a finite aperture width determines the thickness of a particular layer that can be sampled. In practice, the minimum aperture size and, therefore, the best spatial resolution is usually dictated by loss of signal with aperturing and coaddition of scans for a reasonable time rather than by diffraction as wavelength-sized aperture dimensions are approached.

SYNCHROTRON MICROSPECTROSCOPY

The use of synchrotron radiation in place of the conventional globar (thermal) source provides a greater signal to noise with small apertures. In the synchrotron (Fig. 1) of the National Synchrotron Light Source (NSLS) at Brookhaven National Laboratory, electrons from an electron source (A) are accelerated with a linear accelerator (B) to an energy of

CP430, *Fourier Transform Spectroscopy:* 11th International Conference
edited by J.A. de Haseth

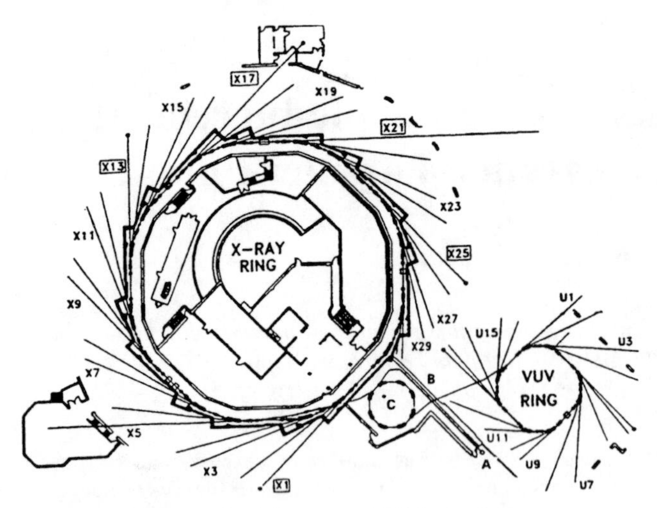

Figure 1. Diagram of synchrotron at NSLS. A=electron source, B=LINAC, C=booster ring

approximately 75 MeV. They subsequently enter into a booster ring (C), where they are accelerated to a higher energy state before being injected into either an X-ray storage ring or, in this situation, a VUV storage ring. Electrons entering the VUV storage ring have an energy of approximately 750 MeV. Electrons that have been accelerated to velocity near the speed of light give off radiation in a relativistic manner. This radiation is highly directional and thus highly concentrated. Synchrotron radiation besides possessing brightness also has no thermal noise. In the National Synchrotron Light Source at Brookhaven National Laboratory, bunches of electrons orbiting in the VUV storage ring emit radiation that is directed from the ring port into the beamlines in proximity to the various bending magnets. At Beamline U2B a special interface at the ring and a long evacuated tube allow introduction of the synchrotron beam into the IRμs™ FT-IR microspectrometer (Spectra-Tech, Inc, Shelton, CT) 14 meters downline. The interface at the ring was designed for other infrared experiments prior to recently establishing the facility for a synchrotron IR illuminated microspectrometer (1,2). The microscpectrometer was modified only slightly by removing the source mirror.

In the 4000-600 cm^{-1} range, the signal (expressed in Watts into a 2 cm^{-1} bandwidth) through a 10 μm pinhole at f/1 has a maximum of slightly more than 10^{-6} for the synchrotron compared to 10^{-9} for a compable black body source. Calculated signal vs. wavenumber profiles (3) were based on a 40 mrad vertical x 40 mrad horizontal NSLS U10 beamline source and a 1200K black body. The noise of a typical liquid nitrogen cooled HgCdTe detector in the same spectral range is estimated at 10^{-11}. Even without achieving the potential theoretical thousandfold signal enhancement, the signal to noise advantage of the synchrotron is obvious.

Besides the brightness of the synchrotron beam and the absence of thermal noise, the most important feature in this experiment for incremental probing is that the beam is highly nondivergent. Thus, the restriction imposed by projected apertures results in very little attenuation, so the infrared beam enters the specimen even at an aperture 6 μm wide with

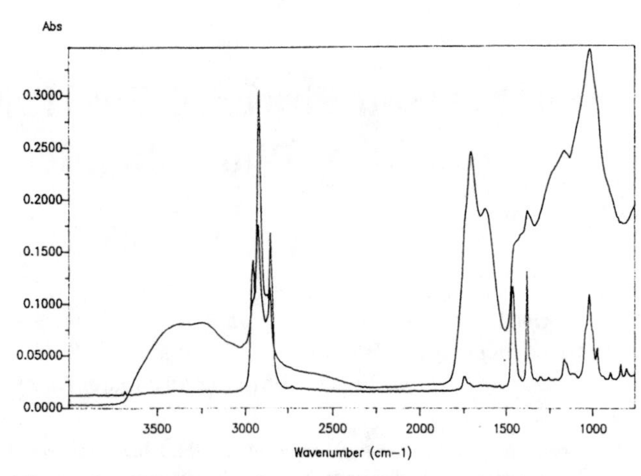

Figure 2. ATR spectra of unexposed (lower) and high exposure (upper) polymers

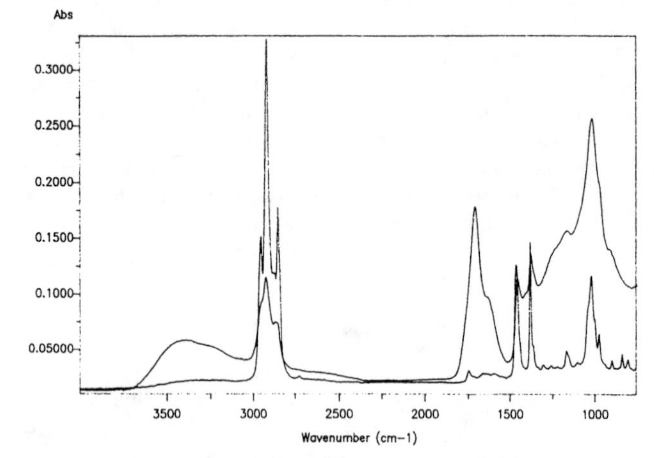

Figure 3. ATR spectra of unexposed (lower) and medium exposure (upper) polymers

sufficient energy to produce good spectroscopic data without coaddition.

EXPERIMENTATION AND RESULTS

Attenuated Total Reflectance

Polypropylene specimens were impacted by O^{+4} ions at different levels of flux. These samples were subjected to ATR microspectroscopy using the ATR objective attachment for the IRμs instrument. Figures 2 and 3 each show the spectrum of the O^{+4} ion impacted surface and the spectrum of the back side of pieces of polypropylene. Figure 2 shows the absorption bands produced from the higher flux bombardment representing OH stretch and carbonyl as well as a broad band of degradation products at frequencies below 1600 cm^{-1}. A comparison of Figs. 2 and 3 show that the intensities of reaction product bands are greater for the surface exposed to a greater flux of O^{+4}. The fluxes for these two were of 2 x 10^{14} and 1 x 10^{14} ions/mm^2. This is a relatively simple experiment; the results are immediately apparent, and at least semi-quantitation is implied.

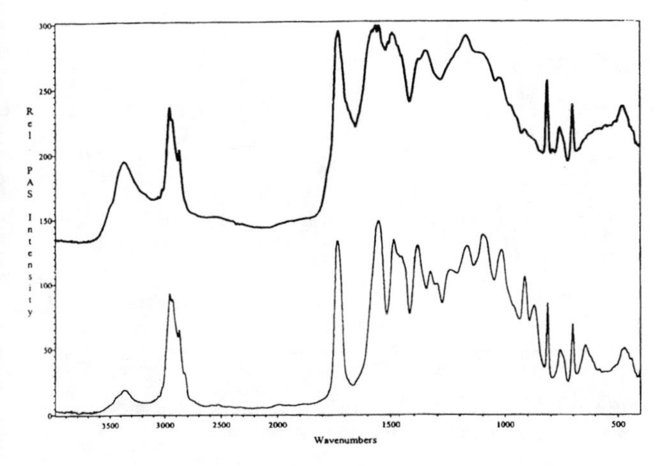

Figure 4. Photoacoustic IR spectrum of control (lower) and accelerated exposure (upper) polymers

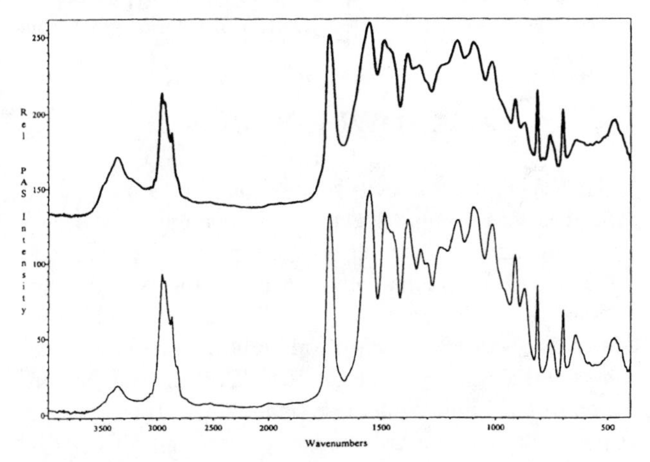

Figure 5. Photoacoustic IR spectrum of control (lower) and natural exposure (upper) polymers

Photoacoustic Spectroscopy

For studying the weathering properties of an acrylic polymer, an acrylic polymer laminate was divided into two parts, one of which was subjected to attack and the other was a retained sample. The acrylic polymer consisted of a clear coat, a base coat, and primer. Of interest was determining the extent to which the clear coat was penetrated with ultraviolet rays that had sufficient energy to cause photochemical degradation. The samples discussed in this paper were subjected to solar photochemical attack and weathering for a period of 3 years in the Florida sun. Subsequently, accelerated photochemical attack was produced by exposure of one specimen to high intensity Xenon lamps. The clear coats of the polymer laminate subjected to natural and accelerated exposure were compared to that of the retained (unexposed) specimen. Figures 4 and 5 compare the photoacoustic spectra of the degraded samples and the control. Disappearance of the characteristic acrylic bands including one at 1550 cm^{-1} and appearance of broad bands from degradation products are observed. From these figures,

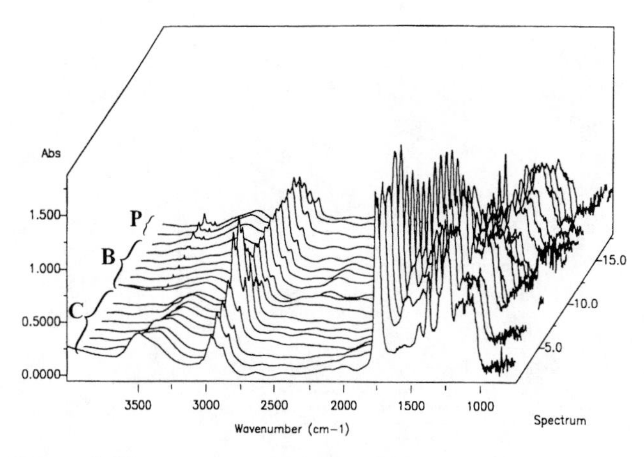

Figure 6. Line mapping experiment of polymer section C=clearcoat, B=basecoat, P=primer

the breakdown of the original polymer and appearance of degradation products is readily apparent. The question that remains is how far into the polymer surface does this occur.

Transmission Cross Sectional Mapping

Cross sections of the polymer (clear, base, primer) laminate were microtomed to thicknesses of 10 μm and 6 μm. Data presented in this report are from 6 μm thick specimens placed between two barium fluoride windows in microcompression cells. Apertures 6 μm wide and 36 μm long were used and aligned with the outer edge and care was taken to avoid sampling any adjacent material. Figure 6 shows a series of spectra taken in sequence in 5 μm steps for a typical line mapping experiment. Spectra shown at the front of the figure were obtained at the edge of the clear coat. Successive spectra go through the clear coat (spectrum 7) and continue on to the base coat and finally into the area of the primer beginning with spectrum 13. The first three spectra exemplify photochemically degraded material and 5th and 6th are pristine. Pigments of the base and primer possibly account for peaks at ~3700 cm^{-1}. Note the missing 1550 cm^{-1} band in the first spectrum. Subsequent to the mapping of the polymer that had been subjected to accelerated photochemical attack, the sample subjected to natural light for 3 years and the retained (unexposed) specimens also were mapped. Mapping of the sun exposed clear coat was done by advancing a 6 μm x 36 μm aperture in 1 μm steps. This provided the opportunity to better define the point where changes occur in the film composition and its spectrum.

DISCUSSION AND CONCLUSIONS

Figures 7 and 8 compare four spectra selected from the line mapping experiment compared to the bottom spectrum of retained acrylic. The middle spectrum in Fig. 7 obtained 7 μm from the surface reduced shows acrylic band intensities at 2960, 1740, and 1550 cm^{-1} and obscuration of spectral

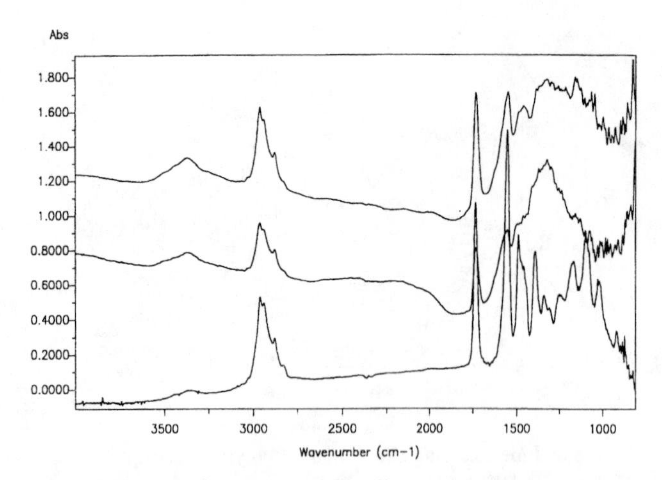

Figure 7. Spectra from exposed film line mapping experiment at 7 μm (middle) and 10 μm (top) and retained film spectrum (bottom).

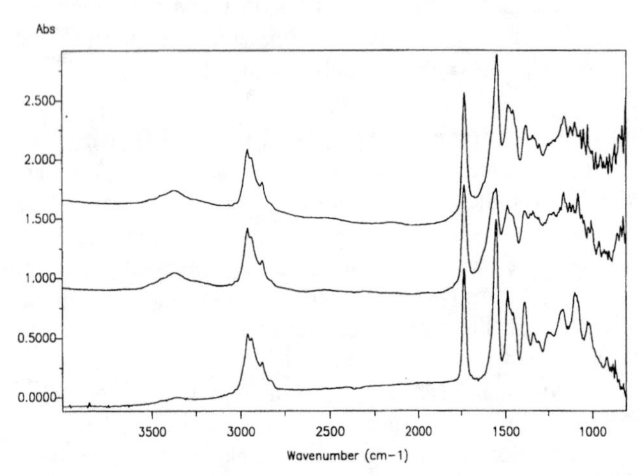

Figure 8. Spectra from exposed film at 15 μm (middle) and 18 μm (top) and retained film spectrum (bottom)

features in the fingerprint region by a broad band centered at 1400 cm^{-1}. The top spectrum in Fig. 7 taken 10 μm from the edge still shows considerable degradation products but the three strong bands are more prominent. The 15 μm point (middle spectrum) in Fig. 8 shows more normal acrylic features and less degradation. By the 18 μm mark, the ratio of the 1550 and 1740 cm^{-1} bands appears normal and further changes were not in evidence beyond that point. Besides finding the depth beyond which degradation did not occur under the conditions used, it was possible to observe signs of interdiffusion of wet clear coat applied directly over wet base coat. Mapping of the clear coat layer of the pristine retained specimen and just part of the base coat interface allowed us to see differences between the outer and center part of the unexposed specimen. Figure 9 shows the interdiffusion between clear and base coats within the 6 μm distance between spectra 27 and 33.

Data from the experiment that we chose to show indicate that determination of the actual depth of penetration is possible with incremental probing. The spatial resolution

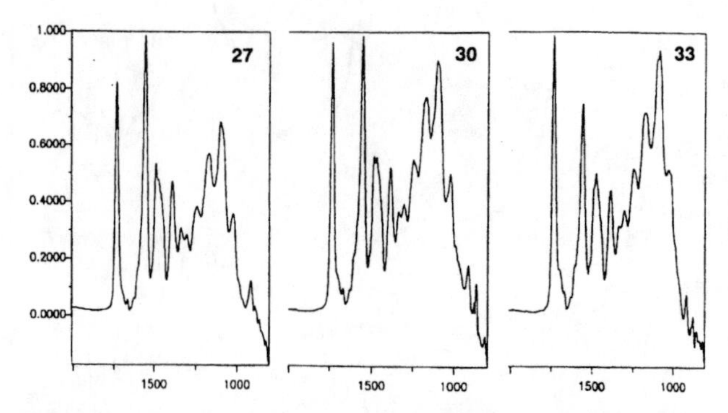

Figure 9. Spectra at 27, 30, and 33 μm into sample showing interdiffusion of layers in the unexposed film.

possible with synchrotron powered FT-IR microspectroscopy allows incremental probing and localization of the damage caused by photochemical degradation of various conditions.

ACKNOWLEDGMENTS

The ATR microspectroscopy was performed at the Microbeam Molecular Spectroscopy Laboratory at Kansas State University. The photoacoustic spectroscopy was done at the Science and Technology Center (Chemistry Laboratory) at Ford Motor Company, and the incremental probing was done at the National Synchrotron Light Source, Brookhaven National Laboratory, Upton, New York. The National Synchrotron Light Source is maintained by the U.S. Department of Energy under project no. DE-AC02-76CH00016. This work was supported by NSF EPSCoR Grant no. OSR 9255223.

REFERENCES

1. Carr, G.L., Reffner, J.A., and Williams, G.P., *Rev. Sci.Instr.*, **66**, 1490, 1995.
2. Wetzel, D.L., "Microbeam molecular spectroscopy of biological materials", in: *Food Flavors: Generation, Analysis, Process Influence*, Charalambous, G. (ed.), Amsterdam, Elsevier Science, 1995, pp. 2039-2108.
3. Wetzel, D.L., Reffner, J.A., and Williams, G.P., *Mikrochim. Acta [Suppl.]* **14**, 353-355, 1997.

Contribution no. 98-52-B, Kansas Agricultural Experiment Station, Manhattan.

FT-IR Specular Reflection and FT-Raman Study of the Structure of PET

K. C. Cole, A. Ajji, and E. Pellerin

Industrial Materials Institute, National Research Council Canada, 75 De Mortagne, Boucherville QC, Canada J4B 6Y4

Samples of PET with different states of crystallinity, prepared by annealing and/or drawing amorphous sheet, were examined by front-surface reflection FT-IR and FT-Raman spectroscopy. The spectra were analyzed in detail by means of spectral subtraction and curve fitting in order to obtain information on the molecular conformation. The strain-induced crystalline (SIC) material produced by drawing has a quite different structure from the true crystalline material produced by thermal annealing. In the latter the glycol groups are known to exist entirely in the form of *trans* conformers, and the carbonyl groups are coplanar with the benzene ring and in the *trans* conformation. The SIC structure possesses an extended *trans* glycol structure, but the carbonyl groups are shown to have the same conformation as in amorphous material, believed to involve a distribution of out-of-plane conformations. As a result, the SIC structure cannot achieve the same degree of molecular packing as the true crystalline structure. Furthermore, it is shown that on thermal annealing the material passes through an intermediate structure similar to the SIC structure.

INTRODUCTION

The manufacture of many products such as fibers, films, and bottles from poly(ethylene terephthalate), or PET, involves both molecular orientation and crystallization during processing. In spite of the numerous studies that have been published, the behavior at the molecular level is still not fully understood. While the conversion of *gauche* glycol conformers to *trans* upon drawing amorphous PET is well established, the conformational changes involving the carbonyl groups and benzene rings are less clear. Recently we have attempted to better understand the phenomena that govern the process of drawing PET, and have shown that external reflection infrared spectroscopy is a useful addition to the battery of tools available for characterizing the changes in molecular structure (1,2). In the present paper we extend this work to a new set of samples that were prepared with the intention of performing more complete characterization under optimal conditions in order to obtain more detailed information.

EXPERIMENTAL

Amorphous PET sheet of thickness 0.45 mm was prepared by compression molding of dried granules of DuPont Selar PT 7086, followed by quenching in ice water. The crystallinity, as measured by DSC, was 5%. Oriented sheet was prepared by drawing this material in an Instron test machine at 2 cm/min and 80°C. Infrared spectra were measured by external reflection at 11° angle of incidence on a Nicolet 170SX instrument equipped with a Model 134 external reflection accessory and zinc selenide wire grid polarizer from Spectra-Tech. To assure the highest quality of spectra, films were mounted in epoxy resin and carefully polished to a finish of 0.05 μm. Although the initial stages of polishing affect the molecular structure of the sample surface, the successive stages remove the disturbed material and the surface left after the final stage is identical in nature to the initial material, except that the measured reflectance is higher because of the better surface quality and planarity (3). Furthermore, mounting in epoxy eliminates any reflection from the back surface of the sample, although this is not a serious problem for the relatively thick samples used here. The spectra were measured with a resolution of 2 cm^{-1}, higher than in our previous work. The reflectance spectra were converted into refractive index (n) and absorption index (k) spectra by means of the Kramers-Kronig transformation, and these were used to calculate the imaginary molecular polarizability ϕ, which is a more precise quantity to use when detailed analysis of band shapes is to be performed (4). To correct for global variations in spectral intensity, all transformed spectra were normalized with respect to the peak at 1410 cm^{-1}, which is reported to be insensitive to orientation and crystallinity effects (5). Finally, for the uniaxially drawn films, spectra were measured with polarization parallel and perpendicular to the draw direction, and from these the structural factor spectrum was calculated as $\phi_0 = (\phi_{\parallel} + 2\phi_{\perp})/3$. Both the Kramers-Kronig transformation and spectral band fitting (with a Pearson VII lineshape, m = 8) were performed with Galactic GRAMS/386 software. FT-Raman conditions were: Nicolet 800 FT-IR with Raman accessory, Nd:YAG laser (1.06 μm), laser power 0.45 W, 180° backscattering mode, resolution 4 cm^{-1}, 200 scans.

RESULTS AND DISCUSSION

Fig. 1 shows the spectrum of amorphous PET (crystal-

CP430, *Fourier Transform Spectroscopy:* 11th International Conference
edited by J.A. de Haseth
 1-56396-746-4/98/$15.00

linity 5%). The positions and widths of the main peaks as obtained by curve fitting are listed in the second column of Table 1. Fig. 2 shows the spectrum of a sheet prepared by slow cooling from the melt in order to develop a relatively high degree of thermally-induced crystallinity (35%). By subtracting the spectra shown in Figs. 1 and 2 with a factor chosen so as to eliminate the peaks known to be associated with the amorphous phase (*gauche* glycol conformers), the spectrum shown in Fig. 3 is obtained. This may be considered to represent the spectrum of fully crystalline PET. The band parameters corresponding to Fig. 3 are listed in the last column of Table 1. The results are in excellent agreement with previous work done by transmission spectroscopy [6], but are somewhat more complete because they include all the strong bands of the spectrum, which are often saturated in transmission spectra.

Fig. 4 is the structural factor spectrum of a film drawn at 80°C and 2 cm/min to a draw ratio λ of 3.8. The drawing induces "strain-induced crystallization" and the crystallinity as measured by DSC was 30%. As was done for the thermally crystallized sample, the spectrum of the sample before drawing was subtracted in order to obtain the hypothetical spectrum of the strain-induced crystalline (SIC) phase, which is shown in Fig. 5. The main peaks are listed in the third column of Table 1 under "Inter" for intermediate phase.

In general appearance Fig. 5 resembles Fig. 3, but closer examination shows that most of the peaks differ in position and/or width. Hence the strain-induced crystalline structure is quite different from the true crystalline structure. The most important differences concern the carbonyl band at 1728 cm^{-1} and the out-of-plane C–H deformation band of the ring at 730 cm^{-1}. In the SIC phase, these have the same position and width as in the amorphous material. In the true crystalline phase they occur at lower frequency and are sharper. The SIC phase thus corresponds to an intermediate structure in which the glycol groups have been converted from *gauche* to *trans*, but the ring-carbonyl conformation is the same as in the amorphous phase. The carbonyls are not entirely *trans* with respect to the ring, and this prevents the molecules from efficiently packing into the true crystalline structure.

Yet another sample was prepared by annealing the amorphous material for 15 min at 200°C; the measured crystallinity was 31%. When the spectrum of the amorphous sample was subtracted, the result in Fig. 6 was obtained. This spectrum is similar to that of the fully crystalline material (Fig. 3) but on close inspection it shows some subtle differences. For instance, some of the peak maxima are slightly lower in frequency. Also, whereas in Figs. 3 and 5 the C–H in-plane deformation band at respectively 1023.4 and 1021.1 cm^{-1} is fairly symmetric, in Fig. 6 it is broader, centred at 1022.6 cm^{-1}, and distinctly asymmetric. It appears to be a composite of the 1023.4 and 1021.1 peaks. A similar observation applies to the *trans* glycol CH_2 wagging band near 1340 cm^{-1}. It can be concluded that annealing under these conditions has produced not only some of the true crystalline structure but also some of an intermediate structure similar to that produced by drawing.

This material was then drawn at 80°C to λ = 3.6. Subtraction of the spectrum of the undrawn (but thermally crystallized) material gives the spectrum of the structure generated by the drawing, which is shown in Fig. 7. This is similar to Fig. 5 (SIC phase), indicating that the main effect of drawing this partially crystalline material is to convert *gauche* conformers in the amorphous phase into extended *trans* units. When the parallel and perpendicular polarized spectra of this sample are subtracted, strong peaks arising from dichroism are observed at 1729, 1341, 1270, 1127, 1107, 1020, and 730 cm^{-1}. These correspond very well to Fig. 7 and show that all the orientation occurs in the amorphous and intermediate phases. No dichroism is observed for the peaks associated with the thermally-induced crystals present in the sample before drawing.

Finally, Figs. 8 and 9 show FT-Raman spectra for the samples just discussed; they confirm the conclusions already derived. Drawing at 80°C produces no detectable change in the carbonyl band, whereas thermal crystallization produces the sharpening at 1725 cm^{-1} that has long been associated with crystallinity in PET (7). Our results show that this sharpening is indicative mainly of the perfect crystalline structure, and not strain-induced crystallinity. The same is true for the bands at 1093 and 995 cm^{-1}, as can be seen from Fig. 9.

It can be concluded that the crystallization of amorphous PET is achieved in at least two fairly distinct stages. The first involves conversion of *gauche* glycol conformers into *trans* to produce linear extended *trans* segments that can align themselves into quasi-crystalline domains. However, the crystal structure is imperfect because the carbonyl groups are not all in the *trans* conformation with respect to the benzene rings. This structure must therefore be considered intermediate. The true crystalline structure is developed only when the carbonyl groups rotate into the proper conformation. Thermal crystallization brings about both stages, but straining at temperatures below those required for thermal annealing brings about only the first stage. Front-surface reflection infrared spectroscopy is a very useful tool for obtaining detailed information on the different structures present in PET processed under different conditions.

REFERENCES

1. Cole, K. C., Guèvremont, J., Ajji, A., and Dumoulin, M. M., *Appl. Spectrosc.*, **48**, 1513-1521 (1994).

2. Ajji, A., Guèvremont, J., Cole, K. C., and Dumoulin, M. M., *Polymer*, **37**, 3707-3714 (1996).

3. Ben Daly, H., Cole, K. C., Sanschagrin, B., and Nguyen, K. T., *Proc. ANTEC '97*, 2201-2205 (1997).

4. Bertie, J. E., Zhang, S. L., and Keefe, C. D., *J. Mol. Struct.*, **324**, 157-176 (1994).

5. Walls, D. J., *Appl. Spectrosc.*, **7**, 1193-1198 (1991).

6. D'Esposito, L., and Koenig, J. L., *J. Polym. Sci.: Polym. Phys. Ed.*, **14**, 1731-1741 (1976).

7. A. J. Melveger, J. Polym. Sci.: Part A-2, 10, 317-322 (1972).

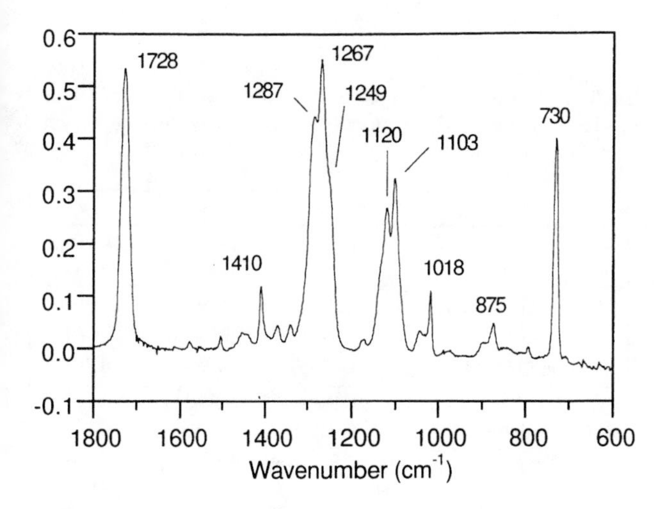

FIGURE 1. Spectrum of amorphous PET sheet.

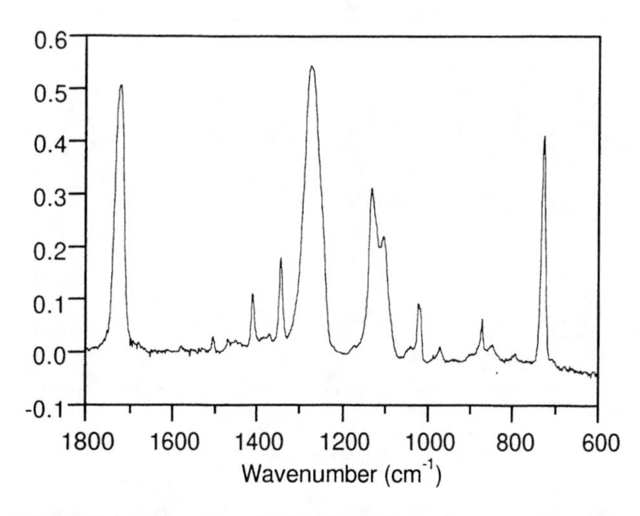

FIGURE 2. Spectrum of PET sheet thermally crystallized by slow cooling from the melt.

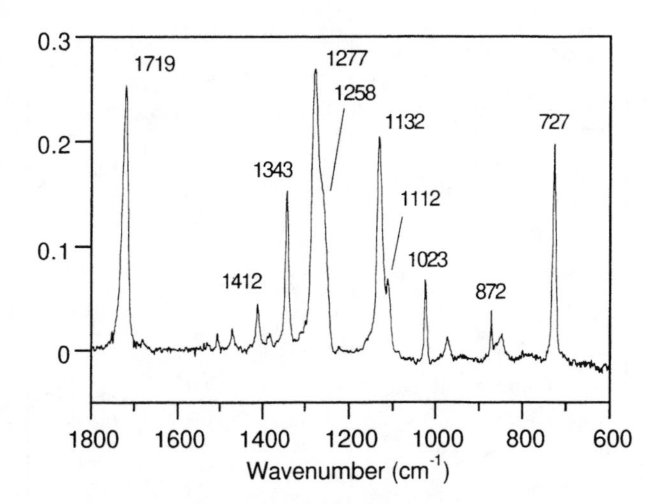

FIGURE 3. Hypothetical spectrum of fully crystalline PET obtained by difference from Figures 1 and 2.

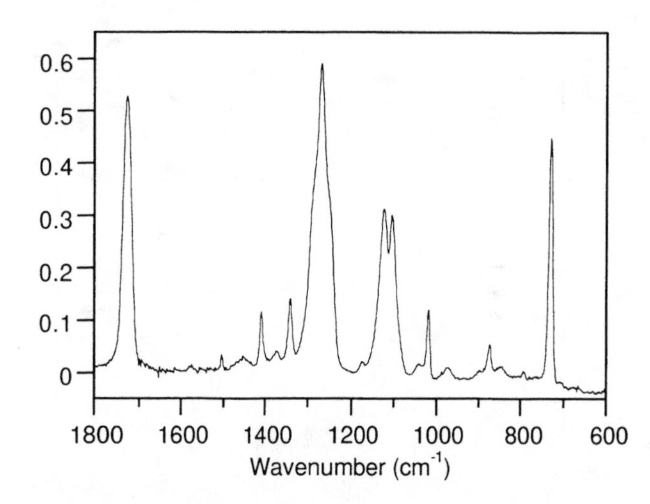

FIGURE 4. Spectrum of film obtained by drawing amorphous PET at 80°C to $\lambda = 3.8$.

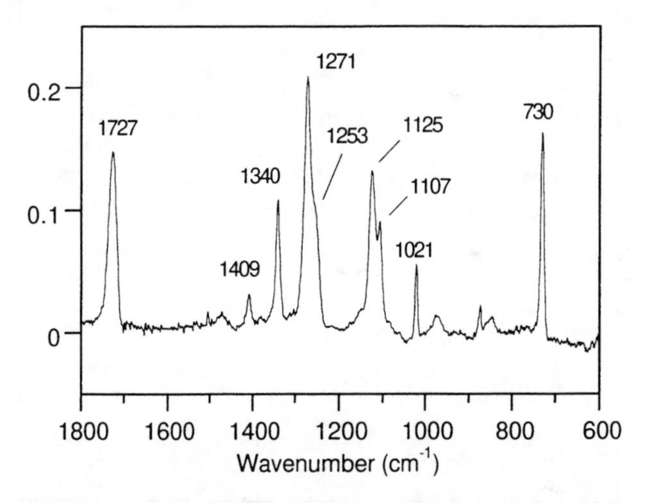

FIGURE 5. Hypothetical spectrum of "strain-induced" crystalline PET obtained by difference from Figures 1 and 4.

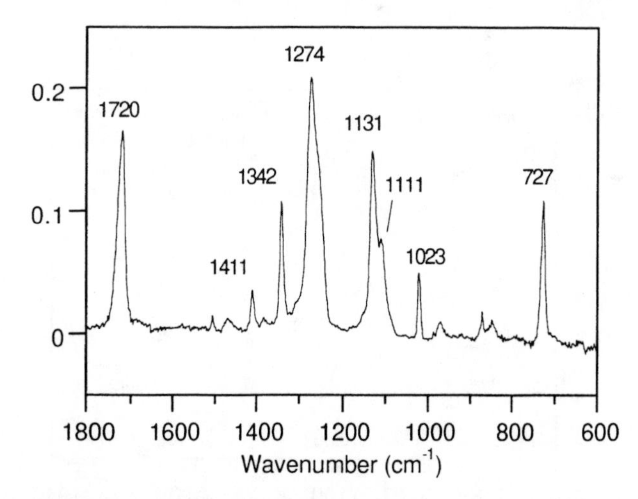

FIGURE 6. Difference spectrum resulting from annealing of amorphous sheet at 200°C for 15 min.

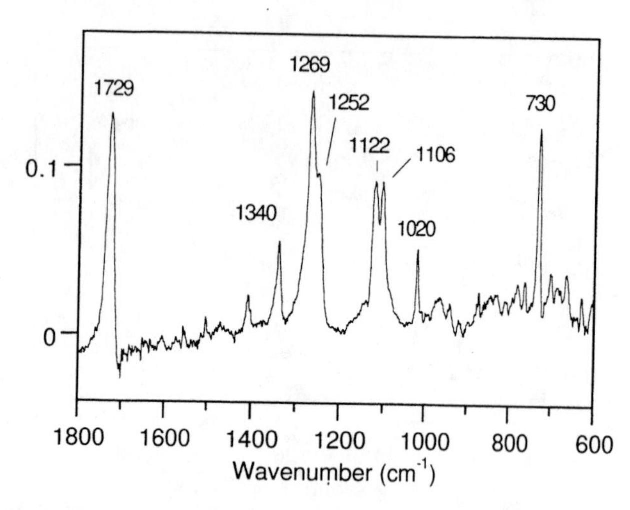

FIGURE 7. Difference spectrum resulting from drawing at 80°C to λ = 3.6 of sheet previously annealed at 200°C for 15 min.

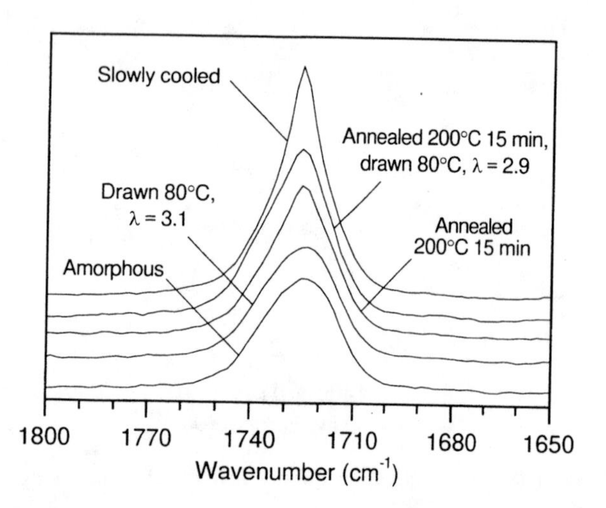

FIGURE 8. Raman spectra of different samples in carbonyl region.

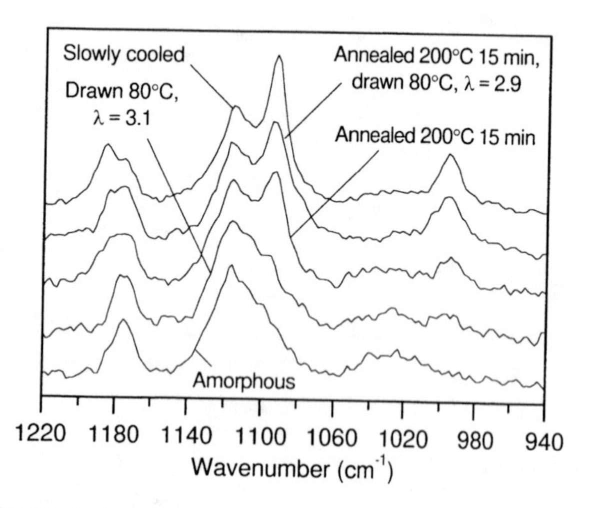

FIGURE 9. Raman spectra of different samples in 1220-940 cm^{-1} region.

TABLE 1. Peak Positions and Widths for the Main Bands Corresponding to the Amorphous, Intermediate, and Crystalline Phases.

Band	Amorph	Inter	Cryst
C=O stretch	1728.2 (21.8)	1727.4 (21.2)	1719.0 (9.3)
C–H in-plane def.	1504.7 (6.8)	1504.5 (4.2)	1506.0 (4.1)
CH_2 bend	1448.7 (30.3)	1473.4 (26.5)	1470.7 (7.6)
C–H in-plane def.	1409.9 (8.2)	1409.0 (12.9)	1411.7 (9.1)
C–H in-plane def.			1385.2 (10.6)
CH_2 wag (*gauche*)	1371.5 (17.0)		
CH_2 wag (*trans*)	1341.3 (16.6)	1340.4 (9.6)	1343.1 (9.7)
Glycol bend mode?	1286.7 (26.0)		
Ring + ester modes	1266.5 (16.5)	1271.0 (15.9)	1277.0 (18.2)
	1248.9 (20.4)	1252.9 (17.7)	1257.9 (18.4)
Glycol stretch and bend modes + in-plane ring modes	1133.6 (25.4)		
	1120.1 (14.5)	1125.1 (17.1)	1132.1 (14.6)
	1102.8 (15.7)	1106.9 (12.4)	1112.4 (14.7)
	1091.6 (21.5)		
C–O stretch (*gauche*)	1043.5 (19.0)		
C–C and C–O stretch?	1025.4 (16.7)		
C–H in-plane def.	1018.1 (6.7)	1021.1 (6.9)	1023.4 (6.4)
C–O stretch (*trans*)	979.2 (25.8)	972.9 (26.9)	972.5 (12.8)
CH_2 rock (*gauche*)	895.8 (24.4)		
C–H out-of-plane def.	875.0 (13.1)	874.5 (7.8)	872.3 (4.5)
CH_2 rock (*trans*)	845.1 (39.1)	849.7 (26.6)	852.3 (22.0)
C–H out-of-plane def.	732.6 (10.3)	731.0 (9.1)	
C–H out-of-plane def.	727.7 (6.7)	727.8 (5.4)	727.0 (8.9)

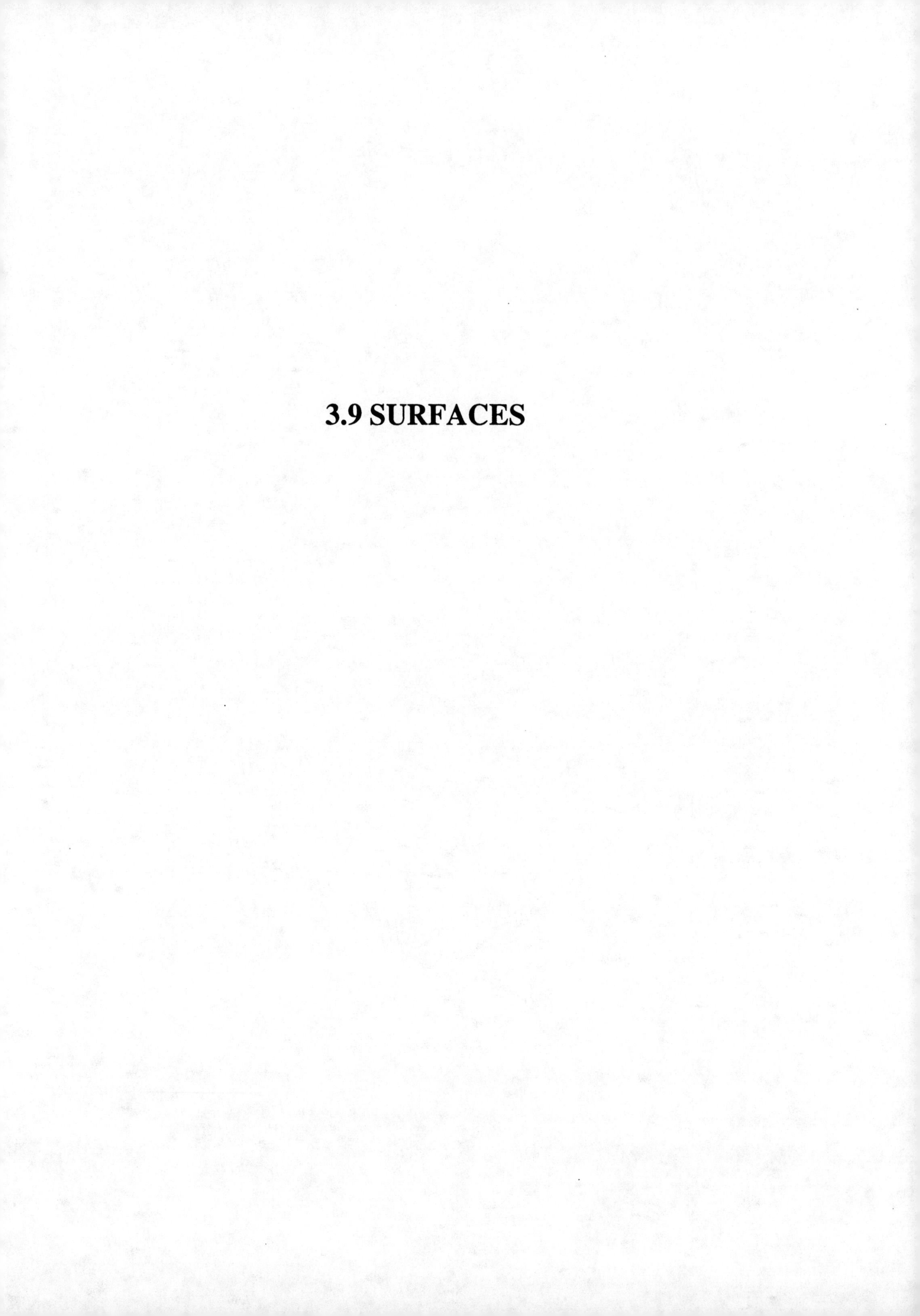

3.9 SURFACES

Surface-Sensitive FTIR Spectral Measurements of Nanogram Samples Using 30–100-μm-Thick Planar Ge Waveguides.

Mark S. Braiman, Susan E. Plunkett, and James J. Stone

Biochemistry Department, University of Virginia, Charlottesville, VA 22908

We have developed techniques for fabricating supported planar Ge waveguides that are useful for measuring broadband mid-IR evanescent-wave absorption spectra from small numbers of molecules. The waveguides are ground and polished as thin as 30 μm, starting from larger pieces of commercially-available single-crystal Ge. Such waveguides are useful as extremely sensitive multiple-internal-reflection elements, when used in conjunction with an FTIR spectrometer with microscope accessory and a small-area detector. With careful polishing and alignment, optical losses can be reduced to those predicted as resulting from ray spreading within the waveguide and from reflection at the air-Ge interface. For example, by using a Spectra-Tech Research IR-Plan™ microscope with a $(100\ \mu m)^2$ detector element, we have been able to demonstrate broadband IR power transmission through a 30-μm-thick, 3-mm-wide waveguide that is 4% of the open-beam throughput focused through a 30 μm × 3 mm slit at the sample focal plane. Supported planar Ge waveguides between 30 and 100 μm in thickness function as internal reflection elements with 10-20 reflections per mm of length. Using such waveguides, we have been able to make surface-sensitive spectral measurements of tiny samples. For example, we can easily observe the FTIR absorption spectrum selectively from the adhesive layer on 3M Scotch™ tape, without interfering absorption from the tape backing. A 0.07 mm^2 piece of tape, carrying only several μg of adhesive, still produces distinct bands with absorbance values as large as 0.18. The great sensitivity to monolayer samples covering small contact areas has allowed us to measure useful FTIR spectra of the plasma membranes of intact cells, e.g. individual 1-mm-diameter frog oocytes, submerged in aqueous media. The protein and lipid components in a patch of such a single-bilayer membrane give absorbances on the order of 0.01.

BACKGROUND

Planar-trapezoidal germanium (Ge) prisms are commonly used as internal reflection elements (IRE's) for measuring attenuated total reflection infrared (ATR-IR) spectra of thin film samples. These IREs can be thought of as thick planar waveguides. When IR light propagates through them, a portion of the energy in each mode is carried as an evanescent wave that penetrates into the external medium, which has a lower refractive index than that of the Ge waveguide itself (4.0). The energy carried in the evanescent wave can be absorbed by IR-active vibrational modes of molecules sitting in the external medium, but close (generally <1 μm) to the Ge surface (1).

It is well known that decreasing the waveguide thickness increases the fraction of energy carried in the evanescent wave, and therefore increases the sensitivity of evanescent-wave absorption (or ATR-IR) measurements (2-4). Ge is a very brittle material, and generally lacks sufficient mechanical strength to allow fabrication of free-standing IREs much less than 1 mm in thickness. To create thinner Ge IREs, we have therefore developed methods of fabricating them as supported planar waveguides 30–100 μm in thickness (5-7).

We have shown that is possible to fabricate rugged planar Ge waveguides as thin as 30 μm that transmit >4% of the open beam intensity measured using an IR microscope with a $(100\ \mu m)^2$ detector (6). These are apparently the thinnest broadband IR-transmitting waveguides ever fabricated. We have demonstrated their utility as IRE's for measuring evanescent-wave absorption spectra of thin films and coatings on objects having a surface area <0.1 mm^2 in contact with the waveguide (6-8).

WAVEGUIDE FABRICATION

Our methods for fabricating waveguides have been described in several recent publications. Standard Ge ATR orthorhombic prisms (Spectral Systems Inc.) are coated on one face with a 1.2-μm-thick CVD coating of ZnS, using a relatively low temperature (<150° C). This proprietary coating method (of Hughes-Santa Barbara Research Corporation, Santa Barbara CA) was found to be necessary to give adequate mechanical adhesion of the ZnS layer for the subsequent grinding and polishing to <50 μm thickness.

Coated prisms are then cemented to quartz substrates (Quartz Scientific, Fairport Harbor, OH) using a UV-curing optical adhesive (optical adhesive 81 from Norland, New Brunswick NJ; or type SK-9 adhesive from Summers

CP430, *Fourier Transform Spectroscopy:* 11th International Conference
edited by J.A. de Haseth

Optical, Fort Washington, PA). The adhesive is cured with a long wave UV lamp (Blak-Ray®, 115 V, 60 Hz) at a distance of 50 mm for a minimum of 3 h.

We use a low-viscosity UV-curing optical adhesive in order to produce a very thin (sub-μm) adhesive layer. This helps keep a consistent stress on the CVD-ZnS coating and Ge waveguide during the grinding and polishing operations. Optical adhesives also give the advantage of allowing repositioning of the prism on the substrate in the event that small air bubbles are trapped in the adhesive layer. Such air bubbles trapped in the epoxy cement typically lead to chipping of the waveguides during grinding/polishing.

After curing, 45° bevels are ground into each end of the waveguide using aluminum oxide grinding powders (Buehler Ltd., Lake Bluff, IL) of 400 grit, 600 grit, 25 μm, and 12.5 μm particle sizes. Beveled pieces of poly(tetrafluoroethylene) are used as guides during grinding/polishing to define the bevel angle, α, i.e. the angle between the long axis of the waveguide and its end faces (see Fig. 1). It is crucial that both faces have nearly-identical bevel angles in order to obtain efficient coupling of light into and out of the waveguide.

The beveled Ge waveguides are then ground to nearly the desired thickness of <100 μm by using the grinding powders and a grinding stone (Harrick Corp., Ossining NY). The bevels and top surface are then polished using aluminum oxide or diamond films (Buehler Ltd., Lake Bluff, IL), mounted on the grinding stone to insure flatness. Careful fine-polishing to a 0.1-μm finish is crucial for minimizing light scattering from imperfections in the planar surface.

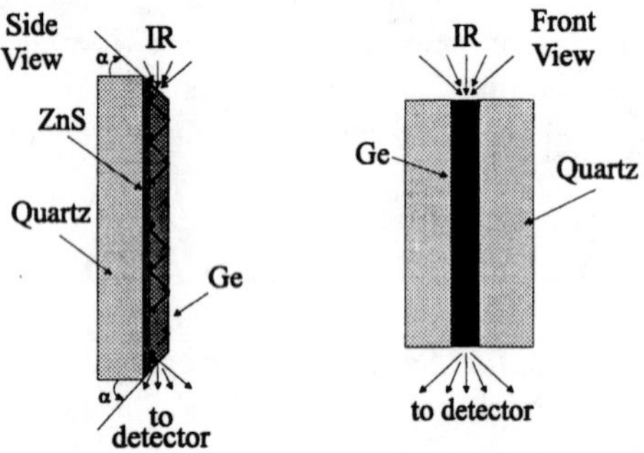

Figure 1: Schematic diagram of planar Ge waveguides and light paths. Left diagram shows a side cross-sectional view of a waveguide, and right diagram shows a view of the top surface of the waveguide. Drawings are not to scale. Typical dimensions of the waveguide are 12×3 mm, with a thickness of 30-100 μm. Typical dimensions of the quartz substrate are 50×12×2-mm. Reproduced from ref. 6.

For spectral transmission and ATR measurements, the waveguide is mounted vertically in an IR microscope (IR-Plan™ microscope, Spectra Tech, Stamford, CT) interfaced to an FTIR interferometer (Illuminator, Midac Corp., Irvine CA). As shown in Fig. 1, broadband mid-IR light from the source (silicon carbide Globar at ~1500 K) is focused through the microscope's objective into the waveguide via the top beveled surface. Light exiting the similarly-beveled surface at the bottom of the waveguide is collected by the condensor and focused onto a 100 μm×100 μm HgCdTe detector (Graseby Infrared, Model FTIR-M16-0.01). Both the condenser (16×) and objective (32×) are reflective Cassegrain optics with numerical aperture of 0.8. The waveguide and focusing optics must be carefully aligned to maximize the transmitted light intensity.

TRANSMISSION THROUGH THE WAVEGUIDE

Figure 2 shows the transmittance spectrum of a typical waveguide of dimensions 12 mm × 3 mm × 30 μm. We observed a transmittance of up to 4.2% in the range from 1500 to 2500 cm^{-1} (see Fig. 2). This is a large fraction of the maximum expected energy throughput for a waveguide of our cross-sectional area, considering that the combined reflection loss from the two 45°-beveled end surfaces of the waveguide is expected to be nearly 50%, and that, furthermore, the beam entering the waveguide is ~100 μm wide, but is expected to spread out to fill the waveguide's 3-mm width nearly uniformly at the output end. Only a fraction of this output light can be collected and focused onto the tiny detector (100 μm × 100 μm).

The relatively high percent transmittance and the shape of the measured spectrum indicate that planar waveguides of this design are capable of propagating light very efficiently across the spectral region transmitted by Ge (λ=2-15 μm). The sharp Ge cutoff at 5200 cm^{-1} shows that light is transmitted directly through the waveguide and not through the surrounding media (air and quartz). We conclude that very little power is lost through the lateral surfaces of the waveguide.

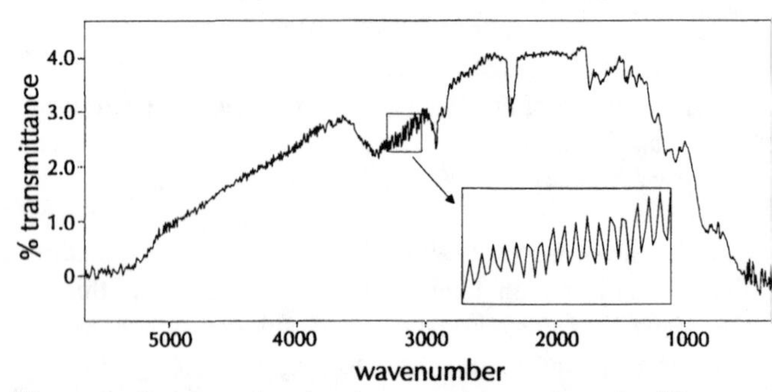

Figure 2: Spectrum showing the transmittance through a 30-μm thick waveguide. As a background, an open beam spectrum was measured using dual apertures set to mask off all but a 3-mm × 30-μm area at the sample plane. Boxed section of the spectrum is enlarged to show oscillation pattern. Reproduced from ref. 6.

In the transmittance spectrum in Fig. 2, much of what appears to be noise are actually regular oscillations, which result from interference within the waveguide. These oscillations can be clearly observed when a small section of the spectrum is enlarged and are expected as a result of interference within the waveguide. For example, the boxed section in Fig. 2 which appears to be the noisiest part of the spectrum can be seen to be a periodic function when enlarged. This periodic function is easily removed from absorbance spectra by careful background ratioing and Fourier filtering.

ATR MEASUREMENTS USING THE WAVEGUIDES

An example of an ATR measurement made with a thin supported Ge waveguide is shown in Fig. 3, which shows an absorbance spectrum of a 1-mm^2 piece of 3M Scotch™ transparent tape. Note the peaks (e.g. 3500 cm^{-1}) due to the tape backing that are found in the transmission spectrum but are not observed using the evanescent-wave absorption method because the evanescent wave penetrates into the adhesive layer only. We estimate the concentration of the adhesive on the tape to be <20 µg/mm^2, so the absorbances in Fig. 3 were detected using <1 µg of sample. In order to observe the dependence of the ATR-IR spectrum on the waveguide thickness, we obtained spectra of a single piece of tape, adhered in separate experiments to the surfaces of 30-µm-thick and 70-µm-thick waveguides.

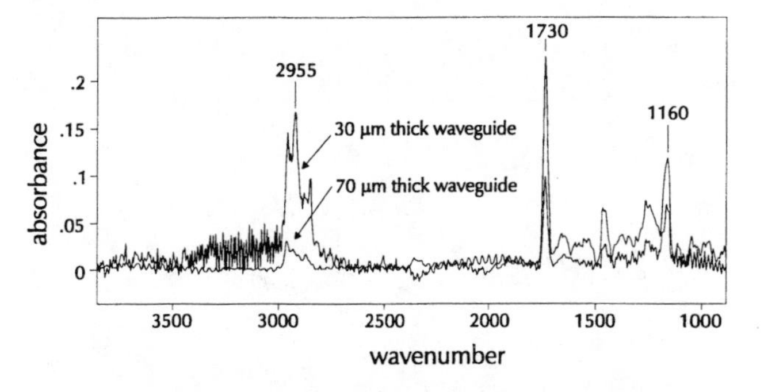

Figure 3: Spectra of the same 1×1-mm piece of 3M Highlands Removable™ tape on the surface of 30-µm thick and 70-µm thick waveguides. Figure reproduced from reference 6.

Since the number of reflections per unit length on the top surface of the waveguide increases linearly with decreasing thickness, the absorption signal for a fixed number of molecules is expected to increase as the reciprocal of the thickness. Therefore our 30-µm-thick waveguide should be ~2.3 times as sensitive as the 70-µm-thick waveguide. This is confirmed by the relative peak heights in Fig. 3.

One of the main difficulties with using traditional ATR methods to measure absorption spectra from thin samples is the requirement of a large sample area (4). However, our miniature supported waveguides have such a small cross-sectional area that they permit such measurements.

This increased sensitivity opens opportunities for experiments that have been limited in the past by the small surface area samples. For example, past experiments measuring IR absorption of supported lipid bilayers have been performed using 50-mm-long × 1-mm-thick Ge ATR plates (8,9). In these experiments, the whole 50×1-mm surface needed to be covered with sample.

Our miniaturized design allows for the sample size to be reduced to study a small surface area such as the membrane of a *Xenopus* oocyte (live unfertilized single cell egg, 1.5 mm in diameter) (10). An example of such a spectrum, obtained using a 50-µm-thick waveguide, is shown in Fig. 4 below. This spectrum demonstrates absorption bands of both protein (amide I, 1650 cm^{-1}) and lipid (ester carbonyl, 1745 cm^{-1}), that are a significant fraction of those previously observed with single-bilayer-thick membranes covering a 1600-mm^2 area of a traditional 1-mm-thick IRE (8,9).

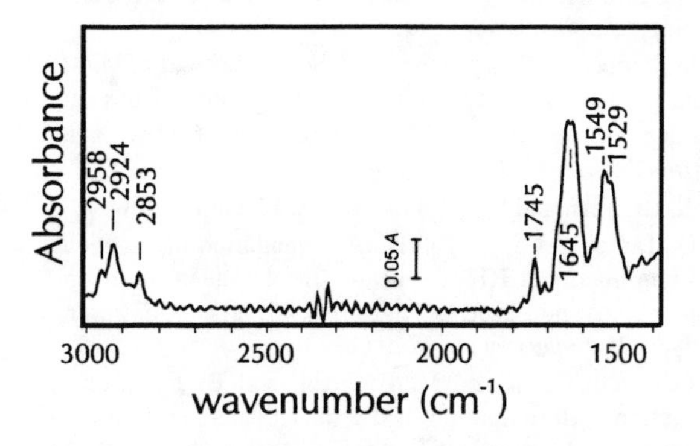

Figure 3: ATR spectrum of an intact single frog egg (*Xenopus laevis* oocyte), fully submerged under buffer. Oocyte was defolliculated and devitellinized prior to placing in contact with a 3-mm-wide, 50-µm-thick supported planar Ge waveguide. Data are reproduced from reference 10.

CONCLUSION

Our results demonstrate that decreasing the thickness of a planar IRE waveguide, even to below 50 µm, can increase the sensitivity of an ATR measurement made with the waveguide, thereby reducing the sample surface area and the total number of molecules needed. Our 30–100-µm-thick waveguides are among the thinnest functioning mid-IR ATR elements ever described. Evanescent-wave absorption measurements using these waveguides should be useful for examining thin films and coatings as well as biological membranes of single cells and vesicles.

ACKNOWLEDGMENTS

This work was supported by NSF grant MCB-9406681 to M.S.B.

REFERENCES

1. Dietrich Marcuse. *Theory of Dielectric Optical Waveguides*, Academic Press, Boston, (1991).
2. N. J. Harrick. *Internal Reflection Spectroscopy*. Ossining, New York, Harrick Scientific Corporation, (1987).
3. D. Ohta and R. Iwamoto. "Lower limit of the thickness of the measurable surface layer by Fourier transform infrared attenuated total reflection spectrometry". *Analytical Chemistry* 57: 2491-2499, (1985).
4. M. S. Braiman and S. E. Plunkett. "Design for supported planar waveguides for obtaining mid-IR evanescent-wave absorption spectra from biomembranes of individual cells," *Applied Spectroscopy*:51:592-597 (1997).
5. S. E. Plunkett, M. S. Braiman, and S. Propst. "Supported planar germanium waveguides for infrared evanescent-wave sensing," *Applied Optics: 36*: 4055-4061 (1997).
6. J. J. Stone, M. S. Braiman, and S. E. Plunkett, "Mid-IR evanescent-wave absorption spectra of thin films and coatings measured with ~50-μm-thick planar Ge waveguide sensors." *Proc. SPIE 3105*: 371-377 (1997).
7. M. S. Braiman and S. E. Plunkett, U.S. Patent Pending.
8. L. D. Tamm and S. A. Tatulian. "Orientation of functional and nonfunctional PTS permease signal sequences in lipid bilayers. A polarized attenuated total reflection infrared study," *Biochemistry* 32: 7720 (1993).
9. S. O. Smith, R. Jonas, M. Braiman, and B. J. Bormann. "Structure and orientation of the transmembrane domain of glycophorin A in lipid bilayers," *Biochemistry*, Vol. 33, 6334-6341 (1994).
10. S. E. Plunkett, M. S. Braiman, R. E. Jonas. "Vibrational spectra of individual millimeter-size membrane patches using miniature IR waveguides," *Biophys. J.* , 1997, in press.

Infrared Surface Analysis of Semiconductors by a Noncontact Air Gap ATR Method

N. Nagai, Y. Izumi, H. Ishida, Y. Suzuki and A. Hatta

Materials Science Laboratories, Toray Research Center Inc., Sonoyama, Otsu, Shiga 520, Japan(N.N., Y.I., H.I.); and Department of Materials Science, Faculty of Engineering, Tohoku University, Aramaki Aoba, Sendai 980, Japan (Y.S., A.H.)

Characterization of chemical bonding structure of semiconductor surface is very important in order to develop a new surface treatment method. The air gap ATR method is presented, which gives an enhanced infrared absorption of surface chemical structure 40 times larger than conventional transmission, and is very useful to study the surface of materials with high refractive indices, such as semiconductors. The enhancement is attributable to the so-called optical cavity effect between the prism and the semiconductor samples. This technique is applied to the study of the mechanism of native oxide growth on a Si(100)substrate, H-terminated Si(111) surfaces, and acid treated GaAs(001) surfaces.

INTRODUCTION

The increasing needs for nondestructive surface characterization of industrial materials to provide structural chemical information have stimulated interest in the use of infrared spectroscopy. Infrared spectroscopy, however, has suffered from a lack of sensitivity compared to other surface analytical techniques, such as XPS (X-ray photoelectron spectroscopy), AES (Auger electron spectroscopy), and SIMS (secondary ion mass spectrometry). Many methods have been proposed to enhance the surface sensitivity of ATR measurements, and ATR is the most useful surface analysis method in infrared spectroscopy. Hjortsberg et al. (1) and Ishino et al. (2) have observed enhanced IR absorption spectra of thin films on metal surface under the Otto ATR configuration (3). Suzuki et al. (4) have recently concluded that the enhanced infrared absorption of thin films on metal surfaces under the Otto ATR configuration can be attributed to an optical cavity effect (5), on the basis of theoretical calculations of electromagnetic fields in the air gap.

In this paper, the air gap ATR method has been successfully applied to surface studies of semiconductor materials such as silicon and GaAs wafers. The present method has shown the enhanced infrared absorption of extremely thin films on Si and GaAs wafers. The usefulness and versatile applicability of the present method will be demonstrated in this paper.

EXPERIMENTAL

Sample Preparation

Silicon wafers used for the native oxide measurements were polished on one side n-type Czochralski (CZ) grown wafers (100) with resistivity of 3.0~8.09 Ωcm. Silicon wafers with 0.525 mm thickness were cut into rectangles of 15×25 mm^2. The film thickness of native oxide on these silicon wafers is 1.5 nm, which was determined by grazing incidence X-ray reflectometry (GLXR). To observe the atomically flat or non-flat H-terminated Si surfaces by air gap ATR, we used 0.4° off Si(111) substrates. The silicon wafers were polished on one side p-type CZ-grown with resistivity of 1 Ωcm. The wafers were cut into 0.45×15×25 mm^3 so that the short sides were perpendicular to the <110> direction.

The atomically flat H-terminated surface was prepared by wet chemical treatment as follows: First, silicon substrates are treated in a 60% NH_4OH solution. Native oxide films are formed by this treatment. Native oxide is removed with 1% HF for 30 sec. This oxide formation and removal process is repeated 3 times. Second, the samples are treated by a 40% NH_4F solution for 15 min and atomically flat H-terminated Si(111) surfaces are obtained. The atomically non-flat surfaces are obtained with only 1% HF solution treatment.

We used liquid encapusulated Czochralski (LEC) grown (001) GaAs wafers with a resistivity of 1 Ωcm for studying the native oxide on GaAs surfaces. The GaAs wafers were polished on one side and cut into 0.5×15×25 mm^3. The samples were treated in an H_2SO_4-H_2O_2-H_2O

CP430, *Fourier Transform Spectroscopy:* 11th International Conference
edited by J.A. de Haseth

solution (H_2SO_4:H_2O_2:H_2O=5:1:1) or an H_3PO_4-H_2O_2-H_2O solution (H_3PO_4:H_2O_2.:H_2O=5:1:1) for 5 min, and rinsed in deionized water following chemical treatment. Native oxides were made by this treatment. We also prepared non-treated samples to compare with the chemical bonding structure of the native oxides.

Measurement Method

A schematic drawing of the air gap ATR geometry used in the present work is shown in Figure 1. A Ge hemisphere prism served as the internal reflection elemental (IRE). The air gap necessary for the measurements is attained with the use of thin spacers between sample and prism, as shown in Fig. 1. We used polypropylene (PP) thin films as spacers and controlled the air gap thickness by using various thicknesses of PP films.

Fourier transform infrared (FT-IR) spectra were obtained with an IFS-120HR spectrometer manufactured by Bruker, and a Harrick ATR illuminator. The instrumental resolution was 4 cm^{-1} and 512 scans were accumulated and averaged. In order to excite the longitudinal-optical (LO) mode of native oxide on silicone (6) and GaAs wafers, p-polarized radiation was employed. The incident angle was changed from 5° to 85° for LO mode measurements of native oxide on silicon surfaces. For measurement of NH_4F (or HF) treated Si and acid treated GaAs, we fixed the incident angle at 15°.

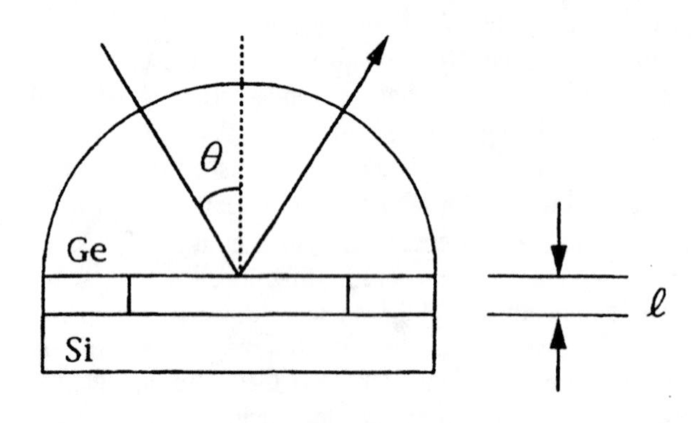

FIGURE 1. Schematic illustration of the air gap ATR geometry. θ: incident angle, *l*: air gap thickness.

RESULTS AND DISCUSSIONS

Air Gap AIR Spectra of Native Oxide Film on a Silicon Wafer

Figure 2 shows the incident angle dependence of the air gap ATR spectra using 5 μm thickness PP films as spacers. The bands near 1230 cm^{-1} are the LO modes of surface native oxides (surface mode), the 1100 cm^{-1} bands are the interstitial oxygen vibrations of silicon substrates, and the other bands are due to the multiphonon modes of Si substrates (bulk mode). The absorption of the LO mode shows a maximum at 15°~20° incident angle, and gradually decreases with increasing incident angle. It was characteristic that the bands of the substrate decreased when the incident angle was large. Figure 3 shows the incident angle dependence of absorption intensity of interstitial vibrational modes. Figure 4 shows the incident angle dependence of peak strength of the LO mode of native oxide. We can observe the LO modes are maximized at 15° incident angle. Different dependencies on incident angle have been observed in the surface and bulk modes.

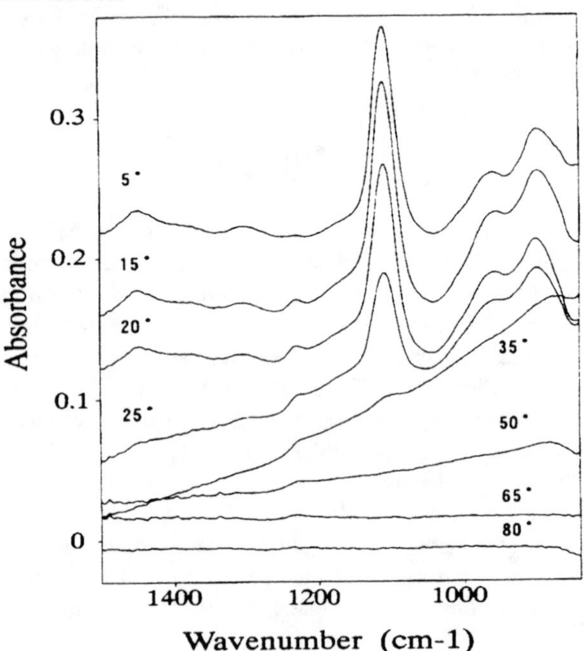

FIGURE 2. Air gap ATR spectra of native oxide film on silicon wafer measured as a function of incident angle. Air gap thickness: 5 μm.

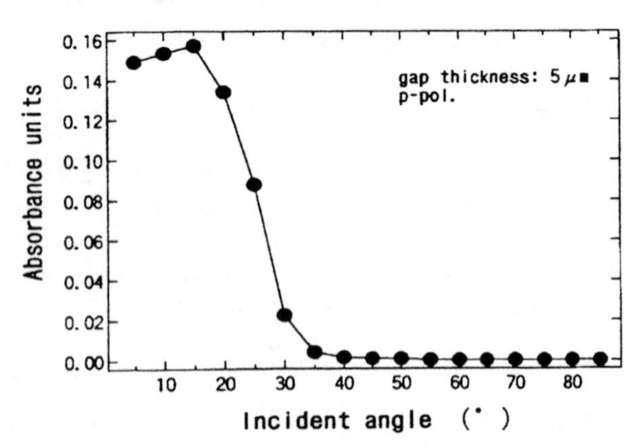

FIGURE 3. Dependence of the absorption intensity of interstitial oxygen (1107 cm^{-1}) in the substrate silicon wafer on the incident angle.

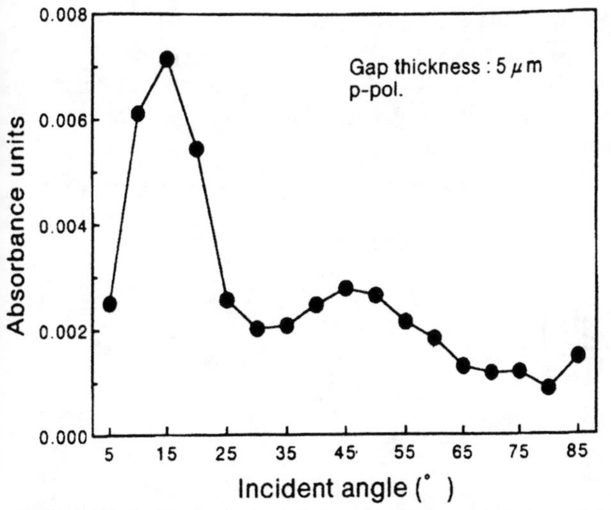

FIGURE 4. Dependence of the absorption intensity of LO mode (1230 cm^{-1}) characteristic of the native oxide film on silicon wafer.

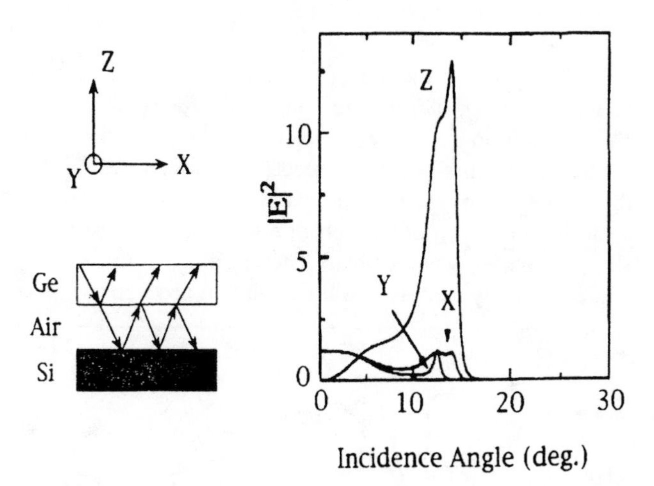

FIGURE 5. Calculated results of electric field amplitude created on silicon surface by multiple reflection between the prism and silicon.

Figure 5 shows the results of the optical cavity effects calculation of electric field amplitude generated on a Si surface by multiple reflection between a Ge prism and a Si surface (5). The amplitude of the electric field perpendicular to the Si surface has a maximum near the critical angle of the Ge/air interface (~14.5°). This result is coincident with the incident angle dependence of the LO mode shown in Fig. 4. Therefore it can be concluded that the detection of very thin films by air gap ATR around 15° incident angle is due to the optical cavity effect.

Figure 6 compares the air gap ATR spectrum of the Si wafer with the transmission spectrum and the specular reflectance spectrum when the incident angle is fixed at 15°. We can clearly detect the 1230 cm^{-1} LO mode of surface oxide by the air gap ATR method, however, the LO mode cannot be observed by transmission and specular reflection methods.

Therefore it demonstrates the air gap ATR is a high sensitive characterization method for the study of semiconductor surfaces. The enhancement of the LO mode is estimated to be 40 times greater than transmission.

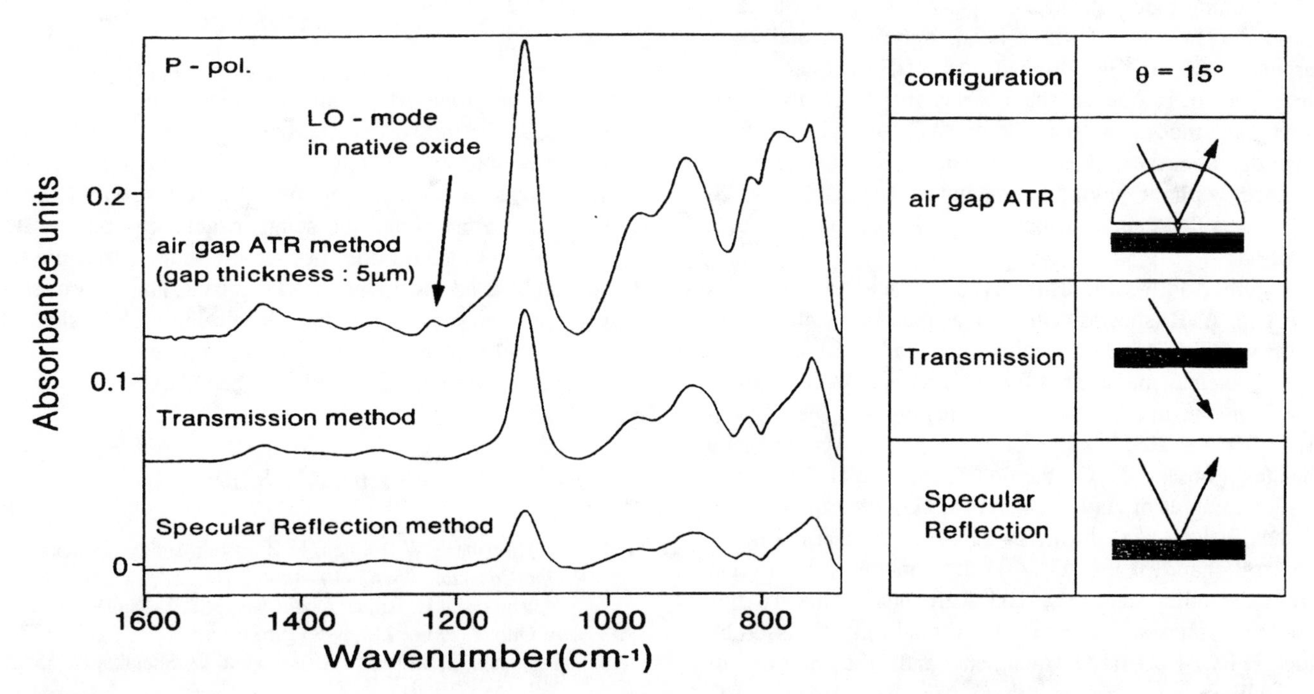

FIGURE 6. Comparison of air the gap ATR spectrum of the silicon wafer covered with native oxide film with those measured by transmission and reflection at the same incident angle (15°).

Application to the Surface Characterization of Treated Si and GaAs

Figure 7 shows a series of air gap ATR spectra applied to the growth of native oxide on Si after being etched by 1% HF solutions. In Fig. 7, p-polarized spectra were subtracted from s-polarized spectra in order to eliminate the interstitial oxygen vibrations and phonons. It can be seen that the LO mode absorption intensity increases and shifts to higher wavenumber. Therefore the Si-O-Si bond angle gradually increases as the oxidation proceeds.

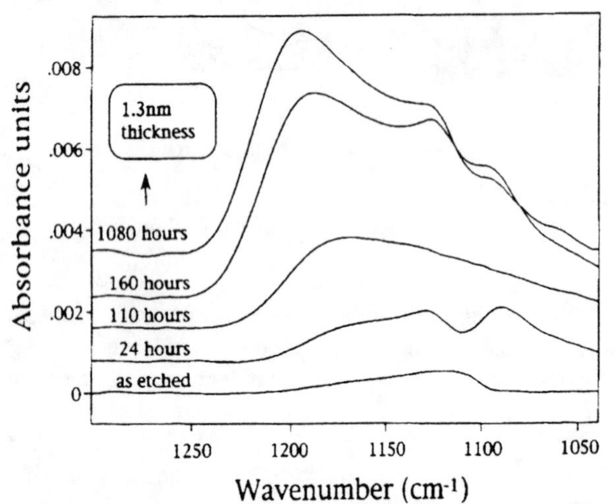

FIGURE 7. Growth of native oxide LO mode on silicon after etching with 1%HF solution.

Figure 8 shows the Si-H stretching mode air gap ATR spectra of HF/NH_4F treated and HF treated Si(111). The sharp 2085 cm^{-1} peak in the HF/NH_4F treatment is the Si-H stretching mode on the atomically flat terrace of the Si surface (7). On the other hand, only the broad band around 2080 cm^{-1} is observed in the HF treatment. The broad band is due to the overlapping Si-H and SiH_2 stretching modes on steps or on the atomically non-flat disordered surface. The surface characterization of the Si surface, without having to manufacture the beveled Si substrates (8), can be achieved by using the air gap ATR method.

Figure 9 shows the as-received and acid treated GaAs air gap ATR spectra. In the as-received sample, we observed the As-O stretching mode at 1050 cm^{-1}, the As_2O_5 skeletal mode at 900 cm^{-1}, and the As_2O_3 skeletal mode in the lower wavenumber region. In the H_2SO_4-H_2O_2-H_2O treated sample, only the As_2O_3 skeletal mode has been observed. On the other hand, the H_3PO_4-H_2O_2-H_2O treatment of As_2O_3, as well as Ga_2O_3 formed on the GaAs surface are shown in Fig. 9. Furthermore, the spectral shape of the As_2O_3 skeletal mode was different from the other samples. The sharp peak shape of this mode indicates an increase in the crystallinity of As_2O_3 by the H_3PO_4-H_2O_2-H_2O treatment and the degree of crystallinity is higher than the H_2SO_4-H_2O_2-H_2O treatment. XPS analysis also indicated the presence of As_2O_3 on the H_3PO_4-H_2O_2-H_2O treated GaAs. However, the XPS analysis does not give us information about the crystallinity of As_2O_3. Therefore, the air gap ATR method is a unique technique for surface studies of semiconductors materials.

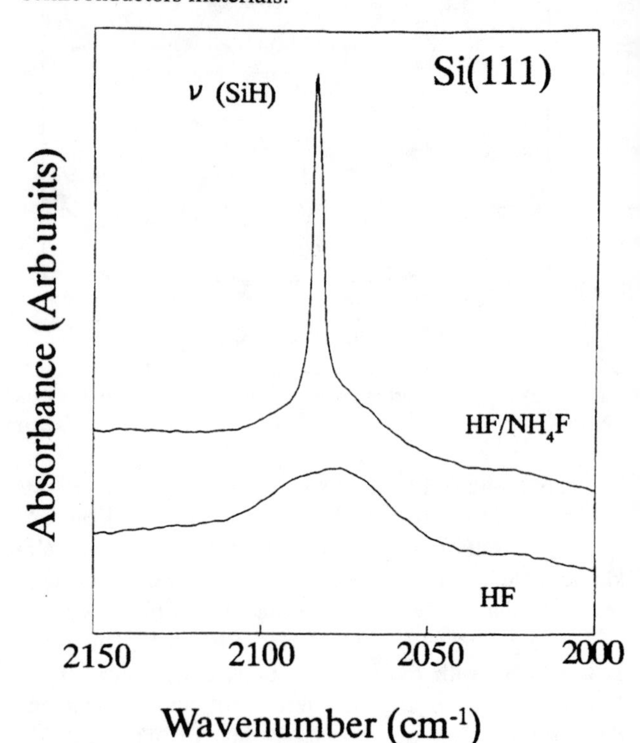

FIGURE 8. Air gap ATR spectra of HF/NH_4F treated silicon wafers.

CONCLUSIONS

We have proposed the air gap ATR method as a new technique for the surface study of semiconductors. The increased sensitivity is due to the electric field created on semiconductor surfaces by multiple reflection of light between the prism and the semiconductor samples. The air gap ATR method does not require the manufacture of special beveled substrates as ATR prisms, and can cover a wide infrared spectral region. The air gap ATR method is very useful for the surface studies of high refractive index semiconductor materials such as Si and GaAs.

REFERENCES

1. A. Hjortsberg, W.P. Chen, E. Burnstein and M. Pomerants, *Opt. Commun.*, 25, 65 (1978).
2. Y. Ishino and H. Ishida, *Anal.Chem.*, 58, 2448 (1986).
3. A. Otto, *Z. Phys.*, 216, 398 (1968).
4. Y. Suzuki, S. Shirnada, A. Hatta and W. Suetaka, *Sur. Sci.*, 219, 1595 (1989)

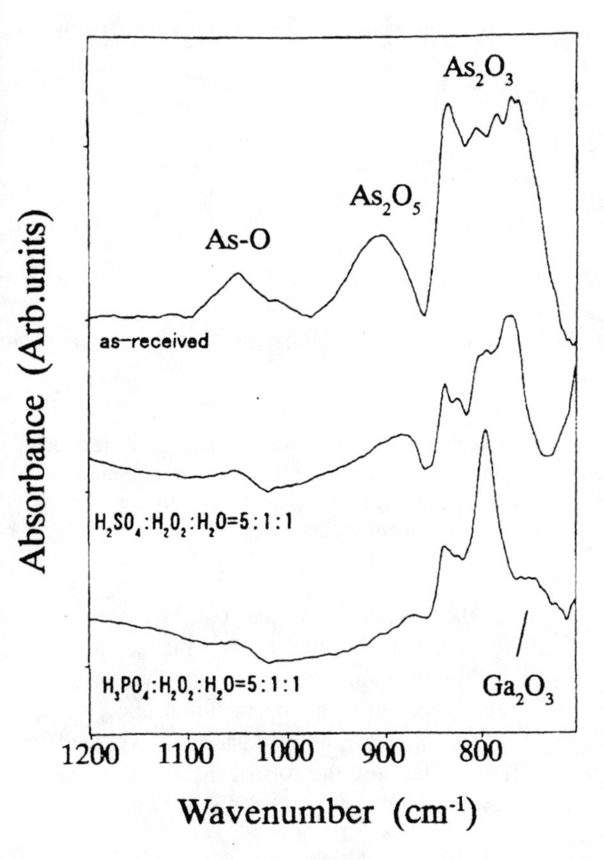

FIGURE 9. Air gap ATR spectra of the as-received and acid treated GaAs wafers.

5. N.J. Harrick, Internal Reflection Spectroscopy, (Wiley-Interscience, NewYork, 1967).
6. B. Harbecke, B. Heinz and P. Grosse, *Appl. Phys.*, A38, 263 (1985). 7. G.S. Higashi, Y.J. Chabal, G.W. Trucks, and K. Raghavachari, *Appl. Phys. Lett.* , 56, 656 (1990).
8. YJ. Chabal, G.S. Higashi, I.C. Raghavachari, and V.A. Burrows, *J. Vac. Sci. Technol.*, A7, 2104 (1989).

Much Without Touch – Conditions for Surface Enhanced Infrared Absorption

W.Theiß [a)], R.Detemple [a)], F.Ozanam [b)]

a) I. Phys. Inst., Aachen University of Technology, D-52056 Aachen, Germany

b) Laboratoire de Physique de la Matière Condensée, CNRS-École Polytechnique, 91128 Palaiseau, Cédex, France

Optical properties of metal-insulator composite materials are discussed with respect to the effect called 'surface enhanced infrared absorption (SEIRA)'. After an introductory discussion of the electric interaction of polarizable particles the Bergman representation is discussed as the basis of effective medium theories. It is applied to describe experimental spectra and leads directly to the SEIRA effect. Finally we suggest what kind of systems should be prepared to obtain efficient absorption enhancement.

INTRODUCTION

Absorption bands may appear enhanced in optical experiments if the material under consideration is placed close to inhomogeneous noble metal films. This well-established experimental fact [1] could be used to lower detection limits in the investigation of small amounts of materials or low concentrated impurities.

Before this so-called 'surface enhanced infrared absorption' (SEIRA) can be utilized in spectroscopic methods such as transmission or attenuated total reflection (ATR) the effect should be understood in detail in order to optimize it.

In this work we first introduce to the topic by an intuitive look on the response of a finite number of polarizable particles exposed to an external electric field. Here the SEIRA effect is already found and explained.

Then we present the Bergman representation as the basis of a proper description of macroscopic experiments. It is applied to investigate inhomogeneous silver layers that are deposited by sputtering on glass substrates. After a successful simulation of the experimental findings we discuss topologies leading to optimized absorption enhancement.

LOCAL FIELD ENHANCEMENT: A MICROSCOPIC VIEW

To understand the effect of surface enhanced infrared absorption it is instructive to have a microscopic view on what happens when a finite number of small particles is exposed to an electric field (of a light wave). We consider small metal spheres and, in addition, 'test spheres' of a different material (the molecules to be identified) somewhere in the scenery. All particles are treated as dipoles in the calculation that takes into account the coupling of the dipole moments due to the electric dipole fields. No retardation effects are included, i.e. we are dealing with electrostatics. The calculations developed for this work [2] are extending the formalism discussed in [3] which were developed for one type of dipoles only.

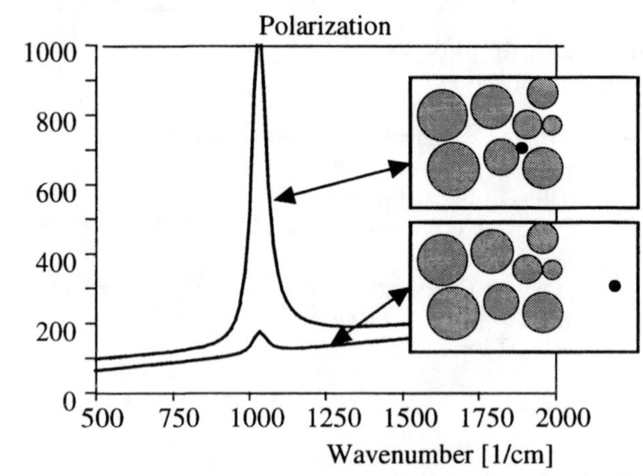

FIGURE 1. Enhancement of an absorption band of a test particle (solid) due to the presence of surrounding silver spheres (dashed). Obviously the strength of the band depends strongly on position which is due to the strong variation of the local electric field. Note the broad polarization 'background' which is due to absorption in the silver clusters.

An example is given in fig.1 where the resulting imaginary part of the total polarization of the sketched configurations of particles is displayed. The absorption band of the test particle appears in very different strengths depending on its position with respect to the surrounding silver clusters. Far away from the metal particles the band is weak whereas in between it is amplified strongly.

CP430, *Fourier Transform Spectroscopy:* 11th International Conference
edited by J.A. de Haseth

The computational algorithm allows also the calculation of the position-dependence of the electric field. It turns out that the internal local fields in between the metal particles are much stronger than the externally applied field. Enhancement factors larger than 200 were found in the simulations. It is noted that the largest internal fields are obtained when the metal clusters are packed densely.

EFFECTIVE MEDIUM THEORIES

The description of macroscopic experiments requires the application of an effective medium theory [3] which gives the response of an extended composite system to an external electric field. Effective medium concepts replace the inhomogeneous composite by a homogeneous effective medium with the same optical properties as the real system (see fig. 2). The characteristic dimensions of the microtopology must be much smaller than the light wavelength which is the case in typical SEIRA systems.

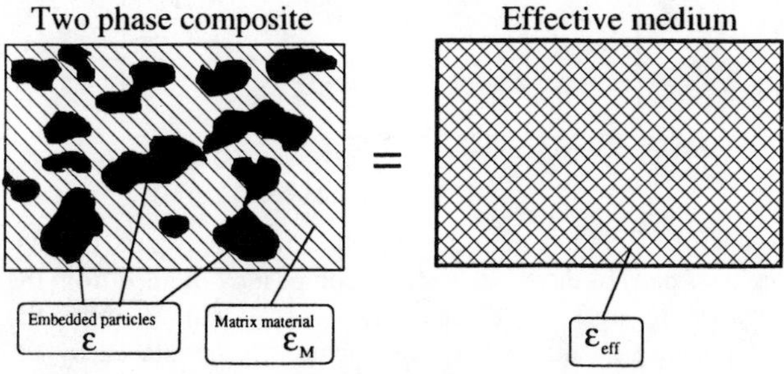

FIGURE 2. Effective medium principle: The details of the microtopology (which are not resolved in optical experiments as long as the light wavelength is much larger than typical particle sizes and distances) are averaged to a quasi-homogeneous effective medium with dielectric function ε_{eff}. The system consists of a host material with dielectric function ε_M and embedded particles with dielectric function ε.

Simple and often applied effective medium concepts are the Maxwell Garnett [4], Bruggeman [5] and Looyenga [6] formulae which use only one parameter to characterize the microtopology, namely the volume fraction of the embedded particles. In many cases more flexible formalisms must be employed (see [3]) which can be derived from the general Bergman representation [7]. The latter gives an integral expression for the effective dielectric function based on the dielectric functions of the materials in the system and the so-called spectral density g(n,f) which is a statistical distribution function of so-called geometric resonances (see the discussion in [3]):

$$\varepsilon_{\text{eff}} = \varepsilon_{\text{M}}\left(1 - f\int_0^1 \frac{g(n,f)}{\dfrac{\varepsilon_{\text{M}}}{\varepsilon_{\text{M}} - \varepsilon} - n}dn\right) .$$

Unfortunately the Bergman representation has been developed only for two-phase composites. SEIRA systems consist of at least three phases (vacuum, metal, molecules), on the other hand, which obviously leads to difficulties. In this work we apply the following approximation. Without test molecules we use silver particles embedded in vacuum. The latter is then replaced by a diluted version of the test material's dielectric function as if it would fill the whole space in between the silver clusters.

As a model system for metal-insulator composites we have prepared sputtered silver films on glass substrates. The deposition was done for various time intervals in a way to produce series of films ranging from isolated metal islands to thick homogeneous silver layers. The obtained layers were investigated in the mid infrared (IR, 500 ... 5000 1/cm, Bruker IFS 48 spectrometer) and the visible-UV (Vis/UV, 1100 ... 200 nm wavelength, Perkin-Elmer λ2 grating spectrometer) spectral range.

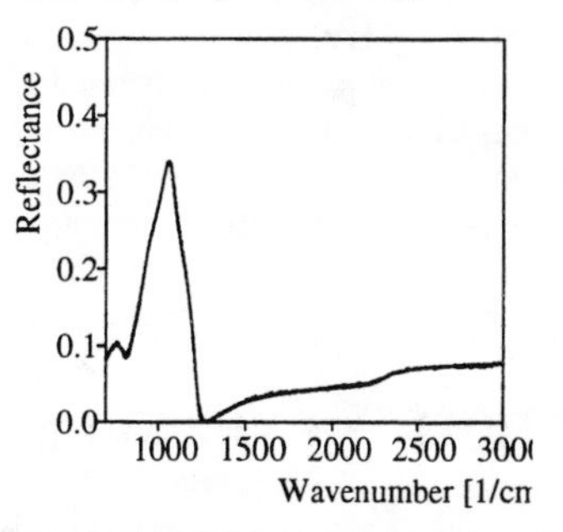

FIGURE 3. Infrared reflectivity (30° angle of incidence, s-polarization) of a thin silver film (2 s sputtering time) on a glass substrate (Solid line: experiment, dashed: simulation). The thickness of the metal-vacuum composite is 7 nm and the metal volume fraction is 50%.

The spectra of the thin island films are an excellent model system to test the performance of effective medium theories. The silver islands are extremely high polarizable leading to strong electric interactions between the individual particles. The simple effective medium formulas severly fail to describe the spectra quantitatively. The complete series of thin film samples ranging from unconnected islands structures to a complete percolated film can be described using a flexible parameterization of the Bergman representation. The form of the spectral density is specified by a set of definition points which are connected by spline interpolated continuous curves.

It can be seen by closer inspection of the dielectric functions that optical spectra in the visible region depend mainly on the spectral density in the range n = 0.05 ... 0.3 whereas in the mid infrared only resonances in the range n = 0 ... 0.005 can be seen. Hence the form of the spectral density in this range is important.

As an example we show results for the sample obtained with 2 s sputtering time. The silver islands are unconnected and the reflectivity in the infrared is quite low. The large structures in the infrared spectrum (see fig. 3) are due to the glass substrate. With a proper parameterization of the spectral density an excellent agreement of simulated and measured spectra is achieved (see also fig.3). The spectral density that is obtained from the fit is shown in fig. 4. The effective dielectric function of the silver-vacuum composite shown in fig. 6 (solid lines) is that of a highly polarizable dielectric material.

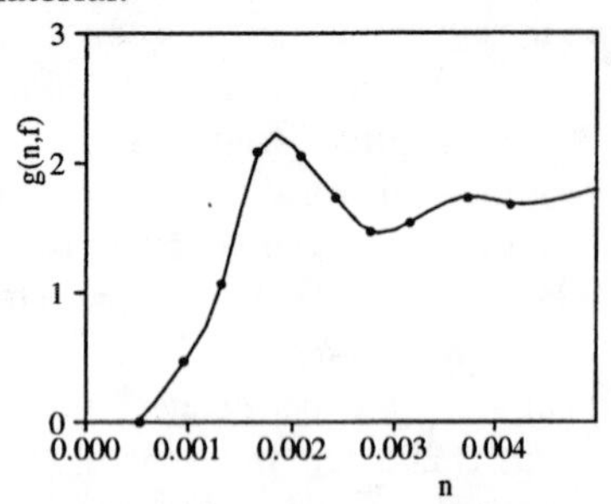

FIGURE 4. Spectral density used for the fit shown in fig.3. This part of the spectral density is responsible for the infrared spectral range in the case of silver particles embedded in vacuum. The form is characteristic for a system of unconnected particles of medium density (volume fraction of silver: 50%). The black dots indicate the definition points whose heights are the fitting parameters.

ABSORPTION ENHANCEMENT

The description of the metal-insulator system obtained this way already features the SEIRA effect. If the vacuum component is replaced in the simulation by a test material with a weak oscillator at 2000 1/cm (see fig. 5) the new effective dielectric function exhibits an amplified version of this absorption band (see fig. 6, dashed line). The peak height in the imaginary part is enhanced by a factor of about 10. Note the asymmetric peak form.

Due to the amplification the absorption band could easily be detected in an ATR experiment as the model calculation shown in fig.7 demonstrates. On the other hand, without the help of the metal clusters almost no absorption band would be seen (see also fig. 7).

With increasing film thickness more and more connections between the metal particles are built up which leads to an increasing dc-conductivity and a negative real part of the effective dielectric function of such systems. The 'test absorption band' also appears enhanced in such systems but it is very 'deformed' and an experimental identification is more difficult.

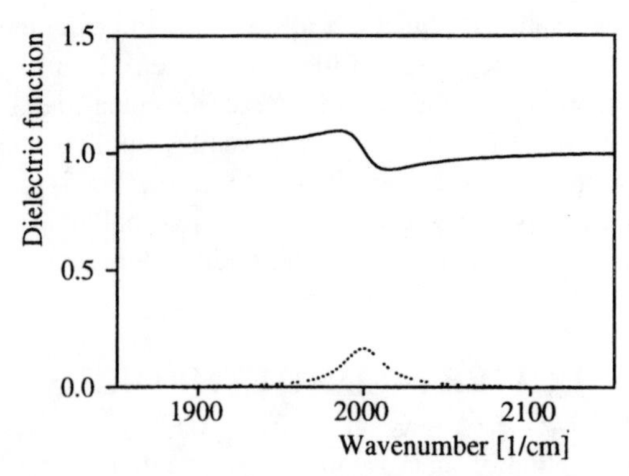

FIGURE 5. Dielectric function of the test material used in this section. As discussed above it is assumed that the whole space between the metal clusters is filled with this material.

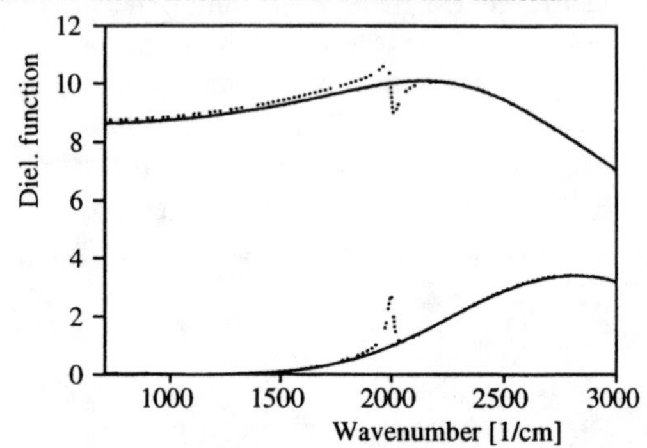

FIGURE 6. Effective dielectric function (top: real part, bottom: imaginary part) of the silver-vacuum composite obtained from the fit shown in fig.3 (solid line). The dashed curves show the corresponding quantities with the vacuum host replaced by the test material.

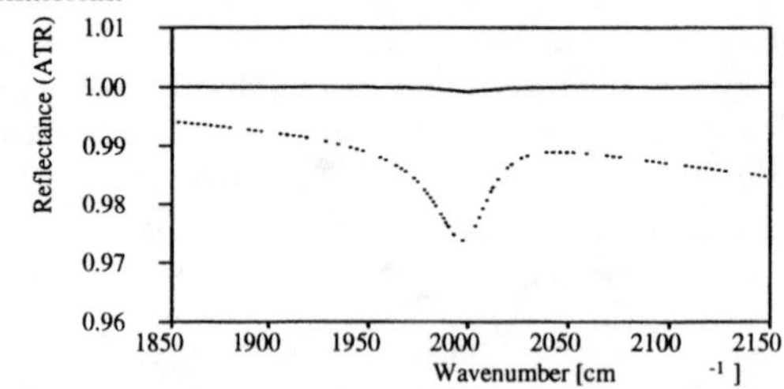

FIGURE 7. Simulated ATR spectra (30°, s-polarized light, silicon prism) of a 3.5 nm layer of the test substance (solid line) and a 7 nm composite layer (50% silver, 50% test substance, plotted dashed).

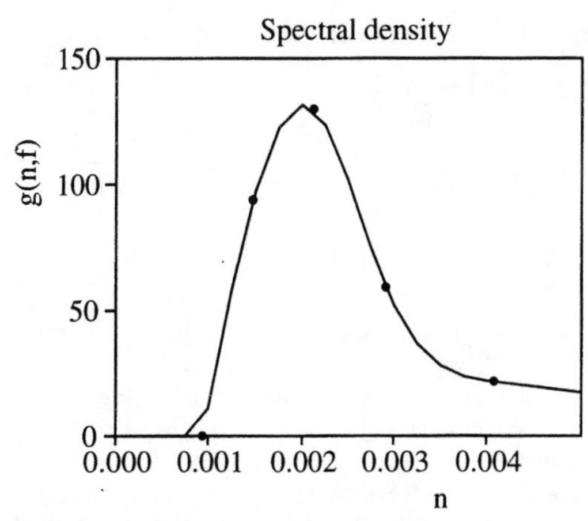

FIGURE 8. Suggested spectral density for very high SEIRA performance.

Now, on the basis of a quantitative spectrum simulation, one can speculate about the best configuration for SEIRA. Certainly the goal is to get an undistorted version of an absorption band which should be amplified as much as possible. From our experience in spectrum simulation the ideal system for SEIRA is one with a high volume fraction of metal particles which are still unconnected. A hypothetical system with a volume fraction of 60% and a spectral density shown in fig. 8 would lead to the effective dielectric function shown in fig. 9 which is much more amplifying the test oscillator as the one displayed in fig. 6. The same amount of test material as considered previously (see fig. 7) could now be observed easily in a single external reflection as shown in fig.10.

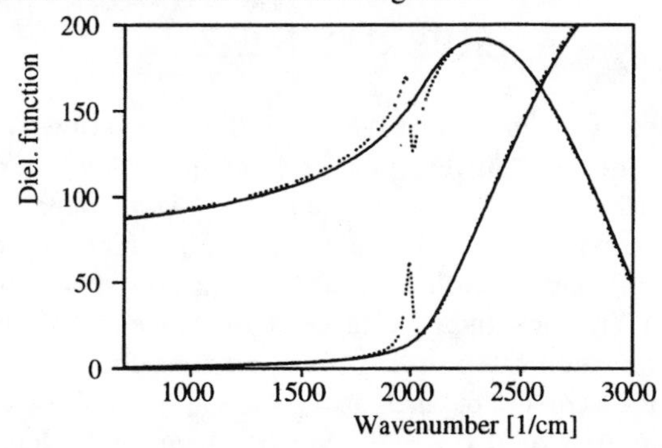

FIGURE 9. The same as fig. 6 but now for a system of larger metal volume fraction (60%) and the spectral density shown in fig. 8.

Unfortunately the relation between spectral density and topology is unknown in general. This means that we do not know how to realize metal films with optimal SEIRA performance. It might help to know that the spectral density shown in fig.8 has been obtained from a spectral density that was adjusted to fit experimental data (3 s sputtering time). The real sample shows dc-conductivity and a large metallic reflection and – as mentioned above – the bands of test materials would appear not in their original form but very distorted. For the suggestion of fig.8 the connections of the silver clusters have been switched off in the spectral density. This could be obtained in reality by tempering the sample which causes in many cases a decrease of the silver network connectivity.

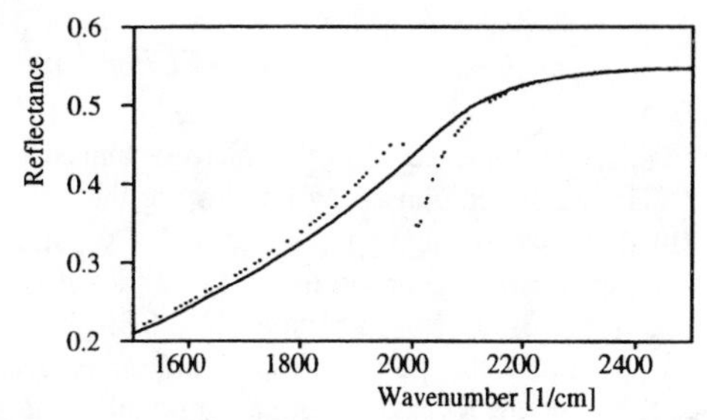

FIGURE 10. External reflectance (30°, s-polarized light) without (solid line) and with (dashed) test substance.

CONCLUSION

Applying the Bergman representation a quantitative description of optical spectra of metal-insulator composites has been achieved. This was used to verify the effect of surface enhanced infrared absorption and to suggest dense systems of unconnected metal clusters for maximum performance.

The suggestion is to be justified by experimental work which will be done in the future. Also – for practical applications in infrared spectroscopy – stable metal films have to be developed which can be used many times, i.e. which can be coated with test molecules and cleaned after the optical measurements.

REFERENCES

[1] Y. Nishikawa, T. Nagasawa, K. Fujiwara, Vibrational Spectroscopy **6** (1993), 43-53
[2] W.Theiß, F.Ozanam, unpublished
[3] W.Theiß, in Festkörperprobleme/Advances in Solid State Physics **33**, ed. by R.Helbig (Vieweg, Braunschweig, Wiesbaden 1994), 149
[4] J.C.Maxwell Garnett, Philos. Trans. R. Soc. London **203** (1904), 385
[5] D.A.G. Bruggeman, Ann. Phys. **24** (1935), 636
[6] H.Looyenga, Physica **31** (1965), 401
[7] D.Bergman, Phys. Rep. **C 43** (1978), 377

Surface-Enhanced Infrared Absorption Studies of p-Nitrothiophenol on Silver

Lin-Tao He and Peter R. Griffiths*

Department of Chemistry, University of Idaho, Moscow, ID 83844-2343

SEIRA spectra measured in the microtransmission mode have been used to investigate the chemistry of p-nitrothiophenol (PNTP) on silver films. By monitoring the S-H stretching band at ~2550 cm^{-1}, we demonstrated that PNTP dimerizes oxidatively when applied to bare ZnSe substrate or Ag surfaces from acetone or methanol solution under ambient conditions. SEIRA spectra of the PNTP dimer were measured on both silver underlayer and overlayer surfaces by casting films of the dimer from solution. The SEIRA spectrum of PNTP dimer on the Ag overlayer is very different from, and more enhanced than, the corresponding underlayer spectrum. We conclude that the S-S bond of the dimer is broken and Ag-S bonds are formed when energetic silver atoms are vapor deposited on a film of the PNTP dimer already on ZnSe plate. PNTP monomer was passed through a GC column and deposited on a 5-nm Ag layer on a ZnSe plate mounted in a direct deposition GC/FT-IR interface. At temperatures above -30°C, the SEIRA spectrum of PNTP monomer measured in this way is similar to that of PNTP dimer measured with an Ag overlayer, indicating that the active proton of the S-H group of the PNTP monomer dissociates and the PNTP chemisorbs on the Ag surface via an S-Ag bond. It was also found that the chemical reaction between PNTP monomer and the Ag surface can occur at temperatures as low as -30°C, but not as low as -40°C.

INTRODUCTION

Since the discovery of the phenomenon of surface-enhanced infrared absorption (SEIRA) by Hartstein et al. [1], the effect has been observed in attenuated total reflection (ATR) [2], external reflection and transmission measurements [3,4]. The SEIRA spectrum of the model compound, p-nitrobenzoic acid, has been reported in several publications [2, 5-11]. Osawa et al. [10,11] investigated the band selection rule for the SEIRA based on their results with various nitrobenzoic acids and proposed that the dipole derivatives of bands due to the adsorbate must have a component in the direction perpendicular to the surface to be infrared active.

Molecules containing functional groups with acidic protons, *e. g.*, -COOH, -OH and -SH, can chemisorb on the surface of metals such as silver, copper and gold. In this work, SEIRA has been used to investigate the chemistry of p-nitrothiophenol (PNTP) on silver films. Thin films of PNTP were applied to the surface of a silver-coated ZnSe plate both from solution, with SEIRA spectra measured in the microtransmission mode with both silver over- and under-layers, and from the vapor phase using the Bio-Rad *Tracer* direct deposition (DD) GC/FT-IR interface.

Aromatic thiols are oxidatively dimerized to disulfides under relatively mild conditions through thiyl radicals.

$$2RSH + 1/2\ O_2 \longrightarrow RSSR + H_2O$$

The kinetics of the oxidative dimerization of some thiols have been investigated by Wakefield and Waring [12] and D'Souza et al. [13] and found to be very fast. Using mass spectrometry, Waring and Wakefield found that the dimerization of p-nitrothiophenol and other aromatic thiophenols during thin-layer chromatography was essentially instantaneous.

EXPERIMENTAL

Underlayer and overlayer SEIRA by casting films from solution. Silver films, 10 nm in thickness, were deposited on ZnSe substrates via physical vapor deposition at a pressure of 2 $\times 10^{-6}$ Torr. The film thickness was monitored and controlled with a quartz crystal thickness monitor (Kronos Inc., Torrance, CA, Model ADS-200). The deposition rate was kept at approximately 1 nm/min. A dilute acetone or methanol solution of PNTP at a concentration of 6.4 x 10^{-4} M (100 ng/μl) was applied either to the bare ZnSe substrate or the Ag-coated substrate with a microsyringe so that 30-50 ng of PNTP was present on the surface after the solvent had evaporated in the air. For overlayer SEIRA spectra, the silver film was deposited after the solution had been applied to the ZnSe plate and dried. PNTP was also adsorbed from the vapor phase using a Bio-Rad/Digilab (Cambridge, MA) *Tracer* direct deposition GC/FT-IR interface [14]. In this case, spectra were measured at a resolution of 4 cm^{-1} with a Bio-Rad

CP430, *Fourier Transform Spectroscopy:* 11th International Conference
edited by J.A. de Haseth

FTS-40 FT-IR spectrometer equipped with Schwarzchild microscope optics and a narrow-band MCT detector.

***Tracer* GC SEIRA.** 0.3μl of a solution of PNTP (30 ng) was injected into an on-column injector and eluted from a 30-m long, 0.32-mm I.D., 0.5-μm thick DB-17 capillary column (J&W Scientific, Folsom, CA) in a Hewlett-Packard (Palo Alto, CA) 5890 gas chromatograph. GC elutes were passed into the *Tracer* GC/FT-IR interface and deposited on a ZnSe window, which had been pre-coated with a 5-nm thick Ag film and cooled to ~-85°C with liquid nitrogen. The vacuum chamber of Tracer interface was maintained at a pressure below 1.0 x 10^{-4} Torr. The infrared spectra were recorded either on-line during the GC run or post-run.

RESULTS AND DISCUSSION

1. The dimerization of PNTP

The spectrum of PNTP powder prepared as a KBr pellet is shown in Fig. 1A; the -S-H stretching band can be seen at ~2550 cm^{-1}. However, when the PNTP solution (acetone or methanol) was deposited on a bare ZnSe window and dried in air, the spectrum lacked the peak at ~2550cm^{-1}, as shown in Fig.1B. This spectrum is similar to a reference spectrum of the 4-nitrophenyl disulfide [15], indicating that PNTP in solution is oxidatively forming the disulfide upon evaporation of solution in air.

2. Ag underlayer and overlayer SEIRA of PTNP dimer

To investigate the difference in the SEIRA spectra of the PNTP dimer with a thin silver under- and over-layer, films of 4-nitrophenyl disulfide were prepared in two ways. For the spectrum of the dimer with a silver underlayer, a solution of PNTP was applied on a ZnSe plate coated with a 5-nm thick film of silver and the film of PNTP was allowed to oxidatively dimerize in the air. To prepare the corresponding spectrum with a silver overlayer, a film of the dimer was prepared on a bare ZnSe plate and a 5-nm film of silver was formed on top of the dimer layer by physical vapor deposition.

The SEIRA spectra of the sample with the Ag overlayer and underlayer are shown in Figure 2, along with the unenhanced spectrum of PNTP dimer on a bare ZnSe plate. The SEIRA spectrum of PNTP dimer on an Ag underlayer is very similar to its unenhanced infrared spectrum. The symmetric NO_2 stretch at 1339 cm^{-1} is only enhanced by a factor of about 8. However, the SEIRA spectrum of PNTP dimer with the Ag overlayer is qualitatively different from, and more enhanced than, the corresponding spectrum with the Ag underlayer. For the SIERA spectrum in the case of the Ag overlayer , the symmetric NO_2 stretching band at 1338 cm^{-1} is enhanced by a factor of ~28. The band from the corresponding antisymmetric stretching vibration at 1514 cm^{-1}, which is very strong in the unenhanced IR spectrum of PNTP dimer, is very weak in the SEIRA spectrum measured with an Ag overlayer. We conclude that the S-S bond of the dimer is broken and Ag-S bonds are formed when energetic silver atoms are vapor deposited on a film of the PNTP dimer already on ZnSe plate. Since the molecules are oriented on the surface of the Ag overlayer, their absorption bands accurately follow the surface selection rule.

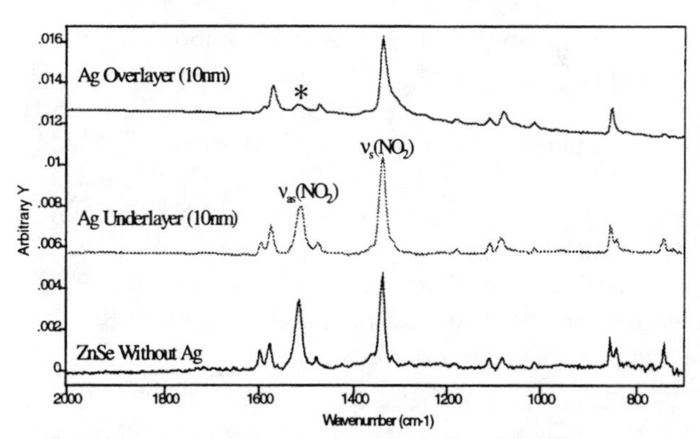

Figure 2. (Above) SEIRA spectra of PNTP dimer measured on an Ag overlayer and (middle) on an Ag underlayer . (Below) Unenhanced IR spectrum of PNTP dimer on ZnSe plate.

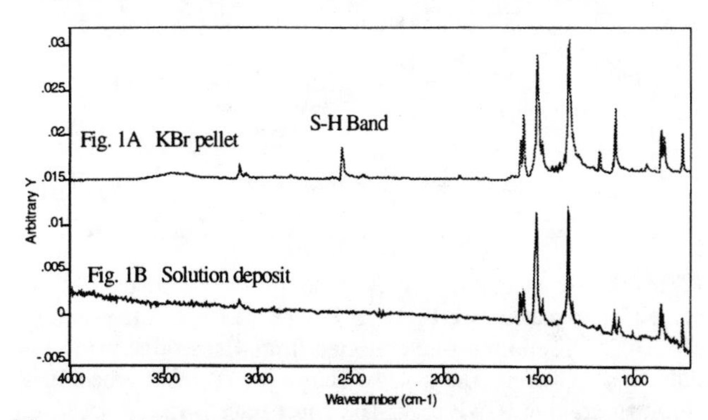

Figure 1. (Above) Spectrum of PNTP powder in a KBr Pellet. (Below) Spectrum of PNTP deposited from solution onto a bare ZnSe window.

3. Direct deposition GC/SEIRA of PNTP monomer

The spectrum of a thin film of the PNTP monomer can be obtained on a ZnSe plate in the complete absence of a solvent by using the Bio-Rad *Tracer* GC/FT-IR interface. Since this interface is evacuated, the conditions for the oxidative dimerization of p-nitrothiophenol are no longer present. The PNTP monomer was passed through the GC column and deposited both on a bare ZnSe window or a ZnSe plate with a 5-nm Ag layer deposited on its surface. The temperature of the plate could be varied. Figure 3 shows the unenhanced spectrum of PNTP monomer on a bare ZnSe plate at -85°C measured in the *Tracer.* The same sample was deposited at -90°C at 1x10^{-4} Torr on a 5-nm Ag layer on ZnSe and spectra were recorded as the plate was allowed to warm up. The spectra recorded at -

10°C and +40°C are also shown in Figure 3. The S-H stretching band is present in the unenhanced spectrum of PNTP monomer. However, the S-H band disappears and the antisymmetric NO_2 band becomes very weak in the SEIRA spectra of PNTP monomer measured in this way. These spectra are similar to the SEIRA spectrum of PNTP dimer measured with an Ag overlayer, indicating that the active proton of the PNTP monomer dissociates and the PNTP chemisorbs on the Ag surface via an S-Ag bond. The fact that the PNTP monomer can chemisorb on the Ag surface even at +40°C and a pressure of less than $1x10^{-4}$ Torr indicates that the strength of the S-Ag bond is quite high.

To investigate the lowest temperature at which a chemical reaction between the PNTP monomer and the Ag surface can take place, depositions of PNTP from GC column were made on a silver-coated ZnSe window at temperatures of -18°, -30°, and -40°C under a vacuum of $1x10^{-4}$ Torr. The resulting spectra are shown in Figures 4 - 6. In the SEIRA spectrum obtained from the deposition at -18°C, the S-H stretching band has disappeared and the antisymmetric NO_2 stretching band is very weak. These observations indicate that a reaction between PNTP monomer and the silver surface to form an S-Ag bond can occur at a temperature as low as -18°C, even though the sample evaporates rapidly from a bare ZnSe plate at this temperature and pressure. In contrast, when the PNTP monomer is deposited on the Ag surface at -40°C, the S-H stretch and the strong antisymmetric NO_2 stretch are readily observed (see the lower trace of Fig. 4). These bands are characteristic of the spectrum of unreacted PNTP monomer and indicate that chemisorption does not occur at -40°C. Thus after elution from the GC column, the PTNP monomer simply condenses on the Ag-coated ZnSe plate without reaction at temperatures below -40°C.

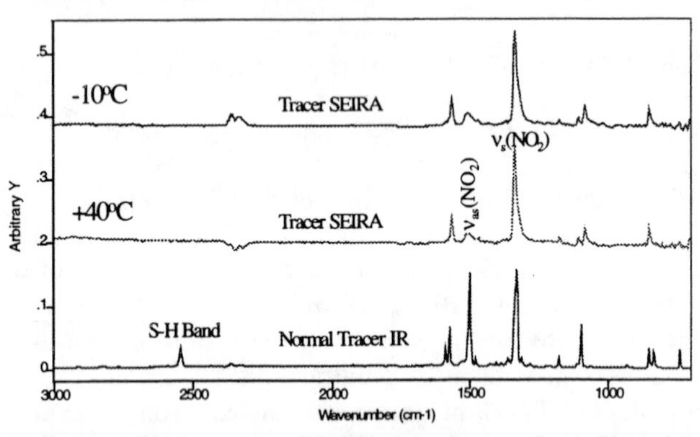

Figure 3. SEIRA spectra of PNTP monomer on 5-nm film of Ag on ZnSe plate mounted in the Bio-Rad *Tracer* at pressure of 9 x 10^{-5} Torr. The sample was deposited at -90°C, and the temperature of the plate was allowed to increase to -10°C (above) and +40°C (middle). (Below) Unenhanced DD GC/FT-IR spectrum of PNTP monomer on ZnSe plate.

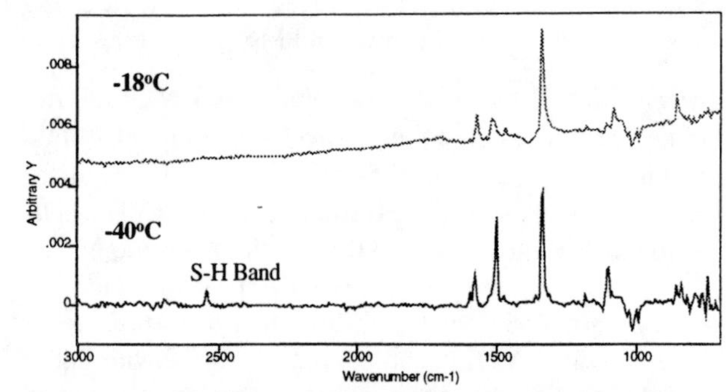

Figure 4. IR spectra of PNTP monomer deposited on the Ag-coated ZnSe plate in the Bio-Rad *Tracer* interface at temperatures of -18°C (above), and -40°C (below). The spectra were collected from 256 co-add scans in the post-run mode after GC deposition.

Figures 5 and 6 illustrate that chemisorption, condensation and evaporation of the PNTP monomer occur when the deposition is made at -30°C. The IR spectra collected from thin layer of sample either on-line during GC run or post-run at -30°C are similar to the spectrum from the deposition at -18°C, indicating that the chemical reaction between PNTP monomer and the Ag surface can occur at a temperature as low as -30°C. The IR spectrum measured at the apex of the chromatographic peak immediately after deposition of the sample, with the plate held at -30°C, is the same as that of the unenhanced IR spectrum of PNTP monomer on a bare ZnSe plate (shown in the lower trace of Fig. 3). However, in the corresponding post-run IR spectrum, the S-H stretching band, characteristic of PTNP monomer, is missing. This result indicates that, with the sample at -30°C under a vacuum of $8.5x10^{-5}$ Torr, the lower layers of PTNP monomer have reacted with the Ag film to form an S-Ag bond and the upper layers have evaporated during the time between the on-line and post-run measurements.

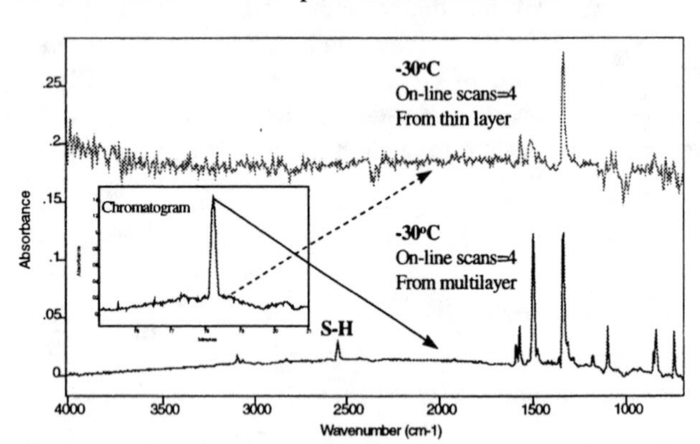

Figure 5. IR spectra of PNTP monomer deposited on the Ag-coated ZnSe plate in the Bio-Rad *Tracer* interface at temperature of -30°C. The spectra were collected from 4 co-add scans in on-line during GC run. The window shows the IR reconstructed gas chromatogram of PTNP peak. The upper trace is the IR spectrum collected from a thin layer of sample in the wings of the GC peak. The lower trace is the corresponding multilayer spectrum from the apex of the GC peak.

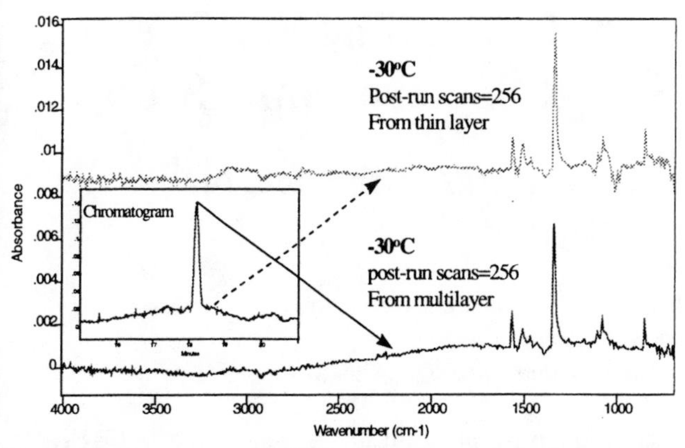

Figure 6. IR spectra of PNTP monomer deposited on the Ag-coated ZnSe plate in the Bio-Rad *Tracer* interface at temperature of -30°C. The spectra were collected from 256 co-add scans in post-run after GC deposition. The window is IR restructured GC chromatogram of PTNP peak. The upper trace is the IR spectrum collected from a thin layer of sample in the wings of the GC peak. The lower trace is the corresponding multilayer spectrum from the apex of the GC peak.

CONCLUSIONS

The oxidative dimerization of p-nitrothiophenol can be readily monitored by FT-IR spectrometry. Surface-enhanced infrared absorption spectrometry has been combined with the direct deposition GC/FT-IR technique to investigate the chemistry of PNTP on silver films. The PNTP disulfide dimer appears to be cleaved by energetic silver atoms and to form an S-Ag bond in the Ag overlayer surface geometry. The GC/SEIRA spectra demonstrate that the S-H bond of the PNTP monomer dissociates and PNTP chemisorbs on the Ag surface via an S-Ag bond. This reaction can occur at a temperature as low as -30°C, but no lower than -40°C.

REFERENCES

1. Hartstein, A., Kirtley, J. R., Tsang, T. C., Phys Rev. Lett. **45,** 201-204 (1980).
2. Hatta, A., Ohshima, T., Suetaka, W., Appl. Phys. A **29**, 71 (1982).
3. Kamata, T., Kato, A., Umemura, J., Takenaka, T., Langmuir **3**, 1150 (1987).
4. Nishikawa, Y., Fujiwara, K., Shima, T., Appl. Spectrosc. **44**, 691 (1990).
5. Hatta, A., Suzuki, Y., Suetaka, W., Appl. Phys. A **35**, 135-140 (1984).
6. Suzuki, Y., Osawa, M., Hatta, A., Suetaka, W., Appl. Surf. Sci. **33/34**, 875 (1988).
7. Badilescu, S., Ashirt, P.V., Truong, V-V., Appl. Phys. Lett. **52**, 1551 (1988).
8. Badilescu, S., Ashirt, P.V., Truong, V-V., Badilescu, I-I., Appl. Spectrosc. **43**, 549-525 (1989).
9. Osawa, M., Ikeda, M., J. Phys. Chem. **95,** 9914 (1991).
10. Osawa, M., Ataka, K-I., Ikeda, M., Uchihara, H., Nanba, R., Anal. Sci., **7**, 503 (1991).
11. Osawa, M., Ataka, K-I., Yoshii, K., Nishikawa, Y., Appl. Spectrosc. **47**, 1497-1502 (1993).
12. Wakefield, C. J., Waring, D. R., J. Chromatogr. Sci., **15**, 82 (1977).
13. D'Souza, V. T., Iyer, V. K., Szmant, H. H., J. Org. Chem. **52**, 1725-1728 (1987).
14. Bourne, S., Haefner, A. M., Norton, K. L., Griffiths, P. R., Anal. Chem. **62**, 2448-2452 (1990).
15. Pouchert, C. J., *The Aldrich Library of FT-IR Spectra*, The Aldrich Chemical Company, Inc., First ed., V.1 p. 1386A (1985).

Vesicles Containing Ion Channels on Crystalline Surfaces - An FTIR and Surface Enhanced FTIR Spectroscopic Study

W. B. Fischer, I. Unverricht, Ch. Kuhne, G. Steiner, A. Schrattenholz[1], A. Maelicke[1], and R. Salzer

Institut für Analytische Chemie, TU Dresden, D-01062 Dresden, Germany
Email: wolfgang.fischer@chemie.tu-dresden.de
[1] *Institut für Physiologische Chemie und Pathobiochemie, Universität Mainz, D-55099 Mainz, Germany*

The kinetics of the adsorption of native vesicles containing the nicotinic acetylcholine receptor (nAChR) is monitored by ATR-FTIR and SEIRA spectroscopy. The membrane vesicles are adsorbed on Ge crystals. Experiments are done with neat Ge and with Ge crystals covered with a thin layer of silver clusters in order to obtain the enhancement effect of infrared adsorption. The nAChR shows β-sheet/turn structures at the interface. These results give evidence for the existence of these structures in the extracellular domains of the receptor. The potential for SEIRA in the investigation of proteins at interfaces and membrane processes is outlined

INTRODUCTION

Ion channels are a special class of integral membrane proteins responsible for the electrochemical signal transduction at the synapses and neuromuscular junctions. One type of ion channel is the nicotinic acetylcholine receptor (nAChR). Activated by the neurotransmitter acetylcholine the receptor allows the diffusion controlled flow of Na^+ and K^+ across the cell membrane (1). The receptor is a 255 kD pentameric protein consisting of subunits in the stoichiometry $2\alpha:\beta:\gamma:\delta$ (2). Each subunit consist of at least four membrane spanning regions called M1 - M4 which are expected to be α helical. Of all subunits the membrane spanning region M2 faces the pore of the channel and guides the ion across the protein. It is still not clear if more membrane spanning regions exist (3). A recent molecular modelling study predicted significant amounts of α-helix and β-sheet elements in the extracellular domains of the receptor (4).

The nAChR is prepared from the electric organ of the ray *Torpedo marmorata*. After this preparation the receptor is embedded in vesicles consisting of the natural lipids, nAChR, and other proteins present in the membranes of the electric organ.

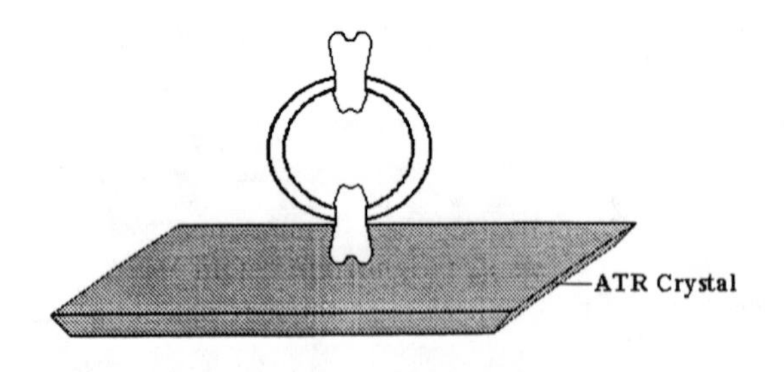

Scheme 1: Prinziple of adsorption of vesicles containing the nAChR

The preparations are assumed to lead preferentially to a so-called right-side out configuration of the receptor, meaning the extracellular domains are facing the outside of the vesicle. To use the nAChR for analytical purposes, i.e. in sensor-actor devices, the vesicles have to be immobilized on artificial materials like crystals or synthetic lipids. The immobilization might either be realized by adsorption of the intact vesicles (see scheme 1) or by fusion of the vesicles on an interface.

The nAChR is extensively studied by FTIR spectroscopy by various groups using purified nAChR membranes (5) and affinity-purified reconstituted nAChR (6-8). Structural changes of the receptor upon ligand binding is monitored by FTIR difference spectroscopy (9,10) and Fluorescence spectroscopy (9,11). There is an ongoing controversy about the contribution of the secondary structural elements observed by spectroscopic methods (for a brief review see (5)).

In this study we used FTIR and, for the first time, surface enhanced FTIR spectroscopy (SEIRA) [see also 12. to monitor the adsorption of vesicles containing the nAChR onto Germanium crystals. This allows us also to illustrate the potential of the SEIRA technique. The data in this study are interpreted on a qualitative basis.

EXPERIMENTAL

nAChR-rich fragments (3 mg/ml; 5 nmol of ACh-sites/mg) where obtained by homogenization of 200 g of *Torpedo marmorata* electric organ, followed by several centrifugation steps and alkaline treatment as described elsewhere (13). Pellet membranes were dissolved in a

CP430, *Fourier Transform Spectroscopy:* 11th International Conference
edited by J.A. de Haseth

buffer containing 250 mM NaCl, 5 mM KCl, 5 mM NaH_2PO_4 (pH 7.4) and stored in 0.5 ml aliquots at -80° C. Samples (200 µl aliquots) were washed twice by spinning down for 3 min at 8000 U/min at 2°C and uptake with buffer containing 150 mM NaCl, 2 mM $CaCl_2$ and $MgCl_2$, 4 mM KCl, and 10 mM HEPES (2-[4-(2-hydroxyethyl)-1-piperazinyl.-ethan-sulfonic acid, pH 7).

Germanium crystals were coated with a 3 nm layer of silver applying a cover rate of 0.5 nm/s prior to vesicle adsorption. 200 µl of solution containing the vesicles where brought onto the crystals. A continuous flow sample holder for ATR spectroscopy (SPECAC, Graseby, UK) was used.

Natural lipids (0.5g) from soybean extract S 100 (>94% Phosphatidylcholine, from LIPOID GmbH, Ludwigshafen, Germany), were resolved in 10 ml methanol. Methanol was evaporated in a rotary evaporator. The lipids were resuspended in 10 ml destilled water forming various sizes of vesicles. 1 ml of this solution was placed on a Ge crystal to form self assembled layers of various consistence. After ca. 2 h the Ge crystal was carefully rinsed (30 min) by a continuous flow of buffer solution. Afterwards vesicles were added.

FTIR spectra where recorded on a BRUKER IFS 88 with MCT detector and purged with dry air (dew point -70°C). The resolution was 2 cm^{-1} using a Happ-Genzel apodization. 1024 interferograms were coadded for a single spectrum representing 3 minutes of data collection. Spectra were compensated for buffer by interactive subtraction of spectra of pure buffer until a flat baseline was obtained in the region around 2300 cm^{-1} (combination bands of water). Contributions of water vapor were eliminated by interactive subtraction of a water vapor spectrum. A multiple point baseline correction was applied to each single spectrum. The spectra in the figures are averaged on 5 consecutive single spectra treated as mentioned. For mathematical data analysis GRAMS 3.0 from Galactic Industries was used. Deconvolution parameters for all spectra were γ= 2.5, 78 % smoothing, and FWHH = 11.73. For the second derivative spectra a 17 point Savitzky-Golay function was used.

RESULTS

In a very early stage of adsorption (after 10 min, Fig. 1 upper panel) on neat Ge we observe protein bands in the amide I region at 1686 cm^{-1}, 1655 cm^{-1} (most intense) 1632 cm^{-1}, and 1619 cm^{-1} (for band assignment see Tab. 1). In the amide II region the most intense band is at 1547 cm^{-1}. After 74 min of adsorption (Fig. 1 lower panel) the band due to α-helix substructure remains the most intense band. Spectra recorded on Ge, wich was precoated with Ag, show a band, of enhanced intensity in the amide I region at 1675 cm^{-1} (Fig. 2 upper panel, 10 min of adsorption). All other bands like those mentioned above are also resolved. The band at 1646 cm^{-1} is only poorly resolved. During further adsorption intensities of the bands at 1675 cm^{-1}/ 1666 cm^{-1} and 1646 cm^{-1} seem to increase most (Fig. 2 lower panel).

The spectra recorded on Ge crystal precoated with lipids show the bands around 1670 cm^{-1} and 1656 cm^{-1} as most intense in the early phase of adsorption (Fig. 3. upper panel). After 80 min of adsorption a slight preference for the 1656 cm^{-1} band is found (Fig. 3 lower panel). If the Ge crystal was coated with Ag cluster prior to vesicle adsorption we find an intense band at 1687 cm^{-1} accompanied by intense bands at 1643 cm^{-1} and 1632 cm^{-1} (Fig. 4. upper panel). After 50 min of adsorption again the band due to α-helical substructure at 1656 cm^{-1} increased intensity (Fig. 4 lower panel).

TABLE 1. Band assignment and band positions for the amide I band envelope. Assignment for this work derived from 2nd derivative spectra. Trans. = Transmission

	(7)	(6)	(5)	This work	
assignments (7)	Trans. CaF_2	Trans. CaF_2	Trans. CaF_2	ATR Ge	SEIRA Ge/Ag)
turns (T)	1693	1691	1692		
turns	1683	1680	1681 T/β	1689	1689
β-sheet	1674	1672		1675	1675
turns	1665		1668	1666	1666
α-helix	1656	1656	1656 α/U	1655	1656
unordered (U)	1647	1644 (loop)	1645 β	1643	1646
β-sheet	1638	1633	1635	1632	1637
β-sheet	1628	1624	1627		1623
side chains	1619			1619	1610
side chains	1609				1601

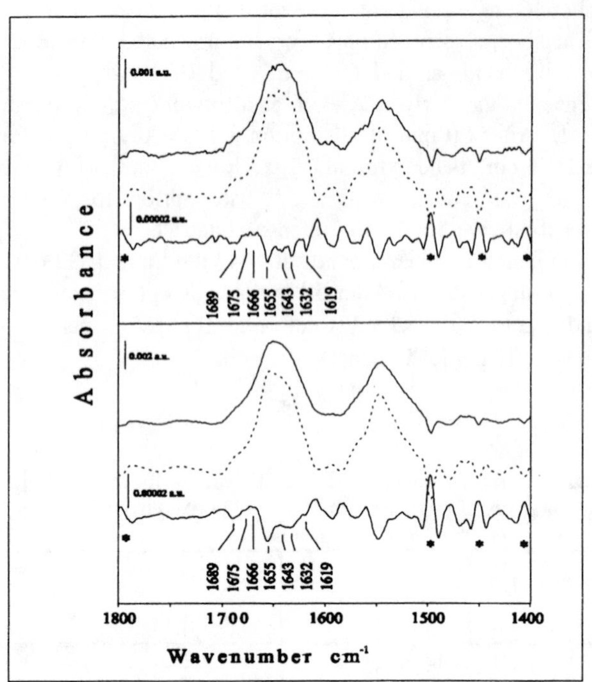

FIGURE 1. FTIR-ATR spectra of nAChR; neat Ge crystal. Upper traces represent 10 min and lower traces 74 min of adsorption. For both panels: absorbance spectrum (solid line), deconvolved spectra (dottet line) and 2nd derivative spectra (thin line, scaled to lower bar)

*Contribution of uncompensated background spectrum.

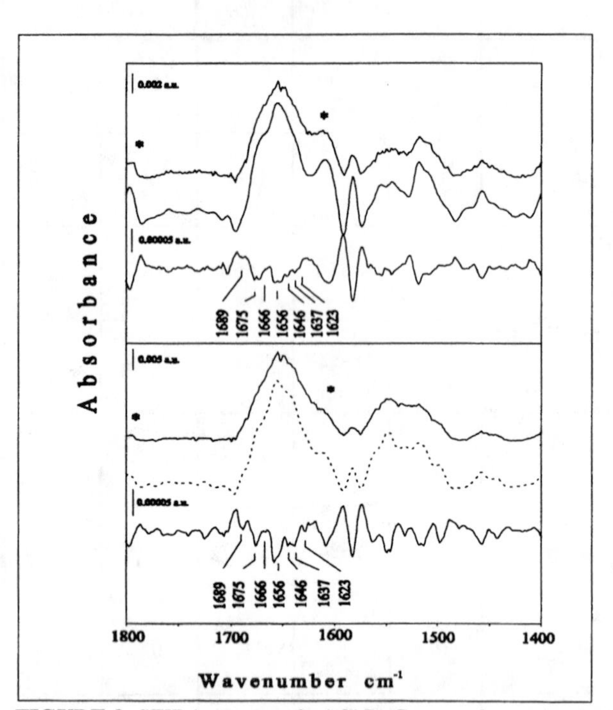

FIGURE 2. SEIRA spectra of nAChR; Ge crystal precoated with Ag. Upper traces represent 10 min and lower traces 58 min of adsorption. For both panels: absorbance spectrum (solid line), deconvolved spectra (dottet line) and 2nd derivative spectra (thin line, scaled to lower bar)

*Contribution of uncompensated background spectrum.

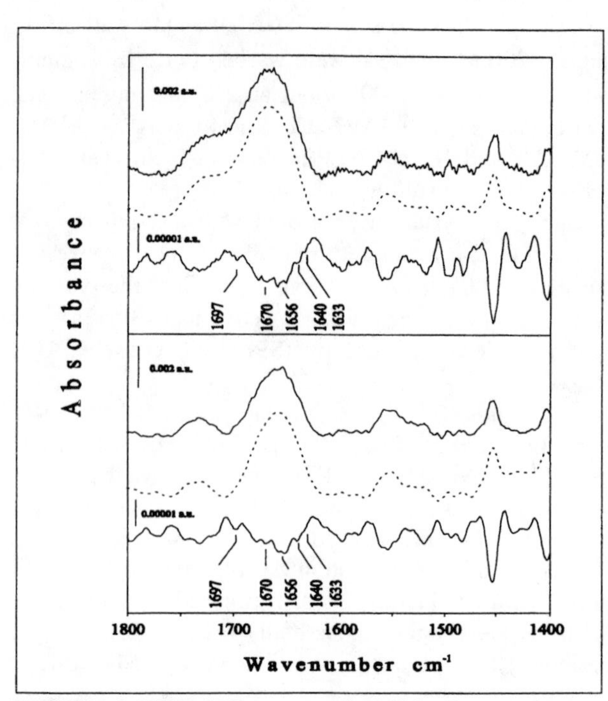

FIGURE 3. FTIR-ATR spectra of nAChR; Ge crystal precoated with lipid. Upper traces represent 25 min and lower traces 80 min of adsorption. For both panels: absorbance spectrum (solid line), deconvolved spectra (dotted line) and 2nd derivative spectra (thin line, scaled to lower bar)

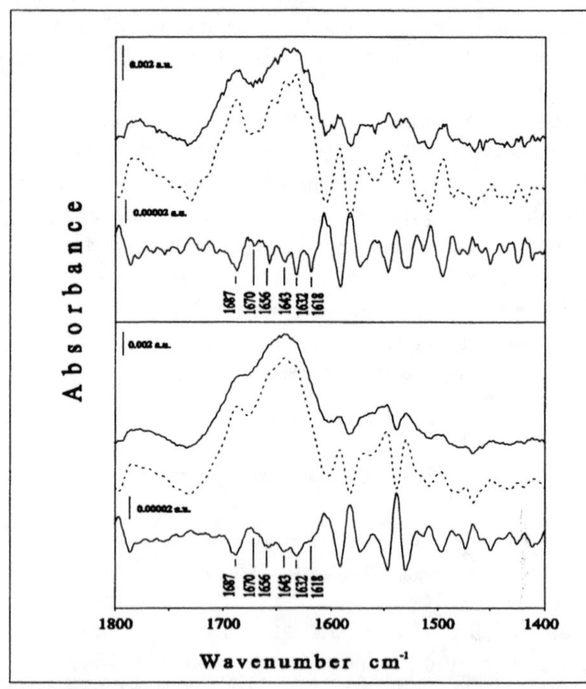

FIGURE 4. SEIRA spectra of nAChR; Ge crystal precoated with Ag and lipid. Upper traces represent 25 min and lower traces 52 min of adsorption. For both panels: absorbance spectrum (solid line), deconvolved spectra (dottet line) and 2nd derivative spectra (thin line, scaled to lower bar)

DISCUSSION

The proteins in the vesicles interact via their β-sheet/turn structure at the protein/crystal interface. The lipid subphase seems to induce β-sheet/turn structure. In the vicinity of the Ag clusters vibrations of these structural units in the proteins are enhanced at the protein/crystal or protein/lipid/Ag-cluster interface. A quantitative estimation of the content of the structural subunits from the spectra recorded on Ge and Ge/Ag (14) reveals similar values as found for pellets of nAChR rich membranes dried on CaF_2 windows and measured in transmission (5). This technique does not overemphasize bands due to protein structures at interfaces. We assume that the internal cohesive forces (15) should prevent the extracellular domains which are in contact with the interface from turning into β-sheet/turn structure. Atomic force microscopic data show that the vesicles do not fuse on the crystal (16). Hydrophobic interfaces might induce stronger forces to the hydrophilic parts of the extracellular domains which lead to turn like formations of these protein regions by repulsive forces.

The enhancement of absorption energy around the Ag cluster decreases by a factor of 1/e within a distance of 3 - 5 nm. This distance is sufficient to enhance bands due to structures of the extracellular domains of the receptor. Thus, the observation of α-helix and β-sheet/turn structures in the early stage of adsorption might represent these elements in the extracellular domains. This is the first spectroscopic evidence, which verifies findings from molecular modelling data on the extracellular domains (4).

For the use of SEIRA spectroscopy one has to take into account the distance of the enhancement effect, if molecules which extend more than 3 - 5 nm apart from the Ag cluster have to be investigated. Assuming intact protein structure at the Ag interface this method could be developed to a powerful method for studying protein structure or cell membrane processes in more detail.

CONCLUSION

The nAChR shows β-sheet/turn and α-helical structures at the protein/crystal interfaces. These substructures are part of the extracellular domain. Hydrophilic surfaces on the other hand induce structural changes on the extracellular domains of the receptor. SEIRA spectroscopy is a powerful method to study proteins at interfaces and to investigate processes at cell membranes.

ACKNOWLEDGEMENT

The authors WBF, IU, GS, and RS thank the BMBF and the Deutsche Forschungsgemeinschaft for financial support. ChK thanks the Free State of Saxony for a stipend. The free gift of lipids from LIPOID GmbH, Ludwigshafen, Germany, is acknowledged. Thanks to Dr. T. Böhme (SENTRONIC GmbH, Dresden) for preparing the silver coated Ge crystals.

REFERENCES

1. Katz B. and Thesleff S., *J. Physiol. Lond.* **138**, 63-80(1957).
2. Neubig R. R., and Cohen J. B., *Biochemistry* **18**, 5464-5475 (1979).
3. Criado M., Hochschwender S., Sarin V., Fox J. L., and Lindstrom J., *Proc. Natl. Acad. Sci. USA* **82,** 2004-2008 (1985).
4. Tsigelny I., Sugiyama N., Sine S. M., and Taylor P., *Biophys. J.* **73,** 52-66 (1997).
5. Naumann D., Schultz Ch., Görne-Tschelnokow U., and Hucho F., *Biochemistry* **32**, 3162-3168 (1993).
6. Méthot N., McCarthy M. P., and Baenziger J. E,*Biochemistry* **33**, 7709-7717 (1994).
7. Butler D. H., and McNamee M. G.; *Biochim. Biophys. Acta* **1150**, 17-24 (1993).
8. Castresana J., Fernandez-Ballester G., Fernandez A. M., Laynez J. L., Arrondo J.-L., Ferragut J. A., and Gonzalez-Ros J. M.; *FEBS Lett.* **314**, 171-175 (1992).
9. Baenziger J. E., Miller K. W., and Rothschild K. J.; *Biophys. J.* **61**, 983-992 (1992).
10. Görne-Tschelnokow U., F. Hucho, D. Naumann, A. Barth, and W. Mäntele; *FEBS Lett.* **309**, 213-217 (1992).
11. Fischer W. B., Schwenke D., Salzer R.; in preparation
12. Kuhne Ch., Steiner G., Fischer W. B., and Salzer R.; *Fres. J. Anal. Chem.* submitted 1997
13. Schrattenholz, A., Godovac-Zimmermann, J., Schäfer H.J., Albuquerque, E. X., and Maelicke A., Eur. J. Biochem. **216,** 671-677 (1993).
14. Fischer W. B., Unverricht I., Kuhne Ch., and Salzer R.; in preparation
15. Norde W, Haynes C. A., *ACS Symp. Ser.* **602**, (Proteins at Interfaces 2), pp. 26-40.
16. Steiner G., Pham , Fischer W. B., and Salzer R.;unpublished results

Surface Enhanced Infrared Absorption (SEIRA) Observed on Different Metal Surface Structures

H. D. Wanzenböck*, N. Weissenbacher, R. Kellner

Institute for Analytical Chemistry, Vienna University of Technology
Getreidemarkt 9/151, A-1060 Vienna, Austria (Europe)

Surface enhanced infrared absorption (SEIRA) spectroscopy is a novel method providing a significantly decreased limit of detection to conventional IR spectroscopy. The extent of spectral enhancement due to the presence of metal island layers is dependent on the surface morphology of the rough metal film. An improved two-staged deposition process is introduced enabling better control of the surface topology of the metal film. By modifying the angle of deposition a directional orientation of the growth process can be achieved approaching a needle-like shape of metal islands. With oblique metal deposition a significantly reduced background absorption of the metal film is obtained resulting in a decreased spectral noise. The strongest surface enhancement effect was obtained with metal layers deposited under a 18° angle.

INTRODUCTION

The absorption of infrared light is increased when molecules are located n or near rough metal surfaces [1,2]. Metal island films are required for obtaining a spectral enhancement effect and the enhancement factor usually increases with higher mass thickness [3]. On the other hand merging of metal islands was experienced to increase the background absorption of the metal film [4] so that no transmission was obtained with multireflection ATR-elements as used during this work. An attempt was made to prevent merging of metal islands to a continuous film and additionally obtain a higher mass thickness. By depositing the metal under a grazing angle a process was chosen promoting the growth of the metal islands in the shape of a needle.

Utilizing grazing angle deposition for surface enhanced Raman scattering (SERS) substrates arrays of sharp needle-like structures have been grown in a stochastic pattern on an inherently rough surface [5,6]. The presence of non-ideal structures promotes bridged needles and side growth [7]. The sharpness of the needle expressed by the aspect ratio is determined by the spacing between the growth sites, the amount of metal deposited and the angle of incidence that is usually chosen near 85°.

A regular pattern with aspect ratios of 5:1 could be obtained by a two-stage process. Wachter et al. [8] first deposited a controlled base layer of polymer microspheres on a clean substrate. In a second step a suitable thickness of metal was deposited by evaporation at a grazing angle (85° - 89° to axis orthogonal to surface) in order to produce metal needles. Needles with 39 nm in diameter and 150 nm to 300 nm length could be observed by means of scanning electron microscopy.

The principle of needle growth benefits from the linear propagation of metal atoms in the low pressure deposition chamber. Any particle in the propagation line of the metal atoms would catch the metal atom and shield off its "shadow" area on the substrate surface. Using a grazing incidence angle the metal clusters formed on the substrate (but also the roughness of the substrate) will act as localized masks preventing metal deposition on certain parts of the substrate's surface. No metal is deposited in the shadow of the metal island and the formation of a continuous film is restrained. By preventing the merging of metal films the background absorption of the metal layer is reduced and the optical throughput can be increased.By varying the incidence angle the shadow area of the particle can be enlarged or reduced. Thus a control over the interparticle distance can be achieved.

EXPERIMENTAL

Metal film deposition

Gold layers with a purity higher than 99.999% were evaporated by an electron beam physical vapor deposition on the surface of trapezoidal ZnSe-ATR-crystals (50x20x2mm/45°) with a deposition rate of 0.06 $nm.s^{-1}$ and at a base pressure below 10^{-6} mbar. The ATR-crystals were polished with 0.3 µm Al_2O_3 suspension and rinsed multiply with distilled water prior to deposition. They were mounted on a sample holder that could be inclined to the incident particle beam with the length axis of the crystal as rotation axis.

CP430, *Fourier Transform Spectroscopy:* 11th International Conference
edited by J.A. de Haseth

The deposition procedure was a two-staged process consisting of a nucleation step and a length growth of needles. The first stage was conventional deposition under an incidence angle of 90°. The metal amount deposited was below 2 nm mass thickness generally providing separated gold islands Such obtained metal islands may act as nuclei for the needle growth.

In the second deposition stage the ATR-crystal was inclined so that the incident particle beam would approach under an angle of 5° to 45° to the substrate surface. Subsequently, further growth of the nuclei commences mainly in the direction of the metal source. By choosing the incidence angle the shadow area and subsequently the interparticle distance was adjusted. With the mass thickness deposited under grazing angle the length of the obtained needle-shaped clusters could be controlled within a certain range.

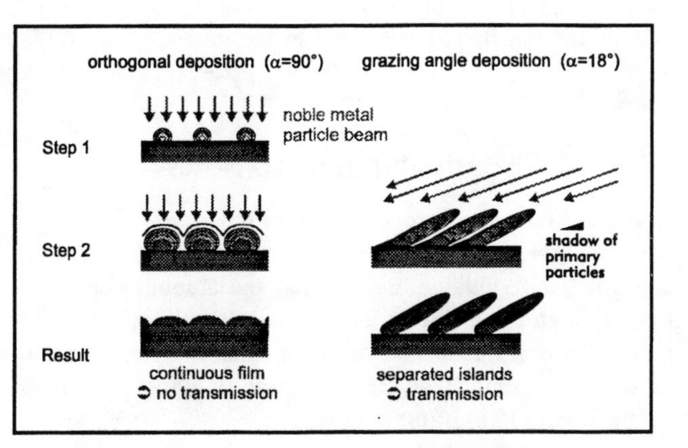

FIGURE 1 Schematic illustration of the metal deposition at orthogonal (left) and grazing deposition angle (right) for fabrication of needle-shaped islands.

Deposited metal island films were investigated by atomic force microscopy (AFM) in tapping mode. With grazing angle deposition separate metal islands could be imaged up to 6 nm mass thickness while gold films produced by orthogonal deposition were observed to be continuous layers above 3 nm mass thickness.

Spectroscopic measurements

The analyte was dissolved in methanol and defined volume was deposited on the metal coated surface of the ATR-crystal. The solvent evaporated off so that the analyte was obtained as a thin film on the metal surface. The surface enhanced spectra were recorded in dry state. The SEIRA-active substrates could be recycled by rinsing with methanol allowing multiple reproducible measurements with the same ATR-crystal.

All spectra were recorded with a Bruker IFS 66 FT-IR spectrophotometer equipped with a MCT detector (Bruker D315) and a Perkin-Elmer ATR-bench. Each spectrum was recorded in the unpolarized beam and represents a coaddition of 1024 scans at a resolution of 4 cm^{-1}. The analyte spectra presented below are ratioed against the spectra of the metal coated ATR-crystals. Contributions of residual water and carbondioxide were eliminated.

RESULTS

Gold layers were deposited by evaporation on a polished ZnSe-ATR covering an area of approximately 3 cm^2. The total amount of metal deposited was larger than the mass thickness of previous metal films that was found to exceed the acceptable loss of transmission of the ATR-crystal due to background absorption of the metal. In order to adjust the optimum needle shape parameters were varied.

Comparison of oblique and orthogonal deposition methods

One objective of this work was the reduction of merging of metal islands for obtaining a reasonable S/N-ratio with multiple internal reflection. With a gold film deposited at an orthogonal angle spectra could only be obtained up to a maximal mass thickness of 2.7 nm providing a surface enhancement factor of 10.

A gold-needle film was produced by orthogonal deposition of 2.0 nm (nucleation) followed by the deposition of 4.0 nm gold for length growth of the needle. The deposition angle was chosen 18° relative to the substrate surface. The gold layer of 6.0 nm mass thickness was obtained that still supplied sufficient transmission of the ATR-crystal. Although the maximum mass thickness in comparison to orthogonal deposition was exceeded by 100 % the background absorption of the metal film was not increased correspondingly. A surface enhanced infrared spectrum of PNBA could be measured but no improvement in the magnitude of the spectral enhancement could be obtained.

It is concluded that merging of metal islands could be prevented by grazing-angle deposition. The features necessary for a surface enhancement could be maintained with this differently structured metal-island film - a surface enhancement by a factor of 5 could be obtained with this needle film. The lower enhancement magnitude observed with the needle-shaped metal clusters might be explained by the changed aspect ratio of the metal islands.

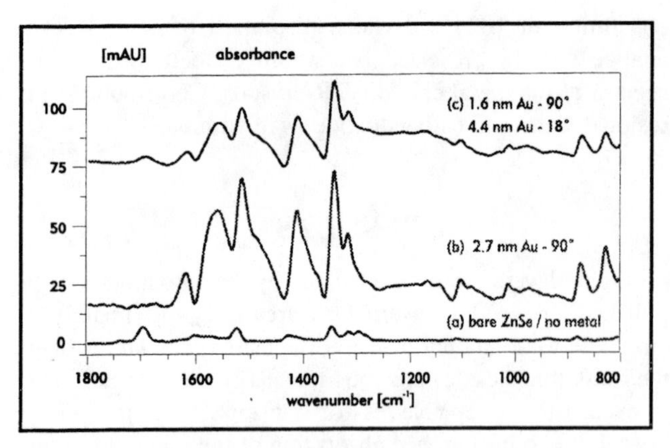

FIGURE 2 Comparison of orthogonal deposited metal islands and directed growth of metal islands. 835 ng p-nitrobenzoic acid (5 µl 0.001M) deposited on (a) a bare ZnSe-crystal (b) a 2.7 nm gold layer deposited under orthogonal angle and (c) a film of needle-shaped metal islands generated by a two-stage process with a grazing deposition angle of 18° in the second stage.

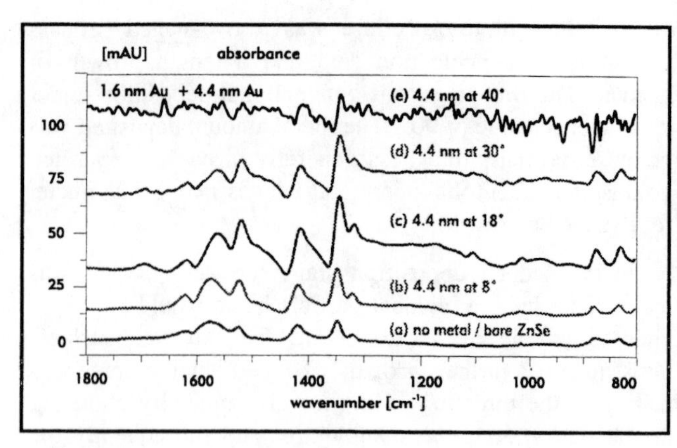

FIGURE 3 835 ng p-nitrobenzoic acid (5 µl 0.001M) deposited on (a) a bare ZnSe-crystal and (b-e) 6.0 nm gold layer generated by a two-stage process with a grazing deposition angle between 8° and 40° in the second stage. A primary gold layer of 1.6 nm mass thickness was deposited under an orthogonal angle and a subsequent layer of 4.4 nm was deposited at an (b) 40°, (c) 30°, (d) 18°, (e) 8° angle inclined to the surface of the ATR-crystal.

Surface topology

In order to obtain information on the surface structure and orientation of the needles the metal layers were imaged by AFM. A gold film with a total thickness of 6.0 nm (2.0 nm at 90° / 4.0 nm at 18°) shows clusters of gold on the surface with a diameter of 30 to 40 nm. The metal island appear to be well separated which is in good correlation with the high optical throughput observed in spectroscopic investigations.

Variation of deposition angle

The influence of the incidence angle of the metal atoms impinging on the surface was varied. In principle this parameter would allow to vary the spacing between the needles by shadowing smaller or larger portions of the surface.

The optimum surface enhancement was achieved for a metal deposition under a grazing angle of 18°. With a deposition angle steeper than 30° no analyte bands could be identified because of the low S/N-ratio. The reason for this behavior could be found in the fact that metal clusters start to gow together similar to coalescence during orthogonal deposition. A high noise level was obtained due to a strong background absorption by the metal film.

Variation of mass thickness

In the two-staged process applied for needle generation the mass thickness of every deposition step could be controlled individually. By varying the amount deposited in the first step the size of islands and thus the diameter of the needle is influenced. The mass thickness deposited in the second process stage could be used to adjust the length and the aspect ratio of the needles.

In the first stage acting as nucleation step the mass thickness of the deposited gold film was varied between 1.2 nm and 2.0 nm. The following layer was deposited in a thickness resulting in a total mass thickness of 6.0 nm for the entire gold film. The spectral enhancement was found to be stronger for less metal deposited in this first stage. This leads to the conclusion that nucleation stage forming a high number of smaller metal clusters is more favorable for a string spectral enhancement.

The length of the inclined needle was varied by depositing metal layers in the range from 3.2 nm to 5.6 nm with a deposition angle of 18 ° relative to the surface. The underlying gold layer (from the first step) was 1.6 nm thick and fabricated by orthogonal metal deposition. All metal layers provided a sufficient transmission through the ATR-crystal, but an increase of metal's background absorption was observed with higher mass thickness.

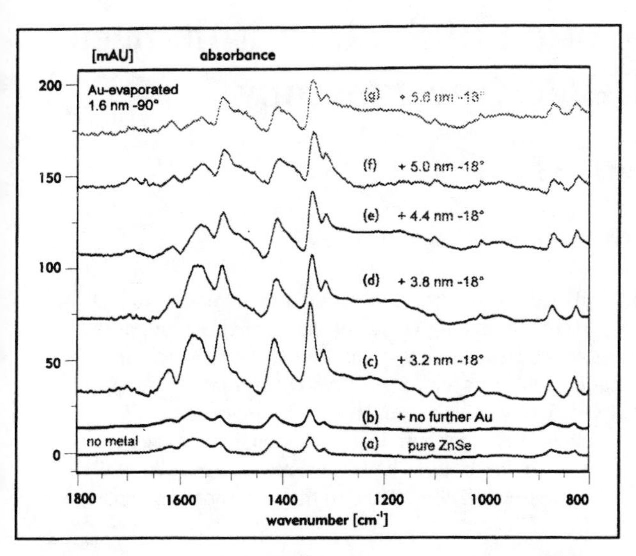

FIGURE 4 835 ng p-nitrobenzoic acid (5 µl 0.001M) deposited on (a) a bare ZnSe-crystal and (b-g) 6.0 nm gold layer generated by a two-stage deposition process. The PNBA is precipitated on a metal film starting from (b) 1.6 nm layer thickness deposited at orthogonal angle and subsequent metal deposition under grazing angle of 18° with a mass thickness of (c) 3.2 nm, (d) 3.8 nm, (e) 4.4 nm (f) 5.0 nm and (g) 5.6 nm of the second layer.

The surface-enhanced spectra shown in Figure 3 were obtained by the deposition of 835 ng PNBA on the metal film. Spectra of PNBA were found surface enhanced proportional for all bands.

Already the first deposition stage alone was found sufficient to cause a small surface enhancement effect. For the needle substrates a higher enhancement magnitude was observed but the magnitude was decreasing with a higher mass thickness deposited in the second process stage. This unexpected trend was confirmed by several measurements. Shorter needles seem to fulfill the requirements for a high surface enhancement better than long needles. In comparison to the ellipsoidal islands typically generated by orthogonal deposition the different orientation of the needles and the low aspect ratio η (=diameter/height) are suggested as explanation for these findings.

CONCLUSION

By applying grazing angle deposition metal films consisting of islands with a orientated structure (needle-like) could be generated. Metal films obtained in such a way showed a reduced background absorption of the metal film. This feature allowed to extend the mass thickness of the used gold films deposited on ZnSe from 3.0 nm with orthogonal deposition up to 7.2 nm.

The maximum signal enhancement was obtained with a 18° deposition angle. However, with the "needle technique" no significant improvement in surface enhancement was obtained in comparison to metal layers produced under orthogonal metal deposition. This correlates with the result that deposition parameters promoting rather flat metal islands instead of long needles were found to effect a stronger spectral enhancement. The shape of the metal islands (with regard of the aspect ratio) is proposed to be an important factor, but a needle-like structure is apparently not substantial for a high surface enhancement effect.

The benefit of grazing angle metal deposition is the prevention of coalescence of metal clusters. With oblique deposition the existence of metal islands necessary for a SEIRA effect can be extended to a higher mass thickness of the film deposited. Further investigations with strictly defined metal structures as achievable with micro-lithography are endorsed.

ACKNOWLEDGEMENT

The financial support of this work by the Austrian Research Council ("Fonds zur Förderung der wissenschaftlichen Forschung") under Project No. FWF 10386 CHE is greatly acknowledged. We also thank the staff of the Institute for Solide State Electronics (Vienna University of Technology) under supervision of Prof. E. Gornik for performing the metal deposition.

REFERENCES

1 Johnson, E., Aroca, R.; *J. Phys. Chem.* **99**, 9325 (1995)

2 Hartstein, A., Kirtley, J.R., Tsang, J.C.; *Phys. Rev. Lett.* **45**, 201 (1980)

3 Wanzenböck, H. D., Edl-Mizaikoff, B., Friedbacher, G., Grasserbauer, M., Kellner, R., Arntzen, M., Luyven, T., Theiss, W., Grosse, P.; *Mikrochim. Acta [Suppl.]* **14**, 665 (1997)

4 Nishikawa, Y., Nagasawa, T., Fujiwara, K., Osawa, M.; *Vibrat. Spectrosc.* **6**, 43 (1993)

5 Dirks A.G., Leamy H.J., *Thin Solid Films* **47**, 219 (1977)

6 Bloemer M.J., Buncick M.C., Warmack R.J., Ferrell T.L.; *J. Opt. Soc. Am.* **B 5**, 2552 (1988)

7 Bloemer M.J., Ferrell T.L., Buncick M.C., Warmack R.J.; *Phys. Rev.* **B 37**, 8015 (1988)

8 Wachter E.A., Moore A.K., Haas J.W.; *Vibr. Spectrosc.* **3**, 73 (1992)

Multilayer Cooperative Chemisorption in Surface-Enhanced Infrared Absorption (SEIRA) Spectroscopy

Lin-Tao He and Peter R. Griffiths*

Department of Chemistry, University of Idaho, Moscow, ID 83844-2343

The cooperative chemisorption of multilayers of 2,4-dinitrobenzoic acid on a silver surface under high vacuum was studied by microtransmission infrared spectrometry. The fact that the antisymmetric stretching modes of the carboxylate and nitro groups are active in the SEIRA spectra of 2,4- and 3,5-dinitrobenzoic acid suggests that the bottom layer is adsorbed perpendicular to the metal surface with higher layers oriented obliquely. Multilayer cooperative chemisorption and the effect of the solvent on the SEIRA spectra of p-nitrobenzoic acid (PNBA) were also observed. The symmetric COO^- stretch in the SEIRA spectra of multilayers of PNBA is split into a doublet. Because there is a weaker interaction between the COO^- group of PNBA in the upper layers with the Ag surface than for the bottom layers, we assign the 1420 cm^{-1} component to the symmetric COO^- stretch of the upper layers of PNBA and the 1387 cm^{-1} band to the corresponding mode of the lower layers.

INTRODUCTION

Surface-enhanced infrared absorption (SEIRA) was first observed and investigated by Harstein et al [1] in 1980. The SEIRA phenomenon has been reported in attenuated total reflection (ATR) [1,2], external reflection and transmission geometries [3,4]

Osawa [5] has suggested that both electromagnetic and chemical mechanisms contribute to the total magnitude of the enhancement of vibrational bands in SEIRA, in a similar manner to surface-enhanced Raman scattering (SERS). For the SEIRA spectra of o, m, and p-nitrobenzoic acid adsorbed on Ag films, Osawa's group [6, 7] found that only those vibrational modes for which the dipole derivative has a component that is perpendicular to the metal surface are SEIRA active, thus demonstrating that the surface selection rule known for reflection-absorption spectroscopy of adsorbates on the smooth metal surface also holds for SEIRA spectroscopy. Figures 1A and 1B show the surface selection rule in SEIRA spectroscopy.

In this report, we propose that the multilayer SEIRA spectra are more complicated than the monolayer SEIRA spectra under the simple surface selection rule. We suggest a multilayer cooperative chemisorption mechanism to explain the spectroscopic properties of multilayer SEIRA spectra.

$\nu_s(NO_2)$ $\nu_{as}(NO_2)$ $\nu_s(COO^-)$ $\nu_{as}(COO^-)$ metal

Only symmetric bands $\nu_s(NO_2)$ and $\nu_s(COO^-)$ are observed

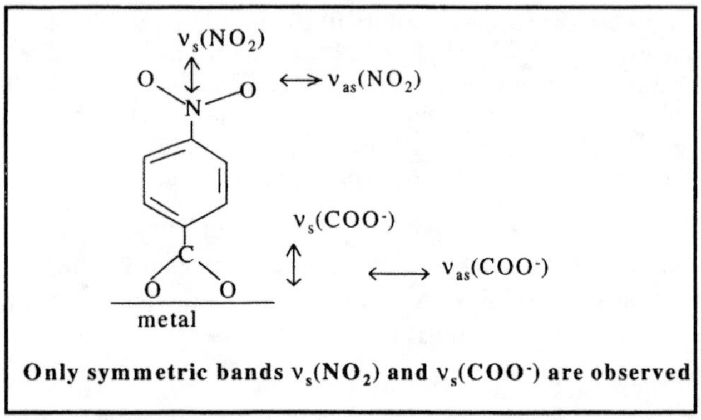

Figure 1A. Surface selection rule for the SEIRA spectrum of PNBA: only the symmetric stretching bands, $\nu_s(NO_2)$ and $\nu_s(COO^-)$, are active. The anti-symmetric stretching bands, $\nu_{as}(NO_2)$ and $\nu_{as}(COO^-)$, are forbidden .

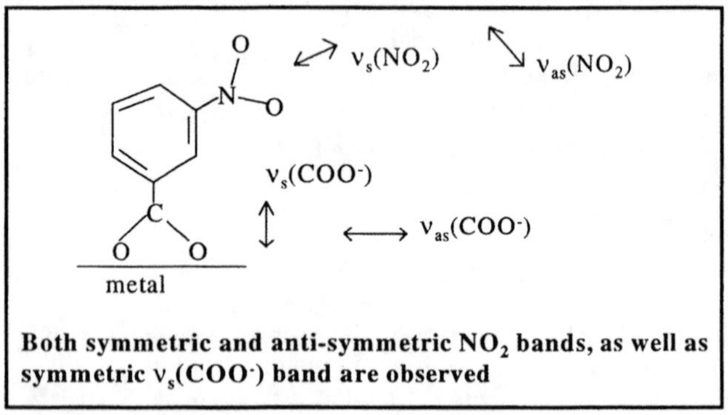

$\nu_s(NO_2)$ $\nu_{as}(NO_2)$ $\nu_s(COO^-)$ $\nu_{as}(COO^-)$ metal

Both symmetric and anti-symmetric NO_2 bands, as well as symmetric $\nu_s(COO^-)$ band are observed

Figure 1B. Surface selection rules for SEIRA spectrum of m-nitrobenzoic acid: $\nu_s(NO_2)$, $\nu_{as}(NO_2)$ and $\nu_s(COO^-)$ are active, but $\nu_{as}(COO^-)$ is forbidden.

CP430, *Fourier Transform Spectroscopy:* 11th International Conference
edited by J.A. de Haseth

EXPERIMENTAL

A 5-nm thick film of silver was deposited onto a ZnSe substrate by physical vapor deposition at a rate of 1 nm per minute under a pressure of 2 x 10^{-6} torr. The thickness of the film was measured by a quartz crystal microbalance. The analytes were dissolved in acetone or methanol at a concentration of 100 ng/μl and deposited on the Ag films with a 1-μl microsyringe. About 30 ng of sample was spotted on the Ag surface and the solvent allowed to evaporate in air. Microtransmission spectra were measured by the Schwartzschild microscope objective and narrow-band MCT detector built into the Bio-Rad *Tracer* direct deposition GC/FT-IR interface (Digilab Division of Bio-Rad Laboratories, Cambridge, MA), which was mounted on a Bio-Rad FTS40 FT-IR spectrometer. Computer-controlled stepper-motors allowed the sample spot to be located on the Ag surface. A 100-μm diameter aperture was used to select the region that was interrogated. All spectra were collected at resolution of 4 cm^{-1}.

RESULTS AND DISCUSSION

1. Dinitrobenzoic acid SEIRA and the observation of multilayer cooperative chemisorption

Figures 2 and 3 show the normal (unenhanced) infrared spectra of 3,5-dinitrobenzoic acid (3,5-DNBA) and 2,4-dinitrobenzoic acid (2,4-DNBA) on a bare ZnSe window along with the microtransmission SEIRA spectra of 3,5- and 2,4-DNBA on 5-nm Ag films. The absence of a carbonyl band at ~1710cm^{-1} in the SEIRA spectra of 3,5-DNBA and 2,4-DNBA indicates that the protons of the COOH group dissociate and the molecules chemisorb on the Ag surface.

By making use of the fact that the *Tracer* GC/FT-IR interface is evacuable, we investigated the changes of 2,4-DNBA molecules absorbed on a 5-nm thick Ag surface under high vacuum (8 x 10^{-5} torr). Figures 4A and B show the SEIRA spectra of 2,4-DNBA collected from two different spots on Ag surface measured at a pressure of 8.2 x 10^{-5} torr at times between 0 and 70 hours after deposition. These spectra demonstrate that several molecular layers of 2,4-DNBA molecules originally adsorb on the Ag surface and that the desorption of the molecules in the upper layers occurs under high vacuum. Silver islands with different size and thickness have different adsorption capacities because the Ag films made by physical vapor deposition are not uniform. Each molecular layer appears to be stable until a particular time, after which all molecules in that layer desorb rapidly, indicating that the desorption is cooperative.

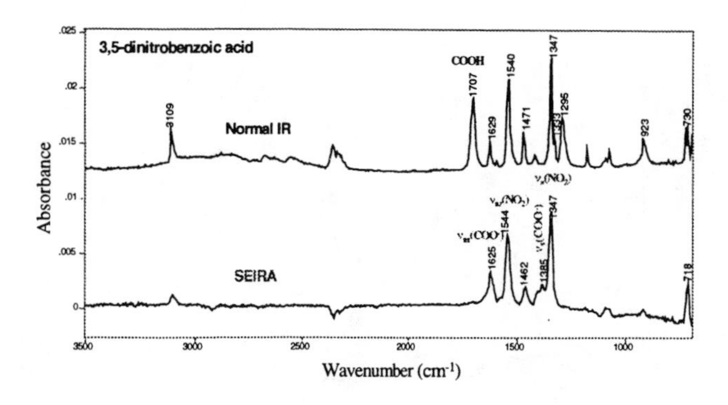

Figure 2. Normal (unenhanced) IR spectrum of 3,5-dinitrobenzoic acid on bare ZnSe and its SEIRA spectrum on 5-nm Ag /ZnSe.

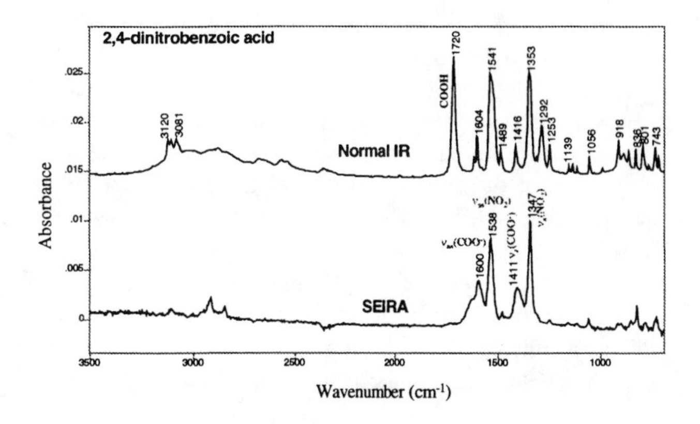

Figure 3. Normal (unenhanced) IR spectrum of 2,4-dinitrobenzoic acid on bare ZnSe and its SEIRA spectrum on 5-nm Ag /ZnSe.

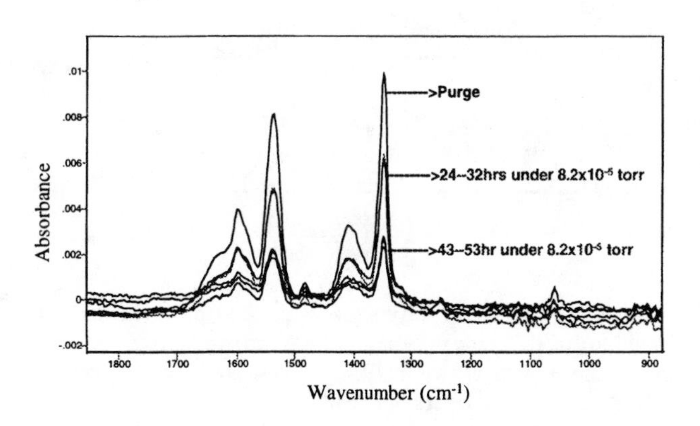

Figure 4A. SEIRA spectra of one spot of 2,4-dinitrobenzoic acid deposited on 5-nm Ag/ZnSe at a pressure of 8.2 x 10^{-5} torr at various times after deposition.

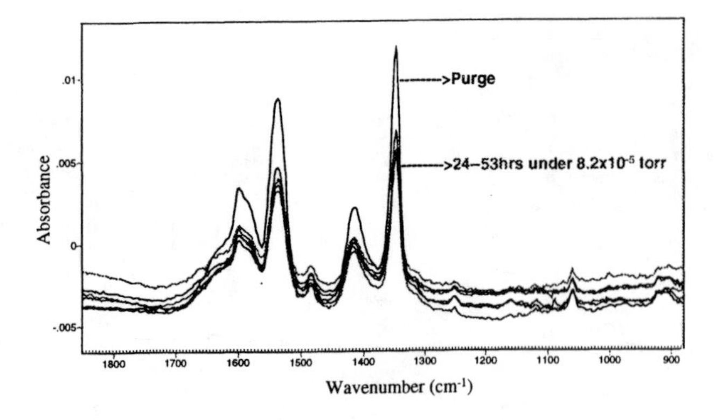

Figure 4B. SEIRA spectra of a different spot of 2,4-DNBA deposited on 5-nm Ag/ZnSe at a pressure of 8.2 x 10^{-5} torr at various times after deposition.

2. Multilayer cooperative chemisorption of p-nitrobenzoic acid on Ag films

When a solution of any nitrobenzoic acid in a volatile solvent is applied dropwise to an Ag surface, it spreads out. Thus after the solvent evaporates, the thickness of the layer of solute near the center part of the sample is greater than that near the edge. The SEIRA spectrum of the thin (approximately one monolayer) film at the outer edge of a spot of p-nitrobenzoic acid (PNBA) deposited from acetone solution onto a 5-nm thick film of Ag surface is shown in Figure 5A. Only the symmetric NO_2 and COO^- stretching bands are enhanced in this spectrum.

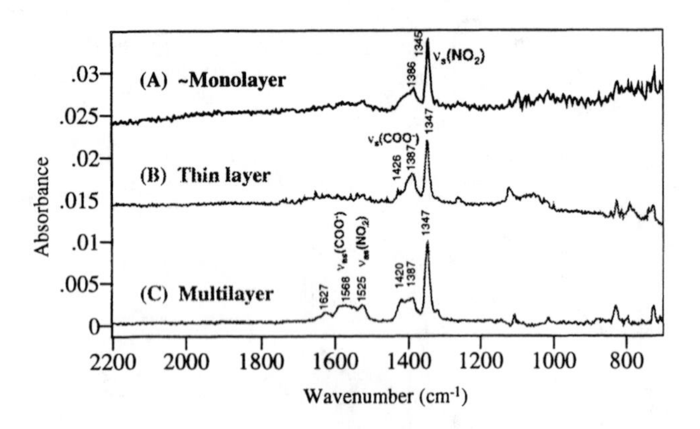

Figure 5. Spectra of layers of p-nitrobenzoic acid (acetone solution) of different thickness on a 5-nm Ag surface. A: about one molecular layer of PNBA; B: a slightly thicker layer of PNBA; C: multiple layers of PNBA.

The SEIRA spectrum of a thicker PNBA film deposited from acetone solution on a 5-nm Ag layer is quite different, as shown in Figure 5C. In this multilayer SEIRA spectrum, not only are the symmetric NO_2 and COO^- stretching bands seen, but also the anti-symmetric NO_2 and COO^- stretching bands are quite apparent in the 1500-1600 cm^{-1} region. The symmetric COO^- stretch in the SEIRA spectrum of multilayers of PNBA is split into a doublet at 1387cm^{-1} and 1420cm^{-1}. Because there is a weaker interaction between COO^- group of PNBA in the upper layers with the Ag surface than for the lowest layer, we assign the 1420 cm^{-1} component to the symmetric COO^- stretch of the upper layers of PNBA and the 1387 cm^{-1} band to the corresponding mode of the lower layer.

The fact that the antisymmetric stretching modes of the carboxylate and nitro groups of PNBA, and the corresponding modes of the carboxylate groups of 2,4-DNBA and 3,5-DNBA, are active in their SEIRA spectra strongly suggests the bottom layer is adsorbed perpendicular to the metal surface with higher layers oriented obliquely, as, they should be forbidden by the surface selection rule if they were perpendicular to the surface.

3. The effect of solvent on multilayer SEIRA spectra

The SEIRA spectra of multilayers of p-nitrobenzoic acid on a 5-nm Ag surface after deposition from acetone and methanol solution are shown in Figure 6. It may be seen that the band at 1420 cm^{-1} is much stronger than the 1387 cm^{-1} band when the film is cast from methanol. We assign this behavior to the effect of hydrogen bonding between methanol and PNBA molecules during the period of deposition, so that as the solvent evaporates, more $PNBA^-$ species from the methanol solution exist at oblique orientations to the Ag surface than when acetone is used as the solvent.

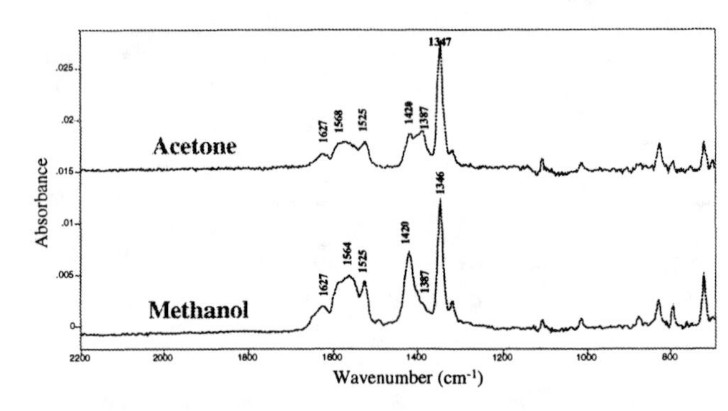

Figure 6. SEIRA spectra of multilayers of p-nitrobenzoic acid on 5-nm Ag/ZnSe after deposition from acetone solution (above) and methanol solution (below).

CONCLUSIONS

Ag island films prepared by conventional physical vapor deposition possess a range of adsorption capacities at Ag islands of different size. These islands can cooperatively chemisorb and desorb multiple layers of nitrobenzoic acid molecules on the surface. The SEIRA spectra of such multilayers can be explained by invoking the basic surface selection rule by assuming cooperative chemisorption with higher layers oriented obliquely to the surface.

REFERENCES

1. Hartstein, A., Kirtley, J. R., Tsang, J. C., Phys. Rev. Lett. **45**, 201-204 (1980).
2. Hatta, A., Ohshima, T., Suetaka, W., Appl. Phys. A **29**, 71-75 (1982).
3. Kamata, T., Kato, A., Umemura, J., Takenaka, T., Langmuir **3**, 1150-1154 (1987).
4. Nishikawa, Y., Fujiwara, K., Shima, T., Appl. Spectrosc. **44**, 691-694 (1990).
5. Osawa, M., and Ikeda, M., J. Phys. Chem. **95**, 9914-9919 (1991).
6. Osawa, M., Ataka, K-I., Ikeda, M., Uchihara, H., Nanba, R., Anal. Sci. **7**, 503-506 (1991).
7. Osawa, M., Ataka, K-I., Yoshii, K., Nishikawa, Y., Appl. Spectrosc. **47**, 1497-1502 (1993).

SEIRA Spectroscopy on Biomembranes

Ch. Kuhne, G. Steiner, W.B. Fischer, and R. Salzer

Institut fuer Analytische Chemie, TU Dresden, D-01062 Dresden, Germany
Email: christian.kuhne@chemie.tu-dresden.de

The infrared absorption of molecules which are adsorbed or near small metal clusters, is enhanced due to an electric field around the clusters. This phenomenon is known as surface enhanced FTIR spectroscopy (SEIRA). The potential of SEIRA consists in the investigation of adsorbates and of thin layers like membranes. We used SEIRA to analyse small and specific changes in the structure of biomembranes. Two SEIRA configurations were used: underlayer and overlayer. An intensity increase in the spectra was observed for both configurations. The SEIRA spectra are discussed in terms of field interactions of clusters and adsorbed material. An effective medium model is used to explain the differences in the excitation levels of the clusters in under- and overlayer configurations.

INTRODUCTION

In metal clusters a local electrical dipole can be excited by incident photons. The local dipole results in an electric field around the clusters. Depending on size and on cluster density the electric fields of neighbouring clusters show strong interactions. The infrared absorption of molecules situated in the resulting alternating field of the clusters will be enhanced. The enhancement effect was described for the first time by Hartstein et al. (1) using an ATR configuration. SEIRA was also used to improve the sensitivity of chemical IR sensors (2). SEIRA has been carried out mainly on Ag or gold clusters (1-9) and was observed in transmission (4-6,8), attenuated total reflection (ATR) (1,3) and external reflection (7).

Surface enhanced optical spectroscopy is an essential tool for the investigation of functionalized membranes. At present, Surface Enhanced Raman Spectroscopy (SERS) (10) and Surface Plasmon Resonance Spectroscopy (SPR) (11) have found wider acceptance than SEIRA. In this study we illustrate the potential of SEIRA to investigate the adsorption of vesicles containing the nicotinic acetylcholine receptor onto Ge and Ag (12). Structural differences between upside and downside of the membranes could be studied by the over- and underlayer configurations of the Ag clusters. The scheme of underlayer and overlayer configuration is given in Fig.1.

THEORETICAL CONSIDERATIONS

In the following the infrared absorption enhancement will be discussed briefly using the effective medium model of Bruggeman (13). Details of the calculation are described in (14). A simple electrostatic model is given in (15).

MIE theory (16) shows that in small metal clusters a collective electron resonance is induced by incident radiation if the clusters are smaller than the wavelength of the photons. A consequence of the electron resonance is a strong electric field around the metal clusters. The field strength is related to the intensity of the incident light and to the morphology of the cluster layer (size, shape and distribution of the metal clusters). The field strengthening leads to an enhanced light absorption by the molecular oscillators in the vicinity of the excited clusters. A further enhancement of the field strength is observed between the clusters because of the interaction of the fields around neighbouring clusters. Since the strength of the induced dipole in the molecules is directly proportional to the strength of the electric field, the resulting effect is a further enhancement of the molecular absorption.

a)

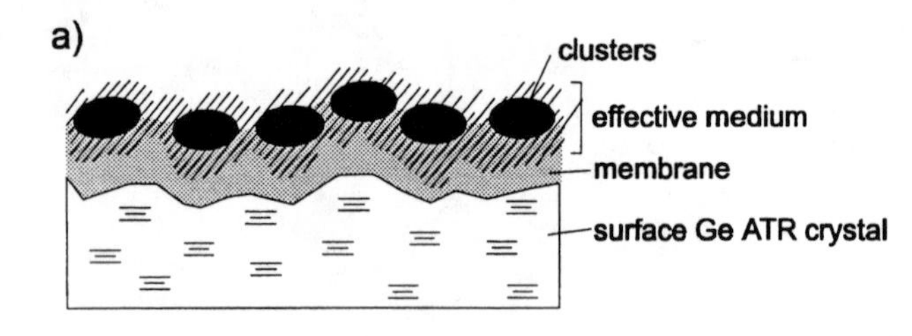

b)

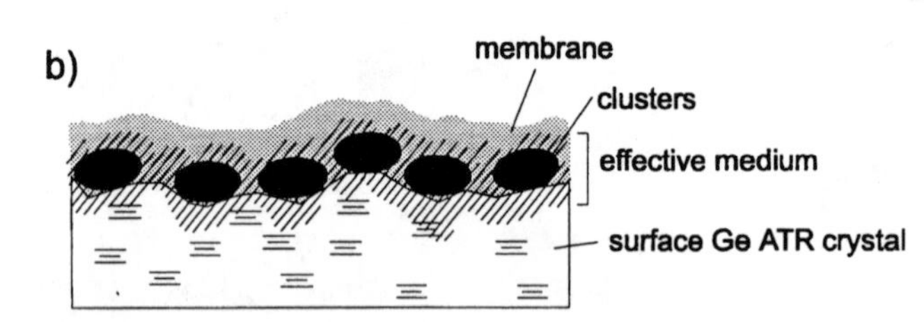

FIGURE 1. Scheme of sample arrangements for SEIRA investigation of membranes: a) overlayer configuration and b) underlayer configuration.

The modeling of SEIRA by calculation of the resulting effective medium with the method of Bruggeman has been suggested by Osawa et al. (8). Because of the large

CP430, *Fourier Transform Spectroscopy:* 11th International Conference
edited by J.A. de Haseth

variation of many factors in the calculations this model can not give good quantitative results, but it shows a very good qualitative validity.

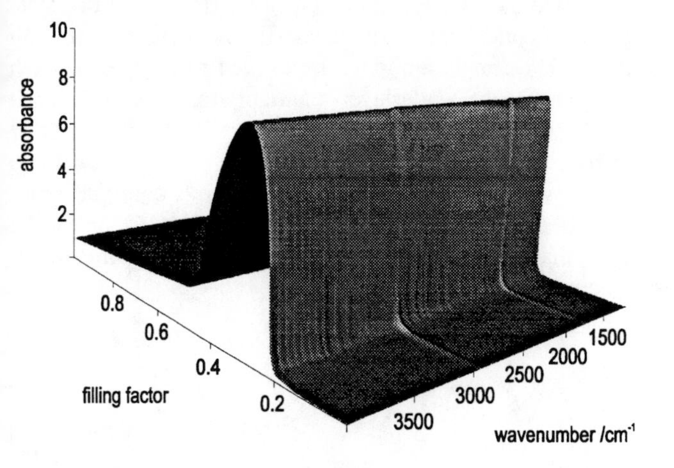

FIGURE 2. 3D plot of calculated spectra for the idealized SEIRA in ATR configuration (Fig.1b).

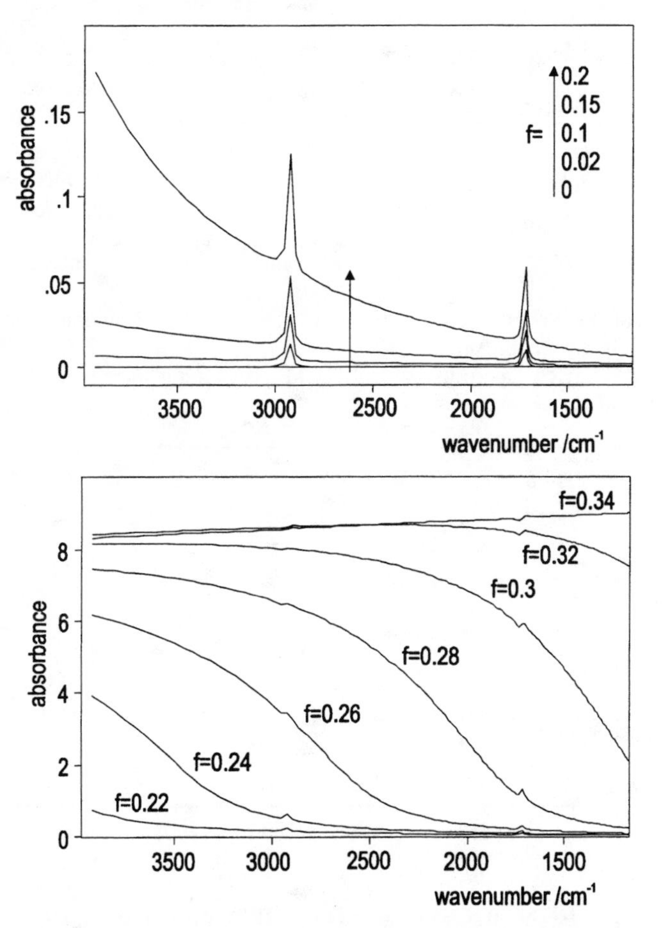

FIGURE 3. Spectra for different filling factors of Fig.2.

The Fig.2 shows the 3D plot of calculated spectra for the idealized SEIRA in ATR configuration (Fig.1b). The filling factor is the ratio between the mass thickness of the cluster layer and the optical thickness of the effective medium. Fig.3 shows spectra for different filling factors. A filling factor equal to zero characterizes an effective medium without any clusters, i.e. the normal case of ATR. The two example absorption bands in Fig.3 will be strongly enhanced due to the collective electron resonance in the metal clusters. The absorption intensities and the baseline level grow if the filling factor increases. This baseline is very typical for the resonance of the metal clusters. In Fig.2 one can see also, that further increase of the filling factor results in a strong metallic absorption and afterwards a strong metallic reflection without any absorption at the wavelengths of the example bands. For the membrane investigations with Ge substrates we found an optimal filling factor respectively an optimal mass thickness of the cluster layer of 4 nm.

The effective medium model takes into account the chemical properties of the cluster material, the substrate and the nature of the analyte between the clusters. The model provides a good qualitative description. The enhancement effects can clearly be demonstrated.

Nicotinic acetylcholine receptor (nAChR) is one type of ion channels. Ion channels are a special class of integral membrane proteins. Activated by the neurotransmitter acetylcholine the receptor allows the diffusion controlled flow of Na^+ and K^+ across the cell membrane. One transmitter molecule causes the flow of several 10^4 ions, which makes these systems good candidates for analytical applications. The functioning of the ion channels may be disturbed by their interactions with the artificial support. Therefore it is very important to investigate the outer functional groups of the membranes and their interactions with the support.

EXPERIMENTAL

The FTIR spectra were recorded with a resolution of 4 cm^{-1}. 256 interferograms were coadded for a single spectrum on a Nicolet 205 equipment with a DTGS detector. The ATR crystals consist of Ge with an incident angle of 45° and 12 active reflections (Spectra Tech, Stanford/CT, USA). Crystals were cleaned with trichlorethylene and water before every usage. Ag clusters were generated by thermal evaporation in a bell jar evaporator at a pressure of 2.10^{-5} Torr. The layer thickness of 4 nm was controlled by a quartz monitor. The deposition rate was kept at »0.1 nm/s. During condensation of the clusters the Ge crystal was heated to a temperature of 50 °C. The Ag purity was 99.9 %. The Ag layers were characterised by atomic force microscopy.

The membranes were spin coated onto the substrates at 4000 rpm. 40µl of methanol solution of nAChR (5nmol ACh sites / mg protein, 4-6 mg protein / ml solution) were applied per test. In the underlayer configuration Ge crystals were precoated by a 4nm Ag layer. In the overlayer configuration a 4 nm Ag top layer on the membranes was produced. The resulting spectra of the membranes are shown in Fig.4. Neat Ge crystals were used for background in reference and overlayer spectra. A Ag coated Ge crystals were used for background in the underlayer configuration. The SEIRA enhancement factor was calculated as ratio of the integral areas of corresponding bands in SEIRA and reference spectra. A tangential baseline has been constructed for every individual band.

A possible influence of the high vacuum on the biomembrane was tested before condensation of the Ag clusters. The comparison of the 2nd derivative spectra of the samples before and after evacuation did not reveal any structural changes.

DISCUSSION AND CONCLUSIONS

The band intensities of the spectra recorded with Ag clusters are up to ten fold enhanced compared to the reference spectrum without the clusters (cf. Fig.4) (15).

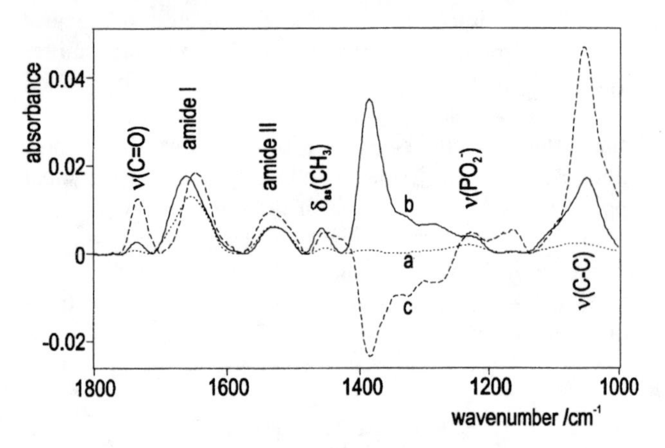

FIGURE 4. a) FTIR ATR spectrum of a nAChR containing membrane. The membrane was adsorbed on a Ge crystal free of Ag clusters. A neat Ge crystal was employed to establish the background spectrum.
b) SEIRA spectrum of a nAChR containing membrane in overlayer configuration. The adsorbed membrane was top-coated by a 4 nm layer of Ag clusters. A neat Ge crystal was employed to establish the background spectrum.
c) SEIRA spectrum of a nAChR containing membrane in underlayer configuration. The Ge crystal was coated by a 4 nm layer of Ag clusters before adsorption of the membrane. An Ag coated Ge crystal was employed to establish the background spectrum.

The total enhancement in overlayer configuration is smaller than in underlayer configuration. The reason is the larger distance of the cluster layer in overlayer configuration to the ATR interface. The larger this distance the smaller is the excitation level of the clusters (14). The excitation level of the clusters corresponds with the strength of the baseline deformation in the spectra. In Fig.5 are shown the reference spectrum and the spectra in over- and underlayer configurations without baseline correction. A neat Ge crystal was employed to establish the background spectrum. For both configurations one can recognize the very typical resonance baselines, but in spite of the identical mass thickness (filling factor) of the Ag layers and sample amounts the excited resonance is much stronger in the underlayer configuration. As mentioned before, the reason is the larger distance of the cluster layer in overlayer configuration to the ATR interface. In Fig.6 are shown the calculated spectra for both configurations. Details of the calculation are described in (14). It can be seen, that the effective medium model is able to mirror the practical results quite well.

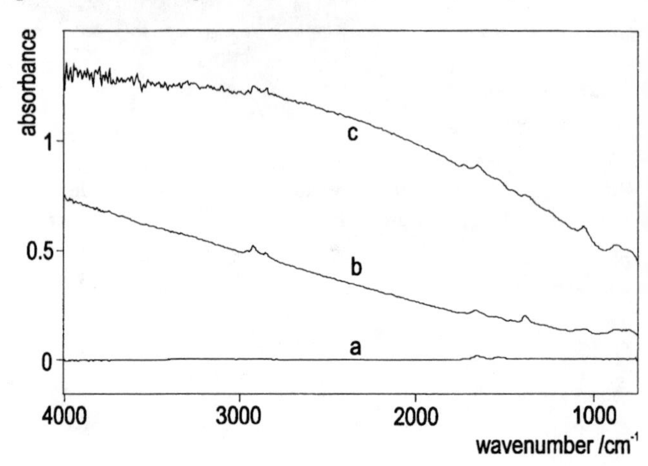

FIGURE 5. a) FTIR ATR spectrum of a nAChR containing membrane (reference).
b) SEIRA spectrum of a nAChR containing membrane in overlayer configuration.
c) SEIRA spectrum of a nAChR containing membrane in underlayer configuration.

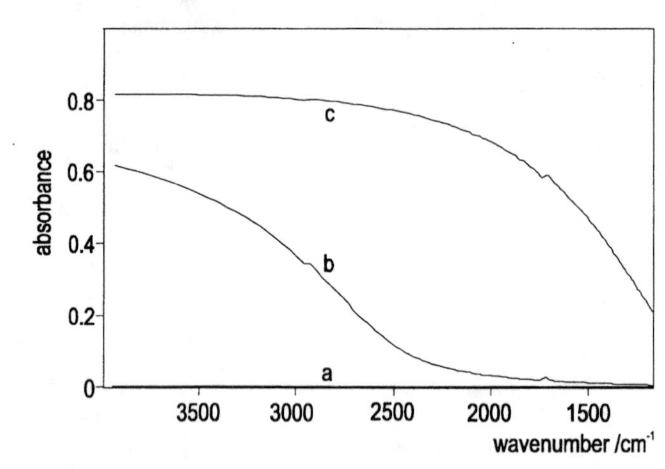

FIGURE 6. a) Caculated FTIR ATR spectrum of a model sample.
b) SEIRA spectrum of the same amout of the model sample in overlayer configuration.
c) SEIRA spectrum of the same amout of the model sample in underlayer configuration.

The differences of the enhancement factors of particular vibrations in Fig.4 are the result of different orientations of the functional groups of the membrane in both layer configurations. The carbonyl ν(C=O) and backbone ν(C-C) vibrations exhibit the strongest enhancement. The amide I band shows only weakly enhanced intensities. This band consists of individual components for the protein substructures: turns, β - sheet, α - helical and unordered structure (17). Structural differences induced in the membrane by contact with the Ag clusters are indicated by a shift of the band envelope compared to the Ag - free reference membrane. We find a downshift of the complete amide I band for the underlayer configuration, whereas an up shift is observed for the overlayer configuration. According to its position at 1653 cm^{-1} the reference spectrum is dominated by the α-helical structure. Moreover, the amide II band at about 1535 cm^{-1} is enhanced in underlayer configuration, while in overlayer configuration this is not noticeable. These differences may have their origin in miscellaneous orientations of the amide groups of the ion channels to the Ag clusters in both configurations. Since the surface electric field of the clusters is oriented perpendicular to the cluster surface, only vibrations with dipole changes perpendicular to the surface are enhanced (8). This surface selection rule in the main should also explain the different enhancement of the lipide structures at about 1230 cm^{-1} of the membrane fragments in under- and overlayer configuration.

At 1385 cm^{-1} the overlayer spectrum shows a strong band that corresponds to background incompensation. A similar effect is observed in the underlayer spectrum, but in this case the background bands are overcompensated. We also observed this bands on ZnSe ATR crystals. Comparison with other spectral regions reveals this band as a CH vibration. Therefore we assume an organic pollutant, which is adsorbed by the active Ag layer immediately after removing the samples from the evaporator.

SEIRA spectroscopy can be used to investigate the outer structure of functionalized membranes. The advantage of this method is the possibility to observe structural differences within the membrane up to depth of several nm. The observation region can be shifted across the membrane simply by shifting the Ag layer. The enhancement factor depends on the orientation of the structural groups to the Ag layer. SEIRA spectroscopy is an essential tool to investigate the interacting groups of proteins towards the interfaces.

ACKNOWLEDGEMENT

The authors thank the BMBF and Deutsche Forschungsgemeinschaft for financial support. ChK acknowledges a stipend of the Free State Saxony. We also acknowledge Prof. A. Maelicke and Dr. A. Schrattenholz for providing us native nAChR membrane vesicles.

REFERENCES

1. Hartstein A, Kirtley JR, Tsang JC (1980) Phys. Rev. Lett. 45: 201-204
2. Kellner R (1997) Appl. Spec. 51: 495-503
3. Hatta A, Suzuki Y, Suetaka W (1984) Appl. Phys. A 35: 135-140
4. Nishikawa Y, Fujiwara K, Shima T (1990) Appl. Spectr. 44: 691-694
5. Osawa M, Ikeda M (1991) J. Phys. Chem. 95: 9914-9919
6. Nishikawa Y, Fujiwara K, Shima T (1991) Appl. Spectr. 45: 747-751
7. Nishikawa Y, Fujiwara K, Ataka K, Osawa M (1993) Anal. Chem. 65: 556-562
8. Osawa M, Ataka K, Yoshii K, Nischikawa Y (1993) Appl. Spectr. 47: 1497-1502
9. Terui Y, Hirokawa K (1994) Vibr. Spectr. 6: 309-314, 315-321
10. Chang RK, Furtak TE (1982) Surface Enhanced Raman Scattering, Plenum Press New York
11. Advincula R, Aust E, Meyer W, Knoll W (1996) Langmuir 12: 3536-3540
12. Fischer WB, Unverricht I, Schwenke D, Kuhne Ch, Steiner G, Schrattenhalz A, Maelicke A, Salzer R, Structural and functional analysis of vesicles containing the nicotinic acetylcholine receptor using ATR-FTIR, SEIRA-, and Fluorescence Spectroscopy, in preparation
13. D.A.G. Bruggeman; Ann. Phys. 24 (1935) 5, 636-651
14. Kuhne Ch, Steiner G, Salzer R, Surface Enhanced Infrared Spectroscopy: Modeling and Spectra, in preparation
15. Kuhne Ch, Steiner G, Fischer WB, Salzer R, Surface Enhanced FTIR Spectroscopy on Membranes, Fres. J. Analyt. Chem. submitted 1997
16. Mie G (1908) Anal. Phys. 25: 377-445
17. Butler DH, McNamee MG (1993) Biochim. Biophys. Acta 1150: 17-24

Comparison of the FT-SERS Behaviour of the Fullerenes C_{60} and C_{70}

P. M. Fredericks

Centre for Instrumental and Developmental Chemistry, Queensland University of Technology, PO Box 2434, Brisbane, Qld 4001, Australia

The surface-enhanced Raman (SER) spectrum of the fullerene C_{60} has been measured on an electrochemically-roughened silver surface using a Fourier transform (FT) Raman spectrometer. The FT-SER spectrum is similar to the normal Raman spectrum of this compound indicating that only the electromagnetic mechanism of enhancement is occurring. Detection limit was found to be 7 ng of C_{60}. Comparison of these data with previously reported SER data for C_{60}, shows distinct spectral differences obtained under different experimental conditions. C_{60} does not appear to show a resonance effect, as has been reported for C_{70}, but it is postulated that C_{60} exhibits phase variation depending on the method of deposition on the surface.

INTRODUCTION

There has been much interest in recent years in the physical and chemical properties of the molecules, C_{60} (1,2) and C_{70}. The Raman spectra were first reported by Bethune *et al* (3) in 1990 and subsequently by a number of other workers. The first reported Fourier transform Raman spectra utilizing near-infrared excitation were by Dennis *et al* in 1991 (4). Initial differences in reported spectra have been explained by variations in sample purity, variations in sample state (powder, film etc.), the presence of oxygen, sample temperature, and variation of the exciting wavelength (5). Recently, it has been proposed that there are two possible phases for C_{60} giving rise to two different Raman spectra (6).

SER Scattering of Fullerenes

There has also been much interest in the surface-enhanced Raman scattering (SERS) of fullerenes. SERS of C_{60} has been studied under a variety of conditions: on a gold electrode roughened by an oxidation-reduction procedure with 676 nm excitation (7), on a silver mirror with 488 nm excitation (8), on tin and gold substrates with 488 nm excitation (9), on a vapour-deposited silver film with 488 nm excitation (10), on a gold electrode roughened by an oxidation-reduction procedure with 647 nm excitation (11), on cold-deposited silver surfaces with excitation lines in the range 514.5 - 457.9 nm (12,13), on a cold-deposited indium surface with 457.9 nm excitation (13), on roughened silver and copper surfaces (14), on a rough silver film (15), on gold, copper and silver surfaces (16), and on gold surfaces in acetonitrile (17). The SERS spectra produced by these differing techniques are all different, particularly in the positions and relative intensities of the bands. In some instances additional bands have been obtained which do not occur in the normal Raman spectrum of C_{60}.

SERS of C_{70} has been less well studied because of the greater difficulty of obtaining pure samples. 488 nm excitation has been used with a vapour deposited silver surface (10) and with a chemically deposited silver surface (9). Mo *et al* (18) used 632.8 nm excitation with both a chemically deposited silver surface and with silver island films. The FT-SERS spectrum of C_{70} on an electrochemically roughened silver surface has been reported (19).

This paper reports the FT-SER spectrum of C_{60} measured on an electrochemically roughened surface using near-IR excitation and compares the spectra with those previously reported in the literature for C_{60} and C_{70}.

EXPERIMENTAL

The pure sample of C_{60} was obtained from CSIRO, Division of Petroleum Resources, Sydney, Australia. The silver surface was prepared as described previously (20), except that the electrolyte was 0.2 M KCl solution. 1 μL of the fullerene solution in toluene was placed on the silver surface and the solvent allowed to evaporate at ambient temperature.

FT-SER spectra were obtained with a Perkin Elmer System 2000 FT-Raman spectrometer equipped with a Spectra Physics diode-pumped Nd-YAG laser emitting at a wavelength of 1064 nm. A laser power of 100 mW was used for all SERS experiments. The optical path difference velocity was 0.1 cm s^{-1} and the resolution was generally 4 cm^{-1}. Strong Beer-Norton apodization was employed. The spectra were corrected for instrument response.

CP430, *Fourier Transform Spectroscopy:* 11th International Conference
edited by J.A. de Haseth

RESULTS AND DISCUSSION

Carbon 60

C_{60} deposited on the silver surface gave rise to a moderately strong SER spectrum which was similar to the normal FT-Raman spectrum of the solid compound.

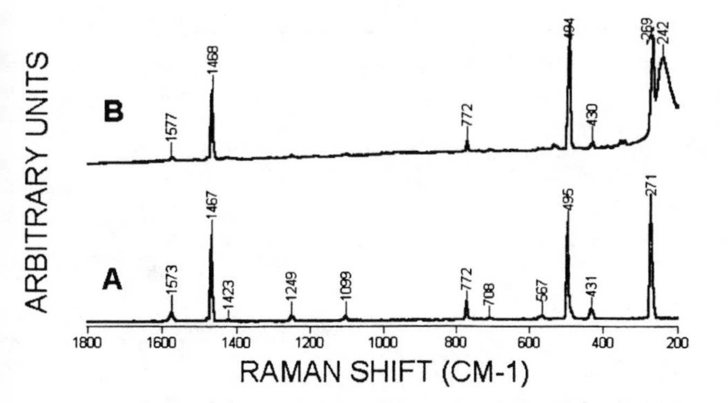

FIGURE 1. Comparison of (A) FT-Raman spectrum of solid C_{60}, and (B) FT-SERS spectrum of 1 μL of 2 x 10^{-3} M solution of C_{60} on a roughened silver surface. The band at 242 cm^{-1} is due to AgCl on the silver surface.

Figure 1 shows the normal FT-Raman spectrum of C_{60}, together with the FT-SER spectrum obtained from 1 μL of 0.8 x 10^{-4} M solution (equivalent to 57 ng of C_{60}) added directly to thesilver surface. The SER spectrum is similar to the solid C_{60} spectrum and shows the same three intense bands near 1468, 493 and 269 cm^{-1}. The less intense bands seen in the solid C_{60} spectrum are also present in the SER spectrum but are considerably weaker. The same amount of C_{60} deposited on a polished silver surface showed only an extremely weak spectrum. This large difference between the polished and roughened silver surfaces consitutes a strong indication that a surface-enhanced Raman scattering process is occurring. As the SER spectrum and the solid C_{60} spectrum are similar, it can be inferred that the molecule is not interacting strongly with the silver surface.

A detailed comparison of the SER spectrum with the solid C_{60} spectrum shows the major differences to be the relative intensities of the bands, and the exact position of the band near 495 cm^{-1}. The three most intense bands in the spectrum of solid C_{60}, at 1467, 495 and 271 cm^{-1}, are also the major bands in the SER spectrum, however in the SER spectrum the most intense band is at 493 cm^{-1}, not 271 cm^{-1}. The three most intense bands appear to experience a greater enhancement than the remaining weaker bands in the spectrum. For example, the ratio of intensities of the 1468 cm^{-1} and the 772 cm^{-1} bands is 4.8 for the solid C_{60} spectrum, but rises to 7.9 for the SER spectrum.

Despite the only moderate enhancement found for C_{60} on the silver surface, the detection limit was found to be 1 μL of 0.98 x 10^{-5} M solution, equivalent to7 ng of C_{60}. As the silver surface is simple and fast to prepare, and the spectrum is quite specific, this form of SERS constitutes a suitable method for the detection of even small quantities of C_{60}. In fact, the actual detection limit is much lower because the drop of solution placed on the surface wets an area about 6-8 mm in diameter. However, the laser beam has a diameter of only 200 μm and thus probes only a small proportion of the sample placed on the surface.

Comparison with Previous C_{60} Work

As mentioned in the introduction, there have been a number of reports dealing with the surface-enhanced Raman scattering, or related phenomena, of the C_{60} molecule (7-17). Of these reports, several discuss the use of silver surfaces (9,10,12-16). Liu *et al* (9) used a silver mirror prepared by chemical deposition onto a glass or quartz substrate. The C_{60} was introduced onto the silver surface in a toluene solution which was allowed to dry in the atmosphere. The SER spectrum obtained with excitation at 488 nm was similar to the one described in this paper using excitation at 1064 nm, with major bands at 1468, 492 and 272 cm^{-1}. However, the intensities are quite different with the band at 1468 cm^{-1} being the most intense and about 8 times more intense than the band at 492 cm^{-1}. In our work the band at 493 cm^{-1} is nearly twice as intense as the band at 1468 cm^{-1}. Also, Liu *et al* find a band at 711 cm^{-1} which is of similar intensity to the band near 770 cm^{-1}. Our SER spectrum shows the band at 770 cm^{-1}, but the band at 711 cm^{-1} is not observed.

Two other reports describe the SERS of vapour deposited C_{60} on silver island films with excitation at 488 nm (10) or 514 nm (16). The spectra obtained are similar to those reported in this paper except that the $A_g(2)$ mode (the pentagonal pinch mode) is very intense compared to all other bands in the spectrum, and occurs near 1436 cm^{-1} instead of 1468 cm^{-1}. Two other reports of enhanced C_{60} spectra on cold-deposited silver films also exist (12,13). In these spectra the $A_g(2)$ mode occurs at 1446 cm^{-1} for spectra measured at 8 K. At higher temperatures the spectral intensity increases and the $A_g(2)$ band becomes a doublet with peak positions 1464 and 1472 cm^{-1}. As the intensity increases at higher temperatures the authors conclude that the spectra are ***not*** SER spectra because cold-deposited films lose their roughness at higher temperatures which would be expected to lead to a diminishing of any SERS effect (13). Our results are consistent with this proposition since we only see a spectrum for the roughened silver surface and almost no signal at all for the polished surface. The Table compares the literature results for silver surfaces with the results given in this paper, and also details the nature of the metal surface, the mode of addition of the C_{60}, the laser excitation wavelength, and the position of the three major bands. It can be seen that the position of the $A_g(2)$ mode has been measured at different positions between 1436 and 1468 cm^{-1}. For silver surfaces there appears to be a correlation between the position of the $A_g(2)$ mode and the method of depositing the C_{60} on the surface. In cases where the C_{60} was sublimed onto the surface (10,16) the $A_g(2)$ mode occurs around 1438 cm^{-1}, however if the C_{60} was deposited from solution (9), as in the work described in this paper, then this band occurs at 1468 cm^{-1}.

TABLE 1. Comparison of this work with some previous work on SERS of C_{60}

	This work	**Reference 9**	**Reference 10**	**Reference 16**
Nature of surface	Electrochemically roughened Ag	Chemically deposited Ag	Vapour deposited Ag	Ag island film
Method of addition of C_{60}	Evaporation of solution	Evaporation of solution	Sublimed onto surface	Sublimed onto surface
Excitation wavelength (nm)	1064	488	488	514
Position of the three main bands [Ag(2), Ag(1) and Hg(1)]	1468 493 269	1468 492 272	1438 482 264	1436 - -

Recently, it has been suggested that solid films of C_{60} can exist in two different phases depending on sample temperature, having different Raman spectra (6). Specifically, the $A_g(2)$ mode occurred either at 1469 or 1459 cm^{-1}. Furthermore, the authors also suggested that exposure to the laser beam, especially at high irradiance, could cause conversion of the phase. It may be that the varied results for SERS and similar phenomena for C_{60} summarised in the Table can also be rationalised by proposing the possibility of different phases for C_{60} layers deposited on surfaces, perhaps depending to some extent on the method of deposition.

Carbon 70

Figure 2 shows the SER spectrum of C_{70} (from 1 µL of 1.3 x 10^{-3} M solution) on the roughened silver surface together with the normal Raman spectrum of the solid compound.

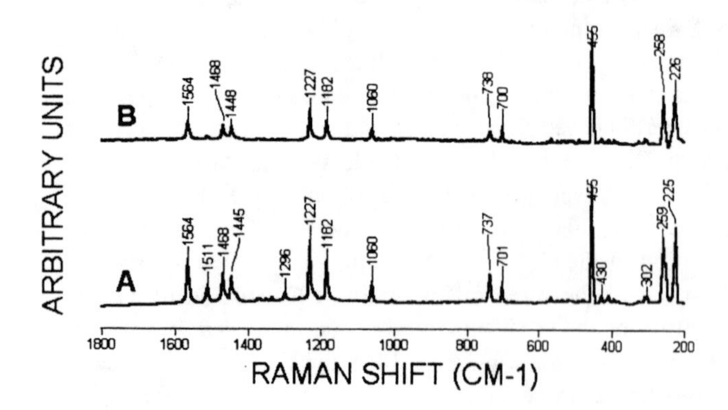

FIGURE 2. Comparison of (A) FT-Raman spectrum of C_{70} and (B) FT-SERS spectrum of 1 µL of 1 X 10^{-5} M solution of C_{70} on a roughened silver surface. A background spectrum of the clean silver surface has been subtracted.

Clearly, the two spectra are very similar indicating little interaction of the C_{70} with the silver surface. Previous work (19) also found large differences between the SERS spectra of C_{70}, mainly in the relative intensities of the bands. However, unlike C_{60}, the differences appeared to be related to the excitation frequency, rather than to the method of deposition of the compound on the surface. It was suggested that a resonance or pre-resonance effect occurred when excitation was in the visible region. The possiblity of a slight resonance effect of some kind is consistent with the electronic absorption spectrum of C_{70} which shows a weak absorption band at 468 nm with a long wavelength shoulder tailing to about 670 nm (21). C_{60} does not show bands at such long wavelength in its electronic spectrum and therefore would not be expected to exhibit resonance effects.

CONCLUSIONS

The SER spectrum of C_{60} on an electrochemically roughened silver surface has been measured using an FT-Raman spectrometer with near-IR excitation. The spectrum is similar to the normal Raman spectrum except that the three major peaks are enhanced by about a factor of two compared with the other spectral features. The similarity of the SER and normal Raman spectrum indicate that the C_{60} experiences only the electromagnetic (EM) mechanism of SERS and therefore it can be inferred that the molecule does not interact strongly with the silver surface.

Although this work was not designed to determine an accurate detection limit, it was found that 1 µL of a solution of concentration 0.98 x 10^{-5} M (equivalent to 7 ng of C_{60} on the surface) could be reliably detected. As the method is simple and fast, and the spectrum is highly specific, it represents a useful method for detecting the presence of small amounts of C_{60}.

Comparison with previous SERS measurements carried out in a variety of different ways, seems to indicate two kinds of spectra which correlate to some extent with the manner of deposition of the C_{60}. The major difference between the two kinds of spectra is the position of the $A_g(2)$ mode (the pentagonal pinch mode) which occurs near 1468 cm^{-1} for most experiments where the C_{60} has been deposited by evaporation of a solution, or near 1438 cm^{-1} for vapour deposited (sublimed) C_{60}. It is suggested that a previous hypothesis of two phases for C_{60} solid films may also be the explanation for the variety of results in SERS experiments. Resonance does not seem to be important in the SER spectra of C_{60}, but has been suggested as significant in the spectra of C_{70}. This is consistent with the differing electronic spectra of the two fullerene molecules.

ACKNOWLEDGMENT

The author is indebted to M. A. Wilson of CSIRO, Division of Petroleum Resources, for a sample of C_{60} and for many helpful discussions.

REFERENCES

1. Kroto, H. W., Allaf, A. W., and Balm, S. P., *Chem. Rev.*, **91**, 1213 (1991).
2. Wilson, M. A., Pang, L. S. K., Willett, G. D., Fisher, K. J., and Dance, I. G., *Carbon*, **30**, 675 (1992).
3. Bethune, D. S., Meijer, G., Tang, W. C., and Rosen, H. J., *Chem. Phys. Lett*, **174**, 219 (1990).
4. Dennis, T. J., Hare, J. P., Kroto, H. W., Taylor, R., Walton, D. W., and Hendra, P. J., *Spectrochim. Acta*, **47A**, 1289 (1991).
5. Lynch, K., Tanke, C., Menzel, F., Brockner, W., Scharff, P., and Stumpp, E., *J. Phys. Chem.*, **99**, 7985 (1995).
6. Akers, K. L., Douketis, C., Haslett, T. L., and Moskovits, M., *J. Phys. Chem.*, **98**, 10824 (1994).
7. Garrell, R. L., Herne, T. M., Szafranski, C. A., Diederich, F., Ettl, F., and Whetton, R. L., *J. Amer. Chem. Soc.*, **113**, 6302 (1991).
8. Zhao, T., Zhao, W., Liu, J., Chen, L., Li, Y., and Zhu, D. *Solid State Comm.*, **83**, 789 (1992).
9. Liu, J., Zhao, T., Mo, Y., Li, T., Liu, Y., Zhu, K., Li, Y., Yao, Y., Yang, D., Wang, X., and Zhu, D. *Solid State Comm.*, **81**, 757 (1992).
10. Akers, K., Cousins, L. M., and Moskovits, M. *Chem. Phys. Lett.*, **190**, 614 (1992).
11. Zhang, Y., Du, Y., Shapley, J. R., and Weaver, M. J. *Chem. Phys. Lett.*, **205**, 508 (1993).
12. Rosenberg, A., and DiLella, D. P. *Chem. Phys. Lett.*, **223**, 76 (1994).
13. Rosenberg, A., and DiLella, D. P. *Solid State Comm.*, **95**, 729 (1995).
14. Li, S. J., Xu, X. P., Qi, Z. P., and Li, W. Z. *Chin. Chem. Lett.*, **4**, 909 (1993).
15. Akers, K. L., and Moskovits, M., *J. Electron Spectrosc. Relat. Phenom.*, **64-65**, 871 (1993).
16. Chase, S. J., Bacsa, W. S., Mitch, M. G., Pillione, L. J., and Lannin, S. J., *Phys. Rev. B: Condens. Matter*, **46**, 7873 (1992).
17. Zhang, Y., Edens, G., and Weaver, M. J., *J. Amer. Chem. Soc.*, **113**, 9395 (1991).
18. Mo, Y., Mattei, G., Pagannone, M., and Xie, S., *Appl. Phys. Letters*, **66**, 2591 (1995).
19. Fredericks, P. M., *Chem. Phys. Letters*, **253**, 251 (1996).
20. Roth, E., Hope, G., Schweinsberg, D. P., Kiefer, W., and Fredericks, P. M., *Appl. Spectrosc.*, **47**, 1794 (1993).
21. Hare, J. P., Kroto, H. W., and Taylor, R., *Chem. Phys. Letters*, **177**, 394 (1991).

Raman and Surface-Enhanced Raman Spectroscopy of Adsorbed Phthalic Acid on Oxidized Aluminum Foil

O. Klug[1], Gy. Parlagh[1] and W. Forsling[2]

[1] *Dept. of Phys. Chem., Technical University of Budapest, H-1521 Budapest, Hungary*
[2] *Division of Inorg. Chem., Luleå University of Technology, S-97187 Luleå, Sweden*

Adsorption of phthalic acid on anodically oxidized aluminium has been investigated at different pHs and ionic strengths by means of FT-Raman and surface-enhanced FT-Raman spectroscopy (SERS). The surface-enhancement was achieved by deposition of silver sol after adsorption. The spectra of the surface species obtained by the two techniques were significantly different. Raman spectra of the adsorbed phthalic compounds were pH and ionic strength dependent, but the surface-enhanced spectra appeared to be identical at each conditions. Supported by further spectroscopic evidences of the phthalic acid and the silver sol interaction (without aluminium oxide), it is plausible that the deposition of the silver sol results in a new surface complex. The evolution of the SER effect is therefore suggested to be a result of two steps: at first an adsorption occurs on aluminium oxide, and when depositing the aqueous silver sol onto the surface the phthalate ligands form complexes rather with the silver than with the aluminium oxide. However, the aluminium oxide surface may contribute to the non-linear spectroscopic effect due to its surface structure and charge.

INTRODUCTION

Anodically oxidized aluminium foil is used by the electrolytic capacitor manufacturers as it constitutes the anodic part in a capacitor winding. As a part of an ongoing research project in capacitor chemistry Raman and surface-enhanced Raman (SER) spectroscopy were applied to investigate the adsorbed or complexed phthalic acid on the surface of capacitor foils. The anodic capacitor foil is an etched, electrochemically oxidized aluminium foil with γ-aluminium oxide on its surface. The high roughness of the oxide surface makes it difficult to apply most of the spectroscopic techniques. However, a novel method of surface-enhanced Raman spectroscopy offered a new approach for investigating adsorbed molecules on oxidized metal surfaces (1,2).

EARLY ATTEMPTS AND CLAIMS

Surface-enhanced Raman spectra of phthalic acid isomers adsorbed on anodic capacitor foils have been reported by us on ICOFTS-10 (3). The enhanced Raman signal was achieved by a silver sol deposition onto the surface after that the adsorption has occurred and the samples were dried. The adsorbed phthalate species became possible to investigate by conventional Raman spectroscopy too if the metal aluminium between the oxide layers of the capacitor foil was dissolved prior to the adsorption. The anodic oxide films were then ground and used as powder samples in the experiments (4). The spectra of the adsorbed species obtained by the two techniques were significantly different.

In the present work Raman and SER spectra of the adsorbed species at different pHs and ionic strengths are compared and a possible explanation is given by assuming different surface complexes to be formed in the two cases.

EXPERIMENTS AND RESULTS

The oxidized aluminium foil was provided by RIFA Electrolytics AB and was produced by Becromal S.P.A. with a forming voltage of 690V. The metal free anodic oxide was prepared as described before (4). The silver sol was a citrate reduced aqueous sol as detailed in (5). Phthalic acid was bought from Fluka Co. with purity of 99.5%. The solid state neutralized phthalates were prepared according to Arenas and Marcos (6). 1 mM aqueous solutions of phthalate were mixed with the appropriate amount of aluminium oxide to get an oxide concentration of 2g/l. Then the suspension was vigorously stirred at room temperature. The equilibration time was different at each pH and ionic strength, but exceeding 100 hours were always sufficient at room temperature(7).

Raman spectra were recorded on a Perkin-Elmer 1760 FT-IR spectrometer equipped with a 1064 nm laser bench. 180° backscattering and 200-800 mW laser power was used. Usually 400-800 scans were accumulated from the solid samples. All spectra in the figures are presented in their original magnitudes and they are only offset.

CP430, *Fourier Transform Spectroscopy:* 11th International Conference
edited by J.A. de Haseth

Conventional Raman spectra of the adsorbed phthalate on aluminium oxide powder at different pHs are shown in Fig. 1. According to the concentration of the phthalic compound remaining in the bulk solution (measured by UV at 280 nm) more phthalate was adsorbed in the acidic range.

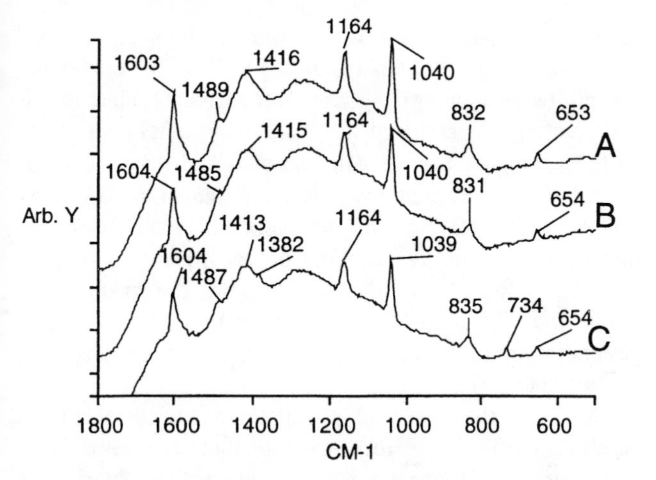

FIGURE 1. Raman spectra of the adsorbed phthalic acid on aluminium oxide at different pHs. The adsorption took place for 116 hs at room temperature from aqueous phthalic solution without ionic strength adjustment. Starting pHs of the solution: A - 3.19 (pure acid), B - 4.05 (mono salt) and C - 6.38 (di-salt).

Figures 2.-4. present the spectra recorded at different ionic strengths for each pH range.

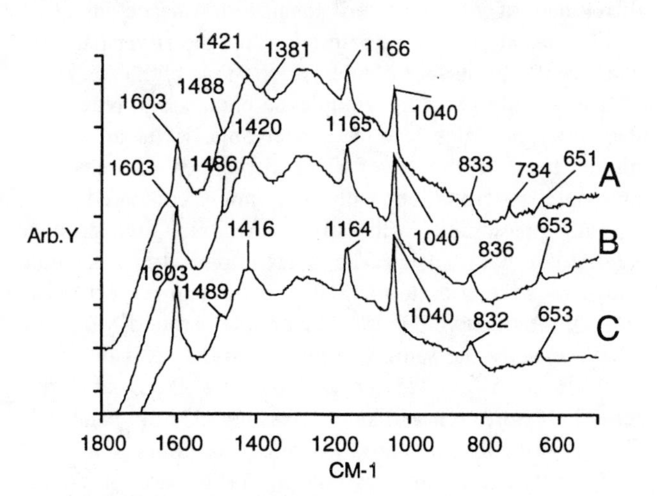

FIGURE 2. Raman spectra of the adsorbed phthalic acid on aluminium oxide at different ionic strengths. The adsorption took place for 116 hs at room temperature from aqueous phthalic solution. The ionic medium: A - 1 M KCl (starting pH: 3.23), B - 0.1 M KCl (starting pH: 3.28) and C - no KCl (starting pH: 3.19).

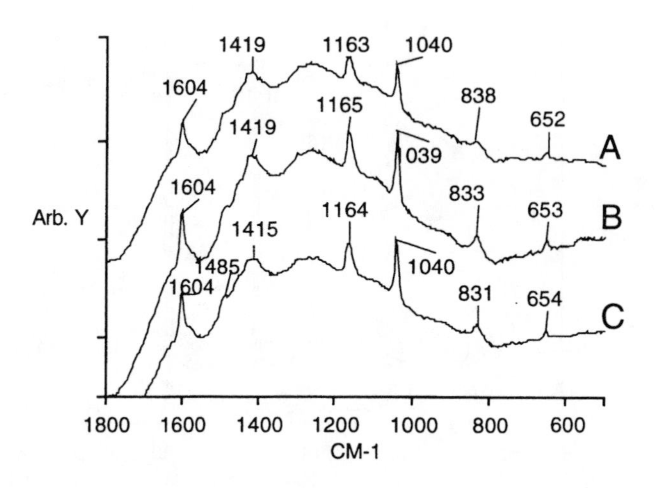

FIGURE 3. Raman spectra of the adsorbed potassium hydrogen phthalate on aluminium oxide at different ionic strengths. The adsorption took place for 116 hs at room temperature from aqueous phthalic solution. The ionic medium: A - 1 M KCl (starting pH: 4.05), B - 0.1 M KCl (starting pH: 4.18) and C - no KCl (starting pH: 4.05).

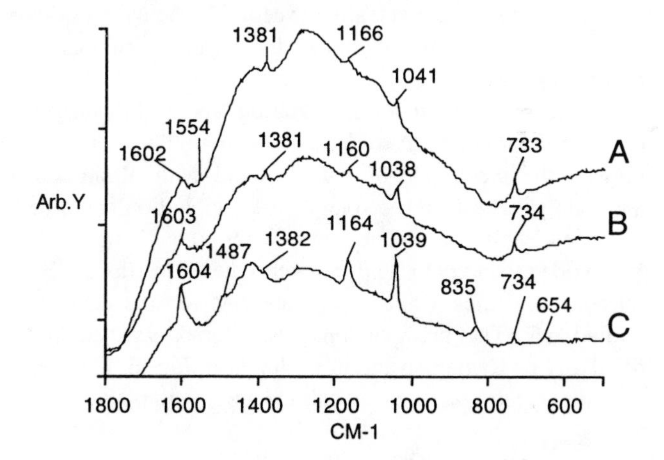

FIGURE 4. Raman spectra of the adsorbed dipotassium phthalate on aluminium oxide at different ionic strengths. The adsorption took place for 116 hs at room temperature from aqueous phthalic solution. The ionic medium: A - 1 M KCl (starting pH: 6.33), B - 0.1 M KCl (starting pH: 6.56) and C - no KCl (starting pH: 6.38).

The adsorbed phthalate at different pHs and different ionic strengths was also studied by SERS. In this case capacitor foils were used as substrates. After the adsorption period a drop of silver sol was deposited on the previously dried surface and the samples were dried once more before the spectra were recorded. Raman spectra obtained from these samples showed high surface-enhancement, but the bands seemed to be independent of adsorption pH as shown in Fig. 5. The only difference was their magnitude which corresponds to the different amount of adsorbed material.

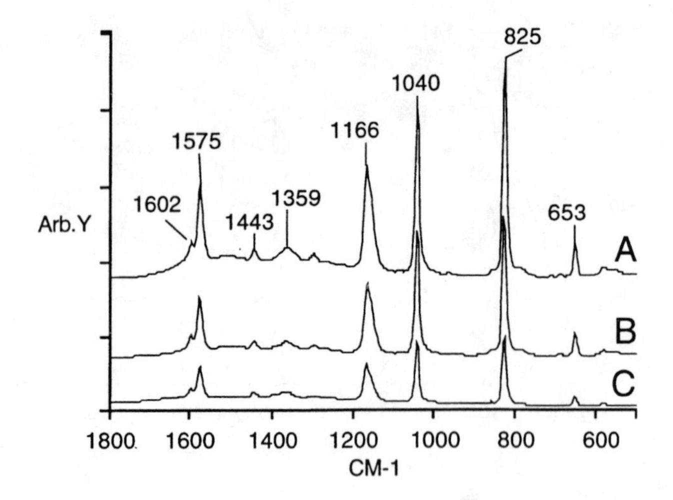

FIGURE 5. SER spectra of the adsorbed phthalic compound on capacitor foils at different pHs. A - from a solution of pure acid, B - from a solution of potassium hydrogen phthalate and C - from a solution of dipotassium phthalate.

Similarly to Fig. 5. SER spectra obtained at different ionic strengths were much alike except that their intensities were somewhat different due to the various amounts of adsorbed phthalate.

In order to get more information about the surface complexes formed in the two cases an attempt was made to prepare the precipitate of phthalic acid with aluminium and with silver ions respectively. No precipitation could be produced with aluminium ions other than aluminium hydroxide around pH=5-6. However, when silver ions were introduced a thick white precipitate was formed at a pH around 6-7. The precipitation was filtered, washed and dried and its Raman spectrum is shown in Fig. 6. together with the spectra of pure acid and the dipotassium salt.

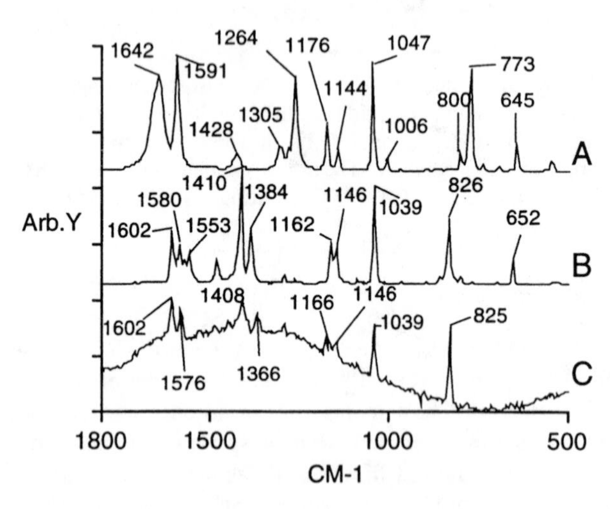

FIGURE 6. Raman spectrum of A - pure phthalic acid, B - dipotassium phthalate and C - the silver precipitate from a phthalate solution.

DISCUSSIONS AND CONCLUSIONS

In the conventional Raman spectra of the adsorbed phthalate two new bands appear at higher pH of the solution: at 1382 cm^{-1} and at 734 cm^{-1}. The first band is also present in the dipotassium phthalate and has been firstly assigned to CC stretching (6). Later the assignment of this band was proposed to be seen as the symmetric O-C-O stretching vibration instead (8). The 734 cm^{-1} band can not be assigned to any vibration of the phthalate ion. It was tentatively assigned to the Al-O stretching in the surface complex formed with the phthalate (7). The ν(Al-O) vibration is normally very weak in Raman, but its range is from 720-760 cm^{-1} in the aluminas. Both bands of interest appear in case of high pH adsorption or at high ionic strength in lower pH. Based on this observation it is likely that different surface complexes can be formed depending on pH and ionic strength as already proposed on the boemite surfaces (9).

Contrary, the adsorbed phthalate in surface-enhanced conditions (Fig. 5.) seemed to be identical at each pH and ionic strength suggesting the same surface species is formed always. However, the Raman spectra of this species were quite distinct from those seen in conventional Raman (Fig. 1.). Many of the bands, especially of the ring vibrations shifted: from 1603 to 1575 cm^{-1}, from 833 to 825 cm^{-1}, etc. In order to achieve SER conditions the surface was covered with a silver sol. The change in those vibrations which originate from the aromatic ring can well be due to the electronic effect of the silver. The distinct difference of the adsorbed species in surface-enhanced conditions supports the assumption that the silver particles can modify the surface complexation of the phthalate.

The possibility of the complexed phthalate to react with the silver particles is only conceivable, if its affinity is higher towards the silver. The difficulties of making a phthalate precipitation with aluminium compared to a similar precipitate with silver, exhibits the stronger interaction with silver. The spectrum of the precipitate shows close relation to the SER spectra with respect to the ring vibrations: two bands of the doublet around 1600 cm^{-1} have exactly the same wavenumbers and so is the other CC vibration at 825 cm^{-1}. However, there are also remarkable differences like the missing 652 cm^{-1} band (O-C-O deformation) and a new unassigned vibration at 1408 cm^{-1}. Certain differences in the spectra are of course expectable since the precipitate was formed with silver ions and not with silver particles as it happens in SER conditions.

The phthalic acid interaction with silver particles has already been studied in solution by FT-Raman and UV spectroscopy (10). It was shown that the silver sol starts to aggregate rapidly when the phthalic acid solution is introduced into the sol. During the first few seconds of the aggregation a SER spectrum of an adsorbed phthalic acid was obtainable, but after that the particles were grown big enough to sink down in the cuvette. The low number of scans in FT-Raman visualize only the more intensive bands, like 1578, 1154, 1043 and 807 cm^{-1} respectively. The 1408 cm^{-1} vibration (Fig. 6.) was missing indicating a different structure from the Ag^{+} precipitate which was rather ionic than co-ordinative complex.

Concluding the results of the different spectroscopic measurements the evolution of the SER effect on the anodically oxidized aluminium foils is suggested to be seen as a result of two steps:

1. Phthalic acid adsorption on aluminium oxide of the capacitor foil (prior to the addition of the silver sol). A pH dependent surface complex is formed between phthalic acid or phthalate and the aluminium oxide.
2. After depositing the aqueous silver sol onto the surface the phthalate ligands seem to form complex rather with the silver particles than with the aluminium oxide.

The reason of this second step is probably related to the different strength of the two complexes as the surface species formed with silver is assumed to be stronger than the one with aluminium oxide. On the other hand the exact way how the ligand exchange occurs is not understood yet and it needs further studies.

The primary function of the aluminium oxide can be to collect the phthalic acid on its surface. However, it may contribute to the non-linear spectroscopic effect due to its rough surface and can influence the aggregation of the silver particles by its texture and surface charge.

ACKNOWLEDGEMENTS

The authors gratefully acknowledge the financial support given by RIFA Electrolytics AB.

REFERENCES

1. Wilson, H. M. M. and Smith, W. E. (Eds. Kiefer, W., Cardona, M., Schaak, G., Schneider, F. W. and Schrötter, H. W.), *Proc. XIIIth Int. Conf. on Raman Spectroscopy*, New York, Wiley, 1992, pp. 698-699.
2. Haigh, J. A., Hendra, P. J. and Forsling, W., *Spectrochim. Acta*, **50A**, 2027-2034 (1994).
3. Klug, O., Száraz, I., Forsling, W., and Ranheimer M., *Microchim. Acta [Suppl.]*, **14**, 649-651 (1997).
4. Klug, O., Parlagh, Gy., and Forsling W., *J. Mol. Structure*, **410/411**, 183-188 (1997).
5. Lee, P. C. and Meisel, D., *J. Phys. Chem.*, **86**, 3391 (1982).
6. Arenas, J. F. and Marcos, J. I., *Spectrochim. Acta*, **35A**, 55 (1979).
7. Klug, O., *Licentiate Thesis*, Luleå, LTU Press, 1997, pp. 27-28.
8. Tejedor-Tejedor, M. I., Yost E. C. and Anderson, M. A., *Langmuir*, **8**, 525-533 (1992).
9. Nordin, J., Persson, P., Laiti, E. and Sjöberg S., Submitted to Langmuir.
10. Klug, O., Gy. Parlagh and Forsling, W. (Eds. Hórvölgyi, Z., Németh, Zs., and Pászli, I.), *Proceedings of the 7th Conference on Colloid Chemistry*, Budapest, HCS Press, 1997, pp.301-305.

Analysis of Thin Film Multilayer Structures by Reflectance Spectroscopy

A. J. Syllaios, D. R. Bessire, P. D. Dreiske, P. -K. Liao, D. Chandra, and H. F. Schaake

Raytheon TI Systems, P. O. Box 655936, MS 154, Dallas, Texas 75265

The optical transfer matrix formulation was applied to the analysis of thin solid film multilayer structures used in infrared detection device technology. Our analysis is similar to optimization techniques applied to optical multilayer coatings, i.e., performing simulation of spectra and obtaining best fits to experimental spectral data (targets). We present, as a example, the application of the transfer matrix method to determine the thickness of HgCdTe films attached with epoxy to CdZnTe carriers, using normal incidence reflectance spectra. MID Infrared reflection spectra were obtained using a NICOLET Magna IR 750 Fourier Transform Infrared Spectrometer, equipped with a SPECTRATECH, Inc., microscope, taking care to operate in the linear response range of the detector. The obtained thickness values of HgCdTe films and epoxy layers agree very well with those measured by photographic techniques.

INTRODUCTION

Multilayer structures are commonly used in modern infrared detection technology. Single layer HgCdTe diode array devices and their associated passivation layers and antireflection coatings form multilayer stacks. More complex multilayers are found in double layer HgCdTe heterostructure arrays and in multicolor devices. Resonant cavity detectors also consist of multilayers. The spectral characteristics of a multilayer are determined by the physical and optical properties of the materials involved such as thickness and complex index of refraction, and the properties of their interfaces. When these parameters are known, the optical response of these multilayer device structures can be analyzed. Conversely from specific spectral responses such as reflectance of a multilayer the optical constants and thickness of its layers can be determined. We have used optical filter design optimization techniques to determine the thickness and thickness uniformity of HgCdTe multilayer structures.

THEORY OF OPTICAL MULTILAYERS

One of the most widely used methods for the optical treatment of multilayers is the amplitude transfer method (1,2). In this formulation each layer is described by a matrix which relates the amplitude of the incident tangential components of the electric, E, and magnetic, H, fields incident on one side of the layer to the amplitudes of the corresponding transmitted and reflected fields after interaction with the layer. Writing the field components in matrix form

$$[A(z)] = \begin{bmatrix} E(z) \\ H(z) \end{bmatrix}$$

where z is the position of the boundary with the incident medium, the matrix describing the components at the next interface located at z+d for a layer of thickness d, the transfer matrix for the layer is

$$[A(z)] = \begin{bmatrix} \cos\delta & \frac{i}{g}\sin\delta \\ ig\sin\delta & \cos\delta \end{bmatrix} [A(z+d)] = \begin{bmatrix} m_{11} & m_{12} \\ m_{21} & m_{22} \end{bmatrix} [A(z+d)] \quad (1)$$

The phase change, δ, upon traversing the layer of thickness d, and index of refraction n, is $\delta = \frac{2\pi n d \cos\theta}{\lambda}$

(2)

where λ is the wavelength of the radiation and θ is the angle of incidence. For absorbing layers n = n-ik, where k is the extinction coefficient. The optical admittance g, is

$$g_p = \frac{n}{\cos\theta} \quad (3)$$

for p-polarization, and

$$g_s = n\cos\theta \quad (4)$$

for s-polarization.

For multilayers, multiplying the matrices for the layers in the proper order produces a system transfer matrix from which amplitude reflection and transmission coefficients can be obtained. As expressed in terms of matrix elements in Eq. (1), the amplitude reflection and transmission coefficients are given by

$$r = \frac{g_0(m_{11} + g_N m_{12}) - (m_{21} + g_N m_{22})}{g_0(m_{11} + g_N m_{12}) + (m_{21} + g_N m_{22})} \quad (5)$$

$$t = \frac{2g_0}{g_0(m_{11} + g_N m_{12}) + (m_{21} + g_N m_{22})} \quad (6)$$

CP430, *Fourier Transform Spectroscopy:* 11th International Conference
edited by J.A. de Haseth

and the measurable power reflectance and transmittance are:

$$R = rr^* \quad (7)$$

$$T = \frac{g_N}{g_0} tt^* \quad (8)$$

where the subscripts 0 and N refer to the incident and final media.

The set of equations (1)-(8) can be used to calculate the spectral response of a multilayer system such as a detector structure when the thickness d and the optical constants n(λ) and k(λ) of each layer are known as a function of incident wavelength.

In general the layer parameters d, n(λ) and k(λ) can be derived from measured reflectance and transmittance spectra by adjusting the layer constants until a good fit between the measured and calculated values is obtained. This is accomplished by minimizing a merit function M defined (3) as

$$M = \left\{ \frac{1}{2} \left(\frac{R_{calc} - R_{meas}}{\varepsilon_R} + \frac{T_{calc} - T_{meas}}{\varepsilon_T} \right) \right\}^{\frac{1}{2}} \quad (9)$$

with respect to the layer parameters. The quantities ε_R and ε_T are the measurement tolerances for reflectance and transmittance.

We used this approach to measured the thickness of HgCdTe films in multilayer structures. Our calculations agree in general with results from commercially available optical filter design software (4,5). We assumed that the optical constants were known and so they were not varied during optimization. In some cases it was possible to determine both the layer thickness and the optical constants n and k for some of the layers of a multilayer structure if the parameters of the remaining layers were known.

HgCdTe THICKNESS MEASUREMENTS

A typical room temperature normal incidence reflectance spectrum of a LWIR HgCdTe film attached with epoxy to a CdZnTe carrier is shown in Figure 1.

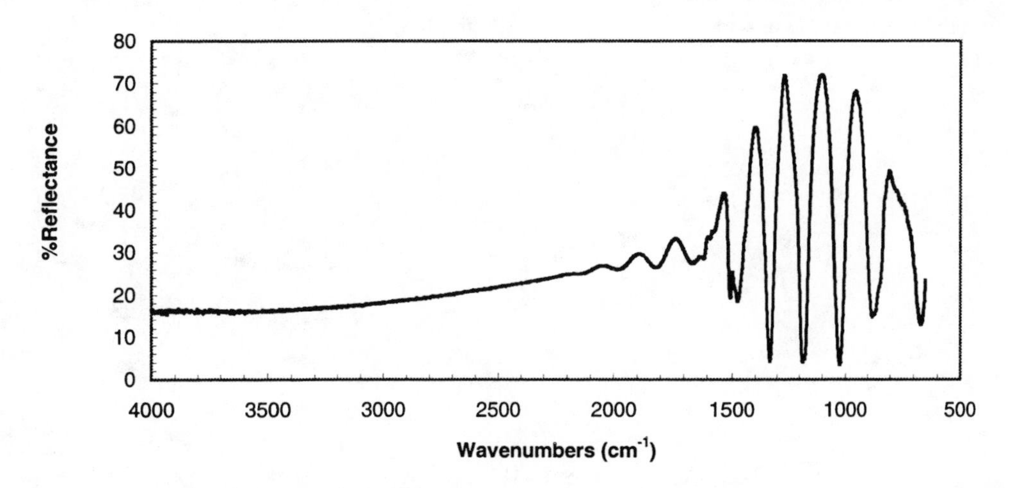

FIGURE 1. Normal incidence Reflectance of a LWIR HgCdTe film attached with epoxy to a CdZnTe carrier.

At normal incidence, for layer j in the stack its characteristic matrix is

$$M_j(t) = \begin{bmatrix} \cos(\frac{2\pi}{\lambda} n\, d) & \frac{i}{n} \sin(\frac{2\pi}{\lambda} n\, d) \\ i\, n \sin(\frac{2\pi}{\lambda} n\, d) & \cos(\frac{2\pi}{\lambda} n\, d) \end{bmatrix},$$

Fitting the reflectance spectrum, we find that the thickness of the HgCdTe film is 7.91 microns, and the epoxy layer thickness is 4.84 microns, using a value of 3.52 for the refractive index of HgCdTe and 1.31 for the epoxy index of refraction. The thickness values of HgCdTe film and epoxy determined from the reflectance spectrum agree very well with the thickness values measured by optical microscopy. The results for two more such film structures are shown in Table 1.

TABLE 1. Comparison of HgCdTe film thickness measurements

Sample	Thickness measured by Reflection (μm)	Directly Measured (μm)
M1482/2-6	7.91	8.250
M1482/1-2	7.00	6.5
M1482/4-2	6.33	6.75

The reflectance spectra were obtained using a NICOLET Magna IR 750 Fourier Transform Infrared Spectrometer, equipped with a SPECTRATECH, Inc., microscope stage for normal incidence reflection and transmission spectroscopic measurements. Various fitting algorithms were used to fit the reflectance spectra, with Monte Carlo or global Simplex converging faster to a solution.

This is a non destructive measurement technique that is easily inserted in the material preparation process flow. Using appropriate apertures the beam diameter of the infrared radiation incident on the top sample surface was selected to be on the order of 100 microns. Thus detailed mapping of the HgCdTe film thickness can be obtained.

SUMMARY

A non destructive method for deteremination of the thickness and optical constants was described that can be applied to a variety of optical multilayer structures. It is based on using optical filter design optimization techniques to fit calculated reflectance and/or transmittance spectra to measured values. This method is easily applicable to reliable non destructive thickness measurements of HgCdTe films in complex multilayer structures.

REFERENCES

1. Born and E. Wolf, Principles of Optics, Pergamon Press, 6th Edition, 1980, p. 58.
2. Abeles, "Optics of Thin Films", in Advanced Optical Techniques, F. Abeles ed. North Holland, 1967, Ch. 5, pp 143-188.
3. A. Dobrowolski, F. C. Ho, and A. Waldorf, Appl. Optics, 22, 3191(1983)
4. Film Wizard Optical Thin Film Design Software, Scientific Computing International.
5. TFCalc, Thin Film Design Software for Windows, Software Spectra, Inc.

A Technique for Measuring the Far-Infrared Radiative Properties Of Metal and Superconducting Thin Films

C.G. Malone and E.G. Cravalho
Department of Mechanical Engineering
Massachusetts Institute of Technology
Cambridge, MA 02139
and
T.J. Johnson and R.S. Jackson
Bruker Optics, Inc.
Billerica, MA 01821

INTRODUCTION

The absolute accuracy of reflectance measurements performed using FT-IR spectroscopy has been limited by standard single-reflection techniques to ±1% at best. This is primarily due to the photometric accuracy of the spectrometer and sample-positioning difficulties. Additionally, the absolute reflectance of the reference required for these techniques is typically not known to better than ±1%. The reflectance of materials such as metals and high-temperature superconductors approaches unity in the infrared, rendering standard techniques useless for the accurate measurement of this parameter.

Genzel et al. [1] developed a technique for measuring the radiative properties of thin films with near-unity reflectances which overcomes these limitations. They applied this method to determine the absorptance of $YBa_2Cu_3O_7$ at cryogenic temperatures with a relative accuracy of approximately ±20%. Thus, for a material with an absorptance of 1%, this method yields a measurement accuracy of approximately ± 0.2%. Malone [2] achieved a similar measurement accuracy for gold and $YBa_2Cu_3O_7$ films using this technique.

EXPERIMENTAL

The method involves measuring the infrared reflectance of a sandwich consisting of a thin, plane–parallel (127 ±0.05 μm), high-resistivity silicon film and a thin film of the material to be investigated. The sandwich, shown in Fig. 1, is referred to as a reflection Fabry-Pérot étalon. Infrared radiation incident upon the stack suffers multiple reflections inside the transparent silicon layer. The resultant reflectance spectrum yields a series of resonance fringes. By applying thin film optics to describe the propagation of the electromagnetic waves inside the stack, the radiative properties of the sample may be determined from the measured reflectance fringe minima with an accuracy approximately one order of magnitude better than with the single-bounce technique. The experiments were carried out on a Bruker IFS 66v in reflectance mode, both at room temperature and at cryogenic temperatures. The 66v is ideally suited to such measurements, in the far-IR because it is a vacuum bench and because of its high stability and photometric accuracy.

THEORY AND DISCUSSION

The thin film optics analysis of this structure assumes that the silicon layer is nonabsorbing, the sample is opaque, and the interfaces are smooth and plane parallel. With these assumptions, the following expression for the absorptance of the sample at the resonance minima, $A_s(\nu_{RFPO})$, is determined

$$A_S(\nu_{RFPO}) = \frac{(1-R_{12})^2(1-R_{RFPO})}{(1+2\sqrt{R_{RFPO}R_{12}}+R_{12})^2} \qquad (1)$$

where R_{12} is the reflectance of the silicon and R_{RFPO} is a reflection Fabry-Pérot étalon reflectance minimum. Malone [2] determined that departures from the thin-film optics solution caused by surface roughness, air gaps at the silicon-sample interface, etc., are small for the experimental data presented here.

Figure 2 shows the absorptance of an evaporated gold film at 300 K and 6 K for the spectral range from 400 to 100 cm^{-1}. The absorptance data error bars (~±20%) are determined by applying an error analysis to Eq. 1 and estimating the uncertainty interval for R_{12} and R_{RFPO}. A prediction for the absorptance made using the anomalous skin effect model with diffuse electron reflections at the metal surface is also shown.

CP430, *Fourier Transform Spectroscopy:* 11th International Conference
edited by J.A. de Haseth

In conclusion, the reflection Fabry-Pérot étalon provides a simple yet powerful technique for determining the radiative properties of thin films with very high reflectances. The experimental absorptance data presented here for a gold film at 300 K and 6 K are less that 1% for the investigated spectral range. These data could not be obtained using standard single-reflection techniques.

References

1. Genzel, L., Bauer, M., Habermeier, H.-U., Brandt, E.H., *'Determination of the Gap Distribution in $YBa_2Cu_3O_7$ Using a Far-Infrared Reflection-Fabry-Pérot Device,'* *Z. Phys. B*, **90**, 3-12, (1993).

2. Malone, C.G., *'A Technique for the Measurement of the Far-Infrared Radiative Properties of Metal and Superconducting Thin Films'*, PhD *Thesis*, Massachusetts Institute of Technology, 1997.

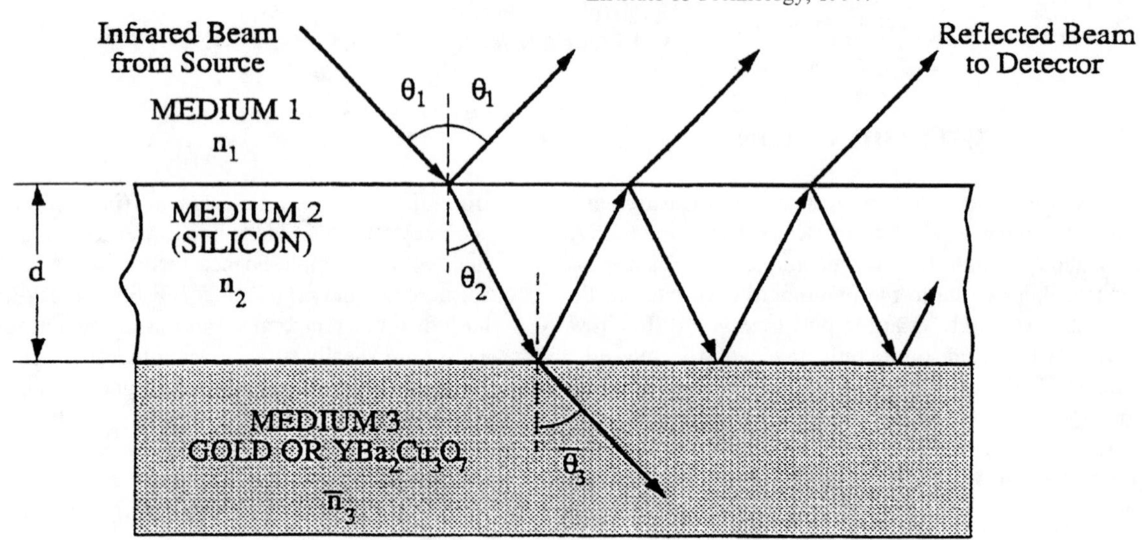

FIGURE 1. The silicon reflection Fabry-Pérot étalon.

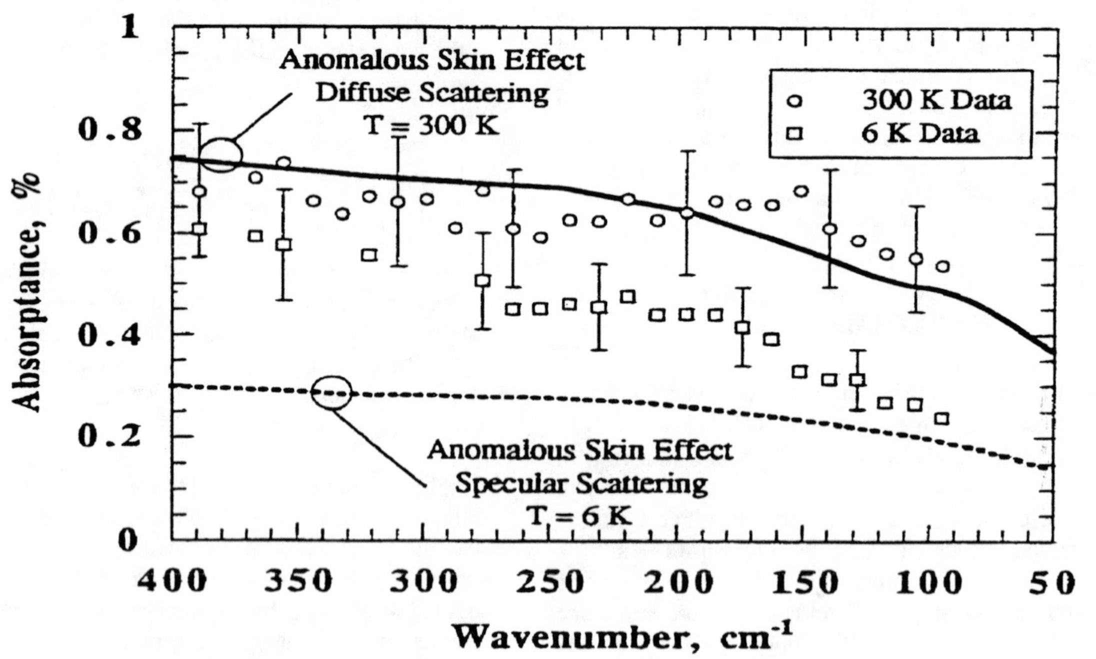

FIGURE 2. Absorptance of a gold film at 300 K and 6 K.

Gold Island Films as Seen by Infrared Ellipsometry

A. Röseler and E.H. Korte

Institut für Spektrochemie und angewandte Spektroskopie, Institutsteil Berlin, 12484 Berlin, Germany

The infrared reflectance of granular films consisting of gold islands of ca. 20 nm diameter and 5 nm thickness has been studied. The ellipsometric technique is outlined. Spectra of such films, also when covered with an organic layer exhibiting Surface Enhanced Infrared Spectroscopy (SEIRA), and their exceptional optical constants are presented and discussed.

INTRODUCTION

The nanophase state is attracting much interest because the increased surface brings about properties which differ essentially from those known for the bulk. Such materials offer unparalled flexibility of adjusting reflection, transmittance and absorption which make possible e.g. to optimize the cutoff wavelength of spectrally selective surfaces for photothermal conversion. The spectra of granular metallic films, also often called metal island films, include a feature in the visible range, often attributed to a surface plasmon or conductivity resonance, and an anomalous absorption in the far infrared [1], while in between no remarkable structure is expected.

However, in this particular interval one observes Surface Enhanced Infrared Absorption (SEIRA) when an adsorbate is deposited on a metal island film [2]. Similarities with Surface Enhanced Raman Scattering (SERS) exist at least in so far that cluster-like metallic particles are necessary to evoke the effects. Regrettable SEIRA does not provide enhancement factors as encountered with SERS, but in the order of 10 to 100. However, a decrease of the detection limit by such a factor would widen the scope of infrared spectroscopy considerably.

Several research groups are testing SEIRA for such purposes. Usually the preparation of the island film is optimized without reference to a physical model. Since commonly transmission measurements are performed, little is known about the infrared-optical properties of the film. Practising spectroscopic infrared ellipsometry [3,4] in order to characterize inorganic and organic layers which are as thin as some nanometers, we decided to apply this technique on metal island films also.

Already the first experimental results showed a remarkable spectral feature and the evaluation revealed striking values of the optical constants [5]. In this contribution ellipsometric results from gold island films and from those when covered with an organic layer are discussed in more detail. To start with, an outline of ellipsometry with experimental aspects is given.

INFRARED ELLIPSOMETRY

The ellipsometer used was constructed in our laboratory, it is mounted as a reflection attachment at a port of a Bruker IFS 55 spectrometer basically equipped for the infrared range of about 5000 cm^{-1} to 400 cm^{-1}. As shown in Fig. 1 the modulated radiation leaving the spectrometer is focussed onto the sample with a maximum angle of convergence of about 4°; the reflected component is collected with an ellipsoidal mirror onto a standard DTGS detector providing linear response. The optical path comprises a polarizer before the sample and an analyzer behind the sample. Both are of wire grid type on KRS5 substrate (SPECAC). Their azimuths (electric vector of transmitted radiation) are adjustable under computer control with a repeatability better than 0.01° (OWIS). The angle of incidence at the sample can be varied up to grazing incidence by rotating the sample and pivoting the detection branch of the optical path through twice the sample angle. The irradiated area upon the sample is about

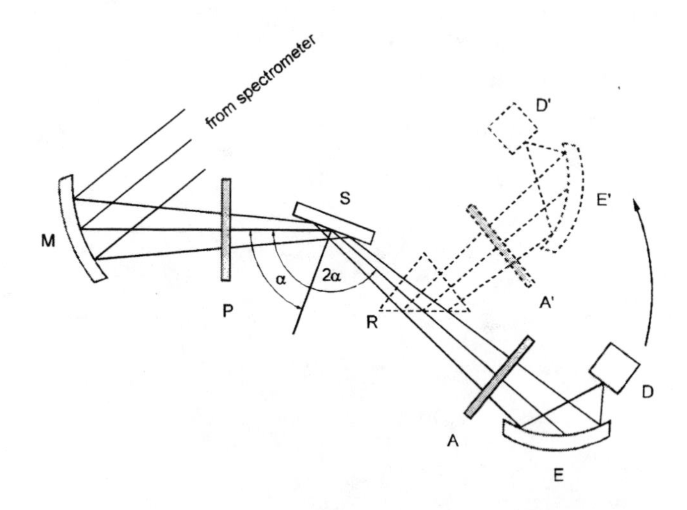

FIGURE 1. Optical path of the ellipsometer attachment. M focussing mirror; P polarizer; S sample; α angle of incidence; A analyzer; E ellipsoidal mirror; D Detector; R retarder prism (optional with A→A', E→E', and D→D').

CP430, *Fourier Transform Spectroscopy:* 11th International Conference
edited by J.A. de Haseth

8 mm high and 8 to 20 mm wide depending on the angle of incidence. The detection path can be folded to accommodate a retarder, preferably a KBr, ZnSe or KRS5 prism for one internal reflection [6]. The ellipsometric attachment is cased and can be purged.

The azimuth of the polarizer is precisely set to 45° so that equal radiation components are supplied parallel (0°) and perpendicular (90°) to the plane of reflection. The process of being reflected effects amplitude and phase of the radiation and this differently for the two components. This is stated by the complex reflection coefficients r_0 and r_{90}. Since these are not directly measurable, one refers to their ratio instead

$$\varrho = r_0/r_{90} = |r_0|/|r_{90}| \exp\{i(\delta_0\text{-}\delta_{90})\} \equiv \tan\Psi\exp\{i\Delta\} \quad (1)$$

leaving us with two real numbers Ψ and Δ which are called ellipsometric parameters and serve as basis for all further evaluation and interpretation. Trigonometric functions of these parameters are related to two experimentally attainable results [3,4], namely

(i) the ratio of the single beam spectra recorded when the polarizer is set first to 0° and then to 90°, and

(ii) the analogous ratio for the polarizer at 45° and 135°.

Equivalent measurements with a retarder, adding an extra phase difference, improve the precision and resolve ambiguity from the trigonometric functions involved. The exclusive use of quotients eliminates not only the spectrometer function and residual atmospheric absorption, but also possible effects of an undersized sample, obscurations, scattering as well as other experimental shortcomings. Even more important - particularly in terms of accuracy - is that no reflectance standard is neccessary when applying ellipsometry.

The ellipsometric parameters are related via Fresnel's equations to the compound-specific optical constants, i.e.

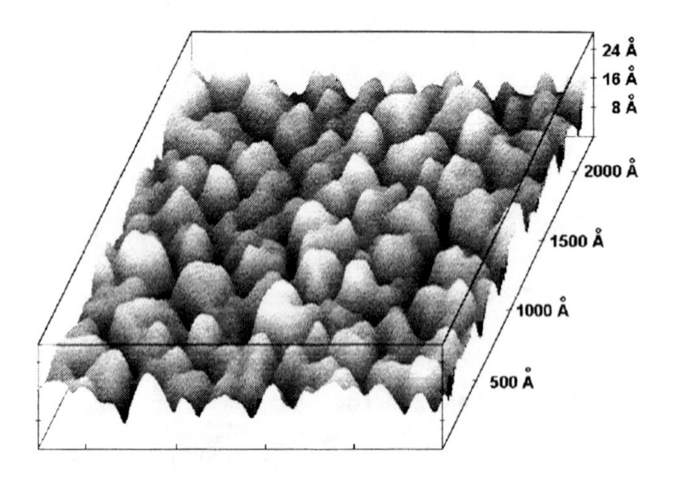

FIGURE 2. Atomic force microscopy scan of a gold island film with expanded ordinate.

the refractive index n and the absorption index k of the material 'behind' the reflecting surface. In the case of a thin layer on a thick substrate five quantities are involved including the optical constants of both materials and the spectrally constant layer thickness. The constants of the substrate are obtained from ellipsometric measurement at an uncoated part (possibly the back surface). It remains to characterize just the layer. From literature appropriate algorithms are known which we have transformed into software according our needs. Granular or otherwise heterogeneous materials, can be treated as being homogeneous as long as the structural dimensions are well below the wavelength. Various Effective Medium Approaches for calculating effective optical constants of a material from those of the constituents are found in the literature [7].

Compared to conventional infrared spectroscopy based on transmission measurements, the sensitivity of ellipsometry with respect to thin layers is remarkable. Generally a large angle of incidence beyond the Brewster angle is used and this is surely advantageous. However of higher importance seems to be that the accuracy of the experimental data, is of such a level that the data can readily be interpreted by the quantitative theory, without adjustment and over a wide spectral range.

GOLD ISLAND FILMS

Gold island films were produced on slides of ordinary glass by thermally evaporating gold. An AFM scan (Zeiss Beetle) is presented in Fig. 2. The picture conveys the impression of densely packed, still separated particles of similar size. Their average diameter can be estimated to be in the order of 20 nm parallel to the plane of the film and definitely less perpendicular. These dimensions are about three orders of magnitude smaller than infrared wavelengths and thus, it is fully justified to use effective optical constants.

SPECTRA OF ISLAND FILMS

In Fig. 3 typical spectra of the ellipsometric parameters are shown which were obtained with the angle of incidence set to 70°. Down to below 1500 cm^{-1} the parameters are almost constant. The island film is thin enough to allow the reststrahlen band of the substrate to structure the reflection response. However, the most pronounced feature occurs where the refractive index of glass equals unity so that the substrate becomes almost invisible. This happens since the dispersion of the strong reststrahlen oscillator forces the refractive index below unity in the spectral interval between the said feature and the resonance frequency. We have several indications that the amplitude

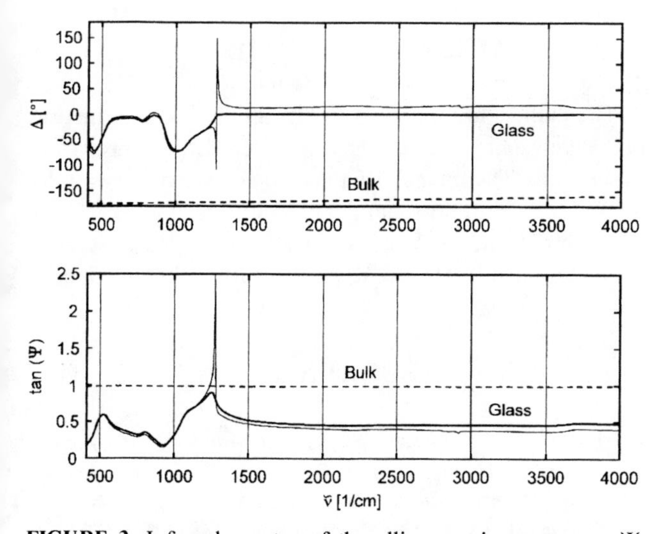

FIGURE 3. Infrared spectra of the ellipsometric parameters Ψ and Δ of a gold island film on glass substrate (thin lines) and of a uncovered glass surface (thick lines); those of bulk gold (broken lines) were calculated. Angle of incidence 70°.

of the features in the Ψ and Δ spectra are correlated to the SEIRA factor the particular layer can produce when covered with an adsorbate [5].

Included in Fig. 3 are the ellipsometric parameter spectra of glass. These are very similar to the ones of the island film on glass as it would have been anticipated when considering the thickness of the island film which is at least three orders of magnitude smaller than the wavelength. The only substantial deviation are the features, which as outlined before are a consequence of the stratified structure of the sample.

For easier reference, in Fig. 3 also the parameters are presented which are to be expected for a thick slab (>50 nm) of dense gold according the optical constants listed [8]. These curves exemplify the behaviour of a metal and the difference to those of the insulator glass is obvious. From this point of view the glass slide covered with a gold island film behaves even less metallic than the neat glass does.

From the measured spectra of the ellipsometric parameters characterizing the reflectance of the coated substrate, the optical constants of the coating, i.e. the gold island film were derived. They are presented in Fig. 4 and can similarly be interpreted. The absorption index being correlated with conductivity, is constantly just above unity, while the absorption index of bulk gold in this spectral range is between 25 and 45. The refractive index of the gold island film is found to be around 6.4 and therefore even higher than the one of bulk gold for most of the infrared and near-infrared range.

The optical constants were evaluated using algorithms based on Fresnel's and Airy's equations. One important side product is the thickness which was found to be about 5 nm. For the calculations the layer was assumed to be isotropic even though Fig. 2 conveys the impression of uniaxiality with the optical axis perpendicular to the plane of the film. However even with different angles of incidence, no influence was observed which could be traced back to anisotropy and thus it was left out of consideration. The reason for this insensitivity is the high refractive index which on the whole is an outstanding property of this material and might bear potential also for other applications.

The evaluated values of the effective optical constants as well as their independence of the wavenumber can be simulated using an effective medium approach [2]. For this purpose the gold islands are assumed to be oblate spheroids with their short axes perpendicular to the glass surface. Such calculations also support the effectively vanishing absorption and thus the proposed interpretation of the results.

ADSORBATE ON GOLD ISLAND FILM

In order to check what can be expected with an organic compound upon the gold island film we applied a Langmuir-Blodgett (LB) monolayer of arachidic acid [9]. Due to the film structure we expect a similar density of molecules on the gold island film and on the neat glass surface. Therefore this should allow us a direct comparison between the analytically usable signals (band height, dispersion amplitude). On the other hand it cannot be taken for granted that this is the best arrangement in view of high SEIRA factors, all the more a close contact throughout must be assumed to be vital. The spectral

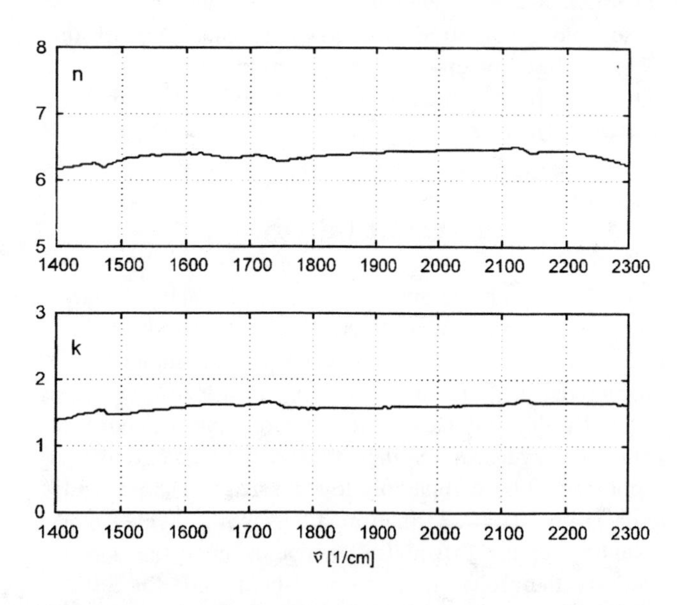

FIGURE 4. Refractive index n and absorption index k of a gold island film in the infrared range as derived from ellipsometric measurements.

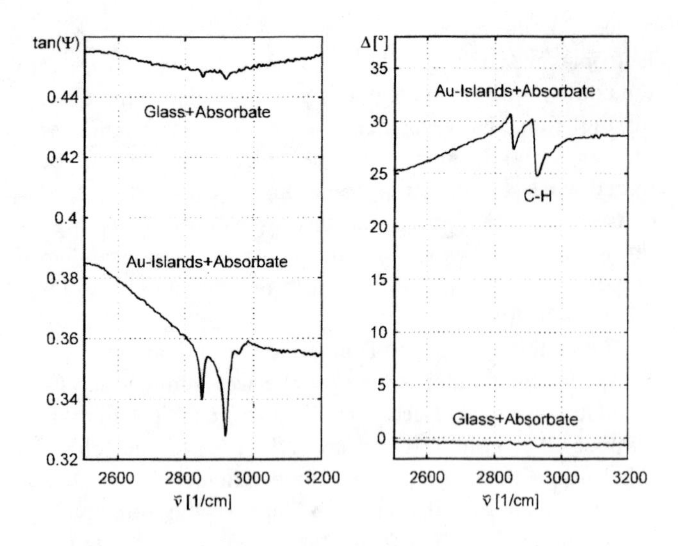

FIGURE 5. Ellipsometric parameters in the C-H streching range for demonstrating SEIRA; experimental results from an arachidic acid monolayer on a gold island film on glass substrate, and an equivalent layer on a glass surface.

sections of the experimentally determined ellipsometric parameters shown in Fig. 5, comprise the spectral range of the C-H streching vibrations. The ordinate scale is expanded in comparison to Fig. 3 to make the effect caused by the LB film deposited on glass at all visible. The enhanced bands and dispersion effects are well resolved without exuberant noise. The enhancement attributed to SEIRA is about 15 and according to what was said before it should be closer to the lower limit of what is attainable with adsorbates.

The effective optical constants of the island film plus adsorbate were simulated. It turned out that under reasonable assumptions the general tendencies of the experimental results can be reproduced [10]. These calculations also indicate that different effects contribute to SEIRA.

CONCLUSIONS

It has been shown that spectroscopic infrared ellipsometry provides an interesting and instructive view on metallic island films. The optical constants found characterize a material of fascinating properties since a spectrally almost constant high refractive index - above ca. 1500 cm^{-1} even higher than the one of solid gold - is accompanied by comparably low absortion. The presently undertaken detailed interpretation will reveal more essentials of the effect itself and thus open the way to optimize the preparation of gold island films for SEIRA applications.

ACKNOWLEDGMENTS

The authors are indebted to Mr. Shiming Geng, Institut für Angewandte Chemie, Berlin-Adlershof, for preparing island films. The financial support by the Senatsverwaltung für Wissenschaft, Forschung und Kultur des Landes Berlin and by the Bundesministerium für Bildung, Wissenschaft, Forschung und Technologie is gratefully acknowledged.

REFERENCES

1. Lafait, J., and Berthier, S., *Optical Properties of Granular Solids*, in: Hadjipanayis, G.C., Siegel, R.W. (eds.), *Nanophase Materials*, Dordrecht: Kluwer, 1994, pp. 449-469.
2. Osawa, M., et al., *Appl. Spectrosc.* **47**, 1497-1502 (1993).
3. Röseler, A., *Infrared Spectroscopic Ellipsometry*, Akademie-Verlag: Berlin, 1980.
4. Röseler, A., Korte, E.H., and Reins, J., *Vib. Spectrosc.* **5**, 275-283 (1993).
5. Röseler, A., and Korte, E.H., *Appl. Spectrosc.* **51**, 902-904 (1997).
6. Korte, E.H., et al., *Appl. Spectrosc.* **42**, 1394-1400 (1988).
7. Aspnes, D.E., in: [8], ch. 5, pp. 89-112.
8. Palik, E.D.(ed.), *Handbook of Optical Constants of Solids*, Academic Press: Orlando, 1985.
9. Röseler, A., Dietel, R., and Korte, E.H., *Mikrochim. Acta [Supplement]* **14**, 657-659 (1997).
10. Röseler, A., and Korte, E.H., *Thin Solid Films*, in press

External Reflection Studies of CO on Platinized Platinum Electrodes

Amy E. Bjerke and Peter R. Griffiths

Department of Chemistry, University of Idaho, Moscow, ID 83844-2343

Several studies have shown a difference in the electrochemical response of bare platinum vs platinized platinum electrodes. We have found that the thickness of the platinum black coverage on a Pt electrode also changes the spectroscopic response of CO adsorbed on such a surface. A polycrystalline electrode was platinized at 2.75 V in a solution of chloroplatinic acid/lead acetate for time periods of 10 to 30 s. External reflection spectra were calculated using a single-beam spectrum collected with a monolayer of CO on the electrode and a single beam spectrum without the adsorbed CO as a reference. At lower platinization times, the CO peak existed as the expected upward-going band. As the platinization time was increased, the band shape became dispersive and the intensity of the band increased. At the highest platinization time, the peak appeared to invert itself into a downward going band, i.e. the reflection increased in the region of the CO absorption. This surprising phenomenon appears to be directly related to the change in the surface features with increased platinization time.

INTRODUCTION

The properties of CO adsorbed on platinum have been studied extensively by electrochemists for over 30 years (1,2). CO adsorbed on platinum has become of great interest because it is a byproduct of methanol electro-oxidation that poisons platinum electrodes, decreasing their activity. Initial spectroelectrochemical studies of CO on platinum determined the types of bonding occurring between CO and platinum. Although previous spectroelectrochemical studies have explored the various aspects and effects of CO bound to platinum, rarely have they been performed on platinized electrodes. A study performed by Christensen, Hamnett and Weeks contrasted CO adsorbed on a platinized glassy carbon electrode with that adsorbed on a polished platinum electrode (3). This study failed to determine any distinctive features that distinguished the absorption bands of platinized glassy carbon CO from those of CO bonded to polished platinum.

The process of platinization has been utilized by scientists since the end of the nineteenth century. Initially developed as heat-absorbing strips in bolometers, platinum black electrodes are now used as electrocatalysts in fuel cells, as components of hydrogen reference electrodes, and in a variety of other electrochemical situations.

There are several variables involved which may be manipulated to form a wide variety of platinized surfaces. Surface roughness of a platinum black electrode may be controlled by the type of platinum solution used, the voltage or current applied, and the time of platinization. (4)

Examination of platinized surfaces by electron microscopy often reveals a surface that is not smooth but is made up of a fairly uniform series of metal islands. The size of these islands can be controlled by varying the voltage or time used for the platinization. It is well known that silver, copper, gold, and iridium thin films comprised of metal islands much smaller than the wavelength of the excitation source can cause the phenomenon of surface-enhanced infrared absorption (SEIRA) in the transmission, internal and external reflection modes (5,6). To the best of our knowledge, no infrared enhancement from Pt islands has ever been noted. The ability to provide enhancement through the platinization of platinum electrodes would be a practical way to improve the sensitivity of standard spectro-electrochemical techniques. This study was performed in the hope that the effect of platinization on CO absorption might be probed, and also to explore the possibility of using platinization for IR enhancement of species adsorbed on standard Pt disk electrodes.

CP430, *Fourier Transform Spectroscopy:* 11th International Conference
edited by J.A. de Haseth

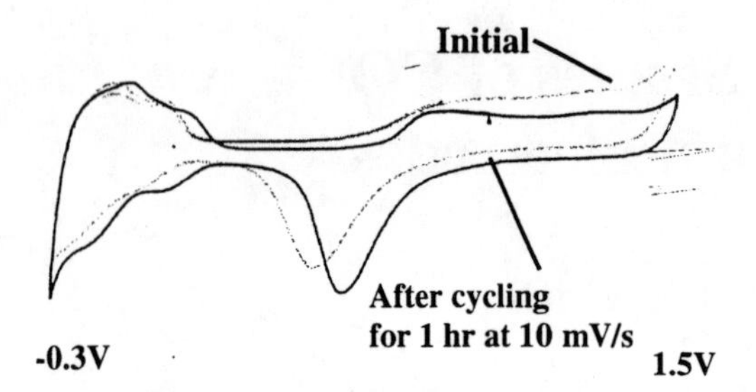

FIGURE 1. Voltammogram of a clean Pt electrode after one hour of cycling between –0.3 and 1.5 V.

EXPERIMENTAL

A 7-mm diameter platinum disk electrode was prepared by polishing with 1 and 0.25 μm Metadi diamond polish, then soaked for ten minutes in dilute aqua regia and rinsed thoroughly with Milli-Q water to ensure a clean electrode surface. The electrode was then platinized at a constant voltage of 2.75 V (vs. SCE) in a solution of 0.072 M chloroplatinic acid and 2.6 x 10^{-4} M lead acetate. The platinum black coverage was controlled by varying the platinization time. Platinization for 5 s yielded a semi-opaque dark gray coverage, compared to the dense black layer formed during a 26-s platinization.

After platinization, the electrode was rinsed with Milli-Q water, and transferred to an electrochemical cell containing 1 M $HClO_4$. The electrochemical cell consisted of a Pyrex cell equipped with a CaF_2 window and a Pt auxiliary electrode. The reference electrode was connected to the main compartment of the cell through a Luggin capillary.

The electrode was then electrochemically cleaned by cycling the electrode voltage from -0.3 to 1.5 V until cyclic voltammograms showed the characteristic hydrogen oxidation and reduction peaks for a clean Pt electrode in $HClO_4$ (Fig. 1). The electrode was held at 0 V for 40 minutes while CO was bubbled through the cell to ensure that the solution was saturated with CO and that a monolayer of CO covered the electrode. A single-beam spectrum was measured at this voltage for use as the reference spectrum. Subsequently the voltage was raised to 1V, at which potential the CO was oxidized off the electrode surface. A single-beam spectrum was also recorded at this voltage. In the single-beam spectra measured at 0 Volts, the presence of CO is evident as a peak at approximately 2080 cm^{-1}. An absorption band at 2345 cm^{-1} in the spectrum collected at 1V is indicative of dissolved CO_2 formed from the oxidized CO.

External reflection spectra were collected at an incidence angle of 60° using a Bio-Rad FTS-60A equipped with a Specac reflection accessory and a mercury cadmium telluride detector. All spectra were measured at 2 cm^{-1} resolution by signal averaging 2048 scans.

TABLE 1. Variation in Particle Size with Platinization Time

Platinization Time (s)	Average Particle Diameter (μm)
5	0.1
10	0.5
15	0.6
20	0.6

RESULTS AND DISCUSSION

For short platinization times, there was little effect on the infrared signal (Fig. 2). With increased platinization time, the CO absorption band shape progressed through a dispersive band shape, until at a 26-s platinization time, the band became fully inverted. The signal reflected from the electrode became weaker with increased platinization times. Beyond 26 s the reflection was too low for the bands to be seen above the noise.

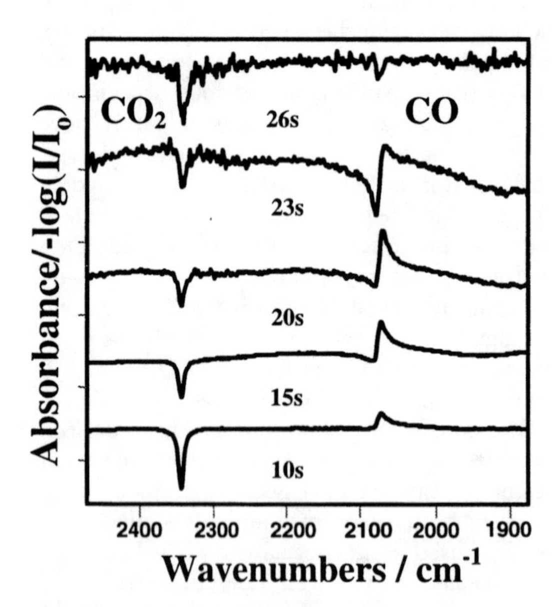

FIGURE 2. External Reflection Spectra of CO on electrodes prepared with different platinization times. The single-beam spectrum (I) was collected at 0.0V, before oxidation of the CO; I_0 was collected at 1.0V.

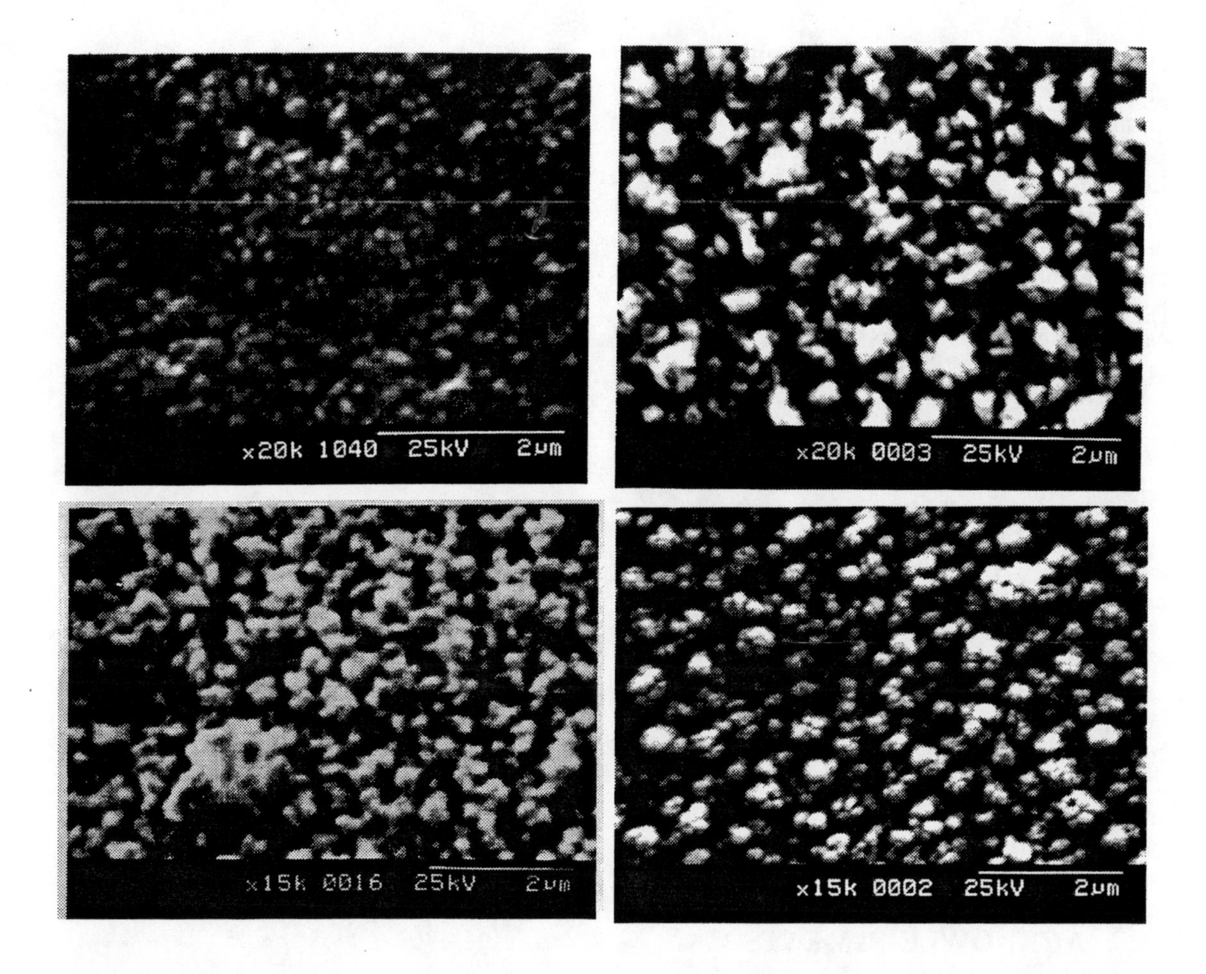

FIGURE 3. SEM micrographs of platinized platinum electrode surfaces, clockwise from upper left: 5s, 10s, 15s, and 20s platinization times.

Scanning electron microscope (SEM) images allowed for examination of the physical changes occurring on the electrode surface, see Fig.3. There is a significant increase in the platinum island size for platinization times between 5 and 10 seconds, but this growth appeared to slow at higher platinization times (see Table 1).

Dispersive band shapes have previously been observed in external reflection spectra of roughened carbon electrodes with low infrared reflectivity (7) but have not previously been reported on metal electrodes. With increased platinization times, there is a reduction in IR reflectivity and an increase in substrate thickness, both of which may contribute to the change in band shape happening in this progression. We are not able to fully explain the complete inversion of the peak at this time, however.

In addition to the change in band shape, there is also a notable enhancement of the signal which reaches a maximum in the when the band shape is most dispersive and then decreases to the original intensity at a 26s platinization time. We found this enhancement by comparing the size of the CO peak to that of the band due to dissolved CO_2 in each spectrum. We estimate that the CO band is enhanced to five times its original size based on this comparison. That this enhancement is not due to the increased surface area of the electrode can be seen by the reduced intensity of the CO band in the lowest spectrum in Fig. 2, for which the surface area is the highest. In SEIRA studies, the metal islands are much smaller than the wavelength of light used for excitation. We do not believe the moderate enhancement seen in these spectra is due to the SIERA effect, since the size of the metal islands in this study is larger than those usually used to give SIERA enhancement, which are commonly on the order of 10 nm in diameter (5). Other factors that are known to effect enhancement of IR bands are the nature of the substrate and thickness of the metal film. Because the size of our platinum islands is significantly larger than the optimum reported sizes for SEIRA, we believe this moderate enhancement may be caused by some other factors than those involved in SEIRA spectra.

CONCLUSIONS

We have observed the unexpected effects that a change in the size and thickness of platinum islands has on the band shape of carbon monoxide adsorbed on platinized platinum electrode surfaces. This change in band shape appears to be accompanied by signal enhancement of the dispersive bands until the thickness of the platinized Pt layer reaches a certain thickness. In the future, we plan to make a closer examination of these effects, possibly creating platinized electrodes with sizes and shapes of the Pt islands leading to even greater band enhancement.

ACKNOWLEDGMENTS

We gratefully acknowledge financial support for this work through the EPA-EPSCOR program. We would also like to express thanks to Dr. William Knowles for his help obtaining the SEM micrographs.

REFERENCES

1. Gelman, S., *J. Phys. Chem.* **66**, 2657 (1962).
2. Giner, *Electrochim. Acta* **8**, 857 (1963).
3. Christensen, P.A., Hamnett, A., and Weeks, S.A., *Electroanal. Chem.* **250**, 127-142 (1988).
4. Feltham, A. M. and Spiro, M., *Chemical Reviews* **71**, 177-193, (1971).
5. Nishikawa, Y., Nagasawa, T., Fujuwara, K, Osawa, M., *Vibrational Spectroscopy* **6**, 43-53 (1993)
6. Nishikawa, Y. and Fujiwara, K., Ataka, K., and Osawa, M., *Anal. Chem.* **65**, 556-562 (1993)
7. Porter, M., *Anal. Chem.* **60,** 1143A-1155A (1988)

FTIR Reflectance Spectroscopy of $SiC_XN_YO_Z(H)$ Ceramic Coatings: Structure Interpretation by Optical Modelling

W. Grählert, V. Hopfe

Fraunhofer Institute Material and Beam Technology, Winterbergstraße 28, 01277 Dresden, Germany

$SiC_XN_YO_Z(H)$ ceramic coatings have been deposited on steel substrate by plasma enhanced chemical vapour deposition (PE-CVD) using hexamethyldisilazane as precursor. The influence of the discharge power (3-90 W) on the network structure and optical properties of the amorphous coatings are discussed. The dielectric function of the layer which has been described by a set of Lorentzian oscillators have been determined from off-normal polarized FTIR reflectance spectra. The calculated spectra are fitted to the experimental measurements by variations of oscillator parameters and layer thickness. Based on dielectric functions the assignment of functional groups and the change of their concentration with deposition discharge power has been derived. A direct dependence of the intensity of the hydrogen modes and the layer thickness on discharge power has been found. An unambiguous influence of the discharge power on lattice modes was not detected.

INTRODUCTION

$SiC_xN_yO_z(H)$ ceramic coatings become increasingly attractive because of their superior properties as hardness, wear resistance, chemical and thermal stability. To deposite such layers, plasma enhanced chemical vapour deposition (PE-CVD) was used. FTIR spectroscopy is an excellent tool for the characterization of the amorphous structure and composition of deposited coatings.

FTIR specular reflection allows the non-destructive investigation of the complex layer systems. However, the interpretation of reflectance spectra is more difficult than the interpretation of usual absorbance spectra because of the superposition of the intrinsic optical properties of layer and substrate with optical phenomena of the layer stack as interference fringes and non-ideal interfaces and surfaces.

An intuitive characterization of the deposited layer based on measured reflectance spectra would be chancy because of the possible misinterpretation of the optical effects. Because of unknown additional phase corrections the common Kramers-Kronig Transformation do not yield the optical properties of the layer. However, spectra simulation based on an optical 3-media model using Maxwell's theory allows the optical properties and the thickness of the layer to be determined.

Structural parameters can be derived from the optical function or the dielectric function and, as a consequence, the properties of the deposited layer are correlated with the deposition parameters.

EXPERIMENTAL

The $SiC_xN_yO_z(H)$ layers were prepared by plasma enhanced chemical vapour deposition (PE-CVD). The layers were deposited at a pressure of 130 Pa on polished steel substrate using a hexamethyldisilazane - nitrogen mixture as precursor. The deposition time was fixed (130 s) but the deposition discharge power was changed from 2.5 W to 90 W [1].

The off-normal polarized infrared reflectance spectra were obtained with a Bruker IFS-48 FT-IR spectrometer equipped with a DTGS detector in conjunction with a specular reflectance attachment (Seagull; Harrick) and complemented by a wire-grid polarizer. The spectra were recorded in a frequency range of 400 cm^{-1} - 5000 cm^{-1} at a resolution of 4 cm^{-1}, with different angles of incidence and with TM- and TE-polarization.

All calculations were carried out by a special developed software package for simulation of spectra of solids, of interfaces and of multiple layered systems

RESULTS AND DISCUSSION

Measurements

The $SiC_xN_yO_z(H)$ layer deposited by 20 W discharge power has been measured under different measuring conditions (10°/TM, 70°/TM, 70°/TE) (Fig. 1). Two significant spectral regions are obvious in all measured spectra: a strong absorption region characterized by typical reflectance bands due to lattice modes (400 - 1400 cm^{-1}) and weak transmission like region due to hydrogen modes

CP430, *Fourier Transform Spectroscopy:* 11th International Conference
edited by J.A. de Haseth

(1400 - 4000 cm^{-1}). The TE spectra are dominated by the interference fringes, and the TM spectra obviously show an intensification of the hydrogen modes. Further investigations are focussed on the 70°/TM spectra.

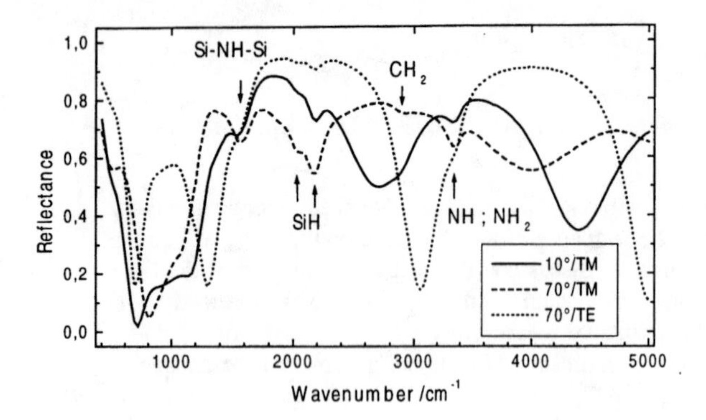

FIGURE 1. Measured reflectance spectra for different measuring conditions of $SiC_xN_yO_z(H)$ layer deposited by 20 W discharge power.

Variations of the discharge power are of strong influence onto reflectance spectra of the $SiC_xN_yO_z(H)$ layers. Band shape, intensity and band position are shifted (Fig. 2). However, as discussed, the change in reflectance spectra does not reflect directly the properties of the layers.

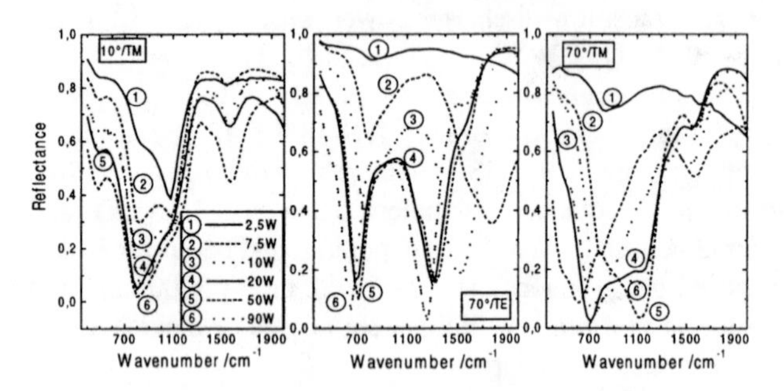

FIGURE 2. Change of spectra shape depending on deposition discharge power for different measuring conditions

Spectra fit procedure

Because measured reflectance spectra represent a superposition of optical properties of layer and substrate with optical interference effects caused by the multilayer configuration, a straightforward interpretation of such spectra is not possible. To extract the optical properties of the $SiC_xN_yO_z(H)$ layer the theoretical reflectance spectra were calculated based on a 3-media optical model (Fig. 3) using the Fresnel equations (1) and were fitted to measured ones. The dielectric function of the layer is described by a set of Lorentz oscillators (2) and the dielectric function of the steel substrate has been determined previously in separate experiments.

$$R(\tilde{\nu}) = f(\hat{\varepsilon}_{Substrate}, \hat{\varepsilon}_{Coating}, d_{Coating}, \theta, Pol) \quad (1)$$

$$\hat{\varepsilon}_{Coating} = \varepsilon' + i\varepsilon'' = \varepsilon_\infty + \sum_{i=1}^{n} \frac{S_i}{\tilde{\nu}_i^2 - \tilde{\nu} + i\gamma_i\tilde{\nu}} \quad (2)$$

$\hat{\varepsilon}$: dielectric function of substrate or coating; d: thickness of layer; θ angle of incidence; Pol: polarization; S_i: oscillator strength; γ_i: oscillator damping; $\tilde{\nu}_i$: oscillator position; $\tilde{\nu}$: wavenumber

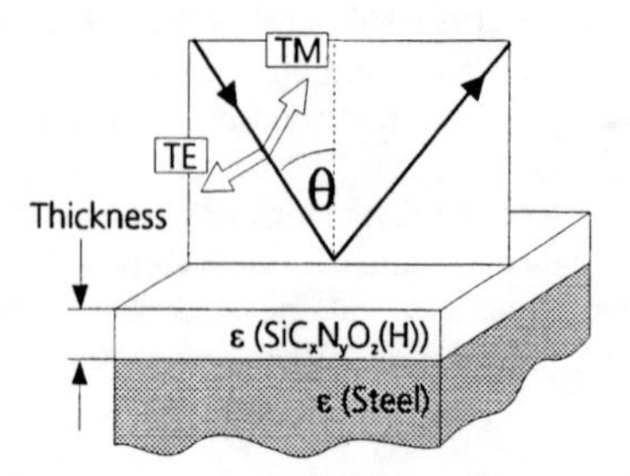

FIGURE 3. Optical model for calculation of polarized off-normal reflectance spectra of layered systems

The calculated spectra are fitted to the measured spectra by variations of the oscillator parameters and layer thickness. The fit procedure is based on the generalized simulated annealing algorithm (GSA). This modified Monte Carlo algorithm is characterized by his robustness concerning the absolute convergence of the minimization procedure [2].

However, the calculated 70°/TM spectra do not coincide well in absolute intensity with the experiment. As a consequence minor spectral features vanish, e.g. the low intensity hydrogen modes. By using the first derivatives of the reflectance spectra in the fit procedure, relevant features of the spectra are made to stand out more clearly and, as a consequence, the hydrogen modes are well described (Fig. 4).

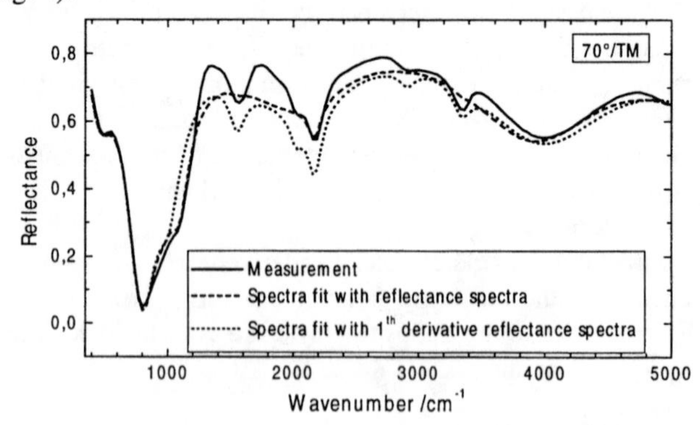

FIGURE 4. Comparison of the measured and fitted reflectance spectra (70°/TM). The reflectance spectra or the first derivative of the reflectance spectra have been used.

A band model which covers 10 Lorentzian oscillators results in a high quality fit. The calculated dielectric function shows a strong broad resonance about 850 cm^{-1} of the lattice modes (Si-N, Si-C) and week resonances caused by the hydrogen modes (Si-H, N-H, C-H) (Fig. 5). The band assignments are shown in Table 1.

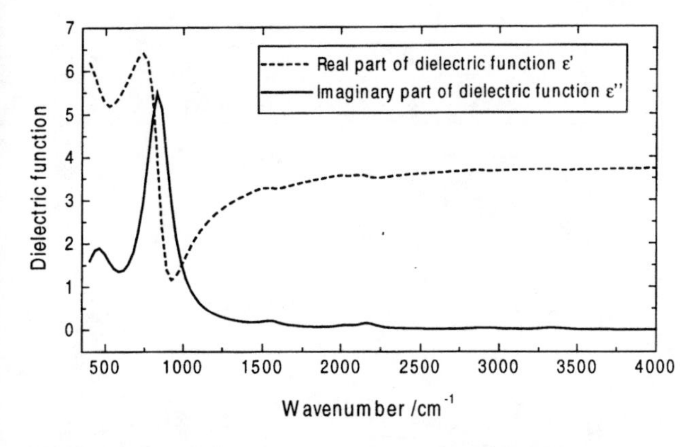

FIGURE 5. Calculated dielectric function of $SiC_xN_yO_z(H)$ layer deposited by 20 W discharge power

TABLE 1. Determined oscillator parameters of the $SiC_xN_yO_z(H)$ layer deposited by 20 W discharge power

Number i of oscillator	ν_i /cm^{-1}	S_i / cm^{-1}	γ_i / cm^{-1}	Band Assignment /3,4/
1	462	0,81	235,0	ν_s Si-N
2	830	0,798	209,6	ν Si-C
3	847	0,154	144,7	ν_{as} Si-N
4	856	0,276	168,7	ω Si/C-$(CH_n)_m$
5	1149	0,00013	94,5	ν_s Si-O
6	1556	0,0102	145,9	δ N-H
7	2029	0,00298	125,8	ν_s (Si-H)Si
8	2170	0,00677	126,0	ν_s $(Si\text{-}H_2)Si$
9	2910	0,00140	155,6	ν C-H
10	3343	0,00189	165,4	ν N-H
e_∞	3,81			
d /nm	1480			

In order to check the validity of the determined dielectric function their transferability to other measuring conditions (70°/TE, 10°/TM) have been tested. The calculated reflectance spectra of different measuring conditions well fit the experimental data in both spectra shape and intensity (Fig. 6).

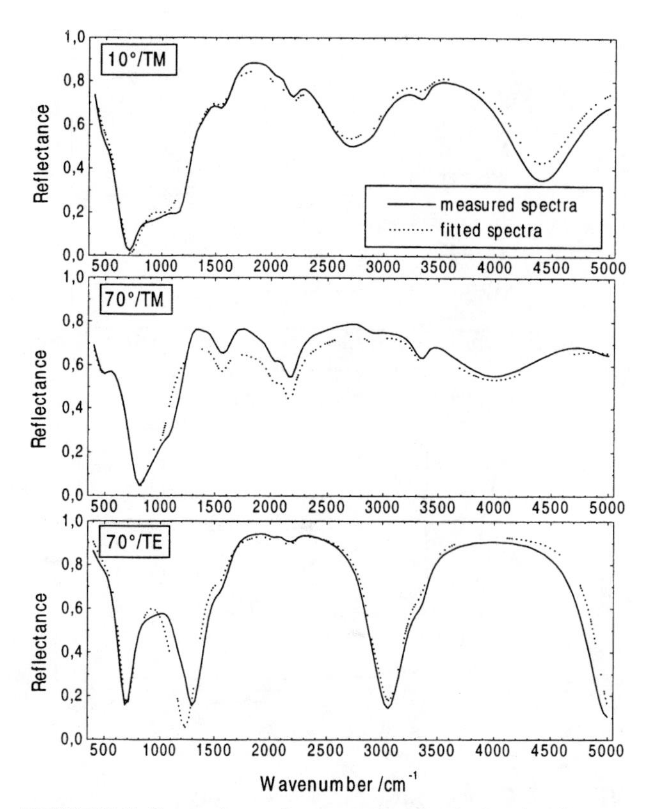

FIGURE 6. Comparison of measured and calculated reflectance spectra for different measuring conditions. Calculation based on oscillator parameters derived from spectra fit of 70°/TM spectra.

Influence of discharge power on the layer properties

In order to evaluate compositional changes of the $SiC_xN_yO_z(H)$ layers the oscillator strength has been used as an direct measure of the concentration of the corresponding functional group. A correlation of the concentration with the discharge power exists in the region of the hydrogen modes. The concentrations of $(SiH_2)Si$ groups (2170 cm^{-1}) and of N-H groups (3343 cm^{-1}) linearly increase with discharge power. There is no significant change in C-H concentration (2910 cm^{-1}). The concentration dependence of the Si(H)Si groups (2040 cm^{-1}) and the N-H groups (represented by δ N-H vibration: 1550 cm^{-1}) exhibit a broad maximum around 50 W discharge power (Fig. 7, 9). Also, the calculated thickness of layers goes through a maximum at the same power (Fig. 8).

In the region of network ("lattice") modes no simple correlation with dicharge power is seen. However using the imaginary part of the dielectric function it seems to be feasible to subdivide the experimental set into a low, a middle, and a high power region (Fig. 7, bottom).

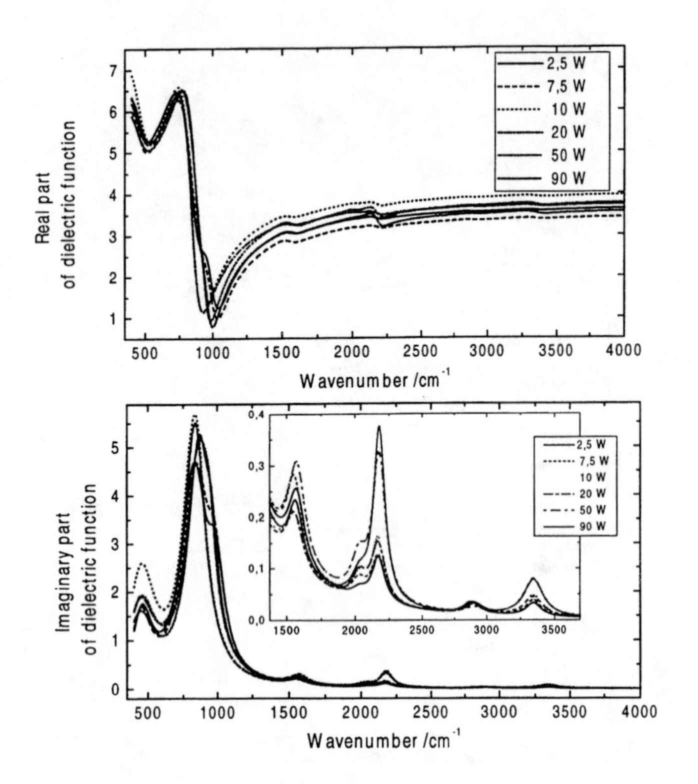

FIGURE 7. Calculated real part and imaginary part of dielectric function of $SiC_xN_yO_z(H)$ layers for various deposition discharge power

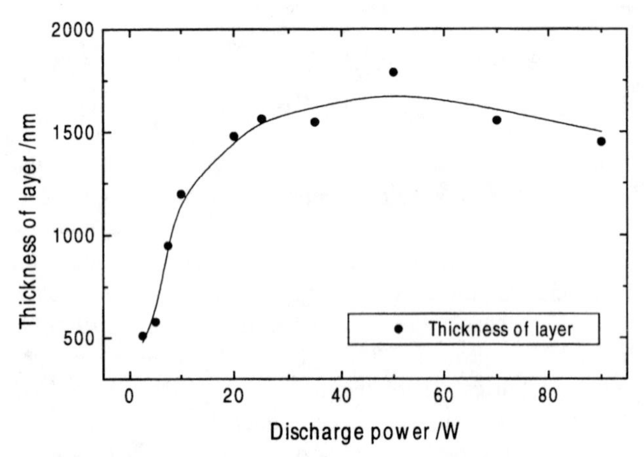

FIGURE 8. Dependence of thickness of $SiC_xN_yO_z(H)$ layer on deposition discharge power

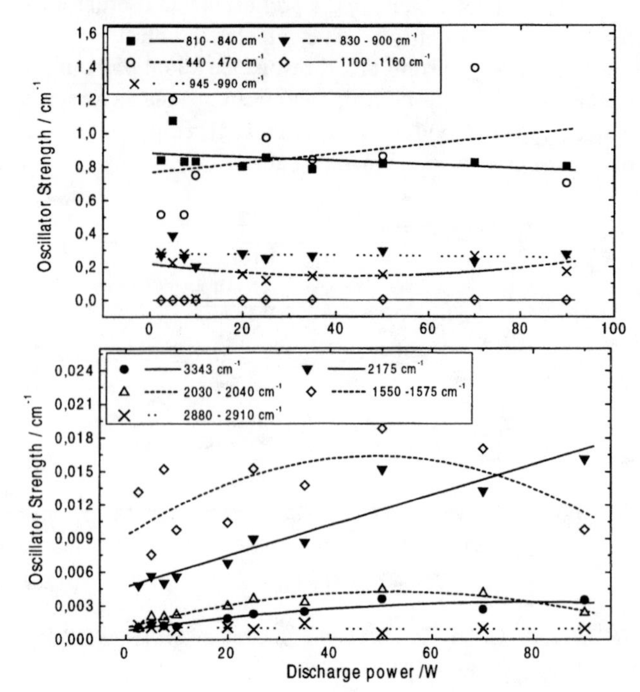

FIGURE 9. Oscillator strength of different modes of the $SiC_xN_yO_z(H)$ layer vs. deposition discharge power. top: „lattice modes“, bottom: hydrogen modes

CONCLUSION

$SiC_xN_yO_z(H)$ layers deposited on a metallic substrate can be investigated by polarized off-normal FTIR reflectance spectroscopy. Using a Lorentz oscillator fit, both the optical properties and thickness of the layer can be determined. The influence of discharge power is apparent in the oscillator strength, mainly of the hydrogen modes, and in the thickness of the layer.

REFERENCES

1. Männel, S., *Diploma Thesis, TU Chemnitz* (1993).
2. Bohachevsky, I. O., Johnson, M. E., Stein, M. L., Technometrics **28** (1986) 3 , 209- 217.
3. Maury, F., Hatim, Z, Biran, C., Birot, M., Dunogues, J., Morancho, R., Proc. 6th European Conference on Chemical Vapour Deposition (1987) 390- 397.
4. Rübel, H., Schröder, B., Fuhs, W., Krauskopf, J., Rupp, T., Bethge, K., *phys. stat. sol. (b)* **131,** 131-143 (1987) .

A RAIRS Study of Propanone Adsorption on a $Pt_{0.25}Rh_{0.75}(111)$ Alloy Single Crystal

S. D. Coomber and M. A. Chesters

Department of Chemistry, University of Nottingham, Nottingham, NG7 2RD, UK.

Monolayer and multilayer films of propanone adsorbed on an alloy metal single crystal have been studied using RAIRS. Analysis of the relative intensities of modes of different symmetry and comparison with the liquid spectrum leads to a determination of the orientation of the molecule. In the monolayer the molecular plane is close to parallel with the surface. On increasing coverage restructuring of the monolayer occurred due to interaction with the second adsorbed layer. Multilayer spectra show the propanone molecule to be orientated with the C=O axis parallel and the molecular plane perpendicular to the surface. IR spectra of considerably thicker films deposited on a KBr plate also show evidence for a strongly preferred molecular orientation with the molecular z axis parallel to the plate and the xy plane perpendicular to the plate.

INTRODUCTION

The adsorption of propanone on metal single crystals has been widely studied. Vibrational spectroscopies such as Reflection Absorption Infrared Spectroscopy (RAIRS) and Electron Energy Loss Spectroscopy (EELS) have been used to characterise the adsorption geometry of propanone on Pt(111), Rh(111), Ru(001), Pd(111) and Cu(110) (1-6). Adsorption on a Pt-Rh alloy single crystal is of interest because propanone exhibits very different chemical behaviour on adsorption at the separate single component surfaces. On Pt(111) the majority monolayer species is η^1(O), that is propanone bonding 'end-on' through the oxygen lone pair. On heating to 200K there is some evidence for a η^2(C,O) species, propanone bonding 'side-on' through π-bonding or di-σ bonding after rehybridisation. Propanone molecularly desorbs on heating above 200K. On Rh(111) the majority monolayer species is η^2(C,O) there is no evidence for a perpendicular species on the clean surface and on heating to 275K propanone decomposes to leave CO, H_2 and surface carbon. The interest in adsorbing on the alloy surface therefore is in determining whether the majority component of the alloy will dominate the chemical behaviour or if contributions will be observed from both metal components.

EXPERIMENTAL

The liquid and solid film propanone spectra were recorded with a Mattson Sirius 100 FT-IR spectrometer using a transmission geometry. The solid film samples were prepared by adsorbing propanone vapour onto a liquid nitrogen cooled KBr plate.

RAIR experiments were carried out in an ultra-high vacuum chamber equipped with standard surface analysis techniques and an ion sputtering gun. The Pt-Rh crystal was cleaned as previously described (7), cooled with liquid nitrogen and could be resistively heated to 1300K. Propanone was dosed onto the crystal by backfilling the chamber or through a metal capillary terminating 3cm from the crystal face. The RAIR spectra were recorded with a Digilab FTS-40 spectrometer equipped with a narrowband MCT detector. Spectra were measured at 4cm^{-1} resolution and resulted from the co-addition of 1024 scans. All RAIR spectra were recorded at a crystal temperature of 100K.

RESULTS AND DISCUSSION

Transmission infrared spectra were obtained for liquid and solid film propanone samples, see Fig. 1. There are a number of important differences between these spectra.

The random molecular motion of the liquid gives rise to broadening of the vibrational bands in the liquid spectrum

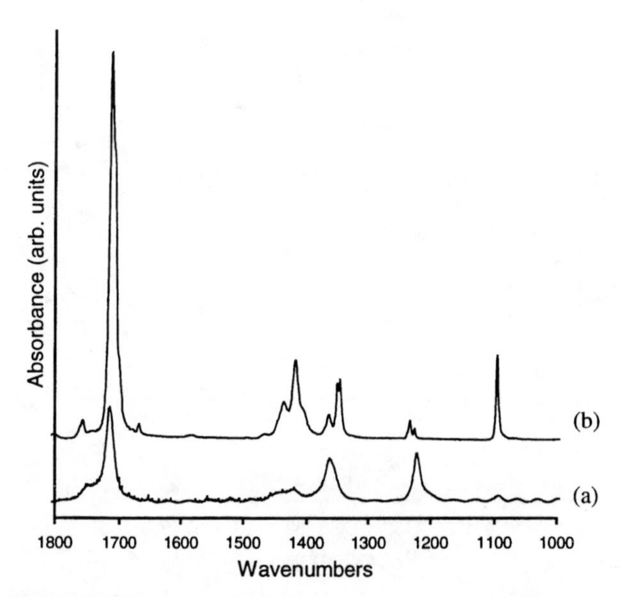

FIGURE 1. Bulk propanone spectra (a) liquid (b) solid film.

CP430, *Fourier Transform Spectroscopy:* 11th International Conference
edited by J.A. de Haseth

and correlation field splitting may be present in some of the bands in the solid film spectrum. The relative intensities of bands of different symmetry also show differences between the spectra. The vibrational mode assignments were made with reference to the work of Dellepiane and Overend (8) and are given in Table 1. Most obviously the B_2 modes show greater intensity than the B_1 modes in the liquid spectrum and vice versa in the solid film spectrum. Since molecules in the liquid phase have a random orientation this suggests that propanone adsorbed in the solid film adopts a preferential orientation. In the transmission experiment the electric field is parallel to the substrate, the relative intensities of the bands in the solid film spectrum therefore show the molecule to be oriented with it's C=O axis parallel and molecular plane perpendicular to the substrate. A small amount of tilt about the molecular x axis accounts for the presence of weak bands associated with B_2 modes.

Figure 2 compares the propanone solid film and multilayer RAIR spectrum. The multilayer spectrum is dominated by B_2 modes and is effectively the inverse of the solid film spectrum where the B_2 modes are the least intense. The differences between the multilayer, solid film and liquid spectra show that propanone adsorbed in the multilayer on a metal surface also preferentially orients. The electric field is polarised perpendicular to the metal surface in the RAIR experiment therefore the molecule is oriented with it's C=O axis parallel and molecular plane perpendicular to the surface i.e. identical to the solid film.

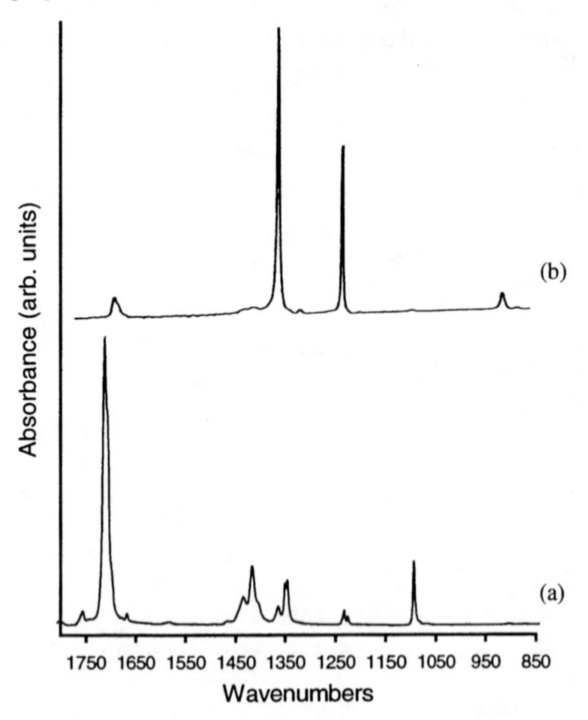

FIGURE 2. (a) Solid film (b) RAIR multilayer

The presence of the weak A_1 and B_1 modes is due to the first few adsorbed layers being tilted and a small amount of tilt of the molecular plane away from the surface normal.

Low coverage RAIR spectra are shown in Fig. 3. The monolayer spectrum contains bands of all IR allowed symmetries. Therefore strict assignment of the orientation to a perpendicular η^1(O) or parallel η^2(C,O) species is not correct. On increasing coverage restructuring occurs due to interaction with the second adsorbed layer. This is characterised by a shift in the frequency of the ν(CO) mode to it's multilayer position. The relative intensities of the B_1 and B_2 modes also change showing a change in the orientation of the molecule. The orientation of the monolayer and restructured species is discussed in greater detail later. The restructured phase grows for approximately ten adsorbed layers before the characteristic multilayer spectrum appears. Growth of the multilayer is identified by the ν(CO) mode reaching a constant intensity while the B_2 modes carry on growing i.e. growth of a phase with it's C=O bond parallel to the surface.

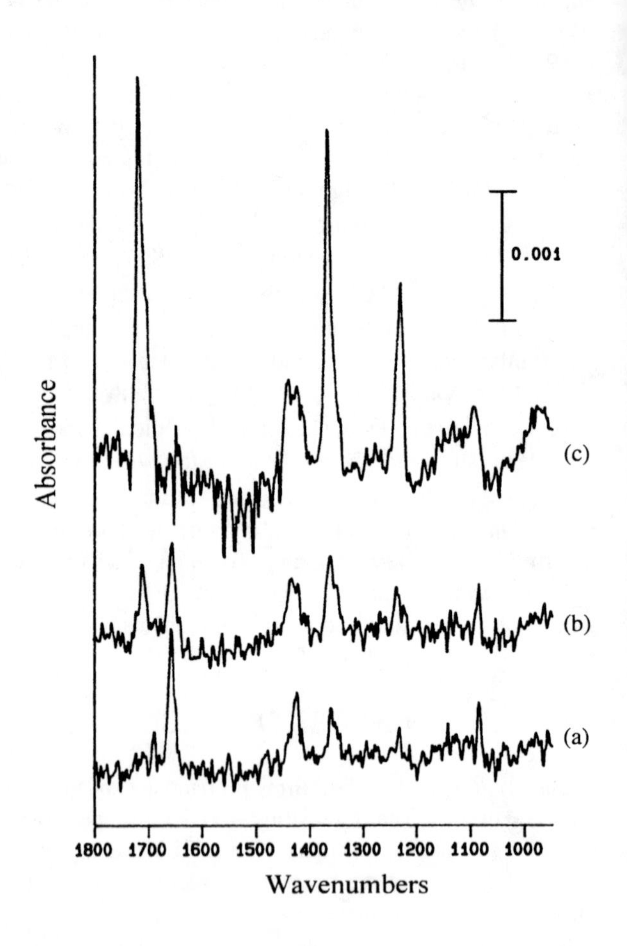

FIGURE 3. Low coverage RAIR spectra (a) monolayer (b) intermediate (c) restructured phase

TABLE 1. Vibrational mode assignments of propanone (all frequencies in cm^{-1}).

Mode	Symmetry	Liquid/Solid Film	Multilayer	Monolayer
$\nu(CO)$	A_1	1713	1715	1658
$\delta_a(CH_3)$	A_1	1440	1440	1440
$\delta_a(CH_3)$	B_1	1420	1423	1425
$\delta_s(CH_3)$	B_2	1362	1373	1361
$\nu_a(CCC)$	B_2	1222	1238	1232
$\rho(CH_3)$	B_1	1094	1095	1085
$\rho(CH_3)$	B_2	-	906	-

Orientation Calculation

The integrated absorbance of bands in a RAIR spectrum is proportional to the square of the component of the dynamic dipole moment in the direction of the electric field. Since the electric field is polarised normal to the surface the intensities of bands of each symmetry can be resolved in terms of two tilt angles of the propanone molecule with respect to the metal surface. The orientation of the molecule in the monolayer and restructured phase can therefore be calculated. The inherent strength of the dipole moment of each vibrational mode is accounted for by normalising the RAIR intensities with respect to those in the liquid spectrum.

In the monolayer the propanone molecule is oriented with it's molecular plane close to parallel with the surface but with a rotation of $17^o \pm 2^o$ about both the y and z molecular axes.

In the restructured layers the tilt angles are $21^o \pm 2^o$ about the y axis and $32^o \pm 3^o$ about the z axis.

CONCLUSIONS

Propanone adsorbed in the first monolayer on a $Pt_{0.25}Rh_{0.75}(111)$ surface bonds with it's molecular plane close to parallel with the surface. On increasing coverage reorientation occurs with the molecular plane becoming more perpendicular to the surface. This restructured phase grows for approximately 10 molecular layers. Propanone adsorbed in the multilayer on the metal surface, a film of between 10 and 100 molecular layers, orients with it's molecular plane perpendicular and C=O axis parallel to the surface. Considerably thicker films of propanone adsorbed on a KBr plate adopt an identical geometry to that of the multilayer.

ACKNOWLEDGEMENTS

The EPSRC and Renishaw Transducer Systems plc are thanked for providing funding.

REFERENCES

1. Avery, N. R., *Surf. Sci.*, **125**, 771, (1983)
2. Vannice, M. A., Erley, W., Ibach, H., *Surf. Sci.*, **254**, 1, (1991)
3. Houtman, C., Barteau, M. A., *J. Phys. Chem.*, **95**, 3755, (1991)
4. Anton, A. B., Avery, N.R., Toby, B.H., Weinberg, W. H., *J. Am. Chem. Soc.*, **108**, 648, (1986)
5. Davis, J. L., Barteau, M. A., *Surf. Sci.*, **208**, 383, (1989)
6. Prabhakaran, K., Rao, C. N. R., *Appl. Surf. Sci.*, **44**, 205, (1990)
7. Rutten, F. J. M., Nieuwenhuys, B. E., McCoustra, M. R. S., Chesters, M. A., Hollins, P., *J. Vac. Sci. & Tech. A*, **15**, 1619, (1997)
8. Dellepiane, G., Overend, J., *Spectrochim. Acta.*, **22**, 593, (1966)

Quantitative Analysis of Molecular Orientation of Structurally Heterogeneous Langmuir-Blodgett Films by Infrared Reflection-Absorption Spectra

T. Hasegawa,[1] D.A. Myrzakozha,[2] T. Imae,[3] J. Nishijo,[1] and Y. Ozaki[2]

Kobe Pharmaceutical University, Motoyama-kita, Higashinada, Kobe 658, Japan[1]; School of Science, Kwansei Gakuin University, Nishinomiya 662, Japan[2]; Faculty of Science, Nagoya University, Chikusa-ku, Nagoya 464-01, Japan[3]

Infrared reflection-absorption (RA) spectra were measured for Langmuir-Blodgett (LB) films of octadecyl-dimethylamine oxide ($C_{18}DAO$) and dioctadecyl-dimethylammonium chloride ($2C_{18}DAC$) on gold-evaporated slide glasses. The RA spectra showed unique relations between the reflection-absorbance and the number of layers of the LB films. Since the RA spectrum of the monolayer LB film showed different properties from the upper layers especially for the $C_{18}DAO$ LB film, this layer was separated from other layers in the molecular orientation analysis. We then proposed a new schematic model comprising a four-phase architecture (IR//air/LB2/LB1/gold) to explain the RA spectra (LB1 refers to the first monolayer only). The results by the new model indicated that the first monolayer directly on the substrate had different film structure from that of the upper layers. These relations could not be figured out by a simple three-phase model (IR//air/LB/gold) that ignored the heterogeneous architecture in the film phase. The present structurally heterogeneous property was also recognized in the temperature-dependent RA spectra of the LB films.

INTRODUCTION

Infrared RA spectrometry (1) has become one of the most powerful measurement techniques of vibrational spectroscopy especially for ultra-thin films on metallic substrates. The application of this technique has been restricted, however, to structurally homogeneous LB films thus far. In other words, all of the layers in the LB films are assumed to have uniform film structure. For the analysis of the layer-location dependent molecular orientation in LB films or heterogeneous LB films, the deuteration method has been the most popular method to investigate the film structure of a specific layer in the stratified multi-layer films (2). The deuteration method is not appropriate for every LB film, especially when deuteration is difficult. In the present study, we have attempted to explore quantitative molecular orientation in a specific layer as well as that in the remaining layers simultaneously from one RA spectrum. The results showed that the first monolayer directly on the metal substrate has a different film structure from that of the remaining layers.

EXPERIMENTAL

$C_{18}DAO$ and $2C_{18}DAC$ were synthesized as described previously (3). $C_{18}DAO$ has one hydrocarbon chain with dimethylamine oxide, while $2C_{18}DAC$ has two hydrocarbon chains with a positively charged dimethylammonium chloride (Fig. 1).

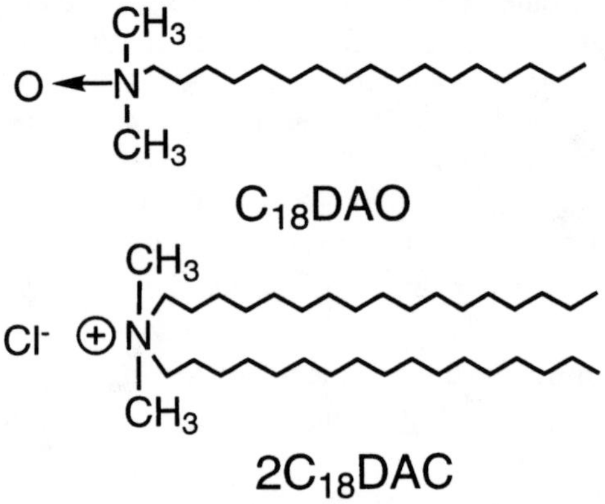

CH3
O←N
CH3
C18DAO
CH3
Cl- ⊕N
CH3
2C18DAC

FIGURE 1. Chemical Structures of $C_{18}DAO$ and $2C_{18}DAC$

Chloroform solutions of the two compounds were individually spread on the water surface to prepare Langmuir (L) films. The L film was compressed by a Teflon® coated aluminum barrier at a compression speed of 15 cm^2 min^{-1} from an area of 900 cm^2. They were transferred to a gold evaporated slide glass by the conventional LB technique at 35 mN m^{-1} at neutral pH. The water subphase was maintained at 25°C by a water circulator.

The infrared *p*-polarized RA measurements of the LB films were performed on a Nicolet Magna 550 FTIR spectrometer with an MCT detector at a resolution of 4 cm^{-1}. The number of scans was 1000 for each measurement. The temperature-dependent RA measurements of the LB films were performed from 35°C to 110°C.

CP430, *Fourier Transform Spectroscopy:* 11th International Conference
edited by J.A. de Haseth

RESULTS AND DISCUSSION

The infrared RA spectra of the $C_{18}DAO$ LB films depended on the number of layers as follows: For the monolayer film, the symmetric CH_3 stretching vibration mode ($\nu_s(CH_3)$) was almost invisible whereas the symmetric CH_2 stretching vibration mode ($\nu_s(CH_2)$) was intense in the CH stretching region (2800~3000 cm^{-1}). For the 3-layer LB film, however, the $\nu_s(CH_3)$ mode was clearly visible, and was comparable to the $\nu_s(CH_2)$ mode. This suggests that the film structure was better organized during the accumulating process. The lower shift of the $\nu_s(CH_2)$ mode from 2853 to 2850 cm^{-1} (respectively, for the monolayer and 3-layer LB films) also supports this organization. For more accumulated LB films, it was remarkable that the band intensity change of the $\nu_s(CH_2)$ mode was not proportional to the number of layers.

In the case of the $2C_{18}DAC$ LB film, the RA spectra depended on the number of layers showed a little different behavior from the previous spectra. The band intensity profile of this LB film was almost linear. However, the fitted straight line did not pass through the zero point.

If an LB film has completely homogeneous film structure from the bottom to the top layer, the band intensity profile should be a simple linearly proportional relation (4). Our results then suggest that the LB films investigated have a heterogeneous film structure. Since the band intensity for the monolayer LB film is different to the accumulated layers, the first layer is considered to have a specific molecular orientation.

In order to understand of the results, we attempted to apply the conventional 3-phase model (IR//air/LB/gold) calculation to the RA spectra. The calculation procedure was the same as that mentioned previously (5). In the calculation, the optical anisotropy reflecting the molecular orientation was taken into account for the film phase. With this calculation procedure, the molecular orientation angles of the $\nu_a(CH_2)$ (antisymmetric stretching) and $\nu_s(CH_2)$ modes (α and β) were quantitatively obtained. These angles were converted to the tilt angle of the hydrocarbon chain (γ) with the following equation (5).

$$\cos^2\alpha + \cos^2\beta + \cos^2\gamma = 1 \qquad (1)$$

The results were, however, insensitive to the change of the band intensity. Although the band intensity itself largely jumped between the monolayer and 3-layer LB films, the calculated molecular orientation almost linearly changed all through the number of layers. The results by the 3-phase model could not clarify the physico-chemical properties of the LB films. This may be because this model ignored the heterogeneous change in the film phase.

We then proposed a 4-phase model. In this model, the film phase was divided into two phases for the first monolayer and the layers named LB1 and LB2 remaining, respectively. The first monolayer was thus separated from other layers. In this model, two unknown molecular orientation angles are introduced for one RA spectrum. Then, the set of the two solutions would not be obtained rigidly. In the present study, however, we attempted to draw the solutions from one RA spectrum. The solutions of the orientation angles have maximum and minimum values to satisfy the border condition restricted by the observed value and the multiple reflection in the 4-phase system. We then obtained two molecular orientations in the two phases simultaneously from an RA spectrum where the result was expressed as an averaged value with a deviation bar.

The averaged orientation angles showed that the first monolayer in accumulated LB films may have similar film structure to the monolayer LB film, and the upper layers have better organized film structure for the two LB films. In particular, the tilt angle from the surface normal of the hydrocarbon chain in the LB2 phase was getting lower with an increase of the number of layers. Since the LB2 phase does not interact with the substrate directly, only the interaction among the film molecules should have played an important role to stabilize the film phase. Regarding the tilt angle in the LB2 phase, the tilt angle for the $2C_{18}DAC$ LB film was better organized than the $C_{18}DAO$ LB film. This may be because the two-leg (two hydrocarbon chains) molecule is likely to form a more stable molecular assembly than the one-leg molecule. Regarding the LB1 phase, the tilt angle for the $2C_{18}DAC$ LB film was also better ordered than the $C_{18}DAO$ LB film. This may reflect the difference of the net charge of the head-group on the film molecules. Since $2C_{18}DAC$ has a positively charged region in the head-group, this region may help the interaction between the head-group and the stronger gold-evaporated substrate ($C_{18}DAO$ has no net charge).

We also measured temperature dependence of the RA spectra for both LB films. The 7-layer LB film of each compound was used for the experiments. They were sufficiently dried in a desiccator in advance. The RA spectra of the $C_{18}DAO$ LB film showed clear melting temperature at ~ 50°C. The band intensity of the CH_2 stretching modes abruptly began to increase at about 50°C, and increased up to 60°C. Above 60°C, the intensity decreased to a lower value, and kept almost constant value for higher temperature.

On the other hand, the $2C_{18}DAC$ LB film showed no melting temperature in the temperature region between 20°C and 110°C. The band intensity began to increase at 50°C, but it was moderate. Above 70°C, the intensity rapidly increased, and continued to increase up to 110°C.

We also applied the 4-phase model to these RA results. The result indicated that the sufficiently dried LB film of $C_{18}DAO$ had quite ordered film structure, giving an homogeneous property. This stability remained until 40°C. At 50°C, however, the tilt angle of the hydrocarbon chain in the LB1 phase was greatly disordered, leaving the tilt angle in the LB2 phase not so disordered. This means the LB film showed a heterogeneous property at this temperature. At 60°C, the tilt angles of the two phases agreed with each other, again indicating the homogeneous property. This should correspond to the complete melt of the film. The 4-phase model showed, in this way, that the first monolayer was firstly disordered before the entire film melted at 60°C.

The results for the $2C_{18}DAC$ LB film were quite different. This film did not show a complete melt, but only the LB1 phase was greatly disordered. The LB film at 35°C after the heating showed fairly good recovery of the film structure.

The difference between the LB films could be explained by the difference of the net charge of the head-groups and the number of the hydrocarbon chains in their molecules. The difference of the net charge should have affected on the property of the LB films more severely than the number-of-layer dependence.

REFERENCES

1. Greenler, R.G., *J. Chem. Phys.*, **44**, 310 (1966).
2. Umemura, J., Takeda, S., Hasegawa, T., Kamata, T., and Takenaka, T., *Spectrochim. Acta*, **50A**, 1563-71 (1994).
3. Imae, T., Tsubota, T., Okamura, H., Mori, O., Takagi, K., Itoh, M., and Sawaki, Y., *J. Phys. Chem.*, **99**, 6046-53 (1995).
4. Hasegawa, T., Umemura, T., and Takenaka, T., *J. Phys. Chem.*, **97**, 9009-12 (1993).
5. Hasegawa, T., Takeda, S., Kawaguchi, A., Umemura, J., *Langmuir*, **11**, 1236-43 (1995).

Use Of Polarized Infrared External Reflection Spectroscopy To Study Phospholipid Monolayers At The Air-Water Interface

Keith M. Faucher, Zhao Ping, and Richard A. Dluhy

Department of Chemistry, University of Georgia, Athens, GA 30602

The pressure dependence of phospholipid monolayers at the air-water interface has been investigated using polarized infrared external reflection spectroscopy under carefully controlled environmental conditions. Results show that a splitting of the methylene symmetric and antisymmetric stretching vibrations is observed in the polarized spectra that has not been previously reported for unpolarized monolayer spectra. The splitting of the C-H bands results in sub-bands identifiable with ordered and disordered conformations. The intensities of these sub-bands correlate the particular thermodynamic state of the monolayer. The splitting of these bands qualitatively tracks the formation of these domains upon compression as previously seen in epiflourescence microscopy. As the surface pressure of the phospholipid monolayer film increases, the band resulting from the liquid domain structures in the film disappears. Curve fitting based on these results indicates that we can now determine the mole fraction of either solid or liquid phase domains as a function of surface pressure. These results have been observered at both 30 and 60 degree angles of incidence. The splitting of the methylene hydrocarbon vibration, and the presence of individual bands corresponding to particular conformational states has not previously been observed in polarized infrared external reflection spectra of monolayers at the air-water interface.

INTRODUCTION

Polarized external reflection-absorbance infrared spectra of phospholipid monolayer films at the air-water interface were acquired with radiation polarized either parallel (R_p) or perpendicular (R_s) to the surface normal. Previous studies have shown that the reflectance of polarized radiation at the air-water (A/W) interface is highly dependent on the angle of incidence employed (1). Calculations using the Fresnel reflectance formulae have previously been made of the reflectance of light *vs.* angle of incidence for both R_s and R_p polarizations at the air-water interface (1,2). These calculations indicate that the absolute reflectance of the radiation at R_p polarization is greater at a 30 degree angle of incidence than at a 60 degree angle, however, the opposite is true for R_s polarized radiation. Thus, orientational as well as structural information on a monolayer film at the air-water interface may be obtained using polarized infrared external reflectance spectroscopy at varying angles of incidence (3).

Recent experiments in our laboratory have focused on using polarized infrared external reflectance spectroscopy to study a monomolecular film of the model membrane dipalmitoylphosphotidylcholine (DPPC). Spectra were obtained at varying surface pressures, angles of incidence, and polarizations at the air-water interface. The results of these experiments show that a band splitting of the CH_2 vibrational modes is observed in the polarized spectra that has not previously been reported for unpolarized monolayer spectra at the A/W interface. In addition, the polarized monolayer spectra are able to distinguish individual sub-bands due to both the ordered and disordered conformational monolayer states. Using these band intensities, we have been able to use these polarized IR spectra to quantitatively track the formation of fractional solid and liquid phases throughout the monolayer phase transition.

EXPERIMENTAL

The phospholipid used throughout these experiments was 1,2-dipalmitoyl-sn-glycero-3-phosphocholine (DPPC) purchased from Avanti Polar Lipids (Alabaster, Alabama) at 99% purity and used without further purification. Samples solutions of DPPC were prepared by dissolving the lipid in a chloroform solution (Baker, HPLC grade) at a concentration of 1.5 mg/ml. The sample was then spread onto a Nima 601M Series Langmuir Film Balance (Coventry, England) and allowed to settle for at least 10-15 min before film compression was begun.

Infrared reflection-absorbance spectra were acquired using a Perkin-Elmer 2000 Fourier Transform infrared spectrometer. The IR beam was brought outside of the instrument using the spectrometer's external beam port and brought to the surface of the Langmuir film balance using custom-designed optics. The IR beam was focused onto the A/W interface using gold-coated, off-axis parabolic mirrors (Janos, Townshend, VT) machined at either 30 or 60 degrees (which resulted in a angle of incidence at the surface of 60 and 30 degrees relative to the surface normal, respectively). A wire grid polarizer was placed in the beam path in order to collect polarized specta; this polarizer was placed on a rotatable mount that allowed the polarization direction to be changed and the polarizer to be placed in and out of the external IR beam without a change in environmental conditions. Spectra were collected at 4

CP430, *Fourier Transform Spectroscopy:* 11th International Conference
edited by J.A. de Haseth

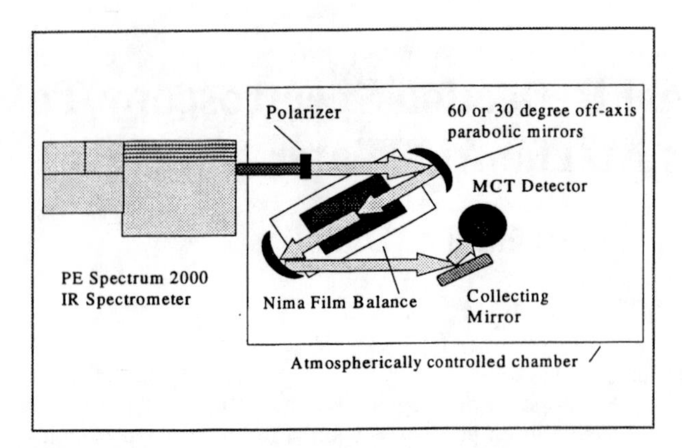

FIGURE 1. Instrumental design used in the experiments reported here.

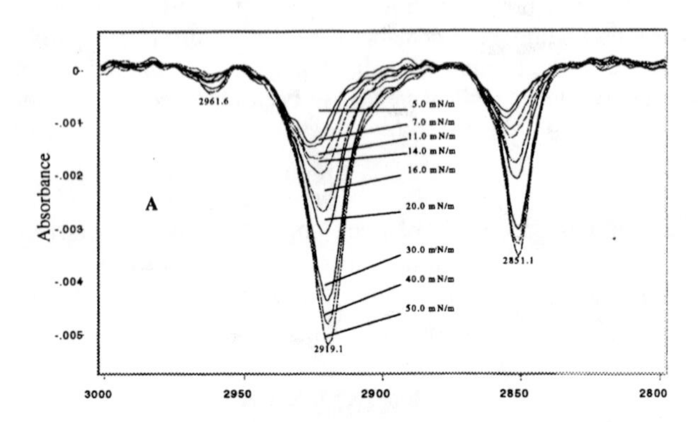

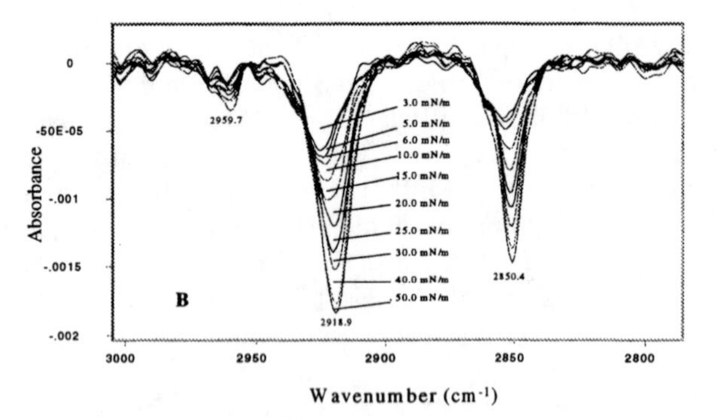

FIGURE 2. Unpolarized IR external-reflection spectra of DPPC monolayers at the A/W interface. (A) 30 degree angle of incidence. (B) 60 degree angle of incidence.

cm^{-1} resolution. At a 30 degree angle of incidence 1024 scans were co-added for both R_p and R_s polarizations. At a 60 degree angle of incidence, 1024 scans were co-added for the R_s polarization, while 2048 scans were coadded for the R_p to obtain an acceptable signal-to-noise ratio. The external IR reflectance setup was enclosed in a custom built chamber that allowed the humidity to be controlled. Figure 1 illustrates the instrumental setup used in these experiments. Spectral manipulations performed on the data such as baseline correction and curve fitting was performed using the Grams/32 software package (Galactic Industries Corporation, Salem, NH).

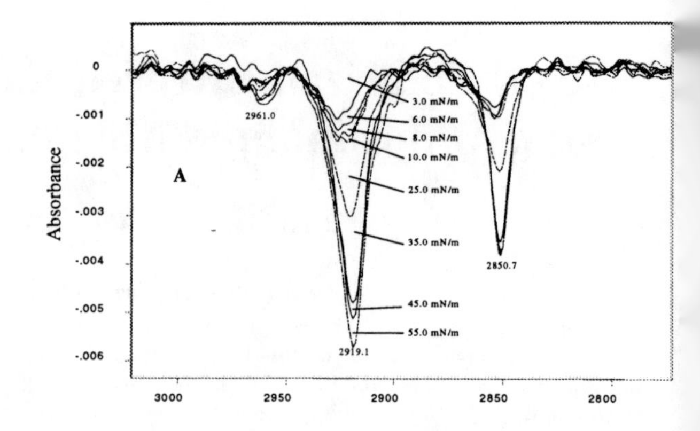

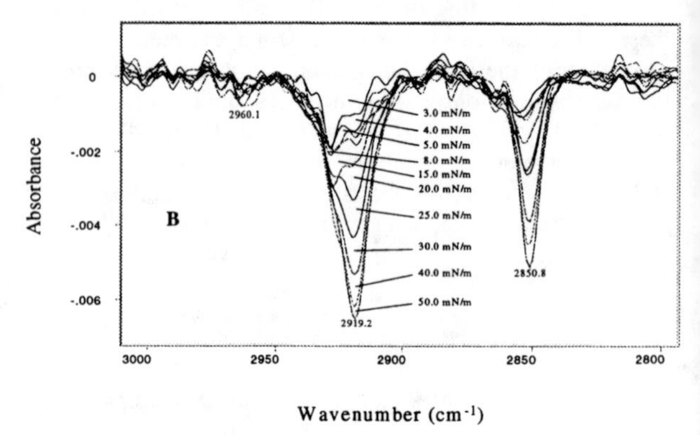

FIGURE 3. IR reflection-absorption spectra at the A/W interface collected as a function of surface pressure for DPPC at a 30 degree angle of incidence. (A) R_s polarized spectra. (B) R_p polarized spectra.

RESULTS AND DISCUSSION

Infrared external reflection-absorption spectra at the air-water (A/W) interface were acquired for DPPC monolayers at three different states of polarization of the incoming radiation: 1) without polarization, 2) with perpendicular (R_s) polarization, and 3) with parallel (R_p) polarization. Each series of spectra collected at a specific polarization setting were also collected at two different (30 and 60 degree) angles of incidence. Finally, each set of spectra was also collected at increasing surface pressures in order to monitor the thermodynamic state of the monolayer. These different types of monolayer spectra are illustrated in Figures 2-4. Figure 2 presents the methyl and methylene stretching bands for DPPC in the C-H stretching region between 3000-2800 cm^{-1} without polarization at 30 and 60 degree angles of incidence. Figure 3 shows reflection-absorbance spectra for DPPC in this same spectral region with R_s and R_p polarization at a 30 degree angle of incidence. Finally, Figure 4 illustrates reflection-absorbance spectra for DPPC in the C-H stretching region with R_s and R_p polarization at a 60 degree angle of incidence.

The polarized reflection-absorbance spectra for the methylene symmetric and antisymmetric stretching bands for DPPC at a 30 degree angle of incidence (Figure 3)

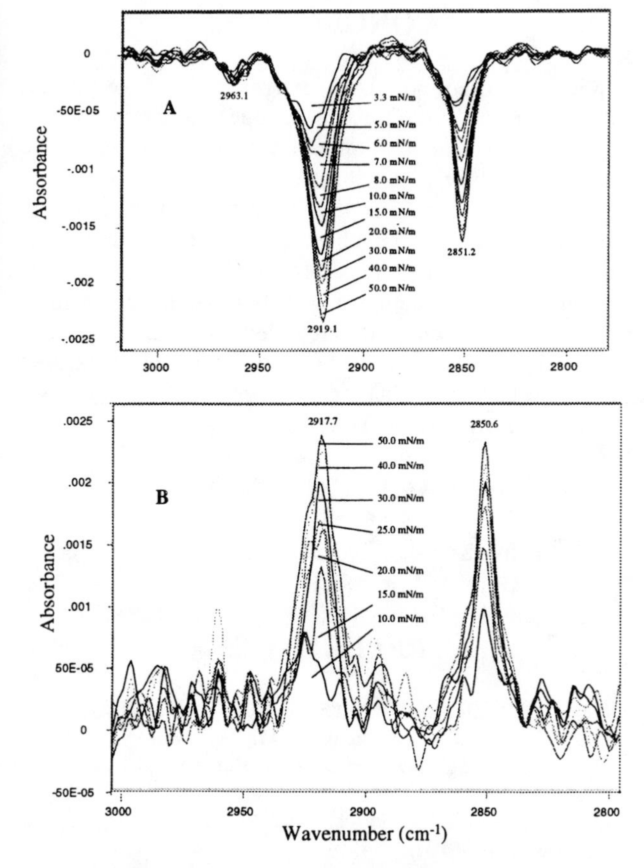

FIGURE 4. IR external reflection-absorption spectra at the A/W interface collected as a function of surface pressure for DPPC at a 60 degree angle of incidence. (A) R_s polarized spectra. (B) R_p polarized spectra.

appear quite different from the unpolarized spectra (Figure 2). In both the R_s and R_p polarized spectra, a splitting of the methylene antisymmetric stretching band at 2920 cm^{-1} is observed. The splitting is most prominent in the R_p polarization. We have also observed such splitting in the R_s and R_p polarized spectra of the antisymmetric methyl stretching vibration (~2960 cm^{-1}), although it is difficult to determine the trends in this band due to their lower intensities.

The polarized reflection-absorbance spectra for the methylene symmetric and antisymmetric streching bands at the 60 degree angle of incidence (Figure 4) also appear quite different from the unpolarized spectra. For the R_s spectra, splitting is observed at low surface pressure which appears to shift to higher wavenumber, similar to the R_p spectra collected at a 30 degree angle of incidence. However, the spectra collected at the R_p polarization at a 60 degree angle of incidence are qualitatively different from the spectra collected at the 30 degree incident angle. The methylene stretching bands have reversed intensity and are now positive rather than negative. This band intensity reversal has been theoretically predicted and previously observed (1,2,3). The DPPC antisymmetric stretching band for the R_p polarization at a 60 degree angle of incidence also appears to exhibit splitting characteristic

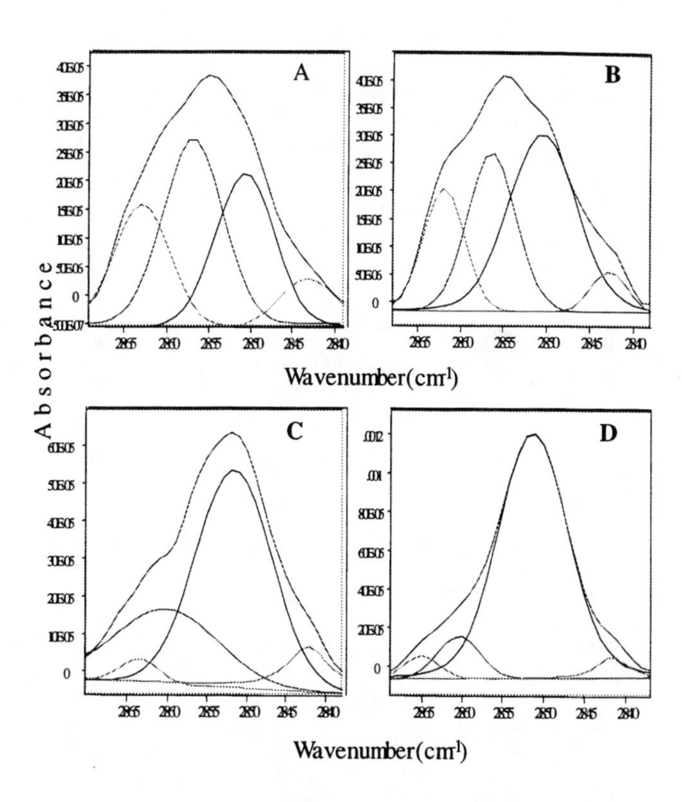

FIGURE 5. Curve fitting results for the 2850 cm^{-1} band of DPPC as a function of increasing surface pressure. Spectra used for curve fitting were collected unpolarized at a 60 degree angle of incidence. (A) 3.0 mN m^{-1}. (B) 5.0 mN m^{-1}. (C) 10.0 mN m^{-1}. (D) 30.0 mN m^{-1}.

of the spectra at a 30 degree angle of incidence. However, the signal to noise level is much lower for the R_p polarization at a 60 degree angle of incidence due to the fact that the absolute reflectivity of the R_p polarization is very near zero at angles close to the Brewster angle (1). It is for these reasons that it is difficult to determine the degree of splitting in the antisymmetric band observed with R_p polarization at 60 degrees.

We have correlated the polarized splitting of the methylene antisymmetric stretching band at low surface pressures with the presence of two different conformational structures, ordered (solid) and disordered (liquid) phases. Curve fitting of the methylene symmetric stretching band results in the ability to determine the amount of either liquid or solid phase lipid as a function of surface pressure. Representative curve-fitting results obtained at varying surface pressures at 60 degree incident angle are presented in Figure 5. These data show a liquid (disordered) phase band at ~2859 cm^{-1} and a solid (ordered) phase band at ~2852 cm^{-1}. At low surface pressures these bands are approximately of equal intensity, and their presence results in the splitting observed in the methylene vibrations. As the surface pressure of the lipid monolayer is increased, the intensity of the liquid-phase band decreases while the solid-phase band increases. This increase in intensity for the solid-phase band corresponds

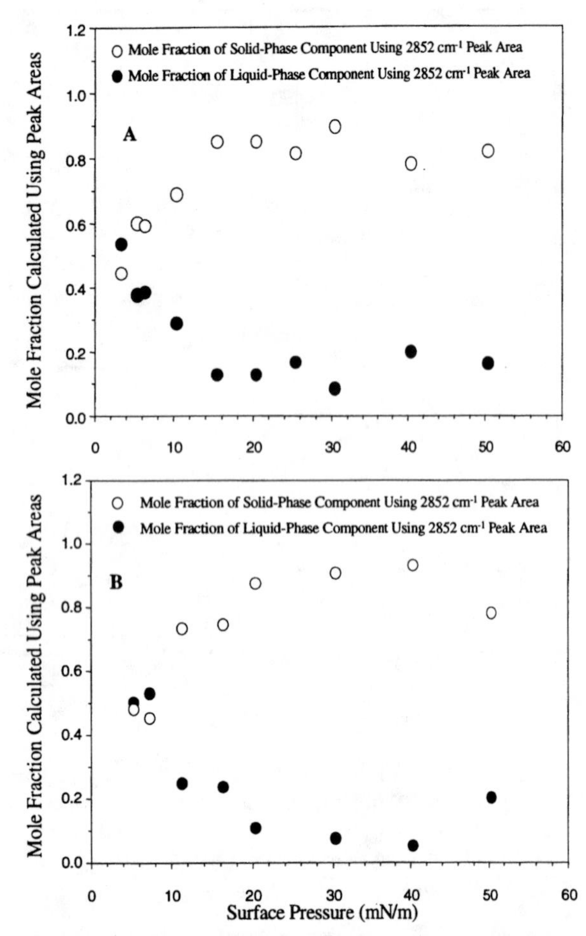

FIGURE 6. Fractional component of solid and liquid phase in the DPPC monolayer calculated using the integrated peak areas obtained from the curve fit results of the 2850 cm^{-1} band. (A) Fractional liquid and solid phase components obtained from 60 degree angle of incidence spectra. (B) Fractional liquid and solid phase components obtained from 30 degree angle of incidence spectra.

to a shift to lower wavenumber for the methylene stretching bands. At high surface pressures the band due to the liquid phase virtually (but not entirely) disappears as the intensity of the solid band predominates.

The curve fitting results can also be used to determine the fractional composition of both the solid and liquid phases as a function of surface pressure. Figure 6 presents plots of mole fraction calculated using peak areas as a function of surface pressure using the 2850 cm^{-1} band at both 30 and 60 degree angles of incidence. Both of these plots show that the mole fraction of both the ordered and disordered phases are approximately equal at low surface pressures. As the surface pressure is increased the mole fraction due to the solid band increases with a concomitant decrease of the liquid phase mole fraction. The plots in Figure 6 assist in explaining the observed splitting of the methylene stretching bands at low pressures and the disappearance of the splitting for these bands at higher surface pressures.

CONCLUSIONS

We have used polarized external-reflectance infrared spectroscopy to moinitor the changes in thermodynamic states of a phospholipid monolayer. Using parallel (R_p) and perpindicular (R_s) polarizations at 30 and 60 degree angles of incidence, disordered and ordered phases of the lipid monolayer can be observered as a splitting of the anitsymmetric methylene stretching band at low surface pressures. This conclusion is supported by curve-fitting results of the symmetric methylene streching band for DPPC. From the curve fitting results it is now possible to determine the mole fraction for both the liquid and solid phases of the monolayer as a function of pressure.

ACKNOWLEDGEMENTS

Support for the research described here was provided by NIH grant GM40117 (R.A.D.)

REFERENCES

(1) Dluhy, R.A. *J. Phys. Chem.* **90**, 1373-1379, 1986.

(2) Dluhy, R.A.; Stephens, S. M.; Widayati, S.; Williams, A. D. *Spectrochimica Acta Part A.* **51**, 1413-1447, 1995.

(3) Flach, C. R.; Gericke, A.; Mendelsohn, R. *J. Phys. Chem. B.* **101**, 58-65, 1997.

An FT-IR Study of Time-Dependent Orientational Changes in One-monolayer Langmuir-Blodgett Films of Alkyl-tetracyanoquinodimethane

S. Morita[1], Y. Wang[1], K. Iriyama[2], and Y. Ozaki[1]

[1]*Department of Chemistry, School of Science, Kwansei Gakuin University, Uegahara, Nishinomiya, 662, Japan*
[2]*The Jikei University School of Medicine, Nishi-sinbashi, Minato-ku, Tokyo 105, Japan*

Time-dependent infrared spectral changes were investigated for one-monolayer Langmuir-Blodgett (LB) films of 2-alkyl-7,7,8,8-tetracyanoquinodimethane (alkyl-TCNQ) on gold-evaporated glass slides and CaF_2 plates. Significant intensity increases were observed with time for bands due to both the alkyl chain and TCNQ plane in the infrared reflection-absorption (RA) spectra of the LB films of dodecyl-, and pentadecyl-TCNQ while most of bands in the infrared transmission spectra of the LB films of the same compounds showed time-dependent intensity decreases. Time-dependent spectral changes in the ultraviolet-visible region were also measured for the one-monolayer LB films of pentadecyl-TCNQ.

INTRODUCTION

During the last decade, great interest has been paid to Langmuir-Blodgett (LB) films of functional organic dyes because such films may exhibit better functionality than inorganic devices. Among the functional LB films, we have been interested in conducting LB films in which a tetracyanoquinodimethane (TCNQ) derivative is used as an electron acceptor. We have also studied the structure of LB films of 2-alkyl-7,7,8,8-tetracyanoquinodimethane (alkyl-TCNQ) and that of its mixed-stack charge transfer complex films by use of infrared and visible spectroscopies and atomic force microscopy (AFM) (1-8). Thus far, we have investigated [1] the molecular orientation and structure in TCNQ LB films and their dependences upon the length of the substituted hydrocarbon chain (1,2), [2] the order-disorder transitions in the TCNQ LB films (3), and [3] the molecular orientation, structure, morphology, and thermal behavior of mixed-stack charge transfer LB films of octadecyl-TCNQ and 3,3',5,5'-tetramethylbenzidine and 5,10-dimethyl-5,10-dihydrophenazine (4-8).

Recently, we found that the molecular orientation in one-monolayer LB films of dodecyl- and pentadecyl-TCNQ changes with time after depositing the films on CaF_2 plate and gold-evaporated glass slide. This communication reports about this time-dependent orientational changes.

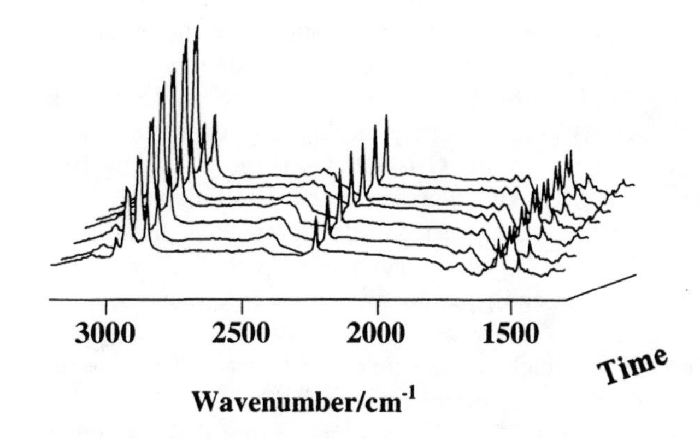

FIGURE 1. Time-dependent infrared RA spectra of a one-monolayer LB film of pentadecyl-TCNQ measured at 5 to 120 minute after the film deposition.

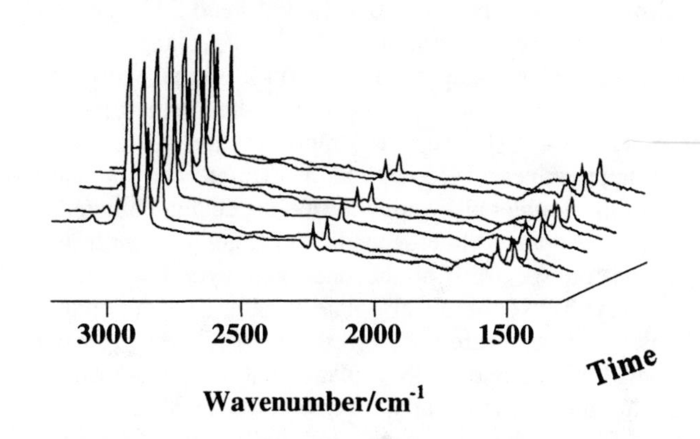

FIGURE 2. Time-dependent infrared transmission spectra of a one-monolayer LB film of pentadecyl-TCNQ measured at 5 to 120 minute after the film deposition.

EXPERIMENTAL

Dodecyl-, pentadecyl-, and octadecyl-TCNQ were purchased from the Japanese Research Institute for Photosensitizing Dyes Co., Ltd., and used without further purification.

The Z-type LB films of alkyl-TCNQ were fabricated by using a Kyowa Kaimen Kagaku Model HBM-AP Langmuir trough with a Whilhelmy balance. A detailed procedure for the fabrication of LB films of alkyl-TCNQ was described in the papers previously reported (1-4). The transfer ratio was found to be nearly unity (0.95± 0.05) throughout the experiments.

CP430, *Fourier Transform Spectroscopy:* 11th International Conference
edited by J.A. de Haseth

1-56396-746-4/98/$15.00

The instrumentation and sample-handling techniques employed for measuring the ultraviolet-visible (UV-vis) and the infrared spectra were the same as those described before (4,8).

RESULTS AND DISCUSSION

Figure 1 shows time-dependent changes in an infrared reflection-absorption (RA) spectrum of a one-monolayer LB film of pentadecyl-TCNQ on a gold-evaporated glass slide. These spectra were measured at 5 to 120 minutes after the film deposition. One can observe the clear time-dependent spectral changes even from the one-monolayer LB film. A band due to a CH_2 antisymmetric stretching mode of the hydrocarbon chain is identified near 2920 cm^{-1} as a doublet. The doublet feature was observed for infrared spectra of dodecyl-, pentadecyl-, and octadecyl-TCNQ in powdered solids, LB films, and evaporated films (1-8). However, the cause of the splitting of the CH_2 antisymmetric stretching band is still not clear. A band at 2850 cm^{-1} is assigned to the CH_2 symmetric stretching mode. Particularly striking is that the intensity ratio of the two band changes largely during the of course time, although not only the CH_2 antisymmetric stretching band but also the CH_2 symmetric stretching band increase.

According to the surface selection rule in infrared RA spectroscopy (9), vibrational modes with their transition moments perpendicular to the surface are enhanced in a RA spectrum. Therefore, the observations in Fig. 1 suggest that the alkyl chain becomes tilted with respect to the surface normal during the time course.

A band at 2222 cm^{-1} is assignable to a C≡N stretching mode of the TCNQ chromophore and those at 1547 and 1529 cm^{-1} are attributed to its C=C stretching mode (1). These bands also increase with time. The reversal of the intensity ratio of the two bands at 1547 and 1529 cm^{-1} is noted. Therefore, it seems that the TCNQ plane is more perpendicular with respect to the surface normal while the molecular axis of the plane becomes tilted with time. A band due to a CH_2 scissoring mode of the hydrocarbon chain appears near 1465 cm^{-1} as a doublet. The intensity ratio of the doublet also shows a time-dependent change.

Figure 2 depicts time-dependent changes in an infrared transmission spectrum of the one-monolayer LB film of pentadecyl-TCNQ on a CaF_2 plate. Again, time-dependent intensity changes were observed for the bands arising from the alkyl chain and TCNQ plane, but the direction of spectral changes are nearly reverse to those observed for the LB film on the gold-evaporated glass slide. The results in Figure 2 also indicate that both the alkyl chain and the molecular axis of the TCNQ plane become tilted with time with respect to the surface normal and that the TCNQ plane gradually stands up.

One-monolayer LB films of dodecyl-TCNQ on a gold-evaporated glass slide and CaF_2 plate underwent similar time-dependent spectral changes while such changes were not observed for those of octadecyl-TCNQ. Probably, the interaction between the substrate and TCNQ chromophore plays an key role in the orientational changes. In the case of octadecyl-TCNQ, the interaction among the long alkyl chains prevents the time-dependent orientational changes.

UV-vis spectra were measured for one-monolayer LB films of pentadecyl-TCNQ at 5 to 120 minutes after the film deposition. The UV-vis spectrum of the LB film of pentadecyl-TCNQ on a CaF_2 plate consists of bands due to the monomeric form and stacked form of the TCNQ chromophore (10). As a function of time, the intensity of the monomer band increases while that of the band due to the stacked form decreases. Therefore, in the reorientation process some of the stacked form change into the monomeric form.

The LB film of pentadecyl-TCNQ on a gold-evaporated glass slide showed gradual base line shift with time in the UV-vis spectra. This may be due to the formation of the domain structure on the order of a few hundred nanometers (4,5,10).

REFERENCES

1. M. Kubota, Y. Ozaki, T. Araki, S. Ohki, and K. Iriyama, Langmuir, **7**, 774 (1991).
2. S. Terashita, K. Nakatsu, T. Mochida, T. Araki, Y. Ozaki, and K. Iriyama, Langmuir, **8**, 3051(1992).
3. S. Terashita, Y. Ozaki, and K. Iriyama, J. Phys. Chem., **97**, 10445 (1993).
4. Y. Wang, K. Nichogi, S. Terashita, K. Iriyama, and Y. Ozaki, J. Phys. Chem., **100**. 368 (1996).
5. Y. Wang, K. Nichogi, K. Iriyama, and Y. Ozaki, J. Phys. Chem., **100**, 374 (1996).
6. Y. Wang, K. Nichogi, K. Iriyama, and Y. Ozaki, J. Phys. Chem., **100**, 17232 (1996).
7. Y. Wang, K. Nichogi, K. Iriyama, and Y. Ozaki, J. Phys. Chem., **100**, 17238 (1996).
8. Y. Wang, K. Nichogi, K. Iriyama, and Y. Ozaki, J. Phys. Chem., in press.
9. R. G. Greenler, J. Chem. Phys., **44**, 310 (1996).
10. Y. Wang, Y. Ozaki, and K. Iriyama, Langmuir, **11**, 705 (1995).

Infrared Analysis of Multilayer Stacks Deposited on Rough and Curved Polymer Substrates

Jacob Valk [a], Sjaak van Enckevort [a] and Wolfgang Theiß [b]

a) Océ-Technologies B.V., P.O.Box 101, 5900 MA Venlo, The Netherlands

b) I. Phys. Inst., Aachen University of Technology, D-52056 Aachen, Germany

We describe a method to analyze multilayer thin film systems deposited on microscopically rough and macroscopically curved polymer substrates by means of optical spectroscopy. As an example SiO_x multilayers are analyzed with respect to layer thicknesses and oxygen content.

INTRODUCTION

The best way to optically analyze thin film systems is to apply a spectrum simulation method [1]. Here we present a method to investigate layers deposited on curved substrates which also have a rough surface.

Basically there are two steps to do. First, the dielectric functions of all the layers in the stack have to be defined. For the materials of interest models for the optical constants are used. Then, in the second step, the propagation of electromagnetic waves through the layer stack has to be computed taking into account reflection and transmission of partial waves at the various interfaces (given by Fresnel's equations) and absorption and phase shift effects inside the individual layers. In our case also the corrections for curvature and roughness effects must be done in this step.

The simulated spectrum is compared to the measured one and the parameters (dielectric function models, layer thicknesses) are adjusted to minimize the deviation of simulation and experiment.

THE DIELECTRIC FUNCTION OF SIO_X VS. OXYGEN CONTENT

The dielectric function $\varepsilon(\tilde{\nu})$ of SiO_x layers deposited on various substrate materials (silicon, KBr and epoxy) can be described in the infrared spectral region by the following model (where $\tilde{\nu}$ denotes the wavenumber):

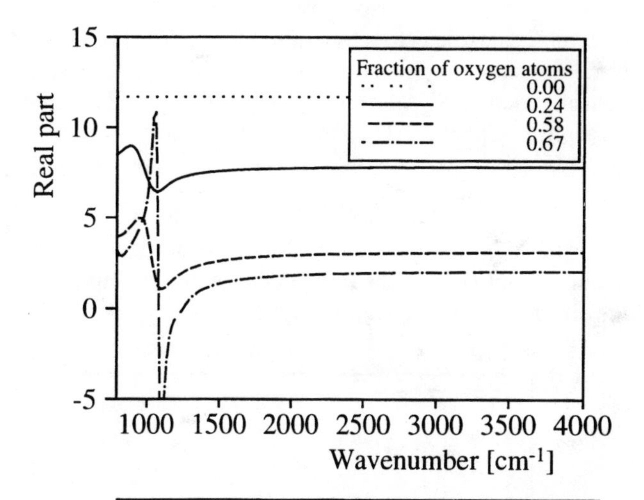

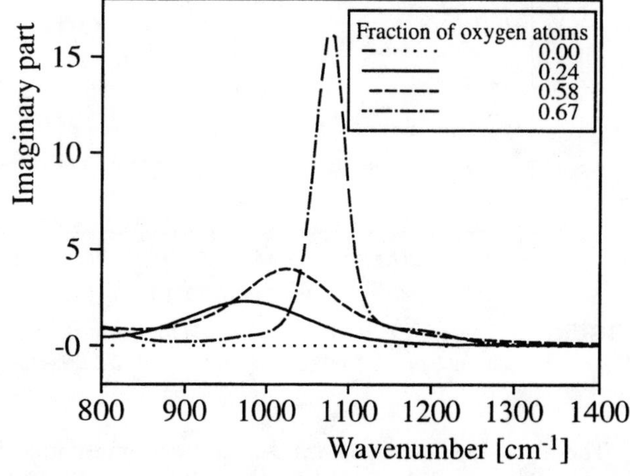

FIGURE 1. Dielectric functions of different SiO_x layers (including pure silicon and SiO_2) obtained from parameter fits on 'flat substrate' spectra. A clear relation between the oxygen content and dielectric function features is observed.

CP430, *Fourier Transform Spectroscopy:* 11th International Conference
edited by J.A. de Haseth

$$\varepsilon(\tilde{\nu}) = \varepsilon_\infty + \sum_j \frac{1}{\sqrt{2\pi}\sigma_j} \int_{-\infty}^{\infty} \exp\left(-\frac{\left(x-\Omega_{0,j}\right)^2}{2\sigma_j^2} \right)$$

$$\otimes \frac{\Omega_{P,j}^2}{x^2 - \tilde{\nu}^2 - i\tilde{\nu}\Omega_{\tau,j}} dx \quad .$$

It is composed of a constant and real dielectric background ε_∞ and up to four oscillator terms (*j=1..4*) which represent Si-O-Si vibrations. The integral sums up harmonic oscillator susceptibilities with Gauss-distributed resonance frequencies. This accounts for local disorder in the atomic configurations which may lead, for example, to variations of bonding angles and atomic distances. These cause frequency shifts of vibrational modes. A more detailed discussion of this model can be found in [2].

Fig.1 shows that the oxygen content of SiO_x (determined separately by energy dispersive x-ray experiments) clearly determines its dielectric function. Hence, in reverse, one can expect to be able to determine the oxygen concentration of SiO_x layers in a stack from proper analysis of optical spectra. The model parameters chosen to determine the oxygen content are ε_∞ and the mean resonance frequency Ω_0 of the strongest Si-O-Si mode. It was verified that there is a direct connection of these quantities not only to the oxygen concentration but also to the electrical properties of the layers.

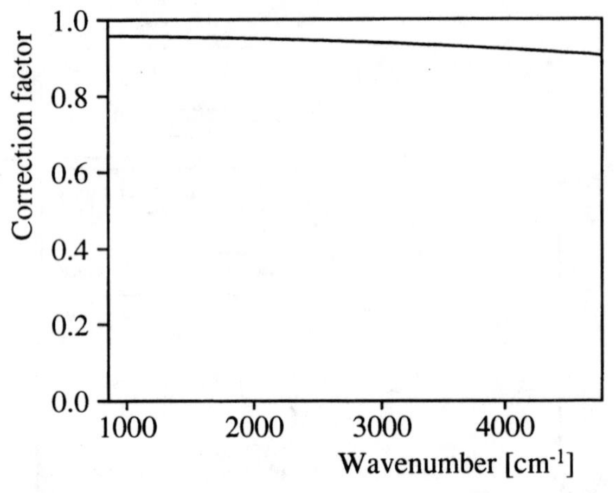

FIGURE 2. Scattering correction factor used for the fit shown in fig.3. The parameters c1 and c2 are 0.96 and 20000 cm-1, respectively.

The algorithm to calculate reflectance and transmittance spectra used in this work is based on the one published in [3]. In contrast to the conventional matrix formalism it allows to switch off in the simulation narrow interference patterns of thick layers (typically substrate layers). In many cases these are not resolved experimentally and hence must also be eliminated in the theory in order to adjust the model spectra to measured data.

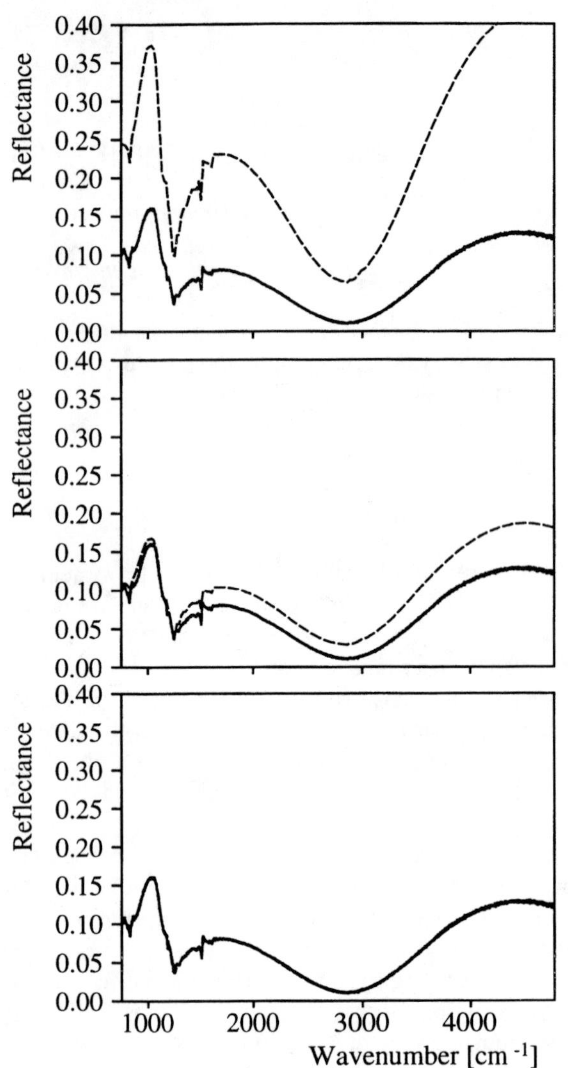

FIGURE 3. Measured (solid line) and simulated (dashed) reflectance spectrum of a SiO_x double layer system on a curved epoxy substrate.
Top: no correction
Center: curvature correction
Bottom: curvature and roughness correction

CURVATURE AND ROUGHNESS CORRECTIONS

The method for the computation of wave propagation was developed for stacks of parallel interfaces and does not include any roughness corrections. For the present case of curved substrates (radius of curvature much larger than the light wavelengths) the reflectance spectra must be corrected in order to account for the different directions into which radiation is directed in the cases of the curved samples and the flat reference mirror. It turns out that a global factor (independent of frequency) should be used which varies with angle of incidence, beam divergence and the size of the illuminated sample spot.

Surface roughness effects depend on the ratio of the typical 'roughness dimension' and light wavelength and hence require a frequency-dependent correction. We introduce losses due to light scattering at rough interfaces by multiplying the complex reflection and transmission Fresnel coefficients of that interface by real correction factors $c(\tilde{\nu})$ according to

$$c(\tilde{\nu}) = c_1 \cdot \exp\left(-\left(\frac{\tilde{\nu}}{c_2}\right)^2\right) .$$

The parameters c_1 and c_2 are adjusted by comparing spectra deposited on flat and curved substrates under the same sputtering conditions. The correction factors are close to 1.0 for low frequencies (no scattering losses) and decrease with increasing frequency (see fig.2).

In order to keep the number of parameters low we assume that the roughness is the same for all interfaces for which consequently the same correction is applied.

As an example a typical fit result for a double layer is shown in fig.3 (bottom graph) which also demonstrates the influence of the spectrum corrections discussed above (top and middle).

RESULTS

The method described above was used to analyze SiO_x single and multilayers on curved and rough epoxy substrates. In this work results are given for double layer samples only. The two layer thicknesses and the dielectric function parameters (with information on the oxygen content) of both layers are obtained from the fit simultaneously.

In fig.4 we compare the total thickness of some double layers obtained from the optical analysis just described to the thickness given by a mechanical profiler. The profiler measurements cannot be done on the curved and rough epoxy substrates but were performed on small pieces of silicon wafers that were placed on the epoxy substrates during layer deposition. The thicknesses that were determined optically turn out to be larger by about 20 nm. This is in agreement with results from SEM pictures which show less dense SiO_x structures on epoxy substrates than on silicon wafers. The good overall agreement gives rise to trust film thickness values from simulation analysis. Note that in contrast to the mechanical profiling the optical method can be done directly with the object of interest. It gives not only the total double layer thickness, but also thickness values of the individual layers.

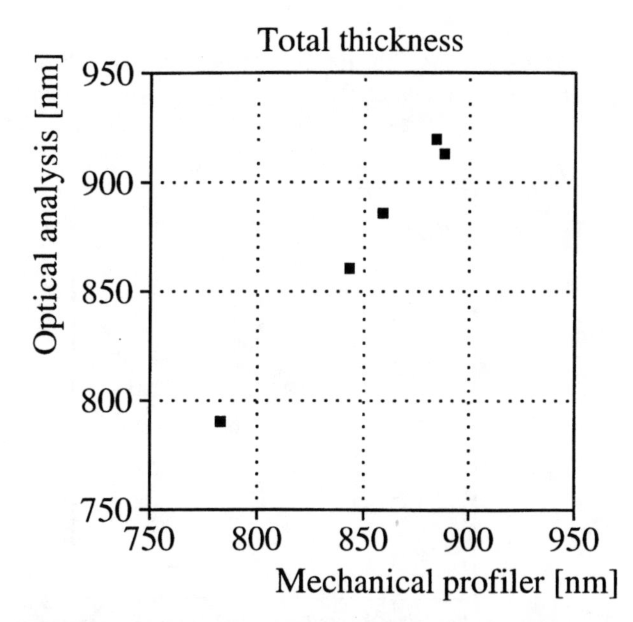

FIGURE 4. Comparison of the total thickness of various double layers obtained by a mechanical profiler and by the optical analysis described in this work. Note that the layers for the profiler analysis were deposited on silicon wafers whereas the optical analysis was done on epoxy samples prepared in the same deposition run.

The fitted parameters of the dielectric function models for both SiO_x layers carry information on their chemical composition. In particular, the oxygen concentration (or the fraction of oxygen atoms) can be determined.

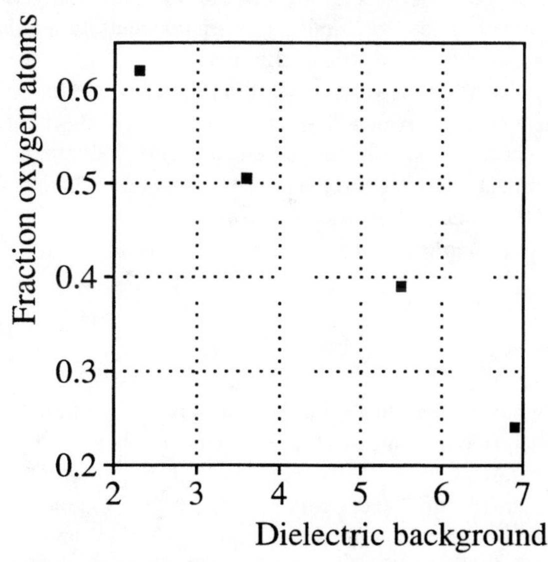

FIGURE 5. Relation of the oxygen content and the dielectric background of the top layer of a double layer system on epoxy.

As can be seen from fig.1, the large wavenumber limit of the real part of the dielectric function (the dielectric background ε_∞) is a parameter with a clear relation to the oxygen content. Fig. 5 shows a monotonic, almost linear dependence of the fraction of oxygen atoms on the dielectric background of the top layer of a SiO_x double layer system.

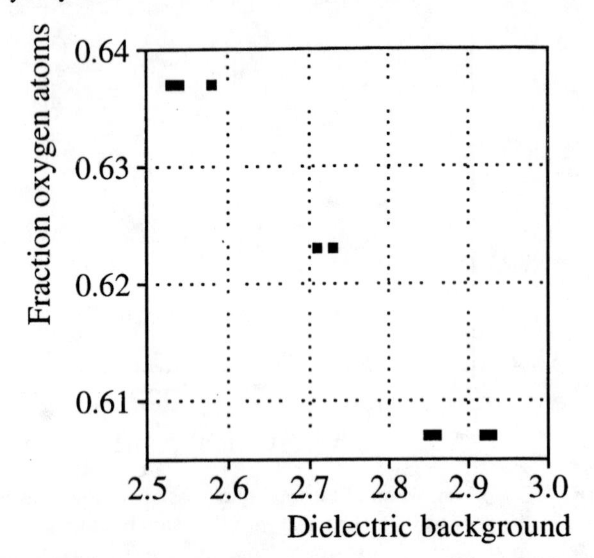

FIGURE 6. Detection of small concentration differences by the analysis of the dielectric background parameter of the top layer of a double layer system.

As already mentioned before, another parameter to determine the oxygen content is the resonance frequency of the main Si-O-Si vibrational mode that shifts from about 970 cm^{-1} for low oxygen content to about 1030 cm^{-1} for large oxygen fractions. It was checked that the analysis of this parameter leads to similar oxygen concentration values as the use of the dielectric background.

Finally fig.6 shows that the optical method can discriminate between rather small oxygen content differences. It should be noted that the agreement of experiment and simulation must be very good to distinguish clearly between dielectric background values of 2.55 and 2.7, for example.

CONCLUSIONS

We have shown that important information on individual layers inside a stack of thin films can be obtained by the analytical method described here. The possibility to consider rough and curved substrates extends the simulation approach to solve many industrial analytical problems, in particular since the method is applicable in cases of absorbing (i.e. non-transparent) substrates.

REFERENCES

[1] P.Grosse, Proc. SPIE Vol. **1575** (1991), 169

[2] R.Brendel, D.Bormann, J. Appl. Phys. **71** (1992), 1

[3] B.Harbecke, Appl. Phys. B **39** (1986), 165

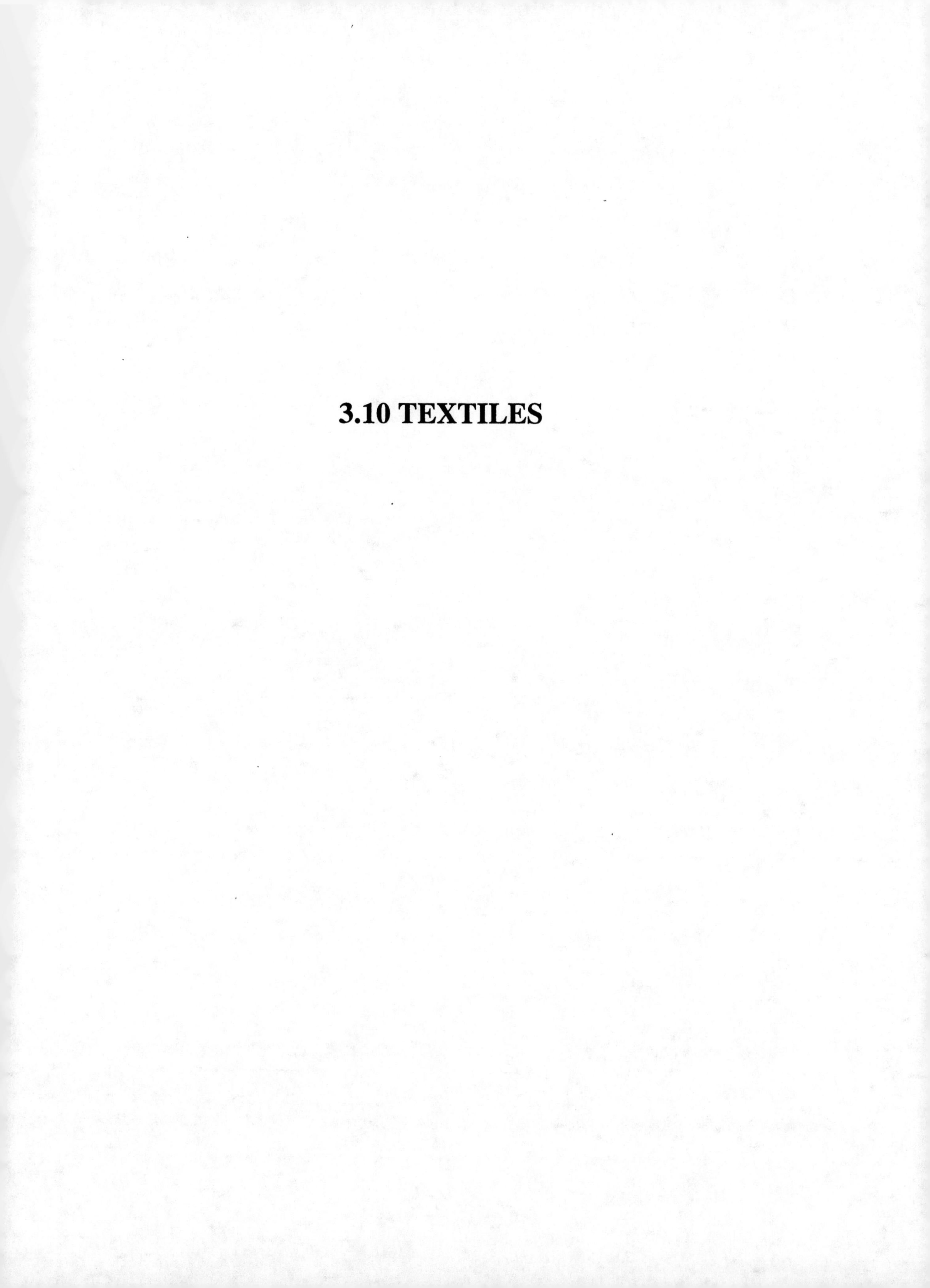

3.10 TEXTILES

Spectroscopic Characterization of Enzymatic Flax Retting: Factor Analysis of FT-IR and FT-Raman Data

D. D. Archibald, [a]G. Henrikssen, D. E. Akin and F. E. Barton

Quality Assessment Research Unit, USDA-ARS-Russell Research Center, Athens, Georgia 30604
and
[a]*Royal Institute of Technology, PMT/Div. of Wood Chemistry, Stockholm, Sweden*

Flax retting is a chemical, microbial or enzymatic process which releases the bast fibers from the stem matrix so they can be suitable for mechanical processing before spinning into linen yarn. This study aims to determine the vibrational spectral features and sampling methods which can be used to evaluate the retting process. Flax stems were retted on a small scale using an enzyme mixture known to yield good retted flax. Processed stems were harvested at various time points in the process and the retting was evaluated by conventional methods including weight loss, color difference and Fried's test, a visual ranking of how the stems disintegrate in hot water. Spectroscopic measurements were performed on either whole stems or powders of the fibers that were mechanically extracted from the stems. Selected regions of spectra were baseline and amplitude corrected using a variant of the multiplicative signal correction method. Principal component regression and partial least-squares regression with full cross-validation were used to determine the spectral features and rate of spectral transformation by regressing the spectra against the retting time in hours. FT-Raman of fiber powders and FT-IR reflectance of whole stems were the simplest and most precise methods for monitoring the retting transformation. Raman tracks the retting by measuring the decrease in aromatic signal and subtle changes in the C-H stretching vibrations. The IR method uses complex spectral features in the fingerprint and carbonyl region, many of which are due to polysaccharide components. Both spectral techniques monitor the retting process with greater precision than the reference method.

INTRODUCTION

Flax stems must undergo microbial, chemical or enzymatic 'retting' in order to release the bast fibers from the stem matrix so the fibers can be mechanically separated and spun to make yarn for linen. The chemical and physical basis of the transformation is not fully characterized, but retting is thought to selectively remove the pectins and hemicellulose which bind the fiber bundles together. Flax stems are composed of three main components: a waxy cuticle (23 % g/g), the cellulosic bast fibers (27 %, g/g), and a lignified shive core (50 % g/g). The quality of linen is highly dependent on the success of the retting process. Under-retted flax fiber processes poorly and over-retted flax fiber is too weak. The retting endpoint is difficult to measure and the conventional methods often require some form of expert judgment of color, texture and the mechanical properties of the stems through visual and hand testing (1). Fried's test is one of the more objective methods for measuring the degree of retting. It involves visual scoring of the degree of fiber release due to mechanical agitation in hot water.

This study examines whether FT-IR and FT-Raman spectroscopy can track the structural transformations of the retting process. An enzymatic retting method is used for the study because it is well characterized and also is a good candidate for commercialization. The objectives are to understand the spectral, chemical and structural changes which occur during the flax retting process; and to determine the best sampling approach for developing a spectral method to measure the optimal endpoint of flax retting. A spectral method for monitoring retting endpoint must measure the product internal structure while ignoring color, which varies substantially with cultivar, growth and harvest conditions, and retting methods. Furthermore, a spectral method must have a simple sampling procedure. Thus we have examined several ways to present the sample to the spectrometer. The most elaborate involves grinding the bast fibers after removing them from the stem, while the simplest involves mounting the desiccated stems in a round clamp.

MATERIALS AND METHODS

Flax cultivar 'Ariane' grown in The Netherlands in 1994 was supplied by Van de Bilt zaden en vlas, b.v., Sluiskil, Holland. The middle 12 cm segments of the stems were used for the study. Enzymatic retting was accomplished in test tubes using the Flaxzyme™ enzyme mixture (Novo, Inc., Chapel Hill, NC) in buffer with chelator and incubation at 50°C with slow agitation (2). A specimen was collected every hour for the first 23 hours and an additional sample was obtained at 37 hours. As 8 hours is recommended, this procedure yields under-retted and over-retted samples. The

CP430, *Fourier Transform Spectroscopy:* 11th International Conference
edited by J.A. de Haseth

reaction was stopped by rinsing specimens with water and drying samples in air at room temperature. Recovery of stems was quantitative so that weight loss could be determined. The Fried's test preparations were scored as coded samples so that the scorer would not be biased. Samples were coded and subsequent analyses were done in a non-sequential order.

Two kinds of samples were prepared for further analysis. Fiber powders were prepared by removing and discarding shive from the desiccated stems, chopping the fibers to less than 1 mm, and grinding in a Wig-L-Bug™ ball mill. Whole stem samples were prepared by securing flattened stems in a clamp composed of two 2" diameter cylindrical rings.

FT-IR spectra of KBr pellets of flax fiber powder were collected with a model 740 Nicolet (Madison, WI) using an MCT detector and a resolution of 4 cm^{-1}. Water and carbon dioxide vapor signals were subtracted from these spectra. FT-IR reflectance spectra of the flattened stem were collected at 4 cm^{-1} resolution using a model 850 Nicolet and a Gemini™ diffuse reflectance accessory (Spectra-Tech, Inc., Shelton, CT). To use this apparatus, the stems were trimmed to size and taped to span across the sample cup position. A DTGS detector was used and the mirror velocity was adjusted to maximize the dynamic range so that peaks did not saturate. Five replicates were averaged.

FT-Raman spectra at 8 cm^{-1} resolution were collected with a Nicolet model Raman 950, with a excitation at 1064 nm and using a Ge detector. The ellipsoidal mirror for a 180 degree collection geometry was employed. Powders were scanned in NMR tubes and the round cells for whole stems were scanned while under slow rotation.

Desired regions of spectra in each set were matched to one another using slope and offset correction and application of multiplicative signal correction (3) to adjacent portions of the spectra showing little variance in spectral features. This was done using subroutines written in the MATLAB™ programming environment (The MathWorks, Inc., Natick, MD). Principal components regression (PCR) or partial least squares regression (PLS) multivariate analyses of these corrected spectra were performed in Unscrambler® 6.1 (Camo, AS, Trondheim, Norway). By correlating to retting time and selecting the early factors and their loadings, it is possible to gauge the extent of the retting transformation. Moreover, one can identify subtle changes in spectral features which are correlated to retting time. All models were calculated by full leave-one-out cross-validation. Spectra were occasionally omitted from the analyses when they were extreme outliers to the multivariate models.

Color analysis of the flattened stem samples was performed using a Minolta Chroma Meter model CR-200. Color values were averages of 4 replicates and were converted to color difference as defined by the L*a*b color system, and using the unretted sample as the reference.

RESULTS AND DISCUSSION

(A)

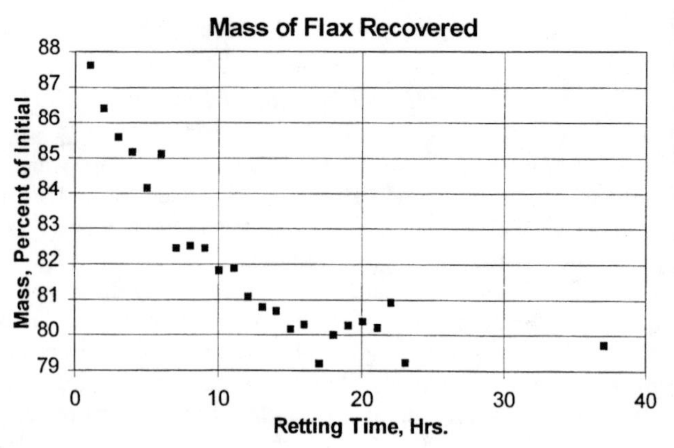

(B)

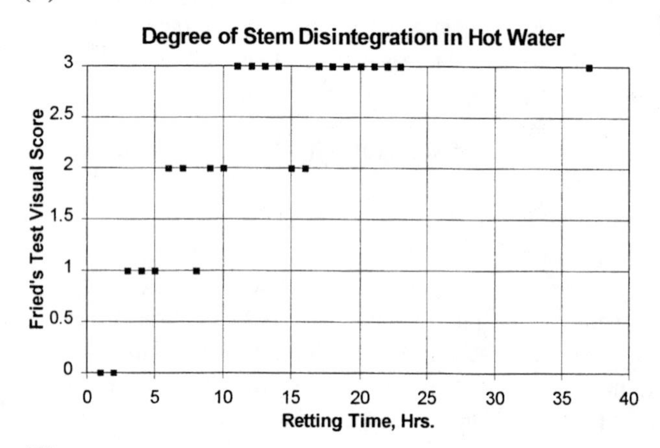

(C)

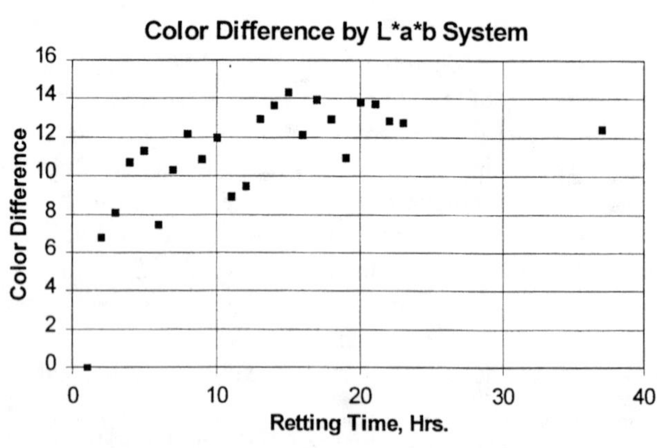

FIGURE 1. Three reference tests which were used to quantify the degree to which the specimens were retted, as plotted versus the actual hours of enzymatic retting that were applied. Note that there is substantial uncertainty in the endpoint, which is believed to be about 8 hours.

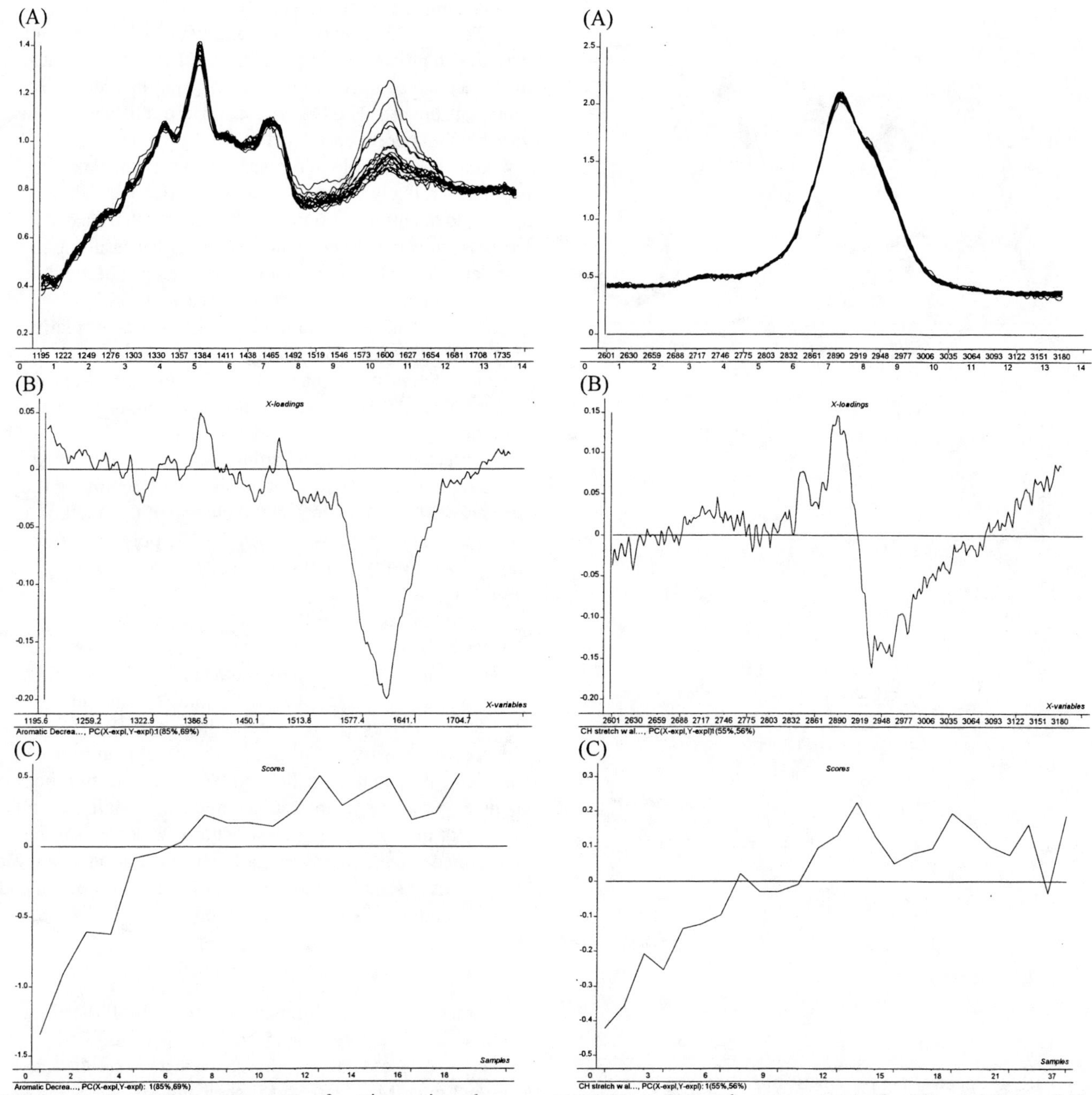

FIGURE 2. Spectral measurement of retting using the middle part of the Raman spectrum of flax fiber powders: (A) corrected spectra for the range 1195-1763 cm^{-1}; (B) variable loading for the first PLS factor versus the spectral range in (A); and (C) PLS sample scores for the first factor versus retting time over the range from 0-18 hours. The decrease in the aromatic band at 1604.4 cm^{-1} indicates that retting is complete after 6 hours.

FIGURE 3. Spectral measurement of retting using the C-H stretching vibrations in the Raman spectrum of flax fiber powders: (A) corrected spectra for the range 2601-3180 cm^{-1}; (B) variable loading for the first PLS factor versus the spectral range in (A); and (C) PLS sample scores for the first factor versus retting time over the range from 0-37 hours. The changes in the bandshape indicate that retting is complete after 7 hours.

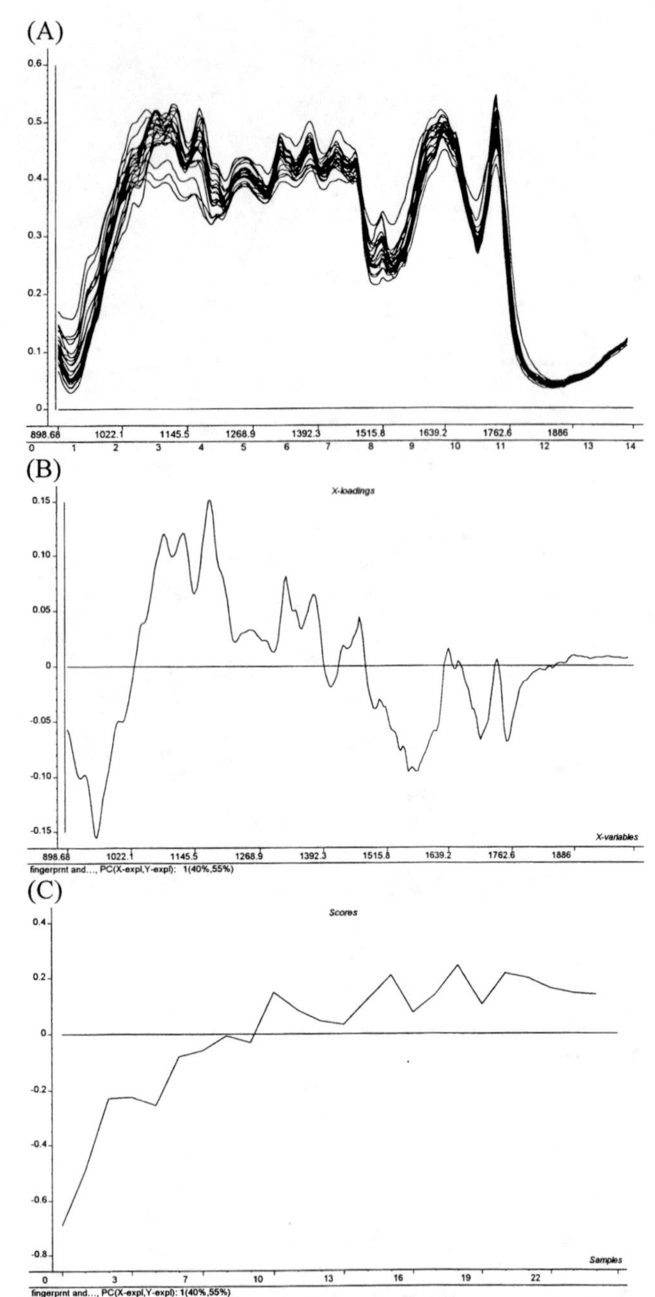

FIGURE 4. Spectral measurement of retting using the IR reflectance of flax stems: (A) corrected spectra for the range 898-2009 cm^{-1}; (B) variable loading for the first PLS factor versus the spectral range in (A); and (C) PLS sample scores for the first factor versus retting time over the range from 0-37 hours. The measurement indicates that the retting is complete after 9 hours. The complex spectral changes include a decrease in the shoulder at 956.5 cm^{-1}, increase in the C-O absorbance at 1168.6 cm^{-1}, and decreases in the carbonyl signals at 1700.9 cm^{-1} and 1754.9 cm^{-1}.

The mass loss of flax stems continues beyond the time when retting is complete (Fig. 1(A)). There is also great uncertainty in the retting endpoint using either Fried's Test or sample color (Fig. 1(B)-(C)).

The carbonyl region of the KBr pellet IR of flax powder was able to monitor retting, but required two factors because of the need to compensate for varying sample hydration (data not shown). The C-H stretch absorptions were highly variable, but poorly correlated to retting (data not shown).

Raman is better able to measure the retting of flax powders. A likely interpretation of the effects in Fig. 2 is decreased quantities of lignin in the better retted flax. Decrease in wax may be an underlying effect resulting in the correlation of retting to the bandshape changes of the C-H region of the flax powder Raman spectrum (Fig. 3).

Raman of flax stems did not have very good correlation to the retting time even though high quality spectra were obtained. Stem spectra had a strong lignin signal from the inner shive and this interfered with the retting measurement (data not shown).

IR reflectance is the most promising method for analysis of the retting of stems without mechanically removing the fibers (Fig. 4). The main spectral change of this method is an increase in the relative intensity of the polysaccharide signal. However wax and lignin signals may also be involved.

Several of the spectral methods track the flax retting process with greater precision than any of the conventional reference methods. Raman of extracted flax fibers and IR reflectance of flax stems are the two simplest and most precise alternatives for spectral monitoring of flax retting. The spectral sensitivities of IR and Raman appear to be complementary in this application. Raman tends to monitor changes in waxes and aromatic components, while IR tends to monitor the polysaccharide structures. It remains to be determined whether these methods can accommodate a more realistic set of samples with variation in cultivar, growth and harvest conditions, and retting methods.

ACKNOWLEDGMENTS

We thank Dr. Dave Himmelsbach for helpful discussions. We also thank the following students who assisted in the collection of the data: Ms Sonja Barber, Ms Jing Lu, Ms Audrey Loper, Ms Meredith Mitchell and Mr. Farris Poole.

REFERENCES

1. Sharma, H.S.S. and Van Sumere, C.F., Eds., *The Biology and Processing of Flax*, Belfast, M Publication, 1986.
2. Henriksson, G., Akin, D.E., Rigsby, L.L., Patel, R. and Eriksson, K.-E.L., "The Influence of Chelating Agents and Mechanical Treatment on Enzymatic Retting of Flax", *Textile Research Journal* (in press).
3. Martens, H. and Næs, T., *Multivariate Calibration*, New York, John Wiley, 1989.
4. Himmelsbach, D.S. and Akin, D.E., "NIR-FT-Raman Spectroscopy of Flax (*Linum usitatissimum* L.) Stems" (manuscript in journal review).

Synchrotron Powered FT-IR Microspectroscopy Permits Small Spot ATR Sampling of Fiber Finish and Other Materials

David L. Wetzel[1], John A. Reffner[2], G. Lawrence Carr[3], and Liling Cho[1]

[1]*Microbeam Molecular Spectroscopy Laboratory, Kansas State University, Shellenberger Hall, Manhattan, KS, 66506, USA*
[2]*Spectra-Tech Inc., 2 Research Drive, Shelton, CT, 06484, USA*
[3]*National Synchrotron Light Source, Beamline U2B, Bldg. 725, Brookhaven National Laboratory, Upton, NY 11973, USA*

The absence of thermal noise, the nondivergence, and brightness of synchrotron radiation for FT-IR microspectroscopy permit aperturing the beam entering the ATR objective to less than the 100 μm diameter pattern usually used. With a ZnSe crystal, the 100 μm aperture produces a 42 μm spot of optical contact with the specimen. With synchrotron radiation, excellent S/N with smaller apertures permits spatially resolved probing of adjacent small spots along a specimen to reveal heterogeneity in the surface chemistry.

INTRODUCTION

Surface preferenced sampling attainable with attenuated total reflectance (ATR) is a useful technique in FT-IR microspectroscopy. The availability of a specialized ATR objective provides the opportunity for routine sampling of a 42 μm spot of optical contact with the specimen when using an appropriate remote projected aperture that allows a light beam of a 100 μm diameter pattern to enter the ATR objective. Using an ATR-equipped FT-IR microspectrometer with a conventional globar (thermal) source, the spatial resolution obtainable is limited by the signal-to-noise that the optical system provides. The object of this work was to improve on the spatial resolution by using the brightness, nondivergence of the synchrotron radiation and the absence of thermal noise.

PREVIOUS ATR WORK WITH SINGLE FIBERS

FT-IR microspectroscopy of single fibers with a specialized ATR objective has been used to find the finish on the surface of a fiber prepared in manufacturing or a finish deliberately applied for end use such as antistatic or antistain. With this technique, a small amount of finish that would not contribute much to the transmission spectrum of the whole fiber shows up better with the surface preferenced sampling of ATR. Subtracting the spectrum of the unfinished fiber from that of a fiber with finish further reveals the spectral features of the finish. This type of single fiber ATR is done routinely in the Microbeam Molecular Spectrosocpy Laboratory at Kansas State University (1).

ADVANTAGE OF SYNCHROTRON SOURCE FOR FT-IR

One advantage that synchrotron radiation has over a thermal source is that it has no thermal noise. Also, its brightness is reportedly 1000 times greater than that of a globar (2). Figure 1, in which the illumination intensity is compared to that of a globar and to the detector noise anticipated from a liquid nitrogen cooled mercury cadmium telluride detector, makes it clear that the signal-to-noise advantage is substantial (3). Perhaps more important but less obvious is that synchrotron light is highly forward directed and concentrated within a narrow angle. Because of this characteristic, the projected aperture used in the IR microscope to limit the part of the field being interrogated

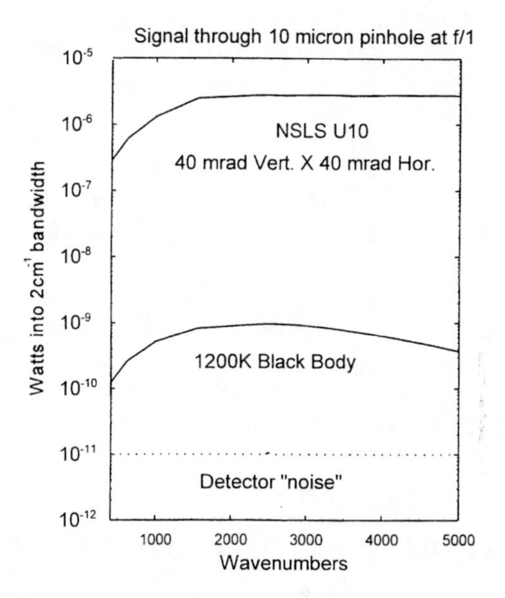

Figure 1. Illumination intensity of synchrotron vs. thermal source (from Ref. 3 with permission)

CP430, *Fourier Transform Spectroscopy:* 11th International Conference
edited by J.A. de Haseth

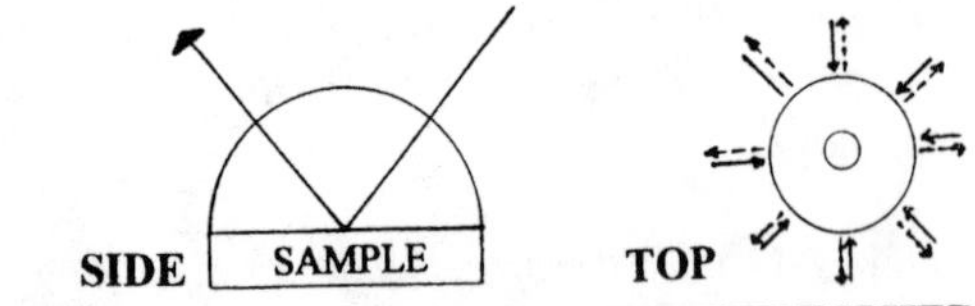

Figure 2. Diagrams for macro and micro ATR experiments

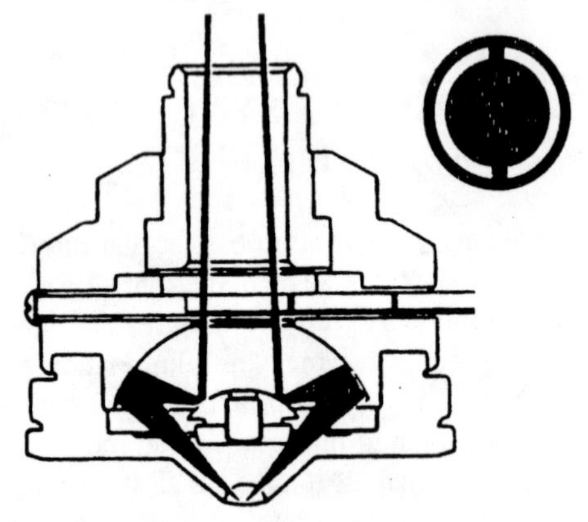

Figure 3. Path of radiation for ATR objective through mask pattern shown (top right)

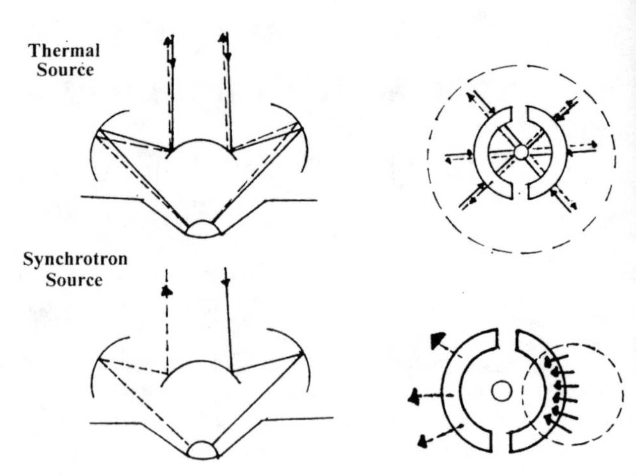

Figure 4. Sketch of illumination of ATR mask for thermal vs. synchrotron source and resulting pathways (left)

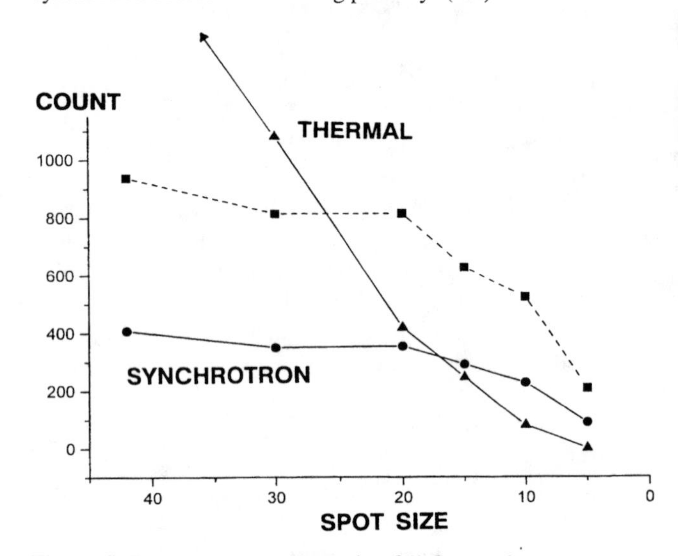

Figure 5. Detector counts (at a gain of 1) for synchrotron vs thermal sources at various ATR spot sizes

does not attenuate the beam as severely as would be the case with a thermal source. Spatial resolution is dependent on high signal-to-noise illumination of a small spot with minimal loss due to aperturing. For illuminating a larger area, there may be no advantage. For the ATR microscope objective when using a large aperture, filling the openings in the mask is more important. When using a small aperture, a small, bright spot works when a broader source will not.

Unlike macro ATR shown at the top of Fig. 2, the mirror lens ATR objective (bottom Fig. 2) uses a single bounce from all directions The Spectra-Tech ATR objective path is shown in Fig. 3 (4). The path results from the mask at the right through which rays enter and exit. With large aperture illumination, all sides of the mask are filled with incident radiation as seen in Fig. 4. A small aperture of nondivergent synchrotron radiation may preferentially illuminate one side, resulting in one-direction operation.

EXPERIMENTATION AND RESULTS

Synchrotron radiation from the vacuum ultraviolet ring of the National Synchrotron Light Source was brought into the optical system of an IRμs microspectrometer (Spectra-Tech, Shelton, CT) as a substitute for the conventional globar (thermal) source. Radiation from a beamline port U2B of the vacuum ultraviolet storage ring impinged upon a water-cooled copper mirror which absorbs the soft x-rays and the vacuum ultraviolet radiation. This interface system was previously described as was the mirror system leading to the 14 m evacuated stainless steel tube and steering mirrors at the other end of the tube to take the radiation to the interferometer of the IRμs instrument (5). The brightness of the synchrotron and its nondivergent character were used to concentrate energy on a small projected aperture directed through the ATR objective. Apertures smaller than the conventional circular 100 micron aperture were used to obtain ATR spectra. The refractive index of the zinc selenide crystal in the ATR objective was 2.4, and this number divided into the size of the projected aperture determines the spot size of contact for the ATR microspectroscopy. For the preliminary attempt to enhance localization of ATR microspectroscopy reported here, the beam line was not adjusted specifically for the ATR

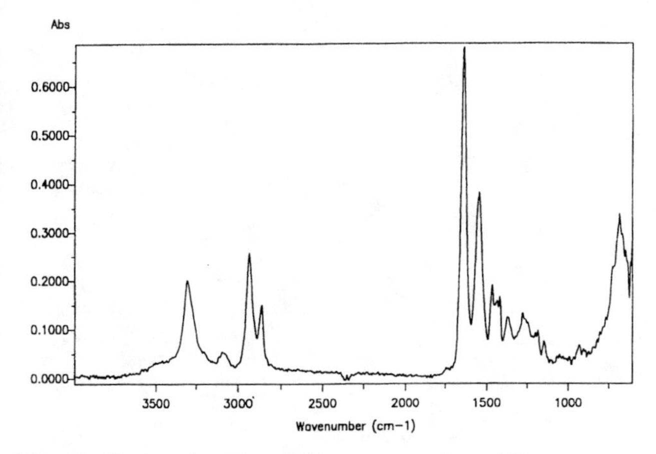

Fiber 6. Single nylon fiber ATR spectrum using a 100 μm aperture and synchrtron source

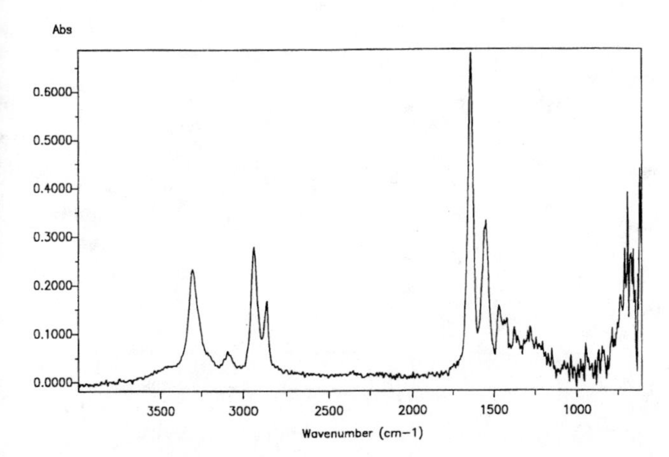

Figure 7a. Single nylon fiber ATR spectrum using a 24 μm aperture and synchrotron source

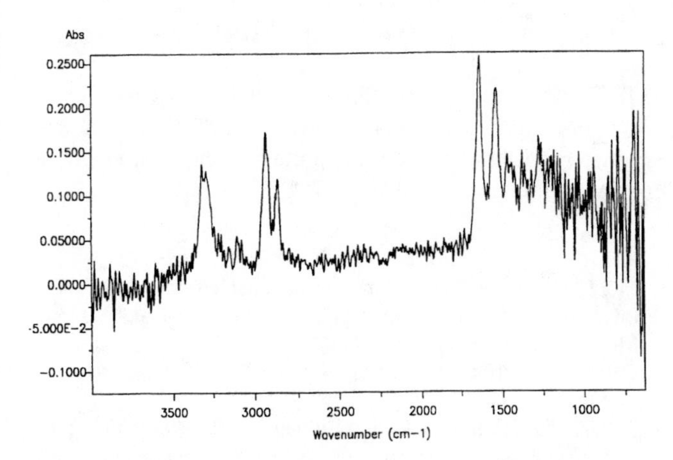

Figure 7b. Single nylon fiber ATR spectrum using a 24 μm aperture and thermal source

but had been adjusted for the microspectrometer equipped with a conventional 15x cassagrainian objective. The illumination pattern striking the ATR objective with the mask in place was not necessarily optimized for that purpose. Nevertheless, an advantage of the synchrotron is evident from

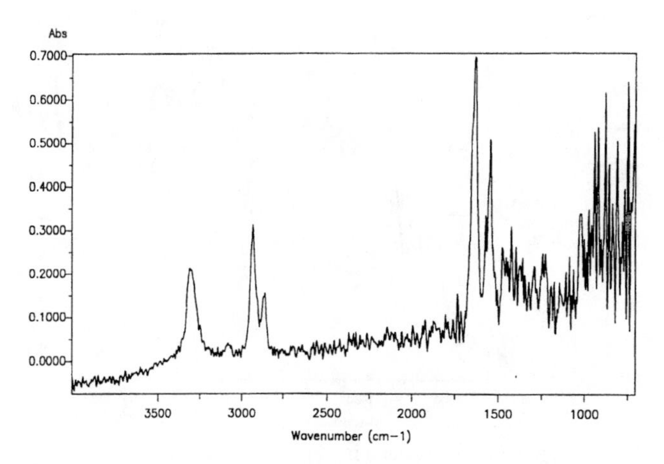

Figure 8. Single nylon fiber ATR spectrum using a 12 μm aperture and synchrotron source

looking at the graph in Fig. 5. Detector counts at a gain of 1 were plotted against spot size of contact.

Using a ZnSe hemisphere-equipped ATR objective, apertures of 100, 72, 48, 36, 24, and 12 μm were used to produce the 42, 30, 20, 15, 10 and 5 μm spot sizes reported. Note that at higher apertures, the synchrotron offers no advantage. The broad beam of the thermal source fully illuminated the crystal from all sides as a cumulative effect of the large spot size, and its actual condition of being spread out enhanced the total input of energy onto the specimen. At apertures below 36 μm, however, the syncrhotron offers a significant advantage because aperturing does not attenuate the nondivergent synchrotron beam as much as that of the globar, as evidenced by the increased count. These data were reported when the synchrotron was in the last half of a fill. At the beginning of the next storage ring fill, the count of the 10 μm spot was repeated, and the dotted line shows the counts that would accompany a higher flux.

DISCUSSION AND CONCLUSIONS

Figure 6 shows the single nylon fiber ATR spectrum obtained at the 100 μm aperture. The spectrum in Fig. 7a shows the benefit of the synchrotron for a 10 μm spot of contact compared to the noisy spectrum (Fig. 7b) for the globar. Figure 8 demonstrates that it is possible to observe spectral features at the lowest value where a 12 μm aperture is used to sample a 5 μm spot.

For routine synchrotron ATR use, we chose the 24 μm aperture producing a 10 μm contact spot with the ZnSe crystal. This ATR configuration was used to find the variation of ink quantity coated on different but closely spaced spots on the surface of a polyester fiber (Fig. 9). In this case, a strong carbonyl band of the fiber substrate was blocked out by the ink as observed in Fig. 10. Antistatic finish on the surface of a commercial nylon carpet fiber also was detected by the presence of carbonyl in the ATR spectrum. The spectra of adjacent fibers in U.S. currency clearly show the spectrum of

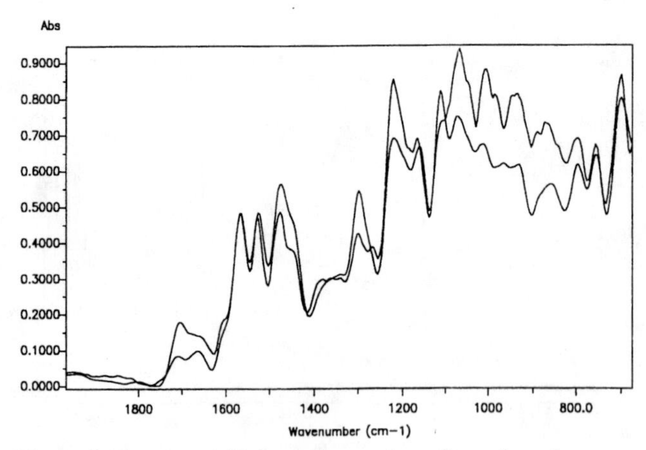

Figure 9. Two spots of ink coating on the surface of a polyester fiber showing variation in quantity of ink

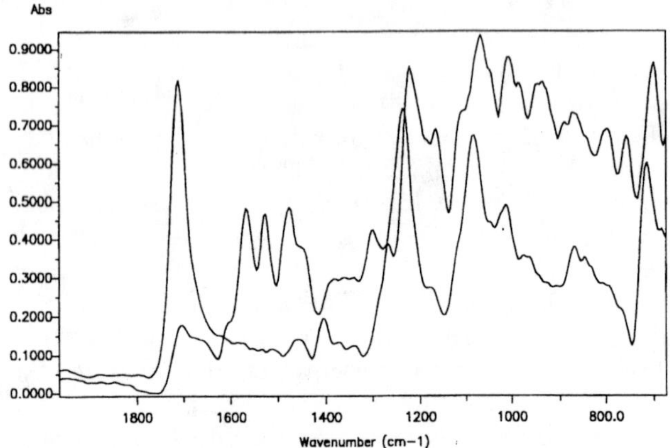

Figure 10. Polyester ATR spectra with (low carbonyl) and without ink

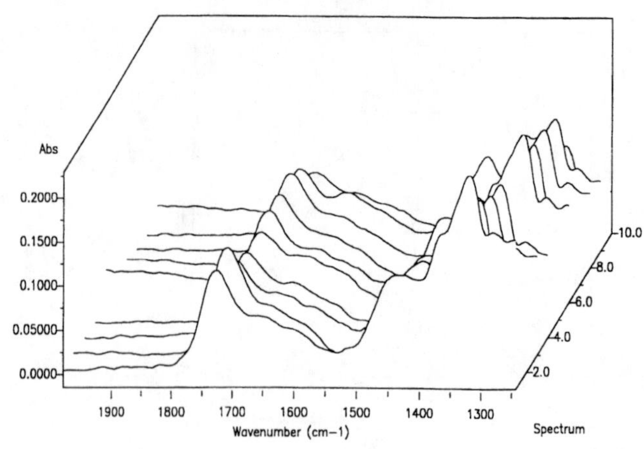

Figure 11. Variation in 1740 cm^{-1} in sequence of adjacent spots 15 μm apart along a single cotton fiber with permanent press finish involving the citrate group

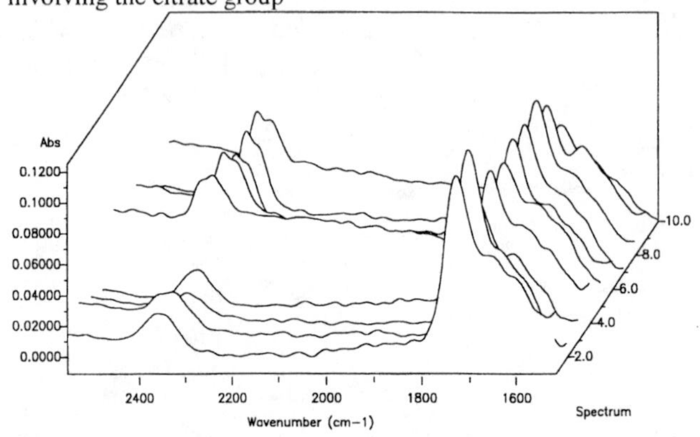

Figure 12. Variation in the residue from the reaction used to bond the permanent press finish to the cotton fiber

nylon for the "marker" fiber in contrast to the cellulose spectrum of fibers in the paper itself. Ten micron spot ATR spectra were obtained at 15 μm intervals along a single cotton fiber. Permanent press finish containing an ester group had been applied to the fiber. The series of ATR spectra in Fig. 11 shows variation in the quantity of permanent press finish by examination of the carbonyl bands at 1740 cm^{-1}. Points 1 and 2 are abnormally high and points 3 and 4 are abnormally low in comparison to the rest.

Surface heterogeneity of the chemically bonded finish (citrate) as observed by the carbonyl absorption band variation at 15 μm intervals along the fiber in Fig. 11 is not the only value of spatially resolved ATR microspectroscopy. Residue (phosphate) from the bonding reaction appears to vary also as evidenced by the absorption intensity in the region of 2400 cm^{-1} (Fig. 12).

Spatially resolved single fiber ATR data of the type presented here demonstrate the potential of localization of surface heterogeneity through the use of synchrotron powered ATR microspectroscopy. There is potential room for improvement of intensity at the smaller spot sizes reported by optimization of the incoming beam for the ATR objective with the appropriate mask in place. Conventional optimization is done with respect to the reflectance from a gold mirror with the 15x objective.

ACKNOWLEDGEMENTS

This work was supported in part by the National Science Foundation, EPSCoR Grant no. OSR 9255223. The National Synchrotron Light Source is supported by the U.S. Dept. of Energy, Contract no. DE-AC02-76CH00016.

REFERENCES

1. Cho, L. And Wetzel, D.L., "Single fiber characterization by ATR infrared microspectroscopy, presented at the FACSS 23rd Meeting, Kansas City, MO, Sept. 29-Oct 4, 1996.
2. Carr, G.L., Reffner, J.A., and Williams, G.P, *Rev. Sci. Intr.*, 66, **1490**, 1995.
3. Wetzel, D.L., Reffner, J.A., and Williams, G.P., *Mikrochim. Acta [Suppl.]* **14**, 353-355, 1997.
4. Bartick, E.G., M.W. Tungol, and Reffner, J.A., *Anal. Chim. Acta*, **288**, 35-42, 1994.
5. Wetzel, D.L., "Microbeam molecular spectroscoy of biological materials", In: Charalambous, G., (ed.), *Food Flavors, Ingredients, and Components*, Elsevier Science, Amsterdam, 1995, 2039-2108.

Contribution no. 98-58-B, Kansas Agricultural Experiment Station, Manhattan.

Applications of FT-IR Spectroscopy to the Studies of Esterification and Crosslinking of Cellulose by Polycarboxylic Acids: Part I. Formation of Cyclic Anhydrides as the Reactive Intermediates

Charles Q. Yang and Xilie Wang

Department of Textiles, Merchandising and Interiors, The University of Georgia, Athens, Georgia 30602, USA

Butanetetracarboxylic acid (BTCA) has been the most effective nonformaldehyde crosslinking agents for cotton cellulose. In this research, we applied FT-IR spectroscopy in combination with thermalgravimetry (TG) and differential scanning calorimetry (DSC) to investigate the formation of carboxylic anhydrides by BTCA in its pure form and as a finish on cotton fabric. The FT-IR spectroscopy and TG/DSC data indicated that BTCA is converted to anhydride when the temperature reaches the vicinity of its melting point. The formation of the anhydride is accelerated above the melting point. The FT-IR spectroscopy data also showed that BTCA forms an anhydride at much lower temperatures when it is applied to cotton fabric as a finish.

INTRODUCTION

Since the identification of formaldehyde as a probable human carcinogen, extensive efforts have been made to find nonformaldehyde durable press finishes to replace the traditional N-methylol reagents (1). Polycarboxylic acids have appeared to be the most promising formaldehyde-free crosslinking agents (1). In 1988, Welch reported that tetracarboxylic acids, with butanetetracarboxylic acid (BTCA) in particular, are able to impart high levels of wrinkle resistance and fabric strength retention to cotton fabrics. BTCA has been proved to be the most efficient crosslinking agent for cellulose.

In our previous research, we used Fourier transform infrared spectroscopy (FT-IR) to investigate the mechanism of esterification of cotton cellulose by a polycarboxylic acid (2-5). We identified the formation of five-member cyclic anhydride intermediates under curing conditions on the cotton fabric treated with various polycarboxylic acid (2,3). We found that a polycarboxylic acid esterifies cotton cellulose through the formation of an 5-membered cyclic anhydride as a reactive intermediate. In this research, we used FT-IR spectroscopy in combination with TG and DSC to study the formation a 5-membered cyclic anhydride intermediate by BTCA both in its pure form and as a finish on cotton fabric.

EXPERIMENTAL

Infrared Spectroscopy, TG and DSC Instrumentation

A Nicolet 510 FT-IR spectrometer with a Specac "selector" diffuse reflectance accessory was used to collect the infrared spectra. All the infrared spectra of cotton fabric samples were collected with the diffuse reflectance accessory and presented at absorbance mode (-log R/R_0). Potassium bromide powder was used as a reference material to produce a background diffuse reflectance spectrum. All the spectra of the BTCA samples, including the residues after the TG isothermal heating, were collected as transmission spectra using a ZnSe window. Resolution for all the infrared spectra was 4 cm^{-1}. No smoothing function or baseline correction was used. The ester carbonyl band intensities in the infrared spectra of the cotton fabric treated with BTCA was normalized against the 1317 cm^{-1} band associated with the C-H bending mode of cellulose.

A Mettler TG50 thermobalance and Mettler DSC20 differential scanning calorimeter with a standard cell were used for the thermal analysis. All the samples were heated from room temperature (25°C) to a specified temperature at a rate of 10°C/min with a continuous nitrogen flow rate of 10 ml/min. The sample size for the DSC and TG experiments was approximately 9 mg.

Fabric Treatment

The cotton fabric was impregnated with a solution containing a polycarboxylic acid. The impregnated fabric was passed through the squeeze rolls of a Cromtax 3-Roll Laboratory Padder to give a wet pickup of 100-110%

CP430, *Fourier Transform Spectroscopy:* 11th International Conference
edited by J.A. de Haseth

based on the original weight of the fabric. The treated fabric was dried at 85°C for 10 min, then cured in an oven at a specified temperature. The cured fabric was also washed in tap water at room temperature for 10 min to remove the unreacted acid, dried at 85°C for 10 min. Because the carboxyl carbonyl band overlaps the ester carbonyl band, the cotton fabric was treated in a 0.1 M NaOH solution for 2 min at room temperature to convert the free carboxyl in the fabric to carboxylate, dried at 85°C for 10 min, and finally ground in a Wiley mill to form a powder before FT-IR analysis.

RESULTS AND DISCUSSION

The DSC and TG curves of the BTCA powder is presented in Figures 1 and 2, respectively. The data indicated that no heat absorption and weight loss was observed when the BTCA powder was heated from room temperature to 180°C. Above 180°C, BTCA started to absorb heat and to lose weight simultaneously as shown in Figures 1 and 2, respectively. The heat absorption reached its maximum around 205°C, whereas the differential TG curve (dM/dT) showed a peak at 196°C, indicating the maximum rate of weight loss at that temperature. The BTCA sample lost approximately 12% of its weight when the temperature reached 250°C (Figure 2).

The BTCA powder was also heated in the thermal balance with a 2 min isothermal heating at different final temperatures. The TG curves of BTCA with 2-min isothermal heating at 160, 180, and 200°C are presented in Figure 3. Little weight loss is observed with the isothermal heating of BTCA at 160°C and 180°C. With the isothermal heating at 200°C, however, BTCA lost approximately 2% of its weight as demonstrated in Figure 3.

The melting point of BTCA is 196°C. The heat absorption observed in the DSC curve of BTCA in the 180-220°C region (Figure 1) is certainly not due to melting because of the weight loss observed in the TG curves in the same temperature region (Figures 2 and 3). To determine the cause of the heat absorption and weight loss with BTCA in this temperature region indicated by the thermal analysis data, we applied FT-IR spectroscopy to analyze the residue of the BTCA sample after the TG isothermal heating described above (Figure 4).

The infrared spectra of BTCA before heat treatment is presented in Figure 4A. No changes are observed in the spectrum of the BTCA after the 2-min isothermal heating at 160 °C (Figure 4B). Weak shoulders around 1800 cm^{-1} emerge in the spectrum of the BTCA after the 2-min isothermal heating at 180°C (Figure 4C). Those observations are consistent with the thermal analysis data

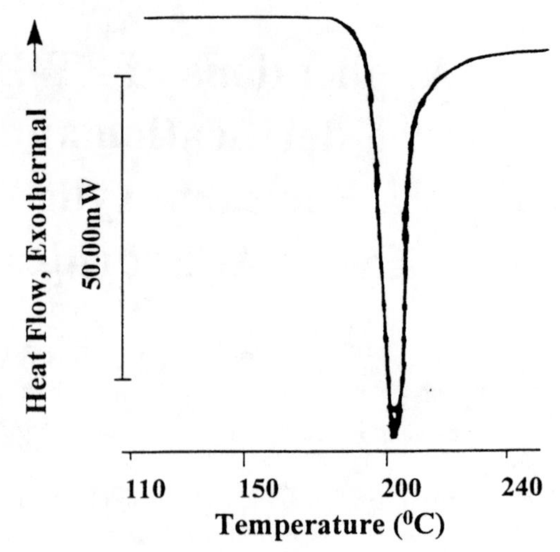

Figure 1. DSC curve of BTCA powder

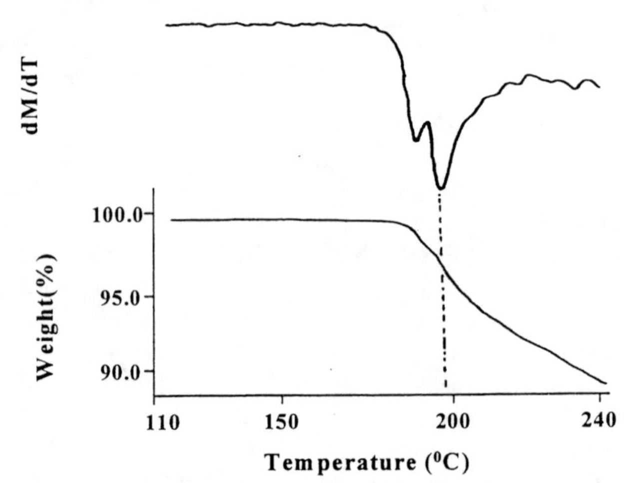

Figure 2. TG and DTG curves of BTCA powder

discussed above. Two distinct bands at 1853 and 1781 cm^{-1} appear in the spectrum after BTCA was exposed to the 2-min isothermal heating at 190°C (Figure 4D). Those two bands become more intense with the isothermal heating at 200°C (Figure 4E).

The two bands at 1853 and 1781 cm^{-1} observed in Figures 4D and 4E are due to the symmetric and asymmetric stretching modes of an anhydride carbonyl. The symmetric stretching band at the high frequency is stronger than the asymmetric stretching band at the low frequency for a noncyclic anhydride carbonyl, whereas the opposite is true for a cyclic anhydride (6). Therefore, we can conclude that the anhydride detected by FT-IR spectroscopy is a 5-member cyclic one formed by the dehydration of two adjacent carboxyl groups in a BTCA molecule. The FT-IR spectroscopy and TG/DSC data indicated that BTCA is converted to anhydride when the temperature reaches the vicinity of its melting point.

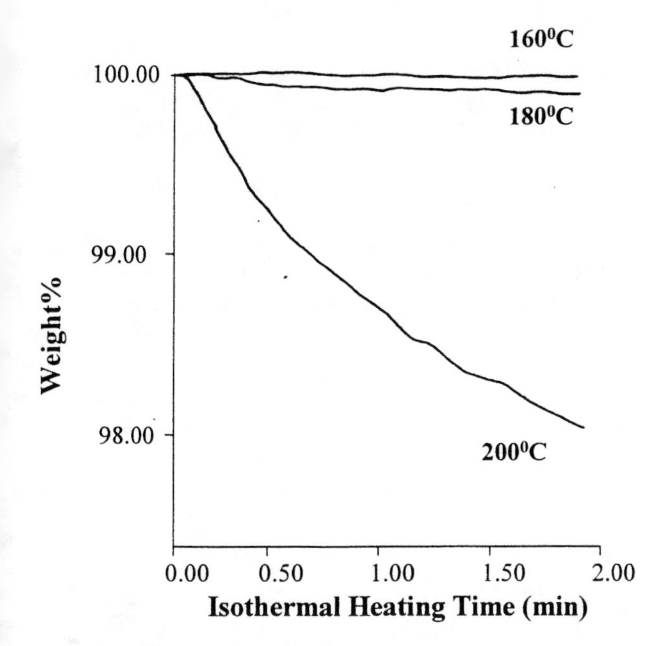

Figure 3. TG curve of BTCA with 2-min isothermal heating at different final temperatures

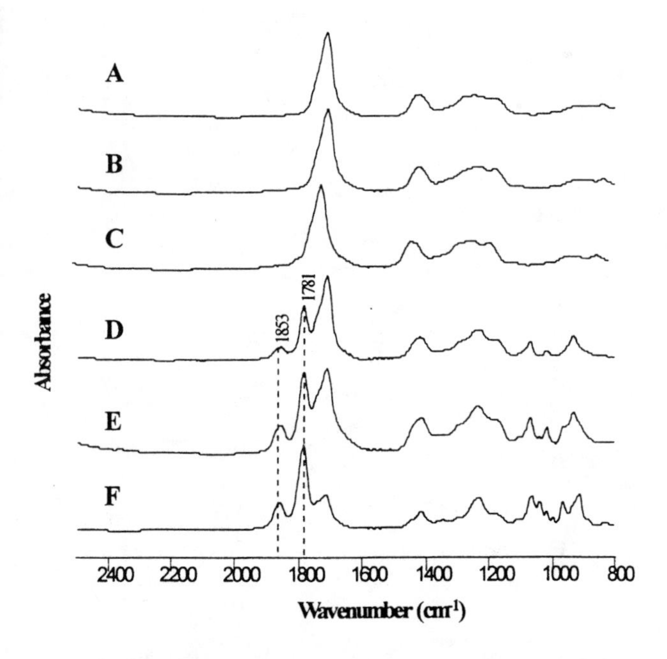

Figure 4. Transmission infrared spectra of (A) BTCA, and (B-F) BTCA with 2-min TG isothermal heating at different temperature (°C): 160, 180, 190, 200, 250.

The two anhydride carbonyl bands become predominant in the spectrum whereas the intensity of the band at 1710 cm^{-1} due to carboxyl is significantly reduced when the BTCA sample was exposed to the 2-min isothermal heating at 250°C (Figure 4F), indicating that majority of carboxyl groups in a BTCA molecule is converted to anhydride. Figure 2 shows that BTCA lost approximately 12% of its weight in the thermal balance when the temperature reached 250°C, which represents the dehydration of approximately 80% of the carboxyl groups in BTCA. Therefore, the FT-IR spectra are consistent with the TG data presented above.

We also studied the formation of 5-membered cyclic anhydride by 1,2,3,4-cyclopentanetetracarboxylic acid and citric acid. All the data showed that the polycarboxylic acids start to form 5-member cyclic anhydride when the temperature reaches the vicinity of their melting points.

We investigated the formation of the cyclic anhydride on cotton fabric under curing conditions. The cotton fabric treated with 12% BTCA is heated in the thermal balance with a isothermal heating at 100°C for 10 min to remove the absorbed moisture, then was continuously heated in the thermal balance at a 10°C/min rate from 100 to 190°C. The TG curve of the treated cotton fabric (100-190°C) is presented in Figure 5. Unlike the TG data of BTCA powder which showed little weight loss under 180°C (Figure 2 and 3), the BTCA-treated cotton fabric gradually lost approximately 1.8% of its weight in the entire temperature region from 100 to 190°C (Figure 5).

The cotton fabric treated with 6% BTCA was exposed to 2-min isothermal heating at different final temperatures. The infrared spectra of the BTCA-treated cotton fabric after the 2-min isothermal heating at different temperatures from 130 to 190 °C are shown in Figure 6. One observes that the two bands due to the 5-membered cyclic anhydride appear in the spectrum when BTCA-treated fabric was exposed to the 2-min isothermal heating at 130°C (Figure 6), indicating the formation of the anhydride on the fabric at that temperature. The intensity of those two bands due to the anhydride increases as the isothermal heating temperature is increased from 130 to 200°C (Figure 6). Both TG data and the FT-IR spectra presented above indicated that BTCA forms anhydride at temperatures much lower than its melting temperature when it is applied to the cotton fabric as finish, which is different from the formation of anhydride by BTCA in its pure form.

CONCLUSIONS

BTCA and other polycarboxylic acids in the form of a powder start to form 5-member cyclic carboxylic anhydrides when the temperature reaches the vicinity of their melting points. When BTCA and other polycarboxylic acids are applied to cotton fabric as finishing agents, they form anhydrides as reactive intermediates on the fabric at much lower temperatures.

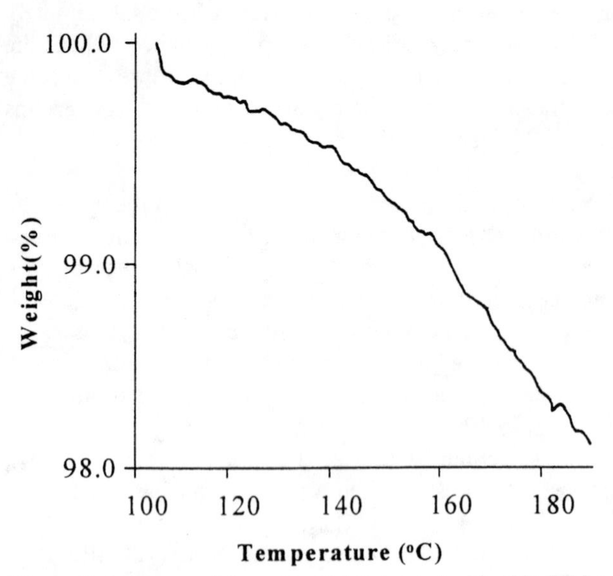

Figure 5. TG curve of the cotton fabric treated with 12% BTCA.

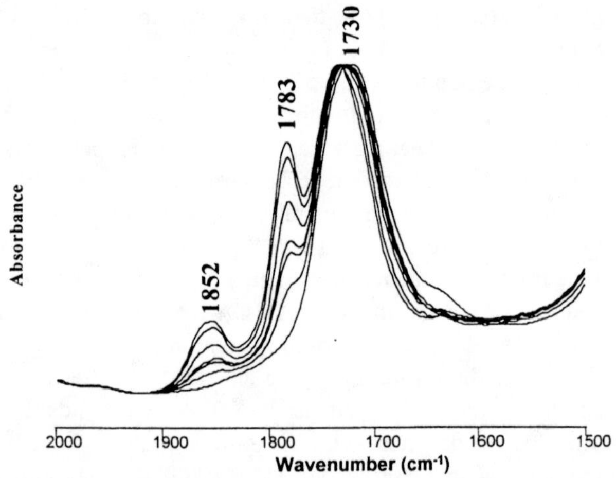

Figure 6. Diffuse reflectance infrared spectra of BTCA-treated cotton fabric with 2-min isothermal heating at different final temperatures (°C): before isothermal heating, 130, 140, 150, 160, 180, and 200 (From bottom to top).

REFERENCES

1. Welch, C. M., *Rev. Prog. Coloration*, **22**, 32-41 (1992).
2. Yang, C. Q., *Textile Res. J.*, **66**, 595-603 (1996).
3. Yang, C. Q., *J. Polym. Sci. Polym. Chem. Ed.*, **31**, 1187 (1993).
4. Yang, C. Q., *J Polym. Sci. Polym. Chem. Ed.*, **34,** 1573-1580 (1996).
5. Yang, C. Q., *J Polym. Sci. Polym. Chem. Ed.*, **35,** 557-564 (1997).
6. Clothup, N. B., Daly, L. H. and Wiberley, S. E., *Introduction to Infrared and Raman Spectroscopy*, 1990, San Diego, CA: Academic Press, Chapter 9, pp310-312.

Applications of FT-IR Spectroscopy to the Studies of Esterification and Crosslinking of Cellulose by Polycarboxylic Acids: Part II. The Performance of the Crosslinked Cotton Fabrics

Weishu Wei and Charles Q. Yang

Department of Textiles, Merchandising and Interiors, The University of Georgia, Athens, Georgia 30602, USA

Durable press finishing processes are commonly used in the textile industry to produce wrinkle-free cotton fabrics and garments. A durable press finishing agent forms covalent bands with cellulosic hydroxyl groups, thus crosslinking the cellulose molecules. The crosslinking of cellulose increases wrinkle resistance of the treated cotton fabric and reduces fabric mechanical strength. Wrinkle recovery angle (WRA) and tensile strength are the two most important parameters used to evaluate the performance of the crosslinked cotton fabrics and garments. In this study, we investigated the correlation between WRA and tensile strength on one hand, and the amount of crosslinkages formed by the crosslinking agents including dimethyloldihydroxylethyleneurea (DMDHEU) and 1,2,3,4-butanetetracarboxylic acid (BTCA) determined by FT-IR spectroscopy on the other hand. Linear regression curves between the carbonyl band absorbance, and WRA and tensile strength of the treated cotton fabric were developed. The data indicated that FT-IR spectroscopy is a reliable technique for predicting the performance of durable press finished cotton fabrics, therefore can be used as a convenient instrumental method for quality control in the textile and garment industry.

INTRODUCTION

Durable press finishing processes are used in the textile industry to produce wrinkle-resistant cotton fabrics and garments. N-methylol reagents, such as dimethylol-dihydroxyethyleneurea (DMDHEU), are the conventional durable press finishing agents used by the textiles industry (1). In recent years, extensive efforts have been made to use polycarboxylic acids, such as 1,2,3,4-butane-tetracarboxylic acid (BTCA), as nonformaldehyde durable press finishes to replace DMDHEU due to the increasing concern with the toxicity and the adverse impact on the environment by formaldehyde (2, 3).

A durable press finishing agent forms covalent bands with cellulosic hydroxyl groups under elevated temperatures, thus forming crosslinkages between the cellulose molecules. The crosslinking of cellulose increases wrinkle resistance and reduces mechanical strength of the treated cotton fabric. Wrinkle recovery angle (WRA) and tensile strength are two important parameters widely used in the textile industry to evaluate the performance of the crosslinked cotton fabrics.

In our previous research, we used Fourier transform infrared (FT-IR) spectroscopy for identification and quantitative measurement of the ester formed on the cotton fabric treated polycarboxylic acids (4-6). In this research, we used carbonyl band absorbance quantified by FT-IR spectroscopy as the basis for the measurement of the "fixed" finishing agents on the cotton fabric, and investigated the correlation between the performance of the finished cotton fabrics, and the carbonyl band absorbance of the "fixed" finishing agents on the fabric so that we can evaluate the feasibility of predicating the performance of the durable press finished cotton fabrics using FT-IR spectroscopy.

EXPERIMENTAL

Materials

The cotton fabric used in this study was a desized, scoured and bleached 40×40 cotton printcloth weighing 3.2 oz/yd^2 (Testfabrics Style 400). Freerez 901 and Catalyst LF, the commercial products supplied by Freedom Chemical, were used as the DMDHEU-based crosslinking agent and its catalyst.

Fabric treatment

The cotton fabric was first impregnated in a solution containing a crosslinking agent and a catalyst. The impregnated fabric was pressed between the squeezing rolls of a Cromax laboratory padder with two dips and

CP430, *Fourier Transform Spectroscopy:* 11th International Conference
edited by J.A. de Haseth

nips. The wet pick-up of the impregnated fabric was in the range of 100-105%. The fabric was then dried at 85°C for 5 min, and cured in a Mathis curing oven at a specified temperature for a specified period of time.

Fabric Performance Evaluation

The cured cotton fabric was evaluated for its performance after one washing/drying cycle to remove the unreacted reagents and the catalyst. The conditioned wrinkle recovery angle and tensile strength of the fabric were measured according to AATCC method 66-1992 and ASTM method D5035-95, respectively.

Infrared Spectroscopy

All the infrared spectra in this research are diffuse reflectance spectra collected with a Nicolet 510 FT-IR spectrometer and a Specac diffuse reflectance accessory, and are presented in absorbance mode (-log R/R_0). Resolution for all the infrared spectra is 4 cm^{-1}, and there were 100 scans for each spectrum. The carbonyl band absorbance in the 1710-1730 cm^{-1} region was normalized against the 1320 cm^{-1} band absorbance associated with the C-H bending mode of cellulose.

RESULTS AND DISCUSSION

Cotton Fabric Treated with A Polycarboxylic Acid

When cotton fabric is treated with a polycarboxylic acid, such as BTCA, esterification and crosslinking of cotton cellulose takes place under elevated temperatures.

Among the various factors effecting the esterification and crosslinking of cotton cellulose, the amount of the crosslinking agent applied to the cotton fabric is probably the most important parameter in determining the degree of crosslinking of cellulose, and therefore the performance of the finished cotton fabric. An increase in the degree of crosslinking between the cellulose molecules in the fabric improves the wrinkle-resistance of the fabric at the expense of its mechanical strength.

In this research, the cotton fabric was treated with the BTCA solutions with concentration ranging from 0.25 to 10.0% and sodium hypophosphite as a catalyst. The treated cotton fabric samples were cured at 180°C for 3 min. The carboxyl concentrations on the treated cotton fabric samples in mmole/g before and after the curing process were determined by acid-base titration. The ester concentration on a treated cotton fabric sample, also in mmole/g, can be calculated by subtracting the carboxyl concentration on the fabric after curing from that before curing (4). In our previous research, we found that the ester carbonyl band can be separated from the overlapping carboxyl band by rinsing the fabric sample in a dilute NaOH solution, thus converting the free carboxyl to carboxylate an anion (4-6). Consequently, the relative amount of ester on the fabric can be measured quantitatively (4-6).

Shown in Figure 1 are the infrared spectra of the cotton fabric samples treated with the BTCA solutions of different concentrations, cured at 180°C for 3 min, and finally rinsed in a 0.1 M NaOH solution for 4 min. The two strong bands at 1730 and 1580 cm^{-1} in Figure 1 are due to the ester and carboxylate carbonyl, respectively.

A polycarboxylic acid crosslinks cotton cellulose by forming ester linkages between cellulosic molecules. Therefore, the effects of BTCA on the performance of the treated cotton fabric is determined by the amount ester linkage formed on the fabric. Presented in Figure 2 is the linear regression curves for the correlation between the WRA and the ester carbonyl band absorbance for the cotton fabric samples treated with the BTCA solutions of different concentrations. One observes an almost perfect linear correlation between the WRA and the ester carbonyl band absorbance with a r^2 value of 0.989. The data indicate that the linear regression model is suitable for predicting the wrinkle recovery angle based on the ester carbonyl band absorbance of the cotton fabric crosslinked by a polycarboxylic acid.

Crosslinking of cotton cellulose molecules causes fiber embrittlement. In addition, acid-catalyzed depolymerization also takes place during a curing process, thus further reducing the strength of the crosslinked cotton fabric. The correlation between the tensile strength and the ester carbonyl band absorbance of the treated cotton fabric is shown in Figure 3. The tensile strength data for the BTCA-treated cotton fabric samples appear to be more scattered than the WRA data, and the linear regression model between the tensile strength and the ester carbonyl band absorbance has a r^2 value of 0.902. Nevertheless, the linear regression model is still valid for the predicting the tensile strength of the cotton fabric crosslinked by a polycarboxylic acid.

Cotton Fabric Treated with DMDHEU

DMDHEU is the most commonly used crosslinking agent today in the textile industry. DMDHEU is the addition product of urea, formaldehyde and glyoxal. The structure of DMDHEU is shown as follows.

```
                    O
                    ‖
HO—CH2—N     N—CH2—OH
        |     |
    HO-CH—CH-OH
```

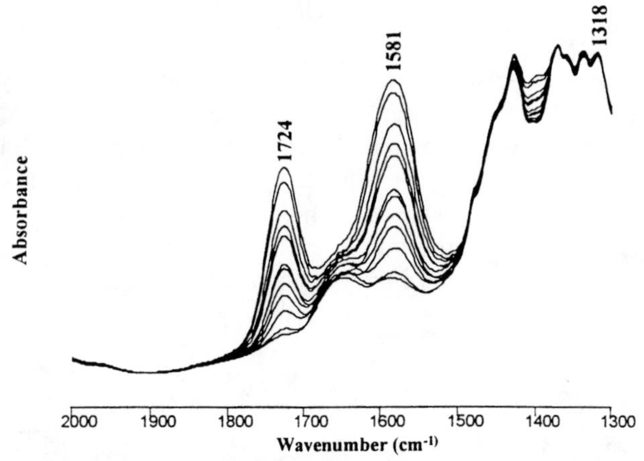

Figure 1. The infrared spectra of the cotton fabric samples treated with the BTCA solutions of different concentration, cured at 180°C for 3 min, and finally rinsed with a NaOH solution. The BTCA concentration (from bottom to top): 0.25, 0.50, 2.00, 2.50, 3.00, 4.00, 6.00, 8.00, and 10.0%.

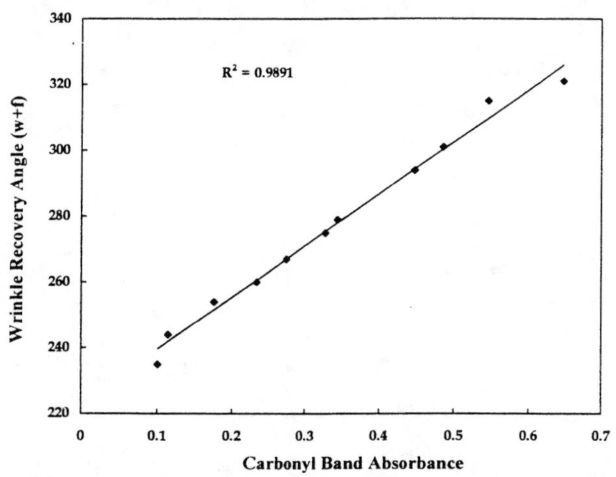

Figure 2. The linear regression curve for the correlation between the wrinkle-recovery angel and the ester carbonyl band absorbance for the BTCA-treated fabric samples.

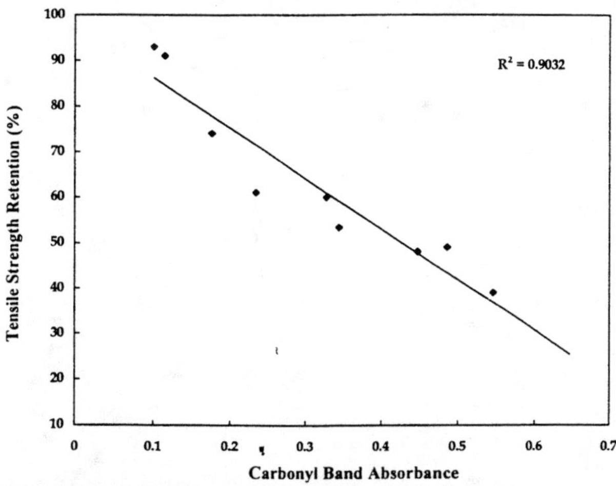

Figure 3. The linear regression curve for the correlation between the tensile strength and the ester carbonyl band absorbance for the BTCA-treated fabric samples.

The two methylol groups and the two hydroxyl groups of DMDHEU react with cellulosic hydroxyl to form crosslinkages between cellulose molecules. The cotton fabric samples were treated with DMDHEU solution which concentration ranges from 2.5 to 5.5%, cured at 170°C for 3 min, and finally rinsed in water to remove the reagent not bound to the cotton fabric. The infrared spectra of the fabric thus treated are presented in Figure 4. The carbonyl band at 1714 cm^{-1} in Figure 4 is due to the DMDHEU which is "fixed" to the cotton fabric after the curing process. The absorbance of the carbonyl band at 1714 cm^{-1} is proportional to the amount of DMDHEU bound to the cotton fabric.

Presented in Figure 5 is the linear regression model of WRA versus the carbonyl band absorbance for the DMDHEU-treated cotton fabric samples. Similar to the BTCA-treated fabric, the cotton fabric treated with DMDHEU show a linear correlation between WRA and the carboxyl band absorbance with a r^2 value of 0.946 (Figure 5). The linear regression curve for the correlation between the tensile strength and the carbonyl band absorbance of the DMDHEU-treated cotton fabric samples has a r^2 value of 0.919 (Figure 6). The crosslinkages formed by DMDHEU increases the wrinkle-resistance whereas it decreases the strength of the treated fabric. Those linear regression models presented in Figure 5 and 6 can be used for predicting the performance of the cotton fabric treated with N-methylol compounds.

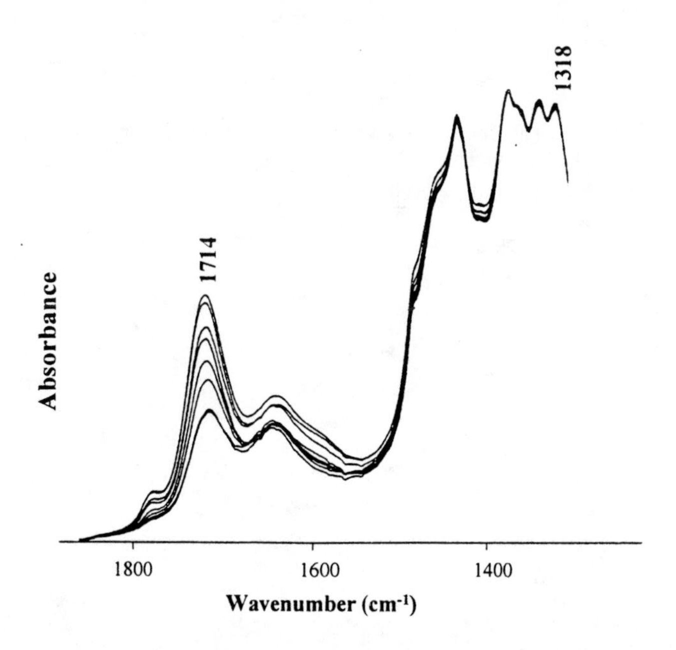

Figure 4. The infrared spectra of the cotton fabric samples treated with the DMDHEU solutions of different concentration, cured at 170°C for 3 min. The DMDHEU (commercial product) concentration (from bottom to top): 5.00, 6.00, 7.00, 8.00, 9.00, 10.0, and 11.0%.

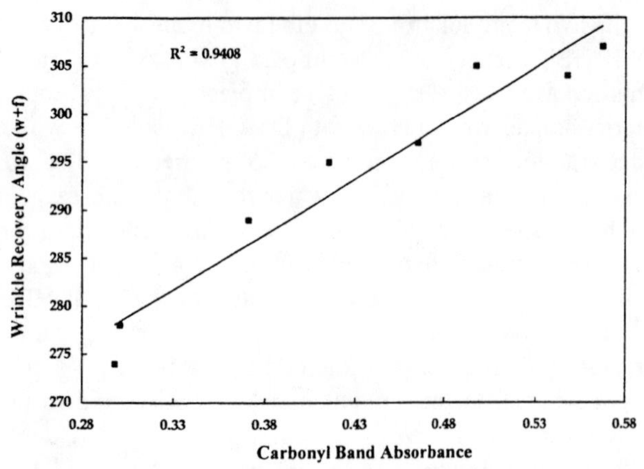

Figure 5. The linear regression curve for the correlation between the wrinkle-recovery angel and the ester carbonyl band absorbance for the DMDHEU-treated fabric samples.

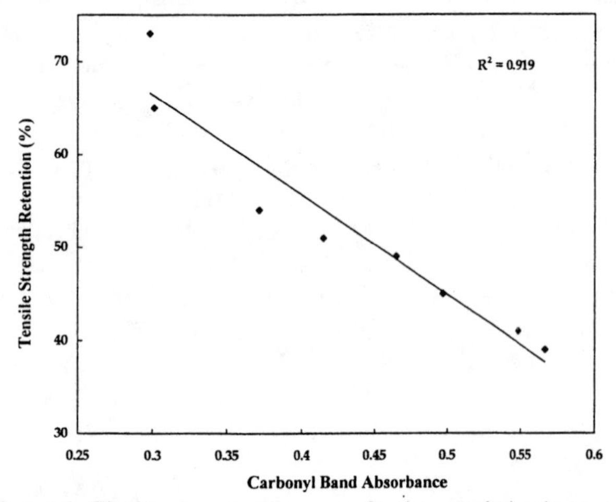

Figure 6. The linear regression curve for the correlation between the tensile strength and the ester carbonyl band absorbance for the DMDHEU-treated fabric samples.

CONCLUSIONS

Linear correlation existed between the wrinkle recovery angle and tensile strength retention, and the carbonyl band absorbance of the cotton fabric treated with BTCA and DMDHEU. Therefore, infrared spectroscopy can be used as a fast and reliable technique for predicting the performance of the durable press finished cotton fabrics and garments when calibration models are established.

REFERENCES

1. Peterson, H., *Rev. Prog. Coloration*, **17**, 7-22 (1987).
2. Welch, C., M., *Rev. Prog. Coloration.*, **22**, 32-41 (1992).
3. Laemmermann, D., *Melliand Textilbericht*, **3**, 274-279 (1992).
4. Yang, C. Q., and Bakshi, G. D., *Textile Res. J.*, **66**, 377-384 (1996).
5. Yang, C. Q., *Textile Res. J.*, **61**, 289-305 (1991).
6. Yang, C. Q., and Andrews, B. A. K., J. *Appl. Polym. Sci.*, **43**, 1609-1616 (1991).

3.11 MINERALS

Monitoring The Sorption of Propanoic Acid By Montmorillonite Using Diffuse Reflectance Fourier Transform Infrared Spectroscopy

R.W. Parker and R.L. Frost

Centre for Instrumental and Developmental Chemistry, Queensland University of Technology, GPO Box 2434, BRISBANE, 4001 Australia

This paper describes how Diffuse Reflectance Fourier Transform Infrared (DRIFT) spectroscopy was used to monitor the sorption behaviour of a short chain fatty acid, propanoic acid, on the clay mineral, montmorillonite. Organic acids bind to montmorillonite in two ways, either by dipole interaction with the oxygens in the interlayer space, or by bonding of the carboxylate anions to exposed aluminium ions. The DRIFT spectra of propanoic acid-montmorillonite complexes have bands at 1728 and 1554 cm^{-1}, which are attributed to the symmetric, and antisymmetric stretching vibrations, respectively, of the C=O, $\nu_{(C=O)s}$, and O-C-O, $\nu_{(O\text{-}C\text{-}O)a}$, bonds of the carboxylic acid group. Each band represents one of the two different binding modes. These bands can be used to monitor the physical and chemical adsorption of the acid by the montmorillonite. When the peak area of each vibration is plotted against increasing acid concentration, both increase to a maximum. However the peak area for the $\nu_{(O\text{-}C\text{-}O)a}$ vibration reaches a maximum at a much lower acid concentration than the $\nu_{(C=O)s}$ vibration. The former maximum corresponds to saturation of the available binding sites on the edge surface aluminium ions. This concentration can be used to calculate the number of binding sites on the clay crystal. Where propanoic acid is allowed to diffuse from the clay, the bound fraction remains on the montmorillonite reducing the available acid that can be desorbed or leached from the clay.

INTRODUCTION

Diffuse Reflectance Fourier Transform Infrared (DRIFT) spectroscopy requires little sample preparation and when used with neat samples can readily detect both minor and major spectral features (1). However DRIFT spectroscopy shows interpretative problems below 1400 cm^{-1}, where peak inversion and *restrahlen* effects occur.

Montmorillonitic clay minerals have a layered structure. Each layer consists of two silica tetrahedral sheets sandwiching a central alumina octahedral sheet (2). The outer surface of the layer consists of a negatively charged oxygen sheet, which is balanced by exchangeable cations and water molecules adsorbed in the region between the layers, the interlayer space (3).

Organic acids bind to montmorillonite in two ways, either by dipole interaction with the oxygens in the interlayer space (4), or by bonding of the carboxylate anion to an exposed aluminium ion on the edge surfaces of the octahedral sheet (5). This latter bonding is quite strong, being resistant to removal by both solvent extraction and heating (4). However the sorption interactions between organic acids, such as propanoic acid, and montmorillonite are expected to be influenced by both types of interaction. This work shows how DRIFT spectroscopy can be used to monitor the sorption of propanoic acid by montmorillonite.

MATERIALS AND METHODS

A sodium montmorillonite obtained from southern Queensland, Australia, was used for this study. Montmorillonite samples were dried at 130 °C. Complexes of propanoic acid (BDH Chemicals) and montmorillonite of varying concentration were prepared by blending the acid with montmorillonite. The resulting complexes were finely ground prior to spectra collection. The DRIFT spectra were obtained using a Perkin Elmer 1600 series FT-IR spectrometer and a Perkin Elmer Diffuse Reflectance attachment. Finely ground potassium bromide was used for the background correction. Spectra were recorded between 4000 and 400 cm^{-1}, over 16 scans, at 4 cm^{-1} resolution. Replicate spectra were obtained on 5 sub-samples of each sample. The DRIFT spectra were related to concentration using the Kubelka-Munk (KM) transformation (6). Peak areas for the vibrations of interest were obtained using integration (Spectra Calc Software) of the Kubelka-Munk transformed spectra. Peak areas were normalised between spectra by dividing the peak area obtained for the vibration of interest, by the peak area of the structural hydroxyl stretching vibration of montmorillonite, which is found near 3630 cm^{-1}. This procedure enables DRIFT spectra to be reliably used for quantitative analysis (7).

Propanoic acid loss from the complex was achieved by surface desorption of the volatile organic. A number of replicate samples were placed in glass vials, with identical surface areas, and exposed to set temperature (25±3 °C), humidity and airflow conditions. Samples were placed in a

CP430, *Fourier Transform Spectroscopy:* 11th International Conference
edited by J.A. de Haseth

grid pattern. From this grid pattern, three samples were randomly selected at 0, 8, 42, 72 and 216 hours.

RESULTS

The infrared spectra of saturated aliphatic carboxylic acids typically have a strong band around 1730 to 1700 cm^{-1} corresponding to the C=O stretching vibration $\nu_{(C=O)s}$. In addition, salts of these acids have bands around 1620 to 1550 cm^{-1} and 1400 to 1300 cm^{-1} corresponding to the asymmetric and symmetric O-C-O stretching, $\nu_{(O-C-O)a}$ and $\nu_{(O-C-O)s}$, of the carboxylate anion (8). Figure 1 shows the neat DRIFT spectra for a sodium montmorillonite, and the complex formed between propanoic acid and that montmorillonite.

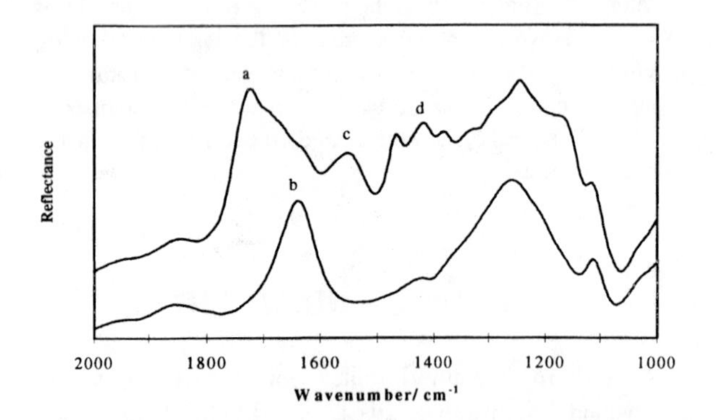

FIGURE 1. Neat DRIFT spectra over the range 2000 to 1000 cm^{-1} for the sodium montmorillonite (lower spectra), and propanoic acid adsorbed on sodium montmorillonite (upper spectra). Positions for the peaks labelled 'a', 'b', 'c', and 'd' are 1728, 1646, 1554, and 1418 cm^{-1} respectively.

Bands marked at 1728 (a), 1646 (b), 1554 (c), and 1418 (d) cm^{-1} , are assigned to the symmetric C=O stretching vibration, $\nu_{(C=O)s}$, the hydration HOH deformation, $\delta_{(H-O-H)}$, and the asymmetric and symmetric O-C-O stretching vibration bands, $\nu_{(O-C-O)a}$ and $\nu_{(O-C-O)s}$, respectively.

Figure 2 shows a plot of the normalised peak areas for the bands at 1726, $\nu_{(C=O)s}$, and 1553 cm^{-1}, $\nu_{(O-C-O)a}$.

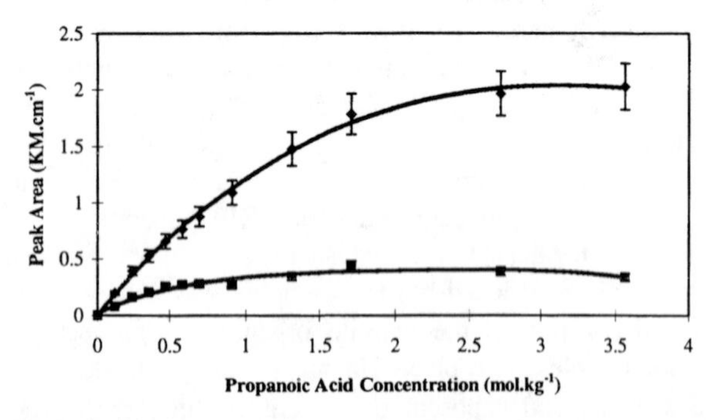

FIGURE 2. Normalised peak areas obtained for the $\nu_{(C=O)s}$ (upper curve) and $\nu_{(O-C-O)a}$ (lower curve) bands, plotted against total concentration of propanoic acid added to montmorillonite.

Both peak areas show an initial linear increase with acid concentration, before reaching a maximum. The relationship between the peak area of the $\nu_{(C=O)s}$ band, '$PA(\nu_{(C=O)s})$', and acid concentration, 'c', can be fitted (R^2 = 0.9974) to a quadratic equation with the following form:

$$PA(\nu_{(C=O)s}) = 0.026c^3 - 0.37c^2 + 1.6c + 0.0062 \quad (1)$$

Similarly, the relationship between the peak area of the $\nu_{(O-C-O)a}$ band, PA ($\nu_{(O-C-O)a}$), and acid concentration, 'c', can also be fitted (R^2 = 0.9513) to a quadratic equation, with the following form:

$$PA(\nu_{(O-C-O)a}) = 0.11x^3 - 0.013x^4 - 0.38x^2 + 0.59x + 0.019 \quad (2)$$

Figure 4 shows the peak areas for the $\nu_{(C=O)s}$ and $\nu_{(O-C-O)a}$ bands when the complex is exposed and propanoic acid is allowed to evaporate.

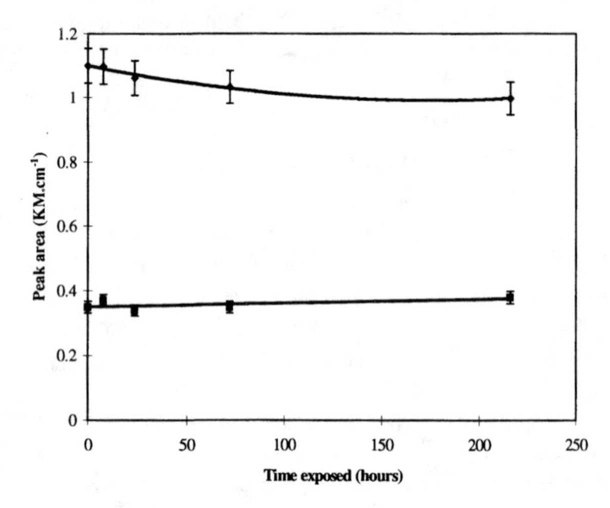

FIGURE 3. Peak areas for the $\nu_{(C=O)s}$ (upper) and $\nu_{(O-C-O)s}$ (lower) bands plotted against time exposed.

DISCUSSION

The $\nu_{(C=O)s}$ bands are due to intercalated propanoic acid, whilst the $\nu_{(O-C-O)a}$ and $\nu_{(O-C-O)s}$ bands are due to the bound propanate anion. The data for the $\nu_{(C=O)s}$ band is fitted to a quadratic equation, equation 1, for which the maximum was calculated to occur at 3.1 mol kg^{-1}. Once the $\upsilon_{(O-C-O)a}$ band reaches a maximum, increases in total acid concentration results only in an increase of the amount of adsorbed acid. This section of the $\nu_{(C=O)s}$ band peak area versus concentration graph is linear (r = 0.9999) and by extrapolating back to zero peak area, a value for the corresponding maximum peak area for the $\upsilon_{(O-C-O)a}$ can be obtained. Substituting this value in equation (2) enables the maximum concentration of the propanate anion to be obtained. In this case the maximum propanate anion concentration is 0.13 mol kg^{-1}. When repeated solvent extraction (3 times) of this complex with methanol was carried out, it was found that a concentration of 0.19 mol kg^{-1} of propanoic acid remained bound to the clay. This figure is in good agreement with the value obtained using the DRIFT determination.

Subtracting the value for the maximum propanate anion concentration (0.13 mol kg^{-1}), from the total acid concentration at which the maximum for the $\nu_{(C=O)s}$ band occurs, 3.1 mol kg^{-1}, gives a value of 3.0 mol kg^{-1}. This value corresponds to the maximum adsorption of propanoic acid as the acid. Because excess acid rapidly evaporates from the complex upon preparation, the adsorbed propanoic acid corresponds to acid trapped in the interlayer space. This value differs from that obtained by determining isotherms using adsorption from solution. For this sample, adsorption from aqueous solution resulted in an adsorption maximum of 1.4 mol kg^{-1}. This value is considerably different from that calculated for adsorption of the concentrated acid. In the case of adsorption from solution, there is a solvent effect with water and acid both entering the interlayer space.

The concentration of 0.13 mol kg^{-1} obtained for the propanate anion corresponds to saturation of the available binding sites, the edge surface aluminium ions. This value can be used to calculate the number of binding sites for a given montmorillonite. The number of binding sites will be dependant on the coordination structure adopted by the anion. Carboxylate anions binding to minerals form three possible coordination structures, unidentate, bidentate and bridging (9). The separation of the $\nu_{(O-C-O)a}$ and $\nu_{(O-C-O)s}$ bands is about 130 cm^{-1} for sodium propanate, and 120 cm^{-1} for the montmorillonite complex. The closeness of these separation distances indicates that the propanate anion forms a bridging complex (9). Consequently the number of number of available binding sites can be calculated by multiplying the concentration of adsorbed propanate anion (0.13 mol kg^{-1}) by Avogadro's number (6.02×10^{23}), by the number of sites occupied by each anion (2), to give a value of 1.6×10^{23} sites per kilogram for this sample.

Once the maximum adsorption for the propanate anion is exceeded, peak area changes for the $\nu_{(C=O)a}$ band correspond to changing propanoic acid concentration, whilst the peak area for the $\nu_{(O-C-O)s}$ band remains constant. Where propanoic acid is allowed to evaporate from the montmorillonite complex, the peak area for the carboxylate anion remains constant, whilst that for the adsorbed propanoic acid declines (see Fig. 3). The rate of diffusion of the acid from the complex will be limited by movement into and out of the interlayer space (4), but the contribution of the anion to this process, if any, is unknown. This work shows there is no relationship between adsorbed acid and anion in the propanoic acid - montmorillonite complex, and that the propanoic acid available for translocation within the clay is limited to the intercalated acid.

CONCLUSIONS

Propanoic acid is adsorbed on montmorillonite as both the intercalated acid and propanate anion. DRIFT spectroscopy can be used to rapidly determine levels of both anion and intercalated acid. Adsorption maximums of 3.0, and 0.13 mol kg^{-1}, for the adsorbed acid, and the anion, respectively, were obtained. The maximum concentration for the adsorbed propanate anion was used to determine the number of anion binding sites on the montmorillonite as 1.6×10^{23} sites per kilogram. Desorption or leeching of propanoic acid from the montmorillonite involves only the intercalated propanoic acid.

ACKNOWLEDGEMENTS

One of the authors (RWP) would like to acknowledge the support of the former Queensland Department of Lands for this research. Both authors would like to acknowledge the support of the Centre for Instrumental and Developmental Chemistry at the Queensland University of Technology.

REFERENCES

1. Parker, R.W. and Frost, R.L., *Clays and Clay Minerals*. **44**(1), 32-40 (1996).
2. Grim, R.E., *Clay Mineralogy*, 2nd Ed. New York: McGraw Hill, 1968, pp. 77-92.
3. Theng, B.K.G., *The Chemistry Of Clay-Organic Reactions*. London: Adam Hilger, 1974, pp. 9-13.
4. Brindley, G.W. and Moll, W.F. *The American Mineralogist*. **50**, 1355-70 (1965).
5. Sieskind, O. and Siffert, B. *C. R. Acad. Sci. Paris*. **274**, 973-76 (1972).
6. Fuller, M.P., and Griffiths, P.R. *American Laboratory*. **10**, 69-80 (1978).
7. Frost, R.L. and Parker, R.W. *Mikrochim. Acta [Suppl.]*. **14**, 691-2 (1997).
8. Kemp, W. *Organic spectroscopy*. 3rd. Ed. London: MacMillan, 1991, pp 58-88.
9. Nakamoto, K. *Infrared and Raman spectra of inorganic and coordination compounds*. 3 rd Ed. NewYork: John Wiley and Sons, 1978 pp 232.

Fourier Transform Raman Spectroscopy of Gels derived from modified Tetra(-n-butoxy)zirconium (IV)

S. M. Dutt[1], R. L. Frost[1] and J. M. Bell[2]

[1] *Centre for Instrumental and Developmental Chemistry, and* [2] *School of Mechanical, Manufacturing and Medical Engineering, Queensland University of Technology, GPO Box 2434, Brisbane, Qld. 4001.*

Reaction products of alkoxides with certain modifying agents and the determination of the structure of zirconia sols and gels can be observed through the use of vibrational spectroscopy. FT-Raman spectroscopy has proved to be a useful technique in observing the gelation process and the resulting structure. Raman spectroscopy has been used to follow the structural evolution of the gel from the mixing the two solvents to the final stages of gelation. The unmodified gel spectra consist of strong bands at 1049 and ~ 260 cm^{-1} which are assigned to the aquated nitrate N-O symmetric stretch and to the Zr-O symmetric stretch respectively. These bands are also present in the acetic and propanoic modified sols and gels, produced by the peptisation of the hydrolysate using nitric acid as the stabilizing agent. The spectra of the acetate and propanoate systems exhibit additional strong peaks at 535, 1350, 1425, 1450, and 1560 cm^{-1}. The latter three bands are assigned to the co-ordinated acetate and propanoate. The band positions indicate that both monodentate and bridging ligands are present within the gel. The spectra of the hydrolysate has a less complex carboxylate region indicating only one type of ligand present. The absence of Zr-O-C bands between 1200 and 1000 cm^{-1} indicate that there is no residual alkoxide present within the gel.

INTRODUCTION

Sol-gel technology has become increasingly important in the production of ceramics that have unique chemical and physical properties. The advantage of this method lies within its ability to produce highly pure crystalline ceramic materials, at lower processing temperatures. It is well known that the types of materials used (alkoxides, salts and modifying agents) and the physical conditions used in the process influence the physical and chemical properties of the final product. FT-Raman and FT-IR spectroscopies are frequently utilized in the characterization of sol-gel materials. These techniques allow for the study of the reactions closely without altering it. However, FT-Raman is particularly important in analyzing sol-gel materials since it has the ability to obtain excellent spectra of colloidal suspensions, wet samples or weak scatterers.

EXPERIMENTAL

Modification of alkoxide with carboxylic acids

The sols, gels and hydrolysates prepared for this study were derived from the modification of Tetra(n-butoxy)zirconium (IV) TBZ with various carboxylic acids. The carboxylic acids used in this research were ethanoic, propanoic and octanoic acids. The alkoxide:acid molar ratio was 1:1 and 1:2. The carboxylic acid was slowly added dropwise to 0.1 moles of TBZ [3.678g]: which was constantly stirred to promote thorough mixing. The hydrolysates of the modified TBZ were obtained by the addition of distilled water at a molar ratio of 100:1 for H_2O:TBZ. The hydrolysate was allowed to settle and the remaining liquid phase was then decanted. The remaining solids were washed with clean distilled water. The solids were allowed once again to settle after stirring had been completed. This procedure of washing the solids was continued for approximately 5-6 washes or until there was no further evidence of the vapor of organic residues present. The hydrolysate was then diluted with clean distilled water to a Zr(IV) concentration of ~0.6M. In the case of the unmodified, propanoic, octanoic and 1:1 ethanoic acid modified gels, a peptizing agent was introduced into the system. The agent selected was concentrated nitric acid, the Zr:HNO_3 molar ratio of 10:1 was employed. The sols, gels and hydrolysates obtained, were submitted to FT-Raman analysis using a Perkin-Elmer 2000 NIR/FT-Raman Spectrometer.

RESULTS AND DISCUSSION

The spectra presented have been assigned a letter, which corresponds to the spectra of a particular gel. The labeling for the spectra is as follows: (A) unmodified gel, (B) 1 mole acetic acid modified gel, (C) 2 moles acetic acid modified gel, (D) 1 mole propanoic acid modified gel, and (E) 1 mole octanoic acid modified hydrolysate.

CP430, *Fourier Transform Spectroscopy:* 11th International Conference
edited by J.A. de Haseth

Region 3600 – 2600 cm^{-1}

All the gels modified with carboxylic acid had the alkyl stretching bands present. Also evident in this region are water bands located at 3377, 3230 and 3120 cm^{-1}, these bands occur only in the unmodified gel and the gels modified with 1 mole of acetic and propanoic acids. These bands are shown in Figures 1 and 2.

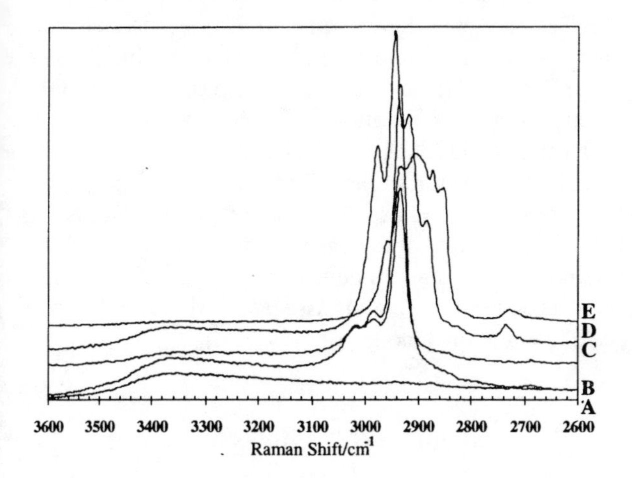

FIGURE 1. FT-Raman spectra of the water and alkyl-CH stretching region of 3600 - 2600 cm^{-1}

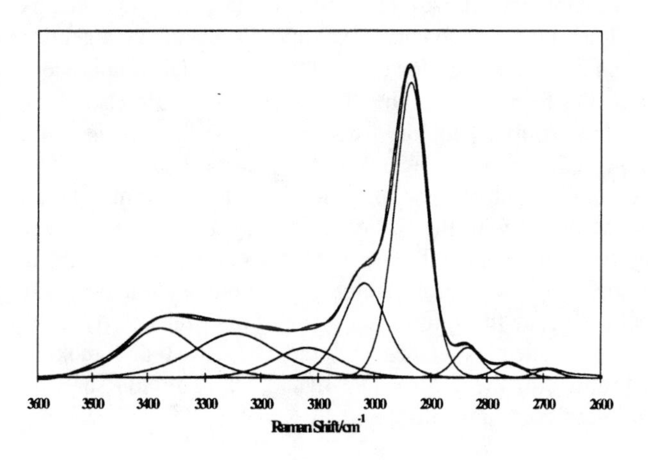

FIGURE 2. Band deconvolution of the FT-Raman spectra of 1 mole acetic acid modified gel: Region 3600 - 2600 cm^{-1}

Region 1800 – 1200 cm^{-1}

The spectral region from 1800 – 1200 cm^{-1} reveals the combination bands of the carboxylate stretching vibration region and the bending vibration of the terminal methyl group. The bands associated with the carboxylate ion consist of the asymmetric and the symmetric vibrational modes. In the Raman spectra the asymmetric vibration is a depolarized band of low intensity whilst the symmetric band shows strong intensity. This region is useful in determining the type of ligand formation between the carboxylate ion and the metal atom. The difference in wavenumbers between the asymmetric and the symmetric stretch of the carboxylate ion indicates whether the ligand is bidentate or monodentate. The bidentate ligand can either be bridging or chelating. It is hypothesized that it is possible to ascertain the types of ligands within a gel using Raman spectroscopy. Figure 3 shows the carboxylate and methyl deformation region of the gels. The spectra of the 1 mole acetic acid modified gel has an additional band located at 1660 cm^{-1}, and is attributed to the bending mode of water which is strongly hydrogen bonded. The asymmetric bands in the modified gels are located at 1625 and ~1565 cm^{-1} (3-5). The assignment of the two asymmetric bands suggests that there are at least two types of ligands present within the gels. In the propanoic acid modified gel there is a further asymmetric band positioned at 1525 cm^{-1} (2). Figure 4 shows that the octanoic acid hydrolysate has only one type of ligand present. The number of bands in the spectra indicates what types of ligands are present in the gel or if a specific ligand undergoes bond angle variations. These bands can be obtained by band deconvolution of the experimental band profile between 1500–1300 cm^{-1} resulting in component bands at 1460, 1420, and 1380 cm^{-1} for the acetic and propanoic acid modified gels. Figures 4, 5 and 6 show the bands obtained by band fitting using the Jandel Scientific 'Peak Fit' program.

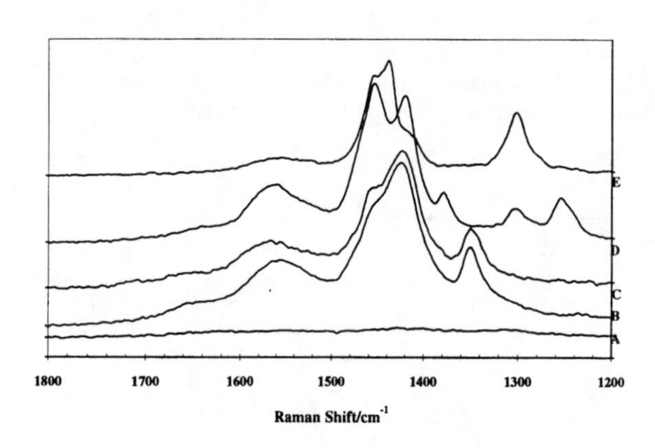

FIGURE 3. FT-Raman Spectra of Gels: Region 1800-1200 cm^{-1}

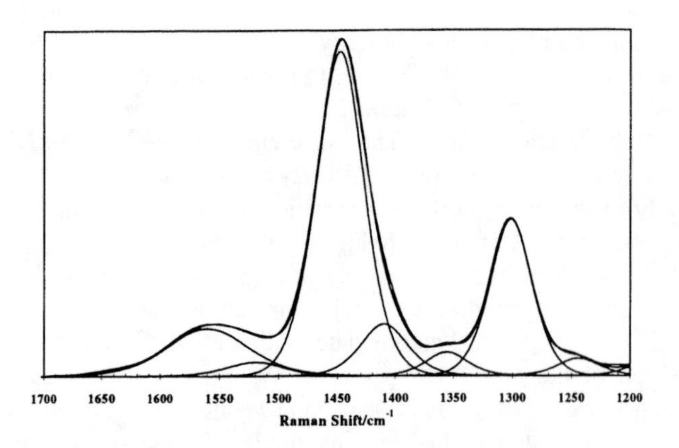

FIGURE 4. Band deconvolution of the FT-Raman spectra of octanoic acid hydrolysate: Region 1700 - 1200 cm^{-1}

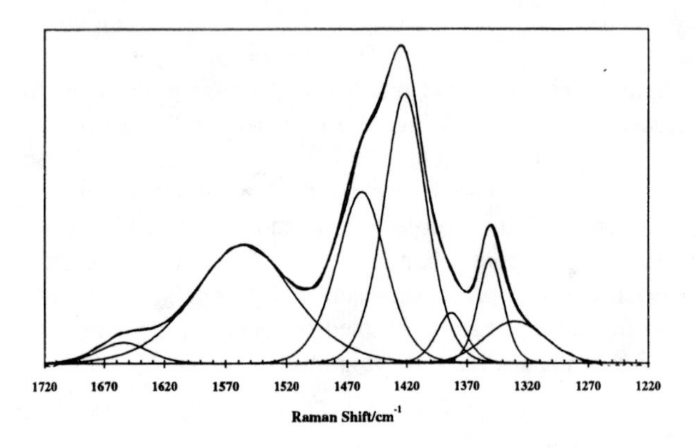

FIGURE 5. Band deconvolution of the FT-Raman spectra of 1 propanoic acid modified gel: Region 1720 - 1200 cm^{-1}

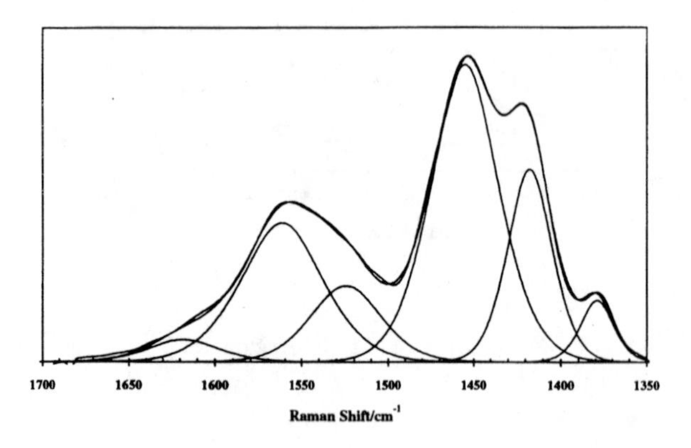

FIGURE 6. Band deconvolution of the FT-Raman spectra of 1 mole acetic acid: Region 1720 - 1200 cm^{-1}

The bands at 1460, 1380 and 1350 cm^{-1} are assigned ν_s (CO_2^-); the 1460 cm^{-1} band has been previously assigned as a symmetric carboxylate (2,6), the 1350 and 1380 cm^{-1} bands have never been previously assigned as carboxylate stretches in a zirconia gel. This leads to the question as to what type of ligands exist based on the spectral data. It is proposed that the 1460 cm^{-1} band belongs to the chelating bridging ligand, whilst the lower energy bands are attributable to the bridging bidentate ligands. The band located at 1420 cm^{-1} is assigned to the asymmetric deformation of the carboxylate methyl group. There are several possibilities for the assignment of the strong band centered at 1420 cm^{-1}, Straughan *et al* (2) assigned the band as the asymmetric deformation of the methyl group, however Paul *et al* (5) indicated that the peak was due to the symmetric stretching vibration of the carboxylate ion. The band located at 1325 cm^{-1} has also been assigned to the symmetric deformation of the methyl group. The propanoic and octanoic hydrolysates have further peaks located at 1317 and 1260 cm^{-1}. It is suggested that the peak at 1317 cm^{-1} band is the symmetric deformation of the alkyl groups, whilst the band at 1260 cm^{-1} is associated with the CH_2 twist or rocking vibrational mode.

Region 1200 – 800 cm^{-1}

This region of the spectra is primarily concerned with the carbon-carbon stretching and the nitrate vibrational region (Figure 7). The common bands within all the spectra are located at 1049 and 945 cm^{-1}. The 1049 cm^{-1} band is the symmetric NO stretch assigned to the aquated nitrate ion that stabilizes the zirconia sol, whilst the 945 cm^{-1} is attributed the carbon-carbon stretch. Propanoic acid gel has an additional ν(C-C) located at 895 cm^{-1} which is consistent with the bidentate ligand. The acetic acid gels also show predominantly a shoulder located at ~ 920 cm^{-1} which is the expected bidentate ligand carbon-carbon stretch. The unmodified gel has a weak band located at 988 cm^{-1}, which is associated with the carbon-carbon stretch of the residual by-product of butanol. The octanoic hydrolysate spectra is complex in this region, which has a combination of bands belonging to the nitrate, carbon-carbon stretches, carbon-carbon skeletal stretches and the CH_2 twist and rocking vibrations of the carboxylate ligand. This region shall be fully assigned in the future.

Region 800 – 75 cm^{-1}

The bands in this spectral region (Figure 8) belong to the deformation bands associated with the carboxylate ion and the zirconium-oxygen stretching and bending vibrations. The unmodified gel illustrates the band positions that are solely associated with the zirconium-oxygen bonds. The broadness of these bands within the region 570 – 75 cm^{-1} suggest that all the gels produced are non-crystalline. The similarity between the spectra of the modified gels, with the unmodified gel, indicates that the bands present, and the assignments made, are due to the influence of the zirconium and oxygen bonds only. The bands positioned at 700, 652 and 617 cm^{-1} are given the following assignments: asymmetric deformation band of the carboxylate ion, the out of phase rocking vibration of the carboxylate ion and the symmetric deformation. The band located at 669 cm^{-1} has been previously assigned as a zirconium oxygen stretch (6). The bands at ~530, ~410, 248 and ~ 210 cm^{-1}: the first two bands are tentatively assigned to the zirconium-oxygen stretch. This produces a total of three different Zr-O bands, this could be due to the various zirconium-oxygen stretches influenced by the different type of ligands available and the Zr-O stretches of the zirconium oxides. The bands at 248 and 210 cm^{-1} are attributed to the O-Zr-O bending vibrations.

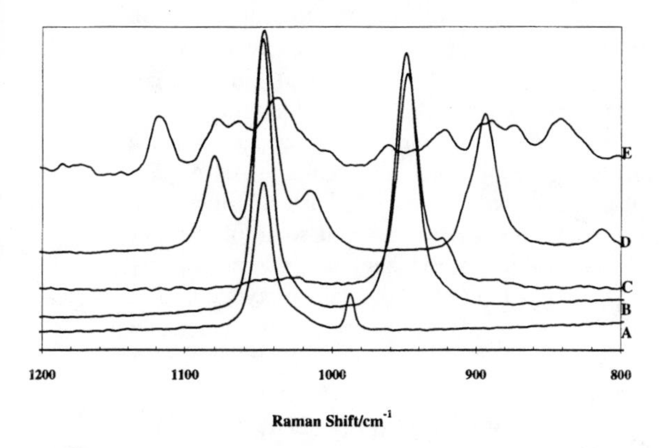

FIGURE 7. FT-Raman sSpectra of the nitrate and $\nu_{(C-C)}$ region of the zirconia gels

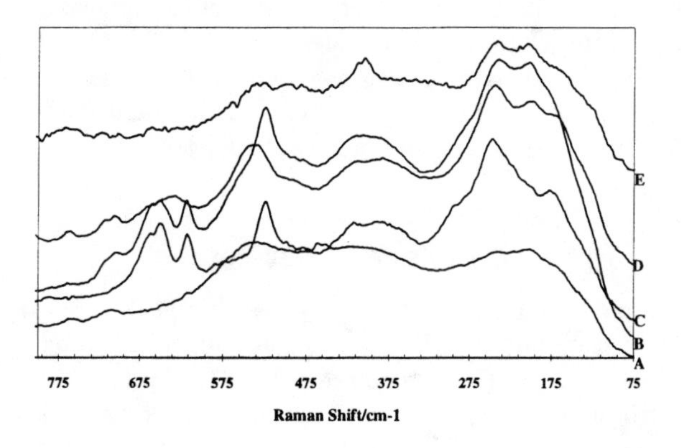

FIGURE 8. FT-Raman sSpectra of the region 800-75 cm^{-1} of the zirconia gels

CONCLUSIONS

FT-Raman spectroscopy is useful in obtaining good quality spectra for aqueous based gels, with the added advantage of acquiring distinct spectra for the carboxylate region, enabling the band assignments for the monodentate, bidentate and bridging ligands to be made. The spectra indicate that the amorphous gel structures contain various carboxylate ligands, and a number of zirconium oxygen stretches. All this information can provide an insight into the structure of the gel which influences the final stages of obtaining a tailored ceramic product.

REFERENCES

1. Ayral, A., Assih, T., Abenzoza, M., Phalippou, J., Lecomte, A., and Dauger, A., *Journal of Materials Science,* **25**, 1268-1274 (1990).
2. Straughan, B.P., Moore, W., and McLaughlin, R., *Spectrochimica Acta,* **42A,** 451-456 (1986).
3. Kaklhana, M., Kotaka, M., and Okamato, M., *J. Phys. Chem.*, **87**, 2526-2535 (1983).
4. Severin, K.G., Ledford, J.S., Torgerson, B.A. and Berglund, K.A., *Chem. Mater,* **6**, 890-898 (1994).
5. Paul, R.C., Baidya, O.M., Kumar, R.C., and Kapoor, R., *Aust. J. Chem.*, **29**, 1605-1607 (1976).
6. Wolf, C., and Russel, C., *Journal of Materials Science,* **27**, 3749-3755 (1992).

Dehydroxylation Of Intercalated Kaolinite An Infrared Emission Spectroscopic Study

Gina N. Paroz, Ray L. Frost, and Greg A. Cash

Centre for Instrumental and Developmental Chemistry
Queensland University of Technology, Brisbane, Australia

Intercalation of kaolinites with organic substances provides a method of expansion for a typically non-expanding clay. The dehydroxylation process is affected by the interaction of the kaolinite with the organic molecules, providing a mechanism for the dehydroxylation to occur at a decreased temperature. The dehydroxylation of the Birdwood kaolinite and its intercalates has been investigated by Fourier Transform *in-situ* infrared emission spectroscopy over the temperature range 200 °C to 600 °C at 25 °C intervals. Dehydroxylation of the intercalated kaolinite was observed by the decrease in intensity of the bands at 3670 cm^{-1}, 3650 and 3600 cm^{-1} (inner surface hydroxyl bands) and 3620 cm^{-1} (inner hydroxyl band). The first two bands represent the out-of-phase vibrations of the hydroxyl stretching region of the kaolinite. The band at 3605 cm^{-1} is assigned to the inner surface hydroxyl hydrogen bonded to the acetate ion. It is proposed that the infrared emission process occurs through the loss of energy through the kaolinite out-of-phase vibrations. Kaolinite and its intercalated complexes lose their inner surface and inner hydroxyl groups simultaneously, except at the final stages of dehydroxylation, where the inner surface hydroxyl group is lost and some of the inner hydroxyl group remains. For the dehydroxylation of the potassium acetate intercalated kaolinite, the process was studied by the decrease in the bands at 3670, 3650, 3620 and 3600 cm^{-1}. It was found that the dehydroxylation of the intercalated kaolinite occurred at 175 °C less than that for the untreated kaolinite, the presence of the intercalating molecule provided a different mechanism by which dehydroxylation could occur.

INTRODUCTION

Fourier transform infrared emission spectroscopy has been identified as an extremely useful tool for following the dehydroxylation of kaolinite and its intercalate complexes. This *in situ* technique shows the temperature at which the process both commences and is complete, and further allows the process to be followed through the loss in intensity of the bands associated with the hydroxyl groups in kaolinite. Dehydroxylation of kaolinite and other clay minerals has been studied using infrared emission spectroscopy (1-3), and this technique has been applied, in this study, to the intercalation compounds of kaolinite.

EXPERIMENTAL

Kaolinite from the Birdwood deposit (South Australia) is a highly ordered, low defect structure, with a Hinckley index of 1.32. Kaolinite was dried in a desiccator to remove adsorbed water and used without further purification. Intercalation with potassium acetate was achieved through mixing 1g of kaolinite with 50 cm^3 of 7M potassium acetate solution (4). The sample was stirred at ambient temperature for 7 days. The excess solution was removed using a 'MSE ultra centrifuge' and the sample was dried in a vacuum desiccator at a controlled humidity.

SPECTROSCOPY

Infrared Absorption Spectroscopy:

Sample of untreated kaolinite and kaolinite-potassium acetate intercalate (1mg) was finely ground and combined with potassium bromide and then pressed into a disc using 8 tonnes of pressure for 5 minutes under vacuum. The spectrum of each sample was recorded in the range 4000 cm^{-1} to 400 cm^{-1} over 64 scans at 4 cm^{-1} resolution using the 'Perkin Elmer 1600 series infrared spectrometer'.

Infrared Emission Spectroscopy:

Powdered samples of submicron size were placed on the platinum plate and the FT-IR emission undertaken using a modified 'Digilab FTS7 spectrometer' (1). Spectra were recorded in the range 4000 cm^{-1} to 400 cm^{-1} over 128 scans at 8 cm^{-1} resolution from 200 °C to 600 °C at 25 °C intervals. Three sets of spectra were obtained: 1. the black body radiation, 2. the Pt plate radiation and 3. the sample plus the Pt plate. Raw emissivity data was analysed by subtracting the emission spectrum of the clean platinum plate and then ratioing the result to the graphite black body spectrum at the relevant temperature.

Data manipulation an interpretation was carried out using the Spectracalc software package GRAMS (Galactic Industries Corporation, NH, USA) and Microsoft Excel.

CP430, *Fourier Transform Spectroscopy:* 11th International Conference
edited by J.A. de Haseth

RESULTS AND DISCUSSION

The unit cell structure of kaolinite contains two types of hydroxyl groups, those situated between successive kaolinite sheets and those situated within the sheets. These are referred to as the inner sheet hydroxyl groups, denoted OuOH and the inner hydroxyl groups, denoted InOH respectively. There are four distinct bands found in the hydroxyl region of the infrared spectrum, associated with the vibrations of these groups.

Extensive study of the infrared spectrum has resulted in the following band assignments (5): the three infrared absorption higher frequency vibrations (ν_1, ν_2 and ν_3) at approximately 3695, 3670 and 3650 cm^{-1} are associated with the inner surface hydroxyls, OuOH. The band at 3620 cm^{-1} is associated with the inner hydroxyl (InOH). Figure 1 shows the FTIR spectra of a low defect kaolinite from the Birdwood deposit of South Australia (A) and the Birdwood kaolinite-potassium acetate intercalate (B).

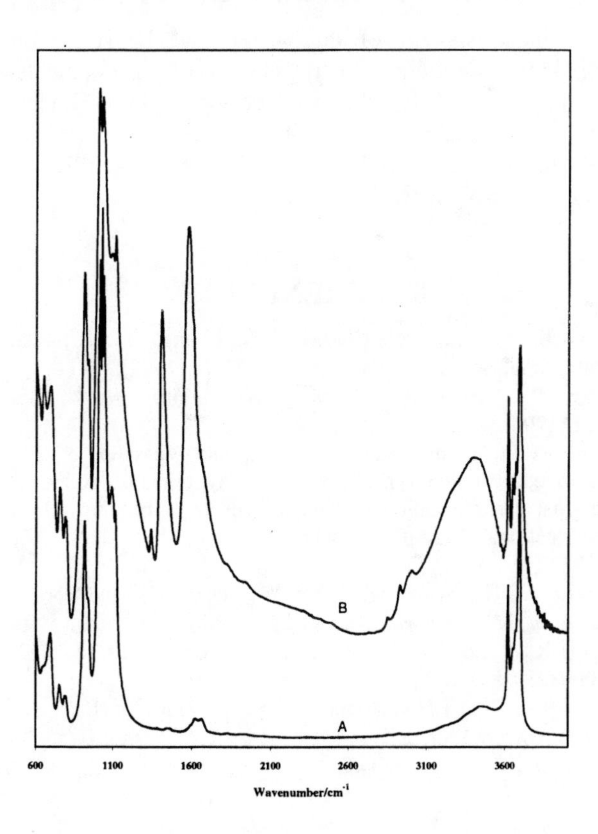

FIGURE 1: Infrared absorption spectrum of (A) Birdwood kaolinite (lower spectrum) and (B) Birdwood kaolinite-potassium acetate intercalate (upper spectrum)

Additional bands observed are the result of the intercalating molecule, potassium acetate. Further Figure 2, spectrum (B) shows the splitting of the 3650 cm^{-1} band into two distinct bands at 3646 and 3656 cm^{-1}.

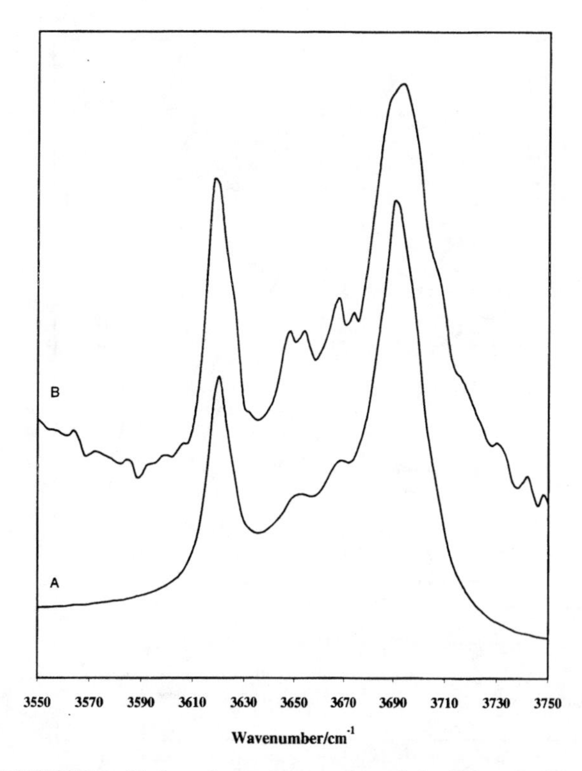

FIGURE 2: Hydroxyl stretching region of the infrared absorption spectrum of (A) Birdwood kaolinite (lower spectrum) and (B) Birdwood kaolinite-potassium acetate intercalate (upper spectrum)

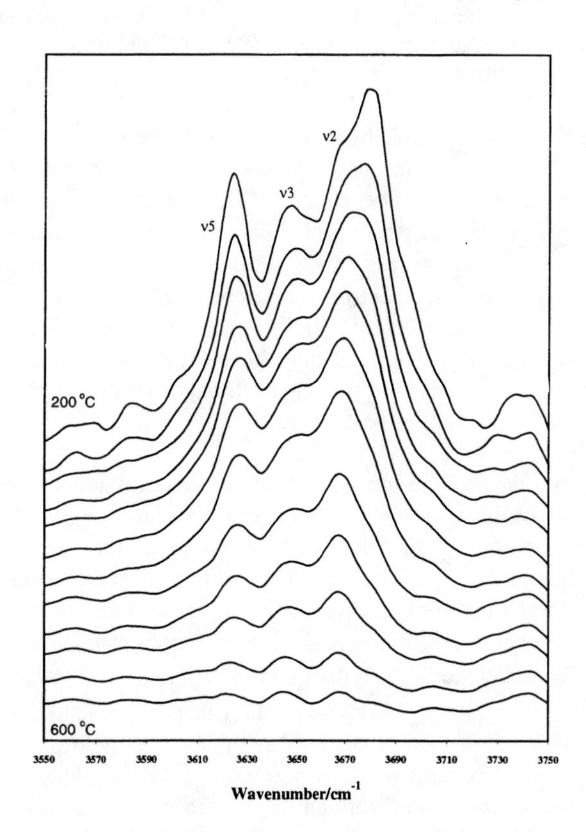

FIGURE 3: Hydroxyl stretching region of the Infrared Emission Spectra of Birdwood kaolinite from 200 °C to 600 °C

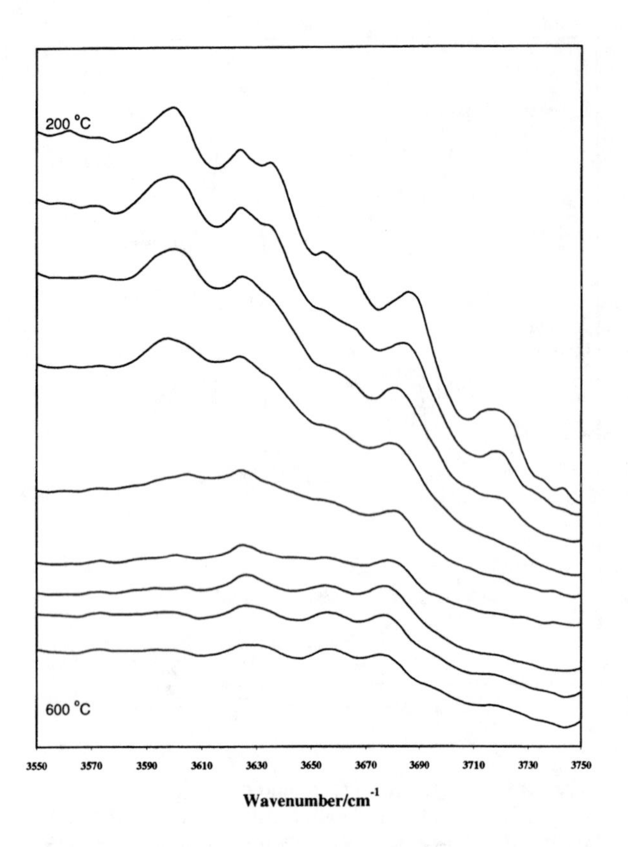

FIGURE 4: Infrared emission spectra of Birdwood kaolinite-potassium acetate intercalate over the hydroxyl stretching region from 200 °C to 600 °

Examination of the hydroxyl stretching regions of both the infrared absorption and infrared emission spectra shows a feature unique to the former. The stretching vibration assigned to the in-phase stretching vibration of the inner surface hydroxyl groups in kaolinite, denoted ν_1, is present in the absorption spectrum at approximately 3695 cm^{-1}, however this band is absent in the emission spectrum (Figure 3). This suggests that the energy loss associated with dehydroxylation is facilitated through the out-of-phase hydroxyl stretching vibrations of the inner surface hydroxyl groups.

A similar observation was found for the emission spectra of the kaolinite-potassium acetate intercalate (Figure 4). Intercalation of the kaolinite with potassium acetate substantially altered the mechanism for dehydroxylation. Adsorbed water was not lost from the intercalate until 350 °C, showing the strength of the hydrophilic nature of potassium actetate. Dehydroxylation commenced at 300 °C for the intercalate, 175 °C less than that of the untreated kaolinite. The spectra also show high concentrations of water and it is proposed that intercalation of the potassium acetate can not occur without the presence of water. It is proposed the water is necessary to hydrate the potassium ion with the acetate hydrogen bonding to the inner surface hydroxyl groups (6).

CONCLUSIONS

Fourier transform infrared emission spectroscopy is the preferred method for the study of the dehydroxylation processes of kaolinite and its intercalates. This technique provides structural information *in situ*, at elevated temperatures, with no sample preparation. Changes occurring in the bands associated with the hydroxyl groups in kaolinite, in the region 3550 cm^{-1} to 3750 cm^{-1}, detail the effect of intercalation upon the dehydroxylation mechanism.

Intercalation is observed in the emission spectra by the presence of the hydroxyl band at 3600 cm^{-1}. Dehydroxylation takes place at 300 °C and it is proposed that intercalation of the potassium acetate provides a different mechanism for the kaolinite dehydroxylation. Further detailed investigation of this concept is required.

ACKNOWLEDGEMENTS

The financial support of the Queensland University of Technology's Centre for Instrumental and Developmental Chemistry is gratefully acknowledged. As is Prof. Graeme George for use of the infrared emission spectrometer. Mr L. Barnes of Commerical Minerals Pty Ltd is thanked for the supply of the clay mineral.

REFERENCES

1. Vassallo, A.M., Cole-Clarke, P.A., Pang, L.S.K., and Palmisano, A., *J. Appl. Spectrosc.* **46,** 73-78 (1992).
2. Frost, R.L., and Vassallo, A.M., *Clay Clay Minerals* **44**, 635-651 (1996).
3. Frost, R.L., Finnie, K, Collins, B., and Vassallo, A. M., "Infrared emission spectroscopy of clay minerals and their thermal transformations". The Proceedings of the 10 th Intertnational Clay Conference, Adelaide, Australia. July 1995.
4. Weiss, A., Thielepape, W., Ritter, W., Schafer, H., and Goring, G., *Anorg. Allg. Chem.* **320,** 183-204 (1963).
5. Frost R.L., and Van der Gaast, S. J., *Clay Minerals* **32,** 293-306 (1997).
6. Frost R.L., Tran T.H. and Janos Kristof J., Vibrational Spectroscopy **13**(2):175-186 (1997).

FT-RAMAN and FTIR Spectroscopy of Intercalated Kaolinites

R.L. Frost[1], G.N. Paroz[1], T.H. Tran[1] and J.Kristof[2]

[1]*Centre for Instrumental and Developmental Chemistry, Queensland University of Technology, 2 George Street, GPO Box 2434, Brisbane Q 4001, Australia, and*
[2] *Department of Analytical Chemistry, University of Veszprem, H 8201 Veszprem, PO Box 158 Hungary.*

Changes in the molecular structure of a low defect structured kaolinite, intercalated with potassium and cesium acetates have been studied using FTIR reflectance and FT-Raman spectroscopy. Additional Raman bands, attributed to the inner surface hydroxyl groups strongly hydrogen bonded to the acetate, are observed at ~3605 cm^{-1} for the potassium and at 3598 and 3606 cm^{-1} for cesium acetate intercalates with the consequential loss of intensity in the bands at 3652, 3670, 3684 and 3693 cm^{-1}. Changes in the position of the band assigned to the inner hydroxyl group are observed upon the formation of the cesium acetate intercalate. DRIFT results are complementary to the Raman microscopic investigations and have proven particularly useful in the study of the hydration sphere of the intercalating cation and the possible effect of the cation on the position of the band assigned to the inner hydroxyl group.

INTRODUCTION

Clay minerals can interact with both organic and inorganic chemicals through a number of mechanisms such as adsorption, intercalation and cation exchange. The basic principles of intercalation reactions have been elucidated for kaolinite (1). One group of intercalating molecules consists of the alkali salts of short chain fatty acids in particular acetic acid (2). FT-Raman spectroscopy has been used to study the structure of kaolinites and kaolinite polymorphs (4-7). The application of FT-Raman and FTIR spectroscopy to the elucidation of the structure of kaolinite hydroxyls in the cesium and potassium acetate intercalates is reported.

EXPERIMENTAL

Clay Mineral Intercalates

The clay minerals used in this study are a highly ordered, low defect structured kaolinite from Kiralyhegy, Hungary, and a disordered, high defect kaolinite from Szeg, Hungary. The air dried kaolinite was intercalated according to Weiss et al. (2,3). A 300 mg portion of the kaolinite was gently stirred in 30 cm^3 of 7M potassium acetate solution or 5 M cesium acetate solution at room temperature for 80 hours. The excess solution was removed by centrifugation. The degree of intercalation was checked by X-ray diffraction and the kaolinites were found to be 90% and 35% intercalated, for the potassium acetate and cesium acetate intercalates, respectively.

FT-Raman and FTIR spectrometry

FT-Raman spectra were obtained using the Perkin-Elmer 2000 series Fourier Transform spectrometer fitted with a Raman accessory comprising a Spectron Laser Systems SL301 Nd-YAG laser operating at a wavelength of 1064 nm, and a Raman sampling compartment incorporating 180 degree optics. The spectrometer contained a quartz beam splitter capable of covering the spectral range 15,000-4000 cm^{-1}. The Raman detector was a highly sensitive indium-gallium-arsenide detector and was operated at room temperatures. Under these conditions Raman shifts would be observed in the spectral range 3000-100 cm^{-1}. Spectra were corrected for the instrumental function and detector response. Measurement times of between 0.5 and 1 hour were used to collect the FT Raman spectra with a signal to noise ratio of better than 100/1 at a resolution of 2 cm^{-1}. A laser power of 100 mW was used. It was found that the best spectra were obtained by simply placing a lump of the raw kaolinite clay mineral in the beam, whereas the powdered clay gave spectra of a lesser quality for an equal number of scans. The spectra of the intercalates were measured in the powdered form. Spectra were calibrated using the 520.5 cm^{-1} line of a silicon wafer. FTIR (DRIFT) analyses were undertaken using a Bio-Rad 60A spectrometer. 512 scans were obtained at a resolution of 2 cm^{-1}. Band fitting was done using a Lorentz-Gauss cross-product function with the minimum number of component bands used for the fitting process. The Gauss-Lorentz ratio was maintained at values greater than 0.7 and fitting was undertaken until reproducible results were obtained with correlations for r^2 greater than 0.995.

CP430, *Fourier Transform Spectroscopy:* 11th International Conference
edited by J.A. de Haseth

TABLE 1. Band positions of the Raman and of the FTIR spectra of the hydroxyl stretching region for both low and high defect kaolinites intercalated with potassium and cesium acetates

Sample	Band centre cm^{-1} Raman	Band centre cm^{-1} Raman	Band centre cm^{-1} FTIR	Band centre cm^{-1} FTIR
	Low defect kaolinite	High defect kaolinite	Low defect kaolinite	High defect kaolinite
ν_8		3555		3500
ν_7		3600	3600	3600
ν_5	3620.3	3621	3619.2	3619.2
ν_6		3635		
ν_3	3651.5	3655	3650	3650
ν_2	3670	3675	3668.5	3668.5
ν_4	3685	3685	3683	3683
ν_1	3693	3696	3693.5	3693.5
cesium acetate intercalate				
ν_8	3580	3580		
ν_7	3598	3598		
ν_6	3606	3606	3606	3605
ν_5	3617	3617	3620	3617
ν_3	3649	3649	3652	3652
ν_2	3670	3670	3668	3668
ν_4	3684	3684	3685	3685
ν_1	3691	3691	3695	3695
potassium acetate intercalate				
ν_7	3555	3585	3580	3580
ν_6	3605	3602.5	3595	3595
ν_5	3620	3619	3607	3607
ν_3	3645 (vw)	3645 (w)	3650	3650
ν_2	3670 (vw)	3670(vw)	3672	3672
ν_4	3686	3686	3684	3684
ν_1	3695	3695	3691	3691

RESULTS AND DISCUSSION

Kaolinite and the other kaolins contain two types of hydroxyl groups: hydroxyl groups at the internal surface, OuOH, often referred to as inner surface hydroxyl groups and hydroxyl groups, InOH, often referred to as inner hydroxyl groups, within the layer. The OuOH groups are situated in the outer, unshared plane whereas the InOH groups are located in the inner-shared plane of the octahedral sheet, often termed the gibbsite sheet. Infrared spectroscopy has been extensively used for the study of kaolinite polymorphs and the four distinct infrared bands are normally assigned as follows: the three higher frequency vibrations (ν_1,ν_2,ν_3) are due to the three inner surface hydroxyls (OuOH) and the band at 3620 cm^{-1} is due to the inner hydroxyl (InOH). Because of the existence of the Raman active band at 3684 cm^{-1}, this InOH band at 3620 cm^{-1} is defined as ν_5 in this research. The hydrogens of the inner hydroxyl groups, are bound to the oxygen below the aluminium atoms and directed towards the intralayer cavity of the kaolinite (8). The InOH hydroxyls essentially lie parallel to the 001 plane. This hydroxyl points towards the vacant dioctahedral site in the kaolinite structure (9). There is considerable debate as to the exact position of this inner hydroxyl group and it is possible that the position of the group depends on whether the Al-O-H bond is linear or bent. Figure 1 shows the Raman spectra of a high defect kaolinite (spectrum (a)), the high defect kaolinite potassium acetate intercalate (spectrum (b)) and the low defect kaolinite potassium acetate intercalate (spectrum (c)). Clearly additional Raman bands are found at ~3605 cm^{-1} upon intercalation. It is apparent that the low defect structured kaolinite is more fully intercalated than the high defect kaolinite. In the first example, spectrum (b), some intensity remains in the ν_1, ν_2, ν_3 bands after intercalation. In the second example (c), almost no intensity remains in these bands.

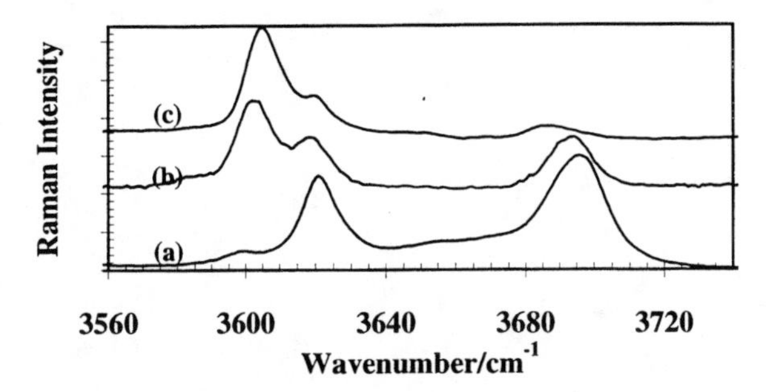

FIGURE 1. Raman spectra of the hydroxyl stretching region of (a) high defect kaolinite (b) potassium acetate intercalated high defect kaolinite (c) potassium acetate intercalated low defect kaolinite

When the low defect structured kaolinite is intercalated with cesium acetate, additional Raman bands are found at 3606, 3598 and 3580 cm^{-1}. Further the band assigned to the inner hydroxyl group has shifted to 3617 cm^{-1}. Upon intercalation with cesium acetate, significant intensity remains in the ν_1 and ν_4 vibrations at 3691 and 3684 cm^{-1}. This shows that the kaolinite could not be fully intercalated with the cesium acetate, in good agreement with our XRD studies. The normalised band areas of the ν_1 and ν_4 vibrations at 3691 and 3684 cm^{-1} are 15.9 and 15.6% respectively. This represents a reduction in area of these bands from 47.7 % for the untreated kaolinite to 31.5 % upon intercalation. The ν_2 and ν_3 bands are reduced to 0 and 5.9 % respectively. The additional bands at 3606, 3598 and 3580 cm^{-1} make up 26.3, 16.6 and 3.7% of the total normalised band intensities. The band assigned to the inner hydroxyl in the untreated kaolinite at 3620 cm^{-1} has shifted to 3617 cm^{-1} but the band width remains unchanged. Clearly there are now two distinct additional Raman bands which are assigned to the inner surface hydroxyl groups which are hydrogen bonded to the acetate. The spectral results of the Raman and the infrared reflectance spectra of the hydroxyl stretching region of the kaolinite and the kaolinites intercalated with cesium acetate and potassium acetates are shown in Table 1. The FTIR spectra of the low defect kaolinites intercalated with potassium and cesium acetates are shown in Figures 2 and 3 respectively. The ν_4 band at 3685 cm^{-1} is infrared inactive but a small component can be observed in the spectra of low defect kaolinite. Additional FTIR band was observed at 3605 cm^{-1} for the cesium acetate intercalate and the inner hydroxyl band, ν_5, was shifted to 3617 cm^{-1}, thus confirming the Raman results. No bands were observed at 3598 or 3580 cm^{-1} in the FTIR spectra. This means that the bands found at 3598 and 3580 cm^{-1} are Raman active and infrared inactive. Such a result arises from a vibration which is totally symmetric.

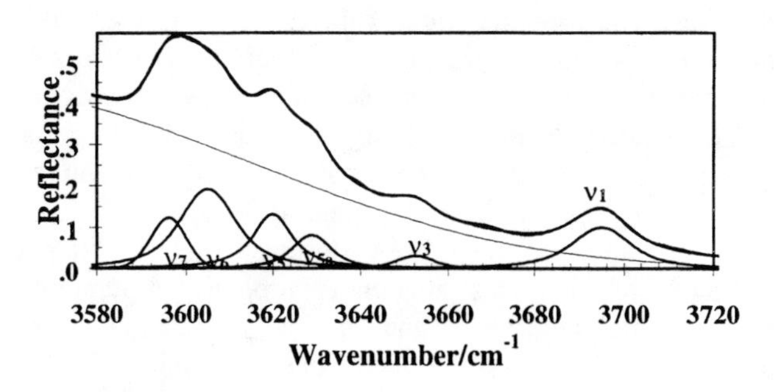

FIGURE 2. FTIR spectra of the hydroxyl stretching region for potassium acetate intercalated low defect kaolinite

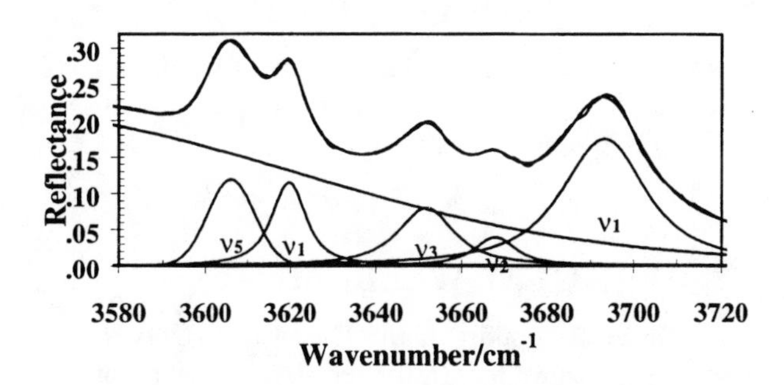

FIGURE 3. FTIR spectra of the hydroxyl stretching region for cesium acetate intercalated low defect kaolinite

When the kaolinite is intercalated with potassium acetate, an additional Raman band is found at 3606 cm^{-1}. The band comprises 62.5% of the total normalised intensity. It is attributed to the formation of strong hydrogen bond between the kaolinite and the acetate ion. The band area of the InOH group remains unchanged although the band is now very broad with a half width of 6.2 cm^{-1} compared to that of the untreated kaolinite of 2.9 cm^{-1}. The normalised band areas of the ν_1, ν_4, ν_2 and ν_3 vibrations are 4.2, 6.4, 2.2 and 3.4%. Essentially, these areas are approaching zero and probably represent the small amount of kaolinite that was not intercalated. Such an observation is supported by our Xray diffraction results. This means that practically all of the OuOH groups are hydrogen bonded to the acetate ion. The effect of intercalation with potassium acetate not only results in the formation of a new Raman band at 3606 cm^{-1} with the concomitant subsequent loss of the ν_1, ν_4, ν_2 and ν_3 bands, but the inner hydroxyl group at 3620 cm^{-1} is considerably broadened. The size of the ditrigonal space is

1.32 Å and the size of the non-hydrated potassium ion is 1.33 Å so it is possible that the potassium ion fits into the ditrigonal hole and thus effects the polarisability of the InOH bond. In the case of the cesium acetate intercalate, this ν_5 band assigned to the inner hydroxyl group, has shifted to 3617 cm^{-1}. The inner hydroxyl can take up two positions in the kaolinite structure: firstly points towards the vacant dioctahedral hole and secondly points towards the ditrigonal space of the siloxane layer. It is now possible that the position of cesium, because of its non-hydrated cationic size of 1.65 Å, is such that it sits above the dioctahedral space and thus forces the inner hydroxyl group to point towards the ditrigonal hole. This then results in position of the inner hydroxyl band at 3617 cm^{-1}. Such a band position has also been observed in the Raman spectrum of kaolinites at liquid nitrogen temperatures. Further both the Raman and infrared spectra show the presence of water in the intercalated kaolinites. It is considered that the presence of water plays a significant role in the intercalation of the kaolinite with cesium acetate.

CONCLUSIONS

Remarkable insights into the structure of kaolinite and its intercalates have been obtained using a combination of FTIR and FT Raman spectroscopy.

ACKNOWLEDGMENTS

The financial support of the Queensland University of Technology Centre for Instrumental and Developmental Chemistry is gratefully acknowledged.

REFERENCES

1. Lagaly, G., *Phil Trans. R. Soc.* Lond. **A311,** 315-332 (1984).
2. Weiss, A., Thielepape, W., Ritter, W., Schafer, H., and Goring, G., *Allg. Chem.* **320,** 183-204 (1963).
3. Weiss, A., Thielepape, W., and Orth, H., *Intercalation into kaolinite minerals.* In Heller L, Weiss A editors. Proc. Int. Clay Conf. Jerusalem I, Jerusalem: Israel University Press, 277-293 (1966).
4. Frost, R.L., Fredericks, P.M., and Bartlett, J.R. *Spectrochim. Acta.* **20**, 667-674 (1993).
5. Frost, R.L., *Clays and Clay Minerals.* **43**, 191-195 (1995).
6. Frost, R.L., *Clay Minerals* **32**, 73-85 (1997).
7. Frost, R.L. , Tran, T. H., and Le, T., ***Mikrochimica Acta.*** **14**, 747-749 (1997).
8. Hess, A.C., and Saunders, V.R., *J. Phys. Chem.* **96**, 4367-4374 (1992).
9. Collins, D.R., and Catlow, C.R.A., *Acta Cryst* **B47**, 678-682 (1991).
10. Ledoux, R.L., and White, J.L., *J. of Colloid and Interface Science* **21**, 127-152 (1964).

Vibrational Spectroscopy of Ring Silicates

M. Sitarz, W. Mozgawa and M. Handke

University of Mining and Metallurgy (AGH),Department of Materials Science and Ceramics,al. Mickiewicza 30 30-059 Kraków, Poland.

IR and Raman spectra of silicates that contain isolated 3-membered ($Ca_3[Si_3O_9]$ and $Sr_3[Si_3O_9]$), 4-membered ($Pb_8[Si_4O_{12}]O_4$), 6-membered ($Na_4Ca_4[Si_6O_{18}]$ and $Na_6Ca_3[Si_6O_{18}]$) and 12-membered ($Na_{16}Ca_4[Si_{12}O_{36}]$) rings in their structure are presented. The bands corresponding to individual ring vibrations have been assigned. The number of members in the silicooxygen rings leads to changes in the shape and the positions of the bands originating from Si-O vibrational modes in the spectra. The influence of the cations outside the ring on the spectra has been analysed as well. Experimental IR spectra have been compared with the appropriate hypothetical ones ($H_6[Si_3O_9]$, $H_8[Si_4O_{12}]$, $H_{12}[Si_6O_{18}]$) calculated using semi-empirical molecular orbital method (PM3). This allowed to confirm the band assignments given by the present authors in their previous work (1).

INTRODUCTION

The main aim of the work is to determine the bands characteristic of different types of ring silicates. The conclusions drawn for ring silicates (cyclosilicates) can be adopted also for more complex silicate structures, such as layered or framework silicates containing in their structure tetrahedra forming silicooxygen rings. The number of papers devoted to spectroscopic studies of ring silicate is rather limited. A paper dealing with the spectroscopic studies of ring silicates was published by Choisnet and Deschanvres (2) who found that in the IR one band occurs at 760 - 700 cm^{-1}, and in the Raman two bands are observed in the 600 - 500 cm^{-1} region. These characteristic bands can be seen when the ring exhibits C_{3h} symmetry. Ring deformations lead to the splitting of these bands. Recently Michailova et al. (3) carried out a detailed analysis of vibrational spectra of various membered silico-oxygen rings. They established that the number of IR and Raman active modes is determined by the topological disorder whereas the number of ring members and the degree of puckering influences only the position and intensity of the normal vibrations. Introduction of boundary conditions results in the shifting of the bands in the high wavenumber region to higher values. Results of a recent literature study (4) indicate that frequencies computed using semi-empirical PM3, AM1 and MNDO methods compare well to values obtained by ab initio, and PM3 showed the closest correspondence to experimental values. Calculations by PM3 included in the MOPAC 6.1 program for siloxane ring molecules have been conducted by West and Hench (5,6). They found that the PM3 model achieved much more realistic values (close to the experimental measurements for amorphous silicates) than ab initio calculations. Moreover, the PM3 model produces infrared vibrational modes in the 1150 cm^{-1} and 464 cm^{-1} regions that match the experimental modes at amorphous silica.

RESULTS AND DISCUSSION

The previous study (1) has presented vibrational spectra (MIR) of synthetic ring silicates exemplifying various groups of cyclosilicates (3-($Ca_3[Si_3O_9]$, $Sr_3[Si_3O_9]$), 4-($Pb_8O_4[Si_4O_{12}]$) and 6-($Na_4Ca_4[Si_6O_{18}]$, $Na_6Ca_3[Si_6O_{18}]$) and carried out a detailed analysis of the compounds. Fig. 1 shows also the spectrum of 12-membered ring silicate, $Na_{16}Ca_4[Si_{12}O_{36}]$, prepared via devitrification of the glass of the same composition.

Configuration of bands within all the presented spectra is similar, however a change in the number of ring members and its deformation, as well as presence of various non-ring cations, affect the shape and location of the bands.

The most characteristic feature is a shift of the "analytical" band typical for a given kind of ring to lower wavenumber with an increase in the number of ring members. The spectra from 719 - 708 cm^{-1} (Fig.1a, Fig.1b) for the 3-membered rings to 654 cm^{-1} (Fig.1c), for the 4-membered rings and down to 627 - 615 cm^{-1} (Fig.1d, Fig.1e) for the 6-membered rings, conform to literature data. (7). In the MIR spectrum of $Na_{16}Ca_4[Si_{12}O_{36}]$ there is no intense band characteristic of ring silicates in this region. One can only observe a number of bands of low intensity at (710, 685, 648, 617, and 587 cm^{-1}) which may

CP430, *Fourier Transform Spectroscopy:* 11th International Conference
edited by J.A. de Haseth

be associated with the deformation of rings present in the structure of this compound. On the other hand, a high number of the bands of low intensity in this spectral region is typical of chain silicates (7,8). It is therefore probable that such large ring structure (12 tetrahedra in the ring) behaves in the IR as the chain silicate. More information of the vibrations of 12-membered rings can be provided only by Raman spectroscopy.

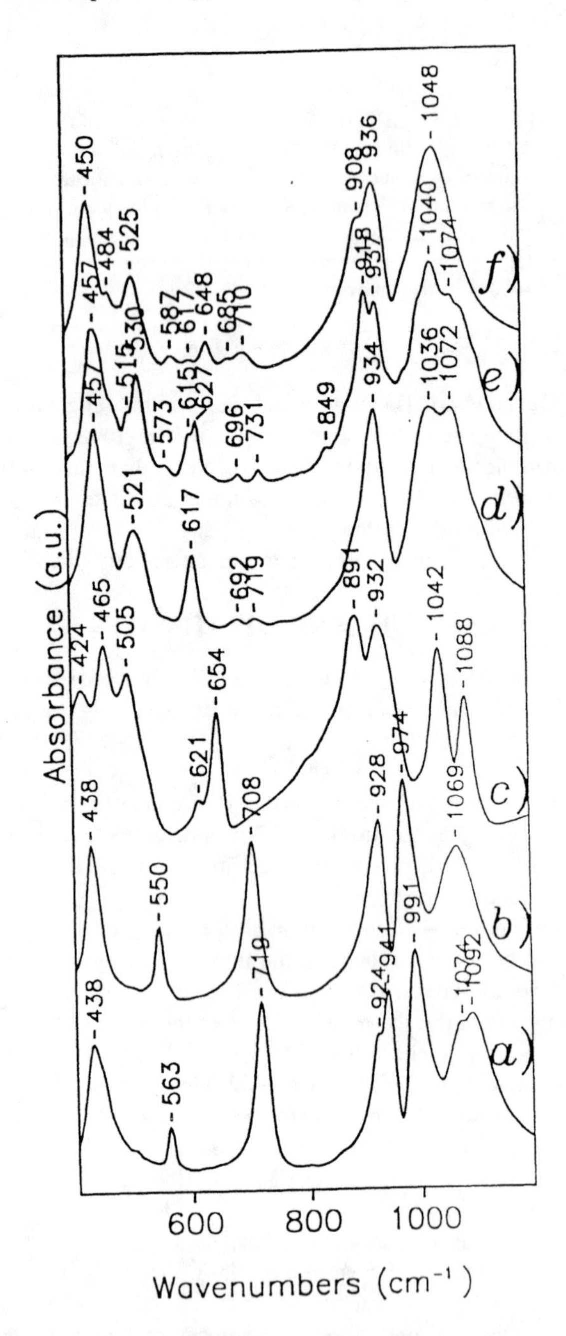

FIGURE . 1. MIR spectra of $Ca_3[Si_3O_9]$ (a), $Sr_3[Si_3O_9]$ (b), $Pb_8O_4[Si_4O_{12}]$ (c), $Na_6Ca_3[Si_6O_{18}]$ (d), $Na_4Ca_4[Si_6O_{18}]$ (e), and $Na_{16}Ca_4[Si_{12}O_{36}]$ (f).

Appearance of low-intense bands in the region of 800 - 500 cm^{-1} in the case of the 6-membered cyclosilicates (Fig.1d, Fig.1e) is most likely connected with a deformation of the rings (for 3 and 4-membered rings any significant deformation has not been noted). The number of the bands is higher in the case of $Na_4Ca_4[Si_6O_{18}]$ (Fig.1e), which is due to the deformation of the rings occurring in the compound structure.

The influence of non-ring cations transformation is most visible in the groups of 3-membered cyclosilicates, where the heavier cation Ca^{+2} (Fig.1a) moves the bands to lower wavenumber and significantly larger splitting than the Sr^{+2} cation (Fig.1b). The splitting may be also influenced by the change of symmetry.

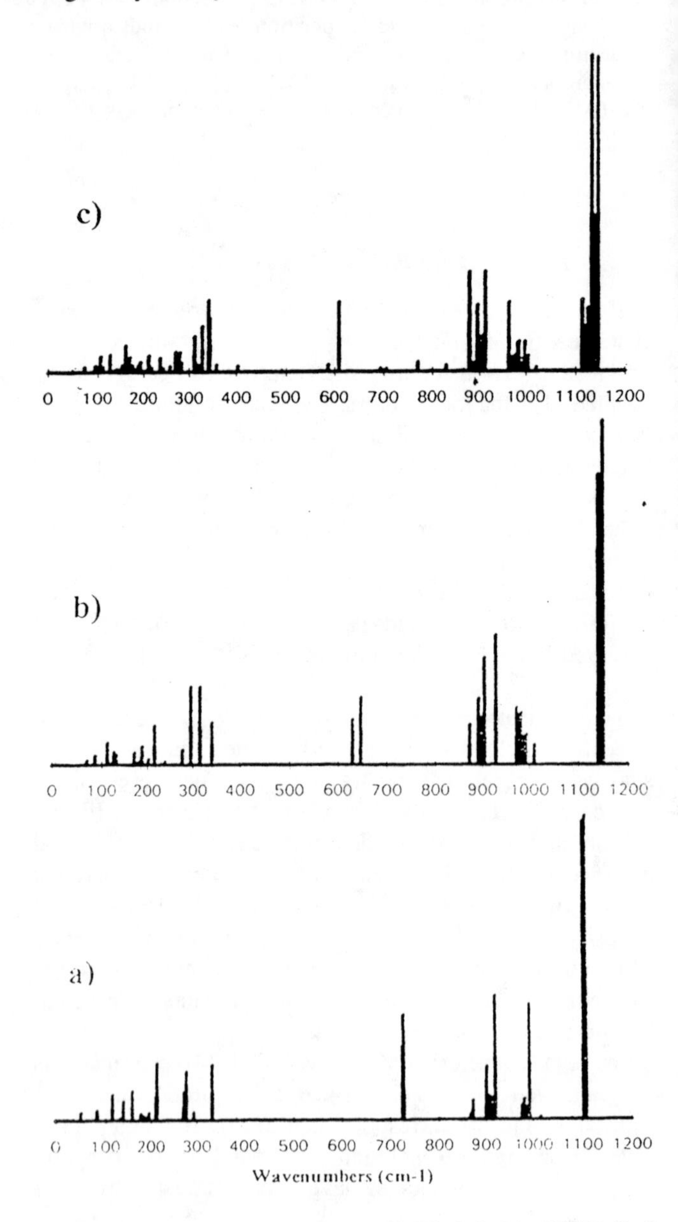

FIGURE 2. Calculated spectra of $H_6[Si_3O_9]$ (a), $H_8[Si_4O_{12}]$ (b), and $H_{12}[Si_6O_{18}]$ (c).

Computer calculations (geometric optimization) were carried out with PM3. Using PM3 (Hyper-Chem software) geometric optimization for the isolated silicooxygen 3-membered rings ($[Si_3O_9]^{-6}$), 4-membered rings ($[Si_4O_{12}]^{-8}$) and 6-membered rings ($[Si_6O_{18}]^{-12}$) were carried out. Since the algorithm requires a neutral molecule, the excess charge has been compensated with protons (H^+) which resulted in the hypothetical molecules $H_6[Si_3O_9]$, $H_8[Si_4O_{12}]$, $H_{12}[Si_6O_{18}]$. The optimization was carried out until minimum energy was reached. Hypothetical IR spectra of individual molecules after geometric optimization were calculated. Fig.#2 shows the calculated spectra. In the figures only the spectral range of 0 - 1200 cm^{-1} is show since above 1200 cm^{-1} there are no bands due to silicooxygen vibrations. According to the literature data (3,5), as well as the obtained results, it has been concluded that in the spectra of $H_6[Si_3O_9]$, the band at 726 cm^{-1} is the characteristic band of silicooxygen rings (Fig.2a). In the ranges of 993 - 874 cm^{-1} and 1109 - 1105 cm^{-1} bands due to nonbridging $Si\text{-}O^-$ stretching and bridging Si-O(Si) vibrations occur respectively.

In the spectra of $H_8[Si_4O_{12}]$ (Fig.2b) and $H_{12}[Si_6O_{18}]$ (Fig.2c) the type of bands in the 993 - 870 cm^{-1} and 1150 - 1010 cm^{-1} is similar as in the spectrum of $H_6[Si_3O_9]$. It is clearly seen that the bands typical of silicooxygen rings vibrations shift to lower wavenumber with the increase in the number of ring members. These bands occur at 650 and 633 cm^{-1} in the spectrum of $H_8[Si_4O_{12}]$ and at 610 cm^{-1} in the spectrum of $H_{12}[Si_6O_{18}]$.

In Fig. 3 Raman spectra of selected ring silicates are presented. It can be seen that the band positions and shapes are similar. Two intense bands at 983 – 966 cm^{-1} and 585 – 563 cm^{-1} are the most characteristic features of these spectra. As it follows from the literature (2,8) the band at 983 - 966 cm^{-1} shows high ^{28}Si - ^{30}Si isotope shift (18 - 10 cm^{-1}) and it originates from the stretching Si-O-Si vibrations. The band at 585 - 563 cm^{-1} shows much lower isotope shift and it is characteristic of ring structures (2,8). It is interesting to note that in the Raman spectrum of 12-membered ring (Fig.3) the band characteristic of the ring structure at 585 cm^{-1} occurs even though in the IR spectrum the band associated with this structure does not appear.

Thus it can be concluded that in the Raman spectra of ring silicates there appears a characteristic intense band at 585 - 563 cm^{-1} associated with the ring structure. The intensity of this band becomes slightly lower (with simultaneous increase in the full width at half maximum) as the number of the ring members increases, which is probably due to the increased ring deformations. It is also noteworthy that the change of the cation influences this band position more significantly than the number of ring members (see Fig.3).

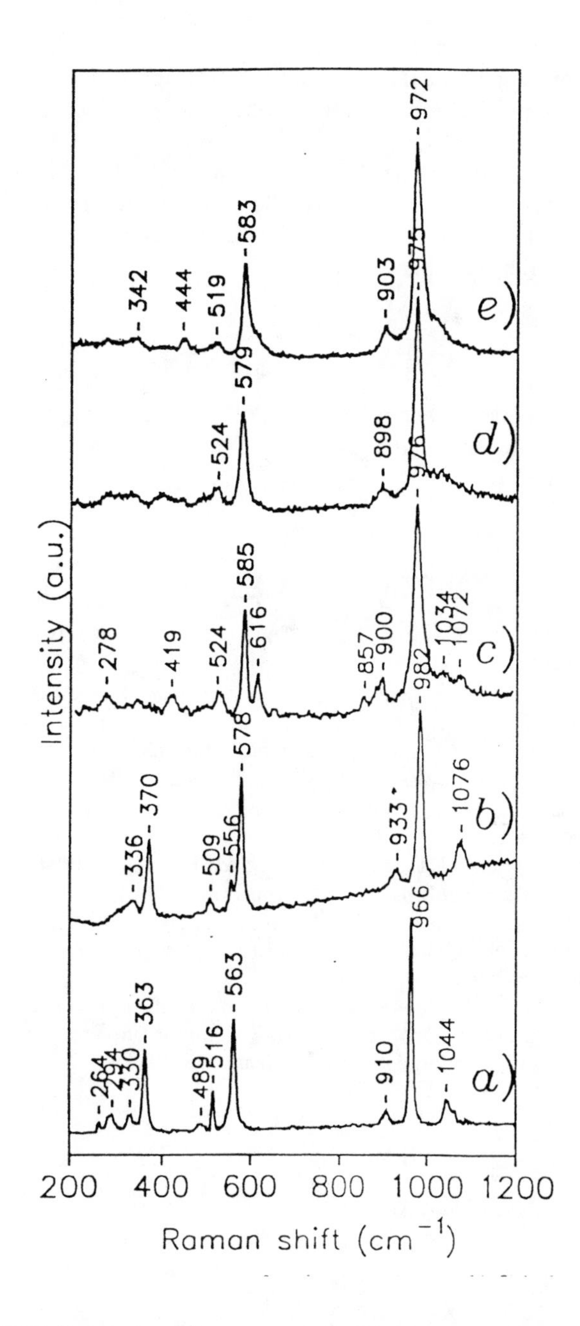

FIGURE 3. Raman spectra of $Sr_3[Si_3O_9]$ (a), $Ca_3[Si_3O_9]$ (b), $Na_4Ca_4[Si_6O_{18}]$ (c), $Na_6Ca_3[Si_6O_{18}]$ (d), and $Na_{16}Ca_4[Si_{12}O_{36}]$ (e).

Spectroscopic studies of ring silicates as well as computer calculation that were carried out made it possible to recognize that bands originating from the vibrations of silicooxygen rings occur in the more complicated silicates structure. The bands typical of ring silicates are located in the range of 730 - 610 cm^{-1} (except for large cyclic systems).

The ring band asssignments can be exemplified by the spectrum of crystalline $K_2[Si_2O_5]$. This compound contains in its structure the rings composed exclusively of 6-

membered silicooxygen rings. The bands due to their vibrations appear in IR spectra in the range of 700 - 600 cm^{-1}. The precise determination of the band positions and other parameters is possible only after the spectral decomposition which has been shown in Fig. 4. The great number of bands in the region of 700 - 600 cm^{-1} (626, 640, and 669 cm^{-1}) may be explained by the ring deformations which results in the band splitting.

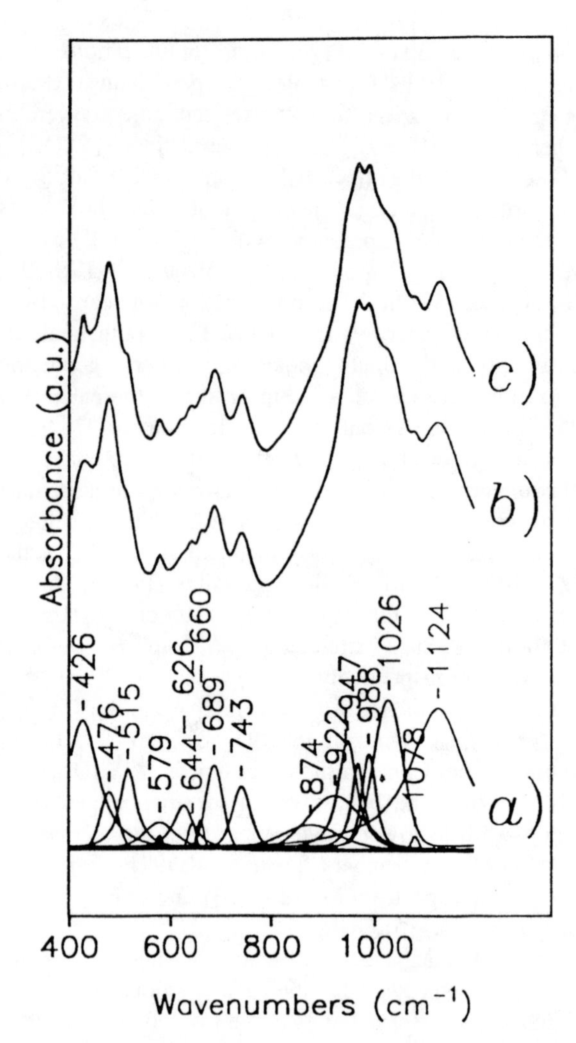

FIGURE 4. Decomposition of IR spectrum of crystalline $K_2[Si_2O_5]$, peaks (a), curvefit (b), and original spectrum (c).

ACKNOWLEDGMENTS

This work is supported by Polish Committee for Scientific Research under grant no. 3 T09A 070 11 and no. 7 T08D 020 10.

REFERENCES

1. Sitarz M., .Mozgawa W., Handke M., *J. Mol. Str.,* **404** 193-197 (1997).
2. Choisnet J., Deschanvres A., *Spectrochim. Acta* **31A** 1023-1034 (1975).
3. Mihailova B., Zotov N., Marinov M., JNikolov J., Konstantinov L., *J.Non-Cryst. Solids* **168** 265-274 (1994).
4. Seeger D.M., Korzeniewski C., KowalczykW., *J. Phys. Chem.*, **95** 68 (1991).
5. West J.K., Hench L.L., *J. Non-Cryst. Solids* **180** 11-16 (1994).
6. West J.K., Hench L.L., *J. Am. Ceram. Soc.*, **78** 1093-1096 (1995).
7. Lazariev A.N., „*Kolebatielnnyje spektry i strojenije silikatov*", Nauka, Leningrad 1968.
8. Handke M., „ *Vibrational Spectroscopy of the Silicates and Si-O Bond Character in Silicates*", Sci. Bulletins of University of Mining and Metallurgy, No. 990, Ceramics 48, 1984.

Charge Transfer in some Lanthanum Transition Metallates with the Perovskite and K_2NiF_4 Structures: an Infrared Spectroscopic Study

Richard Mortimer

Defence Evaluation and Research Agency, Fort Halstead, Sevenoaks, Kent, TN14 7BP UK.

The electronic properties of transition metal-containing systems sometimes indirectly manifest themselves through their infrared spectra. The infrared spectra of some lanthanum and dilanthanum transition metallates have been probed either as a function of composition or applied pressure. It is suggested that there is charge transfer to and from neighbouring transition metal ions. In some cases, La_2CuO_4 for instance, charge separation (valence localisation) is manifest at ambient temperature and pressure. In others, such as $LaV_{0.5}Ni_{0.5}O_3$, it is necessary to apply an external pressure in order to induce charge separation.

INTRODUCTION

Over the last three quarters of a century, transition metal oxides have been recognised to demonstrate some rather remarkable electronic phase transitions, such as the semiconductor-to-metal phase transition in VO_2 and V_2O_3 (1) and charge disproportionation in Fe_2O_3 (2). On going from binary transition metal oxides to ternary oxides, the latter containing both non-transition and transition metal ions, gives rise to a wider variety of materials with the possibility of an even greater range of electronic properties. Two important types of ternary oxide are based either on the structure of barium bismuthate ($BaBiO_3$) -the perovskite structure- or dipotassium nickel tetrafluoride (K_2NiF_4) -the K_2NiF_4 structure. The interest in lanthanide transition metallate perovskites primarily stems from research carried out about thirty years ago with lanthanum cobaltate, $LaCoO_3$, (3). It was found that this material possesses both a progressive thermally-induced low to high spin-state transition as well as a semiconducting-to-metal transition.

Further interest in these types of materials emerged when, in 1986, the first high temperature (T>ca. 30 K) superconductor (4) was discovered and it was based upon alkaline-earth metal substituted dilanthanum cuprate (La_2CuO_4). A more recent discovery has been that of colossal magnetoresistance (CMR) in hole-doped manganese perovskite (5). The lanthanum nickel perovskite ($LaNiO_3$) is metal-like from at least 1.5 to 1200 K. However, on going to other lanthanide nickelates, a semiconductor to metal phase transition (6) is induced, the temperature of which is inversely proportional to the lanthanide ionic radius. This phenomenon has been explained in terms of closing of a charge gap generated by an increase of the electronic bandwidth (7). The analogous cuprate perovskite, however, has been little studied: this is probably associated with the difficulty in its preparation -it requires an oxygen pressure of oxygen about $6 * 10^9$ Pa. However, it is known to be isostructural to $LaNiO_3$ and its synthetic, electrical magnetic properties have been reported (8). Further, Bringley et al. (9) have found that the copper valence can be varied by partial reduction of lanthanum cuprate ($LaCuO_3$) to form the defect perovskite series $LaCuO_{3-\delta}$. It was found that there are three phases, tetragonal, monoclinic and orthorhombic with δ ranging from about 0 to 0.17, 0.21 to 0.38 and 0.41 to 0.5, respectively.

The properties of K_2NiF_4-like transition metallates are rather less well documented. However, Ogita et al. (10) have studied the very different behaviour of La_2CuO_4 and La_2NiO_4 towards partial substitution of lanthanum by strontium. In the former, there are profound changes in the infrared spectrum even with only low levels of strontium substitution (ca. 3%): this is not the case for the nickel analogue. The authors comment upon the significance of the collapse of phonon features in the infrared spectrum of strontium-doped La_2CuO_4 and the onset of sub-ambient temperature superconductivity in these materials.

In this paper, it will be demonstrated that charge transfer manifests itself in the infrared spectra of some lanthanum and dilanthanum first-row transition metallates. In some cases this charge transfer can be manipulated by the application of pressure and in some others it is by composition dependence.

EXPERIMENTAL

All materials were prepared by the thermal decomposition of a solution containing the appropriate quantities of metal nitrates (purity 99.9 % or better) for 16 hours followed by cooling to ambient temperature at a

CP430, *Fourier Transform Spectroscopy:* 11th International Conference
edited by J.A. de Haseth

rate of 1 K min.[-1] All materials were assessed for phase purity using powder X-ray diffraction with a home-built powder diffractometer with a position sensitive detector. For the series $LaNi_xCu_{1-x}O_{3-\delta}$, scanning electron microscopy was used to confirm the ratios of the metal ions; the values of 3-δ were found to be 2.62 with a spread of +/- 0.06 (assuming no impurities). Infrared spectra (KBr pellets) were obtained at ambient temperature and pressure (ca. 295 K and 0.1 $*10^6$ Pa) using a Nicolet 740 infrared spectrometer and a deuterated triglycine sulfate detector. For applied pressure measurements, a liquid nitrogen cooled mercury cadmium telluride detector was used along with a diamond anvil cell using gaskets with approximately 350 μm diameter apertures. For determining the applied pressure, an internal vibrational pressure calibrant was used (11).

RESULTS

Charge Transfer in Dilanthanum Cuprate and Lanthanum Vanadium Nickelate

The infrared spectra of dilanthanum nickelate (La_2NiO_4), dilanthanum cuprate (La_2CuO_4) and dilanthanum chromium cuprate ($La_2Cr_{0.5}Cu_{0.5}O_4$) in the transition metal-oxygen stretching region are shown in Fig. 1 (A), (B) and (C), respectively. Powder X-ray diffractograms of these materials, in the same order, are shown in Fig. 2.

The infrared spectra of La_2CuO_4 and La_2NiO_4 are characterised by two relatively broad bands at 686/516 and 653/505 cm,[-1] respectively. The spectrum of $La_2Cr_{0.5}Cu_{0.5}O_4$, is somewhat different there being a single strong relatively broad band around 619 cm^{-1} and relatively weak shoulders around 680 and 521 cm.[-1] There is also a relatively strong complex family of bands around 900 cm.[-1] The X-ray diffraction pattern of $La_2Cr_{0.5}Cu_{0.5}O_4$ more closely resembles that of La_2NiO_4 than La_2CuO_4 (Fig. 2). The pressure dependencies of La_2CuO_4 and $La_2Cr_{0.5}Cu_{0.5}O_4$ have been investigated but no generation or collapse of bands analogous to pressure cycling in $LaV_{0.5}Ni_{0.5}O_3$ vide infra has been observed to at least 5.3 and 7.1 $*10^9$ Pa, respectively.

The infrared spectrum of $LaV_{0.5}Ni_{0.5}O_3$ (Fig. 3) in this

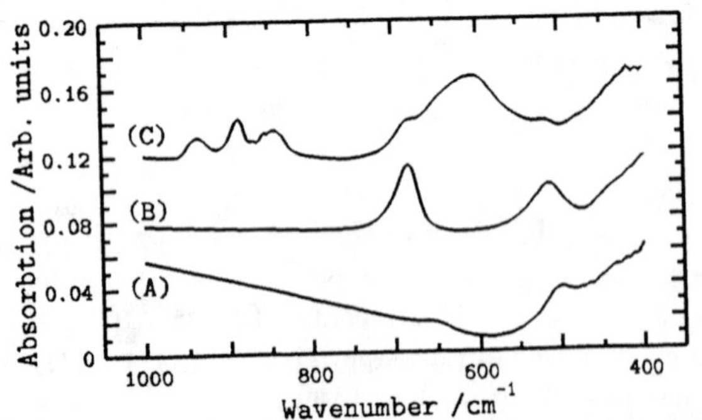

FIGURE 1. Infrared spectra of La_2NiO_4 (A), La_2CuO_4 (B) and $La_2Cr_{0.5}Cu_{0.5}O_4$ (C). (A) and (B) are normalised.

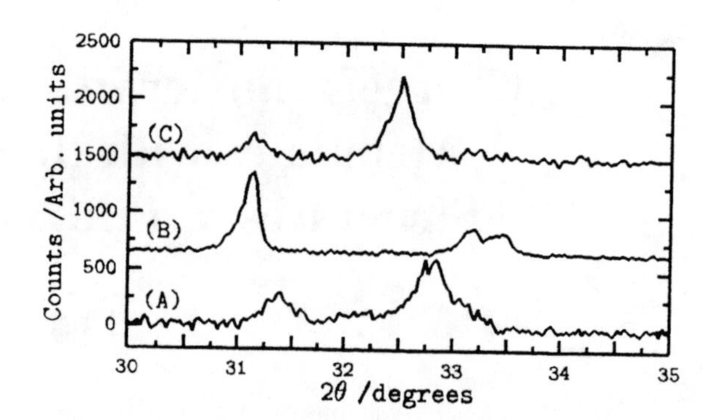

FIGURE 2. X-ray diffractograms for La_2NiO_4 (A), La_2CuO_4 (B) and $La_2Cr_{0.5}Cu_{0.5}O_4$ (C).

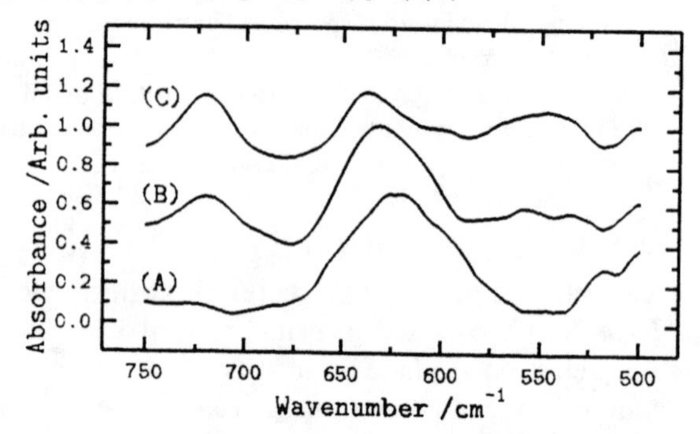

FIGURE 3. Pressure-dependent infared spectra of $LaV_{0.5}Ni_{0.5}O_3$ at 2.1 (A), 2.4 (B) and 2.6 (C) $*10^9$ Pa.

region comprises a single band around 618 cm^{-1} at ambient pressure (ca. $0.1*10^6$ Pa) and this remains the case up to about 2.3 $*10^9$ Pa. Above this pressure, new bands at 717 and 545 cm^{-1} appear, although the parent band around 618 cm^{-1} does not disappear completely. This transition is reversible.

Stabilisation of Lanthanum Cuprate using Nickel

The infrared spectra of some typical lanthanum transition metallates are shown in Fig. 4. These spectra are dominated by a single relatively broad band around 600 cm,[-1] bands to the lower edge of our detector limit around 400 cm^{-1} and, in some, bands around 900 cm^{-1} are also seen. The infrared spectra of "lanthanum cuprate" and some lanthanum nickel cuprates ($LaNi_xCu_{1-x}O_{3-\delta}$, 3-δ~

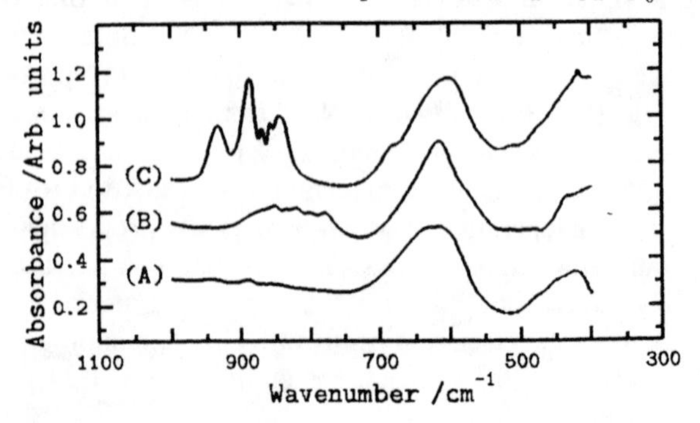

FIGURE 4. Infrared spectra of $LaCrO_3$ (A), $LaMnO_3$ (B) and $LaCr_{0.5}Cu_{0.5}O_3$ (C).

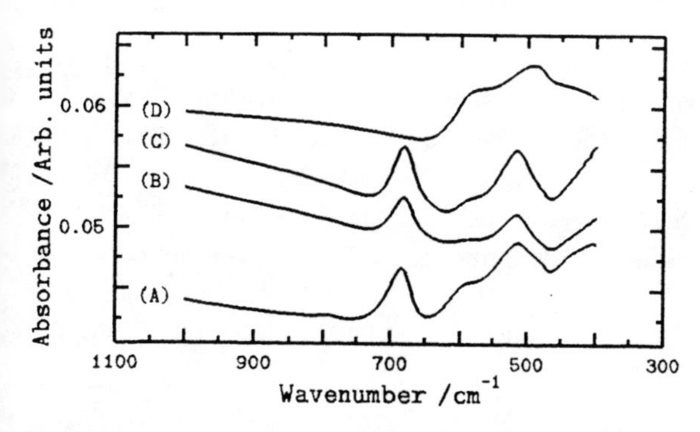

FIGURE 5. Infrared spectra of "LaCuO3" (A) $LaNi0_{.17}Cu_{0.83}O_{3-\delta}$ (B), $LaNi_{0.12}Cu_{0.88}O_{3-\delta}$ (C) and CuO (D).

2.62) are shown in Fig. 5. In contrast to the typical spectra of lanthanum perovskites, there are two bands in the ν_{TM-O} region located at 680 and 516 cm.$^{-1}$ The relative intensities of these two bands does not change with x. At relatively low values of x (ca. 0.17 and below), a shoulder develops on the lower wavenumber band. New peaks arise in the X-ray diffraction patterns for materials with low values of x (ca. 0.17 and below) and correlates with the emergence of the shoulder in the infrared.

Lanthanum Nickel Cuprate - A New Candidate for Superconductor Studies?

The infrared spectruum of a strontium-substituted lanthanum nickel cuprate ($La_{0.971}Sr_{0.029}Ni_{0.3}Cu_{0.7}O_{3-\delta}$) and the superconductor $YBa_2Cu_2O_{7-\delta}$ are shown in Fig. 7. The relatively simple structure of a typical lanthanum nickel cuprate persists with spectral features around 684 and 508 cm.$^{-1}$ There are, however, a number of new spectral bands not found in the lanthanum nickel cuprates around 860, 1370/1470 and 1640 cm.$^{-1}$ In the superconductor $YBa_2Cu_3O_{7-\delta}$, there are similar bands around 574, 630, 667, 1414 and 1634 cm^{-1} and bands near to the low wavenumber cut-off of the detector.

DISCUSSION

Charge Transfer

The infrared spectra of "$LaCuO_3$", $LaNi_xCu_{1-x}O_3$ (x>ca. 0.17) and La_2CuO_4 exhibit two relatively narrow peaks around 683 and 520 cm^{-1} which arise from copper-oxygen stretching (ν_{TM-O}) motion -deformation modes are typically found around 400 cm.$^{-1}$ Many first-row transition metal perovskites exhibit broad features around 600 cm,$^{-1}$ for example $LaTMO_3$ where TM=Cr, Mn, Fe, and Co and thus the spectral behaviour of copper is considered to be anomalous. The spectral similarity between these copper-containing materials, however, suggests a common origin for these features and it is to this which we now turn.

The collapse of spectral features in "$LaCuO_3$" and La_2CuO_4 upon partial chromium substitution of copper is unlikely to have a purely structural origin. La_2CuO_4 has

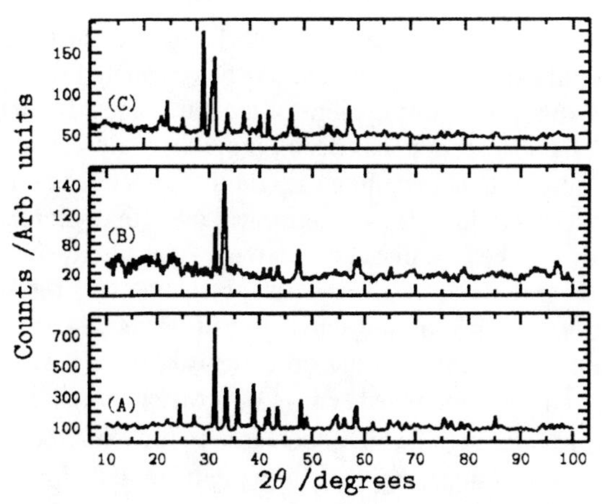

FIGURE 6. X-ray diffractograms of "$LaCuO_3$" (A), $LaNi_{0.17}Cu_{0.83}O_{3-\delta}$ (B), $LaNi_{0.12}Cu_{0.88}O_{3-\delta}$ (C)

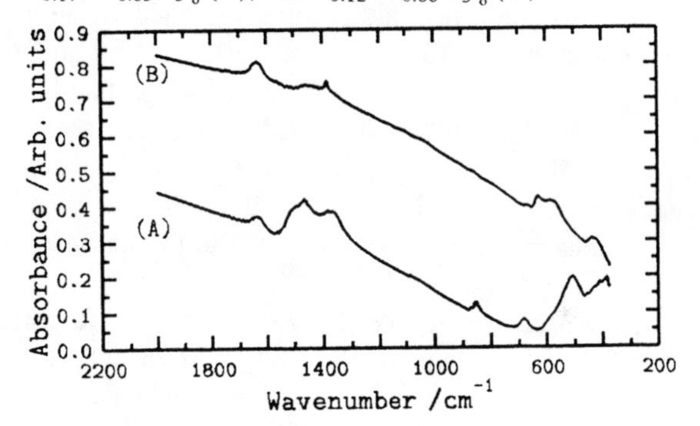

FIGURE 7. Infrared spectra of $La_{0.971}Sr_{0.029}Ni_{0.3}Cu_{0.7}O_{3-\delta}$ (A) and $YBa_2Cu_3O_7$ (B).

tetragonal symmetry as probably does $La_2Cr_{0.5}Cu_{0.5}O_4$, although their infrared spectra are very different. Also, it is unlikely that the substitution of a small amount of lanthanum by strontium in La_2CuO_4 would effect the collapse of the 682 cm^{-1} band because of relatively minor structural changes [Ogita (10)]. Further, in the case of the dilanthanide cuprates, only when the lanthanide ion is lanthanum (15) is there a spectral feature at such a relatively high wavenumber (683 cm.$^{-1}$) It is interesting to note that the crystallographic structure of La_2CuO_4 is orthorhombic whereas other dilanthanide cuprates (La=Pr, Nd and Sm) as well as $LaNiO_4$ are tetragonal. It a series such as this, it is common for the crystallographic structure to distort and adopt a lower symmetry as the lanthanide ion radius decreases: in this case, the reverse is the case. Also, the powder X-ray diffractogram of $La_2Cr_{0.5}Cu_{0.5}O_4$ more closely resemble that of La_2NiO_4 than La_2CuO_4. These data certainly suggests that there is a distortive effect in La_2CuO_4, probably arising from the interplay of subtle electronic effects. Maybe the chromium ions distorts or isolates electronic interactions between copper ions. Certainly in $LaCrO_3$ the electrical resistivity (12) suggests a system in which the "d" transition metal electrons are localised.

A further insight into these systems comes from the pressure dependence study of $LaV_{0.5}Ni_{0.5}O_3$. It is unlikely that any structural change (without change of electronic state) would cause a sufficient change of vibrational force-field to explain the development of the new bands. In fact, since this phenomenon has been previously observed in $LaNi_{0.5}Fe_{0.5}O_3$ (13), a common origin for this behaviour needs to be sought. It is suggested here that there is increased charge localisation between transition metal ions at around the phase transition pressure. This results in the strengthening of one set of transition metal-oxygen bonds and a weakening of the other set, although it is not possible to determine whether it is $\nu_{V\text{-}O}$ which increases in wavenumber and $\nu_{Ni\text{-}O}$ which decreases in wavenumber or vice versa. Transfering this argument to the copper-containing materials, in La_2CuO_4 and the nickel-stabilised copper perovskites this charge transfer must be relatively complete since the spectral features assigned to copper-oxygen stretching are found around 680 and 520 cm^{-1} but no residual feature around 600 cm^{-1} where bands are typically found for lanthanum and dilanthanum transition metallates. The reason for the incomplete transfer of charge in the hetero-transition metal analogues possibly lies in the electronic inequivalence of the transition metal ions.

We have used the term "charge transfer" in a rather generic way. The Zaanen, Sawatzky, Allen (ZSA) (14) classification allows for two general types of gap: the Mott-Hubbard gap due to the correlation energy U and the charge transfer gap associated with an energy Δ. Associated with these two types of gaps, there are two types of insulators: Mott insulators (where the smallest gap is associated with U) and charge-transfer insulators (where the smallest gap is associated with Δ). As a result, there are two types of metal-insulator phase transitions, either as the Mott-Hubbard gap closes or when the charge gap closes. Further work would be required to establish the relative contributions of U and Δ in these systems.

A New Candidate for Superconductivity Studies?

The X-ray powder data indicates that the $LaNi_xCu_{1-x}O_{3-\delta}$ series are not homogeneous, although significant amounts of the imputrities La_2CuO_4 and CuO can only be detected with relatively high copper contents ($x<0.17$). In $LaNi_xCu_{1-x}O_{3-\delta}$ with $x>0.17$ impurities might arise from different perovskite-like phases. Interestingly, the oxygen contents in $LaNi_xCu_{1-x}O_{3-\delta}$ are in the vicinity where there is a transition from a tetragonal to a monoclinic phase in $LaCuO_{3-\delta}$ (δ ca. 0.2).

The infrared spectra and powder X-ray data for $LaNi_xCu_{1-x}O_{3-\delta}$ are similar to La_2CuO_4. Since La_2CuO_4 is a useful precursor to superconducting materials it is of interest to enquire whether the lanthanum nickel cuprates could be potential candidates as superconducting materials. The infrared spectra of $La_{0.971}Sr_{0.029}Ni_{0.3}Cu_{0.7}O_{3\delta}$ and the superconductor $YBa_2Cu_3O_{7-\delta}$ have some similarities, having considerable spectral intensity in the anharmonic region, ca. 1600 cm^{-1}. In the $LaNi_xCu_{1-x}O_{3-\delta}$ materials, there are features in this region but they are extremely weak. The Cu-O stretching region is more complicated in $YBa_2Cu_3O_{7-\delta}$ than in $La_{0.971}Sr_{0.029}Ni_{0.3}Cu_{0.7}O_{3-\delta}$. It is likely that the Cu...Cu interactions in the latter need to be reduced in order to arrive at materials in which the electronic forces acting in this type of material are more like those in the superconductor.

CONCLUSIONS

Infrared spectroscopy has been used to probe charge transfer between transition metal ions in some lanthanum- and dilanthanum transition metallates. Strong charge transfer is found in the copper-containing materials. These interactions may be removed by the partial substitution of copper by electronically insulating ions. It has been found that the partial substitution of copper by nickel will stabilise an oxygen-deficient perovskite series. The new nickel copper materials $LaNi_xCuO_{3-\delta}$ could potentially form the basis of a new class of superconducting materials.

ACKNOWLEDGEMENTS

I wish to thank M. T. Weller for donation of $YBa_2Cu_3O_7$.

REFERENCES

(1) Mott N. F., *Metal Insulator Transitions*, England, Taylor and Francis Ltd., 1974, Chapter 5.
(2) Mott N. F., *Metal Insulator Transitions*, England, Taylor and Francis Ltd., 1974, Chapter 4.
(3) Madhusudan Hanumantharo W., Krishnaswarmy J., Ganguly P., Rao Ramachandra N. C., *J. Chem. Soc. (Dalton)*, 1397-1400, 1980 and references therein.
(4) J. G. Bednorz and K. A. Muller, *Z. Phys.* B, **64**, 189-193 1986.
(5) A. P. Ramirez, *J. Phys.: Condens. Matter*, scheduled October, 1997.
(6) Garcia-Munoz J. L. and Rodriguez-Carvajal J., *Phys. Rev.*, **46**, 4414-4425, 1992.
(7) Torrance J. B., Lacorre P., and Nazzal A. I., *Phys. Rev.*, **45**, 8209-8212, 1992.
(8) Goodenough J. B., Mott N. F., Pouchard M., Demazeau G., and Hagenmuller P., *Mat. Res. Bull.*, **8**, 647-656, 1973.
(9) J. F. Bringley, B. A. Scott, S. J. La Placa, R. F. Boeheme, T. M. Shaw, M. W. McElfresh, S. S. Trail and D. E. Cox, *Nature*, **347**, 263,-265 1990.
(10) Ogita N. Udagawa M., Kojima K. and Ohbayashi K. *J. Phys. Soc. Japan*, **57**, 3932-3940, 1988.
(11) Klugg D. and Whalley E., *Rev. Sci. Instrumen.*, **54**, 1205-1208, 1983.
(12) P. Ganguly and N. Y. Vasanthacharya, *J. Solid State Chem.*, **61**, 164-170, 1986.
(13) R. Mortimer, *Synthetic Metals*, **45**, 2165-2166, 1997.
(14) J. Zaanen, G. A. Sawatzky and J. W. Allen J. Solid State Chem., **88**, 8-27, (1990).
(15) K. K. Singh and P. Ganguly, *Spectrochim. Acta*, **40A**, 539-545, 1984.

3.12 STRUCTURAL ANALYSIS

New Light On An Old Subject : An FT-Raman Study Of Ion Association In Aqueous Solutions Of Cobalt(II) Nitrate

Bradley M. Collins

School of Chemistry, University of Sydney, Sydney NSW 2006, Australia

Raman spectroscopy has been ignored as a tool to investigate ion association in highly coloured salt solutions. With the advent of FT-Raman many of the problems associated with conventional Raman spectroscopy have been eliminated or minimized. This study shows that it is possible to study ion association in highly coloured aqueous cobalt(II) nitrate solutions. Cobalt (II) nitrate has been found to be highly associated in aqueous solution in the concentration range 0.5 m to near saturation. The ion species in solution consist of the free aquated nitrate and cobalt(II) ions and two aquated metal nitrate ion pairs.

INTRODUCTION

There is a long history of probing ion association in salt solutions with the aid of Raman spectroscopy (1,2). High symmetry, multi-atom anions, such as nitrates and perchlorates, have very distinctive vibrational spectra. As these ions participate in ion association and solvation in solution, their vibrational spectra change as a result of a lowering of symmetry. This can manifest itself in the loss of degeneracy in certain bands, the appearance of forbidden bands or the appearance of new bands.

Traditionally this work has been performed using conventional Raman spectroscopy employing visible excitation sources such as krypton and argon ion lasers (1,2). More recently FT-Raman spectroscopy has been used to probe ion association (3-6). In FT-Raman spectroscopy the excitation source is an Nd:Yag laser emitting at 1064 nm. Due to problems associated with fluorescence, self absorption and resonance effects, ion association studies have been confined to colourless solutions.

The advent of FT-Raman spectroscopy has meant that most of these problems are eliminated or minimised. Researchers have not taken advantage of FT-Raman to study coloured salt solutions but have stayed with the familiar colourless solutions. This study looks at ion association in the highly coloured aqueous cobalt(II) nitrate solution assessing the use of FT-Raman as a tool for such studies.

EXPERIMENTAL

Solutions of cobalt(II) nitrate were produced using cobalt(II) nitrate hexahydrate (Merck AR Grade). Solutions were made using the temperature and pressure independent molality scale. To produce solutions, the solid for each solution was weighed and the appropriate weight of water was added to make up the correct molality.

Raman spectra were acquired on a Bruker RFS100 dedicated FT-Raman spectrometer. The excitation source was an Nd:YAG laser delivering an excitation wavelength of 1064 nm. Spectra were run with a resolution of 1 cm^{-1} and zero filled to give an apparent resolution of 0.2 cm^{-1}. A laser power of 350 mW was used. Solutions, held in glass cells at *ca.* 22°C, were presented to the spectrometer in a 180° backscattering arrangement.

Band analyses and spectral manipulations were performed using Grams386 (Galactic Industries). All band analyses were initially constrained but after initial fitting were allowed to vary in both band position, width at half height and band shape. A mixture of Gaussian and Lorentzian band shape was used. Initial band positions were estimated using deconvolution, second derivative and subtraction methods. Consistent trends in band shape, width at half height and band position along with residual spectra, statistical measures and minimum number of bands required were used to determine the goodness of fit.

RESULT AND DISCUSSION

The unperturbed nitrate ion has D_{3h} symmetry. This gives rise to four vibrations, the symmetric stretch ($\nu_1(A_1')$), the antisymmetric stretch ($\nu_3(E')$), the out of plane deformation ($\nu_2(A_2'')$) and the in-plane deformation ($\nu_4(E')$). In the unperturbed nitrate these would occur at approximately 1050, 1380, 830 and 716 cm^{-1}, respectively (1,2). Only ν_1, ν_3 and ν_4 are Raman active although ν_2 can become Raman active if the symmetry of the ion is lowered. A typical nitrate Raman spectrum is given in Fig. 1.

The symmetry of the nitrate ion is lowered through interactions with its environment. Solvation and ion

CP430, *Fourier Transform Spectroscopy:* 11th International Conference
edited by J.A. de Haseth

association will lead to changes in the spectrum. The antisymmetric stretch is very sensitive to symmetric distortions and appears as a broad doublet even in very dilute aqueous solution. The in-plane deformation can also be split by symmetry lowering. The stronger the interaction or distorting electronic field applied to the ion, the larger the splitting of the degenerate band becomes.

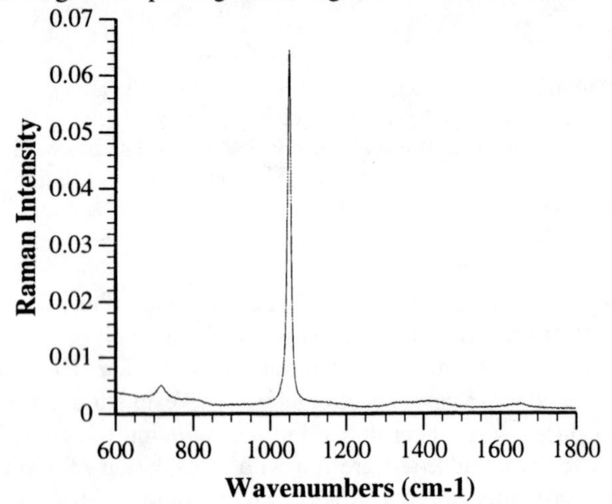

FIGURE 1. A typical nitrate ion Raman spectrum from a 4.0 m aqueous cobalt(II) nitrate solution.

The most important part of the nitrate spectrum is the symmetric stretch, ν_1. It is a nondegenerate band that is very sensitive to changes in the nitrate ion environment. All new bands observed in this region can be assigned to new chemical environments of the nitrate ion. In dilute aqueous nitrate solution, ν_1 is observed at 1047.6 cm^{-1} (1-3). The nitrate ν_1 of the aqueous cobalt(II) nitrate solutions was observed as a band profile asymmetric to higher energy with a band maximum around 1048 cm^{-1}. The band maximum for the nitrate ν_1 showed a movement to higher energy with increasing salt concentration. A plot of band maximum versus concentration for the cobalt(II) nitrate solution (Fig. 2) shows a linear change in the band position with concentration. The *y* intercept of the plot is 1047.65 cm^{-1} and is consistent with the position of the nitrate ν_1 in dilute solution.

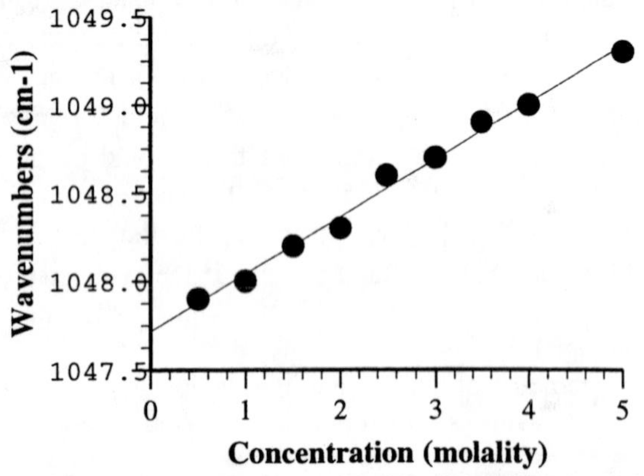

FIGURE 2. A plot of nitrate ν_1 band maximum position versus salt concentration for aqueous cobalt(II) nitrate solutions.

Attempts at deconvolution of the nitrate ν_1 band profiles were inconclusive but second derivative and spectral subtraction pointed to the existence of at least two bands under the band profile. These bands were located at approximately 1050 and 1047 cm^{-1}. Initial band analyses were carried out on all spectra using these two bands. It was found that above a salt concentration of 3 m it was necessary to introduce a third band at 1040 cm^{-1}. The spectra were fitted to obtain consistent trends with band position, band width at half height and band shape. Results of the band analyses are shown in Table 1. A typical band analysis is shown in Fig. 3.

TABLE 1. Normalised band areas calculated from the band analyses of the nitrate ν_1 for aqueous solutions of cobalt(II) nitrate.

Concentration (molality)	1047.6 cm^{-1} Band Area	1050 cm^{-1} Band Area	1040 cm^{-1} Band Area
0.5	0.8827	0.1173	
1.0	0.8107	0.1893	
1.5	0.7533	0.2467	
2.0	0.6762	0.3238	
2.5	0.5934	0.4066	
3.0	0.5551	0.4430	0.0019
3.5	0.4207	0.5437	0.0356
4.0	0.3808	0.5765	0.0427
5.0	0.2184	0.6955	0.0861

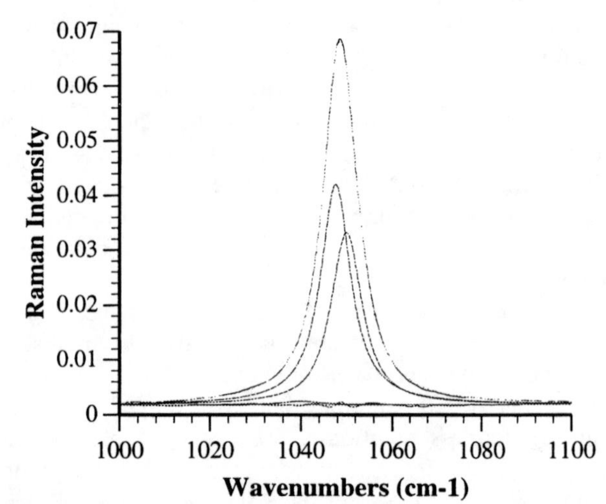

FIGURE 3. A typical band analysis of the nitrate ν_1 band profile. A 2.5 m aqueous cobalt(II) nitrate solution.

A plot of the normalised relative band intensity for the fitted bands versus salt concentration is shown in Fig. 4. Each one of these bands represents a different chemical environment for the nitrate ion. The band at 1047 cm^{-1} is attributed to the free aquated nitrate. The band position of 1050 cm^{-1} is consistent with nitrate involved in an ion pair with a metal ion (1-3). The growth of the band at 1050 cm^{-1}, with increasing salt concentration, at the expense of the free aquated nitrate band at 1047 cm^{-1} is also consistent with ion pair formation. Therefore, the band at 1050 cm^{-1} can be assigned to the symmetric stretch of a $[Co^{2+}\bullet NO_3^-]$ complex.

The appearance of the band at 1040 cm^{-1} indicates the formation of another metal-nitrate complex. This band position is consistent with previously reported metal-nitrate complexes (3). The most likely form of the complex is $[Co^{2+}\bullet(NO_3^-)_2]$. This type of complex would be expected at the very high concentration near the saturated solution (approximately 5 m).

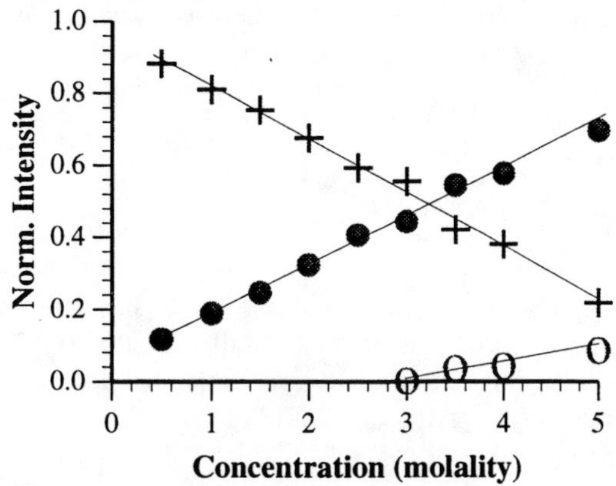

FIGURE 4. A plot of normalised intensity of the component bands under the nitrate ν_1 band profile versus concentration for aqueous cobalt(II) nitrate solutions. + 1047.6 cm^{-1} band, • 1050 cm^{-1} band and o 1040 cm^{-1} band.

Both the nitrate ν_3 and ν_4 band profiles remained unchanged with increasing concentration. The ν_3 vibration was observed as a broad band profile with a maximum at 1412 cm^{-1} and a shoulder at 1339 cm^{-1} and the ν_4 vibration was observed as a band at 717 cm^{-1}. No ν_2 vibration was observed in any of the spectra. The absence of the ν_2 vibration would indicate the interaction of the nitrate ion with the metal ion was similar to that of the nitrate ion with the solvating water.

CONCLUSIONS

The experimental evidence clearly showed the existence of three different chemical environments for the nitrate in aqueous solutions of cobalt(II) nitrate. The three environments consist of the free aquated nitrate, and two aquated metal nitrate complexes. The two complexes are contact ions pairs, most likely of the form $[Co^{2+}\bullet NO_3^-]$ and $[Co^{2+}\bullet(NO_3^-)_2]$. The $[Co^{2+}\bullet NO_3^-]$ complex is seen at all the studied concentrations. Cobalt(II) nitrate is highly associated in aqueous solution. A proposed equilibrium is given below :

$$Co^{2+} + 2NO_3^- \Leftrightarrow [Co^{2+}\bullet NO_3^-] + NO_3^- \Leftrightarrow [Co^{2+}\bullet(NO_3^-)_2]$$

This study has shown that FT-Raman spectroscopy can be used to probe association in highly coloured solutions. A once inaccessible area is now open for investigation.

REFERENCES

1. James D.W., *Prog. Inorg. Chem.*, **33**, 353, 1985.
2. Irish D.E. and Brooker M.H., *Adv Infrared Raman Spectrosc.*, **2**, 212, 1976.
3. Collins B.M., *PhD Dissertation*, University of Queensland, Brisbane, Australia, 1992.
4. Alia J.M., Edwards H.G.M. and Moore J., *Spectrochim. Acta, Part A*, **51**, 2039, 1995.
5. Alia J.M., Demera Y.D., Edwards H.G.M., Garcia F.J. and Lawson E.E., *Z Physik Chemie*, **196**, 209, 1996.
6. Alia J.M., Edwards H.G.M. and Moore J., *Trans. Faraday Soc.*, **92**, 1187, 1996.

The Complex Refractive Index and Dipole-Polarizability of Iodine, I_2, Between 11500 and 17800 cm^{-1}

U.Hohm and D.Goebel

Institut für Physikalische und Theoretische Chemie der TU Braunschweig, Hans-Sommer-Str.10, D-38106 Braunschweig, FRG

We have measured the complex refractive index of molecular iodine in the gas-phase in the wavenumber range between σ=11500 and 17800 cm^{-1} with a resolution of approximately 7 cm^{-1}. The low-resolution absorption spectrum compares favorably well with existing spectra from the literature. The wavenumber-dependence of the polarizability $\alpha(\sigma)$, which is obtained from the refractive index spectrum by means of the Lorenz-Lorentz relation, is recorded for the first time in this wavenumber range and the vibrational energy spacing of the B state of iodine is also resolved in the polarizability spectrum. Fitting the measured polarizability data to a two-term Kramers-Heisenberg dispersion relation yields a static value of $\alpha(0)=10.33(28)\times10^{-30}m^3$.

INTRODUCTION

The complex refractive index $n_C(\sigma,\rho,T)=n(\sigma,\rho,T)-ia(\sigma,\rho,T)/(4\pi\sigma)$ (n = ordinary refractive index, a = absorption coefficient, σ = wavenumber, ρ = density and T = temperature) is a fundamental electro-optical property of matter. Although the absorption coefficient $a(\sigma,\rho,T)$ is known for many substances over broad wavenumber-ranges with exceptional high resolution much less is known about the wavenumber dependence of the real part, $n(\sigma,\rho,T)$, in spectral regions where strong absorption occurs. From a theoretical point of view, $n(\sigma,\rho,T)$ and $a(\sigma,\rho,T)$ are related by the Kramers-Kronig relations. However, if information about $n(\sigma,\rho,T)$ or $a(\sigma,\rho,T)$ is only available in a limited wavenumber region, difficulties may occur in applying the Kramers-Kronig relations. In this case dispersive Fourier transform spectroscopy (DFTS), which is also known as asymmetric or dispersive white-light interferometry, seems to be an appropriate experimental technique which allows for simultaneously recording both, the real and imaginary part of the complex refractive index $n_C(\sigma,\rho,T)$. In order to study the capabilities of DFTS in the visible we have measured the complex refractive index of molecular iodine, which shows strong and well-known absorption in the visible wavenumber range. The complex refractive index is used to obtain the wavenumber dependence of the real part of the complex dipole-polarizability $\alpha_C(\sigma)$ of I_2 according to the well-known Lorenz-Lorentz equation (1).

$$\frac{n_C^2(\sigma,\rho,T)-1}{n_C^2(\sigma,\rho,T)+2}=\frac{4}{3}\pi N_A\rho\alpha_C(\sigma,T) \quad (1)$$

In Eq.(1) $\alpha_C(\sigma)=\alpha(\sigma)-i\alpha_I(\sigma)$ is the complex dipole-polarizability and N_A is Avogadro's constant. In the context of this work, however, we make explicit reference only to the real part $\alpha(\sigma)$. $\alpha(\sigma)$ of iodine has currently attracted a renewed interest especially in the field of intermolecular interactions between molecular iodine and rare-gas atoms (see references in Ref.(1)).

EXPERIMENTAL

The refractive index measurements are carried out with an evacuated Michelson twin-interferometer, which is described in detail in Ref.(2). Two different types of measurements have been applied. First, we made absolute measurements of the refractivity (n-1). To this end 1.1mg of iodine is fed into an evacuated sample cell of length $l/2\approx50$cm and volume $V\approx26.5cm^3$, which is fused off thereafter. This cell is placed into the active arm of the interferometer. The iodine sample is completely evaporated by heating the cell to approximately 345K and the resulting interference fringe shift $N(\sigma)$ is recorded at three different wavenumbers of $\sigma_1=15798.0cm^{-1}$, $\sigma_2=16832.3cm^{-1}$, and $\sigma_3=30756.9cm^{-1}$ (supplied by HeNe- and HeCd-laser radiation sources) during the evaporation. The corresponding change of the refractivity is obtained according to $\Delta(n-1)=N(\sigma)/(\sigma l)$. Absolute values of the refractivity (n-1) (against vacuum) are obtained by extrapolating $\Delta(n-1)$ to zero density. This is necessary because of the finite vapor pressure of I_2 at room-temperature, where the evaporation starts.

These non-isothermal measurements of the refractivity are augmented by isothermal white-light interferometric measurements. These are performed at a temperature of T=296.3K where iodine exhibits a vapor pressure of $\approx$36Pa, according to an amount-of-substance density of $\rho=0.01479$ mol/m^3 in the vapor phase. A 100W halogen lamp is used as white-light source. Due to the very strong

CP430, *Fourier Transform Spectroscopy:* 11th International Conference
edited by J.A. de Haseth

absorption of iodine vapor in the visible the useful spectral range is limited to $11500 cm^{-1} < \sigma < 17800 cm^{-1}$ in our experiments, whereas the resolution is $7 cm^{-1}$. In order to obtain the absolute phase of the Fourier transformed interferogram we use the refractivities measured at σ_1 and σ_2.

RESULTS AND DISCUSSION

In this section we will present the results from the absolute measurements followed by a selection of some experimental results obtained in our white-light interferometric measurements.

Absolute measurements

Our non-isothermal absolute measurements of the refractivity are used to calculate the dipole-polarizability $\alpha(\sigma)$ of iodine at three different wavenumbers according to the Lorenz-Lorentz-equation (1). In particular we obtain $\alpha(\sigma_1)=12.86(33)$, $\alpha(\sigma_2)=13.87(50)$, and $\alpha(\sigma_3)=14.12(28)$ (all in 10^{-30} m^3, uncertainty in parentheses). There are no experimental values determined at these wavenumbers to compare with.

White-light interferometric measurements

Figure 1 shows the real and imaginary part of the complex refractive index $n_C(\sigma)$ of iodine vapor measured at T=296.3K and $\rho=0.01479$ mol/m^3. In the following we will take a closer look at the absorption coefficient $a(\lambda)$ ($\lambda = 1/\sigma$ = vacuum wavelength).

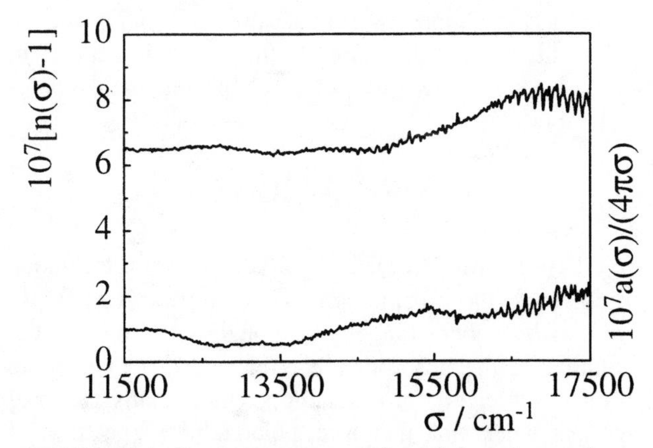

FIGURE 1. Real (upper curve) and imaginary (lower curve) part of the complex refractive index of iodine vapor. For experimental conditions see text.

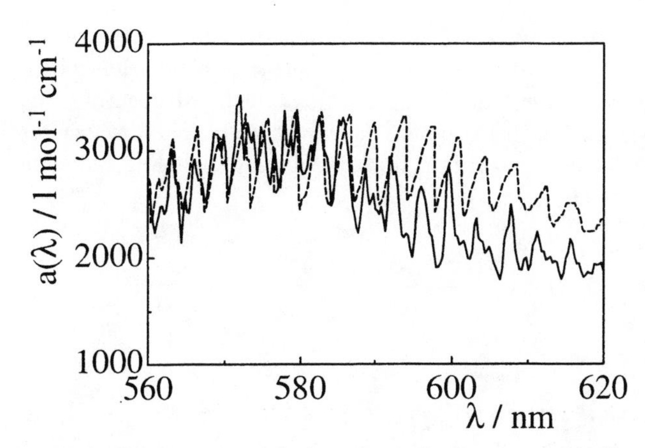

FIGURE 2. Low-resolution absorption spectra of iodine vapor. (——) this work, (- - -) Capelle and Broida (3).

In Fig. 2 the absorption coefficient obtained in our dispersive white-light interferometric experiments is compared to an absorption spectrum with similar resolution reported by Capelle and Broida (3). No scale is given in (3), hence this spectrum has been scaled to fit approximately our experimental result. Overall, quite good agreement can be seen. The absorption in the visible is due to the B $^3\Pi_u^+ \leftarrow$ X $^1\Sigma_g^+$ transition. In our $a(\lambda)$ spectrum progressions from excited vibrational levels v'' can be clearly seen, although a correct assignment of the vibrational levels is not possible in this spectrum, of course.

Figure 3 shows the wavenumber dependence of the dipole-polarizability $\alpha(\sigma)$ obtained in this work together with available data from the literature. Overall we observe good agreement except for the results of Cuthbertson and Cuthbertson (4) above 18400 cm^{-1}. Interestingly, one can recognize the vibrational energy spacing of the B state of iodine in this *polarizability* spectrum, too (see also Fig. 4). In the visible, this feature has never been observed experimentally in a broad wavenumber range. We mention, however, the two rotationally resolved refractive index spectra centered at $\approx 17406 cm^{-1}$ and $\approx 17564 cm^{-1}$ reported by Abraham *et al.* (5), obtained with a scanning wavelength technique in a very narrow interval of approximately $1 cm^{-1}$. As can be seen in Fig.4 we observe a change $\Delta\alpha(\sigma)$ of $\alpha(\sigma)$ of approximately 9% inside a rovibronic absorption band. This causes a change in the real part of the refractive index of $\approx 0.75 \times 10^{-7}$ at the measuring density of 0.01479 mol/m^3 (see Fig.1). Interestingly, Abraham *et al.* (5) have reported in their very high resolution refractive index spectra a change in (n-1) of $\Delta n = 1..2 \times 10^{-7}$ at almost exactly the same thermodynamic conditions in a wavenumber range of $\Delta\sigma = 0.1 cm^{-1}$ at $\approx 17406 cm^{-1}$. In view of the two completely different experimental techniques used, this agreement between the recorded changes of the refractivity and, therefore, of the dipole-polarizability inside a vibronic absorption band is very satisfactory.

In order to obtain the static value of the dipole-polarizability, $\alpha(0)$, we have fitted the global shape of our recorded $\alpha(\sigma)$-spectrum to a two-term Kramers-Heisenberg dispersion-relation, Eq. (2), which disregards the vibrational fine structure. However, we make allowance for the strong absorption in the visible by including radiation damping.

$$\alpha(\sigma) = C\left(\frac{f_1(\sigma_{01}^2 - \sigma^2)}{(\sigma_{01}^2 - \sigma^2)^2 + (\Gamma_1\sigma)^2} + \frac{f_2}{\sigma_{02}^2 - \sigma^2}\right) \quad (2)$$

Here $C=e^2/(16\varepsilon_0\pi^3 mc_0^2)$, where e is the elementary charge, ε_0 is the permittivity of vacuum, m is the electron rest mass, and c_0 is the speed of light in vacuum, respectively. A nonlinear fit of our experimental results yields the effective transition wavenumbers σ_{01} =1.7680(21) and σ_{02} =5.379(46) (both in 10^4 cm^{-1}), with the corresponding oscillator strengths f_1 = 0.02000(88) and f_2 = 4.003(87), and the radiation damping constant Γ_1 = 1.111(46)x10^3 cm^{-1}. Interestingly, σ_{01} corresponds very well to the absorption maximum in the visible, whereas σ_{02} is located in between the continuous absorption band of the D $^1\Sigma_u^+ \leftarrow$ X $^1\Sigma_g^+$ transition of iodine (Cordes bands). However, f_1 is about twice the value of 0.0121 given by Mulliken (6). The static limit of the mean dipole-polarizability of iodine results in $\alpha(0)$ = 10.33(28)x10^{-30}m^3, 4.4% of which is due to the electronic transition in the visible. This experimentally determined static value is close to the best *ab initio* value of 10.51(21)x10^{-30}m^3 obtained from CCSD(T)/MP2 calculations (1). This good accordance is lost, however, if we include the results of Cuthbertson and Cuthbertson (4) measured above 18400 cm^{-1} in our analysis, see Fig.3. In that case we obtain σ_{01}=20242, σ_{02}=43500, and Γ_1=3042 (all in cm^{-1}), whereas the oscillator strengths are f_1=0.098 and f_2=2.08, respectively. The static value of the dipole-polarizability of $\alpha(0)$=9.55x10^{-30}m^3 is also significantly lower than the best *ab initio* result. Nevertheless, the results of the wavenumber dependence of the refractivity reported by Cuthbertson and Cuthbertson (4) clearly indicate that (n-1) has a maximum in the vicinity of 550nm.

In Fig. 4 the polarizability spectrum $\alpha(\sigma)$ is compared to the first derivative of the absorption spectrum, $da(\sigma)/d\sigma$. Although the latter is different in both, physical origin and mathematical form, the maxima of both spectra would nearly coincide if the width of the corresponding absorption line is very small compared to the transition frequency (this means if the relaxation time of the excited level is large). In the case of the B state of iodine, the relaxation time τ is in the order of 1μs (3). Therefore, in view of the limited resolution of our apparatus the maxima should coincide, as can be seen in Fig. 4.

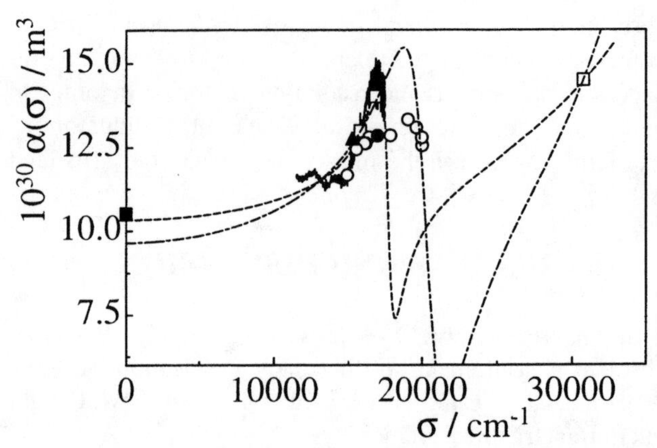

FIGURE 3. Wavenumber-dependence of the dipole-polarizablity of iodine. (——) DFTS-VIS measurements, this work; (□) absolute measurements, this work; (■) CCSD(T)/MP2 *ab initio* calculations (1); (●) Larsson and Folkensson (7); (▲) Braun and Hölemann (8); (O) Cuthbertson and Cuthbertson (4); (- - -) fit according to Eq.(2), only experimental results of this work; (- •-•-) fit according to Eq.(2), all experimental results and large weighting of (4).

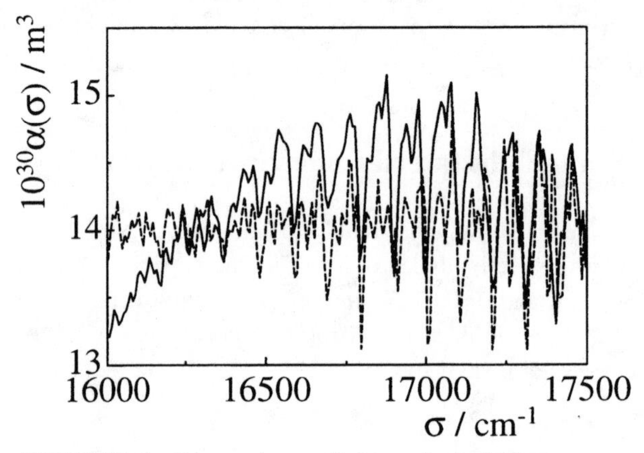

FIGURE 4. Comparison of the polarizability spectrum (——) with the first derivative of the absorption spectrum (- - -), see text. The scale refers to the recorded dipole-polarizability $\alpha(\sigma)$.

CONCLUSIONS

We have shown that DFTS-VIS is a powerful technique for obtaining the complex refractive index $n_C(\sigma,\rho,T)$ in wide wavenumber ranges. Although at present the resolution of the absorption coeffcient $a(\sigma,\rho,T)$ is quite low, it is sufficient for obtaining a vibrationally resolved refractive index and, therefore, polarizability spectrum of iodine. In the case of absorbing substances such as iodine dispersive white-light interferometry is superior to refractivity measurements carried out at discrete wavelengths. As already stated by the pioniers in this field, Cuthbertson and Cuthbertson, inside strong absorption bands the latter technique provides '*numbers (which are), of course, of little individual value, but they show at least the order of magnitude of the variation of refractivity in passing through an absorption band* (4). We notice that our results are in better agreement with the changes reported by Abraham *et al.* (5) than with the few refractivities provided by Cuthbertson and Cuthbertson (4).

ACKNOWLEDGMENTS

Financial support by the Deutsche Forschungsgemeinschaft and Fonds der Chemischen Industrie are gratefully acknowledged.

REFERENCES

1. Maroulis, G., Makris, C., Hohm, U., and Goebel, D., *J.Phys.Chem. A* **101**, 953-956 (1997).
2. Goebel, D., and Hohm, U., *J.Phys.D:Appl.Phys.* **29**, 3132-3136 (1996).
3. Capelle, G.A., and Broida, H.P., *J.Chem.Phys.* **58**, 4212-4222 (1973).
4. Cuthbertson, C., and Cuthbertson, M., *Phil.Trans.* **213**, 1-26 (1913).
5. Abraham, R.G., Booth, J.L., and Dalby, F.W., *Can.J.Phys.* **68**, 81-87 (1990).
6. Mulliken, R.S., *J.Chem.Phys.* **55**, 288-295 (1971).
7. Folkesson, B., and Larsson, R., *J.Electr.Spectr.Rel. Phen.* **50**, 251-266 (1990).
8. Braun, A., and Hölemann, P., *Z.Phys.Chem.* **B34**, 357-380 (1936).

Standard-Oriented Spectroscopic Parameters for the 3-0 Band of $^{12}C^{16}O$ at 1.57 μm.

N. Picqué and G. Guelachvili

Laboratoire de Physique Moléculaire et Applications
Unité Propre du CNRS, Université Paris-Sud, Bâtiment 350
91405 Orsay-Cedex, France

Broadening coefficients, intensities, Herman-Wallis factor, and vibrational transition moment are reported from the analysis of high resolution FTS spectra of 34 lines belonging to the 3-0 band of $^{12}C^{16}O$ centered around 6350 cm^{-1}.

INTRODUCTION

This work, undertaken on the 3-0 band of $^{12}C^{16}O$ centered around 6350 cm^{-1}, is essentially motivated by the lack of high resolution spectroscopic standards in the near infrared range (1) between 1 and 2 μm, due to specific experimental and theoretical difficulties. Indeed, in this spectral range, absorption spectra require high pressure-absorption path products. They are also generally complex (combination and overtone vibrational bands), and therefore difficult to analyse. On the other hand, new sensitive instrumental spectroscopic methods, like diode lasers, have been recently developped in this region. They need reference molecular data (positions, intensities, linewidths) to become efficient in various applications (such as *in situ* trace gas detection, atmospheric remote sensing, optical communications...).

CO has always been a molecule of importance for standard purposes. Its spectrum in the electronic ground state $^1\Sigma$, is well known and its line frequencies are conveniently reproduced using accurate Dunham coefficients (2). Even if the spectrum of the $\Delta v = 3$ transitions is rather simple, curiously the lastest works on the 3-0 band have never been performed with modern instruments. They are about 20 years old. Line positions were measured by Bouanich et Brodbeck (3) in 1974 with an uncertainty of about ± 0.01 cm^{-1}. Individual line intensities measurements are even older (4) and the most recent experimental intensity determination was performed in 1983 by Bouanich et al (5) with no rotational information. To our knowledge, the collisional broadening parameters (self and foreign-gas) were never determined.

In the frame of this work on the 3-0 band of CO, wavenumbers and self induced lineshifts, measured with FTS technique, have been reported (6). The wavenumbers are given within ± 3 10^{-5} cm^{-1}. They agree with the predicted data available from available Dunham coefficients (7). Accuracy on the lineshifts is of the order of 10^{-3} cm^{-1} atm^{-1}. They reach at most - 8 10^{-3} cm^{-1} atm^{-1}. The present paper reports the progress made on the measurement of intensities and of collisional broadening coefficients.

EXPERIMENTAL

Seven FTS absorption spectra have been recorded at LPMA, with CO pressure ranging from about 4 Torr to 1 atmosphere. All spectra cover the range 5895 to 7369 cm^{-1}. Care has been taken to avoid non linearities and ghosts features. Only $^{12}C^{16}O$ was measured from a CO sample in natural abundance. Details concerning the runs are given in Table 1. A small portion of the corresponding spectra in the P-branch is shown on Figure 1.

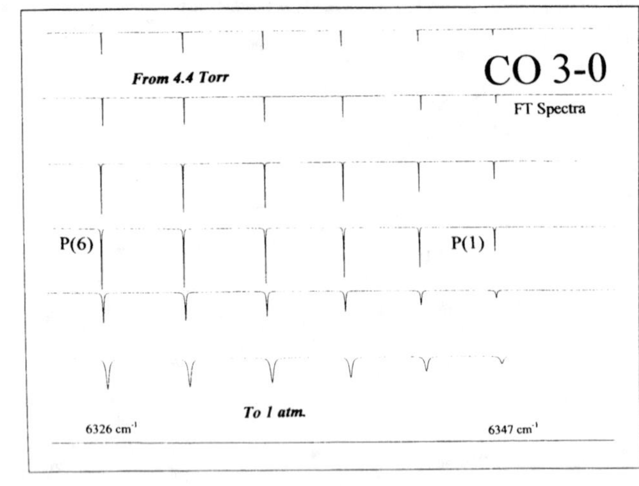

FIGURE 1. CO 3-0 band. From the top to the bottom spectrum nr. 3117, 3118, 3120, 3121, 3122, 3124. See table 1 for the experimental conditions.

Diode laser spectra were also recorded, using the instrumental facilities of the European Laboratory for

CP430, *Fourier Transform Spectroscopy:* 11th International Conference
edited by J.A. de Haseth

TABLE 1. Experimental conditions for the Fourier transform spectra.

Spectrum serial number	Resolution (10^{-3} cm^{-1})	CO pressure (Torr)	Temperature (K)	Absorbing Path (cm)
3117	5.9	4.402	296.65	3218.28
3118	5.9	6.502	297.05	3218.28
3119	5.9	4.402	296.65	3218.28
3120	5.9	40.010	296.85	1618.28
3121	5.9	80.030	296.45	1618.28
3122	5.9	300.20	295.95	418.28
3124	7.3	761.2	295.85	418.28

Non-Linear Spectroscopy (LENS) in Firenze, Italy. The diode emission was centered around 1.578 μm which is convenient for the observation of the $^{12}C^{16}O$ lines P(1) to P(4). The self-broadened P(2) line is shown on Figure 2 where the profile results from the amplification of the difference between a reference and a transmission spectra also given on the figure.These spectra essentially devoted to the determination of foreign-gas broadening are presently under study and no results are given here.

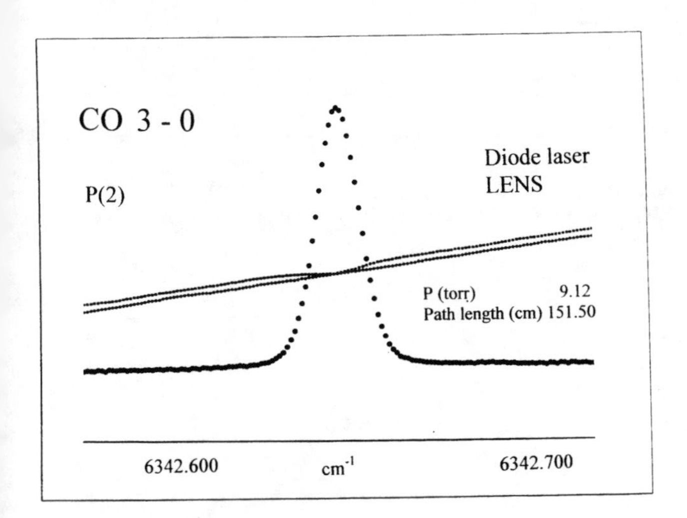

FIGURE 2. CO 3-0 band. Diode laser spectra.

RESULTS

Line intensities and linewidths are measured from the spectra 3120, 3121, 3122, 3124 for which CO is observed at a pressure higher than 40 torr. No instrumental lineshape has been taken into account yet in the fit of the lines. This is justified by the strong influence of the pressure broadening on the line profiles, which is large enough to neglect, in a first step, the apparatus function contribution. Each transmission experimental profile was processed separately without taking into account the negligible far wings of the neighbour profiles.

The position, intensity and Lorentzian halfwidth of each line were simultaneously determined using a non-linear least squares fit of its experimental shape to a Voigt profile.

The transmission of radiation through homogeneous gas sample is described by the Beer-Lambert law :

$$I_t(\nu)=I_0(\nu)\exp(-k(\nu)\,p\,l)$$

where $I_t(\nu)$ and $I_0(\nu)$ are, respectively, the transmitted, and incident, radiation intensities, $k(\nu)$ is the absorption profile, l the optical path length, p the pressure.

The Voigt profile, $k(\nu)$, is defined as:

$$(S\sqrt{\ln 2})/(\gamma_D\sqrt{\pi})\,\mathrm{Re}(W(x,y))$$

where S, which is developped below, is the line intensity in cm^{-2} atm^{-1}, γ_D the Doppler halfwidth at half maximum in cm^{-1}, and W(x,y) is the complex error function given by:

$$W(x,y)=\frac{i}{\pi}\int_{-\infty}^{+\infty}\frac{\exp(-t^2)}{x+iy-t}dt, \qquad (y>0)$$

where $x=\sqrt{\ln 2}\,(\nu-\nu_0)/\gamma_D$, $y=\sqrt{\ln 2}\,\gamma_L/\gamma_D$ and ν_0 is the line center wavenumber in cm^{-1}, γ_L is the collisional (Lorentzian) halfwidth at half maximum.

The numerical form of W(x,y) was from the algorithm given in Hui et al. (8). The parameters left free during the fitting procedure are: baseline level, baseline tilt, and said above, the line position, line intensity and Lorentzian halfwidth. The Doppler halfwidth was held fixed to its value:

$$3.5812\ 10^{-7}\sqrt{T/M}\ \nu_0$$

where M is the the molecular weight, in a.m.u., T is the temperature, in K.

The fits give direcly the Lorentzian halfwidths γ_L which are shown on Figure 3 for four different pressures. The self-broadening coefficients γ_o, for one atmosphere, averaged on the four spectra, are given, for each J, on Table 2. These results are given, within ± 10^{-3} cm^{-1}, estimated from the dispersion of the data. These coefficients should be vibrationally independent and linearly dependent on the pressure.

TABLE 2: CO 3-0 band. γ_0 is the collisional broadening coefficient.

Assignment	Wavenumber (cm^{-1})	γ_0 ($cm^{-1}.atm^{-1}$)	Intensity (10^{-4} cm^{-2} atm^{-1})
P(17)	6270.912 004	0.058	0.79
P(15)	6281.819 777	0.062	1.36
P(14)	6287.119 529	0.062	1.67
P(13)	6292.316 314	0.063	2.07
P(12)	6297.410 001	0.064	2.46
P(11)	6302.400 438	0.065	2.83
P(10)	6307.287 492	0.066	3.21
P(9)	6312.071 007	0.066	3.51
P(8)	6316.750 832	0.068	3.75
P(7)	6321.326 817	0.069	3.86
P(6)	6325.798 824	0.071	3.80
P(5)	6330.166 711	0.073	3.59
P(4)	6334.430 314	0.076	3.18
P(3)	6338.589 402	0.079	2.62
P(2)	6342.644 105	0.083	1.86
P(1)	6346.594 015	0.088	0.98
R(0)	6354.179 033	0.088	1.02
R(1)	6357.813 904	0.083	2.03
R(2)	6361.343 481	0.079	2.97
R(3)	6364.767 593	0.075	3.79
R(4)	6368.086 120	0.072	4.45
R(5)	6371.298 890	0.070	4.93
R(6)	6374.405 766	0.068	5.21
R(7)	6377.406 614	0.067	5.30
R(8)	6380.301 270	0.066	5.21
R(9)	6383.089 606	0.065	4.95
R(10)	6385.771 467	0.064	4.58
R(11)	6388.346 680	0.063	4.12
R(12)	6390.815 138	0.062	3.62
R(13)	6393.176 687	0.062	3.11
R(14)	6395.431 153	0.060	2.59
R(15)	6397.578 415	0.059	2.12
R(16)	6399.618 347	0.058	1.69
R(17)	6401.550 760	0.057	1.32

The intensity of a rotational transition from one level v" = 0, J" ·······> v' = 3, J' is given by:

$$S(0,J";3,J') = (8\,\pi^3/3hc)(N/Q)\,(T_0/T)\,\nu_0\,\exp(-hcE(J")/kT)\,(1-\exp(-hc\nu_0/kT))\,|m|\,F_0^3(m)\,|\,\mu_0^3|^2$$

where J", v", and J', v' are respectively the rotational and vibrational quantum numbers of the lower and upper energy levels of the transitions, h and c are Planck's constant and velocity of light, N is the Loschmidt's number (2.687×10^{19} molecules. cm^{-3}), Q the rovibrational partition function, T_0 the standard temperature (273.15 K), E(J") the energy of the lower level of the transition, k is Boltzman's constant, $|\mu_0^3|$ is the vibrational transition moment, m= -J for P-branch and J + 1 for the R-branch, and $F_0^3(m)$ the Herman Wallis factor:

$$F_0^3(m) = 1 + C\,m + D\,m^2$$

which takes into account the interaction between vibration and rotation.

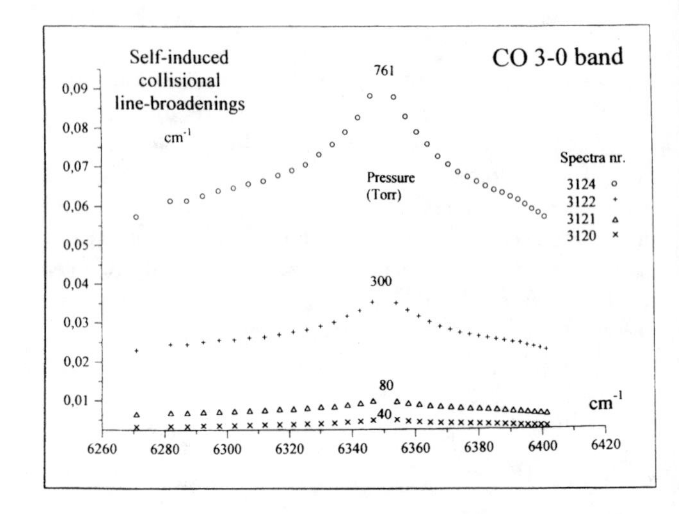

FIGURE 3. CO 3-0 band. Line individual broadenings.

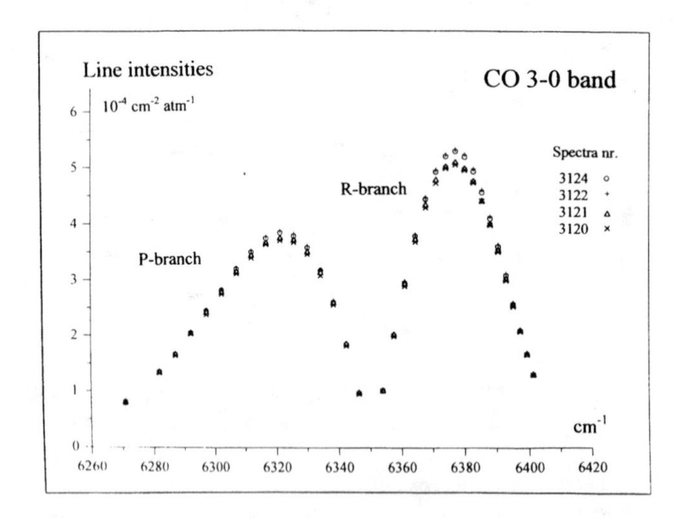

FIGURE 4. CO 3-0 band. Line individual intensities.

The measured intensities S of the lines are shown on Figure 4, for the four different spectra. The two low pressure sets give slightly lower values. This could be due the instrumental function not taken into account in the fits.

The average values of the intensities are given within ± 0.3 10^{-4} $cm^{-2}\ atm^{-1}$, estimated from the dispersion of the data, in Table 2.

The parameters C and D of the F-factor, obtained from a least-squares fit on the S values, are respectively equal to

$$C = 1.1\ (\pm 0.2) \times 10^{-2},$$
$$D = 1.3\ (\pm 0.4) \times 10^{-4}.$$

The squared dipole moment matrix element is equal to

$$|\mu_0^3|^2 = 5.44\ (\pm 0.2)\ 10^{-7}\ \text{Debye}^2.$$

CONCLUSION

Broadening coefficients, intensities, Herman-Wallis factor, and vibrational transition moment are reported from the analysis of FTS spectra of 34 lines belonging to the 3-0 band of $^{12}C^{16}O$. Diode laser spectra of self- and foreign-gas broadening profiles of the P(1) to P(4) lines in the same band have been recorded. Next step is to process these diode laser spectra and to revisit the Fourier transform spectra after including the instrumental function in the fitting procedure of the data.

With the present data and the recent accurate positions and pressure lineshifts (6), the 3-0 band is now appropriate for standard-oriented purposes.

ACKNOWLEDGEMENTS

We warmly acknowledge the help from J.-P Bouanich, R. Farrenq, J.-M Hartmann, and J. Saccani.

REFERENCES

1. Guelachvili, G., Birk, M., Bordé, Ch. J., Brault, J. W., Brown, L. R., Carli, B., Cole, A. R. H., Evenson, K. M., Fayt, A., Hausamann, D., Johns, J. W. C., Kauppinen, J., Kou, Q., Maki, A. G., Rao, K. Narahari, Toth, R. A., Urban, W., Valentin, A., Vergès, J., Wagner, G., Wappelhorst, M. H., Wells, J. S., Winnewisser, B. P., Winnewisser, M., *Pure and Applied Chemistry* **68** (1996) 193-208. *J. Mol. Spectrosc.* **177** (1996) 164-179., *Spectrochimica Acta* **52** (1996) 193-208.
2. George, T., Urban, W., and Le Floch, A. *J. Mol. Spectrosc.* **165** (1994) 500-505.
3. Bouanich, J.-P., and Brodbeck, C. *Revue de Physique Appliquée* **9** (1974) 475-478.
4. Toth, R. A., Hunt, R. H., and Plyler, E. K. *J. Mol. Spectrosc.* **32** (1969) 85-96.
5. Bouanich, J. P., Nguyen-Van-Thanh, and Rossi, I. *J. Quant. Spectrosc. Radiat. Transfer* **30** (1983) 9-15.
6. Picqué, N., and Guelachvili, G. *J. Mol. Spectrosc.* to be published (1997)
7. Farrenq, R., private communication.
8. Hui, A. K., Armstrong, B. H. and Wray, A. A. *J. Quant. Spectrosc. Radiat. Transfer* **19** (1978) 509-516.

Fourier Transform Vibrational Circular Dichroism of Small Pharmaceutical Molecules

Fujin Long, Teresa B. Freedman and Laurence A. Nafie

Department of Chemistry, Syracuse University, Syracuse, New York 13244-4100 USA

Fourier transform vibrational circular dichroism (FT-VCD) spectra of the small pharmaceutical molecules propranolol, ibuprofen and naproxen have been measured in the hydrogen stretching and mid-infrared regions to obtain information on solution conformation and to identify markers for absolute configuration determination. Ab initio molecular orbital calculations of low energy conformations, vibrational frequencies and VCD intensities for fragments of the drugs were utilized in interpreting the spectra. Features characteristic of five conformers of propranolol were identified. The weak positive CH stretching VCD signal in ibuprofen and naproxen is characteristic of the S-configuration of the chiral center common to these two analgesics.

INTRODUCTION

The importance of chirality in the specificity of action of pharmaceuticals is well recognized. The chiroptical technique vibrational circular dichroism (VCD) provides a unique probe of both absolute configuration and solution conformation of chiral drugs (1). In this study, Fourier transform vibrational circular dichroism (FT-VCD) spectra of small pharmaceutical molecules have been investigated in both the mid-infrared and hydrogen-stretching regions to obtain information on predominant solution conformations and to identify VCD marker bands for absolute configuration. The molecules studied include the anti-arrhythmic drug (S)-propranolol, which can assume a number of intramolecularly hydrogen-bonded solution conformations, and the analgesics (S)-ibuprofen [(S)-4-isobutyl-α-methylphenylacetic acid] and (S)-naproxen [(S)-6-methoxy-α-methyl-2-naphthaleneacetic acid], which have similar chiral centers.

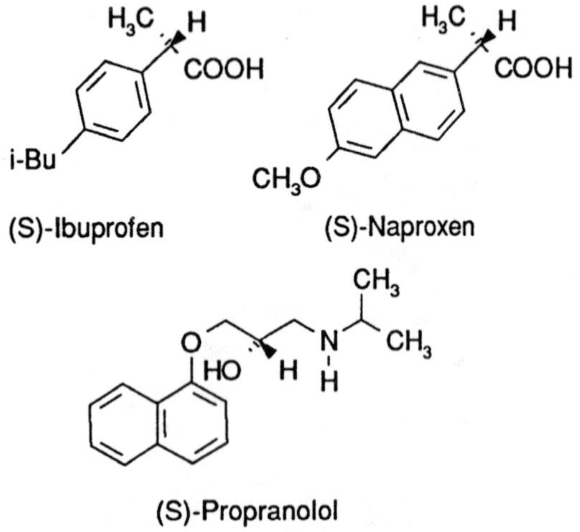

(S)-Ibuprofen (S)-Naproxen

(S)-Propranolol

Figure 1. Structures of the molecules investigated.

EXPERIMENTAL

For FT-VCD measurements in the 800-2000 cm^{-1} region, a Nicolet Magna 850 FTIR spectrometer equipped with an external VCD bench utilizing a HgCdTe detector (3 x 3 mm area) and ZnSe photoelastic modulator (PEM) was employed in rapid-scan mode (2). ZnSe lenses are employed to focus the light emerging from the interferometer at the sample and to focus the beam at the detector. In this upgrade of our previous Nicolet 550 instrument, the IR and VCD signals are measured simultaneously, which improves the quality of the VCD spectra. Measurements in the hydrogen stretching regions were carried out on a Bruker IFS-55 FTIR spectrometer with a VCD accessory bench, in step-scan mode (3). The latter instrument employs a CaF_2 PEM, BaF_2 lenses, and an InSb detector for optimal performance above 2000 cm^{-1}. Step-scan FT-VCD eliminates the high Fourier frequencies present in rapid-scan FTIR that may degrade the VCD signals in this region. Both instruments employ colinear optics on the VCD benches to minimize artifacts in the VCD spectra.

The pharmaceutical samples were obtained from commercial sources. (S)-Propranolol·HCl was converted to the free base. Because of the low solubility of the free acids and salts of ibuprofen and naproxen, the methyl-d_3 esters were also prepared and investigated. Spectra were obtained for $CDCl_3$ solutions. Pure solvent spectra were utilized to obtain VCD baselines.

Complete calculations of structures, IR and VCD spectra of the numerous conformers of these molecules at high levels of theory are not practical. We have therefore considered suitable smaller chiral fragments that retain the same chiral centers and hydrogen-bonding capabilities. *Ab initio* molecular orbital calculations of low energy conformations, vibrational frequencies and VCD intensities for suitable fragments of propranolol and ibuprofen were

CP430, *Fourier Transform Spectroscopy:* 11th International Conference
edited by J.A. de Haseth

utilized in interpreting the VCD spectra, employing either the locally distributed origin gauge (LDO) model (4) or the vibronic coupling VCD theory (5,6).

RESULTS AND DISCUSSION

The chromophores for VCD involve vibrations of groups associated with a chiral center or a chiral conformation of achiral groups. Since the effect occurs on a vibrational time scale, conformations that interchange rapidly, and that are thus averaged on an NMR time scale, each display a distinct VCD spectrum. In addition, vibrations localized on groups that are not in a chiral environment will generate IR, but not VCD intensity. In a region such as the CH-stretching region, where numerous vibrational normal modes overlap, the VCD spectrum will be a selection of features associated only with the chiral chromophores. In the case of naproxen and ibuprofen, the chiral centers are quite similar, and the CH-stretching VCD features arise from the methine and methyl stretches at the chiral carbon. The environment of these two groups is not highly chiral, owing to the similarity in electronic structure of the aromatic ring and carboxyl groups, and the VCD spectra in this region are quite weak, requiring twelve hours of collection time. Nevertheless, distinct positive VCD features are observed in this region for both analgesics, as shown in Fig. 2. *Ab initio* molecular orbital calculations on (*S*)-α-methylphenylacetic acid, a chiral fragment common to naproxen and ibuprofen, indicate the presence of two conformers with similar energies that differ slightly in the orientation of the COOH group. Calculated CH-stretching VCD of both fragments with the vibronic coupling VCD

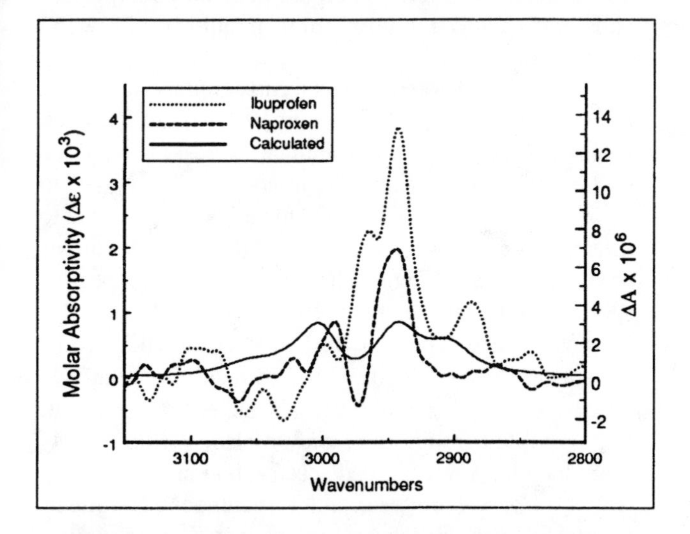

FIGURE 2. CH-stretching VCD spectra of ibuprofen (· · ·) and naproxen (- - -), 0.13 M in $CDCl_3$ 200 μm path length, compared to that calculated for the lowest energy conformer of (*S*)-α-methylphenylacetic acid (—). Spectral acquisition time: 12 hours each for sample and solvent.

theory (6) show primarily positive VCD signal for the methyl and methine stretches. The calculated VCD for the lowest energy conformer is compared to the experimental data in Fig. 2. Although this experimental signal is weak ($\Delta A \sim 5 \times 10^{-6}$), these data demonstrate that positive CH-stretching VCD can serve as a marker for the *S*-configuration of this chiral center, even when other (achiral) CH-stretching modes also occur in this spectral region.

The mid infrared spectra of ibuprofen and naproxen were recorded for the deuteriated methyl esters in CCl_4, since solvent bands obscured the sample features in $CDCl_3$, and the concentration of the saturated solution of the acids in CCl_4 is too low. VCD features appear to be associated with many of the absorbance features in this region, reflecting contributions from motions associated with bonds at the chiral center even to normal modes that are primarily achiral group modes.

For propranolol, conformers with OH--N and OH--O intramolecular hydrogen bonding are prevalent in chloroform solvent, with the NH either free or hydrogen-bonded to one of the oxygen atoms. Features characteristic of at least five conformers have been identified in the VCD spectra in the OH/NH-stretching region of propranolol by comparison to calculations of a fragment containing a phenyl in place of the naphthyl group and a methyl in place of the isopropyl group. For this study involving numerous conformers of a fairly large molecule, the LDO model for VCD (4), which gave good agreement with experiment in a study of the solution conformations of ephedra drugs (7), was employed in the calculations. The VCD features associated with various OH--N and OH--O hydrogen-bonding patterns in propranolol correlate in sign with similar conformations in simple diols and amino alcohols, and also serve as VCD markers for solution conformation.

ACKNOWLEDGMENTS

Support for this work from the National Institutes of Health (GM-23567) is gratefully acknowledged.

REFERENCES

1. Nafie, L .A. *Appl. Spectrosc.* **50** (5), 14A-26A (1996).
2. Long, F., Freedman T. B., Hapanowicz, R., and Nafie, L. A. *Appl. Spectrosc.* **51**, 504-507 (1997).
3. Long, F., Freedman T. B., Tague, T. J., and Nafie, L. A. *Appl. Spectrosc.* **51**, 508-511 (1997).
4. Freedman, T. B., Nafie, L. A., and Yang, D. *Chem. Phys. Lett.* **227**, 419-428 (1994).
5. Nafie, L. A., and Freedman, T. B. *J. Chem. Phys.* **78**, 7108-7116 (1983).
6. Rauk, A., and Yang, D. *J. Phys. Chem.* **96**, 437- (1992).
7. Freedman, T. B., Ragunathan, N., and Alexander, S. *Faraday Discuss.* **99**, 13-34 (1994).

Step-Scan FTIR Absorption Difference Time-Resolved Spectroscopy Studies the Excited State Electronic Structures and Decay Kinetics of d^6 Transition Metal Polypyridine Complexes

G. D. Smith, B. M. Paegel and R. A. Palmer

Department of Chemistry, Duke University, Durham, NC 27708

P. Chen, K. M. Omberg and T. J. Meyer

Department of Chemistry, UNC-Chapel Hill, Chapel Hill, NC 27599-3290

Step-scan FTIR absorption difference time-resolved spectroscopy (S^2 FTIR ΔA TRS) has been used to study the photo-excited states of several low-spin d^6 transition metal polypyridine complexes. Insight into the distribution of electron density in the excited states is obtained by comparing the ground and excited state vibrational frequencies of various bands sensitive to electronic structure. The multiplex, registration, and IR throughput advantages of this interferometric technique are significant in comparison with other methods currently used to probe photo-excited processes on the nanosecond time scale. The S^2 FTIR ΔA TR spectra were obtained by use of a step-scan modified Bruker IFS 88 FTIR spectrometer equipped with an AC/DC-coupled photovoltaic Kolmar Technologies MCT detector with a 20 ns rise time and a 100/200 MHz PAD82a transient digitizer. The complexes were excited with frequency-tripled pulses from a Q-switched Quanta-Ray DCR1A Nd:YAG laser (355 nm, 10 ns, 10 Hz, 3 mJ/pulse). Data were collected with 10 ns time resolution.

INTRODUCTION

As general interest and synthetic efforts in polypyridine complexes of divalent ruthenium continue to grow (1), analytical techniques capable of probing their rich and potentially useful photochemistry have been developing as well (2). Time-resolved infrared (TRIR) methods have proven competitive with the structure sensitive information provided by time-resolved resonance Raman (TR^3) for analyzing the electronic excited states of these potential solar energy conversion complexes. For photochemistry occurring on the nanosecond time scale, the application of step-scan Fourier transform infrared absorption difference time-resolved spectroscopy (S^2 FTIR ΔA TRS) has shown significant advantages over rival techniques in speed, spectral resolution, and free spectral range (2b, 3).

Polypyridine complexes of d^6 transition metals have long been known as promising candidates for solar energy conversion through a photo-initiated reduction-oxidation scheme (4). This exciting property is a direct result of these redox complexes commonly possessing a long-lived triplet metal-to-ligand charge transfer (^{3}MLCT) state as the lowest lying excited state (2c,4,5). In TRIR methods, CO and CN chromophores, both metal-bound and those substituted on the polypyridine ligands, act as IR "reporters" of the electronic excited state structure in these complexes. The sensitivity of these IR active groups to the complexes' symmetry and proximal electron distribution is well established (6). Therefore, continuing progress in the development of structurally sensitive TRIR techniques helps to direct the rational synthesis of important model complexes and the understanding and molding of their photochemical properties.

TRIR methods are currently capable of a wide array of time scales (10^{-15} sec to 10^{-3} sec), and so even the most fundamental chemical processes can be probed. In fact, tunable IR lasers have been used to follow complete reaction mechanisms, from reactants to intermediates to products (7). TRIR measurements have also been demonstrated using a dispersive spectrometer with a time resolution of 10 ns (8). However, associated with monochromatic TRIR systems are several inconveniences including relatively lengthy data collection times for an entire spectrum, large sample consumption, and problems with intensity reproducibility over a broad spectral region.

For the microsecond to nanosecond regime, interferometric approaches are showing encouraging improvements. Inherent to interferometry are high spectral resolution, increased IR throughput, excellent

CP430, *Fourier Transform Spectroscopy:* 11th International Conference
edited by J.A. de Haseth

reproducibility, and the multiplex advantage. Without any prior knowledge of where transient species will absorb, S^2 FTIR ΔA TRS provides a broad spectral range with nanosecond time resolution and fast data acquisition. This paper describes the application of S^2 FTIR ΔA TRS to the study of excited state electronic structure and decay kinetics of low-spin d^6 transition metal complexes.

EXPERIMENTAL

Spectra were measured on a step-scan modified Bruker IFS88 FTIR spectrometer with a standard globar source and dry air purge. All samples were dissolved in CH_3CN to give an IR absorbance of 0.25 to 0.50 for the band of interest. Sparging of the sample for 20 minutes with argon was necessary to remove all dissolved molecular oxygen prior to IR analysis. Approximately 200 µL of sample was loaded into a CaF_2-windowed cell with a 250 µm Teflon spacer under an inert atmosphere.

The sample was repeatedly excited with a 10 ns pulse from a frequency-tripled Q-switched Quanta Ray DCR1A Nd:YAG laser (355nm, 3 mJ/pulse, 10 Hz). Laser excitation and data acquisition were synchronized with a Stanford Research Model DG535 pulse generator. An AC/DC-coupled photovoltaic Kolmar Technologies mercury cadmium telluride (PV MCT) detector with a 50 MHz preamplifier and an effective rise time of ~20 ns was used to sample the transmitted IR probe beam. In order to eliminate extraneous radiation and maximize the detector's digitization range, a Ge low-pass filter and the CaF_2 cell windows limited the free spectral range to 2250-1150 cm^{-1}.

The detector's AC signal was further amplified by a Stanford Research Model SR445 preamplifier (x25) before being directed to a 486 personal computer upgraded with a 100/200 MHz PAD82a transient digitizer board. The DC signal was sent directly to the digitizer to be used for phase correction of the AC signal. Bruker Instrument's Opus 2.2™ software was used to process the recorded data.

The interferogram response both before and after laser excitation was digitized at 10 ns time resolution, and typically, each point on the interferograms was the average of 75-150 laser shots. Spectral resolution was 6 cm^{-1}, and data collection times were on average 30 min. The ΔA spectra were calculated from the single beam ΔI transforms by the relation, $\Delta A = -\log[1 + \Delta I(\nu,t)/I(\nu)]$, where $I(\nu)$ is the static spectrum without excitation and $\Delta I(\nu,t)$ is the change in intensity at time t after the laser shot. For ΔA "snapshots" of the excited states, several post excitation slices were averaged to give a better S/N spectrum. Although one 30 min. experiment is usually adequate for extracting frequency shift data, averaging of several time-resolved spectra further improves S/N for in-depth excited state kinetic decay analysis.

Spectroscopic grade CH_3CN was purchased from Burdick and Jackson and was used without further purification. All low-spin metal complexes were synthesized by standard methods and purified as the PF_6^- salts.

RESULTS AND DISCUSSION

The S^2 FTIR ΔA TRS technique has been used to examine the photo-excited states of several low-spin metal complexes and biological metalloproteins (3c-f). Generally, the transitions observed on the nanosecond time scale are the lowest-lying excited states and can be dπ-pπ (MLCT), pπ-pπ* (ligand centered), or in instances where the two transitions are close in energy, both (9). The energy difference between the ground state absorption of the "reporter ligands" and their excited state absorption contains information about the nature of the excited state being probed at that time and the electron distribution in the excited state structure.

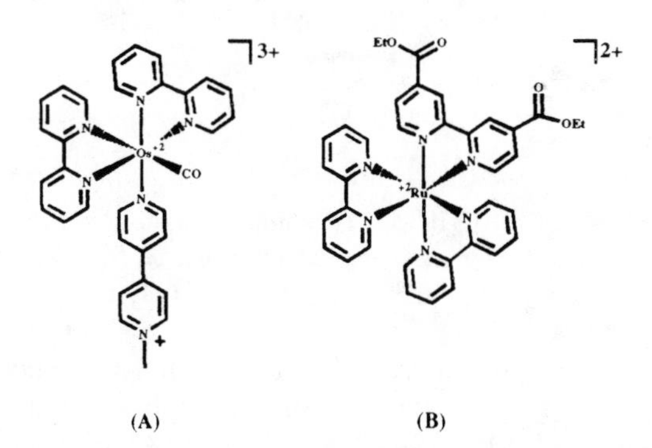

Figure 1. A) $[Os^{II}(bpy)_2(CO)(MQ^+)]^{+3}$. **B)** $[Ru^{II}(bpy)_2(4,4'-(COOEt)_2bpy)]^{2+}$.

Although luminescence experiments often provide valuable information about excited state decay kinetics, TRIR augments that data with structural specificity and excited state symmetry information. Figure 2 shows a 10 ns time resolution ΔA stack plot of $[Os^{II}(bpy)_2(CO)(MQ^+)]^{+3}$ (Fig. 1A) (bpy = 2,2'-bipyridine, MQ^+ = N-methyl-4,4'-bipyridinium cation) both before and after laser excitation. The initial spectral slices are pre-excitation, while the later slices correspond to the decay of the lowest lying excited state, an ^{3}MLCT state, following the laser pulse.

The deep furrow at 1969 cm^{-1} is due to bleaching of the ground state absorption of ν(CO). A corresponding positive peak at 2037 cm^{-1} is associated with ν(CO) in the excited state. The excited state shift of the carbonyl band to higher energy is caused by decreased π-backbonding between the π* orbital of the metal-bound carbonyl and the newly oxidized osmium's dπ orbitals. The extremely intense positive band at 1610 cm^{-1} was not expected, and it might well have been overlooked in a point-by-point TRIR technique. The ability to capture broad spectral ranges without excessively lengthy data collection times allows S^2 FTIR ΔA TRS to potentially monitor numerous "reporter" chromophores widely dispersed in the mid-IR.

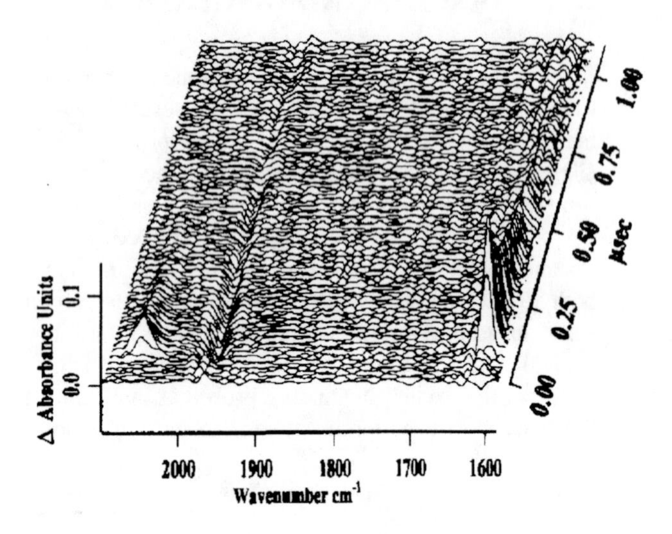

FIGURE 2. S^2 FTIR ΔA TRS time slice stack plot showing the decay of the lowest lying excited state, an MLCT state, of $[Os^{II}(bpy)_2(CO)(MQ^+)]^{+3}$. Time resolution is 10 ns.

The unanticipated peak does not correspond to any strongly absorbing band in the ground state spectrum, and it shows no parallel bleach in the ΔA time slices. Most likely this peak can be attributed to the $MQ^{\bullet}$ radical itself. A future publication from our lab will more fully explore this exceptional compound, its excited state structure, and decay kinetics.

Figure 3A is the ground state IR absorption spectrum of $[Ru^{II}(bpy)_2(4,4'\text{-}(COOEt)_2bpy)]^{2+}$ (Fig. 1B) in the region of the ester ν(CO). Figure 3B shows a ΔA "snapshot" spectrum calculated by averaging several post-excitation time slices from data similar to Fig. 2.

Again, the negative bleach in Fig. 3B corresponds to a loss of the ground state ester species following photo-excitation to the MLCT state. However, in this complex, the ν(CO) originates from a substituted chromophore on the electron *receiving* ligand. The π* orbital of the ester carbonyl mixes heavily into the receiving MO of the ligand, thus weakening the carbonyl bond, and the excited state shift moves to lower energy relative to the ground state.

The excited state frequency shift relative to the ground state for Fig. 1B has been reported by our laboratory to be approximately half of the shift found for the assymetrically substituted monoester analog, $[Ru^{II}(bpy)_2(4\text{-}(COOEt)\text{-}4'\text{-}(CH_3)\ bpy)]^{2+}$ (3f). The excited state shift data indicates enhanced polarization of the excited electron density onto the single ester-substituted ring of the monoester complex, while being delocalized in the mixed-valence sense over both symmetrically substituted rings in the diester complex (Fig. 1B). Reliable frequency shift data can be collected over a broad spectral range for numerous chromophores using S^2 FTIR ΔA TRS. These experiments show that band shift information can be used to elucidate excited state electronic distribution on relatively short lived species.

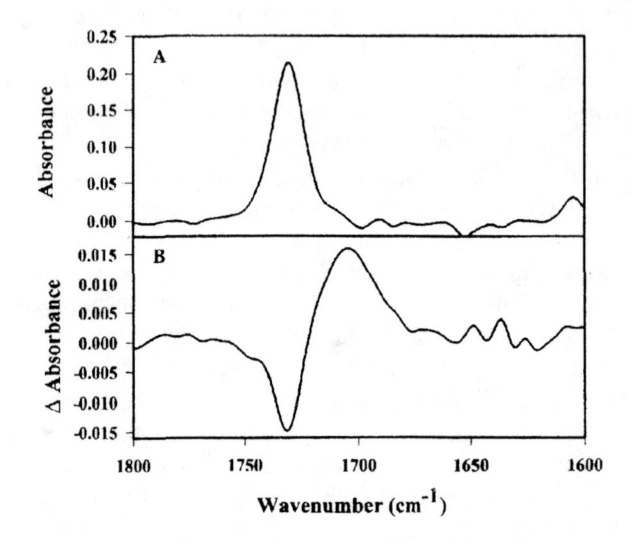

FIGURE 3. A) Ground state IR absorption spectrum of $[Ru^{II}(bpy)_2(4,4'\text{-}(COOEt)_2bpy)]^{2+}$, and **B)** "snapshot" ΔA spectrum in the ester ν(CO) region.

The unchallenged ability of TRIR measurements to access bands not subject to Raman resonance enhancement is one of the strengths of this battery of techniques. Although all TRIR techniques bring increased structural information to the study of photo-excited states, S^2 FTIR ΔA TRS provides unique advantages in spectral breadth, short data collection times, and high spectral resolution in the nanosecond regime. Although state-of-the-art instruments are limited to 10^{-3}-10^{-9} sec time resolution, the technique is fundamentally restrained only by the pulse width of excitation sources and the speed of IR detectors and transient digitizers. As faster instrumentation becomes available, S^2 FTIR ΔA TRS promises to overcome this resolution barrier and compete in the sub-nanosecond time scale.

ACKNOWLEDGEMENTS

Acknowledgements are made to the National Science Foundation, to the Optics Division of Bruker Spectrospin, to Los Alamos National Laboratory, and to Duke University for the support of this work.

REFERENCES

1. (a) Meyer, T. J., *Pure Appl. Chem.* **58**, 1193-1206 (1986). (b) Johnson, S. R.; Westmoreland, T. D.; Caspar, J. V.; Barqawi, K. R.; Meyer, T. J., *Inorg. Chem.* **27**, 3195-3200 (1988). (c) Molnar, S. M.; Neville, K. R.; Jensen, G. E.; Brewer, K. J., *Inorg. Chim. Acta* **206**, 69-76 (1993). (d) Vogler, L. M.; Brewer, K. J., *Inorg. Chem.* **35**, 818-824 (1996).
2. (a) Stoutland, P. O.; Dyer, R. B.; Woodruff, W. H., *Science* **257**, 1913-1917 (1992). (b) Palmer, R. A., *Spectroscopy* **8**, 26-36 (1993). (c) Turner, J. J.; George, M. W.; Johnson, F. P. A.; Westwell, J. R., *Coord. Chem. Rev.* **125**, 101-114 (1993). (d) George, M. W.; Poliakoff, M.; Turner, J. J., *Analyst* **119**, 551-560 (1994). (d). Lian, T.; Bromberg, S. E.; Asplund, M. C.; Yang, H.; Harris, C. B., *Phys. Chem.* **100**, 11994-20001 (1996).
3. (a) Palmer, R. A.; Chao, J.; Dittmar, R.; Gregoriou, V.; Plunkett, S., *Appl. Spectrosc.* **47**, 1297-1310 (1993). (b) Schoonover, J. R.; Strouse, G. F.; Omberg, K. M.; Dyer, R. B., *Comm. Inorg. Chem.* **18**, 165-188 (1996). (c) Palmer, R. A.; Plunkett, S. E.; Chen, P.; Chao, J. L.; Tague, T. J., *Mikrochim. Acta* Suppl. **14**, 603-605 (1997). (d) Palmer, R. A.; Chen, P.; Plunkett, S. E.; Chao, J., *Mikrochim. Acta* Suppl. **14**, 595-597 (1997). (e) Chen, P.; Palmer, R. A., *Appl. Spectrosc.* **51**, 580-583 (1997). (f) Chen, P.; Omberg, K. M.; Kavaliunas, D.; Treadway, J. A.; Palmer, R. A.; Meyer, T. J., *Inorg. Chem.* **36**, 954-955 (1997).
4. Sutin, N.; Creutz, C., *Pure Appl. Chem.* **52**, 2717-2737 (1980).
5. (a) Palmer, R. A.; Piper, T. S., *Inorg. Chem.* **5**, 864-878 (1966). (b) Bradley, P. G.; Kress, N.; Hornberger, B. A.; Dallinger, R. F.; Woodruff, W. H., *J. Am. Chem. Soc.* **103**, 7441-7446 (1981). (c) Caspar, J. V.; Meyer, T. J., *J. Am. Chem. Soc.* **105**, 5583-5590 (1983).
6. Cotton, F. A.; Wilkinson, G., *Advanced Inorganic Chemistry*, New York: John Wiley and Sons, 1988, 5th ed., 61.
7. Glyn, P.; Johnson, F. P. A.; George, M. W.; Lees, A J.; Turner, J. J., *Inorg. Chem.* **30**, 3543-3546 (1991).
8. Yazawa, T.; Kato, C.; George, M. W.; Hamaguchi, H., *Appl. Spectrosc.* **48**, 684-690 (1994).
9. Schoonover, J. R.; Strouse, G. F.; Dyer, R. B.; Bates, W. D.; Chen, P.; Meyer, T. J., *Inorg. Chem.* **35**, 213-214 (1996).

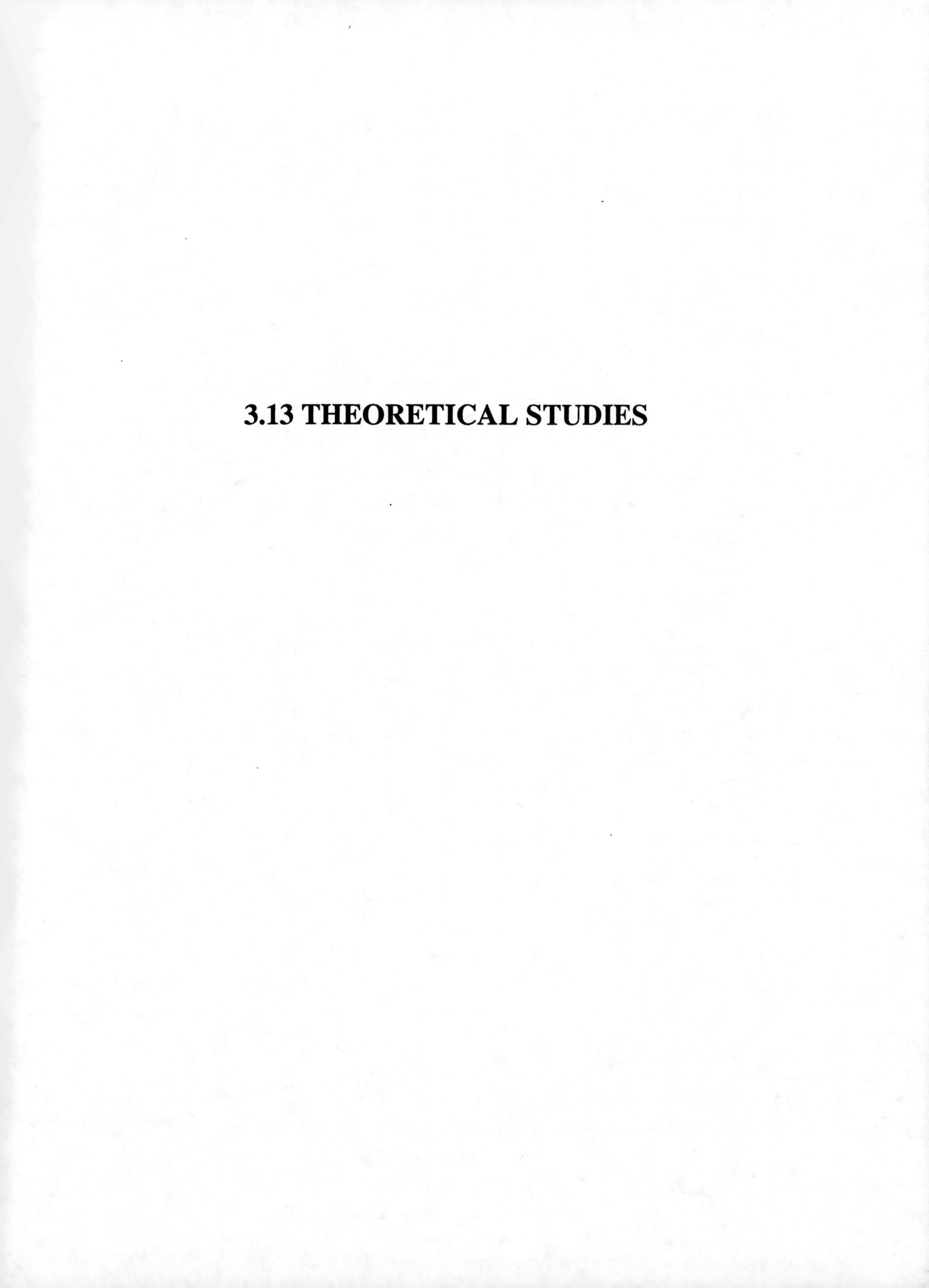

3.13 THEORETICAL STUDIES

Collision Broadening of Spectral Lines in the $X^3\Sigma_g^- \rightarrow b^1\Sigma_g^+$ System of O_2

R.S. Pope, P.J. Wolf, G. P. Perram, and J. J. Cornicelli

Department of Engineering Physics, Air Force Institute of Technology, Wright-Patterson AFB, OH 45433

Spectral line broadening in the (v″, v′) = (0, 0) vibrational band of the O_2 X→b system is studied by Fourier transform spectroscopy. Pressure broadening coefficients are determined by fitting a Voigt function to the measured line profiles for each resolvable rotational transition. Pressure broadening coefficients vs. rotational quantum number are reported for Ne and Xe, and O_2. The pressure broadening coefficients for self-broadening as a function of rotational state agree well with previous grating spectrometer and diode laser measurements. Preliminary results are also discussed for O_2 broadening by He, Ar, Kr, CO, CO_2, and SF_6.

INTRODUCTION

The broadening of molecular absorption lines via collisions with an ambient gas has received much attention in the literature. Accurate line broadening coefficients are useful both for determining interaction potential parameters and testing line broadening theories, and for determining pressures, temperatures, and densities of an environment that surrounds the molecule of interest. Atmospheric remote sensing, for example, requires accurate line broadening coefficients to determine the atmospheric transmission properties. In addition, broadened line profiles can be used as a non-intrusive probe in chemical laser environments, such as the Chemical Oxygen-Iodine Laser, to determine the operating conditions of the device.

The rotational line profiles of the (v′=0 → v″=0) transition in the O_2 (X→b) system, called the "A band," have been studied by several research groups. The earliest measurements were made by absorption spectroscopy with a grating spectrometer and a continuum source (1-3). More recent studies have used diode laser-based instruments (4-5). Researchers have reported the results for broadening with O_2, N_2, and air.

In this paper, we present preliminary results for the broadening of rotational lines in the A band of O_2 measured, we believe, for the first time by Fourier transform spectroscopy. The single advantage of this method over laser-based techniques is the acquisition of line profiles for the entire band system in a single scan. This method, however, does not provide the high resolution obtainable with laser methods. Nevertheless, accurate broadening coefficients can be determined. We report results for self-broadening and broadening in collisions with the noble gases and the molecules CO, CO_2, and SF_6. We compare our self-broadening results to those from previous studies and we comment on resolution and signal-to-noise issues that affect the precision of the results.

EXPERIMENT

The O_2 (X→b) absorption spectrum was recorded using a Bomem DA8.002 Fourier transform spectrometer. A quartz lamp, filtered through an Oriel 57290 notch filter (λ = 766.5 nm, $\Delta\lambda$ = 10 nm), was used as the radiation source. The small absorption cross section for the X→b transition ($\sigma = 2.7 \pm 0.7 \times 10^{-20}$ cm^2) required the use of a 10-meter path length, multi-pass White cell to obtain sufficient absorption. Light absorption was monitored with a Bomem IPH5700L silicon avalanche detector.

The Doppler full width at half maximum (FWHM) of a rotational line in the X→b spectrum was 0.028 cm^{-1}, so initial measurements of the self-broadened spectrum were made at resolutions of 0.020 cm^{-1}. However, the signal-to-noise ratio was very low at this resolution for measurements of practical duration. To improve signal-to-noise, the source aperture was opened from 0.5 mm to 1.0 mm which required that the resolution be decreased to 0.032 cm^{-1}. It is at this resolution that the majority of the measurements were made for this research effort.

Measurements of the self-broadened spectrum were made at 0.020 cm^{-1} and 0.032 cm^{-1} resolution for pressures ranging between 30 Torr and 400 Torr. Each of these measurements was the average of 100 scans of the interferogram. Additionally, a 400-scan measurement at 0.032 cm^{-1} resolution for P = 400 Torr was made to further improve the signal-to-noise ratio at the high-pressure end of the scale. Pressures above 400 Torr could not be measured because the transition became opaque. A typical absorption spectrum is shown in the Figure 1.

Measurements of the X→b transition of O_2 in collisions

CP430, *Fourier Transform Spectroscopy:* 11th International Conference
edited by J.A. de Haseth

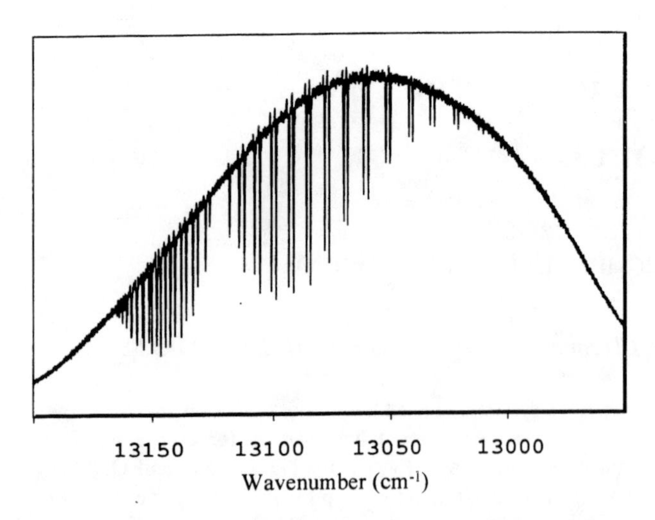

FIGURE 1. Oxygen self-broadening absorption spectrum measured with 400 scans at 0.032 cm^{-1} resolution. The oxygen pressure is 400 Torr. The P-branch features comprise the first set of absorption profiles, followed by R-branch lines.

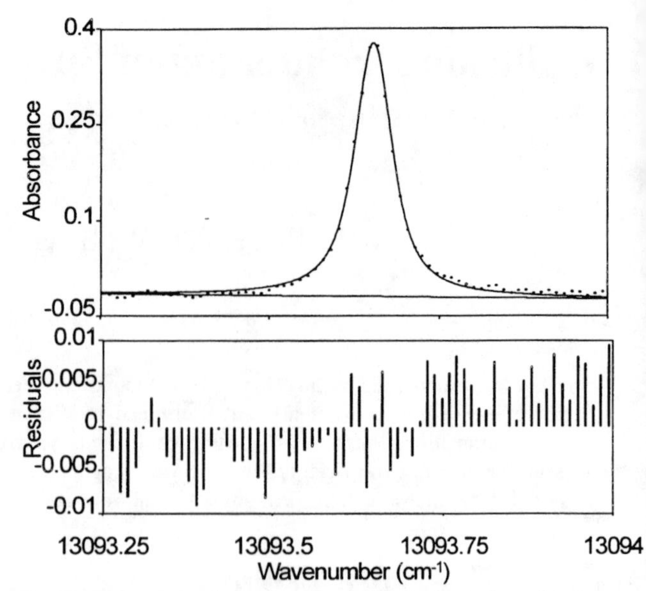

FIGURE 2. Voigt fit to the J″= 8 line in the P branch of the O_2 self-broadened spectrum with 400 scans. The resolution was 0.032 cm^{-1} and the pressure was 400 Torr.

with both the noble gases and the molecules CO, CO_2, and SF_6 were all made with 200 Torr of O_2 in the White cell. Spectra were acquired with broadening gases at pressures ranging between 100 Torr and 400 Torr. The operating limit of the White cell was approximately 600 Torr which restricted the pressure range available in these studies. The spectral resolution was 0.032 cm^{-1} for all the measurements. Initial measurements were all taken with 100 scans, but it became apparent that better signal-to-noise was required at high broadening-gas pressures. Therefore, a series of 400-scan measurements were made to improve the signal-to-noise ratios.

RESULTS AND DISCUSSION

The pressure broadening coefficients have been determined by first correcting the spectra for background. A non-linear least squares fit to the measured spectral feature for each rotational line was then performed using a Voigt profile to extract the linewidth. A fit to a typical peak is shown in Figure 2. Typically, one expects the Gaussian component of the line to remain constant, while the Lorentzian component increases with increasing pressure. Thus, we will follow the standard convention and use the Lorentzian half width at half maximum (HWHM) component as the pressure-broadened line width. Physically, there is a small contribution to the linewidth by Dicke narrowing, which reduces the Gaussian component, but the effect is so small that we were not able to observe it in these studies.

The width of each rotational line was subsequently plotted as a function of collision partner pressure to determine the pressure broadening coefficients. The slope of this plot, calculated using a weighted linear least squares fit to the data, gave the pressure broadening coefficient, γ, for that rotational line. For the self-broadened lines, we expected zero pressure broadening at zero pressure, so we constrained the y-intercept of the plot to be zero. For the foreign-broadened lines, since there was already 200 Torr of O_2 in the cell with the foreign gas, the y-intercept was obviously not zero. Instead, we used the self-broadened slope of each rotational line, multiplied by the pressure of O_2 in the cell, to place a data point at zero pressure of the foreign gas. A typical plot of the self-broadened data is shown in Figure 3.

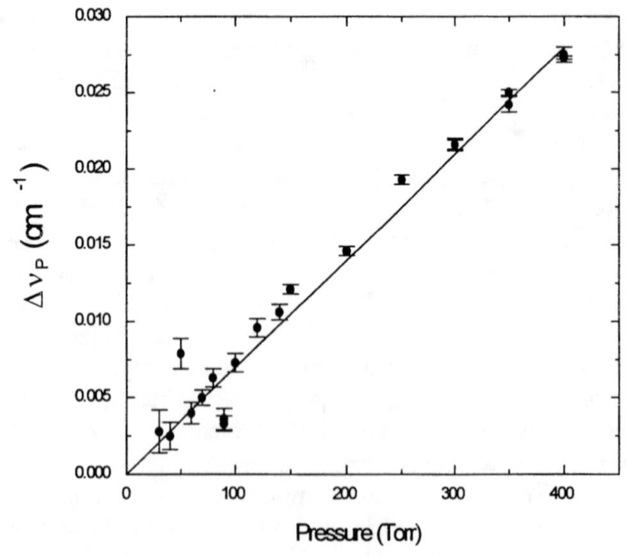

FIGURE 3. Plot of linewidth *vs.* O_2 pressure for the J″ = 8 line in the P-branch. The slope of this line gives the pressure broadening coefficient.

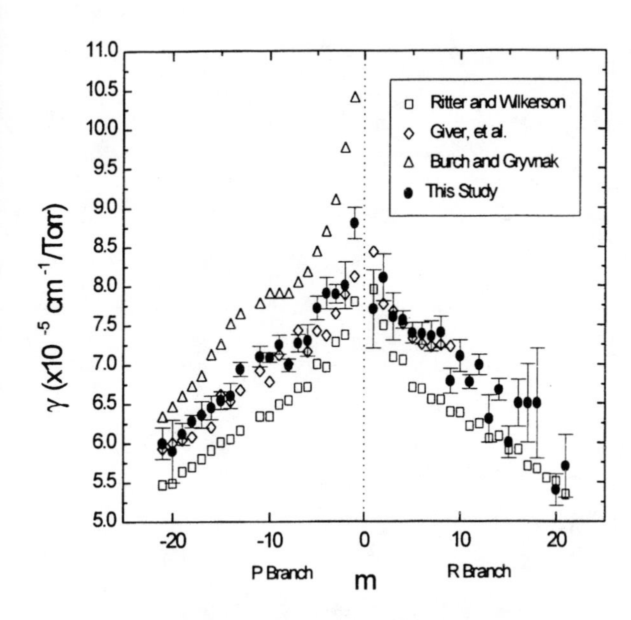

FIGURE 4. Oxygen self-broadened data plotted against m. Data from previous researchers are included for comparison. Previous data is from Ritter and Wilkerson (4), Giver, Boese, and Miller (3), and Burch and Gryvnak (2).

Once the pressure broadening coefficients were determined, we plotted them as a function of rotational quantum number, J″. The self-broadened result are shown in Figure 4, together with the results of several previous researchers. In this figure, γ is plotted against m where m = J″+1 for the R-branch and m = J″- 1 for the P-branch. Clearly, our results by Fourier transform spectroscopy compare favorably with the those determined previously by grating spectroscopy and diode laser spectroscopy.

Results for the noble gases Ne and Xe are shown in Figure 5, and they are in good agreement with an unpublished diode laser study reported by Ritter (7). Ritter, however, only looked at a few rotational lines in each branch, and measured only one line at more than two pressures. This data is, therefore, more precise.

Results showing the behavior of γ with m for He, Ar, Kr and the molecules CO, CO_2, and SF_6 are not yet ready for publication; we are making further measurements with improved signal-to-noise to try to reduce data scatter and the size of the error bars in the plots. However, data for our most intense line, the J″ = 8 line in the P branch, is precise enough that we can use it to demonstrate the trends among the broadening gases. Figure 6 shows a comparison of the J″= 8 broadening coefficients for each of the broadening gases. From the figure, we can see that, with the exception of He, the broadening coefficients for the noble gases increase with the mass of the atom. For the molecules CO, CO_2, and SF_6, the broadening coefficients also appear to increase

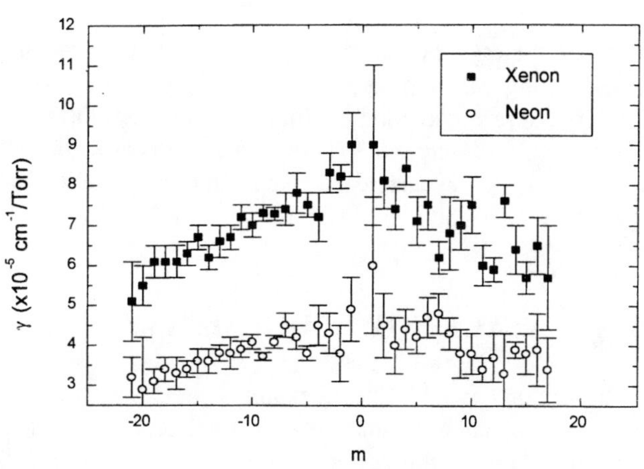

FIGURE 5. Oxygen broadening coefficients plotted against m for the noble gas collision partners neon and xenon.

with the mass or the size of the collision partner.

SUMMARY

The broadening of rotational lines in the (0,0) vibrational transition of the O_2 ($X \rightarrow b$) system by collisions with various ambient gas species was examined using Fourier transform spectroscopy. The self-broadened data presented here compared extremely well in both magnitude and behavior with rotational level with previously published data using high resolution laser diode spectroscopy. Preliminary data detailing O_2 broadening by noble gases and molecular species have also been presented. Currently, we are refining our measurements for several collision partners to increase the signal-to-noise ratio and, subsequently,

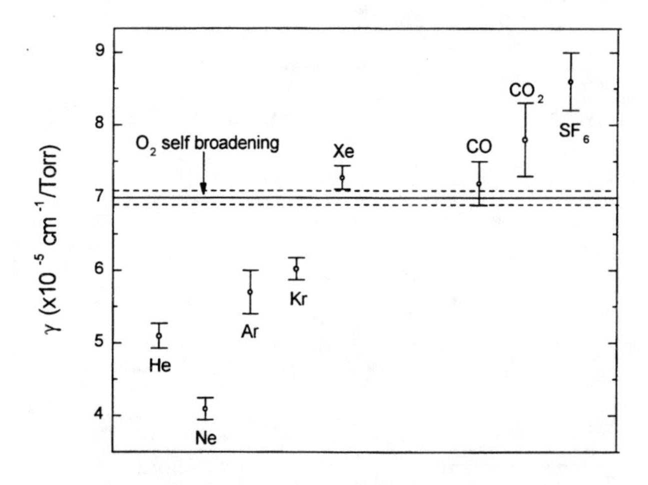

FIGURE 6. Oxygen broadening coefficients for the J″=8 line in the P branch compared for several collision partners. The solid horizontal line is the self-broadening coefficient, while the dashed horizontal lines are the error bars on the self-broadened data.

reduce the error bounds on the data. In addition, we are performing calculations to theoretically determine both the pressure broadening coefficients and intermolecular potential parameters. These calculations will be compared to experimental data and may offer an explanation to the trends we observe.

ACKNOWLEDGEMENTS

The authors would like to thank Tom Niday for his tireless assistance in analyzing the large amount of data generated during this research effort.

REFERENCES

1. J. H. Miller, R. W. Boese, and L. P. Giver, *J. Quant. Spectros. Radiat. Transfer* **9**, 1507-1517 (1969).
2. D. E. Burch and D. A. Gryvnak, *Appl. Opt.* **8**, 1493-1499 (1969).
3. L. P. Giver, R. W. Boese, and J H. Miller, *J. Quant. Spectrosc. Radiat. Transfer* **14**, 793-802 (1974).
4. K. J. Ritter and T. D. Wilkerson, *J. Mol. Spectrosc.* **121**, 1-19 (1987),
5. S. Bhattacharyya, A Hazra, and P. N. Ghosh, *J. Mol. Struct.* **327**, 139-144 (1994).
6. P. L. Varghese and R. K. Hanson, *Appl.Opt.* **23**, 2376-2385 (1984),
7. K. J. Ritter, PhD Dissertation, University of Maryland, 1986.

Coupling between Amide I Vibrations of Neighboring Peptide Groups and Conformation of an Extended Helix

Hajime Torii [1] and Mitsuo Tasumi [2]

[1] *Department of Chemistry, School of Science, The University of Tokyo, Bunkyo-ku, Tokyo 113, Japan*
[2] *Department of Chemistry, Faculty of Science, Saitama University, Urawa, Saitama 338, Japan*

Interrelationship among the coupling between the amide I vibrations of peptide groups, the A–E_1 wavenumber difference of the amide I mode, and the conformation of helical polypeptide chains is examined theoretically. *Ab initio* molecular orbital (MO) calculations are performed for a glycine dipeptide and a glycine tripeptide with various ϕ and ψ angles to obtain the coupling constants between the amide I vibrations of neighboring peptide groups. It is found that the coupling constants between the second nearest peptide groups are reasonably well explained by the transition dipole coupling mechanism, whereas the coupling constants between the nearest peptide groups contain other factors that mainly depend on ψ. The wavenumbers of the A and E_1 components of the amide I mode are calculated for various helices by using these coupling constants. The Raman–infrared wavenumber difference of 10–20 cm^{-1} observed for the amide I bands of poly(L-glutamate) and poly(L-lysine) with charged side chains indicates that these polypeptides, in the so-called "extended helix" state, have conformations giving rise to the E_1 component with a strong IR intensity and the A component with a strong Raman intensity, with the wavenumber difference between them being 10–20 cm^{-1}. The ranges of the ϕ and ψ angles that are consistent with such spectral features are discussed on the basis of the calculated structures and amide I wavenumbers.

INTRODUCTION

Poly(L-glutamic acid) and poly(L-lysine) are known to adopt a few different conformations such as the α-helix, antiparallel β-sheet [for poly (L-lysine) only], and "random-coil", depending on pH and temperature in aqueous solution. These two polypeptides have served as model compounds of proteins in many spectroscopic studies.

In those studies, the character of the so-called "random-coil" conformation, which is adopted in the state with ionized side chains, has been discussed. Tiffany and Krimm [1] have proposed in their CD studies that this conformation is not truly random as was previously conceived but has some helical order similar to that of a left-handed 3_1-helix. This ordered conformation has been referred to as the "extended helix". The existence of helical order has been supported by a large separation (10–20 cm^{-1}) between the wavenumbers of the IR and Raman bands in the amide I region (with the Raman band being higher in wavenumber) [2,3], almost equal integrated intensities of the positive and negative parts in the VCD spectra [4–6] in the amide I region, and some characteristic bands in the IR and Raman spectra in the 1300–900 cm^{-1} region [2,7].

The wavenumbers of the IR and Raman bands in the amide I region are different because they arise from different components of the amide I mode. The A component, in which all the peptide groups stretch in phase, has a strong Raman intensity. By contrast, the E_1 component has a strong IR intensity when the transition dipole of the amide I vibration of each peptide group is almost perpendicular to the helix axis. The A–E_1 wavenumber difference originates from the coupling between the amide I vibrations of peptide groups.

In the present work, the relation among the coupling between the amide I vibrations of neighboring (the nearest and the second nearest) peptide groups, the A–E_1 wavenumber difference of the amide I mode, and the conformation of a helical polypeptide chain is examined theoretically. Coupling constants between the amide I vibrations are calculated by the *ab initio* molecular orbital (MO) method for a glycine dipeptide (CH_3–CONH–CH_2–CONH–CH_3, abbreviated as GLDP hereafter) and a glycine tripeptide (CH_3–(CONH–CH_2–$)_2$CONH–CH_3, GLTP) with various ϕ and ψ angles. The wavenumbers of the A and E_1 components of the amide I mode of regular infinite polypeptide chains are calculated on the basis of these coupling constants. Validity of the transition dipole coupling (TDC) mechanism, which is used in our previous studies on the amide I bands of globular proteins [8], is also examined.

COMPUTATIONAL PROCEDURE

Ab initio MO calculations are carried out at the Hartree–Fock (HF) level by using the Gaussian 94 program [9].

CP430, *Fourier Transform Spectroscopy:* 11th International Conference
edited by J.A. de Haseth

The 6-31G** basis set augmented by one set of diffuse functions each on the oxygen and nitrogen atoms of the peptide groups is used. This augmented basis set is denoted by 6-31(+)G** hereafter.

Coupling constants between the amide I vibrations are calculated as follows. (1) Partial geometry optimization is performed for GLDP with each specified pair of ϕ and ψ values. The ϕ and ψ angles are taken at intervals of 30°, i.e., ϕ, ψ = 180°, 150°, ..., –150°. In the geometry optimization, the structural parameters of the peptide and the methyl groups are fixed to those of the fully optimized structure of *N*-methylacetamide (NMA). Only the structural parameters of the methylene group ($-CH_2-$), including the N–C and C–C bond lengths but excluding the ϕ and ψ angles, are optimized. (2) Both the peptide groups of GLDP (which are denoted by peptide group 1 and 2) are displaced along the coordinate of the amide I mode. The vibrational pattern is taken from the amide I normal mode of NMA. Only the positions of the C, O, N, and H atoms are displaced, since the motions of the α-carbons (the methyl groups in NMA) are negligible in the amide I mode. Four structures are generated by this procedure: (Q_1, Q_2) = (0.03, 0.03), (0.03, –0.03), (–0.03, 0.03), and (–0.03, –0.03) in units of Å $amu^{1/2}$, where Q_1 and Q_2 denote the displacements of peptide group 1 and 2, respectively. The coupling constant is calculated from the energies of these four structures as a finite difference. (3) GLTP with a specified (ϕ, ψ) pair is constructed by using the structural parameters obtained for GLDP. (4) The peptide groups at both ends of GLTP are displaced along the coordinate of the amide I mode. The coupling constant between the amide I vibrations of these peptide groups is calculated in a way similar to that described in (2).

The wavenumbers of the A and E_1 components of the amide I mode of a regular infinite polypeptide chains are calculated on the basis of the coupling constants thus obtained. The diagonal force constant is assumed to be 1.605 mdyn $Å^{-1}$ amu^{-1}. The calculated A–E_1 wavenumber difference is not sensitive to the assumed value of the diagonal force constant.

RESULTS AND DISCUSSION

The coupling constants between the amide I vibrations of the *second nearest* peptide groups calculated by the *ab initio* MO method are shown in Fig. 1 (a), where conformations of polypeptides are numbered by data number n related to the ϕ and ψ angles as $\phi = -180 + 30 \times \mathrm{int}[(n-1)/12]$ and $\psi = 180 - 30 \times \mathrm{mod}[(n-1), 12]$; i.e., conformations are arranged in order of (ϕ, ψ) = (–180°, 180°), (–180°, 150°), ..., (–180°, –150°), (–150°, 180°), ... Here, int(a) stands for the largest integer which is less than or equal to a, and mod(a, b) is calculated as a – int(a/b) × b.

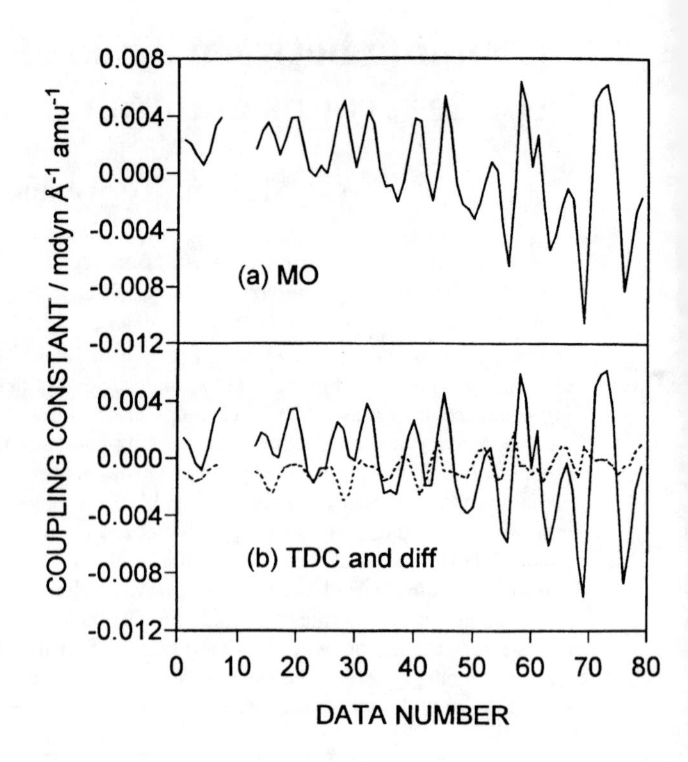

FIGURE 1. Coupling constant between the amide I vibrations of the second nearest peptide groups in a regular polypeptide chain (a) calculated at the HF/6-31(+)G** level, (b) calculated on the basis of the TDC mechanism (solid line) and the difference between the two (broken line).

The coupling constants in Fig. 1 (a) may seem to vary irregularly with data number n. However, since the peptide groups involved in these coupling constants are separated by one peptide group, contributions of interactions other than the electrostatic ones are considered to be small. It is therefore reasonable to consider that the coupling constants (F_{jk}, in mdyn $Å^{-1}$ amu^{-1}) between the amide I vibrations of neighboring peptide groups (j and k) are primarily determined by the TDC mechanism.

Based on the above consideration, the magnitude, direction, and location of the transition dipole (the TDC parameters) are optimized so that the coupling constants between the second nearest peptide groups obtained on the basis of the TDC mechanism reproduce the values obtained by the *ab initio* MO method. The result is shown by the solid line in Fig. 1 (b). The optimized magnitude of the transition dipole $|\partial\mu/\partial Q|$ is 2.73 D $Å^{-1}$ $amu^{-1/2}$ and the angle between $\partial\mu/\partial Q$ and the C=O bond is 10.0°. The optimized location of the transition dipole is represented as $\boldsymbol{r}_C + 0.665\ \boldsymbol{n}_{CO} + 0.258\ \boldsymbol{n}_{CN}$ (in units of Å), where $\boldsymbol{r}_C$ stands for the location of the carbonyl carbon atom, $\boldsymbol{n}_{CO} = (\boldsymbol{r}_O - \boldsymbol{r}_C)/|\boldsymbol{r}_O - \boldsymbol{r}_C|$, and $\boldsymbol{n}_{CN} = (\boldsymbol{r}_N - \boldsymbol{r}_C)/|\boldsymbol{r}_N - \boldsymbol{r}_C|$. All these values are reasonable for the TDC parameters of the amide I vibration of the peptide group. As shown by the broken line in Fig. 1 (b), the difference between the coupling

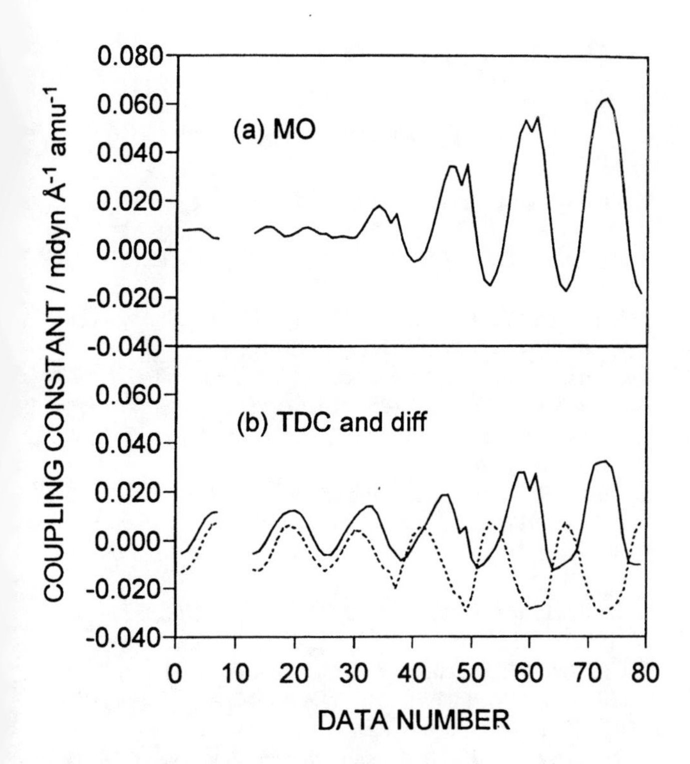

FIGURE 2. Coupling constant between the amide I vibrations of the nearest peptide groups in a regular polypeptide chain (a) calculated at the HF/6-31(+)G** level, (b) calculated on the basis of the TDC mechanism (solid line) and the difference between the two (broken line).

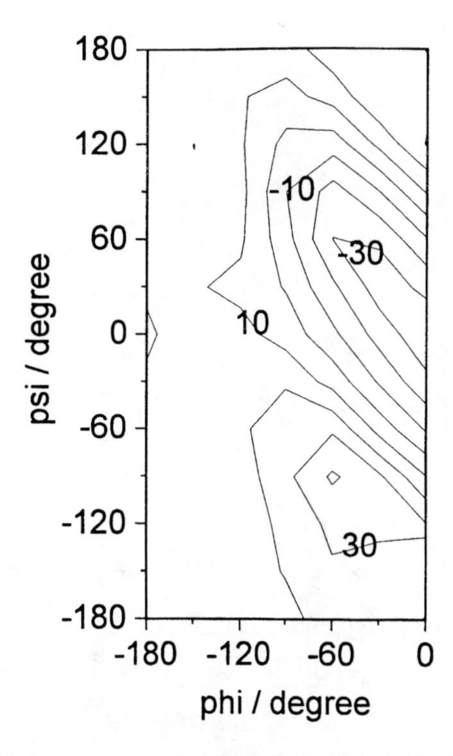

FIGURE 3. Wavenumber difference (in cm-1) between the amide I A and E_1 components of a regular polypeptide chain calculated by using the coupling constants obtained at the HF/6-31(+)G** level.

constants obtained by the two methods is small. It is therefore concluded that the approximation based on the TDC mechanism is valid for describing the coupling between the amide I vibrations of the second nearest peptide groups.

By using these optimized TDC parameters, the coupling constants arising from the TDC mechanism (the TDC constants) between the *nearest* peptide groups are calculated and are compared with the values obtained by the *ab initio* MO method in Fig. 2. The difference between the coupling constants obtained by the two methods varies simusoidally with data number n, indicating that this "residual" coupling constant primarily depends on the ψ angle. We consider that this residual coupling constant originates from the direct (through-bond) interactions between covalently-bonded peptide groups.

The wavenumber difference between the amide I A and E_1 components for each conformation is calculated by using the coupling constants between the amide I vibrations of the nearest peptide groups and between those of the second nearest peptide groups obtained by the *ab initio* MO method. The result is shown as a contour map in Fig. 3.

As explained in Introduction, the Raman band is higher in wavenumber by 10–20 cm-1 than the IR band in the amide I region for polypeptides with ionized side chains [2,3]. This means that the A–E_1 wavenumber difference is 10–20 cm-1 and that the transition dipole of the amide I vibration makes a large angle with the helix axis, generating a strong IR intensity for the E_1 component. In a light-scattering study, it has been shown that a polypeptide with ionized side chains has an extended conformation [10]. The angle (θ) between the transition dipole and the helix axis is generally large for a polypeptide with a pitch (h) larger than 3.0 Å. Although it is not possible to determine the exact values of θ and h, the present results of calculations and the experimental data mentioned above suggest that the likely ranges for these parameters are $\theta \geq 70°$ and $h \geq 3.0$ Å.

Based on the above discussion, the ranges of the ϕ and ψ angles suggested for the extended helix conformation are shown by the hatched area in Fig. 4. The boundary of the hatched area is determined by some of the contours in Fig. 3 and those corresponding to $\theta = 70°$ and $h = 3.0$ Å.

Sengupta and Krimm [7] have observed a band at 943 cm-1 in the Raman spectrum of poly(L-Glu Ca) in the 1000–900 cm-1 region. They have shown that the observed wavenumber of 943 cm-1 corresponds to the ϕ angle of about –100°. According to the result of the present work, the ψ angle would be in the range of $160° \leq \psi \leq 195°$ if the ϕ angle is around –100°. We note that the

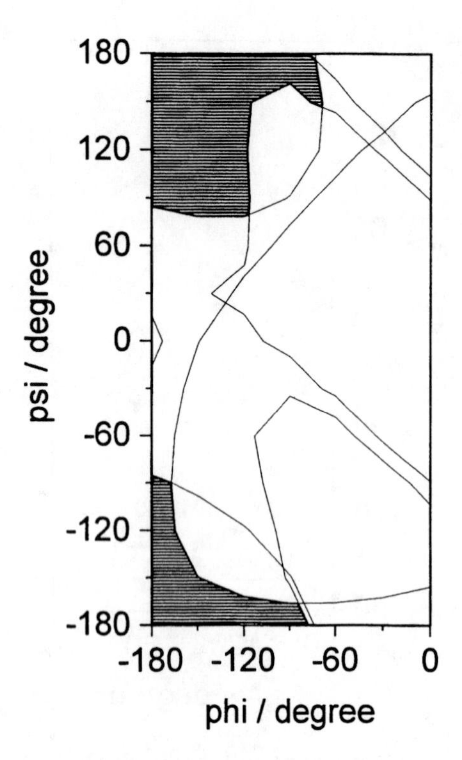

FIGURE 4. Range of the ϕ and ψ angles suggested for the extended helix conformation (hatched area). See text.

conformation with (ϕ, ψ) = (–100°, 170°) is close to that of polyglycine II. Since the VCD spectrum in the amide I region of poly(L-glutamic acid) with ionized side chains is very similar to that of the left-handed 3_1 polyproline II helix [6], the conformation with (ϕ, ψ) = (–100°, 170°) seems to be most likely for the extended helix.

REFERENCES

1. Tiffany, M. L., and Krimm, S., *Biopolymers* **6**, 1379–1382 (1968); **8**, 347–359 (1969).
2. Painter, P. C., and Koenig, J. L., *Biopolymers* **15**, 229–240 (1976).
3. Sugawara, Y., Harada, I., Matsuura, H., and Shimanouchi, T., *Biopolymers* **17**, 1405–1421 (1978).
4. Paterlini, M. G., Freedman, T. B., and Nafie, L. A., *Biopolymers* **25**, 1751–1765 (1986).
5. Yasui, S. C., and Keiderling, T. A., *J. Am. Chem. Soc.* **108**, 5576–5581 (1986).
6. Dukor, R. K. and Keiderling, T. A., *Biopolymers* **31**, 1747–1761 (1991).
7. Sengupta, P. K. and Krimm, S., *Biopolymers* **26**, S 99–107 (1987).
8. Torii, H. and Tasumi, M., *J. Chem. Phys.* **96**, 3379–3387 (1992).
9. Gaussian 94; Frisch, M. J., Trucks, G. W., Schlegel, H. B., Gill, P. M. W., Johnson, B. G., Robb, M. A., Cheeseman, J. R., Keith, T., Petersson, G. A., Montgomery, J. A., Raghavachari, K., Al-Laham, M. A., Zakrzewski, V. G., Ortiz, J. V., Foresman, J. B., Cioslowski, J., Stefanov, B. B., Nanayakkara, A., Challacombe, M., Peng, C. Y., Ayala, P. Y., Chen, W., Wong, M. W., Andres, J. L., Replogle, E. S., Gomperts, R., Martin, R. L., Fox, D. J., Binkley, J. S., Defrees, D. J., Baker, J., Stewart, J. J. P., Head-Gordon, M., Gonzalez, C., and Pople, J. A., Gaussian, Inc., Pittsburgh, PA, 1995
10. Kidera A., and Nakajima, A., *Macromolecules* **17**, 659–663 (1984).

CH-Stretching Vibrational Circular Dichroism of α-Hydroxy Acids and Related Molecules

Teresa B. Freedman, Denise Gigante, Eunah Lee, and Laurence A. Nafie

Department of Chemistry, Syracuse University, Syracuse, New York 13244-4100 USA

Vibrational circular dichroism (VCD) spectra of chiral α-hydroxy acids and related molecules have been investigated in the hydrogen stretching and midinfrared regions to probe factors influencing the VCD intensity of methine and hydroxyl stretching and bending motions. Ab initio calculations were carried out to identify low energy conformers and to calculate VCD intensity with the vibronic coupling VCD theory, utilizing DFT normal modes and geometry. Large methine stretching VCD intensity was correlated with the presence of an oxygen at the chiral center in conjunction with an O=C-C*-O dihedral angle near 0°. Vibrational transition current density plots for the methine stretch in deuteriated methyl lactate reveal angular and circulatory charge flow consistent with the positive rotational strength for the *S*- enantiomer.

INTRODUCTION

Vibrational circular dichroism (VCD) provides information on absolute configuration and solution conformation of chiral molecules (1). In recent years, interpretation of VCD spectra has relied in large part on comparison with ab initio calculations of VCD intensities for a selection of conformers of the molecule. In cases where such calculations are not practical, identification of intense VCD bands that serve as markers for configuration and conformation remains an important interpretational tool. Such markers include the methine-stretching mode in amino acids and α-hydroxy acids (2). This study focuses on factors that influence the intensity of methine stretching VCD and other dominant VCD features in α-hydroxy acid derivatives. In addition, we have utilized transition current density plots (3) to visualize the electron density current, produced by the nuclear motion, that contributes to the magnetic dipole transition moment.

EXPERIMENTAL

Molecules selected for study (Scheme 1) include six with stable hydrogen-bonded rings, 5-membered (**I, III, IV, and VI**) or 6-membered (**VIII and IX**), and three with no hydrogen bonding (**II, V, and VII**). (*S*)-Benzoin, (*S*)-methyl 2-chloropropionate, (*S*)-methyl 3-hydroxy-2-methylpropionate and (*S*)-methyl 3-hydroxybutyrate were obtained from commercial sources. To remove overlapping absorption features, the deuteriated esters (*S*)-methyl-d_3 lactate, (*S*)-methyl-d_3 2-(methoxy-d_3)-propionate, di(methyl-d_3) D-tartrate, (*S*)-methyl-d_3 mandelate, and (*S*)-methyl-d_3 O-(acetyl-d_3)-mandelate were synthesized from the parent acids. Spectra were recorded for 0.01 M CCl_4 solutions.

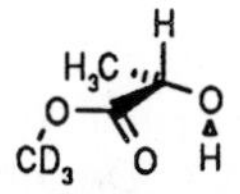

(*S*)-methyl-d_3 lactate (**I**)

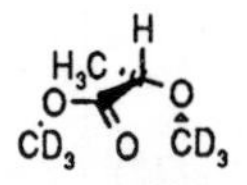

(*S*)-methyl-d_3 2-(methoxy-d_3)- proprionate (**II**)

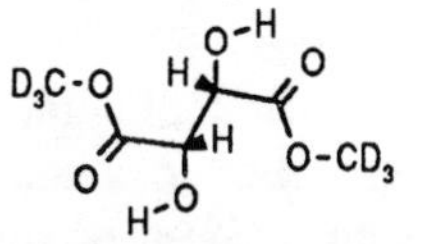

di(methyl-d_3) D-tartrate (**III**)

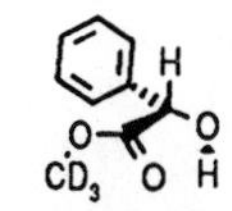

(*S*)-methyl-d_3 mandelate (**IV**)

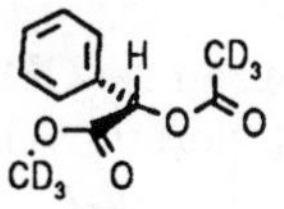

(*S*)-methyl-d_3 *O*-(acetyl-d_3)-mandelate (**V**)

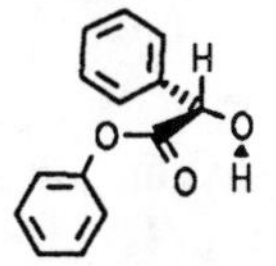

(*S*)-benzoin (**VI**)

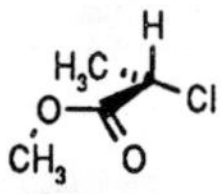

(*S*)-methyl 2-chloroproprionate (**VII**)

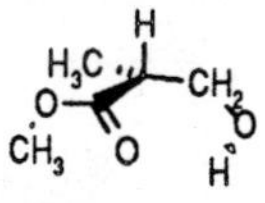

(*S*)-methyl 3-hydroxy-2-methylproprionate (**VIII**)

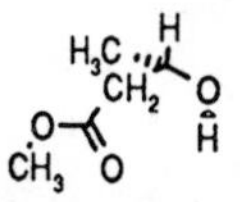

(*S*)-methyl 3-hydroxybutyrate (**IX**)

Scheme 1

CP430, *Fourier Transform Spectroscopy:* 11th International Conference
edited by J.A. de Haseth

Spectra in the mid-infrared region were measured on a Nicolet Magna 850 FTIR equipped with an external VCD bench (4). Spectra in the CH-stretching region were obtained in step-scan mode with a Bruker IFS-55 FTIR adapted for step-scan VCD measurements (5), or on a dispersive VCD instrument.

For some of the molecules studied, *ab initio* molecular orbital calculations of optimized geometries and vibrational frequencies were carried out to identify low energy conformers. VCD intensities were calculated with the vibronic coupling VCD theory (6,7) utilizing normal modes and atomic polar tensors obtained at the HF/6-31G(d) or DFT/6-31G(d) levels. For (S)-methyl-d_3 lactate-Cd_3 vibrational transition current density (TCD) plots (3) were generated for the methine-stretching mode.

RESULTS AND DISCUSSION

(*S*)-Methyl-d_3 lactate (**I**), (*S*)- methyl-d_3 2-(methoxy-d_3)-propionate (**II**), di(methyl-d_3) D-tartrate (**III**), (*S*)-methyl-d_3 mandelate (**IV**), (*S*)-methyl-d_3 O-(acetyl-d_3)-mandelate (**V**) and (*S*)-benzoin (**VI**) all possess oxygen and C=O substituents at the chiral center. For these molecules, large anisotropy ratios ($\Delta A/A$) between $+3 \times 10^{-4}$ and $+5 \times 10^{-4}$ were measured for the methine stretching mode, which dominates the VCD intensity. For example, in the experimental IR and VCD spectra of (*S*)-methyl-d_3 lactate included in Fig. 1, the methine stretch is assigned to the intense broad VCD feature at 2880 cm^{-1}. Replacing the hydroxyl with a chloro group or inserting a methylene group between the chiral carbon and the OH or C=O groups of the methyl lactate structure to form (*S*)-methyl 2-chloro-propionate, (**VII**) (*S*)-methyl 3-hydroxy-2-methylpropionate (**VIII**) and (*S*)-methyl 3-hydroxy-butyrate (**IX**) results in methine stretching VCD intensity that is quite small, relative to the other molecules studied. The VCD spectrum of (*S*)-methyl 3-hydroxybutyrate in Fig. 1 does not exhibit a net VCD intensity bias, in contrast to the large bias to positive intensity for (*S*)-methyl-d_3 lactate, which arises from the methine stretch.

Ab initio molecular orbital calculations were carried out to identify low energy conformers of several of the molecules, and to determine the VCD intensities with the vibronic coupling VCD theory (5,6). Excellent agreement with experimental VCD intensities in the 900-1500 cm^{-1} region of (S)-methyl lactate was found when the DFT (B3LYP) geometry and normal modes were used for the VCD calculation. For the CH-stretching region, the VCD intensity calculations indicate that conformations with an O=C-C*-O dihedral angle near 0° give rise to strong positive methine-stretching VCD. In contrast, the molecules that lack an oxygen heteroatom at the chiral center (**VII-IX**), or conformations with a trans or 90° arrangement of O=C-C*-O exhibit only weak methine stretching VCD, in agreement with experiment. As shown in Fig. 2 for (*S*)-methyl-d_3 lactate, for the lowest energy conformer 1 [OH--O=C hydrogen bonding], intense positive VCD intensity is calculated, whereas for conformer 2 [OH--O(CD_3)-C hydrogen bonding], only weak methine stretching VCD is predicted. Data in the OH-stretching region indicate that conformer 1 is dominant in this solution. Large positive methine-stretching VCD is also observed for II and **V**, which cannot form intramolecular hydrogen bonds. For the lowest energy conformations of **II** and **V**, O=C-C*-O dihedral angles of –11° and O--O distances of ~2.7 Å are calculated, similar to those calculated for I and **IV**.

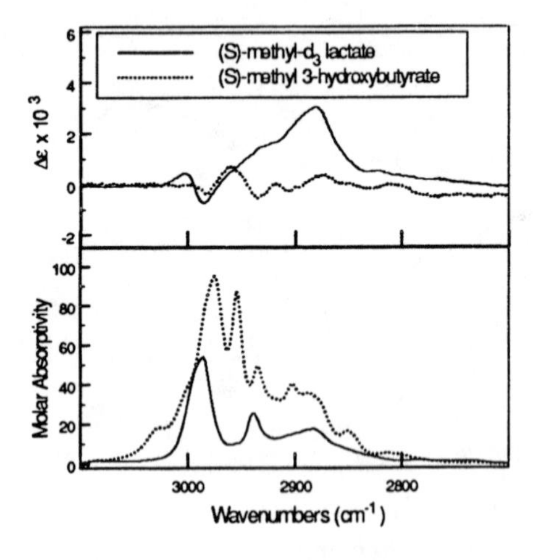

Figure 1. Comparison of IR and VCD spectra of (*S*)-methyl-d_3 lactate and (*S*)-methyl 3-hydroxybutyrate, 0.01 M in CCl_4 solution.

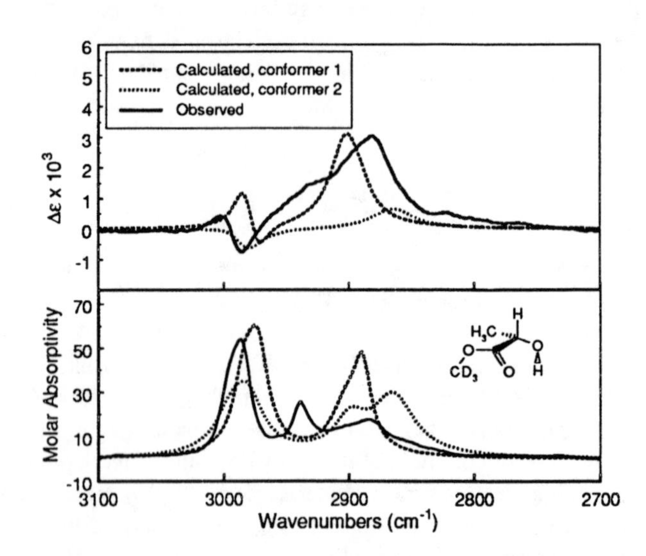

Figure 2. Experimental CH-stretching IR and VCD spectra of (*S*)-methyl-d_3 lactate, 0.01 M in CCl_4 solution (–) compared to calculations for the lowest energy OH--O=C hydrogen-bonded conformer 1 (---) and higher energy OH--O(CD_3)-C hydrogen-bonded conformer 2 (···).

We recently demonstrated the utility of vibrational transition current density (TCD) plots in the visualization of the electron density current that accompanies nuclear motion (3). To gain further insight into the origin of the large positive methine stretching rotational strength, vibrational transition current density calculations were carried out for the deuteriated isotopomer (*S*)-methyl-d_3 lactate-Cd_3 in the most stable, intramolecularly hydrogen-bonded OH--O=C conformation. The TCD calculations allow visualization of the motion of electron density at each point in space for a given phase of vibration, displayed as 2D projections in selected planes. We find that the methine stretch in this isotopomer produces angular charge flow across the O=C-C*-OH bonds and circulatory motion about the oxygen atomic centers, even though these nuclei are nearly stationary. This angular and circulatory charge flow generates a magnetic dipole transition moment with a component parallel to the electric dipole transition moment, consistent with the observed positive rotational strength. An example of the vibrational TCD, shown in Fig. 3 for a plane approximately perpendicular to the methine bond through the ester oxygen atom, illustrates circulatory current about the ester oxygen nucleus. Vibrational TCD along the O=C-C*-OH bonds, accompanied by circulatory motion about the carbonyl oxygen, is evident in a projection plane containing the methyl carbon atom, shown in Fig. 4. Both types of current contribute to rotational strength in the +z-direction, parallel to the electric dipole transition moment for the elongation phase of this mode.

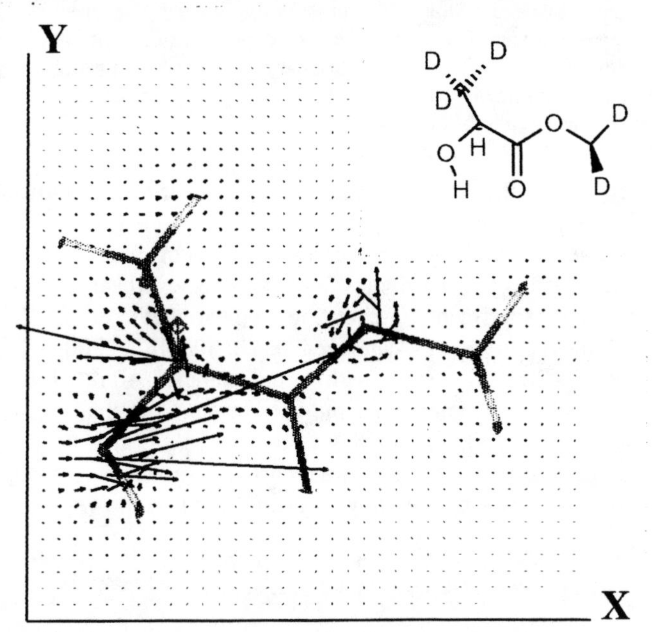

Figure 3. Vibrational TCD plot for the methine elongation in (*S*)-methyl-d_3 lactate-Cd_3 in a plane containing the ester oxygen and perpendicular to the methine bond.

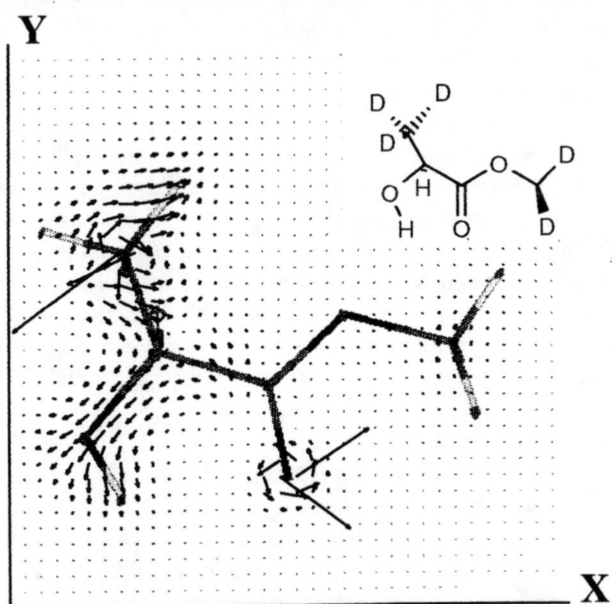

Figure 4. Vibrational TCD plot for the methine elongation in (S)-methyl-d_3 lactate-Cd_3 in a plane containing the methyl carbon and perpendicular to the methine bond.

CONCLUSIONS

Enhanced vibrational circular dichroism intensity for the methine stretch in chiral esters correlates with conformations in which the methine bond is oriented out of the approximate plane of a cis O=C-C*-O structure, independent of the presence or absence of an intramolecularly hydrogen-bonded ring. Intense positive methine-stretching VCD is a marker for the S-configuration for this conformation. The TCD plots for this mode reveal the origin of the angular and circulatory electronic charge flow that accompanies the nuclear motion.

ACKNOWLEDGMENTS

Support for this work from the National Institutes of Health (GM-23567) and the New York State Center for Advanced Technology in Computer Applications and Software Engineering is gratefully acknowledged.

REFERENCES

1. Nafie, L .A. *Appl. Spectrosc.* **50 (5)**, 14A-26A (1996).
2. Freedman, T. B., Balukjian, G. A., and Nafie, L. A. *J. Am. Chem. Soc.* **107**, 6213-6222 (1985).
3. Freedman, T. B., Shih, M.-L., Lee, E., and Nafie, L. A. *J. Am. Chem. Soc.*, in press (1997).
4. Long, F., Freedman T. B., Hapanowicz, R., and Nafie, L. A. *Appl. Spectrosc.* **51**, 504-507 (1997).
5. Long, F., Freedman T. B., Tague, T. J., and Nafie, L. A. *Appl. Spectrosc.* **51**, 508-511 (1997).
6. Nafie, L. A., and Freedman, T. B. *J. Chem. Phys.* **78**, 7108-7116 (1983).
7. Rauk, A., and Yang, D. *J. Phys. Chem.* **96**, 437- (1992).

Anti-Stokes - Stokes Intensity Ratio and Intensity Correction of FT-Raman Spectra

G. Jalsovszky and G. Keresztury

Central Research Institute for Chemistry, Hungarian Academy of Sciences, P.O.Box 17, H-1525 Budapest, Hungary

A relationship between sample temperature and the intensity ratio of anti-Stokes and Stokes Raman scattering intensity has been used to calculate sample temperatures observable during FT-Raman measurements. The results obtained depend strongly on the intensity calibration of the single-beam instrument. When the reference spectrum for intensity calibration is produced by a white light source, the temperature of this source has a significant effect on the accuracy of calibration. Using a relationship between sample temperature and white light source temperature, more accurate intensity calibration is possible.

INTRODUCTION

In the measurement of single-beam spectra the response function of the spectrometer is of primary significance (1). This is the case in FT-Raman experiments, where the response function strongly varies with frequency. Thus, even for the determination of relative intensities, an accurate instrument response function should be known. This function is obtained by measuring the spectrum of a source with known spectral characteristics, and dividing it by the (known) true spectrum of the source. In FT-Raman spectroscopy, where the spectrum is measured in the NIR region, a "known" source may be an ideal black body emitter, the spectrum of which is described (2) by an explicit theoretical expression. In commercial FT-Raman instruments black body source is substituted by a white light source, e.g. tungsten lamp. An important parameter of the source is its temperature, which determines the "known" spectrum used in the calculation. The primary aim of this work was to find a method which can be used to determine this temperature, and thereby to improve the intensity calibration of Raman instruments. The resulting more accurate intensities may find both theoretical and analytical applications.

One of the applications is the determination of sample temperature on the basis of the ratios of Stokes- and anti-Stokes Raman intensities. The relationship between these quantities is well known (3), but its application to temperature measurement (4,5), particularly with FT-Raman technique (6), is limited.

Sample temperature can be expressed from the above cited (3) relationship by a simple transformation:

$$T_{sample} = \frac{1.4388\Delta\nu}{4(\ln\nu_{AS} - \ln\nu_S) - \ln R_{AS/S}} \qquad (1)$$

where T_{sample} is sample temperature in K, $\Delta\nu$ is the (Stokes) Raman shift in cm^{-1}, ν_{AS} and ν_S are the (absolute) frequencies of anti-Stokes and Stokes lines in cm^{-1}, and $R_{AS/S}$ is the ratio of anti-Stokes and Stokes Raman intensities.

Due to their good absorbance and poor internal heat conduction, solid samples are easy to overheat during Raman experiments. According to Marigheto, et al. (6) sample temperatures can even reach 400 K at 300 mW laser power. However, with liquid samples of relatively large volume, temperature may not be far from that of the sample compartment for low laser powers, or, at least, the temperature vs. laser power curve approaches sample compartment temperature with decreasing laser power.

EXPERIMENTAL

Spectra of cyclohexane filled into a glass tube of 5 mm o.d. were recorded on a Nicolet 950 FT-Raman instrument at 4 cm^{-1} resolution in back-scattering mode. Laser source was a Nd^{3+}:YAG laser operating at 1064 nm (9395 cm^{-1}). Laser power was varied between 40 and 940 mW, and 64 scans were co-added for each spectrum. For low laser powers the poor signal-to-noise ratio made it impossible to identify anti-Stokes lines, thus 100 mW was the lowest power for which sample temperatures could be calculated. Sample compartment temperature was 303 K during the measurements. Reference spectra for intensity correction were taken from KBr powder filled into a similar sample tube, illuminated by white light (possible settings: LO and MD) in the same sampling geometry. Correction was performed by the OMNIC software package of Nicolet.

RESULTS AND DISCUSSION

Table 1 shows the temperature values calculated by Equation (1) from the intensity ratios of the 801 cm^{-1} band of cyclohexane. The uncorrected spectra yield obviously

CP430, *Fourier Transform Spectroscopy:* 11th International Conference
edited by J.A. de Haseth

too low temperatures: even with 940 mW excitation the calculated sample temperature is below 0 °C. On the other hand, the corrected spectra yield too high temperatures: 75 or 51 °C for the lowest laser power.

TABLE 1. Sample temperatures of cyclohexane calculated from uncorrected and automatically corrected Raman spectra

Laser power, mW	Sample temperature, K		
	uncorr.	corr., LO	corr., MD
103	251	348	324
241	263	371	344
440	267	379	350
705	270	382	353
940	268	378	369

It is worth noting that the two reference spectra lead to significantly different sample temperatures indicating that either the program of intensity correction is defective, or at least the assumed temperature of white light source, required in the calculation, is far from its true temperature. We have checked both possibilities, arriving at the conclusion that the calibration algorithm was correct, and thus the wrong assumed temperatures must be responsible for the inaccuracy of the correction process.

We have reproduced the correction procedure for the anti-Stokes - Stokes intensity ratios manually by dividing the measured Raman intensity ratio by the measured white light intensity ratio multiplied by the intensity ratio of the theoretical black-body function:

$$R_{AS/S}^{corr.} = R_{AS/S} \frac{R_{BB}}{R_{WL}} \tag{2}$$

where R_{BB} and R_{WL} are the intensity ratios in the black-body curve and the white light reference spectra, respectively, at the frequencies of the anti-Stokes and Stokes Raman lines. We have calculated R_{WL} from the recorded reference spectrum and R_{BB} from the theoretical expression of black-body emission. By substituting corrected *AS/S* ratios into Equation (1), corrected sample temperatures were obtained. As R_{BB} is a function of temperature, corrected sample temperatures depend on black-body temperature. Our calculations have shown (Table 2) that sample temperatures obtained from a reference spectrum with MD white light source setting and source temperatures in the 1300 to 1400 K range fall into the 300 K region for low laser powers, compatible with the measured sample compartment temperature.

It can be stated that the sample temperatures obtained with this correction procedure are much closer to reality than those obtained with the automatic intensity correction. For MD white light setting in the given setup 1400 K is the best choice of source temperature.

TABLE 2. Sample temperatures of cyclohexane calculated from intensity ratios corrected with MD white light setting and assumed white light source temperatures of 1300 and 1400 K

Laser power, mW	Sample temperature, K		
	uncorr.	MD@1300 K	MD@1400 K
103	251	292	301
241	263	308	319
440	267	313	324
705	270	315	326
940	268	314	325

TABLE 3. Sample temperatures of sulphur calculated from intensity ratios corrected with MD white light setting and an assumed white light source temperature of 1400 K

Laser power, mW	Sample temperature, K	
	uncorrected	corrected
200	256	316
300	263	326
395	267	333
510	270	338

It can also be seen from Table 2 that a change of 100 K in white light source temperature involves a change of ca. 10 K in calculated sample temperatures indicating the importance of correct source temperature estimation.

The Raman spectra of sulphur powder were measured under similar experimental conditions, and sample temperatures were calculated by Equations (1) and (2), from the intensity ratio of the 473 cm^{-1} line, using MD white light setting and 1400 K as source temperature (Table 3). A paper on mixtures of sulphur and wheat (6) reported 400 K at 300 mW laser power. Another investigation on sodium molybdate (7) has shown a sample temperature rise of 50 K for a laser power of 375 mW. As sample temperatures strongly depend on exposure time, sample holder, quality of sample, etc., these results are not directly comparable. Nonetheless, our results are closer to those of the latter report (7).

CONCLUSIONS

For the calculation of sample temperatures from anti-Stokes - Stokes intensity ratios, intensity correction of FT-Raman spectra is absolutely necessary. The temperature of white light source must be known for the correction: an error of 100 K in source temperature leads to a difference of ca. 10 K in sample temperature.

ACKNOWLEDGEMENT

The authors are indebted to National Scientific Research Fund for financial support (OTKA T014472).

REFERENCES

1. Petty, C.J., Warnes, G.M., Hendra, P.J., and Judkins, M., *Spectrochim. Acta* **47A,** 1179-1187 (1991).
2. Keresztury, G., and Mink, J., *Appl. Spectrosc.* **46,** 1747-1749 (1992).
3. Long, D.A., *Raman Spectroscopy,* McGraw-Hill, 1977, p84.
4. Kip, B.J., and Meier, R.J., *Appl. Spectrosc.* **44,** 707-711 (1990).
5. Rassat, S.D., and Davis, E.J., *Appl. Spectrosc.* **48,** 1498-1505 (1994).
6. Marigheto, N.A., Kemsley, E.K., Potter, J., Belton, P.S., and Wilson, R.H., *Spectrochim. Acta* **52A,** 1571-1579 (1996).
7. Pope, S.J.A., and West, Y.D., *Spectrochim. Acta* **51A,** 2011-2017 (1995)

Structural Investigations of Oriented Membrane Assemblies by FTIR-ATR Spectroscopy*

Urs Peter Fringeli[1], Jeannette Goette[2], Gerald Reiter[1], Monira Siam[1], and Dieter Baurecht[1]

[1] *Institute of Physical Chemistry, University of Vienna, Althanstrasse 14, A-1090 Vienna, Austria.*
[2] *Medinova AG, Eggbühlstrasse 14, CH-8050 Zurich, Switzerland.*

In situ attenuated total reflection (ATR) Fourier transform (FT) spectroscopy is presented as an adequate tool for studying molecular structure and function of biomembranes. In this article emphasis was directed to the production of suitable model bilayer membranes for optimum mimicking of natural biomembranes, and to special FTIR ATR techniques to achieve enhanced selectivity as well as time resolved information on complex membrane assemblies. In this context, the preparation of supported bilayers according to the LB/vesicle method is presented and the use of such model membranes to build more complex assemblies, e.g. with creatine kinase, a surface bound enzyme, and alkaline phosphatase, a membrane anchored enzyme. A comprehensive summary of equations used for quantitative ATR spectroscopy is given and applied to determine the surface concentration and orientation of membrane bound molecules. The use of supported bilayers for drug membrane interaction studies is demonstrated by the local anesthetic dibucaine. Besides of structural information's, such studies result also thermodynamic date, such as adsorption isotherm and partition coefficient. A special ATR set-up for more precise background compensation is presented enabling the conversion of a single beam spectrometer into a pseudo double beam spectrometer. This optical component may be placed in the sample compartment of the spectrometer, and is referred to as single-beam-sample-reference (SBSR) attachment. Finally, a short theoretical introduction into time resolved modulation spectroscopy is given. Temperature modulated excitation of reversible conformational changes in the polypeptide poly-L-lysine and the enzyme RNase are shown as examples.

INTRODUCTION

Biological model membranes in the form of planar supported bilayer assemblies are of special interest for attenuated total reflection spectroscopy (ATR), since the internal reflection element (IRE) may act as both, the solid support and the wave guide. This set up enables membrane spectroscopy from the part of the solid support, and simultaneous control of the membrane environment from the opposite part. Moreover, the supported membrane is located in the most intense region of the electric field of the evanescent wave which favors the fraction of light absorbed by the membrane with respect to unwanted absorption's of the background (e.g. buffer solution). Oriented regions of the assembly, which should be expected because of intrinsic ordering of lipid bilayers, may be detected and analyzed in a straightforward way by the use of polarized incident light.

Several methods have been described in the literature to prepare symmetric or asymmetric lipid bilayers. Among them the three most important one should be noted: (i) The Langmuir-Schaefer method, adapted for model membrane studies by Tamm and McConnell (1). A hydrophilic plate was coated by a Langmuir-Blodgett (LB) monolayer of a phospholipid. The hydrophobic surface of the supported monolayer was then contacted with the corresponding hydrophobic surface of a compressed monolayer at the air water interface of a film balance. Immersing this plate in vertical direction into the subphase results in the supported bilayer in the aqueous environment of the subphase. (ii) A different way to get supported bilayers was described by Brian and McConnell (2). A dispersion of small unilamellar lipid vesicles is brought into contact with the hydrophobic surface of a solid support, leading to a spontaneous spreading of the vesicles on the solid. (iii) The best characterized procedure leading to supported bilayers is the so called LB/vesicle method (3), (4), (5), (6). As in the first case, the ATR plate (IRE) is coated by a LB lipid monolayer pointing with the hydrocarbon chains towards the air. This plate is now mounted in an ATR flow-through cell where it is contacted by an aqueous solution of lipid vesicles. Since a hydrophobic surface exhibits a large surface energy when in contact with liquid water, a spontaneous oriented adsorption of lipid molecules from the vesicles to the LB monolayer takes place leading to a stable bilayer membrane. Adsorption may be monitored *in situ* by IR ATR spectroscopy.

So far a variety of experiments with different types of supported lipid membranes have been reported for the interaction with proteins, peptides and drugs. First use was made by studying the interaction of local anesthetics (LA) of the tertiary amine type with a dipalmitoylphosphatidic acid (DPPA)/palmitoyl-oleoyl-phosphatidylcholine (POPC) lipid bilayer where POPC was facing the liquid environment (7).

*In memory of Robert Kellner

CP430, *Fourier Transform Spectroscopy:* 11th International Conference
edited by J.A. de Haseth

In order to probe the degree of penetration of the LA dibucaine (DIBU) into the lipid bilayer, as well as the influence on hydrocarbon chain ordering in the membrane, a supported bilayer was used, consisting of DPPA as inner monolayer and dimyristoyl-phosphatidylcholine with deuterated hydrocarbon chains (DMPC-d_{54}) as outer layer facing the aqueous environment (8), (9). In another application, a supported DPPA/POPC bilayer served by itself as a support for natural membrane fragments enriched with the Na^+/K^+-ATPase (sodium-potassium pump) (10). The orientation of the peptide melittin in a POPC/POPG (POPG: Palmitoyl-oleoyl-phosphatidylglycerol) was studied by means of FTIR ATR spectroscopy using polarized incident light (11). This study revealed that the structure of melittin as well as its orientation with respect to the membrane critically depends on the degree of hydration. A similar behavior was observed with the antibiotic alamethicin which was incorporated into dipalmitoyl-phosphatidylcholine (DPPC) mulibilayers (12). Obviously, the structures of oligopeptides are very sensitive to the environment. As a consequence, fully hydrated membrane assemblies must be considered as the most adequate models for ATR spectroscopy. In this respect, it is of great importance to have a good characterization of the supported bilayer from different points of view. In addition to FTIR studies (refs. 3 - 6) valuable information on the nano-structure was obtained by combined application of FTIR and atomic force microscopy (AFM) (13) since nano- and ultra-structure of a sample are of utmost importance for a reliable interpretation of polarized spectra. A further critical point when working with supported bilayers is the lateral mobility of lipid molecules in the inner (LB) monolayer. Fluorescence recovery after photo bleaching (FRAP) has revealed that the lateral mobility of natural phospholipids is preserved (5) giving good evidence for a significant water layer between the solid support (IRE) and the immobilized membrane.

In this presentation we discuss the preparation and characterization of three membrane systems. Two enzyme-membrane assemblies, the mitochondrial creatine kinase (CK) interacting with cardiolipin (CL), and the alkaline phosphatase (AP) interacting with POPC, as well as dibucaine (DIBU), a local anesthetic interacting with a DPPA/POPC bilayer.

Physiologically, the mitochondrial CK is important for the energy metabolism in cells of high and fluctuating energy requirements. The native form of the enzyme is octameric and highly ordered as revealed by x-ray diffraction (14). Typical features are a large channel, connecting the top to the bottom face of the nearly cubic shaped octamer and the accumulation of positively charged residues at these opposite faces. CK was found to adsorb spontaneously to negatively charged phospholipid membranes most probably by one of the positively charged surfaces. First kinetic results have been obtained by plasmon resonance and light scattering experiments (15). *In situ* FTIR ATR spectroscopy with [po]larized light was used to monitor the formation of a membrane assembly consisting of a DPPA/CL bilayer membrane and the binding of CK to this membrane. Quantitative analysis resulted in a surface coverage of about 60% by CK (16). Furthermore, a flow-through system enabled the measurement of enzyme activity simultaneously with spectroscopic data acquisition. Structural changes in the lipid bilayer revealed a predominant electrostatic interaction at the membrane surface (17).

In contrast, the alkaline phosphatase (AP) is a dimeric enzyme with one glycosyl-phosphatidylinositol (GPI) molecule covalently bound to each subunit. The hydrophobic hydrocarbon chains of the phospholipid are responsible for anchoring the enzyme in the lipid bilayer. AP is insoluble in aqueous buffer solution and needs addition of a detergent for solubilization. On the other hand structure and stability of the supported lipid bilayer are influenced, too. Detachment from the IRE was observed as soon as the detergent concentration was equal or higher than the critical micelle concentration (cmc). Obviously, the transfer of AP from solution to the lipid bilayer is much more critical than that of CK. *In situ* measurements of the formation and characterization of the membrane assembly consisting of an inner DPPA LB-layer with an adsorbed POPC layer and bound AP have been performed. A considerable amount POPC molecules from the outer leaflet of the bilayer were found to be replaced by detergent molecules (β-octylglucoside, β-OG) during AP attachment to the bilayer membrane. Moreover, the hydrocarbon chain ordering of the inner membrane leaflet (DPPA) was decreased upon AP binding giving evidence for a hydrophobic interaction, possibly with the GPI anchor (18).

The interaction of DIBU, a tertiary amine local anesthetic (LA) with a supported bilayer membrane consisting of a DPPA inner leaflet and a POPC outer leaflet was studied *in situ* at different DIBU concentrations. Even at bulk pH 5.5 where the amino group of dissolved DIBU, which has a pK value of 8.9, is nearly completely protonated, quantitative FTIR ATR measurements revealed that about 50% of the membrane bound DIBU was in the deprotonated, i.e. in a hydrophobic, water insoluble state. This approximately 1:1 mixture of charged and uncharged DIBU is assumed to be characteristic of aggregated DIBU. A small fraction of the total amount of detected uncharged DIBU is expected to be in the interior of the lipid bilayer. This finding is supported by the observation that the hydrocarbon chain ordering of both, the outer (POPC or DMPC-d_{54}), and the inner (DPPA) m were significantly reduced compared to the ordering in an unperturbed supported bilayer. The majority of uncharged DIBU, however, is expected to be aggregated with the protonated (charged) form, even without influence of the membrane.

A new ATR attachment will be presented, too. It enables the conversion of a single beam (SB) spectrometer into a pseudo double beam instrument in order to achieve more stable background compensation than with the conventional SB technique. This attachment converts convergent light entering the sample compartment into a parallel beam. The latter is focused to the entrance of a multiple internal reflection element (MIRE) by means of a cylindrical mirror thus producing a parallel ray propagation within the MIRE (wave guide). Placing now the sample at the lower half of the plate and the reference at the upper half and switching the beam by a chopper alternatively up and down leads to the so-called single beam sample reference (SBSR) technique (3), featuring two advantages over conventional SB technique: (i) elimination of water vapor incompensation, and (ii) compensation of drifts resulting from the instrument or the sample/reference, respectively. In this article we present a SBSR ATR attachment which makes use of a computer controlled lift in order to move the upper or lower half of the MIRE into the parallel beam. This set up leads to a better base line than the chopper version used earlier (3,4) since the FTIR spectrometer turned out to have inhomogeneous intensity within the cross-section of the IR beam. Subdividing this beam into an upper and lower half and calculating the ratio of the corresponding single channel spectra will therefore not result in a flat 100% line as should be expected for an ideal spectrometer.

Finally, an outlook to the application of more sophisticated dynamic techniques such as modulated excitation (ME) FTIR spectroscopy is given. External periodic stimulation of the membrane assembly may be accomplished via thermodynamic parameters such as temperature (T), concentration (c), electric field (E), light flux (Φ) etc. The power of ME-FTIR spectroscopy is based on the one hand on a significant enhancement of the selectivity of the measurement which is of great importance in the case of macromolecule spectroscopy featuring heavily overlapped absorption bands. Selectivity is achieved by the fact that only molecules or molecular parts (e.g. functional groups) affected by the periodic external perturbation will result in a corresponding modulated absorbance spectrum. Phase sensitive detection (PSD) enables now a separation of modulated and non-modulated parts of the FTIR spectrum with a higher accuracy than achieved by conventional difference spectroscopy. Moreover, if the system response to a sinusoidal excitation with frequency ω contains not only the fundamental ω, but also the first harmonic (2ω) or even higher harmonics, one has unambiguous evidence for a nonlinear chemical system,which may originate in chemical reactions different from first order and/or cooperative phenomena. The latter are of special interest in processes where conformational changes of biological macromolecules are involved. If a time resolved data acquisition of the system response is possible, a kinetic analysis of the process may be performed on the basis of the frequency dependence of both, amplitudes and phases of modulated absorbances of the system response. This information is accessible by phase sensitive detection (PSD) of time resolved FTIR absorbance spectra via analog or digital procedures (19). Variation of the modulation frequency ω results in independent data sets which facilitate the reconstruction of the reaction scheme. This is the main advantage of ME spectroscopy on relaxation (REL) spectroscopy. Although both, ME and REL techniques have the same physico-chemical information content, the read out of this information, however, is facilitated by ME, because more independent experimental data is available due to the additional degree of freedom given by the parameter 'modulation frequency' ω, while the number of unknown kinetic parameters included the reaction scheme are the same for ME and REL. After a theoretical introduction to ME spectroscopy, T-ME spectra of poly-L-lysine (PLL) will be shown. PLL may be converted reversibly from the α-helical to the antiparallel β-pleated sheet structure by T-modulation under adequate conditions. At least three so far unknown transient species have been detected and sequentially assigned based on their phase shift with respect to the T-stimulation (20). Finally, preliminary results from a T-modulation experiment performed with RNase A is shown. This enzyme undergoes a reversible unfolding/ refolding process with a distinct transition temperature at T=64°. Time resolved temperature jump (T-jump) measurements revealed relaxation times in the range of τ_i = 5-10 s (21). A corresponding T-ME experiment has been performed by switching the temperature between 59°C and 69°C with a period τ_{mod} = 25 s.

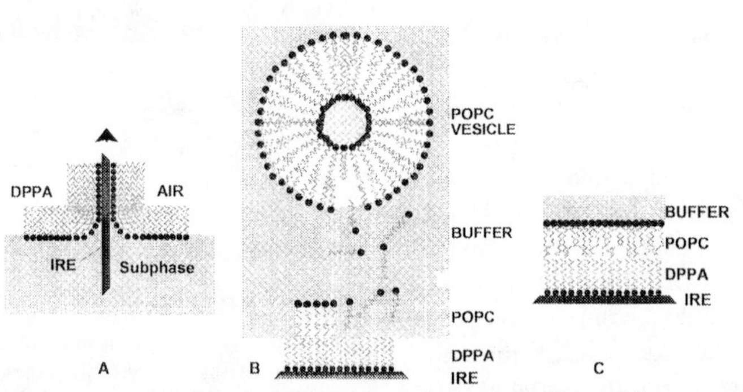

FIGURE 1 . Supported lipid membrane prepared by the LB/vesicle technique. **A**: The solid support (IRE) is coated by a lipid monolayer (e.g. DPPA) transferred in the compressed state from the air-water interface of a film balance by means of the Langmuir-Blodgett technique. **B**: Spontaneous lipid transfer from vesicles (e.g. POPC) to the hydrophobic surface of the LB monolayer occurs by forming a supported bilayer. **C**: Completed asymmetric bilayer (e.g. DPPA/POPC) in contact with buffer solution.

PREPARATION OF SUPPORTED MODEL MEMBRANE ASSEMBLIES

Supported Bilayers

Model bilayers for *in situ* ATR studies have been prepared according to the LB/vesicle method. The principle steps of this procedure are depicted in Fig. 1.

In practice, after LB-layer transfer, see Fig. 1A, the monolayer coated IRE is mounted in an ATR flow through cuvette as depicted by Fig. 2. After spectroscopic examination of quantity and quality (ordering) of the LB monolayer a vesicle solution of any phospholipid is circulated through the cuvette. This set up enables direct monitoring of the state of lipid adsorption shown by Fig. 1B.

After about 30 min the state of Fig. 1C is reached. A special washing procedure turned out to be necessary in order to detach loosely bound vesicle fragments. For details, the reader is referred to refs. (4), (6) and (16).

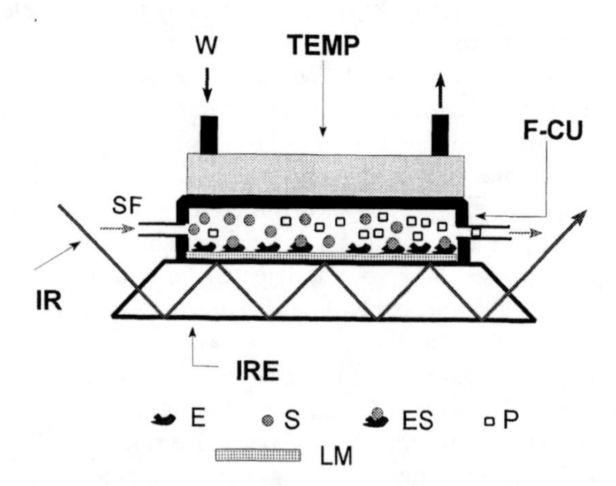

FIGURE 2. Flow-through cuvette (F-CU) for *in situ* FTIR ATR spectroscopy. The IRE is coated by a supported bilayer (LM) to whom the enzyme (E) is immobilized. The substrate (S) flowing through the cuvette (SF) is enzymatically converted into the product (P).

Membrane Bound Creatine Kinase

Mitochondrial creatine kinase (CK) is a highly ordered octamer. It has nearly cubic shape with an edge of 93 Å (14) and features an accumulation of positive charges at two opposite sides of the cube. Probably one of these sides binds to the membrane surface, predominantly by electrostatic interaction. Therefore, a DPPA/CL membrane (CL: cardiolipin) was used where CL formed the outer negatively charged layer of the supported bilayer. Pumping CK through the ATR flow-through cuvette (Fig. 2) lead to a spontaneous adsorption of CK. The observed surface coverage of 60 % and the kinetics of CK adsorption were in accordance with results obtained by a plasmon resonance study (15). For ails the reader is referred to ref. (16).

Membrane Anchored Alkaline Phosphatase

Alkaline phosphatase (AP) is also a membrane bound enzyme. The binding mechanism of AP, however, is quite different from that of CK. Native AP forms a water insoluble dimer where each monomer has one covalently bound glycosyl phosphatidylinositol (GPI) phospholipid molecule. It is assumed that AP is anchored to a bilayer predominantly via the hydrocarbon chains of GPI, penetrating into the hydrophobic region of the bilayer. Consequently, the reconstitution of a membrane assembly with bound AP must be expected to be more delicate than with CK. Indeed, the detergent β-octylglucoside (β-OG) which was added in order to solubilize AP has lead to the fist problem, because it destroyed the POPC layer of a DPPA/POPC bilayer nearly completely. FTIR spectra revealed, however, that the spectra of AP and β-OG appeared at the place of POPC. In a second step it was possible to exchange β-OG by POPC using an aqueous buffer solution with POPC vesicles circulating through the ATR cuvette. For a more detailed discussion, the reader is referred to ref. (16).

QUANTITATIVE ATR SPECTROSCOPY

Evanescent Wave and Penetration Depth

The relevant optical parameters are depicted by Fig. 3, which shows schematically the trace of a ray upon internal reflection at the membrane IRE interface. The goal of this analysis of FTIR ATR spectra is the calculation of the surface concentration and orientation of molecules in an

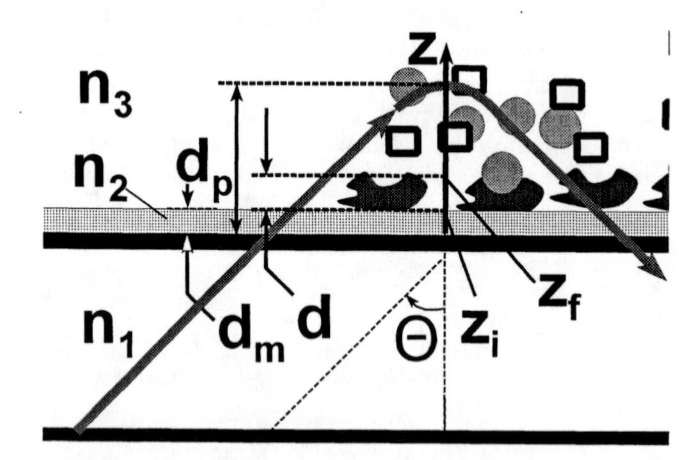

FIGURE 3. Penetration depth d_p of the evanescent wave. The internally reflected light propagates within the IRE (index 1) with an angle of incidence θ. It penetrates the membrane (index 2) of thickness $d_m \approx 50$ Å and the adsorbed enzyme (adlayer of thickness d, ranging from $z = z_i$ to $z = z_f$) and enters the aqueous environment (index 3). The strength of the electric field decreases exponentially with distance z from the 1-2 interface, see eqn. 1. The refractive indices of the media are denoted by n_1, n_2, and n_3.

adsorbed layer as well as the bulk concentration of dissolved substances in the aqueous environment. For that purpose some characteristic properties of internal reflection spectroscopy must be explained. Fig. 4 shows the IRE fixed coordinate system which is relevant for the description of optical and structural features of the system. Straightforward calculation of the propagation of a plane wave from medium 1 (IRE) into a nonabsorbing medium 2 under the conditions of total reflection (22) yields eqn. (1)

$$E_{x,y,z,2}(z) = E_{x,y,z,2}(0)\exp(-z/d_p) \qquad (1)$$

It follows that all electric field components of the so-called evanescent wave decrease exponentially with distance z from the interface. d_p denotes the penetration depth, which according to eqn. (2) is in the order of the wavelength of the light in the medium 1. The relative field components at the interface 1-2, $E^r_{x,y,z,2}(0) = E^r_{0x,0y,0z,2}$ may be calculated by Fresnel's equations (refs. (4),(22),(23), eqn. (7)).

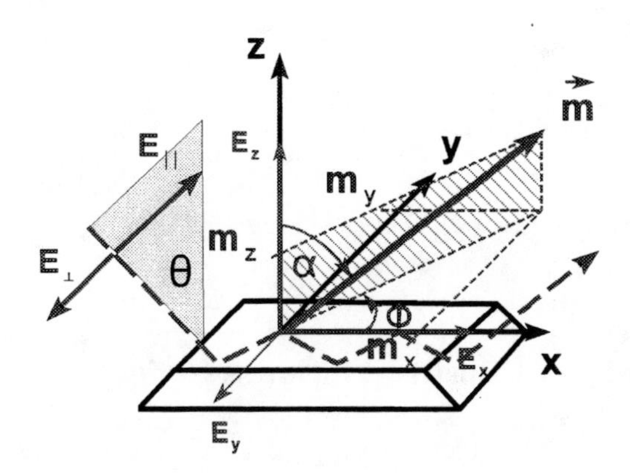

FIGURE 4. ATR set up. Optical and structural features are related to the IRE fixed coordinate system x,y,z. $E_{||}$ and $E_{\perp}$ denote the parallel and perpendicular polarized components of the light incident to the IRE under the angle θ. $E_{||}$ results in the E_x, and E_z components of the evanescent field, while $E_{\perp}$ results in the E_y component. $\vec{m}$ denotes the unit vector in direction of the transition dipole moment vector of a given vibrational mode, and m_x, m_y, m_z are the corresponding components in the IRE coordinate system. $\vec{m}$ goes off at an angle α with respect to the z-axis and the projection of $\vec{m}$ to the xy-plane goes off at an angle ϕ with respect to the x-axis.

$$d_p = \frac{\lambda/n_1}{2\pi(\sin^2\theta - (\frac{n_3}{n_1})^2)^{\frac{1}{2}}} \qquad (2)$$

$\lambda_1 = \lambda/n_1$ denotes the wavelength in medium 1, and λ is the vacuum wavelength. θ is the angle of incidence. Since the thickness of the membrane assembly is in the range of 50 Å to 150 Å, i.e. only about 1 % of the penetration depth, the electric field in the rarer medium is predominantly determined by the bulk medium 3. Therefore, the refractive index n_3, is used instead of n_2, see Fig. 3. This is the basic assumption for the thin layer approximation introduced by Harrick (23). As a consequence the electric field in the membrane (thin medium 2) is then assumed to be determined by the boundary conditions valid for a thin dielectric film in the electric field of the evanescent wave generated by the bulk media 1 and 3, see refs. (4) and (23).

Effective thickness d_e

The concept of effective thickness has been introduced by Harrick, ref. (23). The quantity d_e indicates the thickness of a sample that would result in the same absorbance in a hypothetical transmission experiment, as obtained with the genuine ATR experiment. This concept enables application of Lambert-Beer's law on ATR spectra according to eqn. (3).

$$T = 10^{-\varepsilon c d_e} = 10^{-A} \qquad (3)$$

where $A = \varepsilon c d_e$ denotes the absorbance per internal reflection. For an isotropic layer extended from $z = z_i$ to $z = z_f$ one obtains (ref. (4)).

$$d_e^{iso} = \frac{1}{\cos\theta}\frac{n_2}{n_1}\frac{d_p}{2}E_{02}^{r\,2} \cdot \left(\exp(-\frac{2z_i}{d_p}) - \exp(-\frac{2z_f}{d_p})\right) \qquad (4)$$

According to eqn. (4) d_e turns out to be wavelength dependent via d_p, see eqn. (2). As a consequence, ATR spectra of bulk media generally show increasing intensity with increasing wavelength. However, if the thickness of the layer $d = z_f - z_i$ is small compared to d_p then eqn. (4) reduces to eqn (4a) which is independent of the wavelength.

$$d_e^{iso} = \frac{1}{\cos\theta}\frac{n_2}{n_1}\, d\, E_{02}^{r\,2} \qquad (4a)$$

A further case often encountered is the bulk sample extended from $z_i = 0$ to $z_f = \infty$ resulting in eqn (4b).

$$d_e^{iso} = \frac{1}{\cos\theta}\frac{n_2}{n_1}\frac{d_p}{2}E_{02}^{r\,2} \qquad (4b)$$

$E_{02}^{r\,2}$ denotes the square of the electric field strength in medium 2 which is proportional to the light intensity. For polarized incident light it follows

$$E_{02,||}^{r\ 2} = E_{02,x}^{r\ 2} + E_{02,z}^{r\ 2} \quad and$$

$$E_{02,\perp}^{r\ 2} = E_{02,y}^{r\ 2} \qquad (5)$$

Introducing eqn. (5) into eqn. (4) results in

$$d_{e||}^{iso} = d_{ex}^{iso} + d_{ez}^{iso} \quad and$$

$$d_{e\perp}^{iso} = d_{ey}^{iso} \qquad (6)$$

It should be concluded from eqns. (5) and (6) that straightforward application of Lambert-Beer's law to ATR spectra of isotropic samples needs measurements with either parallel or perpendicular polarized incident light. The effective thickness for unpolarized incident light $d_{e,up}^{iso}$ turns out to be a linear combination of $d_{e||}^{iso}$ and of $d_{e\perp}^{iso}$. The coefficients depend on the polarizing properties of the optical components in the spectrometer (ref. (4)) which must be determined for each instrument.

Relative Electric Field Components

electric field components of the evanescent wave may be calculated by means of Fresnel's equations (refs. (22),(23)). For a 'nonabsorbing' thin film, e.g. a membrane assembly as depicted in Fig. 3 (medium 2), one obtains the following expressions (refs.(4), (23)):

$$E_{0x,2}^{r} = \frac{E_{0x,2}}{E_{0||,1}} = \frac{2\cos\theta\,(\sin^2\theta - n_{31}^2)^{1/2}}{(1-n_{31}^2)^{1/2}\,[(1+n_{31}^2)\sin^2\theta - n_{31}^2]^{1/2}}$$

$$E_{0z,2}^{r} = \frac{E_{0z,2}}{E_{0||,1}} = \frac{2\cos\theta\,\sin\theta\,n_{32}^2}{(1-n_{31}^2)^{1/2}\,[(1+n_{31}^2)\sin^2\theta - n_{31}^2]^{1/2}}$$

and

$$E_{0y,2}^{r} = \frac{E_{0y,2}}{E_{0\perp,1}} = \frac{2\cos\theta}{(1-n_{31}^2)^{1/2}} \qquad (7)$$

The meaning of n_{ik} is the ratio of the refractive indices of media i and k, respectively, i.e. $n_{ik} = n_i/n_k$. According to the thin film approximation by Harrick (23) the corresponding field components of the bulk environment (medium 3) are obtained by replacing index 2 by index 3. This affects only the z-component in accordance with electrostatic boundary conditions.

For intermediate layer thickness, i.e. $d \approx d_p$ the electric field components (eqn. (7)) must be modified, either by accurate treatment of a layered system (22), or by the application of an approximation described by eqn. (8) which is based on the interpolation between the results obtained for a thin layer ($d \ll d_p$) and a bulk medium ($d \gg d_p$) (24). The results obtained by this approximation deviate less than 5% from those got by the much more complicated accurate calculation. Eqn. (8) holds for all field components (x-, y-, and z-direction).

$$E_{02}^{r}(d) = E_{02}^{r}(thin\ layer) + (1-e^{-d/d_p})\cdot(E_{02}^{r}(bulk) - E_{02}^{r}(thin\ layer)) \qquad (8)$$

Validity of Effective Thickness Concept

Since the effective thickness concept enables the application of Lambert-Beer's law to ATR data, experimental validation may be performed easily by comparing spectra of the same sample measured by both, ATR and transmission (T). As long as the results do not differ significantly from each other the analytical approach described above is considered to be justified. ATR and T measurements with aqueous solutions of Na_2SO_4 have shown that at 1 molar concentration Lambert-Beer's law is still fulfilled for the very intense SO_4^{2-} stretching band at 1100 cm^{-1}. Even for the strong H_2O bending ($\delta(H_2O)$) band of liquid water at 1640 cm^{-1} the integral molar absorption coefficients determined by ATR with a germanium MIRE at an angle of incidence of $\theta = 45°$ was found to be equal to T-data within the experimental error (4). However, a few percents of deviation were found when peak values of the absorbance were used to determine the molar absorption coefficient. The latter indicates the onset of band distortion, a phenomenon well known in ATR spectroscopy under extreme conditions (23). This finding is in accordance with calculations by Harrick using Fresnel's equations with complex refractive indices (23). For Ge in contact with liquid water and $\theta = 45°$ the analysis resulted in an upper limit of the absorption coefficient $\alpha_{max} \approx 1000$ cm^{-1}. The concept of effective thickness as described above may be considered to be valid for $\alpha < \alpha_{max}$. For organic compounds this condition is generally fulfilled. In case of $\delta(H_2O)$ of liquid water, however, the absorption coefficient (ref. 4) results in $\alpha = \varepsilon(1640\ \text{cm}^{-1})\cdot c = 1.82\cdot10^4\ \text{cm}^2\cdot\text{mol}^{-1}\cdot5.56\cdot10^{-2}\ \text{mol}\cdot\text{cm}^{-3} = 1011.9\ \text{cm}^{-1}$, which indicates that the limit of validity of the approach is reached, in complete accordance with experimental data mentioned above.

Oriented Samples

Considering a transition dipole moment $\vec{M}$ associated with a vibrational mode of a given molecule and the electric field $\vec{E}$, responsible for vibrational excitation, the intensity of light absorption depends on the mutual orientation of these vectors according to

$$\begin{aligned}\Delta I &\propto (\vec{E}\cdot\vec{M})^2 \\ &= |\vec{E}|^2\cdot|\vec{M}|^2\cdot\cos^2(\vec{E},\vec{M}) \\ &= (E_xM_x + E_yM_y + E_zM_z)^2\end{aligned} \tag{9}$$

Eqn. (9) forms the basis of orientation measurements. M_x, M_y, and M_z denote the components of the transition dipole moment in the IRE fixed coordinate system shown in Fig. 4. It is usual to work with dimensionless relative intensities instead of absolute intensities in order to get rid of physical and molecular constants, e.g. the magnitude of the transition moment. Introducing the so-called dichroic ratio, the absorbance ratio obtained from spectra measured with parallel and perpendicular polarized incident light, i.e.

$$R = \frac{A_{||}}{A_{\perp}} = \frac{d_{e,||}}{d_{e,\perp}} = \frac{\int A_{||}\,d\tilde{\nu}}{\int A_{\perp}\,d\tilde{\nu}} \tag{10}$$

In order to get information on the direction of the transition dipole moment $\vec{M}$, the scalar product notation using vector components (see eqn. (9, 21)) will be used. Taking into account that in the evanescent field pp light is represented by x- and z-components, and vp light by the y-component one obtains for the dichroic ratio:

$$\begin{aligned}R &= \frac{A_{||}}{A_{\perp}} = \frac{d_{e,||}}{d_{e,\perp}} \\ &= \frac{E_x^2\, m_x^2 + E_z^2\, m_z^2 + 2\, E_x\, E_z\, m_x\, m_z}{E_y^2\, m_y^2}\end{aligned} \tag{11}$$

m_x, m_y, and m_z are the unit vector components of $\vec{M}$. Eqn. (11) holds for a single crystalline sample. In a complex non crystalline molecule there are generally many possibilities of molecular arrangements, conformational changes and fluctuations. The experimentally available quantity R is therefore a ensemble mean represented by:

$$\begin{aligned}R &= \frac{A_{||}}{A_{\perp}} = \frac{d_{e,||}}{d_{e,\perp}} \\ &= \frac{E_x^2\,\langle m_x^2\rangle + E_z^2\,\langle m_z^2\rangle + 2\,E_x\,E_z\,\langle m_x\,m_z\rangle}{E_y^2\,\langle m_y^2\rangle}\end{aligned} \tag{12}$$

Uniaxial orientation along the z-axis is often encountered in membrane spectroscopy. In this case $\langle m_x m_z\rangle = 0$, resulting in

$$\begin{aligned}R &= \frac{A_{||}}{A_{\perp}} = \frac{d_{e,||}}{d_{e,\perp}} \\ &= \frac{E_x^2\,\langle m_x^2\rangle + E_z^2\,\langle m_z^2\rangle}{E_y^2\,\langle m_y^2\rangle}\end{aligned} \tag{13}$$

The unit vector components as presented in Fig. 4 are determined by the angles α and Φ

$$\begin{aligned}m_x &= \sin\alpha\,\cos\Phi \\ m_y &= \sin\alpha\,\sin\Phi \\ m_z &= \cos\alpha\end{aligned}$$

with corresponding mean squares

$$\begin{aligned}\langle m_x^2\rangle &= \langle\sin^2\alpha\,\cos^2\Phi\rangle = \frac{1}{2}\,(1 - \langle\cos^2\alpha\rangle) \\ \langle m_y^2\rangle &= \langle\sin^2\alpha\,\sin^2\Phi\rangle = \frac{1}{2}\,(1 - \langle\cos^2\alpha\rangle) \\ \langle m_z^2\rangle &= \langle\cos^2\alpha\rangle\end{aligned} \tag{14}$$

It should be noted that for an isotropic arrangement of transition moments the ensemble mean of any component of eqn. (14) result in 1/3. As a consequence eqn. (13) results in

$$R^{iso} = \frac{E_x^2 + E_z^2}{E_y^2} \tag{15}$$

which, according to eqn. (7) differs from unity. $R^{iso} = 1$ holds only for transmission spectroscopy.

Introducing eqn. (14) into eqn. (13) results in

$$R = \frac{E_x^2}{E_y^2} + 2\,\frac{E_z^2}{E_y^2}\,\frac{\langle\cos^2\alpha\rangle}{1-\langle\cos^2\alpha\rangle} \tag{16}$$

Solving eqn. (16) for $\langle\cos^2\alpha\rangle$ results in

$$\langle\cos^2\alpha\rangle = \frac{(R - \frac{E_x^2}{E_y^2})\,\frac{E_y^2}{E_z^2}}{2 + (R - \frac{E_x^2}{E_y^2})\,\frac{E_y^2}{E_z^2}} \tag{17}$$

This quantity is directly related to the segmental order parameter S_{seg}, which corresponds to the bond order parameter encountered in nuclear magnetic resonance (NMR) spectroscopy.

$$S_{seg} = \frac{3}{2}\langle\cos^2\alpha\rangle - \frac{1}{2} \tag{18}$$

Perfect alignment along the z-axis would result in $\langle\cos^2\alpha\rangle = 1$, and $S_{seg} = 1$, respectively. On the other hand an isotropic arrangement of transition moments would result in the ensemble mean $\langle\cos^2\alpha\rangle = 1/3$, corresponding to $S_{seg} = 0$. Finally, isotropic arrangement of the transition moments in the x,y-plane, i.e. $\alpha = 90°$ would result in $S_{seg} = -1/2$.

Both, $\langle\cos^2\alpha\rangle$ and S_{seg} are experimentally accessible by polarized light measurements.

Effective Thickness of Oriented Samples Surface Concentration

Axial effective thickness' of isotropic samples as introduced by eqns. (4) and (6) must now be weighted by the corresponding ensemble mean of the unit vector components of the transition moment, resulting in

$$\begin{aligned} d_{ex} &= 3\langle m_x^2\rangle d_{ex}^{iso} = \frac{3}{2}(1 - \langle\cos^2\alpha\rangle)\, d_{ex}^{iso} \\ d_{ey} &= 3\langle m_y^2\rangle d_{ey}^{iso} = \frac{3}{2}(1 - \langle\cos^2\alpha\rangle)\, d_{ey}^{iso} \\ d_{ez} &= 3\langle m_z^2\rangle d_{ez}^{iso} = 3\,\langle\cos^2\alpha\rangle\, d_{ez}^{iso} \end{aligned} \tag{19}$$

In analogy to eqn. (6) one obtains for the effective thickness with parallel polarized incident light

$$d_{e||} = d_{ex} + d_{ez} \tag{19a}$$

$$d_{e\perp} = d_{ey} \tag{19b}$$

The surface concentration Γ of a species in a layer of thickness d is considered as projection of the volume concentration c to the surface of the IRE. It follows from eqns. (3) and (19).

$$\begin{aligned} \Gamma = c \cdot d &= \frac{A_{||}\cdot d}{\varepsilon \cdot d_{e||}} = \frac{\int A_{||} d\tilde{v} \cdot d}{\int \varepsilon d\tilde{v} \cdot d_{e||}} \\ &= \frac{A_{\perp}\cdot d}{\varepsilon \cdot d_{e\perp}} = \frac{\int A_{\perp} d\tilde{v} \cdot d}{\int \varepsilon d\tilde{v} \cdot d_{e\perp}} \end{aligned} \tag{20}$$

$A_{||}$ and $A_{\perp}$ denote the absorbances measured with parallel and perpendicular polarized incident light, respectively. ε is the molar absorption coefficient. It should be noted that eqn. (20) holds for integrated absorbance, too, provided that integrated molar absorption coefficients are used.

SBSR SPECTROSCOPY

Most FTIR spectrometers are working in the single beam (SB) mode. As a consequence a single channel reference spectrum has to be stored for later conversion of single channel sample spectra into transmittance and absorbance spectra. This technique favors inaccuracy due to drifts resulting from the instrument or from the sample as well as disturbance by atmospheric absorptions. In order to eliminate these unwanted effects to a great extent a new ATR attachment has been constructed, converting a single beam instrument into a pseudo-double beam instrument. The principle features of this attachment are depicted in Fig. 5.

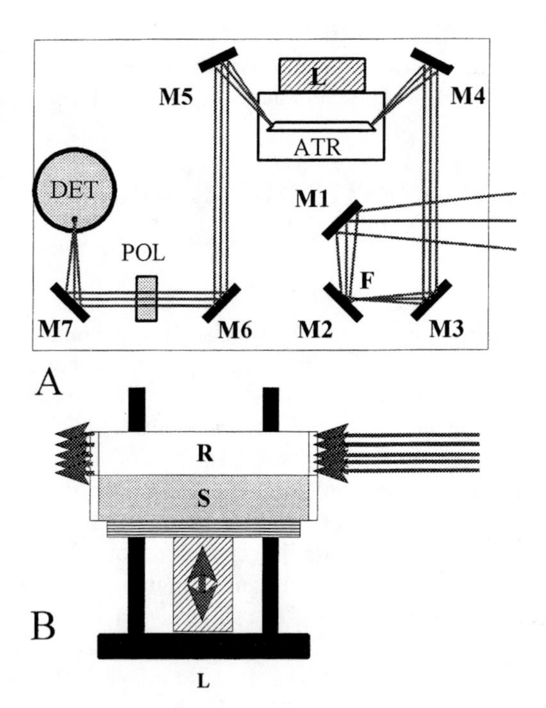

FIGURE 5. Single Beam Sample Reference (SBSR) ATR attachment. (A) The focus in the sample compartment is displaced to the position F by the planar mirrors M1 and M2. The off-axis parabolic mirror M3 produces a parallel beam with a diameter of one centimeter, i.e. half of the height of the IRE. The cylindrical mirror M4 focuses the light to the entrance face of the IRE. M5 which has the same shape as M4 reconverts to parallel light passing via the planar mirror M6 through the polarizer POL and being focused to the detector DET by the off-axis parabolic mirror M7. (B) Alternative change from sample to reference and vice versa is performed by computer controlled lifting and lowering of the ATR cell body.

As usual, a convergent IR beam enters the sample compartment. The focal point is now displaced by the planar mirrors M1 and M2 to the new position F, whereas the off-axis parabolic mirror M3 performs a conversion of the divergent beam into a parallel beam with fourfold reduced cross-section. This beam is focused to the entrance face of a trapezoidal IRE by a cylindrical mirror M4. Therefore, the ray propagation in the IRE is still parallel to the direction of light propagation (x-axis), enabling subdivision of the large IRE surfaces (x,y-plane) in perpendicular direction (y-axis) to the light propagation. One half of the IRE is then used for the sample (S) and the other one for the reference (R). Both, S and R, were encapsulated by flow-through cuvettes, independently accessible by liquid or gaseous flow-through. This principle is referred to as *Single Beam Sample Reference (SBSR)* technique. In a first version (refs. (3), (4)) a computer controlled chopper was used to direct the beam alternatively through the sample and reference. Later on the chopper version has been replaced by the lift version shown in Fig. 5. The cell platform is moved alternatively up and down aligning

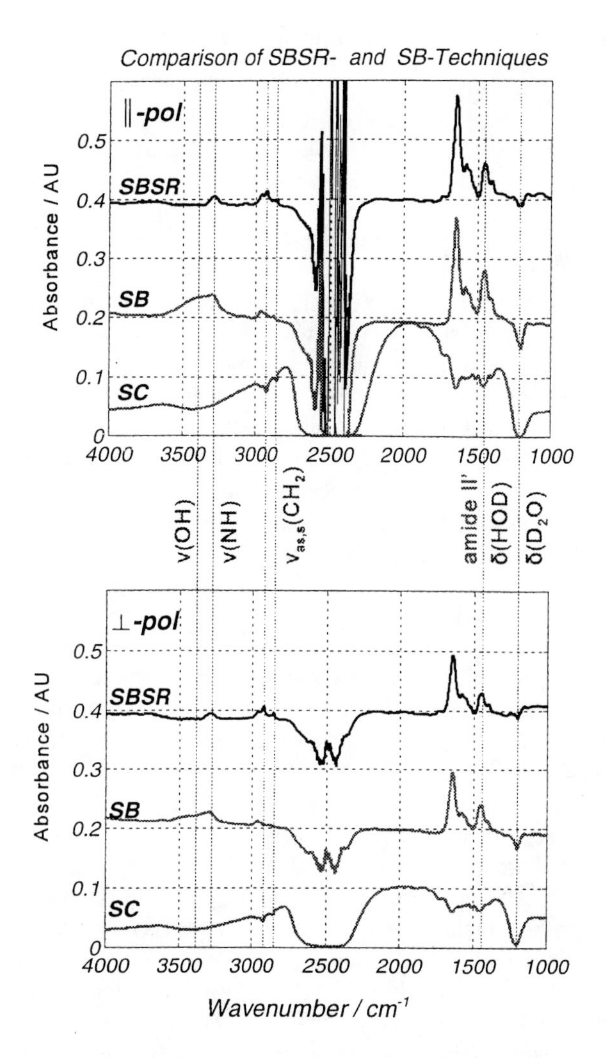

FIGURE 6. Comparison of Single Beam Sample Reference (SBSR) pseudo-double beam technique with conventional single beam (SB) technique. Very low energy in the 2500 cm^{-1} and 1200 cm^{-1} region resulted from stretching $\nu(D_2O)$ and bending $\delta(D_2O)$ absorptions of liquid D_2O, respectively, as shown by a single channel (SC) spectrum. The supported bilayers in the sample and reference cuvettes consisted of a dipalmitoylphosphatidic acid (DPPA) LB monolayer and a cardiolipin (CL) adsorbed monolayer. Both membranes exhibited the same age, since the LB layer covered the whole width of the IRE, and CL adsorption from vesicles occurred synchronously by two independent equal circuits. In a second step creatine kinase (CK) was adsorbed from a circulating solution in the sample channel, ref. (16). Therefore, the absorbance spectra shown in this figure reflect adsorbed CK as well as any other differences between S and R channel. Obviously, there are more detectable differences in the SB mode than in the SBSR mode, because SBSR reflects the actual difference between S and R, while SB shows the difference between the actual sample spectrum and a stored (older) reference spectrum. In this case, the partly different results obtained in the SBSR and SB mode result predominately from a slow uptake of H_2O vapor by the circulating D_2O solutions. This leads to overlapping of ν(NH) and amide II' of CK, as well as an overcompensation of D_2O absorption bands (~ 2500 cm^{-1} and 1200 cm^{-1}). It should be noted that the slight overcompensation of $\delta(D_2O)$ at 1200 cm^{-1} is significant, since it reflects the reduced water

content in the sample cuvette due to displacement by the CK layer. pp and vp absorbances are in accordance with eqns. (3) and (4), using $z_i = 50$ Å, $z_f = 143$ Å, $\varepsilon(\delta(D_2O)) = 1.38 \cdot 10^4$ $cm^2 \cdot mol^{-1}$. Ge IRE, angle of incidence, $\theta = 51°$, number of active internal reflections, N = 36.7. The refractive indices were: $n_1 = 4.0$, $n_2 = 1.45$, and $n_3 = 1.30$, see Fig. 3.

the sample and reference cuvettes with the IR beam, respectively. The lift version has two significant advantages over the chopper version (i) as it makes use of the full beam of the spectrometer resulting in twice the single channel energy of the chopper version, and (ii) still more relevant for most applications, it compensates the inhomogeneous light flux inherent in most IR spectroscopic instruments.

Thus SBSR absorbance spectra are calculated from sample and reference single channel spectra which have been measured with very short mutual time delay. Fig. 6 shows the results of a series of HD exchange measurements performed in the SBSR mode with the enzyme creatine kinase (CK). The enzyme was adsorbed from H_2O buffer to a DPPA/CL supported bilayer as described in ref. (16).

The conventional SB spectrum reflects the whole history of the sample, whereas the SBSR spectrum reflects the sample state when compared with a reference of the same age. Therefore, the SB spectrum contains the HDO produced by slight H_2O contamination during the experiment in addition to the spectrum of CK. The former obscures the shape of ν(NH) and amide II' bands, which is an obvious disadvantage of the SB mode.

For HD exchange a D_2O buffer solution was circulated through the sample and reference cuvette of the ATR cell during three days. As a consequence slight contamination of D_2O by atmospheric H_2O could not be avoided in the course of this long-time experiment. The resulting HDO gave rise to absorption bands near 3400 cm^{-1} and near 1450 cm^{-1} interfering with NH stretching (ν(NH)) of non-exchanged amide protons, and with amid II' of deuterated amide groups of the protein, respectively. Since in sample and reference contamination by hydrogen is approximately the same due to equal treatment of the circulating D_2O buffer solutions, the HDO absorption bands (ν(OH) and δ(HDO)) will be compensated to a major extent, as demonstrated by Fig. 6, trace SBSR. The SBSR trace represents predominately membrane bound CK in a partially deuterated state. The sample consisted of a DPPA/CL/CK assembly, and the reference of a DPPA/CL supported bilayer. Since a sequence of SBSR spectra consists of two independent sequences of single channel spectra, the collected data may by analyzed in the SB mode as well. Doing this by using the single channel spectrum of the sample channel measured in the SBSR mode before CK adsorption (DPPA/CL in D_2O environment in S and R) as reference and a corresponding single channel spectrum after about 12 hours of CK exposure to D_2O buffer as single channel sample spectrum. The resulting SB absorbance spectrum is also presented in Fig. 6, trace SB. Thus SBSR and SB spectra shown in Fig. 6 had exactly the same experimental conditions. To make best use of SBSR data, it is recommended to analyze the data by both modes, SBSR and SB since an unwanted synchronous breakdown of sample and reference assembly, e.g. by hydrolysis of a polymer matrix existing in S- and R-channel, or by equal loss of lipid molecules from a supported bilayer, would be obscured in the SBSR mode, but unambiguously detected in the SB mode.

LOCAL ANESTHETIC MEMBRANE INTERACTION

The mechanism of action of local anesthetics (LA) is still subject to discussion among biophysicists, physicians, and pharmacists, although the teaching opinion, generally accepted among physicians, claims that tertiary amine LA's will react specifically from the interior of the cell with a receptor in the sodium channel (25). This reaction is suggested to block the sodium influx, thus interrupting nerve signal transmission. Of course, FTIR studies with model membranes cannot validate or reject the suggestion of a specific action of LA's. However, significant information on the interaction with lipid bilayers is available with respect to partition coefficient, degree of protonation of bound LA, and influence on structural changes of the lipid bilayer and LA upon binding.

Dibucaine (DIBU) or cinchocaine which is a synonym of DIBU was used as a model LA. Its chemical formula is depicted in Fig. 7. DIBU has a pK_a value of 8.87 in solution. 99.9 % of the molecules in the bulk solution are in the protonated state at pH 5.5 (Fig. 7).

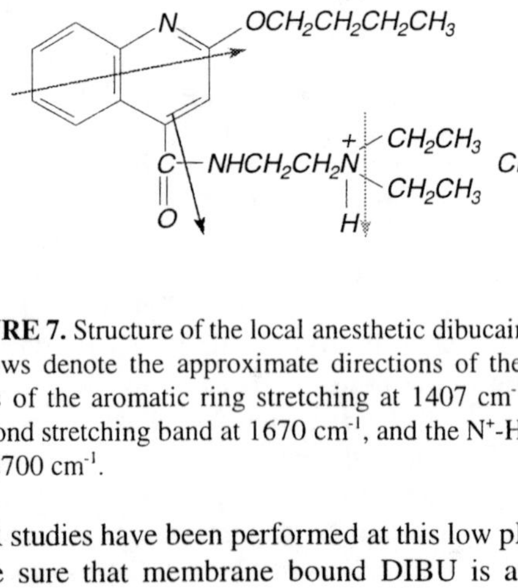

FIGURE 7. Structure of the local anesthetic dibucaine (DIBU). The arrows denote the approximate directions of the transition moments of the aromatic ring stretching at 1407 cm^{-1}, the C=O double bond stretching band at 1670 cm^{-1}, and the N^+-H stretching band at 2700 cm^{-1}.

NMR studies have been performed at this low pH in order to make sure that membrane bound DIBU is also in the protonated state.This would facilitate the interpretation of spectra (26). For the sake of comparison our experiments have been performed under the same conditions. The model membrane consisted of a DPPA/POPC supported bilayer. A series of DIBU concentrations (0.5 mM to 10.0 mM) were prepared in 20 mM phosphate-borate-citrate buffer (pH 5.5) containing 100 mM sodium chloride. The solutions were pumped in a closed cycle through the ATR cuvette (Fig. 2) while ATR spectra were measured at the same time.

Typical polarized absorbance spectra of the membrane interacting with a 10 mM DIBU solution are shown in Fig. 8 trace a. Trace b was obtained from the spectra of a 100 mM solution of DIBU by scaling down with a factor 0.299 for parallel polarization (pp, ||) and 0.325 for perpendicular polarization (vp, ⊥). The aromatic ring stretching vibration at 1407 cm^{-1} was used for scaling, because this band turned out to be insensitive to pH changes. Since the scaling factors were larger than 0.1, there is unambiguous evidence for LA adsorption to the lipid membrane. Moreover, different scaling factors point to a certain ordering of DIBU upon membrane binding.

Structural Changes upon LA Binding

The most prominent spectral changes result from the DPPA/POPC bilayer. First, the negative absorbance at 1740 cm^{-1} resulting from the ester C=O stretching vibrations should be mentioned. At first sight, loss of lipid molecules due to LA interaction could be supposed. Watching closer, however, reveals that in such a case even more negative bands should be expected at the wavenumbers of the more intense $>PO_2^{2-}$ stretching bands at 1230 cm^{-1} and 1090 cm^{-1}. Since there is no evidence for such negative $>PO_2^{2-}$ stretching bands, one may exclude lipid loss upon LA binding. This conclusion is supported by the observation that the original spectrum of the bilayer is restored after washing out of DIBU. A reasonable explanation for the behavior of the ν(C=O) band is a conformational change of one or both fatty acid ester groups induced by LA interaction. This might induce a change of the composition of the corresponding normal mode. As a consequence the associated transition moment $\vec{M}$ (see Fig. 4) must be expected to be changed, too, which according to eqns. (3) and (9) would also influence the magnitude of the molar absorption coefficient ε. In case of ν(C=O) a decrease of ε is expected. Conformational change of the hydrocarbon chain region is indicated by the symmetric and antisymmetric CH_2 stretching bands at 2850 cm^{-1} and 2920 cm^{-1}, respectively. For a perfectly aligned hydrocarbon chain along the z-axis the mean transition moments of both, $\nu_s(CH_2)$ and $\nu_{as}(CH_2)$ are expected to be in the x,y-plane (see Fig. 4), i.e. perpendicular to the z-axis and isotropically distributed around this axis. Since in this case $\langle \cos^2 \alpha \rangle = 0$, it follows from eqn. (19) that $d_{e\perp} = d_{ey} = 1.5\ d_{ey}^{iso}$ is decreased to $d_{ey} = d_{ey}^{iso}$ upon conversion to random conformations. In case of parallel polarized incident light one obtains from eqns. (19) and (19a) $d_{e||} = 1.5\ d_{ex}^{iso}$ for perfect chain alignment along the z-axis. Conversion to random structures increases the effective thickness to $d_{e||} = d_{ex}^{iso} + d_{ez}^{iso}$.

Under the given conditions one obtains for a supported bilayer of thickness d = 50 Å (θ = 45°, $n_1 = 4.0$, $n_2 = 1.45$, n_3 (2900 cm^{-1}) ≈ 1.40 (anomalous dispersion of H_2O)) the axial effective thickness' $d_{ex}^{iso} = 50$ Å, $d_{ey}^{iso} = 58$ Å and $d_{ez}^{iso} = 58$ Å. As a consequence the dichroic response to the above mentioned perturbation of hydrocarbon chain ordering will be: $d_{e||} = 75$ Å $\Rightarrow d_{e||}^{iso} = 108$ Å, and $d_{e\perp} = 88$ Å $\Rightarrow d_{e\perp}^{iso} = 58$ Å thus confirming qualitatively the absorbance decrease in the vp-spectrum and the synchronous increase in the pp-spectrum as a consequence of chain disordering (Fig. 8).

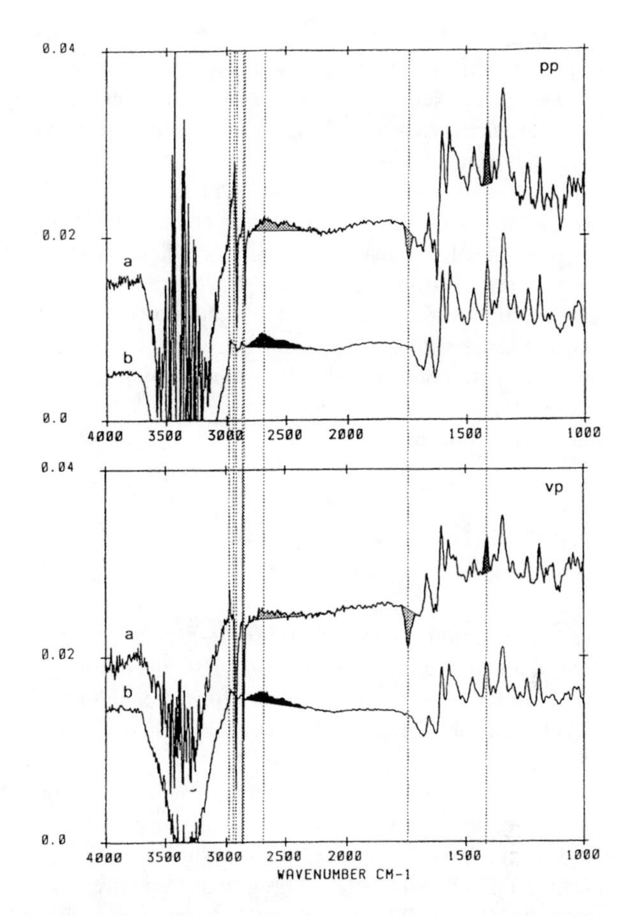

FIGURE 8. SBSR ATR IR spectra of the local anesthetic dibucaine (DIBU). **a:** 10 mM DIBU in contact with a DPPA/POPC supported bilayer. **b:** 100 mM DIBU in direct contact with the IRE, scaled down by 0.299 (pp), and 0.325 (vp). 100 mM NaCl, 20 mM BPC buffer pH 5.5. Parallel (pp) and perpendicular (vp) polarized incident light at θ=45°. Germanium (Ge) IRE, N=19.3 active internal reflections. The reference cuvette contained a DPPA/POPC membrane under the same conditions.

It should be noted that orientational changes of the fatty acid ester groups most probably contribute also to the absorbance of ν(C=O), however, the effect by the expected decrease of ε should be dominant, since negative bandes are observed in both polarizations. On the other hand, a decrease of $\varepsilon(\nu(CH_2))$ should also be taken into account upon conformational changes of the hydrocarbon chain (27).

A quantitative analysis of hydrocarbon chain ordering based on polarized IR ATR spectra and on intensity considerations will be given in ref. (9). The qualitative interpretation of chain order decrease due to LA interaction as given above is strongly supported by a shift of both, $\nu_s(CH_2)$ and $\nu_{as}(CH_2)$ to higher wavenumbers paralleled by a decrease of peak absorbances. This effect is indicative for hydrocarbon chain disordering (28). Finally the behavior of the antisymmetric methyl stretching vibration at 2963 cm^{-1} should be mentioned. In contrast to the CH_2 stretching bands, this band shows an increase in the vp spectrum upon LA interaction. In the pp spectrum, it is only visible as a shoulder. Qualitatively, one has to conclude, that the transition moment of this band ($\nu_{as}(CH_3)$) exhibits a predominant interaction with E_z in the unperturbed state. Reducing the ordering by LA interaction lead to enhanced interaction with E_y, i.e. increasing absorbance in the vp spectrum.

Unambiguously, DIBU interaction with a DPPA/POPC bilayer lead to a significant hydrocarbon chain disordering. As a consequence local anaesthesia might result only from a disturbance of the global membrane structure which could idirectly influence the structure and function of proteins relevant for nerve signal transduction.

Finally, it should be noted, that the original ordering of the supported bilayer was restored, when the LA was detached by circulating pure buffer solution through the ATR cuvette.

Apparent pK_a of Bound DIBU and Partition Coefficient

Several absorption bands in the 1000 - 1800 cm^{-1} region are changing shape and wavenumber upon DIBU interaction. A discussion of some bands reflecting conformational and pH induced changes of DIBU as well as of phospholipids will be given in ref. (9).

Two absorption bands, however, are of special interest in connection with the determination of the degree of protonation of bound DIBU ($DIBU^+$) as well as of the partition coefficients K_p of DIBU, $DIBU^+$ and $DIBU_{tot}$. The aromatic ring stretching vibration $\nu(C{=}C)$ at 1407 cm^{-1} has turned out to be insensitive to pH changes, therefore it can be used to determine the total amount of bound DIBU ($\Gamma(DIBU_{tot})$). On the other hand, the N^+-H stretching vibration $\nu(N^+H)$ at 2700 cm^{-1} gives direct information on the amount of bound $DIBU^+$ ($\Gamma(DIBU^+)$).

The surface concentration of deprotonated DIBU is obtained by the difference $\Gamma(DIBU) = \Gamma(DIBU_{tot}) - \Gamma(DIBU^+)$. The result of a quantitative analysis of an experimental series with measurements at variable bulk DIBU concentration is presented in Fig. 9.

Before going into details, a short explanation shall be given, on how adsorption isotherms are obtained from FTIR ATR raw data. 20 - 40 internal reflections are optimum for *in situ* mono- or submonolayer spectroscopy in aqueous environment when using a germanium IRE with an angle of incidence of $\theta \approx 45°$ (4). Under these conditions, intense bands of dissolved organic compounds become visible in the spectrum at bulk concentrations above about 0.5 mM.

Therefore, in the first step of the analysis, the polarized absorbance spectra have to be corrected for bulk DIBU absorption using eqns. (2 - 4) and eqn. (7). For a bulk medium, index 2 in eqn. (7) must be replaced by index 3 (see text). The following values have been used for calculation: $n_1 = 4.0$, $n_2 = 1.45$ (membrane + DIBU), $n_3 = 1.31$ (H_2O, 1407 cm^{-1}), $\theta = 45°$, $\lambda = 7.11 \cdot 10^{-4}$ cm (1407 cm^{-1}), $\varepsilon(1407\ cm^{-1}) = 1.80 \cdot 10^5$ cm^2/mol, $z_i = 5.0 \cdot 10^{-7}$ cm, $z_f = \infty$, N = 19.3 (number of active internal reflections). Under these conditions, a 10 mM bulk DIBU solution results in at 1407 cm^{-1}: $A_{||}(DIBU) = 2.52$ mAU, and $A_{\perp}(DIBU) = 1.26$ mAU, respectively. For correction of bulk absorption one may subtract the corresponding DIBU spectrum measured without membrane, since according to eqn. (4) there is an attenuation of only about 1% due to the displacement of the solution by the supported membrane of thickness d = 50 Å. The remaining spectrum reflects bound DIBU and conformational changes of the lipid bilayer induced by DIBU interaction. Since most supported membrane assemblies feature uniaxial partial orientation along the membrane normal (z-axis) quantitative analysis must be based on eqns. (17), (19) and (20). The following dichroic ratios were determined from Fig. 8: $R(1407\ cm^{-1}) = 1.60$, and $R(2700\ cm^{-1}) = 1.42$. The former value is very close to that of an isotropic layer which is according to eqn. (15) $R^{iso} = 1.61$, however, since the dichroic ratio of the $\nu(N^+H)$ band deviates significantly from R^{iso} one may conclude that DIBU is not randomly adsorbed to the lipid bilayer but exhibits some ordering. As a consequence, the transition moment $\nu_{ar}(C{=}C)$ at 1407 cm^{-1} is expected to have a mean incline of about 55° (magic angle) with respect to the z-axis.

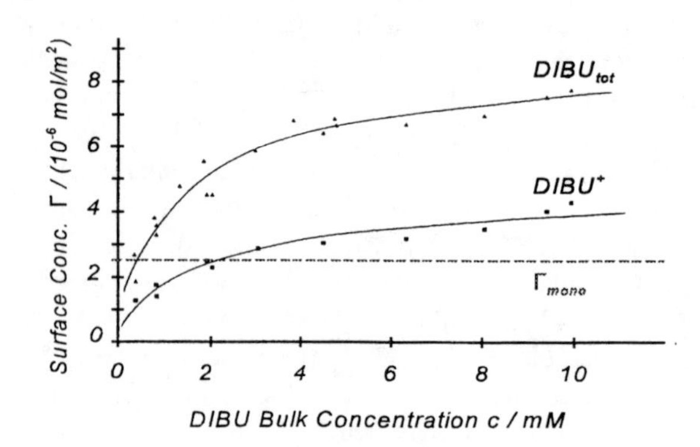

FIGURE 9. Surface concentration of dibucaine (DIBU) determined by means of eqn. (20) as a function of bulk concentration c at bulk pH 5.5. $DIBU_{tot}$: Adsorption isotherm of total amount of membrane bound DIBU. $DIBU^+$: Adsorption isotherm of protonated fraction of membrane bound DIBU. Both adsorption isotherms may be described by the Langmuir model resulting in $\Gamma_{max}(DIBU_{tot}) = 8.22 \cdot 10^{-6}\ mol/m^2$, and $\Gamma_{max}(DIBU^+) = 3.88 \cdot 10^{-6}\ mol/m^2$. On the other hand, $\Gamma_{mono} = 2.5 \cdot 10^{-6}\ mol/m^2$ denotes the calculated surface concentration of a packed monolayer of DIBU from the molecular cross section of 68 $Å^2$, ref. (26). Mean monolayer coverage is already achieved near 1 mM bulk concentration. A possible explanation of the apparent discrepancy with fits according to the Langmuir model is adsorption of preformed DIBU aggregates from bulk solution to the membrane. Membrane bound DIBU consists of about equal amounts of the protonated and deprotonated form.

At first sight Fig. 9 reveals that despite of the low bulk pH of 5.5 membrane bound $DIBU_{tot}$ behaves stoichiometrically like a 1:1 complex of protonated $DIBU^+$ and deprotonated DIBU species. At low bulk concentrations one could argue for DIBU being dissolved in the hydrophobic region of the membrane, and $DIBU^+$ being located at the membrane surface.

DIBU penetrating into the lipid phase would explain the significant disturbance of hydrocarbon chain ordering discussed above. This interpretation may be correct, however, it can't explain the nearly constant ratio of stoichiometric coefficients up to bulk concentrations of 10 mM because already below 1 mM bulk DIBU concentration bound $DIBU_{tot}$ exceeds the quantity required for a monomolecular coverage of the membrane surface (DIBU cross-section 68 $Å^2$, ref. (26)). It is known from earlier work (7) that LA's tend to multilayered adsorption at elevated bulk concentration. Vapor pressure osmometry has revealed that LA form aggregates already in bulk solution (8), which could explain the significantly higher DIBU surface concentrations obtained by fitting experimental data according to the Langmuir isotherm, see Fig. 9. In view of these findings we suggest that the observed stoichiometric ratio DIBU : $DIBU^+$ = 1 : 1 is characteristic of associated DIBU, at least above 1 mM bulk concentration. DIBU multilayer formation at the membrane surface is also supported by the observation that the mean order parameter of the CH_2 groups of hydrocarbon chains of POPC as determined by eqns. (17) and (18) is found to decrease from $S_{mean}(c=0 \text{ mM}) = 0.65$ in the pure DPPA/POPC supported membrane to $S_{mean}(c \geq 3\text{mM}) = 0.20$. Thus hydrocarbon chain ordering was no longer affected by DIBU adsorption above 3 mM bulk DIBU concentration, refs. (8) and (9).

In view of these facts, the use of a partition coefficient is only meaningful for the limiting case c ⇒ 0 mM. Denoting bulk DIBU concentration by c and the concentration of DIBU in the membrane by c_m one obtains for the partition coefficient under consideration of eqn. (20)

$$K_p = \lim_{c \to 0} \frac{c_m}{c} = \frac{\Gamma}{d \cdot c} \tag{21}$$

Taking the experimental data from Fig. 9 to calculate the limits according to eqn. (21) one obtains $K_p(DIBU_{tot}) = 1410$, $K_p(DIBU^+) = 668$, and $K_p(DIBU) = 9.2 \cdot 10^5$, respectively. For details, the reader is referred to refs. (8) and (9).

TIME RESOLVED MODULATED EXCITATION (ME) SPECTROSCOPY

External ME of Lipids, Peptides and Proteins

Change of any external thermodynamic parameter generally excerts a specific influence on the state of a system. The sytem response will be a relaxation from the original state (e.g. an equilibrium) to a new equilibrium state. In case of a periodic change (modulation) of the parameter, the system response will also be periodic, i.e. those absorption bands of the spectrum which result from stimulated molecules or parts of them will be labelled by the same frequency. As a consequence, it will be possible to separate the modulated response of the system, which is correlated with the external stimulation from the stationary response, resulting from parts of the system that were not affected by modulated excitation (ME) and from the background. Moreover, if the kinetics of the stimulated process is in the same time range as the period of external stimulation, phase-lag and amplitude measurements of modulated absorbances give information on the reaction scheme and the kinetics of the stimulated process. Hydration modulation, e.g. was applied to determine the hydration sites of lecithins (29), (30).

Temperature ME of poly-L-lysine was used to study induced periodic secondary structural changes as well as the sequence of transients (20). The classical ATR set-up (see Fig. 2 facilitates the application of electric fields to membrane assemblies, since a Ge ATR plate, supporting the membrane, may be used as one electrode, and the back-wall of the cuvette as counter electrode. First use of electric field ME of immobilized acetyl choline esterase (AChE) was reported in ref. (31). In the view of to-day, the interpretation of a field-dissociation effect (second Wien effect) of carboxylic acid residues of AChE must be qualified. Since Ge decomposes at slightly positive potentials, forming germinic acid, the periodic deprotonation and reprotonation of -COOH groups of AChE may be interpreted as a superposition of electric field and pH-effect. As a consequence, passivation of the Ge surfaces (32) is a prerequisite for electric field ME experiments. The problem of anodic decomposition of germanium does not exist in the case of electric field ME of liquid crystals (33), (34).

Electronic ME of photochemical processes by modulated UV-/VIS-light enables access to a wide range of excitation frequencies. Light flux modulation in the kHz range, which may be easily performed by means of a mechanical chopper, enables kinetic analysis even in the μs region as shown in the case of the photooxidation of pyrocatechol by modulated electronic excitation IR and ESR spectroscopy (35).

Schematic Set-up for ME Spectroscopy

The principle set up for a modulated excitation (ME) experiment is depicted by Fig. 10. In contrast to relaxation experiments where step-excitation (SE) is used ME technique is based on a periodic stimulation. Both techniques give access to the characteristic dynamic quantities of a system, the relaxation constants. The response of SE is a superposition of exponentials, $\exp(-t/\tau_i)$, where τ_i denotes the i-th relaxation time of the system. ME on the other hand results in a superposition of sine waves, $\sin(\omega t+\phi_i)$ where the phase lag ϕ_i is given by $\phi_i=\arctan(-\omega\tau_i)$, see eqn. (28).

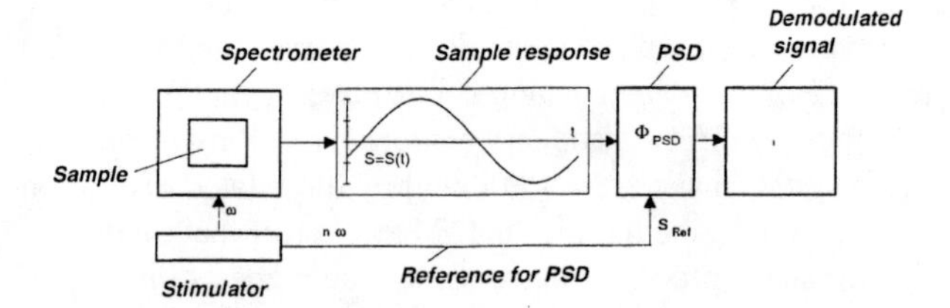

FIGURE 10. Schematic set up for modulated excitation (ME) experiments. A periodic excitation is exerted on the sample with frequency ω. The sample response as sensed e.g. by IR radiation contains the frequency ω at the wavelength's that are significant for those parts of the sample that have been affected by the stimulation. Selective detection of the periodic sample responses is enabled by phase sensitive detection (PSD), resulting in the amplitudes A_n of the fundamentals ω and the harmonics nω (n=2, 3, ...) as well as the phase shifts ϕ_n between the n-th harmonic and the stimulation. This phase shift is indicative of the kinetics of the stimulated process and of the underlaying chemical reaction scheme.

Temperature Modulated Excitation (T-ME) of Chemical Reactions

The principles of T-ME may be elucidated by considering the simple reversible chemical reaction between two species A_1 and A_2 where k_{12} and k_{21} denote the rate constants of forwards and backwards reaction.

$$A_1 \underset{k_{21}}{\overset{k_{12}}{\rightleftarrows}} A_2$$

If the sample is exposed to a periodic temperature stimulation according to

$$T(t)=T_i+\frac{\Delta T}{2}(1-\cos\ \omega t) \tag{22}$$

where T_i, ΔT and ω denote the initial temperature, the peak to peak temperature variation and the angular frequency, respectively. The influence of temperature on rate constants may be described by the Arrhenius equation. Since ΔT is small (≤ 5°C), the linearized form will be used in this context, leading to

$$k_{ik}(t)=k_{ik}(\overline{T})+\frac{\Delta k_{ik}}{2}(1-\cos\omega t)$$
$$with\ \ \Delta k_{ik}=k_{ik}(T_i)\frac{E_{ik}}{RT_i^2}\Delta T \tag{23}$$

k_{ik} denotes the rate constant from species i to species k and E_{ik} is the corresponding activation energy.

Introducing the reaction number ξ as the relevant concentration parameter for the description of turnover

$$[A_1]=[A_1]_0-\xi,\quad [A_2]=[A_2]_0+\xi \tag{24}$$

and inserting eqn. (24) into the rate equations for A_1 and A_2 results in the rate equation for ξ, eqn. (25). Since the peak to peak variation of the rate constants Δk_{12} and Δk_{21} are small compared to the mean values $\overline{k}_{12}$ and $\overline{k}_{21}$, the third term in parantheses of the coefficient of ξ in eqn. (25) may be neglected. The solution of the differential equation with constant coefficients is then given by eqn. (26) which describes the response of the reaction number. Insertion of eqn. (26) into eqn. (24) results in the time dependent behavior of the concentrations of species A_1 and A_2. It contains the relaxation from the initial state to the steady state which is reached for t≥3τ, where τ denotes the relaxation time. In this simplest case of a reversible reaction τ is the inverse of the sum of the two rate constants, i.e. $\tau=(k_{12}+k_{21})^{-1}$. Modulation experiments are generally started after an initial period of 3τ where relaxation is completed to about 95%. The relevant steady state solution is then given by eqn. (27).

$$\begin{aligned}\dot{\xi} =& \\ &-(\overline{k}_{12}+\overline{k}_{21}-\frac{1}{2}(\Delta k_{12}+\Delta k_{21})\cos\omega t)\cdot\xi \\ &-\frac{1}{2}(\Delta k_{12}[A_1]_0-\Delta k_{21}[A_2]_0)\cos\omega t \\ &+\overline{k}_{12}[A_1]_0-\overline{k}_{21}[A_2]_0\end{aligned} \tag{25}$$

It should be noted that the quantity $(k_{12}+k_{21})$ appears as inverse relaxation time τ in the exponentials of the general solution as well as in amplitude and phase angle of the steady state solution eqns. (27) and (28). This fact proves the equivalence of relaxation and modulation techniques.

$$\xi(t) = \frac{\bar{k}_{12}[A_1]_0 - \bar{k}_{21}[A_2]_0}{\bar{k}_{12} + \bar{k}_{21}} (1 - \exp(-(\bar{k}_{12} + \bar{k}_{21})t))$$

$$+ \frac{(\bar{k}_{12} + \bar{k}_{21})(\Delta k_{12}[A_1]_0 - \Delta k_{21}[A_2]_0)}{2((\bar{k}_{12} + \bar{k}_{21})^2 + \omega^2)} \exp(-(\bar{k}_{12} + \bar{k}_{21})t)$$

$$-\frac{1}{2}(\Delta k_{12}[A_1]_0 - \Delta k_{21}[A_2]_0) \sqrt{\frac{1}{1 + (\frac{\omega}{\bar{k}_{12} + \bar{k}_{21}})^2}} \cos(\omega t + \phi) \qquad (26)$$

For t→∞ one obtains the stationary solution

$$\xi(t) = -\frac{1}{2} \frac{\bar{k}_{12}[A_1]_0 - \bar{k}_{21}[A_2]_0}{\bar{k}_{12} + \bar{k}_{21}} \cdot$$

$$\sqrt{\frac{(\Delta k_{12}[A_1]_0 - \Delta k_{21}[A_2]_0)^2}{1 + (\frac{\omega}{\bar{k}_{12} + \bar{k}_{21}})^2}} \cos(\omega t + \phi) \qquad (27)$$

with

$$\phi = \arctan(-\frac{\omega}{\bar{k}_{12} + \bar{k}_{21}}) = \arctan(-\omega\tau) \qquad (28)$$

It follows from eqns. (27) and (28) that the product $\omega\tau$ is the relevant kinetic parameter in ME spectroscopy. For $\omega \to 0$ the system is expected to be able to respond immediately to the external stimulation. $\omega\tau=1$ results in an amplitude damping by a factor of $\sqrt{2}$ (3 dB point), paralleled by a phase shift of ϕ=-45°, whereas for $\omega \to \infty$ the amplitude approachs zero and the phase angle ϕ=-90°. Consequently, the simple chemical reaction under consideration behaves just like an electronic RC low pass filter.

Of course, the amplitude and phase dependence on modulation frequency becomes more complex for more complicated chemical reaction schemes, however, any scheme features a characteristic amplitude/phase-frequency dependence. As soon as phase resolved ME experiments at different modulation frequencies are available, ME technique will enable a more detailed kinetic analysis of the system then SE technique (relaxation technique) because of the additional experimental degree of freedom given by the modulation frequency ω.

T-ME of the α Helix to β Pleated Sheet Conversion of Poly-L-Lysine

A poly-(L)-lysine (PLL) film cast on an ATR plate and hydrated with D_2O (80% rel. humidity, 28°C) was exposed to a periodic temperature variation of $\Delta T/2=\pm 2$°C at the mean value of $\bar{T}$ = 28°C. The results obtained after phase sensitive detection (PSD) are shown in Fig.11. Part (A) shows the stationary spectrum and part (B) phase resolved spectra of the system response with the fundamental frequency ω. The numbers indicated on the spectra denote phase difference between the modulated excitation and phase setting at the phase sensitive detector (PSD). The ME spectra shown in Fig. 11B may be expressed by eqn. (29)[19, 20].

$$\Delta A(\tilde{\nu}, \phi_{PSD}) = \kappa \cdot \sum_{i=1}^{N} \Delta A_{0i}(\tilde{\nu}) \cos(\phi_i - \phi_{PSD}) \qquad (29)$$

$\Delta A_{0i}(\tilde{\nu})$ is the i-th component-spectrum in which each band has the same phase angle ϕ_i. Consequently, this set of bands may be considered to be correlated, i.e. to belong to a population of molecules or functional groups featuring the same kinetic response to the external stimulation. In such a population all absorbance bands exhibit a periodic dependence on the PSD phase setting ϕ_{PSD}. The amplitudes become maximum for $(\phi_i - \phi_{PSD}) = 0°$, minimum (negative) for $(\phi_i - \phi_{PSD}) = 180°$, and zero for $(\phi_i - \phi_{PSD}) = 90°$ *or* 270°. Obviously, ϕ_{PSD} can be used to sense the phase angle ϕ_i of a population of absorption bands, because ϕ_{PSD} is a parameter under experimental control. The most accurate determination of ϕ_i is got by performing a line shape analysis of the phase resolved spectra shown in Fig. 11B, followed by fitting each component according to eqn. (29), see ref. 20.

The first impression on comparing Fig. 11A with Fig. 11B is that modulation spectra are significantly better resolved. The spectral resolution was 4 cm^{-1} for both, stationary and modulation spectra. However, overlap is drastically reduced in the latter, because they contain only anbsorption bands from species that have been affected by the external stimulation. Furthermore, Fig. 11B shows that not only the intensity but also the shape of phase-resolved spectra is changing with ϕ_{PSD}-setting. This is an unambiguous indication of the existance of populations of conformational states featuring different phase angles ϕ_i. Extraction of these populations according to eqn. (29) enabled the assignment of transient species in the amide I' and amide II' regions. For details the reader is referrred to ref. (20). Attention should be drawn to a correlation between CH_2 stretching and the secondary structure of PLL which which has not been reported so far. The weak absorption bands at 2865 cm^{-1} and 2935 cm^{-1} result from symmetric and antisymmetric

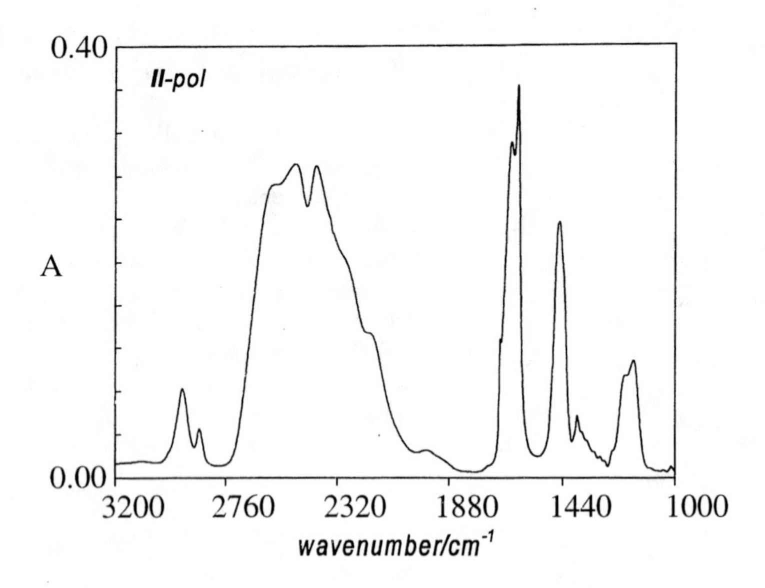

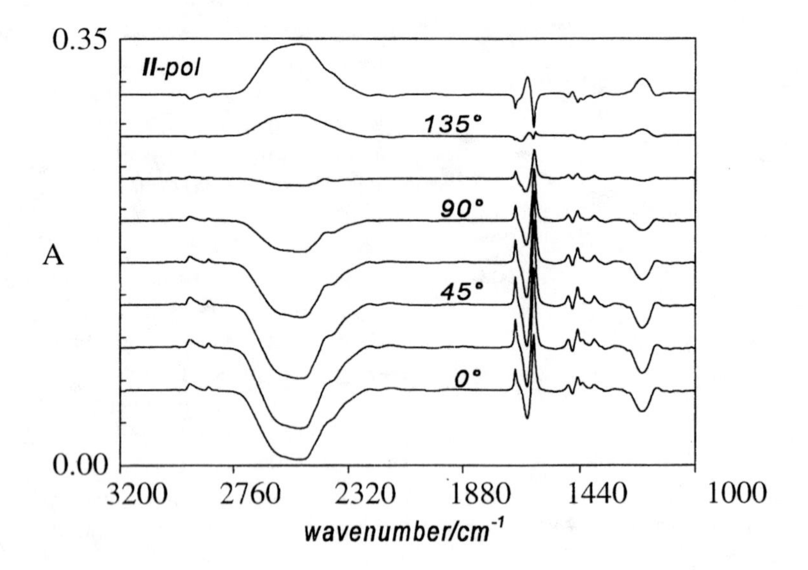

FIGURE 11. Parallel polarized T-ME FTIR spectra of a poly-(L)-lysine deuterobromide film hydrated with D_2O (80% rel. hum., 28°C). The film was deposited on a CdTe ATR plate. A rectangular temperature stimulation was applied with a period of 14.7 min (ω=0.427 min^{-1}) at $\bar{T}$ =28°C ± 2°C. Angle of incidence:θ = 45°, mean number of internal reflections: N=9-10. (A) Stationary part of the T-MEIR spectrum of PLL. (B) Set of phase-resolved T-MEIR spectra after phase sensitive detection (PSD) at phase settings ϕ_{PSD} = 0°-157.5° (phase resolution 22.5°) with respect to the T-stimulation. ϕ_{PSD} = 0° means in-phase with temperature switching from 26°C to 30°C. Heat transfer from the thermostats to the sample resulted in an additional phase lag of ϕ_T = 25°. (From ref. 20).

stretching of the CH_2 groups of the lysine side chains.They displaced by appproximately 3 cm^{-1} towards lower wavenumbers with respect to the corresponding bands in the stationary state (Fig. 11A).

This finding is indicative for a conformational change of a hydrocarbon chain from gauche defects into trans conformations (28). Since these bands are correlated with the formation of antiparallel β-pleated sheet structure (amide I' bands at 1614 cm^{-1} and 1685 cm^{-1}). We conclude therefore, that the conversion of PLL from α helix to β sheet is paralleled by a conformational change of the side chain from a bent to a extended structure.

T-ME of Reversible Unfolding/Folding of RNase A

Understanding of the molecular mechanism of protein folding and unfolding is of increasing interest not least because of molecular biological approaches to protein synthesis and modifications. Nase A is an enzyme that may be unfolded/denatured by heating and refolded/renatured upon cooling. Kinetic FTIR measurements have been reported recently using temperature jump techniques (21), (36). In this article we report the first preliminary T-ME experiments performed in solution. The stimulation amplitude was $\Delta T/2$ = 5°C at $\overline{T}$ = 64°C with a period of τ_m = 25 s. A sequence of time resolved spectra is shown in Fig. 12. Like in case of PLL there are drastic diffrences between the modulation spectra and the stationary spectrum (upper trace). As mentioned above modulation spectra suppress any absorbance which is not labelled by the stimulation frequency. Two interesting observations should be mentioned. (i) the corresponding isosbestic points in the amide I' band at 1667 cm^{-1} and in the amide II' at 1435 cm^{-1}, and (ii) the response of a distinct tyrosin population to T-ME.In the modulation spectrum, the tyrosin band appears at 1517 cm^{-1}, whereas in the stationary spectrum the corresponding band is found to be considerably broader with the peak maximum shifted to lower wavenumbers.

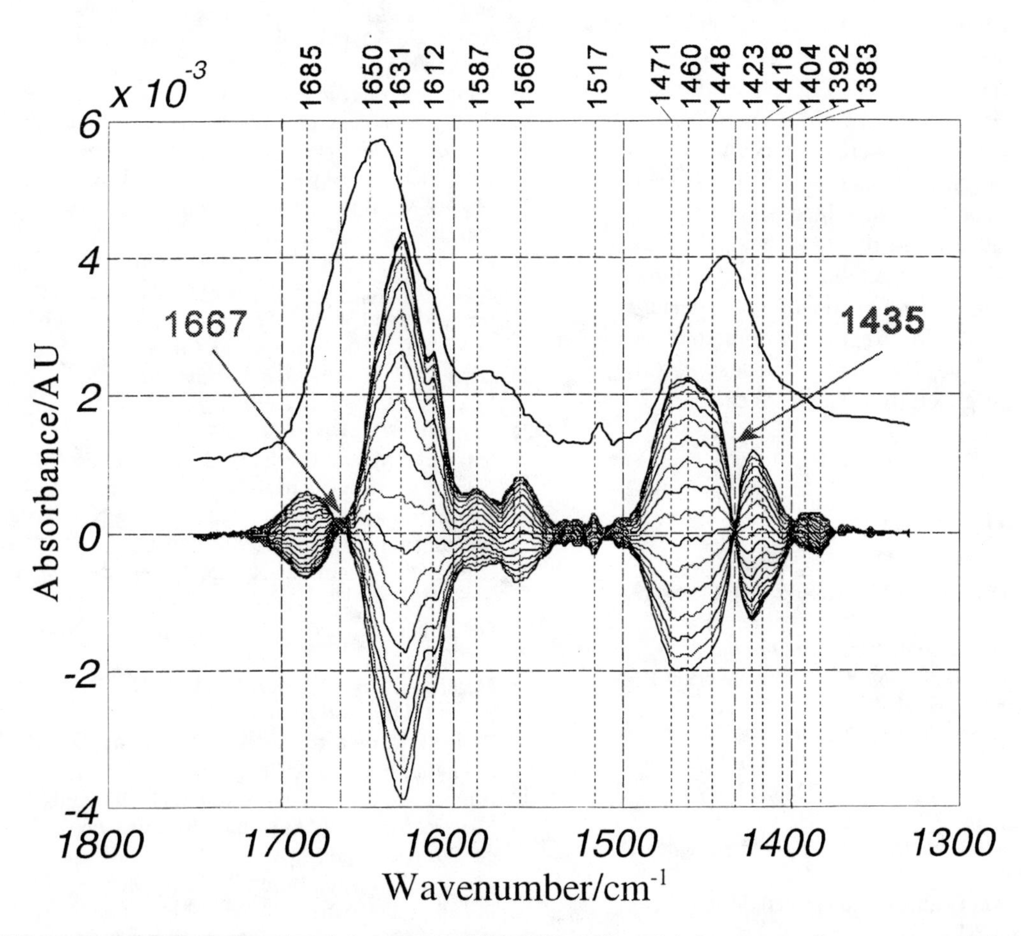

FIGURE 12. Stationary (upper trace) and phase resolved (9°) FTIR ATR modulation spectra of temperature modulated excitation of RNase A dissolved in D_2O buffer, pD 7. Mean temperature $\overline{T}$=64°C, modulation amplitude $\Delta T/2$=5°C. Note tyrosin response T-ME is selective, because the responding population absorbs at 2-3 cm^{-1} higher than the supperposition of all tyrosins of RNase. Secondary structural conversion results in isosbestic points in the amide I' and amide II' regions.

As in the case of PLL the shapes of modulation spectra alter with PSD phase setting ϕ_{PSD}, indicating that some phase resolution is achieved. The effect, however is less distinct than with PLL. Most probably higher modulation frequencies are required in order to get larger differences in the phase angles ϕ_i of different conformational populations. T-ME-experiments at higher frequencies are in progress.

ACKNOWLEDGEMENTS

Grant No. 6106 by the Jubiläumsfonds of the Oesterreichische Nationalbank is kindly acknowledged. It enabled scholarships for G.R. and M.S. We thank the University of Vienna for a preceding research scholarship for G.R. International contacts in the framework of the presented projects were facilitated by the EC research project grant No. BIO4-CT96-0022, and support by the Oesterreichischer Aussendienst (ÖAD). Both grants are kindly acknowledged. U.P.F. thanks gratefully to Prof. R. Hütter, former Vice President for research of the Swiss Federal Institute of Technology (ETH) Zurich, for enabling performance of a substantial part of the presented experimental work at the Technopark ETH. The authors are indebted to J. Gast, Dr. A. Simon, Dr. J. Gronholz from Bruker Analytische Messtechnik, Karlsruhe, Germany, and to Dr. W. Tschopp, Dr. A. Marcuzzi and D. Stierli from Spectrospin AG, Fällanden, Switzerland, for technical support. Supply of Creatine kinase by Prof. Th. Wallimann, O. Stachowiak, and U. Schlattner, Institute for Cellular Biology, ETH Zurich, and supply of Alcaline Phosphatase by Prof. B. Roux and Ms. M. Angrand, Laboratoire de Physico Chimie Biologique, University Claude Bernard, Lyon I, Lyon, France, is gratefully acknowledged.

REFERENCES

1. Tamm, L.K., and McConnell, H.M., *Biophys. J.* **47,** 105-113 (1985).
2. Brian, A., and McConnell, H.M.,*Proceedings of the National Academy of Sciences USA* **81,** 6159-6163 (1984).
3. Fringeli, U. P. in Schlunegger, U.P. (Ed.), *Biologically Active Molecules*; Springer: Berlin, **1989**; pp. 241-252.
4. Fringeli, U. P., in Mirabella, F. M. (Ed.), *Internal Reflection Spectroscopy, Theory and Applications,* Marcel Dekker **1992,** Chpt. 10, 255-324.
5. Kalb, E., Frey, S., and Tamm, L., *Biochim. Biophys, Acta* **1103,** 307-316 (1992).
6. Wenzl, P., Fringeli, M., Goette, J. and Fringeli, U. P., *Langmuir* **10,** 4253-4264 (1994).
7. Schöpflin, M., Perlia, X., and Fringeli, U. P., *J. Am. Chem. Soc.* **109,** 2375-2380 (1987).
8. Jeannette, ATR-IR-spektroskopische Untersuchungen der Lokalanästhetika-Modellmembran-Interaktion und dampfdruckosmometrische Bestimmung der Assoziation von Lokalanästhetika in wässriger Lösung, Zürich: PhD Thesis Nr. 9721, Swiss Federal Institute of Technology (ETH),1992.
9. Goette, J., Fahr, A., and Fringeli, U. P., *Langmuir,* in preparation.
10. Fringeli, U. P., Apell, H.-J., Fringeli, M., and Läuger, P., *Biochim. Biophys. Acta* **984,** 301-312 (1989).
11. Frey, S., and Tamm, *L. K., Biophys. J.* **60,** 922-930 (1991).
12. Fringeli, U. P., and Fringeli, M., *Proc. Natl. Acad. Sci. USA* **76,** 3852-3856 (1979).
13. Stephens, S.M., and Dluhy, *R., Thin Solid Films* **284-285,** 381.-386 (1996).
14. Fritz-Wolf, K., Schnyder, Th., Wallimann, Th., and Kapsch, W., *Nature* **381,** 341-345 (1996).
15. Stachowiak, O., Dolder, M., and Wallimann, Th., *Biochemistry,* **35,** 15522-15528 (1996).
16. Siam, M., Reiter, G., Schwarzott, M., Baurecht, D., and Fringeli, U. P., *"Interaction of two different types of membrane proteins with model membranes investigated with FTIR-ATR spectroscopy",* presented at the 11th International Conference on Fourier Transform Spectroscopy (ICOFTS-11), Athens, Georgia, U.S.A., August 10-15, 1997. See publication in this book.
17. Siam, M., Stachowiak, O., Schlattner, U., Wallimann, Th., and Fringeli, U. P., in preparation.
18. Reiter, G., Angrand, M., Roux, B., and Fringeli, U. P., in preparation.
19. Fringeli, U. P., *Simultaneous phase-sensitive digital detection process for time-resolved, quasi-simultaneously captured data arrays of a periodically stimulated system,* PCT international patent application, WO97/08598, 1997.
20. Müller, M., Buchet, R., and Fringeli, U. P., *J. Phys. Chem.* **100,** 10810-18825 (1996).
21. Reinstädler, D., Fabian,H., Backmann, J., and Naumann D., *Biochemistry* **35,** 15822-15830 (1996).
22. Born, M., and Wolf, E., *Principles of Optics,* Oxford: Pergamon Press, 1983, 6th ed. Ch. 1, pp. 47-51.
23. Harrick, N. J., *Internal Reflection Spectroscopy,* New York: Interscience Publisher, 1967.
24. Fringeli, U. P., *Chimia* **46,** 200-214 (1992).
25. Ritchie, J. M.,and Green, N. M., *Local Anesthetics*, in: Gilman, A. G., Goodman, L. S., and Gilman, A., Eds., *The Pharmacological Basis of Therapeutics,* New York: Macmillan Publishing, 1990, 7th Ed.
26. Seelig, A., Allegrini, P. R., and Seelig, J., *Biochim. Biophys. Acta* **939,** 267-276 (1988).
27. Del Zoppo, M., and Zerbi, G., *Polymer,* **33,** 4667-4672 (1992).
28. Mantsch, H. H., and McElhaney, R. N., *Chem. Phys. Lipids* **57,** 213-226 (1991).
29. Fringeli, U. P., and Günthard, Hs. H., *Biochim. Biophys. Acta* **450,** 101-106 (1976).
30. Fringeli, U. P., and Günthard, Hs. H. "Infrared Membrane Spectroscopy", in: "Membrane Spectroscopy", E.Grell ed.,Springer 1981, pp. 270-332.
31. Fringeli, U. P., and Hofer, P., *Neurochemistry International* **2,** 185-192 (1980).
32. Roberts, *Thin Isolating Films on Semi-Conductors,* Springer Heidelberg, 1981.
33. Fringeli, U. P., Schadt, M., Rihak, P., and Günthard, Hs. H., *Z.Naturforsch.* **31a,** 1098-1107 (1976).

34. Baurecht, D, Neuhäusser, W., and Fringeli, U. P., "Modification of Time-Resolved Step-Scan and Rapid-Scan FTIR Spectroscopy for Modulation Spectroscopy in the Frequency Range from Hz to kHz," *AIP Conference Proceedings no. 430*, pp 367-370, ©1998.
35. Forster, M., Loth, K., Andrist, M., Fringeli, U.P., and Günthard, Hs.H., *Chemical Physics* **17,** 59-85 (1976).
36. RNase A project: Collaboration with Dieter Naumann and Diane Reinstädler, Robert Koch Institute, Berlin, Germany.

Author Index

G

H

I

J

K

L

M

N

O

P

R

S

T

U

V

W

Y

Z